Explorations and Applications

Authors

Senior Authors
Miriam A. Leiva Richard G. Brown

Loring Coes III
Shirley Frazier Cooper
Joan Ferrini-Mundy
Amy T. Herman
Patrick W. Hopfensperger
Celia Lazarski
Stuart J. Murphy
Anthony Scott
Marvin S. Weingarden

McDougal Littell
A HOUGHTON MIFFLIN COMPANY
Evanston, Illinois ◆ Boston ◆ Dallas ◆ Phoenix

Authors

Senior Authors

Richard G. Brown Mathematics Teacher, Phillips Exeter Academy, Exeter, New Hampshire

Miriam A. Leiva Cone Distinguished Professor for Teaching and Professor of Mathematics, University of North Carolina at Charlotte

Loring Coes III Chair of the Mathematics Department, Rocky Hill School, E. Greenwich, Rhode Island

Shirley Frazier Cooper Curriculum Specialist, Secondary Mathematics, Dayton Public Schools, Dayton, Ohio

Joan Ferrini-Mundy Professor of Mathematics and Mathematics Education, University of New Hampshire, Durham, New Hampshire

Amy T. Herman Mathematics Teacher, Atherton High School, Louisville, Kentucky

Patrick W. Hopfensperger Mathematics Teacher, Homestead High School, Mequon, Wisconsin

Celia Lazarski Mathematics Teacher, Glenbard North High School, Carol Stream, Illinois

Stuart J. Murphy Visual Learning Specialist, Evanston, Illinois

Anthony Scott Assistant Principal, Orr Community Academy, Chicago, Illinois

Marvin S. Weingarden Supervisor of Secondary Mathematics, Detroit Public Schools, Detroit, Michigan

The authors wish to thank **Jane Pflughaupt**, Mathematics Teacher, Pioneer High School, San Jose, California, and **Martha E. Wilson**, Preparatory Mathematics Specialist, Mathematical Sciences Teaching and Learning Center, University of Delaware, Newark, Delaware, for their contributions to this Teacher's Edition.

ISBN: 0-395-67136-1

123456789—VH—99 98 97 96 95

Contents of the Teacher's Edition

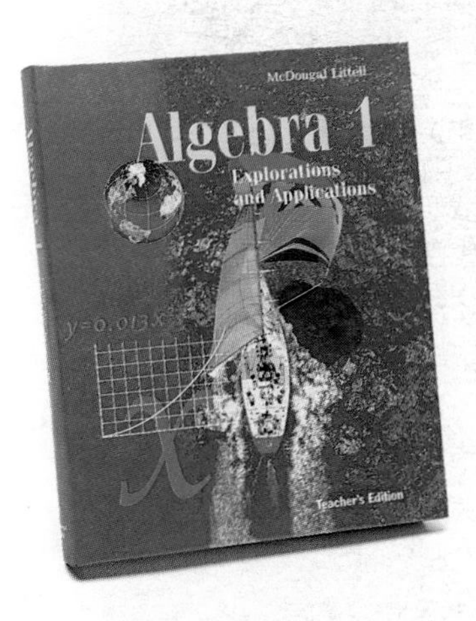

Philosophy of Algebra 1

Explorations and Applications

Goals of the Course

This course has been designed to make mathematics accessible and inviting to the wide range of students who are studying algebra today. It helps you prepare today's students for tomorrow's world by:

- Involving students in exploring and discovering math concepts
- Connecting algebra to the real world and to other subjects and math topics
- Building understanding of the concepts that provide a strong foundation for future courses and careers
- Assessing students' progress in ways that support learning

Mathematical Content

The content and the teaching strategies in the textbook reflect the curriculum, teaching, and assessment standards of the National Council of Teachers of Mathematics. The fresh, new course outline:

- Emphasizes graphing
- Uses functions as a unifying theme
- Integrates technology as a problem-solving tool
- Connects algebra to geometry, data analysis, probability, and discrete mathematics

Teaching Strategies

The flexible course design offers frequent opportunities for you to incorporate:

- Exploratory activities that build conceptual understanding
- Applications that strengthen problem-solving skills
- Discussion and writing questions that develop communication skills

Assessment

You and your students can measure their mathematical growth throughout the course in a variety of ways, including:

- Cooperative learning activities
- Open-ended problems
- Journal writing
- Portfolio projects

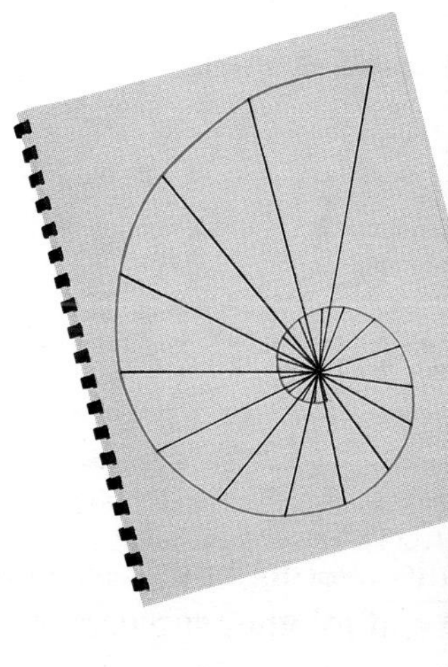

Program Overview

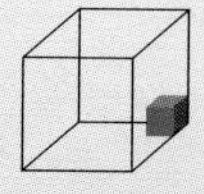
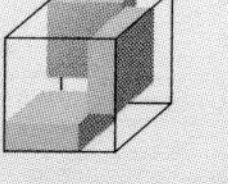
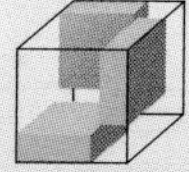
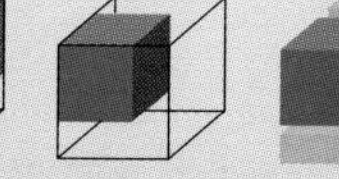

Pages T6-T37 give an overview of *Algebra 1: Explorations and Applications* and the teaching materials that support it. These pages provide information about:

Contents

CHAPTER

1

Using Algebra to Work with Data

Applications

CHAPTER 1

Connections of mathematics to the real world are emphasized throughout this course. In this chapter, variables are used to represent and analyze **real-world data**. Matrices are presented informally as a way to organize data. The use of **technology** (calculator, graphing technology, spreadsheet) as a problem solving tool is introduced.

CHAPTER

2

Equations and Functions

2.2 *Modeling with algebra tiles* 61

Applications

Interview: ***Bob Hall***
Athlete and Racing Wheelchair Designer
57, 60, 82

Additional applications include: zoology, sports, language arts, medicine, history, geometry, personal finance, aviation, biology, consumer economics, energy use, business, recreation, science

CHAPTER 2

In this chapter, students use both **algebra tiles** and **algebraic methods** to solve equations. **Functions** are introduced informally. Students model graphs of linear equations in a **cooperative learning** activity. Then they draw their own coordinate graphs and use graphing technology to solve real-world problems.

CHAPTER

3

Graphing Linear Equations

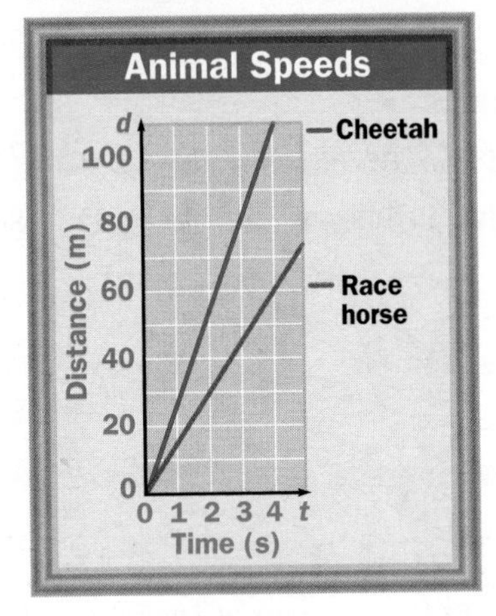

Applications

Interview:
Mae Jemison
Astronaut
99, 100, 104, 110

Additional applications include: aviation, sports, astronomy, weather, employment, consumer economics, physics, business, geometry, energy use, civil engineering, recreation, geology

CHAPTER 3

Modeling linear relationships is the focus of this chapter. Students investigate and apply rates, direct variation, slope, and linear equations. They model **linear data** using scatter plots and lines of fit. The many tables, graphs, and equations in this chapter highlight the **multiple representations** of functions.

CHAPTER

4

Solving Equations and Inequalities

Applications

CHAPTER 4

This chapter continues and expands the work with **solving linear equations** begun in Chapter 2. Through the use of **explorations, manipulatives,** and **technology**, students learn to solve multi-step equations, equations with fractions or decimals, and linear inequalities.

CHAPTER

5 Connecting Algebra and Geometry

Applications

CHAPTER 5

Integrating the branches of mathematics helps students see mathematics as a whole, and enables them to solve a broader range of problems. This chapter connects **algebra**, **geometry**, and **probability** through ratio, proportion, similarity, and area.

CHAPTER

6

Working with Radicals

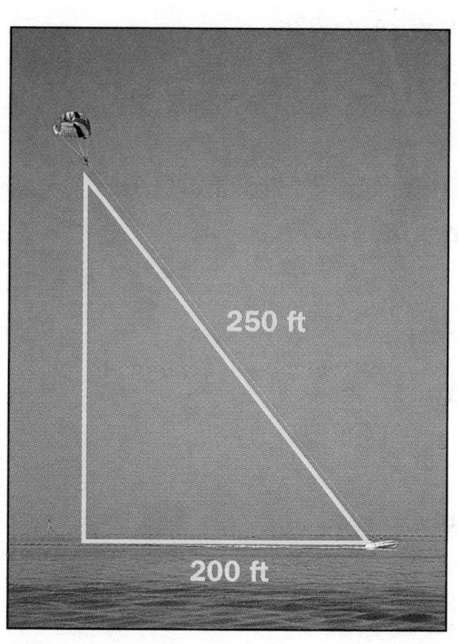

Applications

Additional applications include: recreation, urban planning, aviation, language arts, biology, physics, geometry, medicine, construction

CHAPTER 6

This chapter continues the integration of algebra and geometry by introducing radicals through applications of the Pythagorean theorem. Students learn to calculate with radicals and then extend these numerical procedures to algebraic ones to perform calculations with binomials.

CHAPTER

7 Systems of Equations and Inequalities

Applications

Additional applications include: sports, fitness, aviation, finance, space science, geometry, personal finance, nutrition

CHAPTER 7

This chapter extends the earlier work with linear equations to **systems of linear equations and inequalities** in two variables. Students explore and apply different methods for solving systems of equations, and learn **strategies for choosing** an appropriate method. Throughout the chapter, helpful visuals enhance the lesson presentations.

CHAPTER

8

Quadratic Functions

line of symmetry

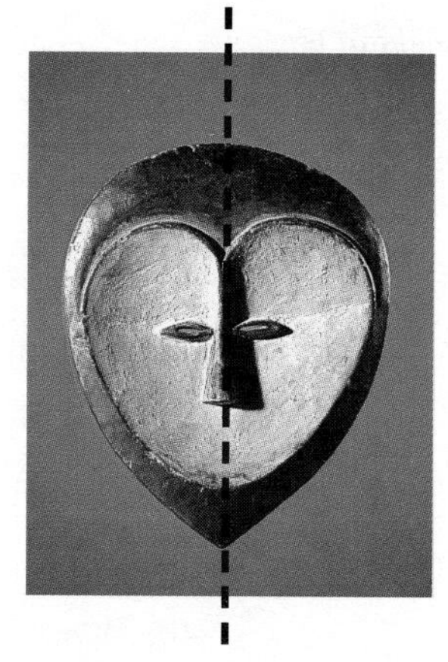

Applications

Additional applications include: geometry, physics, cooking, sports, firefighting, science

CHAPTER 8

In this chapter, students use **graphing calculators** to explore nonlinear relationships, especially quadratic functions. Students **analyze graphs** of second-degree equations using technology and relate the graphs to the solutions of these equations. The usefulness of quadratic functions in **modeling** real-world phenomena is highlighted.

CHAPTER

9 Exponential Functions

9.3 *Tennis* 384

Applications

Interview: ***Lori Arviso Alvord***
Surgeon 380, 387, 399

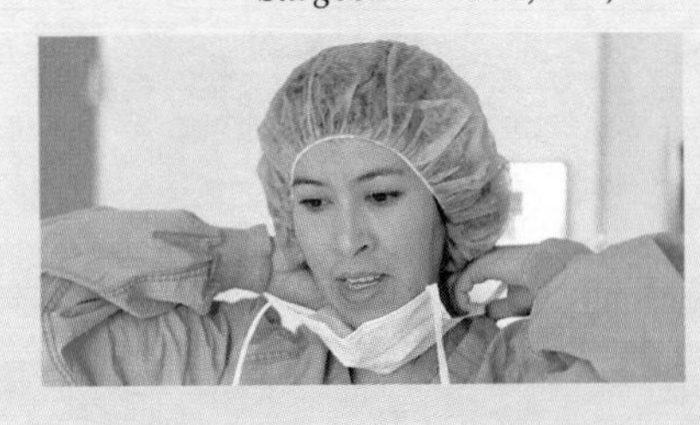

Connection

Computers	374	**Literature**	404
Cooking	388	**History**	410
Consumer Economics	394		

Additional applications include: architecture, geometry, carpentry, number theory, ecology, biology, personal finance, traffic, social studies, physics, meteorology, sports, occupational safety, fuel economy, population, history, astronomy

CHAPTER 9

Technology allows students to explore topics that would be inaccessible otherwise. In this chapter, students use technology to begin to build a base of knowledge about exponents, powers, and exponential functions that will provide a strong foundation for later courses.

CHAPTER

10

Polynomial Functions

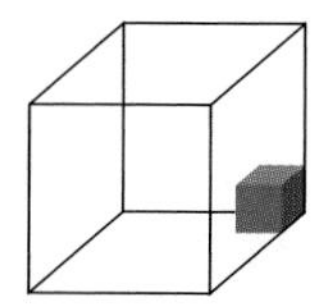

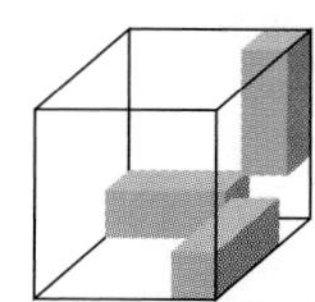

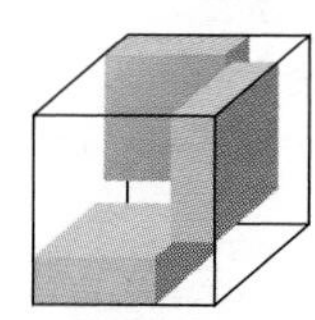

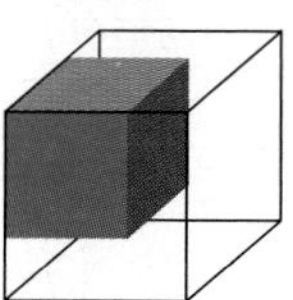

10.2 *Modeling* 431

Applications

Interview:
Linda Pei
Financial Adviser
427, 446

Connection

Crafts	433	**Sports**	452
Business	440		

Additional applications include: geometry, crafts, physics, sports

CHAPTER 10

Manipulatives can provide excellent **visual representations** of operations with polynomials. In this chapter, students use algebra tiles to **explore polynomial operations**, including factoring. Different methods for solving quadratic equations are compared and used to solve real-world problems.

CHAPTER

11 Rational Functions

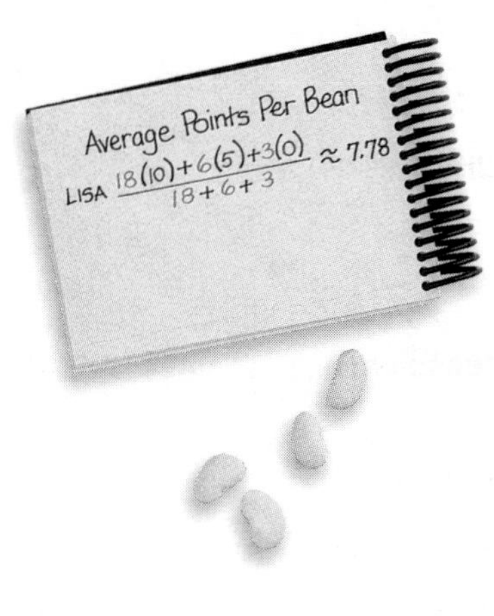

11.2 *Comparing game scores* 469

Applications

Interview: ***Alex MacLean***
Aerial Photographer
467, 486

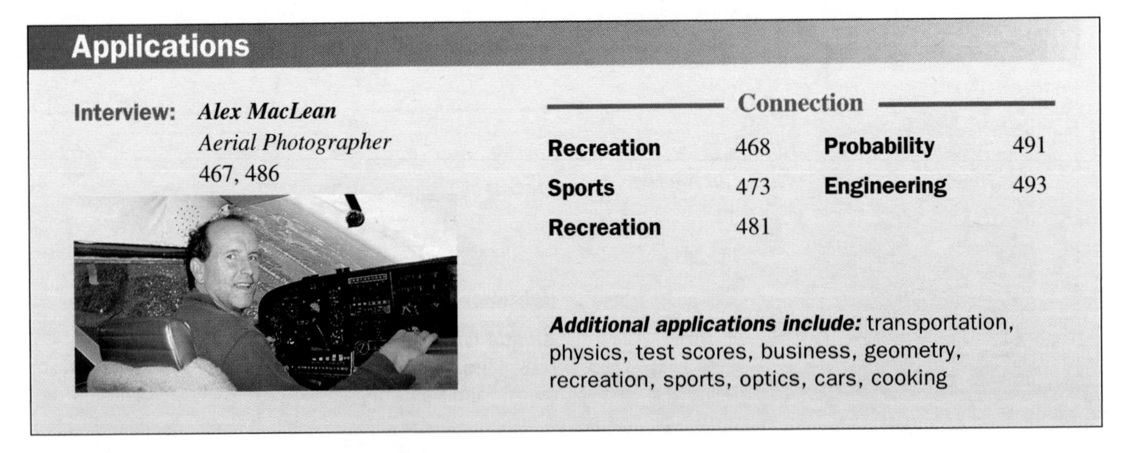

Connection

Recreation	468	**Probability**	491
Sports	473	**Engineering**	493
Recreation	481		

Additional applications include: transportation, physics, test scores, business, geometry, recreation, sports, optics, cars, cooking

CHAPTER 11

New concepts can be learned more easily when they are **connected to prior learning**. In this chapter **inverse variation** is compared to direct variation, **weighted averages** extend students' earlier experiences with averages, and operations with **rational expressions** are related to the corresponding operations with fractions.

CHAPTER 12

Discrete Mathematics

Applications

Interview:
Phil Roman
Animator
509, 513

Additional applications include: fundraising, fine arts, cars, business, fashion design, sports, government, manufacturing, geometry, photography

CHAPTER 12

This chapter demonstrates once again how mathematical topics are **interconnected**. Students explore algorithms from **algebra**, find paths and trees from **graph theory**, study permutations and combinations from **combinatorics**, and connect **probability** and counting techniques.

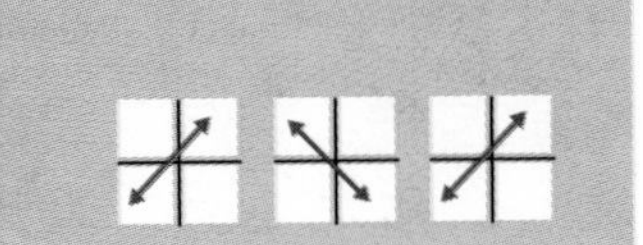

Student Resources

Student Resources

These resources include **practice, review,** and **reference** material to help students with the course. The *Toolbox* provides a quick brush-up on **prerequisite skills** that includes worked examples and practice exercises. The *Technology Handbook* introduces students to the basic features available on most **graphing calculators.**

Complete Teaching Resources

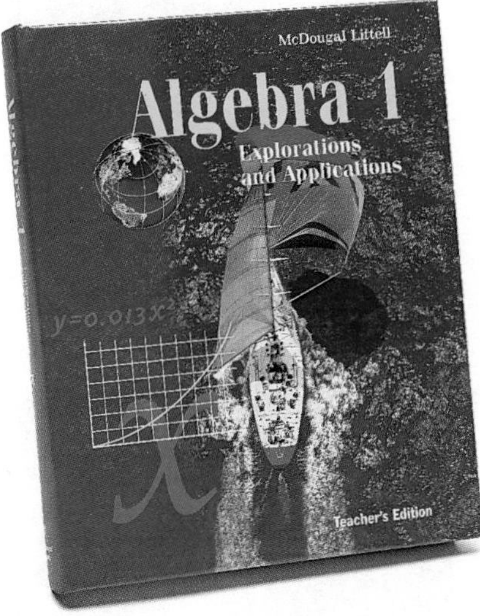

Teacher's Edition includes complete support for planning and teaching your lessons.

- Student pages with complete answers
- Point-of-use teaching and exercise notes
- Planning pages for each chapter, with assignment guides for standard and block scheduling

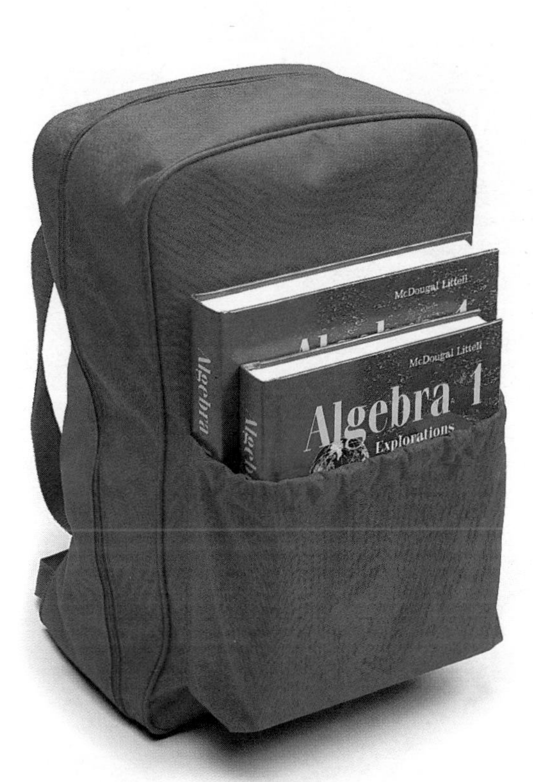

Teacher's Resources
provides a comprehensive selection of support materials, with suggestions for implementing new ideas for teaching and assessment.

- Assessment Book
- Practice Bank
- Study Guide
- Portfolio Project Book
- Professional Handbook
- Course Guides
- Explorations Lab Manual
- Technology Book
- Preparing for College Entrance Tests
- Lesson Plans

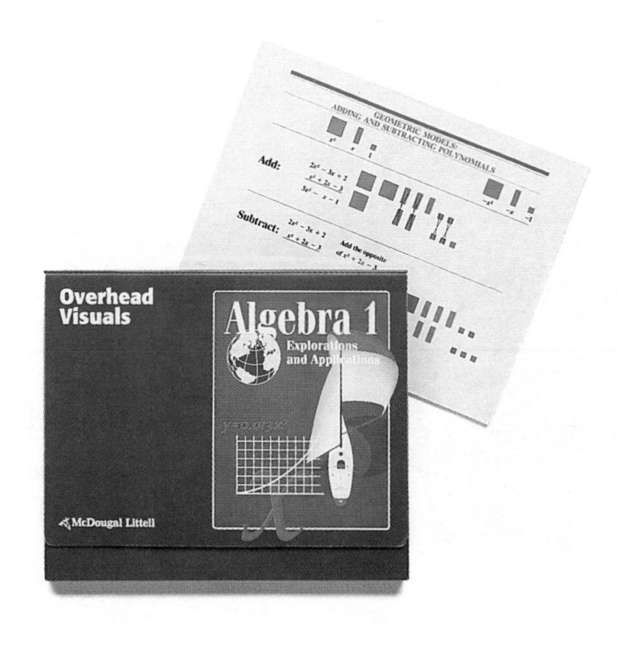

Also available:

- Overhead Visuals
- Warm-Up Transparencies
- Multi-Language Glossary
- Assessment Book, Spanish Edition
- Study Guide, Spanish Edition
- Solution Key
- Test Generator (Macintosh and IBM)
- McDougal Littell Software for exploring, analyzing, and solving problems

Support Materials

for all teaching and learning needs

Of Special Interest ...

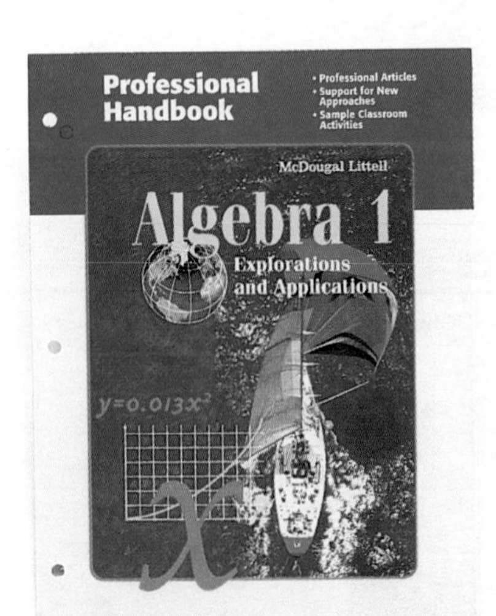

Assessment Book

- Chapter tests — 2 versions
- Performance-based assessment
- Short quizzes and cumulative tests

Also available in Spanish

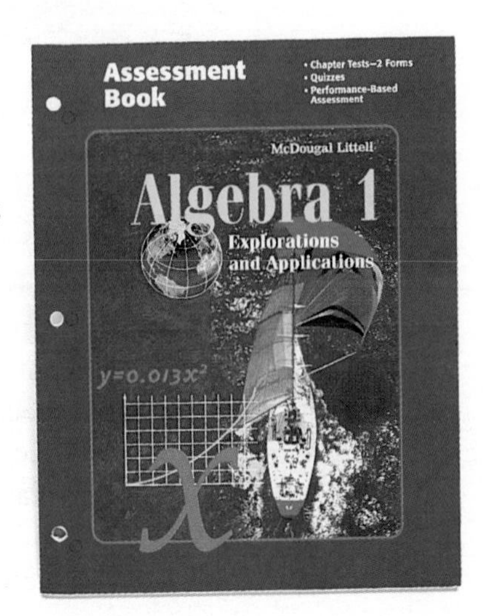

Professional Handbook

- Articles on new teaching approaches
- Practical classroom advice
- Sample activities
- Assessment suggestions

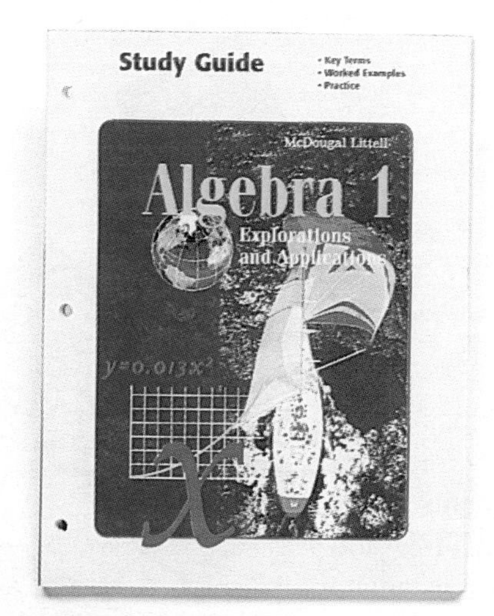

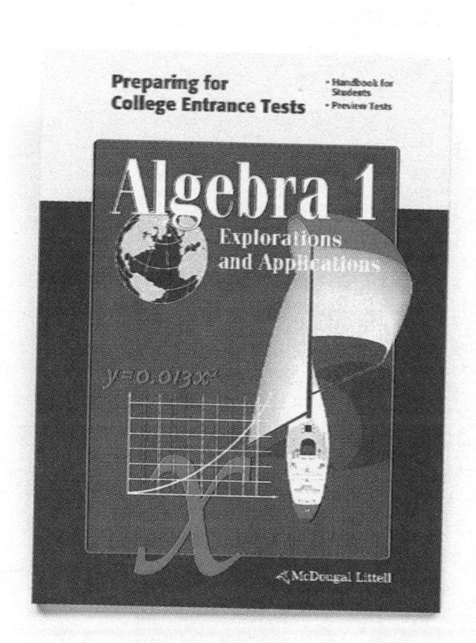

Study Guide

- Study Guide lesson for each section of the textbook
- Key terms
- Worked examples
- Practice and review

Also available in Spanish

Preparing for College Entrance Tests

- Student handbook with test-preparation strategies
- Preview tests with questions in standardized-test formats

Please turn the page for Technology Support ➤

Technology Support

Integrating technology The technology tools and print support materials that you need to integrate technology successfully into your courses are available with this program.

McDougal Littell Software

- Software for exploring, analyzing, and solving problems
- Function and statistical graphing
- Activities linked to the textbook

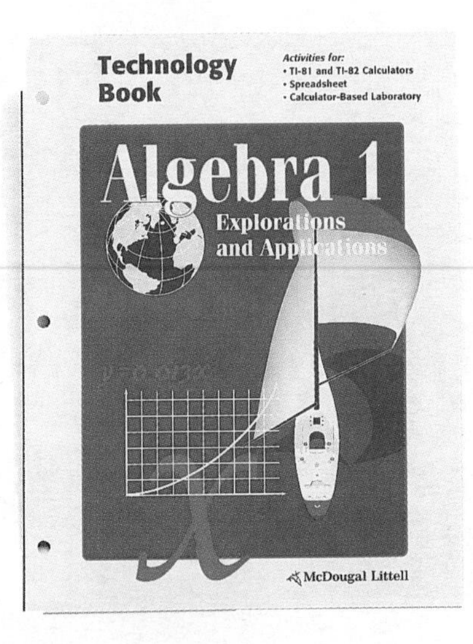

Technology Book

Includes activities for:
- TI-81 and TI-82 graphing calculators
- Spreadsheets
- Calculator-Based Laboratory

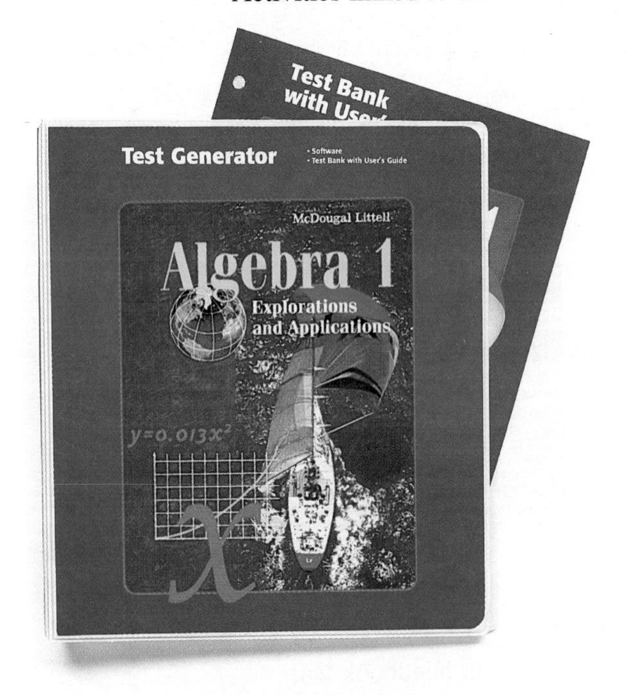

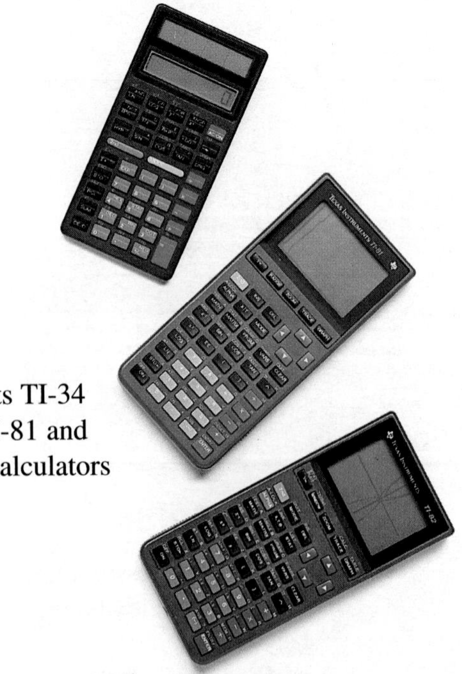

Calculators

- Texas Instruments TI-34 calculator and TI-81 and TI-82 graphing calculators

Test Generator

Technology-based assessment with:
- Test-generating software for Macintosh and IBM
- Test bank with user's guide

Algebra 1 Explorations and Applications

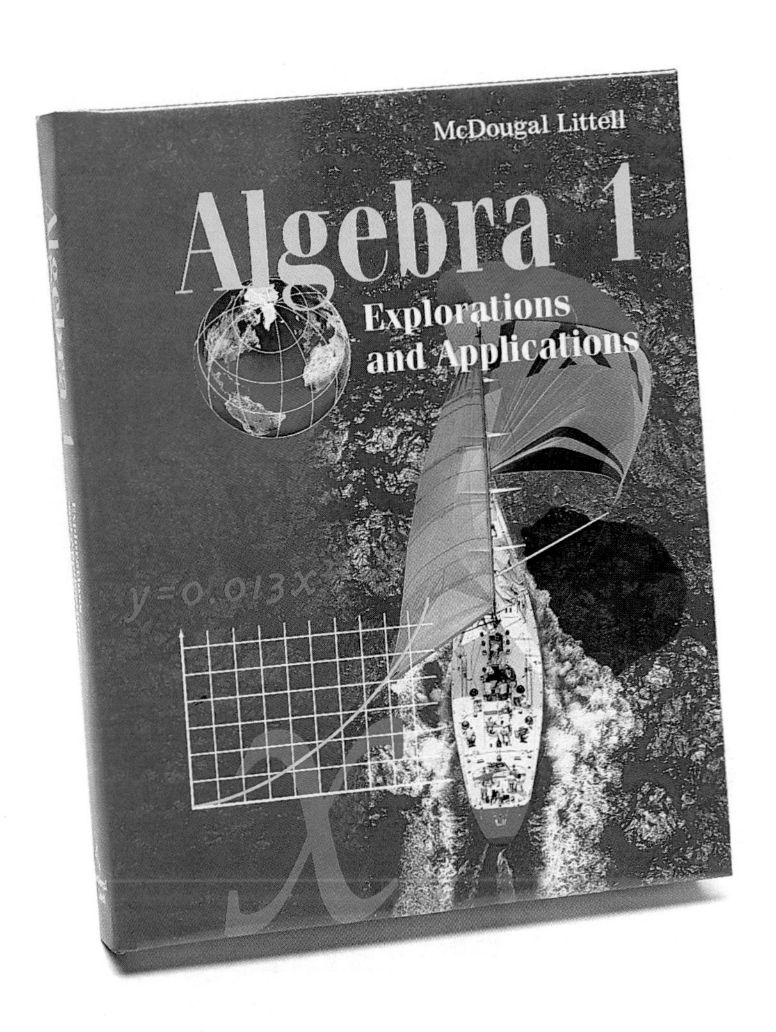

This new program helps you prepare today's students for tomorrow's world.

Involve *students in exploring and discovering math concepts.*

SECTION

4.1

Solving Problems Using Tables and Graphs

Learn how to...

- model situations with tables and graphs

So you can...

- compare options, such as buying CDs from a store or from a CD club

Is it less expensive to buy compact discs (CDs) at a discount store or through a CD club? You can compare options using tables or graphs.

EXPLORATION
COOPERATIVE LEARNING

Comparing Tables and Graphs

Work in 2 teams of 2 students each.
You will need:

- a graphing calculator or graph paper

Discounts!
CDs
$11⁰⁰
Why Pay More?!
Join our
CD Club
and get
8 FREE CDs
(You must buy at least 6 more CDs for $15 each.)

One team should do Steps 1–4 below. The other team should do Steps 5–8 on the next page. The teams should work together on Step 9.

1 Copy and complete the table.

Number of CDs	Cost at store ($)	CD club cost ($)
14	?	?
15	?	?
16	?	?

The CD club minimum is 8 + 6, or 14.

2 For the values in the table, which way of buying CDs is less expensive? Will this always be the case? Explain.

3 Expand your table. Include rows for 20, 30, 40, and 50 CDs.

4 When will the total cost through the CD club equal the total cost at the discount store? How can you tell? When would you choose the CD club? the discount store? Why?

Exploration continued on next page.

4.1 Solving Problems Using Tables and Graphs **147**

Exploration *continued*

5 Write an equation for the cost y of buying x CDs from the discount store.

6 Repeat Step 5 for the CD club.

7 Graph the two equations in the same coordinate plane. Set a good viewing window, like the one shown below.

WINDOW FORMAT
Xmin=0
Xmax=50
Xscl=10
Ymin=0
Ymax=750
Yscl=50

8 When will the total cost through the CD club equal the total cost at the discount store? How can you tell? When would you choose the CD club? the discount store? Why?

9 Discuss your answers to Steps 4 and 8 with the other team. Compare the table method to the graphing method. List some advantages and disadvantages of each method.

BY THE WAY

There are over 7 billion transactions per year at automatic teller machines (ATMs) in the United States.

EXAMPLE Application: Consumer Economics

At Service Bank customers pay a monthly fee of $6.25 plus $.30 per transaction. At Slate Bank customers pay $3.00 per month plus $.40 per transaction. (Checks written, deposits, and visits to automatic teller machines count as transactions.) When will the total monthly cost at Slate Bank be greater than at Service Bank?

SOLUTION

Problem Solving Strategy: Make a table.

You can use a spreadsheet or a graphing calculator, or make a table by hand. Include a column for the number of transactions and a column for the cost at each bank. Use this general formula.

Cost = Monthly fee + Transaction fee × Number of transactions

148 Chapter 4 Solving Equations and Inequalities

Cooperative Learning *A Team Approach*

Mathematics is more meaningful for students when they can explore ideas and discover solutions. Working in cooperative groups makes it easier to investigate alternative methods.

Modeling with Manipulatives *Building Understanding*

Exploring with manipulatives such as algebra tiles helps students develop conceptual understanding and the confidence to move to abstract work.

SECTION

4.3 Solving Multi-Step Equations

Learn how to...
- solve multi-step equations

So you can...
- solve real-world problems, such as finding when one runner catches up to another

Some equations that represent real-world situations have variables on both sides of the equals sign. Algebra tiles can help you understand how to solve this type of equation.

EXPLORATION
COOPERATIVE LEARNING

Getting Variables onto One Side

Work with a partner.
You will need:
- algebra tiles

SET UP Use tiles to represent the equation $3x + 1 = x + 5$.

1 Remove an *x*-tile from each side of your tile equation. What mathematical operation does this model?

2 Finish solving the equation using the algebra tile model. What is the value of one *x*-tile? Check this result in the original equation.

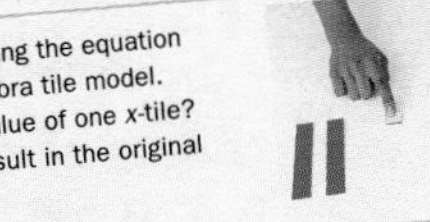

3 Take turns using the algebra tile method introduced in the previous steps to solve each of these equations.

a. $4x + 2 = x + 5$ **b.** $x + 9 = 5x + 1$
c. $2x + 8 = 4x + 2$ **d.** $5x + 3 = 2x + 6$

4 Work together to make a generalization about the first step you should take to solve an equation with variables on both sides of it.

158 Chapter 4 *Solving Equations and Inequalities*

Exploring with Technology *Visualizing Patterns*

With technology, students can explore a wider range of topics than in the past. Graphing calculator activities help students understand the relationship between equations and graphs.

SECTION

8.2 Exploring Parabolas

Learn how to...
- recognize characteristics of parabolas

So you can...
- predict the shape of a parabola
- recognize side views of real objects, such as a radio telescope

Some nonlinear graphs are U-shaped curves called **parabolas**. The graph of the equation $y = x^2$ shown below is one example of a parabola. Every other parabola is a variation of this basic curve.

EXPLORATION
COOPERATIVE LEARNING

Analyzing the Shape of a Parabola

Work with a partner.
You will need:
- a graphing calculator
- graph paper

1 One person should graph the equations below. Set a good viewing window, such as $-9 \le x \le 9$ and $-2 \le y \le 10$. What do you notice about the shapes of the parabolas?

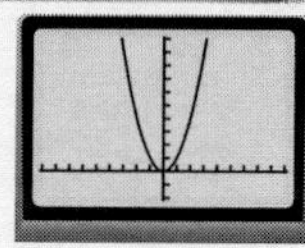

a. $y = x^2$ **b.** $y = 2x^2$ **c.** $y = 3x^2$

2 On graph paper, the other person should sketch what he or she thinks the graph of $y = 4x^2$ will look like in relation to the graphs in Step 1. Use a graphing calculator to check the prediction.

3 Switch roles and repeat Step 1 using the equations below.

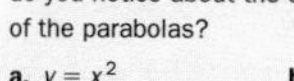

a. $y = x^2$ **b.** $y = \frac{1}{2}x^2$ **c.** $y = \frac{1}{3}x^2$

4 On graph paper, sketch what you think the graph of $y = \frac{1}{4}x^2$ will look like in relation to the graphs in Step 3. Use a graphing calculator to check your prediction.

5 Work together to generalize what happens to the graph of the equation $y = ax^2$ as the value of *a* increases from 0 to 1 and beyond.

6 Work together to generalize what happens to the graph of the equation $y = ax^2$ as the value of *a* decreases from 0 to −1 and beyond. Try some examples. Be sure to set a good viewing window.

8.2 Exploring Parabolas 331

Connect *algebra to the real world and to other subjects and math topics.*

Real-World Applications *Mathematics in Context*

Throughout this course new concepts are introduced, practiced, and extended in the context of real-world applications. Seeing the wide variety of settings in which algebra is useful helps students value mathematics.

SECTION

3.5 Writing an Equation of a Line

Learn how to...

- find the y-intercept of a graph

So you can...

- model real-world situations, such as the altitude of a hot air balloon

Every year at the International Balloon Fiesta in Albuquerque, New Mexico, thousands of hot air balloons take to the sky. The graph below shows the altitude of one balloon as it climbs.

This break in the scale means that the y-axis is not shown for values of y between 0 and 6000.

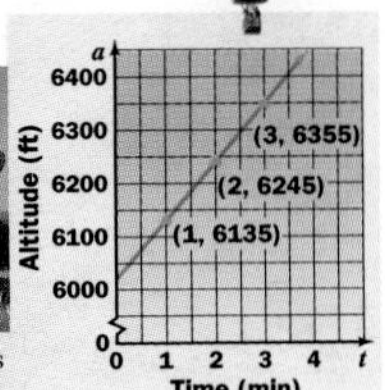

EXAMPLE 1 Application: Recreation

As shown by the slope of the graph, the balloon rises at a rate of 110 ft/min. Write an equation for the balloon's altitude a as a function of time t.

SOLUTION

Step 1 Write the equation in slope-intercept form.

Altitude = Slope × Time + Vertical intercept

$$a = mt + b$$

Step 2 Substitute the slope of the line for m in the equation.

$$a = 110t + b$$

Step 3 To find the vertical intercept, substitute the coordinates of any point on the graph for t and a in the equation. Then solve for b.

$6135 = 110(1) + b$ — Choose the point (1, 6135).

$6135 = 110 + b$

$6025 = b$ — Solve for b.

Step 4 Substitute 6025 for b in the equation $a = 110t + b$.

An equation for the altitude of the balloon as a function of time is $a = 110t + 6025$.

3.5 Writing an Equation of a Line **125**

Connection ECONOMICS

The United States government requires that workers receive a minimum hourly wage. The circle graph shows that most teenagers earn more than the minimum wage.

12. What percent of working teenagers earn more than the minimum wage?

1993 Wages of Teenagers

Above minimum wage 3,713,000

At minimum wage 928,000

Below minimum wage 380,000

Number of teenagers working for hourly wages = 5,021,000

Source: U.S. Bureau of Labor Statistics

13. a. Write a formula for converting the numbers on the circle graph to percents.
    b. Does your formula from part (a) represent a function? Explain.
14. a. In 1994, the minimum wage was \$4.25 per hour. Write an equation for earnings as a function of the number of hours worked at this wage.
    b. Describe in words the domain and the range of the function from part (a).
15. Ana Martinez earns \$500 for working a 40 h week. She also earns time and a half for working overtime. That means her overtime hourly wage is 1.5 times her regular hourly wage.
    a. What is Ana's regular hourly wage?
    b. What is her overtime hourly wage?
    c. How many hours of overtime must she work to earn a total of \$650 in one week?

GEOMETRY For Exercises 16–18, write an equation for the perimeter P of each figure as a function of the side length s. Describe the domain and range of each function.

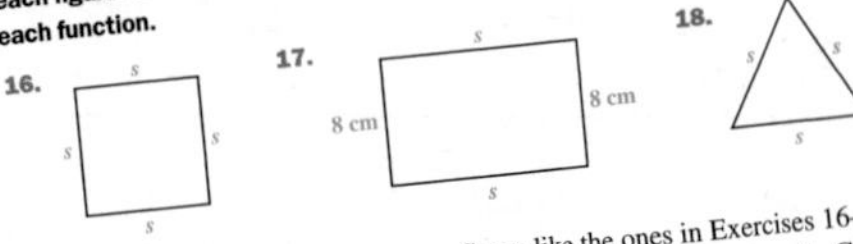

19. **Open-ended Problem** Draw a figure like the ones in Exercises 16–18. Express the perimeter of the figure as a function of the side length. Give the domain and the range of your function.
20. **Challenge** Write an equation for the perimeter P of the triangle at the right as a function of the side length s. The Triangle Inequality says that the sum of the lengths of any two sides of a triangle is greater than the length of the third side. What does this tell you about the domain and the range of the perimeter function for this triangle?

12 12 s

70 Chapter 2 *Equations and Functions*

Interdisciplinary Problems *Connecting Learning*

Each chapter has several clusters of exercises that focus on the connections of algebra to careers and to other subject areas, both within and outside of mathematics.

INTERVIEW Alex MacLean

Look back at the article on pages 460–462.

When Alex MacLean takes aerial photographs, he uses a formula to determine how high he must fly to achieve a desired image size.

$$\text{Altitude} = \frac{k}{\text{image size}}$$ ← The letter k represents a constant.

10. Investigation You will need a ruler. Alex MacLean took picture A from 1500 ft above the ground. At what heights were pictures B and C taken? How do you know?

11. Alex McLean was flying 1500 ft above the water when he took the picture at the right. At what altitude should he fly so the images of the boats are twice as long? half as long?

12. Writing You can use inverse variation to estimate the speed of the waves in the ocean. First estimate the horizontal distance l between the crests of the waves. Then count the time t between wave crests at a fixed point such as a buoy or pier pile. To find the wave speed s, use the formula $s = \frac{l}{t}$.

a. Suppose you estimate the distance between wave crests to be about 7 m and the time between crests to be about 2 s. What is the speed of the wave?

b. Which two variables in the formula are inversely related?

c. If the distance between wave crests is constant, how does the speed of the waves affect the time between crests?

13. Open-ended Problem For each function graphed below, write an equation that could represent the function. Explain your thinking.

a. 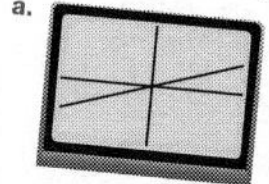
b. 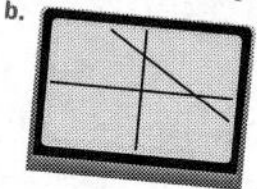
c. 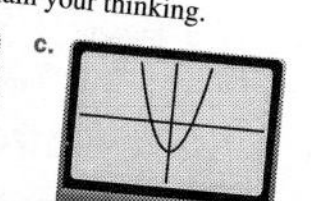
d. 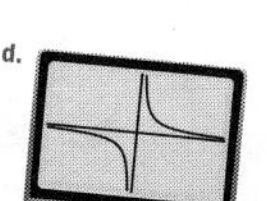

14. To loosen a bolt with a wrench with a 6 in. long handle, David Ruiz must exert a force of 250 lb. (That is like lifting a 250 lb person!) The force needed and the length of the handle vary inversely. If David slips a pipe over the wrench handle to make it 24 in. long, about how much force must he exert?

11.1 Exploring

Careers *Preparing for the Future*

The *Interviews* at the start of each chapter, and the related exercises throughout the chapter, give students a sense of the types of mathematical skills they will need in their future careers.

SECTION

5.6 Geometric Probability

Learn how to...
- find probabilities based on area

So you can...
- predict events such as where meteorites are likely to hit when they fall to Earth

Did you know that over 2000 meteorites weighing 1 kg land on Earth every year? Scientists say most will land in water. But how do they know?

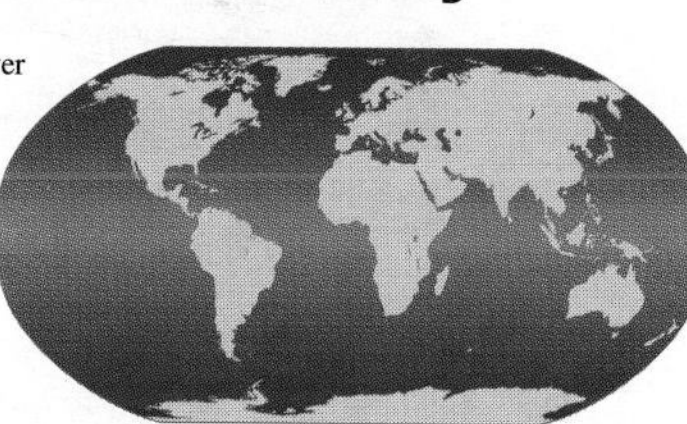

When a meteorite falls to Earth, it can hit either land or water. Because $\frac{7}{10}$ of Earth's surface is covered by water, the probability that an object falling to Earth will land in water is 70%. The chance that a meteorite will land in water is an example of **geometric probability**, a probability based on area.

EXAMPLE 1

Suppose a randomly thrown dart hits the target shown. (A random throw means that you do not aim for any particular part of the target, and each part has an equal chance of being hit.) What is the theoretical probability of the dart hitting the shaded area?

SOLUTION

The square is divided into 8 triangles having equal area, and 2 of them are shaded.

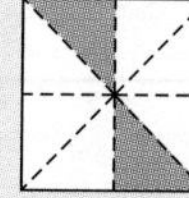

$$\text{Probability of hitting a shaded area} = \frac{\text{Shaded area}}{\text{Total area}} = \frac{\text{Number of shaded triangles}}{\text{Total number of triangles}} = \frac{2}{8} = \frac{1}{4}$$

The probability of a randomly thrown dart hitting the shaded area is 25%.

224 Chapter 5 *Connecting Algebra and Geometry*

Mathematical Connections *Integrated Problem Solving*

In this course students see that integrating algebra with concepts from geometry, statistics, probability, and discrete mathematics enables them to solve a greater variety of problems.

Build *understanding of the concepts that provide a strong foundation for future courses and careers.*

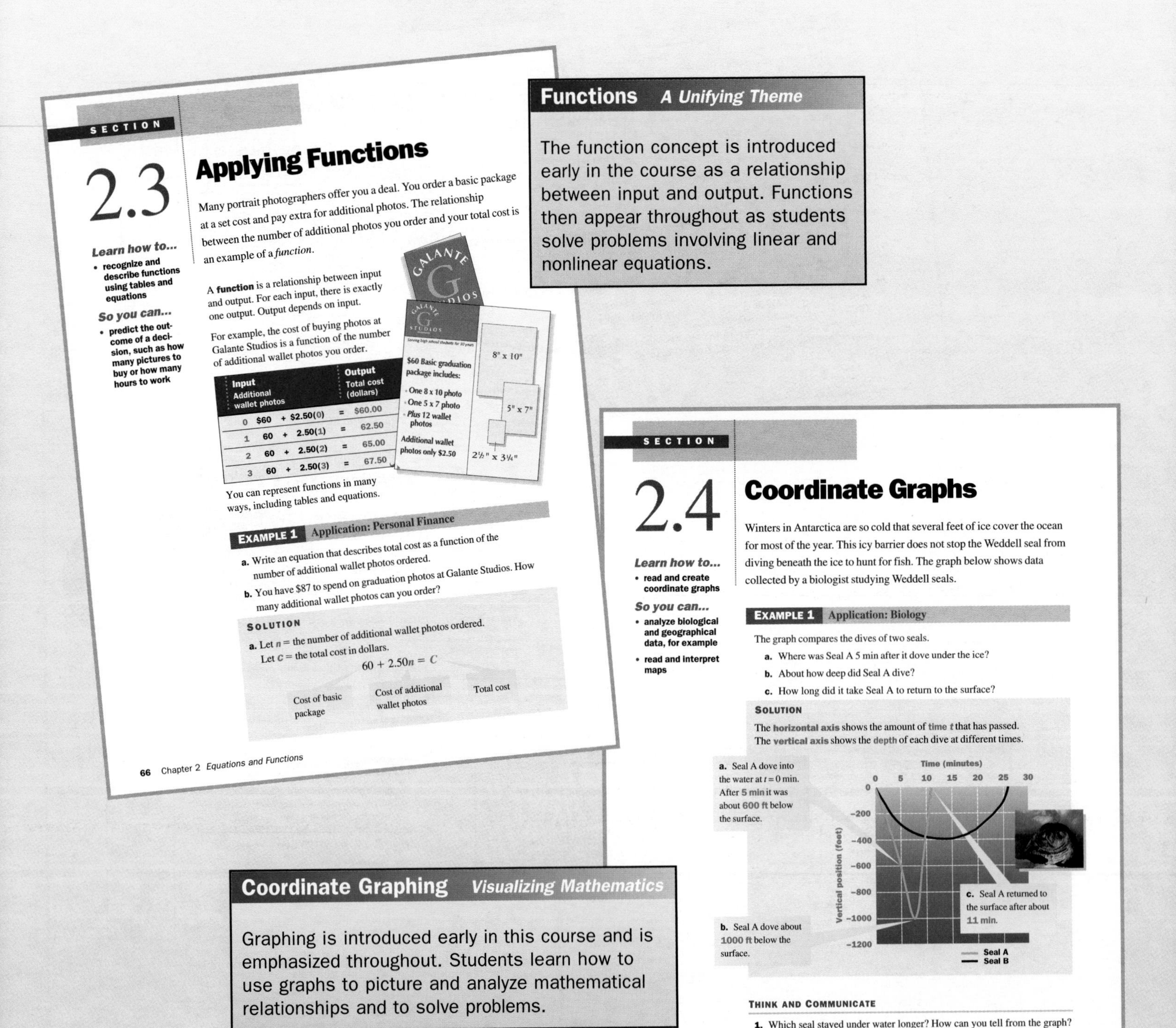

SECTION

2.3

Applying Functions

Learn how to...
- recognize and describe functions using tables and equations

So you can...
- predict the outcome of a decision, such as how many pictures to buy or how many hours to work

Many portrait photographers offer you a deal. You order a basic package at a set cost and pay extra for additional photos. The relationship between the number of additional photos you order and your total cost is an example of a *function*.

A **function** is a relationship between input and output. For each input, there is exactly one output. Output depends on input.

For example, the cost of buying photos at Galante Studios is a function of the number of additional wallet photos you order.

Input Additional wallet photos			Output Total cost (dollars)
0	\$60 + \$2.50(0)	=	\$60.00
1	60 + 2.50(1)	=	62.50
2	60 + 2.50(2)	=	65.00
3	60 + 2.50(3)	=	67.50

GALANTE
G
STUDIOS
\$60 Basic graduation package includes:
One 8 x 10 photo
One 5 x 7 photo
Plus 12 wallet photos
Additional wallet photos only \$2.50
8" x 10"
5" x 7"
2½" x 3¼"

You can represent functions in many ways, including tables and equations.

EXAMPLE 1 Application: Personal Finance

a. Write an equation that describes total cost as a function of the number of additional wallet photos ordered.

b. You have \$87 to spend on graduation photos at Galante Studios. How many additional wallet photos can you order?

SOLUTION

a. Let n = the number of additional wallet photos ordered.
Let C = the total cost in dollars.

$$60 + 2.50n = C$$

Cost of basic package — Cost of additional wallet photos — Total cost

66 Chapter 2 *Equations and Functions*

Functions *A Unifying Theme*

The function concept is introduced early in the course as a relationship between input and output. Functions then appear throughout as students solve problems involving linear and nonlinear equations.

SECTION

2.4

Coordinate Graphs

Learn how to...
- read and create coordinate graphs

So you can...
- analyze biological and geographical data, for example
- read and interpret maps

Winters in Antarctica are so cold that several feet of ice cover the ocean for most of the year. This icy barrier does not stop the Weddell seal from diving beneath the ice to hunt for fish. The graph below shows data collected by a biologist studying Weddell seals.

EXAMPLE 1 Application: Biology

The graph compares the dives of two seals.

a. Where was Seal A 5 min after it dove under the ice?

b. About how deep did Seal A dive?

c. How long did it take Seal A to return to the surface?

SOLUTION

The **horizontal axis** shows the amount of time t that has passed.
The **vertical axis** shows the depth of each dive at different times.

a. Seal A dove into the water at $t = 0$ min. After 5 min it was about 600 ft below the surface.
b. Seal A dove about 1000 ft below the surface.
c. Seal A returned to the surface after about 11 min.
Time (minutes)
0 5 10 15 20 25 30
Vertical position (feet)
0 −200 −400 −600 −800 −1000 −1200
Seal A
Seal B

THINK AND COMMUNICATE

1. Which seal stayed under water longer? How can you tell from the graph?
2. How deep was Seal B's dive? About how many minutes did it take for Seal B to reach the deepest point of its dive?

72 Chapter 2 *Equations and Functions*

Coordinate Graphing *Visualizing Mathematics*

Graphing is introduced early in this course and is emphasized throughout. Students learn how to use graphs to picture and analyze mathematical relationships and to solve problems.

SECTION

3.6 Modeling Linear Data

Learn how to...
- fit a line to data

So you can...
- make predictions about natural occurrences and the life expectancy of future Americans

Old Faithful is a geyser in Yellowstone National Park in Wyoming. For more than 100 years, Old Faithful has erupted every day at intervals of less than two hours. The data in the table suggest that the time between eruptions may be related to the length of the previous eruption.

The table is an example of a *mathematical model.* A mathematical model can be a table, a graph, an equation, a function, or an inequality that describes a real-world situation. Using mathematical models is called modeling.

Length of eruption (min)	Time until next eruption (min)
2.0	57
2.5	62
3.0	68
3.5	75
4.0	83
4.5	89
5.0	92

EXAMPLE Application: Geology

Find an equation that models the data about Old Faithful.

SOLUTION

Make a scatter plot of the data. Draw a line on the scatter plot that is as close as possible to all of the points on the plot. This is called a **line of fit**.

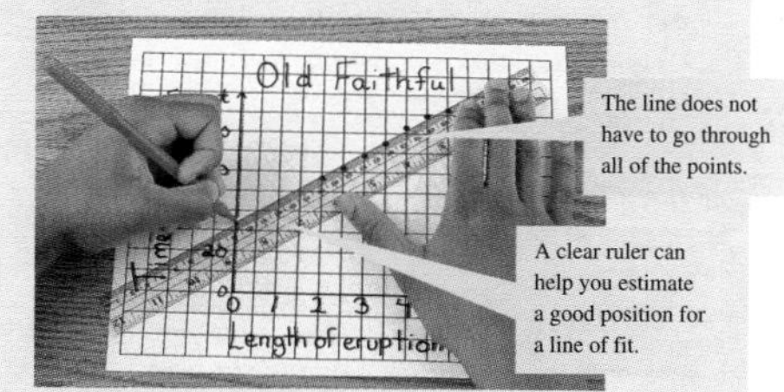

The line does not have to go through all of the points.

A clear ruler can help you estimate a good position for a line of fit.

BY THE WAY

A geyser is a natural hot spring that periodically ejects a column of water and steam into the air. Most geysers are less predictable than Old Faithful.

130 Chapter 3 *Graphing Linear Equations*

Mathematical Modeling *A Tool for Predicting*

Students are introduced to this basic technique for solving real-world problems. They use ideas from statistics, algebra, geometry, and probability as they model a variety of problem situations.

SECTION

5.2 Scale Measurements

Learn how to...
- use scales in drawings and on graphs
- use scale factors to compare sizes of objects and drawings

So you can...
- model large objects with scale drawings
- find large dimensions, such as the size of a whale
- compare maps of different sizes

The blue whale is the largest animal on Earth. To appreciate its size, compare the pictures below. The drawing of the whale has the same *scale* as the photograph of the elephant.

The drawing of the whale and the photograph of the elephant both have a **scale** of 1 in. = 24 ft. This means that each inch in the pictures represents 24 feet in real life.

EXAMPLE 1 Application: Zoology

The drawing of the whale is about 4 in. long. Find the whale's real length.

SOLUTION

Let w = the whale's real length in feet.

$$\frac{\text{Whale's length in drawing in inches}}{\text{Whale's real length in feet}} = \frac{1 \text{ in.}}{24 \text{ ft}}$$

$$\frac{4}{w} = \frac{1}{24}$$

$$4 \cdot 24 = w \cdot 1$$

$$96 = w$$

Use the means-extremes property.

The whale is about 96 ft long.

THINK AND COMMUNICATE

1. The photograph of the elephant is about $\frac{3}{8}$ in. high. Explain how to estimate the real height of the elephant.
2. Suppose a drawing of a whale has a scale of 1 cm = 2 m. To find the whale's real length, what units must you use to measure the drawing? In what units will the whale's length be? How do you know?

198 Chapter 5 *Connecting Algebra and Geometry*

Ratio and Proportion *A Multitude of Applications*

Algebra and geometry are united in these important topics that underlie mathematical applications in daily life and in many different careers. Among the applications that students investigate are estimating wildlife populations and using scale drawings.

Assess students' progress in ways that support learning.

Embedded Assessment *A Part of Learning*

Ongoing Assessment questions in each section ask students to apply concepts and explain their thinking. *Journal* writing gives students a chance to reflect on their own learning process.

ONGOING ASSESSMENT

22. a. Open-ended Problem Use the information in the advertisement to write a system of equations. Explain what each equation represents.

b. Solve your system of equations. Explain what the solution represents.

ARTS

Art's Records
Fabulous 50th
Anniversary Sale!
On May 5th, the 50th customer will receive $50 to spend on tapes and CDs!
Used CDs only $8! Used tapes only $3!

SPIRAL REVIEW

23. GEOMETRY The Cosmoclock 21 in Yokohama City, Japan, is one of the largest Ferris wheels in the world. Its diameter is 328 ft. The 60 arms holding the gondolas serve as second hands for the clock.

a. How many feet would you travel in one revolution of the Ferris wheel? *(Toolbox, page 593)*

b. Find your speed in miles per hour if the wheel makes one revolution per minute. Use the fact that 1 mi = 5280 ft. *(Section 3.1)*

BY THE WAY

The Cosmoclock 21 is $344\frac{1}{2}$ ft high. Each of the 60 gondolas holds eight passengers.

ASSESS YOUR PROGRESS

VOCABULARY

standard form of a linear equation (p. 282)
vertical intercept, y-intercept (p. 282)
horizontal intercept, x-intercept (p. 282)
system of linear equations (p. 288)
solution of a system (p. 288)

Give the x- and y-intercepts of each graph. Graph each equation. *(Section 7.1)*

1. $4x + 3y = -12$ **2.** $-3x + 8y = 12$

Rewrite each equation in slope-intercept form. Then graph the equation. *(Section 7.1)*

3. $5x - y = 5$ **4.** $2x + 3y = 9$

5. Shira gets four points for each correct answer on her biology quiz. She loses a point for each wrong answer. *(Section 7.2)*

a. Write an equation showing that Shira earned 30 points on the quiz.

b. Write an equation showing that Shira answered 20 questions.

c. Solve the system of equations in parts (a) and (b) by graphing.

Use substitution to solve each system of equations. *(Section 7.2)*

6. $6x + 5y = 1$, $y = 1 - 2x$ **7.** $2a + 7b = 24$, $3a - b = 13$

8. Journal You can solve a system of equations by graphing or by using substitution. Describe some advantages and disadvantages of each method.

7.2 Solving Systems of Equations **293**

CHAPTER 4 **Assessment**

VOCABULARY

SECTIONS 4.4, 4.5, *and* 4.6

21. What power of ten would you use to eliminate the decimal numbers in the equation $0.02n + 3.5 = 8.067$ most easily?

22. What is the smallest number that you can multiply by to eliminate the fractions in the equation $\frac{4}{5}x + \frac{2}{9} = \frac{1}{6} - \frac{4}{15}x$?

Solve each equation.

23. $\frac{2}{5}x + 2 = \frac{2}{3} - 2x$ **24.** $\frac{4}{5}m + \frac{3}{4} = \frac{7}{10}m + \frac{1}{4}$

25. $0.2n + 3 = 5.4 - n$ **26.** $0.7 - 0.8x = 0.18 + 0.24x$

The graph shows the fuel economy of an average car as a function of speed. Use the graph for Exercises 27 and 28.

Speed and Fuel Economy
Fuel economy (mi/gal)
Speed (mi/h)

27. At what speeds is a car's fuel economy less than 25 mi/gal? Write your answer as two inequalities.

28. At what speeds is a car's fuel economy more than 30 mi/gal? Write your answer as an inequality.

29. The statement $10 > 2$ is a true inequality. Is the inequality still true if both sides of the inequality are multiplied by 7? by -3? by $\frac{1}{2}$? by $-\frac{5}{2}$? Explain.

30. If $m < r$, which of the following inequalities is not true?

A. $m + 7 < r + 7$ **B.** $-3m < -3r$ **C.** $m - 3 < r - 3$ **D.** $\frac{m}{5} < \frac{r}{5}$

Solve each inequality. Graph each solution on a number line.

31. $c - 3 > 4$ **32.** $2x \leq -8$ **33.** $-3f \leq -27$

34. $7 - 4x \leq -5$ **35.** $-\frac{2}{5}x + 2 \geq 8$ **36.** $5x + 3 < 2x - 6$

PERFORMANCE TASK

37. Find something that can be paid for in two different ways, such as admission to a museum. Compare the costs. Explain the circumstances under which each option is better. Justify your explanation with a table, a graph, or an equation. Use inequalities to express your findings.

PLIMOTH PLANTATION RATES

INDIVIDUALS:	
Adult	$15.00
Child 5–12	$9.00
ANNUAL PASS:	
Valid for one full year from date of purchase for an unlimited number of visits	
Adult	$25.00
Child 5–12	$12.00
1 YEAR MEMBERSHIP:	
Family	$70.00
Individual	$35.00

Assessment **187**

Monitoring Progress *Check for Understanding*

A variety of different types of assessment questions are included throughout the course. A *Performance Task* is included in each chapter assessment.

Portfolio Assessment *Demonstrating Growth*

The *Portfolio Projects* provide opportunities for original student work that can be part of a mathematical portfolio. In each project, students are asked to present their results and assess their work.

PORTFOLIO PROJECT

Using a Cartogram

Most maps of the United States show the size, shape, and location of each state. Some other maps are drawn so that each state's size is distorted to reflect a particular group of data, such as population. This type of map is called a cartogram.

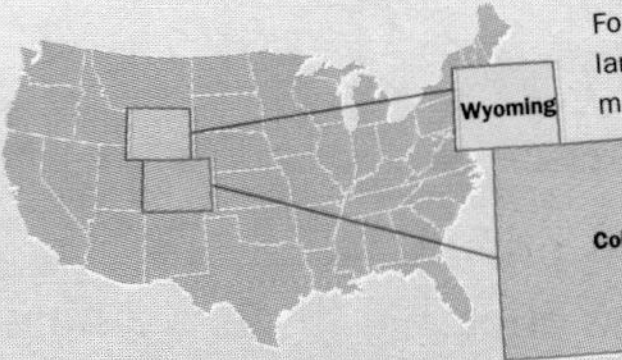

For example, although Colorado is only a little larger than Wyoming, in 1990 it had approximately seven times as many people. The cartogram shows how these states look when their sizes are distorted to represent population instead of area.

PROJECT GOAL **Make a cartogram of a selected group of states. Then make a poster of your cartogram.**

Gathering Data

Work in a group of 2–3 students. First choose a set of statistical data to illustrate with your cartogram. Some possible topics are:

- number of physicians
- amount of land suitable for farming
- amount of a natural resource produced each year, such as oil or coal

Select three to five neighboring states to show on your cartogram, and find the relevant data for each. Organize the information in a table. For example, this table shows the population densities of Nebraska, Iowa, and Illinois.

State	Population density, 1993
Nebraska	20.9
Iowa	50.4
Illinois	210.4

Constructing Your Cartogram

Use a computer drawing program or graph paper to form the shapes of the states in your cartogram with squares.

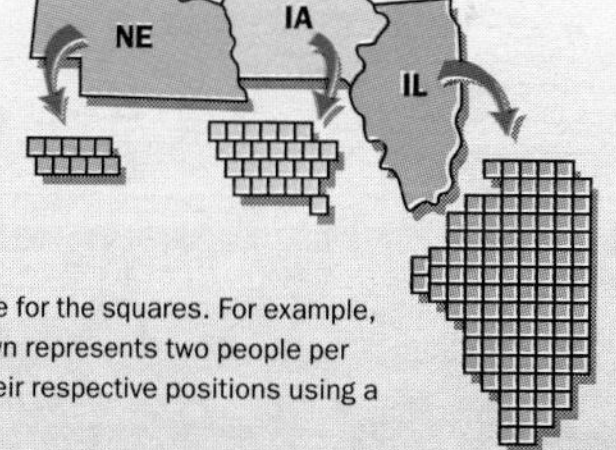

Choose a convenient size and scale for the squares. For example, each square in the cartogram shown represents two people per square mile. Place the states in their respective positions using a geographical map as a guide.

Making a Poster

Display your cartogram on a poster. Label each state and describe the scale you used. Include a labeled geographical map of the states and a table of the data.

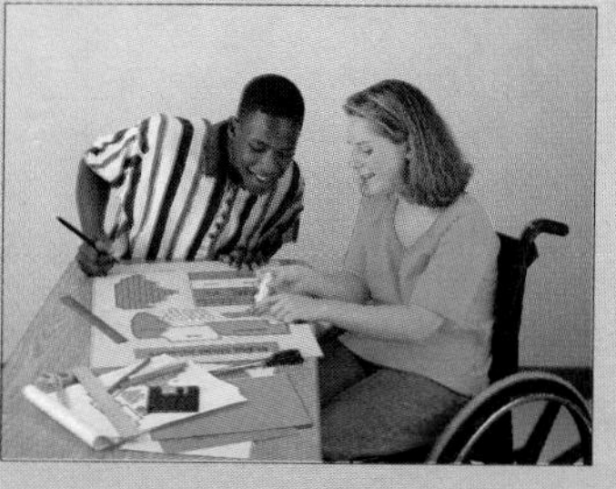

You and your classmates can extend your study of cartograms by exploring some of the ideas below:

- Create a class cartogram of the entire United States.
- Make three or four cartograms that illustrate different data for the same group of states.
- Start a scrapbook of cartograms that you find in newspapers, magazines, or reference books.

Self-Assessment

Write a paragraph explaining your project. Describe how you chose the scale and how you created each state. Why is it important to use the same sized squares for all the states? How does it help to have a geographical map next to your cartogram? Describe any difficulties you had. What would you do differently if you were to do the project over again?

Planning Pages for Every Chapter

CHAPTER

2

Equations and Functions

OVERVIEW

Connecting to Prior and Future Learning

⇔ Students begin Chapter 2 by studying and solving one- and two-step equations. This introduction to equations will help students build a firm foundation for their continued work.

⇔ Functions are introduced in this chapter. In Section 2.3, the previously introduced concepts of equations and tables are used to represent functions. The concept of a function will be used throughout this course and future mathematics courses.

⇔ Students begin their study of graphing functions and using graphs to solve problems in Chapter 2. These basic concepts of graphing and problem solving are used throughout this course and are important in all levels of mathematics.

Chapter Highlights

Interview with Bob Hall: The interview and related exercises and examples on pages 57, 60, and 82 emphasize the variety of contexts in which math occurs. Bob Hall, the first person to compete in the Boston Marathon in a wheelchair, uses both algebra and geometry in his work.

Explorations in Chapter 2 provide students with the opportunity to solve equations using algebra tiles in Section 2.2 and to use the entire class as a coordinate grid to model graphing equations in Section 2.5.

The Portfolio Project: Students collect and graph data in which a pattern of change is obvious. As students measure and record the data, make a graph, and present their ideas, they will learn to work together to utilize many of the mathematical skills they have learned so far.

Technology: Scientific calculators are used in Section 2.2 to solve equations. The use of graphing calculators to add matrices is discussed in Section 2.4. Graphing calculators are also used in Section 2.6 to graph linear equations and to find solutions. Graphin calculators are used to graph linear and a variety of other equations throughout this course Section 2.5, spreadsheet formulas are used to represent and evaluate functions.

OBJECTIVES

Section	Objectives	NCTM Standar
2.1	• Write and solve one-step equations. • Solve real-world problems including those involving distance, rate, and time.	1, 2, 3, 4, 5
2.2	• Write and solve two-step equations. • Solve real-world problems using equations. • Solve geometry problems using equations.	1, 2, 3, 4, 5, 8
2.3	• Recognize and describe functions using tables and equations. • Predict the outcome of a decision.	1, 2, 3, 4, 5, 6
2.4	• Read and create coordinate graphs. • Analyze biological and geographical data. • Read and interpret maps.	1, 2, 3, 4, 5, 8, 12
2.5	• Graph equations. • Visually represent functions.	1, 2, 3, 4, 5, 6, 8, 12
2.6	• Use a graph to solve a problem. • Analyze the costs and profits of running a business.	1, 2, 3, 4, 5, 8

52A

OVERVIEW

The **Overview** provides a summary of connections to prior and future learning and highlights the chapter interview, explorations, portfolio project, and use of technology.

OBJECTIVES

Chapter **Objectives** gives objectives and NCTM Standards for each section.

INTEGRATION

Mathematical Connections	2.1	2.2	2.3	2.4	2.5	2.6
algebra	**55–60***	**61–65**	**66–71**	**72–77**	**78–83**	**84–89**
geometry	60	65	70	**72–77**	**78–83**	**84–89**
data analysis, probability, discrete math			70	72, 74–77	82	**84–89**
patterns and functions			**66–71**	77	**78–83**	89
logic and reasoning	56, 60	62–65	67–69, 71	72–76	79–83	86–89
Interdisciplinary Connections and Applications						
history and geography	59			75, 76		
reading and language arts	58		69			
biology and earth science	55			72		87, 88
business and economics			70		80	84, 85, 86
sports and recreation	58					87
personal finance		63	66, 69	74, 75	81	
medicine, cars, aviation, energy use	59	64	71		81	

* ***Bold page numbers*** *indicate that a topic is used throughout the section.*

TECHNOLOGY

Section	*opportunities for use with* Student Book	Support Material
2.1	scientific calculator	
2.2	scientific calculator	**Technology Book:** Spreadsheet Activity 2
2.3	scientific calculator	
2.4	graphing calculator	**Technology Book:** Calculator Activity 2
2.5	graphing calculator spreadsheet software McDougal Littell Software *Stats!* *Function Investigator*	
2.6	graphing calculator	

52B

INTEGRATION

The **Integration** chart highlights the mathematical and interdisciplinary connections and applications found throughout the chapter.

TECHNOLOGY

The **Technology** chart highlights opportunities to use technology in both the student book and support materials.

Planning Guide

Regular Scheduling (45 min)

Section	Materials Needed	Core Assignment	Extended Assignment	*exercises that feature* Applications	Communication	Technology
2.1		1–31, 33–41	1–41	20–22, 25–29, 31	23, 30, 32, 33	
2.2	algebra tiles, scientific calculator	**Day 1:** 1–21 **Day 2:** 22–33, 35–48	**Day 1:** 1–21 **Day 2:** 22–48	19–21	17, 18 34, 39	17
2.3		1–3, 5–18, 22–35, AYP*	1–35, AYP	4, 10–15	4, 19–21, 23	
2.4	graph paper, graphing calculator	1–15, 17–22, 25–28, 31–39	1–4, 7, 10, 11, 13–39	17–22, 25–28	23, 24, 36	39
2.5	graph paper, spreadsheet software	1–8, 10–24, 27–33	1–33	10–13	9, 14, 25–27	19–23
2.6	graph paper, graphing calculator	1–8, 11–14, 16–29, AYP	1–8, 11–29, AYP	1–7, 10–15	8, 9, 15, 22	16–20
Review/ Assess		**Day 1:** 1–13 **Day 2:** 14–28 **Day 3:** Ch. 2 Test	**Day 1:** 1–13 **Day 2:** 14–28 **Day 3:** Ch. 2 Test	12, 13 27	1, 2 26–28	
Portfolio Project		Allow 2 days.	Allow 2 days.			

Yearly Pacing (with Portfolio Project)	Chapter 2 Total	Chapters 1–2 Total	Remaining	Total
	12 days	28 days	132 days	160 days

Block Scheduling (90 min)

	Day 9	Day 10	Day 11	Day 12	Day 13	Day 14	Day 15
Teach/Interact	Ch. 1 Test 2.1	2.2: Exploration, page 61	2.3 2.4	2.5: Exploration, page 78 2.6	Review Port. Proj.	Review Port. Proj.	Ch. 2 Test 3.1
Apply/Assess	**Ch. 1 Test** **2.1:** 2–6, 10–41	**2.2:** 1–48	**2.3:** 1–18, 22, 24–35, AYP* **2.4:** 1–23, 25–28, 31–39	**2.5:** 1–33 **2.6:** 1–6, 10–14, 16–29, AYP	**Review:** 1–13 **Port. Proj.**	**Review:** 14–28 **Port. Proj.**	**Ch. 2 Test** **3.1:** 1–11, 13–16, 23–25, 27–36

NOTE: A one-day block has been added for the Portfolio Project—timing and placement to be determined by the teacher.

Yearly Pacing (with Portfolio Project)	Chapter 2 Total	Chapters 1–2 Total	Remaining	Total
	6 days	$14\frac{1}{2}$ days	$66\frac{1}{2}$ days	81 days

* ***AYP*** *is Assess Your Progress.*

52C

PLANNING GUIDE

The **Planning Guide** gives materials, pacing, and suggested assignments for each section, and block scheduling assignments.

Lesson Support

Section	Practice Bank	Study Guide*	Assessment Book*	Visuals	Explorations Lab Manual	Lesson Plans	Technology Book
2.1	10	2.1		Warm-Up 2.1		2.1	
2.2	11	2.2		Warm-Up 2.2 Folder B	Master 4	2.2	Spreadsheet Act. 2
2.3	12	2.3	Test 6	Warm-Up 2.3		2.3	
2.4	13	2.4		Warm-Up 2.4 Folder 2	Master 2	2.4	Calculator Act. 2
2.5	14	2.5		Warm-Up 2.5 Folder A	Masters 1, 2 Add. Expl. 3	2.5	
2.6	15	2.6	Test 7	Warm-Up 2.6 Folder A	Masters 1, 2	2.6	
Review Test	16	Chapter Review	Tests 8, 9 Alternative Assessment			Review Test	Calculator Based Lab 2

***Spanish versions** of *Study Guide* and *Assessment Book* are available.

Chapter Support

- Course Guide
- Lesson Plans
- Portfolio Project Book
- Preparing for College Entrance Tests
- Multi-Language Glossary
- *Test Generator* Software
- Professional Handbook

Software Support

McDougal Littell Software
Stats!
Function Investigator

Internet Support

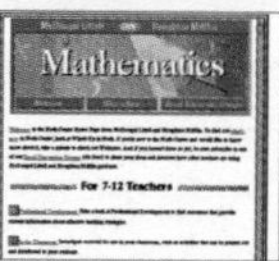

http://www.hmco.com
Next go to McDougal Littell; then the Education Center; then Secondary Math.

Outside Resources

Books, Periodicals

Mercer, Joseph. "Teaching Graphing Concepts with Graphics Calculators." *Mathematics Teacher* (April 1995): pp. 268–273.

Martignette-Boswell, Carol and Albert A. Cuoco. "Say It With Machines." *Mathematics Teacher* (April 1995): pp. 338–341.

Activities, Manipulatives

Algeblocks. Cincinnati, OH: Southwestern Publications.

A Graphing Matter: Activities for Easing into Algebra. Berkeley, CA: Key Curriculum Press.

Software

Harvey, Wayne, Judah Schwartz, and Michal Yerushalmy. *Visualizing Algebra: The Function Analyzer.* Scotts Valley, CA: Sunburst.

IBM Mathematics Exploration Toolkit. Armonk, NY: IBM.

Videos

Visualizing Algebra: A New Vision for Learning and Teaching Algebra. Reston, VA: NCTM, 1993.

Internet

Join the discussions in Kidsphere (excellent mailing list for K–12 education) by sending message to:
kidsphere-request@vms.cis.pitt.edu
In the body of the message, type:
subscribe kidsphere [first name, last name]

52D

LESSON SUPPORT

The **Lesson Support** chart lists all support materials for each section.

OUTSIDE RESOURCES

Outside Resources lists books, periodicals, manipulatives and activities, software, videos, and Internet addresses.

A Teaching Plan for Every Section

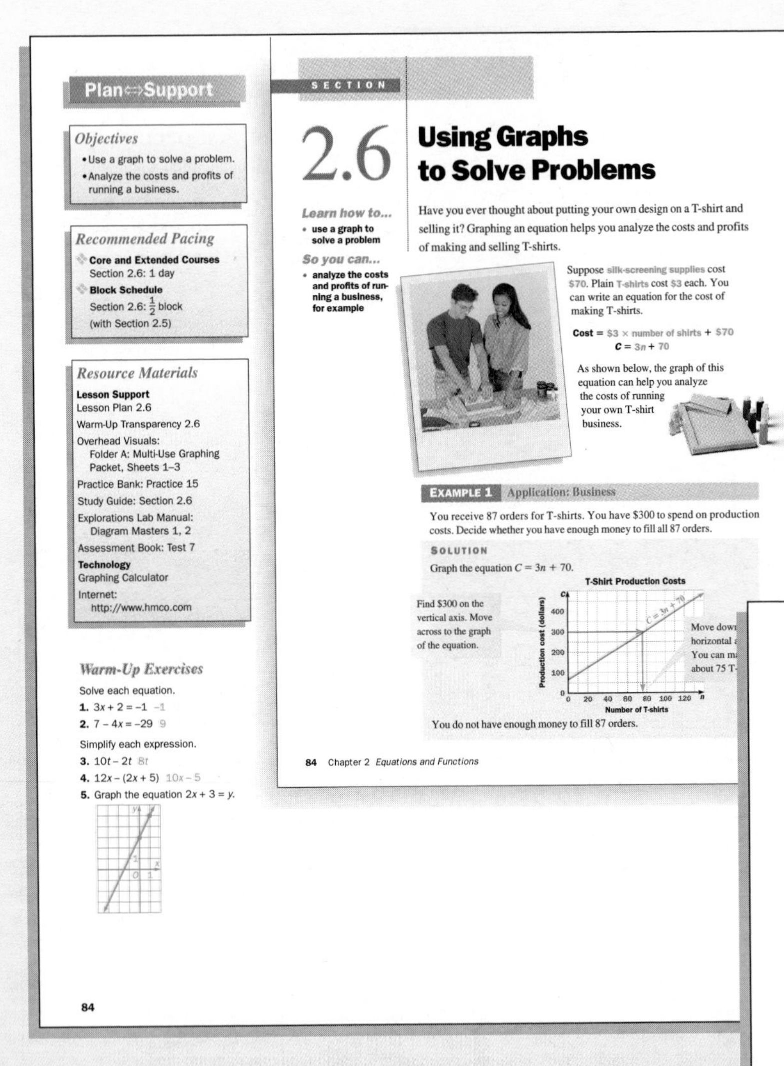

Plan⇔Support

Objectives

- Use a graph to solve a problem.
- Analyze the costs and profits of running a business.

Recommended Pacing

Core and Extended Courses
Section 2.6: 1 day
Block Schedule
Section 2.6: $\frac{1}{2}$ block
(with Section 2.5)

Resource Materials

Lesson Support
Lesson Plan 2.6
Warm-Up Transparency 2.6
Overhead Visuals:
Folder A: Multi-Use Graphing Packet, Sheets 1–3
Practice Bank: Practice 15
Study Guide: Section 2.6
Explorations Lab Manual:
Diagram Masters 1, 2
Assessment Book: Test 7
Technology
Graphing Calculator
Internet:
http://www.hmco.com

Warm-Up Exercises

Solve each equation.
1. $3x + 2 = -1$ -1
2. $7 - 4x = -29$ 9

Simplify each expression.
3. $10t - 2t$ $8t$
4. $12x - (2x + 5)$ $10x - 5$
5. Graph the equation $2x + 3 = y$.

84

SECTION

2.6 Using Graphs to Solve Problems

Learn how to...
- use a graph to solve a problem

So you can...
- analyze the costs and profits of running a business, for example

Have you ever thought about putting your own design on a T-shirt and selling it? Graphing an equation helps you analyze the costs and profits of making and selling T-shirts.

Suppose silk-screening supplies cost \$70. Plain T-shirts cost \$3 each. You can write an equation for the cost of making T-shirts.

Cost = \$3 × number of shirts + \$70
$C = 3n + 70$

As shown below, the graph of this equation can help you analyze the costs of running your own T-shirt business.

EXAMPLE 1 Application: Business

You receive 87 orders for T-shirts. You have \$300 to spend on production costs. Decide whether you have enough money to fill all 87 orders.

SOLUTION

Graph the equation $C = 3n + 70$.

T-Shirt Production Costs

Find \$300 on the vertical axis. Move across to the graph of the equation.

Move down ... horizontal ... You can m... about 75 T-...

You do not have enough money to fill 87 orders.

84 Chapter 2 Equations and Functions

The graph in Example 1 helps you estimate production costs. But a graphing calculator can help you make even closer estimates, as shown in Example 2.

EXAMPLE 2 Application: Business

a. You decide to charge customers \$10 per T-shirt. Your production costs are the same as in Example 1. Write an equation for your profit P from making and selling n T-shirts.

b. What is your profit from making and selling 75 T-shirts?

SOLUTION

a. Profit = Sales Income − Production Cost
\$10 × number of shirts − \$3 × number of shirts + \$70
$P = 10n - (3n + 70)$

$P = 10n - (3n + 70)$

You can leave the equation in this form. Or you can simplify it.

$P = 10n - 3n - 70$
$P = 7n - 70$

b. **Problem Solving Strategy:** Use a graph.
First enter the equation $P = 7n - 70$ as $y = 7x - 70$.

Then set the viewing window by choosing *scales* for the *x*- and *y*-axes.

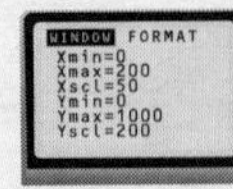

Then graph the equation. Use the *trace* feature to find your profit from selling 75 T-shirts.

The *trace* feature shows you *x*- and *y*-values on the graph.

Use the *zoom* feature to get more accurate values for *x* and *y*.

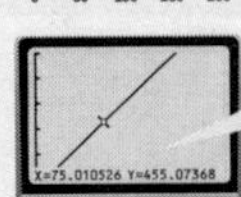

When the *x*-value is close to 75, the *y*-value is close to 455.

Your profit from selling 75 T-shirts is about \$455.

For more information about using a graphing calculator, see the Technology Handbook, p. 603.

Technology Note

Example 2 demonstrates several key aspects of using graphing calculators: setting the viewing window, trace, and zoom. Show students how the range and domain of the function can help choose a viewing window. Also point out that the trace function takes the place of finding a point on the vertical axis and moving across to the graph and then down to the horizontal axis, as was shown in Example 1.

Although this Example shows graphing calculator screens, many software graphing packages work in a similar way. You can show your class this Example with an overhead projector using such software or a graphing calculator.

Teach⇔Interact

Learning Styles: Visual

Students who are visual learners usually prefer to use graphs to solve problems rather than equations. By using graphs, these students can actually see the solution, which helps them to better understand both the problem and its solution.

About Example 1

Reasoning

In this example, students see that they do not have enough money to fill 87 orders. Since a successful business would want to be able to fill as many orders as possible, explore with students how they might possibly fill all 87 orders. For example, if they could buy the plain T-shirts for less than \$3 each, then the cost of each shirt would go down and more orders could be filled. This possibility is explored in Additional Example 1.

Additional Example 1

Suppose you buy plain T-shirts for \$2.50 each and have \$300 to spend on production costs. Silk-screening supplies still cost \$70. Can you now fill an order for 87 shirts?

Graph the equation $C = 2.5n + 70$.

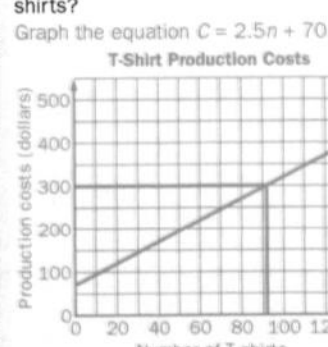

You can make 92 T-shirts, so you can fill an order for 87 shirts.

Additional Example 2

a. You charge customers \$10 per T-shirt and your production costs are the same as in Additional Example 1 ($2.5n + 70$). Write an equation for your profit P from making and selling n T-shirts.
$P = 10n - (2.5n + 70)$
$P = 7.5n - 70$

Example continued on next page.

85

Plan⇔Support

- Section Objectives
- Recommended Pacing
- Resource Materials
- Warm-Up Exercises

Teach⇔Interact

- Additional Examples
- Closure Questions
- Notes on the student lesson, including:

Technology
Explorations
Learning Styles
Communication
Multicultural Information
Second-Language Learners

Apply⇔Assess

- Suggested Assignment
- *Practice Bank* facsimile
- Notes on the exercises, including:

Applications
Problem Solving
Technology
Cooperative Learning
Integrating the Strands
Assessment

RECREATION Exercises 3 and 4 are about Central Park in New York City. Solve at least one of the exercises by writing and graphing an equation.

3. Central Park is rectangular and has a perimeter of about 6 mi. The length of one side is about 2.5 mi. Find the lengths of the other three sides.
4. To celebrate each new year, runners participate in a "Midnight Run" through Central Park. The women's record for this 5 mi race is about 26.2 min. The men's record is about 23.3 min. Estimate the running speed in miles per minute of these record-holding runners.

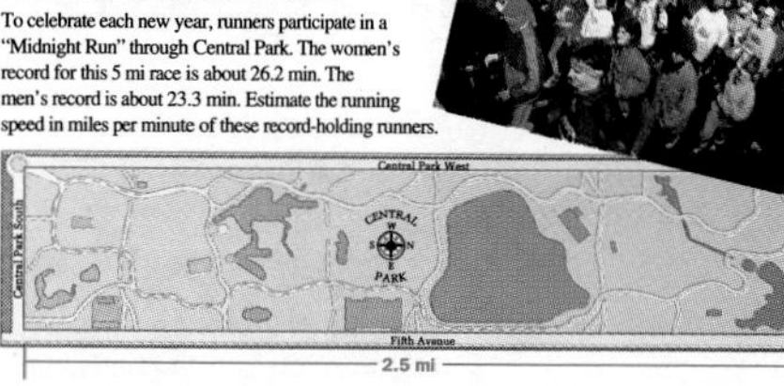

SCIENCE The graph shows the relationship between the Fahrenheit and Celsius temperature scales. Use the graph for Exercises 5–9.

5. Estimate the temperature in degrees Fahrenheit when the temperature is:
   a. 0°C b. 20°C c. −20°C d. −30°C
6. Estimate the temperature in degrees Celsius when the temperature is:
   a. 32°F b. 90°F c. 0°F d. −40°F
7. You can use the formulas below to convert Celsius temperatures to Fahrenheit temperatures and Fahrenheit temperatures to Celsius. Use the formulas to find more exact answers to Exercises 5 and 6.

Let F = the temperature in degrees Fahrenheit. $F = \frac{9}{5}C + 32$
Let C = the temperature in degrees Celsius. $C = \frac{5}{9}(F - 32)$

Temperature Scales (Fahrenheit temperature (°F); Celsius temperature (°C))

8. **Writing** A friend on a trip to Africa writes that the temperature in Nairobi, Kenya, is 25°. Which temperature scale do you think your friend is using? Why?
9. **Open-ended Problem** A friend from a country that uses the Celsius scale wants to visit you. Write a letter to your friend describing the range of temperatures in your area. Give all temperatures in degrees Celsius.
10. The film club pays $500 to rent *Casablanca*. Members sell tickets for $3.50 each. Write and graph an equation for the club's profit P from selling n tickets. How many tickets must be sold to make a $150 profit?

2.6 Using Graphs to Solve Problems 87

Apply⇔Assess

Exercise Notes

Communication: Discussion
Exs. 1, 2 A class discussion of the business concepts involved in this section can help students to understand how they are related. Profit, for example, is affected by both cost and price. Profit can be increased by cutting costs, raising prices, or both. Profit can also be increased by selling more items, thus generating more sales income. Before reaching the break-even point, there is no profit because costs are greater than sales income.

Topic Spiraling: Review
Ex. 7 You may need to review with students how to multiply a whole number by a fraction. Also, remind students to apply the order of operations when using the two temperature formulas.

Challenge
Ex. 7 Each temperature formula expresses one variable in terms of another variable. You may wish to challenge some students to start with either formula and use what they know about inverse operations to derive the other one.

Multicultural Note
Ex. 8 Nairobi, the capital of Kenya, is a very modern city, and is the most important commercial center in eastern Africa. Like many capital cities, Nairobi has office and parliament buildings, conference centers, hotels, restaurants, and museums. Unlike other capitals, however, Nairobi has a national park, which is home to many wild animals, including lions, gazelles, rhinos, and zebras. The income from tourism generated by national parks and game reserves throughout Kenya is vital to Nairobi's economy.

2. a. Sand Costs (Cost (dollars); Number of tons; about 58)
   b. $C = 20 + 19(2) = 58$; $58

Exercises and Applications
1. 10 T-shirts
2. a. $P = 12n - 70$
   b. 48 T-shirts
3. 2.5 mi, 0.5 mi, 0.5 mi
4. women's record speed: about 0.19 mi/min; men's record speed: about 0.21 mi/min
5, 6. Estimates may vary.
5. a. about 30°F
   b. about 70°F
   c. about −5°F
   d. about −20°F
6. a. about 0°C
   b. about 32°C
   c. about −18°C
   d. about −40°C
7. Ex. 5: 32°C; 68°C; −4°C; −22°C
   Ex. 6: 0°F; about 32.2°F; about −17.8°F; −40°F
8. The friend is using the Celsius scale. Kenya is near the equator, so it is generally quite hot there. In degrees Fahrenheit, 25° is below freezing.
9. Answers may vary.
10. $P = 3.5n - 500$
    Film Club Profits (Profit (dollars); Number of tickets sold)
    From the graph, you can estimate the club must sell about 185 tickets. The equation yields a solution of about 186.

87

Teach⇔Interact

Additional Example 2 (continued)

b. What is your profit from making and selling 92 T-shirts?
Use a graphing calculator. First enter the equation $P = 7.5n - 70$ as $y = 7.5x - 70$. Then set the viewing window by choosing scales for the x- and y-axes.

```
WINDOW FORMAT
Xmin=0
Xmax=200
Xscl=50
Ymin=0
Ymax=1000
Yscl=200
```

Then graph the equation. Use the trace feature to find your profit from selling 92 T-shirts.

X=91.489362 Y=616.17021

Use the zoom feature to get more accurate values for x and y.

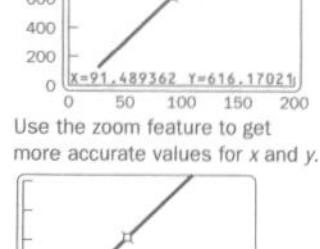

When the x-value is 92, the y-value is 620. Your profit from selling 92 T-shirts is $620.

Closure Question

Does a graph usually give an exact solution or an estimated solution to a problem? Why?
estimated solution; It is not always possible to read exact coordinates on the x- and y-axes, which are the coordinates of the solution.

THINK AND COMMUNICATE

Use the information in Examples 1 and 2 on pages 84 and 85.

1. Suppose you want to make a profit of $1000. Use the graph shown at the right to estimate the number of T-shirts you must sell.
2. Explain how you could use the graph in Example 1 to estimate the cost of making 87 T-shirts. Why is this cost an estimate?

Graph of $y = 7x - 70$

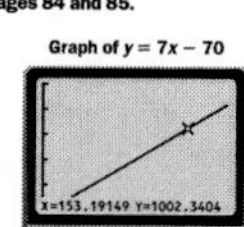

BY THE WAY

During 1989–1990, over 2.6 million teenagers in the United States studied Spanish as a second language. French was the second most studied language among teens, with about 1.1 million students.

CHECKING KEY CONCEPTS

Cooperative Learning Work with a partner. One of you should do part (a) of each question. The other should do part (b). Compare your work. Discuss which way of solving each problem you prefer.

1. Mandy has to read a 126-page book in Spanish. She has already read 19 pages. She estimates she can read about 9 pages per day.
   a. Write and solve an equation to find how many days it will take Mandy to finish the book.
   b. Use a graph to estimate how many days it will take Mandy to finish the book.
2. **BUSINESS** Jon Li buys sand from a building supply company. The sand costs $19 per ton. There is a $20 delivery charge. Jon needs 2 tons of sand delivered to a construction site.
   a. Use a graph to estimate Jon's total cost.
   b. Write and solve an equation to find Jon's total cost.

2.6 Exercises and Applications

Extra Practice exercises on page 561

BUSINESS For Exercises 1 and 2, use the information in Examples 1 and 2 on pages 84 and 85.

1. Suppose you charge $10 per T-shirt. How many T-shirts do you have to sell to break even? (To break even means that you sell just enough T-shirts to cover your production costs.)
2. a. Suppose you charge $15 per T-shirt. Write an equation for your profit P from selling n T-shirts.
   b. How many T-shirts do you have to sell to make $500?

86 Chapter 2 Equations and Functions

Apply⇔Assess

Suggested Assignment

Core Course
Exs. 1–8, 11–14, 16–29, AYP

Extended Course
Exs. 1–8, 11–29, AYP

Block Schedule
Day 12 Exs. 1–6, 10–14, 16–29, AYP

86

ANSWERS Section 2.6

Think and Communicate

1. about 153 T-shirts; Use the zoom function. Create a "zoom box" and trace the new graph. Repeated zooming and tracing gives better estimates.
2. Find 87 on the horizontal axis. Move directly up until you reach the graph of the equation. Then move horizontally to the left until you reach the vertical axis. Estimate the C-coordinate of this point to estimate the cost (about $325). It is only an estimate because the positions of both coordinates are estimated and the vertical axis is marked off in groups of 50.

Checking Key Concepts

1. a. $19 + 9d = 126$; $d \approx 12$; about 12 days
   b.

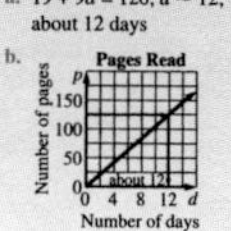

ANSWERS

Answers to Explorations, Think and Communicate questions, Checking Key Concept exercises, and Exercises and Applications are conveniently located at the bottom of each page.

In addition to the section side-column notes, a **Progress Check** is provided for each Assess Your Progress in the student book.

Pacing and Making Assignments

PACING CHART

A yearly Pacing Chart and daily assignments are provided for three courses—a core course, an extended course, and a block-scheduled course. The core and extended courses require 160 days, and the block-scheduled course requires 81 days. These time frames include days for using the Portfolio Projects and time for review and testing. The Pacing Chart below shows the number of days allotted for each of the three courses. Semester and trimester divisions are indicated by red and blue rules, respectively.

Chapter	1	2	3	4	5	6	7	8	9	10	11	12
Core Course	16	12	14	13	13	12	14	16	13	14	12	11
Extended Course	16	12	14	13	13	12	14	16	13	14	12	11
Block Schedule	$8\frac{1}{2}$	6	7	$6\frac{1}{2}$	$6\frac{1}{2}$	6	7	8	$6\frac{1}{2}$	7	6	6

trimester semester trimester

Core Course

The Core Course is intended for students who enter with typical mathematical and problem-solving skills. The course covers all twelve chapters. The daily assignments provide students with substantial work with the skills and concepts presented in each lesson. The exercises assigned range from exercises that involve straightforward application of the new material to exercises involving higher-order thinking skills.

Extended Course

The Extended Course is intended for students who enter with strong mathematical and problem-solving skills and who are able to understand new concepts quickly. The course covers all twelve chapters. The daily assignments include all material in the core course plus additional exercises that focus on higher-order thinking skills.

Block-Scheduled Course

The Block-Scheduled Course is intended for schools that use longer periods, typically 90-minute blocks, for instruction. The course covers all twelve chapters. The exercises assigned range from exercises that involve straightforward application of the new material to exercises involving higher-order thinking skills. All material in the core course is included, plus some additional exercises requiring higher-order thinking skills.

Part of the Block-Scheduled Course for Chapter 2 is shown on the facing page. The entire chart for each chapter is located on the interleaved pages preceding the chapter.

Planning Guide

The Planning Guide for each chapter is located on the interleaved pages preceding the chapter. Part of the Planning Guide for Chapter 2 is shown here.

Regular Scheduling (45 min)

Section	Materials Needed	Core Assignment	Extended Assignment	*exercises that feature* Applications	Communication	Technology
2.1		1–31, 33–41	1–41	20–22, 25–29, 31	23, 30, 32, 33	
2.2	algebra tiles, scientific calculator	**Day 1:** 1–21 **Day 2:** 22–33, 35–48	**Day 1:** 1–21 **Day 2:** 22–48	19–21	17, 18 34, 39	17
2.3		1–3, 5–18, 22–35, AYP*	1–35, AYP	4, 10–15	4, 19–21, 23	
2.4	graph paper, graphing calculator	1–15, 17–22, 25–28, 31–39	1–4, 7, 10, 11, 13–39	17–22, 25–28	23, 24, 36	39

Applications

Each section contains exercises that relate the mathematics of that section to real-world applications. These exercises are usually assigned in the daily assignments and are listed in the Planning Guide under the *Applications* head.

Communication

Each section contains exercises that require students to communicate mathematically. These exercises have students discuss or write about the mathematical concepts presented in the section and are usually assigned in the daily assignments. These exercises are denoted by in-line heads in the Exercises and Applications sets and are listed in the Planning Guide under the *Communication* head.

Technology

Each chapter contains exercises that involve technology, usually graphing calculators, spreadsheets, or computer software. Technology-based exercises are usually assigned in the daily assignments. Exercises that require technology or are especially appropriate for using technology have a logo (shown in the chart above) beside them in the textbook. These exercises are listed in the Planning Guide under the *Technology* head.

Block Scheduling (90 min)

	Day 9	Day 10	Day 11	Day 12	Day 13	Day 14	Day 15
Teach/Interact	Ch. 1 Test 2.1	2.2: Exploration, page 61	2.3 2.4	2.5: Exploration, page 78 2.6	Review Port. Proj.	Review Port. Proj.	Ch. 2 Test 3.1
Apply/Assess	**Ch. 1 Test** **2.1:** 2–6, 10–41	**2.2:** 1–48	**2.3:** 1–18, 22, 24–35, AYP* **2.4:** 1–23, 25–28, 31–39	**2.5:** 1–33 **2.6:** 1–6, 10–14, 16–29, AYP	**Review:** 1–13 **Port. Proj.**	**Review:** 14–28 **Port. Proj.**	**Ch. 2 Test** **3.1:** 1–11, 13–16, 23–25, 27–36

Algebra 1

Explorations and Applications

Authors

Senior Authors

Miriam A. Leiva   Richard G. Brown

Loring Coes III
Shirley Frazier Cooper
Joan Ferrini-Mundy
Amy T. Herman
Patrick W. Hopfensperger
Celia Lazarski
Stuart J. Murphy
Anthony Scott
Marvin S. Weingarden

McDougal Littell
A HOUGHTON MIFFLIN COMPANY
Evanston, Illinois ◆ Boston ◆ Dallas ◆ Phoenix

Authors

Senior Authors

Richard G. Brown Mathematics Teacher, Phillips Exeter Academy, Exeter, New Hampshire

Miriam A. Leiva Cone Distinguished Professor for Teaching and Professor of Mathematics, University of North Carolina at Charlotte

Loring Coes III Chair of the Mathematics Department, Rocky Hill School, E. Greenwich, Rhode Island

Shirley Frazier Cooper Curriculum Specialist, Secondary Mathematics, Dayton Public Schools, Dayton, Ohio

Joan Ferrini-Mundy Professor of Mathematics and Mathematics Education, University of New Hampshire, Durham, New Hampshire

Amy T. Herman Mathematics Teacher, Atherton High School, Louisville, Kentucky

Patrick W. Hopfensperger Mathematics Teacher, Homestead High School, Mequon, Wisconsin

Celia Lazarski Mathematics Teacher, Glenbard North High School, Carol Stream, Illinois

Stuart J. Murphy Visual Learning Specialist, Evanston, Illinois

Anthony Scott Assistant Principal, Orr Community Academy, Chicago, Illinois

Marvin S. Weingarden Supervisor of Secondary Mathematics, Detroit Public Schools, Detroit, Michigan

ISBN: 0-395-67135-3

123456789—VH—99 98 97 96 95

Editorial Advisors

Martha A. Brown Mathematics Supervisor, Prince George's County Public Schools, Capitol Heights, Maryland

Diana Garcia Mathematics Supervisor, Laredo Independent School District, Laredo, Texas

Sue Ann McGraw Mathematics Teacher, Lake Oswego High School, Lake Oswego, Oregon

Editorial Advisors and Reviewers helped plan the concept, teaching approach, and format of the book. They also reviewed draft manuscripts in the early stages of development.

Review Panel

Judy B. Basara Curriculum Chair, St. Hubert's High School, Philadelphia, Pennsylvania

Dane Camp Mathematics Teacher, New Trier High School, Winnetka, Illinois

Pamela W. Coffield Mathematics Teacher, Brookstone School, Columbus, Georgia

Kathleen Curran Mathematics Teacher, Ball High School, Galveston, Texas

Randy Harter Mathematics Specialist, Buncombe County Schools, Asheville, North Carolina

William Leonard Assistant Director of Mathematics, Fresno Unified School District, Fresno, California

Betty McDaniel Coordinator of Mathematics, Florence School District, Florence, South Carolina

Roger O'Brien Mathematics Supervisor, Polk County Schools, Bartow, Florida

Leo Ramirez Mathematics Teacher, McAllen High School, McAllen, Texas

Michelle Rohr Director of Mathematics, Houston Independent School District, Houston, Texas

May Samuels Mathematics Chairperson, Weequahic High School, Newark, New Jersey

Betty Takesuye Mathematics Teacher, Chaparral High School, Scottsdale, Arizona

Members of the review panel read and commented upon outlines, sample lessons, and research versions of the chapters.

Manuscript Reviewers

Elda López Jefferson Davis High School, Houston, Texas

Anita G. Morris Coordinator of Mathematics, Anne Arundel County Schools, Annapolis, Maryland

Michelle Rohr Director of Mathematics, Houston Independent School District, Houston, Texas

May Samuels Mathematics Chairperson, Weequahic High School, Newark Public Schools, Newark, New Jersey

Straightline Editorial Development, Inc. Editorial Consultants, San Francisco, California

Manuscript Reviewers read and reacted to draft manuscript, focusing on its effectiveness from a teaching/learning viewpoint.

The authors wish to thank Pam Coffield and her students at Brookstone School in Columbus, Georgia, for using and providing comments on the Portfolio Projects in this book. Ms. Coffield's students are Kate Baker, Bo Bickerstaff, Lizzie Bowles, Bradford Carmack, Lucy Cartledge, Henry Dunn, Patrick Graffagnino, Charles Haines, Daniel McFall, Sravanthi Meka, Blake Melton, Jason Pease, Ann Phillips, Dorsey Staples, and Jeffrey Usman.

Contents

CHAPTER

1

Using Algebra to Work with Data

Applications

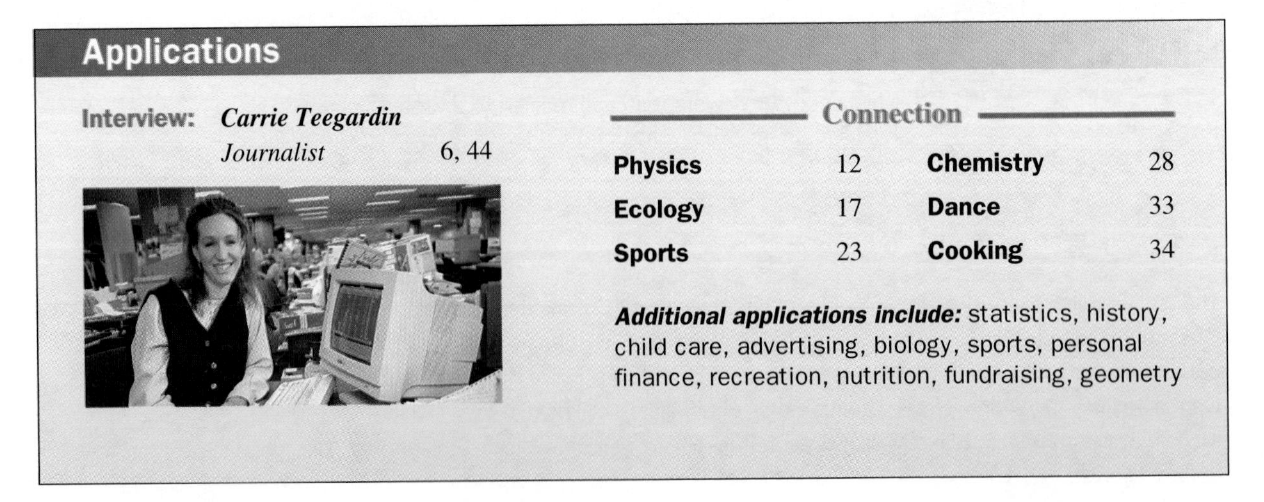

CHAPTER

2 Equations and Functions

Applications

Additional applications include: zoology, sports, language arts, medicine, history, geometry, personal finance, aviation, biology, consumer economics, energy use, business, recreation, science

CHAPTER

3

Graphing Linear Equations

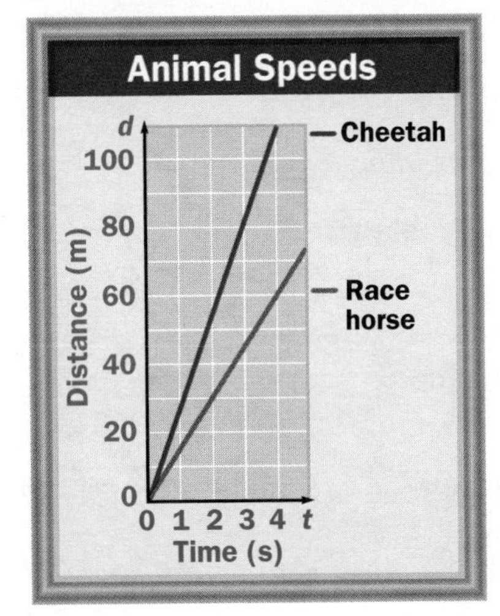

Applications

Additional applications include: aviation, sports, astronomy, weather, employment, consumer economics, physics, business, geometry, energy use, civil engineering, recreation, geology

CHAPTER

4 Solving Equations and Inequalities

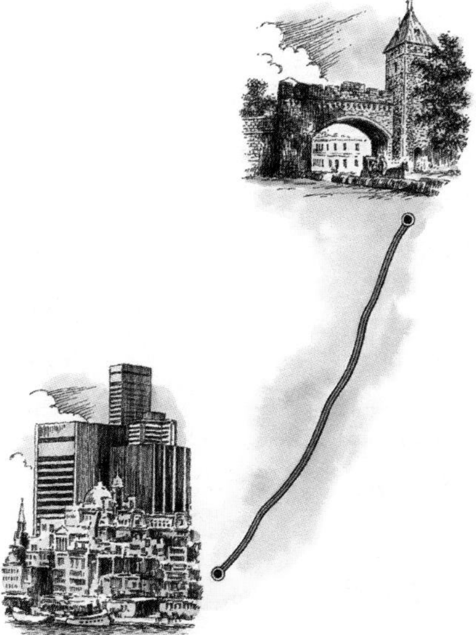

Applications

Additional applications include: consumer economics, personal finance, recreation, geometry, fuel economy, transportation, sports, architecture, communications, business

CHAPTER

5

Connecting Algebra and Geometry

Applications

Additional applications include: history, ecology, forestry, language arts, sports, languages, zoology, business, geometry, architecture, construction, photography

CHAPTER

6 Working with Radicals

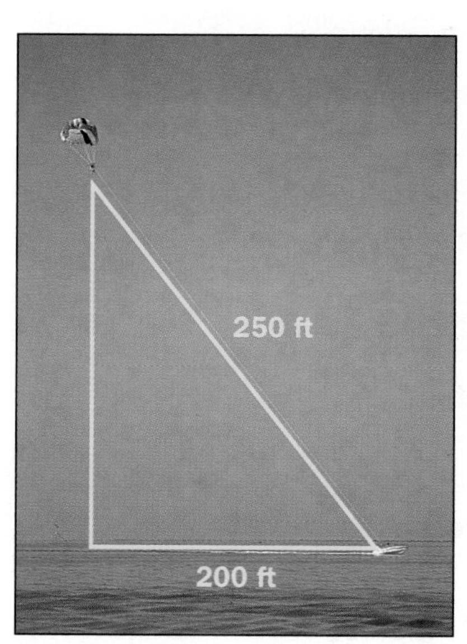

Applications

Additional applications include: recreation, urban planning, aviation, language arts, biology, physics, geometry, medicine, construction

CHAPTER 7

Systems of Equations and Inequalities

Applications

Additional applications include: sports, fitness, aviation, finance, space science, geometry, personal finance, nutrition

CHAPTER

8 Quadratic Functions

Applications

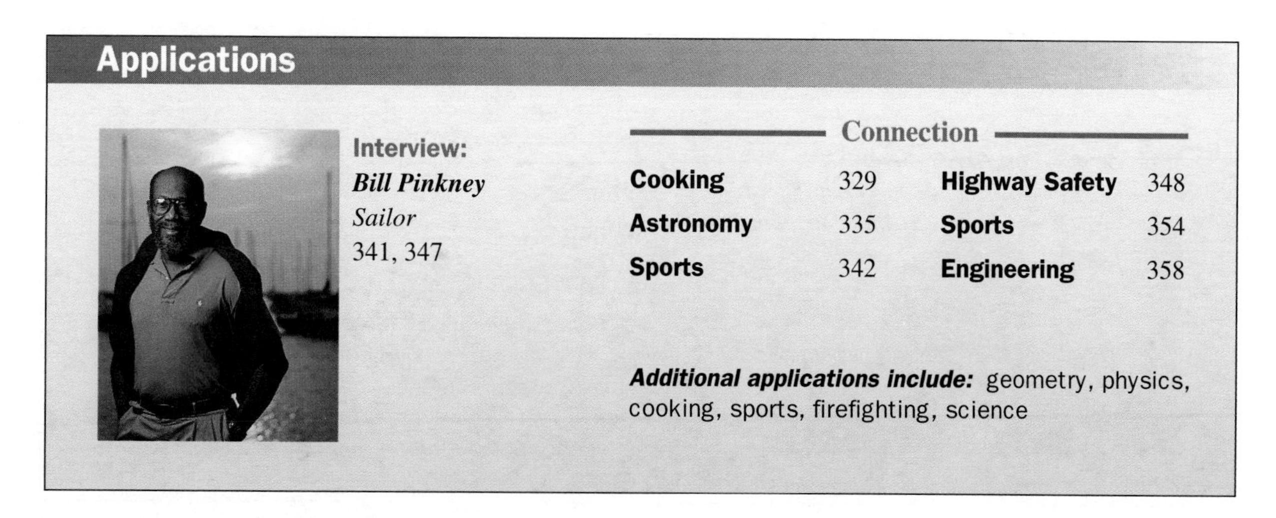

Additional applications include: geometry, physics, cooking, sports, firefighting, science

CHAPTER

9

Exponential Functions

9.3 *Tennis* 384

Applications

Interview: ***Lori Arviso Alvord***
Surgeon 380, 387, 399

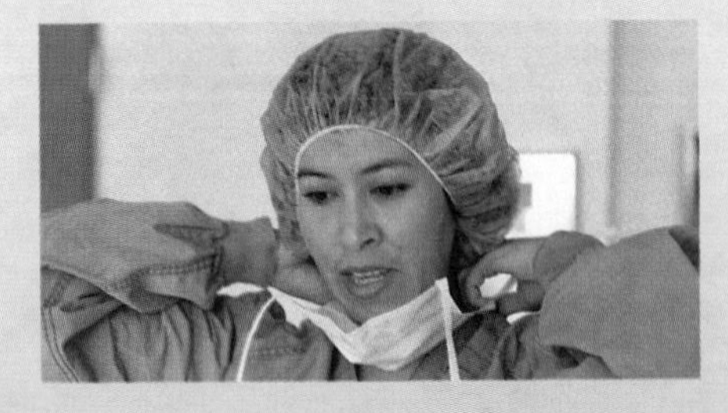

Connection

Additional applications include: architecture, geometry, carpentry, number theory, ecology, biology, personal finance, traffic, social studies, physics, meteorology, sports, occupational safety, fuel economy, population, history, astronomy

CHAPTER

10 Polynomial Functions

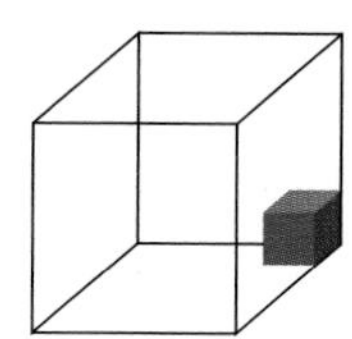

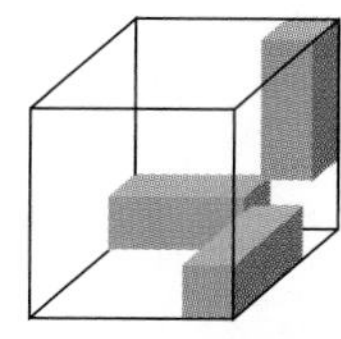

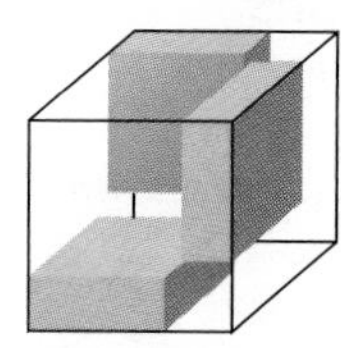

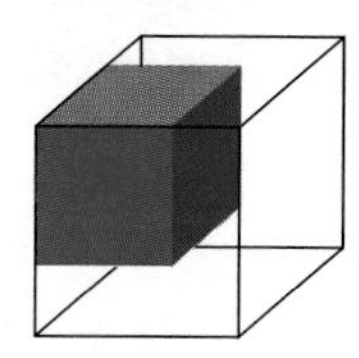

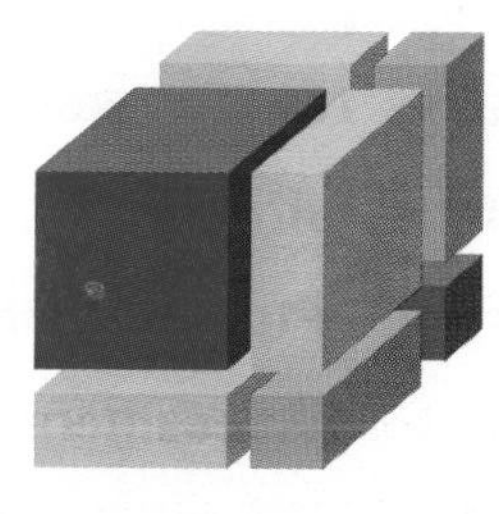

Applications

Additional applications include: geometry, crafts, physics, sports

CHAPTER

11 Rational Functions

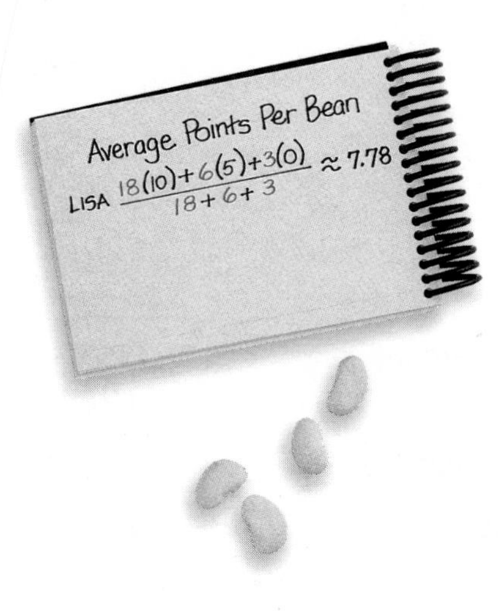

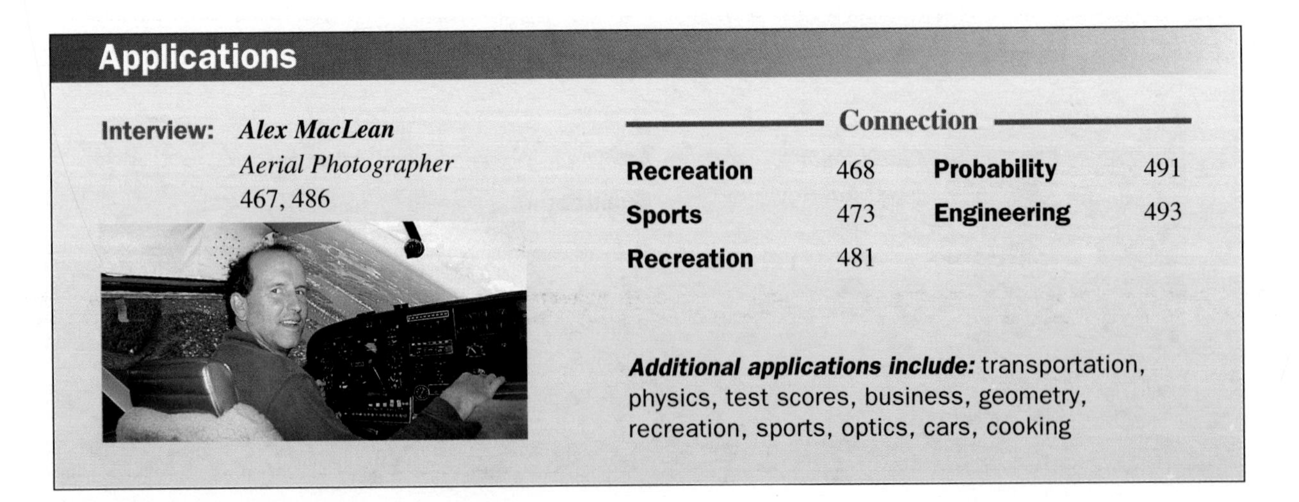

Applications

Additional applications include: transportation, physics, test scores, business, geometry, recreation, sports, optics, cars, cooking

CHAPTER

12 Discrete Mathematics

Applications

Additional applications include: fundraising, fine arts, cars, business, fashion design, sports, government, manufacturing, geometry, photography

Student Resources

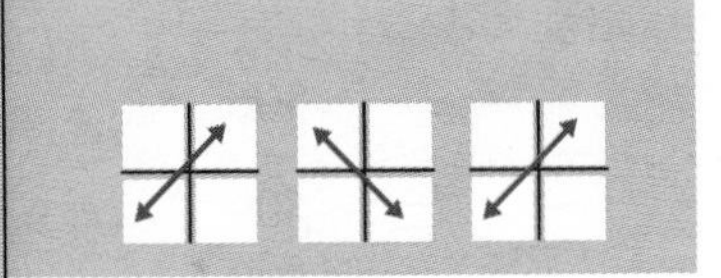

About the Interviews

Each chapter of this book starts with a personal interview with someone who uses mathematics in his or her daily life. Does this mean that all the interviews are with mathematicians or scientists? Not at all. You may be surprised by the wide range of careers that are included. These are the people you will be reading about:

- Journalist ***Carrie Teegardin***
- Racing Wheelchair Designer ***Bob Hall***
- Astronaut ***Mae Jemison***
- Seismologist ***Barbara Romanowicz***
- Artist ***Daniel Galvez***
- Accident Reconstructionist ***Nathan Shigemura***
- Sports Nutritionist ***Nancy Clark***
- Sailor ***Bill Pinkney***
- Surgeon ***Lori Arviso Alvord***
- Financial Adviser ***Linda Pei***
- Aerial Photographer ***Alex MacLean***
- Animator ***Phil Roman***

Phil Roman

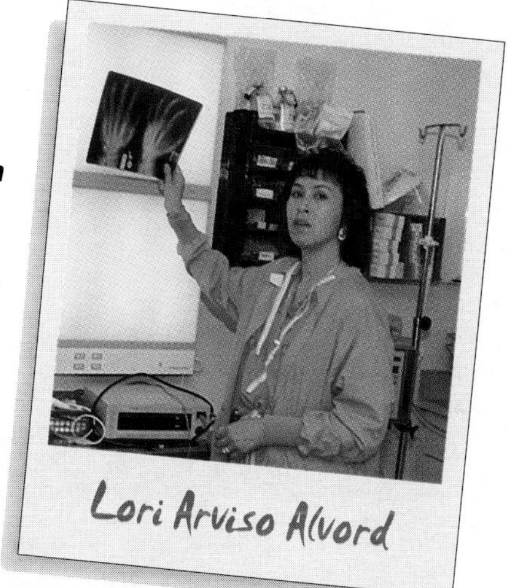

Lori Arviso Alvord

At the end of each interview, there are *Explore and Connect* questions that guide you in learning more about the career being discussed. Some of these questions involve research that is done outside of class. In addition, in each chapter there are ***Related Examples and Exercises*** that show how the mathematics you are learning is directly related to the career highlighted in the interview.

Nathan Shigemura

xvii

Welcome to Algebra 1

Explorations and Applications

Goals of The Course

This book will help you use mathematics in your daily life and prepare you for success in future courses and careers.

In this course you will:

- Study the algebra concepts that are most important for today's students
- Apply these concepts to solve many different types of problems
- Learn how calculators and computers can help you find solutions

You will have a chance to develop your skills in:

- Reasoning and problem solving
- Communicating orally and in writing
- Studying and learning independently and as a team member

xviii

MATHEMATICAL CONTENT

This contemporary algebra course gives you a strong background in the types of mathematical reasoning and problem solving that will be important in your future.

The book emphasizes:

- Using functions, equations, and graphs to model problem situations
- Investigating connections of algebra to geometry, statistics, probability, and discrete mathematics

ACTIVE LEARNING

To learn algebra successfully, you need to get involved!

There will be many opportunities in this course for you to participate in:

- Explorations of mathematical concepts
- Cooperative learning activities
- Small-group and whole-class discussions

So don't sit back and be a spectator. If you join in and share your ideas, everyone will learn more.

Course Overview

To get an overview of your course, turn to pages xx–xxix to see some of the types of problems you will solve and topics you will explore.

xix

"What does algebra have to do with me?"

Applications and Connections

Algebra is about you and the world around you.

In this course you'll learn how algebra can help answer many different types of questions in daily life and in careers.

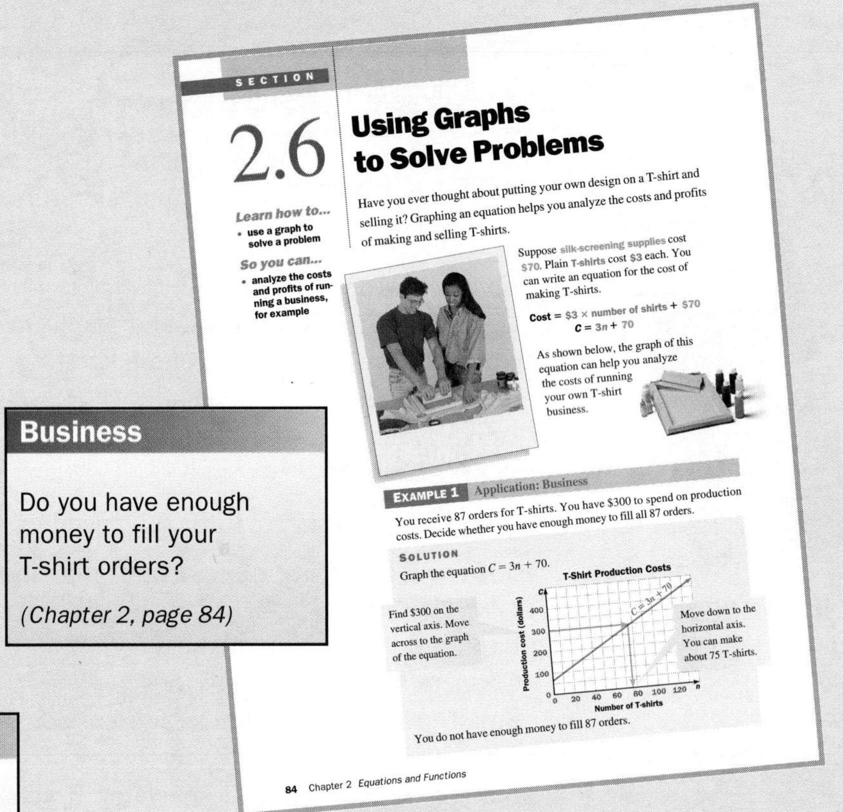

SECTION

2.6 Using Graphs to Solve Problems

Learn how to...
- use a graph to solve a problem

So you can...
- analyze the costs and profits of running a business, for example

Have you ever thought about putting your own design on a T-shirt and selling it? Graphing an equation helps you analyze the costs and profits of making and selling T-shirts.

Suppose silk-screening supplies cost \$70. Plain T-shirts cost \$3 each. You can write an equation for the cost of making T-shirts.

Cost = \$3 × number of shirts + \$70

$C = 3n + 70$

As shown below, the graph of this equation can help you analyze the costs of running your own T-shirt business.

EXAMPLE 1 Application: Business

You receive 87 orders for T-shirts. You have \$300 to spend on production costs. Decide whether you have enough money to fill all 87 orders.

SOLUTION

Graph the equation $C = 3n + 70$.

Find \$300 on the vertical axis. Move across to the graph of the equation.

Move down to the horizontal axis. You can make about 75 T-shirts.

You do not have enough money to fill 87 orders.

84 Chapter 2 *Equations and Functions*

Business

Do you have enough money to fill your T-shirt orders?

(Chapter 2, page 84)

Sports Nutrition

Can a teenage athlete get enough calcium from milk and ice cream without eating too much fat?

(Chapter 7, page 309)

xx

Fuel Economy

At what speeds should a car be driven to get more than 25 mi/gal?

(Chapter 4, page 173)

Speed and Fuel Economy

f; 40, 30, 20, 10, 0 — Fuel economy (mi/gal)

0 10 20 30 40 50 60 70 80 s — Speed (mi/h)

20 40 60 80 100 120 140 MPH

GASOLINE

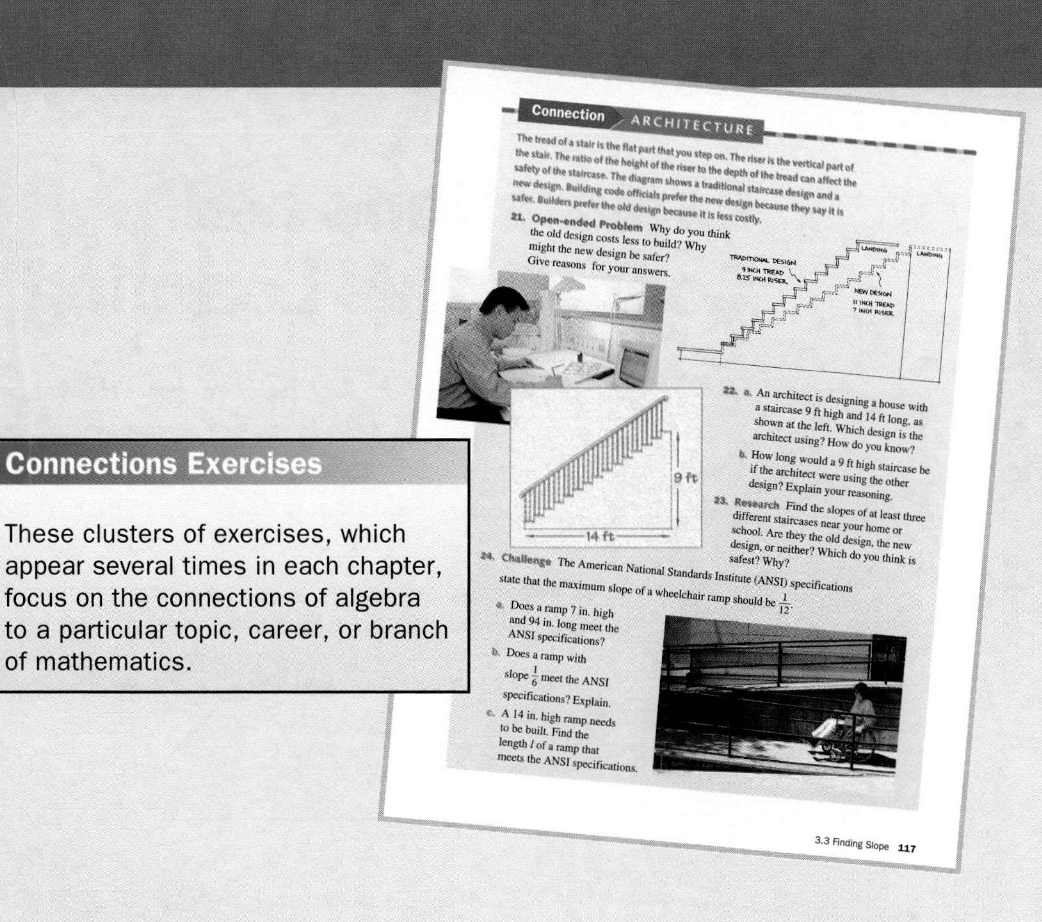

Connection ARCHITECTURE

The tread of a stair is the flat part that you step on. The riser is the vertical part of the stair. The ratio of the height of the riser to the depth of the tread can affect the safety of the staircase. The diagram shows a traditional staircase design and a new design. Building code officials prefer the new design because they say it is safer. Builders prefer the old design because it is less costly.

21. Open-ended Problem Why do you think the old design costs less to build? Why might the new design be safer? Give reasons for your answers.

22. a. An architect is designing a house with a staircase 9 ft high and 14 ft long, as shown at the left. Which design is the architect using? How do you know?

b. How long would a 9 ft high staircase be if the architect were using the other design? Explain your reasoning.

23. Research Find the slopes of at least three different staircases near your home or school. Are they the old design, the new design, or neither? Which do you think is safest? Why?

24. Challenge The American National Standards Institute (ANSI) specifications state that the maximum slope of a wheelchair ramp should be $\frac{1}{12}$.

a. Does a ramp 7 in. high and 94 in. long meet the ANSI specifications?

b. Does a ramp with slope $\frac{1}{6}$ meet the ANSI specifications? Explain.

c. A 14 in. high ramp needs to be built. Find the length l of a ramp that meets the ANSI specifications.

3.3 Finding Slope **117**

Connections Exercises

These clusters of exercises, which appear several times in each chapter, focus on the connections of algebra to a particular topic, career, or branch of mathematics.

Earth Science

How can you predict when Old Faithful geyser will erupt next?

(Chapter 3, pages 130–131)

Old Faithful

Time until next eruption (min) vs. Length of eruption (min)

Social Studies

What is New Jersey's quota of seats in the United States House of Representatives?

(Chapter 5, page 196)

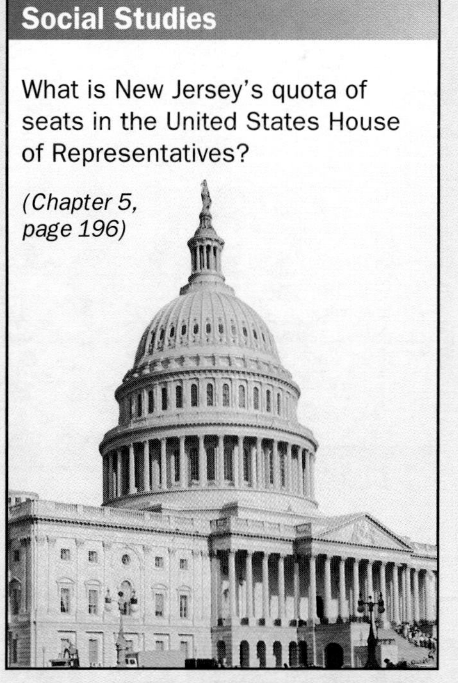

xxi

"Do we just sit back and listen?"

Explorations and Cooperative Learning

In this course you'll be an active learner.

Working individually and in groups, you'll investigate questions and then present and discuss your results. Here are some of the topics you'll explore.

Writing Variable Expressions

How many toothpicks do you need to make n triangles?

(Chapter 1, page 30)

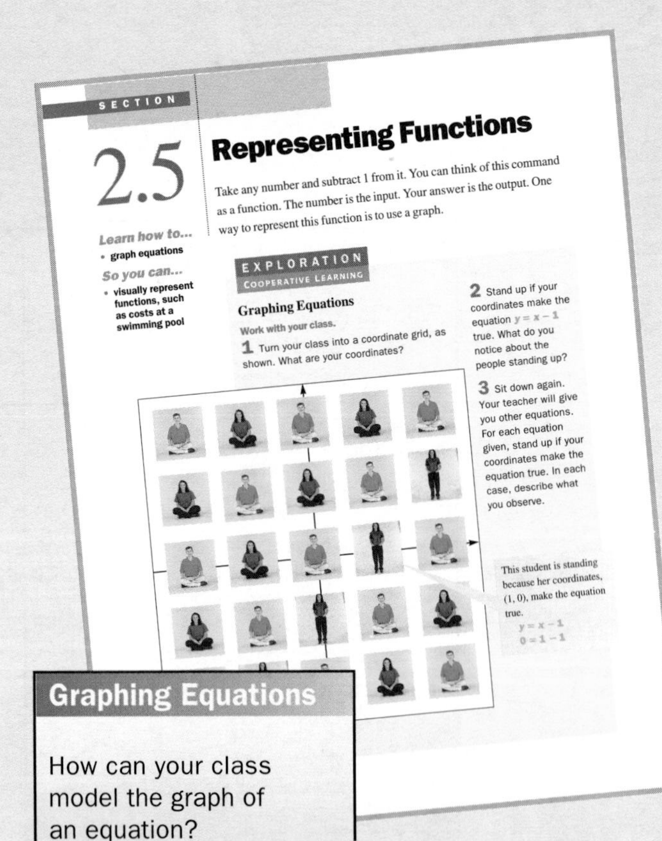

SECTION

2.5 Representing Functions

Take any number and subtract 1 from it. You can think of this command as a function. The number is the input. Your answer is the output. One way to represent this function is to use a graph.

Learn how to...
- graph equations

So you can...
- visually represent functions, such as costs at a swimming pool

EXPLORATION
COOPERATIVE LEARNING

Graphing Equations

Work with your class.

1 Turn your class into a coordinate grid, as shown. What are your coordinates?

2 Stand up if your coordinates make the equation $y = x - 1$ true. What do you notice about the people standing up?

3 Sit down again. Your teacher will give you other equations. For each equation given, stand up if your coordinates make the equation true. In each case, describe what you observe.

This student is standing because her coordinates, (1, 0), make the equation true.

$y = x - 1$
$0 = 1 - 1$

Graphing Equations

How can your class model the graph of an equation?

(Chapter 2, page 78)

Creating a Linear Model

Can you find a linear equation to predict how many marbles are needed to break the pieces of spaghetti?

(Chapter 3, page 132)

Number of pieces of spaghetti	Number of marbles needed to break spaghetti
1	?
2	?
3	?
4	?
5	?

xxii

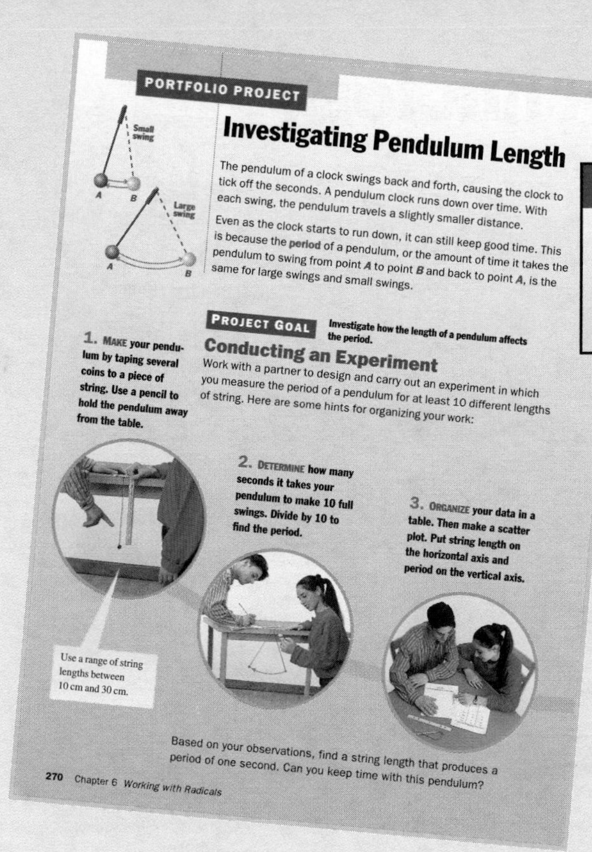
PORTFOLIO PROJECT

Investigating Pendulum Length

The pendulum of a clock swings back and forth, causing the clock to tick off the seconds. A pendulum clock runs down over time. With each swing, the pendulum travels a slightly smaller distance.

Even as the clock starts to run down, it can still keep good time. This is because the **period** of a pendulum, or the amount of time it takes the pendulum to swing from point *A* to point *B* and back to point *A*, is the same for large swings and small swings.

PROJECT GOAL **Investigate how the length of a pendulum affects the period.**

Conducting an Experiment

Work with a partner to design and carry out an experiment in which you measure the period of a pendulum for at least 10 different lengths of string. Here are some hints for organizing your work:

1. **MAKE your pendulum by taping several coins to a piece of string. Use a pencil to hold the pendulum away from the table.**
2. **DETERMINE how many seconds it takes your pendulum to make 10 full swings. Divide by 10 to find the period.**
3. **ORGANIZE your data in a table. Then make a scatter plot. Put string length on the horizontal axis and period on the vertical axis.**

Use a range of string lengths between 10 cm and 30 cm.

Based on your observations, find a string length that produces a period of one second. Can you keep time with this pendulum?

270 Chapter 6 *Working with Radicals*

Portfolio Projects

These open-ended projects give you a chance to explore applications of the topics you have studied.

Exploring Exponents

How many grains of rice will be on the 64th square of the chessboard?

(Chapter 9, page 369)

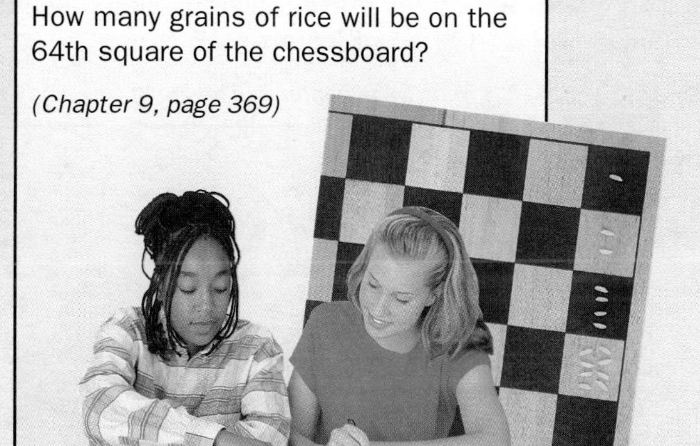

Using Models to Factor Trinomials

How can algebra tiles be used to model the process of factoring?

(Chapter 10, page 442)

$x - 3$

$x - 1$

xxiii

"How can I visualize the problem?"

Using Technology

Calculators and computers can help you see mathematical relationships.

In this course there are many opportunities to use technology to model problem situations, identify patterns, and find solutions.

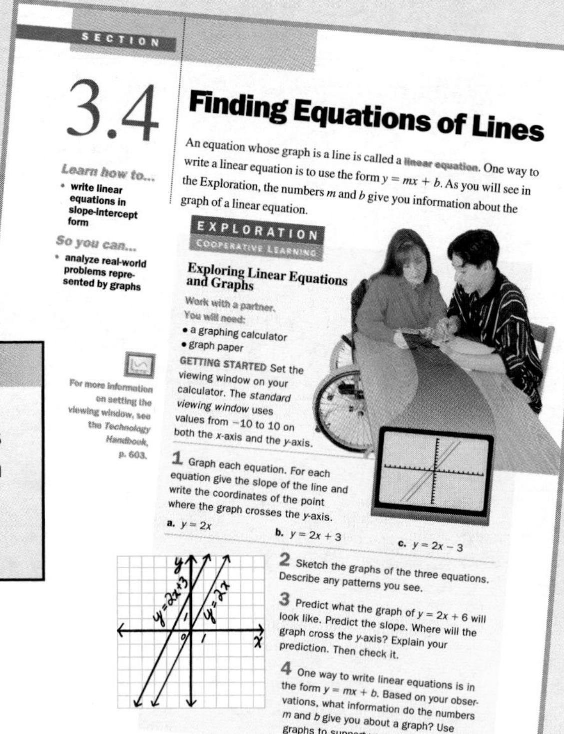

SECTION

3.4

Finding Equations of Lines

Learn how to...
- write linear equations in slope-intercept form

So you can...
- analyze real-world problems represented by graphs

An equation whose graph is a line is called a **linear equation**. One way to write a linear equation is to use the form $y = mx + b$. As you will see in the Exploration, the numbers m and b give you information about the graph of a linear equation.

EXPLORATION
COOPERATIVE LEARNING

Exploring Linear Equations and Graphs

Work with a partner.
You will need:
- a graphing calculator
- graph paper

GETTING STARTED Set the viewing window on your calculator. The *standard viewing window* uses values from −10 to 10 on both the x-axis and the y-axis.

For more information on setting the viewing window, see the *Technology Handbook*, p. 603.

1 Graph each equation. For each equation give the slope of the line and write the coordinates of the point where the graph crosses the y-axis.

a. $y = 2x$ **b.** $y = 2x + 3$ **c.** $y = 2x - 3$

2 Sketch the graphs of the three equations. Describe any patterns you see.

3 Predict what the graph of $y = 2x + 6$ will look like. Predict the slope. Where will the graph cross the y-axis? Explain your prediction. Then check it.

4 One way to write linear equations is in the form $y = mx + b$. Based on your observations, what information do the numbers m and b give you about a graph? Use graphs to support your answer.

3.4 Finding Equations of Lines **119**

Graphing Calculator

What information do the numbers m and b give you about the graph of $y = mx + b$?

(Chapter 3, page 119)

Calculator

Why did Robin and Jeff's calculators give different answers for the amount they owe for the pizza?

(Chapter 1, page 8)

xxiv

Technology Exercises

In these exercises you will be using graphing technology or spreadsheets to practice, apply, and extend what you have learned.

Technology **Use a graphing calculator to solve each equation. If an equation has no solution, write *no solution*.**

9. $x^2 + 2x - 8 = 0$
10. $2x^2 + 7x - 4 = 0$
11. $-x^2 + 9 = 0$
12. $x^2 - 9x + 18 = 0$
13. $x^2 - 8x + 16 = 0$
14. $2x^2 - 3x + 5 = 0$
15. $-4x^2 - 12x = 9$
16. $x^2 - 4x + 9 = 4$
17. $2x^2 + 6x - 4 = 1$
18. $-3x^2 - 6x + 8 = 3$
19. $3x^2 - 15x = 24$
20. $-x^2 + 8 = 12x$
21. **SAT/ACT Preview** If S represents the greater of the two solutions of the equation $x^2 - 4x - 3 = 5$, then:
    A. $S > 3$ B. $S < 3$ C. $S = 3$
    D. relationship cannot be determined

Connection HIGHWAY SAFETY

Speed limits vary from country to country. The speed limit on highways in Mexico is 100 km/h. In France the limits vary from 90 to 130 km/h. In Germany, some sections of the highway have no speed limit. Other sections are limited to either 180 km/h or 130 km/h.

The equation $d = 0.0056s^2 + 0.14s$ models the relationship between a vehicle's stopping distance d in meters and the vehicle's speed s in kilometers per hour.

Germany

France

Technology **For Exercises 22–24, use a graphing calculator.**

22. How many solutions does the equation $0.0056s^2 + 0.14s = 0$ have? How many of the solutions make sense in this situation?
23. About how fast should you be driving in order to be able to stop within 100 m? In which of the above countries is this speed within the speed limit?
24. Suppose a car is traveling at the speed of 130 km/h on a German highway. What distance should the driver of the car leave between the driver's car and the car in front of it?
25. **Research** Find out what is a reasonable length in meters for a car. Then estimate the number of car lengths that corresponds to the stopping distance you found in Exercise 24.

348 Chapter 8 *Quadratic Functions*

Graphing Technology

How has the population of Alaskan peregrine falcons grown since 1980?

(Chapter 9, page 378)

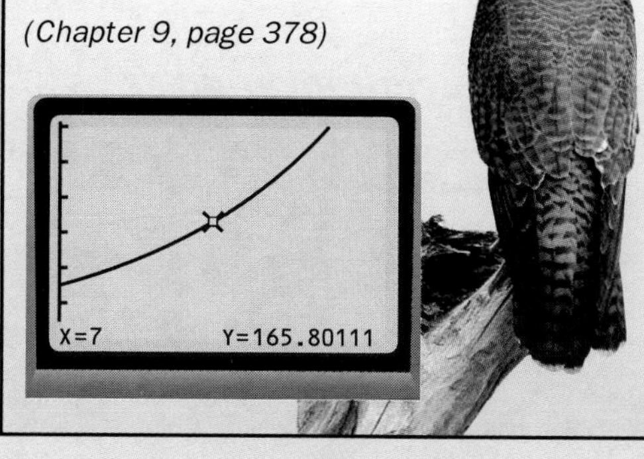

Spreadsheet

Which region of the United States had the greatest increase in the number of in-line skaters?

(Chapter 1, page 46)

	A	B	C	D
1		1992	1993	Percent chan
2	Northeast	1,720,000	2,889,000	67
3	North central	2,646,000	3,963,000	
4	South	2,306,000	2,899,000	
5	West	2,692,000	2,808,000	4.3
6				
7				
8				

XXV

"Can I solve this problem with algebra?"

Integrating Math Topics

Sometimes you need to combine algebra with other math topics in order to find a solution.

In this course you'll see how you can solve problems by integrating algebra with geometry, statistics, probability, and discrete mathematics.

CHAPTER

5 Connecting Algebra and Geometry

Painting a neighborhood

INTERVIEW **Daniel Galvez**

Daniel Galvez always wanted to be an artist, but he didn't know what kind of artist until he learned in college about the great Mexican muralists of the 1900s. Galvez admired the way these artists used colorful paintings on public walls to reach all types of people. "I liked the idea of making a painting that you don't have to go see in a gallery," he says, "the idea that it was part of people's lives whether or not they understood art. It would be something they would see on their way to school or work."

"I liked the idea of making a painting that you don't have to go see in a gallery."

188

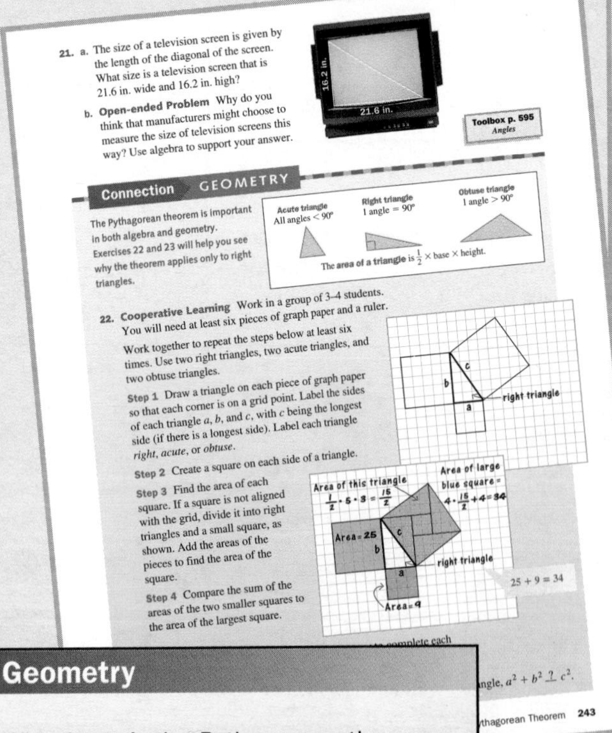

21. a. The size of a television screen is given by the length of the diagonal of the screen. What size is a television screen that is 21.6 in. wide and 16.2 in. high?

    b. **Open-ended Problem** Why do you think that manufacturers might choose to measure the size of television screens this way? Use algebra to support your answer.

Toolbox p. 595
Angles

Connection GEOMETRY

The Pythagorean theorem is important in both algebra and geometry. Exercises 22 and 23 will help you see why the theorem applies only to right triangles.

The area of a triangle is $\frac{1}{2}$ × base × height.

22. **Cooperative Learning** Work in a group of 3–4 students. You will need at least six pieces of graph paper and a ruler.

    Work together to repeat the steps below at least six times. Use two right triangles, two acute triangles, and two obtuse triangles.

    Step 1 Draw a triangle on each piece of graph paper so that each corner is on a grid point. Label the sides of each triangle a, b, and c, with c being the longest side (if there is a longest side). Label each triangle *right*, *acute*, or *obtuse*.

    Step 2 Create a square on each side of a triangle.

    Step 3 Find the area of each square. If a square is not aligned with the grid, divide it into right triangles and a small square, as shown. Add the areas of the pieces to find the area of the square.

    Step 4 Compare the sum of the areas of the two smaller squares to the area of the largest square.

...complete each

...ngle, $a^2 + b^2 \underline{?} c^2$.

...thagorean Theorem 243

Geometry

Why doesn't the Pythagorean theorem work for obtuse triangles?

(Chapter 6, page 243)

68 67 70 66 69 71 67 70 62

xxvi

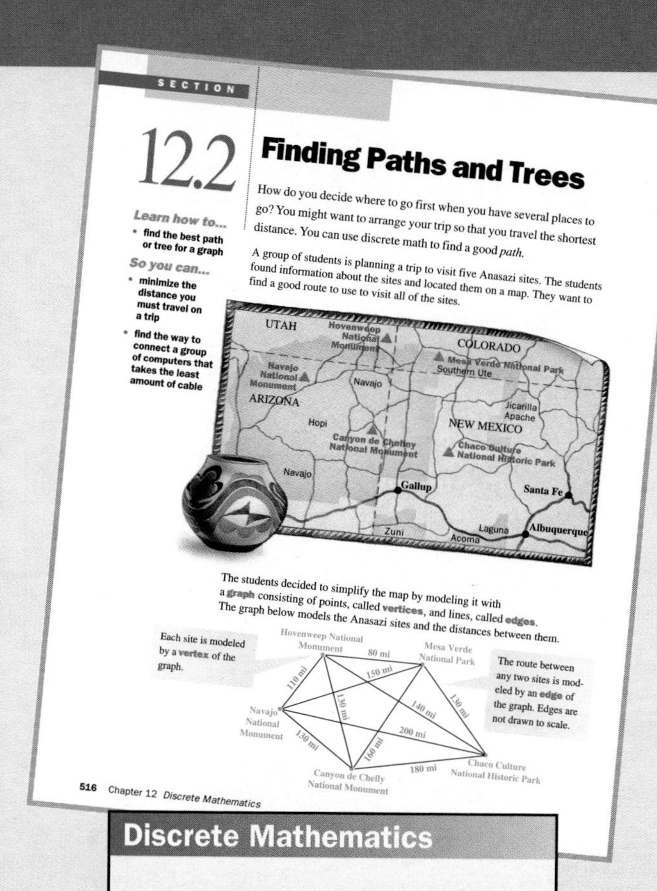
SECTION

12.2 Finding Paths and Trees

Learn how to...
- find the best path or tree for a graph

So you can...
- minimize the distance you must travel on a trip
- find the way to connect a group of computers that takes the least amount of cable

How do you decide where to go first when you have several places to go? You might want to arrange your trip so that you travel the shortest distance. You can use discrete math to find a good *path*.

A group of students is planning a trip to visit five Anasazi sites. The students found information about the sites and located them on a map. They want to find a good route to use to visit all of the sites.

The students decided to simplify the map by modeling it with a **graph** consisting of points, called **vertices**, and lines, called **edges**. The graph below models the Anasazi sites and the distances between them.

516 Chapter 12 *Discrete Mathematics*

Discrete Mathematics

How can you find a short path for visiting all five of the Anasazi sites on the map?

(Chapter 12, pages 516-517)

Probability

How many times will each number on a die occur in 24 rolls?

(Chapter 5, page 218)

Statistics

What is the life expectancy of a person born in the United States in the year 2000?

(Chapter 3, page 133)

Life Expectancy

Year of birth	Life expectancy (years)
1900	47.3
1910	50.0
1920	54.1
1930	59.7
1940	62.9
1950	68.2
1960	69.7
1970	70.8
1980	73.7
1990	75.4

Data Analysis

What two inequalities would you use to describe the heights of these students?

(Chapter 1, page 3)

68 62 66 63 63 69 66 62 66

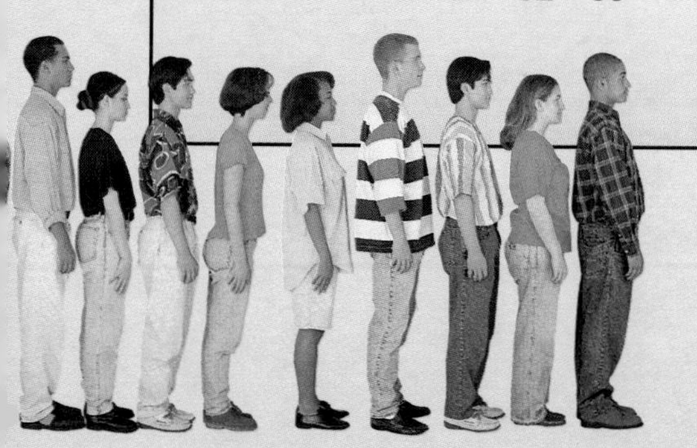

xxvii

"When will I ever use this?"

Building for the Future

The skills you'll learn in this course will form a strong foundation for the future.

They'll prepare you for more advanced courses and increase your career opportunities.

Problem Solving and Communication

There are many exercises to help you develop your problem solving skills and your ability to communicate orally and in writing.

Open-ended Problems give you a chance to share your thinking.

SAT/ACT Preview exercises help you prepare for these college entrance tests.

Writing You can be creative in answering these questions.

WEATHER For Exercises 23–25, use the table showing rainfall amounts.

World Record Rainfall Amounts

Location	Amount	Duration
Unionville, Maryland (U.S.)	3.1 cm	1 min
Curtea-de-Arges (Romania)	20.5 cm	20 min
Holt, Missouri (U.S.)	30.5 cm	42 min
Cilaos, La Réunion	188 cm	24 h
Cherrapunji (India)	930 cm	1 month

23. Write a rate and a unit rate for each location. Use the same units for all the unit rates.
24. In Exercise 23, what information does each rate give you that the unit rate does not? What information does the unit rate give you that the rate does not?
25. Suppose rain continued to fall at a rate of 3.1 cm/min in Unionville. How long would it take to break Cherrapunji's record? Explain.
26. **Open-ended Problem** A child's height is an example of a variable showing a positive rate of change over time. Give two examples of a variable showing a negative rate of change over time. For one of these variables, estimate a rate of change. Explain your estimate.
27. **SAT/ACT Preview** Vera has \$20. The cost of a pair of socks ranges from \$1.75 to \$2.60. What is the greatest number of pairs she can buy?
    A. 7 B. 8 C. 11 D. 9 E. 10

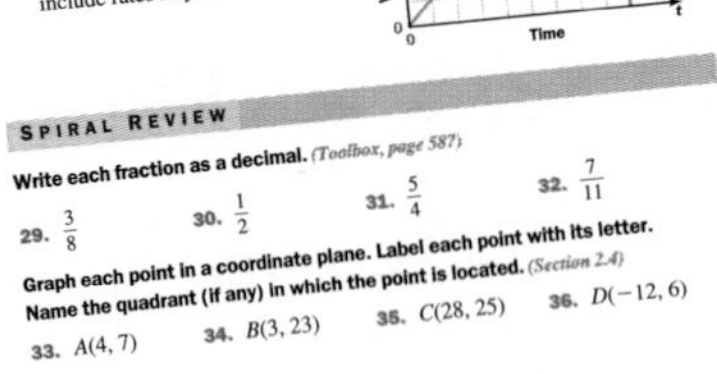

ONGOING ASSESSMENT

28. **Writing** Write a story that relates the graph and the picture shown at the right. Be sure to include rates in your story.

SPIRAL REVIEW

Write each fraction as a decimal. *(Toolbox, page 587)*

29. $\frac{3}{8}$ 30. $\frac{1}{2}$ 31. $\frac{5}{4}$ 32. $\frac{7}{11}$

Graph each point in a coordinate plane. Label each point with its letter. Name the quadrant (if any) in which the point is located. *(Section 2.4)*

33. $A(4, 7)$ 34. $B(3, 23)$ 35. $C(28, 25)$ 36. $D(-12, 6)$

3.1 Applying Rates 105

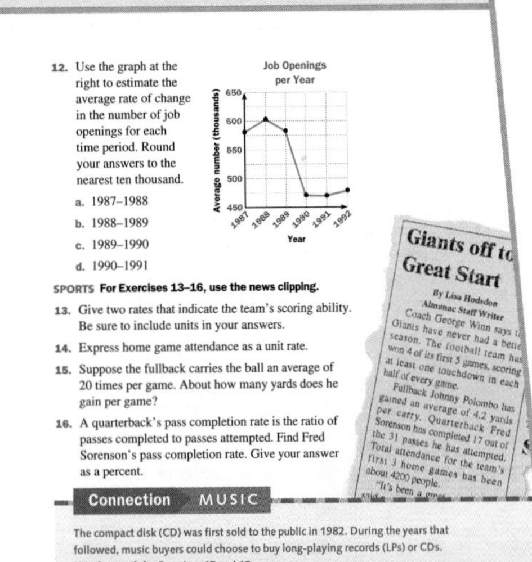

12. Use the graph at the right to estimate the average rate of change in the number of job openings for each time period. Round your answers to the nearest ten thousand.
    a. 1987–1988
    b. 1988–1989
    c. 1989–1990
    d. 1990–1991

SPORTS For Exercises 13–16, use the news clipping.

13. Give two rates that indicate the team's scoring ability. Be sure to include units in your answers.
14. Express home game attendance as a unit rate.
15. Suppose the fullback carries the ball an average of 20 times per game. About how many yards does he gain per game?
16. A quarterback's pass completion rate is the ratio of passes completed to passes attempted. Find Fred Sorenson's pass completion rate. Give your answer as a percent.

Connection MUSIC

The compact disk (CD) was first sold to the public in 1982. During the years that followed, music buyers could choose to buy long-playing records (LPs) or CDs. Use the graph for Exercises 17 and 18.

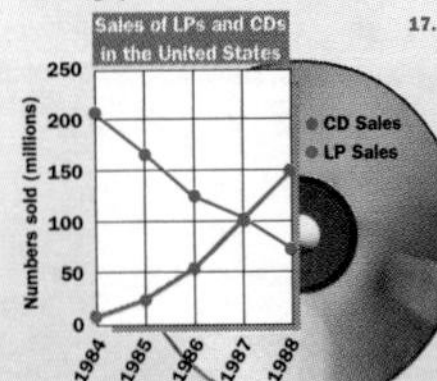

17. For each time period, find the average rate of change for LP sales and the average rate of change for CD sales. Organize your results in a table. Round your answers to the nearest ten million.
    a. 1984–1985 b. 1985–1986
    c. 1986–1987 d. 1987–1988
18. **Writing** Take yourself back in time to the year 1988. A music producer wants to know how many CDs and LPs to produce in 1989. Write a letter describing a good business strategy. Use rates of change to support your suggestions.

3.1 Applying Rates 103

xxviii

Career Preview

The Interview pages at the start of each chapter, and the related exercises throughout the chapter, show how math is used in a variety of different careers.

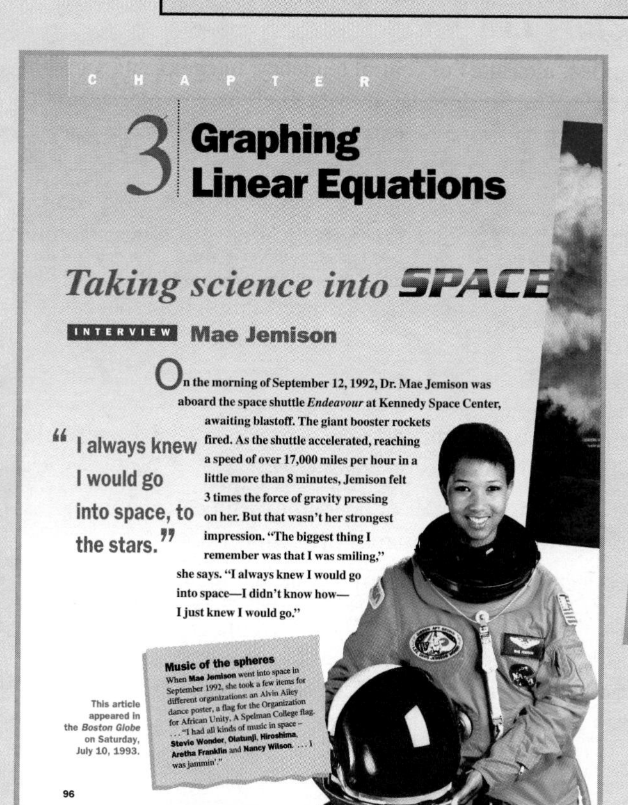

CHAPTER

3 Graphing Linear Equations

Taking science into SPACE

INTERVIEW Mae Jemison

On the morning of September 12, 1992, Dr. Mae Jemison was aboard the space shuttle *Endeavour* at Kennedy Space Center, awaiting blastoff. The giant booster rockets fired. As the shuttle accelerated, reaching a speed of over 17,000 miles per hour in a little more than 8 minutes, Jemison felt 3 times the force of gravity pressing on her. But that wasn't her strongest impression. "The biggest thing I remember was that I was smiling," she says. "I always knew I would go into space—I didn't know how—I just knew I would go."

"I always knew I would go into space, to the stars."

Music of the spheres
When **Mae Jemison** went into space in September 1992, she took a few items for different organizations: an Alvin Ailey dance poster, a flag for the Organization for African Unity, A Spelman College flag. . . . "I had all kinds of music in space – **Stevie Wonder**, **Olatunji**, **Hiroshima**, **Aretha Franklin** and **Nancy Wilson**. . . . I was jammin'."

This article appeared in the *Boston Globe* on Saturday, July 10, 1993.

96

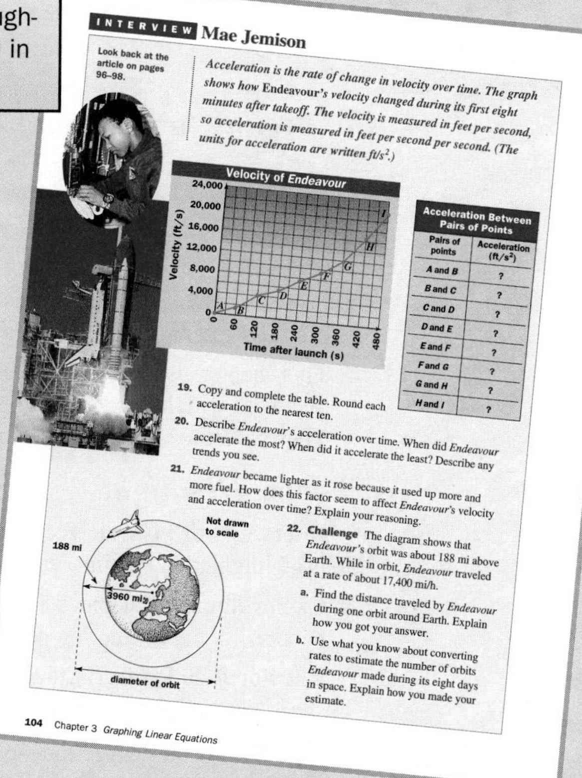

INTERVIEW Mae Jemison

Look back at the article on pages 96–98.

Acceleration is the rate of change in velocity over time. The graph shows how Endeavour's *velocity changed during its first eight minutes after takeoff. The velocity is measured in feet per second, so acceleration is measured in feet per second per second. (The units for acceleration are written ft/s².)*

Acceleration Between Pairs of Points	
Pairs of points	Acceleration (ft/s²)
A and *B*	?
B and *C*	?
C and *D*	?
D and *E*	?
E and *F*	?
F and *G*	?
G and *H*	?
H and *I*	?

19. Copy and complete the table. Round each acceleration to the nearest ten.

20. Describe *Endeavour*'s acceleration over time. When did *Endeavour* accelerate the most? When did it accelerate the least? Describe any trends you see.

21. *Endeavour* became lighter as it rose because it used up more and more fuel. How does this factor seem to affect *Endeavour*'s velocity and acceleration over time? Explain your reasoning.

22. Challenge The diagram shows that *Endeavour*'s orbit was about 188 mi above Earth. While in orbit, *Endeavour* traveled at a rate of about 17,400 mi/h.

a. Find the distance traveled by *Endeavour* during one orbit around Earth. Explain how you got your answer.

b. Use what you know about converting rates to estimate the number of orbits *Endeavour* made during its eight days in space. Explain how you made your estimate.

104 Chapter 3 *Graphing Linear Equations*

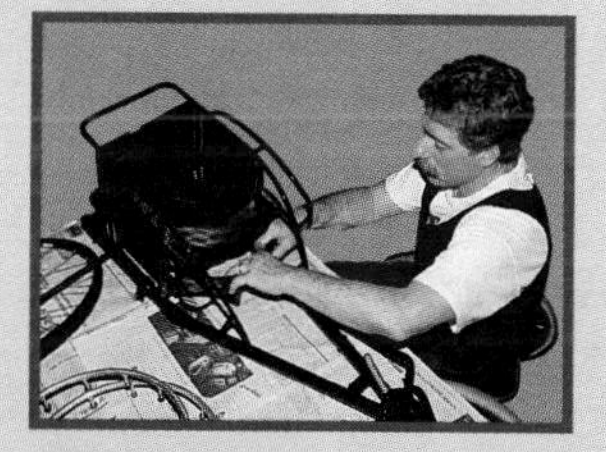

Bob Hall
Athlete and Racing Wheelchair Designer

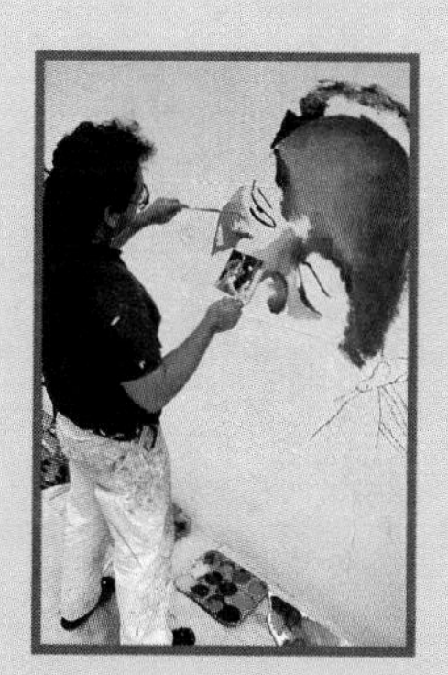

Daniel Galvez
Mural Artist

Carrie Teegardin
Journalist

CHAPTER

1 Using Algebra to Work with Data

OVERVIEW

Connecting to Prior and Future Learning

⇔ In Chapter 1, variables and inequalities, measures of central tendency, integers, and variable expressions are used to work with data. The **Student Resources Toolbox** provides a convenient review of statistical graphs on pages 596–599, inequalities on page 590, and integers on pages 588 and 589.

⇔ Problem solving is also emphasized in Chapter 1. Students can refresh their memory of using problem-solving strategies by studying pages 600 and 601 in the **Student Resources Toolbox**.

⇔ This chapter also includes an introduction to the use of spreadsheets and matrices to organize and work with large amounts of data. Students will use spreadsheets throughout this course. They will study matrices again in Algebra 2.

Chapter Highlights

Interview with Carrie Teegardin: The use of mathematics in newspaper database reporting is highlighted in this interview, with related exercises on pages 6 and 44.

Explorations in Chapter 1 involve exploring the order of operations with calculators in Section 1.2, and exploring patterns and variable expressions in Section 1.6.

The Portfolio Project: Students collect data that change over time and then use a spreadsheet, percents, and a graph to analyze the data.

Technology: Students use a scientific calculator to evaluate expressions in Section 1.2 and to add and subtract integers in Section 1.4. In Section 1.8, spreadsheets are introduced as a way to organize data and graphing calculators are used to add matrices.

OBJECTIVES

Section	Objectives	NCTM Standards
1.1	• Use variables and inequalities. • Analyze real-world data using histograms.	1, 2, 3, 4, 5, 10
1.2	• Simplify an expression using the order of operations. • Solve real-world problems.	1, 2, 3, 4, 5
1.3	• Find the mean, the median, and the mode(s) of a set of data. • Analyze and compare real-world data.	1, 2, 3, 4, 5, 10
1.4	• Find the absolute value of a number. • Add and subtract integers.	1, 2, 3, 4, 5
1.5	• Multiply and divide integers. • Use properties of addition and multiplication. • Solve problems involving negative numbers.	1, 2, 3, 4, 5
1.6	• Write variable expressions. • Solve real-world problems by using variables.	1, 2, 3, 4, 5
1.7	• Simplify variable expressions. • Use the distributive property. • Apply variable expressions in problem-solving situations.	1, 2, 3, 4, 5
1.8	• Organize data using spreadsheets and matrices. • Work conveniently with large amounts of data.	1, 2, 3, 4, 5, 10, 12

INTEGRATION

Mathematical Connections	1.1	1.2	1.3	1.4	1.5	1.6	1.7	1.8
algebra	**3–7***	**8–13**	**14–18**	**19–23**	**24–29**	**30–34**	**35–39**	**40–45**
geometry				23			39	
data analysis, probability, discrete math	**3–7**		**14–18**					**40–45**
patterns and functions						30, 31		
logic and reasoning	6, 7	11–13	**14–18**	21–23	24, 27, 29	**30–34**	38, 39	41, 42, 44
Interdisciplinary Connections and Applications								
history and geography	5			22				
biology and earth science	7		17					
chemistry and physics		12			28			
arts and entertainment						33		
sports and recreation			15	23		31		40, 41
child care, advertising, personal finance, nutrition, cooking, fundraising	5, 6					31, 33, 34	38	

**Bold page numbers* indicate that a topic is used throughout the section.*

TECHNOLOGY

Section	*opportunities for use with* Student Book	Support Material
1.1	scientific calculator	
1.2	scientific calculator	**Technology Book:** Calculator Activity 1
1.3	graphing calculator McDougal Littell Software *Stats!*	
1.4	scientific calculator graphing calculator	
1.5	scientific calculator	
1.6	scientific calculator	
1.7	scientific calculator	
1.8	graphing calculator spreadsheet software McDougal Littell Software *Stats!* *Function Investigator*	**Technology Book:** Spreadsheet Activity 1

PLANNING GUIDE

Regular Scheduling (45 min)

Section	Materials Needed	Core Assignment	Extended Assignment	*exercises that feature* Applications	Communication	Technology
1.1		1–13, 18–22, 24–34	1–34	4–10, 12–15, 17–19	6, 11, 18, 19, 23, 25, 28	
1.2	scientific calculator	**Day 1:** 1–6, 9–20, 22, 23 **Day 2:** 24–38, 41, 43–48	**Day 1:** 1–23 **Day 2:** 24–48	36–38	7 39, 44, 45	7, 21 39, 41, 42
1.3		1–4, 6–15, AYP*	1–15, AYP	3, 5	3, 4, 6–9	
1.4	scientific calculator	1–10, 12–17, 19–38, 41, 42, 46–52	1, 3, 5, 6, 8–17, 19, 21–30, 32, 35, 37–52	13–17, 43–45	18, 45, 46	42
1.5		**Day 1:** 2–11, 14–24 **Day 2:** 26–33, 35–43, AYP	**Day 1:** 1–25 **Day 2:** 26–43, AYP	13, 16	1, 11, 12, 25 35, 37	
1.6	30 toothpicks	1–9, 12, 13, 15–17, 18–26	1–26	1–6, 14–17	2, 5, 8–14, 18	
1.7	ruler	**Day 1:** 1–33 odd **Day 2:** 2–34 even, 35, 38–45	**Day 1:** 1–33 odd, **Day 2:** 2–34 even, 35–45	25	13, 15 35–37, 39	
1.8	spreadsheet, graphing calculator	1–6, 8–12, 15–21, AYP	1–21, AYP	4–6, 8–14	13–15	5, 7, 10, 12
Review/ Assess		**Day 1:** 1–15 **Day 2:** 16–31 **Day 3:** Ch. 1 Test	**Day 1:** 1–15 **Day 2:** 16–31 **Day 3:** Ch. 1 Test	23, 28	1–3, 10 28	
Portfolio Project		Allow 2 days.	Allow 2 days.			

Yearly Pacing (with Portfolio Project)	**Chapter 1 Total** 16 days	**Remaining** 144 days	**Total** 160 days

Block Scheduling (90 min)

	Day 1	Day 2	Day 3	Day 4	Day 5	Day 6	Day 7	Day 8	Day 9
Teach/ Interact	1.1	1.2: Exploration, page 8	1.3 1.4	1.5	1.6: Exploration, page 30 1.7	Continue with 1.7 1.8	Review Port. Proj.	Review Port. Proj.	Ch. 1 Test 2.1
Apply/ Assess	**1.1:** 1–15, 18–22, 24–34	**1.2:** 1–20, 22–48	**1.3:** 1–15, AYP* **1.4:** 1–10, 12–17, 19–37 odd, 39–52	**1.5:** 2–11 14–33, 35–43, AYP	**1.6:** 1–13, 15–26 **1.7:** 1–33 odd	**1.7:** 2–34 even, 35, 38–45 **1.8:** 1–21, AYP	**Review:** 1–15 **Port. Proj.**	**Review:** 16–31 **Port. Proj.**	**Ch. 1 Test** **2.1:** 2–6 10–41

NOTE: A one-day block has been added for the Portfolio Project—timing and placement to be determined by teacher.

Yearly Pacing (with Portfolio Project)	**Chapter 1 Total** $8\frac{1}{2}$ days	**Remaining** $72\frac{1}{2}$ days	**Total** 81 days

***AYP** *is Assess Your Progress.*

Lesson Support

Section	Practice Bank	Study Guide*	Assessment Book*	Visuals	Explorations Lab Manual	Lesson Plans	Technology Book
1.1	1	1.1		Warm-Up 1.1		1.1	
1.2	2	1.2		Warm-Up 1.2		1.2	Calculator Act. 1
1.3	3	1.3	Test 1	Warm-Up 1.3		1.3	
1.4	4	1.4		Warm-Up 1.4	Add. Expl. 2	1.4	
1.5	5	1.5	Test 2	Warm-Up 1.5		1.5	
1.6	6	1.6		Warm-Up 1.6	Master 6	1.6	
1.7	7	1.7		Warm-Up 1.7 Folders 1, B	Master 4	1.7	
1.8	8	1.8	Test 3	Warm-Up 1.8		1.8	Spreadsheet Act. 1
Review Test	9	Chapter Review	Tests 4, 5 Alternative Assessment			Review Test	Calculator Based Lab 1

***Spanish versions** of *Study Guide* and *Assessment Book* are available.

Chapter Support

- Course Guide
- Lesson Plans
- Explorations Lab Manual: Additional Exploration 1
- Portfolio Project Book: Additional Projects 1, 2: The History of Algebra, Absolute Value
- Preparing for College Entrance Tests
- Multi-Language Glossary
- *Test Generator* Software
- Professional Handbook

Software Support

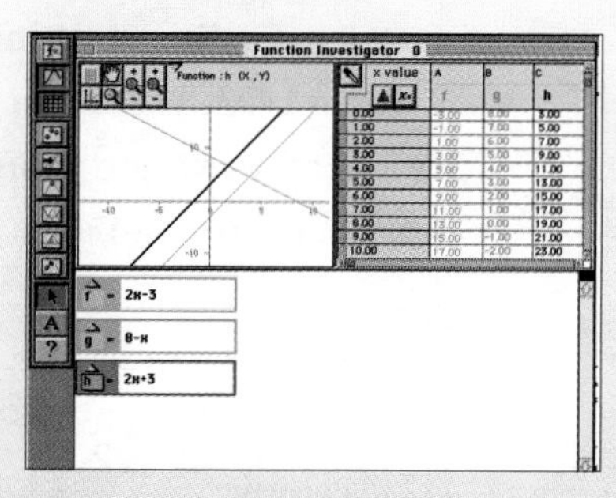

McDougal Littell Software
Stats!
Function Investigator

Internet Support

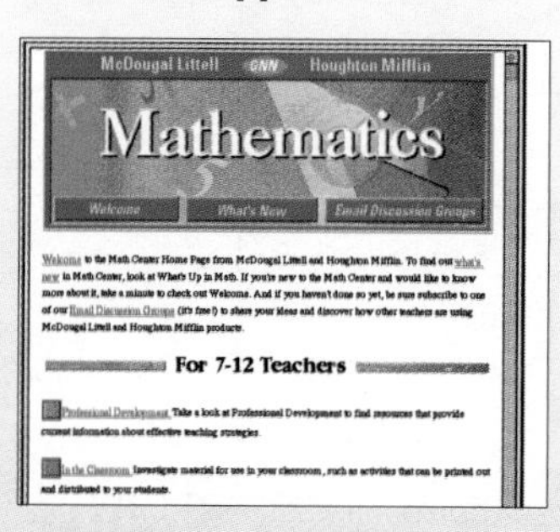

http://www.hmco.com
Next go to McDougal Littell; then the Education Center; then Secondary Math.

Outside Resources

Books, Periodicals

"Exploring Data," from the *Quantitative Literacy Series*; written by members of the Joint Committee on the Curriculum in Statistics and Probability of the American Statistical Assoc. and the NCTM. Dale Seymour publication.

Wilson, Melvin R. (Skip) and Carol M. Krapfl. "Exploring Mean, Median, and Mode with a Spreadsheet." *Mathematics Teaching in the Middle School,* (Sept.–Oct. 1995): pp. 490–495.

Countryman, Joan. *Writing to Learn Mathematics.* Portsmouth, NH: Heinemann Educational Books, 1992.

Driscoll, Mark and Jere Confrey, eds. *Teaching Mathematics: Strategies that Work.* Chelmsford, MA: Northeast Regional Exchange, Inc., 1985.

Activities, Manipulatives

Illingworth, Mark. *A Graphing Matter: Activities for Easing into Algebra.* Berkeley, CA: Key Curriculum Press, 1995.

Algebra Tiles for the Overhead Projector. Instructional booklet and 70-piece set provide materials to model addition and subtraction of polynomials, and so on. Hayward, CA: Activity Resources Co., Inc.

Spikell, Mark A. *Teaching Mathematics with Manipulatives: A Resource of Activites for the K–12 Teacher.* Needham Heights, MA: Allyn and Bacon, 1993.

Software

"How the West was One + Three × Four." Order of operations challenge. Pleasantville, NY: Sunburst/Wings for Learning, 1994.

Videos

Algebra for Everyone: Videotape and discussion guide. Reston, VA: NCTM, 1991.

Internet

For great discussions among mathematics teachers, subscribe to the NCTM list by sending e-mail to:
majordomo@forum.swarthmore.edu
In the body of the message type:
subscribe nctm-l [firstname, lastname]

Interview Notes

Background

Journalism
Journalism is an occupation that is concerned with the daily news. Journalists are engaged in reporting, writing, editing, photographing or broadcasting news, usually for a newspaper, magazine, or television studio. Students can prepare for a career in journalism by enrolling in a school of journalism associated with a college or university. Sometimes journalists develop stories that are based on an in-depth analysis of events or data. Carrie Teegardin turned numbers into news by using the findings of the 1990 census to take a statistical snapshot of every child in the South. She also traveled the South extensively and talked to hundreds of children about growing up Southern. Her statistical data and records of interviews provided the basis for her report *Growing Up Southern.*

Carrie Teegardin
Carrie Teegardin is a staff writer for the *Atlanta Journal and Constitution.* Her specialty is to cover demographics for the newspaper. To write her report, she began with a simple question: Who are the children of the South and what makes them Southern? She started to answer this question by using census data to analyze hundreds of thousands of statistics that covered every aspect of Southern children's lives, from poverty and television to shopping and religion. In so doing, she was able to create a portrait of these children that was rounded out through the use of actual interviews. Her final report was published in a series of eight parts during June 1993.

CHAPTER

1 Using Algebra to Work with Data

Turning numbers into NEWS

INTERVIEW Carrie Teegardin

Carrie Teegardin liked algebra in high school, but she never dreamed it would be so important to her career. As a reporter at the *Atlanta Journal and Constitution* in Georgia, Teegardin does database reporting, a job that requires finding meaning hidden in large sets of numbers. In 1993, Teegardin produced an eight-part series called "Growing Up Southern." She used data and interviews to investigate the lives of American children. On a computer, she analyzed data from 3248 counties.

"Math helps you see the world clearly ... you need math to figure out what's happening around you."

Turning Numbers into Percents

For each county, Teegardin turned numbers into percents. The percents helped her compare aspects of life across the country, from income level to the percent of working mothers. "It's like organizing batting averages," says Teegardin. "You may know somebody's hitting well, but you don't know exactly how well until you quantify their hitting with a number and compare it to others."

Practical Problem Solvers

Teegardin also interviewed people about the real life behind the statistics. She talked to a cross section of teens about their families, their sense of the future, and current problems such as drugs and poverty. "Teens hear people saying they're going to be the first generation that won't be better off than their parents," says Teegardin. "They worry about being able to afford college. They're the first generation not to know life without AIDS. Even so, I got the impression that a lot of kids want to be practical problem solvers for themselves and society." One student she interviewed joined the Marines so he could afford to go to college and become a pediatrician. He wants to help kids like himself who don't get the attention he thinks they deserve from their doctors. "I like that," says Teegardin. "I like the fact that teens are determined to make their lives work and make the world a better place in a very practical way."

> "...teens are determined to make their lives work..."

1

Interview Notes

Background

Journalism Pioneer
Some historians credit Ida Tarbell (1857–1944) with being one of the pioneers of modern journalism. She is best known for her meticulous and scathing "History of the Standard Oil Company," first published in installments in *McClure's* magazine from 1902 to 1904. Tarbell taught science for a time, and the influence of her knowledge of scientific inquiry is evident in her highly detailed, fact-filled reporting.

Second-Language Learners

Some students learning English might need an explanation of the term *cross section* as meaning a sample that is typical of the whole. On the next page, you may need to explain that the phrase *boil thousands of them [numbers] down to four or five* means "taking thousands of numbers, analyzing them, and coming up with fewer percentages that make sense to the reader."

Interview Notes

Mathematical Connection

The analysis of real-world data often involves the use of algebra and graphs, topics that students study in Section 1.1. In this section, students learn to use variables, percents, inequalities, and bar graphs to answer questions about the amount of TV eighth-grade students watch. In Section 1.8, students analyze teenage employment using spreadsheets and matrices.

Explore and Connect

Writing/Research
Discuss questions 1 and 2 as a whole class activity. Make sure students understand the meaning of the percent change in child population shown by the small squares above the map.

Project
Allow students to take a few days to research their articles. You may wish to ask for volunteers to respond with verbal reports, or students can pass them around the class and select one to read at home.

Organizing Data

As Teegardin weaves together personal interviews and data, she has to organize the data to make sense of them. "The power of numbers is that you can combine them, and manipulate them, and boil thousands of them down to four or five," says Teegardin. "Then you can convert them to forms that readers can understand, like maps, charts, and graphs." For example, during the 1980s, families with children tended to move from rural areas to urban areas. The map below gives you a picture of this shift in population.

Where children live
The number of children in the United States remained almost unchanged over the 1980s, while the elderly population grew rapidly. Rural areas saw an exodus of nearly 1 million children, while metro areas – particularly Atlanta, Boston, Dallas, Los Angeles and Washington – saw their under-18 population grow substantially.
Percent change in child population, 1980 to 90
-10% or more
-10% to 0
0 to 10%
+ 10% or more
Missing
Growing Up SOUTHERN
Kids in America
KID DATA
PERSONAL MONEY GUIDE
BUSINESS
The Kids Take Over
Fashioning Identity
Southern teens spend more and save less

Explore and Connect

Carrie Teegardin took six months to gather and analyze the data to create a statistical portrait of Southern children.

1. Writing According to the map above, in what parts of the United States did the largest percent change in child population take place? In what parts did the smallest percent change take place? In what parts did the number of children stay about the same? How does using percents help you compare population changes in different parts of the country?

2. Research What percent of the students in your class have lived in at least one other state? How did you calculate the percent?

3. Project Find a newspaper or magazine article about a topic that interests you. The article should include at least one table, graph, or map. Share the article with your class.

Mathematics & Carrie Teegardin

In this chapter, you will learn more about how mathematics is related to database reporting.

Related Exercises

Section 1.1
- **Exercises 12–16**

Section 1.8
- **Exercises 8–14**

SECTION

1.1 Working with Variables and Data

Learn how to...
- use variables and inequalities

So you can...
- analyze real-world data, such as the heights of different students

Numbers can describe your height, your grades, even where you live. Numbers that give you information are called **data**. For example, the heights of the students in the photo are data.

Notice that the heights vary from student to student. You can use a letter called a **variable** to stand for a quantity that varies.

Let the variable h stand for the heights of the students in inches. Each student's height is called a **value** of h.

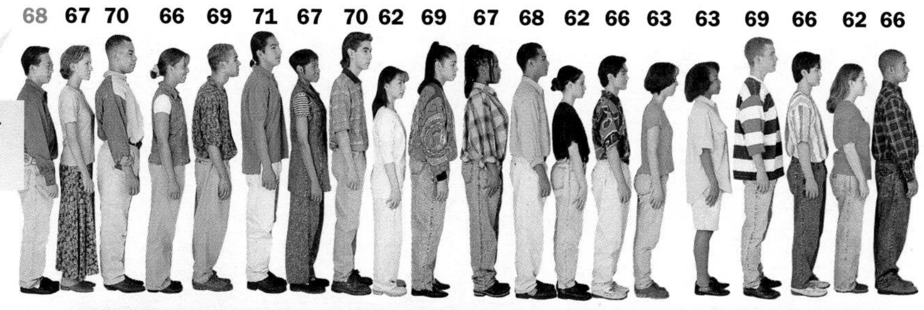

The value of h for this student is 68.

EXAMPLE 1 Application: Statistics

a. What is the smallest value of h for the students above? What is the largest?

b. Write two inequalities that describe the heights of the students.

Toolbox p. 590 *Inequalities*

SOLUTION

a. The shortest students are 62 in. Therefore, the smallest value of h is 62. The tallest student is 71 in. Therefore, the largest value of h is 71.

b. All students are 62 in. or taller. $h \geq 62$
All students are 71 in. or shorter. $h \leq 71$

THINK AND COMMUNICATE

1. Explain why you can write $h > 61$ and $h < 72$ to describe the students' heights. How are these two inequalities different from the ones in part (b) of Example 1?
2. Let d = the distance in miles traveled from home to school. Let n = the number of telephone calls someone made yesterday. Give some reasonable values of each of these variables.

Plan⇔Support

Objectives
- Use variables and inequalities.
- Analyze real-world data using histograms.

Recommended Pacing
Core and Extended Courses
Section 1.1: 1 day
Block Schedule
Section 1.1: 1 block

Resource Materials
Lesson Support
Lesson Plan 1.1
Warm-Up Transparency 1.1
Practice Bank: Practice 1
Study Guide: Section 1.1
Technology
Internet:
http://www.hmco.com

Warm-Up Exercises
1. The time of the day is always changing. If the letter t represents time, what time is it now? Will t be different one second from now? one minute from now? Answers will vary; Yes; Yes.

Read each inequality.

2. $t > 2$ t is greater than two.
3. $t < 4$ t is less than four.
4. $t \geq 7$ t is greater than or equal to seven.
5. $t \leq 9$ t is less than or equal to nine.

ANSWERS Section 1.1

Think and Communicate

1. because all the students' heights are greater than 61 in. and are less than 72 in.; The inequalities in part (b) use symbols that mean "greater than *or equal to*" and "less than *or equal to*," while the inequalities in this question use symbols that mean "greater than" and "less than."
2. Answers may vary. Examples are given. d: $\frac{1}{10}$, 0.5, 2; n: 0, 1, 3

Teach⇔Interact

About Example 1

Communication: Reading The concept of a variable can be reinforced by having individual students each read one of the heights shown in the photo.

Additional Example 1

Suppose the grades, g, on a math quiz were 73, 94, 82, 67, 79, 99, 85, 63, 97, 80, 89, 74, 90, 65, 77.

a. What is the lowest grade? What is the highest grade? 63; 99

b. Write two inequalities that describe the grades of the students. $g \geq 63$; $g \leq 99$

Think and Communicate

Since some students may have difficulty grasping the meaning of a variable, you may wish to continue with additional activities such as those in question 2. One student can suggest a real-world situation that can be represented by a variable and the others can give some reasonable values.

Additional Example 2

Use the histogram from Example 2.

a. How many students are taller than 64 in.? Count the number of students represented by the bars to the right of $h = 64$. Nine students are taller than 64 in.

b. How many students are at most 66 in. tall? Two students are 66 in. tall. Thirteen students are less than 66 in. tall. A total of 15 students are at most 66 in. tall.

Section Note

Integrating the Strands
In working with variables and data, students can see how algebra is used to describe data that are organized in a graph. This approach integrates the strands of algebra and statistics.

Reading Data from Graphs

One way to organize data is in a graph. If you group the students by height, you can make a graph called a *histogram.*

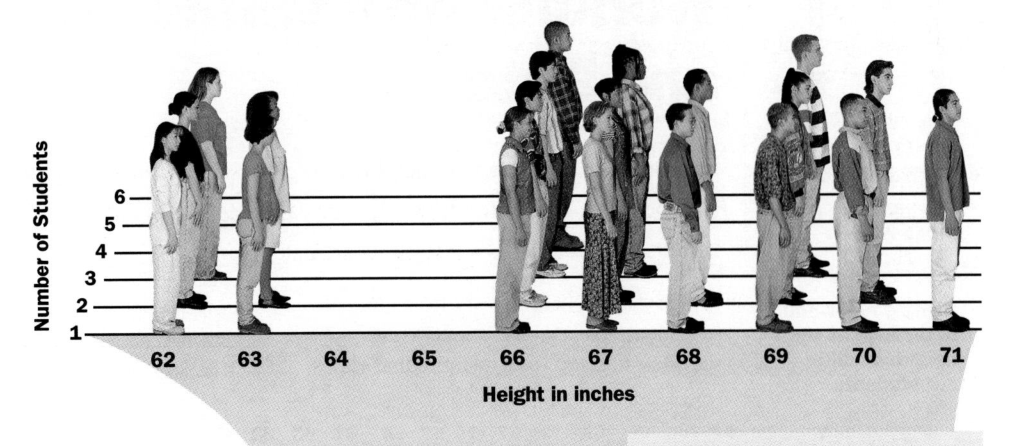

The bars show the number of students for each height. The bar at $h = 69$ is three units high. Therefore, **3** students are **69 in.** tall.

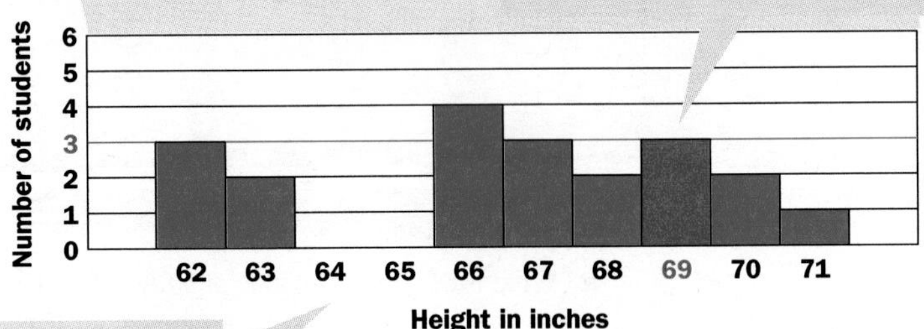

The **values of *h*** are shown along the bottom of the histogram.

EXAMPLE 2 Application: Statistics

Use the histogram.

a. How many students are shorter than 66 in.?

b. How many students are at least 68 in. tall?

SOLUTION

a. Count the number of students represented by the bars to the left of $h = 66$. Five students are shorter than 63 in.

b. Two students are 68 in. tall. Six students are more than 68 in. tall. A total of eight students are at least 68 in. tall.

Checking Key Concepts

1. Answers may vary.
2. Answers may vary.
3. Answers may vary. An example is given. I disagree. It is true that $h > 0$ and $h < 100$, but these inequalities apply to the heights of all human beings and give no useful description of this particular group.
4. 2; The bar at $h = 63$ is 2 units high.
5. 17 students
6. 17; 8; To find the number of students whose heights are less than 70 in., add the number of students with heights of 62, 63, 66, 67, 68, and 69 in. To find the number of students whose heights are 68 in. or greater, add the number of students with heights of 68, 69, 70, and 71 in.
7. that you spend at most 6 h visiting
8. Answers may vary. An example is given. In question 6, h represents a student's height in inches, while in question 7, h represents an interval of time in hours. Also, the values for h in question 6 are integers, while the values for h in question 7 could be fractions or decimals.

CHECKING KEY CONCEPTS

For Questions 1 and 2, let t = outdoor temperature.

1. Give several reasonable values of t for your part of the world.
2. Use the variable t to write two inequalities describing temperatures where you live.
3. Sandra says that you can use the inequalities $h < 100$ and $h > 0$ to describe the heights of the students shown on page 3. Explain why you agree or disagree.

For Questions 4–6, use the histogram on page 4.

4. How many students are 63 in. tall? How do you know?
5. How many students are at least 63 in. tall?
6. For how many students is $h < 70$? For how many students is $h \geq 68$? How can you tell from the histogram?
7. Let h = the number of hours you spend visiting a friend. What does $h \leq 6$ mean?
8. Compare the variables in Questions 6 and 7 above. How are they different?

1.1 Exercises and Applications

Extra Practice exercises on page 557

HISTORY **For Exercises 1–3, use the histogram at the right.**

1. How many states were admitted between 1925 and 1949?
2. How many states were there at the end of 1849?
3. How many states were admitted after 1874?

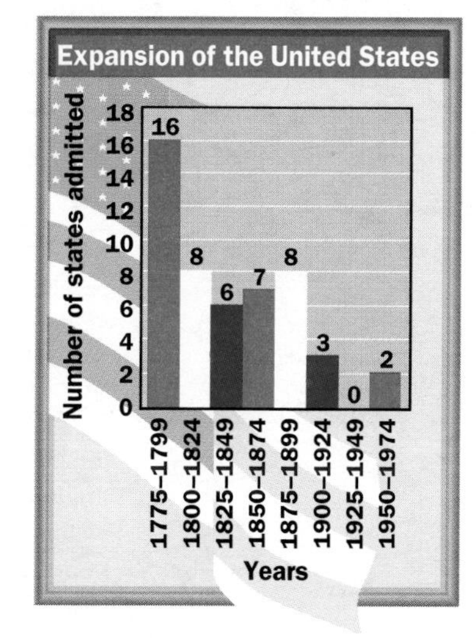

CHILD CARE **For Exercises 4–6, use the data in the chart at the left.**

Age at which Infants First Walk

Name	Age (months)	Name	Age (months)
John	14	Cora	12
Lynne	12	Theo	11
Parker	12	Bryan	14
Andrea	13	Yi Hsa	13
Brad	11	Alison	12

4. Write two inequalities that describe the ages at which the infants began walking.
5. What percent of the infants began walking before 12 months? after 12 months?
6. **Writing** Who might be interested in these data and your answers to Exercises 4 and 5? Explain.

Teach⇔Interact

Checking Key Concepts

Communication: Discussion
Questions 1–8 can be used as a basis for a class discussion to review and strengthen students' understanding of the objectives of this section. Ask students to suggest another way to express phrases such as *at least* (no less than) and *at most* (no greater than).

Closure Question

If the variable n represents any number from 20 to 30, write two inequalities that describe the value of n.
$n \geq 20$; $n \leq 30$

Apply⇔Assess

Suggested Assignment

Core Course
Exs. 1–13, 18–22, 24–34

Extended Course
Exs. 1–34

Block Schedule
Day 1 Exs. 1–15, 18–22, 24–34

Exercise Notes

Application
Exs. 1–11, 18, 19 Applications of algebra to real-world topics and situations help to make the study of mathematics relevant and interesting to students. In these exercises, there are applications to history, child care, advertising, and biology.

Student Study Tip
Ex. 1 Point out to students that the word *between* is used in a way that includes the years 1925 and 1949. This word is also used in mathematics as a technical term having a different meaning than the one used here. For example, the whole numbers between 20 and 30 are the numbers 21, 22, 23, 24, 25, 26, 27, 28, 29; that is, 20 and 30 are *not* included.

Exercises and Applications

1. 0
2. 30 states
3. 13 states
4. Let a = the age in months at which an infant began walking. Then $a > 10$ and $a < 15$, or $a \geq 11$ and $a \leq 14$.
5. 20%; 40%
6. Answers may vary. Examples are given. Parents and doctors might be interested because they might want to be able to compare a particular infant's development to that of other children. Manufacturers of infants' shoes might use the information in planning for production and advertising.

Apply⇔Assess

Exercise Notes

Mathematical Procedures
Exs. 7–10 Students should understand that they can use any letter of the alphabet to represent a variable. A customary procedure is to choose a letter that is descriptive of the actual situation, such as *h* for height, *d* for distance, *t* for time, or *n* for number.

Problem Solving
Ex. 11(b) Students are asked to write their own ad for an imaginary product. Such open-ended problems allow students to think for themselves and to express their originality and creativity. They also reveal a great deal of information about students' understanding of the concepts involved in the situation.

Second-Language Learners
Ex. 11(b) Have students who are learning English work with English-proficient students to write the advertisement.

Interview Note
Exs. 12, 13, 15 You may need to remind some students of the meaning of percent. For Ex. 15, challenge students to write their two inequalities as a single inequality. For example, if $h \le 6$ and $h \ge 2$, then this can be written as $2 \le h \le 6$.

Topic Spiraling: Preview
Ex. 23 Many important ideas and concepts can be introduced to students in an informal and intuitive way. This exercise introduces students to the idea that the value of a variable can depend upon certain factors. Since the factors themselves can also vary (such as pay rate and hours worked), an intuitive basis is being established to build an understanding of the concept of a functional relationship between two variables.

ADVERTISING Choose a variable and write an inequality for each ad. Explain what each variable stands for.

7. 

8. 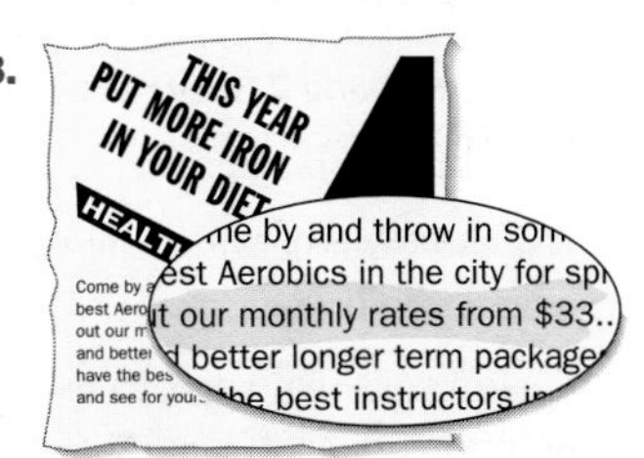

9. 

10. 

11. a. **Writing** Look back at the ads in Exercises 7–10. Why do you think advertisers use phrases that suggest inequalities rather than providing specific prices?
    b. **Open-ended Problem** Write your own ad for an imaginary product. Explain how your ad suggests an inequality.

INTERVIEW Carrie Teegardin

Look back at the article on pages xxx–2.

In "Growing Up Southern," Carrie Teegardin reported on the amount of TV eighth-grade students watch. The graph shows some data she used. Let h = the number of hours spent watching TV each day.

Carrie Teegardin at work in the *Atlanta Journal and Constitution* office.

12. For about what percent of eighth graders is $h = 3$?

13. For about what percent of eighth graders is $h \ge 6$?

14. **Challenge** Tom says that, according to the graph, most eighth graders in the South watch four or five hours of TV daily. Do you agree? Use the graph to support your answer.

15. What is the greatest amount of TV you watch in a day? the least? Write two inequalities that describe your daily viewing habits.

16. **Writing** The graph shown below is a *bar graph*. How is it different from a histogram?

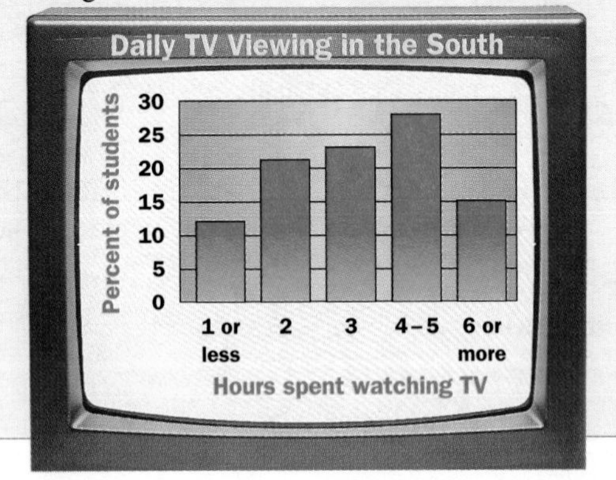

7. Let e = weekly earnings; $e \le 600$.
8. Let r = monthly rate; $r \ge 33$.
9. Let l = price of lunch; $l \ge 3.95$. Let d = price of dinner; $d \ge 4.95$.
10. Let n = the number of hours the movers work; $n \ge 2$.
11. Answers may vary. Examples are given.
    a. Advertisers want to put only the most attractive prices in their ads.
    b. "Own your own home at last! Sunnyside Acres has beautiful condominiums in the desirable Forestvale community starting at just $80,000. Call 555-6543 today for an appointment. You'll be glad you did." The ad suggests the inequality $p \ge 80{,}000$, where p is the price of a condominium.
12. Estimates may vary; about 23%.
13. about 15%
14. Answers may vary. An example is given. I disagree. More than half (about 55%) of the students watch less than 4 h of TV daily.
15. Answers may vary.
16. Answers may vary. An example is given. The bars in a histogram show how many data items are in each category. A bar graph can display any numerical data along the vertical axis, such as percents or actual heights, not just the number in each category. Also, the bars in a histogram are connected, while the bars in a bar graph are not.
17. a, b. Answers may vary.
18. If w = the weight in grams, then $w \ge 2$ and $w \le 500{,}000$; if w = the weight in kilograms, then $w \ge 0.002$ and $w \le 500$.
19. If w = the weight in grams, then $w \ge 2$ and $w \le 30{,}000{,}000$; if w = the

17. a. Research Find out how many pets each student in your class has. Make a histogram of the data.

b. Let $p =$ the number of pets each student has. Write two inequalities to describe the values of p.

BIOLOGY For each statement, choose a variable and write at least two inequalities. Explain what each variable stands for and what units you are using. Remember that 1 kg = 1000 g.

18. Throughout the history of Earth, birds have ranged in size from the bee hummingbird (2 g) to the now extinct elephant bird (about 500 kg).

19. The smallest land mammal is Savi's pygmy shrew (2 g). The largest land mammal ever was the Baluchitherium, a prehistoric giant rhinoceros (about 30,000 kg).

BY THE WAY

The Baluchitherium lived about 35 million years ago. It was so large that a group of six people could have stood under it with plenty of extra room.

For Exercises 20–22, give several reasonable values of each variable.

20. the number of words w a person says in an hour

21. the cost c in dollars of a pair of sneakers

22. the number of people n at a concert

23. Open-ended Problem Many factors can cause the values of a variable to vary. For example, the amount a of your paycheck may depend on your pay rate and how many hours you work. Choose one of the variables in Exercises 20–22. Describe some factors that could cause the values of the variable to vary.

24. SAT/ACT Preview If 2 identical packages contain a total of 12 bagels, how many bagels are there in 5 packages?

A. 12 **B.** 24 **C.** 30 **D.** 36 **E.** 60

ONGOING ASSESSMENT

25. Writing Describe three ways to represent data. Compare the advantages and disadvantages of the three representations. When might each be used?

SPIRAL REVIEW

For Exercises 26–28, use the bar graph at the right. *(Toolbox, page 596)*

26. Which country shown had the highest gasoline price in 1992?

27. Which had the lowest price?

28. Writing How else can the bars on this graph be arranged? Why doesn't it make sense to rearrange the bars of the histogram on page 5?

1992 Gasoline Prices

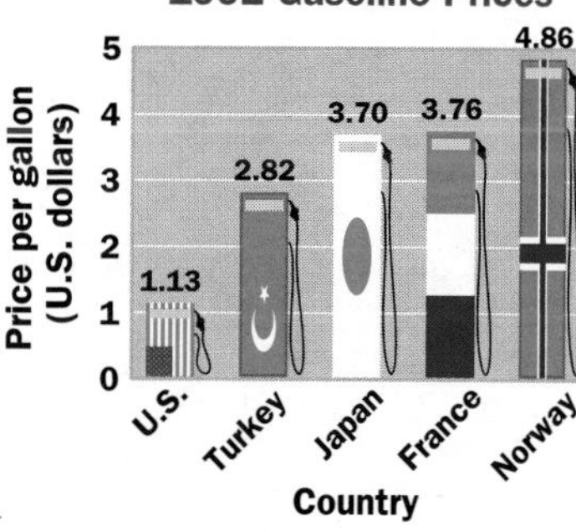

Write each number as a decimal rounded to the nearest hundredth. *(Toolbox, page 587)*

29. $\frac{8}{3}$ **30.** $\frac{3}{7}$ **31.** $\frac{100}{18}$ **32.** $5\frac{7}{9}$ **33.** $\frac{1}{8}$ **34.** $3\frac{1}{3}$

Apply⇔Assess

Exercise Notes

Common Error

Ex. 24 A common error students make when solving problems is to not read the problem carefully. For example, in this exercise, some students may select 60 as the answer because they multiplied 5 and 12. These students did not read carefully that the two packages contain a *total* of 12 bagels, thus there are 6 bagels in a package and $6 \times 5 = 30$.

Assessment Note

Ex. 25 An essential aspect of learning mathematics is to be able to express ideas in written form. Writing activites allow students an opportunity to display their thinking and understanding of mathematics. You might wish to suggest that students' reponses to the Ongoing Assessment exercises be kept in a mathematics journal.

Practice 1 for Section 1.1

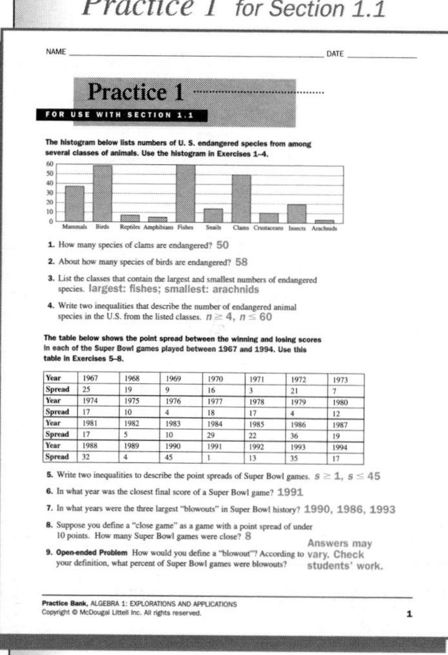

NAME ________________ DATE ________

Practice 1

FOR USE WITH SECTION 1.1

The histogram below lists numbers of U. S. endangered species from among several classes of animals. Use the histogram in Exercises 1–4.

1. How many species of clams are endangered? 50
2. About how many species of birds are endangered? 58
3. List the classes that contain the largest and smallest numbers of endangered species. largest: fishes; smallest: arachnids
4. Write two inequalities that describe the number of endangered animal species in the U.S. from the listed classes. $n \geq 4$, $n \leq 60$

The table below shows the point spread between the winning and losing scores in each of the Super Bowl games played between 1967 and 1994. Use this table in Exercises 5–8.

Year	1967	1968	1969	1970	1971	1972	1973
Spread	25	19	9	16	3	21	7
Year	1974	1975	1976	1977	1978	1979	1980
Spread	17	10	4	18	17	4	12
Year	1981	1982	1983	1984	1985	1986	1987
Spread	17	5	10	29	22	36	19
Year	1988	1989	1990	1991	1992	1993	1994
Spread	32	4	45	1	13	35	17

5. Write two inequalities to describe the point spreads of Super Bowl games. $s \geq 1$, $s \leq 45$
6. In what year was the closest final score of a Super Bowl game? 1991
7. In what years were the three largest "blowouts" in Super Bowl history? 1990, 1986, 1993
8. Suppose you define a "close game" as a game with a point spread of under 10 points. How many Super Bowl games were close? 8
9. **Open-ended Problem** How would you define a "blowout"? According to your definition, what percent of Super Bowl games were blowouts? Answers may vary. Check students' work.

Practice Bank, ALGEBRA 1: EXPLORATIONS AND APPLICATIONS

weight in kilograms, then $w \geq 0.002$ and $w \leq 30{,}000$.

20–22. Answers may vary. Examples are given.

20. 100, 500, 2000

21. 29.95, 49.99, 119.99

22. 3000, 12,000, 20,000

23. Answers may vary. Examples are given.
(20) being alone, being in a class or a meeting, being the speaker at a lecture
(21) the type of shoe, the brand name, whether the shoe is for a young child, a teen, or an adult
(22) the popularity of the performers, the cost of tickets, the location of the concert, the seating capacity of the performance site

24. C

25. Answers may vary. For example, you can represent data by using a table (as in Exs. 4–6), a histogram (as in Exs. 1–3), or inequalities (as in Exs. 18 and 19). An advantage of a table is that it gives exact values. Disadvantages are that a table can require a lot of room and can be difficult to interpret. An advantage of a histogram is that it is a visual summary of data and is easy to interpret. A disadvantage is that it does not provide exact data values and may not include all the available information. Inequalities are compact and give a range of data values, but do not give information about how many data values there are, or what the exact data values are.

26. Norway

27. United States

28–34. See answers in back of book.

Plan⇔Support

Objectives

- Simplify an expression using the order of operations.
- Solve real-world problems.

Recommended Pacing

Core and Extended Courses
Section 1.2: 2 days

Block Schedule
Section 1.2: 1 block

Resource Materials

Lesson Support
Lesson Plan 1.2
Warm-Up Transparency 1.2
Practice Bank: Practice 2
Study Guide: Section 1.2

Technology
Technology Book:
Calculator Activity 1
Calculator
Internet:
http://www.hmco.com

Warm-Up Exercises

Evaluate.

1. 11 + 4 15
2. 15 ÷ 3 5
3. 27 − 13 14
4. 9 • 7 63
5. $\frac{7}{3}+\frac{2}{3}$ 3
6. $\frac{5}{6}-\frac{1}{6}$ $\frac{2}{3}$
7. 9 • 9 81
8. 4 • 4 • 4 64

SECTION

1.2 Using the Order of Operations

Learn how to...

- simplify an expression using the order of operations

So you can...

- solve real-world problems, such as estimating the height of a waterfall or a tall bridge

While studying for their math exam, Robin and Jeff had a pizza delivered. The pizza cost $12.50, and they gave the driver a $2.00 tip. To figure out how much each of them should pay, they both used the keystrokes 12.5 [+] 2 [÷] 2 [=] on their calculators. Why did they get two different answers, as shown below?

EXPLORATION
COOPERATIVE LEARNING

Calculator Order of Operations

Work with a partner.
You will need:

- a calculator

1 Reread the paragraph above about Robin and Jeff. Whose answer makes more sense? Why?

2 Try using Robin's and Jeff's keystrokes on your calculator. Did your calculator add first, or did it divide first? How can you tell?

3 Which operation did Robin's calculator do first? Jeff's? How do you know?

4 Press the following keys on your calculator:

24 [÷] 7 [−] 3 [=]

Which operation did the calculator do first?

Now try: 24 [÷] [(] 7 [−] 3 [)] [=]

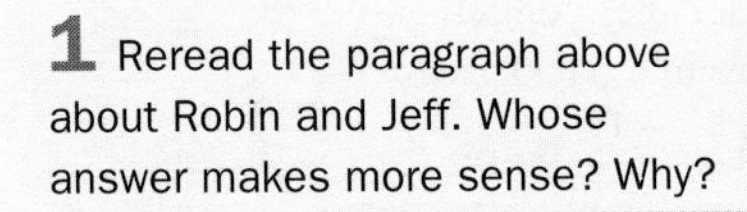

Which operation did the calculator do first this time?

5 Explain how Jeff can use parentheses keys to figure out how much to contribute for pizza. Test your answer on your calculator.

Jeff needs to use parentheses because his calculator follows the *order of operations*, a set of rules people agree to use so that an expression has only one value. You should use the order of operations when you simplify expressions.

Exploration Note

Purpose
The purpose of this Exploration is to have students understand why an order of operations is needed.

Materials/Preparation
Students should have as many different types of calculators available as possible.

Procedure
Before beginning, make sure that each pair of students has two different calculators and that one of the calculators has parentheses keys. Students should record their answers to each step.

Closure
Groups can share their answers and results with each other. All students should arrive at an understanding that when simplifying a numerical expression, an order of operations is essential so that the expression has only one value.

Explorations Lab Manual
See the Manual for more commentary on this Exploration.

Order of Operations

1. First do all work inside parentheses.
2. Then do all multiplications and divisions in order from left to right. Multiplication and division are usually written like this:

   Multiplication: $5 \cdot 3$, $5(3)$, or $(5)(3)$ **Division:** $10 \div 5$ or $\frac{10}{5}$
3. Then do all additions and subtractions in order from left to right.

EXAMPLE 1

Simplify each expression.

a. $7 + 3 \cdot 5 \cdot 4$ **b.** $\frac{9}{4} + 3(6 - 3)$

SOLUTION

a. First do the **multiplications** in order from left to right.

$$7 + 3 \cdot 5 \cdot 4 = 7 + 15 \cdot 4$$
$$= 7 + 60$$
$$= 67$$

Then do the addition.

b. First do the **subtraction** inside the parentheses.

$$\frac{9}{4} + 3(6 - 3) = \frac{9}{4} + 3(3)$$
$$= 2.25 + 9$$
$$= 11.25$$

Then divide and multiply.

Do the addition last.

Variable Expressions

Suppose two students buy a pizza for p dollars and give a tip of t dollars. If they split the total cost, they each pay $\frac{p+t}{2}$ dollars. The fraction $\frac{p+t}{2}$ is an example of a *variable expression.*

A fraction bar groups operations. For example, $\frac{p+t}{2}$ means $(p + t) \div 2$.

A **variable expression** is made up of **variables**, **numbers**, and **operations**. The value of a variable expression depends on what values you choose for the variables. For example, you can **evaluate** the variable expression $s + r$ when $s = 20$ and $r = 3$:

$$s + r = 20 + 3$$
$$= 23$$

Examples of Variable Expressions	
$a + b + c$	$10x - 3$
$\frac{t}{60}$	$s + r$
$2(l + w)$	x^2
$2lw$	$-3v + 1.5$

Teach⇔Interact

Section Notes

Teaching Tip

The order of operations can be illustrated by using a modified version of the expression in Step 4 of the Exploration: $24 \div (7 - 3) + 5 = ?$

Second-Language Learners

Be sure students learning English understand that *order* here means "what is done first, what is done next, and so on," and that *operations* refers to adding, subtracting, multiplying, and dividing numbers.

Additional Example 1

Simplify each expression.

a. $12 - 9 \div 3 + 7$

First do the division in order from left to right. Then do the subtraction and addition.

$12 - 9 \div 3 + 7$
$= 12 - 3 + 7$
$= 9 + 7$
$= 16$

b. $15 + 2 \cdot 9 \div 3 - 4$

First do the multiplication and division in order from left to right. Then do the addition and subtraction.

$15 + 2 \cdot 9 \div 3 - 4$
$= 15 + 18 \div 3 - 4$
$= 15 + 6 - 4$
$= 21 - 4$
$= 17$

Section Note

Communication: Reading

You may wish to have individual students read aloud the examples of variable expressions given on this page. Ask what x^2 means.

ANSWERS Section 1.2

Exploration

1. Robin's answer; They gave the driver \$14.50 in all, so \$7.25 is a reasonable estimate of half and \$13.50 is not.
2. Answers may vary. If your answer is 13.50, your calculator divided first. If your answer is 7.25, your calculator added first.
3. addition; $14.50 \div 2 = 7.25$; division; $12.50 + 1 = 13.50$
4. 0.428571428; division; 6; subtraction
5. If Jeff enters a left parenthesis before he enters 12.50 and a right parenthesis after he enters the first 2, the calculator will add 12.5 and 2 before dividing.

Teach⇔Interact

About Example 2

Problem Solving
By having students write and evaluate a variable expression, this Example not only illustrates these particular skills, but it also prepares students for the first key step in solving real-world problems. The first step necessary to solve a problem is to choose a variable and then use it to represent the conditions of the problem.

Additional Example 2

a. Let m = the miles a person lives from school. Write a variable expression for the distance the person lives from school after moving 7 miles closer to school.
$m - 7$

b. Lucie lives 12 miles from school. Evaluate the expression from part (a) when $m = 12$.
Substitute 12 for m.
$m - 7 = 12 - 7$
$= 5$
Lucie will live 5 miles from school after moving 7 miles closer.

Additional Example 3

Evaluate each variable expression when $c = 15$ and $d = 10$.

a. $5(c - d)$
Substitute 15 for c and 10 for d.
$5(c - d) = 5(15 - 10)$
$= 5(5)$
$= 25$

b. $\frac{c^2 - cd}{5}$
Substitute 15 for c and 10 for d.
$\frac{c^2 - cd}{5} = \frac{15^2 - 15 \cdot 10}{5}$
$= \frac{225 - 150}{5}$
$= \frac{75}{5}$
$= 5$

EXAMPLE 2

a. Let h = a person's height in inches. Write a variable expression for the person's height after the person grows 5 in.

b. James is 63 in. tall. Evaluate the expression from part (a) when $h = 63$.

SOLUTION

a. $h + 5$

b. $h + 5 = 63 + 5$ — Substitute 63 for h.
$= 68$

James will be 68 in. tall after he grows 5 in.

EXAMPLE 3

Evaluate each variable expression when $a = 4$ and $b = 3$.

a. $2(a + b)$ **b.** $\frac{b^2 + ab}{7}$

SOLUTION

a. Substitute 4 for a and 3 for b.

$2(a + b) = 2(4 + 3)$ — Simplify inside the parentheses first.
$= 2(7)$
$= 14$ — Remember, 2(7) means $2 \cdot 7$.

b. The "2" in this expression is an **exponent**. It tells you to use b as a factor two times: $b^2 = b \cdot b$. You read b^2 as "b squared."

Substitute 4 for a and 3 for b. Notice that ab means $a \cdot b$.

$\frac{b^2 + ab}{7} = \frac{3^2 + 4 \cdot 3}{7}$ — **Simplify exponents** before other multiplications and divisions.
$= \frac{9 + 4 \cdot 3}{7}$
$= \frac{9 + 12}{7}$ — Simplify the numerator completely before dividing by 7.
$= \frac{21}{7}$
$= 3$

Checking Key Concepts
1. 9 2. 5
3. 5 4. 8
5. 5.1 6. 10
7. $\frac{1}{2}$ 8. 2
9. 0 10. 21
11. 4 12. 1
13. 10.7 14. 2.8
15. $\frac{7}{17}$

Exercises and Applications
1. a. 0 b. 4
2. a. 0 b. 12
3. a. 42 b. 42
4. a. 14 b. 12
5. a. 8 b. 11.5
6. a. 8 b. 20
7. Answers may vary. An example is given. To simplify the expression in part (a), I used the keystrokes 6 [×] 3 [−] 2 [×] 2 [=]. The calculator used the order of operations and did the multiplications before the subtraction; the result was 14. To simplify the expression in part (b), I used the keystrokes 6 [×] [(] 3 [−] 2 [)] [×] 2 [=]. The calculator did the subtraction in parentheses first and then multiplied the numbers in order from left to right; the result was 12.

✓ Checking Key Concepts

Simplify each expression.

1. $8 + 2 \div 2$
2. $2 \cdot 4 - 6 \div 2$
3. $15 - 5 \cdot 4 \div 2$
4. $2(1 + 3)$
5. $3.4(5 \div 2 - 1)$
6. $1 + 3(5 - 2)$
7. $\frac{1+2}{6}$
8. $\frac{1+3}{2(4-3)}$
9. $2 - \frac{6}{2(1.5)}$

Evaluate each variable expression when $m = 4$, $n = 7$, and $p = 10$.

10. $m + n + p$
11. $m^2 - 3m$
12. $p - (5 + m)$
13. $\frac{10p + n}{p}$
14. $\frac{mn}{p}$
15. $\frac{p - n + m}{n + p}$

1.2 Exercises and Applications

Extra Practice exercises on page 557

Simplify each expression.

1. a. $8 \div 4 - 2$
   b. $8 \div (4 - 2)$
2. a. $6 - 2 \cdot 3$
   b. $(6 - 2) \cdot 3$
3. a. $5 \cdot 8 + 2$
   b. $2 + 5 \cdot 8$
4. a. $6 \cdot 3 - 2 \cdot 2$
   b. $6 \cdot (3 - 2) \cdot 2$
5. a. $\frac{7+9}{2}$
   b. $7 + 9 \div 2$
6. a. $12 - 2(9 - 7)$
   b. $(12 - 2)(9 - 7)$

7. **Writing** Use your calculator to simplify each expression in Exercise 4. Describe your keystrokes and explain what the calculator does.
8. a. Simplify the expression $6 + 9 \div 3 - 2$.
   b. Add parentheses to the expression so that it equals 3 when simplified.
   c. Add parentheses to the expression so that it equals 1 when simplified.

Simplify each expression.

9. $3(7 - 2.5) - 10$
10. $24 \div 6 + 3.2 \div 4$
11. $0.80(7 + 7 \cdot 0.05)$
12. $15 + \frac{95}{100} \cdot 3$
13. $\frac{3}{15} + 2(3)$
14. $\frac{5(7)}{4(3)}$
15. $\frac{5-3}{3+8}$
16. $\frac{4(4)+5}{5}$
17. $\frac{12-2}{4} + 3(10)$
18. $\frac{12.25 - 2 \cdot 5}{3+2}$
19. $2\left(3 \cdot 8 - \frac{15}{16}\right)$
20. $\frac{(5+2)(6-3)}{3+2}$
21. **Open-ended Problem** Describe how to test a calculator to see if it follows the order of operations. Explain why your test works.
22. Let g = the number of gallons of gasoline in a car's tank. Write a variable expression for the amount in the tank if 3 gal are added.

8. a. 7
   b. $(6 + 9) \div 3 - 2$
   c. not possible
9. 3.5
10. 4.8
11. 5.88
12. 17.85
13. $6\frac{1}{5}$
14. $\frac{35}{12}$
15. $\frac{2}{11}$
16. 4.2
17. $32\frac{1}{2}$
18. 0.45
19. $46\frac{1}{8}$
20. 4.2
21. Answers may vary. An example is given. Evaluate $12 - 4 \cdot 2$ by using the keystrokes 12 [−] 4 [×] 2 [=]. If your calculator follows the order of operations, it will do the multiplication before the subtraction and give 4 as the result. If not, it will subtract first and give the result 16.
22. $g + 3$ gal

Teach⇔Interact

Checking Key Concepts

Common Error
Different students can read their answers to questions 1–15 to the class or write them on the board. Incorrect answers will most likely be a result of not using the order of operations correctly. Refer students making errors to the order of operations chart at the top of page 9.

Closure Questions

Why is it necessary to have an agreement about the order of operations when simplifying a numerical expression? How do you evaluate a variable expression?
to avoid having different values for the same expression; Substitute given numbers for the variables, then use the order of operations.

Apply⇔Assess

Suggested Assignment

Core Course
Day 1 Exs. 1–6, 9–20, 22, 23
Day 2 Exs. 24–38, 41, 43–48

Extended Course
Day 1 Exs. 1–23
Day 2 Exs. 24–48

Block Schedule
Day 2 Exs. 1–20, 22–48

Exercise Notes

Student Study Tip
Ex. 8 Mention to students that parentheses are often referred to as *grouping symbols*. As the numbers in the expression $6 + 9 \div 3 - 2$ are grouped differently using parentheses, different answers result: for example, $(6 + 9) \div 3 - 2 = 3$ and $6 + (9 \div 3) - 2 = 7$.

Problem Solving
Ex. 21 Students should be able to support their description and explanation by using a specific numerical expression.

Apply⇔Assess

Exercise Notes

Interdisciplinary Problems
Exs. 36–39 An academic discipline is a branch of instruction or learning. Mathematics is a discipline as is the science of physics. These exercises involve physical laws that govern free-falling objects. Other disciplines included in the first chapter of this book are history, biology, ecology, medicine, chemistry, agriculture, finance, and business. Mathematics is often called the *language of science* because it is used by scientists in all disciplines to express their ideas.

Career Connection
Exs. 36–39 Physics is the branch of science that concerns itself with basic principles of the physical world involving matter and energy in terms of motion and force. A scientist who specializes in the study of physics is called a physicist. Most physicists hold advanced degrees in their field and work in a large variety of jobs. Since physics is the most basic of sciences, its principles are used in industry to manufacture new products and create new technologies. Physicists are also employed by the federal government, most notably in the space program, and by colleges and universities to teach and do basic research.

Multicultural Note
Exs. 38, 39 The three highest waterfalls in the world are in South America, and two of them are in Venezuela. The Angel Falls (2648 ft) in Venezuela is the highest. The Itatinga Falls (2060 ft) in Brazil is the second highest. The Cuquenan Falls (2000 ft) in Venezuela is the third highest. The Ribbon Falls in California is the highest in the United States (1612 ft).

BY THE WAY

Kudzu was brought from Japan in 1876 to decorate the Japanese pavilion at a centennial celebration. It became popular in the South, but its popularity faded as the vine spread. Now most people consider kudzu a weed.

23. Let g = the number of gallons of gasoline in a car's tank. Write a variable expression for the number of miles the car can travel if it gets 25 miles to the gallon.

"If you want to plant kudzu, drop it and run," is a joke you might hear in the southeastern United States. The kudzu vine grows very quickly. A vine can grow up to 12 in. per day. Use this information for Exercises 24–26.

24. Write a variable expression for how much a kudzu vine can grow in d days.

25. How much can kudzu grow in 3 days?

26. How much can kudzu grow in 10 days?

Evaluate each variable expression when $a = 2$, $b = 7$, and $c = 6$.

27. $a^2 + b^2$ **28.** $6c - b$ **29.** $\frac{5b}{3c - 5a}$

30. $\frac{7a + b}{c}$ **31.** $\frac{a^2 + b}{c}$ **32.** $\frac{6b - 2a}{c}$

33. $b^2 + c^2 - 2bc$ **34.** $(b + a)(b - a) + a^2$ **35.** $c^2 - ab + c - 4b$

Connection PHYSICS

You can tell how high a bridge is, how deep a hole is, or how high a cliff is by dropping a pebble and counting the time it takes to hit the bottom. In t seconds, an object falls about $16t^2$ feet.

36. A stone dropped into Mexico's **Sótano de las Golondrinas,** possibly the deepest natural pit in the world, takes about nine seconds to reach the bottom. About how deep is the pit?

37. A stone dropped from the Golden Gate Bridge in San Francisco takes about four seconds to reach the water. About how high above the water is the Golden Gate Bridge?

38. A stick going over **Victoria Falls** on the Zambia-Zimbabwe border in Africa takes about five seconds to reach the bottom. About how high is the waterfall?

39. Writing A log going over Niagara Falls plummets about 170 ft. Use a calculator to estimate how long it takes the log to reach the bottom of the falls. Explain how you found your answer.

23. $25g$ mi 24. $12d$ in.
25. 36 in. 26. 120 in.
27. 53 28. 29
29. $4\frac{3}{8}$ 30. 3.5
31. $\frac{11}{6}$ 32. $6\frac{1}{3}$
33. 1 34. 49
35. 0
36. about 1296 ft
37. about 256 ft
38. about 400 ft
39. about 3.25 seconds; Methods may vary. An example is given. I knew from Ex. 37 that a drop of 4 s indicated a distance of 256 ft. I used guess-and-check and the square key on my calculator: $16 \cdot 3^2 = 144$; $16 \cdot 3.5^2 = 196$; $16 \cdot 3.25^2 = 169$
40. Answers may vary. Example: $5 - 4 \cdot 3 \div (2 + 1)$
41. Answers may vary. Example: [(] 2 [×] 5 [−] 7 [)] [÷] 15 [=]
42. a. No; if the cost before the tip is \$8.40 and each person pays \$9.03, that means the tip is larger than the cost of the pizza. Each person should pay a bit more than half of \$8.40, or \$4.20.

40. Challenge Add parentheses to this expression so that it equals 1 when simplified: $5 - 4 \cdot 3 \div 2 + 1$.

41. **Technology** Write the keystrokes you would use to simplify this expression with a calculator: $\frac{2(5) - 7}{15}$.

42. **Technology** Ben and Diana are splitting the cost of an \$8.40 pizza. They want to leave a 15% tip. Using her calculator, Diana finds that they should each pay \$9.03.

a. Does Diana's answer make sense? Explain.

b. Diana's calculator has parentheses keys. What keystrokes should she use to find her share of the cost?

c. What keystrokes do you think Diana used?

43. SAT/ACT Preview $(8 - 6 \div 2) \cdot (1 + 2 \cdot 2) = \underline{?}$.

A. 5 **B.** 18 **C.** 25 **D.** 6 **E.** 30

"YOUR SPAGHETTI DINNER IS READY, SIR."

44. Open-ended Problem Write a short story or draw a cartoon like the one shown to show what can go wrong if you do something nonmathematical in the wrong order.

ONGOING ASSESSMENT

45. Writing Explain the difference between simplifying a numerical expression and evaluating a variable expression. Give an example of each.

SPIRAL REVIEW

Exercises 46–48 give you information about tropical rain forests. For each statement, choose a variable and write an inequality. Explain what each variable stands for. *(Section 1.1)*

46. At least 60 in. of rain falls in a single year.

47. Scientists have counted up to 80 species of trees per acre in some forests.

48. Trees may be as tall as 150 ft.

Apply⇔Assess

Exercise Notes

Assessment Note

Ex. 45 Two important and effective types of assessment are *student self-assessment* and *peer assessment.* Both types of assessment can be applied to this exercise. Students can work in small groups to exchange their written responses and then evaluate one another's work and their own work. A discussion of the results is essential to provide feedback to each student. In this way, students can monitor their own progress toward the objectives of this section in particular and toward the objectives of the chapter itself.

Practice 2 for Section 1.2

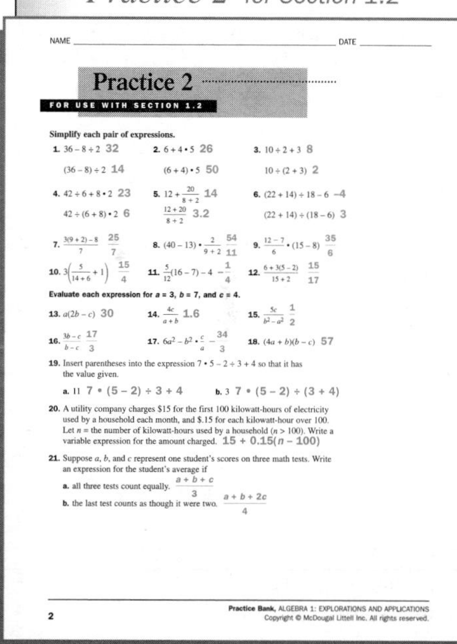

NAME ____________ DATE ________

Practice 2

FOR USE WITH SECTION 1.2

Simplify each pair of expressions.

1. $36 - 8 \div 2$ 32; $(36 - 8) \div 2$ 14
2. $6 + 4 \cdot 5$ 26; $(6 + 4) \cdot 5$ 50
3. $10 \div 2 + 3$ 8; $10 \div (2 + 3)$ 2
4. $42 \div 6 + 8 \cdot 2$ 23; $42 \div (6 + 8) \cdot 2$ 6
5. $12 + \frac{20}{8 + 2}$ 14; $\frac{12 + 20}{8 + 2}$ 3.2
6. $(22 + 14) \div 18 - 6$ −4; $(22 + 14) \div (18 - 6)$ 3
7. $\frac{3(9 + 2) - 8}{7}$ $\frac{25}{7}$
8. $(40 - 13) \cdot \frac{2}{9 + 2}$ $\frac{54}{11}$
9. $\frac{12 - 7}{6} \cdot (15 - 8)$ $\frac{35}{6}$
10. $3\left(\frac{5}{14 + 6} + 1\right)$ $\frac{15}{4}$
11. $\frac{5}{12}(16 - 7) - 4$ $-\frac{1}{4}$
12. $\frac{6 + 3(5 - 2)}{15 + 2}$ $\frac{15}{17}$

Evaluate each expression for $a = 3$, $b = 7$, and $c = 4$.

13. $a(2b - c)$ 30
14. $\frac{4c}{a + b}$ 1.6
15. $\frac{5c}{b^2 - a^2}$ $\frac{1}{2}$
16. $\frac{3b - c}{b - c}$ $\frac{17}{3}$
17. $6a^2 - b^2 \cdot \frac{c}{a}$ $-\frac{34}{3}$
18. $(4a + b)(b - c)$ 57

19. Insert parentheses into the expression $7 \cdot 5 - 2 \div 3 + 4$ so that it has the value given.

a. 11 $7 \cdot (5 - 2) \div 3 + 4$ b. 3 $7 \cdot (5 - 2) \div (3 + 4)$

20. A utility company charges \$15 for the first 100 kilowatt-hours of electricity used by a household each month, and \$.15 for each kilowatt-hour over 100. Let n = the number of kilowatt-hours used by a household ($n > 100$). Write a variable expression for the amount charged. $15 + 0.15(n - 100)$

21. Suppose a, b, and c represent one student's scores on three math tests. Write an expression for the student's average if

a. all three tests count equally. $\frac{a + b + c}{3}$

b. the last test counts as though it were two. $\frac{a + b + 2c}{4}$

2 **Practice Bank,** ALGEBRA 1: EXPLORATIONS AND APPLICATIONS

---

b. Answers may vary. Example: (8.40 + 0.15 × 8.40) ÷ 2 =; This gives \$4.83 as each person's share.

c. 8.4 + 0.15 × 8.4 ÷ 2 =

43. C

44. Answers may vary. Examples might include putting on your socks after your shoes or pouring milk from a carton before getting a glass ready.

45. Answers may vary. An example is given. You simplify a numerical expression by using the order of operations to find the value of the expression. You evaluate a variable expression by substituting a value for each variable and then following the same steps used to simplify an expression.

46–48. Choices of variables may vary. Examples are given.

46. Let r = the annual rainfall in inches; $r \geq 60$.

47. Let s = the number of species counted; $s > 80$.

48. Let h = the height in feet of a rain forest tree; $h \leq 150$.

Plan⇔Support

Objectives

- Find the mean, the median, and the mode(s) of a set of data.
- Analyze real-world data.
- Compare data.

Recommended Pacing

Core and Extended Courses
Section 1.3: 1 day

Block Schedule
Section 1.3: $\frac{1}{2}$ block
(with Section 1.4)

Resource Materials

Lesson Support
Lesson Plan 1.3
Warm-Up Transparency 1.3
Practice Bank: Practice 3
Study Guide: Section 1.3
Assessment Book: Test 1

Technology
Graphing Calculator
McDougal Littell Software
Stats!
Internet:
http://www.hmco.com

Warm-Up Exercises

Find each sum.

1. 6 + 3 + 9 + 14 + 8 + 22 62
2. 4(0) + 6(2) + 7(5) + 3(9) + 8(11) 162
3. Arrange these numbers from smallest to largest: 9, 3, 11, 14, 7, 4, 8, 13, 1, 5. 1, 3, 4, 5, 7, 8, 9, 11, 13, 14
4. In this set of data, which data item has the same number of items both less than it and greater than it? 22, 14, 20, 16, 30, 10, 27, 31, 9, 17, 25 20
5. Simplify this expression. $\frac{7(1) + 2(5) + 9(3) + 6(7)}{2 + 6 + 4 + 8}$ 4.3

SECTION

1.3 Mean, Median, and Mode

Learn how to...

- **find the mean, the median, and the mode(s) of a set of data**

So you can...

- **analyze real-world data, such as the number of hours of TV students watch**
- **compare the ages of athletes in different sports**

According to a national report, people under the age of 18 typically watch between 3 and 4 hours of TV daily. Todd made a histogram showing how much TV his friends watch daily.

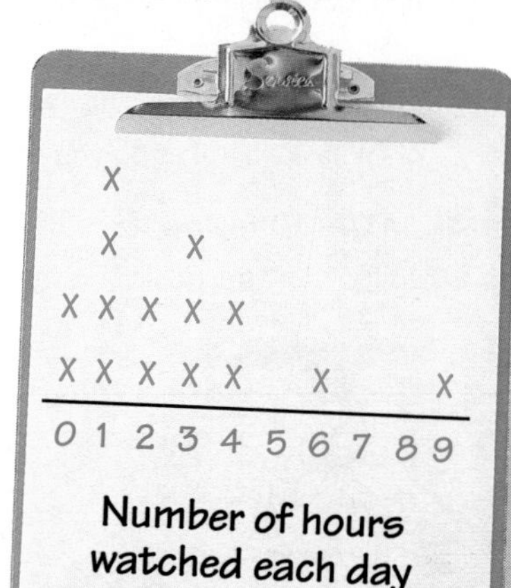

Think and Communicate

1. Each X represents one student. On average, how many hours of TV do Todd's friends watch? Explain your answer.
2. A report says that in 1993, the average family in the United States had 3.16 members. What do you think this statement means?

Three Kinds of Averages

When people talk about averages, they often mean different things.

Mean To find the *mean* of the data, first find the sum of all the data. Then divide by the number of data items.

Median The *median* is the middle number when you put the data in order from smallest to largest.

Mode The *mode* is the data value that appears most often.

Example 1

Use the histogram of Todd's data to find each kind of average.

a. the mean **b. the median** **c. the mode**

Solution

a. Use this formula to find the mean of Todd's data:

$$\text{Mean} = \frac{\text{Total number of hours watched}}{\text{Number of students}}$$

Two students watch 4 h of TV.

$$= \frac{2(0) + 4(1) + 2(2) + 3(3) + 2(4) + 1(6) + 1(9)}{2 + 4 + 2 + 3 + 2 + 1 + 1}$$

$$= \frac{40}{15}$$

≈ 2.7 The mean number of hours watched is about 2.7.

The symbol $\approx$ means *approximately equal to.*

ANSWERS Section 1.3

Think and Communicate

1. Answers may vary. Examples are given.
(1) On average, Todd's friends watch about 2 h of TV per day. I base this answer on the fact that 2 seems to be about in the middle of the data.
(2) Since more students answered 1 than anything else, on average, Todd's friends watch about 1 h of TV daily.
(3) I think on average Todd's friends watch about $2\frac{2}{3}$ h per day. I added the number of hours they watched altogether and divided by the number of friends.
2. Answers may vary. An example is given. I think if you divided the total number of people in the United States in 1993 by the number of families, the result would be about 3.16.

b. To find the median, put the data values in order from smallest to largest. For Todd's data, the median is 2.

0 0 1 1 1 1 2 **2** 3 3 3 4 4 6 9

7 data items below the median | **median** | 7 data items above the median

c. In a histogram, the mode is represented by the tallest bar. For Todd's data, the mode is 1.

THINK AND COMMUNICATE

3. Explain why the tallest bar of a histogram represents the mode.
4. Do you think the mean, the median, or the mode best describes the average hours of TV watched by Todd's class? Explain your choice.

EXAMPLE 2 Application: Sports

Find the median and the mode(s) of the ages shown in the histogram at the right.

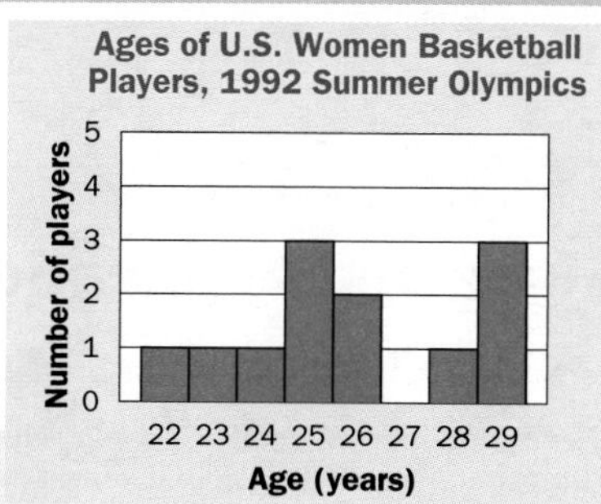

SOLUTION

To find the median, put the ages in order from smallest to largest.

There is no middle number.

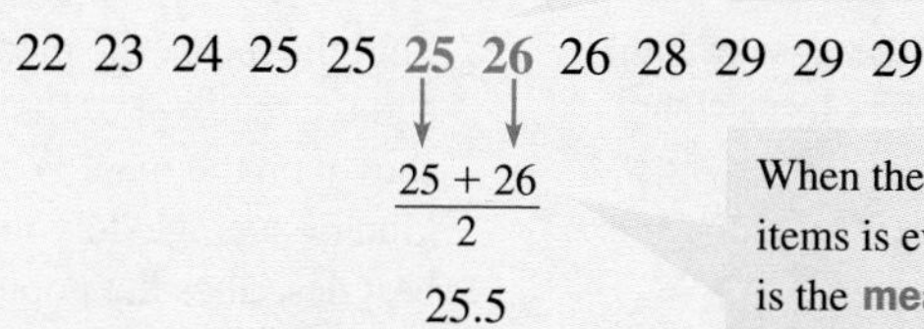

When the number of data items is even, the median is the **mean of the two middle numbers.**

The median age of the players is 25.5.

To find the mode, look for the data values that appear most often.

22 23 24 **25 25 25** 26 26 28 **29 29 29**

The modes are ages 25 and 29.

25 and **29** each appear three times. There can be more than one mode.

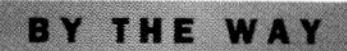

The athletes on the United States women's gymnastics team at the 1992 Summer Olympics were all younger than the women basketball players. The mode of their ages was 15, the median 15, and the mean 15.86.

Think and Communicate

3. The tallest bar shows which data value occurs most often.
4. Answers may vary. An example is given. I think the mode, 1 h, is too low. I think the mean, $2\frac{2}{3}$ h, and the median, 2 h, both represent the average viewing time reasonably well.

Teach⇔Interact

Think and Communicate

Some students may have difficulty answering question 1 because they do not understand how to find the total number of hours watched. This difficulty should be resolved after students discuss Example 1, in which the mean number of hours Todd's friends watch TV is calculated.

Additional Example 1

The histogram shows the number of goals scored by Lauren's soccer team in its 15-game season. Each × represents one goal.

0	1	2	3	4	5
× ×	×	× × × ×	× × × × ×	×	× ×

Number of Goals Scored

Use the histogram of Lauren's data to find each kind of average.

a. the mean

$\text{Mean} = \frac{\text{Total number of goals scored}}{\text{Number of games}}$

$= \frac{2(0) + 1(1) + 4(2) + 5(3) + 1(4) + 2(5)}{2 + 1 + 4 + 5 + 1 + 2}$

$= \frac{38}{15} \approx 2.5$

The mean number of goals scored is about 2.5.

b. the median

To find the median, put the data values in order from smallest to largest.

0 0 1 2 2 2 2 **3** 3 3 3 3 4 5 5

The median is 3.

c. the mode

In a histogram, the mode is represented by the tallest bar. The mode is 3.

Additional Example 2

Find the median and the mode of the number of miles run shown in the histogram.

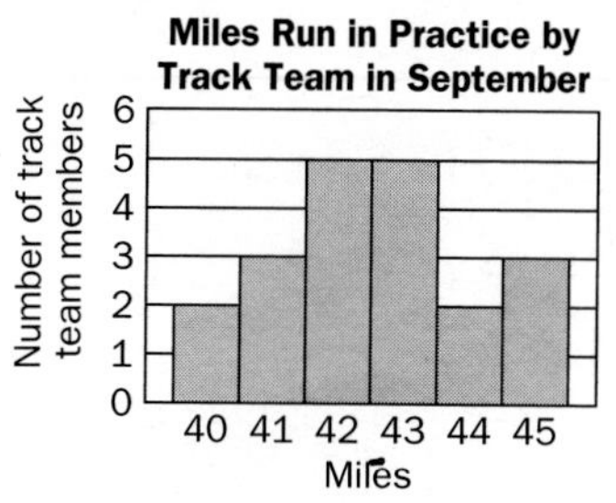

Example continued on next page.

Teach⇔Interact

Additional Example 2 (continued)

To find the median, put the miles in order from smallest to largest.

40 40 41 41 41 42 42 42 42 **42**
43 43 43 43 43 44 44 45 45 45

There is no middle number. The median is the mean of the two middle numbers.

$\frac{42 + 43}{2} = 42.5$

The median number of miles is 42.5.

To find the mode, look for the data values that appear most often. The modes are 42 and 43 miles. Each appears five times.

Think and Communicate

Question 5 brings to students' attention an important fact about the mean of a set of data; that is, it is influenced or changed by an extreme value, either low or high. Ask students if they would prefer to know only one average, mean, median or mode, or all three when analyzing real-world data and why.

Closure Question

Describe how to find the mean, the median, and the mode of a set of data.
See definitions on page 14.

Apply⇔Assess

Suggested Assignment

Core Course
Exs. 1–4, 6–15, AYP

Extended Course
Exs. 1–15, AYP

Block Schedule
Day 3 Exs. 1–15, AYP

THINK AND COMMUNICATE

5. Look back at Example 1. Sharon told Todd that she was joking when she said she watches nine hours of TV each day. Find the mean, the median, and the mode of the data if you ignore Sharon's data. Which average changes? Why?

6. Is the mean always greater than the median and the mode(s)? Give examples to support your answer.

CHECKING KEY CONCEPTS

Find the mean, the median, and the mode(s) of each set of data.

1. Days absent from school
2, 6, 1, 5, 2, 2, 4, 3, 0, 7, 3

2. Ages of students in a cooking class
23, 17, 18, 19, 23, 25, 67, 19, 22, 18

3. **Ages of U.S. Male Gymnasts, 1992 Summer Olympics**

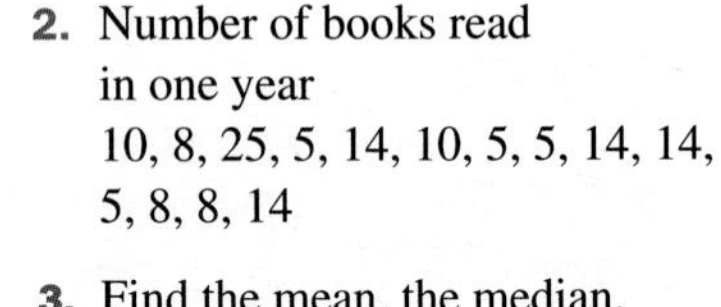

1.3 Exercises and Applications

Extra Practice exercises on page 557

For Exercises 1 and 2, find the mean, the median, and the mode(s) of the data.

1. Number of sitcoms watched in one weekend
3, 0, 3, 2, 1, 0, 0, 4, 4, 8, 2, 2, 2, 1, 0, 0

2. Number of books read in one year
10, 8, 25, 5, 14, 10, 5, 5, 14, 14, 5, 8, 8, 14

Estimated Population Density of the Lesser Antilles, 1991 (people/km²)

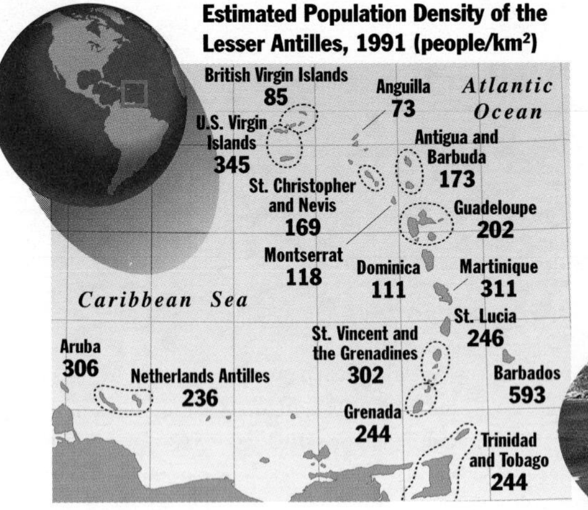

3. Find the mean, the median, and the mode(s) of the data shown on the map. Which kind of average do you think best describes the population density of the islands of the Lesser Antilles? Explain your reasoning.

Think and Communicate

5. 2.2 h; 2 h; 1 h; the mean; The sum of the data values decreases by 9, while the number of data items decreases only by 1.

6. No. For example, for the data values 5, 6, 7, 7, the mean is 6.25, which is less than both the median, 6.5, and the mode, 7. For the data values 5, 6, 6, 7, the mean, median, and mode are all 6.

Checking Key Concepts

1. about 3.2; 3; 2
2. 25.1; 20.5; 18, 19, and 23
3. about 21.9, 22, 22

Exercises and Applications

1. 2; 2; 0
2. about 10.4; 9; 5 and 14
3. about 234.9; 240; 244; Answers may vary. An example is given. The three values are fairly close, so I think no one average describes the data better than the others.
4. a. about 4.2; 3; 2 and 8
   b. Answers may vary. An example is given. According to the graph, it is difficult to describe the typical customer's wait. More than half the customers wait in line 3 min or less. However, for every customer who waits only 2 min, there is another customer who waits 8 min.

4. a. Find the mean, the median, and the mode(s) of the data in the histogram.

   b. **Writing** Paul Pelletier, the restaurant manager, wonders how long a typical customer must wait in line. What answer would you give him? Explain.

5. **Challenge** Find a set of five data values with modes 0 and 2, median 2, and mean 2. Explain how you found your answer.

Time Spent in a Fast Food Restaurant Line
Number of customers
0 1 2 3 4 5 6 7 8
1 2 3 4 5 6 7 8
Time (minutes)

Connection ECOLOGY

Gulf of California

The Gulf of California harbor porpoise, or vaquita (pronounced bah-KEE-tah), is an endangered species. About 30 to 35 vaquitas die each year after being captured in fishing nets. The histogram shows the ages of a group of captured vaquitas. The age groupings of the captured vaquitas may represent the ages of the entire population of vaquitas.

Ages of Captured Vaquitas

Number captured: 0 2 4 6 8 10 12

less than 1 year old

Age (years): 0 2 4 6 8 10 12 14 16 18 20 22 24 26

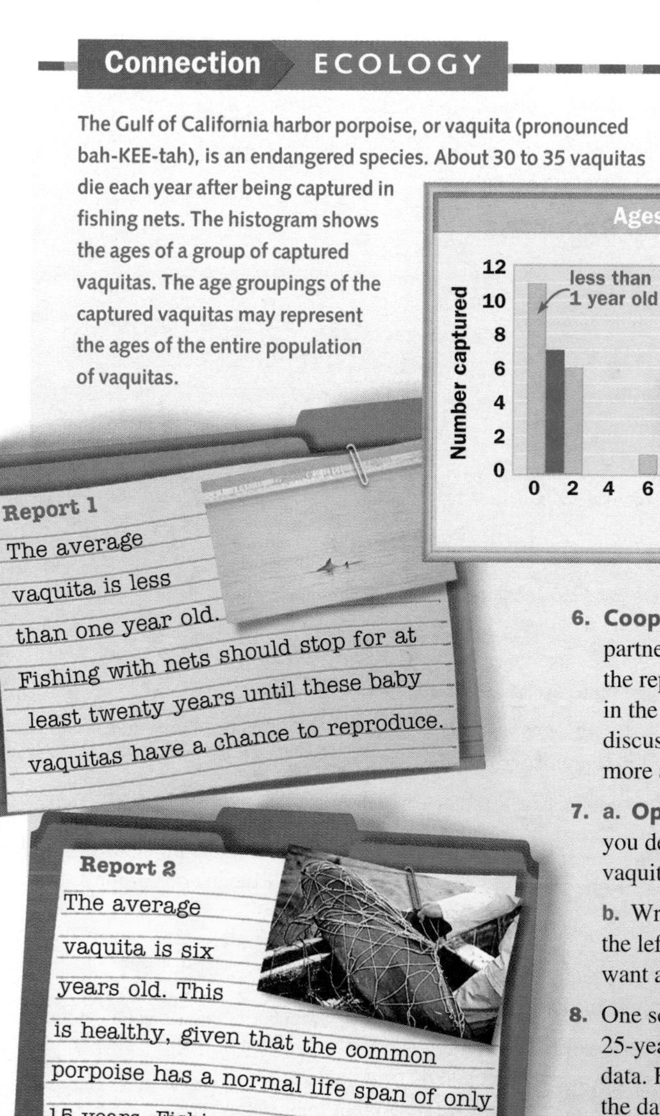

Report 1
The average vaquita is less than one year old. Fishing with nets should stop for at least twenty years until these baby vaquitas have a chance to reproduce.
Report 2
The average vaquita is six years old. This is healthy, given that the common porpoise has a normal life span of only 15 years. Fishing with nets can continue.

6. **Cooperative Learning** Work with a partner. Each of you should choose one of the reports at the left. Explain how the data in the histogram supports the report. Then discuss together which report you think is more accurate. Explain your conclusions.

7. a. **Open-ended Problem** How would you describe the typical age of the captured vaquitas? Give reasons for your answer.

   b. Write your own report like the ones at the left, or write some questions you would want answered before writing a report.

8. One scientist decides not to include the 25-year-old vaquita in an analysis of the data. How does this affect the mean of the data?

5. 0, 0, 2, 2, 6; Since there are five data values and the median is 2, the middle value is 2. Since the modes are 0 and 2, there must be 2 zeros and 2 twos. The sum of these four numbers is 4. Since the mean of the five numbers is 2, their sum must be 10, so the fifth number must be 6.

6. Answers may vary. Examples are given. Report 1: The mode is 0. This supports the claim that the average vaquita is less than one year old. Report 2: The mean age is about 6. This supports the claim that the average vaquita is six years old. Comparisons of the reports may vary. For example, I think that Report 1 is more accurate. I think that the mean is too high because the ages of two vaquitas in the data list are much greater than the others, and this increases the value of the mean. The three data values that appear most often are 0, 1, and 2, so I think that the average in Analysis 1 is probably too low, but is more accurate than the average in Report 2.

7. See answers in back of book.

8. It decreases from about 6 to about 5.65.

Apply⇔Assess

Exercise Notes

Using Technology

Exs. 1, 2 Students can use the *Stats!* software to create and analyze these sets of data. To do these exercises with a TI-82 calculator, enter the data as statistical data in L1. To find the mean and the median, press STAT ▶ and select 1-Var Stats (1-variable statistics) from the CALC menu. The calculator will show 1-Var Stats on the home screen. Type 2nd L1 and press ENTER. The calculator displays 2 lists of one-variable statistics. The mean is the value $\bar{x}$. Press ▼ until you come to the median (Med). To find the mode, sort the data in ascending or descending order. Use the STAT EDIT menu. Choose SortA(or SortD(. Type 2nd L1) and press ENTER. Then view the data in the sorted list. Find the data value that occurs most often.

Reasoning

Ex. 3 In this exercise, students are asked to select the kind of average they think best describes the data and to explain their reasoning. It is important that students understand that the mean is best used when the data are reasonably centered around it and there are no extreme values. If there is an extreme value that distorts the data, the median is the more appropriate average to use. When data cannot be averaged or listed in order, the mode can be used. The best approach, however, is to calculate and use all three averages whenever possible.

Research

Exs. 6–8 You may wish to ask some students to look up the meaning of the word *ecology* in a dictionary and explain its meaning to the class. (Ecology is the branch of biology that deals with the relationships between organisms and their environment.)

Assessment Note

Ex. 9 This exercise provides an excellent opportunity to assess students' understanding of the three averages discussed in this section. The mean, the median, and the mode are numbers that tell where the center of a set of data is located. Thus, students should be able to see from a histogram that if these three numbers are equal, then about half the data are both below them and above them.

Apply⇔Assess

Assess Your Progress

You may wish to review the meanings of the *Terms to Know* verbally as a whole class activity. Students can then note those terms whose meanings they are still not sure of and refer to the page on which the term is introduced for further study.

Journal Entry
Ask students to share their everyday uses of the words mean, median, and mode. They can then enter a list of these everyday meanings in their journals.

Progress Check 1.1–1.3

See page 48.

Practice 3 for Section 1.3

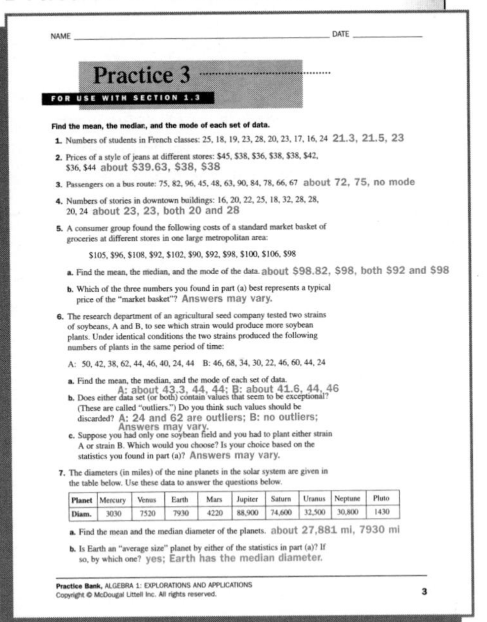

NAME ______________________ DATE __________

Practice 3

FOR USE WITH SECTION 1.3

Find the mean, the median, and the mode of each set of data.

1. Numbers of students in French classes: 25, 18, 19, 23, 28, 20, 23, 17, 16, 24 21.3, 21.5, 23
2. Prices of a style of jeans at different stores: \$45, \$38, \$36, \$38, \$38, \$42, \$36, \$44 about \$39.63, \$38, \$38
3. Passengers on a bus route: 75, 82, 96, 45, 48, 63, 90, 84, 78, 66, 67 about 72, 75, no mode
4. Numbers of stories in downtown buildings: 16, 20, 22, 25, 18, 32, 28, 28, 20, 24 about 23, 23, both 20 and 28
5. A consumer group found the following costs of a standard market basket of groceries at different stores in one large metropolitan area:

   \$105, \$96, \$108, \$92, \$102, \$90, \$92, \$98, \$100, \$106, \$98

   a. Find the mean, the median, and the mode of the data. about \$98.82, \$98, both \$92 and \$98

   b. Which of the three numbers you found in part (a) best represents a typical price of the "market basket"? Answers may vary.
6. The research department of an agricultural seed company tested two strains of soybeans, A and B, to see which strain would produce more soybean plants. Under identical conditions the two strains produced the following numbers of plants in the same period of time:

   A: 50, 42, 38, 62, 44, 46, 40, 24, 44 B: 46, 68, 34, 30, 22, 46, 60, 44, 24

   a. Find the mean, the median, and the mode of each set of data. A: about 43.3, 44, 44; B: about 41.6, 44, 46

   b. Does either data set (or both) contain values that seem to be exceptional? (These are called "outliers.") Do you think such values should be discarded? A: 24 and 62 are outliers; B: no outliers; Answers may vary.

   c. Suppose you had only one soybean field and you had to plant either strain A or strain B. Which would you choose? Is your choice based on the statistics you found in part (a)? Answers may vary.
7. The diameters (in miles) of the nine planets in the solar system are given in the table below. Use these data to answer the questions below.

Planet	Mercury	Venus	Earth	Mars	Jupiter	Saturn	Uranus	Neptune	Pluto
Diam.	3030	7520	7930	4220	88,900	74,600	32,500	30,800	1430

   a. Find the mean and the median diameter of the planets. about 27,881 mi, 7930 mi

   b. Is Earth an "average size" planet by either of the statistics in part (a)? If so, by which one? yes; Earth has the median diameter.

Practice Bank, ALGEBRA 1: EXPLORATIONS AND APPLICATIONS

3

ONGOING ASSESSMENT

9. **Writing** Which histogram represents data for which the mean, the median, and the mode are equal? Explain your reasoning.

A. B.

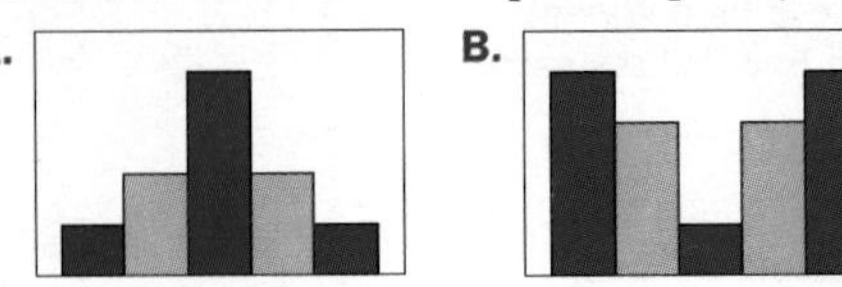

C.

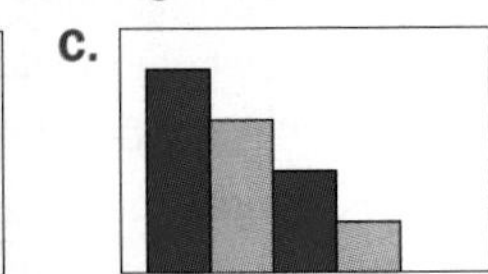

SPIRAL REVIEW

For Exercises 10–15, write each decimal as a fraction in lowest terms. Then write each decimal as a percent. *(Toolbox, page 587)*

10. 0.765 **11.** 0.05 **12.** 0.2
13. 0.25 **14.** 0.125 **15.** 1.75

ASSESS YOUR PROGRESS

VOCABULARY

data (p. 3)
variable (p. 3)
value of a variable (p. 3)
variable expression (p. 9)
evaluate (p. 9)
exponent (p. 10)
mean (p. 14)
median (p. 14)
mode (p. 14)

Simplify each expression. *(Section 1.2)*

1. $12 \div 3 \cdot 2 + 8$ **2.** $26 - 3(6 + 4 \div 2)$ **3.** $\frac{2(5) - 7}{15}$

Evaluate each variable expression when $x = 3$. *(Section 1.2)*

4. $6x + \frac{2}{x}$ **5.** $x^2 - 2x + 7$ **6.** $\frac{x - 2}{8 - x}$

Use the histogram for Exercises 7–10. *(Sections 1.1 and 1.3)*

7. How many students are represented by the histogram?

8. How many students saw three or more movies during the month?

9. Write two inequalities that describe the data in the histogram. What does the variable represent?

10. Find the mean, the median, and the mode(s) of the data.

Movies Seen During One Month
Number of students: 0 1 2 3 4 5 6 7 8
Number of movies seen: 0 1 2 3 4 5 6

11. Journal Review the meanings of the words *mean*, *median*, and *mode*. What are some everyday uses of these words that can help you remember their meanings?

9. histogram A, because the middle value, or median, is the tallest bar, which is also the mode; Since the bars are symmetric about the middle value, this value is also the mean. In the second histogram, the modes are the lowest and greatest values and the median is the middle value. In the third histogram, the mode is the lowest value and the median and mean are greater than the mode.

10. $\frac{153}{200}$; 76.5% 11. $\frac{1}{20}$; 5%
12. $\frac{1}{5}$; 20% 13. $\frac{1}{4}$; 25%
14. $\frac{1}{8}$; 12.5% 15. $\frac{7}{4}$; 175%

Assess Your Progress

1. 16 2. 2
3. $\frac{1}{5}$ 4. $18\frac{2}{3}$
5. 10 6. $\frac{1}{5}$

7. 24 students
8. 15 students
9. For example, $n \geq 0$ and $n \leq 6$; the variable n represents the number of movies seen by each student.
10. 2.75; 3, 3 and 4
11. Answers may vary. Examples are given. The word *mean* describes something midway between two extremes, such as mean temperature or mean score. The word *median* means located near the middle, such as the median strip on a highway. *Mode* can be used to described a customary fashion, the style that is most common. This helps me remember that the mode of a data set is the most common value.

1.4 Working with Integers

Learn how to...

- find the absolute value of a number
- add and subtract integers

So you can...

- use negative numbers to find differences on number scales, for example

In 1978, a group of American women climbed Annapurna, one of the highest mountains in the world. The illustration below shows some of the camps the women made. The scale is labeled with *integers.*

Integers are numbers like these:
... −4, −3, −2, −1, 0, 1, 2, 3, 4 ...

Compared to base camp, Arlene Blum is at 400 ft and Piro Kramar is at −400 ft. Although they are in different places, they are both 400 ft from base camp because distance is never negative.

The **absolute value** of a number is its **distance from zero** on a number line. The symbol $|x|$ means *the absolute value of x.*

The absolute value of a number is never negative.

$|-400| = |400| = 400$

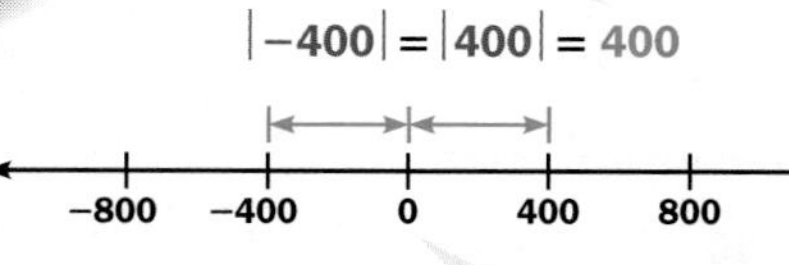

The absolute value of 0 is 0.

EXAMPLE 1

Evaluate each expression when $x = -2$.

a. $|x|$

b. $2|1 + x| + |4|$

SOLUTION

a. $|x| = |-2|$ Substitute −2 for x.

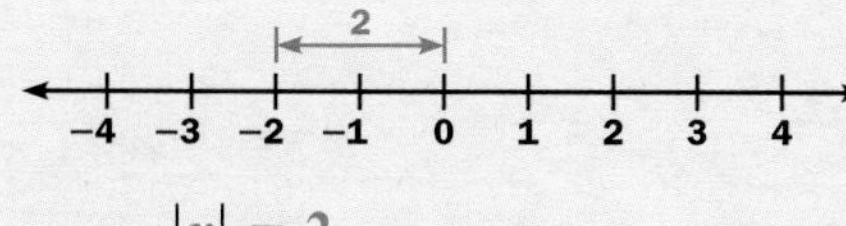

−2 is a **distance of 2** away from zero.

$|x| = 2$

Solution continued on next page.

Plan⇔Support

Objectives

- Find the absolute value of a number.
- Add and subtract integers.
- Use negative numbers to find differences on number scales.

Recommended Pacing

Core and Extended Courses
Section 1.4: 1 day

Block Schedule
Section 1.4: $\frac{1}{2}$ block
(with Section 1.3)

Resource Materials

Lesson Support
Lesson Plan 1.4
Warm-Up Transparency 1.4
Practice Bank: Practice 4
Study Guide: Section 1.4
Explorations Lab Manual: Additional Exploration 2

Technology
Calculator
Graphing Calculator
Internet: http://www.hmco.com

Warm-Up Exercises

Evaluate each expression.

1. 3(39 − 32) 21

2. $5^2 - 2(3)^2$ 7

Evaluate each expression when $a = 5$ and $b = 9$.

3. $2a + 5b$ 55

4. $b^2 - 2a^2$ 31

5. Write an inequality to show that the distance, d, from point A to point B is greater than or equal to zero. $d \geq 0$

6. If a and b are two numbers greater than zero, do you think that $a + b = b + a$ is *always, sometimes,* or *never* true? always true

Teach⇔Interact

Learning Styles: Kinesthetic

The fundamental idea behind the notion of absolute value is that *distance* is always a nonnegative number. This fact can be reinforced by having students use their rulers to make measurements at their desks, or by having two students use a tape measure to find the distance between various objects in the classroom.

Section Note

Student Study Tip
Point out to students that 1, 2, 3, 4, ... are the positive integers and –1, –2, –3, –4, ... are the negative integers. The integer 0 is neither positive nor negative. The numbers 0, 1, 2, 3, 4, ... are also called *whole numbers*. The numbers used for counting are 1, 2, 3, 4, ... and are called the *counting numbers*.

Additional Example 1

Evaluate each expression when $y = -5$.

a. $|y + 2|$
$|y + 2| = |-5 + 2| = |-3| = 3$

b. $3|10 + y| + |-2|$
$3|10 + y| + |-2|$
$= 3|10 + (-5)| + |-2|$
$= 3|5| + |-2| = 15 + 2 = 17$

Section Note

Common Error
A common error that some students make is to think that $-a$ is always a negative number. Stress that if a is a negative number, then $-a$, which is the *opposite* of a, must be a positive number. For example, if $a = -3$, then $-a$ = the opposite of –3, which is –(–3) or 3.

Additional Example 2

Simplify each expression.

a. 11 – 19
11 – 19 = 11 + (–19) = –8

b. –15 – (–9)
–15 – (–9) = –15 + 9 = –6

Toolbox p. 588 *Integers*

SOLUTION *continued*

b. Substitute -2 for x.

$$2|1 + x| + |4| = 2|1 + (-2)| + |4|$$
$$= 2|-1| + |4|$$
$$= 2 \cdot 1 + 4$$
$$= 6$$

The absolute value bars act like parentheses. Do the operations inside them first.

$2|-1|$ means $2 \cdot |-1|$.

Adding and Subtracting Negatives

The integers 400 and -400 are **opposites** because their sum is zero. Every number has an opposite.

Property of Opposites

For every number a, there is a number $-a$ such that $a + (-a) = 0$.
The expression $-a$ means "the opposite of a."

For example: $3 + (-3) = 0$ ← The numbers 3 and -3 are opposites.
$0 + 0 = 0$ ← The opposite of 0 is 0.

Subtracting a **number** is the same as **adding** its **opposite**.

$$5 - 3 = 5 + (-3)$$

EXAMPLE 2

Simplify each expression.

a. $6 - 10$ **b.** $-700 - (-400)$

SOLUTION

a. $6 - 10 = 6 + (-10)$ ← Subtracting 10 is the same as adding -10.

To add a positive number, move to the right. To add a negative number, move to the left.

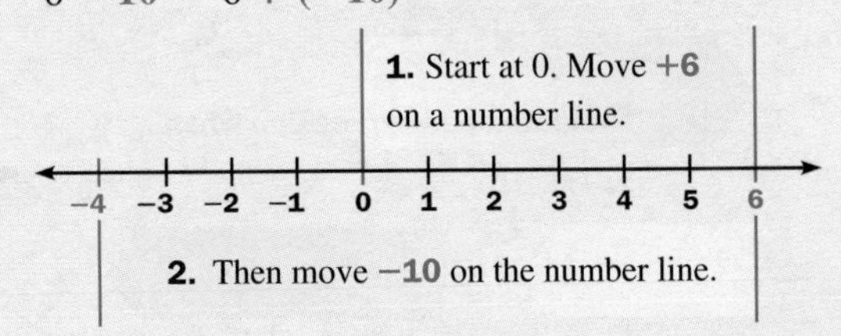

$6 + (-10) = -4$

b. $-700 - (-400) = -700 + 400$ ← Subtracting -400 is the same as adding 400.
$= -300$

ANSWERS Section 1.4

Think and Communicate

1. sometimes; For example, if $a = 2$ and $b = 3$, then $|2 + 3| = |2| + |3|$. If $a = 5$ and $b = -1$, then $|5 + (-1)| = 4$, but $|5| + |-1| = 5 + 1 = 6$.
2. a. 1200 – 400 = 800 ft
   b. 400 – (–400) = 400 + 400 = 800 ft
3. sometimes; If x is a positive number, then $-x$ is a negative number. If x is negative, then $-x$ is positive, and if $x = 0$, then $-x = 0$.

THINK AND COMMUNICATE

1. Decide whether the statement below is *always*, *sometimes*, or *never* true. Give examples to support your answer.

   $|a + b| = |a| + |b|$

2. Look back at the illustration on page 19. Write and simplify an expression to find each distance.
   a. Annie Whitehouse's distance above Arlene Blum
   b. Arlene Blum's distance above Piro Kramar
3. Is $-x$ *always, sometimes,* or *never* a negative number? Explain.

CHECKING KEY CONCEPTS

State the opposite of each number or variable.

1. 24
2. -24
3. $\frac{1}{4}$
4. -12.5
5. n
6. 0

Simplify each expression. Which of your answers are *not* integers?

7. a. $10 - 4$ b. $10 + (-4)$
8. a. $12 - 6$ b. $6 - 12$
9. a. $-4 - 2$ b. $-4 - (-2)$
10. a. $|-2|$ b. $|2|$
11. a. $|3 + (-5)|$ b. $|3| + |-5|$
12. a. $\frac{2|5 - 7|}{3}$ b. $\frac{2}{3}|5 - 7|$

1.4 Exercises and Applications

Extra Practice exercises on page 558

Simplify each expression.

1. $5 - 15$
2. $-2 + 5$
3. $11 + (-24)$
4. $10 + (-2)$
5. $-6 - 2$
6. $-3 + 1$
7. $4 + (-8)$
8. $-6 + 4$
9. $-2 - (-5)$
10. What is the sum of any number and its opposite?
11. **Open-ended Problem** Draw a number line with your home at zero. Label neighboring buildings or landmarks on the number line according to their distance and direction from your home.
12. What two numbers have an absolute value of 18? How many numbers have an absolute value of zero?

Checking Key Concepts

1. –24
2. 24
3. $-\frac{1}{4}$
4. 12.5
5. $-n$
6. 0
7. a. 6 b. 6
8. a. 6 b. –6
9. a. –6 b. –2
10. a. 2 b. 2
11. a. 2 b. 8
12. a. $\frac{4}{3}$ b. $\frac{4}{3}$

7–12. The expression in question 12 is not an integer.

Exercises and Applications

1. –10
2. 3
3. –13
4. 8
5. –8
6. –2
7. –4
8. –2
9. 3
10. 0
11. Answers may vary.
12. 18 and –18; only one, 0

Teach⇔Interact

Think and Communicate

In question 1, some students may not understand what the term *sometimes* means. Point out that this term means *one or more times.* In question 3, students should substitute negative values for *x* to see that *–x* is not necessarily a negative number.

Checking Key Concepts

Alternate Approaches

A variety of approaches can be used by students to do these exercises. For example, students can work on them individually or in small groups. Volunteers can give verbal responses or write answers on the board. An overhead projector can be used to show answers so that students can check their work.

Closure Question

What are the sum and the absolute values of two numbers that are opposites? The sum is zero; the absolute values are the same.

Apply⇔Assess

Suggested Assignment

Core Course
Exs. 1–10, 12–17, 19–38, 41, 42, 46–52

Extended Course
Exs. 1, 3, 5, 6, 8–17, 19, 21–30, 32, 35, 37–52

Block Schedule
Day 3 Exs. 1–10, 12–17, 19–37 odd, 39–52

Exercise Notes

Alternate Approach

Exs. 1–9 The concept of a negative number can be difficult for some students to understand because they have not had sufficient experience with negative numbers in everyday contexts. To help these students, suggest that they think of a negative number in terms of borrowing money. Then, for example, in Ex. 1, if you have \$5 and borrow \$15, then you are \$10 in debt (–10). Thus, $5 - 15 = 5 + (-15) = -10$.

Apply⇔Assess

Exercise Notes

Reasoning

Ex. 12 This question can be generalized to ask: If $x \geq 0$, what two numbers have an absolute value of x? (x and $-x$)

Interdisciplinary Problems

Exs. 13–18 These exercises relate the skills of adding and subtracting integers to the discipline of history. For Ex. 18, suggest that students choose the civilizations for their reports so that each of the five civilizations is represented.

Using Technology

Exs. 19–39 Students can evaluate absolute value expressions on a TI-81 or TI-82 graphing calculator. Press [2nd][ABS] and enter the numerical expression. If the expression is a single number in decimal form, no parentheses are needed. Expressions that contain operation signs require parentheses. The usual written form (with vertical bars) can help students see where parentheses are needed. For example, to enter $|5 - 23|$, type abs(5 – 23). Students can discuss why $\left|\frac{7+3}{4-10}\right|$ should be entered as abs((7 + 3)/(4 – 10)).

Reasoning

Exs. 39, 40 Students can arrive at answers to these questions by examining specific values of a and b. By doing so, they can begin to understand the importance of looking at specific examples in order to arrive at a more general understanding of a mathematical situation.

Student Progress

Exs. 19–27, 30–38 For some students, progress in understanding negative numbers and being able to simplify or evaluate expressions involving them may take some time. Subsequent sections that involve computing with negative numbers and solving equations having negative numbers as solutions should help these students to solidify their understanding of negative numbers.

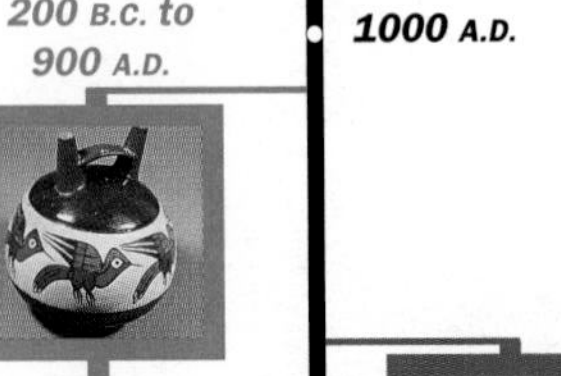

2000 A.D.

1428 A.D. to 1521 A.D.

200 B.C. to 900 A.D.

1000 A.D.

27 B.C. to 476 A.D.

750 B.C. to 300 A.D.

1000 B.C.

1766 B.C. to 1122 B.C.

2000 B.C.

HISTORY The time line shows the approximate beginning and ending dates of some civilizations from the past. Each civilization is matched by color to the time line. About how long did each civilization last?

13. Kingdom of Kush **14.** Shang Dynasty **15.** Roman Empire

16. Aztec Empire **17.** Nacza Civilization

18. Research Write a report about one of the civilizations in Exercises 13–17.

Simplify each expression.

19. $8 - 20$ **20.** $4 + (-5)$ **21.** $72 - (-6)$

22. $-5 - (-14)$ **23.** $1.9 - (-2.4)$ **24.** $|7 - 3 \cdot 3|$

25. $|-14| - |17|$ **26.** $|7 - 2| + 3|-4|$ **27.** $\frac{3|2 - 5|}{4}$

28. Which of the answers to Exercises 19–27 are *not* integers?

29. Find values of x and y for which the value of $x - y$ is:

a. positive b. negative c. zero

Evaluate each variable expression when $t = 4$, $u = 13$, and $v = -5$.

30. $t + (-u)$ **31.** $12 - u$ **32.** $u - v$

33. $u + v$ **34.** $t - (-v) + 1$ **35.** $5 - |v|$

36. $|t - u|$ **37.** $|t + u| + v$ **38.** $|t - v| - u$

39. a. Challenge Simplify each expression. Describe any patterns you see.

$|7 - 4|$ $|4 - 7|$ $|8 - 2|$ $|2 - 8|$

b. Suppose a and b are positive numbers. What can you say about $|a - b|$ and $|b - a|$?

c. Suppose a is a positive number and b is a negative number. Do you think there is a relationship between $|a - b|$ and $|b - a|$? Design and carry out an investigation to find out.

40. Decide whether the statement $a \leq |a|$ is *always*, *sometimes*, or *never* true. Give examples to support your answer.

41. SAT/ACT Preview $-3 - (-4) + 5 - 2 = \underline{\ ?\ }$.

A. 0 **B.** -2 **C.** -4 **D.** 4 **E.** -14

42. Technology Many calculators have a key that changes the sign of a number. On some calculators, this key looks like this: (–). On others, it looks like this: +/–.

a. Which key does your calculator use to change the sign of a number? How can you use the key to enter -6? How do you enter $-(-6)$?

b. Use the change sign key to simplify each expression. Check your work to make sure you are using the key correctly.

$12 + (-6)$ $8 - (-6)$ $3 - (-2)$

c. Write an expression that means "the opposite of the opposite of x."

13–27. See answers in back of book.

28. The answers to Exs. 23 and 27 are not integers.

29. Answers may vary. Examples are given.

a. $x = 5, y = 2$

b. $x = 2, y = 5$

c. $x = 5, y = 5$

30. –9 31. –1

32. 18 33. 8

34. 0 35. 0

36. 9 37. 12

38. –4

39. a. 3; 3; 6; 6; It appears that the absolute value of the difference of two positive numbers is equal no matter which number is taken first.

b. They are equal.

c. Investigations may vary. Students should find that if a is positive and b is negative, then $|a - b| = |b - a|$.

40. The statement is always true. If a is a positive number or zero, then $a = |a|$. For example, $5 = |5|$ and $0 = |0|$. If a is a negative number, then $|a|$ is positive and so $a < |a|$. For example, $-3 < |-3|$. Every number is positive, negative, or zero, so $a \leq |a|$ for every number a.

41. D

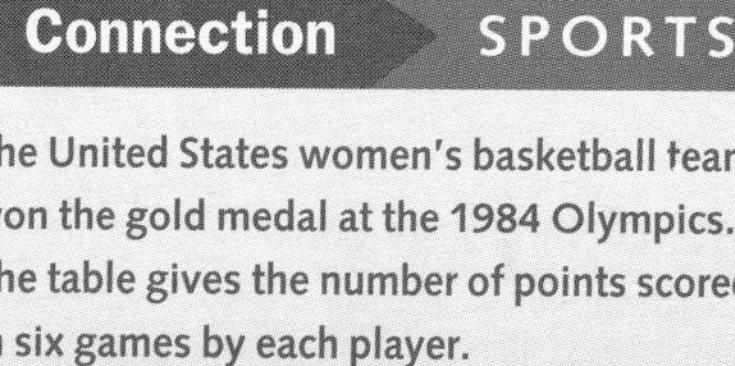

Connection SPORTS

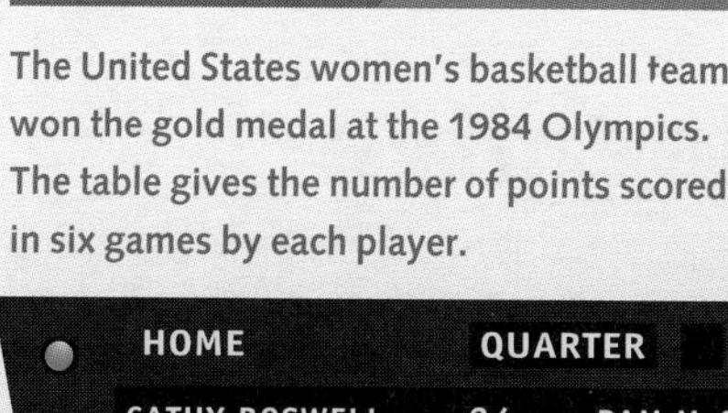

The United States women's basketball team won the gold medal at the 1984 Olympics. The table gives the number of points scored in six games by each player.

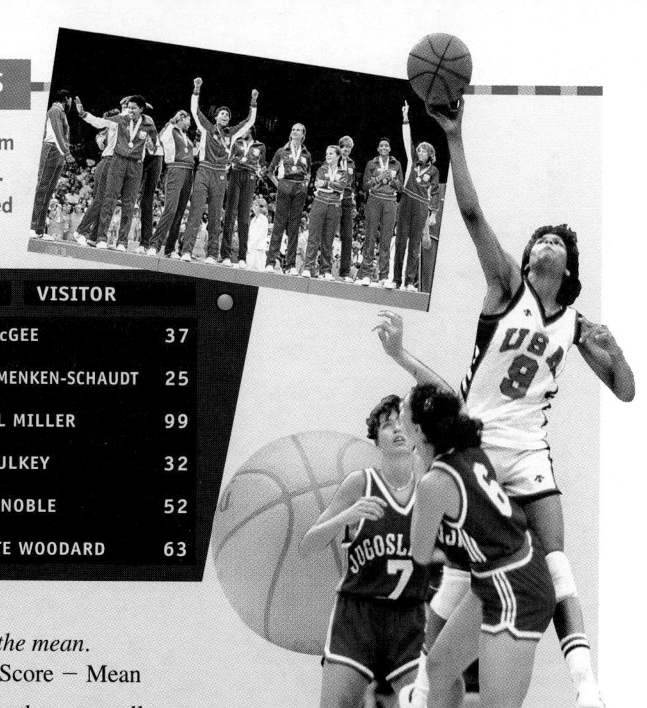

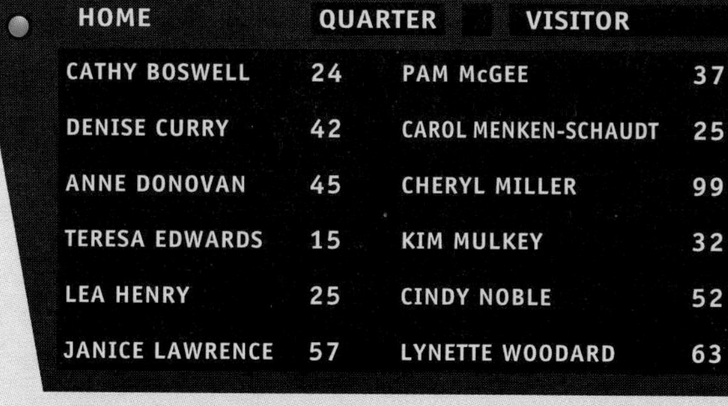

HOME	QUARTER	VISITOR	
CATHY BOSWELL	24	PAM McGEE	37
DENISE CURRY	42	CAROL MENKEN-SCHAUDT	25
ANNE DONOVAN	45	CHERYL MILLER	99
TERESA EDWARDS	15	KIM MULKEY	32
LEA HENRY	25	CINDY NOBLE	52
JANICE LAWRENCE	57	LYNETTE WOODARD	63

43. Find the mean of the scores.

44. Find each player's *deviation from the mean.*
Note: Deviation from the mean = Score − Mean

45. Writing What does deviation from the mean tell you about a player?

ONGOING ASSESSMENT

46. Writing Yael says that $|-x| - x = 0$ for all values of x. Explain why you agree or disagree.

SPIRAL REVIEW

Find the perimeter and area of each figure. *(Toolbox, page 593)*

47. square (side 2)

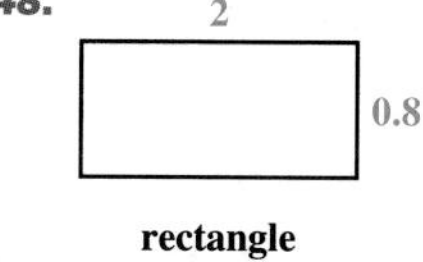

48. rectangle (2 by 0.8)

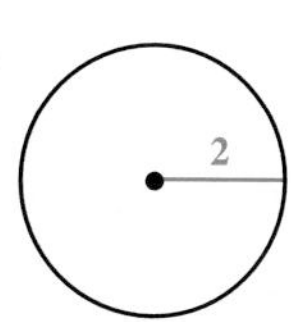

49. circle (radius 2)

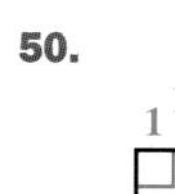

50. (figure with sides 1, 1, 2, 3, 3, 2)

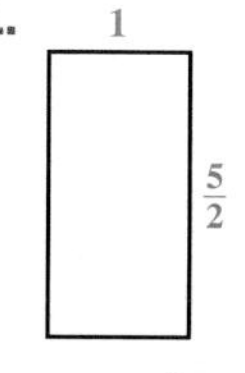

51. rectangle (1 by $\frac{5}{2}$)

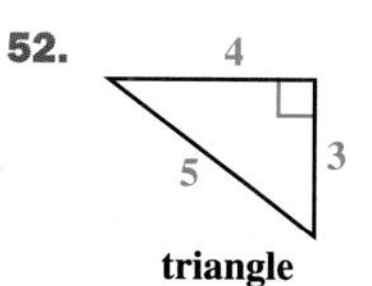

52. triangle (4, 3, 5)

Apply⇔Assess

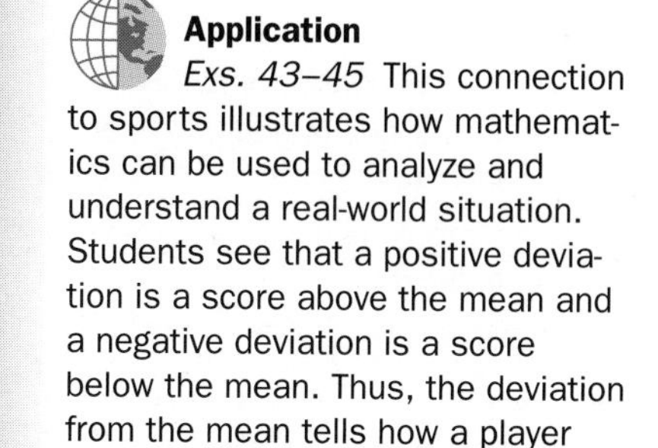

Exercise Notes

Application
Exs. 43–45 This connection to sports illustrates how mathematics can be used to analyze and understand a real-world situation. Students see that a positive deviation is a score above the mean and a negative deviation is a score below the mean. Thus, the deviation from the mean tells how a player scored relative to her teammates.

Assessment Note
Ex. 46 It may help some students to think of this absolute value equation in the equivalent form of $|-x| = x$. This equality is clearly not true if x is a negative number; for example, if $x = -5$, then the equation becomes $|-(-5)| = -5$, which is a false statement.

Practice 4 for Section 1.4

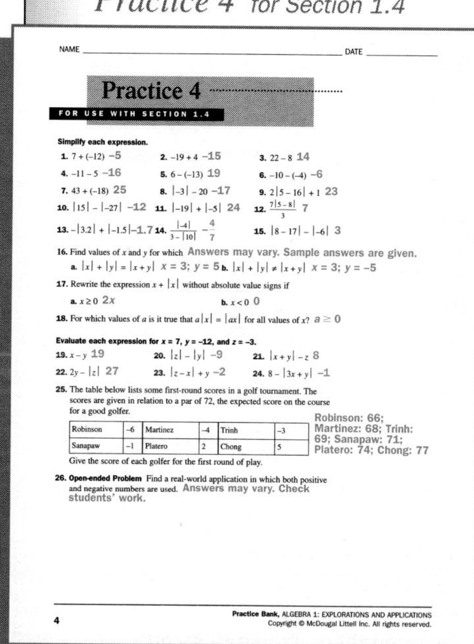

NAME ____ DATE ____

Practice 4

FOR USE WITH SECTION 1.4

Simplify each expression.

1. $7 + (-12)$ −5
2. $-19 + 4$ −15
3. $22 - 8$ 14
4. $-11 - 5$ −16
5. $6 - (-13)$ 19
6. $-10 - (-4)$ −6
7. $43 + (-18)$ 25
8. $|-3| - 20$ −17
9. $2|5 - 16| + 1$ 23
10. $|15| - |-27|$ −12
11. $|-19| + |-5|$ 24
12. $\frac{7|5-8|}{3}$ 7
13. $-|3.2| + |-1.5|$ −1.7
14. $\frac{|-4|}{3 - |10|}$ $-\frac{4}{7}$
15. $|8 - 17| - |-6|$ 3

16. Find values of x and y for which Answers may vary. Sample answers are given.
a. $|x| + |y| = |x + y|$ $x = 3$; $y = 5$ b. $|x| + |y| \neq |x + y|$ $x = 3$; $y = -5$

17. Rewrite the expression $x + |x|$ without absolute value signs if
a. $x \geq 0$ $2x$ b. $x < 0$ 0

18. For which values of a is it true that $a|x| = |ax|$ for all values of x? $a \geq 0$

Evaluate each expression for $x = 7$, $y = -12$, and $z = -3$.

19. $x - y$ 19
20. $|z| - |y|$ −9
21. $|x + y| - z$ 8
22. $2y - |z|$ 27
23. $|z - x| + y$ −2
24. $8 - |3x + y|$ −1

25. The table below lists some first-round scores in a golf tournament. The scores are given in relation to a par of 72, the expected score on the course for a good golfer.

Robinson	−6	Martinez	−4	Trinh	−3
Sanapaw	−1	Platero	2	Chong	5

Give the score of each golfer for the first round of play. Robinson: 66; Martinez: 68; Trinh: 69; Sanapaw: 71; Platero: 74; Chong: 77

26. **Open-ended Problem** Find a real-world application in which both positive and negative numbers are used. Answers may vary. Check students' work.

4 Practice Bank, ALGEBRA 1: EXPLORATIONS AND APPLICATIONS

---

42. a. Answers may vary. For calculators with a [(−)] key, you enter −6 as [(−)] 6 and −(−6) as [(−)] [(−)] 6. For calculators with a [+/−] key, you enter −6 as 6 [+/−] and −(−6) as 6 [+/−] [+/−].

b. 6; 14; 5

c. $-(-x)$

43. Estimates may vary due to rounding; about 43.1 points.

44. The following deviations use 43.1 for the mean.

Player	Deviation from the mean	Player	Deviation from the mean
Boswell	−19.1	McGee	−6.1
Curry	−1.1	Mencken-Schaudt	−18.1
Donovan	1.9	Miller	55.9
Edwards	−27.1	Mulkey	−11.1
Henry	−18.1	Nobel	8.9
Lawrence	13.9	Woodard	19.9

45. Answers may vary. An example is given. The deviation from the mean compares the players scoring to the average. A positive deviation indicates that a player's scoring was better than average. A negative deviation indicates that a player's scoring was worse than average.

46. disagree; It is not true that $|-x| - x = 0$ unless x is positive or 0. For example, if $x = -3$, then $|-(-3)| - (-3) = |3| + 3 = 6$, not 0.

47. 8; 4 **48.** 5.6; 1.6 **49.** $4\pi \approx 12.56$; $4\pi \approx 12.56$

50. 12; 8 **51.** 7; 2.5 **52.** 12; 6

Plan⇔Support

Objectives

- Multiply and divide integers.
- Use properties of addition and multiplication.
- Solve problems involving negative numbers.

Recommended Pacing

Core and Extended Courses
Section 1.5: 2 days

Block Schedule
Section 1.5: 1 block

Resource Materials

Lesson Support
Lesson Plan 1.5
Warm-Up Transparency 1.5
Practice Bank: Practice 5
Study Guide: Section 1.5
Assessment Book: Test 2

Technology
Internet:
http://www.hmco.com

Warm-Up Exercises

Add.

1. (–3) + (–3) –6
2. (–3) + (–3) + (–3) –9
3. Name the factors of 6. 1, 2, 3, 6

Evaluate each expression when $b = 4$.

4. $7b$ 28
5. $2b^2 - 3$ 29
6. Find each sum. What do you observe?
   a. 4 + (–2) 2
   b. (–2) + 4 2
   The sums are the same.

SECTION

1.5 Exploring Negative Numbers

Learn how to...

- **multiply and divide integers**
- **use properties of addition and multiplication**

So you can...

- **solve problems involving negative numbers, such as finding very cold temperatures**

If you multiply a number by 1, nothing happens to it. But what if you multiply the number by -1? You can use the multiplicative property of -1 to explore multiplication by any negative number.

Identity Property of Multiplication

For every number *a*:

$1 \cdot a = a$

For example:

$1 \cdot 23 = 23$

$(1)(-14) = -14$

Multiplicative Property of –1

For every number *a*:

$-1 \cdot a = -a$

For example:

$(-1)(100) = -100$

$(-1)(-6) = 6$

THINK AND COMMUNICATE

For Questions 1–3, use the multiplicative property of –1 to complete each simplification. Then answer Question 4.

1. a. $(-1)(2) = \underline{?}$
   b. $(-1)(-5) = \underline{?}$
   c. $(-1)(-3) = \underline{?}$
2. a. $-3 = (-1)(\underline{?})$
   b. $-6 = (-1)(\underline{?})$
   c. $-4 = (-1)(\underline{?})$
3. a. $(-2)(3) = (-1)(\underline{?})(3) = (-1)(\underline{?}) = \underline{?}$
   b. $(-7)(4) = (-1)(\underline{?})(\underline{?}) = (-1)(\underline{?}) = \underline{?}$
   c. $(-3)(4) = \underline{?}$
4. If you multiply a negative number by a positive number, is the product positive or negative? Use the multiplicative property of -1 to explain your reasoning.

ANSWERS Section 1.5

Think and Communicate

1. a. –2
   b. 5
   c. 3
2. a. 3
   b. 6
   c. 4
3. a. 2; 6; –6
   b. 7; 4; 28; –28
   c. –12
4. negative; Suppose a and b are positive numbers. If you multiply the negative number $-a$ by the positive number b, then $-a \cdot b = (-1 \cdot a)b = -1(ab) = -ab$, which is a negative number, since it is the opposite of the positive number ab.

For Question 5, use the multiplicative property of −1 to complete each simplification. Then answer Question 6.

5. a. $(-3)(-5) = (-1)(\underline{?})(-5)$
$= (-1)(\underline{?})$
$= \underline{?}$

b. $(-7)(-4) = (-1)(\underline{?})(\underline{?})$
$= (-1)(\underline{?})$
$= \underline{?}$

c. $(-3)(-4) = \underline{?}$

6. If you multiply two negative numbers, is the product positive or negative? Use the multiplicative property of −1 to explain your reasoning.

The rules below summarize what you saw in Think and Communicate Questions 1–6.

The product or quotient of two numbers is **positive** if **both** numbers are positive or if **both** are negative.

Same sign → Positive product

$6 \cdot 2 = 12$
$(-6)(-2) = 12$

Same sign → Positive quotient

$\frac{6}{2} = 3 \qquad \frac{-6}{-2} = 3$

The product or quotient of two numbers is **negative** if **one** of the numbers is negative.

Different sign → Negative product

$(-6)(2) = -12$
$(6)(-2) = -12$

Different sign → Negative quotient

$\frac{-6}{2} = -3 \qquad \frac{6}{-2} = -3$

In general, for all numbers a and all nonzero numbers b:
$\frac{-a}{b} = \frac{a}{-b} = -\frac{a}{b}$

EXAMPLE 1

Evaluate each expression when $m = -3$.

a. $-5m + 4$ **b.** $\frac{7}{m}$

SOLUTION

a. Substitute −3 for m. Then follow the order of operations.

$-5m + 4 = (-5)(-3) + 4$
$= 15 + 4$
$= 19$

−5 and −3 are **both** negative, so the product is **positive**.

WATCH OUT!
$-5m$ means $(-5) \cdot m$. Use parentheses when you substitute −3 for m to avoid confusion: $(-5)(-3)$.

b. Substitute −3 for m.

$\frac{7}{m} = \frac{7}{-3}$
$= -\frac{7}{3} \approx -2.3$

7 and −3 have **different** signs, so the quotient is **negative**.

Think and Communicate

5. a. 3; −15; 15
b. 7; −4; −28; 28
c. 12

6. positive; Suppose a and b are positive numbers. If you multiply the negative numbers $-a$ and $-b$, then $(-a)(-b) = (-1 \cdot a)(-b) = (-1)(a \cdot (-b)) = (-1)(-ab) = ab$, since ab is the opposite of $-ab$. Note that ab is a positive number.

Teach⇔Interact

Think and Communicate

In answering questions 1–6, it may help some students to think of the multiplicative property of −1 in terms of taking the opposite of a number. Since for every number a, $-1 \cdot a = -a$, then multiplying a by −1 gives the opposite of a. Thus, if a is positive, the opposite of a is negative; if a is negative, its opposite is positive.

Section Note

Reasoning
Ask students how they can write the division of two numbers, for example, $\frac{-6}{2}$, as a multiplication problem. $\left(-6 \cdot \frac{1}{2}\right)$ Since a division of two numbers can be expressed as a multiplication of the numbers, the quotient must have the same sign as the product.

Additional Example 1

Evaluate each expression when $t = -4$.

a. $7t - 6$
Substitute −4 for t. Then follow the order of operations.
$7t - 6 = 7(-4) - 6$
$= -28 - 6$
$= -34$

b. $\frac{t}{-4}$
Substitute −4 for t.
$\frac{t}{-4} = \frac{-4}{-4} = 1$

Teach⇔Interact

Learning Styles: Kinesthetic

The physical meaning of the terms *commute* and *associate* can be reinforced by having students tell how far they commute to school and what groups they associate with. Then ask students to relate these ideas to the commutative and associative properties of numbers.

Additional Example 2

Simplify each expression.

a. $-4\left(9 \cdot \frac{1}{4}\right)$

$-4\left(9 \cdot \frac{1}{4}\right) = -4\left(\frac{1}{4} \cdot 9\right)$
$= \left(-4 \cdot \frac{1}{4}\right) \cdot 9$
$= -1 \cdot 9$
$= -9$

b. $10 - 3 \cdot 5 + 5$

$10 - 3 \cdot 5 + 5 = 10 - 15 + 5$
$= -5 + 5$
$= 0$

Checking Key Concepts

Reasoning
The commutative and associative properties of numbers can be used to compute sums or products mentally. For example, in question 10, the product (0.12)(100) can be found mentally as 12, which is then multiplied by –3 mentally to get the answer, –36. In question 12, the two fives, 5 and –5, can be grouped as 5 + (–5) = 0. This step simplifies the expression to 3 – 10 + 4 – 9, which can be thought of as 3 + 4 – 10 – 9, or 7 – 19, which is equal to –12.

BY THE WAY

People and numbers are not all that different. People *commute* to work and *associate* with each other. Whether you are talking about people or numbers, *commute* involves movement and *associate* involves groupings.

Properties of Addition and Multiplication

There often are many ways to simplify an expression. You can use the *commutative* and *associative properties* to choose the easiest way for you.

Commutative Property

You can add or multiply numbers in any order.

Addition: $-3 + 7 = 7 + (-3)$

Multiplication: $(-3)(7) = (7)(-3)$

Associative Property

When you add or multiply three or more numbers, you can group the numbers without affecting the result.

Addition: $(7 + 8) + 2 = 7 + (8 + 2)$

Multiplication: $(7 \cdot 8) \cdot 2 = 7 \cdot (8 \cdot 2)$

EXAMPLE 2

Simplify each expression.

a. $\left(\frac{2}{3} \cdot 7\right) \cdot 3$

b. $-9 + 7 + 2 \cdot 5$

SOLUTION

a. It may be easier to find $\frac{2}{3} \cdot 3$ than to find $\frac{2}{3} \cdot 7$.

$\left(\frac{2}{3} \cdot 7\right) \cdot 3 = \left(7 \cdot \frac{2}{3}\right) \cdot 3$ — Use the commutative property.
$= 7 \cdot \left(\frac{2}{3} \cdot 3\right)$ — Use the associative property.
$= 7 \cdot 2$
$= 14$

b. $-9 + 7 + 2 \cdot 5 = -9 + 7 + 10$ — Do the multiplication first.
$= [10 + (-9)] + 7$ — Choose which numbers to add first. Brackets [ ] are grouping symbols like parentheses.
$= 1 + 7$
$= 8$

Think and Communicate

7. Answers may vary. An example is given. You can do the addition –9 + 7 + 10 in order from left to right, or you can add 7 and 10 and then add –9 to the sum. I think the easiest way to simplify the expression is to use the method shown in Example 2 because the sum 10 + (–9) is positive and easy to evaluate and so is the sum 1 + 7.
8. No. Examples may vary. $2 \div 4 = \frac{1}{2}$, but $4 \div 2 = 2$. No. Examples may vary. $(12 \div 3) \div 2 = 4 \div 2 = 2$, but $12 \div (3 \div 2) = 12 \div 1.5 = 8$.

Checking Key Concepts

1. –6 2. 2 3. –6
4. $\frac{1}{2}$ 5. 18 6. 2

7–9. See answers in back of book.

10–12. Answers may vary. Examples are given.

10. You can multiply 0.12 and –3 and then multiply the product by 100, or you can multiply 0.12 by 100 and then multiply the product by –3. I prefer the second way because the product of 0.12 and 100 is an integer rather than a decimal.
11. You can multiply 7 and 8 and then multiply the product by 0, or you can multiply 8 and 0 and then multiply the product by 7. I prefer the second way because the product of 8 and 0 is 0, and the product of 0 and 7 is 0; this is easier than remembering the product of 7 and 8.
12. You can evaluate the expression as written from left to right, or you can rewrite and evaluate the expression as (5 – 5) + (3 – 10) + (4 – 9). I prefer the second way because it is easy to evaluate 0 – 7 – 5 and much harder to evaluate the given expression.
13. Answers may vary. An example is given. (–3 + 11) – 1 = –3 + (11 – 1) = –3 + 10 = 7

THINK AND COMMUNICATE

7. Describe at least two other ways to simplify the expression in part (b) of Example 2. Which way do you find easiest? Explain your choice.
8. Do you think division is commutative? associative? Give some examples to support your answer.

CHECKING KEY CONCEPTS

Evaluate each expression when $t = -2$.

1. $3t$
2. $-1 \cdot t$
3. $\frac{12}{t}$
4. $\frac{t}{-4}$
5. $6(t + 5)$
6. $|4 \cdot 2 + 5t|$

Rewrite each fraction with at most one negative sign. Explain each answer.

7. $\frac{-13}{-16}$
8. $\frac{4}{-15}$
9. $-\left(\frac{-12}{-7}\right)$

Describe two ways to simplify each expression. Which do you prefer? Why?

10. $(0.12)(-3)(100)$
11. $7 \cdot 8 \cdot 0$
12. $3 + 5 - 10 + 4 - 5 - 9$
13. **Open-Ended Problem** Write a numerical expression that you might want to use the associative property to simplify. Simplify the expression.

1.5 Exercises and Applications

Extra Practice exercises on page 558

1. **Writing** If you owe a friend \$2, you can think of yourself as having -2 dollars. Think of a situation when you might have $(5)(-2)$ dollars. Explain why it makes sense that $(5)(-2)$ is negative for the situation.

Simplify each expression.

2. $2(-5)$
3. $(-1)(4)$
4. $(-3)(3)$
5. $(-7)(-10)$
6. $\frac{-24}{6}$
7. $-35 \div (-5)$
8. $2 - 4 + (-2) + 4$
9. $|-18 \div 9 + 9|$
10. $\frac{|(4)(-3)|}{-2}$
11. **Writing** Peter says that $-a \cdot -b$ is always positive because the signs are the same. Explain why you agree or disagree. Use examples to support your answer.
12. **Writing** Explain the difference between a subtraction sign $(13 - 4)$ and a negative sign (-4).

Exercises and Applications

1. Answers may vary. An example is given. Someone has given you five \$2 tickets, and you have not yet paid for them. It makes sense that (5)(–2) is negative because (5)(–2) is the same as (–2) + (–2) + (–2) + (–2) + (–2), which is a sum of negative numbers.
2. –10
3. –4
4. –9
5. 70
6. –4
7. 7
8. 0
9. 7
10. 6
11. It is not true that $-a \cdot -b$ is always positive. The symbol in front of each variable indicates the opposite of the number. For example, if $a = 2$ and $b = -3$, then $-a \cdot -b = -2 \cdot 3 = -6$. Also, if a or b is 0, then $-a \cdot -b = 0$, not a positive number.
12. Answers may vary. An example is given. A subtraction sign indicates an operation to be applied to two numbers: subtract the second from the first. A negative sign can indicate a negative number or the opposite of a number.

Teach⇔Interact

Closure Question

Describe what you have learned in this section about multiplying and dividing positive and negative numbers.

The product or quotient of two numbers is positive if both numbers are positive or if they are both negative. The product or quotient of two numbers is negative if one of the numbers is negative.

Apply⇔Assess

Suggested Assignment

Core Course
Day 1 Exs. 2–11, 14–24
Day 2 Exs. 26–33, 35–43, AYP

Extended Course
Day 1 Exs. 1–25
Day 2 Exs. 26–43, AYP

Block Schedule
Day 4 Exs. 2–11, 14–33, 35–43, AYP

Exercise Notes

Communication: Writing
Ex. 1 If students think of (5)(–2) dollars as a situation in which they borrowed \$2 from a friend on 5 different occasions, then this situation can be expressed as (–2) + (–2) + (–2) + (–2) + (–2), which equals –10. Since multiplication is just a short way to do repeated additions, it makes sense that the product of a positive and negative number is negative.

Student Study Tip
Ex. 12 Point out to students who have difficulty with this exercise that the ideas involved are rather subtle. A subtraction sign is a symbol that represents an *operation* between any two numbers. Thus, 13 – 4 means to subtract positive 4 from positive 13. To indicate that a number is negative, as with –4, the same symbol for the operation of subtraction is used (–), but its meaning is completely different. For practical purposes, since 13 – 4 can be written as 13 + (–4), the result is the same.

Apply⇔Assess

Exercise Notes

Topic Spiraling: Preview
Exs. 13–16 When discussing these exercises, ask students to describe the relationship that salt has on the freezing point of water. In so doing, students will be describing a functional relationship between two variables (amount of salt and freezing point of water). Although the concept and language of functions is introduced in Chapter 2, an intuitive understanding of a functional relationship can be developed at this time.

Historical Connection
Exs. 17–24 Considering the history of negative numbers, it should not be surprising that many students have difficulty understanding and accepting them as numbers. The ancient Greek mathematician Diophantus (circa 250) was the first to reject negative numbers as solutions to equations saying instead that the equations were not solvable. The rejection of negative numbers by many mathematicians continued until the nineteenth century. For about 1600 hundred years, many mathematicians regarded negative numbers as utter nonsense and some even called them *absurd* numbers. As late as the nineteenth century, the famous British mathematician Augustus De Morgan (1806–1871) refused to consider numbers less than zero. Negative numbers were not well understood and accepted by all mathematicians until modern times.

Cooperative Learning
Exs. 26–33 Students may wish to do these exercises together in groups of two or three. The benefit in doing so is for group members to provide assistance to classmates having difficulty understanding how the associative and commutative properties are used.

Connection CHEMISTRY

Pure water freezes and ice melts at 32°F. When you dissolve something in the water, however, the freezing temperature of the water is lowered. Most scientists measure temperature in degrees Celsius (°C). If you are more familiar with degrees Fahrenheit (°F), you can use the conversion formula below. Use the formula to convert the temperatures in Exercises 13–15.

Let F = the temperature in degrees Fahrenheit.

$$F = \frac{9}{5}C + 32$$

Let C = the temperature in degrees Celsius.

13. During the winter, salt is used to melt ice on the streets. A solution of 30 g salt and 70 g water freezes at −21°C.

14. Because salt can harm plants, the substance $(NH_2)_2CO$ is sometimes used to melt ice. A solution of 79 g $(NH_2)_2CO$ and 100 g water freezes at −24°C.

15. Ethylene glycol is used as antifreeze in car radiators to lower the freezing point of the coolant. A solution of 1 kg ethylene glycol and 2 kg water freezes at −15°C.

16. Research Ocean water does not freeze at 32°F. At what temperature does ocean water freeze?

Evaluate each expression when $x = -6$.

17. $2x$ **18.** $-2x$ **19.** $3x - 12$ **20.** $16 + x - 4$

21. $-3x - 12$ **22.** $2 \cdot x \cdot 5$ **23.** $|2x|$ **24.** $\frac{5}{x}(-10)(12)$

25. Open-ended Problem Karen explained how she remembers how to multiply with negative numbers.

a. Write your own positive statement that can be made negative by using the word *not*.

b. Write your own negative statement that can be made positive by using the word *not*.

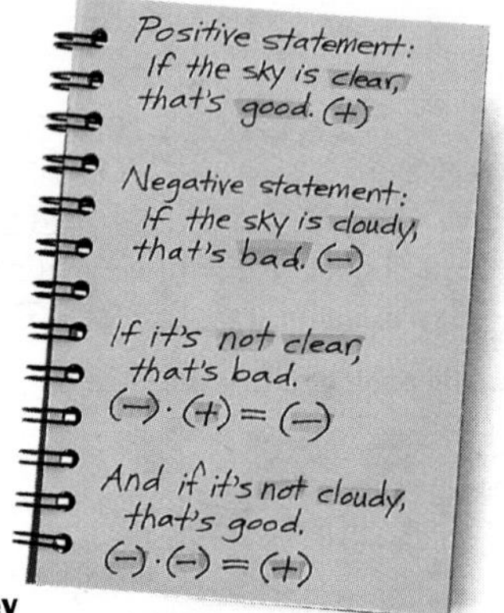

Tell whether the expressions are *equal* or *not equal*. If they are equal, tell whether they demonstrate the *associative property*, the *commutative property*, or *both*.

26. $7 \cdot 10 \stackrel{?}{=} 10 \cdot 7$

27. $-0.2 \cdot (4 \cdot 15) \stackrel{?}{=} (-0.2 \cdot 4) \cdot 15$

28. $6 + (3 \cdot 5) \stackrel{?}{=} (6 + 3) \cdot 5$

29. $(5 + 0.8) + 0.2 \stackrel{?}{=} 5 + (0.8 + 0.2)$

30. $(20 \div 4) \div 2 \stackrel{?}{=} 20 \div (4 \div 2)$

31. $6 - 2 \stackrel{?}{=} 2 - 6$

32. $-1.7 + 13 \stackrel{?}{=} 13 + (-1.7)$

33. $(0.01)(3.42)(100) \stackrel{?}{=} (3.42)(0.01)(100)$

13. –5.8°F 14. –11.2°F
15. 5°F
16. Answers may vary. about 29°F
17. –12 18. 12
19. –30 20. 6
21. 6 22. –60
23. 12 24. 100
25. Answers may vary. Examples are given.
 a. Positive statement: If my team wins the basketball game, that's good. Negative statement: If my team does not win the basketball game, that's bad.
 b. Negative statement: If my team loses the basketball game, that's bad. Positive statement: If my team does not lose the basketball game, that's good.
26. equal; commutative property
27. equal; associative property
28. not equal
29. equal; associative property
30. not equal 31. not equal
32. equal; commutative property
33. equal; both properties
34. Answers may vary. Examples are given.
 $2 + (3 + 7) = 2 + 10 = 12$ and $(2 + 3) + 7 = 5 + 7 = 12$;
 $-5 + [-1 + (-3)] = -5 + (-4) = -9$ and $[-5 + (-1)] + (-3) = -6 + (-3) = -9$;

34. **Open-ended Problem** Write and simplify expressions to show that the associative property of addition works for positive and negative numbers.

35. **Challenge** Are the commutative and associative properties true for subtraction? Give examples to support your answers.

36. **SAT/ACT Preview** Which expression has a positive value?

A. $-3|4 - 9|$ **B.** $3 \cdot 4 - 7 \cdot 9$ **C.** $9(-72)$

D. $-19 + |8 - 3|$ **E.** $|12 - 6 \cdot 3|$

ONGOING ASSESSMENT

37. **Open-ended Problem** Choose the easiest way to simplify $(2)(-4.7)(-5)$ and simplify it. Explain how you made your choice. What properties did you use?

SPIRAL REVIEW

For each pair of numbers, find the greatest common factor and the least common multiple. *(Toolbox, page 584)*

38. 9 and 16 39. 24 and 40 40. 17 and 51

Write each fraction as a decimal and as a percent. *(Toolbox, page 587)*

41. $\frac{6}{5}$ 42. $\frac{4}{3}$ 43. $\frac{8}{9}$

ASSESS YOUR PROGRESS

Vocabulary

integer (p. 19) **opposite** (p. 20)
absolute value (p. 19)

Evaluate each variable expression when $s = -5$ and $t = 7$. *(Section 1.4)*

1. $11 - s - t$ 2. $2 - (t + s)$ 3. $t - 5 + s$
4. $|s|$ 5. $|-t| - s$ 6. $20 - |t - s|$
7. **Writing** A friend tells you that $-x$ is less than x for all values of x. Explain why you agree or disagree.

Simplify each expression. *(Section 1.5)*

8. $(-7)(-6)$ 9. $(-4)(2)$ 10. $-15 \div (-3)$

Evaluate each variable expression when $x = -2$ and $y = 8$. *(Section 1.5)*

11. $x + y + 1$ 12. $y \cdot \frac{1}{6} \cdot x$ 13. $-2y + y + 20$

14. **Journal** Describe some real-world situations in which you might need to perform operations on negative numbers. Explain how to do the operations.

Apply⇔Assess

Assess Your Progress

Journal Entry
For Ex. 14, you may wish to ask students to describe their situations to the class as a whole. Students' knowledge about the uses of these numbers will be enriched by having as many different situations as possible for which operations on negative numbers can be performed.

Progress Check 1.4–1.5

See page 49.

Practice 5 for Section 1.5

NAME ______________________ DATE __________

Practice 5

FOR USE WITH SECTION 1.5

Simplify each expression.

1. $\frac{35}{-7}$ −5 2. $(-8)3$ −24 3. $(-6)(-11)$ 66
4. $4(-12 + 3)$ −36 5. $\frac{-18}{-3}$ 6 6. $13 - 24 \div (-3)$ 21
7. $|40 \div (-5) + 14|$ 6 8. $(-7)(3) + |-8|$ −13 9. $\frac{-30}{3|-5|}$ −2

10. Suppose you multiply several negative numbers $a, b, c, \ldots$ together. Give a rule, based on how many numbers are being multiplied, for determining whether the product will be positive or negative. The product is positive if and only if there is an even number of factors.

11. In an *ionized* solution, such as you find inside a car's battery, there are the same number of negatively charged ions as there are positively charged ions, so that the net charge is 0.

a. Suppose 6 negatively charged ions, each with a charge of −2, are added to the solution. What will the net charge of the solution be? −12

b. Suppose 5 negatively charged ions, with the same charge as before, are removed from the original solution. What will the net charge of the solution be? 10

12. Lillian Ewing sometimes divides her class into 4 groups of 6 students each, and sometimes into 6 groups of 4 students each, for cooperative learning. What property of multiplication is she demonstrating? commutative property

Evaluate each expression for $x = -4$.

13. $5 - 3x$ 17 14. $-2x - 12$ −4 15. $\frac{x - 10}{x + 2}$ 7
16. $\frac{-7x}{-2}$ −14 17. $|8x + 9|$ 23 18. $(-5)\frac{36}{x}$ 45

Tell whether the expressions are *equal* or *not equal*. If they are equal, tell whether they demonstrate the *associative property*, the *commutative property*, or both.

19. $5[3(-4)] \stackrel{?}{=} (5 \cdot 3)(-4)$ equal; associative property
20. $6 \div 2 + 3 \stackrel{?}{=} 6 \div 3 + 2$ not equal
21. $(-8)(3 - 7) \stackrel{?}{=} (3 - 7)(-8)$ equal; commutative property
22. $3 + 7 - 9 \stackrel{?}{=} 9 - 7 + 3$ not equal

Practice Bank, ALGEBRA 1: EXPLORATIONS AND APPLICATIONS

5

$9 + (-7 + 1) = 9 + (-6) = 3$ and $[9 + (-7)] + 1 = 2 + 1 = 3$

35. No. Examples may vary. $5 - 2 = 3$, but $2 - 5 = -3$; $(9 - 3) - 4 = 6 - 4 = 2$, but $9 - (3 - 4) = 9 - (-1) = 10$.

36. E

37. Answers may vary. An example is given. $(2)(-4.7)(-5) = [2(-5)](-4.7) = -10(-4.7) = 47$; I chose this method because when I group 2 and −5, I get −10, and when I multiply −4.7 by −10, I eliminate the decimal. I used the commutative and associative properties of multiplication.

38. 1; 144 39. 8; 120
40. 17; 51 41. 1.2; 120%
42. 1.333; 133.3%
43. 0.888; 88.8%

Assess Your Progress

1. 9 2. 0 3. −3
4. 5 5. 12 6. 8

7. It is not true that $-x < x$ for all values of x. If $x = 0$, then $-x = x$. Also, if x is negative, $-x > x$. For example if $x = -1$, then $-x = 1$ and $-x > x$.

8. 42 9. −8
10. 5 11. 7
12. $-\frac{8}{3}$ 13. 12

14. Answers may vary. Examples are given. If the temperature is 2° below zero and falls 3° more, this can be represented as $-2 - 3 = -5$; that is, the temperature has fallen to 5° below zero. If a solution is chilled during a chemistry experiment and the temperature of the solution falls 4° per hour for 3.5 h, this can be represented as $3.5(-4) = -14$; that is, the temperature has fallen 14° during the 3.5 hour period. Other examples might include finding the difference between two elevations, one above sea level and one below, or calculating the length of time between two dates, one B.C. and one A.D.

Plan⇔Support

Objectives

- Write variable expressions.
- Solve real-world problems by using variables.

Recommended Pacing

Core and Extended Courses
Section 1.6: 1 day

Block Schedule
Section 1.6: $\frac{1}{2}$ block
(with Section 1.7)

Resource Materials

Lesson Support
Lesson Plan 1.6
Warm-Up Transparency 1.6
Practice Bank: Practice 6
Study Guide: Section 1.6
Explorations Lab Manual: Diagram Master 6

Technology
Internet: http://www.hmco.com

Warm-Up Exercises

What is the next number in each pattern shown?

1. 1, 3, 5, 7, 9, ... 11
2. 1, 4, 9, 16, 25, ... 36
3. 0, 1.5, 3, 4.5, 6, ... 7.5
4. If there are n students in a room and 6 leave, write a variable expression for the number of students still in the room. $n - 6$
5. Evaluate the expression $6 + 3y$ when $y = 10$. 36

SECTION

1.6 Exploring Variable Expressions

Learn how to...

- write variable expressions

So you can...

- solve real-world problems about nutrition, for example

How many toothpicks do you need to make 10 triangles in a row? 100 triangles? 1,000,000? Patterns can help you answer questions like these and write variable expressions.

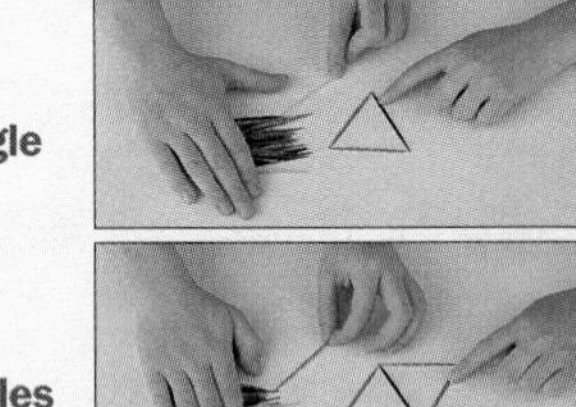

EXPLORATION
COOPERATIVE LEARNING

Patterns and Variable Expressions

Work with a partner.
You will need:
- about 30 toothpicks

SET UP Make as many triangles in a row as you can. Copy and extend the table. Then answer the questions below.

Number of triangles	Number of toothpicks
1	3
2	5
3	7
4	?
5	?
?	?

1 triangle

2 triangles

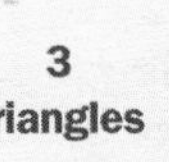

3 triangles

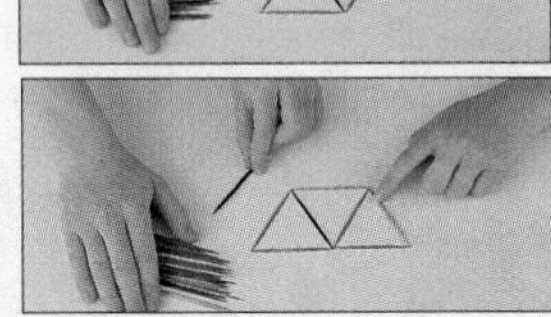

Questions

1. Describe any number patterns you see. What is the relationship between the number of triangles and the number of toothpicks?
2. How many toothpicks do you need to make 10 triangles? 100 triangles? 1,000,000 triangles? How did you get your answers?
3. How many toothpicks do you need to make n triangles? Write your answer as a variable expression. Compare your answer to the answers of other groups in the class.

Exploration Note

Purpose
The purpose of the Exploration is to have students discover a number pattern and then relate that pattern to a variable expression.

Materials/Preparation
Each group of two students should have 30 or more toothpicks.

Procedure
Students should copy the table shown in the Set Up. Using all of their toothpicks, students should make as many triangles as they can. Using their triangles, they can then extend the table.

Closure
Discuss the answers to all three questions. In particular, students should discuss question 3 in detail so they are sure they understand the meaning of the variable expression needed to make n triangles.

Explorations Lab Manual
See the Manual for more commentary on this Exploration.

Diagram Master 6

EXAMPLE 1 Application: Personal Finance

Jason is going dancing. The dance club charges **$8** at the door and **$1.25** for each soft drink. Write a variable expression for the amount of money Jason should bring, based on the number of soft drinks he buys.

SOLUTION

Problem Solving Strategy: Identify a pattern.

Toolbox p. 600 *Problem Solving*

	Number of soft drinks Jason buys	Money needed (dollars)
Step 1 Try different numbers to see if you can find a pattern.	0	$8 + 1.25 \cdot 0 = 8.00$
	1	$8 + 1.25 \cdot 1 = 9.25$
	2	$8 + 1.25 \cdot 2 = 10.50$
	3	$8 + 1.25 \cdot 3 = 11.75$
	4	$8 + 1.25 \cdot 4 = 13.00$
Step 2 After a pattern appears, test a larger number.	10	$8 + 1.25 \cdot 10 = 20.50$

Step 3 Use the pattern to write a variable expression.

Let d = the number of soft drinks Jason buys.

$8 + 1.25d$

Always tell what the variable stands for.

If Jason wants to buy d soft drinks, then he needs to bring $8 + 1.25d$ dollars.

THINK AND COMMUNICATE

1. In Example 1, suppose the dance costs $10 and each soft drink costs $1. How would the variable expression change?

EXAMPLE 2 Application: Recreation

Suppose you are a new customer at the Cactus Video Arcade. Write a variable expression for the amount you pay, based on the number of games you play.

SOLUTION

Problem Solving Strategy: Identify a pattern.

Each game costs **$.50**.

$0.50(3 - 2)$ — You play **3 games** but **2 are free**.

$0.50(4 - 2)$ — You play **4 games** but **2 are free**.

Let g = the number of games you play.

$0.50(g - 2)$ — You play **g games** but **2 are free**.

You pay $0.50(g - 2)$ dollars to play g games.

Teach⇔Interact

About Example 1

Problem Solving

The solution to Example 1 is found by using the problem-solving strategy of identifying a pattern. After students have studied and understood this example, you may wish to discuss what is meant by a problem-solving strategy. Also, students can begin to compile a list of strategies in their journals for future reference.

Additional Example 1

The Sharpes pay Helen $4 an hour plus a $3 tip for babysitting their two children. Write a variable expression for the amount of money Helen earns based on the number of hours she babysits.

Identify a pattern.

Step 1 Try different numbers to see if you can find a pattern.

Number of hours Helen babysits	Money earned (dollars)
1	$3 + 4 \cdot 1 = 7$
2	$3 + 4 \cdot 2 = 11$
3	$3 + 4 \cdot 3 = 15$
4	$3 + 4 \cdot 4 = 19$
5	$3 + 4 \cdot 5 = 23$

Step 2 After a pattern appears, test a larger number.

8	$3 + 4 \cdot 8 = 35$

Step 3 Use the pattern to write a variable expression. Let h = the number of hours Helen babysits.

$3 + 4h$

If Helen babysits h hours, she earns $3 + 4h$ dollars.

Additional Example 2

Suppose that to attract more customers the Cactus Video Arcade lowers the cost of each game to $.40 and offers the first four games free. Write a new variable expression for the amount you pay based on the number of games you play.

Identify a pattern.

Each game costs $.40. If you play 5 games, 4 are free, so you pay $0.40(5 - 4)$. If you play 6 games, you pay $0.40(6 - 4)$.

Let g = the number of games you play.

You pay $0.40(g - 4)$ dollars to play g games.

ANSWERS Section 1.6

Exploration

Number of triangles	Number of toothpicks
4	9
5	11
6	13
7	15
8	17
...	...
14	29

Questions

1. Answers may vary. An example is given. Each time the number of triangles increases by 1, the number of toothpicks increases by 2. In each row, the number of toothpicks is one more than twice the number of triangles.
2. 21; 201; 2,000,001; Answers may vary. An example is given. I used the pattern I described in Question 1.
3. $2n + 1$ toothpicks

Think and Communicate

1. $10 + d$ dollars

Teach⇔Interact

Think and Communicate

After students discuss part (a) of question 3, you may wish to modify it slightly by asking them to write a variable expression for two overdue books. ($0.10d$)

Checking Key Concepts

Student Progress
It is essential that students be able to answer these questions correctly. Translating a mathematical situation into a variable expression is the first key step in solving any mathematics problem that is stated in words.

Closure Question

Describe the steps necessary to identify a pattern.

Step 1 Try different numbers to see if you can find a pattern.

Step 2 After a pattern appears, test a larger number.

Step 3 Use the pattern to write a variable expression.

Apply⇔Assess

Suggested Assignment

Core Course
Exs. 1–9, 12, 13, 15–17, 18–26

Extended Course
Exs. 1–26

Block Schedule
Day 5 Exs. 1–13, 15–26

Exercise Notes

Application
Exs. 1–6, 14–17 The applications in this section cover a wide variety of real-world situations, from pizza toppings to agriculture, dance, personal finance, nutrition, and cooking. As students work through these problems, and others in subsequent chapters, they should begin to appreciate the general nature and power of mathematics as a tool to solve real-world problems. This fact should be brought to students' attention occasionally throughout the course.

THINK AND COMMUNICATE

2. Does the variable expression in Example 2 make sense if you play only one or two games? Explain.
3. a. You pay a library fine of \$.05 a day every day a book is overdue. What is your fine d days after the book is due?
   b. Suppose you pay \$.05 sales tax for every dollar you spend in a store. How much sales tax do you pay when you spend d dollars?
   c. How are the variable expressions in parts (a) and (b) alike? How are they different?

CHECKING KEY CONCEPTS

1. In Example 2, suppose each game costs \$.75 and the first five games are free. How would the variable expression change?
2. Tim buys a water pitcher for \$12 and matching glasses for \$2.50 each. Write a variable expression for the amount he pays, based on the number of glasses he buys.
3. Jill invites her five family members to her birthday party. She also invites her friends. Each friend brings a guest. Use this information to write a variable expression. Tell what the variable stands for. Explain what the expression tells you about the situation.

1.6 Exercises and Applications

Extra Practice exercises on page 559

1. Use the pizza menu. Write a variable expression for the cost of a cheese pizza with t toppings. Based on your own experience, what values of t make sense for your variable expression?

Pizza

Plain Cheese	\$10.00
1 Topping	\$11.50
2 Toppings	\$13.00
Extra Toppings	\$1.50 ea.

2. **Visual Thinking** The table below shows the height of a corn plant each day after it first appeared above ground. Write a variable expression for the height of the corn plant d days after it first appeared above ground.

1.5 cm	3 cm	4.5 cm	...	?
1 day	2 days	3 days	...	d days

Think and Communicate

2. No; if you play fewer than 3 games, then you get a cost that is negative or zero, and this does not make sense.
3. a. $0.05d$ dollars
   b. $0.05d$ dollars
   c. Answers may vary. An example is given. The expressions look exactly alike. The variable d represents different things in the two expressions, days and dollars. Also, in the first expression, d must be a whole number. In the second, d can be 0, 0.01, 0.01, 0.03, and so on.

Checking Key Concepts

1. $0.75(g - 5)$ dollars
2. Let g = the number of glasses he buys; $12 + 2.5g$ dollars.
3. Let f = the number of friends she invites to the party; $5 + 2f$ people. The expression represents the total number of people she invites to the party, which is 5 more than twice the number of friends she invites.

Exercises and Applications

1. $10 + 1.5t$ dollars; Answers may vary. An example is given. whole numbers from 0 to 4
2. $1.5d$ cm

Connection DANCE

Read the article about ballet shoes. Then answer Exercises 3–5.

That's a Lot of Shoes!

Ballet dancing is so strenuous that a professional ballerina's toe shoes usually wear out after about 15 minutes of performing. According to one ballet shoe manufacturer, the New York City Ballet Company orders about 50 pairs of toe shoes for each of its ballerinas four times per year. The American Ballet Theater orders about 60 pairs for each of its ballerinas twice a year.

Westcord ...MUNITY ...LLET ... Events

3. Write a variable expression for the number of pairs of toe shoes the New York City Ballet Company orders in one year, based on the number of ballerinas.

4. Write a variable expression for the number of pairs of toe shoes the American Ballet Theater orders in one year, based on the number of ballerinas.

5. Writing Can you tell which company orders more toe shoes? Explain.

6. PERSONAL FINANCE Use the health club ad. Write a variable expression for the amount you would pay to join Northside Fitness, based on the number of months you belong to the health club.

north SIDE
The Only Health Club That Meets Your Needs!
just $65 per month!
No initiation fee, and your first month is free!

7. Chris says, "I read a lot when I'm on vacation. I bet I read about two books a day. I also like to take a couple of extra books along." Use this information to write a variable expression. Tell what the variable stands for. Explain what the expression tells you about the situation.

Open-ended Problem The same variable expression can represent many situations, depending on what the variable stands for. For example:

Tickets cost \$10. The price of n tickets is $10n$ dollars.

One van holds 10 students. The number of students who fit in n vans is $10n$.

Describe a situation that each variable expression can represent.

8. $50m$ **9.** $16 + d$ **10.** $1.50m + 3.25n$

11. $\frac{p}{25}$ **12.** $20 + 0.15m$ **13.** $0.15(m - 60)$

14. NUTRITION Calories in food come from fat, protein, and carbohydrates. There are nine calories in each gram of fat, four calories in each gram of protein, and four calories in each gram of carbohydrates.

a. Write a variable expression for the total calories in a food, based on its fat, protein, and carbohydrate content.

b. Research Find a food package that gives information about the fat, protein, and carbohydrate content of the food. Show how you can use your variable expression to find the number of calories in the food.

3. Let n = the number of ballerinas; $200n$ pairs of toe shoes are ordered.
4. Let b = the number of ballerinas; $120b$ pairs of toe shoes are ordered.
5. No; the article does not mention how many ballerinas are in each ballet company.
6. Let m = the number of months you belong to the health club; $65(m - 1)$ dollars.
7. Let d = the number of days Chris is on vacation; $2d + 2$ books (if "a couple" of books means two books); the number of books Chris wants to take along is two more than twice the number of vacation days.

8–13. Answers may vary. Examples are given.

8. the value in cents of m half dollars
9. your total savings if you have \$16 and save d dollars more
10. the amount you spend altogether when you buy m pens for \$1.50 each and n notebooks for \$3.25 each
11. the number of quarters you have if your collection of quarters is worth p cents

12–14. See answers in back of book.

Apply⇔Assess

Exercise Notes

Visual Thinking

Ex. 2 Ideas for helping students develop their visual thinking skills are provided at strategic points throughout the side-column notes. Visual thinking skills include:

1. Observation
2. Identification
3. Recognition
4. Recall
5. Interpretation
6. Exploration
7. Correlation
8. Generalization
9. Inference
10. Perception
11. Communication
12. Self-Expression

Communication: Reading

Exs. 3–5 These exercises, and most others in this section, require that students read them carefully in order to answer the questions correctly. Some students tend to rush through the reading process and write down an answer quickly, often making errors in doing so. Encourage all students to read mathematical material slowly and carefully. They must also develop the habit of thinking about the meanings of the terms used and the relationships involved.

Multicultural Note

Exs. 3–5 The New York City Ballet was co-founded in 1948 by George Balanchine. As one of this century's most important ballet choreographers, Balanchine was intent on showcasing world-class ballets that everyone could afford to enjoy. One of the company's principal dancers, Arthur Mitchell, was the only African-American lead dancer in the 1950s.

Alternate Approach

Exs. 8–13 These open-ended problems offer students an opportunity to create their own situations, given different variable expressions. The use of this approach will help to strengthen and solidify students' understanding of how variable expressions are used to describe real-world situations.

Apply⇔Assess

Exercise Notes

Teaching Tip
Exs. 15–17 Students having difficulty with these exercises can use patterns to help in writing the variable expressions involved.

Assessment Note
Ex. 18 Students should use their toothpicks to actually make the squares and pentagons, as they did for the triangles in the Exploration on page 30. Then they can create a table, look for a pattern, and use the pattern to write a variable expression.

Practice 6 for Section 1.6

NAME ______________________ DATE __________

Practice 6

FOR USE WITH SECTION 1.6

1. In California, there are about 650 licensed drivers for every 1000 residents of the state. Write a variable expression for the number of licensed drivers in California, based on the number of residents. $\frac{650n}{1000}$
2. A phone call to Mexico costs \$.23 per minute. Under a special calling plan, your first 30 minutes are free. Write a variable expression for the amount you would pay for calls to Mexico totalling m minutes. (Assume $m > 30$.) $0.23(m - 30)$
3. Lumber for bookshelves costs \$1.30 per linear foot, and there is a \$10 milling charge, no matter how many feet you buy. Write a variable expression for the cost of the lumber for bookshelves, based on the number of linear feet you buy. $10 + 1.3n$
4. A tablespoon of peanut butter contains 5 g of protein. A slice of whole-wheat bread contains 3 g of protein. Write a variable expression for the amount of protein in a peanut butter sandwich made with n tablespoons of peanut butter and two pieces of whole-wheat bread. $6 + 5n$
5. An amusement park ride called the Octopus takes 15 s to extend its "tentacles" and each revolution takes 24 s. Write a variable expression for your total time on the Octopus, based on the number of revolutions. $15 + 24t$
6. Sue Jung has 2300 frequent-flier miles, and her credit card gives her 2 frequent-flier miles for every dollar she charges on the card. Write a variable expression for the number of frequent-flier miles Sue Jung will have after charging d dollars on her credit card. $2300 + 2d$
7. A talk-radio station airs commercials during any time not used for discussions with callers. Commercial time on the station costs \$750 per minute. Write a variable expression for the amount of money the station can make on commercials in one hour, based on the number of minutes in the hour used for discussions with callers. $750(60 - m)$
8. A jetliner cruises at an altitude of 35,000 ft. During the initial phase of its descent, the plane loses 250 ft of altitude per minute. Write a variable expression for the altitude of the plane, based on the number of minutes after the plane begins its descent. $35{,}000 - 250t$
9. A book publisher wants to predict the thickness of a one-volume gardening encyclopedia it plans to publish. Each page is $\frac{1}{400}$ in. thick, and the covers are each $\frac{1}{8}$ in. thick. Write a variable expression for the thickness of the book, based on the number of pages it has. $\frac{1}{4} + \frac{p}{400}$
10. **Open-ended Problem** Describe as many real-life situations as you can in which the variable expression $7.95(10 - x)$ represents a quantity in the situation. Answers may vary. Check students' work.

6

Practice Bank, ALGEBRA 1: EXPLORATIONS AND APPLICATIONS

Connection COOKING

ROAST TURKEY

How long to cook?
You should cook a turkey 20 min for each pound. If the turkey is stuffed, add a half hour.

How many servings per pound?
For a whole turkey, allow 1.5 lb per person. This amount is roughly based on 4 oz servings of cooked turkey with some leftovers for sandwiches the next day.

For Exercises 15–17, refer to the cookbook excerpt.

15. Write a variable expression for the weight of the turkey you should buy, based on the number of people who will be eating it.

16. a. Write a variable expression for the amount of time in minutes you should cook a stuffed turkey, based on its weight.

b. When should you begin cooking a 16 lb stuffed turkey if you want to take it out of the oven at noon?

17. a. Challenge In a pressure cooker, food cooks in about one third of the time it would take otherwise. Write a variable expression for the amount of time it takes to cook a dish in a pressure cooker, based on the normal cooking time.

b. Food in a pressure cooker is usually cooked at a pressure of 15 lb/in.^{2}. You should increase the pressure by 0.5 lb/in.^{2} for every 1000 ft above sea level. What pressure should you use h ft above sea level?

ONGOING ASSESSMENT

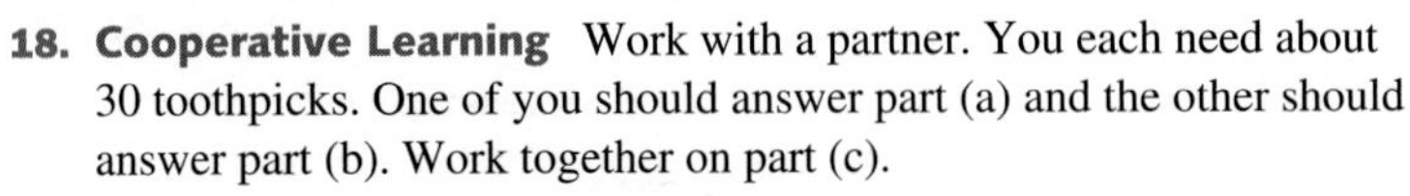

18. Cooperative Learning Work with a partner. You each need about 30 toothpicks. One of you should answer part (a) and the other should answer part (b). Work together on part (c).

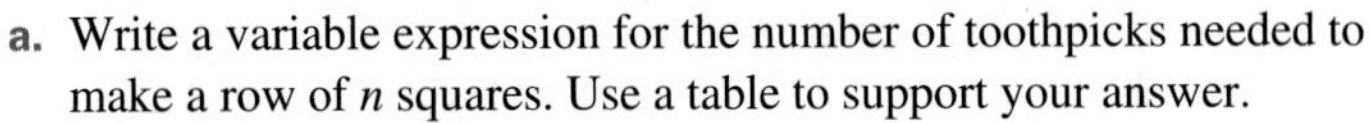

a. Write a variable expression for the number of toothpicks needed to make a row of n squares. Use a table to support your answer.

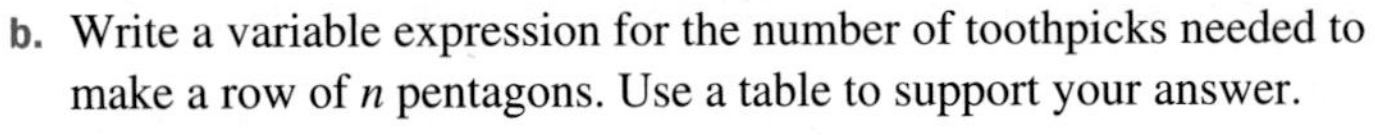

b. Write a variable expression for the number of toothpicks needed to make a row of n pentagons. Use a table to support your answer.

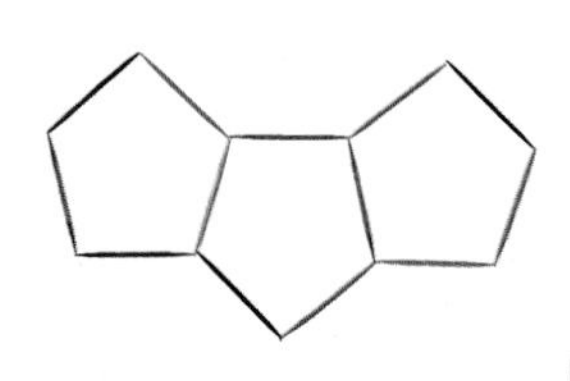

c. Writing Write a paragraph comparing the variable expressions you wrote for squares and for pentagons. Compare these expressions with the ones you wrote for triangles in the Exploration on page 30.

SPIRAL REVIEW

Evaluate each variable expression when $x = 4$. *(Section 1.2)*

19. $0.25x$ **20.** $12(3 + x)$ **21.** $15(2x + 5)$ **22.** $3x + 12$

23. $3(x + 12)$ **24.** $x^2 - 4x$ **25.** $\frac{1}{3}(9x - 9)$ **26.** $\frac{x + 3}{x - 3}$

15. Let n = the number of people to be served; $1.5n$ pounds.
16. a. Let p = the number of pounds; $20p + 30$ minutes.
    b. 6:10 A.M.
17. a. Let t = the amount of time it would take to cook on a stove; $\frac{1}{3}t$.
    b. $15 + 0.5\frac{h}{1000}$ lb/in.^{2} or $15 + 0.0005h$ lb/in.^{2}
18. See answers in back of book.
19. 1
20. 84
21. 195
22. 24
23. 48
24. 0
25. 9
26. 7

SECTION

1.7 Applying Variable Expressions

Learn how to...

- simplify variable expressions
- use the distributive property

So you can...

- apply variable expressions more easily in problem solving situations

The expressions $3(x + 1)$ and $3x + 3$ are called **equivalent expressions**. No matter what value you substitute for x, the expressions are equal.

Algebra tiles show why the expression $3(x + 1)$ is equivalent to $3x + 3$.

The [tile] tile represents the variable x, and the [tile] tile represents 1.

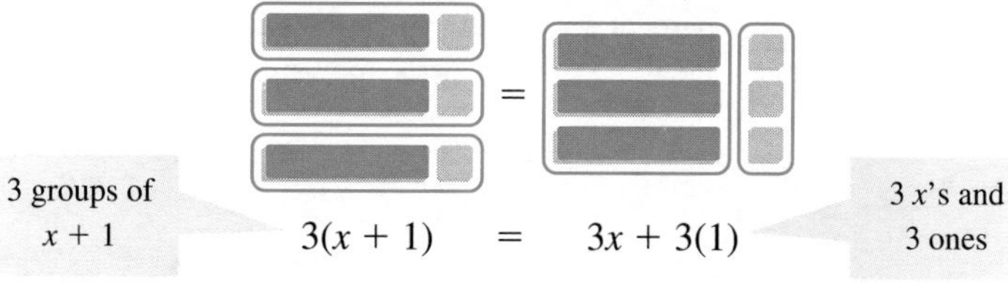

3 groups of $x + 1$ $\quad 3(x + 1) = 3x + 3(1) \quad$ 3 x's and 3 ones

You can use the *distributive property* to write equivalent expressions.

Distributive Property

For all numbers a, b, and c:	For example:
$a(b + c) = ab + ac$	$4(23 + 3) = 4 \cdot 23 + 4 \cdot 3$

EXAMPLE 1

For each expression, use the distributive property to write an equivalent expression.

a. $2(2x + 5)$ **b.** $-(3t - 2)$

SOLUTION

a. $2(2x + 5) = (2 \cdot 2x) + (2 \cdot 5)$

$= 4x + 10$

Use the distributive property. Multiply both terms inside the parentheses by 2.

b. $-(3t - 2) = -1(3t - 2)$

Use the multiplicative property of -1.

$= -1[3t + (-2)]$

$= (-1)(3t) + (-1)(-2)$

$= -3t + 2$

Remember: Subtracting 2 is the same as adding -2. Use the distributive property.

Plan⇔Support

Objectives

- Simplify variable expressions.
- Use the distributive property.
- Apply variable expressions in problem-solving situations.

Recommended Pacing

Core and Extended Courses
Section 1.7: 2 days

Block Schedule
Section 1.7: 2 half-blocks (with Sections 1.6 and 1.8)

Resource Materials

Lesson Support
Lesson Plan 1.7

Warm-Up Transparency 1.7

Overhead Visuals:
Folder 1: Simplifying a Variable Expression
Folder B: Algebra Tiles

Practice Bank: Practice 7

Study Guide: Section 1.7

Explorations Lab Manual: Diagram Master 4

Technology
Internet:
http://www.hmco.com

Warm-Up Exercises

Evaluate each expression when $x = -3$.

1. $2x$ -6
2. $-4x$ 12
3. $2x + 5$ -1
4. $-7 - 3x$ 2

Evaluate each expression when $x = 4$ and $y = -1$.

5. $x + y$ 3
6. $-2y + 3x$ 14
7. $3(2x + 4) - y$ 37
8. $-(3y - x) - y$ 8

Teach⇔Interact

Learning Styles: Visual

The use of algebra tiles makes the understanding of equivalent expressions a visual and physical experience rather than simply an abstract one. This experience can be built upon by substituting some values for x in the expressions $3(x + 1)$ and $3x + 3$, such as 0, 1, and 2, to demonstrate that they are equal.

Section Note

Teaching Tip

The distributive property is one of the most important properties used in algebra. Students need to understand its meaning completely. You may wish to illustrate the property by using numerical examples before discussing Example 1.

Additional Example 1

For each expression, use the distributive property to write an equivalent expression.

a. $3(4x - 2)$

$3(4x - 2) = (3 \cdot 4x) + (3)(-2)$
$= 12x - 6$

b. $-3(-3t + 2)$

$-3(-3t + 2) = (-3)(-3t) + (-3)(2)$
$= 9t - 6$

Section Note

Communication: Writing There are four new ideas presented in the discussion on simplifying variable expressions, namely, *terms, like terms, coefficient,* and *simplest form.* Students would benefit from writing these words in their journals, with an example to illustrate the meaning of each one.

Additional Example 2

Simplify each variable expression by combining like terms.

a. $5x^2 + 2x - 5 + x^2 - 3x + 9$

Group like terms.

$5x^2 + 2x - 5 + x^2 - 3x + 9$
$= (5x^2 + x^2) + (2x - 3x) + (-5 + 9)$
$= 6x^2 - x + 4$

b. $4x^2 - 3x + 9xy + 2y$

There are no like terms. The expression cannot be simplified.

Simplifying Variable Expressions

Algebra tiles can help you see how to *simplify* a variable expression.

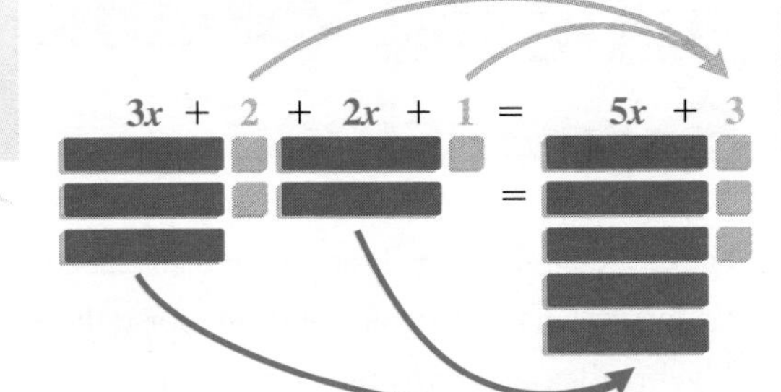

Use tiles to represent the expression 3x + 2 + 2x + 1.
3x + 2 + 2x + 1 = 5x + 3
=
Simplify the expression by grouping the tiles and the tiles.

When you group algebra tiles, you are combining *like terms.* **Terms** are the parts of a variable expression that are added together. When terms have identical variable parts, they are called **like terms**.

$2x + 4 + x^2 + 3x + 6x^2 - 7$

like terms (pointing to x^2 and $6x^2$)

This term is -7 because you are adding -7.

To combine like terms, add their *coefficients*. The numerical part of a variable term is called the **coefficient**. For example, the coefficient of $6x^2$ is **6** and the coefficient of x^2 is **1**.

$$x^2 + 6x^2 = 1 \cdot x^2 + 6 \cdot x^2 = (1 + 6)x^2 = 7x^2$$

An expression is in **simplest form** when there are no parentheses and all like terms are combined.

EXAMPLE 2

Simplify each variable expression by combining like terms.

a. $2x + 4 + x^2 + 3x + 6x^2 - 7$ **b.** $3x + 6xy + 4y$

SOLUTION

a. Group like terms.

$$2x + 4 + x^2 + 3x + 6x^2 - 7 = (2x + 3x) + (x^2 + 6x^2) + (4 - 7)$$
$$= 5x + 7x^2 + (-3)$$
$$= 5x + 7x^2 - 3$$

b. $3x + 6xy + 4y$ There are no like terms. The expression cannot be simplified.

ANSWERS Section 1.7

Checking Key Concepts

1. $2(2x + 1) = 4x + 2$
2. $2x + 2 + x + 1 = 3x + 3$
3. $2x - 8$
4. $-2t - 1$
5. $40x - 16$
6. $y - 3$
7. $13 + 8x$
8. $x^2 + x - 2$
9. $2x - 7$
10. $11t + 1$
11. $-3x - 20$

EXAMPLE 3

Simplify the variable expression $5(2x - 4) - (8x + 3)$.

SOLUTION

Use the distributive property.

$$5(2x - 4) - (8x + 3) = 5 \cdot 2x - 5 \cdot 4 - 8x - 3$$
$$= 10x - 20 - 8x - 3$$
$$= (10x - 8x) + (-20 - 3) \quad \text{Combine like terms.}$$
$$= 2x - 23$$

CHECKING KEY CONCEPTS

Write equivalent expressions for each tile diagram.

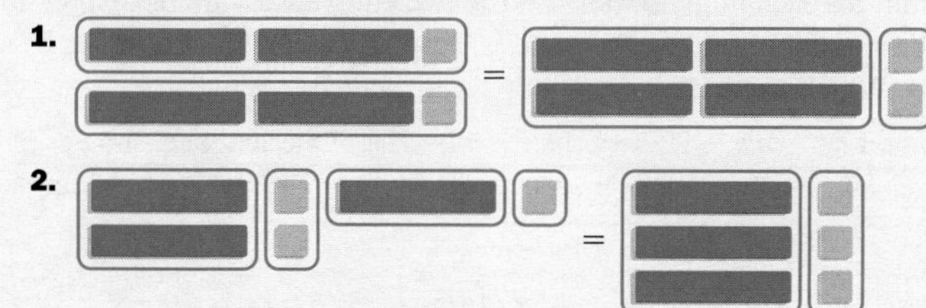

Simplify each variable expression.

3. $2(x - 4)$
4. $-(2t + 1)$
5. $8(5x - 2)$
6. $7 + y - 10$
7. $8 + 7x - 4 + 9 + x$
8. $x^2 + x - 2$
9. $2(x - 1) - 5$
10. $12t - (t - 1)$
11. $-5(x + 4) + 2x$

1.7 Exercises and Applications

Simplify each variable expression.

Extra Practice exercises on page 559

1. $2(x + y)$
2. $4(4 - 4d)$
3. $2(-5t - 7)$
4. $-8(2u + 3v)$
5. $\frac{1}{3}(-5a + 6b)$
6. $\frac{1}{2}(6x + 8y)$
7. $-6(-4 + 2y)$
8. $-6(z - 6)$
9. $-5(-3z - 4)$
10. $0.25(2x + 10)$
11. $-(1.75 - x)$
12. $-(2.5x + 8)$
13. A teacher distributes a textbook and a calculator to each algebra student. How is this situation like using the distributive property?

Exercises and Applications

1. $2x + 2y$
2. $16 - 16d$
3. $-10t - 14$
4. $-16u - 24v$
5. $-\frac{5}{3}a + 2b$
6. $3x + 4y$
7. $24 - 12y$
8. $-6z + 36$
9. $15z + 20$
10. $0.5x + 2.5$
11. $-1.75 + x$
12. $-2.5x - 8$
13. Answers may vary. An example is given. The teacher can give each student a book and a calculator at the same time (which is like adding, then multiplying). The result will be the same if the teacher gives a book to each student and then a calculator to each student (which is like multiplying first). In symbols, $(1 + 1)n = n + n$, where n = number of students.

Teach⇔Interact

Additional Example 3

Simplify the variable expression $2(5 - 3x) - 2(4 + 2x)$.

Use the distributive property and combine like terms.

$2(5 - 3x) - 2(4 + 2x)$
$= 10 - 6x - 8 - 4x$
$= (10 - 8) + (-6x - 4x)$
$= 2 - 10x = -10x + 2$

Checking Key Concepts

Common Error

In those expressions that have a negative sign in front of the parentheses, such as in question 4, some students may not apply the distributive property correctly. They may forget, for example, that $-(2t + 1)$ means to multiply each term inside the parentheses by -1 and write the answer as $-2t + 1$ instead of $-2t - 1$. Refer these students to part (b) of Example 1 on page 35 to study its solution.

Closure Questions

How many terms does $t^2 - 2t + 3 + 5t^2 + 3t - 1$ have? Which are like terms? Combine any like terms to simplify the expression.

6; t^2 and $5t^2$, $-2t$ and $3t$, 3 and -1; $6t^2 + t + 2$

Apply⇔Assess

Suggested Assignment

Core Course
- **Day 1** Exs. 1–33 odd
- **Day 2** Exs. 2–34 even, 35, 38–45

Extended Course
- **Day 1** Exs. 1–33 odd
- **Day 2** Exs. 2–34 even, 35–45

Block Schedule
- **Day 5** Exs. 1–33 odd
- **Day 6** Exs. 2–34 even, 35, 38–45

Exercise Notes

Communication: Discussion

Ex. 13 A discussion of the verb *to distribute* may help some students understand the distributive property on an intuitive level.

Apply⇔Assess

Exercise Notes

Common Error

Ex. 14 The error illustrated in the expression of part (a) shows a lack of understanding about multiplication by a negative number. Remind students that the product of two negative numbers is a positive number. Thus, $-6(-3x) = 18x$, not $-18x$.

Topic Spiraling: Preview

Ex. 14 The expression of part (b) previews the use of the distributive property to *factor* a mathematical expression, which is first discussed in Chapter 10. The property is used to write the sum of two terms in the form $ab + ac$ as the product $a(b + c)$.

Reasoning

Exs. 16–24 Students should be able to explain how the distributive property is used to simplify those expressions that can be simplified. Ex. 18, for example, cannot be simplified. Ask students why this is so.

Mathematical Procedures

Exs. 27–34 Students need to recall that to simplify an expression means to eliminate all parentheses and then to combine like terms. Thus, they should first use the distributive property to eliminate the parentheses and then add or subtract the like terms.

Problem Solving

Ex. 35 The result of this exercise can be used to remind students that when they are solving real-world problems involving variable expressions, they should first simplify the expression for the final answer before attempting to find its numerical value.

Integrating the Strands

Ex. 38 The symbols and methods of algebra are used in all branches of mathematics. Algebra can be thought of as a mathematical language that helps to connect ideas and concepts in geometry, number theory, probability and statistics, discrete mathematics, and other branches of advanced mathematics.

14. Joe used the distributive property to rewrite two variable expressions. Decide whether he rewrote the expressions correctly. If not, correct his work.

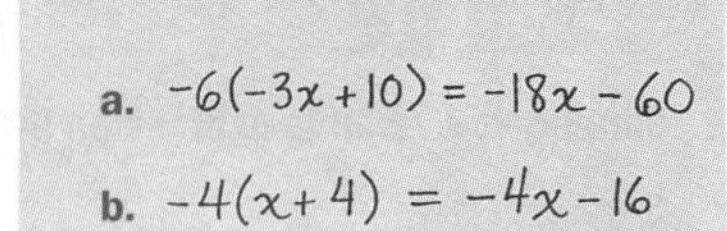

15. Cloth shopping bags cost \$3.95. Liam used mental math to find the total cost of five bags:

$$5(3.95) = 5(4.00 - 0.05) = 20.00 - 0.25 = 19.75$$

a. **Writing** Explain how Liam used the distributive property.

b. **Open-ended Problem** Describe a situation in which you might use the distributive property to do mental math.

Simplify each variable expression.

16. $3m + 5m$ **17.** $10m - 7m + 4$ **18.** $3l - 2w$

19. $4x + 4y^2 - x$ **20.** $8t - 5 - 16t + 12$ **21.** $20 + 6b^2 - 5 - 6b^2$

22. $5x^2 - 13 + 6x + 4$ **23.** $-21u + 21uv - 2v$ **24.** $-n^2 + 2n^2 - 4n^2$

25. FUNDRAISING Dave is raising money in a walk for charity. His friends pay him for each mile he walks. Write two equivalent variable expressions for the amount of money Dave raises, based on the number of miles he walks. Which expression do you prefer to evaluate? Why?

Sponsor's name	Address	City, State, Zip	Telephone number	Pledge per mile	T ple
1. Vera Santos	3 River Street	Westbridge	555-8355	\$1.00	
2. BRUCE SIMMS	10 RIVER STREET	WESTBRIDGE	555-8788	\$1.50	
3. Paul and Sarah Wu	88 Center Street	Westbridge	555-0208	\$2.00	
Danny Wu	88 Center Street	Westbridge	555-0208	\$1.00	
Sam Peterson	1500 8th Street	Westbridge	555-9600	\$1.00	

Sponsor me! WALK FOR HUNGER! SUNDAY, MAY 7

26. Use the book sale sign at the right.

a. A student wrote the variable expression shown for the total cost of buying some hardcover books and some softcover books. Explain why this variable expression does not give the correct cost. Then write a variable expression that does give the correct cost.

Let b = the number of books bought.
1b + 0.50b

b. Sheila bought five softcover books and three hardcover books. Her brother bought five hardcover books and three softcover books. Who paid more money?

THE KING UBLIC LIBRARY
• Annual •
Used Book SALE
This Saturday!
Hardcover books: \$1.00
Softcover books: \$.50

14. No; $-6(-3x + 10) = 18x - 60$.

15. Answers may vary. Examples are given.

a. Liam rewrote 3.95 as $4.00 - 0.05$. Then he used the distributive property to simplify $5(4.00 - 0.05)$.

b. If you want to know the cost of eight 32-cent stamps, you could use the following mental math: $8(32) = 8(30 + 2) = 8(30) + 8(2) = 240 + 16 = 256$.

16. $8m$ 17. $3m + 4$

18. $3l - 2w$ 19. $3x + 4y^2$

20. $-8t + 7$ 21. 15

22. $5x^2 + 6x - 9$

23. $-21u + 21uv - 2v$

24. $-3n^2$

25. If he walks m miles, then the amount of money he raises is $m + 1.5m + 2m + m + m$, or $6.5m$ dollars. I prefer to evaluate the expression $6.5m$ because it requires multiplying once. Evaluating the other expression requires multiplying twice and then adding five numbers.

26. a. The expression does not give the correct cost because it contains only one variable. The expression says he pays \$1 for each book and then \$.50

Simplify each variable expression.

27. $16 + 4(x + 5)$

28. $-11(2 - 2x) + 4$

29. $10 - 3(x - 1)$

30. $4 + 2(6y - 2) + 5y$

31. $2(3x - 2) + 3(5y - 1)$

32. $5 - (n + 3) + 3n$

33. $-3(6x - 6) + 2(9x - 9)$

34. $3(d + 5) - (3 - d)$

35. a. Cooperative Learning Work with a partner to evaluate the variable expression $-5(6 + 12x) + 3(20x + 10)$ when $x = 5$. One of you should evaluate the expression as is. The other should simplify the expression first. Which of you do you think did more work? Why?

b. Work together to write a variable expression with two variables that equals 0 no matter what values you choose for each variable. Simplify the expression to show why it equals 0.

36. Look back at Example 2 on page 31 in Section 1.6. Describe another way you can write the amount you pay for g games at the Cactus Arcade.

37. Challenge Evaluate each expression when $x = 1$ and when $x = -1$.

a. $x^2(x - 2)$

b. $x^3 - 2$ ← *Hint:* x^3 means $x \cdot x \cdot x$.

c. Writing Explain why $x^2(x - 2)$ and $x^3 - 2$ are not equivalent expressions. Support your answer with an example.

38. GEOMETRY The perimeter of a rectangle is the distance around the rectangle. You can use a variable expression to represent the perimeter.

$l + w + l + w$ ← Let l = length. Let w = width.

a. Write two other variable expressions that are equivalent to $l + w + l + w$.

b. Measure the length and the width of a rectangular calculator. Then use the calculator to find its perimeter. Which of the three perimeter expressions describes the key sequence you used? Explain.

ONGOING ASSESSMENT

39. Writing Explain how to simplify the variable expression $10(d - 6) - 3(2d + 8)$. Explain why the original expression and the simplified expression are equivalent.

SPIRAL REVIEW

Evaluate each variable expression when $n = -5$ and $p = 10$. *(Section 1.5)*

40. $8p + 3p - 20$

41. $n - 5n$

42. $\dfrac{-3n - 2p}{5}$

43. $-4(n + 3) + 2n$

44. $2n + 2(6 - n) + 3n$

45. $-11n + 11(n + 2)$

Apply⇔Assess

Exercise Notes

Assessment Note

Ex. 39 Assessment questions need to be evaluated to see if students understand the concepts of the section. They are an integral part of the learning process, providing feedback to both the teacher and students. This feedback should influence the future direction of the instructional process. For example, some students may need to spend more time on the ideas in Ex. 39 to fully understand them, while others are ready to move ahead.

Practice 7 for Section 1.7

NAME ____________ DATE ____________

Practice 7

FOR USE WITH SECTION 1.7

Simplify each variable expression.

1. $-3(a + 5)$ $-3a - 15$
2. $4(-7 - x)$ $-28 - 4x$
3. $\frac{1}{3}(6s + 15t)$ $2s + 5t$
4. $8(2y - 7)$ $16y - 56$
5. $-6(-2 + 3z)$ $12 - 18z$
6. $-2.5(4b - 12c)$ $-10b + 30c$
7. $12(-5p + 9q)$ $-60p + 108q$
8. $-\left(\frac{1}{6} - \frac{1}{2}k\right)$ $-\frac{1}{6} + \frac{k}{2}$
9. $-(-2.7m - 5.4)$ $2.7m + 5.4$
10. $-10w + 3w$ $-7w$
11. $-9 + d - 14d$ $-9 - 13d$
12. $x + 4y - x$ $4y$
13. $7k + 5 - 8 - 4k$ $3k - 3$
14. $-3n + 7 - 11n$ $-14n + 7$
15. $r + 2s - 7s + 5r$ $6r - 5s$
16. $-6a - 5b - 12b + a$ $-5a - 17b$
17. $x^2 - 8x + 4x^2 + x$ $5x^2 - 7x$
18. $-c^2 + 5d - d - 6c^2$ $-7c^2 + 4d$

Simplify each variable expression.

19. $2(-5 + w) - 3w$ $-10 - w$
20. $4 - 5(2n + 7)$ $-31 - 10n$
21. $-7(4 - r) + 9r$ $-28 + 16r$
22. $-6h + 8(-3 + 4h)$ $26h - 24$
23. $10x - 3(-2x + 4) - 12$ $16x - 24$
24. $2(p - q) + 2(-p - q)$ $-4q$
25. $-5(2a - 3b) + 7(a - b)$ $-3a + 8b$
26. $9(4 - 7z) - 3(2 + 5z)$ $30 - 78z$

27. Raoul Ramirez needs an equal number of cans of undercoat and finish paint to waterproof the hull of his boat. Each can of undercoat costs \$8, each can of finish paint costs \$11, and Raoul has a \$3 discount coupon.
    a. Write an unsimplified variable expression for the total cost of the paint, based on the number of cans of each kind that Raoul buys. $8n + 11n - 3$
    b. Simplify the expression you wrote in part (a). $19n - 3$

28. Amanda Hines is driving to the regional sales conference sponsored by her company, which will pay her \$.40 per mile for gas plus \$15 for tolls. Suppose gas for Amanda's car costs her only \$.35 per mile, and she actually spends \$12 on tolls.
    a. Write an unsimplified variable expression for the amount of money Amanda has left over, based on the number of miles she drives. $0.4n + 15 - (0.35n + 12)$
    b. Simplify the expression you wrote in part (a). $0.05n + 3$

Practice Bank, ALGEBRA 1: EXPLORATIONS AND APPLICATIONS

7

more for each book. A variable expression that gives the correct cost would be $h + 0.5s$, where h = the number of hardcover books and s = the number of softcover books.

b. Sheila's brother

27. $36 + 4x$
28. $-18 + 22x$
29. $13 - 3x$
30. $17y$
31. $6x + 15y - 7$
32. $2n + 2$
33. 0
34. $4d + 12$
35. a. evaluating: $-5(6 + 12(5)) + 3(20(5) + 10) = -5(66) + 3(110) = -330 + 330 = 0$; simplifying first: $-30 - 60x + 60x + 30 = 0$; It is easier to simplify the expression first because you can see that the expression is equal to 0.
    b. Answers may vary. An example is given. $4(3x - 6y) - 3(-8y + 4x) = 12x - 24y + 24y - 12x = 0$
36. Since $0.50(g - 2) = 0.5g - 1$, you can use the expression $0.5g - 1$ to represent the amount you pay. Note that g still must be greater than 2.
37. a. -1; -3
    b. -1; -3
    c. $x^2(x - 2) = x^2(x) - x^2(2) = x^3 - 2x^2$; $x^3 - 2x^2$ is not equivalent to $x^3 - 2$. For example, if $x = 3$, then $3^2(3 - 2) = 9(1) = 9$, but $3^3 - 2 = 27 - 2 = 25$.
38. a. $2l + 2w$ and $2(l + w)$
    b. Answers may vary. Any of the expressions can be used to find the perimeter.
39. See answers in back of book.
40. 90
41. 20
42. -1
43. -2
44. -3
45. 22

Plan⇔Support

Objectives

- Organize data using speadsheets and matrices.
- Work conveniently with large amounts of data.

Recommended Pacing

Core and Extended Courses
Section 1.8: 1 day

Block Schedule
Section 1.8: $\frac{1}{2}$ block
(with Section 1.7)

Resource Materials

Lesson Support
Lesson Plan 1.8
Warm-Up Transparency 1.8
Practice Bank: Practice 8
Study Guide: Section 1.8
Assessment Book: Test 3

Technology
Technology Book:
Spreadsheet Activity 1
Graphing Calculator
Spreadsheet Software
McDougal Littell Software
Function Investigator
Stats!
Internet:
http://www.hmco.com

Warm-Up Exercises

Find each sum mentally.

1. 51 + 34 85
2. 26 + 42 68
3. 41 + 39 80
4. 54 + 36 90
5. 17 + 43 60
6. 29 + 11 40
7. 45 + 47 92
8. 48 + 39 87
9. If Bill made 12 of 16 free throws in a basketball game, what percent of free throws did he make? 75%

SECTION

1.8 Organizing Data

Learn how to...

- organize data using spreadsheets and matrices

So you can...

- work conveniently with large amounts of data, such as the win-loss histories of baseball teams

Are you a baseball fan? The tables below compare the win-loss history of each team in the American League East before and after the 1992 All-Star Game.

AMERICAN LEAGUE

Games Before the All-Star Game		
EAST	W	L
Toronto	53	34
Baltimore	49	38
Milwaukee	45	41
Boston	42	43
New York	42	45
Detroit	41	48
Cleveland	36	52
WEST	W	

AMERICAN LEAGUE

Games After the All-Star Game		
EAST	W	L
Toronto	43	32
Baltimore	40	35
Milwaukee	47	29
Boston	31	46
New York	34	41
Detroit	34	39
Cleveland	40	34
WEST	W	

Each table can be written as a *matrix*. A **matrix** is a group of numbers arranged in rows and columns. The plural of matrix is *matrices*.

Matrices are usually named by capital letters and enclosed in brackets.

$$B = \begin{bmatrix} 53 & 34 \\ 49 & 38 \\ 45 & 41 \\ 42 & 43 \\ 42 & 45 \\ 41 & 48 \\ 36 & 52 \end{bmatrix} \qquad A = \begin{bmatrix} 43 & 32 \\ 40 & 35 \\ 47 & 29 \\ 31 & 46 \\ 34 & 41 \\ 34 & 39 \\ 40 & \mathbf{34} \end{bmatrix}$$

Each number is called an **element** of the matrix. This element shows Cleveland's losses after the All-Star Game.

EXAMPLE 1 Application: Sports

Find each team's total wins and losses for the season.

SOLUTION

Add matrices B and A. Add each element in B to the element in the same position in matrix A.

53 + 43 = 96

$$B + A = \begin{bmatrix} 53 & 34 \\ 49 & 38 \\ 45 & 41 \\ 42 & 43 \\ 42 & 45 \\ 41 & 48 \\ 36 & 52 \end{bmatrix} + \begin{bmatrix} 43 & 32 \\ 40 & 35 \\ 47 & 29 \\ 31 & 46 \\ 34 & 41 \\ 34 & 39 \\ 40 & 34 \end{bmatrix} = \begin{bmatrix} 96 & 66 \\ 89 & 73 \\ 92 & 70 \\ 73 & 89 \\ 76 & 86 \\ 75 & 87 \\ 76 & 86 \end{bmatrix}$$

Matrix $B + A$ has 7 rows and 2 columns. Its **dimensions** are 7×2. You can add matrices only when they have the same dimensions.

ANSWERS Section 1.8

Think and Communicate

1. Toronto's number of wins before the All-Star Game; Toronto's number of wins after the All-Star Game; Toronto's total number of wins for the season
2. No; they do not have the same dimensions.

THINK AND COMMUNICATE

1. What is represented by the element in the top left-hand corner of matrix B? of matrix A? of matrix $B + A$?
2. Can the matrices $\begin{bmatrix} 5 & 4 \\ 9 & -1 \end{bmatrix}$ and $\begin{bmatrix} 0 & 7 & 5 \\ 3 & 6 & 1 \end{bmatrix}$ be added? If they can be added, show how. If they cannot, explain why not.

Spreadsheets

For more information about spreadsheets, see the *Technology Handbook*, p. 610.

Who was the top team in the American League East at the end of the 1992 season? To find out, you can compare the winning percentages of the teams. A computer **spreadsheet** can help you perform the calculations. The data in a spreadsheet are arranged in rows and columns. The spreadsheet can also include row and column headings. Each box in the spreadsheet is called a **cell**.

When you highlight a cell, its contents appear in the formula bar.

Wins/Losses

C2 | 66

	A	B	C	D
1		Wins	Losses	
2	Toronto	96	66	
3	Baltimore	89	73	
4	Milwaukee	92	70	
5	Boston	73	89	

Cell C2 shows Toronto's losses for the season.

The origin of baseball is an English game called *rounders*. People in the United States played rounders as early as 1834 under the name *base ball* or *goal ball*.

EXAMPLE 2 Application: Sports

Use a spreadsheet to calculate the winning percentage for each team.

SOLUTION

Step 1 Enter a formula to tell the computer how to find the winning percentage for Toronto. The formula for winning percentage is:

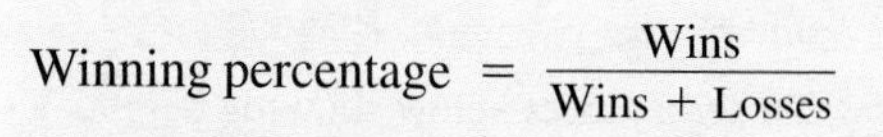

$$\text{Winning percentage} = \frac{\text{Wins}}{\text{Wins} + \text{Losses}}$$

Put parentheses around the denominator and use / for division.

Wins/Losses

D2 | =B2/(B2+C2)

	A	B	C	D
1		Wins	Losses	Winning percentage
2	Toronto	96	66	0.593
3	Baltimore	89	73	
4	Milwaukee	92	70	
5	Boston	73	89	
6	New York	76	86	
7	Detroit	75	87	
8	Cleveland	76	86	

The spreadsheet finds Toronto's winning percentage: $\frac{96}{96 + 66} \approx 0.593$

Solution continued on next page.

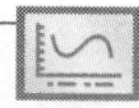

Technology Note

Computer spreadsheet software was the first highly successful commercial software product that was developed for the new small personal computers that began to appear for sale during the 1970s. Spreadsheets are accounting or bookkeeping forms that contain many rows and columns of financial data. The new computer spreadsheet software allowed users to manipulate large quantities of data quickly and accurately and then print out the results. Business people who had to work with spreadsheets found the new software a tremendous labor-saving device that could be used to increase productivity and lower costs.

Teach⇔Interact

Additional Example 1

What are the dimensions of matrices E and F? Find their sum.

$$E = \begin{bmatrix} 3 & 6 & 4 \\ 1 & 7 & 9 \\ 8 & 5 & 2 \\ 6 & 0 & 10 \end{bmatrix}$$

$$F = \begin{bmatrix} 11 & 14 & 12 \\ 16 & 3 & 19 \\ 21 & 17 & 20 \\ 9 & 13 & 8 \end{bmatrix}$$

4×3;

$$E + F = \begin{bmatrix} 3 & 6 & 4 \\ 1 & 7 & 9 \\ 8 & 5 & 2 \\ 6 & 0 & 10 \end{bmatrix} + \begin{bmatrix} 11 & 14 & 12 \\ 16 & 3 & 19 \\ 21 & 17 & 20 \\ 9 & 13 & 8 \end{bmatrix} = \begin{bmatrix} 14 & 20 & 16 \\ 17 & 10 & 28 \\ 29 & 22 & 22 \\ 15 & 13 & 18 \end{bmatrix}$$

About Example 2

Using Technology

Students can use the statistics features and list operations of the TI-82 to do many of the operations performed by spreadsheet technology. For instance, in Example 2, press [STAT] and select Edit from the EDIT menu. Enter the numbers for column B as list L1. Press 96[ENTER]89[ENTER], and so on. In the same way, enter the numbers for column C as list L2. Press [2nd][QUIT] to go to the home screen. Type (L1/(L1 + L2))*100→L3[ENTER] to perform the calculations and store the results in list L3. (To type the → symbol, press [STO▶].) Press [STAT] and select Edit from the EDIT menu to see the results.

Additional Example 2

Can a spreadsheet be used to calculate the losing percentages for each team? Describe two ways this can be done. Yes;

1. The formula for losing percentage is: losing percentage $= \frac{\text{losses}}{\text{wins + losses}}$.
2. This formula can also be used. losing percentage = 1 – winning percentage

Teach⇔Interact

Checking Key Concepts

Mathematical Procedures

Question 2 provides a quick check as to whether students understand the procedure of adding two matrices. Students who find a sum for part (b) need to review the concept of matrix addition.

Closure Question

Describe two ways data can be organized.

Data can be organized in matrices or spreadsheets. A matrix is a group of numbers, called elements, arranged in rows and columns. A spreadsheet also arranges data in rows and columns. Each box in a spreadsheet is called a cell.

SOLUTION *continued*

Step 2 Highlight cells D2 through D8. Then use the *fill down* command.

For more information about the fill down command, see the *Technology Handbook*, p. 610.

Wins/Losses

D2 | =B2/(B2+C2)

	A	B	C	D
1		Wins	Losses	Winning percentage
2	Toronto	96	66	0.593
3	Baltimore	89	73	0.549
4	Milwaukee	92	70	0.568
5	Boston	73	89	0.451
6	New York	76	86	0.469
7	Detroit	75	87	0.463
8	Cleveland	76	86	0.469

The spreadsheet calculates the winning percentages for all the teams.

THINK AND COMMUNICATE

3. Which team was the American League East's top team at the end of the 1992 season? In what order should the teams be ranked? Explain.

4. In Example 2, the formula for cell D4 is B4/(B4 + C4). Compare this formula to the formula for cell D2. What does the fill down command do?

CHECKING KEY CONCEPTS

For Questions 1 and 2, use matrices *A*, *B*, *C*, and *D*.

$$A = [-5\ \ 0.5\ \ 0] \qquad B = \begin{bmatrix} 5 \\ 3.5 \\ 2 \end{bmatrix} \qquad C = [1\ \ 0\ \ 0] \qquad D = \begin{bmatrix} 3 \\ 0 \\ 3 \end{bmatrix}$$

1. What are the dimensions of each matrix?

2. Add each pair of matrices. If they cannot be added, explain why.

a. *A* and *C* **b.** *A* and *B* **c.** *B* and *D*

3.  **Spreadsheets** Use the spreadsheet below showing heights of NBA basketball players.

a. What formula can you enter to find Rik Smits's height in inches?

b. Suppose you fill down to find each athlete's height in inches. Then you enter the formula (D3 + D4 + D5 + D6)/4 in cell D7. What number will you get? How is the number related to the height data?

	A	B	C	D
1		Height		
2		Feet	Inches	Height in inches
3	Rik Smits	7	4	
4	David Robinson	7	1	
5	Grant Hill	6	8	
6	Jason Kidd	6	4	

Think and Communicate

3. Toronto, since that team had the highest winning percentage; The ranking should be in order of winning percentages from greatest to least: Toronto, Milwaukee, Baltimore, New York and Cleveland (tied), Detroit, Boston.

4. The formula for D4 has the same form as the formula for D2, but the formula for D4 is given in terms of the elements in row 4 rather than row 2. The fill down command substitutes the corresponding elements from each row to compute the unknown element for that row.

Checking Key Concepts

1. A: 1×3; B: 3×1; C: 1×3; D: 3×1

2. a. $[-4\ \ 0.5\ \ 0]$

   b. cannot; different dimensions

   c. $\begin{bmatrix} 8 \\ 3.5 \\ 5 \end{bmatrix}$

3. a. 12 * B3 + C3; 88

   b. 82.25; The number is the mean height of the four athletes in inches.

Exercises and Applications

1. A: 2×3; B: 2×3: C: 3×2; D: 3×2

2. a. $\begin{bmatrix} 4 & 8 \\ 11 & 2 \\ 0 & -1 \end{bmatrix}$

   b. $\begin{bmatrix} 15 & 25 & 30 \\ -15 & 30 & 25 \end{bmatrix}$

   c. cannot; different dimensions

3. Answers may vary. An example is given. $A = [1\ \ -2\ \ 3]$ and $B = \begin{bmatrix} 2 & -4 \\ 6 & 3 \end{bmatrix}$; A is a 1×3 matrix and B is a 2×2 matrix.

1.8 Exercises and Applications

For Exercises 1 and 2, use matrices *A*, *B*, *C*, and *D*.

Extra Practice exercises on page 559

$$A = \begin{bmatrix} 12 & 19 & 34 \\ -11 & 25 & 16 \end{bmatrix} \quad B = \begin{bmatrix} 3 & 6 & -4 \\ -4 & 5 & 9 \end{bmatrix} \quad C = \begin{bmatrix} 1 & 4 \\ 9 & 3 \\ -5 & 7 \end{bmatrix} \quad D = \begin{bmatrix} 3 & 4 \\ 2 & -1 \\ 5 & -8 \end{bmatrix}$$

1. Give the dimensions of each matrix.

2. Add each pair of matrices. If the matrices cannot be added, explain why.

a. C and D **b.** A and B **c.** A and C

3. Open-ended Problem Give an example of two matrices that cannot be added. What are the dimensions of each matrix?

Spreadsheets A company asks its traveling salespeople to keep a monthly record of mileage and gasoline expenses. For Exercises 4–6, use the monthly records shown in the spreadsheet below.

Monthly Travel Expenses

D3 | =C3/B3

	A	B	C	D	E	F	G	H
1		June Travel				July Travel		
2	Name	Mileage	Gas expenses (dollars)	Cost per mile		Mileage	Gas expenses (dollars)	Cost per mile
3	Milner	3000	120	0.04		2600	110	
4	Dubois	2000	90	0.05		2200	100	
5	Perez	2600	110	0.04		2800	120	
6	Chen	2400	100	0.04		1800	75	

4. Write a matrix showing the mileage and gas expenses for each time period. Give the dimensions of each matrix.

a. June **b.** July **c.** total for June and July

5. a. The sales manager of the company used the spreadsheet to find the cost per mile for each salesperson during the month of June. Explain what the sales manager did.

b. Use a spreadsheet to find the cost per mile for each salesperson during the month of July.

6. Describe formulas you could use to find the mean mileage and the mean gas expenses for June.

7. Technology You can use a graphing calculator to add matrices. Before you enter a matrix, you need to give its dimensions and name it with a letter.

a. Give the dimensions of matrices A and B.

b. Use a graphing calculator to add matrices A and B. If you do not have a graphing calculator, add them by hand.

$$A = \begin{bmatrix} 2 & 4.5 \\ 0 & 9.1 \\ 40 & 7 \end{bmatrix} \quad B = \begin{bmatrix} -5.4 & 37 \\ -4.8 & 16.6 \\ -4.9 & 23 \end{bmatrix}$$

For more information on adding matrices with a graphing calculator, see the *Technology Handbook*, p. 608.

Apply⇔Assess

Suggested Assignment

Core Course
Exs. 1–6, 8–12, 15–21, AYP

Extended Course
Exs. 1–21, AYP

Block Schedule
Day 6 Exs. 1–21, AYP

Exercise Notes

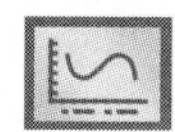

Using Technology
Exs. 2, 5, 7, 10, 12 For Exs. 2 and 7, students can use the *Function Investigator* to add matrices. Students can use the *Stats!* software in their work on the other exercises.

Common Error
Ex. 5(b) Some students may err in identifying each cell by writing the number first and then the letter as, for example, 2B rather than B2. Remind these students that the order is always *column* first, *row* second.

Using Technology
Ex. 7 To enter matrix A on a TI-81 or TI-82 graphing calculator, use the following steps. Press [MATRX] and use the [▶] key to access the EDIT menu. Press 1 to tell the calculator that you want to input the matrix [A]. Enter the matrix dimensions by pressing 3[ENTER] 2[ENTER]. The blinking cursor moves to row 1, column 1. Press 2[ENTER] 4.5[ENTER] to enter the first row of the matrix. Enter the remaining two rows in the same way. Repeat the process to enter matrix [B]. Press [2nd][QUIT] to go to the home screen. On the TI-82, find [A] + [B] by pressing [MATRX]1 +[MATRX] 2[ENTER]. On the TI-81, press [2nd] [[A]]+[2nd][[B]][ENTER].

4. a. $\begin{bmatrix} 3000 & 120 \\ 2000 & 90 \\ 2600 & 110 \\ 2400 & 100 \end{bmatrix}; 4 \times 2$

b. $\begin{bmatrix} 2600 & 110 \\ 2200 & 100 \\ 2800 & 120 \\ 1800 & 75 \end{bmatrix}; 4 \times 2$

c. $\begin{bmatrix} 5600 & 230 \\ 4200 & 190 \\ 5400 & 230 \\ 4200 & 175 \end{bmatrix}; 4 \times 2$

5. a. The sales manager used the fill down command to divide each person's expenses for the month by that person's mileage for the month.

b. Miner, \$.04; Dubois, \$.05; Perez, \$.04; Chen, \$.04

6. To find the mean mileage and gas expenses for June, enter (B3 + B4 + B5 + B6)/4 in cell B7 and enter (C3 + C4 + C5 + C6)/4 in cell C7.

7. a. 3×2; 3×2

b. $\begin{bmatrix} -3.4 & 41.5 \\ -4.8 & 25.7 \\ 35.1 & 30 \end{bmatrix}$

Apply⇔Assess

Exercise Notes

Interview Note

Exs. 8–14 Students are required to know how to read and interpret data in a table for Exs. 8 and 9. For Ex. 10, you might wish to suggest that students actually construct a spreadsheet before attempting to describe it. A concise answer to Ex. 11 depends upon students having a clear understanding of the concept of percent as meaning *so many per hundred*; in this case, *so many students in the work force per hundred.* For Ex. 14, students need to determine how to find the number of 16- and 17-year-olds who were not in the work force to answer the question. (Since total – in the work force = not in the work force, then in the work force + not in the work force = total.)

Second-Language Learners

Ex. 13 Offer students learning English the opportunity to explain their analysis orally to an aide or peer tutor. As an alternative, students learning English might work cooperatively with English-proficient students to complete this writing assignment.

INTERVIEW Carrie Teegardin

Look back at the article on pages xxx–2.

One of the statistics Carrie Teegardin looked at in her report on "Growing Up Southern" was teenage employment. Teens who are working or looking for jobs are part of the work force.

Number of 16- and 17-year-olds in the Work Force (millions)

Year	Males	Females	Total
1960	1.3	0.8	2.1
1970	1.8	1.3	3.1
1980	2.1	1.8	3.9
1990	1.5	1.4	2.9

Total Number of 16- and 17-year-olds (millions)

Year	Males	Females	Total
1960	2.9	2.8	5.7
1970	3.9	3.8	7.7
1980	4.3	4.1	8.4
1990	3.4	3.3	6.7

In 1990, about 44% of working teens in the United States held jobs in wholesale or retail trade.

The Atlanta Journal
THE ATLANTA CONSTITUTION
Growing up Southern
Hutchison win in Texas is big loss for Clinton
Teens working
Southern teens between 16 and 19 are less likely to be working than teens outside the South.
SOUTH
▶ 16 to 19-year olds working or looking for jobs: 48.2%
▶ Unemployment rate: 20.9%
NON-SOUTH
▶ 16 to 19-year olds working or looking for jobs: 52.7%
▶ Percent of the population actually employed: 45.2%
▶ Unemployment rate: 18.0%

Teegardin used data from the 1990 Census to support her report.

8. In which year was the number of males in the work force the highest? In which year was the number of females in the work force the highest?

9. Describe in general what happened to the number of males in the work force and the number of females in the work force from 1960 to 1990.

10. **Spreadsheets** What percent of 16- and 17-year-olds were in the work force during each of the four years? Describe how to use a spreadsheet to make the calculations.

11. Look at the numbers of 16- and 17-year-olds in the work force from year to year and the percents in the work force from year to year. Which figures do you think give you a better sense of the changes in the teen work force from 1960 to 1990? Explain.

12. **Spreadsheets** Find the percent of males in the work force and the percent of females in the work force for the years 1960, 1970, 1980, and 1990. Organize your results in a table.

13. **Writing** Look back at what Carrie Teegardin said about using percents in her work. How does your analysis of teens in the work force support her statements?

14. **Challenge** Write a matrix for each table. Write a matrix showing the number of 16- and 17-year-olds who were not in the work force from 1960 to 1990. How did you get your answer? How can you use matrix addition to check your answer?

8. 1980; 1980

9. The number of males in the work force and the number of females in the work force increased steadily from 1960 to 1980 and then decreased from 1980 to 1990.

10. 1960, about 36.8%; 1970, about 40.3%; 1980, about 46.4%; 1990, about 43.3%; You can compute these percents by entering the data into a spreadsheet and using the fill down command with a formula such as G2 = C2/F2*100.

11. Answers may vary. An example is given. The number and percent of 16- and 17-year-olds in the work force increased between 1960 and 1980 and then declined. I think the percents give a better sense of how many teenagers were in the work force. They tell how likely it is a teenager was in the work force in a given year.

12.

Year	Percent of Males	Percent of Females
1960	44.83	28.57
1970	46.15	34.21
1980	48.84	43.90
1990	44.1	42.42

13. Answers may vary. An example is given. Consider the decline for 16- and 17-year-olds in the work force from 1980 to 1990. You do not know exactly how well teenagers were doing at getting jobs until you consider the percentages. Since population fell significantly from 1980 to 1990, the percentage of teenagers in the work force actually increased.

14. Number of 16- and 17-year-olds in the Work Force (in millions)

$$\begin{bmatrix} 1.3 & 0.8 & 2.1 \\ 1.8 & 1.3 & 3.1 \\ 2.1 & 1.8 & 3.9 \\ 1.5 & 1.4 & 2.9 \end{bmatrix}$$

Total Number of 16- and 17-year-olds (in millions)

$$\begin{bmatrix} 2.9 & 2.8 & 5.7 \\ 3.9 & 3.8 & 7.7 \\ 4.3 & 4.1 & 8.4 \\ 3.4 & 3.3 & 6.7 \end{bmatrix}$$

Number of 16- and 17-year-olds not in the Work Force (in millions)

$$\begin{bmatrix} 1.6 & 2.0 & 3.6 \\ 2.1 & 2.5 & 4.6 \\ 2.2 & 2.3 & 4.5 \\ 1.9 & 1.9 & 3.8 \end{bmatrix}$$

I got my answer by subtracting each element of the first matrix from the element in the same position of the second matrix. To check the elements of the

ONGOING ASSESSMENT

15. Writing Describe a situation in which you might want to add matrices. Describe a situation in which you might want to use a spreadsheet to perform calculations.

SPIRAL REVIEW

Evaluate each expression when $a = 4$, $b = \frac{1}{2}$, and $c = -3$. *(Section 1.5)*

16. $2c + a$ **17.** $6b + c$ **18.** ab

19. $a - c$ **20.** $-c$ **21.** $0.25a - b$

ASSESS YOUR PROGRESS

VOCABULARY

equivalent expressions (p. 35)
terms (p. 36)
like terms (p. 36)
coefficient (p. 36)
simplest form (p. 36)
matrix (p. 40)
element (p. 40)
dimensions (p. 40)
spreadsheet (p. 41)

For Exercises 1 and 2, write a variable expression. *(Section 1.6)*

1. the number of minutes in h hours
2. the current mileage of a car that had m mi before a 250 mi road trip

Simplify each variable expression. *(Section 1.7)*

3. $3(x - 2)$ **4.** $5x^2 + 5y - 3y - x$ **5.** $2\left(x - \frac{1}{2}\right) - x$

6. Give the dimensions of each matrix. Tell which matrices can be added. Then add them. *(Section 1.8)*

$$X = \begin{bmatrix} 5 & 0.25 & -2 \\ 4 & 7 & 0 \\ 0.5 & -1 & 7 \end{bmatrix} \quad Y = \begin{bmatrix} 1.6 & 2 & 0 \\ -6 & 11 & 3 \\ 4.5 & 4 & 4 \end{bmatrix} \quad Z = \begin{bmatrix} -2 & -2 \\ -1.4 & 2.5 \end{bmatrix}$$

7. **Spreadsheets** Use the spreadsheet below. What formula can you enter to find the total number of T-shirts sold from January through April? *(Section 1.8)*

	A	B	C	D
1		T-shirts sold	Sweatshirts sold	Polo shirts sold
2	January	96	129	51
3	February	88	108	46
4	March	72	115	62
5	April	104	92	57

8. **Journal** Describe some advantages of using a spreadsheet to organize data. How does knowing about variable expressions help you work with spreadsheet formulas?

Apply⇔Assess

Exercise Notes

Assessment Note
Ex. 15 Open-ended questions such as this one are one of the best ways to assess students' understanding of a topic. Since many situations are possible, there is no single answer. You may wish to have students share their responses by verbally discussing them in small groups of three or four.

Progress Check 1.6–1.8

See page 49.

Practice 8 for Section 1.8

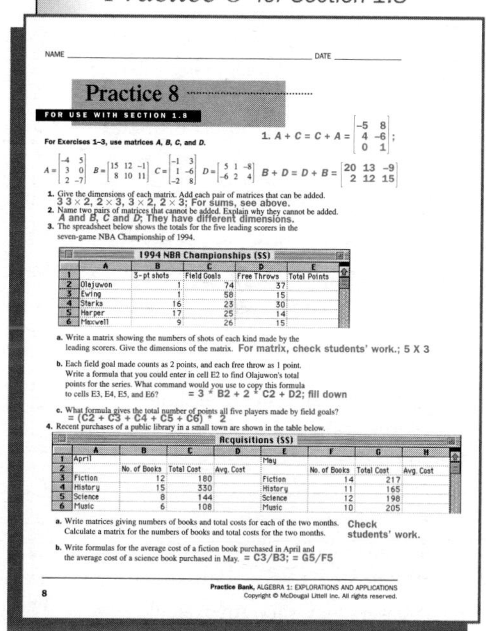

NAME ____________ DATE ________

Practice 8

FOR USE WITH SECTION 1.8

For Exercises 1–3, use matrices *A*, *B*, *C*, and *D*.

$$A = \begin{bmatrix} -4 & 5 \\ 3 & 0 \\ 2 & -7 \end{bmatrix} \quad B = \begin{bmatrix} 15 & 12 & -1 \\ 8 & 10 & 11 \end{bmatrix} \quad C = \begin{bmatrix} -1 & 3 \\ 1 & -6 \\ -2 & 8 \end{bmatrix} \quad D = \begin{bmatrix} 5 & 1 & -8 \\ -6 & 2 & 4 \end{bmatrix}$$

1. $A + C = C + A = \begin{bmatrix} -5 & 8 \\ 4 & -6 \\ 0 & 1 \end{bmatrix}$; $B + D = D + B = \begin{bmatrix} 20 & 13 & -9 \\ 2 & 12 & 15 \end{bmatrix}$

1. Give the dimensions of each matrix. Add each pair of matrices that can be added. 3 3 × 2, 2 × 3, 3 × 2, 2 × 3; For sums, see above.
2. Name two pairs of matrices that cannot be added. Explain why they cannot be added. *A* and *B*, *C* and *D*; They have different dimensions.
3. The spreadsheet below shows the totals for the five leading scorers in the seven-game NBA Championship of 1994.

1994 NBA Championships (SS)

	A	B	C	D	E
1		3-pt shots	Field Goals	Free Throws	Total Points
2	Olajuwon	1	74	37	
3	Ewing	1	58	15	
4	Starks	16	23	30	
5	Harper	17	25	14	
6	Maxwell	9	26	15	

a. Write a matrix showing the numbers of shots of each kind made by the leading scorers. Give the dimensions of the matrix. For matrix, check students' work.; 5 X 3

b. Each field goal made counts as 2 points, and each free throw as 1 point. Write a formula that you could enter in cell E2 to find Olajuwon's total points for the series. What command would you use to copy this formula to cells E3, E4, E5, and E6? = 3 * B2 + 2 * C2 + D2; fill down

c. What formula gives the total number of points all five players made by field goals? = (C2 + C3 + C4 + C5 + C6) * 2

4. Recent purchases of a public library in a small town are shown in the table below.

Acquisitions (SS)

	A	B	C	D	E	F	G	H
1	April				May			
2		No. of Books	Total Cost	Avg. Cost		No. of Books	Total Cost	Avg. Cost
3	Fiction	12	180		Fiction	14	217	
4	History	15	330		History	11	165	
5	Science	8	144		Science	12	198	
6	Music	6	108		Music	10	205	

a. Write matrices giving numbers of books and total costs for each of the two months. Calculate a matrix for the numbers of books and total costs for the two months. Check students' work.

b. Write formulas for the average cost of a fiction book purchased in April and the average cost of a science book purchased in May. = C3/B3; = G5/F5

8 **Practice Bank,** ALGEBRA 1: EXPLORATIONS AND APPLICATIONS

third matrix, add the third matrix to the first matrix and see if the sum is the second matrix.

15. Answers may vary. An example is given. If you have matrices that gave month-by-month sales of various models of computers, you can use matrix addition to calculate total sales for the year. If you have a table listing inventory for a furniture store for each of three consecutive months, you can use a spreadsheet to calculate the number of each item sold per month.

16. –2 **17.** 0
18. 2 **19.** 7
20. 3 **21.** $\frac{1}{2}$

Assess Your Progress

1. $60h$ min
2. $m + 250$ mi
3. $3x - 6$
4. $5x^2 + 2y - x$
5. $x - 1$
6. X: 3 × 3; Y: 3 × 3; Z: 2 × 2; $X + Y = \begin{bmatrix} 6.6 & 2.25 & -2 \\ -2 & 18 & 3 \\ 5 & 3 & 11 \end{bmatrix}$
7. B6 = B2 + B3 + B4 + B5
8. Answers may vary. An example is given. A spreadsheet keeps data well organized, readable, and easy to analyze and compute. Knowing about variable expressions helps you write and interpret spreadsheet formulas correctly. For example, you know when you need to use parentheses to get the correct result.

Project Notes

Mathematical Goals

- Collect and organize data using a spreadsheet or table.
- Calculate a percent of change in the data.
- Display the percent of change results in a graph.

Planning

Materials
- Newspapers, magazines, or almanac
- Graph paper

Project Teams
Students can choose a partner to work with and then discuss how they wish to proceed.

Guiding Students' Work

Suggest that each partner research at least two statistics and calculate their percent of change. Each student should calculate his or her own percent of change. One partner can then make the graph showing their results visually, while the other partner begins to write a report or make a poster. Both partners should work to complete this analysis of their results and participate in explaining their results.

Second-Language Learners
Students learning English may not be familiar with the term "in-line skaters." Explain that in-line skates are a specific kind of rollerskate with three to four wheels, one behind the other, or "in-line."

PORTFOLIO PROJECT

Analyzing Change

Would your school feel more crowded if enrollment increased by 50 students? It depends on the size of your school. If your school has 1000 students, enrollment would increase by 5%. If your school has 100 students, enrollment would increase by 50%. Knowing a percent change often gives you more information than a numerical change.

PROJECT GOAL **Your goal is to investigate the percent change in something that interests you.**

Collecting Data

Work with a partner. Decide on a topic for your investigation. You can collect data yourself, or you can find statistics in a newspaper, a magazine, or an almanac. Your data should include at least four numerical values that change from one time to another.

Some possibilities are:

- the populations of your state and four nearby states in two different years
- the number of cars passing through a green light in the morning and afternoon at various intersections

Use a table or spreadsheet to record and analyze your data. For example, suppose you want to explore a surge in the popularity of in-line skating.

1. USE A SPREADSHEET to record the number of in-line skaters in four regions of the United States for two different years.

2. USE A FORMULA to calculate the percent change from 1992 to 1993 in the Northeast.

$$\text{Percent change} = \frac{\text{New value} - \text{Old value}}{\text{Old value}} \cdot 100$$

Edit Format

Normal

In-Line Skating

D2 =((C2–B2)/B2)*100

	A	B	C	D
1		1992	1993	Percent change
2	Northeast	1,720,000	2,889,000	67.97
3	North central	2,646,000	3,963,000	49.77
4	South	2,306,000	2,899,000	25.72
5	West	2,692,000	2,808,000	4.31
6				
7				
8				

3. FILL DOWN to find the percent change in the other three regions of the country.

General Rubric for Projects
Each project can be evaluated in many possible ways. The following rubric is just one way to evaluate these open-ended projects. It is based on a 4-point scale.

4 The student fully achieves all mathematical and project goals. The presentation demonstrates clear thinking and explanation. All work is complete and correct.

3 The student substantially achieves the mathematical and project goals. The main thrust of the project and the mathematics behind it is understood, but there may be some minor misunderstanding of content, errors in computation, or weakness in presentation.

2 The student partially achieves the mathematical and project goals. A limited grasp of the main mathematical ideas or project requirements is demonstrated. Some of the work may be incomplete, misdirected, or unclear.

1 The student makes little progress toward accomplishing the goals of the project because of lack of understanding or lack of effort.

Making a Graph

Present your results visually. Your display should make it clear which quantities increased and which decreased. You can choose any of these types of displays, or you may have an idea of your own.

- a bar graph
- a colored map
- a histogram

Presenting Your Analysis

Write a report, make a poster, prepare a video, or give a presentation explaining your results.

Remember to:

- Include your original data. Explain how you measured and recorded the data.
- Describe how you did your calculations. (If you used a spreadsheet, include a copy and explain how you used it.)
- Explain your visual display.
- Explain the difference between a positive percent change and a negative percent change.
- Try to draw conclusions from the percent change or from the original data.
- Discuss reasons why these changes may have occurred.

Self-Assessment

What have you and your partner learned about percent change? Describe what you think went well in your project. Describe what you think did not work as smoothly as you would have liked. If you could do the project again, what would you do differently? Why?

Project Notes

Guiding Students' Work

Rubric for Chapter Project

4 Students collect four statistics and calculate their percents of change correctly. They make an accurate graph that clearly shows which quantities increased or decreased. The written analysis includes all six items listed under *Remember to*, and is well organized and easy to read.

3 Students collect their statistics and compute the percents of change accurately. Their graphs present the results correctly. Their presentation, however, has a few flaws and is incomplete in at least one of the ways listed under *Remember to*.

2 Students collect four statistics but make a mathematical mistake in calculating a percent of change. Their graph does not show clearly which quantities increased and which decreased. Their final report is not well organized, contains some errors or omissions, and is difficult to understand.

1 Students fail to collect all four statistics and do not calculate the percents of change correctly. No graphs are made or those made are poorly done. A final analysis is attempted but falls short in explaining the results. Students should be encouraged to speak with the teacher as soon as possible to review their work and to make a new start on the project.

Chapter Support

Course Guide: Chapter 1
Lesson Plans: Chapter 1
Practice Bank: Cumulative Practice 9
Study Guide: Chapter 1 Review
Explorations Lab Manual:
Additional Exploration 1
Assessment Book:
Chapter Tests 4 and 5
Chapter 1 Alternative Assessment
Technology Book:
Calculator Based Lab 1
Portfolio Project Book:
Additional Project 1:
The History of Algebra
Additional Project 2:
Absolute Value
Preparing for College Entrance Tests
Professional Handbook

Progress Check 1.1–1.3

Write an inequality that describes each situation. *(Section 1.1)*

1. There are at most 25 students in each class. $s \le 25$

2. The least number of managers needed to run the company is 6. $6 \le m$

Simplify each expression. *(Section 1.2)*

3. $4 \cdot 5 + 10 \div 2$ 25

4. $9(13 - 1) \div 10$ 10.8

Evaluate each variable expression when $a = 7$ and $b = 3$. *(Section 1.2)*

5. $a + 2b$ 13

6. $\frac{2a^2 - 10b}{4}$ 17

7. Find the mean, the median, and the mode(s) of these data. *(Section 1.3)*
50, 10, 30, 70, 40, 50, 20, 90, 80, 60, 100
mean ≈ 54.5, median = 50, mode = 50

CHAPTER

1 Review

STUDY TECHNIQUE

There are many ways to review what you have learned. Complete the following list of study ideas with some that have worked well for you.

To study for a math test, I usually...

- ***Make an outline of the chapter with worked out sample problems.***
- ***Redo the homework problems that were difficult for me.***
- ...

Compare ideas with other students. As you review this and other chapters, you may want to try new ways of studying.

VOCABULARY

data (p. 3)
variable (p. 3)
value of a variable (p. 3)
variable expression (p. 9)
evaluate (p. 9)
exponent (p. 10)
mean (p. 14)
median (p. 14)
mode (p. 14)
integer (p. 19)
absolute value (p. 19)
opposite (p. 20)
equivalent expressions (p. 35)
terms (p. 36)
like terms (p. 36)
coefficient (p. 36)
simplest form (p. 36)
matrix, matrices (p. 40)
element (p. 40)
dimensions (p. 40)
spreadsheet (p. 41)

SECTIONS 1.1, 1.2, *and* 1.3

To **evaluate** a **variable expression**, substitute a value for each **variable**. Then simplify the numerical expression using the order of operations.

Evaluate the expression $3(a + 2) - 4b + 6$ when $a = 5$ and $b = 2$.

1. Do all work inside parentheses.
2. Then do all multiplications and divisions from left to right.
3. Then do all additions and subtractions from left to right.

$$3(a + 2) - 4b + 6 = 3(5 + 2) - 4 \cdot 2 + 6$$
$$= 3 \cdot 7 - 4 \cdot 2 + 6$$
$$= 21 - 8 + 6$$
$$= 19$$

There are three kinds of averages, as shown in the example below.

The **modes** are 9 and 18 because they are the most frequent values.

9 9 10 11 13 14 16 17 18 18 19 22

15

There is an even number of values, so the **median** is the mean of the two middle values.

The **mean** is the sum of the data items divided by the number of data items.

$$\frac{9 + 9 + 10 + 11 + 13 + 14 + 16 + 17 + 18 + 18 + 19 + 22}{12} \approx \mathbf{14.67}$$

SECTIONS 1.4 *and* 1.5

Subtracting a number is the same as adding its **opposite**.

$$2 - 3 = 2 + (-3) = -1$$

You can use these rules to multiply and divide negative numbers:

Same sign → Positive

$3 \cdot 9 = 27$ and $(-5)(-4) = 20$

$\frac{16}{2} = 8$ and $\frac{-15}{-5} = 3$

Different sign → Negative

$(-3)(9) = -27$ and $(5)(-4) = -20$

$\frac{-16}{2} = -8$ and $\frac{15}{-5} = -3$

SECTIONS 1.6, 1.7, *and* 1.8

You can use patterns to write variable expressions.

To simplify a variable expression:
1. Use the distributive property.
2. Combine like terms.

$$\begin{aligned} 2(5x + 3) - 2x &= 10x + 6 - 2x \\ &= 10x - 2x + 6 \\ &= 8x + 6 \end{aligned}$$

A **matrix** is a group of numbers arranged in rows and columns. You can add matrices that have the same **dimensions**. The dimensions of each of these matrices are 3×2.

$$\begin{bmatrix} 4 & 9 \\ 7 & 1 \\ 0 & -3 \end{bmatrix} + \begin{bmatrix} 1 & 2 \\ 2 & 1 \\ 6 & 3 \end{bmatrix} = \begin{bmatrix} 5 & 11 \\ 9 & 2 \\ 6 & 0 \end{bmatrix}$$

Computer **spreadsheets** contain data arranged in rows and columns. Spreadsheets are useful for performing calculations within large sets of data.

Miles Traveled

	A	B	C
1		Week 1	Week 2
2	Alexa Maglio	75	160
3	Rick Steele	130	95

Progress Check 1.4–1.5

Evaluate each expression when $a = -5$ and $b = 7$. *(Section 1.4)*

1. $2|b - a|$ 24

2. $-3a + b^2$ 64

What is the opposite of each number? *(Section 1.4)*

3. 11 −11

4. −2 2

5. 4.5 −4.5

Simplify each expression. *(Section 1.5)*

6. $-5 - (-16)$ 11

7. $|-3 + 1| - 2|-8|$ −14

Evaluate each expression when $t = -3$ and $w = 10$. *(Section 1.5)*

8. $-9t \div w$ −2.7

9. $-2(t^2 + w)$ −38

Progress Check 1.6–1.8

Write a variable expression for each situation *(Section 1.6)*

1. the number of feet in y yards $3y$

2. the number of cents in d dollars $100d$

3. the cost of an item that sells for d dollars plus 5% tax $d + 0.05d$

Simplify each variable expression. *(Section 1.7)*

4. $3t^2 + 4w - 6t^2 + w$ $-3t^2 + 5w$

5. $-5(2t - 3) - 2(-4t + 8)$ $-2t - 1$

6. Find the sum of the two matrices. *(Section 1.8)*

$$A = \begin{bmatrix} 2 & -1 & 3 \\ 4 & 0 & 5 \\ 1 & 6 & 9 \end{bmatrix}$$

$$B = \begin{bmatrix} -2 & 2 & -2 \\ -3 & 0 & -4 \\ 0 & -5 & -9 \end{bmatrix}$$

$$A + B = \begin{bmatrix} 0 & 1 & 1 \\ 1 & 0 & 1 \\ 1 & 1 & 0 \end{bmatrix}$$

7. In naming a cell in a spreadsheet, do you identify the row or column of the cell first? *(Section 1.8)* column

Chapter 1 Assessment
Form A Chapter Test

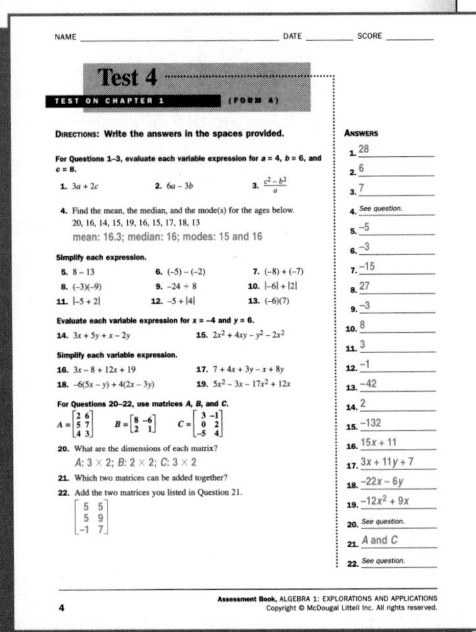

NAME ______ DATE ______ SCORE ______

Test 4

TEST ON CHAPTER 1 (FORM A)

DIRECTIONS: Write the answers in the spaces provided.

For Questions 1–3, evaluate each variable expression for $a = 4$, $b = 6$, and $c = 8$.

1. $3a + 2c$
2. $6a - 3b$
3. $\frac{c^2 - b^2}{a}$
4. Find the mean, the median, and the mode(s) for the ages below.
   20, 16, 14, 15, 19, 16, 15, 17, 18, 13
   mean: 16.3; median: 16; modes: 15 and 16

Simplify each expression.

5. $8 - 13$
6. $(-5) - (-2)$
7. $(-8) + (-7)$
8. $(-3)(-9)$
9. $-24 \div 8$
10. $|-6| + |2|$
11. $|-5 + 2|$
12. $-5 + |4|$
13. $(-6)(7)$

Evaluate each variable expression for $x = -4$ and $y = 6$.

14. $3x + 5y + x - 2y$
15. $2x^2 + 4xy - y^2 - 2x^2$

Simplify each variable expression.

16. $3x - 8 + 12x + 19$
17. $7 + 4x + 3y - x + 8y$
18. $-6(5x - y) + 4(2x - 3y)$
19. $5x^2 - 3x - 17x^2 + 12x$

For Questions 20–22, use matrices A, B, and C.

$A = \begin{bmatrix} 2 & 6 \\ 5 & 7 \\ 4 & 3 \end{bmatrix}$ $B = \begin{bmatrix} 8 & -6 \\ 2 & 1 \end{bmatrix}$ $C = \begin{bmatrix} 3 & -1 \\ 0 & 2 \\ -5 & 4 \end{bmatrix}$

20. What are the dimensions of each matrix?
    A: 3 × 2; B: 2 × 2; C: 3 × 2
21. Which two matrices can be added together?
22. Add the two matrices you listed in Question 21.
    $\begin{bmatrix} 5 & 5 \\ 5 & 9 \\ -1 & 7 \end{bmatrix}$

ANSWERS

1. 28
2. 6
3. 7
4. See question.
5. −5
6. −3
7. −15
8. 27
9. −3
10. 8
11. 3
12. −1
13. −42
14. 2
15. −132
16. $15x + 11$
17. $3x + 11y + 7$
18. $-22x - 6y$
19. $-12x^2 + 9x$
20. See question.
21. A and C
22. See question.

4 **Assessment Book,** ALGEBRA 1: EXPLORATIONS AND APPLICATIONS

Chapter 1 Assessment
Form B Chapter Test

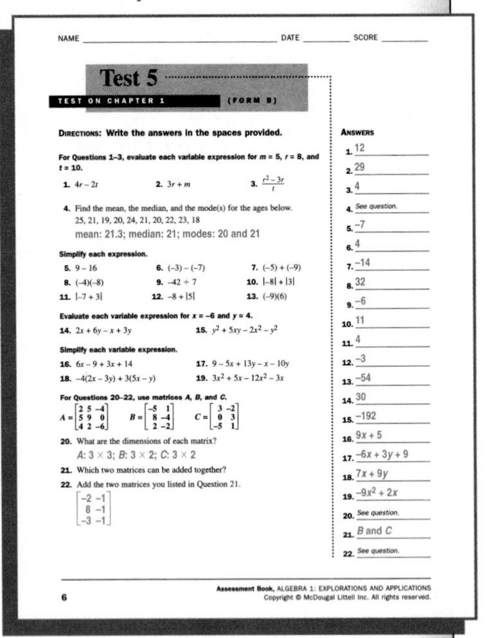

NAME ______ DATE ______ SCORE ______

Test 5

TEST ON CHAPTER 1 (FORM B)

DIRECTIONS: Write the answers in the spaces provided.

For Questions 1–3, evaluate each variable expression for $m = 5$, $r = 8$, and $t = 10$.

1. $4r - 2t$
2. $3r + m$
3. $\frac{r^2 - 3r}{t}$
4. Find the mean, the median, and the mode(s) for the ages below.
   25, 21, 19, 20, 24, 21, 20, 22, 23, 18
   mean: 21.3; median: 21; modes: 20 and 21

Simplify each expression.

5. $9 - 16$
6. $(-3) - (-7)$
7. $(-5) + (-9)$
8. $(-4)(-8)$
9. $-42 \div 7$
10. $|-8| + |3|$
11. $|-7 + 3|$
12. $-8 + |5|$
13. $(-9)(6)$

Evaluate each variable expression for $x = -6$ and $y = 4$.

14. $2x + 6y - x + 3y$
15. $y^2 + 5xy - 2x^2 - y^2$

Simplify each variable expression.

16. $6x - 9 + 3x + 14$
17. $9 - 5x + 13y - x - 10y$
18. $-4(2x - 3y) + 3(5x - y)$
19. $3x^2 + 5x - 12x^2 - 3x$

For Questions 20–22, use matrices A, B, and C.

$A = \begin{bmatrix} 2 & 5 & -4 \\ 5 & 9 & 0 \\ 4 & 2 & -6 \end{bmatrix}$ $B = \begin{bmatrix} -5 & 1 \\ 8 & -4 \\ 2 & -2 \end{bmatrix}$ $C = \begin{bmatrix} 3 & -2 \\ 0 & 3 \\ -5 & 1 \end{bmatrix}$

20. What are the dimensions of each matrix?
    A: 3 × 3; B: 3 × 2; C: 3 × 2
21. Which two matrices can be added together?
22. Add the two matrices you listed in Question 21.
    $\begin{bmatrix} -2 & -1 \\ 8 & -1 \\ -3 & -1 \end{bmatrix}$

ANSWERS

1. 12
2. 29
3. 4
4. See question.
5. −7
6. 4
7. −14
8. 32
9. −6
10. 11
11. 4
12. −3
13. −54
14. 30
15. −192
16. $9x + 5$
17. $-6x + 3y + 9$
18. $7x + 9y$
19. $-9x^2 + 2x$
20. See question.
21. B and C
22. See question.

6 **Assessment Book,** ALGEBRA 1: EXPLORATIONS AND APPLICATIONS

CHAPTER

1 Assessment

VOCABULARY QUESTIONS

1. Give an example of a variable expression that is not in simplest form. Then write the same expression in simplest form.
2. Explain the meaning of *absolute value*.
3. Explain the difference between finding the mean and finding the median of a data set.

SECTIONS 1.1, 1.2, *and* 1.3

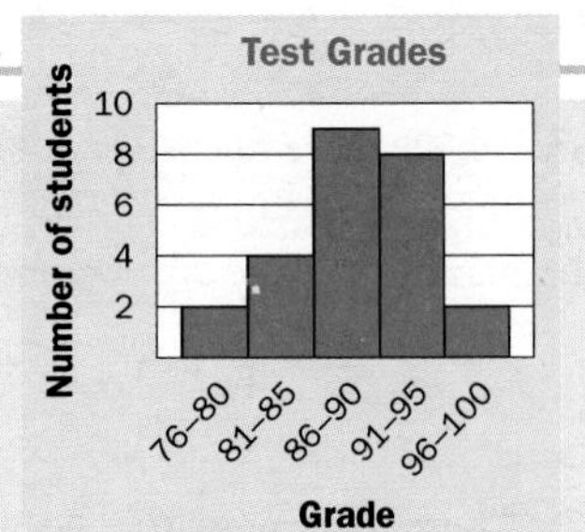

For Questions 4–6, use the histogram. Let g = each student's test grade.

4. For how many students is $g \geq 91$?
5. How many students are represented?
6. Write two inequalities that describe the grades of the students.

Evaluate each variable expression when $a = 3$, $b = 4$, and $c = 5$.

7. $4a + 3b$ 8. $6a - 2c$ 9. $\frac{a^2 - b}{c}$

10. a. The ages of students in a college class are 18, 19, 17, 22, 19, 19, 19, 18, 23, 22. Find the mean, the median, and the mode(s) of the ages.
    b. **Writing** Which of the averages in part (a) will be affected if a 68-year-old person joins the class? Explain.

SECTIONS 1.4 *and* 1.5

Simplify each expression.

11. $7 - 10$ 12. $(-3) - (-4)$ 13. $(-3) + (-4)$ 14. $(-6)(-5)$

15. $-15 \div 3$ 16. $|-2| + |3|$ 17. $|-2 + 3|$ 18. $-12 + \frac{|-3|}{3(3)}$

Evaluate each variable expression when $x = -5$ and $y = 2$.

19. $x + 7y + 2x - 5y$ 20. $-8y + \frac{20}{x}$ 21. $y^2 + xy - 4$

ANSWERS Chapter 1

Assessment

1. Answers may vary. An example is given. $3x + 2y - 5x + 6 = -2x + 2y + 6$
2. The absolute value of a number is its distance from zero on the number line. For example, $|-7| = 7$, $|3| = 3$, and $|0| = 0$.
3. The mean is the sum of the data divided by the number of data items. The median is the middle number when you put the data in order from smallest to largest. When the number of data items is even, the median is the mean of the two middle numbers.
4. 10
5. 25
6. $g \geq 76$ and $g \leq 100$
7. 24
8. 8
9. 1
10. a. 19.6; 19; 19
    b. The mean will be affected because 68 would be added to the sum of the other ages.
11. −3
12. 1
13. −7

SECTIONS 1.6, 1.7, *and* 1.8

22. There are 24 h in one day. Write a variable expression for the number of hours in d days. Use your variable expression to find the number of hours in a week.

23. John is treating his friends to lunch. He pays \$2.25 for each sandwich and \$.65 for each soft drink. Write a variable expression for John's cost if he and his friends each have a sandwich and a soft drink. What is John's cost if he and 7 friends have lunch?

Simplify each variable expression.

24. $2x - 6 + 9x + 17$

25. $5 + 6x + 4y - x + 3y$

26. $-3(2x - y) + 2(3x - y)$

27. $4x^2 - 2x - 10x^2 + 8x$

28. A secretary used a spreadsheet to find the cost of ordering some supplies, as shown. What information is contained in columns B, C, and D? Explain what spreadsheet formula was used for cell D2.

Office Supplies

	A	B	C	D
1	Item	Qty	Price	Cost
2	Paper	6	4.40	26.40
3	Pencils	25	0.69	
4	Envelopes	2	22.90	
5	Disks	5	19.10	

For Questions 29 and 30, use matrices *A*, *B*, *C*, and *D*.

$$A = \begin{bmatrix} 5 & 6 \\ 2 & 8 \\ 7 & 4 \end{bmatrix} \quad B = \begin{bmatrix} 4 & -3 \\ 7 & 0.5 \end{bmatrix} \quad C = \begin{bmatrix} 3 \\ 6 \\ -4 \end{bmatrix} \quad D = \begin{bmatrix} -3 & 2 \\ 5 & -8 \\ -7 & 2 \end{bmatrix}$$

29. What are the dimensions of each matrix?

30. Tell which matrices can be added. Then add them.

PERFORMANCE TASK

31. Work in a group of 2–4 students. Follow the steps below to create a game using variable expressions.

Each group should make a set of 20 cards. Half of the cards contain variable expressions. The other half describe matching situations.

Exchange cards with another group. Shuffle the cards and place them face down. Turn two cards face up. If they are a matching pair, take both cards and go again. If they don't match, put them back and the next person takes a turn. The person with the most pairs of cards wins.

Give comments and suggestions for improvement to the group whose game you played. Make changes to your game if needed.

a - 5
I picked some apples.
I threw out 5 that
had worms.
How many do I have
now?

Assessment **51**

Chapter 1 Assessment
Form C Alternative Assessment

Chapter 1

ALTERNATIVE ASSESSMENT

1. **Open-ended Problem** What is a variable? Include examples with your explanation.

2. **Project** Cut out articles from newspapers and magazines that contain examples of inequalities. Write a mathematical inequality that could represent each example.

3. **Project** Write a play-by-play commentary for part of a football game. Next to each play write a mathematical statement describing the progress of the team. Use positive and negative integers in your description.

4. **Group Activity** Work in a group of four students. In your group, test different types of numbers in the two statements below. As a group, write a general description to answer each of the questions.
   a. Which values for a and b make the statement $|a| + |b| = |a + b|$ true? Which values for a and b make the statement false?
   b. Which values for a and b make the statement $|a| - |b| = |a - b|$ true? Which values for a and b make the statement false?

5. **Performance Task** A set of quiz grades for two different classes is given in the table at the right.
   a. Calculate the mean for each class. What does the mean tell you about each class?
   b. Draw a histogram for each class. What does the histogram tell you about each class?
   c. Which class do you think would be easier for a teacher to teach? Explain why you think so.
   d. Why should a teacher be cautious about making assumptions based on the mean of a set of quiz grades?

Class 1	Class 2
5	8
7	9
10	8
10	6
8	8
7	10
5	8
10	7
8	9
10	7

6. **Research Project** Look up the word *average* in the dictionary and copy its definition. If you can find the definition of *mean* embedded in your definition, underline it. If you can find the definition of *mode* embedded in your definition, circle it. If you can find the definition of *median* embedded in your definition, put a box around it. What does this tell you about the word "average?"

Assessment Book, ALGEBRA 1: EXPLORATIONS AND APPLICATIONS

91

7. **Open-ended Problem** Matrix A below represents the inventory of a music store at the beginning of a month. Matrix B represents the merchandise delivered early in that month.

$$A = \begin{array}{c} \\ \text{Tapes} \\ \text{CDs} \end{array}\begin{array}{c} \begin{matrix} \text{Classical} & \text{Rock} \end{matrix} \\ \begin{bmatrix} 250 & 400 \\ 700 & 650 \end{bmatrix} \end{array} \quad B = \begin{array}{c} \\ \text{Classical} \\ \text{Rock} \end{array}\begin{array}{c} \begin{matrix} \text{Tapes} & \text{CDs} \end{matrix} \\ \begin{bmatrix} 250 & 400 \\ 700 & 650 \end{bmatrix} \end{array}$$

   a. Explain why it does not make sense to add matrix A to matrix B to get a total inventory after the delivery. Rewrite the matrices so that addition is possible.
   b. Add the rewritten matrices from part (a). Explain what you did to add them and what the resulting matrix represents.

8. **Performance Task** Make an integer timeline of your life. Your birth date is represented by 0 on the timeline. Use positive and negative integers to represent the years. Events before your birth will be located along the negative part of the timeline. Mark and label the locations for at least three family events that occurred before your birth. Include photos or drawings of at least five important personal events after your birth. Have a parent or relative help you choose the events and pictures. *Extension:* Use fractions or decimals to represent your events to the nearest day or month.

9. a. Draw a geometric figure that has a perimeter of $2x + 3y$. Label the sides of your figure.
   b. Draw a quadrilateral that has a perimeter of $2x + 3y$. Label the sides of your figure.
   c. Draw a rectangle that has a perimeter of $2x + 3y$. Label the sides of your figure.

92
Assessment Book, ALGEBRA 1: EXPLORATIONS AND APPLICATIONS

14. 30
15. –5
16. 5
17. 1
18. $-11\frac{2}{3}$
19. –11
20. –20
21. –10
22. $24d$; 168 h
23. Let f = the number of friends; $2.90(f + 1)$ dollars; \$23.20.
24. $11x + 11$
25. $5 + 5x + 7y$
26. y
27. $-6x^2 + 6x$
28. amount; price per item; total cost; D2 = B2*C2
29. A: 3×2; B: 2×2; C: 3×1; D: 3×2
30. $A + D = \begin{bmatrix} 2 & 8 \\ 7 & 0 \\ 0 & 6 \end{bmatrix}$
31. Answers may vary. Check students' work.

CHAPTER

2 Equations and Functions

OVERVIEW

Connecting to Prior and Future Learning

⇔ Students begin Chapter 2 by studying and solving one- and two-step equations. This introduction to equations will help students build a firm foundation for their continued work.

⇔ Functions are introduced in this chapter. In Section 2.3, the previously introduced concepts of equations and tables are used to represent functions. The concept of a function will be used throughout this course and future mathematics courses.

⇔ Students begin their study of graphing functions and using graphs to solve problems in Chapter 2. These basic concepts of graphing and problem solving are used throughout this course and are important in all levels of mathematics.

Chapter Highlights

Interview with Bob Hall: The interview and related exercises and examples on pages 57, 60, and 82 emphasize the variety of contexts in which math occurs. Bob Hall, the first person to compete in the Boston Marathon in a wheelchair, uses both algebra and geometry in his work.

Explorations in Chapter 2 provide students with the opportunity to solve equations using algebra tiles in Section 2.2 and to use the entire class as a coordinate grid to model graphing equations in Section 2.5.

The Portfolio Project: Students collect and graph data in which a pattern of change is obvious. As students measure and record the data, make a graph, and present their ideas, they will learn to work together to utilize many of the mathematical skills they have learned so far.

Technology: Scientific calculators are used in Section 2.2 to solve equations. The use of graphing calculators to add matrices is discussed in Section 2.4. Graphing calculators are also used in Section 2.6 to graph linear equations and to find solutions. Graphing calculators are used to graph linear and a variety of other equations throughout this course. In Section 2.5, spreadsheet formulas are used to represent and evaluate functions.

OBJECTIVES

Section	Objectives	NCTM Standards
2.1	• Write and solve one-step equations. • Solve real-world problems including those involving distance, rate, and time.	1, 2, 3, 4, 5
2.2	• Write and solve two-step equations. • Solve real-world problems using equations. • Solve geometry problems using equations.	1, 2, 3, 4, 5, 8
2.3	• Recognize and describe functions using tables and equations. • Predict the outcome of a decision.	1, 2, 3, 4, 5, 6
2.4	• Read and create coordinate graphs. • Analyze biological and geographical data. • Read and interpret maps.	1, 2, 3, 4, 5, 8, 12
2.5	• Graph equations. • Visually represent functions.	1, 2, 3, 4, 5, 6, 8, 12
2.6	• Use a graph to solve a problem. • Analyze the costs and profits of running a business.	1, 2, 3, 4, 5, 8

INTEGRATION

Mathematical Connections	2.1	2.2	2.3	2.4	2.5	2.6
algebra	**55–60***	**61–65**	**66–71**	**72–77**	**78–83**	**84–89**
geometry	60	65	70	**72–77**	**78–83**	**84–89**
data analysis, probability, discrete math			70	72, 74–77	82	**84–89**
patterns and functions			**66–71**	77	**78–83**	89
logic and reasoning	56, 60	62–65	67–69, 71	72–76	79–83	86–89
Interdisciplinary Connections and Applications						
history and geography	59			75, 76		
reading and language arts	58		69			
biology and earth science	55			72		87, 88
business and economics			70		80	84, 85, 86
sports and recreation	58					87
personal finance		63	66, 69	74, 75	81	
medicine, cars, aviation, energy use	59	64	71		81	

* ***Bold page numbers*** *indicate that a topic is used throughout the section.*

TECHNOLOGY

Section	*opportunities for use with* Student Book	Support Material
2.1	scientific calculator	
2.2	scientific calculator	**Technology Book:** Spreadsheet Activity 2
2.3	scientific calculator	
2.4	graphing calculator	**Technology Book:** Calculator Activity 2
2.5	graphing calculator spreadsheet software McDougal Littell Software *Stats!* *Function Investigator*	
2.6	graphing calculator	

PLANNING GUIDE

Regular Scheduling (45 min)

Section	Materials Needed	Core Assignment	Extended Assignment	*exercises that feature* Applications	Communication	Technology
2.1		1–31, 33–41	1–41	20–22, 25–29, 31	23, 30, 32, 33	
2.2	algebra tiles, scientific calculator	**Day 1:** 1–21 **Day 2:** 22–33, 35–48	**Day 1:** 1–21 **Day 2:** 22–48	19–21	17, 18 34, 39	17
2.3		1–3, 5–18, 22–35, AYP*	1–35, AYP	4, 10–15	4, 19–21, 23	
2.4	graph paper, graphing calculator	1–15, 17–22, 25–28, 31–39	1–4, 7, 10, 11, 13–39	17–22, 25–28	23, 24, 36	39
2.5	graph paper, spreadsheet software	1–8, 10–24, 27–33	1–33	10–13	9, 14, 25–27	19–23
2.6	graph paper, graphing calculator	1–8, 11–14, 16–29, AYP	1–8, 11–29, AYP	1–7, 10–15	8, 9, 15, 22	16–20
Review/ Assess		**Day 1:** 1–13 **Day 2:** 14–28 **Day 3:** Ch. 2 Test	**Day 1:** 1–13 **Day 2:** 14–28 **Day 3:** Ch. 2 Test	12, 13 27	1, 2 26–28	
Portfolio Project		Allow 2 days.	Allow 2 days.			

Yearly Pacing (with Portfolio Project)	Chapter 2 Total	Chapters 1–2 Total	Remaining	Total
	12 days	28 days	132 days	160 days

Block Scheduling (90 min)

	Day 9	Day 10	Day 11	Day 12	Day 13	Day 14	Day 15
Teach/Interact	Ch. 1 Test 2.1	2.2: Exploration, page 61	2.3 2.4	2.5: Exploration, page 78 2.6	Review Port. Proj.	Review Port. Proj.	Ch. 2 Test 3.1
Apply/Assess	**Ch. 1 Test** **2.1:** 2–6, 10–41	**2.2:** 1–48	**2.3:** 1–18, 22, 24–35, AYP* **2.4:** 1–23, 25–28, 31–39	**2.5:** 1–33 **2.6:** 1–6, 10–14, 16–29, AYP	**Review:** 1–13 **Port. Proj.**	**Review:** 14–28 **Port. Proj.**	**Ch. 2 Test** **3.1:** 1–11, 13–16, 23–25, 27–36

NOTE: A one-day block has been added for the Portfolio Project—timing and placement to be determined by the teacher.

Yearly Pacing (with Portfolio Project)	Chapter 2 Total	Chapters 1–2 Total	Remaining	Total
	6 days	$14\frac{1}{2}$ days	$66\frac{1}{2}$ days	81 days

* ***AYP*** *is Assess Your Progress.*

Lesson Support

Section	Practice Bank	Study Guide*	Assessment Book*	Visuals	Explorations Lab Manual	Lesson Plans	Technology Book
2.1	10	2.1		Warm-Up 2.1		2.1	
2.2	11	2.2		Warm-Up 2.2 Folder B	Master 4	2.2	Spreadsheet Act. 2
2.3	12	2.3	Test 6	Warm-Up 2.3		2.3	
2.4	13	2.4		Warm-Up 2.4 Folder 2	Master 2	2.4	Calculator Act. 2
2.5	14	2.5		Warm-Up 2.5 Folder A	Masters 1, 2 Add. Expl. 3	2.5	
2.6	15	2.6	Test 7	Warm-Up 2.6 Folder A	Masters 1, 2	2.6	
Review Test	16	Chapter Review	Tests 8, 9 Alternative Assessment			Review Test	Calculator Based Lab 2

***Spanish versions** of *Study Guide* and *Assessment Book* are available.

Chapter Support

- Course Guide
- Lesson Plans
- Portfolio Project Book
- Preparing for College Entrance Tests
- Multi-Language Glossary
- *Test Generator* Software
- Professional Handbook

Software Support

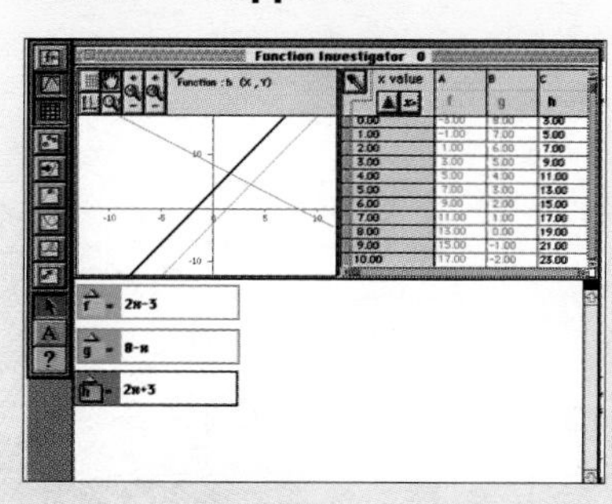

McDougal Littell Software
Stats!
Function Investigator

Internet Support

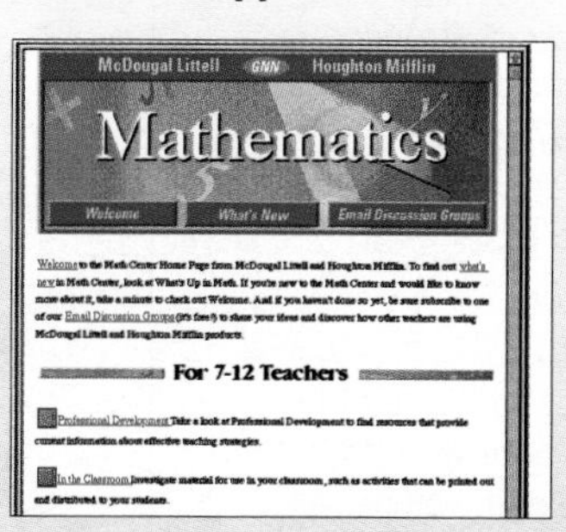

http://www.hmco.com
Next go to McDougal Littell; then the Education Center; then Secondary Math.

Outside Resources

Books, Periodicals

Mercer, Joseph. "Teaching Graphing Concepts with Graphics Calculators." *Mathematics Teacher* (April 1995): pp. 268–273.

Martignette-Boswell, Carol and Albert A. Cuoco. "Say It With Machines." *Mathematics Teacher* (April 1995): pp. 338–341.

Activities, Manipulatives

Algeblocks. Cincinnati, OH: Southwestern Publications.

A Graphing Matter: Activities for Easing into Algebra. Berkeley, CA: Key Curriculum Press.

Software

Harvey, Wayne, Judah Schwartz, and Michal Yerushalmy. *Visualizing Algebra: The Function Analyzer.* Scotts Valley, CA: Sunburst.

IBM Mathematics Exploration Toolkit. Armonk, NY: IBM.

Videos

Visualizing Algebra: A New Vision for Learning and Teaching Algebra. Reston, VA: NCTM, 1993.

Internet

Join the discussions in Kidsphere (excellent mailing list for K–12 education) by sending message to:
kidsphere-request@vms.cis.pitt.edu

In the body of the message, type:
subscribe kidsphere [first name, last name]

Interview Notes

Background

Wheelchair Athletes

Wheelchair athletes are engaged actively in the sports of basketball, softball, tennis, golf, and racing. Competition drives wheelchair athletes as it does all athletes. Their desire to compete in athletic events and to win is strong. In racing, the evolution of the racing wheelchair between 1974 and 1993 has opened up opportunities for many wheelchair athletes to participate in this sport. In 1974, wheelchair racers made some simple modifications to existing chairs to enhance their performance in a race. The first significant changes, however, came in 1978 with the advent of rigid frame chairs that were custom made for racing. These chairs were usually made of aluminum, which was replaced eventually by titanium because it is lighter and stronger. Other advances in racing wheelchair technology since 1978 involved improvements in tires, rims, brakes, steering control, footrests, and forks.

Bob Hall

Bob Hall lost the use of his right leg as an infant after a bout with polio. At Boston State College, where he studied health education and art, Hall learned about wheelchair sports, and started playing basketball. He then became interested in racing and won the National Mile. At that time, in the early 1970s, Hall began to experiment with his equipment, adjusting tire pressure, sitting positions, and wheel alignment. This led him to make his own wheelchair and was the start of his business in 1978 of making racing wheelchairs. Today, Bob Hall is the proprietor of Hall's Wheels, which designs and builds racing chairs for adults and children.

CHAPTER

2 Equations and Functions

Life in the fast lane

INTERVIEW **Bob Hall**

Bob Hall is not slowed down because he uses a wheelchair. In 1975, he became the first wheelchair athlete to win the Boston Marathon. He won again in 1977, setting national records in both races. Hall is also a wheelchair designer. His designs have helped transform the sport of wheelchair racing. "I always thought wheelchairs were the most disabling aspect of being disabled," he says. "They weren't made for speed or comfort. They were made for keeping you still. But I like to keep moving. I needed a chair that would let me do that."

"All of a sudden, you can take a negative and turn it into a positive."

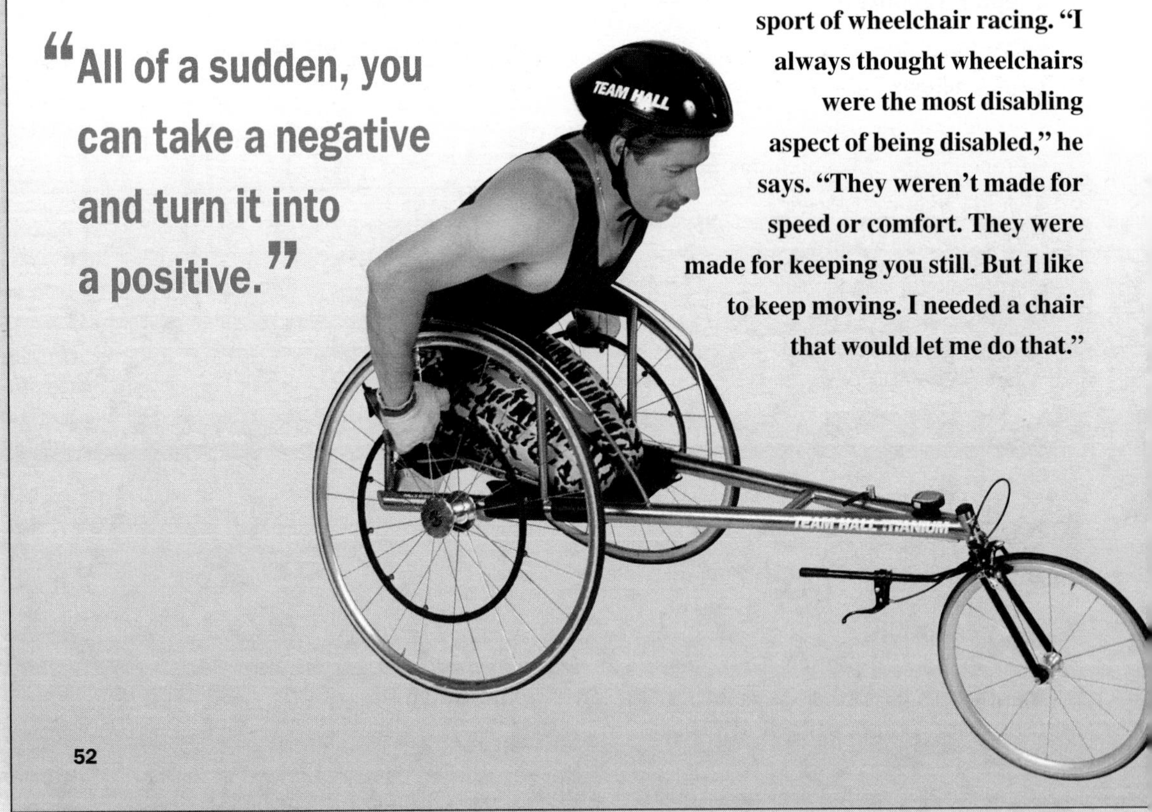

52

The Sky's the Limit

When Hall first began racing, wheelchairs weighed as much as 50 pounds. In 1980, he designed a wheelchair for himself that weighed only 22 pounds. Today Hall owns a company that makes advanced racing wheelchairs for wheelchair athletes all over the world. As chairs have become faster and safer, the races have grown more competitive. "A whole group of people who didn't have an outlet in sports can now compete on an international basis," Hall says.

Revolutionizing a Sport

Probably the most important advances in wheelchair racing come from weight reduction. Through the use of lightweight materials like titanium, Hall's competition chairs now weigh as little as 12.5 pounds. Racers sit lower in their chairs to reduce wind resistance.

Hall also invented a new steering system that makes turning easier. To change course, racers push a "handlebar" connected to the front wheel. After completing the turn, they let go. A powerful spring automatically snaps the wheel back to a centered position. This mechanism keeps the chair straight, so that racers can push with both arms and not worry about steering.

"With the evolution of equipment, suddenly the sky's the limit!"

In 1975, Bob Hall was the only wheelchair entrant in the Boston Marathon. The evolution of equipment has made the race more competitive. In 1994, 90 men and women participated in the race's wheelchair division.

Interview Notes

Background

Wheelchair Technology

The first wheelchairs were made of wood and because they could not be collapsed, they were not practical to transport. The many injured soldiers returning from World War II drew national attention to the plight of those in wheelchairs and increased the demand for better-designed chairs. Herbert A. Everest, an engineer, designed a metal chair for his friend and fellow engineer, Harry C. Jennings. This lighter, more portable chair made many things possible for those in wheelchairs, including wheelchair sports. The first organized competitive wheelchair sport was basketball. In the years from 1948 to 1973, the number of wheelchair basketball teams in the United States grew from eight to over one hundred. Many of those in wheelchairs who do not have the use of their arms use voice-controlled chairs. The technology for many of these chairs is based on a computer program that was created by a French woman when she was just 23 years old. Martine Kempf developed the *Katalavox* (from the Greek *Katal*, "to understand," and the Latin *vox*, meaning "voice") on her computer while she was a student in Germany. In addition to being used to control wheelchairs, the *Katalavox* is also widely used by microsurgeons.

Second-Language Learners

Students learning English may have trouble understanding the double meaning in the title *Life in the fast lane.* If necessary, explain that colloquially it means "living a full, exciting, and busy life," but here in the title it also refers to racing in wheelchairs as a sport. The colloquial phrase *the sky's the limit* may also need to be explained to students learning English.

Interview Notes

Mathematical Connection

A racing wheelchair has a magnetic computer mounted on the fork of the front wheel that calculates speed in miles per hour and distance traveled. This information is relayed to a battery-powered speedometer/odometer in front of the chair that is used by the racer to adjust and set his or her pace. In Section 2.1, students learn that the formula $d = rt$ (distance equals rate times time) can also be used by a racer to make these calculations. In Section 2.5 on representing functions, students draw a graph that shows the turning radius of a racing chair as a function of the turn angle. They do this by using the data given in the table on this page.

Explore and Connect

Research
Students can ask their school librarian how to find information about wheelchair sports. They may also have friends, family members, or neighbors who can help them collect information. Another source would be to locate businesses that make wheelchairs for athletes and write to them for information.

Writing
You may wish to discuss the answer to this question after students have had time to respond in writing.

Project
Since many towns may not have race tracks, students can use drawings of tracks from books or magazines.

Turn Angle and Turning Radius

Racers learn to judge just how much the wheel has to be nudged to make the desired turn. However, mathematics can also provide the answer. There is a precise relationship between the number of degrees the wheel is turned and the turning radius. (The turning radius is the radius of the circular path the wheelchair follows when you turn it.)

When the wheel is turned by one degree, the turning radius is 206 feet. In other words, the wheelchair would need a road wider than the width of a football field to make a U-turn. When the wheel is turned by 15 degrees, the turning radius is 14 feet. The wheelchair could make a U-turn in a small garage. The relationship between the number of degrees and the turning radius is called a *function.*

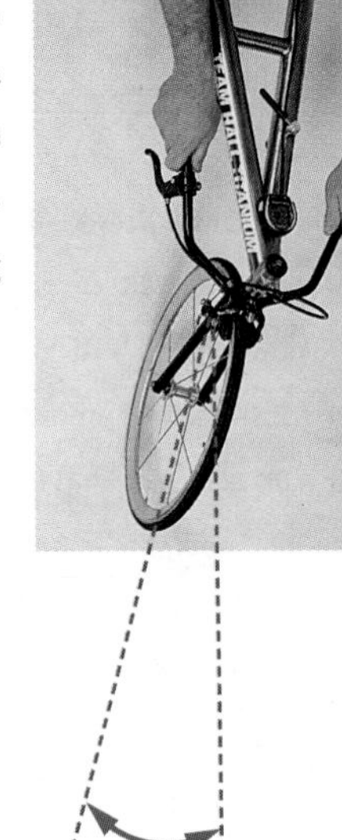

Turn angle	1°	2°	3°	6°	9°	12°	15°
Turning radius	206 ft	103 ft	69 ft	35 ft	23 ft	17 ft	14 ft

EXPLORE AND CONNECT

Bob Hall is shown here in his workshop, modifying one of his racing wheelchairs.

1. **Research** Find out more about wheelchair sports. For example, how are wheelchairs designed for tennis and basketball players different from racing wheelchairs? Why do the chairs have different designs?

2. **Writing** Does it make sense to talk about the turning radius of the chair when the turn angle is 0°? Explain.

3. **Project** Visit a race track in your town and measure its dimensions. Draw a sketch showing your measurements. What turn angle do you think Bob Hall would have to use to round a turn on this track? Explain your thinking.

Mathematics & Bob Hall

In this chapter, you will learn more about how mathematics is related to the sport of wheelchair racing.

Related Examples and Exercises

Section 2.1
- **Example 4**
- **Exercises 31 and 32**

Section 2.5
- **Exercises 11–14**

SECTION

2.1 Solving One-Step Equations

Learn how to...

- write and solve one-step equations

So you can...

- solve real-world problems about distance, rate, and time, for example

It's not every day you see algebra in a newspaper headline. As you can see from this article in the *New York Times*, zoo officials used an *equation* to find the weight of a stubborn baby elephant.

How to Weigh a Baby Elephant, or (A + B) - A = B

Officials at the Bronx Zoo yesterday trying to weigh Astor, an elephant born Aug. 20 at the zoo. When the recalcitrant pachyderm refused to step onto the scale, he was weighed together with his mother (5,396 pounds). The mother was then weighed alone (5,033 pounds), allowing the baby's weight to be calculated.

An **equation** is a mathematical sentence that says two numbers or expressions are equal.

Total weight = Mother's weight + Baby's weight

$5396 = 5033 + w$

To find the unknown weight of the baby elephant, you need to find the value of w that makes the equation true. This is called **solving the equation**.

EXAMPLE 1 Application: Zoology

Find the weight of the baby elephant.

SOLUTION

Solve the equation $5396 = 5033 + w$.

$$\begin{aligned} 5396 &= 5033 + w \\ -5033 & \quad -5033 \\ \hline 363 &= w \end{aligned}$$

The weight of the baby elephant is 363 lb.

Undo the addition. Subtract 5033 from both sides of the equation.

Check
Substitute 363 for w.
$5396 = 5033 + w$
$5396 \stackrel{?}{=} 5033 + 363$
$5396 = 5396$ ✓

Plan⇔Support

Objectives

- Write and solve one-step equations.
- Solve real-world problems including those involving distance, rate, and time.

Recommended Pacing

Core and Extended Courses
Section 2.1: 1 day

Block Schedule
Section 2.1: $\frac{1}{2}$ block
(with Chapter 1 Test)

Resource Materials

Lesson Support
Lesson Plan 2.1
Warm-Up Transparency 2.1
Practice Bank: Practice 10
Study Guide: Section 2.1

Technology
Internet:
http://www.hmco.com

Warm-Up Exercises

Evaluate.

1. $25\left(\frac{1}{25}\right)$ 1
2. $-10 + 10$ 0
3. $13\left(\frac{42}{13}\right)$ 42
4. $\frac{-19}{-19}$ 1
5. $16 + (-16)$ 0
6. Round 0.145 to the nearest hundredth. 0.15
7. Find the value of $-17x$ when $x = -2$. 34

Teach⇔Interact

Section Note

Reasoning

Implicit in the discussion of an equation and the idea that a particular value of a variable can make an equation true is the idea that an equation can also be false. For example, the equation $x + 3 = 7$ is false if x equals any number other than 4. Students may find the idea of a false equation strange because their experience with the concept of equal has always implied a true equation.

Additional Example 1

Suppose the total weight of the two elephants is 5029 pounds and the mother's weight is 4708 pounds. How much does the baby elephant weigh?

Solve the equation $5029 = 4708 + w$.

$$\begin{aligned} 5029 &= 4708 + w \\ -4708 &\quad -4708 \\ 321 &= w \end{aligned}$$

The baby elephant weighs 321 lb.

Check

Substitute 321 for w.

$5029 = 4708 + w$

$5029 \stackrel{?}{=} 4708 + 321$

$5029 = 5029$ ✓

Additional Example 2

Solve each equation.

a. $\frac{x}{-5} = -20$

Undo the division. Multiply both sides of the equation by –5.

$$\frac{x}{-5} = -20$$

$$-5\left(\frac{x}{-5}\right) = (-5)(-20)$$

$$\frac{-5}{-5} \cdot x = (-5)(-20)$$

$$x = 100$$

Check

Substitute 100 for x.

$$\frac{x}{-5} = -20$$

$$\frac{100}{-5} \stackrel{?}{=} -20$$

$$-20 = -20 ✓$$

b. $9x = -63$

Undo the multiplication. Divide both sides by 9.

$$9x = -63$$

$$\frac{9x}{9} = -\frac{63}{9}$$

$$x = -7$$

Check

Substitute –7 for x.

$$9(-7) \stackrel{?}{=} -63$$

$$-63 = -63 ✓$$

Inverse Operations

Any value of a variable that makes an equation true is called a **solution** of the equation. To find a solution, you need to get the variable alone on one side of the equation. You can do this by using *inverse operations*. Addition and subtraction are called **inverse operations** because they undo each other. Multiplication and division are also inverse operations.

EXAMPLE 2

Solve each equation.

a. $\frac{t}{60} = 12.5$ **b.** $-15x = -37$

SOLUTION

a.

$$\frac{t}{60} = 12.5$$

Undo the division. Multiply both sides of the equation by 60.

$$60\left(\frac{t}{60}\right) = 60(12.5)$$

$$\frac{60}{60} \cdot t = 60(12.5)$$

$$t = 750$$

Check

Substitute 750 for t.

$$\frac{t}{60} = 12.5$$

$$\frac{750}{60} \stackrel{?}{=} 12.5$$

$$12.5 = 12.5 ✓$$

b.

$$-15x = -37$$

Undo the multiplication. Divide both sides by −15.

$$\frac{-15x}{-15} = \frac{-37}{-15}$$

$$x = \frac{37}{15}$$

$$\text{or } x \approx 2.47$$

Check

Substitute $\frac{37}{15}$ for x.

$$-15x = -37$$

$$-15\left(\frac{37}{15}\right) \stackrel{?}{=} -37$$

$$-37 = -37 ✓$$

THINK AND COMMUNICATE

1. What inverse operations were used to solve the equation in Example 1 and the equations in Example 2?
2. Explain why you divide by -15 in the solution of part (b) of Example 2.
3. The solution of part (b) of Example 2 says that $x \approx 2.47$. If you try to check this solution, you get $-15(2.47) = -37.05$. Does this mean the solution is incorrect? Explain your reasoning.

ANSWERS Section 2.1

Think and Communicate

1. In Example 1, subtraction was used to undo the addition. In part (a) of Example 2, multiplication was used to undo the division. In part (b) of Example 2, division was used to undo the multiplication.
2. By dividing both sides by –15, you can get x alone on one side of the equation.
3. No; since 2.47 is an approximate value for x, –15(2.47) is approximately equal to –37.

Sometimes it helps to think about opposites when you solve equations. Remember that $-x$ means "the opposite of x."

$-x = 8$ ⟸ The opposite of x is 8.

$x = -8$ ⟸ So x is the opposite of 8.

EXAMPLE 3

Solve the equation $-x - 13 = 9$.

SOLUTION

Undo the subtraction. Add 13 to both sides.

$$\begin{aligned} -x - 13 &= 9 \\ +13 &\quad +13 \\ -x &= 22 \\ x &= -22 \end{aligned}$$

The opposite of x is 22, so $x = -22$.

Check

Substitute -22 for x.

$$-x - 13 = 9$$
$$-(-22) - 13 \stackrel{?}{=} 9$$
$$22 - 13 \stackrel{?}{=} 9$$
$$9 = 9 \checkmark$$

EXAMPLE 4 Interview: Bob Hall

In 1975, Bob Hall became the first wheelchair athlete to win the Boston Marathon. He finished the 26.2 mi race in about 178 min. Use this formula to find his average speed, or rate, in miles per minute:

Distance = Rate × Time

$$d = rt$$

SOLUTION

The distance d is 26.2 mi. The time t is 178 min. Find the rate r.

Substitute the values you know in the formula.

$$d = rt$$
$$26.2 = r \cdot 178$$
$$\frac{26.2}{178} = \frac{178r}{178}$$
$$0.15 \approx r$$

Divide both sides by 178. Round to the nearest hundredth.

Bob Hall's average speed was about 0.15 mi/min.

BY THE WAY

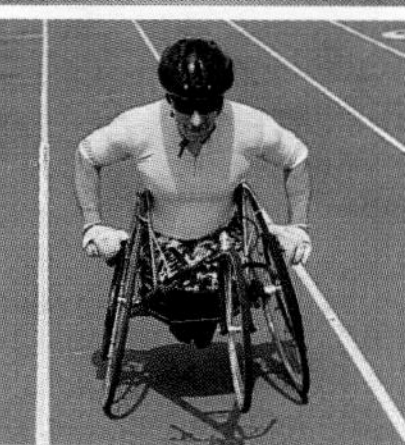

The back wheels of a racing wheelchair are tilted in at the top. This makes them easier for the rider to reach and also makes the chair more stable.

Teach⇔Interact

Additional Example 3

Solve the equation $-t + 17 = -23$.

Undo the addition. Subtract 17 from both sides.

$$\begin{aligned} -t + 17 &= -23 \\ -17 &\quad -17 \\ -t &= -40 \end{aligned}$$

The opposite of t is -40, so $t = 40$.

Check

Substitute 40 for t.

$$-t + 17 = -23$$
$$-(40) + 17 \stackrel{?}{=} -23$$
$$-23 = -23 \checkmark$$

About Example 4

Teaching Tip

After completing Example 4, you may wish to have students convert Bob Hall's speed of 0.15 mi/min to feet per minute and then to feet per second so they can get a better sense of how fast he was actually moving.

(0.15 mi/min $= 0.15 \times 5280$ ft/min $= 792$ ft/min $= 792 \times \frac{1}{60}$ ft/s $= 13.2$ ft/s)

Additional Example 4

In 1992, Hwang Young-jo, from South Korea, won the Marathon in the Olympic Games in about 133 minutes. The distance is exactly the same as the Boston Marathon, 26.2 mi. Find his average speed, or rate in miles per minute.

The distance d is 26.2 mi. The time t is 133 min. Find the rate r.

$$d = rt$$
$$26.2 = r \cdot 133$$
$$\frac{26.2}{133} = \frac{133r}{133}$$
$$0.20 \approx r$$

Hwang Young-jo's average speed was about 0.20 mi/min.

Teach⇔Interact

Checking Key Concepts

Common Error
Some students may make sign errors in questions 5 and 6 because of the negative sign in front of the variable. Remind them that $-m$ and $-r$ can be written as $-1 \cdot m$ and $-1 \cdot r$. Then they can divide both sides of each equation by -1 to get the correct answer.

Closure Question

Describe how to solve a one-step equation.

To solve a one-step equation you need to get the variable alone on one side of the equation. This can be done by using inverse operations.

Apply⇔Assess

Suggested Assignment

Core Course
Exs. 1–31, 33–41

Extended Course
Exs. 1–41

Block Schedule
Day 9 Exs. 2–6, 10–41

Exercise Notes

Communication: Discussion
Exs. 1–18 Students will benefit from discussing with each other how each of these equations is solved.

Interdisciplinary Problems
Exs. 20–23 An English sentence, such as "He invented a steering mechanism for racing wheelchairs" can be related to an equation because the pronoun *He* plays the role of the variable in an equation. Neither the sentence above nor an equation can be true until *He* and the variable are replaced by a name or a number.

BY THE WAY

Short-track speed skates let racers lean into turns. The left skate blade is positioned on the outer edge of the boot. The right skate blade is on the inner edge.

CHECKING KEY CONCEPTS

Solve each equation.

1. $27 + x = 4$ **2.** $8 = y - 4$ **3.** $v + 0.25 = 1$ **4.** $s - \frac{1}{3} = \frac{2}{3}$

5. $16 = -m$ **6.** $-r = -10$ **7.** $28 = \frac{t}{5}$ **8.** $\frac{m}{3} = 11$

9. $-60 = 6t$ **10.** $-2p = 18$ **11.** $13m = 53.3$ **12.** $1.5 = -6y$

13. SPORTS At the 1994 winter Olympics in Lillehammer, Norway, Lee-Kyung Chun of South Korea won the women's 1000 m short-track speed skating race with a time of about 97 seconds. Find her average speed in meters per second.

2.1 Exercises and Applications

Extra Practice exercises on page 559

Solve each equation.

1. $520 + m = 105$ **2.** $-21 = p - 13$ **3.** $-0.5 + n = 6$

4. $-8 = 14 - p$ **5.** $-c + \frac{1}{8} = \frac{1}{2}$ **6.** $-6 - v = 3\frac{1}{2}$

7. $-10 = x - 5$ **8.** $-7 = -x$ **9.** $-s = 0$

10. $10 = \frac{n}{5}$ **11.** $0 = \frac{t}{16}$ **12.** $\frac{n}{12} = -12$

13. $16x = 100$ **14.** $3n = -27$ **15.** $3 = -6h$

16. $3x = 7.5$ **17.** $13.5 = -3z$ **18.** $\frac{x}{2} = 2.5$

19. Challenge Solve for x. Tell what inverse operations you used.

a. $\sqrt{x} = 4$ b. $x^2 = 9$

LANGUAGE ARTS The sentences in Exercises 20–22 are like equations because only certain names make them true. Match each name with a sentence.

A. Mae Jemison **B.** Daniel Galvez **C.** Bob Hall

(*Hint:* Each person is described in an interview in one of the chapters of this book.)

20. __?__ invented a steering mechanism for racing wheelchairs.

21. __?__ is the first African-American woman astronaut.

22. __?__ is an artist who painted a mural called *Crosswinds*.

23. Open-ended Problem Write a word sentence like the ones in Exercises 20–22.

24. SAT/ACT Preview How many positive integers less than 20 are equal to 3 times an even integer?

A. one **B.** two **C.** three **D.** four **E.** five

Checking Key Concepts
1. –23 2. 12
3. 0.75 4. 1
5. –16 6. 10
7. 140 8. 33
9. –10 10. –9
11. 4.1
12. –0.25
13. about 10.31 m/s

Exercises and Applications
1. –415 2. –8
3. 6.5 4. 22
5. $-\frac{3}{8}$ 6. $-9\frac{1}{2}$
7. –5 8. 7
9. 0 10. 50
11. 0 12. –144
13. 6.25 14. –9
15. $-\frac{1}{2}$ 16. 2.5
17. –4.5 18. 5
19. a. $x = 16$; Undo the square root by squaring both sides.
b. $x = \pm 3$; Undo the squaring by taking the square root of both sides.
20. C 21. A 22. B
23. Answers may vary. For example, __?__ was the first president of the United States. (George Washington)
24. C

Connection HISTORY

On May 21, 1932, Amelia Earhart became the first woman to complete a solo west-east flight across the Atlantic Ocean. On September 5, 1936, Beryl Markham became the first woman to complete a solo east-west flight across the Atlantic. The map shows their routes.

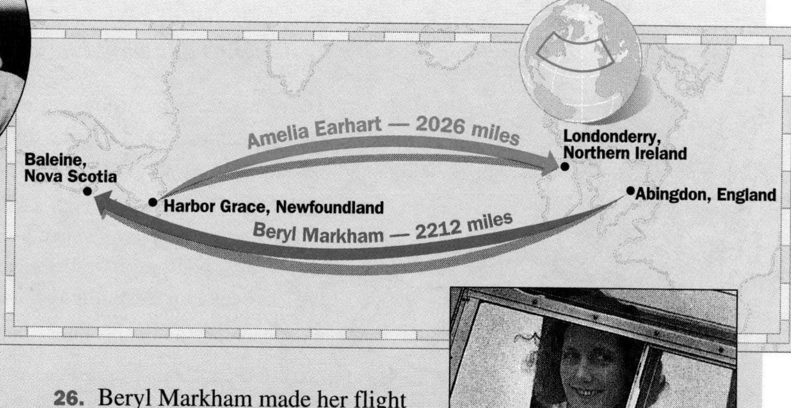

25. Amelia Earhart made her flight in about 14.9 h. Find her average speed in miles per hour.

26. Beryl Markham made her flight in about 21.6 h. Find her average speed in miles per hour.

27. About 100 acres of pizza are eaten each day in the United States. San Francisco, California, has an area of about 29,887 acres. Write and solve an equation to find how many days it would take people in the United States to eat a pizza the size of San Francisco.

28. MEDICINE Health officials predict that by the year 2020, there will be 292,000 women doctors in the United States. These women will represent one third of all doctors. Which equation can you use to find the total number of doctors d in 2020? Explain your choice and solve the equation.

A. $3d = 292{,}000$ **B.** $\frac{292{,}000}{3} = d$ **C.** $\frac{d}{3} = 292{,}000$

29. HISTORY Covered wagons traveled about 15 miles per day on the Oregon Trail. Which equation can you use to find about how many days d the 2000 mi journey took? Explain your choice and solve the equation.

A. $15d = 2000$ **B.** $\frac{d}{15} = 2000$ **C.** $d + 15 = 2000$

30. Investigation How thick is a piece of paper? Make a stack of 100 pieces of paper. Then measure the height h of the stack. Use the value you find for h and one of the formulas below to find the approximate thickness t of one piece of paper. Explain why you chose this formula.

A. $\frac{t}{100} = h$ **B.** $100t = h$ **C.** $100h = t$

Apply⇔Assess

Exercise Notes

Problem Solving

Ex. 24 This is an interesting problem that students can solve by using a variable. If n is any positive integer, then $2n$ is an even integer. Three times an even integer is $3(2n) = 6n$. When $n = 1$, 2, and 3, $6n = 6$, 12, 18, which are the three positive integers less than 20 equal to 3 times an even integer ($6 = 3 \cdot 2$; $12 = 3 \cdot 4$; $18 = 3 \cdot 6$). If $n = 4$, then $6 \cdot 4 = 24$, which is greater than 20.

Teaching Tip

Ex. 27 You may wish to ask students to calculate how many square feet of pizza are eaten each day in the United States. Since 1 acre = 43,560 square feet, then $43{,}560 \cdot 100 = 4{,}356{,}000$ square feet of pizza are eaten.

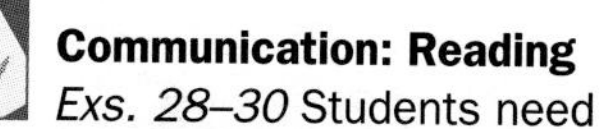

Communication: Reading

Exs. 28–30 Students need to read these exercises carefully in order to identify the key phrases that can be represented by using a variable. In Ex. 28, "one third of all doctors" is $\frac{d}{3}$. In Ex. 29, "15 miles per day" is $15d$. In Ex. 30, careful reading will lead students to see that if t is the thickness of one piece of paper, then $100t$ is the thickness of 100 pieces, which is also the height of the stack. Thus, $100t = h$.

25. about 135.97 mi/h
26. about 102.41 mi/h
27. $29{,}887 = 100d$; about 299 days
28. C; This equation states that the number of women doctors, 292,000, will represent the total number of doctors d divided by 3; $d = 876{,}000$; 876,000 doctors.
29. A; The rate per day, 15 mi, times the number of days d equals the distance traveled, 2000 mi; $d \approx 133.33$; about 133.3 days.
30. B; This equation states that 100 times the thickness of one sheet is equal to the height of a stack of 100 sheets. Actual thicknesses may vary; for example, a sheet of copier paper is about 0.1 mm thick.

Apply⇔Assess

Exercise Notes

Assessment Note

Ex. 33 In part (b), students use the criteria of most interesting and most challenging to evaluate their problems. You may wish to ask students if they want to include other criteria, such as most realistic, for example.

Practice 10 for Section 2.1

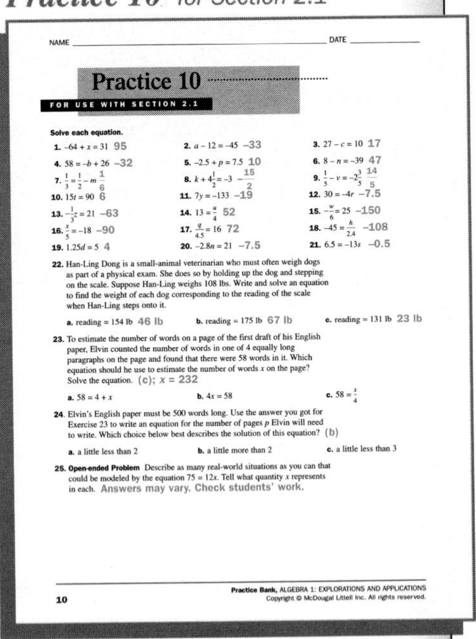

NAME ______________________ DATE ________

Practice 10

FOR USE WITH SECTION 2.1

Solve each equation.

1. $-64 + x = 31$ 95
2. $a - 12 = -45$ −33
3. $27 - c = 10$ 17
4. $58 = -b + 26$ −32
5. $-2.5 + p = 7.5$ 10
6. $8 - n = -39$ 47
7. $\frac{1}{3} = \frac{1}{2} - m$ $\frac{1}{6}$
8. $k + 4\frac{1}{2} = -3$ $-\frac{15}{2}$
9. $\frac{1}{5} - v = -2\frac{3}{5}$ $\frac{14}{5}$
10. $15t = 90$ 6
11. $7y = -133$ −19
12. $30 = -4r$ −7.5
13. $-\frac{1}{3}z = 21$ −63
14. $13 = \frac{u}{4}$ 52
15. $-\frac{w}{6} = 25$ −150
16. $\frac{x}{5} = -18$ −90
17. $\frac{q}{4.5} = 16$ 72
18. $-45 = \frac{h}{2.4}$ −108
19. $1.25d = 5$ 4
20. $-2.8n = 21$ −7.5
21. $6.5 = -13s$ −0.5

22. Han-Ling Dong is a small-animal veterinarian who must often weigh dogs as part of a physical exam. She does so by holding up the dog and stepping on the scale. Suppose Han-Ling weighs 108 lbs. Write and solve an equation to find the weight of each dog corresponding to the reading of the scale when Han-Ling steps onto it.

    a. reading = 154 lb 46 lb
    b. reading = 175 lb 67 lb
    c. reading = 131 lb 23 lb

23. To estimate the number of words on a page of the first draft of his English paper, Elvin counted the number of words in one of 4 equally long paragraphs on the page and found that there were 58 words in it. Which equation should he use to estimate the number of words x on the page? Solve the equation. (c); $x = 232$

    a. $58 = 4 + x$
    b. $4x = 58$
    c. $58 = \frac{x}{4}$

24. Elvin's English paper must be 500 words long. Use the answer you got for Exercise 23 to write an equation for the number of pages p Elvin will need to write. Which choice below best describes the solution of this equation? (b)

    a. a little less than 2
    b. a little more than 2
    c. a little less than 3

25. **Open-ended Problem** Describe as many real-world situations as you can that could be modeled by the equation $75 = 12x$. Tell what quantity x represents in each. Answers may vary. Check students' work.

10

Practice Bank, ALGEBRA 1: EXPLORATIONS AND APPLICATIONS

INTERVIEW **Bob Hall**

Look back at the article on pages 52–54.

In 1975, Bob Hall reached the finish line of the Boston Marathon in the wheelchair shown at the right. His average speed was about 0.15 mi/min. Today Hall uses a racing wheelchair that he designed and built himself.

31. In 1990, Bob Hall finished a marathon (26.2 mi) in a little over 117 min. He used a racing wheelchair like the one shown at the left. Find his average speed in miles per minute.

32. Writing Compare Bob Hall's average speed in the 1990 marathon to his average speed in the 1975 marathon. Write a paragraph describing how improvements in wheelchair design may have affected Hall's speed.

ONGOING ASSESSMENT

33. a. **Open-ended Problem** Write three word problems that can be solved using the formula $d = rt$. Write one problem that asks for a rate, one that asks for a time, and one that asks for a distance. Explain how solving for a distance is different than solving for a rate or a time.

b. **Cooperative Learning** Work with several other students to read and evaluate each other's problems from part (a). Which are the most interesting? Which are the most challenging? Why? Based on your conclusions, rewrite your problems to make them more interesting and more challenging.

SPIRAL REVIEW

Evaluate each expression when $p = -3$. *(Sections 1.4 and 1.5)*

34. $6p + 5$

35. $2 - 8p$

36. $\dfrac{p}{10 + 1}$

37. $\dfrac{|p + 1|}{3}$

38. $|-p + p|$

39. $|-p| + |p|$

GEOMETRY Write a variable expression for the perimeter of each figure. Evaluate each expression when $x = 10$. *(Section 1.6)*

40.

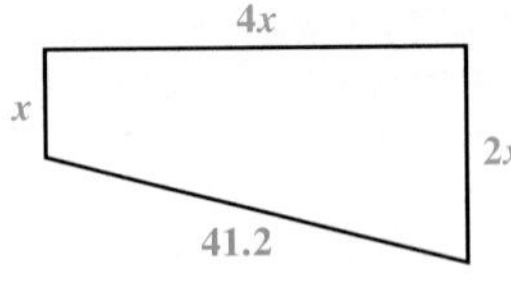

41.

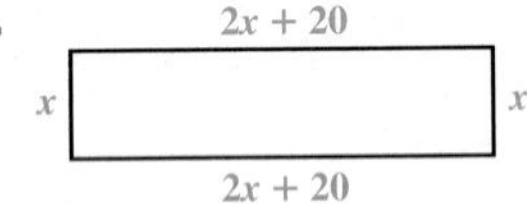

31. about 0.22 mi/min
32. In 1990, Bob Hall's average speed was considerably faster than his average speed in 1975. It is likely that improvements in wheelchair design, such as weight reduction and the improved steering system, allowed Bob Hall to significantly reduce his racing time.
33. a. Answers may vary. Examples are given.

(1) Amelia drove from Albertville to Bakerstown, a distance of 91 mi, in 1.75 h. Find her average speed in miles per hour.

(2) Clarence rides his bike at an average speed of 6 mi/h. How long will it take him to ride to a friend's house 9 mi away?

(3) Sarah drove along the highway for 1.5 h with her cruise control set to 55 mi/h. How far did she travel during this time?

Solving for a distance is different from solving for a rate or a time because you do not need to use inverse operations to find a distance; you just multiply the rate and the time.

b. Answers may vary.

34. −13
35. 26
36. $-\frac{3}{11} \approx -0.27$
37. $\frac{2}{3}$
38. 0
39. 6
40. $7x + 41.2$; 111.2
41. $6x + 40$; 100

SECTION

2.2 Solving Two-Step Equations

Learn how to...

- write and solve two-step equations

So you can...

- make purchasing decisions in real-world situations, such as at a state fair
- solve geometry problems, such as finding a perimeter

Sometimes it takes more than one step to solve an equation. You may need to add or subtract, and then multiply or divide. Algebra tiles help you see the steps involved.

EXPLORATION
COOPERATIVE LEARNING

Solving Equations with Algebra Tiles

Work with a partner.
You will need:
- algebra tiles

The tile represents the variable x.

The tile represents 1.

SET UP Use tiles to represent the equation $2x + 3 = 11$.

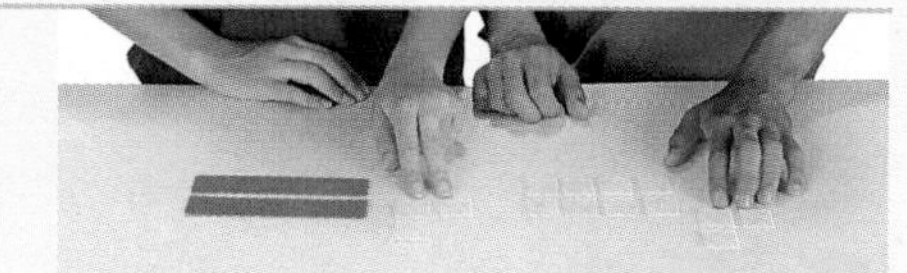

1 **Subtract 3** from both sides of the equation.

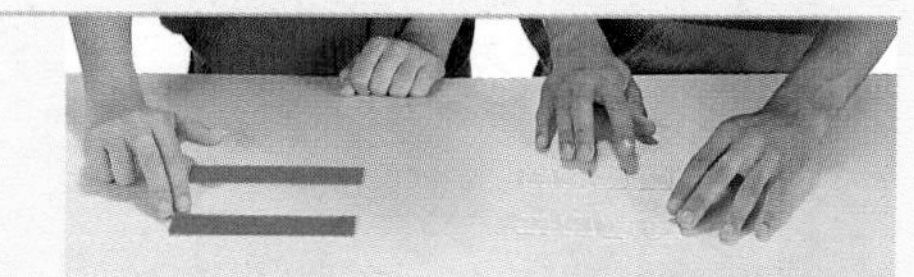

2 **Divide** both sides of the equation into **two** equal parts.

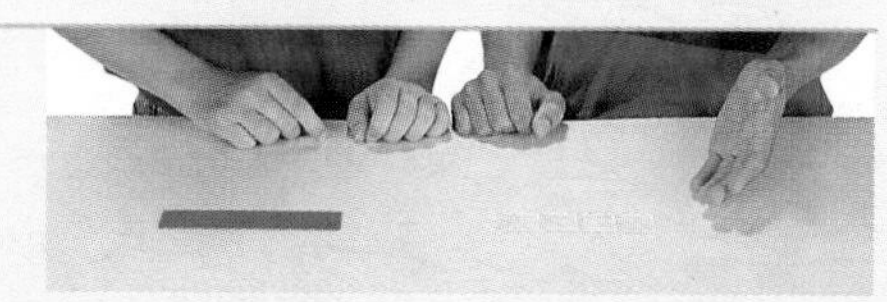

You can see that each tile equals four tiles. So $x = 4$.

Show how to solve each equation using algebra tiles.

1. $5x + 2 = 7$ **2.** $2x + 5 = 15$ **3.** $3x + 5 = 11$ **4.** $13 = 4 + 3x$

Exploration Note

Purpose
The purpose of the Exploration is to use algebra tiles to help students see the steps involved in solving two-step equations.

Materials/Preparation
Each group should have at least five x-tiles and twenty 1-tiles in order to solve equations 1–4.

Procedure
Students should first solve the equation $2x + 3 = 11$ by following the steps shown in the text. Then they can solve equations 1–4.

Closure
Volunteer groups can demonstrate their solutions to equations 1–4 and answer questions from their classmates.

Explorations Lab Manual
See the Manual for more commentary on this Exploration.
Diagram Master 4

For answers to the Exploration, see following page.

Plan⇔Support

Objectives

- Write and solve two-step equations.
- Solve real-world problems using equations.
- Solve geometry problems using equations.

Recommended Pacing

Core and Extended Courses
Section 2.2: 2 days

Block Schedule
Section 2.2: 1 block

Resource Materials

Lesson Support
Lesson Plan 2.2
Warm-Up Transparency 2.2
Overhead Visuals:
Folder B: Algebra Tiles
Practice Bank: Practice 11
Study Guide: Section 2.2
Explorations Lab Manual:
Diagram Master 4

Technology
Technology Book:
Spreadsheet Activity 2
Calculator
Internet:
http://www.hmco.com

Warm-Up Exercises

Evaluate each variable expression.

1. $3x - 7$ when $x = -1$ -10
2. $\frac{4x}{3} + 5$ when $x = 9$ 17
3. $13 - 2.5x$ when $x = -2$ 18
4. $-3 - 7x$ when $x = -4$ 25

Solve each equation.

5. $5x = -30$ -6
6. $x + 8 = -8$ -16
7. $-x - 1 = 0$ -1

Teach⇔Interact

Learning Styles: Kinesthetic

The use of algebra tiles to introduce students to solving two-step equations provides a concrete basis for students to understand an abstract process intuitively. If understanding can be developed on an intuitive and informal level, then students should have little difficulty understanding how to solve equations by using symbols alone.

About Example 1

Using Manipulatives

Students should continue to use their algebra tiles to solve the equation given in this example.

Additional Example 1

Solve the equation $3x + 2 = 8$.

First undo the addition. Subtract 2 from both sides.

$$\begin{aligned} 3x + 2 &= 8 \\ -2 \quad & \quad -2 \\ 3x &= 6 \end{aligned}$$

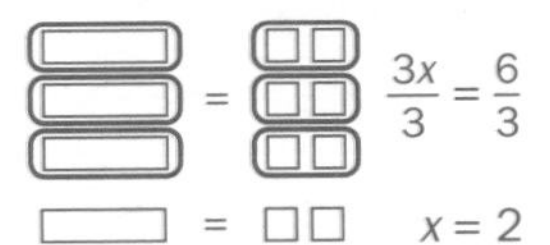

$3x = 6$

Then undo the multiplication. Divide both sides by 3.

$\frac{3x}{3} = \frac{6}{3}$

$x = 2$

Additional Example 2

Solve the equation $\frac{x}{3} + 5 = -3$.

Undo the addition. Subtract 5 from both sides.

$$\begin{aligned} \frac{x}{3} + 5 &= -3 \\ -5 \quad & \quad -5 \\ \frac{x}{3} &= -8 \end{aligned}$$

Undo the division. Multiply both sides by 3.

$$3\left(\frac{x}{3}\right) = 3(-8)$$

$$x = -24$$

Check

Substitute −24 for x.

$\frac{x}{3} + 5 = -3$

$\frac{-24}{3} + 5 \stackrel{?}{=} -3$

$-8 + 5 \stackrel{?}{=} -3$

$-3 = -3$ ✓

EXAMPLE 1

Solve the equation $8 + 2x = 14$.

SOLUTION

First undo the addition. Subtract **8** from both sides.

$$\begin{aligned} 8 + 2x &= 14 \\ -8 \quad & \quad -8 \\ 2x &= 6 \end{aligned}$$

$$2x = 6$$

Then undo the multiplication. Divide both sides by **2**.

$$\frac{2x}{2} = \frac{6}{2}$$

$$x = 3$$

Example 2 shows how to solve a two-step equation using symbols instead of algebra tiles. This method is called *solving algebraically*.

EXAMPLE 2

Solve the equation $\frac{x}{5} - 7 = -3$.

SOLUTION

Undo the subtraction. Add **7** to both sides.

$$\begin{aligned} \frac{x}{5} - 7 &= -3 \\ +7 \quad & \quad +7 \\ \frac{x}{5} &= 4 \end{aligned}$$

Undo the division. Multiply both sides by **5**.

$$5\left(\frac{x}{5}\right) = 5(4)$$

$$x = 20$$

Check

Substitute 20 for x.

$\frac{x}{5} - 7 = -3$

$\frac{20}{5} - 7 \stackrel{?}{=} -3$

$4 - 7 \stackrel{?}{=} -3$

$-3 = -3$ ✓

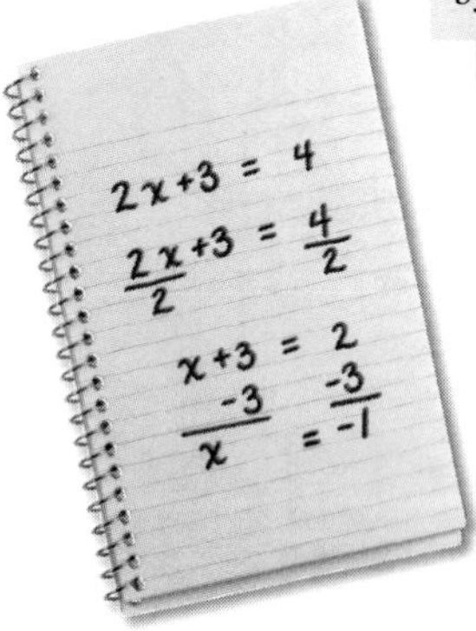

THINK AND COMMUNICATE

1. Describe the steps you would use to solve the equation $8 - 2x = 14$.
2. Does it make sense to solve Example 2 with algebra tiles? Why or why not?
3. Jean tried to solve the equation $2x + 3 = 4$, as shown at the left. What did she do wrong? How would you solve the equation?

ANSWERS Section 2.2

Exploration

1. Subtract two 1-tiles from both sides of the equation. Then divide both sides into five equal parts to find that each x-tile is equal to one 1-tile; $x = 1$.
2. Subtract five 1-tiles from both sides of the equation. Then divide both sides into two equal parts to find that each x-tile is equal to five 1-tiles; $x = 5$.
3. Subtract five 1-tiles from both sides of the equation. Then divide both sides into three equal parts to find that each x-tile is equal to two 1-tiles; $x = 2$.
4. Subtract four 1-tiles from both sides of the equation. Then divide both sides into three equal parts to find that each x-tile is equal to three 1-tiles; $x = 3$.

Think and Communicate

1. Subtract 8 from both sides. Then divide both sides by −2.
2. No; reasons may vary. For example, it would be hard to figure out how to represent the given equation using algebra tiles.

EXAMPLE 3 Application: Personal Finance

The state fair charges $6 admission plus $2.25 for each ride. Surajit has $25 to spend. How many rides can he take?

SOLUTION

Problem Solving Strategy: Use an equation.

Let r = the number of rides Surajit can take.

Admission		Price per ride		Number of rides		Amount Surajit has to spend
$6	+	$2.25	·	r	=	$25

How many rides can Surajit take with $25? Solve the equation for r.

$$6 + 2.25r = 25$$

Subtract 6 from both sides of the equation.

$$\frac{-6 \qquad -6}{2.25r = 19}$$

Divide both sides by 2.25.

$$\frac{2.25r}{2.25} = \frac{19}{2.25}$$

$$r \approx 8.4$$

The solution of the equation is about 8.4. But Surajit must take a whole number of rides. Surajit can take eight rides.

BY THE WAY

The first Ferris wheel in the United States was introduced at the World's Columbian Exposition in Chicago in 1893. Over 27 million people attended the fair.

THINK AND COMMUNICATE

4. In Example 3, suppose Surajit has $40 to spend. How does the equation change? How does the solution change?

CHECKING KEY CONCEPTS

Each diagram represents an equation. Write and solve the equation.

1. 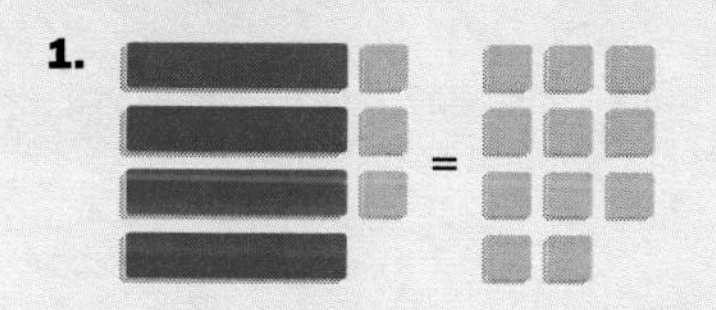

2. 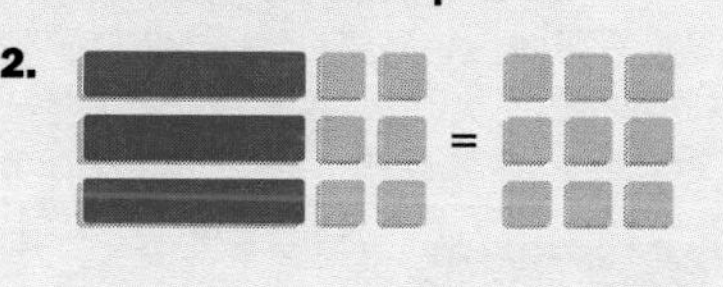

Solve each equation.

3. $5x + 7 = 14$ **4.** $12t - 7 = 19$ **5.** $24 = 3x - 3$

6. $16 - 4c = 20$ **7.** $89 = 11 - 11p$ **8.** $-3 - 9x = -21$

9. $\frac{m}{3} + 6 = 3$ **10.** $20 + \frac{x}{4} = 25$ **11.** $0 = \frac{m}{10} - 1$

3. She did not divide the entire left side by 2, as she should have. I would solve the equation by subtracting 3 from both sides, and then dividing both sides by 2.
4. The equation becomes $6 + 2.25r = 40$; the solution is $r \approx 15.1$, so Surajit can take 15 rides in this case.

Checking Key Concepts

1. $4x + 3 = 11; x = 2$
2. $3x + 6 = 9; x = 1$
3. $\frac{7}{5} = 1.4$
4. $\frac{13}{6} \approx 2.17$
5. 9
6. -1
7. $-\frac{78}{11} \approx -7.09$
8. 2
9. -9
10. 20
11. 10

Teach⇔Interact

About Example 3

Problem Solving

The solution to this example illustrates an important aspect of solving real-world problems; that is, the solution to a problem must be examined in terms of the conditions of the problem. In this case, since the variable r represents the number of rides a person can take, it must be a whole number. The mathematical solution $r \approx 8.4$ is meaningless since a person cannot go on a fractional part of a ride.

Additional Example 3

To park their car at the state fair, Surajit's parents had to pay $5 for the first hour and $1.50 for each additional hour. If they paid $12.50 upon leaving the parking lot, how many hours were they at the fair?

Use an equation. Let h = the number of hours at the fair after the first hour.

First hour		Price per hour		Number of hours		Total cost
$5	+	$1.50	·	h	=	$12.50

Solve the equation for h.

$$5 + 1.50h = 12.50$$

$$\frac{-5 \qquad -5.00}{1.50h = 7.50}$$

$$\frac{1.50h}{1.50} = \frac{7.50}{1.50}$$

$$h = 5$$

The solution of the equation is 5. But this is the number of hours after the first hour. Thus, Surajit and his parents were at the fair for 6 hours.

Checking Key Concepts

Alternate Approach

When solving equations 3–11, some students may ask if they can solve a two-step equation by first undoing a multiplication or division and then undoing an addition or subtraction. The answer, of course, is yes, but the arithmetic can get more involved because fractions may be introduced. For example, dividing by 5 in equation 3 introduces the fractions $\frac{7}{5}$ and $\frac{14}{5}$.

Closure Question

Write an equation that takes two steps to solve and then solve it.

Answers may vary.

Apply⇔Assess

Suggested Assignment

Core Course
Day 1 Exs. 1–21
Day 2 Exs. 22–33, 35–48

Extended Course
Day 1 Exs. 1–21
Day 2 Exs. 22–48

Block Schedule
Day 10 Exs. 1–48

Exercise Notes

Common Error
Exs. 1–16 Students can make a number of different types of errors when solving equations. For example, they can make sign errors when adding, subtracting, multiplying or dividing positive and negative numbers or two negative numbers. They may use the wrong operation as, for example, in Ex. 1 by adding 1200 to each side of the equation instead of subtracting. Or they may make careless errors as in Ex. 16 by writing $2m$ instead of $-2m$. If students organize their work neatly and proceed carefully, most errors can be eliminated. Students should also develop a habit of checking their solutions by substituting them into the original equation.

Reasoning
Ex. 18 This exercise points out that some equations can have many solutions (a) or no solutions (b). You may wish to point out that a *statement* is either true or false, as these equations illustrate. Most equations, however, such as those in Exs. 1–17 and 22–33, are neither true nor false until a number is substituted for the variable. For this reason, equations are often called *open sentences.*

Application
Exs. 19–21 These exercises require that students write and solve two two-step equations in order to analyze the towing situation. Many practical problems in life can be approached mathematically and connections to real-world situations help students to understand this fact.

2.2 Exercises and Applications

Extra Practice *exercises on page 560*

Solve each equation.

1. $1200 + 50x = 2750$
2. $6 + 3h = -3$
3. $10 = 5 - 5m$
4. $12 = 15n - 8$
5. $-4 - 2p = 8$
6. $-17 - 6t = -17$
7. $2m - 2.5 = 1.5$
8. $37 = -4s - 1$
9. $9 + \frac{x}{2} = 7$
10. $\frac{b}{2} - 7 = -11$
11. $10 = \frac{t}{7} + 16$
12. $\frac{n}{4} - 3 = 4$
13. $4 = \frac{p}{8} - 5$
14. $-9 + \frac{w}{4} = 18$
15. $-12 = 10.5 + \frac{r}{3}$

$3 = -7 - 2m$
$+7 \quad +7$
$10 \quad 2m$
$\frac{10}{2} = \frac{2m}{2}$
$5 = m$

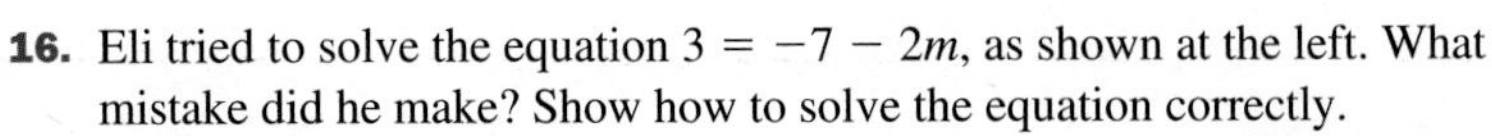

16. Eli tried to solve the equation $3 = -7 - 2m$, as shown at the left. What mistake did he make? Show how to solve the equation correctly.

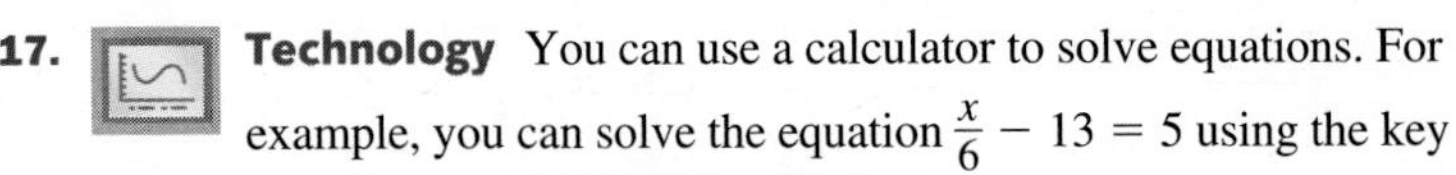

17. **Technology** You can use a calculator to solve equations. For example, you can solve the equation $\frac{x}{6} - 13 = 5$ using the key sequence 5 [+] 13 [ENTER] [×] 6 [ENTER].

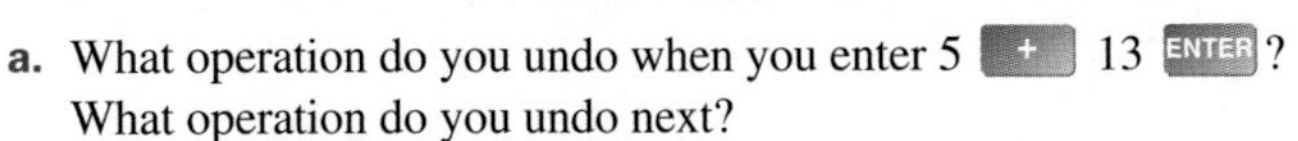

 a. What operation do you undo when you enter 5 [+] 13 [ENTER]? What operation do you undo next?
 b. **Writing** Describe a key sequence you can use to solve the equation $7x + 14 = 42$. How can you use a calculator to check your solution?

18. **Challenge** How many values of x make each statement true? Explain your reasoning.
 a. $3x + 3 = 3(x + 1)$
 b. $x - 4 - x = 0$

Connection CARS

When your car breaks down, you may have to call a tow truck. Towing companies usually charge one fee to hook up the car. They also may charge for each mile the car is towed. The table shows towing charges for two companies.

19. Suppose Company A charges you $70. About how far was your car towed?

Company	Charge to hook up car	Charge per mile
A	$34	$2
B	$28	$3

20. A friend tells you that Company B would have been a better bargain. Do you agree or disagree? Explain.

21. **Open-ended Problem** Under what circumstances would you recommend using Company B? What other information would you like to know before hiring a towing company?

BY THE WAY

The American Automobile Association receives about 67,479 emergency calls per day. About 38.4% of these emergencies involve towing.

Exercises and Applications

1. 31
2. –3
3. –1
4. $\frac{4}{3}$
5. –6
6. 0
7. 2
8. –9.5
9. –4
10. –8
11. –42
12. 28
13. 72
14. 108
15. –67.5
16. When he added 7 to the right side of the equation, the sum should have been $-2m$, not $2m$. Therefore, $10 = -2m$ and so $-5 = m$.
17. a. subtraction of 13; division by 6
 b. Use the key sequence 42 [–] 14 [ENTER] [÷] 7 [ENTER]. This gives the display 4. To check the solution, substitute it in the original equation by multiplying the display value by 7 and adding 14; the result should be 42.
18. a. infinitely many; The two sides are equal for any value of x.
 b. none; Every value of x makes the left side equal to –4, and $-4 \neq 0$.
19. 18 mi

Solve each equation.

22. $5x + 150 = -100$ **23.** $-6x - 3 = -18$ **24.** $-20 = 4 - 12m$

25. $10 - 5x = 25$ **26.** $0 = 12 - 4x$ **27.** $10 = \frac{n}{5} + 4$

28. $17 = \frac{p}{5} - 13$ **29.** $\frac{t}{7} + 3 = -4$ **30.** $-3 + \frac{z}{3} = 3$

31. $\frac{x}{10} - 13 = 0$ **32.** $-3 = \frac{w}{9} - 9$ **33.** $-28 = -8 + \frac{t}{5}$

34. Open-ended Problem Write a two-step equation for which -5 is the solution. Explain how you found your equation.

GEOMETRY For Exercises 35–38, find the value of x. Then find the unknown dimension(s).

35.

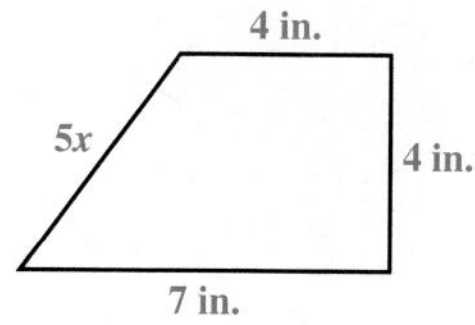

Perimeter = 20 in.

36.

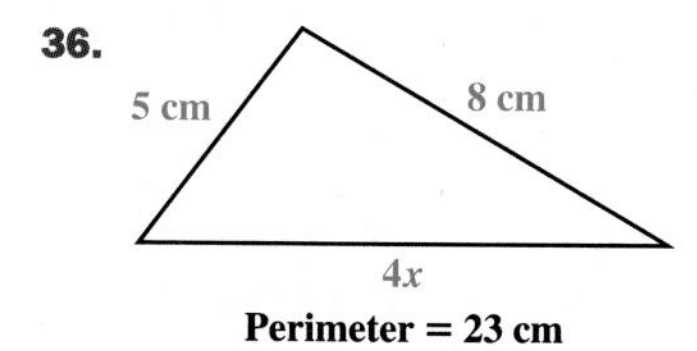

Perimeter = 23 cm

37.

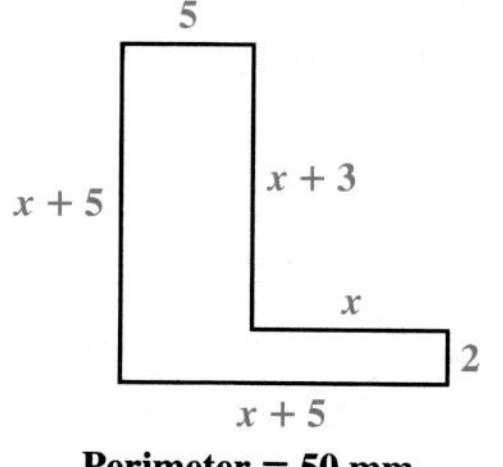

Perimeter = 50 mm

38.

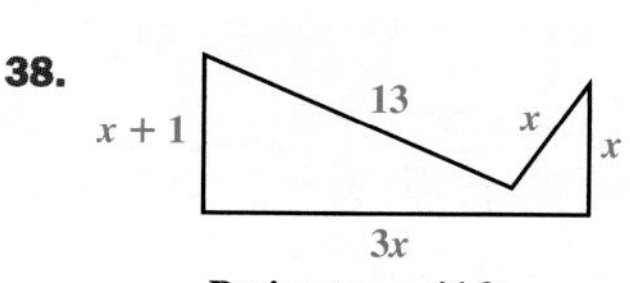

Perimeter = 44 ft

ONGOING ASSESSMENT

39. Open-ended Problem Make a list of the most important steps you should use to solve an equation. Include two examples and solutions that illustrate all four operations.

SPIRAL REVIEW

Write a variable expression to describe each situation. Tell what the variable stands for. *(Section 1.6)*

40. There are about 13 calories in each ounce of a glass of apple juice.

41. Rob is 5 in. taller than his sister.

42. Sandra took $50 out of her savings account.

Simplify each variable expression. *(Section 1.7)*

43. $12t + 13t - 7$ **44.** $3(-6 - p)$ **45.** $-4(2m - 7)$

46. $3 - 5(6 + 2m)$ **47.** $7 - (2x + 7)$ **48.** $4(6t + 3) - 3(t + 3)$

Apply⇔Assess

Exercise Notes

Integrating the Strands

Exs. 35–38 These exercises apply algebra to the solution of some problems in geometry. Geometry can be presented in a way that is completely independent of algebra, but linking these two strands together has many advantages when solving real-world problems.

Assessment Note

Ex. 39 Students' responses to this assessment exercise should be evaluated to make sure they understand how to solve an equation. This can be done as a class activity through discussion, or by means of peer assessment in small groups.

Second-Language Learners

Ex. 39 Second-language learners will benefit from being able to work cooperatively on this activity with students fluent in English.

Practice 11 for Section 2.2

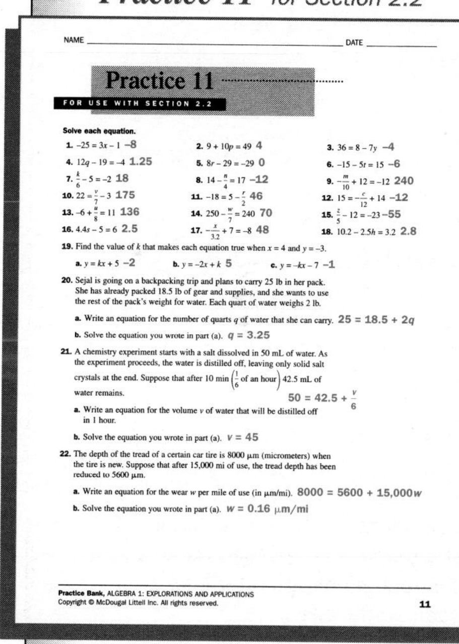

NAME ______ DATE ______

Practice 11

FOR USE WITH SECTION 2.2

Solve each equation.

1. $-25 = 3x - 1$ −8
2. $9 + 10p = 49$ 4
3. $36 = 8 - 7y$ −4
4. $12q - 19 = -4$ 1.25
5. $8r - 29 = -29$ 0
6. $-15 - 5t = 15$ −6
7. $\frac{k}{6} - 5 = -2$ 18
8. $14 - \frac{n}{4} = 17$ −12
9. $-\frac{m}{10} + 12 = -12$ 240
10. $22 = \frac{v}{7} - 3$ 175
11. $-18 = 5 - \frac{r}{2}$ 46
12. $15 = -\frac{c}{12} + 14$ −12
13. $-6 + \frac{u}{8} = 11$ 136
14. $250 - \frac{w}{7} = 240$ 70
15. $\frac{z}{5} - 12 = -23$ −55
16. $4.4s - 5 = 6$ 2.5
17. $-\frac{x}{3.2} + 7 = -8$ 48
18. $10.2 - 2.5h = 3.2$ 2.8

19. Find the value of k that makes each equation true when $x = 4$ and $y = -3$.

a. $y = kx + 5$ −2 b. $y = -2x + k$ 5 c. $y = -kx - 7$ −1

20. Sejal is going on a backpacking trip and plans to carry 25 lb in her pack. She has already packed 18.5 lb of gear and supplies, and she wants to use the rest of the pack's weight for water. Each quart of water weighs 2 lb.

a. Write an equation for the number of quarts q of water that she can carry. $25 = 18.5 + 2q$

b. Solve the equation you wrote in part (a). $q = 3.25$

21. A chemistry experiment starts with a salt dissolved in 50 mL of water. As the experiment proceeds, the water is distilled off, leaving only solid salt crystals at the end. Suppose that after 10 min $\left(\frac{1}{6}\text{ of an hour}\right)$ 42.5 mL of water remains.

a. Write an equation for the volume v of water that will be distilled off in 1 hour. $50 = 42.5 + \frac{v}{6}$

b. Solve the equation you wrote in part (a). $v = 45$

22. The depth of the tread of a certain car tire is 8000 µm (micrometers) when the tire is new. Suppose that after 15,000 mi of use, the tread depth has been reduced to 5600 µm.

a. Write an equation for the wear w per mile of use (in µm/mi). $8000 = 5600 + 15{,}000w$

b. Solve the equation you wrote in part (a). $w = 0.16$ µm/mi

Practice Bank, ALGEBRA 1: EXPLORATIONS AND APPLICATIONS

11

20. I disagree; my car was towed 18 mi, and Company B would have charged me $82, or $12 more than Company A.

21. If the car needs to be towed less than 6 mi, then Company B costs less than Company A. If the car needs to be towed more than 6 mi, then Company A costs less than Company B. A 6-mile tow costs the same at both companies. Before you hire a towing company, you probably want to know where the company is located, if service is prompt, and if the personnel are reliable and honest, in addition to knowing the towing rates.

22. –50
23. 2.5
24. 2
25. –3
26. 3
27. 30
28. 150
29. –49
30. 18
31. 130
32. 54
33. –100

34. Answers may vary. An example is given. $2x - 7 = -17$; To find my equation, I wrote an algebraic expression, $2x - 7$, and evaluated it when $x = -5$. I used that value, –17, as the right side of my equation.

35. $x = 1$; 5 in.

36. $x = 2.5$; 10 cm

37. $x = 7.5$; 7.5 mm, 10.5 mm, 12.5 mm, and 12.5 mm

38. $x = 5$; 5 ft, 5 ft, 6 ft, and 15 ft

39. See answers in back of book.

40. $13x$; the number of ounces in the glass

41. $s + 5$; Rob's sister's height in inches

42. $a - 50$; the amount (in dollars) in Sandra's account before the withdrawal

43. $25t - 7$

44. $-18 - 3p$

45. $-8m + 28$

46. $-27 - 10m$

47. $-2x$

48. $21t + 3$

Plan⇔Support

Objectives

- Recognize and describe functions using tables and equations.
- Predict the outcome of a decision.

Recommended Pacing

Core and Extended Courses
Section 2.3: 1 day

Block Schedule
Section 2.3: $\frac{1}{2}$ block
(with Section 2.4)

Resource Materials

Lesson Support
Lesson Plan 2.3
Warm-Up Transparency 2.3
Practice Bank: Practice 12
Study Guide: Section 2.3
Assessment Book: Test 6

Technology
Internet:
http://www.hmco.com

Warm-Up Exercises

Solve each equation.

1. $10 + 3x = 46$ 12
2. $-2x + 4 = 16$ -6
3. $\frac{n}{3} - 9 = 2$ 33
4. What percent of 75 is 25? $33\frac{1}{3}\%$
5. Evaluate x^2 for $x = -1$. 1
6. Evaluate $-2x + 6$ when $x = -5$. 16

SECTION

2.3 Applying Functions

Learn how to...

- recognize and describe functions using tables and equations

So you can...

- predict the outcome of a decision, such as how many pictures to buy or how many hours to work

Many portrait photographers offer you a deal. You order a basic package at a set cost and pay extra for additional photos. The relationship between the number of additional photos you order and your total cost is an example of a *function*.

A **function** is a relationship between input and output. For each input, there is exactly one output. Output depends on input.

For example, the cost of buying photos at Galante Studios is a function of the number of additional wallet photos you order.

GALANTE
G
STUDIOS
Serving high school students for 30 years
$60 Basic graduation package includes:
One 8 x 10 photo
One 5 x 7 photo
Plus 12 wallet photos
Additional wallet photos only $2.50
8" x 10"
5" x 7"
2½" x 3¼"

Input: Additional wallet photos		Output: Total cost (dollars)
0	$60 + $2.50(0) =	$60.00
1	60 + 2.50(1) =	62.50
2	60 + 2.50(2) =	65.00
3	60 + 2.50(3) =	67.50

You can represent functions in many ways, including tables and equations.

EXAMPLE 1 Application: Personal Finance

a. Write an equation that describes total cost as a function of the number of additional wallet photos ordered.

b. You have $87 to spend on graduation photos at Galante Studios. How many additional wallet photos can you order?

SOLUTION

a. Let n = the number of additional wallet photos ordered.
Let C = the total cost in dollars.

$$60 + 2.50n = C$$

Cost of basic package — Cost of additional wallet photos — Total cost

b. Use the equation from part (a). To find how many photos you can order, substitute **\$87** for **C** and solve for n.

Subtract **60** from both sides.

Divide both sides by **2.50**.

$$\begin{aligned} 60 + 2.50n &= C \\ 60 + 2.50n &= 87 \\ -60 \quad & \quad -60 \\ 2.50n &= 27 \\ \frac{2.50n}{2.50} &= \frac{27}{2.50} \\ n &= 10.8 \end{aligned}$$

You can order 10 additional wallet photos.

THINK AND COMMUNICATE

1. Explain how making a table like the one showing the costs at Galante Studios can help you write an equation for a function.
2. In part (b) of Example 1, the solution of the equation $60 + 2.50n = 87$ is 10.8. Why can you order only 10 additional wallet photos?

Domain and Range

The table below shows only positive whole numbers for n, the number of additional wallet photos you can order from Galante Studios. It doesn't make sense to order -12 photos or $\frac{1}{3}$ of a photo.

The **domain** of a function consists of all input values that make sense for the function.

n	C
0	60.00
1	62.50
2	65.00
3	67.50
...	...

The **range** of a function consists of all possible output values. The range depends on the domain.

These three dots show that the pattern of numbers continues.

EXAMPLE 2

Describe the domain and the range of the cost function in Example 1.

SOLUTION

The domain consists of the whole numbers 0, 1, 2, 3, ...

The range consists of these amounts in dollars and cents: 60.00, 62.50, 65.00, 67.50, ...

Teach⇔Interact

Section Notes

Second-Language Learners
The terms *deal* and *basic package* may confuse second-language learners. Define *deal* as a price that is lower than normal, and *basic package* as items that are included with the lowest-cost group.

Teaching Tip
A key aspect of the definition of a function that students need to understand is that *for each input, there is exactly one output.* This fact will be used by students later in this section to decide whether a table of numbers represents a function or not.

Additional Example 1

After graduating from high school, Susan Murphy took a job working 35 hours per week for \$8 per hour. For each hour over 35 that Susan worked, she was paid \$12 per hour.

a. Write an equation that describes total earnings as a function of the number of hours over 35 that Susan works.
Let n = the number of hours over 35 that Susan works.
Let E = total earnings in dollars.
$280 + 12n = E$

b. One week Susan earned \$364. How many hours over 35 did Susan work?
To find how many hours over 35 Susan worked, substitute \$364 for E and solve for n.

$$\begin{aligned} 280 + 12n &= E \\ 280 + 12n &= 364 \\ -280 \quad & \quad -280 \\ 12n &= 84 \\ \frac{12n}{12} &= \frac{84}{12} \\ n &= 7 \end{aligned}$$

Susan worked 7 hours over 35.

Additional Example 2

In Additional Example 1, Susan's employer required that overtime be worked in increments of $\frac{1}{2}$ hour.

Describe the domain and the range of the earnings function in Additional Example 1.
The domain consists of the numbers $0, \frac{1}{2}, 1, 1\frac{1}{2}, 2, 2\frac{1}{2}, 3, \ldots$.
The range consists of these amounts in dollars: 280, 286, 292, 298, 304, 310, 316, ...

ANSWERS Section 2.3

Think and Communicate

1. In each row, the output is determined using the formula 60 + 2.50(input) = output. By letting n represent the input and C the output, you can write the equation for the function, $C = 60 + 2.50n$.
2. The number of photos must be a whole number. You cannot afford to order 11 additional photos, so the greatest possible number is 10.

Teach⇔Interact

Additional Example 3

Decide whether each table represents a function. Explain your reasoning.

a.

t	$t \pm 1$
5	4, 6
6	5, 7
7	6, 8
8	7, 9
9	8, 10
10	9, 11

No. For each value in the domain, there is more than one value in the range.

b.

y	$2y + 1$
0	1
1	3
−1	−1
2	5
−2	−3
3	7
−3	−5

Yes. For each value in the domain, there is exactly one value in the range.

Checking Key Concepts

Visual Thinking

Ask students to create a table based on the information presented in question 2. For example, they may show the total cost of purchasing one soft drink and different amounts of salad. Ask them to discuss the *input, output, domain,* and *range* represented in their table. This activity involves the visual skills of *identification* and *interpretation.*

Closure Question

Identify one relationship in your life that is an example of a function. Describe your function using a table or an equation.

Answers may vary.

EXAMPLE 3

Decide whether each table represents a function. Explain your reasoning.

a.

Price (dollars)	Sales tax (dollars)
1	0.05
2	0.10
3	0.15
4	0.20
5	0.25
...	...

b.

x	x^2
−2	4
−1	1
0	0
1	1
2	4
...	...

c.

x	Any numbers that equal x when squared
0	0
1	1 and −1
4	2 and −2
9	3 and −3
...	...

SOLUTION

a. Yes. For each value in the domain, there is exactly one value in the range. The table represents a function.

b. Yes. For each value in the domain, there is exactly one value in the range. The table represents a function.

c. No. For some values in the domain, there is more than one value in the range. The table does not represent a function.

THINK AND COMMUNICATE

3. In Example 2 on page 67, how can you find the next three values in the range?

4. a. Use the movie theater sign. Explain why the price of a movie ticket is a function of age. Describe the domain and the range of the function.

b. Can you write an equation for the function? Why or why not?

CHECKING KEY CONCEPTS

1. When you turn the steering wheel of a car to the left, you expect the car to turn to the left. How is the steering wheel like a function? Can you think of another machine that is like a function?

2. a. A soft drink at Cafe Jamaica costs \$.80. If you go to the salad bar, you pay \$.20 per ounce of salad. Write an equation for the cost of a soft drink and a salad at Cafe Jamaica.

b. Explain why the equation you wrote for part (a) represents a function. Describe the domain and the range of the function.

c. You have \$6 to spend on a soft drink and a salad. How many ounces of salad can you buy?

Think and Communicate

3. Answers may vary. For example, continue the pattern of adding 2.50 to the previous value, or evaluate $60 = 2.50n$ for $n = 4, 5,$ and 6. Either method gives 70.00, 72.50, and 75.00.

4. a. For any age, there is exactly one ticket price. The domain is the whole numbers and the range consists of 4.25 and 7.00.

b. You cannot write an equation because no one equation describes the value of the function for all numbers in the domain.

Checking Key Concepts

1. The steering wheel is like a function because the angle through which the car turns depends on how far you turn the steering wheel. For each amount you turn the wheel, the car will turn in a predictable way. A TV channel selector is like a function. For each number you select, one channel will be chosen.

2. a. For example, $C = 0.80 + 0.2x$, where x is the number of ounces of salad and C is the total cost in dollars.

b. This equation represents a function because for every possible salad size (whole-number value for x), there is exactly one cost (value for C). The domain is the whole numbers 0, 1, 2, 3, ... The range is the amounts in dollars and cents 0.80, 1.00, 1.20, 1.40, ...

c. 26 oz

Exercises and Applications

1. No; for some values in the domain (13 and 14), there is more than one value in the range.

2.3 Exercises and Applications

Decide whether each table represents a function. Explain your reasoning.

1.

Age (years)	Height (in.)
13	60
13	63
14	64
14	68
16	68

2.

n	$\lvert n \rvert$
0	0
−3	3
3	3
15	15
−15	15

3.

x	$x - 1$
−1	−2
0	−1
1	0
2	1
3	2

Extra Practice exercises on page 560

4. **LANGUAGE ARTS** A doctor says, "Good health is a function of good nutrition." What does the doctor mean? How is this use of the word *function* like the use of the word *function* in mathematics? How is it different? In what other ways can you use the word *function*?

For Exercises 5–8, match each equation with a situation. Tell what the variables stand for.

A. $C = p - 10$ **B.** $p = \frac{10}{n}$ **C.** $C = 10 - p$ **D.** $C = 10d$

5. Rita buys a book with a $10 bill. The change she gets from the cashier is a function of the price of the book.
6. Matt rents a bicycle for $10 a day. His total cost is a function of the number of days he rents the bicycle.
7. Some friends want to share equally the cost of a $10 pizza. The price each friend pays is a function of the number of friends.
8. Tony buys a slightly damaged pair of shoes. The store clerk subtracts $10 from the original price. Tony's cost is a function of the original price of the shoes.
9. Describe the domain and the range of each function in Exercises 5–8. Explain your reasoning.

PERSONAL FINANCE **The cost of going to a concert depends on the number of tickets you buy and the number of cars you take, as shown in the notebook at the right. Use the information in the notebook for Exercises 10 and 11.**

10. One car seats up to five people. Suppose you take one car and some friends to the concert. Write an equation for your cost as a function of the number of tickets you buy. Give the domain and the range of the function. How does the equation change if you take two cars to the concert? How do the domain and the range change?
11. Janet Neville wants to take some of her cousins to the concert. She has $160 to spend on tickets and parking for one car. How many tickets can she buy?

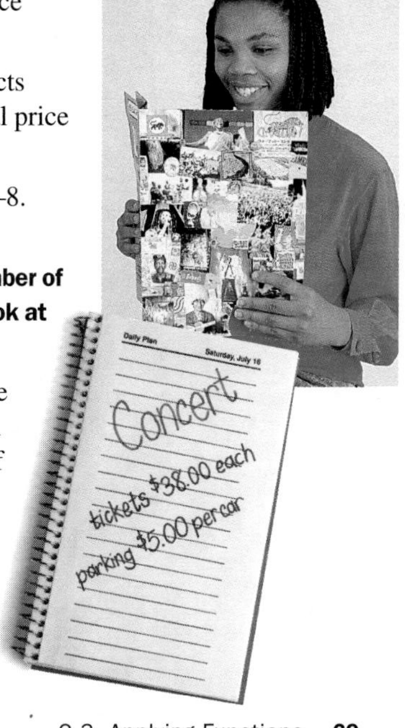

Apply⇔Assess

Suggested Assignment

Core Course
Exs. 1–3, 5–18, 22–35, AYP

Extended Course
Exs. 1–35, AYP

Block Schedule
Day 11 Exs. 1–18, 22, 24–35, AYP

Exercise Notes

Communication: Discussion
Ex. 4 A discussion of this exercise will help students to understand the concept of a function on an informal level as being a relationship between two quantities (variables) such that when one changes, there is a change in the other.

Reasoning
Exs. 5–8 These exercises help students to understand the concept of a function as represented by an equation. It is important that they see how a change in one variable affects the other variable. The notion of one variable being dependent upon the other is a crucial aspect of a functional relationship.

2. Yes; for each value in the domain, there is exactly one value in the range.
3. Yes; for each value in the domain, there is exactly one value in the range.
4. Answers may vary. An example is given. The doctor means that eating right will help keep you healthy. This use of the word *function* in the doctor's expression is like the mathematical one in that it is known that there is a relationship between the kinds and amounts of food a person chooses to eat and that person's general health. In a mathematical function, a given input always guarantees the same output. This is not true for the "function" the doctor describes. A person's diet is a factor, but not the only one, in determining that person's health. Good nutrition does not necessarily protect a person from accidents, some infections, or genetically influenced diseases, for example. The word *function* can also refer to an assigned use (for example, the table *functions* as a desk) or to an event (for example, the public was invited to the *function*).
5. C; p = the price of the book in dollars and C = her change in dollars.
6. D; d = the number of days he rents the bike and C = the total cost in dollars.
7. B; n = the number of friends and p = the price in dollars each friend pays.
8. A; p = the original price of the shoes in dollars and C = the price in dollars Tony pays.
9. Ex. 5: the domain is 0.01, 0.02, 0.03, …, 10.00 and the range is 0, 0.01, 0.02, …, 9.99, since she must pay at least $.01 for the book and can spend no more than $10. Ex. 6: the domain is 1, 2, 3, … and the range is 10, 20, 30, …, since the bike is rented by the day. Ex. 7: the domain is 1, 2, 3, 4, … and the range is 10, 5, 3.33, 2.50, …, since the number of friends is a whole number greater than 0. Ex. 8: the domain is 10.01, 10.02, 10.03, … and the range is 0.01, 0.02, 0.03, since the original price of the shoes must be at least $10.01.
10. $C = 38n + 5$, where n = the number of tickets and C = the total cost in dollars. The domain is the whole numbers from 1 to 5 and the range consists of these amounts in dollars: 43, 81, 119, 157, and 195. The equation becomes $C = 38n + 10$. The domain is the whole numbers from 6 to 10, and the range consists of these amounts in dollars: 238, 276, 314, 352, and 390.
11. 4 tickets

Apply⇔Assess

Exercise Notes

Application
Exs. 12–15 Teenagers are interested in earning money and these exercises provide an opportunity to discuss functions in an economic setting. You may wish to ask their opinions of the minimum wage. Do they think it is high, low, or about right? Ex. 15 introduces the idea of *time and a half* for working overtime. The idea of *double time* (2 times the regular hourly wage) can also be discussed.

Challenge
Ex. 20 You may wish to have students review and explore the Triangle Inequality symbolically when considering this exercise. If a, b, and c are the lengths of the three sides of a triangle, then $a + b > c$, $a + c > b$, and $b + c > a$.

Connection ECONOMICS

The United States government requires that workers receive a minimum hourly wage. The circle graph shows that most teenagers earn more than the minimum wage.

12. What percent of working teenagers earn more than the minimum wage?

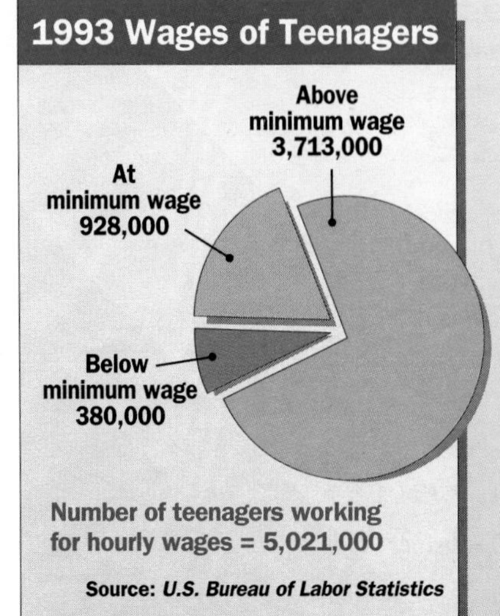

13. a. Write a formula for converting the numbers on the circle graph to percents.

b. Does your formula from part (a) represent a function? Explain.

14. a. In 1994, the minimum wage was \$4.25 per hour. Write an equation for earnings as a function of the number of hours worked at this wage.

b. Describe in words the domain and the range of the function from part (a).

15. Ana Martinez earns \$500 for working a 40 h week. She also earns time and a half for working overtime. That means her overtime hourly wage is 1.5 times her regular hourly wage.

a. What is Ana's regular hourly wage?

b. What is her overtime hourly wage?

c. How many hours of overtime must she work to earn a total of \$650 in one week?

GEOMETRY For Exercises 16–18, write an equation for the perimeter P of each figure as a function of the side length s. Describe the domain and range of each function.

16.

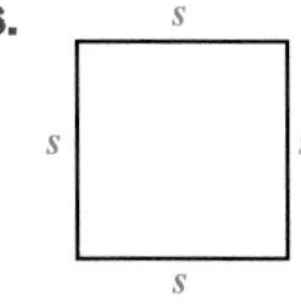

17.

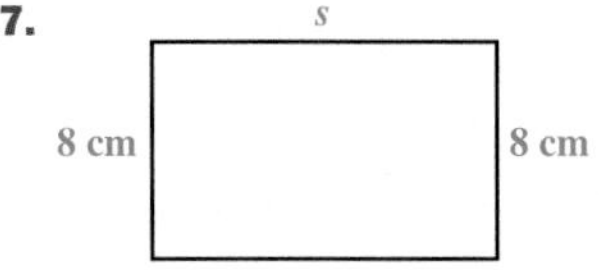

18.

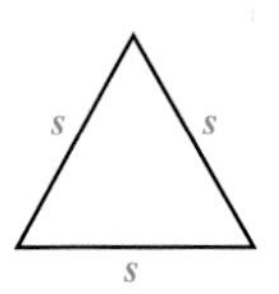

19. Open-ended Problem Draw a figure like the ones in Exercises 16–18. Express the perimeter of the figure as a function of the side length. Give the domain and the range of your function.

20. Challenge Write an equation for the perimeter P of the triangle at the right as a function of the side length s. The Triangle Inequality says that the sum of the lengths of any two sides of a triangle is greater than the length of the third side. What does this tell you about the domain and the range of the perimeter function for this triangle?

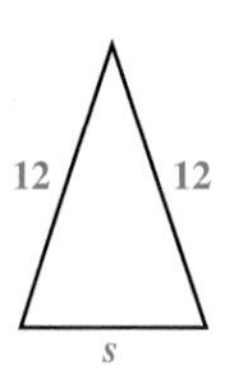

12. about 74%

13. a. $p = \frac{n}{5{,}021{,}000}$, where n = the number of teens associated with a portion of the circle graph and p = the related percent.

b. Yes; for each value of n, there is exactly one value of p.

14. a. $E = 4.25h$, where h = the number of hours worked and E = the earnings.

b. Answers may vary. An example is given. The domain is 0, 1, 2, 3, … and the range is 0, 4.25, 8.50, 12.75, ….

15. a. \$12.50

b. \$18.75

c. 8 h

16. $P = 4s$; the domain and the range are all positive numbers.

17. $P = 2s + 16$; the domain is all positive numbers and the range is all numbers greater than 16.

18. $P = 3s$; the domain and the range are all positive numbers.

19. Answers may vary. An example is given.

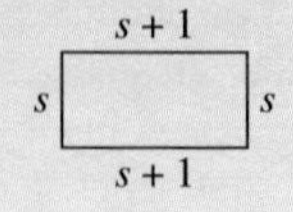

$P = 4s + 2$; the domain is all positive numbers and the range is all numbers greater than 2.

20. $P = s + 24$; this means that $0 < s < 24$ and that $24 < P < 48$.

21. For any number of bottles collected, there is exactly one amount of money earned. The domain is 0, 1, 2, 3, … and the range is 0, 0.05, 0.10, 0.15, ….

22. D

23. Answers may vary. Examples are given. (1) The total cost for gasoline depends on the number of gallons of gas pumped. The domain is the nonnegative numbers and the range is the nonnegative numbers. (2) The

21. **Writing** A hiking club raises money by collecting returnable bottles. The club earns \$.05 for each bottle. Tell how this situation represents a function. Describe the domain and the range of the function.

22. **SAT/ACT Preview** If $2x - 15 = -5$, then $x = \underline{?}$.

A. 25 **B.** –10 **C.** 10 **D.** 5 **E.** 7.5

ONGOING ASSESSMENT

23. **Writing** Give five examples of functions in everyday life. Describe the domain and the range of each one. Then describe a relationship between input and output that is not a function. Explain why it is not a function.

SPIRAL REVIEW

Solve each equation. *(Sections 2.1 and 2.2)*

24. $42 = -b$
25. $-64 + p = -12$
26. $\frac{n}{27} = 3$
27. $57d = 378.5$
28. $72 = 45 + 12h$
29. $-21 = -2w - 16$
30. $2r - 17 = 27$
31. $35 - 2x = 5$
32. $22 - p = -15$
33. $4 + \frac{s}{3} = -13$
34. $\frac{h}{3} - 7 = -19$
35. $\frac{q}{7} + 12 = 15$

ASSESS YOUR PROGRESS

VOCABULARY

equation (p. 55)
solve an equation (p. 55)
solution (p. 56)
inverse operations (p. 56)
function (p. 66)
domain (p. 67)
range (p. 67)

Solve each equation. *(Sections 2.1 and 2.2)*

1. $-3 = 18x$
2. $-r - 7 = 1$
3. $-4 = 37 + s$
4. $24 - 3d = 0$
5. $-n = 20$
6. $20 = 18 + \frac{m}{7}$
7. $\frac{p}{4} = -12$
8. $\frac{t}{6} - 4 = 9$

9. **AVIATION** A plane is flying at an altitude of 10,000 ft. The pilot wants to climb to a higher altitude at a rate of about 1000 ft/min. About how many minutes will it take the pilot to climb to 14,500 ft? *(Section 2.2)*

10. a. Ben buys plant food for \$3 and plants for \$2 each. Write an equation for his total cost as a function of the number of plants he buys. Describe the domain and the range of the function. *(Section 2.3)*

b. Ben has \$45. How many plants can he buy?

11. **Journal** Evaluate $2x + 8$ when $x = 5$. Then solve $2x + 8 = 16$. How is evaluating an expression different from solving an equation?

Apply⇔Assess

Exercise Notes

Assessment Note

Ex. 23 The functions students choose as their examples should involve a relationship between two sets of numbers and not be the type discussed in Ex. 4 on page 69.

Progress Check 2.1–2.3

See page 92.

Practice 12 for Section 2.3

NAME ______ DATE ______

Practice 12

FOR USE WITH SECTION 2.3

1. yes; There is exactly one value of $|n - 2|$ for each value of n.
2. yes; There is exactly one noon temperature for each day.
3. no; Pavement for a street number has more than one width.

Decide whether each table represents a function. Explain your reasoning.

1.

n	$\lvert n - 2 \rvert$
0	2
1	1
2	0
3	1
4	2

2.

Day in May	Temp. at noon (°F)
1	67
2	65
3	65
4	65
5	68

3.

Street number	Pavement width
10	20
11	22
11	24
12	20
12	26

4. At a certain video rental store, you are entitled to 1 free video rental after every 8 rentals at the regular price. Suppose you have rented v videos at the regular price. Describe a way to determine (using a calculator) how many free video rentals f you are entitled to. Is f a function of v? Divide the number of rentals by eight and use the whole-number part of the result.; yes

5. Joon Hee bought some containers of yogurt in a supermarket for \$.65 each and gave the cashier a \$20 bill. His change is a function of the number of containers he bought. Write an equation for this function. What are the domain and range of the function? $c = 20 - 0.65n$; domain = {1, 2, 3, ..., 30}; range = {19.35, 18.70, 18.05, 17.40, ..., 0.50}

6. Write the perimeter of each polygon as a function of x.

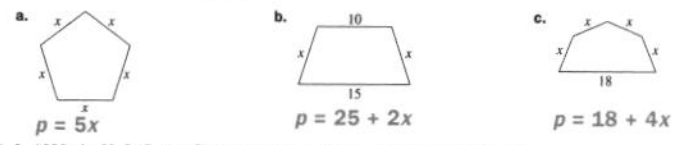

a. $p = 5x$ b. $p = 25 + 2x$ c. $p = 18 + 4x$

7. In 1990, the U. S. Census Bureau estimated that 1 person was added to the U. S. population every 14 seconds. Suppose there were 249,000,000 people in the U. S. at the beginning of 1990. The population t seconds after the beginning of the year is a function of t. Write an equation for this function. What are the domain and range of the function? $p = 249{,}000{,}000 + \frac{t}{14}$; domain = multiples of 14, range = integers ≥ 249,000,000

8. Waneta is at the corner of Elm Avenue and 9th Street. All the numbered streets cross Elm Avenue in order: 1st, 2nd, 3rd, and so on, up to 15th Street. Suppose Waneta wants to walk to nth street. Using absolute value, write the number of blocks b she will have to walk, based on n. Is b a function of n? If so, what are its domain and range? $b = |9 - n|$; yes; domain = {1, 2, 3, ..., 15}; range = {0, 1, 2, 3, ..., 8}

9. **Open-ended Problem** The color of a person's eyes is a function of the person's Social Security number. Make up a table for this function and other similar functions for the members of your family or math class. Answers may vary. Check students' work.

12 **Practice Bank,** ALGEBRA 1: EXPLORATIONS AND APPLICATIONS

distance a car travels at a constant speed of 50 mi/h is a function of the traveling time. The domain is the nonnegative numbers and the range is the nonnegative numbers. (3) Your score on a 20-question true/false quiz is a function of the number of questions you answer correctly. The domain is 0, 1, 2, ..., 20 and the range is 0, 5, 10, ..., 100. (4) The number of Calories burned is not necessarily a function of the number of minutes you exercise. An input of 15 can be matched with more than one output.

24. –42
25. 52
26. 81
27. about 6.64
28. 2.25
29. 2.5
30. 22
31. 15
32. 37
33. –51
34. –36
35. 21

Assess Your Progress

1. $-\frac{1}{6}$
2. –8
3. –41
4. 8
5. –20
6. 14
7. –48
8. 78
9. 4.5 min
10. a. $C = 3 + 2p$, where p is the number of plants bought and C is the total cost in dollars. The domain is the positive whole numbers and the range consists of these amounts in dollars: 5, 7, 9, ...

b. 21 plants

11. 18; 4; To evaluate an expression, you substitute a given value for the variable. To solve an equation, you find the value of the variable that makes the equation true.

Plan⇔Support

Objectives

- Read and create coordinate graphs.
- Analyze biological and geographical data.
- Read and interpret maps.

Recommended Pacing

Core and Extended Courses
Section 2.4: 1 day

Block Schedule
Section 2.4: $\frac{1}{2}$ block
(with Section 2.3)

Resource Materials

Lesson Support
Lesson Plan 2.4
Warm-Up Transparency 2.4
Overhead Visuals:
  Folder 2: Reading Graphs
Practice Bank: Practice 13
Study Guide: Section 2.4
Explorations Lab Manual:
  Diagram Master 2

Technology
Technology Book:
  Calculator Activity 2
Graphing Calculator
Internet:
  http://www.hmco.com

Warm-Up Exercises

1. Would you use a positive number or a negative number to represent 25 feet below sea level? Write the number.
negative number; –25

Complete each blank with < or >.

2. –4 __?__ 7 <
3. 5 __?__ –1 >
4. –6 __?__ –9 >
5. –13 __?__ –4 <
6. 0 __?__ –5 >
7. 4 __?__ 0 >

SECTION

2.4 Coordinate Graphs

Learn how to...
- **read and create coordinate graphs**

So you can...
- **analyze biological and geographical data, for example**
- **read and interpret maps**

Winters in Antarctica are so cold that several feet of ice cover the ocean for most of the year. This icy barrier does not stop the Weddell seal from diving beneath the ice to hunt for fish. The graph below shows data collected by a biologist studying Weddell seals.

EXAMPLE 1 Application: Biology

The graph compares the dives of two seals.

a. Where was Seal A 5 min after it dove under the ice?

b. About how deep did Seal A dive?

c. How long did it take Seal A to return to the surface?

SOLUTION

The **horizontal axis** shows the amount of **time *t*** that has passed.
The **vertical axis** shows the **depth** of each dive at different times.

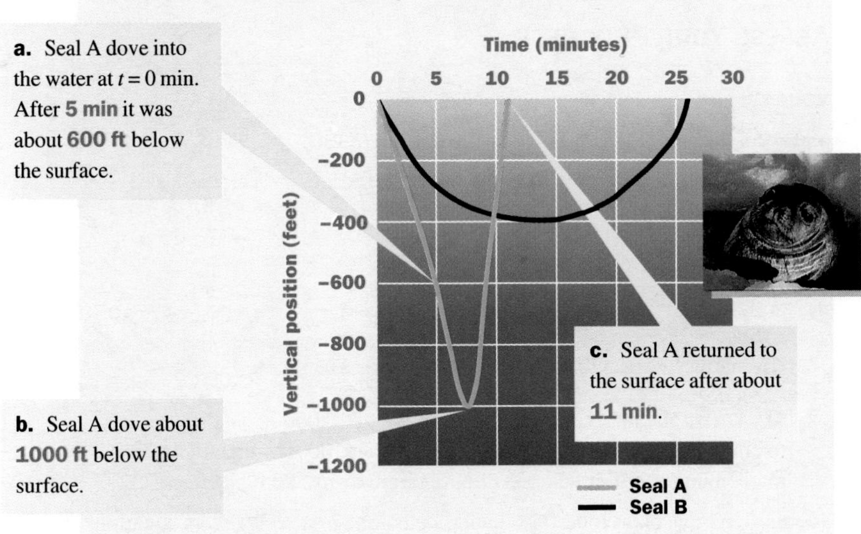

a. Seal A dove into the water at $t = 0$ min. After 5 min it was about 600 ft below the surface.
Time (minutes)
0
5
10
15
20
25
30
0
–200
–400
–600
–800
–1000
–1200
Vertical position (feet)
b. Seal A dove about 1000 ft below the surface.
c. Seal A returned to the surface after about 11 min.
Seal A
Seal B

THINK AND COMMUNICATE

1. Which seal stayed under water longer? How can you tell from the graph?
2. How deep was Seal B's dive? About how many minutes did it take for Seal B to reach the deepest point of its dive?

ANSWERS Section 2.4

Think and Communicate

1. Seal B; both seals dove into the water at time $t = 0$. Seal A returned to the surface after about 11 min, and Seal B returned to the surface after about 26 min.
2. about 400 ft below the surface; about 13 min

Graphing in the Coordinate Plane

You can graph points in a **coordinate plane** like the one shown. You can use an **ordered pair** such as **(–4, 2)** to identify each point.

horizontal coordinate, or ***x*-coordinate**

vertical coordinate, or ***y*-coordinate**

The point with coordinates (0, 0) is called the **origin**.

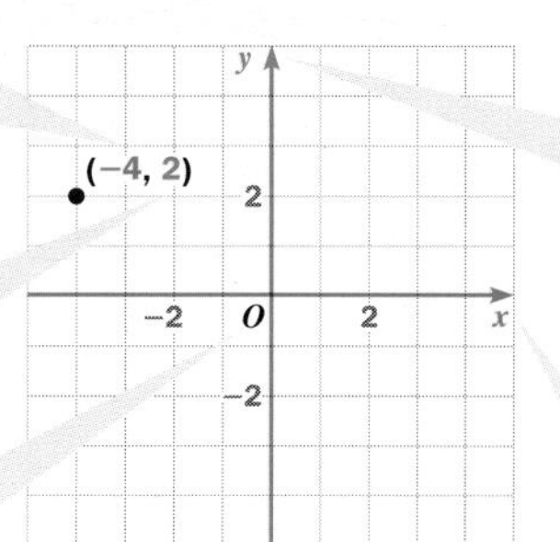

The vertical axis is usually called the ***y*-axis**.

The horizontal axis is usually called the ***x*-axis**.

The *x*- and *y*-axes divide the coordinate plane into four regions called **quadrants**.

This point is in Quadrant III.

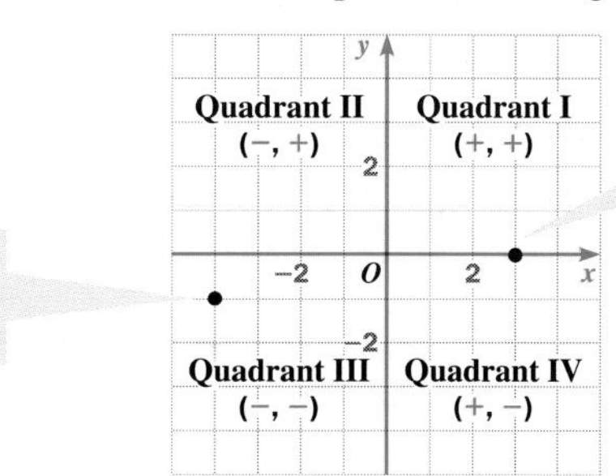

This point is not in any quadrant because it is on an axis.

EXAMPLE 2

Graph each point in a coordinate plane. Name the quadrant (if any) in which the point lies.

a. **(−3, 2)** **b.** **(0, −4)**

SOLUTION

a. Start at the origin. Move **3** units left. Then move **2** units up. Label the point. The point is in Quadrant II.

b. Start at the origin. Move **0** units on the *x*-axis. Move **4** units down. Label the point. The point is not in any quadrant.

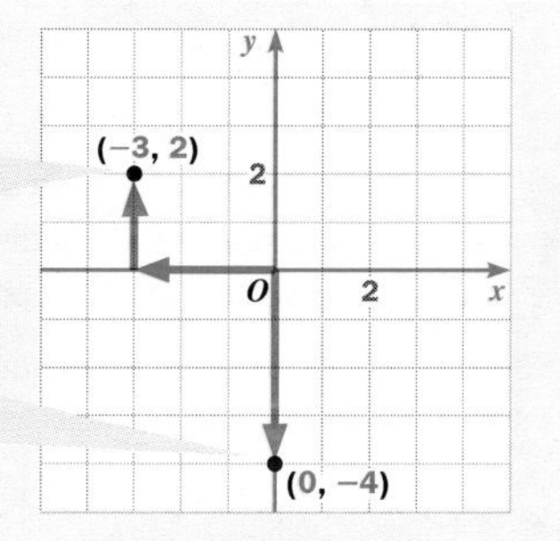

WATCH OUT! The horizontal coordinate always comes first in an ordered pair. So (–3, 2) is *not* the same as (2, –3).

THINK AND COMMUNICATE

3. Which direction is positive on the horizontal axis? on the vertical axis?
4. Which quadrant does the graph on page 72 show? How can you tell?

Think and Communicate

3. right; upward
4. IV; The horizontal coordinates are positive and the vertical coordinates are negative.

Teach⇔Interact

Learning Styles: Visual

Data seem to be most easily understood by students when presented in graphs. Visual images can convey a great deal of information in an interesting and succinct format. The data in Example 1, for example, can also be presented in a table, but the graph shows visually the dives of the two seals, which a table could not do.

Additional Example 1

Use the graph from Example 1.

a. Where was Seal B 5 min after it dove under the ice? After 5 min, Seal B was about 275 ft below the surface.

b. How long did it take Seal B to return to the surface? Seal B returned to the surface after about 26 min.

Section Note

Using Technology
Students can use graphing calculators to explore the four quadrants of the coordinate plane. To set up the coordinate axes, press WINDOW on the TI-82 or RANGE on the TI-81. Choose a "friendly" viewing window by using the following settings.

TI-82: Xmin = –9.4, Xmax = 9.4, Xscl = 1, Ymin = –6.2, Ymax = 6.2, Yscl = 1
TI-81: Xmin = –9.4, Xmax = 9.6, Xscl = 1, Ymin = –6.2, Ymax = 6.4, Yscl = 1
Next, press GRAPH. Use the arrow keys ◀, ▶, ▼, ▲ to move the crosshair cursor to various points in the viewing window. Observe the *x*- and *y*-coordinates shown at the bottom of the screen.

Additional Example 2

Graph each point in a coordinate plane. Name the quadrant (if any) in which the point lies.

a. (3, –2) Quadrant IV
b. (4, 3) Quadrant I
c. (–4, –3) Quadrant III

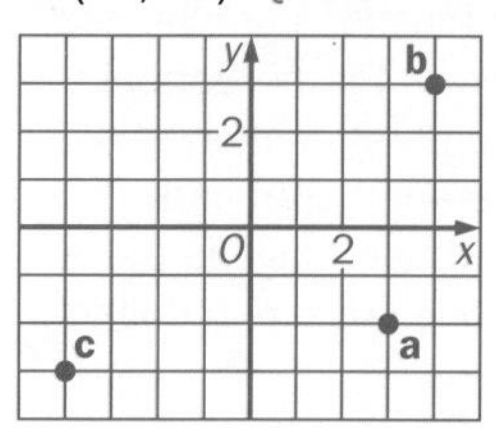

Teach⇔Interact

Checking Key Concepts

Visual Thinking
While responding to questions 1 and 2, ask students to discuss the actual paths of the two seals in relation to the paths shown on the graph. Are they the same? How might they be different? Encourage students to discuss their responses in class. This activity involves the skills of *interpretation* and *generalization.*

Common Error
Students who err in identifying the coordinates of a point in the coordinate plane should study Example 2 again, paying particular attention to the *Watch Out* note.

Closure Question

Identify the parts of a coordinate plane and describe each part.
x-axis, which is the horizontal axis; y-axis, which is the vertical axis; the origin, which is where the axes intersect; four quadrants, which are formed by the intersecting axes

Apply⇔Assess

Suggested Assignment

Core Course
Exs. 1–15, 17–22, 25–28, 31–39

Extended Course
Exs. 1–4, 7, 10, 11, 13–39

Block Schedule
Day 11 1–23, 25–28, 31–39

Exercise Notes

Teaching Tip
Exs. 1–15 Students should use graph paper to draw their coordinate planes. They can use the plane shown at the top of page 73 as a model to get started.

Topic Spiraling: Preview
Exs. 17–19 These exercises preview the work in the next section on representing functions graphically. The graph shown for the Benefit Car Wash represents a function whose domain is the number of cars washed and whose range is the amount of money raised in dollars.

CHECKING KEY CONCEPTS

For Questions 1 and 2, use the graph in Example 1 on page 72.

1. Where was Seal A 10 min after it dove under the ice?

2. About how many minutes did it take Seal A to swim from the deepest point of its dive to the surface?

Match each ordered pair with a point on the graph. Name the quadrant (if any) in which the point lies.

3. (1, 1)
4. (4, 4)
5. (−4, 3)
6. (3, −4)
7. (0, −3)
8. (4, 0)
9. (−3.5, −1)
10. $\left(\frac{7}{3}, -2\right)$

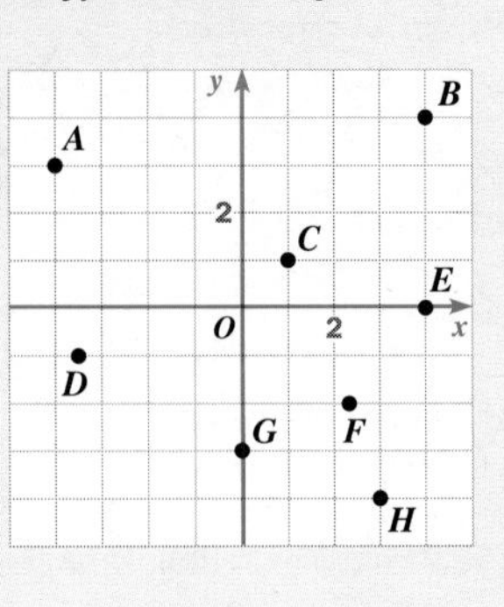

2.4 Exercises and Applications

Extra Practice exercises on page 560

Graph each point in a coordinate plane. Label each point with its letter. Name the quadrant (if any) in which the point lies.

1. $A(4, 1)$
2. $B(4, -1)$
3. $C(-3, -5)$
4. $D(-4, 2)$
5. $E(4, 5)$
6. $F(-5, 4)$
7. $G(0, -2)$
8. $H(-2, 3)$
9. $I(5, -5)$
10. $J(0, 0)$
11. $K(3, 0)$
12. $L(-4, -4)$
13. $M(1.5, 0)$
14. $N\left(-2, 4\frac{1}{2}\right)$
15. $P(2.75, 2.75)$

16. Challenge Point (a, b) is in the second quadrant and point (c, d) is in the fourth quadrant. Complete each blank with < or >. Explain your reasoning.

a. $a \underline{?} b$ **b.** $c \underline{?} d$ **c.** $a \underline{?} c$ **d.** $b \underline{?} d$

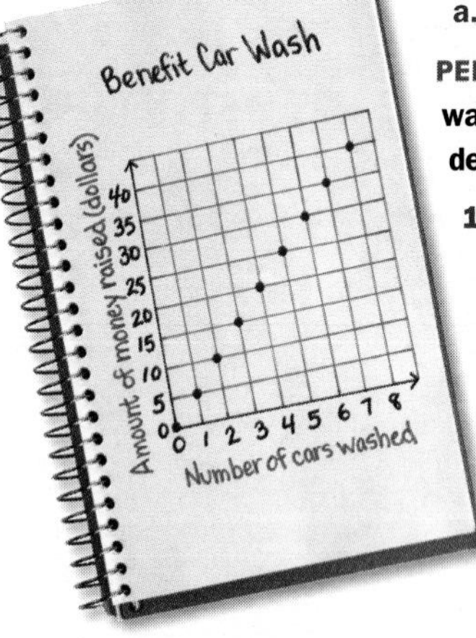

PERSONAL FINANCE Some students decide to raise money by having a car wash. The graph at the left shows how much money they can raise, depending on how many cars they wash.

17. What does the horizontal axis of the graph show? What does the vertical axis show?

18. How much do the students raise per car? How do you know?

19. The students want to raise $200. How many cars do they have to wash?

Checking Key Concepts

1. about 400 ft below the surface
2. about 3.5 min
3. C; I
4. B; I
5. A; II
6. H; IV
7. G; none
8. E; none
9. D; III
10. F; IV

Exercises and Applications

1–15.

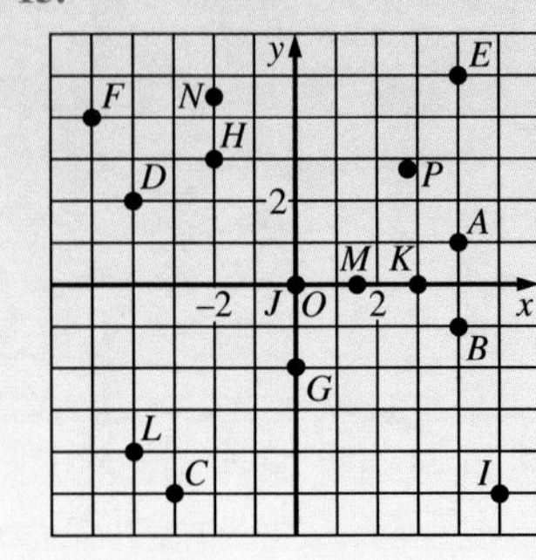

1. I 2. IV 3. III
4. II 5. I 6. II
7. none 8. II 9. IV
10. none 11. none 12. III
13. none 14. II 15. I
16. a. <; $a < 0$ and $b > 0$
b. >; $c > 0$ and $d < 0$
c. <; $a < 0$ and $c > 0$
d. >; $b > 0$ and $d < 0$
17. the number of cars washed; the amount of money raised

Connection GEOGRAPHY

Depending on your location on the globe and the time of year, the number of daylight hours varies, as shown by the graph below. Use the graph for Exercises 20–22.

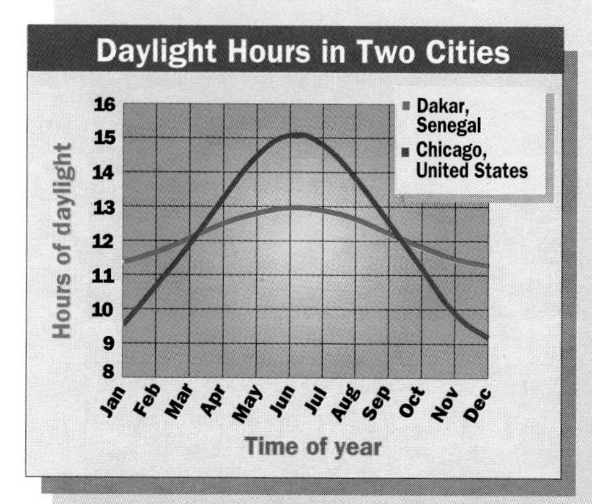

20. a. When does each city receive the fewest hours of daylight? About how many hours of daylight per day does each city receive at this time?

b. When does each city receive the most hours of daylight? About how many hours of daylight per day does each city receive at this time?

BY THE WAY

Dakar and Chicago are both in the Northern Hemisphere. In the Northern and Southern Hemispheres, the day with the greatest number of daylight hours is called the summer solstice. The day with the least number of daylight hours is called the winter solstice.

21. During which months are the hours of daylight in Chicago and Dakar about equal? About how many hours of daylight per day does each city receive during these months?

22. Compare the shapes of the two graphs. How are they the same? How are they different?

For Exercises 23 and 24, use the graph in Example 1 on page 72.

23. Writing About how long do you think a seal would stay under water if it dove to a depth of 700 ft? Explain your answer.

24. Open-ended Problem About how long do you think you can hold your breath under water? How deep do you think you can dive? Answer these questions by drawing a graph. How is your graph different from the seal graph? Give some possible reasons for these differences.

25. PERSONAL FINANCE Suppose you want to borrow money from your parents for a new pair of in-line skates. You plan to pay your parents $10 a week until the loan is paid off.

a. According to the graph at the right, how much money do you borrow? How do you know?

b. How much do you owe after six weeks?

c. How long does it take you to pay back the loan? How can you tell from the graph?

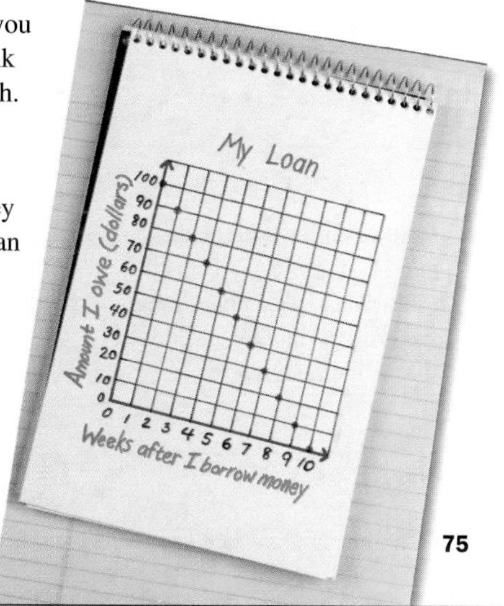

Apply⇔Assess

Exercise Notes

Communication: Reading
Ex. 20–22 Students should come to appreciate the fact that it would be difficult to compare the daylight hours in Dakar and Chicago without using graphs.

Interdisciplinary Problems
Exs. 20–22, 26–30 These exercises connect the mathematical ideas of this section to the disciplines of geography and history, respectively.

Communication: Discussion
Ex. 24 A brief discussion of students' answers to these questions would be interesting from the point of view of seeing how wide or narrow their range of responses is.

18. $5; Answers may vary. An example is given. For every additional car washed, the amount of money raised increases by $5.
19. 40 cars
20. a. Both receive the fewest hours of daylight in December. Dakar receives about 11.25 h and Chicago about 9.25 h.
    b. Both cities receive the most hours of daylight in June. Dakar receives about 13 h and Chicago about 15 h.
21. March and September; about 12 h per day
22. Answers may vary. An example is given. The shapes of the graphs are similar in that both rise steadily from the amount of daylight in January until June, and then both fall steadily back to the original amounts by December. The shapes differ in that the graph for Dakar increases and falls only slightly, whereas the graph for Chicago increases and falls much more steeply.
23. Answers may vary. An example is given. I think a seal might stay under water for about 20 min if it dove to a depth of 700 ft. The graph on page 72 suggests that the deeper a seal dives, the sooner the seal needs to return to the surface. Since 700 ft is halfway between 1000 ft (Seal A's dive) and 400 ft (Seal B's dive), I estimate that the length of the dive would be halfway between 11 min and 26 min (the times for Seals A and B), or 18.5 min.
24. Answers may vary.
25. a. $100; The point (0, 100) is on the graph, so 0 weeks after the money is borrowed, $100 is owed.
    b. $40
    c. 10 weeks; The point (10, 0) is on the graph, so 10 weeks after the money is borrowed, the amount owed is $0.

Apply⇔Assess

Exercise Notes

Application

Exs. 26–30 Students should understand that it would be impossible to locate positions on modern maps precisely without using a coordinate grid. You may wish to ask what coordinate grid is used for the planet Earth itself. (lines of latitude and longitude)

Multicultural Note

In the second century A.D., a brilliant Chinese scientist named Chang Heng invented what is known as quantitative cartography. He was the first person in China to use a grid system for map-making. The calculation and study of distances and positions was made more scientific by laying a system of grids over the maps. This method was so efficient that by the Middle Ages cartographers were doing away with visual images on maps and relying solely on grids to measure and pinpoints distances and locations.

Research

Exs. 29, 30 Students who research these exercises may wish to report their findings to the class. Then all students can share in the knowledge gained by the researchers.

Connection HISTORY

Around the second century A.D., Chinese geographers began using grids on maps. In 1311, the geographer Chu Ssu-Pen created a map of China. The map was printed in book form in about 1555. This section of the map shows part of China's coast.

26. About how many square miles does the map represent? How do you know?

27. The black band in the upper left corner of the map represents the Gobi Desert. According to the map, about how long from east to west is this section of the Gobi Desert? How can you tell?

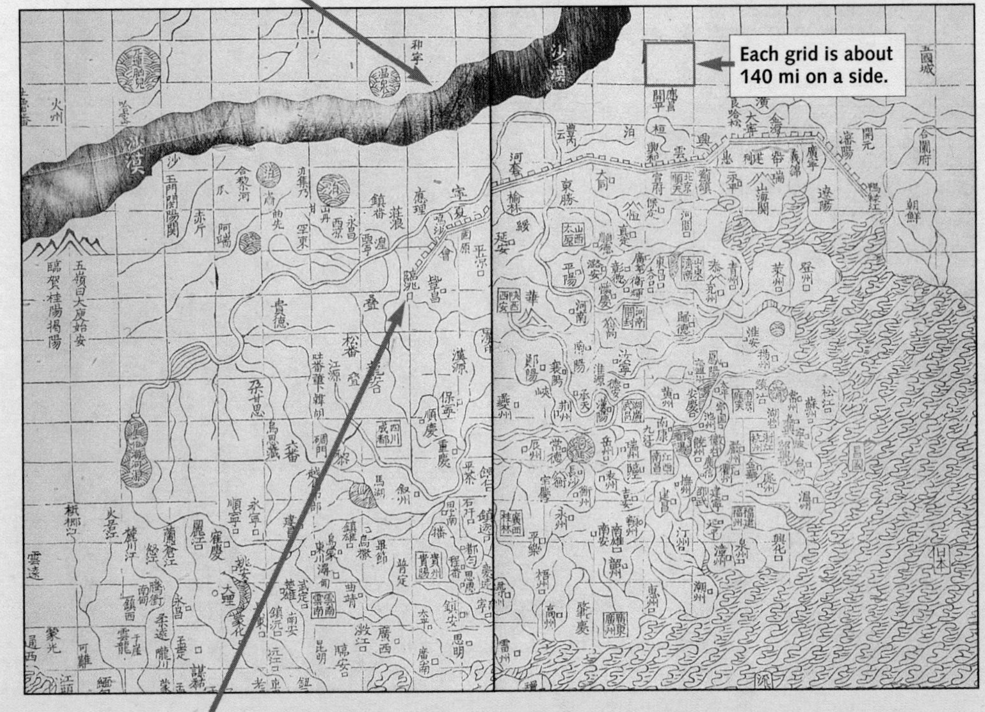

Each grid is about 140 mi on a side.

28. The map shows the Great Wall of China with these special lines (▭▭▭). According to the map, about how long is the Great Wall? How did you make your estimate?

29. Research Use an encyclopedia to find out more about the Great Wall. How long is it? When and why was it built?

30. Research Find a modern map showing roughly the same area as the map above. Does the modern map use a coordinate grid? Can you use the modern map to identify any bodies of water on Chu Ssu-Pen's map? any land features?

26. about 6,272,000 mi^2; The map is 20 grids wide by 16 grids high. Each grid has area about 19,600 mi^2. The total area is about $320 \times 19,600 =$ 6,272,000 mi^2.
27. Estimates may vary; about 1820 mi. The band covers about 13 grids horizontally. Each grid is about 140 mi on a side.
28. Estimates may vary. between 1400 and 1500 mi long; I used the grid to mark a scale on a piece of paper and used the scale to measure the wall to be between 10 and 11 grid marks long.
29. Reports should include the following information. The Wall is about 1500 mi long. It was built in the third century B.C. by the first emperor of China, Ch'in Shih Huang Ti to defend against invaders.
30. Check students' work.

For Exercises 31–35, use the coordinate graph.

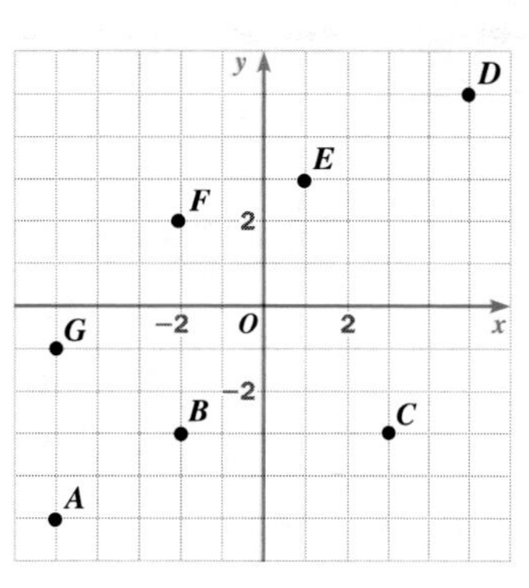

31. Write an ordered pair for each point.

32. Name two points for which the x-coordinate is less than the y-coordinate.

33. Name two points for which the x-coordinate is greater than the y-coordinate.

34. Name two points for which the x-coordinate equals the y-coordinate.

35. Name two points for which the x-coordinate is the opposite of the y-coordinate.

ONGOING ASSESSMENT

36. a. Draw a coordinate plane. Graph at least six points for which the x-coordinate equals the y-coordinate. In which quadrant(s) (if any) do the points lie?

b. Draw another coordinate plane. Graph at least six points for which the x-coordinate is the opposite of the y-coordinate. In which quadrant(s) (if any) do the points lie?

c. Writing Compare your graphs. Write a paragraph describing any patterns you see.

SPIRAL REVIEW

For Exercises 37 and 38, find the mean, the median, and the mode(s) of each set of data. *(Section 1.3)*

37. Movie prices: \$6.00, \$3.50, \$5.50, \$5.75, \$4.50, \$6.75, \$6.75, \$6.50

38. Overtime hours: 7, 1, 10, 5, 0, 3, 1, 1, 2, 2, 0, 0, 0, 9, 0, 4, 2

39. **Technology** Give the dimensions of each matrix. Tell which matrices can be added. Then add them. Use a graphing calculator if you have one. *(Section 1.8)*

$$A = \begin{bmatrix} 2 & 4 \\ 7 & -2 \\ 1 & 16 \end{bmatrix} \qquad B = \begin{bmatrix} 3 & 7 & 11 \\ 3 & 2 & 13 \end{bmatrix} \qquad C = \begin{bmatrix} 5 & 3 & 2 \\ 1 & -1 & -3 \\ 0 & 0 & 3 \end{bmatrix}$$

$$D = \begin{bmatrix} 6 & 0 & 0 & 1 \end{bmatrix} \qquad E = \begin{bmatrix} 1 & 2 \\ 2 & 1 \end{bmatrix} \qquad F = \begin{bmatrix} 1 & 0 & -7 \end{bmatrix}$$

$$G = \begin{bmatrix} 3 & -1 \\ 6 & 17 \\ -12 & 3 \end{bmatrix} \qquad H = \begin{bmatrix} 10 & 5 & 3 & 6 \end{bmatrix} \qquad I = \begin{bmatrix} -3 & -2 & 11 \\ 0 & 1 & -1 \\ 5 & 7 & 1 \end{bmatrix}$$

Apply⇔Assess

Exercise Notes

Cooperative Learning

Exs. 31–36 You may wish to have students organize themselves into small groups to do these exercises. Discussion and shared answers among group members can provide an excellent way of ensuring that all students are successful with coordinate graphs.

Practice 13 for Section 2.4

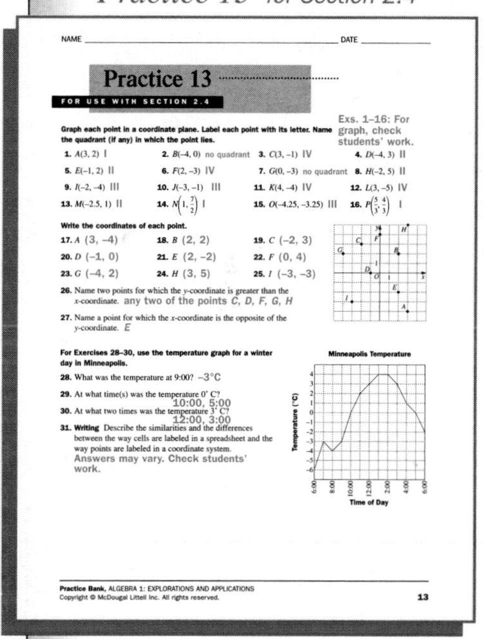
NAME ______________________ DATE __________

Practice 13

FOR USE WITH SECTION 2.4

Graph each point in a coordinate plane. Label each point with its letter. Name the quadrant (if any) in which the point lies. Exs. 1–16: For graph, check students' work.

1. $A(3, 2)$ I	2. $B(-4, 0)$ no quadrant	3. $C(3, -1)$ IV	4. $D(-4, 3)$ II
5. $E(-1, 2)$ II	6. $F(2, -3)$ IV	7. $G(0, -3)$ no quadrant	8. $H(-2, 5)$ II
9. $I(-2, -4)$ III	10. $J(-3, -1)$ III	11. $K(4, -4)$ IV	12. $L(3, -5)$ IV
13. $M(-2.5, 1)$ II	14. $N\left(1, \frac{7}{2}\right)$ I	15. $O(-4.25, -3.25)$ III	16. $P\left(\frac{5}{3}, \frac{4}{3}\right)$ I

Write the coordinates of each point.

17. A (3, −4)	18. B (2, 2)	19. C (−2, 3)
20. D (−1, 0)	21. E (2, −2)	22. F (0, 4)
23. G (−4, 2)	24. H (3, 5)	25. I (−3, −3)

26. Name two points for which the y-coordinate is greater than the x-coordinate. any two of the points C, D, F, G, H

27. Name a point for which the x-coordinate is the opposite of the y-coordinate. E

For Exercises 28–30, use the temperature graph for a winter day in Minneapolis.

28. What was the temperature at 9:00? −3°C

29. At what time(s) was the temperature 0° C? 10:00, 5:00

30. At what two times was the temperature 3° C? 12:00, 3:00

31. **Writing** Describe the similarities and the differences between the way cells are labeled in a spreadsheet and the way points are labeled in a coordinate system. Answers may vary. Check students' work.

Practice Bank, ALGEBRA 1: EXPLORATIONS AND APPLICATIONS

13

31. $A(-5, -5)$; $B(-2, -3)$, $C(3, -3)$; $D(5, 5)$; $E(1, 3)$; $F(-2, 2)$; $G(-5, -1)$

32. any two of E, F, and G

33. B and C

34. A and D

35. C and F

36. a. Choices of points may vary. All such points except for the origin lie in Quadrants I and III.

b. Choices of points may vary. All such points except for the origin lie in Quadrants II and IV.

c. Answers may vary. An example is given. Both sets of points lie on straight lines through the origin and both lines are located "midway" between the axes.

37. about 5.66; about 5.88; 6.75

38. about 2.76; 2; 0

39. A: 3×2; B: 2×3; C: 3×3; D: 1×4; E: 2×2; F: 1×3; G: 3×2; H: 1×4; I: 3×3; A and G, C and I, D and H;

$$A + G = \begin{bmatrix} 5 & 3 \\ 13 & 15 \\ -11 & 5 \end{bmatrix};$$

$$C + I = \begin{bmatrix} 2 & 1 & 13 \\ 1 & 0 & -4 \\ 5 & 7 & 1 \end{bmatrix};$$

$$D + H = \begin{bmatrix} 16 & 5 & 3 & 7 \end{bmatrix}$$

Plan⇔Support

Objectives

- Graph equations.
- Visually represent functions.

Recommended Pacing

Core and Extended Courses
Section 2.5: 1 day

Block Schedule
Section 2.5: $\frac{1}{2}$ block
(with Section 2.6)

Resource Materials

Lesson Support
Lesson Plan 2.5
Warm-Up Transparency 2.5
Overhead Visuals:
Folder A: Multi-Use Graphing Packet, Sheets 1–3
Practice Bank: Practice 14
Study Guide: Section 2.5
Explorations Lab Manual:
Additional Exploration 3, Diagram Masters 1, 2

Technology
Graphing Calculator
Spreadsheet Software
McDougal Littell Software
Function Investigator
Stats!
Internet:
http://www.hmco.com

Warm-Up Exercises

Graph each point in a coordinate plane. Name the quadrant (if any) in which the point lies.

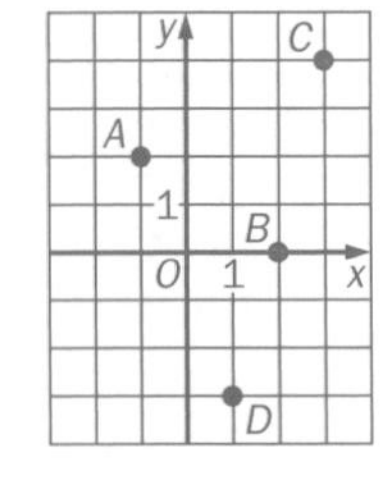

1. $A(-1, 2)$
Quadrant II
2. $B(2, 0)$
none
3. $C(3, 4)$
Quadrant I
4. $D(1, -3)$
Quadrant IV

Complete each sentence.

5. The __?__ axis is the x-axis.
horizontal
6. The __?__ axis is the y-axis.
vertical
7. The point (0, 0) is called the __?__. origin

SECTION

2.5 Representing Functions

Learn how to...
- **graph equations**

So you can...
- **visually represent functions, such as costs at a swimming pool**

Take any number and subtract 1 from it. You can think of this command as a function. The number is the input. Your answer is the output. One way to represent this function is to use a graph.

EXPLORATION
COOPERATIVE LEARNING

Graphing Equations

Work with your class.

1 Turn your class into a coordinate grid, as shown. What are your coordinates?

2 Stand up if your coordinates make the equation $y = x - 1$ true. What do you notice about the people standing up?

3 Sit down again. Your teacher will give you other equations. For each equation given, stand up if your coordinates make the equation true. In each case, describe what you observe.

This student is standing because her coordinates, (1, 0), make the equation true.
$y = x - 1$
$0 = 1 - 1$

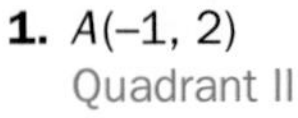

Exploration Note

Purpose
The purpose of this Exploration is to introduce students to the idea of graphing the coordinates that make an equation true. Students see that the solutions to this equation form a straight line.

Materials/Preparation
Students should be arranged in class as shown in the textbook. If any students are absent, or a desk has not been assigned to a student, move other students to fill those positions. In other words, every position in the grid should be occupied by a student.

Procedure
You can begin by asking each student to state verbally his or her coordinates. Then follow Steps 2 and 3. Other equations you can use are $y = x + 1$, $y = -x - 1$, and $y = -x + 1$. After each equation is used, ask students what they see about the people standing up.

Closure
Ask students to describe what they have learned through this Exploration about graphing equations.

Explorations Lab Manual
See the Manual for more commentary on this Exploration.

You can represent the same function in several different ways. As shown in Example 1, you can use an equation, a table of values, or a graph.

EXAMPLE 1

Graph the equation $y = 2x - 1$.

SOLUTION

Step 1 Make a table of values. Choose at least three values of x. Be sure to include both positive and negative values of x.

$y = 2x - 1$	
x	y
−2	−5
0	−1
2	3

Step 2 Draw a coordinate grid. Use the table of values to plot points.

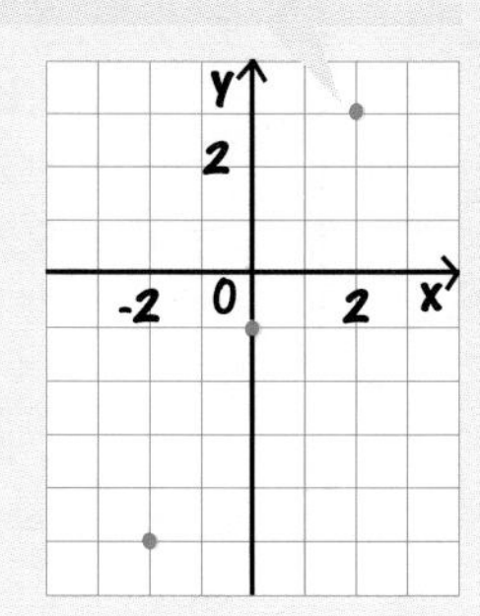

Step 3 Connect the points to make a line. Put an arrowhead on each end to show that the line extends in both directions.

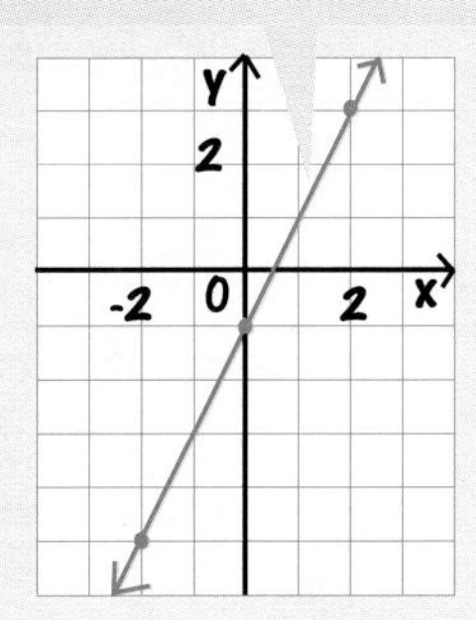

The coordinates of every point on the line make the equation $y = 2x - 1$ true, not just the points plotted in Step 2. For example, the point $\left(-\frac{3}{2}, -4\right)$ is on the graph of the equation, and its coordinates make the equation true:

$$-4 = 2\left(-\frac{3}{2}\right) - 1$$

An ordered pair that makes an equation such as $y = 2x - 1$ true is called a **solution of the equation**. The graph of the equation $y = 2x - 1$ contains all possible solutions of the equation.

THINK AND COMMUNICATE

1. Which of these ordered pairs are solutions of the equation $y = 2x - 1$? How do you know?

   A. $(3, 4)$ **B.** $\left(\frac{1}{2}, 0\right)$ **C.** $\left(\frac{3}{2}, 2\right)$ **D.** $\left(-\frac{5}{2}, -6\right)$

2. Based on what you now know about the graph of an equation, explain what you observed in the Exploration on page 78.

Teach⇔Interact

About Example 1

Reasoning
This example points out that the graph of the equation $y = 2x - 1$ contains all possible solutions of the equation. Ask students: How many solutions does an equation with two variables have? (Since the graph is a straight line, which has infinitely many points, there must be infinitely many solutions to the equation.)

Additional Example 1

Graph the equation $y = -3x + 2$.

Step 1 Make a table of values.

$y = -3x + 2$	
x	y
−1	5
0	2
1	−1

Step 2 Draw a coordinate grid. Use the table of values to plot points.
Step 3 Connect the points to make a line.

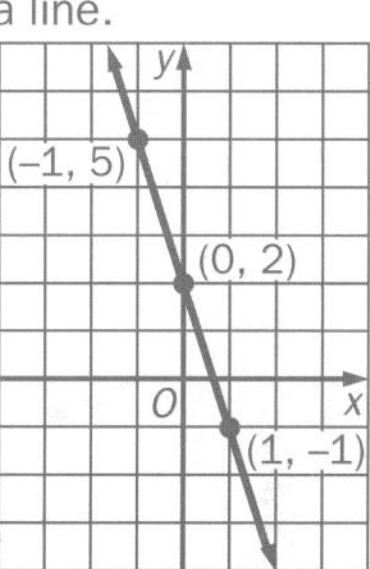

Section Note

Using Technology
Graphing equations is fast and simple with a graphing calculator. Set up a convenient viewing window as described in the Using Technology note on page 73. (You may use other settings for the viewing window if you wish.) Next, press [Y=] and enter the expression in x from the right side of the equation. To type X, press the [X,T,Θ] key on the TI-82 or the [X|T] key on the TI-81. After you have entered the equation, press [GRAPH] to have the calculator display the graph.

ANSWERS Section 2.5

Exploration

1. Answers may vary.
2. Students with coordinates such as (−2, −3), (−1, −2), (0, −1), (1, 0), (2, 1), (3, 2), and so on, should be standing. All the people standing up are located on a straight line.
3. For each equation, all the people will be located on a line.

Think and Communicate

1. *B*, *C*, and *D*; Their coordinates make the equation true.
2. The coordinates of the people standing up make the equation $y = x - 1$ true. The points they represent are on the line that is the graph of the equation.

Teach⇔Interact

Additional Example 2

Suppose an appliance repair technician charges a flat rate of \$68 to come to a home and \$44 per hour for each hour worked, based on incremental charges of 15 minutes each and rounded up for any time over each increment. The cost to the owner to repair an appliance is a function of the number of hours worked.

a. Write a cost equation for the function. Let C = the cost to repair an appliance. Let h = the number of hours worked.
$C = 68 + 44h$

b. Make a table of values for the function. Choose values for h. Use the equation $C = 68 + 44h$.

Number of hours	Cost of repair (\$)
0.25	79
0.5	90
0.75	101
1	112
1.25	123
1.5	134
1.75	145
2	156
...	...
3	200

c. Graph the function. Draw a pair of axes. Use the table to plot points.

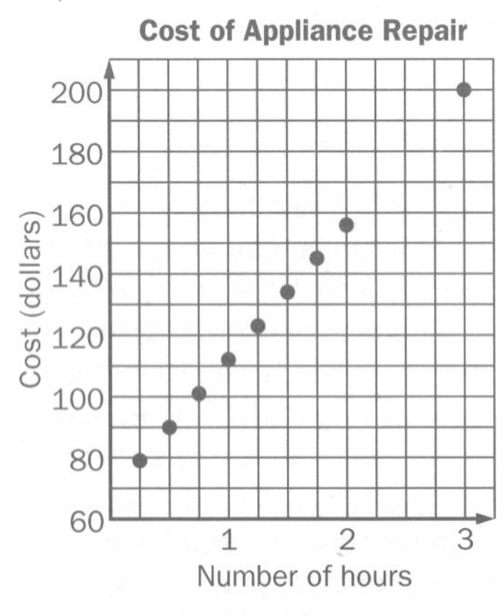

EXAMPLE 2 Application: Consumer Economics

Every year, the Takedas buy a family membership at the Springfield swimming pool. The sign shows that their annual cost is a function of the number of guests they take to the pool.

a. Write an equation for the function.

b. Make a table of values for the function.

c. Graph the function.

SPRINGFIELD MUNICIPAL SWIMMING POOL
MEMBERSHIP FEES PER YEAR:
Individual \$30.00
Family \$60.00
Guests of members \$5.00 per visit

SOLUTION

a. Let C = the annual cost. Let g = the number of guests.

Annual cost = \$60 + \$5 per guest × number of guests

$$C = 60 + 5g$$

b. Choose values for g, the number of guests. It helps to think about the domain of the function.

The domain starts at 0 and includes only whole numbers.

Number of guests	Annual cost (dollars)
0	60
1	65
2	70
3	75
...	...

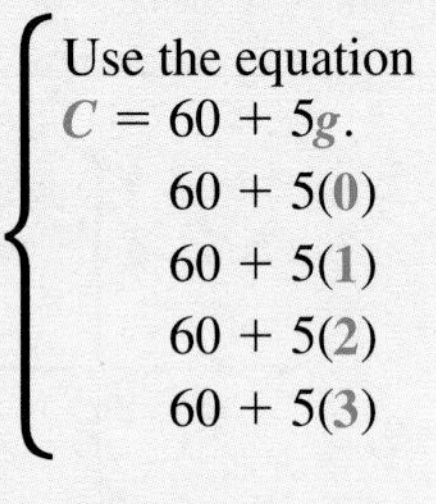

c. Draw a pair of axes. Put the domain values on the horizontal axis and the range values on the vertical axis.

Use the table to plot points. The Takedas can take only a whole number of guests. Do not connect the points because a line would include numbers that are not whole numbers.

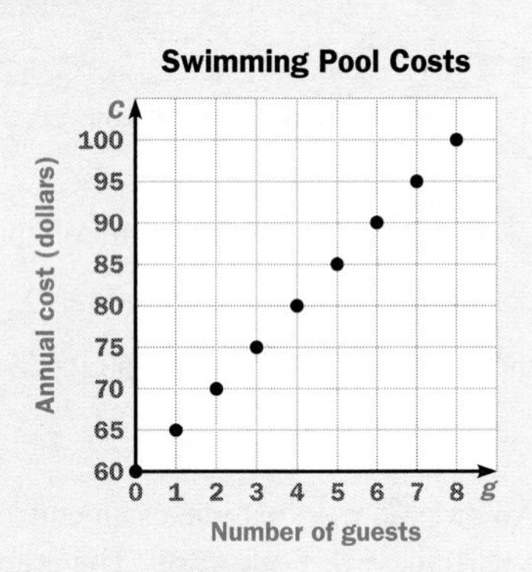

THINK AND COMMUNICATE

3. In Example 2, suppose the points were connected. Then the point (4.5, 82.5) would be on the graph. Explain why these values do not make sense for the function.

4. The cost of an individual membership at the Springfield swimming pool is a function of the number of guests the member takes. How is this function different from the one in Example 2?

Think and Communicate

3. The x-coordinates represent numbers of people and must be whole numbers.
4. The equation is different; $C = 30 + 5g$. The range is multiples of 5 greater than 25.

Checking Key Concepts

1. B; $1 = \frac{1}{2}(2)$ and $-1 = \frac{1}{2}(-2)$
2. C; $-2 = -0 - 2$ and $0 = -(-2) - 2$
3. A; $-2 = -(2)$ and $2 = -(-2)$

4. (1) an equation: $y = 3x$
(2) a table of values:

x	y
1	3
−2	−6
0.4	1.2
$-\frac{4}{3}$	−4

(3) a graph:

5. (1) an equation: $C = 8 + 0.32n$
(2) a table of values:

number of invitations sent	total cost (dollars)
0	8
1	8.32
2	8.64
...	...
15	12.80

(3) a graph:

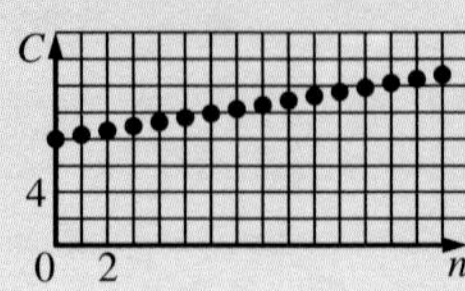

Checking Key Concepts

Match each equation with a graph. Explain how you got your answers.

1. $y = \frac{1}{2}x$
2. $y = -x - 2$
3. $y = -x$

A.

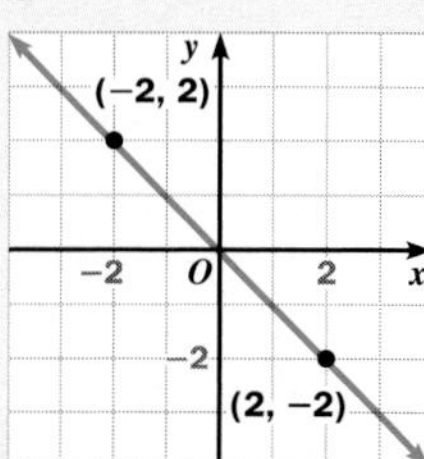

B.

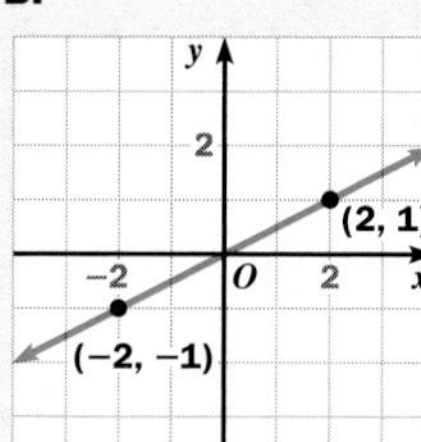

C.

(−2, 0)
(0, −2)

For Questions 4 and 5, show three different ways to represent each function.

4. Think of a number and triple it.
5. **PERSONAL FINANCE** Chandra spends \$8 for 15 graduation party invitations. It costs \$.32 to mail each invitation. Her total cost is a function of the number of invitations she mails.

2.5 Exercises and Applications

Extra Practice exercises on page 560

Graph each equation and each point. Tell whether the ordered pair is a solution of the equation.

1. $y = -3x$; $(-1, 3)$
2. $y = 2 - 2x$; $(2, 4)$
3. $y = 4x - 3$; $(-3, 0)$
4. $y = -4x + 1$; $(6, 21)$
5. $y = 4 - 5x$; $(4, 0)$
6. $y = 2x - 5$; $(3, 4)$
7. $y = -5x - 1$; $(0, -1)$
8. $y = \frac{1}{3}x$; $(3, 1)$
9. **Writing** When you graph an equation, two points are enough to draw a line. How does plotting three points help you graph the equation?
10. **ENERGY USE** Suppose a new refrigerator costs \$1000. Electricity to run the refrigerator costs about \$68 per year. The total cost of the refrigerator is a function of the number of years it is used.
    a. Write an equation for the function.
    b. Make a table of values for the function.
    c. Graph the function.

BY THE WAY

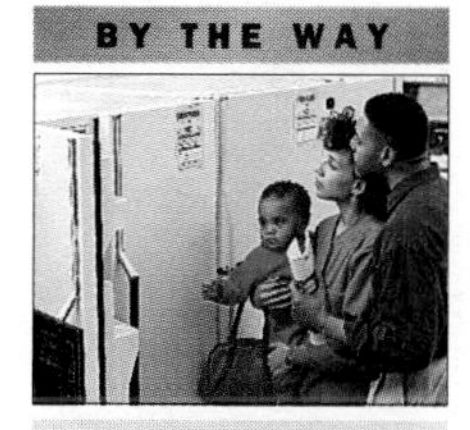

There are an estimated 110 million refrigerators in the United States. They use more than 20% of all household energy.

Teach⇔Interact

Checking Key Concepts

Common Error
To match each equation with a graph, students need to substitute the coordinates of points on the graphs into each of the three equations given. Since the equations all have the variable *y* on the left side of the equals sign, some students may erroneously substitute the *first* number in an ordered pair for *y*. Point out that this is incorrect and that an ordered pair is always (*x*, *y*) and not (*y*, *x*).

Closure Question

What four ways are used in this section to describe functions?
words, equations, tables, graphs

Apply⇔Assess

Suggested Assignment

Core Course
Exs. 1–8, 10–24, 27–33

Extended Course
Exs. 1–33

Block Schedule
Day 12 Exs. 1–33

Exercise Notes

Integrating the Strands
Exs. 1–8 The concept of the coordinate plane made possible the integration of geometry and algebra. Points could be named by ordered pairs of numbers and lines could be described by equations in two variables. If a point is on a line, then the coordinates of the point are a solution to the equation of that line. These exercises highlight this geometric interpretation of the solution of an equation.

Using Technology
Exs. 10–13 Students can use the *Function Investigator* software in their work on these exercises.

Exercises and Applications

1. Yes.

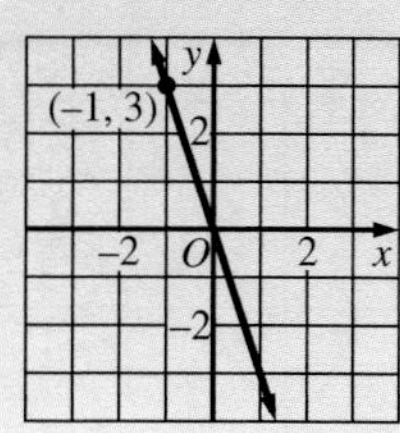

2. No.

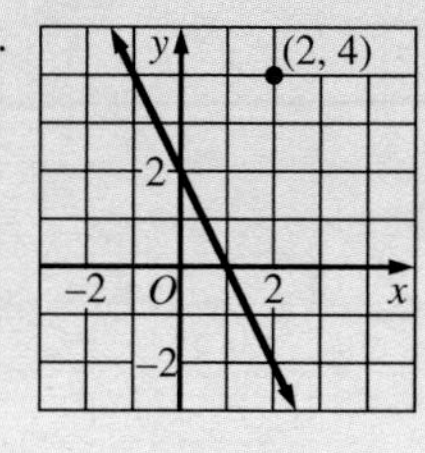

3. No.

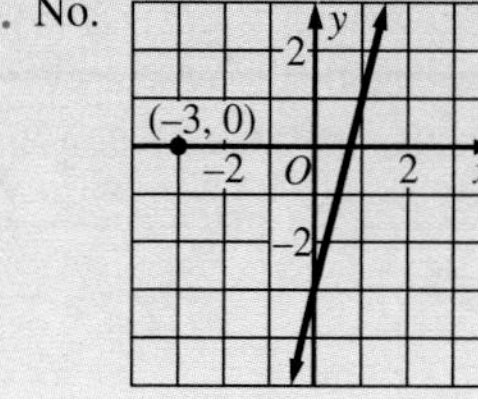

4–10. See answers in back of book.

Apply⇔Assess

Exercise Notes

Interview Note

Ex. 11 Since students' experiences with graphs of functions in this section have been with straight lines only, they may anticipate that the graph they get in part (a) should be a line. The fact that it is a curve can be used to generalize their understanding of the concept of a function. The graph can be used to reinforce the fact that a function is a relationship between input (the domain) and output (the range) and that for each input, there is exactly one output. Students should understand that many lines and curves (in fact, infinitely many) can satisfy the definition of a function.

INTERVIEW Bob Hall

Look back at the article on pages 52–54.

Bob Hall's invention of a new steering system for racing wheelchairs has made it possible for many more athletes to compete in national and international races.

11. The steering mechanism on Bob Hall's racing wheelchair is designed to allow a racer to turn the wheelchair by a maximum angle of 15°.

a. Use the table on page 54 to draw a graph showing turning radius as a function of turn angle. Connect the points on your graph to form a smooth curve.

b. Use your graph to estimate the turning radius when the turn angle is 4°.

c. Describe the domain and the range of the turning radius function.

The table and graph show the number of male wheelchair athletes in the Boston Marathon as a function of the year and the number of female wheelchair athletes as a function of the year. Use the table and graph for Exercises 12–14.

Wheelchair Athletes in the Boston Marathon, 1975–1993

Year	Males	Females
1975	1	0
1976	0	0
1977	6	1
1978	18	2
1979–1983	No official records	
1984	16	3
1985	21	3
1986	30	4
1987	37	4
1988	45	3
1989	34	8
1990	41	5
1991	62	8
1992	52	4
1993	75	9

Wheelchair Athletes in the Boston Marathon

1979–1983 No official records; • Male • Female; Number of Athletes (0–80); Year ('74–'94)

12. Describe the domain and the range of each function.

13. Official records were not kept for the years 1979–1983. Estimate the number of male wheelchair athletes in 1983. Estimate the number of female wheelchair athletes in 1983. How did you make your estimates?

14. Writing What are some of the advantages and disadvantages of seeing these functions displayed in a table? in a graph?

11. a.

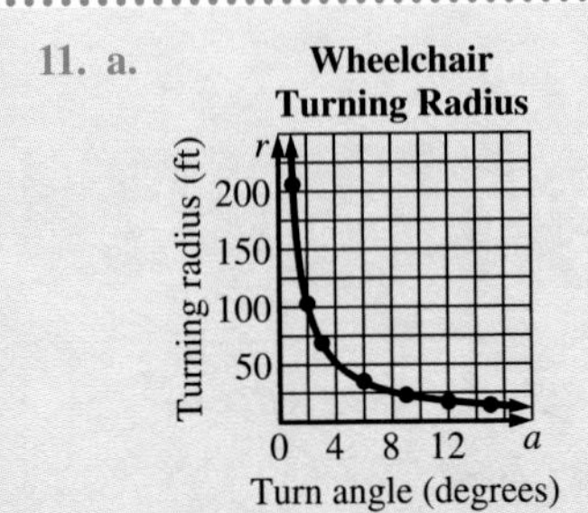

b. Estimates may vary; about 50 ft

c. The domain is the positive numbers less than or equal to 15 and the range is the positive numbers greater than 14.

12. The domain of both functions is the whole numbers from 1975 to 1978 and from 1984 to 1993. The range for the men's function is the whole numbers from 0 to 75. The range for the women's function is the whole numbers from 0 to 9.

13. Estimates and methods may vary. Examples are given. I estimate there were about 15 male wheelchair athletes and about 3 female wheelchair athletes in 1983. I estimated by using the values in the table, giving special attention to the years 1984 to 1993, and following the trends backward from 1984 to 1983.

14. Answers may vary. Examples are given. An advantage of using the table form is that you can see exact values for each year. A disadvantage is that it is not easy to analyze or compare the data when it is displayed in a table. A graph summarizes the data in a visual form that helps you analyze the data, recognize trends, and easily compare two sets of data. Some disadvantages are that you generally must estimate data values and that it may be difficult to identify individual data values quickly.

15. D; –57

16. A; –8

17. B; –28

18. C; 28

19. A; B

20. B1 = A1 + 7, B2 = A2 + 7, and so on; B1 = A1 – 7, B2 = A2 – 7, and so on.

21.

	A	B
1	–5	–24
2	–4	–23
3	–3	–22
4	–0.5	–19.5
5	0	–19
6	1	–18
7	2	–17
8	5	–14

22.

	A	B
1	–5	–16.5
2	–4	–13.5
3	–3	–10.5
4	–0.5	–3
5	0	–1.5
6	1	1.5
7	2	4.5
8	5	13.5

For Exercises 15–18, match each sentence with an equation. For each equation, find the value of y when $x = -6$.

15. Multiply any number by 10 and then add 3.

16. Multiply any number by 3 and then add 10.

17. Subtract 10 from the product of any number and 3.

18. Subtract the product of any number and 3 from 10.

A. $y = 3x + 10$

B. $y = 3x - 10$

C. $y = 10 - 3x$

D. $y = 10x + 3$

Spreadsheets You can use a spreadsheet to make a table of values. You need to enter a formula and use the fill down command. Use the spreadsheet at the right for Exercises 19 and 20.

19. The spreadsheet shows a table of values for the function *Add 7 to a number*. Which column, *A* or *B*, contains the domain values? the range values?

20. What spreadsheet formula is used to represent the function? What would a spreadsheet formula for the function $y = x - 7$ look like?

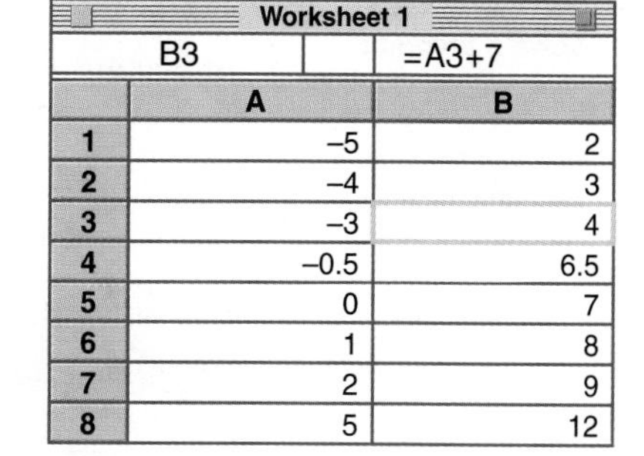

Worksheet 1

B3		=A3+7
	A	**B**
1	–5	2
2	–4	3
3	–3	4
4	–0.5	6.5
5	0	7
6	1	8
7	2	9
8	5	12

Spreadsheets Make a table of values for each function. Use a spreadsheet if you have one.

21. Subtract 19 from any number.

22. Multiply any number by 3 and then subtract 1.5 from the result.

23. Divide any positive integer by 2.

24. SAT/ACT Preview If $-3x + 3 = y$, and $y = 15$, then $x = \underline{?}$.

25. Challenge Describe the graph of each equation.

a. $y = 3$

b. $x = -8$

26. Open-ended Problem Describe a situation in which a graph would be more helpful than a table. Explain why.

ONGOING ASSESSMENT

27. Cooperative Learning Work with a partner. Each of you should make up a function that models something from everyday life, such as buying lunch for a number of friends. Exchange functions and show three ways to represent your partner's function. Check your work together.

SPIRAL REVIEW

Simplify each expression. *(Sections 1.2, 1.4, and 1.5)*

28. $\dfrac{-8(1-8)}{7}$

29. $\dfrac{5}{8} \cdot \dfrac{4}{3} + \dfrac{7}{12}$

30. $\dfrac{3}{-4} - 5 + 6(5)$

Simplify each variable expression. *(Section 1.7)*

31. $25s + 3t - 7 - 6s + t$

32. $10 - 3(2t + 8)$

33. Show that $-6(2m - 3n) + 12m$ is equivalent to $18n$.

Apply⇔Assess

Exercise Notes

Communication: Reading
Exs. 15–18 These matching exercises help students to understand how to express word sentences as algebraic equations. This is a key skill in solving problems whose conditions need to be expressed as equations.

Using Technology
Exs. 19–23 You may want to suggest that students use the spreadsheet features of the *Stats!* software for these exercises. For Exs. 21–23, students can also use a calculator or paper and pencil.

Assessment Note
Ex. 27 Each student should write their everyday life function using words. The partner then has to translate the function into an equation, table, and graph. Students can grade their functions for originality and mathematical correctness.

Practice 14 for Section 2.5

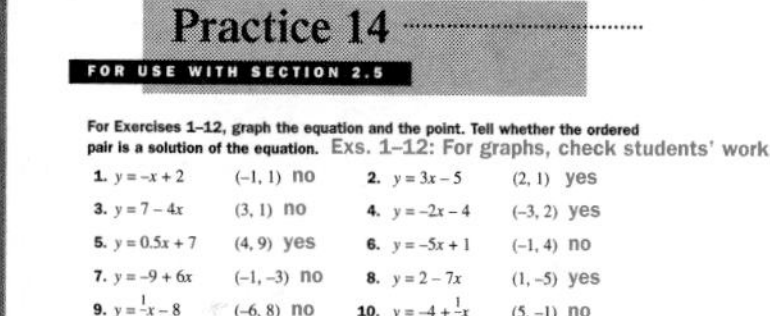

NAME ______________________ DATE __________

Practice 14

FOR USE WITH SECTION 2.5

For Exercises 1–12, graph the equation and the point. Tell whether the ordered pair is a solution of the equation. Exs. 1–12: For graphs, check students' work.

1. $y = -x + 2$ (–1, 1) no

2. $y = 3x - 5$ (2, 1) yes

3. $y = 7 - 4x$ (3, 1) no

4. $y = -2x - 4$ (–3, 2) yes

5. $y = 0.5x + 7$ (4, 9) yes

6. $y = -5x + 1$ (–1, 4) no

7. $y = -9 + 6x$ (–1, –3) no

8. $y = 2 - 7x$ (1, –5) yes

9. $y = \frac{1}{4}x - 8$ (–6, 8) no

10. $y = -4 + \frac{1}{5}x$ (5, –1) no

11. $y = -\frac{2}{3}x + 1$ (0, 1) yes

12. $y = \frac{3}{4}x - 5$ (1, 8) no

For Exercises 13–16, write each sentence as an equation. For each equation, find the value of y when $x = -4$.

13. Divide any number by 4 and then add 8. $y = \frac{x}{4} + 8$; 7

14. Multiply any number by –6 and then subtract 13. $y = -6x - 13$; 11

15. Subtract the product of any number and 7 from –2. $y = -2 - 7x$; 26

16. Add the product of any number and 3 to –10. $y = -10 + 3x$; –22

17. A major appliance repair company charges $38 for a house call and $26 per hour for time spent repairing an appliance. The cost of a repair is a function of the time spent by the repair person.

a. Write an equation for the function. $y = 38 + 26x$

b. Make a table of values for the function. Check students' work.

c. Graph the function. Check students' work.

18. The hiking club will charge each member $2.50 for going on the annual club outing. The club uses the $27 in its treasury to help pay for the outing. The amount available for the outing is a function of the number of members who go on the outing.

a. Write an equation for the function. $y = 27 + 2.5x$

b. Make a table of values for the function. Check students' work.

c. Graph the function. Check students' work.

14

Practice Bank, ALGEBRA 1: EXPLORATIONS AND APPLICATIONS

23.

	A	B
1	1	0.5
2	2	1
3	3	1.5
4	4	2
5	5	2.5
6	6	3
7	7	3.5
8	8	4
…	…	…

24. –4

25. a. a horizontal line that lies 3 units above the x-axis

b. a vertical line that lies 8 units to the left of the y-axis

26. Answers may vary. An example is given. If you were given a complicated function with a complicated graph and you needed to identify intervals for which the function was positive, a graph would be more helpful than a table because you could identify the appropriate intervals at a glance.

27. Answers may vary.

28. 8

29. $\frac{17}{12}$

30. $24\frac{1}{4}$

31. $19s + 4t - 7$

32. $-14 - 6t$

33. $-6(2m - 3n) + 12m = -12m + 18n + 12m = 18n + 0 = 18n$

Plan⇔Support

Objectives

- Use a graph to solve a problem.
- Analyze the costs and profits of running a business.

Recommended Pacing

Core and Extended Courses
Section 2.6: 1 day

Block Schedule
Section 2.6: $\frac{1}{2}$ block
(with Section 2.5)

Resource Materials

Lesson Support
Lesson Plan 2.6
Warm-Up Transparency 2.6
Overhead Visuals:
Folder A: Multi-Use Graphing Packet, Sheets 1–3
Practice Bank: Practice 15
Study Guide: Section 2.6
Explorations Lab Manual:
Diagram Masters 1, 2
Assessment Book: Test 7

Technology
Graphing Calculator
Internet:
http://www.hmco.com

Warm-Up Exercises

Solve each equation.

1. $3x + 2 = -1$ -1

2. $7 - 4x = -29$ 9

Simplify each expression.

3. $10t - 2t$ $8t$

4. $12x - (2x + 5)$ $10x - 5$

5. Graph the equation $2x + 3 = y$.

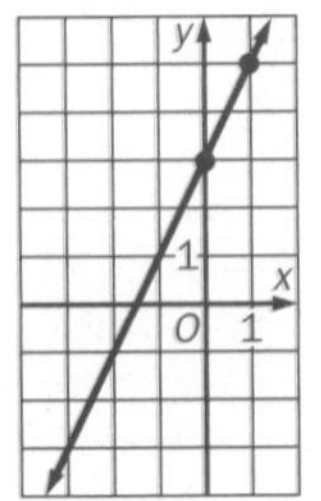

SECTION

2.6 Using Graphs to Solve Problems

Learn how to...

- **use a graph to solve a problem**

So you can...

- **analyze the costs and profits of running a business, for example**

Have you ever thought about putting your own design on a T-shirt and selling it? Graphing an equation helps you analyze the costs and profits of making and selling T-shirts.

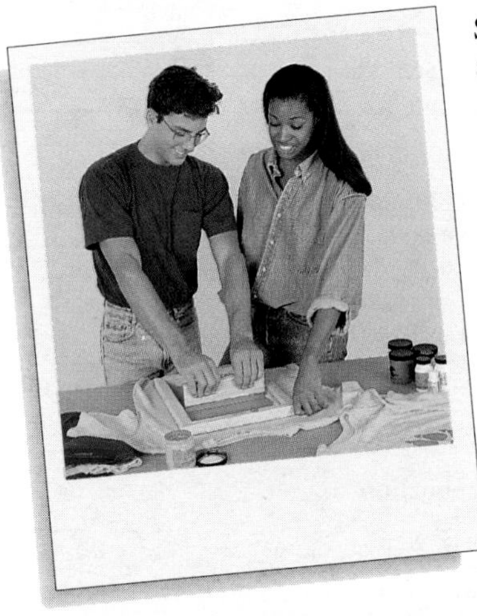

Suppose **silk-screening supplies** cost **\$70**. Plain **T-shirts** cost **\$3** each. You can write an equation for the cost of making T-shirts.

Cost = \$3 × number of shirts + \$70

$$C = 3n + 70$$

As shown below, the graph of this equation can help you analyze the costs of running your own T-shirt business.

EXAMPLE 1 Application: Business

You receive 87 orders for T-shirts. You have \$300 to spend on production costs. Decide whether you have enough money to fill all 87 orders.

SOLUTION

Graph the equation $C = 3n + 70$.

T-Shirt Production Costs

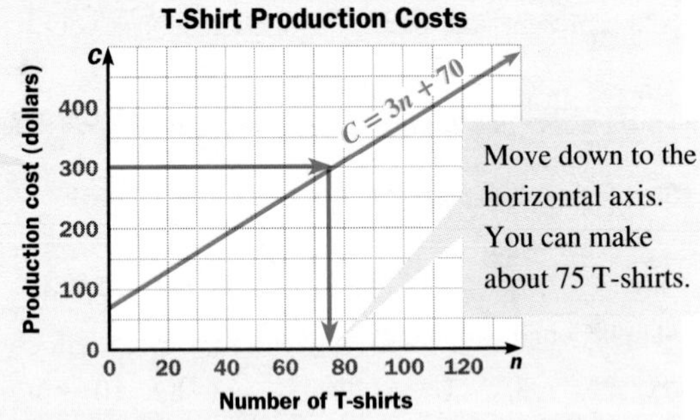

Find \$300 on the vertical axis. Move across to the graph of the equation.

Move down to the horizontal axis. You can make about 75 T-shirts.

You do not have enough money to fill 87 orders.

The graph in Example 1 helps you estimate production costs. But a graphing calculator can help you make even closer estimates, as shown in Example 2.

EXAMPLE 2 Application: Business

a. You decide to charge customers \$10 per T-shirt. Your production costs are the same as in Example 1. Write an equation for your profit P from making and selling n T-shirts.

b. What is your profit from making and selling 75 T-shirts?

SOLUTION

a. Profit = Sales Income − Production Cost

\$10 × number of shirts − \$3 × number of shirts + \$70

$P = 10n - (3n + 70)$

$P = 10n - (3n + 70)$

You can leave the equation in this form. Or you can simplify it.

$P = 10n - 3n - 70$

$P = 7n - 70$

b. **Problem Solving Strategy:** Use a graph.
First enter the equation $P = 7n - 70$ as $y = 7x - 70$.

Then set the viewing window by choosing *scales* for the x- and y-axes.

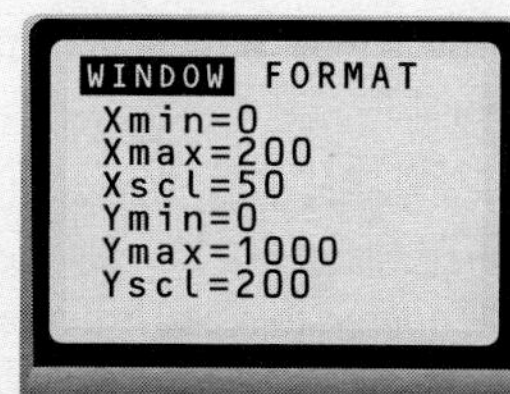

For more information about using a graphing calculator, see the *Technology Handbook*, p. 603.

Then graph the equation. Use the *trace* feature to find your profit from selling 75 T-shirts.

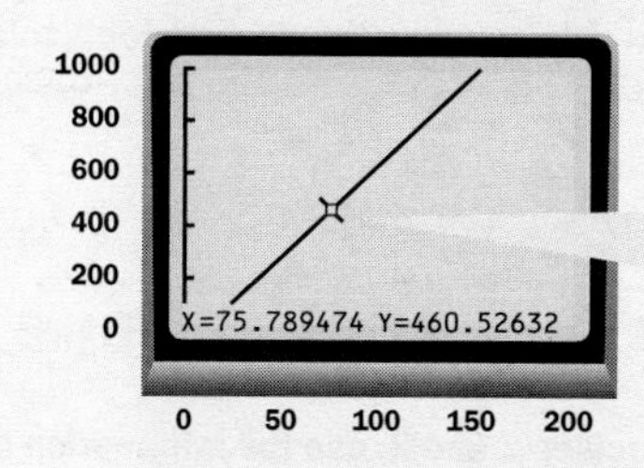

The *trace* feature shows you x- and y-values on the graph.

Use the *zoom* feature to get more accurate values for x and y.

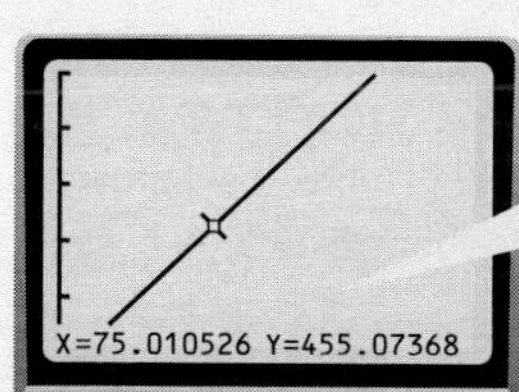

When the x-value is close to 75, the y-value is close to 455.

Your profit from selling 75 T-shirts is about \$455.

Technology Note

Example 2 demonstrates several key aspects of using graphing calculators: setting the viewing window, trace, and zoom. Show students how the range and domain of the function can help choose a viewing window. Also point out that the trace function takes the place of finding a point on the vertical axis and moving across to the graph and then down to the horizontal axis, as was shown in Example 1.

Although this Example shows graphing calculator screens, many software graphing packages work in a similar way. You can show your class this Example with an overhead projector using such software or a graphing calculator.

Teach⇔Interact

Learning Styles: Visual

Students who are visual learners usually prefer to use graphs to solve problems rather than equations. By using graphs, these students can actually see the solution, which helps them to better understand both the problem and its solution.

About Example 1

Reasoning
In this example, students see that they do not have enough money to fill 87 orders. Since a successful business would want to be able to fill as many orders as possible, explore with students how they might possibly fill all 87 orders. For example, if they could buy the plain T-shirts for less than \$3 each, then the cost of each shirt would go down and more orders could be filled. This possibility is explored in Additional Example 1.

Additional Example 1

Suppose you buy plain T-shirts for \$2.50 each and have \$300 to spend on production costs. Silk-screening supplies still cost \$70. Can you now fill an order for 87 shirts?

Graph the equation $C = 2.5n + 70$.

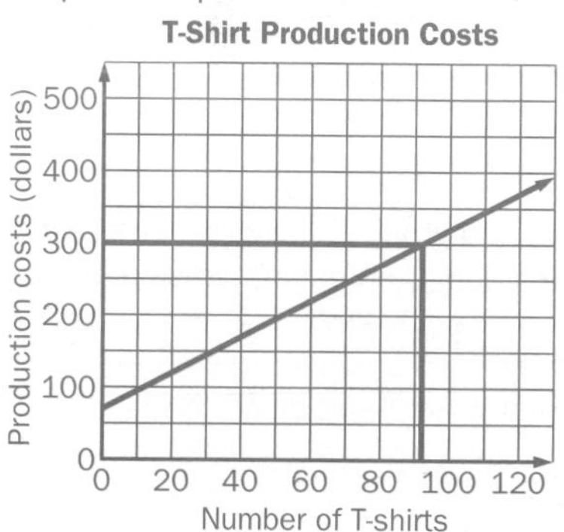

You can make 92 T-shirts, so you can fill an order for 87 shirts.

Additional Example 2

a. You charge customers \$10 per T-shirt and your production costs are the same as in Additional Example 1 ($2.5n + 70$). Write an equation for your profit P from making and selling n T-shirts.

$P = 10n - (2.5n + 70)$

$P = 7.5n - 70$

Example continued on next page.

Teach⇔Interact

Additional Example 2 (continued)

b. What is your profit from making and selling 92 T-shirts?

Use a graphing calculator. First enter the equation $P = 7.5n - 70$ as $y = 7.5x - 70$. Then set the viewing window by choosing scales for the x- and y-axes.

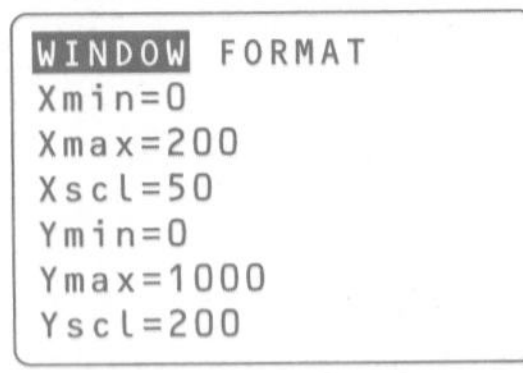

Then graph the equation. Use the trace feature to find your profit from selling 92 T-shirts.

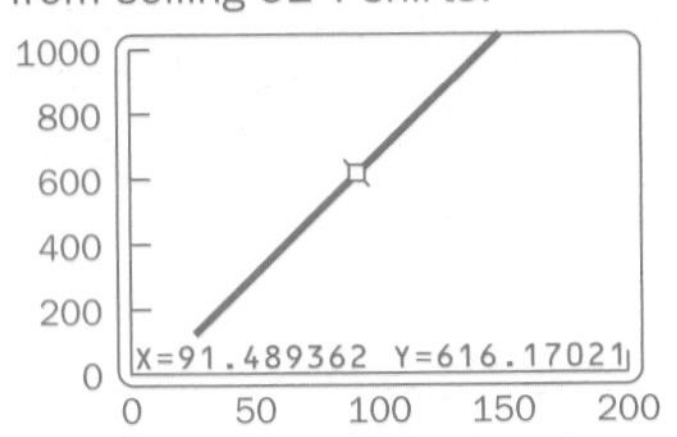

Use the zoom feature to get more accurate values for x and y.

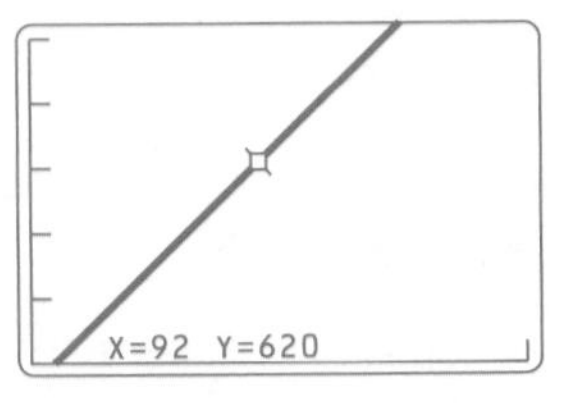

When the x-value is 92, the y-value is 620.
Your profit from selling 92 T-shirts is $620.

Closure Question

Does a graph usually give an exact solution or an estimated solution to a problem? Why?

estimated solution; It is not always possible to read exact coordinates on the x- and y-axes, which are the coordinates of the solution.

Apply⇔Assess

Suggested Assignment

Core Course
Exs. 1–8, 11–14, 16–29, AYP

Extended Course
Exs. 1–8, 11–29, AYP

Block Schedule
Day 12 Exs. 1–6, 10–14, 16–29, AYP

THINK AND COMMUNICATE

Use the information in Examples 1 and 2 on pages 84 and 85.

Graph of $y = 7x - 70$

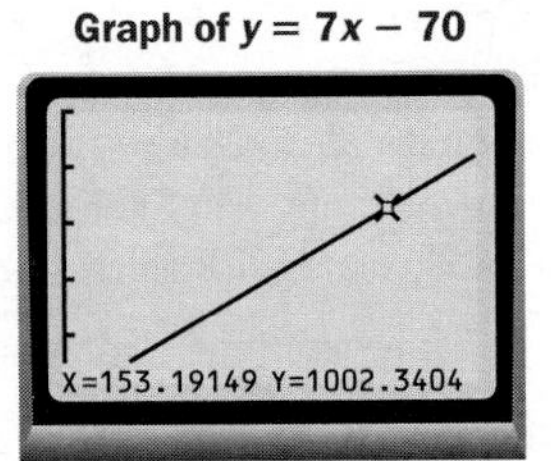

1. Suppose you want to make a profit of $1000. Use the graph shown at the right to estimate the number of T-shirts you must sell.
2. Explain how you could use the graph in Example 1 to estimate the cost of making 87 T-shirts. Why is this cost an estimate?

BY THE WAY

During 1989–1990, over 2.6 million teenagers in the United States studied Spanish as a second language. French was the second most studied language among teens, with about 1.1 million students.

CHECKING KEY CONCEPTS

Cooperative Learning Work with a partner. One of you should do part (a) of each question. The other should do part (b). Compare your work. Discuss which way of solving each problem you prefer.

1. Mandy has to read a 126-page book in Spanish. She has already read 19 pages. She estimates she can read about 9 pages per day.
   a. Write and solve an equation to find how many days it will take Mandy to finish the book.
   b. Use a graph to estimate how many days it will take Mandy to finish the book.
2. **BUSINESS** Jon Li buys sand from a building supply company. The sand costs $19 per ton. There is a $20 delivery charge. Jon needs 2 tons of sand delivered to a construction site.
   a. Use a graph to estimate Jon's total cost.
   b. Write and solve an equation to find Jon's total cost.

2.6 Exercises and Applications

Extra Practice exercises on page 561

BUSINESS For Exercises 1 and 2, use the information in Examples 1 and 2 on pages 84 and 85.

1. Suppose you charge $10 per T-shirt. How many T-shirts do you have to sell to break even? (To break even means that you sell just enough T-shirts to cover your production costs.)
2. a. Suppose you charge $15 per T-shirt. Write an equation for your profit P from selling n T-shirts.
   b. How many T-shirts do you have to sell to make $500?

ANSWERS Section 2.6

Think and Communicate

1. about 153 T-shirts; Use the zoom function. Create a "zoom box" and trace the new graph. Repeated zooming and tracing gives better estimates.
2. Find 87 on the horizontal axis. Move directly up until you reach the graph of the equation. Then move horizontally to the left until you reach the vertical axis. Estimate the C-coordinate of this point to estimate the cost (about $325). It is only an estimate because the positions of both coordinates are estimated and the vertical axis is marked off in groups of 50.

Checking Key Concepts

1. a. $19 + 9d = 126$; $d \approx 12$; about 12 days
   b.

RECREATION Exercises 3 and 4 are about Central Park in New York City. Solve at least one of the exercises by writing and graphing an equation.

3. Central Park is rectangular and has a perimeter of about 6 mi. The length of one side is about 2.5 mi. Find the lengths of the other three sides.

4. To celebrate each new year, runners participate in a "Midnight Run" through Central Park. The women's record for this 5 mi race is about 26.2 min. The men's record is about 23.3 min. Estimate the running speed in miles per minute of these record-holding runners.

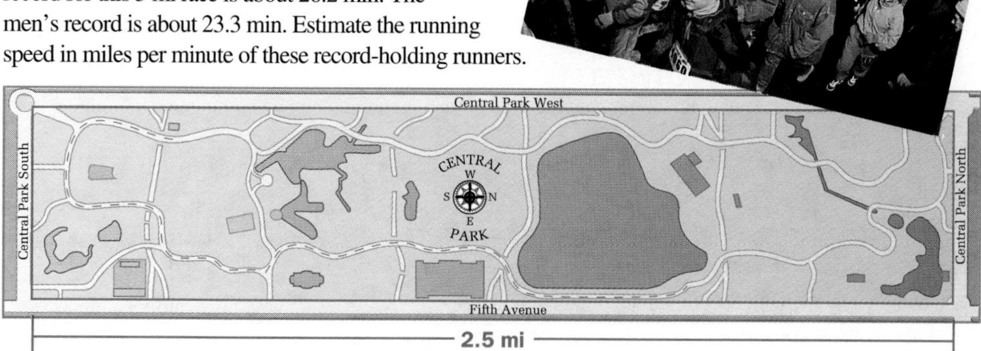

SCIENCE The graph shows the relationship between the Fahrenheit and Celsius temperature scales. Use the graph for Exercises 5–9.

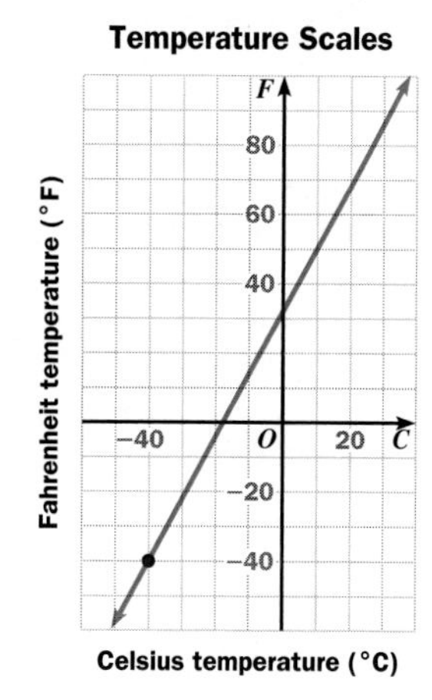

5. Estimate the temperature in degrees Fahrenheit when the temperature is:

a. 0°C **b.** 20°C **c.** −20°C **d.** −30°C

6. Estimate the temperature in degrees Celsius when the temperature is:

a. 32°F **b.** 90°F **c.** 0°F **d.** −40°F

7. You can use the formulas below to convert Celsius temperatures to Fahrenheit temperatures and Fahrenheit temperatures to Celsius. Use the formulas to find more exact answers to Exercises 5 and 6.

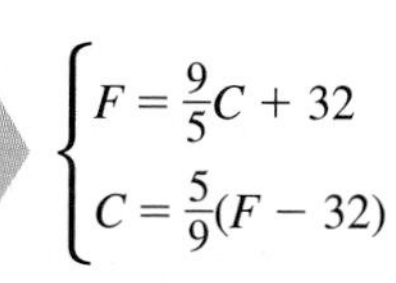

Let F = the temperature in degrees Fahrenheit.

Let C = the temperature in degrees Celsius.

$$F = \frac{9}{5}C + 32$$
$$C = \frac{5}{9}(F - 32)$$

8. Writing A friend on a trip to Africa writes that the temperature in Nairobi, Kenya, is 25°. Which temperature scale do you think your friend is using? Why?

9. Open-ended Problem A friend from a country that uses the Celsius scale wants to visit you. Write a letter to your friend describing the range of temperatures in your area. Give all temperatures in degrees Celsius.

10. The film club pays $500 to rent *Casablanca*. Members sell tickets for $3.50 each. Write and graph an equation for the club's profit P from selling n tickets. How many tickets must be sold to make a $150 profit?

Apply⇔Assess

Exercise Notes

Communication: Discussion

Exs. 1, 2 A class discussion of the business concepts involved in this section can help students to understand how they are related. Profit, for example, is affected by both cost and price. Profit can be increased by cutting costs, raising prices, or both. Profit can also be increased by selling more items, thus generating more sales income. Before reaching the break-even point, there is no profit because costs are greater than sales income.

Topic Spiraling: Review

Ex. 7 You may need to review with students how to multiply a whole number by a fraction. Also, remind students to apply the order of operations when using the two temperature formulas.

Challenge

Ex. 7 Each temperature formula expresses one variable in terms of another variable. You may wish to challenge some students to start with either formula and use what they know about inverse operations to derive the other one.

Multicultural Note

Ex. 8 Nairobi, the capital of Kenya, is a very modern city, and is the most important commercial center in eastern Africa. Like many capital cities, Nairobi has office and parliament buildings, conference centers, hotels, restaurants, and museums. Unlike other capitals, however, Nairobi has a national park, which is home to many wild animals, including lions, gazelles, rhinos, and zebras. The income from tourism generated by national parks and game reserves throughout Kenya is vital to Nairobi's economy.

---

2. a. Sand Costs (graph: Cost (dollars) vs. Number of tons; about 58)

b. $C = 20 + 19(2) = 58$; $58

Exercises and Applications

1. 10 T-shirts
2. a. $P = 12n - 70$
   b. 48 T-shirts
3. 2.5 mi, 0.5 mi, 0.5 mi
4. women's record speed: about 0.19 mi/min; men's record speed: about 0.21 mi/min

5, 6. Estimates may vary.

5. a. about 30°F
   b. about 70°F
   c. about −5°F
   d. about −20°F
6. a. about 0°C
   b. about 32°C
   c. about −18°C
   d. about −40°C
7. Ex. 5: 32°C; 68°C; −4°C; −22°C
   Ex. 6: 0°F; about 32.2°F; about −17.8°F; −40°F
8. The friend is using the Celsius scale. Kenya is near the equator, so it is generally quite hot there. In degrees Fahrenheit, 25° is below freezing.
9. Answers may vary.
10. $P = 3.5n - 500$

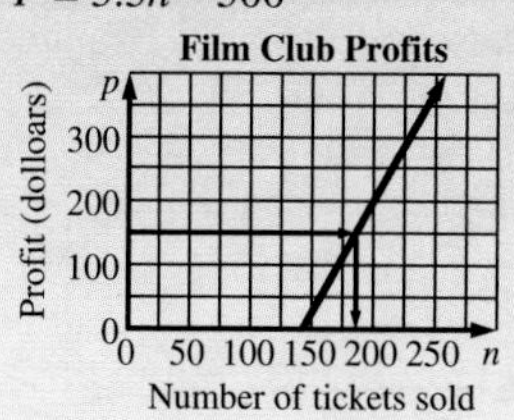

From the graph, you can estimate the club must sell about 185 tickets. The equation yields a solution of about 186.

Apply⇔Assess

Exercise Notes

Career Connection

Exs. 11–15 Students interested in animal behavior may wish to pursue a career in biology. The science of biology covers all living matter, including plant life as well as animal life. A biologist specializes in studying some aspect of biology. To become a biologist, a person would study biology in a college or university. Employment opportunities are diverse and exist in private industry, state governments, the federal government, and educational institutions. A biologist, for example, may be employed by a botanical garden, a zoo, a pharmaceutical company, an environmental agency, or a school.

Assessment Note

Ex. 23 The exchange rate for the dollar and yen changes daily. Students can find the current rate in the financial pages of most large newspapers. After completing this exercise, they may wish to use the current rate to see how much 100,000 to 150,000 yen is worth in dollars.

Connection BIOLOGY

Use the article below for Exercises 11–15.

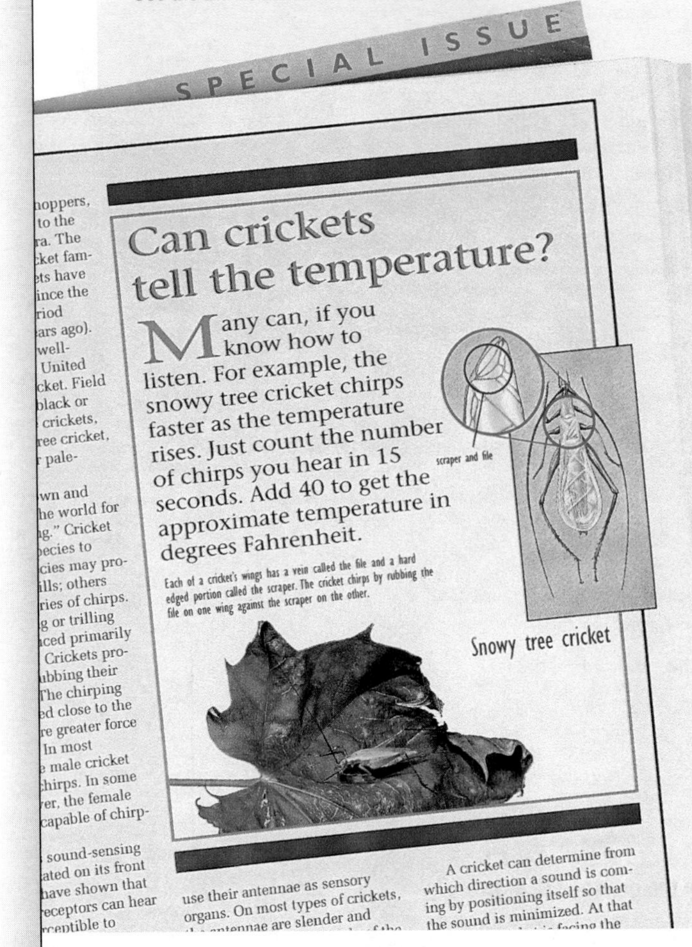
SPECIAL ISSUE

Can crickets tell the temperature?

Many can, if you know how to listen. For example, the snowy tree cricket chirps faster as the temperature rises. Just count the number of chirps you hear in 15 seconds. Add 40 to get the approximate temperature in degrees Fahrenheit.

Each of a cricket's wings has a vein called the file and a hard edged portion called the scraper. The cricket chirps by rubbing the file on one wing against the scraper on the other.

use their antennae as sensory organs. On most types of crickets,

A cricket can determine from which direction a sound is coming by positioning itself so that the sound is minimized. At that

11. Suppose you hear 23 cricket chirps in 15 seconds. What is the approximate temperature?

12. Write and graph an equation for the temperature as a function of the number of chirps you hear in 15 seconds.

13. Suppose you count the number of chirps in 15 seconds. According to your graph, what is the temperature when you hear:

a. 5 chirps **b.** 10 chirps

c. 20 chirps **d.** 25 chirps

14. How many chirps would you expect to hear in 15 seconds when the temperature is:

a. 50°F **b.** 65°F

c. 0°F **d.** 95°F

15. a. Challenge According to your graph, what is a lower limit for the temperature at which a snowy tree cricket starts chirping? Explain.

b. Do you think there is an upper limit for the number of chirps you might hear in 15 seconds? Explain your thinking.

Technology Graph each equation. Use a graphing calculator if you have one. Use the graph to estimate the unknown value to the nearest tenth.

16. $y = 2 - 0.3x$; find the value of x when $y = 0$.

17. $y = 8 - 0.4x$; find the value of x when $y = 6$.

18. $y = -5 + 2.7x$; find the value of x when $y = 1$.

19. $y = 0.5x - 7$; find the value of x when $y = 2$.

20. $y = -9.2 + 3.7x$; find the value of y when $x = 5$.

21. SAT/ACT Preview Which ordered pair is a solution of the equation $y = 3x - 4$?

A. $(-1, 1)$ **B.** $(1, -1)$ **C.** $(-4, 0)$ **D.** $\left(0, \frac{4}{3}\right)$ **E.** $\left(0, \frac{3}{4}\right)$

11. 63°F

12. $T = c + 40$, where c is the number of chirps in 15 s and T is the temperature in degrees Fahrenheit.

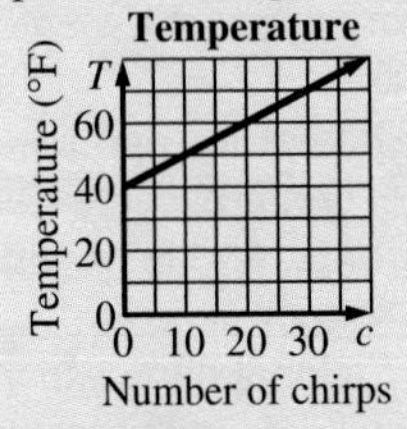

13. a. 45°F b. 50°F

c. 60°F d. 65°F

14. a. 10 chirps b. 25 chirps

c. no chirps d. 55 chirps

15. a. 40°F; This is the temperature at which $c = 0$.

b. Answers may vary. An example is given. It is likely that there are physiological limits to the possible number of chirps per second.

16–20. Answers may vary. Examples are given.

16. $x \approx 6.67$

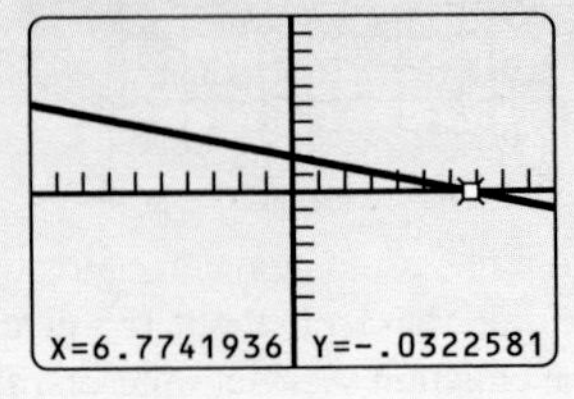

17. $x = 5$

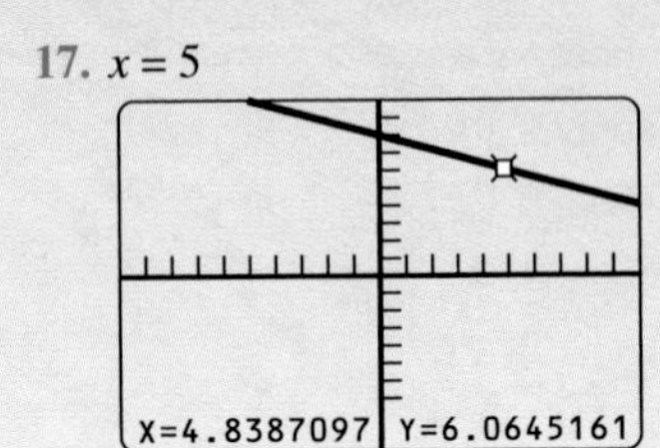

18. $x \approx 2.22$

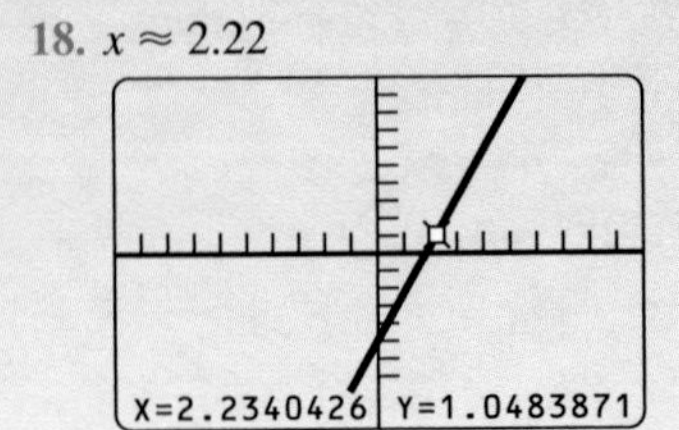

19. $x = 18$

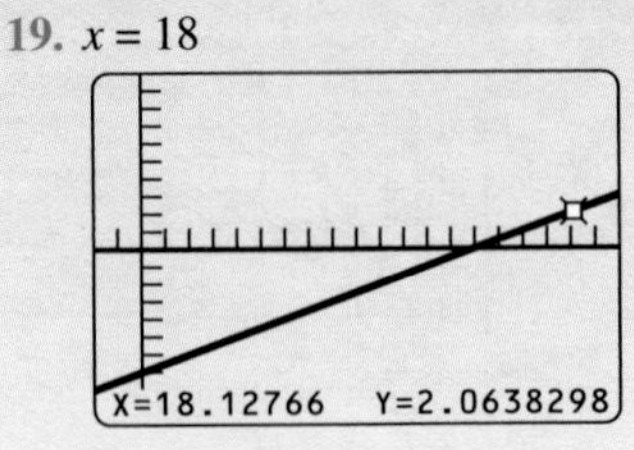

20–29. See answers in back of book.

ONGOING ASSESSMENT

22. **a.** Suppose you read that you can exchange one United States dollar for 100 Japanese yen. Make a graph that shows the number of yen you receive as a function of the number of dollars you exchange.

b. **Writing** Suppose you are going to Japan. You think you will need between 100,000 and 150,000 yen. Explain how can you use the graph from part (a) to find how many dollars you should exchange.

SPIRAL REVIEW

Use the graph. Tell whether each ordered pair is a solution of the equation $y = 6 - 2x$. *(Section 2.5)*

23. (6, 6) **24.** (3, 0) **25.** (−2, 4) **26.** (4, −2)

Graph of $y = 6 - 2x$

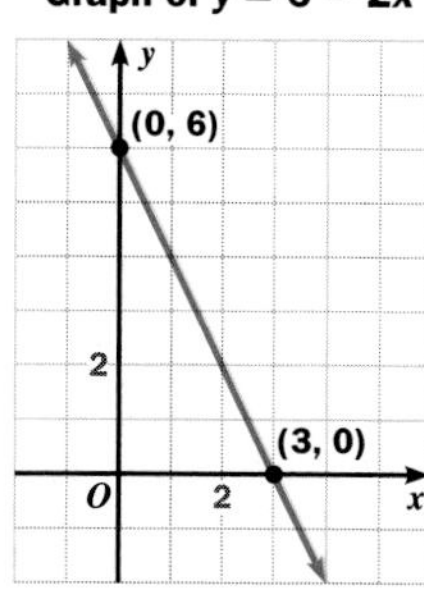

Show three ways to represent each function. *(Section 2.5)*

27. Take any number, double it, and add 3.

28. Take any number and multiply it by 5.

29. Subtract any number from 4.

ASSESS YOUR PROGRESS

VOCABULARY

horizontal axis (p. 72)	**y-coordinate** (p. 73)
vertical axis (p. 72)	**origin** (p. 73)
coordinate plane (p. 73)	**x-axis** (p. 73)
ordered pair (p. 73)	**y-axis** (p. 73)
horizontal coordinate (p. 73)	**quadrant** (p. 73)
x-coordinate (p. 73)	**solution of an equation in two variables** (p. 79)
vertical coordinate (p. 73)	

Graph each point in a coordinate plane. Name the quadrant (if any) in which the point lies. *(Section 2.4)*

1. $A(-5, 7)$ **2.** $B(0, 0)$ **3.** $C(-1, -1)$ **4.** $D(-6, 0)$

5. $E(2, -3)$ **6.** $F(0, 7)$ **7.** $G(4, 1)$ **8.** $H(-3, 3)$

Graph each equation. *(Section 2.5)*

9. $y = -4x - 1$ **10.** $y = 3x + 2$ **11.** $y = -x + 5$

12. The student council pays \$780 for a band and refreshments for a dance. They sell dance tickets for \$8 each. Write and graph an equation for their profit P from selling n tickets. About how many tickets must be sold to make a profit of \$500? *(Section 2.6)*

13. **Journal** Show four different ways to represent the same function. Which do you think is easiest to create? Which do you think is easiest to use? Explain your answers.

Apply⇔Assess

Assess Your Progress

A verbal review in class of each vocabulary term would help all students to assess their understanding of these key concepts.

Journal Entry
Students may need help to come up with a fourth way of describing a function. Often they do not think of the word description. Responses to these questions should be recorded in students' journals.

Progress Check 2.4–2.6

See page 93.

Practice 15 for Section 2.6

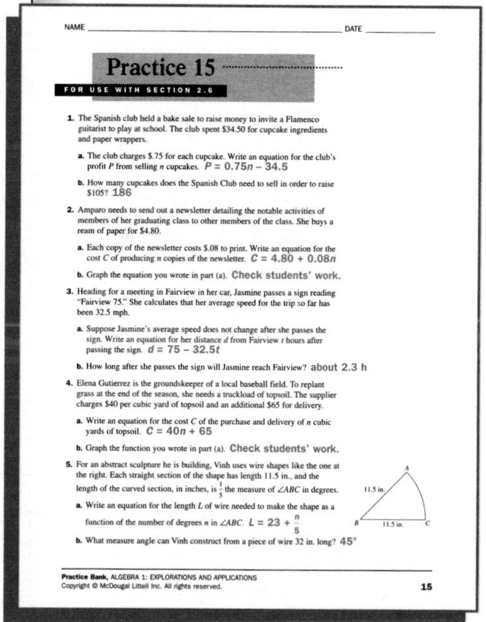

NAME ______________________ DATE __________

Practice 15

FOR USE WITH SECTION 2.6

1. The Spanish club held a bake sale to raise money to invite a Flamenco guitarist to play at school. The club spent \$34.50 for cupcake ingredients and paper wrappers.

a. The club charges \$.75 for each cupcake. Write an equation for the club's profit P from selling n cupcakes. $P = 0.75n - 34.5$

b. How many cupcakes does the Spanish Club need to sell in order to raise \$105? 186

2. Amparo needs to send out a newsletter detailing the notable activities of members of her graduating class to other members of the class. She buys a ream of paper for \$4.80.

a. Each copy of the newsletter costs \$.08 to print. Write an equation for the cost C of producing n copies of the newsletter. $C = 4.80 + 0.08n$

b. Graph the equation you wrote in part (a). Check students' work.

3. Heading for a meeting in Fairview in her car, Jasmine passes a sign reading "Fairview 75." She calculates that her average speed for the trip so far has been 32.5 mph.

a. Suppose Jasmine's average speed does not change after she passes the sign. Write an equation for her distance d from Fairview t hours after passing the sign. $d = 75 - 32.5t$

b. How long after she passes the sign will Jasmine reach Fairview? about 2.3 h

4. Elena Gutierrez is the groundskeeper of a local baseball field. To replant grass at the end of the season, she needs a truckload of topsoil. The supplier charges \$40 per cubic yard of topsoil and an additional \$65 for delivery.

a. Write an equation for the cost C of the purchase and delivery of n cubic yards of topsoil. $C = 40n + 65$

b. Graph the function you wrote in part (a). Check students' work.

5. For an abstract sculpture he is building, Vinh uses wire shapes like the one at the right. Each straight section of the shape has length 11.5 in., and the length of the curved section, in inches, is $\frac{1}{5}$ the measure of $\angle ABC$ in degrees.

a. Write an equation for the length L of wire needed to make the shape as a function of the number of degrees n in $\angle ABC$. $L = 23 + \frac{n}{5}$

b. What measure angle can Vinh construct from a piece of wire 32 in. long? 45°

(Figure: A, B, C; 11.5 in.; 11.5 in.)

Practice Bank, ALGEBRA 1: EXPLORATIONS AND APPLICATIONS

15

Assess Your Progress

1–8.

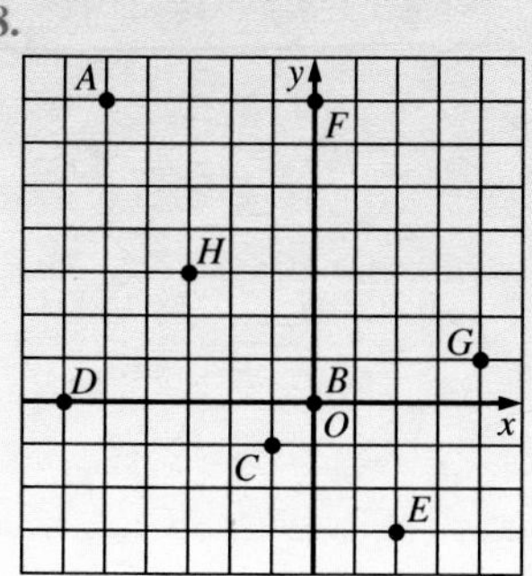

1. II	2. none
3. III	4. none
5. IV	6. none
7. I	8. II

9.

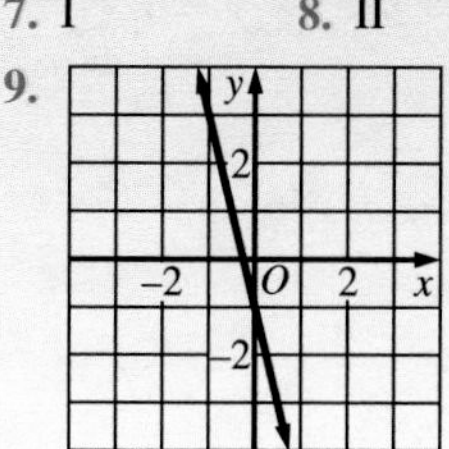

10.

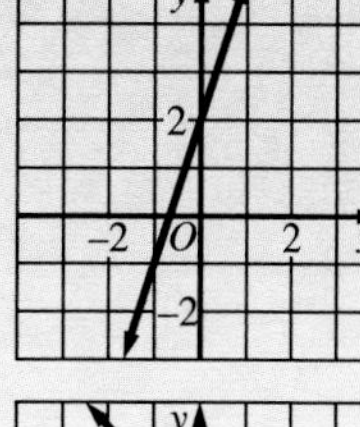

11.

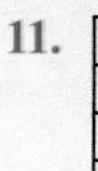

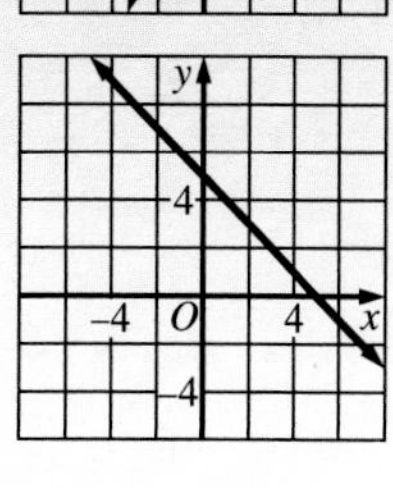

12. $P = 8n - 780$

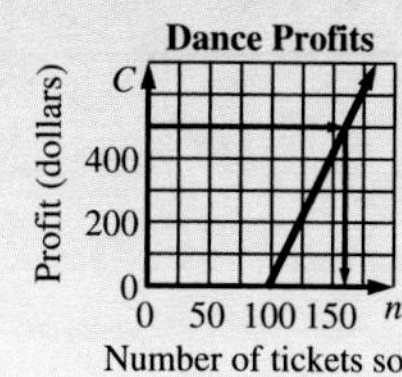

Estimates may vary; the actual number of tickets is 160.

13. Answers may vary. An example is given; words, an equation, a table of values, and a graph. Answers may vary.

Project Notes

Mathematical Goals

- Collect data that represent a function.
- Graph the data.
- Describe the domain and range of the function.

Planning

Materials

- Graph paper
- Books or magazines

Project Teams

Students should choose a partner to work with and discuss which event they would like to investigate and how they can divide up the work.

Guiding Students' Work

You may wish to begin this project by having the whole class brainstorm things that change over time. Then each pair of students can discuss further what they wish to investigate. As students choose an event to investigate, you can check with each group to see that their event is easy to measure. This will ensure that all groups get off to a good start. Students should complete all aspects of the project as a team and decide for themselves how to divide up the work.

Second-Language Learners

A peer tutor or aide might help students learning English discuss and organize their ideas for the presentation or written report.

PORTFOLIO PROJECT

Functions of Time

Many of the changes you see are examples of functions. A graph of a function can help you see a pattern in the change. For example, the graph on page 82 shows how the number of wheelchair athletes in the Boston Marathon has grown over time.

PROJECT GOAL **Collect and graph data about something that changes over time. You will look for patterns in your graph.**

Collecting Data

Work with a partner. Choose an event that occurs in the world around you. Whatever you choose should be easy to measure and should change noticeably within minutes, hours, or days.

Some possibilities:

- the height of a pedal on a moving bicycle
- the number of cars in a fast food restaurant's parking lot at different times of the day
- the temperature at each hour of the day
- the number of people going into a mall on different days or at different hours

Investigate and record what you observe. For example, suppose you decide to investigate how the height of a bicycle pedal changes over time.

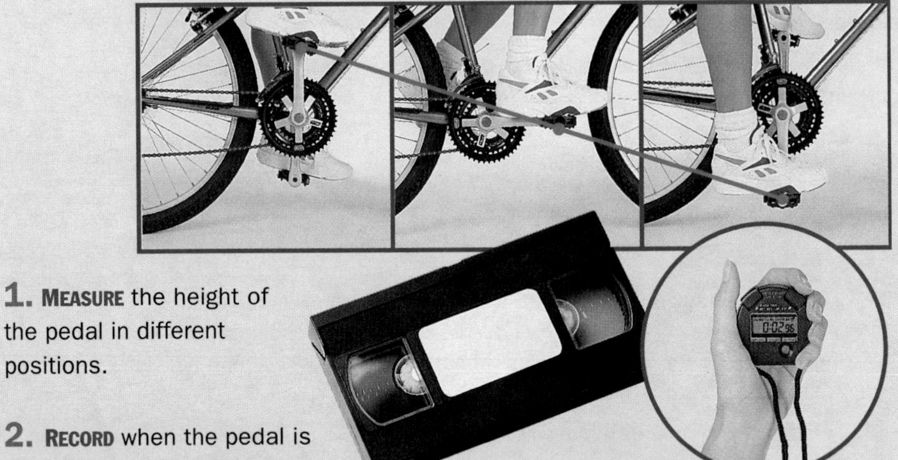

1. **MEASURE** the height of the pedal in different positions.

2. **RECORD** when the pedal is in each position. Making a video can help you see a change that happens quickly.

3. **ORGANIZE** your data in a table or spreadsheet.

Making a Graph

Use your data to make a graph of the function. Do you see a pattern? What do you think happens between the points that you plot? Does it make sense to connect the points?

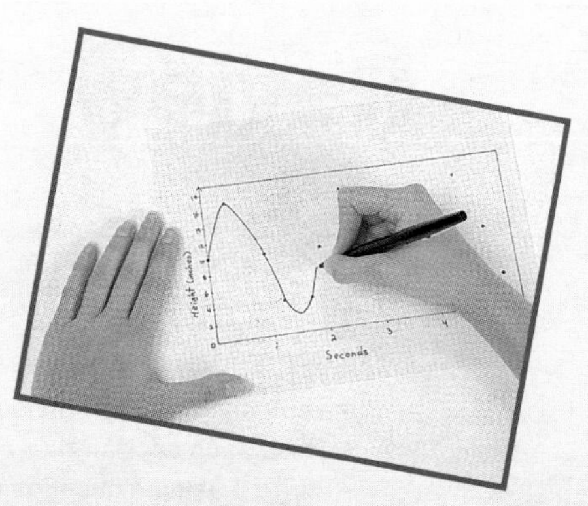

Presenting Your Ideas

Write a report, make a poster, or give an oral presentation explaining your graph. Describe what you observed and discuss any patterns you see. Describe the domain and the range of your function.

You may also want to look through books, magazines, and other sources at the library. Can you find data about functions like the one you investigated? How do the data compare with yours?

Before giving your presentation, you may want to show your classmates your graph without the title or the labels to see if they can guess what you observed.

Self-Assessment

Describe what you and your partner learned about collecting and graphing data. For example, how did you decide what to investigate? How did you decide on values to use for the axes of your graph? Did you discuss ideas together? Did you divide up some of the work? What might you do differently next time?

Project Notes

Guiding Students' Work

Rubric for Chapter Project

4 Students investigate a function and record their data accurately. They make a neatly drawn graph that correctly represents the data. Their presentation of ideas explains the graph, describes what they observed, and correctly describes the domain and range of their function. They use library sources to find other data to compare with the data they investigated. Their final report is interesting and well organized.

3 Students complete their collection of data and present the data accurately in a graph. They identify a pattern and work together to present their ideas. Some details of the final report are not entirely satisfactory. For example, the title of the graph may not be clear or the graph may be labeled incorrectly.

2 Students' data are somewhat incomplete and so is their graph. A pattern is difficult to see. A report is prepared, but it has some serious omissions and errors. For example, the domain and the range of the function are identified incorrectly. The report, poster, or oral presentation is not well organized and shows a lack of serious effort.

1 Students collect some data, but they do not exhibit a pattern. A graph is attempted but is drawn incorrectly. No attempt is made to explain the graph in a final report. Students should be encouraged to speak with the teacher as soon as possible to review their work and to make a new start on the project.

Chapter Support

Course Guide: Chapter 2
Lesson Plans: Chapter 2
Practice Bank:
Cumulative Practice 16
Study Guide: Chapter 2 Review
Assessment Book:
Chapter Tests 8 and 9
Chapter 2 Alternative Assessment
Test Generator software
Technology Book:
Calculator Based Lab 2
Portfolio Project Book
Preparing for College Entrance Tests
Professional Handbook

Progress Check 2.1–2.3

Solve each equation. *(Section 2.1)*

1. $n - 9 = 0$ 9

2. $-3x = 18$ −6

3. $y + 2 = -8$ −10

Solve each equation. *(Section 2.2)*

4. $3t - 6 = 15$ 7

5. $\frac{w}{4} + 9 = 36$ 108

6. $\frac{x}{-2} - 7 = -1$ −12

7. Eleanor earns $6 an hour for a 40-hour workweek and time and a half for overtime. Write an equation to express her weekly earnings, E, as a function of the number of hours, n, of overtime she works. *(Section 2.3)*
$E = 240 + 9n$

CHAPTER

2 Review

STUDY TECHNIQUE

Summarize what you have learned in the chapter by following these steps:

- **Look back through the sections in the chapter.**
- **Read each section title and each subtitle.**
- **Write a sample question or a short summary for each one.**

In your summary be sure to include the topics that gave you the most difficulty. Use your summary as a study sheet.

VOCABULARY

SECTIONS 2.1 *and* 2.2

To **solve** an **equation**, find the value of the variable that makes the equation true. Use **inverse operations** and keep both sides of the equation equal.

First **undo** any addition or subtraction.

$$\begin{aligned} 3x + 29 &= 8 \\ -29 &\quad -29 \\ 3x &= -21 \end{aligned}$$

Then **undo** any multiplication or division.

$$\frac{3x}{3} = \frac{-21}{3}$$

$$x = -7$$

Sections 2.3, 2.4, and 2.5

A **function** is a relationship between input and output. Each input has only one output. You can use equations, tables, and graphs to represent functions.

Dollar value V of n quarters

Equation

$V = 0.25n$

Table

domain values → **range** values

Number of quarters	Value (dollars)
0	0
1	0.25
2	0.50
3	0.75
…	…

Graph

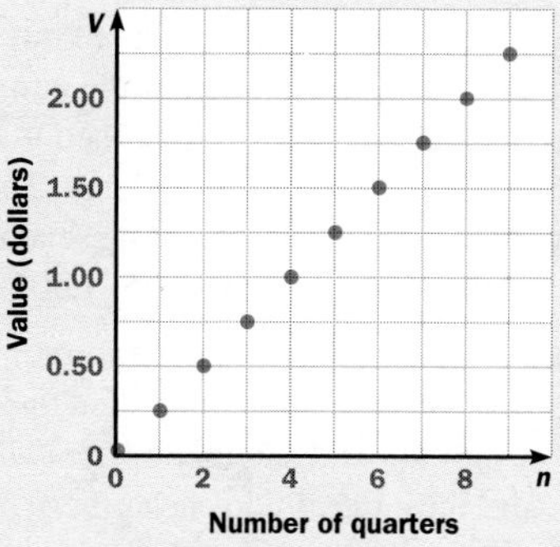

You can graph equations in a **coordinate plane**.

$y = -2x - 3$

x	y
−2	1
−1	−1
0	−3
1	−5
2	−7

Graph of $y = -2x - 3$

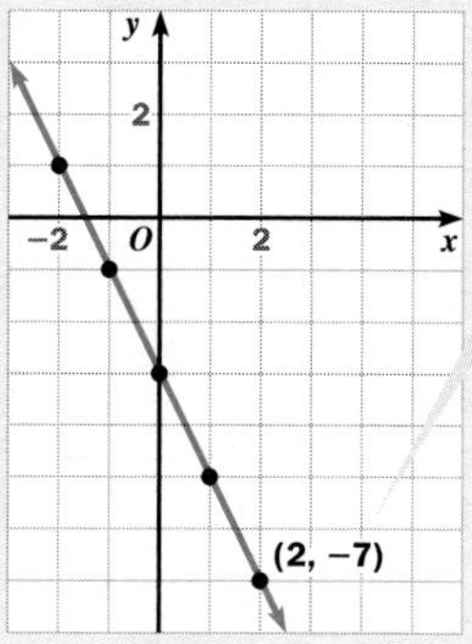

The **ordered pair** (2, −7) is a **solution** of the equation $y = -2x - 3$. The graph contains all solutions of the equation.

Section 2.6

You can use an equation or a graph to solve a problem. For example, to find how many hours h it will take you to drive 500 miles at 55 mi/h, you can solve an equation.

$$500 = 55h$$

$$9.1 \approx h$$

The graph at the right gives you the same answer.

Graph of $y = 55x$

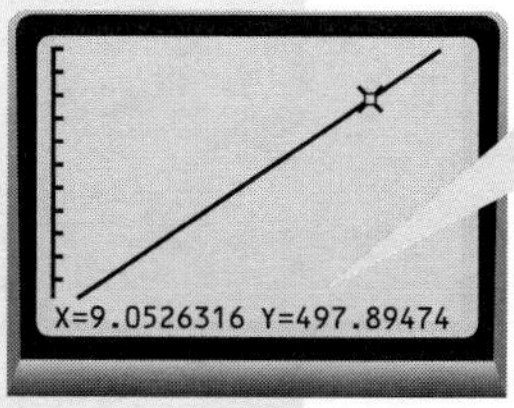

When $y \approx 500$, $x \approx 9.1$.

Progress Check 2.4–2.6

Give the coordinates of each point and name the quadrant (if any) in which the point lies. *(Section 2.4)*

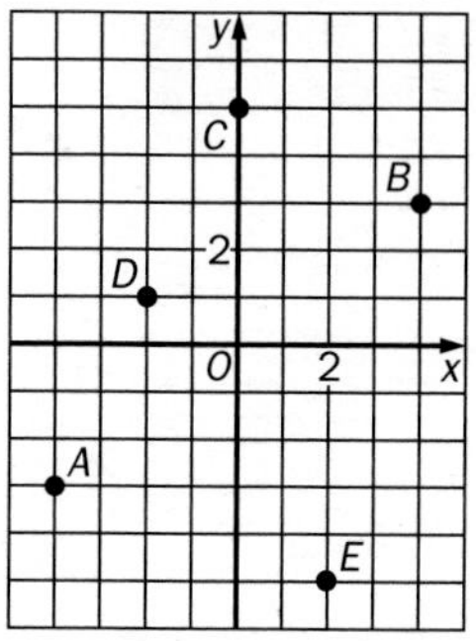

1. A (−4, −3); III
2. B (4, 3); I
3. C (0, 5); none
4. D (−2, 1); II
5. E (2, −5); IV
6. Which of these ordered pairs is a solution of the equation $y = -2x + 6$? *(Section 2.5)*
   a. (2, 10) **b.** (1, 4) **c.** (0, −6)
   (1, 4) is a solution.
7. A manufacturing company has production costs of $10,000 for a new product that it intends to sell for $80 each. Write an equation for their profit P from selling n items. How many items must be sold to break even? *(Section 2.6)*
   $P = 80n - 10{,}000$
   $0 = 80n - 10{,}000$
   $80n = 10{,}000$
   $n = 125$
   To break even, 125 items must be sold.

Chapter 2 Assessment
Form A Chapter Test

NAME ______ DATE ______ SCORE ______

Test 8

TEST ON CHAPTER 2 (FORM A)

DIRECTIONS: **Write the answers in the spaces provided.**

For Questions 1–9, solve each equation.

1. $5t + 125 = 600$
2. $3 = 15 + 12x$
3. $-z = 51$
4. $\frac{x}{2} - 8 = 2$
5. $-1 + \frac{n}{6} = -3$
6. $13 - b = 18$
7. $-p = -43$
8. $-4r + 2.6 = 10.6$
9. $-q - 10 = -20$
10. Last summer, Jack Smith and his family toured the United States for 2 weeks. They traveled a total of 5152 miles during 112 hours of driving time. Find their average speed in miles per hour.

Felicia Johnson paid $125 to join a tennis club. She pays an additional $5 every time she uses one of the club's tennis courts. Use this information for Questions 11–13.

11. Write an equation that describes Felicia's total cost for playing tennis as a function of the number of times she plays. Let C = the total cost and n = the number of times she plays.
12. Describe the domain and range of the function.
    The domain is all whole numbers and the range is 125, 130, 135, 140, ….
13. Felicia does not want to spend more than $275 to play tennis during the summer. What is the maximum number of times that she can play tennis on the club's courts for this amount?

The graph at the right shows the temperature in degrees Fahrenheit in a Minnesota city during a 24-hour period last winter. Use this information for Questions 14 and 15.

Time of day: 12 P.M., 4 P.M., 8 P.M., 12 A.M., 4 A.M., 8 A.M., 12 P.M.
Temperature (°F): 0, −4, −8, −12

14. What was the highest temperature during this 24-hour period?
15. At what hour was the temperature about −12°F?

ANSWERS
1. 95
2. −1
3. −51
4. 20
5. −12
6. −5
7. 43
8. −2
9. 10
10. 46 mi/h
11. $C = 5n + 125$
12. *See question.*
13. 30 times
14. −2°F
15. 4 A.M.

10 **Assessment Book,** ALGEBRA 1: EXPLORATIONS AND APPLICATIONS

Chapter 2 Assessment
Form B Chapter Test

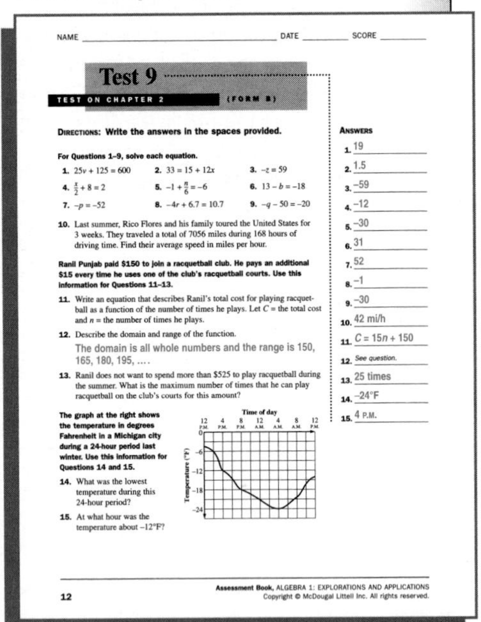
NAME ______ DATE ______ SCORE ______

Test 9

TEST ON CHAPTER 2 (FORM B)

DIRECTIONS: **Write the answers in the spaces provided.**

For Questions 1–9, solve each equation.

1. $25v + 125 = 600$
2. $33 = 15 + 12x$
3. $-z = 59$
4. $\frac{x}{2} + 8 = 2$
5. $-1 + \frac{n}{6} = -6$
6. $13 - b = -18$
7. $-p = -52$
8. $-4r + 6.7 = 10.7$
9. $-q - 50 = -20$
10. Last summer, Rico Flores and his family toured the United States for 3 weeks. They traveled a total of 7056 miles during 168 hours of driving time. Find their average speed in miles per hour.

Ranil Punjab paid $150 to join a racquetball club. He pays an additional $15 every time he uses one of the club's racquetball courts. Use this information for Questions 11–13.

11. Write an equation that describes Ranil's total cost for playing racquetball as a function of the number of times he plays. Let C = the total cost and n = the number of times he plays.
12. Describe the domain and range of the function.
    The domain is all whole numbers and the range is 150, 165, 180, 195, ….
13. Ranil does not want to spend more than $525 to play racquetball during the summer. What is the maximum number of times that he can play racquetball on the club's courts for this amount?

The graph at the right shows the temperature in degrees Fahrenheit in a Michigan city during a 24-hour period last winter. Use this information for Questions 14 and 15.

Time of day: 12 P.M., 4 P.M., 8 P.M., 12 A.M., 4 A.M., 8 A.M., 12 P.M.
Temperature (°F): 0, −6, −12, −18, −24

14. What was the lowest temperature during this 24-hour period?
15. At what hour was the temperature about −12°F?

ANSWERS
1. 19
2. 1.5
3. −59
4. −12
5. −30
6. 31
7. 52
8. −1
9. −30
10. 42 mi/h
11. $C = 15n + 150$
12. *See question.*
13. 25 times
14. −24°F
15. 4 P.M.

12 **Assessment Book,** ALGEBRA 1: EXPLORATIONS AND APPLICATIONS

CHAPTER 2 Assessment

VOCABULARY QUESTIONS

For Questions 1 and 2, complete each paragraph.

1. The graph shows the cost of buying pretzels as a function of the number of pretzels. The _?_ of the function consists of the numbers 0, 1, 2, 3, … . The _?_ of the function consists of these amounts in dollars and cents: 0, 0.75, 1.50, 2.25, … .

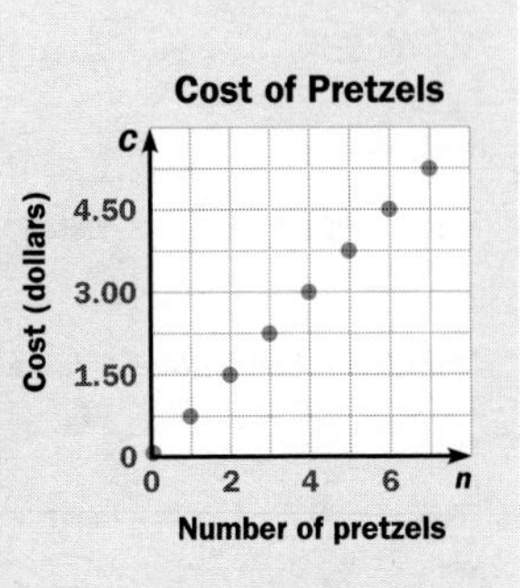

2. When you solve an equation, first undo any addition or subtraction, then undo any _?_. Addition and subtraction are called _?_ because they undo each other.

SECTIONS 2.1 *and* 2.2

Solve each equation.

3. $6t + 122 = 278$ **4.** $3 = 18 + 16w$ **5.** $-x = 23$

6. $-w = -32$ **7.** $11 - m = 15$ **8.** $-h - 21 = -26$

9. $\frac{x}{13} - 7 = 1$ **10.** $-4 + \frac{n}{3} = 14$ **11.** $-3p + 1.8 = 10.8$

12. **HISTORY** On February 20, 1962, John Glenn became the first person from the United States to orbit Earth. He traveled 80,966 mi in about 4.92 h. Find his average speed in miles per hour.

SECTIONS 2.3, 2.4, *and* 2.5

13. a. **PERSONAL FINANCE** Sequina Jackson pays $50 to join a golf club. She pays an additional $12 every time she plays golf. Write an equation that describes her total cost for playing golf as a function of the number of times she plays.

b. Describe the domain and the range of the function.

c. Sequina does not want to spend more than $250 to play golf during the summer. Find the maximum number of times she can play for this amount.

ANSWERS Chapter 2

Assessment

1. domain; range
2. multiplication or division; inverse operations
3. 26
4. $-\frac{5}{16}$
5. −23
6. 32
7. −4
8. 5
9. 104
10. 54
11. −3
12. about 16,456.5 mi/h
13. a. $C = 50 + 12n$, where n is the number of times she plays golf and C is the total cost in dollars
    b. The domain is 0, 1, 2, 3, … and the range is 50, 62, 74, 86, … .
    c. 16 times
14. a. Estimates may vary. highest temperature: about −30°C after 15 h; lowest temperature: about −85°C after 5 h

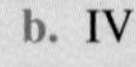

b. IV

15–18.

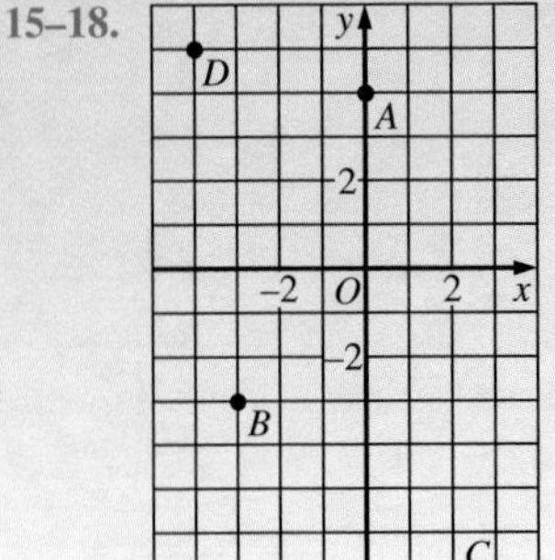

15. none
16. III
17. IV
18. II
19.

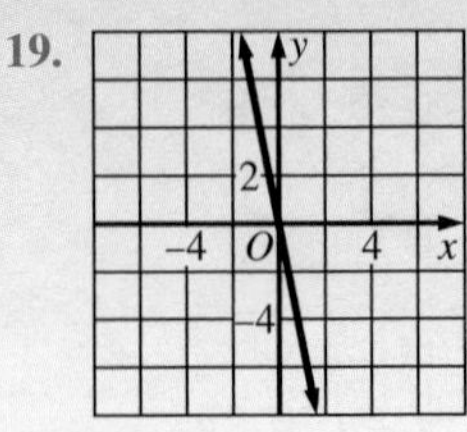

14. ASTRONOMY The graph shows the temperature on Mars in degrees Celsius during a 24-hour period.

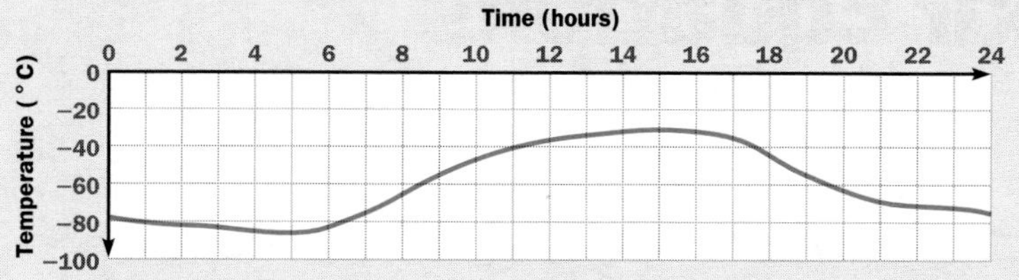

a. Estimate the highest and lowest temperatures during this 24-hour period. At what recorded hours did these temperatures occur?

b. In which quadrant of the coordinate plane does the graph lie?

Graph each point. Name the quadrant (if any) in which the point lies.

15. $A(0, 4)$ **16.** $B(-3, -3)$ **17.** $C(2, -7)$ **18.** $D(-4, 5)$

Graph each equation in a coordinate plane.

19. $y = -5x$ **20.** $y = 5 - 2x$ **21.** $y = 4x - 3$

22. $y = -x - 1$ **23.** $y = \frac{x}{2} + 3$ **24.** $y = 2.5 + x$

25. For which equation in Exercises 19–24 is $(-1, -7)$ a solution?

26. Open-ended Problem Describe the steps you took to graph the equation in Exercise 24. Describe another way to graph the equation.

SECTION 2.6

27. a. CONSUMER ECONOMICS Ted and Elaine Kemp are planning an anniversary party. The disc jockey costs \$450. The caterer charges \$20 per person. Write an equation describing the Kemps' cost for the disc jockey and catering as a function of the number of people at the party.

b. Graph the equation.

c. The Kemps do not want to spend more than \$2000 for a disc jockey and catering. At most, how many people can attend the party?

d. Writing How did you solve part (c)? Describe one other way to solve the problem.

PERFORMANCE TASK

28. Make a poster of a function. Represent your function in at least three ways. Be sure to describe the domain and the range of your function.

Chapter 2 Assessment
Form C Alternative Assessment

Chapter 2
ALTERNATIVE ASSESSMENT

1. **Project** Plumber A charges \$30 per hour plus a \$20 service fee per job. Plumber B charges \$25 per hour plus a \$40 service fee per job. Use tables and graphs to compare the plumbers' rates. Assuming the quality of their work is similar, which plumber's rates are better? Explain your answer.

2. **Open-ended Problem** Use the graph below to describe each student's school performance. Explain the relationship between attendance and grades for each student.

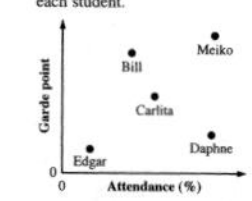

3. **Open-ended Problem** What are some questions that could be asked based on the graph shown below? What are the answers to those questions?

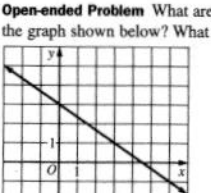

4. **Performance Task** Find an example of a function in your life. Describe the function. Describe the domain and range. Collect some data from your function. Represent your data in a graph or table. Write an equation that can be used to describe your function.

5. Mathematicians call the variable that represents the values in the domain of a function the *independent* variable and the variable that represents the values in the range of a function the *dependent* variable. Choose a real-life example of a function that is modeled by an equation from this chapter. Use the function to discuss why the terms *independent variable* and *dependent variable* might be used.

6. **Group Activity** There is a formula in physics called Hooke's Law. This law defines the relationship between the mass of an object and the amount that a spring will stretch when the object is suspended from it. To study this relationship, follow the directions below.

   a. Gather the following materials: 15 to 20 pennies, paper cup with a wire handle, half of a Slinky™, 2 meter sticks.

   b. Suspend one meter stick across two chairs.

   c. Form a hook on one end of the Slinky™.

   d. Attach the Slinky™ to the meter stick.

   e. Attach the cup to the other end of the Slinky™.

   f. Measure and record the length of the Slinky™.

   g. Place one or more pennies in the cup. Measure and record the new length of the Slinky™.

   h. Repeat this procedure at least 10 times using a different number of pennies for each trial. Record the number of pennies and the length of the Slinky™ for each trial.

   i. Graph your results. What is your domain? What is your range? Justify your choices for domain and range. What other information can you get from your graph?

   j. Repeat the experiment, measuring the distance from the bottom of the cup to the floor. How does this change affect the graph?

   k. If you repeated the experiment with another mass (like dried beans), how do you think the graph would change?

7. **Project** Create a comic strip about solving equations. In your comic strip, give students at least one tip to help them avoid making common mistakes when solving an equation.

8. **Open-ended Problem** Make up a story to go with the function graph shown below.

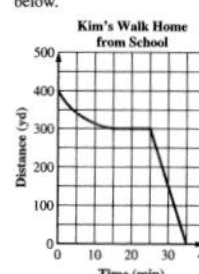

20.
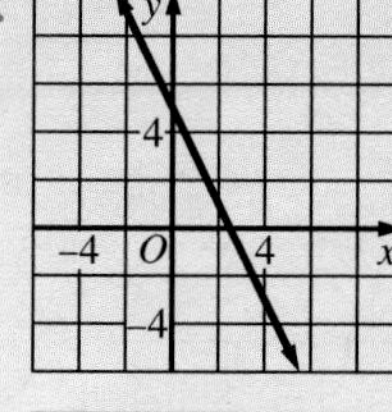

22.
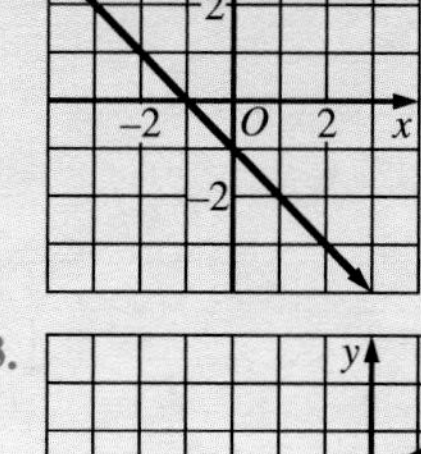

24.
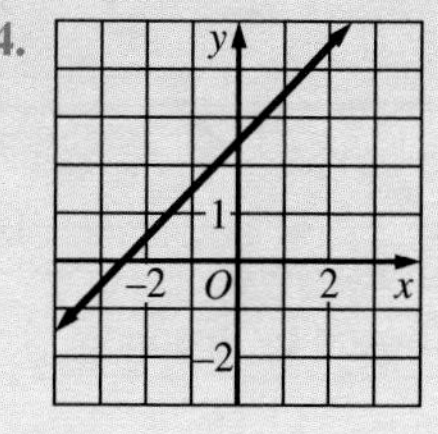

21.
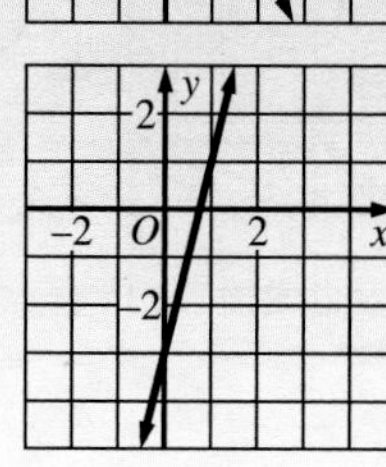

23.

25. the equation $y = 4x - 3$ in Ex. 21

26. Answers may vary. One way to graph the equation is to make a table of values with at least three ordered pairs, plot the points, and draw a line through them. Another way is to use a graphing calculator.

27. See answers in back of book.

28. Answers may vary. Check students' work.

CHAPTER

3 Graphing Linear Equations

OVERVIEW

Connecting to Prior and Future Learning

⇔ The concepts of rates and ratios are reviewed briefly, applied to real-world situations, and then used to introduce direct variations. A more detailed review of rates and ratios, as well as a review of systems of measurement, can be found in the **Student Resources Toolbox** on pages 586 and 592, respectively. Ratios are presented again in Chapter 5.

⇔ Students' work with the coordinate plane continues as they study the slope of a line, the slope-intercept form of a linear equation, and writing the equation of a line with given characteristics. These concepts are basic to algebra and mathematics in general and will be used throughout this and future courses.

⇔ Chapter 3 ends with a section on using mathematical models to describe real-world data. The model created by finding a line of fit is emphasized and will be used in future studies in mathematics.

Chapter Highlights

Interview with Mae Jemison: The relationship between mathematics and space science is emphasized in this interview with Dr. Mae Jemison, a science mission specialist on the space shuttle *Endeavour.* This application is emphasized throughout the chapter in the related examples and exercises on pages 99, 100, 104, and 110.

Explorations focus on studying patterns and direct variation using a meterstick and graph paper or a graphing calculator in Section 3.2, exploring linear equations and graphs using a graphing calculator and graph paper in Section 3.4, and creating a linear model in Section 3.6.

The Portfolio Project: Students use the coordinate plane and linear equations to help them create linear puzzles. Groups of students use the puzzles they have created to make a puzzle book.

Technology: Graphing calculators are used throughout this chapter. In Section 3.2, students use graphing calculators to explore patterns and direct variations. Students continue their study of linear equations and graphing begun in Chapter 2 by using graphing calculators to explore these concepts in Section 3.4. In Section 3.6, graphing calculators are used to find a line of fit.

OBJECTIVES

Section	Objectives	NCTM Standards
3.1	• Find unit rates from words and graphs. • Compare real-world rates.	1, 2, 3, 4, 5
3.2	• Recognize and describe direct variation. • Explore relationships between real-world variables.	1, 2, 3, 4, 5, 12
3.3	• Find the slope of a line. • Analyze a real-world graph.	1, 2, 3, 4, 5, 8
3.4	• Write linear equations in slope-intercept form. • Analyze real-world problems represented by graphs.	1, 2, 3, 4, 5, 8
3.5	• Find the *y*-intercept of a graph. • Model real-world situations.	1, 2, 3, 4, 5, 8
3.6	• Fit a line to data. • Make predictions about real-world situations.	1, 2, 3, 4, 5, 8, 10, 12

INTEGRATION

Mathematical Connections	3.1	3.2	3.3	3.4	3.5	3.6
algebra	**99–105***	**106–112**	**113–118**	**119–124**	**125–129**	**130–135**
geometry	100, 103–105	106, 108–112	**113–118**	**119–124**	125–128	131–135
data analysis, probability, discrete math	103–105	**106–112**	114, 116–118	120, 123, 124	125, 126, 128, 129	**130–135**
patterns and functions		**106–112**	118			
logic and reasoning	99, 101–105	106, 108–112	113, 114, 116–118	**119–124**	126, 128, 129	131–134
Interdisciplinary Connections and Applications						
history and geography				123	128	
biology and earth science	105					130
chemistry and physics		109, 110				
business and economics		107, 108, 111				
arts and entertainment	103	111				
sports and recreation	102, 103				125	
aviation, astronomy, energy use, civil engineering, architecture	99–102, 104		114, 116, 117			

* ***Bold page numbers*** *indicate that a topic is used throughout the section.*

TECHNOLOGY

	opportunities for use with	
Section	**Student Book**	**Support Material**
3.1	McDougal Littell Software *Stats!* *Function Investigator*	**Technology Book:** Spreadsheet Activity 3
3.2	graphing calculator McDougal Littell Software *Stats!*	**Technology Book:** Spreadsheet Activity 3
3.3	graphing calculator	**Technology Book:** Spreadsheet Activity 3
3.4	graphing calculator	
3.5	scientific calculator	**Technology Book:** Calculator Activity 3
3.6	graphing calculator McDougal Littell Software *Stats!*	

PLANNING GUIDE

Regular Scheduling (45 min)

Section	Materials Needed	Core Assignment	Extended Assignment	*exercises that feature* Applications	Communication	Technology
3.1		1–11, 13–17, 23–25, 27–36	1–8, 11–36	11–17, 19–25	11, 18, 22, 26, 28	
3.2	meterstick, graph paper, graphing calculator	**Day 1:** 1–5, 7–12 **Day 2:** 13–21, AYP*	**Day 1:** 1–12 **Day 2:** 13–21, AYP	7–10 13, 14	6, 8, 11, 12 16	
3.3	graph paper	1–19, 22, 24, 27–36	1–14, 19–36	20–22, 24	19, 21, 23, 25, 26, 29	
3.4	graph paper, graphing calculator	**Day 1:** 1–17 **Day 2:** 18–23, 25–34, AYP	**Day 1:** 1–17 **Day 2:** 18–34, AYP	20–23	16, 17 18, 19, 24, 25	
3.5	graph paper	**Day 1:** 1–15 **Day 2:** 16–25, 28–34	**Day 1:** 1–15 **Day 2:** 16–34	10 22–26, 33, 34	11 27, 28	
3.6	paper cup, paper clips, 30 marbles, 25 pieces of spaghetti, 2 chairs, graph paper, graphing calculator	1, 2, 4–19, AYP	1–19, AYP	4–6, 17–19	3, 6, 7, 10	8
Review/ Assess		**Day 1:** 1–11 **Day 2:** 12–24 **Day 3:** Ch. 3 Test	**Day 1:** 1–11 **Day 2:** 12–24 **Day 3:** Ch. 3 Test	8	10	
Portfolio Project		Allow 2 days.	Allow 2 days.			

Yearly Pacing (with Portfolio Project)	**Chapter 3 Total** 14 days	**Chapters 1–3 Total** 42 days	**Remaining** 118 days	**Total** 160 days

Block Scheduling (90 min)

	Day 15	Day 16	Day 17	Day 18	Day 19	Day 20	Day 21	Day 22
Teach/Interact	Ch. 2 Test 3.1	3.2: Exploration, page 106	3.3 3.4: Exploration, page 119	Continue with 3.4 3.5	Continue with 3.5 3.6: Exploration, page 132	Review Port. Proj.	Review Port. Proj.	Ch. 3 Test 4.1: Exploration, page 147
Apply/Assess	**Ch. 2 Test** **3.1:** 1–11, 13–16, 23–25, 27–36	**3.2:** 1–5, 9–21, AYP*	**3.3:** 1–19, 22, 24, 27–36 **3.4:** 1–17	**3.4:** 18–23, 25–34, AYP **3.5:** 1–15	**3.5:** 16–34 **3.6:** 1–19, AYP	**Review:** 1–11 **Port. Proj.**	**Review:** 12–24 **Port. Proj.**	**Ch. 3 Test** **4.1:** 1–11, 13, 15–23

NOTE: A one-day block has been added for the Portfolio Project—timing and placement to be determined by teacher.

Yearly Pacing (with Portfolio Project)	**Chapter 3 Total** 7 days	**Chapters 1–3 Total** $21\frac{1}{2}$ days	**Remaining** $59\frac{1}{2}$ days	**Total** 81 days

* ***AYP*** *is Assess Your Progress.*

Lesson Support

Section	Practice Bank	Study Guide*	Assessment Book*	Visuals	Explorations Lab Manual	Lesson Plans	Technology Book
3.1	17	3.1		Warm-Up 3.1		3.1	Spreadsheet Act. 3
3.2	18	3.2	Test 10	Warm-Up 3.2 Folder A	Masters 1, 7 Add. Expl. 4	3.2	Spreadsheet Act. 3
3.3	19	3.3		Warm-Up 3.3 Folder A	Master 2	3.3	Spreadsheet Act. 3
3.4	20	3.4	Test 11	Warm-Up 3.4 Folders 3, A	Master 2	3.4	
3.5	21	3.5		Warm-Up 3.5 Folder A	Masters 1, 2	3.5	Calculator Act. 3
3.6	22	3.6	Test 12	Warm-Up 3.6 Folder A	Masters 1, 2, 8	3.6	
Review Test	23	Chapter Review	Tests 13, 14 Alternative Assessment			Review Test	Calculator Based Lab 3

*__Spanish versions__ of *Study Guide* and *Assessment Book* are available.

Chapter Support

- Course Guide
- Lesson Plans
- Portfolio Project Book:
  Additional Project 3: Slope
- Preparing for College Entrance Tests
- Multi-Language Glossary
- *Test Generator* Software
- Professional Handbook

Software Support

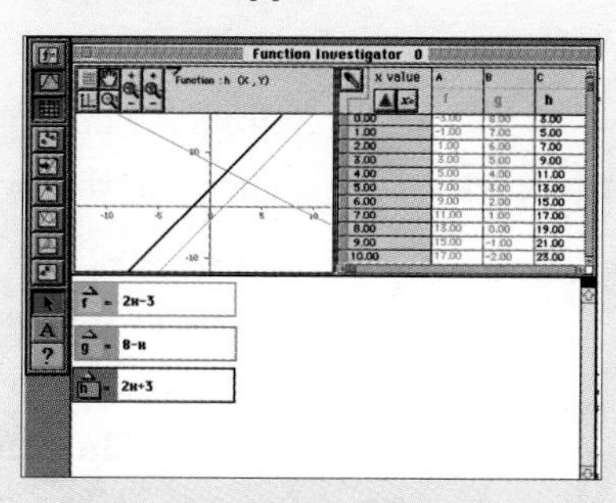

McDougal Littell Software
Stats!
Function Investigator

Internet Support

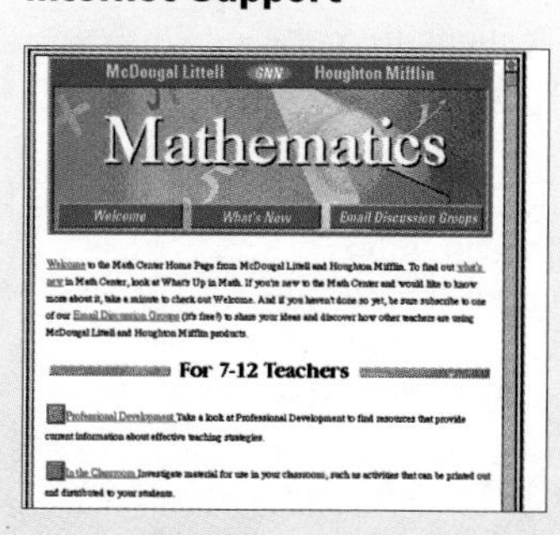

http://www.hmco.com
Next go to McDougal Littell; then the Education Center; then Secondary Math.

Outside Resources

Books, Periodicals

Illingworth, Mark. *A Graphing Matter.* Real-world applications of variables and relationships. Hayward, CA: Activity Resources Co., Inc.

Van Dyke, Frances. "Relating to Graphs in Introductory Algebra." *Mathematics Teacher* (September 1994): pp. 427–432, 438, 439.

Activities, Manipulatives

Algebra Lab Gear. Manipulatives for algebra concepts. Creative Publications, 1990.

Andersen, Edwin D. and Jim Nelson. "An Introduction to the Concept of Slope." *Mathematics Teacher* (January 1994): pp. 27–30, 37–41.

Software

IBM Mathematics Exploration Toolkit. Armonk, NY: IBM.

Math Tools for Algebra. Spring Branch Software. Acton, MA: William K. Bradford Publishing.

Dugdale, Sharon and David Kibbey. *Green Globs and Graphing Equations.* Pleasantville, NY: Sunburst Communications.

Videos

Algebra in Simplest Terms. Linear equations. Burlington, VT: Annenburg/CPB Collection, 1991.

Interview Notes

Background

Space Travel

Before humans can take journeys into space to explore other planets, scientists need to understand the effects of weightlessness on humans, plants, and animals. There have been over fifty space shuttle flights to conduct science experiments aimed at gathering knowledge about the effects of long-term space flight. In these flights, the astronauts do the experiments both as investigators and as test subjects. Scientists have observed that with the influence of gravity removed, many basic physical processes, such as the growth of protein crystals, can be studied more easily. Not only will the knowledge learned in the space shuttle flights be used to keep people healthy and productive on long space missions, but it may also have important applications to medical problems on Earth.

Mae Jamison

Mae Jamison was the first African American woman in space. After graduating from Stanford University in 1977 with a bachelor of science degree in chemical engineering, she earned a doctorate degree in medicine from Cornell in 1981. After completing her medical internship in 1982, she worked as a General Practitioner for a short time before going to Sierra Leone and Liberia in West Africa as the Area Peace Corps Medical Officer. Dr. Jamison was selected for the astronaut program in 1987. She enjoys traveling, reading, graphic arts, photography, sewing, skiing, collecting African art, and speaks Russian, Swahili, and Japanese.

C H A P T E R

3 Graphing Linear Equations

Taking science into SPACE

INTERVIEW **Mae Jemison**

On the morning of September 12, 1992, Dr. Mae Jemison was aboard the space shuttle *Endeavour* at Kennedy Space Center, awaiting blastoff. The giant booster rockets fired. As the shuttle accelerated, reaching a speed of over 17,000 miles per hour in a little more than 8 minutes, Jemison felt 3 times the force of gravity pressing on her. But that wasn't her strongest impression. "The biggest thing I remember was that I was smiling," she says. "I always knew I would go into space—I didn't know how—I just knew I would go."

"I always knew I would go into space, to the stars."

Music of the spheres

When **Mae Jemison** went into space in September 1992, she took a few items for different organizations: an Alvin Ailey dance poster, a flag for the Organization for African Unity, A Spelman College flag. . . . "I had all kinds of music in space – **Stevie Wonder**, **Olatunji**, **Hiroshima**, **Aretha Franklin** and **Nancy Wilson**. . . . I was jammin'."

This article appeared in the *Boston Globe* on Saturday, July 10, 1993.

A Love of Science

Jemison achieved her seat on *Endeavour* by following her love of science. "From as long ago as I can remember, I wanted to be involved in science," Jemison says. In her teens, she wanted to design spacecraft. Later, she discovered biomedical engineering, which "allowed you to do engineering but bring biology into it—both areas that I liked." Jemison earned degrees in chemical engineering and medicine, achievements that helped open the door to NASA's astronaut program.

Hands, Eyes, and Ears

Aboard *Endeavour*, Jemison worked as a science mission specialist, overseeing scientific experiments sent along by earthbound researchers. "The scientists had worked on these studies for years, and they entrusted their work to us," Jemison says. "We were their hands, eyes, and ears on the shuttle." One experiment used weightlessness to grow large protein crystals. Protein crystals are important body substances that do many jobs, such as fighting disease. "In weightlessness, you can grow any size crystals and they won't break. Then you can look at them under microscopes and study their exact structures." By studying protein structures, scientists can build new molecules that copy the protein, Jemison explains. "We took up some proteins that we think are responsible for the immune system responding to AIDS. If we can grow large crystals and see their structure, we might be able to help make drugs to fight the disease."

Interview Notes

Background

Space Pioneers

On August 30, 1983, Guion Stewart Bluford became the first African American to travel in space. Bluford and the four other astronauts aboard the space shuttle *Challenger* launched a communications and weather satellite for India. During their six-day mission, they also conducted medical tests.

Sally Ride was the first American woman to travel in space. In June 1983, she was part of the crew that launched communication satellites for Indonesia and Canada, and tested the shuttle's special *remote manipulator arm*, which was capable of releasing a satellite into space, retrieving it, and placing it in the cargo area of the shuttle.

Second-Language Learners

Students learning English may have trouble understanding the expression "We were their hands, eyes, and ears on the shuttle." If necessary, explain that the *Endeavour* crew was being instructed by the scientists on the ground on how to conduct their scientific experiments.

Interview Notes

Mathematical Connection

Space travel requires an understanding of a number of basic physical ideas such as gravity, velocity, acceleration, mass, and weight. Physicists have expressed these concepts mathematically, using variables and formulas, and have been able to analyze the dynamics of space travel. In this chapter, students learn to calculate the rate of ascent of the space shuttle *Endeavour* and its speed in orbit in feet per second. They also use a graph to analyze acceleration over time, and learn how acceleration affects an astronaut's weight during lift-off. The concept of direct variation is used to relate the weight of a person or an object on Earth with its weight on another planet.

Explore and Connect

Project
Students can work cooperatively in small groups on this project to create a poster. The most interesting and best made posters can be displayed in class.

Research
Students can organize the results of their research into three lists headed Requirements, Reasons, and Science Experiments.

Writing
Students should first research and identify those events they think were important in the 1950s and 1960s for space scientists before creating their time lines.

Experiencing Weightlessness

Some people think there is no gravity outside Earth's atmosphere. In fact, the shuttle is affected by Earth's gravity, though less so than on the planet's surface. So why do astronauts experience weightlessness? The answer has to do with the rate at which the shuttle travels. To maintain its orbit, the shuttle has to travel about 17,400 miles per hour. Even when traveling at this great speed, the shuttle and the astronauts are falling back to Earth. It's just that the shuttle is moving fast enough so that the path of its fall matches the curvature of Earth. The astronauts inside the shuttle are also falling, so gravity does not press them against the shuttle. Because the astronauts are in a constant state of falling, they feel weightless.

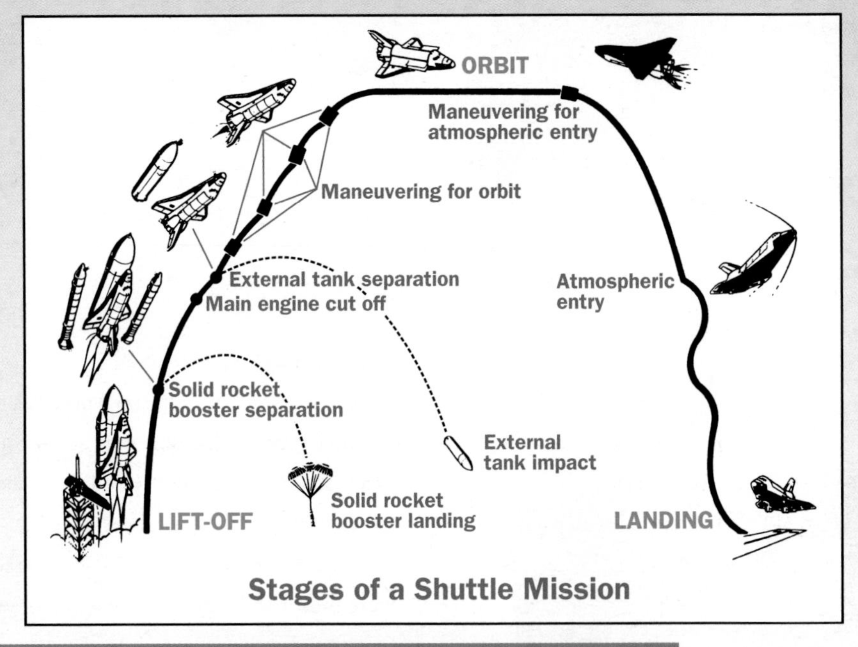

Stages of a Shuttle Mission

EXPLORE AND CONNECT

Mae Jemison floating weightless during the *Endeavour* mission.

1. Project Create a poster that gives an idea of what it would be like to travel at 17,000 miles per hour. For example, you may want to compare this speed to speeds of other fast-moving objects.

2. Research Find out the requirements for becoming an astronaut. What are some reasons astronauts need to study mathematics? What types of science experiments have astronauts performed in space?

3. Writing Mae Jemison began to study Russian when she was in high school. Several events in the 1950s and 1960s may have convinced her that Russian would become an important language for space scientists. Create a time line of events in space travel that support this conclusion.

Mathematics
& Mae Jemison

In this chapter, you will learn more about how mathematics is related to space science.

Related Examples and Exercises

Section 3.1
- **Examples 1 and 2**
- **Exercises 19–22**

Section 3.2
- **Exercises 7 and 8**

SECTION

3.1 Applying Rates

Learn how to...

- find unit rates from words and graphs

So you can...

- compare real-world rates, such as those of the space shuttle and the Concorde jet

It did not take long for astronaut Mae Jemison to leave Earth far behind as she rocketed into space aboard the space shuttle *Endeavour*. The shuttle climbed about 71 mi in about 8.5 min. You can use this information to describe the shuttle's average *rate* of ascent.

A **rate** is a ratio that compares two units. $\frac{71 \text{ mi}}{8.5 \text{ min}}$

Endeavour's rate of ascent is 71 mi in 8.5 min.

Rates are often expressed as *unit rates*. A **unit rate** is a rate per one given unit. $\frac{8.4 \text{ mi}}{1 \text{ min}}$

Find the unit rate in miles per minute. Divide 71 by 8.5 and round to the nearest tenth.

To get an idea of how fast 8.4 mi/min really is, you can convert it to units that are more familiar, such as miles per hour.

EXAMPLE 1 Interview: Mae Jemison

Express *Endeavour's* average rate of ascent, 8.4 mi/min, in miles per hour.

Toolbox p. 586 Ratios and Rates

SOLUTION

$$\text{Rate of ascent} = 8.4 \text{ mi/min}$$

$$= \frac{8.4 \text{ mi}}{1 \text{ min}} \cdot \frac{60 \text{ min}}{1 \text{ h}}$$

Use the fact that 60 min = 1 h. Multiply by $\frac{60 \text{ min}}{1 \text{ h}}$, which is equal to 1.

$$= \frac{8.4 \text{ mi} \cdot 60 \not{\text{min}}}{1 \not{\text{min}} \cdot 1 \text{ h}}$$

The minute units divide out.

$$= \frac{8.4 \text{ mi} \cdot 60}{1 \text{ h}}$$

$$= \frac{504 \text{ mi}}{1 \text{ h}}$$

Simplify.

Endeavour's average rate of ascent was about 504 mi/h.

THINK AND COMMUNICATE

1. Describe how to convert *Endeavour*'s average rate of ascent to miles per second. Which units give you the best idea of the speed of the shuttle?
2. Give some examples of rates you use in everyday life. Are any of these rates unit rates?

Plan⇔Support

Objectives

- Find unit rates from words and graphs.
- Compare real-world rates.

Recommended Pacing

Core and Extended Courses
Section 3.1: 1 day

Block Schedule
Section 3.1: $\frac{1}{2}$ block (with Chapter 2 Test)

Resource Materials

Lesson Support
Lesson Plan 3.1
Warm-Up Transparency 3.1
Practice Bank: Practice 17
Study Guide: Section 3.1

Technology
Technology Book: Spreadsheet Activity 3
McDougal Littell Software *Function Investigator* *Stats!*
Internet: http://www.hmco.com

Warm-Up Exercises

Complete each equation.

1. 32 days = __?__ hours 768
2. 1 second = __?__ minute $\frac{1}{60}$
3. 7 yards = __?__ inches 252
4. Write the fraction $\frac{120}{280}$ in simplest form. $\frac{3}{7}$
5. Find the missing numerator in the equation $\frac{8}{5} = \frac{?}{1}$. 1.6

ANSWERS Section 3.1

Think and Communicate

1. Multiply the rate of ascent, $\frac{8.4 \text{ mi}}{1 \text{ min}}$, by the ratio $\frac{1 \text{ min}}{60 \text{ s}}$, and simplify; about 0.14 mi/s. Answers may vary. An example is given. I think that the rate in miles per hour gives the best idea because I can compare it to the speed of a car or an airplane.
2. Answers may vary. Examples are given. the rate of fuel consumption in miles per gallon, a cost in cents per ounce, a rate of pay in dollars per hour, and the rate of rainfall in inches over a period of time; The first three rates are unit rates.

Teach⇔Interact

About Example 1

Mathematical Procedures
A procedure used in mathematics is to convert information given in a form that may be unfamiliar or difficult to work with to another form that is more understandable or easier to work with. A rate in miles per minute may not be as familiar to students as a rate in miles per hour.

Additional Example 1

Suppose a spacecraft leaves Earth at an average rate of 0.225 km/s. Express its rate of ascent in kilometers per minute.

$$0.225 \text{ km/s} = \frac{0.225 \text{ km}}{1 \text{ s}} \cdot \frac{60 \text{ s}}{1 \text{ min}} = \frac{0.225 \text{ km}}{1 \cancel{\text{s}}} \cdot \frac{60 \cancel{\text{s}}}{1 \text{ min}} = \frac{0.225 \text{ km} \cdot 60}{1 \text{ min}} = \frac{13.5 \text{ km}}{1 \text{ min}}$$

The spacecraft's average rate of ascent is 13.5 km/min.

Additional Example 2

Suppose the speed of the spacecraft in orbit is 27,900 km/h. How fast is this in meters per second?

Use the fact that 1 h = 3600 s and that 1 km = 1000 m.

First change kilometers per *hour* to kilometers per *second*.

$$\frac{27{,}900 \text{ km}}{1 \cancel{\text{h}}} \cdot \frac{1 \cancel{\text{h}}}{3600 \text{ s}} = \frac{7.75 \text{ km}}{1 \text{ s}}$$

Then change *kilometers* per second to *meters* per second.

$$\frac{7.75 \cancel{\text{km}}}{1 \text{ s}} \cdot \frac{1000 \text{ m}}{1 \cancel{\text{km}}} = \frac{7750 \text{ m}}{1 \text{ s}}$$

The speed of the spacecraft is 7750 m/s.

Section Note

Teaching Tip
After discussing Example 2 and Additional Example 2, ask if the calculations could have been done in a different order. Students should see that the distance units could have been changed first and the time units second. The end result is, of course, the same.

Toolbox p. 592
Systems of Measurement

EXAMPLE 2 Interview: Mae Jemison

The interview with Mae Jemison gives the speed of the space shuttle in orbit as about 17,400 mi/h. How fast is this in feet per second?

SOLUTION

First change miles per *hour* to miles per *second*.

$$\frac{17{,}400 \text{ mi}}{1 \cancel{\text{h}}} \cdot \frac{1 \cancel{\text{h}}}{3600 \text{ s}} \approx \frac{4.83 \text{ mi}}{1 \text{ s}}$$

1 h = 3600 s

Then change *miles* per second to *feet* per second.

$$\frac{4.83 \cancel{\text{mi}}}{1 \text{ s}} \cdot \frac{5280 \text{ ft}}{1 \cancel{\text{mi}}} \approx \frac{25{,}502 \text{ ft}}{1 \text{ s}}$$

5280 ft = 1 mi

The speed of the space shuttle in orbit is about 25,502 ft/s.

Finding Rates from Graphs

The graph below shows the altitude of the supersonic Concorde jet during a flight from New York to London. You can use the graph to find the Concorde's rate of change in altitude.

EXAMPLE 3 Application: Aviation

During its first ten minutes of flight, the Concorde climbs from ground level to the altitude shown by point B on the graph. Estimate the Concorde's rate of change in altitude during this time. Round your answer to the nearest hundred.

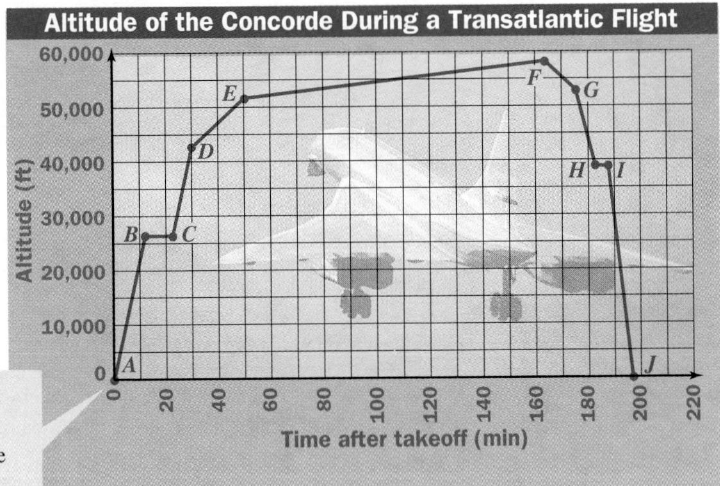

The coordinates of point A are (0, 0). At takeoff, 0 min have elapsed and the Concorde's altitude is 0 ft.

SOLUTION

Estimate the coordinates of point B and point A. Use the coordinates to find the Concorde's rate of change in altitude.

$$B \approx (11 \text{ min}, 27{,}000 \text{ ft}) \text{ and } A = (0 \text{ min}, 0 \text{ ft})$$

$$\begin{aligned}\text{Rate of change} &= \frac{\text{Change in altitude}}{\text{Change in time}} \\ &= \frac{\text{Altitude at point } B - \text{Altitude at point } A}{\text{Time at point } B - \text{Time at point } A} \\ &\approx \frac{27{,}000 \text{ ft} - 0 \text{ ft}}{11 \text{ min} - 0 \text{ min}} \\ &\approx 2455 \text{ ft/min}\end{aligned}$$

The Concorde's rate of change in altitude is about 2500 ft/min.

EXAMPLE 4 Application: Aviation

Between point G and point H on the graph, the Concorde is preparing to descend into London. Estimate the rate of change in altitude during this time.

SOLUTION

Estimate the coordinates of point H and point G.

$$H \approx (183 \text{ min}, 39{,}000 \text{ ft}) \text{ and } G \approx (175 \text{ min}, 53{,}000 \text{ ft})$$

$$\begin{aligned}\text{Rate of change} &= \frac{\text{Altitude at point } H - \text{Altitude at point } G}{\text{Time at point } H - \text{Time at point } G} \\ &\approx \frac{39{,}000 \text{ ft} - 53{,}000 \text{ ft}}{183 \text{ min} - 175 \text{ min}} \\ &\approx \frac{-14{,}000 \text{ ft}}{8 \text{ min}} \\ &\approx -1750 \text{ ft/min}\end{aligned}$$

The Concorde's rate of change in altitude is about −1800 ft/min.

BY THE WAY

One reason the Concorde flies at such a high altitude is to limit the noise pollution caused by sonic booms. Sonic booms occur when jets first cross the sound barrier by traveling faster than the speed of sound. The Concorde does not cross the sound barrier until it is about 100 mi offshore.

THINK AND COMMUNICATE

Use the graph on page 100.

3. Why is the rate in Example 4 negative? Explain how the graph shows whether a rate of change is positive or negative.
4. Find the rate of change in altitude between points E and F on the graph. Is this rate greater than or less than the rate between points A and B? How does the graph visually show that one rate is greater than the other?
5. Can the rate of change between any two points ever be 0? Explain. If possible, give an example from the graph.

Think and Communicate

3. The rate is negative because the altitude of the Concorde is decreasing between points G and H. If the graph rises from left to right, the rate is positive; if the graph falls from left to right, the rate is negative.
4. Estimates may vary. about 45 ft/min; less than the rate between A and B; The segment connecting A and B is much steeper than the segment connecting E and F.
5. Yes; if the graph is horizontal between two points, then the rate of change between the two points is 0. In the graph, the rate of change from B to C and from H to I is approximately 0.

Teach⇔Interact

Additional Example 3

The graph shows how the altitude of a hot air balloon changes during a 15-minute test flight. Estimate the balloon's rate of change in altitude during this time.

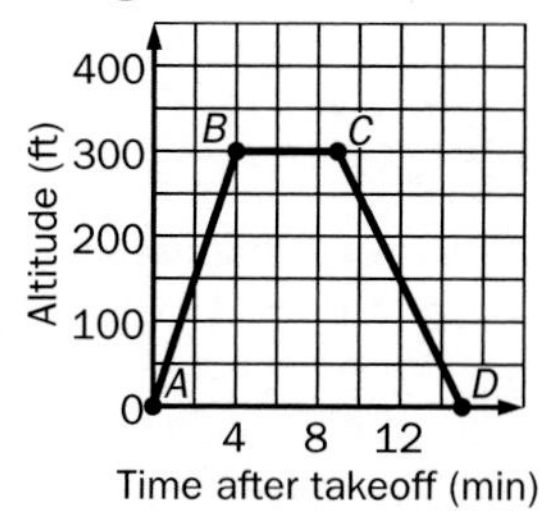

Estimate the coordinates of point A and point B. Use the coordinates to find the balloon's rate of change in altitude.

$A = (0 \text{ min}, 0 \text{ ft})$ and
$B \approx (4 \text{ min}, 300 \text{ ft})$

Rate of change

$$= \frac{\text{Change in altitude}}{\text{Change in time}}$$
$$= \frac{\text{Altitude at } B - \text{Altitude at } A}{\text{Time at } B - \text{Time at } A}$$
$$\approx \frac{300 \text{ ft} - 0 \text{ ft}}{4 \text{ min} - 0 \text{ min}}$$
$$\approx 75 \text{ ft/min}$$

The rate of change in altitude is about 75 ft/min.

Additional Example 4

Use the graph in Additional Example 3. Between points C and D on the graph, the balloon is descending. Estimate the rate of change in altitude during this time.

Estimate the coordinates of point C and point D.

$C \approx (9 \text{ min}, 300 \text{ ft})$ and
$D \approx (15 \text{ min}, 0 \text{ ft})$

Rate of change

$$= \frac{\text{Altitude at } D - \text{Altitude at } C}{\text{Time at } D - \text{Time at } C}$$
$$\approx \frac{0 \text{ ft} - 300 \text{ ft}}{15 \text{ min} - 9 \text{ min}}$$
$$\approx \frac{-300 \text{ ft}}{6 \text{ min}}$$
$$\approx -50 \text{ ft/min}$$

The rate of change in altitude is about −50 ft/min.

Think and Communicate

If students have any hesitation about how to answer question 3, a discussion of Example 4 and the graph for Additional Example 3 should clear up any difficulties. You may wish to ask students to think of other situations that involve negative rates of change.

Teach⇔Interact

Checking Key Concepts

Communication: Discussion
You may wish to suggest that students discuss situations that would lead to each of the calculations in questions 1–4.

Closure Questions

How is a rate different from other ratios? How can real-world rates be compared?
A rate is a ratio that compares two units. Real-world rates can be compared by expressing them as unit rates.

Apply⇔Assess

Suggested Assignment

Core Course
Exs. 1–11, 13–17, 23–25, 27–36

Extended Course
Exs. 1–8, 11–36

Block Schedule
Day 15 Exs. 1–11, 13–16, 23–25, 27–36

Exercise Notes

Reasoning
Ex. 11 This exercise provides a check of students' reasoning in converting rates to unit rates. Students who cannot explain and correct the error should review Example 1.

CHECKING KEY CONCEPTS

Copy and complete each equation.

1. $\frac{\$6.75}{1\text{ h}} \cdot \frac{8\text{ h}}{1\text{ day}} = \underline{?}$
2. $\frac{\$7}{1\text{ h}} \cdot \frac{40\text{ h}}{1\text{ week}} \cdot \frac{52\text{ weeks}}{1\text{ year}} = \underline{?}$
3. $\frac{21\text{ mi}}{1\text{ gal}} \cdot 15\text{ gal} = \underline{?}$
4. $\frac{24\text{ h}}{1\text{ day}} \cdot \underline{?} = \frac{168\text{ h}}{1\text{ week}}$
5. **SPORTS** In 1988, Florence Griffith-Joyner set an Olympic record of 10.54 s in the women's 100 m dash. Find her average speed in kilometers per hour. Then use the fact that 1 km ≈ 0.62 mi to give her average speed in miles per hour.

Use the graph in Example 3 on page 100. Find the rate of change in altitude between the altitudes indicated by each pair of points on the graph.

6. B and C
7. C and D
8. I and J

3.1 Exercises and Applications

Extra Practice exercises on page 561

Copy and complete each equation.

1. $\frac{18\text{ rooms}}{1\text{ dormitory}} \cdot \frac{2\text{ students}}{1\text{ room}} = \underline{?}$
2. $\frac{\$2.12}{8.2\text{ min}} \cdot \underline{?} \approx \frac{\$15.51}{1\text{ h}}$
3. $\frac{1500\text{ mi}}{1\text{ s}} \cdot \underline{?} = \frac{90{,}000\text{ mi}}{1\text{ min}}$
4. $\frac{50\text{ km}}{1\text{ h}} \cdot \frac{1\text{ h}}{3600\text{ s}} \cdot \underline{?} \approx \frac{13.9\text{ m}}{1\text{ s}}$

Express each rate in the given units. For conversion rates, see the Table of Measures on page 612.

5. 25 cm/s = ? km/h
6. 186,282 mi/s = ? mi/h
7. 5280 ft/mi = ? yd/mi
8. \$15,900 per year = ? per week
9. 4.2 mi/h = ? ft/min
10. \$13 per hour = ? cents per minute
11. **ASTRONOMY** During the Perseid meteor shower, Lance counted 9 shooting stars in 15 min. As shown below, he tried to estimate the number of shooting stars per hour.

$$\frac{9\text{ shooting stars}}{15\text{ min}} \cdot \frac{1\text{ h}}{60\text{ min}} = \frac{0.01\text{ shooting stars}}{1\text{ h}}$$

What is wrong with his work? Show how to find the correct rate.

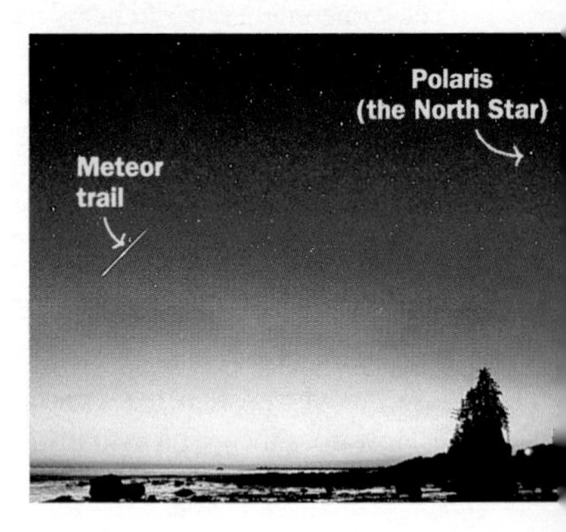

BY THE WAY

The Perseid meteor shower lasts from mid-July through mid-August every year, as Earth passes through an area of space with a large concentration of dust particles. These particles ignite when they pass through Earth's upper atmosphere.

Checking Key Concepts

1. $\frac{\$54}{1\text{ day}}$
2. $\frac{\$14{,}560}{1\text{ year}}$
3. 315 mi
4. $\frac{7\text{ days}}{1\text{ week}}$
5. about 34.2 km/h; about 21.2 mi/h

6–8. Estimates may vary.

6. about 0 ft/min
7. about 2800 ft/min
8. about –2600 ft/min

Exercises and Applications

1. $\frac{36\text{ students}}{1\text{ dormitory}}$
2. $\frac{60\text{ min}}{1\text{ h}}$
3. $\frac{60\text{ s}}{1\text{ min}}$
4. $\frac{1000\text{ m}}{1\text{ km}}$
5. 0.9
6. 51.745
7. 1760
8. about \$305.77
9. 369.6
10. about 21.7
11. He multiplied by $\frac{1\text{ h}}{60\text{ min}}$ when he should have multiplied by $\frac{60\text{ min}}{1\text{ h}}$ to get 36 as the average number of shooting stars per hour.
12. Estimates may vary.
    a. about 20,000 job openings per year
    b. about –20,000 job openings per year
    c. about –110,000 job openings per year
    d. about 0 job openings per year

12. Use the graph at the right to estimate the average rate of change in the number of job openings for each time period. Round your answers to the nearest ten thousand.

   a. 1987–1988
   b. 1988–1989
   c. 1989–1990
   d. 1990–1991

Job Openings per Year

Average number (thousands): 450, 500, 550, 600, 650

Year: 1987, 1988, 1989, 1990, 1991, 1992

SPORTS For Exercises 13–16, use the news clipping.

13. Give two rates that indicate the team's scoring ability. Be sure to include units in your answers.

14. Express home game attendance as a unit rate.

15. Suppose the fullback carries the ball an average of 20 times per game. About how many yards does he gain per game?

16. A quarterback's pass completion rate is the ratio of passes completed to passes attempted. Find Fred Sorenson's pass completion rate. Give your answer as a percent.

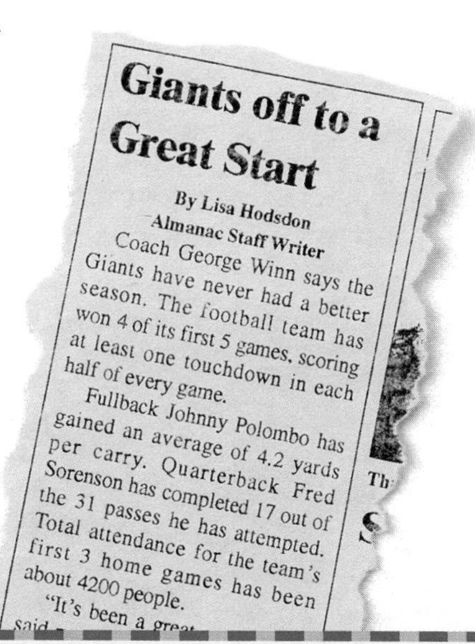

Giants off to a Great Start

By Lisa Hodsdon
Almanac Staff Writer

Coach George Winn says the Giants have never had a better season. The football team has won 4 of its first 5 games, scoring at least one touchdown in each half of every game.

Fullback Johnny Polombo has gained an average of 4.2 yards per carry. Quarterback Fred Sorenson has completed 17 out of the 31 passes he has attempted. Total attendance for the team's first 3 home games has been about 4200 people.

"It's been a great

Connection MUSIC

The compact disk (CD) was first sold to the public in 1982. During the years that followed, music buyers could choose to buy long-playing records (LPs) or CDs. Use the graph for Exercises 17 and 18.

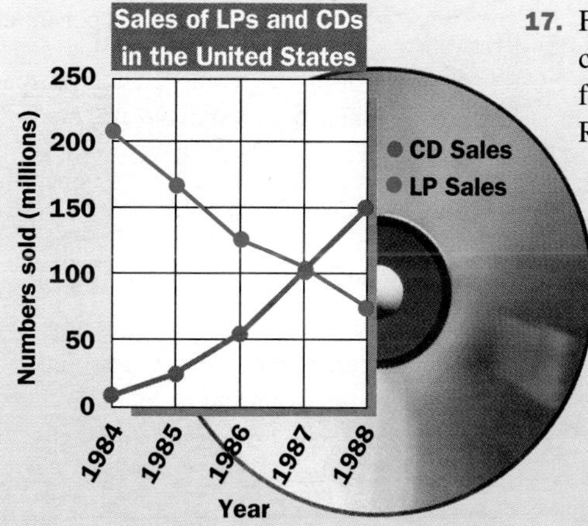

17. For each time period, find the average rate of change for LP sales and the average rate of change for CD sales. Organize your results in a table. Round your answers to the nearest ten million.

   a. 1984–1985
   b. 1985–1986
   c. 1986–1987
   d. 1987–1988

18. **Writing** Take yourself back in time to the year 1988. A music producer wants to know how many CDs and LPs to produce in 1989. Write a letter describing a good business strategy. Use rates of change to support your suggestions.

Apply⇔Assess

Exercise Notes

Application

Exs. 11–25 The concept of a rate has numerous applications in the real world. These exercises apply rates to problems in astronomy, employment, sports, music, space travel, and weather.

Interdisciplinary Problems

Exs. 17, 18 Analyzing rates of change and using this information to make decisions is a crucial aspect of the business world. Decisions that affect the future are often made by relying on data collected in the past and present. In these exercises, a good business strategy would rely upon past sales data.

13. Answers may vary. Examples are given. 4 wins out of the first 5 games, 1 touchdown per half

14. 1400 people per game

15. about 84 yards per game

16. about 75%

17. Estimates may vary.

	Rate of change for LP sales (LPs/y)	Rate of change for CD sales (CDs/y)
a.	about –40 million	about 20 million
b.	about –40 million	about 30 million
c.	about –20 million	about 50 million
d.	about –30 million	about 50 million

18. Answers will vary. An example is given. Sales of LPs declined by a rate of 20 to 40 million per year since 1984. I project a decrease of about 30 million LPs sold in 1989, and would recommend that you reduce your production of LPs by about 25% from 1988 levels. Sales of CDs have been increasing very rapidly since 1984, although the growth did slow somewhat over the last year. I anticipate that overall CD sales will increase by at least 30 million CDs in 1989, and would advise increasing CD production by at least 20% next year.

Apply⇔Assess

Exercise Notes

Interview Note
Exs. 19–22 Students familiar with the concept of gravity may point out correctly that as the distance of *Endeavour* from Earth increases, the force of gravity on the space shuttle decreases, which results in a decrease in weight. (In physics, weight is a function of gravity. Mass, however, is not affected by gravity.)

Using Technology
Exs. 19, 20 Students may wish to use the *Stats!* or *Function Investigator* software to aid their thinking in these exercises.

Integrating the Strands
Ex. 22 This exercise integrates the strands of algebra and geometry. In part (a), some students may need to be reminded of the formula for calculating the circumference of a circle. Part (b) shows how calculations involving geometric ideas can be used in problems related to space travel.

Research
Exs. 23–25 Some students may find it interesting to investigate other ways in which rates are used in reporting weather statistics. Almanacs are one possible source of information. Travel guides often give information about weather conditions for different locations. Students can prepare posters or other bulletin board displays to summarize their findings and explain how rates can have practical applications.

INTERVIEW Mae Jemison

Look back at the article on pages 96–98.

Acceleration is the rate of change in velocity over time. The graph shows how **Endeavour's** *velocity changed during its first eight minutes after takeoff. The velocity is measured in feet per second, so acceleration is measured in feet per second per second. (The units for acceleration are written ft/s².)*

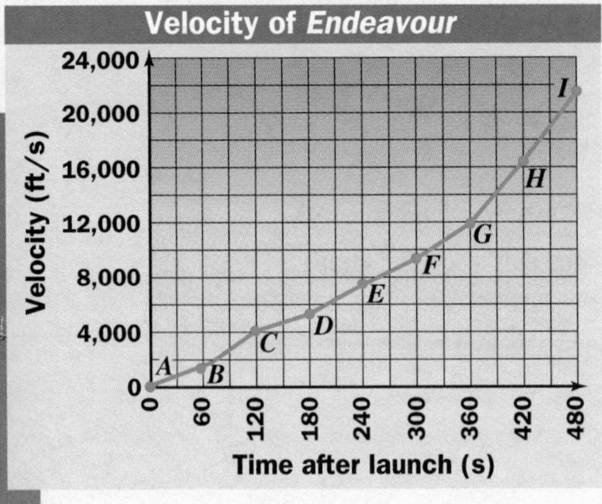

Acceleration Between Pairs of Points	
Pairs of points	Acceleration (ft/s²)
A and *B*	?
B and *C*	?
C and *D*	?
D and *E*	?
E and *F*	?
F and *G*	?
G and *H*	?
H and *I*	?

19. Copy and complete the table. Round each acceleration to the nearest ten.

20. Describe *Endeavour*'s acceleration over time. When did *Endeavour* accelerate the most? When did it accelerate the least? Describe any trends you see.

21. *Endeavour* became lighter as it rose because it used up more and more fuel. How does this factor seem to affect *Endeavour*'s velocity and acceleration over time? Explain your reasoning.

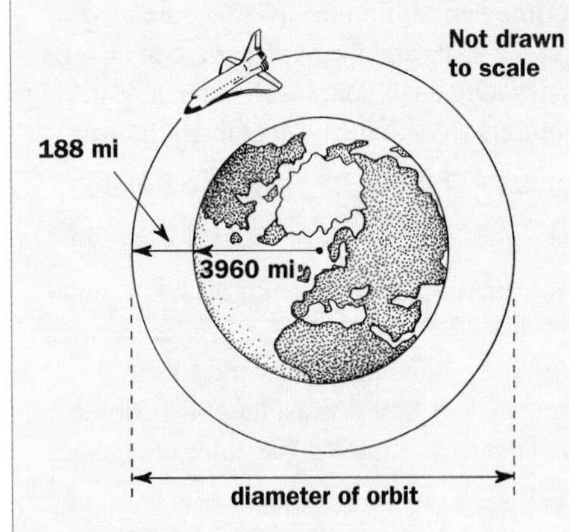

22. Challenge The diagram shows that *Endeavour*'s orbit was about 188 mi above Earth. While in orbit, *Endeavour* traveled at a rate of about 17,400 mi/h.

a. Find the distance traveled by *Endeavour* during one orbit around Earth. Explain how you got your answer.

b. Use what you know about converting rates to estimate the number of orbits *Endeavour* made during its eight days in space. Explain how you made your estimate.

19. Estimates may vary.

Pair of points	Acceleration (ft/s²)
A and *B*	about 25
B and *C*	about 41.7
C and *D*	about 25
D and *E*	about 25
E and *F*	about 41.7
F and *G*	about 41.7
G and *H*	about 66.7
H and *I*	about 100

20. Answers may vary. Examples are given. *Endeavour*'s acceleration averaged about 25 ft/s² during the first minute, increased during the next minute, returned to its previous level during the third and fourth minutes, and then began a steady increase during the last four minutes. *Endeavour* accelerated the most during the last two minutes of the first eight, and accelerated the least during the first four minutes. The acceleration data suggest that *Endeavour*'s acceleration alternately increases and then levels off.

21. The graph shows that the velocity of the *Endeavour* increased steadily over time. The table shows that the acceleration also increased over time. It is reasonable that as *Endeavour* becomes lighter and lighter over time, its velocity and acceleration will both increase and eventually level off.

22. a. $2\pi(3960 + 188) \approx 26{,}062$ mi

b. about 128 orbits; *Endeavour* traveled for 8 days, which is $24 \cdot 8 = 192$ h. At a speed of 17,400 mi/h, that is a total distance of 3,340,800 mi. $3{,}340{,}800 \text{ mi} \cdot \frac{1 \text{ orbit}}{26{,}062 \text{ mi}} \approx 128$

23. Answers may vary. An example is given using centimeters per minute. Unionville: 3.1 cm/min; Curtea-de-Arges: 1.025 cm/min; Holt: about 0.73 cm/min; Cilaos: about 0.13 cm/min; Cherrapunji: about 0.02 cm/min

WEATHER For Exercises 23–25, use the table showing rainfall amounts.

World Record Rainfall Amounts

Location	Amount	Duration
Unionville, Maryland (U.S.)	3.1 cm	1 min
Curtea-de-Arges (Romania)	20.5 cm	20 min
Holt, Missouri (U.S.)	30.5 cm	42 min
Cilaos, La Réunion	188 cm	24 h
Cherrapunji (India)	930 cm	1 month

23. Write a rate and a unit rate for each location. Use the same units for all the unit rates.

24. In Exercise 23, what information does each rate give you that the unit rate does not? What information does the unit rate give you that the rate does not?

25. Suppose rain continued to fall at a rate of 3.1 cm/min in Unionville. How long would it take to break Cherrapunji's record? Explain.

26. Open-ended Problem A child's height is an example of a variable showing a positive rate of change over time. Give two examples of a variable showing a negative rate of change over time. For one of these variables, estimate a rate of change. Explain your estimate.

27. SAT/ACT Preview Vera has \$20. The cost of a pair of socks ranges from \$1.75 to \$2.60. What is the greatest number of pairs she can buy?

A. 7 **B.** 8 **C.** 11 **D.** 9 **E.** 10

ONGOING ASSESSMENT

28. Writing Write a story that relates the graph and the picture shown at the right. Be sure to include rates in your story.

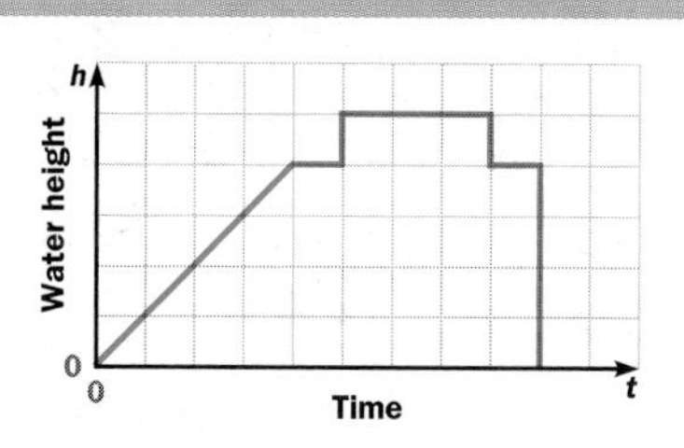

SPIRAL REVIEW

Write each fraction as a decimal. *(Toolbox, page 587)*

29. $\frac{3}{8}$ **30.** $\frac{1}{2}$ **31.** $\frac{5}{4}$ **32.** $\frac{7}{11}$

Graph each point in a coordinate plane. Label each point with its letter. Name the quadrant (if any) in which the point is located. *(Section 2.4)*

33. $A(4, 7)$ **34.** $B(3, 23)$ **35.** $C(28, 25)$ **36.** $D(-12, 6)$

Apply⇔Assess

Exercise Notes

Multicultural Note

Exs. 23–25 Cherrapunji is located in northeast India. This region receives plenty of rainfall as a result of moisture-laden ocean winds from the Bay of Bengal in the south being chilled rapidly as they are forced upwards by the Himalayas to the north. Cherrapunji holds the world record for greatest annual rainfall at 1041.8 inches.

Second-Language Learners

Ex. 28 Most students learning English should be able to complete this writing assignment successfully.

Assessment Note

Ex. 28 It should be interesting to hear how students account for the vertical sections of the graph in this exercise. Have students work together to critique stories for their interest as well as for their correct use of mathematics.

Practice 17 for Section 3.1

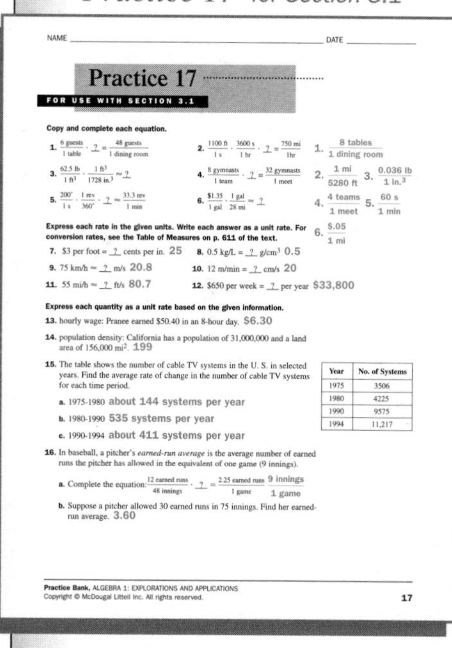

NAME ______ DATE ______

Practice 17

FOR USE WITH SECTION 3.1

Copy and complete each equation.

1. $\frac{6 \text{ guests}}{1 \text{ table}} \cdot \underline{?} = \frac{48 \text{ guests}}{1 \text{ dining room}}$
2. $\frac{1100 \text{ ft}}{1 \text{ s}} \cdot \frac{3600 \text{ s}}{1 \text{ hr}} \cdot \underline{?} = \frac{750 \text{ mi}}{1 \text{hr}}$
3. $\frac{62.5 \text{ lb}}{1 \text{ ft}^3} \cdot \frac{1 \text{ ft}^3}{1728 \text{ in.}^3} \approx \underline{?}$
4. $\frac{8 \text{ gymnasts}}{1 \text{ team}} \cdot \underline{?} = \frac{32 \text{ gymnasts}}{1 \text{ meet}}$
5. $\frac{200°}{1 \text{ s}} \cdot \frac{1 \text{ rev}}{360°} \cdot \underline{?} \approx \frac{33.3 \text{ rev}}{1 \text{ min}}$
6. $\frac{\$1.35}{1 \text{ gal}} \cdot \frac{1 \text{ gal}}{28 \text{ mi}} \approx \underline{?}$

1. $\frac{8 \text{ tables}}{1 \text{ dining room}}$ 2. $\frac{1 \text{ mi}}{5280 \text{ ft}}$ 3. $\frac{0.036 \text{ lb}}{1 \text{ in.}^3}$ 4. $\frac{4 \text{ teams}}{1 \text{ meet}}$ 5. $\frac{60 \text{ s}}{1 \text{ min}}$ 6. $\frac{\$.05}{1 \text{ mi}}$

Express each rate in the given units. Write each answer as a unit rate. For conversion rates, see the Table of Measures on p. 611 of the text.

7. \$3 per foot = ? cents per in. 25
8. 0.5 kg/L = ? g/cm^3 0.5
9. 75 km/h ≈ ? m/s 20.8
10. 12 m/min = ? cm/s 20
11. 55 mi/h ≈ ? ft/s 80.7
12. \$650 per week = ? per year \$33,800

Express each quantity as a unit rate based on the given information.

13. hourly wage: Pranee earned \$50.40 in an 8-hour day. \$6.30
14. population density: California has a population of 31,000,000 and a land area of 156,000 mi^2. 199
15. The table shows the number of cable TV systems in the U. S. in selected years. Find the average rate of change in the number of cable TV systems for each time period.

Year	No. of Systems
1975	3506
1980	4225
1990	9575
1994	11,217

a. 1975-1980 about 144 systems per year
b. 1980-1990 535 systems per year
c. 1990-1994 about 411 systems per year

16. In baseball, a pitcher's *earned-run average* is the average number of earned runs the pitcher has allowed in the equivalent of one game (9 innings).

a. Complete the equation: $\frac{12 \text{ earned runs}}{48 \text{ innings}} \cdot \underline{?} = \frac{2.25 \text{ earned runs}}{1 \text{ game}}$ $\frac{9 \text{ innings}}{1 \text{ game}}$

b. Suppose a pitcher allowed 30 earned runs in 75 innings. Find her earned-run average. 3.60

Practice Bank, ALGEBRA 1: EXPLORATIONS AND APPLICATIONS

17

---

24. The rate gives you the amount of rainfall and its duration; the unit rate gives the average amount of rainfall per minute.

25. 300 min, or 5 h; $930 \text{ cm} \cdot \frac{1 \text{ min}}{3.1 \text{ cm}} = 300 \text{ min}$

26. Answers may vary. Examples are given. the number of leaves on a maple tree from August to November; the amount of fuel in an automobile's gas tank during a trip; Suppose a car's gas tank holds 15 gal, and the car can travel about 8 h on a full tank, then the rate of change in the amount of gas during an 8 h drive is $-\frac{15}{8} = -1.875$ gal/h.

27. C

28. Answers may vary. Suppose each unit on the horizontal axis represents 3 min and each unit on the vertical axis represents 3 in. The family decided to give Rover a bath. They let the tub fill at a rate of 1 in./min. Rover sat and watched but ran off as soon as the tub was filled. It took 3 min to catch him and put him in the tub. After about 9 min, Rover jumped out. It took 3 min to catch him again. Once he was caught, the tub was overturned and it emptied all at once.

29. 0.38 30. 0.5

31. 1.25 32. 0.64

33–36.

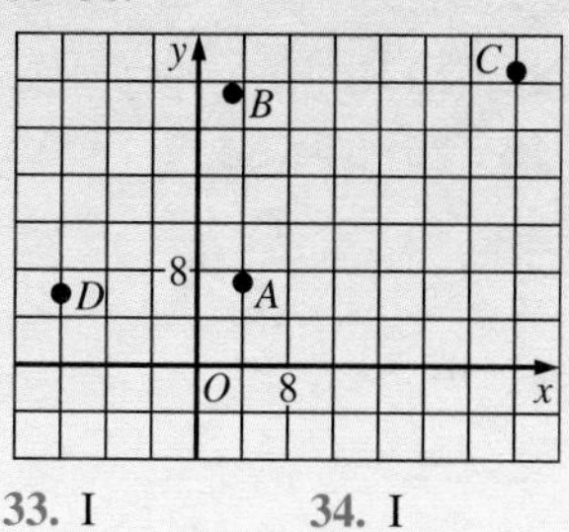

33. I 34. I
35. I 36. II

Plan⇔Support

Objectives

- Recognize and describe direct variation.
- Explore relationships between real-world variables.

Recommended Pacing

Core and Extended Courses
Section 3.2: 2 days

Block Schedule
Section 3.2: 1 block

Resource Materials

Lesson Support
Lesson Plan 3.2
Warm-Up Transparency 3.2
Overhead Visuals:
Folder A: Multi-Use Graphing Packet, Sheets 1–3
Practice Bank: Practice 18
Study Guide: Section 3.2
Explorations Lab Manual:
Additional Exploration 4, Diagram Masters 1, 7
Assessment Book: Test 10

Technology
Technology Book:
Spreadsheet Activity 3
Graphing Calculator
McDougal Littell Software
Stats!
Internet:
http://www.hmco.com

Warm-Up Exercises

Write each ratio as a decimal.

1. 75 km to 50 km 1.5
2. 18 dollars : 4.5 dollars 4
3. $\frac{10 \text{ gallons}}{4 \text{ gallons}}$ 2.5

Tell whether each equation is *True* or *False*.

4. 14 : 28 = 36 : 72 True.
5. 21 : 35 = 15 : 25 True.
6. $\frac{6}{15} = \frac{18}{20}$ False.

SECTION

3.2 Exploring Direct Variation

Learn how to...

- recognize and describe direct variation

So you can...

- explore relationships between real-world variables, such as standing height and kneeling height

The artist Leonardo da Vinci (1452–1519) studied human proportions in order to make more accurate drawings. He observed that the kneeling height of a person is $\frac{3}{4}$ of the person's standing height. In the Exploration, you will test Leonardo da Vinci's observation.

EXPLORATION
COOPERATIVE LEARNING

Patterns and Direct Variation

Work in a group of at least 6 students.
You will need:

- a meterstick
- graph paper or a graphing calculator

1 Measure and record the kneeling and standing heights of each person in your group.

2 For each person, find the ratio of kneeling height to standing height. Compare your group's results with results from other groups.

Name	Standing height (cm)	Kneeling height (cm)	$\frac{\text{Kneeling height}}{\text{Standing height}}$
Tim	156	119	0.76
?	?	?	?

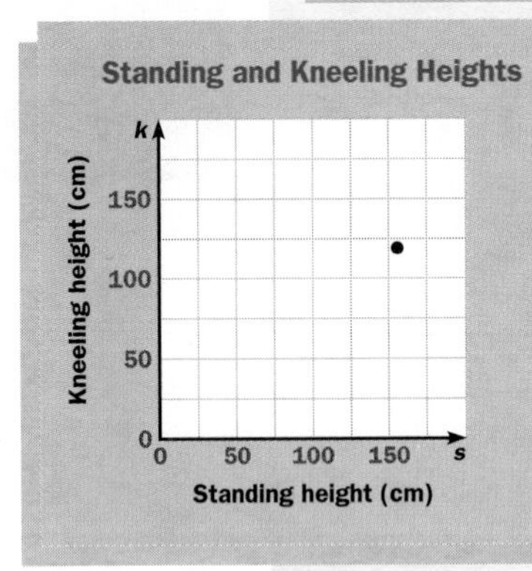

3 Make a graph of your data. Plot a point for each student. Do not connect the points. This **scatter plot** shows the relationship between the two sets of data.

4 Based on Leonardo da Vinci's observation, write an equation for kneeling height as a function of standing height. Graph the equation on your scatter plot.

5 Describe any patterns you see. Do you agree with Leonardo da Vinci's observation? Explain.

Exploration Note

Purpose
The purpose of this Exploration is to have students explore a direct variation situation and to write an equation to describe it.

Materials/Preparation
Each group needs one meterstick and graph paper or a graphing calculator.

Procedure
Students should record their data in a table such as the one on this page. They may use calculators to express ratios as decimals. Suggest that all decimals be rounded to the nearest hundredth. Groups should compare their results and discuss what they have found.

Closure
Ask students if they think that regardless of how many people were used, the ratio of kneeling height of a person is $\frac{3}{4}$ the person's standing height.

Explorations Lab Manual
See the Manual for more commentary on this Exploration.

Diagram Master 7

Whenever two variables have a constant ratio, they are said to show **direct variation**. The constant ratio is called the **constant of variation**.

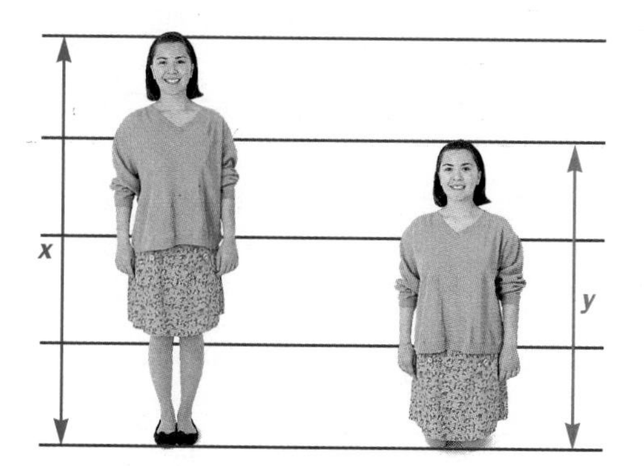

Let x = standing height.

Let y = kneeling height.

$\frac{y}{x} \approx 0.75$ — The constant of variation is 0.75.

You can rewrite the equation $\frac{y}{x} = 0.75$ to show that kneeling height y is a function of standing height x.

$x \cdot \frac{y}{x} = 0.75x$ — Multiply both sides of the equation by x.

$y = 0.75x$ — Kneeling height is a function of standing height.

Direct Variation

A direct variation can be described by an equation in this form:

$y = kx$, or $\frac{y}{x} = k$, where $k \neq 0$ ← The letter k represents the constant of variation.

You say that y *varies directly* with x.

EXAMPLE 1 Application: Employment

Do earnings vary directly with the number of hours worked? If so, give the constant of variation and write an equation that describes the situation.

Hours worked	2	3	4	5
Earnings (dollars)	11.50	17.25	23.00	28.75

SOLUTION

Check to see whether the data pairs have a constant ratio. The data show direct variation because the data pairs have a constant ratio.

Earnings / Hours worked
$\frac{11.50}{2} = 5.75$
$\frac{17.25}{3} = 5.75$
$\frac{23.00}{4} = 5.75$
$\frac{28.75}{5} = 5.75$

The constant of variation is 5.75.

Let E = earnings in dollars, and h = hours worked.

$E = 5.75h$

Teach⇔Interact

Section Notes

Second-Language Learners

Students learning English will benefit from a preliminary discussion of the concept of *human proportions*, particularly a demonstration of *kneeling height vs. standing height.*

Student Study Tip

Mention to students that ratios often use two quantities that involve the same unit. Since ratios use division to compare the quantities, the units "cancel out." For example, 2 feet to 3 feet = $\frac{2 \text{ feet}}{3 \text{ feet}} = \frac{2}{3}$.

A ratio that compares two unlike quantities is called a rate.

Additional Example 1

In the state where Dorothy lives, there is a state income tax. Does the income tax she pays on weekly earnings vary directly with her income? If so, give the constant of variation and write an equation that describes the situation.

Earnings (dollars)	State Tax (dollars)
220	17.60
226	18.08
235	18.80
244	19.52

Check to see whether the data pairs have a common ratio.

$\frac{17.60}{220} = 0.08$ $\frac{18.08}{226} = 0.08$

$\frac{18.80}{235} = 0.08$ $\frac{19.52}{244} = 0.08$

The data show direct variation because the data pairs have a constant ratio. The constant of variation is 0.08.

Let E = earnings and T = tax.

$T = 0.08E$

ANSWERS Section 3.2

Exploration

1, 2. Results may vary. Ratios should be close to 0.75.

3. Graphs may vary. Check students' work.

4. $y = 0.75x$; Check students' work.

5. The points plotted in Step 3 should lie close to the line graphed in Step 4. This result would support Leonardo da Vinci's observation.

Teach⇔Interact

Additional Example 2

Does the sales tax on different sizes of juice vary directly with size? If so, give the constant of variation and write an equation that describes the situation.

Size (oz)	Sale Tax (dollars)
10	0.02
48	0.10
64	0.15

Check the ratio of sales tax to size for each data pair.

$\frac{0.02}{10} = 0.002$

$\frac{0.10}{48} \approx 0.0021$

$\frac{0.15}{64} \approx 0.0023$

The data pairs do not have a constant ratio. The data do not show direct variation.

Section Notes

Application

Although the graph of a direct variation equation is always a line through the origin, real-world data may not lie exactly on a line but still show direct variation. This illustrates an important aspect about applications of mathematics to real-world problems. Real data may not fit a mathematical model exactly but may come close enough for the mathematical model to be a useful tool in understanding the data.

Reasoning

You may wish to ask students this question: If quantity *A* varies directly with quantity *B*, does it follow that quantity *B* varies directly with quantity *A*? (Yes.) Why? (If $\frac{y}{x} = k$, then $\frac{x}{y} = \frac{1}{k}$.)

EXAMPLE 2 Application: Consumer Economics

Do the prices of different containers of juice vary directly with their sizes? If so, give the constant of variation. Write an equation.

Size (oz)	10	48	64	64
Price (dollars)	0.40	1.99	1.69	2.99

SOLUTION

The data pairs do not have a constant ratio. The data do not show direct variation.

Juice Prices

C2 | =B2/A2

	A	B	C
1	Size (oz)	Price (dollars)	Ratio
2	10	0.40	0.04
3	48	1.99	0.04
4	64	1.69	0.03
5	64	2.99	0.05

THINK AND COMMUNICATE

1. In Example 1, what real-world information does the constant of variation give you? Explain why the constant of variation is a unit rate.
2. A person buys bananas for $.63 per pound. How does this situation show direct variation? What are the variables? What is the constant of variation? Describe another situation that shows direct variation.

Direct Variation Graphs

The graph of a direct variation equation is always a line through the origin. You can use this fact to see whether data show direct variation. For example, the scatter plot below shows data collected by one group of students in the Exploration on page 106. The data points on their scatter plot lie on or close to the graph of the equation $y = 0.75x$, so the data show direct variation.

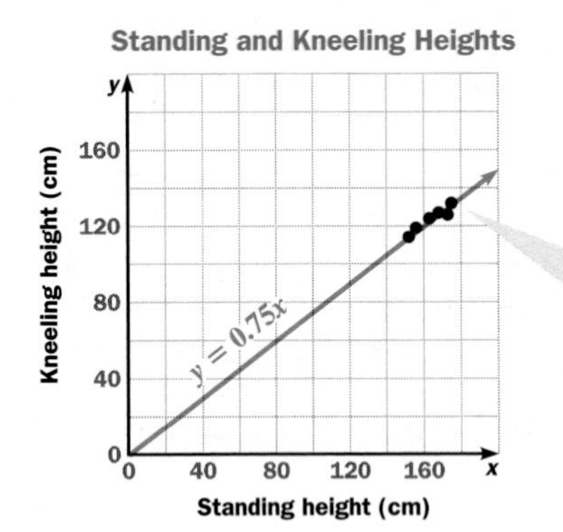

The points on the scatter plot lie on or close to a line through the origin.

Think and Communicate

1. The rate of pay is $5.75/h; the rate is in dollars per hour, so it is a unit rate.
2. Answers may vary. Examples are given. If x represents the number of pounds of bananas bought and y represents the price in dollars paid for the bananas, then $y = 0.63x$. The variables are the number of pounds bought and the price paid. The constant of variation is 0.63. If i represents a length measured in inches and c represents the same length measured in centimeters, then $c = 2.54i$.
3. Example 1; the data show direct variation. Each unit on the horizontal axis represents 1 h worked. Each unit on the vertical axis represents $5 in earnings.

THINK AND COMMUNICATE

3. The graph at the right shows data from one of the Examples in this section. Which data does the graph represent?

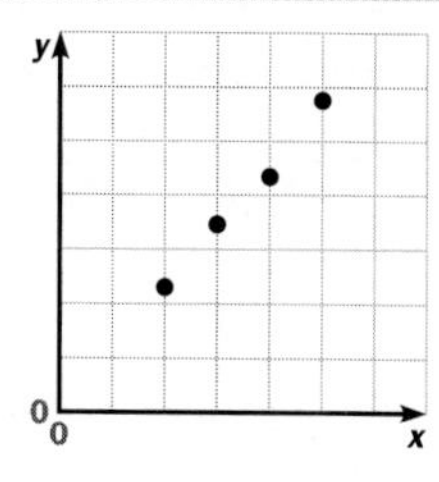

CHECKING KEY CONCEPTS

Tell whether the data show direct variation. If they do, give the constant of variation and write an equation that describes the situation.

1. 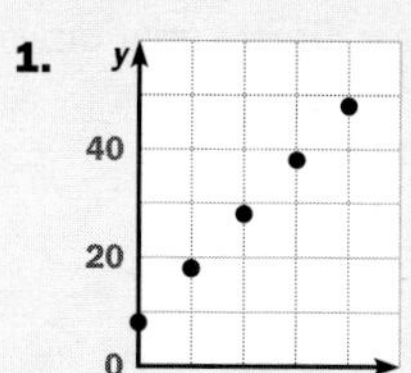

2. 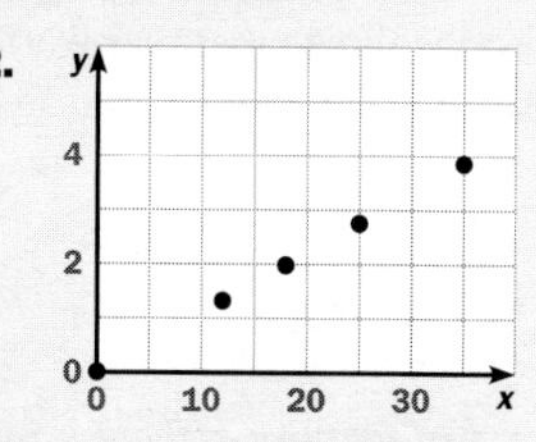

PHYSICS A group of students dropped a ball from different heights and measured the bounce heights. The table shows their data.

3. Bounce height varies directly with drop height. Give the constant of variation and write an equation that describes the situation.

4. Make a scatter plot of the data. Graph your equation from Question 3 in the same coordinate plane. What does your scatter plot show?

Drop height (in.)	Bounce height (in.)
20	15.0
30	24.0
40	32.5
50	41.0
60	48.5

3.2 Exercises and Applications

Tell whether the data show direct variation. If they do, give the constant of variation and write an equation that describes the situation.

Extra Practice exercises on page 561

1.

Time (h)	Distance (mi)
0.5	15.0
1.5	24.0
3.0	32.5
4.5	41.0

2.

Weight (lb)	Weight (kg)
50	22.7
75	34.0
120	54.4
180	81.6

3.

Year	Cable subscribers
1970	5,100,000
1980	17,500,000
1985	35,430,000
1990	50,500,000

Checking Key Concepts

1. No; the graph is not a line through the origin.
2. Yes; the data points lie on a line that goes through the origin.
3. about 0.8; If x = drop height in inches and y = bounce height in inches, then $y = 0.8x$.
4. Ball Heights

Bounce height (in.) vs. Drop height (in.); line $y = 0.8x$

The scatter plot shows that the points in the scatter plot lie close to the line $y = 0.8x$, which verifies that the data show direct variation.

Exercises and Applications

1. No.
2. Yes; about 0.453; if x = weight in pounds and y = weight in kilograms, then $y = 0.45x$.
3. No.

Teach Interact

Think and Com...

When discussing question 3, ... students to explain how they dec... ed which Example the scatter plot goes with.

Checking Key Concepts

Assessment Note
For question 1, ask students how the graph could be altered to show direct variation. Responses to this question can help you determine whether students have a good grasp of the definition of direct variation.

Visual Thinking
After responding to questions 3 and 4, ask students to use their graphs to predict the bounce height of the ball at a drop height of 100 in., 500 in., and 5 in. Ask them what height the ball would have to be dropped from to bounce 100 in., 500 in., and 5 in. This activity involves the visual skills of *interpretation* and *generalization.*

Closure Question

Describe how you can use tables and graphs to decide whether data pairs show direct variation.
For tables, check whether the data pairs have a constant ratio. For graphs, see whether the points of the graph lie on a (nonvertical/nonhorizontal) line that goes through the origin.

Apply⇔Assess

Suggested Assignment

Core Course
Day 1 Exs. 1–5, 7–12
Day 2 Exs. 13–21, AYP

Extended Course
Day 1 Exs. 1–12
Day 2 Exs. 13–21, AYP

Block Schedule
Day 16 Exs. 1–5, 9–21, AYP

Exercise Notes

Using Technology
Exs. 1–3 Students should use calculators to check for a constant ratio.

Exerci-ation

Exs. 6 Ask students to share answers with their classmates. An excellent in-class activity would be to have students write equations and create graphs for any situations where this was not done originally.

Interview Note

Exs. 7, 8 Some students may use the numbers in the second column of the chart as the constants of variation. You can ask these students how much a person weighing 200 lb would weigh on Mercury according to their equation and according to the chart. Students should soon see that the constant of variation is obtained by using $\frac{\text{weight of 100 lb object on planet}}{\text{100 lb}}$.

Reasoning

Ex. 8 Ask students if they can determine which planets are larger and smaller than Earth by using the chart. Ask them to explain their reasoning. (On a planet larger than Earth, a person would weigh more than on Earth because of its greater force of gravity. The opposite is true for a smaller planet.)

For Exercises 4 and 5, decide whether each scatter plot suggests that the data show direct variation. Give reasons for your answers.

4. **Jane Tanner's Height**

h; 150; 100; 50; 0; Height (cm); 0 4 8 12 16 a; Age (years)

5. 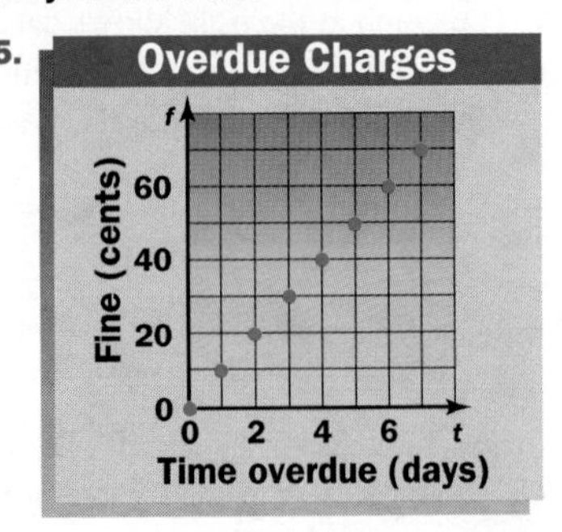

6. **Open-Ended Problem** Suppose you buy cans of juice from a vending machine. The cost varies directly with the number of cans you buy. Describe three other real-world situations that show direct variation. For one of the situations, write an equation and draw a graph.

INTERVIEW Mae Jemison

Look back at the article on pages 96–98.

The force of gravity affects how much you weigh. Astronauts use the letter g to stand for the pull of Earth's gravity on an object at sea level. Accelerating away from Earth increases the "g-force," making you weigh more than you would ordinarily.

BY THE WAY

NASA used this centrifuge to give astronauts the experience of high g-forces. A force of 3 g's is considered the maximum force under which astronauts can function easily. During liftoff for the Apollo missions to the Moon, astronauts experienced g-forces as high as 8 g's.

7. During *Endeavour*'s launch, the astronauts experienced a maximum force of 3 g's. This made them weigh 3 times their normal weight. Use a graph to show that maximum weight varied directly with normal weight.

8. The weight of an object on another planet varies directly with its weight on Earth. Use the table to answer the following questions.

   a. Let w = the weight in pounds of an object on Earth. Write an equation for the weight of the object on each planet.

   b. **Writing** Explain how weight varies on different planets. Be sure to use equations and talk about direct variation in your answer. Include graphs.

Weight of a 100 lb Object on Various Planets	
Mercury	36.5 lb
Venus	90.5 lb
Mars	38 lb
Jupiter	266 lb
Saturn	114 lb
Uranus	107 lb
Neptune	135 lb
Pluto	22.5 lb

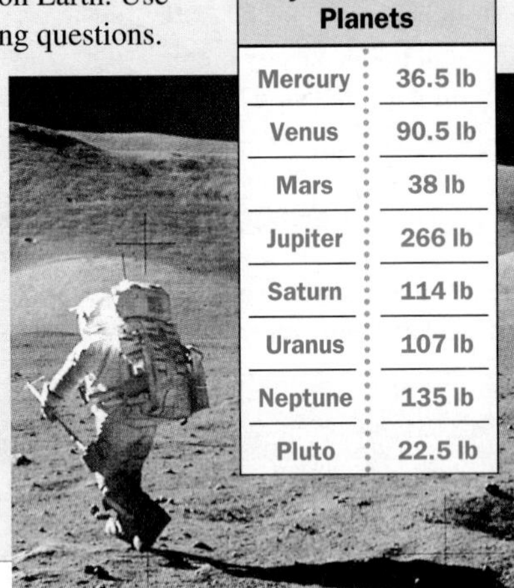

4. No; the points appear to lie on a line, but the line does not go through the origin.
5. Yes; the points lie on a line that goes through the origin.
6. Answers may vary. Check students' work.
7. y: 400, 300, 200, 100, 0; x: 0 50 100

   The graph is a line through the origin, so the data show direct variation.

8. a. Let W = the weight in pounds of the object on the given planet.
   Mercury: $W = 0.365w$;
   Venus: $W = 0.905w$;
   Mars: $W = 0.38w$;
   Jupiter: $W = 2.66w$;
   Saturn: $W = 1.14w$;
   Uranus: $W = 1.07w$;
   Neptune: $W = 1.35w$;
   Pluto: $W = 0.225w$

   b. Answers may vary. An example is given. Since the weight W of an object on any planet varies directly with its weight w on Earth, the weights are related by the equation $W = kw$, where $k > 0$. For Mercury, Venus, Mars, and Pluto, the weight of an object is less than its weight on Earth and $0 < k < 1$. For Jupiter,

Connection ART

In his notebooks, Leonardo da Vinci described how proportions can be used to draw figures accurately. For example, he observed that "... from the top to the bottom of the chin ... is the sixth part of the face, and it is the fifty-fourth part of the man."

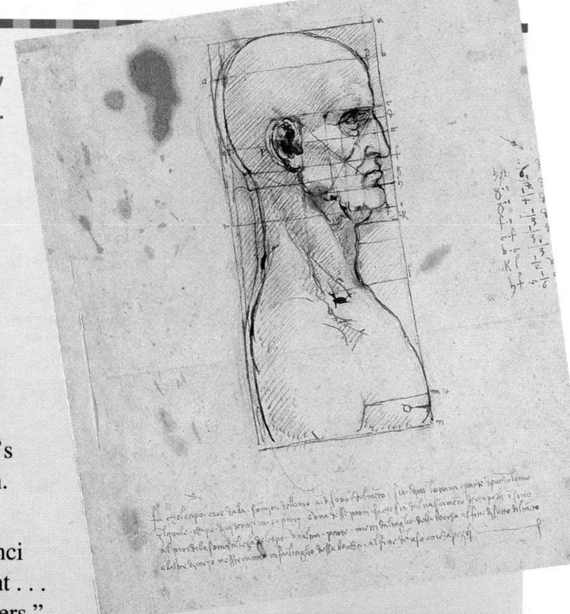

9. Write an equation describing chin length l as a function of face length f.

10. Write an equation describing chin length l as a function of height h.

11. Challenge Based on Leonardo's observation, do you think a person's face length varies directly with the person's height? If so, find the constant of variation. Explain how you got your answer.

12. Cooperative Learning Leonardo da Vinci wrote ". . . the fourth part of a man's height . . . is equal to the greatest width of the shoulders." Work in a group of 4–6 students to test this observation.

a. Measure and record the distance across each person's shoulders. Measure and record each person's height.

b. Make a scatter plot of your group's data. Put height on the horizontal axis and shoulder width on the vertical axis.

c. Do your data show direct variation? Do your data support Leonardo da Vinci's observation? Explain why or why not.

BY THE WAY

Leonardo da Vinci often used mirror writing in his notebooks. He was left-handed and may have found writing backward as easy as writing forward.

BUSINESS Use the graphs at the right for Exercises 13 and 14.

13. In her new sales job, Mina Joshi will not make any commissions until she has sold \$1000 of merchandise. Then she will receive a 5% commission on every sale. Which graph shows the amount Mina will receive in commissions? Explain how you know.

14. Does the amount Mina receives in commissions vary directly with her sales? Give a reason for your answer.

A.

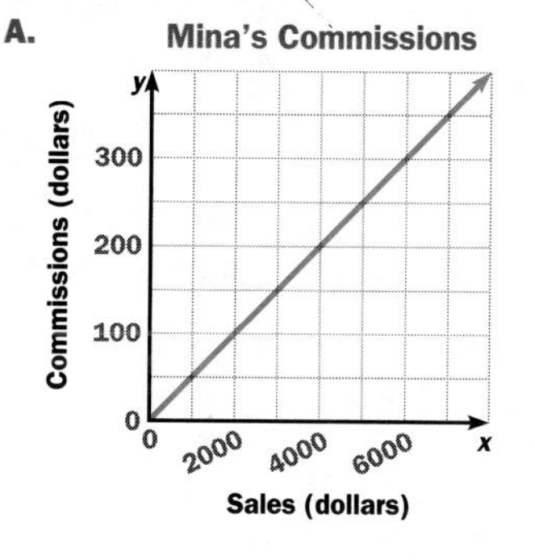

B.

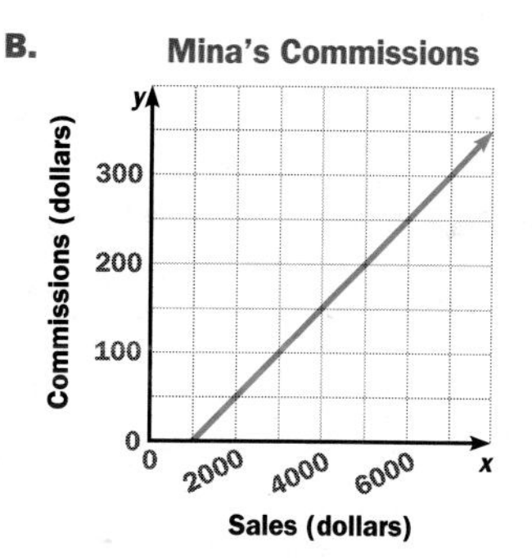

Apply⇔Assess

Exercise Notes

Research

Exs. 9–11 These exercises, based on the quote from Leonardo da Vinci's notebook, assume that the human subjects are adults. Students may find it interesting to gather information on body proportions for people their own age.

Second-Language Learners

Exs. 9–12 Students will benefit from a preliminary discussion of chin length as a function of face length. Also, paraphrase for students the expressions *the sixth part* and *the fifty-fourth part* (one-sixth; one fifty-fourth); or write these on the board as fractions.

Using Technology

Ex. 12 Students can use the *Stats!* software to help make the scatter plot for part (b) of this exercise.

Challenge

Exs. 13, 14 Ask students if they can write an equation for the graph they selected for Ex. 13. They should be able to explain how their equation supports their answer for Ex. 14.

Saturn, Uranus, and Neptune, the weight of an object is greater than its weight on Earth and $k > 1$. Examples:

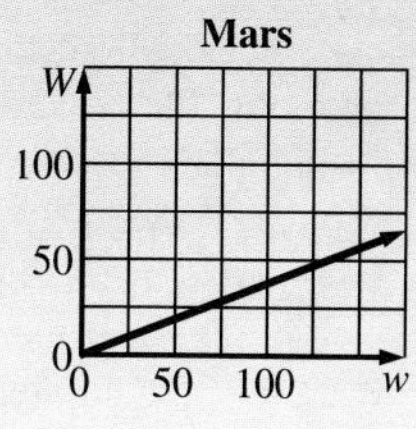

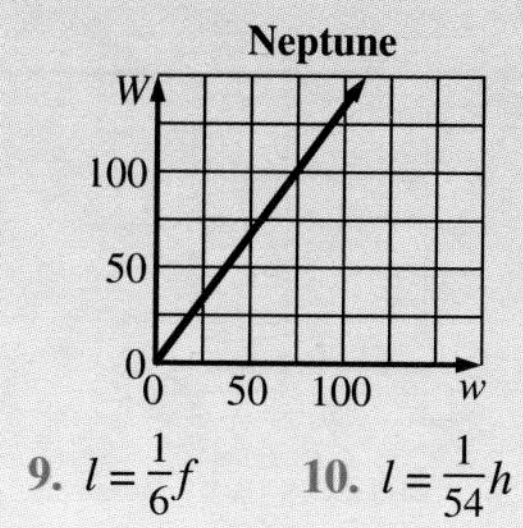

9. $l = \frac{1}{6}f$

10. $l = \frac{1}{54}h$

11. Yes; since $l = \frac{1}{6}f$ and $l = \frac{1}{54}h$, $\frac{1}{6}f = \frac{1}{54}h$, and so $f = \frac{1}{9}h$. The constant of variation is $\frac{1}{9}$.

12. a. Answers may vary.

b. Scatter plots may vary. If the observation is correct, the points in the scatter plot should lie close to the line $y = \frac{1}{4}h$, where h is a person's height and y is the shoulder width.

c. Answers may vary.

13. the graph on the right; This graph shows that Mina receives no commissions on sales at or below \$1000.

14. No; the graph of the commission amount is not a line through the origin.

Apply⇔Assess

Exercise Notes

Integrating the Strands
Ex. 15 This exercise provides a good opportunity to relate direct variation to geometric concepts.

Assess Your Progress

You can provide a review of the vocabulary by asking students to illustrate the meaning of each term using specific instances in which it does or does not apply. Students can deepen their understanding by making up situations to illustrate each term.

Journal Entry
Review with students the examples they have chosen to ensure they are correct.

Progress Check 3.1–3.2

See page 138.

Practice 18 for Section 3.2

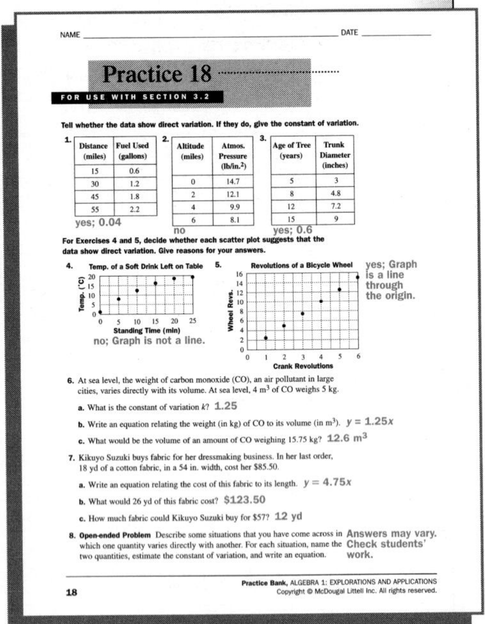
NAME ________________ DATE ________

Practice 18

FOR USE WITH SECTION 3.2

Tell whether the data show direct variation. If they do, give the constant of variation.

1.

Distance (miles)	Fuel Used (gallons)
15	0.6
30	1.2
45	1.8
55	2.2

yes; 0.04

2.

Altitude (miles)	Atmos. Pressure (lb/in.^{2})
0	14.7
2	12.1
4	9.9
6	8.1

no

3.

Age of Tree (years)	Trunk Diameter (inches)
5	3
8	4.8
12	7.2
15	9

yes; 0.6

For Exercises 4 and 5, decide whether each scatter plot suggests that the data show direct variation. Give reasons for your answers.

4. Temp. of a Soft Drink Left on Table (Temp. (°C) vs. Standing Time (min))
no; Graph is not a line.

5. Revolutions of a Bicycle Wheel (Wheel Revs. vs. Crank Revolutions)
yes; Graph is a line through the origin.

6. At sea level, the weight of carbon monoxide (CO), an air pollutant in large cities, varies directly with its volume. At sea level, 4 m^3 of CO weighs 5 kg.
   a. What is the constant of variation k? 1.25
   b. Write an equation relating the weight (in kg) of CO to its volume (in m^3). $y = 1.25x$
   c. What would be the volume of an amount of CO weighing 15.75 kg? 12.6 m^3

7. Kikuyo Suzuki buys fabric for her dressmaking business. In her last order, 18 yd of a cotton fabric, in a 54 in. width, cost her $85.50.
   a. Write an equation relating the cost of this fabric to its length. $y = 4.75x$
   b. What would 26 yd of this fabric cost? $123.50
   c. How much fabric could Kikuyo Suzuki buy for $57? 12 yd

8. **Open-ended Problem** Describe some situations that you have come across in which one quantity varies directly with another. For each situation, name the two quantities, estimate the constant of variation, and write an equation. Answers may vary. Check students' work.

18

Practice Bank, ALGEBRA 1: EXPLORATIONS AND APPLICATIONS

15. GEOMETRY Write an equation for the perimeter of each of the figures shown at the right. For each figure, decide whether the perimeter varies directly with x. Give reasons for your answers.

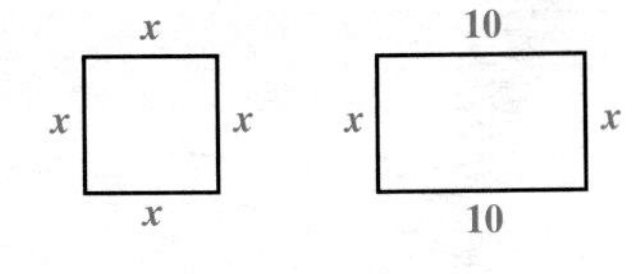

ONGOING ASSESSMENT

16. **Writing** Explain why the graph of an equation in the form $y = kx$ always goes through the origin. Give an example of a graph that shows direct variation and one that does not show direct variation.

SPIRAL REVIEW

17. Find the mean, the median, and the mode(s) of the data shown in the table, if each of the three secretaries has a $25,000 salary. *(Section 1.3)*

Salaries at a Small Company	
President	$135,000
Vice-president	$90,000
Office manager	$60,000
3 Secretaries	$25,000

Graph each equation and each point.
Tell whether the ordered pair is a solution of the equation. *(Section 2.5)*

18. $y = 2x + 2$; (2, 6)
19. $y = \frac{2}{3}x + 1$; (3, 4)
20. $y = 5 - 2x$; (−1, 3)
21. $y = 6x - 2$; (1, 4)

ASSESS YOUR PROGRESS

VOCABULARY

rate (p. 99)
unit rate (p. 99)
scatter plot (p. 106)
direct variation (p. 107)
constant of variation (p. 107)

Express each rate as a unit rate in kilometers per hour. *(Section 3.1)*

1. 0.7 km in 2 min
2. 500 km in 5 days
3. 1 m/s

Tell whether the data show direct variation. If they do, give the constant of variation and write an equation that describes the situation. *(Section 3.2)*

4.

Age of car (years)	Mileage
1	10,020
2	30,023
3	50,072
4	90,365

5.

Tickets sold	Profit (dollars)
50	150
100	300
150	450
200	600

6. **Journal** Give some examples of direct variation that can help you remember what direct variation means.

15. square: $P = 4x$; rectangle: $P = 2x + 20$; The perimeter of the square varies directly with x because the equation $P = 4x$ is in the form of a direct variation equation. The perimeter of the rectangle does not vary directly with x because the equation for the perimeter cannot be written in the form $P = kx$ and because the graph of $P = 2x + 20$ does not pass through the origin.

16. Answers may vary. An example is given. Since $0 = k \cdot 0$, (0, 0) is a point on the graph of $y = kx$. The graph of $y = 0.5x$, below, shows direct variation; the graph of $y = 2x + 1$ does not.

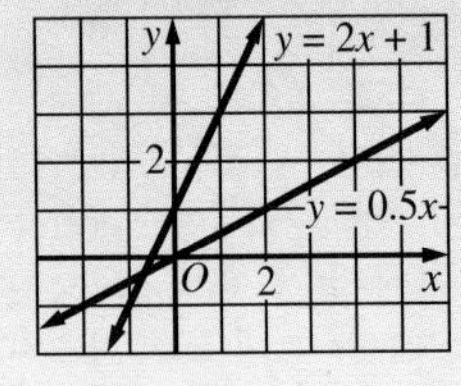

17. mean, $60,000; median, $42,500; mode, $25,000

18. Yes.

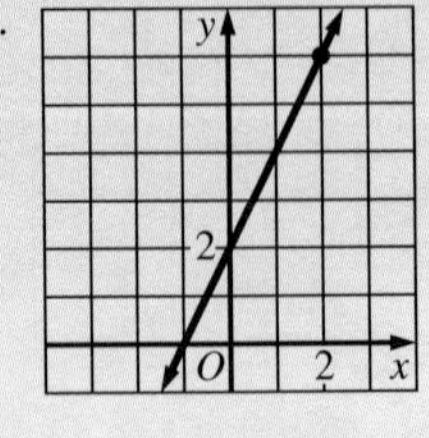

19–21. See answers in back of book.

Assess Your Progress

1. 21 km/h
2. about 4.17 km/h
3. 3.6 km/h
4. No.
5. Yes; 3; $y = 3x$, where x = the number of tickets sold and y = profit in dollars.
6. See answers in back of book.

SECTION

3.3 Finding Slope

Learn how to...
- find the slope of a line

So you can...
- analyze a real-world graph showing fuel consumption, for example

The cheetah is the fastest animal on land. Race horses are among the fastest domestic animals. The graph shows the distances these animals can cover when traveling at top speeds.

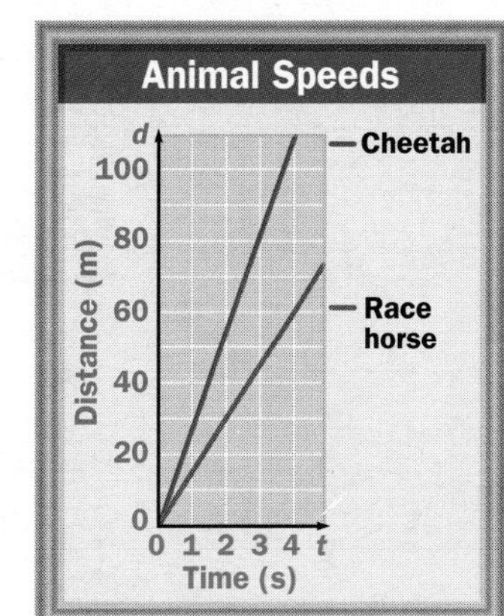

The steepness of a line is called its **slope**. The slope of a line is a rate of change.

$$\text{Slope} = \frac{\text{Vertical change}}{\text{Horizontal change}} = \frac{27}{1}$$

The slope of the cheetah's line is 27.

THINK AND COMMUNICATE

1. Estimate the slope of the horse's line. What information does the slope of the line give you?
2. Which line is steeper? What does this tell you about the speeds?

EXAMPLE 1

Graph the equation $y = 2x + 3$. Find the slope of the line.

SOLUTION

Step 1 To graph the equation on a coordinate plane, make a table of values and plot the points.

x	y
1	5
0	3
−1	1

Step 2 To find the slope of the line, choose two points and count the vertical change and the horizontal change between them.

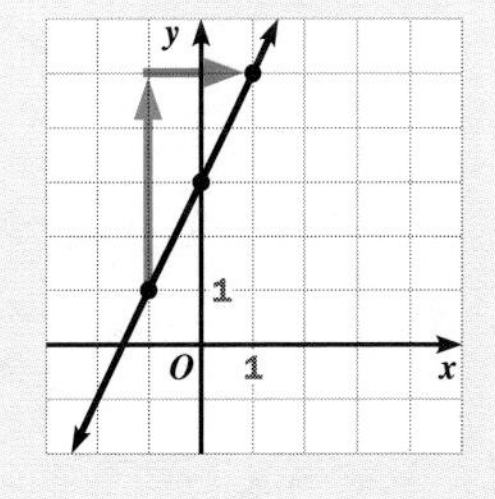

$$\text{Slope} = \frac{\text{Vertical change}}{\text{Horizontal change}} = \frac{4}{2} = 2$$

The slope of the line is 2.

BY THE WAY

Cheetahs can maintain top speeds for about 300 m. They are an endangered species because of the destruction of their habitat and the desirability of their fur.

Plan⇔Support

Objectives
- Find the slope of a line.
- Analyze a real-world graph.

Recommended Pacing
Core and Extended Courses
Section 3.3: 1 day
Block Schedule
Section 3.3: $\frac{1}{2}$ block
(with Section 3.4)

Resource Materials
Lesson Support
Lesson Plan 3.3
Warm-Up Transparency 3.3
Overhead Visuals:
Folder A: Multi-Use Graphing Packet, Sheets 1–3
Practice Bank: Practice 19
Study Guide: Section 3.3
Explorations Lab Manual:
Diagram Master 2
Technology
Technology Book:
Spreadsheet Activity 3
Graphing Calculator
Internet:
http://www.hmco.com

Warm-Up Exercises
Find the change in the x-coordinates as you go from the first point to the second. Then find the change in the y-coordinates.

1. $A(2, 5)$ $B(4, 10)$ 2; 5
2. $C(-1, 0)$ $D(0, -6)$ 1; −6
3. $E(3, 7)$ $F(5.5, -1.2)$ 2.5; −8.2
4. $L(-6.2, -0.8)$ $M(1, 0.2)$ 7.2; 1
5. $P(0, 0)$ $Q(12, -4)$ 12; −4

ANSWERS Section 3.3

Think and Communicate
1. Estimates may vary; about 15; The slope gives the horse's top speed in meters per second.
2. the line for the cheetah; The cheetah's top speed is greater than the horse's top speed.

Teach⇔Interact

Learning Styles: Kinesthetic

Students may find it helpful to relate slope to uphill, downhill, or horizontal motion. If you travel from left to right along a nonvertical line in the coordinate plane, then positive slope corresponds to uphill motion. Negative slope corresponds to downhill motion. Zero slope corresponds to horizontal motion.

Additional Example 1

Graph the equation $y = \frac{2}{3}x + 4$. Find the slope of the line.

Step 1 Make a table of values and plot the points.

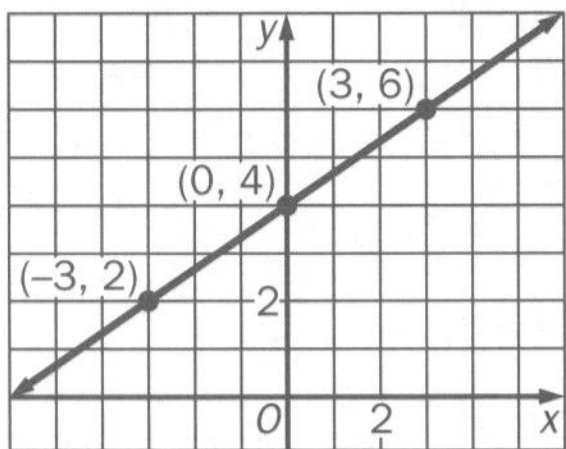

Step 2 Use the points (0, 4) and (3, 6). Find the ratio of vertical change to horizontal change.

$$\text{Slope} = \frac{\text{vertical change}}{\text{horizontal change}} = \frac{2}{3}$$

The slope of the line is $\frac{2}{3}$.

Additional Example 2

In one month, Mary Aragon's fuel oil gauge showed 120 gal at the beginning of the month but only 18 gal 30 days later. The graph shows the amount of fuel oil in the tank as a function of the number of days since the beginning of the month.

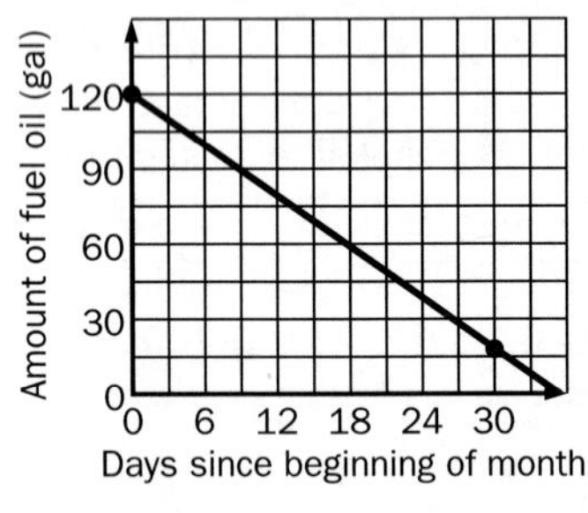

Find the slope. What does the slope tell you about the situation?

Use the points (0, 120) and (30, 18).

$$\text{Slope} = \frac{\text{vertical change}}{\text{horizontal change}} = \frac{18 - 120}{30 - 0} = -3.4$$

Mary's home uses about 3.4 gallons of fuel oil per day. The slope is negative because the amount of fuel oil is decreasing.

Positive and Negative Slope

A line that rises as you move from left to right has a **positive slope**. A line that falls as you move from left to right has a **negative slope**.

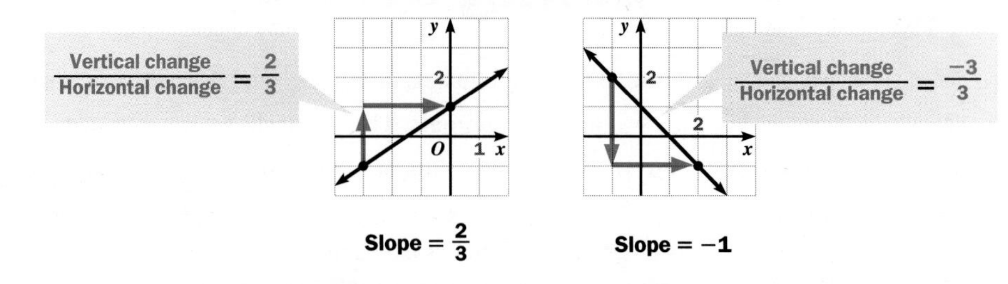

Slope = $\frac{2}{3}$ Slope = -1

EXAMPLE 2 Application: Energy Use

Rob Mueller had 12 gal of gasoline in his car. After he drove 160 mi on the highway, there were 6 gal left. The graph shows the amount of gasoline left in his car as a function of the number of miles driven. Find the slope of the line. What does the slope tell you about the situation?

SOLUTION

$$\text{Slope} = \frac{\text{Vertical change}}{\text{Horizontal change}} = \frac{6 - 12}{160 - 0} = \frac{-6}{160} = -0.0375$$

In many real-world situations, it may make sense to write the slope as a decimal.

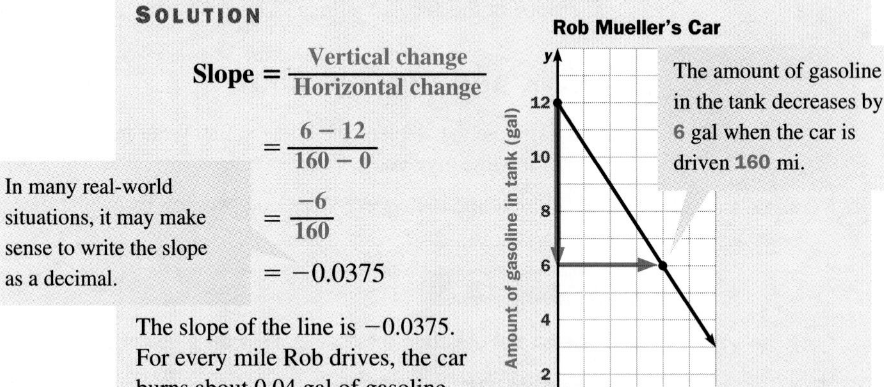

The slope of the line is -0.0375. For every mile Rob drives, the car burns about 0.04 gal of gasoline. The slope is negative because the amount of gasoline in the car is decreasing.

THINK AND COMMUNICATE

Use the information in Example 2.

3. Suppose you graphed the distance Rob's car travels as a function of time. Would the slope of the graph be positive or negative? Explain.
4. Rob's car uses more gasoline when he drives on city streets than it does on the highway. Suppose you graphed the amount of gasoline left in Rob's car after he drove 160 mi in the city. How would the graph be different from the graph shown in Example 2? How would the slope be different?

Think and Communicate

3. positive; As the amount of traveling time increases, the distance traveled increases, so the line rises from left to right.
4. Since the car uses more gasoline in the city than on the highway, there would be less than 6 gal in the tank after Rob drove 160 mi in the city. The graph would be steeper, and the slope of the line would be less than –0.0375.

Example 2 shows that you can calculate slope if you know the coordinates of two points on a line.

$$\textbf{Slope} = \frac{\text{Vertical change}}{\text{Horizontal change}} = \frac{y_2 - y_1}{x_2 - x_1}$$

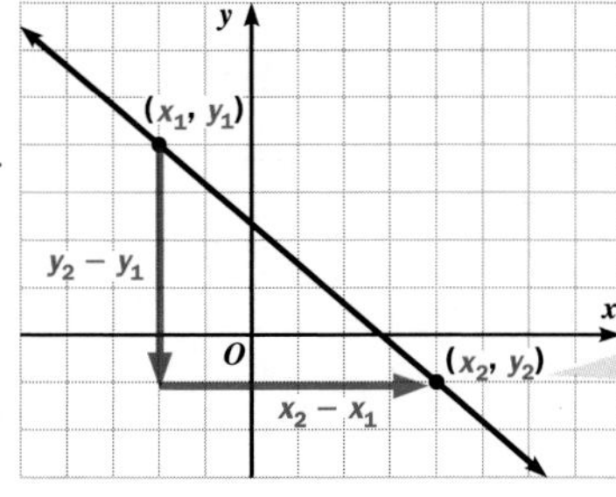

Subscripts are used to designate the coordinates of the **first** point (x_1, y_1) and the coordinates of the **second** point (x_2, y_2).

If you use this formula to find the slope of a horizontal line, you will see that the slope is zero. If you use this formula with a vertical line, you will see that the slope of a vertical line is **undefined**.

Horizontal line

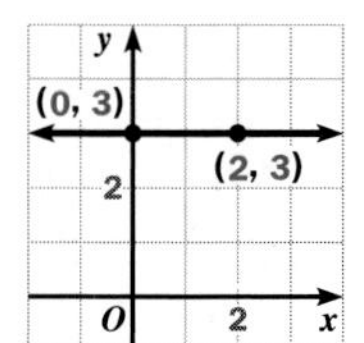

$$\frac{\text{Vertical change}}{\text{Horizontal change}} = \frac{3-3}{2-0} = \frac{0}{2}$$

The slope of a horizontal line is always zero.

Vertical line

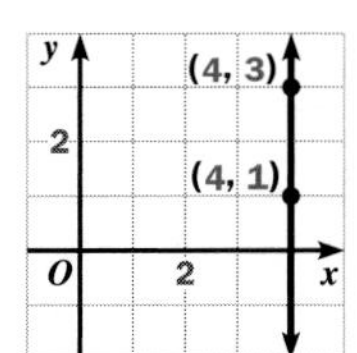

$$\frac{\text{Vertical change}}{\text{Horizontal change}} = \frac{3-1}{4-4} = \frac{2}{0}$$

Division by zero is not possible. The slope of a vertical line is always undefined.

CHECKING KEY CONCEPTS

For Questions 1–5, use the coordinate grid.

1. Which lines have a positive slope?
2. Which lines have a negative slope?
3. Which line has a slope of zero?
4. Which line has an undefined slope?
5. Which line has the greatest slope?
6. Calculate the slope of the line through the points $(-4, 5)$ and $(1, -4)$.

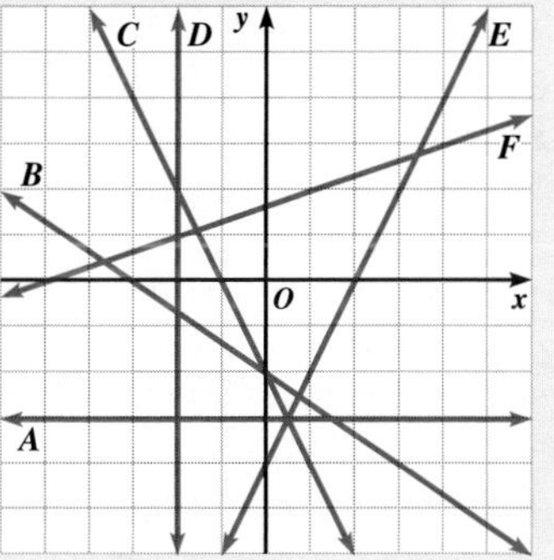

Checking Key Concepts

1. *E* and *F*
2. *B* and *C*
3. *A*
4. *D*
5. *E*
6. –1.8

Teach⇔Interact

Section Notes

Student Study Tip
Subscript notation may be new to some students. If so, point out that this notation is used when talking about two or more values of the same variable. The small numbers tell which value is being used as the first value, second value, and so on. The symbol "x_1" is read "*x* sub 1," "y_2" is read "*y* sub 2," and so on.

Communication: Writing/Listening
If any students feel that subscript notation is complicated, suggest that they try using words alone to convey the meanings of expressions and equations. Once a situation goes beyond the simplest level, students will see that subscript notation is brief, easy to write, and easier to understand in oral communication than words alone.

Mathematical Procedures
The question may arise whether it makes any difference which point is (x_1, y_1) and which is (x_2, y_2) when the formula for slope is used. Students will learn in Ex. 19 that it does not matter which point is labeled (x_1, y_1) and which is labeled (x_2, y_2).

Checking Key Concepts

Teaching Tip
You may wish to ask students whether they can find two lines in the diagram that appear to have the same absolute value for their slopes, but appear to go in different directions. What is the relationship between the slopes of these lines? (Lines *C* and *E* appear to have the same absolute value for their slopes, but appear to go in different directions. Their slopes are opposites.)

Closure Question

How can you tell by looking at a line in the coordinate plane whether it has positive, negative, or zero slope?
The slope is positive if the line rises from left to right. The slope is negative if the line falls from left to right. The slope is zero if the line is horizontal.

Apply⇔Assess

Suggested Assignment

Core Course
Exs. 1–19, 22, 24, 27–36

Extended Course
Exs. 1–14, 19–36

Block Schedule
Day 17 Exs. 1–19, 22, 24, 27–36

Exercise Notes

Challenge
Exs. 4–9 Ask students if they notice anything interesting about the slopes of the lines and the numbers in the equations. Some students may notice that for each equation, the slope is the number by which the variable has been multiplied. Ask what number x is multiplied by in Ex. 6. Someone may see that $y = -5$ can be thought of as $y = 0x - 5$.

Common Error
Exs. 10–18 Some students may use the slope formula incorrectly when finding the slope of a line. For example, they may find $y_2 - y_1$, but then use $x_1 - x_2$. Reinforce the fact that the subscripts must appear in the same order.

Using Technology
Exs. 10–18 Students who are already well acquainted with slope may want to study the manual for a TI-81 or TI-82 graphing calculator to see if they can write a program for calculating slopes. Here is a simple program for the TI-82. It calculates the slope of the line through points (A, B) and (P, Q).

```
PROGRAM:SLOPE
:Input A
:Input B
:Input P
:Input Q
:Disp ((Q–B)/(P–A))
```

If $A = P$, any attempt to execute the program will give an error message.

Application
Ex. 20 Students should think about how the concept of a road's grade is related to the concept of slope. If a line is drawn in the coordinate plane to represent a straight road, the grade of the road equals the *absolute value* of its slope.

3.3 Exercises and Applications

Extra Practice *exercises on page 562*

Find the slope of each line.

1.

2. 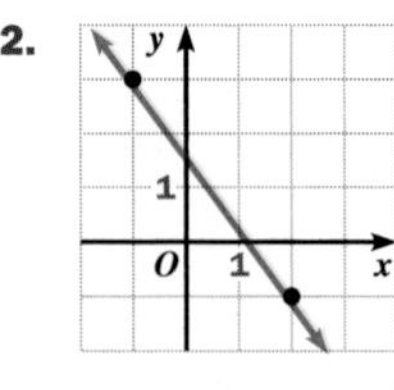

3. 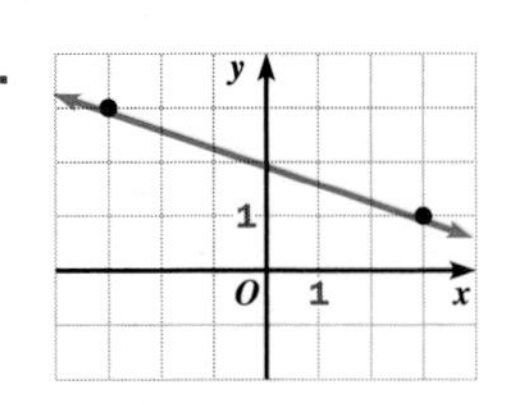

Graph each equation. Find the slope of the line.

4. $y = 3x - 2$ **5.** $y = 10 - x$ **6.** $y = -5$

7. $y = \frac{1}{2}x + 7$ **8.** $y = \frac{1}{4}x$ **9.** $y = -2x - \frac{1}{2}$

Find the slope of the line through each pair of points.

10. (1, 3) and (−2, 1) **11.** (0, 0) and (−5, 3) **12.** (5, −1) and (1, 6)

13. (2, 12) and (2, −9) **14.** (3, 3) and (−7, −7) **15.** (45, 3) and (30, 35)

16. (−12, 5) and (23, 5) **17.** (5, 22) and (126, 7) **18.** (7, −3) and (−11, 22)

19. Mona Jhaveri asked two students to find the slope of the line through the points (3, −1) and (−2, 2).

a. **Writing** Describe each student's method. Explain why both methods work.

b. Look back at the formula $\frac{y_2 - y_1}{x_2 - x_1}$ on page 115. Does it matter which coordinates are labeled (x_1, y_1) and which are labeled (x_2, y_2)? Explain. Give examples to support your answer.

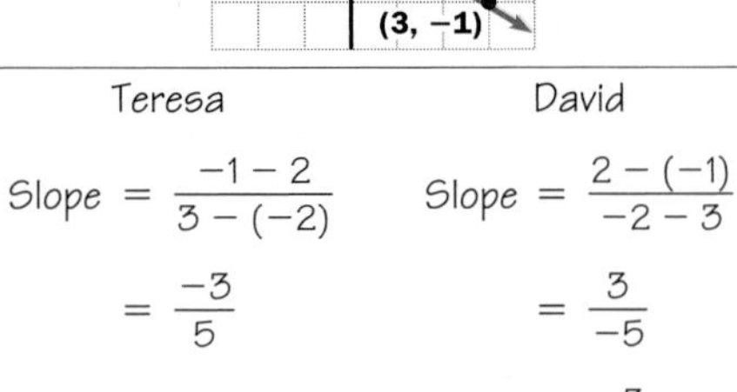

Teresa

$\text{Slope} = \frac{-1 - 2}{3 - (-2)}$

$= \frac{-3}{5}$

David

$\text{Slope} = \frac{2 - (-1)}{-2 - 3}$

$= \frac{3}{-5}$

$= \frac{-3}{5}$

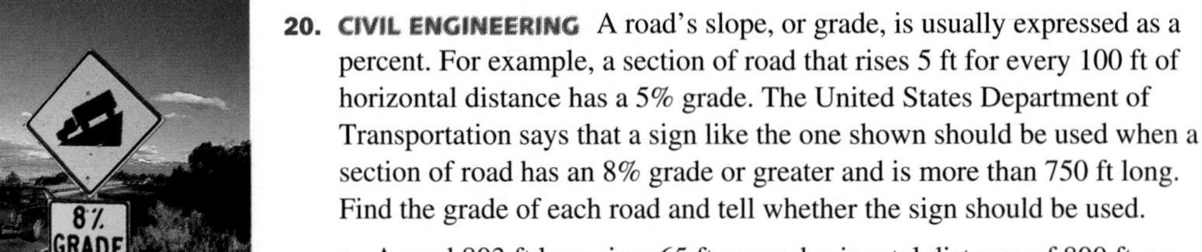

20. **CIVIL ENGINEERING** A road's slope, or grade, is usually expressed as a percent. For example, a section of road that rises 5 ft for every 100 ft of horizontal distance has a 5% grade. The United States Department of Transportation says that a sign like the one shown should be used when a section of road has an 8% grade or greater and is more than 750 ft long. Find the grade of each road and tell whether the sign should be used.

a. A road 803 ft long rises 65 ft over a horizontal distance of 800 ft, as shown in the diagram.

b. A road 975 ft long rises 75 ft over a horizontal distance of 972 ft.

c. A road 650 ft long rises 57 ft over a horizontal distance of 647 ft.

Exercises and Applications

1. $\frac{1}{2}$ 2. $-\frac{4}{3}$ 3. $-\frac{1}{3}$

4. 3

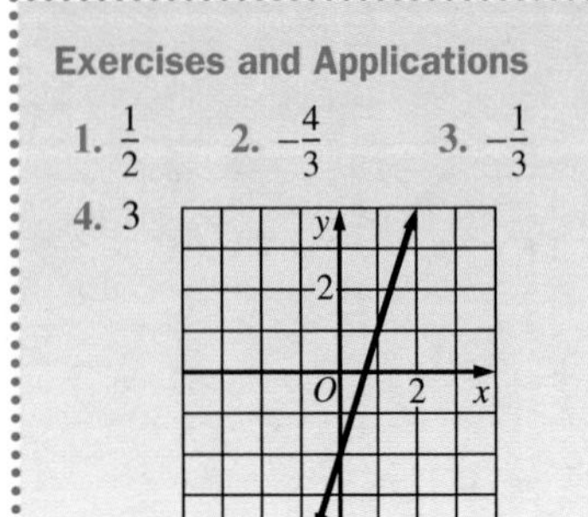

5. −1

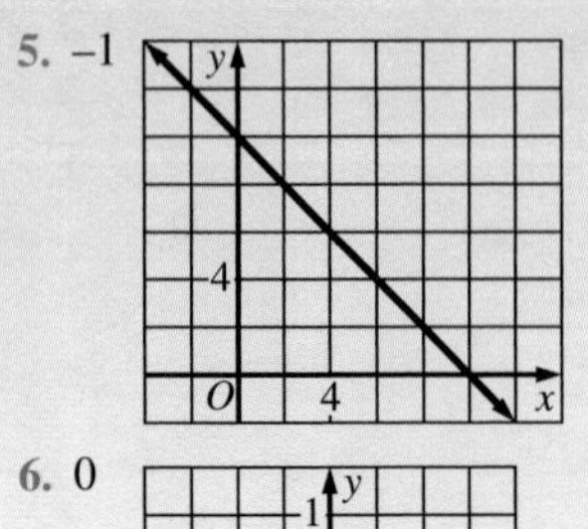

6. 0

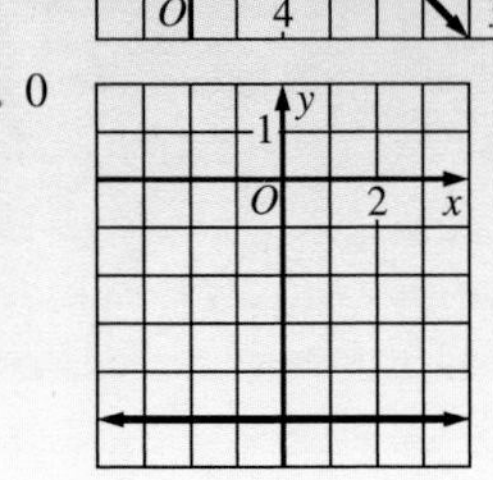

7. $\frac{1}{2}$

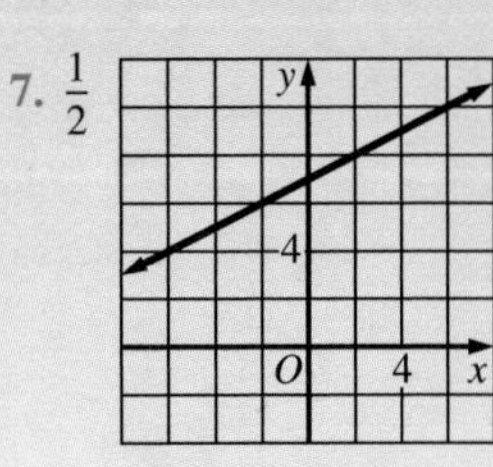

8. $\frac{1}{4}$

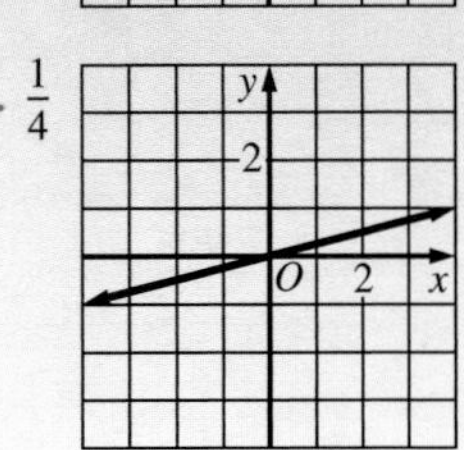

Connection ARCHITECTURE

The tread of a stair is the flat part that you step on. The riser is the vertical part of the stair. The ratio of the height of the riser to the depth of the tread can affect the safety of the staircase. The diagram shows a traditional staircase design and a new design. Building code officials prefer the new design because they say it is safer. Builders prefer the old design because it is less costly.

21. Open-ended Problem Why do you think the old design costs less to build? Why might the new design be safer? Give reasons for your answers.

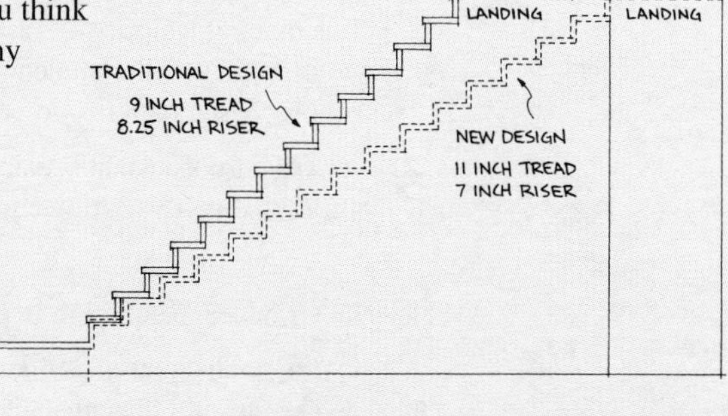

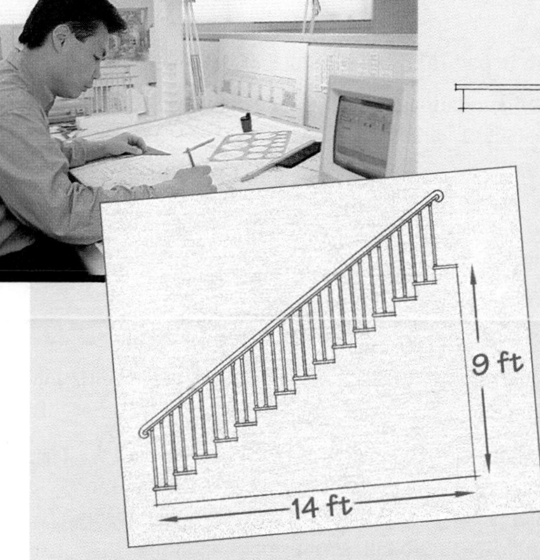

22. a. An architect is designing a house with a staircase 9 ft high and 14 ft long, as shown at the left. Which design is the architect using? How do you know?

b. How long would a 9 ft high staircase be if the architect were using the other design? Explain your reasoning.

23. Research Find the slopes of at least three different staircases near your home or school. Are they the old design, the new design, or neither? Which do you think is safest? Why?

24. Challenge The American National Standards Institute (ANSI) specifications state that the maximum slope of a wheelchair ramp should be $\frac{1}{12}$.

a. Does a ramp 7 in. high and 94 in. long meet the ANSI specifications?

b. Does a ramp with slope $\frac{1}{6}$ meet the ANSI specifications? Explain.

c. A 14 in. high ramp needs to be built. Find the length l of a ramp that meets the ANSI specifications.

Apply⇔Assess

Exercise Notes

Interdisciplinary Problems

Exs. 20–24 Road construction and the design and the construction of buildings are both dependent upon the concept of slope. Civil engineers and architects need to be able to calculate slopes accurately to meet safety requirements for roads and buildings.

Career Connection

Exs. 21–24 Architects play an essential role in helping to create a physical environment that meets peoples' needs for living and working. Their work also has a large impact on the aesthetics of a residential neighborhood or a large city, which is populated with many different kinds of office buildings and other structures. Engaging students in a discussion of what architects do, how they affect peoples' lives, and what training and education they receive will heighten their awareness of this career. You can prepare students for this discussion by asking them to bring to class some examples of what they think architects do.

Multicultural Note

Ex. 24 The Amercian National Standard Institute requirements for the construction of new buildings and facilities state that the slope of a ramp cannot exceed one inch for every 12 inches it is off the ground. The minimum width of a ramp is set at 36 inches. Level landings as wide as the ramp are required at the top and bottom. If the rise of a ramp exceeds six inches or if the horizontal projection of a ramp (that is, the entire length of the ramp without the landing) exceeds 72 inches, handrails must be installed on either side.

9. −2

10. $\frac{2}{3}$ 11. $-\frac{3}{5}$

12. $-\frac{7}{4}$ 13. undefined

14. 1 15. $-\frac{32}{15}$

16. 0 17. $-\frac{15}{121}$

18. $-\frac{25}{18}$

19. Answers may vary. Examples are given.

a. David calculated the horizontal change and the vertical change from (3, −1) to (−2, 2). Teresa calculated the horizontal and vertical change from (−2, 2) to (3, −1). Both methods work because $\frac{-1-2}{3-(-2)} = -\frac{3}{5}$ and $\frac{2-(-1)}{-2-3} = -\frac{3}{5}$.

b. No; examples and explanations may vary. Since $\frac{-(y_2-y_1)}{-(x_2-x_1)} = \frac{y_1-y_2}{x_1-x_2}$, either point can be taken as (x_1, y_1). For example, if you use the points (1, 3) and (−2, 1) from Ex. 10, you get $\frac{2}{3}$ for the slope no matter in which order you write the points in the formula.

20. a. 0.08125 = 8.125%; Yes.

b. about 0.077 ≈ 7.7%; No.

c. about 0.088 ≈ 8.8%; No, since the road is not more than 750 ft long.

21, 22. See answers in back of book.

23. Answers may vary. Staircases with low ratios for rise to tread and with deep treads are safer than staircases with high rise to tread ratios and with narrow treads.

24. a. Yes. $\frac{7}{94} < \frac{1}{12}$ b. No. $\frac{1}{6} > \frac{1}{12}$

c. a ramp that is at least 168 in. or 14 ft long

Apply⇔Assess

Exercise Notes

Cooperative Learning

Ex. 26 Students can work in groups of four. Each student should graph his or her own line. After students graph a line with slope –2, ask them to graph another line with slope 2 and then discuss any patterns.

Topic Spiraling: Review

Ex. 27 This exercise affords a good opportunity to review ideas about direct variation from Section 3.2. You may also review ideas from Chapter 2 by asking whether these equations represent functions.

Assessment Note

Ex. 29 This exercise can help to check students' understanding of how graphs can be used to find slope. You may wish to ask students what points they used to find the slopes and whether there was any special reason for choosing those particular points.

Practice 19 for Section 3.3

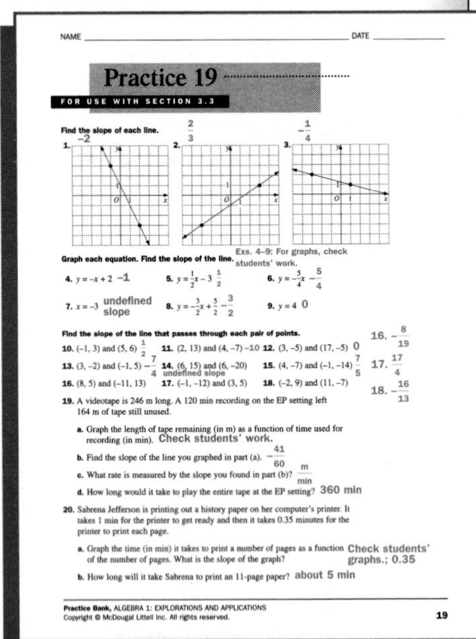

NAME ______ DATE ______

Practice 19

FOR USE WITH SECTION 3.3

Find the slope of each line.

1. -2 2. $\frac{2}{3}$ 3. $-\frac{1}{4}$

Graph each equation. Find the slope of the line. Exs. 4–9: For graphs, check students' work.

4. $y = -x + 2$ -1
5. $y = \frac{1}{2}x - 3$ $\frac{1}{2}$
6. $y = -\frac{5}{4}x$ $-\frac{5}{4}$
7. $x = -3$ undefined slope
8. $y = -\frac{3}{2}x + \frac{5}{2}$ $-\frac{3}{2}$
9. $y = 4$ 0

Find the slope of the line that passes through each pair of points.

10. $(-1, 3)$ and $(5, 6)$ $\frac{1}{2}$
11. $(2, 13)$ and $(4, -7)$ -10
12. $(3, -5)$ and $(17, -5)$ 0
13. $(3, -2)$ and $(-1, 5)$ $-\frac{7}{4}$
14. $(6, 15)$ and $(6, -20)$ undefined slope
15. $(4, -7)$ and $(-1, -14)$ $\frac{7}{5}$
16. $(8, 5)$ and $(-11, 13)$
17. $(-1, -12)$ and $(3, 5)$
18. $(-2, 9)$ and $(11, -7)$

16. $-\frac{8}{19}$ 17. $\frac{17}{4}$ 18. $-\frac{16}{13}$

19. A videotape is 246 m long. A 120 min recording on the EP setting left 164 m of tape still unused.
    a. Graph the length of tape remaining (in m) as a function of time used for recording (in min). Check students' work.
    b. Find the slope of the line you graphed in part (a). $-\frac{41}{60}$
    c. What rate is measured by the slope you found in part (b)? $\frac{\text{m}}{\text{min}}$
    d. How long would it take to play the entire tape at the EP setting? 360 min

20. Sabrena Jefferson is printing out a history paper on her computer's printer. It takes 1 min for the printer to get ready and then it takes 0.35 minutes for the printer to print each page.
    a. Graph the time (in min) it takes to print a number of pages as a function of the number of pages. What is the slope of the graph? Check students' graphs.; 0.35
    b. How long will it take Sabrena to print an 11-page paper? about 5 min

Practice Bank, ALGEBRA 1: EXPLORATIONS AND APPLICATIONS

19

25. Visual Thinking Tanya filled two jars with water. She poured the water into each jar at the same rate. Then she made a graph showing the depth of the water in each jar over time. On her graph, which line represents the water level in which jar? Explain your reasoning.

A B

Depth d Time t

26. Cooperative Learning Graph a line that has a slope of -2. Compare your graph to other students' graphs. What patterns do you see?

27. a. Give the constant of variation for each direct variation equation.

$$y = 4x \qquad y = -7x \qquad y = \frac{3}{5}x$$

b. Graph each equation in part (a). Find the slope of the line.

c. For each equation, what relationship do you see between the slope of the line and the constant of variation?

28. SAT/ACT Preview What is the slope of a line through the points (0, 3) and (3, 3)?

A. 3 **B.** -3 **C.** 1 **D.** 0 **E.** undefined

Typing Speed

Words w: 0, 40, 80, 120; Time (min) t: 0, 1, 2, 3, 4; Bob; Paul

ONGOING ASSESSMENT

29. a. Use the graph at the left. Give the slope of each line. What information do the slopes give you about the typists?

b. Open-ended Problem The amount each typist charges varies directly with the person's typing speed. Suppose you want to hire someone to type a paper for you. What is a good reason for hiring Bob? for hiring Paul? Be sure to talk about slopes and rates.

SPIRAL REVIEW

Decide whether each table represents a function. Explain your reasoning. *(Section 2.3)*

30.

n	Number with absolute value equal to n
2	2 and –2
3	3 and –3
5	5 and –5

31.

Height (in.)	Shoe size
69	9
69	9.5
70	9.5
70	10.5

32.

x	$5x$
–2	–10
–1	–5
0	0
1	5
2	10

Express each rate in the given units. For conversion rates, see the Table of Measures on page 612. *(Section 3.1)*

33. 50 cm/h = _?_ km/h

34. 6160 ft/min = _?_ mi/h

35. 2 km/min ≈ _?_ m/h

36. \$20 per hour ≈ _?_ cents per minute

25. The red line represents jar A. The depth of the water in jar A will increase more quickly than the depth in jar B because the area of the base of jar A is smaller than the area of the base of jar B.

26. Check students' work. All the lines with slope –2 are parallel.

27. a. $4; -7; \frac{3}{5}$ b. $4; -7; \frac{3}{5}$

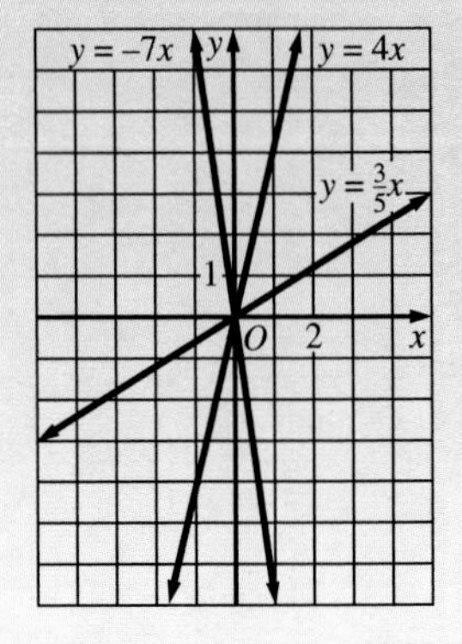

c. The constant of variation in each direct variation equation is equal to the slope of the graph of the equation.

28. D

29. a. The steeper line has slope 50; Bob's rate is 50 words/min. The other line has slope 25; Paul's rate is 25 words/min.

b. The slope of Bob's graph is steeper than that of Paul's, so Bob has greater speed and will finish more quickly. Paul will charge less because he is slower and the amount charged varies directly with typing speed.

30. No; for each value in the domain, there are two values in the range.

31. No; for each value in the domain, there are two values in the range.

32. Yes; for each value in the domain, there is exactly one value in the range.

33. 1.8 34. 70

35. 74.4 36. 33.3

3.4 Finding Equations of Lines

Learn how to...
- write linear equations in slope-intercept form

So you can...
- analyze real-world problems represented by graphs

An equation whose graph is a line is called a **linear equation**. One way to write a linear equation is to use the form $y = mx + b$. As you will see in the Exploration, the numbers m and b give you information about the graph of a linear equation.

EXPLORATION
COOPERATIVE LEARNING

Exploring Linear Equations and Graphs

Work with a partner.
You will need:
- a graphing calculator
- graph paper

For more information on setting the viewing window, see the *Technology Handbook*, p. 603.

GETTING STARTED Set the viewing window on your calculator. The *standard viewing window* uses values from −10 to 10 on both the x-axis and the y-axis.

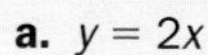
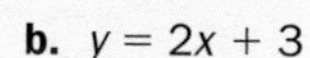
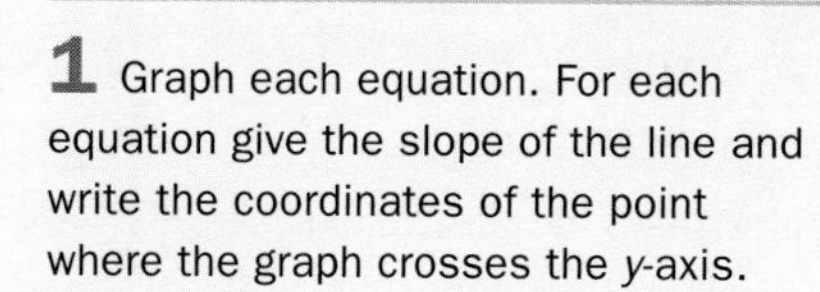

1 Graph each equation. For each equation give the slope of the line and write the coordinates of the point where the graph crosses the y-axis.

a. $y = 2x$ **b.** $y = 2x + 3$ **c.** $y = 2x - 3$

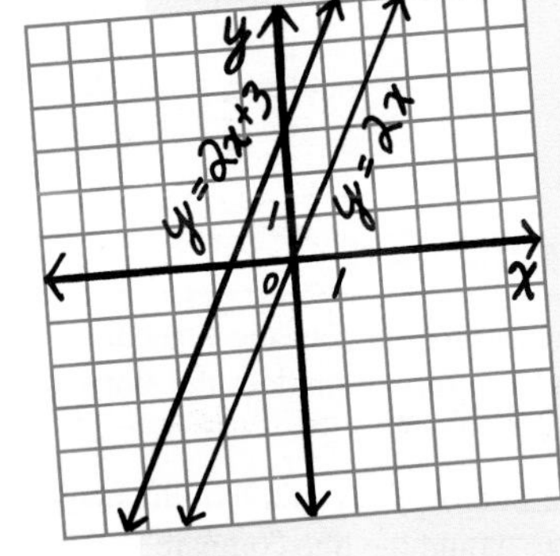

2 Sketch the graphs of the three equations. Describe any patterns you see.

3 Predict what the graph of $y = 2x + 6$ will look like. Predict the slope. Where will the graph cross the y-axis? Explain your prediction. Then check it.

4 One way to write linear equations is in the form $y = mx + b$. Based on your observations, what information do the numbers m and b give you about a graph? Use graphs to support your answer.

Exploration Note

Purpose
The purpose of this Exploration is to help students discover that the graph of an equation in the form $y = mx + b$ has slope m and y-intercept b.

Materials/Preparation
Before students begin graphing, help them choose the *standard window* from the ZOOM menu.

Procedure
As students work with their partners, one person should write down the answers to each activity. In Step 3, students can check their predictions by drawing a graph of $y = 2x + 6$. In Step 4, they should make up an equation in the form $y = mx + b$ and graph it.

Closure
Ask pairs of partners to share their results. You can then have a whole-class discussion of the fact that the graph of $y = mx + b$ has slope m and y-intercept b. Some students may also discover that equations with the same value for m but different values for b have graphs that are parallel lines.

Explorations Lab Manual
See the Manual for more commentary on this Exploration.
Diagram Master 2

For answers to the Exploration, see following page.

Plan⇔Support

Objectives
- Write linear equations in slope-intercept form.
- Analyze real-world problems represented by graphs.

Recommended Pacing
Core and Extended Courses
Section 3.4: 2 days

Block Schedule
Section 3.4: 2 half-blocks (with Sections 3.3 and 3.5)

Resource Materials
Lesson Support
Lesson Plan 3.4
Warm-Up Transparency 3.4
Overhead Visuals:
Folder 3: Graphing $y = mx + b$
Folder A: Multi-Use Graphing Packet, Sheets 1–3
Practice Bank: Practice 20
Study Guide: Section 3.4
Explorations Lab Manual:
Diagram Master 2
Assessment Book: Test 11

Technology
Graphing Calculator
Internet:
http://www.hmco.com

Warm-Up Exercises
Find the slope of each line.

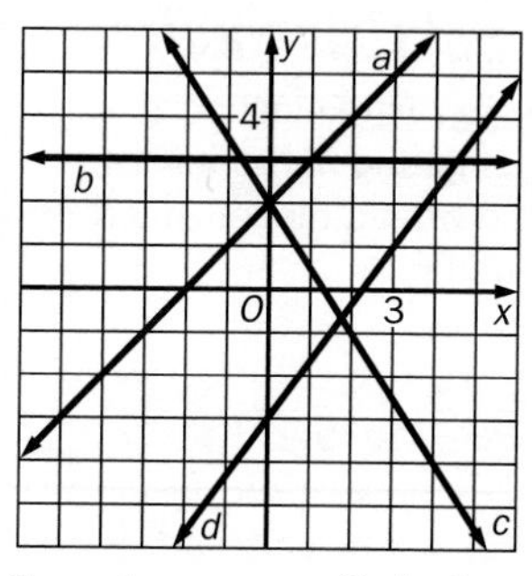

1. a 1 **2.** b 0

3. c $\frac{-3}{2}$ **4.** d $\frac{4}{3}$

Find the slope of the line through each pair of points.

5. (−3, 7) and (0, −2) −3

6. (4, −1) and (4, 7) undefined

Teach⇔Interact

Section Note

Communication: Reading The Exploration examines three equations in which the coefficient of x stays the same, while the constant term varies. This approach helps reveal the effect that m and b have on the graph of $y = mx + b$. After students have completed the Exploration, they should become familiar with the terms presented on this page.

Additional Example 1

The graph shows the weight of a candle as it burns. Find the slope and the vertical intercept of the graph. Write an equation for the weight of the candle in slope-intercept form.

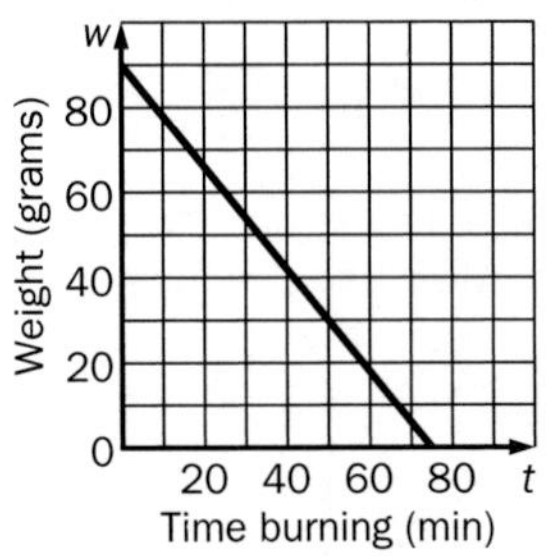

Choose two convenient points to find the slope.

Slope $= \frac{0 - 90}{75 - 0} = -1.2$

The slope is -1.2.

The graph crosses the vertical axis at (0, 90). The vertical intercept is 90.

Let w be the weight of the candle in grams after it has been burning t minutes. An equation for the weight of the candle is $w = -1.2t + 90$.

Think and Communicate

In connection with question 1, ask students what the horizontal intercept is and what it means in terms of the situation.

Slope-Intercept Form

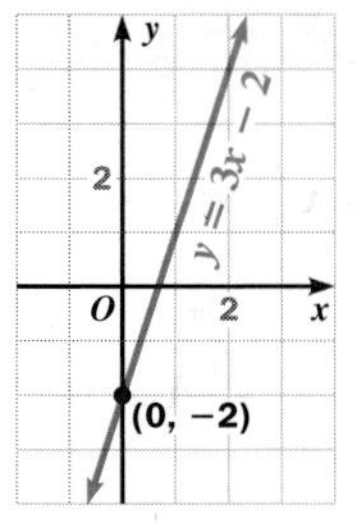

As you can see, the 3 in the linear equation $y = 3x - 2$ is the slope of the graph. The -2 is the y-value of the point where the graph crosses the y-axis. This value is called the **y-intercept**, or the **vertical intercept**.

One way to write a linear equation is in **slope-intercept form**.

$$y = mx + b$$

The letter m represents the slope of the line.

The letter b represents the vertical intercept.

EXAMPLE 1

The graph shows the height of a candle as it burns. Find the slope and the vertical intercept of the graph. Write an equation for the height of the candle in slope-intercept form.

SOLUTION

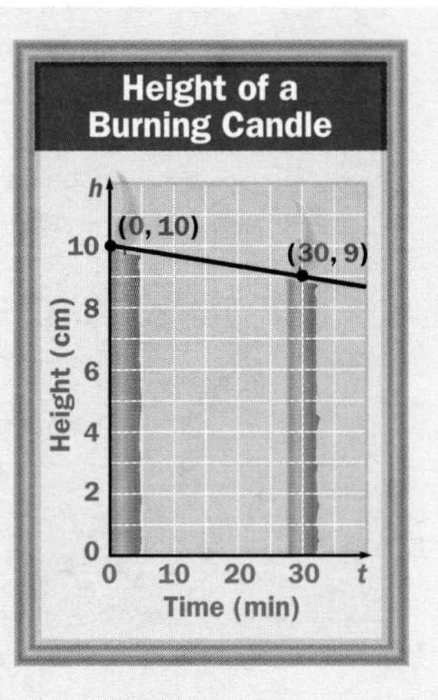

Choose two convenient points to find the slope.

$$\text{Slope} = \frac{9 - 10}{30 - 0} = \frac{-1}{30} \approx -0.03$$

The slope is about -0.03.

The graph crosses the vertical axis at (0, **10**). The vertical intercept is **10**.

Let h = the height of the candle in centimeters after it has been burning t minutes.

An equation for the height of the candle is $h = -0.03t + 10$.

BY THE WAY

People in ancient China and Japan used slowly burning ropes to keep track of time. The rope was marked with glue or knots to show equal periods of time.

THINK AND COMMUNICATE

1. In Example 1, what information does the vertical intercept give you about the candle? What information does the slope give you? Why is the slope negative?
2. Estimate the vertical intercept of a graph of the height of a burning birthday candle. How would the line compare to the graph in Example 1? Would the slope be greater or less? Explain.

ANSWERS Section 3.4

Exploration

1. a. 2; (0, 0)

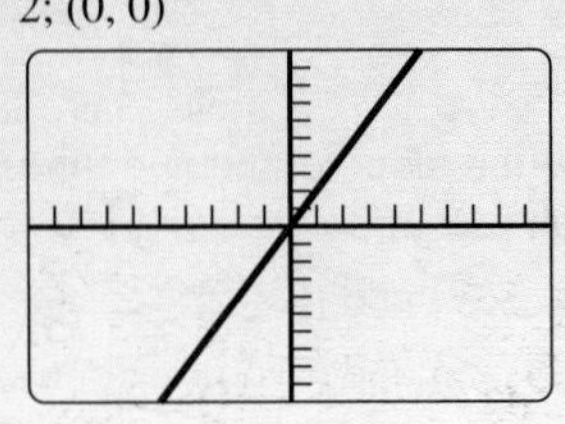

b. 2; (0, 3)

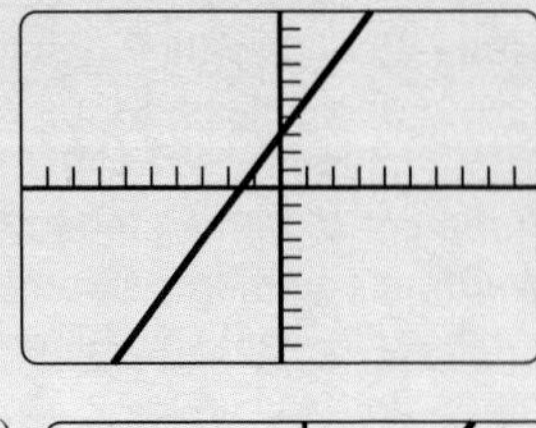

c. 2; (0, –3)

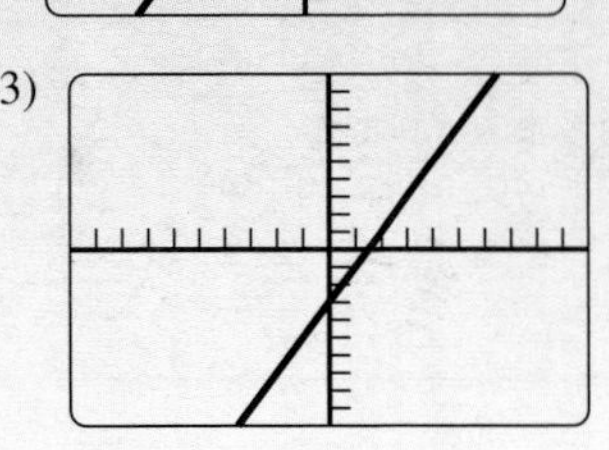

2.

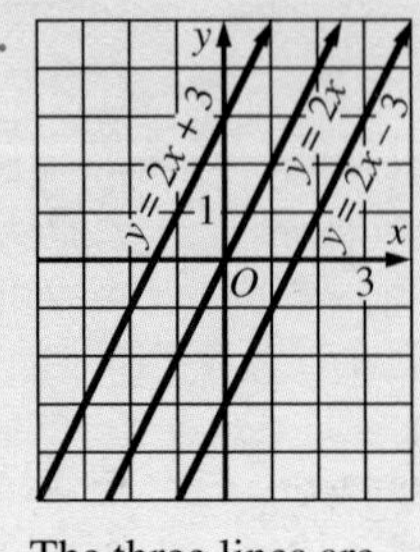

The three lines are parallel. Each line has slope 2, which is the coefficient of x in each equation. The y-coordinate of the point where each graph crosses the y-axis is equal to the number that is added to or subtracted from $2x$ in the equation.

3. The graph will be a line that is parallel to the lines in Step 1. The slope will be 2, since the coefficient of x is 2, and the graph will cross the y-axis at (0, 6), since 6 is added to $2x$ in the equation.

EXAMPLE 2

Graph the equation $y = 6 - \frac{3}{2}x$.

SOLUTION

Method 1 Use graph paper.

Step 1 Rewrite the equation in slope-intercept form to find the slope and y-intercept.

$$y = 6 - \frac{3}{2}x$$

$$y = -\frac{3}{2}x + 6$$

Step 2 The y-intercept is 6. Graph the point (0, 6).

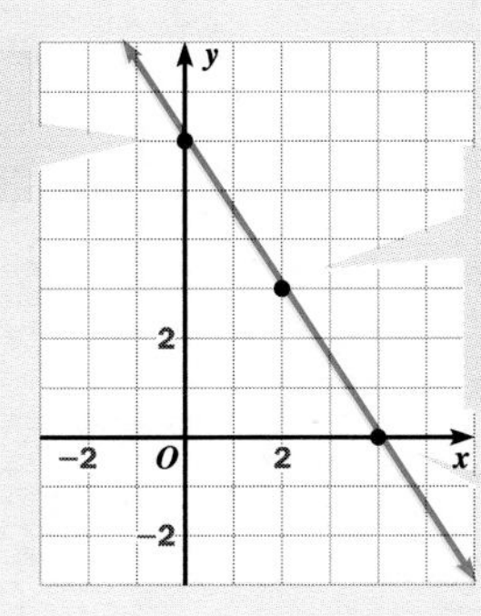

Step 3 Start at the y-intercept. Count 3 units down and 2 units to the right. Graph a second point.

Step 4 Count down 3 and 2 to the right again to plot a third point. Connect the points with a line.

Method 2 Use a graphing calculator.

Enter the equation $y = 6 - \frac{3}{2}x$ as $y = 6 - (3/2)x$.

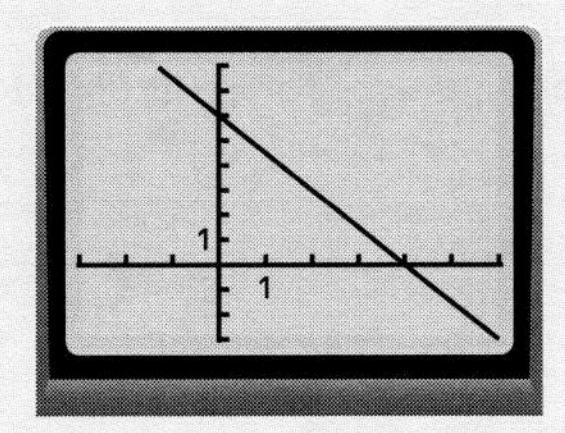

WATCH OUT!

Use parentheses to enter $y = 6 - (3/2)x$, so your calculator graphs $y = 6 - \frac{3}{2}x$, not $y = 6 - \frac{3}{2x}$.

THINK AND COMMUNICATE

3. Check that the three points plotted in method 1 of Example 2 are solutions of the equation $y = 6 - \frac{3}{2}x$. How can checking help you make sure a graph is accurate? How can you check a graph on a graphing calculator?

4. In Section 2.5, you learned to graph equations by creating a table of values and plotting points. Compare that method with method 1 of Example 2. What are some advantages and disadvantages of each of the two methods?

Teach⇔Interact

Additional Example 2

Graph the equation $y = -4 + \frac{3}{5}x$.

Method 1 Use graph paper.

Step 1 Rewrite the equation in slope-intercept form:
$y = \frac{3}{5}x + (-4)$.

Step 2 The y-intercept is −4. Graph the point (0, −4).

Step 3 Start at the y-intercept. Count 3 units up and 5 units to the right. Graph a second point.

Step 4 Count up 3 and 5 to the right to plot a third point. Connect the points with a line.

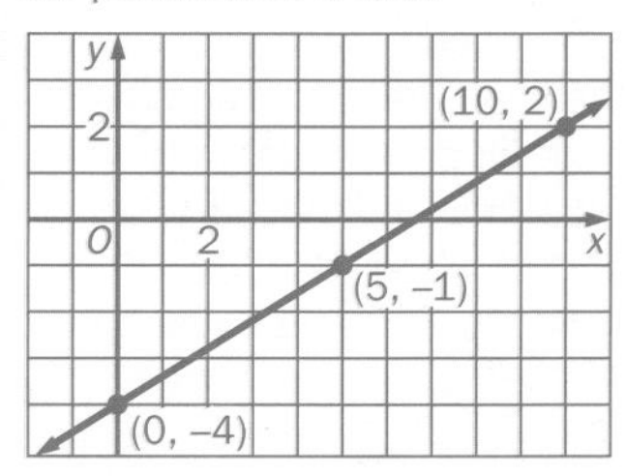

Method 2 Use a graphing calculator. Enter the equation as $y = (3/5)x - 4$ or as $y = -4 + (3/5)x$. (Students who enter the equation $y = -4 + \frac{3}{5}x$ on the graphing calculator should be careful to use the [(−)] key rather than the subtraction key. Otherwise, the calculator will show an error message.)

Think and Communicate

For question 3, students need to be aware that line graphs on graphing calculators may appear to have different slopes than the same graphs drawn on graph paper. This happens when the scales on the axes are different. They can remedy this by pressing [ZOOM] and selecting menu item 5. This will "square up" the graph.

---

4. Steps 1–3 show that the graph of $y = mx + b$ is a line with slope m and with b as the y-coordinate of the point where the line crosses the y-axis.

Think and Communicate

1. the candle's height in centimeters before the candle was lit; the rate at which the candle burns in centimeters per minute; because the candle's height decreases as it burns

2. Estimates may vary; about 6. The line would be steeper, that is, the slope would be less than −0.03, since birthday candles burn very quickly.

3. $6 = 6 - \frac{3}{2}(0)$; $3 = 6 - \frac{3}{2}(2)$; $0 = 6 - \frac{3}{2}(4)$; By checking that plotted points are solutions to the equation, you can be sure that the graph is correct. Answers may vary; for example, use the trace feature to find the coordinates of a few points of the graph, and check that these points are solutions of the original equation.

4. Answers may vary. An example is given. Using a table of values to graph an equation takes longer, I think, than using the slope and the y-intercept, but I am less likely to make an error in graphing using a table of values. Using the slope and y-intercept is a quick way to graph a linear equation, but I have to be careful to rewrite the equation correctly. Also, the method in Example 2 may be more difficult for an equation such as $y = \frac{1}{2}x + \frac{2}{3}$, where the slope and the y-intercept are not whole numbers.

Teach⇔Interact

Checking Key Concepts

Communication: Discussion You can have an interesting discussion by asking students to point out similarities and differences between the situation in question 4 and the candle-burning situation in Example 1.

Closure Question

If you have the graph of a line but not its equation, how can you find the slope, y-intercept, and an equation for the line?

Find the coordinates of two points on the graph and use them to find the slope (m). Find the y-intercept (b) from the graph. Use the values for m and b to write an equation in the form $y = mx + b$.

Apply⇔Assess

Suggested Assignment

Core Course
Day 1 Exs. 1–17
Day 2 Exs. 18–23, 25–34, AYP

Extended Course
Day 1 Exs. 1–17
Day 2 Exs. 18–34, AYP

Block Schedule
Day 17 Exs. 1–17
Day 18 Exs. 18–23, 25–34, AYP

Exercise Notes

Common Error
Exs. 1–6 Students can find the slope of each line by using the graph paper to count the vertical change and horizontal change for the two points shown. For lines having a negative slope, as in Exs. 3 and 5, some students may forget to use a negative sign and treat the slope as positive. Remind these students that to get from one point to the other point, they have to move either left or down, which are negative directions.

Teaching Tip
Exs. 7–15 You may wish to have students first draw the graphs on graph paper and then check them using a graphing calculator.

CHECKING KEY CONCEPTS

For Questions 1–3, give the slope and vertical intercept of the graph of each equation. Then graph the equation.

1. $y = x + 4$ **2.** $y = \frac{1}{2}x - 3$ **3.** $y = x$

4. Some students performed an experiment by dropping marbles into a jar of water. They measured the height of the water after every five marbles. Then they graphed their data.

a. Write an equation for the height of the water as a function of the number of marbles.

b. What do the vertical intercept and the slope of the graph of your equation tell you about the experiment?

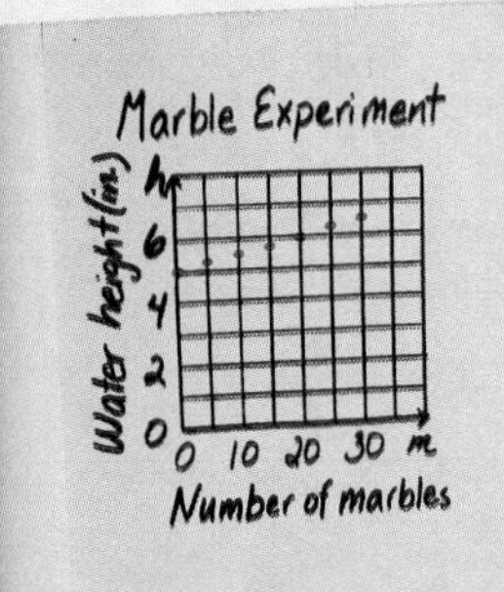

3.4 Exercises and Applications

Extra Practice exercises on page 562

Write an equation in slope-intercept form for each graph.

1.
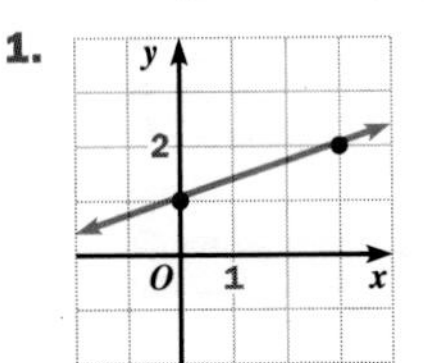

2.
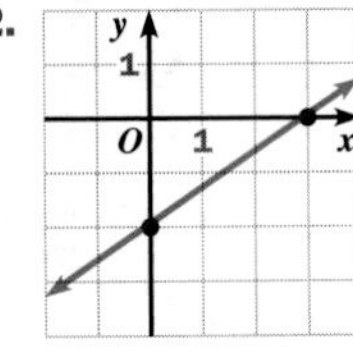

3.
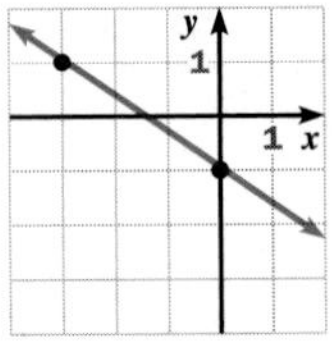

4.
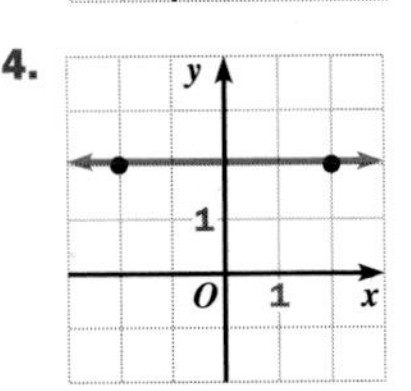

5.
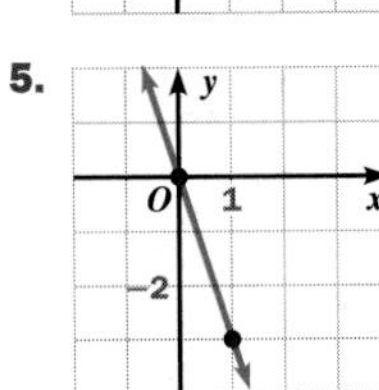

6.
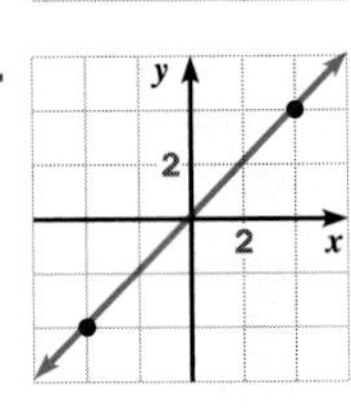

Graph each equation.

7. $y = x - 3$ **8.** $y = 3x + 1$ **9.** $y = 5x - 2$

10. $y = -x$ **11.** $y = -5$ **12.** $x = 3$

13. $x = -\frac{1}{2}$ **14.** $y = -\frac{2}{3}x + 4$ **15.** $y = \frac{5}{3}x + 1$

Checking Key Concepts

1. 1; 4

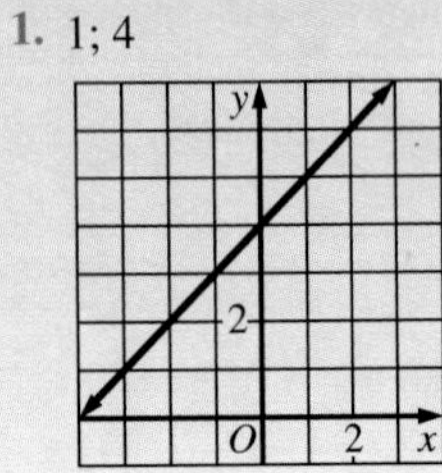

2. $\frac{1}{2}$; −3

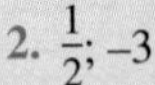

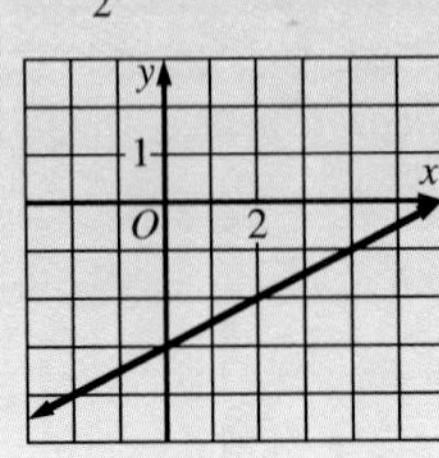

3. 1; 0

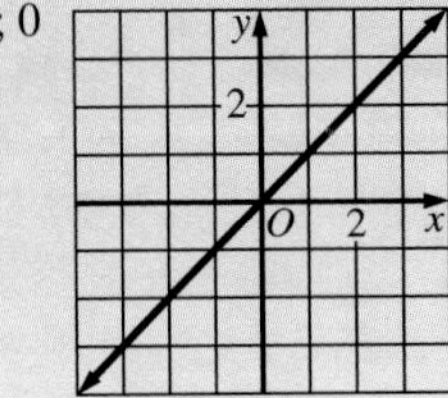

4. a. $h = 0.05m + 5$

b. The vertical intercept 5 is the water level in inches in the jar before any marbles are put in the water. The slope 0.05 is the increase in the water height in inches for each marble added.

16. Open-ended Problem Describe two different ways to graph the equation $y = \frac{1}{4}x + 2$ by hand. Which method do you prefer? Why?

17. a. Explain why the equations in Exercises 10 and 11 are in slope-intercept form. Give the values of m and b for the equation.

b. Writing What is the y-intercept of the graph of a direct variation? Explain your answer, using both graphs and equations.

18. Cooperative Learning Work with a partner.

a. Write three linear equations in slope-intercept form. On a separate sheet of paper, graph each of your equations. Keep your graphs secret. Trade equations with your partner.

b. Graph the three equations that you received from your partner.

c. Compare your graph of each equation with your partner's. If they are different, work together to make an accurate graph of the equation.

19. Challenge What is true about an equation of any line through the point $(0, -2)$? What is true about any line with a slope of 3? How many lines have slope 3 and pass through the point $(0, -2)$? Explain.

Connection HISTORY

Alfred the Great (849–899) was king of Wessex in southwestern England. He is said to have invented a candle clock. A 12 in. candle was divided into twelve equal parts colored alternately black and white. Three parts burned each hour.

20. Graph the height h of one candle as a function of the time t it has burned.

21. Give the slope and vertical intercept of the graph. Explain what each tells you about the situation.

22. How long did it take for one candle to burn? How do you know?

23. Write an equation of the graph in slope-intercept form.

Modern reconstruction of King Alfred's candleclock

Egyptian water clock

24. Research Another type of clock used before modern mechanical clocks was the clepsydra, or water clock. Find out more about the clepsydra. Write a report and draw a graph that illustrates how the clepsydra worked.

BY THE WAY

King Alfred also is said to have invented the "lant-horn" or lantern to protect the candles from drafts that would affect the rate at which the candle burned.

Apply⇔Assess

Exercise Notes

Journal Entry

Ex. 16 Students may wish to include their answers for this exercise in their journals.

Cooperative Learning

Ex. 18 Students will need to work separately from their partners when drawing their graphs. Partners should also use the same kind of graph paper. Any groups having difficulty in making accurate graphs can be assigned a peer tutor for further instruction on graphing linear equations.

Challenge

Ex. 19 You may wish to ask how ideas from this exercise can be used to decide whether three points whose coordinates are given lie on the same straight line.

Exercises and Applications

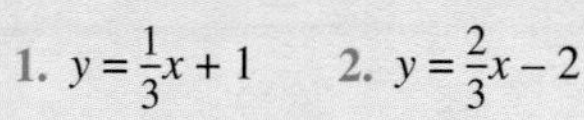

1. $y = \frac{1}{3}x + 1$ 2. $y = \frac{2}{3}x - 2$

3. $y = -\frac{2}{3}x - 1$ 4. $y = 2$

5. $y = -3x$ 6. $y = x$

7.

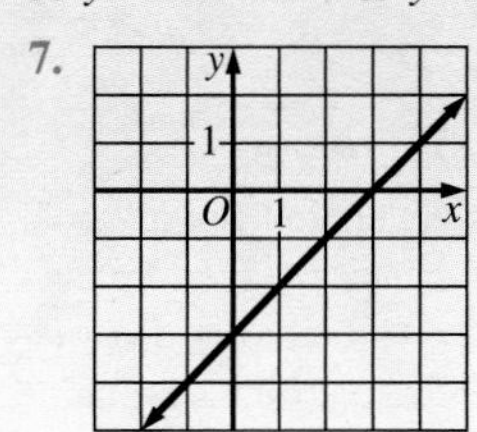

8.

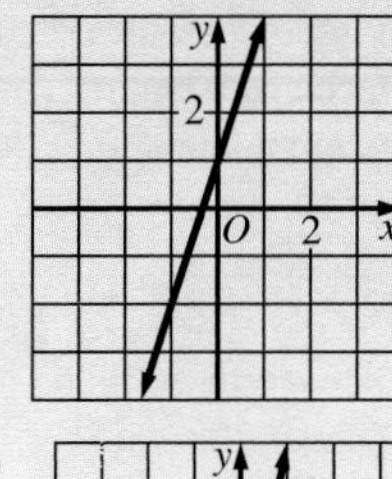

9.

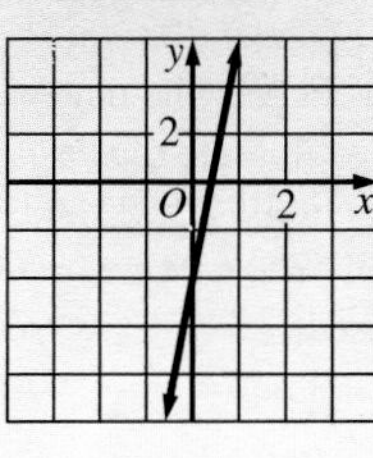

10.

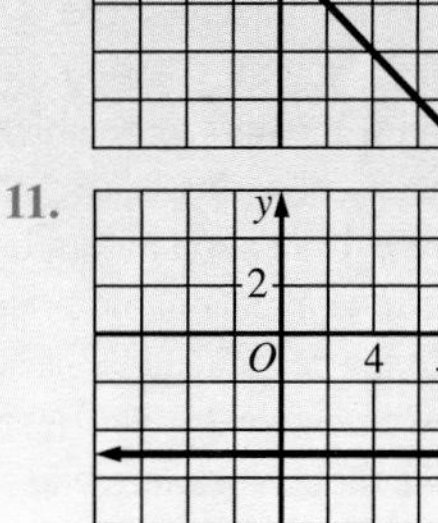

11.

12.

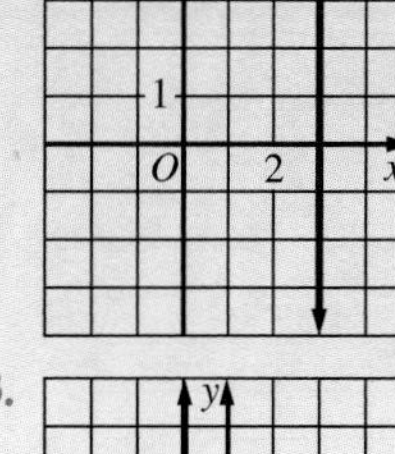

13.

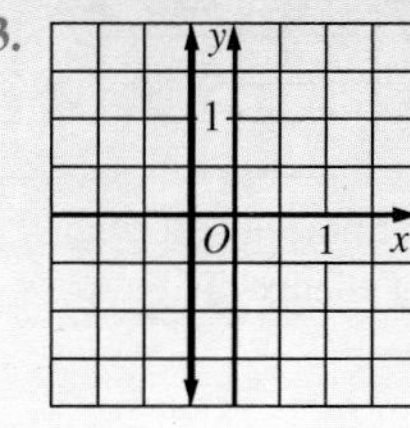

14.

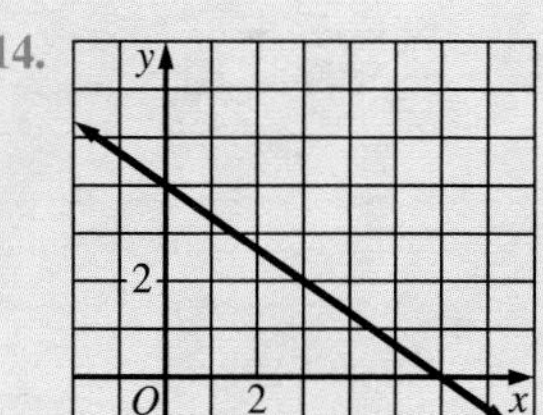

15.

16. Answers may vary. An example is given. One way to graph the equation is to prepare and graph a table of values. Another way is to graph the point (0, 2), since the y-intercept is 2 and then use the slope, $\frac{1}{4}$, to find additional points on the graph. I prefer the slope-intercept method because the equation is already written in slope-intercept form.

17. a. Both equations are in slope-intercept form because they can be expressed as $y = mx + b$. For Ex. 10, $m = -1$ and $b = 0$. For Ex. 11, $m = 0$ and $b = -5$.

b. 0; A direct variation equation has the form $y = kx$, or $y = kx + 0$. Comparing this with $y = mx + b$, you see that $b = 0$. Also $0 = k(0)$, so 0 is the vertical intercept of the graph of $y = kx$. The graph of a direct variation is a line through the origin.

18–24. See answers in back of book.

Apply⇔Assess

Exercise Notes

Assessment Note

Ex. 25 Students will benefit by comparing their answers to this exercise. They can do this in small groups or through class discussion.

Assess Your Progress

When students review the vocabulary terms, ask them to accompany verbal definitions with appropriate diagrams. Many students may find it easier to explain their ideas with a diagram rather than verbally.

Progress Check 3.3–3.4

See page 139.

Practice 20 for Section 3.4

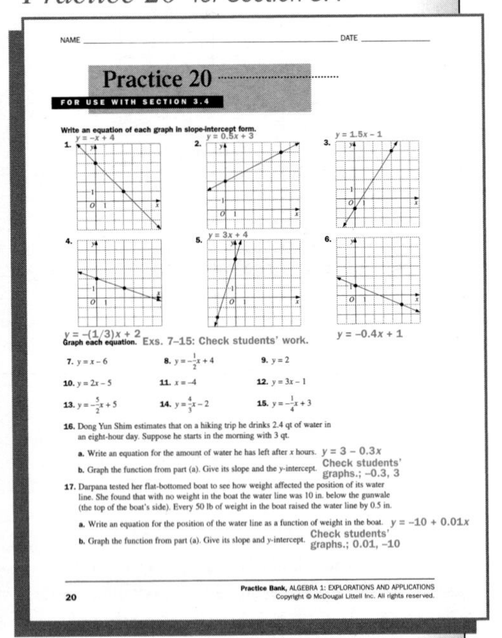

NAME ______________________ DATE __________

Practice 20

FOR USE WITH SECTION 3.4

Write an equation of each graph in slope-intercept form.

1. $y = -x + 4$
2. $y = 0.5x + 3$
3. $y = 1.5x - 1$
4. $y = -(1/3)x + 2$
5. $y = 3x + 4$
6. $y = -0.4x + 1$

Graph each equation. Exs. 7–15: Check students' work.

7. $y = x - 6$
8. $y = -\frac{1}{2}x + 4$
9. $y = 2$
10. $y = 2x - 5$
11. $x = -4$
12. $y = 3x - 1$
13. $y = -\frac{5}{2}x + 5$
14. $y = \frac{4}{3}x - 2$
15. $y = -\frac{1}{4}x + 3$

16. Dong Yun Shim estimates that on a hiking trip he drinks 2.4 qt of water in an eight-hour day. Suppose he starts in the morning with 3 qt.
    a. Write an equation for the amount of water he has left after x hours. $y = 3 - 0.3x$
    b. Graph the function from part (a). Give its slope and the y-intercept. Check students' graphs.; –0.3, 3

17. Darpana tested her flat-bottomed boat to see how weight affected the position of its water line. She found that with no weight in the boat the water line was 10 in. below the gunwale (the top of the boat's side). Every 50 lb of weight in the boat raised the water line by 0.5 in.
    a. Write an equation for the position of the water line as a function of weight in the boat. $y = -10 + 0.01x$
    b. Graph the function from part (a). Give its slope and y-intercept. Check students' graphs.; 0.01, –10

20 **Practice Bank,** ALGEBRA 1: EXPLORATIONS AND APPLICATIONS

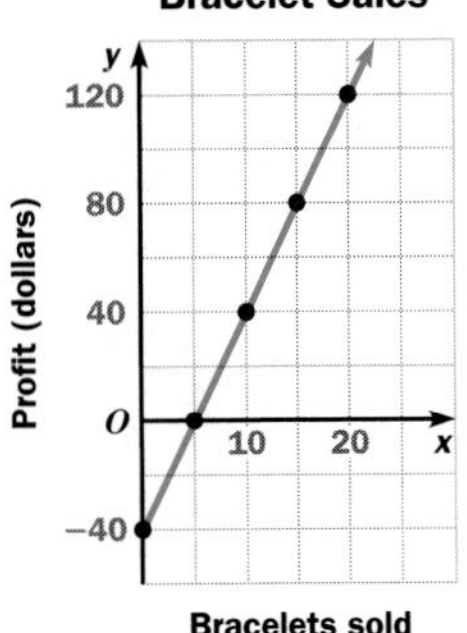

ONGOING ASSESSMENT

25. a. **Writing** Alicia Hartwell owns a jewelry business. Every week she buys materials and sells the bracelets she makes at a crafts market. The graph shows her weekly profit from bracelet sales. What does the slope of the line tell you about Alicia's business? What does the vertical intercept of the line tell you?

    b. The crafts market has decided to charge a weekly membership fee of \$15. How does this affect the graph of Alicia's weekly profit? Do you think she should increase the price of her bracelets? Explain your suggestion using graphs and equations.

SPIRAL REVIEW

Solve each equation. *(Sections 2.1 and 2.2)*

26. $4x - 2 = 10$
27. $6t + 1 = 17$
28. $-2m - 9 = 11$
29. $-s + 4s = 45$
30. $2x + 18 = 38$
31. $35y + 17y = 104$
32. $\frac{1}{3}x + 2 = 5$
33. $\frac{5}{6}r + 9 = 4$
34. $\frac{2}{7}z + 5 = 9$

ASSESS YOUR PROGRESS

VOCABULARY

slope (pp. 113, 115)
positive slope (p. 114)
negative slope (p. 114)
undefined slope (p. 115)
linear equation (p. 119)
y-intercept, vertical intercept (p. 120)
slope-intercept form (p. 120)

Find the slope of the line through each pair of points. *(Section 3.3)*

1. (2, 7) and (1, 5)
2. (−1, 0) and (−1, 3)
3. (12, 9) and (9, 12)
4. Sue Hill drove at a constant speed from her home to Tampa. The graph at the right shows her distance from Tampa as a function of time. *(Sections 3.3 and 3.4)*
    a. Write an equation for the line in slope-intercept form.
    b. What information do the slope and the vertical intercept give you about the situation?

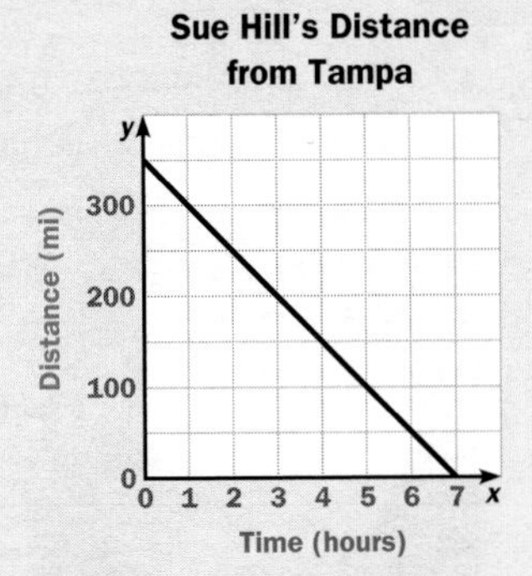

Graph each equation. Give the slope and y-intercept of the graph. *(Section 3.4)*

5. $y = 3x - 4$
6. $y = -x + 2$
7. $y = \frac{1}{3}x - \frac{4}{3}$
8. **Journal** Describe two ways to find the slope of the graph of a linear equation. Which do you prefer? Why?

---

25. a. The slope 8 tells the profit in dollars for each bracelet sold; the vertical intercept –40 shows that Alicia has expenses every week of \$40, whether or not she sells any bracelets.

    b.

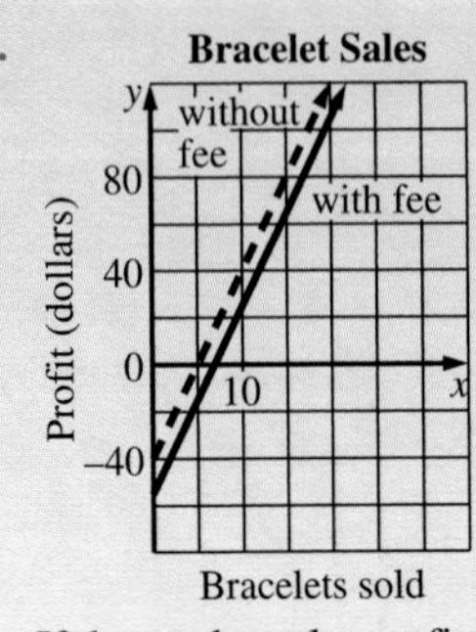

If the per bracelet profit does not change but the costs change to \$55, the horizontal intercept increases, that is, Alicia must sell more bracelets to begin making a profit. With the new vertical intercept, the slope must be greater to equal the profit for any given number of bracelets sold. The equation for the old situation is $y = 8x - 40$ and for the new is $y = 8x - 55$. If Alicia still wants to be able to break even by selling 5 bracelets, she must earn \$11 per bracelet. ($0 = m(5) - 55$; $m = 11$) That may not be reasonable. She must determine a reasonable price by evaluating the demand for her bracelets, her competitors' prices, and the average weekly profit she had been earning before the new charge was assessed.

26. 3
27. $\frac{8}{3}$
28. –10
29. 15
30. 10
31. 2
32. 9
33. –6
34. 14

Assess Your Progress

1–8. See answers in back of book.

SECTION

3.5 Writing an Equation of a Line

Learn how to...

- **find the *y*-intercept of a graph**

So you can...

- **model real-world situations, such as the altitude of a hot air balloon**

Every year at the International Balloon Fiesta in Albuquerque, New Mexico, thousands of hot air balloons take to the sky. The graph below shows the altitude of one balloon as it climbs.

This break in the scale means that the *y*-axis is not shown for values of *y* between 0 and 6000.

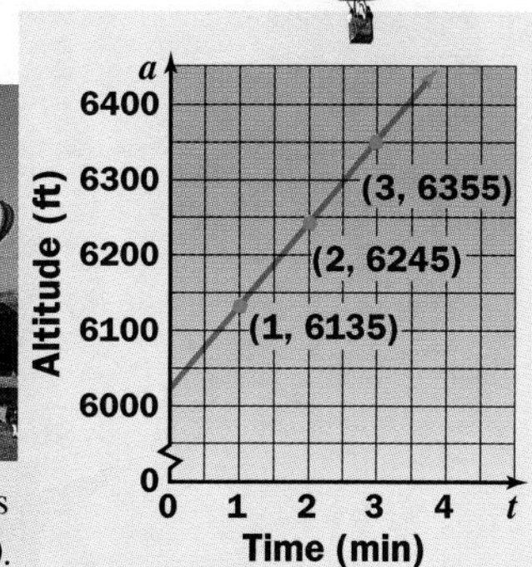

EXAMPLE 1 Application: Recreation

As shown by the slope of the graph, the balloon rises at a rate of 110 ft/min. Write an equation for the balloon's altitude a as a function of time t.

SOLUTION

Step 1 Write the equation in slope-intercept form.

Altitude = Slope × Time + Vertical intercept

$$a = mt + b$$

Step 2 Substitute the **slope** of the line for m in the equation.

$$a = 110t + b$$

Step 3 To find the **vertical intercept**, substitute the coordinates of any point on the graph for t and a in the equation. Then solve for b.

$$6135 = 110(1) + b$$ Choose the point (1, 6135).

$$6135 = 110 + b$$

$$6025 = b$$ Solve for b.

Step 4 Substitute **6025** for b in the equation $a = 110t + b$.

An equation for the altitude of the balloon as a function of time is $a = 110t + 6025$.

Plan⇔Support

Objectives

- Find the *y*-intercept of a graph.
- Model real-world situations.

Recommended Pacing

Core and Extended Courses
Section 3.5: 2 days

Block Schedule
Section 3.5: 2 half-blocks (with Sections 3.4 and 3.6)

Resource Materials

Lesson Support
Lesson Plan 3.5
Warm-Up Transparency 3.5
Overhead Visuals:
Folder A: Multi-Use Graphing Packet, Sheets 1–3
Practice Bank: Practice 21
Study Guide: Section 3.5
Explorations Lab Manual:
Diagram Masters 1, 2

Technology
Technology Book:
Calculator Activity 3
Internet:
http://www.hmco.com

Warm-Up Exercises

Solve each equation.

1. $5 = 14 + k$ −9

2. $-8 = 7(-3) + n$ 13

3. $14 = -0.8(6) + b$ 18.8

Find the slope of the line through each pair of points.

4. $P(0, -2)$ and $Q(4, 9)$ 2.75

5. $M(-1, -2.5)$ and $N(3, -6.5)$ −1

6. Is $\left(\frac{1}{2}, 1\right)$ on the graph of $y = -62x + 32$? How do you know?
Yes; Replacing x with $\frac{1}{2}$ and y with 1 gives $1 = -62\left(\frac{1}{2}\right) + 32$, which is a true equation.

Teach⇔Interact

Additional Example 1

Suppose a hot air balloon has been descending at a rate of 55 ft/min for some time. Write an equation for the balloon's altitude as a function of the time t it has been descending.

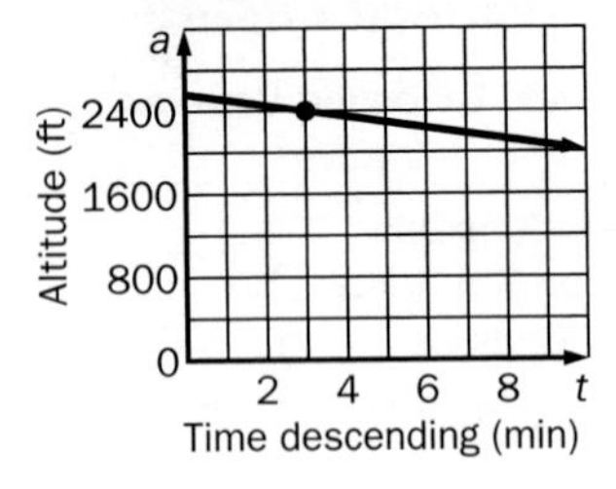

Step 1 Write the equation in slope-intercept form.
$a = mt + b$
Step 2 Substitute the slope of the line for m in the equation.
$a = -55t + b$
Step 3 To find the vertical intercept, substitute the coordinates of any point on the graph for t and a in the equation. Choose the point (3, 2400). Then solve for b.

$2400 = -55(3) + b$
$2400 = -165 + b$
$2565 = b$

Step 4 Substitute 2565 for b in the equation $a = -55t + b$.
An equation for the altitude of the balloon as a function of descending time is $a = -55t + 2565$.

Additional Example 2

Find an equation of the line through the points (–1 ,2) and (4, –0.5).
Draw a graph.

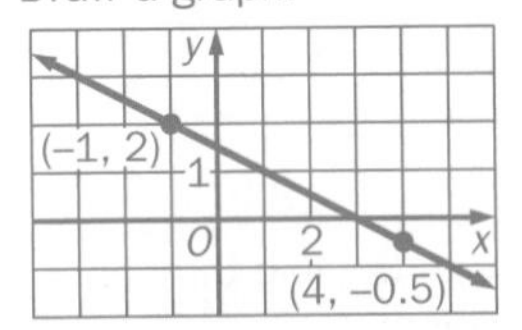

Step 1 The equation will be in slope-intercept form, $y = mx + b$.
Step 2 Find the slope.

$\text{slope} = \frac{-0.5 - 2}{4 - (-1)} = -0.5$

Substitute –0.5 for m in the equation. $y = -0.5x + b$
Step 3 Find the y-intercept. Substitute the coordinates of (–1, 2) in $y = -0.5x + b$ and solve for b.

$2 = -0.5(-1) + b$
$1.5 = b$

Step 4 Substitute 1.5 for b in the equation.
An equation of the line is $y = -0.5x + 1.5$.

BY THE WAY

Because hot air is lighter than cold air, hot air balloons rise when the air inside the balloon is warmer than the air outside. The hottest air is at the top of the balloon. The air ranges in temperature from 180°F to 225°F.

THINK AND COMMUNICATE

1. In Example 1, why can you substitute the coordinates of the point (1, 6135) into the equation and solve for b? Can you use another point? Explain why or why not.

2. Federal aviation regulations require that balloons at the International Balloon Fiesta fly at an altitude of at least 6500 ft while over the Fiesta grounds. How long will it take the balloon in Example 1 to reach this altitude?

EXAMPLE 2

Find an equation of the line through the points (2, 6) and (3, 1).

SOLUTION

Draw a graph.

Step 1 The equation will be in slope-intercept form:

$$y = mx + b$$

Step 2 Find the slope.

$$\text{slope} = \frac{1 - 6}{3 - 2} = \frac{-5}{1} = -5$$

Substitute −5 for m in the equation.

$$y = -5x + b$$

Step 3 Find the y-intercept. Substitute the coordinates of any point on the line for x and y in the equation. Then solve for b.

$6 = -5(2) + b$ — Choose the point (2, 6).
$6 = -10 + b$
$16 = b$ — Solve for b.

Step 4 Substitute 16 for b in the equation.

An equation of the line is $y = -5x + 16$.

THINK AND COMMUNICATE

3. In Example 2, substitute the coordinates of the second point for x and y in the equation $y = -5x + b$. Why do you get the same value for b?

4. a. Graph the horizontal line through (1, 2) and (5, 2). Write an equation for this line.

b. What is the slope of any horizontal line?

c. What do all equations of horizontal lines have in common?

ANSWERS Section 3.5

Think and Communicate

1. Since the point (1, 6135) lies on the line, its coordinates must make the equation $a = 110t + b$ true. Yes; you can use any point on the line to solve for b, since the coordinates of any point make the equation true.

2. about 4.3 min after its ascent begins

3. $1 = -5(3) + b$; $b = 1 + 15 = 16$; You get the same value for b because every point on the line makes the same slope-intercept equation true.

4. a. $y = 2$

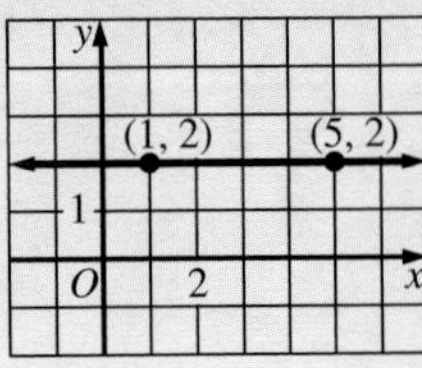

b. 0

c. It has the form $y = k$, where k is a constant.

EXAMPLE 3

Writing **Find an equation of the vertical line through the point (−2, 3). Explain your reasoning.**

SOLUTION

Here is one student's solution.

The student included a diagram.

First I drew a sketch of the line.

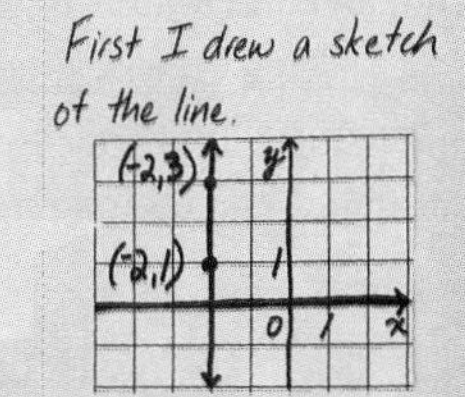

Next I picked a second point on the line (−2, 1), and tried to find the slope.

There was a zero in the denominator, so I knew the slope was undefined. I started looking at some other points. The coordinates of every point I chose looked like (−2, y). If I plot every point with x = −2, I get the vertical line, so an equation of the line is x = −2.

The student explained each step in reaching a solution.

CHECKING KEY CONCEPTS

Find an equation of the line with the given slope and through the given point.

1. slope = 2; (0, −3)
2. slope = −2; (5, 7)
3. slope = 0; (3, 4)
4. slope is undefined; (4, 6)

Find an equation of each line.

5. 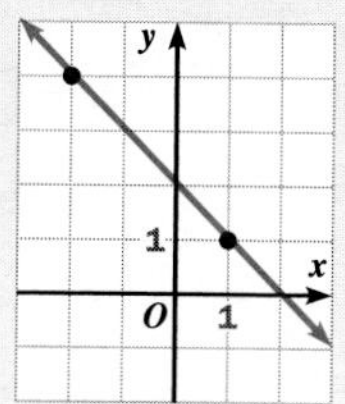

6. 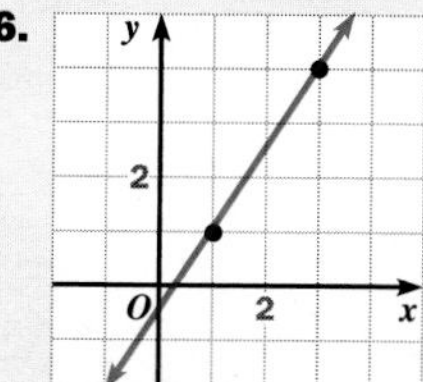

3.5 Exercises and Applications

Find an equation of the line with the given slope and through the given point.

1. slope = 1; (5, 8)
2. slope = 4; (5, 9)
3. slope = −3; (7, 10)
4. slope = $\frac{1}{3}$; (3, 9)
5. slope = −2; (−4, 8)
6. slope = $-\frac{1}{2}$; (1, 6)
7. slope = 0; (−2, 5)
8. slope = −0.5; (3, −2)
9. undefined slope; (1, 1)

Extra Practice *exercises on page 563*

Checking Key Concepts

1. $y = 2x - 3$
2. $y = -2x + 17$
3. $y = 4$
4. $x = -2$
5. $y = -x + 2$
6. $y = \frac{3}{2}x - \frac{1}{2}$

Exercises and Applications

1. $y = x + 3$
2. $y = 4x - 11$
3. $y = -3x + 31$
4. $y = \frac{1}{3}x + 8$
5. $y = -2x$
6. $y = -\frac{1}{2}x + \frac{13}{2}$
7. $y = 5$
8. $y = -0.5x - 0.5$
9. $x = 1$

Teach⇔Interact

Think and Communicate

Question 4 can be approached as a special case of Example 2. You may wish to list several other points on the board that have *y*-coordinates equal to 2. Then ask students if they see a pattern in the coordinates of the points. Ask how they can express this pattern using the variable *x*. (*x*, 2) This approach may help students understand the solution to Example 3.

Additional Example 3

Find an equation of the vertical line through the point (7, −2). Explain your reasoning.

A possible solution is to draw a diagram of the line.

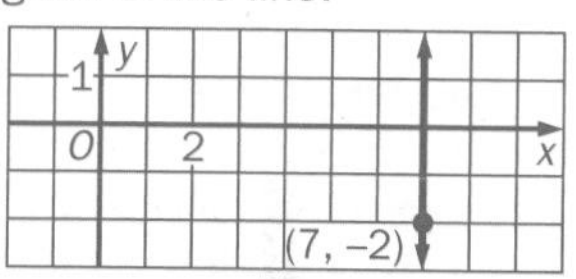

List several other points that appear to be on the line; for example, (7, −3), (7, 0), (7, 1), and (7, 8). Look for a pattern. It appears that the *y*-coordinate does not matter, but that all of the *x*-coordinates are 7. An equation for the line is *x* = 7.

Checking Key Concepts

Visual Thinking

Ask students to sketch each of the lines given in questions 1–4. This activity involves the visual skills of *interpretation* and *communication.*

Closure Question

Describe how you can find an equation for a line if you know the coordinates of two points on the line.

Plot the points and draw the line. If the line is not vertical, use the two points to find its slope. Substitute the slope for *m* in the equation *y* = *mx* + *b*. Then substitute the coordinates of one of the given points in the equation just obtained. Solve for *b*. Write an equation in the form *y* = *mx* + *b* using the values found for *m* and *b*. If the line is vertical, use the equation *x* = *a*, where *a* is the *x*-coordinate of one of the given points.

Apply⇔Assess

Suggested Assignment

Core Course
Day 1 Exs. 1–15
Day 2 Exs. 16–25, 28–34

Extended Course
Day 1 Exs. 1–15
Day 2 Exs. 16–34

Block Schedule
Day 18 Exs. 1–15
Day 19 Exs. 16–34

Exercise Notes

Communication: Writing
Exs. 1–9 You may wish to have students write their solutions to these exercises on the board in order to review the procedures involved and to check for accuracy.

Integrating the Strands
Ex. 11 This exercise illustrates the fact that the geometric concept of parallel lines can be related to the algebraic concept of lines that have equal slope (or undefined slopes).

Student Progress
Exs. 12–20 Some students have serious difficulties finding an equation of the line through two given points. The techniques used involve a number of steps and students may need time to assimilate and understand the pertinent ideas. These students can be helped by doing one or two of these exercises over a period of a week or more.

Teaching Tip
Exs. 12–20 Some students may need to sketch a diagram for each exercise. Others may enjoy seeing if they can mentally visualize the locations of the points. They can then find the equations by using the techniques of Example 2 or Example 3.

10. A balloon at the International Balloon Fiesta lifted off from a different location from the one in Example 1. It reached an altitude of 6125 ft above sea level 1 min into its flight. After 3 min, it reached an altitude of 6245 ft. Write an equation for the balloon's altitude a after t min.

11. A line passes through the points (0, 4) and (5, 0).

a. Graph the line and find an equation for it.

b. A line passes through the point (0, 6) and has the same slope as the line in part (a). Graph this line and find an equation for it.

c. **Open-ended Problem** Write an equation for another line with the same slope as the lines in parts (a) and (b). Explain how you found this line. What can you say about these lines and their equations?

Find an equation of the line through the given points.

12. (8, 4), (10, 2) **13.** (12, 18), (−22, 18) **14.** (7, 1), (7, 4)

15. (0, −3), (6, 9) **16.** (0.5, −2), (2.5, −2) **17.** (−2, 5), (4, −5)

18. (15, −3), (15, 2) **19.** (−1, 12), (6, −7) **20.** (−5, −7), (7, 6)

21. **Challenge** In the graph at the right, an equation of line j is $y = x$.

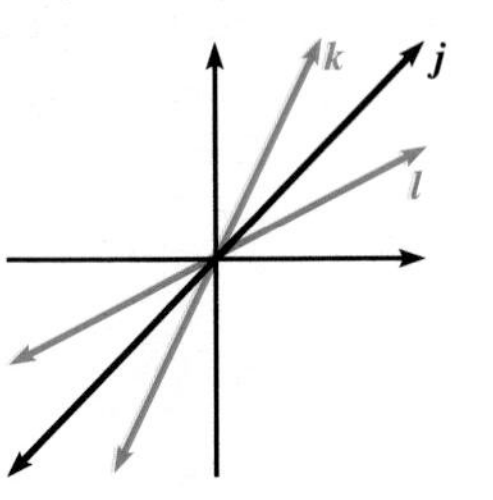

a. Write an equation that could be an equation of line l. Write another for line k.

b. Write an inequality that describes all possible slopes for line k. Write inequalities that describe all possible slopes for line l.

Connection GEOGRAPHY

The Grand Canyon has a large range in temperatures. Park rangers suggest that if you visit the park in June, you should plan on temperatures like those in the table.

South Rim 7050 ft
North Rim 8340 ft
Inner Gorge 2450 ft
8000 ft
6000 ft
4000 ft
2000 ft
Sea Level

	Temperature (°F) High	Temperature (°F) Low
South Rim	81	47
Inner Gorge	101	72
North Rim	73	40

22. Make a scatter plot with elevation on the horizontal axis and high temperature on the vertical axis.

23. Use the temperatures at the South Rim and Inner Gorge to find an equation for high temperature as a function of elevation. Use your equation to find the expected temperature at the North Rim.

24. Repeat Exercise 23 using the low temperature data.

25. What high and low temperatures would you expect to find at an elevation of 4600 ft on the trail down to the Inner Gorge? at 5200 ft?

The Colorado River runs through the Inner Gorge.

10. $a = 60t + 6065$

11. a. $y = -\frac{4}{5}x + 4$

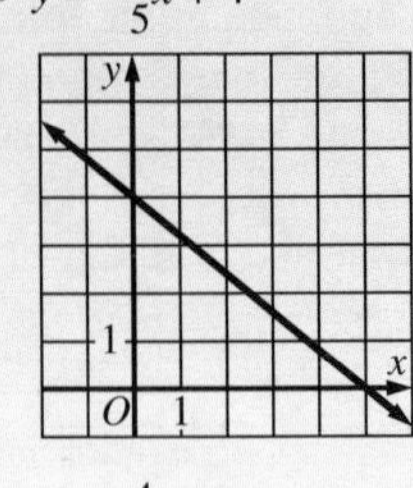

b. $y = -\frac{4}{5}x + 6$

c. Answers may vary. An example is given. $y = -\frac{4}{5}x - 2$; I know that any line with slope $-\frac{4}{5}$ has an equation of the form $y = -\frac{4}{5}x + b$, and I can pick any value for b. The graphs in parts (a) and (b) suggest that all the lines with slope $-\frac{4}{5}$ are parallel.

12. $y = -x + 12$
13. $y = 18$
14. $x = 7$
15. $y = 2x - 3$
16. $y = -2$
17. $y = -\frac{5}{3}x + \frac{5}{3}$
18. $x = 15$
19. $y = -\frac{19}{7}x + \frac{65}{7}$
20. $y = \frac{13}{12}x - \frac{19}{12}$

21. a. Answers may vary. Examples are given. l: $y = 0.5x$; k: $y = 2x$

b. Let m_k be the slope of line k and m_l the slope of line l; $m_k > 1$; $0 < m_l < 1$.

22.

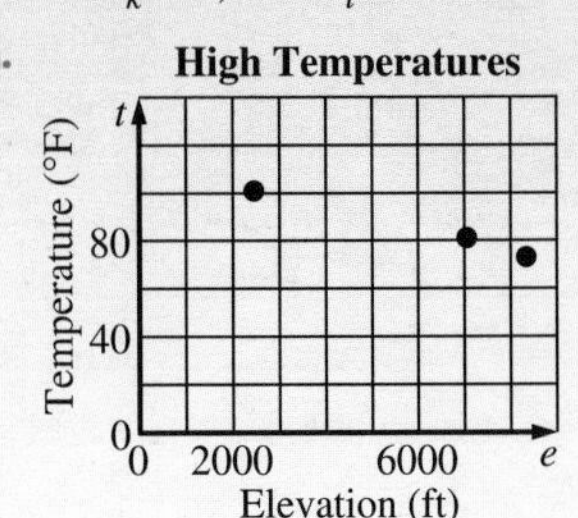

Ruth Emerson thought that her oven was not heating to the temperature on the dial. She recorded the oven temperature at several dial settings. For Exercises 26 and 27, use her results as shown in the table below.

26. Make a scatter plot of Ruth's results. Write an equation for the temperature of Ruth's oven as a function of the dial setting. What assumptions did you make?

27. Writing Write instructions that explain how to find the correct dial setting for Ruth's oven.

Dial Setting (°F)	Oven Temperature (°F)
300	325
320	350
400	450

ONGOING ASSESSMENT

28. Writing Cynthia Wong frequently visits the science museum with her two children. She has three options for paying admission:

(a) Pay admission for one adult and two children each time she visits.

(b) Buy a family membership.

(c) Buy an individual membership, and pay admission for two children.

For each of the three options, write an equation for cost as a function of the number of visits. Graph each equation. What do the slope and the vertical intercept of each line tell you about that option?

THE MUSEUM OF SCIENCE

■ *General admission fees*

Adults **$8.00**

Children (3–14) **$6.00**

Seniors (65+) **$6.00**

(Members free)

■ *Annual membership fees*

Individual **$50.00**

Free admission for member presenting card

Family **$80.00**

Free admission for 2 adults and their children aged 18 and under

SPIRAL REVIEW

For Exercises 29–32, use the histogram at the right. *(Section 1.1)*

29. How many students are at least 15 years old?

30. How many students are younger than 18?

31. What percent of the students are over 16 years old?

32. Write two inequalities that describe the ages of the students.

33. Nikki buys 15 post card stamps and 5 airmail stamps. Write a variable expression for her total cost, based on the cost of each type of stamp. *(Section 1.6)*

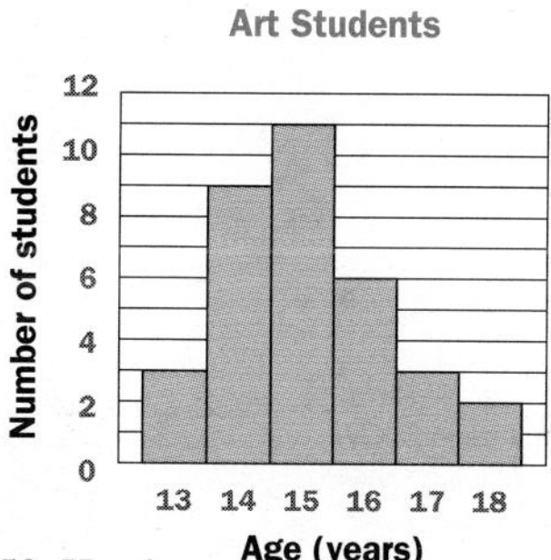

34. Bryan Noland uses his car for his job. His weekly salary is $550. He also gets paid $.29 per mile. Write a variable expression for his total weekly pay, based on the number of miles he drives. *(Section 1.6)*

Apply⇔Assess

Exercise Notes

Research

Exs. 22–25 Students can look in reference works for data on how temperatures vary with altitude on the slopes of mountain ranges.

Problem Solving

Exs. 26, 27 You may wish to ask students if they can think of other ways to solve these problems. If so, students should explain their methods to the class.

Assessment Note

Ex. 28 This exercise affords a good opportunity to see how well students are assimilating ideas about slope, intercepts, and equations of linear functions.

Practice 21 for Section 3.5

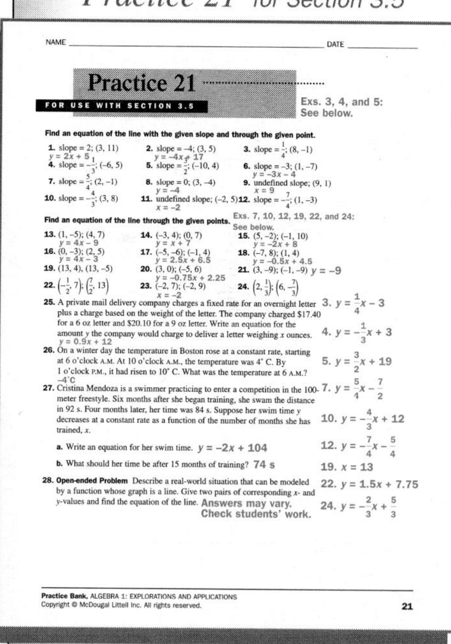

NAME ____________ DATE ____________

Practice 21

FOR USE WITH SECTION 3.5

Exs. 3, 4, and 5: See below.

Find an equation of the line with the given slope and through the given point.

1. slope = 2; (3, 11) $y = 2x + 5$ **2.** slope = –4; (3, 5) $y = -4x + 17$ **3.** slope = $\frac{1}{4}$; (8, –1)

4. slope = $-\frac{1}{3}$; (–6, 5) **5.** slope = $\frac{3}{2}$; (–10, 4) **6.** slope = –3; (1, –7) $y = -3x - 4$

7. slope = $\frac{5}{4}$; (2, –1) **8.** slope = 0; (3, –4) $y = -4$ **9.** undefined slope; (9, 1) $x = 9$

10. slope = $-\frac{4}{3}$; (3, 8) **11.** undefined slope; (–2, 5) $x = -2$ **12.** slope = $-\frac{7}{4}$; (1, –3)

Find an equation of the line through the given points. Exs. 7, 10, 12, 19, 22, and 24: See below.

13. (1, –5); (4, 7) $y = 4x - 9$ **14.** (–3, 4); (0, 7) $y = x + 7$ **15.** (5, –2); (–1, 10) $y = -2x + 8$

16. (0, –3); (2, 5) $y = 4x - 3$ **17.** (–5, –6); (–1, 4) $y = 2.5x + 6.5$ **18.** (–7, 8); (1, 4) $y = -0.5x + 4.5$

19. (13, 4), (13, –5) **20.** (3, 0); (–5, 6) $y = -0.75x + 2.25$ **21.** (3, –9); (–1, –9) $y = -9$

22. $\left(-\frac{1}{2}, 7\right)$; $\left(\frac{7}{2}, 13\right)$ **23.** (–2, 7); (–2, 9) $x = -2$ **24.** $\left(2, \frac{1}{3}\right)$; $\left(6, -\frac{7}{3}\right)$

25. A private mail delivery company charges a fixed rate for an overnight letter plus a charge based on the weight of the letter. The company charged $17.40 for a 6 oz letter and $20.10 for a 9 oz letter. Write an equation for the amount y the company would charge to deliver a letter weighing x ounces. $y = 0.9x + 12$

26. On a winter day the temperature in Boston rose at a constant rate, starting at 6 o'clock A.M. At 10 o'clock A.M., the temperature was 4° C. By 1 o'clock P.M., it had risen to 10° C. What was the temperature at 6 A.M.? –4°C

27. Cristina Mendoza is a swimmer practicing to enter a competition in the 100-meter freestyle. Six months after she began training, she swam the distance in 92 s. Four months later, her time was 84 s. Suppose her swim time y decreases at a constant rate as a function of the number of months she has trained, x.

a. Write an equation for her swim time. $y = -2x + 104$

b. What should her time be after 15 months of training? 74 s

28. Open-ended Problem Describe a real-world situation that can be modeled by a function whose graph is a line. Give two pairs of corresponding x- and y-values and find the equation of the line. Answers may vary. Check students' work.

3. $y = \frac{1}{4}x - 3$

4. $y = -\frac{1}{3}x + 3$

5. $y = \frac{3}{2}x + 19$

7. $y = \frac{5}{4}x - \frac{7}{2}$

10. $y = -\frac{4}{3}x + 12$

12. $y = -\frac{7}{4}x - \frac{5}{4}$

19. $x = 13$

22. $y = 1.5x + 7.75$

24. $y = -\frac{2}{3}x + \frac{5}{3}$

Practice Bank, ALGEBRA 1: EXPLORATIONS AND APPLICATIONS

21

---

23. Answers may vary, depending on rounding. $t = -0.004e + 109.2$; about 76°

24. Answers may vary, depending on rounding. $t = -0.005e + 82.25$; about 40.55°

25. Answers may vary. Examples are given. about 91°, about 60°; about 88°, about 56°

26.

Oven Temperatures

Oven temperature (°F): 300, 400

Dial setting (°F): 0, 300, 400, x

Answers may vary. Examples are given. $y = 1.25x - 50$, where x is the dial setting in degrees Fahrenheit and y is the oven temperature in degrees Fahrenheit. I assumed that the line that passes closest to the three given data points gives a good estimate of the true relationship between the dial setting and the oven temperature and that this relationship is linear.

27, 28. See answers in back of book.

29. 22 students **30.** 32 students

31. about 14.7%

32. Answers may vary. Examples are given. Let a = the age of a student; $a > 12$ and $a < 19$.

33. $15p + 5a$, where p is the cost of a postcard stamp and a is the cost of an airmail stamp

34. $550 + 0.29m$, where m is the number of miles he drives in a week

Plan⇔Support

Objectives

- Fit a line to data.
- Make predictions about real-world situations.

Recommended Pacing

Core and Extended Courses
Section 3.6: 1 day

Block Schedule
Section 3.6: $\frac{1}{2}$ block
(with Section 3.5)

Resource Materials

Lesson Support
Lesson Plan 3.6
Warm-Up Transparency 3.6
Overhead Visuals:
Folder A: Multi-Use Graphing Packet, Sheets 1–3
Practice Bank: Practice 22
Study Guide: Section 3.6
Explorations Lab Manual:
Diagram Masters 1, 2, 8
Assessment Book: Test 12

Technology
Graphing Calculator
McDougal Littell Software
Stats!
Internet:
http://www.hmco.com

Warm-Up Exercises

Find an equation for the line through each pair of points.

1. (3, 10) and (15, 12.4)
$y = 0.2x + 9.4$

2. (4.8, 17) and (14.3, –59)
$y = -8x + 55.4$

For each equation, find the value of y when $x = 55$.

3. $y = 12x - 38$ 622

4. $y = 5.4x + 430$ 727

5. $y = -0.7x + 8.6$ –29.9

SECTION

3.6 Modeling Linear Data

Learn how to...
- **fit a line to data**

So you can...
- **make predictions about natural occurrences and the life expectancy of future Americans**

Old Faithful is a geyser in Yellowstone National Park in Wyoming. For more than 100 years, Old Faithful has erupted every day at intervals of less than two hours. The data in the table suggest that the time between eruptions may be related to the length of the previous eruption.

The table is an example of a *mathematical model.* A mathematical model can be a table, a graph, an equation, a function, or an inequality that describes a real-world situation. Using mathematical models is called modeling.

Length of eruption (min)	Time until next eruption (min)
2.0	57
2.5	62
3.0	68
3.5	75
4.0	83
4.5	89
5.0	92

EXAMPLE Application: Geology

Find an equation that models the data about Old Faithful.

SOLUTION

Make a scatter plot of the data. Draw a line on the scatter plot that is as close as possible to all of the points on the plot. This is called a **line of fit**.

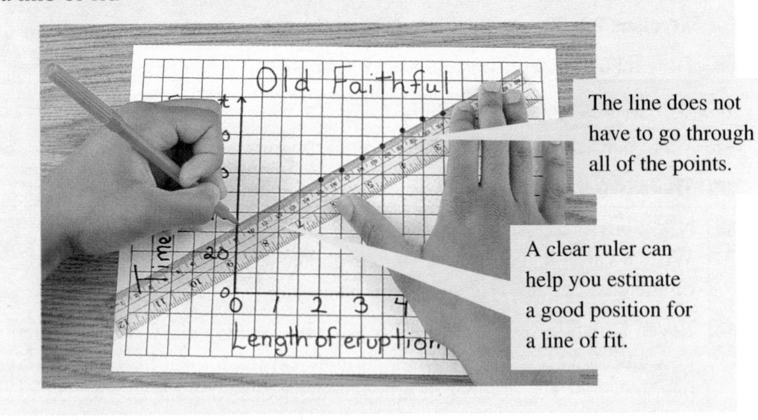
Old Faithful
Time
Length of eruption
The line does not have to go through all of the points.
A clear ruler can help you estimate a good position for a line of fit.

BY THE WAY

A geyser is a natural hot spring that periodically ejects a column of water and steam into the air. Most geysers are less predictable than Old Faithful.

Write an equation for the line of fit by choosing two points on the line.

Step 1 Let l = the length of the eruption in minutes.
Let t = the number of minutes until the next eruption.
The equation will be in the form

$$t = ml + b$$

Step 2 Find the **slope** m of the line of fit.

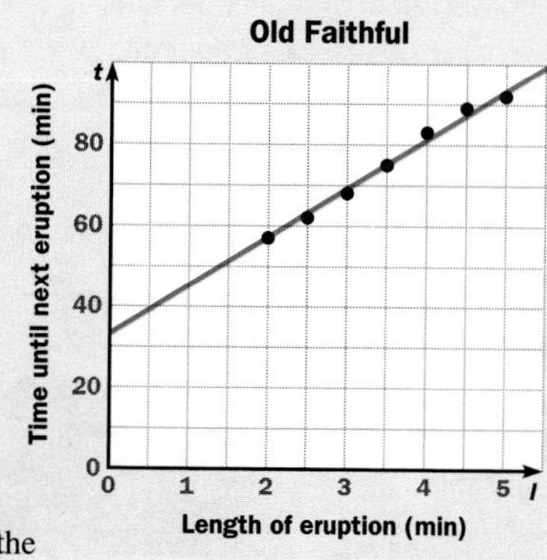

Choose the points (2, 57) and (5, 92).

$$m = \frac{92 - 57}{5 - 2} = \frac{35}{3} \approx 12$$

The slope of the line is 12.

Step 3 Find the **vertical intercept** b of the line.

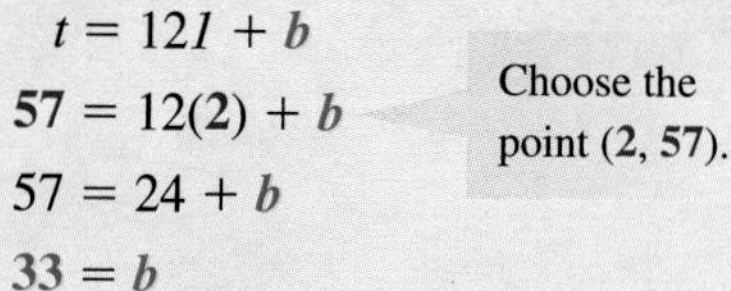

$$t = 12l + b$$
$$57 = 12(2) + b$$
$$57 = 24 + b$$
$$33 = b$$

Choose the point (2, 57).

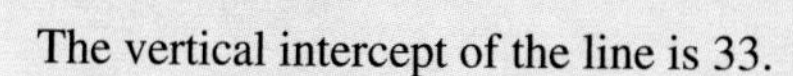

The vertical intercept of the line is 33.

An equation for the line of fit is $t = 12l + 33$.

THINK AND COMMUNICATE

1. In the Example, what information does the slope of the line of fit give you about the situation?
2. Suppose an eruption of Old Faithful lasts 4.2 min. Use the equation for the line of fit to predict the time until the next eruption.
3. In the Example, why does it make sense in this situation to round the slope to 12?

When the points on a scatter plot come close to forming a line, you can say that the data are linear. You can use a line of fit to model linear data.

Even though a few of the points are far from the line, you can say the data are linear.

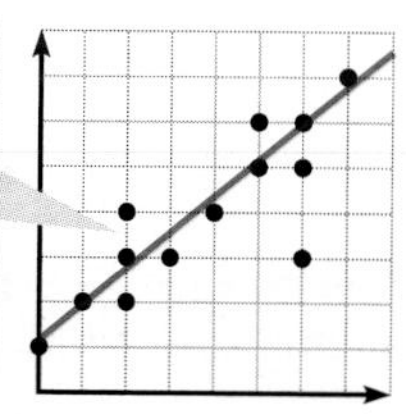

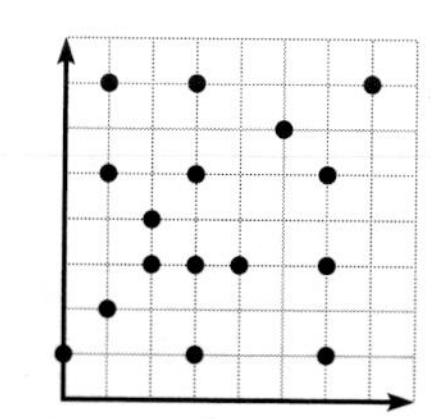

These data are not linear.

Teach⇔Interact

Additional Example 1

Suppose that an observer recorded the following data for another geyser in Yellowstone.

Length of eruption (min)	Time until next eruption (min)
1	60
2	50
2.5	80
3	72
3.5	70
4	89

Find an equation that models the data.

Make a scatter plot of the data. Draw a line of fit.

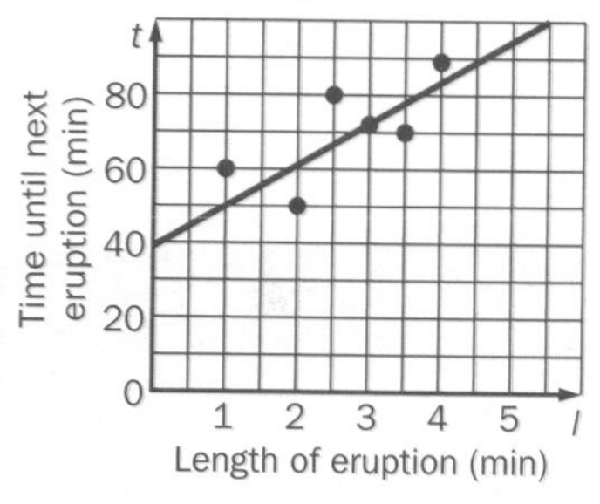

Write an equation for the line of fit by choosing two points on the line.

Step 1 Let l = the length of the eruption in minutes. Let t = the number of minutes until the next eruption. The equation will be in the form $t = ml + b$.

Step 2 Find the slope m of the line of fit. Choose the points (1, 50) and (3, 72).

$m = \frac{72 - 50}{3 - 1} = 11$

The slope of the line is 11.

Step 3 Find the vertical intercept b of the line.

$t = 11l + b$

$50 = 11(1) + b$

$39 = b$

The vertical intercept of the line is 39.

An equation for the line of fit is $t = 11l + 39$.

Think and Communicate

When discussing question 2, ask students whether there is another way to arrive at a prediction. (They could use the graph for the line of fit to estimate the second coordinate of the point whose first coordinate is 4.2.)

ANSWERS Section 3.6

Think and Communicate

1. For each increase of 1 min in the length of an eruption, the time until the next eruption increases by 12 min.
2. about 82.4 min

Teach⇔Interact

Checking Key Concepts

Common Error
Some students may think erroneously that some data points must be on a line of fit or that all points must be very close to the line before they can say that the data fit a linear model. Stress that most of the points should be near the line and that no points actually have to be on the line.

Closure Question

When can you say that a set of data points can be described by a linear equation?
Make a scatter plot and see whether a line can be drawn that is near most of the points.

Apply⇔Assess

Suggested Assignment

Core Course
Exs. 1, 2, 4–19, AYP

Extended Course
Exs. 1–19, AYP

Block Schedule
Day 19 Exs. 1–19, AYP

EXPLORATION
COOPERATIVE LEARNING

Creating a Linear Model

Work in a group of 3–4 students:
You will need:
- a paper cup
- a paper clip
- 30 marbles
- 25 pieces of uncooked spaghetti
- 2 chairs of the same height

SET UP Arrange the chairs, spaghetti, and cup as shown.

1 Place marbles in the cup, one at a time, until the spaghetti breaks. Copy and complete the table.

Number of pieces of spaghetti	Number of marbles needed to break spaghetti
1	?
2	?
3	?
4	?
5	?

2 Make a scatter plot of the data. Draw a line of fit. Find an equation for your line of fit.

3 Use your equation to predict how many marbles are needed to break six pieces of spaghetti. Check your prediction using spaghetti.

4 Do you think that moving the chairs farther apart or closer together would affect your results? If so, how? Describe and carry out an experiment to test your theory.

CHECKING KEY CONCEPTS

Are the data linear? If so, use an equation to model the data.

1.

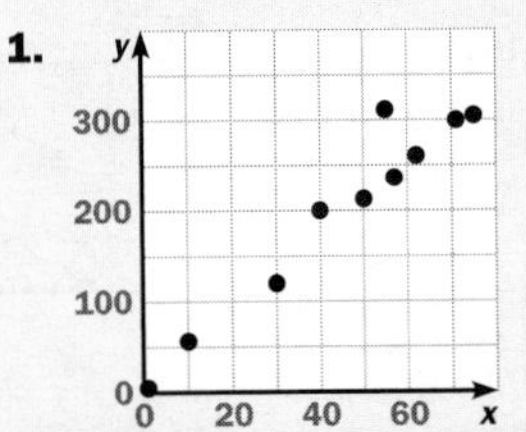

2.

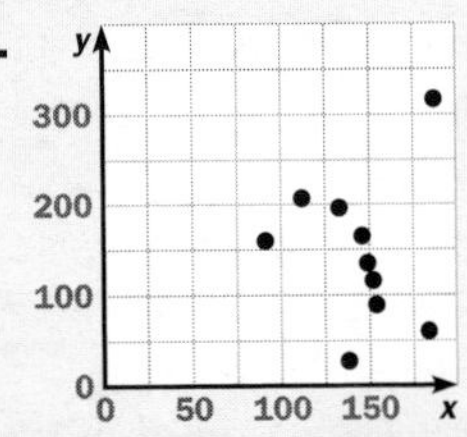

For Questions 3–5, use the equation from the Example to predict the time until the next eruption for the given length of an eruption.

3. 5.8 min **4.** 2.3 min **5.** 6.1 min

Exploration Note

Purpose
The purpose of this Exploration is to have students gather data from a simple experiment and model it by a linear equation. The equation is then used to make a prediction.

Materials/Preparation
After students are organized into groups, provide each group with the materials listed.

Procedure
Discuss briefly the sequence of steps students are to follow. For best results, suggest that students add the marbles to the cup gently. Discuss why dropping the marbles into the cup from too great a height is not a good idea.

Closure
Groups can compare their data and the equations they obtained. Ask if their results are different in any major way. If so, they should try to explain why this is the case.

Explorations Lab Manual
Diagram Masters 1 and 8

For answers to the Exploration, see answers in back of book.

3.6 Exercises and Applications

Extra Practice exercises on page 563

Find an equation that models the data in each scatter plot.

1. 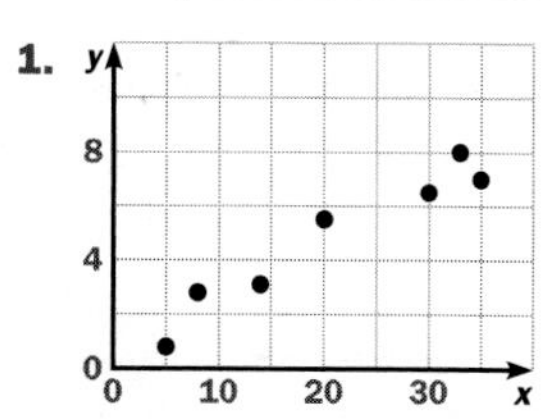

2. 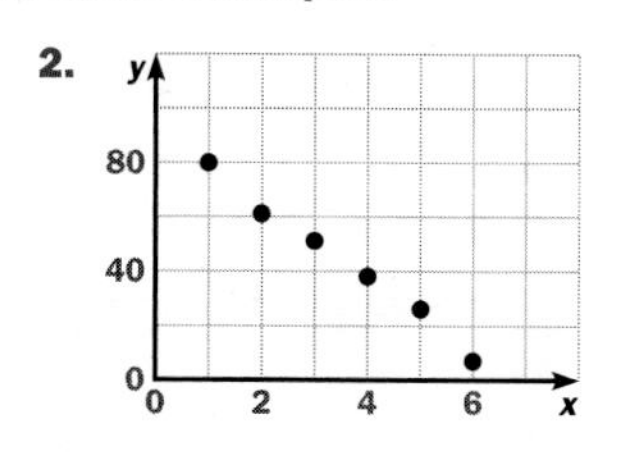

3. **Writing** Comment on the trends suggested by the data shown below. Support your comments with graphs, equations, and predictions.

Garbage Generated and Recycled in the United States
(pounds per person per day)

Year	1960	1965	1970	1975	1980	1985	1990
Garbage generated	2.66	3.00	3.27	3.26	3.65	3.77	4.30
Garbage recycled	0.18	0.19	0.23	0.25	0.35	0.38	0.70

Connection STATISTICS

The table shows the life expectancy of people born in the United States of America between 1900 and 1990. Use these data for Exercises 4–7.

Life Expectancy

Year of birth	Life expectancy (years)
1900	47.3
1910	50.0
1920	54.1
1930	59.7
1940	62.9
1950	68.2
1960	69.7
1970	70.8
1980	73.7
1990	75.4

4. Make a scatter plot of the data. Draw a line of fit on your scatter plot. Find an equation for your line of fit.
5. Predict the life expectancy of a person born in each year.
   a. 1978 b. 1994 c. 1942
6. a. Use your equation to predict the life expectancy of a person born in the United States in the year 2000.
   b. **Writing** The United States Census Bureau predicts that the average life expectancy of a person born in the United States in the year 2000 will be 76.7 years. Compare this with your prediction. Why do you think the predictions might be different?
7. **Challenge** Use your equation to predict the life expectancies of people born in 1800 and in 2100. How accurate do you think your predictions are? What do your answers say about the accuracy of using a linear equation to model the data?

Apply⇔Assess

Exercise Notes

Integrating the Strands
Exs. 1, 2 Students will find that modeling data with equations is an important application of algebra to statistics.

Second-Language Learners
Ex. 3 Students learning English might benefit from the help of a peer tutor as they write their explanations.

Cooperative Learning
Exs. 3–7 Students can work productively in groups of three to complete these exercises. Groups can be asked to report their answers to the class to check for accuracy of answers and understanding of the concepts involved.

Using Technology
Exs. 4–7 You may want to suggest that students use the *Stats!* software for this group of exercises.

Checking Key Concepts

1. Yes. Equations may vary. An example is given. $y = 5x$
2. No.
3. about 102.6 min
4. about 60.6 min
5. about 106.2 min

Exercises and Applications

Equations based on line of fit may vary throughout. Examples are given.

1. $y = \frac{3}{16}x$
2. $y = -14x + 1720$
3. See answers in back of book.

4–7. Answers may vary. Examples are given.

4. 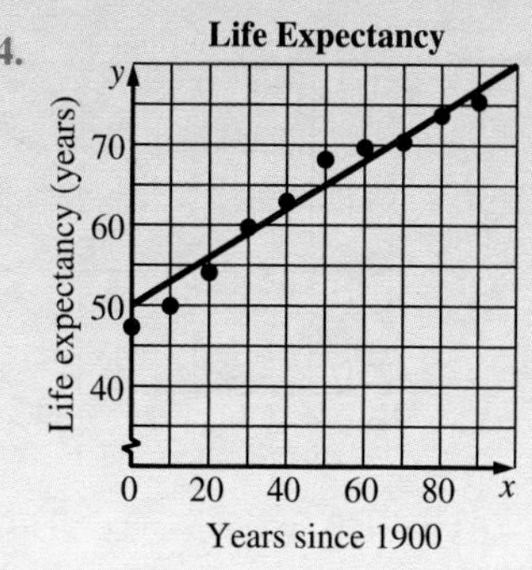

$y = 0.3x + 50$, where x is the number of years since 1900 and y is the life expectancy in years.

5. a. about 73.4 years
   b. about 78.2 years
   c. about 62.6 years
6. a. about 80 years
   b. My estimate is higher than the estimate of the United States Census Bureau. I may have used a different line of fit. Also, my estimate assumes that the rate of increase in life expectancy will be about the same for the entire period 1900 to 2000. This may not be the case. If there are not as many opportunities to increase life expectancy from 1990 to 2000 as in earlier years, then the rate of increase will become smaller and the life expectancy in the year 2000 will be less than my model would predict.
7. about 20 years; about 110 years; The estimate for 1980 is probably too low and the estimate for 2100 is probably too high. This suggests to me that sometimes a linear model is useful for predicting values in a certain range of values, but not very accurate outside of that range.

Apply⇔Assess

Exercise Notes

Using Technology
Ex. 8 When using the TI-82 for completing this exercise, data should be entered as statistical lists L1 and L2. To first clear these lists, either of two techniques can be used. For example, to clear list L1, press [STAT] and select item 1 from the EDIT menu. With the rectangular cursor on the first item of L1, press [DEL] and keep it held down until all items have been deleted. If L1 is a list with many numbers, it is faster to clear L1 by pressing [STAT] and selecting item 4 (ClrList) from the EDIT menu. When you see ClrList on the home screen, press [2nd][L1][ENTER]. The calculator will clear the list of all data and display "Done" on the home screen.

Assessment Note
Ex. 10 Students can give evidence that they understand the three kinds of models by discussing which model best fits the data and why. They can also discuss whether one kind of model is better than another for making predictions.

8. Technology Choose a data set from the Example on page 130 or from the Exercises in this lesson.

a. Follow the steps below to use a graphing calculator to find a line of fit for the data. Your graphing calculator manual gives more complete instructions.

Step 1 Clear the calculator's statistical memory. Enter the data as *xy*-pairs.

Step 2 Choose a good viewing window. Then draw a scatter plot.

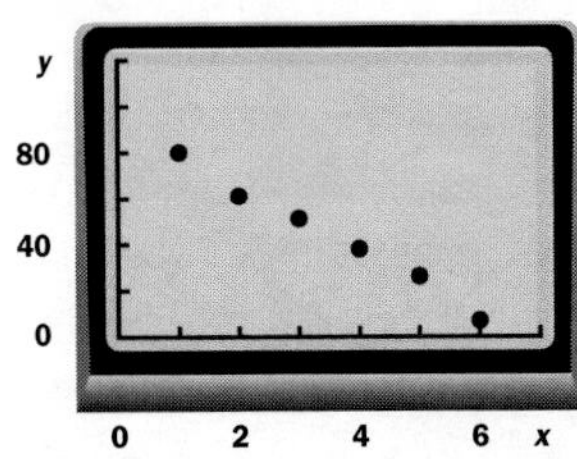

These numbers are the result of performing a linear regression on the data from Exercise 2. The calculator's line of fit is $y = -13.8x + 92.13$.

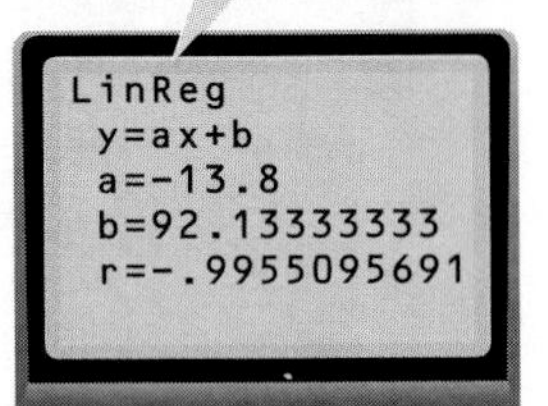

Step 3 Have the calculator perform a linear regression. Many calculators use the letter a for the vertical intercept and b for the slope. The number r is called the correlation coefficient. The closer $|r|$ is to 1, the more linear the data.

Step 4 Graph the line of fit on the scatter plot from Step 2.

b. Does the calculator give a line of fit that is close to the one you found?

9. SAT/ACT Preview If $2x + 3 = 9$, then $x - 2 = \underline{\ ?\ }$.

A. 4 **B.** -1 **C.** 3 **D.** 1 **E.** -2

ONGOING ASSESSMENT

10. a. Model the data with a scatter plot, a line of fit, and an equation.

b. Open-ended Problem What predictions can be made from each of your models? Who might be interested in using these predictions?

Energy Consumption

State	1990 Population (thousands)	Energy consumption (trillion Btu)
Alaska	550	582
Iowa	2,777	899
Tennessee	4,877	1753
Massachusetts	6,016	1341
Ohio	10,847	3698
Florida	12,938	3059
New York	17,990	3584
California	29,760	7307

8. Answers may vary. Examples are given, based on the data in the Example on page 130.

a. $t = 12.4l + 31.6$, where l is the length of the eruption in minutes and t is the number of minutes until the next eruption.

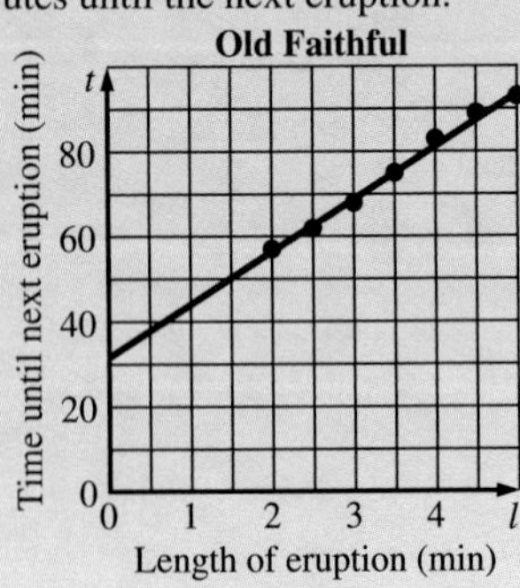

b. Yes, the equation for the fitted line, found on page 131, is $t = 12l + 33$.

9. D

10. Answers may vary. Examples are given.

a.

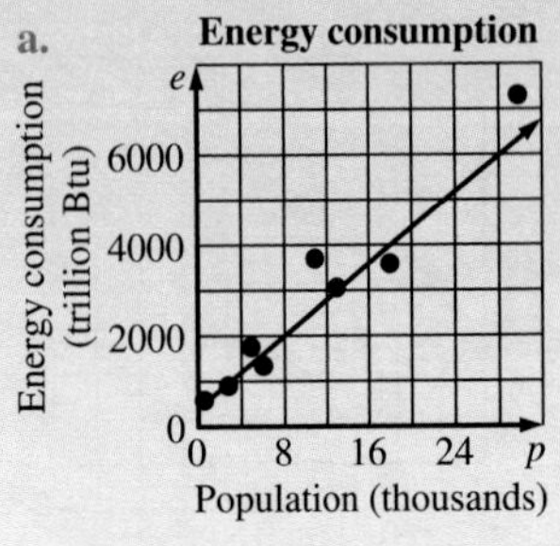

$e = 0.2p + 400$, where p is the 1990 population in thousands and e is the energy consumption in trillion Btu.

b. The line of fit and the equation can both be used to predict energy consumption for a given 1990 population and to predict 1990 population for a given energy consumption. Public utility companies, governmental agencies, and investors might all be interested in using these predictions.

11. $14b + 3$ 12. $2 + 5b$

13. $9 - 9z$ 14. $7x - 14$

15. $-7n + 24$ 16. $4a + 32$

17.

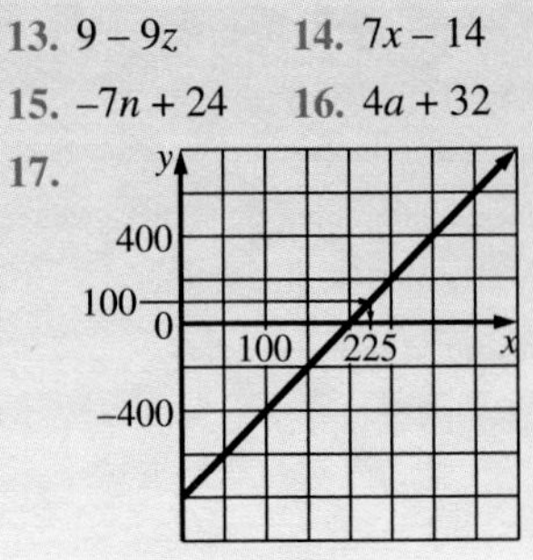

at least 225 tickets

SPIRAL REVIEW

Simplify each variable expression. *(Section 1.7)*

11. $17b + 5 - 3b - 2$ **12.** $0.25(8 + 20b)$ **13.** $3(3 - 3z)$

14. $9x - 2(x + 7)$ **15.** $-6(2n - 4) + 5n$ **16.** $16(0.25a + 2)$

For Exercises 17–19, use a graph to find the solution. *(Section 2.6)*

17. The drama club spent \$1000 on their production of *Fiddler on the Roof.* The club's goal is to make a profit of at least \$300. If tickets are sold at \$5 each, how many does the club need to sell to reach this goal?

18. The club is considering increasing the ticket price to \$10. How many tickets must be sold at this price to reach the club's goal of \$300?

19. Suppose the club earns \$200 by selling advertising space in their program. To reach their goal of \$300, how many tickets must the club sell at \$5 each? at \$10 each?

ASSESS YOUR PROGRESS

VOCABULARY

line of fit (p. 130)

Find an equation of the line through the given points. *(Section 3.5)*

1. (7, 2), (4, −1) **2.** (2, 3), (−1, 3) **3.** (6, 1), (6, −3)

4. (−4, 6), (1, −4) **5.** (3, 7), (−9, −1) **6.** (5, −3), (7, 8)

7. a. Graph the line through the points (−3, 0) and (0, 2). Write an equation of this line. *(Section 3.5)*

b. A line through the point (2, 0) has the same slope as the line in part (a). Graph this line and find an equation for it.

The table lists the number of cable-television systems in the United States from 1980 to 1990. (The number of cable systems has been rounded to the nearest hundred.) Use the table for Exercises 8–10. *(Section 3.6)*

Year	Number of cable systems
1980	4200
1982	4800
1984	6200
1986	7600
1988	8500
1990	9600

8. Make a scatter plot of the data.

9. Draw a line of fit for the data. Find an equation for your line.

10. Predict the number of cable systems in the United States in 2005.

11. Journal How are mathematical models different from other kinds of models you have seen or used? How are they the same?

Apply⇔Assess

Assess Your Progress

Journal Entry

You can assign Ex. 11 for homework and then have students compare and discuss their responses in class. Each student can later revise his or her response as needed to arrive at a well thought-out journal entry.

Progress Check 3.5–3.6

See page 139.

Practice 22 for Section 3.6

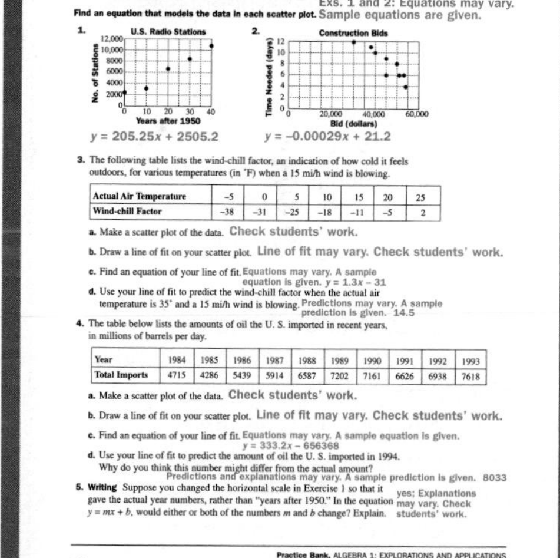
NAME ______________________ DATE ____________

Practice 22

FOR USE WITH SECTION 3.6

Find an equation that models the data in each scatter plot. Exs. 1 and 2: Equations may vary. Sample equations are given.

1. U.S. Radio Stations (No. of Stations vs. Years after 1950) $y = 205.25x + 2505.2$

2. Construction Bids (Time Needed (days) vs. Bid (dollars)) $y = -0.00029x + 21.2$

3. The following table lists the wind-chill factor, an indication of how cold it feels outdoors, for various temperatures (in °F) when a 15 mi/h wind is blowing.

Actual Air Temperature	−5	0	5	10	15	20	25
Wind-chill Factor	−38	−31	−25	−18	−11	−5	2

a. Make a scatter plot of the data. Check students' work.

b. Draw a line of fit on your scatter plot. Line of fit may vary. Check students' work.

c. Find an equation of your line of fit. Equations may vary. A sample equation is given. $y = 1.3x - 31$

d. Use your line of fit to predict the wind-chill factor when the actual air temperature is 35° and a 15 mi/h wind is blowing. Predictions may vary. A sample prediction is given. 14.5

4. The table below lists the amounts of oil the U. S. imported in recent years, in millions of barrels per day.

Year	1984	1985	1986	1987	1988	1989	1990	1991	1992	1993
Total Imports	4715	4286	5439	5914	6587	7202	7161	6626	6938	7618

a. Make a scatter plot of the data. Check students' work.

b. Draw a line of fit on your scatter plot. Line of fit may vary. Check students' work.

c. Find an equation of your line of fit. Equations may vary. A sample equation is given. $y = 333.2x - 656368$

d. Use your line of fit to predict the amount of oil the U. S. imported in 1994. Why do you think this number might differ from the actual amount? Predictions and explanations may vary. A sample prediction is given. 8033

5. Writing Suppose you changed the horizontal scale in Exercise 1 so that it gave the actual year numbers, rather than "years after 1950." In the equation $y = mx + b$, would either or both of the numbers m and b change? Explain. yes; Explanations may vary. Check students' work.

22

Practice Bank, ALGEBRA 1: EXPLORATIONS AND APPLICATIONS

18.

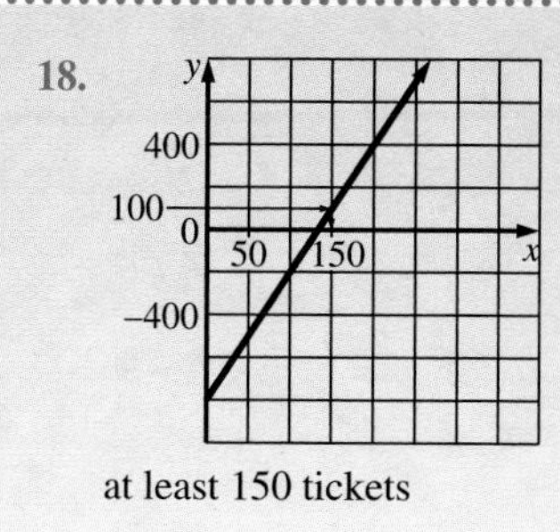

at least 150 tickets

19.

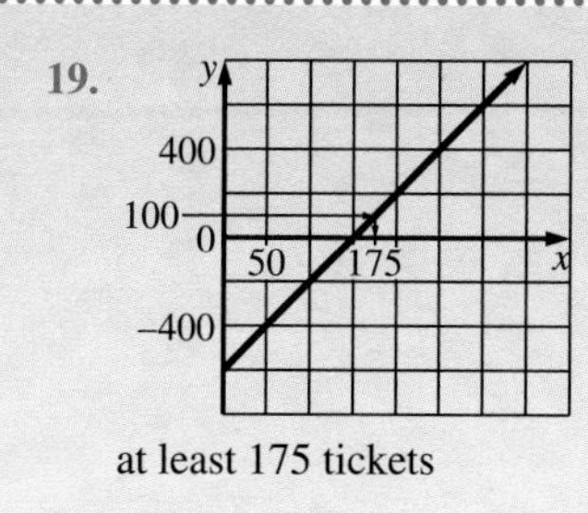

at least 175 tickets

Assess Your Progress

1. $y = x - 5$ 2. $y = 3$

3. $x = 6$ 4. $y = -2x - 2$

5. $y = \frac{2}{3}x + 5$

6. $y = 5.5x - 30.5$

7. a.

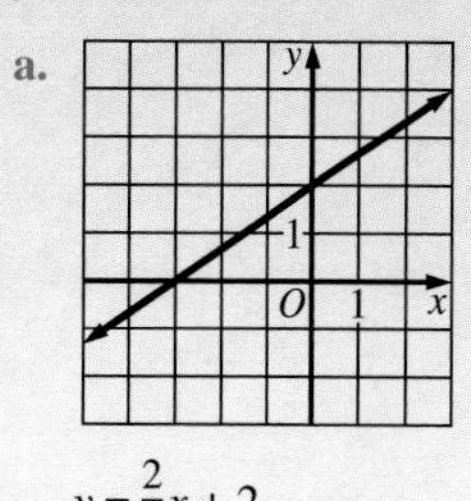

$y = \frac{2}{3}x + 2$

b.

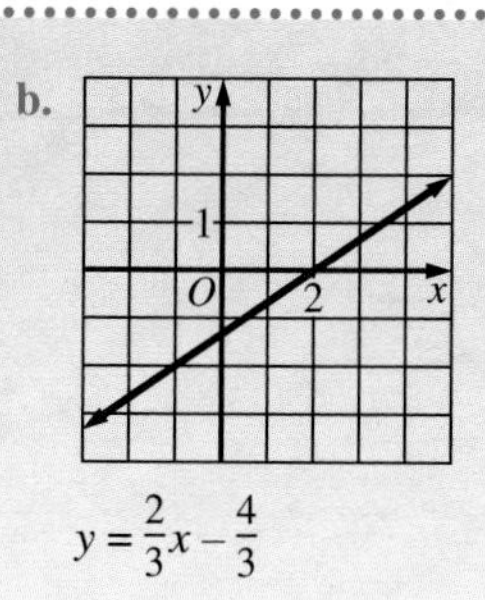

$y = \frac{2}{3}x - \frac{4}{3}$

8–11. See answers in back of book.

Project Notes

Mathematical Goals

- Write an equation of a line in slope-intercept form.
- Write an inequality for a line segment that identifies its endpoints.

Planning

Materials
- Graph paper
- Loose-leaf notebook

Project Teams
For teams as large as five or six students, you should work with the class to help organize the groups. Students having a mix of abilities working together can be advantageous to the development of the project goal.

Guiding Students' Work

All students should participate in each activity involved in creating a book of linear puzzles, following Steps 1–3 on this page. You can circulate among the groups and provide help to individual students as needed. As students check their work, they should feel free to discuss what they are doing with other group members. To assemble their puzzle book, each group should discuss who will do the different tasks and then organize the work accordingly.

Second-Language Learners
Students learning English may benefit from working with an English-proficient partner to plan their instructions for solving and creating puzzles.

PORTFOLIO PROJECT

Creating Linear Puzzles

Computer graphics programs use equations to determine each shape in a picture. Displaying a line on a computer starts with finding an equation of the line and specifying its endpoints. You can use this idea to create a puzzle with linear equations.

PROJECT GOAL **Design and create a book of linear puzzles for other students in your class to solve.**

Designing the Puzzles

Work in groups of five or six to create a book of linear puzzles. Here are some hints for designing your puzzles:

1. CHOOSE A SHAPE OR LETTER for your puzzle that you can draw using straight lines. **DRAW** your design on a coordinate plane. Make sure that the endpoints of the line segments are on grid intersections. Keep your design secret from the other members of your group.

Choose a shape that has line segments with at least four different slopes.

2. FIND an equation in slope-intercept form of each line in your design. Write an equation for each line segment on a separate sheet of paper.

3. PROVIDE INFORMATION about the endpoints of each line segment. Write inequalities that describe the smallest and largest *x*- or *y*-values for each line. Add this information to your set of equations. This list is your puzzle. Your drawing is the solution.

Puzzle

Equation	Endpoints
$y = \frac{3}{2}x + \frac{1}{2}$	$x \geq 1$ $x \leq 5$
$y = -2x + 18$	$x \geq 5$ $x \leq 8$
$y = -\frac{1}{2}x + \frac{5}{2}$	$x \geq 1$ $x \leq 5$
$y = \frac{2}{3}x - \frac{10}{3}$	$x \geq 5$ $x \leq 8$
$x = 5$	$y \geq 0$ $y \leq 8$

Checking Your Work

- Exchange puzzles within your group and solve them by graphing the lines described by the equations and inequalities.
- Check to see if your drawing matches the solution.
- Write an evaluation of the puzzle you solve. Give suggestions for improvements if you have any.

Assembling Your Puzzle Book

Collect a copy of the puzzle and solution made by each member of your group. Use these puzzles to make a book with instructions for solving and creating puzzles.

Remember to include:

- names of puzzle designers
- instructions for solving
- an answer key

You may want to extend your report to explore some of the ideas below:

- Images on a computer monitor are made up of tiny dots called pixels. Each pixel has a coordinate based on its distance from one corner of the monitor. How do computer programmers use pixel coordinates to generate images on a computer screen?
- Chess players move pieces according to a specific set of rules. Find out what these rules are. Then write a set of equations to describe the paths of different pieces as they are moved from one end of the chess board to the other.

Self-Assessment

In your report, describe which puzzles were the most challenging to solve. Why were they challenging? How could you have made your puzzle more challenging? What general advice would you give someone who is trying to create a puzzle?

Project Notes

Guiding Students' Work

Rubric for Chapter Project

4 Students create shapes and write correct equations in slope-intercept form for each line in their design. The inequalities that describe the endpoints of each line segment are also correct. Students check their work and assemble their puzzle book. The book is clearly organized, the puzzles and their solutions are accurate, and the names of the puzzle designers, instructions for solving, and an answer key are included.

3 Students proceed to create shapes for their puzzles, write equations for each line segment, and write inequalities that describe the endpoints. In checking their work, not all mathematics errors are found and corrected. A puzzle book is assembled, but there are some mistakes in the answer key.

2 Students have some difficulties getting started. Not all members of the group can write correct equations in slope-intercept form or correct inequalities. Group members work together to help one another, but the final puzzle book has some serious mathematical errors.

1 Students do not always follow the instructions for designing their puzzles. Some drawings do not have endpoints that are on grid intersections. Many equations and inequalities are wrong. The checking procedure is incomplete and the puzzle book does not include instructions for solving the puzzles or an answer key. Students should be encouraged to speak with the teacher as soon as possible to review their work and to make a new start on the project.

Chapter Support

Course Guide: Chapter 3
Lesson Plans: Chapter 3
Practice Bank:
Cumulative Practice 23
Study Guide: Chapter 3 Review
Assessment Book:
Chapter Tests 13 and 14
Chapter 3 Alternative Assessment
Test Generator software
Technology Book:
Calculator Based Lab 3
Portfolio Project Book:
Additional Project 3: Slope
Preparing for College Entrance Tests
Professional Handbook

Progress Check 3.1–3.2

Express each in the given units. *(Section 3.1)*

1. 54 yd/h = __?__ ft/h 162
2. $12.00 per hour = __?__ cents per minute 20

Use the graph for Exs. 3–5. *(Sections 3.1 and 3.2)*

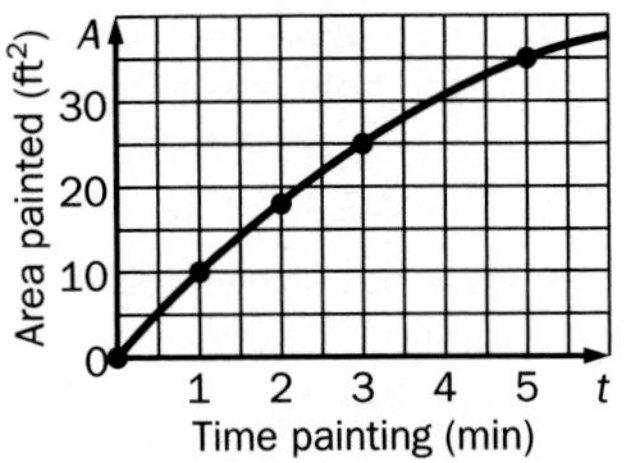

3. Find the average rate of change for area painted from minute 1 to minute 3. 7.5 ft^2/min
4. Find the average rate of change for painted during the first 5 minutes. 7 ft^2/min
5. Does the graph show direct variation? How do you know? No; it is not a straight line.

CHAPTER 3

Review

STUDY TECHNIQUE

What did you just study? Without looking at the book, take five or ten minutes and list all the words and important concepts related to this chapter. When you are finished, compare your list with the vocabulary and with the lesson titles in the table of contents. Did you miss anything important?

VOCABULARY

rate (p. 99)
unit rate (p. 99)
scatter plot (p. 106)
direct variation (p. 107)
constant of variation (p. 107)
slope (pp. 113, 115)
positive slope (p. 114)
negative slope (p. 114)
undefined slope (p. 115)
linear equation (p. 119)
***y*-intercept, vertical intercept** (p. 120)
slope-intercept form (p. 120)
line of fit (p. 130)

SECTIONS 3.1 *and* 3.2

A **unit rate** is a rate per one given unit.

A **rate** is a ratio that compares two units. $\frac{220 \text{ miles}}{4 \text{ hours}} = \frac{55 \text{ miles}}{1 \text{ hour}} = 55$ mi/h

Fuel Economy

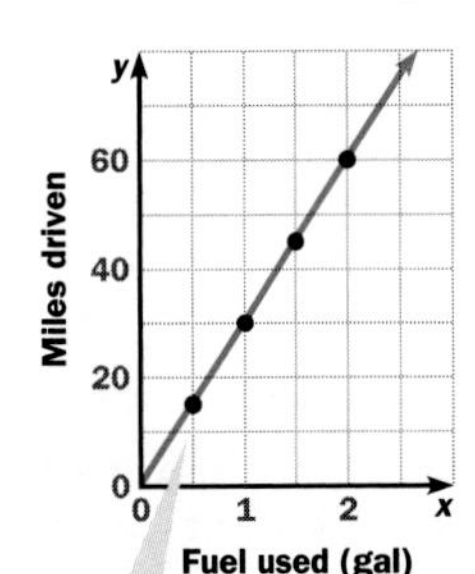

The points on the **scatter plot** shown at the left lie on a line through the origin. An equation for this **direct variation** is $y = 30x$.

The **constant of variation** is 30.

SECTIONS 3.3 *and* 3.4

The **slope** or steepness of a line is a rate of change.

$$\text{Slope} = \frac{y_2 - y_1}{x_2 - x_1} = \frac{5-1}{4-1} = \frac{4}{3}$$

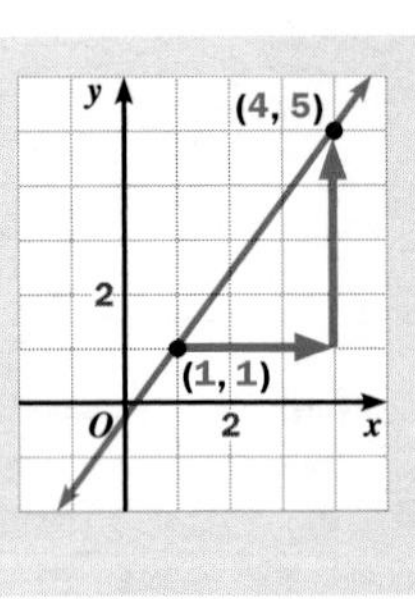

A line that rises to the right has **positive slope**.

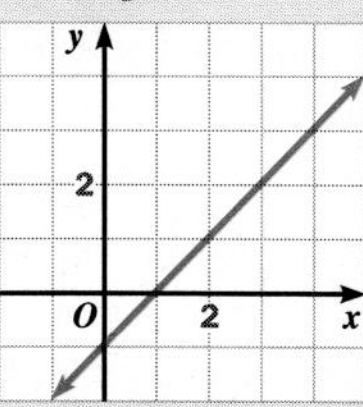

A line that falls to the right has **negative slope**.

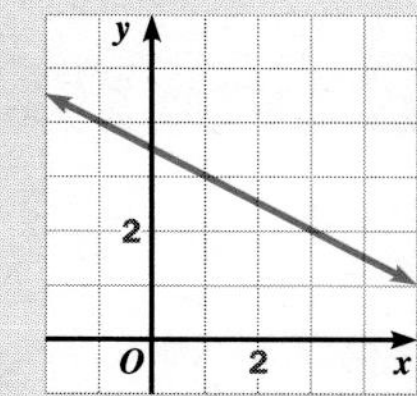

A horizontal line has **0 slope**.

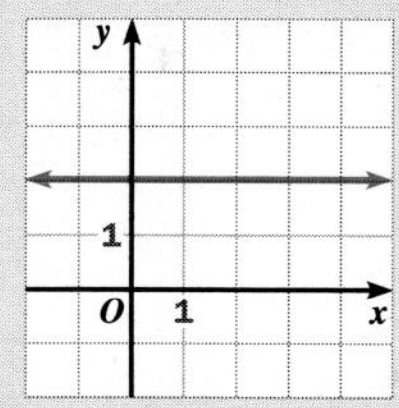

A vertical line has **undefined slope**.

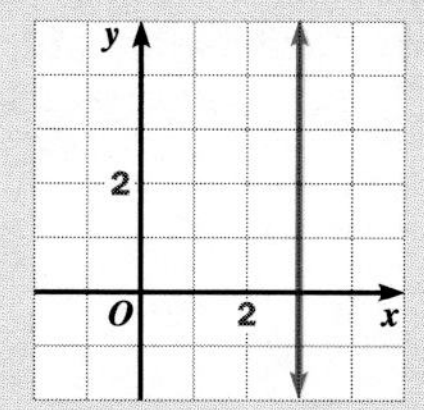

A linear equation can be written in **slope-intercept form**: $y = mx + b$, where m is the slope and b is the y-intercept.

SECTIONS 3.5 *and* 3.6

Mathematical models are used to describe real-world situations.

A **line of fit** is a line drawn as close as possible to all the points on a scatter plot. You can write an equation for the line in slope-intercept form using two of the points on the line.

Find the slope using the points (2, 1) and (8, 5).

$$m = \frac{5-1}{8-2} = \frac{4}{6} = \frac{2}{3}$$

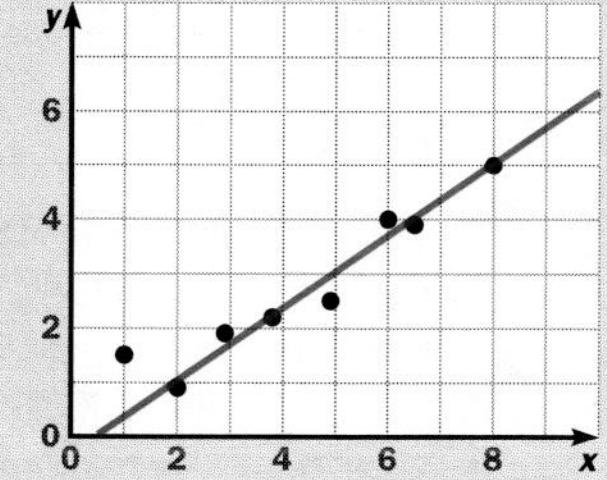

Find the vertical intercept by substituting the slope and the coordinates of one point in the equation $y = mx + b$. Then solve for b.

$$1 = \frac{2}{3}(2) + b$$ Choose the point (**2**, **1**).

$$-\frac{1}{3} = b$$

An equation for the line of fit is $y = \frac{2}{3}x - \frac{1}{3}$.

Progress Check 3.3–3.4

Find the slope of the line through each pair of points. *(Section 3.3)*

1. $A(3, 8)$ and $B(5, 6)$ -1

2. $C(-1, 4)$ and $D(10, 7)$ $\frac{3}{11}$

3. $E(-3, -3)$ and $F(-1, -10)$ -3.5

Give the slope and y-intercept for the graph of each equation. *(Section 3.4)*

4. $y = \frac{3}{5}x + 9$ $\frac{3}{5}$, 9

5. $y = -4x - \frac{1}{2}$ $-4, -\frac{1}{2}$

6. Write an equation for a line that goes through (0, 0) and has slope $\frac{7}{5}$. *(Section 3.4)* $y = \frac{7}{5}x$

Progress Check 3.5–3.6

Find the equation of the line through the given points. *(Section 3.5)*

1. (−7, 7), (5, −5) $y = -x$

2. (4, 12), (14, 23) $y = 1.1x + 7.6$

3. (5.8, 0), (5.8, −4) $x = 5.8$

4. (3, 1.2), (9, 0.6) $y = -0.1x + 1.5$

The scatter plot shows how many minutes eight students spent on a make-up math test and what scores they got. Use the scatter plot for Exs. 5 and 6. *(Section 3.6)*

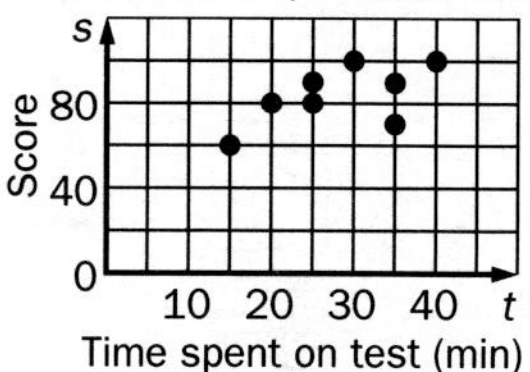

5. Draw a line of fit for the data and write an equation for the line of fit.
A possible line of fit is shown. The equation for this line of fit is $y = x + 55$.

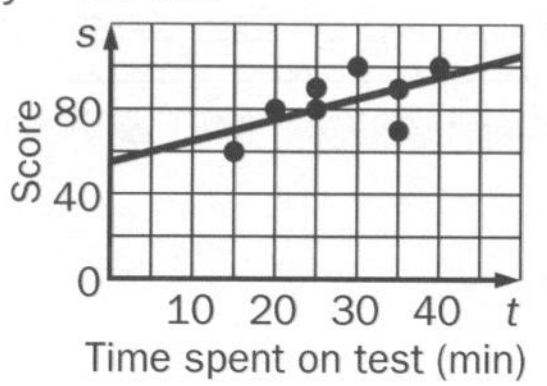

6. Suppose a ninth student has to take the make-up test. If the student spends 30 minutes on the test, predict the score the student will get.
Answers will vary. An example is given. 85

Chapter 3 Assessment
Form A Chapter Test

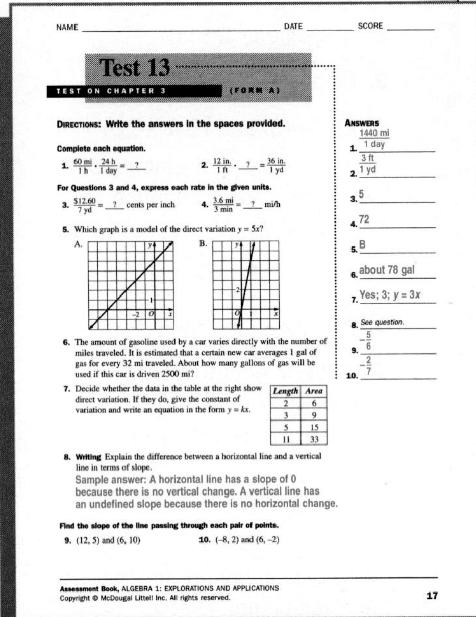

NAME ______ DATE ______ SCORE ______

Test 13

TEST ON CHAPTER 3 (FORM A)

DIRECTIONS: Write the answers in the spaces provided.

Complete each equation.

1. $\frac{60\text{ mi}}{1\text{ h}} \cdot \frac{24\text{ h}}{1\text{ day}} = \underline{\ ?\ }$
2. $\frac{12\text{ in.}}{1\text{ ft}} \cdot \underline{\ ?\ } = \frac{36\text{ in.}}{1\text{ yd}}$

For Questions 3 and 4, express each rate in the given units.

3. $\frac{\$12.60}{7\text{ yd}} = \underline{\ ?\ }$ cents per inch
4. $\frac{3.6\text{ mi}}{3\text{ min}} = \underline{\ ?\ }$ mi/h
5. Which graph is a model of the direct variation $y = 5x$?
   A. B.
6. The amount of gasoline used by a car varies directly with the number of miles traveled. It is estimated that a certain new car averages 1 gal of gas for every 32 mi traveled. About how many gallons of gas will be used if this car is driven 2500 mi?
7. Decide whether the data in the table at the right show direct variation. If they do, give the constant of variation and write an equation in the form $y = kx$.

Length	*Area*
2	6
3	9
5	15
11	33

8. **Writing** Explain the difference between a horizontal line and a vertical line in terms of slope.
   Sample answer: A horizontal line has a slope of 0 because there is no vertical change. A vertical line has an undefined slope because there is no horizontal change.

Find the slope of the line passing through each pair of points.

9. (12, 5) and (6, 10)
10. (–8, 2) and (6, –2)

ANSWERS
1. $\frac{1440\text{ mi}}{1\text{ day}}$
2. $\frac{3\text{ ft}}{1\text{ yd}}$
3. 5
4. 72
5. B
6. about 78 gal
7. Yes; 3; $y = 3x$
8. *See question.*
9. $-\frac{5}{6}$
10. $-\frac{2}{7}$

Assessment Book, ALGEBRA 1: EXPLORATIONS AND APPLICATIONS

17

Chapter 3 Assessment
Form B Chapter Test

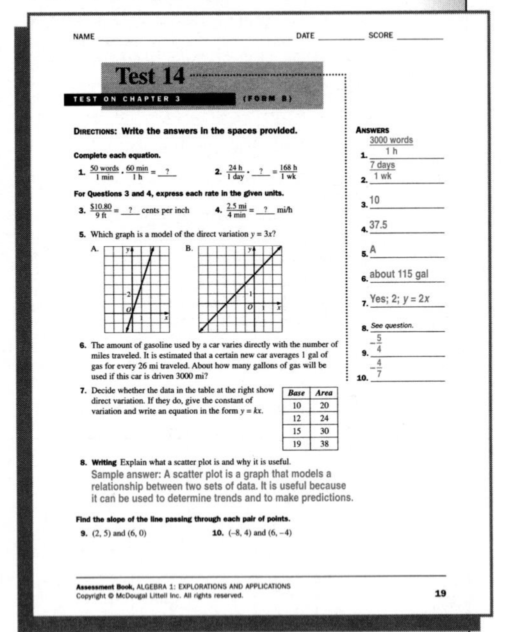

NAME ______ DATE ______ SCORE ______

Test 14

TEST ON CHAPTER 3 (FORM B)

DIRECTIONS: Write the answers in the spaces provided.

Complete each equation.

1. $\frac{50\text{ words}}{1\text{ min}} \cdot \frac{60\text{ min}}{1\text{ h}} = \underline{\ ?\ }$
2. $\frac{24\text{ h}}{1\text{ day}} \cdot \underline{\ ?\ } = \frac{168\text{ h}}{1\text{ wk}}$

For Questions 3 and 4, express each rate in the given units.

3. $\frac{\$10.80}{9\text{ ft}} = \underline{\ ?\ }$ cents per inch
4. $\frac{2.5\text{ mi}}{4\text{ min}} = \underline{\ ?\ }$ mi/h
5. Which graph is a model of the direct variation $y = 3x$?
   A. B.
6. The amount of gasoline used by a car varies directly with the number of miles traveled. It is estimated that a certain new car averages 1 gal of gas for every 26 mi traveled. About how many gallons of gas will be used if this car is driven 3000 mi?
7. Decide whether the data in the table at the right show direct variation. If they do, give the constant of variation and write an equation in the form $y = kx$.

Base	*Area*
10	20
12	24
15	30
19	38

8. **Writing** Explain what a scatter plot is and why it is useful.
   Sample answer: A scatter plot is a graph that models a relationship between two sets of data. It is useful because it can be used to determine trends and to make predictions.

Find the slope of the line passing through each pair of points.

9. (2, 5) and (6, 0)
10. (–8, 4) and (6, –4)

ANSWERS
1. $\frac{3000\text{ words}}{1\text{ h}}$
2. $\frac{7\text{ days}}{1\text{ wk}}$
3. 10
4. 37.5
5. A
6. about 115 gal
7. Yes; 2; $y = 2x$
8. *See question.*
9. $-\frac{5}{4}$
10. $-\frac{4}{7}$

Assessment Book, ALGEBRA 1: EXPLORATIONS AND APPLICATIONS

19

CHAPTER

3 Assessment

VOCABULARY QUESTIONS

1. Draw a scatter plot of data that show direct variation. Give the constant of variation for the data.
2. Draw a graph of a line with each type of slope.
   a. positive **b.** negative **c.** zero **d.** undefined

SECTIONS 3.1 *and* 3.2

For Questions 3 and 4, copy and complete each equation.

3. $\frac{420\text{ mi}}{1\text{ day}} \cdot \underline{\ ?\ } = \frac{2940\text{ mi}}{1\text{ week}}$
4. $\frac{\$7.25}{1\text{ h}} \cdot \frac{8\text{ h}}{\text{day}} = \underline{\ ?\ }$

Express each rate in the given units.

5. $\frac{\$30.82}{23\text{ gal}} \approx \underline{\ ?\ }$ cents/gal
6. $\frac{0.75\text{ mi}}{2\text{ min}} = \underline{\ ?\ }$ mi/h
7. Which graph represents the direct variation $C = 1.25g$?

A.

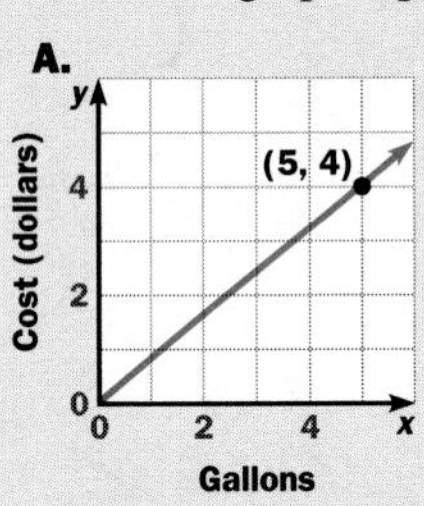

B.

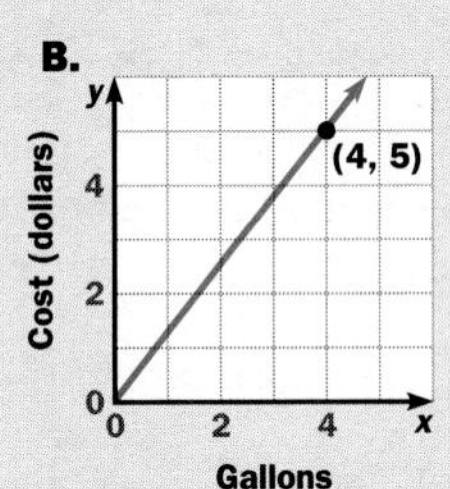

C.

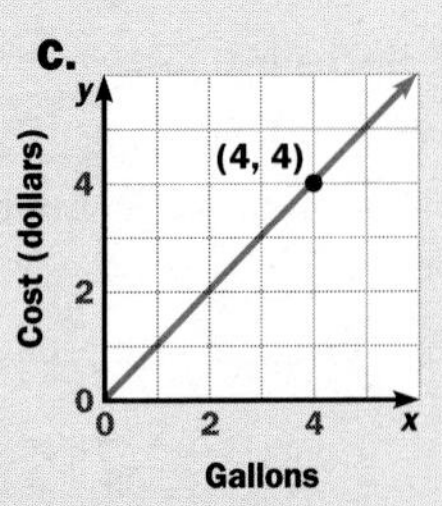

8. The amount of garbage produced in the United States varies directly with the number of people who produce it. It is estimated that 200 people produce 50 tons of garbage annually. How many tons of garbage are produced each year in a city with 100,000 inhabitants?
9. Decide whether the data in the table show direct variation. If they do, give the constant of variation and write an equation that describes the situation.

Circumference (cm)	Radius (cm)
12.56	2
31.4	5
43.96	7
62.8	10

ANSWERS Chapter 3

Assessment

1. Answers may vary. An example is given. The data are from Example 1 on page 107. The constant variation is 5.75.

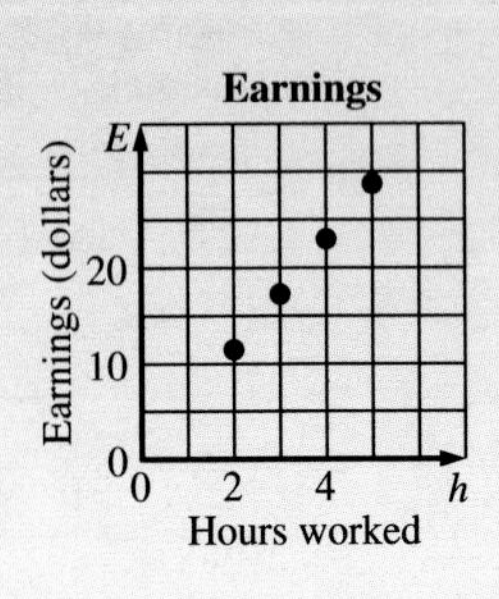

2. Answers may vary. Examples are given.

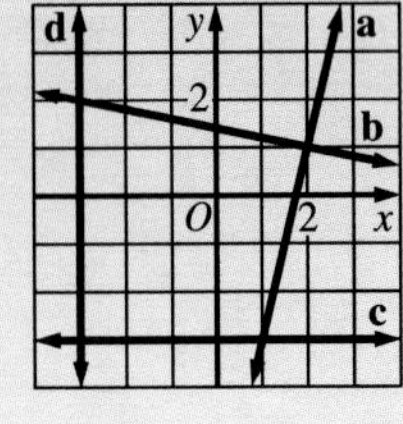

3. $\frac{7\text{ days}}{1\text{ week}}$
4. $\frac{\$58}{1\text{ day}}$
5. 134
6. 22.5
7. B
8. about 25,000 tons of garbage
9. Yes; the constant of variation is about 0.159; $r = 0.159C$, where C is the circumference in centimeters and r is the radius in centimeters.
10. Answers may vary. An example is given. The constant of variation, k, gives the constant rate of change between x and y and is equal to the slope of the graph of the equation $y = kx$. A direct variation equation is also a linear equation because its graph is a line.
11. Let m = number of miles towed and C = cost in dollars; $C = 2m + 34$; the slope indicates the rate of change of the cost, \$2/mi. The vertical intercept indicates the basic charge of the trip excluding the cost per mile. The tow costs \$34 plus the cost per mile.

SECTIONS 3.3 *and* 3.4

10. Writing Explain how the constant of variation in a direct variation equation is related to the slope of the graph of the equation. Explain why a direct variation equation is also a linear equation.

11. Write an equation in slope-intercept form for the graph shown at the right. What information do the slope and vertical intercept give you about the situation?

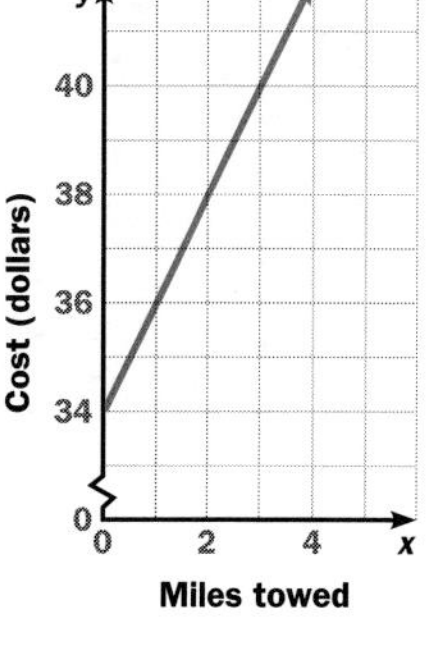

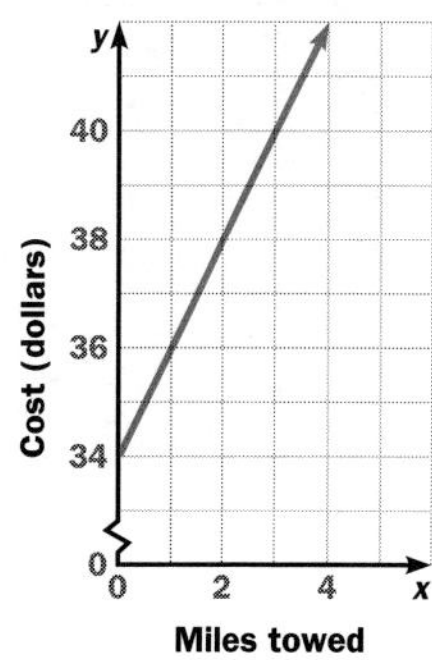

Find the slope of the line through each pair of points.

12. (7, 5) and (8, 4) **13.** (−6, 3) and (2, −3)

Write an equation in slope-intercept form for each graph.

14.

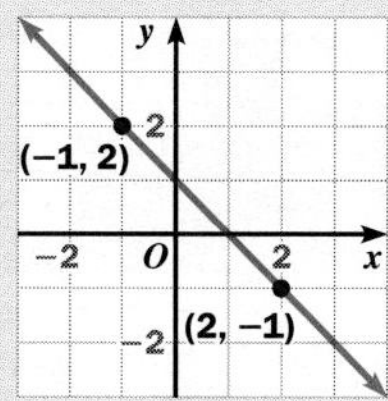

15.

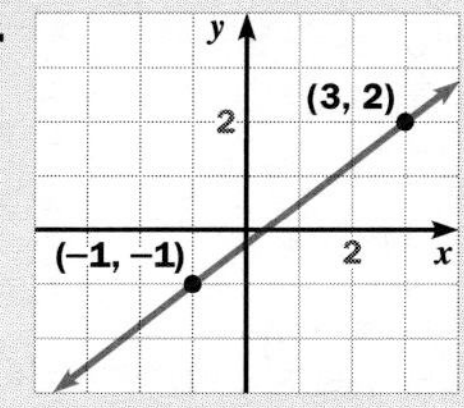

Graph each equation.

16. $y = -3x - 3$ **17.** $y = \frac{2}{5}x$ **18.** $y = \frac{3}{2}x - 4$

SECTIONS 3.5 *and* 3.6

Find an equation of the line through the given points.

19. (2, 3), (4, 6) **20.** (−5, −6), (6, 5) **21.** (2, 7), (−3, 7)

22. Find an equation of a line with slope 3 and through the point (4, 7).

23. a. Make a scatter plot of the data in the table at the right.

b. Draw a line of fit on your scatter plot.

c. Write an equation for the line of fit.

d. Use your equation to predict a weight for a 120 cm tall male who is nine years old.

Average Weight and Height of American Males Aged 6 to 10 Years

Weight (kg)	Height (cm)
20.9	114.2
23.1	119.3
25.9	127.0
28.6	132.0
31.3	137.1

PERFORMANCE TASK

24. Write a story about a trip you have taken or might want to take. Include at least two rates and one linear equation. Illustrate your story with at least one graph. Include as many vocabulary words from the chapter as you can.

Chapter 3 Assessment
Form C Alternative Assessment

Chapter 3
ALTERNATIVE ASSESSMENT

1. a. Some people think of ski slopes when they read about the mathematical term *slope*. The drawings below represent three different "ski" slopes.

If you were rating these ski slopes in terms of difficulty, which would get the highest rating? Use the grids shown to estimate the numerical slope of each figure. Discuss how the numerical slopes confirm your ratings.

b. Find the numerical slope of a horizontal line. What would be an appropriate ski rating for a "hill" with a similar slope? Discuss how you might use the "ski" slope to help remember the numerical slope.

c. Find the numerical slope of a vertical line. What would be an appropriate ski rating for a "hill" with a similar slope? Discuss how you might use the "ski" slope to help remember the numerical slope.

2. Open-ended Problem William needed to calculate the slope of a line containing the points (1, 4) and (−2, −5). He wrote the following calculation: $\frac{4-(-5)}{-2-1} = \frac{9}{-3} = -3$. Then William drew the graph of the line and realized the slope of the line containing (1, 4) and (−2, −5) should be positive. Find and discuss William's error. What can William do to avoid repeating this error?

3. Group Activity Discuss the following questions in your group.

a. Find a real-life function that has a graph with a positive slope. Graph the function.

b. Explain how you knew that the slope of the line would be positive.

c. Can a function whose graph has a positive slope be called an *increasing* function? Explain.

d. Find a real-life function that has a graph with a negative slope. Graph the function.

e. Explain how you knew that the slope of the line would be negative.

f. Can a function whose graph has a negative slope be called an *decreasing* function? Explain.

4. Draw a line of fit for each scatter plot. Estimate the equation for your line of fit.

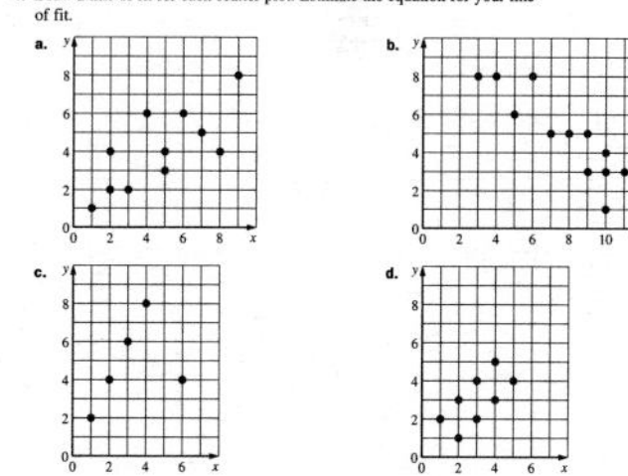

5. Performance Task Suppose a line contains the points (0, 0) and (q, r).

a. What happens to the slope of the line if q and r are both tripled?

b. Suppose the horizontal and vertical distances from the origin to (q, r) were originally measured in inches. What happens to the slope of the line if these distances are measured again in centimeters? Make a generalization about the effect of the unit of measurement on the slope of a line.

6. a. Find two points on a graph that model a rate of change of $\frac{3}{4}$.

b. Is there more than one correct answer to part (a)? If so, find other examples of pairs of points for which the rate of change is $\frac{3}{4}$.

c. Is there a generalization that can be made about all the pairs of points that have a rate of change of $\frac{3}{4}$?

7. Open-ended Problem What are some questions that could be asked based on the graph shown at the right? What are the answers to your questions?

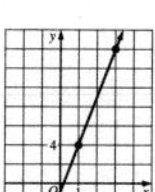

12. −1

13. $-\frac{3}{4}$

14. $y = -x + 1$

15. $y = \frac{3}{4}x - \frac{1}{4}$

16.

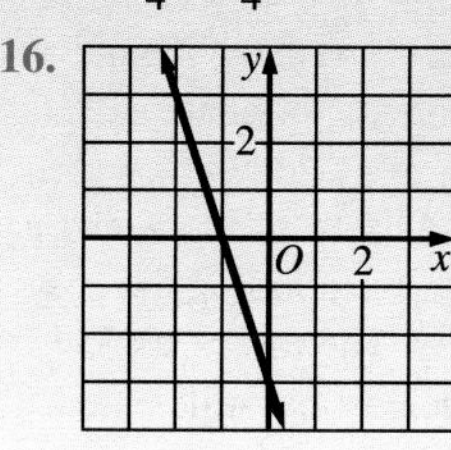

17.

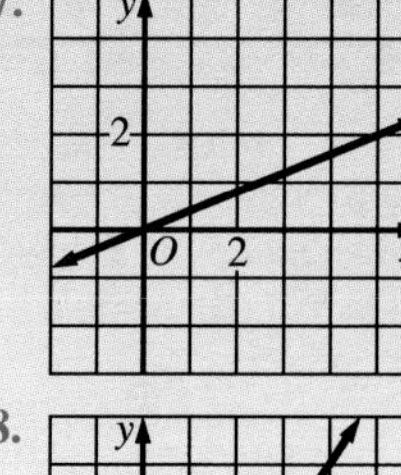

18.

19. $y = \frac{3}{2}x$

20. $y = x - 1$

21. $y = 7$

22. $y = 3x - 5$

23. Answers may vary. Examples are given.

a, b.

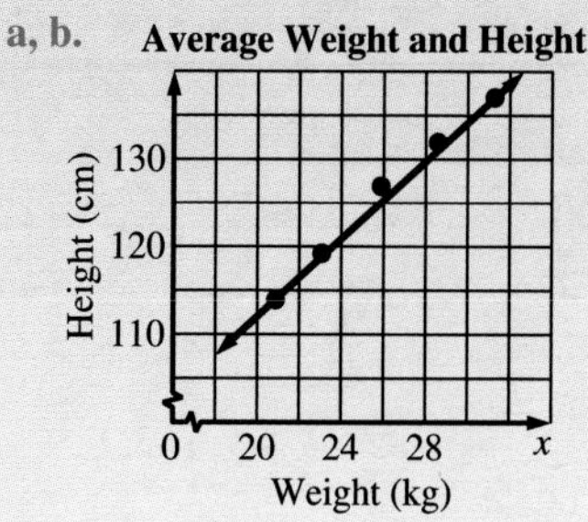

c. Answers may vary. $h = 2.2w + 68$

d. about 23.6 kg

24. Answers may vary. Check students' work.

Cumulative Assessment

CHAPTERS 1–3

CHAPTER 1

Use the histogram at the right.

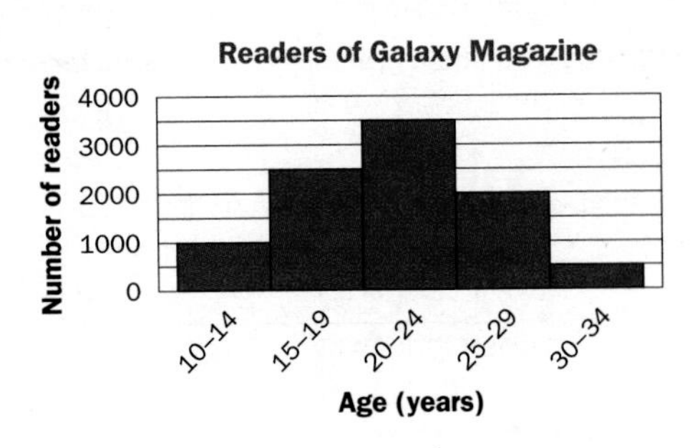

1. How many readers of *Galaxy Magazine* are less than 20 years old?

2. How many readers are at least 25 years old?

3. What interval is the mode of the readers' ages?

Simplify each expression.

4. $6 + 2 \cdot 8$ **5.** $-8 - (-3)$ **6.** $-44 \div (-4)$

Evaluate each expression when $a = 3$ and $b = -5$.

7. $5a - b^2$ **8.** $\frac{a^2 + 3b}{6}$ **9.** $\frac{a - b^2}{b}$

For Questions 10 and 11, find the mean, the median, and the mode(s) of the data.

10. Hours studied
3, 3, 3, 4, 4, 5, 8, 16

11. Meals served
20, 21, 21, 25, 30

12. Writing Uma says that $|n^2| = n^2$ for all values of n. Do you agree or disagree? Explain.

Open-ended Problem Describe two real-world situations that each variable expression can represent.

13. $t + 2$ **14.** $\frac{p}{4}$ **15.** $5n$

CHAPTER 2

Solve each equation.

16. $-7p = 105$ **17.** $\frac{t}{4} = -52$ **18.** $15 = 7 + x$

19. $2n - 6 = 16$ **20.** $3 - 6x = -19$ **21.** $7 + \frac{x}{4} = 15$

ANSWERS Chapters 1–3

Cumulative Assessment

1. 3500 2. 2500
3. 20–24 4. 22
5. –5 6. 11
7. –10 8. –1
9. $\frac{22}{5}$, or 4.4
10. mean: 5.75; median: 4; mode: 3
11. mean 23.4; median: 21; mode: 21
12. Agree. Any number squared is greater than or equal to 0. The absolute value of any number is greater than or equal to 0. Therefore, for all values of n, $|n^2| = n^2$.
13–15. Answers may vary. An example is given.
13. Travel time will include two hours driving to and from the airport plus the time for the flight.
14. Four friends decided to split their profits from a plant sale equally.
15. A bank customer has to write on the deposit slip the value of the nickels being deposited.
16. –15 17. –208
18. 8 19. 11
20. 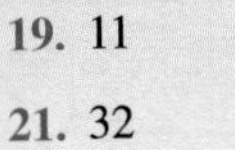 $3\frac{2}{3}$ 21. 32

22–25.

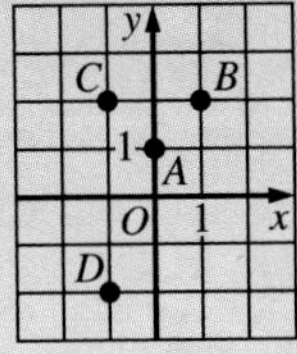

22. A: none 23. B: I
24. C: II 25. D: III
26. about 32 mi/h
27. about 2 s

Graph each point on a coordinate grid. Name the quadrant (if any) in which the point lies.

22. $A(0, 1)$ **23.** $B(1, 2)$ **24.** $C(-1, 2)$ **25.** $D(-1, -2)$

The graph below shows the velocity *v* of a car undergoing constant acceleration for *t* seconds. Use the graph for Questions 26 and 27.

26. Estimate the velocity v when the time is 4 s.

27. Estimate the time t when the velocity is 15 mi/h.

Car Acceleration

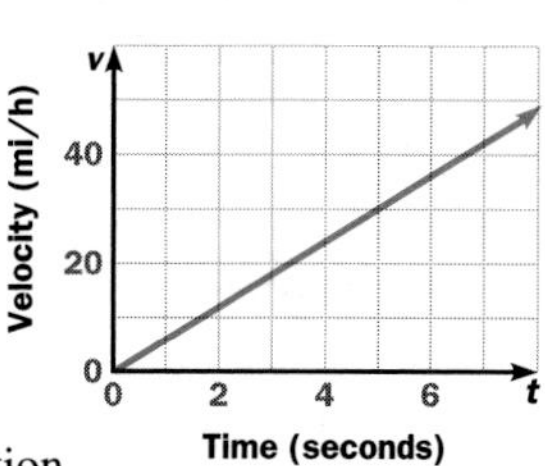

28. Writing Each pizza ordered for the Ski Club gathering costs \$7. Tell how this situation represents a function. Describe the domain and the range of the function.

29. a. Open-ended Problem Describe a situation that can be represented by the equation $y = 4x$. Graph the equation.

b. Tell how your graph can be used to solve a problem relating to the situation in part (a).

CHAPTER 3

For Questions 30 and 31, copy and complete each equation.

30. $\frac{\$4.25}{1\text{ h}} \cdot \frac{20\text{ h}}{1\text{ week}} = \underline{?}$ **31.** $\frac{55\text{ mi}}{1\text{ h}} \cdot \underline{?} = \frac{440\text{ mi}}{\text{day}}$

32. Decide whether the data in the table at the right show direct variation. If they do, give the constant of variation and write an equation that describes the situation.

Size (oz)	Price (dollars)
2.4	2.88
3.0	3.60
3.5	4.20

33. A line contains the two points $A(-2, 3)$ and $B(4, 12)$.

a. Find the slope of the line.

b. Write an equation of the line in slope-intercept form.

c. What is the y-coordinate when $x = 0$?

Use the scatter plot at the right.

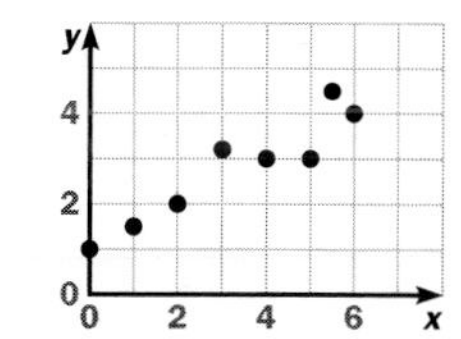

34. Which equation is a better model for the data?

A. $y = -\frac{1}{2}x + 1$ **B.** $y = \frac{1}{2}x + 1$

35. Writing Can a line with a steep slope have a negative slope? Write a few sentences to explain your answer.

36. Open-ended Problem Choose two points in different quadrants of a grid. Find the equation of the line through the points.

28. The total cost of pizzas is a function of the number of pizzas ordered. The domain is the number of pizzas ordered, which will be a whole number. The range is total cost, which will also be a whole number.

29. Answers may vary. An example is given.

a. A photographer is buying rolls of film. The cost of each roll of film is \$4.

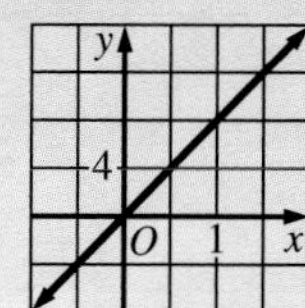

b. You can determine from the graph the total cost of 3 rolls of film.

30. $\frac{\$85}{\text{week}}$ 31. $\frac{8\text{ h}}{1\text{ day}}$

32. Yes; 1.2; $p = 1.2s$.

33. a. $\frac{3}{2}$

b. $y = \frac{3}{2}x + 6$

c. $y = 6$

34. B

35. Answers may vary. An example is given. A steep line can have a slope which is either positive or negative. If the slope is positive, then the line goes up steeply from left to right. If the slope is negative, then the line goes up steeply from right to left.

36. Answers may vary. An example is given. (1, 4) and (–1, –2); $y = 3x + 1$

CHAPTER

4 Solving Equations and Inequalities

OVERVIEW

Connecting to Prior and Future Learning

⇔ Students use tables and graphs to solve problems in Section 4.1. This strategy continues throughout the chapter and the course.

⇔ Students learn to solve multi-step equations and equations containing fractions or decimals. Students use their knowledge of reciprocals, area, factors, multiples, and powers of ten as they solve these equations. The **Student Resources Toolbox** reviews area on page 594, factors and multiples on page 584, and powers of ten on page 585.

⇔ An exploration of inequalities and their graphs introduces students to the skills necessary to solve inequalities. A study of graphing linear inequalities on a coordinate plane is presented in Chapter 7, where students also study systems of inequalities.

Chapter Highlights

Interview with Barbara Romanowicz: The relationship between mathematics and seismology is emphasized in this interview, with related exercises on page 169.

Explorations involve using a graphing calculator or graph paper to compare tables and graphs in Section 4.1, using algebra tiles to explore solving equations in Section 4.3, and working with inequalities in Section 4.6.

The Portfolio Project: The mathematics and decision-making processes that are necessary to produce and market a product are emphasized in this project.

Technology: In this chapter, students use graphing calculators to study linear equations. In Section 4.1, students use a graphing calculator to compare the strategy of using a table versus using a graph to solve a problem. Students graph and compare equations in Section 4.4. In Section 4.5, students find the point of intersection of two lines. The use of spreadsheets as a tool to analyze data is also continued in this chapter.

OBJECTIVES

Section	Objectives	NCTM Standards
4.1	• Model situations with tables and graphs. • Compare options using tables and graphs.	1, 2, 3, 4, 5, 10
4.2	• Solve one-step and two-step equations using reciprocals. • Solve problems involving fractions.	1, 2, 3, 4, 5
4.3	• Solve multi-step equations. • Use multi-step equations to solve real-world problems.	1, 2, 3, 4, 5
4.4	• Solve equations that involve more than one fraction. • Use equations to solve real-world problems.	1, 2, 3, 4, 5
4.5	• Use inequalities to represent intervals on a graph. • Use inequalities to represent a group of solutions to real-world problems.	1, 2, 3, 4, 5, 8
4.6	• Solve inequalities. • Tell when one quantity is greater than another.	1, 2, 3, 4, 5, 8

INTEGRATION

Mathematical Connections	4.1	4.2	4.3	4.4	4.5	4.6
algebra	**147–152***	**153–157**	**158–164**	**165–170**	**171–175**	**176–181**
geometry	**147–152**	157	162		172, 173, 175	
data analysis, probability, discrete math	**147–152**			169, 170	**171–175**	176
logic and reasoning	**147–152**	153–155, 157	159, 161–164	166, 169, 170	171, 173–175	176, 178–181
Interdisciplinary Connections and Applications						
history and geography			163			
biology and earth science				165, 169	174	
business and economics	148					180
arts and entertainment		156				
sports and recreation	152		162	168	172	177, 181
personal finance, energy use, fuel economy, transportation, architecture, communications	150, 151		159, 162	167	172, 173, 175	179

Bold page numbers *indicate that a topic is used throughout the section.*

TECHNOLOGY

	opportunities for use with	
Section	**Student Book**	**Support Material**
4.1	graphing calculator spreadsheet software McDougal Littell Software *Stats!* *Function Investigator*	**Technology Book:** Spreadsheet Activity 4
4.2	graphing calculator	
4.3	scientific calculator	
4.4	graphing calculator spreadsheet software McDougal Littell Software *Stats!*	
4.5	graphing calculator	
4.6	graphing calculator	**Technology Book:** Calculator Activity 4

PLANNING GUIDE

Regular Scheduling (45 min)

Section	Materials Needed	Core Assignment	Extended Assignment	*exercises that feature* Applications	Communication	Technology
4.1	graph paper, graphing calculator, spreadsheet software	1–11, 13, 15–23	1–23	5–11, 13, 14	1–4, 12, 15	
4.2		1–19, 21–23, 25–33, 35, 37–49	1, 3–6, 8, 9, 11–13, 16–49	18, 19, 35	20, 24, 36, 37	
4.3	algebra tiles, graph paper	**Day 1:** 2–13, 15–23 **Day 2:** 25–30, 34, 36–41, AYP*	**Day 1:** 1–24 **Day 2:** 25–41, AYP	1, 15–23	14, 24 36	
4.4	graphing calculator, spreadsheet software	1–25, 27–30, 32–34, 36–40	4–6, 9–11, 13–40	17, 28–31	26, 35, 36	13–16, 33, 37–39
4.5	graph paper, graphing calculator	1–9, 11–22	1–22	1, 5–8, 11, 13	1, 9, 10, 14	12
4.6	graph paper	**Day 1:** 1–21 **Day 2:** 22–33, 35, 37–44, AYP	**Day 1:** 1–21 **Day 2:** 22–44, AYP	13, 15–17 35	18 34, 36, 37	
Review/ Assess		**Day 1:** 1–19 **Day 2:** 20–37 **Day 3:** Ch. 4 Test	**Day 1:** 1–19 **Day 2:** 20–37 **Day 3:** Ch. 4 Test	3 20, 27, 28	1, 2 29	
Portfolio Project		Allow 2 days.	Allow 2 days.			

Yearly Pacing (with Portfolio Project)	**Chapter 4 Total** 13 days	**Chapters 1–4 Total** 55 days	**Remaining** 105 days	**Total** 160 days

Block Scheduling (90 min)

	Day 22	Day 23	Day 24	Day 25	Day 26	Day 27	Day 28
Teach/Interact	Ch. 3 Test 4.1: Exploration, page 147	4.2 4.3: Exploration, page 158	Continue with 4.3 4.4	4.5 4.6: Exploration, page 176	Continue with 4.6 Review	Review Port. Proj.	Port. Proj. Ch. 4 Test
Apply/Assess	**Ch. 3 Test** **4.1:** 1–11, 13, 15–23	**4.2:** 1–19, 21–23, 25–33, 37–49 **4.3:** 2–13, 16–23	**4.3:** 25–32, 34, 36–41, AYP* **4.4:** 1–25, 27–30, 36–40	**4.5:** 1–19, 11–22 **4.6:** 1–21	**4.6:** 22–44, AYP **Review:** 1–19	**Review:** 20–37 **Port. Proj.**	**Port. Proj.** **Ch. 4 Test**

NOTE: A one-day block has been added for the Portfolio Project—timing and placement to be determined by teacher.

Yearly Pacing (with Portfolio Project)	**Chapter 4 Total** $6\frac{1}{2}$ days	**Chapters 1–4 Total** 28 days	**Remaining** 53 days	**Total** 81 days

**AYP* is *Assess Your Progress.*

Lesson Support

Section	Practice Bank	Study Guide*	Assessment Book*	Visuals	Explorations Lab Manual	Lesson Plans	Technology Book
4.1	24	4.1		Warm-Up 4.1 Folder A	Masters 1, 9	4.1	Spreadsheet Act. 4
4.2	25	4.2		Warm-Up 4.2	Add. Expl. 5	4.2	
4.3	26	4.3	Test 15	Warm-Up 4.3 Folders 4, B	Masters 2, 4	4.3	
4.4	27	4.4		Warm-Up 4.4		4.4	
4.5	28	4.5		Warm-Up 4.5 Folder A	Master 2	4.5	
4.6	29	4.6	Test 16	Warm-Up 4.6	Master 2	4.6	Calculator Act. 4
Review Test	30	Chapter Review	Tests 17, 18 Alternative Assessment			Review Test	

***Spanish versions** of *Study Guide* and *Assessment Book* are available.

Chapter Support

- Course Guide
- Lesson Plans
- Portfolio Project Book:
  Additional Project 4: Statistics
- Preparing for College Entrance Tests
- Multi-Language Glossary
- *Test Generator* Software
- Professional Handbook

Software Support

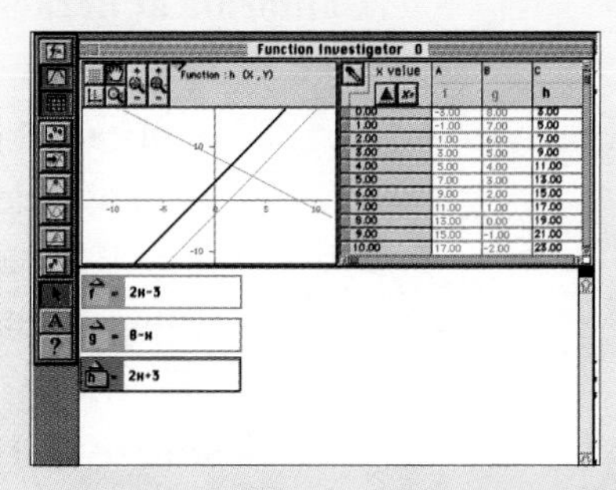

McDougal Littell Software
Stats!
Function Investigator

Internet Support

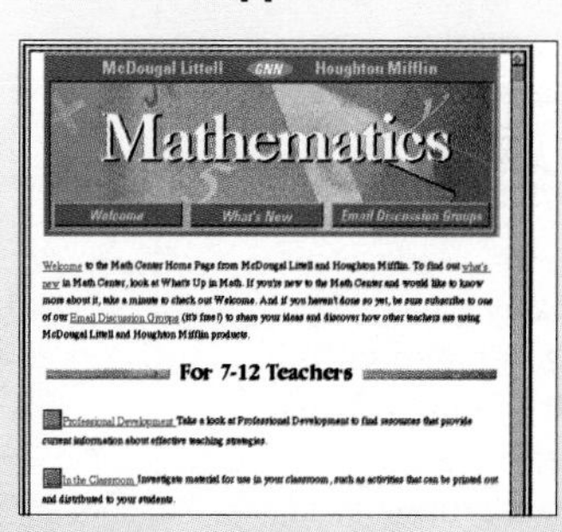

http://www.hmco.com
Next go to McDougal Littell; then the Education Center; then Secondary Math.

Outside Resources

Books, Periodicals

Friel, James O. and Gerald E. Gannon. "'What If...?' A Case in Point." *Mathematics Teacher* (April 1995): pp. 320–322.

Nord, Gail D. and John Nord. "An Example of Algebra in Lake Roosevelt." *Mathematics Teacher* (February 1995): pp. 116–120.

Stewart, Carolyn and Lucinda Chance. "Making Connections: Journal Writing and the Professional Teaching Standards." *Mathematics Teacher* (February 1995): pp. 92–95.

Dickey, Edwin M. "The Golden Ratio: A Golden Opportunity to Investigate Multiple Representations of a Problem." *Mathematics Teacher* (October 1993): pp. 554–557.

Activities, Manipulatives

Algebra Tiles for the Overhead Projector. Booklet and materials. Hayward, CA: Activity Resources Co., Inc.

Leiva, Miriam, Joan Ferrini-Mundy, and Loren P. Johnson. "Playing With Blocks: Visualizing Functions." *Mathematics Teacher* (November 1992): pp. 641–646.

Software

UCSMP. *StatExplorer.* Scott Foresman, 1993.

Videos

Futures with Jaime Escalante. Program No. 1: Statistics. PBS, 1990.

Interview Notes

Background

Earthquakes
The surface of Earth is called its crust. Below the crust is the mantle, which surrounds the outer core. Within the outer core is the inner core. Modern earthquake science is based upon a concept called plate tectonics. This concept describes the crust as being supported by underlying large, broad, and thick plates that are in the upper part of the mantle. In essence, these plates float on an underlying viscous material in the mantle. As the plates move, they can bump one another and one plate may then move over or below the other plate. The forces created by these collisions build up slowly and steadily until so much energy is involved that it cannot be contained by the mantle any longer. When this occurs, the burst of energy that is released causes earthquakes that can result in fractures of Earth's surface.

Barbara Romanowicz
Barbara Romanowicz is a geophysicist and director of the University of California Seismographic Station in Berkeley. She is also a professor of geology and geophysics at Berkeley. Geophysics is the science that deals with the physics of Earth and includes subspecialties such as oceanography, volcanology, and seismology. Romanowicz's special field is seismology, which is the study of earthquakes and their phenomenona. Her study of earthquakes that have occurred over the past 80 years has revealed a 20-year to 30-year cycle in which earthquake energy is transferred around the globe. This is a much greater time and distance than was previously thought to be true. Romanowicz thinks that large Pacific earthquakes in the early 1960s appear to have been responsible for other major quakes in the 1980s and early 1990s. Her work may help scientists forecast more accurately when future earthquakes will occur.

CHAPTER

4 Solving Equations and Inequalities

Shaking up SCIENCE

INTERVIEW **Barbara Romanowicz**

When the ground in Northern California starts to rumble, Barbara Romanowicz goes to work. A seismologist at the University of California at Berkeley, Romanowicz studies the sudden, violent upheavals of the land called earthquakes. "Although we can't predict earthquakes, we can save lives and money by promptly telling people where an earthquake is," she says. "Armed with this information, officials will be able to direct emergency aid much more effectively."

"It's important to know where earthquakes are,..."

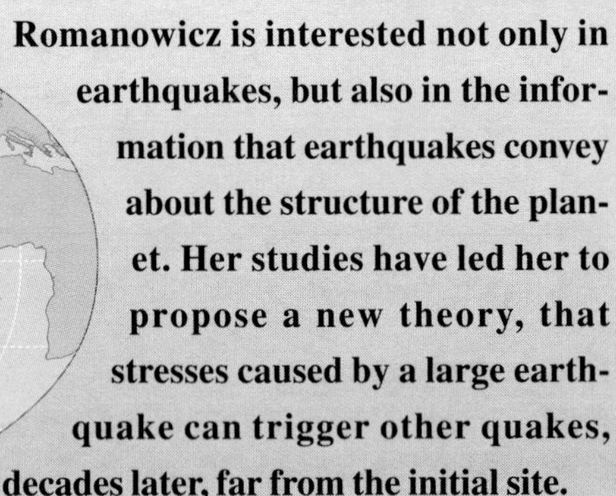

Romanowicz is interested not only in earthquakes, but also in the information that earthquakes convey about the structure of the planet. Her studies have led her to propose a new theory, that stresses caused by a large earthquake can trigger other quakes, decades later, far from the initial site.

Collecting Data

Romanowicz and other seismologists collect data about earthquakes from seismology stations located throughout the country. The stations are equipped with seismographs, machines that measure the movement of the ground.

In an earthquake, the ground moves in waves that spread out from the epicenter. (The epicenter is the point on Earth's surface closest to the source of the earthquake. Large earthquakes originate at great depths, sometimes 10 miles below the surface.) Earthquakes produce different kinds of waves. For example, pressure waves (called *P waves*) move the ground back and forth as they travel away from the epicenter. Shear waves (called *S waves*) move the ground sideways as they travel away from the epicenter. P waves and S waves show up as different patterns when recorded by a seismograph. P waves appear first on the seismograph record, because P waves travel faster than S waves.

Locating the Epicenter

To find out how far a particular seismology station is from an earthquake's epicenter, seismologists find the difference between the arrival times of

"...but it's also important to know how big they are and the time at which they occurred."

Rescuers on the roof of a collapsed house try to free people trapped inside following a major earthquake in western Japan, January, 1995.

Interview Notes

Background

Ancient Chinese Seismograph
The world's first seismograph was built in A.D. 132 by a Chinese scientist named Zhang Heng. The instrument consisted of a bronze cylindrical object from which protruded several dragons' heads. Each dragon's mouth held a bronze ball which, in the event of an earthquake, would fall into the open mouth of a bronze toad standing directly below. The noise created by this action would alert people to an earthquake even if it was hundreds of miles away. The seismograph contained a device which, after one ball was released, prevented the release of the others so that the direction of the epicenter of the earthquake could be determined.

Second-Language Learners

Students learning English may have trouble understanding the meaning of the phrase *armed with information* on page 144. If necessary, explain that it means "to have all the necessary information you need before making a decision."

Interview Notes

Mathematical Connection

An important fact that seismologists need to know about an earthquake is the location of the epicenter. By using the difference between the arrival times of P waves and S waves at a seismograph, they can write an equation that involves fractions to calculate the distance from the seismograph to the epicenter. Students learn how to solve equations involving fractions in Section 4.4.

Explore and Connect

Research
There is a great deal of information available about earthquakes in reference books and articles. Students should take notes of their research data for use during a class report.

Writing
You may wish to suggest that students also complete a list of the five strongest earthquakes since 1900, as measured by the Richter scale.

Project
This project would make an excellent activity for groups of four students. After finding out about the different kinds of seismographs, students can choose one model to build. Each group can demonstrate how its model works.

In January, 1995, this section of the elevated Hanshin Expressway collapsed during an earthquake in western Japan.

P waves and S waves. They use this value to solve an equation that gives the distance of the station from the epicenter. (You will learn more about this equation in Section 4.4.)

A modern model of Chang Heng's seismograph of 132 A.D.

Predicting Earthquakes

"It's important to know where earthquakes are, but it's also important to know how big they are and the time at which they occurred," Romanowicz says. This information is collected in catalogs that record the location, size, and date of major earthquakes all over the world. After reviewing catalogs that cover the past 80 years, Romanowicz was able to propose her theory about the long-term effects of earthquakes. "If I'm correct," she says, "soon we might be able to state with confidence that there will be a large earthquake in a certain region sometime in the next 20 years. Of course, that doesn't help people living there much, but it's a step towards our ultimate goal—reliable, long-range earthquake predictions."

EXPLORE AND CONNECT

Despite Japan's strict building codes, the earthquake caused severe damage.

1. Research The map on page 144 shows that some regions of the world are more susceptible to earthquakes than others. Find out why earthquakes tend to occur in these regions. Are earthquakes common in the area where you live? Report your findings to your class.

2. Writing To measure the size of an earthquake, seismologists commonly use the *Richter scale*. Write a report explaining how the Richter scale works and how seismologists use it.

3. Project Find out how modern seismographs work. Draw a diagram or build a model of one. Show how it works.

Mathematics
& Barbara Romanowicz

In this chapter, you will learn more about how mathematics is related to seismology.

Related Examples and Exercises

Section 4.4
- **Example 1**
- **Exercises 28–31**

SECTION

4.1 Solving Problems Using Tables and Graphs

Learn how to...

- **model situations with tables and graphs**

So you can...

- **compare options, such as buying CDs from a store or from a CD club**

Is it less expensive to buy compact discs (CDs) at a discount store or through a CD club? You can compare options using tables or graphs.

EXPLORATION
COOPERATIVE LEARNING

Comparing Tables and Graphs

Work in 2 teams of 2 students each.
You will need:

- a graphing calculator or graph paper

Discounts!
CDs
$11.00
Why Pay More !?!
Join our
CD Club
and get
8 FREE
CDs
(You must buy at least 6 more CDs for $15 each.)

One team should do Steps 1–4 below. The other team should do Steps 5–8 on the next page. The teams should work together on Step 9.

1 Copy and complete the table.

The CD club minimum is 8 + 6, or 14.

Number of CDs	Cost at store ($)	CD club cost ($)
14	?	?
15	?	?
16	?	?

2 For the values in the table, which way of buying CDs is less expensive? Will this always be the case? Explain.

3 Expand your table. Include rows for 20, 30, 40, and 50 CDs.

4 When will the total cost through the CD club equal the total cost at the discount store? How can you tell? When would you choose the CD club? the discount store? Why?

Exploration continued on next page.

Plan⇔Support

Objectives

- Model situations with tables and graphs.
- Compare options using tables and graphs.

Recommended Pacing

Core and Extended Courses
Section 4.1: 1 day

Block Schedule
Section 4.1: $\frac{1}{2}$ block
(with Chapter 3 Test)

Resource Materials

Lesson Support
Lesson Plan 4.1
Warm-Up Transparency 4.1
Overhead Visuals:
Folder A: Multi-Use Graphing Packet, Sheets 1–3
Practice Bank: Practice 24
Study Guide: Section 4.1
Explorations Lab Manual:
Diagram Masters 1, 9

Technology
Technology Book:
Spreadsheet Activity 4
Graphing Calculator
Spreadsheet Software
McDougal Littell Software
Function Investigator
Stats!
Internet:
http://www.hmco.com

Exploration Note

Purpose
The purpose of this Exploration is to have students use tables and graphs to explore and solve the same problem in order to have them compare and contrast the two methods.

Materials/Preparation
Have graphing calculators or graph paper available for the groups to use.

Procedure
Each team uses a table or a graph to decide whether it is more economical to buy CDs through a club or at a discount store. Each team should work independently on its table or graph. Some teams may need help in writing the equation for Step 5.

Closure
After Step 9 has been completed, all groups should participate in a general discussion about using tables or graphs to compare cost options. Students should conclude that either a table or a graph may be used as a problem-solving strategy and understand the advantages and disadvantages of each method.

Explorations Lab Manual
See the Manual for more commentary on this Exploration.

Diagram Masters 1 and 9

For answers to the Exploration, see answers in back of book.

Warm-Up Exercises

1. Complete the table of values for the equation $y = 4x + 1$. 1; 5; 9; 13

x	y
0	?
1	?
2	?
3	?

2. Find the slope of the line through points (14, 10) and (20, 22). 2

3. Find an equation of the line in Ex. 2. $y = 2x - 18$

4. Find the salary of a person earning \$10.50 an hour for 40 hours of work, plus \$15.75 an hour for 3 hours of overtime. \$467.25

5. For Ex. 4, write an expression to calculate the salary in terms of h, the number of hours of overtime. $10.50(40) + 15.75h$

Teach⇔Interact

Learning Styles: Visual

Tables and graphs help students to perceive the solution to a problem that requires a comparison of options. The visual aspect of these problem-solving techniques help students to learn and understand abstract ideas at a more concrete level.

Section Note

Teaching Tip
In Step 7 of the Exploration, remind students that a good viewing window for two graphs in the same coordinate plane is probably one in which their intersection point shows.

About the Example

Using Technology Students may understand the "relative" nature of a spreadsheet if they enter the formulas into a spreadsheet and view it as a list of formulas. Switching back to the table values may help them see that variables really do vary.

Additional Example

U-Park Garage charges customers a daily fee of \$3.50 plus \$.75 for each hour a car is parked. Self-Park Garage charges customers a daily fee of \$5.75 plus \$.50 for each hour a car is parked. When will it cost more to park at the U-Park Garage?

Make a table. You can use a spreadsheet or a graphing calculator, or make a table by hand. Include a column for the number of hours the car is parked and a column for the cost at each garage. Use this general formula.

Cost = Daily Fee + Hourly Fee × Number of Hours

Step 1 Enter an initial value for the number of hours in cell A2. Enter a formula in cell A3 to add a fixed amount to the initial value. Fill down to get a series of hourly values.

A3		=A2+4	
	A	B	C
1	Hours	U-Park Garage	Self-Park Garage
2	4	6.50	7.75
3	8	9.50	9.75
4	12	12.50	11.75
5	16	15.50	13.75

Exploration *continued*

5 Write an equation for the cost y of buying x CDs from the discount store.

6 Repeat Step 5 for the CD club.

7 Graph the two equations in the same coordinate plane. Set a good viewing window, like the one shown below.

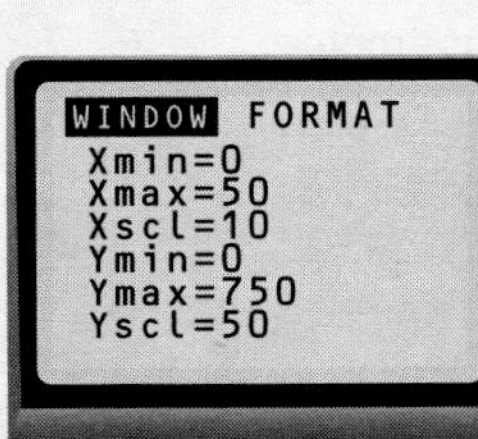

8 When will the total cost through the CD club equal the total cost at the discount store? How can you tell? When would you choose the CD club? the discount store? Why?

9 Discuss your answers to Steps 4 and 8 with the other team. Compare the table method to the graphing method. List some advantages and disadvantages of each method.

BY THE WAY

There are over 7 billion transactions per year at automatic teller machines (ATMs) in the United States.

EXAMPLE Application: Consumer Economics

At Service Bank customers pay a monthly fee of \$6.25 plus \$.30 per transaction. At Slate Bank customers pay \$3.00 per month plus \$.40 per transaction. (Checks written, deposits, and visits to automatic teller machines count as transactions.) When will the total monthly cost at Slate Bank be greater than at Service Bank?

SOLUTION

Problem Solving Strategy: Make a table.

You can use a spreadsheet or a graphing calculator, or make a table by hand. Include a column for the number of transactions and a column for the cost at each bank. Use this general formula.

Cost = Monthly fee + Transaction fee × Number of transactions

Technology Note

Students who are new to the TI-82 graphing calculator may not realize that they can use the CALCULATE menu to find where two graphs intersect. (This is usually a much faster approach than using the TRACE and ZOOM features.) From the CALCULATE menu, students should select 5:intersect. See the TI-82 manual for details.

Step 1 Enter an initial value for the number of transactions in cell A2. Enter a formula in cell A3 to add a fixed amount to the initial value. Fill down to get a series of transaction values.

Bank Options

A3 =A2+5

	A	B	C
1	Transactions	Service Bank	Slate Bank
2	5	7.75	5.00
3	10	9.25	7.00
4	15	10.75	9.00
5	20	12.25	11.00
6	25	13.75	13.00
7	30	15.25	15.00
8	35	16.75	17.00
9	40	18.25	19.00

Step 2 Enter a formula and fill down to find the **cost** at Service Bank.
For example, **18.25 = 6.25 + (0.3)(40)**.

Step 3 Enter a formula and fill down to find the **cost** at Slate Bank.
For example, **19 = 3 + (0.4)(40)**.

The total monthly cost at Slate Bank becomes greater than at Service Bank somewhere between 30 and 35 transactions. Try more values between 30 and 35 transactions until you find more precisely when the cost at Slate Bank becomes greater.

Bank Options

A8 =A7+1

	A	B	C
7	30	15.25	15.00
8	31	15.55	15.40
9	32	15.85	15.80
10	33	16.15	16.20
11	34	16.45	16.60
12	35	16.75	17.00

When the number of transactions in a month reaches 33, the total monthly cost at Slate Bank becomes greater than at Service Bank.

THINK AND COMMUNICATE

1. **a.** For each bank in the Example, write an equation for the monthly cost of a checking account as a function of the number of transactions.
   b. Describe how you could use a graph of the equations to solve the problem.
2. Suppose Jennifer expects to write fewer than 30 checks in a month. Based on the Example, what advice would you give her? Why?

Teach⇔Interact

Additional Example (continued)

Step 2 Enter a formula and fill down to find the cost at U-Park Garage. For example: 3.50 + 0.75(4) = 6.50.
Step 3 Enter a formula and fill down to find the cost at Self-Park Garage. For example: 5.75 + 0.50(4) = 7.75.
The total daily cost becomes greater at U-Park Garage somewhere between 8 and 12 hours. Try more values between 8 and 12 until you find more precisely when the cost at U-Park Garage becomes greater.

A3 =A2+1

	A	B	C
1	Hours	U-Park Garage	Self-Park Garage
2	8	9.50	9.75
3	9	10.25	10.25
4	10	11.00	10.75

When the number of hours is over 9, the daily cost at U-Park Garage becomes greater than at Self-Park Garage.

Think and Communicate

When discussing question 1, you may need to remind students of the definition of a function. The input values for the function are the domain, and the output values are the range.

ANSWERS Section 4.1

Think and Communicate

1. a. Service Bank, $c = 6.25 + 0.3t$ and Slate Bank, $c = 3 + 0.4t$, where t is the number of transactions in a month and c is the total monthly cost in dollars.
   b. You could use a graphing calculator or graph paper to graph both equations in the same coordinate plane. The t-coordinate of the intersection point tells you the number of monthly transactions for which the costs at the two banks are equal.
2. I would suggest that she choose Slate Bank rather than Service Bank because her total costs will be less with that bank.

Teach⇔Interact

Checking Key Concepts

Teaching Tip
Have students interpret the graph for questions 1–3 before they give the equations of the lines. Include a discussion of how the graph is a model of the pool options and what the intersection point of the two lines represents.

Closure Question

If you want to compare two options in order to make a buying decision, how can you model the situation to make the best decision?
Use either a table or a graph to model the situation.

Apply⇔Assess

Suggested Assignment

Core Course
Exs. 1–11, 13, 15–23

Extended Course
Exs. 1–23

Block Schedule
Day 22 1–11, 13, 15–23

Exercise Notes

Cooperative Learning
Exs. 1–4 Allow students to share their answers to these questions. Students can share their personal knowledge about the factors important in joining a fitness club. Each group should decide what factors are important and to what extent they are important.

Topic Spiraling: Preview
Ex. 4 Students should understand that each factor they consider may be a variable in a real decision for joining a fitness club. This is an intuitive underpinning for a function of more than one variable. You can extend the exercise by listing and ranking the other factors and then using the ranking to help choose a fitness club. This can be done by assigning a weight to each ranking.

CHECKING KEY CONCEPTS

People who want to use a community pool during the summer have the two options modeled by the graph shown. Use the graph for Questions 1–3.

1. Write an equation for the red line. Describe the relationship between the cost of a season pass and the number of visits to the pool.
2. Write an equation for the green line. What is the cost without a season pass of one visit to the pool?
3. Is it worth buying a season pass if you expect to use the pool only 10 times during the summer? Explain.

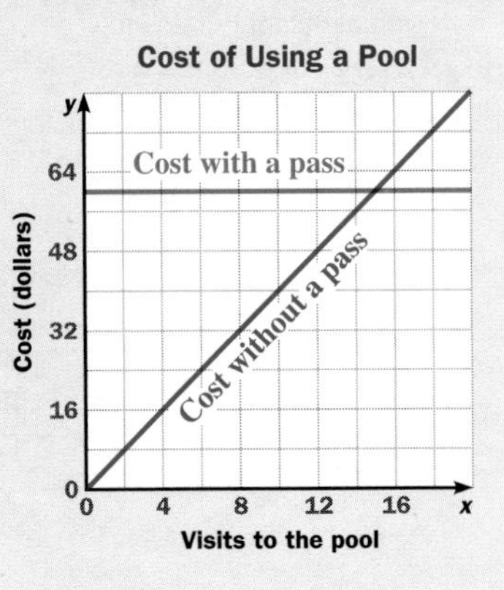

4.1 Exercises and Applications

Extra Practice exercises on page 563

Cooperative Learning For Exercises 1–4, work with a partner. Use the table below.

1. Which club has a greater initiation fee? a greater monthly fee? How can you tell?

Fitness Club Membership Costs		
Number of months	Cost at Club A ($)	Cost at Club B ($)
0	100	0
1	140	50
2	180	100
3	220	150

2. One person should copy and expand the table to cover a total of 18 months. The other person should write and graph two equations to model the membership costs at the two clubs.
3. When is the membership cost the same for both fitness clubs?
4. **Open-ended Problem** Suppose you decide to join a fitness club. What factors will you consider besides cost? Explain.

PERSONAL FINANCE Exercises 5–8 should be done together.

5. Use both a table and a graph to model the earnings from these two jobs as a function of the number of overtime hours worked.

   Job A: $30,000 per year for full-time work, with no pay for overtime

   Job B: $24,000 per year for full-time work, plus $12 per hour for each hour of overtime

Checking Key Concepts

1. $y = 60$; The cost is constant no matter how many visits a person makes to the pool.
2. $y = 4x$; $4
3. No; the cost for a pass is $60 and the cost for 10 visits without a pass is $40.

Exercises and Applications

1. Club A, since at 0 months Club A has a cost of $100 and Club B has a cost of $0; Club B, since the cost of the first month is 50 – 0 or $50 and the cost at Club A is 140 – 100 or $40
2. See answers in back of book.
3. during the tenth month of membership
4. Answers may vary. Examples are given. You might want to know the location of the fitness clubs, their convenience with respect to public transportation or availability of parking, their hours of operation, the equipment that is available at each club, and the expertise of the staff.
5. See answers in back of book.
6. 500 h; In the table, look for the number of hours of overtime for which the entries in the second and third columns are the same. In the graph, find the x-coordinate of the intersection point.

6. How many hours of overtime will you have to work to earn as much in a year from Job B as from Job A? How can you tell from the table? from the graph?

7. Based on earnings alone, under what circumstances would you be better off with Job A? with Job B?

8. Choose the letter of the equation you could use to solve Exercise 6.

   A. $12h = 30{,}000$

   B. $24{,}000 + 12h = 30{,}000$

   C. $24{,}000 + 12h = 30{,}000 + 12h$

Connection ENERGY USE

An advertisement for a fluorescent light bulb says that it lasts 9 times longer than a comparable standard light bulb—up to 9000 hours. It also uses less energy. The fluorescent bulb uses only 15 watts. A typical standard bulb costs about $.75 and uses 60 watts. This graph models the cost of buying and using each kind of bulb.

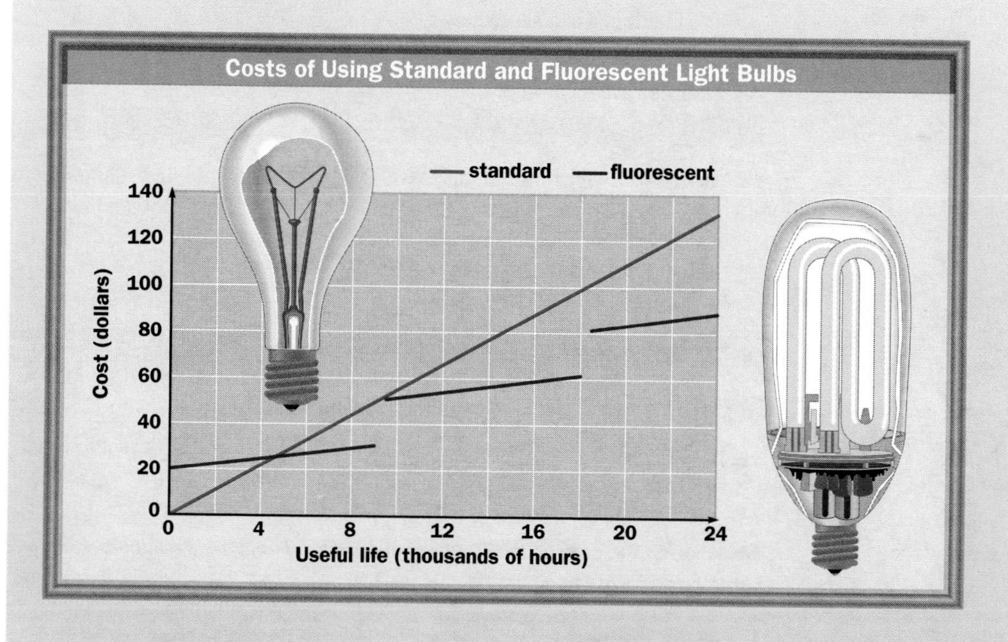

9. How does the graph show that after using a fluorescent light bulb for 9000 hours, you must buy a new one? Estimate the price of one fluorescent bulb.

10. Why aren't there noticeable jumps in the graph for standard bulbs?

11. How many hours each month do you think you have the light on in your bedroom? How many hours do you think the lights are on in your school each month?

12. **Writing** Do you think you should use fluorescent lights at home? at school? Explain.

7. If you work no more than 500 h of overtime per year, then Job A would pay more than Job B. If you work more than 500 h of overtime per year, then Job B would pay more.

8. B

9. The cost of using a fluorescent bulb increases sharply after each 9000 hour period; about $20.

10. because the cost of a standard bulb is very small compared to the units on the vertical axis

11. Answers may vary.

12. Answers may vary. An example is given. It may not be economical to use fluorescent bulbs at home because the lights are on for fairly short periods of time and people may not want to pay the high cost of fluorescent bulbs. However, at school, there are very many bulbs which are in use for long periods of time; the cumulative savings resulting from using fluorescent rather than standard bulbs would make it worthwhile to use them despite the higher initial cost. There may be other considerations as well. For example, some people experience eyestrain from fluorescent lighting and would not choose to use it at home.

Apply⇔Assess

Exercise Notes

Application
Exs. 5–13 Applications of algebra to real-world topics and situations makes mathematics more authentic for students. In these exercises, there are applications to personal finance, energy use, and recreation.

Using Technology
Exs. 5–7 Students can use the *Stats!* and *Function Investigator* software to generate a spreadsheet and graph for these exercises. On a TI-82 graphing calculator, students can use the TABLE feature once they have modeled the situation with two equations. Enter Y=30000 and Y=24000+12X on the Y= list. Next, press [2nd][TblSet]. Use TblMin = 0, ΔTbl = 20 to tell the calculator what value of x to start with and how much x will increase from one line of the table to the next. Press [2nd][TABLE] to view the table. Use the [▼] key to move down through the table. Look for the line where the numbers in the Y_1- and Y_2-columns are equal. Students can display graphs for these exercises on the TI-82. Follow the usual procedure after selecting an appropriate viewing window.

Communication: Discussion
Ex. 7 Ask students if they would consider earnings alone when choosing a job. List some of the other factors that students think are important. Then ask students to rank the factors from most important to least important.

Research
Ex. 11 Students are asked to estimate how many hours each month they use a bedroom light. This exercise can be extended by having them record the actual number of hours for one month, compare it with their estimate, and calculate how much of the family's electric bill is due to their bedroom light. Doing this will help students understand how to read and interpret an electric bill.

Apply⇔Assess

Exercise Notes

Alternate Approach

Ex. 14(a) This exercise can be done with a spreadsheet or a calculator with TABLE features. In either case, have students decrease the time intervals used to generate the table until they see the solution to the problem.

Student Progress

Ex. 14(b) At this point, students should be comfortable with thinking of an equation as a possible model for a situation. Ask them to look at their table or graph from part (a) and use it to write an equation as a model.

Assessment Note

Ex. 15 Students' responses to this exercise should be evaluated on the clarity and organization of their letter as well as on the correct use of a table or graph to support their advice.

Practice 24 for Section 4.1

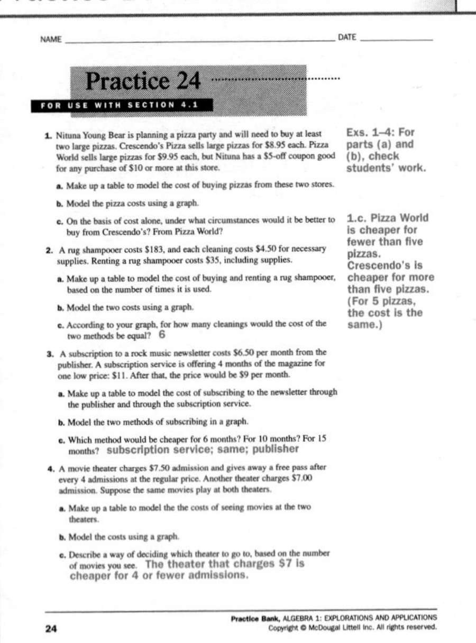

NAME ______________________ DATE __________

Practice 24

FOR USE WITH SECTION 4.1

1. Nituna Young Bear is planning a pizza party and will need to buy at least two large pizzas. Crescendo's Pizza sells large pizzas for $8.95 each. Pizza World sells large pizzas for $9.95 each, but Nituna has a $5-off coupon good for any purchase of $10 or more at this store.
   - a. Make up a table to model the cost of buying pizzas from these two stores.
   - b. Model the pizza costs using a graph.
   - c. On the basis of cost alone, under what circumstances would it be better to buy from Crescendo's? From Pizza World?
2. A rug shampooer costs $183, and each cleaning costs $4.50 for necessary supplies. Renting a rug shampooer costs $35, including supplies.
   - a. Make up a table to model the cost of buying and renting a rug shampooer, based on the number of times it is used.
   - b. Model the two costs using a graph.
   - c. According to your graph, for how many cleanings would the cost of the two methods be equal? 6
3. A subscription to a rock music newsletter costs $6.50 per month from the publisher. A subscription service is offering 4 months of the magazine for one low price: $11. After that, the price would be $9 per month.
   - a. Make up a table to model the cost of subscribing to the newsletter through the publisher and through the subscription service.
   - b. Model the two methods of subscribing in a graph.
   - c. Which method would be cheaper for 6 months? For 10 months? For 15 months? subscription service; same; publisher
4. A movie theater charges $7.50 admission and gives away a free pass after every 4 admissions at the regular price. Another theater charges $7.00 admission. Suppose the same movies play at both theaters.
   - a. Make up a table to model the the costs of seeing movies at the two theaters.
   - b. Model the costs using a graph.
   - c. Describe a way of deciding which theater to go to, based on the number of movies you see. The theater that charges $7 is cheaper for 4 or fewer admissions.

Exs. 1–4: For parts (a) and (b), check students' work.

1.c. Pizza World is cheaper for fewer than five pizzas. Crescendo's is cheaper for more than five pizzas. (For 5 pizzas, the cost is the same.)

24

Practice Bank, ALGEBRA 1: EXPLORATIONS AND APPLICATIONS

13. RECREATION A small movie theater has two admissions options.

Members: $40 annual membership fee, plus $4.50 per show
Nonmembers: $7.00 per show

Under what circumstances do you recommend buying a membership?

14. Challenge Montreal and Quebec City are about 260 km apart along Route 20 in Canada. Sylvie Potvin leaves Montreal and drives toward Quebec City on Route 20 at the same time that Anne Durand leaves Quebec City headed for Montreal. Sylvie drives at a steady speed of 90 km/h and Anne drives at a steady speed of 105 km/h.

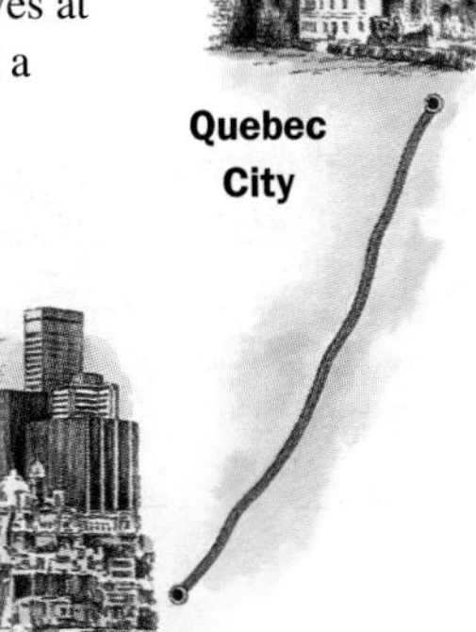

a. When and where do Sylvie and Anne pass each other? (Assume that neither one stops along the way.) Include a table or graph with your answer. (*Hint:* When are they the same distance from Montreal?)

b. Write an equation to model this situation. Then show how to solve the equation.

ONGOING ASSESSMENT

15. Writing Every time Mr. and Mrs. Dalton go to the movies, it costs them $13. They wonder whether they should buy a VCR for $350 and rent videotapes for $3 each instead. Write a letter to the Daltons to give them your advice. Include a table or graph in your letter.

SPIRAL REVIEW

Solve each equation. *(Sections 2.1 and 2.2)*

16. $m - 17 = -3$ **17.** $\frac{n}{5} = 2$ **18.** $15 = \frac{c}{2} + 1$

19. $5a - 3 = 7$ **20.** $-6 = 3x + 21$ **21.** $13 - c = 18$

Tell whether the data show direct variation. If they do, give the constant of variation and write an equation. *(Section 3.2)*

22.

Time (minutes)	Distance (meters)
3	5.25
8	14
14	24.5
21	36.75

23.

Age (years)	Height (inches)
2	32
10	60
25	66
45	66

13. If a customer plans to attend up to 15 shows, then it is less expensive to pay the nonmember rate. If a customer plans to attend more than 16 shows, buying a membership is the better choice. At 16 shows, the costs are equal.
14. See answers in back of book.
15. Answers may vary. An example is given. Dear Mr. and Mrs. Dalton: If you plan to see more than 35 movies, it may make sense for you to buy the VCR because then the cost of buying a VCR and renting videotapes is cheaper than going to the movies. (Take a look at the graph below.) If not, it is cheaper to go to the movies. Cost may not be the only consideration, however. For example, are the movies you want to see available on video? Do you prefer to go out or stay home? It might be possible to borrow videotapes from the library. You might also use the VCR to show your own videotapes if you own a video camera.

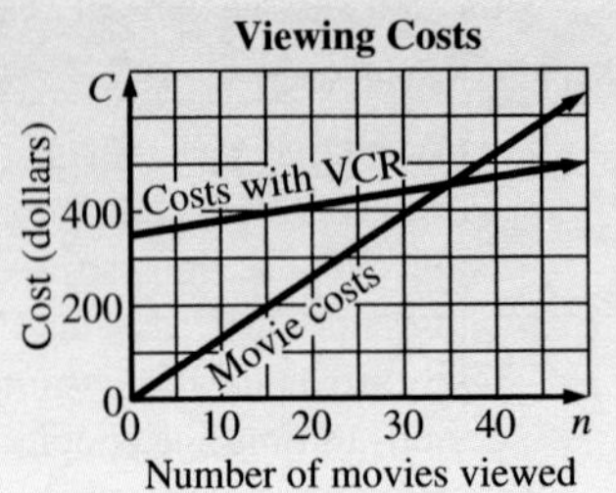

16. 14
17. 10
18. 28
19. 2
20. –9
21. –5
22. Yes; 1.75; $d = 1.75t$.
23. No.

SECTION

4.2 Using Reciprocals

Learn how to...

- solve one-step and two-step equations using reciprocals

So you can...

- solve problems involving fractions

How would you solve an equation like $\frac{3}{4}x = 12$? You could divide both sides by $\frac{3}{4}$ or multiply both sides by the reciprocal of $\frac{3}{4}$, which is $\frac{4}{3}$. Two numbers are **reciprocals** if their product is 1.

$$\frac{3}{4} \cdot \frac{4}{3} = \frac{3 \cdot 4}{4 \cdot 3} = \frac{12}{12} = 1$$

THINK AND COMMUNICATE

1. What is the reciprocal of $\frac{2}{3}$? of $\frac{1}{3}$? of 3? of 1? Give some other examples of reciprocals.
2. Do you think 0 has a reciprocal? Explain.

Property of Reciprocals

For every nonzero number a, there is exactly one number $\frac{1}{a}$ such that

$$a \cdot \frac{1}{a} = 1 \text{ and } \frac{1}{a} \cdot a = 1.$$

EXAMPLE 1

Solve the equation $\frac{3}{4}x = 12$.

SOLUTION

In order to solve an equation algebraically, you must get the variable alone on one side of the equation.

Multiply both sides of the equation by the reciprocal of $\frac{3}{4}$.

$$\frac{3}{4}x = 12$$

$$\frac{4}{3} \cdot \frac{3}{4}x = \frac{4}{3} \cdot 12$$

$$1 \cdot x = \frac{4 \cdot 12}{3}$$

$$x = 16$$

Plan⇔Support

Objectives

- Solve one-step and two-step equations using reciprocals.
- Solve problems involving fractions.

Recommended Pacing

Core and Extended Courses
Section 4.2: 1 day

Block Schedule
Section 4.2: $\frac{1}{2}$ block (with Section 4.3)

Resource Materials

Lesson Support
Lesson Plan 4.2
Warm-Up Transparency 4.2
Practice Bank: Practice 25
Study Guide: Section 4.2
Explorations Lab Manual: Additional Exploration 5

Technology
Graphing Calculator
Internet: http://www.hmco.com

Warm-Up Exercises

1. Find $\frac{5}{3} \cdot \frac{3}{5}$. 1

Solve each equation.

2. $3x = 12$ 4
3. $-4t = 20$ −5
4. $5s + 9 = 19$ 2
5. $-2z + 8 = 16$ −4

ANSWERS Section 4.2

Think and Communicate

1. $\frac{3}{2}$; 3; $\frac{1}{3}$; 1; Answers may vary. Examples are given. The following pairs of numbers are reciprocals: $\frac{1}{4}$ and 4, $\frac{4}{7}$ and $\frac{7}{4}$, and 0.1 and 10.
2. No. Explanations may vary. The product of 0 and any number is 0, so there is no number whose product with 0 is 1.

Teach⇔Interact

Think and Communicate

For question 1, you may want to generalize the reciprocal of a nonzero number n as $\frac{1}{n}$. This generalization should help students answer question 2 also.

Section Notes

Using Technology
Both the TI-81 and TI-82 graphing calculators have a $\boxed{x^{-1}}$ key for finding reciprocals. Students should be alert to situations where parentheses are needed to obtain correct results.

Topic Spiraling: Review
The property of reciprocals can be reinforced by having students always use the property as shown in the second step of Example 1. Remind students that it is the identity property of multiplication which states that $1 \cdot x = x$.

Additional Example 1

Solve the equation $\frac{2}{3}x = 16$.

$$\frac{2}{3}x = 16$$
$$\frac{3}{2} \cdot \frac{2}{3}x = \frac{3}{2} \cdot 16$$
$$1 \cdot x = \frac{3 \cdot 16}{2}$$
$$x = 24$$

Additional Example 2

Solve each equation.

a. $-\frac{1}{3}x = -16$

$$-\frac{1}{3}x = -16$$
$$(-3)\left(-\frac{1}{3}\right)x = (-3)(-16)$$
$$1 \cdot x = 48$$
$$x = 48$$

b. $-\frac{4s}{5} = 24$

$$-\frac{4s}{5} = 24$$
$$-\frac{4}{5}s = 24$$
$$\left(-\frac{5}{4}\right)\left(-\frac{4}{5}\right)s = \left(-\frac{5}{4}\right)(24)$$
$$1 \cdot s = -30$$
$$s = -30$$

When you solve an equation, remember to keep it balanced. If you apply an operation like multiplication to one side of the equation, then you must apply the same operation to the other side of the equation.

BY THE WAY

The Babylonians used reciprocals for division. Instead of dividing by a number, they multiplied by its reciprocal. They created tables of reciprocal values to simplify this work.

Reciprocals and Negative Numbers

The reciprocal of a negative number is negative.

For example, the reciprocal of $-\frac{5}{4}$ is $-\frac{4}{5}$.

The product of two negative numbers is positive.

$$-\frac{5}{4} \cdot -\frac{4}{5} = +\frac{5 \cdot 4}{4 \cdot 5} = \frac{20}{20} = 1$$

EXAMPLE 2

Solve each equation.

a. $-\frac{1}{6}y = -18$ **b.** $\frac{-3n}{10} = 21$

SOLUTION

a.
$$-\frac{1}{6}y = -18$$
$$(-6)\left(-\frac{1}{6}\right)y = (-6)(-18)$$
Multiply both sides by the **reciprocal**.
$$1 \cdot y = 108$$
$$y = 108$$

b.
$$\frac{-3n}{10} = 21$$
$$-\frac{3}{10}n = 21$$
Rewrite $\frac{-3n}{10}$ with a fraction as the coefficient.
$$\left(-\frac{10}{3}\right)\left(-\frac{3}{10}\right)n = \left(-\frac{10}{3}\right)(21)$$
Multiply both sides by the **reciprocal**.
$$n = -70$$

WATCH OUT!

When you work with negative fractions, remember these rules:

$$-\frac{a}{b} \neq \frac{-a}{-b}$$

$$-\frac{a}{b} = \frac{-a}{b} = \frac{a}{-b}$$

THINK AND COMMUNICATE

3. Explain how a reciprocal was used in part (a) of Example 2 to get y alone on one side of the equation.

4. How could you solve part (b) of Example 2 using inverse operations, as you learned in Chapter 2?

5. Describe how you would solve the equation $\frac{-a}{3} = 5$.

Think and Communicate

3. By multiplying both sides of the equation by –6, which is the reciprocal of $-\frac{1}{6}$, the coefficient of y becomes 1, and so the left side has y alone.

4. Divide both sides by $-\frac{3}{10}$.

5. Rewrite $\frac{-a}{3}$ as $-\frac{1}{3}a$. Multiply both sides of the equation $\frac{-a}{3} = 5$ by –3 to get $a = -15$.

EXAMPLE 3

Sheila has \$20 available to spend on materials to make a vest. The pattern, thread, and buttons cost \$10 altogether. The pattern calls for $\frac{7}{8}$ yd of fabric. What is the most Sheila can afford to pay per yard of fabric?

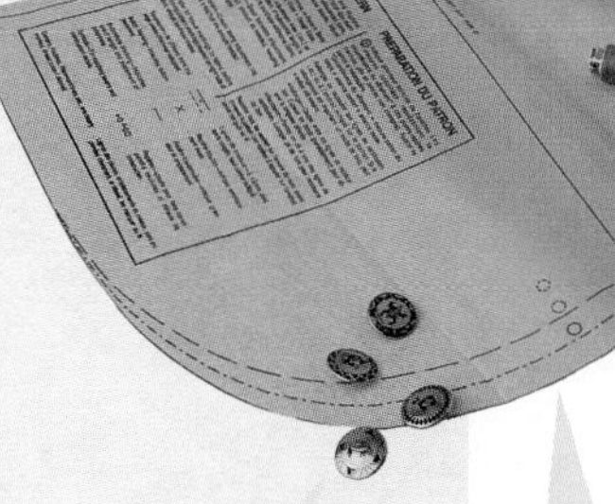

SOLUTION

Problem Solving Strategy: Use an equation.

Let p = the maximum price per yard.

Cost of fabric + Cost of other materials = Money available

$$\frac{7}{8}p + 10 = 20$$

$$\frac{7}{8}p + 10 - 10 = 20 - 10$$

$$\frac{7}{8}p = 10$$

Multiply both sides by the reciprocal.

$$\frac{8}{7} \cdot \frac{7}{8}p = \frac{8}{7} \cdot 10$$

$$p = \frac{80}{7} \approx 11.43$$

The most Sheila can afford to pay per yard of fabric is \$11.43.

THINK AND COMMUNICATE

6. Suppose someone wrote the equation in Example 3 as $20 - \frac{7}{8}p = 10$. Describe the steps you would take to solve the equation.
7. Solve Example 3 assuming that the pattern calls for only $\frac{3}{4}$ yd of fabric.

CHECKING KEY CONCEPTS

Find the reciprocal of each number.

1. $\frac{9}{10}$
2. $\frac{1}{11}$
3. 8
4. $-\frac{7}{12}$

Rewrite each expression with a fraction as the coefficient.

5. $\frac{2n}{3}$
6. $\frac{-4a}{7}$
7. $\frac{6c}{-5}$
8. $\frac{-x}{2}$

Solve each equation.

9. $\frac{4}{5}x = 16$
10. $\frac{-4x}{3} = 5$
11. $\frac{3}{5}n - 11 = 19$

Teach⇔Interact

Additional Example 3

Jorge has \$8 to spend on the ingredients to make an apple pie. The spices, dry ingredients, and dough cost \$5.30 altogether. If the recipe calls for $2\frac{1}{2}$ or $\frac{5}{2}$ pounds of apples, how much can Jorge afford to pay for each pound of apples?

Use an equation.

Let p = the maximum price per pound.

$$\frac{5}{2}p + 5.30 = 8$$

$$\frac{5}{2}p + 5.30 - 5.30 = 8 - 5.30$$

$$\frac{5}{2}p = 2.70$$

$$\frac{2}{5} \cdot \frac{5}{2}p = \frac{2}{5} \cdot 2.70$$

$$p = 1.08$$

Jorge can afford to pay \$1.08 per pound for the apples.

Think and Communicate

Question 6 points out that different, but related, equations can be used to solve the same problem. Have students compare their answers to this question with the steps in Example 3.

Checking Key Concepts

Common Error

In questions 9 and 11, some students may confuse $\frac{4}{5}x$ with $\frac{4}{5x}$ and $\frac{3}{5}n$ with $\frac{3}{5n}$. Emphasize that $\frac{4}{5}x = \frac{4x}{5}$ and that $\frac{3}{5}n = \frac{3n}{5}$, that is, the variable goes in the numerator and not in the denominator.

Closure Question

Explain how to solve the equation $-\frac{3}{4}x + 12 = 18$.

First, subtract 12 from both sides of the equation to get $-\frac{3}{4}x = 6$. Then multiply both sides by the reciprocal of $-\frac{3}{4}$ to get $1 \cdot x = -8$. State the solution as $x = -8$.

Think and Communicate

6. Answers may vary. An example is given. Subtract 20 from both sides. Then multiply both sides by the reciprocal of $-\frac{7}{8}$, which is $-\frac{8}{7}$.
7. $\frac{3}{4}p + 10 = 20$; $\frac{3}{4}p = 10$; $\frac{4}{3} \cdot \frac{3}{4}p = \frac{4}{3} \cdot 10$; $p \approx 13.33$; She can pay up to \$13.33 per yard.

Checking Key Concepts

1. $\frac{10}{9}$
2. 11
3. $\frac{1}{8}$
4. $-\frac{12}{7}$
5. $\frac{2}{3}n$
6. $-\frac{4}{7}a$
7. $-\frac{6}{5}c$
8. $-\frac{1}{2}x$
9. 20
10. $-\frac{15}{4}$
11. 50

Apply⇔Assess

Suggested Assignment

Core Course
Exs. 1–19, 21–23, 25–33, 35, 37–49

Extended Course
Exs. 1, 3–6, 8, 9, 11–13, 16–49

Block Schedule
Day 23 Exs. 1–19, 21–23, 25–33, 37–49

Exercise Notes

Communication: Writing/Discussion
Exs. 1–17 To assess students' understanding of how to solve equations using reciprocals, you can ask some students to write their solutions to these exercises on the board. Each solution should be explained by the writer.

Teaching Tip
Exs. 9–16 Encourage students to check their answers for each equation. If they use calculators to check their answers, remind them to use parentheses when entering fractions.

Interdisciplinary Problems
Exs. 18–20 Mathematics can be connected to music directly in the way notes are positioned on a musical scale. Mathematics can be connected to music indirectly when it is used to analyze sound waves in physics or to study how notes are related to frequencies.

Second-Language Learners
Ex. 24 Students learning English should not have any difficulty with this writing assignment, but they may benefit from working with an English-proficient partner.

4.2 Exercises and Applications

Extra Practice exercises on page 564

Identify the reciprocal you would use to solve each equation. Then solve.

1. $\frac{2}{9}x = 8$ **2.** $\frac{1}{3}n = 21$ **3.** $12 = \frac{5}{6}a$ **4.** $\frac{2}{5}b = -3$

Rewrite each expression with a fraction as the coefficient.

5. $\frac{7c}{16}$ **6.** $\frac{n}{15}$ **7.** $\frac{-5x}{11}$ **8.** $\frac{-b}{20}$

Solve each equation.

9. $-\frac{4}{9}c = 20$ **10.** $2 = -\frac{1}{15}n$ **11.** $-\frac{2}{3}x = -7$ **12.** $\frac{10c}{3} = -20$

13. $9 = \frac{5a}{2}$ **14.** $\frac{-12x}{5} = 36$ **15.** $10 = \frac{-n}{8}$ **16.** $\frac{-4a}{5} = 30$

17. SAT/ACT Preview Which number is the solution of $-\frac{3}{5}x = -15$?

A. -25 **B.** -9 **C.** $-\frac{5}{3}$ **D.** 9 **E.** 25

Connection MUSIC

Musicians generate sound waves when they sing or play an instrument. The rate at which a sound wave vibrates determines the musical note, or *pitch*, that you hear. For example, the A above middle C has a frequency of 440 vibrations per second and is referred to as *international*, or *concert*, pitch.

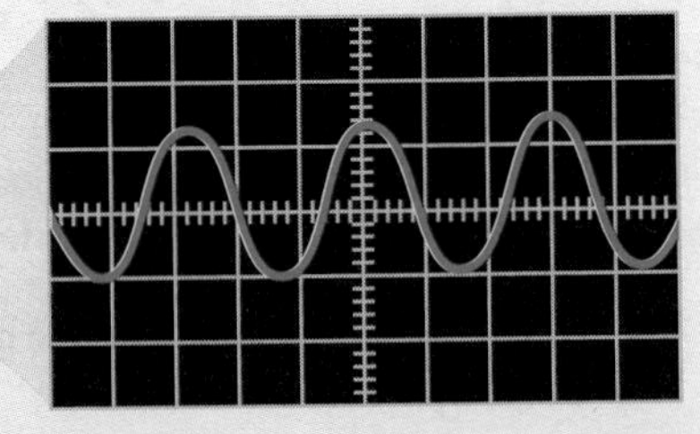

18. When two notes are a "fifth" apart, the frequency of the lower note is $\frac{2}{3}$ the frequency of the higher note.

a. Write an equation to model this relationship.

b. Find the frequency of the note that is a fifth above concert pitch.

c. Find the frequency of the note that is a fifth below concert pitch.

19. When two notes are a "fourth" apart, the frequency of the lower note is $\frac{3}{4}$ the frequency of the higher note. Repeat Exercise 18 using a "fourth" instead of a "fifth."

20. Research What is the letter of the note a fifth above concert pitch? a fourth below concert pitch? Compare the notes and their frequencies.

BY THE WAY

Orchestra concerts begin with the musicians tuning their instruments to a single note. That note is the A above middle C.

Exercises and Applications

1. $\frac{9}{2}$; 36
2. 3; 63
3. $\frac{6}{5}$; $\frac{72}{5} = 14\frac{2}{5}$
4. $\frac{5}{2}$; $-\frac{15}{2} = -7\frac{1}{2}$
5. $\frac{7}{16}c$
6. $\frac{1}{15}n$
7. $-\frac{5}{11}x$
8. $-\frac{1}{20}b$
9. -45
10. -30
11. $\frac{21}{2} = 10\frac{1}{2}$
12. -6
13. $\frac{18}{5} = 3\frac{3}{5}$
14. -15
15. -80
16. $-\frac{75}{2} = -37\frac{1}{2}$
17. E
18. a. $l = \frac{2}{3}h$, where l = the frequency of the lower note and h = the frequency of the higher note
 b. 660 vibrations per second
 c. $293\frac{1}{3}$ vibrations per second
19. a. $l = \frac{3}{4}h$, where l = the frequency of the lower note and h = the frequency of the higher note
 b. $586\frac{2}{3}$ vibrations per second
 c. 330 vibrations per second
20. E; E; The notes are both E, but they are an octave apart. The frequency of the higher note is double the frequency of the lower note.
21. $2\frac{2}{3}$ yd
22. 4 in.
23. $1\frac{1}{3}$ ft

GEOMETRY Find each unknown measurement.

Toolbox p. 594
Area and Volume

21. parallelogram
Area = 2 yd^2

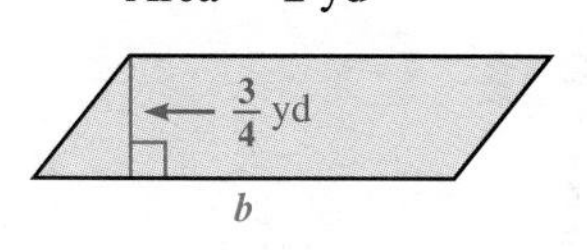

22. triangle
Area = 10 in.^{2}

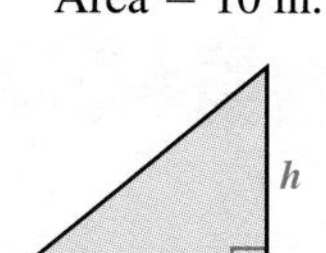

23. rectangle
Area = 3 ft^2

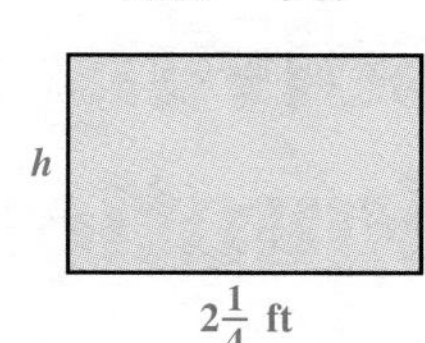

24. Writing Explain how you found the unknown measurement for one of the figures in Exercises 21–23.

Solve each equation.

25. $\frac{2}{3}x + 7 = 15$ **26.** $\frac{1}{10}n - 5 = 15$ **27.** $60 = \frac{4}{5}a + 16$

28. $\frac{9}{10}c - 8 = 4$ **29.** $-\frac{3}{4}x - 12 = 15$ **30.** $14 - \frac{7}{8}n = 56$

31. $\frac{-2a}{5} + 9 = 5$ **32.** $32 = \frac{-n}{8} + 8$ **33.** $-2 - \frac{5c}{6} = 7$

34. Open-ended Problem Write an equation that can be solved using inverse operations or using reciprocals. Solve the equation both ways.

35. FUEL ECONOMY The gasoline tank on Lorenzo Shepard's car holds 13 gal. The car uses $\frac{1}{35}$ gal of gasoline for every mile that Lorenzo drives it. The car's fuel light comes on when there are 3 gal of gasoline left in the tank. How many miles can Lorenzo drive the car before the fuel light comes on?

36. Challenge How is the equation $\frac{3}{5x} + 2 = 8$ different from the others in this section? How would you solve this equation?

ONGOING ASSESSMENT

37. Open-ended Problem Write a quiz with five questions that could be used to check your understanding of the material in this section. Then make a solution key for your quiz.

SPIRAL REVIEW

Simplify each expression. *(Section 1.2)*

38. $4 + 10 \div 2$ **39.** $18 - 2(12 - 4)$ **40.** $35 \cdot 4 + 20 \div 4$

41. $\frac{5 + 3(5)}{4}$ **42.** $\frac{16 + 8}{2 + 4}$ **43.** $\frac{15 - 3}{5} + 7(2)$

Simplify each variable expression. *(Section 1.7)*

44. $12x + 1 - 3x$ **45.** $6a^2 - 2 + 5a - a^2$ **46.** $0.35(3q - 4)$

47. $-4(2m - 3n)$ **48.** $3(c + 5) + c - 7$ **49.** $5(p + 7) + 3(p + 7)$

Apply⇔Assess

Exercise Notes

Research
Ex. 35 New car manufacturers include an estimated annual fuel cost based on city and highway mileage ratings for new cars. Some students may wish to research how new cars receive their ratings.

Mathematical Procedures
Ex. 36 This exercise can be used to point out the importance of learning mathematical procedures. Even though the equation may not look familiar to students, they can still use the procedures involving reciprocals to solve it.

Assessment Note
Ex. 37 After students write their quizzes and solutions, you may have them proceed further by exchanging quizzes. Students should do one another's quizzes and critique them regarding their appropriateness for the material in the section.

Practice 25 for Section 4.2

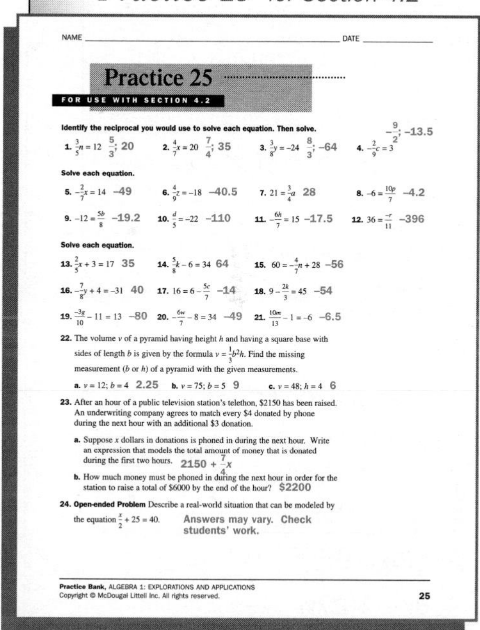

NAME ______________________ DATE __________

Practice 25

FOR USE WITH SECTION 4.2

Identify the reciprocal you would use to solve each equation. Then solve.

1. $\frac{3}{5}n = 12$ $\frac{5}{3}$; 20 **2.** $\frac{4}{7}x = 20$ $\frac{7}{4}$; 35 **3.** $\frac{3}{8}y = -24$ $\frac{8}{3}$; −64 **4.** $-\frac{2}{9}c = 3$ $-\frac{9}{2}$; −13.5

Solve each equation.

5. $-\frac{2}{7}x = 14$ −49 **6.** $\frac{4}{9}z = -18$ −40.5 **7.** $21 = \frac{3}{4}a$ 28 **8.** $-6 = \frac{10p}{7}$ −4.2

9. $-12 = \frac{5b}{8}$ −19.2 **10.** $\frac{d}{5} = -22$ −110 **11.** $-\frac{6h}{7} = 15$ −17.5 **12.** $36 = \frac{-r}{11}$ −396

Solve each equation.

13. $\frac{2}{5}x + 3 = 17$ 35 **14.** $\frac{5}{8}k - 6 = 34$ 64 **15.** $60 = -\frac{4}{7}n + 28$ −56

16. $-\frac{7}{8}y + 4 = -31$ 40 **17.** $16 = 6 - \frac{5c}{7}$ −14 **18.** $9 - \frac{2k}{3} = 45$ −54

19. $\frac{-3g}{10} - 11 = 13$ −80 **20.** $-\frac{6w}{7} - 8 = 34$ −49 **21.** $\frac{10m}{13} - 1 = -6$ −6.5

22. The volume v of a pyramid having height h and having a square base with sides of length b is given by the formula $v = \frac{1}{3}b^2h$. Find the missing measurement (b or h) of a pyramid with the given measurements.

a. $v = 12$; $b = 4$ 2.25 **b.** $v = 75$; $b = 5$ 9 **c.** $v = 48$; $h = 4$ 6

23. After an hour of a public television station's telethon, $2150 has been raised. An underwriting company agrees to match every $4 donated by phone during the next hour with an additional $3 donation.

a. Suppose x dollars in donations is phoned in during the next hour. Write an expression that models the total amount of money that is donated during the first two hours. $2150 + \frac{7}{4}x$

b. How much money must be phoned in during the next hour in order for the station to raise a total of $6000 by the end of the hour? $2200

24. Open-ended Problem Describe a real-world situation that can be modeled by the equation $\frac{x}{2} + 25 = 40$. Answers may vary. Check students' work.

Practice Bank, ALGEBRA 1: EXPLORATIONS AND APPLICATIONS

25

---

24. Answers may vary. Examples are given. In Ex. 21, I solved $\frac{3}{4}b = 2$ by multiplying both sides by $\frac{4}{3}$. In Ex. 22, I solved $\frac{1}{2} \cdot 5h = 10$ by multiplying both sides by $\frac{2}{5}$. In Ex. 23, I rewrote $2\frac{1}{4}h$ as $\frac{9}{4}h$ and solved $\frac{9}{4}h = 3$ by multiplying both sides by $\frac{4}{9}$.

25. 12 26. 200
27. 55 28. $13\frac{1}{3}$
29. −36 30. −48
31. 10 32. −192
33. $-10\frac{4}{5}$

34. Answers may vary. An example is given. $-4x + 2 = 40$; using inverse operations: $-4x = 38$; $\frac{-4x}{-4} = \frac{38}{-4}$; $x = -9.5$; using a reciprocal: $-4x = 38$; $-\frac{1}{4}(-4x) = -\frac{1}{4} \cdot 38$; $x = -9.5$

35. 350 mi

36, 37. See answers in back of book.

38. 9 39. 2
40. 145 41. 5
42. 4 43. $16\frac{2}{5}$
44. $9x + 1$ 45. $5a^2 + 5a - 2$
46. $1.05q - 1.4$ 47. $-8m + 12n$
48. $4c + 8$ 49. $8p + 56$

Plan⇔Support

Objectives

- Solve multi-step equations.
- Use multi-step equations to solve real-world problems.

Recommended Pacing

Core and Extended Courses
Section 4.3: 2 days

Block Schedule
Section 4.3: 2 half-blocks
(with Sections 4.2 and 4.4)

Resource Materials

Lesson Support
Lesson Plan 4.3
Warm-Up Transparency 4.3
Overhead Visuals:
  Folder 4: Solving Multi-Step Equations
  Folder B: Algebra Tiles
Practice Bank: Practice 26
Study Guide: Section 4.3
Explorations Lab Manual:
  Diagram Masters 2, 4
Assessment Book: Test 15

Technology
Internet:
  http://www.hmco.com

Warm-Up Exercises

Simplify each expression.

1. $4(x-5)$ $4x-20$
2. $9n+(-12n)$ $-3n$
3. $\frac{1}{2}(6y-18)$ $3y-9$
4. $8y-(3+2y)$ $6y-3$

Solve each equation.

5. $4x=20$ 5
6. $3x+18=24$ 2

SECTION

4.3 Solving Multi-Step Equations

Learn how to...

- solve multi-step equations

So you can...

- solve real-world problems, such as finding when one runner catches up to another

Some equations that represent real-world situations have variables on both sides of the equals sign. Algebra tiles can help you understand how to solve this type of equation.

EXPLORATION
COOPERATIVE LEARNING

Getting Variables onto One Side

Work with a partner.
You will need:
- algebra tiles

SET UP Use tiles to represent the equation

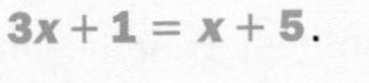

$3x+1=x+5$.

1 Remove an x-tile from each side of your tile equation. What mathematical operation does this model?

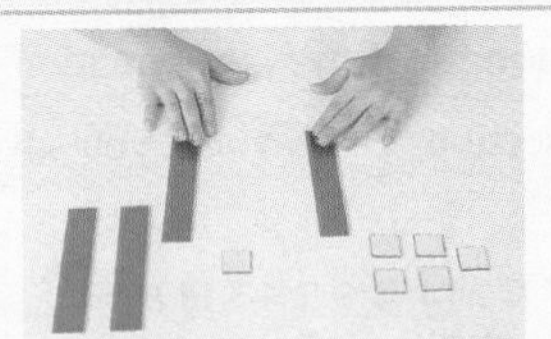

2 Finish solving the equation using the algebra tile model. What is the value of one x-tile? Check this result in the original equation.

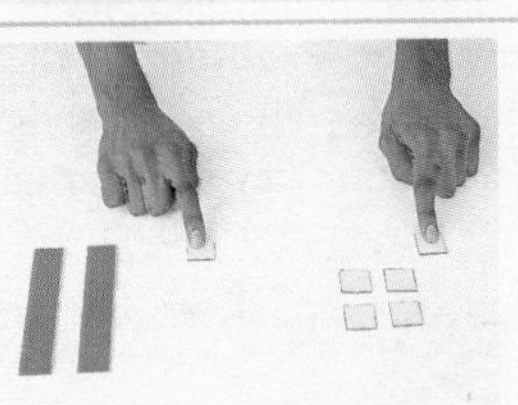

3 Take turns using the algebra tile method introduced in the previous steps to solve each of these equations.

a. $4x+2=x+5$ **b.** $x+9=5x+1$
c. $2x+8=4x+2$ **d.** $5x+3=2x+6$

4 Work together to make a generalization about the first step you should take to solve an equation with variables on both sides of it.

Exploration Note

Purpose
The purpose of this Exploration is to help students understand how to solve an equation with variables on both sides.

Materials/Preparation
Have sets of algebra tiles available. Review Section 2.2 if necessary.

Procedure
Students begin by representing the equation $3x+1=x+5$ with algebra tiles. They then remove an x-tile from both sides to simulate subtracting a variable term. Students should discuss how they arrived at the solutions to each equation in Step 3. Some students may need to record their steps as they go along.

The class should reach a consensus about the generalization in Step 4.

Closure
Groups can compare results with one another. Students should understand that removing an x-tile from each side of an algebra-tile model for an equation is equivalent to subtracting x from both sides of an equation. This method gives an equivalent equation with the variable on one side.

Explorations Lab Manual
See the Manual for more commentary on this Exploration.

Diagram Master 4

In Section 4.1 you used a table and a graph to analyze options for buying CDs. The example below shows how to solve the same problem using an equation with variables on both sides.

EXAMPLE 1 Application: Personal Finance

A discount store charges \$11 per CD. A CD club gives members 8 free CDs, then charges \$15 per CD. When does the cost at the discount store equal the cost through the CD club?

SOLUTION

Let x = the number of CDs you buy.

Write the cost for each option as a function of the number of CDs you buy.

Discount store: Cost = $11x$
CD club: Cost = $15(x - 8)$

WATCH OUT! The second function does not apply to the first 8 CDs.

To find when the two costs are equal, write and solve an equation.

Cost at discount store = Cost through CD club

Use the distributive property to simplify $15(x - 8)$.

Get the x-terms on just one side by subtracting $15x$ from both sides.

$$11x = 15(x - 8)$$
$$11x = 15x - 120$$
$$11x - 15x = 15x - 120 - 15x$$
$$-4x = -120$$
$$\frac{-4x}{-4} = \frac{-120}{-4}$$
$$x = 30$$

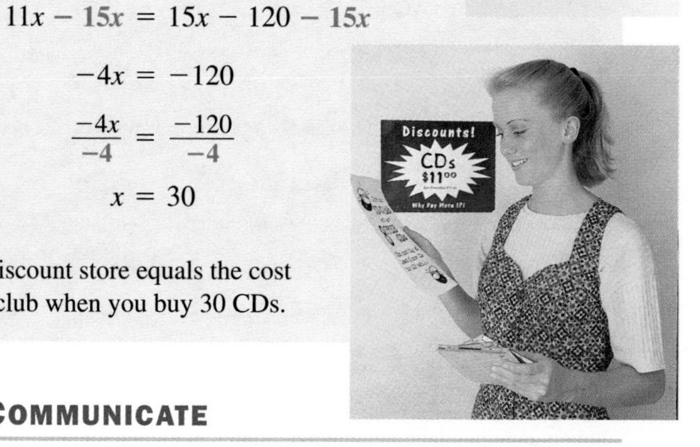

The cost at the discount store equals the cost through the CD club when you buy 30 CDs.

THINK AND COMMUNICATE

1. What are the three key steps used to solve the equation in Example 1? In what order are the steps done?
2. How can you check the solution of Example 1?
3. What would you do first to solve the equation $5x = 3(x + 10)$? What would you do second? What would you do third?
4. a. What would you do first to solve the equation $5x = 3x + 10$? What would you do second?
   b. Are your answers to part (a) the same as your answers to Question 3? Why or why not?

Teach⇔Interact

Learning Styles: Kinesthetic

Using algebra tiles to simulate solving equations with a variable on both sides provides a kinesthetic and visual experience for students that can help them to construct a mental image of how to solve such an equation.

About Example 1

Communication: Reading Some students may have difficulty knowing how to write the equation after they write the cost for each option. Have these students read the last sentence of the example. Stress the fact that the costs should be *equal*.

Additional Example 1

A hotel charges \$12 plus \$2 per hour to park in its underground garage. A parking lot across the street charges \$5 plus \$3 per hour to park. When does the cost of parking at the hotel equal the cost of parking in the lot?

Let h = the number of hours you park.

Write the cost for each option as a function of the number of hours parked.

Hotel: Cost = $12 + 2h$
Lot: Cost = $5 + 3h$

To find when the two costs are equal, write and solve an equation.

Cost at hotel = Cost at parking lot

$$12 + 2h = 5 + 3h$$
$$12 + 2h - 2h = 5 + 3h - 2h$$
$$12 = 5 + h$$
$$12 - 5 = 5 + h - 5$$
$$7 = h$$

The cost of parking at the hotel equals the cost of parking in the lot when you park for 7 hours.

Think and Communicate

These questions can be used for a class discussion to summarize the procedures for solving a multi-step equation.

ANSWERS Section 4.3

Exploration

1.

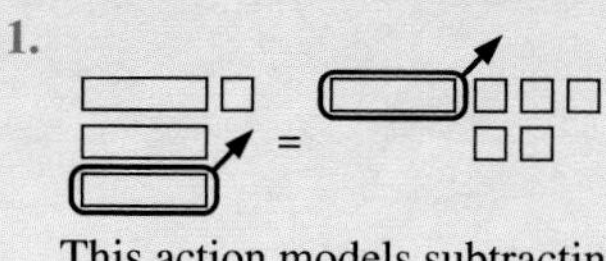

This action models subtracting x from both sides.

2.

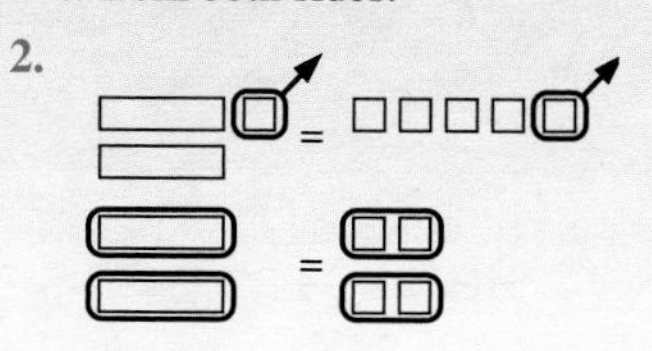

$x = 2$; check: $3x + 1 = x + 5$; $3(2) + 1 \stackrel{?}{=} 2 + 5$; $7 = 7$ ✓

3. a. $x = 1$ b. $x = 2$ c. $x = 3$ d. $x = 1$

4. Answers may vary. An example is given. If there are variables on both sides, add the same variable expression to both sides or subtract the same variable expression from both sides to get all the variable terms on the same side of the equation.

Think and Communicate

1. Use the distributive property to simplify $15(x - 8)$. Subtract $15x$ from both sides. Divide both sides by -4. These steps are done in the order listed.
2. Substitute $x = 30$ in the original equation to see if that value makes the equation true.

3, 4. See answers in back of book.

Teach⇔Interact

About Example 2

Problem Solving

The problem-solving strategy of using a diagram is essential in helping students to see how to express each runner's distance using the variable t for time. Without a diagram, many students would have difficulties solving this problem.

Additional Example 2

Carrie is a swimmer on the school swim team. She can swim at the rate of 5 ft/s. Her younger brother Larry can swim at a rate of 3 ft/s, so she gives him a 50 ft head start in a 500 ft race.

a. After how many seconds will Carrie catch up to her brother?

Use a diagram.

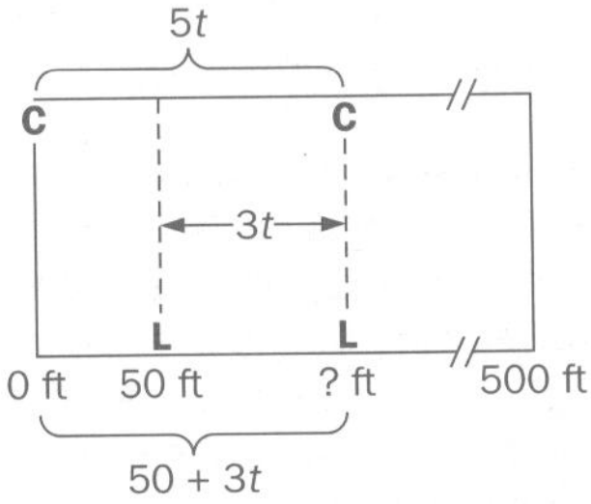

Let t = the number of seconds they swim.

Use the formula $d = rt$ to find the distance d each person can swim in t seconds.

Carrie's distance from the start after t seconds is $5t$.

Larry's distance from the start after t seconds is $50 + 3t$.

To find when the distances are equal, write and solve an equation.

Carrie's distance = Larry's distance.

$$5t = 50 + 3t$$
$$5t - 3t = 50 + 3t - 3t$$
$$2t = 50$$
$$t = 25$$

Carrie catches up to Larry after swimmimg 25 seconds.

b. How far did Carrie swim before she caught Larry?

$5(25) = 125$ ft

EXAMPLE 2

Al is a sprinter on a track team. He can run at a rate of 9 m/s. His brother Bill can run only 6 m/s, so Al gives him a 25 m head start in a 100 m race.

a. After how many seconds will Al catch up to Bill?

b. How far will they be from the starting line when Al catches up to Bill?

Toolbox p. 600 *Problem Solving*

SOLUTION

Problem Solving Strategy: Use a diagram.

Let t = the number of seconds they run.

You know each runner's rate r. Use the formula $d = rt$ to find the distance d each person can run in t seconds.

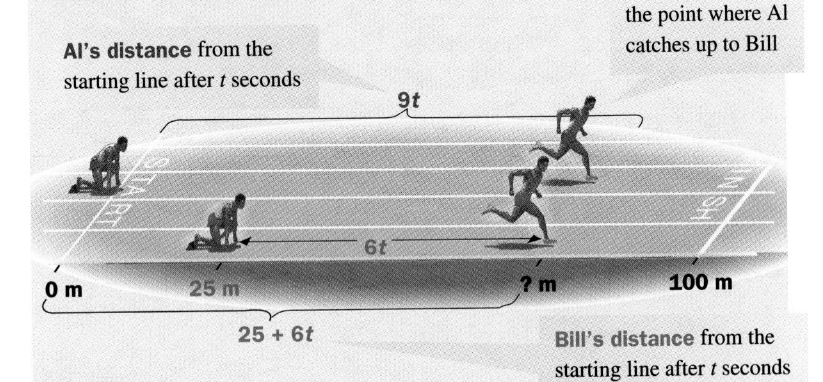

a. To find out when the distances are equal, write and solve an equation.

$$\text{Al's distance} = \text{Bill's distance}$$
$$9t = 25 + 6t$$
$$9t - 6t = 25 + 6t - 6t \quad \text{Subtract } 6t \text{ from both sides.}$$
$$3t = 25$$
$$\frac{3t}{3} = \frac{25}{3}$$
$$t = 8\frac{1}{3} = 8.\overline{3}$$

Al catches up to Bill after running about 8.3 seconds.

b. Al's distance $= 9t$

$$= 9 \cdot \frac{25}{3} \quad \text{Substitute } 8\tfrac{1}{3}, \text{ or } \tfrac{25}{3}, \text{ for } t.$$
$$= 75$$

Al catches up to Bill when they are 75 m from the starting line.

Solving Multi-Step Equations

You can follow these steps to solve an equation with a variable term on both sides of the equals sign.

1. Simplify each side of the equation.
2. Get all the variable terms on one side of the equation. Remember to keep it balanced by doing the same operation on both sides.
3. Get the variable alone on one side of the equation.

Sometimes an equation has *no solution*, because no value of the variable will make the equation true. Sometimes *all numbers* are solutions of an equation. An equation that is true for all numbers is called an **identity**. The following questions give examples of these types of equations.

THINK AND COMMUNICATE

Identify the steps used to solve each equation and interpret the final statement. If all numbers are solutions, write *identity*. If there is no solution, write *no solution*. Explain your reasoning.

5.
$$7x - 4 = 7(x - 4)$$
$$7x - 4 = 7x - 28$$
$$7x - 4 - 7x = 7x - 28 - 7x$$
$$-4 = -28$$

This statement is never true.

6.
$$8x - 12 = 2x + 2(3x - 6)$$
$$8x - 12 = 2x + 6x - 12$$
$$8x - 12 = 8x - 12$$

This statement is always true.

CHECKING KEY CONCEPTS

Copy and complete each solution by replacing each ? with the correct expression.

1.
$$3x + 3 = 2x + 5$$
$$3x + 3 - \underline{?} = 2x + 5 - 2x$$
$$x + 3 = 5$$
$$x + 3 - \underline{?} = 5 - \underline{?}$$
$$x = \underline{?}$$

2.
$$3(2x - 4) = 5x - 6$$
$$\underline{?} - 12 = 5x - 6$$
$$6x - 12 - 5x = 5x - 6 - \underline{?}$$
$$x - 12 = -6$$
$$x - 12 + \underline{?} = -6 + \underline{?}$$
$$x = \underline{?}$$

Tell what you would do first to solve each equation. If an equation is an *identity* or there is *no solution*, say so.

3. $4x = 2x + 9$

4. $3y = 20 - y$

5. $5n - 17 = 7n + 3$

6. $9a - 60 = 12a + 3$

7. $2(6x + 4) = 12x - 6$

8. $2(5c + 3) = 6 + 10c$

Think and Communicate

5. First the distributive property was used to rewrite $7(x - 4)$ as $7x - 28$. Then $7x$ was subtracted from both sides, and both sides were simplified. Since the final statement, $-4 = -28$, is a false statement, the original equation has no solution.
6. First the distributive property was used to rewrite $2(3x - 6)$ as $6x - 12$. Then the right side was simplified. The final statement, $8x - 12 = 8x - 12$, is true for every value of x, so all numbers are solutions and the equation is an identity.

Checking Key Concepts

1. $2x$; 3; 3; 2
2. $6x$; $5x$; 12; 12; 6
3. Subtract $2x$ from both sides; $x = \frac{9}{2}$.
4. Add y to both sides; $y = 5$.
5. Subtract $5n$ from both sides; $n = -10$.
6. Subtract $9a$ from both sides; $a = -21$.
7. Rewrite $2(6x + 4)$ as $12x + 8$; no solution.
8. Rewrite $2(5c + 3)$ as $10c + 6$; identity.

Teach⇔Interact

Section Notes

Journal Entry
Students may want to write the steps in solving multi-step equations in a journal. Suggest they include an example for each step.

Teaching Tip
Using examples from real-life may make it easier for students to understand how the solution to an equation can be "all numbers" or "no numbers." Ask students to think of real-life situations that can have the answers *always* or *never*. For example, if two towns start with equal populations and the first town continues to grow at a faster rate, then the population of the second town can never catch up to the population of the first town.

Using Manipulatives
If students have difficulty understanding the idea of no solutions, you may wish to have them use algebra tiles to represent an equation such as $3x + 4 = 3(x + 1)$. They should try to solve the equation using the tiles.

Think and Communicate

In question 5, students should understand that the equation $7x - 4 = 7x - 28$ can never be true for any value of x and, therefore, has no solution. Question 6 may include the steps $-12 = -12$ or $0 = 0$. If a statement of equality is between a number and itself, then all numbers are solutions to the *original* equation.

Checking Key Concepts

Teaching Tip
For questions 3–8, encourage students to check their solution to each equation by substituting it in the original equation.

Closure Question

Describe how to solve a multi-step equation.

See the chart at the top of this page.

Apply⇔Assess

Suggested Assignment

Core Course
Day 1 Exs. 2–13, 15–23
Day 2 Exs. 25–30, 34, 36–41, AYP

Extended Course
Day 1 Exs. 1–24
Day 2 Exs. 25–41, AYP

Block Schedule
Day 23 Exs. 2–13, 16–23
Day 24 Exs. 25–32, 34, 36–41, AYP

Exercise Notes

Communication: Writing
Ex. 14 Writing exercises provide students with an opportunity to express the mathematics they know in an alternate form. This particular exercise requires that students use what they have learned about solving certain types of equations to write their letter. A clear understanding of the mathematics involved provides a firm foundation for learning more about solving equations.

Second-Language Learners
Ex. 14 Some second-language learners may prefer first to create examples that show how to move a variable term, and then work with a peer tutor to write an explanation to accompany the examples.

4.3 Exercises and Applications

Extra Practice ***exercises on page 564***

1. RECREATION Look back at Exercises 1–4 on page 150. You can use the equations below to model the membership costs for the two fitness clubs.

Let n = the number of months you belong to the club.

Club A: Cost = $100 + 40n$ Club B: Cost = $50n$

Write and solve an equation to answer this question: When is the membership cost the same for both fitness clubs?

Solve each equation.

2. $7x = 3x + 12$ **3.** $3n = n - 2$ **4.** $8b - 4 = 5b + 8$

5. $4a - 5 = 6a - 11$ **6.** $12c - 5 = 7c + 30$ **7.** $3n - 11 = 29 - 5n$

8. $2y - 1 = 5y$ **9.** $4n + 4 = 16n$ **10.** $-7x + 5 = 4x + 5$

11. $3y - 9 = 9y - 3$ **12.** $7x + 8 = -18 - 6x$ **13.** $2r - 15 = -9r + 18$

14. Writing Write a letter to a friend explaining how to get a variable term from one side of an equation onto the other side. Include examples.

15. TRANSPORTATION Suppose the McCall family wants to rent a truck for a day. They have a choice of two truck rental companies.

Company A: \$30 per day plus \$.60 per mile
Company B: \$55 per day plus \$.35 per mile

Write and solve an equation to answer this question: After how many miles are the total costs at the two companies equal?

Visual Thinking For Exercises 16–19, use the graph below.

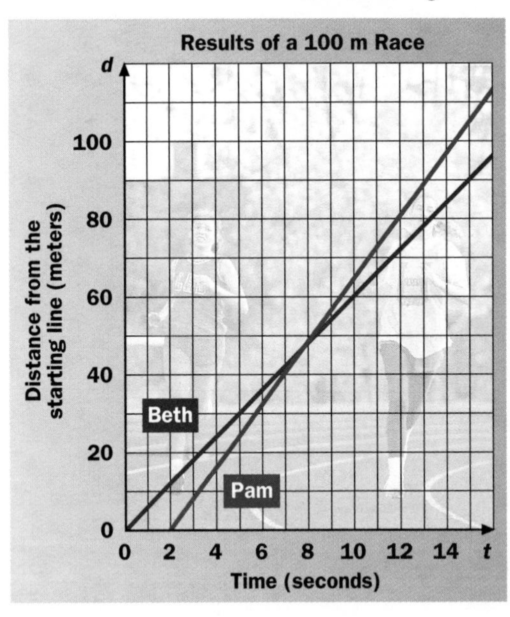

In Example 2, Al and Bill started running at the same time, but Bill had a *distance* advantage. Another way to give someone a head start is to give a *time* advantage. Both runners leave from the starting line, but the slower runner leaves earlier. The graph gives information about another 100 m race in which one person was given a time advantage.

16. Who is the slower runner? How can you tell?

17. Estimate how far the runners were from the starting line when one of them caught up to the other.

18. Write an equation for the distance covered by each runner as a function of time t.

19. Write and solve an equation to find the time t at which the faster runner caught up to the slower runner. Calculate how far the runners were from the starting line at that time.

Exercises and Applications

1. $100 + 40n = 50n$; $n = 10$; The membership cost is the same for 10 months of membership.
2. 3
3. –1
4. 4
5. 3
6. 7
7. 5
8. $-\frac{1}{3}$
9. $\frac{1}{3}$
10. 0
11. –1
12. –2
13. 3
14. Answers may vary. An example is given. Dear Luis, Suppose you have an equation with variable terms on both sides. You need to combine these two terms by moving one of the terms from one side to the other. To do this, follow these suggestions. (1) If a variable term is added or has a positive coefficient, subtract that term from both sides of the equation. For example, to solve $3y - 8 = 10y + 20$, subtract $3y$ from both sides to get $-8 = 7y + 20$. (2) If a variable term is subtracted or has a negative coefficient, add the variable term to both sides of the equation. For example, to solve $5 - 8k = -2k - 1$, add $2k$ or $8k$ to both sides of the equation. I hope this helps! Your friend, Maura
15. Let m = the number of miles they must drive; $30 + 0.6m = 55 + 0.35m$; $m = 100$; after 100 miles.
16. Beth; She started running before Pam.
17. about 50 m
18. Beth: $d = 6t$; Pam: $d = 8(t - 2)$
19. $8(t - 2) = 6t$; $t = 8$; The faster runner, Pam, caught up to the slower runner, Beth, 8 s after the race began. At that time, they were 48 m from the starting line.
20. Assume the quantity is 4 because $\frac{1}{4}$ of 4 is an integer. The result is 5. It should be 25. The assumption is false. Notice that $5 \cdot 5 = 25$. Work backward. $5 \cdot 4 = 20$, $5 \cdot 1 = 5$, and $20 + 5 = 25$. The unknown quantity is 20.
21. Assume the quantity is 3 because $\frac{1}{3}$ of 3 is an integer: $3 + \frac{1}{3} \cdot 3 = 4$, not 36. If you

Connection HISTORY

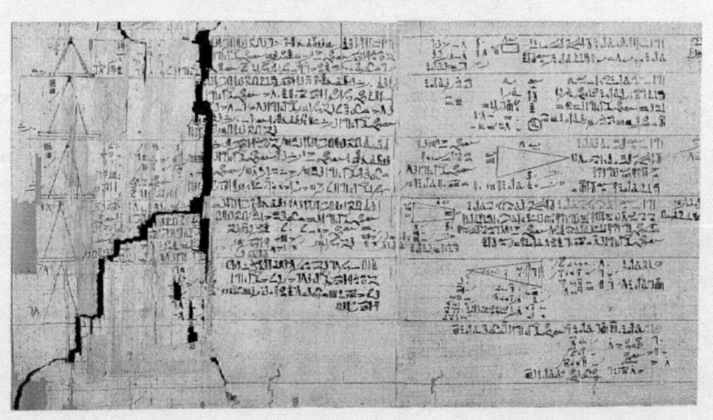

This may not look like a math problem to you, but it is. It was written in Africa over 3000 years ago by a scribe named Ahmes. Scholars have studied the Ahmes papyrus to understand how the Egyptians used mathematics.

The diagram at the right shows the reasoning used in the Ahmes papyrus to solve this problem:

A quantity and one quarter of the quantity added together equal 15. What is the quantity?

This method of reasoning is often called *false assumption.*

For Exercises 20–22, use the method of false assumption to find each unknown number.

Assume the quantity is 4. **Choose 4 because $\frac{1}{4}$ of 4 is an integer.** **The result is 5. It should be 15. The assumption is false.**

▲▲▲▲ + △ = ▲▲▲▲△ **Notice that $3 \cdot 5 = 15$.**

Work backward. Multiply each term by 3.

$3 \cdot 4 = 12$ $3 \cdot 1 = 3$ $3 \cdot 5 = 15$

The unknown quantity is 12.

20. A number plus $\frac{1}{4}$ of the number is 25.

21. A number plus $\frac{1}{3}$ of the number is 36.

22. A number plus $\frac{1}{12}$ of the number is 65.

23. Use an equation to find each unknown number in Exercises 20–22.

24. Research Find out about the Rosetta stone. How did it enable scholars to read Egyptian writing?

Solve. If an equation is an *identity* or there is *no solution*, say so.

25. $4(x - 3) = 7x$

26. $4(x - 3) = 4x - 3$

27. $4(x - 3) = 2(2x - 6)$

28. $7(-6 - 2t) = 9t + 4$

29. $10q - 36 = 2(3q + 4)$

30. $6n - 5 = 2(3n - 4)$

31. $2(4a - 2) = \frac{2}{3}(9 - 48a)$

32. $12x - (6 + 8x) = \frac{1}{2}(8x - 12)$

33. Open-ended Problem Write an equation that has one solution, another equation that has no solution, and another equation that is an identity.

Apply⇔Assess

Exercise Notes

Cooperative Learning

Exs. 16–19 You may wish to have students work with a partner on these exercises. Then they can help each other interpret the graph.

Historical Connection

Exs. 20–22 The Ahmes papyrus was written in about 1650 B.C. and is the chief source of information that exists today concerning ancient Egyptian mathematics. Ahmes was a scribe who copied the mathematical text of an earlier work that was essentially a practical handbook containing 85 problems. The Ahmes papyrus is also known as the Rhind papyrus, named after A. Henry Rhind, a British Egyptologist who purchased it in Egypt.

Problem Solving

Exs. 20–23 Ask students to explain how they solved each of the problems in Exs. 20–22. Then have them compare their work to the equations in Ex. 23.

Reasoning

Ex. 33 This exercise will help students to solidify their understanding of how to determine the number of solutions to an equation.

multiply both sides by 9, you get $3 \cdot 9 + \frac{1}{3} \cdot 3 \cdot 9 = 4 \cdot 9$, or $27 + 9 = 36$. Since $9 = \frac{1}{3} \cdot 27$, the unknown quantity is 27.

22. Assume the number is 12 because $\frac{1}{12}$ of 12 is an integer: $12 + \frac{1}{12} \cdot 12 = 13$, not 65. If you multiply both sides by 5, you get $5 \cdot 12 + 5 \cdot \frac{1}{12} \cdot 12 = 5 \cdot 13$, or $60 + 5 = 65$. Since $5 = \frac{1}{12} \cdot 60$, the unknown quantity is 60.

23. Ex. 20: $x + \frac{1}{4}x = 25$; $\frac{5}{4}x = 25$; $x = \frac{4}{5} \cdot 25 = 20$. Ex. 21: $x + \frac{1}{3}x = 36$; $\frac{4}{3}x = 36$; $x = \frac{3}{4}x \cdot 36 = 27$. Ex. 22: $x + \frac{1}{12}x = 65$; $\frac{13}{12}x = 65$; $x = \frac{12}{13} \cdot 65 = 60$

24. The Rosetta stone was discovered near Rosetta, Egypt, in 1799. Believed to have been carved between 203 B.C. and 181 B.C., it contained the same inscription in three languages, hieroglyphics, Demotic (which was the popular language of Egypt at the time of the carving), and Greek. A French scholar, Jean François Champollion, used his knowledge of Greek and Egyptian to determine the meanings of the hieroglyphic characters.

25. –4 26. no solution
27. identity 28. –2
29. 11 30. no solution
31. $\frac{1}{4}$ 32. identity

33. Answers may vary. Examples are given. The equation $2x - 5 = -4x + 13$ has one solution (3), the equation $-2(2x - 5) = -4x + 13$ has no solution, and the equation $-2(2x - 5) = -4x + 10$ is an identity.

Apply⇔Assess

Exercise Notes

Integrating the Strands

Ex. 35 The graph of the equations in part (a) illustrates very clearly why the equation $x + 2 = x - 2$ has no solution. A geometric interpretation of an algebraic situation can often help students to understand the algebra better.

Assess Your Progress

Journal Entry

Ask students to share their preferences for solving problems using equations or tables or graphs. Encourage them to list situations for which they may prefer one method to another in their journal.

Progress Check 4.1–4.3

See page 184.

Practice 26 for Section 4.3

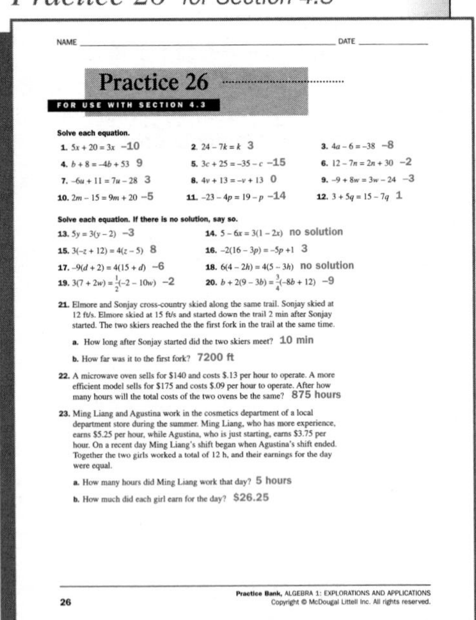

NAME ______________________ DATE ____________

Practice 26

FOR USE WITH SECTION 4.3

Solve each equation.

1. $5x + 20 = 3x$ −10
2. $24 - 7k = k$ 3
3. $4a - 6 = -38$ −8
4. $b + 8 = -4b + 53$ 9
5. $3c + 25 = -35 - c$ −15
6. $12 - 7n = 2n + 30$ −2
7. $-6u + 11 = 7u - 28$ 3
8. $4v + 13 = -v + 13$ 0
9. $-9 + 8w = 3w - 24$ −3
10. $2m - 15 = 9m + 20$ −5
11. $-23 - 4p = 19 - p$ −14
12. $3 + 5q = 15 - 7q$ 1

Solve each equation. If there is no solution, say so.

13. $5y = 3(y - 2)$ −3
14. $5 - 6x = 3(1 - 2x)$ no solution
15. $3(-z + 12) = 4(z - 5)$ 8
16. $-2(16 - 3p) = -5p + 1$ 3
17. $-9(d + 2) = 4(15 + d)$ −6
18. $6(4 - 2h) = 4(5 - 3h)$ no solution
19. $3(7 + 2w) = \frac{1}{2}(-2 - 10w)$ −2
20. $b + 2(9 - 3b) = \frac{3}{4}(-8b + 12)$ −9

21. Elmore and Sonjay cross-country skied along the same trail. Sonjay skied at 12 ft/s. Elmore skied at 15 ft/s and started down the trail 2 min after Sonjay started. The two skiers reached the the first fork in the trail at the same time.
    a. How long after Sonjay started did the two skiers meet? 10 min
    b. How far was it to the first fork? 7200 ft
22. A microwave oven sells for \$140 and costs \$.13 per hour to operate. A more efficient model sells for \$175 and costs \$.09 per hour to operate. After how many hours will the total costs of the two ovens be the same? 875 hours
23. Ming Liang and Agustina work in the cosmetics department of a local department store during the summer. Ming Liang, who has more experience, earns \$5.25 per hour, while Agustina, who is just starting, earns \$3.75 per hour. On a recent day Ming Liang's shift began when Agustina's shift ended. Together the two girls worked a total of 12 h, and their earnings for the day were equal.
    a. How many hours did Ming Liang work that day? 5 hours
    b. How much did each girl earn for the day? \$26.25

26

Practice Bank, ALGEBRA 1: EXPLORATIONS AND APPLICATIONS

34. SAT/ACT Preview Which of these equations is not an identity?

A. $5(n + 2) = 5n + 10$ **B.** $8(n - 2) = 8n - 16$

C. $3n + 8 = \frac{1}{4}(12n + 32)$ **D.** $2(n + 6) - 5n = 14 - 3n$

35. a. Challenge Graph the equations $y = x + 2$ and $y = x - 2$ in the same coordinate plane. How does the graph help you to see that there is no solution to the equation $x + 2 = x - 2$?

b. When will an equation in the form $ax + b = cx + d$ have no solution? only one solution? When will all numbers be solutions?

ONGOING ASSESSMENT

36. Cooperative Learning Work with a partner. Take turns using algebra tiles to solve each equation. Record the steps with algebraic symbols.

a. $x + 6 = 6x + 1$ **b.** $4x + 8 = 5x + 5$ **c.** $6x + 2 = 4x + 10$

SPIRAL REVIEW

Find the least common multiple of each set of numbers. *(Toolbox, page 584)*

37. 2, 3, 5 **38.** 2, 6, 12 **39.** 3, 12, 15 **40.** 4, 14, 21

41. Gerry makes a phone call at a rate of \$.20 per minute or fraction thereof. Her total cost is a function of the call's length. Find the domain and range of the function. *(Section 2.3)*

ASSESS YOUR PROGRESS

VOCABULARY

reciprocals (p. 153) **identity** (p. 161)

1. A bank has two basic checking accounts. For the *Minimum Account* the monthly fee is \$2.50 and the first ten checks you write each month are free. After that you pay \$.75 per check. The *Maximum Account* fee is \$7.00 no matter how many checks you write each month. *(Section 4.1)*

a. Model this situation with a table.

b. When will the monthly cost be the same for both accounts?

Solve. If the equation is an *identity* or there is *no solution*, say so. *(Sections 4.2 and 4.3)*

2. $\frac{2}{3}a = 6$ **3.** $-\frac{2}{7}b + 4 = 6$ **4.** $\frac{5t}{3} = -20$

5. $15 = -\frac{1}{5}b + 2$ **6.** $6x + 72 = 15x$ **7.** $-9c - 8 = 6c + 7$

8. $4n - 21 = 15 - 8n$ **9.** $2(2a + 4) = 4a + 1$ **10.** $8x + 6 = 2(4x + 3)$

11. Journal Compare solving problems by using equations to solving problems by using tables and graphs. Which method do you prefer? Why?

34. D

35. a.

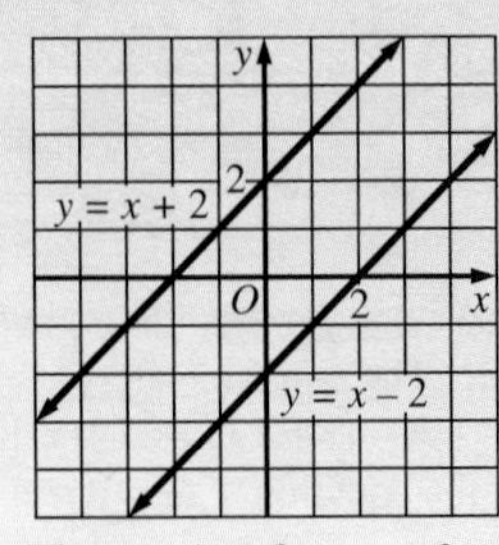

The graphs of $y = x + 2$ and $y = x - 2$ do not intersect. This shows that there is no common solution. There is no value of x for which $x + 2 = x - 2$.

b. An equation in the form $ax + b = cx + d$ will have no solution when $a = c$ and $b \neq d$; it will have one solution when $a \neq c$; it will be an identity when $a = c$ and $b = d$.

36. a. $x = 1$ b. $x = 3$

c. $x = 4$

37. 30 38. 12

39. 60 40. 84

41. The domain consists of the nonnegative real numbers. The range consists of the numbers 0, 0.20, 0.40, 0.60, … .

Assess Your Progress

1. See answers in back of book.

2. 9 3. −7

4. −12 5. −65

6. 8 7. −1

8. 3 9. no solution

10. identity

11. Answers may vary. An example is given. I prefer using equations to solve problems, because it is fast and gives an exact solution. Using tables can be very time-consuming and sometimes it is hard to identify the solution. Graphs take a long time to draw if I do not have a graphing calculator, and sometimes I cannot get an exact solution using a graph.

SECTION

4.4 Equations with Fractions or Decimals

Learn how to...

- solve equations that involve more than one fraction or decimal

So you can...

- solve real-world problems, such as finding the distance to the epicenter of an earthquake

A catastrophic earthquake struck Kobe, Japan, in January, 1995. A seismograph in Western Samoa recorded the wave patterns shown.

For the same earthquake, a seismograph near Tulsa, Oklahoma, recorded a difference of 700 seconds between the arrival times of pressure waves (P waves) and shear waves (S waves). Both waves traveled the same distance from the epicenter to Tulsa, but the average speed of the P waves was about 13 km/s, while the average speed of the S waves was only about 7 km/s.

EXAMPLE 1 Interview: Barbara Romanowicz

Seismologists like Barbara Romanowicz calculate distances to epicenters. Estimate the distance from the epicenter to the seismograph near Tulsa, to the nearest ten thousand kilometers.

SOLUTION

Let d = the distance in kilometers from the epicenter to the seismograph.

Travel time for the S waves − Travel time for the P waves = Time lapse between S waves and P waves

$$\frac{d}{7} - \frac{d}{13} = 700$$

Remember: $d = rt$, or $t = \frac{d}{r}$.

$$91\left(\frac{d}{7} - \frac{d}{13}\right) = 91 \cdot 700$$

Multiply both sides by **91**, the least common multiple of 7 and 13.

$$\frac{91d}{7} - \frac{91d}{13} = 63{,}700$$

$$13d - 7d = 63{,}700$$

$$6d = 63{,}700$$

Divide both sides of the equation by 6.

$$d = \frac{63{,}700}{6} \approx 10{,}617$$

The epicenter was about 10,000 km from the seismograph near Tulsa.

Plan⇔Support

Objectives

- Solve equations that involve more than one fraction or decimal.
- Use equations to solve real-world problems.

Recommended Pacing

Core and Extended Courses
Section 4.4: 1 day

Block Schedule
Section 4.4: $\frac{1}{2}$ block (with Section 4.3)

Resource Materials

Lesson Support
Lesson Plan 4.4
Warm-Up Transparency 4.4
Practice Bank: Practice 27
Study Guide: Section 4.4

Technology
Graphing Calculator
Spreadsheet Software
McDougal Littell Software *Stats!*
Internet: http://www.hmco.com

Warm-Up Exercises

1. Find the least common denominator of $\frac{1}{6}$ and $\frac{3}{8}$. 24
2. Simplify $\frac{9}{5} + \frac{2}{3}$. $\frac{37}{15}$

Solve each equation.

3. $\frac{5}{3}n = 20$ 12
4. $\frac{-4y}{7} = 2$ $-\frac{7}{2}$
5. $2c = 8 + 5c$ $-\frac{8}{3}$
6. $0.5x + 0.8 = 6.3$ 11

Teach⇔Interact

Additional Example 1

The average speed of an airplane traveling over the Atlantic Ocean is 540 mi/h, while the average speed of another airplane traveling over land is 345 mi/h. If the difference in their travel times for the same distance is 2 hours, how far did each airplane travel?

Let d = the distance each airplane traveled.

Travel time for the plane over land – Travel time for the plane over the ocean = Difference in travel time

$$\frac{d}{345} - \frac{d}{540} = 2$$

$$345 \cdot 540\left(\frac{d}{345} - \frac{d}{540}\right) = 2 \cdot 345 \cdot 240$$

$$\frac{345 \cdot 540d}{345} - \frac{345 \cdot 540d}{540} = 2 \cdot 345 \cdot 540$$

$$540d - 345d = 372{,}600$$

$$195d = 372{,}600$$

$$d = \frac{372{,}600}{195} \approx 1911$$

Each airplane traveled about 1911 miles.

Section Note

Student Study Tip

Mention to students that the LCM of two numbers is not always the product of the numbers, as is the case with 7 and 13. For example, the LCM of 4 and 6 is 12, not 24. The LCM of 4 and 6 is found by taking the product of the greatest power of each factor in 4 and 6, which is $2^2 \cdot 3 = 12$.

Additional Example 2

Solve the equation $\frac{2}{3}x + \frac{3}{4} = \frac{1}{2}x$.

$$\frac{2}{3}x + \frac{3}{4} = \frac{1}{2}x$$

$$12\left(\frac{2}{3}x + \frac{3}{4}\right) = 12\left(\frac{1}{2}x\right)$$

$$\frac{24}{3}x + \frac{36}{4} = \frac{12}{2}x$$

$$8x + 9 = 6x$$

$$8x + 9 - 8x = 6x - 8x$$

$$9 = -2x$$

$$-\frac{9}{2} = x$$

Toolbox p. 584
Factors and Multiples

You can make an equation with fractions easier to solve. You multiply both sides of the equation by the least common denominator of the fractions. As you saw in Example 1, the least common denominator of a group of fractions is the least common multiple (LCM) of the denominators.

$$91\left(\frac{d}{7} - \frac{d}{13}\right) = 91 \cdot 700$$

The least common denominator is the LCM of 7 and 13: $7 \cdot 13 = 91$

EXAMPLE 2

Solve the equation $\frac{3}{2}x - \frac{1}{5} = \frac{1}{6}x$.

SOLUTION

$$\frac{3}{2}x - \frac{1}{5} = \frac{1}{6}x$$

$$30\left(\frac{3}{2}x - \frac{1}{5}\right) = 30 \cdot \frac{1}{6}x$$

Multiply both sides of the equation by **30**, the LCM of 2, 5, and 6.

$$\frac{90}{2}x - \frac{30}{5} = \frac{30}{6}x$$

$$45x - 6 = 5x$$

$$45x - 6 - 45x = 5x - 45x$$

Subtract **45x** from both sides of the equation.

$$-6 = -40x$$

$$\frac{-6}{-40} = \frac{-40x}{-40}$$

$$\frac{3}{20} = x$$

THINK AND COMMUNICATE

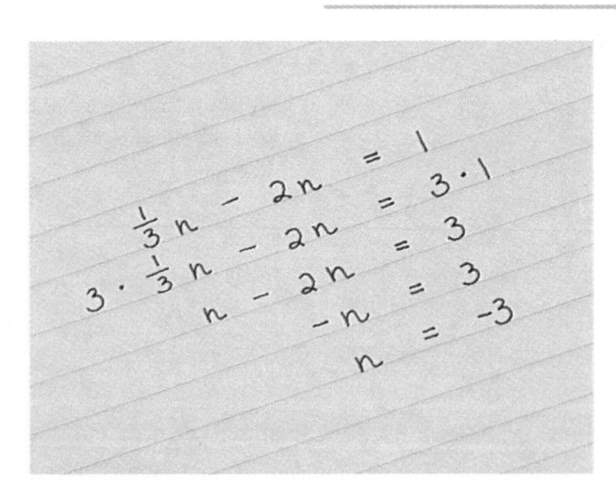

1. A student tried to solve the equation $\frac{1}{3}n - 2n = 1$ as shown. What mistake did the student make? What is the correct solution?
2. In Example 2, why can't you just multiply the right side of the equation by 6 and the left side of the equation by the least common denominator of $\frac{3}{2}$ and $\frac{1}{5}$?
3. Why might someone try to solve the equation in Example 2 by multiplying both sides by 60 instead of 30? Will that method work?

ANSWERS Section 4.4

Think and Communicate

1. The student did not multiply both terms of the left side of the equation by 3; the correct solution is $n - 6n = 3$; $-5n = 3$; $n = -\frac{3}{5}$.
2. because the resulting equation would not have the same solution as the original equation; It is important to apply the same operation with the same number to both sides of an equation when solving.
3. because 60 is the product of all the denominators; Yes.

Equations with Decimals

You can also make an equation with decimals easier to solve. You multiply both sides of the equation by a power of ten that will eliminate the decimals.

EXAMPLE 3 Application: Personal Finance

It costs \$6.50 to see a movie at the theater near where Nina Ramirez lives. It costs only \$2.99 to rent a video at the local video store. Nina wants to buy a VCR for \$325.95. How many movies will Nina have to watch at home instead of at the theater before she recovers the cost of the VCR?

SOLUTION

Let n = the number of movies Nina watches.

Write and solve an equation to find the number of movies that will make the costs equal.

Cost of seeing n movies at the theater = Cost of VCR + Cost of renting n videos

$$6.50n = 325.95 + 2.99n$$

$$100(6.50n) = 100(325.95 + 2.99n)$$

Multiply both sides of the equation by 100 to eliminate the decimals.

$$650n = 32{,}595 + 299n$$

$$650n - 299n = 32{,}595 + 299n - 299n$$

$$351n = 32{,}595$$

Divide both sides of the equation by 351.

$$n \approx 92.86$$

Nina will have to watch 93 movies at home instead of at the theater before she recovers the cost of the VCR.

BY THE WAY

Hollywood studios earn almost half of their revenues from video rentals and sales.

CHECKING KEY CONCEPTS

Find the least common denominator of each group of fractions.

1. $\frac{1}{2}, \frac{3}{5}, \frac{1}{3}$ **2.** $\frac{1}{10}, \frac{1}{5}, \frac{7}{2}$ **3.** $\frac{5}{9}, \frac{5}{6}, \frac{3}{8}$

Tell what power of ten you would multiply by to make each equation easier to solve.

4. $10.5x + 1.1 = 43.1$ **5.** $1.5n = 11.25 - 6n$ **6.** $5.15x - 29 = -2.1x$

Solve each equation.

7. $\frac{x}{2} + \frac{x}{3} = 1$ **8.** $\frac{d}{8} + \frac{d}{12} = \frac{5}{6}$ **9.** $\frac{3}{8}n = 10 - \frac{1}{4}n$

10. $8y + 40.8 = 1.2y$ **11.** $0.34c + 1.2 = 0.74c$ **12.** $1.5a = 15 - 2.25a$

Toolbox p. 585 *Powers of Ten*

Checking Key Concepts

1. 30
2. 10
3. 72
4. $10^1 = 10$
5. $10^2 = 100$
6. $10^2 = 100$
7. $\frac{6}{5}$
8. 4
9. 16
10. −6
11. 3
12. 4

Teach⇔Interact

Think and Communicate

Question 1 points out a common error that some students make in solving equations with fractional coefficients. Students should verify that −3 is not the correct solution by substituting −3 for n in the original equation.

About Example 3

Visual Thinking

Ask students to graph $y = 6.5x$ and $y = 325.95 + 2.99x$. Encourage a discussion of the graphs and how they can be used to answer the question in this Example. This activity involves the visual skills of *interpretation* and *exploration.*

Additional Example 3

Toby sells bread for \$1.30 a loaf. It costs \$.45 for the ingredients and electricity to bake each loaf in an oven that costs \$199.95. How many loaves of bread will Toby have to sell before he recovers the cost of the oven?

Let n = the number of loaves Toby has to sell.

Income from sales = Cost of baking loaves + Cost of oven

$$1.30n = 0.45n + 199.95$$

$$100(1.30n) = 100(0.45n + 199.95)$$

$$130n = 45n + 19{,}995$$

$$130n - 45n = 45n + 19{,}995 - 45n$$

$$85n = 19{,}995$$

$$n \approx 235$$

Toby will have to sell about 235 loaves of bread.

Checking Key Concepts

Topic Spiraling: Review

For questions 1–3, you may need to review the procedure for finding the least common denominator of a group of fractions.

Closure Question

When solving an equation that contains fractions or decimals, what would you do first and how would you do it?

Eliminate the fractions or decimals by multiplying both sides of the equation by the least common denominator of the fractions or by a power of ten that will eliminate the decimals.

Apply⇔Assess

Suggested Assignment

Core Course
Exs. 1–25, 27–30, 32–34, 36–40

Extended Course
Exs. 4–6, 9–11, 13–40

Block Schedule
Day 24 Exs. 1–25, 27–30, 36–40

Exercise Notes

Reasoning
Exs. 13–16 These exercises afford students an opportunity to use technology to relate two seemingly unrelated sets of equations and a single equation. You may wish to lead students in a discussion of how the concepts presented in these exercises can be used to solve equations such as those in Exs. 4–12 graphically.

Communication: Writing
Ex. 26 You may wish to extend this exercise to include a discussion of how to solve equations such as those in Exs. 4–12.

4.4 Exercises and Applications

Extra Practice *exercises on page 564*

Find the least common denominator of each group of numbers.

1. $\frac{1}{3}, \frac{5}{6}, \frac{1}{2}$ **2.** $\frac{3}{2}, \frac{3}{4}, \frac{1}{3}$ **3.** $\frac{2}{18}, \frac{4}{15}, 8$

Solve each equation.

4. $\frac{m}{3} + \frac{m}{4} = 14$ **5.** $\frac{x}{2} + \frac{x}{3} = \frac{25}{3}$ **6.** $\frac{7y}{9} + 2 = \frac{2y}{3}$

7. $\frac{2}{3}a - \frac{1}{2}a = \frac{7}{2}$ **8.** $\frac{3}{4}a - 6 = \frac{1}{4}$ **9.** $\frac{3}{5}p - \frac{7}{10} = \frac{7}{10}p$

10. $\frac{2y}{3} + \frac{1}{3} = \frac{y}{2} + \frac{1}{6}$ **11.** $\frac{3x}{4} - \frac{5}{2} = \frac{x}{8} + 10$ **12.** $\frac{1}{8}c - \frac{1}{5}c + 1 = \frac{5}{2}$

Technology For Exercises 13–16, use a graphing calculator to graph and compare the equations.

13. Graph the equations $y = \frac{x}{2} + \frac{x}{3}$ and $y = \frac{25}{3}$ in the same coordinate plane.

14. Graph the equations $y = 3x + 2x$ and $y = 50$ in the same coordinate plane.

15. How are the equations in Exercise 14 related to the equations in Exercise 13? What do their graphs have in common?

16. How is the solution of the equation in Exercise 5 related to the graphs?

17. SPORTS With the wind at his back, Russell can skate across a lake at a speed of 14 km/h. When skating back into the wind, he can skate only 12 km/h. It takes Russell half an hour to skate across the lake and back.

a. Copy and complete the table.

b. Write and solve an equation to find the distance d across the lake.

Skating Across a Lake and Back

	Distance across (km)	Rate (km/h)	Time (hours)
Wind at his back	d	14	$\frac{d}{14}$
Into the wind	?	?	?

Solve each equation.

18. $0.2x + 5.3 = 6.1$ **19.** $0.05a + 2.33 = 1.08$

20. $4.8 - 0.3n = 0.3n$ **21.** $2.54a - 0.008 = 0.5$

22. $3.25c = 6 + 2.5c$ **23.** $464 - 0.018n = -0.25n$

24. $3k - 0.75 = k - 0.5$ **25.** $8 - 0.02x = 2.23x + 3.5$

26. Writing Write a letter to a friend explaining how you decide what power of ten to multiply by when you solve equations like those in Exercises 18–25. Include examples.

27. SAT/ACT Preview If $0.3x - 14 = 0.02x$, what is the value of $2x + 9$?

A. 0.5 **B.** 14 **C.** 50 **D.** 59 **E.** 109

Exercises and Applications

1. 6 2. 12
3. 90 4. 24
5. 10 6. –18
7. 21 8. $\frac{25}{3}$
9. –7 10. –1
11. 20 12. –20

13. [graph]

14. [graph]

15. The equations in Ex. 14 are obtained by multiplying the right-hand side of the equations in Ex. 13 by the LCD, 6, of the fractions in the equations in Ex. 13. The intersections of both graphs have the same x-coordinate, 10.

16. The solution, 10, of the equation in Ex. 5 is the x-coordinate of the intersection points of the graphs in Exs. 13 and 14. That is the value of x for which $\frac{x}{2} + \frac{x}{3} = \frac{25}{3}$ (Ex. 13) and $3x + 2x = 50$ (Ex. 14).

17. See answers in back of book.

18. 4 19. –25
20. 8 21. 0.2
22. 8 23. –2000
24. 0.125 25. 2

26. Answers may vary. An example is given. Dear Fumihiro, If you want to solve an equation that contains decimals, decide

INTERVIEW Barbara Romanowicz

Look back at the article on pages 144–146.

Barbara Romanowicz is shown here reading a seismogram.

Seismologists like Barbara Romanowicz need data from at least three seismographs at three different locations in order to locate the epicenter of an earthquake.

Readings for the Northridge Earthquake January 17, 1994

	Average S wave speed (km/s)	Average P wave speed (km/s)	Difference in arrival time (seconds)
Topopah Spring, Nevada	3.3	6.0	49
San Andreas Observatory, California	3.75	7.5	51
Albuquerque, New Mexico	3.5	7.7	169

For Exercises 28 and 29, round each answer to the tens' place.
For Exercise 30, round the answer to the hundreds' place.

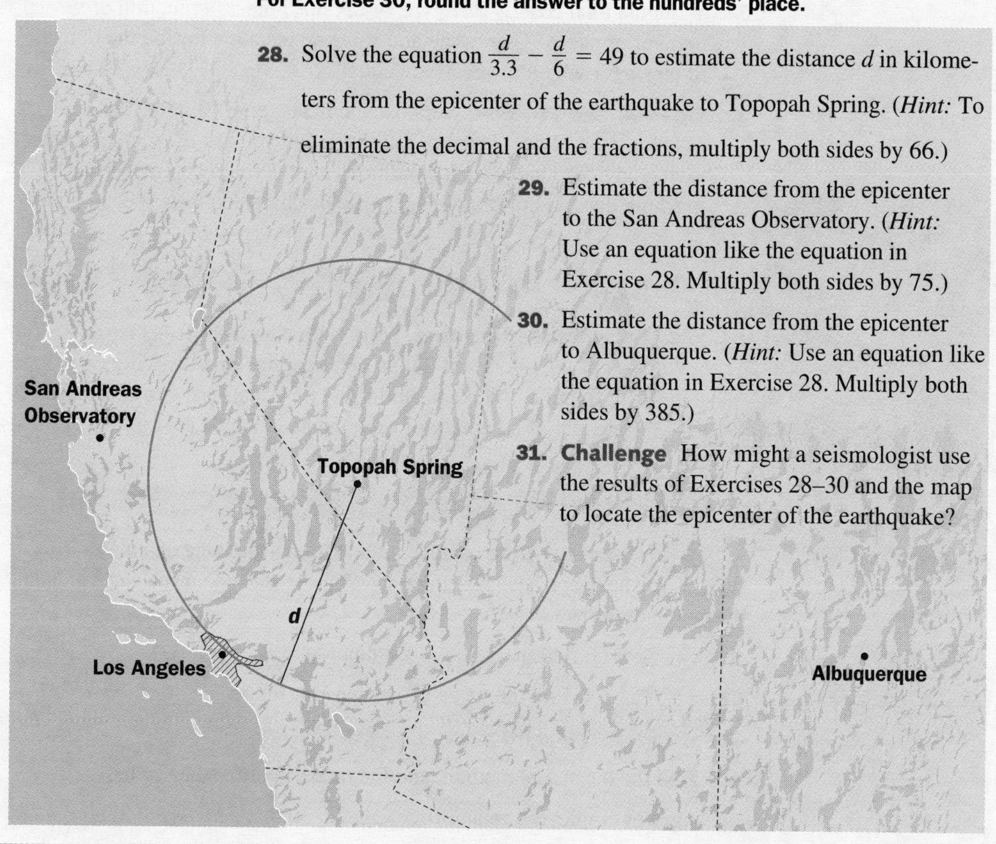

28. Solve the equation $\frac{d}{3.3} - \frac{d}{6} = 49$ to estimate the distance d in kilometers from the epicenter of the earthquake to Topopah Spring. (*Hint:* To eliminate the decimal and the fractions, multiply both sides by 66.)

29. Estimate the distance from the epicenter to the San Andreas Observatory. (*Hint:* Use an equation like the equation in Exercise 28. Multiply both sides by 75.)

30. Estimate the distance from the epicenter to Albuquerque. (*Hint:* Use an equation like the equation in Exercise 28. Multiply both sides by 385.)

31. Challenge How might a seismologist use the results of Exercises 28–30 and the map to locate the epicenter of the earthquake?

Apply⇔Assess

Exercise Notes

Interview Note

Exs. 29–31 You might need to remind students that the time lapse between the S waves and the P waves is the difference between the travel times for the S waves and the P waves.

Challenge

Ex. 31 When discussing students' answers to this exercise, examine why data from one or two seismograph locations are insufficient to locate the epicenter of an earthquake. Students should use their compasses to draw three circles to simulate how to locate the epicenter.

what the greatest number of decimal places is in any of the numbers in the equation. For example, the equation $5 = 1.2x - 1$ involves tenths, and the equation $1.008 - 3y = 0.15y$ involves thousandths. Multiply both sides of an equation like this by 10^n, where n is the greatest number of decimal places in the equation. This will give you an equation that involves integers. To solve $5 = 1.2x - 1$, multiply both sides by 10^1, or 10, to get $50 = 12x - 10$; $60 = 12x$; $x = 5$. To solve $1.008 - 3y = 0.15y$, multiply both sides by 10^3, or 1000, to get $1008 - 3000y = 150y$; $1008 = 3150y$; $y = 0.32$. I hope this helps you solve the equations in this lesson. Sincerely, Rosa

27. E

28. about 360 km

29. about 380 km

30. about 1100 km

31. A seismologist would draw arcs on the map to locate the point that is about 360 km from Topopah Spring, about 380 km from the San Andreas Observatory, and about 1100 km from Albuquerque. The intersection of the three arcs is the epicenter of the earthquake.

Apply⇔Assess

Exercise Notes

Using Technology
Exs. 33–35, 37–39 For Exs. 33–35, students can use the *Stats!* or the TABLE feature of the TI-82 for these exercises. For Exs. 37–39, students can use the Zoom 8:ZInteger feature on the TI-82 to change the trace display to integral values of *x*. Trace to find the desired *y*-values. As an alternate, students can graph each equation in the same coordinate plane and then use the CALC 5:Intersect function to find the intersection point of the lines.

Practice 27 for Section 4.4

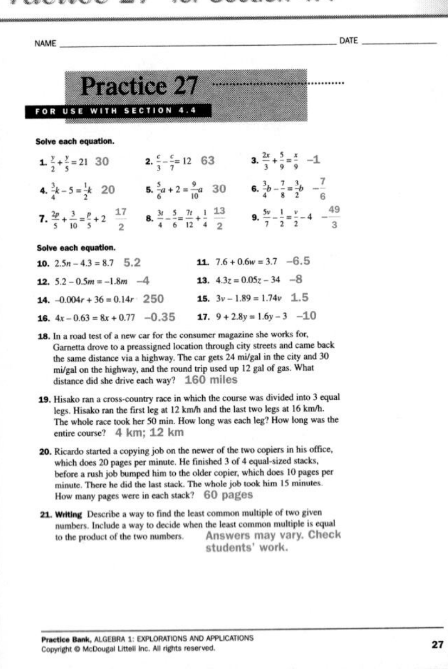
NAME ______________ DATE ________

Practice 27

FOR USE WITH SECTION 4.4

Solve each equation.

1. $\frac{y}{2}+\frac{y}{5}=21$ 30
2. $\frac{c}{3}-\frac{c}{7}=12$ 63
3. $\frac{2x}{3}+\frac{5}{9}=\frac{x}{9}$ −1
4. $\frac{3}{4}k-5=\frac{1}{2}k$ 20
5. $\frac{5}{6}a+2=\frac{9}{10}a$ 30
6. $\frac{3}{4}b-\frac{7}{8}=\frac{3}{2}b$ $-\frac{7}{6}$
7. $\frac{2p}{5}+\frac{3}{10}=\frac{p}{5}+2$ $\frac{17}{2}$
8. $\frac{3t}{4}-\frac{5}{6}=\frac{7t}{12}+\frac{1}{4}$ $\frac{13}{2}$
9. $\frac{5v}{7}-\frac{1}{2}=\frac{v}{2}-4$ $-\frac{49}{3}$

Solve each equation.

10. $2.5n-4.3=8.7$ 5.2
11. $7.6+0.6w=3.7$ −6.5
12. $5.2-0.5m=-1.8m$ −4
13. $4.3z=0.05z-34$ −8
14. $-0.004r+36=0.14r$ 250
15. $3v-1.89=1.74v$ 1.5
16. $4x-0.63=8x+0.77$ −0.35
17. $9+2.8y=1.6y-3$ −10
18. In a road test of a new car for the consumer magazine she works for, Garnetta drove to a preassigned location through city streets and came back the same distance via a highway. The car gets 24 mi/gal in the city and 30 mi/gal on the highway, and the round trip used up 12 gal of gas. What distance did she drive each way? 160 miles
19. Hisako ran a cross-country race in which the course was divided into 3 equal legs. Hisako ran the first leg at 12 km/h and the last two legs at 16 km/h. The whole race took her 50 min. How long was each leg? How long was the entire course? 4 km; 12 km
20. Ricardo started a copying job on the newer of the two copiers in his office, which does 20 pages per minute. He finished 3 of 4 equal-sized stacks, before a rush job bumped him to the older copier, which does 10 pages per minute. There he did the last stack. The whole job took him 15 minutes. How many pages were in each stack? 60 pages
21. **Writing** Describe a way to find the least common multiple of two given numbers. Include a way to decide when the least common multiple is equal to the product of the two numbers. Answers may vary. Check students' work.

Practice Bank, ALGEBRA 1: EXPLORATIONS AND APPLICATIONS 27

Spreadsheets For Exercises 32–35, use a spreadsheet and the information below.

Kellem Company wants to buy a fax machine. The company will buy either a machine that uses ordinary paper or one that uses special paper.

Cost (ordinary paper): $1274 plus $2.41 per package of 500 sheets of paper

Cost (special paper): $622 plus $8.45 per roll of 350 sheets of paper

Kellem Company is strongly considering buying the fax machine for $622, but faxes on ordinary paper are easier to read.

32. Explain how to find the cost per page of each type of paper.

33. Complete the spreadsheet shown.

Fax Machine Costs

B2		=1274+(2.41/500)*A2	
	A	B	C
1	Number of pages	Cost (ordinary)	Cost (special)
2	1,000	1278.82	
3	10,000		
4	100,000		

34. Write the total cost of each type of fax machine as a function of the number of pages. Then write and solve an equation to find the number of pages for which the costs are equal.

35. **Open-ended Problem** Suppose the manager of Kellem Company asks for your advice. What questions would you ask the manager before giving advice? Under what conditions would you recommend the machine that uses ordinary paper? the machine that uses special paper?

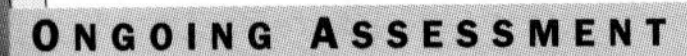

ONGOING ASSESSMENT

36. **Cooperative Learning** Work with a partner. Together make a list of the important things to remember about solving equations with fractions. Make a similar list for equations with decimals. Write a brief summary of each list and include examples.

SPIRAL REVIEW

Technology Graph each equation. Use a graphing calculator if you have one. Use the graph to estimate the unknown value to the nearest tenth. *(Section 2.6)*

37. $y = 2x + 3$; find the value of x when $y = 7$.

38. $y = -x + 4$; find the value of x when $y = 6$.

39. $y = \frac{2}{3}x + 1$; find the value of x when $y = -1$.

40. Find the mean, the median, and the mode(s) of the data. *(Section 1.3)*

Ages of employees in an office: 21, 25, 30, 45, 52, 24, 38, 27, 35, 48

32. Divide the cost per package by the number of sheets per package.

33–35. See answers in back of book.

36. Answers may vary. Students should realize that they can eliminate fractions from an equation by multiplying both sides by the least common denominator of all the fractions in the equation. (The LCD is the LCM of the denominators.) They can eliminate decimals in an equation by multiplying both sides by an appropriate power of ten (10, 100, 1000, and so on). In either case, the resulting equation involves integers only and can be solved using techniques presented in earlier chapters.

37.

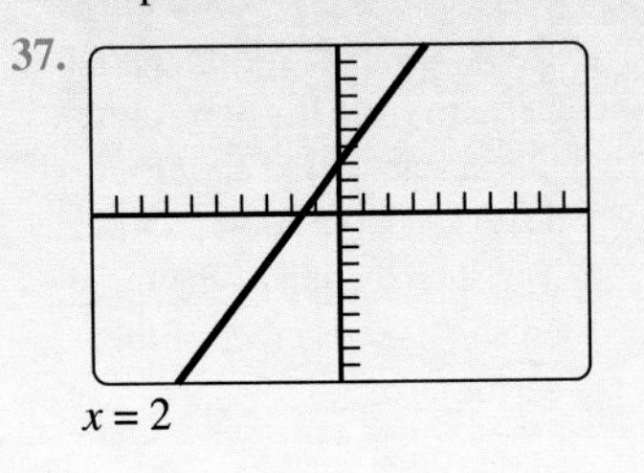

$x = 2$

38.

$x = -2$

39.

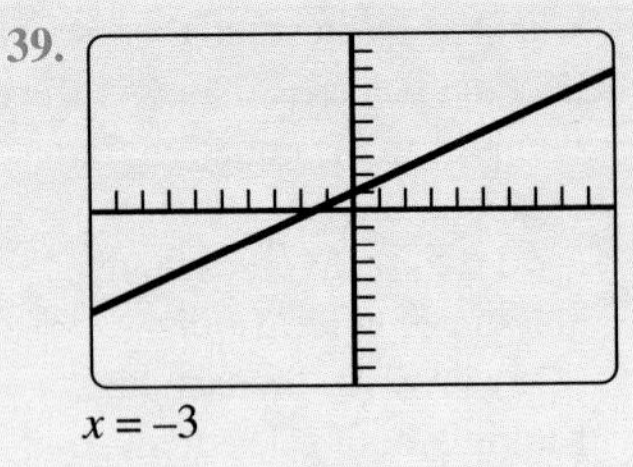

$x = -3$

40. 34.5; 32.5; none

SECTION

4.5 Writing Inequalities from Graphs

Learn how to...

- use inequalities to represent intervals on a graph

So you can...

- represent a group of solutions, such as times of day when temperatures are more than 100°F

You can be comfortable inside an adobe house even when the outside temperature is more than 100°F. The clay building materials absorb heat during the day and give off heat during the night. The graph below models this situation.

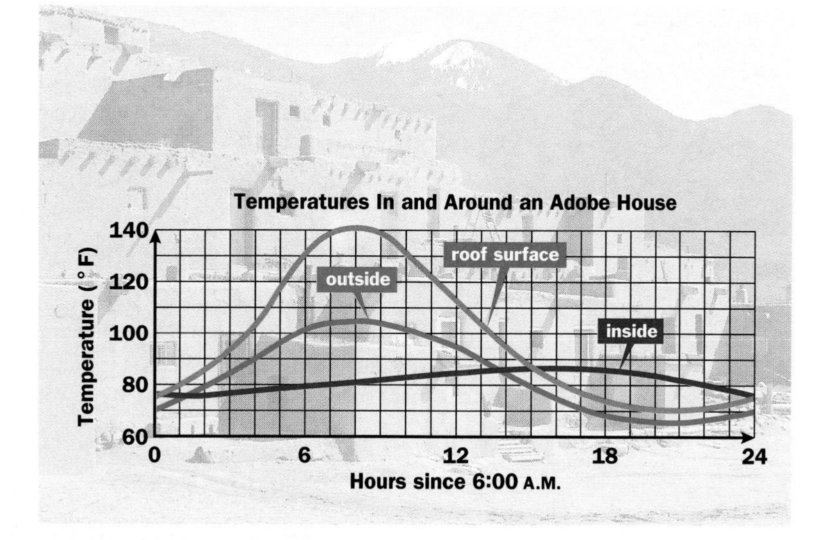

THINK AND COMMUNICATE

Use the graph above.

1. Where will you find the hottest temperature: *inside*, *outside*, or *on the roof*? Where will you find the coolest temperature?
2. What is the range of inside temperatures? of outside temperatures?
3. Where do the temperatures vary least: *inside*, *outside*, or *on the roof*?
4. When is the outside temperature hotter than the inside temperature? When is the outside temperature cooler than the inside temperature?
5. How do the outside temperature and the temperature of the roof surface compare?

An **inequality** is a mathematical statement formed by placing an inequality symbol between numerical or variable expressions. You can use an inequality to represent the solution of a problem.

Plan⇔Support

Objectives

- Use inequalities to represent intervals on a graph.
- Use inequalities to represent a group of solutions to real-world problems.

Recommended Pacing

Core and Extended Courses
Section 4.5: 1 day

Block Schedule
Section 4.5: $\frac{1}{2}$ block
(with Section 4.6)

Resource Materials

Lesson Support
Lesson Plan 4.5
Warm-Up Transparency 4.5
Overhead Visuals:
Folder A: Multi-Use Graphing Packet, Sheets 1–3
Practice Bank: Practice 28
Study Guide: Section 4.5
Explorations Lab Manual:
Diagram Master 2

Technology
Graphing Calculator
Internet:
http://www.hmco.com

Warm-Up Exercises

Write an inequality that describes each sentence.

1. n is a number less than –3. $n < -3$
2. t is a temperature greater than 20°F. $t > 20$
3. p is a price less than \$10 and more than \$5. $5 < p < 10$
4. w is a weight greater than or equal to 120 lb. $w \geq 120$
5. d is a distance less than or equal to 75 mi. $d \leq 75$

ANSWERS Section 4.5

Think and Communicate

1. on the roof; outside
2. about 10°; about 40°
3. inside
4. Estimates may vary; from about 7:30 A.M. until about 7:30 P.M.; from about 7:30 P.M. until about 7:30 A.M.
5. Answers may vary. An example is given. The temperature on the roof surface rises and falls with the outside temperature. During the afternoon and night, the roof temperature is slightly greater than the outside temperature. During the morning, the temperature of the roof is substantially greater than the outside temperature.

Teach⇔Interact

Think and Communicate

These questions lead students to the idea of an inequality and to writing inequalities. For question 2, remind students that the range of a set of numbers is the highest number minus the lowest number in the set. On the graph, the range can be found by using the vertical temperature axis.

Section Note

Multicultural Note

Adobe structures have been used as dwellings for thousands of years. Most adobe is made from clay and water, plus a bit of grass or straw to stabilize the mixture. The block shape is created by placing this mixture into wooden forms and allowing it to dry. Once dry, the blocks are removed and left to bake in the sun for up to fourteen days. The finished blocks are joined with mortar to form walls. Because adobe structures cannot withstand long periods of rain, they are usually found in desert regions such as the southwestern part of the United States.

About Example 1

Communication: Reading
Remind students when they read the inequality $5.5 < t < 10.5$ that it means t is greater than 5.5 ($t > 5.5$) and t is less than 10.5 ($t < 10.5$). Since t is *between* these two values, it does not include either one of them.

Additional Example 1

Use the graph from page 171. Write an inequality to represent the temperature of the roof surface between 6:00 A.M. and 12:00 noon.
Let T = temperature of roof surface. Then T is between 75° and 130°, or $75° < T < 130°$.

Additional Example 2

The park district charges $3.50 to use the pool for nonmembers and $2 to use the pool for members. It costs $30 to join the park district. When is the total cost of using the pool for a nonmember more than the total cost for a member? Write your answer as an inequality.

BY THE WAY

The word *adobe* is of Arabic origin. It means "earth from which unburnt bricks are made." Adobe houses are common in Mexico and in the southwestern United States. Adobe was also used by the ancient Egyptians and Babylonians.

EXAMPLE 1 Application: Architecture

Use the graph on page 171. When is the outside temperature more than 100°F? Write your answer as an inequality.

SOLUTION

Here the **outside** temperature is more than 100°F.

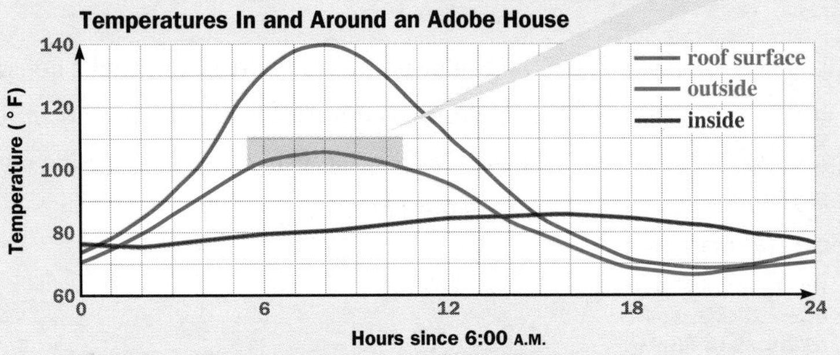

Let t = hours since 6:00 A.M. Then t is between 5.5 and 10.5, or $5.5 < t < 10.5$.

EXAMPLE 2 Application: Recreation

A dance organization sponsors a few dances every week. The admission price to each dance is $4 for members and $6 for nonmembers. It costs $50 for an annual membership. When is the total cost for a nonmember more than the total cost for a member? Write your answer as an inequality.

SOLUTION

Problem Solving Strategy: Use a graph.

Let x = the number of dances a person attends in a year.

Step 1 Write and graph two equations that model this situation.

Cost for a member: $y = 4x + 50$

Cost for a nonmember: $y = 6x$

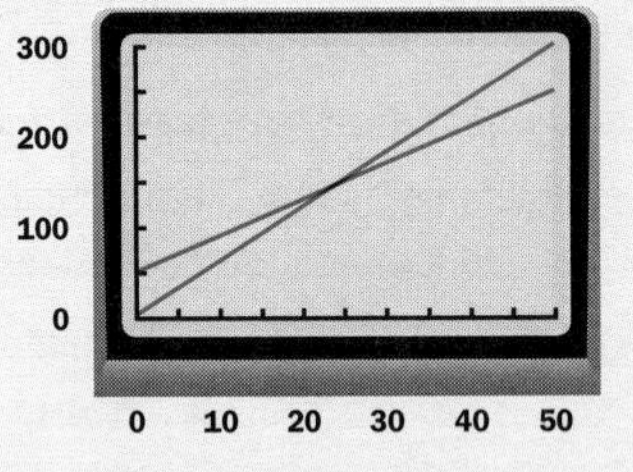

Step 2 Find the coordinates of the intersection point.

Nonmember costs are more than **member** costs beyond this point.

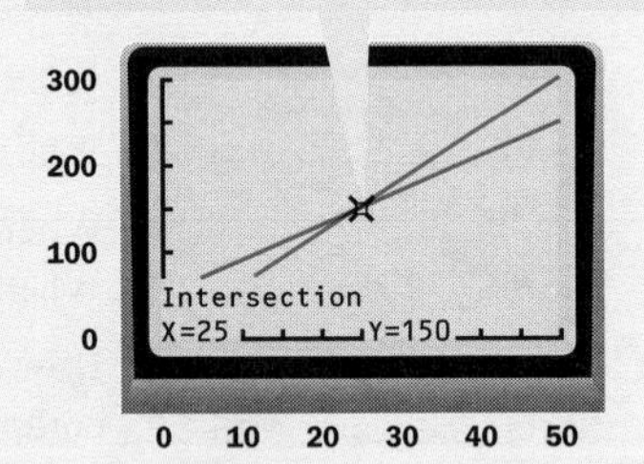

For more information about finding where graphs intersect, see the *Technology Handbook*, p. 607.

The total cost for a nonmember is more than the total cost for a member when a person attends more than 25 dances, or when $x > 25$.

Think and Communicate

6. because the question asks when the total cost for a nonmember is more than the total cost for a member; When $x = 25$, the costs are equal.
7. $x < 25$

Checking Key Concepts

1. $x < 8$
2. $x \geq 8$
3. $x > 2$
4. $2 < x < 8$
5. $x > 60$, where x = the number of dances attended in a year

Exercises and Applications

1. a. $20 < s < 65$
   b. Answers may vary. For example, the fuel economy of an average car increases as the speed increases up to about 40 mi/h, levels off at about 45 mi/h, and then decreases. The fuel economy is maximum, about 33 mi/gal, at a speed of about 40 mi/h.
   c. Answers may vary. An example is given. (1) At what speeds is the fuel economy at least 30 mi/gal? (2) If you drive over 40 mi/h, what fuel economy can be expected? (3) At what speeds is the fuel economy between 25 mi/gal and 30 mi/gal?

THINK AND COMMUNICATE

6. In Example 2, why is the inequality $x > 25$ and not $x \geq 25$?
7. In Example 2, when is the total cost for a nonmember *less than* the total cost for a member? Write your answer as an inequality.

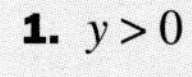

CHECKING KEY CONCEPTS

For Questions 1–4, use the graph. Write an inequality to represent the x-values that correspond to each range of y-values.

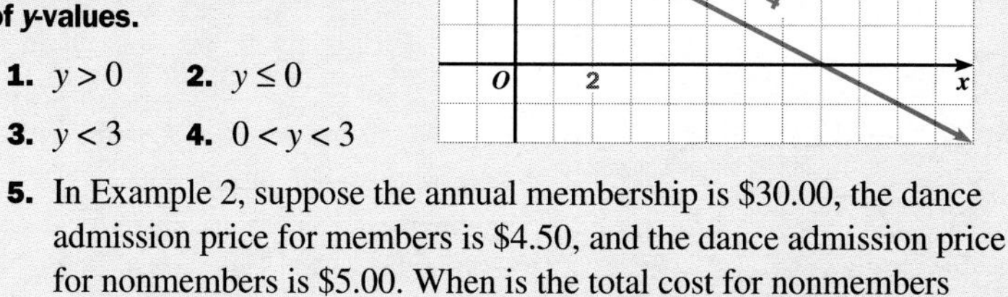

1. $y > 0$
2. $y \leq 0$
3. $y < 3$
4. $0 < y < 3$
5. In Example 2, suppose the annual membership is \$30.00, the dance admission price for members is \$4.50, and the dance admission price for nonmembers is \$5.00. When is the total cost for nonmembers greater than the total cost for members? Write your answer as an inequality.

4.5 Exercises and Applications

Extra Practice *exercises on page 565*

1. **FUEL ECONOMY** The graph shows the fuel economy of an average car as a function of speed.
   a. At what speeds should the car be driven so that it gets more than 25 mi/gal? Write your answer as an inequality.
   b. **Writing** What does the graph tell you about fuel economy at different speeds?
   c. **Open-ended Problem** Write three problems that can be answered using the graph and an inequality.

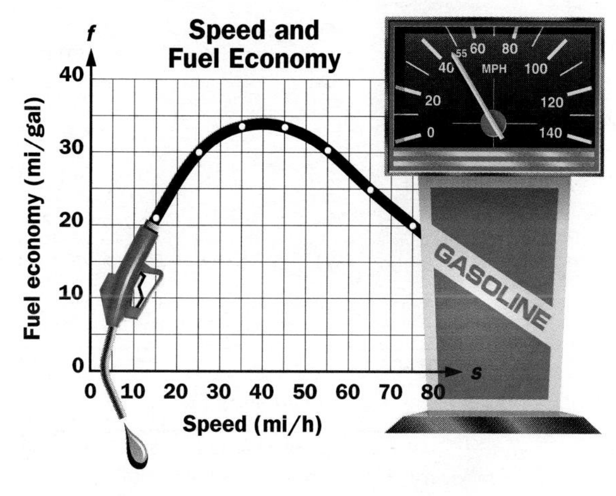

For Exercises 2–4:

a. Graph each equation.

b. Write an inequality to represent the x-values when $y > 1$.

2. $y = 2x + 1$
3. $y = -2x + 1$
4. $y = 2x$

Teach⇔Interact

Additional Example 2 (continued)

Use a graph. Let x = the number of visits to the pool.

Step 1 Write and graph two equations that model this situation.

Cost for a member: $y = 30 + 2x$

Cost for a nonmember: $y = 3.50x$

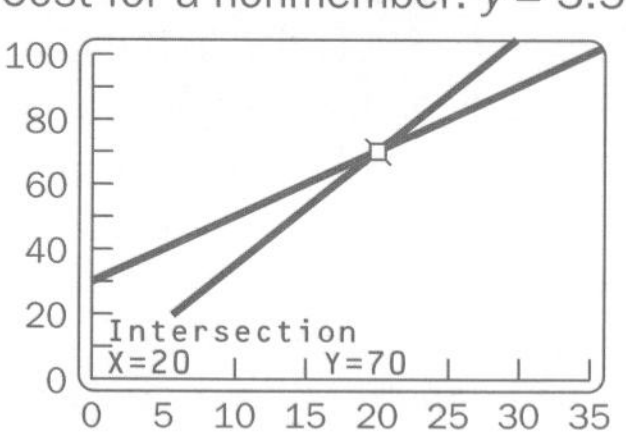

Step 2 Find the coordinates of the intersection point. The total cost for a nonmember is more than the total cost for a member when there are more than 20 visits to the pool, or when $x > 20$.

Checking Key Concepts

Topic Spiraling: Review

Questions 1–4 can be used to review the concepts of domain and range introduced in Section 2.3. Point out that the possible range values for y are restricted and therefore the domain values for x are also restricted.

Closure Question

Compare the inequalities $7 \leq x \leq 15$ and $7 < x < 15$. How are they alike? How are they different? Both inequalities include all numbers between 7 and 15. However, the first inequality includes the numbers 7 and 15 as values for x, while the second inequality does not.

Apply⇔Assess

Suggested Assignment

- **Core Course**
  Exs. 1–9, 11–22
- **Extended Course**
  Exs. 1–22
- **Block Schedule**
  Day 25 Exs. 1–9, 11–22

2. a. [graph]
   b. $x > 0$

3. a. [graph]
   b. $x < 0$

4. a. [graph]
   b. $x > \frac{1}{2}$

Apply⇔Assess

Exercise Notes

Communication: Discussion
Ex. 1 Discuss with students what it means to say that fuel economy is a function of speed.

Application
Exs. 1, 5–11, 13 In their previous study of mathematics, students were exposed to the concept of equality more than the concept of inequality. In real-life applications and in advanced mathematics, inequalities appear rather frequently. The applications in these exercises to fuel economy, meteorology, communications, and personal finance will help students to better appreciate the role of inequalities in solving problems.

Interdisciplinary Problems
Ex. 5 The troposphere and the stratosphere are the lower layers of the atmosphere in which most weather phenomena occur. Therefore, they are the chief focus of most meteorological forecasting.

Career Connection
Exs. 5–10 A meteorologist is a scientist who studies Earth's atmosphere and the variations in conditions that affect weather patterns. A modern meteorologist draws upon many fields of science, including physics, chemistry, and engineering. Mathematical modeling and computer technology are used extensively to study weather patterns and to make predictions and display information. The scientific study of weather and weather forecasting is very difficult because the mathematical models developed so far are only rough approximations to what really happens. Most meteorologists work for government agencies, corporations, or in universities.

Connection METEOROLOGY

Earth's atmosphere is divided into layers. Temperatures vary from layer to layer. The black jagged line on the graph shows a representative temperature for each altitude. For Exercises 5–7, write each answer as an inequality. Use the variable *a* to represent altitude, and the variable *t* to represent temperature.

5. What is the range of temperatures in the troposphere? in the mesosphere?

6. At which altitudes is the temperature above 30°C?

7. At which altitudes is the temperature below −70°C?

8. Which layer has the greatest range of temperatures? the smallest range?

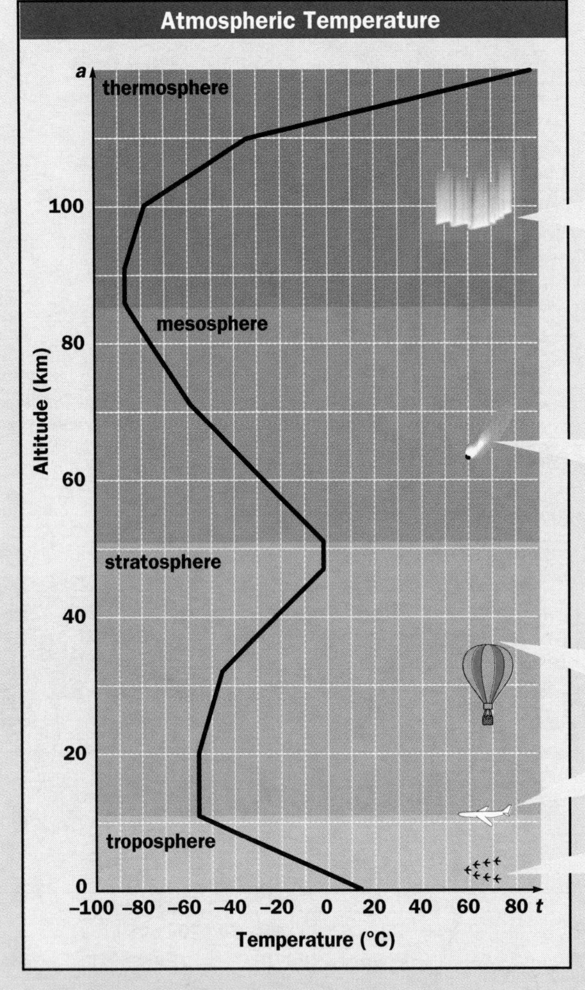

Auroras are bright displays of light in the sky. They usually occur at least 97 km above Earth's surface.

Meteors burn up at heights around 65 km.

A record-setting balloon flight in 1961 reached 34.668 km.

Airplanes generally fly between 8.8 km and 12.5 km.

Pilots have observed geese in migratory flight at 1.05 km.

9. Challenge Between which two altitudes is the rate of decrease in temperature the greatest? How can you tell?

10. Research Temperatures do not follow the pattern in the graph during an *inversion*. Find out what an *inversion* is.

5. Estimates may vary. $-55 \le t \le 15$; $-85 \le t \le 0$
6. Estimates may vary. $a > 115$
7. Estimates may vary. $78 < a < 102$
8. the thermosphere; the stratosphere
9. Answers may vary. An example is given. $50 < a < 85$; Of all the intervals in which the temperature is decreasing, in that interval the graph is steepest.
10. Answers may vary. An example is given. A temperature inversion occurs when the normal decrease of atmospheric temperature changes over a relatively short vertical distance. If an inversion is strong enough, there will be a layer in the atmosphere where temperature actually increases with height. Causes of inversions include night cooling of the lowest layers of the atmosphere, stagnant high-pressure systems in the middle and upper latitudes, and cold air blowing over warmer ocean water.
11. a. Company A charges $12 for the first 60 min and $.20 for each additional minute. Company B charges $10 for the first 60 min and $.25 for each additional minute.
    b. $t > 100$, where t = the number of minutes of calling per month

11. **COMMUNICATIONS** The graph models the cost of using two different long distance companies. Cost is a function of the total number of minutes of your long distance calls each month.

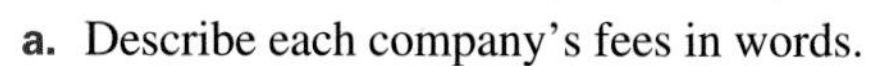

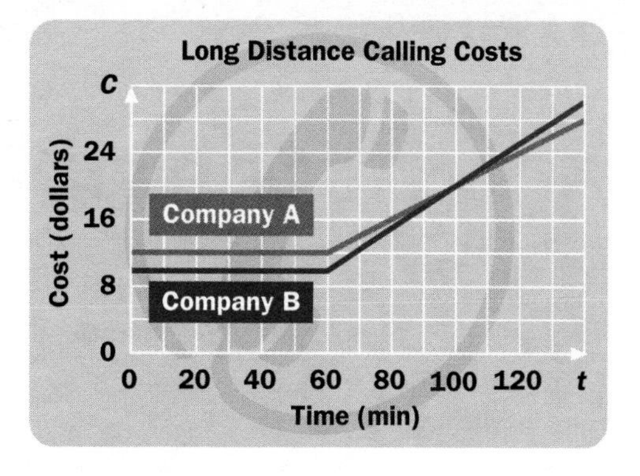

a. Describe each company's fees in words.

b. When does it cost less to use Company A? Write your answer as an inequality.

12. 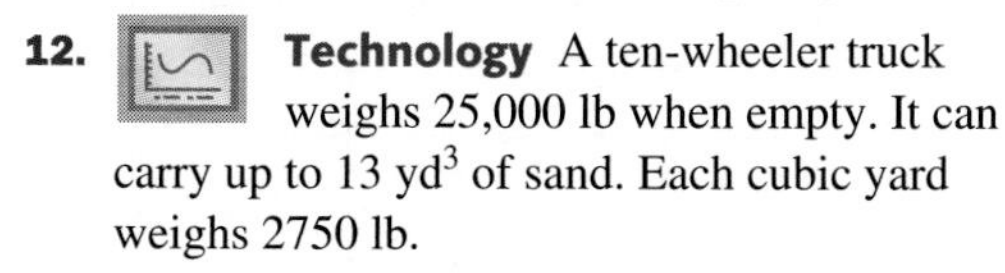 **Technology** A ten-wheeler truck weighs 25,000 lb when empty. It can carry up to 13 yd^3 of sand. Each cubic yard weighs 2750 lb.

a. Write an equation for the weight of the truck as a function of the number of cubic yards of sand it carries.

b. Graph the equation from part (a). Set a good viewing window.

c. For what amounts of sand is it safe for the truck to cross a bridge with a 20 ton load limit? Write your answer as an inequality. (*Hint:* Use the fact that 1 ton = 2000 lb.)

13. **PERSONAL FINANCE** The camcorder that Yoshi Minami is considering costs \$750 to buy. A similar model costs \$40 per day to rent. When is the cost to rent a camcorder greater than the cost to buy a camcorder? Write your answer as an inequality.

ONGOING ASSESSMENT

14. **Open-ended Problem** The graph models car trips for Theresa Jones and Jason Wills between Dallas and Houston. Write a story based on the graph and the point where the lines intersect. Be sure to include speeds and inequalities in your story.

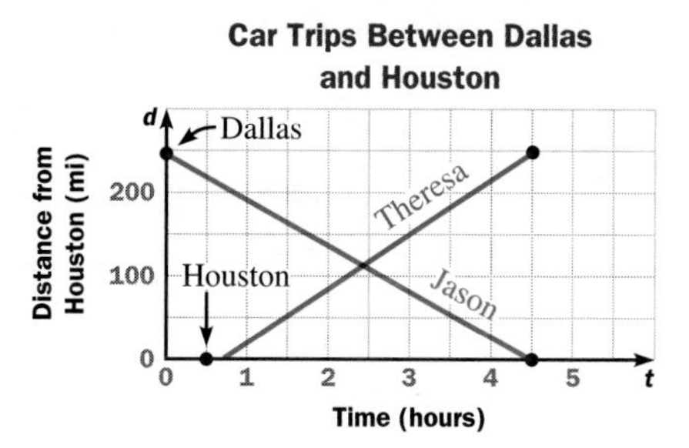

SPIRAL REVIEW

Find an equation of the line with the given slope and through the given point. *(Section 3.5)*

15. slope = 2; (5, 3)

16. slope = $-\frac{3}{5}$; (−15, 9)

17. slope = 0; (12, −12)

18. undefined slope; (2, 2)

Graph each inequality on a number line. *(Toolbox, page 591)*

19. $x > -4$

20. $a \le 3$

21. $n \ge 0$

22. $y < -1$

Apply⇔Assess

Exercise Notes

Assessment Note

Ex. 14 Using open-ended problems for assessment gives students an opportunity to be original and to communicate the mathematics they know in more than one way. An analysis of students' stories will show clearly how well they understand not only the graph given in this exercise, but also the complete concept of reading and interpreting information presented in graphs.

Practice 28 for Section 4.5

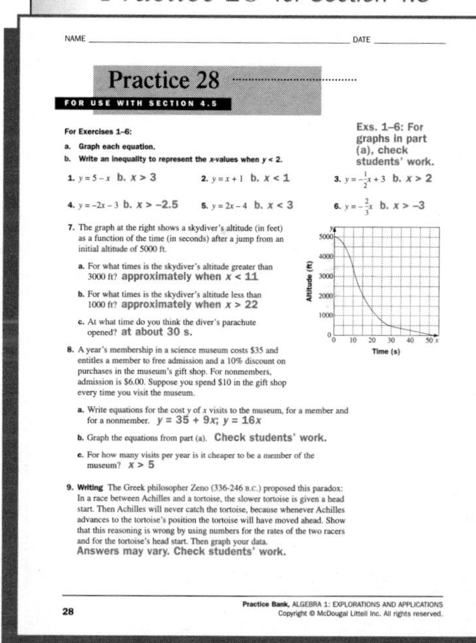

NAME ______ DATE ______

Practice 28

FOR USE WITH SECTION 4.5

For Exercises 1–6:

a. **Graph each equation.**

b. **Write an inequality to represent the x-values when y < 2.**

Exs. 1–6: For graphs in part (a), check students' work.

1. $y = 5 - x$ b. $x > 3$
2. $y = x + 1$ b. $x < 1$
3. $y = -\frac{1}{2}x + 3$ b. $x > 2$
4. $y = -2x - 3$ b. $x > -2.5$
5. $y = 2x - 4$ b. $x < 3$
6. $y = -\frac{2}{3}x$ b. $x > -3$

7. The graph at the right shows a skydiver's altitude (in feet) as a function of the time (in seconds) after a jump from an initial altitude of 5000 ft.

a. For what times is the skydiver's altitude greater than 3000 ft? approximately when $x < 11$

b. For what times is the skydiver's altitude less than 1000 ft? approximately when $x > 22$

c. At what time do you think the diver's parachute opened? at about 30 s.

8. A year's membership in a science museum costs \$35 and entitles a member to free admission and a 10% discount on purchases in the museum's gift shop. For nonmembers, admission is \$6.00. Suppose you spend \$10 in the gift shop every time you visit the museum.

a. Write equations for the cost y of x visits to the museum, for a member and for a nonmember. $y = 35 + 9x$; $y = 16x$

b. Graph the equations from part (a). Check students' work.

c. For how many visits per year is it cheaper to be a member of the museum? $x > 5$

9. **Writing** The Greek philosopher Zeno (336-246 B.C.) proposed this paradox: In a race between Achilles and a tortoise, the slower tortoise is given a head start. Then Achilles will never catch the tortoise, because whenever Achilles advances to the tortoise's position the tortoise will have moved ahead. Show that this reasoning is wrong by using numbers for the rates of the two racers and for the tortoise's head start. Then graph your data.
Answers may vary. Check students' work.

28 **Practice Bank,** ALGEBRA 1: EXPLORATIONS AND APPLICATIONS

---

12. a. $w = 25{,}000 + 2750s$, where w = the weight in pounds and s = the number of cubic yards of sand

b.

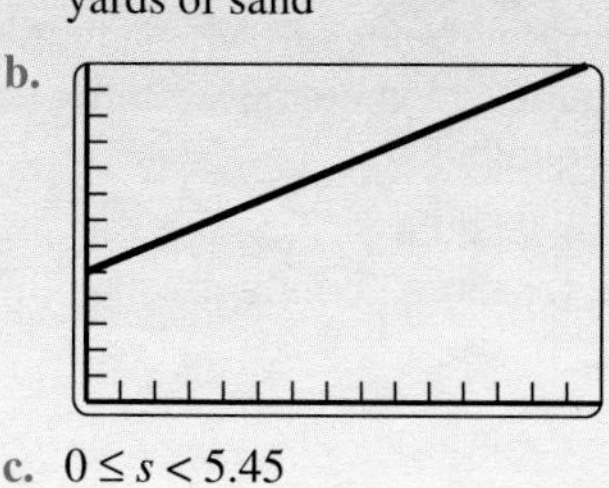

c. $0 \le s < 5.45$

13. $d > 18$ or $d \ge 19$, where d is the number of days

14. Answers may vary. An example is given. Jason left Dallas heading for Houston at an average speed of about 55.6 mi/h. One half hour later, Theresa left Houston heading for Dallas at an average speed of about 62.5 mi/h. The two passed each other about 2 h 20 min after Jason left. If t is the time in hours since Jason left, $0 < t < 4.5$ for Jason and $0.5 < t < 4.5$ for Theresa. If d is the distance in miles from Houston, $0 \le d \le 250$ for both Jason and Theresa.

15. $y = 2x - 7$

16. $y = -\frac{3}{5}x$

17. $y = -12$

18. $x = 2$

19.–22.

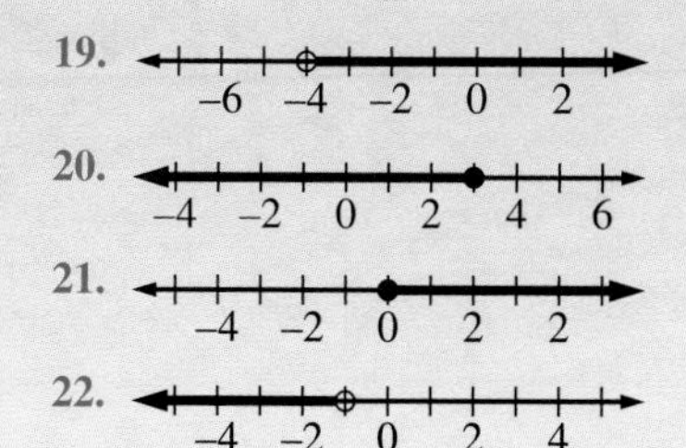

Plan⇔Support

Objectives

- Solve inequalities.
- Tell when one quantity is greater than another.

Recommended Pacing

Core and Extended Courses
Section 4.6: 2 days

Block Schedule
Section 4.6: 2 half-blocks
(with Section 4.5 and Review)

Resource Materials

Lesson Support
Lesson Plan 4.6
Warm-Up Transparency 4.6
Practice Bank: Practice 29
Study Guide: Section 4.6
Explorations Lab Manual:
Diagram Master 2
Assessment Book: Test 16

Technology
Technology Book:
Calculator Activity 4
Graphing Calculator
Internet:
http://www.hmco.com

Warm-Up Exercises

Solve each equation.

1. $-4x = 24$ -6
2. $7n + 2 = 9n - 8$ 5
3. $2(5y - 3) = 4y$ 1
4. $4x + 3 = 2(2x + 6)$ no solution
5. Use a number line to graph the inequality $x > -2$.

SECTION

4.6 Solving Inequalities

Learn how to...

- solve inequalities

So you can...

- tell when one quantity is greater than another, such as when renting is more expensive than buying

Suppose you start with an inequality that is true. What happens when you add the same number to both sides? The example below should give you some idea.

$$2 < 3$$
$$2 + 4 < 3 + 4$$
$$6 < 7$$

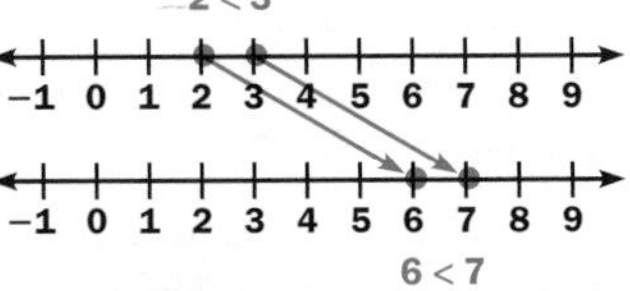

EXPLORATION
COOPERATIVE LEARNING

Working with Inequalities

Work with a partner.

For each of Steps 1–4, draw pairs of number lines like those shown above. One of you should work with the inequality 2 < 3 and one of you should work with the inequality 1 < 5. Use these numbers for each step:

a. 2 **b.** 10 **c.** −2 **d.** −10

1 **Add** each number to both sides of your inequality. Is each new inequality *True* or *False*?

2 **Subtract** each number from both sides of your inequality. Is each new inequality *True* or *False*?

3 **Multiply** both sides of your inequality by each number. Is each new inequality *True* or *False*?

4 **Divide** both sides of your inequality by each number. Is each new inequality *True* or *False*?

Questions

1. Work together to generalize how an inequality is affected by each of the four basic operations: addition, subtraction, multiplication, and division. Test your generalization with some inequalities.
2. Assume that $x > 0$, $y > 0$, and $x > y$. Replace each __?__ with > or <.
   a. $x - 3$ __?__ $y - 3$ **b.** $3x$ __?__ $3y$ **c.** $\frac{x}{-3}$ __?__ $\frac{y}{-3}$

Exploration Note

Purpose
The purpose of this Exploration is to have students discover how an inequality is affected by the four basic operations.

Materials/Preparation
Paper and pencil

Procedure
Students draw number lines and express an inequality between pairs of numbers on the lines. They then use the four basic operations and the numbers 2, 10, −2, and −10 to develop specific examples from which generalizations about inequalities can be made. Encourage students to test their generalizations with different inequalities.

Closure
Students should conclude that adding or subtracting positive or negative numbers to both sides of an inequality preserves the direction of the inequality. Multiplying or dividing both sides of an inequality by a negative number reverses the inequality.

Explorations Lab Manual
See the Manual for more commentary on this Exploration.

Solving Inequalities

When you add, subtract, multiply by, or divide by the same positive number on both sides of an inequality, the inequality symbol stays the same. It also stays the same when you add or subtract the same negative number. When you multiply by or divide by the same **negative** number on both sides, the inequality symbol is **reversed**. For example:

$1 < 2$	$1 < 2$	$1 < 2$
$1 - 5 < 2 - 5$	$5 \cdot 1 < 5 \cdot 2$	$-5 \cdot 1 > -5 \cdot 2$
$-4 < -3$	$5 < 10$	$-5 > -10$

When you solve an inequality that involves a variable, the **solution of the inequality** is the set of numbers that makes the inequality true.

EXAMPLE 1 Application: Recreation

Brian pays $4 to rent skates every time he goes roller skating. He thinks he can save money if he buys his own skates for $45. He wonders, "When is the cost of renting skates more expensive than buying a new pair?"

SOLUTION

Here is how Brian solved the problem.

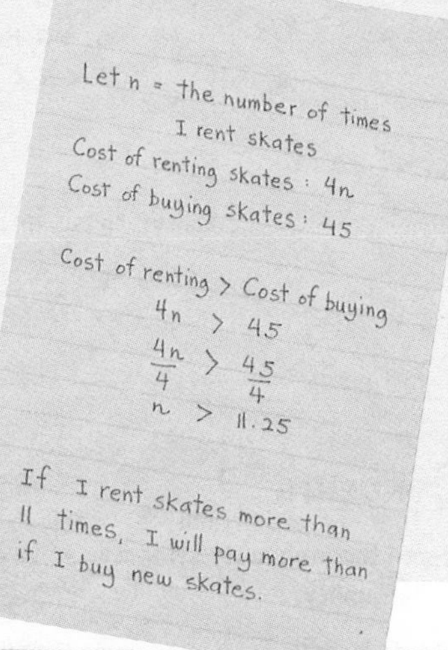

Check

Substitute a number greater than 11 **for** n.

$4n > 45$

$4(12) \overset{?}{>} 45$

$48 > 45$ ✓

EXAMPLE 2

Solve the inequality $-2x \leq 10$. Graph the solution on a number line.

SOLUTION

Divide both sides by -2 and **reverse** the inequality symbol.

$$-2x \leq 10$$

$$\frac{-2x}{-2} \geq \frac{10}{-2}$$

$$x \geq -5$$

Graph -5 and all numbers greater than -5.

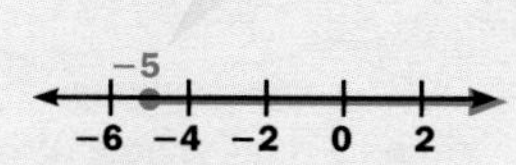

Toolbox p. 591 *Graphing Inequalities*

Teach⇔Interact

Section Note

Common Error

Some students tend to forget to reverse the inequality symbol when they multiply or divide by a negative number. Encourage all students to check at least one number they think is a solution in the original inequality. This will help students to correct their errors and eventually to avoid making them altogether.

Additional Example 1

Tisha pays $2.50 to rent bowling shoes every time she goes bowling. A new pair of bowling shoes costs $37.50. When is the cost of renting shoes more expensive than buying a new pair?

Let n = the number of times she goes bowling.

Cost of renting shoes: $2.50n$

Cost of buying shoes: 37.50

Cost of renting > Cost of buying

$2.50n > 37.50$

$100(2.50n) > 100(37.50)$

$250n > 3750$

$\frac{250n}{250} > \frac{3750}{250}$

$n > 15$

If Tisha rents shoes more than 15 times, she will pay more than if she bought them.

Additional Example 2

Solve the inequality $-3x \geq 15$. Graph the solution on a number line.

$-3x \geq 15$

$\frac{-3x}{-3} \leq \frac{15}{-3}$

$x \leq -5$

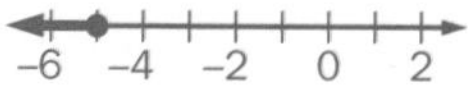

(number line: -6 -4 -2 0 2)

ANSWERS Section 4.6

Exploration

1–4. See answers in back of book.

Questions

1. Students should discover the following generalizations.

   Any number can be added to or subtracted from each side of a true inequality to get another true inequality.

   Each side of a true inequality can be multiplied or divided by the same positive number to get another true inequality.

   If each side of a true inequality is multiplied or divided by the same negative number, the resulting inequality is false. To get a true inequality, reverse the inequality sign. Examples will vary.

2. a. >

   b. >

   c. <

Teach⇔Interact

Think and Communicate

For question 2, point out that using the numbers 0, 1, or –1 to check the solution to an inequality often makes the check quick and easy.

About Example 3

Mathematical Procedures
When discussing this example, relate the procedures for solving the equation $6 + x = 3 - 4x$ to those for solving the inequality $6 + x < 3 - 4x$. Students should see that the procedures are exactly the same.

Additional Example 3

Solve the inequality $5 + y < 2 - 3y$. Graph the solution on a number line.

$$5 + y < 2 - 3y$$
$$5 + y - 5 < 2 - 3y - 5$$
$$y < -3 - 3y$$
$$y + 3y < -3 - 3y + 3y$$
$$4y < -3$$
$$y < -\frac{3}{4}$$

$-\frac{3}{4}$

–4 –2 0 2 4

Checking Key Concepts

Communication: Discussion
Discuss with students the fact that the graph of the solutions of an inequality shows that there are infinitely many solutions to all inequalities.

Closure Questions

When solving an inequality, when is it necessary to reverse the inequality symbol? How is the solution to an inequality graphed?
when you multiply or divide both sides by the same negative number; The solution is graphed on a number line.

THINK AND COMMUNICATE

1. Describe how Brian could have used a graph to solve the problem in Example 1 on page 177.

2. In Example 2 on page 177, can you use -5 to check the solution? Why is it better to use a number greater than -5 to check the solution? Check the solution.

EXAMPLE 3

Solve the inequality $6 + x < 3 - 4x$. Graph the solution on a number line.

SOLUTION

$$6 + x < 3 - 4x$$
$$6 + x - 6 < 3 - 4x - 6$$
$$x < -3 - 4x$$
$$x + 4x < -3 - 4x + 4x$$
$$5x < -3$$
$$\frac{5x}{5} < \frac{-3}{5}$$
$$x < -\frac{3}{5}$$

WATCH OUT!
The number you are dividing by is positive 5, so you do not reverse the inequality symbol.

Check
Substitute a number less than $-\frac{3}{5}$ for x, such as -1.

$$6 + x < 3 - 4x$$
$$6 + (-1) \overset{?}{<} 3 - 4(-1)$$
$$5 \overset{?}{<} 3 + 4$$
$$5 < 7 \checkmark$$

Graph all numbers less than $-\frac{3}{5}$.

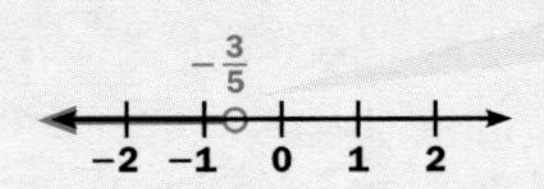

x is less than but not equal to $-\frac{3}{5}$, so you use an open circle.

☑ CHECKING KEY CONCEPTS

Tell whether you would reverse the inequality symbol to solve each inequality. Then solve each inequality.

1. $-3x > 6$ **2.** $4n < -20$ **3.** $a + 4 \geq 5$

4. $\frac{m}{2} \leq 3$ **5.** $\frac{x}{-4} \geq 5$ **6.** $-\frac{c}{3} > 1$

Solve each inequality. Graph each solution on a number line.

7. $x + 9 < 4$ **8.** $\frac{n}{-5} \leq 7$ **9.** $-5t + 6 > 46$

10. $15 \geq 3a - 6$ **11.** $-3x > x - 9$ **12.** $7n + 4 < 9n - 3$

13. In Example 2 on page 177, suppose that Brian pays only \$3 each time he rents skates. When is the cost of renting skates more expensive than buying a new pair?

Think and Communicate

1. Answers may vary. An example is given. Brian could have graphed $y = 4x$ on a graphing calculator and used the trace feature to find the approximate x-value when $y = 45$. From the graph, the solution would be that x-value and all the numbers greater than that value.
2. No; –5 is the boundary value. You need to check which side of the number line to graph: the numbers greater than –5 or the numbers less than –5. It is better to use a number greater than –5 since the algebraic solution is the numbers greater than or equal to –5. Choose 0; $0 \geq -5$.

Checking Key Concepts

1. Yes; $x < -2$. 2. No; $n < -5$. 3. No; $a \geq 1$. 4. No; $m \leq 6$. 5. Yes; $x \leq -20$. 6. Yes; $c < -3$.

7. $x < -5$

–10 –8 –6 –4 –2 0

8. $n \geq -35$

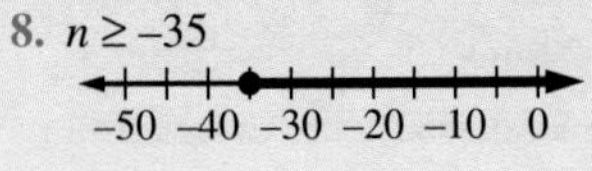

9. $t < -8$

–16 –12 –8 –4 0 4

10. $a \leq 7$

0 2 4 6 8 10

4.6 Exercises and Applications

Solve each inequality. Graph each solution on a number line.

Extra Practice exercises on page 565

1. $x - 5 > 0$ **2.** $-4n \leq 16$ **3.** $10 > k + 9$

4. $10x \geq 7$ **5.** $\frac{c}{-3} > 2$ **6.** $\frac{3}{4}a \leq 21$

7. $8 - x > 8$ **8.** $-7k + 4 < 18$ **9.** $3n + 4 \leq -8$

10. $6 + 4c > -10$ **11.** $15 < 3a - 5$ **12.** $1 - 4a > 21$

13. Malcolm has already invited some people to a party. Now he wants to invite 8 more. His father says, "If you add 8 more people to the guest list, that's over 40 people." Write and solve an inequality based on this situation.

14. **GEOMETRY** A rectangle is 7 cm longer than it is wide. The perimeter is between 30 cm and 50 cm. Find the range of possible widths. Write your answer as an inequality.

Connection TRANSPORTATION

Many people work in areas where public transportation is available. They often have a number of commuting options to choose from. The options shown are for a 36 mi commute.

For Exercises 15–17, use the information with each photograph. Explain your reasoning.

15. For what number of days is it less expensive to commute by train than by subway?

16. When is it less expensive to commute by train than by bus?

17. When is it less expensive to commute by subway than by bus?

18. **Open-ended Problem** Is the least expensive option always the best option? What other things besides cost should you consider when you choose a commuting option?

Walk to the commuter bus stop and pay \$7 per day.

Drive to the subway and pay \$13.50 per day for subway, parking, and car operating costs.

Walk to the train station and pay \$120 per month.

Apply⇔Assess

Suggested Assignment

Core Course
Day 1 Exs. 1–21
Day 2 Exs. 22–33, 35, 37–44, AYP

Extended Course
Day 1 Exs. 1–21
Day 2 Exs. 22–44, AYP

Block Schedule
Day 25 Exs. 1–21
Day 26 Exs. 22–44, AYP

Exercise Notes

Cooperative Learning
Exs. 1–12 Working in groups of three or four, students can share their ideas about how to solve each inequality and draw its graph. Group members can take turns doing each exercise.

Integrating the Strands
Ex. 14 Inequalities are often very helpful in analyzing geometric situations, as is illustrated by this problem involving the range of possible widths of a rectangle given certain conditions placed upon its length and perimeter.

Second-Language Learners
Ex. 18 Students should prepare written examples of when particular options are most beneficial, and then orally explain each one.

11. $x < 3$

0 1 2 3 4 5

12. $n > \frac{7}{2}$

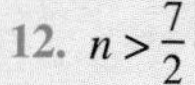

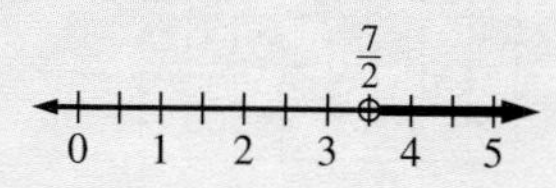

13. if he rents skates 16 or more times

Exercises and Applications

1. $x > 5$

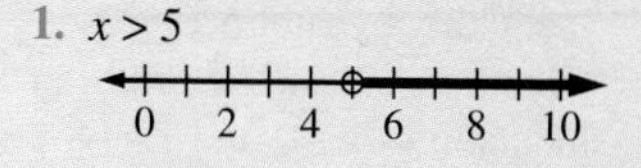

2. $n \geq -4$

−8 −6 −4 −2 0 2

3. $k < 1$

−4 −2 0 2 4 6

4. $x \geq 0.7$

0 0.2 0.4 0.6 0.8 1

5. $c < -6$

−8 −6 −4 −2 0 2

6. $a \leq 28$

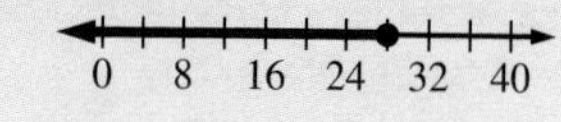

7. $x < 0$

8. $k > -2$

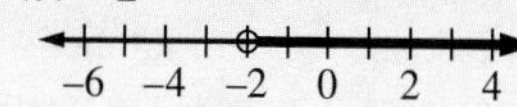

9. $n \leq -4$

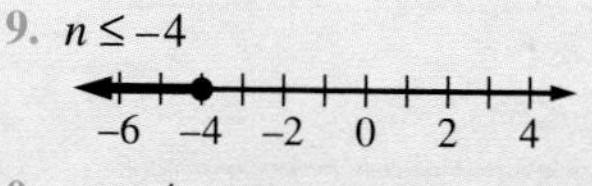

10. $c > -4$

−6 −4 −2 0 2 4

11. $a > \frac{20}{3}$

0 2 4 6 8 10

12. $a < -5$

−10 −8 −6 −4 −2 0

13. Let g be the number of guests already invited; $g + 8 > 40$; $g > 32$; more than 32 people have already been invited.

14. $4 < w < 9$, where w = the width in centimeters

15. 9 or more days

16. 18 or more days

17. never

18. See answers in back of book.

Apply⇔Assess

Exercise Notes

Using Technology

Ex. 36 Students can see a geometric interpretation for the solution to $x + 3 \le 5$ by using a graphing calculator. They should first graph $y = x + 3 \le 5$. The graph is the ray corresponding to $x \le 2$ at the $y = 1$ line. The graph $y = 1$ for an inequality graph represents the logical condition *true*. When $x + 3 > 5$, $y = 0$ is graphed (the logical condition *false*), but the graph may not show on the screen because it is on the x-axis.

Match each inequality with its solution.

19. $3x + 4 < 7$ **20.** $3x + 4 < 3x$ **21.** $3x + 4 < 3(x + 4)$

A. all numbers

B. $x < 1$

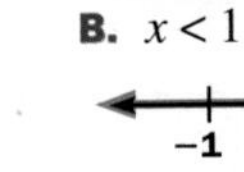

C. no solution

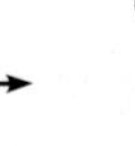

Solve each inequality. Graph each solution on a number line. If there is no solution, write *no solution*.

22. $12x - 36 < 3x$ **23.** $7n > 12n - 5$ **24.** $2.5n - 14 < -12.75$

25. $-\frac{2}{5}c \ge \frac{3}{10}$ **26.** $\frac{5}{2}x - 1 > 14$ **27.** $-\frac{3}{2}t - 1 \le 5$

28. $-6a - 16 \le 6a + 8$ **29.** $5a < 2(a - 9)$ **30.** $3(5x - 4) < 15x$

31. $2(1 - x) \ge 2x$ **32.** $5(n + 1) < 5n$ **33.** $-6k > -2(3k + 3)$

34. Writing How is solving an inequality like solving an equation? How is it different?

35. BUSINESS Veronica Kelly is a salesperson in a large department store. She receives a yearly salary of \$19,500. In addition, she makes a 5% commission on all sales. Last year, Veronica made more than \$43,000. Write and solve an inequality based on this situation.

36. Challenge Solve the inequalities $x + 3 \le 5$ and $x + 3 \ge -5$. Explain how you can use the solutions to solve the inequality $|x + 3| \le 5$.

ONGOING ASSESSMENT

37. Cooperative Learning Work with a partner. Discuss how you could solve the inequality $5 - \frac{1}{2}x > 4 - \frac{1}{4}x$ by graphing the equations $y = 5 - \frac{1}{2}x$ and $y = 4 - \frac{1}{4}x$ in the same coordinate plane. Also discuss how you would solve the inequality algebraically. Then solve the inequality both ways.

SPIRAL REVIEW

Copy and complete each equation. *(Section 3.1)*

38. $\frac{\$10.25}{1 \text{ page}} \cdot \frac{5 \text{ pages}}{1 \text{ h}} = \underline{?}$

39. $\frac{2 \text{ tokens}}{1 \text{ trip}} \cdot \underline{?} = 56 \text{ tokens}$

40. $\frac{4 \text{ quarters}}{1 \text{ wash}} \cdot \frac{\$1}{4 \text{ quarters}} \cdot \frac{6 \text{ washes}}{1 \text{ day}} = \underline{?}$

41. $\frac{60 \text{ mi}}{1 \text{ h}} \cdot \underline{?} \cdot \frac{1 \text{ h}}{3600 \text{ s}} = \underline{?} \text{ ft/s}$

Find the slope of the line through each pair of points. *(Section 3.3)*

42. (5, 1) and (4, 4) **43.** (−2, 5) and (7, 1) **44.** (1, 8) and (1, 0)

19. B 20. C 21. A

22. $x < 4$

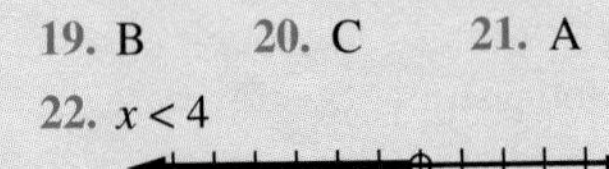

23. $n < 1$

24. $n < 0.5$

25. $c \le -\frac{3}{4}$

26. $x > 6$

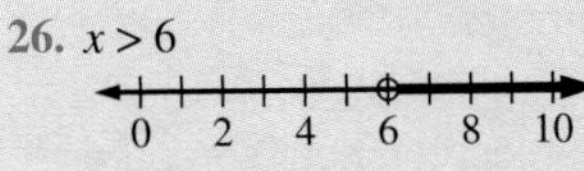

27. $t \ge -4$

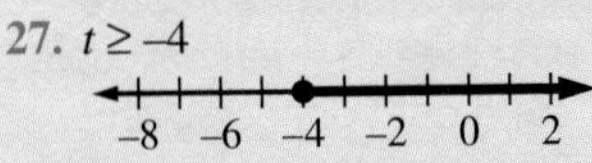

28. $a \ge -2$

29. $a < -6$

30. identity

31. $x \le \frac{1}{2}$

32. no solution

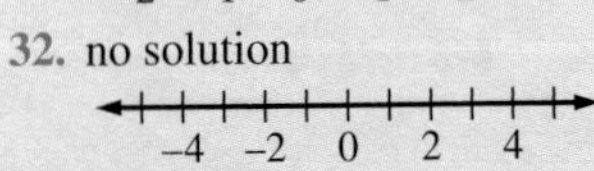

33. identity

34. Answers may vary. An example is given. When solving an equation or an inequality, you can add the same number to both sides or subtract the same number from both sides without changing the solution. You can also multiply or divide both sides by the same positive number. However, you can multiply or divide both sides of an equation by the same negative number without changing the solution, but you cannot do so with an inequality. In this case, you must reverse the direction of the inequality.

35. Let s be Veronica's annual sales. $19,500 + 0.05s > 43,000$; $s > 470,000$; Her annual sales last year were more than \$470,000.

36. To solve $x + 3 \le 5$, subtract 3 from both sides to get $x \le 2$. To solve $|x + 3| \le 5$, use the definition of absolute value to rewrite the inequality as a pair of inequalities, $x + 3 \le 5$ and $x + 3 \ge -5$. Solve each inequality by subtracting 3 from both sides. The resulting inequalities $x \le 2$ and $x \ge -8$ can be combined in a single inequality, $-8 \le x \le 2$. The solution of $x + 3 \le 5$ is all numbers less than or equal to 2; the solution of $|x + 3| \le 5$ is all numbers greater than or equal to −8 and less than or equal to 2.

37.

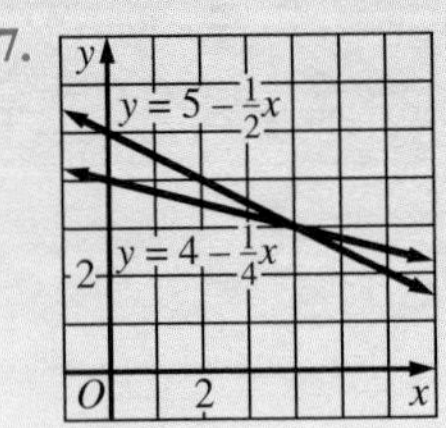

The graphs intersect at (4, 3), and the graph of $y = 5 - \frac{1}{2}x$ is above the graph of $y = 4 - \frac{1}{4}x$ when $x < 4$, so the solution is

ASSESS YOUR PROGRESS

VOCABULARY

inequality (p. 172) **solution of an inequality** (p. 177)

Solve each equation. *(Section 4.4)*

1. $\frac{x}{5} = \frac{3}{2}$ **2.** $\frac{3}{4}a - \frac{2}{3} = \frac{5}{6}$ **3.** $\frac{3}{8} - \frac{n}{2} = 1$

4. $0.75t + 0.15 = 1.65$ **5.** $5x - 0.5 = 5.75$ **6.** $2.208 - 0.09n = 0.6n$

7. SPORTS When you hit a baseball near either end of a bat, it vibrates. It stings your hands and the ball does not travel far. But if you hit the ball near the bat's *node*, the bat vibrates less and the ball travels farther.

Suppose a pitcher throws a ball at 85 mi/h and you swing the bat at about 70 mi/h. The graph shows the distance the ball will travel, based on where the bat hits the ball. *(Section 4.5)*

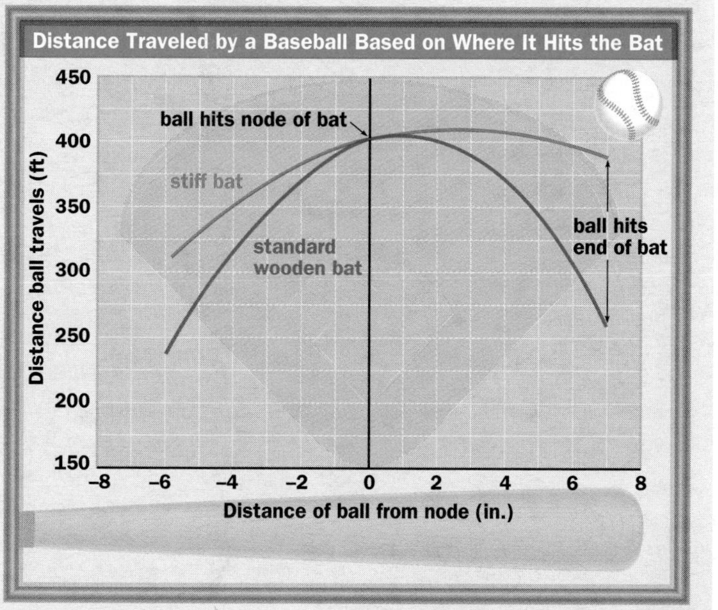

a. With a stiff bat, how far from the node should you hit the ball so that it travels at least 350 ft? Write your answer as an inequality.

b. Repeat part (a) for a standard wooden bat.

Solve each inequality. Graph each solution on a number line. *(Section 4.6)*

8. $n + 8 > 5$ **9.** $\frac{x}{-5} \geq 8$ **10.** $3 - 2a < 7$

11. $10t - 4 \leq 2t$ **12.** $6k + 9 \geq k - 1$ **13.** $4(c + 2) < 4c + 10$

14. Journal Write a summary of the various methods you have learned for solving equations and inequalities.

BY THE WAY

Aluminum bats, used almost exclusively in amateur baseball, were first made in the early 1970s. They weigh 2–4 oz less than wooden bats and are usually filled with a sound-deadening insert such as cork.

Apply⇔Assess

Assess Your Progress

The following class activity can be used to assess students' understanding of the steps involved in solving equations and inequalities. Have one student write a problem on the board; another student can describe the steps needed to solve the problem; a third student can write out the solution step by step; the next student can check the solution. Start again with a new problem. Continue this pattern until all students have had a chance to participate.

Progress Check 4.4–4.6

See page 185.

Practice 29 for Section 4.6

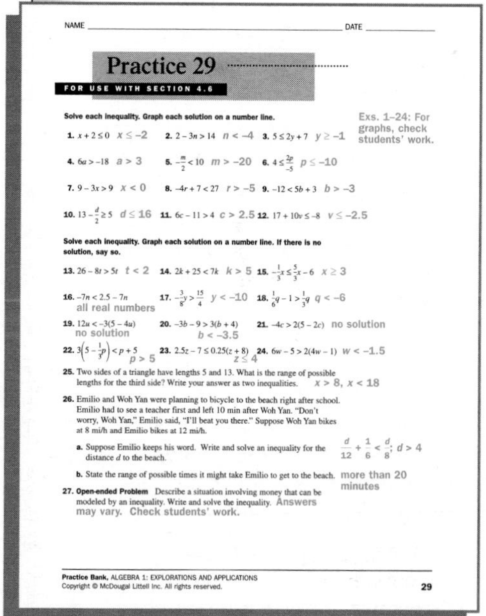

NAME ______ DATE ______

Practice 29

FOR USE WITH SECTION 4.6

Solve each inequality. Graph each solution on a number line. Exs. 1–24: For graphs, check students' work.

1. $x + 2 \leq 0$ $x \leq -2$ **2.** $2 - 3n > 14$ $n < -4$ **3.** $5 \leq 2y + 7$ $y \geq -1$

4. $6a > -18$ $a > 3$ **5.** $-\frac{m}{2} < 10$ $m > -20$ **6.** $4 \leq \frac{2p}{-5}$ $p \leq -10$

7. $9 - 3x > 9$ $x < 0$ **8.** $-4r + 7 < 27$ $r > -5$ **9.** $-12 < 5b + 3$ $b > -3$

10. $13 - \frac{d}{2} \geq 5$ $d \leq 16$ **11.** $6c - 11 > 4$ $c > 2.5$ **12.** $17 + 10v \leq -8$ $v \leq -2.5$

Solve each inequality. Graph each solution on a number line. If there is no solution, say so.

13. $26 - 8t > 5t$ $t < 2$ **14.** $2k + 25 < 7k$ $k > 5$ **15.** $-\frac{1}{3}x \leq \frac{5}{3}x - 6$ $x \geq 3$

16. $-7n < 2.5 - 7n$ all real numbers **17.** $-\frac{3}{8}y > \frac{15}{4}$ $y < -10$ **18.** $\frac{1}{6}q - 1 > \frac{1}{3}q$ $q < -6$

19. $12u < -3(5 - 4u)$ no solution **20.** $-3b - 9 > 3(b + 4)$ $b < -3.5$ **21.** $-4c > 2(5 - 2c)$ no solution

22. $3\left(5 - \frac{1}{3}p\right) < p + 5$ $p > 5$ **23.** $2.5z - 7 \leq 0.25(z + 8)$ $z \leq 4$ **24.** $6w - 5 > 2(4w - 1)$ $w < -1.5$

25. Two sides of a triangle have lengths 5 and 13. What is the range of possible lengths for the third side? Write your answer as two inequalities. $x > 8$, $x < 18$

26. Emilio and Woh Yan were planning to bicycle to the beach right after school. Emilio had to see a teacher first and left 10 min after Woh Yan. "Don't worry, Woh Yan," Emilio said, "I'll beat you there." Suppose Woh Yan bikes at 8 mi/h and Emilio bikes at 12 mi/h.

a. Suppose Emilio keeps his word. Write and solve an inequality for the distance d to the beach. $\frac{d}{12} + \frac{1}{6} < \frac{d}{8}$; $d > 4$

b. State the range of possible times it might take Emilio to get to the beach. more than 20 minutes

27. Open-ended Problem Describe a situation involving money that can be modeled by an inequality. Write and solve the inequality. Answers may vary. Check students' work.

Practice Bank, ALGEBRA 1: EXPLORATIONS AND APPLICATIONS

29

$x < 4$. To solve algebraically, multiply both sides by 4 to get $20 - 2x > 16 - x$; $4 > x$, or $x < 4$.

38. \$51.25/h

39. 28 trips

40. \$6/day

41. 5280 ft/mi; 88

42. –3

43. $-\frac{4}{9}$

44. undefined

Assess Your Progress

1. $\frac{15}{2}$ 2. 2

3. $-\frac{5}{4}$ 4. 2

5. 1.25 6. 3.2

7. Let d be the distance of the ball from the node.

a. $-4 \leq d \leq 7$

b. $-2.8 \leq d \leq 4.8$

8. $n > -3$

9. $x \leq -40$

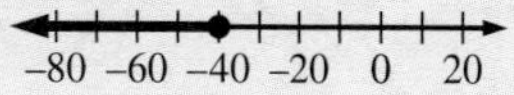

10. $a > -2$

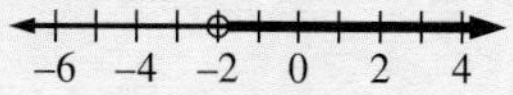

11. $t \leq \frac{1}{2}$

12. $k \geq -2$

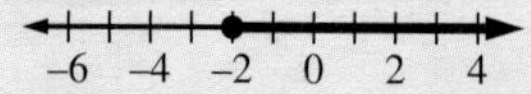

13. identity

14. Answers may vary. An example is given. You can solve an equation by using algebra tiles, inverse operations, tables, and graphs. You can solve inequalities by using graphs and algebraic methods.

Project Notes

Mathematical Goals

- Calculate expenses and income for making a batch of cookies.
- Use a table and graph to model expenses and income as functions of the number of batches sold.
- Use the table and graph to find the point at which a profit is made.
- Write and solve an inequality to find the point at which a profit is made.

Planning

Materials
- Cookie recipe
- Graph paper
- Writing paper

Project Teams
Allow students to choose their own partners. They should then read the project's goal and decide how they want to proceed.

Guiding Students' Work

Suggest that each student bring one cookie recipe to class. You may also wish to have a few recipes available for use by students who cannot find their own. Students can share the cost of a bag of sugar, but they should determine themselves the number of cups per bag. Students should share evenly the other tasks involved in doing the project. For example, one student can make a table while the partner creates the graph. Both students should write parts of the final report.

Second-Language Learners
Students learning English may benefit from working with a peer tutor to plan, write, and refine their report and their self-assessment paragraph.

PORTFOLIO PROJECT

Making a Profit

Have you ever wondered how prices are determined? People who make a product must determine how much money to charge so that they make a *profit*. Profit is the money remaining after expenses are subtracted from income.

PROJECT GOAL **Determine the expenses of making cookies and decide on a selling price so that you earn a profit.**

Financial Plans

Work with a partner. Choose a cookie recipe you want to make.

1. CALCULATE EXPENSES

- **INGREDIENTS**
  Calculate or estimate how much the ingredients in one batch of cookies cost. For example:

  Cost of 1 cup of sugar =

  $$\frac{\text{Cost of a bag}}{\text{Cups per bag}} = \frac{\$1.89}{10} \approx \$.19$$

- **TIME**
  Calculate the cost of your time to make one batch.

  $$\text{Labor costs} = \text{Hours worked} \times \text{Hourly pay}$$

- **FIXED EXPENSES**
  Assume that you must pay $25 for a permit to sell cookies.

2. CALCULATE INCOME Research a reasonable price to charge for selling your cookies individually. How much income can you make per batch?

Analyzing Your Data

Use a table and a graph to model expenses and income as functions of the number of batches sold. What is the point at which you make a profit? Write and solve an inequality to show another method of finding the point at which you make a profit.

Making a Report

Write a report about your cookie-making venture. Describe the decisions that you made concerning ingredients, labor costs, and pricing. Include your table, your graph, and a discussion of income and expenses.

You may want to extend your report by examining some of the ideas below:

- Survey friends and neighbors to find the price they would be willing to pay for a cookie. Display that data and any conclusions you can make from the data.
- How do cookie manufacturers' prices differ from your price? Why do you think this is so? How are their expenses different from yours?

Self-Assessment

Write a paragraph explaining your decisions. Which method, making a table, making a graph, or writing an inequality, works best for you?

Describe how you did research. For example, how did you find out what the minimum wage is or what local businesses pay workers for similar work?

What was the most difficult aspect of the project? the easiest? Why?

Project Notes

Guiding Students' Work

Rubric for Chapter Project

4 Students calculate their expenses and income accurately and charge a reasonable price for their cookies. They construct correct tables and graphs and determine their profit-making point. They are successful in writing and solving an inequality that also finds the profit-making point. The final report includes all the features listed in the text and is clearly organized, well written, and interesting to read.

3 Students complete the initial activities involving the financial plans and analysis of data correctly. Their final report, however, has a few minor errors and is incomplete in some way, for example, the discussion of the decisions made concerning ingredients, labor costs, and pricing.

2 Students complete all of the activities, but some errors and misconceptions permeate the entire project. Their table or graph does not clearly show the profit-making point. They do not write an inequality to show another method of finding the profit-making point. The final written report is hastily written and has some serious omissions.

1 Students start the project but are confused about how to set a price to make a profit. Their table and graph reflect this confusion and no profit-making point is determined. They cannot write a correct inequality and no report is attempted. Students should be encouraged to speak with the teacher as soon as possible to review their work and to make a new start on the project.

Chapter Support

Course Guide: Chapter 4
Lesson Plans: Chapter 4
Practice Bank:
Cumulative Practice 30
Study Guide: Chapter 4 Review
Assessment Book:
Chapter Tests 17 and 18
Chapter 4 Alternative Assessment
Test Generator software
Portfolio Project Book:
Additional Project 4: Statistics
Preparing for College Entrance Tests
Professional Handbook

Progress Check 4.1–4.3

Use a table or graph. *(Section 4.1)*

1. At Central Bank, customers pay a monthly fee of \$5.25 plus \$.25 per transaction. At West Side Bank, customers pay \$4.00 per month plus \$.30 per transaction. When will the total monthly cost at West Side Bank be higher than at Central Bank?
after 25 transactions

Solve each equation. *(Section 4.2)*

2. $\frac{3}{5}y = 12$ 20

3. $-\frac{3}{4}x + 6 = -18$ 32

4. $24 = \frac{-2n}{9} - 2$ −117

Solve each equation. If there is no solution, write *no solution.* *(Section 4.3)*

5. $4a = 2a - 12$ −6

6. $-3x + 8 = 7x + 2$ $\frac{3}{5}$

7. $8y - 6 = 2(4y - 5)$ no solution

CHAPTER

4 Review

STUDY TECHNIQUE

The diagram at the right is the start of a **concept map**. A map like this can help you review new words and ideas. Keep these things in mind when you make a concept map:

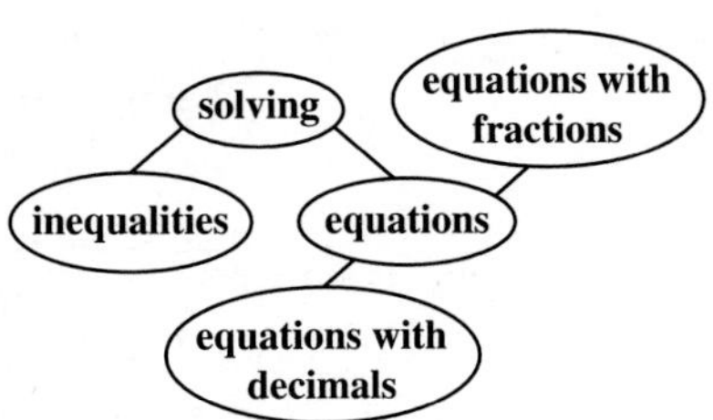

- **A group of ideas can be summarized in more than one way.**
- **Lines should connect related concepts.**
- **Concepts should be arranged to emphasize relationships.**

Expand this concept map to review the key ideas in this chapter.

VOCABULARY

reciprocals (p. 153)
identity (p. 161)
inequality (p. 171)
solution of an inequality (p. 177)

SECTIONS 4.1, 4.2, *and* 4.3

Tables, graphs, and equations can be used to compare different situations.

For example, should you pay \$7 for each visit to a museum or pay \$50 for an annual membership? The models below show that 8 or more visits in a year cost more than the annual \$50 fee.

Number of visits	Total cost (\$) With membership	Total cost (\$) Without membership
1	50	7
2	50	14
3	50	21
4	50	28
5	50	35
6	50	42
7	50	49
8	50	56

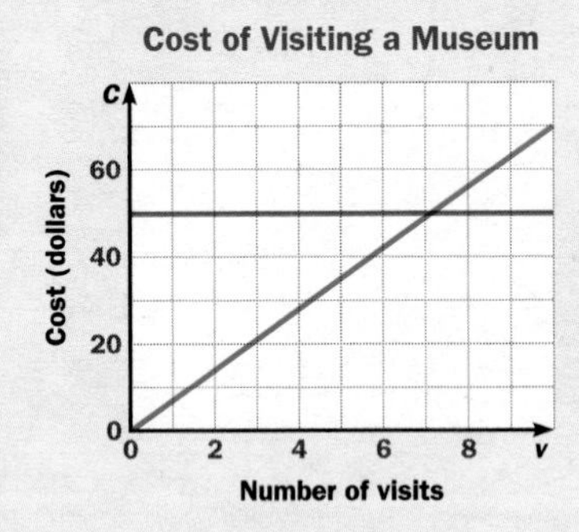

Let v = the number of visits.

$7v = 50$

$v \approx 7.14$

You can use *reciprocals* to solve equations with fractional coefficients.

$$\frac{5}{8}x = 10$$

The numbers x and $\frac{1}{x}$ are **reciprocals** because their product is one. Zero does not have a reciprocal.

$$\frac{8}{5} \cdot \frac{5}{8}x = \frac{8}{5} \cdot 10$$

$$x = 16$$

Multiply both sides of the equation by the **reciprocal** of $\frac{5}{8}$.

To solve multi-step equations with parentheses:

1. Simplify each side of the equation.
$$5(x + 3) = 3 + 7x - 14$$
$$5x + 15 = -11 + 7x$$
2. Get all the variable terms on one side.
$$5x + 15 - 5x = -11 + 7x - 5x$$
$$15 = -11 + 2x$$
3. Get the variable alone on one side.
$$15 + 11 = -11 + 2x + 11$$
$$26 = 2x$$
$$13 = x$$

An **identity** is an equation that is true for all values of the variable. An equation has no solution when no value of the variable will make the equation true.

SECTIONS 4.4, 4.5, *and* 4.6

You can solve equations with fractions by **multiplying both sides by the least common denominator**.

$$\frac{3}{4}x + 5 = \frac{2}{3}$$
$$12\left(\frac{3}{4}x + 5\right) = 12\left(\frac{2}{3}\right)$$
$$9x + 60 = 8$$
$$9x = -52$$
$$x = -5\frac{7}{9}$$

You can eliminate decimals in an equation by **multiplying both sides by a power of ten**.

$$3.65x - 0.6 = 3.45x$$
$$100(3.65x - 0.6) = 100(3.45x)$$
$$365x - 60 = 345x$$
$$-60 = -20x$$
$$3 = x$$

Subtract $365x$ from both sides.

An **inequality** is a statement formed by placing an inequality symbol between mathematical expressions. For the museum example on page 184, you can use the inequality $v \geq 8$ to represent the interval where the total cost for a nonmember is more than the cost for a member.

When you find the **solution of an inequality**, remember to **reverse** the inequality symbol when you multiply or divide both sides by a **negative** number.

$$-3x < 12$$
$$\frac{-3x}{-3} > \frac{12}{-3}$$
$$x > -4$$

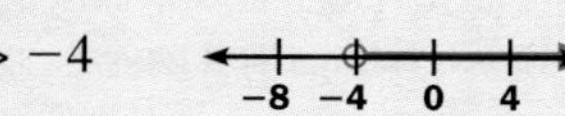

Progress Check 4.4–4.6

Solve each equation. *(Section 4.4)*

1. $\frac{2x}{3} - \frac{x}{4} = 10$ 24
2. $3.5y + 5.4 = 45.16 - 2.1y$ 7.1
3. $\frac{4}{5}z - \frac{7}{2} = \frac{1}{10}z + 3$ $\frac{65}{7}$

Use the graph below for Exs. 4–6. *(Section 4.5)*

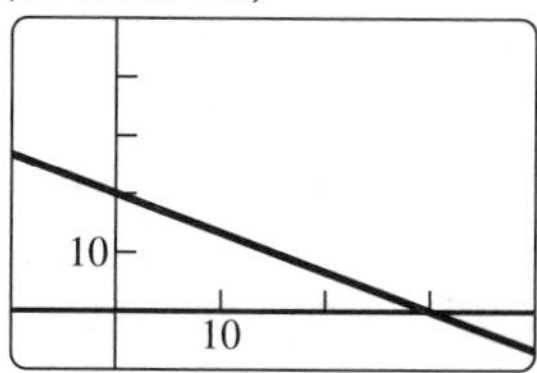

4. If $y > 0$, what are possible values of x? $x < 40$
5. If $0 < y < 20$, what are possible values of x? $0 < x < 40$
6. If $y \leq 0$, what are possible values of x? $x \geq 40$

Solve each inequality. Graph each solution on a number line. *(Section 4.6)*

7. $3a \leq 5a - 10$ $a \geq 5$

0 2 4 6 8

8. $-2x + 8 > 7x - 10$ $x < 2$

-4 -2 0 2 4

9. $4y - 6 < 2(2y - 5)$ no solution

Chapter 4 Assessment
Form A Chapter Test

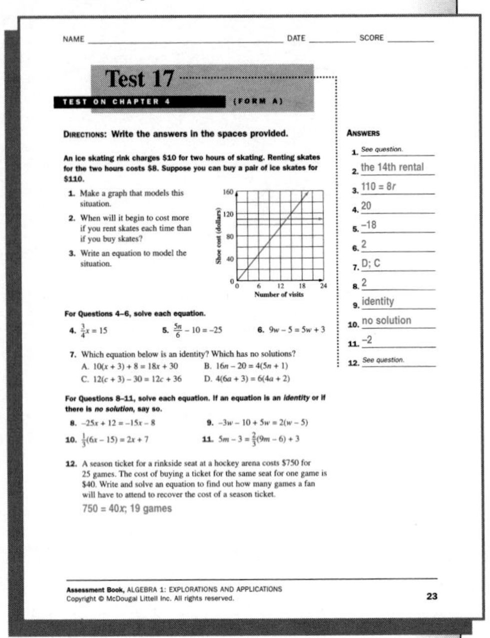
NAME ______ DATE ______ SCORE ______

Test 17
TEST ON CHAPTER 4 (FORM A)

DIRECTIONS: Write the answers in the spaces provided.

An ice skating rink charges $10 for two hours of skating. Renting skates for the two hours costs $8. Suppose you can buy a pair of ice skates for $110.

1. Make a graph that models this situation.
2. When will it begin to cost more if you rent skates each time than if you buy skates?
3. Write an equation to model the situation.

For Questions 4–6, solve each equation.

4. $\frac{3}{4}x = 15$ 5. $\frac{5n}{6} - 10 = -25$ 6. $9w - 5 = 5w + 3$

7. Which equation below is an identity? Which has no solutions?
A. $10(x + 3) + 8 = 18x + 30$ B. $16n - 20 = 4(5n + 1)$
C. $12(c + 3) - 30 = 12c + 36$ D. $4(6a + 3) = 6(4a + 2)$

For Questions 8–11, solve each equation. If an equation is an *identity* or if there is *no solution*, say so.

8. $-25x + 12 = -15x - 8$ 9. $-3w - 10 + 5w = 2(w - 5)$
10. $\frac{1}{3}(6x - 15) = 2x + 7$ 11. $5m - 3 = \frac{2}{3}(9m - 6) + 3$

12. A season ticket for a rinkside seat at a hockey arena costs $750 for 25 games. The cost of buying a ticket for the same seat for one game is $40. Write and solve an equation to find out how many games a fan will have to attend to recover the cost of a season ticket.
$750 = 40x$; 19 games

ANSWERS
1. See question.
2. the 14th rental
3. $110 = 8r$
4. 20
5. –18
6. 2
7. D; C
8. 2
9. identity
10. no solution
11. –2
12. See question.

Assessment Book, ALGEBRA 1: EXPLORATIONS AND APPLICATIONS

23

Chapter 4 Assessment
Form B Chapter Test

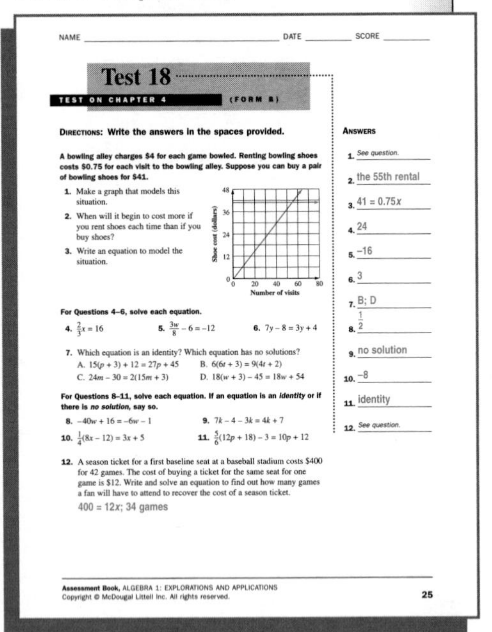
NAME ______ DATE ______ SCORE ______

Test 18
TEST ON CHAPTER 4 (FORM B)

DIRECTIONS: Write the answers in the spaces provided.

A bowling alley charges $4 for each game bowled. Renting bowling shoes costs $0.75 for each visit to the bowling alley. Suppose you can buy a pair of bowling shoes for $41.

1. Make a graph that models this situation.
2. When will it begin to cost more if you rent shoes each time than if you buy shoes?
3. Write an equation to model the situation.

For Questions 4–6, solve each equation.

4. $\frac{2}{3}x = 16$ 5. $\frac{3w}{8} - 6 = -12$ 6. $7y - 8 = 3y + 4$

7. Which equation is an identity? Which equation has no solutions?
A. $15(p + 3) + 12 = 27p + 45$ B. $6(6t + 3) = 9(4t + 2)$
C. $24m - 30 = 2(15m + 3)$ D. $18(w + 3) - 45 = 18w + 54$

For Questions 8–11, solve each equation. If an equation is an *identity* or if there is *no solution*, say so.

8. $-40w + 16 = -6w - 1$ 9. $7k - 4 - 3k = 4k + 7$
10. $\frac{1}{4}(8x - 12) = 3x + 5$ 11. $\frac{5}{6}(12p + 18) - 3 = 10p + 12$

12. A season ticket for a first baseline seat at a baseball stadium costs $400 for 42 games. The cost of buying a ticket for the same seat for one game is $12. Write and solve an equation to find out how many games a fan will have to attend to recover the cost of a season ticket.
$400 = 12x$; 34 games

ANSWERS
1. See question.
2. the 55th rental
3. $41 = 0.75x$
4. 24
5. –16
6. 3
7. B; D
8. $\frac{1}{2}$
9. no solution
10. –8
11. identity
12. See question.

Assessment Book, ALGEBRA 1: EXPLORATIONS AND APPLICATIONS

25

CHAPTER

4 Assessment

VOCABULARY

Complete each statement.

1. The product of a number and its _?_ is 1.
2. All numbers are solutions of a(n) _?_.

SECTIONS 4.1, 4.2, *and* 4.3

3. A public golf course charges $25 for an 18-hole round of golf. Renting clubs costs an additional $10 per round. Suppose you can buy a set of golf clubs for $525. Make a graph that models this situation. When will it begin to cost more if you rent clubs each time than if you buy clubs?
4. Write an equation to model the point at which the two options in Question 3 cost the same amount.

Find the reciprocal of each number.

5. $\frac{3}{5}$ 6. $-\frac{7}{8}$ 7. $\frac{3}{2}$

Solve each equation.

8. $\frac{3}{5}m = 15$ 9. $\frac{2x}{3} - 8 = 4$ 10. $-17 = 19 - \frac{6}{7}a$
11. $8a = 2a + 54$ 12. $4y - 3 = 6y + 7$ 13. $2 - 7n = 8n - 3$

14. Which equation is an identity? Which equation has no solution?
A. $2(6a + 3) = 3(4a + 2)$ **B.** $5(x + 3) + 4 = 9x + 15$
C. $8x - 10 = \frac{2}{3}(15x + 3)$ **D.** $6(c + 3) - 15 = 6c + 18$
15. Solve the equation $3(x + 2) = 5(x - 3) + 13$. Explain each step.

Solve. If an equation is an *identity* or there is *no solution*, say so.

16. $-30x + 15 = -6x + 31$ 17. $5t - 6 - 8t = -3(t + 2)$
18. $\frac{1}{2}(6x - 8) = 3x - 2$ 19. $4a - 2 = \frac{4}{5}(10a - 5) - 6$

20. A season ticket for an orchestra seat at a theater costs $140 for six performances. The cost of a ticket for the same seat for one performance is $35. Write and solve an equation to find out how many performances Rafael Santos will have to attend to recover the cost of a season ticket.

ANSWERS Chapter 4

Assessment

1. reciprocal 2. identity

3. Golf Costs

C; Cost (dollars): 0, 800, 1600; Number of rounds: 0, 20, 40, 60, n; Buy clubs; Rent clubs

Estimates may vary; after about 55 rounds.

4. Let n = the number of rounds and C = the cost in dollars; $35n = 25n + 525$.

5. $\frac{5}{3}$ 6. $-\frac{8}{7}$
7. $\frac{2}{3}$ 8. 25
9. 18 10. 42
11. 9 12. –5
13. $\frac{1}{3}$ 14. A; D

15. Answers may vary. An example is given. First use the distributive property to simplify both sides: $3x + 6 = 5x - 15 + 13$; $3x + 6 = 5x - 2$. Next subtract $3x$ from both sides to get the variables on one side of the equation: $3x + 6 - 3x = 5x - 2 - 3x$; $6 = 2x - 2$. Add 2 to both sides: $6 + 2 = 2x - 2 + 2$; $8 = 2x$. Finally, divide both sides by 2: $\frac{8}{2} = \frac{2x}{2}$; $x = 4$.

SECTIONS 4.4, 4.5, *and* 4.6

21. What power of ten would you use to eliminate the decimal numbers in the equation $0.02n + 3.5 = 8.067$ most easily?

22. What is the smallest number that you can multiply by to eliminate the fractions in the equation $\frac{4}{5}x + \frac{2}{9} = \frac{1}{6} - \frac{4}{15}x$?

Solve each equation.

23. $\frac{2}{5}x + 2 = \frac{2}{3} - 2x$ **24.** $\frac{4}{5}m + \frac{3}{4} = \frac{7}{10}m + \frac{1}{4}$

25. $0.2n + 3 = 5.4 - n$ **26.** $0.7 - 0.8x = 0.18 + 0.24x$

The graph shows the fuel economy of an average car as a function of speed. Use the graph for Exercises 27 and 28.

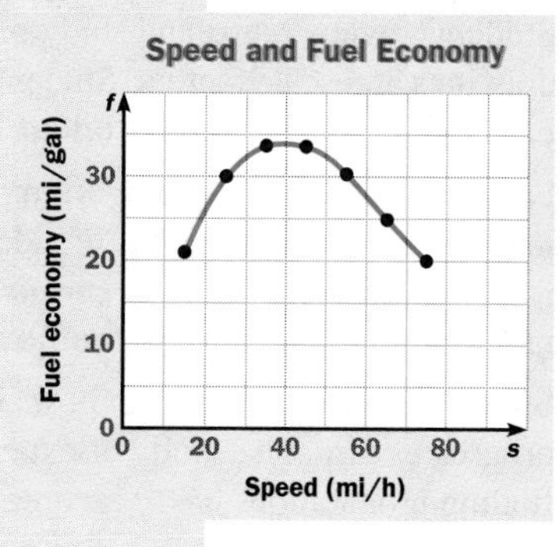

27. At what speeds is a car's fuel economy less than 25 mi/gal? Write your answer as two inequalities.

28. At what speeds is a car's fuel economy more than 30 mi/gal? Write your answer as an inequality.

29. The statement $10 > 2$ is a true inequality. Is the inequality still true if both sides of the inequality are multiplied by 7? by -3? by $\frac{1}{2}$? by $-\frac{5}{2}$? Explain.

30. If $m < r$, which of the following inequalities is not true?

A. $m + 7 < r + 7$ **B.** $-3m < -3r$ **C.** $m - 3 < r - 3$ **D.** $\frac{m}{5} < \frac{r}{5}$

Solve each inequality. Graph each solution on a number line.

31. $c - 3 > 4$ **32.** $2x \leq -8$ **33.** $-3f \leq -27$

34. $7 - 4x \leq -5$ **35.** $-\frac{2}{5}x + 2 \geq 8$ **36.** $5x + 3 < 2x - 6$

PERFORMANCE TASK

37. Find something that can be paid for in two different ways, such as admission to a museum. Compare the costs. Explain the circumstances under which each option is better. Justify your explanation with a table, a graph, or an equation. Use inequalities to express your findings.

PLIMOTH PLANTATION RATES

INDIVIDUALS:	
Adult	\$15.00
Child 5–12	\$9.00
ANNUAL PASS: *Valid for one full year from date of purchase for an unlimited number of visits*	
Adult	\$25.00
Child 5–12	\$12.00
1 YEAR MEMBERSHIP:	
Family	\$70.00
Individual	\$35.00

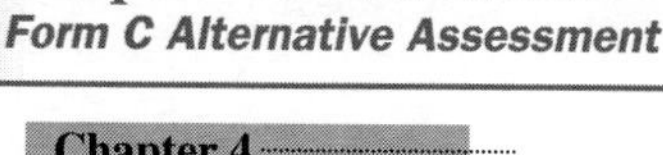

Chapter 4 Assessment
Form C Alternative Assessment

Chapter 4
ALTERNATIVE ASSESSMENT

1. a. Solve the inequality $4x + 2 \leq 2(x - 1)$. Graph the solution on a number line.
   b. Explain how to check your answer.
   c. Name three values that would be appropriate to use when checking your solution.
   d. Compare the procedure for checking the answer to an inequality with the procedure for checking the answer to an equation.
2. a. Write an inequality that has one solution.
   b. Write an inequality that has no solutions.
   c. Write an inequality for which all numbers are solutions.
3. a. Create a table of x- and y-values for the function $y = 3x - 1$. Create a table of x- and y-values for the function $y = -2x + 24$.
   b. Use the two tables from part (a) to discuss the solution of the equation $3x - 1 = -2x + 24$.
   c. Use the two tables from part (a) to discuss the solutions of the inequality $3x - 1 \leq -2x + 24$.
4. **Open-ended Problem** Write three other inequalities that are equivalent to $x > -5$.
5. **Project** Call a nearby amusement park or skating rink that offers a season pass. Using a graph or table, compare the cost of the season pass to the single admission charge. Write an article for the school newspaper discussing the advantages and disadvantages of buying a season pass. In your article, include the number of visits you would need to make in order for the season pass to be cost-effective.
6. **Group Activity** Use tables to solve the equation $4x + \frac{1}{3} = 3x + 1$. (*Hint:* If the solution is not immediately evident, sometimes a graph will help.) What decisions did your group make in trying to find a solution? How can your group be assured that the solution is correct? How do you know that there is only one solution?

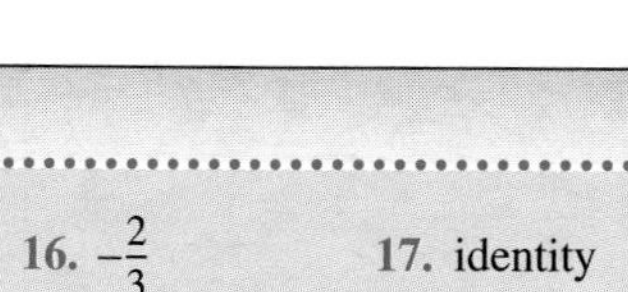

16. $-\frac{2}{3}$ 17. identity
18. no solution 19. 2
20. Let n = the number of performances; $35n = 140$; $n = 4$.
21. $10^3 = 1000$ 22. 90
23. $-\frac{5}{9}$ 24. -5
25. 2 26. 0.5
27. Let s = the speed of the car in mi/h. Estimates may vary; $15 \leq s < 20$ and $65 < s \leq 75$.
28. Let s = the speed of the car in mi/h. Estimates may vary; $25 < s < 55$.
29. Yes; No; Yes; No; if both sides of a true inequality are multiplied by the same positive number, the result is a true inequality. If both sides of a true inequality are multiplied by the same negative number, the result is not a true inequality. The order of the inequality sign must be reversed to make the resulting inequality true.

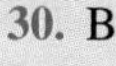

30. B

31.
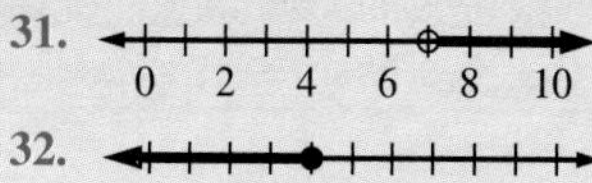

32.

33.
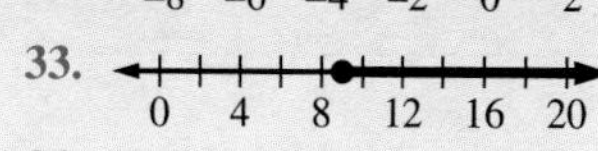

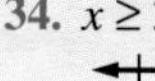

34. $x \geq 3$
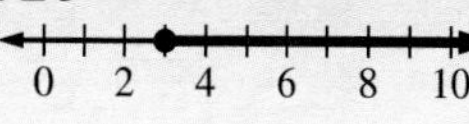

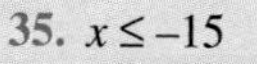

35. $x \leq -15$
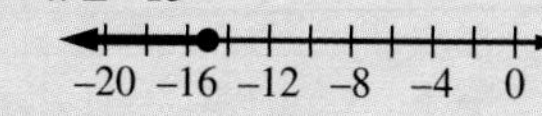

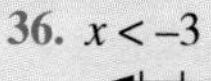

36. $x < -3$
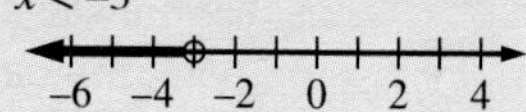

37. Answers may vary. Check students' work.

CHAPTER

5 Connecting Algebra and Geometry

OVERVIEW

Connecting to Prior and Future Learning

⇔ Students review ratios and study proportions and then use proportions to work with scale drawings and scale factors. Students can review ratios, fractions, decimals, and percents on pages 586 and 587 in the **Student Resources Toolbox**.

⇔ Proportions are applied to working with similar geometric figures. Students find missing measures, including lengths of sides, measures of angles, perimeters, and areas. Students can review angles on page 595, perimeter and circumference on page 593, and area on page 594 in the **Student Resources Toolbox**.

⇔ Students begin to work with probability. The work begins with a die rolling experiment, progresses through a study of experimental and theoretical probabilities, and concludes with finding probabilities based on area. This topic is encountered again in Chapter 12.

Chapter Highlights

Interview with Daniel Galvez: This interview highlights the relationship between art, algebra, and geometry, with related exercises on pages 195 and 203.

Explorations in Chapter 5 begin in Section 5.1, with students exploring the capture-recapture method of sampling. In Section 5.3, students construct similar triangles. Students roll a die to explore probability in Section 5.5.

The Portfolio Project: Pairs of students collect data about a selected group of states and then use the data to make a cartogram.

Technology: Graphing calculators are used in Section 5.2 to study the effects of changing the scale of a scatter plot. Random numbers generated by a calculator are used in Section 5.6 to determine experimental probability.

OBJECTIVES

Section	Objectives	NCTM Standards
5.1	• Use ratios to compare quantities. • Solve proportions. • Estimate quantities that are difficult to count.	1, 2, 3, 4, 5
5.2	• Use scales in drawings and on graphs. • Use scale factors to compare sizes of objects and drawings. • Model large objects with scale drawings.	1, 2, 3, 4, 5
5.3	• Recognize and draw similar figures. • Use similar figures to measure large distances.	1, 2, 3, 4, 5, 7
5.4	• Find the ratios of the perimeters and areas of similar figures. • Use scale drawings to find the perimeters and areas of large objects. • Create accurate pictographs that represent statistics.	1, 2, 3, 4, 5, 7, 10
5.5	• Calculate theoretical and experimental probability. • Predict the outcome of events.	1, 2, 3, 4, 5, 11
5.6	• Find probabilities based on area. • Predict events based on geometric probability.	1, 2, 3, 4, 5, 7, 11

INTEGRATION

Mathematical Connections	5.1	5.2	5.3	5.4	5.5	5.6
algebra	**191–197***	**198–204**	**205–211**	**212–217**	**218–223**	**224–229**
geometry		198–203	**205–211**	**212–217**		**224–229**
data analysis, probability, discrete math	191, 193, 195–197	200, 202, 203	211	216	**218–223**	**224–229**
patterns and functions				212		
logic and reasoning	191, 192, 195–197	198, 200–204	205, 206, 209–211	212, 215–217	218, 221–223	225–229
Interdisciplinary Connections and Applications						
history and geography	192, 196		209, 210			
reading and language arts	195, 197					
biology and earth science	193, 195	198				
business and economics		202				
sports and recreation	197				219	227
forestry, measuring heights, architecture, construction, photography, social studies	194		207	213–216	222	

* ***Bold page numbers*** *indicate that a topic is used throughout the section.*

TECHNOLOGY

	opportunities for use with	
Section	**Student Book**	**Support Material**
5.1	graphing calculator	**Technology Book:** Spreadsheet Activity 5
5.2	graphing calculator McDougal Littell Software *Stats!*	
5.3	scientific calculator	
5.4	scientific calculator	**Technology Book:** Spreadsheet Activity 5
5.5	graphing calculator	**Technology Book:** Calculator Activity 5
5.6	scientific calculator graphing calculator	

PLANNING GUIDE

Regular Scheduling (45 min)

Section	Materials Needed	Core Assignment	Extended Assignment	*exercises that feature* Applications	Communication	Technology
5.1	large coffee can, bag of dried beans, marker	**Day 1:** 1–20 **Day 2:** 23–33a, 34, 35, 37–46	**Day 1:** 1–22 **Day 2:** 23–46	14, 16–20, 22 33, 35	16, 21 33, 36	
5.2	graphing calculator, graph paper, overhead projector, transparencies, marker, ruler	1–6, 10–15, 17–24, AYP*	1–24, AYP	8	8, 9, 12–16, 18	12
5.3	scissors, protractor, ruler, yardstick, tape measure, mirror	**Day 1:** 1–10 **Day 2:** 11–17, 19–27	**Day 1:** 1–10 **Day 2:** 11–27	8 13–18	8, 10 11, 12, 15, 18, 20, 21	
5.4	ruler	1–8, 10–13, 15–22, AYP	1–22, AYP	6–8, 11–13	7, 14, 16	
5.5	dice, pennies, graph paper	1–10, 12–31	1–31	8–11	6, 7, 18, 19	
5.6	scientific or graphing calculator, dried bean	1–6, 8–10, 12–24, AYP	1–24, AYP	11, 12	7, 13, 18	17
Review/ Assess		**Day 1:** 1–10 **Day 2:** 11–25 **Day 3:** Ch. 5 Test	**Day 1:** 1–10 **Day 2:** 11–25 **Day 3:** Ch. 5 Test		14, 16	
Portfolio Project		Allow 2 days.	Allow 2 days.			

Yearly Pacing (with Portfolio Project)	Chapter 5 Total	Chapters 1–5 Total	Remaining	Total
	13 days	68 days	92 days	160 days

Block Scheduling (90 min)

	Day 29	Day 30	Day 31	Day 32	Day 33	Day 34	Day 35
Teach/Interact	5.1: Exploration, page 191	5.2 5.3: Exploration, page 205	Continue with 5.3 5.4	5.5: Exploration, page 218 5.6	Review Port. Proj.	Review Port. Proj.	Ch. 5 Test 6.1
Apply/Assess	**5.1:** 1–46	**5.2:** 1–6, 10–15, 17–24, AYP* **5.3:** 1–10	**5.3:** 11–27 **5.4:** 1–22, AYP	**5.5:** 1–31 **5.6:** 1–10, 12–24, AYP	**Review:** 1–10 **Port. Proj.**	**Review:** 11–25 **Port. Proj.**	**Ch. 5 Test** **6.1:** 1–11, 14, 17–31

NOTE: A one-day block has been added for the Portfolio Project—timing and placement to be determined by teacher.

Yearly Pacing (with Portfolio Project)	Chapter 5 Total	Chapters 1–5 Total	Remaining	Total
	$6\frac{1}{2}$ days	$34\frac{1}{2}$ days	$46\frac{1}{2}$ days	81 days

* ***AYP*** *is Assess Your Progress.*

LESSON SUPPORT

Section	Practice Bank	Study Guide*	Assessment Book*	Visuals	Explorations Lab Manual	Lesson Plans	Technology Book
5.1	31	5.1		Warm-Up 5.1	Add. Expl. 7	5.1	Spreadsheet Act. 5
5.2	32	5.2	Test 19	Warm-Up 5.2	Master 1	5.2	
5.3	33	5.3		Warm-Up 5.3	Master 10	5.3	
5.4	34	5.4	Test 20	Warm-Up 5.4		5.4	Spreadsheet Act. 5
5.5	35	5.5		Warm-Up 5.5 Folder 5	Masters 2, 11	5.5	Calculator Act. 5
5.6	36	5.6	Test 21	Warm-Up 5.6		5.6	
Review Test	37	Chapter Review	Tests 22, 23 Alternative Assessment			Review Test	

*__Spanish versions__ of *Study Guide* and *Assessment Book* are available.

Chapter Support

- Course Guide
- Lesson Plans
- Explorations Lab Manual: Additional Exploration 6
- Portfolio Project Book
- Preparing for College Entrance Tests
- Multi-Language Glossary
- *Test Generator* Software
- Professional Handbook

Software Support

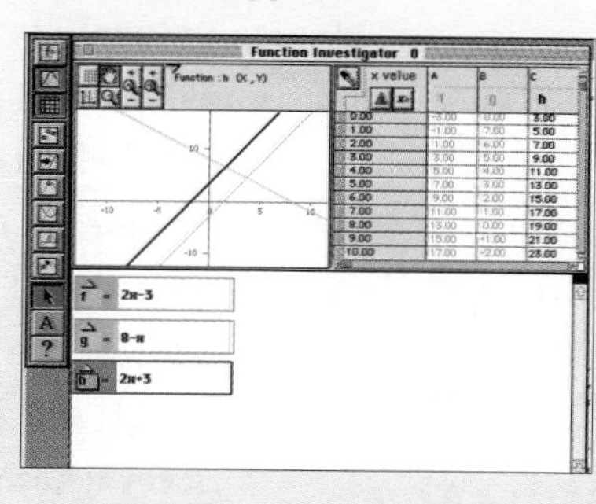

McDougal Littell Software
Stats!

Internet Support

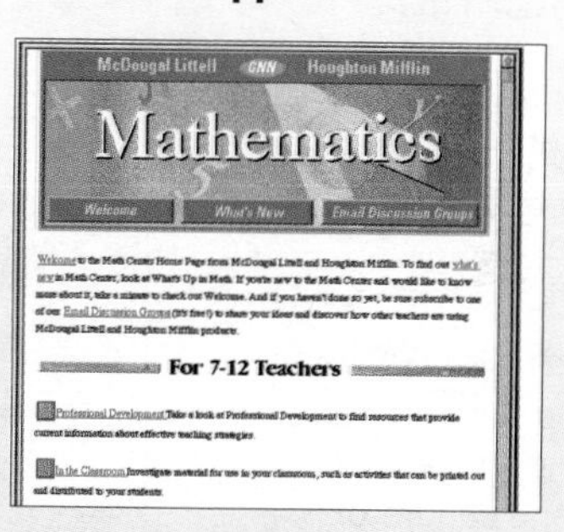

http://www.hmco.com
Next go to McDougal Littell; then the Education Center; then Secondary Math.

OUTSIDE RESOURCES

Books, Periodicals

Brutlag, Dan. "Making Connections: Beyond the Surface." *Mathematics Teacher* (March 1992): pp. 230–235.

Wills, Herbert, III. *Leonardo's Dessert: No Pi.* Reston, VA: NCTM, 1985.

Walton, Karen Doyle. "Albrecht Durer's Renaissance Connections between Mathematics and Art." *Mathematics Teacher* (April 1994): pp. 278–282.

Addenda Series. *Geometry From Multiple Perspectives (Grades 9–12).* Reston, VA: NCTM.

Froelich, Gary, Kevin Bartkovich, and Paul Foerster. *Connecting Mathematics: Addenda Series, Grades 9–12.* Reston, VA: NCTM.

Software

Schwartz, Judah L. and Michal Yerushalmy. *The Geometric superSupposer.* Pleasantville, NY: Sunburst Communications.

Videos

Apostol, Tom. *Similarity.* Video plus program guide. Reston, VA: NCTM.

Internet

Join the Geometry Forum at Swarthmore by connecting to:

gopher forum.swarthmore.edu

Seven newsgroups are available.

Interview Notes

Background Notes

Mural Painting
Mural painting is done on a wall and has a tradition that goes back many centuries. Prior to 1922, when Mexican mural painting appeared, this great art form had been neglected for a long time. The two great contributors to Mexican mural painting were José Clemente Orozco (1883–1949) and Diego Rivera (1886–1957), who is considered one of the great creators in contemporary art. Rivera painted historical, social, and philosophical subjects and, in particular, depicted the history of Mexico from its indigenous past, through the Spanish conquest and colonization and into the modern world. Other Mexican mural artists have continued in the tradition of Orozco and Rivera and have enriched the production of mural art since the rebirth of this art form in the early 1920s.

Daniel Galvez
Mural artist Daniel Galvez follows the tradition of the great Mexican muralists by painting scenes that depict the history and culture of people living in an area. Although most of Galvez's murals are in the Oakland area of California, he also has done work in other states, most notably in Cambridge and Pittsfield, Massachusetts. His mural *Crosswinds* in Cambridge shows the richness of the different people who live in the Central Square area of Cambridge. His mural in Pittsfield honors Vietnam War veterans.

CHAPTER

5 Connecting Algebra and Geometry

Painting a neighborhood

INTERVIEW **Daniel Galvez**

Daniel Galvez always wanted to be an artist, but he didn't know what kind of artist until he learned in college about the great Mexican muralists of the 1900s. Galvez admired the way these artists used colorful paintings on public walls to reach all types of people. "I liked the idea of making a painting that you don't have to go see in a gallery," he says, "the idea that it was part of people's lives whether or not they understood art. It would be something they would see on their way to school or work."

"I liked the idea of making a painting that you don't have to go see in a gallery."

188

Viva la Raza

Galvez began to look for places to paint murals and for money to pay for his time and supplies. When he was 24, he painted *Viva la Raza*, a history of Mexican Americans, on the wall of a building in Berkeley, California. Over the next 16 years, he painted more than 15 murals in California and Massachusetts.

Painting a Neighborhood

In 1992, Galvez was hired to paint a mural on the wall of a restaurant in Cambridge, Massachusetts. He wanted the mural, called *Crosswinds*, to show the diverse cultures of people living near the restaurant. "The more I can make people aware of their differences, the more they become aware of their similarities," he says. "The more you know of your neighbors, the less threatening they are." He hired a photographer who took hundreds of photos of people from the neighborhood. Galvez chose his favorite photos to include in the mural.

Interview Notes

Background

City Murals
San Francisco and Los Angeles are two cities having the most murals per person in the world. There are over 580 murals in San Francisco and over 2000 in Los Angeles.

Interview Notes

Mathematical Connection

Artists who paint very large pictures often use scale drawings to get the proportions of the final painting correct. In Section 5.1, students learn about ratios and proportions and then use those concepts to make estimates of the size of real-life objects based upon the measurements in a photograph. In Section 5.2, students learn to use scales in drawings and graphs and then apply this knowledge to create a poster. In so doing, they follow a method similar to the one used by Daniel Galvez to paint his *Crosswinds* mural.

Explore and Connect

Project
Students may wish to work in groups of three on this project. One student can create the scale drawing, another can write the letter, and the third can play the role of the artist by describing the contents of the mural.

Research
By checking students' selections, you can be sure that each artist is the subject of a report.

Writing
You may wish to have three students use the same scale model of an object but work independently to estimate the scale. This procedure would provide a check of each student's work.

Using a Scale Drawing

The wall of the restaurant was 24 ft high and 41 ft long. To create his mural, Galvez first made a scale drawing. He used a scale of 1 in. to 1 ft, which made the drawing 24 in. high and 41 in. long. Using slides, he projected the photos onto his scale drawing to find the right size and position for each figure. Then he took the same slides and projected them onto sheets of paper at the actual size he wanted the figures to appear on the wall. After drawing the figures on the sheets of paper, Galvez coated the back of each sheet with chalk and traced the outlines of the figures onto the wall of the restaurant.

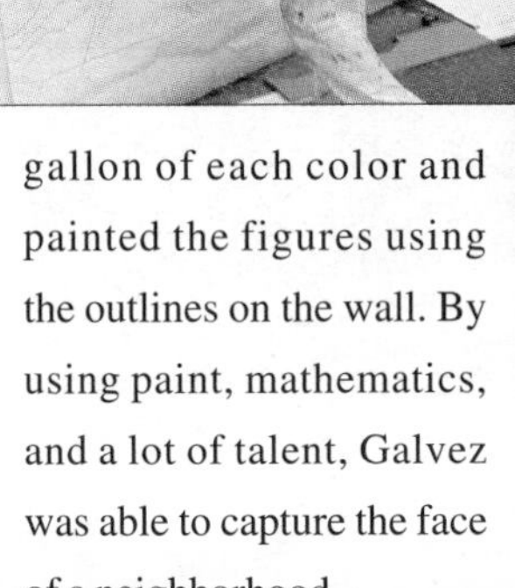

He estimated he needed at least a half gallon of each color of paint for every 1000 ft^2 of wall. He multiplied the height of the restaurant wall by the length and got 984 ft^2, so he bought a half gallon of each color and painted the figures using the outlines on the wall. By using paint, mathematics, and a lot of talent, Galvez was able to capture the face of a neighborhood.

EXPLORE AND CONNECT

Cambridge residents stand in front of their portraits with artist Daniel Galvez.

1. Project Measure a bulletin board or wall in your school or community and create a scale drawing of a mural for the space. Include the scale with your drawing. Write a letter to the school board or the owner of the space proposing that your mural be painted.

2. Research José Clemente Orozco, Diego Rivera, David Alfaro Siqueiros, and Rufino Tamayo are four Mexican artists who painted murals. Find out more about one of these artists and write a report about his work.

3. Writing Find a scale model of an object, such as a toy car or doll. Estimate the scale that was used to create the model. How accurate is the model? Be sure to explain your reasoning.

Mathematics & Daniel Galvez

In this chapter, you will learn more about how mathematics is related to art.

Related Exercises

Section 5.1
- Exercises 17–19

Section 5.2
- Exercises 13–16

SECTION

5.1 Ratios and Proportions

How do wildlife biologists find the number of fish in a lake or birds in a forest? They don't count the animals one by one. Instead, they use an estimation method called *capture-recapture.*

Learn how to...

- use ratios to compare quantities
- solve proportions

So you can...

- estimate quantities that are difficult to count, such as wildlife populations

EXPLORATION
COOPERATIVE LEARNING

The Capture-Recapture Method

Work with a partner.
You will need:

- a large coffee can
- a bag of dried beans
- a marker

Follow Steps 1–4. Then answer the questions below.

1 Pour the bag of beans into the can.

2 Take out a handful of beans, mark them, and record how many you marked.

3 Put the marked beans back into the can. Mix the beans thoroughly.

4 Take out another handful, or *sample*, of beans. Record the number of marked beans and the total number of beans in your sample.

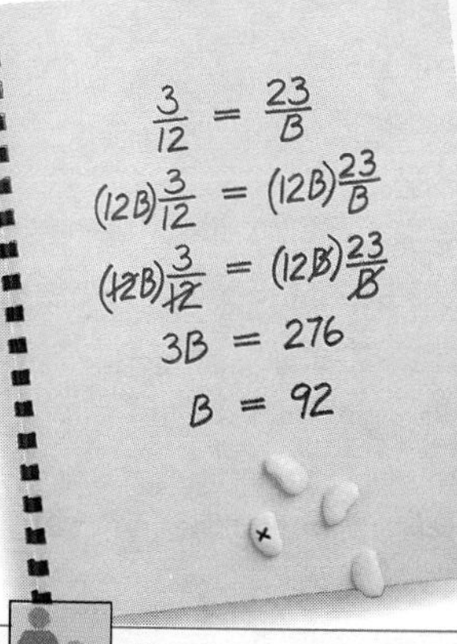

Questions

1. Let B = the unknown number of beans in the can. What is the fraction of beans that are marked? (Your answer will involve B.)
2. What fraction of beans in your sample were marked? How do you think this fraction relates to the fraction from Question 1?
3. Ayita marked 23 beans. Then she took a sample of 12 beans and found 3 marked. Use Ayita's method and the fractions you found in Questions 1 and 2 to estimate the number of beans in your coffee can.

Exploration Note

Purpose
The purpose of this Exploration is to have students discover how to use a pair of ratios to estimate quantities that are difficult to count.

Materials/Preparation
You may wish to ask students to bring their own cans and dried beans to school if they are not available in the classroom. All groups should begin their work at the same time.

Procedure
As students work through Steps 1–4, one partner should have the responsibility of recording the number of beans marked in Steps 2 and 4. The other partner can record the answers to the questions.

Closure
Ask students to write a brief description of the procedure they used to estimate the number of beans in the can. Students should understand how a sample of beans allows them to estimate the total number of beans in the can. Ask students why they think this method is called the capture-recapture method.

Explorations Lab Manual
See the Manual for more commentary on this Exploration.

For answers to the Exploration, see following page.

Plan⇔Support

Objectives

- Use ratios to compare quantities.
- Solve proportions.
- Estimate quantities that are difficult to count.

Recommended Pacing

Core and Extended Courses
Section 5.1: 2 days
Block Schedule
Section 5.1: 1 block

Resource Materials

Lesson Support
Lesson Plan 5.1
Warm-Up Transparency 5.1
Practice Bank: Practice 31
Study Guide: Section 5.1
Explorations Lab Manual: Additional Exploration 7
Technology
Technology Book: Spreadsheet Activity 5
Graphing Calculator
Internet: http://www.hmco.com

Warm-Up Exercises

1. Convert 2560 yards to miles. (1 mi = 1760 yd) 2560 yd ≈ 1.45 mi

Write each fraction in lowest terms.

2. $\frac{10}{12}$ $\frac{5}{6}$
3. $\frac{5}{25}$ $\frac{1}{5}$
4. $\frac{7\frac{1}{2}}{3}$ $\frac{5}{2}$
5. $\frac{\frac{5}{2}}{3}$ $\frac{5}{6}$
6. Express the fraction $\frac{27}{12}$ as a decimal. 2.25

Teach⇔Interact

Section Note

Topic Spiraling: Review
Review with students the fact that a rate is a ratio in which the units of the two quantities are different. A rate is a special type of ratio.

About Example 1

Mathematical Procedures
The solution to this Example uses a conversion factor of 0.305 m = 1 ft. As a general procedure, students can find conversion factors in books containing conversion tables. Or, if students remember that 1 m = 3.28 ft, then dividing both sides of the equality by 3.28 yields 1 ft = 0.305 m.

Additional Example 1

The height of the Egyptian pyramid at Maydum is about 90 m high. The Pyramid of the Sun in Mexico is about 210 ft high. Find the ratio of the Egyptian pyramid to the Mexican pyramid, using feet.

Convert the height of the pyramid at Maydum to feet.

$$90 \text{ m} \cdot \frac{3.28 \text{ ft}}{1 \text{ m}} \approx 295 \text{ ft}$$

$$\frac{\text{Height of Pyramid at Maydum}}{\text{Height of Pyramid of the Sun}} = \frac{295 \text{ ft}}{210 \text{ ft}} = \frac{59}{42}$$

Think and Communicate

You can use question 3 to introduce the idea of the reciprocal of a ratio with numerator 1. Ask students how a fraction with denominator 1 is different from a ratio with denominator 1. (Since a ratio is a comparison of two numbers, the denominator of 1 should not be omitted when writing a ratio. It can be omitted, however, in a fraction.)

Ratios

A **ratio** is a quotient that compares two quantities. For example, in the Exploration on page 191, there were 3 marked beans in a sample of 12 beans. The ratio of marked beans to all beans in the sample is 3 to 12, or 1 to 4. You can write this ratio in three ways:

$$\frac{1}{4} \qquad 1:4 \qquad 1 \text{ to } 4$$

EXAMPLE 1 Application: History

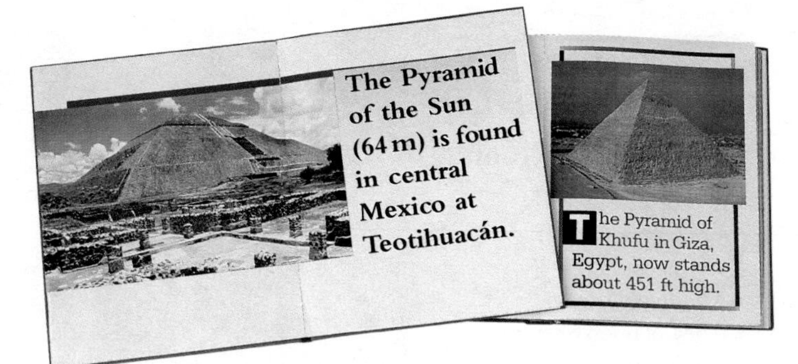

The Pyramid of the Sun (64 m) is found in central Mexico at Teotihuacán.
The Pyramid of Khufu in Giza, Egypt, now stands about 451 ft high.

Grace wants to compare the heights of two pyramids, but the books she has use different units of measure. Use the information shown to find the ratio of the height of the Pyramid of Khufu to the height of the Pyramid of the Sun.

WATCH OUT!
When you write a ratio, the two quantities must be measured in the same units.

SOLUTION

Convert the height of the Pyramid of Khufu to meters.

$$451 \text{ ft} \cdot \frac{0.305 \text{ m}}{1 \text{ ft}} \approx 138 \text{ m}$$

Use the fact that 1 ft = 0.305 m.

$$\frac{\text{Height of the Pyramid of Khufu}}{\text{Height of the Pyramid of the Sun}} = \frac{138 \text{ m}}{64 \text{ m}} = \frac{69}{32}$$

Write the fraction in lowest terms.

The ratio of the height of the Pyramid of Khufu to the height of the Pyramid of the Sun is $\frac{69}{32}$.

THINK AND COMMUNICATE

1. Convert $\frac{69}{32}$ to a decimal rounded to the nearest hundredth. What does this number tell you about the heights of the pyramids in Example 1?
2. Suppose you found the ratio in Example 1 by first converting the height of the Pyramid of the Sun to feet. What do you think the ratio would be?
3. Find the ratio of the height of the Pyramid of the Sun to the height of the Pyramid of Khufu. How is this ratio related to the ratio found in Example 1?

ANSWERS Section 5.1

Exploration

1–3. Answers may vary. Let m = the total number of marked beans, s = the number of beans in the sample, and n = the number marked in the sample.

1. $\frac{m}{B}$

2. $\frac{n}{s}$; The two fractions should be approximately equal.

3. If, for example, 22 beans are marked in Step 2 and 3 of 15 beans are marked in the sample drawn in Step 4, then $\frac{22}{B} = \frac{3}{15}$. Solving for B gives: $(15B)\left(\frac{22}{B}\right) = (15B)\left(\frac{3}{15}\right)$; $330 = 3B$; $B = 110$. There are about 110 beans in the coffee can.

Think and Communicate

1. 2.16; The Pyramid of Khufu is about twice as tall as the Pyramid of the Sun.
2. $64 \text{ m} \cdot \frac{1 \text{ ft}}{0.305 \text{ m}} \approx 210 \text{ ft}$; $\frac{451 \text{ ft}}{210 \text{ ft}}$
3. $\frac{32}{69}$; It is the reciprocal of the ratio found in Example 1.

Proportions

A **proportion** is an equation that shows two ratios to be equal. Each of the proportions below can be read as "a is to b as c is to d."

You call a and d the **extremes** of the proportion.

$\frac{a}{b} = \frac{c}{d}$ $\quad a : b = c : d \quad$ a to b = c to d

You call b and c the **means** of the proportion.

Means-Extremes Property

For any proportion, the product of the **extremes** is equal to the product of the **means**.

If $\frac{a}{b} = \frac{c}{d}$, then $ad = bc$.

For example, you know that $\frac{4}{6} = \frac{2}{3}$.

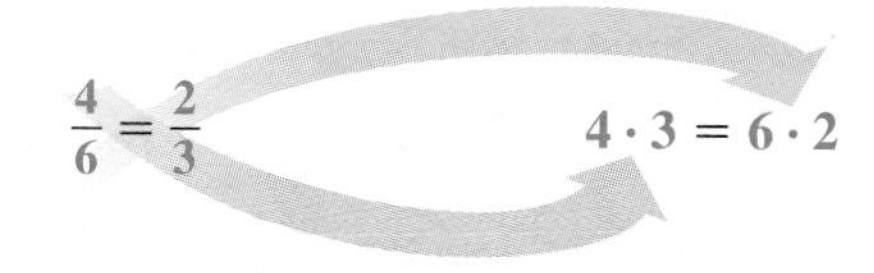

EXAMPLE 2 Application: Ecology

At Dryden Lake in New York, researchers caught, marked, and released 213 largemouth bass. Later, they took a sample of 104 bass and found that 13 were marked. Estimate the number of largemouth bass in Dryden Lake.

SOLUTION

Let x = the total number of largemouth bass in the lake.

$$\frac{\text{Marked bass in sample}}{\text{Total bass in sample}} = \frac{\text{Marked bass in lake}}{\text{Total bass in lake}}$$

$$\frac{13}{104} = \frac{213}{x}$$ Use the means-extremes property.

$$13x = 104 \cdot 213$$
$$13x = 22{,}152$$
$$x = 1704$$

There are about 1700 largemouth bass in Dryden Lake.

Teach⇔Interact

Section Note

Teaching Tip
Ask students how they might use a familiar meaning of the word *extreme*, such as extreme temperatures, to help them remember the positions of the extremes in a proportion.

About Example 2

Application
The strategy of using a proportion to solve a problem is one of the most useful and powerful ways to solve problems that are related to real-world situations. The application to ecology here is only one of many applications in this section.

Additional Example 2

Umbagog Lake lies on the border of New Hampshire and Maine. Assume that New Hampshire Fish and Wildlife employees caught, tagged, and released 105 chain pickerel. If a sample of 43 chain pickerel taken later contained 10 marked pickerel, estimate the number of chain pickerel in Umbagog Lake.

Let p = the total number of chain pickerel in the lake.

$$\frac{\text{Marked pickerel in sample}}{\text{Total pickerel in sample}} = \frac{\text{Marked pickerel in lake}}{\text{Total pickerel in lake}}$$

$$\frac{10}{43} = \frac{105}{p}$$
$$10p = 43 \cdot 105$$
$$p = 451.5$$

There are about 450 chain pickerel in Umbagog Lake.

Teach⇔Interact

Additional Example 3

Julie Renaud owns a 200 acre wood lot that contains a spruce forest. Julie estimates that the total number of Norway spruce trees and red spruce trees is 1000. If the ratio of Norway spruce trees to red spruce trees is 5 to 3, how many Norway spruce trees are there in the forest?

Let n = the number of Norway spruce trees. Then $(1000 - n)$ = the number of red spruce trees.

$$\frac{5}{3} = \frac{n}{1000 - n}$$
$$5(1000 - n) = 3n$$
$$5000 - 5n = 3n$$
$$5000 = 8n$$
$$625 = n$$

There are about 625 Norway spruce trees in the forest.

Checking Key Concepts

Common Error

For questions 2 and 3, some students may not write the ratios using common units. For example, in question 3, students may use the ratio 5 to 15, neglecting to convert 5 feet to inches or 15 inches to feet. Remind these students that when they write a ratio, the units must be the same.

Closure Question

Describe a procedure for showing that two ratios are equal.

Apply the means-extremes property to show that the product of the extremes is equal to the product of the means.

Apply⇔Assess

Suggested Assignment

Core Course
Day 1 Exs. 1–20
Day 2 23–33a, 34, 35, 37–46

Extended Course
Day 1 Exs. 1–22
Day 2 23–46

Block Schedule
Day 29 Exs. 1–46

BY THE WAY

Zoe Emily Schnabel developed a form of the capture-recapture method in the 1930s. The Schnabel Estimate allowed scientists to estimate populations more accurately.

EXAMPLE 3 Application: Forestry

In a large state forest, the ratio of white pine trees to red pine trees is 7 to 4. A forester estimates that the total number of white and red pine trees is 5000. How many white pine trees are in the forest?

SOLUTION

Let w = the number of white pine trees. Then $5000 - w$ = the number of red pine trees.

$$\frac{7}{4} = \frac{w}{5000 - w}$$
$$7(5000 - w) = 4w$$
$$35{,}000 - 7w = 4w$$
$$35{,}000 = 11w$$
$$3182 \approx w$$

Use the distributive property.

There are about 3200 white pine trees in the state forest.

CHECKING KEY CONCEPTS

Find the ratio of the first quantity to the second quantity. Write your answer as a fraction in lowest terms.

1. $4\frac{1}{2}$ lb; $13\frac{1}{2}$ lb **2.** 6 days; 3 weeks **3.** 5 ft; 15 in.

Solve each proportion.

4. $\frac{12}{35} = \frac{p}{100}$ **5.** $\frac{x}{x + 12} = \frac{2}{3}$ **6.** $9:11 = 6:m$

7. **NUTRITION** A 10.5 oz can of clam chowder contains 220 Cal. Find the number of calories in an 18.5 oz can of the same brand of clam chowder.

5.1 Exercises and Applications

Extra Practice exercises on page 566

Find the ratio of the first quantity to the second quantity. Write your answer as a fraction in lowest terms.

1. length of a football field: 360 ft; width of a football field: 160 ft
2. number of gray whales in 1970: 21,000; number of gray whales in 1946: 2000
3. weight of a mouse: 20 g; weight of an African elephant: 7500 kg
4. speed of light: 186,282 mi/s; speed of sound: 1100 ft/s

Checking Key Concepts

1. $\frac{1}{3}$
2. $\frac{2}{7}$
3. $\frac{4}{1}$
4. $\frac{240}{7} \approx 34.29$
5. 24
6. $\frac{22}{3} \approx 7.33$
7. about 387.62 Cal

Exercises and Applications

1. $\frac{9}{4}$
2. $\frac{21}{2}$
3. $\frac{1}{375{,}000}$
4. $\frac{98{,}356{,}896}{110}$
5. 2
6. 35
7. 12
8. 10.5
9. 24
10. 3.5
11. 4
12. 60
13. $\frac{33}{7} \approx 4.7$
14. about 1190 wood ducks

Solve each proportion.

5. $\frac{x}{5} = \frac{2}{5}$ **6.** $\frac{5}{1} = \frac{d}{7}$ **7.** $\frac{9}{y} = \frac{3}{4}$

8. $\frac{7}{4} = \frac{v}{6}$ **9.** $\frac{1.5}{18} = \frac{2}{k}$ **10.** $\frac{2}{5.08} = \frac{h}{8.89}$

11. 144 to 18 = 32 to a **12.** $6:1 = w:12$ **13.** $3:n = 7:11$

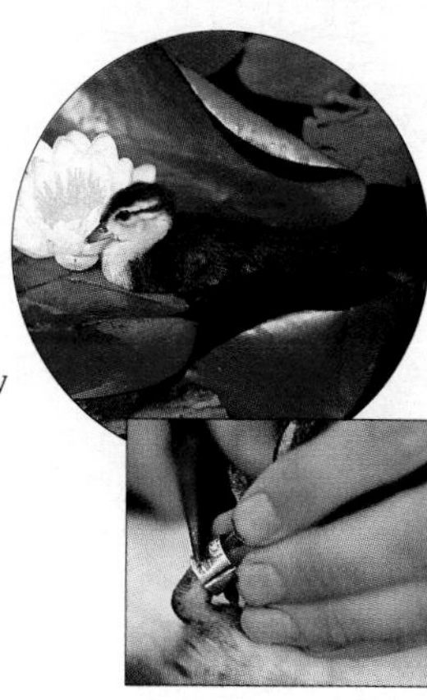

14. ECOLOGY In 1974, researchers at the Montezuma Wildlife Refuge in New York captured, marked, and released 364 young wood ducks. A few weeks later, the researchers took a sample of 196 young wood ducks and found that 60 were marked. Estimate the number of young wood ducks in the refuge. Round your answer to the nearest ten.

15. Open-ended Problem Write a real-life problem that can be solved using the proportion $\frac{6.5}{180} = \frac{g}{325}$.

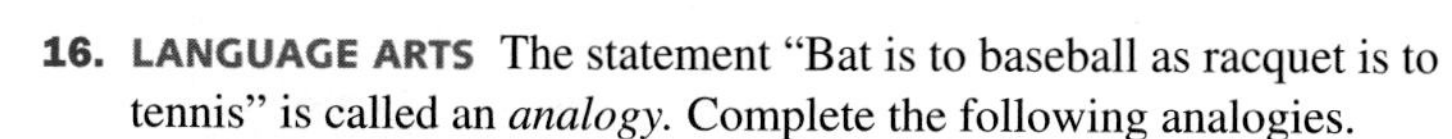

16. LANGUAGE ARTS The statement "Bat is to baseball as racquet is to tennis" is called an *analogy*. Complete the following analogies.

a. Kitten is to cat as _?_ is to dog.

b. Red is to _?_ as green is to go.

c. Driver is to car as pilot is to _?_.

d. _?_ is to Texas as Sacramento is to California.

e. **Writing** Explain how analogies are like proportions.

INTERVIEW Daniel Galvez

Look back at the article on pages 188–190.

The photo below at the right shows a part of Daniel Galvez's mural, Crosswinds, in Cambridge, Massachusetts. The girl is standing in front of her portrait that Galvez painted as part of the mural.

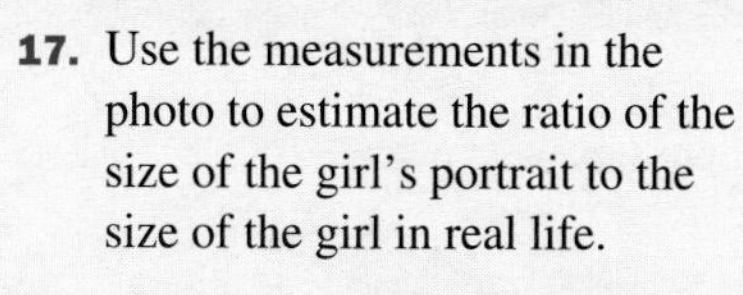

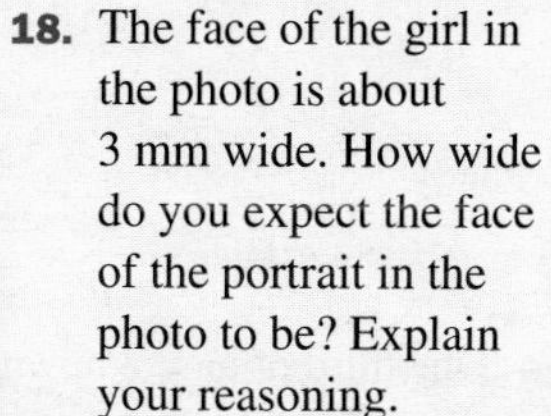

17. Use the measurements in the photo to estimate the ratio of the size of the girl's portrait to the size of the girl in real life.

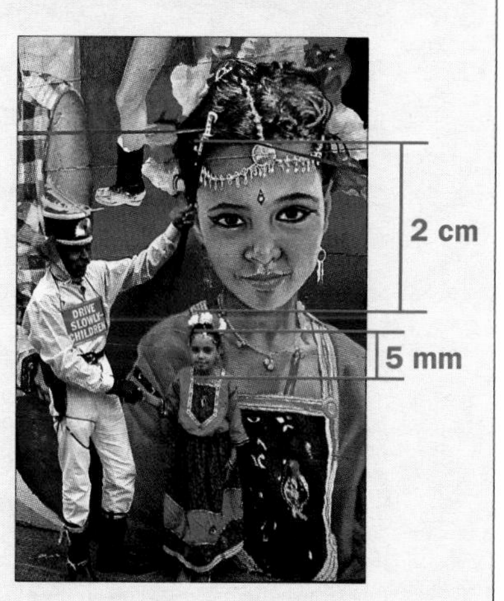

18. The face of the girl in the photo is about 3 mm wide. How wide do you expect the face of the portrait in the photo to be? Explain your reasoning.

19. How many times as large as the girl is the portrait? How can you tell from the photo?

15. Answers may vary. An example is given. The Wasiuks used 6.5 gal of gasoline on a 180-mi trip. How many gallons of gasoline should they expect to use on a similar trip of 325 mi?

16. a. puppy
b. stop
c. airplane
d. Austin
e. Answers may vary. An example is given. The words in an analogy are related to each other in the same way that the numbers in a proportion are related to each other.

17. Estimates may vary; about $\frac{4}{1}$.

18. 12 mm; The ratio of the size of the girl's portrait to the actual size of the girl is about 4 to 1.

19. The portrait is about 4 times as large as the girl. The ratio in the photo is about 4 to 1, so the ratio of the actual sizes is about 4 to 1.

Apply⇔Assess

Exercise Notes

Mathematical Procedures

Exs. 5–13 Students should understand that the position of the variable in the proportion does not affect the way in which the proportion is solved. Applying the means-extremes property will always result in an equation that can be solved by dividing both sides of the equation by the coefficient of the variable.

Research

Ex. 14 The male wood duck is a strikingly crested, multicolored duck. Ask students to research the wood duck's habitat, nesting habits, and range. Students can prepare a bulletin board display of the information they gather. If any students live in areas inhabited by wood ducks, those students may be able to provide photographs of the ducks for the display.

Problem Solving

Ex. 15 This exercise allows students to use their creativity and writing skills to demonstrate their understanding of the strategy of using proportions to solve real-world problems.

Teaching Tip

Ex. 16 Explain to students that one way to solve an analogy is to determine what the relationship is between the two terms that are separated by the words *is to* and then to apply that relationship to the remaining terms of the analogy. Challenging students to write their own analogies may help them to see the similarity between analogies and proportions.

Multicultural Note

Exs. 17–19 Urban mural art became popular in the 1960s. Some social analysts believe that this art form grew out of a community's need to express itself and speak out on social and political issues. The longest urban mural in the world is in Los Angeles and is called "The Great Wall." The mural, which depicts California's rich and diverse history from prehistoric times to the present, is part of a unique collaboration of historians, artists, and local youth. Each of the 350-foot sections represents a decade of California's history and has taken about a year to complete.

Apply⇔Assess

Exercise Notes

Using Technology
Exs. 20–22 Students can use a programmable calculator to find the quotas. Instructions are given for the TI-82 calculator.

```
PROGRAM:QUOTAS
: 435→T
: Lbl 1
: 248710000→P
: Prompt S
: S*T/P→Q
: Disp Q
: Goto 1
```

For Ex. 22(a), edit the program QUOTAS so that T = 14 and P = 110,000. Use the program to find the quotas.

Reasoning
Ex. 22(b) Asking students to analyze the results of the exercise helps them to apply critical thinking skills to real-world problems. You can point out to students that the study of statistics requires the careful analyzing of data.

Integrating the Strands
Ex. 24 Mention that this type of proportion was first written to describe a situation in geometry involving the lengths of line segments and, thus, the variable g is called the *geometric mean* of the proportion.

Problem Solving
Ex. 32 Since the original number of beans is unknown, the number after 20 beans have been added, and one third the original number are also unknown quantities. Because one equation is used to solve the problem, all the unknown quantities must be expressed in terms of a single variable.

Connection GOVERNMENT

There are 435 seats in the United States House of Representatives. Every 10 years, these seats are apportioned, or divided, among the states. You can find a state's quota, or share, of House seats by solving the proportion below.

$$\frac{\text{State's quota of House seats}}{\text{Total number of House seats}} = \frac{\text{Population of the state}}{\text{Population of United States}}$$

20. Copy and complete the table at the right. Use 248,709,873 for the population of the United States in 1990. Round your answers to the nearest hundredth.

State	Population in 1990	Quota
Arizona	3,665,228	?
Massachusetts	6,016,425	?
New Jersey	7,730,188	?
Oklahoma	3,145,585	?

21. Research Find the number of House seats each state in the table actually had in 1990. How do these numbers compare to the states' quotas?

22. Alexander Hamilton proposed one method for apportioning the House seats among the states. He suggested the government give each state the whole-number part of its quota, and then distribute the extra seats to the states whose decimal parts of their quotas are largest. (For example, 8 is the whole-number part of 8.53.)

District	Population	Quota	Seats
A	35,000	?	?
B	27,000	?	?
C	12,000	?	?
D	36,000	?	?

a. Consider a city having 14 seats on its city council. Copy and complete the table at the left using the Hamilton method.

b. **Open-ended Problem** Suppose the city council increases the number of seats to 15. How does the number of seats change for each district? Do you think the new apportionment is fair? Explain.

Solve each proportion.

23. $\frac{2}{9} = \frac{4a}{13}$ **24.** $\frac{3}{g} = \frac{g}{12}$ **25.** $\frac{y+1}{12} = \frac{2}{3}$

26. $\frac{3}{7} = \frac{t-4}{21}$ **27.** $\frac{11}{8} = \frac{1}{3r+2}$ **28.** $40 = \frac{80}{x}$

29. $\frac{b}{6} = \frac{2+b}{12}$ **30.** $\frac{p}{36-p} = \frac{4}{5}$ **31.** $\frac{6x-2}{7} = \frac{5x+7}{8}$

32. Challenge One third of an unknown number of beans is marked. Twenty more unmarked beans are added. Then a sample of 25 beans is taken. Six beans have marks. Estimate the total number of beans there were before the twenty beans were added.

20. Arizona: 6.41; Massachusetts: 10.52; New Jersey: 13.52; Oklahoma: 5.50

21. Arizona: 6; Massachusetts: 10; New Jersey: 13; Oklahoma: 6; All are approximately equal to the quota.

22. a.

District	Population	Quota	Seats
A	35,000	$4.\overline{45}$	4
B	27,000	$3.4\overline{36}$	3
C	12,000	$1.5\overline{27}$	2
D	36,000	$4.5\overline{81}$	5

b. Districts A and B gain an additional seat, while District C loses a seat, and District D retains its 5 seats. Answers may vary. An example is given. I do not think it is fair to District C, since it loses a seat with no change in population. Also, the ratio of seats to population is about the same for Districts A, B, and D, but much lower for District C.

23. $\frac{13}{18}$ 24. ± 6 25. 7

26. 13 27. $-\frac{14}{33}$ 28. 2

29. 2 30. 16 31. 5

32. Estimates may vary due to rounding; about 71 beans.

33. a. Estimates may vary due to rounding; about 61 home runs.

b. The record, 61 home runs, was set by former New York Yankee Roger Maris in 1961. Answers may vary. An example is given. I do not think he would have.

33. SPORTS Major league baseball teams play 162 games during a typical season. Due to the baseball strike in 1994, the San Francisco Giants played only 115 games during that season. In these 115 games, Matt Williams hit 43 home runs for the Giants.

a. Predict how many home runs Matt Williams would have hit during the 1994 season if it had been a typical season.

b. **Research** Find out the record for the number of home runs hit in a single season by one player. Do you think Matt Williams would have broken this record if it had been a typical season? Explain your answer.

34. SAT/ACT Preview Suppose $0 < x < y$. Which of the following is true?

A. $\frac{x}{y} > 1$ **B.** $\frac{x}{y} < 1$ **C.** $\frac{x}{y} = 1$

D. relationship cannot be determined

35. LANGUAGES The bar graph shows the six most widely spoken languages in the world in 1993. To put these numbers in perspective, scale the numbers in the graph to the number of students in your school. Use a 1993 world population estimate of 5.5 billion people. How many students would speak each of the six languages?

ONGOING ASSESSMENT

36. Writing Explain how to solve a proportion without using the means-extremes property. Illustrate your method with an example.

SPIRAL REVIEW

Find an equation of the line through the given points. *(Section 3.5)*

37. $(-3, -16)$ and $(4, 5)$

38. $(-6, 3)$ and $(12, -15)$

39. $(0, 0)$ and $\left(\frac{1}{3}, \frac{1}{4}\right)$

40. $\left(-\frac{1}{2}, -2\right)$ and $(1.5, -6)$

41. $(0, 1)$ and $(0, -1)$

42. $(0, b)$ and $(1, b + m)$

Solve each equation. Round answers to the nearest hundredth. *(Section 4.4)*

43. $3.57u - 4.73 = -8.07$

44. $\frac{13.42(t - 7.53)}{12.56} = -4.79$

45. $12.71v - 13.07 = 1.12v$

46. $0.1w + 0.7w = 0.24$

Apply⇔Assess

Exercise Notes

Communication: Drawing
Ex. 35 Ask students to draw graphs using the data obtained in the exercise. Students can discuss how representative the drawings are of their actual school population. Students who have access to the Internet might contact students in other schools, send them their graphs, and ask them to return their results for a similar exercise.

Assessment Note
Ex. 36 If students have difficulty with this exercise, refer them to Sections 2.1 and 2.2, where they solved equations such as $\frac{m}{2} = 14$. Students should understand that any proportion can be solved by applying what they have already learned about solving equations.

Practice 31 for Section 5.1

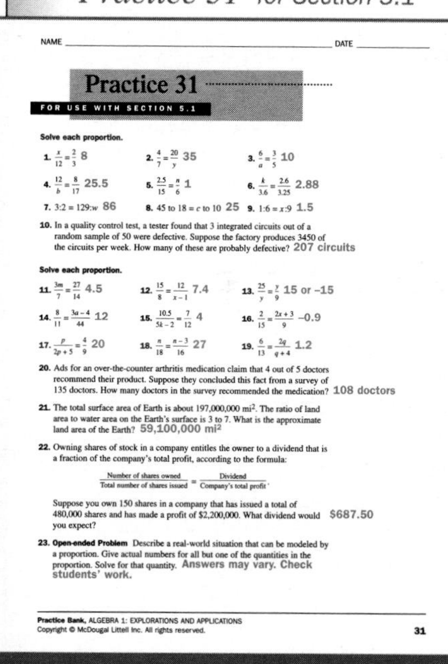

NAME ______________________ DATE __________

Practice 31

FOR USE WITH SECTION 5.1

Solve each proportion.

1. $\frac{x}{12} = \frac{2}{3}$ 8
2. $\frac{4}{7} = \frac{20}{y}$ 35
3. $\frac{6}{a} = \frac{3}{5}$ 10
4. $\frac{12}{b} = \frac{8}{17}$ 25.5
5. $\frac{2.5}{15} = \frac{n}{6}$ 1
6. $\frac{k}{3.6} = \frac{2.6}{3.25}$ 2.88
7. $3{:}2 = 129{:}w$ 86
8. 45 to 18 = c to 10 25
9. $1{:}6 = x{:}9$ 1.5
10. In a quality control test, a tester found that 3 integrated circuits out of a random sample of 50 were defective. Suppose the factory produces 3450 of the circuits per week. How many of these are probably defective? 207 circuits

Solve each proportion.

11. $\frac{3m}{7} = \frac{27}{14}$ 4.5
12. $\frac{15}{8} = \frac{12}{x - 1}$ 7.4
13. $\frac{25}{y} = \frac{y}{9}$ 15 or −15
14. $\frac{8}{11} = \frac{3a - 4}{44}$ 12
15. $\frac{10.5}{5k - 2} = \frac{7}{12}$ 4
16. $\frac{2}{15} = \frac{2x + 3}{9}$ −0.9
17. $\frac{p}{2p + 5} = \frac{4}{9}$ 20
18. $\frac{n}{18} = \frac{n - 3}{16}$ 27
19. $\frac{6}{13} = \frac{2q}{q + 4}$ 1.2
20. Ads for an over-the-counter arthritis medication claim that 4 out of 5 doctors recommend their product. Suppose they concluded this fact from a survey of 135 doctors. How many doctors in the survey recommended the medication? 108 doctors
21. The total surface area of Earth is about 197,000,000 mi^2. The ratio of land area to water area on the Earth's surface is 3 to 7. What is the approximate land area of the Earth? 59,100,000 mi^2
22. Owning shares of stock in a company entitles the owner to a dividend that is a fraction of the company's total profit, according to the formula:

$$\frac{\text{Number of shares owned}}{\text{Total number of shares issued}} = \frac{\text{Dividend}}{\text{Company's total profit}}.$$

Suppose you own 150 shares in a company that has issued a total of 480,000 shares and has made a profit of $2,200,000. What dividend would you expect? $687.50

23. **Open-ended Problem** Describe a real-world situation that can be modeled by a proportion. Give actual numbers for all but one of the quantities in the proportion. Solve for that quantity. Answers may vary. Check students' work.

Practice Bank, ALGEBRA 1: EXPLORATIONS AND APPLICATIONS

31

Williams would have had to hit 19 home runs in the remaining 47 games. That is about 0.40 home runs per game. His ratio of home runs to games for the shortened season was about 0.37.

34. B

35. Answers may vary. An example is given for a school with a population of 625 students. Mandarin: about 95; Hindustani: about 38; English: about 37; Spanish: about 38; Bengali: about 21; Arabic: about 21

36. Multiply both sides of the equation by the least common denominator of the ratios. Then solve the resulting equation.

37. $y = 3x - 7$

38. $y = -x - 3$

39. $y = \frac{3}{4}x$

40. $y = -2x - 3$

41. $x = 0$

42. $y = mx + b$

43. −0.94

44. 3.05

45. 1.13

46. 0.3

Plan⇔Support

Objectives

- Use scales in drawings and on graphs.
- Use scale factors to compare sizes of objects and drawings.
- Model large objects with scale drawings.

Recommended Pacing

Core and Extended Courses
Section 5.2: 1 day

Block Schedule
Section 5.2: $\frac{1}{2}$ block
(with Section 5.3)

Resource Materials

Lesson Support
Lesson Plan 5.2
Warm-Up Transparency 5.2
Practice Bank: Practice 32
Study Guide: Section 5.2
Explorations Lab Manual: Diagram Master 1
Assessment Book: Test 19

Technology
Graphing Calculator
McDougal Littell Software *Stats!*
Internet: http://www.hmco.com

Warm-Up Exercises

Solve each proportion.

1. $\frac{5}{12} = \frac{x}{24}$ 10
2. $6 : y = 18 : 21$ 7

Express each fraction as a decimal.

3. $\frac{3}{4}$ 0.75
4. $\frac{5}{8}$ 0.625

Express each fraction as a percent.

5. $\frac{3.4}{2}$ 170%
6. $\frac{0.5}{2.5}$ 20%

SECTION

5.2 Scale Measurements

Learn how to...

- use scales in drawings and on graphs
- use scale factors to compare sizes of objects and drawings

So you can...

- model large objects with scale drawings
- find large dimensions, such as the size of a whale
- compare maps of different sizes

The blue whale is the largest animal on Earth. To appreciate its size, compare the pictures below. The drawing of the whale has the same *scale* as the photograph of the elephant.

The drawing of the whale and the photograph of the elephant both have a **scale** of 1 in. = 24 ft. This means that each inch in the pictures represents 24 feet in real life.

EXAMPLE 1 Application: Zoology

The drawing of the whale is about 4 in. long. Find the whale's real length.

SOLUTION

Let w = the whale's real length in feet.

$$\frac{\text{Whale's length in drawing in inches}}{\text{Whale's real length in feet}} = \frac{1 \text{ in.}}{24 \text{ ft}}$$

$$\frac{4}{w} = \frac{1}{24}$$

Use the means-extremes property.

$$4 \cdot 24 = w \cdot 1$$

$$96 = w$$

The whale is about 96 ft long.

THINK AND COMMUNICATE

1. The photograph of the elephant is about $\frac{3}{8}$ in. high. Explain how to estimate the real height of the elephant.
2. Suppose a drawing of a whale has a scale of 1 cm = 2 m. To find the whale's real length, what units must you use to measure the drawing? In what units will the whale's length be? How do you know?

ANSWERS Section 5.2

Think and Communicate

1. Let h = the elephant's actual height in feet. Write and solve the proportion $\frac{\frac{3}{8}}{h} = \frac{1}{24}$. (The elephant is about 9 ft tall.)
2. centimeters; meters; The units must be the same in both ratios.

Scale Factors

1 in. = 48 ft

The drawing here has a different scale from the one on page 198. In this drawing the whale is 2 in. long.

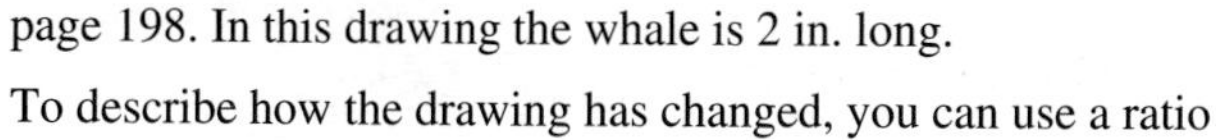

To describe how the drawing has changed, you can use a ratio called the *scale factor*.

$$\textbf{Scale factor} = \frac{\text{Whale's length in second drawing}}{\text{Whale's length in first drawing}} = \frac{2 \text{ in.}}{4 \text{ in.}} = \frac{1}{2}$$

The second drawing is half the size of the original drawing. You can write the scale factor as a fraction $\left(\frac{1}{2}\right)$, as a decimal (0.5), or as a percent (50%).

WATCH OUT!

The scale factor is the ratio of the new size to the old size, not the old size to the new size. Be sure to write the ratio in the correct order.

EXAMPLE 2

Nathan plans to include a map in his report on Africa. The map he has is 6 in. wide and 8 in. long. He wants to use a photocopier to enlarge the map so that it fills the width of the page.

a. What scale factor should he use to make the map $7\frac{1}{2}$ in. wide? Give your answer as a percent.

b. How long will the enlarged map be?

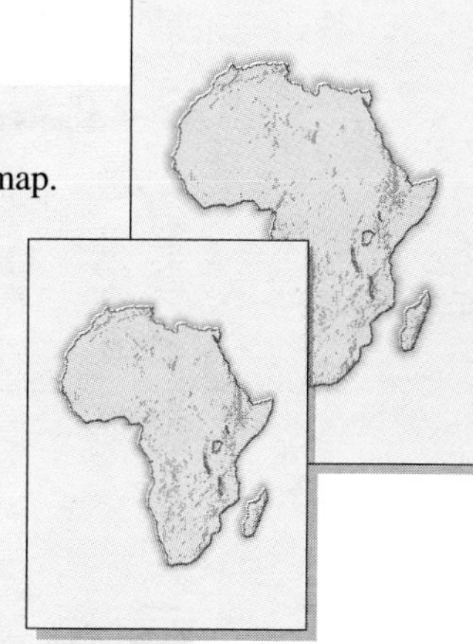

SOLUTION

a. Let p = the scale factor at which Nathan should copy the map.

$$\frac{\text{Width of enlargement}}{\text{Width of original}} = p$$

$$\frac{7.5}{6} = p$$

$$1.25 = p$$

Nathan should copy the map at 125%.

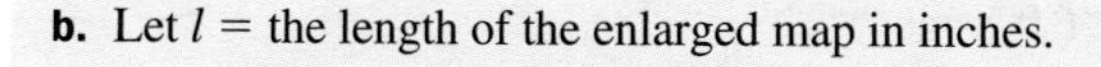

b. Let l = the length of the enlarged map in inches.

$$\frac{\text{Length of enlarged map}}{\text{Length of original map}} = 1.25$$

Set the ratio of the maps' lengths equal to the scale factor.

$$\frac{l}{8} = 1.25$$

$$l = 8(1.25)$$

$$l = 10$$

The new map will be 10 in. long.

Teach⇔Interact

Learning Styles: Visual

Use a dimension of the classroom to help students visualize 24 ft. Students then can get a visual image of the scale 1 in. = 24 ft. Students can see how a scale allows them to model large objects.

Additional Example 1

The whale made famous in Herman Melville's novel, *Moby Dick*, was a sperm whale. Using the same scale as in Example 1, the drawing of a sperm whale would be about $2\frac{1}{2}$ in. long. Find the actual length of a sperm whale.

Let l = the whale's length in feet. To find l, write a proportion.

$$\frac{\text{Whale's length in drawing}}{\text{Whale's actual length}} = \frac{1 \text{ in.}}{24 \text{ ft}}$$

$$\frac{2\frac{1}{2}}{l} = \frac{1}{24}$$

$$2\tfrac{1}{2} \cdot 24 = l \cdot 1$$

$$60 = l$$

The whale is about 60 ft long.

Additional Example 2

Jill has just developed a print 5 in. wide by 7 in. long in her dark room. She wants to reduce the print on a photocopier.

a. What scale factor should she use to make the print 3 inches wide?

Let p = the percent at which Jill should copy the print.

$$\frac{\text{Width of reduction}}{\text{Width of original}} = p$$

$$\frac{3}{5} = p$$

$$0.6 = p$$

Jill should copy the print at 60%.

b. How long will the reduced print be?

Let l = the length of the reduced print.

$$\frac{\text{Length of reduced print}}{\text{Length of original print}} = 0.60$$

$$\frac{l}{7} = 0.60$$

$$l = 7 \cdot 0.60$$

$$l = 4.2$$

The new print will be 4.2 in. long.

Teach⇔Interact

Think and Communicate

Questions 3–5 should be discussed thoroughly in class so that all students understand the concepts involved. Questions 6 and 7 alert students to two important facts: (1) graphs can be used to mislead readers and (2) readers must determine if the data presented in a graph are relevant to their situations. Discuss ways in which graphs like those used for questions 6 and 7 might be used incorrectly. (For example, an employee might use the graph on the right to suggest to his employer that there has been a dramatic increase in the number of foreign tourists.)

Second-Language Learners
When discussing question 7, emphasize the value of being able to speak two or more languages. Consider having students brainstorm other businesses or organizations that would benefit by having bilingual and multilingual individuals available to serve customers.

Section Note

Common Error
Choosing a scale for the vertical axis of a graph is often a difficult process for some students. Not only do students make errors in choosing scales that are completely inappropriate for the data being graphed, but they also make errors in actually labeling the axis. These aspects of placing scales on graphs should be discussed with students for all graphs they draw. Stress that careful thought needs to be given to the choice of a scale and how it is written on the axis.

THINK AND COMMUNICATE

3. In Example 2, the original map has a scale of 1 in. = 1000 mi. Nathan did the calculation shown to find the scale of the enlarged map. Explain why he multiplied 1 in. by 125% to find the new scale.

4. How many times as small as the original is a picture changed by a scale factor of $\frac{1}{20}$? How many times as small as the building is a blueprint drawn to a scale of 1 in. = 20 ft? Use these examples to describe the difference between a scale and a scale factor.

5. How does the size of a copy compare to the size of the original when the scale factor is greater than 1? less than 1? equal to 1?

Scales on Graphs

The table shows an increase in tourism in the United States from 1970 through 1990.

Year	Visitors from other countries
1970	12,362,000
1975	15,698,000
1980	22,326,000
1985	25,399,000
1990	39,539,000

Even though the graphs below both represent these data, they give different impressions of the increase in tourism. The choice of the scale on the vertical axis of each graph affects the appearance of the graph.

Tourism in the United States

Visitors (millions): 0, 20, 40, 60, 80, 100
Year: 1970, 1975, 1980, 1985, 1990

Each interval on the vertical axis represents 20 million visitors.

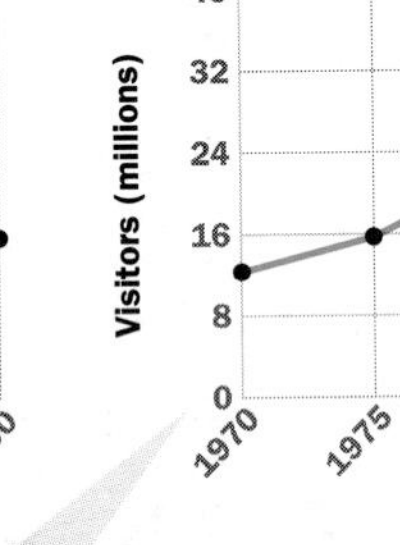

Each interval on the vertical axis represents 8 million visitors.

THINK AND COMMUNICATE

6. Compare and contrast the graphs. How does the scale on the vertical axis of each graph affect your impression of the data?

7. Suppose you are trying to encourage a hotel manager to hire more bilingual employees. Which graph above would you show him? How might you change the graph to make your point even more dramatic?

Think and Communicate

3. A distance of 1 in. on the original map corresponds to a distance of 1.25 in. on the new map.
4. 20 times; 240 times; A scale factor compares two quantities measured in the same units. A scale compares two quantities that may or may not be measured in the same units.
5. It is larger; it is smaller; it is the same size.
6. The graphs represent the same data. However, the graph on the right has smaller intervals and makes the increase look much more dramatic than it does in the graph on the left.
7. Answers may vary. An example is given. I would show the manager the second graph, which emphasizes the growth in tourism more than the first graph. I might change the intervals on the vertical axis to represent 1 or 2 million visitors to make the increase in tourism look even more pronounced.

CHECKING KEY CONCEPTS

1. The floor plan of a 90 ft by 200 ft building has a scale of 1 in. = 8 ft. Give the dimensions of the floor plan.
2. A portrait of a man about 6 ft tall is drawn 2.5 times life size in one of Daniel Galvez's murals. About how tall is the portrait?

For Questions 3 and 4, find the scale factor used to create each object.

3. a 5 in. by $7\frac{1}{2}$ in. postcard that pictures a 2 ft by 3 ft painting
4. a photograph 15 cm wide made from a negative 35 mm wide
5. Look back at the graphs on page 200. How can you change the scale on the *horizontal* axis to make the graph seem even more dramatic?

5.2 Exercises and Applications

Extra Practice exercises on page 566

1. This photograph of the space shuttle *Columbia* has a scale of 1 in. = 48 ft. Use a ruler to measure the photograph. Then use a proportion to find the real length of *Columbia*.

Tonya, the editor of the school yearbook, has to scale each picture in Exercises 2–6 to fill a column 2 in. wide. Give the scale factor she should use for each picture. Then find the height of each scaled picture.

2.

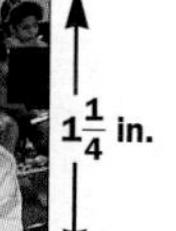

3.

4.

5.

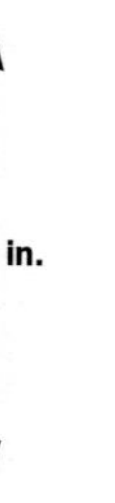

6.

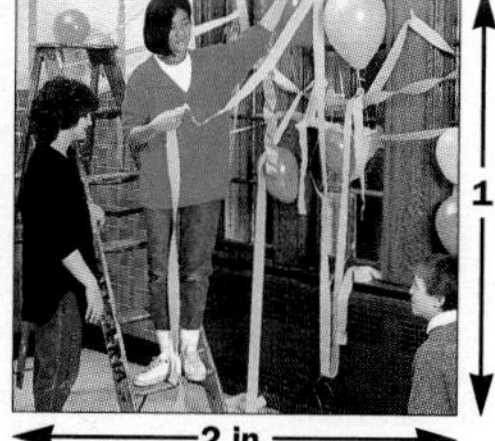

Teach⇔Interact

Checking Key Concepts

Student Progress
To ensure that students can work successfully with scales and scale factors, you may wish to have the solutions to these questions placed on the board for discussion.

Visual Thinking
Encourage students to work in small groups to collect magazine photos of people, buildings, and other objects. Have them estimate the height or width of the real person or object and then determine the approximate scale of the photo. Then, ask them to consider one portion of the photo, for example, the arm of the person or the door of the building, and to estimate its real size. This activity involves the visual skills of *recognition* and *generalization.*

Closure Question

How is a scale factor related to a scale and how are the two different?
A scale factor tells how a scale has changed. A scale uses units of measure; a scale factor does not.

Apply⇔Assess

Suggested Assignment

Core Course
Exs. 1–6, 10–15, 17–24, AYP

Extended Course
Exs. 1–24, AYP

Block Schedule
Day 30 Exs. 1–6, 10–15, 17–24, AYP

Exercise Notes

Reasoning
Exs. 2–6 Ask students to state whether each of these exercises will have a scale factor greater than 1, less than 1, or equal to 1. This activity will help students to set up the ratio for a scale factor.

Checking Key Concepts

1. 11.25 in. by 25 in.
2. about 15 ft
3. $\frac{5}{24}$
4. $\frac{30}{7}$
5. Answers may vary. An example is given. Let each of the given intervals on the horizontal axis represent 10 years instead of 5.

Exercise and Applications

1. Estimates may vary; about 120 ft.
2. $\frac{8}{5}$; 2 in.
3. $\frac{1}{2}$; $\frac{3}{4}$ in.
4. $\frac{4}{3}$; $3\frac{1}{3}$ in.
5. $\frac{2}{3}$; $1\frac{1}{3}$ in.
6. $\frac{1}{1}$; $1\frac{7}{8}$ in.

Apply⇔Assess

Exercise Notes

Application

Ex. 7 Architects use scales and scale drawings when designing buildings. Their scale drawings are called blueprints because the process used to create copies of the original design shows the drawings in white against a blue background.

Cooperative Learning

Ex. 9 Pairs of students can exchange diagrams and find the actual measurements of each other's home or school.

Communication: Discussion

Ex. 11 It may be helpful for students to discuss how the map of Montgomery is related to the map of Alabama. (The map of Montgomery is an enlargement of the same area on the map of Alabama.)

7. **Challenge** Kathryn Tate is designing a building in Germany. Her drawing has a scale of 1 cm = 1200 cm. She wants to compare the drawing to the blueprint of a building in the United States, which has a scale of 1 in. = 200 ft. At what percent should she copy the drawing of the building in Germany so that its scale becomes 1 in. = 200 ft? (*Hint:* First convert 200 ft to inches. Then use the fact that the scale 1 cm = 1200 cm is equivalent to the scale 1 in. = 1200 in.)

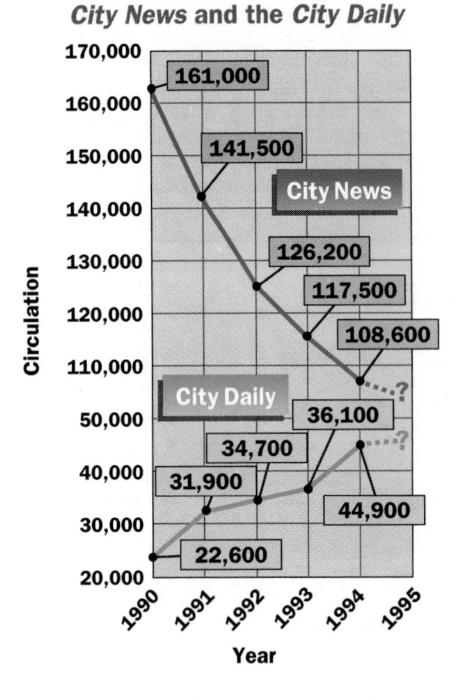

8. **BUSINESS** The graph shows estimated circulation figures for two competing newspapers, the *City News* and the *City Daily.*
   a. What does the graph suggest will happen in 1995?
   b. Find the average rate of change per year for each paper from 1990 to 1994. Use this rate to predict each paper's circulation for 1995.
   c. **Writing** What is deceptive about this graph? Which newspaper do you think is more likely to have created the graph? Why?
9. **Open-ended Problem** Draw a floor plan of a room in your home. Show the scale you used and explain how you calculated each measurement.
10. Troy wants to reduce a 12 in. by 12 in. drawing to 3 in. by 3 in. Unfortunately, the smallest scale factor the photocopier allows for is 50%. What scale factor can he use twice to reduce the drawing to the size he wants?
11. A map of Alabama has a scale of 1 in. = 20 mi. The close-up of Montgomery has a scale of 1 in. = 4 mi. What scale factor was used to change the size of the map of Alabama? Explain your reasoning.

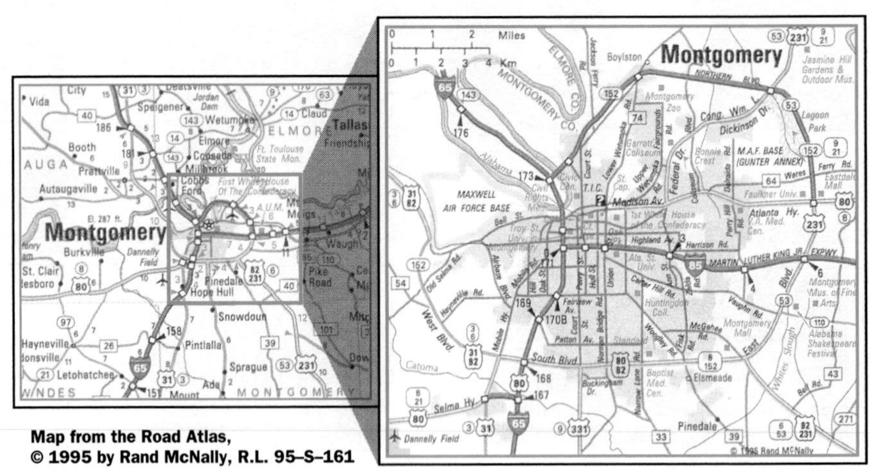

Map from the Road Atlas, © 1995 by Rand McNally, R.L. 95-S-161

7. 50%
8. a. The graph suggests at first glance that the circulation of the two papers will be the same, and the *City Daily*'s circulation will eventually surpass that of the *City News*.
   b. *City News*: –13,100 papers per year; 95,500
   City Daily: 5,575 papers per year; 50,475
   c. The scale on the vertical axis is not uniform. The labels indicate that seven of the eight intervals represent 10,000 papers. The fourth interval up from the origin represents 60,000 papers. The *City Daily* probably created the graph to improve the comparison.
9. Answers may vary. Check students' work.
10. 50%
11. 5 to 1; The scale factor is $\frac{\text{distance on Montgomery map}}{\text{distance on Alabama map}}$. Consider a distance of 4 mi. The corresponding distance on the Montgomery map is 1 in. and on the Alabama map is $\frac{4}{20}$ in. $= \frac{1}{5}$ in. $\frac{1}{\frac{1}{5}} = 1 \cdot \frac{5}{1} = \frac{5}{1}$
12. a. b.

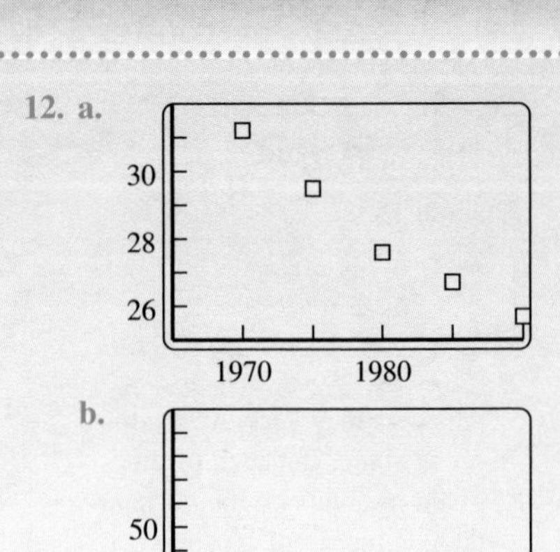

12. **Technology** The table gives the average number of gallons of milk consumed per person in the United States for various years from 1970–1990.

a. Use a graphing calculator to make a scatter plot of the data in the table. Set the viewing window as shown below. Make a sketch of the graph.

b. Change the scale on the vertical axis so that $0 \le y \le 100$. Make a sketch of the new graph.

c. Compare the two graphs. How does the scale affect the visual impression given by the data?

d. **Open-ended Problem** Find a viewing window that you think gives the "best" graph. Sketch your graph and show the scale you used. Explain why you think your graph is better than the graphs in parts (a) and (b).

Milk Consumption in the U.S. (per capita)

Year	Gallons consumed
1970	31.2
1975	29.5
1980	27.6
1985	26.7
1990	25.7

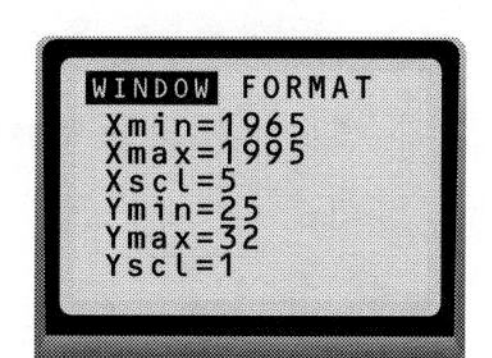

INTERVIEW Daniel Galvez

Look back at the article on pages 188–190.

Before Daniel Galvez painted the mural Crosswinds, he made a drawing of it by projecting slides onto a piece of paper. In this activity, your class will create a poster using a similar method.

Cooperative Learning **You will need an overhead projector, overhead transparencies, markers, a ruler, and a large piece of paper.**

13. Draw a picture of a person on a transparency. Measure the height of the person. What scale factor will make the person appear 12 in. tall when projected on a piece of paper taped to the wall?

14. Put your transparency on the overhead projector and measure the height of the person as it appears on the paper. By what scale factor did the image change? Without moving the projector, project the other students' transparencies onto the paper. Did they all change by the same scale factor? How do you know?

15. Will you have to move the projector forward or backward to make the person appear 12 in. tall? How do you know? Find this position and trace the outline of the person on the piece of paper. Then draw the details of your picture to the new scale. Repeat these steps for each student's picture.

16. **Writing** Describe a way to change this activity so that each person in the poster appears a different size.

c. Answers may vary. An example is given. The first graph appears to show a sharp drop, with milk consumption approaching 0 gallons per year. The second appears to show a low rate of consumption holding nearly steady over the twenty-year period.

d. Answers may vary. An example is given. Change the given scale so that the vertical axis has range $20 \le y \le 35$. I think the resulting graph indicates the decrease in the rate of consumption without exaggerating it.

13. Answers may vary.

14. Answers may vary, but the scale factor for each transparency within a project will be the same, because the overhead projector is not moved.

15. Answers may vary. Moving the projector forward will make an image smaller. Moving the projector back will make the image larger.

16. Answers may vary. An example is given. Move the projector to a different position for each drawing to make each image a different size.

Apply⇔Assess

Exercise Notes

Using Technology

Ex. 12 Students using a TI-82 should first clear all equations from the Y= list and clear statistical data lists L1 and L2. They should then enter the data from the table in these lists (years in L1, milk consumption data in L2). Next, they should enter appropriate window settings. To draw the scatter plot, press [2nd][STAT PLOT]4[ENTER]. Next, press [STAT PLOT] and select On from the second line of the Plot 1 screen. On the next line, select the first logo (the scatter plot logo). On the line for Xlist, select L1 and on the line for Ylist, select L2. On the last line, select the kind of mark desired for points of the scatter plot.

The TI-81 can also display scatter plots. The procedures for entering data and displaying the scatter plot are, however, quite different from those for the TI-82. Students should consult the TI-81 manual for details.

The *Stats!* software can also be used for this exercise. If students are working in small groups, there may be significant advantages to displaying the scatter plot on the larger computer screen.

Reasoning

Ex. 12(d) This exercise requires that students think about how the visual representation of the data is altered when the scales on the axes are changed. Students must apply critical thinking skills to make judgments about the value of each graph.

Interview Note

Ex. 15 The second question in this exercise (How do you know?) may be difficult for some students to answer without experimenting with the position of the projector. Refer these students to their answer for Ex. 13.

Journal Entry

Ex. 16 Have students record their answers to this exercise in their journals. Then have students organize themselves into groups of four to read their journal entries to each other, discussing which method seems the most efficient. By reading and listening to the journal entries, students can assess their own methods and the methods of the other students in their group.

Apply⇔Assess

Exercise Notes

Problem Solving

Ex. 17 Discuss with students how they solved this problem. One approach would be to observe that $1\frac{3}{4}$ min has seven 15-second intervals. After 1 min $\left(\frac{4}{4}\right)$, 4 of the 7 intervals are used and thus $\frac{4}{7}$ of the tank has been filled.

Progress Check 5.1–5.2

See page 232.

Practice 32 for Section 5.2

NAME ______________________ DATE __________

Practice 32

FOR USE WITH SECTION 5.2

For Exercises 1–6: A rectangle has the given dimensions: length b and width h.

a. Find the scale factor that should be used to transform each rectangle so that it has length 3 in.

b. Find the width of each scaled rectangle.

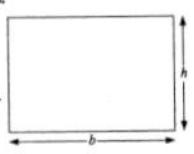

1. $b = 10$ in.; $h = 5$ in.
2. $b = 12$ in.; $h = 8$ in.
3. $b = 15$ in.; $h = 9$ in.
4. $b = 4.5$ in.; $h = 3.9$ in.
5. $b = 7.5$ in.; $h = 10$ in.
6. $b = 5$ in.; $h = 4$ in.

1.a. 0.3 b. 1.5 in.
2.a. 0.25 b. 2 in.
3.a. 0.2 b. 1.8 in.
4.a. $\frac{2}{3}$ b. 2.6 in.
5.a. 0.4 b. 4 in.
6.a. 0.6 b. 2.4 in.

7. A classroom in an architect's model of a proposed school building measures 8 in. by 11 in. The scale of the model is 1 in. = 3 ft. Find the dimensions of the actual classroom that will be built from the model. 24 ft by 33 ft

8. Bernadina wants to photocopy two pages that she has written, each of which measures 8.5 in. × 11 in. At the same time, she wants to reduce the two pages so that they fit side by side on one page of 8.5 in. × 14 in. paper, turned sideways, as shown.

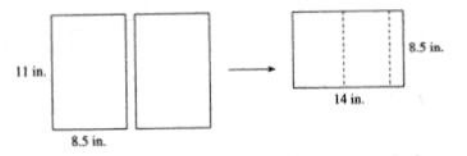

a. What is the largest scale factor Bernadina can use to reduce the height of each page? about 0.77

b. Bernadina used this scale factor and found that a narrow blank strip remained on the right side of the new paper. How wide was this strip? about 0.9 in.

9. The earnings of a small company are shown in the table at the right.

a. Construct a histogram from the table, using a vertical scale that runs from 0 to 20 (in millions of dollars). Check students' work.

b. Construct a histogram from the table, using a vertical scale that runs from 0 to 10. Check students' work.

c. If you were the chairman of the board of the company, which histogram would you show at a stockholders' meeting? probably the second histogram

Year	Earnings (million \$)
1989	1.5
1990	1.2
1991	2.6
1992	3.8
1993	4.7
1994	6.9

32 **Practice Bank,** ALGEBRA 1: EXPLORATIONS AND APPLICATIONS

17. SAT/ACT Preview A person spent $1\frac{3}{4}$ min filling a car's gasoline tank completely. If the person had stopped after 1 min, what part of the tank would have been filled?

A. $\frac{2}{7}$ **B.** $\frac{3}{7}$ **C.** $\frac{4}{7}$ **D.** $\frac{3}{4}$ **E.** $\frac{5}{7}$

ONGOING ASSESSMENT

18. Writing Describe the difference between a scale and a scale factor. Explain why one specifies units and the other does not.

SPIRAL REVIEW

Solve each inequality. Graph the solution on a number line. *(Section 4.6)*

19. $2x + 1 < 7$ **20.** $5(x - 4) > 0$ **21.** $3 - 5x \le 5(x + 3)$

22. $\frac{3}{5}x \ge 8$ **23.** $4(-x + 1) \ge 3x + 9$ **24.** $3x - 5 \le 6\left(\frac{1}{2}x + 1\right)$

ASSESS YOUR PROGRESS

VOCABULARY

ratio (p. 192)
proportion (p. 193)
means (p. 193)
extremes (p. 193)
scale (p. 198)
scale factor (p. 199)

Find the ratio of the first quantity to the second quantity. Write your answer as a fraction in lowest terms. *(Section 5.1)*

1. 5 min; 2 h **2.** 3 kg; 200 g

Solve each proportion. *(Section 5.1)*

3. $4:x = 6:15$ **4.** $\frac{3.6}{10.8} = \frac{p}{100}$ **5.** $\frac{x}{20 - x} = \frac{2}{3}$

6. A week before Election Day, 276 out of 450 people said that they would vote for Jean Ferreira. If 15,240 registered voters plan to vote, how many votes should Jean Ferreira expect to get? *(Section 5.1)*

7. Gary Leeds wants to build a basement playroom for his children. He makes a drawing using a scale of $\frac{3}{4}$ in. = 1 ft. *(Section 5.2)*

a. What width in the scale drawing should Gary use to represent a doorway 30 in. wide?

b. The room in the scale drawing is 15 in. by 22 in. Find the actual dimensions of the playroom.

8. Journal Explain the difference between a ratio and a scale. Which was easier for you to understand? Why?

17. C

18. Answers may vary. A scale factor compares two quantities measured in the same units. A scale compares two quantities that may or may not be measured in the same units, so the units must be specified.

19. $x < 3$

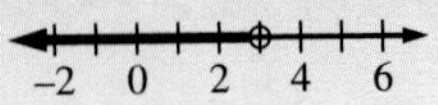

20. $x > 4$

21. $x \ge -1.2$

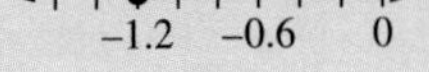

22. $x \ge \frac{40}{3}$

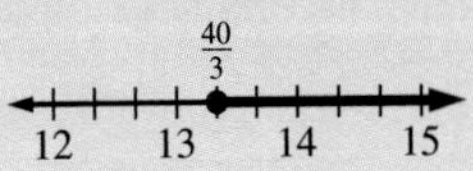

23. $x \le -\frac{5}{7}$

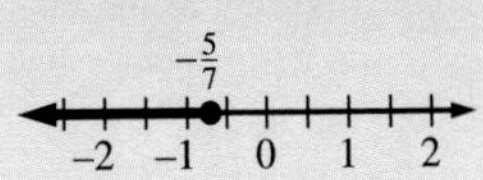

24. identity

Assess Your Progress

1. $\frac{1}{24}$
2. $\frac{15}{1}$
3. 10
4. $33.\overline{3}$
5. 8
6. Estimates may vary due to rounding; about 9347 votes.
7. a. $1\frac{7}{8}$ in.
   b. 20 ft by $29\frac{1}{3}$ ft
8. A ratio is a quotient that compares two quantities. A scale is one type of ratio; it compares two quantities that may or may not be measured in the same units. Answers may vary. An example is given. I found ratios easier to understand, because sometimes I confused scale with scale factor and did not know whether or not I needed to be concerned about units.

SECTION

5.3 Working with Similarity

Learn how to...

- recognize and draw similar figures

So you can...

- measure large distances in real-life situations

What do triangles have to do with totem poles, trees, and other tall objects? *Similar* triangles can help you measure the heights of tall objects indirectly.

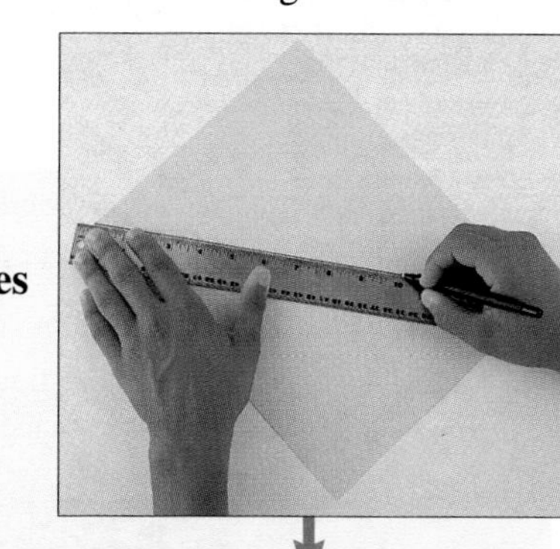

EXPLORATION
COOPERATIVE LEARNING

Constructing Similar Triangles

Work with a partner.
You will need:

- a rectangular sheet of paper
- scissors
- a protractor
- a ruler

SET UP Cut along the diagonal of a piece of paper to make two triangles. Label the corners of one triangle *A*, *B*, and *C* as shown. Fold the unlabeled triangle to make a line parallel to the shortest side. Cut along this line and label the corners of the smaller triangle *D*, *E*, and *F* as shown. Then answer the questions below.

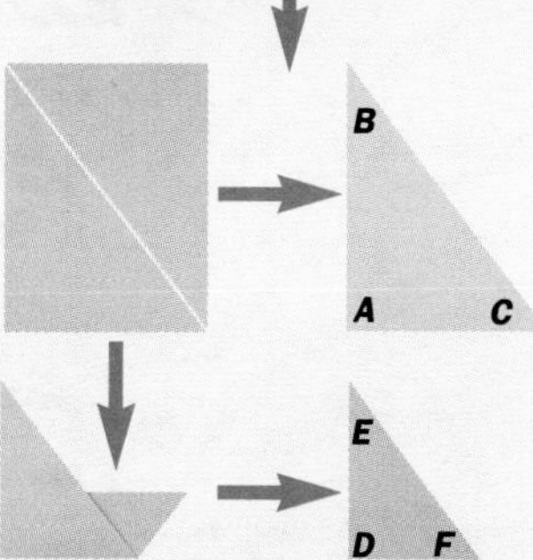

Questions

1. Copy and complete the tables. Give lengths to the nearest millimeter.
2. What do you notice about the angles of $\triangle ABC$ and $\triangle DEF$?

$\triangle ABC$	
Measure of $\angle A$	?
Measure of $\angle B$	?
Measure of $\angle C$	?
Length of $\overline{AB}$	?
Length of $\overline{BC}$	?
Length of $\overline{AC}$	?

$\triangle DEF$	
Measure of $\angle D$	?
Measure of $\angle E$	?
Measure of $\angle F$	?
Length of $\overline{DE}$	?
Length of $\overline{EF}$	?
Length of $\overline{DF}$	?

3. Find each ratio. What do you notice? (AB stands for the length of $\overline{AB}$.)

   a. $\frac{AB}{DE}$ **b.** $\frac{BC}{EF}$ **c.** $\frac{AC}{DF}$

4. The triangles you made are called *similar* triangles. What do you think it means for two triangles to be similar?

Plan⇔Support

Objectives

- Recognize and draw similar figures.
- Use similar figures to measure large distances.

Recommended Pacing

Core and Extended Courses
Section 5.3: 2 days

Block Schedule
Section 5.3: 2 half-blocks
(with Sections 5.2 and 5.4)

Resource Materials

Lesson Support
Lesson Plan 5.3
Warm-Up Transparency 5.3
Practice Bank: Practice 33
Study Guide: Section 5.3
Explorations Lab Manual:
Diagram Master 10

Technology
Internet:
http://www.hmco.com

Warm-Up Exercises

Solve each proportion.

1. $\frac{1}{x} = \frac{4}{160}$ 40
2. $\frac{x}{19} = \frac{10}{7}$ ≈27.14
3. $\frac{2}{8} = \frac{x}{4+8}$ 3
4. $\frac{7}{16} = \frac{5 \cdot 2}{x}$ ≈22.86

Exploration Note

Purpose
The purpose of this Exploration is to have students discover the properties of similar triangles.

Materials/Preparation
Each group should have all the materials listed in the text. Students may need a review of measuring angles with a protractor.

Procedure
Make sure students label their triangles correctly. Suggest that different groups fold along different parallel lines.

Closure
Have students discuss their results for questions 1–4. Because of measuring inaccuracies, not all students will have equal measures for corresponding angles or equal ratios for corresponding sides. By examining all the results, however, students should understand that corresponding angles of similar triangles have the same measure and that the three ratios of the pairs of corresponding sides are all equal.

Explorations Lab Manual
See the Manual for more commentary on this Exploration.

Diagram Master 10

For answers to the Exploration, see following page.

Teach⇔Interact

Section Notes

Second-Language Learners
It may be helpful to students learning English to have *similar* defined as being the same in some important ways.

Alternate Approach
You may wish to relate the ratio of corresponding sides of similar figures to students' understanding of scale factor. In a pair of similar figures, one figure is either an enlargement, a reduction, or a copy of the other figure. Therefore, the second figure is related to the first figure by a scale factor that is either greater than 1, less than 1, or equal to 1. Since the ratio of any side of $\triangle KLM$ to the corresponding side of $\triangle RST$ is 1 : 2, $\triangle KLM$ is related to $\triangle RST$ by a scale factor of 0.5.

Additional Example 1

Triangle $ABC \sim$ triangle XYZ.

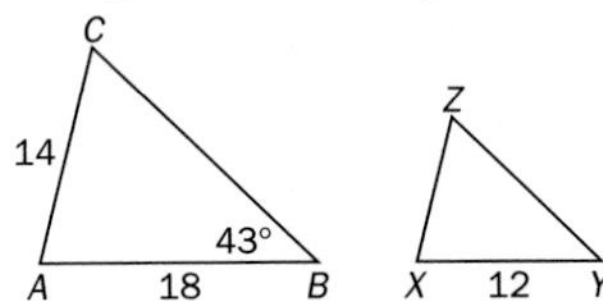

a. Find the measure of $\angle Y$.
$\angle Y$ and $\angle B$ are corresponding angles, so $\angle Y = \angle B = 43°$.

b. Find XZ.
Find the ratio of the lengths of corresponding sides.

$$\frac{AB}{XY} = \frac{18}{12}$$
$$\frac{AC}{XZ} = \frac{18}{12}$$
$$\frac{14}{XZ} = \frac{18}{12}$$
$$168 = 18XZ$$
$$9\frac{1}{3} = XZ$$

Similar Figures

The triangles you made in the Exploration are called *similar figures.* Two figures are **similar** if corresponding angles have equal measure and the ratios of the lengths of corresponding sides are equal.

WATCH OUT!
When you name similar figures, always name the corresponding angles in the same order. That is, if $\angle A = \angle D$ and $\angle B = \angle E$ and $\angle C = \angle F$, then write $\triangle ABC \sim \triangle DEF$.

The figures below are similar. To show this you write $\triangle KLM \sim \triangle RST$, which you read as "triangle KLM is similar to triangle RST."

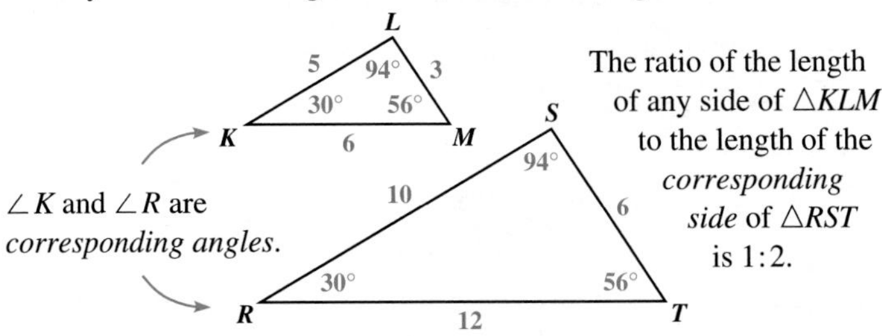

Similar figures can be the same size or different sizes, but they are always the same shape.

EXAMPLE 1

Figure $ABCD \sim$ figure $EFGH$.

a. Find the measure of $\angle F$.

b. Find FG.

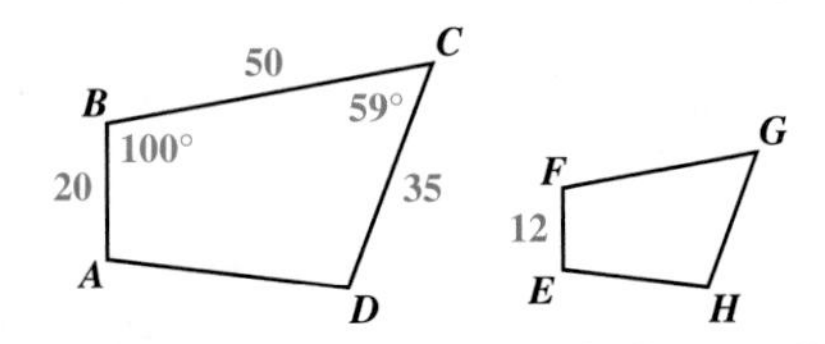

SOLUTION

a. $\angle F$ and $\angle B$ are corresponding angles, so $\angle F = \angle B = 100°$.

b. Find the ratio of the lengths of corresponding sides.

$$\frac{AB}{EF} = \frac{20}{12}$$

$\overline{BC}$ and $\overline{FG}$ are also corresponding sides. Set the ratio of their lengths equal to $\frac{AB}{EF}$.

$$\frac{BC}{FG} = \frac{20}{12}$$

Substitute 50 for BC. Solve the proportion.

$$\frac{50}{FG} = \frac{20}{12}$$
$$600 = FG \cdot 20$$
$$30 = FG$$

THINK AND COMMUNICATE

Questions 1 and 2 refer to figures *ABCD* and *EFGH* in Example 1.

1. How would you use similarity to find GH?
2. Can you use similarity to find EH? Explain.

ANSWERS Section 5.3

Exploration

1–4. Answers may vary. An example is given for a sheet of typing paper, about 280 mm by 216 mm, oriented vertically, with the fold in Step 2 made 51 mm up from the bottom. Angle measures are rounded to the nearest degree and lengths to the nearest millimeter.

1.

$\triangle ABC$	
Measure of $\angle A$	90°
Measure of $\angle B$	38°
Measure of $\angle C$	52°
Length of $\overline{AB}$	280 mm
Length of $\overline{BC}$	353 mm
Length of $\overline{AC}$	216 mm

$\triangle DEF$	
Measure of $\angle D$	90°
Measure of $\angle E$	38°
Measure of $\angle F$	52°
Length of $\overline{DE}$	229 mm
Length of $\overline{EF}$	290 mm
Length of $\overline{DF}$	179 mm

2. $\angle A$ and $\angle D$ have the same measure, $\angle B$ and $\angle E$ have the same measure, and $\angle C$ and $\angle F$ have the same measure.
3. a. about 1.2
   b. about 1.2
   c. about 1.2
   The three ratios are the same.
4. I think it means the triangles have angles with equal measure and three pairs of sides whose lengths have the same ratio.

Similar Triangles

To see whether two figures are similar, you need to check that all corresponding angles are equal in measure and that the ratios of lengths of corresponding sides are equal. There is a simpler test for similar triangles.

A Test for Similar Triangles

Two triangles are similar if two angles of one triangle have the same measures as two angles of the other triangle.

EXAMPLE 2 Application: Measuring Height Indirectly

Jane wants to estimate the height of the Thunderbird House Post totem pole. She stands so that the end of her shadow meets the end of the totem pole's shadow (A). Use similar triangles and the measurements given to find the height of the totem pole.

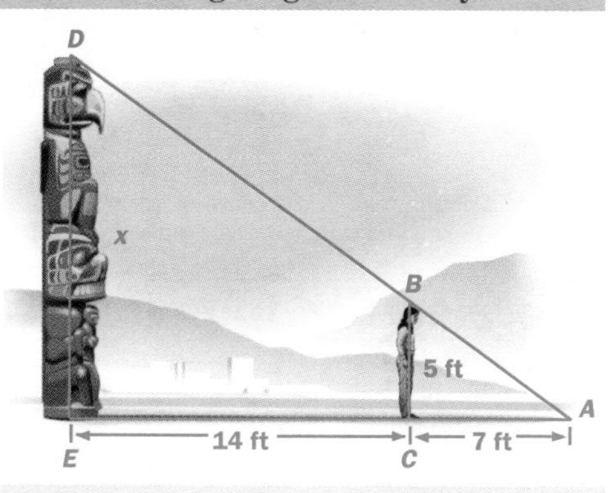

SOLUTION

First, find a pair of similar triangles. $\triangle ABC$ and $\triangle ADE$ share $\angle A$. Since Jane and the totem pole are both perpendicular to the ground, $\angle C = \angle E = 90°$. Two angles of $\triangle ABC$ have the same measures as two angles of $\triangle ADE$, so the triangles are similar.

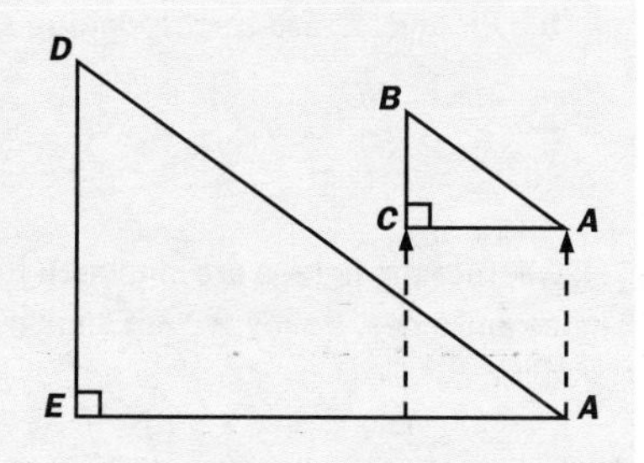

Let $x =$ the height of the totem pole. Write and solve a proportion involving x.

$$\frac{BC}{DE} = \frac{AC}{AE}$$

Since $\triangle ABC \sim \triangle ADE$, the ratios of the lengths of corresponding sides are equal.

$$\frac{5}{x} = \frac{7}{7 + 14}$$

$$5(7 + 14) = x \cdot 7$$

$$105 = 7x$$

$$15 = x$$

The totem pole is about 15 ft tall.

BY THE WAY

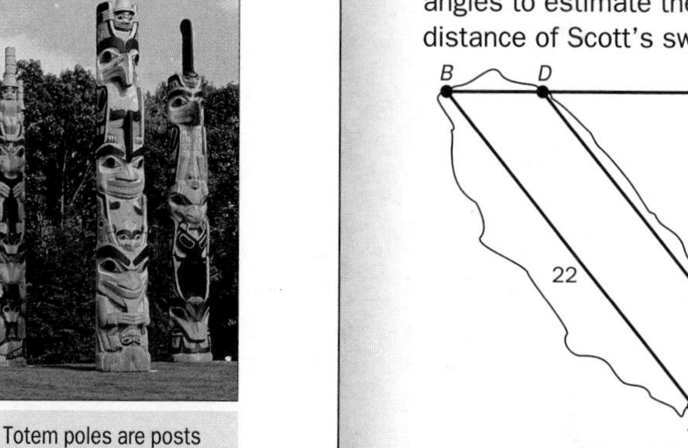

Totem poles are posts carved and painted by Native Americans of the Northwest Coast. These poles support houses, serve as memorials, and preserve family histories.

Teach⇔Interact

Learning Styles: Visual

Working in pairs, students can draw pairs of triangles to demonstrate the test for similar triangles. Students in each group should first choose the measures for two angles of a triangle. They should then each draw a triangle with two angles having the measures chosen. Have students test the triangles for similarity.

Additional Example 2

Scott and his family vacationed at a large lake 22 miles long. One summer, Scott swam close to the shore of the lake from point C to point D. Scott's brother watched him through binoculars from a fire tower. The base of the fire tower, at point X is about 12.4 mi from point C and about 17.6 mi from point A. Assume that $\triangle XBA \sim \triangle XDC$. Use similar triangles to estimate the straight line distance of Scott's swim.

$$\frac{CD}{AB} = \frac{XC}{XA}$$

$$\frac{CD}{22} = \frac{12.4}{17.6}$$

$$CD \cdot 17.6 = 272.8$$

$$CD = 15.5$$

Scott swam about 15.5 mi.

Think and Communicate

1. Solve the proportion $\frac{35}{GH} = \frac{20}{12}$; $GH = 21$.
2. No; you would need to know AD.

Teach⇔Interact

Checking Key Concepts

Communication: Discussion Point out to students that the notation *ABCD* ~ *EFGH* identifies not only the corresponding angles of the two figures, but also their corresponding sides. For example, because *AB* and *EF* occupy the same position when the figures are named, *AB* and *EF* are corresponding sides. Ask students to name each pair of corresponding sides and corresponding angles in figures *ABCD* and *EFGH*.

Closure Question

If two figures are similar, what two facts do you know about the figures? Corresponding angles are equal in measure; the ratios of the lengths of corresponding sides are equal.

Apply⇔Assess

Suggested Assignment

Core Course
Day 1 Exs. 1–10
Day 2 Exs. 11–17, 19–27

Extended Course
Day 1 Exs. 1–10
Day 2 Exs. 11–27

Block Schedule
Day 30 Exs. 1–10
Day 31 Exs. 11–27

Checking Key Concepts

For Questions 1–5, use the fact that *ABCD* ~ *EFGH*.

1. Complete each statement.
   a. $\angle B$ and __?__ are corresponding angles.
   b. __?__ and $\overline{FG}$ are corresponding sides.
2. Find the measure of $\angle A$.
3. Find the measure of $\angle H$.
4. Find *AB*.
5. Find *HG*.

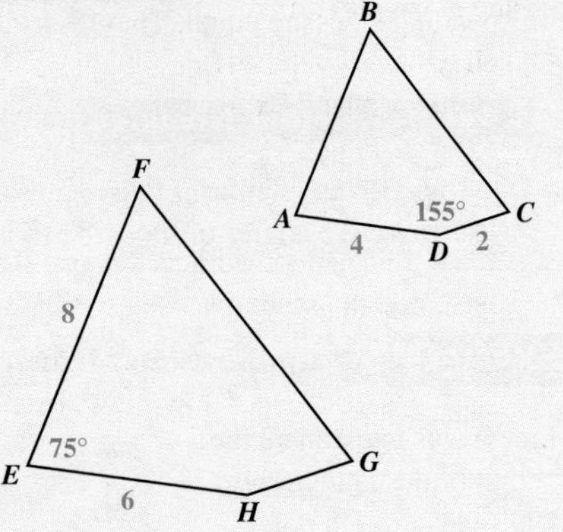

5.3 Exercises and Applications

Extra Practice exercises on page 567

1. Use the fact that $\triangle MNO \sim \triangle PQR$ to complete each statement.
   a. $\angle N =$ __?__
   b. $\overline{PR}$ and __?__ are corresponding sides.
   c. $\frac{MN}{?} = \frac{NO}{?}$

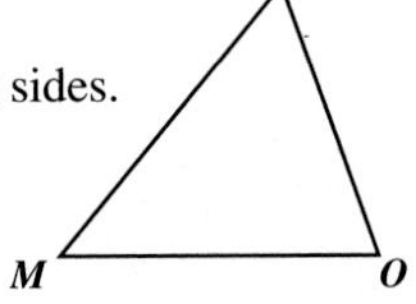

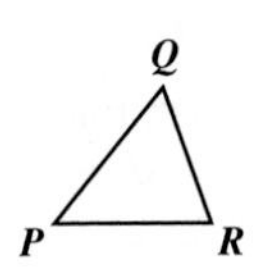

Tell whether the figures are similar. If they are, list each pair of corresponding angles and sides. If they are not similar, explain why.

2.

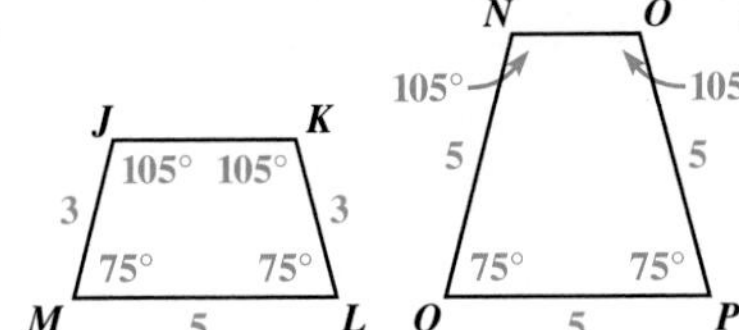

3.

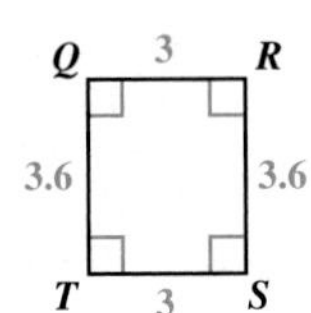

4.

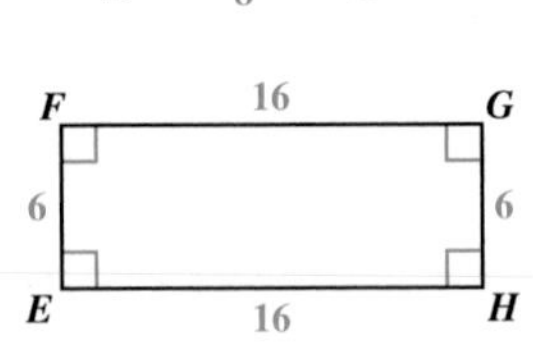

5.

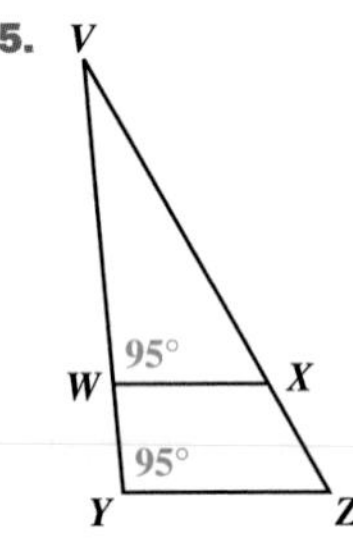

Checking Key Concepts

1. a. $\angle F$ b. $\overline{BC}$
2. 75°
3. 155°
4. $\frac{16}{3}$
5. 3

Exercises and Applications

1. a. $\angle Q$
   b. $\overline{MO}$
   c. *PQ*; *QR*
2. No.
3. Answers may vary. Because the two figures are rectangles, there are several ways to write the similarity. For example, *MNOP* ~ *QRST* and *MNOP* ~ *RSTQ*. The examples are based on the similarity *MNOP* ~ *QRST*; $\angle M$ and $\angle Q$, $\angle N$ and $\angle R$, $\angle O$ and $\angle S$, $\angle P$ and $\angle T$, $\overline{MN}$ and $\overline{QR}$, $\overline{NO}$ and $\overline{RS}$, $\overline{PO}$ and $\overline{TS}$, $\overline{MP}$ and $\overline{QT}$.
4. No.
5. $\angle V$ and $\angle V$, $\angle VWX$ and $\angle Y$, and $\angle VXW$ and $\angle Z$, and $\overline{VW}$ and $\overline{VY}$, $\overline{WX}$ and $\overline{YZ}$, $\overline{VX}$ and $\overline{VZ}$
6. $\angle J = 66°$, $\angle L = 121°$, $\angle O = 95°$, $\angle Q = 78°$, $JK = 3$, $OP = 3.75$, $NQ = 6$
7. a. Yes; the corresponding angles have equal measure and the ratios of the lengths of corresponding sides are equal.
   b. Yes; the corresponding angles have equal measure and the ratios of the lengths of corresponding sides are equal.
   c. No; the corresponding angles do not have equal measure and the ratios of the lengths of corresponding sides are not equal.
8. a. Answers may vary.
   b. Answers may vary. An example is given. I would use a protractor to measure the angles of two patches I thought might be similar. If the patches were triangles,

For Exercises 6 and 7, use the quadrilaterals *JKLM* and *NOPQ*.

6. Figure $JKLM \sim$ figure $NOPQ$. Find each missing side length and angle measure.

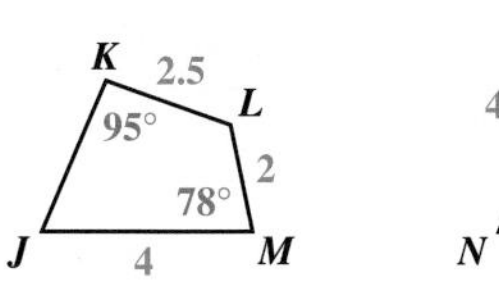

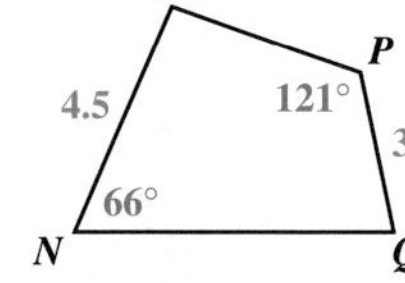

7. Tell whether each statement is *True* or *False*. Explain your reasoning.

   a. $KLMJ \sim OPQN$ b. $MLKJ \sim QPON$ c. $JKLM \sim QPON$

8. **HISTORY** The photo shows a patchwork coat that belonged to Uesugi Kenshin, a famous Japanese general. The coat is over 400 years old.

   a. Look carefully at the shapes of the patches. Find a pair of quadrilaterals that appear to be similar. Explain your reasoning.

   b. **Writing** Explain precisely what you would do to decide whether two patches in the coat are actually similar.

9. **GEOMETRY** A *regular polygon* is a figure whose angles are all of equal measure and whose side lengths are all equal.

   a. Are all squares similar to each other? How do you know?

   b. An equilateral triangle has three sides of equal length and three 60° angles. Are all equilateral triangles similar to each other? How do you know?

   c. Draw a regular pentagon whose angles all measure 108°. Draw another one the same way, but use a different length for the sides. Are these pentagons similar? Are all regular pentagons similar to each other? Explain.

   d. Make a conjecture about regular polygons and similarity based on your answers to parts (a)–(c).

10. **Open-ended Problem** Draw and label two similar triangles so that the ratio of the lengths of corresponding sides is 3:5.

11. **Writing** Is a figure similar to itself? Explain.

12. **Visual Thinking** Draw a triangle and label its corners *A*, *B*, and *C*. Use a ruler to find the midpoint of each side. Draw $\triangle DEF$ where *D* is the midpoint of $\overline{AB}$, *E* is the midpoint of $\overline{BC}$, and *F* is the midpoint of $\overline{AC}$.

    a. How many triangles are inside the original triangle?

    b. Name all the similar triangles. Give the ratio of corresponding sides.

    c. Draw another triangle and repeat the exercise. Did you get the same answers?

Apply⇔Assess

Exercise Notes

Student Progress
Ex. 7 This exercise asks students to explain why the pairs of figures do or do not fit the definition of similar figures. Explaining why a mathematical definition cannot be applied to a certain situation is an excellent means of assessing students' understanding of the definition.

Student Study Tip
Ex. 9(b) This exercise gives rise to the statement "All equilateral triangles are similar triangles." Ask students to provide one example that shows the statement "All similar triangles are equilateral triangles" is false. Point out that such an example is called a *counterexample*.

Second-Language Learners
Ex. 9(d) If necessary, paraphrase *make a conjecture* as *state a rule that you think may be true.*

Cooperative Learning
Ex. 12 This exercise can be done by having students work in groups, sharing their results with each other and thereby verifying their answer for part (c). You may wish to challenge students to examine the relationship of the sides of the *middle* triangle to the sides of the original triangle. Ask them to complete this statement: "A segment joining the midpoints of two sides of a triangle is __?__ to the third side of the triangle and __?__ the measure of the third side." (parallel; one-half)

I would need to determine only whether the measures of two angles of one patch were equal to the measures of two angles of the other. If the patches were quadrilaterals, I would first have to make sure corresponding angles had equal measures. Then I would have to compare the ratios of lengths of corresponding sides to see if they were equal.

9. a. Yes; all four angles of a square are right angles, so corresponding angles of the two squares have equal measure. Let x be the length of a side of one square and y the length of a side of the other. The ratio of the length of any side of the first square to any side of the second is $\frac{x}{y}$.

   b. Yes; the measure of each angle of an equilateral triangle is 60°, so corresponding angles of the two triangles have equal measure. Let x be the length of a side of one equilateral triangle and y the length of a side of the other. The ratio of the length of any side of the first triangle to any side of the second is $\frac{x}{y}$.

   c. Check students' work; Yes; Yes. Corresponding angles of the two pentagons have equal measure. Let x be the length of a side of one pentagon and y the length of a side of the other. The ratio of the length of any side of the first pentagon to any side of the second is $\frac{x}{y}$.

   d. Regular polygons with the same number of sides are similar.

10. Check students' work.

11. Yes, the corresponding angles have equal measures and the ratios of the lengths of corresponding sides is 1:1. (Each side corresponds to itself.)

12. See answers in back of book.

Apply⇔Assess

Exercise Notes

Multicultural Note

Exs. 13–18 Ancient Chinese mathematicians developed some interesting and useful mathematical concepts and theories. Chinese mathematicians understood the importance of the number zero and the idea of a negative number. The decimal system may have been developed in China before it began to be used in Europe. The mathematician Liu Hui refined the value of π (originally computed by the Greeks), accurately, calculating it to the value 3.14159.

Challenge

Ex. 15 If students think about the difference between a ratio and a rate, it may help them to better understand how to answer this question.

Cooperative Learning

Ex. 20 Once students have completed this exercise and know the height of the wall, they can use the same activity to find the heights of students who were not measured. These students can walk backward from the mirror until they can see the top of the wall. Then similar triangles and the known height of the wall can be used to find students' heights.

Visual Thinking

Ex. 20 When students have completed this activity, you may wish to ask them to create a sketch that shows how they might estimate the height of a tall building, cliff, or tree by using similar triangles. Encourage them to explain their sketches to the class. This activity involves the visual skills of *correlation* and *communication.*

Connection HISTORY

One of the earliest known mathematics textbooks is the *Nine Chapters on the Mathematical Art*, the *Jiŭzhāng suànshù* of China. The following problem is from this 2000-year-old book.

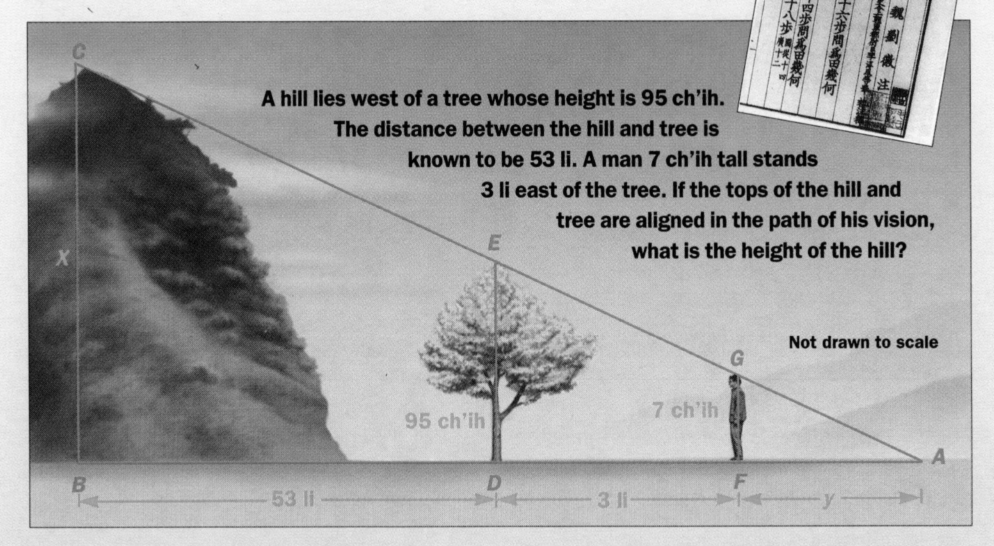

A hill lies west of a tree whose height is 95 ch'ih. The distance between the hill and tree is known to be 53 li. A man 7 ch'ih tall stands 3 li east of the tree. If the tops of the hill and tree are aligned in the path of his vision, what is the height of the hill?
Not drawn to scale
C
E
G
x
95 ch'ih
7 ch'ih
A
B
D
F
53 li
3 li
y

13. Explain why $\triangle AFG \sim \triangle ADE$.

14. Solve the proportion below:
$$\frac{y \text{ li}}{y + 3 \text{ li}} = \frac{7 \text{ ch'ih}}{95 \text{ ch'ih}}$$

15. **Challenge** Explain why you can use the proportion in Exercise 14 to find AF, even though both li and ch'ih are used.

16. Explain why $\triangle AFG \sim \triangle ABC$.

17. Find the length of $\overline{BC}$. Explain how you got your answer.

18. **Writing** Based on the measurements in the illustration, about how long do you think a li is? How long do you think a ch'ih is? Explain your reasoning.

19. **SAT/ACT Preview** For which of the following values of x is rectangle $ABCD$ similar to rectangle $EFGH$?

I. 5 **II.** 17 **III.** 7

A. I only

B. II only

C. III only

D. I and II only

E. I, II, and III

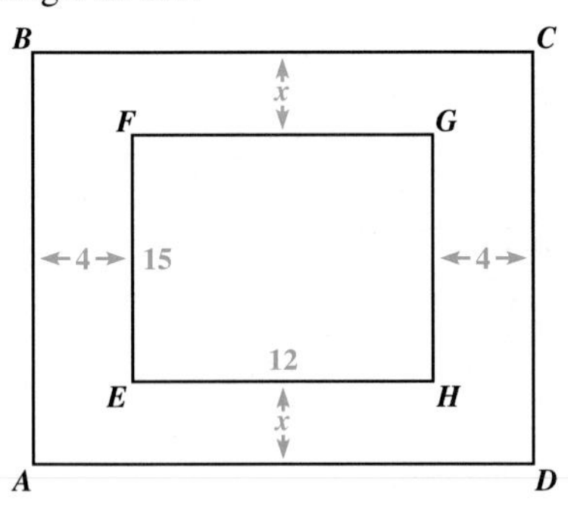

13. $\triangle AFG$ and $\triangle ADE$ share $\angle A$ and $\overline{ED}$ and $\overline{GF}$ are both perpendicular to the ground, so $\angle D$ and $\angle F$ are both right angles. Two angles of $\triangle AFG$ have the same measures as two angles of $\triangle ADE$, so the triangles are similar.

14. $\frac{21}{88} \approx 0.24$

15. The units are the same in the numerator and denominator of each ratio; $\frac{y \text{ li}}{(y+3) \text{ li}}$ is the same as $\frac{y}{y+3}$, and $\frac{7 \text{ ch'ih}}{95 \text{ ch'ih}}$ is the same as $\frac{7}{95}$.

16. $\triangle AFG$ and $\triangle ABC$ share $\angle A$ and $\overline{ED}$ and $\overline{CB}$ are both perpendicular to the ground, so $\angle B$ and $\angle F$ are both right angles. Two angles of $\triangle AFG$ have the same measures as two angles of $\triangle ABC$, so the triangles are similar.

17. about 1635 ch'ih; Use the fact that $\triangle AFG \sim \triangle ABC$ to write the proportion $\frac{y}{y+56} = \frac{7}{x}$. Substitute the value of y found in Ex. 14 and solve for x.

18. Answers may vary. An example is given. about 1 yd; about 1 ft; Explanations may vary.

20. **Cooperative Learning** Work with a partner to follow the steps below. You will need a yardstick or a tape measure and a small mirror.

- Place the mirror some distance from a wall. Measure the distance between the mirror and the wall.
- Facing the wall, back up along the line perpendicular to the wall and passing through the mirror. Stop when you can see the top of the wall in the mirror. Have your partner measure the distance between the mirror and your feet.

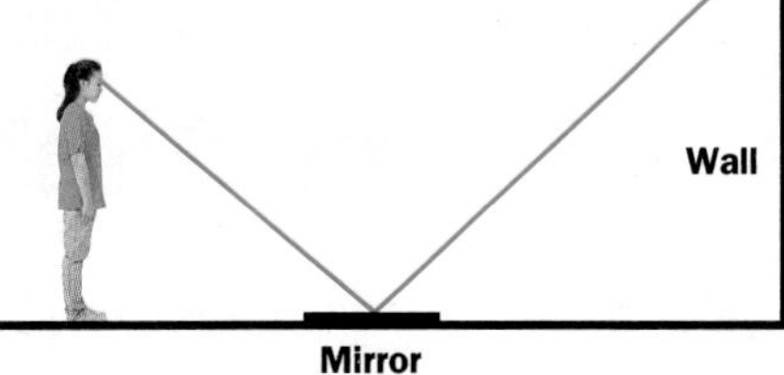

- Have your partner measure your height.

a. Draw a diagram like the one shown. Include all measurements.

b. The two triangles in your drawing are similar. Use this fact to calculate the height of the wall.

c. Exchange roles with your partner and repeat the activity with the mirror in a different position. Do you get the same answer for the height of the wall?

ONGOING ASSESSMENT

21. **Writing** If $\triangle ABC \sim \triangle DEF$ and $\triangle DEF \sim \triangle GHI$, is it true that $\triangle ABC \sim \triangle GHI$? Explain your reasoning.

SPIRAL REVIEW

Express each rate in the given units. Write each answer as a unit rate. *(Section 3.1)*

22. $21,000 per year = __?__ per month

23. 160 pages per hour = __?__ pages per day

24. 186,000 miles per second = __?__ feet per minute

Tell whether the data show direct variation. If they do, give the constant of variation and write an equation. *(Section 3.2)*

25.

Capacity (liters)	Distance (meters)
2.5	13.125
2.6	14.976
2.7	16.983
2.8	19.152
2.9	21.489
3.0	24.000

26.

Time (min)	Distance (cm)
1.3	5.59
2.5	10.75
4.0	17.20
6.1	26.23
7.2	30.96
7.7	33.11

27.

Age (years)	Weight (lb)
5	7.5
10	15.0
15	22.5
30	45.0
50	75.0
70	112.5

Apply⇔Assess

Exercise Notes

Assessment Note

Ex. 21 Most students intuitively understand the concept of transitivity of similar figures. For students who have difficulty with this idea, the use of a few real-world examples of transitivity should help them. For example, If I am older than Dillon and Dillon is older than Kate, then I am older than Kate. Ask volunteers to suggest other examples.

Practice 33 for Section 5.3

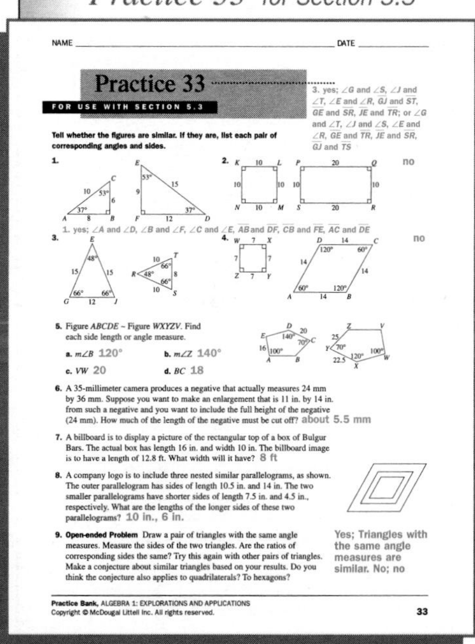
NAME ______ DATE ______

Practice 33

FOR USE WITH SECTION 5.3

Tell whether the figures are similar. If they are, list each pair of corresponding angles and sides.

1. 1. yes; $\angle A$ and $\angle D$, $\angle B$ and $\angle F$, $\angle C$ and $\angle E$, $\overline{AB}$ and $\overline{DF}$, $\overline{CB}$ and $\overline{FE}$, $\overline{AC}$ and $\overline{DE}$

2. no

3. 3. yes; $\angle G$ and $\angle S$, $\angle J$ and $\angle T$, $\angle E$ and $\angle R$, $\overline{GJ}$ and $\overline{ST}$, $\overline{GE}$ and $\overline{SR}$, $\overline{JE}$ and $\overline{TR}$; or $\angle G$ and $\angle T$, $\angle J$ and $\angle S$, $\angle E$ and $\angle R$, $\overline{GE}$ and $\overline{TR}$, $\overline{JE}$ and $\overline{SR}$, $\overline{GJ}$ and $\overline{TS}$

4. no

5. Figure $ABCDE \sim$ Figure $WXYZV$. Find each side length or angle measure.
   a. $m\angle B$ 120° b. $m\angle Z$ 140°
   c. VW 20 d. BC 18

6. A 35-millimeter camera produces a negative that actually measures 24 mm by 36 mm. Suppose you want to make an enlargement that is 11 in. by 14 in. from such a negative and you want to include the full height of the negative (24 mm). How much of the length of the negative must be cut off? about 5.5 mm

7. A billboard is to display a picture of the rectangular top of a box of Bulgur Bars. The actual box has length 16 in. and width 10 in. The billboard image is to have a length of 12.8 ft. What width will it have? 8 ft

8. A company logo is to include three nested similar parallelograms, as shown. The outer parallelogram has sides of length 10.5 in. and 14 in. The two smaller parallelograms have shorter sides of length 7.5 in. and 4.5 in., respectively. What are the lengths of the longer sides of these two parallelograms? 10 in., 6 in.

9. **Open-ended Problem** Draw a pair of triangles with the same angle measures. Measure the sides of the two triangles. Are the ratios of corresponding sides the same? Try this again with other pairs of triangles. Make a conjecture about similar triangles based on your results. Do you think the conjecture also applies to quadrilaterals? To hexagons? Yes; Triangles with the same angle measures are similar. No; no

Practice Bank, ALGEBRA 1: EXPLORATIONS AND APPLICATIONS

33

19. A
20. a, b. Check students' work.
    c. The results should be about the same but may vary slightly due to measurement and rounding differences.
21. Yes. Reasoning may vary. An example is given. Since $\triangle ABC \sim \triangle DEF$, the measure of $\angle A$ = the measure of $\angle D$ and the measure of $\angle B$ = the measure of $\angle E$. Since $\triangle DEF \sim \triangle GHI$, the measure of $\angle D$ = the measure of $\angle G$ and the measure of $\angle E$ = the measure of $\angle H$. Then the measure of $\angle A$ = the measure of $\angle G$ and the measure of $\angle B$ = the measure of $\angle H$, so $\triangle DEF \sim \triangle GHI$.
22. $1750
23. 3840
24. 58,924,800,000
25. No.
26. Yes; 4.3; $y = 4.3x$.
27. No.

Plan⇔Support

Objectives

- Find the ratios of the perimeters and areas of similar figures.
- Use scale drawings to find the perimeters and areas of large objects.
- Create accurate pictographs that represent statistics.

Recommended Pacing

Core and Extended Courses
Section 5.4: 1 day

Block Schedule
Section 5.4: $\frac{1}{2}$ block
(with Section 5.3)

Resource Materials

Lesson Support
Lesson Plan 5.4
Warm-Up Transparency 5.4
Practice Bank: Practice 34
Study Guide: Section 5.4
Assessment Book: Test 20

Technology
Technology Book:
Spreadsheet Activity 5
Internet:
http://www.hmco.com

Warm-Up Exercises

Find the perimeter of each figure.

1. a rectangle whose length is 2.5 m and whose width is 1.3 m 7.6 m
2. a square whose side is $9\frac{1}{2}$ in. long 38 in.
3. a circle whose radius is 2 cm about 12.57 cm

Find the area of each figure.

4. a square whose side is $15\frac{1}{4}$ ft long $232\frac{9}{16}$ ft^2
5. a triangle whose base is 20 yd long and whose height is 20 yd 200 yd^2

SECTION

5.4 Perimeters and Areas of Similar Figures

Learn how to...

- **find the ratios of the perimeters and areas of similar figures**

So you can...

- **use scale drawings to find the perimeters and areas of houses, for example**
- **create accurate pictographs that represent statistics such as the price of stamps**

What does it mean for a square to double in size? Look at the sets of tiles below and see whether you can find a pair of figures in which one set appears to be twice as large as the other.

Figure *A*

Figure *B*

Figure *C*

Figure *D*

THINK AND COMMUNICATE

Use the figures above to answer Questions 1–3.

1. Explain how you know that figures *A, B, C,* and *D* are similar.
2. For each pair of figures, find the ratio of the perimeters. Then find the ratio of the areas. Describe any patterns you see.
   a. *A* and *B* **b.** *B* and *D* **c.** *A* and *C* **d.** *B* and *C*
3. How do the perimeter and area of a square change when its side length is doubled? when it is tripled? Explain your answers.
4. Look at the diagrams below. All the triangles shown are similar equilateral triangles. How do the perimeter and area of a triangle change when its side length is doubled? when it is tripled?

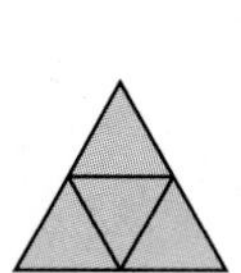

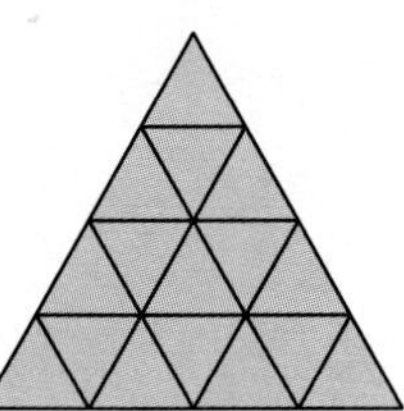

As you saw in Questions 1–4, there is a relationship between the side lengths, perimeters, and areas of similar figures. One way to express this relationship is to use ratios.

ANSWERS Section 5.4

Think and Communicate

1. The measures of all the angles are equal; the ratios of the lengths of corresponding sides are equal.
2. a. $\frac{1}{2}$; $\frac{1}{4}$ b. $\frac{1}{2}$; $\frac{1}{4}$ c. $\frac{1}{3}$; $\frac{1}{9}$ d. $\frac{2}{3}$; $\frac{4}{9}$
   In each case, the ratio of the areas is the square of the ratio of the perimeters.
3. The perimeter doubles. Let the length of a side of the original square be s. The length of a side of the larger square is $2s$, so its perimeter is $4(2s)$ or $8s$, which is double the perimeter of the original square. The area is multiplied by 4, since $(2s)^2 = 4s^2$. If the length of a side of the larger square is $3s$, its perimeter is $4(3s)$ or $12s$, which is triple the perimeter of the original square. The area is multiplied by 9, since $(3s)^2 = 9s^2$.
4. Perimeter is doubled and area is multiplied by 4; perimeter is tripled and area is multiplied by 9.

Suppose the ratio of the lengths of corresponding sides of two similar figures is $a:b$. The ratio of the perimeters is $a:b$, and the ratio of the areas is $a^2:b^2$.

For example, the rectangles shown are similar.

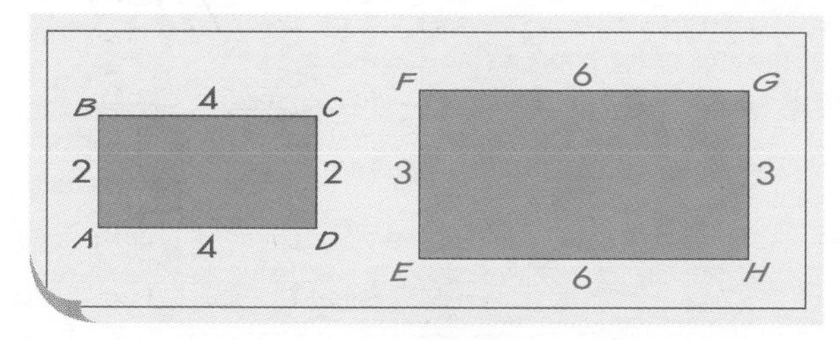

Side length of **ABCD** : Side length of **EFGH** = **2:3**

Perimeter of **ABCD** : Perimeter of **EFGH** = **12:18** = **2:3**

Area of **ABCD** : Area of **EFGH** = **8:18** = **4:9**

EXAMPLE 1 Application: Architecture

The drawing below shows the floor plan of the Manuél Atencio house in Trampas, New Mexico. The ratio of the length of a side of the house in the diagram to the length of the corresponding side of the actual house is 1 cm: 300 cm. Estimate the perimeter of the actual house.

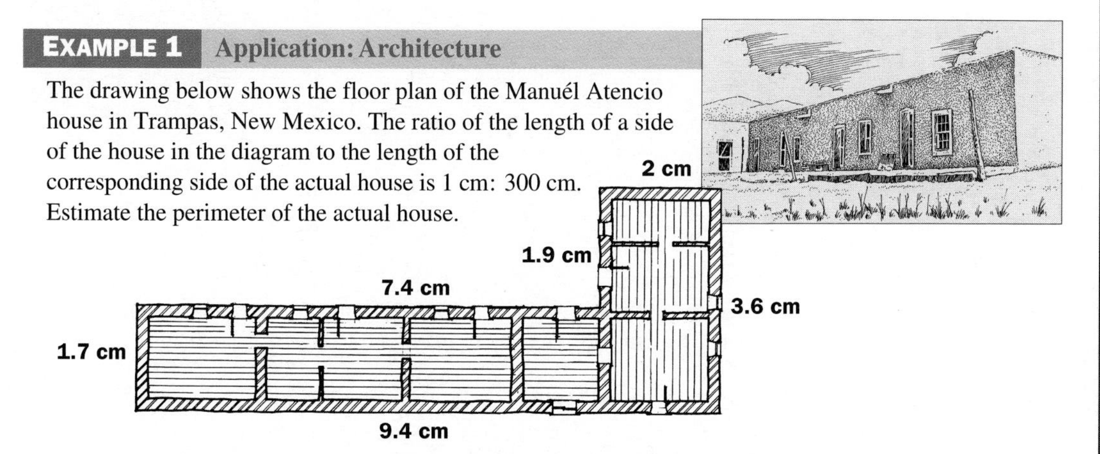

SOLUTION

Find the perimeter of the house in the diagram by adding the side lengths.

$$1.7 + 7.4 + 1.9 + 2 + 3.6 + 9.4 = 26 \text{ cm}$$

Then use what you know about the perimeters of similar figures to estimate the perimeter of the actual house. Let P = the perimeter of the house.

$$\frac{\text{Side length of house in diagram}}{\text{Side length of actual house}} = \frac{\text{Perimeter of house in diagram}}{\text{Perimeter of actual house}}$$

$$\frac{1}{300} = \frac{26}{P}$$

Use the means-extremes property.

$$P = 7800$$

The perimeter of the house is about 7800 cm, or 78 m.

Teach⇔Interact

Additional Example 1

The drawing below shows a design for one of the gardens at Flower Farm. Each side of the drawing of the rhombus-shaped garden is 3 cm. The diameter of the circle is 2.5 cm. The ratio of a length in the drawing to a corresponding length of the actual garden is 1 cm : 120 cm. Find the perimeter of the garden.

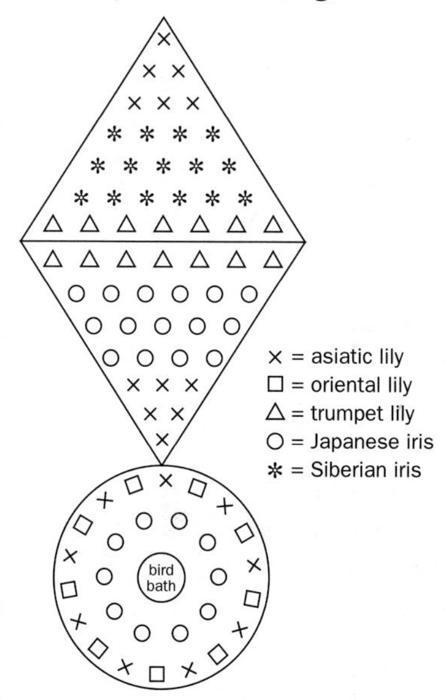

Find the perimeter of the garden in the diagram by adding the side lengths. Find the circumference of the circle by using the formula $C = \pi d$. Then add the results.

$4(3) + 3.14(2.5) = 19.85$ cm

Let P = the perimeter of the garden.

$$\frac{1}{120} = \frac{19.85}{P}$$

$P = 2382$

The perimeter of the garden is about 2382 cm, or 23.82 m.

Section Note

Multicultural Note

In the 1820s, when Manuél Atencio was building his house, the land that is today New Mexico became a province of Mexico, At that time, Mexico had just gained independence from Spain and changed its name from New Spain to Mexico. When New Mexico gained statehood, it adopted its current name to distinguish itself from Mexico.

Teach⇔Interact

Additional Example 2

Find the area of the garden given in Additional Example 1. Assume the base of each triangle is 3.3 cm and the height is 2.5 cm.

First find the area of the garden in the diagram. Use $A = \pi r^2$ with $r = \frac{1}{2}d = \frac{1}{2}(2.5) = 1.25$ for the area of the circle.

$2\left[\frac{1}{2}(3.3 \cdot 2.5)\right] + (3.14 \cdot 1.25^2) = 13.15625$

The area is about 13.16 cm^2.

Now use what you know about the areas of similar figures to find the area of the actual garden.

Let A = the area of the actual garden.

$\frac{1^2}{120^2} = \frac{13.16}{A}$

$\frac{1}{14{,}400} = \frac{13.16}{A}$

$A = 189{,}504$

The area of the garden is about 189,504 cm^2. Convert this to square meters by using the fact that $(1\text{ m})^2 = (100\text{ cm})^2 = 10{,}000\text{ cm}^2$.

$189{,}500\text{ cm}^2 \cdot \frac{1\text{ m}^2}{10{,}000\text{ cm}^2} \approx 19\text{ m}^2$

The area of the garden is about 19 m^2.

Checking Key Concepts

Visual Thinking

Check students' understanding of the concept of perimeter by asking them to clip floor plans of a house or apartment from a home or architectural magazine, or from the home section of a newspaper. Ask them to estimate the length of one side of the real building. Then, ask them to determine the perimeter of the real building based on the perimeter of the floor plan. This activity involves the visual skills of *correlation* and *interpretation.*

Closure Question

How are scale drawings used to find perimeters and areas of actual objects?

Use the scale of the drawing to write a proportion in which one ratio is the scale itself and the other ratio is the perimeter or area of the drawing to the unknown perimeter or area of the actual object.

EXAMPLE 2 Application: Architecture

Find the area of the Manuél Atencio house.

SOLUTION

First find the area of the house in the diagram by dividing the house into two rectangles. Add the areas.

$(1.7)(7.4) + (3.6)(2) = 19.78\text{ cm}^2$

2 cm
7.4 cm
3.6 cm
1.7 cm

Now use what you know about the areas of similar figures to find the area of the actual house.

The ratio of the squares of the side lengths is equal to the ratio of the areas.

$$\frac{(\text{Side length of house in diagram})^2}{(\text{Side length of actual house})^2} = \frac{\text{Area of house in diagram}}{\text{Area of actual house}}$$

$$\frac{1^2}{300^2} = \frac{19.78}{A}$$

$$\frac{1}{90{,}000} = \frac{19.78}{A}$$

$$A = 1{,}780{,}200$$

The area of the house is about 1,780,200 cm^2. Convert this to square meters by using the fact that $(1\text{ m})^2 = (100\text{ cm})^2 = 10{,}000\text{ cm}^2$.

$$1{,}780{,}200\text{ cm}^2 \cdot \frac{1\text{ m}^2}{10{,}000\text{ cm}^2} \approx 178\text{ m}^2$$

The area of the house is about 178 m^2.

CHECKING KEY CONCEPTS

Find the missing perimeter and area for each pair of similar figures.

1. 10, $P = 44$, $A = 104$; 8, $P = ?$, $A = ?$

2. 12, $P = ?$, $A = ?$; 9, $P = 24$, $A = 27$

Checking Key Concepts

1. 35.2; 66.56
2. 32; 48

Exercises and Applications

1. $\frac{4}{9}$; $\frac{16}{81}$
2. $\frac{7}{6}$; $\frac{7}{6}$
3. 132.3; 36
4. 25; 6.25
5. 225π; 15; 5
6. $50; The cost of the fence should depend on the perimeter of the area enclosed. The perimeter of the smaller area is half that of the larger area, so the fence should cost half as much as for the larger area.
7. a. $\frac{25}{49}$
   b. $14.70; The area of the large pizza is $\frac{49}{25}$ times that of the small pizza, so the large pizza should cost $\frac{49}{25}$ times as much as the small pizza.

5.4 Exercises and Applications

Extra Practice exercises on page 567

1. The perimeters of two similar pentagons are 20 in. and 45 in. What is the ratio of lengths of corresponding sides? What is the ratio of the areas?
2. Two similar triangles have areas of 49 cm^2 and 36 cm^2. What is the ratio of lengths of corresponding sides? What is the ratio of the perimeters?

Find the missing values for each pair of similar figures.

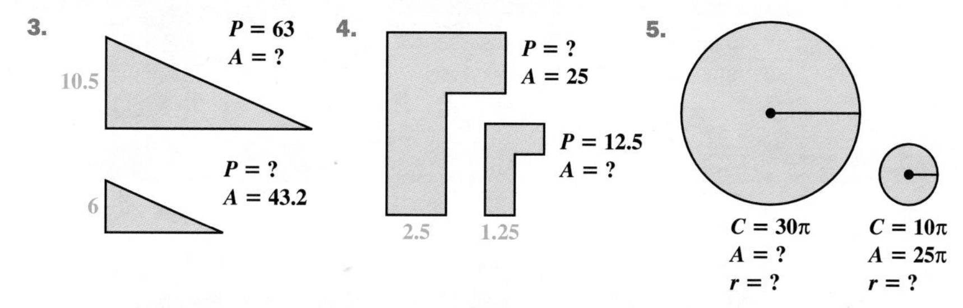

6. Caitlin wants to enclose a 20 ft by 24 ft rectangular area with a fence. She is told that the fence will cost $100. Since this is more than she can afford, she decides to enclose a smaller area, a 10 ft by 12 ft rectangle. What price should she expect to pay for the fence now? Explain.
7. Dee's Pizza has 10 in. and 14 in. diameter pizzas on the menu.
   a. What is the ratio of the areas of the two pizzas?
   b. The price of a 10 in. pizza is $5. What do you think the price of a 14 in. pizza should be? Explain your reasoning.
   c. **Writing** Dee is thinking of adding a 20 in. pizza to the menu. Should she charge twice as much as she does for the 10 in. pizza? Explain your thinking.
8. **CONSTRUCTION** Carpenters refer to a blueprint when building a home. The one shown has a scale of 1 in. = 15 ft, or 1 in. = 180 in. Use a ruler to measure the dimensions of the family room. Find the actual area of the family room in *square yards.* How did you get your answer?

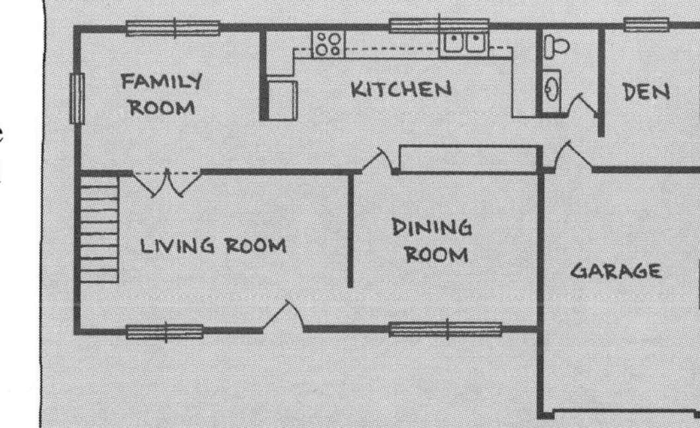

9. **Challenge** Suppose the ratio of the side lengths of two cubes is 1 : 3.
   a. The surface area of a cube is $6x^2$ where x is the length of one side of the cube. What is the ratio of the surface areas of the cubes?
   b. What is the ratio of the volumes of the cubes? How do you know?
   c. Suppose you have two spheres with diameters of 3 m and 4 m. What do you think is the ratio of the volumes of the spheres? Explain.

$\frac{x}{3x}$ $\frac{x^2}{9x^2} = \frac{1}{9}$ $\frac{x^3}{(3x)^3} = \frac{1}{27}x^3$

c. No; the area of the large pizza is four times that of the small pizza, so the large pizza should cost four times as much as the small pizza, or $40.

8. 18.75 yd^2; Answers may vary. I measured the area of the family room on the diagram. It is $\frac{3}{4}$ $in.^2$. The scale is 1 in. to 15 ft, so the ratio of the areas is 1 $in.^2$ to 225 ft^2. Then $\frac{3}{4}$ $in.^2$ represents 168.75 ft^2. Since 3 ft = 1 yd, 9 ft^2 = 1 yd^2. The actual area of the family room is $\frac{168.75}{9}$ yd^2 or 18.75 yd^2.

9. a. $\frac{1}{9}$
   b. $\frac{1}{27}$; Let x and $3x$ represent the lengths of the sides; the volumes are x^3 and $(3x)^3$ or $27x^3$.
   c. $\frac{27}{64}$; $\frac{\frac{4}{3}\pi \cdot 3^3}{\frac{4}{3}\pi \cdot 4^3} = \frac{3^3}{4^3} = \frac{27}{64}$

Apply⇔Assess

Suggested Assignment

Core Course
Exs. 1–8, 10–13, 15–22, AYP

Extended Course
Exs. 1–22, AYP

Block Schedule
Day 31 Exs. 1–22, AYP

Exercise Notes

Common Error
Ex. 2 Some students mistakenly take half a number when they intend to take the square root of a number. Remind these students that the area of a square is found by multiplying a number by itself and not by two.

Application
Ex. 7(c) When students can relate the mathematics they are learning to a real-world problem, they see the value of mathematics. Students can discuss the considerations Dee must take into account when determining a price for her goods. She needs to consider, for example, how much her food and labor costs will increase for a 20 in. pizza.

Teaching Tip
Ex. 8 Although house and room measurements are generally given in square feet, carpeting, wood flooring, and linoleum are sold by the square yard.

Second-Language Learners
Ex. 8 Be sure students know that a *blueprint* is a diagram that shows how a building is to be built. Point out that the blueprint shown here is a *floor plan.* Explain that a floor plan shows the size of the floor of each room on one level of a house. A floor plan would exist for each level of a house. So, when you see a floor plan, it is as if you are looking down from the sky at a house whose roof and ceilings have been removed.

Using Manipulatives
Ex. 9 The use of actual cubes and spheres may be helpful to students who are visual and kinesthetic learners. Also, you may need to review cubic measures with some students.

Apply⇔Assess

Exercise Notes

Integrating the Strands
Exs. 11–14 These exercises integrate concepts from the mathematical strands of statistics, geometry, measurement, and number. Mention this to students and ask them to identify one concept from each strand used in the exercises.

Career Connection
Exs. 11–14 You may wish to explore with students the types of tasks that a graphic designer may be involved in. A few suggestions, such as magazine design, book design, or commercial product design, can get students thinking about other types of graphic designs. Ask students what may motivate a person to pursue a career in graphic design. (artistic talent, strong visual skills) Point out that there are special schools and courses of study that teach graphic design.

10. PHOTOGRAPHY Suppose you have a 5 in. by 7 in. photo framed. A framing store charges you $7 for the wood you select, $3 for the glass to cover the photo, and $20 for the labor to assemble the parts. What should you expect to pay for having a 10 in. by 14 in. photo framed? Explain your reasoning.

Connection STATISTICS

Graphic designers are often hired to create interesting pictographs. A *pictograph* is a bar graph that uses pictures or symbols to represent quantities. In Exercises 11–14, you will explore the accuracy of pictographs.

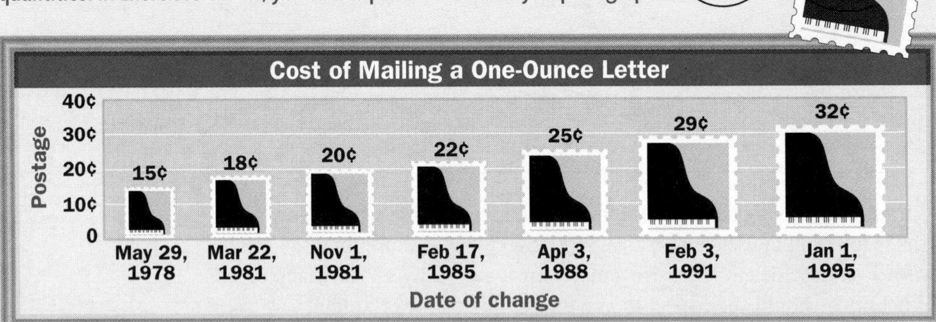

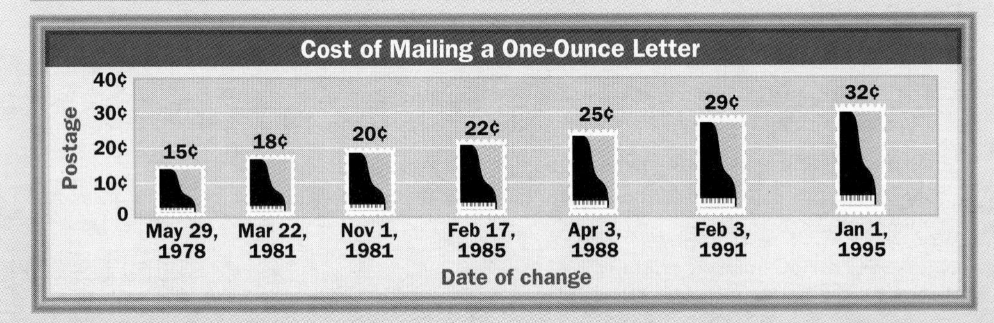

11. Without measuring, choose a stamp on each pictograph that seems to be twice the size of the May 29, 1978, stamp. Explain your choice.

12. Use a ruler to measure the height of each May 29, 1978, postage stamp to the nearest millimeter. Find the stamp on each pictograph that is about twice the height of the May 29, 1978, stamp on the same pictograph. Does each stamp represent about twice as much postage?

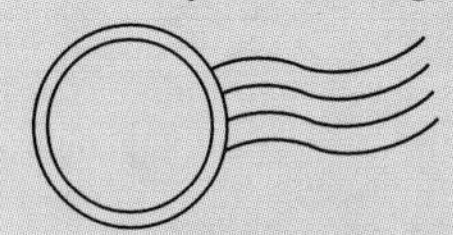

13. Now find the area of each May 29, 1978, postage stamp. Find the stamp on each pictograph that has about twice the area of the May 29, 1978, stamp of the same pictograph. Does this stamp represent about twice as much postage? Are these the same stamps you found for Exercise 12?

14. Open-ended Problem Which pictograph do you think is more accurate? Why? Describe another way to use a stamp to illustrate the data accurately.

10. $46; The perimeter of the new frame is twice that of the old, so the wood should cost twice as much, $14. The glass for the new frame is 4 times the area of the glass for the old frame, so it should cost 4 times as much, $12. The parts to be assembled require about the same amount of work, so the labor cost should still be $20, although that cost could vary.

11. Answers may vary. An example is given. In the upper pictograph, I would choose the 22¢ stamp. In the lower pictograph, I would choose the 29¢ stamp. Both appear to have area about twice that of the May 29, 1978 stamp.

12. about 8 mm; the 29¢ stamp (upper pictograph), 29¢ stamp (lower pictograph); Yes.

13. about 64 mm^2; the 22¢ stamp (upper pictograph), 32¢ stamp (lower pictograph); No, Yes; No, No.

14. Answers may vary. Examples are given. I think the lower pictograph is more accurate, because the ratios of the areas of the stamps reflect the ratios of the costs. I would use square stamps. I would determine the length of each side by finding the appropriate area and taking the square root. For example, if the cost of the stamp doubles, the ratio of the areas should be 1 : 2, so the ratio of the side lengths should be $1 : \sqrt{2}$. To represent doubling the cost of the May 29, 1978 stamp, I would draw a square stamp with sides about $8(\sqrt{2})$ or about 11 mm long.

15. D

15. SAT/ACT Preview If the side of a square is increased by 20%, the area of the square is increased by:

A. 20% **B.** 40% **C.** 37.5% **D.** 44%

ONGOING ASSESSMENT

16. Writing Suppose the lengths of the sides of rectangle $ABCD$ are x and y, and the lengths of the sides of rectangle $EFGH$ are kx and ky. Show that the ratio of the perimeters is k:1 and the ratio of the areas is k^2:1.

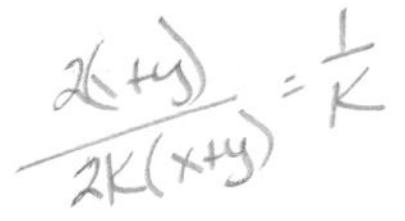

SPIRAL REVIEW

Solve. If an equation is an *identity* or there is *no solution*, say so.
(Sections 4.2 and 4.3)

17. $\frac{2}{3}t - 7 = -\frac{1}{3}$ **18.** $-4(3 + 5p) = 48$ **19.** $-1 = \frac{5d}{7} + 1$

20. $\frac{3m - 7}{5} = -1.2$ **21.** $29 = \frac{3}{5}n + 2$ **22.** $13(3 - 4s) = 0$

ASSESS YOUR PROGRESS

VOCABULARY

similar (p. 206)

1. Writing A tabletop air hockey game (top right) is $18\frac{1}{2}$ in. by 40 in. An ice hockey rink (bottom right) is 100 ft by 200 ft. Is the tabletop game an accurate representation of a real hockey rink? Explain your reasoning. *(Section 5.3)*

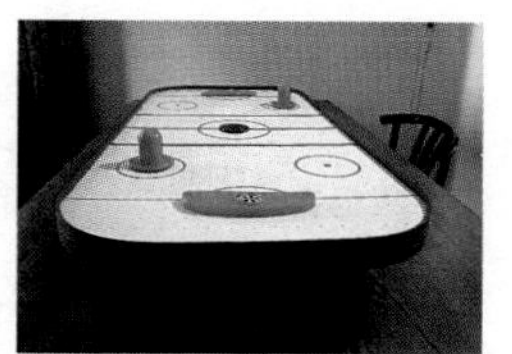

In the drawing, $\triangle ABC \sim \triangle DEF$. Use this drawing for Exercises 2–5.

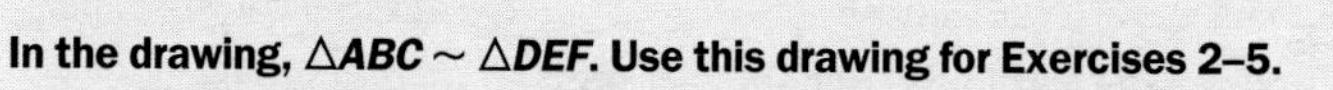

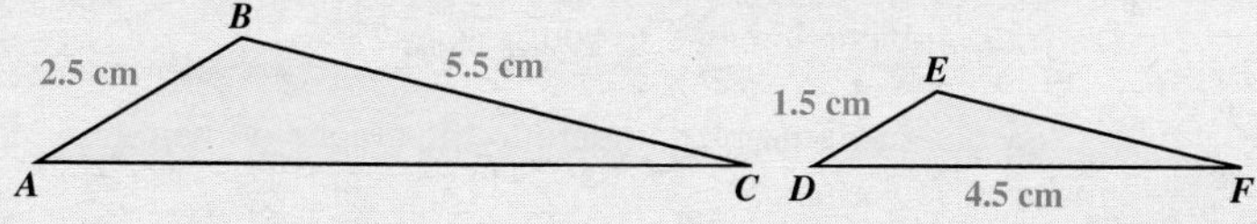

2. Find the length of $\overline{AC}$. *(Section 5.3)*

3. Find the length of $\overline{EF}$. *(Section 5.3)*

4. Find the ratio of the perimeters of $\triangle ABC$ and $\triangle DEF$. *(Section 5.4)*

5. Find the ratio of the areas of $\triangle ABC$ and $\triangle DEF$. *(Section 5.4)*

6. Draw two squares of appropriate size to compare the areas of Alaska (591,004 mi^2) and Texas (266,807 mi^2). *(Section 5.4)*

7. Journal Describe an occupation that might require knowledge of similar figures and their perimeters and areas. How would someone with this occupation use the skills taught in Sections 5.3 and 5.4?

Apply⇔Assess

Assess Your Progress

Journal Entry
For Ex. 7, ask students to brainstorm a list of occupations. They can then choose one occupation to discuss in their journal entry.

Progress Check 5.3–5.4

See page 233.

Practice 34 for Section 5.4

NAME ______________________ DATE __________

Practice 34

FOR USE WITH SECTION 5.4

For Exercises 1–3, the perimeters of two similar polygons are given. Find
a. The ratio of the corresponding side lengths.
b. The ratio of the areas.

1. 11 in.; 22 in. a. 1:2 b. 1:4 **2.** 24 ft; 16 ft a. 3:2 b. 9:4 **3.** 9 cm; 15 cm a. 3:5 b. 9:25

For Exercises 4–6, the areas of two similar polygons are given. Find
a. The ratio of the corresponding side lengths.
b. The ratio of the perimeters.

4. 64 in.^{2}; 25 in.^{2} a. 8:5 b. 8:5 **5.** 5 m^2; 45 m^2 a. 1:3 b. 1:3 **6.** 63 yd^2; 28 yd^2 a. 3:2 b. 3:2

7. It costs $375 for carpeting to cover a rectangular floor that measures 15 ft by 20 ft. How much should it cost for the carpeting needed to cover a floor measuring 12 ft by 16 ft? $240

8. Suppose enough ceiling molding to go around the larger room in Exercise 7 costs $37.50. What should it cost to buy enough ceiling molding to go around the smaller room? $30

9. An overhead projector produces a screen image 40 in. high when it is placed 10 ft from the screen. The image is 56 in. high when the projector is placed 14 ft from the screen.

a. What is the ratio of the perimeters of the two screen images? 5:7

b. What is the ratio of the areas of the two screen images? 25:49

c. How do you think the ratio of projector-to-screen distances is related to the ratio of the areas? Its square is in the ratio of the areas.

10. A ceramic floor tile has the shape of a regular hexagon with side length 1.5 in. A larger version of the same tile has a side length of 2.5 in.

a. What is the ratio of the areas of the two tiles? 9:25

b. Suppose it takes 3600 of the smaller tiles to cover a bathroom floor. How many of the larger tiles would be needed for the same floor? (Hint: If one tile has twice the area of another, it takes half as many of the larger tiles to cover the same area.) 1296

11. Writing Suppose a and b are the bases of two similar rectangles and h and k are the heights of the same rectangles. Explain how you know that the ratio of the areas of the two rectangles is $\frac{a^2}{b^2}$. Start by writing the ratio of the areas, and then substitute using the area formula A = (base)(height). Remember that the ratios of corresponding sides are equal. Check students' work.

34 **Practice Bank,** ALGEBRA 1: EXPLORATIONS AND APPLICATIONS

---

16. perimeter of $EFGH = 2kx + 2ky = 2k(x + y)$; perimeter of $ABCD = 2x + 2y = 2(x + y)$; ratio of perimeters: $\frac{2k(x + y)}{2(x + y)} = \frac{k}{1}$; area of $EFGH$: $(kx)(ky) = k^2xy$; area of $ABCD$: xy; ratio of areas: $\frac{k^2xy}{xy} = \frac{k^2}{1}$

17. 10
18. –3
19. $-\frac{14}{5}$
20. $\frac{1}{3}$
21. 45
22. $\frac{3}{4}$

Assess Your Progress

1. No; $\frac{18.5 \text{ in.}}{100 \text{ ft}} \neq \frac{40 \text{ in.}}{200 \text{ ft}}$, so the figures are not similar.
2. 7.5 cm
3. 3.3 cm
4. $\frac{5}{3}$
5. $\frac{25}{9}$
6. Check students' work. The ratio of the lengths of the sides should be about $\frac{3}{2}$.
7. Answers will vary. Examples are given. architects, engineers, and interior decorators; An interior decorator, for example, could draw sketches and use the skills taught in these sections to determine the amounts of materials needed for actual projects.

Plan⇔Support

Objectives

- Calculate theoretical and experimental probability.
- Predict the outcome of events.

Recommended Pacing

Core and Extended Courses
Section 5.5: 1 day

Block Schedule
Section 5.5: $\frac{1}{2}$ block
(with Section 5.6)

Resource Materials

Lesson Support
Lesson Plan 5.5
Warm-Up Transparency 5.5
Overhead Visuals:
Folder 5: Probability Spinners
Practice Bank: Practice 35
Study Guide: Section 5.5
Explorations Lab Manual:
Diagram Masters 2, 11

Technology
Technology Book:
Calculator Activity 5
Graphing Calculator
Internet:
http://www.hmco.com

Warm-Up Exercises

Write each ratio as a decimal and as a percent.

1. $\frac{8}{36}$ about 0.22; about 22%
2. $\frac{6}{6}$ 1.0; 100%
3. $\frac{12}{25}$ 0.48; 48%
4. Find the sum. $\frac{1}{4}+\frac{1}{4}+\frac{1}{4}$ $\frac{3}{4}$
5. What is 50% of 1? $\frac{1}{2}$

SECTION

5.5 Exploring Probability

Learn how to...

- calculate theoretical and experimental probability

So you can...

- predict the outcome of events, such as whether a basketball player will make a free throw

There's a 70% chance of rain tomorrow. Melissa has a 50-50 chance of making a basket in basketball. You hear statements about chance, or *probability*, every day. Probability can help you to understand sports, politics, science, and other aspects of your life.

Probability is a ratio that measures how likely an **event** is to happen. When you roll a die, there are six possible results or **outcomes**. How often do you think you will roll a 2?

EXPLORATION
COOPERATIVE LEARNING

Rolling a Die

Work with a partner.
You will need:
- one die

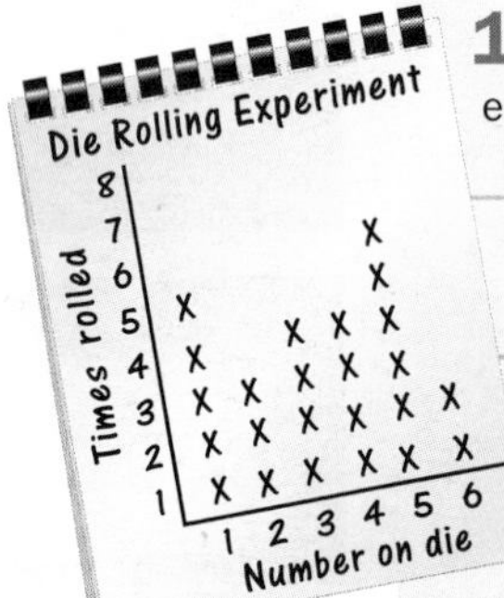

1 How many times do you think each number will occur in 24 rolls? Explain your reasoning.

2 Roll the die 24 times. Record your data on a graph as shown. Were your predictions close to your results?

3 How does your graph compare to the graphs of the other groups in your class?

4 Find the ratio of the number of times each number occurs to the total number of rolls. For example, if 2 occurs 3 times in 24 rolls:

$$\frac{\text{Number of 2's rolled}}{\text{Total number of rolls}} = \frac{3}{24} = \frac{1}{8}$$

5 What do you think the ratios found in Step 4 will be if you combine the results from all the groups in your class? Explain.

6 Combine your results with the rest of the class. Create a class graph and calculate the ratios in Step 4 for each number. Are these numbers close to your prediction in Step 5?

Exploration Note

Purpose
The purpose of this Exploration is to have students explore the concept of probability by conducting an experiment of rolling a die 24 times.

Materials/Preparation
Each group of students should have one die and a place where they can roll it and record their results. You may wish to organize students into groups of six.

Procedure
Students in each group should take turns rolling the die. One student can be chosen to draw the group's graph and show it to the other groups. If students are working in groups of six, each student can be responsible for finding the ratio for a particular number on the face of the die.

Closure
Students should understand that each number has the same probability of occurring on any one roll of the die and that this probability is approached as more rolls are made.

Explorations Lab Manual
See the Manual for more commentary on this Exploration.

Diagram Master 11

Experimental Probability

In Step 4 of the Exploration, you found the *experimental probability* of rolling each number. **Experimental probability** is the ratio of the number of times an event actually occurs to the number of times the experiment is done.

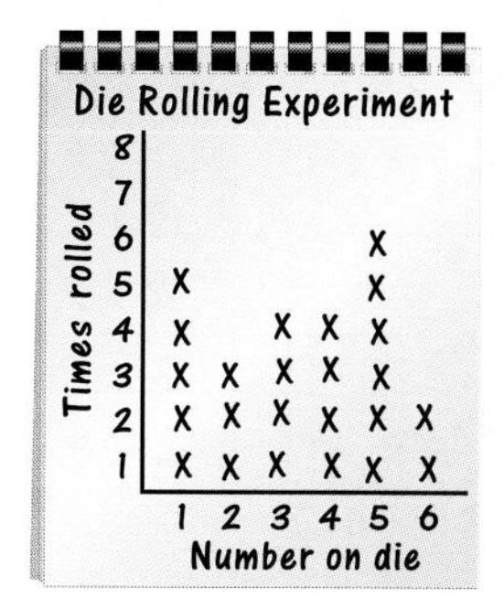

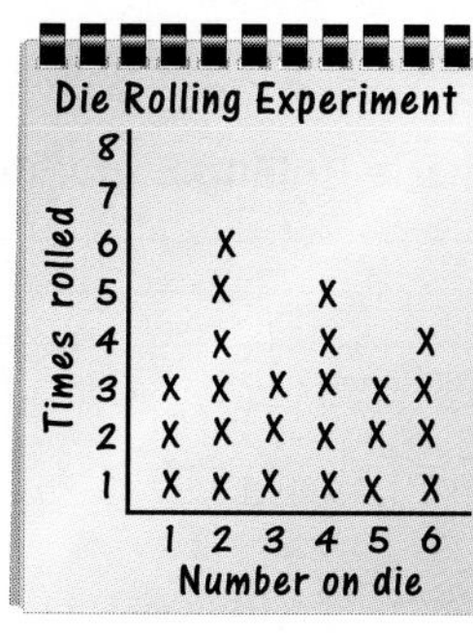

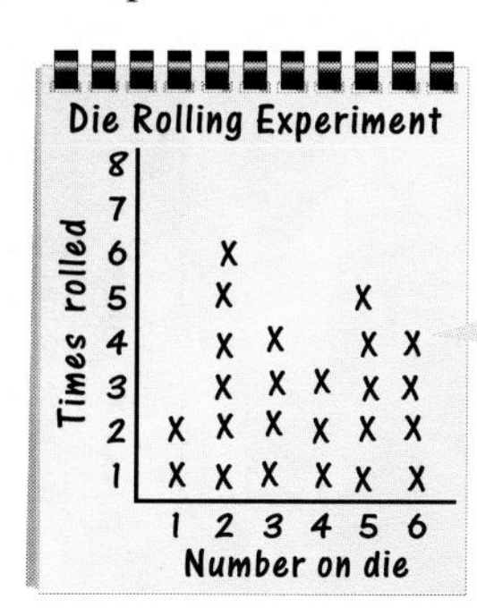

This student's experimental probability of rolling a 6 was $\frac{4}{24}$, or about 17%.

One way to think of experimental probability is as the following ratio:

$$\text{Experimental probability} = \frac{\text{Number of successes}}{\text{Number of tries}}$$

EXAMPLE 1 Application: Sports

Basketball player Jerry West of the Los Angeles Lakers attempted 1507 free throws during playoff games from 1961–1974. He made 1213 of these. What was the experimental probability of Jerry West making a free throw during a playoff game? Give your answer as a percent.

SOLUTION

$$\frac{\text{Number of free throws made}}{\text{Number of free throws attempted}} = \frac{1213}{1507} \approx 0.80$$

The probability that Jerry West would make a free throw was about 80%.

BY THE WAY

Jerry West holds the NBA playoff record for the most free throws made throughout his career. He also holds the record for the most free throws made during a single series, making 86 free throws in 6 games in 1965.

Probabilities are always between 0 and 1 and can be expressed as fractions, decimals, or percents.

The closer a probability is to 0, the less likely the event. An event with a probability of 0 can never happen. The closer a probability is to 1, the more likely the event. An event with a probability of 1 is certain to happen.

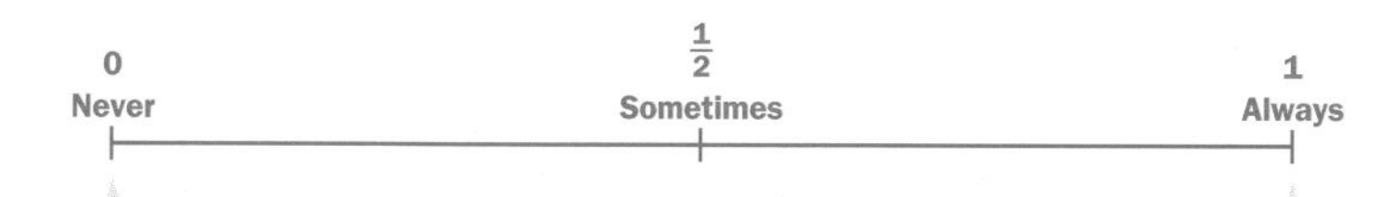

The probability of rolling a 7 is 0.

The probability of rolling a number less than 7 is 1.

Teach⇔Interact

Learning Styles: Verbal

Students can discuss what they mean by expressions such as, "I probably will," "I might," "It's not likely," and "I probably won't." Ask them to assign probabilities to each expression and place them on a number line between 0 and 1.

Additional Example 1

Many Health Maintenance Organizations (HMO) require patients to get prior approval before they can obtain health care. One hospital reports that of 30,000 HMO patients seeking medical care last year, 600 had not received prior approval for their treatment. What was the experimental probability that an HMO patient seeking care last year at that hospital had not received prior approval?

$$\frac{\text{Patients who did not receive approval}}{\text{Total patients}} = \frac{600}{30{,}000} = 2\%$$

The probability that an HMO patient did not receive prior approval was 2%.

Section Note

Reasoning

If students examine the ratio for finding an experimental probability, number of success ÷ number of tries, they can see that if the number of successes is 0, the probability is 0. If the number of successes equals the number of tries, the probability is 1.

ANSWERS Section 5.5

Exploration

1. If the die is fair, each outcome is equally likely to happen, so each number should occur about 4 times.

2–6. Answers may vary. Since the number of rolls is small, results may vary greatly.

2. Predictions should be reasonably close.
3. Most graphs should be similar.
4. Each fraction should be close to $\frac{1}{4}$.
5. An example is given. I think the more rolls there are, the closer the ratios will be to $\frac{1}{4}$.
6. It should be reasonably close.

Teach⇔Interact

Section Note

Communication: Discussion
When discussing theoretical and experimental probabilities, mention that in many real-world situations, it may not be possible to determine a theoretical probability. For example, in a new manufacturing process, it may not be possible to predict the number of defects produced until after the process is in use for a certain period of time.

Additional Example 2

A die is rolled. What is the theoretical probability of rolling a number that is not 1, 4, or 6?

The three favorable outcomes are 2, 3, or 5.

Probability of a number not equal to 1, 4, or 6
= Probability of 2
+ Probability of 3
+ Probability of 5

$= \frac{1}{6} + \frac{1}{6} + \frac{1}{6}$

$= \frac{3}{6} = \frac{1}{2}$

The probability of rolling a number that is not 1, 4, or 6 is $\frac{1}{2}$.

Checking Key Concepts

Student Progress
Students should be able to answer these questions correctly and also be able to discuss the difference between experimental and theoretical probability. They should understand that experimental probability is concerned with the actual results of an experiment. Theoretical probability is concerned with the possible results of an experiment.

Using Technology
For question 3, if time permits, students can simulate the tossing of a coin with a TI-81 or TI-82 graphing calculator. They should use the random number feature of the calculator. Press [MATH] [◄] and select menu item 1 (Rand on the TI-81, rand on the TI-82). Next, press [ENTER]. The calculator will display a randomly selected number between 0 and 1. Think of getting a number from 0 to 0.5 as getting heads on a single toss of a coin. Think of getting a number from 0.5 3to 1 as getting tails. Press [ENTER]

Continued on following page.

Theoretical Probability

The more data you collect in a probability experiment, the less the results will vary. When students combined their data from the Exploration shown on page 219, the experimental probability came much closer to the *theoretical probability*.

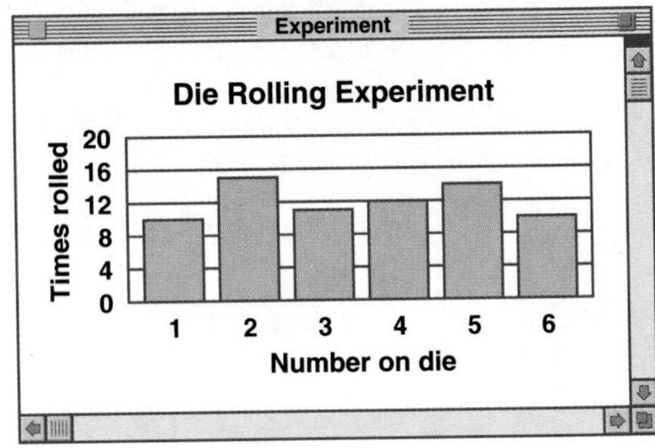

Theoretical probability is the ratio of the number of *favorable outcomes* to the total number of possible outcomes. An outcome is **favorable** if it makes the event happen.

You may have found a theoretical probability in Step 1 of the Exploration. The die has six numbers on it. Each number is equally likely to occur, so you should have expected a favorable outcome of 2 about $\frac{1}{6}$ of the time. The theoretical probability of rolling a 2 is $\frac{1}{6}$.

EXAMPLE 2

A die is rolled. What is the theoretical probability of rolling a number greater than 4?

SOLUTION

The two favorable outcomes are 5 and 6 because both of these numbers are greater than 4.

$$\text{Probability of rolling a number greater than 4} = \text{Probability of rolling a 5} + \text{Probability of rolling a 6}$$

$$= \frac{1}{6} + \frac{1}{6}$$

$$= \frac{2}{6}$$

$$= \frac{1}{3}$$

The probability of rolling a number greater than 4 is $\frac{1}{3}$.

Think and Communicate

1. You cannot have more successes than tries, so the fraction $\frac{\text{number of successes}}{\text{number of tries}}$ cannot be greater than 1.
2. It was determined by dividing the number of times the event actually occurred by the total number of times the experiment was done.
3. Answers may vary. An example is given. Experimental probability takes into account the skill of the player. Theoretical probability would take into account only the number of possible outcomes (2) and the number of outcomes that make the event happen (1).

Checking Key Concepts

1. a. 0 b. 1 c. $\frac{1}{3}$
2. $\frac{2}{17}$
3. theoretical

Exercises and Applications

1–4. Answers may vary. Examples are given.

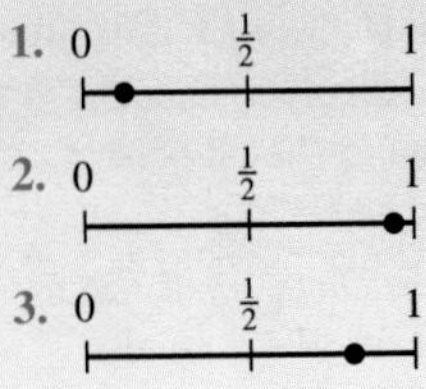

THINK AND COMMUNICATE

1. Explain why a probability cannot be greater than 1. (*Hint:* Think of the ratio of the number of successes to the number of tries.)
2. How do you know that the probability in Example 1 is experimental?
3. In Example 1, what does experimental probability take into account that theoretical probability does not?

BY THE WAY

Dice are used in many cultures and have been made of many materials, such as peach stones, walnut shells, and animal horns. Six-sided dice have been found in ancient Egyptian tombs nearly 4000 years old.

CHECKING KEY CONCEPTS

1. Suppose you roll a die like the one used in the Exploration on page 218. Find the theoretical probability of rolling:
   a. a number greater than 6.
   b. a number greater than 0.
   c. a 1 or a 2.
2. You roll a 3 twice in 17 rolls of a die. What is the experimental probability of rolling a 3?
3. There are two sides to a coin, so the probability of a coin landing heads up is 50%. Is this an experimental or theoretical probability?

5.5 Exercises and Applications

Extra Practice exercises on page 568

Make a copy of this number line. On your number line mark your estimate of the probability of each event described in Exercises 1–5.

0 ——— $\frac{1}{2}$ ——— 1

1. It will rain sometime during the next two days where you live.
2. You will have breakfast tomorrow morning.
3. You will finish your homework before 9 P.M. tonight.
4. You will receive a Nobel Prize next week.
5. You will talk to a friend today.
6. **Writing** Choose one of the events from Exercises 1–5 and explain your estimate.
7. **Open-ended Problem** Describe an event that has the given theoretical probability. Will the experimental probability of each event necessarily be the same as the theoretical probability? Explain your reasoning.

   a. 1 b. 0 c. $\frac{1}{3}$ d. $\frac{3}{4}$

Teach⇔Interact

Using Technology (continued)

repeatedly. Tally heads and tails (numbers from 0 to 0.5, numbers from 0.5 to 1). Students can compare results of the simulation experiment with those obtained by actually tossing a coin. This procedure can also be used with Ex. 19 on page 223.

Closure Question

Explain the difference between experimental and theoretical probability.

Experimental probability compares the number of times an event actually occurs to the total number of times the experiment is done. Theoretical probability compares the number of ways an event could happen to the total number of possible outcomes.

Apply⇔Assess

Suggested Assignment

Core Course
Exs. 1–10, 12–31

Extended Course
Exs. 1–31

Block Schedule
Day 32 Exs. 1–31

Exercise Notes

Cooperative Learning

Ex. 7 Students can share their descriptions with a partner. Each student can repeat the exercise using his or her partner's descriptions. By comparing their answers, students can assess and strengthen their understanding of the relationship between theoretical and experimental probability.

5. 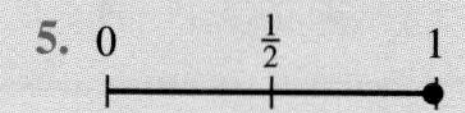

6. Answers may vary. Examples are given. Ex. 1: It rarely rains where I live. Ex. 2: I almost never miss breakfast. Ex. 5: It is almost certain that I will talk to at least one friend today.

7. Answers may vary. Examples are given.
   a. getting a number less than 7 when a standard die is rolled
   b. getting a number greater than 7 when a standard die is rolled
   c. getting a multiple of 3 when a standard die is rolled
   d. getting a red marble when you take a marble from a bag containing 3 red marbles and 1 white marble

   The experimental probability will be the same as the theoretical probability if the probability is 1 or 0 because the event either always occurs or never occurs. This is not necessarily the case for other probabilities like the one in parts (c) and (d).

Exercise Notes

Interdisciplinary Problems

Exs. 8–11 The mathematics of statistics and probability plays a large role in helping social scientists collect, analyze, present, and interpret data. Probability not only helps scientists to understand data, but it is also extremely helpful in making predictions.

Topic Spiraling: Review

Ex. 11 Have students review the difference between a mean and a median. Ask them why they think the government chose the median rather than the mean as a measure of *middle* income. (A mean is affected by extreme values in a data set and a very small percent of Americans make very large salaries. These high incomes would pull the mean up but would not affect the median. The median, therefore, is a more reliable measure of a typical income.) Challenge students to consider whether the probability that a head of household would earn more than the mean income for his or her group would be higher or lower than the probability calculated in Ex. 11.

Student Study Tip

Exs. 13–15 Mention to students that probability theory belongs to the branch of mathematics called *discrete* mathematics. Discrete mathematics deals with distinct, separate objects. Counting is a familiar discrete mathematics operation. Measurement, on the other hand, deals with an operation of *continuous* mathematics. Ask students: Which set of numbers do they think form the basic number system for discrete mathematics? (the natural numbers) for continuous mathematics? (the real numbers)

Reasoning

Ex. 16 This exercise introduces students informally to the idea of a complement. If the probability of rolling an even number is *p*, then the probability of *not* rolling an even number, which is the complement of *p*, is 1 – *p*. Ask students to explain why *p* is subtracted from 1. (The sum of the probabilities of an event and its complement is 1 because either the event or its complement will occur.)

Connection SOCIAL STUDIES

BY THE WAY

In 1992, there were over 14 million people enrolled in higher education in the United States. The average student spent $12,250 for school and received $745 in financial support.

According to the United States Census, these were the median yearly income levels for heads of households with different levels of education in 1992. Probability can help you understand statistics like these.

Education and Income of Heads of Households in the United States

Education level attained	Number of people	Median income
Less than 9th grade	9,060,000	$13,383
9th to 12th grade, no diploma	9,933,000	$17,375
High school graduate	30,103,000	$29,006
Some college, no degree	15,387,000	$35,327
Associate degree	5,502,000	$38,382
Bachelor's degree or more	21,382,000	$54,117

8. How many people were "heads of households" in 1992? How do you know?

9. What is the probability that a head of a household had less than a 9th grade education in 1992? How did you get your answer?

10. What is the probability that a head of a household had some college education in 1992, but no degree?

11. Challenge What is the probability that a head of a household earned more than the median income for his or her group?

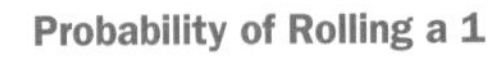

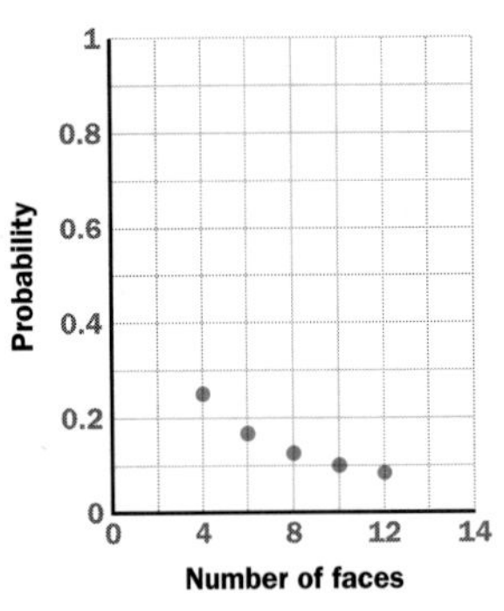

12. a. A 3 occurs 23 times when a die is rolled 180 times. What is the experimental probability of rolling a 3?

b. What is the theoretical probability of rolling a 3?

Most dice come with six faces. You can also find dice with more than six faces. The graph shows the probability of rolling a 1 as a function of the number of faces on the die. Use the graph for Exercises 13–15.

13. Describe the domain and the range of the function. Write an equation describing the function. How did you get your answer?

14. Would you connect the points on the graph? Why or why not?

15. Suppose the graph showed the probability of rolling a 2 as a function of the number of faces on the die. Would the graph look different? Why or why not?

8. 91,367,000; Add the numbers in the middle column.
9. about 9.9%; Divide the number of people with less than a ninth-grade education (9,060,000) by the total number (91,367,000).
10. about 16.8%
11. 50%
12. a. $\frac{23}{180} \approx 13\%$ b. $\frac{1}{6} \approx 17\%$
13. The domain is the positive even numbers greater than 2, and the range is the reciprocals of numbers in the domain. Let x = the number of faces on the die and let y = the probability of rolling a 1; $y = \frac{1}{x}$; for any number x of faces, there are x possible outcomes, one of which is the desired outcome.
14. No; the number of faces must be a whole number.
15. No; for any number of faces, the probability or rolling a 1 is the same as the probability of rolling a 2.
16. a. 50%
    b. 30%
    c. 0.671
    d. $1 - x$; The two events represent all the possible outcomes, so the sum of their probabilities must be 1.

16. a. The probability that you will get an even number when you roll a die is 50%. What is the probability that you will get an odd number?

b. If there is a 70% chance that it will rain tomorrow, what is the probability that it will not rain?

c. If Kirby Puckett's *batting average*, the probability that he would get a hit, was .329 in 1992, what was the probability that he would not get a hit?

d. Suppose x represents the probability that something will happen. Write an expression for the probability that it will not happen. Explain your reasoning.

17. SAT/ACT Preview Which one of the following equals 7.5%?

A. 0.75 **B.** $\frac{3}{4}$ **C.** $\frac{75}{100}$ **D.** 0.075

18. Writing Look back at the Exploration on page 191 in Section 5.1. How does the capture-recapture method rely on probability? Explain.

ONGOING ASSESSMENT

19. Cooperative Learning Work with a partner to toss a pair of pennies 50 times. Record the number of times you get two heads, two tails, or a head and a tail.

a. Before you begin, discuss the possible outcomes and favorable outcomes for each event with your partner.

b. One of you should calculate the theoretical probability for each event. The other person should calculate the experimental probability for each event. (*Hint:* Think of the events as head and head, tail and tail, head and tail, tail and head.)

c. Are your predictions close to your results?

SPIRAL REVIEW

Graph each point in a coordinate plane. Name the quadrant (if any) in which the point lies. *(Section 2.4)*

20. $A(-1, 1)$ **21.** $B(4, 0)$ **22.** $C(0, -5)$

23. $D(5, -1)$ **24.** $E(-5, -4)$ **25.** $F\left(5, \frac{1}{2}\right)$

Solve each equation. *(Section 4.4)*

26. $\frac{a}{8} = \frac{5}{56}$ **27.** $\frac{7}{3}b - 9 = -10\frac{2}{3}$ **28.** $0.3x - 1.33 = 6.5$

29. $d + \frac{1}{2}d + \frac{1}{3}d = 1$ **30.** $7 - 0.5x = 7.5x - 9$ **31.** $f - 1 = 2(f - 2) - f$

Apply⇔Assess

Exercise Notes

Assessment Note

Ex. 19 A tree diagram that shows how the two tosses of the coins are related to the outcomes is usually helpful to students.

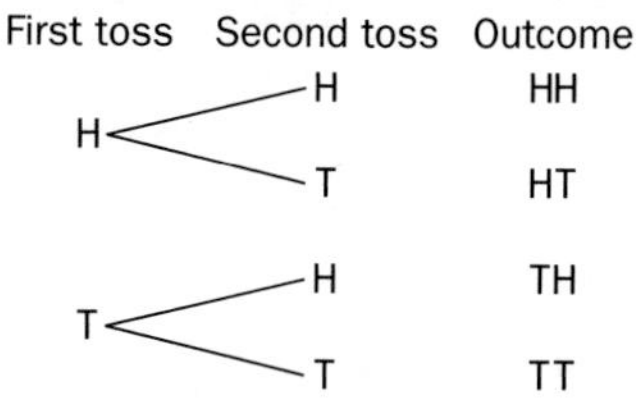

You may wish to suggest that one student record each event in a table having four columns, using as headings the events in part (b). Each time an event occurs, a tally mark is placed in the appropriate column. The other partner can toss the coins.

Practice 35 for Section 5.5

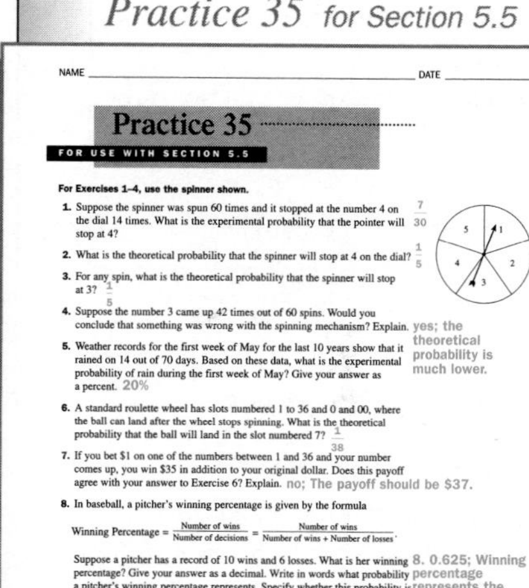

NAME ______________________ DATE __________

Practice 35

FOR USE WITH SECTION 5.5

For Exercises 1–4, use the spinner shown.

1. Suppose the spinner was spun 60 times and it stopped at the number 4 on the dial 14 times. What is the experimental probability that the pointer will stop at 4? $\frac{7}{30}$
2. What is the theoretical probability that the spinner will stop at 4 on the dial? $\frac{1}{5}$
3. For any spin, what is the theoretical probability that the spinner will stop at 3? $\frac{1}{5}$
4. Suppose the number 3 came up 42 times out of 60 spins. Would you conclude that something was wrong with the spinning mechanism? Explain. yes; the theoretical probability is much lower.
5. Weather records for the first week of May for the last 10 years show that it rained on 14 out of 70 days. Based on these data, what is the experimental probability of rain during the first week of May? Give your answer as a percent. 20%
6. A standard roulette wheel has slots numbered 1 to 36 and 0 and 00, where the ball can land after the wheel stops spinning. What is the theoretical probability that the ball will land in the slot numbered 7? $\frac{1}{38}$
7. If you bet $1 on one of the numbers between 1 and 36 and your number comes up, you win $35 in addition to your original dollar. Does this payoff agree with your answer to Exercise 6? Explain. no; The payoff should be $37.
8. In baseball, a pitcher's winning percentage is given by the formula

$$\text{Winning Percentage} = \frac{\text{Number of wins}}{\text{Number of decisions}} = \frac{\text{Number of wins}}{\text{Number of wins + Number of losses}}.$$

Suppose a pitcher has a record of 10 wins and 6 losses. What is her winning percentage? Give your answer as a decimal. Write in words what probability a pitcher's winning percentage represents. Specify whether this probability is experimental or theoretical. 8. 0.625; Winning percentage represents the probability that a pitcher will win a game in which he or she is involved in the decision.; experimental

9. A poll of seniors showed that 110 favored Thu Pham for senior class president, 100 favored Aurora Martinez, and 40 were undecided.

   a. What is the experimental probability that a randomly chosen senior will vote for Thu Pham? 0.44

   b. What is the experimental probability that a randomly chosen senior will vote for Aurora Martinez? 0.4

Practice Bank, ALGEBRA 1: EXPLORATIONS AND APPLICATIONS

35

17. D

18. The ratio of the number of marked items to total items in the sample should be the same as the ratio in the general population. That is, the probability of selecting a marked item in the sample should be about the same as the probability of selecting a marked item from the general population.

19. a, b. Each of the four possible outcomes (head & head, tail & tail, head & tail, tail & head) is equally likely and has probability 25%. The theoretical probability of two heads is 25%, of two tails is 25%, and of one head and one tail is 50%. Experimental probabilities may vary.

c. Answers may vary.

20–25.

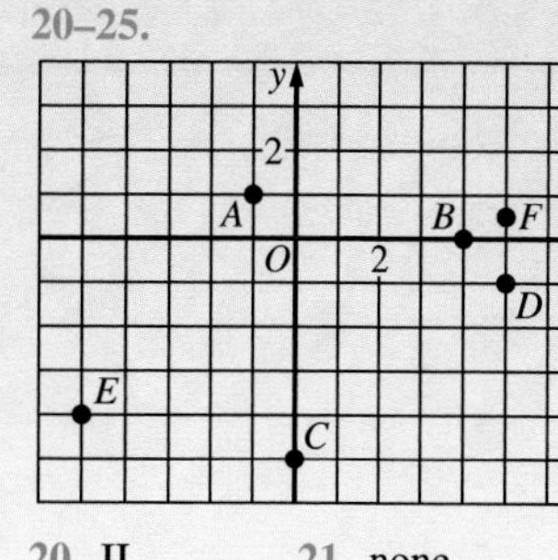

20. II 21. none
22. none 23. IV
24. III 25. I

26. $\frac{5}{7}$

27. $-\frac{5}{7}$

28. 26.1

29. $\frac{6}{11}$

30. 2

31. no solution

Plan⇔Support

Objectives

- Find probabilities based on area.
- Predict events based on geometric probability.

Recommended Pacing

Core and Extended Courses
Section 5.6: 1 day

Block Schedule
Section 5.6: $\frac{1}{2}$ block
(with Section 5.5)

Resource Materials

Lesson Support
Lesson Plan 5.6
Warm-Up Transparency 5.6
Practice Bank: Practice 36
Study Guide: Section 5.6
Assessment Book: Test 21

Technology
Scientific Calculator
Graphing Calculator
Internet:
http://www.hmco.com

Warm-Up Exercises

Find the area of each figure.

1. a rectangle whose length is 0.5 m and whose width is 2.3 m 1.15 m^2
2. a circle whose radius is 4 yd about 50.27 yd^2

Find each probability.

3. rolling at least a 4 on one roll of a fair die 50%
4. tossing two tails on two tosses of a coin 25%

Find the ratio of the areas of two similar figures given the ratio of the lengths of corresponding sides.

5. 3 : 5 9 : 25
6. 9 : 4 81 : 16

SECTION

5.6 Geometric Probability

Learn how to...

- **find probabilities based on area**

So you can...

- **predict events such as where meteorites are likely to hit when they fall to Earth**

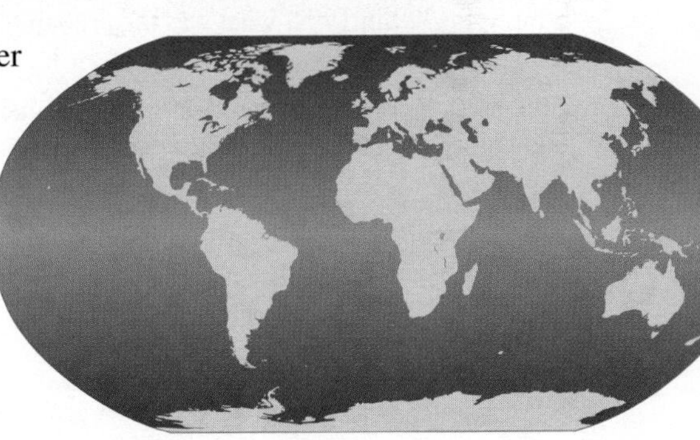

Did you know that over 2000 meteorites weighing 1 kg land on Earth every year? Scientists say most will land in water. But how do they know?

When a meteorite falls to Earth, it can hit either land or water. Because $\frac{7}{10}$ of Earth's surface is covered by water, the probability that an object falling to Earth will land in water is 70%. The chance that a meteorite will land in water is an example of **geometric probability**, a probability based on area.

EXAMPLE 1

Suppose a randomly thrown dart hits the target shown. (A random throw means that you do not aim for any particular part of the target, and each part has an equal chance of being hit.) What is the theoretical probability of the dart hitting the shaded area?

SOLUTION

The square is divided into 8 triangles having equal area, and 2 of them are shaded.

$$\text{Probability of hitting a shaded area} = \frac{\text{Shaded area}}{\text{Total area}}$$

$$= \frac{\text{Number of shaded triangles}}{\text{Total number of triangles}}$$

$$= \frac{2}{8}$$

$$= \frac{1}{4}$$

The probability of a randomly thrown dart hitting the shaded area is 25%.

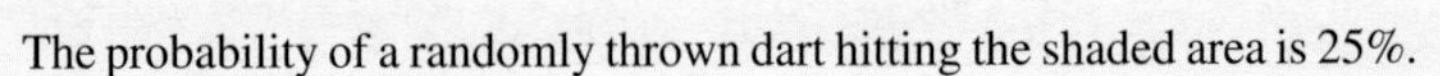

EXAMPLE 2

Suppose you have the chance to win a prize by throwing a dart at either of the two targets shown without hitting a balloon. Which target gives you a better chance of winning a prize?

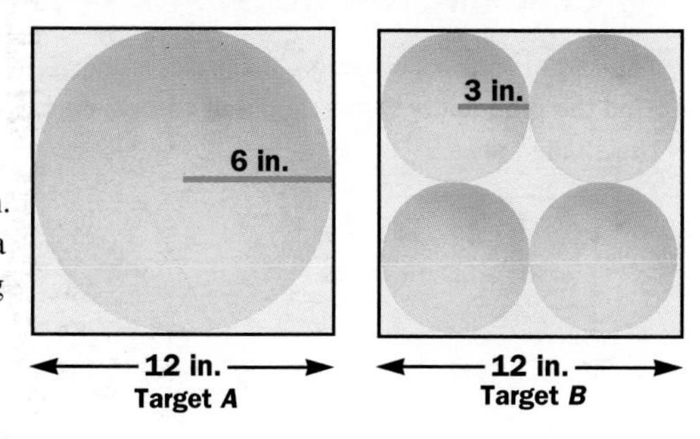

SOLUTION

Compare the probabilities of hitting a balloon on each of the two targets.

$$\text{Probability of hitting the balloon on Target } A = \frac{\text{Area covered by large balloon}}{\text{Area of Target } A}$$
$$= \frac{\pi(6)^2}{12^2}$$
$$= \frac{36\pi}{144}$$
$$\approx 0.785$$

$$\text{Probability of hitting a balloon on Target } B = \frac{4 \times (\text{Area covered by one small balloon})}{\text{Area of Target } B}$$
$$= \frac{4\pi(3)^2}{12^2}$$
$$= \frac{36\pi}{144}$$
$$\approx 0.785$$

Since the probability of hitting a balloon is the same for both targets, neither one gives you a better chance of winning a prize.

THINK AND COMMUNICATE

1. In Section 5.4, you learned that if the ratio of the lengths of corresponding sides of two similar figures is $a:b$, then the ratio of the areas is $a^2:b^2$. How can you use this fact to answer the question in Example 2?
2. Suppose a square board with side length 12 in. has nine balloons attached to it. If the radius of each balloon is 2 in., what is the probability that a randomly thrown dart that hits the target hits a balloon?

Teach⇔Interact

Section Note

Integrating the Strands
The topic of geometric probability integrates concepts from geometry and probability as well as number concepts, such as percents, and algebraic concepts, such as using formulas.

About Example 1

Communication: Reading
Have the students read the example to themselves. Then ask one student to read it aloud. Ask students how they might rephrase the question. (How much of the total area is shaded?)

Additional Example 1

A forester is marking trees for cutting with a can of orange spray paint. Assume he randomly sprays at the rectangular target area, which is 6 in. wide and 15 in. long. The diameter of the circle is 3 in. What is the theoretical probability that he hits the "bull's-eye" circle?

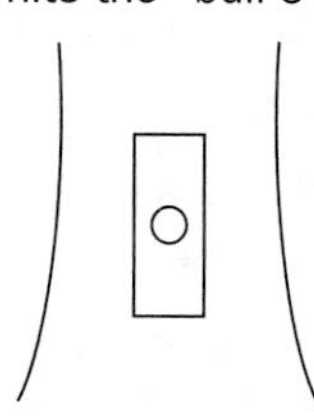

Probability of hitting the circle

$$= \frac{\text{Area of circle}}{\text{Area of rectangle}} = \frac{\pi(1.5)^2}{90}$$
$$= \frac{2.25\pi}{90}$$
$$\approx 0.08$$

The probability that the paint sprayed randomly hits the circle is about 8%.

ANSWERS Section 5.6

Think and Communicate

1. The ratio of the radii of the large and small circular areas (corresponding lengths of similar figures) is $\frac{1}{2}$, so the ratio of the areas is $\frac{1}{4}$. Then the total area of the four small circular areas is the same as the area of the large one. Since the area of the targets is also the same, the probability of hitting the target without hitting a balloon is the same for both targets.
2. $\frac{9\pi \cdot 2^2}{144} \approx 0.785$

Teach⇔Interact

Additional Example 2

Refer to Example 2 on page 225. Suppose the radius of the larger balloon is 4 in. Which target gives you a better chance of winning?

Probability of hitting the balloon Target A

$= \frac{\text{Area of large balloon}}{\text{Area of Target A}} = \frac{\pi(4)^2}{144}$

$= \frac{16\pi}{144}$

≈ 0.349

Since the probability of hitting a balloon on Target A is now about 35% and the probability of hitting a balloon on Target B is still about 79%, Target B gives you a better chance of winning a prize.

Closure Question

What kinds of ratios are involved in finding a geometric probability?

ratios based upon areas of figures or lengths of segments

Apply⇔Assess

Suggested Assignment

Core Course
Exs. 1–6, 8–10, 12–24, AYP

Extended Course
Exs. 1–24, AYP

Block Schedule
Day 32 Exs. 1–10, 12–24, AYP

Exercise Notes

Second-Language Learners

Ex. 7 Make sure students learning English have access to a United States map that shows the names of the states.

Communication: Writing

Ex. 7 This exercise offers an excellent opportunity to assess students' understanding of the fact that geometric probability is based on area.

CHECKING KEY CONCEPTS

The targets shown below are a square, an equilateral triangle, and a circle. Find the probability that a randomly thrown dart that hits each target hits the shaded area.

1. 

2. 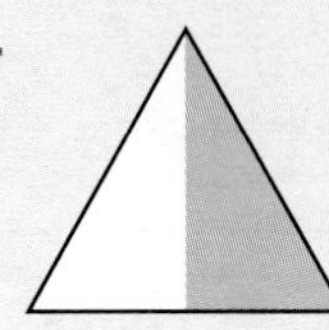

3. 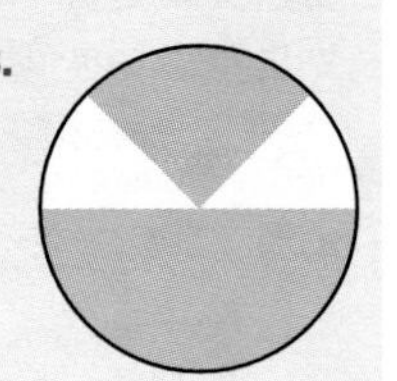

5.6 Exercises and Applications

Extra Practice exercises on page 568

The targets shown below are squares, equilateral triangles, and circles. Find the probability that a randomly thrown dart that hits each target hits the shaded area.

1. 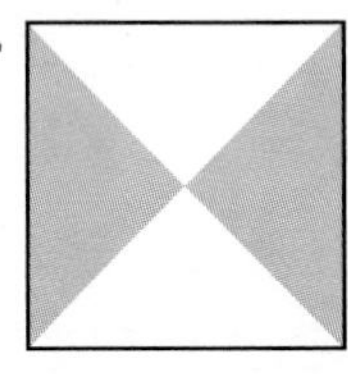

2. 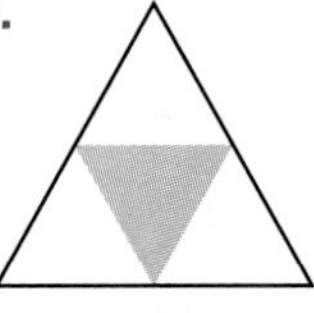

3. 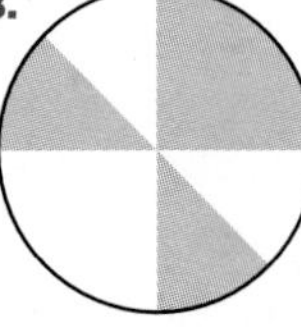

4. 

5. 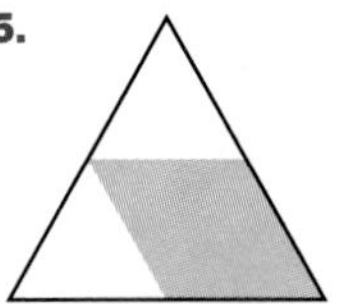

6. 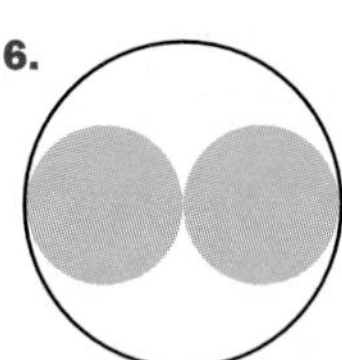

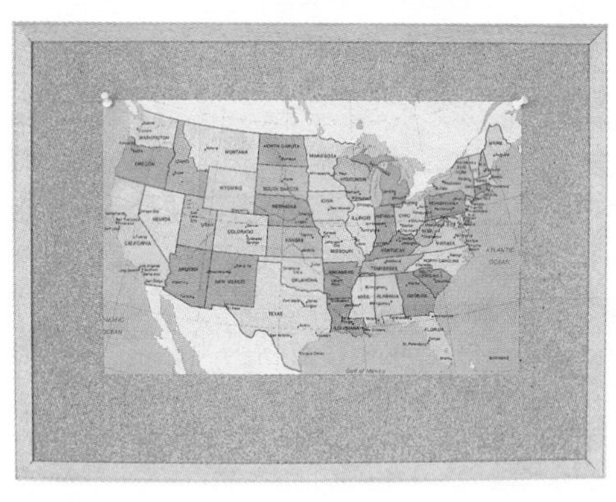

7. Suppose you and a friend want to take a vacation somewhere in the 48 states shown in the map. You can't decide where to go, so you agree to throw a dart randomly at the map and go where the dart lands.

a. Which state do you have the highest probability of visiting? Why?

b. **Writing** Explain why your method is *not* equivalent to writing the names of the 48 states on slips of paper and picking one slip of paper at random.

Checking Key Concepts

1. $\frac{5}{9} \approx 56\%$
2. 50%
3. 75%

Exercises and Applications

1. 50% 2. 25%
3. 50% 4. 50%
5. 50% 6. 50%
7. a. Texas; It has the largest area of the states shown on the map.
   b. If the names of the 48 states are put on slips of paper and one is randomly drawn, then each state has a probability of being drawn of $\frac{1}{48} \approx 2\%$. When the dart method is used, the probability for each state varies with its area.
8. $\frac{1}{36} \approx 2.8\%$
9. $\frac{4}{9} \approx 44\%$
10. $\frac{25}{144} \approx 17\%$

In Exercises 8–10, the targets and the shaded areas are all equilateral triangles. Find the probability that a randomly thrown dart that hits each target hits the shaded area.

8.

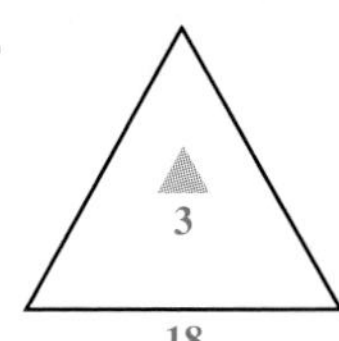

9.

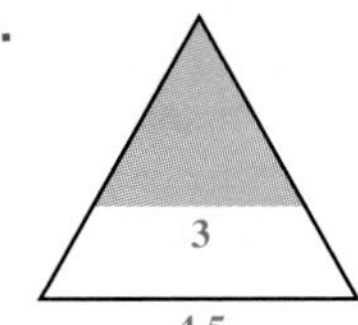

10.

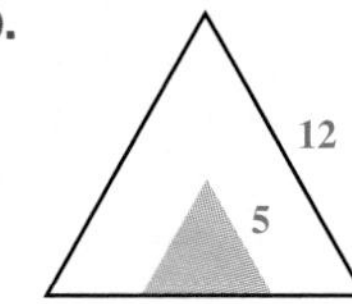

11. Challenge The map shows the location of sunken ships in a 28.5 mi by 34.5 mi section of the Atlantic Ocean just off the North Carolina coast. Suppose you want to search for a sunken ship in these waters, but you don't know the location of any of these ten ships. If you select a point in this area at random and search within a 0.5 mi radius of that point, what is the probability that you will find one of the ships? Show how you got your answer.

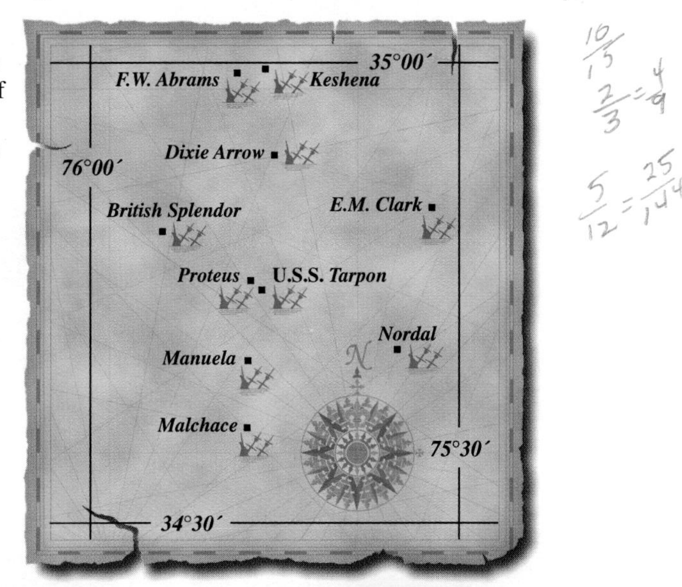

Connection SPORTS

At the Olympics, archers shoot at a target like the one shown. They earn the most points when they hit the center, or the bull's-eye, of the target. You can use geometric probability to see just how difficult it is to hit the bull's-eye.

12.2 cm

122 cm

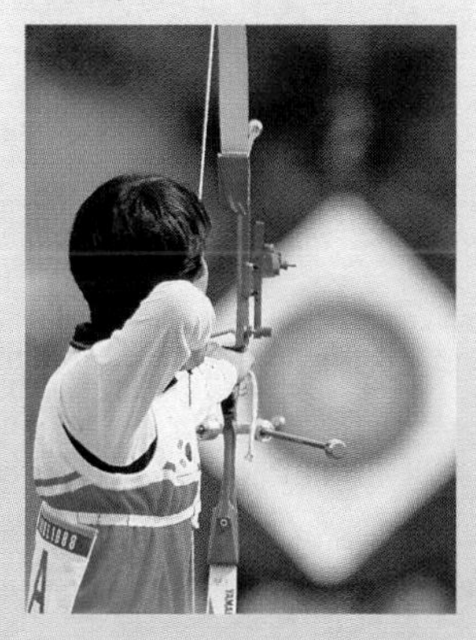

12. The target has a diameter of 122 cm. The bull's-eye has a diameter of only 12.2 cm. What is the probability that a randomly shot arrow that hits the target hits the bull's-eye? Is this probability experimental or theoretical?

13. Open-ended Problem Do you think the probability you found in Exercise 12 takes into account the skill of the archer? Explain your reasoning.

BY THE WAY

In 1984 at age 17, Seo Hyang-Soon of Korea became the youngest gold medalist in archery when she won the women's title. Galen Spencer of the United States was the oldest gold medalist in the sport, winning at age 64 in 1904.

Apply⇔Assess

Exercise Notes

Problem Solving

Exs. 8–10 Students are expected to use the fact that similar triangles with corresponding sides in the ratio $\frac{a}{b}$ have areas in the ratio $\frac{a^2}{b^2}$. Because the geometric probabilities in these exercises are based on area, this property of similar triangles allows the probabilities to be determined quickly.

Research

Ex. 12 Students may like to research other sports that involve targets. They could find the target and bull's-eye sizes and then find the probability of hitting the bull's-eye on those targets.

11. about 0.80%; To find a ship, the point chosen must be within 0.5 mi of one of the 10 ships. That is, the point must be within one of 10 circles with radius 0.5 mi and center at one of the ships. The probability is $\frac{\text{total area of the 10 circles}}{\text{area of the map region}} = \frac{10\pi(0.5)^2}{28.5(34.5)}$.

12. 1%; theoretical

13. Answers may vary. An example is given. I think the archer's skill is taken into account only to the degree that we have to assume the archer hits the target. A skilled archer should have a better probability than the theoretical probability.

Apply⇔Assess

Exercise Notes

Student Study Tip

Ex. 15 Mention to students that the pattern found in part (b) can be used to write a geometric series. Each of the terms of a geometric series is multiplied by the same number, called the *common ratio,* to produce the next number in the series. Ask students to find the common ratio of the series in part (b). $\left(\frac{8}{9}\right)$

Topic Spiraling: Review

Ex. 16 Students can solve this problem by writing a proportion and then solving it using methods learned in Section 5.1.

Cooperative Learning

Ex. 17 Students should take turns using the calculator and recording where the randomly generated point lies so that both students experience the full intent of this exercise.

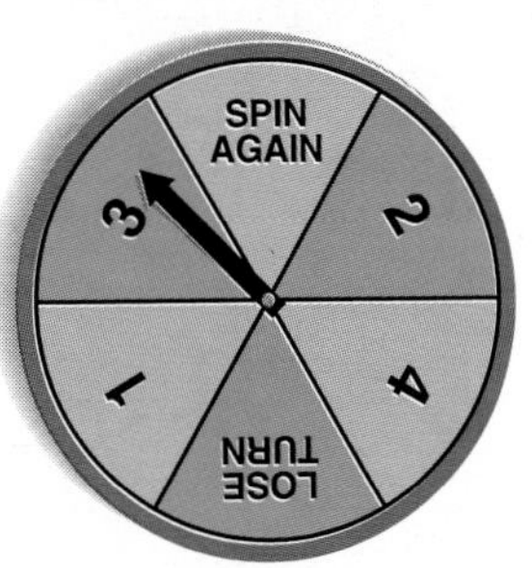

14. Suppose you are playing a board game that comes with the spinner shown. Find the probability that you will:

a. lose your turn. **b.** spin a number greater than 2.

15. Suppose you create an elaborate target in stages, as shown below. For Stage 0, you have a square, blank target. For Stage 1, you divide the target into 9 squares and shade the center one. For Stage 2, you divide each of the unshaded squares into 9 squares and shade the center one.

Stage 0

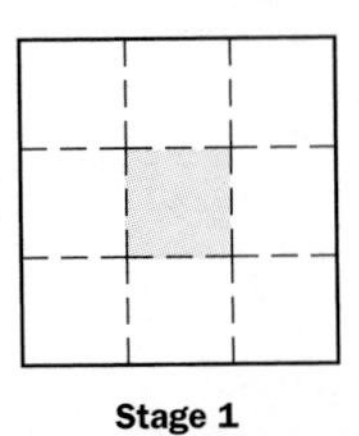
Stage 1

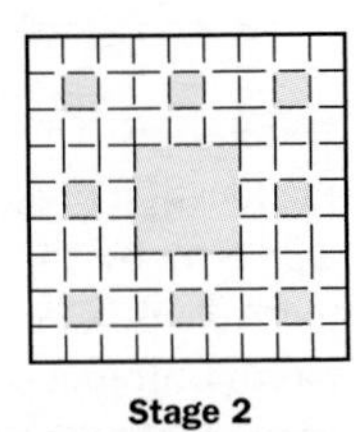
Stage 2

a. At each stage, suppose you randomly throw a dart at the target. What is the probability of hitting the target's unshaded area? Answer this question by copying and completing the table. Express the probabilities as fractions.

Stage	Probability of hitting the target's unshaded area
0	?
1	?
2	?

b. What do you think is the probability for a Stage 3 target?

16. The probability that a randomly thrown dart that hits the square target shown hits the shaded area is $\frac{5}{9}$. The area of the inner square is 16 in.^{2}. Find the area of the target.

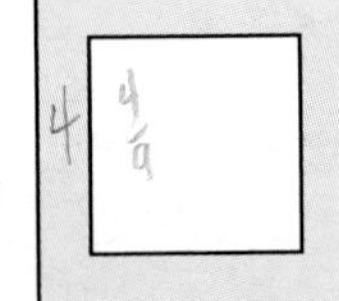

17. **Technology** Work with a partner. You need a scientific or graphing calculator with a random-number feature. This feature will give you randomly chosen decimals between 0 and 1. You will use this feature to simulate the random throwing of darts at a dart board. Repeat the following steps 40 times:

- Get a pair of random numbers from your calculator. Treat these as the coordinates of a point in the 1-by-1 square shown.
- Have your partner record whether the point lies in the shaded area of the square shown. Assume the point (0.5, 0.5) lies in the shaded area.

a. What must be true about x and y if the point (x, y) lies in the shaded area?

b. What is the theoretical probability of hitting the shaded area?

c. What experimental probability do you get as a result of this activity? Is this close to the theoretical probability?

14. a. $\frac{1}{6} \approx 17\%$

b. $\frac{1}{3} \approx 33\%$

15. a.

Stage	Probability of hitting the target's unshaded area
0	1
1	$\frac{8}{9}$
2	$\frac{64}{81}$

b. $\frac{512}{729}$

16. 36 in.^{2}

17. a. $x \le 0.5$ and $y \le 0.5$

b. 25%

c. Answers may vary.

18. a. theoretical; The target was designed and the probability determined without performing an experiment.

b. Check students' work.

ONGOING ASSESSMENT

18. **Cooperative Learning** Work in a group of 3–5 students. Each student needs notebook paper and a small dried bean.
    a. Design a target to cover a piece of notebook paper. Shade in part of it so that the probability that a bean hits the shaded area when dropped on the target is 75%. What kind of probability is this? Explain.
    b. Place another student's target on the floor and drop the bean on the target twenty times.
    c. Find the ratio of the number of times the bean lands clearly inside the shaded area to the number of times the bean hit the target. What kind of probability is this? Explain.

SPIRAL REVIEW

Evaluate each variable expression when $a = -2$, $b = 10$, and $c = \frac{3}{4}$. *(Section 1.5)*

19. $a^2 + b^2$
20. $c^2 - b^2$
21. $9a^2 + 16b^2$
22. $a(bc + a)$
23. $\left(\frac{a+b}{2}\right) + 5c$
24. $b + c(a + b) - a$

ASSESS YOUR PROGRESS

VOCABULARY

probability (p. 218)
event (p. 218)
outcome (p. 218)
experimental probability (p. 219)
theoretical probability (p. 220)
favorable outcome (p. 220)
geometric probability (p. 224)

Find the theoretical probability of each event. Then choose one event and find its experimental probability. *(Section 5.5)*

1. A die is rolled and lands showing a number less than 5.
2. A calendar is opened randomly to a month that begins with "M."

For Exercises 3–5, find the theoretical probability that a randomly thrown dart that hits each target hits the shaded area. *(Section 5.6)*

3. 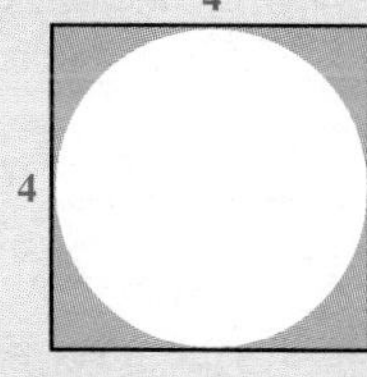

4. 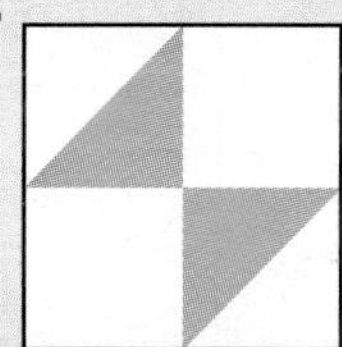

5. 

6. **Journal** List five real-world situations that involve probability. One of them should involve geometric probability.

Apply⇔Assess

Exercise Notes

Assessment Note
Ex. 18 After groups complete their work, you may wish to call upon some of them to give their answers to the class.

Progress Check 5.5–5.6

See page 233.

Practice 36 for Section 5.6

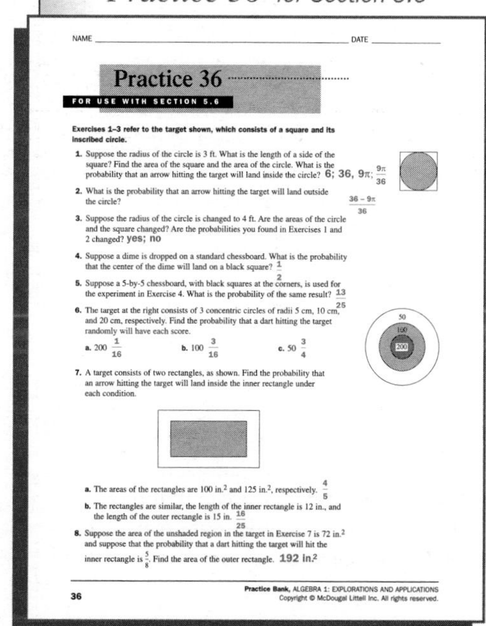

NAME ______ DATE ______

Practice 36
FOR USE WITH SECTION 5.6

Exercises 1–3 refer to the target shown, which consists of a square and its inscribed circle.

1. Suppose the radius of the circle is 3 ft. What is the length of a side of the square? Find the area of the square and the area of the circle. What is the probability that an arrow hitting the target will land inside the circle? **6; 36, 9π; $\frac{9\pi}{36}$**
2. What is the probability that an arrow hitting the target will land outside the circle? **$\frac{36 - 9\pi}{36}$**
3. Suppose the radius of the circle is changed to 4 ft. Are the areas of the circle and the square changed? Are the probabilities you found in Exercises 1 and 2 changed? **yes; no**
4. Suppose a dime is dropped on a standard chessboard. What is the probability that the center of the dime will land on a black square? **$\frac{1}{2}$**
5. Suppose a 5-by-5 chessboard, with black squares at the corners, is used for the experiment in Exercise 4. What is the probability of the same result? **$\frac{13}{25}$**
6. The target at the right consists of 3 concentric circles of radii 5 cm, 10 cm, and 20 cm, respectively. Find the probability that a dart hitting the target randomly will have each score.
   a. 200 **$\frac{1}{16}$** b. 100 **$\frac{3}{16}$** c. 50 **$\frac{3}{4}$**
7. A target consists of two rectangles, as shown. Find the probability that an arrow hitting the target will land inside the inner rectangle under each condition.
   a. The areas of the rectangles are 100 in.^{2} and 125 in.^{2}, respectively. **$\frac{4}{5}$**
   b. The rectangles are similar, the length of the inner rectangle is 12 in., and the length of the outer rectangle is 15 in. **$\frac{16}{25}$**
8. Suppose the area of the unshaded region in the target in Exercise 7 is 72 in.^{2} and suppose that the probability that a dart hitting the target will hit the inner rectangle is $\frac{5}{8}$. Find the area of the outer rectangle. **192 in.^{2}**

36 **Practice Bank,** ALGEBRA 1: EXPLORATIONS AND APPLICATIONS

---

c. Answers may vary. An example is given. experimental; An experiment was performed and the probability was determined by the results.

19. 104
20. $-99\frac{7}{16}$
21. 1636
22. –11
23. $7\frac{3}{4}$
24. 18

Assess Your Progress

1. $\frac{2}{3}$
2. $\frac{1}{6}$

1, 2. Answers may vary. An example is given. A die is rolled 30 times. It lands with a number less than 5 showing 25 times. The experimental probability of rolling a die and having it land with a number less than 5 showing is $\frac{5}{6}$.

3. about 21.5%
4. 37.5%
5. 50%
6. Answers may vary.

Project Notes

Mathematical Goals

- Choose a set of statistical data to illustrate with a cartogram.
- Select three to five neighboring states with relevant data and organize the data in a table.
- Choose a convenient size and scale to make squares that can be used to form the shapes of the states chosen.

Planning

Materials
- Computer drawing program
- Graph paper
- Poster board

Project Teams
Work with students to organize the class into groups of two or three students. Each team should discuss the goal of the project and decide how to proceed.

Guiding Students' Work

You may wish to discuss the project with the whole class before the groups are organized. Ask the class for suggestions about possible topics for gathering data. This would provide a source of ideas for students to get started. Suggest that the groups explore reference materials in their school or public libraries or use reference books they may have at home.

Second-Language Learners
Students learning English may not be familiar with the expression *neighboring states.* Explain that this phrase refers to states that have borders in common.

PORTFOLIO PROJECT

Using a Cartogram

Most maps of the United States show the size, shape, and location of each state. Some other maps are drawn so that each state's size is distorted to reflect a particular group of data, such as population. This type of map is called a cartogram.

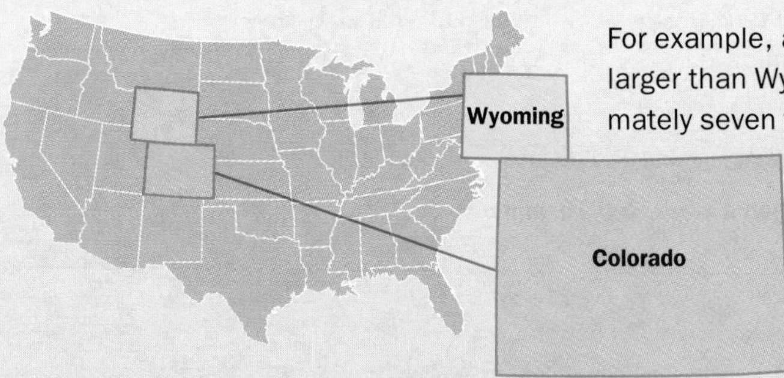

For example, although Colorado is only a little larger than Wyoming, in 1990 it had approximately seven times as many people. The cartogram shows how these states look when their sizes are distorted to represent population instead of area.

PROJECT GOAL **Make a cartogram of a selected group of states. Then make a poster of your cartogram.**

Gathering Data

Work in a group of 2–3 students. First choose a set of statistical data to illustrate with your cartogram. Some possible topics are:

- number of physicians
- amount of land suitable for farming
- amount of a natural resource produced each year, such as oil or coal

Select three to five neighboring states to show on your cartogram, and find the relevant data for each. Organize the information in a table. For example, this table shows the population densities of Nebraska, Iowa, and Illinois.

State	Population density, 1993
Nebraska	20.9
Iowa	50.4
Illinois	210.4

Constructing Your Cartogram

Use a computer drawing program or graph paper to form the shapes of the states in your cartogram with squares.

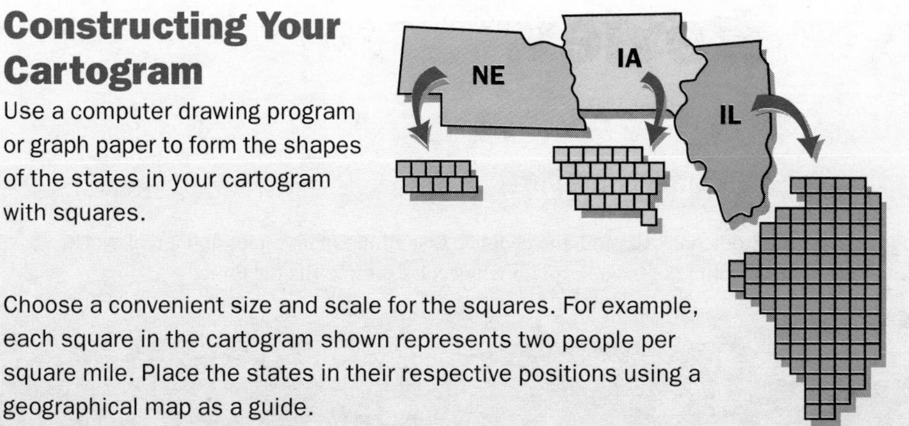

Choose a convenient size and scale for the squares. For example, each square in the cartogram shown represents two people per square mile. Place the states in their respective positions using a geographical map as a guide.

Making a Poster

Display your cartogram on a poster. Label each state and describe the scale you used. Include a labeled geographical map of the states and a table of the data.

You and your classmates can extend your study of cartograms by exploring some of the ideas below:

- Create a class cartogram of the entire United States.
- Make three or four cartograms that illustrate different data for the same group of states.
- Start a scrapbook of cartograms that you find in newspapers, magazines, or reference books.

Self-Assessment

Write a paragraph explaining your project. Describe how you chose the scale and how you created each state. Why is it important to use the same sized squares for all the states? How does it help to have a geographical map next to your cartogram? Describe any difficulties you had. What would you do differently if you were to do the project over again?

Project Notes

Guiding Students' Work

Rubric for Chapter Project

4 Students work cooperatively in their group to choose an interesting set of statistical data for their cartogram and find relevant data for three to five states. They choose an appropriate scale for the squares and are successful in forming the shapes of their states. The states are placed in their correct respective positions, a poster is made, and the correct information is placed on the poster.

3 Students are successful in collecting the necessary data, and they organize it properly to make their cartogram. The size and scale chosen for the squares could have been better, but it works. A poster is made and labeled, but a little more effort would have made it visually more attractive.

2 The statistical data students choose are difficult to work with and are somewhat incomplete. States are chosen and relevant data collected. The size and scale chosen for the squares is not a convenient one and the resulting cartogram does not come out very well and is misleading. A poster is made, but labels are incorrect or missing altogether.

1 Students have difficulty getting started and cannot find appropriate data to begin constructing their cartograms correctly. Squares are drawn, but they are not representative of the data. No attempt is made to construct a poster. Students should be encouraged to speak with the teacher as soon as possible to review their work and to make a new start on the project.

Chapter Support

Course Guide: Chapter 5
Lesson Plans: Chapter 5
Practice Bank:
Cumulative Practice 37
Study Guide: Chapter 5 Review
Explorations Lab Manual:
Additional Exploration 6
Assessment Book:
Chapter Tests 22 and 23
Chapter 5 Alternative Assessment
Test Generator software
Portfolio Project Book
Preparing for College Entrance Tests
Professional Handbook

Progress Check 5.1–5.2

1. Identify the means and the extremes in the proportion $\frac{2}{7} = \frac{3}{10.5}$. *(Section 5.1)*
   means: 7 and 3;
   extremes: 2 and 10.5

Solve each proportion. *(Section 5.1)*

2. $\frac{5}{x} = \frac{2}{3}$ 7.5
3. $\frac{3}{x+3} = \frac{2}{9-2x}$ 2.625

Solve using a proportion. *(Section 5.1)*

4. The ratio of spruce trees to fir trees on a wood lot is 3 : 5. If there are 750 spruce trees on the lot, how many fir trees are on the lot? 1250 fir trees
5. The ratio of cars to pickup trucks that pass through an intersection from 5:00 P.M. to 6:00 P.M. is 7 : 2. Of the 225 cars and trucks that passed through the intersection during this hour one afternoon, how many were cars? 175 cars

Jeremy's father built a 20 ft sailboat. Jeremy is building a model of his father's boat. The scale of Jeremy's model is 1 in. = 2 ft. *(Section 5.2)*

6. How long must Jeremy's model be? 10 in.
7. What is the scale factor of Jeremy's model? $\frac{1}{24}$

CHAPTER

5 Review

STUDY TECHNIQUE

Look back through the chapter. List all the main topics and a real-world example of each. Explain why each example fits the topic.

VOCABULARY

ratio (p. 192)
proportion (p. 193)
means (p. 193)
extremes (p. 193)
scale (p. 198)
scale factor (p. 199)
similar (p. 206)
probability (p. 218)
event (p. 218)
outcome (p. 218)
experimental probability (p. 219)
theoretical probability (p. 220)
favorable outcome (p. 220)
geometric probability (p. 224)

SECTIONS 5.1 *and* 5.2

A **ratio** is a quotient that compares two numbers. The ratio of 1 to 2 can be written in three ways:

$$\frac{1}{2} \qquad 1:2 \qquad 1 \text{ to } 2$$

A **proportion** is an equation that shows two ratios to be equal. For any proportion, the product of the **extremes** is equal to the product of the **means**.

$$\frac{9}{12} = \frac{6}{8}$$
$$9 \cdot 8 = 12 \cdot 6$$
$$72 = 72$$

A **scale** is a ratio of two different measurements. The photograph of *Columbia* has a scale of about $\frac{1 \text{ cm}}{6 \text{ m}}$. Each centimeter in the picture represents 6 meters in real life.

A **scale factor** is the ratio of a new size to an old size. You can write the scale factor of the photograph as $\frac{1}{600}$, 0.0017, or 0.17%.

SECTIONS 5.3 *and* 5.4

Two figures are **similar** when corresponding angles have equal measure and the ratios of the lengths of corresponding sides are equal. Similar figures are always the same shape.

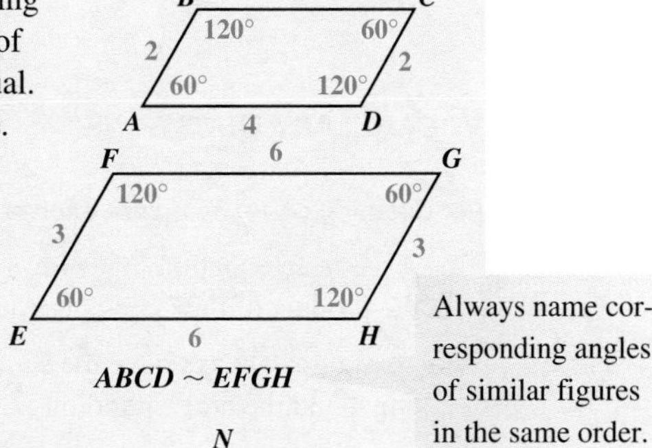

Always name corresponding angles of similar figures in the same order.

Triangles are similar when two angles of one triangle have the same measures as two angles of the other triangle.

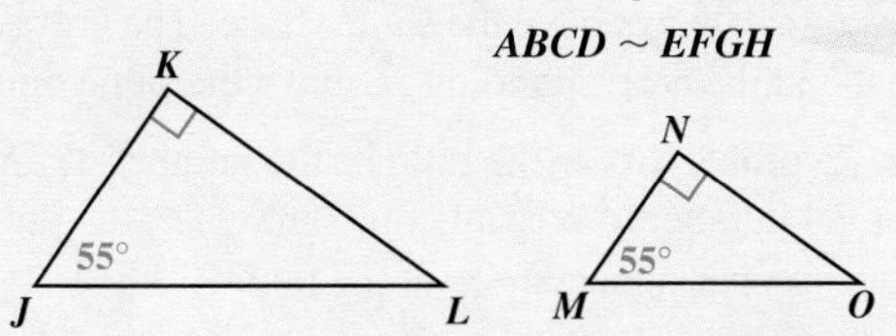

If the ratio of corresponding sides of similar figures is $x : y$, then the ratio of the perimeters of these figures is $x : y$, and the ratio of the areas is $x^2 : y^2$.

For figure *ABCD* and figure *EFGH* above, the ratio of the lengths of corresponding sides is 2 : 3. The ratio of the perimeters is 2 : 3, and the ratio of the areas is $2^2 : 3^2$ or 4 : 9.

SECTIONS 5.5 *and* 5.6

Probability is a ratio that measures how likely an **event** is to happen.

- The probability that water will freeze at 100°F is 0.
- The probability that the sun will rise tomorrow is 1.

Experimental probability is the ratio of the number of times an event actually happens to the number of times the experiment is performed.

- June oversleeps an average of once every four days, so the probability that she will oversleep tomorrow is 25%.

Theoretical probability is the ratio of the number of **favorable outcomes** to the total number of outcomes.

- Three of the six numbers on a die are prime, so the probability of rolling a prime number is 0.5.

Geometric probability is a probability based on area.

- One ninth of the target is shaded, so the probability that a randomly thrown dart will hit the shaded area is $\frac{1}{9}$.

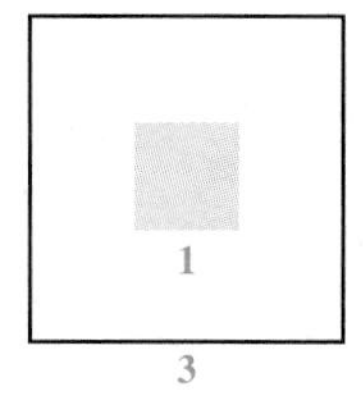

Progress Check 5.3–5.4

1. Suppose triangle *ABC* and triangle *RST* are both isosceles triangles with vertex angles *A* and *R*, respectively. If $\angle R = \angle A$, are the two triangles similar? Explain. *(Section 5.3)* Yes; the other corresponding angles have to be equal in measure.
2. What is the ratio of any pair of corresponding sides in two equilateral triangles if both triangles have sides of length *a*. *(Section 5.3)* 1 : 1
3. Complete. *(Section 5.4)* 1 m = __?__ cm; 1 m^2 = __?__ cm^2 100; 10,000

In the drawing, rectangle *MNOP* ~ rectangle *UVWX*.

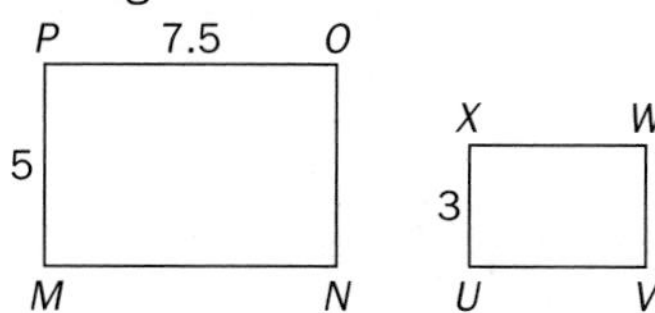

4. Find the length of side *WX*. *(Section 5.3)* 4.5
5. Find the ratio of the perimeter of rectangle *MNOP* to the perimeter of rectangle *UVWX*. *(Section 5.4)* 5 : 3
6. Find the ratio of the area of rectangle *UVWX* to the area of rectangle *MNOP*. *(Section 5.4)* 9 : 25

Progress Check 5.5–5.6

Tell whether each probability described is experimental or theoretical. *(Section 5.5)*

1. The probability of drawing a club card from a deck of cards is 0.25. theoretical
2. A coin was tossed 50 times and the results were used to calculate the probability of tossing a head on this coin. experimental

Find the theoretical probability of each event. *(Section 5.5)*

3. A number less than 5 is randomly chosen from the digits 0 to 9. 0.5
4. A coin is flipped and lands with tails showing. 0.5

Determine whether each probability is geometric. *(Section 5.6)*

5. the probability of pointing to a West Coast state when randomly pointing to a map of the United States Yes.
6. the probability that at least one head will show when two coins are tossed No.

Chapter 5 Assessment
Form A Chapter Test

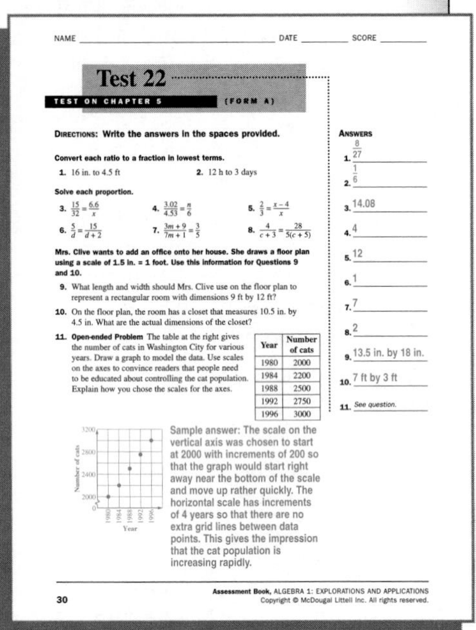

NAME ______ DATE ______ SCORE ______

Test 22

TEST ON CHAPTER 5 (FORM A)

DIRECTIONS: Write the answers in the spaces provided.

Convert each ratio to a fraction in lowest terms.

1. 16 in. to 4.5 ft
2. 12 h to 3 days

Solve each proportion.

3. $\frac{15}{32} = \frac{6.6}{x}$
4. $\frac{3.02}{4.53} = \frac{n}{6}$
5. $\frac{2}{3} = \frac{x-4}{x}$
6. $\frac{5}{d} = \frac{15}{d+2}$
7. $\frac{3m+9}{7m+1} = \frac{3}{5}$
8. $\frac{4}{c+3} = \frac{28}{5(c+5)}$

Mrs. Clive wants to add an office onto her house. She draws a floor plan using a scale of 1.5 in. = 1 foot. Use this information for Questions 9 and 10.

9. What length and width should Mrs. Clive use on the floor plan to represent a rectangular room with dimensions 9 ft by 12 ft?
10. On the floor plan, the room has a closet that measures 10.5 in. by 4.5 in. What are the actual dimensions of the closet?
11. **Open-ended Problem** The table at the right gives the number of cats in Washington City for various years. Draw a graph to model the data. Use scales on the axes to convince readers that people need to be educated about controlling the cat population. Explain how you chose the scales for the axes.

Year	Number of cats
1980	2000
1984	2200
1988	2500
1992	2750
1996	3000

Sample answer: The scale on the vertical axis was chosen to start at 2000 with increments of 200 so that the graph would start right away near the bottom of the scale and move up rather quickly. The horizontal scale has increments of 4 years so that there are no extra grid lines between data points. This gives the impression that the cat population is increasing rapidly.

ANSWERS

1. $\frac{8}{27}$
2. $\frac{1}{6}$
3. 14.08
4. 4
5. 12
6. 1
7. 7
8. 2
9. 13.5 in. by 18 in.
10. 7 ft by 3 ft
11. *See question.*

30 **Assessment Book,** ALGEBRA 1: EXPLORATIONS AND APPLICATIONS

Chapter 5 Assessment
Form B Chapter Test

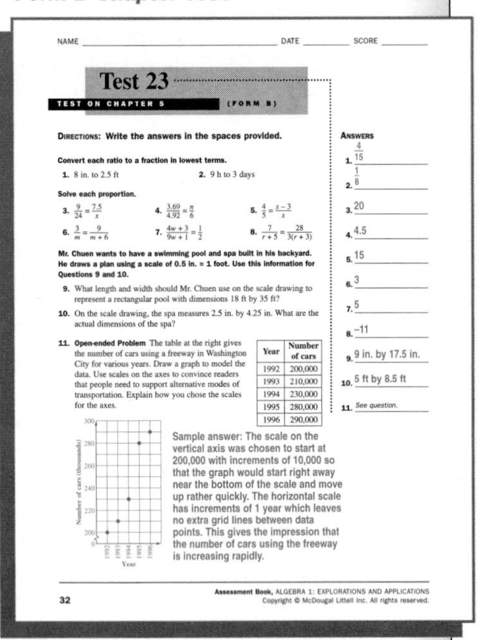

NAME ______ DATE ______ SCORE ______

Test 23

TEST ON CHAPTER 5 (FORM B)

DIRECTIONS: Write the answers in the spaces provided.

Convert each ratio to a fraction in lowest terms.

1. 8 in. to 2.5 ft
2. 9 h to 3 days

Solve each proportion.

3. $\frac{9}{24} = \frac{7.5}{x}$
4. $\frac{3.69}{4.92} = \frac{n}{6}$
5. $\frac{4}{5} = \frac{x-3}{x}$
6. $\frac{3}{m} = \frac{9}{m+6}$
7. $\frac{4w+3}{9w+1} = \frac{1}{2}$
8. $\frac{7}{r+5} = \frac{28}{3(r+3)}$

Mr. Chuen wants to have a swimming pool and spa built in his backyard. He draws a plan using a scale of 0.5 in. = 1 foot. Use this information for Questions 9 and 10.

9. What length and width should Mr. Chuen use on the scale drawing to represent a rectangular pool with dimensions 18 ft by 35 ft?
10. On the scale drawing, the spa measures 2.5 in. by 4.25 in. What are the actual dimensions of the spa?
11. **Open-ended Problem** The table at the right gives the number of cars using a freeway in Washington City for various years. Draw a graph to model the data. Use scales on the axes to convince readers that people need to support alternative modes of transportation. Explain how you chose the scales for the axes.

Year	Number of cars
1992	200,000
1993	210,000
1994	230,000
1995	280,000
1996	290,000

Sample answer: The scale on the vertical axis was chosen to start at 200,000 with increments of 10,000 so that the graph would start right away near the bottom of the scale and move up rather quickly. The horizontal scale has increments of 1 year which leaves no extra grid lines between data points. This gives the impression that the number of cars using the freeway is increasing rapidly.

ANSWERS

1. $\frac{4}{15}$
2. $\frac{1}{8}$
3. 20
4. 4.5
5. 15
6. 3
7. 5
8. −11
9. 9 in. by 17.5 in.
10. 5 ft by 8.5 ft
11. *See question.*

32 **Assessment Book,** ALGEBRA 1: EXPLORATIONS AND APPLICATIONS

CHAPTER

5 Assessment

VOCABULARY QUESTIONS

For Questions 1–4, complete each statement.

1. A __?__ is an equation that shows two ratios are equal. In this equation, the product of the __?__ equals the product of the __?__.
2. __?__ figures always have the same shape. The corresponding __?__ are equal and the corresponding __?__ have the same ratio of lengths.
3. The __?__ probability is the ratio of the number of favorable outcomes to the total number of possible outcomes. The __?__ probability is the ratio of the number of successes to the number of tries.
4. An event with a probability of __?__ will certainly occur. An event with a probability of __?__ will never occur.

SECTIONS 5.1 *and* 5.2

Write each ratio as a fraction in lowest terms.

5. 6 in. to 5.5 ft
6. \$42.50 to \$5.00
7. A 1.25 acre school sits on a 7 acre campus. One acre of the grounds is devoted to parking and 4 acres are devoted to playing fields. For parts (a)–(c), find the ratio of the first quantity to the second. Write each answer as a fraction in lowest terms.
   a. school area to campus area
   b. school area to playing field area
   c. unused area to campus area

Solve each proportion.

8. $\frac{15}{35} = \frac{82.5}{x}$
9. $\frac{3.14}{50.24} = \frac{n}{16}$
10. $\frac{3}{5} = \frac{n-12}{n}$
11. $4 : p = 6 : p + 9$
12. $a : 8 = 7 : 1$
13. $5 : c = 9 : 8$
14. **Writing** Explain how to solve the proportion $5 : 8 = 30 : n$ without using the means-extremes property.
15. A 12 ft by 20 ft room is being remodeled. What scale should be used in order to fit the largest possible drawing on an $8\frac{1}{2}$ in. by 11 in. paper? What is the scale factor of the drawing?

ANSWERS Chapter 5

Assessment

1. proportion; means; extremes
2. similar; angle measures; sides
3. theoretical; experimental
4. 1; 0
5. $\frac{1}{11}$
6. $\frac{17}{2}$
7. a. $\frac{5}{28}$ b. $\frac{5}{16}$ c. $\frac{3}{28}$
8. 192.5
9. 1
10. 30
11. 18
12. 56
13. $4\frac{4}{9}$
14. Answers may vary. Examples are given. One way is to notice that because $30 = 6 \cdot 5$, $n = 6 \cdot 8 = 48$. Another way is to rewrite the proportion as $\frac{5}{8} = \frac{30}{n}$. Multiply both sides of the equation by the least common denominator of the fractions, $8n$, to get $5n = 240$ or $n = 48$.
15. 11 in. = 20 ft, or 1 in. ≈ 1 ft 10 in.; $\frac{11}{240}$

16. Writing The graph shows the number of school-age children in Hopedale. Suppose you want to use the graph to convince the town council that Hopedale needs a new school. Redraw the graph using a different scale on the vertical axis. Explain how the scale you chose changes the visual effect of the graph.

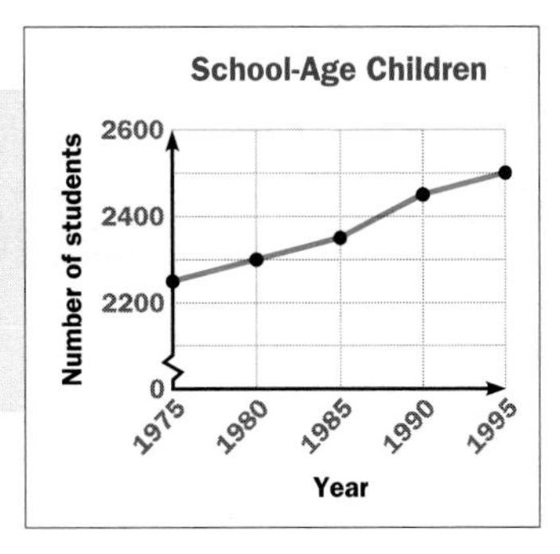

SECTIONS 5.3 *and* 5.4

17. Open-ended Problem Draw a pair of similar figures so that the ratio of lengths of corresponding sides is 7 : 2.

For Questions 18–22, use the fact that *ABCD* ~ *EFGH*.

18. Find *AD* and *FG*.

19. Find $\angle B$, $\angle C$, $\angle E$, and $\angle H$.

20. The perimeter of *ABCD* is 19.2. Find the perimeter of *EFGH*.

21. The area of *EFGH* is 31.5. Find the area of *ABCD*.

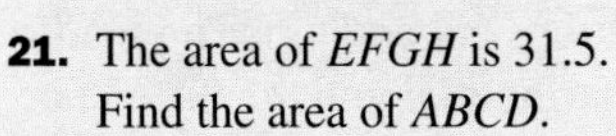

22. Tell whether each statement is *True* or *False*.

a. *BCDA* ~ *FGHE*

b. *CBAD* ~ *GHEF*

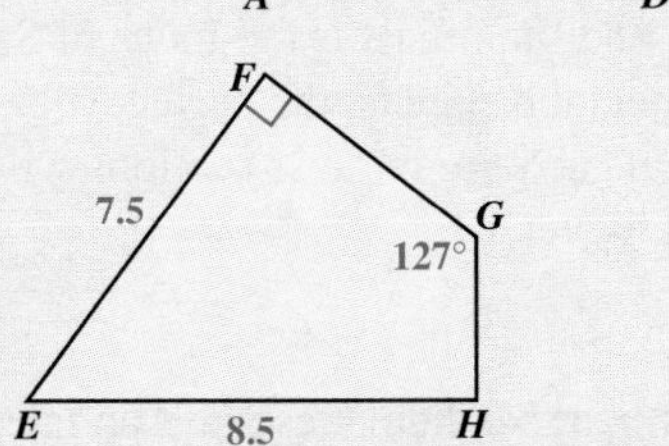

SECTIONS 5.5 *and* 5.6

23. The table shows results of 50 rolls of a die. Find the theoretical and experimental probabilities of rolling either a 2 or a 5 on one roll.

Number rolled	1	2	3	4	5	6
Times rolled	8	12	5	9	10	6

24. Darts are tossed at the square board pictured. What is the geometric probability that a dart will land inside the circle?

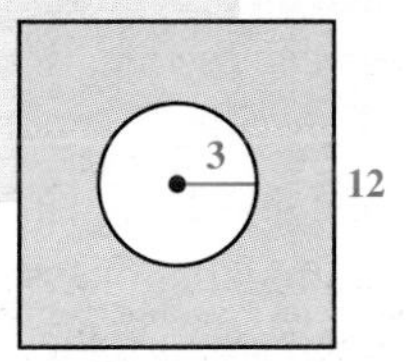

PERFORMANCE TASK

25. Choose an exercise or quiz question you solved incorrectly. Explain what you did wrong. Correct it. Explain your correction.

Chapter 5 Assessment

Form C Alternative Assessment

Chapter 5

ALTERNATIVE ASSESSMENT

1. a. Are the parallelograms below similar? How do you know?

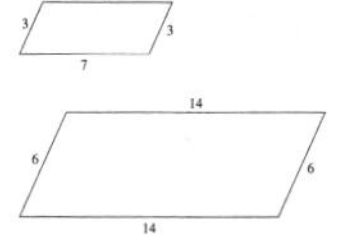

b. How are the areas of the parallelograms related? Explain your answer.

2. Group Activity Work in a group of four. (This activity could also be done with the whole class.) Do the following activity to estimate the percent of Earth that is covered by water. Randomly toss an inflatable globe to different group members 20 times. Count how many times the receiving member's right thumb touches water on the globe when it is caught. Use this information to make the estimation. Repeat the process for 40 tosses and 60 tosses. Use the new information to estimate again. Look up the exact percent in an encyclopedia. As you increased the number of tosses, did the results of the activity get closer to the actual percent? Why do you think that happened?

3. Group Activity Discuss the the following questions in groups. Answer each question in as many different ways as you can. Include examples with your answers.

a. If rectangle *ABCD* is similar to rectangle *EFGH*, what would be the minimum amount of information that you would need in order to calculate the perimeter and area of each rectangle?

b. If you knew the areas, what other information would you need in order to find the perimeters of the rectangles?

4. a. Graph $y = -x^2 + 5$ using x-values from -10 to 10 and y-values from -10 to 10.

b. Graph $y = -x^2 + 5$ using x-values from 0 to 0.1 and y-values from 0 to 6.

c. Use your graphs from parts (a) and (b) to help you answer this question: Can the scale of a graph affect your impression of that graph?

98 **Assessment Book,** ALGEBRA 1: EXPLORATIONS AND APPLICATIONS

5. Open-ended Problem Solve the proportion $\frac{x}{15} = \frac{10}{6}$ without using the means-extremes property. (*Hint:* Use what you know about solving equations.)

6. Performance Task Make a scale model of your favorite room in your house. Be sure to indicate the scale of your model.

7. Research Project Compare the size of a doll or action figure to an average-sized adult. Consider the lengths of the arms and the legs, the circumference of the head, and so on. What scale was used to reduce the doll or action figure from the size of an average adult? Are the proportions accurate? The Golden Rectangle was used by many ancient artists to keep the images of human bodies in proportion. Investigate and report on the Golden Rectangle.

Assessment Book, ALGEBRA 1: EXPLORATIONS AND APPLICATIONS 99

16. Answers may vary. An example is given.

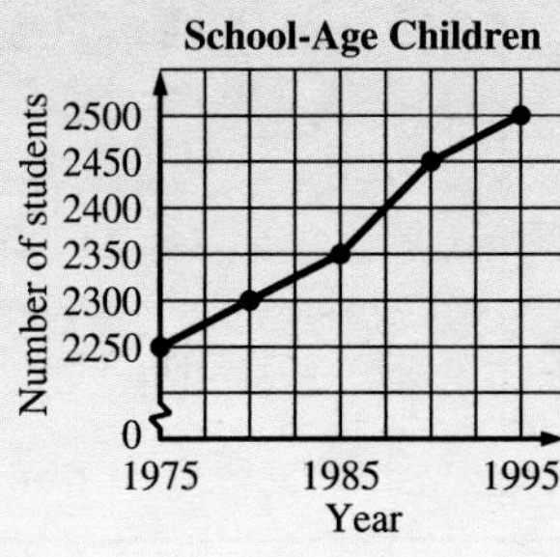

I changed the vertical scale to give a steeper graph by increasing the space between units and adding new units. This makes the growth rate seem higher.

17. Check students' work.

18. 6.8; 5

19. 90°; 127°; 53°; 90°

20. 24

21. 20.16

22. a. True.

b. False.

23. $\frac{1}{3}$; $\frac{11}{25}$

24. $\frac{\pi}{16} \approx 0.2$

25. Answers may vary. Check students' work.

CHAPTER

6 Working with Radicals

OVERVIEW

Connecting to Prior and Future Learning

⇔ Students begin their work with radicals by studying square roots as they relate to the Pythagorean theorem. Section 6.1 includes exercises that help students understand why this important theorem applies only to right triangles. Students may find it helpful to study page 595 in the **Student Resources Toolbox**, which presents a review of the concept of angles.

⇔ Irrational numbers and calculating with radicals are presented in Sections 6.2 and 6.3, respectively. As students work with these concepts, they will be preparing for future work with quadratic functions and other applications that require the use of radicals.

⇔ Finding products of monomials and binomials is the focus of Sections 6.4 and 6.5. The work in Section 6.5 focuses on two special products, the perfect square trinomial and the difference of two squares. The skills students acquire in these sections will be used again when they study polynomial functions in Chapter 10.

Chapter Highlights

Interview with Nathan Shigemura: Mathematics plays an integral role in investigating traffic accidents. This fact is emphasized in the interview with Nathan Shigemura, an accident investigator with the Illinois State Police. Students have the opportunity to study this relationship in more detail in the related examples and exercises found on pages 241, 245, 249, and 255.

Explorations in Chapter 6 ask students to use graph paper to draw right triangles that can be used to compare radicals, and to find products with algebra tiles.

The Portfolio Project: This project guides students through an experiment with a pendulum. As students make their pendulum, and collect and organize their data, they learn about the relationship between the length of a pendulum and its period. Students summarize their work on this project by writing a report describing their work.

Technology: Calculators are used to find decimal approximations for both rational and irrational numbers. Students also use calculators and spreadsheets throughout the chapter to evaluate formulas. As in previous chapters, graphing calculators are used to graph and compare equations. These typical uses of calculators will continue to be used throughout this and future mathematics courses.

OBJECTIVES

Section	Objectives	NCTM Standards
6.1	• Identify right triangles using the Pythagorean theorem. • Use the Pythagorean theorem to solve real-world problems.	1, 2, 3, 4, 5, 7, 8
6.2	• Find noninteger square roots. • Estimate square roots to solve real-world problems.	1, 2, 3, 4, 5
6.3	• Multiply and divide radicals. • Simplify radicals. • Use skills with radicals to find values in real-world problems.	1, 2, 3, 4, 5
6.4	• Find products of monomials and binomials. • Use products of monomials and binomials to find areas of complex figures.	1, 2, 3, 4, 5
6.5	• Recognize some special products of binomials. • Find products of some variable expressions more quickly.	1, 2, 3, 4, 5

INTEGRATION

Mathematical Connections	6.1	6.2	6.3	6.4	6.5
algebra	**239–244***	**245–250**	**251–257**	**258–263**	**264–269**
geometry	**239–244**	250	251, 254, 256, 257	262, 263	267
data analysis, probability, discrete math		249	256		268
patterns and functions	244				
logic and reasoning	240–244	246, 247, 249, 250	251, 254	259, 260, 262, 263	264, 266–269
Interdisciplinary Connections and Applications					
history and geography				262	
reading and language arts		249			
biology and earth science		250			
chemistry and physics			253		
sports and recreation	240				
urban planning, aviation, medicine, construction, genetics	244	248	255	262	267, 268

****Bold page numbers** indicate that a topic is used throughout the section.*

TECHNOLOGY

	opportunities for use with	
Section	**Student Book**	**Support Material**
6.1	graphing calculator	**Technology Book:** Spreadsheet Activity 6
6.2	graphing calculator spreadsheet software McDougal Littell Software *Stats!*	**Technology Book:** Spreadsheet Activity 6
6.3	graphing calculator McDougal Littell Software *Stats!*	**Technology Book:** Calculator Activity 6 Spreadsheet Activity 6
6.4	scientific calculator	
6.5	scientific calculator	

PLANNING GUIDE

Regular Scheduling (45 min)

Section	Materials Needed	Core Assignment	Extended Assignment	*exercises that feature* Applications	Communication	Technology
6.1	calculator, graph paper, ruler	1–20, 22, 24, 26–31	1–11, 14, 18–31	21, 24	21, 23	
6.2	spreadsheet software	1–25, 31, 33–39	1–7, 11–22, 26–39	11, 28–31, 33	26, 27, 30, 33	28
6.3	graph paper, graphing calculator, yardstick	**Day 1:** 1–17 **Day 2:** 19–23, 25–31, AYP*	**Day 1:** 1–18 **Day 2:** 19–31, AYP	17 19, 20	19, 24	14 22
6.4	algebra tiles	1–19, 21, 22, 25–30	1–5, 8–30	20–22	20, 23–25	
6.5		**Day 1:** 1–23 **Day 2:** 24–27, 32–44, AYP	**Day 1:** 1–23 **Day 2:** 24–44, AYP	11 29, 30	10 31, 32	
Review/ Assess		**Day 1:** 1–17 **Day 2:** 18–40 **Day 3:** Ch. 6 Test	**Day 1:** 1–17 **Day 2:** 18–40 **Day 3:** Ch. 6 Test	9	1–3 39	
Portfolio Project		Allow 2 days.	Allow 2 days.			

Yearly Pacing (with Portfolio Project)	**Chapter 6 Total** 12 days	**Chapters 1–6 Total** 80 days	**Remaining** 80 days	**Total** 160 days

Block Scheduling (90 min)

	Day 35	Day 36	Day 37	Day 38	Day 39	Day 40	Day 41
Teach/Interact	Ch. 5 Test 6.1	6.2 6.3: Exploration, page 251	Continue with 6.3 6.4: Exploration, page 259	6.5	Review Port. Proj.	Review Port. Proj.	Ch. 6 Test 7.1
Apply/Assess	**Ch. 5 Test** **6.1:** 1–11, 14, 17–31	**6.2:** 1–39 **6.3:** 1–18	**6.3:** 19–31, AYP* **6.4:** 1–19, 21–23, 25–30	**6.5:** 1–28, 32–44, AYP	**Review:** 1–17 **Port. Proj.**	**Review:** 18–40 **Port. Proj.**	**Ch. 6 Test** **7.1:** 1–22

NOTE: A one-day block has been added for the Portfolio Project—timing and placement to be determined by teacher.

Yearly Pacing (with Portfolio Project)	**Chapter 6 Total** 6 days	**Chapters 1–6 Total** $40\frac{1}{2}$ days	**Remaining** $40\frac{1}{2}$ days	**Total** 81 days

****AYP*** *is Assess Your Progress.*

Lesson Support

Section	Practice Bank	Study Guide*	Assessment Book*	Visuals	Explorations Lab Manual	Lesson Plans	Technology Book
6.1	38	6.1		Warm-Up 6.1 Folder 6	Master 1 Add. Expl. 8	6.1	Spreadsheet Act. 6
6.2	39	6.2		Warm-Up 6.2		6.2	Spreadsheet Act. 6
6.3	40	6.3	Test 24	Warm-Up 6.3	Master 1 Add. Expl. 9	6.3	Calculator Act. 6 Spreadsheet Act. 6
6.4	41	6.4		Warm-Up 6.4 Folder B	Masters 4, 5	6.4	
6.5	42	6.5	Test 25	Warm-Up 6.5 Folder B		6.5	
Review Test	43	Chapter Review	Tests 26, 27, 28 Alternative Assessment			Review Test	

***Spanish versions** of *Study Guide* and *Assessment Book* are available.

Chapter Support

- Course Guide
- Lesson Plans
- Portfolio Project Book
- Preparing for College Entrance Tests
- Multi-Language Glossary
- *Test Generator* Software
- Professional Handbook

Software Support

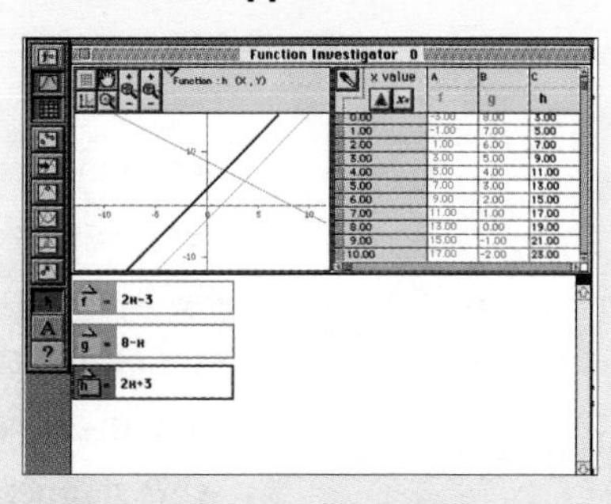

McDougal Littell Software
Stats!

Internet Support

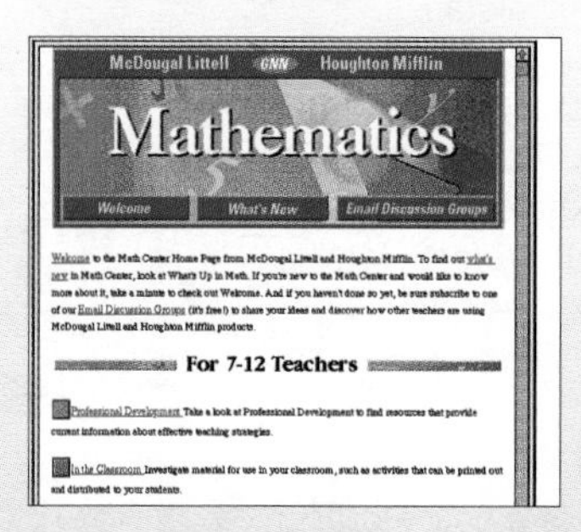

http://www.hmco.com
Next go to McDougal Littell; then the Education Center; then Secondary Math.

Outside Resources

Books, Periodicals

Miller, William and Linda Wagner. "Pythagorean Dissection Puzzles." *Mathematics Teacher* (April 1993): pp. 302–314.

Kolpas, Sidney J. *The Pythagorean Theorem.* Poster set and book. Palo Alto, CA: Dale Seymour Publications.

Activities, Manipulatives

Wenninger, Magnus J. *Polyhedron Models for the Classroom.* Reston, VA: NCTM, 1975.

Naraine, Bishnu. "If Pythagoras Had a Geoboard ..." *Mathematics Teacher* (February 1993): pp. 137–140.

Software

Geometer's Sketchpad. Macintosh and MS-DOS. Berkeley, CA: Key Curriculum Press.

Videos

Apostol, Tom. *The Story of Pi.* Reston, VA: NCTM.

Apostol, Tom. *The Theorem of Pythagoras.* Reston, VA: NCTM. Features narration and computer graphics to demonstrate why the theorem works.

Internet

From an internet account, access lesson plans for K–12 mathematics, compiled by the Eisenhower Network, using the command:

gopher enc.org

Interview Notes

Background

Speed and Safety
While riding in an automobile, most people understand intuitively that safety is related to speed. Ask students if they would feel safer in a car traveling 40 mi/h or in one traveling 90 or 100 mi/h. Many responses to this question would probably involve the ability of the driver to stop the car if something unexpected happened. It is a fact that excessive speed is a major cause of traffic accidents. The reason for this involves the time it takes to stop a fast-moving car and the distance the car travels during this time, which is called the *total stopping distance.* Total stopping distance is a function of speed and involves three quantities: perception distance, reaction distance, and braking distance. When a driver sees a hazard, the car travels a certain distance before the driver recognizes it. Then the driver's foot must move to the brake. Then the car must stop. Therefore, Total stopping distance = Perception distance + Reaction distance + Braking distance.

Nathan Shigemura
Nathan Shigemura is a police officer whose job is to investigate accidents to determine what caused them. The results of his investigations are important in assigning responsibility for the accident and can affect the outcome of trials and insurance payments to the people involved. Because of the potential serious consequences of his investigations, Trooper Shigemura employs numerous formulas from science and mathematics to ensure that the results of his investigations are correct.

CHAPTER

6 Working with Radicals

On the scene of the CRASH

INTERVIEW **Nathan Shigemura**

Many of us think we know from television shows and movies what a police investigation is all about. The detective arrives at the scene of the crime, searching for clues and questioning witnesses until a suspect finally breaks down and confesses. But few people realize the importance of mathematics in an investigation, particularly an investigation of a traffic accident. "When I get to an accident scene, it's hectic, with rescue personnel, police officers, survivors, witnesses, and victims all intermingled," explains Trooper Nathan Shigemura, an accident reconstructionist with the Illinois State Police. "It is often dark and always chaotic. Out of that chaos, we try to figure out what happened and why."

"Math enables people to understand the world around them."

> **"All accidents obey the same laws of physics and math."**

Trooper Shigemura shown measuring the road surface's coefficient of friction.

Solving a Puzzle

An accident is like a puzzle. Shigemura arrives after the collision and tries to put the pieces together. Mathematics plays a big role in this process because "everything happens according to the laws of physics and many of these can be represented by mathematical equations." By examining the position of the vehicles, the nature and extent of the damage they have sustained, and other physical evidence, he works backwards to identify the factors leading up to the crash.

Making Cars and Roads Safer

Knowledge gained from these investigations can highlight safety problems that weren't recognized before. "Based on what we see, we might learn that a particular make of car has an engineering defect or that the roadway itself is dangerous," says Shigemura. He is pleased that his studies of tragic accidents can eventually pave the way for safer roads and cars. He also finds satisfaction in the task of taking a complex situation like an accident scene and gradually piecing together what happened. "Math enables people to understand the world around them. When you can understand how things work—rather than just looking around in wonder and scratching your head—it opens up a whole new world."

Interview Notes

Background

German Motorways
There is no maximum speed limit on motorways in Germany. The average speed of passenger cars is about 117 km/h (about 73 mi/h) and more than 30% of passenger cars on German roadways travel at speeds over 130 km/h (about 81 mi/h).

Second-Language Learners

You may need to explain that the expression *pave the way* means "to make things easier for the future and to set standards," in this case, for motor vehicles and highway laws.

Interview Notes

Mathematical Connection

In Section 6.1, students learn about square roots of numbers and are introduced to the radical sign. These concepts are then applied in Section 6.2, where the formula to find the speed of a car involved in an accident, $S = \sqrt{30df}$, is given. In this formula, d is the length of the skid mark and f is the coefficient of friction for the road. In Section 6.3, students continue to work with formulas involving radicals that relate braking distance and reaction time.

Explore and Connect

Writing
When discussing question 1, you may wish to point out that the coefficient of friction is the force required to drag a car with all four wheels locked divided by the mass of the car. These coefficients have been calculated for all types of road surfaces and are available for police investigations.

Research
Many community colleges offer courses in police science or law enforcement. Their catalogs would contain information pertinent to this project.

Project
Ask students to relate what they have noticed about how far the car travels to the concept of a coefficient of friction.

The Physical Evidence

Shigemura uses physics and math to reconstruct an accident, charting the precise path the vehicles must have taken before the crash to end up where they did. For example, he uses the formula $S = \sqrt{30df}$ to calculate a car's minimum speed S in miles per hour at the moment the car begins to skid. In the formula, d = the length in feet of the skid mark left by the car and f = the coefficient of friction. The coefficient of friction is a measure of the slipperiness or stickiness of the road. Shigemura finds d with a tape measure, and determines f with a special device. The matrix at the right shows speeds calculated for several different combinations of d and f.

Calculated Speed *S*

f \ d	76	98	123
0.40	30.19	34.29	38.41
0.80	42.70	48.49	54.33
1.20	52.30	59.39	66.54

$\sqrt{30 \cdot 123 \cdot 0.40} = 38.41$

If $d = 123$ and $f = 0.40$, the car was traveling about 38 mi/h when it began to skid.

EXPLORE AND CONNECT

Nathan Shigemura standing in the dent left in the front of a truck by its impact with a telephone pole.

1. Writing Use the formula in the article. If the value of f does not change, what happens to S as d increases? If d does not change, what happens to S as f increases? If a car is traveling at 55 mi/h, will it skid farther on a road with a high coefficient of friction or a low one?

2. Research Find out the requirements for becoming a police officer. What mathematics do officers need to know? What are the course requirements for a degree in law enforcement? Report your findings to your class.

3. Project Use a toy car and a ramp. Let the car roll down the ramp on to various surfaces: rough concrete, a smooth floor, and so on. Observe how far the car travels before it stops. What do you notice?

Mathematics
& Nathan Shigemura

In this chapter, you will learn more about how mathematics is related to police investigations.

Related Examples and Exercises

Section 6.1
- **Example 2**

Section 6.2
- **Example 1**
- **Exercises 28–30**

Section 6.3
- **Exercises 19 and 20**

SECTION

6.1 The Pythagorean Theorem

Learn how to...

- identify right triangles using the Pythagorean theorem

So you can...

- solve real-world problems, such as checking whether a corner is square

Surveyors in ancient Egypt may have used stretched ropes like the one shown to make right angles when laying out land boundaries. The surveyors knew that a triangle with side lengths of 3, 4, and 5 units is a *right triangle*.

A **right triangle** has one right, or 90°, angle. As shown on the map below, people have been aware for centuries of a special relationship among the sides of a right triangle. The *Pythagorean theorem* describes this relationship. The theorem is named after Pythagoras, a Greek mathematician.

The Pythagorean Theorem

In a right triangle, the square of the length of the **hypotenuse** is equal to the sum of the squares of the lengths of the **legs**.

$a^2 + b^2 = c^2$ $\qquad$ $3^2 + 4^2 = 5^2$

leg a, leg b, hypotenuse c $\qquad$ 3, 4, 5

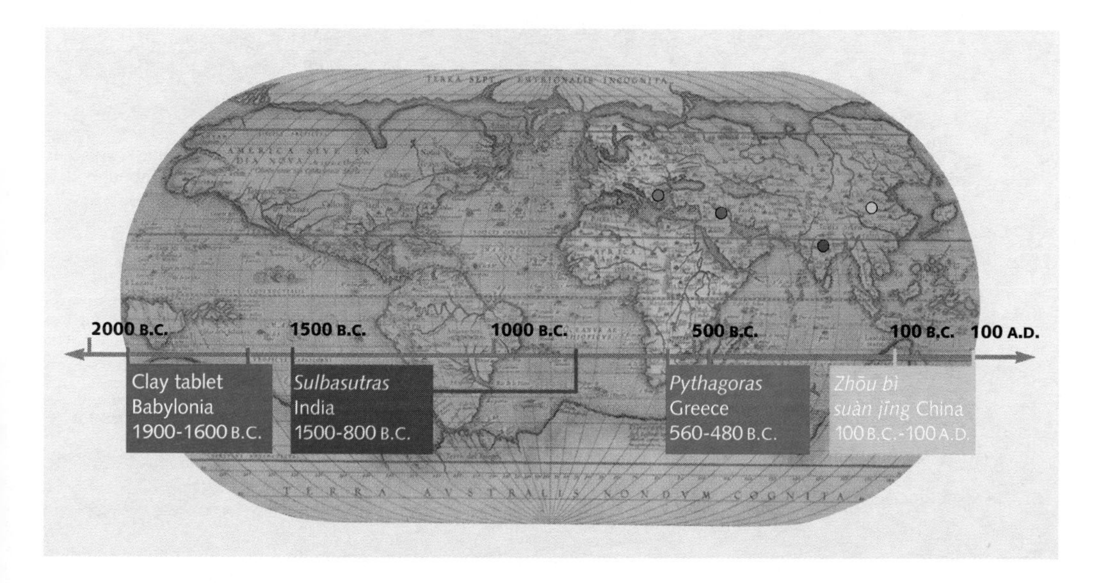

Plan⇔Support

Objectives

- Identify right triangles using the Pythagorean theorem.
- Use the Pythagorean theorem to solve real-world problems.

Recommended Pacing

Core and Extended Courses
Section 6.1: 1 day

Block Schedule
Section 6.1: $\frac{1}{2}$ block
(with Chapter 5 Test)

Resource Materials

Lesson Support
Lesson Plan 6.1
Warm-Up Transparency 6.1
Overhead Visuals:
Folder 6: Pythagorean Triples
Practice Bank: Practice 38
Study Guide: Section 6.1
Explorations Lab Manual:
Additional Exploration 8,
Diagram Master 1

Technology
Technology Book:
Spreadsheet Activity 6
Graphing Calculator
Internet:
http://www.hmco.com

Warm-Up Exercises

Simplify.

1. $5^2 + 6^2$ 61
2. $12^2 + 7^2$ 193
3. $15^2 - 13^2$ 56
4. $30^2 - 20^2$ 500
5. Name two integers that have squares equal to 81. 9, –9
6. In right triangle *ABC*, what are the lengths of the legs? of the hypotenuse?

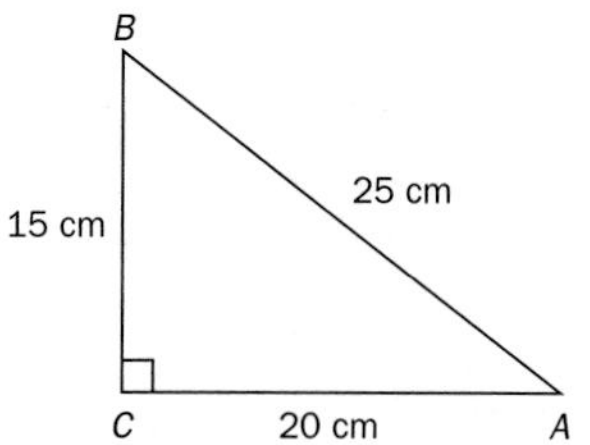

15 cm and 20 cm; 25 cm

Teach⟺Interact

Learning Styles: Visual

Visual learners may find it helpful to have a geometric picture to reinforce the verbal definition of square root. Use the relationship between the length of a side of a square and the area of the square.

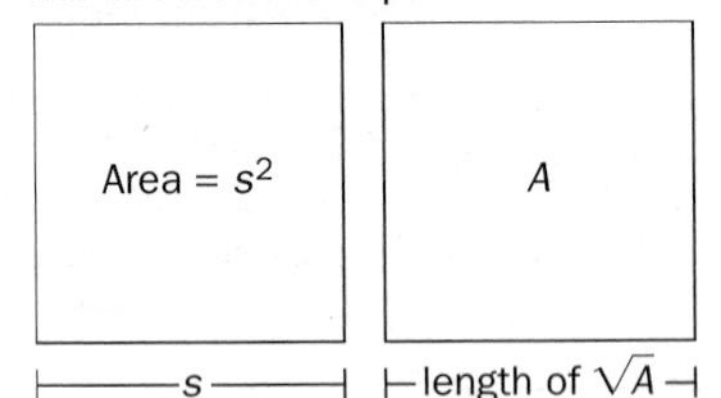

Point out that the geometric interpretation of square root accounts for only one of the square roots of a positive number. The negative square root cannot be pictured, since only positive numbers can represent lengths.

Additional Example 1

A hot-air balloon is attached to a 75 foot rope that is tied to a post. A breeze blows the balloon to one side of the post, so that when the rope is taut the balloon is over the edge of a flower bed 21 feet from the post.

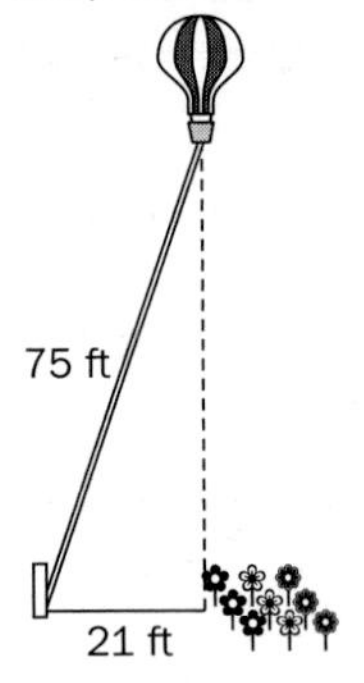

How high above the flower bed is the balloon?

Use the Pythagorean theorem. Let h = the height of the balloon.

$$h^2 + 21^2 = 75^2$$
$$h^2 + 441 = 5625$$
$$h^2 = 5184$$
$$h = \pm\sqrt{5184}$$
$$h = \pm 72$$

Only the positive solution makes sense. The balloon is 72 ft above the flower bed.

Section Note

Using Technology Students can use a TI-82 or TI-81 calculator to find the square root in Example 1. Press [2nd][$\sqrt{\ }$] 22500[ENTER]. The calculator will display the positive square root.

BY THE WAY

A German book published in 1525 included the first known use of the symbol $\sqrt{\ }$ in print. The symbol eventually replaced R̷x, an older notation. R̷x came from the first and last letters of the word *radix*, a Latin translation of an Arabic word for *root*.

Working with the Pythagorean theorem involves using *squares* and *square roots*. The number 9 is called a **perfect square** because it is the square of the integer 3. The number 3 is called a **square root** of 9 because $3^2 = 9$. The symbol $\sqrt{\ }$ is called a **radical sign**. The radical sign indicates the positive square root of a number. All positive numbers have two square roots. Negative numbers do not have square roots.

The positive square root is written as $\sqrt{9} = 3$.

The negative square root is written as $-\sqrt{9} = -3$.

Both square roots are written as $\pm\sqrt{9} = 3$ **or** -3.

The symbol $\pm$ is read as *plus or minus*.

EXAMPLE 1 Application: Recreation

Kevin is parasailing. He is wearing a special parachute and is connected to the boat by a 250 ft rope. As the speed of the boat increases, he is lifted into the air. Kevin is 200 ft behind the boat when the rope is taut. What is his height in feet above the water?

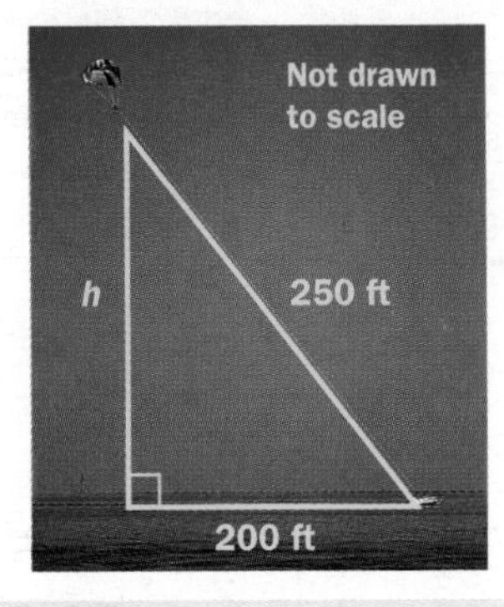

SOLUTION

Use the Pythagorean theorem. Let h = Kevin's height in feet above the water.

$$h^2 + 200^2 = 250^2$$
$$h^2 + 40{,}000 = 62{,}500$$
$$h^2 = 22{,}500$$
$$h = \pm\sqrt{22{,}500}$$
$$h = \pm 150$$

Find the *square root* of both sides of the equation.

The value of h can be either the positive square root or the negative square root of 22,500. Only the positive solution makes sense in this situation. Kevin is 150 ft above the water.

THINK AND COMMUNICATE

Find each square root. You may be able to use mental math. You can also use a calculator or the Table of Squares and Square Roots on page 614.

1. $\sqrt{1}$ **2.** $-\sqrt{4}$ **3.** $\pm\sqrt{289}$ **4.** $\sqrt{0}$

ANSWERS Section 6.1

Think and Communicate

1. 1
2. –2
3. ±17
4. 0

As you will see in Example 2 below, Nathan Shigemura uses the *converse* of the Pythagorean theorem in his work as an accident reconstructionist.

Converse of the Pythagorean Theorem

If the lengths of the sides of a triangle satisfy the relationship $a^2 + b^2 = c^2$, then the triangle is a right triangle.

EXAMPLE 2 Interview: Nathan Shigemura

To measure the extent of damage to a car, Nathan Shigemura creates a rectangle around the damaged car to represent its original shape. By comparing the shape of the damaged car to its original shape, he can see how badly the car is bent.

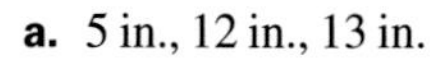

To make sure that the corners of his rectangle are right angles, Shigemura uses measuring tapes to form a triangle at each corner. He uses the converse of the Pythagorean theorem to check whether each triangle is a right triangle. Tell whether the given lengths can be the sides of a right triangle.

a. 5 in., 12 in., 13 in.

b. 16 in., 11 in., 19 in.

SOLUTION

Use the converse of the Pythagorean theorem.

a. $5^2 + 12^2 \stackrel{?}{=} 13^2$
$25 + 144 = 169$
These lengths *can* be the sides of a right triangle.

b. $16^2 + 11^2 \stackrel{?}{=} 19^2$
$256 + 121 \neq 361$
These lengths *cannot* be the sides of a right triangle.

THINK AND COMMUNICATE

5. Choose a corner of something. Use the converse of the Pythagorean theorem and a ruler to check whether the corner is a right angle.

6. To check the angles of his rectangle, Nathan Shigemura always uses a right triangle with side lengths of 3, 4, and 5 units. Why do you think he chooses to use this right triangle?

Teach⇔Interact

Section Notes

Multicultural Note
The development of mathematics in ancient Babylon progressed at a rate that many historians find impressive. By 1800 B.C., Babylonian mathematics reflected a knowledge of the properties of arithmetic and geometrical progressions, and an understanding of some geometrical relationships. While historians believe the Babylonians knew the relationships contained in the Pythagorean theorem, there is no formal Babylonian proof of it. Babylonian mathematics generally relied on knowledge of properties as practical facts rather than on formal proofs.

Teaching Tip
Some students may not recall what the word *converse* means. Explain that the converse of an if-then statement of the form "if *p*, then *q*" is the statement "if *q*, then *p*." It will probably help to discuss two or three instances of such statements.

Additional Example 2

A building contractor measured the sides and diagonals of two door frames to be sure that the door frames were "squared off" at the floor. Tell whether the given measurements can be the sides of a right triangle.

a. 3 ft, 7 ft, 7.5 ft

Use the converse of the Pythagorean theorem.
$3^2 + 7^2 \stackrel{?}{=} 7.5^2$
$9 + 49 \neq 56.25$
These measurements *cannot* be the sides of a right triangle.

b. 6 ft, 8 ft, 10 ft

$6^2 + 8^2 \stackrel{?}{=} 10^2$
$36 + 64 = 100$
These measurements can be the sides of a right triangle.

Think and Communicate

5. Answers may vary.
6. Answers may vary. An example is given. This right triangle has the shortest side lengths that are whole numbers, which makes his measuring easier.

Teach⇔Interact

Checking Key Concepts

Teaching Tip
Ask students which questions require finding a square root? (questions 1–3) Ask whether the square roots will be positive or negative square roots and why. (positive; Lengths are always positive.)

Closure Question

Suppose you know the lengths of all three sides of a triangle. How would you check to see that the triangle is a right triangle and, if it is not, how could you redraw the shortest side to make it a right triangle?
Check whether the sum of the squares of the two smaller lengths equals the square of the length of the longest side. If not, find the difference of the squares of the two longer sides. Change the length of the shorter side to be equal to the square root of the difference.

Apply⇔Assess

Suggested Assignment

Core Course
Exs. 1–20, 22, 24, 26–31

Extended Course
Exs. 1–11, 14, 18–31

Block Schedule
Day 35 Exs. 1–11, 14, 17–31

Exercise Notes

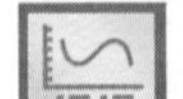
Using Technology
Exs. 10–18 Allow students to use calculators for these exercises. A good way to use a calculator for an exercise such as Ex. 12 is to set up an equation but not square any of the numbers until there is a single expression that can be entered. After finding that $b^2 = 3541^2 - 2291^2$, students can get the final answer by entering [2nd] [√][(]3541[x²][−]2291[x²][)] [ENTER]. The parentheses are important. Other keystroke sequences are possible, but this sequence has the advantage of displaying a single expression that can be related easily to key steps used in solving the problem.

CHECKING KEY CONCEPTS

For Questions 1–3, refer to the right triangle below. Find the unknown length for each right triangle.

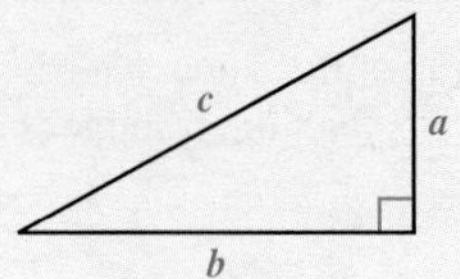

1. $a = 8, b = 15, c = \underline{?}$
2. $a = \underline{?}, b = 120, c = 169$
3. $a = 1.5, b = \underline{?}, c = 2.5$

For Questions 4–7, tell whether the given lengths can be the sides of a right triangle.

4. 3 cm, 4 cm, 5 cm
5. 7 in., 9 in., 15 in.
6. 6 ft, 8 ft, 12 ft
7. 1 m, 2.4 m, 2.6 m

6.1 Exercises and Applications

Extra Practice exercises on page 568

Find each square root.

1. $\sqrt{25}$
2. $-\sqrt{64}$
3. $\pm\sqrt{100}$
4. $\sqrt{36}$
5. $\pm\sqrt{400}$
6. $-\sqrt{144}$

Find the unknown length for each right triangle.

7. 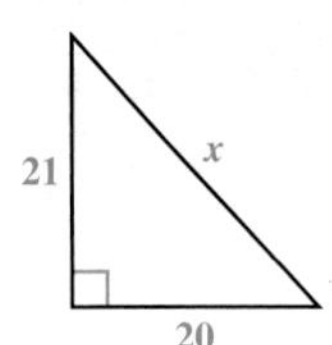

8. 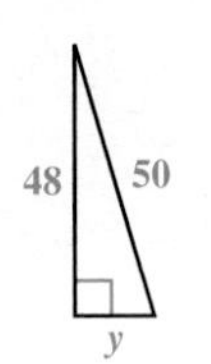

9. 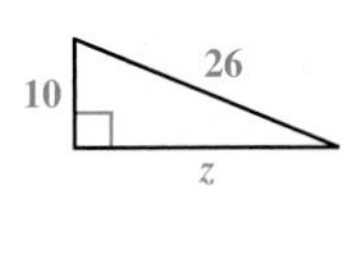

The legs of a right triangle have lengths *a* and *b*. The hypotenuse has length *c*. Find the unknown length for each right triangle.

10. $a = 60, b = 11$
11. $b = 18, c = 82$
12. $a = 2291, c = 3541$
13. $a = 12, c = 37$

For Exercises 14–19, tell whether the given lengths can be the sides of a right triangle.

14. 2.4 cm, 1.8 cm, 3.0 cm
15. 0.6 in., 1.0 in., 1.2 in.
16. 48 ft, 74 ft, 55 ft
17. 91 m, 84 m, 35 m
18. 33 mm, 19 mm, 27 mm
19. **Challenge** $2x, 3x, 4x$
20. An equilateral triangle has sides of equal length. Use algebra to show why an equilateral triangle cannot be a right triangle.

Checking Key Concepts

1. 17
2. 119
3. 2
4. Yes.
5. No.
6. No.
7. Yes.

Exercises and Applications

1. 5
2. −8
3. ±10
4. 6
5. ±20
6. −12
7. 29
8. 14
9. 24
10. $c = 61$
11. $a = 14$
12. $b = 2700$
13. $b = 35$
14. Yes.
15. No.
16. No.
17. Yes.
18. No.
19. No.
20. Answers may vary. An example is given. Let the length of each side be x. By the converse of the Pythagorean theorem, an equilateral triangle is a right triangle only if $x^2 + x^2 = x^2$. However, the only value of x for which $2x^2 = x^2$ is 0. A triangle cannot have a side of length 0, so an equilateral triangle cannot be a right triangle.

21. a. The size of a television screen is given by the length of the diagonal of the screen. What size is a television screen that is 21.6 in. wide and 16.2 in. high?

b. Open-ended Problem Why do you think that manufacturers might choose to measure the size of television screens this way? Use algebra to support your answer.

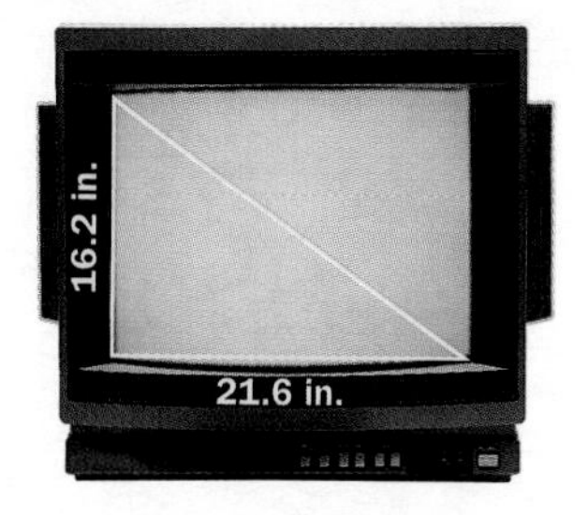

Toolbox p. 595
Angles

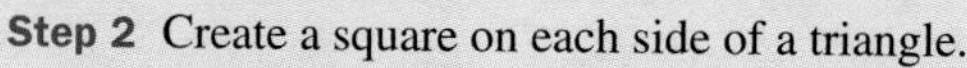

The Pythagorean theorem is important in both algebra and geometry. Exercises 22 and 23 will help you see why the theorem applies only to right triangles.

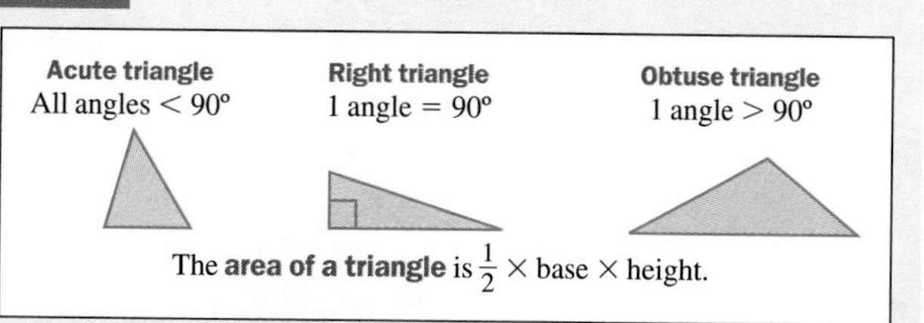

The **area of a triangle** is $\frac{1}{2}$ × base × height.

22. Cooperative Learning Work in a group of 3–4 students. You will need at least six pieces of graph paper and a ruler.

Work together to repeat the steps below at least six times. Use two right triangles, two acute triangles, and two obtuse triangles.

Step 1 Draw a triangle on each piece of graph paper so that each corner is on a grid point. Label the sides of each triangle a, b, and c, with c being the longest side (if there is a longest side). Label each triangle *right*, *acute*, or *obtuse*.

Step 2 Create a square on each side of a triangle.

Step 3 Find the area of each square. If a square is not aligned with the grid, divide it into right triangles and a small square, as shown. Add the areas of the pieces to find the area of the square.

Step 4 Compare the sum of the areas of the two smaller squares to the area of the largest square.

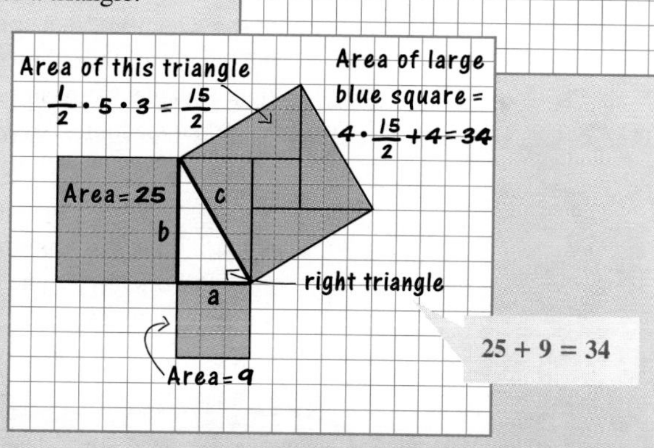

23. Writing Based on your observations, use > or < to complete each statement. Explain how your observations show that the Pythagorean theorem applies only to right triangles.

a. In an obtuse triangle, $a^2 + b^2 \underline{?} c^2$.

b. In an acute triangle, $a^2 + b^2 \underline{?} c^2$.

21. a. 27 in.

b. Answers may vary. An example is given. The diagonal of the screen is the hypotenuse of a right triangle, so it is longer than either the width or the height. Consumers probably feel that they are getting a bigger screen when the screen size is given in terms of the diagonal.

22. See answers in back of book.

23. a. <

b. >

Every triangle is obtuse, acute, or right. My observation from Ex. 22 shows that the Pythagorean theorem does not apply to obtuse triangles or acute triangles. Since it does apply to right triangles, it applies only to right triangles.

Apply⇔Assess

Exercise Notes

Mathematical Procedures

Exs. 10–19 At this stage of their mathematical development, many students may be inclined to perform all calculations for a problem without first thinking of a procedure to follow. This often makes it difficult to see quickly what steps were used. It also makes it difficult for students to check their work. For these exercises, suggest that students set up procedures to organize their work.

Topic Spiraling: Review

Ex. 19 This exercise can be used to illustrate how ideas about simplifying expressions can be applied in new situations.

Ex. 20 Some students may answer by stating that an equilateral triangle has all of its angles equal to 60°, and this is a perfectly acceptable answer. Ask these students, however, if they can provide an explanation that involves the Pythagorean theorem or its converse.

Research

Ex. 21 Students may find it interesting to research how screens for movies are measured. They can then compare movie screens with television screens to see what problems may be encountered by people who want to adapt material from a movie format to a TV format.

Challenge

Exs. 22, 23 Ask students whether their results for these exercises suggest a way to tell how close a triangle is to being a right triangle. (If the sum of the squares of the lengths of the two shortest sides is very close to the square of the length of the third side, then the triangle is close to being a right triangle. If the sum is *equal* to the square of the length of the third side, the triangle is a right triangle.)

Apply⇔Assess

Exercise Notes

Integrating the Strands

Ex. 26 This exercise involves concepts from number theory, geometry, and algebra.

Historical Connection

Ex. 26 The French mathematician Blaise Pascal conjectured in the 1600s that when the exponent n is greater than 2, there are no positive integers a, b, and c for which $a^n + b^n = c^n$. Many of the world's best mathematicians tried for centuries to show that the conjecture was true or false. In 1993, the British mathematician Andrew Wiles presented a proof of Pascal's conjecture. So this famous conjecture, known as Pascal's Last Theorem, is indeed a theorem and not a conjecture.

Practice 38 for Section 6.1

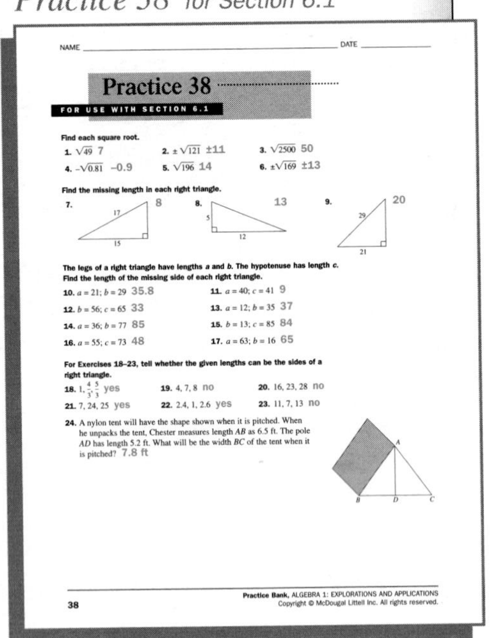

NAME ______________________ DATE ____________

Practice 38

FOR USE WITH SECTION 6.1

Find each square root.

1. $\sqrt{49}$ 7
2. $\pm\sqrt{121}$ ±11
3. $\sqrt{2500}$ 50
4. $-\sqrt{0.81}$ −0.9
5. $\sqrt{196}$ 14
6. $\pm\sqrt{169}$ ±13

Find the missing length in each right triangle.

7. 17, 15 — 8
8. 5, 12 — 13
9. 29, 21 — 20

The legs of a right triangle have lengths *a* and *b*. The hypotenuse has length *c*. Find the length of the missing side of each right triangle.

10. $a = 21$; $b = 29$ 35.8
11. $a = 40$; $c = 41$ 9
12. $b = 56$; $c = 65$ 33
13. $a = 12$; $b = 35$ 37
14. $a = 36$; $b = 77$ 85
15. $b = 13$; $c = 85$ 84
16. $a = 55$; $c = 73$ 48
17. $a = 63$; $b = 16$ 65

For Exercises 18–23, tell whether the given lengths can be the sides of a right triangle.

18. $1, \frac{4}{3}, \frac{5}{3}$ yes
19. 4, 7, 8 no
20. 16, 23, 28 no
21. 7, 24, 25 yes
22. 2.4, 1, 2.6 yes
23. 11, 7, 13 no

24. A nylon tent will have the shape shown when it is pitched. When he unpacks the tent, Chester measures length AB as 6.5 ft. The pole AD has length 5.2 ft. What will be the width BC of the tent when it is pitched? 7.8 ft

38 Practice Bank, ALGEBRA 1: EXPLORATIONS AND APPLICATIONS

BY THE WAY

Belo Horizonte, the capital of Minas Gerais, was the first planned city in Brazil. The layout of the city was modeled after Washington, D.C., and La Plata, Argentina.

24. URBAN PLANNING Choose two locations on the map below. Copy part of the map to show two different routes between the points. One route should use the diagonal streets. The other route should not. Estimate the length of each route. Use a city block as one unit. Round your answer to the nearest tenth. Explain why city planners would choose to include the diagonal streets.

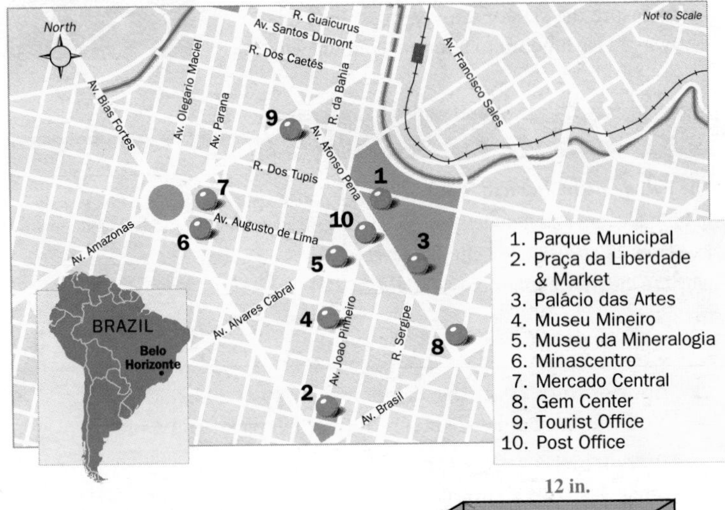

25. Challenge Find the length d of the diagonal of the rectangular box.

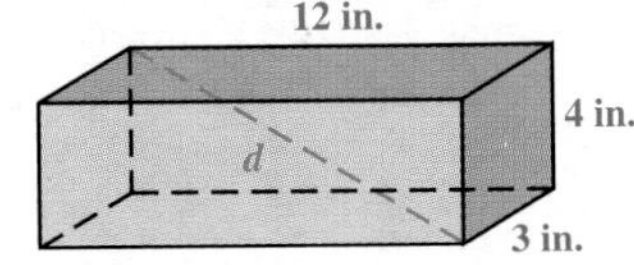

ONGOING ASSESSMENT

26. a. A *Pythagorean triple* is a group of three numbers that could be the lengths of the sides of a right triangle. Copy and complete the table. Are the three numbers in each row a Pythagorean triple?

n	$2n$	$n^2 - 1$	$n^2 + 1$
3	6	8	10
4	?	?	?
5	?	?	?
6	?	?	?

b. Do you think this pattern will always work? Try some other values of n to test your theory.

SPIRAL REVIEW

27. In 1990, the average price for gasoline was \$1.22 per gallon. Write and graph a direct variation equation based on this fact. *(Section 3.2)*

Find the reciprocal of each number. *(Section 4.2)*

28. $\frac{23}{5}$ **29.** $-\frac{13}{36}$ **30.** 15 **31.** −11

24. See answers in back of book.

25. 13 in.

26. a.

n	$2n$	$n^2 - 1$	$n^2 + 1$
3	6	8	10
4	8	15	17
5	10	24	26
6	12	35	37

Yes.

b. Yes. Answers may vary. Examples are given.
(1) $n = 9$; $2n = 18$, $n^2 - 1 = 80$, $n^2 + 1 = 82$, and $18^2 + 80^2 = 82^2$
(2) $n = 13$; $2n = 26$, $n^2 - 1 = 168$, $n^2 + 1 = 170$, and $26^2 + 168^2 = 170^2$

27. $y = 1.22x$

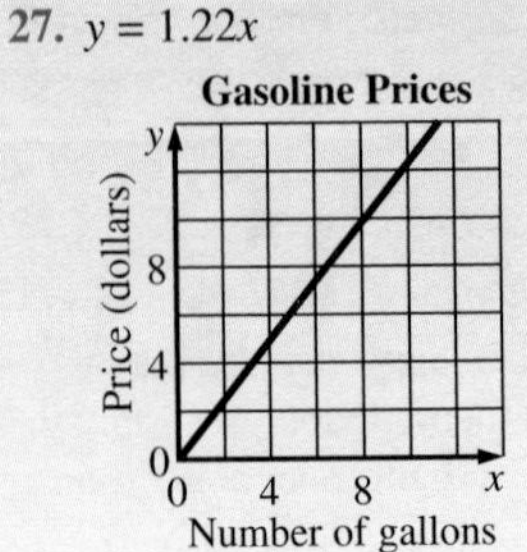

28. $\frac{5}{23} \approx 0.22$

29. $-\frac{36}{13} \approx -2.77$

30. $\frac{1}{15} \approx 0.07$

31. $-\frac{1}{11} \approx 0.09$

SECTION

6.2 Irrational Numbers

Learn how to...

- find noninteger square roots

So you can...

- estimate the speed of a car before it started to skid

Nathan Shigemura can estimate the minimum speed a car was traveling before a crash. He measures the distance the car skidded and finds the coefficient of friction, or slipperiness, for the road surface. He uses a formula that involves finding a square root.

EXAMPLE 1 Interview: Nathan Shigemura

An accident reconstructionist measures the distance a car skidded before coming to a stop and finds the coefficient of friction for the road. Was the car exceeding the 25 mi/h speed limit when the driver applied the brakes?

length of the skid mark is 83 ft

coefficient of friction is 0.5

$S = \sqrt{30df}$

SOLUTION

Use the formula from the article on pages 236–238. Let S = the speed of the car in miles per hour when it started to skid. Let d = the length of the skid mark in feet. Let f = the coefficient of friction for the road.

$$S = \sqrt{30df}$$

Substitute the measured values for d and f into the formula.

$S = \sqrt{(30)(83)(0.5)}$ — The radical sign is a grouping symbol. First do the operations under the radical sign.

$= \sqrt{1245}$

≈ 35.28 — A calculator gives an approximate value for $\sqrt{1245}$.

The car was traveling about 35 mi/h. The car was exceeding the 25 mi/h speed limit.

Plan⇔Support

Objectives

- Find noninteger square roots.
- Estimate square roots to solve real-world problems.

Recommended Pacing

Core and Extended Courses
Section 6.2: 1 day

Block Schedule
Section 6.2: $\frac{1}{2}$ block (with Section 6.3)

Resource Materials

Lesson Support
Lesson Plan 6.2
Warm-Up Transparency 6.2
Practice Bank: Practice 39
Study Guide: Section 6.2

Technology
Technology Book: Spreadsheet Activity 6
Graphing Calculator
McDougal Littell Software *Stats!*
Internet: http://www.hmco.com

Warm-Up Exercises

Write each fraction as a decimal.

1. $\frac{3}{4}$ 0.75

2. $\frac{16}{25}$ 0.64

Write each decimal as a fraction in lowest terms.

3. $0.\overline{3}$ $\frac{1}{3}$

4. 1.25 $\frac{5}{4}$

Write >, =, or < to make a true statement.

5. 28.099 __?__ 28.9 <

6. $\frac{2}{3}$ __?__ 0.6666... =

Teach⇔Interact

Additional Example 1

In court, a motorist claimed he had been traveling no more than 30 mi/h when he put on his brakes to avoid hitting a truck. A highway patrol officer said she measured the car's skid marks to be 88 feet long. Photos confirmed the officer's information. The coefficient of friction for the road was 0.55. Was the motorist reporting his speed correctly?

Use the formula from the article on pages 236–238. Let S = the speed of the car in miles per hour when it started to skid. Let d = the length of the skid marks in feet. Let f = the coefficient of friction for the road.

$S = \sqrt{30df}$

Substitute the measured values for d and f into the formula.

$S = \sqrt{(30)(88)(0.55)}$

$= \sqrt{1452}$

≈ 38.11

The car was traveling about 38 mi/h. The motorist was not correct.

Section Note

Historical Connection

The ancient Greeks appear to have been the first to discover that some numbers cannot be expressed as ratios of whole numbers. The discovery was probably made around 400 B.C. It destroyed a fundamental tenet of Pythagorean philosophy, which held that all quantities were expressible by such ratios.

Think and Communicate

For question 5, students may wish to explore ways to use their calculators to find when the decimal for $\frac{1}{23}$ begins to repeat. They will need to use a combination of pencil and paper and a calculator.

BY THE WAY

The word *rational* comes from the Latin word *ratio*, which means *calculation*. The prefix *ir-* in the word *irrational* means *not*.

Rational and Irrational Numbers

Any number that can be written as a ratio of two integers $\frac{a}{b}$, when $b \neq 0$, is a **rational number**. For example 7.5 and $0.\overline{3}$ are rational because $7.5 = \frac{15}{2}$ and $0.\overline{3} = \frac{1}{3}$. When rational numbers are written in decimal form, they either terminate or repeat.

The number $\sqrt{1245}$ cannot be written as a ratio of two integers. When written as a decimal, it never terminates or repeats. It is an **irrational number**. The square root of any integer that is not a perfect square is irrational.

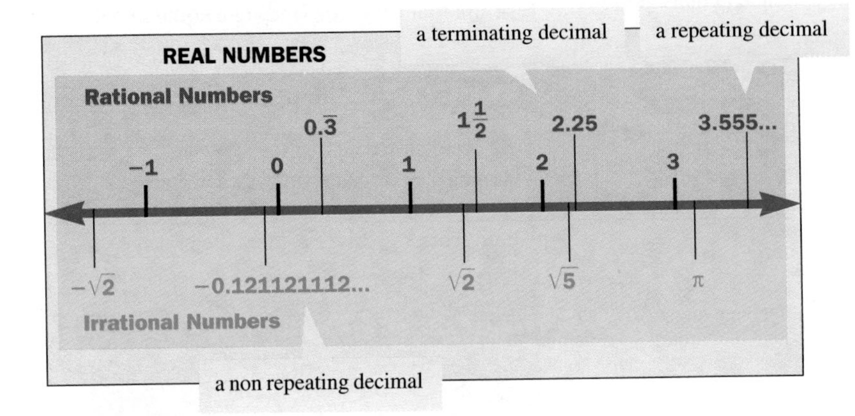

All **real numbers** are either *rational* or *irrational* and can be located on the real number line. The real numbers are **dense**. This means you can always find another real number between any two real numbers.

Unless you are given different directions, in this book you should round square roots to the nearest hundredth.

THINK AND COMMUNICATE

Tell whether each number is *rational* or *irrational*. Explain how you know.

1. $\frac{1}{23}$ **2.** 0.37 **3.** $\sqrt{3}$ **4.** $\sqrt{16}$

5. Use a calculator to find a decimal approximation for $\frac{1}{23}$. Does the decimal appear to terminate or repeat? Why can't you use a calculator to tell the difference between a rational number and an irrational number?

6. Find a real number between 0.3 and 0.4. Now find another real number between 0.4 and the number you found. How many times can you repeat this process?

7. Explain how to write 0.37 as a ratio of two integers.

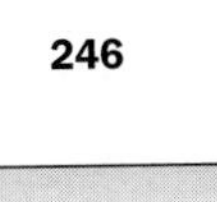

ANSWERS Section 6.2

Think and Communicate

1. rational; It is a ratio of two integers, 1 and 23.
2. rational; It is a terminating decimal.
3. irrational; 3 is not a perfect square.
4. rational; It can be written as the ratio of two integers; 4 and 1.
5. about 0.04347826; No, the decimal does not seem to terminate or repeat. You cannot use a calculator to tell whether a number is rational or irrational because the digits may terminate or begin to repeat beyond the digits that are displayed on the calculator screen. The number may also have a repeating block that has more digits than the display can show.
6. Answers may vary. Examples are given. 0.35; 0.375; infinitely many times
7. 0.37 = 37 hundredths = $\frac{37}{100}$

EXAMPLE 2

Write the repeating decimal 0.272727... as a fraction in lowest terms.

Toolbox p. 587 *Fractions, Decimals, and Percents*

SOLUTION

Write 0.272727... as $0.\overline{27}$. Let $b = 0.\overline{27}$. Because there are 2 digits in the repeating block, multiply both sides of the equation $b = 0.\overline{27}$ by 100.

$$100b = 100(0.\overline{27})$$
$$100b = 27.\overline{27}$$

Subtract the original equation from the new equation to eliminate the decimal part of $27.\overline{27}$.

$$\begin{aligned} 100b &= 27.\overline{27} \\ -\ b &= -\ 0.\overline{27} \\ \hline 99b &= 27 \\ b &= \frac{27}{99} \\ b &= \frac{3}{11} \end{aligned}$$

The decimal 0.272727... can be written as $\frac{3}{11}$.

THINK AND COMMUNICATE

8. Describe the steps you would use to write $0.\overline{2}$ as a fraction in lowest terms. What steps would you use to write $0.\overline{2397}$ as a fraction in lowest terms?

CHECKING KEY CONCEPTS

For Questions 1–4, decide which numbers below belong in each category.

$-5, -4.24, -\sqrt{29}, -\sqrt{16}, 0.0, 3, 4.1, 0.\overline{72}, \sqrt{11}$

1. integer
2. rational
3. irrational
4. real

Write each decimal as a fraction in lowest terms.

5. 0.4
6. 0.77777...
7. 1.26
8. $0.\overline{37}$
9. At the scene of a car accident, an accident reconstructionist found a 112 ft skid mark. The coefficient of friction for the road was 0.6. Use the formula $S = \sqrt{30df}$ to find the speed of the car when the driver applied the brakes.

Teach⇔Interact

About Example 2

Teaching Tip

Some students may have trouble understanding the second equation in the solution. Point out that if a decimal is multiplied by 100, the decimal point shifts 2 places to the right.

Additional Example 2

Write the repeating decimal 0.162162162... as a fraction in lowest terms.

Write 0.162162162... as $0.\overline{162}$. Let $b = 0.\overline{162}$. Because there are 3 digits in the repeating block, multiply both sides of the equation $b = 0.\overline{162}$ by 1000.

$$1000b = 1000(0.\overline{162})$$

Subtract the original equation from the new equation.

$$\begin{aligned} 1000b &= 162.\overline{162} \\ -\ b &= -\ 0.\overline{162} \\ \hline 999b &= 162 \\ b &= \frac{162}{999} \\ b &= \frac{6}{37} \end{aligned}$$

Checking Key Concepts

For questions 1–4, it may be helpful to use a Venn diagram to show how the various categories of numbers are related.

Closure Question

How do the decimal forms for rational numbers differ from those for irrational numbers?

Decimals for rational numbers terminate or repeat. Decimals for irrational numbers do not terminate and do not repeat.

Think and Communicate

8. Let $b = 0.\overline{2}$. Multiply both sides of the equation by 10, then subtract the original equation from the new one to get $9b = 2$, or $b = \frac{2}{9}$. Let $b = 0.\overline{2397}$. Multiply both sides of the equation by 10,000, then subtract the original equation from the new one to get $9999b = 2397$, or $b = \frac{2397}{9999}$. Simplify the fraction to get $b = \frac{799}{3333}$.

Checking Key Concepts

1. $-5, -\sqrt{16}, 0, 3$
2. $-5, -4.24, -\sqrt{16}, 0, 3, 4.1, 0.\overline{72}$
3. $-\sqrt{29}, \sqrt{11}$
4. all
5. $\frac{2}{5}$
6. $\frac{7}{9}$
7. $\frac{63}{50}$
8. $\frac{37}{99}$
9. about 44.90 mi/h

Apply⇔Assess

Suggested Assignment

Core Course
Exs. 1–25, 31, 33–39

Extended Course
Exs. 1–7, 11–22, 26–39

Block Schedule
Day 36 Exs. 1–39

Exercise Notes

Common Error
Exs. 5–10 Some students may get the wrong result when they multiply decimals written with bar notation by a power of 10. For example, in Ex. 7, students may multiply $0.\overline{63}$ by 100 as $100(0.\overline{63}) = 63$.

Help students correct this error by having them write the decimal without bar notation. After multiplying, they can then switch back to bar notation.
$0.\overline{63} = 0.636363...$
$100(0.636363...) = 63.\overline{63}$

Using Technology
Exs. 5–10 The TI-82 will convert many repeating decimals to fractions in simplest form. The repeating block must contain 4 digits or fewer. For example, to change $0.\overline{63}$ to a fraction, type .6363636363636363 [MATH] 1 [ENTER]. The calculator will display 7/11. Students can experiment to see what kinds of repeating decimals the calculator can convert. They should always type enough digits to fill one complete line of the screen before they press [MATH] 1 [ENTER].

Reasoning
Exs. 5–10 After students are confident that they understand how to do these exercises, ask them how decimals such as $0.72\overline{53}$ and $0.824\overline{9}$ can be written as fractions. You may wish to have two students write their results on the board, including a check that converts their fraction back to a decimal. Ask if they notice anything interesting about the second number. (A possible response for the second number is the following calculation.

$$\begin{aligned} 0.824\overline{9} &= 0.824 + 0.000\overline{9} \\ &= 0.824 + (0.001)(0.\overline{9}) \\ &= \frac{824}{1000} + \frac{1}{1000}(0.\overline{9}) \end{aligned}$$

Continued on following page.

6.2 Exercises and Applications

Extra Practice *exercises on page 569*

For Exercises 1–4, decide which numbers below belong in each category.

$-6, -4.3, -\sqrt{25}, -\sqrt{29}, -\frac{3}{5}, 0, 0.42, 0.45, 8, \sqrt{961}, \sqrt{19}, 0.\overline{32}$

1. integer **2.** rational **3.** irrational **4.** real

Write each decimal as a fraction in lowest terms.

5. 0.35 **6.** 0.768 **7.** $0.\overline{63}$

8. $0.\overline{423}$ **9.** 0.4545... **10.** $0.\overline{504}$

AVIATION For Exercises 11–13, use the article below from an airline magazine.

You can figure out how far you can see to the horizon on a clear day if you know how high your plane is flying. Your viewing distance v in miles is related to your altitude a in feet by the formula below:

$$v = 1.22\sqrt{a}$$

You can use this formula to calculate how far away the horizon is no matter where you are. For example, suppose the plane you are in is flying above Hawaii at an altitude of ... feet. A...

Honolulu, Hawaii, from the air. Using the formula at the left and the height at which this photo was taken, you can determine the distance to the horizon.

11. Suppose you are in a plane that is cruising at an altitude of 30,000 ft. How far away is a lake that you see on the horizon?

12. As the plane starts to descend, a city comes into view on the horizon. If the plane is at a height of 20,000 ft when you see the city, how far is the plane from the city?

13. Suppose you visit the observation tower at the top of a 1127 ft high skyscraper. How far can you see from the top of the building?

For Exercises 14–18, name a real number, if possible, that fits each description.

14. not rational **15.** not irrational

16. rational but not an integer **17.** not rational and not irrational

18. an integer that is not positive and not negative

19. Arrange the numbers below in order from smallest to largest.

$0.26, 0.\overline{265}, 0.265, 0.\overline{26}, 0.2\overline{6}, 0.2, 0.\overline{2}$

Exercises and Applications

1. $-6, -\sqrt{25}, 0, 8, \sqrt{961}$
2. $-6, -4.3, -\sqrt{25}, -\frac{3}{5}, 0, 0.42, 0.45, 8, \sqrt{961}, 0.\overline{32}$
3. $-\sqrt{29}, \sqrt{19}$
4. all
5. $\frac{7}{20}$
6. $\frac{96}{125}$
7. $\frac{7}{11}$
8. $\frac{47}{111}$
9. $\frac{5}{11}$
10. $\frac{56}{111}$

11–13. Estimates may vary. Answers are given to the nearest hundredth.

11. about 211.31 mi
12. about 172.53 mi
13. about 40.96 mi

14–16. Answers may vary. Examples are given.

14. $\sqrt{2}$
15. 1.07
16. $\frac{4}{5}$
17. not possible
18. 0
19. $0.2, 0.\overline{2}, 0.26, 0.\overline{26}, 0.265, 0.\overline{265}, 0.2\overline{6}$
20. $c \approx 9.43$
21. $a \approx 10.95$
22. $a \approx 20.12$
23. $b \approx 17.66$
24. $c \approx 59.94$
25. $b \approx 73.97$
26. Answers may vary. Examples are given. $\frac{17}{24}$, 0.68, and 0.72; infinitely many; Between any

The legs of a right triangle have lengths *a* and *b*. The hypotenuse has length *c*. Estimate the unknown length for each right triangle.

20. $a = 5, b = 8$ **21.** $b = 13, c = 17$ **22.** $b = 18, c = 27$

23. $a = 7, c = 19$ **24.** $a = 53, b = 28$ **25.** $a = 53, c = 91$

26. Writing Find three rational numbers between $\frac{2}{3}$ and $\frac{3}{4}$. How many rational numbers are there between $\frac{2}{3}$ and $\frac{3}{4}$? Explain your reasoning.

27. LANGUAGE ARTS Look up the word *dense* in a dictionary. Which meanings help you understand the mathematical meaning of the word?

INTERVIEW Nathan Shigemura

Look back at the interview on pages 236–238.

A driver applies the brakes of a car and the car comes to a stop, leaving a skid mark on the pavement. The greater the initial speed of the car, the longer the skid mark, or stopping distance. Road conditions also affect stopping distance.

Nathan Shigemura finds the coefficient of friction of a road.

Stopping Distance

C2 =SQRT(30*A2*B2)

	A	B	C
1	*d*	*f*	*S*
2	70	0.4	28.98
3	70	0.8	
4	70	1.2	
5	100	0.4	
6	100	0.8	
7	100	1.2	
8	130	0.4	
9	130	0.8	
10	130	1.2	

28. Spreadsheet Look back at the formula $S = \sqrt{30df}$ in Example 1 on page 245. The value of the coefficient of friction f depends on road conditions. Copy the table and use the formula for car speed to complete it. Use a spreadsheet if you have one.

29. When the skid distance remains constant, how does the value of f affect the calculated speed of the car? Why is it important to take road conditions into account when estimating the minimum speed of a car before an accident?

30. Open-ended Problem What types of road conditions would you expect to have a high coefficient of friction? a low coefficient of friction? Explain. What other factors might affect stopping distance?

two rational numbers there is another rational number, so if you choose a rational number between $\frac{2}{3}$ and $\frac{3}{4}$, there will be another number between $\frac{2}{3}$ and the number you have chosen, and so on.

27. Answers may vary. An example is given. The meanings that help me are "thick" and "crowded closely together."

28.

	A	B	C
1	*d*	*f*	*S*
2	70	0.4	28.98
3	70	0.8	40.99
4	70	1.2	50.20
5	100	0.4	34.64
6	100	0.8	48.99
7	100	1.2	60.00
8	130	0.4	39.50
9	130	0.8	55.86
10	130	1.2	68.41

29, 30. See answers in back of book.

Apply⇔Assess

Exercise Notes

Reasoning (continued)
Changing $0.\overline{9}$ to a fraction in simplest form gives 1. So the result is $\frac{825}{1000}$ or $\frac{33}{40}$. To check, change $\frac{33}{40}$ to a decimal. The result is 0.825. This shows that $0.824\overline{9} = 0.825$. In fact, every repeating decimal that has the single digit 9 as the repeating digit is equal to a terminating decimal.)

Application
Exs. 11–13 Students may wish to see if they can use the formula $v = 1.22\sqrt{a}$ to approximate the observer's altitude if the viewing distance is given. They should see that this can be done by substituting the given value of v into the formula and squaring both sides.

Assessment Note
Exs. 14–19 These exercises can afford an excellent insight into how well students have grasped basic concepts about rational and irrational numbers. A discussion of Ex. 19 can provide valuable feedback for students on how well they understand bar notation for repeating decimals.

Using Manipulatives
Exs. 20–23 In these exercises, the values of *a*, *b*, and *c* are relatively small. Many students may benefit from checking their answers with a drawing. They can use graph paper diagrams of the right triangles. Have them cut off one long strip of squares from the graph paper and use it as a ruler to measure the hypotenuses.

Using Technology
Ex. 28 Students can use the *Stats!* software to generate spreadsheets for this exercise.

Interview Note
Exs. 28–30 Some students may be interested in knowing what additional information they can find about coefficients of friction and how they are determined. Suggest that they consider incorporating their findings in their answers to Ex. 30.

Apply⇔Assess

Exercise Notes

Interdisciplinary Problems

Ex. 31 Mathematical formulas are used in a large number of scientific disciplines, as is illustrated by this problem from biology.

Using Technology

Ex. 31 This exercise lends itself well to small group work and using technology. Students may want to measure actual walking speeds. With data on 20 people or more from various groups, they could use the *Stats!* software to analyze the data and report their findings.

Assessment Note

Ex. 33 In addition to providing insight into students' problem-solving skills, this exercise also provides an opportunity to assess their understanding of the Pythagorean theorem.

Practice 39 for Section 6.2

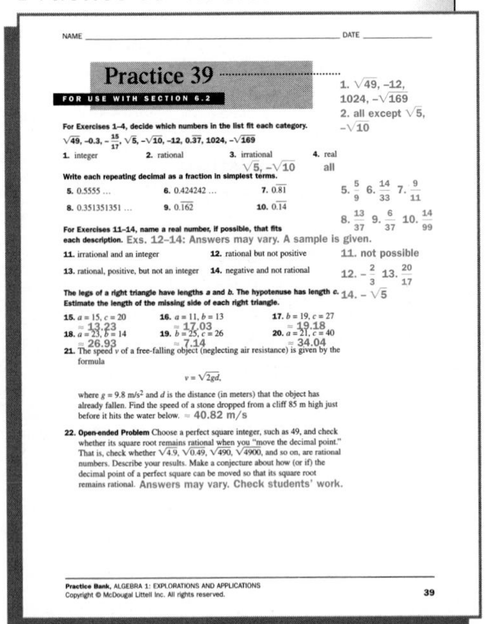

NAME ______________ DATE ______

Practice 39

FOR USE WITH SECTION 6.2

For Exercises 1–4, decide which numbers in the list fit each category.

$\sqrt{49}, -0.3, -\frac{15}{17}, \sqrt{5}, -\sqrt{10}, -12, 0.\overline{37}, 1024, -\sqrt{169}$

1. integer 2. rational 3. irrational $\sqrt{5}, -\sqrt{10}$ 4. real all

1. $\sqrt{49}, -12, 1024, -\sqrt{169}$
2. all except $\sqrt{5}, -\sqrt{10}$

Write each repeating decimal as a fraction in simplest terms.

5. 0.5555 … 6. 0.424242 … 7. $0.\overline{81}$

8. 0.351351351 … 9. $0.\overline{162}$ 10. $0.\overline{14}$

5. $\frac{5}{9}$ 6. $\frac{14}{33}$ 7. $\frac{9}{11}$ 8. $\frac{13}{37}$ 9. $\frac{6}{37}$ 10. $\frac{14}{99}$

For Exercises 11–14, name a real number, if possible, that fits each description. Exs. 12–14: Answers may vary. A sample is given.

11. irrational and an integer 12. rational but not positive

13. rational, positive, but not an integer 14. negative and not rational

11. not possible 12. $-\frac{2}{3}$ 13. $\frac{20}{17}$ 14. $-\sqrt{5}$

The legs of a right triangle have lengths *a* and *b*. The hypotenuse has length *c*. Estimate the length of the missing side of each right triangle.

15. $a = 15, c = 20$ ≈ 13.23 16. $a = 11, b = 13$ ≈ 17.03 17. $b = 19, c = 27$ ≈ 19.18

18. $a = 23, b = 14$ ≈ 26.93 19. $b = 25, c = 26$ ≈ 7.14 20. $a = 21, c = 40$ ≈ 34.04

21. The speed v of a free-falling object (neglecting air resistance) is given by the formula

$$v = \sqrt{2gd},$$

where $g = 9.8$ m/s^2 and d is the distance (in meters) that the object has already fallen. Find the speed of a stone dropped from a cliff 85 m high just before it hits the water below. ≈ 40.82 m/s

22. **Open-ended Problem** Choose a perfect square integer, such as 49, and check whether its square root remains rational when you "move the decimal point." That is, check whether $\sqrt{4.9}, \sqrt{0.49}, \sqrt{490}, \sqrt{4900}$, and so on, are rational numbers. Describe your results. Make a conjecture about how (or if) the decimal point of a perfect square can be moved so that its square root remains rational. Answers may vary. Check students' work.

Practice Bank, ALGEBRA 1: EXPLORATIONS AND APPLICATIONS

39

31. **BIOLOGY** The maximum walking speed S in centimeters per second of any animal is related to the animal's leg length l in centimeters and the force of gravity g of the planet. You can use the formula $S = \sqrt{gl}$ to estimate an animal's maximum walking speed.
   a. Measure your own leg length. Use the formula to estimate your maximum walking speed. Does this speed seem reasonable? Explain how you decided. (*Note:* On Earth, $g \approx 981$ cm/s^2.)
   b. Use the formula to estimate your maximum walking speed on the moon. (*Note:* On the moon, $g \approx 162$ cm/s^2.)

32. **Challenge** Write $0.4\overline{5}$ as a fraction in lowest terms. Explain how you got your answer.

ONGOING ASSESSMENT

33. **Writing** Sue Boswell purchased a king-size mattress and box spring. Use the Pythagorean theorem to explain how she was able to fit the 76 in. by 80 in. mattress through her bedroom door. Why do you think king-size box springs are sold as two smaller pieces, as shown in the picture at the right?

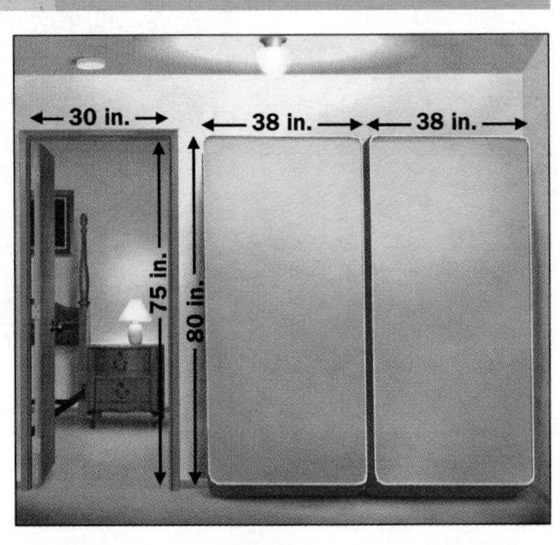

SPIRAL REVIEW

For each pair of similar figures, find the ratio of the perimeters. Find the ratio of the areas. *(Section 5.4)*

34.

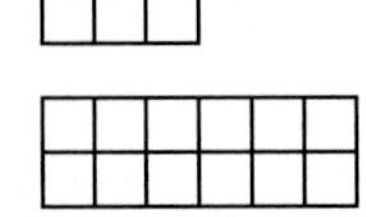

35.

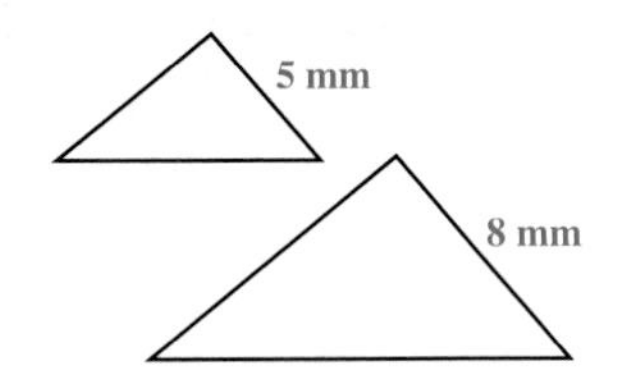

The legs of a right triangle have lengths *a* and *b*. The hypotenuse has length *c*. Find the unknown length for each right triangle. *(Section 6.1)*

36. $a = 1.5$ in., $b = 2$ in.
37. $a = 5$ cm, $b = 12$ cm
38. $a = 15$ km, $c = 39$ km
39. $a = 5$ ft, $c = 13$ ft

31. Answers may vary. An example is given for a person whose leg length is 86 cm.
   a. about 290 cm/s; That is reasonable.
   b. about 118 cm/s

32. $\frac{41}{90}$; Answers may vary. An example is given. Let $b = 0.4\overline{5}$; then $10b = 4.\overline{5} = 4.5\overline{5}$; subtracting the first equation from the second gives $9b = 4.1$; $b = \frac{4.1}{9} = \frac{41}{90}$.

33. Answers may vary. An example is given. The diagonal of the door is about 80.8 in., so the mattress will fit through the doorway. I think the box spring is sold as two smaller pieces to make it lighter and easier to handle.

34. 1:2; 1:4
35. 5:8; 25:64
36. $c = 2.5$ in.
37. $c = 13$ cm
38. $b = 36$ km
39. $c = 12$ ft

SECTION

6.3 Calculating with Radicals

Learn how to...

- multiply and divide radicals
- simplify radicals

So you can...

- find your reaction time or the escape velocity for a planet

There are many ways to write the square root of eight. You can use a calculator to find a decimal approximation. You can write the number in radical form as $\sqrt{8}$. In the Exploration, you will find other ways to write the square root of 8.

EXPLORATION
COOPERATIVE LEARNING

Comparing Radicals

Work with a partner.

You will need:

- graph paper

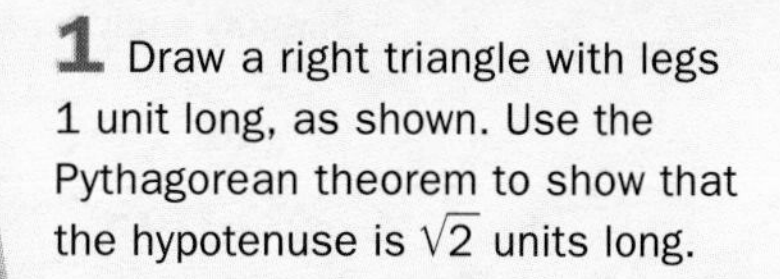

1 Draw a right triangle with legs 1 unit long, as shown. Use the Pythagorean theorem to show that the hypotenuse is $\sqrt{2}$ units long.

2 Draw a right triangle with legs 2 units long. Show that the hypotenuse is $\sqrt{8}$ units long. Divide the triangle into smaller triangles like the one in Step 1 to show that the hypotenuse is also $2\sqrt{2}$ units long.

3 Draw a right triangle with legs 3 units long, one with legs 4 units long, and one with legs 5 units long. For each triangle, write the length of the hypotenuse two different ways. Describe any patterns you see.

4 The number 8 can also be written as the product $4 \cdot 2$. You can write $\sqrt{8}$ as $\sqrt{4 \cdot 2}$. How does writing $\sqrt{8}$ this way help explain what you saw in Step 2? How does it help to explain what you saw in Step 3?

Exploration Note

Purpose
The purpose of this Exploration is to have students understand how certain radical expressions can be written in simpler form.

Materials/Preparation
Each pair of students should have a sheet of graph paper with squares that are large enough for the drawing activity. Centimeter grid paper would serve this purpose.

Procedure
Suggest that students make a table to show the two ways of expressing the length of the hypotenuse of each triangle.

Closure
Call on several partners to describe their results to the class. Discuss the patterns observed in Step 3 of the activity. All students should be able to answer the questions in Step 4 correctly.

Explorations Lab Manual
See the Manual for more commentary on this Exploration.

Diagram Master 1

For answers to the Exploration, see following page.

Plan⇔Support

Objectives

- Multiply and divide radicals.
- Simplify radicals.
- Use skills with radicals to find values in real-world problems.

Recommended Pacing

Core and Extended Courses
Section 6.3: 2 days

Block Schedule
Section 6.3: 2 half-blocks
(with Sections 6.2 and 6.4)

Resource Materials

Lesson Support
Lesson Plan 6.3
Warm-Up Transparency 6.3
Practice Bank: Practice 40
Study Guide: Section 6.3
Explorations Lab Manual:
Additional Exploration 9,
Diagram Master 1
Assessment Book: Test 24

Technology
Technology Book:
Calculator Activity 6,
Spreadsheet Activity 6
Graphing Calculator
McDougal Littell Software
Stats!
Internet:
http://www.hmco.com

Warm-Up Exercises

Find the length c of the hypotenuse of a right triangle whose legs have the given lengths. If the answer is not a whole number, leave the square root symbol in the answer.

1. $a = 6$, $b = 8$ $c = 10$
2. $a = 5$, $b = 5$ $c = \sqrt{50}$
3. $a = 9$, $b = 9$ $c = \sqrt{162}$
4. $a = 4$, $b = 12$ $c = \sqrt{160}$

Simplify.

5. $\frac{\sqrt{100}}{\sqrt{25}}$ 2

6. $\frac{\sqrt{49}}{\sqrt{121}}$ $\frac{7}{11}$

Teach⇔Interact

Section Notes

Second-Language Learners
In the Exploration and other activities in this section, students need to understand and use a comparatively large number of math terms. Working cooperatively with a partner or in small groups should help students learning English become comfortable with new terms.

Visual Thinking
Ask students to create their own chart demonstrating the properties of square roots. Ask them to select two positive numbers, one of which has a perfect square factor. Then, have them explain the different ways that the square root of the product of these numbers can be written. This activity involves the visual skills of *interpretation* and *inference*.

Additional Example 1

Simplify each expression.

a. $\sqrt{\frac{100}{49}}$

Use the property $\sqrt{\frac{a}{b}} = \frac{\sqrt{a}}{\sqrt{b}}$.

$\sqrt{\frac{100}{49}} = \frac{\sqrt{100}}{\sqrt{49}}$

$= \frac{10}{7}$

b. $\frac{1}{3}\sqrt{5} + \sqrt{125}$

Simplify each term in the expression. Look for a factor of 125 that is a perfect square.

$\frac{1}{3}\sqrt{5} + \sqrt{125}$

$= \frac{1}{3}\sqrt{5} + \sqrt{25 \cdot 5}$

$= \frac{1}{3}\sqrt{5} + \sqrt{25} \cdot \sqrt{5}$

$= \frac{1}{3}\sqrt{5} + 5\sqrt{5}$

$= \frac{1}{3}\sqrt{5} + \frac{15}{3}\sqrt{5}$

$= \frac{16}{3}\sqrt{5}$

Properties of Square Roots

For all nonnegative numbers a and b:	**For example:**
$\sqrt{a^2} = a$	$\sqrt{3^2} = 3$
$\sqrt{ab} = \sqrt{a} \cdot \sqrt{b}$	$\sqrt{4 \cdot 3} = \sqrt{4} \cdot \sqrt{3} = 2\sqrt{3}$
$\sqrt{\frac{a}{b}} = \frac{\sqrt{a}}{\sqrt{b}}, b \neq 0$	$\sqrt{\frac{3}{4}} = \frac{\sqrt{3}}{\sqrt{4}} = \frac{\sqrt{3}}{2}$

An expression that is written using the symbol $\sqrt{\ }$ is in **radical form**. You can use the properties of square roots to simplify an expression in radical form. An expression in radical form is in *simplest form* when:

1. there is no integer under the radical sign with a perfect square factor,
2. there are no fractions under the radical sign, and
3. there are no radicals in the denominator.

For example, $\sqrt{8}$ can be simplified because it has a perfect square factor.

$$\sqrt{8} = \sqrt{4 \cdot 2} = \sqrt{4} \cdot \sqrt{2} = 2\sqrt{2}$$

EXAMPLE 1

Simplify each expression.

a. $\sqrt{\frac{15}{81}}$ **b.** $\sqrt{24} + 5\sqrt{6}$

SOLUTION

a. Use the property $\sqrt{\frac{a}{b}} = \frac{\sqrt{a}}{\sqrt{b}}$.

$$\sqrt{\frac{15}{81}} = \frac{\sqrt{15}}{\sqrt{81}} = \frac{\sqrt{15}}{9}$$

b. Simplify each term in the expression.

Look for a factor of 24 that is a perfect square.

$$\sqrt{24} + 5\sqrt{6} = \sqrt{4 \cdot 6} + 5\sqrt{6}$$
$$= \sqrt{4} \cdot \sqrt{6} + 5\sqrt{6}$$
$$= 2\sqrt{6} + 5\sqrt{6}$$
$$= 7\sqrt{6}$$

You can combine terms that have the same radical part. $2 \cdot \sqrt{6} + 5 \cdot \sqrt{6} = (2 + 5)\sqrt{6}$

ANSWERS Section 6.3

Exploration

1–3. Check students' work.

1. Let h = the length of the hypotenuse; $h^2 = 1^2 + 1^2 = 2$, so $h = \sqrt{2}$.
2. Let h = the length of the hypotenuse; $h^2 = 2^2 + 2^2 = 4 + 4 = 8$, so $h = \sqrt{8}$. If you draw a line from the vertex of the right angle to the midpoint of the hypotenuse, you get two identical isosceles right triangles, each with a hypotenuse of length 2 units. Let x = length of each leg of the isosceles triangles. Since $x^2 + x^2 = 2^2$, $2x^2 = 4$, $x^2 = 2$, and $x = \sqrt{2}$. Therefore, $h = 2x = 2\sqrt{2}$.
3. 3 units: $\sqrt{18}$ units and $3\sqrt{2}$ units; 4 units: $\sqrt{32}$ units and $4\sqrt{2}$ units; 5 units: $\sqrt{50}$ units and $5\sqrt{2}$ units; The pattern seems to be that for any number x, $\sqrt{x^2 + x^2} = x\sqrt{2}$.
4. Since $\sqrt{8} = \sqrt{4 \cdot 2}$, it makes sense that $\sqrt{8} = \sqrt{4} \cdot \sqrt{2} = 2\sqrt{2}$. Since $\sqrt{18} = \sqrt{9 \cdot 2}$, it makes sense that $\sqrt{18} = \sqrt{9} \cdot \sqrt{2} = 3\sqrt{2}$. Similarly, it makes sense that $\sqrt{32} = \sqrt{16} \cdot \sqrt{2} = 4\sqrt{2}$ and $\sqrt{50} = \sqrt{25 \cdot 2} = 5\sqrt{2}$.

Estimating Square Roots

Sometimes you may want to estimate the square root of an integer. To estimate $\sqrt{59}$, find the two perfect squares closest to 59 on either side of it. They are 49 and 64.

$$\sqrt{49} < \sqrt{59} < \sqrt{64}$$

$$7 < \sqrt{59} < 8$$

The square root of 59 is between 7 and 8.

EXAMPLE 2 Application: Physics

Elisa and Zack are testing their reaction time in physics lab. Elisa holds a yardstick above Zack's hand. He tries to catch it as quickly as he can after Elisa drops it. The yardstick falls 9 in. before Zack catches it. Use the formula in the lab notebook to find his reaction time.

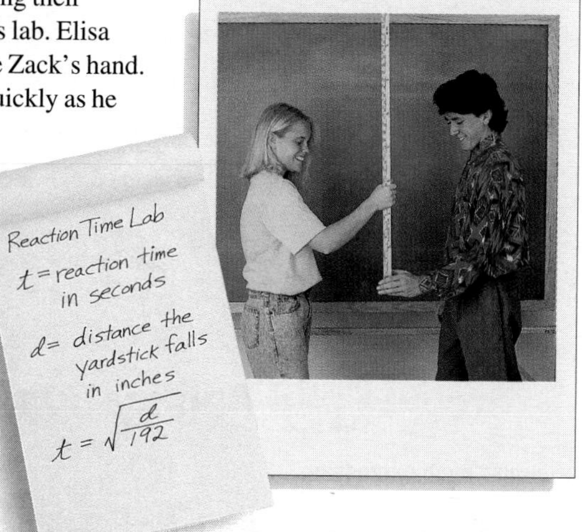

Reaction Time Lab
t = reaction time in seconds
d = distance the yardstick falls in inches
$t = \sqrt{\frac{d}{192}}$

SOLUTION

Method 1 Use estimation.
Substitute 9 for d in the formula.

$$t = \sqrt{\frac{d}{192}}$$
$$= \sqrt{\frac{9}{192}}$$
$$= \frac{\sqrt{9}}{\sqrt{192}}$$
$$\approx \frac{3}{14}$$
$$\approx 0.21$$

$13^2 < 192 < 14^2$
Because 192 is very close to 14^2, $\sqrt{192} \approx 14$.

Zack's reaction time is about 0.2 s.

Method 2 Use a calculator.
Substitute 9 for d in the formula.

$$t = \sqrt{\frac{d}{192}}$$
$$= \sqrt{\frac{9}{192}}$$
$$= \sqrt{0.046875}$$
$$\approx 0.22$$

Divide before you find the square root.

Zack's reaction time is about 0.2 s.

Teach⇔Interact

Section Note

Mathematical Procedures
Students should understand that the procedure for estimating square roots can be used repeatedly to refine estimates. For example, $\sqrt{4} < \sqrt{5} < \sqrt{9}$ means that $\sqrt{5}$ is between 2 and 3. Next, point out that since $\sqrt{4.84} < \sqrt{5} < \sqrt{5.29}$, then $\sqrt{5}$ is between 2.2 and 2.3.

Additional Example 2

Refer to the situation described in Example 2. When Elisa dropped the yardstick, it fell 24 cm before Zack caught it. Use the formula in the lab notebook to find his reaction time.

Method 1 Use estimation.
Substitute 24 for d in the formula.

$$t = \sqrt{\frac{d}{192}}$$
$$t = \sqrt{\frac{24}{192}}$$
$$= \sqrt{\frac{1}{8}}$$
$$= \frac{\sqrt{1}}{\sqrt{8}}$$

$2^2 < 8 < 3^2$. Because 8 is closer to 3^2, use $\sqrt{8} \approx 2.8$.

$$\frac{\sqrt{1}}{\sqrt{8}} \approx \frac{1}{2.8}$$
$$\approx 0.36$$

Zack's reaction time is about 0.4 s.

Method 2 Use a calculator.
Substitute 24 for d in the formula.

$$t = \sqrt{\frac{d}{192}}$$
$$= \sqrt{\frac{24}{192}}$$
$$= \sqrt{0.125}$$
$$\approx 0.35$$

Zack's reaction is about 0.4 s.

Teach⇔Interact

Think and Communicate

For question 1, it may help to discuss in detail what it means to simplify an expression containing radicals. In particular, point out that calculator estimates of the value of a radical expression are not simplified forms of the expression. The simplified form must have exactly the same value as the original expression. In question 4, it is very important for students to remember that negative numbers do not have square roots.

Closure Question

How can you decide whether $\sqrt{a}$ can be simplified when a is a positive whole number?

See if a is a perfect square. If it is not, see whether it is divisible by a perfect square greater than 1.

Apply⇔Assess

Suggested Assignment

Core Course
- Day 1 Exs. 1–17
- Day 2 Exs. 19–23, 25–31, AYP

Extended Course
- Day 1 Exs. 1–18
- Day 2 Exs. 19–31, AYP

Block Schedule
- Day 36 Exs. 1–18
- Day 37 Exs. 19–31, AYP

Exercise Notes

Using Technology

Ex. 14 Students should graph these equations together by entering Y1 = $2\sqrt{2}$X and Y2 = $\sqrt{8}$X on the Y= list. They should then press TRACE. They can move left and right using ◀ and ▶. They can switch from Y1 to Y2 and back by using the ▼ and ▲ keys. Students should note that when they use the up and down arrow keys without moving left or right, the coordinate readouts at the bottom of the screen do not change.

THINK AND COMMUNICATE

1. Can $\sqrt{2} + \sqrt{3}$ be simplified? Can $11 + 2\sqrt{3}$? Explain why or why not.
2. The property $\sqrt{ab} = \sqrt{a} \cdot \sqrt{b}$ also works in reverse. Use this fact to simplify each expression.
   a. $\sqrt{2} \cdot \sqrt{3}$ b. $\sqrt{5} \cdot \sqrt{7}$ c. $3\sqrt{2} \cdot \sqrt{8}$
3. Explain how to use the prime factors of 720 to simplify $\sqrt{720}$.
4. Explain why the properties of square roots are true only for nonnegative numbers.

CHECKING KEY CONCEPTS

Simplify each expression.

1. $\sqrt{12}$
2. $\sqrt{18}$
3. $\sqrt{\frac{13}{36}}$
4. $5\sqrt{6} \cdot \sqrt{3}$
5. $\sqrt{20} + \sqrt{5}$
6. $\sqrt{14} + 2\sqrt{6}$

Estimate each square root within a range of two integers.

7. $\sqrt{5}$
8. $\sqrt{33}$
9. $\sqrt{84}$

6.3 Exercises and Applications

Extra Practice exercises on page 569

Simplify each expression.

1. $\sqrt{40}$
2. $\sqrt{75}$
3. $4 + 2\sqrt{10}$
4. $\sqrt{128}$
5. $3\sqrt{20}$
6. $8\sqrt{2} + 5\sqrt{2}$
7. $\sqrt{28} + \sqrt{7}$
8. $\sqrt{\frac{27}{49}}$
9. $(5\sqrt{3})^2$
10. $5\sqrt{3} - \sqrt{75}$
11. $\sqrt{\frac{5}{16}}$
12. $4\sqrt{3} \cdot \sqrt{18}$
13. Nina says that $11 + 6\sqrt{2}$ cannot be simplified. Ellen thinks it can be simplified to $17\sqrt{2}$. Who is correct? Explain how you know.
14. **Technology** Graph the equations $y = (2\sqrt{2})x$ and $y = (\sqrt{8})x$ on a graphing calculator. What do you observe about the two graphs? Explain what you see.

GEOMETRY Find the perimeter and area of each rectangle.

15. Rectangle with sides $3\sqrt{2}$ and 10
16. Rectangle with sides $2\sqrt{3}$ and $5\sqrt{2}$

Think and Communicate

1. No; No; the terms have no common factor.
2. a. $\sqrt{6}$ b. $\sqrt{35}$ c. 12
3. Write 720 as the product of its prime factors: $\sqrt{720} = \sqrt{2 \cdot 2 \cdot 2 \cdot 2 \cdot 3 \cdot 3 \cdot 5} = \sqrt{16 \cdot 9 \cdot 5} = \sqrt{16} \cdot \sqrt{9} \cdot \sqrt{5} = 4 \cdot 3 \cdot \sqrt{5} = 12\sqrt{5}$
4. Negative numbers do not have square roots that are real numbers.

Checking Key Concepts

1. $2\sqrt{3}$
2. $3\sqrt{2}$
3. $\frac{\sqrt{13}}{6}$
4. $15\sqrt{6}$
5. $3\sqrt{5}$
6. $\sqrt{14} + 2\sqrt{6}$
7. $2 < \sqrt{5} < 3$
8. $5 < \sqrt{33} < 6$
9. $9 < \sqrt{84} < 10$

Exercises and Applications

1. $2\sqrt{10}$
2. $5\sqrt{3}$
3. $4 + 2\sqrt{10}$
4. $8\sqrt{2}$
5. $6\sqrt{5}$
6. $13\sqrt{2}$
7. $3\sqrt{7}$
8. $\frac{3\sqrt{3}}{7}$
9. 75
10. 0
11. $\frac{\sqrt{5}}{4}$
12. $12\sqrt{6}$
13. Nina; Answers may vary. An example is given. You could evaluate $11 + 6\sqrt{2}$ and $17\sqrt{2}$

17. **MEDICINE** For making medical decisions, such as how much medicine to prescribe, a doctor may need to know a person's body surface area. You can use a formula to find a person's body surface area A in square meters. Let h = height in inches. Let w = weight in pounds.

$$A = \sqrt{\frac{h \cdot w}{3131}}$$

Find the body surface area of a person who is 5 ft 4 in. tall and weighs 160 lb. Round your answer to the nearest hundredth.

18. **Challenge** A square and a rectangle have the same area. The side lengths of the rectangle are integers, and the perimeter is 28 cm. Which of the areas below cannot be the area of the square? Explain.

A. 33 cm^2 **B.** 40 cm^2 **C.** 42 cm^2 **D.** 45 cm^2

INTERVIEW Nathan Shigemura

Look back at the interview on pages 236–238.

In addition to investigating accidents that have already happened, Nathan Shigemura is interested in preventing future accidents. One way drivers can avoid getting into accidents is by considering braking distance and reaction time.

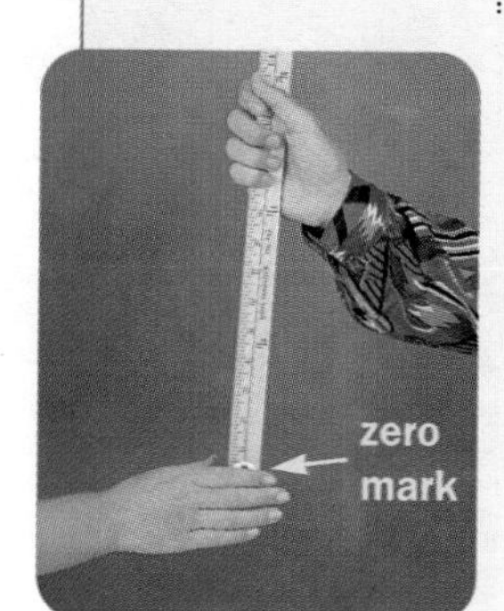

19. **Cooperative Learning** Work with a partner. Try the experiment described in Example 2 on page 253. You will need a yardstick.
    a. Hold the yardstick so that the zero mark is just between your partner's fingers. Without warning, drop the yardstick. Record your partner's "reaction distance" in inches. Then have your partner measure your reaction distance.
    b. Let d = your reaction distance from part (a). Let t = your reaction time in seconds. Use the formula $t = \sqrt{\frac{d}{192}}$ to estimate your reaction time to the nearest tenth of a second.

20. a. A study found that a typical driver's reaction time to brake lights is about 0.66 s. Suppose a driver is traveling at 55 mi/h. Suddenly, the car directly ahead slows down. About how far will the driver travel during the time it takes to react and apply the brake?
    b. **Open-ended Problem** Based on your answer to part (a), what advice would you give a new driver about preventing accidents?

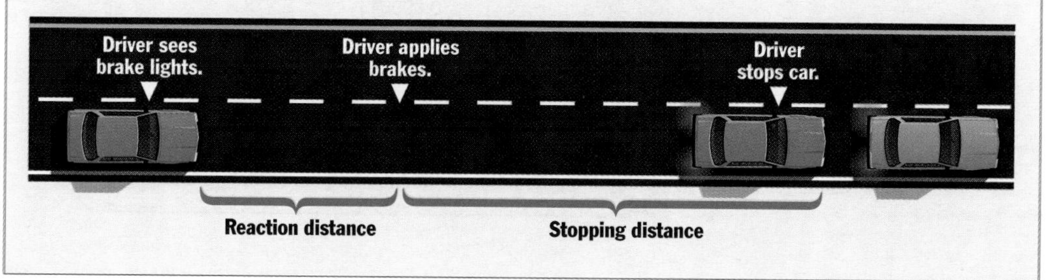

Apply⇔Assess

Exercise Notes

Career Connection

Ex. 17 It is very likely that most, if not all, students in your class have had some contact with a medical doctor during their lives. Even though students are familiar with doctors in general, they may have little understanding of what it takes to become a doctor. A brief discussion of this topic should bring out the fact that to earn a medical degree requires four years of pre-medical study at a college or university and then four more years of medical school. Internships and residency study may add one to four more years of work in a hospital to qualify for a license to practice medicine in a special field, such as surgery, for example.

Research

Ex. 17 Students can look in books on medicine and physiology for other mathematical formulas that model data about the human body and body chemistry. They should take special note of formulas that involve radical expressions.

Challenge

Ex. 18 A good strategy for students to use in this exercise is to make an organized list of the dimensions of all rectangles that match the given description.

Cooperative Learning

Ex. 19 Students can work with a partner to compile enough data for a scatter plot. They can then graph $t = \sqrt{\frac{d}{192}}$ on the same coordinate grid with the scatter plot to see how close the data points are to the curved graph. To do this, students can use the *Stats!* software or a TI-82 graphing calculator.

using a calculator to show they are not equal, or you could point out that $11 + 6\sqrt{2}$ cannot be simplified because only one term contains $\sqrt{2}$.

14. 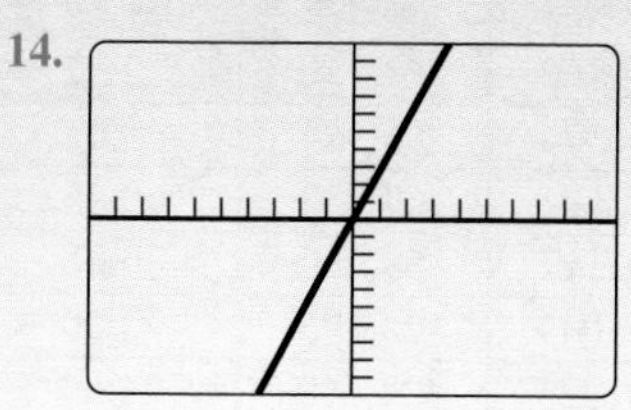

Since only one graph appears on the screen, the two equations have the same graph. This is another way of showing that $\sqrt{8} = 2\sqrt{2}$.

15. $20 + 6\sqrt{2}$; $30\sqrt{2}$
16. $4\sqrt{3} + 10\sqrt{2}$; $10\sqrt{6}$
17. about 1.81 m^2
18. C; If you make a list of all the possibilities, you will see that no rectangle with side lengths that are integers and perimeter 28 cm has area 42 cm^2. (The rectangle in A is 11 cm by 3 cm, the rectangle in B is 10 cm by 4 cm, and the rectangle in D is 9 cm by 5 cm.)
19. a. Answers may vary.
    b. Answers may vary.
20. a. about 0.01 mi or about 53 ft
    b. Answers may vary. For example, I would tell a new driver, "A car can travel quite a distance between the time you notice a danger ahead and the time you can bring your car to a complete stop. At 55 mi/h, you can travel about 53 ft before you even apply the brakes. The car will travel even farther before you can bring it to a complete stop. Be sure to leave plenty of room between your car and the car in front of you to allow time for you to react and for the car to brake."

Apply⇔Assess

Exercise Notes

Integrating the Strands

Ex. 21 This exercise illustrates an interesting and beautiful connection between algebra and geometry. The spiral generated in this manner is known as the spiral of Archimedes, after the Greek mathematician who was one of the first to study it (about 250 B.C.).

Using Technology

Ex. 22 Students can use the *Stats!* software to generate their spreadsheets. Students who are proficient on the TI-82 can explore ways to use the graphing calculator to produce tables for the situation.

Second-Language Learners

Ex. 24 Most students learning English should be able to answer this exercise by presenting examples. They may need some help writing a supporting explanation.

Communication: Writing

Ex. 24 Students may want to compare this situation with that for $\sqrt{a \cdot b} = \sqrt{a} \cdot \sqrt{b}$.

BY THE WAY

Spirals often occur in nature, as shown in this picture of a hurricane. Individual thunderstorms extend from the eye of the hurricane in two spiral bands.

21. GEOMETRY Make a spiral by following these steps.

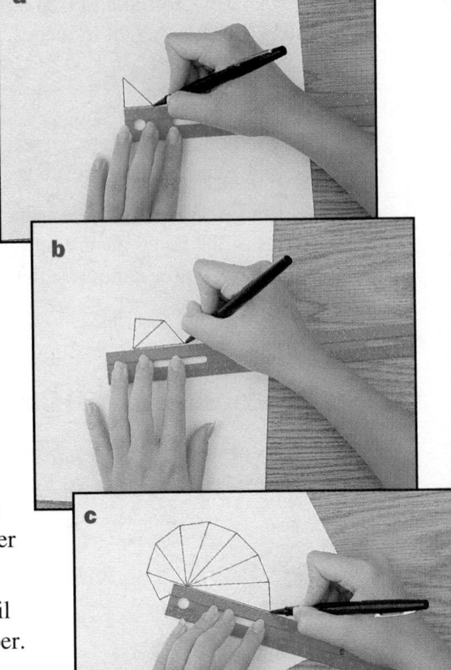

a. Start close to the center of a clean sheet of paper. Draw a right triangle with legs 1 unit long. The hypotenuse has length $\sqrt{2}$. Use the hypotenuse as one leg of a second right triangle. Make the other leg 1 unit long. Use algebra to find the length of the hypotenuse of the second triangle. Write your answer as a radical in simplest form.

b. Use the hypotenuse of the second triangle as one leg of a third right triangle. Make the other leg 1 unit long. Find the length of the hypotenuse of the third triangle. Write your answer as a radical in simplest form.

c. Continue to make triangles until you run out of room on the paper. What pattern do you see in the hypotenuse lengths?

22. **Spreadsheets** A spacecraft must reach *escape velocity* to escape from the gravitational pull of a planet or moon. Let v = the escape velocity, g = the force of gravity on the planet or moon, and r = the radius of the planet or moon. Use the formula $v = \sqrt{2gr}$ to find each escape velocity in the chart below. Use a spreadsheet if you have one.

Escape Velocity

D2 =SQRT(2*B2*C2)

	A	B	C	D
1		g (km/s^2)	r (km)	v (km/s)
2	Mercury	0.00358	2,436	4.18
3	Venus	0.00887	6,059	
4	Earth	0.00981	6,378	
5	Moon	0.00162	1,740	
6	Mars	0.00374	3,384	
7	Jupiter	0.02601	69,833	
8	Saturn	0.01117	58,280	
9	Uranus	0.01049	23,495	
10	Neptune	0.01325	22,740	
11	Pluto	0.00221	5,706	

21. a. The length of the hypotenuse of the second triangle is $\sqrt{3}$ units.

b. The length of the hypotenuse of the third triangle is $\sqrt{4} = 2$ units.

c. The lengths of the hypotenuses of the succeeding triangles are $\sqrt{5}$ units, $\sqrt{6}$ units, $\sqrt{7}$ units, and so on. The pattern is that the nth triangle has a hypotenuse that is $\sqrt{n+1}$ units long.

22.

	A	B	C	D
1		g (km/s^2)	r (km)	v (km/s)
2	Mercury	0.00358	2,436	4.18
3	Venus	0.00887	6,059	10.37
4	Earth	0.00981	6,378	11.19
5	Moon	0.00162	1,740	2.37
6	Mars	0.00374	3,384	5.03
7	Jupiter	0.02601	69,833	60.27
8	Saturn	0.01117	58,280	36.08
9	Uranus	0.01049	23,495	22.20
10	Neptune	0.01325	22,740	24.55
11	Pluto	0.00221	5,706	5.02

23. SAT/ACT Preview $\frac{900}{10} + \frac{90}{100} + \frac{9}{1000} = \underline{\ ?\ }$.

A. 99.09 **B.** 90.99 **C.** 90.909 **D.** 99.009 **E.** 90.09

ONGOING ASSESSMENT

24. Writing Is the equation $\sqrt{a+b} = \sqrt{a} + \sqrt{b}$ true for all values of a and b? Explain how you reached your conclusion. Describe a quick check you can use if you forget your conclusion.

SPIRAL REVIEW

Solve each equation. *(Section 4.3)*

25. $7x = 35 + 2x$ **26.** $4 + 4x = 20x$ **27.** $3x = 24 - 3x$

28. $4x - 15 = 16 - 8x$ **29.** $3(x - 2) = 9$ **30.** $4(x - 8) = 2x - 12$

31. $\triangle ABC \sim \triangle PQR$. Find each missing side length and angle measure. *(Section 5.3)*

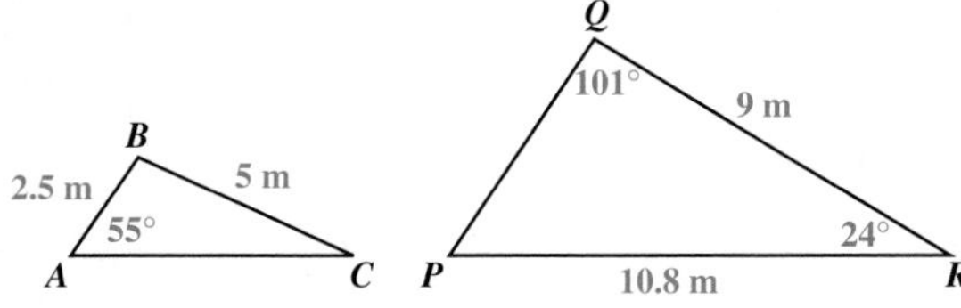

ASSESS YOUR PROGRESS

VOCABULARY

right triangle (p. 239)
Pythagorean theorem (p. 239)
hypotenuse (p. 239)
leg (p. 239)
perfect square (p. 240)
square root (p. 240)
radical sign (p. 240)
converse of the Pythagorean theorem (p. 241)
real number (p. 246)
rational number (p. 246)
irrational number (p. 246)
dense (p. 246)
radical form (p. 252)

The legs of a right triangle have lengths a and b. The hypotenuse has length c. Find the unknown length for each right triangle. *(Sections 6.1 and 6.2)*

1. $a = 9, c = 15$ **2.** $a = 6, b = 15$ **3.** $a = 15, c = 22$

Write each decimal as a fraction in lowest terms. *(Section 6.2)*

4. 3.65 **5.** 3.5757... **6.** $2.\overline{5}$

Simplify each expression. *(Section 6.3)*

7. $\sqrt{45}$ **8.** $\sqrt{5} + 2\sqrt{20}$ **9.** $\sqrt{\frac{28}{121}}$

10. Journal What do you find hard about working with radical numbers? What do you find easy? Why?

Apply⇔Assess

Assess Your Progress

Students should be able to give examples to illustrate each vocabulary term.

Journal Entry
For Ex. 10, encourage students to give specific examples to illustrate what they find hard or easy to do.

Progress Check 6.1–6.3

See page 272.

Practice 40 for Section 6.3

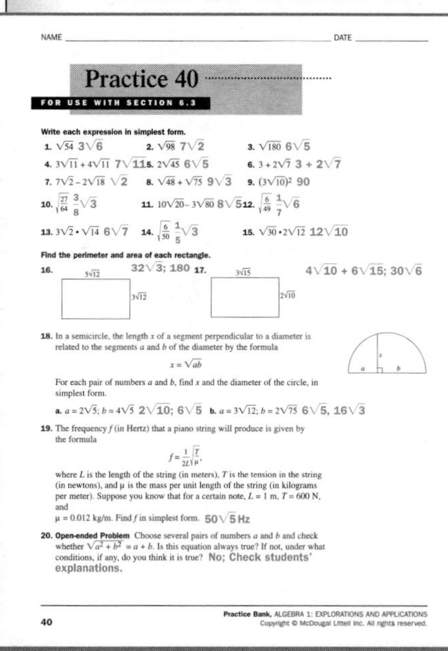

NAME ______________________ DATE __________

Practice 40

FOR USE WITH SECTION 6.3

Write each expression in simplest form.

1. $\sqrt{54}$ $3\sqrt{6}$ **2.** $\sqrt{98}$ $7\sqrt{2}$ **3.** $\sqrt{180}$ $6\sqrt{5}$

4. $3\sqrt{11} + 4\sqrt{11}$ $7\sqrt{11}$ **5.** $2\sqrt{45}$ $6\sqrt{5}$ **6.** $3 + 2\sqrt{7}$ $3 + 2\sqrt{7}$

7. $7\sqrt{2} - 2\sqrt{18}$ $\sqrt{2}$ **8.** $\sqrt{48} + \sqrt{75}$ $9\sqrt{3}$ **9.** $(3\sqrt{10})^2$ 90

10. $\sqrt{\frac{27}{64}}$ $\frac{3}{8}\sqrt{3}$ **11.** $10\sqrt{20} - 3\sqrt{80}$ $8\sqrt{5}$ **12.** $\sqrt{\frac{6}{49}}$ $\frac{1}{7}\sqrt{6}$

13. $3\sqrt{2} \cdot \sqrt{14}$ $6\sqrt{7}$ **14.** $\sqrt{\frac{6}{50}}$ $\frac{1}{5}\sqrt{3}$ **15.** $\sqrt{30} \cdot 2\sqrt{12}$ $12\sqrt{10}$

Find the perimeter and area of each rectangle.

16. $5\sqrt{12}$, $3\sqrt{12}$ — $32\sqrt{3}$; 180 **17.** $3\sqrt{15}$, $2\sqrt{10}$ — $4\sqrt{10} + 6\sqrt{15}$; $30\sqrt{6}$

18. In a semicircle, the length x of a segment perpendicular to a diameter is related to the segments a and b of the diameter by the formula

$$x = \sqrt{ab}$$

For each pair of numbers a and b, find x and the diameter of the circle, in simplest form.

a. $a = 2\sqrt{5}$; $b = 4\sqrt{5}$ $2\sqrt{10}$; $6\sqrt{5}$ **b.** $a = 3\sqrt{12}$; $b = 2\sqrt{75}$ $6\sqrt{5}$, $16\sqrt{3}$

19. The frequency f (in Hertz) that a piano string will produce is given by the formula

$$f = \frac{1}{2L}\sqrt{\frac{T}{\mu}},$$

where L is the length of the string (in meters), T is the tension in the string (in newtons), and μ is the mass per unit length of the string (in kilograms per meter). Suppose you know that for a certain note, $L = 1$ m, $T = 600$ N, and $\mu = 0.012$ kg/m. Find f in simplest form. $50\sqrt{5}$ Hz

20. Open-ended Problem Choose several pairs of numbers a and b and check whether $\sqrt{a^2 + b^2} = a + b$. Is this equation always true? If not, under what conditions, if any, do you think it is true? No; Check students' explanations.

40 Practice Bank, ALGEBRA 1: EXPLORATIONS AND APPLICATIONS

---

23. C

24. No. Answers may vary. An example is given. $\sqrt{9+16} = \sqrt{25} = 5$, but $\sqrt{9} + \sqrt{16} = 3 + 4 = 7$. A quick check might be the following: $\sqrt{1+1} = \sqrt{2}$, but $\sqrt{1} + \sqrt{1} = 1 + 1 = 2$.

25. 7 26. $\frac{1}{4}$

27. 4 28. $\frac{31}{12}$

29. 5 30. 10

31. $\angle B = 101°$; $\angle C = 24°$; $\angle P = 55°$; $PQ = 4.5$ m; $AC = 6$ m

Assess Your Progress

1. $b = 12$
2. $c = \sqrt{261} \approx 16.16$
3. $b = \sqrt{259} \approx 16.09$
4. $\frac{73}{20}$
5. $\frac{118}{33}$
6. $\frac{23}{9}$
7. $3\sqrt{5}$
8. $5 + 4\sqrt{5}$
9. $\frac{2\sqrt{7}}{11}$
10. Answers may vary. An example is given. I find the hardest part of working with radical numbers is remembering how to combine them. You have to remember rules for adding, subtracting, multiplying, and dividing. You also need to remember that you cannot add or subtract numbers with different radical parts. The easiest part is simplifying or evaluating a single radical number because you can use a calculator or look for perfect-square factors of the number.

Plan⇔Support

Objectives

- Find products of monomials and binomials.
- Use products of monomials and binomials to find areas of complex figures.

Recommended Pacing

Core and Extended Courses
Section 6.4: 1 day

Block Schedule
Section 6.4: $\frac{1}{2}$ block
(with Section 6.3)

Resource Materials

Lesson Support
Lesson Plan 6.4
Warm-Up Transparency 6.4
Overhead Visuals:
Folder B: Algebra Tiles
Practice Bank: Practice 41
Study Guide: Section 6.4
Explorations Lab Manual:
Diagram Masters 4, 5

Technology
Internet:
http://www.hmco.com

Warm-Up Exercises

Find the area of each rectangle whose length and width are given.

1. length = 15 cm, width = 6 cm
90 cm^2

2. length = 23 in., width = 17 in.
391 in.^2

Simplify.

3. $7x + 9x$ $16x$

4. $3x + 10y + 5x - y$ $8x + 9y$

5. $12x + 3(x + y)$ $15x + 3y$

SECTION

6.4 Multiplying Monomials and Binomials

Learn how to...

- **find products of monomials and binomials**

So you can...

- **find the area of a complex figure, such as the sidewalk around a fountain**

You can visualize the product $20 \cdot 40$ by thinking of the surface area of a 20 ft by 40 ft pool. To help visualize a product that involves a variable, you can use algebra tiles.

	1-tile	x-tile	x^2-tile
Dimensions	1 by 1	1 by x	x by x
Area	$1 \cdot 1 = 1$	$1 \cdot x = x$	$x \cdot x = x^2$

EXAMPLE 1

Write an equation modeled by the algebra tiles.

SOLUTION

The length of one side of the rectangle is $3x$. The length of the other side is $2x$. Therefore this model shows the product $(3x)(2x)$.

There are 6 x^2-tiles.

The tiles model the equation $(3x)(2x) = 6x^2$.

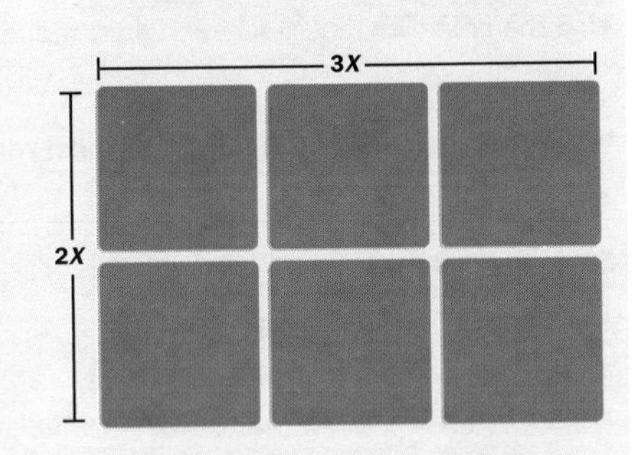

EXPLORATION
COOPERATIVE LEARNING

Finding Products with Algebra Tiles

Work with a partner.
You will need:
- algebra tiles
- ruler
- paper

SET UP Use tiles to find the product $2x(x + 3)$.

1 Draw two perpendicular lines on your paper.

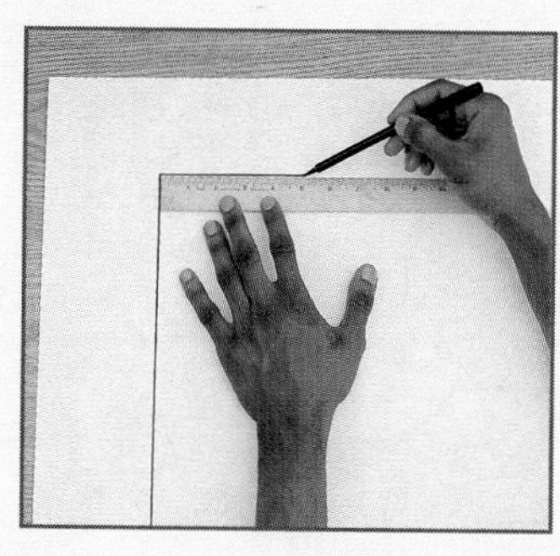

2 Using tiles, mark off the dimensions $2x$ and $(x + 3)$.

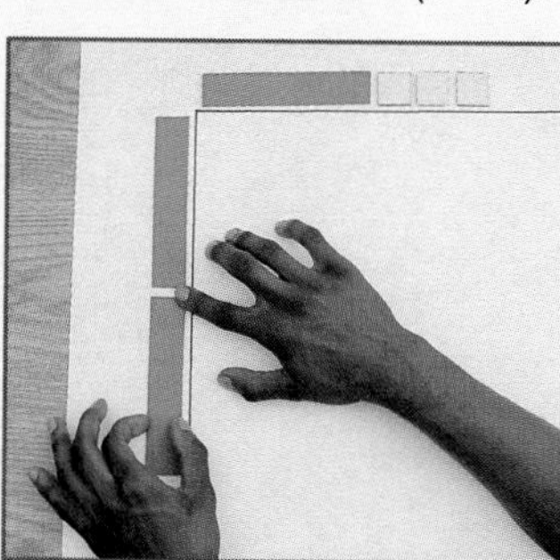

3 Create and completely fill in a rectangle using x^2-tiles, x-tiles, and 1-tiles.

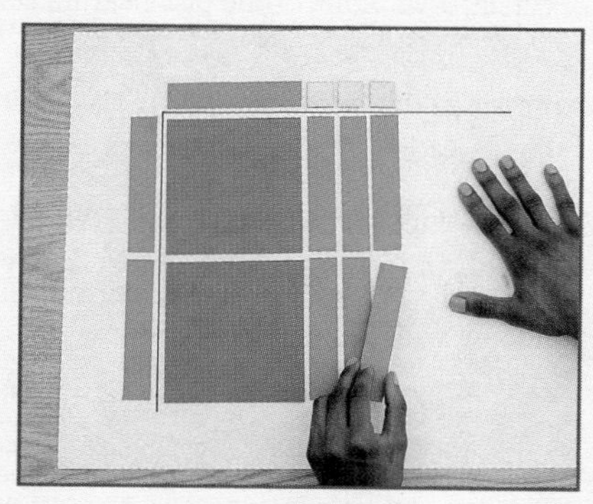

4 Count the tiles to see that $2x(x + 3) = 2x^2 + 6x$.

Use algebra tiles to find each product. Make a drawing of each tile model and show its dimensions.

1. $3x(x + 2)$ **2.** $(x + 2)(x + 3)$ **3.** $(2x + 1)(2x + 3)$

4. Choose your own product to model. Make a drawing of your tile model and show its dimensions.

Exploration Note

Purpose
The purpose of this Exploration is to have students use algebra tiles to find the product of two algebraic expressions.

Materials/Preparation
Each pair of partners will need at least 4 x^2-tiles, 8 x-tiles, and 6 1-tiles.

Procedure
Mention to students that it is important to make sure the lines they draw are perpendicular. Also, they should be careful in marking off the dimensions $2x$ and $x + 3$.

Closure
Have students share with the class their ideas about how to find a correct arrangement of the tiles. Explore with students how they might find the products $2x(x + 3)$ and $3x(x + 2)$ without using algebra tiles.

Explorations Lab Manual
See the Manual for more commentary on this Exploration.

Diagram Masters 4 and 5

For answers to the Exploration, see following page.

Teach⇔Interact

Learning Styles: Visual

The use of algebra tiles to visualize a product that involves variables and to find a product of variable expressions will help all students learn how to multiply monomials and binomials.

Section Note

Integrating the Strands
The use of algebra tiles to develop an understanding of algebraic concepts illustrates how mathematical ideas from geometry and algebra can be integrated.

About Example 1

Mathematical Procedures
After studying Example 1, make sure students see that the product $6x^2$ can be found by multiplying the numbers 3 and 2 and the two x-variables.

Additional Example 1

Write an equation modeled by the algebra tiles.

The length of one side of the rectangle is $4x$. The length of the other side is $3x$. Therefore, this model shows the product $(4x)(3x)$.
There are 12 x^2-tiles.
The tiles model the equation $(4x)(3x) = 12x^2$.

Teach⇔Interact

Additional Example 2

Find each product.

a. $5x(x + 8)$

Use the distributive property.

$5x(x + 8) = 5x^2 + 5x(8)$
$= 5x^2 + 40x$

b. $(3 + \sqrt{7})(4 - 2\sqrt{7})$

Use the distributive property twice.

$(3 + \sqrt{7})(4 - 2\sqrt{7})$
$= (3 + \sqrt{7})4 - (3 + \sqrt{7})(2\sqrt{7})$
$= 12 + 4\sqrt{7} - 6\sqrt{7} - 2(\sqrt{7})(\sqrt{7})$
$= 12 - 2\sqrt{7} - 14$
$= -2 - 2\sqrt{7}$

Section Notes

Second-Language Learners

Check to see that all students learning English have read the By the Way note. It can help them gain an understanding of the key terms *monomial*, *binomial*, and *trinomial* more quickly.

Student Study Tip

Mention to students that there are no special words for expressions with four or more terms. The term polynomial can be used, but it does not tell how many terms are involved.

Think and Communicate

For question 2, suggest that students use a dictionary if they have difficulty thinking of examples of words that use the given prefixes.

Not all products can be modeled conveniently with algebra tiles. You can use the distributive property to find many products, as shown in Example 2.

EXAMPLE 2

Find each product.

a. $x(x - 5)$ **b.** $(4 - \sqrt{3})(2 + \sqrt{3})$

SOLUTION

a. $x(x - 5) = x \cdot x + x(-5)$
$= x^2 - 5x$

Use the distributive property.

b. Use the distributive property twice.

$(4 - \sqrt{3})(2 + \sqrt{3}) = (4 - \sqrt{3})2 + (4 - \sqrt{3})\sqrt{3}$
$= (4 - \sqrt{3})2 + (4 - \sqrt{3})\sqrt{3}$
$= 8 - 2\sqrt{3} + 4\sqrt{3} - (\sqrt{3})(\sqrt{3})$
$= 8 + 2\sqrt{3} - 3$
$= 5 + 2\sqrt{3}$

Notice that $(\sqrt{3})(\sqrt{3}) = \sqrt{9} = 3$. In general, $\sqrt{a} \cdot \sqrt{a} = \sqrt{a^2} = a$.

WATCH OUT!

Pay attention to the negative signs when you use the distributive property.

BY THE WAY

You may recognize these prefixes.

mono-	*one*
bi-	*two*
tri-	*three*

The ending *-nomial* comes from the Greek word *nomos*, which means *part.*

You can name a variable expression by the number of terms in the expression. Each term can be a constant, a variable, or the product of a constant and one or more variables.

A **monomial** is an expression with *one* term.

-2 x $6x^2$ $-3xy$

A **binomial** is an expression with *two* unlike terms.

$3x + 1$ $2x^2 - 6x$ $x + y$ $y - y^2$

A **trinomial** is an expression with *three* unlike terms.

$x^2 + 5x - 6$ $3x + 2y + 1$ $4a - 3a^2 - 2$ $x - xy + y$

THINK AND COMMUNICATE

1. When do you think it is easier to model a product using algebra tiles? When do you think it is easier to use the distributive property?
2. What are some other words with the prefixes *mono-*, *bi-*, and *tri-*? Talk to someone who knows another language. Are these prefixes used in other languages?

ANSWERS Section 6.4

Exploration

1. $x + 2$; $3x$; tiles: three rows of x^2, x, x

$3x(x + 2) = 3x^2 + 6x$

2. $x + 3$; $x + 2$; tiles: x^2, x, x, x; x, 1, 1, 1; x, 1, 1, 1

$(x + 2)(x + 3) = x^2 + 5x + 6$

3. $2x + 3$; $2x + 1$; tiles

$(2x + 1)(2x + 3) = 4x^2 + 8x + 3$

4. Answers may vary. Check students' work.

Think and Communicate

1. Answers may vary. An example is given. I think it is easier to model a product using algebra tiles when the constant terms are positive and the numbers are small. For negative numbers and large numbers, I use the distributive property.
2. Answers may vary. Examples are given. monorail, monotone, bicycle, binoculars, tricycle, and triangle.

Checking Key Concepts

1. $x(2x) = 2x^2$
2. $2(2x) = 4x$
3. $(x + 2)x = x^2 + 2x$
4. $(x + 2)2x = 2x^2 + 4x$
5. $5x$; x; tiles: x^2, x^2, x^2, x^2, x^2
6. $x + 2$; 3; tiles: three rows of x, 1, 1

CHECKING KEY CONCEPTS

For Exercises 1–4, write an equation for each tile model.

1. 

2. 

3. 

4. 

Use algebra tiles to find each product. Make a drawing of your model.

5. $x(5x)$ **6.** $3(x + 2)$ **7.** $(x + 1)(x + 1)$

8. Look back at the Exploration on page 259. Use the distributive property to find each of the products you found using algebra tiles.

Tell whether each expression is a *monomial*, a *binomial*, a *trinomial*, or *none of these*.

9. $3p + 4$ **10.** $x^2 + xy + y^2$ **11.** 0 **12.** $a + b + c + d$

6.4 Exercises and Applications

For Exercises 1–4, write an equation for each tile model.

Extra Practice exercises on page 569

1. 

2. 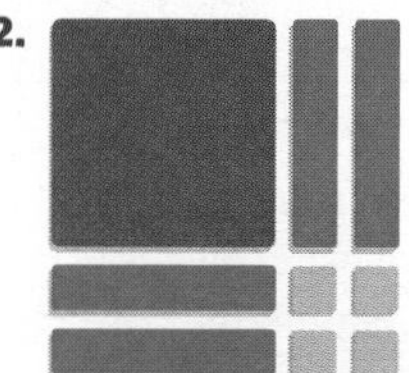

3. 

4. 

Use algebra tiles to find each product. Make a drawing of your model.

5. $x(3x)$ **6.** $2(4x)$ **7.** $3(x + 4)$

8. $3x(x + 3)$ **9.** $(x + 1)(2x + 5)$ **10.** $(2x + 1)(2x + 1)$

Teach⇔Interact

Checking Key Concepts

Student Progress

In questions 1–4, students should be able to write an equation using as few terms as possible for length, width, and total area. Students who write sums that contain a term for each individual tile will need additional help.

A desirable goal for students is to learn how to sketch a diagram for products like those in questions 5–7 *without always* beginning with the concrete model.

Closure Question

Explain how to find the product of a monomial and a trinomial.

Use the distributive property to multiply each term of the trinomial by the monomial.

Apply⇔Assess

Suggested Assignment

Core Course
Exs. 1–19, 21, 22, 25–30

Extended Course
Exs. 1–5, 8–30

Block Schedule
Day 37 Exs. 1–19, 21–23, 25–30

Exercise Notes

Common Error

Exs. 5–10 Some students may think that they can add the terms in two factors to tell what tiles they will need to find a product. They may think, for instance, that to find $3x(x + 3)$, they will need 4 x-tiles and 3 1-tiles. To correct this error, urge students to use the procedure they learned in the Exploration on page 259.

7. 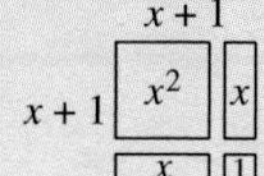

8. $2x(x + 3) = 2x \cdot x + 2x \cdot 3 = 2x^2 + 6x$; $3x(x + 2) = 3x \cdot x + 3x \cdot 2 = 3x^2 + 6x$; $(x + 2)(x + 3) = (x + 2)x + (x + 2)3 = x \cdot x + 2 \cdot x + x \cdot 3 + 2 \cdot 3 = x^2 + 5x + 6$; $(2x + 1)(2x + 3) = (2x + 1)2x + (2x + 1)3 = 2x \cdot 2x + 1 \cdot 2x + 2x \cdot 3 + 1 \cdot 3 = 4x^2 + 8x + 3$

9. binomial
10. trinomial
11. monomial
12. none of these

Exercises and Applications

1. $(x + 1)2x = 2x^2 + 2x$
2. $(x + 2)(x + 2) = x^2 + 4x + 4$
3. $(x + 2)(x + 4) = x^2 + 6x + 8$
4. $(x + 2)(2x + 1) = 2x^2 + 5x + 2$

5. 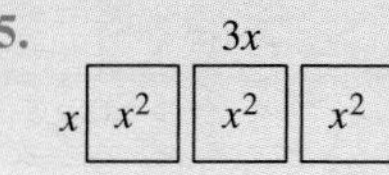

6. 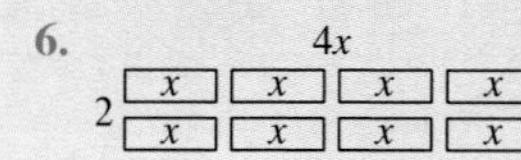

7. 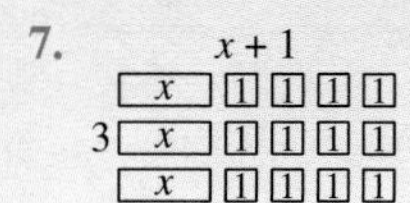

8–10. See answers in back of book.

Apply⇔Assess

Exercise Notes

Cooperative Learning
Exs. 11–16 Students can work in pairs on these exercises. One student can find the product symbolically while the other draws a diagram or makes a model with algebra tiles.

Application
Ex. 20 Students may find it interesting to find and report on actual examples of urban planning in their local environments.

Research
Exs. 21, 22 Many proofs for the Pythagorean theorem have been discovered during the past centuries. Some students may be interested in researching just how many proofs of this theorem are known today.

Journal Entry
Exs. 21, 22 Students who do a good job on these exercises may want to include their responses in their journals.

Using Manipulatives
Ex. 23 Students who find it difficult to think about this exercise in symbolic terms can use algebra tiles to aid their thinking.

Multicultural Note
Ex. 24 The NAMES Project AIDS Memorial Quilt was started by people who believed that the individual experiences of the men and women living with AIDS, and their friends and families, could be brought together to create something beautiful. Friends and families of the victims of AIDS created quilted panels to represent the lives of their loved ones. Volunteers helped sew the quilt together, as hundreds of panels were sent in each week. After initially working out of their own apartments, the volunteers moved to an empty storefront, where the first month's rent was donated by local merchants. When the quilt was initially displayed in Washington, D.C. in 1987, it contained 1920 panels. Since then, tens of thousands of new panels have been created and added to the quilt.

Find each product.

11. $4(x + 1)$ **12.** $3(5 - 2x)$ **13.** $4x(x + 3)$

14. $(x + 2)(x + 4)$ **15.** $(x + 4)(2x - 1)$ **16.** $(5x + 1)(3x + 2)$

17. $3(5 + \sqrt{2})$ **18.** $(6 - \sqrt{5})(4 - \sqrt{3})$ **19.** $(3\sqrt{3} + 2)(\sqrt{2} + 1)$

20. URBAN PLANNING A town council is considering a proposal to create a rectangular fountain as part of a new park. The fountain will be surrounded by a concrete walkway that is the same width on each side.

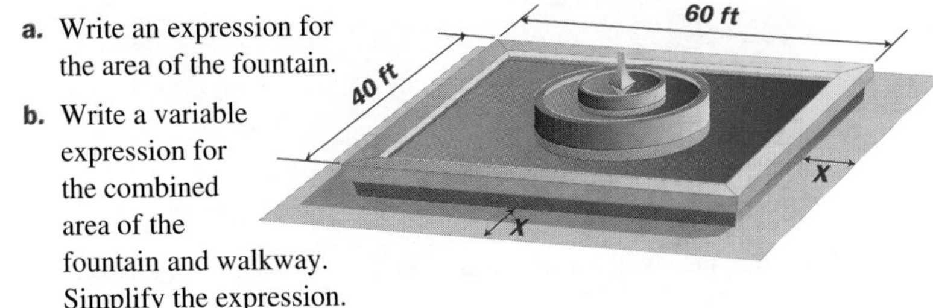

a. Write an expression for the area of the fountain.

b. Write a variable expression for the combined area of the fountain and walkway. Simplify the expression.

c. Use your answers to parts (a) and (b) to write a variable expression for the area of the walkway.

d. Open-ended Problem The walkway must be at least 5 ft wide to allow two wheelchairs to pass. Compare the area of a 5 ft wide walkway and a 6 ft wide walkway. What are some reasons for choosing one width over the other?

Connection HISTORY

To prove a theorem means to show that it is always true. Historians believe that around 275 A.D., the commentator Zhao Shuang added the diagram at the right to the *Zhōu bi suàn jīng* (100 B.C. – 100 A.D.). This diagram can be used to prove the Pythagorean theorem.

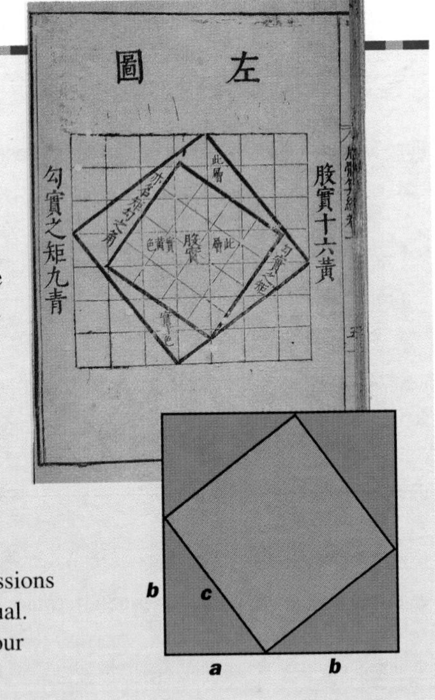

21. a. The length of each side of the large square can be represented as $a + b$. Write a variable expression for the area of the large square. Simplify the expression.

b. The large square can be divided into four identical right triangles shown in blue and a smaller red square. Write a variable expression for the area of the large square as the sum of the areas of the blue triangles and the red square.

22. Write an equation that states that the variable expressions you wrote in parts (a) and (b) of Exercise 21 are equal. Simplify both sides of the equation. Explain how your results prove the Pythagorean theorem.

11. $4x + 4$ 12. $15 - 6x$

13. $4x^2 + 12x$ 14. $x^2 + 6x + 8$

15. $2x^2 + 7x - 4$

16. $15x^2 + 13x + 2$

17. $15 + 3\sqrt{2}$

18. $24 - 4\sqrt{5} - 6\sqrt{3} + \sqrt{15}$

19. $3\sqrt{6} + 2\sqrt{2} + 3\sqrt{3} + 2$

20. a. 2400 ft^2

b. $(40 + 2x)(60 + 2x) = (2400 + 200x + 4x^2)$ ft^2

c. $(4x^2 + 200x)$ ft^2

d. 5-ft walkway: 1000 ft^2; 6-ft walkway: 1344 ft^2; Reasons for choosing a width may vary, and could include the following: the funds available for the park, the average and maximum number of people expected to use the park, including the expected number of people in wheelchairs, the land available for the park, and other design requirements.

21. a. $(a + b)(a + b) = a^2 + 2ab + b^2$

b. $4\left(\frac{1}{2}ab\right) + c^2 = 2ab + c^2$

22. $a^2 + 2ab + b^2 = 2ab + c^2$; Subtracting $2ab$ from both sides gives $a^2 + b^2 = c^2$. Since each triangle is a right triangle with legs of length a and b and hypotenuse of length c, this proves the Pythagorean theorem.

23. **Writing** A student tried to find the product $2x(x + 5)$, as shown. Explain what is wrong with the student's work. Find the correct product.

$$2x(x + 5) = 2(x + 5) + x(x + 5)$$
$$= 2x + 10 + x^2 + 5x$$
$$= x^2 + 7x + 10$$

24. **Challenge** The NAMES Project AIDS Memorial Quilt is displayed in 24 ft square sections. Each section contains 32 panels. A 5 ft walkway may be left between the sections. Draw a sketch of part of the displayed quilt. Write an expression for the area of the quilt and walkways. Explain how much space is required to display 900 sections in a square.

BY THE WAY

In February 1995, the AIDS Memorial Quilt contained 28,972 panels. To display the entire quilt with walkways would require an area equal to the area of 17 football fields.

ONGOING ASSESSMENT

25. **Cooperative Learning** Work in a group of 3–4 students. Make a list of examples of the product of a monomial and a binomial and of the product of two binomials. Discuss how you would find each product. Would you use algebra tiles or the distributive property? Why?

SPIRAL REVIEW

Exercises 26–28 refer to a six-sided die. *(Section 5.5)*

26. What is the theoretical probability of rolling a number less than 3?

27. What is the theoretical probability of rolling a number greater than 3?

28. Add your values from Exercises 26 and 27. Is the sum equal to 1? Explain why or why not.

Use an equation to model each set of data. *(Section 3.6)*

29.

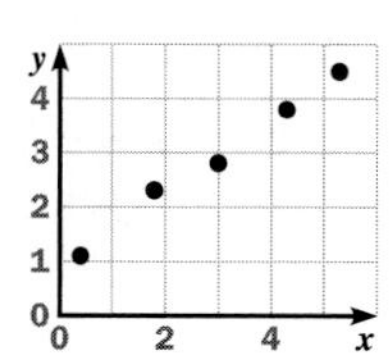

30.

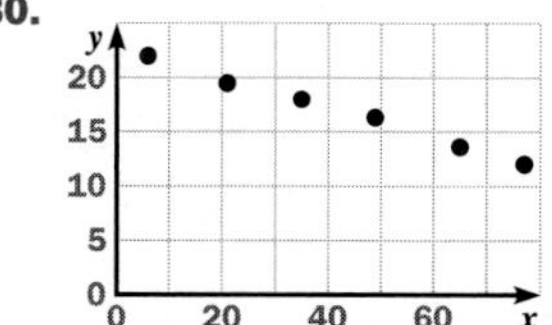

Apply⇔Assess

Exercise Notes

Assessment Note

Ex. 25 Students can examine the work of other groups and modify their responses if necessary. Groups that do an especially thorough job may want to include their responses in their journals.

Practice 41 for Section 6.4

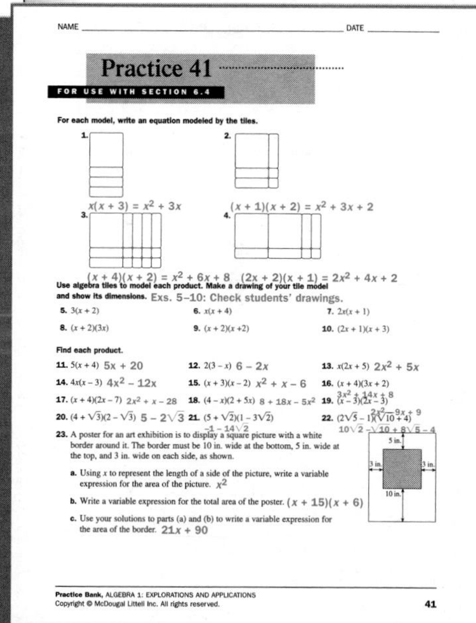

NAME ________________ DATE ________

Practice 41

FOR USE WITH SECTION 6.4

For each model, write an equation modeled by the tiles.

1. $x(x + 3) = x^2 + 3x$
2. $(x + 1)(x + 2) = x^2 + 3x + 2$
3. $(x + 4)(x + 2) = x^2 + 6x + 8$
4. $(2x + 2)(x + 1) = 2x^2 + 4x + 2$

Use algebra tiles to model each product. Make a drawing of your tile model and show its dimensions. Exs. 5–10: Check students' drawings.

5. $3(x + 2)$ 6. $x(x + 4)$ 7. $2x(x + 1)$
8. $(x + 2)(3x)$ 9. $(x + 2)(x + 2)$ 10. $(2x + 1)(x + 3)$

Find each product.

11. $5(x + 4)$ $5x + 20$
12. $2(3 - x)$ $6 - 2x$
13. $x(2x + 5)$ $2x^2 + 5x$
14. $4x(x - 3)$ $4x^2 - 12x$
15. $(x + 3)(x - 2)$ $x^2 + x - 6$
16. $(x + 4)(3x + 2)$ $3x^2 + 14x + 8$
17. $(x + 4)(2x - 7)$ $2x^2 + x - 28$
18. $(4 - x)(2 + 5x)$ $8 + 18x - 5x^2$
19. $(x - 3)(2x - 3)$ $2x^2 - 9x + 9$
20. $(4 + \sqrt{3})(2 - \sqrt{3})$ $5 - 2\sqrt{3}$
21. $(5 + \sqrt{2})(1 - 3\sqrt{2})$ $-1 - 14\sqrt{2}$
22. $(2\sqrt{5} - 1)(\sqrt{10} + 4)$ $10\sqrt{2} - \sqrt{10} + 8\sqrt{5} - 4$

23. A poster for an art exhibition is to display a square picture with a white border around it. The border must be 10 in. wide at the bottom, 5 in. wide at the top, and 3 in. wide on each side, as shown.
   a. Using x to represent the length of a side of the picture, write a variable expression for the area of the picture. x^2
   b. Write a variable expression for the total area of the poster. $(x + 15)(x + 6)$
   c. Use your solutions to parts (a) and (b) to write a variable expression for the area of the border. $21x + 90$

5 in. 3 in. 3 in. 10 in.

Practice Bank, ALGEBRA 1: EXPLORATIONS AND APPLICATIONS

41

23. The student used the distributive property incorrectly, as though the problem were $(2 + x)(x + 5)$. The correct product is $2x(x) + 2x(5) = 2x^2 + 10x$.

24.

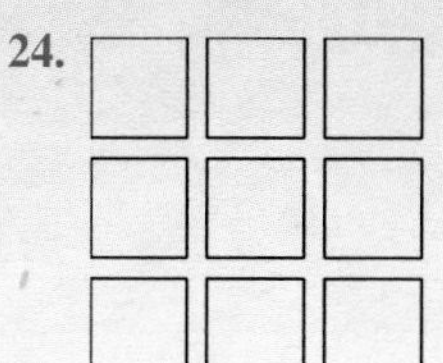

If there are n sections by n sections, the area is $(24n + 5(n - 1))^2 = 841n^2 - 290n + 25$ ft^2. For a display of 900 sections in a square, there would be 30 sections on a side and the necessary area would be 748,225 ft^2.

25. Answers may vary.

26. $\frac{1}{3}$ 27. $\frac{1}{2}$

28. No, the sum is $\frac{5}{6}$. The two events do not consider all the possible outcomes. If you add the probability of rolling a 3, $\frac{1}{6}$, then the sum is 1.

29, 30. Answers may vary. Examples are given.

29. $y = 0.72x + 0.75$

30. $y = -0.13x + 22.55$

Plan⇔Support

Objectives

- Recognize some special products of binomials.
- Find products of some variable expressions more quickly.

Recommended Pacing

Core and Extended Courses
Section 6.5: 2 days

Block Schedule
Section 6.5: 1 block

Resource Materials

Lesson Support
Lesson Plan 6.5
Warm-Up Transparency 6.5
Overhead Visuals:
Folder B: Algebra Tiles
Practice Bank: Practice 42
Study Guide: Section 6.5
Assessment Book: Test 25

Technology
Internet:
http://www.hmco.com

Warm-Up Exercises

Find each product.

1. $5x(2 + 7x)$ $10x + 35x^2$
2. $-3x(12x - 3)$ $-36x^2 + 9x$
3. $(x + 3)(x + 8)$ $x^2 + 11x + 24$
4. $(x + 4)(2x - 5)$ $2x^2 + 3x - 20$
5. $(x - 5)(x - 5)$ $x^2 - 10x + 25$
6. $(x + 5)(x - 5)$ $x^2 - 25$

SECTION

6.5 Finding Special Products

Learn how to...

- **recognize some special products of binomials**

So you can...

- **find products of some variable expressions more quickly**

The previous section showed how to multiply binomials. Sometimes it is helpful to know about special cases, such as when a binomial is multiplied by itself.

The square of a binomial follows a special pattern. You may see a pattern in the diagrams below that will help you remember the product.

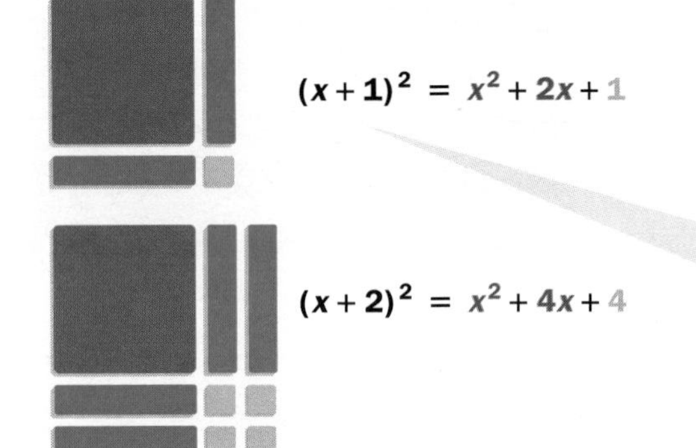

Read this as "the square of the quantity x plus 1," or as "x plus 1, quantity squared."

THINK AND COMMUNICATE

1. Use the diagram below to find the product $(x + 3)^2$.

2. What pattern do you see in the second term of all three products shown on this page?
3. What pattern do you see in the third term?
4. Use the patterns you saw in Questions 2 and 3 to find the product $(x + 4)^2$. Draw a diagram or use the distributive property to check your answer.
5. Find the product $(x + 10)^2$ using only mental math.

ANSWERS Section 6.5

Think and Communicate

1. $x^2 + 6x + 9$
2. In each case, the second term is twice the product of the terms of the binomial.
3. In each case, the third term is the square of the constant.
4. $(x + 4)^2 = x^2 + 2(4)(x) + 4^2 = x^2 + 8x + 16$; $(x + 4)^2 = (x + 4)(x + 4) = (x + 4)x + (x + 4)4 = x^2 + 4x + 4x + 16 = x^2 + 8x + 16$
5. $x^2 + 20x + 100$

Squaring a Binomial

The diagrams on page 264 show you a pattern that can help you find the square of a binomial. Remembering this pattern can help you square binomials quickly. For any numbers a and b:

$$(a+b)^2 = a^2 + 2ab + b^2$$
$$(a-b)^2 = a^2 - 2ab + b^2$$

The square of a binomial is called a **perfect square trinomial**.

EXAMPLE 1

Find each product.

a. $(x+9)^2$ **b.** $(x-15)^2$

SOLUTION

a. $(x+9)^2 = x^2 + 2(9)x + 9^2$

$= x^2 + 18x + 81$

b. $(x-15)^2 = x^2 - 2(15)x + 15^2$

$= x^2 - 30x + 225$

Check

Use the distributive property.

$(x+9)(x+9) = (x+9)x + (x+9)9$

$= x^2 + 9x + 9x + 81$

$= x^2 + 18x + 81$ ✓

EXAMPLE 2

Find each product.

a. $\left(3^2 - \sqrt{5}\right)^2$ **b.** $(2t+3)^2$

SOLUTION

a. $\left(3-\sqrt{5}\right)^2 = 3^2 - 2(3)\left(\sqrt{5}\right) + \left(\sqrt{5}\right)^2$

$= 9 - 6\sqrt{5} + 5$

$= 14 - 6\sqrt{5}$

b. $(2t+3)^2 = (2t)^2 + 2(2t)(3) + 3^2$

$= 4t^2 + 12t + 9$

Teach⇔Interact

Think and Communicate

For question 4, you may want to have algebra tiles available for students who have difficulty drawing diagrams.

Section Notes

Reasoning

Some students may find it helpful to observe that the pattern for finding $(a-b)^2$ can be thought of as a special case of finding $(a+b)^2$. For example, if $a - b$ is written as $a + (-b)$, then

$(a-b)^2 = (a+(-b))^2$
$= a^2 + 2a(-b) + (-b)^2$
$= a^2 - 2ab + b^2.$

Problem Solving

Challenge students to use their experiences with algebra tiles and area to show a geometric interpretation of $(a+b)^2 = a^2 + 2ab + b^2$. Some students may discover the following diagram.

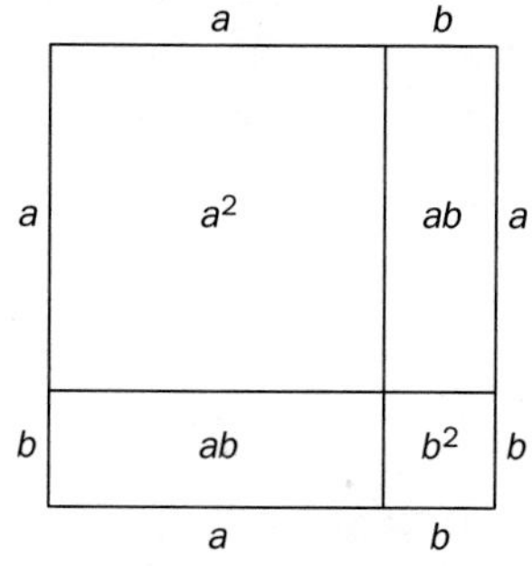

Additional Example 1

Find each product.

a. $(x+5)^2$

$(x+5)^2 = x^2 + 2(5)x + 5^2$
$= x^2 + 10x + 25$

b. $(y-9)^2$

$(y-9)^2 = y^2 - 2(9)y + 9^2$
$= y^2 - 18y + 81$

Additional Example 2

Find each product.

a. $(7-\sqrt{13})^2$

$(7-\sqrt{13})^2$
$= 7^2 - 2 \cdot 7\sqrt{13} + (\sqrt{13})^2$
$= 49 - 14\sqrt{13} + 13$
$= 62 - 14\sqrt{13}$

b. $(5k+4)^2$

$(5k+4)^2$
$= (5k)^2 + 2(5k)(4) + 4^2$
$= 25k^2 + 40k + 16$

Teach⇔Interact

Section Note

Integrating the Strands

The product of the sum and difference of two terms has a geometric interpretation. Draw a smaller square inside one corner of a large square. Then shade the area between the squares as shown below.

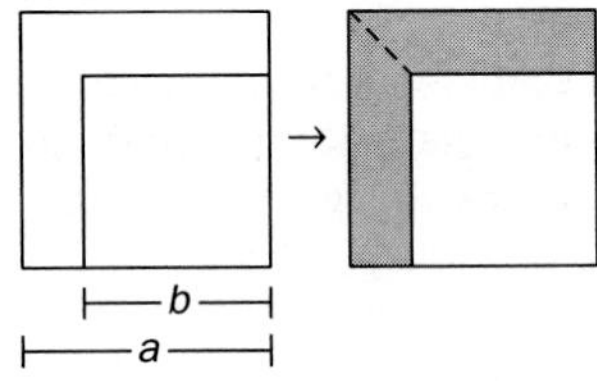

The shaded area represents $a^2 - b^2$. One can cut apart the figure on the right and arrange the shaded pieces to form a rectangle.

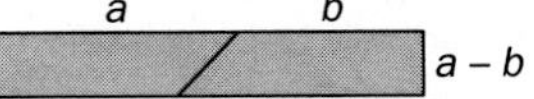

This rectangle has dimensions $a + b$ by $a - b$, and an area of $(a + b)(a - b)$. Hence, $(a + b)(a - b) = a^2 - b^2$.

Additional Example 3

Find the product $(x + 11)(x - 11)$.

$(a + b)(a - b) = a^2 - b^2$
$(x + 11)(x - 11) = x^2 - 11^2$
$= x^2 - 121$

Think and Communicate

In discussing question 7, point out that it is easy to find products such as $14 \cdot 26$ by using $(a + b)(a - b) = a^2 - b^2$ when the mean of the numbers being multiplied is a multiple of 10, 100, 1000, and so on. The mean of 14 and 26 is $\frac{14 + 26}{2} = 20$.

The Difference of Two Squares

Joelle Rozene designs office spaces. In estimating the cost of doing a job, she often uses mental math for multiplying numbers. For example, to find the area of a room that is 42 ft by 38 ft, she does this calculation in her head:

$$42 \cdot 38 = (40 + 2)(40 - 2) = 40^2 - 2^2$$

Joelle finds the product of the sum of the numbers 40 and 2 and the difference of the numbers 40 and 2.

The product of the sum and difference of two terms is called the **difference of two squares**.

$$(a + b)(a - b) = a^2 - b^2$$

EXAMPLE 3

Find the product $(x + 3)(x - 3)$.

SOLUTION

Notice that you are finding the product of the sum of two terms (x and 3) and the difference of the same two terms.

$(a + b)(a - b) = a^2 - b^2$
$(x + 3)(x - 3) = x^2 - 3^2$
$= x^2 - 9$

THINK AND COMMUNICATE

6. Use the distributive property to show that the solution found in Example 3 is correct. Explain why the pattern $(a + b)(a - b) = a^2 - b^2$ always works. (*Hint:* What happens to two of the terms when you use the distributive property?)
7. Explain how to use the difference of two squares to find the product $14 \cdot 26$.
8. Explain how you could find 42^2 by using mental math and the pattern for squaring binomials.
9. Why is $(2 + \sqrt{3})(2 - \sqrt{3})$ the product of the sum and difference of two terms? Use mental math to find the product.

Think and Communicate

6. $(x + 3)(x - 3) = x(x - 3) + 3(x - 3) = x^2 - 3x + 3x - 9 = x^2 - 9$
When you use the distributive property, the terms ab and $-ab$ are opposites of each other, so there are no ab terms in the product.
7. $14 \cdot 26 = (20 - 6)(20 + 6) = 20^2 - 6^2 = 400 - 36 = 364$
8. $42^2 = (40 + 2)^2 = 40^2 + 2(40)(2) + 2^2 = 1600 + 160 + 4 = 1764$
9. because the product has the form $(a + b)(a - b)$; The product is equal to $2^2 - (\sqrt{3})^2 = 4 - 3 = 1$.

Checking Key Concepts

1. $x^2 + 16x + 64$
2. $x^2 - 10x + 25$
3. $x^2 - 81$
4. $14 + 6\sqrt{5}$
5. $52 - 14\sqrt{3}$
6. -2
7. $(50 + 4)^2 = 50^2 + 2(50)(4) + 4^2 = 2500 + 400 + 16 = 2916$
8. $53 \cdot 47 = (50 + 3)(50 - 3) = 50^2 - 3^2 = 2500 - 9 = 2491$

Exercises and Applications

1. $w^2 - 4w + 4$
2. $x^2 + 10x + 25$
3. $x^2 - 6x + 9$
4. $x^2 - 64$
5. $y^2 - 16$
6. $c^2 - 0.04$
7. $4r^2 + 24r + 36$
8. $16x^2 - 16x + 4$
9. $100t^2 - 36$
10. a. In each case, the result in Step 2 is 1 more than the result in Step 3. That is, the square of the middle number is 1 more than the product of the first and third numbers.

Checking Key Concepts

Find each product.

1. $(x + 8)^2$
2. $(x - 5)^2$
3. $(x + 9)(x - 9)$
4. $(3 + \sqrt{5})^2$
5. $(7 - \sqrt{3})^2$
6. $(2 + \sqrt{6})(2 - \sqrt{6})$
7. Find 54^2 by writing it as the square of a binomial.
8. Look back at page 266. Use Joelle Rozene's method to find the product $53 \cdot 47$.

6.5 Exercises and Applications

Find each product.

1. $(w - 2)^2$
2. $(x + 5)^2$
3. $(x - 3)^2$
4. $(x + 8)(x - 8)$
5. $(y - 4)(y + 4)$
6. $(c + 0.2)(c - 0.2)$
7. $(2r + 6)^2$
8. $(4x - 2)^2$
9. $(10t + 6)(10t - 6)$

Extra Practice exercises on page 570

10. **Investigation** Repeat the following steps several times using different groups of integers:

    Step 1 Write down three consecutive integers, such as 3, 4, and 5.
    Step 2 Square the middle number.
    Step 3 Multiply the first and third numbers together.

    a. Describe the relationship between the results of Steps 2 and 3.

    b. Let n = an integer. Then $n - 1$, n, and $n + 1$ are three consecutive integers. Use these variable expressions to do Steps 2 and 3. How do your results explain the relationship you saw in part (a)?

    c. **Open-ended Problem** Design and carry out an investigation of other groups of integers, such as four consecutive integers or three consecutive even integers. Can you identify any patterns? Try to find an algebraic explanation for your observations.

11. **CONSTRUCTION** A contractor needs to estimate the cost of building a stone path around a square garden, as shown. Explain why the cost of building the path depends on the width of the path. Use a variable expression to support your answer.

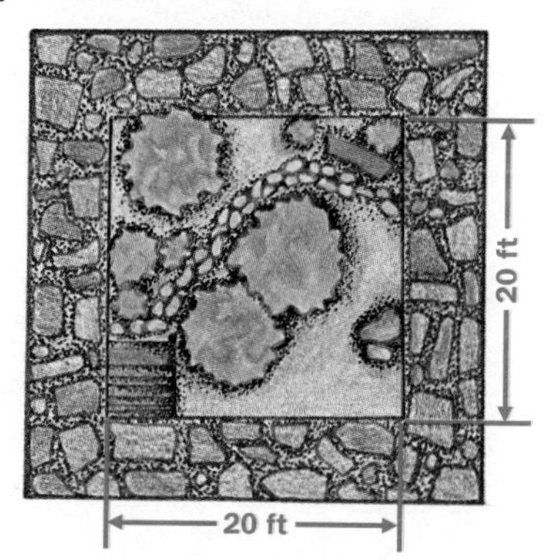

b. The middle number is $n + 1$; $(n + 1)^2 = n^2 + 2n + 1$. The first and third numbers are n and $n + 2$; $n(n + 2) = n^2 + 2n$. Since $n^2 + 2n + 1$ is 1 more than $n^2 + 2n$, this explains the relationship found in part (a).

c. Answers may vary. Examples are given. For any four consecutive integers, the product of the second and third is 2 more than the product of the first and fourth because $n(n + 3) = n^2 + 3n$ and $(n + 1)(n + 2) = n^2 + 3n + 2$. For any three consecutive even integers, the square of the second integer is 4 more than the product of the first and the third because $(n + 2)^2 = n^2 + 4n + 4$ and $n(n + 4) = n^2 + 4n$.

11. The greater the width, the greater the value of x and the greater the area of the path; the cost depends on the area. The area of the path is equal to the combined area of the path and garden minus the area of the garden alone. Then the area of the path $= (2x + 20)^2 - 20^2 = 4x^2 + 80x$.

Teach⇔Interact

Checking Key Concepts

Communication: Listening Some students may find it easy to respond to questions 1–3 by having the expression presented orally.

Closure Question

What special products have you learned to recognize in this section? Give the formulas for these special products.
squaring a binomial and finding the difference of two squares; $(a + b)^2 = a^2 + 2ab + b^2$
$(a - b)^2 = a^2 - 2ab + b^2$
$(a + b)(a - b) = a^2 - b^2$

Apply⇔Assess

Suggested Assignment

Core Course
Day 1 Exs. 1–23
Day 2 Exs. 24–27, 32–44, AYP

Extended Course
Day 1 Exs. 1–23
Day 2 Exs. 24–44, AYP

Block Schedule
Day 38 Exs. 1–28, 32–44, AYP

Exercise Notes

Cooperative Learning
Exs. 1–9 Students can work with a partner to complete these exercises and to check their understanding by creating more examples like these. This will help students recognize special products quickly.

Challenge
Exs. 1–9 Students who understand special products may be interested in an example such as the following.
$(2x + 3 - y)(2x + 3 + y)$
$= [(2x + 3) - y][(2x + 3) + y]$
$= (2x + 3)^2 - y^2$
$= (4x^2 + 12x + 9) - y^2$
$= 4x^2 + 12x - y^2 + 9$
Challenge students to use this idea to find the product $(3 + \sqrt{2} + \sqrt{5})(3 + \sqrt{2} - \sqrt{5})$.
$(6 + 6\sqrt{2})$

Apply⇔Assess

Exercise Notes

Teaching Tip
Exs. 12–17 Suggest that students estimate the products before doing the mental computation. They can check their final answers by using a calculator.

Topic Spiraling: Review
Exs. 24–26 Have students simplify each radical expression to provide a brief review of topics from Section 6.3.

Topic Spiraling: Preview
Ex. 28 This is a good exercise to prepare students for factoring.

Second-Language Learners
Exs. 29–31 It may be helpful for second-language learners to hear a student fluent in English read this material aloud before they begin working on it.

Interdisciplinary Problems
Exs. 29–31 The diagrams used here are also called Punnett squares. Students interested in applications of genetics to biology can consult a biology textbook for more information on this topic.

Find each product using mental math.

12. $18 \cdot 22$ **13.** 39^2 **14.** $25 \cdot 35$

15. 2.1^2 **16.** $\left(6\frac{1}{2}\right)^2$ **17.** $\left(9\frac{1}{2}\right)\left(10\frac{1}{2}\right)$

Find each product.

18. $(x + y)^2$ **19.** $(x - y)^2$ **20.** $(3x - 5)^2$

21. $(x + y)(x - y)$ **22.** $(3x + 6y)(3x - 6y)$ **23.** $(4x - 7)(4x + 7)$

24. $\left(1 - \sqrt{6}\right)^2$ **25.** $\left(4 + 2\sqrt{3}\right)^2$ **26.** $\left(6 + \sqrt{13}\right)^2$

27. SAT/ACT Preview For which value of x does a rectangle with side lengths $x + 4$ and $x - 4$ have an area of 84?

A. 5 **B.** 10 **C.** 15 **D.** 20 **E.** 25

28. Challenge Rewrite each polynomial as the product of two binomials.

a. $y^2 + 6y + 9$ b. $25x^2 - 10x + 1$ c. $36 - y^2$

Connection GENETICS

Chicken breeders use *genetics* to breed chickens for desired characteristics. For example, the shape of a chicken's comb is determined by a combination of two genes, one inherited from each of its parents. Let G_R represent the gene for a "rose comb." Let G_S represent the gene for a "single comb." For Exercises 29–31, suppose a hen and rooster each have a mixed pair of comb-shape genes.

This geneticist is inspecting chicken genes that have been separated so that they can be identified.

29. The table at the right is called an *inheritance table*. The table is used to show possible gene combinations for the offspring of the hen and rooster. Copy and complete the table to find which gene combinations are possible.

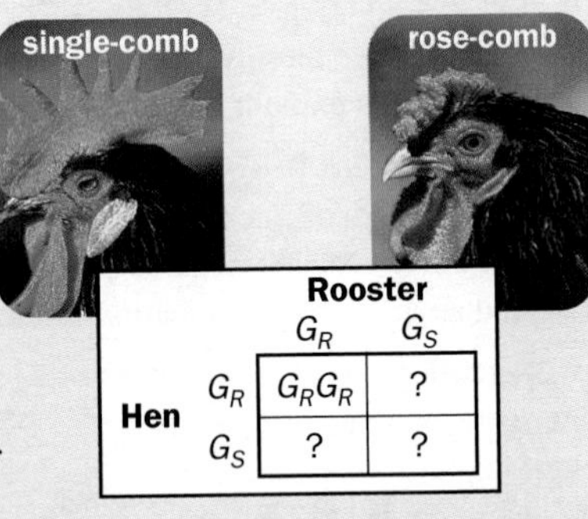

		Rooster	
		G_R	G_S
Hen	G_R	G_RG_R	?
	G_S	?	?

BY THE WAY

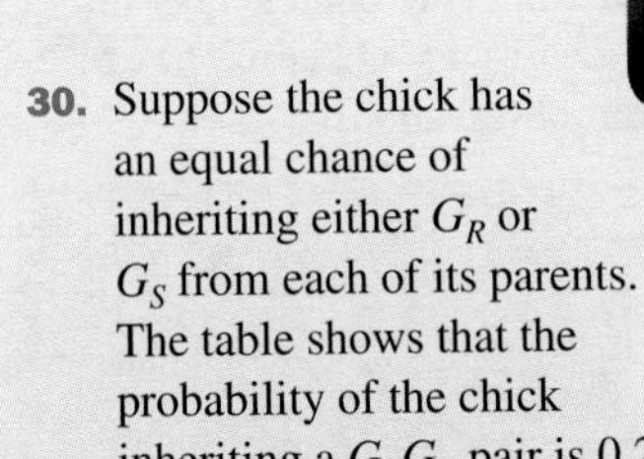

As early as 4000 years ago, chickens were domesticated. The picture shown is from a Roman mosaic found in North Africa.

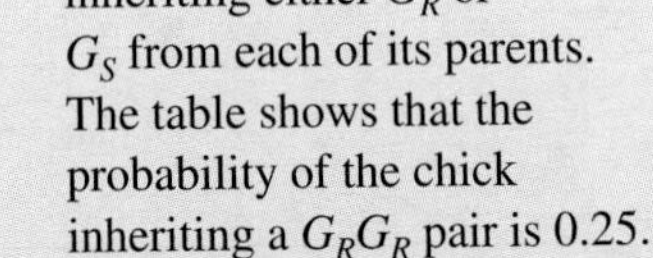

30. Suppose the chick has an equal chance of inheriting either G_R or G_S from each of its parents. The table shows that the probability of the chick inheriting a G_RG_R pair is 0.25. What is the probability of inheriting a G_SG_S pair? a mixed pair?

31. Writing Square the binomial $0.5G_R + 0.5G_S$. Explain how the product is related to your answer to Exercise 30.

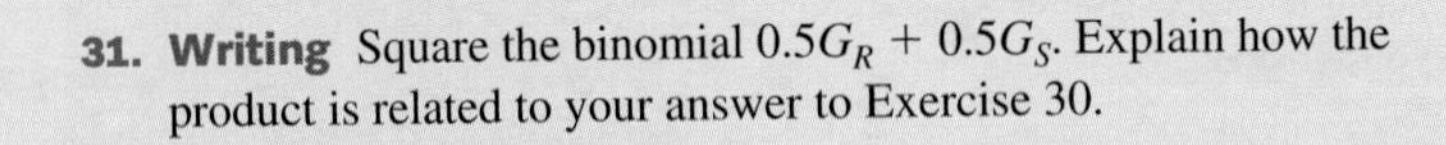

12. $(20 - 2)(20 + 2) = 20^2 - 2^2 = 400 - 4 = 396$
13. $(40 - 1)^2 = 40^2 - 2(40)(1) + 1^2 = 1600 - 80 + 1 = 1521$
14. $(30 - 5)(30 + 5) = 30^2 - 5^2 = 900 - 25 = 875$
15. $(2 + 0.1)^2 = 2^2 + 2(2)(0.1) + (0.1)^2 = 4 + 0.4 + 0.01 = 4.41$
16. $\left(6 + \frac{1}{2}\right)^2 = 6^2 + 2(6)\left(\frac{1}{2}\right) + \left(\frac{1}{2}\right)^2 = 36 + 6 + \frac{1}{4} = 42\frac{1}{4}$
17. $\left(10 - \frac{1}{2}\right)\left(10 + \frac{1}{2}\right) = 10^2 - \left(\frac{1}{2}\right)^2 = 100 - \frac{1}{4} = 99\frac{3}{4}$
18. $x^2 + 2xy + y^2$
19. $x^2 - 2xy + y^2$
20. $9x^2 - 30x + 25$
21. $x^2 - y^2$
22. $9x^2 - 36y^2$
23. $16x^2 - 49$
24. $7 - 2\sqrt{6}$
25. $28 + 16\sqrt{3}$
26. $49 + 12\sqrt{13}$
27. B
28. a. $(y + 3)^2$
 b. $(5x - 1)^2$
 c. $(6 + y)(6 - y)$
29.

		Rooster	
		G_R	G_S
Hen	G_R	G_RG_R	G_RG_S
	G_S	G_SG_R	G_SG_S

30. 0.25; 0.5
31. $(0.5G_R + 0.5G_S)^2 = 0.25(G_R)^2 + 0.5G_RG_S + 0.25(G_S)^2$. The coefficients of the terms give the probability of inheriting the associated gene pair. That is, the coefficient of the first term gives the probability of inheriting a G_RG_R pair, the coefficient of the second term gives the probability of inheriting a mixed pair, and the coefficient of the third term gives the probability of inheriting a G_SG_S pair.
32. Answers may vary. An example is given. Subtract the number given by the audience member from the nearest multiple of 10. Add the result to the same multiple of 10. This new result is the second number. The number pair is the sum and difference of two

ONGOING ASSESSMENT

32. **Writing** A magician asks a member of the audience for a number less than 100. The magician then gives a second number. The audience is amazed that the magician can find the product of the two numbers mentally before the audience member can find it with a calculator. Use what you have learned in this section to write instructions for performing this trick.

SPIRAL REVIEW

Find an equation of the line through the given points. *(Section 3.5)*

33. (1, 4) and (3, 12) **34.** (0, 3) and (−2, 7) **35.** (4, −2) and (−8, −8)

Find an equation of the line with the given slope and through the given point. *(Section 3.5)*

36. slope = 3; (0, 5) **37.** slope = $\frac{2}{3}$; (3, 6) **38.** slope = −1.4; (−5, 4)

Find each product. *(Section 6.4)*

39. $3m(2m - 5)$ **40.** $(x + 1)(2x + 3)$ **41.** $(5x - 1)(1 - 5x)$

42. $(y - 1)(y + 3)$ **43.** $5s(3s - 4)$ **44.** $(3x + 5)(5x - 6)$

ASSESS YOUR PROGRESS

VOCABULARY

monomial (p. 260)
binomial (p. 260)
trinomial (p. 260)
perfect square trinomial (p. 265)
difference of two squares (p. 266)

Find each product. *(Sections 6.4 and 6.5)*

1. $5x(x + 3)$ **2.** $5(2 + \sqrt{3})$ **3.** $(x - 5)(x + 5)$

4. $(6 + \sqrt{10})^2$ **5.** $3x(2x + 3)$ **6.** $(x - 4)^2$

7. $5x(3x)$ **8.** $(2x + 1)(x + 5)$ **9.** $(x + \sqrt{8})^2$

10. Write a variable expression for the area of a square with sides of length $(n + 8)$ units. Write your expression as a trinomial. *(Section 6.5)*

11. Write variable expressions for the perimeter and the area of the rectangle shown. *(Section 6.5)*

$x + 7$

$x - 7$

12. **Journal** How do you keep track of terms when you use the distributive property to multiply binomials? Do you think your method works well?

Apply⇔Assess

Exercise Notes

Assessment Note
Ex. 32 This exercise will demonstrate a student's ability to analyze a situation, model it algebraically, and apply knowledge of special products.

Assess Your Progress

Journal Entry
Students should be encouraged to provide examples to illustrate their methods for Ex. 12.

Progress Check 6.4–6.5

See page 273.

Practice 42 for Section 6.5

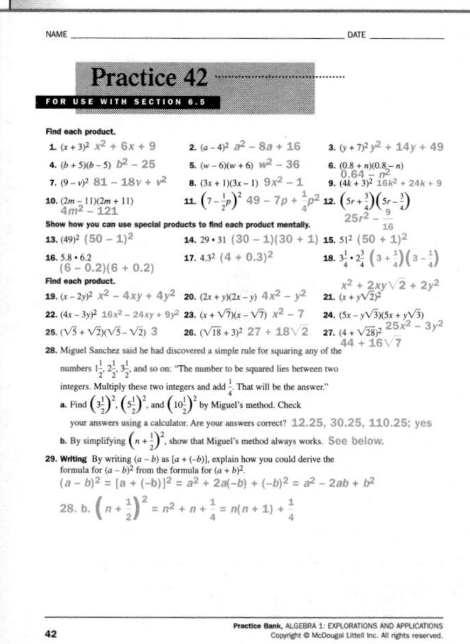

NAME ______ DATE ______

Practice 42

FOR USE WITH SECTION 6.5

Find each product.

1. $(x + 3)^2$ $x^2 + 6x + 9$
2. $(a - 4)^2$ $a^2 - 8a + 16$
3. $(y + 7)^2$ $y^2 + 14y + 49$
4. $(b + 5)(b - 5)$ $b^2 - 25$
5. $(w - 6)(w + 6)$ $w^2 - 36$
6. $(0.8 + n)(0.8 - n)$ $0.64 - n^2$
7. $(9 - v)^2$ $81 - 18v + v^2$
8. $(3x + 1)(3x - 1)$ $9x^2 - 1$
9. $(4k + 3)^2$ $16k^2 + 24k + 9$
10. $(2m - 11)(2m + 11)$ $4m^2 - 121$
11. $\left(7 - \frac{1}{2}p\right)^2$ $49 - 7p + \frac{1}{4}p^2$
12. $\left(5r + \frac{3}{4}\right)\left(5r - \frac{3}{4}\right)$ $25r^2 - \frac{9}{16}$

Show how you can use special products to find each product mentally.

13. $(49)^2$ $(50 - 1)^2$
14. $29 \cdot 31$ $(30 - 1)(30 + 1)$
15. 51^2 $(50 + 1)^2$
16. $5.8 \cdot 6.2$ $(6 - 0.2)(6 + 0.2)$
17. 4.3^2 $(4 + 0.3)^2$
18. $3\frac{1}{4} \cdot 2\frac{3}{4}$ $\left(3 + \frac{1}{4}\right)\left(3 - \frac{1}{4}\right)$

Find each product.

19. $(x - 2y)^2$ $x^2 - 4xy + 4y^2$
20. $(2x + y)(2x - y)$ $4x^2 - y^2$
21. $(x + y\sqrt{2})^2$ $x^2 + 2xy\sqrt{2} + 2y^2$
22. $(4x - 3y)^2$ $16x^2 - 24xy + 9y^2$
23. $(x + \sqrt{7})(x - \sqrt{7})$ $x^2 - 7$
24. $(5x - y\sqrt{3})(5x + y\sqrt{3})$ $25x^2 - 3y^2$
25. $(\sqrt{5} + \sqrt{2})(\sqrt{5} - \sqrt{2})$ 3
26. $(\sqrt{18} + 3)^2$ $27 + 18\sqrt{2}$
27. $(4 + \sqrt{28})^2$ $44 + 16\sqrt{7}$

28. Miguel Sanchez said he had discovered a simple rule for squaring any of the numbers $1\frac{1}{2}, 2\frac{1}{2}, 3\frac{1}{2}$, and so on: "The number to be squared lies between two integers. Multiply these two integers and add $\frac{1}{4}$. That will be the answer."
    a. Find $\left(3\frac{1}{2}\right)^2$, $\left(5\frac{1}{2}\right)^2$, and $\left(10\frac{1}{2}\right)^2$ by Miguel's method. Check your answers using a calculator. Are your answers correct? 12.25, 30.25, 110.25; yes
    b. By simplifying $\left(n + \frac{1}{2}\right)^2$, show that Miguel's method always works. See below.

29. **Writing** By writing $(a - b)$ as $[a + (-b)]$, explain how you could derive the formula for $(a - b)^2$ from the formula for $(a + b)^2$.
$(a - b)^2 = [a + (-b)]^2 = a^2 + 2a(-b) + (-b)^2 = a^2 - 2ab + b^2$

28. b. $\left(n + \frac{1}{2}\right)^2 = n^2 + n + \frac{1}{4} = n(n + 1) + \frac{1}{4}$

42 **Practice Bank,** ALGEBRA 1: EXPLORATIONS AND APPLICATIONS

terms, a product of the form $(10n + k)(10n - k)$. Find the product, $100n^2 - k^2$. For example, suppose the audience member chooses 54. Since 50 − 54 = −4, choose 50 + (−4) = 46 as the other number. The product is (50 + 4)(50 − 4) = 2500 − 16 = 2484.

33. $y = 4x$
34. $y = -2x + 3$
35. $y = \frac{1}{2}x - 4$
36. $y = 3x + 5$
37. $y = \frac{2}{3}x + 4$
38. $y = -1.4x - 3$
39. $6m^2 - 15m$
40. $2x^2 + 5x + 3$
41. $-25x^2 + 10x - 1$
42. $y^2 + 2y - 3$
43. $20s - 15s^2$
44. $15x^2 + 7x - 30$

Assess Your Progress

1. $5x^2 + 15x$
2. $10 + 5\sqrt{3}$
3. $x^2 - 25$
4. $46 + 12\sqrt{10}$
5. $6x^2 + 9x$
6. $x^2 - 8x + 16$
7. $15x^2$
8. $2x^2 + 11x + 5$
9. $x^2 + 4\sqrt{2}x + 8$
10. $(n + 8)^2 = n^2 + 16n + 64$
11. $4x$; $x^2 - 49$
12. Answers may vary. An example is given. I draw arrows like the ones shown in the examples. This method works well for me.

Project Notes

Mathematical Goals

- Collect data relating the period of a pendulum to its length and make a scatter plot of the data.
- Use the formula $P = 0.2\sqrt{L}$ that relates the period P of a pendulum to its length L to calculate P for different values of L.
- Graph $P = 0.2\sqrt{L}$ on a graphing calculator.
- Use the formula $P = 0.2\sqrt{L}$ to find L when P is one second.

Planning

Materials
- String
- Coins
- Pencil
- Graph paper
- Ruler
- Graphing calculator
- Watch

Project Teams
Students can choose a partner to work with and then discuss how they wish to proceed. They should choose a convenient place to conduct the experiment and record their results.

Guiding Students' Work

Make sure students have a sufficient number of tables to use for their experiments. Suggest that they share the work of timing the pendulum for 10 swings and recording the results. Both students in each group should do the work stated under *Using a Formula* on the following page. They should also share in writing the report about their experiment.

Second-Language Learners
Make sure students learning English recognize that the term *period* (of a pendulum) is defined as *the amount of time it takes the pendulum to swing from point A to point B and back again.*

PORTFOLIO PROJECT

Investigating Pendulum Length

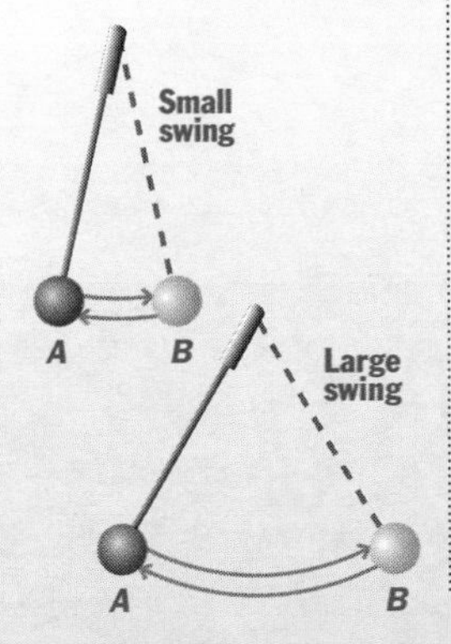

The pendulum of a clock swings back and forth, causing the clock to tick off the seconds. A pendulum clock runs down over time. With each swing, the pendulum travels a slightly smaller distance.

Even as the clock starts to run down, it can still keep good time. This is because the **period** of a pendulum, or the amount of time it takes the pendulum to swing from point **A** to point **B** and back to point **A**, is the same for large swings and small swings.

PROJECT GOAL **Investigate how the length of a pendulum affects the period.**

Conducting an Experiment

Work with a partner to design and carry out an experiment in which you measure the period of a pendulum for at least 10 different lengths of string. Here are some hints for organizing your work:

1. MAKE your pendulum by taping several coins to a piece of string. Use a pencil to hold the pendulum away from the table.

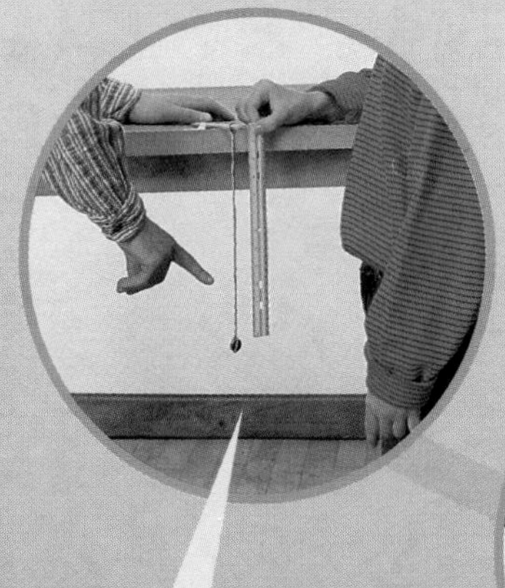

Use a range of string lengths between 10 cm and 30 cm.

2. DETERMINE how many seconds it takes your pendulum to make 10 full swings. Divide by 10 to find the period.

3. ORGANIZE your data in a table. Then make a scatter plot. Put string length on the horizontal axis and period on the vertical axis.

Based on your observations, find a string length that produces a period of one second. Can you keep time with this pendulum?

Using a Formula

The Italian scientist Galileo Galilei (1564 – 1642) showed that the period of a pendulum is proportional to the square root of the length of the pendulum. You can use a formula to describe how the period P of a pendulum is related to its length L.

$$P = 0.2\sqrt{L}$$ **P is in seconds and L is in centimeters.**

- Calculate P for the L-values you used in your experiment. (Make sure you express the L-values in centimeters.) Compare your calculated P-values to your measured values. What could cause differences?
- Graph the equation $P = 0.2\sqrt{L}$ on a graphing calculator. How does the graph compare with your scatter plot?
- Use the formula to find L when the period is one second. Compare this L-value to the length you found from experimenting.

Writing A Report

Write a report about your experiment. Describe your procedure and tell what conclusions you made about the relationship between the length of a pendulum and its period. Include pictures, graphs, and tables to support your ideas.

You may want to extend your report to explore some of the ideas below:

- The first pendulum clock was invented in 1657. How does a pendulum clock work?
- Pendulum clocks have a weight that can be raised and lowered. Why would this come in handy?
- How does the weight of a pendulum affect the period? Design and carry out an experiment to find out.

Self-Assessment

In your report, be sure to describe any difficulties you had. Tell what went well in your experiment and why. Is there anything that you would like to remember for a future experiment, such as a useful way to organize data, or a way to use your data to set up a good viewing window on a graphing calculator?

Project Notes

Guiding Students' Work

Rubric for Chapter Project

4 Students complete all aspects of the experiment successfully. Their report describes all procedures clearly and states the relationship between the length of a pendulum and its period correctly. Pictures, graphs, and tables are used to support statements made in the report. The report is well organized, neat, and does not contain any mathematical errors.

3 Students are successful in conducting their experiment. They collect the results and write a report. There are a few problems with the report, but none of them are serious. For example, the pictures and graphs could have been drawn better. Students' conclusions are correct and the report is complete.

2 Students do not use 10 different lengths of string and their measurements of 10 full swings are sometimes inaccurate. A report is written, but it does not contain a full description of the procedures followed. The conclusions reached are correct but not fully supported by pictures, graphs, and tables.

1 Students have difficulty setting up their experiments. They make a few measurements of 10 full swings but do not collect sufficient data to make a meaningful scatter plot. Students should be encouraged to speak with the teacher as soon as possible to review their work and to make a new start on the project.

Chapter Support

Course Guide: Chapter 6

Lesson Plans: Chapter 6

Practice Bank:
  Cumulative Practice 43

Study Guide: Chapter 6 Review

Assessment Book:
  Chapter Tests 26 and 27
  Chapter 6 Alternative Assessment
  Cumulative Test 28

Test Generator software

Portfolio Project Book

Preparing for College Entrance Tests

Professional Handbook

Progress Check 6.1–6.3

1. Write an equation that describes the relationship between the known lengths and the unknown lengths for the right triangle shown. *(Section 6.1)*

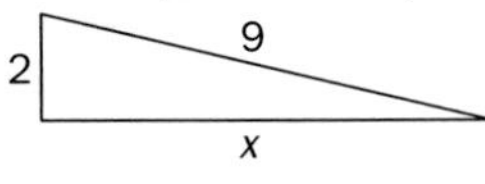

$2^2 + x^2 = 9^2$ (or any equivalent equation)

2. Find the length of the hypotenuse of a right triangle whose legs have lengths 9 and 12. *(Section 6.1)* 15

3. Is it possible for 5, 6, and 10 to be the lengths of the sides of a right triangle? Why or why not? *(Section 6.1)*
No; $5^2 + 6^2 \neq 10^2$.

4. Tell which of the terms listed below apply to the number $\sqrt{17}$. *(Section 6.2)*

integer	rational
irrational	real

irrational, real

5. Write $0.\overline{93}$ as a fraction in lowest terms. *(Section 6.2)* $\frac{31}{33}$

6. *True* or *False*? $1 = 0.\overline{9}$ *(Section 6.2)* True.

7. Simplify $\sqrt{200} + \sqrt{3}$. *(Section 6.3)* $10\sqrt{2} + \sqrt{3}$

8. Between what two consecutive whole numbers is $\sqrt{50}$? *(Section 6.3)* between 7 and 8

CHAPTER

6 Review

STUDY TECHNIQUE

Find the exercises in each lesson that you had the most difficulty answering. Do these exercises again without looking at your previous work. Check your new work against your previous work.

VOCABULARY

right triangle (p. 239)
Pythagorean theorem (p. 239)
hypotenuse (p. 239)
leg (p. 239)
perfect square (p. 240)
square root (p. 240)
radical sign (p. 240)
converse of the Pythagorean theorem (p. 241)
rational number (p. 246)
irrational number (p. 246)
real number (p. 246)
dense (p. 246)
radical form (p. 252)
monomial (p. 260)
binomial (p. 260)
trinomial (p. 260)
perfect square trinomial (p. 265)
difference of two squares (p. 266)

SECTION 6.1

A **right triangle** has one 90° angle.

Pythagorean theorem: In a right triangle, the square of the length of the hypotenuse is equal to the sum of the squares of the lengths of the legs.

$$a^2 + b^2 = c^2$$

leg a, hypotenuse c, leg b

Converse of the Pythagorean theorem: If the lengths of the sides of a triangle satisfy the relationship $a^2 + b^2 = c^2$, then it is a right triangle.

A **perfect square** is a number whose **square root** is an integer. A positive number has two square roots. A negative number has no square roots.

$\sqrt{25} = 5$ $\qquad$ $-\sqrt{25} = -5$

SECTIONS 6.2 *and* 6.3

Every **real number** is either a rational number or an irrational number.

A **rational number** can be written as the ratio of two integers. When a rational number is written as a decimal it either repeats or terminates.

An **irrational number** cannot be written as the ratio of two integers. When an irrational number is written as a decimal it never repeats or terminates.

Properties of Square Roots

For positive numbers *a* and *b*:	For example:
$\sqrt{a^2} = a$	$\sqrt{4^2} = 4$
$\sqrt{ab} = \sqrt{a} \cdot \sqrt{b}$	$\sqrt{4 \cdot 6} = \sqrt{4} \cdot \sqrt{6} = 2\sqrt{6}$
$\sqrt{\frac{a}{b}} = \frac{\sqrt{a}}{\sqrt{b}}, b \neq 0$	$\sqrt{\frac{64}{49}} = \frac{\sqrt{64}}{\sqrt{49}} = \frac{8}{7}$

SECTIONS 6.4 *and* 6.5

A variable expression can be named by the number of terms in the expression.

monomial	*binomial*	*trinomial*
$2x$	$2r + 3$	$3xy - 5y + 7$
1 term	2 terms	3 terms

You can use the distibutive property to multiply binomials. For example, find the product $(x + 26)(2x - 5)$.

$$\begin{aligned}(x + 26)(2x - 5) &= (x + 26)2x - (x + 26)5 \\ &= 2x^2 + 52x - 5x - 130 \\ &= 2x^2 + 47x - 130\end{aligned}$$

The product $(2x + 8)^2$ is a **perfect square trinomial**.

$$\begin{aligned}(2x + 8)^2 &= (2x)^2 + 2(2x)(8) + 8^2 \\ &= 4x^2 + 32x + 64\end{aligned}$$

The product $(5 + \sqrt{7})(5 - \sqrt{7})$ is the **difference of two squares**.

$$\begin{aligned}(5 + \sqrt{7})(5 - \sqrt{7}) &= 5^2 - (\sqrt{7})^2 \\ &= 25 - 7 \\ &= 18\end{aligned}$$

Progress Check 6.4–6.5

Find each product. *(Sections 6.4 and 6.5)*

1. $(x - 3)(x + 7)$ $x^2 + 4x - 21$
2. $(3x + 5)(2x + 8)$ $6x^2 + 34x + 40$
3. $(\sqrt{2} + 5)(\sqrt{2} - 1)$ $4\sqrt{2} - 3$
4. $(x + 2y)^2$ $x^2 + 4xy + 4y^2$
5. $(2k + 3)(2k - 3)$ $4k^2 - 9$
6. A flower bed will be built around the square base of a sculpture in a park, as shown by the shaded area in the diagram.

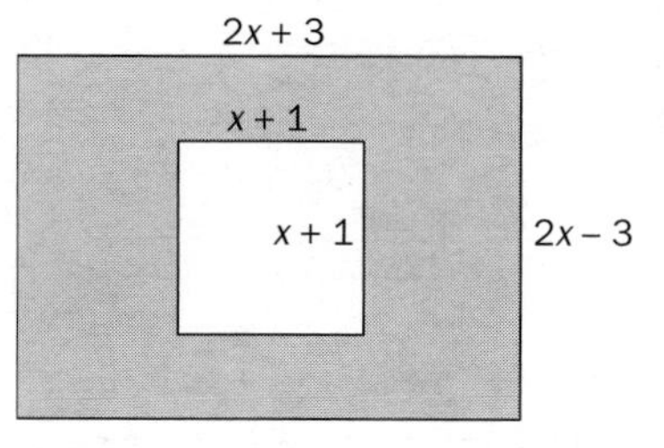

Write a simplified expression for the area of the flower bed. *(Section 6.5)*
$3x^2 - 2x - 10$

Chapter 6 Assessment
Form A Chapter Test

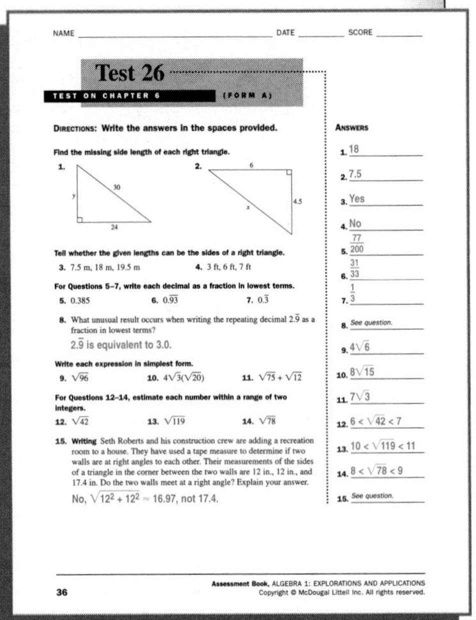

NAME ______ DATE ______ SCORE ______

Test 26

TEST ON CHAPTER 6 (FORM A)

DIRECTIONS: Write the answers in the spaces provided.

Find the missing side length of each right triangle.

1. (triangle: legs y and 24, hypotenuse 30)
2. (triangle: legs 6 and 4.5, hypotenuse x)

Tell whether the given lengths can be the sides of a right triangle.

3. 7.5 m, 18 m, 19.5 m
4. 3 ft, 6 ft, 7 ft

For Questions 5–7, write each decimal as a fraction in lowest terms.

5. 0.385
6. $0.\overline{93}$
7. $0.\overline{3}$
8. What unusual result occurs when writing the repeating decimal $2.\overline{9}$ as a fraction in lowest terms?
   $2.\overline{9}$ is equivalent to 3.0.

Write each expression in simplest form.

9. $\sqrt{96}$
10. $4\sqrt{3}(\sqrt{20})$
11. $\sqrt{75} + \sqrt{12}$

For Questions 12–14, estimate each number within a range of two integers.

12. $\sqrt{42}$
13. $\sqrt{119}$
14. $\sqrt{78}$
15. **Writing** Seth Roberts and his construction crew are adding a recreation room to a house. They have used a tape measure to determine if two walls are at right angles to each other. Their measurements of the sides of a triangle in the corner between the two walls are 12 in., 12 in., and 17.4 in. Do the two walls meet at a right angle? Explain your answer.
    No, $\sqrt{12^2 + 12^2} \approx 16.97$, not 17.4.

ANSWERS

1. 18
2. 7.5
3. Yes
4. No
5. $\frac{77}{200}$
6. $\frac{31}{33}$
7. $\frac{1}{3}$
8. *See question.*
9. $4\sqrt{6}$
10. $8\sqrt{15}$
11. $7\sqrt{3}$
12. $6 < \sqrt{42} < 7$
13. $10 < \sqrt{119} < 11$
14. $8 < \sqrt{78} < 9$
15. *See question.*

36

Assessment Book, ALGEBRA 1: EXPLORATIONS AND APPLICATIONS

Chapter 6 Assessment
Form B Chapter Test

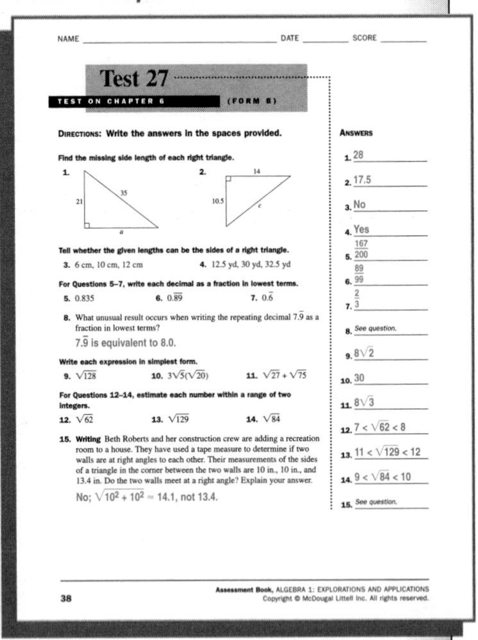

NAME ______ DATE ______ SCORE ______

Test 27

TEST ON CHAPTER 6 (FORM B)

DIRECTIONS: Write the answers in the spaces provided.

Find the missing side length of each right triangle.

1. (triangle: legs 21 and a, hypotenuse 35)
2. (triangle: legs 14 and 10.5, hypotenuse c)

Tell whether the given lengths can be the sides of a right triangle.

3. 6 cm, 10 cm, 12 cm
4. 12.5 yd, 30 yd, 32.5 yd

For Questions 5–7, write each decimal as a fraction in lowest terms.

5. 0.835
6. $0.\overline{89}$
7. $0.\overline{6}$
8. What unusual result occurs when writing the repeating decimal $7.\overline{9}$ as a fraction in lowest terms?
   $7.\overline{9}$ is equivalent to 8.0.

Write each expression in simplest form.

9. $\sqrt{128}$
10. $3\sqrt{5}(\sqrt{20})$
11. $\sqrt{27} + \sqrt{75}$

For Questions 12–14, estimate each number within a range of two integers.

12. $\sqrt{62}$
13. $\sqrt{129}$
14. $\sqrt{84}$
15. **Writing** Beth Roberts and her construction crew are adding a recreation room to a house. They have used a tape measure to determine if two walls are at right angles to each other. Their measurements of the sides of a triangle in the corner between the two walls are 10 in., 10 in., and 13.4 in. Do the two walls meet at a right angle? Explain your answer.
    No; $\sqrt{10^2 + 10^2} \approx 14.1$, not 13.4.

ANSWERS

1. 28
2. 17.5
3. No
4. Yes
5. $\frac{167}{200}$
6. $\frac{89}{99}$
7. $\frac{2}{3}$
8. *See question.*
9. $8\sqrt{2}$
10. 30
11. $8\sqrt{3}$
12. $7 < \sqrt{62} < 8$
13. $11 < \sqrt{129} < 12$
14. $9 < \sqrt{84} < 10$
15. *See question.*

38

Assessment Book, ALGEBRA 1: EXPLORATIONS AND APPLICATIONS

CHAPTER

6 Assessment

VOCABULARY REVIEW

1. Complete each sentence. In triangle ABC, $\angle C$ measures $\underline{?}$ degrees. Side $\overline{AB}$ is called the $\underline{?}$ and sides $\overline{AC}$ and $\overline{BC}$ are called $\underline{?}$. Describe what the Pythagorean theorem says about the side lengths of $\triangle ABC$.

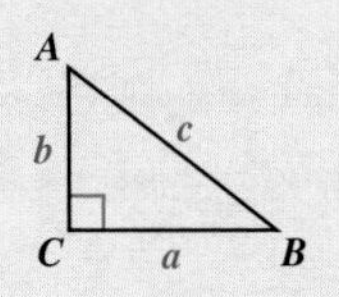

2. What is a *perfect square*? Give an example.
3. Define *rational number*. Give three examples of rational numbers between 1 and 3. Define *irrational number*. Give three examples of irrational numbers between 1 and 3.

SECTIONS 6.1, 6.2, *and* 6.3

For Questions 4–6, find the unknown length for each right triangle.

4. 22, 122, m
5. 20, p, 15
6.

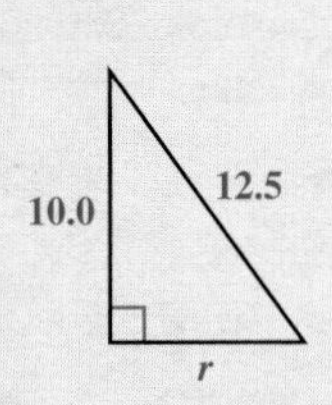

For Questions 7 and 8, tell whether the given lengths can be the sides of a right triangle.

7. 5 ft, 3.75 ft, 6.25 ft
8. 13 cm, 14 cm, 15 cm
9. **ART** Rosalie is making a frame for a drawing with the dimensions shown. To check whether the corners of her frame are right angles, she measures the lengths of both diagonals. If the corners are right angles, how long are the diagonals?

8.25 in.

11 in.

Write each decimal as a fraction in lowest terms.

10. 0.684
11. 0.777 . . .
12. $0.\overline{95}$

Write each expression in simplest form.

13. $\sqrt{48}$
14. $3 + \sqrt{200}$
15. $\sqrt{50} + \sqrt{18}$
16. $\sqrt{\frac{36}{100}}$
17. $2\sqrt{3} \cdot \sqrt{240}$
18. $\sqrt{\frac{192}{121}}$

ANSWERS Chapter 6

Assessment

1. 90; hypotenuse; legs; $a^2 + b^2 = c^2$, that is, the square of the length of the hypotenuse is equal to the sum of the squares of the lengths of the legs.
2. a number whose square root is an integer; examples: 100, 12^2, 0
3. any number that can be written as a ratio of two integers $\frac{a}{b}$, where $b \neq 0$; examples: 1.5, $\frac{7}{6}$, 2; any number that cannot be written as a ratio of two integers, and whose decimal form never terminates or repeats; examples: $\sqrt{7}$, 1.010010001..., $\sqrt{3}$
4. $m = 120$
5. $p = 25$
6. $r = 7.5$
7. Yes.
8. No.
9. 13.75 in.
10. $\frac{171}{250}$
11. $\frac{7}{9}$
12. $\frac{95}{99}$
13. $4\sqrt{3}$
14. $3 + 10\sqrt{2}$
15. $8\sqrt{2}$
16. $\frac{3}{5}$
17. $24\sqrt{5}$
18. $\frac{8\sqrt{3}}{11}$
19. $7 < \sqrt{54} < 8$
20. $10 < \sqrt{107} < 11$

Estimate each number within a range of two integers.

19. $\sqrt{54}$ **20.** $\sqrt{107}$ **21.** $\sqrt{72}$

22. Find the perimeter and area of the rectangle. Find the length of a diagonal of the rectangle. Write each dimension in simplest form.

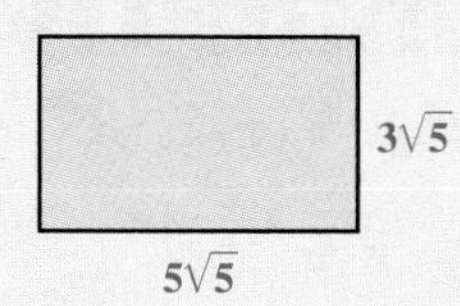

SECTIONS 6.4 *and* 6.5

Find each product.

23. $7(2x + 5)$ **24.** $3y(y - 9)$

25. $(2m - 7)(3m + 5)$ **26.** $\sqrt{3}(8 + \sqrt{3})$

27. $(5 + 2\sqrt{3})(5 - 2\sqrt{3})$ **28.** $(3 + 2\sqrt{20})(5 + 3\sqrt{45})$

29. a. Write a trinomial that represents the area of the rectangle.

$x - 5$

$2x - 6$

b. What is the smallest integer value of x that will give positive values for the length and the width of the rectangle?

Find each product.

30. $(n - 8)^2$ **31.** $(2a + 3)^2$ **32.** $(7 + 2\sqrt{3})^2$

33. $(c - 9)(c + 9)$ **34.** $\left(6x - \frac{1}{5}\right)\left(6x + \frac{1}{5}\right)$ **35.** $(5 + \sqrt{5})(5 - \sqrt{5})$

Show how to use special products to find each product.

36. $(5.5)(4.5)$ **37.** $(23)(37)$ **38.** $(28)^2$

39. Writing Explain the error in the product below. Show how to find the correct product.

$$(w - 7)^2 = w^2 + 14w + 49$$

PERFORMANCE TASK

40. Use the Pythagorean theorem to approximate the distance between two places that cannot be connected directly by a straight line. Explain how you approximated the distance. Include a sketch. An example is shown.

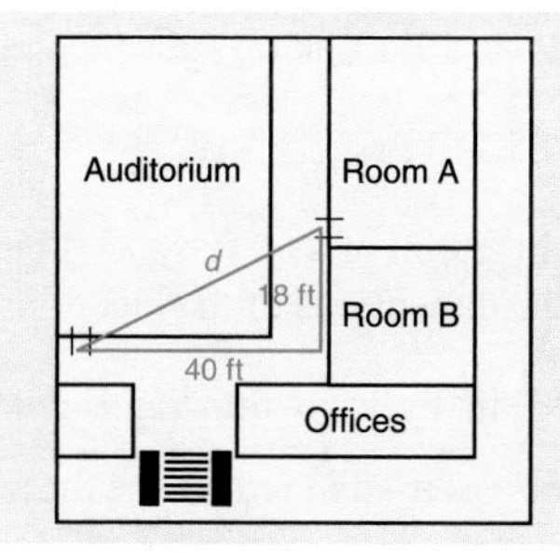

Chapter 6 Assessment
Form C Alternative Assessment

Chapter 6
ALTERNATIVE ASSESSMENT

1. a. Name two irrational numbers between 9 and 10.

b. Name two rational numbers between 9 and 10.

2. Open-ended Problem The following question appeared on a quiz. Tyler's work is shown to the right of the figure.

Find the length of $\overline{CD}$.

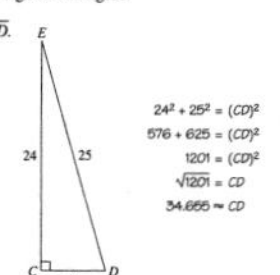

As you can see from his work, Tyler found the length of the missing side of this triangle to be $\sqrt{1201}$. He knows something is wrong because $\sqrt{1201} \approx 34.66$ and the length of $\overline{CD}$ appears to be much less than 35 units. Find the error in Tyler's work. Write a paragraph explaining his error. Think of a helpful hint to give Tyler so that he can remember to avoid this error next time.

3. a. Find several values of a and b such that $\sqrt{a^2 + b^2} = a + b$.

b. Find several values of a and b such that $\sqrt{a^2 + b^2} \neq a + b$.

c. Explain why the statement in part (a) is not always true.

4. Performance Task Describe several ways that you could verify the equality $\sqrt{18} + \sqrt{50} = 8\sqrt{2}$.

5. Research Project Look up the definitions for *scalene triangle*, *isosceles triangle*, and *equilateral triangle* in a dictionary. Write the definitions in your own words. Using graph paper, draw (if possible) a scalene *right* triangle, an isosceles *right* triangle, and an equilateral *right* triangle.

6. Open-ended Problem Describe the method you would use to put the following values in order.

$\sqrt{2} - \sqrt{3}$, $\sqrt{2} + \sqrt{3}$, $\sqrt{2} + 3$, $\sqrt{2} - 3$, $2 + \sqrt{3}$

7. Project A lifeguard must be able to quickly swim the longest distance in a swimming pool.

a. Visit a rectangular pool in or near your community. Describe the longest distance that a lifeguard might have to swim.

b. Measure the length and width of the swimming pool.

c. Calculate the diagonal length of the swimming pool.

d. Ask the lifeguard how many seconds it would take her or him to swim the longest distance in the swimming pool.

8. Performance Task Follow the directions below to verify the Pythagorean theorem.

1. Draw a right triangle.
2. Measure one side of the triangle.
3. On a sheet of heavy paper (like tagboard), draw a square that has a side length equal to the length you measured in Step 2.
4. Repeat Steps 2 and 3 for the other two sides of the triangle you drew.
5. Cut out your three squares and place the second largest square on the largest square. Now cut the smallest square into pieces that can be arranged so that the largest square is completely covered. (*Note:* There is more than one way to do this.)

Explain how this procedure verifies the Pythagorean theorem for your triangle. Does this example prove that the Pythagorean theorem is always true?

21. $8 < \sqrt{72} < 9$
22. $16\sqrt{5}$; 75; $\sqrt{170}$
23. $14x + 35$
24. $3y^2 - 27y$
25. $6m^2 - 11m - 35$
26. $8\sqrt{3} + 3$
27. 13
28. $195 + 47\sqrt{5}$
29. a. $(2x - 6)(x - 5) = 2x^2 - 16x + 30$
    b. 6
30. $n^2 - 16n + 64$
31. $4a^2 + 12a + 9$
32. $61 + 28\sqrt{3}$
33. $c^2 - 81$
34. $36x^2 - \frac{1}{25}$
35. 20
36. $24\frac{3}{4}$
37. 851
38. 784
39. $\left(5\frac{1}{2}\right)\left(4\frac{1}{2}\right) = \left(5 + \frac{1}{2}\right)\left(5 - \frac{1}{2}\right) = 5^2 - \left(\frac{1}{2}\right)^2 = 25 - \frac{1}{4} = 24\frac{3}{4}$; $(23)(37) = (30 - 7)(30 + 7) = 30^2 - 7^2 = 900 - 49 = 851$; $(28)^2 = (20 + 8)^2 = 20^2 + 2(20)(8) + 8^2 = 400 + 320 + 64 = 784$
40. Answers may vary. Check students' work.

Cumulative Assessment

CHAPTERS 4–6

CHAPTER 4

Solve. If an equation is an *identity* or there is *no solution*, say so.

1. $-\frac{2}{5}z = 8$

2. $\frac{2}{3}c - \frac{1}{2}c = 1$

3. $7x = 15 - 3x$

4. $1.3t - 0.7 = 2.7t$

5. $6(2u - 3) = -8 + 8u$

6. $3(n + 4) - n = 2(n + 5)$

Solve each inequality. Graph each solution on a number line.

7. $5 - x > 1$

8. $3m - 6 \geq 3$

9. $8p + 5 \geq 23p$

PERSONAL FINANCE One car rental company charges customers $225.00 per week with unlimited mileage to rent a mid-sized car. Another company charges $200.00 per week plus $.25 per mile. A person needs to rent a car for one week.

10. Model this situation with a table and with a graph.

11. Let m = the number of miles. For what number of miles is it less expensive to rent from the company with unlimited mileage? Write your answer as an inequality.

12. Write and solve an equation to find when the costs are equal.

13. **Writing** Do you find a table, a graph, or an equation most helpful when you compare car-renting options? Explain.

14. **Open-ended Problem** What factors might you consider besides cost when deciding which company to rent from?

CHAPTER 5

For Questions 15–17, solve each proportion.

15. $13 : 65 = 3 : s$

16. $\frac{7}{8.4} = \frac{r}{1.8}$

17. $\frac{15}{p} = \frac{20}{p + 3}$

18. The blueprint of a 60 ft by 75 ft building has a scale of 1 in. = 8 ft. What are the dimensions of the blueprint?

19. An $8\frac{1}{2}$ in. by 11 in. drawing is reduced to $6\frac{3}{8}$ in. by $8\frac{1}{4}$ in. What scale factor was used to reduce the drawing?

ANSWERS Chapters 4–6

Cumulative Assessment

1. –20 2. 6 3. 1.5

4. –0.5 5. 2.5 6. no solution

7. $x < 4$

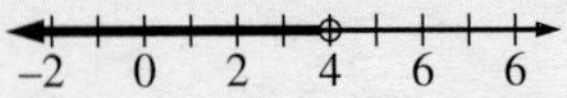

8. $m \geq 3$

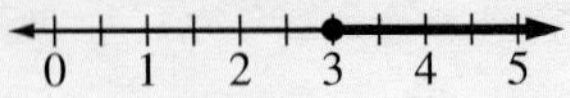

9. $p \leq \frac{1}{3}$

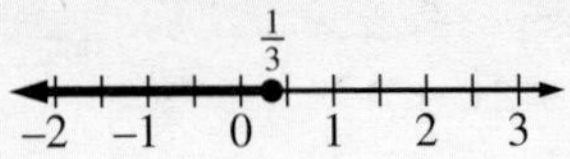

10.

miles	cost with unlimited mileage	cost with a mileage charge
0	225	200
20	225	205
40	225	210
60	225	215
80	225	220
100	225	225
120	225	230

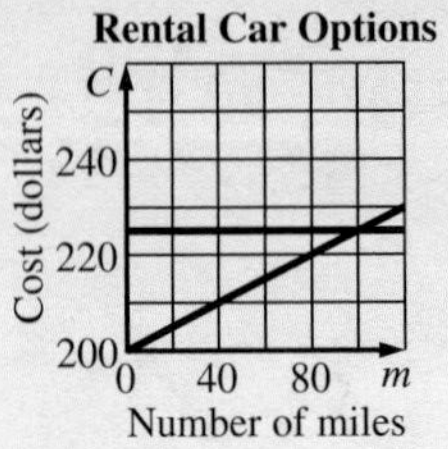

11. $m > 100$

12. $0.25m + 200 = 225$; 100 mi

13. Answers may vary.

14. Answers may vary. Examples are given. the types of cars available, pick-up and delivery services

15. 15 16. 1.5

17. 9 18. $7\frac{1}{2}$ in. by $9\frac{3}{8}$ in.

19. $\frac{3}{4}$ 20. 12

21. 22.5 22. $\angle X$

23. 4:5 24. 16:25

For Questions 20–24, use the fact that $\triangle ABC \sim \triangle XYZ$. For Questions 20–22, complete each statement.

20. $\overline{AC} = \underline{?}$ **21.** $\overline{YZ} = \underline{?}$ **22.** $\angle A = \underline{?}$

23. Find the ratio of the perimeter of $\triangle ABC$ to the perimeter of $\triangle XYZ$.

24. Find the ratio of the area of $\triangle ABC$ to the area of $\triangle XYZ$.

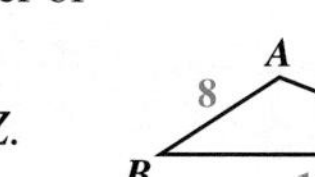

25. Writing Is every rectangle similar to every other rectangle? Explain.

26. What is the probability that someone born in September was born on or before September 12?

27. Open-ended Problem Give an example of experimental probability.

28. A dart lands on a square of a checkerboard at random. What is the probability that the dart lands on one of the four corner squares?

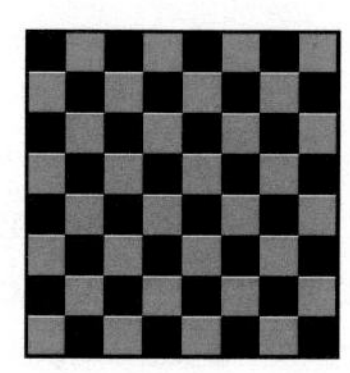

CHAPTER 6

Tell whether the given lengths can be sides of a right triangle. Explain your reasoning.

29. 15, 36, 39 **30.** 160, 340, 300 **31.** 8.2, 9.1, 11.9

32. Open-ended Problem Use a right triangle with lengths 3, 4, and 5.

a. Multiply the length of each side of the triangle by 3. Is the new shape a right triangle?

b. Add 3 to the length of each side. Is the new shape a right triangle?

c. Choose a different right triangle. Repeat parts (a) and (b). Compare your answers for both triangles. Do you think your results would be the same for any right triangle you chose?

Write each repeating decimal as a fraction.

33. 0.030303. . . **34.** $17.\overline{04}$ **35.** $0.\overline{961}$

Simplify each expression.

36. $\sqrt{63}$ **37.** $7\sqrt{2} + 12\sqrt{2}$ **38.** $\sqrt{\frac{144}{9}}$

39. $4\sqrt{5} \cdot \sqrt{15}$ **40.** $\sqrt{\frac{6}{49}}$ **41.** $10\sqrt{7} - \sqrt{28}$

Find each product.

42. $3q(2q - 5)$ **43.** $(5\sqrt{3} - 1)(6\sqrt{3} + 4)$

44. $(5 - 2\sqrt{3})^2$ **45.** $(6x + \sqrt{5})(6x - \sqrt{5})$

46. Writing Explain the error in the product below. Show how to find the correct product.

$$(m + 4)(m - 4) = m^2 + 16$$

25. No. All the corresponding angles are equal in measure, but corresponding sides are not necessarily proportional, unless the rectangle is a square.
26. $\frac{2}{5}$
27. Answers may vary. An example is given. throwing darts onto a circular dart board with different colored concentric circles and using the results to determine the probability of hitting a specific region of the board
28. $\frac{1}{16}$
29. Yes; $15^2 + 36^2 = 39^2$.
30. Yes; $160^2 + 300^2 = 340^2$.
31. No; $8.2^2 + 9.1^2 \neq 11.9^2$.
32. a. Yes.
 b. No.
 c. Answers may vary. An example is given. 5, 12, 13; The results are the same. The results will be the same for any right triangle.
33. $\frac{1}{33}$
34. $\frac{1687}{99}$
35. $\frac{961}{999}$
36. $3\sqrt{7}$
37. $19\sqrt{2}$
38. 4
39. $20\sqrt{3}$
40. $\frac{\sqrt{6}}{7}$
41. $8\sqrt{7}$
42. $6q^2 - 15q$
43. $86 + 14\sqrt{3}$
44. $37 - 20\sqrt{3}$
45. $36x^2 - 5$
46. The sign of the constant term is wrong. Use the special products rule, $(a + b)(a - b) = a^2 - b^2$: $(m + 4)(m - 4) = m^2 - (4)^2 = m^2 - 16$.

CHAPTER

7 Systems of Equations and Inequalities

OVERVIEW

Connecting to Prior and Future Learning

⇔ Students continue their work from Chapter 3 as they write and graph linear equations in standard form. This concept is then expanded as students use their graphing skills to solve systems of linear equations. Students will continue to encounter these concepts during their future studies in mathematics.

⇔ Algebraic solutions for systems of linear equations are presented in Sections 7.3 and 7.4. Students add or subtract a system of equations in Section 7.3 in order solve the system. Using multiplication before adding or subtracting is discussed in Section 7.4.

⇔ Chapter 7 concludes with two sections that deal with graphing linear inequalities and systems of inequalities. Students use the skills they acquired in the first two sections of this chapter to complete these last two sections. As with all of the skills in this chapter, these skills are ones that students will use many times while studying mathematics.

Chapter Highlights

Interview with Nancy Clark: Sports and mathematics come together in this interview with Nancy Clark, M.S., R.D., a sports nutritionist. Examples and exercises relating to this interview can be found on pages 281, 285, 309, and 314.

Explorations can be found in Section 7.4, where students use graph paper to graph and solve linear systems of equations, and in Section 7.5, where students use a ruler and colored pencils to graph linear inequalities.

The Portfolio Project: Codes that increase the safety of stairs are used by pairs of students to design a stairway. Students present their results in a report that includes graphs, tables, and scales.

Technology: Graphing calculators are used throughout this chapter to graph linear equations, and systems of equations and inequalities.

OBJECTIVES

Section	Objectives	NCTM Standards
7.1	• Write and graph equations in standard form. • Solve problems with two variables.	1, 2, 3, 4, 5, 8
7.2	• Solve systems of equations. • Solve problems with two variables.	1, 2, 3, 4, 5, 8
7.3	• Solve systems of equations by adding or subtracting. • Solve problems with two variables.	1, 2, 3, 4, 5
7.4	• Write equivalent systems of equations. • Solve equivalent systems of equations that model real-world situations.	1, 2, 3, 4, 5
7.5	• Graph linear inequalities. • Visualize problems involving inequalities.	1, 2, 3, 4, 5, 8
7.6	• Graph systems of inequalities. • Graph systems of inequalities to model real-world situations when there are restrictions.	1, 2, 3, 4, 5, 8

INTEGRATION

Mathematical Connections	7.1	7.2	7.3	7.4	7.5	7.6
algebra	**281–286***	**287–293**	**294–298**	**299–305**	**306–310**	**311–315**
geometry	**281–286**	288–293	297	305	**306–310**	311–314
logic and reasoning	**281–286**	287, 291–293	294, 295, 298	299, 302–305	**306–310**	312–315
Interdisciplinary Connections and Applications						
history and geography				303		
reading and language arts				304		
business and economics		292				
sports and recreation	284					
fitness, finance, space science, cooking, aviation, crafts, nutrition	286	291	296–298	301		311, 312

* ***Bold page numbers*** *indicate that a topic is used throughout the section.*

TECHNOLOGY

Section	Student Book (opportunities for use with)	Support Material
7.1	graphing calculator McDougal Littell Software *Function Investigator*	**Technology Book:** Spreadsheet Activity 7
7.2	graphing calculator McDougal Littell Software *Function Investigator*	**Technology Book:** Spreadsheet Activity 7
7.3	graphing calculator McDougal Littell Software *Function Investigator*	**Technology Book:** Calculator Activity 7
7.4	graphing calculator	
7.5	scientific calculator	
7.6	graphing calculator	

Planning Guide

Regular Scheduling (45 min)

Section	Materials Needed	Core Assignment	Extended Assignment	*exercises that feature* Applications	Communication	Technology
7.1	graph paper, graphing calculator	**Day 1:** 1–22 **Day 2:** 23–50	**Day 1:** 1–22 **Day 2:** 23–50	34, 35, 38	33, 36, 37, 39, 40	20–22 23–28, 41–44
7.2	graph paper, graphing calculator	1–14, 16–19, 22, 23, AYP*	1–7, 11–23, AYP	15, 17–19	21, 22	1–4, 16
7.3	graph paper, graphing calculator	1–22, 24, 26–33	1–33	11, 21–24	1, 20, 25, 26	18, 19
7.4	graph paper, graphing calculator	**Day 1:** 1–16 **Day 2:** 20–29, AYP	**Day 1:** 1–16 **Day 2:** 17–29, AYP	7–10	17, 19, 22, 23	
7.5	ruler, colored pencils, graph paper	**Day 1:** 1–22 **Day 2:** 23–34, 36–45	**Day 1:** 1–22 **Day 2:** 23–45	29–32	17–19 34, 36	
7.6	graph paper	1–15, 21, 23–33, AYP	1, 2, 5, 6, 8–33, AYP	15, 16, 18	17, 24	20
Review/ Assess		**Day 1:** 1–14 **Day 2:** 15–29 **Day 3:** Ch. 7 Test	**Day 1:** 1–14 **Day 2:** 15–29 **Day 3:** Ch. 7 Test		1–3	6, 7
Portfolio Project		Allow 2 days.	Allow 2 days.			

Yearly Pacing (with Portfolio Project)	**Chapter 7 Total** 14 days	**Chapters 1–7 Total** 94 days	**Remaining** 66 days	**Total** 160 days

Block Scheduling (90 min)

	Day 41	Day 42	Day 43	Day 44	Day 45	Day 46	Day 47	Day 48
Teach/Interact	Ch. 6 Test 7.1	Continue with 7.1 7.2	7.3 7.4: Exploration, page 299	Continue with 7.4 7.5: Exploration, page 306	Continue with 7.5 7.6	Review Port. Proj.	Review Port. Proj.	Ch. 7 Test 8.1: Exploration, page 325
Apply/Assess	**Ch. 6 Test** **7.1:** 1–22	**7.1:** 23–50 **7.2:** 1–14, 16–20, 22, 23, AYP*	**7.3:** 1–24, 26–33 **7.4:** 1–16	**7.4:** 17–29, AYP **7.5:** 1–22	**7.5:** 23–34, 36–45 **7.6:** 1–15, 20–33, AYP	**Review:** 1–14 **Port. Proj.**	**Review:** 15–29 **Port. Proj.**	**Ch. 7 Test** **8.1:** 1–13, 15–18

NOTE: A one-day block has been added for the Portfolio Project—timing and placement to be determined by teacher.

Yearly Pacing (with Portfolio Project)	**Chapter 7 Total** 7 days	**Chapters 1–7 Total** $47\frac{1}{2}$ days	**Remaining** $33\frac{1}{2}$ days	**Total** 81 days

****AYP*** *is Assess Your Progress.*

Lesson Support

Section	Practice Bank	Study Guide*	Assessment Book*	Visuals	Explorations Lab Manual	Lesson Plans	Technology Book
7.1	44	7.1		Warm-Up 7.1 Folder A	Masters 1, 2 Add. Expl. 10	7.1	Spreadsheet Act. 7
7.2	45	7.2	Test 29	Warm-Up 7.2 Folders 7, A	Master 1	7.2	Spreadsheet Act. 7
7.3	46	7.3		Warm-Up 7.3	Master 2	7.3	Calculator Act. 7
7.4	47	7.4	Test 30	Warm-Up 7.4 Folder A	Masters 1, 2	7.4	
7.5	48	7.5		Warm-Up 7.5 Folder A	Masters 2, 12 Add. Expl. 11	7.5	
7.6	49	7.6	Test 31	Warm-Up 7.6 Folder A	Masters 1, 2	7.6	
Review Test	50	Chapter Review	Tests 32, 33 Alternative Assessment			Review Test	Calculator Based Lab 4

***Spanish versions** of *Study Guide* and *Assessment Book* are available.

Chapter Support

- Course Guide
- Lesson Plans
- Portfolio Project Book
- Preparing for College Entrance Tests
- Multi-Language Glossary
- *Test Generator* Software
- Professional Handbook

Software Support

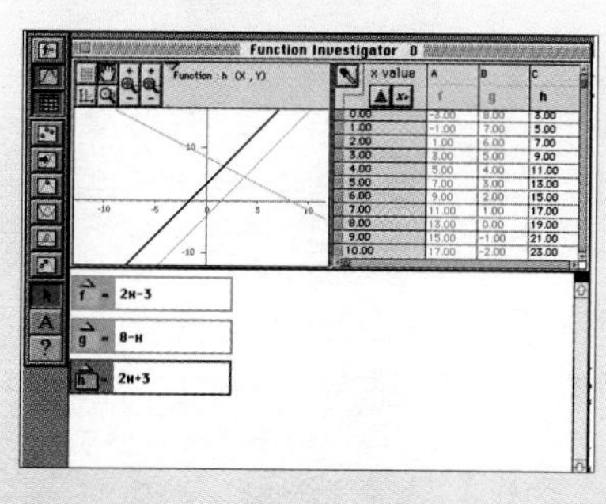

McDougal Littell Software
Function Investigator

Internet Support

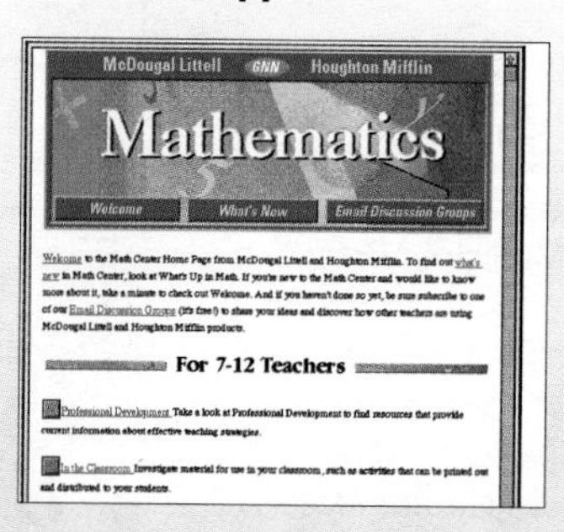

http://www.hmco.com
Next go to McDougal Littell; then the Education Center; then Secondary Math.

Outside Resources

Books, Periodicals

McNally, Ann. "The Point-Slope Equation Revisited." *Mathematics Teacher* (November 1990): pp. 639, 640.

Eisner, Gail A., CPA. "Using Algebra in an Accounting Practice." *Mathematics Teacher* (April 1994): pp. 272–276.

Activities, Manipulatives

Baker, Patricia Cooper. "Supply and Demand—An Application of Linear Equations." *Mathematics Teacher* (October 1991): pp. 554–556.

Wood, Eric. "Gas-Bill Mathematics." *Mathematics Teacher* (March 1995): pp. 214–217.

Software

Tools of Mathematics: Supplementary Topics from MacNumerics II. Includes curve fitting, linear programming, linear systems, with a solve option. Acton, MA: Bradford Publishing Co., 1993.

Videos

Desktop Realities: Mathematical Modeling to Solve Real-World Problems, Gr. 9–12. Videocassette plus teacher's resource book. Includes systems of inequalities and linear programming. Pleasantville, NY: HRM Video.

Internet

Subscribe to a newsgroup dedicated to discussions, questions, and answers regarding Texas Instruments graphics calculators by sending a message to majordomo@lists.ppp.ti.com
Include no subject; in the body of the message, type
subscribe graph-ti [your e-mail address]

Interview Notes

Background

Sports Nutrition and Training
Recent research findings about sports nutrition have brought forth a number of interesting facts concerning the fitness of athletes. One study found that many runners do not eat enough carbohydrates, averaging 52% instead of the recommended 60–70%. Another study suggested that fluids are important to all athletes, including swimmers who showed signs of dehydration during swimming despite the cooling effect of the water. Researchers studying the performance of cyclists looked at the effects of tapering off exercise prior to a cycling event. They found that after a six-week, high-intensity training program, a one- to two-week taper resulted in a 8–9% strength gain and enhanced performance in collegiate cyclists.

Nancy Clark
Nancy Clark is an expert in sports nutrition and is a nutrition counselor at Sports Medicine Brookline. She is also a fellow of the American College of Sports Medicine. Clark has written many journal articles for *American Fitness* and is the author of *Nancy Clark's Sports Nutrition Guidebook, The Athlete's Kitchen,* and *The New York City Marathon Cookbook.* She has also made an audiocassette on dieting tips called *How to Lose Weight and Have Energy to Exercise.*

CHAPTER

7 Systems of Equations and Inequalities

Feeding a need for athletes

INTERVIEW **Nancy Clark**

Just about everyone knows that being a top athlete takes talent and hard work. But fewer people appreciate the importance of diet in helping athletes perform at their best. Sports nutritionist Nancy Clark, M.S., R.D., is trying hard to change this situation. "Many people train hard but neglect their training diet; this is a big mistake," Clark says. "They're shooting themselves in the foot." Clark meets with athletes of all ages, sizes, and shapes — from high school and college students to Olympic athletes and members of the Boston Celtics and Red Sox. Though each of her clients has different needs, Clark is guided by one general principle: finding the right balance of training, nutrition, and rest.

"You can't compete at your best if you don't train at your best."

278

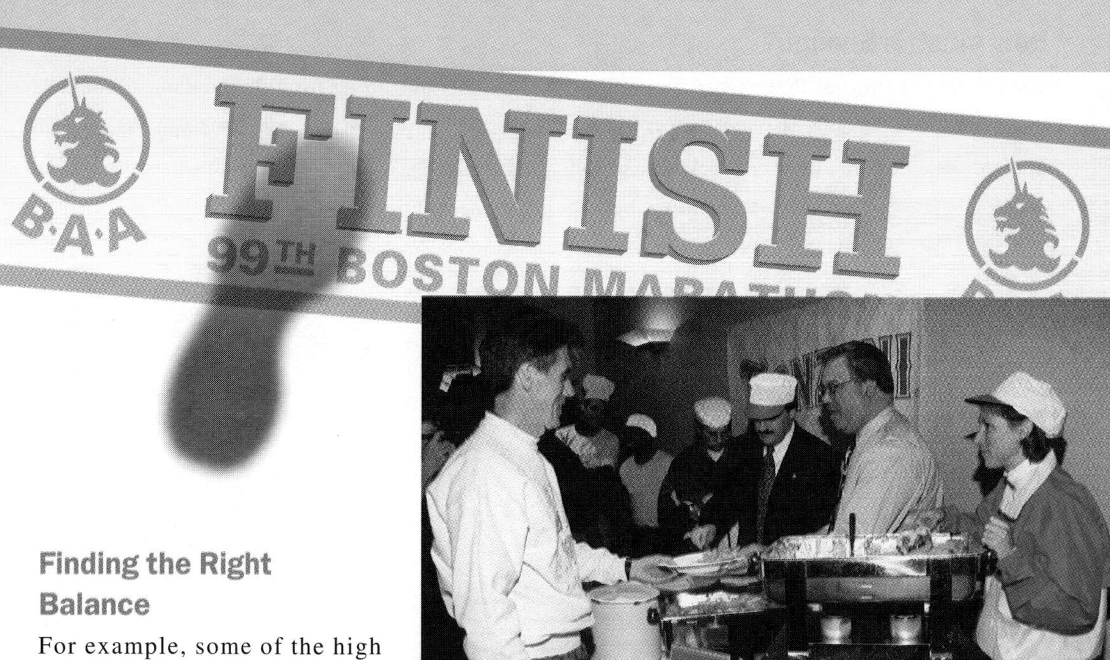

Nancy Clark serving food at the annual pasta fest on the eve of the Boston Marathon.

Finding the Right Balance

For example, some of the high school students Clark works with tend to skip both breakfast and lunch. When they show up for after-school sports, they are running on empty. "They come to me looking for a quick energy fix, but no amount of fix can make up for a poor breakfast and lunch," she says. "Together we figure out some smart eating choices—a combination of foods that they like to eat which will provide them with energy they need throughout the day."

Fueling Up

"Food is really just fuel, not so different from the gasoline you put in a car," she says. "You need to fill up before strenuous exercise and refuel afterwards." Food provides energy from three sources. Proteins build muscles but get used for energy in emergency situations, carbohydrates fuel muscles, and fats contain energy reserves. Of the three, carbohydrates are the most critical for athletes because they provide the body with glycogen, a form of sugar that is stored within muscles as a source of energy. Foods that are high in carbohydrates include bread, pasta, potatoes, and rice. Athletes should load up on carbohydrates while preparing for and recovering from demanding events such as triathlons, marathons, or long-distance ski races. Between 60 and 70 percent of the calories they consume daily should come from carbohydrates, according to Clark. She stresses that these carbohydrate guidelines make sense for everybody, not just athletes. "It's what we call a healthy diet," she says.

"You can't train at your best if you don't eat right on a daily basis."

Interview Notes

Background

Distance Runners

One of Japan's top women marathoners in the 1980s was Chie Matsuda. During training, Matsuda ate three meals a day containing vegetables, a meat or fish protein, rice, and bread. For a high-energy snack, she ate sushi. New Zealand's Rod Dixon was one of the world's top distance runners throughout the 1980s and the winner of the New York Marathon in 1983. Aside from an occasional pizza or hamburger, his diet consisted of a fairly traditional fare including fresh vegetables and fiber.

Second-Language Learners

Some students learning English may be unfamiliar with the expressions *running on empty* and *load up on carbohydrates.* If necessary, explain that these mean "to run out of energy because of lack of food" and "to eat foods with plenty of carbohydrates," respectively.

Interview Notes

Mathematical Connection

In Section 7.1, students write and graph equations that show how various combinations of food can provide a given amount of carbohydrates. Possible solutions can then be selected that best meet an individual's needs. In Section 7.5, inequalities are used to model and visualize nutritional situations that involve determining the least or most amounts of certain foods that can be eaten to achieve particular goals involving calcium and fat consumption. In Section 7.6, systems of inequalities are graphed to find an individual's ideal carbohydrate intake.

Explore and Connect

Writing
When discussing question 1, point out that the solutions to a system of inequalities lie in a region of the coordinate plane. Every point within the region is a solution to each inequality and thus is a solution to the system.

Research
Students can read about the meaning of a calorie in a high school physics book. As students understand that a calorie is a unit of heat and that heat is a form of energy, the connection to food calories should become clearer.

Project
Students can work in groups to list foods high in proteins and fats and to create their posters. A discussion of the posters made by each group will help all students understand how carbohydrates, fats, and proteins help athletes perform at their best.

How Much is Enough?

The number of calories that a particular athlete should consume depends on many factors including age, height, and activity level. The graph shown below can help an athlete figure out how many calories from carbohydrates to include in a daily diet. Once the athlete's daily calorie needs are known, the number of calories from carbohydrates should be within the shaded region of the graph. As you will learn in this chapter, the graph represents a system of inequalities.

Nutrition labeling can help create daily meals with appropriate amounts of proteins, carbohydrates, and fats.

foods
proteins
carbohydrates
fats
build muscles
provides fuel
bread
pasta
potatoes
rice

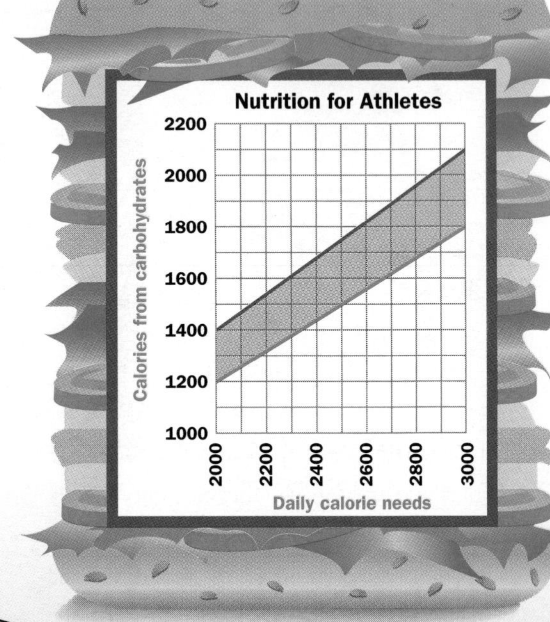

EXPLORE AND CONNECT

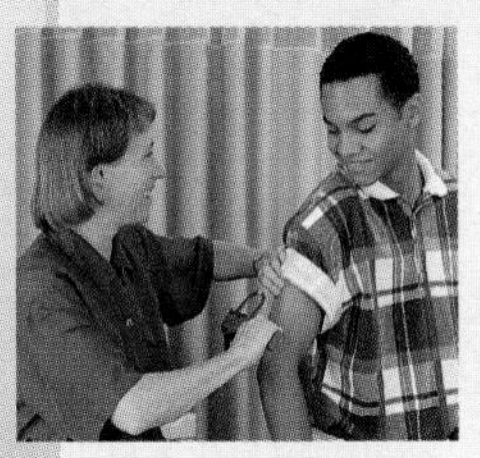

Nancy Clark gathering data, to better advise on nutrition issues.

1. Writing According to the graph, if an athlete needs to consume 2200 Cal each day, about how many should come from carbohydrates? If another athlete needs to consume 1700 Cal from carbohydrates each day, about how many total calories does that athlete need?

2. Research In physics, a calorie is defined as a unit of heat. What is the relationship between calories used in physics and food calories? Find out how the numbers of calories in foods are determined.

3. Project The other 30–40% of an athlete's daily calorie intake should come from proteins and fats. What kinds of foods are high in proteins and fats? Create a poster showing how carbohydrates, fats, and proteins help the body to function.

Mathematics & Nancy Clark

In this chapter, you will learn more about how mathematics is related to sports nutrition.

Related Examples and Exercises

Section 7.1
- **Example 1**
- **Exercises 34–36, 38**

Section 7.5
- **Exercises 29–31**

Section 7.6
- **Exercises 15–18**

SECTION

7.1 Using Linear Equations in Standard Form

Learn how to...

- write and graph equations in standard form

So you can...

- solve problems with two variables, such as nutritional planning or telecommunications problems

Nancy Clark recommends that runners replace their muscles' energy stores after a marathon. A runner should eat at least 0.5 g of foods containing carbohydrates for each pound of body weight within 2 h after the race. The athlete should eat at least this amount again 2 h later.

EXAMPLE 1 Interview: Nancy Clark

Leon should eat at least 70 g of carbohydrates after running a marathon. One cup of apple juice contains **30 g** of carbohydrates, and 1 oz of pretzels contains **23 g**. Write and graph an equation showing combinations of pretzels and juice that provide 70 g of carbohydrates.

SOLUTION

Step 1 Write an equation.

Let j = number of cups of juice. Let p = number of ounces of pretzels.

Carbohydrates from juice + Carbohydrates from pretzels = 70

$$30j + 23p = 70$$

Step 2 Graph the equation.

First make a table of values.

j	p
0	3.0
1.0	1.7
2.3	0

Then graph each point and draw a line through the points.

Combinations of Pretzels and Juice Containing 70 g of Carbohydrates

Pretzels (oz)

Juice (cups)

BY THE WAY

After the Boston Marathon, runners eat snacks such as pretzels and bread supplied by their sponsors. Ten small pretzels weigh about one ounce.

THINK AND COMMUNICATE

1. Why does the graph in Example 1 only make sense in the first quadrant?
2. How would the graph in Example 1 be different if you put cups of juice on the vertical axis?

Plan⇔Support

Objectives

- Write and graph equations in standard form.
- Solve problems with two variables.

Recommended Pacing

Core and Extended Courses
Section 7.1: 2 days

Block Schedule
Section 7.1: 2 half-blocks (with Chapter 6 Test and Section 7.2)

Resource Materials

Lesson Support

Lesson Plan 7.1

Warm-Up Transparency 7.1

Overhead Visuals:
Folder A: Multi-Use Graphing Packet, Sheets 1–3

Practice Bank: Practice 44

Study Guide: Section 7.1

Explorations Lab Manual:
Additional Exploration 10, Diagram Masters 1, 2

Technology

Technology Book:
Spreadsheet Activity 7

Graphing Calculator

McDougal Littell Software
Function Investigator

Internet:
http://www.hmco.com

Warm-Up Exercises

Evaluate each expression when $x = 15$ and $y = 8$.

1. $10x + 14y$ 262
2. $3x - 7y$ −11

Tell whether each equation is true or false when $x = -2$ and $y = 6$.

3. $5x + 2y = -2$ False.
4. $-4x - y = 2$ True.
5. $3x = -4y$ False.
6. $10x + 20y = 100$ True.

ANSWERS Section 7.1

Think and Communicate

1. because both the x-coordinates and y-coordinates represent physical quantities and must be positive
2. The graph would look similar. It would be the first-quadrant part of a line, but the horizontal and vertical intercepts would be reversed from those shown in Example 1.

Teach⇔Interact

Additional Example 1

Before an afternoon workout, Leon plans a lunch of pasta and skim milk that will provide about 200 g of carbohydrates. One cup of dry pasta has 41 g of carbohydrates and 1 cup of skim milk has 12 g. Write and graph an equation showing combinations of pasta and skim milk that provide 200 g of carbohydrates.

Step 1 Write an equation.
Let p = number of cups of pasta.
Let m = number of cups of milk.
Carbohydrates from pasta + Carbohydrates from milk = 200
$41p + 12m = 200$

Step 2 Graph the equation.
First make a table of values.

p	m
0	16.7
4	3
4.9	0

Then graph each point and draw a line through the points.

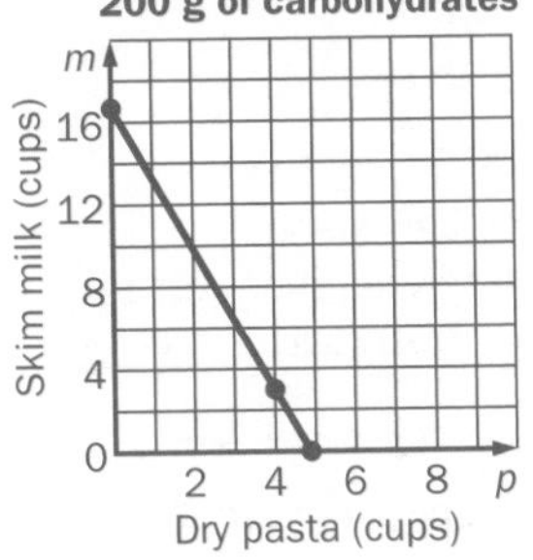

Additional Example 2

Graph the equation $30x + 18y = 450$.

Step 1 The graph is a line. Use the equation to find two points on the line. Find y when $x = 0$.

$30(0) + 18y = 450$
$18y = 450$
$y = 25$

One point on the line is (0, 25).
Find x when $y = 0$.

$30x + 18(0) = 450$
$30x = 450$
$x = 15$

Another point on the line is (15, 0).

Step 2 Graph the two points and draw the line through them.

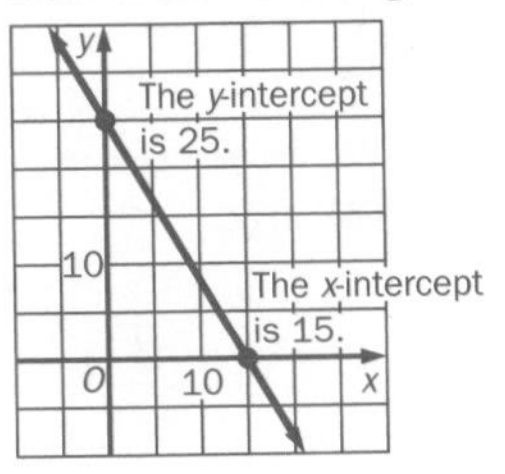

As Example 1 shows, the graph of $30j + 23p = 70$ is a line. All linear equations can be written in the form $ax + by = c$, where a and b are both real numbers.

Linear equations written like this are in **standard form**, where a, b, and c are integers.

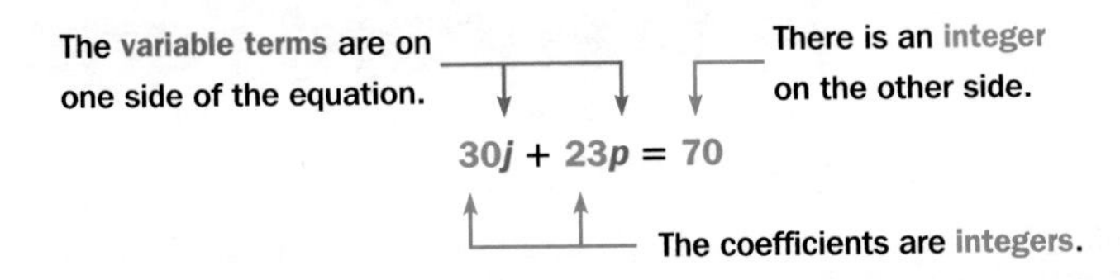

EXAMPLE 2

Graph the equation $2x + 5y = 1720$.

SOLUTION

Step 1 The graph is a line. Use the equation to find two points on the line. Choose two points that are easy to find.

Find y when $x = 0$.

$2(0) + 5y = 1720$
$5y = 1720$
$y = 344$

One point on the line is (0, 344).

Find x when $y = 0$.

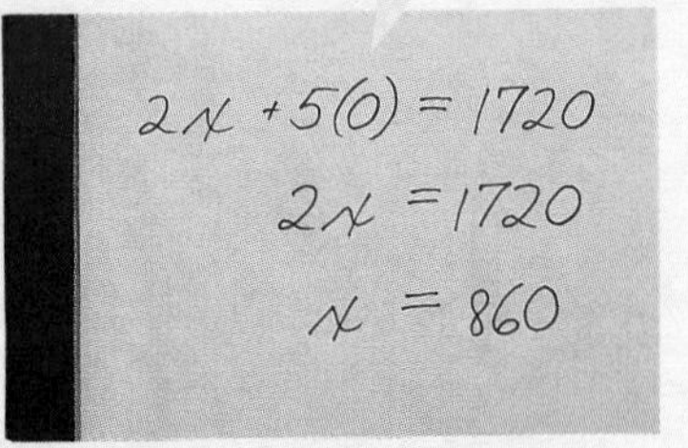

Another point on the line is (860, 0).

Step 2 Graph the two points and draw the line through them.

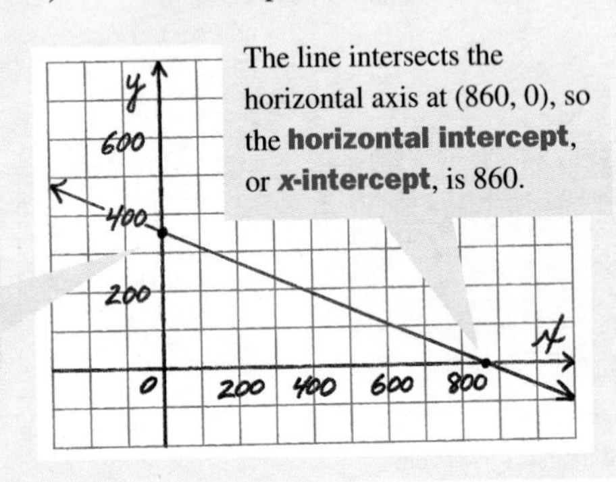

The line intersects the horizontal axis at (860, 0), so the **horizontal intercept**, or **x-intercept**, is 860.

The **vertical intercept**, or **y-intercept**, is 344.

THINK AND COMMUNICATE

3. Explain how to find two points to graph when graphing an equation in standard form.

Think and Communicate

3. Find the value of each variable when the value of the other variable is zero. Substituting 0 for a variable produces an equation that can be solved for the other variable in one step.
4. Answers may vary. Examples are given. One advantage is that writing the equation in slope-intercept form allows you to graph it on a graphing calculator. One disadvantage is that the coefficients of the variable terms may not be integers.
5. Solve for y by subtracting x from both sides and then dividing both sides by 2; $y = -\frac{1}{2}x + 2$.

If you are given a linear equation in standard form, you can rewrite it in slope-intercept form. Many graphing calculators cannot graph equations in standard form. Rewriting the equation can help you to graph the equation on a graphing calculator.

EXAMPLE 3

Graph the equation $-5x + 2y = 4$ on a graphing calculator.

SOLUTION

First rewrite the equation in slope-intercept form.

Solve for y.

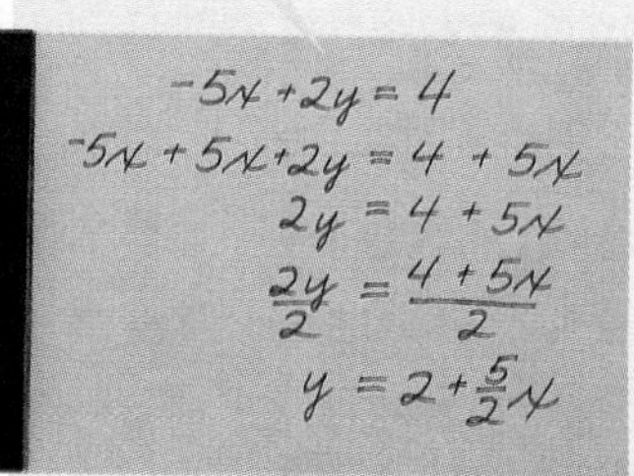

Now you can graph the equation on your calculator.

$y = 2 + \frac{5}{2}x$

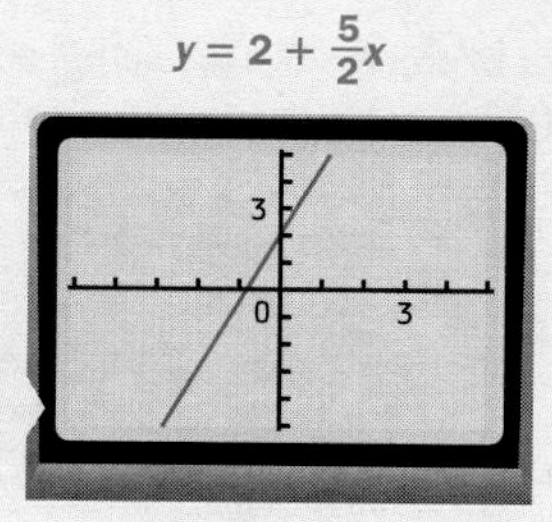

THINK AND COMMUNICATE

4. Example 3 shows an equation written in standard form and slope-intercept form. Describe an advantage and a disadvantage of writing an equation in slope-intercept form.
5. Explain how to rewrite $x + 2y = 4$ in slope-intercept form.

CHECKING KEY CONCEPTS

1. Adult tickets to a museum cost $8 each and tickets for children cost $5 each.
   a. What is the cost of tickets for a adults and c children?
   b. Write an equation showing combinations of adults' tickets and children's tickets that cost $40.

Tell whether each equation is in standard form. If it is not, explain why not.

2. $3x + 5y = 30$
3. $y = 2x - 5$
4. $8x - 9y = 36$
5. Find the x-intercept and the y-intercept of the graph of $3x + 5y = 30$.
6. Explain how to rewrite $3x + 5y = 30$ in slope-intercept form.
7. Describe two ways to graph the equation $3x + 5y = 30$.

Checking Key Concepts

1. a. $8a + 5c$
   b. $8a + 5c = 40$
2. Yes.
3. No; the variable terms are not on one side.
4. Yes.
5. 10; 6
6. Solve for y by subtracting $3x$ from both sides and then dividing both sides by 5; $y = -\frac{3}{5}x + 10$.
7. Answers may vary. Examples are given. One method is to graph the x- and y-intercepts and then draw the line through the points. Another method is to rewrite the equation in slope-intercept form and enter this equation into a graphing calculator.

Teach⇔Interact

Section Note

Teaching Tip

You may wish to point out what happens to the graph of $ax + by = c$ when $b = 0$. Use a specific example such as $2x + 0 \cdot y = 10$. The equation simplifies to $x = 5$. Students should understand that this equation cannot be written in slope-intercept form. However, it is easy to draw its graph. The graph is the vertical line 5 units to the right of the y-axis.

Additional Example 3

Graph the equation $7x - 3y = 21$ on a graphing calculator.

First rewrite the equation in slope-intercept form. Solve for y.

$-7x - 3y = 21$

$-3y = 7x + 21$

$y = -\frac{7}{3}x - 7$

Graph the equation on a calculator.

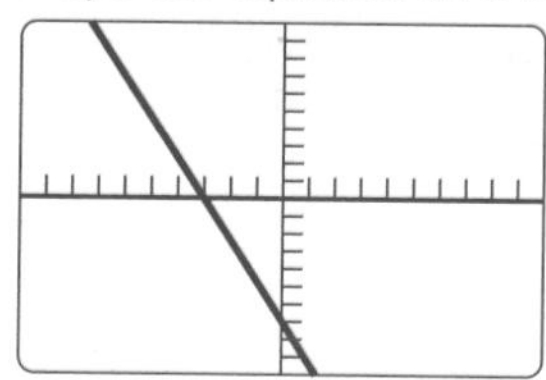

Think and Communicate

In discussing question 4, students can use specific examples to help explain the advantages and disadvantages of writing an equation in slope-intercept form.

Checking Key Concepts

Cooperative Learning

Students can work on these questions in pairs. Call on different pairs of partners to explain their answers to individual questions.

Closure Question

What is the standard form of a linear equation? Explain how you would use an equation in this form to graph the equation.

$ax + by = c$; Choose two points that are easy to find on the line, such as $(0, y)$ and $(x, 0)$, graph the points and draw a line through them or use a graphing calculator.

Apply⇔Assess

Suggested Assignment

Core Course
Day 1 Exs. 1–22
Day 2 Exs. 23–50

Extended Course
Day 1 Exs. 1–22
Day 2 Exs. 23–50

Block Schedule
Day 41 Exs. 1–22
Day 42 Exs. 23–50

Exercise Notes

Problem Solving
Exs. 1–5, 6–10 In each of these two groups of exercises, students use a series of particular instances to arrive at a general pattern. Some students may be able to answer Exs. 5 and 10 without going through this process, but many students will find the particular instances helpful.

Mathematical Procedures
Exs. 11–19 Students should note that two of the points used in the table for Example 1 are the *x*- and *y*-intercepts. These two points are sufficient for drawing the graph, but having a third point provides a useful check on the correctness of the graph.

Using Technology
Exs. 20–28 The exercises can also be done using the *Function Investigator* software. Some students may observe that it is possible to graph many linear equations without using slope-intercept form. For example, the equation $27x - 43y = 72$ could be graphed by entering the equation $y = \frac{(72 - 27x)}{-43}$.

7.1 Exercises and Applications

Extra Practice exercises on page 570

SPORTS Exercises 1–5 refer to a soccer league in which a team gets 2 points for each win and 1 point for each tie game. How many points does a team with each record have?

1. 6 wins and 3 ties
2. 8 wins and 2 ties
3. no wins and 5 ties
4. w wins and t ties
5. a. Write an equation showing combinations of wins and ties that result in a total of 13 points.
b. Graph your equation. What do the intercepts of your graph represent?

For Exercises 6–10, use the advertisement. How much does each combination of compact disks and tapes cost?

6. 3 compact disks and 2 tapes
7. 4 compact disks and 7 tapes
8. 5 compact disks and no tapes
9. c compact disks and t tapes
10. a. Write an equation that shows how many compact disks and tapes you can buy with exactly \$60.
b. Graph your equation. Should the graph be a line? Explain.

Graph each equation by using the method in Example 1 on page 281.

11. $3x + 2y = 24$
12. $4x + 3y = 18$
13. $8x - 3y = 12$
14. $-5x + 2y = 20$
15. $2x - 4y = -5$
16. $-3s + 2t = -1$
17. $1.5a - 0.5b = 3.0$
18. $\frac{x}{2} - \frac{y}{3} = 1$
19. $3x + 4y = 1.2$

Technology Rewrite each equation in slope-intercept form. Then graph it. Use a graphing calculator if you have one.

20. $2x + y = 8$
21. $3x + 2y = 10$
22. $2x - 7y = 21$
23. $-4x + 5y = 12$
24. $2x + 2y = 7$
25. $8 - 3y = 4x$
26. $50 = 5x + 4y$
27. $0.3x + 0.2y = 1.2$
28. $\frac{2}{3}x - \frac{1}{2}y = 2$

29. Which equations in Exercises 11–28 are not in standard form? Explain your answers.

30. SAT/ACT Preview A line crosses the axes at (15, 0) and (0, 12). Which of the equations below could be an equation of the line?

A. $y = 15 + x$ **B.** $y = 12 + x$ **C.** $x + y = 27$
D. $y = 12 - \frac{4}{5}x$ **E.** $y = \frac{5}{4}x$

Exercises and Applications

1. 15 points 2. 18 points
3. 5 points 4. $2w + t$ points
5. a. $2w + t = 13$
b.

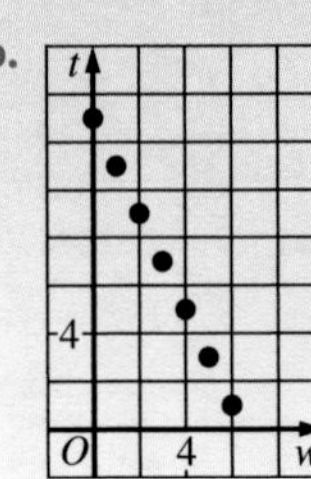

The t-intercept 13 represents the number of ties it would take to have 13 points without any wins. The graph has no w-intercept since the domain and range of this function include only whole numbers. If the graph did intersect the w-axis, the w-intercept would represent the number of wins it would take to have 13 points without any ties.

6. \$63 7. \$123
8. \$75 9. $15c + 9t$
10. a. $15c + 9t = 60$
b.

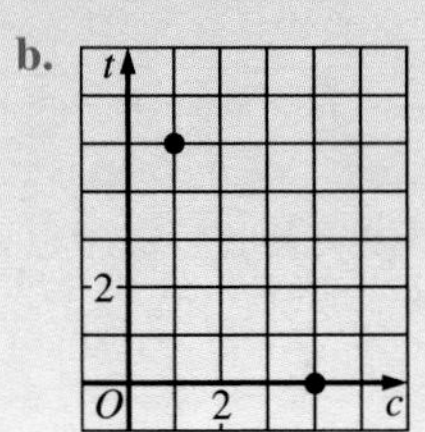

No, you cannot buy part of a compact disc or part of a tape. Only points for which both coordinates are whole numbers make sense.

11.

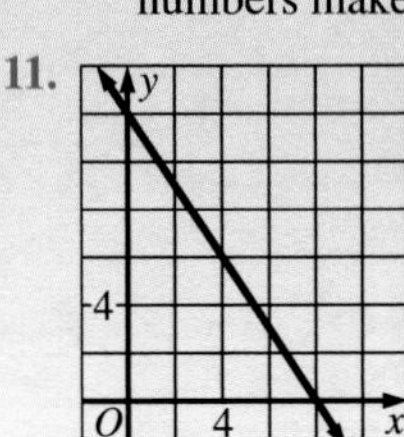

Cooperative Learning For Exercises 31–33, work with a partner. One of you should write an equation for the situation in Exercise 31. The other should write an equation for the situation in Exercise 32. Work together on Exercise 33.

31. The roller coaster at the state fair costs \$1.50 to ride and the bumper cars cost \$1.00. Ben has \$10.00 left and decides to spend it all on x roller coaster rides and y bumper car rides.

32. At Macomber's Market, you can buy grapes for \$1.50 per pound and mushrooms for \$1.00 per pound. Suppose you spend \$10.00 on x pounds of grapes and y pounds of mushrooms.

33. Writing Explain to your partner how your equation represents your situation. Which of the graphs describes the situation in Exercise 31? Which describes the situation in Exercise 32? Explain your answers.

A.

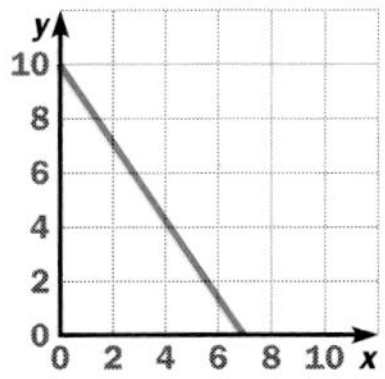

B.

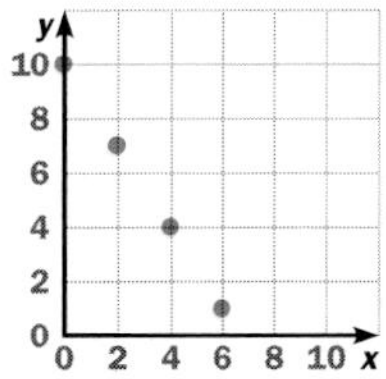

INTERVIEW Nancy Clark

Look back at the article on pages 278–280.

In an article that appeared in the journal American Fitness, Nancy Clark discusses sources of carbohydrates.

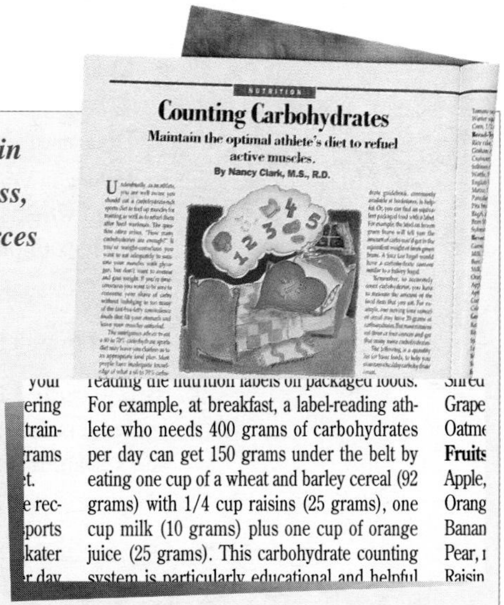
NUTRITION

Counting Carbohydrates

Maintain the optimal athlete's diet to refuel active muscles.

By Nancy Clark, M.S., R.D.

reading the nutrition labels on packaged foods. For example, at breakfast, a label-reading athlete who needs 400 grams of carbohydrates per day can get 150 grams under the belt by eating one cup of a wheat and barley cereal (92 grams) with 1/4 cup raisins (25 grams), one cup milk (10 grams) plus one cup of orange juice (25 grams). This carbohydrate counting system is particularly educational and helpful

34. a. How many grams of carbohydrates are in c cups of a wheat and barley cereal?

b. How many grams of carbohydrates are in m cups of milk?

c. Write an equation showing combinations of cereal and milk that provide 100 g of carbohydrates.

d. List three possible solutions of your equation. Which solution describes the combination of cereal and milk that you like best?

Oatmeal Nutrition Information	
Serving size	$\frac{1}{2}$ cup dry (40 g)
Calories	150
Fat	3 g
Carbohydrates	27 g

35. How many cups of oatmeal and how many cups of raisins do you need to eat to get 100 g of carbohydrates? Write an equation and list three possible solutions.

36. Research What do you eat for breakfast? Look at nutrition labels and calculate the total amount of carbohydrates. If some foods are not labeled, you may want to look at a nutrition guidebook from the library.

12.

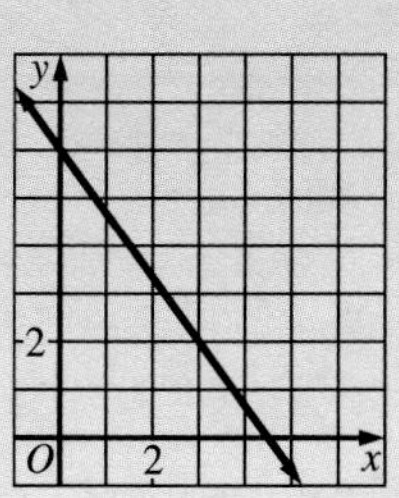

13–33. See answers in back of book.

34. a. $92c$ grams

b. $10m$ grams

c. $92c + 10m = 100$

d. Answers may vary. Examples are given. 10 cups of milk alone, about 1.1 cups of cereal alone, and 1 cup of cereal with 0.8 cups of milk; I would prefer the last choice, cereal and milk together.

35. $54c + 100r = 100$, where c = the number of cups of oatmeal and r = the number of cups of raisins; Solutions may vary. Examples are given. 1 cup of raisins alone, about 1.85 cups of oatmeal alone, and 1 cup of oatmeal with 0.46 cup of raisins

36. Answers may vary.

Apply⇔Assess

Exercise Notes

Second-Language Learners

Exs. 31–33 Students learning English should benefit from working cooperatively on these exercises with a peer tutor.

Interview Note

Exs. 34, 35 Students may find it interesting to research recommended daily allowances of vitamins, carbohydrates, fiber, and so on. They can then write problems similar to those in these exercises. Students can do the research with a partner, write problems individually, then trade and solve the problems.

Career Connection

Ex. 36 Students are generally aware that good nutrition is essential to good health. Individuals should know their nutritional needs. Specific knowledge and training in this area is required of all professional dietitians and nutritionists. There are many job opportunities for such professionals. Clinical dietitians provide nutritional services in hospitals and nursing homes. Community dietitians work in places such as health clinics, home health agencies, and health maintenance organizations. Administrative dietitians are responsible for large-scale meal planning in health care facilities, company cafeterias, prisons, schools, and other large institutions.

Multicultural Note

Ex. 36 People around the world enjoy a variety of foods for breakfast. For instance, in India, small cakes made from fermented rice flour called *idlis* are eaten with coriander chutney or a lentil stew called *sambar*. In Brazil, people might eat a simple mini baguette with butter and cheese or jam. In Japan, some people begin the day with a clear soup made with soybean paste called *misoshiru*, or a bowl of rice sprinkled with dried seaweed called *nori*.

Apply⇔Assess

Exercise Notes

Topic Spiraling: Review
Ex. 39 This exercise provides a good opportunity to review direct variation concepts studied in Chapter 3.

Assessment Note
Ex. 40 Students can use this exercise to demonstrate their understanding of the main ideas of this section. Students may wish to share how they related the mathematical concepts to real-life situations.

Reasoning
Exs. 41–44 Ask students what they can conclude in each case about the x- and y-values in relation to the graph of the given equation. (The x- and y-values are the x- and y-coordinates of a point that lies on the graph.)

Practice 44 for Section 7.1

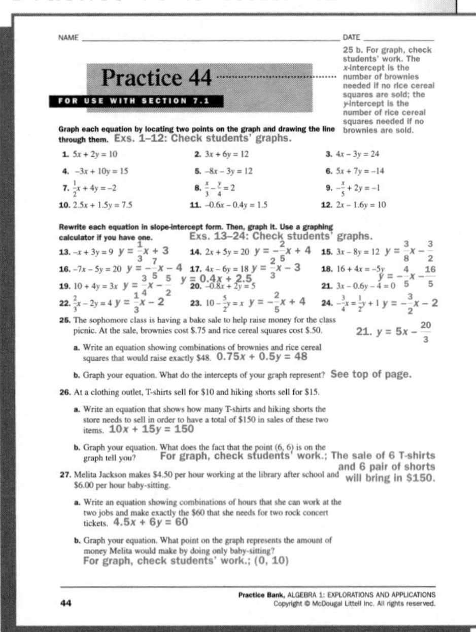

NAME ______________________ DATE __________

Practice 44

FOR USE WITH SECTION 7.1

25 b. For graph, check students' work. The x-intercept is the number of brownies needed if no rice cereal squares are sold; the y-intercept is the number of rice cereal squares needed if no brownies are sold.

Graph each equation by locating two points on the graph and drawing the line through them. Exs. 1–12: Check students' graphs.

1. $5x + 2y = 10$
2. $3x + 6y = 12$
3. $4x - 3y = 24$
4. $-3x + 10y = 15$
5. $-8x - 3y = 12$
6. $5x + 7y = -14$
7. $\frac{1}{2}x + 4y = -2$
8. $\frac{x}{3} - \frac{y}{4} = 2$
9. $-\frac{x}{5} + 2y = -1$
10. $2.5x + 1.5y = 7.5$
11. $-0.6x - 0.4y = 1.5$
12. $2x - 1.6y = 10$

Rewrite each equation in slope-intercept form. Then, graph it. Use a graphing calculator if you have one. Exs. 13–24: Check students' graphs.

13. $-x + 3y = 9$ $y = \frac{1}{3}x + 3$
14. $2x + 5y = 20$ $y = -\frac{2}{5}x + 4$
15. $3x - 8y = 12$ $y = \frac{3}{8}x - \frac{3}{2}$
16. $-7x - 5y = 20$ $y = -\frac{7}{5}x - 4$
17. $4x - 6y = 18$ $y = \frac{2}{3}x - 3$
18. $16 + 4x = -5y$ $y = -\frac{4}{5}x - \frac{16}{5}$
19. $10 + 4y = 3x$ $y = \frac{3}{4}x - \frac{5}{2}$
20. $-0.8x + 2y = 5$ $y = 0.4x + 2.5$
21. $3x - 0.6y - 4 = 0$
22. $\frac{2}{3}x - 2y = 4$ $y = \frac{1}{3}x - 2$
23. $10 - \frac{5}{2}y = x$ $y = -\frac{2}{5}x + 4$
24. $-\frac{3}{4}x = \frac{1}{2}y + 1$ $y = -\frac{3}{2}x - 2$

21. $y = 5x - \frac{20}{3}$

25. The sophomore class is having a bake sale to help raise money for the class picnic. At the sale, brownies cost \$.75 and rice cereal squares cost \$.50.
    a. Write an equation showing combinations of brownies and rice cereal squares that would raise exactly \$48. $0.75x + 0.5y = 48$
    b. Graph your equation. What do the intercepts of your graph represent? See top of page.
26. At a clothing outlet, T-shirts sell for \$10 and hiking shorts sell for \$15.
    a. Write an equation that shows how many T-shirts and hiking shorts the store needs to sell in order to have a total of \$150 in sales of these two items. $10x + 15y = 150$
    b. Graph your equation. What does the fact that the point (6, 6) is on the graph tell you? For graph, check students' work.; The sale of 6 T-shirts and 6 pair of shorts will bring in \$150.
27. Melita Jackson makes \$4.50 per hour working at the library after school and \$6.00 per hour baby-sitting.
    a. Write an equation showing combinations of hours that she can work at the two jobs and make exactly the \$60 that she needs for two rock concert tickets. $4.5x + 6y = 60$
    b. Graph your equation. What point on the graph represents the amount of money Melita would make by doing only baby-sitting? For graph, check students' work.; (0, 10)

44 — **Practice Bank,** ALGEBRA 1: EXPLORATIONS AND APPLICATIONS

BY THE WAY

When people talk about calories in food, they usually mean *kilocalories.* A kilocalorie is the amount of heat it takes to raise the temperature of one kilogram of water 1°C.

37. Open-ended Problem Write an equation in standard form. Graph your equation and describe the method you used to graph it.

38. FITNESS The energy you expend while exercising can be measured in calories (Cal). Based on a table in *Nancy Clark's Sports Nutrition Guidebook,* John finds that he may use 6 Cal/min while doing calisthenics and 8.5 Cal/min while jogging. At football practice, John does calisthenics and jogs for a total of 30 min.

a. Copy and complete the table.

Minutes of calisthenics (x)	Minutes of jogging (y)	Total calories used (c)
0	30	?
5	25	?
10	?	?
18	?	?
?	0	?

b. Write an equation relating x and y. Graph the equation.

c. Give the x- and y-intercepts of the graph. What do they tell you about the situation?

39. a. Challenge Graph the equations $3x - 2y = 12$, $3x - 2y = 6$, and $3x - 2y = 0$ in the same coordinate plane.

b. Do any of these equations describe direct variation?

c. How can you tell by looking at a linear equation in standard form whether it describes direct variation? Support your answer with examples.

ONGOING ASSESSMENT

40. Open-ended Problem Make up an equation in standard form that represents some situation in your life. Explain what your equation and the variables represent. Give at least three solutions of your equation.

SPIRAL REVIEW

Technology Graph each equation. Use a graphing calculator if you have one. Use the graph to find the missing value. *(Section 2.6)*

41. $y = 3.2x - 1.5$; find the value of x when $y = 0$.

42. $y = 10 + 1.75x$; find the value of x when $y = -1$.

43. $y = -3 - 4.7x$; find the value of x when $y = -2$.

44. $y = -0.1x - 0.6$; find the value of x when $y = 0.2$.

Solve each equation. *(Section 2.2)*

45. $4x - 3 = 7$ **46.** $2y - 10 = 1$ **47.** $1.5 + 2z = 2.5$

48. $9 = 14 - 3t$ **49.** $2r + 5 = \frac{3}{7}$ **50.** $4s - \frac{3}{2} = 1$

37. Answers may vary. Check students' work.

38. a.

Minutes of calisthenics (x)	Minutes of jogging (y)	Total calories used (c)
0	30	255.0
5	25	242.5
10	20	230.0
18	12	210.0
30	0	180.0

b. $x + y = 30$

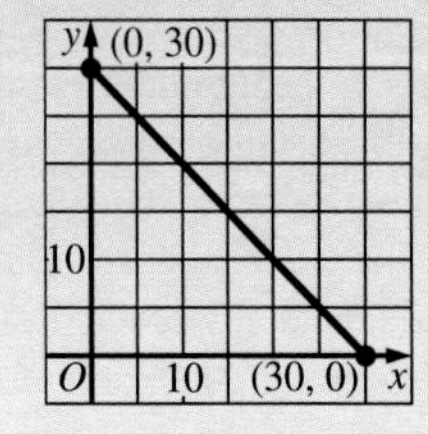

c. The x-intercept 30 indicates the number of minutes spent doing calisthenics if no time is spent jogging. The y-intercept 30 indicates the number of minutes spent jogging if no time is spent doing calisthenics.

39. a.

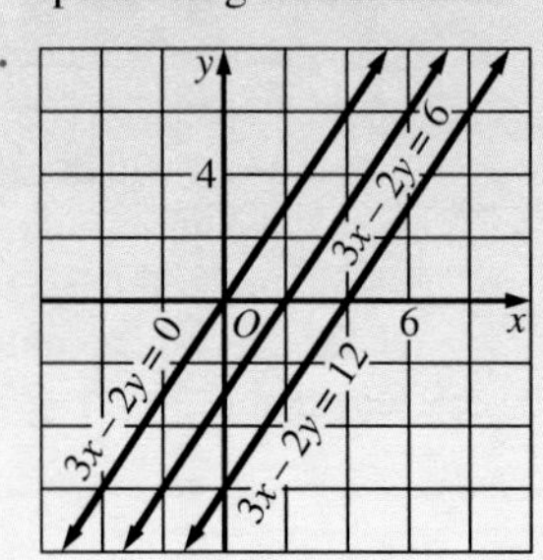

b. Yes; $3x - 2y = 0$.

c. If a linear equation in standard form has 0 as one side, then it describes direct variation. If $ax + by = 0$, then the graph is a line through the origin, since $a(0) + b(0) = 0$. So any such equation describes direct variation.

40–50. See answers in back of book.

SECTION

7.2 Solving Systems of Equations

Learn how to...

- solve systems of equations

So you can...

- solve problems with two variables, such as planning flight training

How would you like a career flying a plane? In addition to flying between cities, commercial pilots bring mail and food to remote towns and even stock lakes with fish.

An experienced private pilot must train for fifty hours to get a commercial license. Pilots can train in airplanes and on the ground in a flight simulator.

Plane Instruction = $98/h

Simulator = $46/h

EXAMPLE 1 Application: Aviation

Nancy Loesch is training for her commercial pilot's license. She can afford to spend only $3500 on the required 50 hours of instruction. Nancy would like to spend as much time as possible training in an airplane, but airplane instruction costs $98 per hour. Training in the simulator costs only $46 per hour.

Let a = number of hours spent training in an airplane.

Let s = number of hours spent training in a flight simulator.

Write two equations relating a and s.

SOLUTION

Hours spent training in an airplane	+	Hours spent training in a simulator	=	Hours that Nancy must train
a	+	s	=	50
Cost of a hours in an airplane	+	Cost of s hours in a simulator	=	Amount that Nancy can spend
$98a$	+	$46s$	=	3500

THINK AND COMMUNICATE

1. Give two solutions of the equation $a + s = 50$. Give two solutions of the equation $98a + 46s = 3500$. Do you think that there is an ordered pair that is a solution of both equations?

Plan⇔Support

Objectives

- Solve systems of equations.
- Solve problems with two variables.

Recommended Pacing

Core and Extended Courses
Section 7.2: 1 day

Block Schedule
Section 7.2: $\frac{1}{2}$ block (with Section 7.1)

Resource Materials

Lesson Support
Lesson Plan 7.2
Warm-Up Transparency 7.2
Overhead Visuals:
  Folder 7: A System of Equations
  Folder A: Multi-Use Graphing Packet, Sheets 1–3
Practice Bank: Practice 45
Study Guide: Section 7.2
Explorations Lab Manual:
  Diagram Master 1
Assessment Book: Test 29

Technology
Technology Book:
  Spreadsheet Activity 7
Graphing Calculator
McDougal Littell Software
  Function Investigator
Internet:
  http://www.hmco.com

Warm-Up Exercises

Tell whether the given point is on the graph of the given equation. Answer *Yes* or *No.*

1. (5, 8); $8x + 5y = 80$ Yes.
2. (5, 8); $-2x + 3y = 14$ Yes.
3. (–3, 4); $6x - 7y = -48$ No.

Solve each equation for y.

4. $-4x + y = 16$ $y = 4x + 16$
5. $4x + 9y = -12$ $y = \frac{-4x - 12}{9}$

ANSWERS Section 7.2

Think and Communicate

1. Answers may vary. Examples are given. (50, 0), (20, 30), and (0, 50); about (35.7, 0) and about (0, 76); Yes, I think so.

Teach⇔Interact

Additional Example 1

Pedro Torres pilots small planes for various jobs in a farming region. He earns \$30 per hour for dusting crops and \$45 per hour transporting cargo. He works a 40-hour week and earns \$1600 per week. Let d = number of hours dusting crops. Let c = number of hours transporting cargo. Write two equations relating d and c.

Hours dusting crops + Hours transporting cargo = Hours Pedro works

$d + c = 40$

Earnings dusting crops + Earnings transporting cargo = Amount Pedro earns

$30d + 45c = 1600$

Think and Communicate

In discussing question 1, ask students to explain why they think there is or is not a solution. If students try various ordered pairs at random, you can suggest a more organized approach of trying ordered pairs in the second equation that are known to be solutions of $a + s = 50$.

Section Note

Visual Thinking

Check students' understanding of the graphs of a system of linear equations by asking them to discuss the meaning of the point at which the two lines intersect. Ask them to select other points on the graph and describe their meaning. This activity involves the visual skills of *recognition* and *perception.*

Additional Example 2

Use a graphing calculator to solve the system of equations.

$d + c = 40$

$30d + 45c = 1600$

Step 1 Rewrite each equation using x and y as variables.

$x + y = 40$

$30x + 45y = 1600$

Step 2 Solve each equation for y so it can be entered into the calculator.

$x + y = 40$

$y = 40 - x$

$30x + 45y = 1600$

$45y = 1600 - 30x$

$y = \dfrac{1600 - 30x}{45}$

Example continues on following page.

BY THE WAY

In June of 1994, 12-year-old Vicki Van Meter flew a plane across the Atlantic Ocean from Augusta, Maine, to Glasgow, Scotland.

The two equations in Example 1 are a **system of linear equations**. Any pair of numbers that satisfies both equations is a **solution** of the system. If you graph the equations together, you can see that they have one common solution.

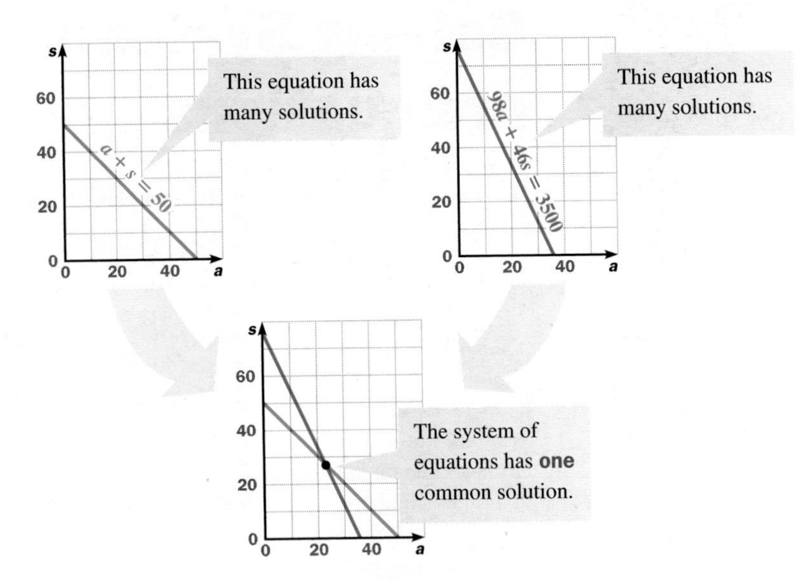

EXAMPLE 2 Method: Graphing

Use a graphing calculator to solve the system of equations.

$\mathbf{a + s = 50}$

$\mathbf{98a + 46s = 3500}$

SOLUTION

Step 1 Rewrite each equation using x and y as variables.

$x + y = 50$ $\qquad 98x + 46y = 3500$

Step 2 Solve each equation for y so you can enter it into the calculator.

$y = 50 - x$ $\qquad 46y = 3500 - 98x$

$y = \dfrac{3500 - 98x}{46}$

Step 3 Graph each equation and find the point where the graphs intersect.

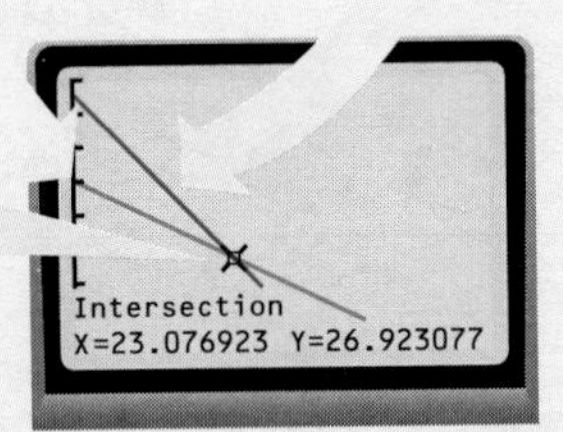

Write your answer in terms of the original variables. The solution (a, s) is about (23.1, 26.9).

Solving Systems by Substitution

You can substitute variable expressions into equations just as you substitute numbers. Substitution gives you another method for solving systems of equations.

For example, to solve the system

$$y = 2x$$
$$3x - y = 5$$

you can substitute **2x** for y in the **second equation**.

$$3x - 2x = 5$$
$$x = 5$$

Now that you have a value for x, you can find a value for y.

$$y = 2x$$
$$y = 2 \cdot 5$$
$$y = 10$$

Substitute 5 for x in the **first equation**.

The ordered pair (5, 10) is a solution of the system of equations.

EXAMPLE 3 Method: Substitution

Use substitution to solve the system of equations from Example 1. Explain what your solution tells you about the situation.

$$a + s = 50$$
$$98a + 46s = 3500$$

SOLUTION

Step 1 Solve the **first equation** for a.

$$a + s = 50$$
$$a = 50 - s$$

Step 2 Substitute **50 − s** for a in the **second equation** and solve for s.

$$98a + 46s = 3500$$
$$98(50 - s) + 46s = 3500$$
$$4900 - 98s + 46s = 3500$$
$$1400 = 52s$$
$$s \approx 26.9$$

Step 3 To find a value for a, substitute 26.9 for s in either of the original equations.

$$a + s = 50$$
$$a + 26.9 \approx 50$$
$$a \approx 23.1$$

The solution (a, s) is about (23.1, 26.9). Nancy should spend about 23 h training in an airplane and about 27 h training in a flight simulator.

Check

Substitute 26.9 for s and 23.1 for a in each equation.

$$a + s = 50$$
$$23.1 + 26.9 \stackrel{?}{=} 50$$
$$50 = 50 \checkmark$$
$$98a + 46s = 3500$$
$$98(23.1) + 46(26.9) \stackrel{?}{=} 3500$$
$$2263.8 + 1237.4 \stackrel{?}{=} 3500$$
$$3501.2 \approx 3500 \checkmark$$

Teach⇔Interact

Additional Example 2 (continued)

Step 3 Graph each equation and find the point where the graphs intersect.

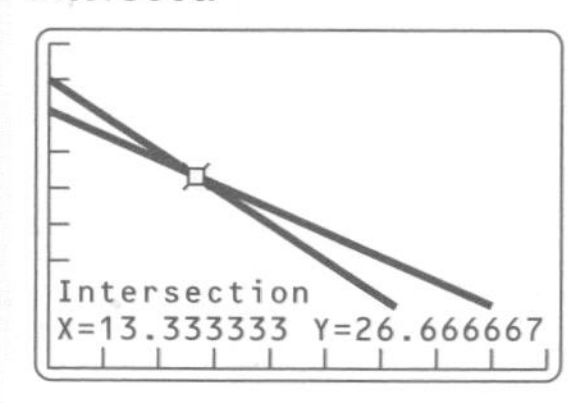

Write the answer in terms of the original variables. The solution (d, c) is about (13.3, 26.7).

Section Note

Reasoning

After discussing solving systems by substitution, ask students how they can check the solution (5, 10) to see that it is correct.

Additional Example 3

Use substitution to solve the system of equations from Additional Example 1. Explain what your solution tells you about the situation.

$$d + c = 40$$
$$30d + 45c = 1600$$

Step 1 Solve the first equation for d.

$$d + c = 40$$
$$d = 40 - c$$

Step 2 Substitute $40 - c$ for d in the second equation and solve for c.

$$30d + 45c = 1600$$
$$30(40 - c) + 45c = 1600$$
$$1200 - 30c + 45c = 1600$$
$$15c = 400$$
$$c \approx 26.7$$

Step 3 To find d, substitute 26.7 for c in either of the original equations.

$$d + c = 40$$
$$d + 26.7 \approx 40$$
$$d \approx 13.3$$

The solution (d, c) is about (13.3, 26.7). Pedro Torres spends about 13 hours dusting crops and about 27 hours transporting cargo.

Check

Substitute 13.3 for d and 26.7 for c in each equation.

$$d + c = 40$$
$$13.3 + 26.7 \stackrel{?}{=} 40$$
$$40 = 40 \checkmark$$
$$30d + 45c = 1600$$
$$30(13.3) + 45(26.7) \stackrel{?}{=} 1600$$
$$399 + 1201.5 \stackrel{?}{=} 1600$$
$$1600.5 \approx 1600 \checkmark$$

Teach⇔Interact

Think and Communicate

You may wish to discuss why certain systems of linear equations can be solved easily by substitution. All of the systems in this section have an equation in which at least one variable has a coefficient of 1 or –1. These systems are especially easy to solve by substitution. This fact can be reinforced by using a system that is not easy to solve by substitution, such as
$2x + 3y = 5$
$-6x + 4y = 8$.

Checking Key Concepts

Mathematical Procedures
Be sure students understand that they should always look for a variable in a system that it is easy to solve for. This will ensure that the entire solution process is as simple as possible.

Closure Question

Describe two methods that can be used to solve a system of linear equations. Graph the equations together on a coordinate grid or by using a graphing calculator or solve the system using substitution.

Apply⇔Assess

Suggested Assignment

Core Course
Exs. 1–14, 16–19, 22, 23, AYP

Extended Course
Exs. 1–7, 11–23, AYP

Block Schedule
Day 42 Exs. 1–14, 16–20, 22, 23, AYP

Exercise Notes

Using Technology
Exs. 1–4 These exercises can be done with a graphing calculator or with the *Function Investigator* software.

Teaching Tip
Exs. 5–13 Suggest that students check their solutions by substituting them into the given systems or by using a graphing calculator.

THINK AND COMMUNICATE

Tell what you would do first to solve each system by substitution.

2. $3x + 5y = 18$
$2x + y = 4$

3. $x + 3y = 15$
$5x - 6y = 19$

4. $x + y = 1$
$4x - 3y = 18$

5. In step 3 of Example 3, what happens if you substitute $s = 26.9$ into $98a + 46s = 3500$ instead of into $a + s = 50$?

6. Which method do you prefer for solving the system of equations in Example 1? Why?

CHECKING KEY CONCEPTS

1. Use the substitution method to solve each system in Think and Communicate Questions 2–4.

2. Match each of the systems in Think and Communicate Questions 2–4 with one of the graphs below.

A.
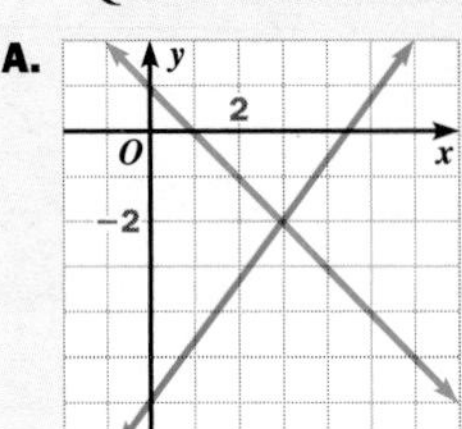

B.
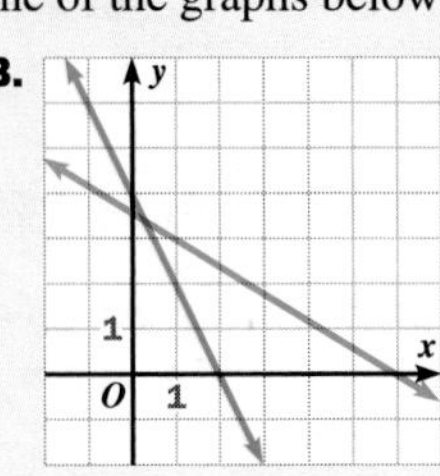

C.
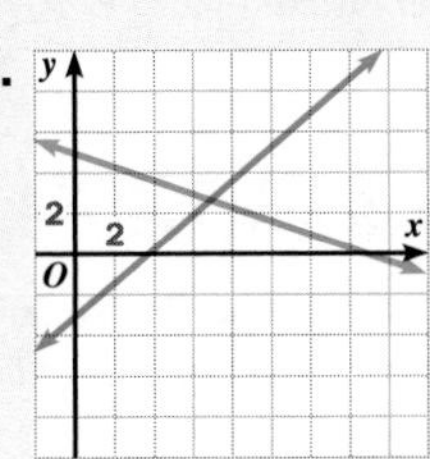

7.2 Exercises and Applications

Extra Practice exercises on page 570

Technology For Exercises 1–4, solve each system of equations by graphing. Use a graphing calculator if you have one.

1. $2x - 3y = 24$
$2x + y = 8$

2. $3x - y = 9$
$x + 2y = 10$

3. $6x + 9y = -5$
$x + y = 1$

4. $7x - 14y = 2$
$y - \frac{3}{5}x = 0$

Use substitution to solve each system of equations.

5. $3x + 4y = 4$
$y = x - 6$

6. $2x + 7y = 3$
$x = 1 - 4y$

7. $x + 4y = 1$
$2x - 3y = -9$

8. $3x + 2y = 13$
$2x + y = 7$

9. $x - 3y = 8$
$4x + 5y = -2$

10. $2a + 3b = 12$
$a + b = 5$

11. $4c + 5d = 11$
$3c - d = 13$

12. $5r - s = 5$
$5s - 4r = 17$

13. $5x - 4y = -6$
$7x - y = 10$

Think and Communicate

2. Solve the second equation for *y*.
3. Solve the first equation for *x*.
4. Solve the first equation for either *x* or *y*.
5. You get the same result, but with more calculating needed.
6. Answers may vary. An example is given. I think the substitution method is simplest, because the first equation can so easily be solved for one of the variables.

Checking Key Concepts

1. $\left(\frac{2}{7}, \frac{24}{7}\right)$; $\left(7, \frac{8}{3}\right)$; $(3, -2)$
2. A: Question 4; B: Question 2; C: Question 3

Exercises and Applications

1.
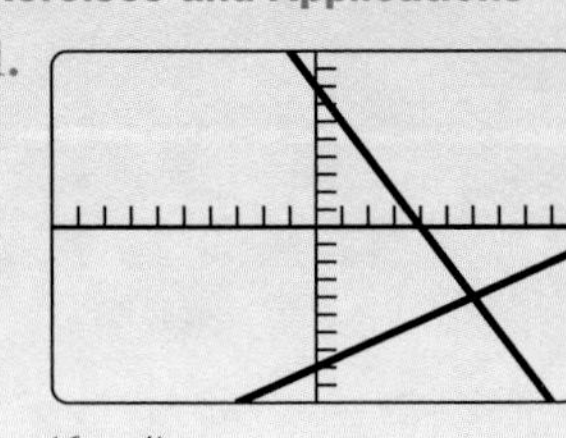
$(6, -4)$

14. SAT/ACT Preview Suppose a section of the test has 50 multiple-choice questions. Trevor earns one point for each right answer and loses $\frac{1}{4}$ point for each wrong answer. For example, if he gives 32 right answers, 8 wrong answers, and leaves 10 questions blank, his score is $32 - \frac{1}{4}(8) = 30$.

a. Let r and w represent the number of questions Trevor got right and wrong. Write an equation stating that Trevor answered 41 questions.

b. Write an equation stating that Trevor's score for answering 41 questions was $34\frac{3}{4}$.

c. Solve your system of equations from parts (a) and (b).

15. a. FINANCE On March 1, 1995, one French franc was worth \$.19 and one German mark was worth \$.68. Write two equations expressing these facts.

b. Write one equation relating the value of a German mark to a French franc. Explain how to use your equation to find the value in marks of 2000 francs.

16. **Technology** Use algebra or a graphing calculator to decide whether or not these three equations have a common solution.

$$2x - y = -10$$
$$-x + y = 7$$
$$-3x - y = 6$$

17. SPACE SCIENCE Use the information in the article to write and solve a system of equations.

Let $N =$ the heart's rate of pumping blood under normal conditions.

Let $L =$ the heart's rate of pumping blood during launch.

As a space shuttle or rocket is launched, the astronauts inside experience a great deal of stress. An astronaut's heart pumps twice as much blood per minute during rocket launch as under normal conditions—the heart pumps 10 pints more blood each minute during launch than when the astronaut is at rest on the ground.

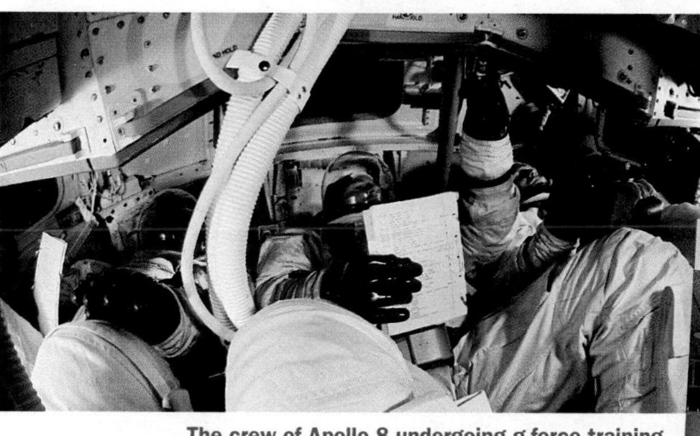

The crew of Apollo 8 undergoing g-force training. The Apollo program used a centrifuge, like the one shown at the right, in g-force training.

Apply⇔Assess

Exercise Notes

Second-Language Learners

Ex. 14 Students learning English will benefit from working on this activity with an English-proficient student. Suggest that the English-proficient student read the paragraph of information aloud and that the partners then discuss the information before working to solve the system of equations.

Problem Solving

Ex. 15 A useful variation of this financial problem would be to have students use the given exchange rates to find the value of one dollar in terms of francs and marks on March 1, 1995. (\$1.00 = 5.26 francs; \$1.00 = 1.47 marks)

Using Technology

Ex. 16 This exercise can be used to stress the need for caution when solving systems by graphing. Students who use the standard viewing window to display the graphs on a graphing calculator may decide quickly that the system does have a common solution. However, if they trace and zoom in, they will see that the graphs intersect in three points and therefore the system does *not* have a common solution.

Application

Ex. 17 Point out to students that a major goal of scientists for the first flights of astronauts into space was to monitor their body reactions to the conditions of stress and weightlessness in space.

2.

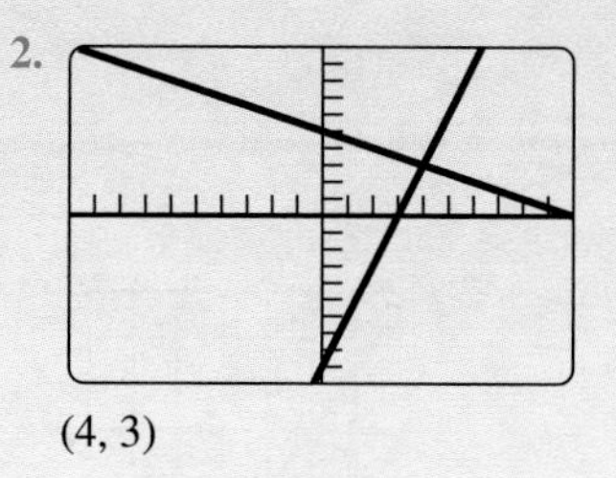

(4, 3)

3.

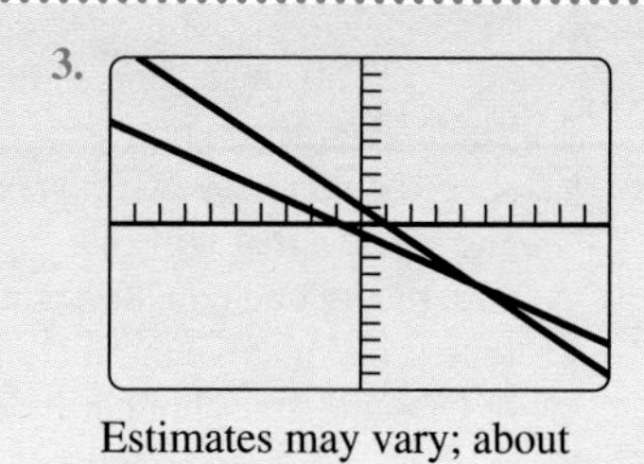

Estimates may vary; about (4.7, –3.7).

4. Estimates may vary; about (–1.4, –0.9).

5. (4, –2) 6. (5, –1)

7. (–3, 1) 8. (1, 5)

9. (2, –2) 10. (3, 2)

11. (4, –1) 12. (2, 5)

13. (2, 4)

14. a. $r + w = 41$

b. $r - \frac{1}{4}w = 34\frac{3}{4}$

c. 36 right answers and 5 wrong ones

15–17. See answers in back of book.

Exercise Notes

Interdisciplinary Problems
Ex. 18 System of equations are used frequently to analyze and solve problems in business because two or more variables are usually involved. In fact, you can point out to students that it is not uncommon to have systems with three or more variables that require special techniques or computers to solve.

Mathematical Procedures
Ex. 20 This exercise illustrates a useful mathematical procedure. By examining the *form* of a mathematical expression or equation, it may be possible to solve a problem by using a relationship that involves the form only, without doing any additional algebra.

Problem Solving
Ex. 21 You may wish to have a few students explain how they solved this problem and demonstrate their approaches at the board.

Connection BUSINESS

Many airlines provide meals or snacks on their flights. These meals are prepared in the huge kitchens of catering firms. The airlines must pay different amounts for the meals based on what they contain.

18. Mario Clementi is in charge of ordering meals for a flight on a jet plane. There will be 164 coach-class passengers and 24 first-class passengers. From past experience, Mario knows that first-class meals cost about twice as much as coach-class meals. How much can he spend on each type of meal if his budget is \$1300? Write and solve a system of equations.

19. Mario decides to spend a total of \$1000 on the coach-class meals. He orders 164 coach-class meals. 60% of the meals will contain beef, and 40% will contain chicken. Chicken meals cost about 80% as much as beef meals. How much can he spend on each type of meal? Write and solve a system of equations.

20. a. Challenge Verify that $(2, -4)$ is a solution of the first system of equations at the right.

$$3x - 5y = 26$$
$$6x + y = 8$$

b. Compare the two systems of equations. How are they alike? How are they different?

$$3\left(\frac{1}{x}\right) - 5\left(\frac{1}{y}\right) = 26$$
$$6\left(\frac{1}{x}\right) + \left(\frac{1}{y}\right) = 8$$

c. Use your answers to parts (a) and (b) to find a solution of the second system of equations. Explain your reasoning.

21. Open-ended Problem Find equations of at least three lines that intersect each other at the point $(6, 3)$. Explain why there are other possible answers.

18. Let c = the cost in dollars of a coach-class meal and f = the cost in dollars of a first-class meal. $164c + 24f = 1300$; $f = 2c$; about \$6.13 for a coach-class meal and \$12.26 for a first-class meal

19. Let b = the cost in dollars of a beef meal and c = the cost in dollars of a chicken meal. $98b + 66c = 1000$; $c = 0.8b$; about \$6.63 for a beef meal and \$5.31 for a chicken meal

20. a. $3(2) - 5(-4) = 26$ and $4(2) + (-4) = 4$

b. The systems have the same general form, that is, the same coefficients and constants, and they contain the same two variables, x and y. They differ only in the form of the variables.

c. $\left(\frac{1}{2}, -\frac{1}{4}\right)$; If (x, y) is a solution of the first system, then $\left(\frac{1}{x}, \frac{1}{y}\right)$ is a solution of the second.

21. Answers may vary. Examples are given. $x + y = 9$, $y = 3$, and $y = 0.5x$; There are many lines through $(6, 3)$. You can see this by thinking of the point-slope form, $y = mx + b$, or $3 = 6m + b$. For every different value of m and of b, you get a different line through $(6, 3)$.

22. a. Answers may vary. An example is given.

$8d + 3t = 50$ and $d + t = 15$; The first equation represents the number of CDs, d, and tapes, t, that will cost \$50; the second equation represents the total number of CDs and tapes bought.

b. $d = 1$, $t = 14$; the number of CDs and tapes that can be bought for \$50

23. a. about 1030 ft

b. about 11.7 mi/h

Assess Your Progress

1. $-3, -4$

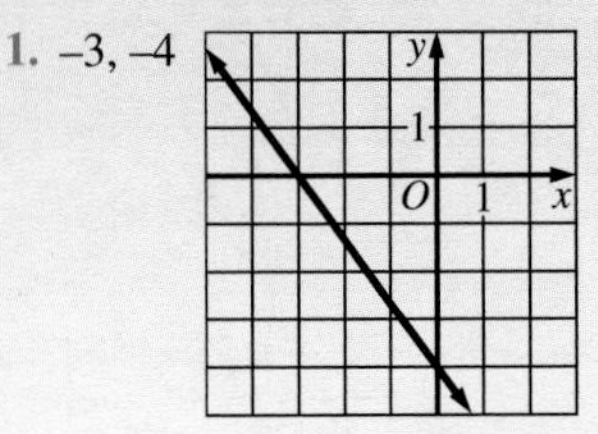

ONGOING ASSESSMENT

22. a. **Open-ended Problem** Use the information in the advertisement to write a system of equations. Explain what each equation represents.

    b. Solve your system of equations. Explain what the solution represents.

A R T S
Art's Records
Fabulous 50th
Anniversary Sale!
On May 5th, the 50th customer will receive $50 to spend on tapes and CDs!
Used CDs only $8!
Used tapes only $3!

SPIRAL REVIEW

23. **GEOMETRY** The Cosmoclock 21 in Yokohama City, Japan, is one of the largest Ferris wheels in the world. Its diameter is 328 ft. The 60 arms holding the gondolas serve as second hands for the clock.

    a. How many feet would you travel in one revolution of the Ferris wheel? *(Toolbox, page 593)*

    b. Find your speed in miles per hour if the wheel makes one revolution per minute. Use the fact that 1 mi = 5280 ft. *(Section 3.1)*

BY THE WAY

The Cosmoclock 21 is $344\frac{1}{2}$ ft high. Each of the 60 gondolas holds eight passengers.

ASSESS YOUR PROGRESS

VOCABULARY

standard form of a linear equation (p. 282)
vertical intercept, *y*-intercept (p. 282)
horizontal intercept, *x*-intercept (p. 282)
system of linear equations (p. 288)
solution of a system (p. 288)

Give the *x*- and *y*-intercepts of each graph. Graph each equation. *(Section 7.1)*

1. $4x + 3y = -12$
2. $-3x + 8y = 12$

Rewrite each equation in slope-intercept form. Then graph the equation. *(Section 7.1)*

3. $5x - y = 5$
4. $2x + 3y = 9$
5. Shira gets four points for each correct answer on her biology quiz. She loses a point for each wrong answer. *(Section 7.2)*

   a. Write an equation showing that Shira earned 30 points on the quiz.

   b. Write an equation showing that Shira answered 20 questions.

   c. Solve the system of equations in parts (a) and (b) by graphing.

Use substitution to solve each system of equations. *(Section 7.2)*

6. $6x + 5y = 1$
   $y = 1 - 2x$
7. $2a + 7b = 24$
   $3a - b = 13$
8. **Journal** You can solve a system of equations by graphing or by using substitution. Describe some advantages and disadvantages of each method.

Apply⇔Assess

Assess Your Progress

Journal Entry
Groups of students may want to compare their responses to Ex. 8 and revise them before recording them in their journals.

Progress Check 7.1–7.2

See page 318.

Practice 45 for Section 7.2

NAME ______________________ DATE ____________

Practice 45
FOR USE WITH SECTION 7.2

Solve each system of equations by graphing. Use a graphing calculator if you have one.

1. $x + y = 8$, $2x - y = 7$ (5, 3)
2. $2x + y = -3$, $x + 4y = 2$ (−2, 1)
3. $-x + y = 4$, $5x - 2y = -8$ (0, 4)
4. $3x + y = 6$, $4x - y = 1$ (1, 3)
5. $2x + 3y = 5$, $x - 3y = 7$ (4, −1)
6. $4x - 5y = 7$, $6x - y = -9$ (−2, −3)

Use substitution to solve each system of equations.

7. $y = x + 3$, $x + 5y = 3$ (−2, 1)
8. $-4x + y = 3$, $x = 2y + 1$ (−1, −1)
9. $6x + 5y = -7$, $y = -x - 2$ (3, −5)
10. $x - 4y = 15$, $2x + 5y = 4$ (7, −2)
11. $6p + 5q = 9$, $3p - q = 15$ (4, −3)
12. $-x + 4y = 3$, $7x - 9y = 17$ (5, 2)
13. $3a - b = 11$, $2a - 7b = 1$ (4, 1)
14. $-m + 6n = 14$, $5m - 3n = -16$ (−2, 2)
15. $4x - 7y = 20$, $x - 3y = 5$ (5, 0)
16. $\frac{1}{2}x + y = 0$, $\frac{5}{2}x + 2y = -9$ (−6, 3)
17. $0.4x - y = 1$, $0.6x + 3y = 15$ (10, 3)
18. $5r - \frac{1}{4}s = -13$, $r + \frac{3}{4}s = 7$ (−2, 12)
19. During one week of playing basketball, Abdul Williams hit 30 shots from the floor and scored 68 points. Each shot was a 2-point shot or a 3-point shot.

    a. Write a system of equations that expresses the conditions on the numbers of 2-point and 3-point shots that Abdul made. $2x + 3y = 68$; $x + y = 30$

    b. How many of each kind of shot did he make? 2-point shots: 22; 3-point shots: 8
20. Kichi Hiyoshi is a landscape architect who needs to buy 17 shrubs for a foundation planting. Azalea plants cost $20 each, rhododendrons cost $25 each, and Kichi wants to spend exactly $400.

    a. Write a system of equations that models the situation. $x + y = 17$; $20x + 25y = 400$

    b. Solve the system you wrote in part (a) for the number of each kind of plant Kichi should buy. 5 azeleas, 12 rhododendrons
21. **Writing** Must any pair of straight-line graphs that you draw in a coordinate plane intersect? Explain your answer in terms of the equations of the lines and the solution, if any, of the corresponding system. no; Check students' explanations.

Practice Bank, ALGEBRA 1: EXPLORATIONS AND APPLICATIONS

45

---

2. $-4, \frac{3}{2}$

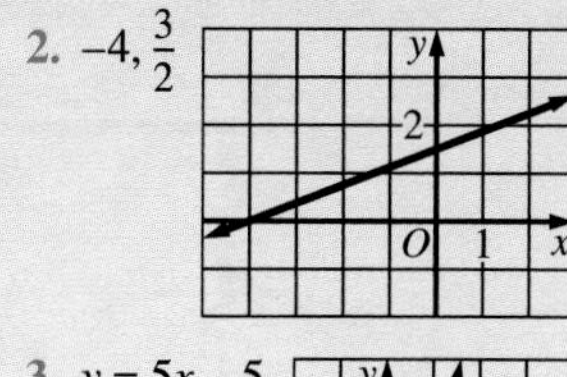

3. $y = 5x - 5$

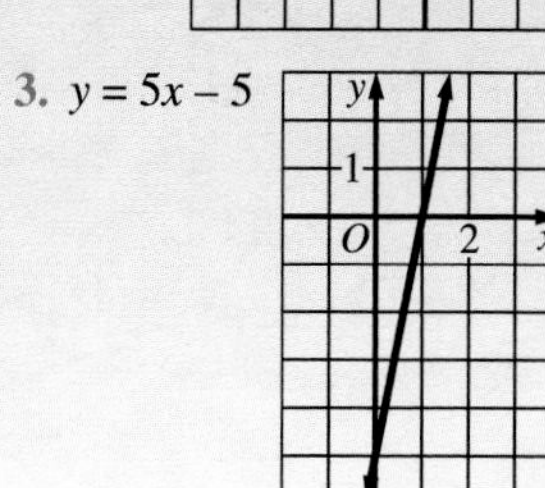

4. $y = -\frac{2}{3}x + 3$

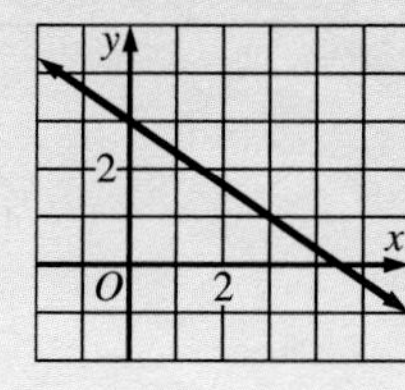

5. Let c = the number of correct answers and w = the number of wrong answers.

   a. $4c - w = 30$

   b. $c + w = 20$

   c.

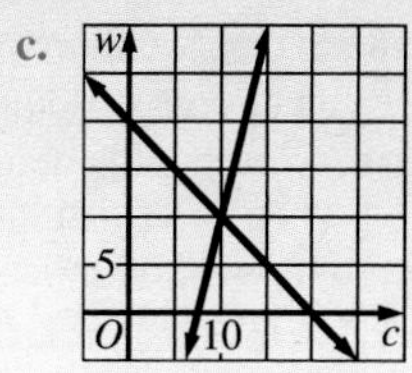

10 correct answers and 10 wrong ones

6. (1, −1)
7. (5, 2)
8. Answers will vary. An example is given. Graphing method: An advantage is that I can use my graphing calculator and get an answer pretty quickly. A disadvantage is that I may need to rewrite the equations in slope-intercept form. Substitution method: An advantage is that I can get an exact solution. A disadvantage is that it is easy to make a mistake while I am solving the system.

Plan⇔Support

Objectives

- Solve systems of equations by adding or subtracting.
- Solve problems with two variables.

Recommended Pacing

Core and Extended Courses
Section 7.3: 1 day

Block Schedule
Section 7.3: $\frac{1}{2}$ block
(with Section 7.4)

Resource Materials

Lesson Support
Lesson Plan 7.3
Warm-Up Transparency 7.3
Practice Bank: Practice 46
Study Guide: Section 7.3
Explorations Lab Manual: Diagram Master 2

Technology
Technology Book: Calculator Activity 7
Graphing Calculator
McDougal Littell Software *Function Investigator*
Internet: http://www.hmco.com

Warm-Up Exercises

Add the expressions.

1. $\begin{array}{r} 5x - 2y \\ \underline{-3x + 4y} \end{array}$ $2x + 2y$
2. $\begin{array}{r} 17x + 11y \\ \underline{3x - 11y} \end{array}$ $20x$
3. $\begin{array}{r} -8x + 6y \\ \underline{8x - 3y} \end{array}$ $3y$

Subtract the second expression from the first expression.

4. $\begin{array}{r} 7x - 12y \\ \underline{-3x - 12y} \end{array}$ $10x$
5. $\begin{array}{r} -5x + y \\ \underline{-7x + y} \end{array}$ $2x$
6. $\begin{array}{r} 4x - 9y \\ \underline{4x + 18y} \end{array}$ $-27y$

SECTION

7.3 Solving Linear Systems by Adding or Subtracting

Learn how to...

- solve systems of equations by adding or subtracting

So you can...

- solve problems in two variables, such as understanding a restaurant bill

Many Chinese restaurants serve *dim sum*. Instead of ordering from a menu, you choose items from carts of appetizers and desserts that circulate through the restaurant. Some items are more expensive than others. To distinguish between them, some restaurants serve the more expensive items from larger plates.

The character 小 means *small* and 大 means *large*.

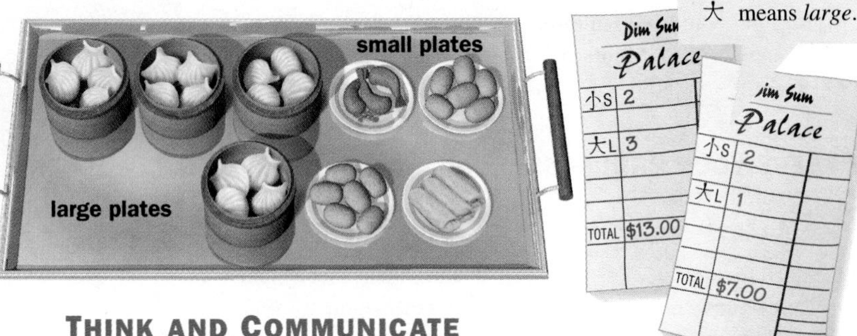

THINK AND COMMUNICATE

1. Can you figure out how much a large *dim sum* plate costs?
2. How much does a small *dim sum* plate cost? Explain your reasoning.

EXAMPLE 1 Method: Subtraction

Write and solve a system of equations to find the cost of a large *dim sum* plate.

SOLUTION

Let l = the cost of a large *dim sum* plate.

Let s = the cost of a small *dim sum* plate.

According to the bills above, $3l + 2s = \$13$ and $l + 2s = \$7$.

If you **subtract** each side of the second equation from the corresponding side of the first, you can solve for l.

$$\begin{aligned} 3l + 2s &= 13 \\ -(l + 2s) &= -(7) \\ \hline 2l &= 6 \\ l &= 3 \end{aligned}$$

Notice that the coefficients of s are the same.

A large *dim sum* plate costs \$3.

ANSWERS Section 7.3

Think and Communicate

1. \$3
2. \$2; Comparing the two bills, you can see that two more of the large items cost an additional \$6, so each large item costs \$3. Then two small items cost \$4, so each costs \$2.

THINK AND COMMUNICATE

3. To keep both sides of an equation balanced, you must always perform the same operation on both sides. In Example 1, was the same amount subtracted from both sides of the first equation? Explain.
4. In Example 1, how can you find how much an item on a small *dim sum* plate costs?

BY THE WAY

You can use any shape or character to represent a variable. For example, you could write the system of equations in Example 1 using Chinese characters for *large* and *small*.

3 小 + 2 大 = 13

小 + 2 大 = 7

In Example 1, the coefficients of s in each equation are the same, so you can eliminate s from the system by subtracting. In Example 2, the coefficients of y are opposites. If you add corresponding sides of each equation, you can eliminate y.

EXAMPLE 2 Method: Addition

Solve the system of equations.

$$8x - 9y = 13$$
$$3x + 9y = 9$$

SOLUTION

Step 1 Add corresponding sides of each equation to eliminate one variable.

$$\begin{aligned} 8x - 9y &= \quad 13 \\ +\ (3x + 9y) &= +\ (9) \\ \hline 11x \quad &= \quad 22 \\ x &= 2 \end{aligned}$$

Step 2 Now you can solve for x.

Step 3 Substitute 2 for x in either equation and solve for y.

$$\begin{aligned} 8x - 9y &= 13 \\ 8(2) - 9y &= 13 \\ -9y &= -3 \\ y &= \frac{1}{3} \end{aligned}$$

The solution (x, y) is $\left(2, \frac{1}{3}\right)$.

Check

Substitute 2 for x and $\frac{1}{3}$ for y in each equation.

$$8x - 9y = 13 \qquad 3x + 9y = 9$$
$$8(2) - 9\left(\frac{1}{3}\right) \stackrel{?}{=} 13 \qquad 3(2) + 9\left(\frac{1}{3}\right) \stackrel{?}{=} 9$$
$$16 - 3 \stackrel{?}{=} 13 \qquad 6 + 3 \stackrel{?}{=} 9$$
$$13 = 13 \checkmark \qquad 9 = 9 \checkmark$$

THINK AND COMMUNICATE

5. When can you solve a system by subtracting? When can you solve a system by adding?

Think and Communicate

3. Yes; $l + 2s$ was subtracted from one side and 7 from the other. But $l + 2s = 7$.
4. Substitute 3 for l in either equation, and solve for s.
5. when the coefficients of a variable are the same in both equations; when the coefficients of a variable are opposites in the two equations

Teach⇔Interact

Additional Example 1

The Montrose Cafeteria charges one price for all main courses and one price for all vegetable dishes. Lynn took her parents to lunch and had a bill of $8.65 for 3 main courses and 2 vegetable dishes. Rolando took a friend there to lunch and had a bill of $6.30 for 2 main courses and 2 vegetable dishes. Write and solve a system of equations to find the cost of a main course.

Let m = the cost of a main course.
Let v = the cost of a vegetable dish.
$3m + 2v = \$8.65$ and
$2m + 2v = \$6.30$.
Subtract each side of the second equation from the corresponding side of the first equation to solve for m.

$$\begin{aligned} 3m + 2v &= \quad 8.65 \\ -\ (2m + 2v) &= -\ (6.30) \\ \hline m \quad &= \quad 2.35 \end{aligned}$$

A main course costs $2.35.

Think and Communicate

After discussing question 3, refer to Additional Example 1. Ask whether Lynn and her parents had the same number of vegetable dishes as Rolando and his friend. (Yes.) What was different about their two bills? (Lynn and her parents had 1 more main course on their bill.) Is it reasonable that the difference between the two bills represents the cost of 1 main course? (Yes.)

Additional Example 2

Solve the system of equations.
$5x - 6y = 2$
$-5x + 9y = -1$

Step 1 Add corresponding sides of each equation to eliminate one variable.

$$\begin{aligned} 5x - 6y &= \quad 2 \\ +\ (-5x + 9y) &= +\ (-1) \\ \hline 3y &= \quad 1 \end{aligned}$$

Step 2 Solve for y.
$3y = 1$
$y = \frac{1}{3}$

Step 3 Substitute $\frac{1}{3}$ for y in either equation and solve for x.

$$\begin{aligned} 5x - 6y &= 2 \\ 5x - 6\left(\frac{1}{3}\right) &= 2 \\ 5x - 2 &= 2 \\ 5x &= 4 \\ x &= \frac{4}{5} \end{aligned}$$

The solution (x, y) is $\left(\frac{4}{5}, \frac{1}{3}\right)$.

Teach⇔Interact

Checking Key Concepts

Reasoning
Ask students how they decided whether to use addition or subtraction in questions 1–6.

Closure Question

Describe how you would decide to use addition or subtraction to solve a linear system of equations.
If the coefficients of one of the two variables are the same, then add. If the coefficients are opposites, then subtract.

Apply⇔Assess

Suggested Assignment

Core Course
Exs. 1–22, 24, 26–33

Extended Course
Exs. 1–33

Block Schedule
Day 43 Exs. 1–24, 26–33

Exercise Notes

Communication: Writing
Ex. 1 Encourage students to show all of their work and to explain their thinking, including how they decided whether to add or subtract in order to solve the system of equations.

Mathematical Procedures
Exs. 2–10 Students should write their solutions by using ordered pairs to indicate which variable corresponds to which number. For instance, in Ex. 2, the solution (x, y) is $(6, 2)$. Point out to students that solutions are usually written in alphabetical order such as (x, y), (a, b), (r, s), and so on. Remind students to check their solutions by substitution.

CHECKING KEY CONCEPTS

Tell whether you would use addition or subtraction to solve each system of equations. Then solve the system.

1. $3x + y = 10$
$x - y = 2$

2. $4x + 3y = 14$
$-4x + 5y = 2$

3. $4a - 2b = -6$
$3a - 2b = 8$

4. $4x + 5y = 55$
$2x + 5y = 35$

5. $2a + 3b = -12$
$2a - 3b = 0$

6. $8x + 3y = 20$
$5x + 3y = 17$

7.3 Exercises and Applications

Extra Practice *exercises on page 570*

1. Writing At the Peking Sky restaurant, you can order small, medium, or large plates. The waiters keep track of how much you spend by stamping the appropriate section of your bill each time you ask for a plate. Susan Kim went to Peking Sky with a large group of friends for *dim sum*. They sat at two tables and each table got a separate bill.

a. At Peking Sky, how much does a medium plate cost? How do you know?

b. How much does a large plate cost? How do you know?

Solve each system of equations by adding or subtracting.

2. $x + y = 8$
$x - y = 4$

3. $2a + b = 5$
$a - b = 1$

4. $6 = 5n + 2p$
$22 = 9n + 2p$

5. $c - 3d = 7$
$c + 2d = 2$

6. $3r - 5s = -35$
$-2r + 5s = 30$

7. $13a + 5b = -11$
$13a + 11b = 7$

8. $-4x = -7 + 5y$
$8x = 9 - 5y$

9. $-13 = 3x + 5y$
$-5 = 3x + y$

10. $a - 2b - 9 = 0$
$-2b + 3a = 9$

11. PERSONAL FINANCE John Haag knows that the Thacker Tree Company charges its customers a certain amount per cord of fire wood, plus a fixed price for delivery. The checks that John wrote for the last two deliveries were \$260 for 1 cord of wood and \$377.50 for $1\frac{1}{2}$ cords of wood. How much will it cost to have 2 cords of wood delivered?

Checking Key Concepts

1. addition; (3, 1)
2. addition; (2, 2)
3. subtraction; (–14, –25)
4. subtraction; (10, 3)
5. addition; (–3, –2)
6. subtraction; $\left(\frac{37}{3}, -\frac{236}{9}\right)$

Exercises and Applications

1. a. \$2.15; From the two bills, you can write the system of equations $2m + 3l = 15.55$ and $3m + 3l = 17.70$; solving the system by subtraction yields $m = 2.15$.
b. \$3.75; Substituting 2.15 for m in $2m + 3l = 15.55$ yields $l = 3.75$.
2. (6, 2)
3. (2, 1)
4. (4, –7)
5. (4, –1)
6. (–5, 4)
7. (–2, 3)
8. $\left(\frac{1}{2}, 1\right)$
9. (–1, –2)
10. (0, –4.5)
11. \$495
12. (2, 1)
13. $\left(-3, -\frac{2}{5}\right)$
14. $\left(\frac{5}{3}, -\frac{2}{3}\right)$
15. (1, –2)
16. (–3, 12)
17. (7, 5)

Solve each system of equations by graphing, substituting, adding, or subtracting.

12. $y = 5x - 9$
$3x + 4y = 10$

13. $3a + 5b = -11$
$6a - 5b = -16$

14. $6c + 9d = 4$
$6c + 3d = 8$

15. $4y - 3z = 10$
$y + 4z = -7$

16. $3y = 21 - 5x$
$x + y = 9$

17. $2y + x = 17$
$3y = x + 8$

Technology **Graph each system of equations. Add corresponding sides of the equations and graph the resulting equation with the original system. What do you notice about the graph of the third equation? Use a graphing calculator if you have one.**

18. $2x + 3y = 6$
$-2x + 5y = 10$

19. $2x + 3y = 6$
$3x - 2y = 24$

20. Open-ended Problem How can you tell if these three equations have a common solution?

$3x - 5y = 1$ $\quad$ $4x - y = 24$ $\quad$ $2x - 3y = 2$

Connection COOKING

Some recipes, like the ones shown for bread pudding and quiche, are excellent for using leftover ingredients. If you don't have enough ingredients to follow a recipe exactly, you can make smaller batches by scaling down the amounts.

Bread Pudding
5 cups milk
3 eggs
3 cups bread pieces
Quiche
1 cup milk
4 eggs
3/4 cup mozzarella cheese
Flan
2 cups milk
6 eggs
1/2 cup sugar

21. If you want to make bread pudding but only have two eggs, how many cups of bread should you use?

22. Sandi Gubin wants to use up 14 eggs and 12 cups of milk. How many quiches and bread pudding recipes can she make? (*Hint:* Write a system of equations that involves the number of quiches, x, and number of bread pudding recipes, y, and the ingredients.)

23. Challenge Paul Gill wants to use up one dozen eggs and a half gallon of milk. (*Note:* 1 gal = 16 cups) He decides to make some flan and bread pudding. How many recipes of each can he make?

Apply⇔Assess

Exercise Notes

Student Progress
Ex. 11 Solving problems in two variables that require writing a system of equations, such as in this exercise, is not easy for some students. First they have to be able to translate the conditions of the problem into equations, and then they have to apply newly learned skills to solve the system. Although progress may be slow at first for some students, understanding and skill will increase through experience.

Cooperative Learning
Exs. 12–17 These exercises can be done by students working in groups of three. Different students can solve the same system by different methods. They should then compare their results and discuss which method seems best for each system. All solutions should routinely be checked.

Using Technology
Exs. 18, 19 Students can use a graphing calculator or the *Function Investigator* software.

18. 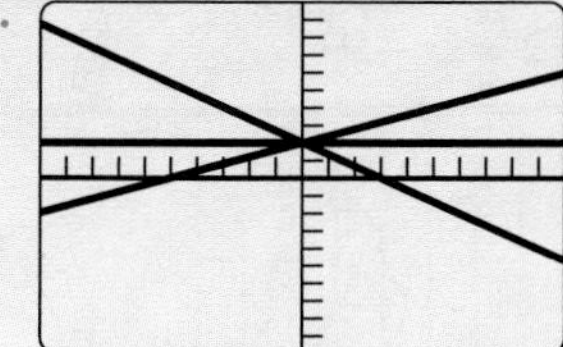

The graph of the third equation contains the intersection point of the system.

19. 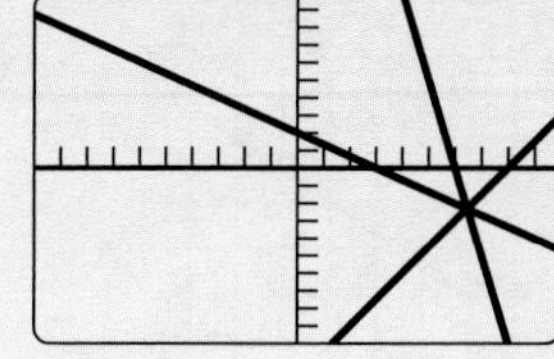

The graph of the third equation contains the intersection point of the system.

20. Answers may vary. Examples are given. Graph all three equations in the same coordinate plane and see if there is one point that is on all three graphs. Or, solve a system of two of the equations, and see if the coordinates of the intersection point make the third equation true.

21. 2 cups

22. two bread pudding recipes and two quiches

23. one bread pudding recipe and one and one-half flans

Apply⇔Assess

Exercise Notes

Communication: Discussion
Ex. 24 After discussing this exercise with the class, call on volunteers to make up similar problems for traveling upstream and downstream on a river. Students should be able to do this with little difficulty and can present their problems to the class orally. The current of a calm river is 3 or 4 mi/h.

Assessment Note
Ex. 26 Students should describe each method carefully and illustrate it with an example. A good clue to how well students understand solving systems of equations is their awareness that most systems can be solved by all four methods. In order to decide when one particular method is best requires using individual judgment.

Practice 46 for Section 7.3

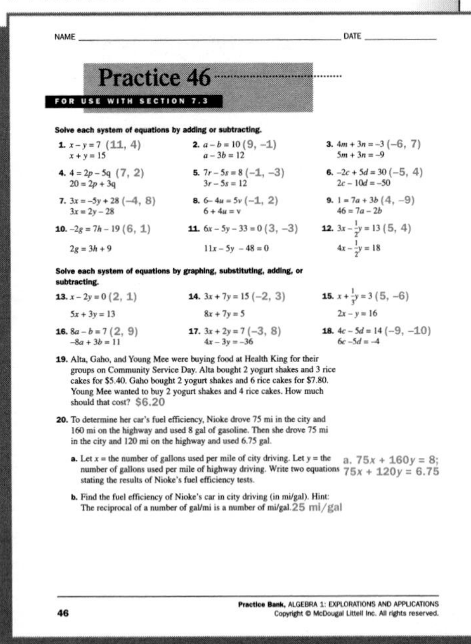
NAME ______________________ DATE __________

Practice 46

FOR USE WITH SECTION 7.3

Solve each system of equations by adding or subtracting.

1. $x - y = 7$, $x + y = 15$ (11, 4)
2. $a - b = 10$, $a - 3b = 12$ (9, −1)
3. $4m + 3n = -3$, $5m + 3n = -9$ (−6, 7)
4. $4 = 2p - 5q$, $20 = 2p + 3q$ (7, 2)
5. $7r - 5s = 8$, $3r - 5s = 12$ (−1, −3)
6. $-2c + 5d = 30$, $2c - 10d = -50$ (−5, 4)
7. $3x = -5y + 28$, $3x = 2y - 28$ (−4, 8)
8. $6 - 4u = 5v$, $6 + 4u = v$ (−1, 2)
9. $1 = 7a + 3b$, $46 = 7a - 2b$ (4, −9)
10. $-2g = 7h - 19$, $2g = 3h + 9$ (6, 1)
11. $6x - 5y - 33 = 0$, $11x - 5y - 48 = 0$ (3, −3)
12. $3x - \frac{1}{2}y = 13$, $4x - \frac{1}{2}y = 18$ (5, 4)

Solve each system of equations by graphing, substituting, adding, or subtracting.

13. $x - 2y = 0$, $5x + 3y = 13$ (2, 1)
14. $3x + 7y = 15$, $8x + 7y = 5$ (−2, 3)
15. $x + \frac{1}{3}y = 3$, $2x - y = 16$ (5, −6)
16. $8a - b = 7$, $-8a + 3b = 11$ (2, 9)
17. $3x + 2y = 7$, $4x - 3y = -36$ (−3, 8)
18. $4c - 5d = 14$, $6c - 5d = -4$ (−9, −10)

19. Alta, Gaho, and Young Mee were buying food at Health King for their groups on Community Service Day. Alta bought 2 yogurt shakes and 3 rice cakes for \$5.40. Gaho bought 2 yogurt shakes and 6 rice cakes for \$7.80. Young Mee wanted to buy 2 yogurt shakes and 4 rice cakes. How much should that cost? \$6.20

20. To determine her car's fuel efficiency, Nioke drove 75 mi in the city and 160 mi on the highway and used 8 gal of gasoline. Then she drove 75 mi in the city and 120 mi on the highway and used 6.75 gal.
   a. Let x = the number of gallons used per mile of city driving. Let y = the number of gallons used per mile of highway driving. Write two equations stating the results of Nioke's fuel efficiency tests. a. $75x + 160y = 8$; $75x + 120y = 6.75$
   b. Find the fuel efficiency of Nioke's car in city driving (in mi/gal). Hint: The reciprocal of a number of gal/mi is a number of mi/gal. 25 mi/gal

46 **Practice Bank,** ALGEBRA 1: EXPLORATIONS AND APPLICATIONS

Los Angeles

New York City

24. AVIATION On a typical day with mild winds, the 2459 mi flight from Los Angeles to New York City takes about $5\frac{1}{4}$ h. The return flight takes about $\frac{3}{4}$ h longer because the plane is flying into the wind.

a. What is the plane's average speed on the way to New York City? on the return flight?

b. Let s = the speed of the plane when there is no wind.
Let w = the speed of the wind.
On the way to New York City, the plane's total speed is $s + w$.
Write a variable expression for the plane's speed on the return trip.

c. Use your answers to parts (a) and (b) to write and solve a system of equations. How fast would the plane fly if there were no wind?

25. Cooperative Learning Work with a partner. You and your partner will explore two methods for finding an equation in the form $y = mx + b$ for the line through the points (3, 2) and (5, 10). One of you should do part (a). The other should do part (b). Work together on part (c).

a. Look back at Section 3.5. Use the method described in Example 2 on page 126.

b. Substitute 3 for x and 2 for y in the equation $y = mx + b$. Then substitute 5 for x and 10 for y in the equation $y = mx + b$. Solve this system of equations to find m and b. Then write an equation.

c. Explain the method you used to your partner. Make sure that you understand both methods. Which method do you prefer? Why?

ONGOING ASSESSMENT

26. Writing Describe four methods for solving systems of equations. When can you use each method?

SPIRAL REVIEW

Solve each equation. *(Section 4.4)*

27. $3.2x + 8.7 = 2.3$

28. $0.4 - 2.54n = 0.108$

29. $\frac{2}{3}x - \frac{1}{3} = 7$

30. $\frac{a}{3} + \frac{a}{4} = 5$

For Exercises 31–33, graph each equation. Then write an inequality to represent the x-values when $y \le 2$. *(Section 4.5)*

31. $y = 2x - 4$

32. $y = \frac{2}{3}x + 2$

33. $y = -2x + 3$

24. a. about 468.4 mi/h; about 409.8 mi/h
b. $s - w$
c. $s + w = 468.4$; $s - w = 409.8$; about 439.1 mi/h

25, 26. See answers in back of book.

27. −2

28. about 0.11

29. 11

30. $\frac{60}{7} \approx 8.57$

31. $x \le 3$

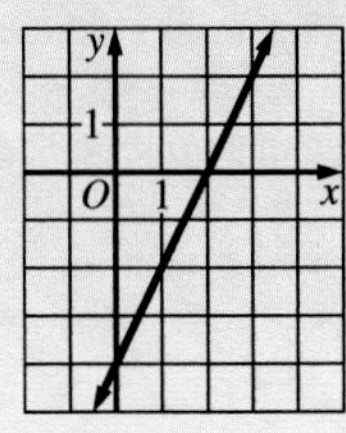

32. $x \le 0$

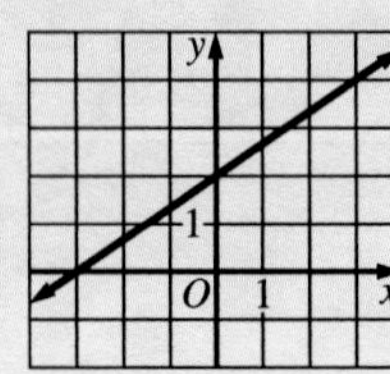

33. $x \ge \frac{1}{2}$

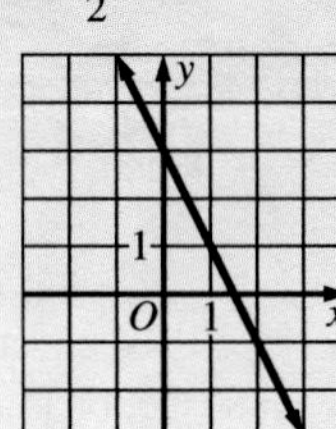

SECTION

7.4 Solving Linear Systems Using Multiplication

Learn how to...
- write equivalent systems of equations

So you can...
- make purchasing decisions, such as deciding how much fabric to buy

You can solve any system of linear equations by using multiplication to rewrite the system so that addition or subtraction will eliminate a variable. Complete the Exploration to discover how solving by multiplication works.

EXPLORATION
COOPERATIVE LEARNING

Graphing Multiple Equations

Work in a group of 3 students.
You will need:
- graph paper

1 Choose one of the equations below and graph it. Each of you should choose a different equation.

a. $2x + 2y = 5$

b. $4x + 4y = 10$

c. $6x + 6y = 15$

2 Compare your graphs. What do you notice?

3 Compare the three equations. How are they related?

4 Choose one of the systems of equations below and solve it by graphing. Each of you should choose a different system.

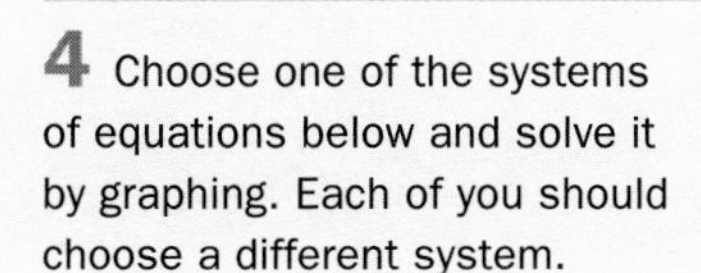

a. $2x + 2y = 5$
$5x - 6y = 7$

b. $4x + 4y = 10$
$5x - 6y = 7$

c. $6x + 6y = 15$
$5x - 6y = 7$

5 Compare your solutions. What do you notice? Explain your results.

6 Which of the systems in step 4 is easiest to solve algebraically? Why?

Exploration Note

The purpose of this Exploration is to have students discover how to solve a system of equations by using multiplication.

Materials/Preparation
Each member of a group should have two or three pieces of graph paper and a ruler. Allow students to organize themselves into groups.

Procedure
Group members should agree in advance on the scales for the *x*- and *y*-axes. This will make it easier to compare their graphs.

Closure
Groups should compare their results and discuss how they can be used to solve systems of equations.

Explorations Lab Manual
See the Manual for more commentary on this Exploration.

Diagram Master 2

For answers to the Exploration, see following page.

Plan⇔Support

Objectives

- Write equivalent systems of equations.
- Solve equivalent systems of equations that model real-world situations.

Recommended Pacing

Core and Extended Courses
Section 7.4: 2 days

Block Schedule
Section 7.4: 2 half-blocks (with Sections 7.3 and 7.5)

Resource Materials

Lesson Support
Lesson Plan 7.4
Warm-Up Transparency 7.4
Overhead Visuals:
Folder A: Multi-Use Graphing Packet, Sheets 1–3
Practice Bank: Practice 47
Study Guide: Section 7.4
Explorations Lab Manual:
Diagram Masters 1, 2
Assessment Book: Test 30

Technology
Graphing Calculator
Internet:
http://www.hmco.com

Warm-Up Exercises

Simplify.

1. $2(-7x + 5y)$ $-14x + 10y$

2. $9(2x + 8y)$ $18x + 72y$

3. $-6(4x - 3y)$ $-24x + 18y$

4. Add.
$15x - 18y$
$7x + 18y$
$22x$

5. Subtract.
$9x - 6y$
$9x - 2y$
$-4y$

Teach⇔Interact

Section Note

Student Study Tip

Point out to students that equivalent equations have the same solution. For example, $3x = 9$ and $6x = 18$ are equivalent equations because they both have the solution $x = 3$. Similarly, two systems are equivalent if the systems have the same solution.

Additional Example 1

Use multiplication to solve the system of equations from Example 1 in a different way.

$98a + 46s = 3500$

$a + s = 50$

Multiply one equation by 98 so the coefficients of a in both equations are the same.

$98a + 46s = 3500$

$a + s = 50 \quad \times 98$

Then use subtraction to solve the system.

$$\begin{aligned} 98a + 46s &= 3500 \\ -(98a + 98s) &= -(4900) \\ -52s &= -1400 \\ s &\approx 26.9 \end{aligned}$$

$$\begin{aligned} a + s &= 50 \\ a + 26.9 &\approx 50 \\ a &\approx 23.1 \end{aligned}$$

The solution (a, s) is about (23.1, 26.9).

Additional Example 2

Solve the system of equations.

$6x + 11y = 13$

$-5x + 7y = 70$

Transform both equations so the coefficients of x are opposites.

$6x + 11y = 13 \quad \times 5$

$-5x + 7y = 70 \quad \times 6$

Solve by addition.

$$\begin{aligned} 30x + 55y &= 65 \\ +(-30x + 42y) &= 420 \\ 97y &= 485 \\ y &= 5 \end{aligned}$$

$$\begin{aligned} 6x + 11y &= 13 \\ 6x + 11(5) &= 13 \\ 6x + 55 &= 13 \\ 6x &= -42 \\ x &= -7 \end{aligned}$$

The solution (x, y) is (−7, 5).

You can use multiplication to rewrite a system of equations as an equivalent system that can be solved by adding or subtracting.

EXAMPLE 1 Method: Multiplication

Use multiplication to solve the system of equations from Example 1 on page 287.

$\mathbf{98a + 46s = 3500}$

$\mathbf{a + s = 50}$

SOLUTION

Multiply one equation by **46**, so the coefficients of s in both equations are the same.

$98a + 46s = 3500$

$a + s = 50 \quad \times 46$

$$\begin{aligned} 98a + 46s &= 3500 \\ -(46a + 46s) &= -(2300) \\ 52a &= 1200 \\ a &\approx 23.1 \end{aligned}$$

Then use subtraction to solve the system.

$$\begin{aligned} a + s &= 50 \\ 23.1 + s &\approx 50 \\ s &\approx 26.9 \end{aligned}$$

The solution (a, s) is about (23.1, 26.9).

EXAMPLE 2

Solve the system of equations.

$\mathbf{5x + 3y = 2}$

$\mathbf{3x - 2y = 5}$

SOLUTION

Transform both equations so the coefficients of y are opposites.

$5x + 3y = 2 \quad \times 2$

$3x - 2y = 5 \quad \times 3$

$$\begin{aligned} 10x + 6y &= 4 \\ +(9x - 6y) &= +(15) \\ 19x &= 19 \\ x &= 1 \end{aligned}$$

Solve by addition.

$$\begin{aligned} 5x + 3y &= 2 \\ 5(1) + 3y &= 2 \\ 5 + 3y &= 2 \\ y &= -1 \end{aligned}$$

The solution (x, y) is $(1, -1)$.

WATCH OUT!

Always check your solutions by substituting the values of x and y in both equations.

ANSWERS Section 7.4

Exploration

1. a–c. The graph is the graph of all three equations.

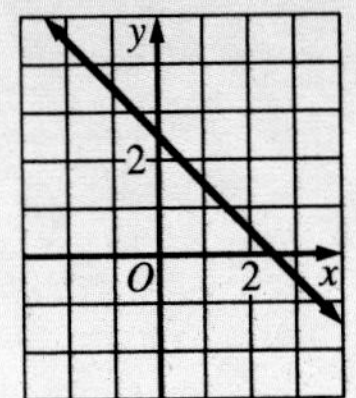

2. All three graphs are identical.

3. Given any two of the equations, one can be obtained by multiplying both sides of the other by some number.

4. a–c. The graph is the graph of all three systems.

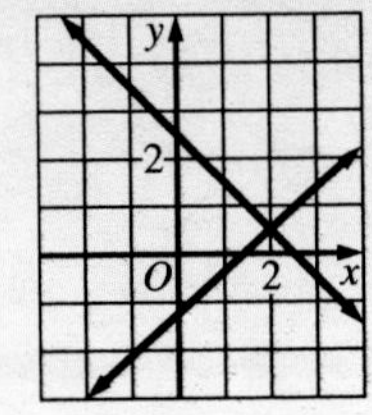

5. Each system has the solution $\left(2, \frac{1}{2}\right)$. This makes sense because each system consists of the equation $5x - 6y = 7$ and an equation with the same graph as $2x + 2y = 5$.

6. the system in part (c); Since the coefficients of y are opposites, this system can be solved by adding.

A system of equations can have one solution, many solutions, or no solutions. When you solve a system with **many solutions**, you get a statement that is **always true**. When you solve a system with **no solutions**, you get a statement that is **never true**.

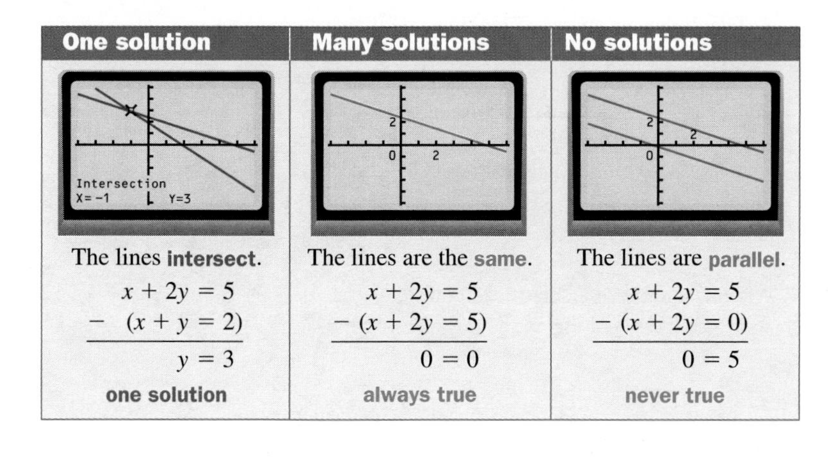

One solution	Many solutions	No solutions
The lines **intersect**.	The lines are the **same**.	The lines are **parallel**.
$x + 2y = 5$ $-\ (x + y = 2)$ $y = 3$	$x + 2y = 5$ $-\ (x + 2y = 5)$ $0 = 0$	$x + 2y = 5$ $-\ (x + 2y = 0)$ $0 = 5$
one solution	**always true**	**never true**

EXAMPLE 3 Application: Crafts

Ella Robinson needs 20 yd of fabric for a quilt. She has \$80 to spend and wants to buy two different fabrics. She likes the four fabrics shown.

A. 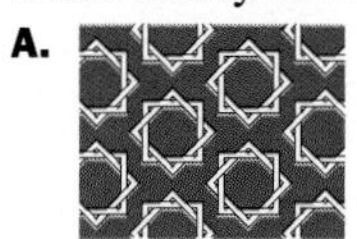 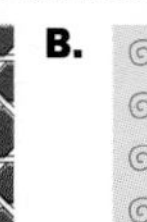 **$10/yd**

B. 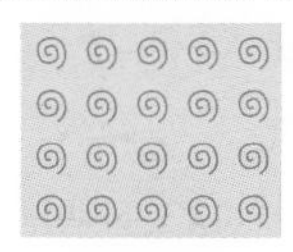 **$10/yd**

C. 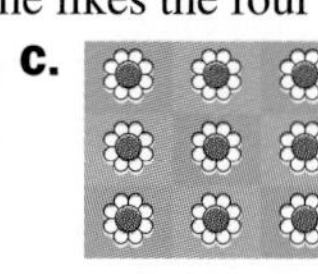 **$4/yd**

D. 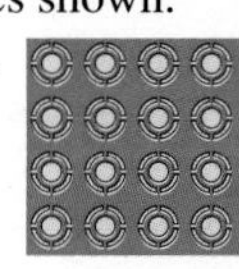 **$4/yd**

a. Ella thinks A and B will look the best in her quilt. How many yards of each can she buy?

b. How many yards each of fabrics C and D can Ella buy?

SOLUTION

a. Problem Solving Strategy: Use a graph.

Let a = amount of fabric A (in yards). $\quad a + b = 20$ (Ella wants 20 yd of fabric.)

Let b = amount of fabric B (in yards). $\quad 10a + 10b = 80$ (Ella has \$80 to spend.)

Graph the system on a graphing calculator.

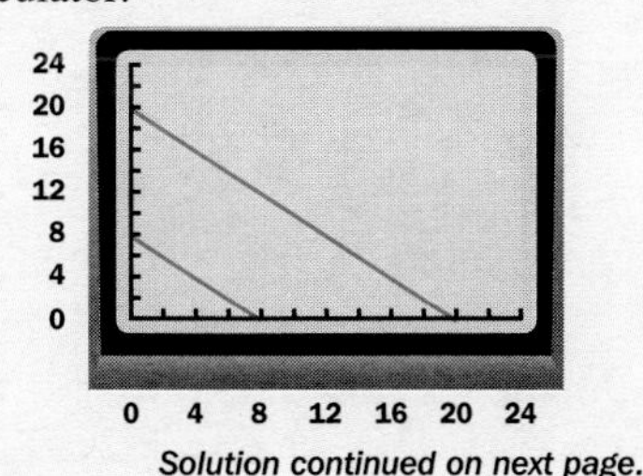

The lines are parallel, so they will never intersect. This means that the two equations do not have any common solutions.

Ella cannot buy 20 yd of fabrics A and B with \$80.

Solution continued on next page.

***Pyramid*, pieced by Gussie Wells, who began quilting at age nine and continued to quilt into her 80's.**

Teach⇔Interact

Section Notes

Teaching Tip

When discussing how many solutions a system of equations can have, emphasize what happens with the algebraic approach. When there are many solutions or no solutions, this approach results in equations that contain no variables.

Alternate Approach

Another way to predict the number of solutions of a system of equations is to write each equation in slope-intercept form. If the slopes are the same but the y-intercepts are different, the graphs are parallel lines and there are no solutions. If the slopes are the same and the y-intercepts are equal, there are many solutions since the graphs coincide. If the slopes are different, the graphs intersect in one point. In this case, there is exactly one solution.

Additional Example 3

Suppose Ella Robinson decided to find a shop where fabric was less expensive. She wanted to buy 30 yd of fabric and spend \$60. She found a shop where there were two fabrics she wanted, fabrics E and F, each costing \$8 per yard. There were two discontinued fabrics she also wanted, G and H, each costing just \$2 per yard.

a. How many yards of fabrics E and F can she buy?

Let e = yards of fabric E.
Let f = yards of fabric F.
The system of equations is:
$e + f = 30$
$8e + 8f = 60$
Graph the system on a graphing calculator.

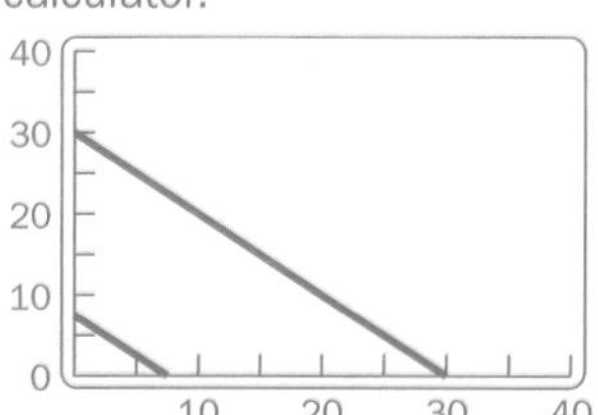

The lines are parallel, so there are no common solutions. She cannot buy 30 yd of fabric E and F with \$60.

Example continued on following page.

Teach⇔Interact

Additional Example 3 (continued)

b. How many yards of fabrics G and H can she buy?

Let g = yards of fabric G.
Let h = yards of fabric H.
The system of equations is:
$g + h = 30$
$2g + 2h = 60$
Graph the system on a graphing calculator.

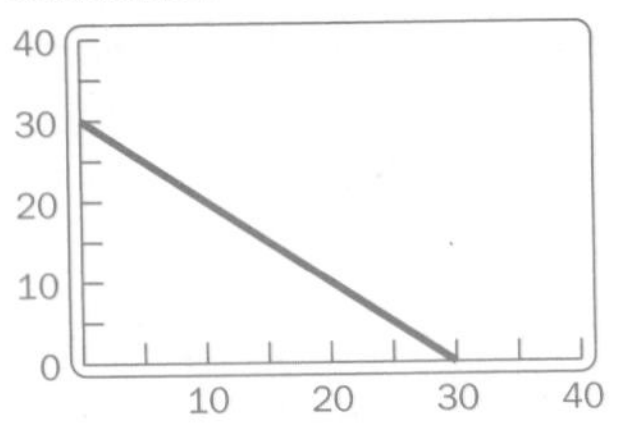

The equations have the same graph, so every solution of $g + h = 30$ is also a solution of $2g + 2h = 60$. She can buy as much of each fabric as she likes as long as she buys 60 yd total.

Think and Communicate

Discuss and compare the graphing approach and the algebraic approach. Ask students which approach they prefer and why.

Section Note

Top Spiraling: Review
Use the chart shown to help students review all the methods they have learned for solving systems of equations.

BY THE WAY

In Links and Lineage (1986), artist Paul Goodnight adapted African tribal designs from the Yoruba, the Dogon, the Ashanti, and other African tribes for the quilt design.

SOLUTION *continued*

b. Let c = the amount of fabric C (in yards). $c + d = 20$

Let d = the amount of fabric D (in yards). $4c + 4d = 80$

Graph the system on a graphing calculator.

The equations have the same graph, so every solution of $c + d = 20$ is also a solution of $4c + 4d = 80$.

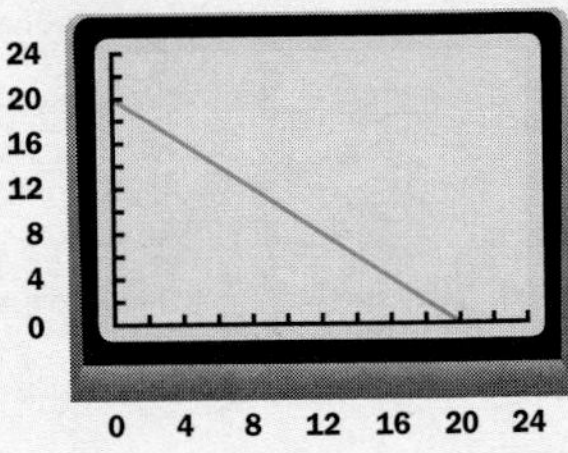

Ella can buy as much of each fabric as she likes, as long as she buys 20 yd total.

THINK AND COMMUNICATE

1. Solve each system in Example 3 by multiplying. Explain your results.
2. How does the system in part (b) of Example 3 change if Fabric C costs \$8/yd?

Choosing Strategies

You have learned many methods for solving systems of equations. When should you use each strategy?

Method	Example	When to use
Graphing	Intersection X=23.076923 Y=26.923077	When you want a visual understanding of the problem
Substitution	$y = 4 + 2x$ $2x + 7y = 3$	When one equation is already solved for x or y
Addition	$3x - 5y = 23$ $4x + 5y = 14$	When coefficients of one variable are opposites
Subtraction	$3x + 5y = 23$ $3x + 2y = 51$	When coefficients of one variable are the same
Multiplication	$5x + 3y = 2$ $3x + 2y = 5$	When no corresponding coefficients are the same

Think and Communicate

1. The first system results in the false statement $0 = 120$. The system has no solution. The second system results in the true statement $0 = 0$. Any solution of the equation $c + d = 20$ is a solution of the system.
2. The second equation in the system becomes $8c + 4d = 80$. The new system has a single solution, (0, 20). That is, Ella must buy none of fabric C and 20 yd of fabric D.

✓ CHECKING KEY CONCEPTS

Describe at least two ways to solve each system of equations.

1. $2x + 3y = -2$
$2x - 3y = -16$

2. $x - 5y = 15$
$7x + 9y = -2$

3. $2x + 5y = 13$
$9x + 4y = 3$

Solve each system of equations.

4. $2x + y = 4$
$-4x - 2y = -8$

5. $3x + 4y = 11$
$4x - 3y = 23$

6. $6 = 21x - 15y$
$3 = 7x - 5y$

7.4 Exercises and Applications

For each system of equations, give the solution or state whether there are *no solutions* or *many solutions*.

Extra Practice exercises on page 570

1. $3x + 8y = 1$
$7x + 4y = 17$

2. $-3x + 5y = -7$
$9x + 2y = 38$

3. $7r + 3s = 13$
$14r - 15s = -16$

4. $2x - 7y = 42$
$4x - 14y = 64$

5. $9x + 12y = 15$
$3x + 4y = 5$

6. $2 = 4s - 3t$
$2 = 12s - 9t$

Connection HISTORY

Mathematicians in ancient China used counting rods to solve systems of equations. The illustration at the right shows how these rods were used.

The illustration shows this system of equations:
$4x - 3y = 11 \quad -5x + 2y = -12$

A vertical rod stands for 1.
A horizontal rod stands for 10. 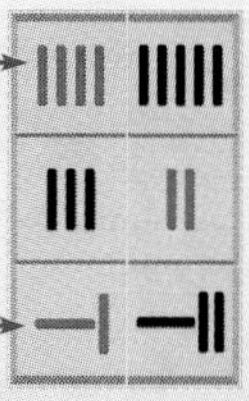
Black represents **negative**.
Red represents positive.

Use the illustration at the right to solve the system of equations in each rod diagram.

7.

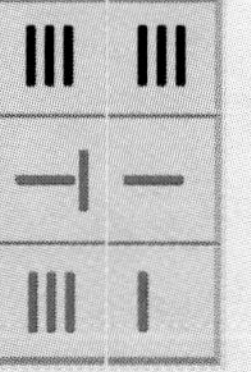

8.

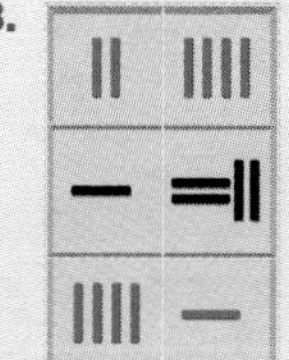

9.

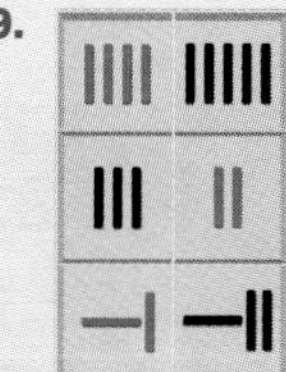

10. Challenge The problem at the right is from the *Jiŭzhāng suànshù,* a mathematical text described on page 210.

a. Draw a rod diagram to represent the problem.

b. What is the solution?

"Five large containers and one small container have a total capacity of 3 hu. One large container and five small containers have a capacity of 2 hu. Find the capacities of one large container and one small container."

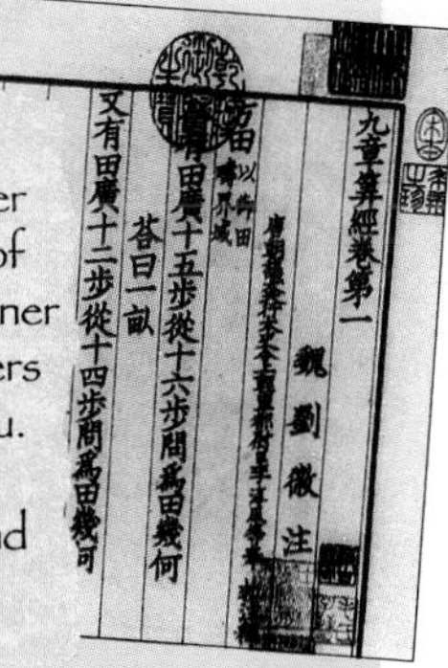

Teach⇔Interact

Checking Key Concepts

Teaching Tip
You may wish to ask students to solve the equations in questions 4–6 in two ways. Ask why they chose the methods they used.

Closure Question

Name five different methods for solving systems of equations and when you would use any one of them.
See the chart on page 302.

Apply⇔Assess

Suggested Assignment

Core Course
Day 1 Exs. 1–16
Day 2 Exs. 20–29, AYP

Extended Course
Day 1 Exs. 1–16
Day 2 Exs. 17–29, AYP

Block Schedule
Day 43 Exs. 1–16
Day 44 Exs. 17–29, AYP

Exercise Notes

Using Manipulatives
Exs. 1–6, 10 You may wish to have students do these exercises using counting rods as outlined in the history connection. The *x*-tiles from a set of algebra tiles would make an excellent set of counting rods.

Checking Key Concepts

1–3. Answers may vary. Examples are given.
1. addition or subtraction
2. graphing or substitution
3. graphing or multiplication
4. all (x, y) such that $2x + y = 4$
5. $(5, -1)$
6. no solution

Exercises and Applications

1. $(3, -1)$
2. $(4, 1)$
3. $(1, 2)$
4. no solution
5. many solutions (all (x, y) such that $3x + 4y = 5$)
6. no solution
7. $\left(6\frac{1}{3}, 2\right)$
8. $(-3, -1)$
9. $(2, -1)$
10. a.

b. large container, $\frac{13}{24}$ *hu*; small container, $\frac{7}{24}$ *hu*

Apply⇔Assess

Exercise Notes

Teaching Tip

Exs. 11–16 Ask students to predict in advance how many solutions each system has. Students should check their solutions by graphing.

Using Technology

Ex. 11 For this and many other systems of equations, the TI-82 can calculate exact answers from a graph. The procedure is illustrated here for this exercise.

Graph $2x + 5y = 4$ and $3x - 7y = -10$ in the usual way. For convenience, use the standard viewing window. Press [2nd][CALC]<intersect>. Use the left and right arrow keys ([◀] and [▶]) to move the blinking cursor as close as possible to the point where the graphs intersect. When the cursor is at this point, press [ENTER] *three* times. Approximate coordinates for the intersection point are automatically displayed. Next press [2nd][QUIT] to go to the home screen. Press [X,T,θ] [MATH] < ▶Frac>[ENTER]. The result is $-\frac{22}{29}$. Next press [ALPHA][Y] [MATH]< ▶Frac>[ENTER]. The result is $\frac{32}{29}$. The exact solution (x, y) of the system is $\left(-\frac{22}{29}, \frac{32}{29}\right)$. To check the solution by hand, substitute the x- and y-coordinates in the original system and simplify.

Second-Language Learners

Exs. 17–19 The English text is likely to prove very challenging to second-language learners because it deals with abstract ideas and includes difficult vocabulary, a considerable amount of figurative language, and unconventional sentence structures. It will be helpful to students to have this passage read aloud and then paraphrased by a native speaker of English before attempting these exercises.

Assessment Note

Ex. 23 This exercise will show how well students understand basic ideas about solutions of systems of linear equations. It may also reveal how well students understand the concepts of slope and intercepts.

Solve each system of equations.

11. $2x + 5y = 4$
$3x - 7y = -10$

12. $2x - 3y = 6$
$3x + 4y = -2$

13. $3a - 6b = 18$
$4a + 8b = 24$

14. $4x + 8y = 5$
$5x + 10y = 13$

15. $13 = 8a - 8b$
$6 = 4a - 5b$

16. $160 = 2x - 10y$
$120 = 3x - 15y$

Connection LITERATURE

Christian Morgenstern first published *Songs from the Gallows* in Germany in March 1905. Some of his verses were set to music, and some inspired the artist Paul Klee. Morgenstern's artistic strength is not so surprising, considering that his father and both of his grandfathers had been artists.

Die zwei Parallelen
by Christian Morgenstern

Es gingen zwei Parallelen
ins Endlose hinaus,
zwei kerzengerade Seelen
und aus solidem Haus.

Sie wolten sich nicht schneiden
bis an ihr seliges Grab:
Das var nun einmal der beiden
geheimer Stolz und Stab.

Doch als sie zehn Lichtjahre
gewandert neben sich hin,
da ward's dem einsamen Paare
nicht irdish nehr zu Sinn.

War'n sie noch Parallelen?
See wusstens selber nicht,
sie flossen nur wie zwei Seelen
zusammen durch ewiges Licht.

Das ewige Licht durchdrang sie,
da wurden sie eins in ihm;
die Ewigkeit verschlang sie,
als wie zwei Seraphim.

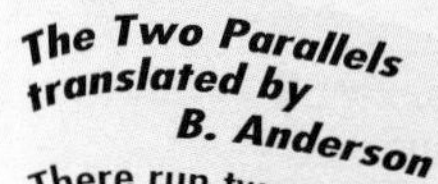

The Two Parallels
translated by
B. Anderson

There run two parallels
out into endlessness,
two candle-straight souls
of the best families.

They won't let themselves be cut
until their quiet grave:
This was actually once
their secret pride and staff.

But when they had wandered
ten light years side by side,
then the lonesome pair was
no longer of earthly mind.

Were they still parallel?
They didn't know themselves,
they only flow like two souls
together through eternal light.

The eternal light transfused them,
they were made one with it;
eternity had twisted them,
as would two seraphim.

17. Open-ended Problem Give an example of a system of equations whose graph could be described by this poem.

18. Writing Graph a system of equations and write a poem or story about the graph.

19. a. Research Find someone who can read German and ask him or her to read the original poem to you. Compare the original and its translation. What are the advantages and disadvantages of each?

b. When you graph a system of equations, you can think of it as translating from the language of symbols to the language of pictures. What are the advantages and disadvantages of representing a linear system by its equations and by its graph?

11. $\left(-\frac{22}{29}, \frac{32}{29}\right)$ 12. $\left(\frac{18}{17}, -\frac{22}{17}\right)$

13. (6, 0) 14. no solution

15. $\left(\frac{17}{8}, \frac{1}{2}\right)$ 16. no solution

17. Answers may vary. An example is given. $x + 4y = 8$ and $3x + 12y = 9$

18. Answers may vary.

19. a. The advantage of reading the poem in German is that you read exactly what the author intended without having changes forced by the use of English. Often a word exists in one language which has no exact translation in another. The obvious disadvantage is that you can read the original only if you read German. The advantage of reading the translation is that, even if you are unable to read German, you get some understanding of what the author was trying to say. The disadvantage is that something may be lost or altered in the translation.

b. A graph visually indicates how the equations are related. It is immediately obvious from the graph how many solutions the system has, along with their approximate values. It is more difficult to obtain the

20. An exam will have 20 questions worth a total of 100 points. There will be True-False questions worth 3 points each and short essay questions worth 11 points each. Several students ask how many essay questions there will be, but the teacher won't tell. How many essay questions will there be?

21. Sarah wants to buy 15 flower bulbs. Freesia bulbs cost \$3 and dahlia bulbs cost \$3. She has \$45 to spend. How many of each type of bulb can she buy?

22. a. **Visual Thinking** Match lines (a), (b), and (c) with equations (1), (2), and (3). Explain your reasoning.

(1) $x + y = 6$ (2) $x + 2y = 12$ (3) $x + 2y = 6$

b. What are the coordinates of points P and Q?

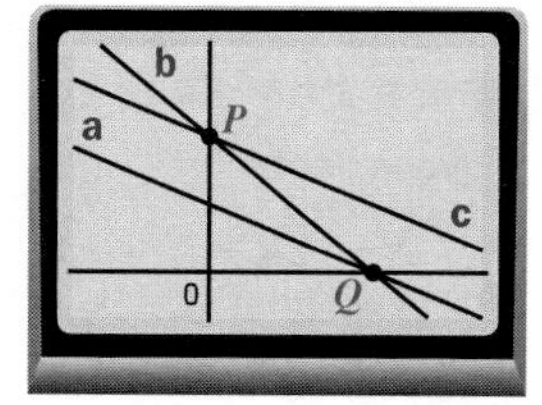

ONGOING ASSESSMENT

23. **Writing** Describe two ways to tell how many solutions a system of equations has.

SPIRAL REVIEW

Graph each equation and each point. Tell whether the ordered pair is a solution of the equation. *(Section 2.5)*

24. $y = 2x + 3$; $(0, 3)$ 25. $y = 6 - \frac{2}{3}x$; $(1, 5)$ 26. $y = 5x - 10$; $(2, 0)$

Graph each equation. *(Section 3.4)*

27. $y = 0$ 28. $y = 3x + 3$ 29. $x = 5$

ASSESS YOUR PROGRESS

Solve each system of equations. *(Sections 7.3 and 7.4)*

1. $7c + 9d = 13$
   $5c + 9d = 17$
2. $-4x + 11y = 93$
   $4x - 7y = -65$
3. $-4r + 5s = 27$
   $r - 6s = -2$
4. $5 = 2t - s$
   $20 = 8t - 4s$
5. $4t - 6s = 2$
   $-6t + 9s = -3$
6. $12a + 20b = 15$
   $9a + 15b = 8$

7. Emily Burnes's health plan has a monthly cost of \$35 and a health benefit for which she pays \$5 per visit to a fitness facility. One month she paid \$75 total. Adam Katz has the same health plan and spent \$135 one month. Between the two of them, they went to the health facility a total of 28 times. How many times did each person visit the health facility?

8. **Journal** What is the purpose of using multiplication as the first step when solving a system of equations?

Apply⇔Assess

Assess Your Progress

If there is any indication that students are having trouble with the correct use of terminology, use the vocabulary list in the Assess Your Progress on page 293.

When discussing Exs. 1–6, have students describe all the methods they know for solving systems of linear equations. They can demonstrate their understanding of these methods by using each method with at least one of the exercises.

Journal Entry
Students might also be asked whether they think multiplication would ever be necessary if they were going to solve a system by graphing.

Progress Check 7.3–7.4

See page 319.

Practice 47 for Section 7.4

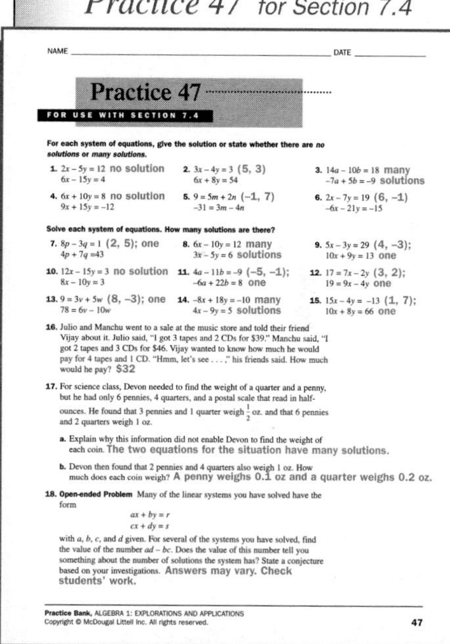

NAME ____________________ DATE __________

Practice 47

FOR USE WITH SECTION 7.4

For each system of equations, give the solution or state whether there are *no solutions* or *many solutions*.

1. $2x - 5y = 12$, $6x - 15y = 4$ no solution
2. $3x - 4y = 3$, $6x + 8y = 54$ (5, 3)
3. $14a - 10b = 18$, $-7a + 5b = -9$ many solutions
4. $6x + 10y = 8$, $9x + 15y = -12$ no solution
5. $9 = 5m + 2n$, $-31 = 3m - 4n$ (−1, 7)
6. $2x - 7y = 19$, $-6x - 21y = -15$ (6, −1)

Solve each system of equations. How many solutions are there?

7. $8p - 3q = 1$, $4p + 7q = 43$ (2, 5); one
8. $6x - 10y = 12$, $3x - 5y = 6$ many solutions
9. $5x - 3y = 29$, $10x + 9y = 13$ (4, −3); one
10. $12x - 15y = 3$, $8x - 10y = 3$ no solution
11. $4a - 11b = -9$, $-6a + 22b = 8$ (−5, −1); one
12. $17 = 7x - 2y$, $19 = 9x - 4y$ (3, 2); one
13. $9 = 3v + 5w$, $78 = 6v - 10w$ (8, −3); one
14. $-8x + 18y = -10$, $4x - 9y = 5$ many solutions
15. $15x - 4y = -13$, $10x + 8y = 66$ (1, 7); one

16. Julio and Manchu went to a sale at the music store and told their friend Vijay about it. Julio said, "I got 3 tapes and 2 CDs for \$39." Manchu said, "I got 2 tapes and 3 CDs for \$46. Vijay wanted to know how much he would pay for 4 tapes and 1 CD. "Hmm, let's see . . . ," his friends said. How much would he pay? \$32

17. For science class, Devon needed to find the weight of a quarter and a penny, but he had only 6 pennies, 4 quarters, and a postal scale that read in half-ounces. He found that 3 pennies and 1 quarter weigh $\frac{1}{2}$ oz. and that 6 pennies and 2 quarters weigh 1 oz.

a. Explain why this information did not enable Devon to find the weight of each coin. The two equations for the situation have many solutions.

b. Devon then found that 2 pennies and 4 quarters also weigh 1 oz. How much does each coin weigh? A penny weighs 0.1 oz and a quarter weighs 0.2 oz.

18. **Open-ended Problem** Many of the linear systems you have solved have the form

$$ax + by = r$$
$$cx + dy = s$$

with a, b, c, and d given. For several of the systems you have solved, find the value of the number $ad - bc$. Does the value of this number tell you something about the number of solutions the system has? State a conjecture based on your investigations. Answers may vary. Check students' work.

Practice Bank, ALGEBRA 1: EXPLORATIONS AND APPLICATIONS

47

equations from the graph than vice versa. The system of equations can be worked with algebraically to obtain numerical solutions readily. It is more difficult to tell from the equations alone how many solutions the system has.

20. five essay questions

21. She can buy any number of Freesia bulbs from 0 to 15 in combination with any number of dahlia bulbs from 15 to 0, just so the number of bulbs totals 15.

22. a. (1): (b); (2): (c); (3): (a); Reasoning may vary. An example is given. The graphs of equations (2) and (3) are parallel lines, so these are lines (a) and (c). This leaves line (b) as the graph of equation (1). Since the y-intercept of the graph of equation (2) is greater than that of equation (3), line (c) is the graph of equation (2).

b. $P(0, 6)$; $Q(6, 0)$

23. Answers may vary. An example is given. One way is to solve the system algebraically and see if you get a single solution, a statement that is always true, or a false statement. Another way is to graph the equations and see if you get intersecting lines, a single line, or two parallel lines.

24. Yes.

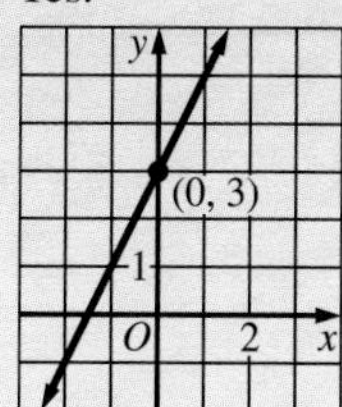

25–29. See answers in back of book.

Assess Your Progress

1–8. See answers in back of book.

Plan⇔Support

Objectives

- Graph linear inequalities.
- Visualize problems involving inequalities.

Recommended Pacing

Core and Extended Courses
Section 7.5: 2 days

Block Schedule
Section 7.5: 2 half-blocks
(with Sections 7.4 and 7.6)

Resource Materials

Lesson Support
Lesson Plan 7.5
Warm-Up Transparency 7.5
Overhead Visuals:
Folder A: Multi-Use Graphing Packet, Sheets 1–7
Practice Bank: Practice 48
Study Guide: Section 7.5
Explorations Lab Manual:
Additional Exploration 11,
Diagram Masters 2, 12

Technology
Internet:
http://www.hmco.com

Warm-Up Exercises

Find the value of each expression when $x = -4$ and $y = 7$.

1. $7x + 8y$ 28
2. $-3x + 2y$ 26
3. $5x - 13y$ −111

Tell whether the value of each expression is *greater than* 10, *less than* 10, or *equal to* 10 when $a = 5$ and $b = 2$.

4. $2a + 5b$ greater than
5. $-9a + 13b$ less than
6. $a - 6b$ less than

SECTION

7.5 Linear Inequalities

Learn how to...

- graph linear inequalities

So you can...

- visualize problems involving inequalities

You can tell that (5, 3) and (5, 4) are both solutions of the inequality $x - y \geq 1$ because they both make the inequality true. What other solutions does the inequality have? What does the graph of all of the solutions of the inequality look like?

EXPLORATION
COOPERATIVE LEARNING

Graphing $x - y \geq 1$

Work with a partner.
You will need:

- a ruler
- colored pencils

1 Make a large coordinate grid like the one in the picture. Both axes should extend from −4 to 4. Draw the grid lines.

2 At each point where two grid lines cross, write the value of $x - y$ for that point. For example, at the point (3, −2), $x - y = 3 - (-2) = 5$.

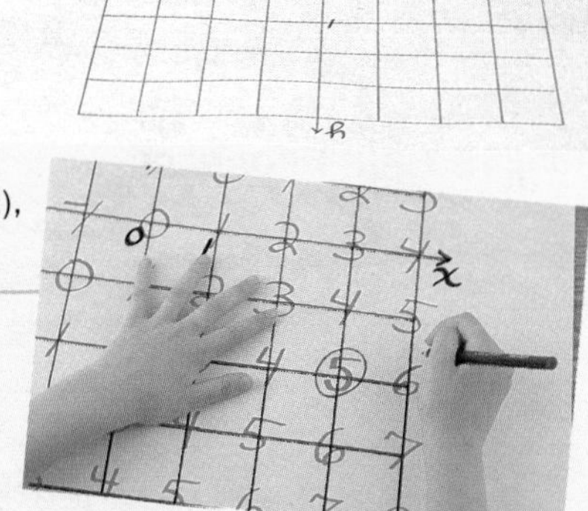

3 With a colored pencil, circle each point that represents a solution of the inequality $x - y \geq 1$. What do you notice?

4 Find the point (2.5, 0.5) on your grid. Do you think this is a solution of the inequality? Do you think (2.5, 1.5) is a solution of the inequality? Check your answers.

5 Shade your grid to show all of the solutions of the inequality $x - y \geq 1$. How is the border between the shaded and the unshaded regions related to the inequality $x - y \geq 1$? Explain.

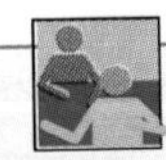

Exploration Note

Purpose
The purpose of this Exploration is to have students discover how to find and graph the solutions to an inequality.

Materials/Preparation
Each pair of students should have colored pencils and a ruler.

Procedure
Students should circle *all* points where grid lines intersect that are solutions of the inequality and discuss what they notice.

Closure
Have a class discussion of the results of this Exploration. Students should notice that the region of the coordinate plane shaded in color shows all points on the line $x - y = 1$ and all points *below* the line.

Explorations Lab Manual
See the Manual for more commentary on this Exploration.

Diagram Master 12

The inequality $x - y \geq 1$ is an example of a **linear inequality**. As you saw in the Exploration, the graph of a linear inequality is a region on a coordinate plane whose edge is a line.

EXAMPLE 1

Graph the inequality $y \leq 4 - \frac{x}{2}$.

SOLUTION

Step 1 Graph the equation $y = 4 - \frac{x}{2}$. The line shows all ordered pairs for which y equals $4 - \frac{x}{2}$.

Step 2 Now shade all points for which y is less than $4 - \frac{x}{2}$.

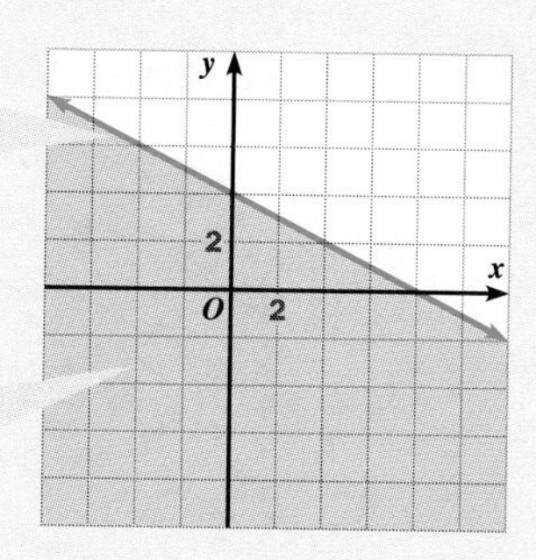

Check

Test any point in the shaded region, such as (0, 0).

$y \leq 4 - \frac{x}{2}$

$0 \overset{?}{\leq} 4 - \frac{0}{2}$

$0 \leq 4$ ✓

THINK AND COMMUNICATE

1. In Example 1, why should you shade all points *below* the line? What does *below* the line mean?

EXAMPLE 2

Graph the inequality $3x - 2y < 12$.

SOLUTION

Step 1 Solve $3x - 2y < 12$ for y.

$$3x - 2y < 12$$
$$-2y < 12 - 3x$$
$$y > -6 + \frac{3}{2}x$$

WATCH OUT!

When you multiply or divide both sides of an inequality by a negative number, remember to switch the direction of the inequality sign.

Step 2 Graph the equation $y = -6 + \frac{3}{2}x$.

Make the line dashed, to show that the points on the line are not solutions of the inequality.

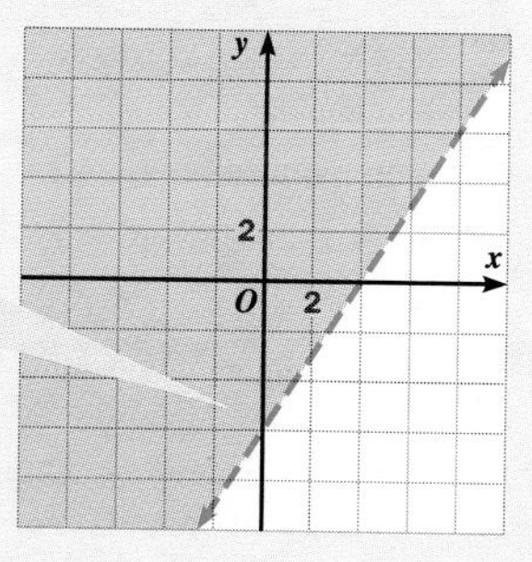

Step 3 Shade all points above the line.

Teach⇔Interact

About Example 1

Visual Thinking

Ask students to state the coordinates of some points along the line in the graph of the Solution. Ask them to discuss the meaning of the shaded area. This activity involves the visual skills of *correlation* and *communication.*

Reasoning

In the check of the Solution, ask students why (0, 0) is a good point to test. (Since the terms involving x and y become 0, the calculations are simple.)

Additional Example 1

Graph the inequality $y \geq 2x - 6$.

Step 1 Graph the equation $y = 2x - 6$. The line shows all ordered pairs for which y equals $2x - 6$.

Step 2 Now shade all points for which y is greater than $2x - 6$.

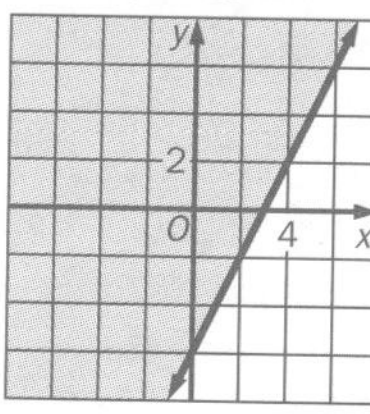

Check

Test the point (0, 0).

$y \geq 2x - 6$

$0 \overset{?}{\geq} 2(0) - 6$

$0 \geq -6$ ✓

Additional Example 2

Graph the inequality $0.5x - 1.5y > 6$.

Step 1 Solve $0.5x - 1.5y > 6$ for y.

$$0.5x - 1.5y > 6$$
$$-1.5y > 6 - 0.5x$$
$$y < -4 + \frac{1}{3}x$$

Step 2 Graph the equation $y = -4 + \frac{1}{3}x$. Make the line dashed to show that the points on the line are not solutions of the inequality.

Step 3 Shade all points below the line.

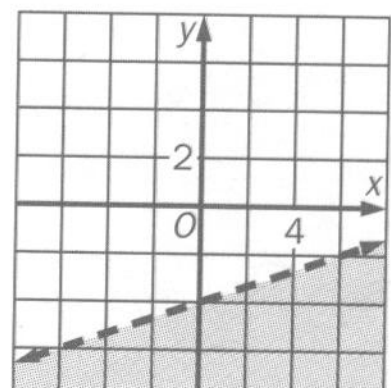

ANSWERS Section 7.5

Exploration

1–3.

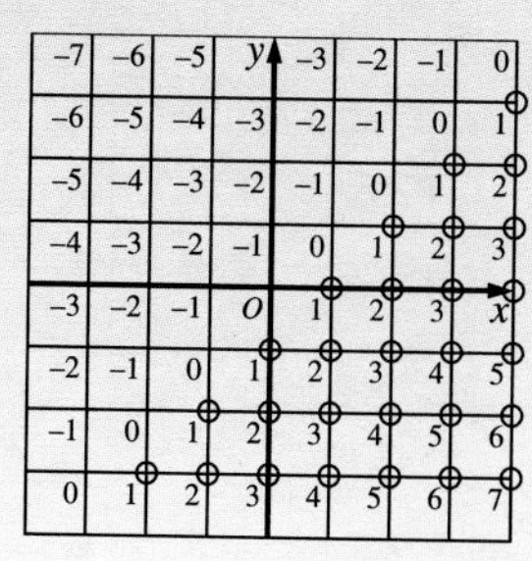

3. Answers may vary. An example is given. All the circled points lie below and to the right of the points labeled with zeros

4. Yes; Yes.

5.

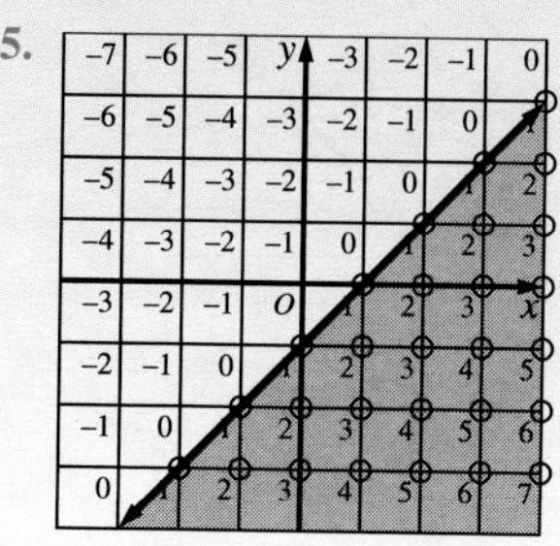

The border is the graph of the equation $x - y = 1$. Points on the line make the sides $x - y$ and 1 of the inequality $x - y \geq 1$ equal.

Think and Communicate

1. because the inequality $y \leq 4 - \frac{x}{2}$ contains the points for which y is *less than* or equal to $4 + \frac{x}{2}$; in the lower of the two half-planes determined by the line

Teach⇔Interact

Think and Communicate

Students can easily remember when to include the boundary line in the graph of an inequality by examining the inequality symbol. When the symbol includes the "equal to" part, a solid line is used to include the boundary solutions.

Checking Key Concepts

Teaching Tip
Students should associate "$y >$" and "$y \geq$" with points above the boundary and associate "$y <$" and "$y \leq$" with points below the boundary.

Section Note

Challenge
After discussing the questions in Checking Key Concepts, students might consider whether it would be possible to draw the graph of a linear inequality by solving for x instead of y. If so, how would one decide which region to shade. (Yes, it is possible. Decide the boundary line question in the same way as before. If the inequality has been solved for x and has "$x <$" or "$x \leq$," then shade points to the left of the boundary. If the inequality has "$x >$" or "$x \geq$," then shade points to the right of the boundary.)

Closure Question

Describe the graph of the solutions of a linear inequality.
The graph is a region on a coordinate plane whose edge is a line.

Apply⇔Assess

Suggested Assignment

Core Course
Day 1 Exs. 1–22
Day 2 Exs. 23–34, 36–45

Extended Course
Day 1 Exs. 1–22
Day 2 Exs. 23–45

Block Schedule
Day 44 Exs. 1–22
Day 45 Exs. 23–34, 36–45

THINK AND COMMUNICATE

2. In Example 1, explain why the solutions of the equation $y = 4 - \frac{x}{2}$ are also solutions of the inequality $y \leq 4 - \frac{x}{2}$. In Example 2, explain why the solutions of $y = -6 + \frac{3}{2}x$ are not solutions of $y > -6 + \frac{3}{2}x$.
3. When should you use a dashed line to graph an inequality? When should you use a solid line?

CHECKING KEY CONCEPTS

Match each inequality with its graph.

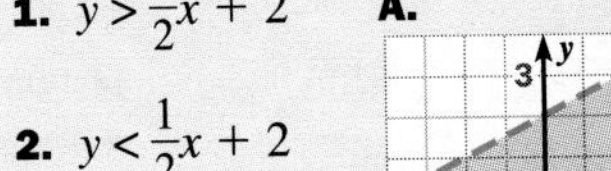

1. $y > \frac{1}{2}x + 2$
2. $y < \frac{1}{2}x + 2$
3. $y \geq \frac{1}{2}x + 2$

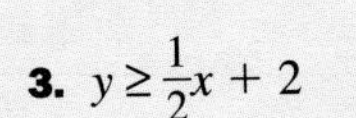

A.

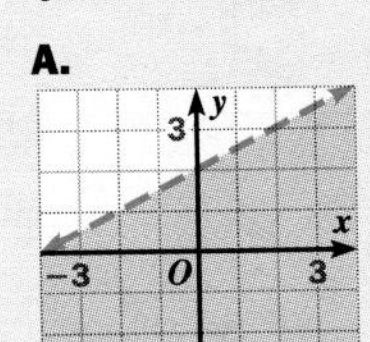

B.

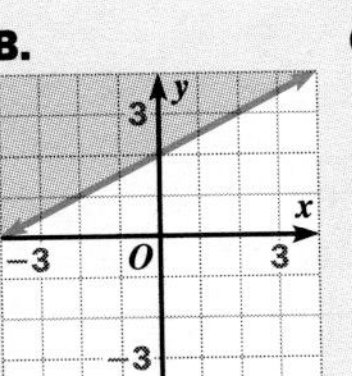

C.

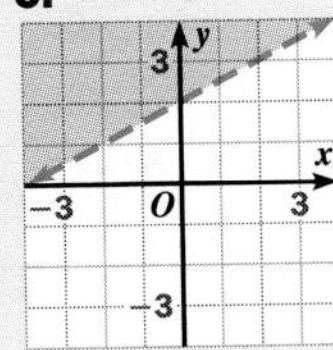

Graph each inequality.

4. $y \geq x + 1$
5. $x + y \leq 3$
6. $2x - 5y < 20$

7.5 Exercises and Applications

Extra Practice exercises on page 570

Match each inequality with its graph.

1. $y < -x + 2$
2. $y > -x + 2$
3. $y \leq 2$
4. $x \leq 2$

A.

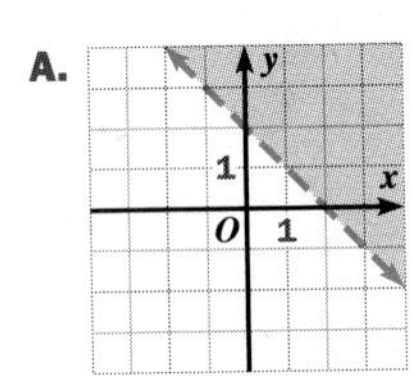

B.

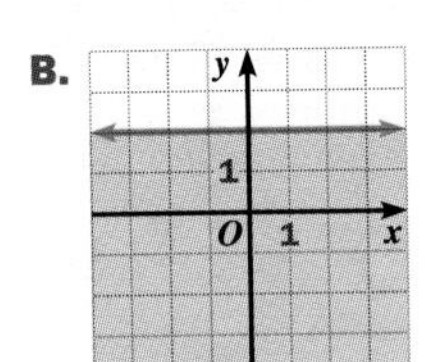

C.

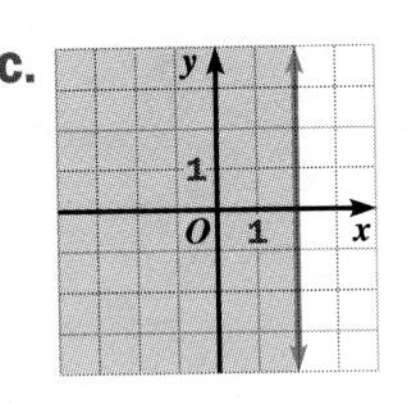

D.

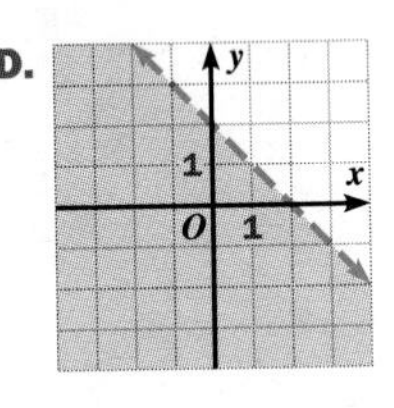

Graph each inequality.

5. $y > 2$
6. $x \geq -1$
7. $y \geq 2x$
8. $y < 3x$
9. $y \leq 2x - 1$
10. $y > 5 - \frac{1}{2}x$
11. $y < -\frac{1}{3}x + 3$
12. $x + 2y \leq 4$
13. $2x + 3y \geq 18$
14. $3x - y > 0$
15. $2x - y \geq 5$
16. $4x - 4y \leq 9$

Think and Communicate

2. because the symbol $\leq$ means *less than or equal to*; because the symbol $>$ means *greater than* and does not include points where the two sides of the inequality are equal
3. when the inequality symbol is $<$ or $>$; when the inequality symbol is $\leq$ or $\geq$

Checking Key Concepts

1. C
2. A
3. B
4.

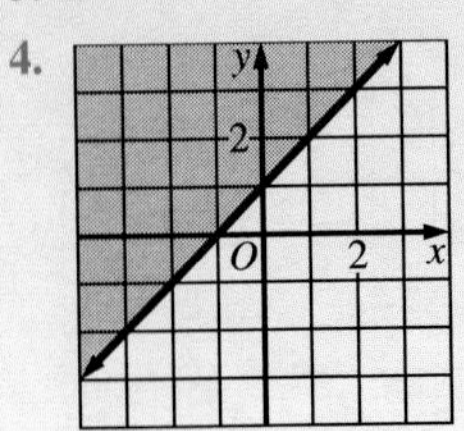

5.

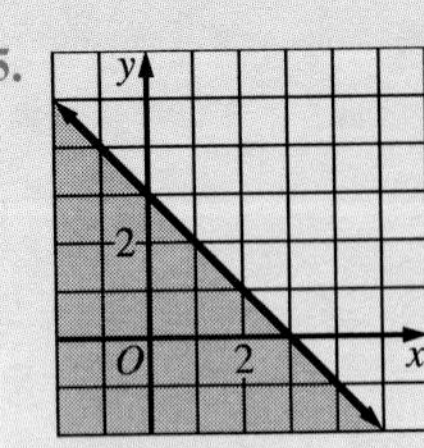

6.

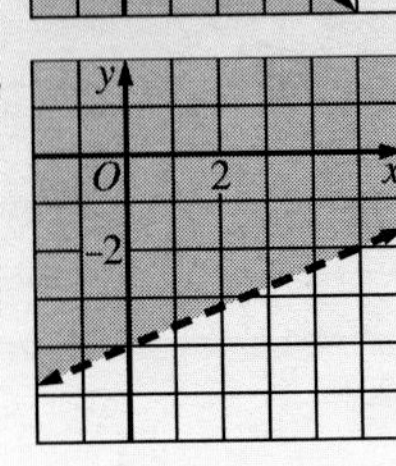

Cooperative Learning **Work with a partner. One of you should do Exercise 17. The other should do Exercise 18. Work together on Exercise 19.**

17. Use the method described in Example 2 on page 307 to graph the inequality $4x - 3y \geq 12$.

18. a. Graph the equation $4x - 3y = 12$ by finding the x- and y-intercepts and drawing a line through them.

b. Pick a point not on the line. Do the coordinates of the point make the inequality $4x - 3y \geq 12$ true? What does this tell you about the graph of the inequality $4x - 3y \geq 12$?

c. Graph the inequality $4x - 3y \geq 12$.

19. Explain to your partner the method you used. Make sure that you both understand both methods. Which method do you prefer? Why?

Graph each inequality.

20. $2x + 4y > 5$ **21.** $x + 2y \leq 7$ **22.** $-3x + 2y < 9$

23. $5x - 3y \leq 10$ **24.** $3x - 8y > 10$ **25.** $-4x - 7y \geq 18$

26. $-\frac{2}{3}x + 4y \geq \frac{5}{6}$ **27.** $2.3x - 4.9y < 5.2$ **28.** $\frac{x}{2} - \frac{y}{3} < \frac{1}{6}$

INTERVIEW Nancy Clark

Look back at the article on pages 278–280.

In her Sports Nutrition Guidebook, Nancy Clark advises athletes to choose low-fat sources of calcium. Teenagers need at least 1200 mg of calcium each day to help build strong bones.

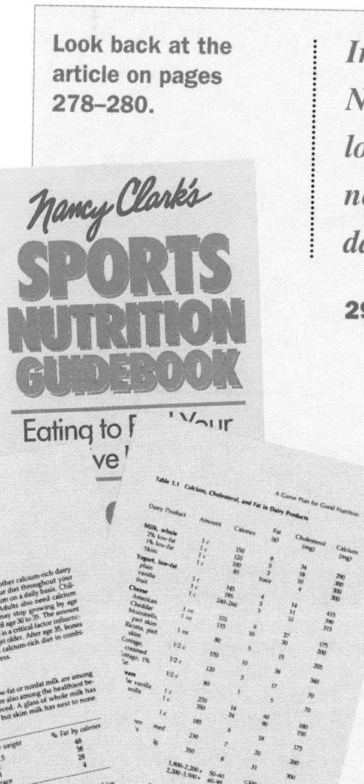

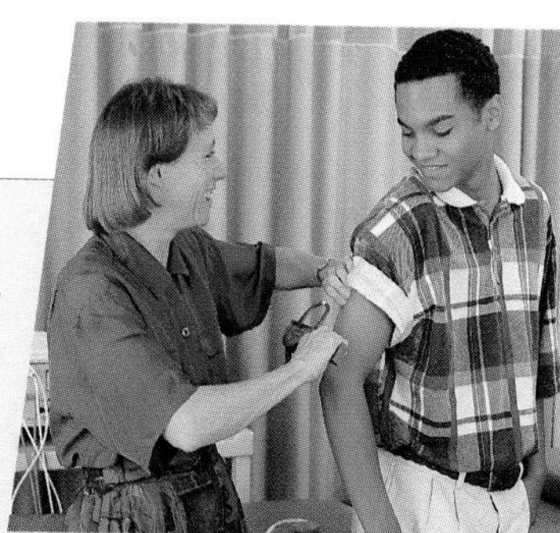

Nancy Clark uses calipers to help determine a healthy level of body fat for young adults.

29. One cup of low-fat milk supplies 300 mg of calcium. One cup of vanilla ice cream provides 180 mg of calcium.

a. Write an inequality stating that x cups of milk and y cups of ice cream supply more than 1200 mg of calcium.

b. Graph the inequality and list two different solutions.

30. One cup of 2% low-fat milk contains 5 g of fat. One cup of vanilla ice cream contains 14 g of fat. If you eat 2000 calories per day, you should eat between about 55 g and 65 g of fat.

a. Write an inequality stating that x cups of milk and y cups of ice cream contain less than 65 g of fat.

b. Graph your inequality. List two different solutions of the inequality.

31. Challenge Can you get 1200 mg of calcium from milk and ice cream without eating more than 65 g of fat? Explain.

Apply⇔Assess

Exercise Notes

Communication: Discussion
Ex. 4 Be sure students understand that the value of the y-coordinate is irrelevant when deciding which ordered pairs are solutions of $x \leq 2$. Every pair with an x-coordinate less than 2 or equal to 2 is a solution and corresponds to a point of the graph.

Common Error
Exs. 5–16 Students often shade the wrong region when graphing linear inequalities. They can correct their errors by using a test point, as in Example 1. If the boundary line goes through (0, 0), they will need to use a point other than (0, 0) to be sure the correct region is shaded. Other easy test points are (0, 1), (1, 0), (0, –1), and (–1, 0).

Reasoning
Exs. 20–28 Ask students how they can quickly decide whether they will shade points above the boundary line or below it. They should be able to anticipate the direction of the inequality sign once they have solved it for y. (If the y-term is on the left side of the inequality and has a positive coefficient, the inequality sign will not change direction after the inequality has been solved for y. If the coefficient is negative, then the direction of the inequality will be reversed.)

Research
Exs. 29–31 Some students may be interested in doing a brief research report on the proper nutrition for teenagers. Such a report could be created in the form of a bulletin board display.

Exercises and Applications

1. D 2. A
3. B 4. C

5.
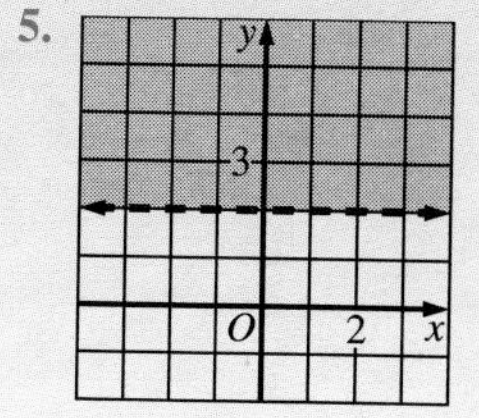

6.
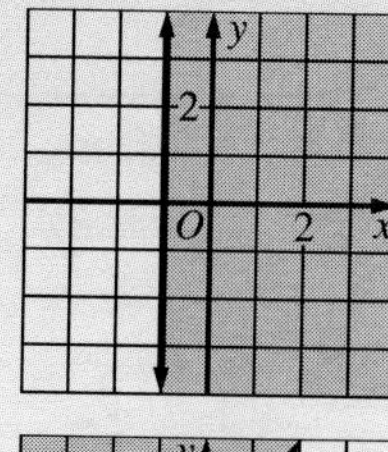

7.
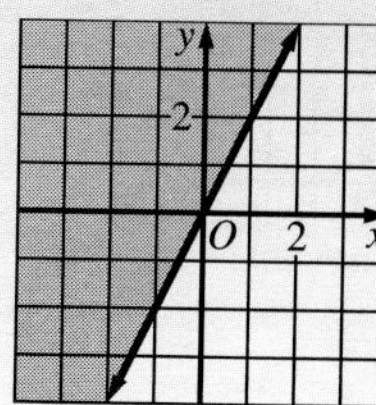

8.
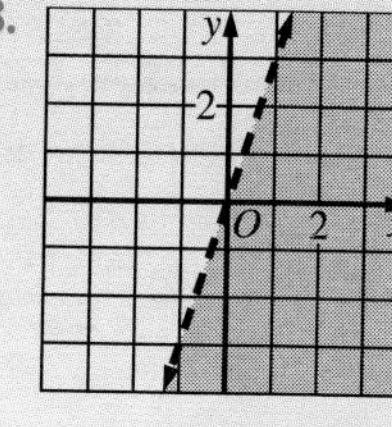

9.
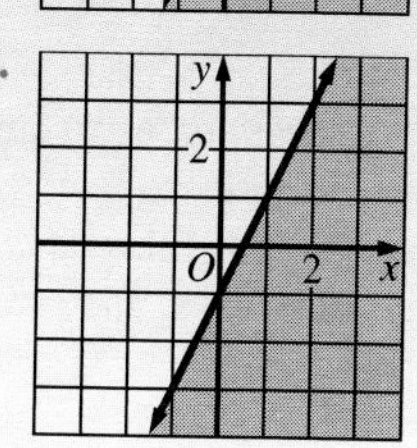

10., 11.
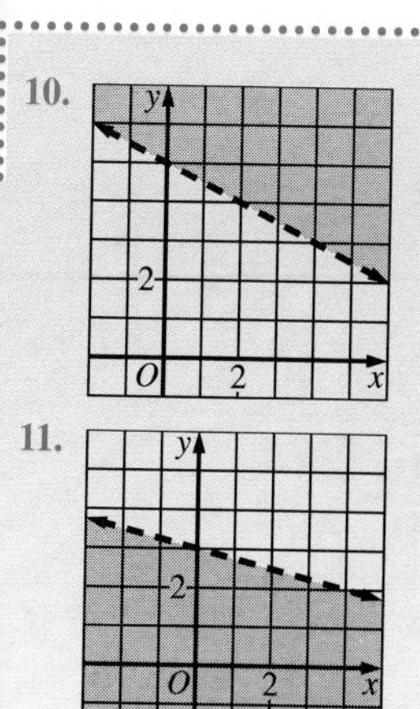

12–31. See answers in back of book.

Apply⇔Assess

Exercise Notes

Cooperative Learning

Ex. 34 This writing exercise can be assigned as a small group activity. Students can discuss each part of the exercise, write their responses, and review one another's work.

Assessment Note

Ex. 36 Check to see how clearly students formulate their situations. Each situation should result in a graph that includes the boundary line.

Topic Spiraling: Review

Exs. 37–42 Use these exercises to maintain skills that students will need when they study factoring in Chapter 10.

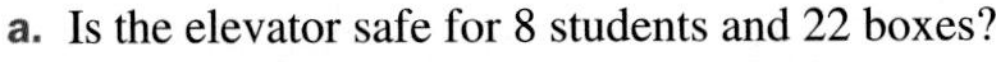

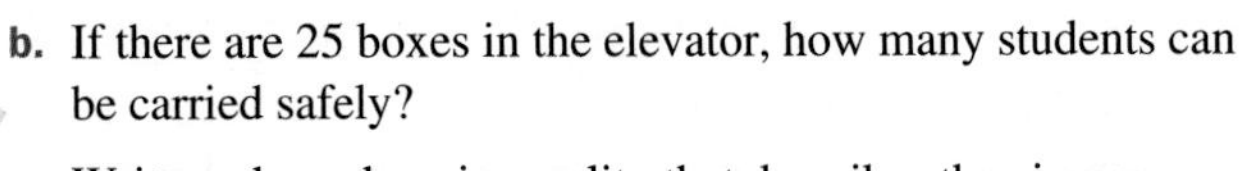

32. When students move into a college dormitory, the elevators get a great deal of use. Suppose an elevator has a 3000 lb weight limit, that a typical student weighs 150 lb, and a typical box weighs 60 lb.

a. Is the elevator safe for 8 students and 22 boxes?

b. If there are 25 boxes in the elevator, how many students can be carried safely?

c. Write and graph an inequality that describes the circumstances under which it is safe to load s students and b boxes into the elevator at the same time.

33. SAT/ACT Preview If $A = -y$ and $B = 2 - y$, then:

A. $A > B$ **B.** $B > A$ **C.** $A = B$ **D.** relationship cannot be determined

34. Writing For his birthday, Jacob got a movie gift certificate worth \$25. Matinees cost \$4 and evening shows cost \$7.

a. Write an equation stating that Jacob saw m matinees and n evening shows for \$25.

b. Write an inequality stating that Jacob saw m matinees and n evening shows for at most \$25.

c. Compare the equation from part (a) with the inequality from part (b). Which do you think best represents the situation? Why?

35. Open-ended Problem Choose a problem from Sections 7.1–7.4 that can be represented by either an equation or an inequality. Write and graph an inequality that describes the problem.

Practice 48 *for Section 7.5*

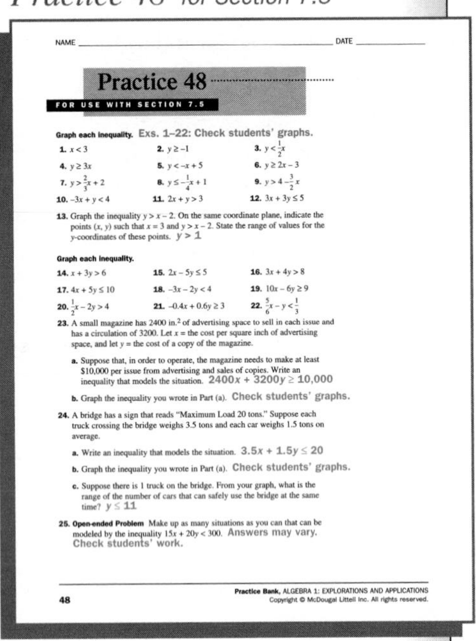

NAME ______________________ DATE __________

Practice 48

FOR USE WITH SECTION 7.5

Graph each inequality. Exs. 1–22: Check students' graphs.

1. $x < 3$
2. $y \geq -1$
3. $y < \frac{1}{2}x$
4. $y \geq 3x$
5. $y < -x + 5$
6. $y \geq 2x - 3$
7. $y > \frac{2}{3}x + 2$
8. $y \leq -\frac{1}{4}x + 1$
9. $y > 4 - \frac{3}{2}x$
10. $-3x + y < 4$
11. $2x + y > 3$
12. $3x + 3y \leq 5$
13. Graph the inequality $y > x - 2$. On the same coordinate plane, indicate the points (x, y) such that $x = 3$ and $y > x - 2$. State the range of values for the y-coordinates of these points. $y > 1$

Graph each inequality.

14. $x + 3y > 6$
15. $2x - 5y \leq 5$
16. $3x + 4y > 8$
17. $4x + 5y \leq 10$
18. $-3x - 2y < 4$
19. $10x - 6y \geq 9$
20. $\frac{1}{2}x - 2y > 4$
21. $-0.4x + 0.6y \geq 3$
22. $\frac{5}{6}x - y < \frac{1}{3}$
23. A small magazine has 2400 in.^{2} of advertising space to sell in each issue and has a circulation of 3200. Let x = the cost per square inch of advertising space, and let y = the cost of a copy of the magazine.
    a. Suppose that, in order to operate, the magazine needs to make at least \$10,000 per issue from advertising and sales of copies. Write an inequality that models the situation. $2400x + 3200y \geq 10,000$
    b. Graph the inequality you wrote in Part (a). Check students' graphs.
24. A bridge has a sign that reads "Maximum Load 20 tons." Suppose each truck crossing the bridge weighs 3.5 tons and each car weighs 1.5 tons on average.
    a. Write an inequality that models the situation. $3.5x + 1.5y \leq 20$
    b. Graph the inequality you wrote in Part (a). Check students' graphs.
    c. Suppose there is 1 truck on the bridge. From your graph, what is the range of the number of cars that can safely use the bridge at the same time? $y \leq 11$
25. **Open-ended Problem** Make up as many situations as you can that can be modeled by the inequality $15x + 20y < 300$. Answers may vary. Check students' work.

48 **Practice Bank,** ALGEBRA 1: EXPLORATIONS AND APPLICATIONS

ONGOING ASSESSMENT

36. Open-ended Problem Describe an everyday situation that this graph could represent.

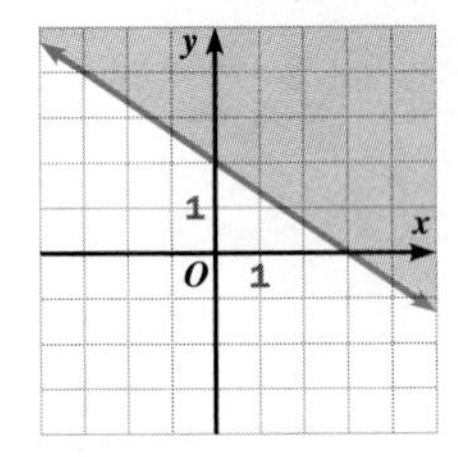

SPIRAL REVIEW

Find each product. *(Section 6.4)*

37. $2x(x + 7)$ **38.** $4x(5 - x)$ **39.** $(a + 7)(a - 5)$

40. $(3b + 5)(b + 3)$ **41.** $-2(3 - \sqrt{5})$ **42.** $(4 + \sqrt{2})(1 + \sqrt{2})$

Use substitution to solve each system of equations. *(Section 7.2)*

43. $x = y$; $3x - 2y = 1$

44. $a = b - 2$; $3a + 2b = 29$

45. $3x - y = 1$; $2x + 5y = 63$

32. a. Yes. b. 10 students
    c. $150s + 60b \leq 3000$

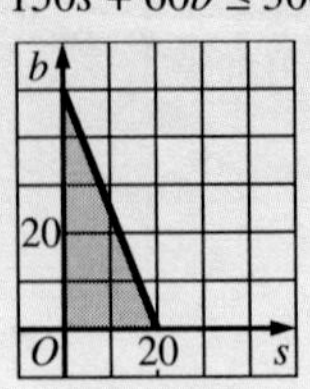

33. B

34. a. $4m + 7n = 25$
    b. $4m + 7n \leq 25$
    c. the inequality; The only possible way to spend exactly \$25 is to see 1 matinee and 3 evening shows. The inequality covers every possible way to use \$25 on matinees and evening shows.

35. Answers may vary. An example is given. Refer to Example 3 on page 301. If Ella wants to buy fabrics A and C and can spend up to \$80, then $10x + 4y \leq 80$, where x is the number of yards of fabric A and y is the number of yards of fabric C.

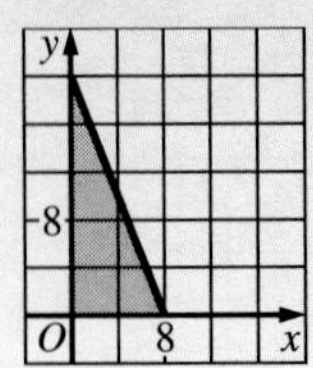

36. Answers may vary.
37. $2x^2 + 14x$
38. $20x - 4x^2$
39. $a^2 + 2a - 35$
40. $3b^2 + 14b + 15$
41. $-6 + 2\sqrt{5}$
42. $6 + 5\sqrt{2}$
43. (1, 1)
44. (5, 7)
45. (4, 11)

SECTION

7.6 Systems of Inequalities

Learn how to...

- graph systems of inequalities

So you can...

- make decisions when there are many restrictions, such as when making food for a hiking trip

Because backpackers need large amounts of energy and must carry their food, they must plan meals carefully. Many hikers make a mixture of peanuts and chocolate chips. The mixture, sometimes called *gorp*, provides quick energy from carbohydrates.

One ounce of chocolate chips contains 20.9 g of carbohydrates.

One ounce of peanuts contains 7.8 g of carbohydrates.

BY THE WAY

Backpackers often eat high-fat diets when hiking, because one ounce of fat provides more than twice as much energy as one ounce of carbohydrates.

EXAMPLE 1 Application: Nutrition

Andrea wants her gorp to contain at least 150 g of carbohydrates and weigh less than 16 oz. Write and graph two inequalities to show how much of each ingredient she should use.

SOLUTION

Let p = the number of ounces of peanuts.

Let c = the number of ounces of chocolate chips.

The gorp should weigh less than 16 oz.	Carbohydrates from peanuts	+	Carbohydrates from chocolate	≥ 150
$p + c < 16$	$7.8p$	+	$20.9c$	≥ 150

Graph the inequalities in the same coordinate plane.

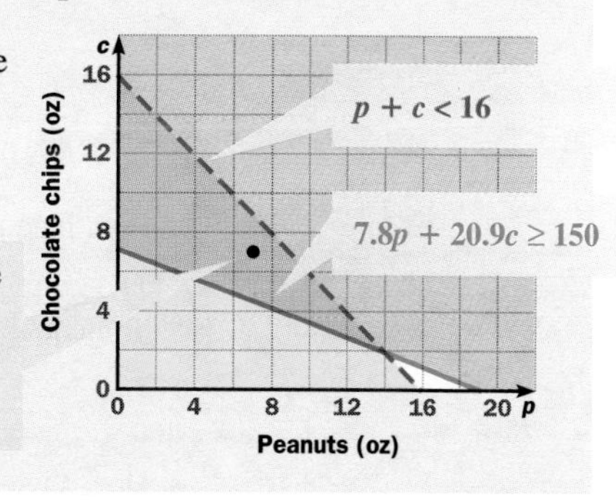

The points that are in both the blue and the red regions satisfy both inequalities. For example, Andrea could use 7 oz of peanuts and 7 oz of chocolate chips.

Plan⇔Support

Objectives

- Graph systems of inequalities.
- Graph systems of inequalities to model real-world situations when there are restrictions.

Recommended Pacing

Core and Extended Courses
Section 7.6: 1 day

Block Schedule
Section 7.6: $\frac{1}{2}$ block
(with Section 7.5)

Resource Materials

Lesson Support
Lesson Plan 7.6
Warm-Up Transparency 7.6
Overhead Visuals:
Folder A: Multi-Use Graphing Packet, Sheets 1–7
Practice Bank: Practice 49
Study Guide: Section 7.6
Explorations Lab Manual:
Diagram Masters 1, 2
Assessment Book: Test 31

Technology
Graphing Calculator
Internet:
http://www.hmco.com

Warm-Up Exercises

Tell whether each ordered pair is or is not a solution of the given inequality. Answer *Yes* or *No*.

1. (4, 5); $y \le 0.5x + 4$ Yes.
2. (4, 5); $y > 2x + 5$ No.
3. (4, 5); $y < -x + 6$ No.
4. (5, –1); $y < -x + 6$ Yes.
5. (–2, 6); $y \ge 0.5x + 4$ Yes.
6. (–2, 6); $y \le 2x + 5$ No.

Teach⇔Interact

Additional Example 1

Katrina plans to sell her handmade wooden toys at a local crafts show. Airplanes sell for \$19.50 and trucks for \$11.00. She can fit no more than 40 toys in her car and she wants to sell at least \$500 worth of toys. Write and graph two inequalities to show how many of each toy she should bring.

Let a = the number of airplanes. Let t = the number of trucks. Katrina can bring no more than 40 toys, so $a + t \leq 40$. She wants to sell at least \$500 worth, so $19.5a + 11t \geq 500$. Graph the inequalities.

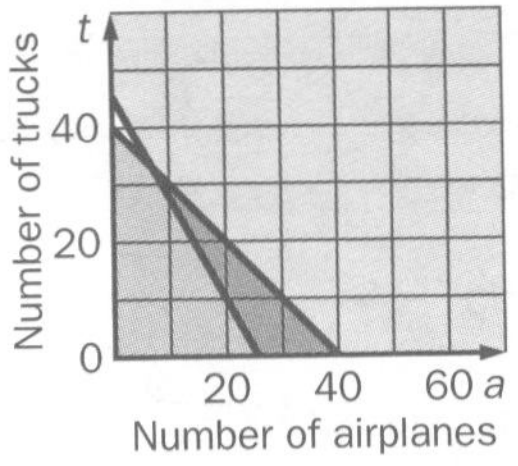

The points in the double-shaded region satisfy both inequalities. For example, Katrina could bring 30 airplanes and 10 trucks.

Additional Example 2

Use the information from Additional Example 1. Katrina wants to package each toy but has only 80 ft^2 of cardboard.

a. Airplane boxes are made with 4 ft^2 of cardboard and truck boxes with 2.5 ft^2. Write an inequality for the numbers of each toy she can package with only 80 ft^2 of cardboard.

cardboard for airplanes	+	cardboard for trucks	≤ 80
$4a$	+	$2.5t$	≤ 80

b. Graph your inequality along with the inequalities from Additional Example 1. Can Katrina package enough toys to sell at least \$500 worth?

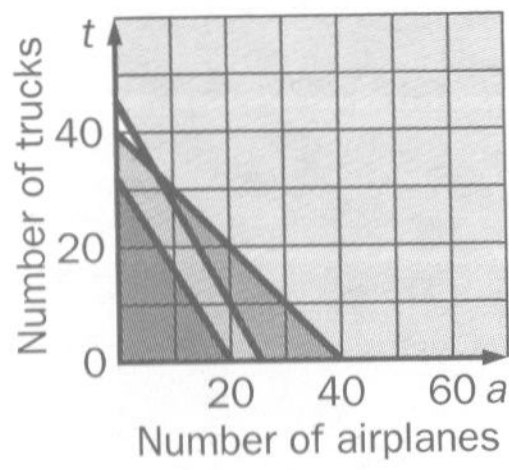

There are no points that satisfy all three inequalities. It is not possible for Katrina to package and transport enough toys to sell for at least \$500 with only 80 ft^2 of cardboard.

The inequalities in Example 1 are called a **system of inequalities**. The **solution of a system of inequalities** consists of all ordered pairs of numbers that satisfy both inequalities.

EXAMPLE 2 Application: Nutrition

The American Heart Association and the American Cancer Society recommend that you should eat at most 1 g of fat for every 5 g of carbohydrates you eat.

a. One ounce of peanuts contains **9.1 g of fat**. One ounce of chocolate chips contains **5.3 g of fat**. Write an inequality for the amount of each ingredient Andrea should use so that the gorp will contain **less than 30 g of fat**.

b. Graph your inequality along with the inequalities from Example 1. How much of each ingredient should Andrea use?

SOLUTION

a.

Fat from peanuts	+	Fat from chocolate	< 30
$9.1p$	+	$5.3c$	< 30

b.

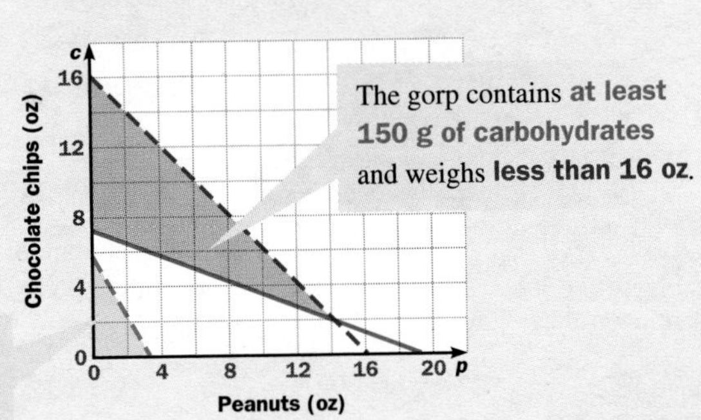

There are no points that satisfy all three inequalities.

It is impossible to make a gorp of peanuts and chocolate chips that contains more than 150 g of carbohydrates, contains less than 30 g of fat, and weighs less than 16 oz.

THINK AND COMMUNICATE

1. Is an inequality a function? Explain your answer.
2. The graph in Example 2 is the first quadrant of the graph of the system of inequalities
$$p + c < 16$$
$$7.8p + 20.9c \geq 150$$
$$9.1p + 5.3c < 30.$$
Does this system have solutions in any other quadrants? Explain.

ANSWERS Section 7.6

Think and Communicate

1. No; each value in the domain corresponds to more than one value in the range. Note that the graph of an inequality does not satisfy the vertical-line test.
2. Yes; there are solutions in Quadrant II. An example is (–6, 10).

Checking Key Concepts

1–5. See answers in back of book.

Exercises and Applications

1.

2.

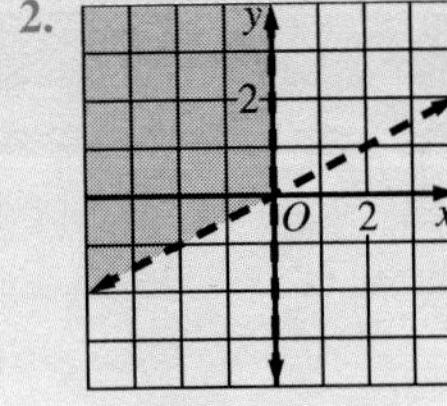

3.

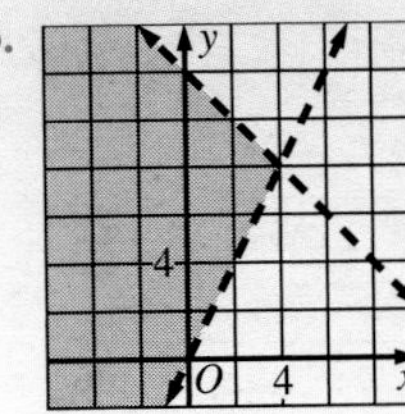

☑ CHECKING KEY CONCEPTS

Graph each system of inequalities.

1. $x \geq 3$
$y \leq 5$

2. $y > x$
$y > 0$

3. $y < x + 4$
$y \geq -2x + 2$

4. $2x - y \geq 4$
$x + y \leq 3$

5. The graphs of the equations $y = x - 3$ and $y = 7 - x$ divide the coordinate plane shown at the right into four regions. Give a system of two inequalities that describes each region. Region A is done for you.

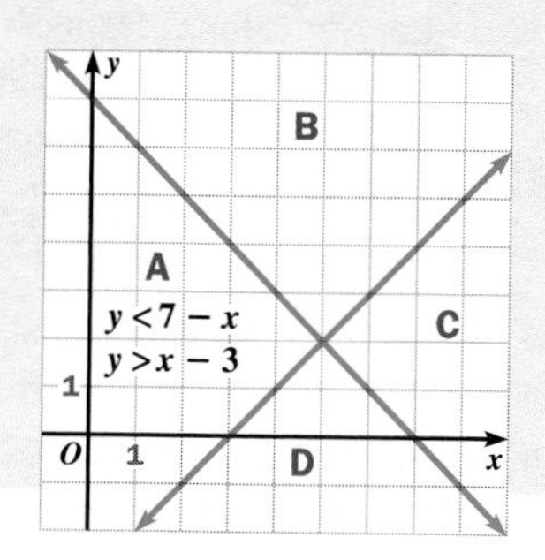

7.6 Exercises and Applications

Graph each system of inequalities.

Extra Practice exercises on page 570

1. $y \geq 0$
$x \geq -2$

2. $x < 0$
$y > \frac{1}{2}x$

3. $y > 2x$
$y < 12 - x$

4. $y \leq 4 - 2x$
$y \geq -2$

5. $y > 4x$
$y \leq 7 + x$

6. $x + y < 8$
$x + y > 2$

7. $2x + 3y < 6$
$3x + 2y > 6$

8. $5x - 2y \leq 12$
$-3x - y \leq 5$

9. $2 < x - 3y$
$x - 2y \leq 2$

Use the information about cat food for Exercises 10–12.

	Canned cat food	Dry cat food
Cost	$.50	$2.50
Ounces	6 oz	48 oz

10. Write an inequality for each statement.

a. Erik Mueller can spend at most $5 for cat food each week.

b. Erik's cat eats at least 72 oz of cat food each week.

11. Graph the system of inequalities from Exercise 10. List some solutions of the system.

12. Erik prefers to give his cat more canned food than dry food. Express this preference as an inequality. Graph the inequality with your graph from Exercise 11. How much of each kind of food should Erik buy?

Graph each pair of equations. The graphs will divide the coordinate plane into four regions. Write a system of inequalities that describes each region. Label the region with these inequalities.

13. $x = 6$
$y = -2$

14. $y = x - 3$
$y = 9 - x$

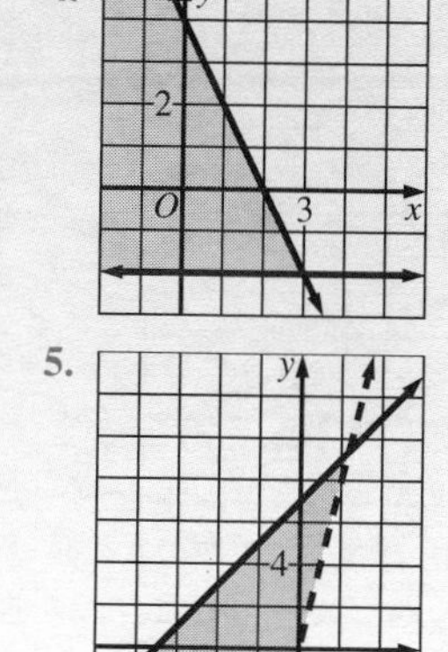

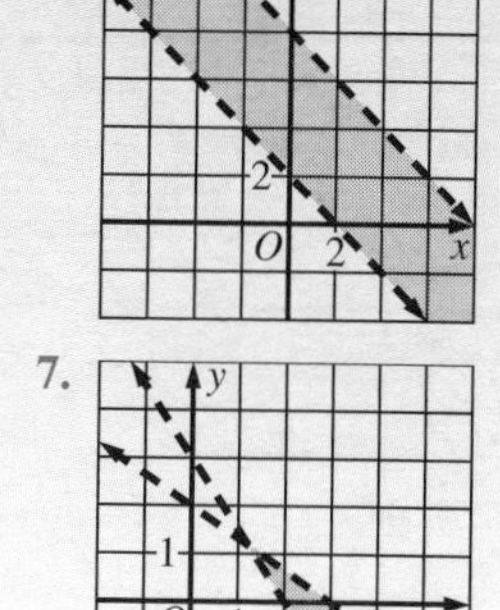

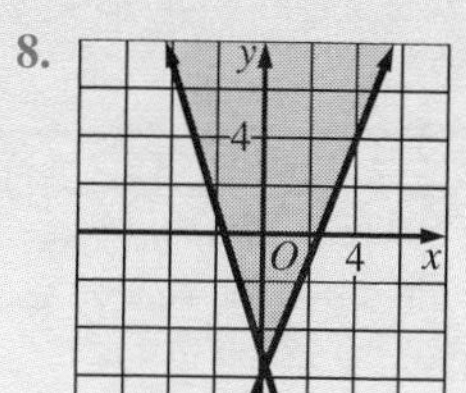

9–14. See answers in back of book.

Teach⇔Interact

Think and Communicate

For question 1, students should understand immediately that since the graph of an inequality in two variables is a region of the plane, it fails the vertical-line test and cannot be a function.

Checking Key Concepts

Teaching Tip

Some students may have difficulty with question 1. Since each of the inequalities contains just one variable, students may wonder how it is possible to substitute the coordinates of a point into an inequality when there is a variable missing. Point out that the missing variable can be written with coefficient 0. For example, $x \geq 3$ can be rewritten as $x + 0 \cdot y \geq 3$. The inequality $y \leq 5$ can be rewritten as $0 \cdot x + y \leq 5$.

Apply⇔Assess

Suggested Assignment

Core Course
Exs. 1–15, 21, 23–33, AYP

Extended Course
Exs. 1, 2, 5, 6, 8, 9–33, AYP

Block Schedule
Day 45 Exs. 1–15, 20–33, AYP

Exercise Notes

Mathematical Procedures

Exs. 1–9 Students can use different kinds of shading for the inequalities in each system. They may choose to use different colors to shade the regions or striped shading with the stripes going in different directions.

Common Error

Exs. 1–9 Some students may shade the wrong region when they graph an inequality. They may also err in showing the boundary of a region with a solid line or a dotted line. The first type of error can be corrected by having students use check points. The second type of error can be corrected by calling attention to the inequality symbols in the system. The boundary is included if the symbol $\geq$ or $\leq$ is used, but it is not included if the symbol $>$ or $<$ is used.

Apply⇔Assess

Exercise Notes

Interview Note

Exs. 15–18 Students can benefit from studying situations like this to see if they can describe other situations that might result in a different solution region. The boundary lines for the individual inequalities should be the same as those in the original situation.

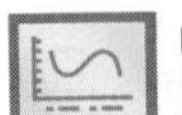

Using Technology

Ex. 20 Different keystroke sequences are possible for graphing systems like these. The graphs are easiest when one of the inequalities has ≥ or > and the other has ≤ or <. You can use the TI-82 for the system in part (b) by using the following procedure. (Note that the inequalities have already been solved for *y*.)

Enter an equation for the inequality that has "*y* ≥" or "*y* >" as Y1 on the Y= list. Enter the equation for the other inequality as Y2. In this case, the equations will be

Y1=2X–8

Y2=X+1

Press the following keys to display boundary lines and the solution graph for the system of inequalities. [2nd][QUIT][CLEAR][2nd][DRAW] <Shade(>[2nd][Y-VARS]<Function...> <Y1>[,][2nd][Y-VARS]<Function...> <Y2>[)][ENTER].

The instruction Shade(Y1,Y2) tells the calculator to shade the region containing points that lie *above* the line for Y1 and *below* the line for Y2. (See the answer for Ex. 20(c) for another approach.)

Problem Solving

Ex. 22 Students should be able to extrapolate from their experience with systems of two linear inequalities to solve this problem. Students who are skillful with a graphing calculator may see that they can obtain a quadrilateral and its interior rather easily by using absolute value inequalities. For example, they could enter Y1=absX–3 and Y2=–absX+5. Then use the instruction Shade (Y1,Y2). (See the Using Technology note above.)

INTERVIEW Nancy Clark

Look back at the article on pages 278–280.

Nancy Clark stresses that a sports diet is healthy for all people, not just athletes. She says that, regardless of your sport or activity level, 60–70% of the calories you eat should come from carbohydrates.

15. a. Let c = calories from carbohydrates. Let t = total calories. Write an inequality that states that at least 60% of total calories should come from carbohydrates.

b. Write an inequality that states that at most 70% of calories should come from carbohydrates.

16. Graph the system of inequalities from Exercise 15. Compare your graph to the graph on page 280. How are they alike? How are they different?

17. Writing How can you use your graph to find your ideal carbohydrate intake? What do you need to know?

18. Athletes in training require 70–80% of their calories to be from carbohydrates. Suppose a cyclist in the Tour DuPont, an international cycling stage race, uses about 963 Cal during one stage of the race. Write an inequality that represents the number of calories that should come from carbohydrates.

In 1995, the Tour DuPont followed a 1130 mile-long route from Wilmington, DE, to Piedmont Triad, NC, and took 12 days to complete. Cyclists often have to eat their meals while riding.

19. a. You can write $2 \le y \le 4$ as a short way of writing the system $2 \le y$ and $y \le 4$. Graph this system of inequalities.

b. What system of inequalities does $0 \le x \le 5$ represent? Graph the system.

c. Graph this system of four inequalities: $2 \le y \le 4$; $0 \le x \le 5$

20. Technology Find out how to graph a system of inequalities on your graphing calculator. You may need to use features designed for drawing and shading. Graph each system of inequalities.

a. $y \ge 3$
$y \le 2x$

b. $y \ge 2x - 8$
$y \le x + 1$

c. Describe how to graph a system of inequalities on your graphing calculator. Be specific enough so that your directions can be followed accurately by someone who is familiar with your calculator but has never used it to graph inequalities before.

15. a. $c \ge 0.6t$
 b. $c \le 0.7t$

16. See answers in back of book.

17. You can look in the shaded region of the graph that contains the solutions of the system to find a range of amounts of carbohydrates you should eat for a particular calorie total. You need to know what your total calorie consumption per day is.

18. The number of calories can be represented by two inequalities, $c > 674.1$ and $c < 770.4$. They can be combined and written $674.1 < c < 770.4$.

19. a.

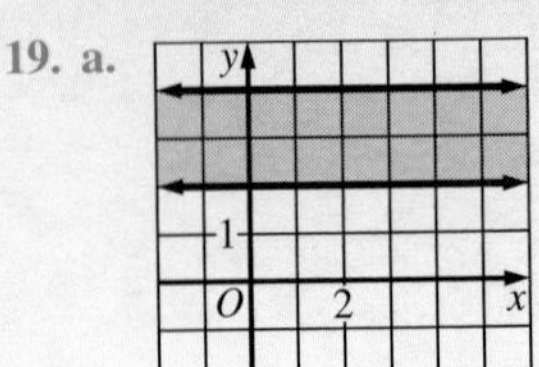

b. $0 \le x$ and $x \le 5$

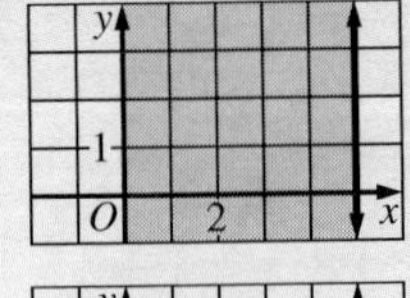

c.

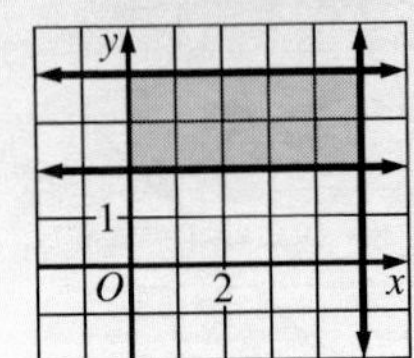

21. **SAT/ACT Preview** What is the area of the region described by the inequalities $x > 0$, $y > 0$, and $y < 8 - 2x$?

A. 8 **B.** 16 **C.** 24 **D.** 32 **E.** 64

22. **Open-ended Problem** Create a system of inequalities whose solution region is a polygon.

23. **Challenge** How many pairs of positive integers x and y satisfy the inequality $2x + 5y \leq 10$?

ONGOING ASSESSMENT

24. **Writing** How is solving a system of inequalities like solving a system of equations? How is it different?

SPIRAL REVIEW

Solve each proportion. *(Section 5.1)*

25. $\frac{3}{7} = \frac{x}{2}$

26. $\frac{y}{3} = \frac{7}{8}$

27. $\frac{12}{n} = \frac{3}{5}$

28. $\frac{a}{2} = \frac{18}{a}$

29. $\frac{t-2}{2} = \frac{4}{3}$

30. $\frac{2n+1}{2} = \frac{3}{2}$

Find each product. *(Section 6.5)*

31. $(x + 2)^2$

32. $\left(4 + \sqrt{3}\right)^2$

33. $(x + 5)(x - 5)$

ASSESS YOUR PROGRESS

VOCABULARY

linear inequality (p. 307)
system of inequalities (p. 312)
solution of a system of inequalities (p. 312)

Graph each inequality or system of inequalities. *(Sections 7.5 and 7.6)*

1. $y \leq 4 - \frac{1}{3}x$

2. $3x + 2y \leq -12$

3. $5x - 2y > 10$

4. $y > 2$
   $x < -1$

5. $4x + y \geq 0$
   $4x + y \leq 6$

6. $3y - x < 3$
   $2x + 7y \geq 7$

7. **Business** The members of the math team are selling T-shirts and sweatshirts. They need to sell a minimum of 10 T-shirts and 10 sweatshirts. They earn \$5 for each T-shirt and \$10 for each sweatshirt sold, and would like to raise at least \$200. Write and graph a system of inequalities that shows how many T-shirts and sweatshirts the team needs to sell.

8. **Journal** How is graphing the inequality $3x - y < -6$ like graphing the equation $3x - y = -6$? How is it different?

Apply⇔Assess

Exercise Notes

Assessment Note

Ex. 24 This exercise provides a good review of major ideas of this chapter. Students who do a good job with this exercise may wish to include their responses in their journals.

Assess Your Progress

Review the vocabulary terms. Students can use the inequalities and systems of inequalities in Exs. 1–6 to illustrate the meaning of each term.

Journal Entry

Students can draw on their response to Ex. 24 of the Exercises and Applications when they work on Ex. 8.

Progress Check 7.5–7.6

See page 319.

Practice 49 for Section 7.6

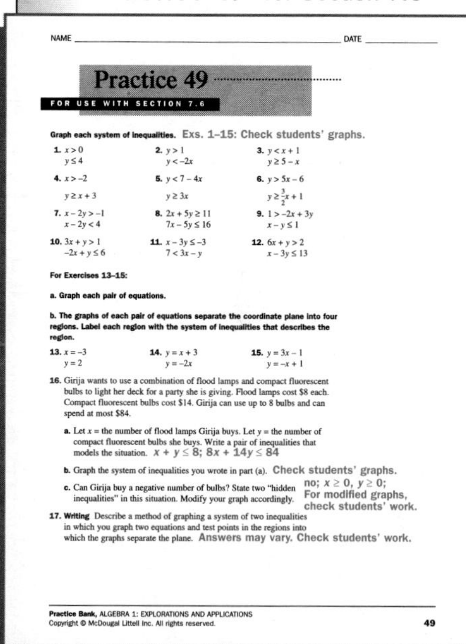
NAME ____________ DATE ______

Practice 49

FOR USE WITH SECTION 7.6

Graph each system of inequalities. Exs. 1–15: Check students' graphs.

1. $x > 0$, $y \leq 4$
2. $y > 1$, $y < -2x$
3. $y < x + 1$, $y \geq 5 - x$
4. $x > -2$, $y \geq x + 3$
5. $y < 7 - 4x$, $y \geq 3x$
6. $y > 5x - 6$, $y \geq \frac{3}{2}x + 1$
7. $x - 2y > -1$, $x - 2y < 4$
8. $2x + 5y \geq 11$, $7x - 5y \leq 16$
9. $1 > -2x + 3y$, $x - y \leq 1$
10. $3x + y > 1$, $-2x + y \leq 6$
11. $x - 3y \leq -3$, $7 < 3x - y$
12. $6x + y > 2$, $x - 3y \leq 13$

For Exercises 13–15:

a. Graph each pair of equations.

b. The graphs of each pair of equations separate the coordinate plane into four regions. Label each region with the system of inequalities that describes the region.

13. $x = -3$, $y = 2$
14. $y = x + 3$, $y = -2x$
15. $y = 3x - 1$, $y = -x + 1$

16. Girija wants to use a combination of flood lamps and compact fluorescent bulbs to light her deck for a party she is giving. Flood lamps cost \$8 each. Compact fluorescent bulbs cost \$14. Girija can use up to 8 bulbs and can spend at most \$84.

   a. Let x = the number of flood lamps Girija buys. Let y = the number of compact fluorescent bulbs she buys. Write a pair of inequalities that models the situation. $x + y \leq 8$; $8x + 14y \leq 84$

   b. Graph the system of inequalities you wrote in part (a). Check students' graphs.

   c. Can Girija buy a negative number of bulbs? State two "hidden inequalities" in this situation. Modify your graph accordingly. no; $x \geq 0$, $y \geq 0$; For modified graphs, check students' work.

17. **Writing** Describe a method of graphing a system of two inequalities in which you graph two equations and test points in the regions into which the graphs separate the plane. Answers may vary. Check students' work.

Practice Bank, ALGEBRA 1: EXPLORATIONS AND APPLICATIONS

49

20. a.

b.

c. Answers may vary. An example is given for the TI-82. First write both inequalities with y alone on one side. Press [2nd] [DRAW] and then press 7 to select Shade. After "Shade(" appears on the screen, you enter *lowerfunction, upperfunction,* where lowerfunction is an expression in terms of x for the lower boundary of the shaded region, and upperfunction is an expression in terms of x for the upper boundary of the shaded region. Then you enter the closing parenthesis, [ )], and press [ENTER].

21. B

22. Answers may vary. An example is given. $x \geq 0$, $0 \leq y \leq 6$, and $y \leq -x + 8$

23. two pairs: (1, 1) and (2, 1)

24–34. See answer in back of book.

Assess Your Progress

1–7. See answer in back of book.

8. Answers may vary. An example is given. Both require you to graph the line $3x - y = -6$. However, to graph the inequality $3x - y < -6$ you need to use a dashed, not solid, line and you need to determine whether to shade the region of the plane above or below the dashed line. In this case, since $y > 3x + 6$, you shade the region above the line.

Project Notes

Mathematical Goals

- Collect riser and tread data for at least five different stairways and decide on an ideal riser/tread ratio.
- Design a stairway and make a scale drawing of it.
- Make a graph that shows all the riser/tread data collected and that meets the two rules stated on the following page.

Planning

Materials

- Ruler
- Graph paper

Project Teams
Students can choose a partner to work with and then discuss how they wish to proceed.

Guiding Students' Work

Suggest that students collect their riser and tread data in different public buildings, apartment houses, schools, homes, and so on. This will ensure a variety of data. Students should then work together to carry out the goals of the project.

Second-Language Learners
Students learning English may benefit from a conference with an aide or peer tutor after they complete the first draft of their report.

PORTFOLIO PROJECT

Designing a Stairway

Have you ever found it difficult to climb a narrow or steep stairway? Stairs may appear to be all the same, but slight changes in the riser/tread ratio can greatly affect the safety and the cost of stairs.

Iowa City

PROJECT GOAL **Compare the dimensions of stairways in your community with some accepted rules for stairs, so that you can better design your own stairway.**

Designing a Stairway

Work with a partner.

1. **COLLECT** **riser** and **tread** data for at least five different stairways. Take notes about how easy or hard it is to climb each stairway.

Number	Location	Tread Length(in)	Riser Length(i
1	Near Gym	10	7

2. **DECIDE** on an ideal riser/tread ratio. Explain your decision.

Chicago

3. **DESIGN** your own stairway. Suppose the total distance from one floor to the next is 105 in. Determine lengths for the risers for a 9 in. tread and an 11 in. tread. Then make a scale drawing of your stairway.

Cycle
by M.C. Escher
(1898 – 1972)

Analyzing Stairway Standards

Codes have been written to increase stair safety. Codes often concern the riser/tread ratio, lighting around stairs, and handrail height. Here are two generally accepted rules for stairway construction:

RULE 1: The sum of one riser and one tread should be from 17 in. to 18 in.

RULE 2: The sum of two risers and one tread should be from 24 in. to 25 in.

New Mexico

1. Make a graph showing all the riser/tread data that meet the conditions of these rules. (*Hint:* You will graph four inequalities.)

2. On the same coordinate grid, graph the data you collected about stairs as ordered pairs. Also graph the ordered pair for the stairway you designed.

3. Describe any patterns you see. Did all the stairways follow the generally accepted rules? Did any not follow the rules?

Presenting Your Results

Give a report that includes your graph, your table, and your scale drawing. Write a paragraph comparing your local stairways' dimensions with the accepted rules for stairs. Give your recommendations for riser/tread ratios.

You may want to extend your report by examining some of the ideas below:

- Learn more about constructing stairs, including how a carpenter's square is used.
- Collect photographs of stairways from other periods of time and from other countries.
- Find the slope for each stairway on your list. Then find the range of slopes for the stairways.
- Interview architects and/or carpenters in your area. Ask what they would consider the best riser/tread ratio for stairs.

Self-Assessment

What method(s) did you use to create your graph? How did you and your partner decide on an ideal riser/tread ratio? Overall, was there enough variety in the stairs you studied to give you a sense of how the riser/tread ratio affects stairway use? What was the most difficult aspect of the project? the easiest? Why?

Questions

1. *What was the most interesting stairway you've designed?*
2. *What's the difference between designing a stairway for a home and for an office building?*
3. *What different types of stairways have you designed?*

Project Notes

Guiding Students' Work

Rubric for Chapter Project

4 Students collect their data, decide on an ideal riser/tread ratio, and design their own stairway. They graph the data that meet the generally accepted rules for stairway construction and the data collected for the project. They present their results in a report that includes their graph, table, and scale drawing. They give their recommendations for riser/tread ratios. All work is accurate and the report is well organized, clearly written, and interesting to read. There are no mathematical errors in the report.

3 Students complete all aspects of the project and present their results in a report. Overall, the work is well done with some minor exceptions. The graphs, for example, are not entirely clear or the scale drawing could have been drawn better. The paragraph comparing local stairways' dimensions with the accepted rules for stairs is somewhat hard to understand.

2 Students design a stairway, analyze stairway standards, and present their results in a report. Most of the work is carelessly done and shows little effort. There are errors in the scale drawing and the graphs of the four inequalities. The report is incomplete and poorly written.

1 Students collect some data and decide on an ideal riser/tread ratio but cannot satisfactorily design their own stairway nor graph the data they collected. A report is attempted but is not completed. Students should be encouraged to speak with the teacher as soon as possible to review their work and to make a new start on the project.

Chapter Support

Course Guide: Chapter 7
Lesson Plans: Chapter 7
Practice Bank:
Cumulative Practice 50
Study Guide: Chapter 7 Review
Assessment Book:
Chapter Tests 32 and 33
Chapter 7 Alternative Assessment
Test Generator software
Technology Book:
Calculator Based Lab 4
Portfolio Project Book
Preparing for College Entrance Tests
Professional Handbook

Progress Check 7.1–7.2

1. Rewrite $3x + 2y = 10$ in slope-intercept form. *(Section 7.1)*
   $y = -\frac{3}{2}x + 5$
2. Graph the equation $3x + 2y = 10$. *(Section 7.1)*

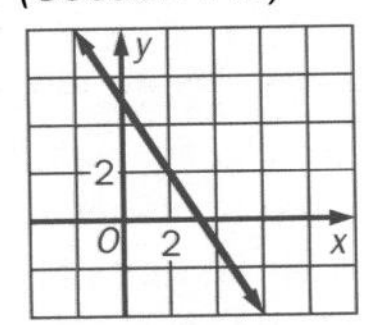

3. Use substitution to solve the system. *(Section 7.2)*
   $5x - 9y = 10$
   $x - y = 3$ (4.25, 1.25)
4. Ming needs to sell 22 tickets to a school play to take in \$80. Children's tickets cost \$2 and adult tickets cost \$5. How many of each kind of ticket does Ming need to sell? *(Section 7.2)*
   Let c = number of children's tickets, and let a = number of adult tickets. Solve the system.
   $2c + 5a = 80$
   $c + a = 22$
   The solution (c, a) is (10, 12). Ming needs to sell 10 children's tickets and 12 adult tickets.
5. Is (2, 7) a common solution of the equations $3x + y = -1$ and $-4x - y = 1$? *(Section 7.2)* No.

CHAPTER

7 Review

STUDY TECHNIQUE

For each vocabulary word or phrase listed below, give an example or draw a graph to illustrate.

VOCABULARY

standard form of a linear equation (p. 282)
vertical intercept, *y*-intercept (p. 282)
horizontal intercept, *x*-intercept (p. 282)
system of linear equations (p. 288)
solution of a system of linear equations (p. 288)
linear inequality (p. 307)
system of linear inequalities (p. 312)
solution of a system of inequalities (p. 312)

SECTIONS 7.1 *and* 7.2

You can solve a system of equations by graphing or by substitution.

By graphing

Graph the intercepts, or solve for y and use a graphing calculator.

The **red line** contains all solutions of the equation $2x + 5y = 11$.

The **green line** contains all solutions of the equation $x - 4y = -1$.

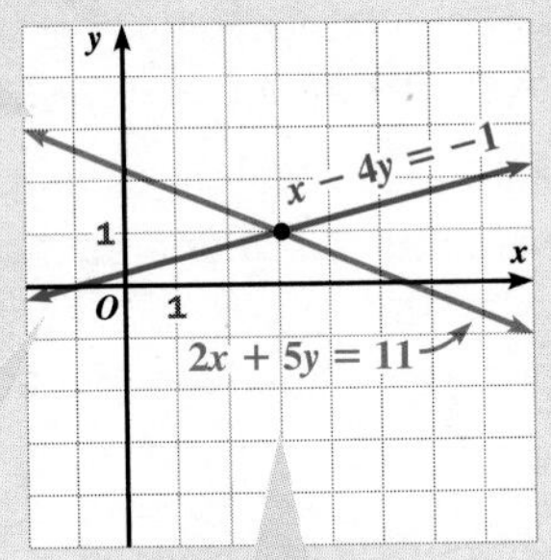

The solution (x, y) is **(3, 1)**.

By substitution

$$2x + 5y = 11$$
$$x - 4y = -1$$
$$x = 4y - 1$$

$$2(4y - 1) + 5y = 11$$
$$13y - 2 = 11$$
$$y = 1$$

$$x - 4y = -1$$
$$x - 4(1) = -1$$
$$x = 3$$

The solution (x, y) is **(3, 1)**.

SECTIONS 7.3 *and* 7.4

You can solve a system of equations by **adding**, **subtracting**, or **multiplying**.

$$\begin{aligned} 2x + 3y &= 1 \\ +(-2x + 4y) &= +(-8) \\ \hline 0 + 7y &= -7 \\ y &= -1 \\ x &= 2 \end{aligned} \qquad \begin{aligned} 3x - 4y &= -6 \\ -(5x - 4y) &= -(-2) \\ \hline -2x + 0 &= -4 \\ x &= 2 \\ y &= 3 \end{aligned}$$

$$\begin{aligned} 4x + 2y = 3 \quad \times 3 \qquad 12x + 6y &= 9 \\ 6x + 3y = 1 \quad \times 2 \qquad -(12x + 6y) &= -(2) \\ \hline 0 &= 7 \end{aligned}$$

no solution

A graph can show whether a system of equations has one solution, many solutions, or no solution.

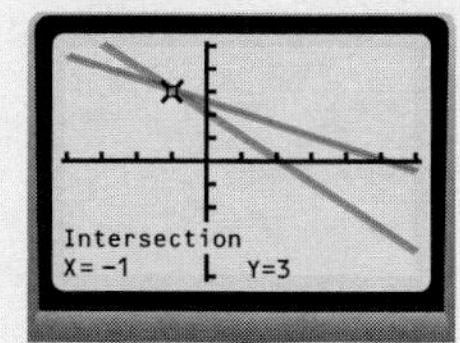

The lines **intersect.**
one solution

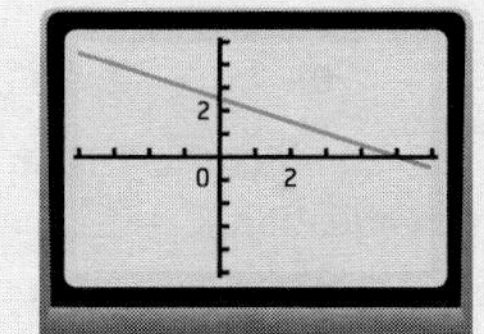

The lines are the **same.**
many solutions

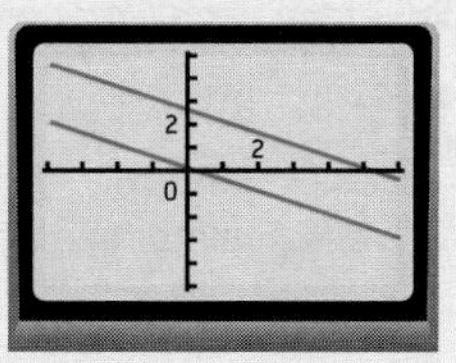

The lines are **parallel.**
no solutions

SECTIONS 7.5 *and* 7.6

The graph of a linear inequality is a region on the coordinate plane whose edge is a line. The region shows all ordered pairs that satisfy the inequality.

The **purple region** contains all points that satisfy **both** inequalities.

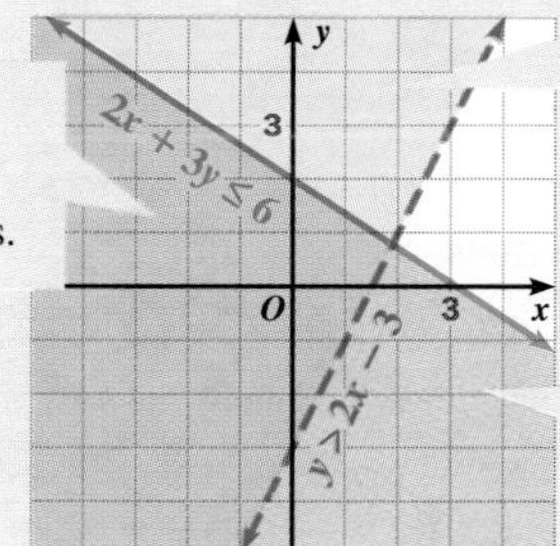

The **blue region** contains all points that satisfy the inequality $y > 2x - 3$.

The **pink region** and the **red line** contain all points that satisfy the inequality $2x + 3y \le 6$.

Progress Check 7.3–7.4

Solve each system by addition or subtraction. *(Section 7.3)*

1. $3x - 5y = 8$
   $3x - 7y = 18$ $\left(-\frac{17}{3}, -5\right)$
2. $5x - 7y = 23$
   $4x + 7y = -32$ (−1, −4)
3. Does the following system have one solution, many solutions, or no solutions? *(Section 7.4)*
   $-7x + 3y = 5$
   $14x - 6y = -10$ many solutions

Use multiplication to solve each system of equations. *(Section 7.4)*

4. $3x + 5y = -11$
   $8x - 2y = 78$ (8, −7)
5. $7x + 9y = 15$
   $5x - 8y = -47$ (−3, 4)

Progress Check 7.5–7.6

1. Consider the following points: (5, 1), (5, 10), (5, 7). Which of these points are solutions of $y \ge 2x - 3$? *(Section 7.5)*
   (5, 10) and (5, 7)
2. Is the line $y = -7x + 1$ part of the graph of $y < -7x + 1$? Is the line $y = -7x + 1$ part of the graph of $y \le -7x + 1$? *(Section 7.5)*
   No; Yes.
3. Write an inequality whose graph consists of the points above the graph of $y = -2x - 3$. *(Section 7.5)* $y > -2x - 3$
4. Graph the system of inequalities. *(Section 7.6)*
   $2x + y > 3$
   $x - y > 8$

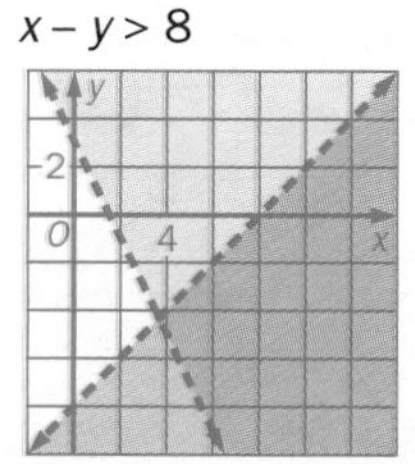

5. Suppose one ounce of juice A has 200 mg of vitamin C and one ounce of juice B has 120 mg of vitamin C. Max wants to have at least 700 mg of vitamin C, but does not want to drink more than 4 ounces of juice. Write a system of inequalities that models this situation. Use *a* for the number of ounces of juice A and *b* for the number of ounces of juice B. *(Sections 7.5 and 7.6)*
   $a + b \le 4$
   $200a + 120b \ge 700$

Chapter 7 Assessment
Form A Chapter Test

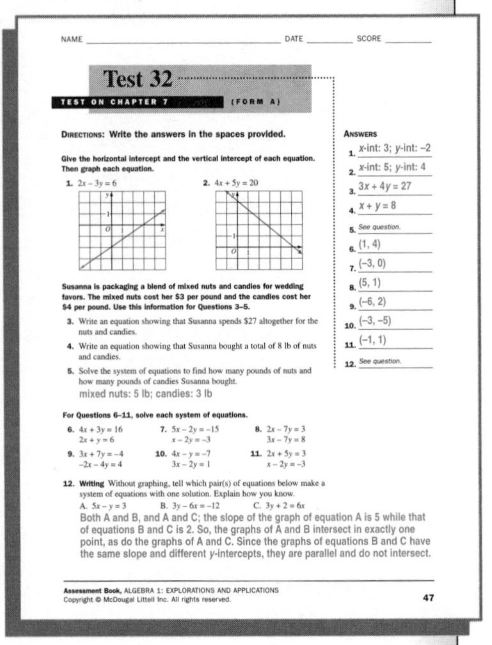

NAME ______ DATE ______ SCORE ______

Test 32

TEST ON CHAPTER 7 (FORM A)

DIRECTIONS: Write the answers in the spaces provided.

Give the horizontal intercept and the vertical intercept of each equation. Then graph each equation.

1. $2x - 3y = 6$
2. $4x + 5y = 20$

Susanna is packaging a blend of mixed nuts and candies for wedding favors. The mixed nuts cost her \$3 per pound and the candies cost her \$4 per pound. Use this information for Questions 3–5.

3. Write an equation showing that Susanna spends \$27 altogether for the nuts and candies.
4. Write an equation showing that Susanna bought a total of 8 lb of nuts and candies.
5. Solve the system of equations to find how many pounds of nuts and how many pounds of candies Susanna bought.
   mixed nuts: 5 lb; candies: 3 lb

For Questions 6–11, solve each system of equations.

6. $4x + 3y = 16$, $2x + y = 6$
7. $5x - 2y = -15$, $x - 2y = -3$
8. $2x - 7y = 3$, $3x - 7y = 8$
9. $3x + 7y = -4$, $-2x - 4y = 4$
10. $4x - y = -7$, $3x - 2y = 1$
11. $2x + 5y = 3$, $x - 2y = -3$

12. **Writing** Without graphing, tell which pair(s) of equations below make a system of equations with one solution. Explain how you know.
    A. $5x - y = 3$ B. $3y - 6x = -12$ C. $3y + 2 = 6x$
    Both A and B, and A and C; the slope of the graph of equation A is 5 while that of equations B and C is 2. So, the graphs of A and B intersect in exactly one point, as do the graphs of A and C. Since the graphs of equations B and C have the same slope and different y-intercepts, they are parallel and do not intersect.

ANSWERS

1. x-int: 3; y-int: –2
2. x-int: 5; y-int: 4
3. $3x + 4y = 27$
4. $x + y = 8$
5. See question.
6. (1, 4)
7. (–3, 0)
8. (5, 1)
9. (–6, 2)
10. (–3, –5)
11. (–1, 1)
12. See question.

Assessment Book, ALGEBRA 1: EXPLORATIONS AND APPLICATIONS

47

Chapter 7 Assessment
Form B Chapter Test

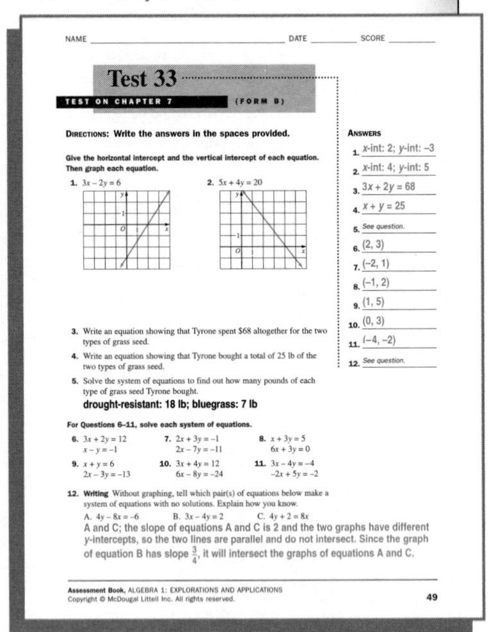

NAME ______ DATE ______ SCORE ______

Test 33

TEST ON CHAPTER 7 (FORM B)

DIRECTIONS: Write the answers in the spaces provided.

Give the horizontal intercept and the vertical intercept of each equation. Then graph each equation.

1. $3x - 2y = 6$
2. $5x + 4y = 20$

3. Write an equation showing that Tyrone spent \$68 altogether for the two types of grass seed.
4. Write an equation showing that Tyrone bought a total of 25 lb of the two types of grass seed.
5. Solve the system of equations to find out how many pounds of each type of grass seed Tyrone bought.
   drought-resistant: 18 lb; bluegrass: 7 lb

For Questions 6–11, solve each system of equations.

6. $3x + 2y = 12$, $x - y = -1$
7. $2x + 3y = -1$, $2x - 7y = -11$
8. $x + 3y = 5$, $6x + 3y = 0$
9. $x + y = 6$, $2x - 3y = -13$
10. $3x + 4y = 12$, $6x - 8y = -24$
11. $3x - 4y = -4$, $-2x + 5y = -2$

12. **Writing** Without graphing, tell which pair(s) of equations below make a system of equations with no solutions. Explain how you know.
    A. $4y - 8x = -6$ B. $3x - 4y = 2$ C. $4y + 2 = 8x$
    A and C; the slope of equations A and C is 2 and the two graphs have different y-intercepts, so the two lines are parallel and do not intersect. Since the graph of equation B has slope $\frac{3}{4}$, it will intersect the graphs of equations A and C.

ANSWERS

1. x-int: 2; y-int: –3
2. x-int: 4; y-int: 5
3. $3x + 2y = 68$
4. $x + y = 25$
5. See question.
6. (2, 3)
7. (–2, 1)
8. (–1, 2)
9. (1, 5)
10. (0, 3)
11. (–4, –2)
12. See question.

Assessment Book, ALGEBRA 1: EXPLORATIONS AND APPLICATIONS

49

CHAPTER

7 Assessment

VOCABULARY REVIEW

1. Give an example of a linear equation in standard form. Give an example of a linear equation in slope-intercept form.
2. Give an example of a linear inequality.
3. What is the difference between a horizontal intercept and a vertical intercept?

SECTIONS 7.1 and 7.2

Give the horizontal intercept and the vertical intercept of each equation. Then graph each equation.

4. $4x - 3y = 24$
5. $2x + 3y = 17$

Technology Solve each equation for *y*. Then use a graphing calculator to graph each equation.

6. $5x - 2y = 11$
7. $2x - 3y = 12$
8. Anita makes a blend of teas for a party. Sunrise Tea costs \$6/lb and Great Taste Tea costs \$4/lb.
   a. Write an equation showing that Anita spends \$20 altogether for tea.
   b. Write an equation showing that Anita buys 4 lb of tea.
   c. Solve the system of equations to find how much of each kind of tea Anita purchases.

Solve each system of equations by graphing or by substitution.

9. $4x + 3y = 24$
   $x - 2y = -5$
10. $3x + 3y = 7$
    $4x - y = 6$

SECTIONS 7.3 and 7.4

Solve each system of equations.

11. $4x + 2y = 14$
    $x = 1 + 2y$
12. $4x + 3y = -2$
    $4x - 2y = 8$
13. $4x + 5y = 14$
    $10x + 5y = 20$
14. $4x + 3y = 10$
    $8x - 6y = -20$
15. $2x - 3y = 11$
    $6y = -22 + 4x$
16. $3y = 5 - 2x$
    $3x + 4y = 2$

ANSWERS Chapter 7

Assessment

1. Any equation $ax + bx = c$, where a, b, and c are integers and a and b are not both zero is in standard form. For example, $2x + 5y = 7$ is in standard form. Any equation $y = mx + b$ is in slope-intercept form. For example, $y = \frac{3}{4}x - \frac{1}{8}$ is in slope-intercept form.
2. Answers may vary. Examples are given. $y \le 2x - 1$; $y < 7$; $y \ge 3x + \frac{1}{5}$; $x > 5$
3. A horizontal intercept is on the x-axis and has coordinates $(x, 0)$ for some number x. A vertical intercept is on the y-axis and has coordinates $(0, y)$ for some number y.

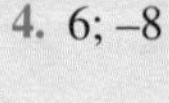

4. 6; –8

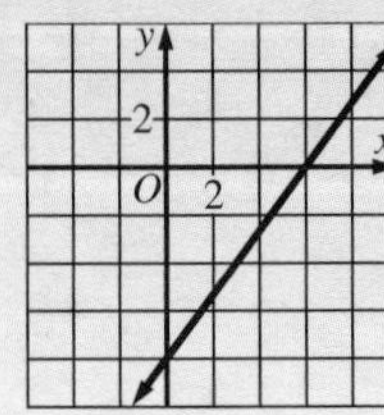

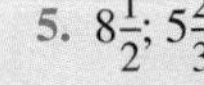

5. $8\frac{1}{2}$; $5\frac{2}{3}$

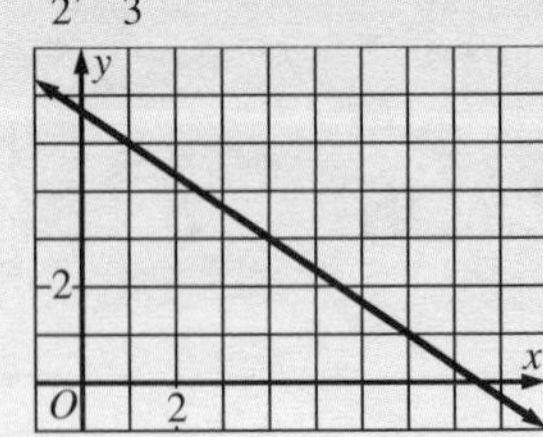

17. At a county fair one day, 600 tickets were sold. Adult tickets cost \$10 each and tickets for children cost \$5 each. If the total money collected from selling tickets was \$4750, how many children's tickets were sold?

18. Writing Use the graphs at the right. Tell what you know about the equation of each line and about the solution(s) of each system of equations.

a.

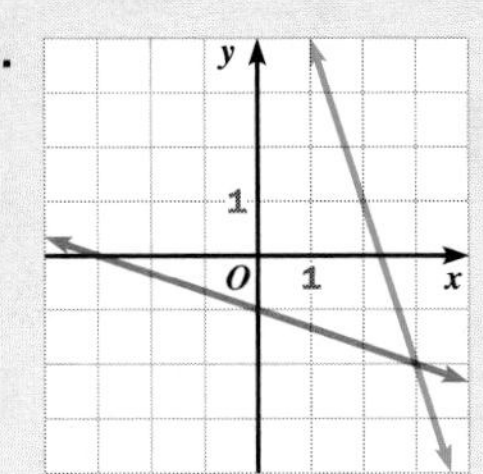

b.

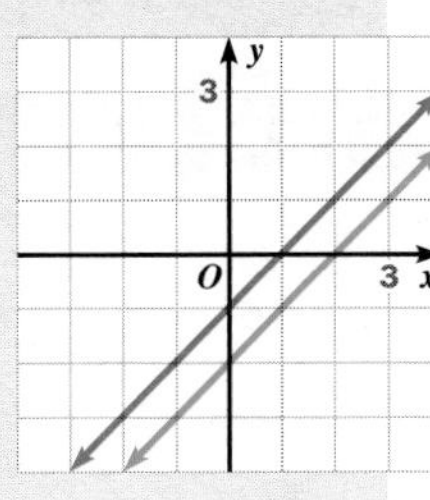

SECTIONS 7.5 *and* 7.6

Graph each inequality or system of inequalities.

19. $y < x + 6$

20. $y > \frac{2}{3}x - 2$

21. $2x + 4y \leq 17$

22. $3y < 16 - 3x$

23. $y \geq 3$
$y \leq -4x$

24. $2x + 3y > 5$
$6y < 1 - 4x$

25. $2x + 3y < 11$
$y \geq \frac{4}{5}x - 3$

26. $y < \frac{3}{4}x + 1$
$y \geq \frac{1}{3}x - 3$

27. $4x + 5y > 6$
$1 \geq x + y$

28. For the opening of an art gallery, Gina DiFazio and Eric Kornfeld need to buy at least 10 lb of nuts. Peanuts cost \$2/lb and deluxe mixed nuts cost \$5/lb. Gina and Eric can spend up to \$25 on nuts.

a. Write an inequality stating that x pounds of peanuts and y pounds of mixed nuts come to a total of at least 10 lb.

b. Write an inequality stating that Gina and Eric can spend up to \$25 on x pounds of peanuts and y pounds of mixed nuts.

c. They cannot buy negative amounts of nuts, so $x \geq 0$ and $y \geq 0$. Graph these inequalities, as well as those you wrote in parts (a) and (b), in the same coordinate plane.

d. What is the greatest amount of mixed nuts Gina and Eric can buy? Explain.

PERFORMANCE TASK

29. Draw three diagrams that show the possibilities for a solution of a system of equations. Give the equations for the lines and solve each system.

Chapter 7

ALTERNATIVE ASSESSMENT

1. Open-ended Problem Write two equations whose graphs intersect at the point (1, 2).

2. Open-ended Problem Use the graph shown below to find the slope-intercept form of the equation $2x + 4y = 8$.

3. Open-ended Problem Find the equations of three lines that are parallel. Write the equations in both slope-intercept form and in standard form.

4. a. Write a system of two equations in standard form that has one solution.

b. Write a system of two equations in standard form that has no solutions.

c. Write a system of two equations in standard form that has an infinite number of solutions.

5. Open-ended Problem Write a system of four inequalities whose solution region is a trapezoid.

6. a. Find the vertices of the triangular solution region formed by the following system.
$y \leq \frac{1}{5}x + 2$
$y \geq x - 2$
$x \geq 0$

b. Find three ordered pairs that are solutions of the system given in part (a).

7. a. Describe and sketch the possible solution regions for a system of two linear inequalities.

b. Describe and sketch the possible solution regions for a system of three linear inequalities.

102

Assessment Book, ALGEBRA 1: EXPLORATIONS AND APPLICATIONS

6. $y = \frac{5}{2}x - \frac{11}{2}$

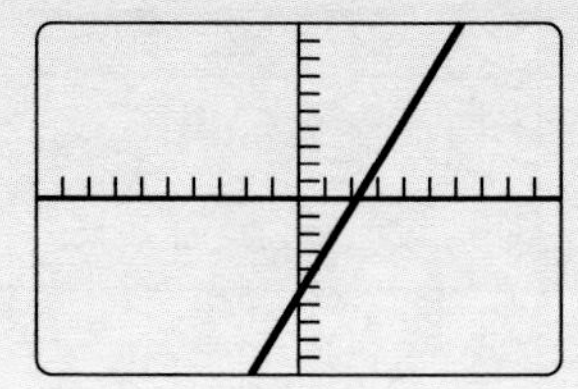

7. $y = \frac{2}{3}x - 4$

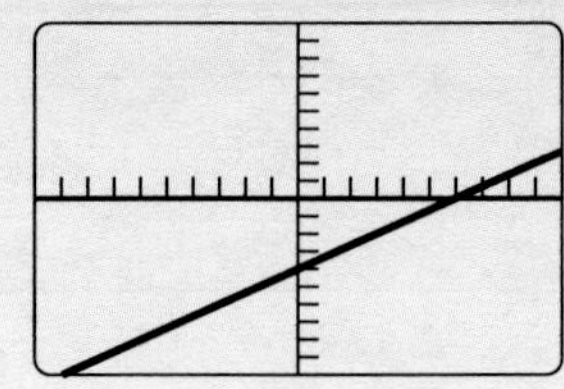

8. Let x = the number of pounds of Heavenly Brew and y = the number of pounds of Great Taste Tea.

a. $6x + 4y = 20$

b. $x + y = 4$

c. (2, 2); Anita buys two pounds of each kind of tea.

9. (3, 4)

10. $\left(1\frac{2}{3}, \frac{2}{3}\right)$

11. (3, 1)

12. (1, –2)

13. (1, 2)

14. $\left(0, 3\frac{1}{3}\right)$

15. all (x, y) such that $2x - 3y = 11$

16. (–14, 11)

17. 250 children's tickets

18–29. See answers in back of book.

CHAPTER

8 Quadratic Functions

OVERVIEW

Connecting to Prior and Future Learning

⇔ Work with graphs, begun in Chapter 3, is continued as students explore nonlinear relationships in Section 8.1. A detailed study of parabolas is presented in Section 8.2.

⇔ In Section 8.3, two ways of solving quadratic equations are presented; algebraic solutions using square roots and graphical solutions utilizing parabolas. These skills are applied to solving problems in Section 8.4.

⇔ This chapter concludes with the quadratic formula and the discriminant. This method of solving quadratic equations is one that students will use in future mathematics courses.

Chapter Highlights

Interview with Bill Pinkney: The feat of sailing around the world illustrates uses of mathematics, with related exercises on pages 341 and 347.

Explorations in Chapter 8 use graph paper and graphing calculators to explore nonlinear graphs and analyze the shape of a parabola. Students also use a graphing calculator to explore the solutions of quadratic equations.

The Portfolio Project: Exploring the effects of gravity on the time it takes a ball to roll down a ramp is the focus of this project. Students conduct an experiment, make a scatter plot of the data they collect, and compare their results to Galileo's.

Technology: Students use graphing calculators to explore, graph, and analyze nonlinear relationships including absolute value and quadratic functions. Students also use this technology to solve quadratic equations.

OBJECTIVES

Section	Objectives	NCTM Standards
8.1	• Analyze the shape of a graph. • Decide whether a relationship is linear.	1, 2, 3, 4, 5, 6, 8
8.2	• Recognize characteristics of parabolas. • Predict the shape of a parabola. • Recognize side views of real objects.	1, 2, 3, 4, 5, 6, 8
8.3	• Use square roots and graphs to solve simple quadratic equations. • Relate x-intercepts and solutions of quadratic equations. • Solve simple motion problems involving acceleration.	1, 2, 3, 4, 5, 6, 8
8.4	• Use graphs to solve more complicated quadratic equations. • Recognize quadratic equations with no solutions. • Solve any quadratic equation and find heights of thrown objects.	1, 2, 3, 4, 5, 6, 8
8.5	• Use the quadratic formula. • Solve quadratic equations without using a graph. • Find more precise solutions to real-world problems.	1, 2, 3, 4, 5, 8
8.6	• Find the discriminant of a quadratic equation. • Find the number of solutions of a quadratic equation.	1, 2, 3, 4, 5, 8

INTEGRATION

Mathematical Connections	8.1	8.2	8.3	8.4	8.5	8.6
algebra	**325–330***	**331–336**	**337–343**	**344–349**	**350–354**	**355–359**
geometry	**325–330**	336	343			357
data analysis, probability, discrete math	328–330		342			
patterns and functions				349		
logic and reasoning	**325–330**	331, 333–336	**337–343**	**344–349**	351, 353, 354	355, 357–359
Interdisciplinary Connections and Applications						
chemistry and physics			337			359
sports and recreation			342	346	351, 354	
cooking, astronomy, highway safety, fire fighting engineering	329	335	340	348	353	358

* ***Bold page numbers*** *indicate that a topic is used throughout the section.*

TECHNOLOGY

	opportunities for use with	
Section	**Student Book**	**Support Material**
8.1	graphing calculator McDougal Littell Software *Stats!* *Function Investigator*	**Techonology Book:** Calculator Activity 8 Spreadsheet Activity 8
8.2	graphing calculator McDougal Littell Software *Function Investigator*	
8.3	graphing calculator McDougal Littell Software *Function Investigator*	
8.4	graphing calculator McDougal Littell Software *Function Investigator*	
8.5	graphing calculator	**Technology Book:** Spreadsheet Activity 8
8.6	graphing calculator	

Regular Scheduling (45 min)

Section	Materials Needed	Core Assignment	Extended Assignment	*exercises that feature* Applications	Communication	Technology
8.1	graph paper, graphing calculator	**Day 1:** 1–13, 15–18 **Day 2:** 19–21, 23–25, 28–39	**Day 1:** 1–18 **Day 2:** 19–39	19–21	10, 14 23–27	16 19, 21, 22c, 23–25
8.2	graph paper, graphing calculator, six circular objects	**Day 1:** 1–17 **Day 2:** 18–28, 30–39	**Day 1:** 1–17 **Day 2:** 18–39	21–23	6, 17 24–30	7–15
8.3	graph paper, graphing calculator	**Day 1:** 1–20 **Day 2:** 21–29, 32, 33, 35–40, AYP*	**Day 1:** 1–20 **Day 2:** 21–40, AYP	10, 14–16 32–34	30, 31, 34, 35	17–20
8.4	graphing calculator	**Day 1:** 1–16 **Day 2:** 17–24, 32–41	**Day 1:** 1–16 **Day 2:** 17–41	5–7 25	8 26, 32	6, 9–16 17–20, 22–24
8.5	graph paper, graphing calculator	1–6, 8–18, 20–33	1–10, 14–33	17, 21–23	7, 18, 24	8–16
8.6	graph paper, graphing calculator	**Day 1:** 1–13, 15–20 **Day 2:** 21–23, 27–38, AYP	**Day 1:** 1–20 **Day 2:** 21–38, AYP	11–13	14	
Review/ Assess		**Day 1:** 1–17 **Day 2:** 18–34 **Day 3:** Ch. 8 Test	**Day 1:** 1–17 **Day 2:** 18–34 **Day 3:** Ch. 8 Test		1–4	24–26
Portfolio Project		Allow 2 days.	Allow 2 days.			

Yearly Pacing (with Portfolio Project)	**Chapter 8 Total** 16 days	**Chapters 1–8 Total** 110 days	**Remaining** 50 days	**Total** 160 days

Block Scheduling (90 min)

	Day 48	Day 49	Day 50	Day 51	Day 52	Day 53	Day 54	Day 55	Day 56
Teach/ Interact	Ch. 7 Test 8.1: Exploration, page 325	Continue with 8.1 8.2: Exploration, page 331	Continue with 8.2 8.3	Continue with 8.3 8.4	Continue with 8.4 8.5	8.6: Exploration, page 355	Review Port. Proj.	Review Port. Proj.	Ch. 8 Test 9.1: Exploration, page 369
Apply/ Assess	**Ch. 7 Test** **8.1:** 1–13, 15–18	**8.1:** 19–21, 23–39 **8.2:** 1–17	**8.2:** 18–39 **8.3:** 1–20	**8.3:** 21–29, 32, 33, 35–40, AYP* **8.4:** 1–16	**8.4:** 17–24, 27–41 **8.5:** 1–6, 8–33	**8.6:** 1–13, 15–23, 27–38, AYP	**Review:** 1–17 **Port. Proj.**	**Review:** 18–34 **Port. Proj.**	**Ch. 8 Test** **9.1:** 1–20

NOTE: A one-day block has been added for the Portfolio Project—timing and placement to be determined by teacher.

Yearly Pacing (with Portfolio Project)	**Chapter 8 Total** 8 days	**Chapters 1–8 Total** $55\frac{1}{2}$ days	**Remaining** $25\frac{1}{2}$ days	**Total** 81 days

**AYP* is *Assess Your Progress.*

Lesson Support

Section	Practice Bank	Study Guide*	Assessment Book*	Visuals	Explorations Lab Manual	Lesson Plans	Technology Book
8.1	51	8.1		Warm-Up 8.1 Folder A	Masters 1, 2, 13	8.1	Calculator Act. 8 Spreadsheet Act. 8
8.2	52	8.2		Warm-Up 8.2 Folder 8	Master 2 Add. Expl. 13	8.2	
8.3	53	8.3	Test 34	Warm-Up 8.3	Masters 1, 2	8.3	
8.4	54	8.4		Warm-Up 8.4		8.4	
8.5	55	8.5		Warm-Up 8.5	Master 2	8.5	Spreadsheet Act. 8
8.6	56	8.6	Test 35	Warm-Up 8.6	Masters 2, 14	8.6	
Review Test	57	Chapter Review	Tests 36, 37 Alternative Assessment			Review Test	Calculator Based Lab 5

*__Spanish versions__ of *Study Guide* and *Assessment Book* are available.

Chapter Support

- Course Guide
- Lesson Plans
- Explorations Lab Manual: Additional Exploration 12
- Portfolio Project Book: Additional Project 5: Quadratic Functions
- Preparing for College Entrance Tests
- Multi-Language Glossary
- *Test Generator* Software
- Professional Handbook

Software Support

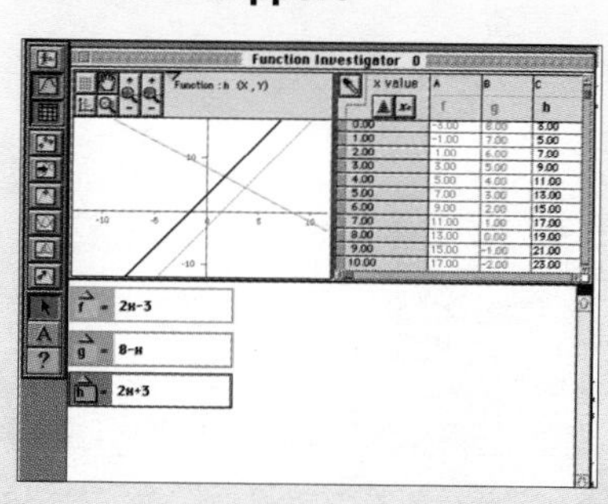

McDougal Littell Software
Stats!
Function Investigator

Internet Support

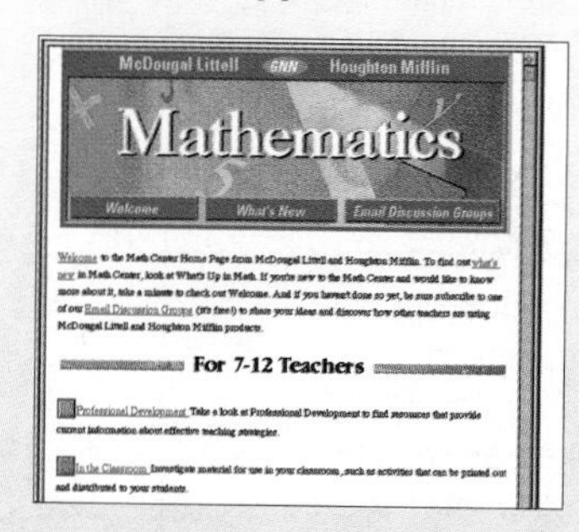

http://www.hmco.com
Next go to McDougal Littell; then the Education Center; then Secondary Math.

Outside Resources

Books, Periodicals

Forringer, Richard. "If the Product of Two Numbers is Zero..." *Mathematics Teacher* (February 1994): p. 89.

Amick, H. Louise. "A Unique Slope for a Parabola." *Mathematics Teacher* (January 1995): p. 38.

Activities, Manipulatives

Disher, Fan. "Graphing Art." *Mathematics Teacher* (February 1995): pp. 124–128, 134–138.

Consortium for Mathematics and its Applications. *Technology in the Classroom, Volume 1.* Lexington, MA: COMAP.

Software

Harvey, Wayne and Judah L. Schwartz. *The Function Supposer: Explorations in Algebra.* Newton, MA: Education Development Center, 1992.

Harvey, Wayne, Judah L. Schwartz, and Michal Yerushalmy. *Visualizing Algebra: The Function Analyzer.* Pleasantville, NY: Sunburst Communications.

Algebra Animator. Models phenomena whose motions are defined by functions. Pleasantville, NY: Sunburst Communications.

Videos

Apostel, Tom. *Polynomials.* Reston, VA: NCTM.

Interview Notes

Background

Pinkney's Route

Bill Pinkney began his voyage around the world in Boston's historic harbor in August 1990. Leaving Boston, he sailed down the coast of North America and part of the coast of South America to Salvador, Brazil and then headed east, across the Atlantic Ocean to Capetown, South Africa. From Capetown, he continued to proceed east through the Indian Ocean to Hobart, Tasmania, which is an island off the southern coast of Australia. Leaving Tasmania, he sailed across the Pacific Ocean to the southern tip of Chile around Cape Horn and up the coasts of South America and North America to Boston. He ended his voyage 22 months later, including 16 months at sea, at the same point he began in Boston Harbor.

Bill Pinkney

Bill Pinkney is the first African American to attempt and complete a solo sail around the world. After graduating from high school, Pinkney was trained as an x-ray technician and worked at Provident Hospital in Chicago. He later served in the Navy as a hospital corpsman and after being discharged, moved to San Juan, Puerto Rico. It was while living in Puerto Rico that Pinkney developed a great love for sailing. When he returned to the United States, Pinkney became a licensed makeup artist in order to work in the film industry. While doing promotional work for cosmetic companies, he was recruited by Revlon to help develop a new cosmetic line and in a few years became the first African American marketing executive in the company's history. Pinkney continued to sail as his business career advanced and in 1983 he became a licensed captain for power and sail vessels up to 100 tons.

CHAPTER

8 Quadratic Functions

Meeting the CHALLENGE

INTERVIEW **Bill Pinkney**

On June 9, 1992, former businessman Bill Pinkney sailed into Boston Harbor—and into history books. After 16 months and 32,000 miles of battling weather, loneliness, and equipment failures, Pinkney became the first African American to sail solo around the world. His route made his achievement even more remarkable. Pinkney circled much of the globe 40 to 60 degrees south of the equator. Here the Pacific, Indian, and Atlantic Oceans merge around Antarctica, creating extreme conditions for sailors: strong winds, high waves, icebergs, and snowstorms, even in summer.

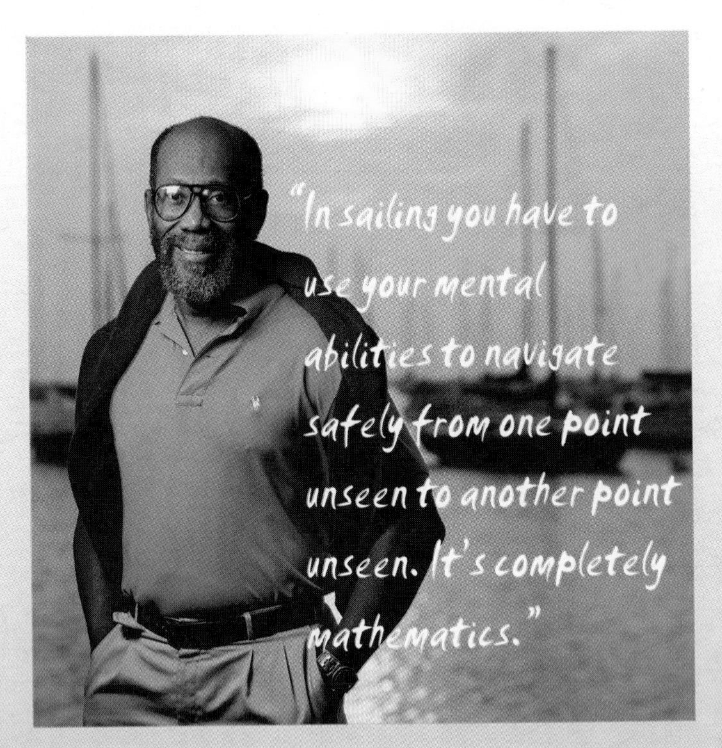

"In sailing you have to use your mental abilities to navigate safely from one point unseen to another point unseen. It's completely mathematics."

Using these winds required many calculations. "You want to match wind speed with the amount of surface area of sail you have up," Pinkney explains. "Too much wind hitting too much sail can knock you flat on your side, which happened to me twice."

Using the Wind

Strong west-to-east winds powered *Commitment*, Pinkney's 47 foot boat. "Cruising sailors like to sail in trade winds of 20 to 22 knots. But ideal sailing weather for me was just below true gale-force winds," Pinkney says. "That meant a 35 knot wind and moderate waves, anywhere from 15 to 25 feet." (The word knot means nautical miles per hour. One knot is about 1.15 miles per hour.)

Storms and Squalls

What did Pinkney do when he encountered storms and squalls during his voyage? "You evaluate a storm as bad, terrible, or life threatening," he says, "and if it's bad, you go on and do other things. I'd take care of my housekeeping, maintenance, and navigation duties, and I'd read, sleep, and eat."

Pinkney experienced the worst storm of the trip as he rounded Cape Horn, the southern tip of

Interview Notes

Background

Ancient Egyptian Boats
Some of the first boats were made by the ancient Egyptians more than 6000 years ago. Those early boats were made of bundles of reeds secured together. They were first powered by poles, and then by rows of oarsmen. About one thousand years later, the Egyptians found that they could harness the power of the wind to propel their boats more quickly and efficiently, and in so doing, they invented the first sails. The design and construction of boats continued to evolve over the centuries, with planks of wood replacing the bundles of reeds and wider sails replacing long, narrow ones. In addition, strong ropes began to be used to link the front and rear of the boat, making boats stronger.

Second-Language Learners

Students learning English may be unfamiliar with the term *squall.* Explain that a squall is "a brief and sudden storm."

Interview Notes

Mathematical Connection

In Section 8.3, students learn to solve simple quadratic equations by using square roots and graphs. These skills can then be used to solve problems involving the force F on a sail by a wind blowing perpendicular to the sail with speed v. The formula relating F and v is $F = 0.00432v^2$. In Section 8.4, students continue to apply quadratic equations to sailing problems using Pinkney's ideal wave height and the equation $h = 0.013s^2$, where h is wave height and s is wind speed.

Explore and Connect

Writing
Students' examples can be ordered pairs of numbers from which they can form a conclusion about how wave height is related to wind speed.

Project
The report students prepare should include the time periods that the civilization existed and the range of their sailing exploits.

Research
Students can trace Pinkney's route by identifying the key turning points of his voyage.

South America. Cold, 65 to 70 knot winds forced Pinkney to pull down his sails. High waves crashed in from two directions. During this 40 hour ordeal, Pinkney called on all his sailing and navigating skill to clear Cape Horn's rocky coast.

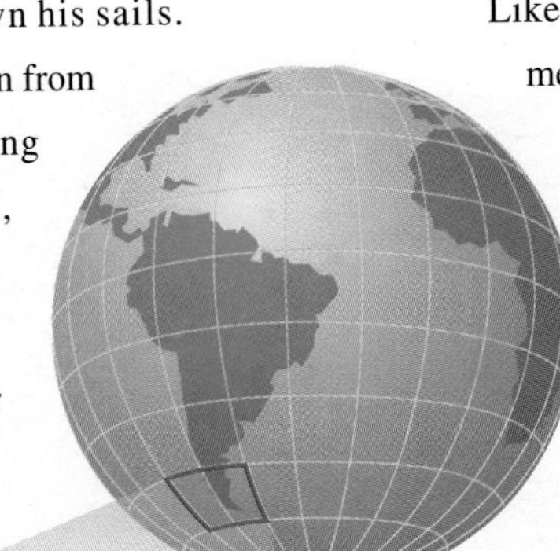

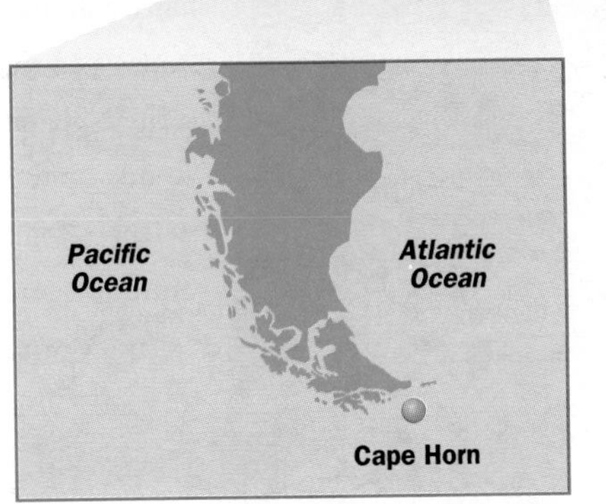

Wind Speed and Wave Height

Like all sailors, Pinkney knows that strong winds mean big waves. The duration of high winds, the amount of open water, and the strength of water currents affect wave height, but wind speed is one of the most important factors. The graph shows that as the wind speed increases, the height of ocean waves increases dramatically. The dependence of wave height on wind speed is an example of a *quadratic function.*

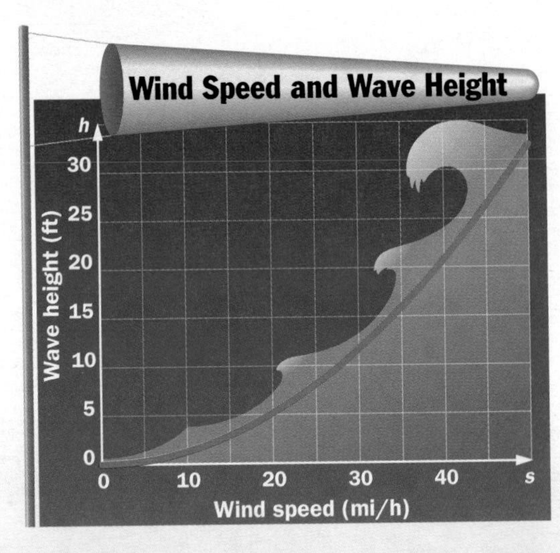

EXPLORE AND CONNECT

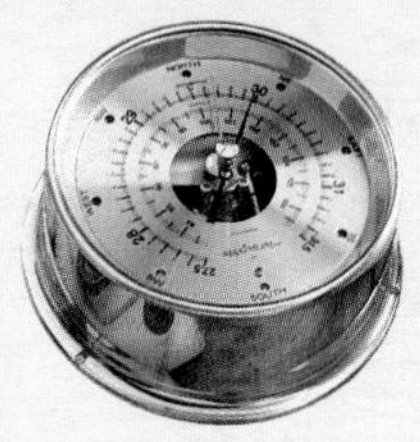
Navigational compass

1. Writing Use the graph to determine what happens to wave height when wind speed doubles. Give examples to justify your conclusion.

2. Project The Egyptians first began using sails for boats about 5000 years ago. The Phoenicians, the Polynesians, and the Vikings also developed a high degree of sailing skill. Prepare a report on the seafaring accomplishments of one of these civilizations.

3. Research In addition to sailing around Cape Horn, Bill Pinkney sailed around four other "great capes." Locate these five great capes on a map or a globe. Roughly trace Bill Pinkney's route.

Mathematics & Bill Pinkney

In this chapter, you will learn more about how mathematics is related to sailing.

Related Exercises

Section 8.3
- Exercises 14–16

Section 8.4
- Exercises 5–8

SECTION

8.1 Nonlinear Relationships

Graphs are not always lines. Often the relationship between two variables is *nonlinear,* as you will see in the Exploration.

Learn how to...

- analyze the shape of a graph

So you can...

- decide whether a relationship is linear

EXPLORATION
COOPERATIVE LEARNING

Exploring Nonlinear Graphs

Work in a group of 2–4 students.
You will need:

- graph paper
- a graphing calculator

You want to design a rectangular pen for a dog, using 30 ft of fencing. You decide that the length of each side must be an integer.

1 Draw all the pens you can build. Then copy and complete the table. Add rows until you have listed all possibilities.

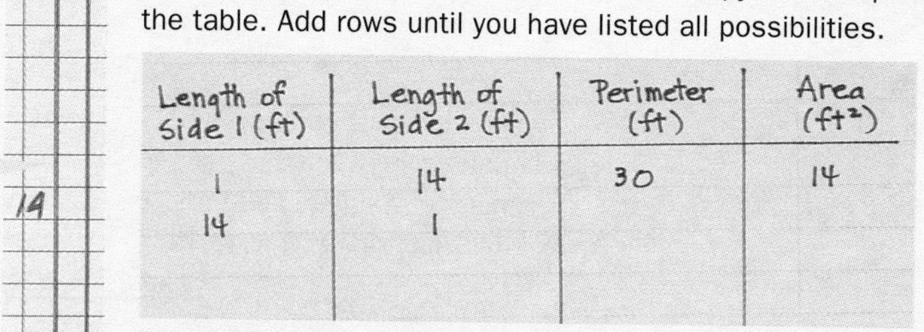

Length of Side 1 (ft)	Length of Side 2 (ft)	Perimeter (ft)	Area (ft²)
1	14	30	14
14	1		

Perimeter = 30 ft
Area = 14 ft²

2 Which dimensions give the dog the most area? Which dimensions give the dog the longest run? Discuss which dimensions you would recommend. Explain your decision.

3 Make a scatter plot of the data. Put the length of Side 1 on the horizontal axis and the area of the pen on the vertical axis. Describe the shape of the graph.

4 Let x = the length of Side 1 of the pen. Write an expression in terms of x for the length of Side 2. Then write and graph an equation for the area of the pen as a function of the length of Side 1. Use a graphing calculator.

5 Discuss with your group how the graphs in Steps 3 and 4 compare.

Exploration Note

Purpose
The purpose of this Exploration is to have students discover that a relationship between two variables can be nonlinear.

Materials/Preparation
Each group should have graph paper and a graphing calculator.

Procedure
Groups should complete the table in Step 1 and answer the questions in Step 2. Then one student in each group can be in charge of drawing the graphs for Steps 3 and 4.

Closure
After completing Step 4, groups can share their results with each other. The class should conclude that the graph of the area of a rectangle as a function of the length of a side is not a straight line.

Explorations Lab Manual
See the Manual for more commentary on this Exploration.

Diagram Masters 1 and 13

For answers to the Exploration, see answers in back of book.

Plan⇔Support

Objectives

- Analyze the shape of a graph.
- Decide whether a relationship is linear.

Recommended Pacing

Core and Extended Courses
Section 8.1: 2 days

Block Schedule
Section 8.1: 2 half-blocks (with Chapter 7 Test and Section 8.2)

Resource Materials

Lesson Support
Lesson Plan 8.1
Warm-Up Transparency 8.1
Overhead Visuals:
Folder A: Multi-Use Graphing Packet, Sheets 1–3
Practice Bank: Practice 51
Study Guide: Section 8.1
Explorations Lab Manual:
Diagram Masters 1, 2, 13

Technology
Technology Book:
Calculator Activity 8,
Spreadsheet Activity 8
Graphing Calculator
McDougal Littell Software
Function Investigator
Stats!
Internet:
http://www.hmco.com

Warm-Up Exercises

1. Find an ordered pair that is a solution of the equation $y = 3x$. Sample: (4, 12)
2. Find an ordered pair that is a solution of the equation $y = 3x^2$. Sample: (5, 75)
3. What is the domain of the function $y = -2x + 1$? all numbers
4. Find $|x|$ if $x = -8$. 8
5. If $P = 6x$ represents the profit on x tickets sold for a concert, what is a reasonable domain for the function P? $x \geq 0$, where x is an integer

Teach⟺Interact

About Example 1

Teaching Tip
After discussing Example 1, you may want to remind students that for the equation $P = 4s$, the number 4 represents the slope of the line.

Additional Example 1

Tell whether each relationship is *linear* or *nonlinear.*

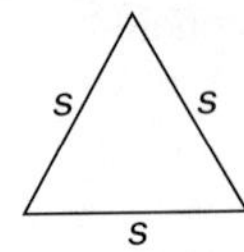

a. the relationship between the perimeter P of an equilateral triangle and the length s of one side

Write and graph an equation for P as a function of s.

$P = 3s,\ s \geq 0$

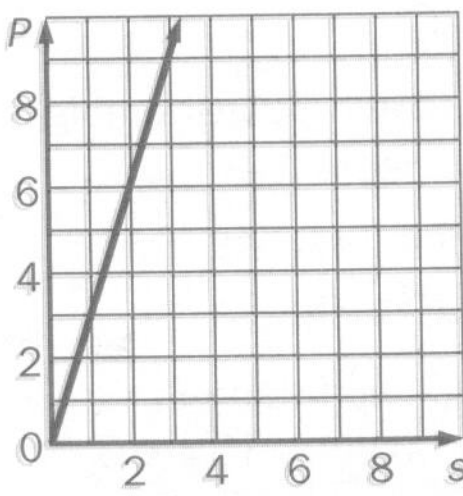

The graph is part of a line. The relationship between the perimeter of an equilateral triangle and the length of one side is linear.

b. the relationship between the area A of an equilateral triangle and the length s of one side

Write and graph an equation for A as a function of s.

$A = \frac{\sqrt{3}}{4}s^2,\ s \geq 0$

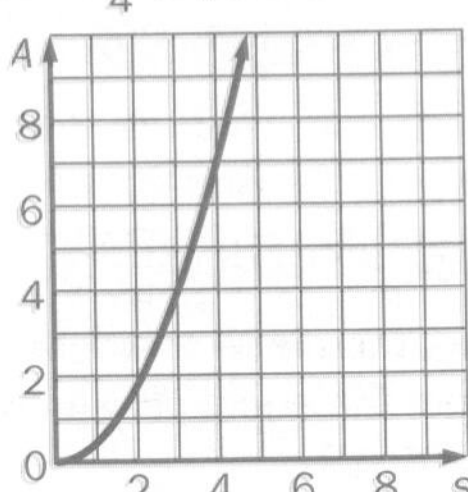

The graph is not a line or part of a line. The relationship between the area of an equilateral triangle and the length of one side is nonlinear.

Think and Communicate

When discussing question 1, students should understand that the domain for the function $V = s^3$ does not have to be restricted to integral values of s.

If the graph of an equation is a line, then the equation shows constant change and the relationship that it models is linear. Otherwise, the equation and the relationship that it models are both **nonlinear**.

EXAMPLE 1

Tell whether each relationship is *linear* or *nonlinear*.

a. the relationship between the perimeter P of a square and the length s of one side

b. the relationship between the area A of a square and the length s of one side

SOLUTION

a. Write and graph an equation for P as a function of s.

$P = 4s,\ s \geq 0$

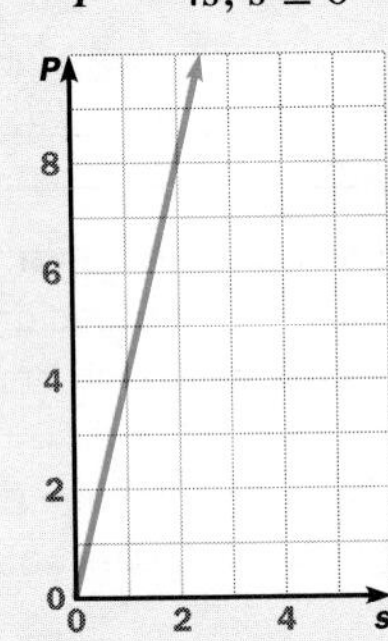

The graph is part of a line. The relationship between the perimeter of a square and the length of one side is linear.

b. Write and graph an equation for A as a function of s.

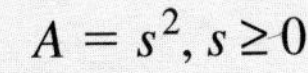

$A = s^2,\ s \geq 0$

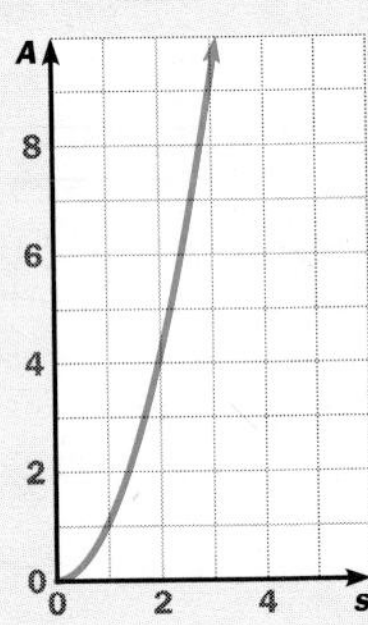

The graph is not a line or part of a line. The relationship between the area of a square and the length of one side is nonlinear.

THINK AND COMMUNICATE

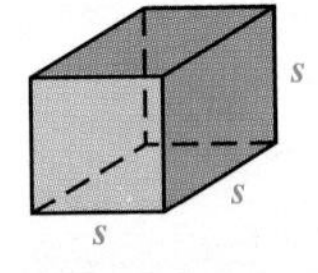

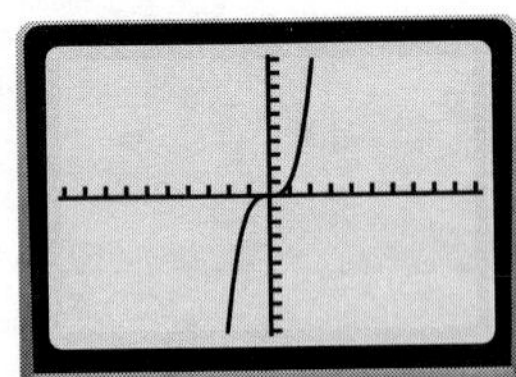

1. An equation for the volume V of a cube with edge length s is $V = s^3$. A graph of the equation is shown.

a. What is a reasonable domain for this function?

b. Is the relationship between volume and edge length *linear* or *nonlinear*? Explain.

2. In the Exploration, is the relationship between the area of the pen and the length of Side 1 *linear* or *nonlinear*? Explain.

ANSWERS Section 8.1

Think and Communicate

1. a. positive real numbers
   b. nonlinear; The graph of the equation is not a line.
2. nonlinear; The graph of the equation is not a line.
3. The graph never falls below the x-axis; that is, the value of y is never less than zero.
4. For $x \geq 0$, the graphs are identical. For negative values of x, the graph of $y = |x|$ is the mirror image of the graph of $y = x$ through the x-axis. $y = x$ is linear; $y = |x|$ is nonlinear.
5. The graph in Example 1(a) is part of a line through the origin. The graph in Example 2 is made up of parts of two different lines through the origin.

EXAMPLE 2

Is the equation $y = |x|$ *linear* or *nonlinear*?

SOLUTION

Graph the equation.

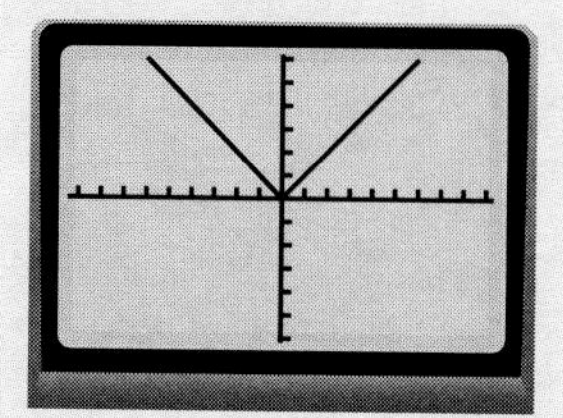

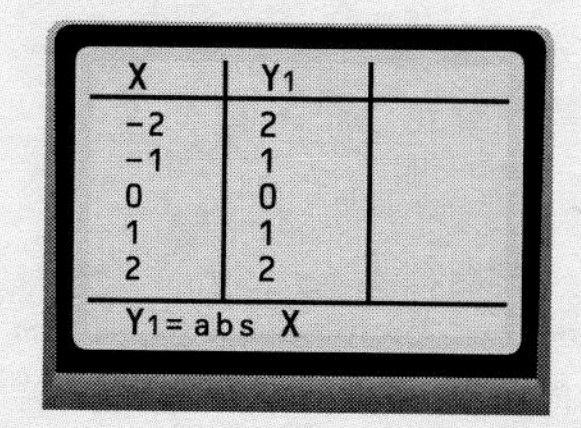

X	Y1
−2	2
−1	1
0	0
1	1
2	2

Y1= abs X

A table of values can also help you to see what's happening.

The graph is not one continuous line. The equation $y = |x|$ is nonlinear.

THINK AND COMMUNICATE

3. In Example 2, how does the graph of the equation $y = |x|$ show that the absolute value of a number x is never negative?
4. How is the graph of the equation $y = |x|$ different from the graph of the equation $y = x$? Which equation is linear and which is nonlinear?
5. The relationship in Example 1(a) is linear, but the relationship in Example 2 is nonlinear. How are the graphs alike? How are they different?

CHECKING KEY CONCEPTS

Tell whether each graph represents a *linear* or a *nonlinear* relationship.

1. 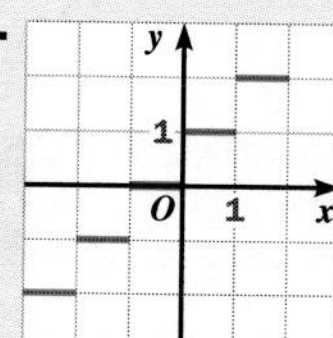

2. 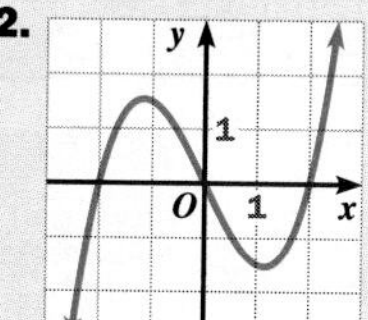

3. 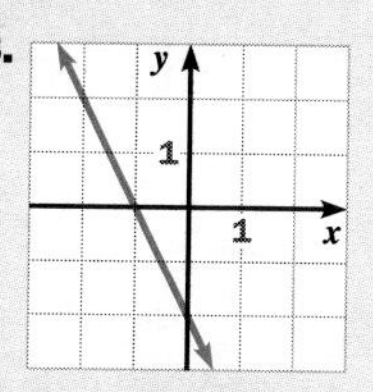

Graph each equation. Tell whether each equation is *linear* or *nonlinear*.

4. $y = 3x + 2$
5. $y = 3x^2$
6. $y = 3|x|$
7. A convenience store has a self-service copier. The rates for using the copier are shown in the table. Is the relationship between the cost of making copies and the number of copies *linear* or *nonlinear*? Explain.

Number of copies	Cost per copy (cents)
1–10	15
11–25	10
26 +	7

Checking Key Concepts

1. nonlinear
2. nonlinear
3. linear
4. linear

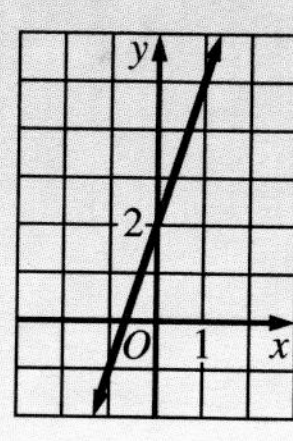

5. nonlinear

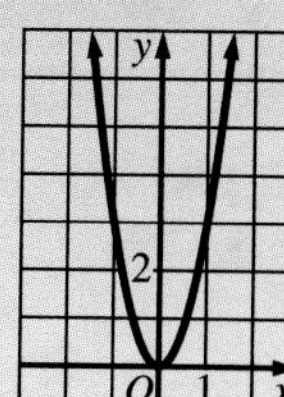

6. nonlinear

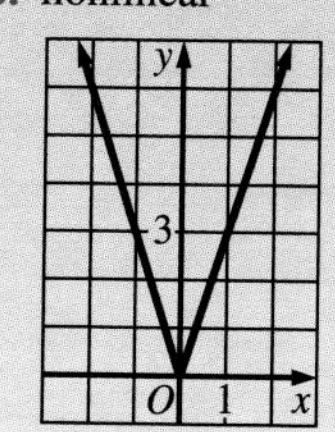

7. nonlinear; The graph is not a line.

Teach⇔Interact

About Example 2

Topic Spiraling: Review
Remind students that the absolute value of a number is either zero or positive. If $y = |x|$, then y is either 0 or a positive number.

Additional Example 2

Is the equation $y = \sqrt{x}$ *linear* or *nonlinear*?
Graph the equation.

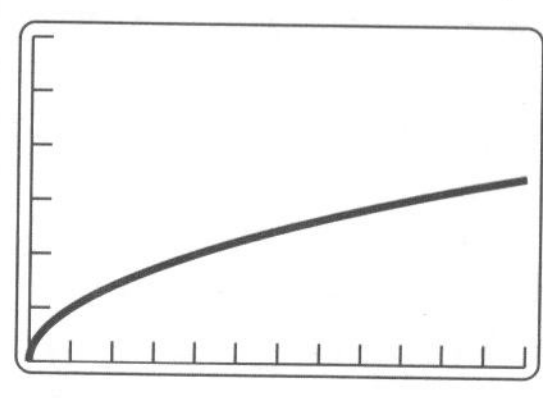

A table of values can also help you see what's happening.

X	Y1
0	0
4	2
8	2.8284
12	3.4641
16	4
20	4.4721
24	4.899

Y1=√X

The graph is not a line or part of a line. The equation $y = \sqrt{x}$ is nonlinear.

Think and Communicate

A class discussion of questions 4 and 5 will help to ensure that all students can differentiate between linear and nonlinear relationships.

Checking Key Concepts

Common Error
Some students may think that the graph in question 1 is linear because it is made up of line segments. Emphasize that a linear graph shows the same constant change between any two points of the graph and is one continuous line.

Closure Question

How can you determine if an equation is linear or nonlinear?
Graph the equation. If the graph is a straight line, the equation is linear. If not, the equation is nonlinear.

Apply⇔Assess

Suggested Assignment

Core Course
Day 1 Exs. 1–13, 15–18
Day 2 Exs. 19–21, 23–25, 28–39

Extended Course
Day 1 Exs. 1–18
Day 2 Exs. 19–39

Block Schedule
Day 48 Exs. 1–13, 15–18
Day 49 Exs. 19–21, 23–39

Exercise Notes

Reasoning
Ex. 10 Students should write additional equations to verify their generalizations. They should understand that their generalizations have to apply to all equations.

Second-Language Learners
Ex. 10 Many students learning English will be able to respond to this exercise independently, but some may need the help of a peer tutor.

Integrating the Strands
Exs. 11–13 These exercises help students to see connections between algebra and geometry. Understanding the algebraic concepts of linear and nonlinear can provide students with a deeper understanding of these common geometric formulas.

Cooperative Learning
Exs. 15–18 Working in small groups will allow students to share their personal knowledge of the ticket structure for theater productions. Each group should decide what price it would charge for a ticket and why.

Using Technology
Ex. 16 Students may use the ZOOM 9:ZoomStat feature of the TI-82 calculator to automatically set the viewing window for data points.

8.1 Exercises and Applications

Extra Practice exercises on page 572

Graph each equation. Tell whether each equation is *linear* or *nonlinear*.

1. $y = 5x$ **2.** $y = 5|x|$ **3.** $y = 5x^2$

4. $y = -2|x|$ **5.** $y = -3x + 1$ **6.** $y = |x| + 3$

7. $y = \frac{1}{2}x + 4$ **8.** $y = 4x^2 - 3$ **9.** $y = x^3 + 2$

10. Writing Based on the results of Exercises 1–9, generalize how you can tell from an equation whether its graph will be a line.

GEOMETRY Tell whether the relationship between the radius *r* and each given variable is *linear* or *nonlinear*. Explain your reasoning.

11. Circumference C

$C = 2\pi r$

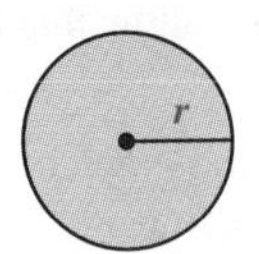

12. Area A

$A = \pi r^2$

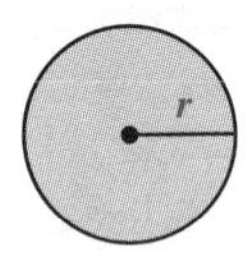

13. Volume V

$V = \frac{4}{3}\pi r^3$

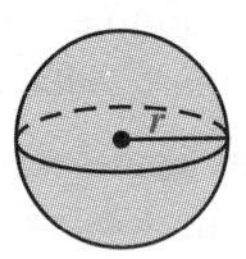

14. Open-ended Problem Find two different examples of real-world relationships that can be modeled with a graph, one linear and one nonlinear. Model each relationship with a graph.

For Exercises 15–18, use the table.

The drama club needs to decide on the ticket price for a school play. They estimate that about 500 people will buy tickets if the ticket price is \$3. Based on previous performances, they estimate that for every \$.50 increase in the ticket price, they will sell 50 fewer tickets.

15. Copy and complete the table.

16. Technology Make a scatter plot of the data. Put ticket price on the horizontal axis and income on the vertical axis.

17. Do you think the relationship between ticket price and income is *linear* or *nonlinear*? Explain.

18. Assume that the estimates are accurate. What price would you charge for a ticket? Give a reason for your answer.

Expected Income from Ticket Sales

Ticket price (dollars)	Number of tickets sold	Income (dollars)
3.00	500	1500
3.50	450	1575
4.00	?	?
4.50	?	?
5.00	?	?

BY THE WAY

Oklahoma! celebrated its 50th anniversary in 1993. According to one survey, *Oklahoma!* was the play most often performed by American high schools that year.

Exercises and Applications

1. linear

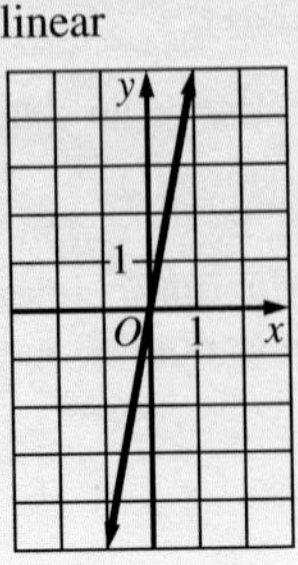

2. nonlinear

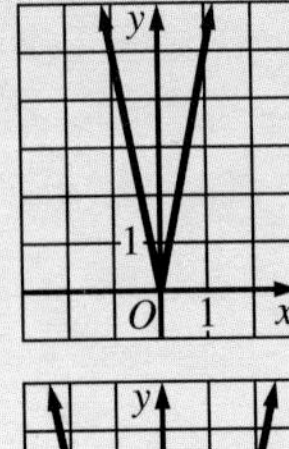

3. nonlinear

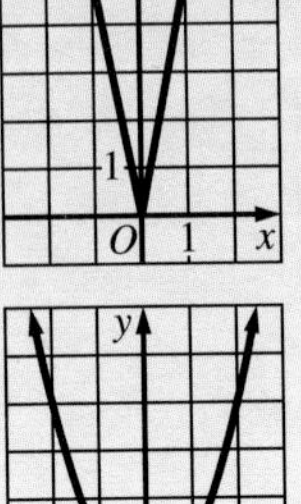

4. nonlinear

5. linear

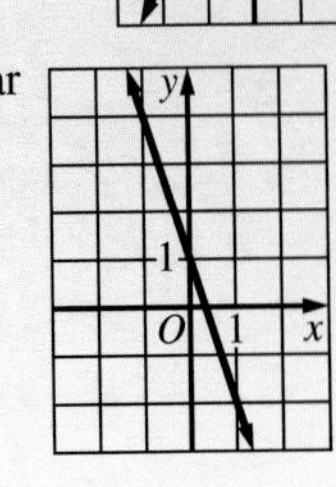

6. nonlinear

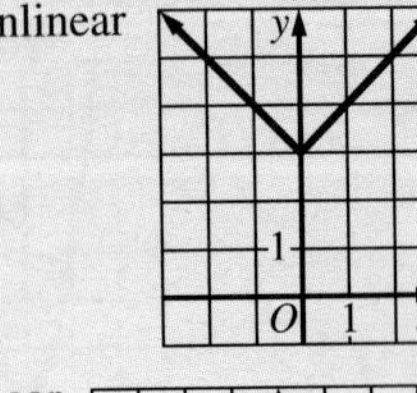

7. linear

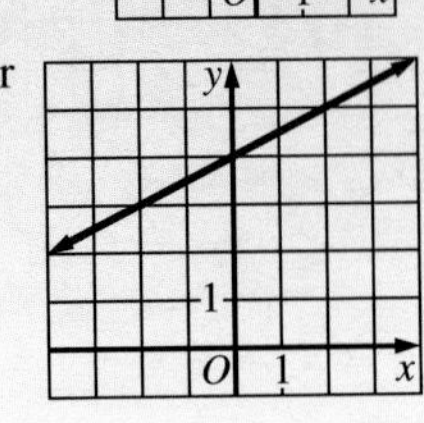

8. nonlinear

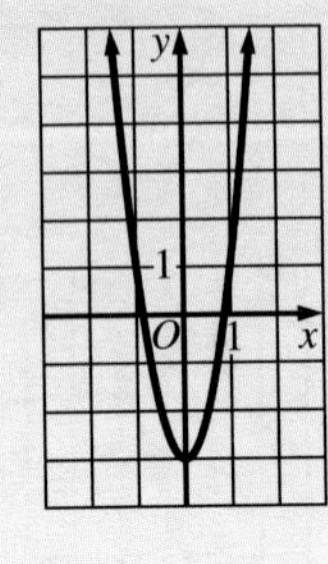

Connection COOKING

A professional baker uses the table at the right as a guide. It shows the greatest number of servings that you can expect from a four-layer cake of a given diameter.

19. **Technology** Make a scatter plot of the data. Put diameter on the horizontal axis and number of servings on the vertical axis.

20. Do you think the relationship between the diameter and the number of servings is *linear* or *nonlinear*? Explain.

21. **Technology** Use the formula below for the volume of a cylinder. Assume that each layer is 1.5 in. high. Decide whether the relationship between number of servings and volume is *linear* or *nonlinear*. Use a spreadsheet or a graphing calculator to make a scatter plot to support your answer.

Servings from a Four-Layer Cake

Diameter (in.)	Number of servings
6	20
8	30
10	45
12	65
14	95
16	125

$$\text{Volume} = \pi r^2 h = 6\pi r^2$$

r

$h = 6$ in.

22. The numbers in the sequence below are called *triangular numbers* because you can arrange each number of dots to form a triangle.

1st	2nd	3rd	...	nth
1	3	6	...	?

The triangular numbers follow a pattern.

$1 = \frac{1 \times 2}{2}$ $\quad 3 = \frac{2 \times 3}{2}$ $\quad 6 = \frac{3 \times 4}{2}$ $\quad \ldots \quad ?$

a. What are the 4th, 5th, 6th, and 10th triangular numbers?

b. Write a formula for the nth triangular number.

c. **Technology** Graph your formula from part (b) on a graphing calculator. Is the formula *linear* or *nonlinear*?

Apply⇔Assess

Exercise Notes

Problem Solving

Ex. 20 You may wish to ask students how they can determine the type of relationship by examining only the data in the table. (The increase in diameter is constantly 2, but the increase in the number of servings is not constant. This implies that the graph of the data cannot be a straight line.)

Using Technology

Exs. 19–25 Students can use the *Stats!* or *Function Investigator* software or a TI-82 graphing calculator for these exercises.

In part (b) of Ex. 22, students should be encouraged to check their formulas by verifying that they give the known triangular numbers for $n = 1, n = 2, \ldots, n = 6$, and $n = 10$. On the TI-82, this can be done by first entering the formula on the Y= list (preparatory to displaying the graph for part (c)). Next press [2nd] [TblSet] and use the settings TblMin = 1, △Tbl = 1. Finally, press [2nd][TABLE] and use the table entries to check the values given by the formula against the known triangular numbers.

In part (b) of Ex. 23, caution students to be careful to use parentheses when they enter their equations. For example, on the TI-82, the equation $y = \frac{1}{2}|x|$ should be entered on the Y= list as follows: Y1=(1/2)absX.

Historical Connection

Ex. 22 Triangular numbers were first investigated more than 2000 years ago by the followers of Pythagoras, who pictured numbers as having orderly designs. The triangular number 10 was special and called the *holy tetractys* by his followers.

9. nonlinear

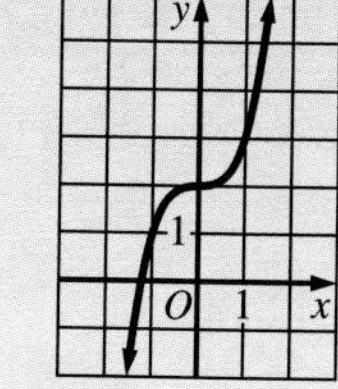

10. An equation whose graph is a line is of the form $y = mx + b$.

11. linear; The graph of the equation is a line; also, the equation is of the form $y = mx + b$, where $m = 2\pi$ and $b = 0$.

12. nonlinear; The graph of the equation is not a line; also, the equation cannot be written in the form $y = mx + b$.

13. nonlinear; The graph of the equation is not a line; also, the equation cannot be written in the form $y = mx + b$.

14. Answers may vary. Examples are given. The relationship between the cost of gasoline at \$1.45/gal and the number of gallons purchased is linear. The cost in dollars of sending a first-class letter and its weight in ounces is nonlinear.

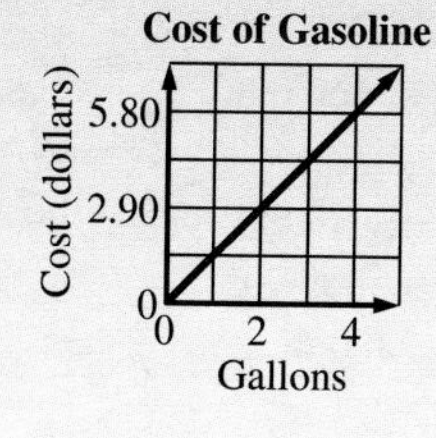

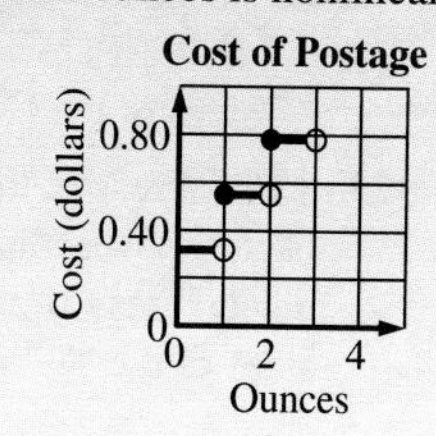

15–21. See answers in back of book.

22. a. 10, 15, 21, and 55

b. $t = \frac{n(n+1)}{2}$

c. nonlinear

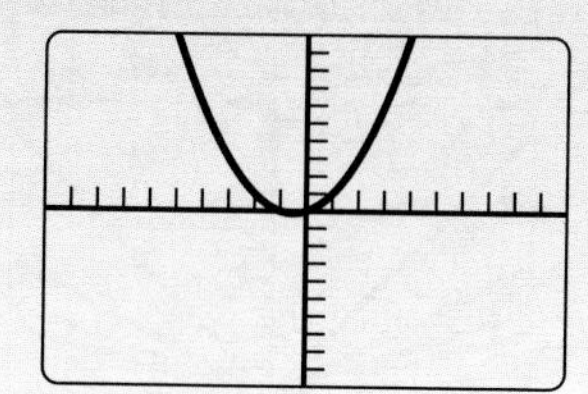

Apply⇔Assess

Exercise Notes

Cooperative Learning
Exs. 23–25 These exercises provide students with valuable experiences in examining specific equations that can be used to form generalizations.

Mathematical Procedures
Ex. 25 At this point, students should be comfortable with describing a procedure for carrying out an exploration. Since technology is involved in this exercise, they may want to include specific keystrokes.

Teaching Tip
Ex. 26 Remind students to use parentheses for the numerator and for the denominator if they use a graphing calculator.

Assessment Note
Ex. 27 Making a concept map will appeal to students having a visual learning style.

Practice 51 for Section 8.1

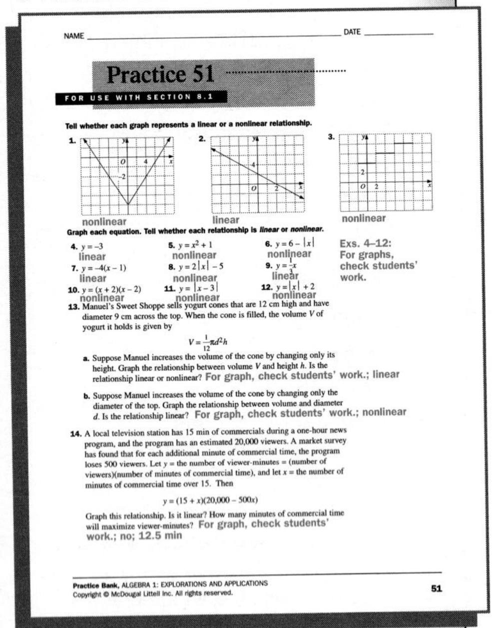

NAME ____________________ DATE __________

Practice 51

FOR USE WITH SECTION 8.1

Tell whether each graph represents a linear or a nonlinear relationship.

1. nonlinear
2. linear
3. nonlinear

Graph each equation. Tell whether each relationship is *linear* or *nonlinear*.

4. $y = -3$ linear
5. $y = x^2 + 1$ nonlinear
6. $y = 6 - |x|$ nonlinear
7. $y = -4(x - 1)$ linear
8. $y = 2|x| - 5$ nonlinear
9. $y = \frac{1}{3}x$ linear
10. $y = (x + 2)(x - 2)$ nonlinear
11. $y = |x - 3|$ nonlinear
12. $y = |x| + 2$ nonlinear

Exs. 4–12: For graphs, check students' work.

13. Manuel's Sweet Shoppe sells yogurt cones that are 12 cm high and have diameter 9 cm across the top. When the cone is filled, the volume V of yogurt it holds is given by

$$V = \frac{1}{12}\pi d^2 h$$

a. Suppose Manuel increases the volume of the cone by changing only its height. Graph the relationship between volume V and height h. Is the relationship linear or nonlinear? For graph, check students' work.; linear

b. Suppose Manuel increases the volume of the cone by changing only the diameter of the top. Graph the relationship between volume and diameter d. Is the relationship linear? For graph, check students' work.; nonlinear

14. A local television station has 15 min of commercials during a one-hour news program, and the program has an estimated 20,000 viewers. A market survey has found that for each additional minute of commercial time, the program loses 500 viewers. Let y = the number of viewer-minutes = (number of viewers)(number of minutes of commercial time), and let x = the number of minutes of commercial time over 15. Then

$$y = (15 + x)(20,000 - 500x)$$

Graph this relationship. Is it linear? How many minutes of commercial time will maximize viewer-minutes? For graph, check students' work.; no; 12.5 min

Practice Bank, ALGEBRA 1: EXPLORATIONS AND APPLICATIONS

51

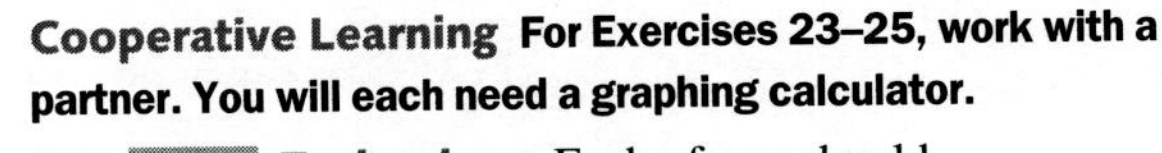

Cooperative Learning For Exercises 23–25, work with a partner. You will each need a graphing calculator.

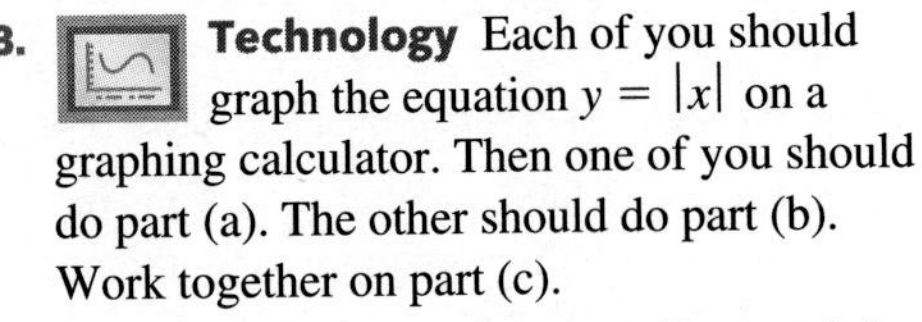

23. **Technology** Each of you should graph the equation $y = |x|$ on a graphing calculator. Then one of you should do part (a). The other should do part (b). Work together on part (c).

a. Graph the equations $y = 2|x|$, $y = 3|x|$, and $y = 4|x|$ along with the graph of $y = |x|$.

b. Graph the equations $y = \frac{1}{2}|x|$, $y = \frac{1}{3}|x|$, and $y = \frac{1}{4}|x|$ along with the graph of $y = |x|$.

c. Describe how the coefficients affect your original graphs.

24. **Technology** Work together to generalize how the graph of an equation in the form $y = a|x|$ changes as the coefficient a changes in value.

25. **Technology** Work together to write a procedure for exploring how the graph of an equation in the form $y = a|x|$ changes as the coefficient a becomes negative. Carry out the exploration and generalize the results.

26. **Challenge** Predict whether the equation $y = \frac{x^2 - 4}{x + 2}$ is *linear* or *nonlinear*. Make a graph to test your prediction. Explain the results.

ONGOING ASSESSMENT

27. **Open-ended Problem** Make a concept map that summarizes the distinction between linear and nonlinear relationships. Be sure to include examples. (For an example of a concept map, see the study technique on page 184 of Chapter 4.)

SPIRAL REVIEW

Find each product. *(Section 6.4)*

28. $5t(15t - 2)$
29. $(p - 3)(p + 7)$
30. $(u + 4)(2u - 3)$
31. $(2r + 5)(3r + 3)$
32. $(5 + \sqrt{3})(4 - \sqrt{3})$
33. $(2 + \sqrt{5})(1 + \sqrt{3})$

Rewrite each equation in slope-intercept form. Then graph it. Use a graphing calculator if you have one. *(Section 7.1)*

34. $3x - 5y = 7$
35. $3x + 5y = 7$
36. $5y - 3x = 7$
37. $7y + 5x = 3$
38. $3x + 7y = 5$
39. $3x - 7y = 5$

23. a., b.

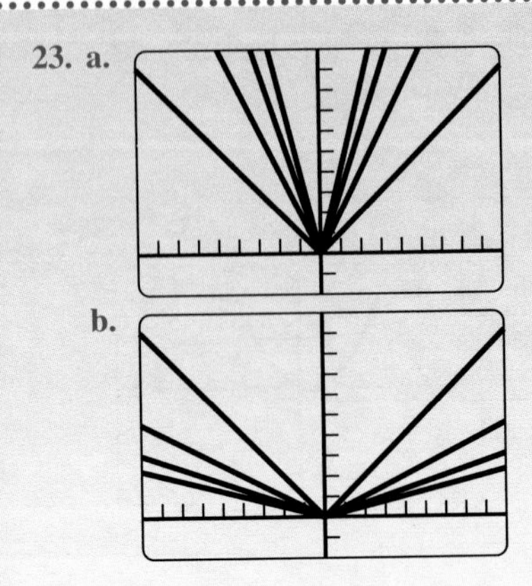

c. Answers may vary. An example is given. As the coefficient increases, the sides of the "V" become steeper. As the coefficient decreases toward 0, the sides of the "V" flatten out.

24. As a increases, the two sides of the "V" become steeper. As a decreases toward 0, the two sides of the "V" flatten out.

25. Graph $y = -2|x|$, $y = -3|x|$, $y = -0.5|x|$, and $y = -0.25|x|$ on your graphing calculator along with the graph of $y = |x|$. The "V" opens in the downward direction for negative values of a, and as $|a|$ increases, the sides of the "V" become steeper.

26. Answers may vary. An example is given. linear; The graph is identical to the graph of $y = x - 2$, except for a break in the line at $x = -2$, where the denominator of the expression $\frac{x^2 - 4}{x + 2}$ is 0. Because of the "hole" in the graph at $x = -2$, the equation is technically nonlinear.

27–39. See answers in back of book.

SECTION

8.2 Exploring Parabolas

Learn how to...

- recognize characteristics of parabolas

So you can...

- predict the shape of a parabola
- recognize side views of real objects, such as a radio telescope

Some nonlinear graphs are U-shaped curves called **parabolas**. The graph of the equation $y = x^2$ shown below is one example of a parabola. Every other parabola is a variation of this basic curve.

EXPLORATION
COOPERATIVE LEARNING

Analyzing the Shape of a Parabola

Work with a partner.
You will need:

- a graphing calculator
- graph paper

1 One person should graph the equations below. Set a good viewing window, such as $-9 \le x \le 9$ and $-2 \le y \le 10$. What do you notice about the shapes of the parabolas?

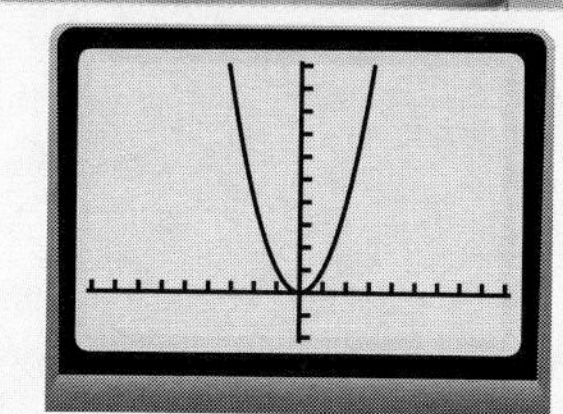

a. $y = x^2$ **b.** $y = 2x^2$ **c.** $y = 3x^2$

2 On graph paper, the other person should sketch what he or she thinks the graph of $y = 4x^2$ will look like in relation to the graphs in Step 1. Use a graphing calculator to check the prediction.

3 Switch roles and repeat Step 1 using the equations below.

a. $y = x^2$ **b.** $y = \frac{1}{2}x^2$ **c.** $y = \frac{1}{3}x^2$

4 On graph paper, sketch what you think the graph of $y = \frac{1}{4}x^2$ will look like in relation to the graphs in Step 3. Use a graphing calculator to check your prediction.

5 Work together to generalize what happens to the graph of the equation $y = ax^2$ as the value of a increases from 0 to 1 and beyond.

6 Work together to generalize what happens to the graph of the equation $y = ax^2$ as the value of a decreases from 0 to -1 and beyond. Try some examples. Be sure to set a good viewing window.

Exploration Note

Purpose
The purpose of this Exploration is to have students explore equations of the form $y = ax^2$ to see what happens to the graphs as a increases from 0 to 1 and beyond and decreases from 0 to –1 and beyond.

Materials/Preparation
Each group needs graph paper and a graphing calculator.

Procedure
Students graph parabolas and make predictions about the graphs of $y = ax^2$. Students should discuss their generalizations in Steps 5 and 6 to verify them. Students should understand that changing the viewing window on a calculator changes the appearance of the parabola being graphed but not the actual parabola.

Closure
Groups can share results with each other. They should conclude that the graph of $y = ax^2$ gets wider as a decreases and narrower as a increases. If a is positive, the graph opens up. If a is negative, the graph opens down.

Explorations Lab Manual
See the Manual for more commentary on this Exploration.

Diagram Master 2

For answers to the Exploration, see following page.

Plan⇔Support

Objectives

- Recognize characteristics of parabolas.
- Predict the shape of a parabola.
- Recognize side views of real objects.

Recommended Pacing

Core and Extended Courses
Section 8.2: 2 days

Block Schedule
Section 8.2: 2 half-blocks
(with Sections 8.1 and 8.3)

Resource Materials

Lesson Support
Lesson Plan 8.2
Warm-Up Transparency 8.2
Overhead Visuals:
Folder 8: Quadratic Functions
Practice Bank: Practice 52
Study Guide: Section 8.2
Explorations Lab Manual:
Additional Exploration 13,
Diagram Master 2

Technology
Graphing Calculator
McDougal Littell Software
Function Investigator
Internet:
http://www.hmco.com

Warm-Up Exercises

1. Find the ordered pairs that are solutions of $y = \frac{1}{2}x^2$ for domain values –4, –2, 0, 2, 4.
(–4, 8); (–2, 2); (0, 0); (2, 2); (4, 8)
2. Find the ordered pairs that are solutions of $y = -\frac{1}{3}x^2$ for domain values –6, –3, 0, 3, 6.
(–6, –12); (–3, –3); (0, 0); (3, –3); (6, –12)

Tell whether each equation is *linear* or *nonlinear.*

3. $y = 2x^2$ nonlinear
4. $y = -5x + 1$ linear
5. $y = -\frac{1}{3}x^2$ nonlinear

Teach⇔Interact

Section Notes

Using Technology
Students should know that a good viewing window for a graph shows all the important features of the graph. For a parabola, a good viewing window usually shows its vertex and its intercepts.

Communication: Discussion
Point out that all parabolas have a minimum or maximum point but not both. That is, if a parabola has a minimum point, then it has no maximum point because the *y*-values increase as *x*-values increase or decrease. Likewise, if a parabola has a maximum point, then it has no minimum point because the *y*-values get smaller as *x*-values increase or decrease.

About the Example

Mathematical Procedures
Making and testing predictions is an important procedure because it reflects a natural way of learning and doing mathematics. Writing the solution as shown helps students to organize their thoughts in a clear and precise way.

Additional Example

Predict how the graph of the equation $y = -2x^2$ will compare with the graph of the equation $y = x^2$. Then sketch the graph.

Possible solution:
First look at the sign of the coefficient of x^2. The graph of $y = -2x^2$ will open down instead of up because the coefficient is negative. Now ignore the sign. The graph will be narrower and steeper than the graph of $y = x^2$ because 2 is greater than 1. The graph will go through the point (1, –2) instead of (1, 1).

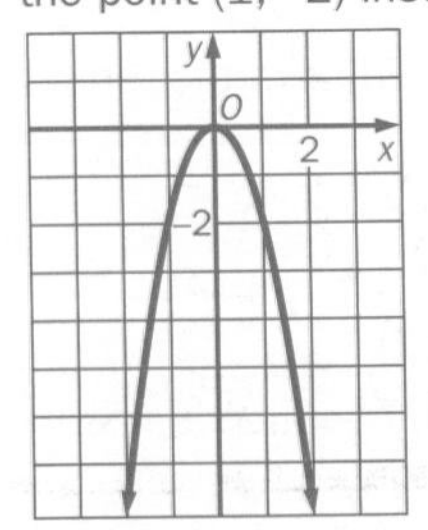

y
O
2
x
-2

Quadratic Functions

A function or an equation whose graph is a parabola is called **quadratic**. The word *quadratic* comes from a Latin word that means *to make square*. All quadratic equations have a squared term in them.

The **vertex of a parabola** is the point where the function reaches a minimum or a maximum value. The plural of *vertex* is *vertices*.

Some examples of quadratic functions are shown below.

$y = 2x^2$

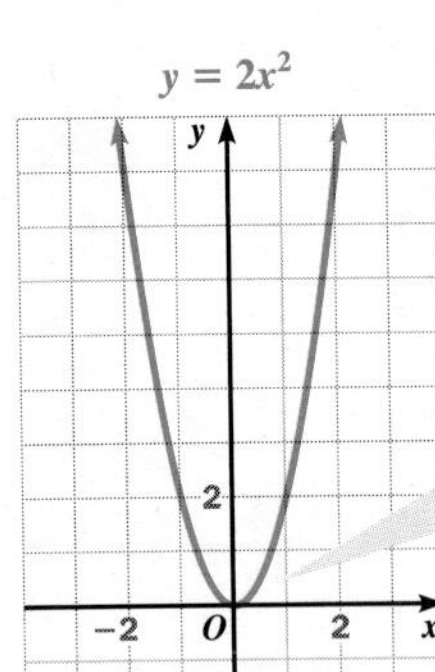

y
2
-2
O
2
x

The function reaches a **maximum** value of $y = 0$ when $x = 0$.

The function reaches a **minimum** value of $y = 0$ when $x = 0$.

$y = -x^2$

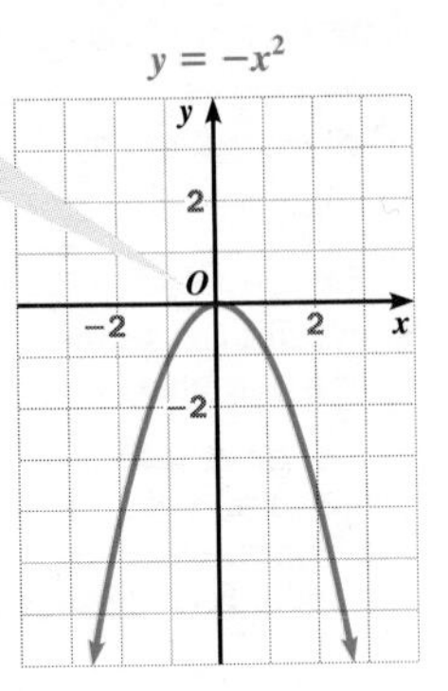

y
2
O
-2
2
x
-2

Notice that the graph of the function with a positive coefficient opens up, while the graph of the function with a negative coefficient opens down.

EXAMPLE

Predict how the graph of the equation $y = -3x^2$ will compare with the graph of the equation $y = x^2$. Then sketch the graph.

SOLUTION

Here is one student's solution of the Example.

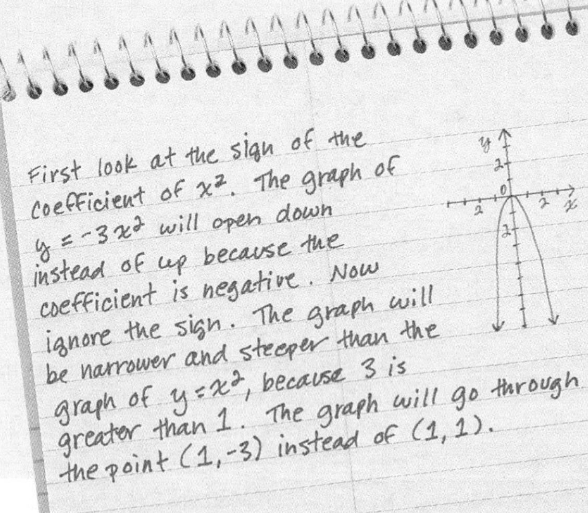

First look at the sign of the
coefficient of x². The graph of
y = -3x² will open down
instead of up because the
coefficient is negative. Now
ignore the sign. The graph will
be narrower and steeper than the
graph of y = x², because 3 is
greater than 1. The graph will go through
the point (1, -3) instead of (1, 1).
y
2
O
2
2
x
2

ANSWERS Section 8.2

Exploration

1. The greater the value of the coefficient of x^2, the narrower and steeper the parabola.
2. Check students' work. The graph of $y = 4x^2$ will be even narrower and steeper than the graph of $y = 3x^2$.
3. The smaller the coefficient of x^2, the wider and flatter the parabola.
4. Check students' work. The graph of $y = \frac{1}{4}x^2$ will be even wider and flatter than the graph of $y = \frac{1}{3}x^2$.
5. As the value of a increases from 0 to 1 and beyond, the graph of $y = ax^2$ gets narrower and steeper.
6. As the value of a decreases from 0 to –1 and beyond, the graph of $y = ax^2$ gets narrower and steeper, and the "U" opens downward.

Think and Communicate

1. No; as $|x|$ increases, y increases without limit.
2. The graph would still open down and would be wider and flatter than the graph of $y = x^2$ because $\frac{1}{3}$ is less than 1. The graph would go through the point $\left(1, -\frac{1}{3}\right)$ instead of (1, –3).
3. $y = \frac{1}{4}x^2$

THINK AND COMMUNICATE

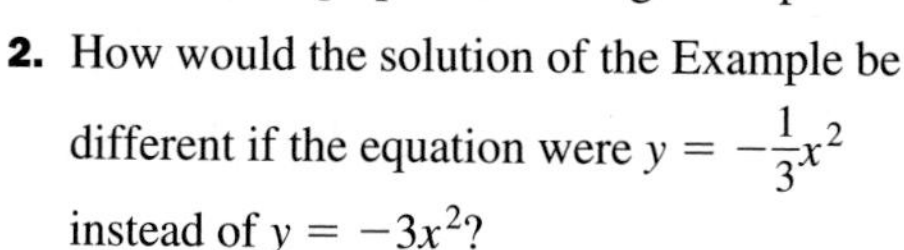

1. Do you think there is a maximum value for the function graphed at the right? Explain.
2. How would the solution of the Example be different if the equation were $y = -\frac{1}{3}x^2$ instead of $y = -3x^2$?

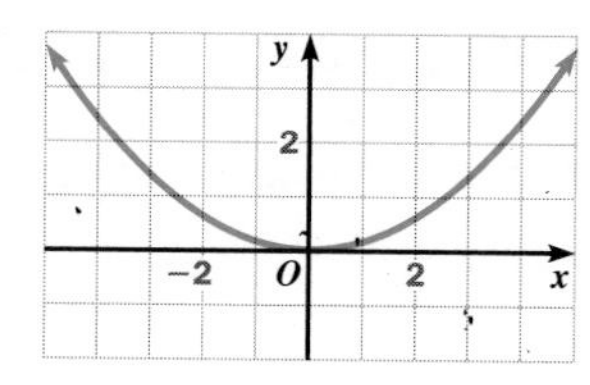

3. Which equation has the widest graph: $y = \frac{1}{4}x^2$, $y = -\frac{1}{2}x^2$, or $y = 4x^2$?

Symmetry

If you fold the graph of an equation in the form $y = ax^2$ along the y-axis, the two sides of the parabola line up. We say that the parabola is **symmetric** about the y-axis. The y-axis is the **line of symmetry**.

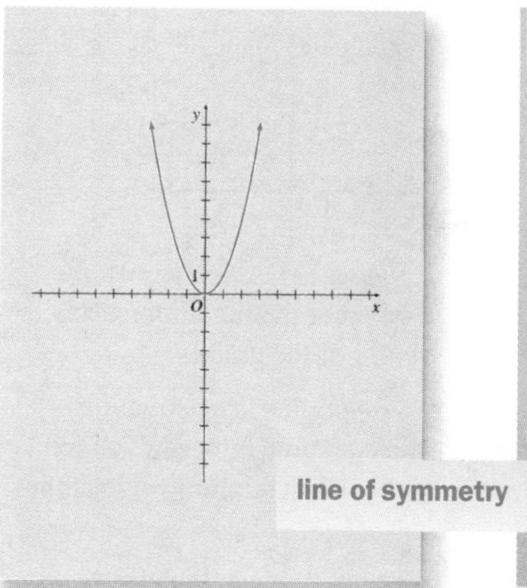

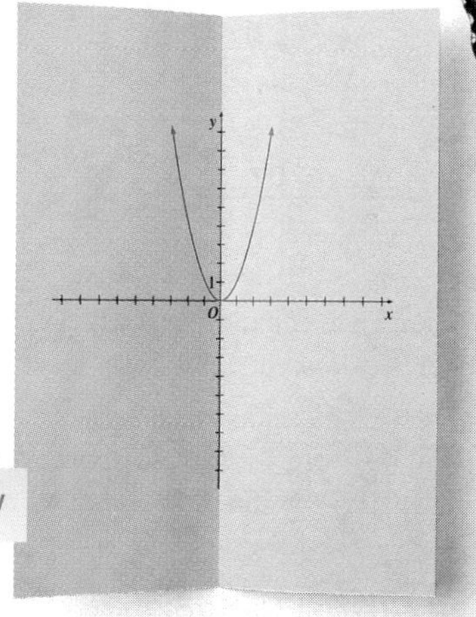

line of symmetry

As a rule, if you draw a vertical line through the vertex of any parabola, the parabola will be symmetric about that line.

CHECKING KEY CONCEPTS

Is the graph of each equation *narrower* or *wider* than the graph of $y = x^2$? Does the graph open *up* or *down*? What does this tell you about the vertex?

1. $y = 5x^2$
2. $y = -5x^2$
3. $y = \frac{1}{5}x^2$
4. $y = -\frac{1}{5}x^2$
5. $y = 0.75x^2$
6. $y = -0.8x^2$

Sketch the graph of each equation.

7. $y = 5x^2$
8. $y = -\frac{1}{5}x^2$
9. $y = 0.75x^2$

BY THE WAY

Many cultures use symmetry in their art, as in this African mask. Archaeologists can sometimes discover how cultures influenced each other by studying changes in their art forms.

Teach⇔Interact

Think and Communicate

To answer questions 2 and 3, students can apply the generalizations they made for the Exploration on page 331.

Section Notes

Teaching Tip
Point out to students that they can use the line of symmetry for a parabola to help locate additional points on the parabola by reflecting known points across this line.

Visual Thinking
Ask students to clip magazine photos of subjects that demonstrate symmetry. Ask them to draw the line of symmetry on each photo and to show the class how the photo is symmetric about that line. This activity involves the visual skills of *exploration* and *communication.*

Checking Key Concepts

Reasoning
Students should understand that the sign of a in $y = ax^2$ does not affect how wide or narrow the graph is. The sign of a only affects whether the orientation of the graph is up or down.

Closure Question

Explain the effect of the value of a on the graph of $y = ax^2$ as compared to the graph of $y = x^2$.
As the absolute value of a increases, the graph of $y = ax^2$ gets narrower and steeper than the graph of $y = x^2$. As the absolute value of a decreases, the graph of $y = ax^2$ gets wider than the graph of $y = x^2$. If a is positive, the graph opens up. If a is negative, the graph opens down.

Checking Key Concepts

1. narrower; up; The function reaches a minimum at the vertex.
2. narrower; down; The function reaches a maximum at the vertex.
3. wider; up; The function reaches a minimum at the vertex.
4. wider; down; The function reaches a maximum at the vertex.
5. wider; up; The function reaches a minimum at the vertex.
6. wider; down; The function reaches a maximum at the vertex.
7. 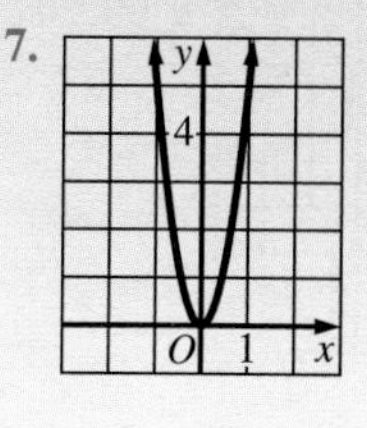
8. 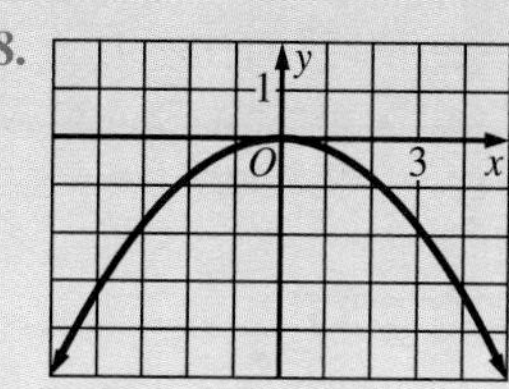
9. 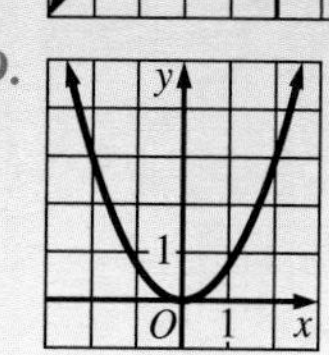

Apply⇔Assess

Suggested Assignment

Core Course
Day 1 Exs. 1–17
Day 2 Exs. 18–28, 30–39

Extended Course
Day 1 Exs. 1–17
Day 2 Exs. 18–39

Block Schedule
Day 49 Exs. 1–17
Day 50 Exs. 18–39

Exercise Notes

Communication: Drawing
Exs. 1–4 These exercises provide further opportunities for students to connect the equation of a parabola with its graph. This will be especially useful to visual learners.

Reasoning
Ex. 5 Students should be able to explain why the maximum point or the minimum point of a parabola is on its line of symmetry.

Cooperative Learning
Exs. 7–15 Students can work on these exercises in small groups. They can then discuss their predictions with one another.

Using Technology
Exs. 7–15 Students who are using graphing calculators may find it helpful to use "friendly" settings for the graphing window. On the TI-82, they can use Xmin = –9.4, Xmax = 9.4. On the TI-81, use Xmin = –9.4, Xmax = 9.6. All of the graphs can also be displayed with the *Function Investigator* software.

8.2 Exercises and Applications

Extra Practice exercises on page 572

Match each equation with its graph.

1. $y = x^2$ **2.** $y = -4x^2$ **3.** $y = \frac{1}{4}x^2$ **4.** $y = 4x^2$

A.
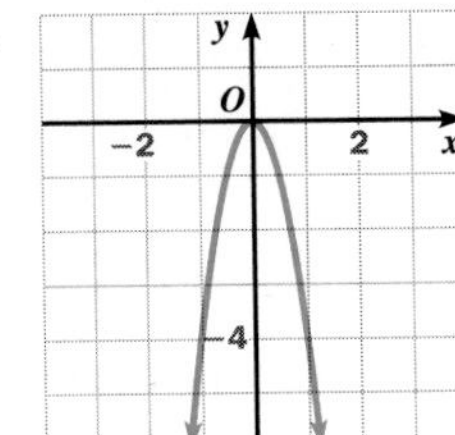

B.
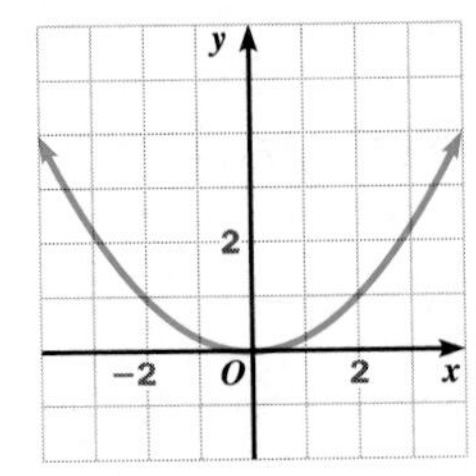

C.
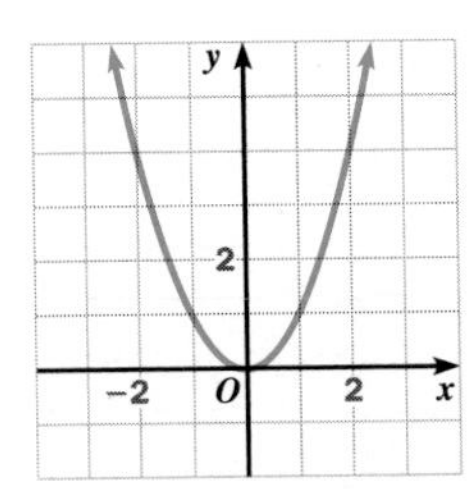

D.
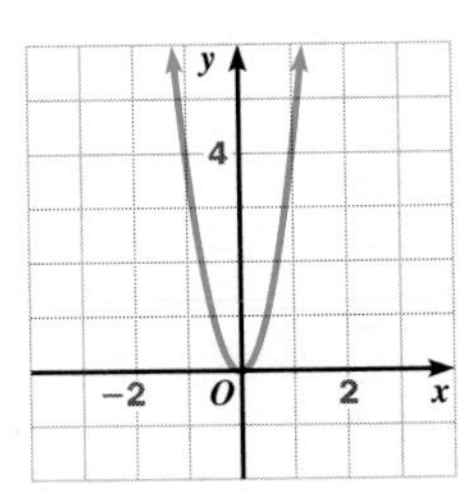

5. For each equation in Exercises 1–4, tell whether the function reaches a *maximum* or a *minimum* value at the vertex of the graph.

6. Writing Suppose someone shows you a function whose equation is in the form $y = ax^2$. Without graphing the equation, how can you tell whether the function will reach a minimum or a maximum value at the vertex of the graph?

Technology For Exercises 7–15, use a graphing calculator to check your predictions.

a. Predict how the graph of each equation will compare with the graph of $y = x^2$. Explain how you made your prediction.

b. Sketch the graph of each equation.

7. $y = 6x^2$ **8.** $y = -2x^2$ **9.** $y = -10x^2$

10. $y = \frac{1}{6}x^2$ **11.** $y = -\frac{1}{3}x^2$ **12.** $y = \frac{2}{3}x^2$

13. $y = 0.5x^2$ **14.** $y = 1.25x^2$ **15.** $y = -2.5x^2$

16. Open-ended Problem Write an equation for a parabola that opens down and is wider than the graph of $y = x^2$. Then write an equation for a parabola that opens up and is narrower than the graph of $y = x^2$. Sketch all three equations in the same coordinate plane.

17. Challenge Is it possible for a quadratic equation in the form $y = ax^2$ to have a vertex at a point other than the origin? Why or why not?

Exercises and Applications

1. C 2. A 3. B 4. D

5. Ex. 1: minimum; Ex. 2: maximum; Ex. 3: minimum; Ex. 4: minimum

6. Consider whether a is positive or negative. If a is positive, the function reaches a minimum at the vertex. If a is negative, the function reaches a maximum at the vertex.

7. a. Since $6 > 1$, the graph will be narrower. Since $6 > 0$, the graph will also open up.

b.
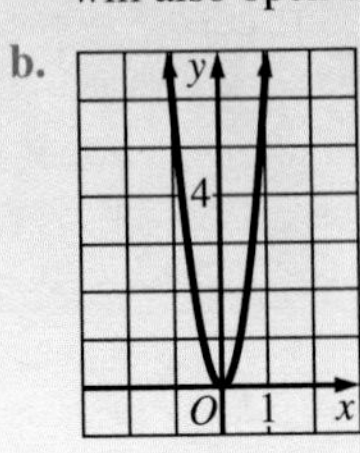

8. a. Since $|-2| > 1$, the graph will be narrower. Since $-2 < 0$, the graph will open down.

b.
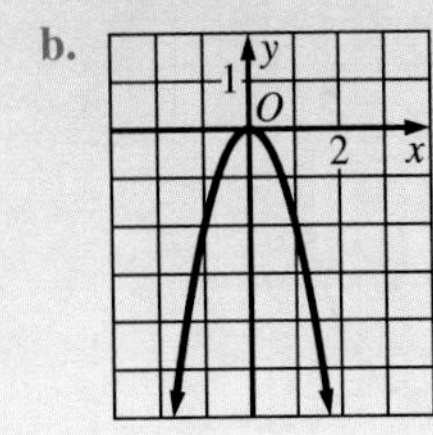

9. a. Since $|-10| > 1$, the graph will be narrower. Since $-10 < 0$, the graph will open down.

b.
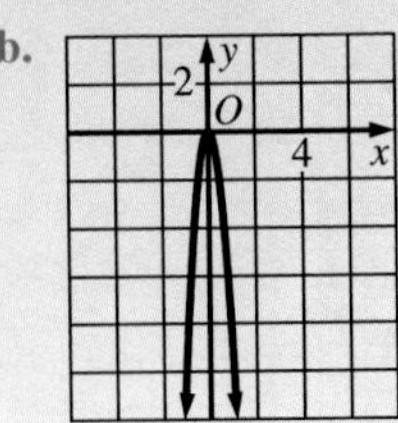

10. a. Since $\frac{1}{6} < 1$, the graph will be wider. Since $\frac{1}{6} > 0$, the graph will also open up.

b.
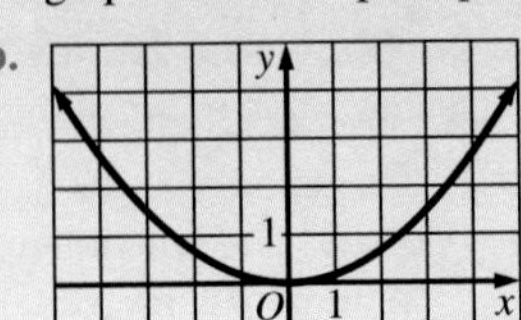

Tell whether each graph is symmetric about the *y*-axis. Explain.

18. 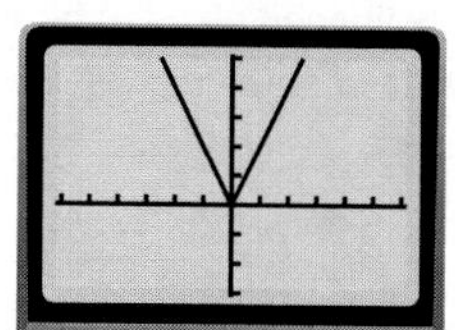

19. 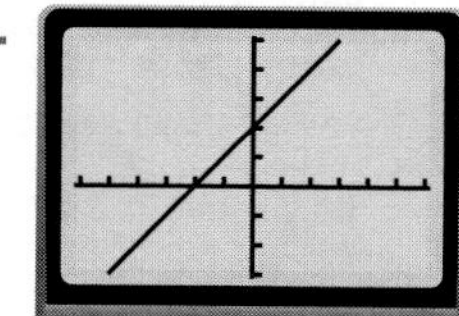

20. 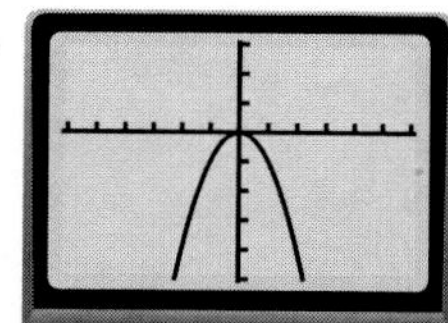

Connection ASTRONOMY

Radio telescopes are designed to monitor radio waves. A common type of radio telescope has an antenna in the shape of a parabolic dish. This means that if you look at a cross section of the antenna, you will see a parabola.

The radio telescopes pictured below are part of a worldwide network. Data collected from these telescopes is pooled and then analyzed by computer to provide high-resolution images.

Each equation approximates the shape of the given telescope dish. Match each equation with its graph. Explain how you made each choice. (*Note:* Assume that the scale is the same on all three graphs.)

21. Parkes telescope, Australia: $y = 0.003x^2$

22. Fortaleza telescope, Brazil: $y = 0.013x^2$

23. Hartebeesthoek telescope, South Africa: $y = 0.007x^2$

A. 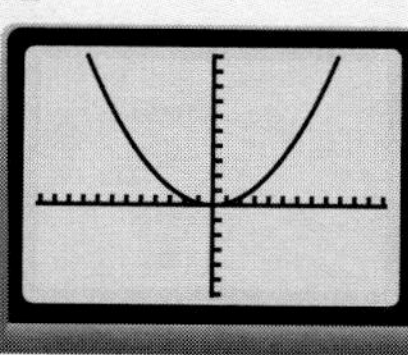

B. 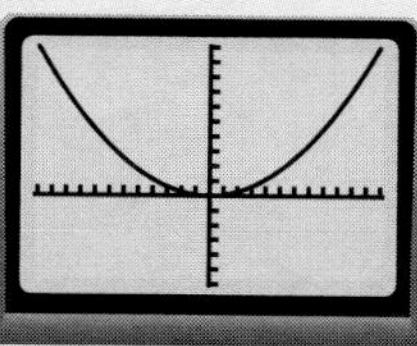

C. 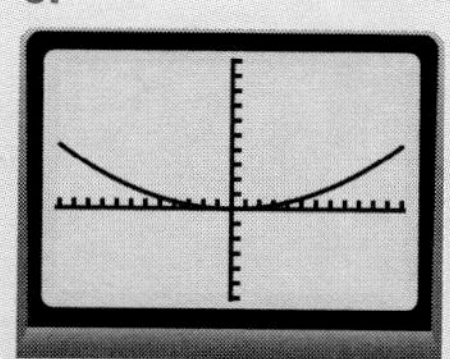

BY THE WAY

At the University of California at Berkeley, scientists use a network of radio telescopes to look for signs of extra-terrestrial civilizations.

Apply⇔Assess

Exercise Notes

Application
Exs. 21–23 Radio telescopes are designed in the shape of a parabolic dish because parabolas have a property that makes them ideal to collect radio waves or light waves. On the line of symmetry of a parabola there is a single point, called the *focus* of the parabola. As radio waves strike a parabolic dish, they are all directed to its focus and then to a receiving mechanism (most likely a computer) to record them.

Interdisciplinary Problems
Exs. 21–23 The first radio telescope was built in 1940 by a ham radio operator named Grote Reber in Wheaton, Illinois. Since then, astronomers have built very large and sophisticated radio telescopes that can construct a radio map or a picture of the sky.

Using Technology
Exs. 21–23 Students can use a graphing calculator to explore the differences in the graphs for these exercises. Stress the importance of keeping the viewing window the same for each graph.

11. a. Since $\left|-\frac{1}{3}\right| < 1$, the graph will be wider. Since $-\frac{1}{3} < 0$, the graph will open down.

b. 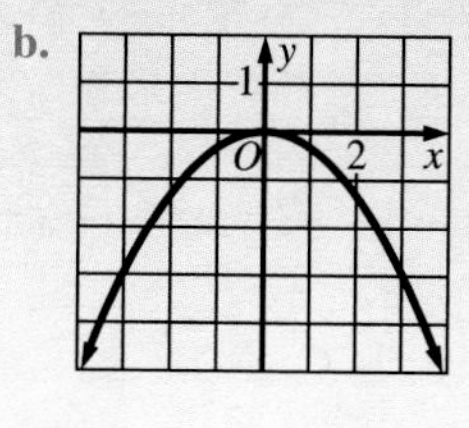

12. a. Since $\frac{2}{3} < 1$, the graph will be wider. Since $\frac{2}{3} > 0$, the graph will also open up.

b. 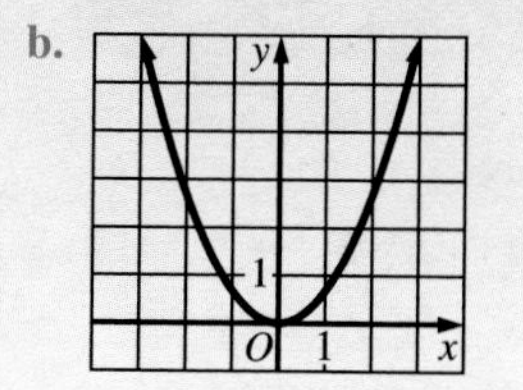

13. a. Since $0.5 < 1$, the graph will be wider. Since $0.5 > 0$, the graph will also open up.

b. 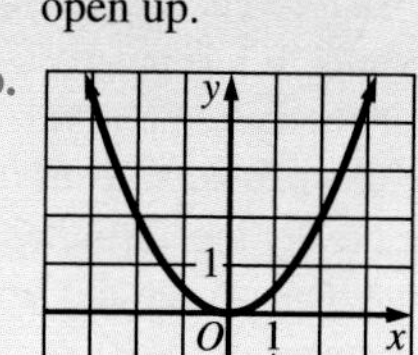

14. a. Since $1.25 > 1$, the graph will be narrower. Since $1.25 > 0$, the graph will also open up.

b. 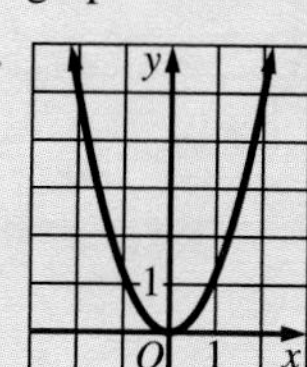

15. a. Since $|-2.5| > 1$, the graph will be narrower. Since $-1.25 < 0$, the graph will open down.

b. 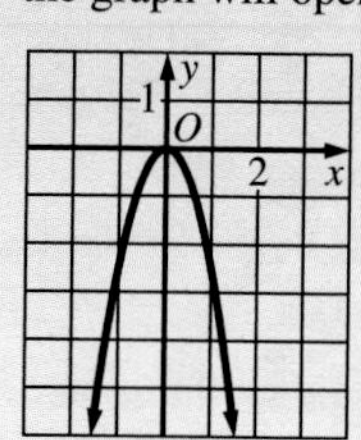

16. Answers will vary. Check students' work.

17–23. See answers in back of book.

Apply⇔Assess

Exercise Notes

Topic Spiraling: Review
Exs. 24–27 These exercises review nonlinear graphing concepts from Section 8.1.

Using Manipulatives
Exs. 24–27 Students can share their circular objects for these exercises to reduce the need to have a large variety of objects available in class.

Challenge
Ex. 29 Students should support their explanations with mathematical equations.

Assessment Note
Ex. 30 Asking students to summarize the lesson in writing is a good way to identify any misconceptions they may have.

Practice 52 for Section 8.2

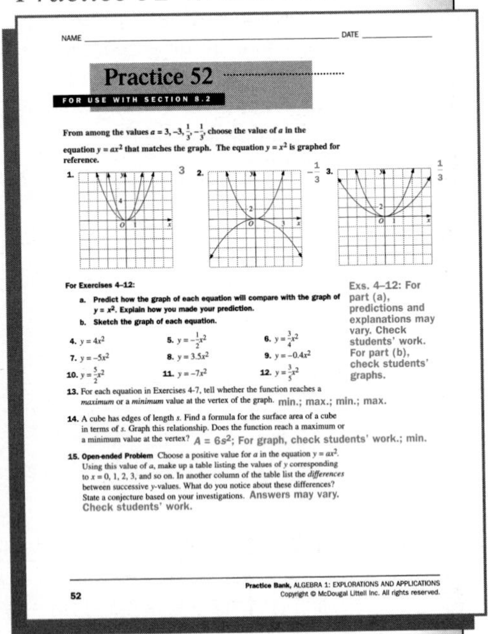
NAME ______ DATE ______

Practice 52

FOR USE WITH SECTION 8.2

From among the values $a = 3, -3, \frac{1}{3}, -\frac{1}{3}$, choose the value of a in the equation $y = ax^2$ that matches the graph. The equation $y = x^2$ is graphed for reference.

1. [graph] 3
2. [graph] $-\frac{1}{3}$
3. [graph] $\frac{1}{3}$

For Exercises 4–12:

a. **Predict how the graph of each equation will compare with the graph of $y = x^2$. Explain how you made your prediction.**
b. **Sketch the graph of each equation.**

Exs. 4–12: For part (a), predictions and explanations may vary. Check students' work. For part (b), check students' graphs.

4. $y = 4x^2$
5. $y = -\frac{1}{2}x^2$
6. $y = \frac{3}{4}x^2$
7. $y = -5x^2$
8. $y = 3.5x^2$
9. $y = -0.4x^2$
10. $y = \frac{5}{2}x^2$
11. $y = -7x^2$
12. $y = \frac{3}{5}x^2$
13. For each equation in Exercises 4-7, tell whether the function reaches a *maximum* or a *minimum* value at the vertex of the graph. min.; max.; min.; max.
14. A cube has edges of length s. Find a formula for the surface area of a cube in terms of s. Graph this relationship. Does the function reach a maximum or a minimum value at the vertex? $A = 6s^2$; For graph, check students' work.; min.
15. **Open-ended Problem** Choose a positive value for a in the equation $y = ax^2$. Using this value of a, make up a table listing the values of y corresponding to $x = 0, 1, 2, 3$, and so on. In another column of the table list the *differences* between successive y-values. What do you notice about these differences? State a conjecture based on your investigations. Answers may vary. Check students' work.

52 **Practice Bank,** ALGEBRA 1: EXPLORATIONS AND APPLICATIONS

Cooperative Learning **For Exercises 24–27, work with a partner. You will need six sheets of graph paper and six objects of various sizes that you can use to trace circles.**

24. Take turns tracing around each object on graph paper. Measure the radius of each circle drawn. Use the width of one square on the graph paper as your unit of measure.

25. Estimate the area of each circle drawn by counting the number of squares inside the circle. Include parts of squares to get as close an estimate as possible.

26. Make a scatter plot of your combined data. Put radius on the horizontal axis and area on the vertical axis.

27. Graph the equation for the area of a circle $A = \pi r^2$, $r \geq 0$, along with your scatter plot. Compare the graph and the scatter plot.

28. SAT/ACT Preview Which of the following equations has a graph that is a parabola?

A. $y = 3x$ **B.** $y = 3|x|$ **C.** $y = 3x^2$ **D.** $y = 2x^3$

29. Challenge Suppose x represents both the radius of a circle and the length of a side of a square. As x increases, which area increases more quickly, the area of the circle or the area of the square? Explain.

ONGOING ASSESSMENT

30. Writing Suppose a friend has missed this lesson. Summarize how to predict what the graph of an equation in the form $y = ax^2$ will look like. Be sure to include examples.

SPIRAL REVIEW

Find each product. *(Section 6.5)*

31. $(y + 3)^2$ **32.** $(r - 4)^2$ **33.** $(2p + 1)^2$

34. $(3 + \sqrt{5})^2$ **35.** $(x + 6)(x - 6)$ **36.** $(3n - 2)(3n + 2)$

Find the unknown length in each right triangle. *(Section 6.1)*

37.

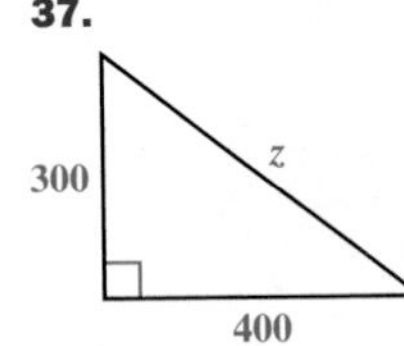

38.

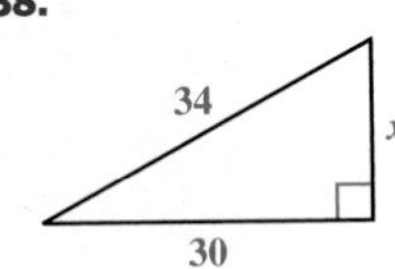

39.

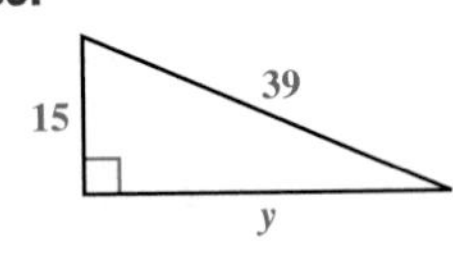

24. Check students' work.
25. The area of each circle should be a rough approximation of πr^2.
26. Check students' work.
27. The graphs should be similar.
28. C
29. The equation for the circle is $y = \pi x^2$ and for the square is $y = x^2$. Since $\pi > 1$, the graph of $y = \pi x^2$ is narrower and steeper than the graph of $y = x^2$. The area of the circle increases more quickly.
30. Answers may vary. An example is given. The graph of any equation of the form $y = ax^2$ is a U-shaped curve called a parabola. If a is positive, the parabola opens up; if a is negative, the parabola opens down. The greater the absolute value of a, the narrower and steeper the parabola is. For example, $y = 3x^2$ is narrower and steeper than $y = x^2$ and it opens up; $y = 0.5x^2$ is wider and flatter and it opens up; $y = -2x^2$ is steeper and narrower than $y = x^2$, and it opens down.
31. $y^2 + 6y + 9$
32. $r^2 - 8r + 16$
33. $4p^2 + 4p + 1$
34. $14 + 6\sqrt{5}$
35. $x^2 - 36$
36. $9n^2 - 4$
37. 500
38. 16
39. 36

SECTION

8.3 Solving Quadratic Equations

Learn how to...

- use square roots and graphs to solve simple quadratic equations

So you can...

- relate x-intercepts and solutions of quadratic equations
- solve simple motion problems involving acceleration

Have you ever considered dropping an egg out of a window? At some schools this is an annual event! Art, physics, and engineering students test their skills by designing containers to prevent eggs from breaking when dropped from heights of 30 ft or more.

When an egg is dropped out of a window, the egg *accelerates*. That means it falls faster and faster over time. The formula $d = 16t^2$ can be used to model the distance d in feet that an egg falls in t seconds.

BY THE WAY

Students use all kinds of materials to cushion their eggs, including marshmallows, popcorn, gelatin, and peanut butter.

EXAMPLE 1 Application: Physics

Suppose an egg is dropped from a height of 80 ft. About how long will it take for the egg to reach the ground?

SOLUTION

Problem Solving Strategy: Use a formula.

Substitute 80 for d in the formula $d = 16t^2$ and solve for t.

$$80 = 16t^2$$
$$5 = t^2$$
$$\pm\sqrt{5} = t$$
$$\pm 2.24 \approx t$$

Find the square root of both sides.

A negative time value does not make sense in this situation. Therefore, there is only one solution. It will take about 2.2 seconds for an egg to reach the ground when dropped from a height of 80 ft.

Plan⇔Support

Objectives

- Use square roots and graphs to solve simple quadratic equations.
- Relate x-intercepts and solutions of quadratic equations.
- Solve simple motion problems involving acceleration.

Recommended Pacing

Core and Extended Courses
Section 8.3: 2 days

Block Schedule
Section 8.3: 2 half-blocks (with Sections 8.2 and 8.4)

Resource Materials

Lesson Support
Lesson Plan 8.3
Warm-Up Transparency 8.3
Practice Bank: Practice 53
Study Guide: Section 8.3
Explorations Lab Manual: Diagram Masters 1, 2
Assessment Book: Test 34

Technology
Graphing Calculator
McDougal Littell Software *Function Investigator*
Internet: http://www.hmco.com

Warm-Up Exercises

1. Find two numbers whose square is 16. 4; –4
2. Give the vertex of the parabola $y = 2x^2$. (0, 0)
3. What is the x-intercept of the graph of a line? the x-coordinate of the point where the line crosses the x-axis
4. Solve the equation $2x - 4 = 20$. 12
5. What is the approximate value of $\sqrt{5}$? 2.24
6. What is the approximate value of $-\sqrt{5}$? –2.24

Teach⇔Interact

About Example 1

Problem Solving

In Example 1, students employ the problem-solving strategy of using a formula. Remind students that formulas are useful because they serve as models for many different kinds of situations and because they can be solved by equation-solving procedures.

Additional Example 1

A ball is dropped from a height of 48 ft. How long will it take for the ball to reach the ground?

Use the formula $d = 16t^2$. Substitute 48 for d in the formula and solve for t.

$$48 = 16t^2$$
$$3 = t^2$$
$$\pm\sqrt{3} = t$$
$$\pm 1.73 \approx t$$

A negative time value does not make sense in this situation. It will take about 1.7 seconds for the ball to reach the ground.

Section Note

Communication: Discussion

A quadratic equation may have 0, 1, or 2 solutions. Stress the importance of considering each solution in terms of the original problem. Mathematical solutions that are unrealistic in terms of the problem should not be used.

Additional Example 2

Solve the equation $x^2 - 16 = 0$.

$$x^2 - 16 = 0$$
$$x^2 = 16$$
$$x = \pm\sqrt{16}$$
$$x = \pm 4$$

The solutions are 4 and −4.

Think and Communicate

For question 3, point out the similarity to solving the equation $3x = 48$. You first divide both sides of the equation by 3.

Section Note

Alternate Approach

On a TI-82, an alternate way to find the x-intercepts of a function, such as $y = ax^2 + c$ entered in the Y= menu, is to use the CALC 2:root function from the CALCULATE menu.

Quadratic equations can have as many as two solutions. You must decide which solutions apply to a given situation and ignore any solution that doesn't make sense, such as the negative time value in Example 1.

EXAMPLE 2

Solve the equation $x^2 - 9 = 0$.

SOLUTION

When you solve quadratic equations algebraically, you must first get x^2 alone on one side of the equation.

$$x^2 - 9 = 0$$
$$x^2 = 9$$
$$x = \pm\sqrt{9}$$
$$x = \pm 3$$

Find the square root of both sides.

The solutions are 3 and −3.

THINK AND COMMUNICATE

1. Do Example 1 and Example 2 have the same number of solutions? Why or why not?
2. What is the first step you would take to solve the equation $x^2 - 25 = 0$?
3. What is the first step you would take to solve the equation $3x^2 = 48$?

Using Graphs to Solve Equations

A graph of the equation $y = x^2 - 9$ is shown below.

When $x = -3$, $y = 0$.
$y = x^2 - 9$
$0 = (-3)^2 - 9$

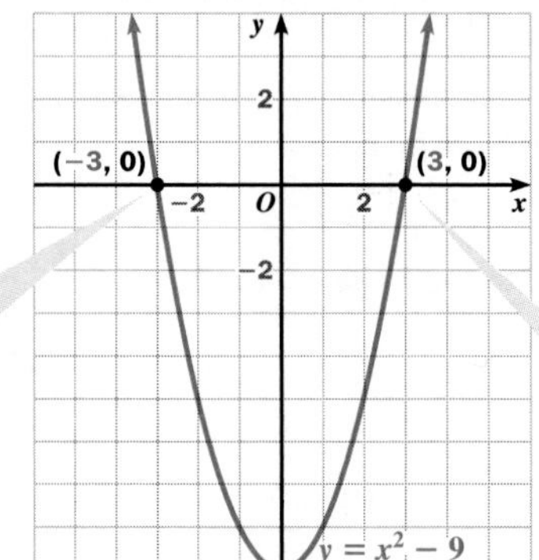

Notice that the x-intercepts −3 and 3 are the solutions of the equation $x^2 - 9 = 0$ in Example 2. This makes sense, because the x-intercepts of $y = x^2 - 9$ occur when $y = 0$.

When $x = 3$, $y = 0$.
$y = x^2 - 9$
$0 = 3^2 - 9$

In general, the solutions of a quadratic equation in the form $ax^2 + c = 0$ are the x-intercepts of the graph of the related equation $y = ax^2 + c$. This is also true for quadratic equations in other forms.

ANSWERS Section 8.3

Think and Communicate

1. No; although the equations in both examples have two solutions, one of the solutions in Example 1 does not make sense as a solution to the real-world problem.
2. Add 25 to both sides.
3. Divide both sides by 3.
4. In Method 1, the first step is to get the x^2 term alone on one side of the equation. In Method 2, the first step is to get 0 alone on one side of the equation.
5. Subtract 25 from both sides of the equation. Subtract 75 from both sides of the equation.
6. No; they would be the solutions of $2x^2 - 4 = 0$.
7. $\sqrt{14}$ and $-\sqrt{14}$; Method 1; the solutions are not rational numbers so they cannot be obtained exactly by graphing.

EXAMPLE 3

Solve the equation $2x^2 - 4 = 24$.

SOLUTION

Method 1 Solve the equation algebraically.

Work to get x^2 alone on one side of the equation.

$$2x^2 - 4 = 24$$
$$2x^2 = 28$$
$$x^2 = 14$$
$$x = \pm\sqrt{14}$$
$$x \approx \pm 3.74$$

Find the square root of both sides.

The solutions are about -3.74 and about 3.74.

Method 2 Solve the equation using a graph.

Step 1 Rewrite the equation in the form $ax^2 + c = 0$.

$$2x^2 - 4 = 24$$
$$2x^2 - 28 = 0$$

Step 2 Graph the related equation $y = 2x^2 - 28$.

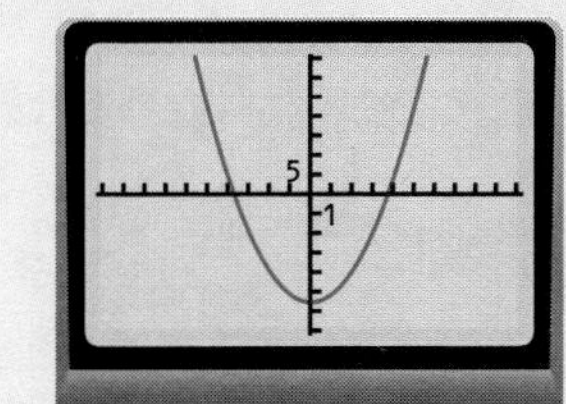

Step 3 Find the x-intercepts.

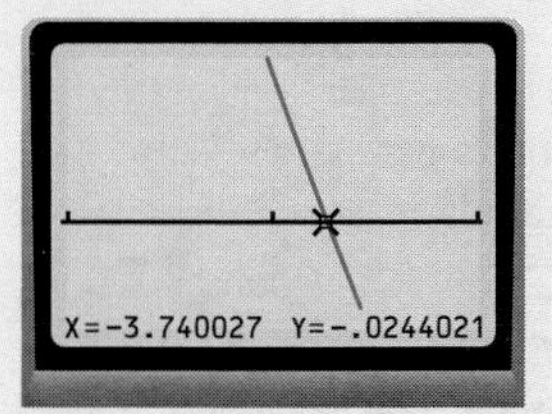

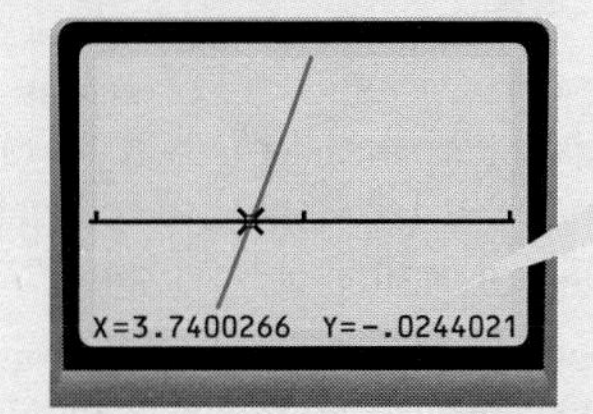

Use the trace and zoom features to get more accurate values.

The solutions are about -3.74 and about 3.74.

THINK AND COMMUNICATE

4. Contrast the very first steps of the two methods shown in Example 3.
5. What would you do first to solve the equation $x^2 + 25 = 75$ algebraically? What would you do first to solve it using a graph?
6. In Example 3, suppose you graphed the equation $y = 2x^2 - 4$. Would the x-intercepts be the solutions of $2x^2 - 4 = 24$? Explain.
7. What are the exact solutions of the equation $2x^2 - 4 = 24$ in Example 3? Which method will give you exact solutions? Explain.

Teach⇔Interact

Additional Example 3

Solve the equation $3x^2 - 8 = 10$.

Method 1 Solve the equation algebraically. First get x^2 alone on one side of the equation.

$$3x^2 - 8 = 10$$
$$3x^2 = 18$$
$$x^2 = 6$$
$$x = \pm\sqrt{6}$$
$$x \approx 2.45$$

The solutions are about -2.45 and about 2.45.

Method 2 Solve the equation using a graph.

Step 1 Rewrite the equation in the form $ax^2 + c = 0$.

$$3x^2 - 8 = 10$$
$$3x^2 - 18 = 0$$

Step 2 Graph the related equation $y = 3x^2 - 18$.

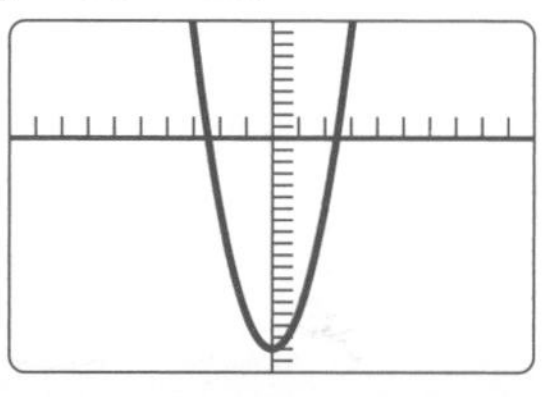

Step 3 Find the x-intercepts.

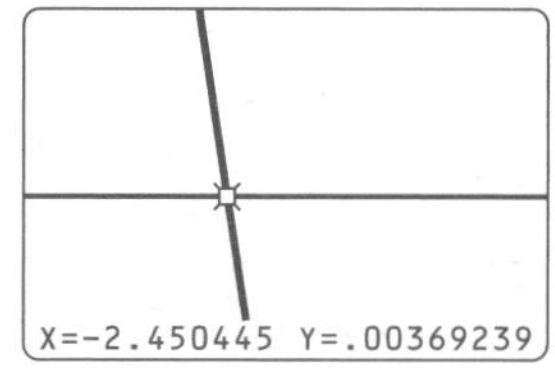

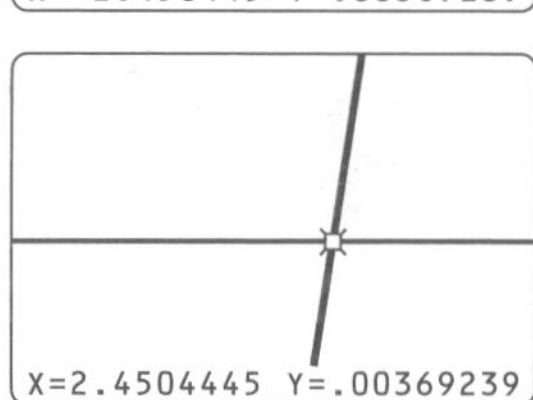

The solutions are about -2.45 and about 2.45.

Think and Communicate

When discussing questions 4–6, point out that Method 1 uses equivalent equations (which are equations that all have the same solution) to find the solution to the given equation. Method 2 gets a 0 on one side of the equation. In this method, 0 is replaced with y, which helps make the connection that $y = 0$ when students are finding the x-intercepts of the related function $y = ax^2 + c$.

Teach⇔Interact

Checking Key Concepts

Teaching Tip
Encourage students to check their answers to questions 4–9 by substituting each solution for the variable in the equation. Have them describe how to check their answers by graphing.

Closure Question

Describe two methods to solve a quadratic equation of the form $ax^2 + c = 0$.
Solve the equation algebraically by working to get x^2 alone on one side of the equation and then taking the square root of both sides. Solve the equation by graphing the related equation $y = ax^2 + c$. Find the x-intercepts by using the trace and zoom features of a calculator.

Apply⇔Assess

Suggested Assignment

Core Course
Day 1 Exs. 1–20
Day 2 Exs. 21–29, 32, 33, 35–40, AYP

Extended Course
Day 1 Exs. 1–20
Day 2 Exs. 21–40, AYP

Block Schedule
Day 50 Exs. 1–20
Day 51 Exs. 21–29, 32, 33, 35–40, AYP

Exercise Notes

Using Technology
Exs. 1–9 Students can verify their solutions by graphing the related equation of the form $y = ax^2 + c$. Have them use the trace and zoom features to get accurate values for the x-intercepts. However, students may get slightly different answers when using a calculator. If this occurs, explain that the answers depend on the choice of window settings and can vary slightly.

CHECKING KEY CONCEPTS

Solve each equation using the given graph.

1. $2x^2 = 0$

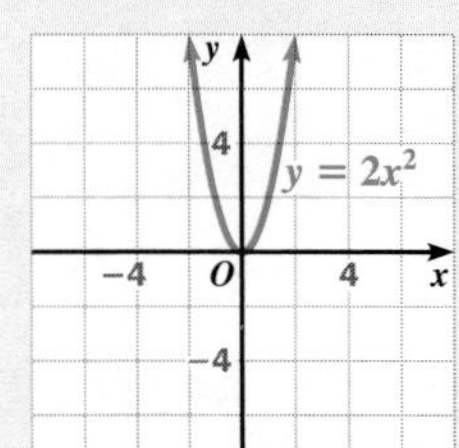

2. $-x^2 + 4 = 0$

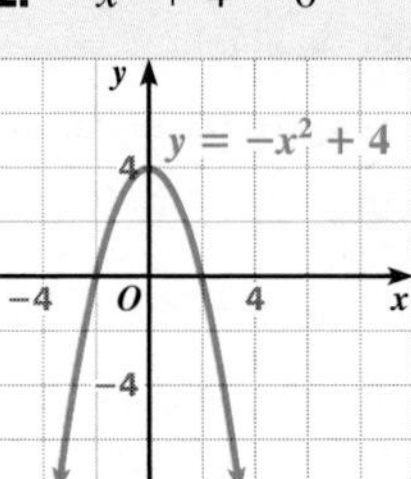

3. $\frac{1}{4}x^2 - 4 = 0$

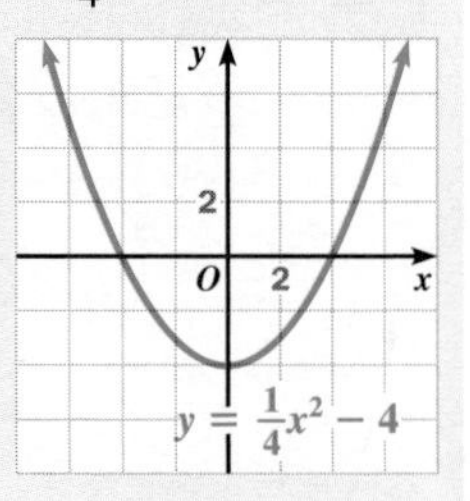

Solve each equation.

4. $x^2 = 100$
5. $2n^2 = 32$
6. $p^2 - 49 = 0$
7. $d^2 - 25 = 16$
8. $4x^2 + 18 = 27$
9. $70 = 5n^2 + 10$

8.3 Exercises and Applications

Extra Practice exercises on page 573

Solve each equation algebraically.

1. $x^2 = 81$
2. $121 = n^2$
3. $3p^2 = 27$
4. $-10k^2 = -1000$
5. $\frac{1}{2}r^2 = 2$
6. $\frac{3}{4}x^2 = 12$
7. $k^2 - 144 = 0$
8. $0 = p^2 - 64$
9. $n^2 - 0.25 = 0$

10. a. COOKING A recipe for pizza dough says to use a 10 in. by 14 in. rectangular pan. Jake only has a circular pizza pan with a 12 in. diameter. If Jake makes the pizza in his circular pan, will the crust be thicker or thinner? Explain.

b. Suppose Jake wants to make a round pizza that is exactly the same thickness as the 10 in. by 14 in. pizza in the recipe. What will the diameter of the pan have to be? Is Jake likely to find such a pan? Why or why not?

Estimate the solutions of each equation using the given graph.

11. $-x^2 + 3 = 0$

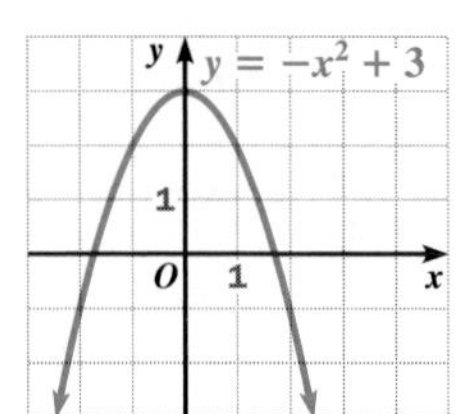

12. $\frac{1}{2}x^2 - 3 = 0$

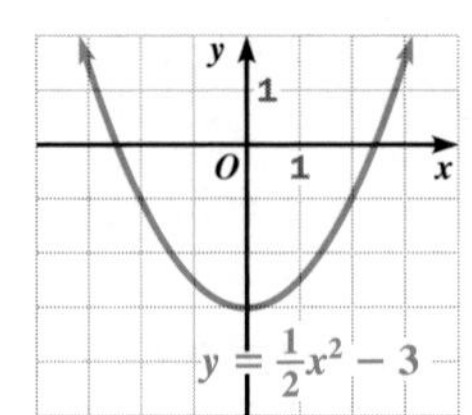

13. $-\frac{1}{4}x^2 + 5 = 0$

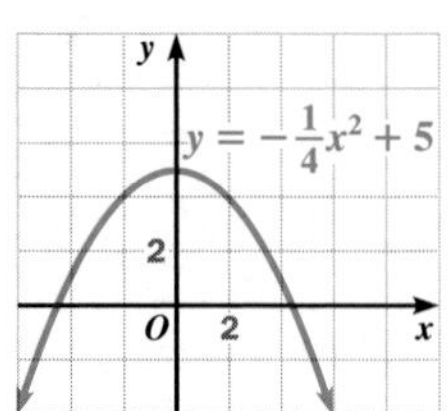

Checking Key Concepts

1. 0
2. ±2
3. ±4
4. ±10
5. ±4
6. ±7
7. $\pm\sqrt{41}$, or about 6.4 and about –6.4
8. $\pm\frac{3}{2}$
9. $\pm 2\sqrt{3}$, or about 3.46 and about –3.46

Exercises and Applications

1. ±9
2. ±11
3. ±3
4. ±10
5. ±2
6. ±4
7. ±12
8. ±8
9. ±0.5
10. a. thicker; The area of the bottom of the circular pan is less than that of the bottom of the rectangular pan.
    b. $2\sqrt{\frac{140}{\pi}}$, or about 13.35 in.; No; manufacturers are not likely to produce a circular pan with an irrational diameter.

11–13. Estimates may vary. Examples are given.
11. about –1.75, about 1.75
12. about –2.5, about 2.5
13. about –4.5, about 4.5

INTERVIEW Bill Pinkney

Look back at the article on pages 322–324.

Sailors like Bill Pinkney have to adjust their sails according to the speed of the wind in order to keep from being blown over. A mathematical formula that models the force F in lb/ft² on a sail when the wind is blowing perpendicular to the sail is $F = 0.00431v^2$, where v is the speed of the wind in knots (1 knot ≈ 1.15 mi/h).

14. According to the article, Pinkney's ideal wind speed was 35 knots. To the nearest tenth, what is the force on a sail of a 35-knot wind blowing perpendicular to the sail?

15. Estimate the wind speed that will produce a force of 4 lb/ft² when the wind is blowing perpendicular to the sail.

16. Challenge How could you use a graph of the equation $F = 0.00431v^2$ to answer Exercises 14 and 15? Are the intercepts of that graph the solutions of either exercise? Why or why not?

Cooperative Learning For Exercises 17–20, work with a partner. You will each need a graphing calculator.

17. **Technology** Each of you should graph the equation $y = x^2$ on a graphing calculator. Then one of you should do part (a). The other should do part (b). Work together on part (c).

a. Graph the equations $y = x^2 + 1$, $y = x^2 + 2$, and $y = x^2 + 3$ along with the graph of $y = x^2$. Record the location of the vertex of each graph.

b. Graph the equations $y = x^2 - 1$, $y = x^2 - 2$, and $y = x^2 - 3$ along with the graph of $y = x^2$. Record the location of the vertex of each graph.

c. Describe how each graph compares with the original graph.

18. **Technology** Work together to generalize how the graph of an equation in the form $y = x^2 + c$ changes as the value of c changes.

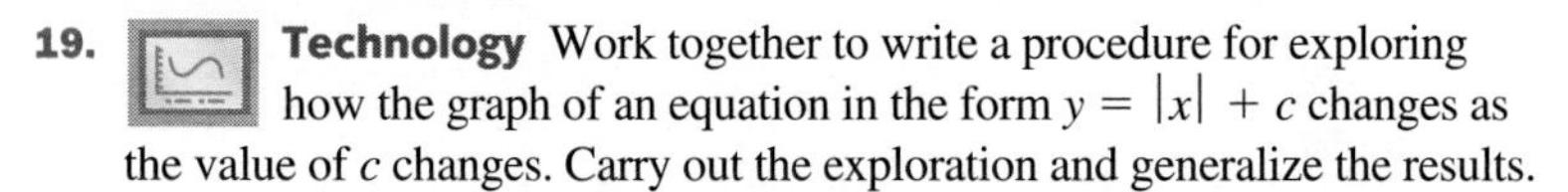

19. **Technology** Work together to write a procedure for exploring how the graph of an equation in the form $y = |x| + c$ changes as the value of c changes. Carry out the exploration and generalize the results.

20. Predict how the graph of each equation will compare with the graph of $y = 2x^2$. Then sketch the graph of each equation. Check your predictions.

a. $y = 2x^2 - 4$ b. $y = 2x^2 + 4$ c. $y = 2x^2 + 0.25$

Apply⇔Assess

Exercise Notes

Interview Note

Ex. 16 You may need to remind students that as the value of a in the equation $y = ax^2$ gets very small, the graph of its related parabola gets very wide. In this case, it may be a challenge for students to find a good viewing window for the graph.

Cooperative Learning

Exs. 17–20 Working with a partner, students can learn how the graphs of $y = x^2 + c$ and $y = |x| + c$ change as the value of c changes. For Exs. 18 and 19, students should be able to explain how they arrived at their generalizations.

Using Technology

Exs. 17–20 These exercises can be done with either a TI-82 or TI-81 graphing calculator, or with the *Function Investigator* software.

Challenge

Ex. 20 Ask students to think about how the graph of $y = ax^2 + c$ changes as both values a and c change. Have students verify their answers using equations they make up on their own.

14. about 5.3 lb/ft²

15. about 30 knots

16. To answer Ex. 14, replace F with y and v with x and graph $y = 0.00431x^2$ on a graphing calculator. Use the trace feature to locate the value of y when $x = 35$. To solve Ex. 15, use the trace feature to locate the value of x when $y = 4$. For Ex. 15, you could also graph $y = 4$ on the same screen and use the trace feature to locate the intersection of the graphs. No; the graph of $y = 0.00431x^2$ has only one intercept, 0, which is not a solution of either exercise.

17. a. Vertices: (0, 0), (0, 1), (0, 2), (0, 3)

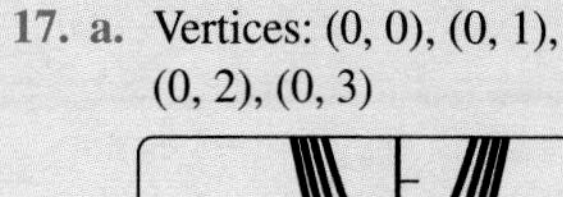

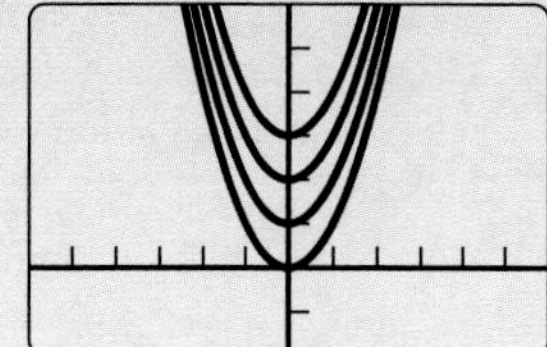

b. Vertices: (0, 0), (0, –1), (0, –2), (0, –3)

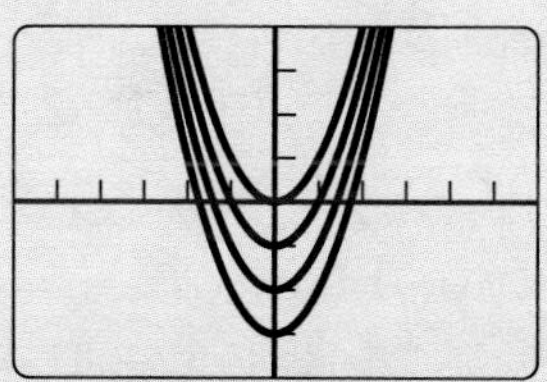

c. The graphs of $y = x^2 + 1$, $y = x^2 + 2$, and $y = x^2 + 3$ are the graph of $y = x^2$ shifted up 1, 2, and 3 units, respectively. The graphs of $y = x^2 - 1$, $y = x^2 - 2$, and $y = x^2 - 3$ are the graph of $y = x^2$ shifted down 1, 2, and 3 units, respectively.

18. The graph of $y = x^2 + c$ is the graph of $y = x^2$ shifted up c units if c is positive and down $|c|$ units if c is negative.

19. Answers may vary. Examples are given. Graph equations using different values for c and compare the graphs to the original graph. Use both positive and negative values. The graph of $y = |x| + c$ is the graph of $y = |x|$ shifted up c units if c is positive, and shifted down $|c|$ units if c is negative.

20. See answers in back of book.

Apply⇔Assess

Exercise Notes

Student Progress

Exs. 21–29 It is essential that students be able to do these exercises correctly before they are presented with more complicated quadratic equations in the next two sections of this chapter.

Communication: Writing

Ex. 30 This exercise provides students with an opportunity to relate previously learned skills to a new but similar situation. This should help to strengthen students understanding of how to solve quadratic equations.

Using Technology

Ex. 32 Students can use the STAT features of a graphing calculator to make a scatter plot of the data. Have them use ZOOM 9:ZoomStat to set an appropriate viewing window.

Solve each equation algebraically or by using a graph.

21. $x^2 + 8 = 12$ **22.** $-n^2 + 10 = -15$ **23.** $4r^2 = 12$

24. $30 = 6p^2$ **25.** $0 = \frac{1}{4}x^2 - 1$ **26.** $4k^2 - 30 = 10$

27. $55 = 45n^2 - 25$ **28.** $-3r^2 + 10 = 4$ **29.** $\frac{2}{3}x^2 - 9 = 17$

30. Writing Solve the equations $3x - 12 = 0$ and $3x^2 - 12 = 0$. How is solving a quadratic equation in the form $ax^2 + c = 0$ like solving a linear equation in the form $ax + c = 0$? How is it different?

31. Open-ended Problem Write a quiz with five questions that could be used to check your understanding of the material in this section. Show how to answer all the questions on your quiz.

Connection SPORTS

Skydivers refer to the period before they open their parachutes as *free fall*. If there were no air resistance, the formula $d = 16t^2$ could be used to model the distance d in feet a skydiver falls in t seconds of free fall. Because of air resistance, a skydiver eventually stops accelerating and reaches a terminal velocity.

32. The table shows the distances a skydiver falls in t seconds of free fall with air resistance. Make a scatter plot of the data in the table. Put time on the horizontal axis and distance on the vertical axis.

Time after jump (seconds)	Distance fallen (feet)
1	16
3	138
5	366
7	652
9	970
11	1,310
13	1,660
15	2,010
25	3,770
35	5,530
45	7,290
55	9,050
65	10,810
75	12,570

33. Graph the equation $d = 16t^2$ in the same coordinate plane as your scatter plot. Compare the graphs. How does comparing the graphs show you that the skydiver's acceleration is slowed down?

34. Challenge About how long does it take for the skydiver to reach terminal velocity? Estimate the terminal velocity. Give reasons for your answers.

21. ± 2
22. ± 5
23. $\pm\sqrt{3}$, or about 1.73 and about -1.73
24. $\pm\sqrt{5}$, or about 2.24 and about -2.24
25. ± 2
26. $\pm\sqrt{10}$, or about 3.16 and about -3.16
27. $\pm\frac{4}{3}$
28. $\pm\sqrt{2}$, or about 1.41 and about -1.41
29. $\pm\sqrt{39}$, or about 6.24 and about -6.24
30. $x = 4$; $x = \pm 2$; In both cases, you solve the equation by getting the variable alone on one side. To solve the linear equation, you may need to add, subtract, multiply, divide, or use some combination of these. To solve the quadratic equation, you also need to take square roots. The linear equation has exactly one solution. The quadratic equation can have as many as two solutions.
31. Answers may vary.

32, 33.

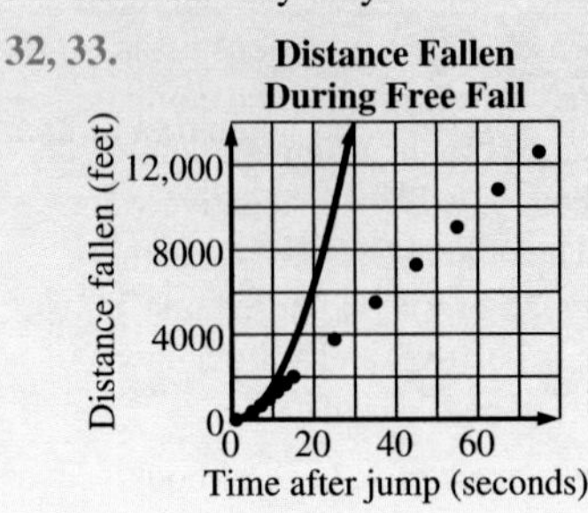

The scatter plot falls below the graph of the equation. That means the skydiver does not fall as far during each time interval as the equation would indicate.

34. See answers in back of book.
35. Check students' work.
36. 12 in. 37. $13\frac{1}{3}$ in.

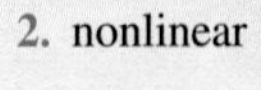

38. 30 in. long and 18 in. wide
39. 128; 36 40. 24; 12

Assess Your Progress

1. linear
2. nonlinear

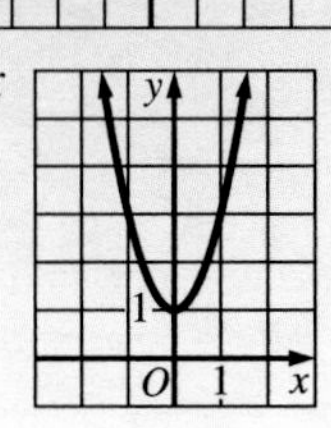

3. nonlinear

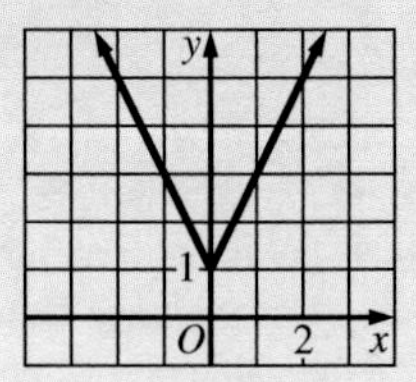

4. Since $7(0) = 0$, $7 > 0$, and $7 > 1$, the graph will have the same vertex, will open in the same direction but will be narrower and steeper.

ONGOING ASSESSMENT

35. Cooperative Learning Work in a group of three students. Take turns listing the key concepts of this section and examples that illustrate the concepts. Each of you should write a summary of the list.

SPIRAL REVIEW

A building is 150 ft long and 90 ft wide. Find each value. *(Section 5.2)*

36. A floor plan of the building is 20 in. long. How wide is the floor plan?

37. A cardboard model of the building is 8 in. wide. How long is the model?

38. An architect uses the scale 1 in. = 5 ft for the plans. What are the dimensions of the architect's plans?

Find the missing values for each pair of similar figures. *P* represents perimeter and *A* represents area. *(Section 5.4)*

39.

8 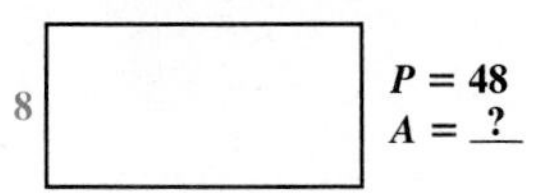 $P = 48$, $A = \underline{?}$

6 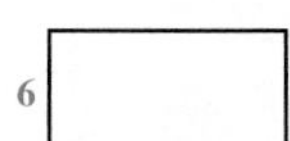 $P = \underline{?}$, $A = 72$

40.

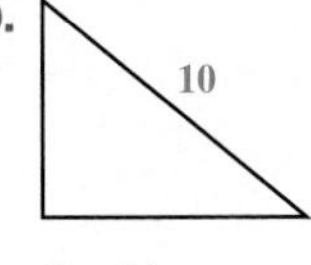

5

$P = 24$, $A = \underline{?}$ $P = \underline{?}$, $A = 6$

ASSESS YOUR PROGRESS

VOCABULARY

nonlinear equation (p. 326)
parabola (p. 331)
quadratic (p. 332)
vertex of a parabola (p. 332)
symmetric (p. 333)
line of symmetry (p. 333)

Graph each equation. Tell whether each equation is *linear* or *nonlinear*. *(Section 8.1)*

1. $y = 2x + 1$ **2.** $y = 2x^2 + 1$ **3.** $y = 2|x| + 1$

Predict how the graph of each equation will compare with the graph of $y = x^2$. Explain your prediction, then sketch the graph. *(Section 8.2)*

4. $y = 7x^2$ **5.** $y = -7x^2$ **6.** $y = 0.7x^2$

Solve each equation algebraically or by using a graph. *(Section 8.3)*

7. $x^2 = 144$ **8.** $\frac{1}{3}n^2 = 3$ **9.** $0 = k^2 - 4$

10. $-5p^2 + 20 = 0$ **11.** $2n^2 - 3 = 13$ **12.** $4x^2 + 25 = 49$

13. Journal Explain how you can solve the equation $3x^2 + 6 = 25$ algebraically. Then explain why the x-intercepts of the graph of the equation $y = 3x^2 - 19$ are the solutions of the equation $3x^2 + 6 = 25$.

Apply⇔Assess

Exercise Notes

Assessment Note
Ex. 35 You may wish to have each group make a brief presentation of their summaries to the class.

Second-Language Learners
Ex. 35 This cooperative summarizing activity should be of considerable benefit to students learning English.

Assess Your Progress

Journal Entry
For Ex. 13, a review may help those students who are having difficulty relating the algebraic and graphic approaches to solving quadratic equations.

Progress Check 8.1–8.3

See page 362.

Practice 53 for Section 8.3

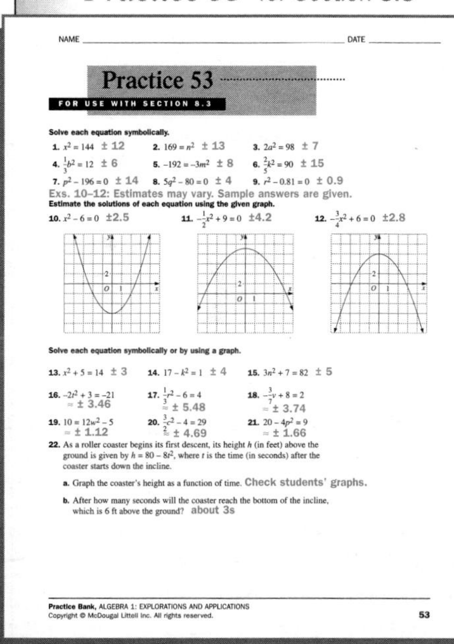
NAME ______ DATE ______

Practice 53

FOR USE WITH SECTION 8.3

Solve each equation symbolically.

1. $x^2 = 144$ ± 12 2. $169 = n^2$ ± 13 3. $2a^2 = 98$ ± 7
4. $\frac{1}{3}b^2 = 12$ ± 6 5. $-192 = -3m^2$ ± 8 6. $\frac{2}{5}k^2 = 90$ ± 15
7. $p^2 - 196 = 0$ ± 14 8. $5q^2 - 80 = 0$ ± 4 9. $r^2 - 0.81 = 0$ ± 0.9

Exs. 10–12: Estimates may vary. Sample answers are given.

Estimate the solutions of each equation using the given graph.

10. $x^2 - 6 = 0$ ±2.5 11. $-\frac{1}{2}x^2 + 9 = 0$ ±4.2 12. $-\frac{3}{4}x^2 + 6 = 0$ ±2.8

Solve each equation symbolically or by using a graph.

13. $x^2 + 5 = 14$ ± 3 14. $17 - k^2 = 1$ ± 4 15. $3n^2 + 7 = 82$ ± 5
16. $-2t^2 + 3 = -21$ ≈ ± 3.46 17. $\frac{1}{3}r^2 - 6 = 4$ ≈ ± 5.48 18. $-\frac{3}{7}v + 8 = 2$ ≈ ± 3.74
19. $10 = 12w^2 - 5$ ≈ ± 1.12 20. $\frac{3}{2}c^2 - 4 = 29$ ≈ ± 4.69 21. $20 - 4p^2 = 9$ ≈ ± 1.66

22. As a roller coaster begins its first descent, its height h (in feet) above the ground is given by $h = 80 - 8t^2$, where t is the time (in seconds) after the coaster starts down the incline.

a. Graph the coaster's height as a function of time. Check students' graphs.

b. After how many seconds will the coaster reach the bottom of the incline, which is 6 ft above the ground? about 3s

Practice Bank, ALGEBRA 1: EXPLORATIONS AND APPLICATIONS

53

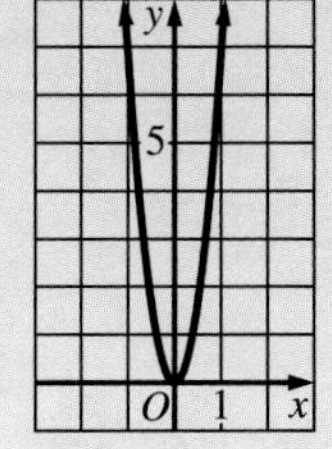

5. Since $-7(0) = 0$, $-7 < 0$, and $|-7| > 1$, the graph will have the same vertex, will open down, and will be narrower and steeper.

6. Since $0.7(0) = 0$, $0.7 < 1$, and $0.7 > 0$, the graph will have the same vertex, will open up but will be wider and flatter.

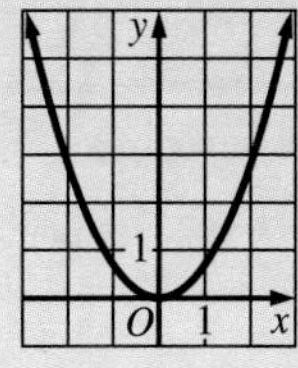

7. ± 12 8. ± 3

9. ± 2 10. ± 2

11. $\pm 2\sqrt{2}$, or about 2.83 and about –2.83

12. $\pm\sqrt{6}$, or about 2.45 and about –2.45

13. First, subtract 6 from both sides. Then divide both sides by 3. Finally, take the square root of both sides; the x-intercepts of the graph of the equation $y = 3x^2 - 19$ are the points with y-coordinate 0, that is, the points for which $3x^2 - 19 = 0$. This equation is equivalent to $3x^2 + 6 = 25$. The solutions are about 2.52 and about –2.52.

Plan⇔Support

Objectives

- Use graphs to solve more complicated quadratic equations.
- Recognize quadratic equations with no solutions.
- Solve any quadratic equation and find heights of thrown objects.

Recommended Pacing

Core and Extended Courses
Section 8.4: 2 days

Block Schedule
Section 8.4: 2 half-blocks (with Sections 8.3 and 8.5)

Resource Materials

Lesson Support
Lesson Plan 8.4
Warm-Up Transparency 8.4
Practice Bank: Practice 54
Study Guide: Section 8.4

Technology
Graphing Calculator
McDougal Littell Software *Function Investigator*

Internet:
http://www.hmco.com

Warm-Up Exercises

1. Tell whether the graph of $y = -3x^2$ opens up or down. down

Solve each equation.

2. $3x^2 - 12 = 0$ ± 2
3. $x^2 + 10 = 18$ $\pm 2\sqrt{2}$ or about ± 2.83
4. Give the related equation to solve Exercise 2 by graphing. $y = 3x^2 - 12$
5. What are the x-intercepts of the graph of $y = x^2 - 4$. ± 2

SECTION

8.4 Applying Quadratic Equations

Learn how to...

- **use graphs to solve more complicated quadratic equations**

So you can...

- **recognize quadratic equations with no solution**
- **solve any quadratic equation and find heights of thrown objects, for example**

Have you ever played or watched a game of table tennis? You may not have realized that the path the ball followed through the air was a parabola. There is a quadratic relationship between the height of the ball above the table and the ball's horizontal distance from where it bounced.

There is a *different* quadratic relationship between the height of the ball above the table and the *time* since the ball bounced. For the lower parabola in the photograph, the height-time relationship is modeled by this equation:

$$h = -4.9t^2 + 2.1t$$

In the equation, h stands for height in meters and t stands for time in seconds.

BY THE WAY

The celluloid ball used in table tennis was introduced in 1902. The diameter is about $1\frac{1}{2}$ in., and the ball weighs less than $\frac{1}{10}$ oz.

THINK AND COMMUNICATE

Use the graph of the equation $h = -4.9t^2 + 2.1t$ below.

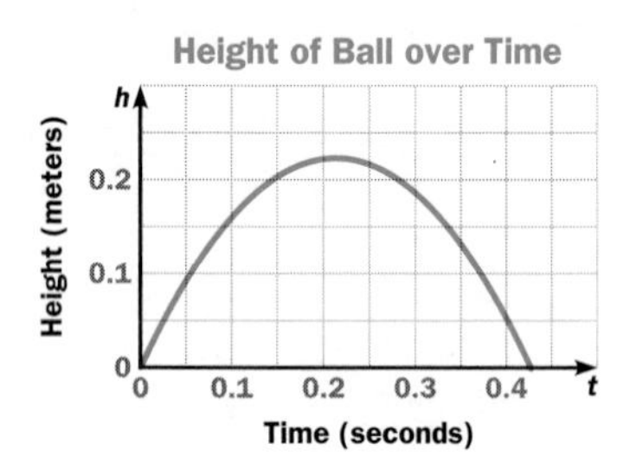

1. Estimate the solutions of the equation $-4.9t^2 + 2.1t = 0$. What do the solutions tell you about the situation?
2. Estimate the maximum height the ball reached. About how much time had gone by when the ball reached this height?

ANSWERS Section 8.4

Think and Communicate

1. 0 and about 0.425; The ball left the table at time $t = 0$ and landed back on the table after about 0.425 s.
2. Estimates may vary; about 0.225 m, about 0.21 s.
3. Determine whether the graph of the related equation has any x-intercepts. If so, these intercepts are the solutions. If not, the original equation has no solutions.
4. Use the trace and zoom features to find the x-coordinates of points with y-coordinates as close to 0 as possible.
5. one solution
6. zero, one, or two solutions

All **quadratic equations** can be written in the form $ax^2 + bx + c = 0$, when $a \neq 0$. You can find the solutions of an equation in this form by using a graph. In fact, finding solutions of a quadratic equation is so important in mathematics that some graphing calculators have a *root* feature that will find the solutions for you. *Root* is another word for the solution of an equation.

EXAMPLE 1

Solve each equation.

a. $x^2 - 2x - 5 = 0$ **b.** $3x^2 + x = -2$

SOLUTION

a. Graph the related equation $y = x^2 - 2x - 5$. Then find the x-intercepts, which are where $y = 0$.

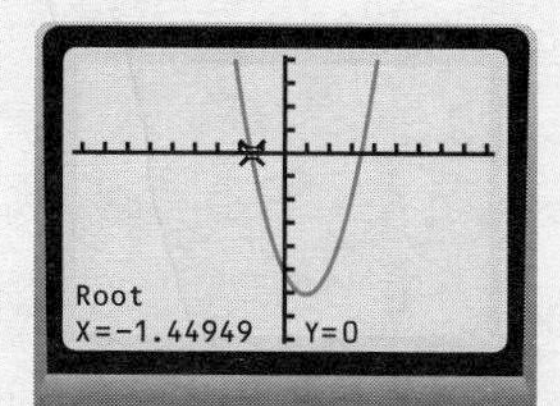

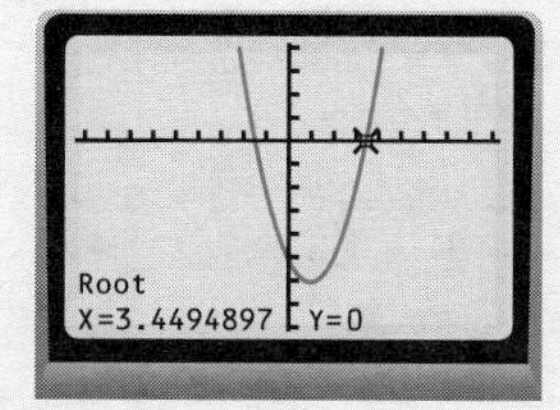

The solutions are about −1.45 and about 3.45.

b. Rewrite the equation in the form $ax^2 + bx + c = 0$.

$$3x^2 + x = -2$$
$$3x^2 + x + 2 = 0$$

Graph the related equation $y = 3x^2 + x + 2$. Notice that there are no x-intercepts. There are no values of x that make the equation $3x^2 + x + 2 = 0$ true.

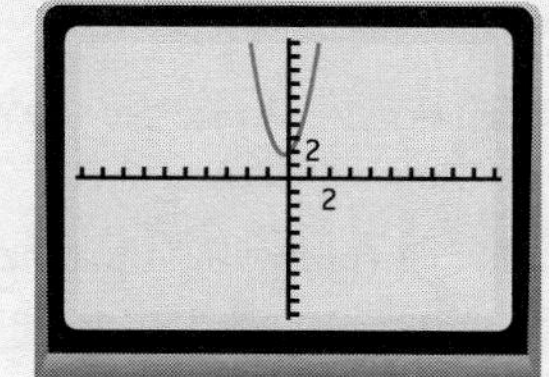

The equation has no solution.

For more information about using the root feature, see the *Technology Handbook*, p. 607.

THINK AND COMMUNICATE

3. How can you use a graph to tell whether a quadratic equation has any solutions?
4. How could you use the trace and zoom features of a graphing calculator to find the solutions in Example 1(a)?
5. The vertex of the parabola $y = x^2 - 4x + 4$ lies on the x-axis. How many solutions does the related equation $x^2 - 4x + 4 = 0$ have?
6. Suppose a is negative. What are the possible numbers of solutions of the equation $ax^2 + bx + c = 0$?

Technology Note

Students may not be able to get a y-coordinate of 0 when they find the x-intercepts by using the root key or the trace and zoom features. Ask them to interpret results like Y = 3E – 13 or Y = 1E – 12 as 0. These numbers are so small that they essentially equal 0. Remind students that a good viewing window for a quadratic equation normally shows the vertex and x-intercepts of the graph.

Teach⇔Interact

Section Note

Teaching Tip

When discussing the formula $h = -4.9t^2 + 2.1t$, it should be clear to students that the height of the ball, h, is a function of the time, t, since the ball has bounced. Make sure students understand that this functional relationship between h and t means that as time changes, the height of the ball also changes.

Think and Communicate

For question 1, have students check their estimated solutions by substituting them into the equation $-4.9t^2 + 2.1t = 0$. For question 2, ask students to describe two ways to determine how much time had gone by when the ball reached its maximum height.

Additional Example 1

Solve each equation.

a. $x^2 - 6x + 9 = 0$

Graph the related equation $y = x^2 - 6x + 9$. Then find the x-intercepts.

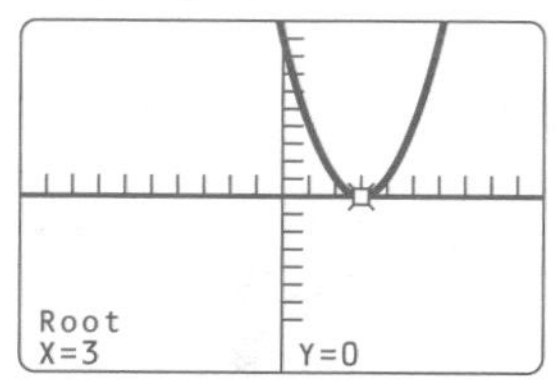

The solution is 3.

b. $4x^2 - x = -3$

Rewrite the equation in the form $ax^2 + bx + c = 0$.

$4x^2 - x + 3 = 0$

Graph the related equation $y = 4x^2 - x + 3$.

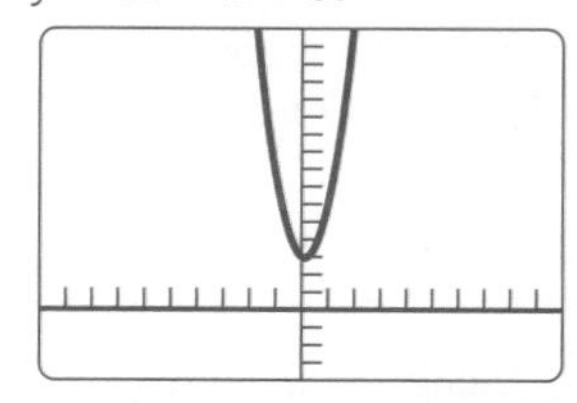

Notice that there are no x-intercepts. There are no values of x that make the equation $4x^2 - x + 3 = 0$ true. The equation has no solution.

Think and Communicate

Questions 3 and 5 can help students summarize their understanding of the relationship between the solutions of a quadratic equation and the x-intercepts of its graph.

Teach⇔Interact

About Example 2

Approximate solutions given by a graphing calculator are appropriate for application problems dealing with time measurements. Estimating an answer to the nearest tenth of a second is close enough for most applications.

Additional Example 2

Juan kicked a football. The equation $h = -16t^2 + 4t + 1$ describes the height h of the ball in feet t seconds after he kicked it. About how many seconds went by before the ball hit the ground?

Use a graph.

To solve the equation using a graph, you must first substitute y for h and x for t in the equation $h = -16t^2 + 4t + 1$.

$y = -16x^2 + 4x + 1$

When the ball hit the ground, the height was 0. Solve the equation $0 = -16x^2 + 4x + 1$ by graphing the related equation $y = -16x^2 + 4x + 1$ and finding the x-intercepts.

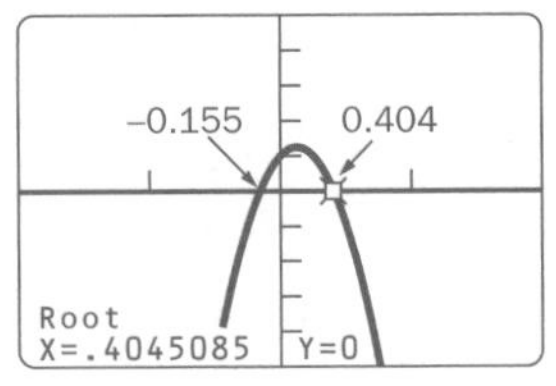

Time cannot be negative, so ignore the intercept –0.155. The ball hit the ground after about 0.4 seconds.

Think and Communicate

Question 7 asks students to interpret the graph in Example 2. It is important that students be able to locate points on a graph and describe what they represent.

Checking Key Concepts

Using Technology
Students can use the intersect function under the CALCULATE menu of a TI-82 to answer questions 4–9 if they graph each side of the equation as separate graphs. For example, for question 8, students can graph $y = x^2 + 10x + 34$ and $y = 9$ and then use the intersect function to find the intersection of the two graphs.

EXAMPLE 2 Application: Sports

Janet threw a basketball. The equation $h = -16t^2 + 5t + 5$ describes the height h of the ball in feet t seconds after she threw it. About how many seconds went by before the ball hit the ground?

BY THE WAY

In early women's basketball, each player had to stay within one of three zones for the entire game and was allowed only two or three dribbles at a time.

SOLUTION

Problem Solving Strategy: Use a graph.

To solve the equation using a graph, you must first substitute y for h and x for t in the equation $h = -16t^2 + 5t + 5$.

$$y = -16x^2 + 5x + 5$$

When the ball hit the ground, the height was 0. Solve the equation $0 = -16x^2 + 5x + 5$ by graphing the related equation $y = -16x^2 + 5x + 5$ and finding the x-intercepts.

Time cannot be negative, so ignore this intercept.

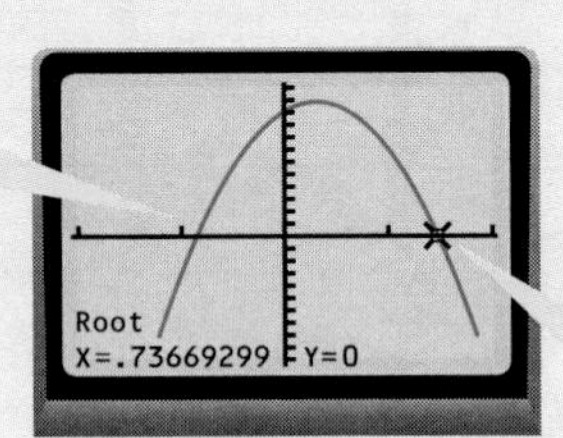

Locate the positive x-intercept.

The ball hit the ground after about 0.7 seconds.

THINK AND COMMUNICATE

7. What does the vertex of the parabola in Example 2 tell you about the situation?
8. What equation would you have to solve to find out when the ball in Example 2 was 2 ft off the ground?

CHECKING KEY CONCEPTS

Rewrite each equation in the form $ax^2 + bx + c = 0$.

1. $x^2 + 3x = 8$
2. $x^2 - 6x + 4 = 7$
3. $5x^2 + x - 8 = 6$

Technology Use a graphing calculator to solve each equation. If an equation has no solution, write *no solution*.

4. $x^2 - 2x - 3 = 0$
5. $6x^2 + 3x + 2 = 0$
6. $-x^2 + 4x = 0$
7. $3x^2 - x = 2$
8. $x^2 + 10x + 34 = 9$
9. $-4x^2 + 16x - 7 = 9$

Think and Communicate

7. The x-coordinate of the vertex indicates how many seconds went by before the ball reached its maximum height. The y-coordinate indicates the maximum height.
8. $0 = -16x^2 + 5x + 3$

Checking Key Concepts

1. $x^2 + 3x - 8 = 0$
2. $x^2 - 6x - 3 = 0$
3. $5x^2 + x - 14 = 0$
4. 3, –1
5. no solution
6. 0, 4
7. $1, -\frac{2}{3}$
8. –5
9. 2

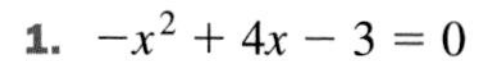

8.4 Exercises and Applications

Estimate the solutions of each equation using the given graph. If an equation has no solution, write *no solution.*

Extra Practice *exercises on page 573*

1. $-x^2 + 4x - 3 = 0$

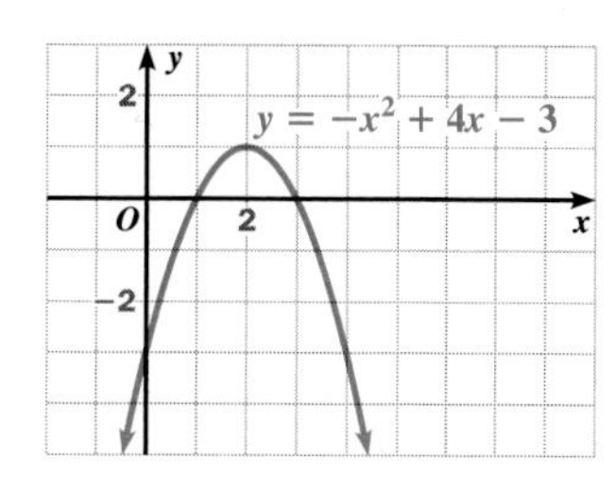

2. $x^2 + 2x + 1 = 0$

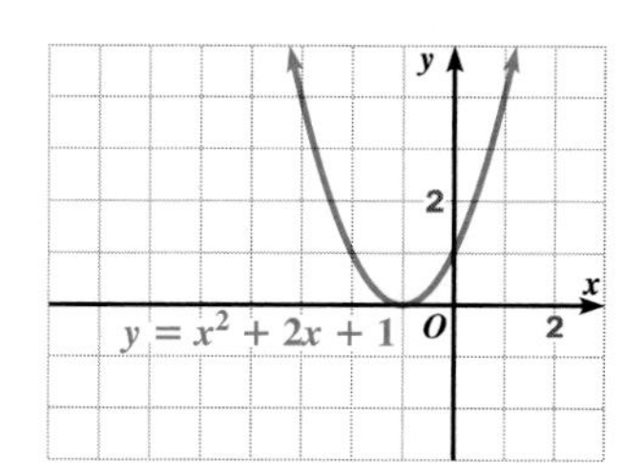

3. $\frac{1}{2}x^2 + 2x - \frac{1}{2} = 0$

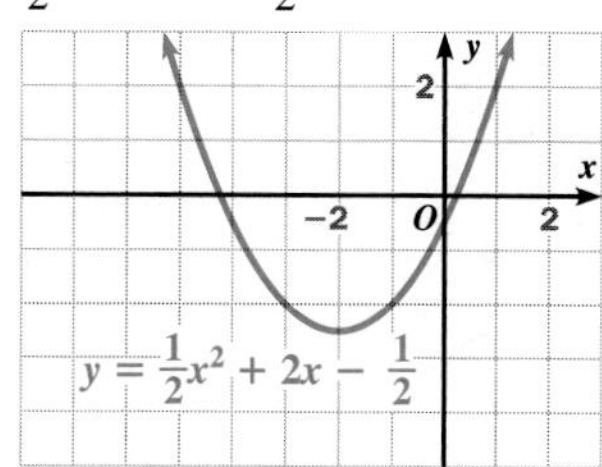

4. $-x^2 + 3x - \frac{15}{4} = 0$

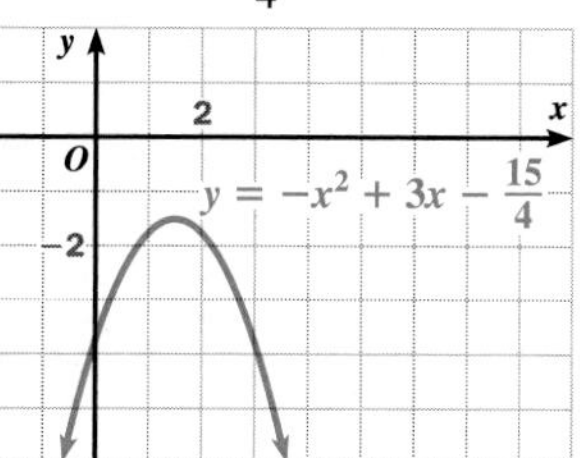

INTERVIEW Bill Pinkney

Look back at the article on pages 322–324.

According to the article, Pinkney's ideal wave height was "anywhere from 15 to 25 feet." An equation for the graph in the article is $h = 0.013s^2$, where h is wave height in feet and s is wind speed in miles per hour.

5. Use the graph on page 324 in the article to estimate the wind speeds that correspond to Pinkney's ideal wave heights.

6. **Technology** Use graphs of the equations $h = 0.013s^2 - 15$ and $h = 0.013s^2 - 25$ to estimate the wind speeds in Exercise 5.

7. Use equation solving to estimate the wind speeds in Exercise 5.

8. **Writing** Compare the methods used in Exercises 5–7. Which do you prefer? Why?

Teach⇔Interact

Closure Question

Explain how to solve the quadratic equation $ax^2 + bx + c = 0$ by graphing.

Graph the related equation $y = ax^2 + bx + c$. Then find the coordinates of the x-intercepts.

Apply⇔Assess

Suggested Assignment

Core Course
Day 1 Exs. 1–16
Day 2 Exs. 17–24, 32–41

Extended Course
Day 1 Exs. 1–16
Day 2 Exs. 17–41

Block Schedule
Day 51 Exs. 1–16
Day 52 Exs. 17–24, 27–41

Exercise Notes

Common Error

Ex. 4 Some students may incorrectly think that a solution is shown where the graph intersects the y-axis. Remind these students that solutions to a quadratic equation are points on its graph where $y = 0$, that is, where the graph intersects the x-axis.

Interview Note

Ex. 8 Students can use this exercise to review what they have learned about solving quadratic equations.

Exercises and Applications

1. 1, 3
2. −1
3. about $-4\frac{1}{4}$, about $\frac{1}{4}$
4. no solution

5–7. Estimates may vary. Examples are given.

5. about 34 mi/h, about 44 mi/h
6. about 34 mi/h, about 43.85 mi/h
7. $\sqrt{\frac{15}{0.013}}$ or about 33.97 mi/h, $\sqrt{\frac{25}{0.013}}$ or about 43.85 mi/h
8. Answers may vary. An example is given. The methods in Exs. 5 and 6 involve reading a graph, so the answers are estimates. The result in Ex. 6 (from a graphing calculator) is probably closer than the result in Ex. 5. Ex. 7 involves adding 15 (or 25) to both sides of the equation, dividing both sides by 0.013, and taking the square root of both sides. If you leave the solutions in radical form, then this method gives the exact answer. Personal preferences may vary.

Apply⇔Assess

Exercise Notes

Using Technology

Exs. 9–20, 22–24 These exercises are appropriate for TI-82 or TI-81 calculators or the *Function Investigator* software. For Exs. 22–24, students may need to experiment to find a good viewing window.

Students who use graphing calculators and the zoom-and-trace approach may wonder how accurately they have approximated the roots of the equations. One way to gauge this is to zoom and trace a few times and then press WINDOW (for the TI-82) or RANGE (for the TI-81). Compare the values displayed for Xmin and Xmax.

Reasoning

Ex. 22 Ask students to describe the situation when $s = 0$ for the equation $0.0056s^2 + 0.14s = 0$. (When the speed of a car is 0, its stopping distance is 0.)

Technology Use a graphing calculator to solve each equation. If an equation has no solution, write *no solution*.

9. $x^2 + 2x - 8 = 0$ **10.** $2x^2 + 7x - 4 = 0$ **11.** $-x^2 + 9 = 0$

12. $x^2 - 9x + 18 = 0$ **13.** $x^2 - 8x + 16 = 0$ **14.** $2x^2 - 3x + 5 = 0$

15. $-4x^2 - 12x = 9$ **16.** $x^2 - 4x + 9 = 4$ **17.** $2x^2 + 6x - 4 = 1$

18. $-3x^2 - 6x + 8 = 3$ **19.** $3x^2 - 15x = 24$ **20.** $-x^2 + 8 = 12x$

21. SAT/ACT Preview If S represents the greater of the two solutions of the equation $x^2 - 4x - 3 = 5$, then:

A. $S > 3$ **B.** $S < 3$ **C.** $S = 3$

D. relationship cannot be determined

Connection HIGHWAY SAFETY

Speed limits vary from country to country. The speed limit on highways in Mexico is 100 km/h. In France the limits vary from 90 to 130 km/h. In Germany, some sections of the highway have no speed limit. Other sections are limited to either 180 km/h or 130 km/h.

The equation $d = 0.0056s^2 + 0.14s$ models the relationship between a vehicle's stopping distance d in meters and the vehicle's speed s in kilometers per hour.

Technology For Exercises 22–24, use a graphing calculator.

22. How many solutions does the equation $0.0056s^2 + 0.14s = 0$ have? How many of the solutions make sense in this situation?

23. About how fast should you be driving in order to be able to stop within 100 m? In which of the above countries is this speed within the speed limit?

24. Suppose a car is traveling at the speed of 130 km/h on a German highway. What distance should the driver of the car leave between the driver's car and the car in front of it?

25. Research Find out what is a reasonable length in meters for a car. Then estimate the number of car lengths that corresponds to the stopping distance you found in Exercise 24.

9. −4, 2
10. $-4, \frac{1}{2}$
11. ±3
12. 3, 6
13. 4
14. no solution
15. $-\frac{3}{2}$
16. no solution
17–20. Estimates may vary.
17. about −3.68, about 0.68
18. about −2.63, about 0.63
19. about −1.27, about 6.27
20. about −12.63, about 0.63
21. A
22. two solutions; one solution
23. about 122 km/h; Germany and some places in France
24. at least 112.84 m
25. Answers may vary. An example is given. A car that is 4.5 m long requires about 25 car lengths at that speed.
26. Answers may vary.
27. a. 0, 2; (1, −1)
 b. 0, 4; (2, −4)
 c. −3, 1; (−1, −4)
28. The x-coordinate of the vertex is the average of the x-intercepts.

26. **Open-ended Problem** Pick one of the real-world equations in this section. Write your own set of problems about it. Solve each problem.

Challenge For Exercises 27 and 28, use the three graphs below. Then do Exercises 29–31.

a. $y = x^2 - 2x$ **b.** $y = x^2 - 4x$ **c.** $y = x^2 + 2x - 3$

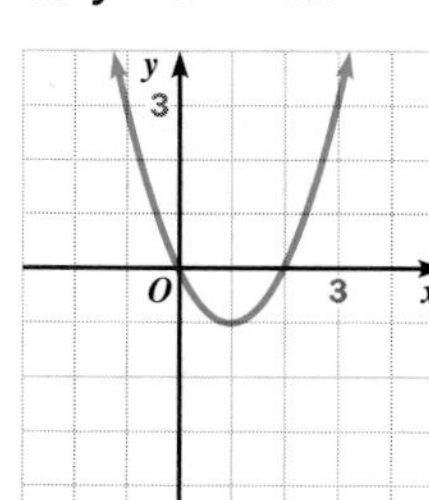

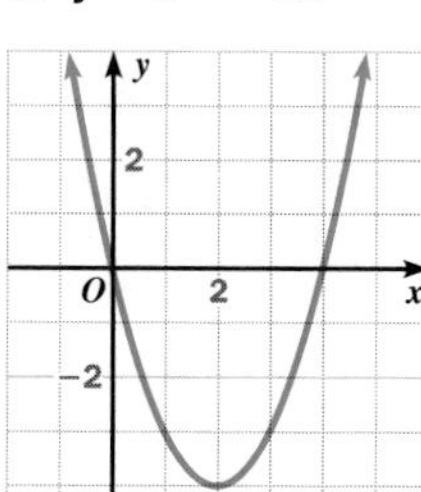

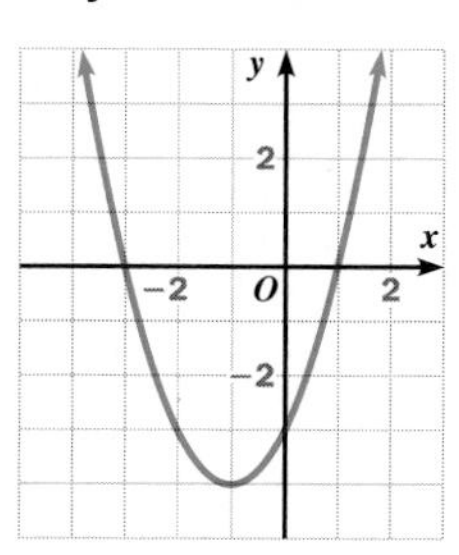

27. Identify the x-intercepts and the coordinates of the vertex of each graph.

28. Explain how the x-intercepts of each graph are related to the x-coordinate of the vertex.

29. The x-intercepts of the graph of the equation $y = x^2 - 12x + 32$ are 4 and 8. Explain how to find the x-coordinate of the vertex without graphing the equation.

30. For a graph of the equation $y = x^2 + bx + c$, the x-coordinate of the vertex is a function of b. Write an equation for the function.

31. Use the function you wrote in Exercise 30. For each equation, predict the x-coordinate of the vertex of the graph.
   a. $y = x^2 - 4x + 2$ **b.** $y = x^2 + 8x - 3$ **c.** $y = x^2 + 5x + 6$

ONGOING ASSESSMENT

32. **Open-ended Problem** A graph of the equation $y = -x^2 + 2x + 8$ is shown. Make a list of all the things you can think of to say about the graph based on what you have learned so far in this chapter.

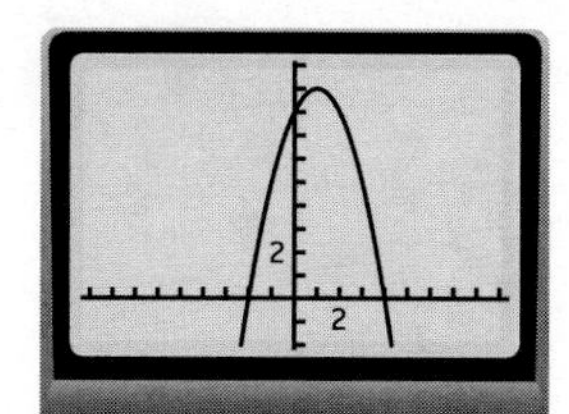

SPIRAL REVIEW

Write each repeating decimal as a fraction in lowest terms. *(Section 6.2)*

33. 0.6666 . . . 34. 0.7171 . . . 35. $0.\overline{001}$

Write each expression in simplest form. *(Section 6.3)*

36. $\sqrt{50}$ 37. $\sqrt{\frac{49}{9}}$ 38. $\sqrt{\frac{7}{16}}$

39. $\sqrt{48}$ 40. $7\sqrt{5} + 3\sqrt{5}$ 41. $\sqrt{8} + 3\sqrt{2}$

Apply⇔Assess

Exercise Notes

Challenge

Ex. 30 This exercise asks students to use graphs to find how the value of b in $y = x^2 + bx + c$ is related to the x-coordinate of the vertex. If the coefficient of the square term is 1, the average of the x-intercepts, which is $-\frac{1}{2}b$, is the x-coordinate of the vertex. In more advanced algebra courses, students use the quadratic formula to show that the x-coordinate of the vertex of $y = ax^2 + bx + c$ is $-\frac{b}{2a}$.

Assessment Note

Ex. 32 You may wish to call upon some students to read their lists aloud to ensure that all the essential ideas have been included.

Topic Spiraling: Preview

Ex. 36–41 These exercises will help prepare students for the next section on using the quadratic formula.

Practice 54 for Section 8.4

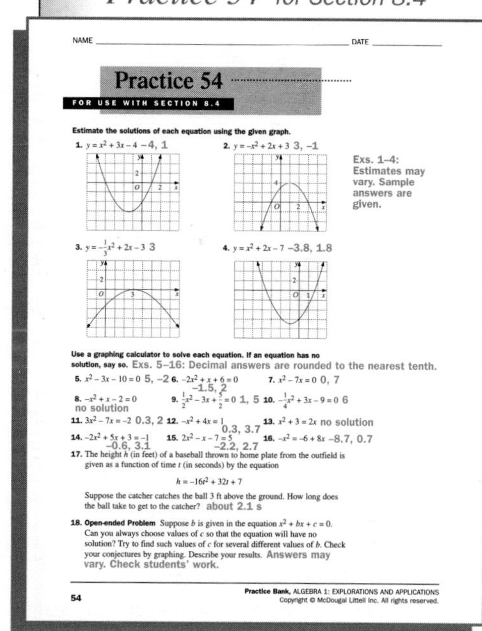
NAME ____________ DATE ____________

Practice 54

FOR USE WITH SECTION 8.4

Estimate the solutions of each equation using the given graph.

1. $y = x^2 + 3x - 4$ −4, 1
2. $y = -x^2 + 2x + 3$ 3, −1

Exs. 1–4: Estimates may vary. Sample answers are given.

3. $y = -\frac{1}{3}x^2 + 2x - 3$ 3
4. $y = x^2 + 2x - 7$ −3.8, 1.8

Use a graphing calculator to solve each equation. If an equation has no solution, say so. Exs. 5–16: Decimal answers are rounded to the nearest tenth.

5. $x^2 - 3x - 10 = 0$ 5, −2
6. $-2x^2 + x + 6 = 0$ −1.5, 2
7. $x^2 - 7x = 0$ 0, 7
8. $-x^2 + x - 2 = 0$ no solution
9. $\frac{1}{2}x^2 - 3x + \frac{5}{2} = 0$ 1, 5
10. $-\frac{1}{4}x^2 + 3x - 9 = 0$ 6
11. $3x^2 - 7x = -2$ 0.3, 2
12. $-x^2 + 4x = 1$ 0.3, 3.7
13. $x^2 + 3 = 2x$ no solution
14. $-2x^2 + 5x + 3 = -1$ −0.6, 3.1
15. $2x^2 - x - 7 = 5$ −2.2, 2.7
16. $-x^2 = -6 + 8x$ −8.7, 0.7
17. The height h (in feet) of a baseball thrown to home plate from the outfield is given as a function of time t (in seconds) by the equation

$$h = -16t^2 + 32t + 7$$

Suppose the catcher catches the ball 3 ft above the ground. How long does the ball take to get to the catcher? about 2.1 s

18. **Open-ended Problem** Suppose b is given in the equation $x^2 + bx + c = 0$. Can you always choose values of c so that the equation will have no solution? Try to find such values of c for several different values of b. Check your conjectures by graphing. Describe your results. Answers may vary. Check students' work.

54 **Practice Bank,** ALGEBRA 1: EXPLORATIONS AND APPLICATIONS

29. The x-coordinate of the vertex is the average of the x-intercepts; $\frac{4+8}{2} = 6$.

30. $x = -\frac{b}{2}$

31. a. 2
   b. −4
   c. $-\frac{5}{2}$

32. Answers may vary. An example is given. The graph of the equation is a parabola that opens downward because the coefficient of x^2 is negative. The vertex is the highest point on the graph of the equation. The x-intercepts of the graph are −2 and 4, so those are the solutions of the equation $-x^2 + 2x + 8 = 0$.

33. $\frac{2}{3}$

34. $\frac{71}{99}$

35. $\frac{1}{999}$

36. $5\sqrt{2}$

37. $\frac{7}{3}$

38. $\frac{\sqrt{7}}{4}$

39. $4\sqrt{3}$

40. $10\sqrt{5}$

41. $5\sqrt{2}$

Plan⟺Support

Objectives

- Use the quadratic formula.
- Solve quadratic equations without using a graph.
- Find more precise solutions to real-world problems.

Recommended Pacing

Core and Extended Courses
Section 8.5: 1 day

Block Schedule
Section 8.5: $\frac{1}{2}$ block
(with Section 8.4)

Resource Materials

Lesson Support
Lesson Plan 8.5
Warm-Up Transparency 8.5
Practice Bank: Practice 55
Study Guide: Section 8.5
Explorations Lab Manual:
Diagram Master 2

Technology
Technology Book:
Spreadsheet Activity 8
Graphing Calculator
Internet:
http://www.hmco.com

Warm-Up Exercises

1. Solve the equation $x^2 - 6x + 9 = 0$. 3
2. Estimate the solutions of $-2x^2 + 3x + 1 = 0$. −0.28; 1.78
3. Verify that −4 is a solution of the equation $x^2 + 5x + 4 = 0$. $(-4)^2 + 5(-4) + 4 = 16 - 20 + 4 = 0$
4. Rewrite the equation $-6x^2 + 3x = -2$ in the form $ax^2 + bx + c = 0$. $-6x^2 + 3x + 2 = 0$
5. Give the x-intercepts of the related graph for Ex. 3. −4; −1

SECTION

8.5 The Quadratic Formula

Learn how to...

- **use the quadratic formula**

So you can...

- **solve quadratic equations without using a graph**
- **find more precise solutions to real-world problems**

Sometimes you want a more precise solution of a quadratic equation than you can get from a graph. You can use the *quadratic formula* to find exact solutions of quadratic equations in the form $ax^2 + bx + c = 0$, when $a \neq 0$:

$$x = \frac{-b + \sqrt{b^2 - 4ac}}{2a} \quad \text{or} \quad x = \frac{-b - \sqrt{b^2 - 4ac}}{2a}$$

The **quadratic formula** is usually written in this form:

$$x = \frac{-b \pm \sqrt{b^2 - 4ac}}{2a}$$

Example 1 shows you how to use the quadratic formula to find the values of x that make an equation in the form $ax^2 + bx + c = 0$ true.

EXAMPLE 1

Solve the equation $2x^2 + 6x - 8 = 0$.

SOLUTION

Find the values of a, b, and c: $2x^2 + 6x - 8 = 0$.

Substitute 2 for a, 6 for b, and −8 for c in the quadratic formula.

$$\begin{aligned} x &= \frac{-b \pm \sqrt{b^2 - 4ac}}{2a} \\ &= \frac{-6 \pm \sqrt{6^2 - 4(2)(-8)}}{2(2)} \\ &= \frac{-6 \pm \sqrt{100}}{4} \\ &= \frac{-6 \pm 10}{4} \end{aligned}$$

There are two exact solutions.

$$x = \frac{-6 + 10}{4} = \frac{4}{4} = 1 \quad \text{or} \quad x = \frac{-6 - 10}{4} = \frac{-16}{4} = -4$$

The solutions are 1 and −4.

Check

Substitute each solution, 1 and −4, in the equation.

$$2x^2 + 6x - 8 = 0$$
$$2(1)^2 + 6(1) - 8 \stackrel{?}{=} 0$$
$$2 + 6 - 8 \stackrel{?}{=} 0$$
$$0 = 0 \checkmark$$

$$2(-4)^2 + 6(-4) - 8 \stackrel{?}{=} 0$$
$$32 - 24 - 8 \stackrel{?}{=} 0$$
$$0 = 0 \checkmark$$

THINK AND COMMUNICATE

1. What do you notice about the two solutions of the equation $2x^2 + 6x - 8 = 0$ from Example 1 and the graph of $y = 2x^2 + 6x - 8$ at the right?

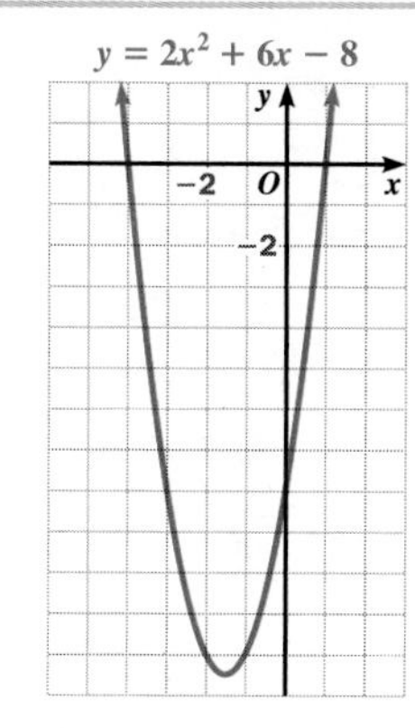

2. How would the answer to Example 1 be different if the equation were $2x^2 - 6x - 8 = 0$ instead of $2x^2 + 6x - 8 = 0$?

When you use the quadratic formula, you can write your answer with a radical sign or find a decimal approximation.

EXAMPLE 2 Application: Sports

A diver jumps from a platform which is 10 m above the water. Her height h in meters above the water t seconds after she jumps is modeled by the equation $h = -4.9t^2 + 3.5t + 10.8$. About how long will it take her to reach the water?

SOLUTION

The diver's height when she reaches the water is 0 m. Use the quadratic formula to solve the equation $0 = -4.9t^2 + 3.5t + 10.8$.

$$t = \frac{-b \pm \sqrt{b^2 - 4ac}}{2a}$$

$$= \frac{-3.5 \pm \sqrt{(3.5)^2 - 4(-4.9)(10.8)}}{2(-4.9)}$$

Substitute **−4.9** for *a*, **3.5** for *b*, and **10.8** for *c*.

$$= \frac{-3.5 \pm \sqrt{223.93}}{-9.8}$$

There are two solutions.

$$t = \frac{-3.5 + \sqrt{223.93}}{-9.8} \approx -1.17 \quad \text{or} \quad t = \frac{-3.5 - \sqrt{223.93}}{-9.8} \approx 1.88$$

Since the time cannot be negative, −1.17 is not a realistic answer.

It will take the diver about 1.9 seconds to reach the water.

BY THE WAY

Fu Mingxia, a member of the Chinese Olympic team, won the gold medal in women's platform diving at the 1992 summer Olympics. She won another gold medal at a world competition in 1994.

ANSWERS Section 8.5

Think and Communicate

1. The x-intercepts of the graph are the solutions of the equation.
2. The only thing that changes in the quadratic formula is the sign of b; the solutions are $\frac{6+10}{4} = 4$ and $\frac{6-10}{4} = -1$. These solutions are the opposites of the solutions of Example 1.

Teach⇔Interact

Section Notes

Communication: Listening
Read the short form of the quadratic formula out loud. Mention that the symbol "±" is read "plus or minus." Emphasize that the short form is used for convenience only. The formula still represents two solutions, not one.

Teaching Tip
Emphasize that regardless of the order of the terms in a quadratic equation, in the quadratic formula the coefficient of the square term is always *a* and the coefficient of the first power term is always *b*.

Additional Example 1

Solve the equation $3x^2 + 6x - 24 = 0$.

Substitute 3 for *a*, 6 for *b*, and −24 for *c* in the quadratic formula.

$$x = \frac{-b \pm \sqrt{b^2 - 4ac}}{2a} = \frac{-6 \pm \sqrt{(6)^2 - 4(3)(-24)}}{2(3)} = \frac{-6 \pm \sqrt{324}}{6}$$

There are two exact solutions.

$$x = \frac{-6 + 18}{6} = \frac{12}{6} = 2 \text{ or}$$

$$x = \frac{-6 - 18}{6} = \frac{-24}{6} = -4$$

The solutions are 2 and −4.

Additional Example 2

A ball is dropped from a building 30 m tall. The height h of the ball in meters above the ground t seconds after it is dropped is modeled by the equation $h = -4.9t^2 + 2.8t + 35$. About how long will it take the ball to reach the ground?

The ball's height when it reaches the ground is 0 m. Use the quadratic formula to solve the equation $0 = -4.9t^2 + 2.8t + 35$.

$$t = \frac{-b \pm \sqrt{b^2 - 4ac}}{2a} = \frac{-2.8 \pm \sqrt{(2.8)^2 - 4(-4.9)(35)}}{2(-4.9)} = \frac{-2.8 \pm \sqrt{693.84}}{-9.8}$$

$$t = \frac{-2.8 \pm \sqrt{693.84}}{-9.8} \approx -2.40 \text{ or}$$

$$t = \frac{-2.8 \pm \sqrt{693.84}}{-9.8} \approx 2.97$$

Since the time cannot be negative, −2.40 is not a realistic answer. It will take the ball about 3 seconds to reach the ground.

Teach⇔Interact

Additional Example 3

Solve the equation
$x^2 + 6x + 2 = -7$.
Rewrite the equation in the form
$ax^2 + bx + c = 0$.
$x^2 + 6x + 2 = -7$
$x^2 + 6x + 9 = 0$
Substitute 1 for a, 6 for b, and 9 for c in the quadratic formula.

$$x = \frac{-b \pm \sqrt{b^2 - 4ac}}{2a}$$
$$= \frac{-6 \pm \sqrt{(6)^2 - 4(1)(9)}}{2(1)}$$
$$= \frac{-6 \pm \sqrt{36 - 36}}{2}$$
$$= \frac{-6 \pm \sqrt{0}}{2}$$

There is only one solution because $\sqrt{0} = 0$.
$\frac{-6}{2} = -3$
The solution is -3.

Checking Key Concepts

Integrating the Strands
Question 1 helps students relate an algebraic solution to a quadratic equation to its geometric representation as the x-intercept of a graph in the coordinate plane.

Closure Question

What steps would you follow to solve a quadratic equation using the quadratic formula?
If necessary, first rewrite the quadratic equation in the form $ax^2 + bx + c = 0$. Then identify the values of a, b, and c. Substitute these values into the formula $x = \frac{-b \pm \sqrt{b^2 - 4ac}}{2a}$ and simplify.

EXAMPLE 3

Solve the equation $x^2 - 8x + 18 = 2$.

SOLUTION

Rewrite the equation in the form $ax^2 + bx + c = 0$.

$$x^2 - 8x + 18 = 2$$
$$x^2 - 8x + 16 = 0$$

Substitute 1 for a, -8 for b, and 16 for c in the quadratic formula.

$$x = \frac{-b \pm \sqrt{b^2 - 4ac}}{2a}$$
$$= \frac{-(-8) \pm \sqrt{(-8)^2 - 4(1)(16)}}{2(1)}$$
$$= \frac{8 \pm \sqrt{64 - 64}}{2}$$
$$= \frac{8 \pm \sqrt{0}}{2}$$
$$= \frac{8}{2}$$
$$= 4$$

There is only one solution because $\sqrt{0} = 0$.

The solution is 4.

CHECKING KEY CONCEPTS

1. What does the graph of the equation $y = x^2 - 8x + 16$ at the right tell you about the solution of the equation $x^2 - 8x + 16 = 0$ from Example 3? Explain.

$y = x^2 - 8x + 16$

For each equation in the form $ax^2 + bx + c = 0$, identify the values of a, b, and c.

2. $x^2 + 10x + 25 = 0$
3. $x^2 - 3x - 10 = 0$
4. $5n^2 - n + 1 = 0$
5. $35t^2 + 3t = 0$

Rewrite each equation in the form $ax^2 + bx + c = 0$. Then solve.

6. $k^2 + 4k - 2 = 3$
7. $4x^2 - 2x = 1$
8. $n^2 - 6 = 4n$
9. $4.9t^2 = 7t$

Checking Key Concepts

1. There is only one x-intercept so the equation has only one solution.
2. 1, 10, 25
3. 1, −3, −10
4. 5, −1, 1
5. 35, 3, 0
6. $k^2 + 4k - 5 = 0$; −5, 1
7. $4x^2 - 2x - 1 = 0$; $\frac{1 \pm \sqrt{5}}{4}$, or about 0.81 and about −0.31
8. $n^2 - 4n - 6 = 0$; $2 \pm \sqrt{10}$, or about 5.16 and about −1.16
9. $4.9t^2 - 7t = 0$; 0, $\frac{10}{7}$

Exercises and Applications

1. −3, 6
2. 1
3. $\frac{1}{3}$, 3
4. −1, $\frac{1}{5}$
5. 2, $-\frac{1}{5}$
6. 0, $\frac{5}{2}$
7. Answers may vary. An example is given.
Dear Jessie, The symbol "±" means "plus or minus." So you actually found two solutions, $\frac{6 + \sqrt{40}}{2}$ and $\frac{6 - \sqrt{40}}{2}$. If you draw the graph of $y = x^2 - 6x - 1$, you will see that it has two x-intercepts, so the related equation $x^2 - 6x - 1 = 0$ has two solutions.

8.5 Exercises and Applications

Solve each equation using the quadratic formula.

1. $x^2 - 3x - 18 = 0$
2. $y^2 - 2y + 1 = 0$
3. $3n^2 - 10n + 3 = 0$
4. $5p^2 + 4p - 1 = 0$
5. $-5x^2 + 9x + 2 = 0$
6. $2t^2 - 5t = 0$

Extra Practice *exercises on page 573*

7. **Writing** Jessie solved the equation $x^2 - 6x - 1 = 0$. When someone asked her how many solutions she found, she said there was only one: $\frac{6 \pm \sqrt{40}}{2}$. Write a letter to Jessie explaining why there are two solutions. Use a graph to make your explanation clearer.

Technology For Exercises 8–16, solve each equation using the quadratic formula. Then check each solution using a graphing calculator.

8. $y^2 + 12y + 10 = 0$
9. $4t^2 - 16t = 0$
10. $3n^2 - 9 = 0$
11. $-x^2 + 3x + 3 = 5$
12. $2n^2 + 9n + 4 = 1$
13. $-6y^2 + 2y = -3$
14. $3t^2 - 4t = 6$
15. $8y^2 + 2 = 10y$
16. $3x^2 = 4x + 5$

17. **FIREFIGHTING** A firefighter aims a hose at a window 25 ft above the ground. The equation $h = -0.01d^2 + 1.06d + 5$ models the path of the water, when h equals height in feet. Estimate, to the nearest foot, the horizontal distance d in feet between the firefighter and the building.

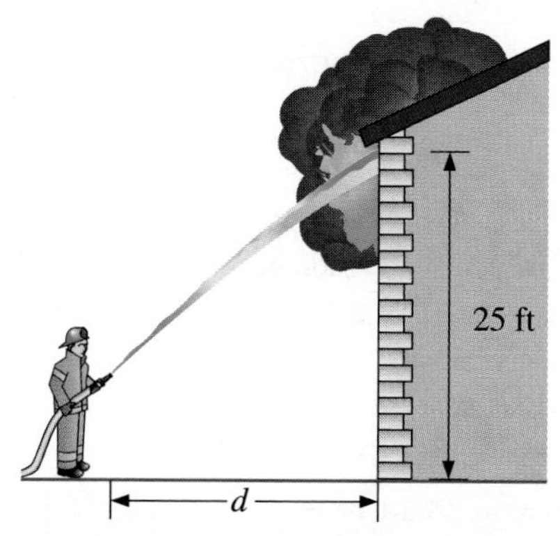

BY THE WAY

The first volunteer fire department in the American Colonies was founded in 1736 by Benjamin Franklin.

18. **Open-ended Problem** You know how to solve quadratic equations using algebra, graphs, and the quadratic formula. Sometimes one method of solving is more convenient than another method. Describe how you would solve each equation. Give reasons for your answers.
    a. $16x^2 = 1600$
    b. $x^2 - 350 = 0$
    c. $x^2 + 10x = 0$
19. **Challenge** Solve the equation $(x + 2)(x - 3) = 14$.
20. **SAT/ACT Preview** Which two numbers are solutions of the equation $x^2 - 6x + 8 = 0$?
    A. 2 and -3
    B. 2 and -4
    C. -2 and 4
    D. 2 and 4

Apply⇔Assess

Suggested Assignment

Core Course
Exs. 1–6, 8–18, 20–33

Extended Course
Exs. 1–10, 14–33

Block Schedule
Day 52 Exs. 1–6, 8–33

Exercise Notes

Common Error
Ex. 6 Some students may not realize that they should use 0 as a replacement for c when they apply the quadratic formula in this exercise. If they use some other value, such as 1 for example, have them relate the equation $2t^2 - 5t = 0$ to $ax^2 + bx + c = 0$ and identify the values for a, b, and c.

Communication: Writing
Ex. 7 This exercise will help students to understand the shortened notation used with the quadratic formula. You may wish to have one student read his or her letter to the class.

Using Technology
Exs. 8–16 Here is a program for the TI-82 that calculates real number roots of quadratic equations. The values A, B, and C correspond to the coefficients a, b, and c in $ax^2 + bx + c = 0$. If there are no real solutions, the program will display a message to that effect.

```
PROGRAM:QROOT
:Lbl 1
:Disp "A"
:Input A
:Disp "B"
:Input B
:Disp "C"
:Input C
:B²–4AC→D
:IF D< 0
:Goto 2
:√D→M
:(–B+M)/2A→E
:(–B–M)/2A→F
:Disp "SOLUTIONS ARE"
:Disp E
:Disp F
:Goto 1
:Lbl 2
:Disp "NO REAL SOLUTIONS"
```

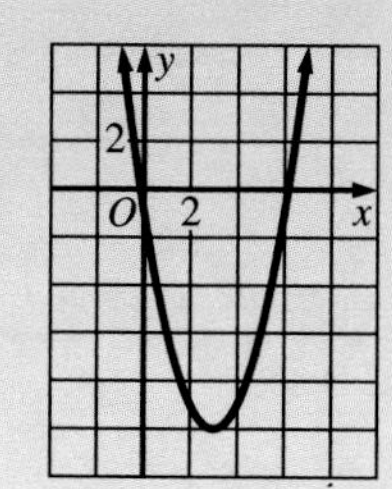

8. $-6 \pm \sqrt{26}$, or about -0.9 and about -11.1
9. 0, 4
10. $\pm\sqrt{3}$, or about 1.73 and about -1.73
11. 1, 2
12. $\frac{-9 \pm \sqrt{57}}{4}$, or about -0.36 and about -4.14
13. $\frac{1 \pm \sqrt{19}}{6}$, or about 0.89 and about -0.56
14. $\frac{2 \pm \sqrt{22}}{3}$, or about -0.90 and about 2.23
15. $1, \frac{1}{4}$
16. $\frac{2 \pm \sqrt{19}}{3}$, or about -0.79 and about 2.12
17. about 20 ft
18. Answers may vary. Examples are given.
    a. algebra
    b. graph
    c. graph
19. 5, -4
20. D

Apply⇔Assess

Exercise Notes

Multicultural Note

Exs. 21–23 The national soccer team of Brazil is usually considered the most formidable in the world. The only country to win four World Cup championships, Brazil achieved two of its victories back to back in 1958 and 1962. Until recently South American and European teams dominated world soccer. Now Asian and African teams are proving strong opponents. Nigeria, Saudi Arabia, and South Korea are some of the teams that qualified for the first time ever for the World Cup in 1994.

Assessment Note

Ex. 24 This exercise will show clearly how well students have understood the ideas in this section. You may wish to call upon individual students to explain their answers to the class.

Practice 55 for Section 8.5

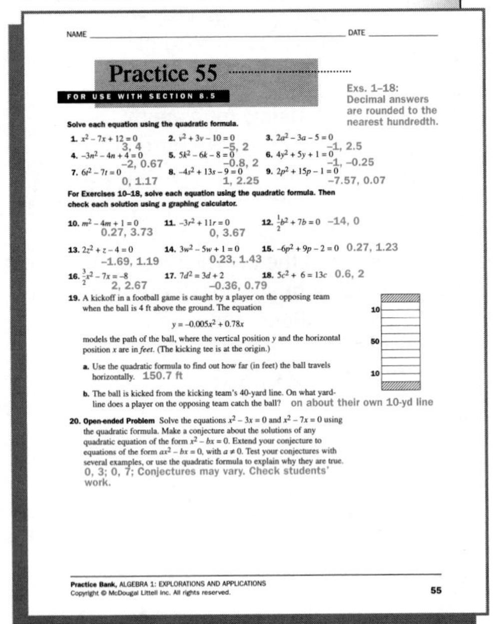

NAME ____________ DATE ______

Practice 55

FOR USE WITH SECTION 8.5

Exs. 1–18: Decimal answers are rounded to the nearest hundredth.

Solve each equation using the quadratic formula.

1. $x^2 - 7x + 12 = 0$ 3, 4
2. $v^2 + 3v - 10 = 0$ −5, 2
3. $2a^2 - 3a - 5 = 0$ −1, 2.5
4. $-3n^2 - 4n + 4 = 0$ −2, 0.67
5. $5k^2 - 6k - 8 = 0$ −0.8, 2
6. $4y^2 + 5y + 1 = 0$ −1, −0.25
7. $6t^2 - 7t = 0$ 0, 1.17
8. $-4s^2 + 13s - 9 = 0$ 1, 2.25
9. $2p^2 + 15p - 1 = 0$ −7.57, 0.07

For Exercises 10–18, solve each equation using the quadratic formula. Then check each solution using a graphing calculator.

10. $m^2 - 4m + 1 = 0$ 0.27, 3.73
11. $-3r^2 + 11r = 0$ 0, 3.67
12. $\frac{1}{2}b^2 + 7b = 0$ −14, 0
13. $2z^2 + z - 4 = 0$ −1.69, 1.19
14. $3w^2 - 5w + 1 = 0$ 0.23, 1.43
15. $-6p^2 + 9p - 2 = 0$ 0.27, 1.23
16. $\frac{3}{2}x^2 - 7x = -8$ 2, 2.67
17. $7d^2 = 3d + 2$ −0.36, 0.79
18. $5c^2 + 6 = 13c$ 0.6, 2
19. A kickoff in a football game is caught by a player on the opposing team when the ball is 4 ft above the ground. The equation

$$y = -0.005x^2 + 0.78x$$

models the path of the ball, where the vertical position y and the horizontal position x are in *feet*. (The kicking tee is at the origin.)

a. Use the quadratic formula to find out how far (in feet) the ball travels horizontally. 150.7 ft

b. The ball is kicked from the kicking team's 40-yard line. On what yard-line does a player on the opposing team catch the ball? on about their own 10-yd line

20. **Open-ended Problem** Solve the equations $x^2 - 3x = 0$ and $x^2 - 7x = 0$ using the quadratic formula. Make a conjecture about the solutions of any quadratic equation of the form $x^2 - bx = 0$. Extend your conjecture to equations of the form $ax^2 - bx = 0$, with $a \neq 0$. Test your conjectures with several examples, or use the quadratic formula to explain why they are true. 0, 3; 0, 7; Conjectures may vary. Check students' work.

Practice Bank, ALGEBRA 1: EXPLORATIONS AND APPLICATIONS

55

Connection SPORTS

In 1994, Brazil won the World Cup soccer tournament by defeating the Italian team 3 to 2. The championship game was decided by a round of penalty kicks, since neither team had scored during 90 min of regulation play and 30 min of overtime. The Italian team overshot the goal twice, sending the ball several feet over the goal's crossbar.

A player takes a penalty kick. The equation $h = -0.002d^2 + 0.36d$ models the height h in feet of the ball at distance d in feet along the ground from where the ball is kicked. Use this equation for Exercises 21–23.

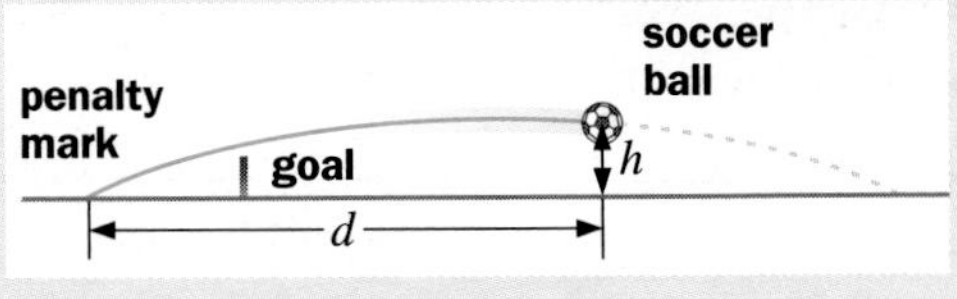

21. Assume that the ball overshoots the goal and doesn't hit any obstacles before it lands on the ground. About how far will the ball go horizontally before it lands?

22. The penalty mark is 12 yd from the goal. The goal's crossbar is 8 ft high. By about how many feet does the ball go over the crossbar? (Remember that 3 ft = 1 yd.)

23. To the nearest foot, what is the ball's horizontal distance from the goal when it is 8 ft high?

BY THE WAY

As of the 1994 tournament, Brazil was the only team to win four World Cups and to qualify for every World Cup tournament.

ONGOING ASSESSMENT

24. Writing Explain why it is possible to solve the equation $3x^2 + 12 = 0$ algebraically, without using a graph or the quadratic formula, but not the equation $3x^2 - 14x + 12 = 0$.

SPIRAL REVIEW

Tell whether each number is *rational* or *irrational*. (Section 6.2)

25. $\sqrt{161}$ **26.** $\sqrt{121}$ **27.** $\sqrt{109}$

Use substitution to solve each system of equations. (Section 7.2)

28. $y = 2x - 1$
$3x + 5y = 34$

29. $y = 5x - 6$
$2x - 7y = 9$

30. $x = 3y - 6$
$-7x - 5y = -10$

31. $a = \frac{4}{3}b$
$15a - 4b = 4$

32. $p = 1$
$7p - q = 6$

33. $y = 5x - 32$
$3x - 4y = -59$

21. 180 ft
22. The ball goes about 2.4 ft over the crossbar.
23. The ball achieves a height of 8 ft twice, once about 10 ft before reaching the goal, and again about 118 ft on the far side of the goal.
24. You can solve the equation $3x^2 + 12 = 0$ algebraically because it is possible to isolate the x^2-term and then find the square root of both sides. This is not the case with the equation $3x^2 - 14x + 12 = 0$ because of the x-term.
25. irrational
26. rational
27. irrational
28. (3, 5)
29. (1, −1)
30. (0, 2)
31. $\left(\frac{1}{3}, \frac{1}{4}\right)$
32. (1, 1)
33. (11, 23)

SECTION

8.6 Using the Discriminant

Learn how to...
- find the discriminant of a quadratic equation

So you can...
- find the number of solutions of a quadratic equation

The value of the expression $b^2 - 4ac$ in the quadratic formula is called the **discriminant**.

$$x = \frac{-b \pm \sqrt{b^2 - 4ac}}{2a}$$

The discriminant gives you information about the number of possible solutions when you solve a quadratic equation.

EXPLORATION
COOPERATIVE LEARNING

Solutions of Quadratic Equations

Work in a group of 2 or 3 students.
You will need:
- a graphing calculator

1 Take turns graphing the following equations. Use a table like the one shown to record the number of solutions of each equation when $y = 0$.

a. $y = -3x^2 + 5x + 5$
b. $y = 5x^2 + 3x + 2$
c. $y = 6x^2 + 12x + 6$
d. $y = x^2 + 8x - 3$
e. $y = x^2 - 2x + 4$
f. $y = x^2 + 4x + 4$

Equation	Number of solutions	$b^2 - 4ac$
$0 = -3x^2 + 5x + 5$	?	?
$0 = 5x^2 + 3x + 2$	?	?
$0 = 6x^2 + 12x + 6$	?	?
$0 = x^2 + 8x - 3$	?	?
$0 = x^2 - 2x + 4$	?	?
$0 = x^2 + 4x + 4$	?	?

2 Find the value of $b^2 - 4ac$ for each equation in Step 1. Record the results in your table.

3 Work together to make a generalization about the value of the discriminant and the number of solutions of an equation.

Exploration Note

Purpose
The purpose of this Exploration is to have students discover how the value of the discriminant of a quadratic equation can be used to predict the number of solutions to the equation.

Materials/Preparation
Graphing calculators are needed by each group.

Procedure
Students graph various quadratic equations to find the number of solutions (x-intercepts) of each equation. They then find the value of the discriminant for each equation. Students record their results in a table and use them to make a generalization about how the value of the discriminant is related to the number of solutions to the equation.

Closure
Students should conclude that if the discriminant is positive, there are 2 solutions; if the discriminant is 0, there is one solution; and if it is negative, there are no solutions.

Explorations Lab Manual
See the Manual for more commentary on this Exploration.

Diagram Master 14

For answers to the Exploration, see following page.

Plan⇔Support

Objectives

- Find the discriminant of a quadratic equation.
- Find the number of solutions of a quadratic equation.

Recommended Pacing

Core and Extended Courses
Section 8.6: 2 days

Block Schedule
Section 8.6: 1 block

Resource Materials

Lesson Support
Lesson Plan 8.6
Warm-Up Transparency 8.6
Practice Bank: Practice 56
Study Guide: Section 8.6
Explorations Lab Manual: Diagram Masters 2, 14
Assessment Book: Test 35

Technology
Graphing Calculator

Internet: http://www.hmco.com

Warm-Up Exercises

Solve each equation.

1. $x^2 + 6x + 10 = 1$ -3
2. $x^2 + 6x + 5 = 0$ $-1; -5$
3. How many x-intercepts does the related graph for Exercise 2 have? 2
4. Find the number of solutions of the equation $x^2 + 6x + 3 = 0$. 2
5. Find the number of solutions of the equation $x^2 + 4x + 5 = 1$. 1

Teach⇔Interact

Learning Styles: Visual

The use of graphs to illustrate algebraic concepts continues to support those students who may be primarily visual learners. However, all learners can also benefit from this approach.

Section Notes

Visual Thinking
Ask students to examine the three graphs and discuss them in relation to the x-axis. Have them answer questions such as, Which one has two x-intercepts? What are they? Which one has no x-intercepts? What does this mean? Is it possible to have no y-intercepts? This activity involves the visual skills of *exploration* and *interpretation*.

Second-Language Learners
Discuss the meaning of *discrimination* with students in terms of making a distinction. Have students develop an understanding of the term *discriminant* in algebra as an expression that makes a distinction among the number of roots of a quadratic equation.

Additional Example

How many solutions are there of the equation $-3x^2 - 4x + 7 = 4$?

In order to use the discriminant, the equation must be in the form $ax^2 + bx + c = 0$.

$-3x^2 - 4x + 7 = 4$

$-3x^2 - 4x + 3 = 0$

Now you can identify a as -3, b as -4, and c as 3. To find the discriminant, substitute these values in the expression $b^2 - 4ac$.

$b^2 - 4ac = (-4)^2 - 4(-3)(3) = 52$

The discriminant is positive, so there are two solutions.

Section Notes

Teaching Tip
Students can use calculators to verify that the square root of a negative number does not exist. The calculator will give them an error message.

Reasoning
Ask students why it would be worthwhile to find the discriminant of a quadratic equation before attempting to solve the equation. (If the discriminant is negative, there is no further work to do. If it is not negative, you apply the formula but do not need to recalculate the value of $b^2 - 4ac$.)

The Discriminant

The discriminant tells you how many real number solutions there are of an equation in the form $ax^2 + bx + c = 0$, when $a \neq 0$.

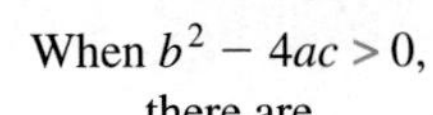

When $b^2 - 4ac = 0$, there is exactly **one** solution.	When $b^2 - 4ac > 0$, there are **two** solutions.	When $b^2 - 4ac < 0$, there are **no** solutions.
The graph of $y = ax^2 + bx + c$ has **one** x-intercept.	The graph of $y = ax^2 + bx + c$ has **two** x-intercepts.	The graph of $y = ax^2 + bx + c$ has **no** x-intercepts.

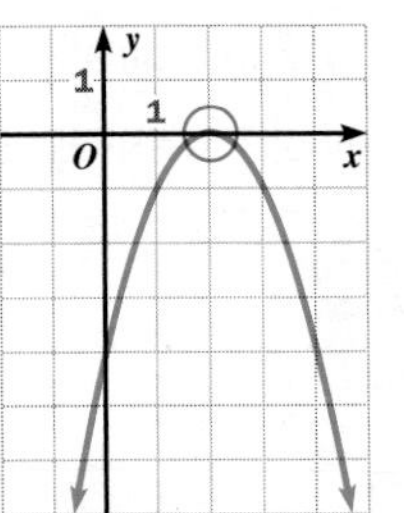

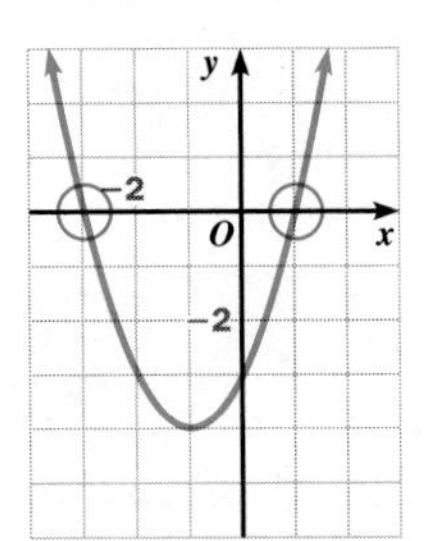

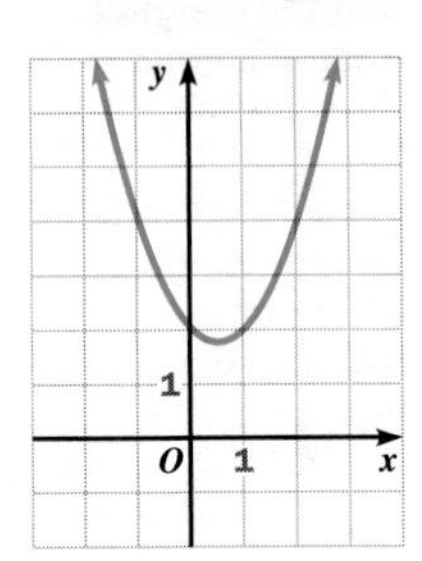

EXAMPLE

How many solutions are there of the equation $-3x^2 - 6x - 1 = 3$?

SOLUTION

In order to use the discriminant, the equation must be in the form $ax^2 + bx + c = 0$.

$$-3x^2 - 6x - 1 = 3$$
$$-3x^2 - 6x - 4 = 0$$

Subtract 3 from both sides of the equation.

Now you can identify a as -3, b as -6, and c as -4. To find the discriminant, substitute these values in the expression $b^2 - 4ac$.

$$b^2 - 4ac = (-6)^2 - 4(-3)(-4) = -12$$

The discriminant is negative, so there are no solutions.

When you solve the equation $-3x^2 - 6x - 4 = 0$ using the quadratic formula, this is what you get:

$$x = \frac{-(-6) \pm \sqrt{(-6)^2 - 4(-3)(-4)}}{2(-3)} = \frac{6 \pm \sqrt{-12}}{-6}$$

Because the square root of a negative number does not exist in the real numbers, there are no solutions.

ANSWERS Section 8.6

Exploration

1, 2.

Equation	Number of solutions	$b^2 - 4ac$
$0 = -3x^2 + 5x + 5$	2	85
$0 = 5x^2 + 3x + 2$	0	–31
$0 = 6x^2 + 12x + 6$	1	0
$0 = x^2 + 8x - 3$	2	76
$0 = x^2 - 2x + 4$	0	–12
$0 = x^2 + 4x + 4$	1	0

3. If the discriminant is positive, the equation has two solutions. If the discriminant is zero, the equation has one solution. If the discriminant is negative, the equation has no solutions.

Think and Communicate

1. Since the graph lies entirely below the x-axis, it has no x-intercepts. There are no solutions when $y = 0$.

THINK AND COMMUNICATE

1. **Technology** Graph the equation $y = -3x^2 - 6x - 4$. What does the graph tell you about the number of solutions when $y = 0$?
2. Describe how you would find the number of solutions of $-3x^2 - 6x + 4 = 7$. How could you check your answer?

CHECKING KEY CONCEPTS

Find the discriminant for each equation. What does it tell you about the number of solutions?

1. $x^2 - 3x + 2 = 0$
2. $5n^2 + 9n + 6 = 0$
3. $3y^2 + 12 = 12y$
4. $-12k^2 + 5k - 9 = 0$
5. $4x^2 - 4x + 1 = 0$
6. $2t^2 - 7t - 4 = 0$

The density of water changes depending on its temperature. The equation $d = -0.0069T^2 + 0.0549T + 999.8266$ models the relationship between the density of water d in mg/cm^3 and its temperature T in °C.

7. For the equation $-0.0069T^2 + 0.0549T + 999.8266 = 0$, identify the values of a, b, and c that you would use to find the discriminant.
8. How many solutions are there when $d = 999.90$? when $d = 1000$?
9. Water freezes at a temperature of 0°C. What is the density of water at this temperature?

BY THE WAY

Water is one of the few substances which is lighter in its solid form than in its liquid form. This is why ice floats.

8.6 Exercises and Applications

Extra Practice exercises on page 574

For each equation, find the number of solutions.

1. $x^2 - 9x + 14 = 0$
2. $6y^2 - 8y + 3 = 0$
3. $2p^2 - 13p - 3 = 0$
4. $0 = 6n^2 - n + 5$
5. $-3x^2 - 5x + 1 = 0$
6. $25k^2 - 10k = 0$
7. $2y^2 + 10 = 0$
8. $n^2 + 49 = 14n$
9. $x^2 = 64$
10. **GEOMETRY** Use the figures shown.
    a. Write an equation for the combined area of the two figures as a function of s.
    b. Find the discriminant of the equation you wrote in part (a) if the combined area is 48. What does this tell you about the number of possible solutions?
    c. Solve your equation using the quadratic formula. How many of the solutions make sense in this situation?

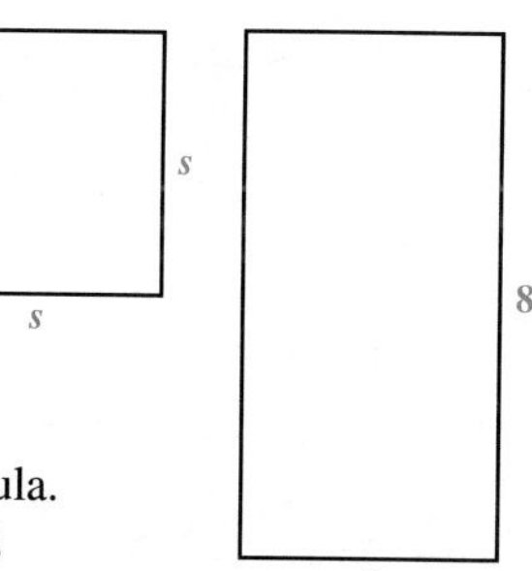

Teach⇔Interact

Checking Key Concepts

Using Technology
Students may check their answers to questions 1–6 by graphing. Remind them to write the equations in the form $y = ax^2 + bx + c$ first.

Closure Question

Describe how to determine the number of solutions to a quadratic equation.

Determine the value of a, b, and c for a quadratic equation in the form $ax^2 + bx + c = 0$. Substitute a, b, and c into the discriminant, $b^2 - 4ac$. If the discriminant is negative, there are no real solutions. If it is 0, there is one solution, and if it is positive, there are two real solutions.

Apply⇔Assess

Suggested Assignment

Core Course
Day 1 Exs. 1–13, 15–20
Day 2 Exs. 21–23, 27–38, AYP

Extended Course
Day 1 Exs. 1–20
Day 2 Exs. 21–38, AYP

Block Schedule
Day 53 Exs. 1–13, 15–23, 27–38, AYP

Exercise Notes

Using Technology
Exs. 1–9 Students can run the following TI-82 program to calculate the discriminant of a quadratic equation in the form $ax^2 + bx + c = 0$.

```
PROGRAM:DISC
:Input "INPUT A", A
:Input "INPUT B", B
:Input "INPUT C", C
:B²–4AC→D
:Disp "DISCRIMINANT"
:Disp D
```

2. First put the equation in the form $ax^2 + bx + c = 0$. Then evaluate the discriminant. Since the discriminant is 0, the equation has one solution. Checking methods may vary. An example is given. Graph the related equation $y = -3x^2 - 6x - 3$ to find the number of x-intercepts.

Checking Key Concepts

1–9. See answers in back of book.

Exercises and Applications

1. two solutions
2. no solutions
3. two solutions
4. no solutions
5. two solutions
6. two solutions
7. no solutions
8. one solution
9. two solutions
10. See answers in back of book.

Apply⇔Assess

Exercise Notes

Interdisciplinary Problems

Exs. 11–13 Engineering is an applied science that depends heavily upon mathematics to express its concepts. Students who major in engineering in college take courses in calculus and very often in differential equations. Many engineering problems, however, such as those given here, can be analyzed and solved using concepts from elementary algebra, geometry, and trigonometry.

Career Connection

Exs. 11–13 An engineer applies science to designing, creating, and using structures and machines. Civil engineers design static structures like bridges, buildings, and dams. Mechanical engineers build dynamic systems like engines and machinery. Other types of engineers include electrical, chemical, and computer engineers. Engineers are employed throughout the business world and many of them eventually are promoted to executive positions with responsibilities for running major parts of a company.

Research

Ex. 24 Some of the more sophisticated calculators do not give an error message for the square root of a negative number. They give the answer as a complex number in coordinate form. Some students may be interested in researching the meaning of a complex number.

Communication: Discussion

Ex. 26 A class discussion can be used to help students learn how other students interpret the solutions to real-world problems.

Connection ENGINEERING

The Water Arc is a fountain that shoots recirculated water across the Chicago River from a water cannon. The path of the Water Arc can be modeled by the equation $y = -0.006x^2 + 1.2x + 10$, where x is the horizontal distance in feet and y is the vertical distance in feet.

The water cannon is about 10 ft above the river.

11. How many solutions of the equation are there when $y = 0$? Find each solution using the quadratic formula.

12. How many solutions of the equation are there when $y = 70$? Find each solution using the quadratic formula.

13. Use the results of Exercises 11 and 12 to sketch the path of the Water Arc in a coordinate plane. Label your graph with key points, including the location of the water cannon.

14. Research The Water Arc was built to commemorate an engineering project which reversed the direction of the Chicago River. Find out about the history of this project.

For each equation, find the number of solutions. Then solve using the quadratic formula.

15. $x^2 + 7x + 6 = 0$ **16.** $y^2 - 5y + 6 = 0$ **17.** $x^2 - 2x + 1 = 0$

18. $n^2 - 8n + 5 = 15$ **19.** $x^2 = 2x + 2$ **20.** $3k^2 - 8k + 5 = 4$

21. $-x^2 - 16 = -8x$ **22.** $3n^2 + 11n = 4$ **23.** $-4x^2 = -12x - 9$

24. Writing Negative numbers do not have square roots in the real numbers. Use what you know about multiplying negative numbers to explain why.

25. Challenge What happens to the equation $0 = ax^2 + bx + c$ when $a = 0$, but $b \neq 0$ and $c \neq 0$? Can you use the discriminant to find the number of solutions? Can you use the quadratic formula to solve the equation? Give an example to support your answer.

26. Open-ended Problem Explain why some real-world problems involving quadratic equations have only one solution, even when the discriminant tells you that there are two. Include examples.

27. SAT/ACT Preview How many solutions does the equation $32 + 2x^2 = 16x$ have?

A. none **B.** one **C.** two **D.** three

11. two solutions; about −8, about 208 (Note that while both are solutions of the equation and can be used to sketch its graph, only the positive solution makes sense as a horizontal distance.)
12. one solution; 100
13.

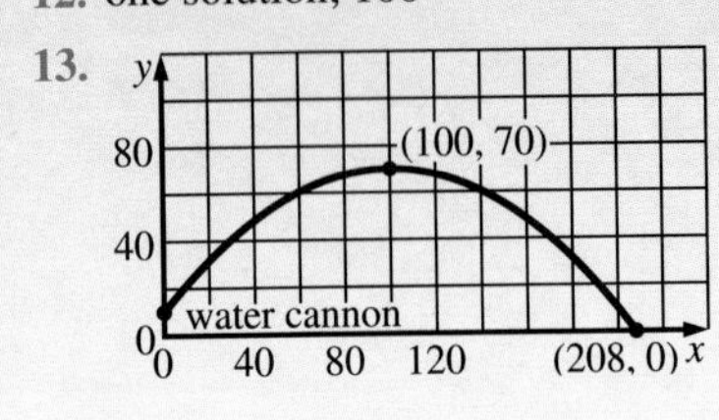

14. The Chicago River originally flowed eastward toward Lake Michigan. The engineering project involved building a series of locks that cause the river to flow westward toward the Des Plaines River.
15. two solutions; −6, −1
16. two solutions; 2, 3
17. one solution; 1
18. two solutions; $4 \pm \sqrt{26}$, or about −1.1 and about 9.1
19. two solutions; $1 \pm \sqrt{3}$, or about −0.73 and about 2.73
20. two solutions; $\frac{4 \pm \sqrt{13}}{3}$, or about 0.13 and about 2.54
21. one solution; 4
22. two solutions; $-4, \frac{1}{3}$
23. two solutions; $\frac{3 \pm 3\sqrt{2}}{2}$, or about −0.62 and about 3.62
24. In the real numbers, when you multiply a positive or a negative number by itself you get a positive number. Therefore, a negative number cannot have a square root in the real numbers because no real number times itself is negative.
25. The equation becomes a linear equation. You cannot use the discriminant. Since $b^2 - 4(0)c = b^2$, the equation would have to have

28. SCIENCE An experiment shows that the relationship between a time t that it takes a person to react to a sound and the person's age a in years can be modeled by the equation $t = 0.0051a^2 - 0.3185a + 15.0008$.

(Each unit of time equals $\frac{1}{77}$ second.)

a. Do you think that the graph of this equation has a horizontal intercept? Why or why not? Check using the discriminant.

b. How many solutions are there to the equation when $t = 12$?

ONGOING ASSESSMENT

29. Cooperative Learning Work with a partner.

a. Each of you should make six flash cards with questions on the front and answers on the back. Your questions should cover the discriminant, the quadratic formula, and solutions of a quadratic equation.

b. Take turns holding up your flash cards and quizzing each other. Give each other helpful hints and go over any mistakes together.

SPIRAL REVIEW

Solve each system of equations. *(Section 7.4)*

30. $2x - 3y = -5$; $3x - y = 3$

31. $5x - y = -1$; $3x - 2y = 5$

32. $4x + 9y = 8$; $-10x - 3y = -7$

Solve each equation. *(Section 8.3)*

33. $x^2 - 4 = 0$ **34.** $4y^2 - 16 = 0$ **35.** $z^2 - 1 = 0$

36. $2 = r^2 - 7$ **37.** $z^2 - 5 = 0$ **38.** $2x^2 - 3 = 0$

ASSESS YOUR PROGRESS

VOCABULARY

quadratic equation (p. 345)
quadratic formula (p. 350)
discriminant (p. 355)

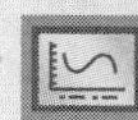

Technology Solve each equation using a graphing calculator. *(Section 8.4)*

1. $x^2 - 5x - 4 = 0$ **2.** $-x^2 + 3x - 2 = 0$ **3.** $x^2 + x + 2 = 0$

Solve each equation using the quadratic formula. *(Section 8.5)*

4. $x^2 - 5x - 24 = 0$ **5.** $2x^2 + x - 15 = 0$ **6.** $2x^2 - x = 2$

Without graphing or solving each equation, find the number of *x*-intercepts of the graph of each equation. *(Section 8.6)*

7. $y = 3x^2 - 4x + 1$ **8.** $y = 3x^2 - 4x + 2$ **9.** $y = 9x^2 - 12x + 4$

10. Journal Summarize the ways you know to solve a quadratic equation.

Apply⇔Assess

Exercise Notes

Assessment Note
Ex. 29 This exercise combines visual and kinesthetic learning styles to help students review this section.

Assess Your Progress

Journal Entry
For Ex. 10, you may wish to ask students to include some real-world examples of the various applications of quadratic equations with their summaries. This will reinforce connections between algebra and real-world problems.

Progress Check 8.4–8.6

See page 363.

Practice 56 for Section 8.6

NAME ______________________ DATE __________

Practice 56

FOR USE WITH SECTION 8.6

For each equation, find the number of solutions.

1. $x^2 + 6x - 3 = 0$ two
2. $3a^2 + 30a + 90 = 0$ none
3. $-2b^2 + 4b - 5 = 0$ none
4. $4n^2 - 7n + 3 = 0$ two
5. $-k^2 + 12k - 36 = 0$ one
6. $0 = 5y^2 + 3y - 10$ two
7. $3p^2 = 30p - 75$ one
8. $6t^2 + 14 = 0$ none
9. $2r^2 + 5 = 12r$ two

10. a. Simplify the equation $(x - k)^2 = 0$ by using a formula that you have learned. $x^2 - 2kx + k^2 = 0$

b. Using the discriminant, explain how you know that, for any given value of k, the equation you got in part (a) has exactly 1 solution. $D = (2k)^2 - 4(1)(k^2) = 4k^2 - 4k^2 = 0$

For each equation, find the number of solutions. Then solve using the quadratic formula.

11. $x^2 - 7x + 10 = 0$ two; 2, 5
12. $2w^2 + 20w + 50 = 0$ one; −5
13. $8c^2 = 17c - 9$ two; 1, 1.13
14. $5y^2 + 10y + 5 = 0$ one; −1
15. $-m^2 + 12m = 0$ two; 0, 12
16. $3x^2 - 7x + 4 = 0$ two; 1, 1.33
17. $-3q^2 + 9q = 5$ two; 0.74, 2.26
18. $7v^2 = 11v + 3$ two; −0.24, 1.81
19. $25x^2 + 1 = 10x$ one; 0.2

20. Tiwa wants to build an enclosure for a rectangular garden. She has 120 ft of fencing. Let x = the width of the garden in feet. Let y = the length of the garden in feet.

a. Suppose Tiwa uses all the fencing she has. Write an equation that relates x and y. Use it to express y in terms of x. $y = 60 - x$

b. Tiwa wants the area of the garden to be 1000 ft². Is it possible to accomplish this with the fencing she has? If it is, find x. Otherwise, explain why it is not possible. no; The discriminant is negative.

c. Can Tiwa enclose a garden with area 850 ft²? Explain. yes; The discriminant is positive.

21. Kyung-Wha is practicing for the shotput event. She has calculated that when she releases the shot at 40 ft/s, the path of the shot is modeled by the equation

$$y = -0.01(1 + m^2)x^2 + mx + 5,$$

where m is the slope of the line along which she aims.

a. Kyung-Wha is trying for a distance of 50 ft. That is, she wants $x = 50$ when $y = 0$. Is there a slope m that will achieve this? yes

b. Can Kyung-Wha find m so that her distance will equal 60 ft? no

56 **Practice Bank,** ALGEBRA 1: EXPLORATIONS AND APPLICATIONS

two solutions, while a linear equation has only one. The quadratic formula cannot be used because it would involve division by 0. Consider the equation $0 = 0x^2 + x + 1$. The "discriminant" is positive, but the equation has only one solution, −1. The quadratic formula yields $x = \frac{-1 \pm 1}{0}$, which is not defined.

26. Some quadratic equations that model real-world situations have both a positive and a negative solution. However, only the positive solution makes sense if the solutions represent quantities that must be positive. Examples include length, width, area, and time.

27. B

28. a. No; it does not make sense that it would take no time at all to react. The discriminant is about −0.205.

b. two solutions (about 11.6, about 50.9)

29. Answers may vary.

30. (2, 3) 31. (−1, −4)

32. $\left(\frac{1}{2}, \frac{2}{3}\right)$ 33. ±2

34. ±2 35. ±1

36. ±3

37. $\pm\sqrt{5}$, or about −2.24 and about 2.24

38. $\pm\sqrt{\frac{3}{2}}$, or about −1.22 and about 1.22

Assess Your Progress

1. about −0.7, about 5.7
2. 1, 2
3. no solutions
4. −3, 8
5. $-3, \frac{5}{2}$
6. $\frac{1 \pm \sqrt{17}}{4}$, or about 1.28 and about −0.7807

7–10. See answers in back of book.

Project Notes

Mathematical Goals

- Collect data on how long it takes a ball to reach the end of a ramp that forms a 10° angle with the floor.
- Make a scatter plot of the data.
- Use the equation $d = 0.61t^2$ to calculate how long it would take a ball to roll 55 cm.

Planning

Materials

- Two metersticks
- Tape
- Solid heavy ball or marble
- Stopwatch
- Graph paper

Project Teams
Students should choose their own partners and collect the materials they need to do the project.

Guiding Students' Work

Suggest that students share all aspects of collecting their data, drawing graphs, and writing the report. Help groups to locate places in the classroom where they can work without interfering with other groups. As the ball rolls down the ramp, one student in each group should be responsible to see that it is stopped after leaving the ramp.

Second-Language Learners
Students learning English may have difficulty understanding the sentence *Since Earth is the biggest thing around, it pulls objects (like skiers) toward it.* Explain to students that since Earth has a greater mass than any other object on it, all other objects are drawn to its surface.

PORTFOLIO PROJECT

Investigating Gravity

Downhill skiers take advantage of steep slopes to ski faster. Gravity enables them to accelerate as they go down the slope. Objects are pulled toward the ground because of gravity. The bigger the object, the greater its pull on other objects. Since Earth is the biggest thing around, it pulls objects (like skiers) toward it.

PROJECT GOAL **Explore how gravity affects the time it takes a ball to roll down a ramp.**

Collecting Data

Work with a partner to recreate an experiment first performed by Galileo Galilei over four hundred years ago. Measure the time it takes a solid heavy ball or marble to roll down a ramp for at least ten different distances. Here are some key steps you should follow:

1. MAKE a ramp by taping two meter sticks together side by side in a "V" shape.

2. PLACE one end of the ramp on a stack of books so that the other end forms a 10° angle with the floor. Check the angle with a protractor.

3. USE a stopwatch to measure the time it takes a ball to reach the end of the ramp. Record the time and the distance for ten starting points.

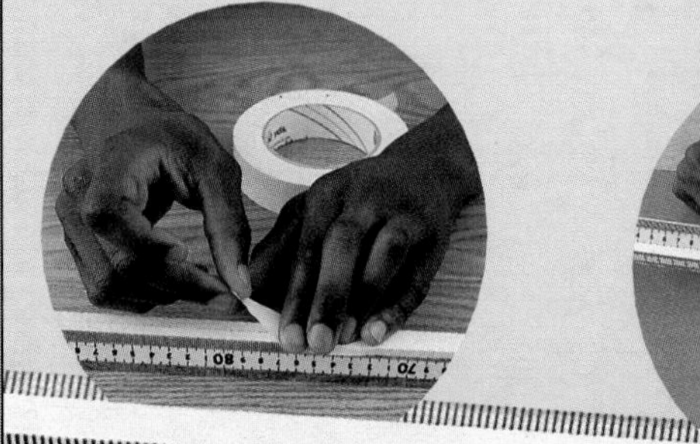

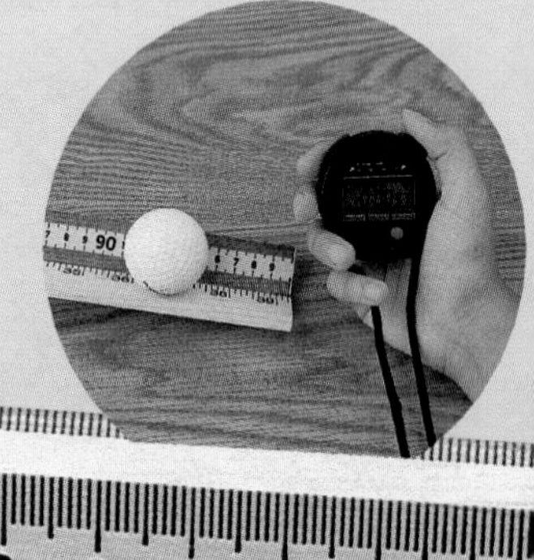

Based on your observations, how long do you think it would take the ball to roll 55 cm? How far would the ball roll in 0.75 s?

Comparing Graphs

Galileo showed that a ball rolling down an inclined plane accelerates at a constant rate. The equation below shows the relationship between the amount of time t in seconds that a solid ball rolls and the distance d it travels along a ramp inclined at a 10° angle:

$$d = 0.61t^2$$

- Calculate how long it would take a ball to roll 55 cm. Compare this value to the value you estimated using your scatter plot. What might explain any differences?

- Sketch a graph of the equation $d = 0.61t^2$ in the same coordinate plane as your scatter plot. How do the graph and the scatter plot compare?

Writing a Report

Write a report about your experiment. Include a brief description of your procedure. Describe the relationship between the distance a ball travels and the time it takes to travel that far.

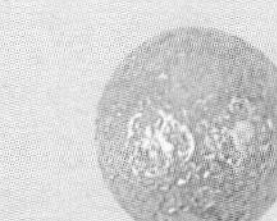

4. MAKE a scatter plot of the data. Put time in seconds on the horizontal axis and distance in meters on the vertical axis.

You may want to extend your report to explore some of the ideas below:

- What affect does doubling the time a ball rolls have on the distance it travels?

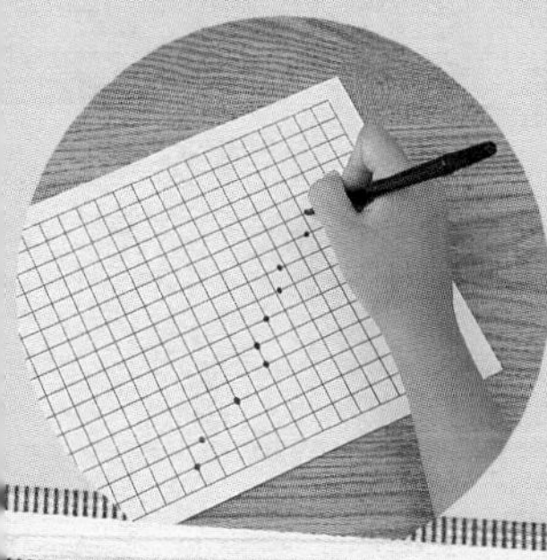

- Will the results of the experiment change if you change the angle of the ramp? If so, how?
- Will the results of the experiment change if you use a lighter or a heavier solid ball? Why or why not?

Self-Assessment

Identify the mathematical skills you used to analyze the relationship you explored in this experiment. What are some possible sources of error in the way you collected your data? Describe how you and your partner worked together or could work together in the future to minimize errors.

Project Notes

Guiding Students' Work

Rubric for Chapter Project

4 Students follow the key steps stated in the text to collect their data. They make an accurate scatter plot of the data, use the formula $d = 0.61t^2$ to calculate how long it would take a ball to roll 55 cm, and compare this value to the one estimated by using their scatter plot. They then sketch a graph of the equation $d = 0.61t^2$ and compare it to their scatter plot. All calculations and graphs are accurate and explanations are clear and correct. A complete and well-written report summarizes the experiment and states the relationship between distance and time correctly.

3 Students collect their data, make the necessary plots, calculations, and graphs, and write a report of their work. Each aspect of the work is done correctly, but the graphs could have been drawn more neatly and the description of the procedure in the report is too brief. Students describe the relationship between distance and time correctly.

2 Students complete the experiment by collecting their data, drawing and comparing graphs, and writing a report. The differences in the estimate value from the scatter plot and the calculated value from the equation $d = 0.61t^2$ for a ball to roll 55 cm are hard to reconcile.

1 Students set up their ramps but do not record the time and distance for ten starting points accurately. The scatter plot of the data is not representative of the real situation. Students have difficulty comparing their results to those given by the formula $d = 0.61t^2$. A report is not written. Students should be encouraged to speak with the teacher as soon as possible to review their work and to make a new start on the project.

Chapter Support

Course Guide: Chapter 8
Lesson Plans: Chapter 8
Practice Bank:
Cumulative Practice 57
Study Guide: Chapter 8 Review
Explorations Lab Manual:
Additional Exploration 12
Assessment Book:
Chapter Tests 36 and 37
Chapter 8 Alternative Assessment
Test Generator software
Technology Book:
Calculator Based Lab 5
Portfolio Project Book:
Additional Project 5:
Quadratic Functions
Preparing for College Entrance Tests
Professional Handbook

Progress Check 8.1–8.3

Graph each equation. Tell whether each equation is *linear* or *nonlinear*. *(Section 8.1)*

1. $y = 4|x|$

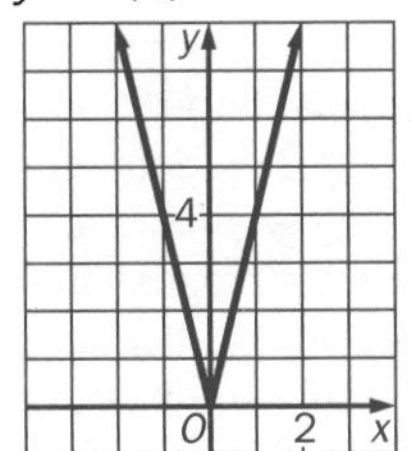

nonlinear

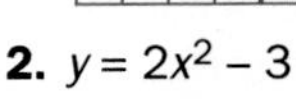

2. $y = 2x^2 - 3$

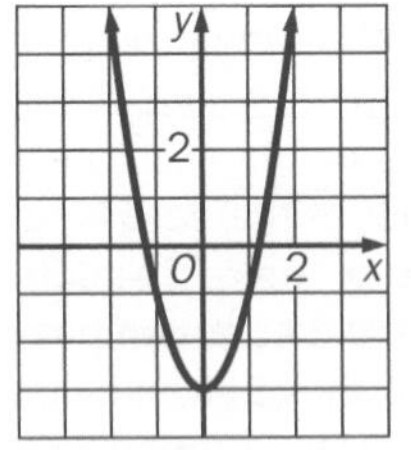

nonlinear

Sketch the graph of each equation. *(Section 8.2)*

3. $y = -3x^2$

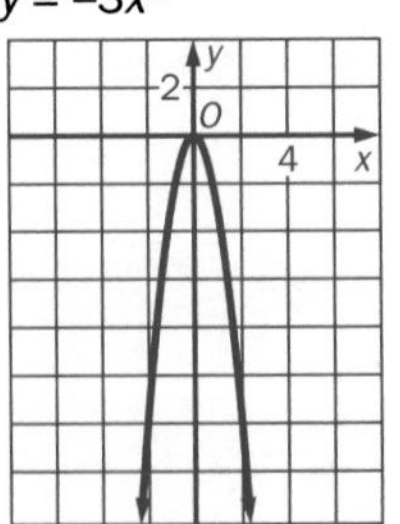

Continued on following page.

CHAPTER

8 Review

STUDY TECHNIQUE

Draw a concept map, as on page 184, to show what you learned in this chapter.

VOCABULARY

nonlinear equation (p. 326)
parabola (p. 331)
quadratic (p. 332)
vertex of a parabola (p. 332)
symmetric (p. 333)
line of symmetry (p. 333)
quadratic equation (p. 345)
quadratic formula (p. 350)
discriminant (p. 355)

SECTIONS 8.1 *and* 8.2

When $a > 0$, the parabola opens up.

The vertex can be a maximum or a minimum.

When $a < 0$, the parabola opens down.

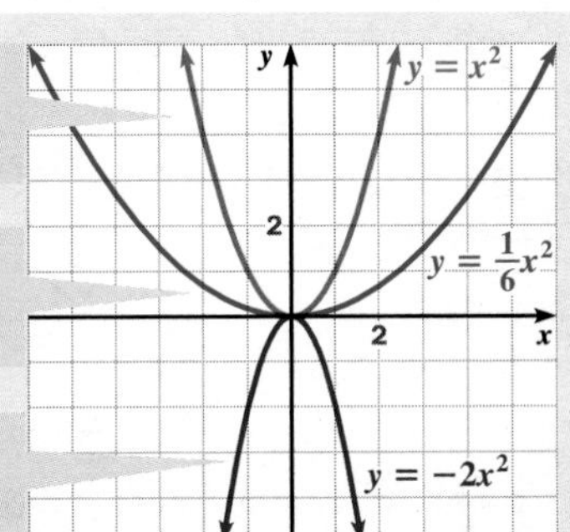

If the graph of an equation is not a line, then the equation and the relationship it models are both **nonlinear**. *Quadratic* functions and absolute value functions are nonlinear.

The graph of a **quadratic** equation is a **parabola**. A parabola that has an equation in the form $y = ax^2$ is **symmetric** about the y-axis. The shape of the parabola depends on the value of $|a|$.

SECTIONS 8.3 *and* 8.4

To solve an equation like $2x^2 - 2 = 6$ algebraically, get the x^2 term alone on one side of the equation. Then find the square root of both sides.

$$2x^2 - 2 = 6$$
$$2x^2 = 8$$
$$x^2 = 4$$
$$x = \pm\sqrt{4} = \pm 2$$

The solutions are 2 and −2.

A quadratic equation can be written in the form $ax^2 + bx + c = 0$, when $a \neq 0$. The solutions are the x-intercepts of the graph of $y = ax^2 + bx + c$.

For example, you can find the solutions of the equation $3x^2 - 2x = 5$ by using a graph. First rewrite the equation in the form $3x^2 - 2x - 5 = 0$. Then graph the related equation $y = 3x^2 - 2x - 5$. The x-intercepts are the solutions.

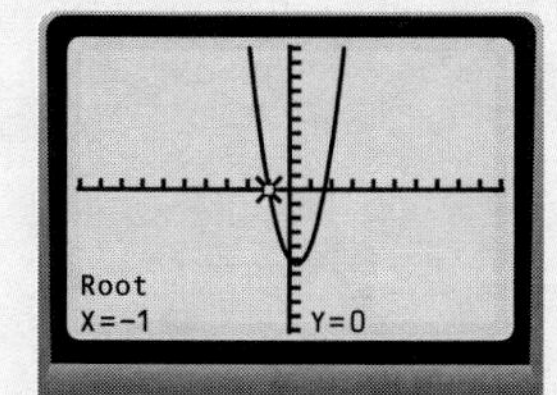

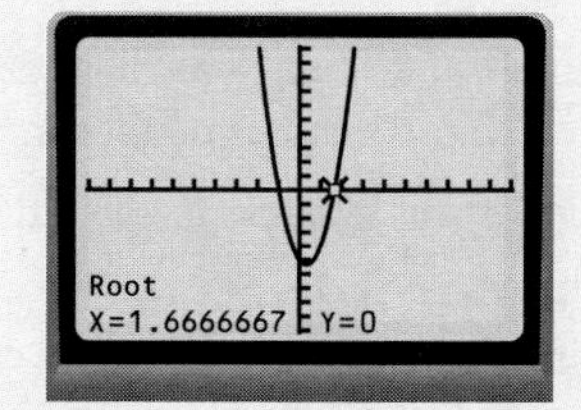

You can use the root feature of a graphing calculator to find the solutions. They are -1 and $1\frac{2}{3}$.

SECTIONS 8.5 *and* 8.6

You can use the **quadratic formula** to solve a quadratic equation in the form $ax^2 + bx + c = 0$, when $a \neq 0$.

$$x = \frac{-b \pm \sqrt{b^2 - 4ac}}{2a}$$

For example, in the equation $8x^2 - 9x + 1 = 0$, $a = 8$, $b = -9$, and $c = 1$.

$$x = \frac{-(-9) \pm \sqrt{(-9)^2 - 4(8)(1)}}{2(8)}$$

$$= \frac{9 \pm \sqrt{49}}{16}$$

$$= \frac{9 \pm 7}{16}$$

There are two solutions.

$$x = \frac{9+7}{16} \quad \text{or} \quad x = \frac{9-7}{16}$$

$$= 1 \qquad\qquad = \frac{1}{8}$$

The solutions are 1 and $\frac{1}{8}$.

The value of the expression $b^2 - 4ac$ in the quadratic formula is called the **discriminant**. You can use the discriminant to find the number of solutions of a quadratic equation.

When $b^2 - 4ac = 0$, the quadratic equation has **one** solution.

When $b^2 - 4ac > 0$, the quadratic equation has **two** solutions.

When $b^2 - 4ac < 0$, the quadratic equation has **no** solutions.

Progress Check 8.1–8.3 (continued)

4. $y = \frac{1}{2}x^2$

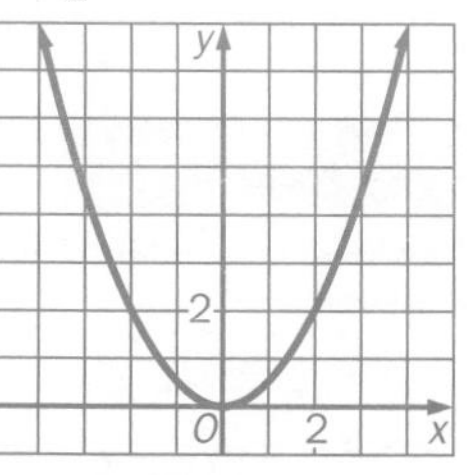

Solve each equation algebraically. *(Section 8.3)*

5. $x^2 - 25 = 0$ ± 5

6. $x^2 + 9 = 25$ ± 4

7. $\frac{1}{6}x^2 - 2 = 4$ ± 6

8. Estimate the x-intercepts of the graph of $y = \frac{1}{2}x^2 - 4$. ± 2.83

Progress Check 8.4–8.6

Use a graphing calculator to solve each equation. *(Section 8.4)*

1. $x^2 + 3x - 4 = 0$ $-4; 1$

2. $2g^2 + 6g + 1 = 0$ $-0.18; -2.82$

3. $-3y^2 + 3y - 1 = 0$ no solution

Solve each equation using the quadratic formula. *(Section 8.5)*

4. $2x^2 + 5x - 3 = 0$ $-3; \frac{1}{2}$

5. $4k^2 - 24k = 0$ $0; 6$

6. $4d^2 = -7d + 2$ $\frac{1}{4}; -2$

For each equation, find the number of solutions. Then solve using the quadratic formula. *(Section 8.6)*

7. $8g^2 + 15g - 2 = 0$
2 solutions; $-2; \frac{1}{8}$

8. $m^2 + 2m + 4 = 0$
no solutions

Chapter 8 Assessment
Form A Chapter Test

NAME ______ DATE ______ SCORE ______

Test 36

TEST ON CHAPTER 8 (FORM A)

DIRECTIONS: Write the answers in the spaces provided.

Tell whether each relationship is *linear* or *nonlinear*.

1. $y = 3x + 4$ 2. $y = 2x^2 - 1$ 3. $y = 3|x| - 2$
4. the relationship between the radius and the area of a circle
5. the relationship between the length of a side and the perimeter of a square

Match each equation with its graph.

6. $y = 2x^2$
7. $y = -3x^2$
8. $y = \frac{1}{5}x^2$

Describe how the graph of each equation compares to the graph of $y = x^2$.

9. $y = 3x^2$ The graph is narrower than that of $y = x^2$, passing through (1, 3) rather than (1, 1).
10. $y = -4x^2$ The graph opens downward rather than upward and is narrower than that of $y = x^2$, passing through (1, –4) rather than (1, 1).
11. $y = -\frac{1}{6}x^2$ The graph opens downward rather than upward and is wider than that of $y = x^2$, passing through (6, –6) rather than (6, 36).

Solve each equation algebraically.

12. $x^2 + 9 = 18$ 13. $2x^2 = 72$ 14. $4x^2 - 4 = 60$
15. $\frac{1}{3}x^2 + 4 = 7$ 16. $\frac{3}{4}x^2 - 9 = 39$ 17. $3x^2 - 42 = 201$

ANSWERS
1. linear
2. nonlinear
3. nonlinear
4. nonlinear
5. linear
6. A
7. C
8. B
9. See question.
10. See question.
11. See question.
12. ±3
13. ±6
14. ±4
15. ±3
16. ±8
17. ±9

Assessment Book, ALGEBRA 1: EXPLORATIONS AND APPLICATIONS

53

Chapter 8 Assessment
Form B Chapter Test

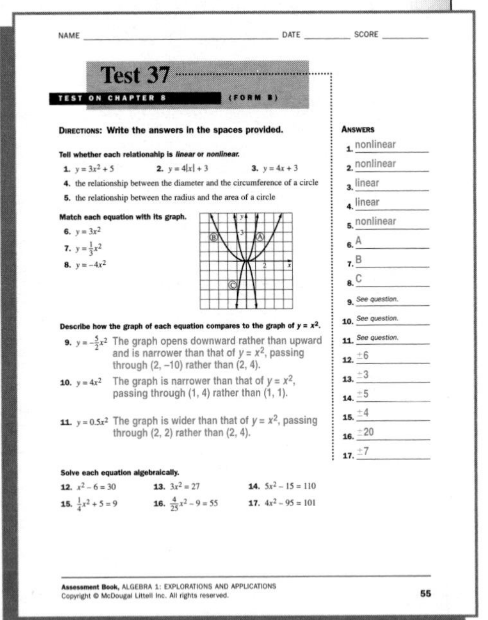
NAME ______ DATE ______ SCORE ______

Test 37

TEST ON CHAPTER 8 (FORM B)

DIRECTIONS: Write the answers in the spaces provided.

Tell whether each relationship is *linear* or *nonlinear*.

1. $y = 3x^2 + 5$ 2. $y = 4|x| + 3$ 3. $y = 4x + 3$
4. the relationship between the diameter and the circumference of a circle
5. the relationship between the radius and the area of a circle

Match each equation with its graph.

6. $y = 3x^2$
7. $y = \frac{1}{3}x^2$
8. $y = -4x^2$

Describe how the graph of each equation compares to the graph of $y = x^2$.

9. $y = -\frac{5}{2}x^2$ The graph opens downward rather than upward and is narrower than that of $y = x^2$, passing through (2, –10) rather than (2, 4).
10. $y = 4x^2$ The graph is narrower than that of $y = x^2$, passing through (1, 4) rather than (1, 1).
11. $y = 0.5x^2$ The graph is wider than that of $y = x^2$, passing through (2, 2) rather than (2, 4).

Solve each equation algebraically.

12. $x^2 - 6 = 30$ 13. $3x^2 = 27$ 14. $5x^2 - 15 = 110$
15. $\frac{1}{4}x^2 + 5 = 9$ 16. $\frac{4}{25}x^2 - 9 = 55$ 17. $4x^2 - 95 = 101$

ANSWERS
1. nonlinear
2. nonlinear
3. linear
4. linear
5. nonlinear
6. A
7. B
8. C
9. See question.
10. See question.
11. See question.
12. ±6
13. ±3
14. ±5
15. ±4
16. ±20
17. ±7

Assessment Book, ALGEBRA 1: EXPLORATIONS AND APPLICATIONS

55

CHAPTER 8

Assessment

VOCABULARY REVIEW

1. Give an example of an equation whose graph is linear. Give an example of an equation whose graph is nonlinear.
2. Sketch a graph that is symmetric about the y-axis. What does the word *symmetry* mean?
3. What do all quadratic equations have in common? What do the graphs of all quadratic equations have in common?
4. What is the discriminant? What information does it give you about a quadratic equation?

SECTIONS 8.1 and 8.2

Tell whether each relationship is *linear* or *nonlinear*.

5. $y = 2|x| - 6$ 6. $y = 5x + 1$ 7. $y = 2x^2 + 10$

8. the relationship between s and the volume V of the box, where $V = 2s^3$

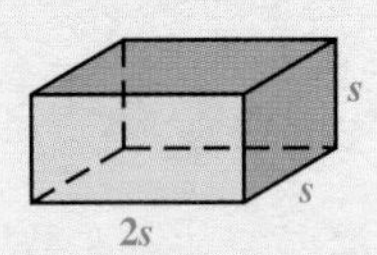

9. the relationship between x and the perimeter P of the equilateral triangle, where $P = 3x$

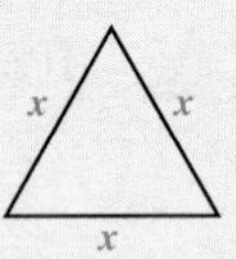

Describe how the graph of each equation compares to the graph of $y = x^2$.

10. $y = 2x^2$ 11. $y = -6x^2$ 12. $y = -\frac{1}{2}x^2$

Match each equation with its graph.

13. $y = -x^2$
14. $y = \frac{1}{3}x^2$
15. $y = 1.5x^2$

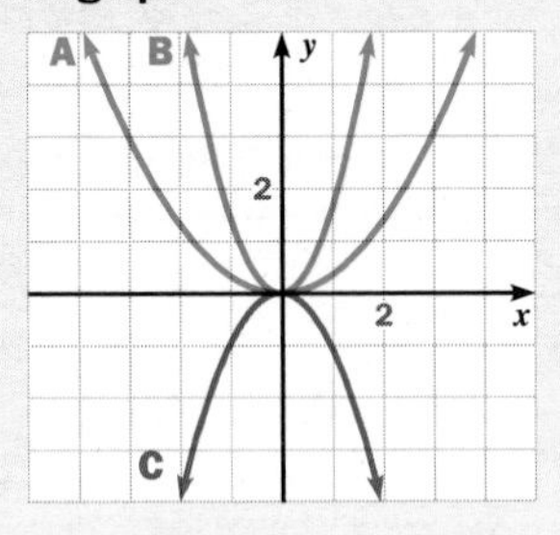

ANSWERS Chapter 8

Assessment

1. Answers may vary. Examples are given; $y = 3x$; $y = 3x^2$.
2. Sketches may vary. An example is given; $y = x^2$.

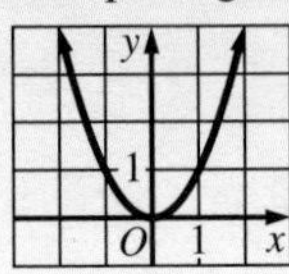

Symmetry is a correspondence in size, shape, and position of parts on opposite sides of a dividing line, plane, or point.

3. All quadratic equations must contain an x^2-term whose coefficient is not equal to zero. All graphs of quadratic equations are parabolas with a vertex and an axis of symmetry.
4. The discriminant is the value of $b^2 - 4ac$ in the quadratic formula. The discriminant gives you information about the number of solutions when you solve a quadratic equation.
5. nonlinear
6. linear
7. nonlinear
8. nonlinear
9. linear
10. The graph opens in the same direction, but it is narrower and steeper.
11. The graph opens in the opposite direction and is narrower and steeper.
12. The graph opens in the opposite direction and is wider and flatter.
13. C
14. A
15. B
16. –4, 4
17. –5, 5
18. –3, 3
19. –4, 4
20. –6, 6
21. $\pm\sqrt{26}$, or about –5.1 and about 5.1
22. about –1.15, about 1.15
23. about –1.6, about 1.6
24. –0.5, 3
25. about –2.76, about –0.24

SECTIONS 8.3 *and* 8.4

Solve each equation algebraically.

16. $x^2 + 4 = 20$ **17.** $2x^2 = 50$ **18.** $3x^2 - 11 = 16$

19. $\frac{1}{2}x^2 + 5 = 13$ **20.** $\frac{2}{3}x^2 - 7 = 17$ **21.** $5x^2 - 43 = 87$

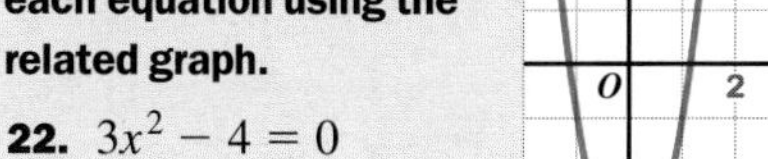

Estimate the solutions of each equation using the related graph.

22. $3x^2 - 4 = 0$

23. $-2x^2 + 5 = 0$

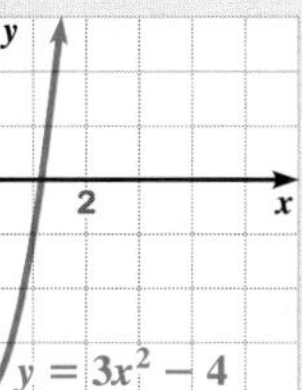

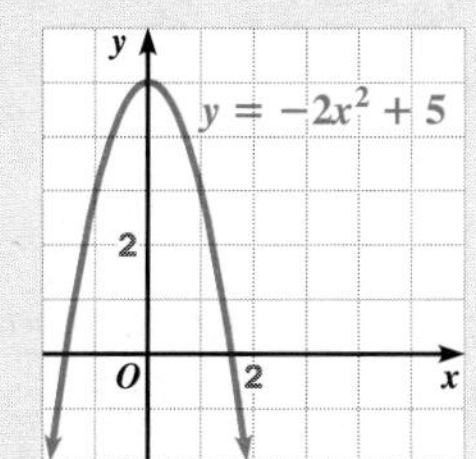

Technology
Solve each equation using a graphing calculator.

24. $2x^2 - 5x = 3$ **25.** $3x^2 + 9x + 2 = 0$ **26.** $6x^2 + 5x + 7 = 0$

SECTIONS 8.5 *and* 8.6

Solve each equation using the quadratic formula.

27. $x^2 + 12x + 11 = 0$ **28.** $16x^2 - 8x + 1 = 0$ **29.** $5x^2 - 3x = 1$

For each equation, find the number of solutions.

30. $2x^2 + 9x - 2 = 0$ **31.** $3x^2 - x + 3 = 0$ **32.** $4x^2 + 4x + 1 = 0$

33. a. The square and the rectangle are equal in area. Write an equation that models this relationship.

b. How many solutions does your equation have?

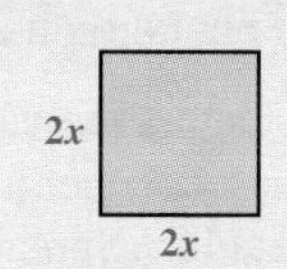

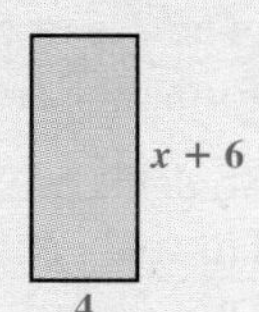

c. Writing How many of the solutions make sense? Explain.

PERFORMANCE TASK

34. Write a quadratic equation that has each combination of solutions. Use what you have learned in this chapter to support each answer.

a. no solutions
b. exactly one positive solution
c. exactly one negative solution
d. two positive solutions
e. two negative solutions
f. one solution of each sign

Chapter 8 Assessment
Form C Alternative Assessment

Chapter 8
ALTERNATIVE ASSESSMENT

1. **Open-ended Problem** What do you think is the maximum number of solutions of (a) a linear equation, (b) a quadratic equation, and (c) a cubic equation? Explain why.

2. **Open-ended Problem** Estimate the equations for each of the parabolas shown in the graph at the right. Use your graphing calculator to check your guess.

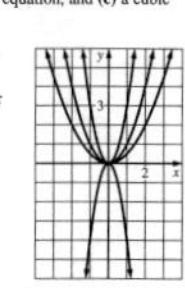

3. **Group Activity** Work in groups of four students. (This activity can also be done with the whole class.) One member of the group is chosen as the "equation reader," who reads off a list of related equations. The rest of the group members demonstrate the shape of the graph of each equation with their arms. The equation reader checks for the accuracy of the "arm graphs." The equation reader should use a list of equations like the ones given below. (Notice that the group members return to the basic graph after the graph of each related equation.)

   a. $y = x^2$ ← This is the basic graph.
   $y = 2x^2$
   $y = x^2$
   $y = 3x^2$
   $y = x^2$
   $y = 0.5x^2$
   $y = x^2$
   $y = -x^2$

   b. $y = x$ ← This is the basic graph.
   $y = 2x$
   $y = x$
   $y = 3x$
   $y = x$
   $y = 0.5x$
   $y = x$
   $y = -x$

Extension: Equations such as $y = 4x^2 + 3$ and $y = 4x + 3$ might also be used.

Assessment Book, ALGEBRA 1: EXPLORATIONS AND APPLICATIONS

103

4. **a.** Find an example of a real-life linear function that can be represented by an equation. Write the equation and describe the domain and range of the function.

   b. Find an example of a real-life nonlinear function that can be represented by an equation. Write the equation and describe the domain and range of the function.

5. **a.** Complete the tables below for the quadratic equation $y = x^2 - x - 20$.

x	y
−6	?
−5	?
−4	?
−3	?
−2	?
−1	?
0	?

x	y
1	?
2	?
3	?
4	?
5	?
6	?

   Use your tables to find the solutions of the equation $0 = x^2 - x - 20$. How do you identify the solutions in the tables?

   b. Generate your own table of values for the equation $y = x^2 + 6x - 7$. Use your table to find the solutions of the equation $0 = x^2 + 6x - 7$.

   c. Generate your own table of values for the equation $y = 2x^2 + 3x + 1$. Use your table to find the solutions of the equation $0 = 2x^2 + 3x + 1$. (*Hint:* Using integer values in your table will only reveal one of the solutions.)

6. **Project** Use the quadratic formula to solve the equation $0 = x^2 - 2x - 5$. Describe how you can use the graph of the equation $y = x^2 - 2x - 5$ to verify your solutions to the equation $0 = x^2 - 2x - 5$. Sketch this graph and verify your solutions.

104
Assessment Book, ALGEBRA 1: EXPLORATIONS AND APPLICATIONS

---

26. no solutions

27. −11, −1 28. $\frac{1}{4}$

29. $\frac{3 \pm \sqrt{29}}{10}$, or about 0.84 and about −0.24

30. two solutions

31. no solutions

32. one solution

33. a. $4x^2 = 4x + 24$

    b. two solutions

    c. One solution makes sense; the other solution is negative and there cannot be a negative length for a side of a square or rectangle.

34. Equations may vary. Examples are given.

    a. $2x^2 + 3x + 2 = 0$; The discriminant is less than 0.

    b. $x^2 - 6x + 9 = 0$; The discriminant is 0 and the solution is 3.

    c. $x^2 + 10x + 25 = 0$; The discriminant is 0 and the solution is −5.

    d. $x^2 - 5x + 6 = 0$; The discriminant is greater than 0 and the solutions are 2, 3.

    e. $x^2 + 7x + 12 = 0$; The discriminant is greater than 0 and the solutions are −3, −4.

    f. $x^2 + 2x - 15 = 0$; The discriminant is greater than 0 and the solutions are −5, 3.

CHAPTER

9 Exponential Functions

OVERVIEW

Connecting to Prior and Future Learning

⇔ Students begin by using the definition of exponents to evaluate powers. Order of operations is reviewed and the step for evaluating powers is added.

⇔ Work with exponents continues as students study exponential growth and decay. Students may review fractions, decimals, and percents on page 587 in the **Student Resources Toolbox**.

⇔ Rules for working with zero and negative exponents, scientific notation, multiplying and dividing powers, finding the power of products and quotients, and the power of a power are discussed.

Chapter Highlights

Interview with Lori Arviso Alvord: The role mathematics plays in medicine is highlighted in this interview, with related exercises on pages 380, 387, and 399.

Explorations in Chapter 9 guide students through activities that investigate various properties and uses of exponents. Students explore exponents, model exponential decay, and multiply and divide powers of ten.

The Portfolio Project: The use of exponential formulas to model spirals is the focus of this project. Students use information from a photograph of a nautilus to create their own spiral.

Technology: Students use calculators to evaluate powers with positive and negative exponents and to study and graph exponential equations. Exercises in Section 9.5 explore how calculators use scientific notation. Spreadsheets are used to compare linear, quadratic, and exponential functions.

OBJECTIVES

Section	Objectives	NCTM Standards
9.1	• Show repeated multiplication with exponents. • Evaluate expressions that involve powers. • Find volumes of rectangular solids and spheres.	1, 2, 3, 4, 5, 8
9.2	• Interpret and evaluate exponential functions. • Model real-world situations with exponential functions. • Make predictions using exponential functions.	1, 2, 3, 4, 5, 6, 8
9.3	• Use equations to model real-world decay situations. • Make predictions about quantities that decrease.	1, 2, 3, 4, 5, 6, 8
9.4	• Evaluate powers with negative and zero exponents. • Use negative exponents to express quotients. • Find past values using negative exponents. • Use formulas to find information.	1, 2, 3, 4, 5, 6
9.5	• Write numbers in scientific notation.	1, 2, 3, 4, 5
9.6	• Multiply and divide expressions involving powers.	1, 2, 3, 4, 5
9.7	• Find the power of products and quotients. • Find the power of a power. • Use numbers written in scientific notation in formulas.	1, 2, 3, 4, 5

INTEGRATION

Mathematical Connections	9.1	9.2	9.3	9.4	9.5	9.6	9.7
algebra	**369–375***	**376–382**	**383–389**	**390–395**	**396–400**	**401–405**	**406–411**
geometry	372, 374, 375		388			403	406–408, 410
data analysis, probability, discrete math	369, 375	**376–382**	**383–389**	392, 394		405	
patterns and functions	369, 370		383, 384				
logic and reasoning	369, 371, 374, 375	377–382	**383–389**	**390–395**	**396–400**	401, 403–405	408–410
Interdisciplinary Connections and Applications							
history and geography				394		403, 405	410
reading and language arts						404	
biology and earth science		376, 377, 379	385	395	399		406, 407, 410
chemistry and physics			384				409, 411
business and economics				394			
sports and recreation			386				
social studies		382		392		405	
personal finance		379	389				
architecture, computers, carpentry, transportation, occupational safety, cooking, astronomy	372, 374 375	382	387–389		398	403	411

* ***Bold page numbers** indicate that a topic is used throughout the section.*

TECHNOLOGY

	opportunities for use with	
Section	**Student Book**	**Support Material**
9.1	scientific calculator graphing calculator	
9.2	graphing calculator spreadsheet software McDougal Littell Software *Stats!* *Function Investigator*	
9.3	graphing calculator McDougal Littell Software *Function Investigator*	**Technology Book:** Calculator Activity 9 Spreadsheet Activity 9
9.4	scientific calculator	**Technology Book:** Spreadsheet Activity 9
9.5	scientific calculator graphing calculator	
9.6	scientific calculator	
9.7	scientific calculator	

PLANNING GUIDE

Regular Scheduling (45 min)

Section	Materials Needed	Core Assignment	Extended Assignment	*exercises that feature* Applications	Communication	Technology
9.1	chessboard, rice, scientific calculator	**Day 1:** 1–20 **Day 2:** 25–40, 43–49	**Day 1:** 1–20 **Day 2:** 21–49	21–23, 41	21, 24, 39, 41–43	19 21–23
9.2	graphing calculator, spreadsheet software	1–18, 22–26, 29, 31–43	7–12, 17–43	10–21, 29, 30	19, 21, 27, 31	22, 23, 28
9.3	100 pennies, cup, graphing calculator, clear drinking glass, water, measuring cup, food coloring	1–12, 14–23, 27–30, 32–41, AYP*	6, 8, 9–16, 19–41, AYP	12, 22–27, 29–31	13, 24–26, 32	28
9.4	scientific calculator	1–28, 30, 31, 34, 35, 37–43	1, 3, 4, 6, 7, 9–11, 15–20, 24–43	29–34	27, 29, 33, 34, 37	28
9.5	ruler, scientific calculator	1–16, 18–20, 22–26, 29, 31–37, AYP	1–6, 10–37, AYP	16, 18–21, 25	17, 21, 27, 28, 31	26
9.6	scientific calculator	1–17, 19–28	1–28	16–21	18, 22	
9.7	scientific calculator	1–15, 21, 23, 25–36, AYP	2–5, 8, 9, 12–36, AYP	17–21	16, 20b, 24	
Review/ Assess		**Day 1:** 1–19 **Day 2:** 20–39 **Day 3:** Ch. 9 Test	**Day 1:** 1–19 **Day 2:** 20–39 **Day 3:** Ch. 9 Test	14, 15 24, 39	1–4	
Portfolio Project		Allow 2 days.	Allow 2 days.			

Yearly Pacing (with Portfolio Project)	Chapter 9 Total	Chapters 1–9 Total	Remaining	Total
	13 days	123 days	37 days	160 days

Block Scheduling (90 min)

	Day 56	Day 57	Day 58	Day 59	Day 60	Day 61	Day 62
Teach/Interact	Ch. 8 Test 9.1: Exploration, page 369	Continue with 9.1 9.2	9.3: Exploration, page 383 9.4	9.5 9.6: Exploration, page 401	9.7 Review	Review Port. Proj.	Port. Proj. Ch. 9 Test
Apply/Assess	**Ch. 8 Test** **9.1:** 1–20	**9.1:** 25–40, 43–49 **9.2:** 1–18, 22–26, 29, 31–43	**9.3:** 1–12, 14–23, 27–30, 32–41, AYP* **9.4:** 1–28, 30, 31, 34, 35, 37–43	**9.5:** 1–6, 10–16, 18–20, 22–26, 29–37, AYP **9.6:** 1–17, 19–28	**9.7:** 1–16, 18, 19, 21, 24–36, AYP **Review:** 1–19	**Review:** 20–39 **Port. Proj.**	**Port. Proj.** **Ch. 9 Test**

NOTE: A one-day block has been added for the Portfolio Project—timing and placement to be determined by teacher.

Yearly Pacing (with Portfolio Project)	Chapter 9 Total	Chapters 1–9 Total	Remaining	Total
	$6\frac{1}{2}$ days	62 days	19 days	81 days

**AYP* is *Assess Your Progress.*

Lesson Support

Section	Practice Bank	Study Guide*	Assessment Book*	Visuals	Explorations Lab Manual	Lesson Plans	Technology Book
9.1	58	9.1		Warm-Up 9.1	Master 15	9.1	
9.2	59	9.2		Warm-Up 9.2		9.2	
9.3	60	9.3	Test 38	Warm-Up 9.3	Masters 1, 2, 16 Add. Expl. 14	9.3	Calculator Act. 9 Spreadsheet Act. 9
9.4	61	9.4		Warm-Up 9.4 Folder 9		9.4	Spreadsheet Act. 9
9.5	62	9.5	Test 39	Warm-Up 9.5		9.5	
9.6	63	9.6		Warm-Up 9.6	Master 17	9.6	
9.7	64	9.7	Test 40	Warm-Up 9.7		9.7	
Review Test	65	Chapter Review	Tests 41, 42 Alternative Assessment			Review Test	Calculator Based Lab 6

***Spanish versions** of *Study Guide* and *Assessment Book* are available.

Chapter Support

- Course Guide
- Lesson Plans
- Portfolio Project Book
- Preparing for College Entrance Tests
- Multi-Language Glossary
- *Test Generator* Software
- Professional Handbook

Software Support

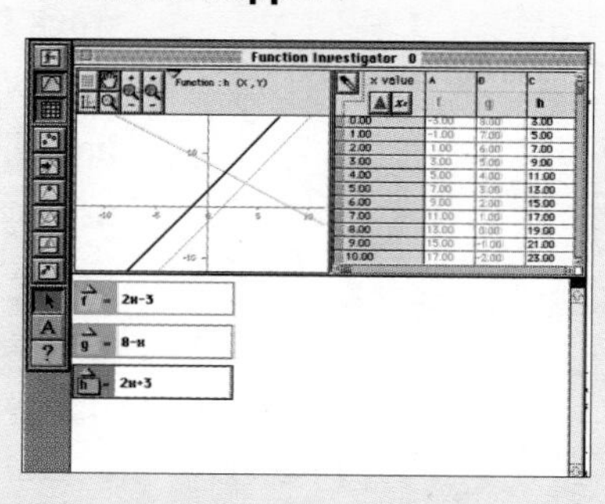

McDougal Littell Software
Stats!
Function Investigator

Internet Support

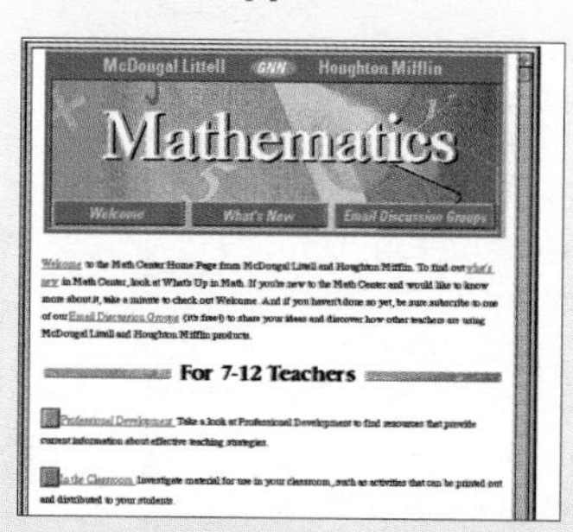

http://www.hmco.com
Next go to McDougal Littell; then the Education Center; then Secondary Math.

Outside Resources

Books, Periodicals

Sandefur, James T. "Drugs and Pollution in the Algebra Class." *Mathematics Teacher* (February 1992): pp. 139–145.

Jones, Graham A. "Mathematical Modeling in a Feast of Rabbits." *Mathematics Teacher* (December 1993): pp. 770–773.

Shell. *The Language of Functions and Graphs,* Unit B3: "Looking at Exponential Functions": pp. 120–125. The Shell Centre for Mathematical Education at the University of Nottingham in the UK.

Activities, Manipulatives

Rahn, James R. and Barry A. Berndes. "Using Logarithms to Explore Power and Exponential Functions." *Mathematics Teacher* (March 1994): pp. 161–170.

Kincaid, Charlene, Guy Mauldin, and Deanna Mauldin. "The Marble Sifter: A Half-Life Simulation." *Mathematics Teacher* (December 1993): pp. 748–759.

Masalski, William J. "Topic: Compound Interest." *How to Use the Spreadsheet as a Tool in the Secondary Mathematics Classroom.* NCTM, 1990. pp. 16–19.

Videos

Southern Illinois University at Carbondale. *World Population Review.* 1990.

Interview Notes

Background

Navajo Healers
In the past, the Navajo people turned to their own spiritual healers when they became ill. Navajo healers used prayers and songs to bring back a person's health, and their rituals tended to be long and complicated. On the average, healers memorized about 300 prayers and songs. Navajo medicine was practiced in a patient's home, where the whole family could unite emotionally behind the patient. A positive atmosphere was essential because the Navajo people believe that bad winds or a negative attitude could destroy the healing process. Navajo healers strived to bring their patients into harmony with their environment.

Lori Arviso Alvord
Dr. Lori Arviso Alvord is a graduate of Dartmouth College and Stanford University Medical School. At the Gallup Indian Medical Center, where she is a staff general surgeon, Dr. Alvord has had to adapt what she learned in medical school to her own knowledge of Navajo ways. She thinks that most doctors could benefit from an understanding of the Navajo culture in the way they view their patients and how they interact with them.

CHAPTER

9 Exponential Functions

Hands that heal

INTERVIEW **Lori Arviso Alvord**

"The patient and I have to work together toward the same goal: his wellness."

As a young girl growing up on the Navajo reservation in Crownpoint, New Mexico, Dr. Lori Arviso Alvord loved working with her hands. She learned to do beadwork from a Sioux friend and studied classical piano for six years. Both skills required fine work with her hands. "I always wanted to do something where I could use my hands," she says. "Surgery was the most appropriate field in medicine to be doing that." According to the Association of American Indian Physicians, Alvord is the first Navajo woman to become a surgeon. After completing her residency, she returned to New Mexico to work among the Navajo. Now, as a general surgeon at the Gallup Indian Medical Center, Alvord uses her hands to heal.

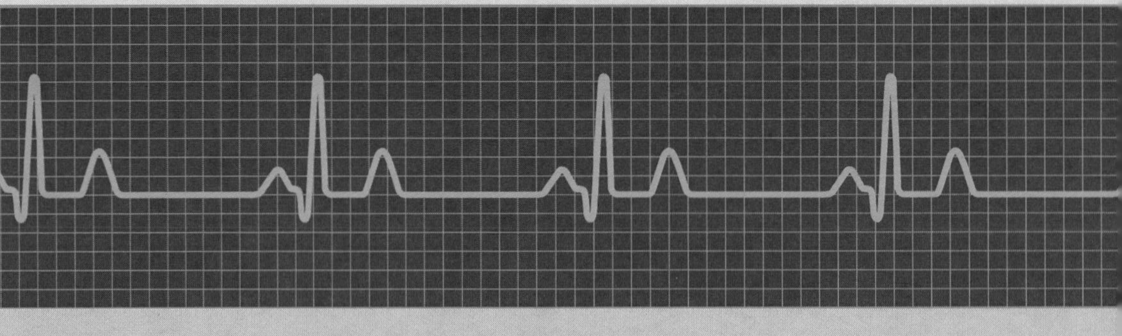

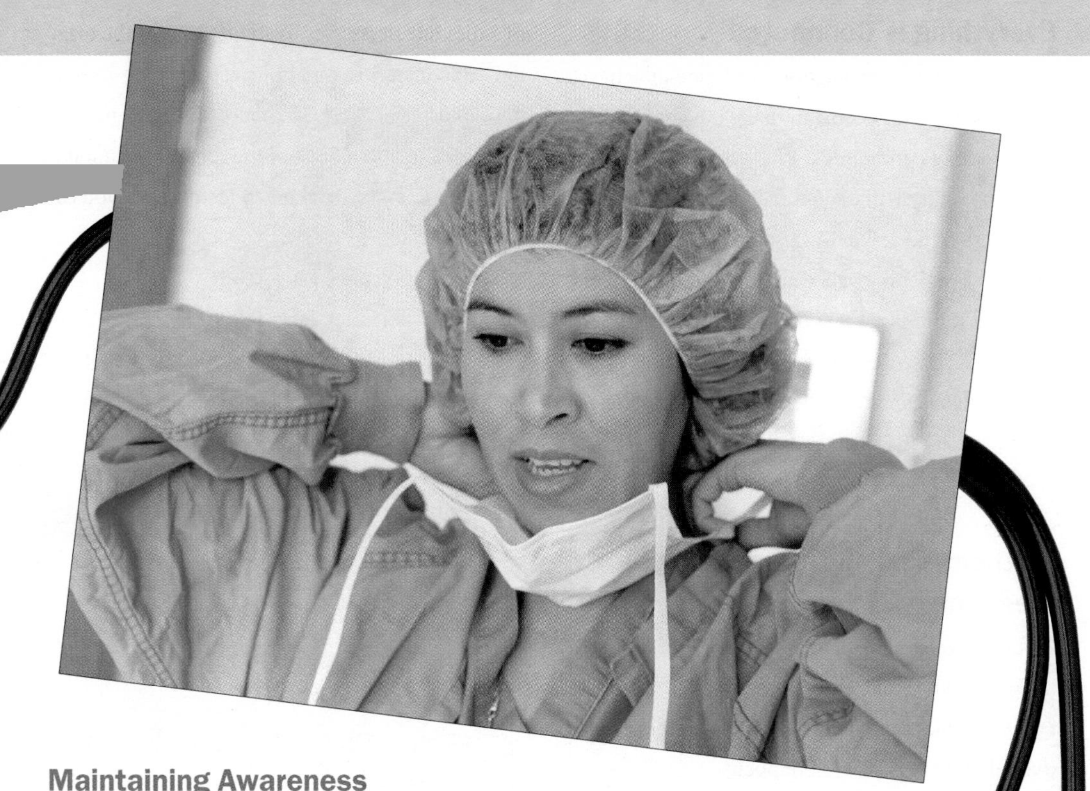

Maintaining Awareness

The challenge of working on the reservation is unique. Navajo people continue to turn to healers from within their own community when there is a need for medical care. But when so-called modern medicine is the only option, an awareness of Navajo customs is a must. For this reason, Alvord applies what she knows of her people's culture to her work.

For example, looking directly into another person's eyes or touching the body without asking permission is considered an insult. And although she does speak some Navajo, Alvord sometimes uses an interpreter to communicate with her patients. "If you don't relate well with Navajo people, they won't allow you to operate on them," she says. "They might just not show up."

Interview Notes

Background

The Navajo Nation

The Navajo are one of the largest Native American nations living within the United States. Their 16-million acre reservation, which includes land in Arizona, New Mexico, and Utah, is rich in mineral deposits, such as coal, gas, and uranium. The mining and sale of these minerals make the Navajo one of the most prosperous tribes as well. Navajo occupations range from engineers, technicians, and teachers to ranchers, crafts people, and farmers. The first college owned and operated by Native Americans is on the Navajo reservation in Arizona.

Second-Language Learners

You may need to explain to students that a *residency* is a specialized clinical training that medical students receive in a hospital after they graduate from medical school. A residency prepares a doctor to practice a particular field of medicine, such as surgery, for example.

Interview Notes

Mathematical Connection

In Section 9.2, students learn about exponential growth and see that the growth of bacteria can be modeled with an exponential function. In Section 9.3, the concept of exponential decay is introduced and used to understand how long certain medications are present in a patient's bloodstream. In Section 9.5, students write numbers in scientific notation to express the measurements of extremely small objects such as blood cells.

Explore and Connect

Writing
For question 1, students should understand from the graph that the concentration of salicylate in a person's bloodstream decreases over time.

Research
Answering the first part of this research activity should be fairly easy for most students. The second question, however, may pose some difficulties now. You may wish to return to this question after students have completed this chapter, at which time they should be able to respond knowledgeably.

Project
Students' posters should be displayed in class. Ask if any students who wrote a report would like to tell the class what they learned.

Everything is Connected

At Alvord's recommendation, medicine men have been encouraged to come in to the hospital to perform healing ceremonies. Plans for an improved medical facility will undoubtedly include her suggestion that the new building have a round room for such ceremonies, called sings. Alvord says that, for her, "there is a great deal of spirituality" in surgery, because the Navajo philosophy holds that everything is connected. "The patient and I have to work together toward the same goal: his wellness."

Mathematics and Medicine

Part of Alvord's medical training involves understanding how long a particular medicine will be effective. In other words, what happens to the concentration of medicine in a person's bloodstream over time? Consider the case of salicylate, an active ingredient in aspirin. The graph shows that the concentration of salicylate decreases to about 80% of the original concentration one hour after taking a small dose of aspirin. After two hours, the concentration decreases to about 80% of the remaining amount. If you let C = the original concentration, you can see a pattern:

Concentration after 1 h $= 0.80C$

Concentration after 2 h $= 0.80(0.80C)$

Concentration after 3 h $= 0.80[0.80(0.80C)]$

In this chapter, you will learn that this pattern can be modeled by an *exponential function*.

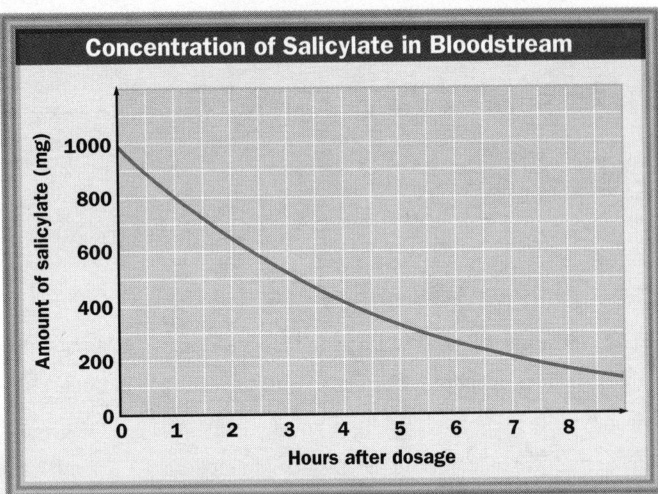

EXPLORE AND CONNECT

Lori Arviso Alvord examines the x-rays of one of her patients.

1. Writing Based on what you read in the interview, describe how to find the concentration of salicylate in the body after four hours. After how many hours do you think the concentration of salicylate will be close to zero? Explain.

2. Research Find out some of the requirements for becoming a medical doctor. What are some of the reasons that doctors might need to know mathematics?

3. Project Many medicines used today are based on chemicals found in plants. Some examples are aspirin, ephedrine, digitalis, quinine, and reserpine. Find out about one of these medicines or some other medicine. Make a poster or write a report on its discovery and use.

Mathematics & Lori Arviso Alvord

In this chapter, you will learn more about how mathematics is related to medicine.

Related Exercises

Section 9.2
- **Exercises 19–21**

Section 9.3
- **Exercises 22–26**

Section 9.5
- **Exercises 18–21**

9.1 Exponents and Powers

Learn how to...

- show repeated multiplication with exponents
- evaluate expressions that involve powers

So you can...

- find the volumes of buildings and figures
- understand computer terms, such as kilobyte and megabyte

According to an ancient folktale from India, a man named Sissa Ben Dahir invented the game of chess. The king of India liked the game so much that he wanted to reward Sissa with 64 gold pieces, one for each square on a chessboard. Instead, Sissa asked for grains of wheat arranged on the chessboard in the pattern shown below.

EXPLORATION
COOPERATIVE LEARNING

Exploring Exponents

Work with a partner.
You will need:

- a chessboard
- a handful of rice

SET UP Use rice to model Sissa Ben Dahir's pattern. Place the grains of rice on a chessboard as shown. Each square should contain twice as many grains as the square before it. Continue this pattern until you run out of rice. Then answer the questions below.

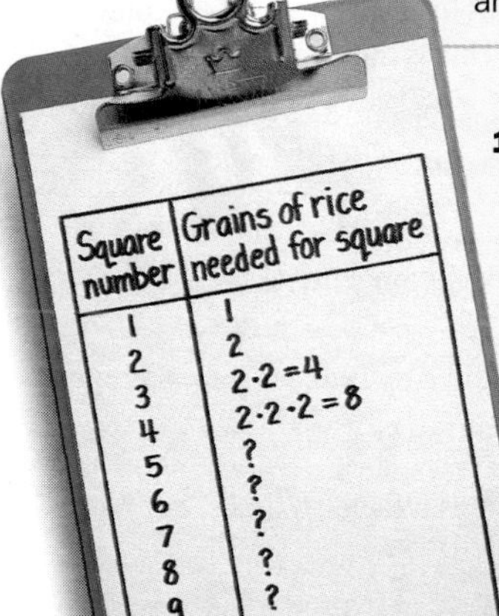

Square number	Grains of rice needed for square
1	1
2	2
3	2·2 = 4
4	2·2·2 = 8
5	?
6	?
7	?
8	?
9	?
10	?

Questions

1. Copy and complete the table. Express each number in the second column as a product of 2's.
2. How many factors of 2 would be in the product for the number of grains on the 30th square? on the 64th square? on the *n*th square? How do you know?
3. A 50 lb bag of wheat contains about 2,500,000 grains. Use a calculator to estimate the number of 50 lb bags of wheat that must be placed on the 30th square to satisfy Sissa Ben Dahir's request. Why do you think he asked for wheat instead of gold?

Exploration Note

Purpose
The purpose of this Exploration is to have students discover what happens to a number as it is multiplied by itself repeatedly.

Materials/Preparation
Each group should have a sufficient amount of rice to fill at least the first five squares of a chessboard.

Procedure
Before beginning, make sure that each pair of students knows how to use the exponent key on a calculator. Students should record their answers to each step.

Closure
Groups can share their results with each other. All students should arrive at an understanding that a whole number multiplied by itself repeatedly increases very rapidly.

Explorations Lab Manual
See the Manual for more commentary on this Exploration.

Diagram Master 15

For answers to the Exploration, see following page.

Plan⇔Support

Objectives

- Show repeated multiplication with exponents.
- Evaluate expressions that involve powers.
- Find volumes of rectangular solids and spheres.

Recommended Pacing

Core and Extended Courses
Section 9.2: 2 days

Block Schedule
Section 9.1: 2 half-blocks (with Chapter 8 Test and Section 9.2)

Resource Materials

Lesson Support
Lesson Plan 9.1
Warm-Up Transparency 9.1
Practice Bank: Practice 58
Study Guide: Section 9.1
Explorations Lab Manual: Diagram Master 15

Technology
Scientific Calculator
Graphing Calculator
Internet: http://www.hmco.com

Warm-Up Exercises

Evaluate each expression.

1. $3 \cdot 3 \cdot 3$ 27
2. $5\left(\frac{1}{2} \cdot \frac{1}{2} \cdot \frac{1}{2}\right)$ $\frac{5}{8}$

State whether each product is positive or negative.

3. $(-4)(-4)(-4)$ negative
4. $(-2)(-2)(-5)(-5)$ positive
5. $(-2 \cdot 3)(-2 \cdot 3)$ positive

Teach⇔Interact

Section Note

Communication: Reading
You may wish to write additional expressions involving powers on the board and call upon individual students to read them.

About Example 1

Common Error
In the solution for part (b), some students may think the answer should be xy^5. Show them that $xy^5 = x \cdot y \cdot y \cdot y \cdot y \cdot y$, which is not the same as the original product.

Additional Example 1

Rewrite each product using exponents.

a. $(-5)(-5)(-5)(-5)(-5)(-5)$
Find the factor repeated in the expression. Then count the number of times it is used as a factor. Parentheses are needed to show that -5 rather than 5 is the base. The number -5 is used as a factor 6 times.
$(-5)(-5)(-5)(-5)(-5)(-5) = (-5)^6$

b. $abc \cdot abc \cdot abc$
Use parentheses to show that a, b, and c are used as factors 3 times.
$abc \cdot abc \cdot abc = (abc)^3$

c. $-2\left(\frac{1}{5} \cdot \frac{1}{5} \cdot \frac{1}{5} \cdot \frac{1}{5}\right)$
Use parentheses to show that $\frac{1}{5}$ is used as a factor 4 times, but -2 is used as a factor only once.
$-2\left(\frac{1}{5} \cdot \frac{1}{5} \cdot \frac{1}{5} \cdot \frac{1}{5}\right) = -2\left(\frac{1}{5}\right)^4$

BY THE WAY

The game of chess is over 1400 years old. It originated in China or India and was brought to Europe during the seventh century.

Using Powers

In the Exploration on page 369, the fifth square of the chessboard had $2 \cdot 2 \cdot 2 \cdot 2$ grains of rice. You can use a *power* to represent repeated multiplication by the same number.

$$\underbrace{2 \cdot 2 \cdot 2 \cdot 2}_{4 \text{ factors}} = 2^4$$

exponent

base

The **base** is the number being used as a factor. The **exponent** is the number of times the base is used as a factor.

The whole expression is called a **power**. You read 2^4 as "two to the fourth power" or "the fourth power of two." It means you use 2 as a factor 4 times.

EXAMPLE 1

Rewrite each product using exponents.

a. $-(9 \cdot 9 \cdot 9 \cdot 9 \cdot 9 \cdot 9 \cdot 9 \cdot 9 \cdot 9 \cdot 9)$

b. $xy \cdot xy \cdot xy \cdot xy \cdot xy$

c. $3\left(\frac{1}{4} \cdot \frac{1}{4} \cdot \frac{1}{4} \cdot \frac{1}{4} \cdot \frac{1}{4} \cdot \frac{1}{4} \cdot \frac{1}{4} \cdot \frac{1}{4}\right)$

SOLUTION

Find the factor repeated in the expression. This is the base. Then count the number of times it is used as a factor. This is the exponent.

a. The number 9 is used as a factor 10 times.

$$-(9 \cdot 9 \cdot 9 \cdot 9 \cdot 9 \cdot 9 \cdot 9 \cdot 9 \cdot 9 \cdot 9) = -(9^{10})$$
$$= -9^{10}$$

Note that the expression -9^{10} means you exponentiate the 9 before you apply the negative.

WATCH OUT!
Notice that $-9^{10} = -(9^{10})$. The expression -9^{10} is *not* equal to $(-9)^{10}$.

b. Use parentheses to show that both x and y are used as factors 5 times.

$$xy \cdot xy \cdot xy \cdot xy \cdot xy = (xy)^5$$

c. Use parentheses to show that $\frac{1}{4}$ is used as a factor 8 times, but 3 is used as a factor only once.

$$3\left(\frac{1}{4} \cdot \frac{1}{4} \cdot \frac{1}{4} \cdot \frac{1}{4} \cdot \frac{1}{4} \cdot \frac{1}{4} \cdot \frac{1}{4} \cdot \frac{1}{4}\right) = 3\left(\frac{1}{4}\right)^8$$

ANSWERS Section 9.1

Exploration

1.

Square number	Grains of rice needed for square
1	1
2	2
3	$2 \cdot 2 = 4$
4	$2 \cdot 2 \cdot 2 = 8$
5	$2 \cdot 2 \cdot 2 \cdot 2 = 16$
6	$2 \cdot 2 \cdot 2 \cdot 2 \cdot 2 = 32$
7	$2 \cdot 2 \cdot 2 \cdot 2 \cdot 2 \cdot 2 = 64$
8	$2 \cdot 2 \cdot 2 \cdot 2 \cdot 2 \cdot 2 \cdot 2 = 128$
9	$2 \cdot 2 \cdot 2 \cdot 2 \cdot 2 \cdot 2 \cdot 2 \cdot 2 = 256$
10	$2 \cdot 2 \cdot 2 \cdot 2 \cdot 2 \cdot 2 \cdot 2 \cdot 2 \cdot 2 = 512$

2. 29; 63; $n - 1$; On each square, the number of factors of 2 is one less than the number of the square.

3. about 215 bags of wheat; Answers may vary. An example is given. The total value of the wheat would be greater than the value of 64 gold pieces.

EXAMPLE 2

Evaluate each power.

a. 2^5 **b.** $\left(\frac{1}{2}\right)^3$ **c.** $(-3)^4$

SOLUTION

a. $2^5 = 2 \cdot 2 \cdot 2 \cdot 2 \cdot 2$
$= 32$

b. $\left(\frac{1}{2}\right)^3 = \frac{1}{2} \cdot \frac{1}{2} \cdot \frac{1}{2}$
$= \frac{1}{8}$

c. $(-3)^4 = (-3)(-3)(-3)(-3)$
$= 9 \cdot 9$
$= 81$

The parentheses show that the base is -3, not 3.

THINK AND COMMUNICATE

1. Look back at the Exploration on page 369. Write a power representing the number of grains that must be placed on the nth square ($1 < n \le 64$) to satisfy Sissa Ben Dahir's request. Explain your reasoning.
2. Let b = any real number and let n = any positive integer. Complete each equation. Explain your answers.
   a. $b^1 = \underline{?}$ **b.** $1^n = \underline{?}$ **c.** $0^n = \underline{?}$
3. **a.** Evaluate each of the powers below when $b = -2$.
   b^1 b^2 b^3 b^4
   b. Let b = any negative number. When is b^n positive? When is b^n negative? Explain your reasoning.

Powers and Order of Operations

The list below tells you where evaluating powers fits into the order of operations you learned about in Section 1.2.

Order of Operations

1. Simplify expressions inside parentheses.
2. Evaluate powers.
3. Do multiplication and division from left to right.
4. Do addition and subtraction from left to right.

Teach⇔Interact

About Example 2

Teaching Tip
When discussing part (b), you may need to review the general procedure for multiplying two fractions, that is, $\frac{a}{b} \cdot \frac{c}{d} = \frac{ac}{bd}$.

Additional Example 2

Evaluate each power.

a. $(3 \cdot 5)^2$
$(3 \cdot 5)^2 = 3 \cdot 5 \cdot 3 \cdot 5 = 225$

b. $\left(\frac{2}{5}\right)^3$ $\left(\frac{2}{5}\right)^3 = \frac{2}{5} \cdot \frac{2}{5} \cdot \frac{2}{5} = \frac{8}{125}$

c. -3^3 $-3^3 = -(3 \cdot 3 \cdot 3) = -27$

Think and Communicate

For question 3, students should understand that when they raise a negative number to a power, the sign of the product is determined by the number of negative factors.

Section Note

Topic Spiraling: Review
Reviewing the order of operations, first stated on page 9, will help students to see the modification that includes evaluating powers.

Think and Communicate

1. 2^{n-1}; On each square, the number of factors of 2 is one less than the number of the square.
2. a. $b^1 = b$; The exponent tells you the base is multiplied by itself one time.
   b. $1^n = 1$; The product of 1 and any number is 1, so 1 multiplied by itself any number of times is 1.
   c. $0^n = 0$; The product of 0 and any number is 0, so 0 multiplied by itself any number of times is 0.
3. a. –2; 4; –8; 16
   b. when n is even; The product of an even number of negative numbers is positive. when n is odd; The product of an odd number of negative numbers is negative.

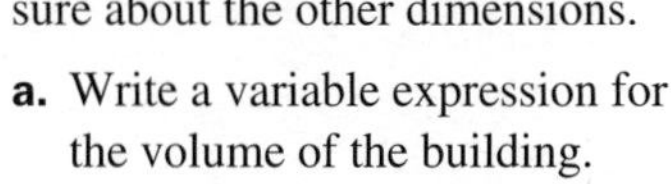

About Example 3

Common Error

In part (b) of the Solution, some students may make the mistake of evaluating $50x^2$ when $x = 60$ as $(50 \cdot 60)^2$. Remind these students that $50x^2 = 50 \cdot x \cdot x = 50 \cdot 60 \cdot 60$.

Additional Example 3

Refer to Example 3. Suppose the building is to be 30 ft tall and cover a square piece of land.

a. Write a variable expression for the volume of the building.
Use a formula.
Let x = the length of one side of the area that the building covers.
Let y = the length of one side of the entrance.
You can think of the building as a box with a cube cut out of it.
Volume = Length × Width × Height
Volume of box $= x \cdot x \cdot 30$
$= 30x^2$
Volume of cube $= y \cdot y \cdot y$
$= y^3$
The building's volume is the volume of the box minus the volume of the cube.
Volume of building
= (Volume of box) – (Volume of cube)
$= 30x^2 - y^3$

b. Find the volume of the building if the area the building covers is 45 ft by 45 ft, and the entrance is 20 ft by 20 ft by 20 ft.
Evaluate the expression when $x = 45$ and $y = 20$.
$30x^2 - y^3 = 30(45)^2 - (20)^3$
$= 30 \cdot 2025 - 8000$
$= 60{,}750 - 8000$
$= 52{,}750$
The volume of the building is $52{,}750\ \text{ft}^3$.

EXAMPLE 3 Application: Architecture

Linda Savas is designing a building with a cube-shaped entrance, as shown by the model. The building is to be 50 ft tall and cover a square piece of land, but she is not sure about the other dimensions.

a. Write a variable expression for the volume of the building.

b. Find the volume of the building if the area the building covers is 60 ft by 60 ft, and the entrance is 25 ft by 25 ft by 25 ft.

SOLUTION

a. Problem Solving Strategy: Use a formula.

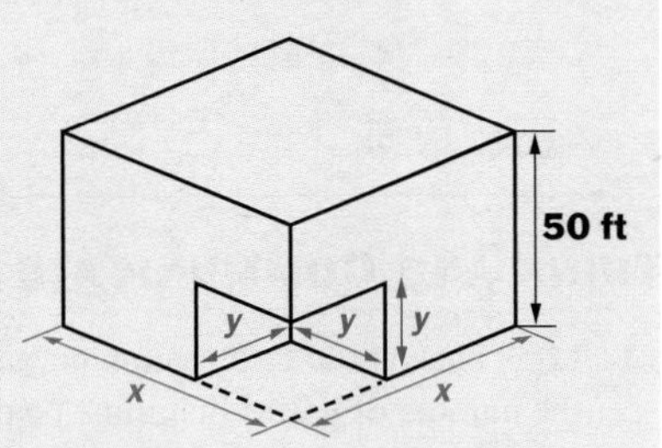

Let x = the length of one side of the area that the building covers.

Let y = the length of one side of the entrance.

You can think of the building as a box with a cube cut out of it.

$$\text{Volume} = \text{Length} \times \text{Width} \times \text{Height}$$

$$\text{Volume of box} = x \cdot x \cdot 50$$
$$= 50x^2$$

$$\text{Volume of cube} = y \cdot y \cdot y$$
$$= y^3$$

The building's volume is the volume of the box minus the volume of the cube.

$$\text{Volume of building} = \text{Volume of box} - \text{Volume of cube}$$
$$= 50x^2 - y^3$$

b. Evaluate the expression when $x = 60$ and $y = 25$.

$$50x^2 - y^3 = 50(60)^2 - (25)^3$$

Evaluate the powers first.

$$= 50 \cdot 3600 - 15{,}625$$
$$= 180{,}000 - 15{,}625$$
$$= 164{,}375$$

The volume of the building is $164{,}375\ \text{ft}^3$.

CHECKING KEY CONCEPTS

Rewrite each product using exponents.

1. $5 \cdot 5 \cdot 5 \cdot 5$ **2.** $4 \cdot 17 \cdot 17$ **3.** $2(y \cdot y \cdot y)$

Evaluate each power.

4. 4^6 **5.** $\left(\frac{1}{7}\right)^3$ **6.** $(-5)^3$

Evaluate each expression when $x = 3$.

7. $6x^2$ **8.** $4x^3 - 5$ **9.** $2(x + 1)^4$

9.1 Exercises and Applications

Extra Practice exercises on page 574

Rewrite each product using exponents.

1. $4.1 \cdot 4.1 \cdot 4.1 \cdot 4.1 \cdot 4.1$ **2.** $(-1.5)(-1.5)(-1.5)$

3. $\left(-\frac{3}{8}\right)\left(-\frac{3}{8}\right)\left(-\frac{3}{8}\right)$ **4.** $-a \cdot a \cdot a \cdot a \cdot a \cdot a \cdot a$

5. $pq(4r \cdot 4r \cdot 4r \cdot 4r)$ **6.** $(8 \cdot 8 \cdot 8 \cdot 8 \cdot 8 \cdot 8)(c - d)$

Evaluate each power.

7. 4^3 **8.** 3^4 **9.** 11^3 **10.** 6^4

11. $\left(\frac{1}{2}\right)^5$ **12.** $\left(-\frac{1}{2}\right)^5$ **13.** $(-2.8)^2$ **14.** $(1729.38)^1$

Write an algebraic expression for each phrase.

15. six to the seventh power **16.** the sixth power of seven

17. the tenth power of y **18.** one fourth to the nth power

19. **Technology** Most calculators have a key for evaluating powers. This key is usually labeled ^, x^y, or y^x. For example, to evaluate 3^5 you might use this key sequence:

Use a calculator to evaluate each power. Round decimal answers to the nearest hundredth.

a. 5^8 b. $(1.65)^{15}$ c. $(-7)^9$

20. Michelle showed her classmates how to evaluate expressions like -3^2 and $(-3)^2$. Use Michelle's work shown here to help you evaluate the expressions in parts (a)–(d).

$-3^2 = -(3 \cdot 3) = -9$

$(-3)^2 = (-3)(-3) = 9$

a. -2^5 and $(-2)^5$ b. -7^3 and $(-7)^3$

c. $-\left(\frac{1}{6}\right)^2$ and $\left(-\frac{1}{6}\right)^2$ d. $-x^9$ and $(-x)^9$

Teach⇔Interact

Checking Key Concepts

Assessment Note
Students should work on these activities independently. A brief verbal review of the answers in class will help to identify those students who need additional instruction on the procedures for using exponents and evaluating expressions containing powers.

Closure Question

How would you read the expression a^b? Identify the exponent and the base.
a to the b power; b is the exponent and a is the base.

Apply⇔Assess

Suggested Assignment

Core Course
Day 1 Exs. 1–20
Day 2 Exs. 25–40, 43–49

Extended Course
Day 1 Exs. 1–20
Day 2 Exs. 21–49

Block Schedule
Day 56 Exs. 1–20
Day 57 Exs. 25–40, 43–49

Exercise Notes

Student Progress
Exs. 1–14 A realistic expectation is that all students be able to complete these exercises successfully. Ask students who do not have correct answers to show their work to you so misunderstandings can be corrected.

Communication: Writing
Exs. 15–18 These exercises provide an excellent experience for interpreting word phrases as mathematical expressions.

Using Technology
Ex. 19 Students should be careful to use parentheses in accordance with the way the calculator will interpret expressions involving powers. For example, on the TI-82, the sixth power of –2 must be evaluated by using (–2)^6. If students enter –2^6, the result will be –64, which is not the correct answer.

Checking Key Concepts

1. 5^4 2. $4^1 \cdot 17^2$
3. $2^1 \cdot y^3$ 4. 4096
5. $\frac{1}{343}$ 6. –125
7. 54 8. 103
9. 512

Exercises and Applications

1. $(4.1)^5$ 2. $(-1.5)^3$
3. $\left(-\frac{3}{8}\right)^3$
4. $(-a)^1 \cdot (a)^6 = -a^7$
5. $pq(4r)^4$ 6. $8^6(c-d)^1$
7. 64 8. 81
9. 1331 10. 1296
11. $\frac{1}{32}$ 12. $-\frac{1}{32}$
13. 7.84 14. 1729.38
15. 6^7 16. 7^6
17. y^{10} 18. $\left(\frac{1}{4}\right)^n$
19. a. 390,625
b. 1829.19
c. –40,353,607
20. a. –32; –32
b. –343; –343
c. $-\frac{1}{36}; \frac{1}{36}$
d. $-x^9; -x^9$

Apply⇔Assess

Exercise Notes

Challenge
Ex. 20 After students complete this exercise, ask them to formulate a general rule for determining when $(-x)^n = -x^n$ if n is a whole number. (If n is an odd whole number, then $(-x)^n = -x^n$.)

Application
Exs. 21–23 These exercises illustrate the utility of exponents in expressing large numbers as well as introducing some useful computer terminology.

Interdisciplinary Problems
Exs. 21–23 Computer science and computer technology are intrinsically dependent upon modern developments in mathematics, physics, and electrical engineering. Mathematically, the binary number system, which uses only the digits 0 and 1, serves as a model for representing the positive or negative states of an electric current. Computers are built from two-state electronic components. The logic of George Boule (1815–1864), an English mathematician, was used to link a two-valued numerical algebra involving the symbols 0 and 1 to the electrical circuits of computers and thus to calculation processes.

Teaching Tip
Exs. 25, 26 Remind students that powers should be evaluated before multiplication is performed.

Reasoning
Ex. 42 Ask students how the two sentences "A Mersenne prime is a Mersenne number" and "A Mersenne number is a Mersenne prime" are related to one another. (They are converses.) Remind students that the converse of a true statement may be true or false.

Connection COMPUTERS

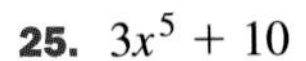

When computers were first invented, they filled entire rooms. Now some computers are so small that you can hold them on your lap.

A *byte* is the amount of computer memory needed to store a single character, such as "a" or "1." A computer's total memory is usually measured in *kilobytes* or *megabytes*. A kilobyte (K) is 2^{10} bytes. A megabyte (MB) is 2^{20} bytes.

***Note:* Your answers to Exercises 21–23 should not involve powers.**

21. a. Find the number of bytes in a kilobyte and in a megabyte.

b. Research Look up the prefixes "kilo-" and "mega-" in a dictionary. Explain why it makes sense to use the words "kilobyte" and "megabyte" for 2^{10} and 2^{20} bytes, respectively.

22. One graphing program requires 250 K of computer memory to run. How many bytes are needed to run the program?

23. A certain brand of computer has 8 MB of memory. Express the computer's memory in bytes.

24. Challenge How many times larger than 2^n is 2^{n+1}? Explain.

Evaluate each expression when $x = 2$ and $y = 5$.

25. $3x^5 + 10$ **26.** $6y^2 - x^4$ **27.** $4(2 - y)^3$ **28.** x^4y

29. xy^3 **30.** $(xy)^3$ **31.** $x^4 + y^4$ **32.** $(x + y)^4$

33. y^x **34.** x^y **35.** $10(x - y)^3$ **36.** $(x - y)^2y^3$

GEOMETRY **Formulas for the surface area S.A. and volume *V* of a sphere with radius *r* are given below. Use the formulas for Exercises 37–39.**

$$\text{S.A.} = 4\pi r^2 \qquad V = \frac{4}{3}\pi r^3$$

r

Radius	Surface Area	Volume
1	?	?
2	?	?
4	?	?
8	?	?
n	?	?
2*n*	?	?

37. Copy and complete the table. Leave your answers in terms of π.

38. How do the surface area and volume of a sphere change when the radius of the sphere is doubled? Use the table to support your answer.

39. Open-ended Problem Suppose a toy store sells rubber balls in two different sizes. The radius of the larger balls is twice the radius of the smaller balls. How do you think the store owner should price the two different kinds of rubber balls? Explain your reasoning.

40. SAT/ACT Preview If $A = 2^n$ and $B = 2n$ for a positive integer n, then:

A. A < B **B.** A > B **C.** A = B

D. relationship cannot be determined

---

21. a. 1024 bytes; 1,048,576 bytes

b. "Kilo-" means "thousand" and 2^{10} is the power of 2 closest to one thousand. "Mega-" means "million" and 2^{20} is the power of 2 closest to one million.

22. 256,000 bytes

23. 8,388,608 bytes

24. 2 times larger; Answers may vary. An example is given. $2^2 = 4$ and $2^1 = 2$, so $2^2 = 2 \cdot 2^1$; $2^3 = 8$, so $2^3 = 2 \cdot 2^2$; $2^4 = 16$, so $2^4 = 2 \cdot 2^3$; it appears that for every n, 2^{n+1} is 2 times 2^n.

25. 106 **26.** 134

27. −108 **28.** 80

29. 250 **30.** 1000

31. 641 **32.** 2401

33. 25 **34.** 32

35. −270 **36.** 1125

37.

Radius	Surface Area	Volume
1	4π	$\frac{4}{3}\pi$
2	16π	$\frac{32}{3}\pi$
4	64π	$\frac{256}{3}\pi$
8	256π	$\frac{2048}{3}\pi$
n	$4\pi n^2$	$\frac{4}{3}\pi n^3$
$2n$	$16\pi n^2$	$\frac{32}{3}\pi n^3$

38. When the radius of a sphere doubles, the surface area is multiplied by 4 and the volume is multiplied by 8.

39. Answers may vary. An example is given. If the balls are hollow, the store owner should base the price comparison on surface area. In that case, the larger ball has surface area four times that of the smaller ball and should cost four times as much. If the

41. CARPENTRY Chris Taylor is building a storage unit like the one shown. The small cabinet portion is a cube with length, width, and height x. The tall shelf portion has length and width x and height y.

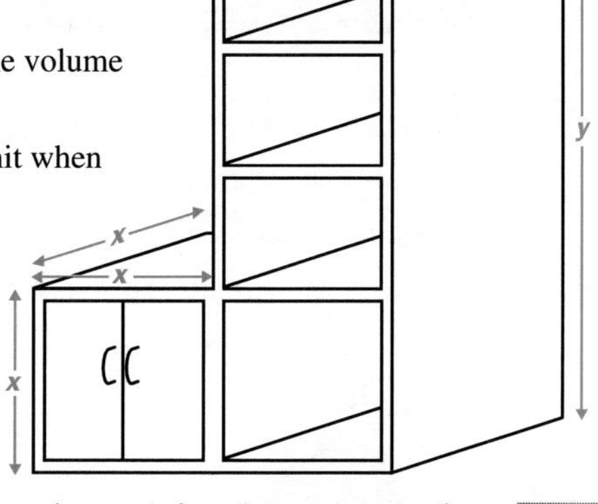

a. Write a variable expression for the volume of the storage unit.

b. Find the volume of the storage unit when $x = 2$ ft and $y = 6$ ft.

c. **Open-ended Problem** Substitute another pair of values for x and y and find the new volume of the storage unit. Be sure to choose reasonable values for each dimension.

42. NUMBER THEORY The largest known primes are found among a set of numbers named for the 17th-century French mathematician Marin Mersenne. A Mersenne number has the form $2^p - 1$, where p is a prime number. A Mersenne prime is a Mersenne number that is also prime.

a. Copy the table at the right. List the five smallest primes in the first column of your table. Then complete the rest of the table.

p	2^p-1	Is 2^p-1 prime?
2	3	yes

b. **Writing** Is every Mersenne number a Mersenne prime? If so, explain how you can tell. If not, give an example of a Mersenne number that is not prime.

BY THE WAY

In 1978, two California students, Laura Nickel and Curt Noll, showed that the Mersenne number $2^{21,701} - 1$ is prime. This number has 6533 digits and was the largest known prime number at the time.

ONGOING ASSESSMENT

43. Writing When is the expression b^n negative? When is it zero? Explain your reasoning.

SPIRAL REVIEW

Describe a situation that can be represented by each expression. Then evaluate the expression when $m = 7$ and $p = 2$. *(Section 1.6)*

44. $32m$ **45.** $4.5p$ **46.** $\frac{2p}{3}$

Find each product. Then check the product by evaluating both sides of your equation when $r = -2$ and $s = 3$. *(Section 6.5)*

47. $(r + 5)^2$ **48.** $(2s + 3)(2s - 3)$ **49.** $(7 - 3r)^2$

Apply⇔Assess

Exercise Notes

Research

Ex. 42 You may wish to ask some students to research the life of Marin Mersenne. Some other students could research what the Greeks called *perfect numbers* (numbers whose proper divisors add up to the number itself) and Mersenne's contribution to the study of those numbers.

Assessment Note

Ex. 43 Students can begin this exercise by recording examples of the value of b^n for positive and negative values of b and positive whole number values of n. Ask students to record their observations in their journals.

Practice 58 for Section 9.1

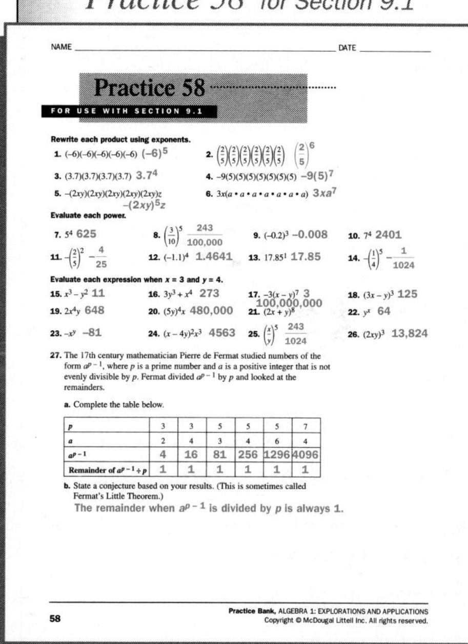
NAME ______ DATE ______

Practice 58

FOR USE WITH SECTION 9.1

Rewrite each product using exponents.

1. $(-6)(-6)(-6)(-6)(-6)$ $(-6)^5$
2. $\left(\frac{2}{5}\right)\left(\frac{2}{5}\right)\left(\frac{2}{5}\right)\left(\frac{2}{5}\right)\left(\frac{2}{5}\right)\left(\frac{2}{5}\right)$ $\left(\frac{2}{5}\right)^6$
3. $(3.7)(3.7)(3.7)(3.7)$ 3.7^4
4. $-9(5)(5)(5)(5)(5)(5)(5)$ $-9(5)^7$
5. $-(2xy)(2xy)(2xy)(2xy)(2xy)z$ $-(2xy)^5z$
6. $3x(a \cdot a \cdot a \cdot a \cdot a \cdot a \cdot a)$ $3xa^7$

Evaluate each power.

7. 5^4 625
8. $\left(\frac{3}{10}\right)^5$ $\frac{243}{100{,}000}$
9. $(-0.2)^3$ -0.008
10. 7^4 2401
11. $-\left(\frac{2}{5}\right)^2$ $-\frac{4}{25}$
12. $(-1.1)^4$ 1.4641
13. 17.85^1 17.85
14. $-\left(\frac{1}{4}\right)^5$ $-\frac{1}{1024}$

Evaluate each expression when $x = 3$ and $y = 4$.

15. $x^3 - y^2$ 11
16. $3y^3 + x^4$ 273
17. $-3(x - y)^7$ 3
18. $(3x - y)^3$ 125
19. $2x^4y$ 648
20. $(5y)^4x$ 480,000
21. $(2x + y)^8$ 100,000,000
22. y^x 64
23. $-x^y$ -81
24. $(x - 4y)^2x^3$ 4563
25. $\left(\frac{x}{y}\right)^5$ $\frac{243}{1024}$
26. $(2xy)^3$ 13,824

27. The 17th century mathematician Pierre de Fermat studied numbers of the form a^{p-1}, where p is a prime number and a is a positive integer that is not evenly divisible by p. Fermat divided a^{p-1} by p and looked at the remainders.

a. Complete the table below.

p	3	3	5	5	5	7
a	2	4	3	4	6	4
a^{p-1}	4	16	81	256	1296	4096
Remainder of $a^{p-1} \div p$	1	1	1	1	1	1

b. State a conjecture based on your results. (This is sometimes called Fermat's Little Theorem.)
The remainder when a^{p-1} is divided by p is always 1.

58

Practice Bank, ALGEBRA 1: EXPLORATIONS AND APPLICATIONS

balls are solid, the owner should base the price comparison on volume. The larger ball has volume eight times that of the smaller ball and should cost eight times as much.

40. D

41. a. $x^3 + x^2y$ b. $32\ \text{ft}^3$
c. Answers may vary. An example is given. If $x = 1.5$ ft and $y = 5$ ft, then the volume is about $14.6\ \text{ft}^3$.

42. a.

p	$2^p - 1$	Is $2^p - 1$ prime?
2	3	Yes.
3	7	Yes.
5	31	Yes.
7	127	Yes.
11	2047	No.

b. No; the Mersenne number $2^{11} - 1 = 2047$ is not prime, since $2047 = 89 \cdot 23$.

43. when b is negative and n is odd, because the product of an odd number of negative terms is negative; when $b = 0$ and n is positive, because a product is zero only when one of the factors is 0 (b^n is not defined when $b = 0$ and n is negative.)

44–46. Answers may vary. Examples are given.

44. the number of students in a school in which there are 32 students in each of m classes; 224

45. the total price of 4.5 lb of a commodity that costs p dollars per pound; $9

46. the cost in dollars per person of an item shared by 3 people when the previous total cost, p, is doubled; $1.33

47. $(r + 5)^2 = r^2 + 10r + 25$; $9 = 9$

48. $(2s + 3)(2s - 3) = 4s^2 - 9$; $27 = 27$

49. $(7 - 3r)^2 = 49 - 42r + 9r^2$; $169 = 169$

Plan⇔Support

Objectives

- Interpret and evaluate exponential functions.
- Model real-world situations with exponential functions.
- Make predictions using exponential functions.

Recommended Pacing

Core and Extended Courses
Section 9.2: 1 day

Block Schedule
Section 9.2: $\frac{1}{2}$ block
(with Section 9.1)

Resource Materials

Lesson Support
Lesson Plan 9.2
Warm-Up Transparency 9.2
Practice Bank: Practice 59
Study Guide: Section 9.2

Technology
Graphing Calculator
Spreadsheet Software
McDougal Littell Software
Function Investigator
Stats!

Internet:
http://www.hmco.com

Warm-Up Exercises

Evaluate each expression.

1. 25% of 150 37.5
2. 125% of 150 187.5
3. $(1.4)^3$ 2.744
4. $30(2.05)^2$ 126.075
5. $5(1 + 0.5)^4$ 25.3125
6. $27(1 + 0.15)^2$ 35.7075

SECTION

9.2 Exponential Growth

Learn how to...

- interpret and evaluate exponential functions
- model real-world situations with exponential functions

So you can...

- make predictions about things that grow, such as population

Toolbox p. 587 *Fractions, Decimals, and Percents*

In 1980, there were only 75 peregrine falcons in Alaska. People decided to try to save the endangered bird. Because of their efforts, the population grew by about 12% per year from 1980 to 1992. You can use these facts to estimate the Alaskan peregrine population after one year.

$$\begin{pmatrix}\text{Population}\\ \text{after 1 year}\end{pmatrix} = \begin{pmatrix}\text{Initial}\\ \text{population}\end{pmatrix} + \begin{pmatrix}\text{Growth in}\\ \text{population}\end{pmatrix}$$

$$\begin{aligned}\begin{pmatrix}\text{Population}\\ \text{after 1 year}\end{pmatrix} &= 75 + (12\% \text{ of } 75)\\ &= 75 + 0.12(75) \quad \leftarrow \text{0.12 is the same as 12\%.}\\ &= 75(1 + 0.12)\\ &= 75(1.12)^1\\ &= 84 \quad \leftarrow \text{population in 1981}\end{aligned}$$

The number 1.12 in the above calculations is called the *growth factor*. You can use the growth factor to find the peregrine population in other years.

EXAMPLE 1 Application: Ecology

Estimate the Alaskan population of peregrine falcons in 1983.

SOLUTION

Use the growth factor 1.12 to find the population in 1982 and in 1983.

$$\begin{aligned}\text{Population after 2 years} &= \begin{pmatrix}\text{Population}\\ \text{after 1 year}\end{pmatrix}(1.12)\\ &= [75(1.12)](1.12)\\ &= 75(1.12)^2 \quad \leftarrow \text{population in 1982}\end{aligned}$$

$$\begin{aligned}\text{Population after 3 years} &= \begin{pmatrix}\text{Population}\\ \text{after 2 years}\end{pmatrix}(1.12)\\ &= [75(1.12)(1.12)](1.12)\\ &= 75(1.12)^3 \quad \leftarrow \text{population in 1983}\\ &\approx 105\end{aligned}$$

In 1983, there were about 105 peregrine falcons in Alaska.

Exponential Equations

You can see from Example 1 that

Population after x years = $75(1.12)^x$.

The number of peregrine falcons increased by a fixed percent each year. This kind of increase is called **exponential growth**. Any quantity y that grows by a fixed percent at regular intervals can be modeled by an equation in the form below. The x in the equation represents the number of intervals.

$$y = a(1 + r)^x$$

The constant a represents the initial amount before any growth occurs.

The constant r represents the **growth rate**. The quantity $1 + r$ represents the **growth factor**.

BY THE WAY

A pesticide referred to as DDT caused peregrine eggs to have thin shells, so that developing chicks were less likely to survive. Since the use of DDT was restricted, the shells have been thicker, and the chemical no longer inhibits peregrine population growth.

EXAMPLE 2 Application: Ecology

Use the information on page 376 to estimate the Alaskan population of peregrine falcons in 1991.

SOLUTION

Let y = the peregrine population x years after 1980.

$$y = a(1 + r)^x$$

The initial population in 1980 was 75 peregrine falcons, so $a = 75$.

The population grows by 12% each year after 1980, so $r = 0.12$.

Substitute these values into the equation.

$$y = 75(1 + 0.12)^x$$
$$= 75(1.12)^{11}$$
$$\approx 261$$

1991 is 11 years after 1980, so $x = 11$.

In 1991, there were about 261 peregrine falcons in Alaska.

THINK AND COMMUNICATE

The equation $y = 84(1.12)^x$ models the number of peregrine falcons in Alaska x years after 1981.

1. What does the number 84 represent in this equation?
2. What does the variable x represent in this equation?
3. What value of x would you substitute into this equation to estimate the number of peregrines in 1991? Explain your reasoning.

ANSWERS Section 9.2

Think and Communicate

1. the peregrine falcon population in 1981
2. the number of years after 1981
3. 10; The number of years from 1981 to 1991 is 10.

Teach⇔Interact

Section Note

Communication: Discussion
When discussing the term *growth factor*, remind students that factor implies multiplication.

Additional Example 1

Estimate the Alaskan population of peregrine falcons in 1984.

Use the growth factor 1.12 to find the population in 1984.
(Population after 4 years)
= (Population after 3 years)(1.12)
= [75(1.12)(1.12)(1.12)](1.12)
= $75(1.12)^4$
≈ 118

In 1984, there were about 118 peregrine falcons in Alaska.

Section Note

Teaching Tip
In the equation $y = a(1 + r)^x$, students need to understand that in Example 1 the growth rate r is 12%; that is, $r = 0.12$ and the growth factor is $1 + r = 1 + 0.12$.

Additional Example 2

Estimate the Alaskan population of peregrine falcons in 1988.

Let y = the population x years after 1980.
$y = a(1 + r)^x$
The initial population in 1980 was 75 peregrine falcons, so $a = 75$.
The population grows by 12% each year after 1980, so $r = 0.12$. 1988 is 8 years after 1980, so $x = 8$.
Substitute these values into the formula.
$y = 75(1 + 0.12)^x$
$= 75(1.12)^8$
≈ 186
In 1988, there were about 186 peregrine falcons in Alaska.

Think and Communicate

After discussing questions 1–3, you may wish to point out that the year in which the initial amount, a, occurs can be thought of as year zero. For the equation $y = 84(1.12)^x$, if $x = 0$, then $y = 84(1.12)^0 = 84 \cdot 1 = 84$.

Teach⇔Interact

Learning Styles: Visual

The graph of the equation $y = 75(1.12)^x$ shows very clearly that exponential growth is a very rapid form of growth. This may not have been apparent to visual learners from the purely algebraic analysis of exponential equations.

Think and Communicate

Question 4 highlights a common error in reasoning that some students make. The constant growth rate of 12% each year is applied to an ever increasing number of peregrine falcons each year, which results in larger numbers from year to year.

Checking Key Concepts

Student Progress
Questions 4–6 should provide a good assessment of how well students understand the concept of exponential growth.

Closure Question

Explain what a, r, and $1 + r$ represent in the exponential growth function $y = a(1 + r)^x$.
a: the initial amount before any growth occurs; r: the growth rate; $1 + r$: the growth factor

Graphing Exponential Equations

In Example 2 you saw that the Alaskan population of the peregrine falcon more than tripled in just 11 years. Another way to show the population increase over time is to graph the equation $y = 75(1.12)^x$.

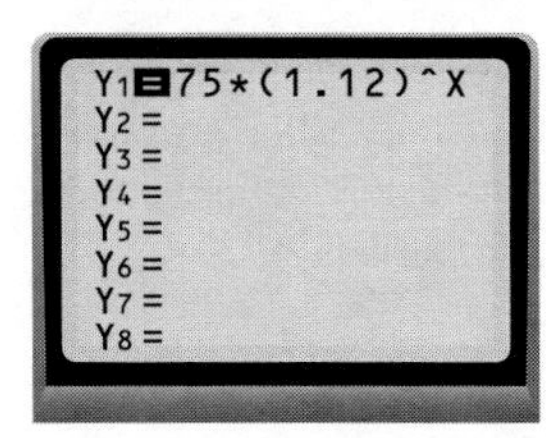

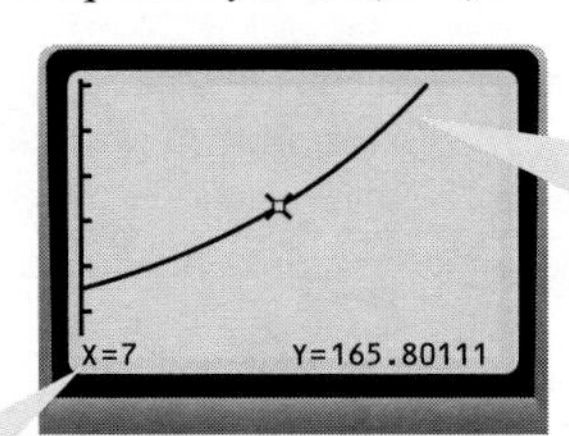

As x becomes larger, y grows faster and the graph becomes steeper.

Seven years after 1980, the population was about 166.

For any graph that represents exponential growth, the value of y increases as x becomes larger.

THINK AND COMMUNICATE

4. Ming says that since the population increases by 12% each year, the population should increase by the same number of peregrines each year. Use the shape of the graph to show Ming his mistake.

5. The population y of New Zealand in thousands x years after 1980 can be modeled by the equation $y = 3113(1.006)^x$.

a. What was the population of New Zealand in 1980? Explain.

b. Predict where the graph of $y = 3113(1.006)^x$ crosses the vertical axis. Check your prediction by graphing the equation on a graphing calculator. Set a good viewing window, such as $0 \leq x \leq 50$ and $2500 \leq y \leq 4500$.

CHECKING KEY CONCEPTS

Find the value of y when $x = 8$.

1. $y = 450(1.07)^x$ **2.** $y = 12(1.5)^x$ **3.** $y = 57(2)^x$

The exponential equation $y = 91.6(1 + 0.77)^x$ models the number of cellular telephone subscribers y in thousands in the United States x years after 1983.

4. What does the number 91.6 represent?

5. What does the number 0.77 represent?

6. Find the number of cellular telephone subscribers in the United States in 1992. Explain how you got your answer.

Think and Communicate

4. Answers may vary. An example is given. If the increase each year were constant, the graph would be a line not a curve. Its slope would be the annual increase.

5. a. 3,113,000; 3113 represents the initial population in thousands.

b. at (0, 3113)

Checking Key Concepts

1, 2. Answers are given to the nearest hundredth.

1. 773.18 2. 307.55

3. 14,592

4. the number of cellular telephone subscribers in thousands in 1983

5. the growth rate of the number of cellular telephone subscribers after 1983

6. about 15, 619, 059; I substituted 9 for x and evaluated $91.6(1.77)^9$; then I multiplied by 1000 to get the number of subscribers.

Exercises and Applications

1–9. When necessary, answers are rounded to the nearest hundredth.

1. 2430 2. 1751.78

3. 32,659.2 4. 335.54

9.2 Exercises and Applications

Find the value of y when $x = 5$.

1. $y = 10(3)^x$ **2.** $y = 22(2.4)^x$ **3.** $y = 4.2(6)^x$

4. $y = (3.2)^x$ **5.** $y = 23.1(1.05)^x$ **6.** $y = 540(1.17)^x$

7. $y = 19(2.07)^x$ **8.** $y = 38(8.2)^x$ **9.** $y = 71(10)^x$

Extra Practice exercises on page 574

Each equation in Exercises 10–15 models the population y in thousands of the given country x years after 1980. For each country:

a. Give the population in 1980.

b. Give the annual growth rate.

c. Estimate the population in 1989 to the nearest thousand.

Explain how you found your answers.

10. Greece: $y = 9643(1.005)^x$ **11.** Austria: $y = 7549(1.002)^x$

12. Canada: $y = 24{,}070(1.01)^x$ **13.** Finland: $y = 4780(1.004)^x$

14. Japan: $y = 116{,}807(1.006)^x$ **15.** Italy: $y = 56{,}451(1.002)^x$

16. BIOLOGY A newly hatched channel catfish typically weighs about 0.3 g. This weight increases by an average of 10% per day for the first six weeks of a catfish's life.

a. Write an equation that models the weight y of a catfish in grams x days after hatching. How did you find your equation?

b. Estimate a catfish's weight 12 days after hatching.

PERSONAL FINANCE When you deposit money into a savings account, the money earns interest. For example, suppose you open a savings account that pays 2% annual interest. The money in your account will increase by 2% each year. For Exercises 17 and 18, assume no other deposits or withdrawals are made.

17. Suppose you use $200 to open an account paying 3.5% annual interest.

a. Write an equation that gives the value y of the account in dollars x years from now.

b. How much money will you have in the account after five years?

18. The value of an investment t years after you make an initial deposit of $500 depends on the interest rate I. Find the value of each account.

a. $t = 10, I = 3\%$ b. $t = 20, I = 3\%$

c. $t = 5, I = 6\%$ d. $t = 5, I = 12\%$

Apply⇔Assess

Suggested Assignment

Core Course
Exs. 1–18, 22–26, 29, 31–43

Extended Course
Exs. 7–12, 17–43

Block Schedule
Day 57 Exs. 1–18, 22–26, 29, 31–43

Exercise Notes

Application
Exs. 10–21, 29, 30 These exercises provide varied applications of exponential growth functions. Students now have an opportunity to apply the skills learned in this section to real-world problems involving population growth, biology, personal finance, traffic analysis, and social studies.

Second-Language Learners
Exs. 17, 18 Some second-language learners may not be familiar with the different types of accounts a person may have at a bank. Consider having a small group conduct research by visiting financial institutions in your community, finding out what kinds of accounts each offers, and the interest and restrictions associated with each. Have students report their findings to the class.

5. 29.48 6. 1183.92
7. 722.11 8. 1,408,811.40
9. 7,100,000

10–15. Answers are obtained from the equation $y = a(1 + r)^x$, where y = the population in thousands, x = the number of years after 1980, a is the initial population, and r is the growth rate.

a. Multiply a by 1000, since a represents the initial population in thousands.
b. Subtract 1 from the quantity in parentheses, since r is the growth rate.
c. Substitute 9 for x in the given equation, since 1989 is 9 years after 1980. Round the result to the nearest thousand.

10. a. 9,643,000 b. 0.005 c. 10,086,000
11. a. 7,549,000 b. 0.002 c. 7,686,000
12. a. 24,070,000 b. 0.01 c. 26,325,000
13. a. 4,780,000 b. 0.004 c. 4,955,000
14. a. 116,807,000 b. 0.006 c. 123,268,000
15. a. 56,451,000 b. 0.002 c. 57,475,000
16. a. $y = 0.3(1.1)^x, 0 \le x \le 42$, where y = the weight in grams and x = the number of days after birth; The weight of the catfish increases exponentially for the first six weeks of life, 0.3 g is the initial weight, and 10%, or 0.1, is the growth rate per day.
b. about 0.9 g
17. a. $y = 200(1.035)^x$, where y = the amount in dollars and x = the number of years from now.
b. $237.54
18. a. $671.96 b. $903.06 c. $669.11 d. $881.17

Apply⇔Assess

Exercise Notes

Interview Note

Exs. 19–21 Body temperature influences the growth of bacteria that enter the body through a wound. When Louis Pasteur, who founded the science of bacteriology, was attempting to infect ducks with anthrax bacteria taken from diseased sheep, he discovered the bacteria would not grow in the ducks because their body temperature is about 5 degrees higher than the body temperature of sheep. After Pasteur lowered the temperature of one duck by immersing its feet in cold water, the anthrax bacteria started growing in the duck and produced the infection caused by the bacteria.

Historical Connection

Ex. 21 In 1915, an English scientist, F. W. Twort, observed a staphylococcus bacterium being destroyed in a culture by an unknown agent. Two years later, a French biologist, F. H. d'Herelle, named the invisible agent of bacterial destruction *bacteriophage* or "bacteria eater." Today, as then, phages can play a disastrous role in industries such as cheese and yogurt industries that use certain bacteria in their manufacturing process.

INTERVIEW Lori Arviso Alvord

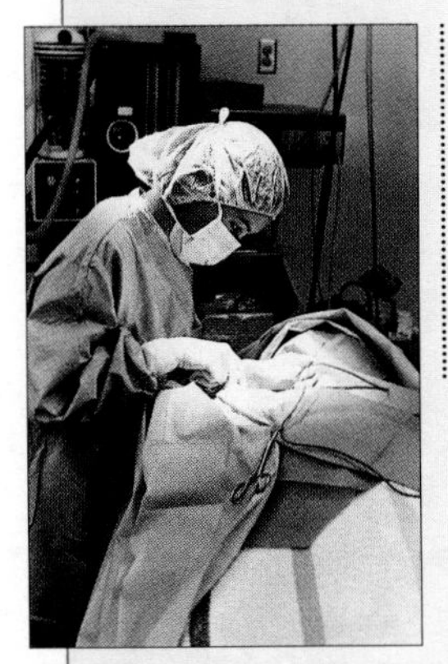

Look back at the article on pages 366–368.

Surgeons like Lori Arviso Alvord need to protect patients from bacteria that can enter a surgical wound and reproduce. Most single-cell bacteria reproduce by splitting into two new cells. The pictures below show the increase in a population of bacteria. The amount of time that elapses between each split is called the generation time.

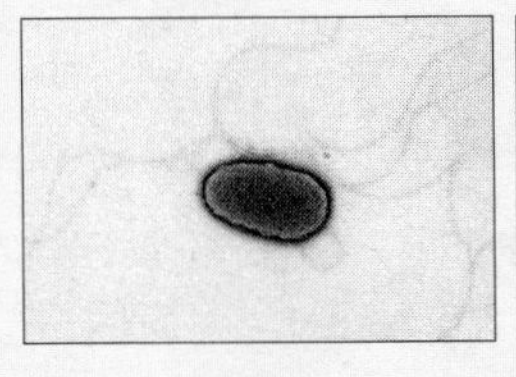
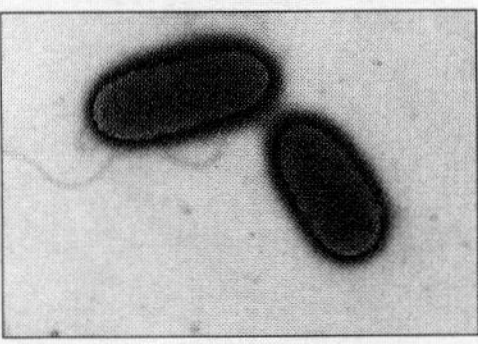
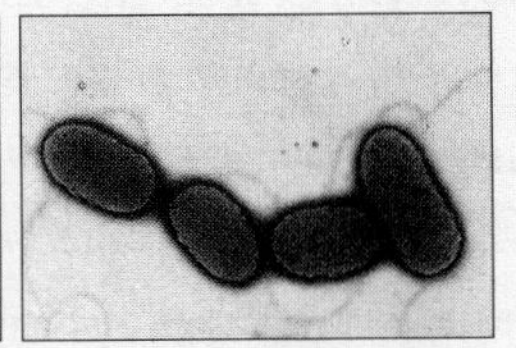

19. a. By what percent does the population increase during each generation?

b. Suppose one cell of bacteria begins to reproduce by splitting into two. Write an equation that models the bacteria population y after x generations.

c. Writing Look back at your answer to part (b). How does your equation show that the population doubles in one generation?

20. One species of bacteria often found in infected surgical wounds is the *Pseudomonas aeruginosa*. It has a generation time of about 33 min.

a. Suppose one *P. aeruginosa* cell starts to reproduce. How many cells are there after 33 min? after 66 min? after 99 min?

b. Write an equation that models the number y of *P. aeruginosa* cells after $33x$ minutes have passed. Explain your reasoning.

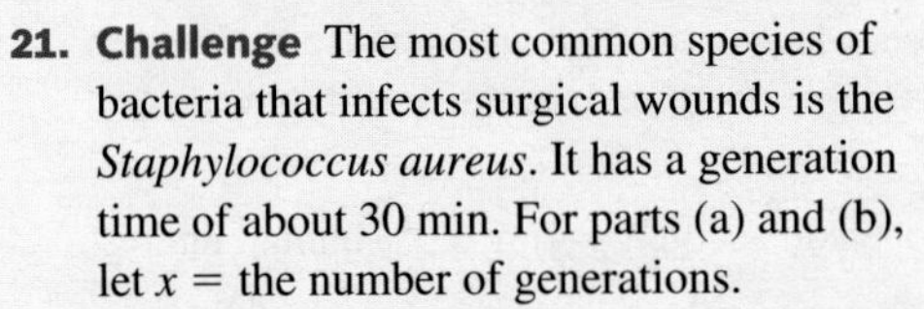

21. Challenge The most common species of bacteria that infects surgical wounds is the *Staphylococcus aureus*. It has a generation time of about 30 min. For parts (a) and (b), let x = the number of generations.

a. Suppose two *S. aureus* cells have just split into four cells. Write and solve an equation to find the number y of *S. aureus* cells $15\frac{1}{2}$ h later.

b. Explain how you chose a value for x in part(a).

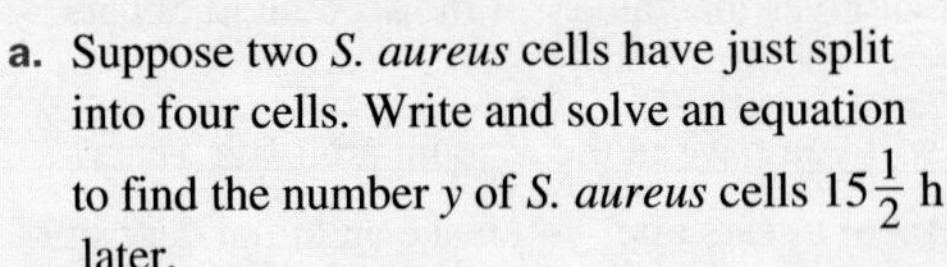

BY THE WAY

Most of the bacteria that infect wounds actually live on normal healthy skin. They become harmful only when they enter the body through a wound.

19. a. 100%

b. $y = (1 + 1)^x$ or $y = 2^x$, where y = the population and x = the number of generations.

c. Answers may vary. An example is given. When x increases by 1, y doubles. Recall that for every positive integer x, $2^{x+1} = 2(2^x)$.

20. a. 2; 4; 8

b. $y = 2^x$; The equation is the same as that in Ex. 19(b), because $33x$ min represents x generations.

21. a. $y = 4(2)^x$, where y = the population and x = the number of generations; $y = 4(2)^{31}$; about 8.6 billion cells

b. $15\frac{1}{2}$ h $= \frac{31}{2}$ h; There are 31 periods of 30 min in $15\frac{1}{2}$ h.

22. a–c.

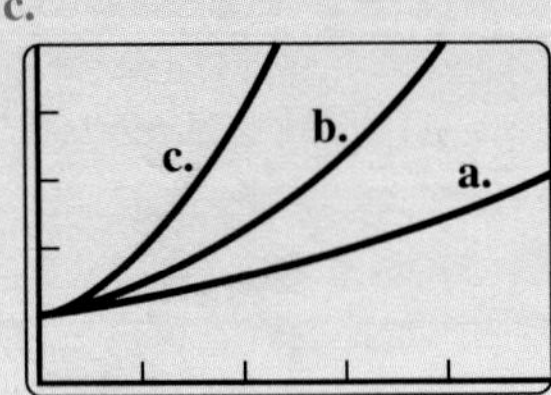

The graphs all show exponential growth and have the same initial value, but the steepness increases as b increases.

Technology How is the graph of $y = ab^x$ affected when you change the value of a? the value of b? For Exercises 22 and 23, use a graphing calculator. Set a good viewing window, such as $0 \le x \le 5$ and $0 \le y \le 25$.

22. Graph each equation on the same screen. How are the graphs similar? How are they different?

a. $y = 5(1.25)^x$ **b.** $y = 5(1.5)^x$ **c.** $y = 5(2)^x$

23. Graph each equation on the same screen. How are the graphs similar? How are they different?

a. $y = 5(1.2)^x$ **b.** $y = 10(1.2)^x$ **c.** $y = 15(1.2)^x$

For Exercises 24–26, match each function with its graph. Explain how you found your answers.

24. $y = 2(1.5)^x$ **25.** $y = 2(1.2)^x$ **26.** $y = 5(1.2)^x$

A.

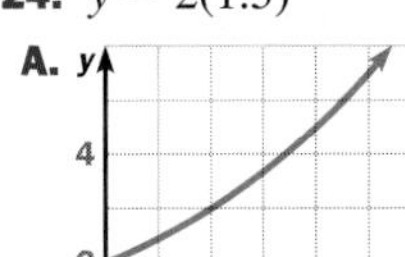

B.

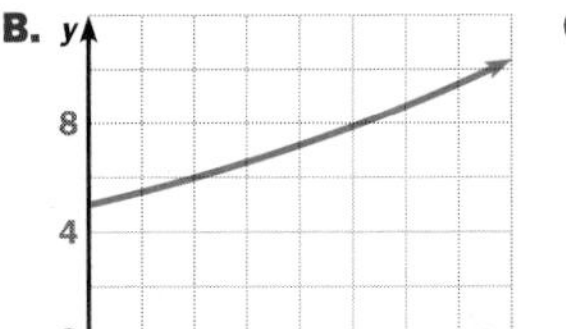

C.

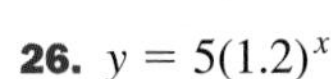

27. Open-ended Problem Describe a real-world situation that can be modeled by the equation $y = 150(1.06)^x$.

28. Spreadsheets In this exercise, you will compare the growth of three functions: the linear function $y = 2x$, the quadratic function $y = x^2$, and the exponential function $y = 2^x$.

a. Use a spreadsheet to find the values of the three functions when $x = 1, 2, \ldots, 20$.

b. Which function grows most quickly when $x > 4$? Use your spreadsheet to justify your answers.

c. Graph the three functions in the same coordinate plane. Do the graphs support your answer to part (b)? If so, how?

Comparing Growth

B2 | =2*A2

	A	B	C	D
1	x	2x	x^2	2^x
2	1	2	1	2
3	2	4	4	4
4	3			
5	4			
6	5			
7	6			
8	7			
9	8			
10	9			

In many spreadsheet programs, the symbol ^ is used for exponentiation.

Apply⇔Assess

Exercise Notes

Cooperative Learning

Exs. 22–27 Have students work in pairs, taking turns to graph the equations. They should discuss the shape and the position of the graphs. For Ex. 27, groups can be combined to share their descriptions of real-world situations.

Reasoning

Ex. 28 The spreadsheet allows students to see that as x increases, the value of y increases most rapidly for the exponential function. Ask students how their graphs show them that the exponential function does not always give the greatest y value. (The graph of $y = 2^x$ is below the graph of $y = x^2$ for $2 \le x < 4$.) For what values of x is 2^x equal to or greater than x^2? ($x \ge 4$)

Using Technology

Ex. 28 Students may wish to explore ways of using the *Stats!* or *Function Investigator* software for this exercise. On the TI-82, students can use the following procedure. Enter the equations on the Y= list, press [2nd][TblSet], and enter the settings TblMin = 1, ΔTbl = 1. Finally, press [2nd][Table] to view the tables of function values. This procedure will generate the tables quickly, but the values of the third function can only be viewed by using the ▶ key.

23. a–c.

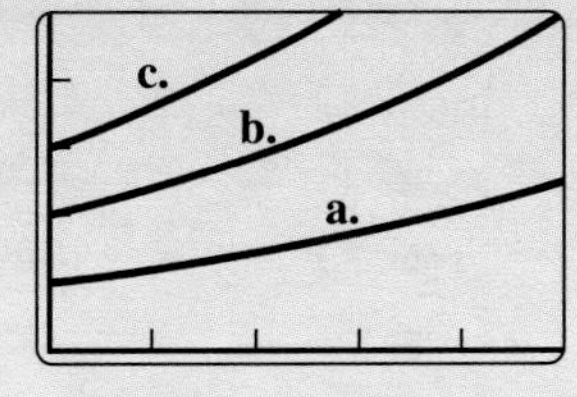

The graphs all show exponential growth with the same growth factor but different initial values.

24–26. Explanations may vary. Examples are given.

24. A; The initial value (at $x = 0$) is 2, so I know the correct graph is either A or C. From Ex. 22, I know that if $b_2 > b_1$, the graph of $y = a(b_2)^x$ is steeper than the graph of $y = a(b_1)^x$.

25. C; The initial value (at $x = 0$) is 2; the only such remaining graph is C.

26. B; The initial value (at $x = 0$) is 5; the only such graph is B.

27. Answers may vary. An example is given. the amount (in dollars) in a savings account after x years if the initial deposit was \$150 and the annual interest rate is 6% (if no deposits or withdrawals are made)

28. See answers in back of book.

Apply⇔Assess

Exercise Notes

Career Connection

Ex. 30 Economics is a social science that has grown in importance as the world's economy has become increasingly interrelated. Economists are highly trained individuals who are involved in every aspect of a nation's financial life. This includes the production, distribution, and consumption of goods and services, which affect the material well being of every person in a society.

Journal Entry

Ex. 31 It will be helpful for students to use examples to explain why a given situation can or cannot be modeled by an exponential equation. Ask students to record their responses to this exercise in their journals.

Practice 59 for Section 9.2

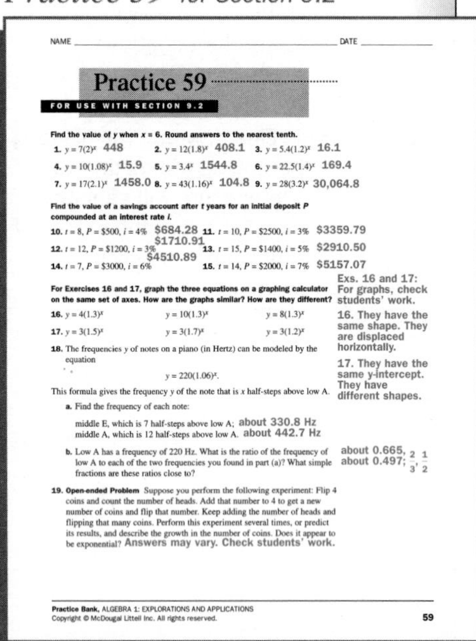

NAME ________________ DATE ________

Practice 59

FOR USE WITH SECTION 9.2

Find the value of y when $x = 6$. Round answers to the nearest tenth.

1. $y = 7(2)^x$ 448
2. $y = 12(1.8)^x$ 408.1
3. $y = 5.4(1.2)^x$ 16.1
4. $y = 10(1.08)^x$ 15.9
5. $y = 3.4^x$ 1544.8
6. $y = 22.5(1.4)^x$ 169.4
7. $y = 17(2.1)^x$ 1458.0
8. $y = 43(1.16)^x$ 104.8
9. $y = 28(3.2)^x$ 30,064.8

Find the value of a savings account after t years for an initial deposit P compounded at an interest rate i.

10. $t = 8$, $P = \$500$, $i = 4\%$ \$684.28
11. $t = 10$, $P = \$2500$, $i = 3\%$ \$3359.79
12. $t = 12$, $P = \$1200$, $i = 3\%$ \$1710.91
13. $t = 15$, $P = \$1400$, $i = 5\%$ \$2910.50
14. $t = 7$, $P = \$3000$, $i = 6\%$ \$4510.89
15. $t = 14$, $P = \$2000$, $i = 7\%$ \$5157.07

For Exercises 16 and 17, graph the three equations on a graphing calculator on the same set of axes. How are the graphs similar? How are they different? Exs. 16 and 17: For graphs, check students' work.

16. $y = 4(1.3)^x$ $y = 10(1.3)^x$ $y = 8(1.3)^x$ 16. They have the same shape. They are displaced horizontally.

17. $y = 3(1.5)^x$ $y = 3(1.7)^x$ $y = 3(1.2)^x$ 17. They have the same y-intercept. They have different shapes.

18. The frequencies y of notes on a piano (in Hertz) can be modeled by the equation

$$y = 220(1.06)^x.$$

This formula gives the frequency y of the note that is x half-steps above low A.

a. Find the frequency of each note:

middle E, which is 7 half-steps above low A; about 330.8 Hz
middle A, which is 12 half-steps above low A. about 442.7 Hz

b. Low A has a frequency of 220 Hz. What is the ratio of the frequency of low A to each of the two frequencies you found in part (a)? What simple fractions are these ratios close to? about 0.665, about 0.497; $\frac{2}{3}, \frac{1}{2}$

19. **Open-ended Problem** Suppose you perform the following experiment: Flip 4 coins and count the number of heads. Add that number to 4 to get a new number of coins and flip that number. Keep adding the number of heads and flipping that many coins. Perform this experiment several times, or predict its results, and describe the growth in the number of coins. Does it appear to be exponential? Answers may vary. Check students' work.

Practice Bank, ALGEBRA 1: EXPLORATIONS AND APPLICATIONS

59

BY THE WAY

It has been estimated that the traffic level in the United States in the year 2020 will be up to 4 trillion vehicle-miles. This rise in traffic will increase congestion delays by 400% unless the highway systems improve.

29. TRAFFIC In 1983, the traffic level in the United States was about 1653 billion vehicle-miles. (That means that all the vehicles in the country drove a total of 1653 billion miles that year.) The traffic level continued growing by about 4% per year for over a decade after that.

a. Write an equation that models the traffic level y in billions of vehicle-miles x years after 1983.

b. Estimate to the nearest billion the number of vehicle-miles traveled in 1989. What value did you substitute for x? Explain your reasoning.

30. SOCIAL STUDIES The British economist Thomas R. Malthus (1766–1834) observed that a human population tends to increase exponentially. A population model that assumes exponential growth is called a *Malthusian model*.

a. The population of Portugal was about 9,778,000 in 1980 and about 10,365,000 in 1990. By what percent did the population increase from 1980 to 1990? Give your answer to the nearest percent.

$$\left(\textit{Hint: } \text{Percent increase} = \frac{\text{1990 population} - \text{1980 population}}{\text{1980 population}} \cdot 100\right)$$

b. Create a Malthusian model by writing an exponential function that predicts Portugal's population y in millions x decades after 1980.

c. Predict Portugal's population in 2020 assuming the growth rate remains the same. What value did you substitute for x? Explain.

ONGOING ASSESSMENT

31. Writing How do you know whether a quantity that grows can be modeled by an exponential equation? What information do you need in order to write the equation?

SPIRAL REVIEW

Solve each system of equations by adding or subtracting. *(Section 7.3)*

32. $x - y = 5$, $x + y = 1$

33. $15p - 7q = 11$, $3p + 7q = -11$

34. $m - 5n = 10$, $m + 7n = 4$

35. $3a - 5b = -12$, $3a + b = 6$

36. $\frac{3}{4}x + \frac{2}{3}y = 25$, $x - \frac{2}{3}y = -10$

37. $5x - 2y = 6$, $5x + 6y = 2$

Solve each equation by using the quadratic formula. *(Section 8.5)*

38. $x^2 + 2x - 2 = 0$

39. $y^2 - y - 1 = 0$

40. $z^2 - 5z + 3 = 0$

41. $2x^2 - 5x - 7 = 0$

42. $3x^2 + 4x = 1$

43. $1 + y^2 = 5y^2 - 2y$

29. a. $y = 1653(1.04)^x$, where y is the number of vehicle-miles in billions x years after 1983.
b. about 2092 billion vehicle-miles to the nearest billion; $x = 6$ since 1989 is 6 years after 1983.

30. a. about 6%
b. $y = 9.778(1.06)^x$, where $y =$ the population in millions and $x =$ the number of decades after 1980.
c. about 11.646 million; $x = 3$ since the number of decades from 1990 to 2020 is 3.

31. Determine whether the quantity increases by a fixed rate over each time period; the initial amount, the growth rate, and the time period.

32. $(3, -2)$

33. $\left(0, -\frac{11}{7}\right)$

34. $\left(\frac{15}{2}, -\frac{1}{2}\right)$

35. $(1, 3)$

36. $\left(\frac{60}{7}, \frac{195}{7}\right)$

37. $\left(1, -\frac{1}{2}\right)$

38. $-1 \pm \sqrt{3}$

39. $\frac{1}{2} \pm \frac{\sqrt{5}}{2}$

40. $\frac{5}{2} \pm \frac{\sqrt{13}}{2}$

41. $-1; 3\frac{1}{2}$

42. $-\frac{2}{3} \pm \frac{\sqrt{7}}{3}$

43. $\frac{1}{4} \pm \frac{\sqrt{5}}{4}$

SECTION

9.3 Exponential Decay

If you have 100 pennies and take away half, how many do you have left? What if you take away half of those? What if you keep taking away half? This kind of decrease is called *exponential decay*.

Learn how to...

- use equations to model real-world decay situations

So you can...

- make predictions about quantities that decrease, such as the height of a bouncing ball

Number of times cup is emptied	Number of pennies left
0	100
1	?
2	?
3	?
4	?
5	?
6	?

EXPLORATION
COOPERATIVE LEARNING

Modeling Exponential Decay

Work with a partner.
You will need:

- 100 pennies
- a cup
- a graphing calculator

Follow Steps 1–3. Then answer the questions below.

1 Put the pennies in the cup. Then empty them onto a desk.

2 Remove all pennies that land heads up. Record the number of pennies remaining.

3 Repeat Steps 1 and 2 until no pennies remain.

Questions

1. Make a scatter plot of the data. Put the number of times the cup is emptied on the horizontal axis. Put the number of pennies remaining on the vertical axis.
2. You can expect to remove about half of the pennies each time. Use this fact to copy and complete the table at the right.
3. Write an equation that models the expected number y of pennies remaining after the cup is emptied x times. Graph the equation in the same coordinate plane as your scatter plot. What do you notice?

x = number of times cup is emptied	y = expected number of pennies remaining
0	100
1	$100(\frac{1}{2})$
2	$(100\cdot\frac{1}{2})(\frac{1}{2})=100(\frac{1}{2})^2$
3	?
4	?
5	?
6	?

Plan⇔Support

Objectives

- Use equations to model real-world decay situations.
- Make predictions about quantities that decrease.

Recommended Pacing

Core and Extended Courses
Section 9.3: 1 day

Block Schedule
Section 9.3: $\frac{1}{2}$ block
(with Section 9.4)

Resource Materials

Lesson Support
Lesson Plan 9.3
Warm-Up Transparency 9.3
Practice Bank: Practice 60
Study Guide: Section 9.3
Explorations Lab Manual:
Additional Exploration 14,
Diagram Masters 1, 2, 16
Assessment Book: Test 38

Technology
Technology Book:
Calculator Activity 9
Spreadsheet Activity 9
Graphing Calculator
McDougal Littell Software
Function Investigator
Internet:
http://www.hmco.com

Warm-Up Exercises

Evaluate each expression.

1. $10(0.25)^3$ 0.15625
2. $312(1 - 0.8)^2$ 12.48
3. $500\left(\frac{2}{3}\right)^4$ $\frac{8000}{81}$

Find the value of y when $x = 2$.

4. $y = 3(0.85)^x$ 2.1675
5. $y = 1982(1 - 0.17)^x$ 1365.3998
6. Write 3(0.4)(0.4)(0.4)(0.4) using exponents. $3(0.4)^4$

Exploration Note

Purpose
The purpose of this Exploration is to have students explore a real-world situation that exhibits a pattern of exponential decay.

Materials/Preparation
Each group will need 100 pennies, a cup, and a graphing calculator. You may wish to review the procedure for drawing a scatter plot.

Procedure
Students should record the number of times the cup is emptied as well as the number of pennies remaining. Then they will have the data needed to make their scatter plots.

Closure
Groups can share their equations and graphs with each other. All students should arrive at an understanding that a quantity that decreases regularly by a fixed percent (in this case, about 50%) will approach 0 rapidly.

Explorations Lab Manual
See the Manual for more commentary on this Exploration.

Diagram Master 16

For answers to the Exploration, see following page.

Teach⇔Interact

Learning Styles: Visual

Students can get a sense of how rapidly a number decreases when it is multiplied repeatedly by a fixed percent by having individual students read and then record on a wall of the classroom the original height of the ball in Example 1 and each new height after each bounce of the ball.

Additional Example 1

A nail is sticking out 15 mm from a board. Each time the nail is hit by a hammer, the height of the nail is 80% of its previous height. After 3 hits, how many millimeters is the nail sticking out from the board?

$\left(\begin{array}{c}\text{Height after}\\ \text{the first hit}\end{array}\right) = (15)(0.8)$

$\left(\begin{array}{c}\text{Height after}\\ \text{the second hit}\end{array}\right) = [(15)(0.8)]0.8$

$\left(\begin{array}{c}\text{Height after}\\ \text{the third hit}\end{array}\right) = [(15)(0.8)(0.8)]0.8$

$= 15(0.8)^3$

$= 7.68$

After the third hit, the nail is sticking out 7.68 mm from the board.

Section Note

Topic Spiraling: Review

Students should recognize the relationship of exponential decay to exponential growth. Make sure that they understand that a decay factor is less than 1 and a growth factor is greater than 1.

In the Exploration, you could expect there to be about half as many pennies left after you cleared away the heads as there were before you emptied the cup.

$$\text{Pennies left after the cup was emptied } \mathbf{1} \text{ time} = 100(0.5)^{\mathbf{1}}$$

$$\text{Pennies left after the cup was emptied } \mathbf{2} \text{ times} = 100(0.5)^{\mathbf{2}}$$

$$\vdots$$

$$\text{Pennies left after the cup was emptied } \boldsymbol{n} \text{ times} = 100(0.5)^{\boldsymbol{n}}$$

Patterns like the one above appear in many situations, as you will see in Example 1.

EXAMPLE 1 Application: Physics

Each time a certain rubber ball bounces, it reaches about 60% of the height from which it fell. Suppose this ball is dropped from a height of 300 cm. What height does the ball reach after the third bounce?

SOLUTION

$$\text{Height after the first bounce} = (300)(0.6)$$

$$\text{Height after the second bounce} = [(300)(0.6)]0.6$$

The bounce height of the first bounce becomes the drop height of the second bounce.

$$\text{Height after the third bounce} = [(300)(0.6)(0.6)]0.6$$

$$= 300(0.6)^3$$

$$= 64.8$$

After the third bounce, the ball reaches a height of about 64.8 cm.

The height reached by the ball in Example 1 decreases by 40% after each bounce. Any quantity y that decreases by a fixed percent at regular intervals is an example of **exponential decay**. Exponential decay can be modeled by an equation in the form below.

$$y = a(1 - r)^x$$

The constant a represents the initial amount before any decay occurs.

The constant r represents the **rate of decrease**. The quantity $1 - r$ is called the **decay factor**.

The x in the equation represents the number of intervals that have passed.

ANSWERS Section 9.3

Exploration

1. Results may vary. Check students' work.

2. a.

x = number of times cup is emptied	y = expected number of pennies remaining
0	100
1	$100\left(\frac{1}{2}\right)$
2	$100\left(\frac{1}{2}\right)\cdot\left(\frac{1}{2}\right) = 100\left(\frac{1}{2}\right)^2$
3	$100\left(\frac{1}{2}\right)^2\cdot\left(\frac{1}{2}\right) = 100\left(\frac{1}{2}\right)^3$
4	$100\left(\frac{1}{2}\right)^3\cdot\left(\frac{1}{2}\right) = 100\left(\frac{1}{2}\right)^4$
5	$100\left(\frac{1}{2}\right)^4\cdot\left(\frac{1}{2}\right) = 100\left(\frac{1}{2}\right)^5$
6	$100\left(\frac{1}{2}\right)^5\cdot\left(\frac{1}{2}\right) = 100\left(\frac{1}{2}\right)^6$

3. $y = 100\left(\frac{1}{2}\right)^x$, where y = the number of pennies remaining and x = the number of times the cup is emptied; the graph of the equation should be close to the points on the scatter plot.

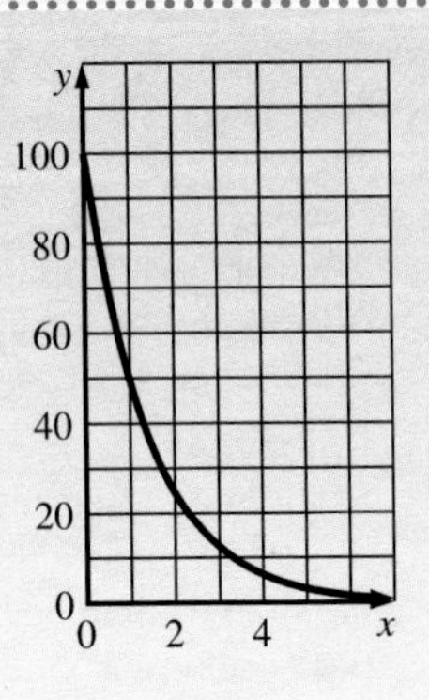

EXAMPLE 2 Application: Meteorology

At sea level, the barometric pressure is 29.92 in. of mercury. For each mile above sea level, the barometric pressure decreases by 18%.

a. Write an equation for the barometric pressure y in inches of mercury x miles above sea level.

b. Find the barometric pressure five miles above sea level.

SOLUTION

a. The initial pressure is 29.92, so $a = 29.92$.

The pressure decreases by 18% for each mile, so $r = 0.18$.

Substitute these values in the equation $y = a(1 - r)^x$.

$$y = 29.92(1 - 0.18)^x$$
$$y = 29.92(0.82)^x$$

b. Substitute **5** for x in the equation from part (a).

$$y = 29.92(0.82)^5$$
$$\approx 11.09$$

The barometric pressure five miles above sea level is about 11.09 in. of mercury.

THINK AND COMMUNICATE

1. How can you tell whether an exponential equation models decay?
2. a. Write an equation for the height y of the ball in Example 1 after x bounces. Graph the equation on a graphing calculator.
   b. Use the graph from part (a) to estimate the number of times the ball bounces before it stops. How did you get your answer?

CHECKING KEY CONCEPTS

Find the value of y when $x = 3$.

1. $y = 150(0.82)^x$
2. $y = 24(0.9)^x$
3. $y = \frac{9}{2}\left(\frac{2}{3}\right)^x$

The percent y of people living on farms in the United States can be modeled by the equation $y = 18(0.95)^x$, where x = the number of years after 1946. Use this information for Questions 4–6.

4. What percent of people lived on farms in 1946? How do you know?
5. By how much did the percent decrease each year? Explain.
6. Estimate the percent of people who lived on farms in 1973.

Teach⇔Interact

Section Note

Visual Thinking

Ask students to find large pieces of paper and to then fold them in half. Have them continue to fold them in half until they are no longer able to do so. Ask them to identify whether or not this activity represents exponential decay. Encourage them to create a graph that shows this decay, for example, by graphing the areas of the rectangles. Have them write an equation for their model. This activity involves the visual skills of *identification* and *communication.*

Additional Example 2

A woman has a 500-acre parcel of land that she wants to leave to her future grandchildren. She decides to give each grandchild 15% of the land remaining when the child is born.

a. Write an equation for the amount y of land remaining in the 500-acre plot after x grandchildren are born.

The original plot contains 500 acres, so $a = 500$.

The remaining acreage decreases by 15% for each grandchild, so $r = 0.15$.

Substitute these values in the equation $y = a(1 - r)^x$.

$y = 500(1 - 0.15)^x$

$y = 500(0.85)^x$

b. Find the amount of land remaining in the original plot after the 6th grandchild is born.

Substitute 6 for the x in the equation from part (a).

$y = 500(0.85)^6$

≈ 189

About 189 acres remain in the original plot after the 6th grandchild is born.

Closure Question

Explain what a, r, and $1 - r$ represent in the exponential decay equation $y = a(1 - r)^x$.

a: the initial amount; r: the decay rate; $1 - r$: the decay factor

Think and Communicate

1. Every exponential function can be written in the form $y = a(b)^x$. If $0 < b < 1$, the equation models decay. If $b > 1$, the equation models growth.
2. a. $y = 300(0.6)^x$, where y = the height in centimeters and x = the number of bounces.

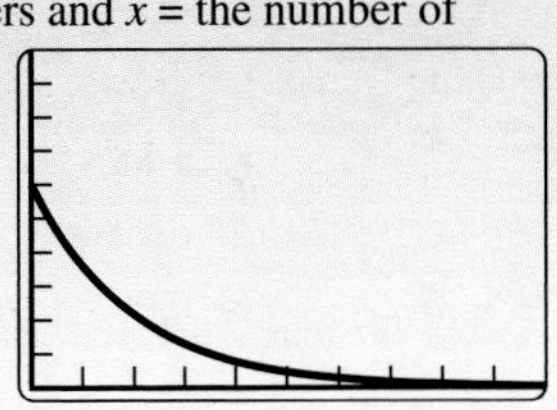

   b. Estimates may vary; after 11 bounces, the bounce height is only about 1 cm. I used TRACE to determine the value of x as y gets close to 0. I think, though, that the ball would stop bouncing before $y = 0$, so I estimated about 11 bounces.

Checking Key Concepts

Where necessary, answers are rounded to the nearest hundredth.

1. 82.71
2. 17.50
3. $\frac{4}{3}$
4. 18%; 18 is the initial value.
5. 5%; The decay factor, $1 - r$, is 0.95, so $r = 0.05$, or 5%.
6. about 4.51%

Apply⇔Assess

Suggested Assignment

- **Core Course**
  Exs. 1–12, 14–23, 27–30, 32–41, AYP
- **Extended Course**
  Exs. 6, 8, 9–16, 19–41, AYP
- **Block Schedule**
  Day 58 Exs. 1–12, 14–23, 27–30, 32–41, AYP

Exercise Notes

Communication: Writing
Exs. 10, 11 These exercises require that students be able to distinguish between the rate of decrease and the decay factor in order to write the correct equation.

Problem Solving
Ex. 13 This exercise allows students to be creative in writing a word problem that may be relevant to them. Students can profit from hearing their classmates' word problems.

Reasoning
Exs. 16–21 Students should be able to identify quickly an exponential growth or decay situation by looking at the number in parentheses. If this number is greater than 1, growth is involved; if it is less than 1, decay is involved. Ask students to explain why this is true.

Using Technology
Exs. 16–21 Students may find it helpful to use the *Function Investigator* software or a graphing calculator to display graphs for those functions. Discuss how the graphs can help in answering the exercises.

9.3 Exercises and Applications

Extra Practice *exercises on page 574*

Find the value of y when $x = 6$. Round decimal answers to the nearest hundredth.

1. $y = 120(0.25)^x$ **2.** $y = 278(0.93)^x$ **3.** $y = 7168(0.97)^x$

4. $y = 402(0.78)^x$ **5.** $y = 3012(0.875)^x$ **6.** $y = 0.8(0.6)^x$

7. $y = 48\left(\frac{1}{2}\right)^x$ **8.** $y = 117\left(\frac{1}{3}\right)^x$ **9.** $y = 1000\left(\frac{4}{5}\right)^x$

For Exercises 10 and 11, write an equation that models the situation described. Tell what each variable in the equation represents.

10. In 1990, the population of a city was 850,000. Each year since, the population has been about 97% what it was the previous year.

11. The amount y of a substance in grams decreases by 17% each hour after 3:00 P.M. At 3:00 P.M., $y = 100$.

12. SPORTS The women's United States Open tennis tournament is held each year. There are 128 players at the start of the tournament. During each round of play, half of the remaining players are eliminated.

a. Write an equation that gives the number y of players remaining after x rounds.

b. Find the number of players remaining after 4 rounds.

c. A player competes in one match during each round. How many matches must a player win in order to win the tournament?

13. Open-ended Problem Write a word problem whose solution is the equation $y = 34{,}000(0.94)^x$.

For Exercises 14 and 15, tell whether each exponential graph represents *growth* or *decay*. Explain your answer.

14.

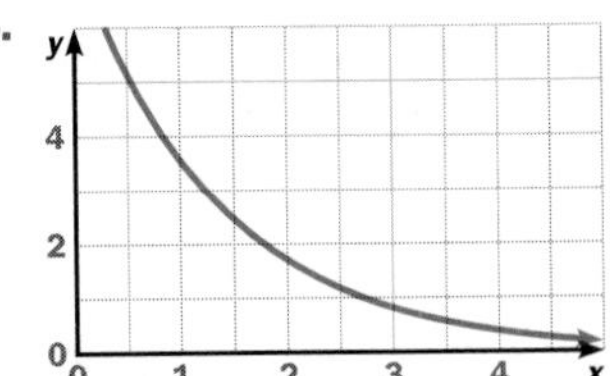

15.

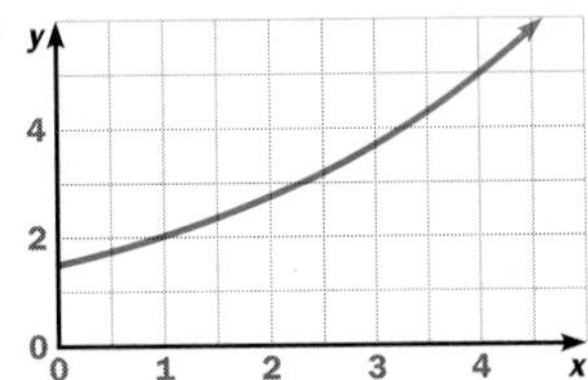

The equations in Exercises 16–21 describe quantities that grow or decay exponentially. For each equation:

a. Tell whether the quantity described *grows* or *decays*.

b. Give the initial amount of the quantity.

c. Give the growth or decay factor for the quantity.

16. $y = 300(1.15)^x$ **17.** $y = 14(0.23)^x$ **18.** $y = 20{,}000(0.72)^x$

19. $y = 0.25(1.88)^x$ **20.** $y = 1.01(0.99)^x$ **21.** $y = 0.99(1.01)^x$

Exercises and Applications

1. 0.29
2. 179.86
3. 5970.74
4. 90.53
5. 1351.77
6. 0.04
7. 0.75
8. 0.16
9. 262.14
10. $y = 850(0.97)^x$, where y = the population in thousands and x = the number of years after 1990.
11. $y = 100(0.83)^x$, where y = the number of grams remaining and x = the number of hours after 3:00 P.M.
12. a. $y = 128(0.5)^x$
    b. 8 players
    c. 7 matches
13. Answers may vary. An example is given. The Washingtons bought a luxury car costing $34,000. Its value will decrease at a rate of 6% per year. Write an equation that gives the value of the car x years from now.
14. exponential decay; The value of y decreases exponentially as x increases.
15. exponential growth; The value of y increases exponentially as x increases.
16. a. grows b. 300 c. 1.15
17. a. decays b. 14 c. 0.23
18. a. decays b. 20,000 c. 0.72
19. a. grows b. 0.25 c. 1.88
20. a. decays b. 1.01 c. 0.99

INTERVIEW Lori Arviso Alvord

Look back at the article on pages 366–368.

Doctors like Lori Arviso Alvord need to know how long certain medications are present in a patient's bloodstream. For example, when a person takes a 325 mg aspirin tablet, the amount of aspirin in the person's bloodstream decreases by about 20% per hour.

22. Write an equation that gives the amount of aspirin y in milligrams in a person's bloodstream x hours after taking a 325 mg aspirin tablet.

23. The directions on many aspirin bottles say to wait 4 hours between doses. How much aspirin is left in the bloodstream 4 hours after taking a 325 mg tablet?

Cooperative Learning In this exploration, you will perform an experiment that models exponential decay. Work in a group of 2–3 students. You will need a clear drinking glass, water, a measuring cup, and food coloring.

24. Pour 10 fluid ounces of water into the drinking glass. Add five drops of food coloring to the water. Dilute the mixture by pouring 2 oz of colored water out of the glass and adding 2 oz of clear water. How does the mixture change?

25. Continue diluting the mixture as in Exercise 24 until the water appears to be clear. How many times did you need to repeat the steps?

26. Writing Explain how this exploration is similar to the decrease of aspirin in a person's bloodstream over time.

OCCUPATIONAL SAFETY The equation $y = 2{,}046{,}322(0.87)^x$ models the maximum number of hours per day y that a worker is allowed to be exposed to a noise of x decibels (dB) according to federal standards. Use this information for Exercises 27 and 28.

27. Find the maximum number of hours per day that a person can work in each location at the given noise level.

a. room with amplified rock music: 110 dB

b. machine shop: 100 dB

c. subway train station: 90 dB

28. Technology Graph the equation $y = 2{,}046{,}322(0.87)^x$. Set a good viewing window, such as $80 \le x \le 120$ and $0 \le y \le 24$.

a. Based on the graph, what noise levels do you think are unhealthy during any amount of time? How did you get your answer?

b. Based on the graph, what noise levels do you think have no effect on an individual's hearing? Explain your reasoning.

c. Explain why it makes sense to set the window so that $0 \le y \le 24$.

Apply⇔Assess

Exercise Notes

Interview Note

Exs. 22–26 Biopharmaceutics and pharmacokinetics are two developing sciences that study how drugs reach certain parts of the body and how they are removed from the body. Computer-aided design (CAD) is being increasingly used to aid pharmaceutical chemists and pharmacologists in drug research and development.

Research

Exs. 27, 28 The Occupational Safety and Health Administration (OSHA) was established by the U.S. Congress in 1970. Its function is to ensure safe and healthful working conditions for American workers. Ask students to research the kinds of jobs OSHA employees perform.

21. a. grows b. 0.99
c. 1.01

22. $y = 325(0.8)^x$

23. 133.12 mg

24, 25. Check students' work. Answers may vary.

26. Every time 2 oz of colored water is replaced with clear water, the amount of food coloring in the water decreases by 20%, just as with every hour that passes, the aspirin in a person's bloodstream decreases by 20%.

27. a. about 0.46 h
b. about 1.83 h
c. about 7.37 h

28.

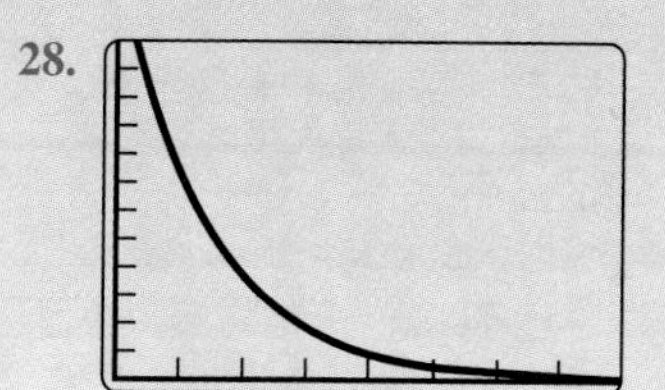

a, b. Answers may vary. Examples are given.

a. about 117 dB and above; I used TRACE to find values of x for which y is very close to 0. For $x > 117$, the maximum safe exposure time is 10 min or less.

b. about 81.5 dB and below; At those levels, the maximum safe exposure time is more than 24 hours per day.

c. Since y = maximum hours of exposure per day, $0 \le y \le 24$.

Apply⇔Assess

Exercise Notes

Problem Solving

Exs. 29–31 An important problem-solving strategy is to look for a pattern. To complete Exs. 29 and 30, students can look for a pattern by setting up a table with the headings *number of folds, number of noodles,* and *total number of feet of noodles.* For Ex. 31, students can use another problem-solving strategy, that of simplifying a problem to arrive at an answer. Ask students to compare the volume of a noodle after 1 fold and stretch to the volume of the beginning noodle with $r = 1$ (found in part (a)). Then they can determine what the radius of the smaller noodle must be after one stretch. Using this information, they should be able to make a general statement about the relationship of the two radii and complete part (c).

Connection COOKING

To make the Chinese dish "dragon's beard noodles," fold a 5 ft strand of dough in half. Stretch the dough back to its original length so that two thinner strands are formed. Repeat this folding and stretching process over and over to produce increasing numbers of progressively thinner noodles.

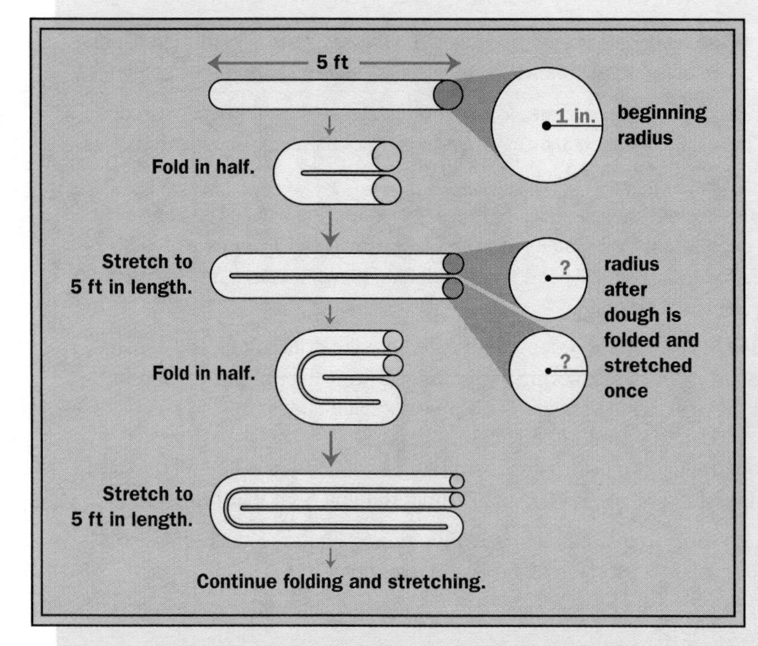

29. Write an equation that gives the number of noodles n formed by folding and stretching the dough x times.

30. Write an equation that gives the combined length l in feet of all the noodles after folding and stretching the dough x times. How did you get your answer?

BY THE WAY

The world's best chefs are able to complete 12 or 13 doublings of dragons's beard noodles. These noodles are only two or three times as thick as human hair.

31. a. The volume V of a noodle in cubic inches is given by the formula $V = 60\pi r^2$, where r is the radius of the noodle in inches. Find the volume of a noodle with a radius of 1 in. Leave your answer in terms of π.

b. Folding and stretching the dough does not change the total volume of all the noodles. Use this fact to find the value of x in the equation below. Explain your answer.

$$x \cdot \left(\begin{array}{c}\text{Volume of a noodle}\\\text{after 1 fold and stretch}\end{array}\right) = \left(\begin{array}{c}\text{Volume of the}\\\text{beginning noodle}\end{array}\right)$$

c. Challenge Use your completed equation from part (b) to complete the equation below.

$$\left(\begin{array}{c}\text{Radius of a noodle}\\\text{after 1 fold and stretch}\end{array}\right) = \left(\begin{array}{c}\text{Radius of the}\\\text{beginning noodle}\end{array}\right) \cdot \underline{\ ?\ }$$

d. Use the relationship in part (c) to write an equation giving the radius r of a noodle after x folds and stretches. By about what percent does the radius decrease after each fold and stretch?

29. $n = 2^x$

30. $l = 5(2)^x$; Answers may vary. Examples are given. (1) The length of the original noodle is 5 ft. After each fold, the number of noodles doubles, with each noodle the same length as the original noodle, so the total length doubles. Then the total length shows exponential growth with an initial length of 5 ft and a growth factor of 2. (2) There will be 2^x noodles, each of which is 5 ft long.

31. a. 60π in.^{2}

b. 2; After 1 fold and stretch, there are 2 noodles. Their total volume is equal to the volume of the original noodle.

c. $\sqrt{\frac{1}{2}} \approx 0.71$

d. $r = (0.71)^x$; about 29%

32. In most situations involving exponential growth or decay, the rate of growth or decay is given as a percent. For example, suppose $1000 is deposited in a savings account that pays 4% interest per year, and no further deposits or withdrawals are made. The amount in the account grows exponentially over time with a growth rate of 4%. The equation that

ONGOING ASSESSMENT

32. **Writing** Explain why you need to know about percent to understand exponential growth and decay. Support your answer with an example.

SPIRAL REVIEW

Simplify each expression. *(Section 1.5)*

33. $\frac{-14+3+2\cdot 7}{5}$
34. $-10\cdot 11\cdot \frac{3}{5}$
35. $\left|-6+\frac{1}{2}\cdot\frac{1}{2}\right|$

Identify the reciprocal you would use to solve each equation. Then solve. *(Section 4.2)*

36. $\frac{4}{5}x = 7$
37. $-\frac{2}{3}a = 4$
38. $\frac{1}{2}r = 3$
39. $-\frac{3}{7}m = -21$
40. $\frac{5}{9}c = \frac{-1}{5}$
41. $\frac{1}{8}p = 8$

ASSESS YOUR PROGRESS

VOCABULARY

base (p. 370)
exponent (p. 370)
power (p. 370)
exponential growth (p. 377)
growth rate (p. 377)
growth factor (p. 377)
exponential decay (p. 384)
rate of decrease (p. 384)
decay factor (p. 384)

Evaluate each power. *(Section 9.1)*

1. 4^3
2. $(-2)^6$
3. $\left(\frac{1}{5}\right)^4$

Evaluate each expression when $x = 5$ and $y = -2$. *(Section 9.1)*

4. xy^4
5. $(x-y)^3$
6. $2x^2 + y^3$
7. $5y - x^2$
8. $y^3 + (x^2y - 1)$
9. $y^5 + (6x + 2)$

10. **PERSONAL FINANCE** Eric Janning has \$250 in a savings account that pays 5% annual interest. *(Section 9.2)*
    a. Write an equation giving the amount y in dollars in Eric's account x years from now, assuming he makes no deposits or withdrawals.
    b. How much money will Eric have in his account 3 years from now?

11. **FUEL ECONOMY** The average annual fuel consumption per car in the United States has decreased by about 2% per year since 1980, when it was about 591 gallons per car. *(Section 9.3)*
    a. Write an exponential function to model this situation.
    b. Sketch the graph of the equation you wrote for part (a).

12. **Journal** Describe some real-world examples of exponential growth and decay. Sketch the graph that models one of these situations.

Apply⇔Assess

Assess Your Progress

Call upon individual students to explain the meaning of at least one vocabulary term.

Journal Entry
Have students share their examples with the class by reading their journal entries aloud. Graphs can be passed around the class as students describe their examples.

Progress Check 9.1–9.3

See page 414.

Practice 60 for Section 9.3

NAME ____________ DATE ________

Practice 60
FOR USE WITH SECTION 9.3

Find the value of y when $x = 5$.

1. $y = 1000(0.5)^x$ 31.25
2. $y = 12.5(0.8)^x$ 4.096
3. $y = 128(0.75)^x$ 30.375
4. $y = 150(0.64)^x$ ≈ 16.1
5. $y = 27{,}000(0.45)^x$ ≈ 498.2
6. $y = 35.4(0.83)^x$ 13.9
7. $y = 640\left(\frac{1}{2}\right)^x$ 20
8. $y = 1875\left(\frac{2}{5}\right)^x$ 19.2
9. $y = 1536\left(\frac{1}{4}\right)^x$ 1.5

The equations in Exercises 10–15 describe quantities that either grow or decay exponentially. For each equation:
a. **Tell whether the quantity described *grows* or *decays*.**
b. **Give the initial amount of the quantity.**
c. **Give the growth or decay factor for the quantity.**

10. $y = 15.2(0.95)^x$
11. $y = 12{,}000(1.25)^x$
12. $y = 0.18(1.04)^x$
13. $y = 1.64(0.73)^x$
14. $y = 25.9\left(\frac{3}{2}\right)^x$
15. $y = 26\left(\frac{3}{4}\right)^x$

10.a. decays b. 15.2 c. 0.95
11.a. grows b. 12,000 c. 1.25
12.a. grows b. 0.18 c. 1.04
13.a. decays b. 1.64 c. 0.73
14.a. grows b. 25.9 c. 1.5
15.a. decays b. 26 c. 0.75

16. The population of Akron, Ohio decreased from 237,177 in 1980 to 223,019 in 1990.
    a. By what percent did the population of Akron drop between 1980 and 1990? about 6%
    b. Suppose the population of Akron continues to drop by the same percentage in future decades. Write an equation that models the population of Akron x decades after 1980. $y = 237{,}177\,(0.94)^x$
    c. According to your model, what will the population of Akron be in 2030? about 174,065

17. A laboratory-produced isotope of silver decays radioactively with a half-life of 20 minutes. This means that at the end of every 20-minute interval, half the amount of the isotope that was present at the beginning of the interval has decayed.
    a. Suppose 10 g of the isotope is present initially. Write an equation that models the amount y of the isotope that is present after x 20-minute intervals. $y = 10(0.5)^x$
    b. Find the amount of the isotope that remains after 1 h 20 min. 0.625 g

18. **Writing** An ancient paradox says that, when traveling, you can never reach your destination because you must first cover half the distance, then half the remaining distance, and so on. Describe the change in the remaining distance in terms of exponential decay. What is wrong with the reasoning in this paradox? Answers may vary. Check students' work.

60 **Practice Bank,** ALGEBRA 1: EXPLORATIONS AND APPLICATIONS

describes the amount (in dollars) in the account after x years is $y = 1000(1.04)^x$.

33. $\frac{3}{5}$
34. –66
35. $5\frac{3}{4}$
36. $\frac{5}{4}; \frac{35}{4}$
37. $-\frac{3}{2}; -6$
38. 2; 6
39. $-\frac{7}{3}; 49$
40. $\frac{9}{5}; -\frac{9}{25}$
41. 8; 64

Assess Your Progress

1. 64
2. 64
3. $\frac{1}{625}$
4. 80
5. 343
6. 42
7. –35
8. –59
9. 0
10. a. $y = 250(1.05)^x$
    b. \$289.41

11. a. $y = 591(0.98)^x$, where y = fuel consumption in gallons and x = the number of years since 1980.
    b.

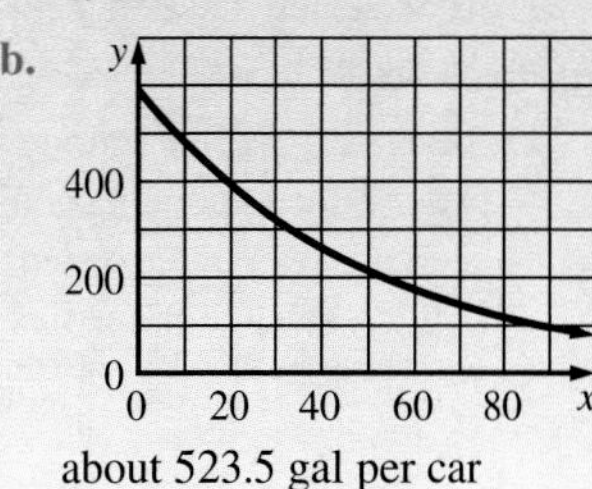

about 523.5 gal per car

12. Answers may vary. Examples are given. exponential growth: the growth of bacteria in a culture; exponential decay: radioactive decay

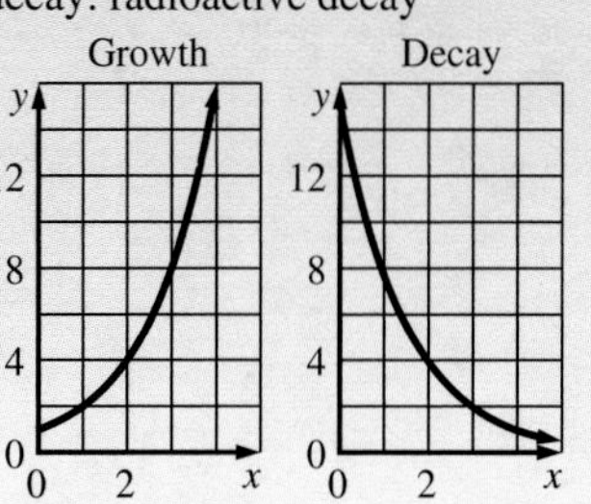

Plan⇔Support

Objectives

- Evaluate powers with negative and zero exponents.
- Use negative exponents to express quotients.
- Find past values using negative exponents.
- Use formulas to find information.

Recommended Pacing

Core and Extended Courses
Section 9.4: 1 day

Block Schedule
Section 9.4: $\frac{1}{2}$ block
(with Section 9.3)

Resource Materials

Lesson Support
Lesson Plan 9.4
Warm-Up Transparency 9.4
Overhead Visuals:
Folder 9: Exponential Growth and Decay
Practice Bank: Practice 61
Study Guide: Section 9.4

Technology
Technology Book:
Spreadsheet Activity 9
Scientific Calculator
Internet:
http://www.hmco.com

Warm-Up Exercises

Evaluate each expression.

1. 5^3 125
2. $(-2)^5$ −32
3. $\frac{1}{3^3}$ $\frac{1}{27}$
4. $\frac{1}{7^2}$ $\frac{1}{49}$
5. $4^1 \div 4$ 1

SECTION

9.4 Zero and Negative Exponents

Learn how to...

- evaluate powers with negative and zero exponents
- use negative exponents to express quotients

So you can...

- find past values, such as the population of Morocco years ago
- use formulas to find information, such as the amount of a car loan payment

You have already seen how to use exponents to predict future values, such as population figures. You can use exponents to look into the past as well as into the future.

Look at the diagram below. Each bar is half as long as the previous bar. The diagram is a visual representation of several powers of 2.

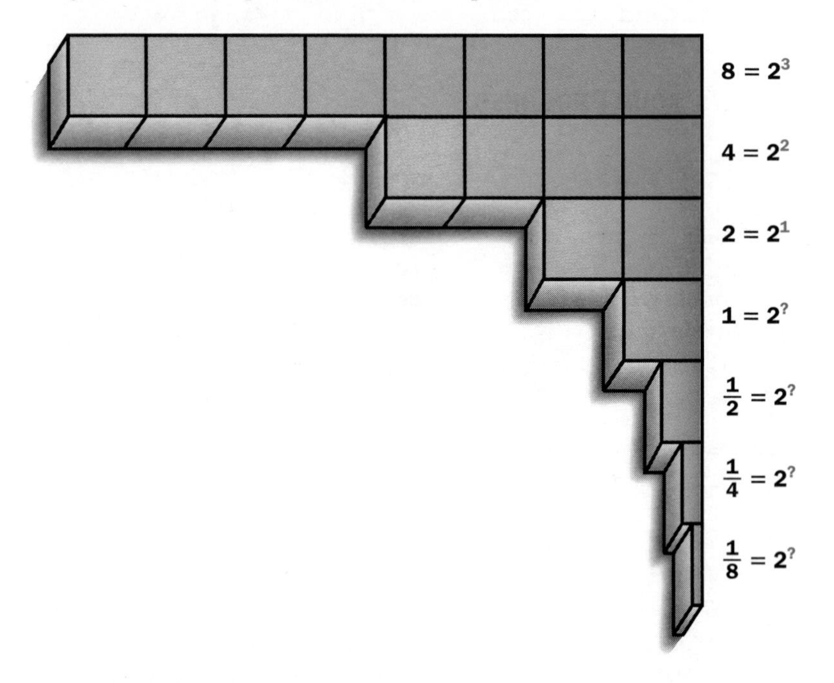

Think and Communicate

1. As the bars in the diagram become shorter, how do the exponents change?
2. What do you think the last four exponents in the diagram should be? Explain your reasoning. Then use a calculator to check your answers.
3. Copy and complete the equations below. How did you get your answers?

$9 = 3^?$ $\quad 3 = 3^?$ $\quad 1 = 3^?$ $\quad \frac{1}{3} = 3^?$ $\quad \frac{1}{9} = 3^?$

ANSWERS Section 9.4

Think and Communicate

1. At each step, the exponent decreases by 1.
2. 0; −1; −2; −3; To continue the pattern, the exponent should decrease by 1 at each step. Using a calculator verifies that $2^0 = 1$, $2^{-1} = \frac{1}{2}$, $2^{-2} = \frac{1}{4}$, and $2^{-3} = \frac{1}{8}$.
3. 2; 1; 0; −1; −2; Answers may vary. An example is given. I used the pattern given by the bars.

Zero and Negative Exponents

For any nonzero number a and any integer n:

$$a^0 = 1 \quad \text{and} \quad a^{-n} = \frac{1}{a^n}$$

EXAMPLE 1

Evaluate each power.

a. $(7.2)^0$ **b.** 6^{-2}

SOLUTION

a. $(7.2)^0 = 1$

b. $6^{-2} = \frac{1}{6^2} = \frac{1}{36}$

EXAMPLE 2

Rewrite each expression using only positive exponents.

a. a^0b^{-4} **b.** $\frac{ac^{-8}}{b^5}$

SOLUTION

a. $a^0b^{-4} = 1 \cdot b^{-4} = \frac{1}{b^4}$

b. $\frac{ac^{-8}}{b^5} = \frac{a}{b^5} \cdot c^{-8} = \frac{a}{b^5} \cdot \frac{1}{c^8} = \frac{a}{b^5c^8}$

THINK AND COMMUNICATE

4. Chayla evaluated the power $\left(\frac{1}{4}\right)^{-2}$ as shown. Use her method to evaluate each of the following powers.

a. $\left(\frac{1}{3}\right)^{-2}$ **b.** $\left(\frac{1}{2}\right)^{-5}$ **c.** $\left(\frac{1}{5}\right)^{-3}$

5. Based on your answers to Question 4, complete the equation below. Explain your reasoning.

$$\left(\frac{1}{a}\right)^{-n} = \underline{?}$$

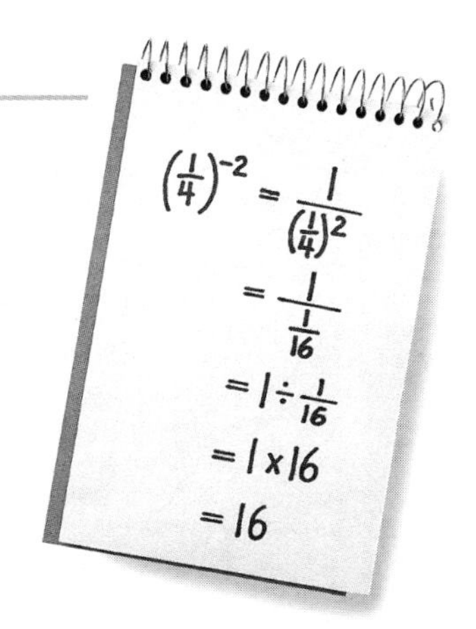

$(\frac{1}{4})^{-2} = \frac{1}{(\frac{1}{4})^2}$
$= \frac{1}{\frac{1}{16}}$
$= 1 \div \frac{1}{16}$
$= 1 \times 16$
$= 16$

Teach⇔Interact

Section Note

Alternate Approach

Many students incorrectly think $a^0 = a$. It may be helpful for these students to compare the powers of two in decreasing order by completing the following. Ask them what they need to do to get the next power of 2. (Divide by 2.)

$2^3 = 2^4 \div 2 = \underline{?}$ (8)
$2^2 = 2^3 \div 2 = \underline{?}$ (4)
$2^1 = 2^2 \div 2 = \underline{?}$ (2)
$2^0 = 2^1 \div 2 = \underline{?}$ (1)

Additional Example 1

Evaluate each power.

a. $\left(\frac{1}{2}\right)^0$ $\left(\frac{1}{2}\right)^0 = 1$

b. 4^{-3}

$4^{-3} = \frac{1}{4^3} = \frac{1}{64}$

Additional Example 2

Rewrite each expression using only positive exponents.

a. $x^{-2}y$

$x^{-2}y = \frac{1}{x^2} \cdot y = \frac{y}{x^2}$

b. $\frac{x^{-5}y^6}{z^0}$

$\frac{x^{-5}y^6}{z^0} = x^{-5} \cdot \frac{y^6}{z^0} = \frac{1}{x^5} \cdot \frac{y^6}{1} = \frac{y^6}{x^5}$

Think and Communicate

4. Chayla's work is correct.
a. 9 b. 32 c. 125

5. a^n; Explanations may vary. An example is given. I noticed in question 4 that $\left(\frac{1}{3}\right)^{-2} = 3^2$, $\left(\frac{1}{2}\right)^{-5} = 2^5$, and $\left(\frac{1}{5}\right)^{-3} = 5^3$. Then I reasoned that if $a \neq 0$, $\left(\frac{1}{a}\right)^{-n}$ must be equal to a^n.

Teach⇔Interact

Section Note

Visual Thinking
Ask students to think of other examples of negative exponents and to create sketches of their ideas to discuss with the class. Examples might include a runner three times behind the leader and another three times farther behind; a submarine four times deeper than another; temperature statements such as: It's ten times colder than last year. This activity involves the visual skills of *recognition* and *communication.*

Additional Example 3

Use the equation given on this page to estimate the population of Morocco in 1972.

Since 1972 is 8 years before 1980, substitute −8 for x.

$y = 20.5(1.025)^{-8}$

≈ 16.8

The population of Morocco was about 16.8 million in 1972.

Think and Communicate

Before answering these questions, it may be helpful for students to review the exponential growth equation, $y = a(1 + r)^x$, identifying the initial amount, the growth rate, the growth factor, and the meaning of the exponent x.

Applying Negative Exponents

The population y of Morocco in millions x years after 1980 can be modeled by this equation:

$$y = 20.5(1.025)^x$$

You already know how to use this equation to estimate the population of Morocco in years after 1980. You can also use it to estimate the population in years *before* 1980.

When $x = 1$, y = the population **1 year after 1980**, in 1981.

When $x = 0$, y = the population **in 1980**.

When $x = -1$, y = the population **1 year before 1980**, in 1979.

EXAMPLE 3 Application: Population

Use the equation above to estimate the population of Morocco in 1975.

SOLUTION

Since 1975 is 5 years *before* 1980, substitute −5 for x.

$$y = 20.5(1.025)^{-5}$$
$$\approx 18.1$$

The population of Morocco was about 18.1 million in 1975.

THINK AND COMMUNICATE

6. a. How can you find the population of Morocco in 1980 just by looking at the equation $y = 20.5(1.025)^x$?

b. Explain how to calculate the population of Morocco in 1980 by substituting a value for x in the equation. Tell what x value to use, and explain why you would choose this value.

c. Describe how you can use the graph of the equation shown at the right to find the population of Morocco in 1980.

d. How do the methods you described in parts (a), (b), and (c) compare? Does each method give you the same results?

7. In Example 3, what is assumed about the annual growth rate of Morocco's population from 1975 to 1980? How can you find out whether this assumption is reasonable?

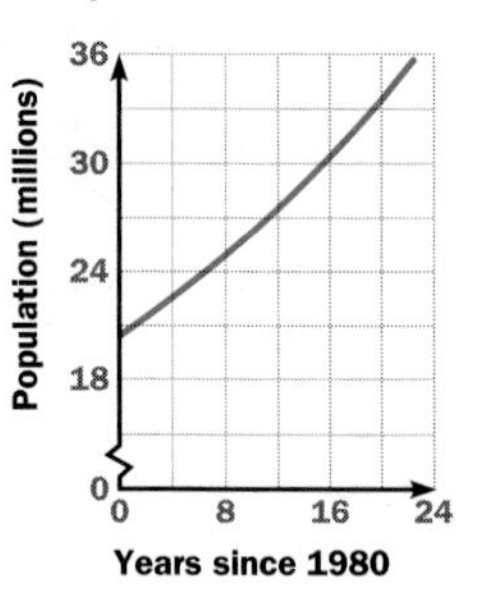

Think and Communicate

6. a. by looking at the initial amount, 20.5; The population was 20.5 million.

b. For 1980, $x = 0$, so $y = 20.5(1.025)^0 = 20.5 \cdot 1 = 20.5$; 20.5 million. Since x represents years before and after 1980 for negative and positive values of x, $x = 0$ represents the year 1980.

c. Find the vertical intercept. It is about 21 million.

d. Answers may vary. An example is given. The method in part (a) is easiest. It is based on the method used in part (b). Both of these methods give the same result. The method in part (c) gives a less accurate result because you have to estimate from a graph. Yes.

7. that the growth rate was the same between 1975 and 1980 as it was after 1980; Answers may vary. An example is given. Use an almanac or encyclopedia to find the growth rate for that period.

Checking Key Concepts

1. $\frac{1}{121}$
2. 1
3. $\frac{1}{49}$
4. 25
5. $\frac{7}{m^3}$
6. $\frac{1}{t^6}$
7. $\frac{1}{a^4c^6}$
8. y^2
9. 1
10. $\frac{25}{27}$
11. $-\frac{1}{8}$
12. $\frac{1}{243}$

CHECKING KEY CONCEPTS

Simplify each expression.

1. 11^{-2} **2.** 0.747966^0 **3.** $(9^0)(7^{-2})$ **4.** $\left(\frac{1}{5}\right)^{-2}$

Rewrite each expression using only positive exponents.

5. $7m^{-3}$ **6.** $t^{-6}r^0$ **7.** $\frac{a^{-4}}{c^6}$ **8.** $\frac{x^0}{y^{-2}}$

Evaluate each expression when $a = 3$, $b = 5$, and $c = -2$.

9. $(ac)^0$ **10.** b^2a^{-3} **11.** c^{-3} **12.** 3^{-b}

9.4 Exercises and Applications

Simplify each expression.

1. 13^0 **2.** 5^{-6} **3.** 2^{-5} **4.** 8^{-2}

5. $\left(\frac{1}{5}\right)^0$ **6.** $\left(\frac{3}{5}\right)^{-3}$ **7.** $\left(\frac{2}{3}\right)^{-4}$ **8.** $\frac{1}{4^0}$

Extra Practice exercises on page 574

Rewrite each expression using only positive exponents.

9. a^{-2} **10.** $x^{-4}y^{-9}$ **11.** $4b^{-6}$

12. ab^2c^{-1} **13.** $\left(\frac{a}{b}\right)^{-3}$ **14.** $\left(\frac{k^{-7}}{12}\right)$

15. $2w^{-8}$ **16.** $\left(\frac{q^{-12}}{r^3}\right)$ **17.** $(yz)^{-2}$

Evaluate each expression when $x = 2$, $y = 3$, and $z = 7$.

18. $(x + z)^{-3}$ **19.** $x^{-2}z^0$ **20.** yz^{-2}

21. $(5x^0 - z)^{-4}$ **22.** $x^{-1} + y^{-2}$ **23.** $xz^0 - y^{-3}$

24. $\left(\frac{z}{x + y}\right)^{-2}$ **25.** 9^{x-y} **26.** yz^{-x}

27. Writing Tony was asked to evaluate the expression $3x^{-4}$ when $x = 2$. His work is shown at the right. Write a sentence or two explaining Tony's mistake. Then evaluate the expression correctly.

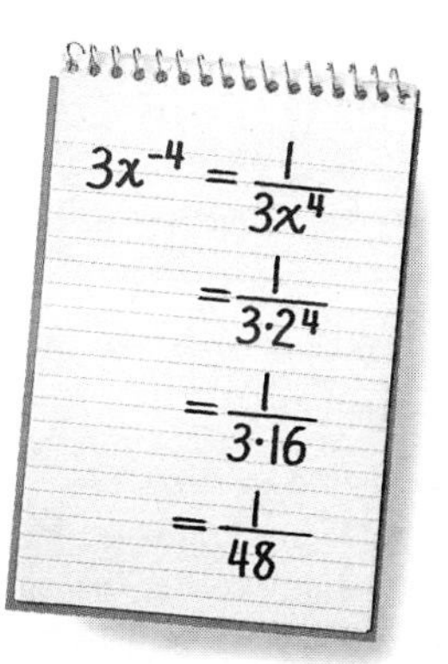

28. 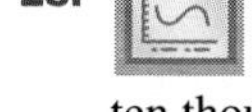 **Technology** Use a calculator to evaluate each power. Round your answers to the nearest ten thousandth.

a. 25^{-1} **b.** 17^{-2} **c.** 1.42^{-3}

Teach⇔Interact

Checking Key Concepts

Common Error

It is not uncommon for many students to make a variety of errors when working with negative exponents because fractions are involved. A thorough review of the answers to these problems would help to identify errors and provide a basis for additional instruction.

Closure Question

Explain the relationship between a^{-n} and $\frac{1}{a^n}$.

They are equal.

Apply⇔Assess

Suggested Assignment

Core Course
Exs. 1–28, 30, 31, 34, 35, 37–43

Extended Course
Exs. 1, 3, 4, 6, 7, 9–11, 15–20, 24–43

Block Schedule
Day 58 Exs. 1–28, 30, 31, 34, 35, 37–43

Exercise Notes

Communication: Reading
Exs. 1–8 Have students read these exercises aloud. Then ask them to read each one as a fraction with a positive exponent.

Assessment Note

Exs. 1–26 Each of these exercises should be discussed in class to assess students' progress in working with negative exponents.

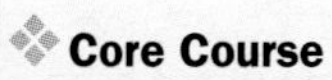

Exercises and Applications

1. 1
2. $\frac{1}{15{,}625}$
3. $\frac{1}{32}$
4. $\frac{1}{64}$
5. 1
6. $\frac{125}{27}$
7. $\frac{81}{16}$
8. 1
9. $\frac{1}{a^2}$
10. $\frac{1}{x^4y^9}$
11. $\frac{4}{b^6}$
12. $\frac{ab^2}{c}$
13. $\frac{b^3}{a^3}$
14. $\frac{1}{12k^7}$
15. $\frac{2}{w^8}$
16. $\frac{1}{r^3q^{12}}$
17. $\frac{1}{y^2z^2}$
18. $\frac{1}{729}$
19. $\frac{1}{4}$
20. $\frac{3}{49}$
21. $\frac{1}{16}$
22. $\frac{11}{18}$
23. $\frac{53}{27}$
24. $\frac{25}{49}$
25. $\frac{1}{9}$
26. $\frac{3}{49}$
27. Tony evaluated the expression $(3x)^{-4}$ instead of $3 \cdot (x^{-4})$; in the given expression, only x is raised to the power -4. The correct solution is $3x^{-4} = \frac{3}{x^4} = \frac{3}{2^4} = \frac{3}{16}$.
28. a. 0.04
b. 0.003460207
c. 0.349248835

Apply⇔Assess

Exercise Notes

Topic Spiraling: Review
Ex. 29 (a), (b) Students should be able to identify these exercises as examples of exponential growth. They provide a good opportunity for students to review the values represented by each of the variables in the exponential growth equation.

Application
Exs. 30–33 Students will find these exercises appealing because they are usually interested in owning and driving a car. Discuss the formula for finding the monthly payment *P* thoroughly so students understand what each variable represents.

Second-Language Learners
Exs. 30–33 Second-language learners can gain a better understanding of the meanings of *loan, interest, installments,* and other terms related to lending through role playing the following activities with other students: purchasing a car, shopping for a loan, making monthly payments (principal and interest), and paying off the loan.

29. HISTORY The first census of the United States population was conducted in 1790. The population was found to be about 3.929 million and continued to increase by roughly 34% per decade until 1860.

a. Write an equation that models the population y of the United States in millions x decades after 1790.

b. Use your equation from part (a) to estimate the United States population in 1780, before the first census was taken.

c. **Research** Use your equation from part (a) to estimate the United States population in 1990. Find out the actual 1990 population of the United States and compare it to your answer. What do you conclude about the growth rate of the population from 1860 to 1990?

Connection CONSUMER ECONOMICS

When you take out a loan to buy a car, you must repay a portion of the loan with interest each month. The monthly payment P required to repay a loan of A dollars in n monthly installments is given by the formula

$$P = \frac{Am}{1 - (1 + m)^{-n}}$$

where m is the monthly interest rate expressed as a decimal. You can find the monthly interest rate by using the formula

$$\text{Monthly interest rate} = \frac{\text{Annual rate as a decimal}}{12}$$

For Exercises 30–32, suppose you take out a loan for the entire cost of the car in the advertisement.

JOE'S NEW AND USED CARS

Low 9% annual financing on loans!

Used '95 Sedan *for only* $11,000!

30. What monthly interest rate is charged for a loan from the car dealership?

31. What is your monthly payment if you repay the loan over each time period?

a. 24 months b. 36 months
c. 48 months d. 60 months

32. Challenge What is the total amount of interest charged if you repay the loan over each time period?

a. 24 months b. 36 months
c. 48 months d. 60 months

33. Open-ended Problem Create your own ad that gives the price of a car and annual interest rate on a loan. Suppose you take out a loan with the dealership for the entire cost of the car. Calculate the monthly payment and total interest charged if you repay the loan over three years.

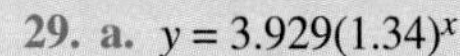

29. a. $y = 3.929(1.34)^x$
b. about 2.932 million
c. about 1.369 billion; 248,709,873; The rate of population growth slowed substantially from 1860 to 1990.

30. $0.75\% = 0.0075$

31. a. $502.53
b. $349.80
c. $273.74
d. $228.34

32. a. $1060.72
b. $1592.80
c. $2139.52
d. $2700.40

33. Answers may vary.

34. a. $y = 2.10(1.3)^x$, where $y =$ the number of bacteria and $x =$ the number of hours after 3:00 P.M.; Yes; for $x = 2$, $y = 3.549$, and for $x = 6$, $y \approx 10.136$.
b. about 1.243 g
c. Answers will vary. An example is given.

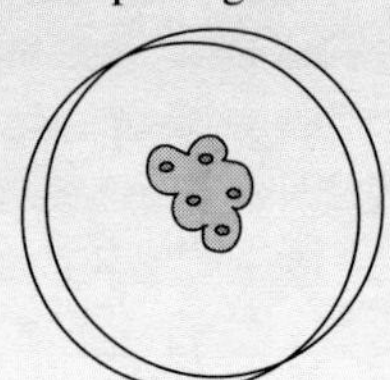

34. BIOLOGY The diagrams below show a growing bacteria colony at three different times. The mass of the colony is increasing by 30% per hour.

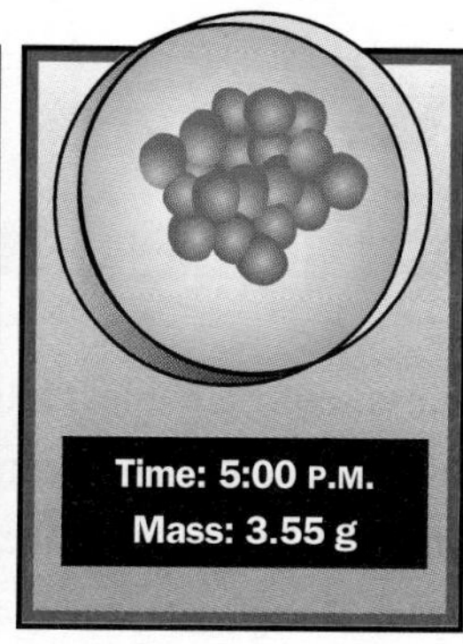

a. Write an equation that models the mass y of the bacteria colony in grams x hours after 3:00 P.M. Does your equation give the correct mass at 5:00 P.M.? at 9:00 P.M.?

b. Estimate the mass of the colony at 1:00 P.M.

c. **Visual Thinking** Draw a diagram showing the approximate size of the colony at 1:00 P.M.

35. SAT/ACT Preview If $A = 3^{-n}$ and $B = 3^n$ for a positive integer n, then:

A. $A < B$ **B.** $A > B$ **C.** $A = B$

D. relationship cannot be determined

36. Any rational number can be written as the sum of multiples of powers of 10. For example, you can write 268.43 like this:

$$268.43 = 2 \cdot 10^2 + 6 \cdot 10^1 + 8 \cdot 10^0 + 4 \cdot 10^{-1} + 3 \cdot 10^{-2}$$

Write each of the following numbers as the sum of multiples of powers of 10.

a. 139.58 **b.** 4234.686 **c.** 73,009.5 **d.** 0.12345

ONGOING ASSESSMENT

37. Writing How would you use division to teach someone about negative and zero exponents? Use examples to support your answer.

SPIRAL REVIEW

Solve each inequality. *(Section 4.6)*

38. $2 + x \leq 7$ **39.** $3m \geq m - 6$ **40.** $p + 25 < 7 - p$

41. $3q + 5 > -9$ **42.** $4 \leq 3x - 2$ **43.** $2 - 2y \leq 10y$

Apply⇔Assess

Exercise Notes

Interdisciplinary Problems

Ex. 34 An essential part of the study of mathematics is applying mathematical techniques to solve real-world problems in areas outside of mathematics. This exercise, like others in this chapter, shows students very important applications of the algebra skills they are learning. Students should understand that medical researchers actually use methods like this one to predict how a colony of bacteria will respond to certain laboratory situations.

Assessment Note

Ex. 37 Students should appreciate that writing an explanation of a mathematical concept or procedure can help them assess whether they understand it well enough to be able to explain it to another student.

Practice 61 for Section 9.4

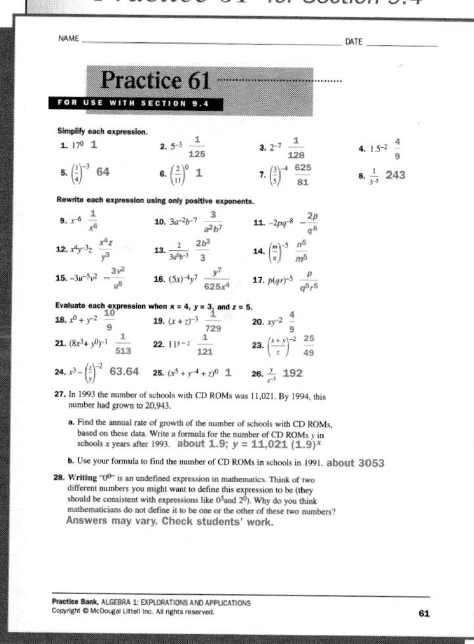

NAME ______ DATE ______

Practice 61

FOR USE WITH SECTION 9.4

Simplify each expression.

1. 17^0 1
2. 5^{-3} $\frac{1}{125}$
3. 2^{-7} $\frac{1}{128}$
4. 1.5^{-2} $\frac{4}{9}$
5. $\left(\frac{1}{4}\right)^{-3}$ 64
6. $\left(\frac{2}{11}\right)^0$ 1
7. $\left(\frac{3}{5}\right)^{-4}$ $\frac{625}{81}$
8. $\frac{1}{3^{-5}}$ 243

Rewrite each expression using only positive exponents.

9. x^{-6} $\frac{1}{x^6}$
10. $3a^{-2}b^{-7}$ $\frac{3}{a^2b^7}$
11. $-2pq^{-8}$ $-\frac{2p}{q^8}$
12. $x^4y^{-3}z$ $\frac{x^4z}{y^3}$
13. $\frac{2}{3a^0b^{-3}}$ $\frac{2b^3}{3}$
14. $\left(\frac{m}{n}\right)^{-5}$ $\frac{n^5}{m^5}$
15. $-3u^{-5}v^2$ $-\frac{3v^2}{u^5}$
16. $(5x)^{-4}y^7$ $\frac{y^7}{625x^4}$
17. $p(qr)^{-5}$ $\frac{p}{q^5r^5}$

Evaluate each expression when $x = 4$, $y = 3$, and $z = 5$.

18. $x^0 + y^{-2}$ $\frac{10}{9}$
19. $(x + z)^{-3}$ $\frac{1}{729}$
20. xy^{-2} $\frac{4}{9}$
21. $(8x^3 + y^0)^{-1}$ $\frac{1}{513}$
22. 11^{y-z} $\frac{1}{121}$
23. $\left(\frac{x+y}{z}\right)^{-2}$ $\frac{25}{49}$
24. $x^3 - \left(\frac{z}{y}\right)^{-2}$ 63.64
25. $(x^5 + y^{-4} + z)^0$ 1
26. $\frac{y}{x^{-3}}$ 192

27. In 1993 the number of schools with CD ROMs was 11,021. By 1994, this number had grown to 20,943.

a. Find the annual rate of growth of the number of schools with CD ROMs, based on these data. Write a formula for the number of CD ROMs y in schools x years after 1993. about 1.9; $y = 11{,}021\ (1.9)^x$

b. Use your formula to find the number of CD ROMs in schools in 1991. about 3053

28. **Writing** "0^0" is an undefined expression in mathematics. Think of two different numbers you might want to define this expression to be (they should be consistent with expressions like 0^3 and 2^0). Why do you think mathematicians do not define it to be one or the other of these two numbers? Answers may vary. Check students' work.

Practice Bank, ALGEBRA 1: EXPLORATIONS AND APPLICATIONS

61

35. A

36. a. $1 \cdot 10^2 + 3 \cdot 10^1 + 9 \cdot 10^0 + 5 \cdot 10^{-1} + 8 \cdot 10^{-2}$

b. $4 \cdot 10^3 + 2 \cdot 10^2 + 3 \cdot 10^1 + 4 \cdot 10^0 + 6 \cdot 10^{-1} + 8 \cdot 10^{-2} + 6 \cdot 10^{-3}$

c. $7 \cdot 10^4 + 3 \cdot 10^3 + 0 \cdot 10^2 + 0 \cdot 10^1 + 9 \cdot 10^0 + 5 \cdot 10^{-1}$

d. $0 \cdot 10^0 + 1 \cdot 10^{-1} + 2 \cdot 10^{-2} + 3 \cdot 10^{-3} + 4 \cdot 10^{-4} + 5 \cdot 10^{-5}$

37. Summaries may vary. An example is given. To raise a number b to the zero or a negative power m, start with 1 and divide by b m times. For example, 3^0 means 1 divided by 3 zero times; $3^0 = 1$. 2^{-3} means 1 divided by 2 three times, that is, $2^{-3} = 1 \div 2 \div 2 \div 2 = \frac{1}{2} \cdot \frac{1}{2} \cdot \frac{1}{2} = \frac{1}{8}$.

38. $x \leq 5$

39. $m \geq -3$

40. $p < -9$

41. $q > -\frac{14}{3}$

42. $x \geq 2$

43. $y \geq \frac{1}{6}$

Plan⇔Support

Objective

- Write numbers in scientific notation.

Recommended Pacing

Core and Extended Courses
Section 9.5: 1 day

Block Schedule
Section 9.5: $\frac{1}{2}$ block
(with Section 9.6)

Resource Materials

Lesson Support
Lesson Plan 9.5
Warm-Up Transparency 9.5
Practice Bank: Practice 62
Study Guide: Section 9.5
Assessment Book: Test 39

Technology
Scientific Calculator
Graphing Calculator
Internet:
http://www.hmco.com

Warm-Up Exercises

Evaluate each expression.

1. 10^3 1000
2. 100^2 10,000
3. 10^7 10,000,000
4. 10^{-1} 0.1
5. 10^{-4} 0.0001
6. 10^{-9} 0.000000001

SECTION

9.5 Working with Scientific Notation

Learn how to...

- **write numbers in scientific notation**

So you can...

- **conveniently express numbers such as the weight of Earth or the diameter of a cell**

Have you ever had trouble entering a number into a calculator because it had too many digits? Mathematicians and scientists use a form of shorthand based on powers of 10 to write numbers with many digits.

The weights of the sun and the nine planets in tons are written in *scientific notation* in the diagram below. A number is expressed in **scientific notation** if it is in the form $a \times 10^n$ where $1 \le a < 10$ and n is an integer.

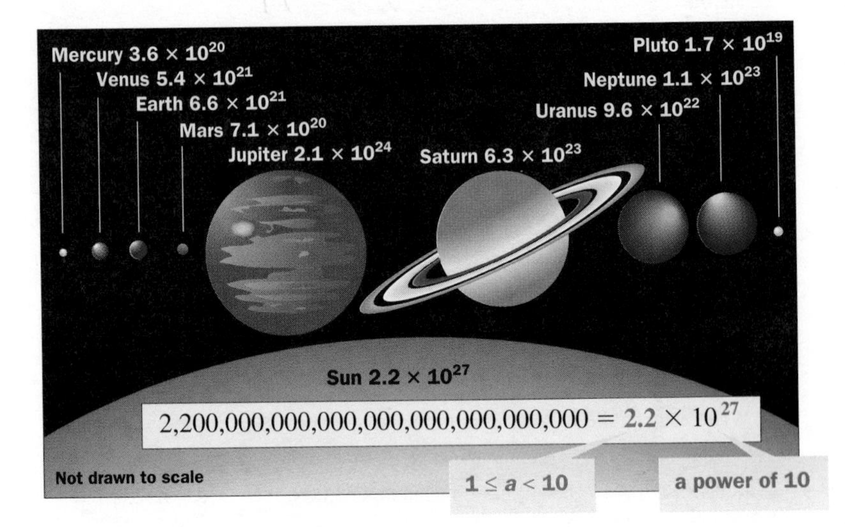

EXAMPLE 1

Write the weight of Earth in decimal notation.

SOLUTION

Follow the order of operations. Simplify the power before you multiply.

Notice that the power 10^{21} has **21** zeros.

$$6.6 \times 10^{21} = 6.6 \times 1{,}000{,}000{,}000{,}000{,}000{,}000{,}000$$
$$= 6{,}600{,}000{,}000{,}000{,}000{,}000{,}000$$

Earth weighs 6,600,000,000,000,000,000,000 tons.

You can also write numbers less than 1 in scientific notation.

$$0.000002 = \frac{2}{1{,}000{,}000} = \frac{2}{10^6} = 2 \cdot \frac{1}{10^6} = 2 \times 10^{-6}$$

Multiplying by 10^{-6} is the same as dividing by 10^6.

When $n > 0$, factoring $\mathbf{10^n}$ out of a number moves the decimal point n places to the **left**. Factoring out $\mathbf{10^{-n}}$ moves the decimal point n places to the **right**.

EXAMPLE 2

Write each number in scientific notation.

a. 72,000 **b. 0.0000006093**

SOLUTION

a. $72{,}000 = 7.2 \times 10^4$

Factoring out 10^4 moves the decimal point **4** places to the **left**.

b. $0.0000006093 = 6.093 \times 10^{-7}$

Factoring out 10^{-7} moves the decimal point **7** places to the **right**.

THINK AND COMMUNICATE

1. Look back at the diagram on page 396. How does the power of 10 show that Saturn weighs more than Earth?
2. Write each number in scientific notation.
   a. 0.1 **b.** 1 **c.** 10
3. Describe how to write a number a in scientific notation when $1 \le a < 10$.

EXAMPLE 3

Write each number in decimal notation.

a. $\mathbf{8.959 \times 10^{-6}}$ **b.** $\mathbf{3.24 \times 10^{-5}}$

SOLUTION

a. $8.959 \times 10^{-6} = 8.959 \times 0.000001 = 0.000008959$

Notice that the power 10^{-6} has **6** decimal places.

b. $3.24 \times 10^{-5} = 3.24 \times 0.00001 = 0.0000324$

ANSWERS Section 9.5

Think and Communicate

1. The power of 10 in 6.3×10^{23} is greater than the power of 10 in 6.6×10^{21}.
2. a. 1×10^{-1}
   b. 1×10^{0}
   c. 1×10^{1}
3. Multiply a by 10^0.

Teach⇔Interact

Additional Example 1

The sun generates energy at the rate of about 3.9×10^{33} ergs/s. Write this in decimal notation.

Follow the order of operations. Simplify the power before you multiply. Notice that 10^{33} has 33 zeros.
$3.9 \times 10^{33} = 3.9 \times$ 1,000,000,000,000,000,000,000,000,000,000,000 = 3,900,000,000,000,000,000,000,000,000,000,000
The sun generates energy at the rate of about 3,900,000,000,000,000,000,000,000,000,000,000 ergs/s.

Section Note

Reasoning
Students should recognize that for any positive integer n, 10^n has n zeros when written in decimal form, but 10^{-n} has $n - 1$ zeros. Ask students to explain why this is true.

About Example 2

Common Error
When attempting to write a number in scientific notation, some students incorrectly choose a number greater than 10 as the first factor. Remind these students that only one digit should remain to the left of the decimal point.

Additional Example 2

Write in scientific notation.

a. 350,000
Factoring out 10^5 moves the decimal point 5 places to the left.
$350{,}000 = 3.5 \times 10^5$

b. 0.0000000775
Factoring out 10^{-8} moves the decimal point 8 places to the right.
$0.0000000775 = 7.75 \times 10^{-8}$

Additional Example 3

Write in decimal notation.

a. 6.43×10^{-4}
$6.43 \times 10^{-4} = 6.43 \times 0.0001 = 0.000643$

b. 5.421×10^{-7}
$5.421 \times 10^{-7} = 5.421 \times 0.0000001 = 0.0000005421$

Teach⇔Interact

Think and Communicate

For question 5, ask one student to describe how to write a number in scientific notation verbally. Then call upon another student to accept or correct the description. Repeat this process until a correct description is given. Then have the student who gave a correct description illustrate his or her method with an example.

Closure Questions

What positive numbers have positive exponents when written in scientific notation? What positive numbers have negative exponents when written in scientific notation? numbers that are greater than or equal to 1; numbers that are between 0 and 1

Apply⇔Assess

Suggested Assignment

Core Course
Exs. 1–16, 18–20, 22–26, 29, 31–37, AYP

Extended Course
Exs. 1–6, 10–37, AYP

Block Schedule
Day 59 Exs. 1–6, 10–16, 18–20, 22–26, 29–37, AYP

Exercise Notes

Using Technology
Exs. 1–15 Students using TI-82 or TI-81 calculators should probably experiment with these exercises to see how their calculators can be used to display numbers in scientific notation. They should also try going from scientific notation to decimal notation.

Students should understand that in a display such as 3.17 E 12 or 5.84 E –9, the integer that follows the letter E is the exponent needed for the scientific notation. Thus, 3.17 E 12 = 3.17×10^{12} and 5.84 E –9 = 5.84×10^{-9}. The calculator can be set to display all results in scientific notation. Press MODE and select Sci on the first line of the MODE screen.

THINK AND COMMUNICATE

4. a. Write each number in decimal notation and in scientific notation.

7 thousand   4.3 billion   6 millionths

b. How do the phrases in part (a) resemble numbers written in scientific notation?

5. Describe how to write a number in scientific notation. Illustrate your method with a number less than 1 and another number greater than 10.

CHECKING KEY CONCEPTS

Write each number in decimal notation.

1. 6×10^5 **2.** 1.2×10^7 **3.** 5.03×10^6

4. 3×10^{-2} **5.** 7.72×10^{-4} **6.** 8.0430×10^{-3}

Write each number in scientific notation.

7. 8,000,000   **8.** 419,500   **9.** 34,005,600,000

10. 0.00001   **11.** 0.00074   **12.** 0.0000208

9.5 Exercises and Applications

Extra Practice exercises on page 574

Write each number in decimal notation.

1. 2×10^3 **2.** 7×10^{-4} **3.** 4.9×10^6

4. 5.12×10^{-5} **5.** 6.761×10^8 **6.** 8.35×10^0

Write each number in scientific notation.

7. 780   **8.** 0.000000046   **9.** 21,000

10. 0.0000003803   **11.** 9,700,000,000,000   **12.** 1.5

13. 3.2 million   **14.** 4 thousandths   **15.** 8.023 hundredths

16. ASTRONOMY The table gives each planet's approximate distance from the sun in scientific notation.

Earth: 9.3×10^7 mi	Mars: 1.4×10^8 mi	Jupiter: 4.8×10^8 mi
Uranus: 1.8×10^9 mi	Venus: 6.7×10^7 mi	Mercury: 3.6×10^7 mi
Pluto: 3.7×10^9 mi	Saturn: 8.9×10^8 mi	Neptune: 2.8×10^9 mi

a. Which planet is farthest from the sun? Which is closest to the sun?

b. Write each distance in decimal notation.

17. Research Create a diagram like the one on page 396 that gives a set of data in scientific notation.

Think and Communicate

4\. a. $7000 = 7 \times 10^3$; $4{,}300{,}000{,}000 = 4.3 \times 10^9$; $0.000006 = 6 \times 10^{-6}$

b. Each consists of a number between 1 and 10 and a word description for a power of 10.

5\. To write a number in scientific notation, move the decimal point to obtain a number between 1 and 10 and multiply the resulting number by the power of 10 determined as follows. If you moved the decimal n places to the left, multiply the number by 10^n; for example, $3729 = 3.729 \times 10^3$. If you moved the decimal point n places to the right, multiply the number by 10^{-n}; for example, $0.003 = 3 \times 10^{-3}$.

Checking Key Concepts

1\. 600,000   2. 12,000,000
3\. 5,030,000   4. 0.03
5\. 0.000772   6. 0.0080430
7\. 8×10^6
8\. 4.195×10^5
9\. 3.40056×10^{10}
10\. 1×10^{-5}
11\. 7.4×10^{-4}
12\. 2.08×10^{-5}

INTERVIEW Lori Arviso Alvord

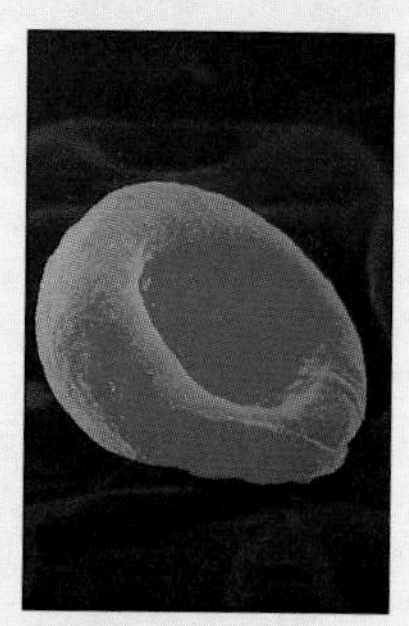

Look back at the article on pages 366-368.

Doctors like Lori Arviso Alvord need to be aware of blood clotting when performing surgery. Blood clotting involves a series of chemical reactions that lead to the formation of a mesh of fibers that are too large for blood cells to pass through.

A micron is a unit of measurement 10^{-6} m long. Extremely small objects, such as blood cells, are measured in microns. For Exercises 18–20, give each measurement in microns. Explain how you got your answers.

18. A typical red blood cell is 0.0000072 m wide.

19. A white blood cell is 0.000007 m to 0.000012 m wide.

20. A platelet is about 0.000003 m wide.

21. Visual Thinking Use the measurements given in Exercises 18–20 to make a scale drawing of the three kinds of blood cells.

For Exercises 22 and 23, write the number in red in scientific notation.

22. A certain brand of "supercomputer" can perform about **2,500,000,000** arithmetic operations per second.

23. The probability of flipping a coin 20 times and getting all heads is about **0.000000954**.

24. Human hair grows at a rate of 6 in./year. How fast does hair grow in miles per hour? Write your answer in scientific notation.

25. EARTH SCIENCE The Amazon River in South America discharges about 4,200,000 ft^3 of water into the Atlantic Ocean each second. Find the volume of water discharged from the Amazon River each day. Write your answer in scientific notation.

26. **Technology** In this exercise, you will explore how your calculator uses scientific notation.

a. Use a calculator to multiply a positive one-digit integer by 10. Then multiply the result by 10. Repeat this until the calculator gives the answer in scientific notation. What is the exponent of the power of 10?

b. Repeat part (a), but start with a two-digit integer. Is the exponent of the power of 10 the same as in part (a)?

c. Repeat parts (a) and (b), but divide by 10 instead of multiplying. How do your answers compare to your answers to parts (a) and (b)?

27. Writing Raj entered the following key sequence on his calculator:

Explain why he got the answer shown.

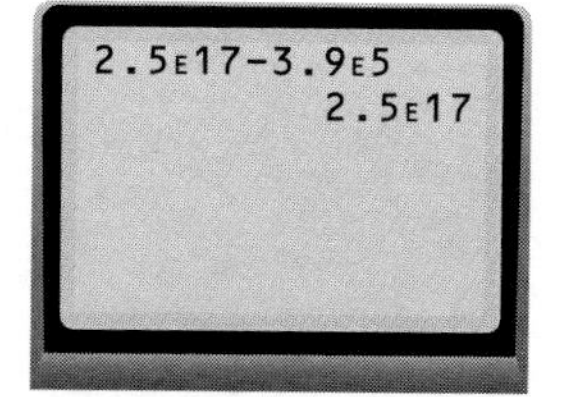

28. Open-ended Problem Describe three situations in which scientific notation might be more useful than decimal notation. Explain.

Exercises and Applications

1. 2000
2. 0.0007
3. 4,900,000
4. 0.0000512
5. 676,100,000
6. 8.35
7. 7.8×10^2
8. 4.6×10^{-8}
9. 2.1×10^4
10. 3.803×10^{-7}
11. 9.7×10^{12}
12. 1.5×10^0
13. 3.2×10^6
14. 4×10^{-3}
15. 8.023×10^{-2}

16–21. See answers in back of book.

22. 2.5×10^9

23. 9.54×10^{-7}

24. about 1.08×10^{-8} mi/h

25. 3.6288×10^{11} ft^3

26. a. 10 b. Yes.
c. (a) –10; (b) –10; The answers to parts (a), (b), and (c) all have the same absolute value.

27. 3.9×10^5 is so small compared to 2.5×10^{17} that it is insignificant.

28. Answers may vary. Examples are given. measuring the diameters of cells, astronomical distances, and describing the weights of extremely large or extremely small objects

Apply⇔Assess

Exercise Notes

Common Error

Exs. 7–15 Some students make errors with the sign of the exponent when writing a number in scientific notation. In Ex. 7, for example, they write $780 = 7.80 \times 10^{-2}$ because, from past experience, they had associated a move to the left on a number line as a negative move. Similarly, they think about moving the decimal point 8 places to the right in Ex. 8 and write $0.000000046 = 4.6 \times 10^8$. Students should always check their answers by converting back to decimal notation to verify that they are correct.

Application

Ex. 16 The science of astronomy makes extensive use of scientific notation because distances in the universe are so vast. Some students may wish to peruse a book on astronomy for examples of other distances or measures expressed in scientific notation. They can then share their findings with the class.

Multicultural Note

Ex. 25 One of the most numerous native peoples living in the Amazon rainforest is the Yanomami. In addition to hunting and gathering, the Yanomami make small clearings in the forest to cultivate a variety of crops used for ceremonies, food, and medicine. Unlike the large-scale slash and burn techniques that tend to leave the land temporarily ravaged, these gardens do not damage the surrounding forest. The Yanomami let the small clearings grow over and, within a decade, their former gardens are indistinguishable from the forest around them.

Using Technology

Ex. 26 To use the TI-82 or TI-81 for this exercise, it is convenient to use the following keystrokes. (Assume the one-digit integer for part (a) is 3.)
3 [×] 10 [ENTER] [×] 10. After typing 10 the second time, press [ENTER] repeatedly to display the successive products. The largest product that the calculator will display in standard form is 3,000,000,000. The next product will be displayed as 3 E 10, meaning 3×10^{10}. Succeeding products will be displayed as 3 E 11, 3 E 12, and so on.

Apply⇔Assess

Exercise Notes

Cooperative Learning

Ex. 28 You may wish to have students discuss this exercise in small groups. Students can choose one discipline such as physics, biology, social studies, or economics from which to gather data. The groups can then share their findings with each other.

Challenge

Ex. 30 Students should see that they can apply the commutative property for multiplication of real numbers to these exercises. Then the multiplication can be done mentally.

Progress Check 9.4–9.5

See page 415.

Practice 62 for Section 9.5

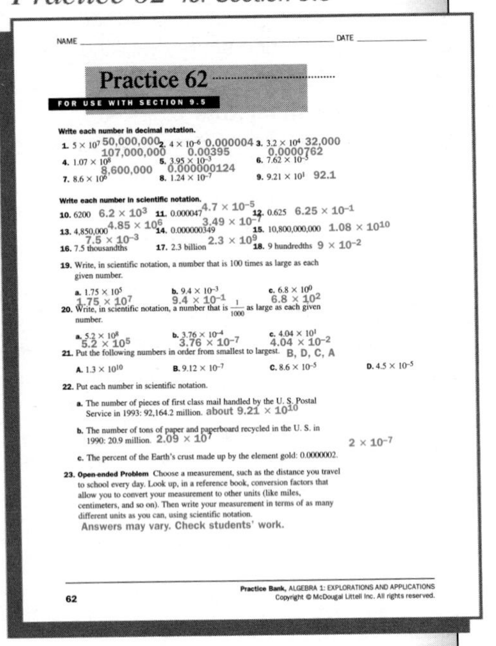

NAME ______ DATE ______

Practice 62

FOR USE WITH SECTION 9.5

Write each number in decimal notation.

1. 5×10^7 50,000,000
2. 4×10^{-6} 0.000004
3. 3.2×10^4 32,000
4. 1.07×10^8 107,000,000
5. 3.95×10^{-3} 0.00395
6. 7.62×10^{-5} 0.0000762
7. 8.6×10^6 8,600,000
8. 1.24×10^{-7} 0.000000124
9. 9.21×10^1 92.1

Write each number in scientific notation.

10. 6200 6.2×10^3
11. 0.000047 4.7×10^{-5}
12. 0.625 6.25×10^{-1}
13. 4,850,000 4.85×10^6
14. 0.000000349 3.49×10^{-7}
15. 10,800,000,000 1.08×10^{10}
16. 7.5 thousandths 7.5×10^{-3}
17. 2.3 billion 2.3×10^9
18. 9 hundredths 9×10^{-2}
19. Write, in scientific notation, a number that is 100 times as large as each given number.
    a. 1.75×10^5 1.75×10^7 b. 9.4×10^{-3} 9.4×10^{-1} c. 6.8×10^0 6.8×10^2
20. Write, in scientific notation, a number that is $\frac{1}{1000}$ as large as each given number.
    a. 5.2×10^8 5.2×10^5 b. 3.76×10^{-4} 3.76×10^{-7} c. 4.04×10^1 4.04×10^{-2}
21. Put the following numbers in order from smallest to largest. B, D, C, A
    A. 1.3×10^{10} B. 9.12×10^{-7} C. 8.6×10^{-5} D. 4.5×10^{-5}
22. Put each number in scientific notation.
    a. The number of pieces of first class mail handled by the U. S. Postal Service in 1993: 92,164.2 million. about 9.21×10^{10}
    b. The number of tons of paper and paperboard recycled in the U. S. in 1990: 20.9 million. 2.09×10^7
    c. The percent of the Earth's crust made up by the element gold: 0.0000002. 2×10^{-7}
23. **Open-ended Problem** Choose a measurement, such as the distance you travel to school every day. Look up, in a reference book, conversion factors that allow you to convert your measurement to other units (like miles, centimeters, and so on). Then write your measurement in terms of as many different units as you can, using scientific notation. Answers may vary. Check students' work.

62 **Practice Bank,** ALGEBRA 1: EXPLORATIONS AND APPLICATIONS

29. SAT/ACT Preview Which of the following numbers is (are) written in scientific notation?

I. 21.8×10^7 II. 9.9005×10^{-2} III. 0.16×10^{-8}

A. I only **B.** II only **C.** III only **D.** I and II **E.** I and III

30. Challenge Write each product in scientific notation. Explain how you found your answers.

a. $(1 \times 10^2)(1 \times 10^2)$ b. $(4 \times 10^2)(1 \times 10^3)$
c. $(1 \times 10^2)(2 \times 10^4)$ d. $(2 \times 10^3)(2 \times 10^4)$

ONGOING ASSESSMENT

31. Writing Describe a method for determining which of two numbers written in scientific notation is larger than the other. Show that your method works for two very large numbers and two very small numbers.

SPIRAL REVIEW

Write each expression in simplest form. *(Section 6.3)*

32. $2^4 \cdot 3^4$ **33.** $(2 \cdot 3)^4$ **34.** $\frac{6^5}{3^5}$

Simplify each power. *(Section 9.1)*

35. $(-5)^3$ **36.** $3\left(\frac{2}{3}\right)^4$ **37.** $\left(3 \cdot \frac{2}{3}\right)^4$

ASSESS YOUR PROGRESS

VOCABULARY

scientific notation (p. 396)

Evaluate each power. *(Section 9.4)*

1. 3.125^0 **2.** 4^{-5} **3.** $\left(\frac{4}{3}\right)^{-3}$

Evaluate each expression when $x = 4$, $y = -3$, and $z = 2$. *(Section 9.4)*

4. $z^{-5}x^0$ **5.** $x^{-2}y^2$ **6.** $\left(\frac{z}{x}\right)^{-2}$

7. The population of Australia in millions x years after 1980 can be modeled by the function $y = 14.6(1.014)^x$. Estimate the population of Australia for each year. *(Section 9.4)*

a. 1976 b. 1980 c. 1972

Write each number in decimal notation. *(Section 9.5)*

8. 7.003×10^9 **9.** 4.21×10^{-5} **10.** 8×10^{-6}

Write each number in scientific notation. *(Section 9.5)*

11. 54,000,000,000 **12.** 3.28 **13.** 0.00000709

14. Journal How can you tell whether a number written in scientific notation is greater than or less than 1? Explain your reasoning.

29. B

30. Methods may vary. I used the commutative property to multiply the "a" parts of the numbers in scientific notation. Then I multiplied the powers of 10 in decimal form and rewrote their product as a power of 10. For example, I multiplied the expressions in part (b) like this: $(4 \times 10^2)(1 \times 10^3) = (4 \times 1)(10^2 \times 10^3) = 4 \times 100 \times 1000 = 4 \times 100{,}000 = 4 \times 10^5$.

a. 1×10^4 b. 4×10^5
c. 2×10^6 d. 4×10^7

31. See answers in back of book.

32. 1296 33. 1296

34. 32 35. –125

36. $\frac{14,641}{81}$ 37. 16

Assess Your Progress

1. 1 2. $\frac{1}{1024}$

3. $\frac{27}{64}$ 4. $\frac{1}{32}$

5. $\frac{9}{16}$ 6. 4

7. a. about 13.81 million
b. 14.6 million
c. about 13.06 million

8. 7,003,000,000

9. 0.0000421

10. 0.000008

11. 5.4×10^{10}

12. 3.28×10^0

13. 7.09×10^{-6}

14. If the exponent in a number written in scientific notation is negative, the number is less than 1. If the exponent is positive, the number is greater than 1.

SECTION

9.6 Exploring Powers

A *light-year*, the distance that light travels in a year, is about 5.88×10^{12} mi. Astronomers use light-years to express large distances in space. To convert light-years to miles, you must often multiply two powers of 10.

Learn how to...

- **multiply and divide expressions involving powers**

So you can...

- **perform calculations such as converting miles to light-years**

EXPLORATION
COOPERATIVE LEARNING

Multiplying and Dividing Powers of 10

Work with a partner.
You will need:

- a calculator

Product of powers	Expanded product	Product as a single power
$10^2 \cdot 10^1$	(10·10)(10)	10^3
$10^4 \cdot 10^2$	?	?
$10^7 \cdot 10^1$	?	?
$10^5 \cdot 10^6$	?	?
$10^0 \cdot 10^3$	?	?

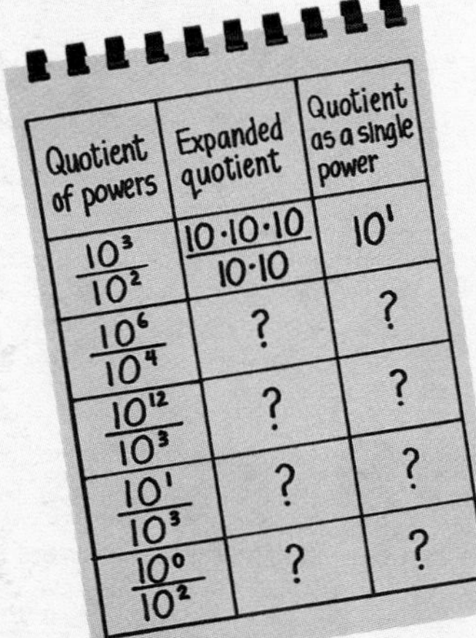

Quotient of powers	Expanded quotient	Quotient as a single power
$\frac{10^3}{10^2}$	$\frac{10 \cdot 10 \cdot 10}{10 \cdot 10}$	10^1
$\frac{10^6}{10^4}$	?	?
$\frac{10^{12}}{10^3}$	?	?
$\frac{10^1}{10^3}$	?	?
$\frac{10^0}{10^2}$	?	?

Copy and complete the tables. Then answer the questions below.

Questions

1. Use any patterns you see in the tables to complete these rules for working with powers of 10:

a. $10^m \cdot 10^n = 10^?$ **b.** $\frac{10^m}{10^n} = 10^?$

2. Use the rules you wrote in Question 1 to simplify each expression. Use a calculator to check your answers.

a. $10^4 \cdot 10^7$ **b.** $10^{10} \cdot 10^3$ **c.** $\frac{10^9}{10^2}$ **d.** $\frac{10^8}{10^{14}}$

Exploration Note

Purpose
The purpose of this Exploration is to have students study a pattern in order to determine a rule for multiplying and dividing expressions involving powers.

Materials/Preparation
Each group should have a calculator. You may wish to have students review the procedures for entering exponents and for displaying numbers in scientific notation on their calculators.

Procedure
Students should look for a relationship between the exponents of the numbers being multiplied or divided and the exponent of the corresponding product or quotient.

Closure
Students should see that the rules they wrote are a shortcut method for multiplying and dividing powers having the same base.

Explorations Lab Manual
See the Manual for more commentary on this Exploration.

Diagram Master 17

For answers to the Exploration, see following page.

Plan⇔Support

Objective

- Multiply and divide expressions involving powers.

Recommended Pacing

Core and Extended Courses
Section 9.6: 1 day

Block Schedule
Section 9.6: $\frac{1}{2}$ block
(with Section 9.5)

Resource Materials

Lesson Support
Lesson Plan 9.6
Warm-Up Transparency 9.6
Practice Bank: Practice 63
Study Guide: Section 9.6
Explorations Lab Manual: Diagram Master 17

Technology
Scientific Calculator
Internet: http://www.hmco.com

Warm-Up Exercises

Write each expression as a power of 2.

1. $(2 \cdot 2 \cdot 2)(2 \cdot 2)$ 2^5
2. $2^2 \cdot 2$ 2^3
3. $\frac{2 \cdot 2 \cdot 2 \cdot 2}{2 \cdot 2}$ 2^2
4. $\frac{2 \cdot 2 \cdot 2}{2 \cdot 2 \cdot 2 \cdot 2 \cdot 2 \cdot 2}$ 2^{-3}
5. $(2 \cdot 2 \cdot 2)(2 \cdot 2 \cdot 2)(2 \cdot 2)$ 2^8

Teach⇔Interact

Learning Styles: Visual

Students with visual learning styles may need to write the exponential form of numbers or variables in factored form before they are comfortable using the power rules. This approach will allow them to see and count the factors being multiplied or divided in order to arrive at an answer. They can then check their answers using the appropriate rule.

Additional Example 1

Simplify each expression.

a. $a^3 \cdot a^9$

$a^3 \cdot a^9 = a^{3+9}$
$= a^{12}$

b. $(a^{-2}b^9)(a^{-1}b^{-5})$

$(a^{-2}b^9)(a^{-1}b^{-5}) = (a^{-2}a^{-1})(b^9b^{-5})$
$= a^{-2+(-1)}b^{9+(-5)}$
$= a^{-3}b^4$
$= \frac{1}{a^3} \cdot b^4$
$= \frac{b^4}{a^3}$

Additional Example 2

Simplify each expression.

a. $\frac{c^6}{c^9}$

$\frac{c^6}{c^9} = c^{6-9}$
$= c^{-3}$
$= \frac{1}{c^3}$

b. $\frac{c^{-2}d^{10}}{c^8d^2}$

$\frac{c^{-2}d^{10}}{c^8d^2} = \frac{c^{-2}}{c^8} \cdot \frac{d^{10}}{d^2}$
$= c^{-2-8}d^{10-2}$
$= c^{-10}d^8$
$= \frac{1}{c^{10}} \cdot d^8$
$= \frac{d^8}{c^{10}}$

Rules for Products and Quotients of Powers

These rules are true for any nonzero number a and any integers m and n:

Product of Powers Rule	Quotient of Powers Rule
$a^m \cdot a^n = a^{m+n}$	$\frac{a^m}{a^n} = a^{m-n}$
$2^4 \cdot 2^5 = 2^{4+5} = 2^9$	$\frac{5^8}{5^2} = 5^{8-2} = 5^6$

EXAMPLE 1

Simplify each expression. Write your answers using positive exponents.

a. $a^3 \cdot a^7$

b. $(a^6b^{-9})(a^8b^4)$

SOLUTION

a. $a^3 \cdot a^7 = a^{3+7}$
$= a^{10}$

b. $(a^6b^{-9})(a^8b^4) = (a^6a^8)(b^{-9}b^4)$
$= a^{6+8}b^{-9+4}$
$= a^{14}b^{-5}$
$= a^{14} \cdot \frac{1}{b^5}$
$= \frac{a^{14}}{b^5}$

Use the associative and commutative properties to group powers with the same base.

EXAMPLE 2

Simplify each expression. Write your answers using positive exponents.

a. $\frac{c^{11}}{c^5}$

b. $\frac{c^5d^3}{c^6d}$

SOLUTION

a. $\frac{c^{11}}{c^5} = c^{11-5}$
$= c^6$

b. $\frac{c^5d^3}{c^6d} = \frac{c^5}{c^6} \cdot \frac{d^3}{d}$
$= c^{5-6}d^{3-1}$
$= c^{-1}d^2$
$= \frac{1}{c} \cdot d^2$
$= \frac{d^2}{c}$

Group powers with the same base.

ANSWERS Section 9.6

Exploration

Product of powers	Expanded product	Product as a single power
$10^2 \cdot 10^1$	$(10 \cdot 10)(10)$	10^3
$10^4 \cdot 10^2$	$(10 \cdot 10 \cdot 10 \cdot 10)(10 \cdot 10)$	10^6
$10^7 \cdot 10^1$	$(10 \cdot 10 \cdot 10 \cdot 10 \cdot 10 \cdot 10 \cdot 10)(10)$	10^8
$10^5 \cdot 10^6$	$(10 \cdot 10 \cdot 10 \cdot 10 \cdot 10)(10 \cdot 10 \cdot 10 \cdot 10 \cdot 10 \cdot 10)$	10^{11}
$10^0 \cdot 10^3$	$(1) \cdot (10 \cdot 10 \cdot 10)$	10^3

Quotient of powers	Expanded quotient	Quotient as a single power
$\frac{10^3}{10^2}$	$\frac{10 \cdot 10 \cdot 10}{10 \cdot 10}$	10
$\frac{10^6}{10^4}$	$\frac{10 \cdot 10 \cdot 10 \cdot 10 \cdot 10 \cdot 10}{10 \cdot 10 \cdot 10 \cdot 10}$	10^2
$\frac{10^{12}}{10^3}$	$\frac{10 \cdot 10 \cdot 10 \cdot 10 \cdot 10 \cdot 10 \cdot 10 \cdot 10 \cdot 10 \cdot 10 \cdot 10 \cdot 10}{10 \cdot 10 \cdot 10}$	10^9
$\frac{10^1}{10^3}$	$\frac{10}{10 \cdot 10 \cdot 10}$	10^{-2}
$\frac{10^0}{10^2}$	$\frac{1}{10 \cdot 10}$	10^{-2}

EXAMPLE 3 Application: Astronomy

The Milky Way galaxy has a diameter of about 5.88×10^{18} mi. The diameter of our solar system is about 7.29×10^{9} mi. How many times as big as the diameter of our solar system is the diameter of the Milky Way?

SOLUTION

$$\frac{\text{Diameter of Milky Way}}{\text{Diameter of solar system}} = \frac{5.88 \times 10^{18}\text{ mi}}{7.29 \times 10^{9}\text{ mi}}$$

$$= \frac{5.88}{7.29} \times \frac{10^{18}}{10^{9}}$$

$$\approx 0.8 \times 10^{18-9}$$

$$\approx 0.8 \times 10^{9}$$

$$\approx (8 \times 10^{-1}) \times 10^{9}$$

Rewrite 0.8 in scientific notation.

$$\approx 8 \times 10^{-1+9}$$

$$\approx 8 \times 10^{8}$$

The diameter of the Milky Way is about 8×10^{8} times as big as the diameter of our solar system.

THINK AND COMMUNICATE

1. The diameter of the circle at the right is 1 in. Suppose this circle represents the size of our solar system. Based on the solution of Example 3, what would be the diameter of a circle that represents the size of the Milky Way? Give your answer in miles.

Solar System

2. Compare the diameter you found in Question 1 with the diameter of Earth, which is about 8000 mi.

CHECKING KEY CONCEPTS

Write each expression as a single power.

1. $4^{5} \cdot 4^{3}$
2. $\frac{7^{12}}{7^{6}}$
3. $\frac{2^{3} \cdot 2}{2^{4}}$

Simplify each expression. Write your answers using positive exponents.

4. $m^{4}m^{2}$
5. $t^{6}t^{-2}$
6. $\frac{k^{5}}{k^{9}}$
7. $\frac{4a^{2}}{6a^{2}}$
8. $\frac{c^{9}d^{7}}{c^{3}d^{2}}$
9. $(3p^{5}q^{4})(5pq^{2})$
10. **GEOGRAPHY** Lake Superior contains about 2.94×10^{3} mi^{3} of water. Find the number of gallons of water in Lake Superior. Use the fact that 1 mi$^{3} \approx 1.10 \times 10^{12}$ gal.

Teach⇔Interact

Additional Example 3

Our solar system has a diameter of about 7.29×10^{9} mi. The diameter of Earth is about 8×10^{3} mi. How many times bigger than the diameter of Earth is the diameter of our solar system?

$$\frac{\text{Diameter of solar system}}{\text{Diameter of Earth}} = \frac{7.29 \times 10^{9}}{8 \times 10^{3}}$$

$$= \frac{7.29}{8} \times \frac{10^{9}}{10^{3}}$$

$$\approx 0.9 \times 10^{9-3}$$

$$\approx 0.9 \times 10^{6}$$

$$\approx (9 \times 10^{-1}) \times 10^{6}$$

$$\approx 9 \times 10^{-1+6}$$

$$\approx 9 \times 10^{5}$$

The diameter of our solar system is about 9×10^{5} times bigger than that of Earth.

Think and Communicate

Comparing very large numbers expressed in scientific notation has little real meaning to most students because they have no intuitive understanding of these numbers. The answer to question 1 will give students an excellent visual comparison between the size of our solar system and the size of the Milky Way.

Checking Key Concepts

Problem Solving

When solving real-world problems, it is often necessary to use conversion factors to express an answer in units that are familiar, meaningful, and understandable.

Closure Question

Explain the product of powers rule and the quotient of powers rule.

Apply the product of powers rule to multiply two powers with the same base. The exponent of the product is the sum of the exponents of the factors. Apply the quotient of powers rule when dividing two powers with the same base. The exponent of the quotient is the difference of the exponent of the dividend and the exponent of the divisor.

1. a. $m + n$ b. $m - n$
2. a. 10^{11} b. 10^{13} c. 10^{7} d. $\frac{1}{10^{6}} = 10^{-6}$

Think and Communicate

1. about 12,626 mi
2. The diameter of the circle depicting the Milky Way would be about 1.58 times the actual diameter of Earth.

Checking Key Concepts

1. 4^{8}
2. 7^{6}
3. 2^{0}
4. m^{6}
5. t^{4}
6. $\frac{1}{k^{4}} = k^{-4}$
7. $\frac{2}{3}$
8. $c^{6}d^{5}$
9. $15p^{6}q^{6}$
10. about 3.234×10^{15} gal

Apply⇔Assess

Suggested Assignment

Core Course
Exs. 1–17, 19–28

Extended Course
Exs. 1–28

Block Schedule
Day 59 Exs. 1–17, 19–28

Exercise Notes

Common Error
Ex. 5 To show students who simplify this expression as 9^{-7} why their answer is incorrect, write the fraction as the product of two fractions and use the product of powers rule.

$$\frac{9^3}{9^{-10}} = 9^3 \cdot \frac{1}{9^{-10}} = 9^3 \cdot \frac{1}{\frac{1}{9^{-10}}} = 9^3 \cdot 9^{10} = 9^{3+10} = 9^{13}$$

Teaching Tip
Exs. 7–15 You may wish to suggest that students assign some simple numerical values to the variables for some of these exercises, and simplify both the variable and numerical expressions. This approach may help some students to feel more comfortable working with the variable expressions.

Second-Language Learners
Exs. 16–18 Second-language learners may find the passage from Heinlein challenging. Consider reading the passage aloud to the class and then having these students work with English-proficient partners to complete the exercises.

9.6 Exercises and Applications

Extra Practice exercises on page 574

Write each expression as a single power.

1. $2^3 \cdot 2^4$ **2.** $\frac{10^6}{10^9}$ **3.** $3^5 \cdot 3^1$

4. $5^7 \cdot 5^{-4}$ **5.** $\frac{9^3}{9^{-10}}$ **6.** $\frac{4^{11}}{4^8 \cdot 4^{-2}}$

Simplify each expression. Write your answers using positive exponents.

7. $a^2 \cdot a^3$ **8.** $\frac{b^{-9}}{b^4}$ **9.** $c^{-5} \cdot c^{-2}$

10. $\frac{a^6}{4a^2}$ **11.** $(5d^4)(8d^6)$ **12.** $\frac{7x^5}{x^{-8}}$

13. $\frac{p^8q^3}{p^6q}$ **14.** $(5pq^3)(10p^{-3}q^2)$ **15.** $\frac{3a^{-2}b^5}{9a^4b^4}$

Connection LITERATURE

The passage below is from *The Rolling Stones* by Robert A. Heinlein. In the story, the Stone family buys a spaceship and leaves their home on the moon to travel around the solar system.

Rock City

The Asteroid Belt is a flattened torus ring or doughnut in space encompassing thirteen thousand five hundred thousand million trillion cubic miles....

13,500,000,000,000,000,000,000,000 cubic miles of space.

Yet the entire human race could be tucked into one corner of a single cubic mile; the average human body is about two cubic feet in bulk....

Write the figure as 1.35 × 10^{25} cubic miles; that makes it easier to see if no easier to grasp. At the time the *Rolling Stone* arrived among the rolling stones of Rock City the Belt had a population density of one human soul for every two billion trillion cubic miles—read 2 × 10^{21}....

16. Challenge When *The Rolling Stones* was written in 1952, the world population was about 2.5 billion. According to the passage, what fraction of a cubic mile would the human race have filled?

17. Use the formula for population density given below to find the population of the Asteroid Belt.

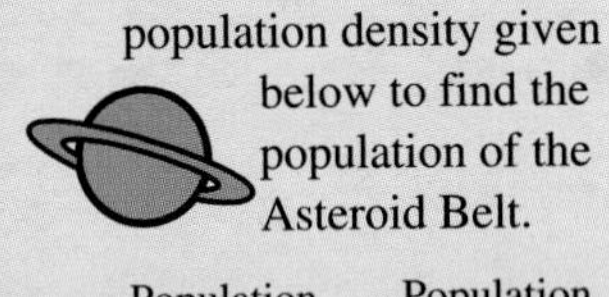

$$\text{Population density} = \frac{\text{Population}}{\text{Volume}}$$

18. Open-ended Problem Write a short science fiction passage that includes calculations with numbers in scientific notation.

Exercises and Applications

1. 2^7 2. $\frac{1}{10^3}$ 3. 3^6
4. 5^3 5. 9^{13} 6. 4^5
7. a^5 8. $\frac{1}{b^{13}}$ 9. $\frac{1}{c^7}$
10. $\frac{a^4}{4}$ 11. $40d^{10}$ 12. $7x^{13}$
13. p^2q^2 14. $\frac{50q^5}{p^2}$ 15. $\frac{b}{3a^6}$
16. Estimates may vary; about $\frac{3}{100}$.
17. 6750 people 18. Answers may vary.

19.

Country	Population	Area (square miles)
China	1.192×10^9	3.696×10^6
France	5.8×10^7	2.10×10^5
Iran	6.0×10^7	6.32×10^5
United States	2.61×10^8	3.679×10^6

20. a. China: about 323 persons/mi²; France: about 276 persons/mi²; Iran: about 95 persons/mi²; United States: about 71 persons/mi²
b. because the population figures are given in millions and the area figures are given in thousands of square miles

POPULATION **Use the graphs below for Exercises 19 and 20.**

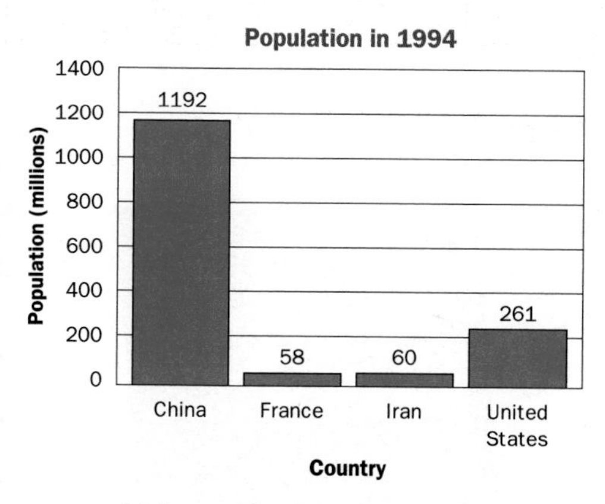

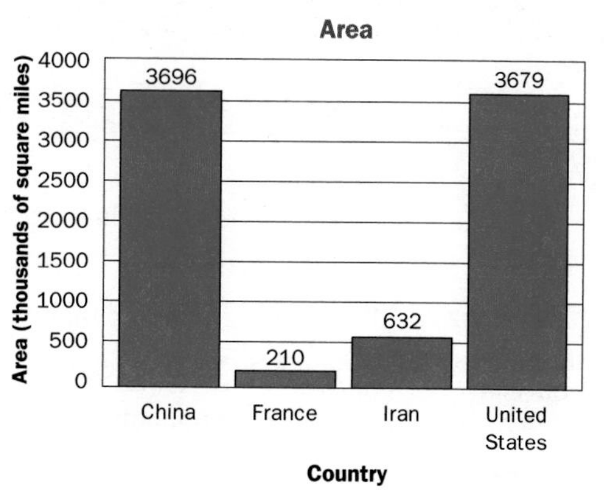

19. Make a table of the data above. Write all numbers in scientific notation.

20. a. Use the formula below to find each country's population density in persons per square mile.

$$\text{Population density} = \frac{\text{Population}}{\text{Area}}$$

b. Writing Explain why the population density of China is not $\frac{1192}{3696}$.

21. a. Recall that a light-year is the distance light travels in a year. Use the fact that light travels at 186,282 mi/s to show that a light-year is about 5.88×10^{12} mi. (Use 365.25 for the number of days in one year.)

b. HISTORY In 1054 A.D., Chinese and Japanese astronomers recorded a *supernova*. A supernova is the explosion of a star. The supernova produced a huge cloud of gas and dust known as the Crab Nebula about 4×10^3 light-years from Earth. Express this distance in miles.

ONGOING ASSESSMENT

22. Writing Explain the mistake that each student made. Then simplify the expression correctly.

$2^3 \cdot 3^4 = (2 \cdot 3)^{3+4}$
$= 6^7$
$= 279{,}936$

$\frac{x^{-9}}{x^{-3}} = x^{-9-3}$
$= x^{-12}$
$= \frac{1}{x^{-12}}$

SPIRAL REVIEW

Simplify each expression. *(Section 9.4)*

23. $\left(-\frac{8}{5}\right)^0$ **24.** $\left(-5\frac{12}{17}\right)^0$ **25.** $\left(\frac{4}{7}\right)^{-3}$

26. $\left(-\frac{4}{7}\right)^{-3}$ **27.** $-\left(\frac{2}{3}\right)^{11}$ **28.** $\frac{17.3^0}{6^{-2}}$

Apply⇔Assess

Exercise Notes

Integrating the Strands

Exs. 19, 20 These exercises integrate algebra and statistics. In so doing, students can see an important real-world application of two mathematical topics.

Assessment Note

Ex. 22 By recognizing the incorrect solution in this exercise, students' attention is focused on the error. Because they can explain the error, students are less apt to repeat it themselves.

Mathematical Procedures

Ex. 28 Students who are inclined to apply the quotient of powers rule to this expression should be reminded that the rule can be applied only when the bases of the powers are the same. The dividend and the divisor in this exercise have different bases.

Practice 63 for Section 9.6

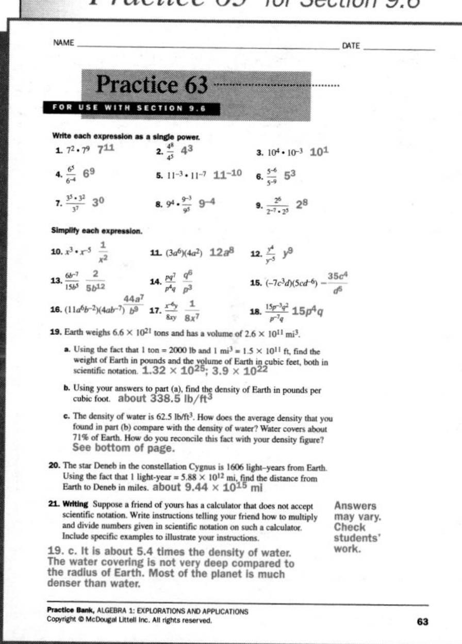

NAME ______ DATE ______

Practice 63

FOR USE WITH SECTION 9.6

Write each expression as a single power.

1. $7^2 \cdot 7^9$ 7^{11} **2.** $\frac{4^8}{4^5}$ 4^3 **3.** $10^4 \cdot 10^{-3}$ 10^1

4. $\frac{6^5}{6^{-4}}$ 6^9 **5.** $11^{-3} \cdot 11^{-7}$ 11^{-10} **6.** $\frac{5^{-6}}{5^{-9}}$ 5^3

7. $\frac{3^5 \cdot 3^2}{3^7}$ 3^0 **8.** $9^4 \cdot \frac{9^{-3}}{9^5}$ 9^{-4} **9.** $\frac{2^6}{2^{-7} \cdot 2^5}$ 2^8

Simplify each expression.

10. $x^3 \cdot x^{-5}$ $\frac{1}{x^2}$ **11.** $(3a^6)(4a^2)$ $12a^8$ **12.** $\frac{y^4}{y^{-5}}$ y^9

13. $\frac{6b^{-7}}{15b^5}$ $\frac{2}{5b^{12}}$ **14.** $\frac{pq^7}{p^4q}$ $\frac{q^6}{p^3}$ **15.** $(-7c^3d)(5cd^{-6})$ $-\frac{35c^4}{d^5}$

16. $(11a^6b^{-2})(4ab^{-7})$ $\frac{44a^7}{b^9}$ **17.** $\frac{x^{-6}y}{8xy}$ $\frac{1}{8x^7}$ **18.** $\frac{15p^{-3}q^2}{p^{-7}q}$ $15p^4q$

19. Earth weighs 6.6×10^{21} tons and has a volume of 2.6×10^{11} mi³.

a. Using the fact that 1 ton = 2000 lb and 1 mi³ = 1.5×10^{11} ft, find the weight of Earth in pounds and the volume of Earth in cubic feet, both in scientific notation. 1.32×10^{25}; 3.9×10^{22}

b. Using your answers to part (a), find the density of Earth in pounds per cubic foot. about 338.5 lb/ft³

c. The density of water is 62.5 lb/ft³. How does the average density that you found in part (b) compare with the density of water? Water covers about 71% of Earth. How do you reconcile this fact with your density figure? See bottom of page.

20. The star Deneb in the constellation Cygnus is 1606 light-years from Earth. Using the fact that 1 light-year = 5.88×10^{12} mi, find the distance from Earth to Deneb in miles. about 9.44×10^{15} mi

21. Writing Suppose a friend of yours has a calculator that does not accept scientific notation. Write instructions telling your friend how to multiply and divide numbers given in scientific notation on such a calculator. Include specific examples to illustrate your instructions. Answers may vary. Check students' work.

19. c. It is about 5.4 times the density of water. The water covering is not very deep compared to the radius of Earth. Most of the planet is much denser than water.

Practice Bank, ALGEBRA 1: EXPLORATIONS AND APPLICATIONS

63

21. a. 1 light-year = $\frac{186{,}282\text{ mi}}{\text{s}} \cdot \frac{60\text{ s}}{\text{min}} \cdot \frac{60\text{ min}}{\text{h}} \cdot \frac{24\text{ h}}{\text{d}} \cdot \frac{365.25\text{ d}}{\text{y}} \cdot 1\text{ y} \approx$ 5.88×10^{12} mi

b. 2.352×10^{16} mi

22. The first student used the product of powers rule incorrectly. The rule applies only if the bases are the same; $2^3 \cdot 3^4 = 8 \cdot 81 = 648$. The second student made an arithmetic mistake in applying the quotient of powers rule.
$\frac{x^{-9}}{x^{-3}} = x^{-9-(-3)} = x^{-9+3} = x^{-6} = \frac{1}{x^6}$

23. 1

24. 1

25. $\frac{343}{64}$

26. $-\frac{343}{64}$

27. $-\frac{2048}{177{,}147}$

28. 36

Plan⇔Support

Objectives

- Find the power of products and quotients.
- Find the power of a power.
- Use numbers written in scientific notation in formulas.

Recommended Pacing

Core and Extended Courses
Section 9.7: 1 day

Block Schedule
Section 9.7: $\frac{1}{2}$ block
(with Review)

Resource Materials

Lesson Support
Lesson Plan 9.7
Warm-Up Transparency 9.7
Practice Bank: Practice 64
Study Guide: Section 9.7
Assessment Book: Test 40

Technology
Scientific Calculator
Internet:
http://www.hmco.com

Warm-Up Exercises

Simplify each expression.

1. 8^3 512
2. $(5-10)^3$ −125
3. $\left(\frac{1}{3}\right)^4$ $\frac{1}{81}$
4. $\left(-\frac{1}{2}\right)^5$ $-\frac{1}{32}$
5. $(a^4)(a^4)$ a^8

SECTION

9.7 Working with Powers

Learn how to...

- **find the power of products and quotients**
- **find the power of a power**

So you can...

- **use numbers written in scientific notation in formulas**

The globe shown has a radius of 8 in. The radius of Earth is about 2×10^7 ft. To find their volumes, you may want to learn a few more rules of exponents. For example, to find Earth's volume, you can use the formula $V = \frac{4}{3}\pi r^3$, where r is the radius of Earth. If you enter $(2 \times 10^7)^3$ into a calculator, you get 8×10^{21}. Why?

EXAMPLE 1 Application: Geology

Show that the volume of Earth is $\frac{4}{3}\pi(8 \times 10^{21})$, or about 3.35×10^{22} ft³.

SOLUTION

Problem Solving Strategy: Use a formula.

Use the formula $V = \frac{4}{3}\pi r^3$. Substitute 2×10^7 for r.

$$\frac{4}{3}\pi r^3 = \frac{4}{3}\pi(2 \times 10^7)^3$$
$$= \frac{4}{3}\pi(2 \times 10^7)(2 \times 10^7)(2 \times 10^7)$$
$$= \frac{4}{3}\pi(2 \cdot 2 \cdot 2)(10^7 \cdot 10^7 \cdot 10^7)$$

Group the factors.

$$= \frac{4}{3}\pi(8)(10^{7+7+7})$$

Use the product of powers rule.

$$= \frac{4}{3}\pi(8 \times 10^{21})$$
$$\approx 3.35 \times 10^{22}$$

In Example 1, you saw that $(2 \times 10^7)^3 = 2^3 \times (10^7)^3$. The exponent **3** tells you to multiply each factor inside the parentheses **3** times. This result illustrates the *power of a product rule.*

To find how the ratio of the radii of two spheres relates to the ratio of the volumes of the spheres, you may need to find the power of a quotient, as in the Example below.

EXAMPLE 2 Application: Geology

The center region of Earth is an extremely hot sphere called the *inner core*. The radius of the inner core is one fifth the radius R of Earth. Write the volume of the inner core as a multiple of the volume of Earth.

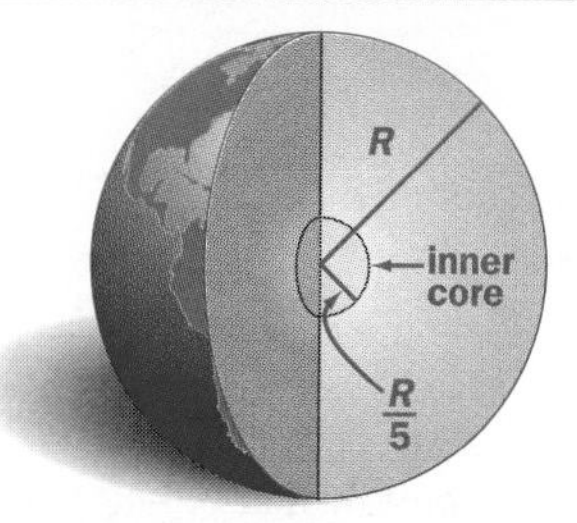

SOLUTION

Let R = the radius of Earth. Then $\frac{R}{5}$ = **the radius of the inner core.**

$$\text{Volume of inner core} = \frac{4}{3}\pi\,(\textbf{Radius of inner core})^3$$
$$= \frac{4}{3}\pi\left(\frac{R}{5}\right)^3$$
$$= \frac{4}{3}\pi\left(\frac{R}{5}\right)\left(\frac{R}{5}\right)\left(\frac{R}{5}\right)$$
$$= \frac{4}{3}\pi\left(\frac{R \cdot R \cdot R}{5 \cdot 5 \cdot 5}\right)$$

Rewrite the numerator and the denominator as powers.

$$= \frac{4}{3}\pi\left(\frac{R^3}{5^3}\right)$$
$$= \frac{4}{3}\pi\left(R^3 \cdot \frac{1}{5^3}\right)$$

Group the factors. Since R = the radius of Earth, $\frac{4}{3}\pi R^3$ = **the volume of Earth.**

$$= \frac{1}{5^3} \cdot \left(\frac{4}{3}\pi R^3\right)$$
$$= \frac{1}{125} \cdot (\textbf{Volume of Earth})$$

The volume of the inner core is $\frac{1}{125}$ the volume of Earth.

In Example 2, you saw that $\left(\frac{R}{5}\right)^3 = \left(\frac{R}{5}\right)\left(\frac{R}{5}\right)\left(\frac{R}{5}\right) = \frac{R^3}{5^3}$. The exponent **3** tells you to use both the numerator and the denominator as factors **3** times. This result illustrates the *power of a quotient rule*.

Teach⇔Interact

Section Note

Topic Spiraling: Preview
Examples 1 and 2 introduce students in an informal way to properties of powers that involve products and quotients. After developing an intuitive sense of these properties, students are introduced to their formal statements on page 408.

Additional Example 1

The radius of the sun is about 4.3×10^5 mi. Show that the volume of the sun is $\frac{4}{3}\pi(79.507 \times 10^{15})$, or about 3.33×10^{17} mi^3.

Use the formula $V = \frac{4}{3}\pi r^3$.

Substitute 4×10^5 for r.

$\frac{4}{3}\pi r^3 = \frac{4}{3}\pi(4.3 \times 10^5)^3$

$= \frac{4}{3}\pi(4.3 \times 10^5)(4.3 \times 10^5) \times (4.3 \times 10^5)$

$= \frac{4}{3}\pi(4.3 \times 4.3 \times 4.3) \times (10^5 \times 10^5 \times 10^5)$

$= \frac{4}{3}\pi(79.507)(10^{5+5+5})$

$= \frac{4}{3}\pi(79.507 \times 10^{15})$

$\approx 3.33 \times 10^{17}$

Additional Example 2

The radius of Earth's moon is about one-fourth the radius R of Earth. Compare the moon's volume to the volume of Earth.

Let R = the radius of Earth.

Then $\frac{R}{4}$ = the radius of the moon.

Volume of moon

$= \frac{4}{3}\pi(\text{Radius of moon})^3$

$= \frac{4}{3}\pi\left(\frac{R}{4}\right)^3$

$= \frac{4}{3}\pi\left(\frac{R}{4}\right)\left(\frac{R}{4}\right)\left(\frac{R}{4}\right)$

$= \frac{4}{3}\pi\left(\frac{R \cdot R \cdot R}{4 \cdot 4 \cdot 4}\right)$

$= \frac{4}{3}\pi\left(R^3 \cdot \frac{1}{4^3}\right)$

$= \frac{1}{4^3} \cdot \left(\frac{4}{3}\pi R^3\right)$

$= \frac{1}{64} \cdot (\text{Volume of Earth})$

The volume of the moon is about $\frac{1}{64}$ the volume of Earth.

Teach⇔Interact

Think and Communicate

Since finding the volume of a sphere involves cubing the radius, ask students to first make an estimate of the answer for question 2.

Checking Key Concepts

Common Error

If some students have difficulty remembering and applying the rules for exponents given on this page, suggest that they apply the basic definition of exponent and write the factors of the expression down on paper.

Closure Question

State the power of a product rule, the power of a quotient rule, and the power of a power rule.

See the rules at the top of the page.

Rules for Powers of Products and Quotients

These rules are true for any nonzero numbers a and b and any integer m:

Power of a Product Rule	Power of a Quotient Rule
$(ab)^m = a^m b^m$	$\left(\frac{a}{b}\right)^m = \frac{a^m}{b^m}$
$(4 \cdot 3)^2 = 4^2 \cdot 3^2$	$\left(\frac{4}{3}\right)^2 = \frac{4^2}{3^2}$
$(12)^2 = 16 \cdot 9$	$\frac{16}{9} = \frac{16}{9}$
$144 = 144$	

In Example 1, you saw that $\left(10^7\right)^3 = 10^7 \cdot 10^7 \cdot 10^7 = 10^{7+7+7} = 10^{21}$. This result illustrates the *power of a power rule.*

Power of a Power Rule

For any nonzero number a and any integers m and n:

$\left(a^m\right)^n = a^{m \cdot n}$

For example: $\left(2^2\right)^3 = 2^{2 \cdot 3}$

$(4)^3 = 2^6$

$64 = 64$

THINK AND COMMUNICATE

1. This globe from page 406 has radius 8 in., or $\frac{2}{3}$ ft. Use the power of a quotient rule to find the volume of the globe in cubic feet.
2. The radius of Earth is about 3.75×10^8 times the radius of the globe. How many times greater than the globe's volume is Earth's volume?

CHECKING KEY CONCEPTS

Evaluate each power.

1. $(3 \cdot 2)^4$
2. $(3 \cdot 5)^3$
3. $\left(2^3\right)^2$
4. $\left(10^5\right)^2$
5. $\left(\frac{7}{8}\right)^2$
6. $\left(\frac{5}{4}\right)^3$

Simplify each expression.

7. $(ab)^9$
8. $\left(x^5\right)^6$
9. $\left(z^4\right)^4$
10. $\left(\frac{2}{d}\right)^3$
11. $(5c)^2$
12. $\left(\frac{m}{n}\right)^7$

ANSWERS Section 9.7

Think and Communicate

1. $\frac{4}{3}\pi\left(\frac{2}{3}\right)^3 = \frac{4}{3}\left(\frac{2}{3}\right)^3\pi = \frac{4}{3} \cdot \frac{2^3}{3^3}\pi = \frac{2^5}{3^4}\pi = \frac{32}{81}\pi$; $\frac{32}{81}\pi$ ft³
2. about 6.54×10^{25} times greater

Checking Key Concepts

1. 1296
2. 3375
3. 64
4. 10,000,000,000
5. $\frac{49}{64}$
6. $\frac{125}{64}$
7. a^9b^9
8. x^{30}
9. z^{16}
10. $\frac{8}{d^3}$
11. $25c^2$
12. $\frac{m^7}{n^7}$

9.7 Exercises and Applications

Extra Practice exercises on page 574

Evaluate each power. Write your answers in scientific notation.

1. $(3 \times 10^4)^2$
2. $(2 \times 10^9)^3$
3. $(5 \times 10^7)^4$

Evaluate each power.

4. $\left(\frac{1}{2}\right)^4$
5. $\left(\frac{4}{7}\right)^3$
6. $\left(\frac{5}{3}\right)^5$

Simplify each expression.

7. $(ab)^5$
8. $(a^4)^8$
9. $\left(\frac{a}{b}\right)^6(2c^3d^2)^4$
10. $(2a)^3$
11. $(3ab)^4$
12. $(5c^7d^8)^2$
13. $(p^{-2}q^5)^7$
14. $\left(\frac{a^2}{b^3}\right)^9$
15. $\left(\frac{b^6c}{d^3}\right)^5$

16. **Open-ended Problem** Leon and Ed were asked to find the product $(x + 3)^2$. Identify any mistakes in their solutions and assign each student a score from 1 to 10, where 10 represents a completely correct solution.

Ed
$(x+3)^2 = x^2 + 3^2$
$= x^2 + 9$

Leon
$(x+3)^2 = (x+3)(x+3)$
$= x^2+3x+3x+9$
$= x^2+6x+9$

17. **PHYSICS** If you stand in a river flowing at s mi/h, the pressure p that the water exerts on your legs is given by the equation $p = 4.125s^2$, where p is measured in pounds per square inch.
    a. Suppose you stand in a river flowing at $\frac{4}{3}$ mi/h. Find the pressure that the water exerts on your legs.
    b. Suppose you stand in a river flowing at 4 mi/h. Find the pressure that the water exerts on your legs.
    c. Refer to the diagrams at the right. Complete this equation:

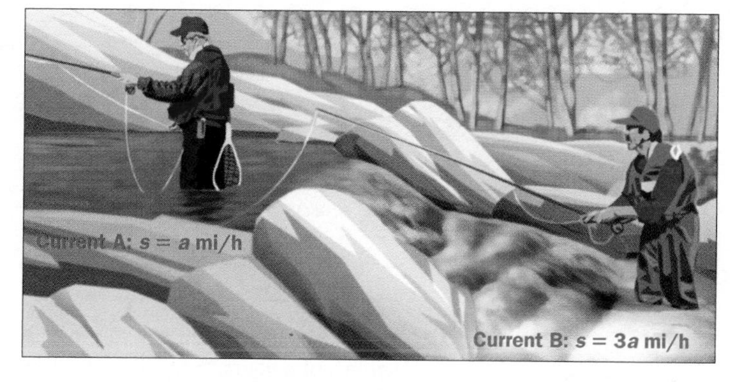

$$\text{Pressure exerted by River B} = \underline{?} \times \left(\text{Pressure exerted by River A}\right)$$

Apply⇔Assess

Suggested Assignment

Core Course
Exs. 1–15, 21, 23, 25–36, AYP

Extended Course
Exs. 2–5, 8, 9, 12–36, AYP

Block Schedule
Day 60 Exs. 1–16, 18, 19, 21, 24–36, AYP

Exercise Notes

Common Error
Exs. 1–3 Some students may err in raising only the power of ten to the power indicated outside the parentheses. Make sure students understand that each factor inside the parentheses is raised to the given power.

Student Progress
Exs. 1–15 Students should be able to achieve complete mastery of these exercises before moving on to the next chapter.

Problem Solving
Ex. 16 For this exercise, students are asked to assume the role of evaluator. The way in which students assign a grade can be revealing because it demonstrates how well they understand the concept involved.

Exercises and Applications

1. 9×10^8
2. 8×10^{27}
3. 6.25×10^{30}
4. $\frac{1}{16}$
5. $\frac{64}{343}$
6. $\frac{3125}{243}$
7. a^5b^5
8. a^{32}
9. $\frac{16a^6c^{12}d^8}{b^6}$
10. $8a^3$
11. $81a^4b^4$
12. $25c^{14}d^{16}$
13. $\frac{q^{35}}{p^{14}}$
14. $\frac{a^{18}}{b^{27}}$
15. $\frac{b^{30}c^5}{d^{15}}$
16. Answers may vary. An example is given. Ed made a very basic mistake, demonstrating that he does not know how to square a binomial. I would give him 2 points for squaring x and 3 correctly. Leon's solution is completely correct. I would give him 10 points.
17. a. $\frac{22}{3}$ lb/in.$^2 \approx 7.33$ lb/in.^{2}
    b. 66 lb/in.^{2}
    c. $\frac{9}{a^2}$

Apply⇔Assess

Exercise Notes

Integrating the Strands
Exs. 18–20 These exercises illustrate that as early as 1850 B.C., algebraic methods were used to solve geometry problems.

Communication: Discussion
Ex. 19 Before students begin any computations, have them discuss how the relationship between the dimensions of the two pyramids will affect the relationship between their volumes.

Problem Solving
Ex. 22 This exercise illustrates that an expression often must be rewritten in a different form before a relationship between two unknowns becomes obvious. Once students have rewritten the quotient as a fraction involving a negative exponent, they should see that the power of a quotient rule is a special case of the power of a product rule.

Connection HISTORY

The plant shown is called the papyrus plant. The ancient Egyptians used this plant to make the papyruses they wrote on. One papyrus written in 1850 B.C. gives evidence that the ancient Egyptians knew the formula for the volume of a pyramid. This papyrus can now be seen at the Museum of Fine Arts in Moscow.

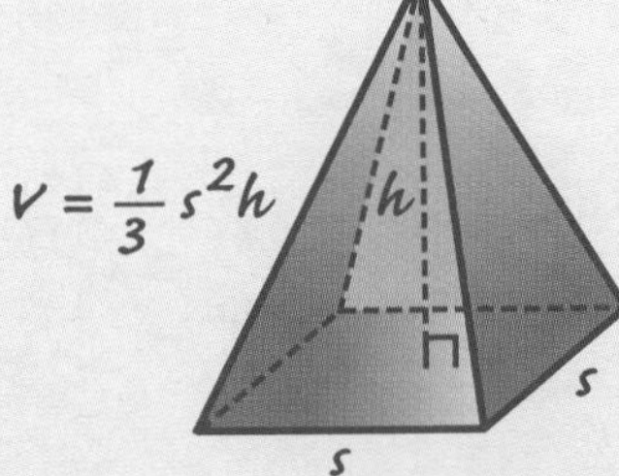

18. For the Great Pyramid in Egypt, $h = 481$ ft and $s = 755$ ft. Find the volume of the Great Pyramid.

19. Show that the volume of Pyramid *A* is twice the volume of Pyramid *B*.

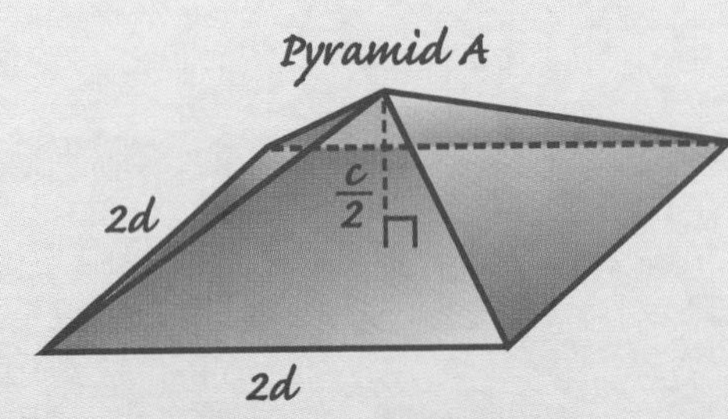

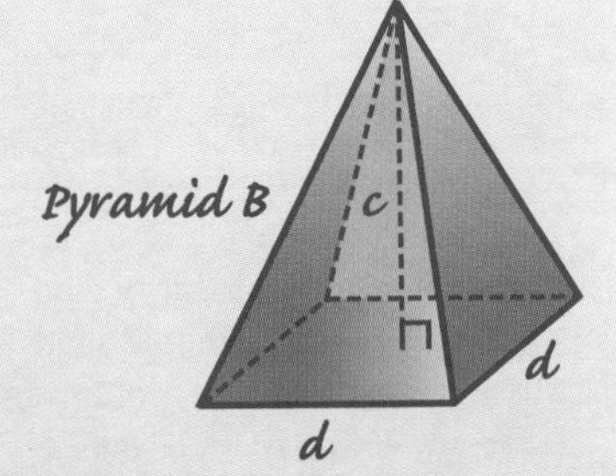

20. The Moscow papyrus also gives evidence that the ancient Egyptians knew how to find the volume of a partially built pyramid, like the one in the diagram. Let f = the fraction of the pyramid that has been built. Then f is given by this equation:

$$f = 1 - \left(\frac{x}{s}\right)^3$$

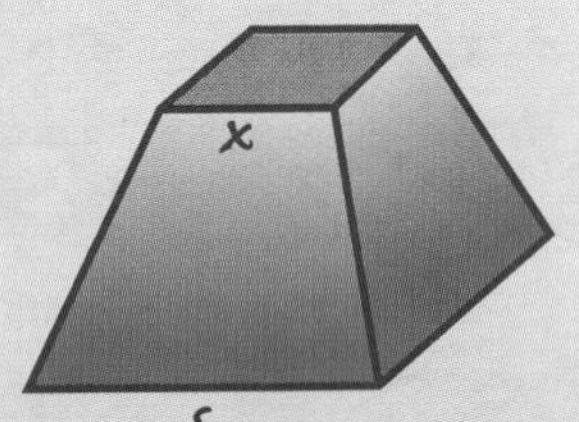

a. Suppose $s = 90$ ft and $x = 60$ ft. What fraction of the pyramid has been built?

b. Writing Suppose $x = \frac{1}{2}s$. Is it true that the pyramid is half built? Explain your reasoning.

21. BIOLOGY The formula $w = 446l^3$ gives the approximate weight w in grams of a hognose snake whose length is l meters.

a. Find the weight of a hognose snake whose length is 0.4 m.

b. One snake is twice as long as another. Show that the weight of the longer snake is eight times the weight of the shorter snake.

22. Challenge Use the power of a product rule and the power of a power rule to prove the power of a quotient rule. (*Hint:* Rewrite the quotient as a product involving a negative exponent.)

18. $91{,}394{,}008\frac{1}{3}$ ft^3

19. volume of pyramid B $= \frac{1}{3}d^2c$; volume of pyramid A $= \frac{1}{3}(2d)^2\left(\frac{c}{2}\right) = \frac{1}{3}(4d^2)\left(\frac{c}{2}\right) = \frac{2d^2c}{3} = 2 \cdot$ volume of pyramid B

20. a. $\frac{19}{27}$

b. No; $f = 1 - \left(\frac{\frac{1}{2}s}{s}\right)^3 = 1 - \left(\frac{1}{2}\right)^3 = 1 - \frac{1}{8} = \frac{7}{8}$; the pyramid is $\frac{7}{8}$ built.

21. a. about 28.544 g

b. Let x = length of short snake, then $2x$ = length of longer snake; weight of longer snake $= 446(2x)^3 = 446 \cdot 8x^3 = 8(446x^3) =$ 8(weight of shorter snake).

22. $\left(\frac{a}{b}\right)^m = \left(a \cdot \frac{1}{b}\right)^m = (a \cdot b^{-1})^m = a^m \cdot (b^{-1})^m = a^m \cdot b^{-m} = a^m \cdot \frac{1}{b^m} = \frac{a^m}{b^m}$

23. B

24. Answers may vary. Examples are given. $a^2b^6c^{-4}$; definition of negative exponents; $\left(\frac{ab^3}{c^2}\right)^2$; power of a power rule

23. SAT/ACT Preview Which of these expressions equals $\left(\frac{x^3y^2}{z^4}\right)^3$?

A. $\frac{x^6y^5}{z^7}$ **B.** $\frac{x^9y^6}{z^{12}}$ **C.** $\frac{x^{15}y^6}{z^{12}}$

D. $\frac{x^9y^6}{z^7}$ **E.** none of these

ONGOING ASSESSMENT

24. Open-ended Problem Use the rules you learned in this section to write the expression $\frac{a^2b^6}{c^4}$ in two other ways. Explain how you used each rule.

SPIRAL REVIEW

Simplify each variable expression. *(Section 1.7)*

25. $-3(5u + 3v) + u$ **26.** $6 + b^2 - (3 - b^2)$ **27.** $2(5 - m) - (4 - m)$

28. $8(4 - p) - (p + 1)$ **29.** $\frac{1}{2}(6q + 10)$ **30.** $(d + 5) + 2(3 - d)$

Solve each proportion. *(Section 5.1)*

31. $\frac{f}{12} = \frac{3}{1}$ **32.** $\frac{7}{2} = \frac{5}{s}$ **33.** $\frac{15}{35} = \frac{p}{100}$

34. $\frac{m}{m+6} = \frac{2}{5}$ **35.** $\frac{10}{x} = \frac{5}{x-1}$ **36.** $\frac{3c}{c+2} = \frac{9}{4}$

ASSESS YOUR PROGRESS

Simplify each expression. *(Sections 9.6 and 9.7)*

1. $a^9 \cdot a^7$ **2.** $\frac{b^{12}}{b^8}$ **3.** $\frac{a^{10}b^{-1}}{a^3b^3}$ **4.** $(ab)^4$

5. $(c^7)^8$ **6.** $b\left(\frac{c}{d}\right)^{10}$ **7.** $\left(\frac{p^2q^6}{r^4}\right)^3$ **8.** $\left(\frac{x^2y}{y^3z^0}\right)^4$

9. TRANSPORTATION In 1990, there were about 1.44×10^8 cars in the United States, and the average car was driven about 1.06×10^4 mi. Estimate the total number of miles traveled by cars in the United States during 1990. Write your answer in scientific notation. *(Section 9.6)*

10. PHYSICS The equation $d = 16t^2$ gives the distance d in feet traveled by a falling object in t seconds. Suppose Rock B has been falling for twice as long as Rock A. Complete the following equation. Be sure to justify your answer algebraically. *(Section 9.7)*

Distance traveled by Rock B = __?__ × Distance traveled by Rock A

11. Journal List the five rules of exponents you learned in the last two sections. Tell whether each was easy or difficult for you to learn.

Apply⇔Assess

Exercise Notes

Teaching Tip

Ex. 23 One of the techniques for answering multiple choice questions is to first eliminate the most obvious incorrect answers. Students should see that choices A and D can be eliminated because the exponent of the denominator is incorrect.

Cooperative Learning

Ex. 24 Students can benefit from each other's insights and explanations by working this exercise in small groups.

Progress Check 9.6–9.7

See page 415.

Practice 64 for Section 9.7

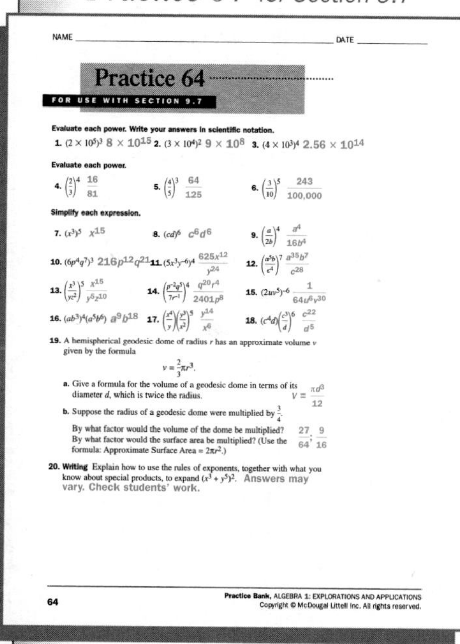

NAME ______________________ DATE __________

Practice 64

FOR USE WITH SECTION 9.7

Evaluate each power. Write your answers in scientific notation.

1. $(2 \times 10^5)^3$ 8×10^{15} 2. $(3 \times 10^4)^2$ 9×10^8 3. $(4 \times 10^3)^4$ 2.56×10^{14}

Evaluate each power.

4. $\left(\frac{2}{3}\right)^4$ $\frac{16}{81}$ 5. $\left(\frac{4}{5}\right)^3$ $\frac{64}{125}$ 6. $\left(\frac{3}{10}\right)^5$ $\frac{243}{100,000}$

Simplify each expression.

7. $(x^3)^5$ x^{15} 8. $(cd)^6$ c^6d^6 9. $\left(\frac{a}{2b}\right)^4$ $\frac{a^4}{16b^4}$

10. $(6p^4q^7)^3$ $216p^{12}q^{21}$ 11. $(5x^3y^{-6})^4$ $\frac{625x^{12}}{y^{24}}$ 12. $\left(\frac{a^5b}{c^4}\right)^7$ $\frac{a^{35}b^7}{c^{28}}$

13. $\left(\frac{x^3}{yz^2}\right)^5$ $\frac{x^{15}}{y^5z^{10}}$ 14. $\left(\frac{p^{-2}q^5}{7r^{-1}}\right)^4$ $\frac{q^{20}r^4}{2401p^8}$ 15. $(2uv^5)^{-6}$ $\frac{1}{64u^6v^{30}}$

16. $(ab^3)^4(a^5b^6)$ a^9b^{18} 17. $\left(\frac{x^4}{y}\right)\left(\frac{y^3}{x^2}\right)^5$ $\frac{y^{14}}{x^6}$ 18. $(c^4d)\left(\frac{c^3}{d}\right)^6$ $\frac{c^{22}}{d^5}$

19. A hemispherical geodesic dome of radius r has an approximate volume v given by the formula

$$v = \frac{2}{3}\pi r^3.$$

a. Give a formula for the volume of a geodesic dome in terms of its diameter d, which is twice the radius. $V = \frac{\pi d^3}{12}$

b. Suppose the radius of a geodesic dome were multiplied by $\frac{3}{4}$.

By what factor would the volume of the dome be multiplied? By what factor would the surface area be multiplied? (Use the formula: Approximate Surface Area = $2\pi r^2$.) $\frac{27}{64}$; $\frac{9}{16}$

20. **Writing** Explain how to use the rules of exponents, together with what you know about special products, to expand $(x^3 + y^5)^2$. Answers may vary. Check students' work.

64

Practice Bank, ALGEBRA 1: EXPLORATIONS AND APPLICATIONS

---

25. $-14u - 9v$ 26. $3 + 2b^2$

27. $6 - m$ 28. $31 - 9p$

29. $3q + 5$ 30. $11 - d$

31. $f = 36$ 32. $s = \frac{10}{7}$

33. $p = 42\frac{6}{7}$ 34. $m = 4$

35. $x = 2$ 36. $c = 6$

Assess Your Progress

1. a^{16} 2. b^4

3. $\frac{a^7}{b^4}$ 4. a^4b^4

5. c^{56} 6. $\frac{bc^{10}}{d^{10}}$

7. $\frac{p^6q^{18}}{r^{12}}$ 8. $\frac{x^8}{y^8}$

9. about 1.53×10^{12} mi

10. 4; Let n_A = the number of seconds that Rock A has been falling; then $2n_A$ = the time in seconds that Rock B has been falling. Then the distance traveled by Rock $B = 16(2n_A)^2 = 16(2^2)(n_A{}^2) = 16(4)n_A{}^2 = 4(16n_A{}^2) = 4$ times the distance traveled by Rock A.

11. Product of Powers Rule: $a^m \cdot a^n = a^{m+n}$

Quotient of Powers Rule: $\frac{a^m}{a^n} = a^{m-n}$

Power of a Product Rule: $(ab)^m = a^m \cdot b^m$

Power of a Quotient Rule: $\left(\frac{a}{b}\right)^m = \frac{a^m}{b^m}$

Power of a Power Rule: $(a^m)^n = a^{mn}$

Answers may vary.

Project Notes

Mathematical Goals

- Measure the distance d_1 in millimeters from the center of the spiral in a nautilus shell to the outside corner of the first chamber.
- Calculate the ratio of the distance for each chamber to the distance for the previous chamber.
- Calculate the mean, r, of the ratios found.
- Substitute values found for d_1 and r into the equation $d_n = d_1 \cdot r^n$ to create an equation for d_n.

Planning

Materials

- Photograph of a nautilus shell
- Ruler
- Graph paper

Project Teams

Students can choose partners to work with and then discuss how they wish to proceed.

Guiding Students' Work

Suggest that students try to find a photograph of a nautilus shell to bring to class. Make sure they understand what measurements they should make and what ratios they are to calculate, and that they are to organize their results in a table or spreadsheet. You may wish to show students having difficulties creating their equation for d_n how to do this. Each student should finish the project by creating his or her own spiral.

PORTFOLIO PROJECT

Models for Spirals

Many patterns found in nature can be modeled by mathematical equations. One pattern that often appears in flowers, spider webs, and sea shells is the spiral. In this project, you will use exponential equations to create a spiral.

PROJECT GOAL **Find an exponential formula that models a spiral. Then create your own spiral.**

Measuring the Nautilus Shell

Work with a partner. Find a photograph of a nautilus shell, or use the one shown here. Measure the distance in millimeters from the center of the spiral to the outside corner of the first chamber. Label this distance d_1 as shown. Measure the corresponding distances for the next seven chambers. Label these distances d_2, d_3, and so on.

Calculating the Ratios

As you move inward along the spiral, the distances you measure become shorter. They are decreasing exponentially. Organize the distances you measured in a table or spreadsheet. Find the ratio of the distance for each chamber to the distance for the previous chamber.

Spiral Project

	A	B	C
1	n	dn	dn+1/dn
2	1	10	0.95
3	2	9.5	
4	3	8.6	
5	4	7.7	
6	5	7.2	

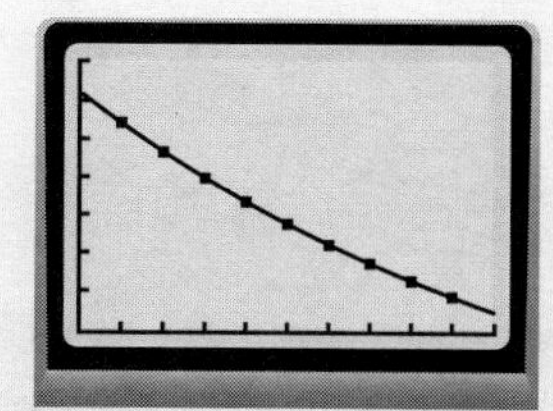

Modeling the Spiral with an Equation

Let r = the mean of the ratios you found. Substitute your values for d_1 and r into the formula below to create an equation for d_n, the distance for the nth chamber.

$$d_n = d_1 \cdot r^n$$

Calculate values for d_n by substituting values of n into your equation. How close do the calculated values come to the distances you measured? Make a scatter plot of your data and graph your equation in the same coordinate plane. Describe your results.

Creating Your Own Spiral

1. **CHOOSE** values for r and d_1. Make sure your r-value is between 0 and 1.
2. **DRAW** a line segment of length d_1.
3. **MULTIPLY** the length of the line segment you drew by r. Draw a line segment of this new length at a 25° angle to the previous line.
4. **REPEAT** Step 3 until the lengths become very close to zero.
5. **CONNECT** the endpoints of the line segments with a smooth curve.

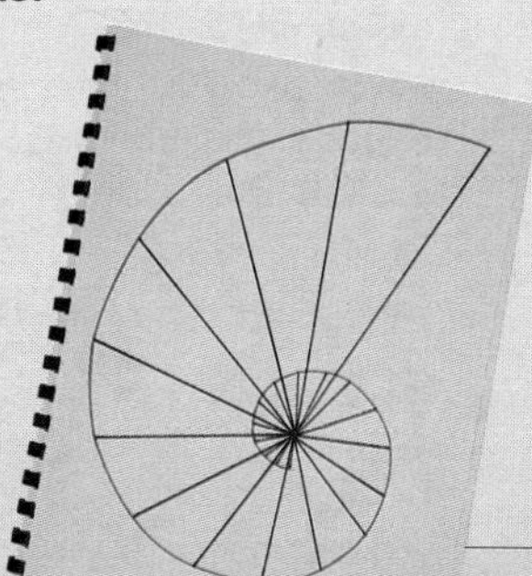

Find an equation to model your partner's spiral. How does the spiral change when you increase the value of r? when you increase the value of d_1? Write a paragraph answering these questions. Include sketches of your work.

Self-Assessment

Tell what you liked and didn't like about this project. How did you and your partner divide up the work in the project? Describe any difficulties you had and discuss how you solved problems. What would you do differently if you were to do the project over again? List some ways you could extend your study of spirals.

Project Notes

Guiding Students' Work

Rubric for Chapter Project

4 Students bring a photograph of a nautilus shell to class. They make the appropriate measurements and accurately calculate the necessary ratios. They also create an equation for d_n, make a scatter plot of their data, and graph their equation with the scatter plot. They accurately describe their results. Each partner creates a spiral following Steps 1–5 on this page. The spiral looks natural. Students answer the questions following Step 5 fully and correctly.

3 Students work through all aspects of the project and their work is generally accurate and well done. Some students have difficulty in actually creating their own spirals. Their spirals do not look fully natural with a smooth curve connecting all the points. The questions following Step 5 are answered correctly.

2 Students' measurements are not as accurate as they could be. Their ratios are not representative of the shell measured. The equation they create gives calculated values for d_n that are not very close to the distances measured. Students create their own spirals, but they are not very natural looking. The curve connecting all the points is a series of line segments that are not smooth.

1 Students start the project by making measurements and calculating ratios. Many errors are made in both the measurements and the calculations. An equation is created, but it does not give values close to those measured. Students cannot create their own spirals. Students should be encouraged to speak with the teacher as soon as possible to review their work and to make a new start on the project.

Chapter Support

Course Guide: Chapter 9
Lesson Plans: Chapter 9
Practice Bank:
Cumulative Practice 65
Study Guide: Chapter 9 Review
Assessment Book:
Chapter Tests 41 and 42
Chapter 9 Alternative Assessment
Test Generator software
Technology Book:
Calculator Based Lab 6
Portfolio Project Book
Preparing for College Entrance Tests
Professional Handbook

Progress Check 9.1–9.3

Evaluate each power. *(Section 9.1)*

1. $(-1)^5$ −1

2. $\left(\frac{2}{3}\right)^4$ $\frac{16}{81}$

Evaluate each expression when $x = 2$ and $y = -3$. *(Section 9.1)*

3. x^3y −24

4. $(2x - y)^2$ 49

5. Lisa Marois has $550 in a savings account that pays 5.5% annual interest. Assuming that she makes no deposits or withdrawals, how much money will she have in her account 4 years from now? *(Section 9.2)*
$681.35

6. DeCosta's Environmental Testing Company was conducting a test to determine the number of gallons N of a chemical remaining x days after a spill into the harbor. They found that the number of gallons decreased 28% each day. If 500 gallons were spilled initially, find the number of gallons still in the water after 3 days. *(Section 9.3)*
about 187 gallons

CHAPTER

9 Review

STUDY TECHNIQUE

Write at least six questions about this chapter. Focus on concepts you had the most trouble learning. Three should be short answer questions about specific situations or details. Three should be more involved questions, concentrating on more general ideas. Then answer the questions.

VOCABULARY

base (p. 370)
exponent (p. 370)
power (p. 370)
exponential growth (p. 377)
growth rate (p. 377)
growth factor (p. 377)
exponential decay (p. 384)
rate of decrease (p. 384)
decay factor (p. 384)
scientific notation (p. 396)

SECTIONS 9.1, 9.2, *and* 9.3

The expression $(-3)^6$ is called a **power**. The number −3 is the **base**, and the number 6 is the **exponent**. The power can be evaluated by using the number −3 as a factor 6 times.

$$(-3)^6 = \underbrace{(-3)\cdot(-3)\cdot(-3)\cdot(-3)\cdot(-3)\cdot(-3)}_{\text{6 factors}} = 729$$

The order of operations for evaluating expressions that include powers is as follows:

1. Simplify expressions inside parentheses.
2. Evaluate powers.
3. Do multiplication and division from left to right.
4. Do addition and subtraction from left to right.

For example:

$$\begin{aligned} 8 - (3+4) + 2\cdot 6^2 &= 8 - (7) + 2\cdot 6^2 \\ &= 8 - 7 + 2\cdot 36 \\ &= 8 - 7 + 72 \\ &= 73 \end{aligned}$$

Evaluate the powers.

You can use equations and graphs to model **exponential growth** and **exponential decay**.

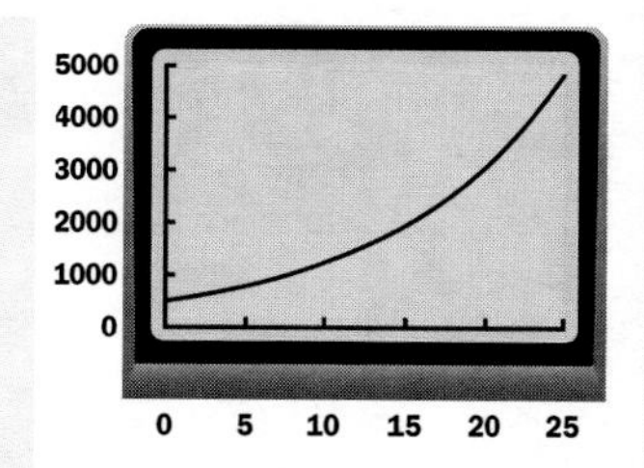

The graph shows that money earning interest *grows* exponentially. The value y of \$500 earning 9.5% interest per year for x years is modeled by the equation $y = 500(1 + 0.095)^x$. The constant 0.095 is called the **growth rate**. The quantity $(1 + 0.095)$ is called the **growth factor**.

The number of teams left in a single-elimination tournament *decreases* exponentially by 50% each round of the tournament. If there are 64 teams at the start of the tournament, the number of teams y left at the end of x rounds is given by $y = 64(1 - 0.50)^x$. The constant 0.50 is called the **rate of decrease**. The quantity $(1 - 0.50)$ is called the **decay factor.**

SECTIONS 9.4 *and* 9.5

For any nonzero number a and any integer n, $a^0 = 1$ and $a^{-n} = \frac{1}{a^n}$. For example, $4^{-3} = \frac{1}{4^3}$ and $4^0 = 1$.

$(-2)^0 = 1$ $\qquad 6^{-2} = \frac{1}{6^2} = \frac{1}{36}$ $\qquad a^5b^{-6} = a^5 \cdot \frac{1}{b^6} = \frac{a^5}{b^6}$

A number is expressed in **scientific notation** if it is in the form $a \times 10^n$, where $1 \le a < 10$ and n is an integer.

$6{,}270{,}000{,}000{,}000 = 6.27 \times 10^{12}$ $\qquad 0.00000000013 = 1.3 \times 10^{-10}$

Move the decimal point 12 places to the left. **Move the decimal point 10 places to the right.**

SECTIONS 9.6 *and* 9.7

Five basic rules for working with powers are:

Product of Powers: $a^m \cdot a^n = a^{m+n}$ $\qquad 3^2 \cdot 3^3 = 3^{2+3} = 3^5 = 243$

Quotient of Powers: $\frac{a^m}{a^n} = a^{m-n}, a \neq 0$ $\qquad \frac{3^5}{3^3} = 3^{5-3} = 3^2 = 9$

Power of a Product: $(ab)^m = a^mb^m$ $\qquad (-2x)^5 = (-2)^5x^5 = -32x^5$

Power of a Quotient: $\left(\frac{a}{b}\right)^m = \frac{a^m}{b^m}, b \neq 0$ $\qquad \left(\frac{5}{2}\right)^3 = \frac{5^3}{2^3} = \frac{125}{8}$

Power of a Power: $(a^m)^n = a^{m \cdot n}$ $\qquad (4^2)^3 = 4^{2 \cdot 3} = 4^6 = 4096$

Progress Check 9.4–9.5

Evaluate each power. *(Section 9.4)*

1. 6^{-3} $\frac{1}{216}$
2. $\left(\frac{1}{4}\right)^0$ 1

Evaluate each expression when $x = 1$ and $y = 5$. *(Section 9.4)*

3. $x^{-3}y^3$ 125
4. $\left(\frac{x}{y}\right)^4$ $\frac{1}{625}$

Write each number in decimal notation. *(Section 9.5)*

5. 3.62×10^6 3,620,000
6. 8.015×10^{-5} 0.00008015

Write each number in scientific notation. *(Section 9.5)*

7. 0.0000000005 5×10^{-10}
8. 890,000,000,000 8.9×10^{11}

Progress Check 9.6–9.7

Simplify each expression. *(Sections 9.6 and 9.7)*

1. $w^5 \cdot w^{-1}$ w^4
2. $\frac{c^7}{c^9}$ $\frac{1}{c^2}$
3. $10(ab)^3$ $10a^3b^3$
4. $\frac{r^5s^{12}t^2}{r^{-5}st^9}$ $\frac{r^{10}s^{11}}{t^7}$
5. $\left(\frac{m}{n}\right)^6$ $\frac{m^6}{n^6}$
6. $\left(\frac{c^4d^3e^{-2}}{d^8}\right)^5$ $\frac{c^{20}}{d^{25}e^{10}}$
7. The diameter of the planet Saturn is about 3.96×10^8 ft. The diameter of Earth is about 4×10^7 ft. How many times bigger is the diameter of Saturn than the diameter of Earth? *(Section 9.6)* about 10 times bigger
8. The radius of Earth is about one-tenth the radius of Saturn. Compare Earth's volume to the volume of Saturn. *(Section 9.7)* The volume of Earth is about $\frac{1}{1000}$ the volume of Saturn.

Chapter 9 Assessment
Form A Chapter Test

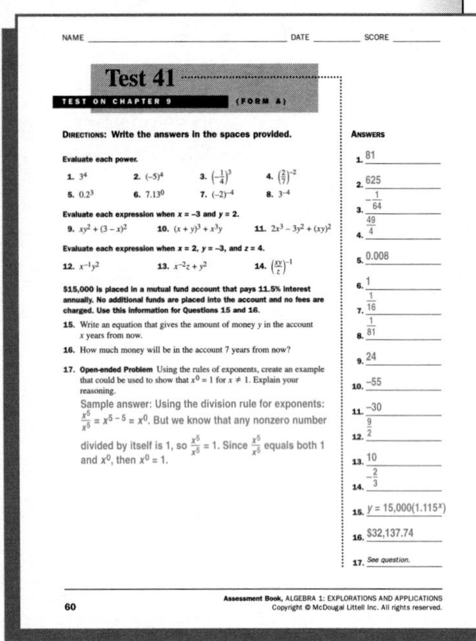
NAME ______ DATE ______ SCORE ______

Test 41

TEST ON CHAPTER 9 (FORM A)

DIRECTIONS: Write the answers in the spaces provided.

Evaluate each power.

1. 3^4 2. $(-5)^4$ 3. $\left(-\frac{1}{4}\right)^3$ 4. $\left(\frac{2}{7}\right)^{-2}$
5. 0.2^3 6. 7.13^0 7. $(-2)^{-4}$ 8. 3^{-4}

Evaluate each expression when $x = -3$ and $y = 2$.

9. $xy^2 + (3 - x)^2$ 10. $(x + y)^3 + x^3y$ 11. $2x^3 - 3y^2 + (xy)^2$

Evaluate each expression when $x = 2$, $y = -3$, and $z = 4$.

12. $x^{-1}y^2$ 13. $x^{-2}z + y^2$ 14. $\left(\frac{xy}{z}\right)^{-1}$

$15,000 is placed in a mutual fund account that pays 11.5% interest annually. No additional funds are placed into the account and no fees are charged. Use this information for Questions 15 and 16.

15. Write an equation that gives the amount of money y in the account x years from now.
16. How much money will be in the account 7 years from now?
17. **Open-ended Problem** Using the rules of exponents, create an example that could be used to show that $x^0 = 1$ for $x \neq 1$. Explain your reasoning.
Sample answer: Using the division rule for exponents: $\frac{x^5}{x^5} = x^{5-5} = x^0$. But we know that any nonzero number divided by itself is 1, so $\frac{x^5}{x^5} = 1$. Since $\frac{x^5}{x^5}$ equals both 1 and x^0, then $x^0 = 1$.

ANSWERS

1. 81
2. 625
3. $-\frac{1}{64}$
4. $\frac{49}{4}$
5. 0.008
6. 1
7. $\frac{1}{16}$
8. $\frac{1}{81}$
9. 24
10. −55
11. −30
12. $\frac{9}{2}$
13. 10
14. $-\frac{2}{3}$
15. $y = 15{,}000(1.115^x)$
16. $32,137.74
17. *See question.*

60 **Assessment Book,** ALGEBRA 1: EXPLORATIONS AND APPLICATIONS

Chapter 9 Assessment
Form B Chapter Test

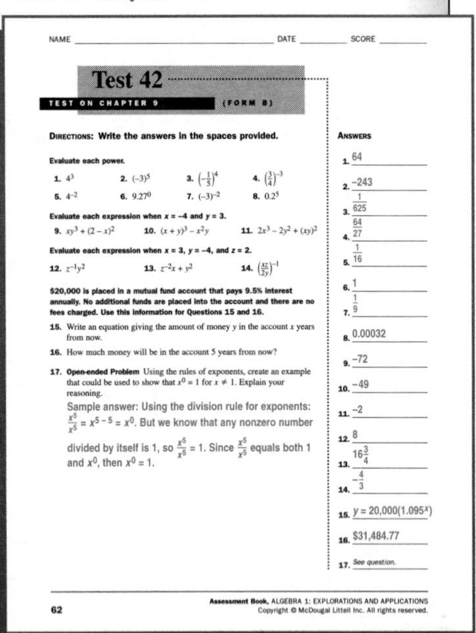
NAME ______ DATE ______ SCORE ______

Test 42

TEST ON CHAPTER 9 (FORM B)

DIRECTIONS: Write the answers in the spaces provided.

Evaluate each power.

1. 4^3 2. $(-3)^5$ 3. $\left(-\frac{1}{5}\right)^4$ 4. $\left(\frac{3}{4}\right)^{-3}$
5. 4^{-2} 6. 9.27^0 7. $(-3)^{-2}$ 8. 0.2^5

Evaluate each expression when $x = -4$ and $y = 3$.

9. $xy^3 + (2 - x)^2$ 10. $(x + y)^3 - x^2y$ 11. $2x^3 - 2y^2 + (xy)^2$

Evaluate each expression when $x = 3$, $y = -4$, and $z = 2$.

12. $z^{-1}y^2$ 13. $z^{-2}x + y^2$ 14. $\left(\frac{xz}{2y}\right)^{-1}$

$20,000 is placed in a mutual fund account that pays 9.5% interest annually. No additional funds are placed into the account and there are no fees charged. Use this information for Questions 15 and 16.

15. Write an equation giving the amount of money y in the account x years from now.
16. How much money will be in the account 5 years from now?
17. **Open-ended Problem** Using the rules of exponents, create an example that could be used to show that $x^0 = 1$ for $x \neq 1$. Explain your reasoning.
Sample answer: Using the division rule for exponents: $\frac{x^5}{x^5} = x^{5-5} = x^0$. But we know that any nonzero number divided by itself is 1, so $\frac{x^5}{x^5} = 1$. Since $\frac{x^5}{x^5}$ equals both 1 and x^0, then $x^0 = 1$.

ANSWERS

1. 64
2. −243
3. $\frac{1}{625}$
4. $\frac{64}{27}$
5. $\frac{1}{16}$
6. 1
7. $\frac{1}{9}$
8. 0.00032
9. −72
10. −49
11. −2
12. 8
13. $16\frac{3}{4}$
14. $-\frac{4}{3}$
15. $y = 20{,}000(1.095^x)$
16. $31,484.77
17. *See question.*

62 **Assessment Book,** ALGEBRA 1: EXPLORATIONS AND APPLICATIONS

CHAPTER 9

Assessment

VOCABULARY REVIEW

For Exercises 1–3, complete each sentence.

1. In the expression 7^6, 7 is called the _?_, 6 is called the _?_, and the whole expression is called a _?_.
2. The population of Sweden increased by 0.3% per year throughout the 1980s. This situation is an example of _?_. The constant 0.3% is the _?_, and the quantity 1.003 is the _?_.
3. The amount of aspirin in a person's blood stream decreases by 20% per hour. This situation is an example of _?_. The quantity 0.80 is the _?_.
4. **Writing** Explain what is meant by *scientific notation*. Give an example of a number written in scientific notation.

SECTIONS 9.1, 9.2, *and* 9.3

Evaluate each power.

5. 9^3 6. $(-3)^4$ 7. $\left(\frac{1}{4}\right)^5$ 8. 0.5^3

Write each expression as a power. Then evaluate the power.

9. the third power of two 10. three to the second power

Evaluate each expression when $x = 4$ and $y = -2$.

11. $x^2y^3 + (5 - y)^2$ 12. $(y - x)^3 + xy^2$ 13. $3y^3 - 2x^2 + (xy)^2$

14. $2000 is placed in an Individual Retirement Account that earns 7.5% interest annually. No other deposits or withdrawals are made.
    a. Write an equation giving the amount of money y in the account x years from now.
    b. How much money will be in the account 40 years from now?
15. A new car was valued at $13,000 in 1995. Each year, the car's value decreases by 12%.
    a. Write an equation giving the value y of the car x years after 1995.
    b. How much will the car be worth in the year 2005?
    c. Graph the equation you wrote in part (a). Where does the graph cross the vertical axis? What does this point represent?

ANSWERS Chapter 9

Assessment

1. base; exponent; power
2. exponential growth; growth rate; growth factor
3. exponential decay; decay factor
4. Scientific notation is a method for writing any number as a product $a \times 10^x$, where $1 \le a < 10$ and x is an integer. For example, $3{,}000{,}000{,}000{,}000{,}000 = 3 \times 10^{15}$ and $0.00000000000072 = 7.2 \times 10^{13}$.

5. 729 6. 81 7. $\frac{1}{1024}$

8. 0.125 9. 2^3; 8 10. 3^2; 9

11. −79 12. −200 13. 8

14. a. $y = 2000(1.075)^x$
    b. $36,088.48
15. a. $y = 13{,}000(0.88)^x$
    b. $3620.51
    c.

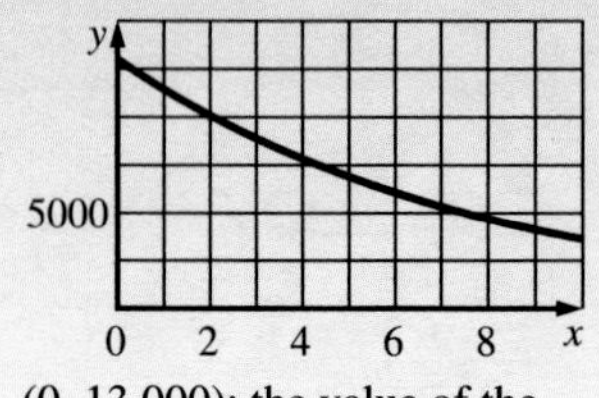

(0, 13,000); the value of the car when it was new

SECTIONS 9.4 *and* 9.5

Evaluate each power.

16. 4.37^0 **17.** 2^{-5} **18.** $(-3)^{-6}$ **19.** $\left(\frac{2}{5}\right)^{-3}$

Evaluate each expression when $x = -5$, $y = 2$, and $z = 3$.

20. $x^{-3}z^0$ **21.** x^2y^{-3} **22.** $\left(\frac{yz}{x}\right)^{-1}$ **23.** $y^{-2}z + x^2$

24. BIOLOGY The mass y in grams of a colony of bacteria x hours after 3:00 P.M., May 2, is given by the equation $y = 2.1(1.35)^x$. Estimate the mass at each of the times below.

a. 1:00 P.M., May 2 **b.** 3:00 P.M., May 2 **c.** 3:00 A.M., May 2

Write each number in decimal notation.

25. 4.57×10^7 **26.** 3.8×10^{-6} **27.** 7.023×10^4

Write each number in scientific notation.

28. 0.00000000678 **29.** 23,000,000,000 **30.** 5.0004

31. Open-ended Problem Give a value of x that satisfies the given inequality. Write your answers in scientific notation.

a. $x < 1$ **b.** $1 \leq x < 10$ **c.** $x > 10$

SECTIONS 9.6 *and* 9.7

Simplify each expression.

32. $\frac{d^4m^5}{d^7m^3}$ **33.** $(2b^2)^4$ **34.** $(3c^5d^7)(-2c^2d^8)$

35. $\frac{x^{-3}y^4}{y^3z^{-1}}$ **36.** $\left(\frac{-2a^2b^{-1}}{3ab^0}\right)^3$ **37.** $\left(\frac{s^3t}{t^2s^0}\right)^4$

38. HISTORY According to the 1990 Census, the population of the United States was 2.49×10^8. The cost of taking the census in 1990 was 2.6×10^9 dollars. How much did the census cost per person in the United States?

PERFORMANCE TASK

39. Open-ended Problem Find an ad for a car you would like to buy. Call a local bank and find out what its interest rate is for a car loan. Set up a plan for buying the car, assuming you make \$100 a week at a part-time job. Use the formula given on page 392 to find your monthly payment. How long would it take you to pay for the car? How would your payment plan change if you made \$75 a week? \$150 a week?

Chapter 9 Assessment
Form C Alternative Assessment

Chapter 9

ALTERNATIVE ASSESSMENT

1. a. Compare $\left(\frac{3}{4}\right)^2$ to $\frac{3^2}{4}$.

b. Compare $(-3)^2$ to -3^2.

c. Compare $(5x)^2$ to $5x^2$.

2. Open-ended Problem Use the rules of exponents to prove that $(27)^2 = (9)^3$.

3. Open-ended Problem Some students confuse $x^2 \cdot x^3$ with $(x^2)^3$. Think of a way to help students remember that $x^2 \cdot x^3 = x^5$ and $(x^2)^3 = x^6$.

4. a. If you double the length of all the edges of a cube, how do the surface area and the volume of the cube change?

b. If you double the length of the four vertical edges of a cube, how do the surface area and the volume of the cube change?

5. Performance Task Describe a real-life situation that can be represented by the equation $y = 250(1.08^x)$. Make a table of ten values for x and y. Explain what these x- and y-values mean in your situation.

6. Open-ended Problem Compare and contrast the linear function $y = 3x + 2$ with the exponential function $y = 2(3^x)$.

7. Performance Task Identify each of the following functions as a model of *exponential growth* or *exponential decay*.

a. The value of a certain new car depreciates by 18% every year for the first 10 years.

b. The half-life of carbon-14 is 5730 years.

c. The cost of living in a certain city has increased by an average of 2.5% per year for the last 5 years.

d. If has been predicted that the amount of information will be doubling every 73 days by the year 2020.

e. The population of a bacteria colony is doubling every 40 min.

Assessment Book, ALGEBRA 1: EXPLORATIONS AND APPLICATIONS

16. 1
17. $\frac{1}{32}$
18. $\frac{1}{729}$
19. $\frac{125}{8}$
20. $-\frac{1}{125}$
21. $\frac{25}{8}$
22. $-\frac{5}{6}$
23. $\frac{103}{4}$
24. a. about 1.15 g
 b. 2.1 g
 c. about 0.057 g
25. 45,700,000
26. 0.0000038
27. 70,230
28. 6.78×10^{-9}
29. 2.3×10^{10}
30. 5.0004×10^0
31. Answers may vary. Examples are given.
 a. 2×10^{-4}
 b. 5×10^0
 c. 6×10^5
32. $\frac{m^2}{d^3}$
33. $16b^8$
34. $-6c^7d^{15}$
35. $\frac{yz}{x^3}$
36. $-\frac{8a^3}{27b^3}$
37. $\frac{s^{12}}{t^4}$
38. about \$10.44
39. Answers may vary. Check students' work.

Cumulative Assessment

CHAPTERS 7–9

CHAPTER 7

1. Give the horizontal intercept and the vertical intercept of the graph of $5x - y = -10$. Then solve the equation for y.

For each system of equations, give the solution or state whether there are *no solutions* or *many solutions*.

2. $x + 3y = 15$
$2x - 3y = 3$

3. $y = 5x + 60$
$10x - 2y = -120$

4. $5x - 2y = 4$
$3x + 7y = 27$

5. Writing Tell which method you used to solve each equation in Questions 2–4. Explain why you chose each method.

RECREATION Sharon Page wants to plant a border of impatiens and marigolds along a walkway. A tray of impatiens costs \$2.50. A tray of marigolds costs \$2.00. Each tray holds six plants.

6. Write an equation showing that Sharon buys a total of 54 plants.

7. Write an equation showing that Sharon spends \$20 altogether for plants.

8. Solve the system of equations from Questions 6 and 7 to find how many trays of each type of plant Sharon buys.

Graph each inequality or system of inequalities.

9. $y \leq \frac{4}{3}x - 3$

10. $x < 2$
$y > -x + 2$

11. $x - y \leq 2$
$x - y > -1$

12. Open-ended Problem In a coordinate plane, draw a simple four-sided figure and shade the interior. Using the endpoints of each side of the figure, write a system of four inequalities that describes your shaded figure.

CHAPTER 8

13. Graph the equation $y = 3|x| - 1$. Tell whether the relationship between x and y is *linear* or *nonlinear*.

Solve each equation algebraically.

14. $4x^2 = 25$

15. $n^2 - 100 = 300$

16. $3a^2 - 8 = 19$

ANSWERS Chapters 7–9

Cumulative Assessment

1. horizontal intercept: –2; vertical intercept: 10; $y = 5x + 10$
2. (6, 3)
3. many solutions
4. (2, 3)
5. Answers may vary. Methods may be graphical or algebraic.
6. Since each tray holds six plants, then Sharon needs a total of nine trays ($54 \div 6 = 9$). If i = the number of trays of impatiens, and m = the number of trays of marigolds, then $i + m = 9$
7. $2.5i + 2m = 20$
8. Sharon buys four trays of impatiens and five trays of marigolds.
9. 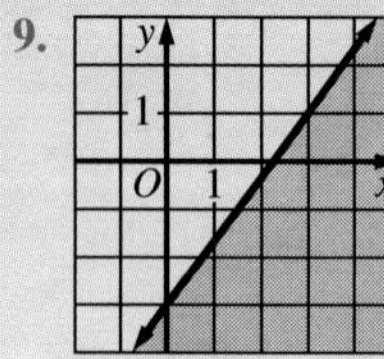

10. 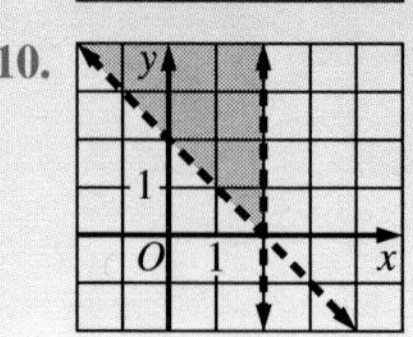

11. 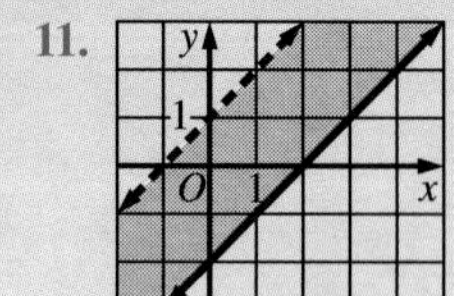

12. Answers may vary. An example is given. $x \geq 0$, $y \leq 2$, $y \geq 0$, $y \leq -x + 5$

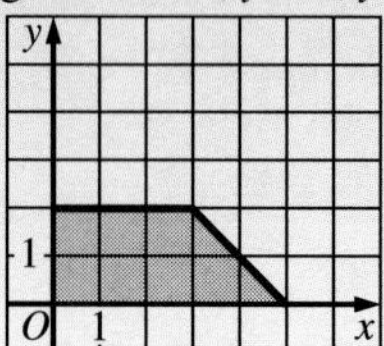

Match each equation with its graph.

17. $y = 3x^2$ **18.** $y = -3x^2$ **19.** $y = -\frac{1}{3}x^2$

A.

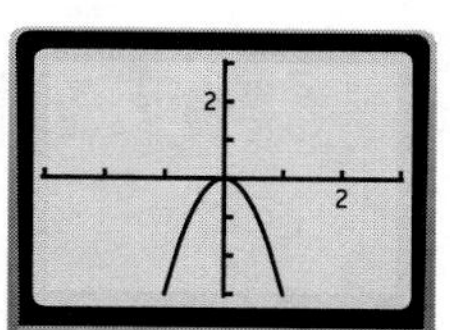

B.

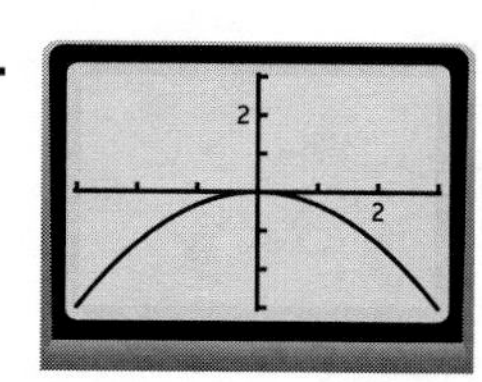

C.

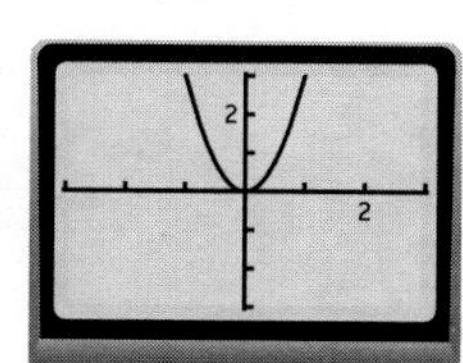

Technology Estimate the solution(s) of each equation using a graphing calculator. Then find the exact solution(s) using the quadratic formula.

20. $x^2 - 3x - 7 = 0$ **21.** $-3x^2 + 5x + 6 = 3$

22. Writing Find the discriminant of the equation $4x^2 - 8x + 4 = 0$. What does it tell you about the number of solutions of the equation? In general, what does the discriminant tell you about the number of solutions of a quadratic equation?

23. Open-ended Problem Write a quadratic equation that has no solution. Use a graph to show why the equation has no solution.

CHAPTER 9

Evaluate each expression when $a = -3$ and $b = 2$.

24. $(ab)^3 - 4b^5$ **25.** $(2a + b)^2 + ab^3$ **26.** $a^0 + b^{-4}$

Simplify each expression. Write each answer using positive exponents.

27. $(ab)^2(a^3b)^4$ **28.** $(x^0y^5)^3$ **29.** $(4m)^2(2m^3)^3$

30. $\frac{(2c)^3}{c^2d}$ **31.** $\left(\frac{2r}{3s}\right)^5$ **32.** $\left(\frac{5k^6}{k^8}\right)^3$

POPULATION The population of the Netherlands in 1990 was about 14,952,000. The annual growth rate for the period from 1980 through the year 2000 is about 0.6%. Use this information for Questions 33–35.

33. Write an equation that models the size y of the population x years after 1990.

34. Estimate the population in 1980 to the nearest thousand.

35. Predict the population in the year 2000 to the nearest thousand.

36. Writing Write *thirty-five billion, thirty million* in decimal notation and in scientific notation. Using your answer as an example, explain how to write a number in scientific notation.

37. Open-ended Problem Give an example of an equation that models exponential decay. Graph the equation. Does the graph ever cross the horizontal axis? Explain.

24. –344 25. –8

26. $1\frac{1}{16}$ 27. $a^{14}b^6$

28. y^{15} 29. $128m^{11}$

30. $\frac{8c}{d}$ 31. $\frac{32r^5}{243s^5}$

32. $\frac{125}{k^6}$

33. $y = 14,952,000(1.006)^x$

34. 14,084,000

35. 15,874,000

36. 35,030,000,000; 3.503×10^{10}; Answers may vary. An example is given. To write a number in scientific notation, write it as the product of a number from 1 up to, but not including, 10 and a power of 10. For 35,030,000,000, the decimal point must be moved to the left until it is behind the leading digit, which is 3. The power of ten is found by counting the number of places that the decimal point was moved, which was 10.

13. 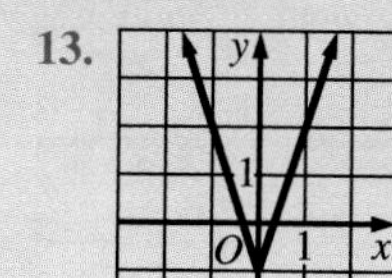 nonlinear

14. ±2.5 15. ±20

16. ±3 17. C

18. A 19. B

20. Check students' work. about –1.54 and about 4.54; $\frac{3 \pm \sqrt{37}}{2}$

21. Check students' work. about –0.47 and about 2.14; $\frac{-5 \pm \sqrt{61}}{-6}$

22. 0; There is exactly one solution. If the discriminant is greater than 0, then there are two solutions; if the discriminant is equal to 0, then there is exactly one solution; if the discriminant is less than 0, then there are no solutions.

23. Answers may vary. An example is given. $x^2 - 3x + 4 = 0$; The graph has no x-intercepts, so the equation has no solutions.

$y = x^2 - 3x + 4$

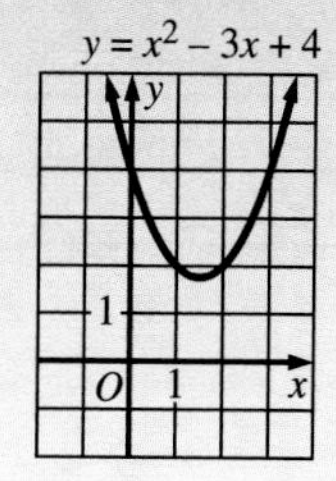

37. Answers may vary. An example is given. $y = 10(0.75)^x$; The graph does not intersect the horizontal axis because y will never equal zero.

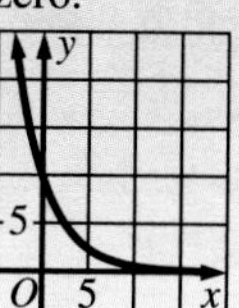

CHAPTER

10 Polynomial Functions

Overview

Connecting to Prior and Future Learning

⇔ Students use skills first acquired in Chapter 1 to add and subtract polynomials. Topics from Chapter 6 are used as students learn to multiply polynomials. Other topics include recognizing and classifying polynomials and using the *FOIL* method to multiply binomials. These skills will be used later in this chapter and in Chapter 11.

⇔ Students begin to explore polynomial equations in Section 10.3. Previous knowledge of greatest common factors and the distributive property are used to introduce the concept of factoring. Students can review greatest common factors on page 584 in the **Student Resources Toolbox**. The zero-product property is introduced as a way to solve equations.

⇔ Factoring trinomials and the application of this skill to solving quadratic equations are presented. These skills will be used again in Chapter 11.

Chapter Highlights

Interview with Linda Pei: Financial advisors use mathematics in many aspects of their work. Exercises relating to this interview can be found on pages 427 and 446.

Explorations: In Section 10.2, students model dimensions with polynomials. Section 10.3 begins with an exploration of factors, intercepts, and solutions using graphing calculators. Students use algebra tiles in Section 10.4 to explore factoring trinomials.

The Portfolio Project: Students measure a doll or action figure and use a scale factor to determine what the figure would weigh if it were a real person.

Technology: Graphing calculators are used to graph volume functions and various nonlinear equations, and to explore factors, intercepts, and solutions. Spreadsheets are used in Section 10.5 to explore data about the relationship between the height of a ball and time.

Objectives

Section	Objectives	NCTM Standards
10.1	• Recognize and classify polynomials. • Add and subtract polynomials. • Use polynomials to model real-world situations.	1, 2, 3, 4, 5
10.2	• Use the *FOIL* method to multiply binomials. • Multiply polynomials. • Use polynomials to model lengths, areas, and volumes.	1, 2, 3, 4, 5
10.3	• Factor out a linear factor from a polynomial. • Use the zero-product property. • Use factoring to solve equations.	1, 2, 3, 4, 5
10.4	• Factor trinomials with positive quadratic terms and positive constant terms. • Use factoring to explore the value of an investment.	1, 2, 3, 4, 5
10.5	• Factor quadratic polynomials with a negative constant term. • Solve quadratic equations by factoring.	1, 2, 3, 4, 5

Integration

Mathematical Connections	10.1	10.2	10.3	10.4	10.5
algebra	**423–428***	**429–435**	**436–441**	**442–446**	**447–453**
geometry	427	**429–435**	437	442, 443, 445	451
data analysis, probability, discrete math	427		440	446	452, 453
logic and reasoning	423, 425–428	429, 430, 432–435	**436–441**	442, 444–446	**447–453**
Interdisciplinary Connections and Applications					
chemistry and physics			438		
business and economics			440		
arts and entertainment		431, 433			
sports and recreation					449, 451, 452

****Bold page numbers*** *indicate that a topic is used throughout the section.*

Technology

	opportunities for use with	
Section	**Student Book**	**Support Material**
10.1	spreadsheet software graphing calculator McDougal Littell Software *Stats!*	
10.2	graphing calculator McDougal Littell Software *Function Investigator*	
10.3	graphing calculator	**Technology Book:** Calculator Activity 10 Spreadsheet Activity 10
10.4	graphing calculator McDougal Littell Software *Function Investigator*	
10.5	graphing calculator spreadsheet software McDougal Littell Software *Stats!* *Function Investigator*	

PLANNING GUIDE

Regular Scheduling (45 min)

Section	Materials Needed	Core Assignment	Extended Assignment	*exercises that feature* Applications	Communication	Technology
10.1	algebra tiles, spreadsheet software	1–22, 25–35, 37–44	1–12, 16–29, 34–44	22–24	24, 36, 37	44
10.2	nonsquare rectangles of paper, ruler, scissors, graphing calculator	**Day 1:** 1–12, 14–17, 19–24 **Day 2:** 25–29, 32–43, AYP*	**Day 1:** 1, 2, 4, 5, 7, 8, 10–24 **Day 2:** 25–43, AYP	15–18	13–18 30, 32–34, 37	35
10.3	graphing calculator	**Day 1:** 1–12, 15–23 **Day 2:** 24–36, 38–56	**Day 1:** 1–23 **Day 2:** 24–56	24–26	13, 14 27, 37, 39–44	41, 43
10.4	algebra tiles	**Day 1:** 1–22 **Day 2:** 23–28, 30–39, 41–52	**Day 1:** 1–22 **Day 2:** 23–52	30–33	13 29, 40	
10.5	spreadsheet software, graph paper	**Day 1:** 1–20, 22–31 **Day 2:** 32–39, 46–54, AYP	**Day 1:** 1–31 **Day 2:** 32–54, AYP	31 38, 39	21 46	38
Review/ Assess		**Day 1:** 1–25 **Day 2:** 26–49 **Day 3:** Ch. 10 Test	**Day 1:** 1–25 **Day 2:** 26–49 **Day 3:** Ch. 10 Test		1–3 48, 49	
Portfolio Project		Allow 2 days.	Allow 2 days.			

Yearly Pacing (with Portfolio Project)	**Chapter 10 Total** 14 days	**Chapters 1–10 Total** 137 days	**Remaining** 23 days	**Total** 160 days

Block Scheduling (90 min)

	Day 63	Day 64	Day 65	Day 66	Day 67	Day 68	Day 69
Teach/Interact	10.1 10.2: Exploration, page 429	Continue with 10.2 10.3: Exploration, page 436	Continue with 10.3 10.4: Exploration, page 442	Continue with 10.4 10.5	Continue with 10.5 Review	Review Port. Proj.	Port. Proj. Ch. 10 Test
Apply/Assess	**10.1:** 1–22, 25–44 **10.2:** 1–12, 14–17, 19–24	**10.2:** 25–29, 32–43, AYP* **10.3:** 1–12, 15–23	**10.3:** 24–36, 38–56 **10.4:** 1–22	**10.4:** 23–28, 30–52 **10.5:** 1–20, 22–31	**10.5:** 32–54, AYP **Review:** 1–25	**Review:** 26–49 **Port. Proj.**	**Port. Proj.** **Ch. 10 Test**

NOTE: A one-day block has been added for the Portfolio Project—timing and placement to be determined by teacher.

Yearly Pacing (with Portfolio Project)	**Chapter 10 Total** 7 days	**Chapters 1–10 Total** 69 days	**Remaining** 12 days	**Total** 81 days

***AYP** *is Assess Your Progress.*

Lesson Support

Section	Practice Bank	Study Guide*	Assessment Book*	Visuals	Explorations Lab Manual	Lesson Plans	Technology Book
10.1	66	10.1		Warm-Up 10.1 Folders 10, B	Masters 4, 5 Add. Expl. 15	10.1	
10.2	67	10.2	Test 43	Warm-Up 10.2	Add. Expl. 16	10.2	
10.3	68	10.3		Warm-Up 10.3		10.3	Calculator Act. 10 Spreadsheet Act. 10
10.4	69	10.4		Warm-Up 10.4 Folders 10, B	Masters 4, 5	10.4	
10.5	70	10.5	Test 44	Warm-Up 10.5	Master 2	10.5	
Review Test	71	Chapter Review	Tests 45, 46 Alternative Assessment			Review Test	

*__Spanish versions__ of *Study Guide* and *Assessment Book* are available.

Chapter Support

- Course Guide
- Lesson Plans
- Portfolio Project Book
- Preparing for College Entrance Tests
- Multi-Language Glossary
- *Test Generator* Software
- Professional Handbook

Software Support

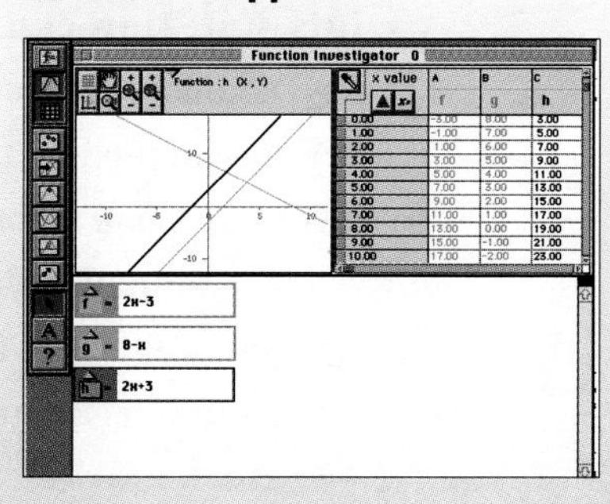

McDougal Littell Software
Stats!
Function Investigator

Internet Support

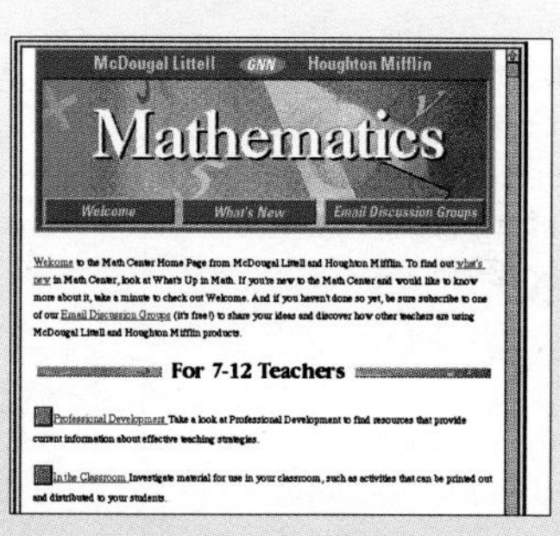

http://www.hmco.com
Next go to McDougal Littell; then the Education Center; then Secondary Math.

Outside Resources

Books, Periodicals

Mayes, Robert L. "Computer Use in Algebra: And Now, the Rest of the Story." *Mathematics Teacher* (October 1993): pp. 538–541.

Spikell, Mark A. *Teaching Mathematics with Manipulatives: A Resource of Activities for the K–12 Teacher.* Needham Heights, MA: Allyn & Bacon, 1993.

Binder, Margery. "A Calculator Investigation of an Interesting Polynomial." *Mathematics Teacher* (October 1995): pp. 558–560.

Software

Tools of Mathematics: Algebra from MacNumerics II. Acton, MA: Bradford Publishing Co., 1993. For MacIntosh II or LC.

Cuoco, Albert A. "Visualizing the Behavior of Functions." (Technology Tips) *Mathematics Teacher* (October 1995): pp. 604–607.

Videos

Apostol, Tom. *Polynomials.* Workbook accompanies video. Reston, VA: NCTM.

Internet

The American Institute of Small Business (AISB) has released a video: *The Internet—What to Know and How to Get On.* Minneapolis, MN. (800) 328-2906.

Interview Notes

Background

Mutual Funds

Mutual funds are a convenient and fairly safe way (risk usually decreases over a period of 5 to 10 years) to invest money for the future. Basically, all mutual funds fall into three categories: stock funds, bond funds, and funds that invest in both stocks and bonds. Among stock funds, there are many varieties to choose from. Some invest only in domestic companies and others invest in foreign corporations. Stock funds also specialize in their objectives, with some looking for capital growth and others for income or growth and income. Some stock funds invest only in large blue-chip companies and others invest only in small companies. Bond funds can also be grouped into three major categories: taxable bond, tax-exempt bond, and global bond. Bond funds are concerned with generating income through the interest paid on the bonds.

Linda Pei

Linda Pei was born in Sichnan, China and raised in Japan. She was educated in the United States, graduating with a bachelor's degree from Brown University and masters degrees from Wesleyan and Stanford Universities. With her MBA in finance from Stanford, she began her career in investment banking in 1975 and later entered the field of consulting. As an independent consultant, she worked in Japan, Great Britain, and Hong Kong. In 1993, Pei started the San Francisco-based Pro-Conscience Mutual Fund Company. Her goal is to provide investors with a means to build and preserve their financial assets, while encouraging corporations to support the needs of their women employees.

CHAPTER

10 Polynomial Functions

INVESTING in Equality

INTERVIEW **Linda Pei**

What are your dreams for the future? Going to college? Traveling the world? Making dreams come true often takes money, sometimes lots of it. How can you save money and make it work for you? You could open a savings account at a bank and make regular deposits. The bank would pay you a small amount of interest on the money in your account. Or you might ask financial adviser Linda Pei about investing in the stock market. She will tell you that some stocks earn 20 percent or more each year, but they can lose that much, too. You can avoid losing money by investing in a variety of carefully chosen companies. "You need to spread your risk around—if one investment does not perform well, hopefully another one will do better," Pei says. "In the end you won't lose everything because you've invested in many different companies."

> "When I was a child growing up in Japan, we got red envelopes at New Year's. ..."

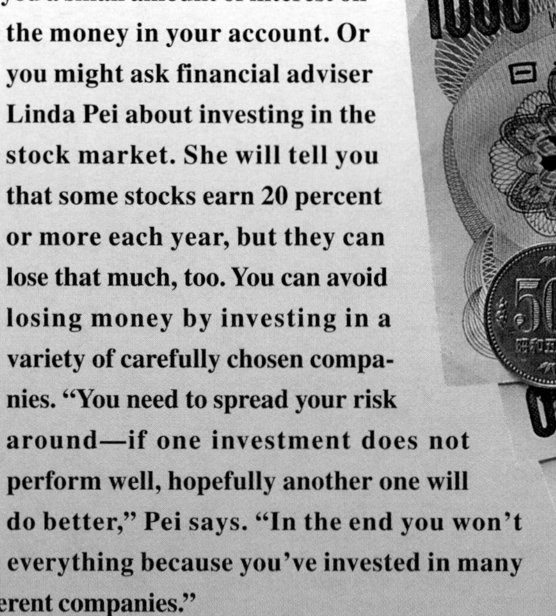

Investing in Change

Pei founded a *mutual fund* to help people invest wisely and support companies that promote equality at the same time. A mutual fund is a company that employs experts to invest money in stocks they think will rise in value. The money comes from thousands of people who share the profits and losses on the stocks owned by the fund.

Pei's experience in banking and consulting gave her a chance to see how different companies operate. "That view made me think about how people manage other people and how senior management directs and treats employees." All too often, Pei noticed women being treated in an unfair way.

"...Whatever money you got in them, you put some away for the future."

Determined to fight for equal treatment of all employees, Pei believed she could become a force for change by supporting companies that promote equality. She worked with others to design a mutual fund to do just that. They developed a set of standards to identify companies actively working to promote equality. They invest only in companies that meet these standards.

Interview Notes

Background

Stock Exchanges
Millions of people invest in stocks that are bought and sold at numerous stock exchanges around the world. Some of the most important exchanges are in Tokyo, New York, Hong Kong, Buenos Aires, and Mexico City. The *Bolsa,* the Mexican Stock Exchange, was founded in 1894; it is now one of the fastest growing in the world. Mexico is also becoming a major player among the world's emerging markets. Many of its mutual funds, government bonds, and companies are listed on stock exchanges in other nations.

Interview Notes

Mathematical Connection

Financial advisors are usually very good at doing mathematics. This is a must in order to advise people on how to invest their money to make it grow for the future. In Section 10.1, students learn what a polynomial is and then write a polynomial to model the growth of a one-time investment. In Section 10.4, students learn how to factor trinomials and then use factoring to explore the value of an annual investment, assuming a fixed rate of growth.

Explore and Connect

Writing
This question can be explored in class with all students doing the calculations. You may wish to ask for volunteers to explain their results.

Research
There are many topics that students can report on about stocks and the stock market. You may wish to have some students investigate mutual funds. Ask them how they would compare the risks of investing in stocks of individual companies as opposed to investing in a mutual fund.

Project
Students should share their findings with their classmates so that a complete picture can emerge regarding the existence of those people and organizations that support equal treatment of people at work.

Calculating Your Returns

Financial advisers like Linda Pei encourage people to save and invest on a regular basis. One consequence of this is that the amount of money earning interest grows. For instance, suppose you invest $1000 at the start of each year over a period of three years in a mutual fund that earns a consistent return of 10% annually. You can calculate the total value of your investment after three years by writing an expression like this one:

$$\text{Total value of investment after three years} = 1000(1 + 0.1)^3 + 1000(1 + 0.1)^2 + 1000(1 + 0.1)$$

value of investment made the first year | value of investment made the second year | value of investment made the third year

If the annual growth rate is r, written as a decimal, the expression above becomes:

$$1000(1 + r)^3 + 1000(1 + r)^2 + 1000(1 + r)$$

This is an example of a *polynomial*, the topic of this chapter.

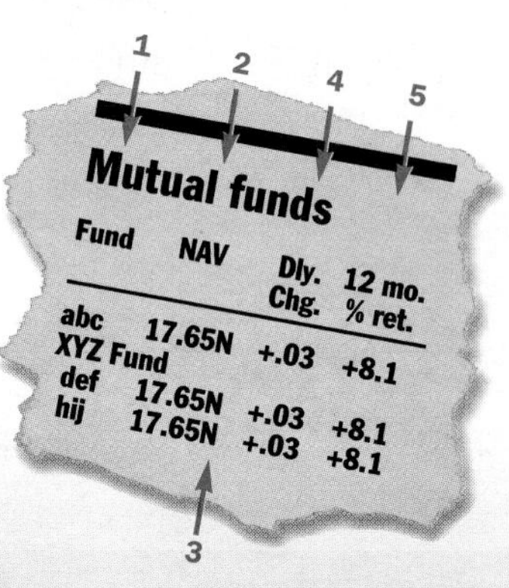

1
2
4
5
Mutual funds
Fund NAV Dly. Chg. 12 mo. % ret.
abc 17.65N +.03 +8.1
XYZ Fund
def 17.65N +.03 +8.1
hij 17.65N +.03 +8.1
3

Many newspapers publish information about the performance of mutual funds.

1. **Fund** Names of fund groups.
2. **NAV** Net asset value.
3. **N** Signifies no load. The fund does not charge any sales fees.
4. **Dly. Chg.** Change in **NAV** from previous session.
5. **12 mo. % ret.** Percent return of fund for last 12 month period.

EXPLORE AND CONNECT

Much of Pei's business is done over the phone and via computer.

1. **Writing** Calculate the total value of the investment represented by the first expression above. What will be the value after four years? Explain.

2. **Research** What are stocks? How do they make money for investors? How are they different from savings accounts at banks? Learn about an aspect of stocks or the stock market. Report your findings to your class.

3. **Project** Linda Pei hopes that her mutual fund will encourage businesses to treat all employees fairly. Find out about another person or organization that supports the equal treatment of people at work. Report your findings to your class.

Mathematics & Linda Pei

In this chapter, you will learn more about how mathematics is related to investing.

Related Exercises

Section 10.1
- **Exercises 22–24**

Section 10.4
- **Exercises 30–33**

SECTION

10.1 Adding and Subtracting Polynomials

Learn how to...
- recognize and classify polynomials
- add and subtract polynomials

So you can...
- use polynomials to model real-world situations, such as the total value of an investment

Financial advisers like Linda Pei sometimes write formulas like the one on page 420 in terms of an annual growth factor g. If you substitute g for $1 + r$, you get the *polynomial* shown below.

A **polynomial** is an expression that can be written as a monomial or as a sum of monomials with exponents that are whole numbers. The polynomial below is in **standard form**, which means it is simplified and the exponents of the terms are in order from largest to smallest.

The **degree of a term** of a polynomial is the exponent of the variable. The degree of this term is 2.

$$1000g^3 + 1000g^2 + 1000g$$

The **degree of a polynomial** in standard form is the largest degree of all the terms. The degree of this polynomial is 3.

For a term with more than one variable, the degree is the *sum* of the exponents of the variables. For example, the degree of $5xy$, or $5x^1y^1$, is 2.

Classifying Polynomials

You can classify a polynomial by its degree.

Degree	Example	Classification
1	$3x + 2y$	**linear polynomial**
2	$-2x^2 + 3x + 1$	**quadratic polynomial**
3	$5x^3 - x^2 + 6$	**cubic polynomial**

THINK AND COMMUNICATE

1. Use what you know about exponents to explain why $\frac{4}{x}$ is not a polynomial.
2. The expression $8x^2 + \frac{1}{2}x^3 - 4$ is a polynomial. Write it in standard form and identify the degree of each term. Then identify the degree of the polynomial and its classification.
3. Explain why the polynomial $4x - 2 - 4x$ has degree 0.

ANSWERS Section 10.1

Think and Communicate

1. $\frac{4}{x} = 4x^{-1}$. Since the exponent of x, -1, is not a whole number, $\frac{4}{x}$ is not a polynomial.
2. $8x^2 + \frac{1}{2}x^3 - 4 = \frac{1}{2}x^3 + 8x^2 - 4$. The degrees of the terms are 3, 2, and 0, respectively. The degree of the polynomial is 3; it is a cubic.
3. Simplifying $4x - 2 - 4x$ gives -2. This is a degree 0 polynomial.

Plan⇔Support

Objectives
- Recognize and classify polynomials.
- Add and subtract polynomials.
- Use polynomials to model real-world situations.

Recommended Pacing
Core and Extended Courses
Section 10.1: 1 day
Block Schedule
Section 10.1: $\frac{1}{2}$ block
(with Section 10.2)

Resource Materials
Lesson Support
Lesson Plan 10.1
Warm-Up Transparency 10.1
Overhead Visuals:
Folder 10: Modeling Polynomial Operations, Sheets 1 and 2
Folder B: Algebra Tiles
Practice Bank: Practice 66
Study Guide: Section 10.1
Explorations Lab Manual:
Additional Exploration 15, Diagram Masters 4, 5
Technology
Graphing Calculator
Spreadsheet Software
McDougal Littell Software
Stats!
Internet:
http://www.hmco.com

Warm-Up Exercises
1. Rewrite $x \cdot x \cdot x \cdot x \cdot x$ using an exponent. x^5
2. Identify the base and exponent of y^8. base: y; exponent: 8

Simplify.

3. $3x + 4 - 3 - 5x$ $-2x + 1$
4. $6v - (3v + 4)$ $3v - 4$
5. $(8z - 2y + 4) - (2z - 7 + 5y)$ $6z - 7y + 11$

Teach⇔Interact

Section Notes

Cooperative Learning
Students can work in groups to model polynomials with algebra tiles. Have them practice several examples together until each student can explain how to use the tiles as a model.

Multicultural Note
In mathematics today, we use the Hindu-Arabic numeral system that was developed in India during the third and second centuries B.C. This system is based upon the concept of place and the use of zero and is called a base 10 system.

Additional Example 1

Add $2x^2 + 3x + 4$ and $4x^2 - x - 1$.
Method 1 Use algebra tiles.
Model each polynomial with algebra tiles.

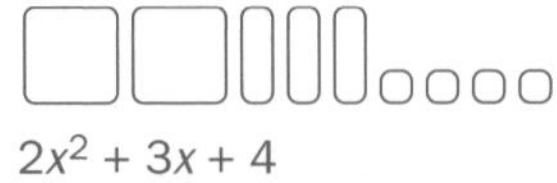

$2x^2 + 3x + 4$

$4x^2 - x - 1$

Group tiles of the same size.

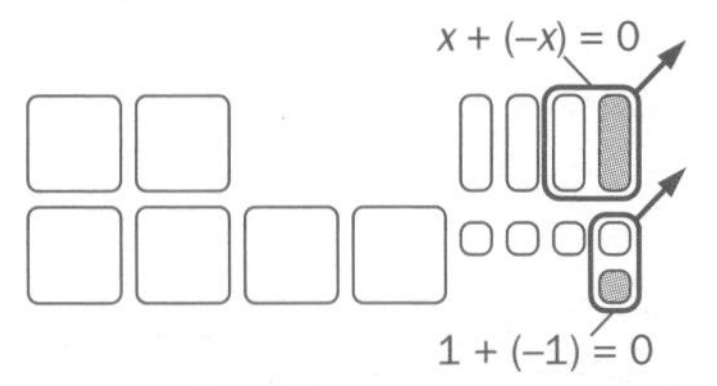

Combine the remaining tiles. The sum is $6x^2 + 2x + 3$.
Method 2 Add vertically.
Line up the like terms. Then add the coefficients.

$$\begin{array}{r} 2x^2 + 3x + 4 \\ \underline{4x^2 - 1x - 1} \\ 6x^2 + 2x + 3 \end{array}$$

Method 3 Add horizontally. Group like terms.

$(2x^2 + 3x + 4) + (4x^2 - x - 1) =$
$(2x^2 + 4x^2) + (3x - 1x) + (4 - 1) =$
$6x^2 + 2x + 3$

BY THE WAY

Hindu scholars were among the first to represent unknown values with symbols or letters. In a 9th century text, the abbreviation *ya* means "an unknown quantity," and *yava* means "the square of the unknown quantity." The quantity $x^2 + 5x$ would be written as *yava* 1 *ya* 5.

Algebra tiles can help you understand how to add polynomials. You can represent positive and negative terms with algebra tiles as shown below.

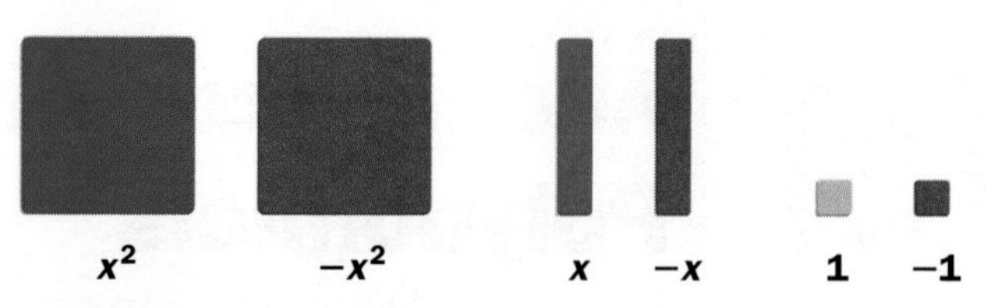

Positive and negative tiles of the same size cancel each other out.

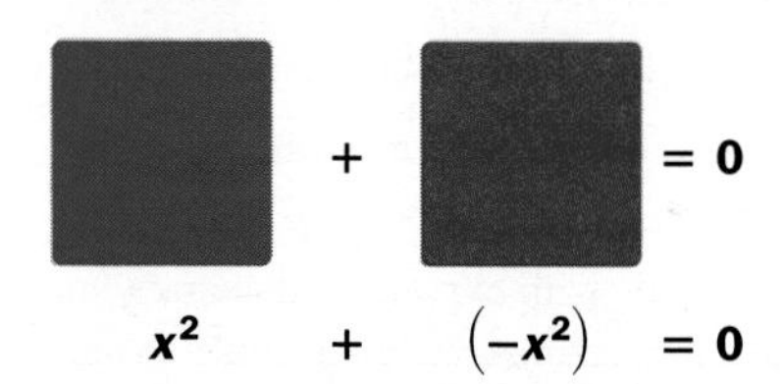

EXAMPLE 1

Add $3x^2 + 2x + 3$ and $2x^2 - x - 2$.

SOLUTION

Method 1 Use algebra tiles.

Model each polynomial with algebra tiles.

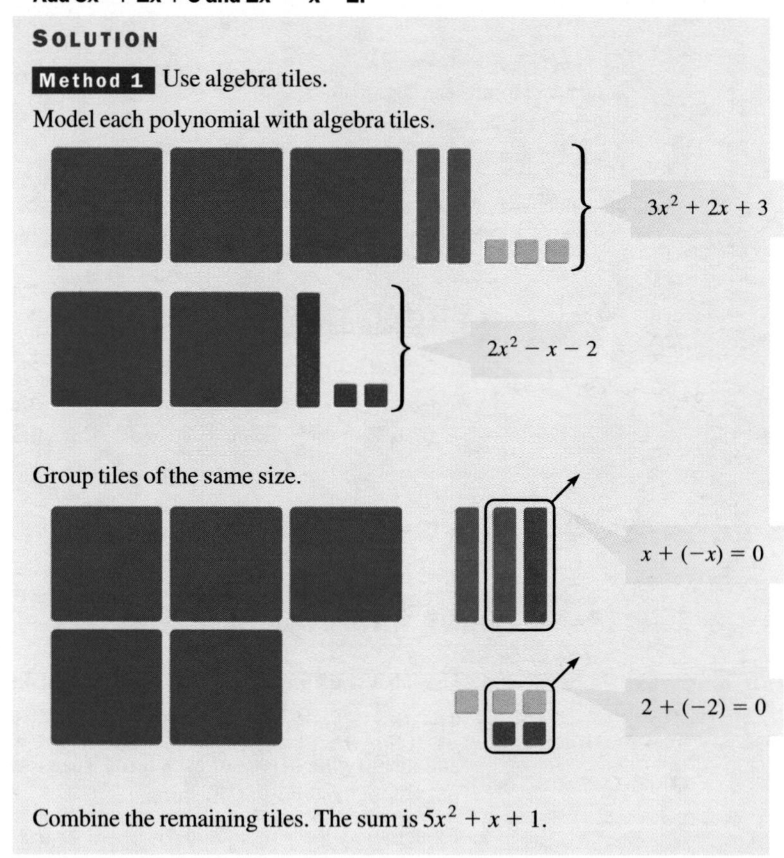

Combine the remaining tiles. The sum is $5x^2 + x + 1$.

Method 2 Add vertically.

Line up the like terms. Then add the coefficients.

$$\begin{array}{r} 3x^2 + 2x + 3 \\ \underline{2x^2 - 1x - 2} \\ 5x^2 + x + 1 \end{array}$$

$x = 1x$

Method 3 Add horizontally.

$$\begin{aligned}(3x^2 + 2x + 3) + (2x^2 - x - 2) &= (3x^2 + 2x^2) + (2x - 1x) + (3 - 2) \\ &= 5x^2 + x + 1\end{aligned}$$

Group like terms.

You can subtract polynomials by using a method similar to the one you used in Chapter 1 to simplify expressions like $5x - (3x + 2)$.

EXAMPLE 2

Subtract $(a^3 - a + 7) - (4a^2 - 3a + 1)$.

SOLUTION

Method 1 Subtract vertically.

Line up the like terms. You may need to insert placeholders, like $0a^3$ and $0a^2$.

$$\begin{array}{r} 1a^3 + 0a^2 - 1a + 7 \\ \underline{0a^3 + 4a^2 - 3a + 1} \end{array}$$

Change each term being subtracted to its opposite. Then *add* like terms.

$$\begin{array}{r} 1a^3 + 0a^2 - 1a + 7 \\ \underline{-0a^3 - 4a^2 + 3a - 1} \\ a^3 - 4a^2 + 2a + 6 \end{array}$$

Method 2 Subtract horizontally.

Change each term being subtracted to its opposite and then add.

$$\begin{aligned}(a^3 - a + 7) - (4a^2 - 3a + 1) &= (a^3 - a + 7) + (-4a^2 + 3a - 1) \\ &= a^3 + (-4a^2) + (-a + 3a) + (7 - 1) \\ &= a^3 - 4a^2 + 2a + 6\end{aligned}$$

THINK AND COMMUNICATE

4. Suppose the polynomials being added in Example 1 were $3x^2 + 2xy + 3$ and $2x^2 - xy - 2$. How would the answer be different?
5. To simplify $(-n^3 + 7n^2 - 3n) - (8n^2 - 5n - 2)$, you must find the *sum* of $-n^3 + 7n^2 - 3n$ and what polynomial?

Teach⇔Interact

About Example 2

Teaching Tip
Method 1 reminds students that when they subtract a quantity, they should change each term being subtracted to its opposite and then add. You may wish to show this fact symbolically as $a - b = a + (-b)$.

Additional Example 2

Subtract $(3a^2 + 7a + 1) - (2a^3 + a^2 - 8)$.

Method 1 Subtract vertically.
Line up the like terms. You may need to insert placeholders, like $0a^3$ and $0a$.

$$\begin{array}{r} 0a^3 + 3a^2 + 7a + 1 \\ \underline{2a^3 + 1a^2 + 0a - 8} \end{array}$$

Change each term being subtracted to its opposite. Then *add* like terms.

$$\begin{array}{r} 0a^3 + 3a^2 + 7a + 1 \\ \underline{-2a^3 - 1a^2 - 0a + 8} \\ -2a^3 + 2a^2 + 7a + 9 \end{array}$$

Method 2 Subtract horizontally.
Change each term being subtracted to its opposite and then add.

$(3a^2 + 7a + 1) - (2a^3 + a^2 - 8) =$
$(3a^2 + 7a + 1) + (-2a^3 - a^2 + 8) =$
$-2a^3 + (3a^2 - a^2) + (7a) + (1 + 8) =$
$-2a^3 + 2a^2 + 7a + 9$

Think and Communicate

For question 5, some students may have difficulty giving the correct polynomial because of sign errors. Make sure students change each term being subtracted to its opposite.

Think and Communicate

4. $5x^2 + xy + 1$; The middle term would be an xy-term instead of an x-term.
5. $-8n^2 + 5n + 2$

Teach⇔Interact

Checking Key Concepts

Common Error
Students may think that the negative number in question 4 is not a polynomial. Emphasize that since -8 may be rewritten as $-8x^0$, it is a monomial.

Closure Question

Explain how to add or subtract two polynomials.
You can add two polynomials vertically by lining up the like terms and then adding the coefficients. Or you can add horizontally by grouping like terms and then adding the coefficients. You can subtract two polynomials vertically by lining up the like terms and then changing each term being subtracted to its opposite. Then add. Or you can subtract horizontally by changing each term being subtracted to its opposite and then adding.

Apply⇔Assess

Suggested Assignment

Core Course
Exs. 1–22, 25–35, 37–44

Extended Course
Exs. 1–12, 16–29, 34–44

Block Schedule
Day 63 Exs. 1–22, 25–44

Exercise Notes

Communication: Discussion
Exs. 1–6 A brief discussion of these exercises would reinforce the meanings of the given terms.

☑ CHECKING KEY CONCEPTS

Tell whether each expression is a polynomial. Write *Yes* or *No*. If not, explain why not.

1. $3a + 4b + 7c$ **2.** $2x^4 - x^6 + x^3$ **3.** $6n^4 + n^{-3}$

4. -8 **5.** $0.7xyz$ **6.** $\frac{3}{x^2} + \frac{4}{x} + \frac{1}{2}$

Write each polynomial in standard form. Then classify the polynomial as *linear, quadratic, cubic,* or *none of these.*

7. $1 - 2x + 5x^2$ **8.** $0.9t + 0.1t^2$ **9.** $1 - x^4$

10. $x - 7x^3 + 4x^2 - 3$ **11.** $14 - \frac{3}{4}h$ **12.** $k^3 + 3k$

Add or subtract each pair of polynomials as indicated. You may want to use algebra tiles.

13. $(7x^2 + 8x - 6) + (-x^2 - 8x - 3)$ **14.** $(9t^2 - 3t + 1) + (12t^2 + 5)$

15. $(9x + 3xy + 6y) - (x - 8xy + 4y)$ **16.** $(2x^3 + 5x^2 - 7x) - (6x^2 - 7x)$

10.1 Exercises and Applications

Extra Practice exercises on page 576

Match each phrase with an algebraic expression.

1. linear binomial **A.** $3a^2 - 4a + 2$

2. quadratic monomial **B.** $\frac{2}{m} + 3$

3. quadratic trinomial **C.** $z^3 - z^2 + z$

4. cubic trinomial **D.** $\frac{1}{2}y^2$

5. not a polynomial **E.** $2x + 3y$

6. Writing Explain why $\frac{x^4}{3}$ is a polynomial, but $\frac{3}{x^4}$ is not.

Add. Give the degree of each sum.

7. $\begin{array}{r} 5x^2 - 3x - 7 \\ x^2 + 2x + 3 \\ \hline \end{array}$ **8.** $\begin{array}{r} -y^2 + 2y - 8 \\ y^2 + 3y - 2 \\ \hline \end{array}$ **9.** $\begin{array}{r} 2r^3 - 7r^2 - r + 2 \\ 4r^2 - r + 1 \\ \hline \end{array}$

10. Open-ended Problem Write the polynomial modeled by the algebra tiles below. Find two polynomials whose sum is that polynomial.

Checking Key Concepts

1. Yes.
2. Yes.
3. No; the exponent of n^{-3} is not a whole number.
4. Yes.
5. Yes.
6. No; $\frac{3}{x^2} = 3x^{-2}$ and $\frac{4}{x} = 4x^{-1}$. The exponents of these terms are not whole numbers.
7. $5x^2 - 2x + 1$; quadratic
8. $0.1t^2 + 0.9t$; quadratic
9. $-x^4 + 1$; none of these
10. $-7x^3 + 4x^2 + x - 3$; cubic
11. $-\frac{3}{4}h + 14$; linear
12. $k^3 + 3k$; cubic
13. $6x^2 - 9$
14. $21t^2 - 3t + 6$
15. $8x - 11xy + 2y$
16. $2x^3 - x^2$

Exercises and Applications

1. E
2. D
3. A
4. C
5. B
6. The exponent of the variable in $\frac{x^4}{3} = \frac{1}{3}x^4$ is 4, a whole number. The exponent of x in $\frac{3}{x^4} = 3x^{-4}$ is -4, which is not a whole number.

7–24. See answers in back of book.

Add.

11. $(9n^2 + 3n - 5) + (8n^2 - n + 2)$

12. $(4x^2 - x - 5) + (6x^2 + x)$

13. $(y^2 + 3y + 2) + (-7y - 9)$

14. $(3k - 5) + (k^2 - 8k)$

15. $(a - 3b + 7) + (-2a - b)$

16. $(9x^2 + 6xy + y^2) + (x^2 - y^2)$

17. $(b^3 + b^2 - 8) + (b^3 - 6b + 9)$

18. $(a^3 + 4a - 1) + (6a^2 - 7a + 2)$

GEOMETRY Express the perimeter of each triangle as a polynomial in standard form.

19.

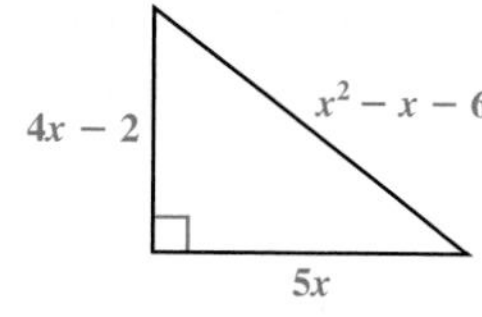

20.

$7x - 10$, $2x^2 + 4$, $x^2 + 8x$

21. Suppose the perimeter of the triangle in Exercise 19 is 120 cm. Find the length of each side of the triangle.

INTERVIEW Linda Pei

Look back at the article on pages 420–422.

Financial advisers like Linda Pei sometimes handle one-time investments. The growth of a one-time investment can be modeled by an exponential function of the annual growth factor g.

22. Does the polynomial in the article on page 422 represent a one-time investment of \$1000? Why or why not?

23. a. Write a polynomial in terms of g to represent the value of a one-time investment of \$2000 at the end of four years.

b. Suppose you invest \$2000 at the beginning of every year for four years. Write a polynomial in terms of g to represent the total value of the investment at the end of four years.

c. Find the sum of the polynomials in parts (a) and (b). What is the degree of the new polynomial?

24. Writing One aspect of a company that Pei might consider when making investments is the number of women in management positions. Use the graph to describe trends in management from 1983 to 1991.

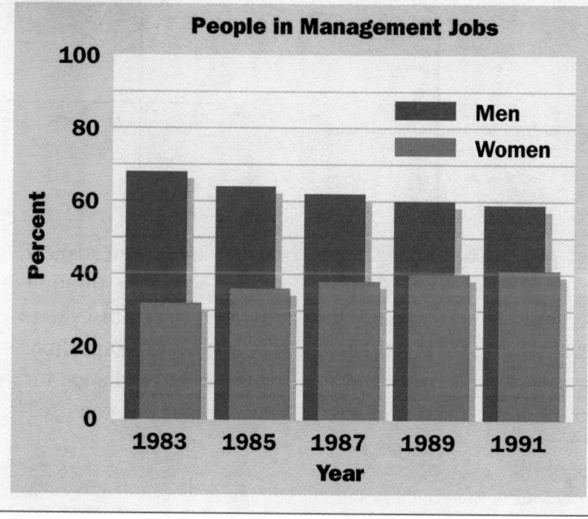

Technology Note

Students can use the list features of the TI-82 to add or subtract polynomials in one variable. To do this, they can construct a list of coefficients for each polynomial and add or subtract the lists. The lists of coefficients must contain the same number of elements. The length of each list should be one more than the greatest exponent that occurs in the polynomials being added or subtracted. Assume that the powers within the polynomials are used in decreasing order and that 0 is used for "missing" powers.

Consider, for example, Exercise 17. To add the two polynomials, add the lists {1, 1, 0, –8} and {1, 0, –6, 9}. Do this by pressing [2nd] [{] 1 [,] 1 [,] 0 [,] [(–)] 8 [2nd] [}] + [2nd] [{] 1 [,] 0 [,] [(–)] 6 [,] 9 [2nd] [}] [ENTER]. The calculator will display the result {2 1 –6 1}, which students should interpret as $2b^3 + b^2 - 6b + 1$. (Note: When the calculator displays a list as an answer, it omits commas between elements in order to save space.) The method described here is especially useful when it is necessary to add or subtract several polynomials at a time.

Apply⇔Assess

Exercise Notes

Student Progress

Exs. 11–18 It is essential that all students be able to do these exercises successfully before attempting to do the subtraction exercises (Exs. 25–35) on the next page.

Interview Note

Exs. 22, 23 These exercises review exponential growth models from Chapter 9.

Career Connection

Exs. 22–24 Financial advisers are professionals who are trained to develop investment portfolios that can help people meet their long-term investment goals. The most common forms of investments are stocks, bonds, and mutual funds. Mutual funds are investments in stocks, bonds, or both, and have become the investment of choice for most small investors. Financial advisers have an expert understanding of financial markets and the mathematics of finance. Many of them have completed a course of study leading to a professional certificate called a Certified Financial Planner, or CFP.

Apply⇔Assess

Exercise Notes

Mathematical Procedures

Exs. 25–35 These exercises allow students to practice the procedure of subtracting two polynomials. Some students may have to be reminded to change each term being subtracted to its opposite before adding.

Assessment Note

Ex. 37 These activities will help students to evaluate their own skill and understanding of how to add and subtract polynomials.

Using Technology

Ex. 45 Students can do this exercises by using the *Stats!* software or the List features of the TI-82 graphing calculator.

Practice 66 for Section 10.1

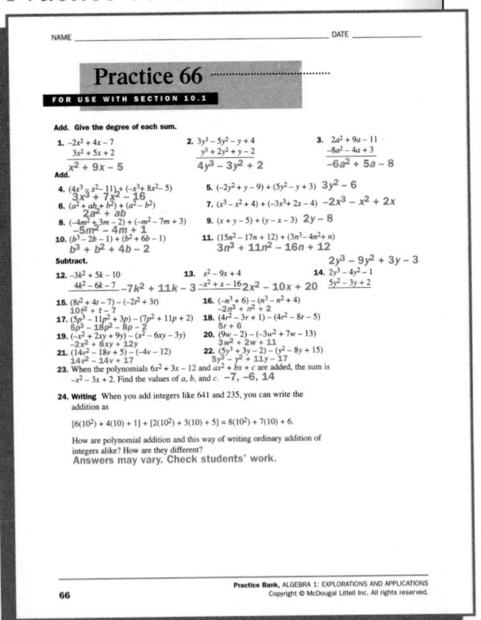

NAME ______ DATE ______

Practice 66

FOR USE WITH SECTION 10.1

Add. Give the degree of each sum.

1. $-2x^2 + 4x - 7$ / $3x^2 + 5x + 2$ — $x^2 + 9x - 5$
2. $3y^3 - 5y^2 - y + 4$ / $y^3 + 2y^2 + y - 2$ — $4y^3 - 3y^2 + 2$
3. $2a^2 + 9a - 11$ / $-8a^2 - 4a + 3$ — $-6a^2 + 5a - 8$

Add.

4. $(4x^3 - x^2 - 11) + (-x^3 + 8x^2 - 5)$ $3x^3 + 7x^2 - 16$
5. $(-2y^2 + y - 9) + (5y^2 - y + 3)$ $3y^2 - 6$
6. $(a^2 + ab + b^2) + (a^2 - b^2)$ $2a^2 + ab$
7. $(x^3 - x^2 + 4) + (-3x^3 + 2x - 4)$ $-2x^3 - x^2 + 2x$
8. $(-4m^2 + 3m - 2) + (-m^2 - 7m + 3)$ $-5m^2 - 4m + 1$
9. $(x + y - 5) + (y - x - 3)$ $2y - 8$
10. $(b^3 - 2b - 1) + (b^2 + 6b - 1)$ $b^3 + b^2 + 4b - 2$
11. $(15n^2 - 17n + 12) + (3n^3 - 4n^2 + n)$ $3n^3 + 11n^2 - 16n + 12$

Subtract.

12. $-3k^2 + 5k - 10$ / $4k^2 - 6k - 7$ — $-7k^2 + 11k - 3$
13. $x^2 - 9x + 4$ / $-x^2 + x - 16$ — $2x^2 - 10x + 20$
14. $2y^3 - 4y^2 - 1$ / $5y^2 - 3y + 2$ — $2y^3 - 9y^2 + 3y - 3$
15. $(8t^2 + 4t - 7) - (-2t^2 + 3t)$ $10t^2 + t - 7$
16. $(-n^3 + 6) - (n^3 - n^2 + 4)$ $-2n^3 + n^2 + 2$
17. $(5p^3 - 11p^2 + 3p) - (7p^2 + 11p + 2)$ $5p^3 - 18p^2 - 8p - 2$
18. $(4r^2 - 3r + 1) - (4r^2 - 8r - 5)$ $5r + 6$
19. $(-x^2 + 2xy + 9y) - (x^2 - 6xy - 3y)$ $-2x^2 + 8xy + 12y$
20. $(9w - 2) - (-3w^2 + 7w - 13)$ $3w^2 + 2w + 11$
21. $(14v^2 - 18v + 5) - (-4v - 12)$ $14v^2 - 14v + 17$
22. $(5y^3 + 3y - 2) - (y^2 - 8y + 15)$ $5y^3 - y^2 + 11y - 17$
23. When the polynomials $6x^2 + 3x - 12$ and $ax^2 + bx + c$ are added, the sum is $-x^2 - 3x + 2$. Find the values of a, b, and c. $-7, -6, 14$
24. **Writing** When you add integers like 641 and 235, you can write the addition as

$$[6(10^2) + 4(10) + 1] + [2(10^2) + 3(10) + 5] = 8(10^2) + 7(10) + 6.$$

How are polynomial addition and this way of writing ordinary addition of integers alike? How are they different?
Answers may vary. Check students' work.

66

Practice Bank, ALGEBRA 1: EXPLORATIONS AND APPLICATIONS

Subtract.

25. $\begin{array}{r} 3c^2 + c + 8 \\ \underline{2c^2 - 2c + 4} \end{array}$

26. $\begin{array}{r} 4x^2 - 5x - 7 \\ \underline{-x^2 - 5x + 1} \end{array}$

27. $\begin{array}{r} 4z^3 - 6z^2 + z + 4 \\ \underline{2z^2 + 9z - 4} \end{array}$

28. $(3a^2 - 5a + 3) - (a^2 + a - 2)$
29. $(-6m^2 - 5m + 1) - (2m^2 - 3)$
30. $(5n^2 - 4n + 8) - (-4n + 2)$
31. $(x^2 + 8) - (3x^2 - x + 1)$
32. $(8r + s + 5) - (r + 8s - 5)$
33. $(2y^3 - 2y + 9) - (5y^3 + y^2 - 2y)$
34. $(n^3 - n^2 + 5n - 8) - (9n^2 - 2)$
35. $(4a^2 - 9b^2) - (5a^2 - 12ab + 9b^2)$
36. **Challenge** Luis claimed that the sum of any four consecutive integers is an even integer. Use algebra to show that Luis is correct. Then use a logical argument to show that Luis is correct.

ONGOING ASSESSMENT

37. **Cooperative Learning** Work in a group of three students. Check each other's work at every stage.
    a. First each of you should write a polynomial on the same piece of paper using the variable x.
    b. Then you should take turns adding your teammates' polynomials.
    c. Finally you should take turns subtracting your original polynomial from the sum of your teammates' polynomials.

SPIRAL REVIEW

Find each product. *(Section 6.4)*

38. $2t(t + 2)$
39. $(x + 1)(x - 1)$
40. $(2n + 1)(n - 4)$

Estimate the solutions of each equation using the given graph. *(Section 8.3)*

41. $x^2 - 2 = 0$
42. $-2x^2 + \frac{5}{2} = 0$
43. $\frac{2}{3}x^2 - 3 = 0$

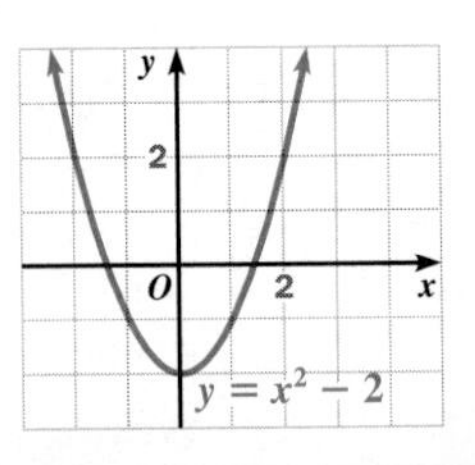

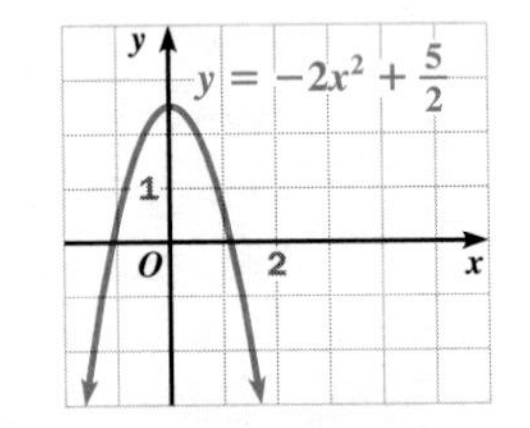

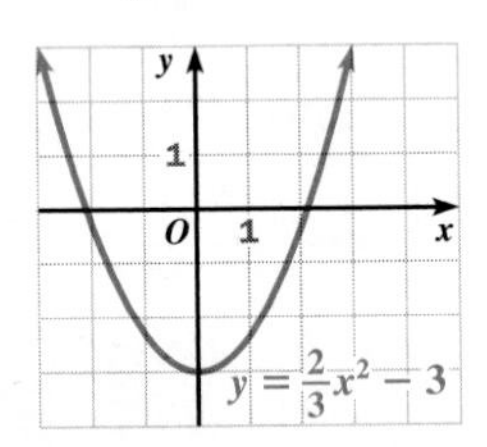

Distribution of Federal Funds

	A	B	C	D
1	Region	1988	1993	Change
2	Northwest	175,419	250,215	
3	Midwest	182,666	257,590	
4	South	303,828	448,269	
5	West	187,577	268,615	

44. **Spreadsheets** Use a spreadsheet. Find the change in the distribution of federal funds for each region from 1988 to 1993. (The numbers shown are in millions of dollars.) Then find the average change for the four regions. What formulas did you use? *(Section 1.8)*

25. $c^2 + 3c + 4$
26. $5x^2 - 8$
27. $4z^3 - 8z^2 - 8z + 8$
28. $2a^2 - 6a + 5$
29. $-8m^2 - 5m + 4$
30. $5n^2 + 6$
31. $-2x^2 + x + 7$
32. $7r - 7s + 10$
33. $-3y^3 - y^2 + 9$
34. $n^3 - 2n^2 + 5n - 6$
35. $-a^2 + 12ab - 18b^2$
36. Suppose the four consecutive integers are n, $n + 1$, $n + 2$, and $n + 3$. Their sum is $4n + 6 = 2(2n + 3)$. Since 2 is a factor of the sum, the sum is even. A logical argument is this: Two of the four integers are even and two are odd. The sum of the evens is even; the sum of the odds is even. The overall sum is the sum of two evens, which is even.
37. Answers may vary.
38. $2t^2 + 4t$
39. $x^2 - 1$
40. $2n^2 - 7n - 4$

41–43. Estimates may vary. Examples are given.

41. about 1.4, about –1.4
42. about 1.1, about –1.1
43. about 2.1, about –2.1
44.

	A	D
1	Region	Change
2	Northeast	74,796
3	Midwest	74,924
4	South	144,441
5	West	81,038

The average change is 93,799.75 million dollars. The formula in cell D2 is "=C2–B2." The formula for the average change in cell D6 is "= Sum(D2:D5)/4."

SECTION

10.2 Multiplying Polynomials

Learn how to...

- use the FOIL method to multiply binomials
- multiply polynomials

So you can...

- use polynomials to model lengths, areas, and volumes, such as the volume of an origami box

People from many different cultures enjoy the art of paper folding, often referred to as *origami*. The word comes from the Japanese *ori*, "to fold," and *kami*, "paper." In traditional origami, you fold a square piece of paper into an object without cutting or pasting.

EXPLORATION
COOPERATIVE LEARNING

Modeling Dimensions with Polynomials

Work in a group of 4 students.
You will need:

- rectangular pieces of paper that are not square
- a ruler

Each of you should build a box and write expressions for its dimensions as outlined in Steps 1–8.

1 Fold a rectangular piece of paper into 16 equal parts as shown. Always fold toward the front of the piece of paper.

Fold in half...

then in quarters...

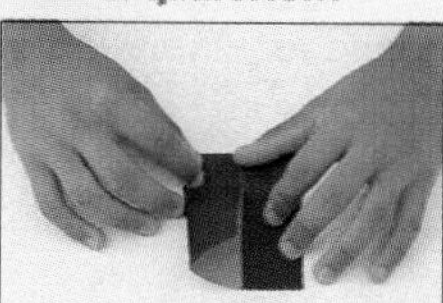

then turn and repeat.

Exploration continued on next page.

Exploration Note

Purpose
The purpose of this Exploration is to have students build a box from a rectangular piece of paper and write polynomial expressions for its dimensions.

Materials/Preparation
Each group should have a ruler and rectangular pieces of paper that are not square.

Procedure
Each group of students first folds a rectangular piece of paper into 16 equal parts and then folds it again in different ways to form a box. Groups then unfold their boxes and write expressions for the length, width, and height of the box in terms of the variable x. Students then compare their box dimensions within their groups and discuss how they are alike and how they are different.

Closure
Groups can share their results with each other. Students should conclude that if the corners of an a in. by b in. piece of paper are folded up x units to make a box, its new dimensions are x, $a - 2x$, and $b - 2x$.

For answers to the Exploration, see following page.

Plan⇔Support

Objectives

- Use the *FOIL* method to multiply binomials.
- Multiply polynomials.
- Use polynomials to model lengths, areas, and volumes.

Recommended Pacing

Core and Extended Courses
Section 10.2: 2 days

Block Schedule
Section 10.2: 2 half-blocks (with Sections 10.1 and 10.3)

Resource Materials

Lesson Support
Lesson Plan 10.2
Warm-Up Transparency 10.2
Practice Bank: Practice 67
Study Guide: Section 10.2
Explorations Lab Manual: Additional Exploration 16
Assessment Book: Test 43

Technology
Graphing Calculator
McDougal Littell Software *Function Investigator*
Internet: http://www.hmco.com

Warm-Up Exercises

Simplify.

1. $(7z - 2y - 4) + (2z - 7 + 5y)$ $9z + 3y - 11$
2. $(3x)(4x)$ $12x^2$
3. $3(4x - 1)$ $12x - 3$
4. $6x^2 - 8x^2$ $-2x^2$
5. $x(3y)$ $3xy$
6. $-4(3x - 2)$ $-12x + 8$

Teach⟺Interact

Section Note

Using Manipulatives
Manipulative activities, such as the paper-folding activity in the Exploration, give concrete meaning to abstract ideas. In this case, they also relate geometric concepts that students understand, namely, area and volume, to the concept of a polynomial. Thus, this approach helps to build mathematical intuition and understanding that forms a foundation for the purely algebraic approach that follows in Example 1 on the next page.

BY THE WAY

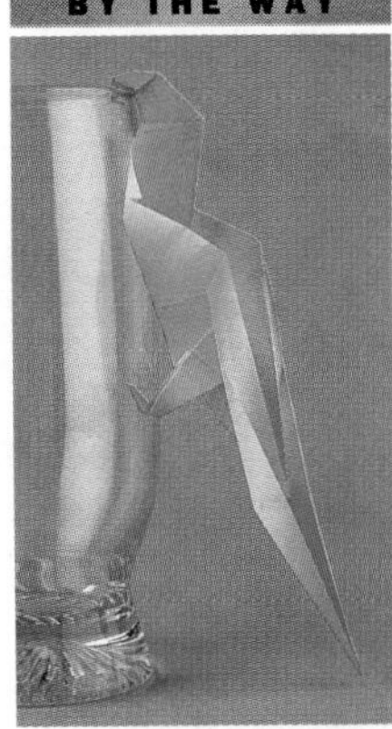

Origami flourished in South America as well as in Japan. This model of a parrot perched on a glass was designed by Ligia Montoya of Argentina.

Exploration *continued*

2 Hold your paper with the longer side facing you and fold the outer flaps back in as shown.

3 Fold in the four corners so the edges line up with the horizontal fold lines as shown.

4 Fold back two vertical strips in the middle, one to the left and the other to the right, to "lock" the corners.

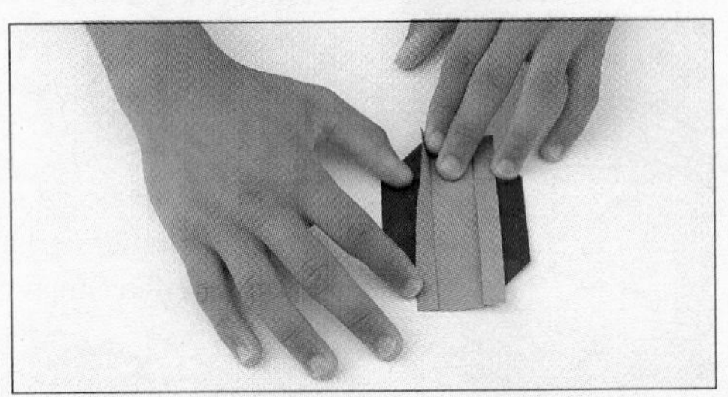

5 Lift up gently as shown to form your box.

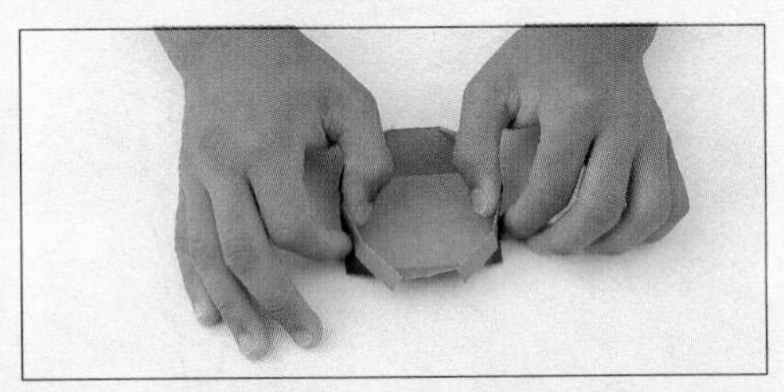

6 Unfold your box as shown. Measure the length and the width of the original piece of paper. Also measure the width of the vertical strips that you folded back.

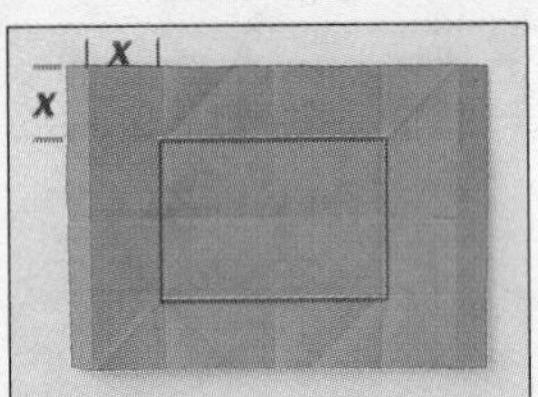

7 What do x and the region outlined in red represent in terms of the box?

8 Use your measurements from Step 6 to write expressions for the length, width, and height of the box in terms of x.

9 Compare expressions with those of the members of your group. How are the expressions alike? How are they different?

ANSWERS Section 10.2

Exploration

1–6. Check students' work.

7. x represents the height of the box. The rectangle outlined in red is the base of the box.

8. Answers may vary. An example is given. For an $8\frac{1}{2}$ in. by 11 in. rectangle, the length is $11 - 2x - 2\left(\frac{5}{8}\right) = 9.75 - 2x$; the width is $8\frac{1}{2} - 2x = 8.5 - 2x$, and the height is x.

9. Answers may vary.

EXAMPLE 1 Application: Crafts

Manuel used a 6 in. by 4 in. piece of paper to do the Exploration on pages 429 and 430. His group expressed the dimensions of the folded box as shown.

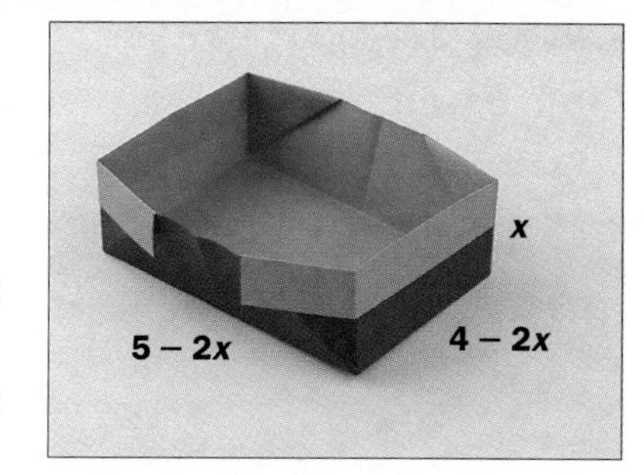

a. Write a polynomial to represent the area of the base of the box.

b. Write a polynomial to represent the volume of the box.

SOLUTION

a. Area of base = Length × Width = $(5 - 2x)(4 - 2x)$

You can use the *FOIL* method to multiply binomials. *FOIL* stands for "**First terms** in the binomials," "**Outer terms** in the expression," "**Inner terms** in the expression," and "**Last terms** in the binomials." You will get the same result as if you used the distributive property.

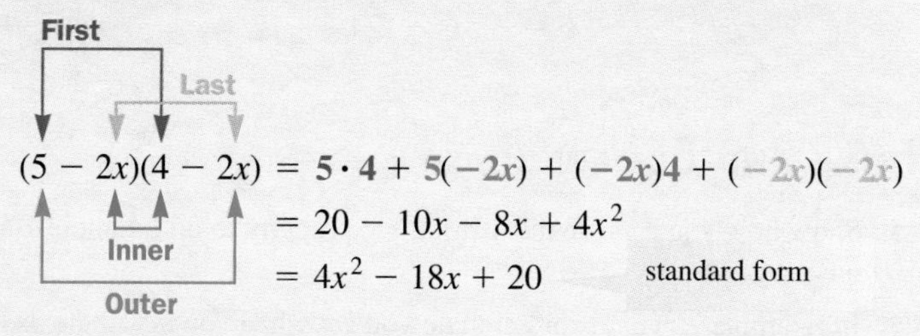

$$(5 - 2x)(4 - 2x) = 5 \cdot 4 + 5(-2x) + (-2x)4 + (-2x)(-2x)$$
$$= 20 - 10x - 8x + 4x^2$$
$$= 4x^2 - 18x + 20 \quad \text{standard form}$$

b. Volume = Area of base × Height

$$= (4x^2 - 18x + 20)x \quad \text{Use the area of the base from part (a).}$$
$$= 4x^2 \cdot x - 18x \cdot x + 20 \cdot x$$
$$= 4x^3 - 18x^2 + 20x \quad \text{Use the distributive property.}$$

You saw on page 259 that area models can be used to represent the product of two binomials. Volume models can be used to represent the product of three binomials, as shown below. The length of each side of the cube is represented by the expression $x + 1$.

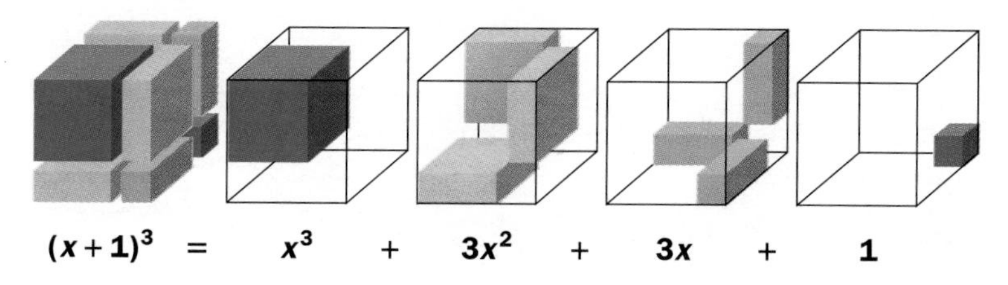

$$(x + 1)^3 = x^3 + 3x^2 + 3x + 1$$

Teach⇔Interact

About Example 1

Communication: Discussion
Point out that the *FOIL* method is a quick way to multiply binomials and only binomials. The distributive property can be used to multiply any two polynomials. Part (b) uses the general volume formula for a prism. This formula applies to a prism of any shape.

Visual Thinking
Ask students to select a pair of binomials and create their own diagram demonstrating how to multiply the pair using the *FOIL* method. They might use color, arrows, dotted lines, and other devices to show the order of operations. Encourage them to share their models with the class. This activity involves the visual skills of *generalization* and *communication.*

Additional Example 1

Katerina used an 8 in. by 11 in. piece of paper to do the Exploration on pages 429 and 430. Her group expressed the dimensions of the folded box as shown.

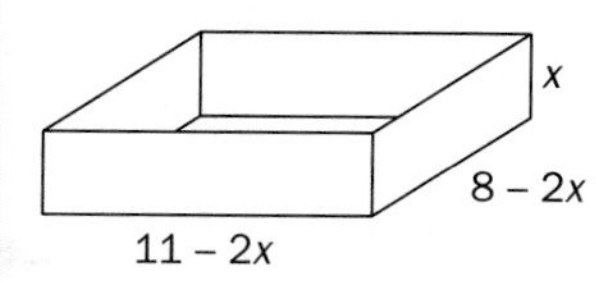

a. Write a polynomial to represent the area of the base of the box.

Area of base = Length × Width = $(8 - 2x)(11 - 2x)$

Use the *FOIL* method to multiply binomials.

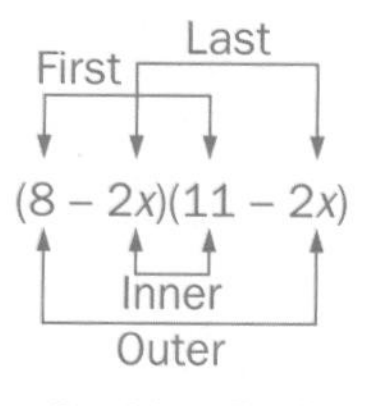

$= 8 \cdot 11 + 8(-2x) + (-2x)11 + (-2x)(-2x)$
$= 88 - 16x - 22x + 4x^2$
$= 4x^2 - 38x + 88$

b. Write a polynomial to represent the volume of the box.

Use the area of the base from part (a).

Volume = Area of base × Height
$= (4x^2 - 38x + 88)x$
$= 4x^2 \cdot x - 38x \cdot x + 88 \cdot x$
$= 4x^3 - 38x^2 + 88x$

Teach⇔Interact

Additional Example 2

Multiply $(2x + 4)(3x^2 - x - 5)$.
Use the distributive property.

Method 1 Multiply horizontally.
$(2x + 4)(3x^2 - x - 5)$
$= 2x(3x^2 - x - 5) + 4(3x^2 - x - 5)$
$= 6x^3 - 2x^2 - 10x + 12x^2 - 4x - 20$
$= 6x^3 + 10x^2 - 14x - 20$

Method 2 Multiply vertically.

$$\begin{array}{r} 3x^2 - x - 5 \\ \underline{2x + 4} \\ 6x^3 - 2x^2 - 10x \quad\quad \\ \underline{12x^2 - 4x - 20} \\ 6x^3 + 10x^2 - 14x - 20 \end{array}$$

Think and Communicate

Question 1 helps students to relate the distributive property to the *FOIL* method.

Checking Key Concepts

Common Error
Various kinds of errors can occur when students multiply polynomials. For example, they can make sign errors, exponent errors, or errors in combining like terms. Actual samples of students' work can be used to illustrate and correct each type of error.

Closure Question

Explain how to multiply two binomials by using the *FOIL* method and the distributive property.
To use the *FOIL* method, multiply the first terms of each binomial, the outer terms, the inner terms, and then the last terms. Add like terms to simplify. To use the distributive property, multiply the entire second expression by each of the terms in the first expression. Then add like terms to simplify.

EXAMPLE 2

Multiply $(3x - 2)(2x^2 + 3x - 6)$.

SOLUTION

Use the distributive property.

Method 1 Multiply horizontally.

WATCH OUT! Be sure to distribute the negative sign with the 2.

$$\begin{aligned}(3x-2)(2x^2+3x-6) &= 3x(2x^2+3x-6) - 2(2x^2+3x-6) \\ &= 6x^3 + 9x^2 - 18x - 4x^2 - 6x + 12 \\ &= 6x^3 + 5x^2 - 24x + 12\end{aligned}$$

Method 2 Multiply vertically.

$$\begin{array}{r} 2x^2 + 3x - 6 \\ \underline{3x - 2} \\ 6x^3 + 9x^2 - 18x \quad\quad \\ \underline{-4x^2 - 6x + 12} \\ 6x^3 + 5x^2 - 24x + 12 \end{array}$$

Multiply each term by $3x$: $3x \cdot 2x^2 = 6x^3$.

Multiply each term by -2. Line up like terms.

Add.

THINK AND COMMUNICATE

1. Show how you can use the distributive property to do Example 1(a) on page 431.
2. In Example 2, what expression do you get when you distribute $3x$? when you distribute -2?
3. Why do you think a polynomial with degree three is called a *cubic* polynomial? (*Hint:* Look at the diagram at the bottom of page 431.)

CHECKING KEY CONCEPTS

Identify the *First*, *Outer*, *Inner*, and *Last* terms of each expression.

1. $(x + 4)(x + 8)$
2. $(a + 7)(a - 9)$
3. $(2n + 4)(n - 8)$

Replace each ? with the correct expression.

4. $8a(4a^2 + 5a - 6) = 8a(\underline{?}) + 8a(\underline{?}) - 8a(\underline{?})$
5. $(a + 4b)(a + 3b) = a^2 + 3ab + \underline{?} + 12b^2$
6. $(2x^2 - 5x + 1)(4x + 5) = (\underline{?})4x + (\underline{?})5$

Multiply.

7. $(m + 4)(m + 5)$
8. $(2z - 1)(z - 7)$
9. $k(k^2 + 2k - 24)$
10. $2x(3x^2 - 5x + 9)$
11. $(y - 1)(3y^2 - 8y - 4)$
12. $(2a + 6)(2a^2 - 3a + 9)$

Think and Communicate

1. $(5 - 2x)(4 - 2x) = 5(4 - 2x) - 2x(4 - 2x) = 20 - 10x - 8x + 4x^2 = 4x^2 - 18x + 20$
2. $6x^3 + 9x^2 - 18x$; $-4x^2 - 6x + 12$
3. The volume of a cube with side length x is x^3.

Checking Key Concepts

1. $x, x; x, 8; 4, x; 4, 8$
2. $a, a; a, -9; 7, a; 7, -9$
3. $2n, n; 2n, -8; 4, n; 4, -8$
4. $4a^2$; $5a$; 6
5. $4ab$
6. $2x^2 - 5x + 1$; $2x^2 - 5x + 1$
7. $m^2 + 9m + 20$
8. $2z^2 - 15z + 7$
9. $k^3 + 2k^2 - 24k$
10. $6x^3 - 10x^2 + 18x$
11. $3y^3 - 11y^2 + 4y + 4$
12. $4a^3 + 6a^2 + 54$

Exercises and Applications

1. $c^2 + 12c + 35$
2. $r^2 - 3r - 40$
3. $s^2 - 6s + 8$
4. $3t^2 - 10t - 8$
5. $12a^2 + 38a + 6$
6. $22m^2 + 43m - 30$
7. $n^3 + 10n$
8. $4p^3 - 9p$
9. $45b^3 + 30b$
10. $6x^3 - 3x^2 + 18x$
11. $-z^3 + 5z^2 - 4z$
12. $4y^3 + 2y^2 - y$
13. Answers may vary.
14–16. See answers in back of book.
17. Substituting $x = 12$ and $y = 3$ in the expressions for perimeter and area gives $P = 422$ and $A = 11{,}074$. Since $P = 2(98) + 2(113) = 422$ and $A = (98)(113) = 11{,}074$, the polynomials check.
18. Answers may vary. An example is given. Let $x = 11$ and $y = 3$. Make the border 5 in. on the sides and 10 in. on the top and bottom.

10.2 Exercises and Applications

Multiply.

1. $(c + 5)(c + 7)$
2. $(r + 5)(r - 8)$
3. $(s - 4)(s - 2)$
4. $(3t + 2)(t - 4)$
5. $(6a + 1)(2a + 6)$
6. $(11m - 6)(2m + 5)$
7. $n(n^2 + 10)$
8. $p(4p^2 - 9)$
9. $5b(9b^2 + 6)$
10. $3x(2x^2 - x + 6)$
11. $-z(z^2 - 5z + 4)$
12. $\frac{1}{2}y(8y^2 + 4y - 2)$
13. **Writing** Write a note to a friend explaining how to do exercises like Exercises 1–12. Include some worked-out examples.
14. Write each product in standard form. Then compare parts (a) and (b). What do you notice? Does the same apply to parts (c) and (d)? Explain.
    a. $(x + 2)(x + 3)$
    b. $(x + 2y)(x + 3y)$
    c. $(4a + 5)(a - 7)$
    d. $(4a + 5b)(a - 7b)$

Extra Practice exercises on page 577

Connection CRAFTS

When quilters design a quilt, they have to decide how wide to make the squares, the lattice, and the border so that the quilt fits nicely on a bed of a given size.

For the quilt pattern shown, suppose the width of each square is x, the width of the lattice is y, and the width of the border is 10 in. all around.

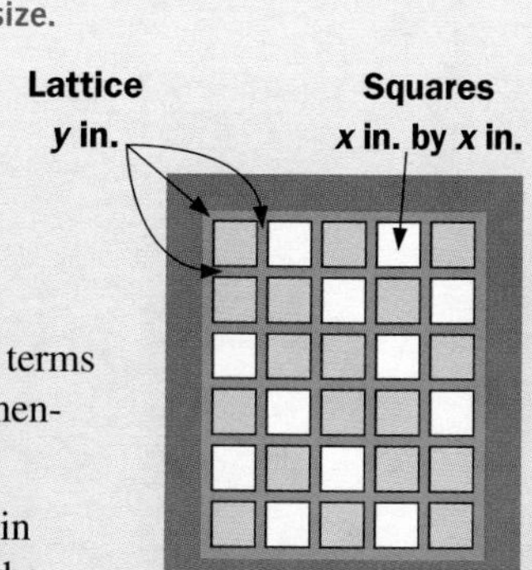

15. a. Write expressions in terms of x and y for the dimensions of the quilt.
    b. Write an expression in terms of x and y for the perimeter of the quilt. Do the same for the area of the quilt.
16. Suppose you want your quilt to be 98 in. wide by 113 in. long (king-size). Write and solve a system of equations to find the widths of the squares and the lattice used in the quilt pattern.
17. Use your answer from Exercise 16 to check the polynomials you wrote in Exercise 15(b).
18. **Open-ended Problem** Suppose you want your quilt to be 83 in. by 107 in. (queen-size). How might you adjust the pattern to fit?

Technology Note

Students can use a TI-82 or TI-81 graphing calculator to check their results after multiplying two polynomials. Consider, for example, Exercise 5. Enter Y1 = (6X + 1)(2X + 6) on the Y= list. (Notice that the variable *a* has been changes to *x*.) Next, enter an equation for the product Y2 = 12X² + 38X + 6. Use an appropriate window to display the graphs. Since $(6x + 1)(2x + 6)$ and $12x^2 + 38x + 6$ are indeed equal, the graphs will be identical.

If an incorrect product, such as $12x^2 + 14x + 6$, had been used for Y2, the graphs will be different.

A similar procedure can be used to check answers with the *Function Investigator* software.

These ideas will be used in the Exploration on page 436.

Apply⇔Assess

Suggested Assignment

Core Course
Day 1 Exs. 1–12, 14–17, 19–24
Day 2 Exs. 25–29, 32–43, AYP

Extended Course
Day 1 Exs. 1, 2, 4, 5, 7, 8, 10–24
Day 2 Exs. 25–43, AYP

Block Schedule
Day 63 Exs. 1–12, 14–17, 19–24
Day 64 Exs. 25–29, 32–43, AYP

Exercise Notes

Mathematical Procedures
Exs. 1–6 Most students will be able to remember the *FOIL* method and apply it correctly. A review of the answers to these exercises would help students to identify and correct any errors.

Application
Exs. 15–18 These exercises illustrate how algebra can be applied to a situation—the design of a quilt—that may not at first appear to be related in any way to mathematics. However, quilt design involves geometric squares which have dimensions that can be expressed in terms of variables. Perimeters and areas can then be calculated as polynomials in standard form.

Multicultural Note
Exs. 15–18 Among tribes in the Sudan and Niger, quilts are used as protection against enemy weapons. Several layers of cotton cloth are stitched together to make quilted armor for both warriors and their horses. Designs are achieved by sewing together patches of different colored cloth.

Apply⇔Assess

Exercise Notes

Integrating the Strands
Exs. 19–21 Although the dimensions of these figures are represented by variable expressions, remind students that the area formulas that they are familiar with when the dimensions are numbers still apply here.

Communication: Writing
Ex. 30 Students should be able to illustrate their explanations by using examples. In fact, an analysis of a few examples is an excellent way to begin thinking about how to write an explanation.

Challenge
Ex. 31 Students may attempt to do this exercise either by a double use of the distibutive property or by using the *FOIL* method first, followed by the distributive property. A demonstration of both solutions will allow students to compare and contrast these two methods.

Cooperative Learning
Exs. 32–34 In these exercises, students make a box and explore the volume function associated with it. If any group appears to be having difficulties, you may wish to provide hints as to how to proceed with these exercises or have the group confer with another, more successful group.

Using Technology
Ex. 35 You may want to discuss a possible domain for the volume function graph. Only first quadrant values make sense in this application. On the TI-82, use the CALC:maximum option to find the value of x that gives the maximum volume.

GEOMETRY Write a polynomial in standard form to represent the area of each figure.

19.
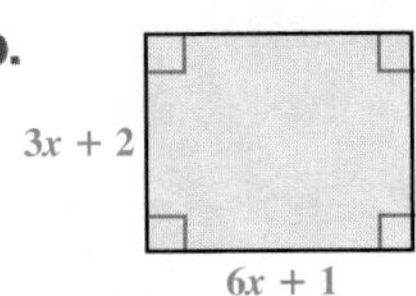

20.
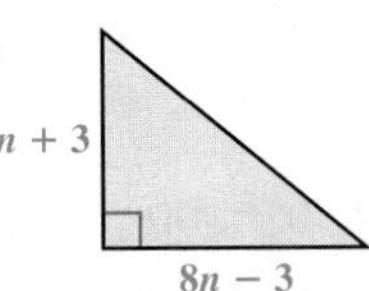

21.
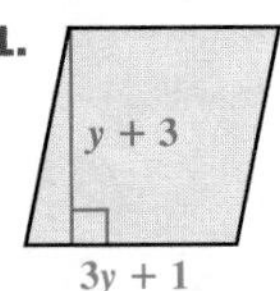

Multiply.

22. $(x + 4)(x^2 + 2x - 5)$
23. $(n - 3)(6n^2 + 8n + 9)$
24. $(3x + 2)(x^2 + 7x - 6)$
25. $(5x - 1)(x^2 - 3x - 10)$
26. $(2x^2 - 5x + 10)(x + 4)$
27. $(7j^2 - j + 4)(6j - 1)$
28. $(3t^2 - 2t - 4)(5t + 9)$
29. $(2x^2 + 9x - 4)(3x - 2)$

30. Writing Explain why the product of a quadratic polynomial and a linear polynomial must be a cubic polynomial.

31. Challenge Show how to multiply $(2y - 3)^3$.

Cooperative Learning Work with a partner. In Exercises 32–35, you will each make a box with an attached lid and explore its volume.

32. Each of you will need a 6 in. by 4 in. piece of paper. Follow the steps below to form two boxes, using two different values for x.

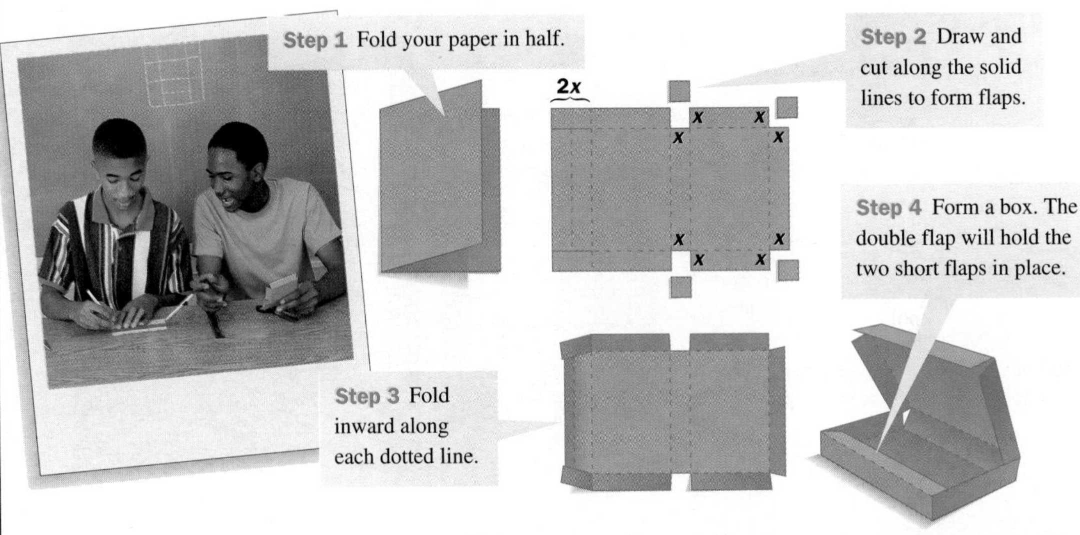

Step 1 Fold your paper in half.
Step 2 Draw and cut along the solid lines to form flaps.
2x
x
Step 3 Fold inward along each dotted line.
Step 4 Form a box. The double flap will hold the two short flaps in place.

33. Write polynomials in terms of x to represent the length, width, and height of your box. Compare results with your partner.

34. Write a function in terms of x to represent the volume of your box. Write the function as a polynomial in standard form. What is a reasonable domain for the function? Compare results with your partner.

35. Technology Graph your volume function from Exercise 34 on a graphing calculator. Find the value of x that gives the maximum volume.

19. $18x^2 + 15x + 2$
20. $4n^2 + 10.5n - 4.5$
21. $3y^2 + 10y + 3$
22. $x^3 + 6x^2 + 3x - 20$
23. $6n^3 - 10n^2 - 15n - 27$
24. $3x^3 + 23x^2 - 4x - 12$
25. $5x^3 - 16x^2 - 47x + 10$
26. $2x^3 + 3x^2 - 10x + 40$
27. $42j^3 - 13j^2 + 25j - 4$
28. $15t^3 + 17t^2 - 38t - 36$
29. $6x^3 + 23x^2 - 30x + 8$
30. Answers may vary. An example is given. The degree of the linear polynomial is one and the degree of the quadratic polynomial is two. The degree of the product of these two polynomials will be three, which is the degree of a cubic polynomial.
31. $(2y - 3)^3 = (2y - 3)(2y - 3)^2 = (2y - 3)(4y^2 - 12y + 9) = 8y^3 - 36y^2 + 54y - 27$
32. Check students' work.
33. length = $(4 - 2x)$ in.; width = $(3 - 2x)$ in.; height = x in.
34. $V = 4x^3 - 14x^2 + 12x$; $0 < x < 1.5$
35. $y = 4x^3 - 14x^2 + 12x$; The maximum volume, y, is created when x is about 0.57 in.

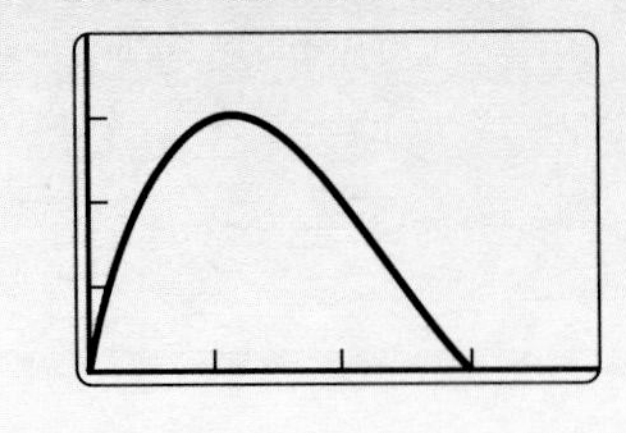

36. A
37. $8x^3 + 14x^2 + 23x + 5$; 9635; The process is similar. Each term is multiplied by each term in the polynomials and the results are added. Each digit is multiplied by each digit in the two numbers and the results are added. There is no carrying involved when multiplying polynomials. The result of the numbers being multiplied can be found using the corresponding polynomial multiplication if 10 is substituted for x.
38. (–1, 3)

36. SAT/ACT Preview If A = $(x + 1)^2$ and B = $(x + 3)(x - 1)$, then:

A. A > B **B.** B > A **C.** A = B

D. relationship cannot be determined

ONGOING ASSESSMENT

37. Writing Use a vertical multiplication method to find the products $(4x + 1)(2x^2 + 3x + 5)$ and $41 \cdot 235$. How is multiplying polynomials like multiplying numbers? How is it different?

SPIRAL REVIEW

For each system of equations, give the solution or state whether there are *no solutions* or *many solutions*. *(Section 7.4)*

38. $3x + 2y = 3$
$4x - y = -7$

39. $-2 = 5m - 4n$
$5 = 7m - 3n$

40. $2x - 3y = -18$
$6x - 9y = 18$

Predict how the graph of each equation will compare with the graph of $y = x^2$. Explain how you made your prediction. *(Section 8.2)*

41. $y = -4x^2$ **42.** $y = \frac{3}{4}x^2$ **43.** $y = -1.2x^2$

ASSESS YOUR PROGRESS

VOCABULARY

polynomial (p. 423)
standard form of a polynomial (p. 423)
degree of a term (p. 423)
degree of a polynomial (p. 423)
linear polynomial (p. 423)
quadratic polynomial (p. 423)
cubic polynomial (p. 423)

1. Is $x - \frac{1}{x}$ a polynomial? Explain your answer. *(Section 10.1)*

Add or subtract each pair of polynomials as indicated. Write each answer in standard form and give its degree.

2. $(3x^2 - 4x + 2) + (2x^2 - x - 5)$ **3.** $(5y^2 + 7y - 9) + (-8y^2 + 6)$

4. $(6a^2 + a - 1) - (6a^2 - a - 1)$ **5.** $(z^3 + 2z - 1) - (2z^2 - 5z - 4)$

Multiply. *(Section 10.2)*

6. $(4y - 7)(2y - 5)$ **7.** $7n(n^2 + 3n - 8)$ **8.** $(5j + 3)(j^2 - 4j + 6)$

9. Show two different ways to find a polynomial that represents the area of the border of the square textile at the right. Then write the polynomial in standard form.

10. Journal What do you find most tricky about multiplying two polynomials such as $7k - 4$ and $3k^2 - k + 5$? Can you think of a way to avoid making mistakes with the part you find tricky?

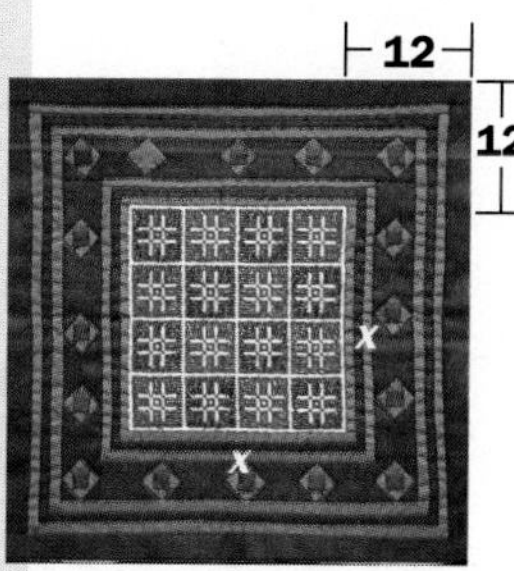

Hmong textile

Apply⇔Assess

Exercise Notes

Assessment Note
Ex. 37 Asking students to compare polynomial multiplication with number multiplication may help them understand how to multiply vertically.

Second-Language Learners
Ex. 37 Most second-language learners should be able to complete this writing task independently. Some might need help from a peer tutor to correctly phrase sentences comparing and contrasting the processes.

Assess Your Progress

Journal Entry
This exercise will allow students insight into what they find difficult about multiplying polynomials.

Progress Check 10.1–10.2

See page 456.

Practice 67 for Section 10.2

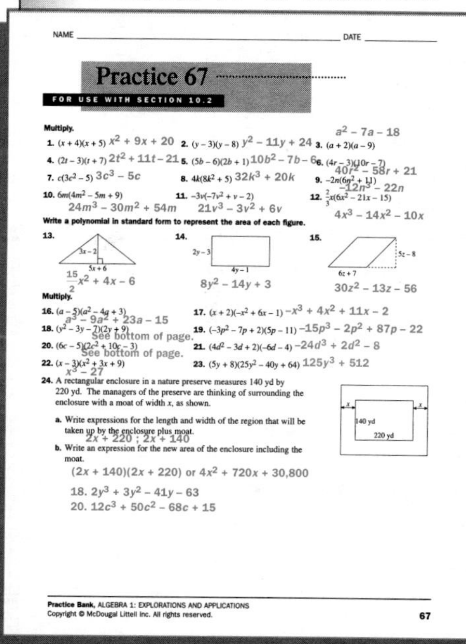

NAME ____________ DATE ________

Practice 67

FOR USE WITH SECTION 10.2

Multiply.

1. $(x + 4)(x + 5)$ $x^2 + 9x + 20$
2. $(y - 3)(y - 8)$ $y^2 - 11y + 24$
3. $(a + 2)(a - 9)$ $a^2 - 7a - 18$
4. $(2t - 3)(t + 7)$ $2t^2 + 11t - 21$
5. $(5b - 6)(2b + 1)$ $10b^2 - 7b - 6$
6. $(4r - 3)(10r - 7)$ $40r^2 - 58r + 21$
7. $c(3c^2 - 5)$ $3c^3 - 5c$
8. $4k(8k^2 + 5)$ $32k^3 + 20k$
9. $-2n(6n^2 + 11)$ $-12n^3 - 22n$
10. $6m(4m^2 - 5m + 9)$ $24m^3 - 30m^2 + 54m$
11. $-3v(-7v^2 + v - 2)$ $21v^3 - 3v^2 + 6v$
12. $\frac{2}{3}x(6x^2 - 21x - 15)$ $4x^3 - 14x^2 - 10x$

Write a polynomial in standard form to represent the area of each figure.

13. $3x - 2$; $5x + 6$ — $\frac{15}{2}x^2 + 4x - 6$
14. $2y - 3$; $4y - 1$ — $8y^2 - 14y + 3$
15. $5z - 8$; $6z + 7$ — $30z^2 - 13z - 56$

Multiply.

16. $(a - 5)(a^2 - 4a + 3)$ $a^3 - 9a^2 + 23a - 15$
17. $(x + 2)(-x^2 + 6x - 1)$ $-x^3 + 4x^2 + 11x - 2$
18. $(y^2 - 3y - 7)(2y + 9)$ See bottom of page.
19. $(-3p^2 - 7p + 2)(5p - 11)$ $-15p^3 - 2p^2 + 87p - 22$
20. $(6c - 5)(2c^2 + 10c - 3)$ See bottom of page.
21. $(4d^2 - 3d + 2)(-6d - 4)$ $-24d^3 + 2d^2 - 8$
22. $(x - 3)(x^2 + 3x + 9)$ $x^3 - 27$
23. $(5y + 8)(25y^2 - 40y + 64)$ $125y^3 + 512$

24. A rectangular enclosure in a nature preserve measures 140 yd by 220 yd. The managers of the preserve are thinking of surrounding the enclosure with a moat of width x, as shown.

a. Write expressions for the length and width of the region that will be taken up by the enclosure plus moat. $2x + 220$; $2x + 140$

b. Write an expression for the new area of the enclosure including the moat.
$(2x + 140)(2x + 220)$ or $4x^2 + 720x + 30,800$

18. $2y^3 + 3y^2 - 41y - 63$
20. $12c^3 + 50c^2 - 68c + 15$

Practice Bank, ALGEBRA 1: EXPLORATIONS AND APPLICATIONS

67

---

39. (2, 3)

40. no solutions

41. The graph will open down and be narrower than the graph of $y = x^2$. The prediction results from the fact that $-4 < 0$ and $|-4| > 1$.

42. The graph will open up but be wider than the graph of $y = x^2$. The prediction results from the fact that $\frac{3}{4} > 0$ and $\left|\frac{3}{4}\right| < 1$.

43. The graph will open down and be narrower than the graph of $y = x^2$. The prediction results from the fact that $-1.2 < 0$ and $|-1.2| > 1$.

Assess Your Progress

1. No. The term $\frac{-1}{x}$, or $-1x^{-1}$, is not a monomial with a whole-number exponent.

2. $5x^2 - 5x - 3$; 2

3. $-3y^2 + 7y - 3$; 2

4. $2a$; 1

5. $z^3 - 2z^2 + 7z + 3$; 3

6. $8y^2 - 34y + 35$

7. $7n^3 + 21n^2 - 56n$

8. $5j^3 - 17j^2 + 18j + 18$

9. Methods may vary. Examples are given. Subtract the area of the center square from the total area: $(x + 24)^2 - x^2 = 48x + 576$. Divide the border into eight parts—four that are 12 by x and four that are 12 by 12. The add the areas: $4(12x) + 4(12)^2 = 48x + 576$.

10. Answers may vary.

Plan⇔Support

Objectives

- Factor out a linear factor from a polynomial.
- Use the zero-product property.
- Use factoring to solve equations.

Recommended Pacing

Core and Extended Courses
Section 10.3: 2 days

Block Schedule
Section 10.3: 2 half-blocks
(with Sections 10.2 and 10.4)

Resource Materials

Lesson Support
Lesson Plan 10.3
Warm-Up Transparency 10.3
Practice Bank: Practice 68
Study Guide: Section 10.3

Technology
Technology Book:
Calculator Activity 10
Spreadsheet Activity 10
Graphing Calculator
Internet:
http://www.hmco.com

Warm-Up Exercises

1. Write $12x^2 - 3x^3 - 8 + 8x$ in standard form. $-3x^3 + 12x^2 + 8x - 8$

Multiply.

2. $x(x - 5)$ $x^2 - 5x$
3. $x(12x^2 + x + 1)$ $12x^3 + x^2 + x$
4. $(x + 2)(x - 5)$ $x^2 - 3x - 10$
5. Solve $4x - 8 = 0$. 2
6. Find the greatest common factor of 24 and 56. 8

SECTION

10.3 Exploring Polynomial Equations

Learn how to...
- factor out a linear factor from a polynomial
- use the zero-product property

So you can...
- begin to use factoring to solve equations

A **polynomial equation** is an equation in which both sides are polynomials. In Chapter 8 you learned how to use a graph to visualize the solutions of a polynomial equation. Another way that you can easily identify the solutions involves *factoring*.

EXPLORATION
COOPERATIVE LEARNING

Factors, Intercepts, and Solutions

Work with a partner.
You will need:
- graphing calculators

1 One of you should graph equations 1–4 while the other graphs equations A–D. Match the equations that have the same graph.

STANDARD FORM	FACTORED FORM
1. $y = x^2 + 2x - 3$	**A.** $y = x(x - 4)$
2. $y = x^3 - 12x^2 + 44x - 48$	**B.** $y = (x + 3)(x - 1)$
3. $y = x^2 - 4x$	**C.** $y = x(x + 3)(x - 1)$
4. $y = x^3 + 4x^2 + 3x$	**D.** $y = (x - 2)(x - 4)(x - 6)$

2 Find the x-intercepts of each graph in Step 1. Then look at the factored form of the equation for each graph. How are the x-intercepts and the factored form related?

3 Suppose you want to solve the polynomial equation $x^2 - 6x = 0$. How can the graph of $y = x^2 - 6x$ help you solve the equation? How can the factored form $x(x - 6) = 0$ help you solve the equation?

Exploration Note

Purpose
The purpose of this Exploration is to have students discover the relationship between the factored form of a polynomial equation and the x-intercepts of its graph in terms of finding solutions when the equation equals 0.

Materials/Preparation
Students should have a graphing calculator.

Procedure
Students graph polynomials in standard form and in factored form. They then match the equations that have the same graph. Students look at the factored form of an equation and tell how the x-intercepts of the graph are related to the factors. Have students use the Zoom 8:ZInteger feature for each graph so that tracing gives exact values for the x-intercepts.

Closure
Students should realize the x-intercepts of the graph of the polynomial function are the numbers that make each factor of the factored form equal to 0 and that these numbers represent the solutions of the equation when it equals 0.

Explorations Lab Manual
See the Manual for more commentary on this Exploration.

For answers to the Exploration, see answers in back of book.

Rewriting a polynomial as a product of polynomial factors is called **factoring a polynomial**. For example:

standard form $\quad x^3 + 2x^2 - 3x = x(x+3)(x-1) \quad$ factored form

EXAMPLE 1

Factor $x^2 + 2x$.

SOLUTION

Method 1 Use algebra tiles.

Arrange one x^2-tile and two x-tiles to form a rectangle. The total area of the rectangle is $x^2 + 2x$. The dimensions of the rectangle will be the factors.

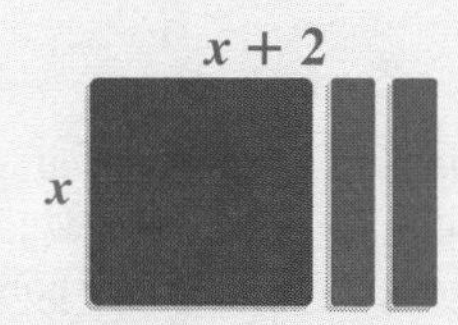

$$\begin{aligned} \text{Area} &= \text{Length} \times \text{Width} \\ x^2 + 2x &= (x+2) \cdot x \\ &= x(x+2) \end{aligned}$$

Method 2 Factor algebraically.

Identify the greatest common factor (GCF) and write each term as a product.

Use the distributive property.

$$\begin{aligned} x^2 + 2x &= x \cdot x + 2 \cdot x \\ &= x(x+2) \end{aligned}$$

The GCF is x.

THINK AND COMMUNICATE

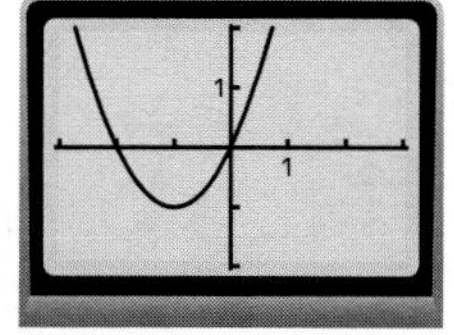

1. a. Use the graph of $y = x^2 + 2x$ at the right. Explain the relationship between the factors of $x^2 + 2x$, found in Example 1, and the x-intercepts of the graph.
   b. How can you use the graph to solve the equation $x^2 + 2x = 0$?
2. Replace each $\underline{?}$ with a monomial to factor each polynomial.
   a. $2x^3 - 6x^2 = 2x^2(\underline{?} - \underline{?})$ b. $6n^3 + 3n^2 + 9n = \underline{?}(2n^2 + n + 3)$

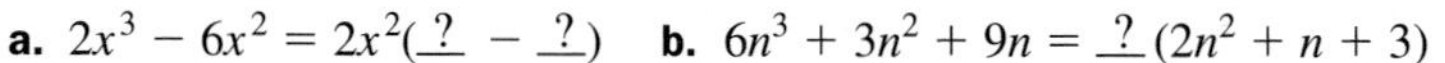

The property below can help you solve equations in factored form. It is based on the fact that zero times any number is zero. For example, $0 \cdot 3 = 0$ and $3 \cdot 0 = 0$.

Zero-Product Property

If a product of factors is equal to zero, then one or more of the factors is equal to zero.

Technology Note

Point out to students that if they want to solve a polynomial equation with a graphing calculator, there are several different options after graphing the related polynomial.

(1) Use the root key.

(2) Use the trace and zoom features to find the x-intercepts.

(3) Also graph $y = 0$ and use the intersect key.

Teach⇔Interact

Learning Styles: Visual

The use of a graphing calculator helps students to connect the x-intercepts of the graph of a polynomial equation with the factors of the equation. This approach provides visual learners with a clear mental image of a solution and motivates the significance of being able to factor a polynomial.

About Example 1

Teaching Tip

When discussing Method 2 of the Solution, a brief review of both the concept of a GCF and the distributive property may be necessary for some students.

Additional Example 1

Factor $x^2 + 4x$.

Method 1 Use algebra tiles. Arrange one x^2-tile and four x-tiles to form a rectangle. The total area of the rectangle is $x^2 + 4x$. The dimensions of the rectangle will be the factors.

x + 4

x

Area = Length × Width

$x^2 + 4x = (x + 4) \cdot x$

$= x(x + 4)$

Method 2 Factor algebraically. Identify the greatest common factor (GCF) and write each term as a product. Then use the distributive property. The GCF is x.

$x^2 + 4x = x \cdot x + 4 \cdot x$

$= x(x + 4)$

Think and Communicate

Question 2 can be used to review and reinforce the meaning of GCF and how to find it.

For answers to Think and Communicate, see following page.

Section Note

Visual Thinking

After students have discussed Example 2 on the next page, ask them to sketch the graph of the equation used in the Example. Ask them to explain the *zero-product property* using their sketches. This activity involves the visual skills of *interpretation* and *communication.*

Teach⇔Interact

Additional Example 2

Solve $(3x + 1)(x - 7) = 0$.

Use the zero-product property. One of the factors must equal zero.

$(3x + 1)(x - 7) = 0$

$3x + 1 = 0$ or $x - 7 = 0$

$x = -\frac{1}{3}$ or $x = 7$

The solutions are $-\frac{1}{3}$ and 7.

Additional Example 3

As part of a science project, Erin launched a small toy rocket straight up into the air. The equation $h = -16t^2 + 28t$ can be used to model the height h in feet of the rocket t seconds after it is launched. According to the model, how long will it take for the rocket to hit the ground?

The rocket hits the ground when the height h is zero.

$h = -16t^2 + 28t$

$0 = -16t^2 + 28t$

$-1 \cdot 0 = -1(-16t^2 + 28t)$

$0 = 16t^2 - 28t$

$0 = 4t(4t - 7)$

$4t = 0$ or $4t - 7 = 0$

$t = 0$ $\quad t = \frac{7}{4} = 1.75$

It takes 1.75 seconds for the rocket to hit the ground.

EXAMPLE 2

Solve $(x + 3)(4x - 5) = 0$.

SOLUTION

Use the zero-product property. One of the factors must equal zero.

$$(x + 3)(4x - 5) = 0$$

$x + 3 = 0$ or $4x - 5 = 0$

$x = -3$ $\qquad x = \frac{5}{4} = 1.25$

The solutions are -3 and 1.25.

WATCH OUT!

Five is not a solution. When $x = 5$, $4x - 5 \neq 0$.

BY THE WAY

The dimples on golf balls have a lot to do with their performance. Other factors that affect the flight of a golf ball are air conditions, ball mass, ball diameter, spin, velocity, and launch angle.

EXAMPLE 3 Application: Physics

Many forces affect the height of a golf ball in flight. Suppose you ignore all forces except gravity. Then under certain initial conditions the equation $h = -16t^2 + 40t$ can be used to model the height h in feet of a golf ball t seconds after the ball has been hit. Using this model, about how long does it take for the golf ball to hit the ground if it is hit on level ground.

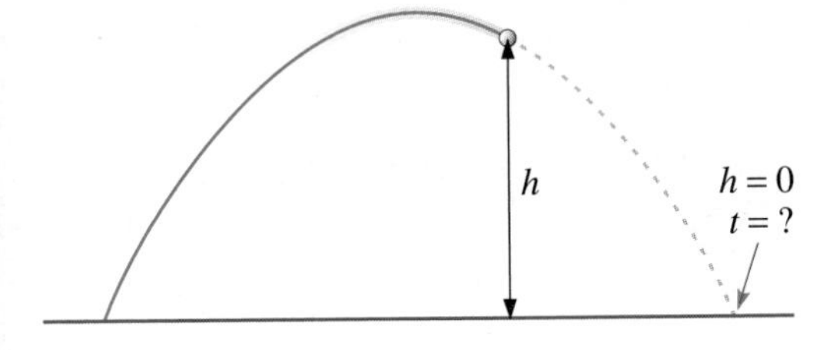

SOLUTION

The ball hits the ground when the height h is zero.

$h = -16t^2 + 40t$

Substitute **0** for h. $\quad 0 = -16t^2 + 40t$

$-1 \cdot 0 = -1(-16t^2 + 40t)$ $\quad$ Multiply both sides by **−1**.

$0 = 16t^2 - 40t$

$0 = 8t(2t - 5)$ $\quad$ Factor.

Use the zero-product property.

$8t = 0$ or $2t - 5 = 0$

$t = 0$ when the ball is hit. $\quad t = 0$ $\qquad t = \frac{5}{2} = 2.5$

It takes the golf ball about 2.5 s to hit the ground.

ANSWERS Section 10.3

Think and Communicate

1. a. The factors of $x^2 + 2x$ can be written as $(x - 0)$ and $(x - (-2))$. The x-intercepts are 0 and -2.
   b. The solutions are the same as the x-intercepts of the graph.
2. a. x; 3 $\quad$ b. $3n$
3. No. The product of the two factors $2x$ and $(x - 3)$ is 4, not 0.
4. No. If $x = 0$, then $2(x - 3) = -6$.
5. Use the zero-product property and solve the equations $x + 3 = 0$ and $9x + 45 = 0$.

Checking Key Concepts

1. x
2. $4n$
3. $2y$
4. $m(m + 5)$
5. $7x(2x + 1)$
6. $3y(2y - 3)$
7. $15n(2n - 3)$
8. $4(z^2 + 5z + 2)$
9. $3t(t^2 - 2t + 3)$
10. $-6, 4$
11. $\frac{2}{5}, -\frac{5}{2}$
12. 0, 2
13. $0, -\frac{3}{4}$

THINK AND COMMUNICATE

3. Can you solve the equation $2x(x - 3) = 4$ using the zero-product property without rewriting the equation first? Why or why not?
4. Is zero a solution of the equation $2(x - 3) = 0$? Why or why not?
5. What would you do to solve the equation $(x + 3)(9x - 45) = 0$?

CHECKING KEY CONCEPTS

Find the greatest common factor of each group of monomials.

1. $5x^2, 3x$
2. $8n^2, 12n$
3. $4y^3, 6y^2, 2y$

Factor each polynomial.

4. $m^2 + 5m$
5. $14x^2 + 7x$
6. $6y^2 - 9y$
7. $30n^2 - 45n$
8. $4z^2 + 20z + 8$
9. $3t^3 - 6t^2 + 9t$

Solve each equation.

10. $(x + 6)(x - 4) = 0$
11. $(5n - 2)(2n + 5) = 0$
12. $y^2 - 2y = 0$
13. $16k^2 + 12k = 0$

10.3 Exercises and Applications

Extra Practice exercises on page 577

Factor each polynomial.

1. $x^2 - 4x$
2. $2m^2 + 11m$
3. $4y^2 + 4$
4. $8j + 10$
5. $2x^2 + 6x$
6. $21k^2 - 3k$
7. $15n^2 + 10n$
8. $36v^2 - 30v$
9. $a^3 - 4a^2 + a$
10. $9t^3 + 6t^2 + 9t + 3$
11. $54x^3 + 18x^2 + 2x$
12. $12r^3 - 4r^2 - 28r$
13. **Visual Thinking** Which polynomial from Exercises 1–12 is modeled by the diagram at the right? How does the diagram show how to factor that polynomial?

14. **Writing** What do you find the most difficult about factoring polynomials? Why? How might you overcome this difficulty?

Solve each equation.

15. $x(x - 2) = 0$
16. $2t(t + 7) = 0$
17. $0 = 3n(n - 4)$
18. $k(5k + 6) = 0$
19. $0 = 7m(3m - 1)$
20. $(k + 3)(k - 3) = 0$
21. $(a + 4)(a - 7) = 0$
22. $(8y - 1)(y - 4) = 0$
23. $0 = (4z + 3)(2z + 1)$

Teach⇔Interact

Think and Communicate

When discussing question 3, ask students if they can use the zero-product property to solve $2x(x - 3) - 4 = 0$. Ask what they think has to be done to solve this equation.

Checking Key Concepts

Reasoning
For questions 4–9, ask students to explain what happens if the GCF is not used when factoring a polynomial. (A polynomial is not completely factored unless the greatest common factor is used.)

Closure Question

Describe how to solve a polynomial equation using the zero-product property.
Simplify the polynomial equation so that one side is 0. Identify the greatest common factor and factor the polynomial. Use the zero-product property to set each factor equal to zero and solve.

Apply⇔Assess

Suggested Assignment

Core Course
Day 1 Exs. 1–12, 15–23
Day 2 Exs. 24–36, 38–56

Extended Course
Day 1 Exs. 1–23
Day 2 Exs. 24–56

Block Schedule
Day 64 Exs. 1–12, 15–23
Day 65 Exs. 24–36, 38–56

Exercise Notes

Using Technology
Exs. 15–23 Students can verify their solutions by graphing the related equation of the form $y = ab$, where a and b are linear factors. The solutions should correspond to the x-intercepts of the graph.

Exercises and Applications

1. $x(x - 4)$
2. $m(2m + 11)$
3. $4(y^2 + 1)$
4. $2(4j + 5)$
5. $2x(x + 3)$
6. $3k(7k - 1)$
7. $5n(3n + 2)$
8. $2v(18v - 15)$
9. $a(a^2 - 4a + 1)$
10. $3(3t^3 + 2t^2 + 3t + 1)$
11. $2x(27x^2 + 9x + 1)$
12. $4r(3r^2 - r - 7)$
13. The tiles model $2x^2 + 6x$. The rectangle is $2x$ high and $x + 3$ wide, so $2x^2 + 6x = 2x(x + 3)$
14. Answers may vary.
15. 0, 2
16. 0, −7
17. 0, 4
18. $0, -\frac{6}{5}$
19. $0, \frac{1}{3}$
20. 3, −3
21. −4, 7
22. $\frac{1}{8}, 4$
23. $-\frac{3}{4}, -\frac{1}{2}$

Apply⇔Assess

Exercise Notes

Interdisciplinary Problems
Exs. 24–26 Economics provides a rational way to analyze business problems involving the variables of price and sales. The development of economics during the past fifty years has become highly dependent upon the creation and use of mathematical models to study and understand complex economic phenomena.

Second-Language Learners
Exs. 24–26 Second-language learners may benefit from a discussion of profit (and the difference between net sales and profit) before beginning this series of exercises.

Teaching Tip
Exs. 33–36 Remind students that they have to get zero on one side of the equation before trying to apply the zero-product property.

Connection BUSINESS

A team of economists must find the most profitable retail price for their company's new laptop computer. The equation $n = -100p^2 + 300{,}000p$ models the expected relationship between net sales n and the retail price p of the laptop computer, in dollars.

24. Show how to solve the equation $-100p^2 + 300{,}000p = 0$ by factoring. What information does the solution give you about the expected sales?

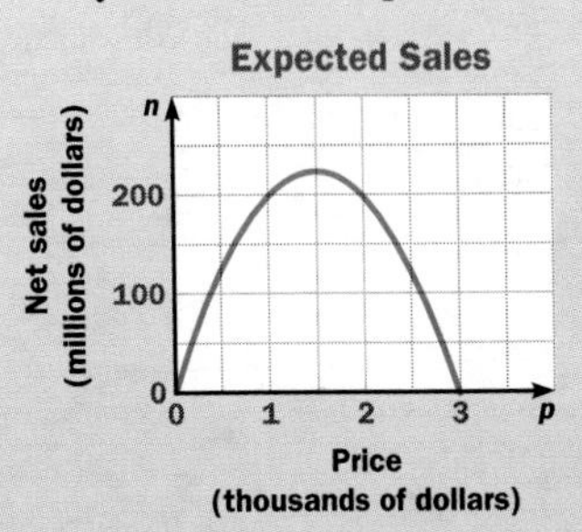

25. At the left is a graph of the equation $n = -100p^2 + 300{,}000p$. What does the graph tell you about the relationship between price and net sales?

26. What price should the team propose for the laptop computer? Explain your reasoning.

BY THE WAY

The Altair 8800 was the first personal computer widely available. The kit cost $397 when it was introduced in 1975, but that price did not include a terminal, floppy disks, or a printer.

27. Writing Jasmine and Patty solved the equation $5x^2 = 15x$. Who got the correct answer? What happens when you use the other student's method?

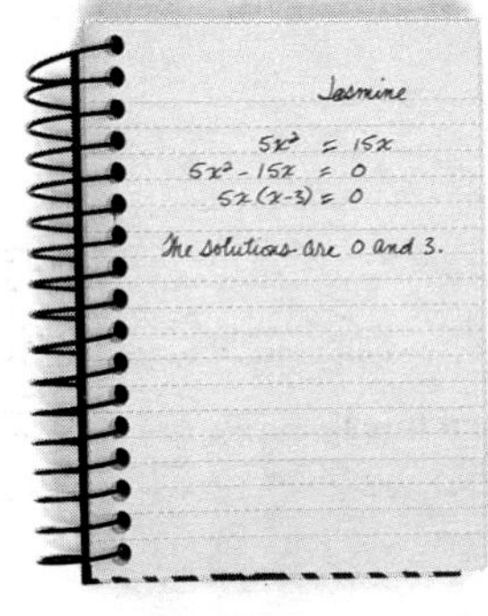

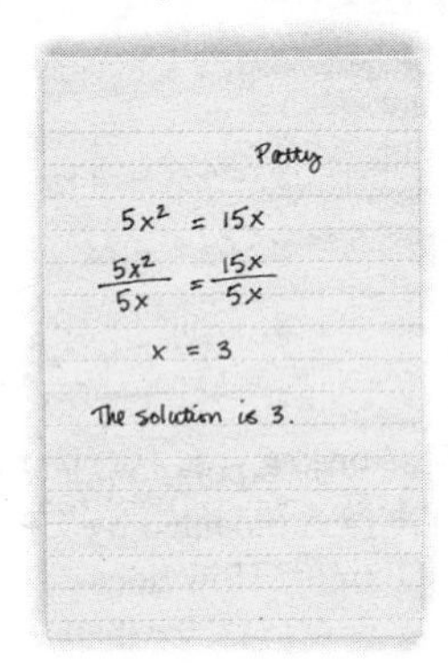

Solve each equation.

28. $r^2 + 10r = 0$

29. $3s^2 + 7s = 0$

30. $2x^2 - 18x = 0$

31. $0 = 3y^2 + 3y$

32. $4b^2 - 6b = 0$

33. $n^2 = 5n$

34. $14x^2 = -x$

35. $12z^2 = 24z$

36. $18p = -15p^2$

37. Open-ended Problem Find at least two different possible quadratic equations for the graph shown at the right. Explain why each is possible.

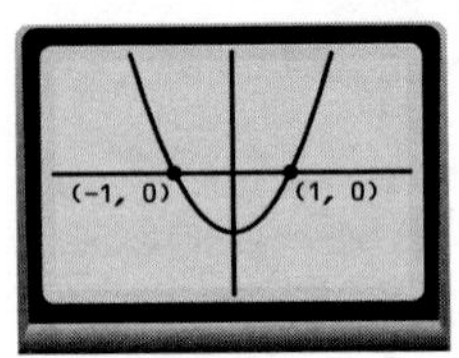

24. Factoring the left side gives $-100p(p - 3000) = 0$. The solutions are 0 and 3000. This tells you that net sales will be 0 if the price is \$0 or \$3000.

25. As the price increases from \$0 to \$1500, net sales increase from \$0 to \$225 million. As the price continues to increase to \$3000, net sales decrease back to \$0.

26. \$1500, because this yields the highest net sales

27. Jasmine is correct. When Patty divided both sides by $5x$, she lost the solution $x = 0$. Division by zero is not defined.

28. $0, -10$ 29. $0, -\frac{7}{3}$

30. $0, 9$ 31. $0, -1$

32. $0, 1.5$ 33. $0, 5$

34. $0, -\frac{1}{14}$ 35. $0, 2$

36. $0, -\frac{6}{5}$

37. Answers will vary. Examples are given. $(x + 1)(x - 1) = y$, or $4(x + 1)(x - 1) = y$; The x-intercepts are -1 and 1, which are the solutions to the equations when $y = 0$.

38. $-1, 2, \frac{5}{2}$; The graph must have -1, 2, and 2.5 as its x-intercepts. The solutions of the equation are the values of x when $y = 0$, that is, the points where the graph intersects the x-axis.

39. a. 2 b. -4 c. -5 d. 3

40. Each graph lies above the x-axis except that its vertex is on the x-axis. The one point where the graph intersects the x-axis is the solution of the corresponding equation in Ex. 39.

41. Check students' work. The graph is above the x-axis except that its vertex is $(-k, 0)$.

38. Use the zero-product property to solve the equation $(x + 1)(x - 2)(2x - 5) = 0$. What must be true about the graph of $y = (x + 1)(x - 2)(2x - 5)$? Explain your reasoning.

Cooperative Learning For Exercises 39–43, work with a partner.

39. Take turns using the zero-product property to solve each equation.

a. $(x - 2)^2 = 0$ b. $(x + 4)^2 = 0$

c. $(x + 5)^2 = 0$ d. $(x - 3)^2 = 0$

40. Use the results of Exercise 39 and what you know about factors and x-intercepts to make a conjecture about the graph of each equation.

a. $y = (x - 2)^2$ b. $y = (x + 4)^2$

c. $y = (x + 5)^2$ d. $y = (x - 3)^2$

41. **Technology** Test your conjectures by graphing each equation in Exercise 40. Generalize your results by describing the graph of an equation in the form $y = (x + k)^2$ in relation to the graph of $y = x^2$.

42. **Challenge** Use what you learned in Chapter 8 and in Exercises 39–41 to describe how to transform the graph of $y = x^2$ to obtain the graph of $y = 2(x + 3)^2 - 4$.

43. **Technology** Make a conjecture about how the graphs of $y = |x - 2|$ and $y = |x + 2|$ are related to the graph of $y = |x|$. Test your conjectures using a graphing calculator. Generalize your results by describing the graph of an equation in the form $y = |x + k|$ in relation to the graph of $y = |x|$.

ONGOING ASSESSMENT

44. **Writing** Suppose one of your friends missed this lesson. Write a description of at least two ways you can solve an equation in the form $0 = ax^2 + bx$. Be sure to include examples.

SPIRAL REVIEW

Simplify each expression. *(Section 9.4)*

45. $(0.17)^0$ 46. 2^{-3} 47. 165^0

48. $\left(\frac{1}{3}\right)^{-4}$ 49. $\left(\frac{5}{11}\right)^{-2}$ 50. $\frac{1}{5^{-2}}$

Multiply. *(Section 10.2)*

51. $(3x + 1)(x + 2)$ 52. $(2y + 5)(y + 3)$ 53. $(z - 7)(2z - 3)$

54. $(2a + 1)(2a + 1)$ 55. $(2a + 1)(2a - 1)$ 56. $(3b - 8)^2$

Apply⇔Assess

Exercise Notes

Cooperative Learning

Exs. 39–41 Using groups for these exercises allows students to share and compare their conjectures about how the graph of $y = (x + k)^2$ is related to the graph of $y = x^2$. Students should summarize their results in a written statement.

Assessment Note

Ex. 44 This exercise provides an excellent opportunity for self-assessment. Students who have difficulty writing their descriptions should review the section to solidify their understanding of how to solve equations of the form $0 = ax^2 + bx$.

Practice 68 for Section 10.3

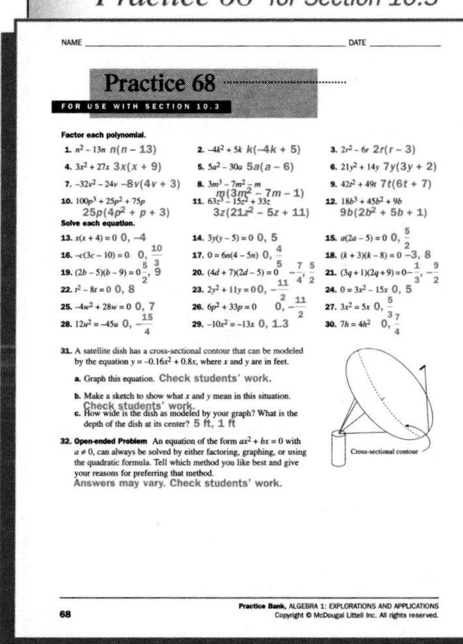

NAME ______ DATE ______

Practice 68

FOR USE WITH SECTION 10.3

Factor each polynomial.

1. $n^2 - 13n$ $n(n - 13)$
2. $-4k^2 + 5k$ $k(-4k + 5)$
3. $2r^2 - 6r$ $2r(r - 3)$
4. $3x^2 + 27x$ $3x(x + 9)$
5. $5a^2 - 30a$ $5a(a - 6)$
6. $21y^2 + 14y$ $7y(3y + 2)$
7. $-32v^2 - 24v$ $-8v(4v + 3)$
8. $3m^3 - 7m^2 - m$ $m(3m^2 - 7m - 1)$
9. $42t^2 + 49t$ $7t(6t + 7)$
10. $100p^3 + 25p^2 + 75p$ $25p(4p^2 + p + 3)$
11. $63z^3 - 15z^2 + 33z$ $3z(21z^2 - 5z + 11)$
12. $18b^3 + 45b^2 + 9b$ $9b(2b^2 + 5b + 1)$

Solve each equation.

13. $x(x + 4) = 0$ $0, -4$
14. $3y(y - 5) = 0$ $0, 5$
15. $a(2a - 5) = 0$ $0, \frac{5}{2}$
16. $-c(3c - 10) = 0$ $0, \frac{10}{3}$
17. $0 = 6n(4 - 5n)$ $0, \frac{4}{5}$
18. $(k + 3)(k - 8) = 0$ $-3, 8$
19. $(2b - 5)(b - 9) = 0$ $\frac{5}{2}, 9$
20. $(4d + 7)(2d - 5) = 0$ $-\frac{7}{4}, \frac{5}{2}$
21. $(3q + 1)(2q + 9) = 0$ $-\frac{1}{3}, -\frac{9}{2}$
22. $t^2 - 8t = 0$ $0, 8$
23. $2y^2 + 11y = 0$ $0, -\frac{11}{2}$
24. $0 = 3x^2 - 15x$ $0, 5$
25. $-4w^2 + 28w = 0$ $0, 7$
26. $6p^2 + 33p = 0$ $0, -\frac{11}{2}$
27. $3x^2 = 5x$ $0, \frac{5}{3}$
28. $12u^2 = -45u$ $0, -\frac{15}{4}$
29. $-10x^2 = -13x$ $0, 1.3$
30. $7h = 4h^2$ $0, \frac{7}{4}$

31. A satellite dish has a cross-sectional contour that can be modeled by the equation $y = -0.16x^2 + 0.8x$, where x and y are in feet.

a. Graph this equation. Check students' work.

b. Make a sketch to show what x and y mean in this situation. Check students' work.

c. How wide is the dish as modeled by your graph? What is the depth of the dish at its center? 5 ft, 1 ft

32. **Open-ended Problem** An equation of the form $ax^2 + bx = 0$ with $a \neq 0$, can always be solved by either factoring, graphing, or using the quadratic formula. Tell which method you like best and give your reasons for preferring that method.
Answers may vary. Check students' work.

68 **Practice Bank,** ALGEBRA 1: EXPLORATIONS AND APPLICATIONS

The new equation is shifted $|k|$ units to the left if k is positive and $|k|$ units to the right if k is negative.

42. The graph of $y = x^2$ is stretched vertically by a factor of 2, then shifted 3 units to the left and 4 units down.

43. The graph of $y = |x - 2|$ is obtained by shifting the graph of $y = |x|$ two units to the right. The graph of $y = |x + 2|$ is obtained by shifting the graph of $y = |x|$ two units to the left. Check students' work. The graph of $y = |x + k|$ is the graph of $y = |x|$ shifted horizontally so that its vertex is $(-k, 0)$.

44. Answers may vary. An example is given. The description should include factoring $ax^2 + bx$ and solving the equation $x(ax + b) = 0$ using the zero-product property. Another way is to use a graphing calculator to graph $y = ax^2 + bx$. The x-intercepts of the graph are the solutions of the original equation.

45. 1

46. $\frac{1}{8}$

47. 1

48. 81

49. $\frac{121}{25}$, or $4\frac{21}{25}$

50. 25

51. $3x^2 + 7x + 2$

52. $2y^2 + 11y + 15$

53. $2z^2 - 17z + 21$

54. $4a^2 + 4a + 1$

55. $4a^2 - 1$

56. $9b^2 - 48b + 64$

Plan⇔Support

Objectives

- Factor trinomials with positive quadratic terms and positive constant terms.
- Use factoring to explore the value of an investment.

Recommended Pacing

Core and Extended Courses
Section 10.4: 2 days

Block Schedule
Section 10.4: 2 half-blocks (with Sections 10.3 and 10.5)

Resource Materials

Lesson Support
Lesson Plan 10.4
Warm-Up Transparency 10.4
Overhead Visuals:
  Folder 10: Modeling Polynomial Operations, Sheets 3–5
  Folder B: Algebra Tiles
Practice Bank: Practice 69
Study Guide: Section 10.4
Explorations Lab Manual:
  Diagram Masters 4, 5

Technology
Graphing Calculator
McDougal Littell Software
  Function Investigator
Internet:
  http://www.hmco.com

Warm-Up Exercises

1. Factor $3x^2 - 12x$. $3x(x - 4)$
2. Multiply $(x - 4)(x - 1)$. $x^2 - 5x + 4$
3. Multiply $(x + 3)(x + 1)$. $x^2 + 4x + 3$
4. Which factors of 6 have a sum of 5? 3 and 2, –1 and 6
5. *True* or *False*? A polynomial with three terms is called a trinomial. True.

SECTION

10.4 Exploring Factoring

In Section 10.2 you learned the FOIL method for multiplying binomials. Can you reverse that process? The Exploration will show you how.

Learn how to...

- **factor trinomials with positive quadratic terms and positive constant terms**

So you can...

- **use factoring to explore the value of an investment, for example**

EXPLORATION
COOPERATIVE LEARNING

Using Models to Factor Trinomials

Work with a partner.
You will need:
- algebra tiles

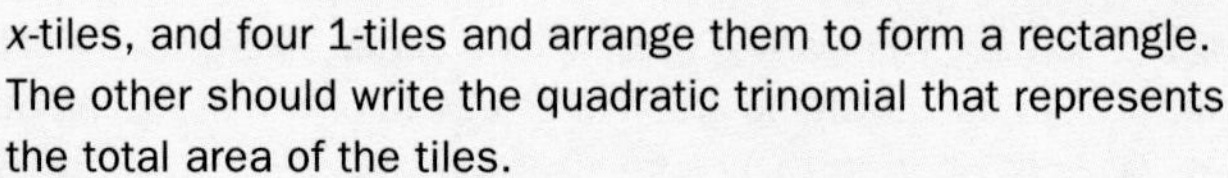

1 One of you should take one x^2-tile, five x-tiles, and four 1-tiles and arrange them to form a rectangle. The other should write the quadratic trinomial that represents the total area of the tiles.

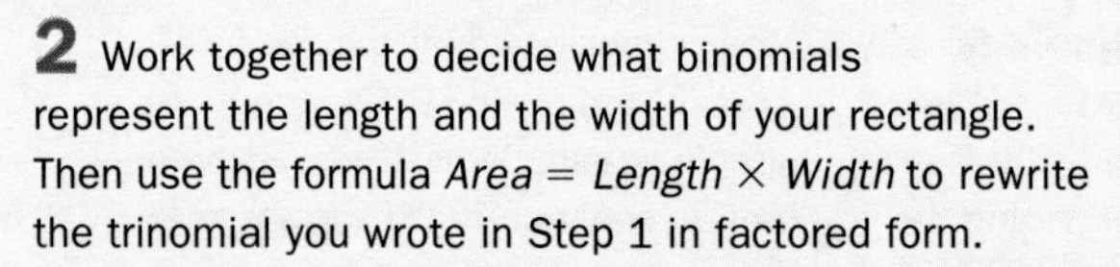

2 Work together to decide what binomials represent the length and the width of your rectangle. Then use the formula *Area* = *Length* × *Width* to rewrite the trinomial you wrote in Step 1 in factored form.

3 Switch roles and repeat Steps 1 and 2 using one x^2-tile, six x-tiles, and eight 1-tiles.

4 Without overlapping tiles, each of you try to build a rectangle using one x^2-tile, two x-tiles, and three 1-tiles. What do you notice? What does this suggest about the trinomial $x^2 + 2x + 3$?

5 You may recall that a [tile] tile represents $-x$. The tiles at the right show how you can factor a trinomial with a negative linear term:

$$x^2 - 4x + 3 = (x - 3)(x - 1)$$

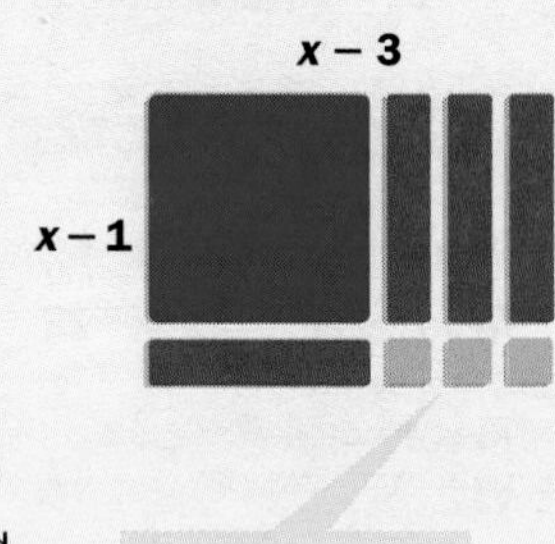

One of you should use tiles to factor $x^2 - 2x + 1$. The other should use tiles to factor $x^2 - 3x + 2$. Use the FOIL method to check each other's results.

Exploration Note

Purpose
The purpose of this Exploration is to have students use algebra tiles to discover how to factor a quadratic trinomial.

Materials/Preparation
Each group should have at least one x^2-tile, 6 x-tiles, 4 negative x-tiles, and 8 1-tiles.

Procedure
Students arrange algebra tiles to form a rectangle and write the quadratic trinomial that represents the total area of the tiles. They then find the binomials that represent the length and width of the rectangle. The formula for the area of a rectangle is used to write the trinomial in factored form. In Step 4, students work with a trinomial that is not factorable.

Closure
Students should understand how to use algebra tiles to factor a quadratic trinomial and that not all quadratic trinomials can be factored. Ask stduents to explain how they would use tiles to factor $x^2 + 3x + 2$.

Explorations Lab Manual
See the Manual for more commentary on this Exploration.
Diagram Masters 4 and 5

EXAMPLE 1

Factor $x^2 + 5x + 6$.

SOLUTION

Method 1 Use algebra tiles.

Step 1 Try to build a rectangle with one x^2-tile, five x-tiles, and six 1-tiles. Begin by finding the possible arrangements of the x^2-tile and the six 1-tiles.

Step 2 Try to complete each rectangle with five x-tiles.

Five x-tiles are not enough to complete this rectangle.

Step 3 Find the dimensions of the completed rectangle: $x + 3$ and $x + 2$.

$$x^2 + 5x + 6 = (x + 3)(x + 2)$$

Method 2 Guess, check, and revise.

Your answer will have this form:

$$(x \pm \underline{?})(x \pm \underline{?})$$

The orange terms will be factors of 6:
6, 1   −6, −1   3, 2   −3, −2

Try different combinations. Focus on getting the correct linear term.

$(x + 6)(x + 1)$	$(x - 6)(x - 1)$	$(x + 3)(x + 2)$
$6x$	$-6x$	$3x$
x	$-x$	$2x$
$7x$	$-7x$	$5x$

Stop when you find the correct linear term.

$$x^2 + 5x + 6 = (x + 3)(x + 2)$$

Teach⇔Interact

Learning Styles: Visual

Using algebra tiles to factor trinomials provides the visual connection that some students need to understand factoring. Students also see that if it is not possible to build a rectangle, then the trinomial cannot be factored.

Section Note

Reasoning
The word *prime* is sometimes used to describe a trinomial that is not factorable. Have students relate prime trinomials to prime numbers.

About Example 1

Reasoning
Some pairs of factors can be eliminated immediately because they do not add to 5. Then students can guess, check, and revise using the remaining possibilities. In general, this method can be used for quadratic trinomials whose leading coefficient is 1.

Additional Example 1

Factor $x^2 - 6x + 8$.

Method 1 Use algebra tiles.
Step 1 Try to build a rectangle with one x^2-tile, six $-x$-tiles, and eight 1-tiles. Begin by finding the possible arrangements of the x^2-tile and the eight 1-tiles.

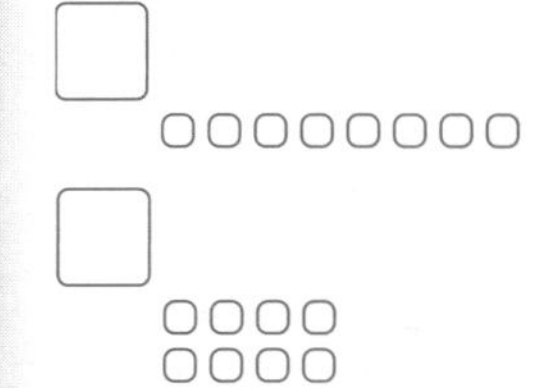

Step 2 Try to complete each rectangle with six $-x$-tiles.

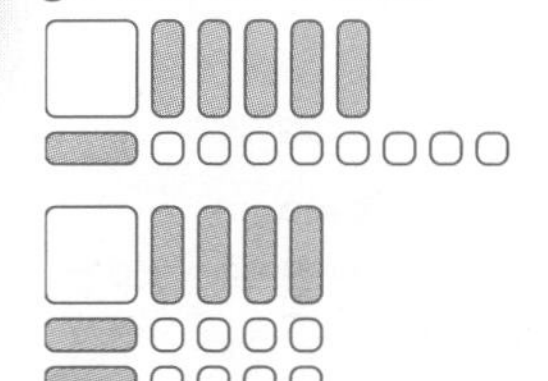

Step 3 Find the dimensions of the completed rectangle: $x - 4$ and $x - 2$.
$x^2 - 6x + 8 = (x - 4)(x - 2)$

Example continued on next page.

ANSWERS Section 10.4

Exploration

1. $x^2 + 5x + 4$
2. $(x + 4)(x + 1)$
3. $x^2 + 6x + 8$; $(x + 4)(x + 2)$
4. You cannot build a rectangle with these tiles. This implies that the polynomial cannot be factored.
5. $(x - 1)(x - 1)$; $(x - 2)(x - 1)$

Teach⇔Interact

Additional Example 1 (continued)

Method 2 Guess, check, and revise.
Your answer will have this form:
$(x \pm \underline{\ ?\ })(x \pm \underline{\ ?\ })$
Try different combinations. Focus on getting the correct linear term.

Guess	Linear term
$(x+8)(x+1)$	$8x + x = 9x$
$(x-8)(x-1)$	$-8x - x = -9x$
$(x-4)(x-2)$	$4x + 2x = 6x$

Stop when you find the correct linear term.
$x^2 - 6x + 8 = (x-4)(x-2)$

About Example 2

Teaching Tip
Point out to students the differences between the guess, check, and revise methods used in Examples 1 and 2. This will help to prevent errors when the leading coefficient is a number other than 1.

Additional Example 2

Factor $3x^2 - 14x + 8$.
Guess, check, and revise.
Your answer will have this form:
$(\underline{\ ?\ }x \pm \underline{\ ?\ })(\underline{\ ?\ }x \pm \underline{\ ?\ })$
Try different combinations. Focus on getting the correct linear term.

Guess	Linear term
$(3x+4)(x+2)$	$4x + 6x = 10x$
$(3x+2)(x+4)$	$2x + 12x = 14x$
$(3x-2)(x-4)$	$-2x - 12x = -14x$

$3x^2 - 14x + 8 = (3x-2)(x-4)$

Think and Communicate

When discussing question 4, you may wish to mention that a perfect square trinomial has a middle term whose coefficient is twice the square root of the leading coefficient times the square root of the constant term.

EXAMPLE 2

Factor $2x^2 - 11x + 15$.

SOLUTION

Toolbox p. 600
Problem Solving

Problem Solving Strategy: Guess, check, and revise.

Your answer will have this form:

$$(\underline{\ ?\ }x \pm \underline{\ ?\ })(\underline{\ ?\ }x \pm \underline{\ ?\ })$$

The blue terms will be factors of $2x^2$: $2x$, x

The orange terms will be factors of 15: 3, 5   −3, −5   1, 15   −1, −15

Try different combinations. Focus on getting the correct linear term.

$(2x+3)(x+5)$	$(2x+5)(x+3)$	$(2x-5)(x-3)$
$3x$	$5x$	$-5x$
$10x$	$6x$	$-6x$
$13x$	$11x$	$-11x$

$2x^2 - 11x + 15 = (2x-5)(x-3)$ the correct linear term

Examples 1 and 2 demonstrate the key things to look for when you factor a trinomial in the form $ax^2 + bx + c$ when c is positive.

When c is positive and b is positive... $\mathbf{x^2 + 5x + 6 = (x+3)(x+2)}$ ...then these terms are both positive.

When c is positive and b is negative... $\mathbf{2x^2 - 11x + 15 = (2x-5)(x-3)}$ ...then these terms are both negative.

THINK AND COMMUNICATE

1. Show how to use multiplication to check the result of Example 1.
2. Look back at Example 1 to figure out how to factor each trinomial.
   a. $x^2 + 7x + 6$ b. $x^2 - 7x + 6$ c. $x^2 - 5x + 6$
3. Look back at Example 2 to figure out how to factor each trinomial.
   a. $2x^2 + 11x + 15$ b. $2x^2 + 13x + 15$ c. $2x^2 - 13x + 15$
4. Factor $9x^2 - 6x + 1$ and $9x^2 - 10x + 1$. Which is a perfect square trinomial?

Think and Communicate

1. $(x+3)(x+2) = x^2 + 2x + 3x + 6 = x^2 + 5x + 6$
2. a. $(x+6)(x+1)$ b. $(x-6)(x-1)$ c. $(x-3)(x-2)$
3. a. $(2x+5)(x+3)$ b. $(2x+3)(x+5)$ c. $(2x-3)(x-5)$
4. $9x^2 - 6x + 1 = (3x-1)^2$; $9x^2 - 10x + 1 = (9x-1)(x-1)$; $9x^2 - 6x + 1$

Checking Key Concepts

1. 10, 1; −10, −1; 2, 5; −2, −5; $2 + 5 = 7$
2. 6, 1; −6, −1; 2, 3; −2, −3; $6 + 1 = 7$
3. 12, 1; −12, −1; 6, 2; −6, −2; 4, 3; −4, −3; $4 + 3 = 7$
4. $(x+1)(x+7)$
5. $(y-2)^2$
6. $(t-5)(t-2)$
7. $(a+9)(a+2)$
8. $(2y-3)(y-1)$
9. $(3x-2)(x-4)$
10. $(4z+1)(z+1)$
11. $(5p+3)^2$
12. $(5b-9)(5b-1)$
13. The factors of 4 are 1 and 4, −1 and −4, 2 and 2, or −2 and −2. None of these pairs add to −3.

Exercises and Applications

1. $(x+3)(x+1)$
2. $(n-1)(n-5)$
3. $(k-4)(k-2)$
4. $(x+5)(x+3)$
5. $(x+2)(x+10)$
6. $(y-6)^2$
7. $(y-9)(y-4)$
8. not factorable
9. $(s-5)(s-4)$
10. $(n+6)(n+4)$
11. $(t-8)(t-3)$
12. not factorable

13–29. See answers in back of book.

CHECKING KEY CONCEPTS

List all the pairs of factors of each number. Then tell which pair of factors has a sum of 7.

1. 10 **2.** 6 **3.** 12

Factor each trinomial. You may want to use algebra tiles.

4. $x^2 + 8x + 7$ **5.** $y^2 - 4y + 4$ **6.** $t^2 - 7t + 10$

7. $a^2 + 11a + 18$ **8.** $2y^2 - 5y + 3$ **9.** $3x^2 - 14x + 8$

10. $4z^2 + 5z + 1$ **11.** $25p^2 + 30p + 9$ **12.** $25b^2 - 50b + 9$

13. Show that $k^2 - 3k + 4$ is not factorable.

10.4 Exercises and Applications

Factor, if possible. If not, write *not factorable*.

Extra Practice exercises on page 578

1. $x^2 + 4x + 3$ **2.** $n^2 - 6n + 5$ **3.** $k^2 - 6k + 8$

4. $x^2 + 8x + 15$ **5.** $x^2 + 12x + 20$ **6.** $y^2 - 12y + 36$

7. $y^2 - 13y + 36$ **8.** $r^2 - 7r + 49$ **9.** $s^2 - 9s + 20$

10. $n^2 + 10n + 24$ **11.** $t^2 - 11t + 24$ **12.** $z^2 + 12z + 24$

13. Writing From Exercises 1–12, choose some trinomials that are factorable and find the value of the discriminant for each one. Then do the same for some trinomials that are not factorable. Explain how you can use the discriminant to decide whether a trinomial is factorable.

14. SAT/ACT Preview Which binomial is a factor of $x^2 - 10x + 9$?

A. $(x - 1)$ **B.** $(x - 2)$ **C.** $(x - 3)$ **D.** $(x - 5)$

15. Express the total area of the tiles shown at the right as a trinomial. Then express the total area as the product of two binomials.

16. Use algebra tiles or draw a diagram to show how to factor $4x^2 + 12x + 5$.

Factor each trinomial.

17. $5k^2 + 6k + 1$ **18.** $2d^2 - 7d + 6$ **19.** $3n^2 + 8n + 4$

20. $4a^2 + 12a + 9$ **21.** $4a^2 - 13a + 9$ **22.** $2m^2 + 15m + 18$

23. $5x^2 - 13x + 8$ **24.** $6n^2 + 17n + 7$ **25.** $10g^2 - 11g + 3$

26. $6q^2 - 11q + 4$ **27.** $9p^2 - 18p + 8$ **28.** $15x^2 + 19x - 10$

29. Open-ended Problem Find five different ways to complete the following trinomial so that it can be factored: $\underline{?}\,x^2 + 5x + \underline{?}$. Show how to factor each of your five trinomials.

Technology Note

There are various ways to use graphing calculators or the *Function Investigator* software to factor polynomials. For example, either type of technology can be used in Exercise 18 to discover that (after changing the variable from d to x) the graph of $y = 2x^2 - 7x + 6$ intersects the x-axis where $x = \frac{3}{2}$ and where $x = 2$. Multiply both sides of $x = \frac{3}{2}$ by 2 and subtract 3 from both sides of the resulting equation to get $2x - 3 = 0$. Subtract 2 from both sides of $x = 2$ to get $x - 2 = 0$. The factors of $2x^2 - 7x + 6$ are $2x - 3$ and $x - 2$. So, for the original exercise, $2d^2 - 7d + 6 = (2d - 3)(d - 2)$. This result can be checked by using the *FOIL* method to multiply the two binomials. Alternatively, the result can be checked by the method described in the Technology Note on page 433.

Be sure that students understand the reasoning used in these procedures.

Teach⇔Interact

Checking Key Concepts

Common Error

Student errors in factoring trinomials can be eliminated if they check the factors by multiplying them to see if the product is the original trinomial. If it is not, then the process of looking for the correct factors and checking them must be continued.

Closure Question

Explain how to factor a trinomial of the form $x^2 + bx + c$, where c is positive.

Use either algebra tiles or the guess, check, and revise strategy. With algebra tiles, build a rectangle with x^2, x, and 1-algebra tiles and find the dimensions of the completed rectangle. The dimensions will be the factors. For the guess, check, and revise method, find all the factors of c. Try different combinations of factors in the form $(x \pm \underline{?})(x \pm \underline{?})$. Focus on getting the correct linear term.

Apply⇔Assess

Suggested Assignment

Core Course
Day 1 Exs. 1–22
Day 2 Exs. 23–28, 30–39, 41–52

Extended Course
Day 1 Exs. 1–22
Day 2 Exs. 23–52

Block Schedule
Day 65 Exs. 1–22
Day 66 Exs. 23–28, 30–52

Exercise Notes

Topic Spiraling: Review

Ex. 13 This exercise makes an important connection to solving quadratic equations in Chapter 8. A trinomial is factorable if the discriminant of the related quadratic equation is a perfect square or zero.

Problem Solving

Ex. 29 Students who can complete this exercise successfully understand factoring. You may wish to have one student demonstrate his or her answer at the board.

Apply⇔Assess

Exercise Notes

Second-Language Learners

Exs. 30–33 When reading the *By the Way* connection to these exercises, some students learning English may be unfamiliar with the concept of a *pension plan.* If necessary, explain that a person pays money into a pension plan while he or she works; then when the person retires, the plan sends money to him or her every month.

Reasoning

Exs. 37–39 When the largest exponent of a trinomial is 3, the polynomial is a cubic. In this case, ask students to explain why they should first look for a common monomial factor.

Assessment Note

Ex. 40 Students may want to work in groups of 2 or 3 to do this activity. In so doing, they can have an opportunity to factor many exercises they have not seen before.

Practice 69 for Section 10.4

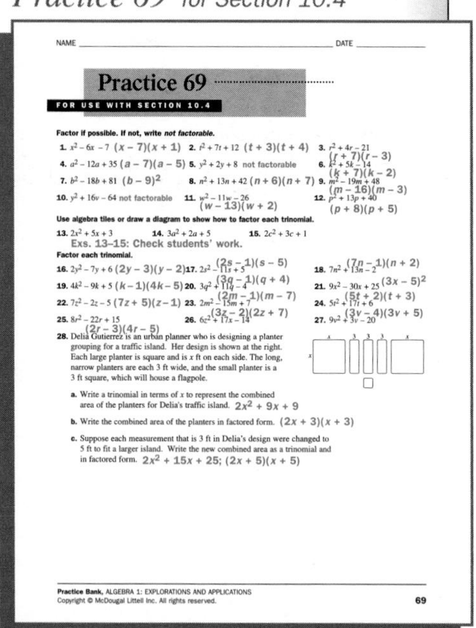

NAME ______________________ DATE __________

Practice 69

FOR USE WITH SECTION 10.4

Factor if possible. If not, write *not factorable*.

1. $x^2 - 6x - 7$ $(x - 7)(x + 1)$
2. $t^2 + 7t + 12$ $(t + 3)(t + 4)$
3. $r^2 + 4r - 21$ $(r + 7)(r - 3)$
4. $a^2 - 12a + 35$ $(a - 7)(a - 5)$
5. $y^2 + 2y + 8$ not factorable
6. $k^2 + 5k - 14$ $(k + 7)(k - 2)$
7. $b^2 - 18b + 81$ $(b - 9)^2$
8. $n^2 + 13n + 42$ $(n + 6)(n + 7)$
9. $m^2 - 19m + 48$ $(m - 16)(m - 3)$
10. $v^2 + 16v - 64$ not factorable
11. $w^2 - 11w - 26$ $(w - 13)(w + 2)$
12. $p^2 + 13p + 40$ $(p + 8)(p + 5)$

Use algebra tiles or draw a diagram to show how to factor each trinomial.

13. $2x^2 + 5x + 3$
14. $3a^2 + 2a + 5$
15. $2c^2 + 3c + 1$

Exs. 13–15: Check students' work.

Factor each trinomial.

16. $2y^2 - 7y + 6$ $(2y - 3)(y - 2)$
17. $2s^2 - 11s + 5$ $(2s - 1)(s - 5)$
18. $7n^2 + 13n - 2$ $(7n - 1)(n + 2)$
19. $4k^2 - 9k + 5$ $(k - 1)(4k - 5)$
20. $3q^2 + 11q - 4$ $(3q - 1)(q + 4)$
21. $9x^2 - 30x + 25$ $(3x - 5)^2$
22. $7z^2 - 2z - 5$ $(7z + 5)(z - 1)$
23. $2m^2 - 15m + 7$ $(2m - 1)(m - 7)$
24. $5t^2 + 17t + 6$ $(5t + 2)(t + 3)$
25. $8r^2 - 22r + 15$ $(2r - 3)(4r - 5)$
26. $6z^2 + 17z - 14$ $(3z - 2)(2z + 7)$
27. $9v^2 + 3v - 20$ $(3v - 4)(3v + 5)$

28. Delia Gutierrez is an urban planner who is designing a planter grouping for a traffic island. Her design is shown at the right. Each large planter is square and is x ft on each side. The long, narrow planters are each 3 ft wide, and the small planter is a 3 ft square, which will house a flagpole.

 a. Write a trinomial in terms of x to represent the combined area of the planters for Delia's traffic island. $2x^2 + 9x + 9$

 b. Write the combined area of the planters in factored form. $(2x + 3)(x + 3)$

 c. Suppose each measurement that is 3 ft in Delia's design were changed to 5 ft to fit a larger island. Write the new combined area as a trinomial and in factored form. $2x^2 + 15x + 25$; $(2x + 5)(x + 5)$

Practice Bank, ALGEBRA 1: EXPLORATIONS AND APPLICATIONS

69

INTERVIEW Linda Pei

Look back at the article on pages 420–422.

Suppose a financial adviser like Linda Pei invests \$1000 for someone at an annual growth rate r, where r is a decimal. The value of the investment at the end of two years can be represented by the polynomial $1000r^2 + 2000r + 1000$.

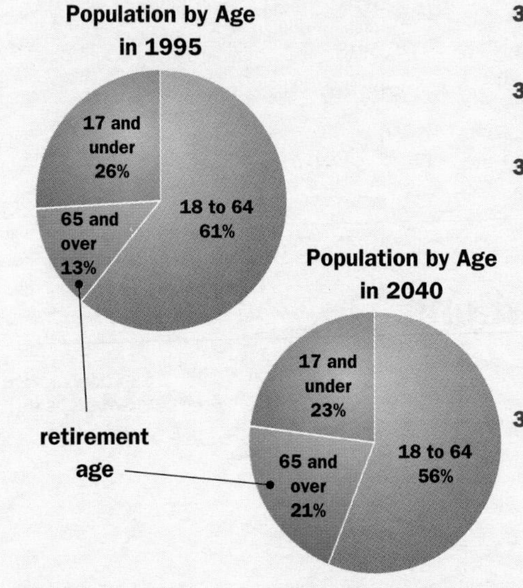

30. Write the polynomial in factored form. (*Hint:* First factor out a common monomial factor.)

31. Explain why the polynomial represents the value of the investment at the end of two years.

32. a. Write a trinomial in terms of r to represent the value of a \$5000 investment at the end of two years. Write the trinomial in standard form and in factored form.

b. Evaluate each trinomial from part (a) when $r = 0.07$. You should get the same result both times. If not, check your work.

33. a. What projected trend in the retirement-age population is shown by the graphs?

b. Suppose you invest \$5000 at an annual growth rate of 7%. What will your investment be worth when you retire?

BY THE WAY

A pension plan is one way to save money for retirement. The first private pension plan in the United States began in 1875, but it wasn't until 1974 that Congress set standards for private pension plans.

Challenge Factor completely. (*Hint:* First look for a common monomial factor.)

34. $2x^2 + 8x + 6$ **35.** $3a^2 + 9a + 6$ **36.** $16k^2 - 80k + 100$

37. $n^3 + 12n^2 + 27n$ **38.** $4y^3 - 5y^2 + y$ **39.** $9x^3 - 51x^2 + 30x$

ONGOING ASSESSMENT

40. Open-ended Problem Make a set of ten flash cards to help you practice your factoring skills. Include five exercises that you have already done in this section, and make up five exercises of your own.

SPIRAL REVIEW

Solve each equation algebraically. *(Section 8.3)*

41. $3y^2 = 147$ **42.** $\frac{2}{3}a^2 = 24$ **43.** $0 = k^2 - 225$

44. $91 = z^2 + 10$ **45.** $25x^2 - 9 = 0$ **46.** $-2n^2 + 63 = 11$

Solve each equation using the quadratic formula. *(Section 8.5)*

47. $n^2 + 7n + 12 = 0$ **48.** $2x^2 + 3x - 1 = 0$ **49.** $-2a^2 + 5a + 1 = 0$

50. $3x^2 - 5x = 0$ **51.** $p^2 - 2p - 1 = 2$ **52.** $3m^2 - 7 = 2m$

30. $1000(r + 1)^2$

31. \$1000 is the amount invested. The growth factor is $(1 + r)$. At the end of year one, the value is $1000(1 + r)$. At the end of year two, the value is $1000(1 + r) \cdot (1 + r)$ or $1000(1 + r)^2$

32. a. $5000(r + 1)^2 = 5000r^2 + 10{,}000r + 5000$

 b. 5724.5

33. a. Answers may vary. The population of persons 65 and over will increase substantially between 1995 and 2040.

 b. Estimates may vary. An example is given: In 50 years, the investment will be worth $5000(1 + 0.07)^{50}$, or about \$147,285.

34. $2(x + 3)(x + 1)$ 35. $3(a + 2)(a + 1)$

36. $4(2k - 5)^2$ 37. $n(n + 9)(n + 3)$

38. $y(4y - 1)(y - 1)$ 39. $3x(3x - 2)(x - 5)$

40. Answers may vary.

41. ± 7 42. ± 6

43. ± 15 44. ± 9

45. $\pm\frac{3}{5}$

46. $\pm\sqrt{26}$, or about –5.1 and about 5.1

47. –3, –4

48. $\frac{-3 \pm \sqrt{17}}{4}$, or about 0.28 and about –1.78

49. $\frac{5 \pm \sqrt{33}}{4}$, or about 2.69 and about –0.19

50. $0, \frac{5}{3}$

51. 3, –1

52. $\frac{1 \pm \sqrt{22}}{3}$, or about 1.9 and about –1.23

SECTION

10.5 Applying Factoring

The pattern you use to factor a quadratic trinomial with a negative constant term is different than the pattern when the constant term is positive.

Learn how to...

- factor quadratic polynomials with a negative constant term

So you can...

- solve quadratic equations by factoring

THINK AND COMMUNICATE

1. Use the FOIL method to multiply each pair of binomials.
   a. $(x + 6)(x + 2)$ b. $(x - 6)(x - 2)$
   c. $(x + 6)(x - 2)$ d. $(x - 6)(x + 2)$
2. Use the results of Question 1 to help you answer these questions.
   a. When the signs in the binomials are the same, what is the sign of the constant term of the product?
   b. When the signs in the binomials are different, what is the sign of the constant term of the product?

EXAMPLE 1

Factor $x^2 - 2x - 8$.

SOLUTION

Problem Solving Strategy: Guess, check, and revise.

Your answer will have this form:

$(x \pm \underline{?})(x \pm \underline{?})$

The orange terms will be factors of −8.
1, −8 −1, 8 2, −4 −2, 4

Try different combinations. Focus on getting the correct linear term.

$(x + 1)(x - 8)$	$(x - 1)(x + 8)$	$(x + 2)(x - 4)$
x	$-x$	$2x$
$-8x$	$8x$	$-4x$
$-7x$	$7x$	$-2x$ (the correct linear term)

$x^2 - 2x - 8 = (x + 2)(x - 4)$

Check

$(x + 2)(x - 4) \stackrel{?}{=} x^2 - 2x - 8$

$x^2 - 4x + 2x - 8 \stackrel{?}{=} x^2 - 2x - 8$

$x^2 - 2x - 8 = x^2 - 2x - 8$ ✓

Plan⇔Support

Objectives

- Factor quadratic polynomials with a negative constant term.
- Solve quadratic equations by factoring.

Recommended Pacing

Core and Extended Courses
Section 10.5: 2 days

Block Schedule
Section 10.5: 2 half-blocks (with Section 10.4 and Review)

Resource Materials

Lesson Support
Lesson Plan 10.5
Warm-Up Transparency 10.5
Practice Bank: Practice 70
Study Guide: Section 10.5
Explorations Lab Manual: Diagram Master 2
Assessment Book: Test 44

Technology
Graphing Calculator
Spreadsheet Software
McDougal Littell Software *Function Investigator* *Stats!*
Internet: http://www.hmco.com

Warm-Up Exercises

Multiply each pair of binomials.

1. $(x + 3)(x + 2)$ $x^2 + 5x + 6$
2. $(x + 3)(x - 2)$ $x^2 + x - 6$
3. $(x - 3)(x - 2)$ $x^2 - 5x + 6$
4. Factor $x^2 - 8x + 15$. $(x - 5)(x - 3)$
5. Give all combinations of factors of −8. −4 and 2; 4 and −2, −8 and 1; 8 and −1

ANSWERS Section 10.5

Think and Communicate

1. a. $x^2 + 8x + 12$
   b. $x^2 - 8x + 12$
   c. $x^2 + 4x - 12$
   d. $x^2 - 4x - 12$
2. a. positive
   b. negative

Teach⇔Interact

Think and Communicate

The answers to questions 1 and 2 point out that the signs in the binomials affect the sign of the constant term of the product.

Additional Example 1

Factor $x^2 - 5x - 6$.

Guess, check, and revise.

Your answer will have this form:

$(x \pm \underline{\ ?\ })(x \pm \underline{\ ?\ })$

Try different combinations. Focus on getting the correct linear term.

$(x + 3)(x - 2)$	$(x - 3)(x + 2)$
$3x$	$-3x$
$\underline{-2x}$	$\underline{2x}$
$1x$	$-1x$

$(x + 1)(x - 6)$

x

$\underline{-6x}$

$-5x$

$x^2 - 5x - 6 = (x + 1)(x - 6)$

Check

$(x + 1)(x - 6) \stackrel{?}{=} x^2 - 5x - 6$

$x^2 + 1x - 6x - 6 \stackrel{?}{=} x^2 - 5x - 6$

$x^2 - 5x - 6 = x^2 - 5x - 6$ ✓

Additional Example 2

Factor $2x^2 + 11x - 6$.

Guess, check, and revise.

Your answer will have this form:

$(\underline{\ ?\ }x \pm \underline{\ ?\ })(\underline{\ ?\ }x \pm \underline{\ ?\ })$

Try different combinations. Focus on getting the correct linear term.

$(2x + 6)(x - 1)$	$(2x - 6)(x + 1)$
$6x$	$-6x$
$\underline{-2x}$	$\underline{2x}$
$4x$	$-4x$

$(2x + 1)(x - 6)$	$(2x - 1)(x + 6)$
x	$-x$
$\underline{-12x}$	$\underline{12x}$
$-11x$	$11x$

$2x^2 + 11x - 6 = (2x - 1)(x + 6)$

Think and Communicate

Questions 3–6 provide an opportunity for students to use the key ideas presented after Example 2 on page 444 in Section 10.4 and after Example 2 on this page. You may wish to have students make a combined list of these ideas in their journals so that they have the complete list available at a glance.

EXAMPLE 2

Factor $3x^2 + 5x - 8$.

SOLUTION

Problem Solving Strategy: Guess, check, and revise.

Your answer will have this form:

$$(\underline{\ ?\ }x \pm \underline{\ ?\ })(\underline{\ ?\ }x \pm \underline{\ ?\ })$$

The blue terms will be factors of $3x^2$: $3x$, x

The orange terms will be factors of -8.
1, −8 −1, 8 2, −4 −2, 4

Try different combinations. Focus on getting the correct linear term.

$(3x + 1)(x - 8)$	$(3x - 1)(x + 8)$
x	$-x$
$\underline{-24x}$	$\underline{24x}$
$-23x$	$23x$

$(3x - 8)(x + 1)$	$(3x + 8)(x - 1)$
$-8x$	$8x$
$\underline{3x}$	$\underline{-3x}$
$-5x$	$5x$ ← the correct linear term

$3x^2 + 5x - 8 = (3x + 8)(x - 1)$

You may have noticed that the key to factoring trinomials of the form $ax^2 + bx + c$ is to position the signs and the factors of the constant term correctly. In Section 10.4 you saw how to do this when c is positive. Example 2 demonstrates the key things to do when c is negative.

When c is negative... $\mathbf{3x^2 + 5x - 8 = (3x + 8)(x - 1)}$ ...one factor of c is positive and the other is negative.

The factored form is either $(\underline{\ ?\ }x + \underline{\ ?\ })(\underline{\ ?\ }x - \underline{\ ?\ })$ or $(\underline{\ ?\ }x - \underline{\ ?\ })(\underline{\ ?\ }x + \underline{\ ?\ })$.

THINK AND COMMUNICATE

Decide whether each trinomial is factored correctly. If it is not, what changes are needed to make the factors correct?

3. $x^2 + 6x + 5 = (x - 1)(x + 5)$ **4.** $x^2 - 4x + 3 = (x + 3)(x + 1)$

5. $x^2 + 2x - 8 = (x - 2)(x + 4)$ **6.** $5x^2 - 13x + 6 = (5x - 2)(x - 3)$

Think and Communicate

3. No. The factoring should be $(x + 1)(x + 5)$.
4. No. The factoring should be $(x - 1)(x - 3)$.
5. Yes.
6. No. The factoring should be $(5x - 3)(x - 2)$.

EXAMPLE 3 Application: Sports

A pitcher throws a warm-up pitch in a softball game. The equation $h = -16t^2 + 24t + 1$ gives the height h in feet of the ball t seconds after the pitcher releases it. How long does it take for the ball to reach a height of 6 ft?

SOLUTION

Substitute **6** for ***h***, and rewrite the equation in the form $ax^2 + bx + c = 0$, with $a > 0$.

$$h = -16t^2 + 24t + 1$$
$$6 = -16t^2 + 24t + 1$$
$$0 = -16t^2 + 24t - 5$$
$$-1 \cdot 0 = -1(-16t^2 + 24t - 5)$$

Multiply both sides by **−1**.

$$0 = 16t^2 - 24t + 5$$
$$0 = (4t - 1)(4t - 5)$$

Factor and use the zero-product property.

$$4t - 1 = 0 \quad \text{or} \quad 4t - 5 = 0$$
$$t = \frac{1}{4} \qquad\qquad t = \frac{5}{4}$$

The ball reaches a height of 6 ft after 0.25 s and after 1.25 s.

BY THE WAY

Softball was first played in the late nineteenth century as an indoor version of baseball.

The diagram below illustrates how it is possible for the ball discussed in Example 3 to be 6 ft off the ground at two different times.

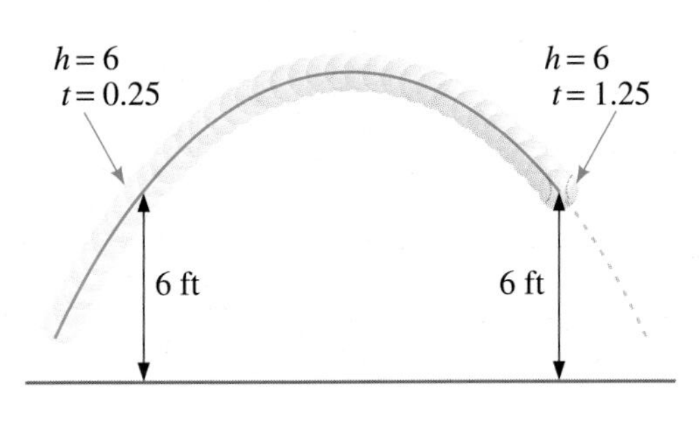

THINK AND COMMUNICATE

7. The notebook at the right shows how Kyle factored the expression $x^2 - 25$. When he was done, he recognized a pattern. What is it?
8. How could you use the pattern you identified in Question 7 to factor $x^2 - 4$? to factor $49x^2 - 16$?

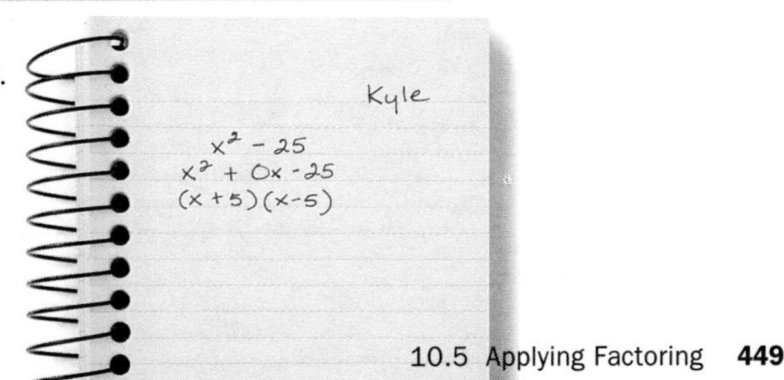

Teach⇔Interact

About Example 3

Reasoning

Ask students to explain why the coefficient of t^2 in the equation $0 = -16t^2 + 24t - 5$ is changed to 16 by multiplying both sides of the equation by −1.

Additional Example 3

A quarterback throws a football in a practice game. The equation $h = -16t^2 + 28t + 6$ gives the height h in feet of the ball t seconds after the quarterback releases it. How long does it take for the ball to reach a height of 12 feet?

Substitute 12 for h and rewrite the equation in the form $ax^2 + bx + c = 0$, with $a > 0$.

Get a zero on one side of the equation. Multiply both sides by −1, then factor and use the zero-product property.

$$h = -16t^2 + 28t + 6$$
$$12 = -16t^2 + 28t + 6$$
$$0 = -16t^2 + 28t - 6$$
$$-1 \cdot 0 = -1(-16t^2 + 28t - 6)$$
$$0 = 16t^2 - 28t + 6$$
$$0 = (4t - 1)(4t - 6)$$
$$4t - 1 = 0 \quad \text{or} \quad 4t - 6 = 0$$
$$t = \frac{1}{4} \qquad\qquad t = \frac{6}{4} = \frac{3}{2}$$

The ball reaches a height of 12 ft after 0.25 s and after 1.5 s.

Think and Communicate

Students learned to find the difference of two squares in Section 6.5. Question 7 reviews this procedure, albeit in reverse, and relates it to factoring.

Think and Communicate

7. the difference of two squares: $(a^2 - b^2) = (a + b)(a - b)$
8. $x^2 - 4 = (x + 2)(x - 2)$; $49x^2 - 16 = (7x + 4)(7x - 4)$

Teach⇔Interact

Section Notes

Communication: Discussion A thorough discussion of the three methods for solving quadratic equations can be done as a whole class activity or by having students work in small groups.

Cooperative Learning
Groups of four can be used to discuss the three methods for solving quadratic equations. Have one student make up a quadratic for each of the other three members to solve using a different method. Then reverse the roles.

Historical Connection
The search for general formulas to solve equations of degree two and higher, such as cubic and quartic equations, was the focus of mathematicians during the sixteenth century. In the early 1800s, a proof was given by P. Ruffini, an Italian physician, that the roots of a general fifth or higher degree equation cannot be expressed by means of radicals in terms of the coefficients of the equation. This fact was later independently established in 1824 by the famous Norwegian mathematician Niels Henrik Abel.

Checking Key Concepts

Questions 1–7 can be discussed verbally with students. Then have the class work independently on solving the equations in 8–10.

Closure Question

Describe how to solve a quadratic equation by factoring.
First rewrite the equation in the form $ax^2 + bx + c = 0$, where a, b, and c are integers and a is positive. Then factor $ax^2 + bx + c$ and use the zero-product property.

You now know three methods for solving quadratic equations. You can use a graph or the quadratic formula to solve any quadratic equation. You can use factoring to solve any quadratic equation whose discriminant is a perfect square, which tells you that the equation can be written in factored form.

Methods to Solve Quadratic Equations

$$x^2 - x - 6 = 0$$

Factor and use the zero-product property.

$$x^2 - x - 6 = 0$$
$$(x + 2)(x - 3) = 0$$
$$x + 2 = 0 \quad \text{or} \quad x - 3 = 0$$
$$x = -2 \qquad x = 3$$

The solutions are -2 and 3.

Graph the related equation $y = x^2 - x - 6$. The x-intercepts are the solutions.

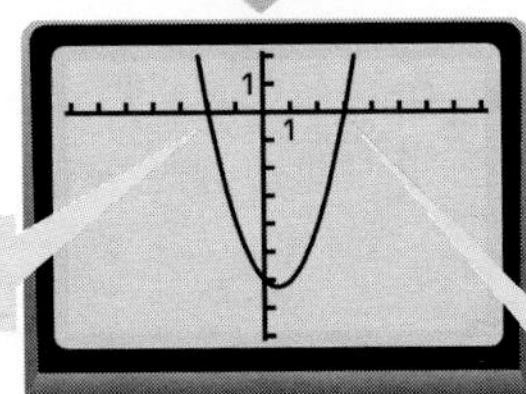

One x-intercept is -2.

One x-intercept is 3.

The solutions are -2 and 3.

Use the quadratic formula.

$$x = \frac{-b \pm \sqrt{b^2 - 4ac}}{2a}$$
$$= \frac{1 \pm \sqrt{1 - 4(1)(-6)}}{2(1)}$$
$$x = \frac{1+5}{2} \quad \text{or} \quad x = \frac{1-5}{2}$$
$$x = 3 \qquad x = -2$$

The solutions are -2 and 3.

CHECKING KEY CONCEPTS

1. Is it possible to solve $2x^2 + x + 1 = 0$ by factoring? Explain.

Choose the letter of the factored form of each polynomial. Then factor the polynomial.

A. (? + ?)(? + ?) **B.** (? − ?)(? − ?) **C.** (? + ?)(? − ?)

2. $x^2 + 7x - 8$ **3.** $3y^2 - 7y + 2$ **4.** $5t^2 + 6t + 1$

5. $4r^2 + 12r + 9$ **6.** $7z^2 + 2z - 5$ **7.** $p^2 - 100$

Solve each equation. Explain why you chose the method you used.

8. $x^2 - 4x - 5 = 0$ **9.** $x^2 - 4x + 1 = 0$ **10.** $4x^2 - 9 = 0$

Checking Key Concepts

1. No. The factors of $2x^2$ are $2x$ and x; the factors of 1 are 1 and 1. To get the right quadratic and constant terms, the product would have to be $(2x + 1)(x + 1)$. But this product gives the wrong linear term.
2. C; $(x + 8)(x - 1)$
3. B; $(3y - 1)(y - 2)$
4. A; $(5t + 1)(t + 1)$
5. A; $(2r + 3)^2$
6. C; $(7z - 5)(z + 1)$
7. C; $(p + 10)(p - 10)$

8–10. Methods may vary. Examples are given.

8. 5, –1; The polynomial is factorable, so use the factored form $(x + 1)(x - 5)$ and the zero-product property.
9. $2 \pm \sqrt{3}$, or about 3.73 and about 0.27; The polynomial is not factorable, so use the quadratic formula.
10. 1.5, –1.5; The polynomial is factorable, so use the factored form $(2x + 3)(2x - 3)$ and the zero-product property.

Exercises and Applications

1. +; –
2. 5; 1
3. –; –
4. $(r - 3)(r + 1)$
5. $(t - 8)(t + 1)$
6. $(y - 4)(y - 3)$
7. $(z - 9)(z + 4)$
8. $(x + 6)(x - 4)$
9. $(y + 6)(y - 6)$
10. $(3k - 7)(k + 1)$
11. $(2d + 11)(d - 1)$
12. $(3v - 2)(v + 2)$
13. $(2x + 3)(x - 2)$
14. $(5w + 4)(w + 1)$
15. $(2y - 1)(3y - 4)$
16. $(x + 10)$ and $(x - 4)$
17. C

10.5 Exercises and Applications

Replace each ? with the correct symbol or number.

1. $3m^2 - 20m - 7 = (3m \underline{?} 1)(m \underline{?} 7)$
2. $2x^2 + 3x - 5 = (2x + \underline{?})(x - \underline{?})$
3. $6r^2 - 17r + 12 = (3r \underline{?} 4)(2r \underline{?} 3)$

Factor each polynomial.

4. $r^2 - 2r - 3$
5. $t^2 - 7t - 8$
6. $y^2 - 7y + 12$
7. $z^2 - 5z - 36$
8. $x^2 + 2x - 24$
9. $y^2 - 36$
10. $3k^2 - 4k - 7$
11. $2d^2 + 9d - 11$
12. $3v^2 + 4v - 4$
13. $2x^2 - x - 6$
14. $5w^2 + 9w + 4$
15. $6y^2 - 11y + 4$
16. **GEOMETRY** If $x^2 + 6x - 40$ represents the area of a rectangle, find possible expressions for its length and width.
17. **SAT/ACT Preview** Which is the correct factorization of $12x^2 - 5x - 2$?

 A. $(6x + 2)(2x - 1)$ **B.** $(6x - 2)(2x + 1)$

 C. $(3x - 2)(4x + 1)$ **D.** $(3x + 2)(4x - 1)$

Find all integer values of *n* for which each trinomial can be factored.

18. $c^2 + nc - 6$
19. $3s^2 + ns - 14$
20. $4p^2 + np - 15$
21. **Writing** Describe how the diagram represents a factoring pattern.

x
x
–
a
a
=
x
x – a
a
a
a
x – a
=
x + a
x – a

Solve each equation by factoring.

22. $x^2 - 8x + 12 = 0$
23. $x^2 - x - 2 = 0$
24. $m^2 - 4m + 1 = -3$
25. $2r^2 - 5r - 12 = 0$
26. $2y^2 + 15y + 10 = 3$
27. $3q^2 + 7q - 4 = 2$
28. $6k^2 - k = 5$
29. $9t^2 - 4 = 0$
30. $16x^2 = 49$
31. **SPORTS** The equation $h = -16t^2 + 24t + 1$ models the height h in feet of a pitched softball t seconds after it is released.

 a. When is the ball 10 ft high?

 b. Is the ball ever 12 ft high? Explain.

Solve each equation using whichever method you prefer.

32. $(3x + 1)(x + 1) = 0$
33. $2d^2 - 6d + 1 = 0$
34. $4r^2 + r = 0$
35. $2y^2 + 5y = 3$
36. $k^2 - 4k = 4$
37. $3x^2 + 2 = 7x + 8$

Extra Practice exercises on page 578

Apply⇔Assess

Suggested Assignment

Core Course
Day 1 Exs. 1–20, 22–31
Day 2 Exs. 32–39, 46–54, AYP

Extended Course
Day 1 Exs. 1–31
Day 2 Exs. 32–54, AYP

Block Schedule
Day 66 Exs. 1–20, 22–31
Day 67 Exs. 32–54, AYP

Exercise Notes

Teaching Tip
Exs. 4–15 Remind students of the key ideas presented on pages 444 and 448 before they begin these exercises.

Using Manipulatives
Ex. 21 Students may wish to actually model the diagram using a paper square in order to better describe the process taking place.

Using Technology
Exs. 22–30 Students can use graphing calculators to check their solutions to these equations. Have them find the *x*-intercepts with the trace and zoom features.

Communication: Writing
Exs. 32–37 You may wish to have students explain why they chose the method they did.

18. –1, 1, –5, 5
19. 1, –1, 11, –11, 19, –19, 41, –41
20. 4, –4, 7, –7, 11, –11, 17, –17, 28, –28, 59, –59
21. Answers may vary. An example is given. Begin with an area of x^2 and remove an area of a^2. This changes the dimensions of the original square so that one dimension is now $(x - a)$. Remove the small bottom rectangle and add it to the end making a rectangle of dimensions $(x + a)(x - a)$. The area is $(x - a)(x + a) = x^2 - a^2$.
22. 6, 2
23. 2, –1
24. 2
25. 4, –1.5
26. $-\frac{1}{2}, -7$
27. $\frac{2}{3}, -3$
28. $1, -\frac{5}{6}$
29. $\frac{2}{3}, -\frac{2}{3}$
30. $\frac{7}{4}, -\frac{7}{4}$
31. a. 0.75 s after release
 b. No. Explanations may vary. An example is given. The graph of the equation shows that 10 ft is the maximum height.
32. $-\frac{1}{3}, -1$
33. $\frac{3 \pm \sqrt{7}}{3}$, or about 2.82 and about 0.18
34. $0, -\frac{1}{4}$
35. $\frac{1}{2}, -3$
36. $2 \pm 2\sqrt{2}$, or about 4.83 and about –0.83
37. $3, -\frac{2}{3}$

Apply⇔Assess

Exercise Notes

Application
Exs. 38, 39 Problems that involve moving objects, such as baseballs, golf balls, footballs, or other projectiles, are studied in detail in physics. The motion of a projectile always can be modeled and analyzed by using a quadratic equation.

Using Technology
Exs. 38, 39 Students can use the *Stats!* software to generate a spreadsheet for these exercises. Students using TI-82 graphing calculators can do the same by using the Table features of the calculator. Some students may also wish to use graphs. They may do so by using a TI-82 or TI-81 calculator or the *Function Investigator* software.

Assessment Note
Ex. 46 Have students compare their concept maps in groups. This will allow students to see alternate versions of these maps.

Connection SPORTS

The table at the right gives data about the outfield walls of Boston's Fenway Park. Suppose the equation $h = -16t^2 + 64t + 4$ models the height h in feet of a ball t seconds after being hit. The equation $d = 77t$ gives the horizontal distance that the ball has traveled in feet from home plate after t seconds.

Outfield wall	Height (ft)	Distance from home plate (ft)
left	37	315
center	17	390
right	3–6	302

Position of Ball Over Time

	A	B	C
1	time	height	horizontal distance
2			
3			
4			
5			
6			

38. **Spreadsheets** Use a spreadsheet to track the height of the ball as a function of time.

a. Find the time interval when the ball is at least as high as the left field wall of Fenway Park. Write your answer as an inequality.

b. Add a column to your spreadsheet to calculate the horizontal distance at time t. Compare the height values with the distance values. If a batter hits the ball down the left field line, will it go over the wall?

39. Another hit is modeled by the equations $h = -16t^2 + 74t + 4$ and $d = 88t$. Use the distance equation to find t when the ball is 390 ft from home plate. How high is the ball at this time? Could it go over one of the outfield walls?

Challenge **Factor completely. (*Hint:* First look for a common monomial factor.)**

40. $3x^3 - 6x^2 - 24x$

41. $20x^3 + 38x^2 + 12x$

42. $10t^3 - 45t^2 + 45t$

43. $3v^2 - vw - 2w^2$

44. $4x^2 - 100y^2$

45. $6n^2 - nq - 2q^2$

ONGOING ASSESSMENT

46. **Open-ended Problem** Make a concept map about factoring. Be sure to include examples.

38. a. to the nearest tenth, $0.7 < t < 3.3$

b. No. At 3.3 s, the ball is just over 37 ft high, but it is only about 254 ft from home plate. The ball is farther from home plate after 3.3 s, but it is no longer 37 ft high. In fact, the ball hits the ground about 313 ft from home plate.

39. about 4.43 s; 17.30 ft; Yes, it could clear the centerfield wall or the rightfield wall.

40. $3x(x - 4)(x + 2)$

41. $2x(2x + 3)(5x + 2)$

42. $5t(t - 3)(2t - 3)$

43. $(3v + 2w)(v - w)$

44. $4(x + 5y)(x - 5y)$

45. $(3n - 2q)(2n + q)$

46. Answers may vary.

47–52.

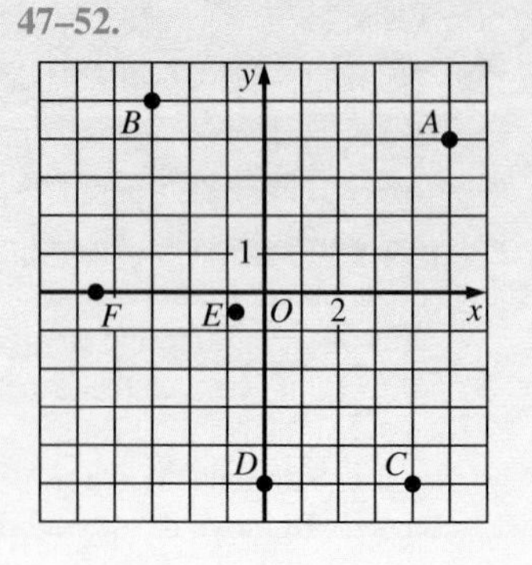

SPIRAL REVIEW

Graph each point in a coordinate plane. Label each point with its letter. Name the quadrant (if any) in which the point lies. *(Section 2.4)*

47. $A(5, 4)$ **48.** $B(-3, 5)$ **49.** $C(4, -5)$

50. $D(0, -5)$ **51.** $E\left(-\frac{3}{4}, -\frac{1}{2}\right)$ **52.** $F(-4.5, 0)$

Tell whether the data show direct variation. Give reasons for your answer. *(Section 3.2)*

53. Air Mail Rates to Asia in 1961

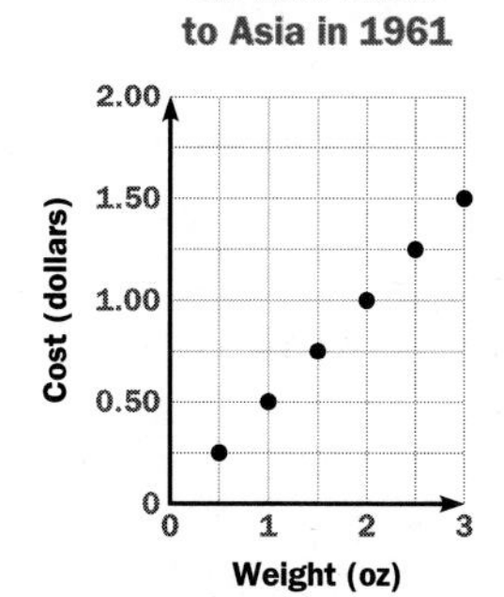

54. Growth of an Investment Earning 10% Interest

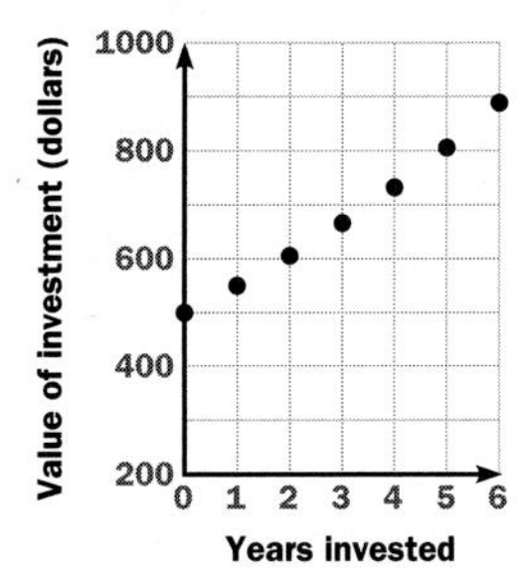

ASSESS YOUR PROGRESS

VOCABULARY

polynomial equation (p. 436) **factoring a polynomial** (p. 437)

Solve each equation. *(Section 10.3)*

1. $(2x - 3)(4x + 5) = 0$ 2. $x^2 - 3x = 0$ 3. $2y^2 = -8y$
4. The equation $d = 0.05s^2 + s$ models the distance d in feet traveled by a car moving s miles per hour once the driver has decided to stop the car. Find the values of s for which $d = 0$. Interpret the solutions.

Factor each polynomial. *(Sections 10.4 and 10.5)*

5. $x^2 + 7x + 10$ **6.** $k^2 + 3k - 28$ **7.** $k^2 - 12k - 28$

8. $2a^2 - 11a + 5$ **9.** $25z^2 + 10z + 1$ **10.** $4m^2 - 100m + 49$

11. $9c^2 - 21c + 10$ **12.** $81t^2 - 64$ **13.** $9d^2 - d - 10$

Solve each equation. *(Section 10.5)*

14. $n^2 + 3n - 28 = 0$ **15.** $4x^2 + 15x + 14 = 0$ **16.** $24p^2 - 5p = 1$

17. Use the stopping-distance formula from Exercise 4. How fast can a car travel if it needs at least 15 ft to come to a complete stop?

18. Journal Discuss how you can use factoring to solve certain polynomial equations, and how you know when this method is appropriate.

Apply⇔Assess

Assess Your Progress

Journal Entry
You may wish to suggest that students include examples of polynomial equations that are appropriate for factoring as well as ones that are not.

Progress Check 10.3–10.5

See page 457.

Practice 70 for Section 10.5

NAME ______ DATE ______

Practice 70

FOR USE WITH SECTION 10.5

Factor each trinomial.

1. $t^2 - 15t + 56$ $(t - 7)(t - 8)$
2. $a^2 + 10a - 24$ $(a + 12)(a - 2)$
3. $x^2 + 2x - 63$ $(x + 9)(x - 7)$
4. $b^2 + 11x + 30$ $(b + 5)(b + 6)$
5. $y^2 - 225$ $(y + 15)(y - 15)$
6. $k^2 - 5k - 66$ $(k - 11)(k + 6)$
7. $2n^2 - 17n + 21$ $(2n - 3)(n - 7)$
8. $3c^2 - 7c - 40$ $(3c + 8)(c - 5)$
9. $2p^2 + 19p + 35$ $(2p + 5)(p + 7)$
10. $10r^2 - r - 3$ $(5r - 3)(2r + 1)$
11. $6m^2 - 37m + 45$ $(2m - 9)(3m - 5)$
12. $15v^2 + 23v - 22$ $(5v + 11)(3v - 2)$

Solve each equation by factoring.

13. $x^2 - 7x - 18 = 0$ $9, -2$
14. $y^2 + y - 110 = 0$ $-11, 10$
15. $n^2 - 11n + 28 = 0$ $4, 7$
16. $3t^2 + 22t + 7 = 0$ $-\frac{1}{3}, -7$
17. $2a^2 - 17a + 30 = 0$ $\frac{5}{2}, 6$
18. $5c^2 + 18c - 8 = 0$ $-4, \frac{2}{5}$
19. $4k^2 - 24k + 27 = 0$ $\frac{3}{2}, \frac{9}{2}$
20. $6w^2 + 5w - 21 = 0$ $-\frac{7}{3}, \frac{3}{2}$
21. $12u^2 - 19u + 5 = 0$ $\frac{1}{3}, \frac{5}{4}$
22. $8x^2 + 5x = 0$ $0, -\frac{5}{8}$
23. $36y^2 - 25 = 0$ $\pm\frac{5}{6}$
24. $9 = 121p^2$ $\pm\frac{3}{11}$

25. The height (in feet) of the ball t seconds after being released for a jump shot in a basketball game can be modeled by the equation

$$y = -16t^2 + 12t + 8$$

a. The rim of the basket is 10 ft high. At what times will the ball reach this height? Explain in terms of the physics of the situation why there are two answers. Which answer represents a moment when the ball has a chance of going in the basket?

b. Suppose the horizontal distance x (in feet) that the ball travels in t seconds is modeled by the equation $x = 40t$. How far away from the basket should the shooter be in order to make the shot (without hitting the backboard)? 20 ft

25. a. $\frac{1}{4}$ s, $\frac{1}{2}$ s; The ball reaches this height on the way up and again on the way down.; $\frac{1}{2}$ s

26. **Writing** Suppose that you are given two rational numbers, like $\frac{2}{3}$ and $-\frac{1}{5}$. Describe how to find a quadratic equation, with integer coefficients, that has those numbers as its roots. Does your method work even if the two given rational numbers are the same? Does it work if the two rational numbers are integers? Answers may vary. Check students' work.

70 **Practice Bank,** ALGEBRA 1: EXPLORATIONS AND APPLICATIONS

---

47. I
48. II
49. IV
50. none
51. III
52. none
53. Yes; the points lie on a straight line that passes through the origin.
54. No; the points are not on a straight line, and the graph does not pass through the origin.

Assess Your Progress

1. $\frac{3}{2}, -\frac{5}{4}$
2. 0, 3
3. 0, –4
4. 0, –20; The car must be traveling at 0 mi/h in order to stop in 0 ft. The negative solution does not make sense in this real-world problem. Speed cannot be negative.
5. $(x + 2)(x + 5)$
6. $(k + 7)(k - 4)$
7. $(k - 14)(k + 2)$
8. $(2a - 1)(a - 5)$
9. $(5z + 1)^2$
10. $(2m - 1)(2m - 49)$
11. $(3c - 2)(3c - 5)$
12. $(9t + 8)(9t - 8)$
13. $(9d - 10)(d + 1)$
14. 4, –7
15. $-2, -\frac{7}{4}$
16. $\frac{1}{3}, -\frac{1}{8}$
17. 10 mi/h
18. Answers may vary.

Project Notes

Mathematical Goals

- Measure the volume of water displaced by a model of a doll or an action figure and convert the volume to liters.
- Calculate the scale factor s of a person's height to the model's height.
- Calculate the volume of the person using the formula $V_p = V_m s^3$.
- Calculate how much the person would weigh using the formula Weight = Volume × Weight Density.

Planning

Materials

- Doll or action figure
- Large pitcher
- Large pot
- Water
- Tape
- Measuring cups and spoons

Project Teams
Students can choose partners to work with and then discuss how they wish to proceed.

Guiding Students' Work

Students can share the use of pitchers and pots to do this project. Have a large pot of water available that can be reused by each pair of students. Make sure that hollow models do not have any openings that would allow water to leak inside them. Provide students with tape to cover any openings.

Second-Language Learners
Students learning English may need help from a peer tutor or aide to write an analysis of results.

PORTFOLIO PROJECT

Volume and Weight

Dolls and action figures are scale models of human beings, but not all such models are shaped like real people. You can learn something about the accuracy of a doll's proportions by measuring its volume.

PROJECT GOAL **Find out how much a doll or an action figure would weigh if it were a real person.**

Taking Measurements

Work with a partner to measure the volume of a model (either a doll or an action figure) by measuring the amount of water it displaces. You may want to do the experiment several times to be sure you are getting reliable results.

1. **MEASURE** the height of your model. Place a pitcher large enough to contain your model into a large pot. Fill the pitcher to the very top with water, being careful not to spill any into the pot.
2. **TAPE** over any places where water might leak in if your model is hollow. Then completely submerge your model in the water without letting your fingers go beneath the surface.
3. **REMOVE** your model and pitcher from the pot. Then use measuring cups and spoons to measure the volume of the water that was displaced, or spilled over, into the pot. This is the volume of your model.
4. **CONVERT** the volume of your model to liters using the conversion chart at the left.

U.S. Customary Units	Metric Units
1 cup	0.24 liters
1 tablespoon	0.015 liters
1 teaspoon	0.005 liters

Calculating Weight

The polynomial equation below shows the relationship between the volume of a real person V_p and the volume of a scale model V_m, in liters.

$$V_p = V_m s^3$$

The variable s is a scale factor.

$$s = \frac{\text{Person's height (in.)}}{\text{Model's height (in.)}}$$

Suppose the doll or action figure that you measured were a real person.

- About how tall do you think the person would be? What is the scale factor s for your model?
- What would be the person's volume?
- How much would the person weigh? Use the formula below to help you find out. The average weight density for males is 2.34 pounds per liter. The average weight density for females is 2.31 pounds per liter.

$$\text{Weight} = \text{Volume} \times \text{Weight density}$$

Writing a Report

Summarize your experiment. Be sure to include:

- a brief description of the procedure you used
- the data you collected
- your calculations
- a discussion of some possible sources of error

Analyze your results. Is the weight you calculated realistic? Why or why not? If not, what could the designers of the model do to make the model more realistic? Explain your answer in terms of the volume relationship at the top of this page.

Self-Assessment

Conclude your report with a discussion of what went well for you during this experiment. Also discuss the aspect of the experiment that you and your partner found the most challenging and how you dealt with it. Describe what you have learned about the relationship between the volume of a scale model and the weight of the object it represents.

Project Notes

Guiding Students' Work

Rubric for Chapter Project

4 Students measure the volume of water accurately and convert the volume to liters. All subsequent calculations are done correctly and all questions are answered. A report is written that includes a complete and accurate summary of the experiment. The results are analyzed and reasonable answers are given to the questions asked under *Writing a Report.*

3 Students perform all aspects of the experiment in an orderly manner and make the required calculations correctly. The written report covers all the points required, but some parts are too brief or not entirely clear. The results are analyzed and the questions answered, but more thought would have produced a better analysis overall.

2 Students take the necessary measurements but are a little careless in their measurements of the volume of the water. They also have errors in their calculations of the volume and weight of the real person. A report is written, but there are shortcomings in both the summary and the analysis.

1 Students attempt to start the experiment but are very careless in measuring the volume of the water. Subsequent calculations are incorrect and no report is written. Students should be encouraged to speak with the teacher as soon as possible to review their work and to make a new start on the project.

Chapter Support

Course Guide: Chapter 10
Lesson Plans: Chapter 10
Practice Bank:
Cumulative Practice 71
Study Guide: Chapter 10 Review
Assessment Book:
Chapter Tests 45 and 46
Chapter 10 Alternative Assessment
Test Generator software
Portfolio Project Book
Preparing for College Entrance Tests
Professional Handbook

Progress Check 10.1–10.2

1. Tell which of the following is not a polynomial. *(Section 10.1)* c

a. $4x - 3y$

b. $6x^3 - 4x + 3$

c. $\frac{5}{3y}$

Add or subtract each pair of polynomials. Write each answer in standard form and give its degree. *(Section 10.1)*

2. $(4x^2 + 5x - 2) + (6x^2 - 2x + 1)$ $10x^2 + 3x - 1$; 2

3. $(3x^2 - x - 2) - (3x^2 + 5x + 4)$ $-6x - 6$; 1

Multiply. *(Section 10.2)*

4. $(s - 7)(s + 2)$ $s^2 - 5s - 14$

5. $(3r - 2)(5r + 1)$ $15r^2 - 7r - 2$

6. $3x(4x^2 - x + 6)$ $12x^3 - 3x^2 + 18x$

7. $(6x + 1)(3x^2 - x + 4)$ $18x^3 - 3x^2 + 23x + 4$

8. Find the area of the rectangle below. *(Section 10.2)*

10 – 2x

4 – 2x

$4x^2 - 28x + 40$

CHAPTER

10 Review

STUDY TECHNIQUE

Without looking at the book, outline this chapter's content. Use the book to check the outline. Fill in any important ideas that you missed.

VOCABULARY

polynomial (p. 423)
standard form of a polynomial (p. 423)
degree of a term (p. 423)
degree of a polynomial (p. 423)
linear polynomial (p. 423)
quadratic polynomial (p. 423)
cubic polynomial (p. 423)
polynomial equation (p. 436)
factoring a polynomial (p. 437)

SECTIONS 10.1 *and* 10.2

Polynomials can be added or subtracted by combining like terms.

$$\begin{array}{r} 2x^2 - 5x + 14 \\ \underline{7x^2 - 9x + 24} \end{array}$$

To subtract, add the **opposite**.

$$\begin{array}{r} 2x^2 - 5x + 14 \\ \underline{-7x^2 + 9x - 24} \\ -5x^2 + 4x - 10 \end{array}$$

You can use one of several methods to multiply polynomials.

Horizontal method

$$\begin{aligned} -2x(x^2 + 3x - 1) &= -2x(x^2) + (-2x)(3x) - (-2x)(1) \\ &= -2x^3 - 6x^2 + 2x \end{aligned}$$

Vertical method

$$\begin{array}{r} x^2 + 5x + 6 \\ \underline{2x - 3} \\ 2x^3 + 10x^2 + 12x \\ \underline{-3x^2 - 15x - 18} \\ 2x^3 + 7x^2 - 3x - 18 \end{array}$$

FOIL method

$$\begin{aligned} (x + 2)(3x - 4) &= x(3x) + x(-4) + 2(3x) + 2(-4) \\ &= 3x^2 - 4x + 6x - 8 \\ &= 3x^2 + 2x - 8 \end{aligned}$$

SECTIONS 10.3, 10.4, *and* 10.5

Some polynomials can be **factored** by finding the greatest common factor (GCF).

$$4x^2 + 10x = 2x \cdot 2x + 2x \cdot 5$$
$$= 2x(2x + 5)$$

The GCF of $4x^2$ and $10x$ is $2x$.

Polynomials can also be factored using a guess and check method.

To factor $2k^2 - 13k + 21$, guess, check, and revise.

$$(\underline{?}\,k \pm \underline{?})(\underline{?}\,k \pm \underline{?})$$

The blue terms will be factors of $2k^2$: $2k$, k.

The orange terms will be factors of 21: 3, 7   −3, −7   1, 21   −1, −21

Try different combinations. Focus on getting the correct linear term.

$(2k + 3)(k + 7)$	$(2k + 7)(k + 3)$	$(2k - 7)(k - 3)$
$3k$	$7k$	$-7k$
$14k$	$6k$	$-6k$
$17k$	$13k$	$-13k$

So, $2k^2 - 13k + 21 = (2k - 7)(k - 3)$. — the correct linear term

You can use the FOIL method to check your results.

To factor a quadratic polynomial with a negative constant term, such as $x^2 + x - 6$, remember that the signs of the binomials will be different.

$$(x \pm \underline{?})(x \pm \underline{?})$$

The orange terms will be factors of −6: 1, −6   −1, 6   3, −2   −3, 2

$(x + 1)(x - 6)$	$(x - 1)(x + 6)$	$(x + 3)(x - 2)$
x	$-x$	$3x$
$-6x$	$6x$	$-2x$
$-5x$	$5x$	x

the correct linear term

So, $x^2 + x - 6 = (x + 3)(x - 2)$.

You have seen three methods for solving quadratic equations.

1. Factor and use the zero-product property.
2. Use the quadratic formula: $x = \dfrac{-b \pm \sqrt{b^2 - 4ac}}{2a}$
3. Graph the related equation. The x-intercepts are the solutions.

Progress Check 10.3–10.5

Factor each polynomial. *(Section 10.3)*

1. $3x^2 + 12x$ $3x(x + 4)$
2. $12g^3 + 16g^2 - 8g$ $4g(3g^2 + 4g - 2)$

Solve each equation. *(Section 10.3)*

3. $3n(n - 5) = 0$ 0; 5
4. $(a + 3)(2a - 9) = 0$ $-3; \frac{9}{2}$

Factor, if possible. If not, write *not factorable*. *(Section 10.4)*

5. $k^2 - 12k + 32$ $(k - 8)(k - 4)$
6. $3d^2 - 13d + 4$ $(3d - 1)(d - 4)$
7. $s^2 - 10s + 8$ not factorable

Factor each trinomial. *(Section 10.5)*

8. $m^2 - 4m - 12$ $(m - 6)(m + 2)$
9. $3g^2 - 10g - 8$ $(3g + 2)(g - 4)$

Solve by factoring. *(Section 10.5)*

10. $w^2 - 2w - 3 = 0$ 3; −1
11. $2k^2 + k = 15$ $-3; \frac{5}{2}$

Chapter 10 Assessment

Form A Chapter Test

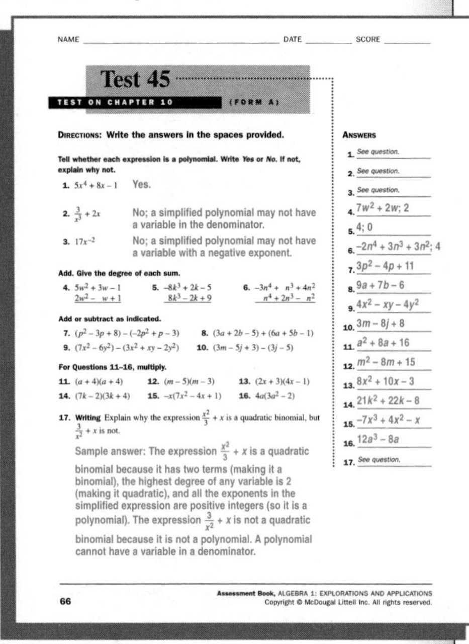

NAME ______________ DATE ________ SCORE ________

Test 45

TEST ON CHAPTER 10 (FORM A)

DIRECTIONS: Write the answers in the spaces provided.

Tell whether each expression is a polynomial. Write *Yes* or *No*. If not, explain why not.

1. $5x^4 + 8x - 1$ Yes.
2. $\frac{3}{x^3} + 2x$ No; a simplified polynomial may not have a variable in the denominator.
3. $17x^{-2}$ No; a simplified polynomial may not have a variable with a negative exponent.

Add. Give the degree of each sum.

4. $5w^2 + 3w - 1$ / $2w^2 - w + 1$
5. $-8k^3 + 2k - 5$ / $8k^3 - 2k + 9$
6. $-3n^4 + n^3 + 4n^2$ / $n^4 + 2n^3 - n^2$

Add or subtract as indicated.

7. $(p^2 - 3p + 8) - (-2p^2 + p - 3)$
8. $(3a + 2b - 5) + (6a + 5b - 1)$
9. $(7x^2 - 6y^2) - (3x^2 + xy - 2y^2)$
10. $(3m - 5j + 3) - (3j - 5)$

For Questions 11–16, multiply.

11. $(a + 4)(a + 4)$
12. $(m - 5)(m - 3)$
13. $(2x + 3)(4x - 1)$
14. $(7k - 2)(3k + 4)$
15. $-x(7x^2 - 4x + 1)$
16. $4a(3a^2 - 2)$
17. **Writing** Explain why the expression $\frac{x^2}{3} + x$ is a quadratic binomial, but $\frac{3}{x^2} + x$ is not.

Sample answer: The expression $\frac{x^2}{3} + x$ is a quadratic binomial because it has two terms (making it a binomial), the highest degree of any variable is 2 (making it quadratic), and all the exponents in the simplified expression are positive integers (so it is a polynomial). The expression $\frac{3}{x^2} + x$ is not a quadratic binomial because it is not a polynomial. A polynomial cannot have a variable in a denominator.

ANSWERS

1. *See question.*
2. *See question.*
3. *See question.*
4. $7w^2 + 2w$; 2
5. 4; 0
6. $-2n^4 + 3n^3 + 3n^2$; 4
7. $3p^2 - 4p + 11$
8. $9a + 7b - 6$
9. $4x^2 - xy - 4y^2$
10. $3m - 8j + 8$
11. $a^2 + 8a + 16$
12. $m^2 - 8m + 15$
13. $8x^2 + 10x - 3$
14. $21k^2 + 22k - 8$
15. $-7x^3 + 4x^2 - x$
16. $12a^3 - 8a$
17. *See question.*

66 **Assessment Book**, ALGEBRA 1: EXPLORATIONS AND APPLICATIONS

Chapter 10 Assessment

Form B Chapter Test

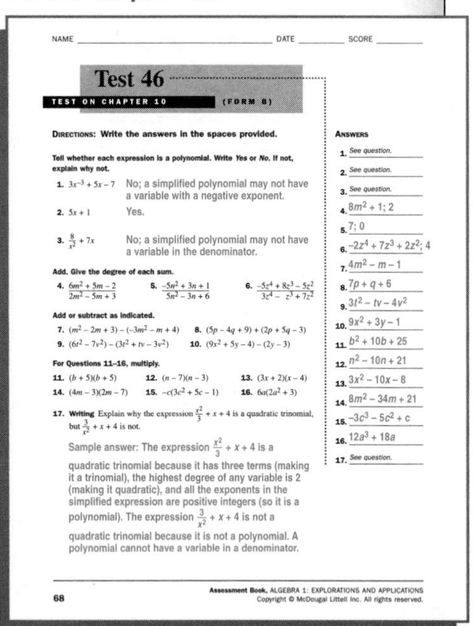

NAME ______________ DATE ________ SCORE ________

Test 46

TEST ON CHAPTER 10 (FORM B)

DIRECTIONS: Write the answers in the spaces provided.

Tell whether each expression is a polynomial. Write *Yes* or *No*. If not, explain why not.

1. $3x^{-3} + 5x - 7$ No; a simplified polynomial may not have a variable with a negative exponent.
2. $5x + 1$ Yes.
3. $\frac{8}{x^2} + 7x$ No; a simplified polynomial may not have a variable in the denominator.

Add. Give the degree of each sum.

4. $6m^2 + 5m - 2$ / $2m^2 - 5m + 3$
5. $-5n^2 + 3n + 1$ / $5n^2 - 3n + 6$
6. $-5z^4 + 8z^3 - 5z^2$ / $3z^4 - z^3 + 7z^2$

Add or subtract as indicated.

7. $(m^2 - 2m + 3) - (-3m^2 - m + 4)$
8. $(5p - 4q + 9) + (2p + 5q - 3)$
9. $(6t^2 - 7v^2) - (3t^2 + tv - 3v^2)$
10. $(9x^2 + 5y - 4) - (2y - 3)$

For Questions 11–16, multiply.

11. $(b + 5)(b + 5)$
12. $(n - 7)(n - 3)$
13. $(3x + 2)(x - 4)$
14. $(4m - 3)(2m - 7)$
15. $-c(3c^2 + 5c - 1)$
16. $6a(2a^2 + 3)$
17. **Writing** Explain why the expression $\frac{x^2}{3} + x + 4$ is a quadratic trinomial, but $\frac{3}{x^2} + x + 4$ is not.

Sample answer: The expression $\frac{x^2}{3} + x + 4$ is a quadratic trinomial because it has three terms (making it a trinomial), the highest degree of any variable is 2 (making it quadratic), and all the exponents in the simplified expression are positive integers (so it is a polynomial). The expression $\frac{3}{x^2} + x + 4$ is not a quadratic trinomial because it is not a polynomial. A polynomial cannot have a variable in a denominator.

ANSWERS

1. *See question.*
2. *See question.*
3. *See question.*
4. $8m^2 + 1$; 2
5. 7; 0
6. $-2z^4 + 7z^3 + 2z^2$; 4
7. $4m^2 - m - 1$
8. $7p + q + 6$
9. $3t^2 - tv - 4v^2$
10. $9x^2 + 3y - 1$
11. $b^2 + 10b + 25$
12. $n^2 - 10n + 21$
13. $3x^2 - 10x - 8$
14. $8m^2 - 34m + 21$
15. $-3c^3 - 5c^2 + c$
16. $12a^3 + 18a$
17. *See question.*

68 **Assessment Book**, ALGEBRA 1: EXPLORATIONS AND APPLICATIONS

CHAPTER

10 Assessment

VOCABULARY REVIEW

1. Give an example of a polynomial with degree 3. Write the polynomial in standard form.
2. Give examples of a linear polynomial, a quadratic polynomial, and a cubic polynomial.
3. Explain what it means to factor a polynomial.

SECTION 10.1

Tell whether each expression is a polynomial. Write *Yes* or *No*. If not, explain why not.

4. $2m + 3n$
5. $5a^3 + 6$
6. 15
7. $7x^2 - 7x - 7$
8. $6x^{-2}$
9. $3x^2 - \frac{2}{x^2}$

Add. Give the degree of each sum.

10. $-3x^2 + 15x + 6$ / $3x^2 - 7x + 4$
11. $6a^2 - 6a - 7$ / $a^2 + 5a + 9$
12. $n^3 + 3n^2 - 2n + 2$ / $3n^2 - 2n + 2$

Add or subtract as indicated.

13. $(9x^2 - 2x + 5) + (-x^2 + 4x + 6)$
14. $(r^2 - 7r + 8) - (-3r^2 + 7r + 6)$
15. $(6x^3 - 9x^2 + 14) - (5x^2 + 10x)$
16. $(b + 2c - 6) - (-2b + 4c - 1)$

SECTION 10.2

Multiply.

17. $(x - 3)(x - 4)$
18. $(b - 9)(b + 5)$
19. $(r - 3)(5r - 6)$
20. $(3c + 8)(2c + 2)$
21. $3a(2a^2 + 9)$
22. $-d(d^2 - 3d - 6)$
23. $(2h + 7)(h^2 - 4h + 6)$
24. $(4x^2 + 3x + 2)(x - 8)$

ANSWERS Chapter 10

Assessment

1. Answers may vary. An example is given. $2x^3 - 4x + 6$
2. Answers may vary. Examples are given. linear: $4x - 6$; quadratic: $x^2 - 2$; cubic: $4x^3 + 8x^2 + x - 5$
3. To factor a polynomial means to write it as a product of two or more polynomials.
4. Yes.
5. Yes.
6. Yes.
7. Yes.
8. No; the power of x is not a whole number.
9. No; the power of x in the term $-\frac{2}{x^2} = -2x^{-2}$ is not a whole number.
10. $8x + 10$; 1
11. $7a^2 - a + 2$; 2
12. $n^3 + 6n^2 - 4n + 4$; 3
13. $8x^2 + 2x + 11$
14. $4r^2 - 14r + 2$
15. $6x^3 - 14x^2 - 10x + 14$
16. $3b - 2c - 5$
17. $x^2 - 7x + 12$
18. $b^2 - 4b - 45$
19. $5r^2 - 21r + 18$
20. $6c^2 + 22c + 16$
21. $6a^3 + 27a$
22. $-d^3 + 3d^2 + 6d$
23. $2h^3 - h^2 - 16h + 42$

25. Write a polynomial in standard form that represents the area of the blue region in the figure at the right.

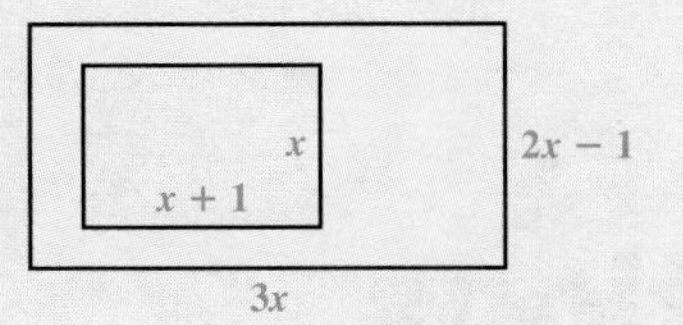

SECTION 10.3

Factor each polynomial.

26. $13y^2 - 4y$ **27.** $6a^2 - 10a$ **28.** $6x^3 + 4x^2 - 8x$

Solve each equation.

29. $7y(y - 7) = 0$ **30.** $0 = 2x(4x + 1)$ **31.** $(3a - 3)(5a + 8) = 0$

32. $x^2 - 5x = 0$ **33.** $6m^2 + 6m = 0$ **34.** $8c = 4c^2$

35. Suppose the area of the shaded region in Question 25 is 12 in.^{2}. Find the value of x.

SECTIONS 10.4 *and* 10.5

Factor, if possible. If not, write *not factorable*.

36. $x^2 + 11x + 10$ **37.** $y^2 - 11y + 24$ **38.** $a^2 - 6a + 6$

Factor each polynomial.

39. $q^2 + 6q - 7$ **40.** $y^2 - 10y + 16$ **41.** $a^2 + a - 12$

42. $m^2 - 9$ **43.** $2x^2 - 3x + 1$ **44.** $4n^2 + 12n + 5$

Solve each equation by factoring.

45. $3x^2 - 8x - 11 = 0$ **46.** $7n^2 + n - 6 = 2$ **47.** $4a^2 - 3a = 10$

48. Writing Explain how to use the graph of the equation $y = 6x^2 + 30x + 36$ at the right to solve the equation $6x^2 + 30x = -36$. Describe two other methods you could use to solve the equation $6x^2 + 30x = -36$.

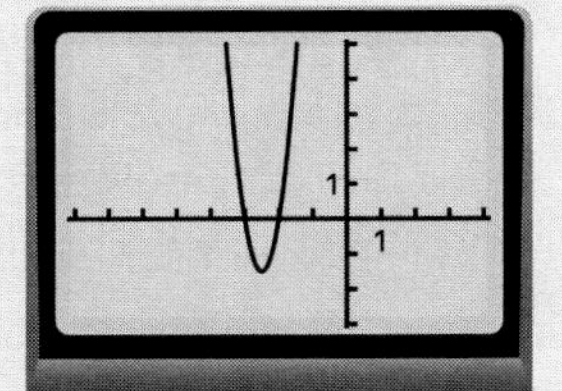

PERFORMANCE TASK

49. Suppose the equation $h = -16t^2 + 40t$ models the height h in feet of a ball t seconds after the ball has been hit. Will the ball reach a height of 24 ft? If so, when? Use methods you have learned in this chapter to support your answer.

Chapter 10 Assessment
Form C Alternative Assessment

Chapter 10

ALTERNATIVE ASSESSMENT

1. a. Give an example of a linear monomial, a quadratic trinomial, and a cubic binomial.

b. Explain the difference between classifying a polynomial by its degree and classifying a polynomial by its number of terms.

2. a. Draw a tile model that represents the product $(x + 2)(x - 4)$ as an area. Then represent the area as a polynomial in standard form.

b. Use tiles to factor the following expressions. Describe your procedure.
$x^2 + 5x + 4$ $2x^2 + 7x + 6$ $x^2 + 2x - 3$

3. a. Make up three polynomials that each have a factor of $x - 3$.

b. Use each polynomial from part (a) to write an equation in the form $y = \underline{\ ?\ }$.

c. What would be true about the graphs of all three equations you wrote in part (b)? Explain your conclusions.

4. Performance Task Explain how you decide which method (factoring, graphing, or the quadratic formula) you will use to solve a quadratic equation.

5. a. How could you use graphing to prove that $(x + 3)(x - 4)$ is the factored form of $x^2 - x - 12$?

b. Is $(3x + 1)(-x - 5)$ the factored form of $-3x^2 + 16x - 5$? Use graphing to check.

6. a. How is the degree of the sum of two polynomials related to the degrees of the two polynomials being added?

b. How is the degree of the product of two polynomials related to the degrees of the two polynomials being multiplied?

106 **Assessment Book,** ALGEBRA 1: EXPLORATIONS AND APPLICATIONS

7. Project Suppose you have been hired to design a rectangular box for a new brand of cereal. You are to use a 16 in. by 20 in. piece of cardboard to do the job. You need to design your box so that it will hold the greatest possible amount. *Note:* Do not include the lid in your design.

a. Cut a square out of each corner. Fold up the sides to form a box.

b. Find the surface area and volume of your box.

c. Repeat parts (a) and (b) for two more boxes.

d. Let x = the length of a side of the square cut from each corner. Write an equation to represent the volume of the box in terms of x.

e. Graph the equation with a graphing calculator. Find the maximum volume. What happens when $x > 8$?

f. Why is the shape of the box with the maximum volume not a cube?

g. If the container did not have to be a rectangular box, how might the problem change?

h. What problems might occur with some of the containers discussed in part (g)?

Assessment Book, ALGEBRA 1: EXPLORATIONS AND APPLICATIONS 107

24. $4x^3 - 29x^2 - 22x - 16$
25. $5x^2 - 4x$
26. $y(13y - 4)$
27. $2a(3a - 5)$
28. $2x(3x^2 + 2x - 4)$
29. 0, 7
30. $0, -\frac{1}{4}$
31. $1, -\frac{8}{5}$
32. 0, 5
33. 0, −1
34. 0, 2
35. 2 in.
36. $(x + 10)(x + 1)$
37. $(y - 8)(y - 3)$
38. not factorable
39. $(q + 7)(q - 1)$
40. $(y - 8)(y - 2)$
41. $(a + 4)(a - 3)$
42. $(m + 3)(m - 3)$
43. $(x - 1)(2x - 1)$
44. $(2n + 1)(2n + 5)$
45. $-1, \frac{11}{3}$
46. $-\frac{8}{7}, 1$
47. $-\frac{5}{4}, 2$
48. The equation is equivalent to $6x^2 + 30x + 36 = 0$. The solutions of this equation are the x-intercepts of the graph. Two other ways to solve the equation are: factor $6x^2 + 30x + 36$ and use the zero product property; use the quadratic formula.
49. Yes, at 1 s and again at 1.5 s. Solve the equation $-16t^2 + 40t = 24$ or $-16t^2 + 40t - 24 = 0$. Factor $-8(2t - 3)(t - 1) = 0$. Solve using the zero-product property.

CHAPTER

11 Rational Functions

OVERVIEW

Connecting to Prior and Future Learning

⇔ The chapter opens by presenting two different applications of rational functions, namely inverse variations and weighted averages. Students study these topics by graphing functions and exploring problems that can be solved by using these concepts.

⇔ The previously learned skills of finding the least common denominator and solving equations are used in Section 11.3 to solve rational equations. Students' equation-solving skills are again put to work as they learn to solve a formula for one of its variables. Both of these topics will be presented in more depth in Algebra 2.

⇔ Learning to multiply, divide, add, and subtract rational expressions is the focus of the last two sections of Chapter 11. The concepts presented here are an extension of skills students have already acquired. These skills will be used in Algebra 2.

Chapter Highlights

Interview with Alex MacLean: The relationship between photography and mathematics is emphasized in this interview, with related exercises on pages 467 and 486.

Explorations in Chapter 11 involve exploring rectangles with equal areas using algebra tiles and graph paper, exploring weighted averages by comparing game scores, and exploring subtracting rational expressions using a ruler and graph paper.

The Portfolio Project: Students explore inverse variations and rational expressions by making a balance and then studying the relationship between weights and their distances from the center of the balance. Students summarize the results of their experiments in a written report.

Technology: The use of technology in Chapter 11 includes studying a rational formula with spreadsheets, graphing rational functions, and solving rational equations with graphing calculators. The method used to solve rational equations with a graph is similar to the method used to solve quadratic equations with a graph in Chapter 8.

OBJECTIVES

Section	Objectives	NCTM Standards
11.1	• Recognize and describe data that show inverse variation. • Solve real-world problems involving inverse variation.	1, 2, 3, 4, 5, 8
11.2	• Find weighted averages. • Use weighted averages to solve real-world problems.	1, 2, 3, 4, 5, 8, 10
11.3	• Solve rational equations. • Solve real-world problems involving rational equations.	1, 2, 3, 4, 5, 8
11.4	• Solve a formula for one of its variables. • Perform multiple calculations with the same formula.	1, 2, 3, 4, 5, 8
11.5	• Multiply and divide rational expressions. • Use multiplication and division of rational expressions to solve mathematical and real-world problems.	1, 2, 3, 4, 5, 8
11.6	• Add and subtract rational expressions. • Use addition and subtraction of rational expressions to solve real-world problems.	1, 2, 3, 4, 5, 6, 8

INTEGRATION

Mathematical Connections	11.1	11.2	11.3	11.4	11.5	11.6
algebra	**463–468***	**469–475**	**476–481**	**482–487**	**488–493**	**494–499**
geometry	463	475	480	485	488, 492, 493	498
data analysis, probability, discrete math	467, 468	**469–475**			491	
patterns and functions						494, 497
logic and reasoning	463, 465–468	**469–475**	476, 478–481	483, 485–487	488, 490–493	494, 495, 497–499
Interdisciplinary Connections and Applications						
chemistry and physics	465		479			
business and economics		471–473		485		
sports and recreation	468	473	478, 480, 481	482, 485		
transportation	464, 466					498
test scores, optics, cars, cooking, engineering		470, 472		484–486	488, 493	498

****Bold page numbers*** *indicate that a topic is used throughout the section.*

TECHNOLOGY

Section	*opportunities for use with* Student Book	Support Material
11.1	graphing calculator	**Technology Book:** Calculator Activity 11
11.2	graphing calculator McDougal Littell Software *Stats!* *Function Investigator*	**Technology Book:** Spreadsheet Activity 11
11.3	graphing calculator McDougal Littell Software *Function Investigator*	
11.4	graphing calculator spreadsheet software McDougal Littell Software *Stats!*	
11.5	graphing calculator	
11.6	graphing calculator	

Planning Guide

Regular Scheduling (45 min)

Section	Materials Needed	Core Assignment	Extended Assignment	*exercises that feature* Applications	Communication	Technology
11.1	36 algebra unit tiles, graph paper, ruler, graphing calculator	1–11, 14–17, 19–25	1–25	4, 10, 11, 14–18	12, 13, 18–20	5–8
11.2	bowl, narrow cup or glass, dried beans, graphing calculator, dowel, string, identical objects heavier than dowel	1–3, 5–7, 9–12, 14, 15, 19–25, AYP*	1–12, 14–25, AYP	1–3, 5–7, 12, 16	4, 16–19	8–11, 13
11.3	graphing calculator	1–10, 12, 13, 15–24, 27, 31–37	1–37	11, 19, 20, 27–29	12–14, 25, 26, 29–31	4, 24
11.4	graphing calculator, spreadsheet software	1–17, 20, 22, 25–31, AYP	1–22, 24–31, AYP	2, 19–22	18, 25	12–17, 24
11.5		1–24, 26, 27, 33–39	1–39	10, 11, 31, 32	9, 18, 29	
11.6	ruler, graph paper	**Day 1:** 1–16, 18–20 **Day 2:** 21–26, 28, 30–36, AYP	**Day 1:** 1–20 **Day 2:** 21–36, AYP	17, 25, 26	15 29, 30	
Review/ Assess		**Day 1:** 1–14 **Day 2:** 15–29 **Day 3:** Ch. 11 Test	**Day 1:** 1–14 **Day 2:** 15–29 **Day 3:** Ch. 11 Test	4–8	1–3, 5	10 17
Portfolio Project		Allow 2 days.	Allow 2 days.			

Yearly Pacing (with Portfolio Project)	**Chapter 11 Total** 12 days	**Chapters 1–11 Total** 149 days	**Remaining** 11 days	**Total** 160 days

Block Scheduling (90 min)

	Day 70	Day 71	Day 72	Day 73	Day 74	Day 75
Teach/Interact	11.1: Exploration, page 463 11.2: Exploration, page 469	11.3 11.4	11.5 11.6: Exploration, page 494	Continue with 11.6 Review	Review Port. Proj.	Port. Proj. Ch. 11 Test
Apply/Assess	**11.1:** 1–11, 14–17, 19–25 **11.2:** 1–3, 5–7, 9–12, 14, 15, 19–25, AYP*	**11.3:** 1–10, 12, 13, 15–24, 27, 31–37 **11.4:** 1–17, 20, 22, 25–31, AYP	**11.5:** 1–24, 26, 27, 33–39 **11.6:** 1–16, 18–20	**11.6:** 21–26, 28, 30–36, AYP **Review:** 1–14	**Review:** 15–29 **Port. Proj.**	**Port. Proj.** **Ch. 11 Test**

NOTE: A one-day block has been added for the Portfolio Project—timing and placement to be determined by teacher.

Yearly Pacing (with Portfolio Project)	**Chapter 11 Total** 6 days	**Chapters 1–11 Total** 75 days	**Remaining** 6 days	**Total** 81 days

**AYP* *is Assess Your Progress.*

LESSON SUPPORT

Section	Practice Bank	Study Guide*	Assessment Book*	Visuals	Explorations Lab Manual	Lesson Plans	Technology Book
11.1	72	11.1		Warm-Up 11.1 Folder 11	Masters 2, 4, 18	11.1	Calculator Act. 11
11.2	73	11.2	Test 47	Warm-Up 11.2	Master 19	11.2	Spreadsheet Act. 11
11.3	74	11.3		Warm-Up 11.3	Master 2	11.3	
11.4	75	11.4	Test 48	Warm-Up 11.4		11.4	
11.5	76	11.5		Warm-Up 11.5	Add. Expl. 17	11.5	
11.6	77	11.6	Test 49	Warm-Up 11.6	Master 1	11.6	
Review Test	78	Chapter Review	Tests 50, 51 Alternative Assessment			Review Test	

*__Spanish versions__ of *Study Guide* and *Assessment Book* are available.

Chapter Support

- Course Guide
- Lesson Plans
- Portfolio Project Book
- Preparing for College Entrance Tests
- Multi-Language Glossary
- *Test Generator* Software
- Professional Handbook

Software Support

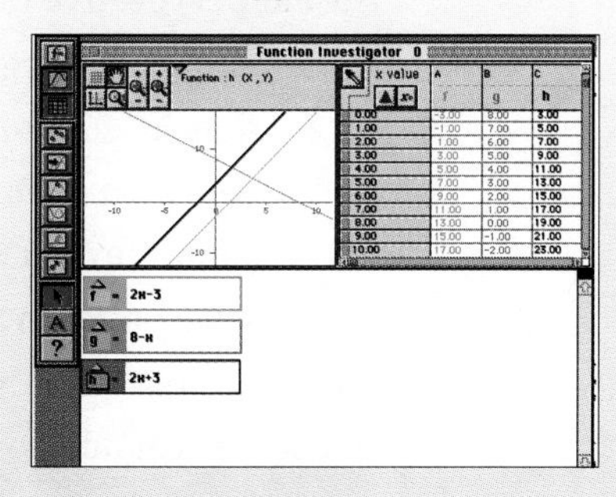

McDougal Littell Software
Stats!
Function Investigator

Internet Support

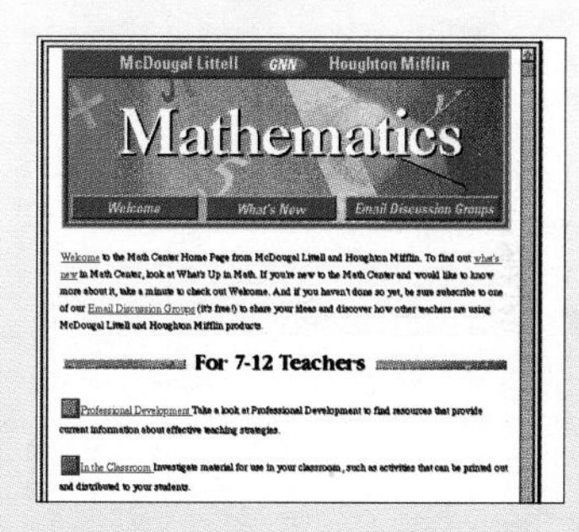

http://www.hmco.com
Next go to McDougal Littell; then the Education Center; then Secondary Math.

OUTSIDE RESOURCES

Books, Periodicals

Dion, Gloria S. "Fibonacci Meets the TI-82." *Mathematics Teacher* (February 1995): pp. 101–105.

Swetz, Frank J. "The Volume of a Sphere: A Chinese Derivation." *Mathematics Teacher* (February 1995): pp. 142–145.

Stone, Michael E. "Teaching Relationships between Area and Perimeter with The Geometer's Sketchpad." *Mathematics Teacher* (November 1994): pp. 590–594.

Activities, Manipulatives

Kennedy, Jane B. "Area and Perimeter Connections." *Mathematics Teacher* (March 1993): pp. 218–221, 231, 232.

Kincaid, Charlene, Guy Mauldin, and Deanna Mauldin. "Perimeters, Patterns, and Conjectures." *Mathematics Teacher* (February 1994): pp. 98–101, 107, 108.

Software

Jackiw, Nicholas. *The Geometer's Sketchpad.* Berkeley, CA: Key Curriculum Press, 1991.

Internet

With a modem, access the FAA Aviation Education menu for information about AOPA's aviation education program. Aircraft Owners and Pilots Association (AOPA) distributes *A Teacher's Guide to Aviation.*

Interview Notes

Background

Aerial Photography
Aerial photography can provide either vertical photos or oblique photos. To take a vertical photo, the camera is pointed straight down so that the photographic plane of the camera is parallel to the horizontal plane of the ground. Vertical photos are most useful for mapping purposes. Oblique photos are taken with the camera intentionally tilted away from a vertical position. A reasonable range for taking an aerial photo is between 500 and 14,000 feet. Aerial photos are used by architects, land developers, environmentalists, urban planners, and others who have a need or an interest to see the land from above.

Alex MacLean
Alex MacLean grew up on a small farm in Maryland. Shortly after earning a master's degree in architecture from Harvard University, he founded Landslides, a Boston-based aerial photography firm, in 1975. Although much of his work has been done for commercial clients, MacLean has taken many photos for himself to capture an image of the American landscape. His book, *Look at the Land: Aerial Reflections on America* shows the work he has done in nearly all of the fifty states. MacLean says, "I'm interested in both man-made and natural patterns as seen from the air—patterns which are curious and telling reflections of underlying cultural and physical forces."

CHAPTER

11 Rational Functions

SIZING things up in the air

INTERVIEW **Alex MacLean**

Sometimes you have to look at things from a different perspective to understand them better. Alex MacLean sees things from a different perspective every time he goes to work. As an aerial photographer, he gets a bird's-eye view of his subjects as he flies over them. "From the air, it's possible to observe large-scale patterns and the connectedness of things," he explains. "That's a perspective you just can't get from the ground."

"It takes some luck to capture the perfect shot, but I also think I'm getting a little better at being lucky."

"From the air, it's possible to observe large-scale patterns and the connectedness of things."

MacLean's photos show interesting patterns of human development. Some show the massive scarring of the land caused by the "clear-cutting" of trees in the Pacific Northwest. Others show vast landfills overflowing with paper trash. His photos do not just illustrate damage to the environment, however. "A lot of the things you see are really gorgeous—natural treasures that ought to be preserved," he says. "Still, by highlighting the negative things, the pictures can help you figure out what the positive things really are."

Patterns of Human Development

MacLean learned to fly while studying architecture and city planning. Instead of becoming an architect, he started an aerial photography company. At first, he provided photos to architects and city planners. Soon he ventured into other areas, taking pictures of such things as an archaeological site in New Mexico, wheat fields in Montana, and New York City's Central Park.

Getting the Right Angle

MacLean almost always flies alone. When he's ready to shoot, he banks his plane on its side and points the camera. For some shots, he points straight down. "If the camera is tilted at even a slight angle, the shot will come out distorted," he notes. He has to be at just the right height so that the subject completely fills the picture. "If you fly too low, you won't get it all in. If you fly too high, you'll have a lot of empty space on the edge of the picture."

Interview Notes

Background

Aerial Photography Firsts
The first successful attempts at aerial photography were made by a French photographer named Gaspard Félix Tournachon in 1858. He took the first aerial photographs while flying in a balloon over Paris. His best-known pictures were aerial photographs of the famous Paris landmark the Arc de Triomphe and its surrounding boulevards, taken in the late 1860s. The first aerial photographs of sites in the United States were taken around 1860 by James Wallace Black. He took the photographs from a balloon which was secured at an altitude of 1200 feet over the Boston Common.

Second-Language Learners

Students learning English may be unfamiliar with the idiomatic expression used in the interview title. If necessary, explain that to *size something up* is to form an opinion about something by judging what you see before you.

Interview Notes

Mathematical Connection

In Section 11.1, students explore the concept of inverse variation and then apply what they have learned to a problem in aerial photography to determine how high a photographer must fly to achieve a desired image size. In Section 11.4, students learn to solve a formula for one of its variables and then use this skill to solve problems that relate the focal length of a lens to flying height, image size, and actual size.

Explore and Connect

Writing
After students complete this activity, you may want to review their answers to the questions to ensure understanding of the concept of inverse variation.

Research
If any students in your class are especially interested in photography, you may suggest that they explain to their classmates what they know about camera lenses.

Project
This project lends itself to working in groups. All groups should include at least one student who is good at drawing. Have the groups work together to lay out their sketches to arrange them in order.

Altitude and Image Size

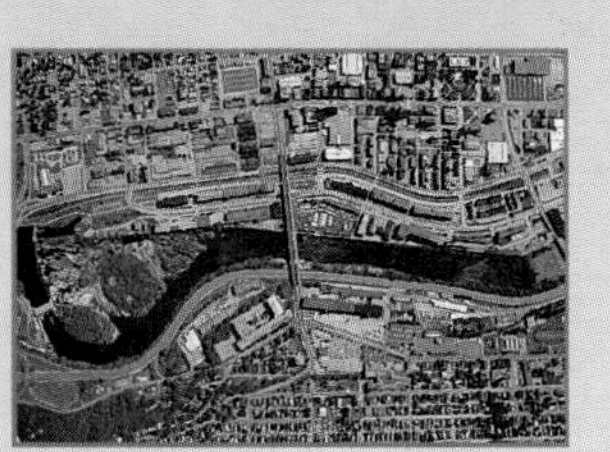

MacLean can use a formula to determine how large the image of an object will be when he flies at a certain altitude. The formula says that flying height and image size show *inverse variation*. In other words, the higher MacLean flies, the smaller the image size. In the photos of the river front in Manchester, New Hampshire, MacLean made the photographic image of the group of trees along the river appear to be large by flying the airplane at a low height. He made the image of the trees smaller and was able to include more of the city by flying higher. The graph shows the relationship between flying height and image size.

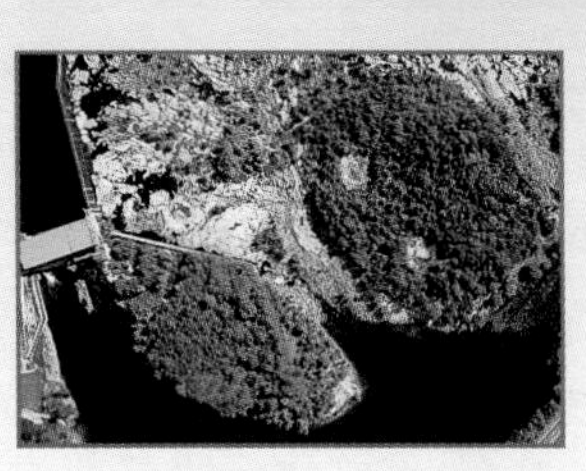

Image Size of the Group of Trees on a 35 mm Slide

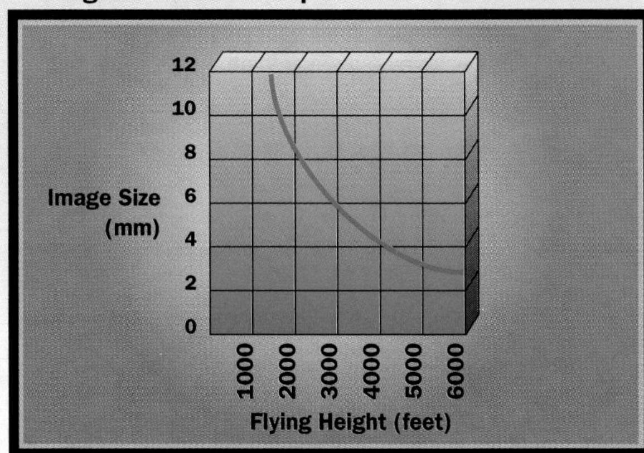

You can use the formula

$$\text{Image size} = \frac{k}{\text{Flying height}}$$

to help determine image size.

EXPLORE AND CONNECT

Alex MacLean takes his photographs from a four-seater Cessna 182.

1. **Writing** Use the graph to describe what it means for flying height and image size to show inverse variation. When flying height doubles, what happens to image size? How is this different from direct variation?

2. **Research** MacLean controls the size of the photographic image by changing his altitude. Some photographers use a telephoto lens to give the impression that the subject was close to them. Find out how a telephoto lens is different from a regular camera lens. Prepare a report of your findings.

3. **Project** Sketch a picture of your school as you think it would look from a plane. Compare your sketch with others made by students in your class. Which sketches look like they were made at the lowest heights? the greatest heights? Work together to put the sketches in order from lowest to highest.

Mathematics & Alex MacLean

In this chapter, you will learn more about how mathematics is related to photography.

Related Exercises

Section 11.1
- **Exercises 10 and 11**

Section 11.4
- **Exercises 20 and 21**

SECTION

11.1 Exploring Inverse Variation

Learn how to...

- recognize and describe data that show inverse variation

So you can...

- solve problems involving inverse variation, such as calculating driving times or planning aerial photographs

These two rectangles have the same area, but different lengths and widths. What are some examples of other rectangles with the same area? If area is constant, how does the length of a rectangle affect its width?

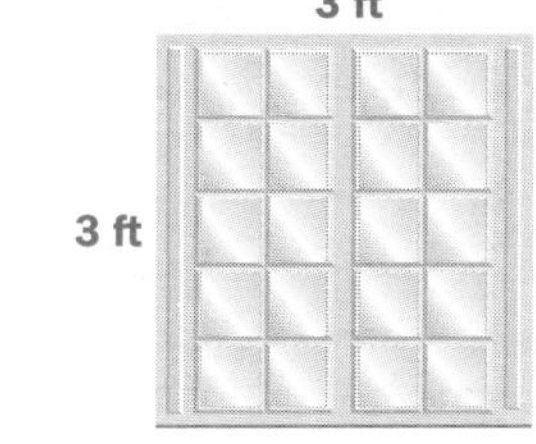

EXPLORATION
COOPERATIVE LEARNING

Exploring Rectangles with Equal Areas

Work with a partner.
You will need:

- 36 algebra unit tiles
- graph paper

1 Arrange the 36 tiles to form a rectangle. What are the length and the width of the rectangle?

2 Use the algebra tiles to form as many rectangles as you can with area 36. Copy and extend the table to record the length l and the width w of each rectangle.

l	w
12	3
3	12
2	?
?	?

3 Graph the data pairs (l, w) from your table in a coordinate plane.

4 **a.** Could a rectangle with area 36 be 4.5 units wide? If so, what would the length be?
b. Draw a smooth curve passing through the points you graphed. Explain why it makes sense to connect the points.

5 Can a rectangle be 0 units long? What happens as the length of the rectangle gets very small? Adjust your graph if necessary.

6 Write an equation relating l and w. Describe how the length of a rectangle affects its width if the area is constant.

Exploration Note

Purpose
The purpose of this Exploration is to have students use rectangles with a constant area to discover the graph and equation for inverse variation.

Materials/Preparation
Each pair of partners needs graph paper and 36 algebra unit tiles.

Procedure
Students should record the dimensions of their rectangles as they form them. Students should then graph the data pairs and answer questions 4–6, adjusting their graphs as necessary.

Closure
Students should understand that if the area of a rectangle is a constant, then the length and width have an inverse relationship, that is, as one increases, the other decreases. Students should also come to the conclusion that this relationship is not linear.

Explorations Lab Manual
See the Manual for more commentary on this Exploration.

Diagram Masters 2, 4, 18

Plan⇔Support

Objectives

- Recognize and describe data that show inverse variation.
- Solve real-world problems involving inverse variation.

Recommended Pacing

Core and Extended Courses
Section 11.1: 1 day

Block Schedule
Section 11.1: $\frac{1}{2}$ block
(with Section 11.2)

Resource Materials

Lesson Support
Lesson Plan 11.1
Warm-Up Transparency 11.1
Overhead Visuals:
Folder 11: Inverse Variation
Practice Bank: Practice 72
Study Guide: Section 11.1
Explorations Lab Manual:
Diagram Masters 2, 4, 18

Technology
Technology Book:
Calculator Activity 11
Graphing Calculator
Internet:
http://www.hmco.com

Warm-Up Exercises

Evaluate the expression xy for each pair of values of x and y.

1. $x = 4$, $y = 25$ 100
2. $x = 50$, $y = 2$ 100
3. $x = 3$, $y = 33\frac{1}{3}$ 100
4. *True* or *False*? If the graph for a set of data values is a straight line, then the data show direct variation. False.

Solve for y.

5. $21y = 7$ $\frac{1}{3}$
6. $3 = \frac{726}{y}$ 242

Teach⇔Interact

Section Note

Teaching Tip
Have students compare the definition of inverse variation with that of direct variation from Section 3.2. Students should point out any differences and similarities in the two definitions.

Additional Example 1

The time t in minutes that it takes Leona Franklin to read one chapter of a children's book (in which all chapters are equal in length) varies inversely with her average reading speed s in pages per minute. If Leona reads at an average speed of 3 pages per minute, it takes her 18 minutes to read one chapter.

a. Write and graph an equation expressing t in terms of s.
Since t varies inversely with s, find an equation in the form $t = \frac{k}{s}$. You know that $t = 18$ when $s = 3$. Use this information to find k.
$18 = \frac{k}{3}$
$k = 54$
An equation expressing t in terms of s is $t = \frac{54}{s}$.
Graph the equation as $y = \frac{54}{x}$. Use the viewing window $-30 \le x \le 30$, $-30 \le y \le 30$.

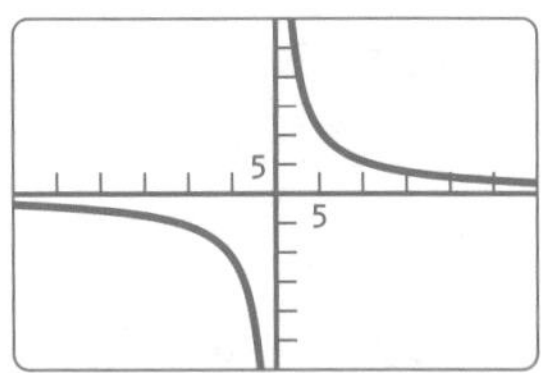

Time and speed are always positive. Only the first-quadrant portion of the graph makes sense in this problem.

b. If Leona wants to read one chapter in 12 minutes, how fast must she read?
Method 1 Substitute 12 for t in the equation from part (a).
$12 = \frac{54}{s}$
$12s = 54$
$s = 4.5$
Method 2 Use the graph.

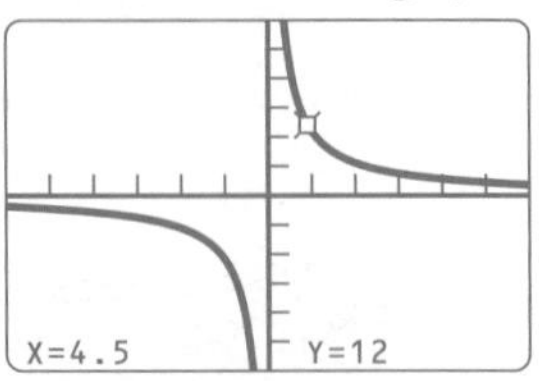

Leona must read at an average speed of 4.5 pages per minute.

Inverse Variation

In the Exploration, the product of the rectangle's length and width was always constant. Whenever two variables have a constant, nonzero product, they are said to show **inverse variation**.

The constant product is called the **constant of variation**.

Inverse Variation

Inverse variation can be described by an equation in the form

$xy = k$ or $y = \frac{k}{x}$, where $k \neq 0$ ← The letter k represents the constant of variation.

BY THE WAY

Between 1974 and 1987, states were required to enforce a 55 mi/h speed limit on all highways in order to receive federal aid for highways.

EXAMPLE 1 Application: Transportation

The time t in hours that it takes Derrick Le Marr to drive home from college varies inversely with his average speed s in miles per hour. If Derrick drives at an average speed of 50 mi/h, it takes him 6 h to get home.

a. Write and graph an equation expressing t in terms of s.

b. If Derrick wants to get home in 5 h, what must his average speed be?

SOLUTION

a. Since t varies inversely with s, find an equation in the form $t = \frac{k}{s}$.

You know that $t = 6$ when $s = 50$. Use this information to find k.

$$6 = \frac{k}{50}$$
$$k = 300$$

An equation expressing t in terms of s is $t = \frac{300}{s}$.

Graph the equation as $y = \frac{300}{x}$ on a graphing calculator.

Inverse variation graphs have two parts. To see both parts, view all four quadrants. One solution is (50, 6), so try the viewing window $-75 \le x \le 75$, $-75 \le y \le 75$.

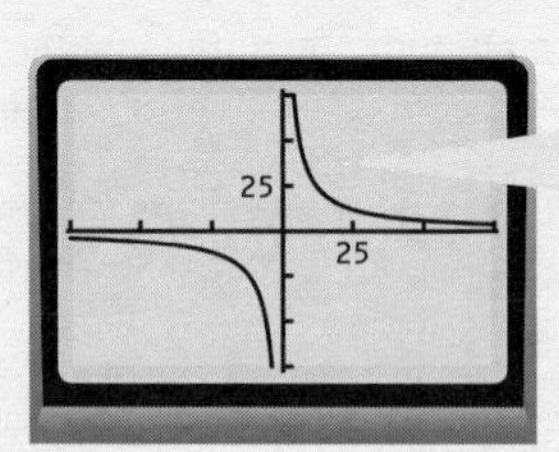

Time and speed are always positive. Only the first-quadrant portion of the graph makes sense in this problem.

ANSWERS Section 11.1

Exploration

1. Answers may vary. An example is given. 12 and 3

2.

l	w
12	3
3	12
2	18
18	2
1	36
36	1
4	9
9	4
6	6

3.

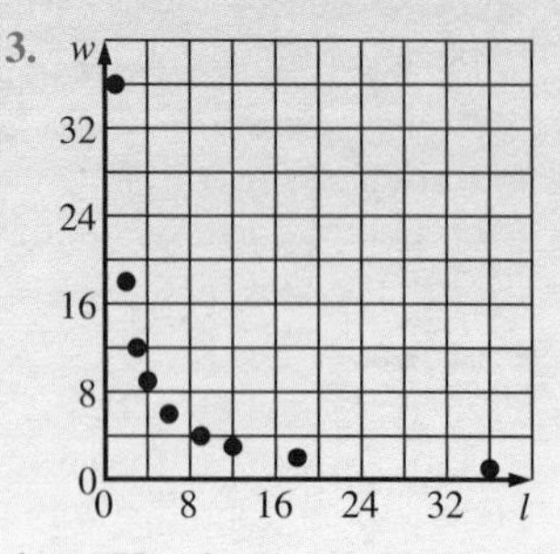

4. a. Yes; 8 units.

b. **Method 1** Substitute 5 for t in the equation from part (a). Solve for s.

$$5 = \frac{300}{s}$$
$$5s = 300$$
$$s = 60$$

Method 2 Use the graph.

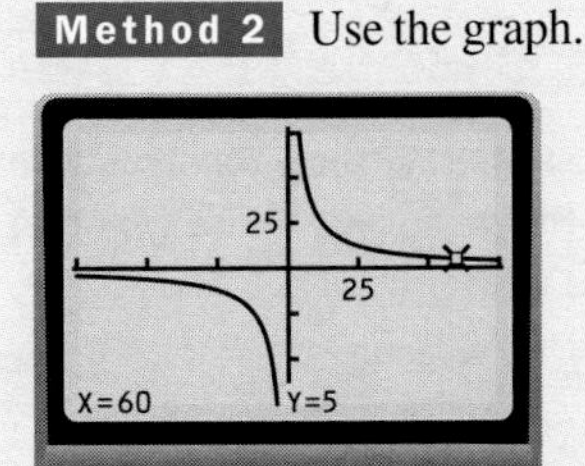

Derrick must drive at an average speed of 60 mi/h to get home in 5 h.

THINK AND COMMUNICATE

For Questions 1 and 2, use the information in Example 1.

1. The graph of $t = \frac{300}{s}$ never crosses either axis. Why?
2. What happens to t as s increases? as s decreases? How does the relationship between t and s compare with direct variation?

EXAMPLE 2 Application: Physics

A scuba diver 99 ft below the surface of the ocean exhales a small air bubble. As the bubble rises to the surface, the pressure on it decreases. The bubble grows in volume. The table describes the changes in pressure and volume. Does **volume** vary inversely with **pressure**?

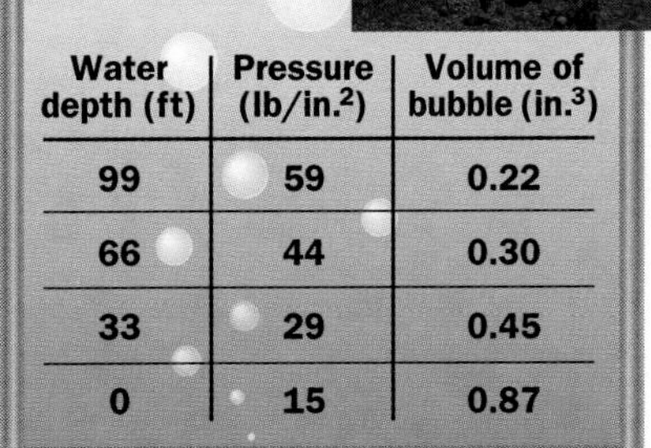

Size of Rising Air Bubble

Water depth (ft)	Pressure (lb/in.^{2})	Volume of bubble (in.^{3})
99	59	0.22
66	44	0.30
33	29	0.45
0	15	0.87

SOLUTION

Find the product of pressure and volume at each depth. Pressure and volume vary inversely if the product is always the same.

$59 \cdot 0.22 = 12.98$ $44 \cdot 0.30 = 13.2$

$29 \cdot 0.45 = 13.05$ $15 \cdot 0.87 = 13.05$

The products are all very close to 13.

Volume varies inversely with pressure.

BY THE WAY

Scuba divers must exhale as they rise to the surface. Otherwise, expanding air in the lungs could cause the lungs to tear.

THINK AND COMMUNICATE

3. In Example 2, does bubble volume vary inversely with water depth?

Do the quantities described show inverse variation? Explain your thinking.

4. the length of a side of a square and the area of the square
5. the number of people sharing a pizza equally and the number of slices per person

b.

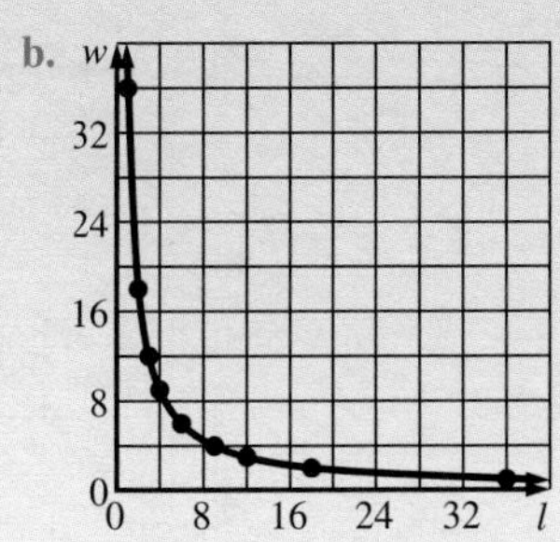

It makes sense to connect the points because any positive number can be the length or width of a rectangle. The length and width need not be integers.

5. No; a length must be positive. As the length gets very small, the width gets very large.
6. $lw = 36$; Answers may vary. An example is given. As the length increases, the width decreases, and as the length decreases, the width increases.

Think and Communicate

1. Answers may vary. An example is given. In the given equation, neither s nor t can ever be 0. No matter how fast Derrick drives, he cannot get home in no time at all, so t is never 0. He will never get home if he drives at a rate of 0 mi/h, so s is never 0.

2–5. See answers in back of book.

Teach⇔Interact

Additional Example 2

A scientist is measuring the velocity of air flowing in a tube that will be used in a new piece of diving equipment. When the tube is bent, its cross-sectional area is decreased and the velocity of air through that section of the tube is changed. The table shows some of the data that the scientist collected.

Cross-sectional area (cm^2)	Air-flow velocity (cm/s)
1.5	26.4
1.0	40.8
0.8	49.7
0.5	81.0

Does air-flow velocity vary inversely with cross-sectional area?

Find the product of cross-sectional area and air-flow velocity. Cross-sectional area and velocity vary inversely if the product is always the same.

$1.5 \cdot 26.4 = 39.6$
$1.0 \cdot 40.8 = 40.8$
$0.8 \cdot 49.7 = 39.76$
$0.5 \cdot 81.0 = 40.5$

The products are all very close to 40. Air-flow velocity varies inversely with cross-sectional area.

Section Notes

Research
In connection with Example 2, some students may find it interesting to research data on how the water pressure varies with the depth of the water.

Historical Connection
If temperature is held constant, then the equation $pV = k$ describes the relationship between the pressure and volume of a gas. This is known as Boyle's law, after Robert Boyle (1627–1691), a British scientist.

Interdisciplinary Problems
Physics is an excellent source of problems involving direct variation and inverse variation. In more advanced courses, students will also encounter many situations in which one quantity varies directly or inversely as the *square* of another quantity.

Teach⇔Interact

Checking Key Concepts

Reasoning
When discussing question 5, ask whether the equation describes an inverse variation. (Yes.) Then ask if, in this case, it is true that when the value of one variable increases, the value of the other variable decreases? Have students explain their answers. (No; if $xy = -150$, then $(5, -30)$ and $(10, -15)$ are both on the graph. The value of the x-coordinate has increased from 5 to 10 and the value of the y-coordinate has increased from -30 to -15.)

Closure Question

How can you tell if two variables vary inversely?
If the product of the variables is a nonzero constant, then they vary inversely.

Apply⇔Assess

Suggested Assignment

Core Course
Exs. 1–11, 14–17, 19–25

Extended Course
Exs. 1–25

Block Schedule
Day 70 Exs. 1–11, 14–17, 19–25

Exercise Notes

Teaching Tip
Exs. 1–3 Ask students to identify the constant of variation for each exercise that shows inverse variation.

Application
Ex. 4 When discussing this exercise, point out that real-world data that show inverse variation seldom have exactly the same product for all data pairs. If the products are *very close* to the same number, it is common practice to say that the data vary inversely. In the case of purely mathematical problems, the products have to be *exactly* the same.

CHECKING KEY CONCEPTS

Tell whether the data show *inverse variation, direct variation,* or *neither.* If the data show direct variation or inverse variation, write an equation for *y* in terms of *x*.

1.

x	y
15	4
12	5
10	6

2.

x	y
6	18
4	12
3	9

3.

x	y
7	10
9	12
12	15

4. The time it takes each editor to proofread a chapter varies inversely with the editor's average rate of proofreading. Julia Poriet usually proofreads at a rate of about six pages per hour. She finished Chapter 11 in about 4.5 h. If Joan Kaplan proofreads at a rate of about five pages per hour, how long will it take her to finish Chapter 11?

5. Write and graph an equation showing that the product of two variables is always -150.

11.1 Exercises and Applications

Extra Practice exercises on page 579

Decide whether the data show inverse variation. If they do, find the missing value.

1.

x	y
12	4
15	5
18	6
?	7

2.

x	y
−15	−8
12	10
$\frac{2}{3}$	?
−8	−15

3.

x	y
12	?
−9	−16
8	18
−1	−144

4. TRANSPORTATION It takes Laurie Grandmaison $3\frac{1}{2}$ h driving at an average speed of 50 mi/h to go from home to her friend's house. How long will it take her to return if she drives at an average speed of 45 mi/h?

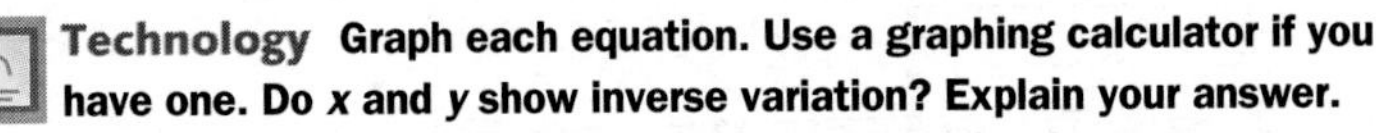
Technology Graph each equation. Use a graphing calculator if you have one. Do *x* and *y* show inverse variation? Explain your answer.

5. $xy = 24$ **6.** $y = \frac{1}{x}$ **7.** $y = \frac{-24}{x}$ **8.** $\frac{y}{x} = 1$

9. SAT/ACT Preview If you want to earn a fixed amount of simple interest I in one year, the yearly interest rate r and the principal P you invest vary inversely. Which equation does *not* show this relationship?

A. $P = \frac{I}{r}$ **B.** $r = \frac{P}{I}$ **C.** $r = \frac{I}{P}$ **D.** $Pr = I$ **E.** none of these

Checking Key Concepts

1. inverse variation; $y = \frac{60}{x}$
2. direct variation; $y = 3x$
3. neither
4. 5.4 h
5. $y = -\frac{150}{x}$ or $xy = -150$

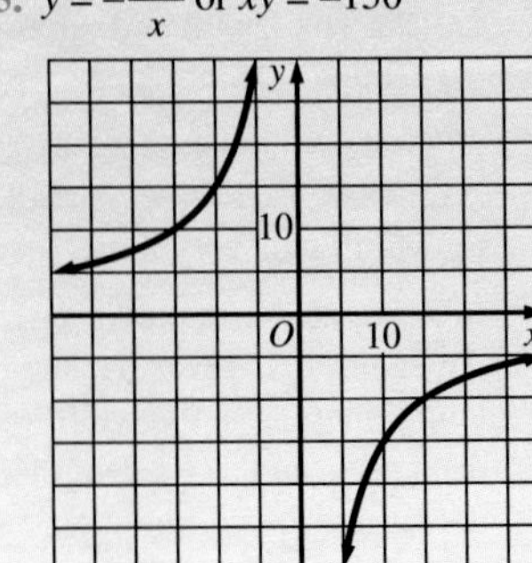

Exercises and Applications

1. Yes. 21
2. Yes. 180
3. Yes. 12
4. about 3 h 53 min
5–8. See answers in back of book.
9. B
10. Picture B was taken from 3000 ft above the ground, and picture C was taken from 6000 ft above the ground. If you measure a particular feature in all three photographs,

INTERVIEW Alex MacLean

Look back at the article on pages 460–462.

When Alex MacLean takes aerial photographs, he uses a formula to determine how high he must fly to achieve a desired image size.

$$Altitude = \frac{k}{image\ size}$$ ← The letter k represents a constant.

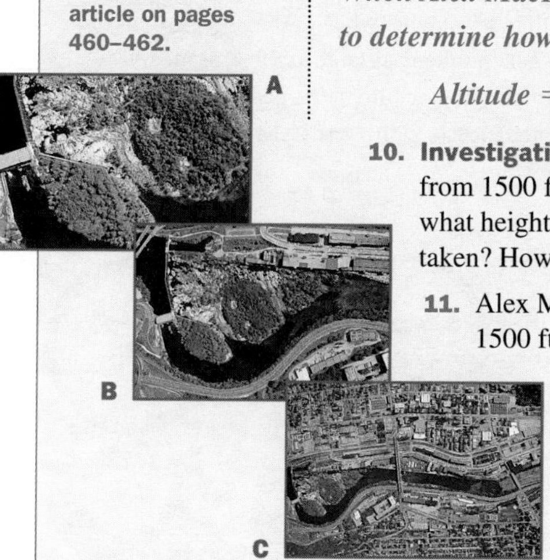

10. Investigation You will need a ruler. Alex MacLean took picture A from 1500 ft above the ground. At what heights were pictures B and C taken? How do you know?

11. Alex McLean was flying 1500 ft above the water when he took the picture at the right. At what altitude should he fly so the images of the boats are twice as long? half as long?

12. Writing You can use inverse variation to estimate the speed of the waves in the ocean. First estimate the horizontal distance l between the crests of the waves. Then count the time t between wave crests at a fixed point such as a buoy or pier pile. To find the wave speed s, use the formula $s = \frac{l}{t}$.

crest
l

a. Suppose you estimate the distance between wave crests to be about 7 m and the time between crests to be about 2 s. What is the speed of the wave?

b. Which two variables in the formula are inversely related?

c. If the distance between wave crests is constant, how does the speed of the waves affect the time between crests?

13. Open-ended Problem For each function graphed below, write an equation that could represent the function. Explain your thinking.

a. 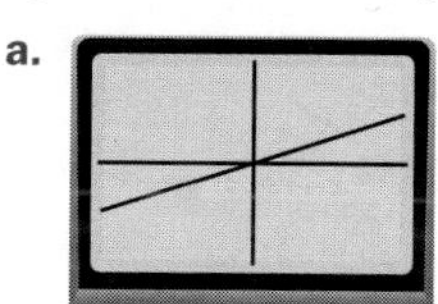
b. 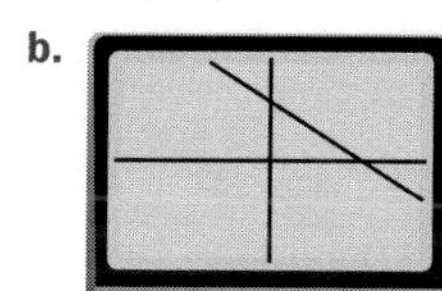
c. 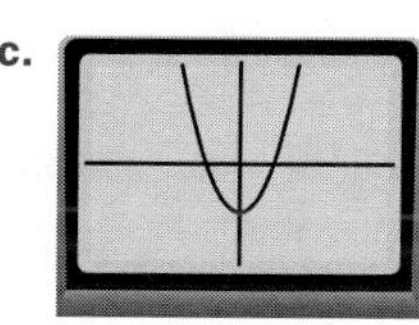
d. 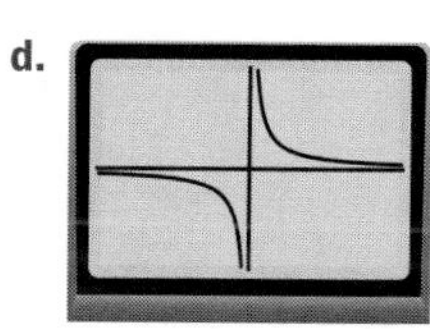

14. To loosen a bolt with a wrench with a 6 in. long handle, David Ruiz must exert a force of 250 lb. (That is like lifting a 250 lb person!) The force needed and the length of the handle vary inversely. If David slips a pipe over the wrench handle to make it 24 in. long, about how much force must he exert?

Apply⇔Assess

Exercise Notes

Interview Note

Exs. 10, 11 Discuss the fact that the formula the photographer uses relates *lengths*, not areas. If "image size" were the area of the image, then another formula is required because altitude varies inversely as the *square* of image area.

Application

Ex. 12 Waves have two basic dimensions. The length of a wave is the horizontal distance between two crests. The height of a wave is the vertical distance between the crest and the trough. In this exercise, students see that the variables s and t are inversely related. Tidal waves, also called tsunamis, have a low wave height at sea, which increases as the wave loses speed when it approaches land and the water becomes more shallow. Thus, in a sense, the height of a tsunamis wave is inversely related to the depth of the ocean.

Challenge

Ex. 13 Ask students how is it possible to decide whether a graph shows inverse variation. (Find the coordinates of several points on the graph and see if the product of each pair of coordinates gives the same number, or very nearly the same number).

Reasoning

Ex. 13 Ask students this question: If you know that a graph shows inverse variation, how many coordinate pairs are needed from the graph to write an equation for the graph? (one pair, since their product gives the constant of variation)

---

you find that the image size in photo B is half the image size in photo A, and the image size in photo C is one-fourth the image size in photo A. By using these facts and the formula relating altitude to image size, you can find the unknown heights.

11. 500 ft; 2000 ft

12. a. 3.5 m/s

b. s and t

c. As the speed increases, the time between crests decreases. As the speed decreases, the time between crests increases.

13. Answers may vary. Examples are given.

a. $y = 0.5x$; The graph shows direct variation with a positive constant of variation.

b. $y = -0.5x + 2$; The graph is a line with negative slope and positive y-intercept.

c. $y = x^2 - 2$; The graph is a parabola that opens up and has a negative y-intercept.

d. $y = \frac{2}{x}$; The graph shows inverse variation with a positive constant of variation.

14. 62.5 lb

Apply⇔Assess

Exercise Notes

Cooperative Learning
Exs. 15–18 You may wish to have students work on these exercises in small groups.

Multicultural Note
Exs. 15–18 Bicycles are a popular means of transportation all over the world. In China and the Netherlands, for instance, many people use bicycles as their primary means of transportation. In China, it is estimated that there are more than 210 million bicycles. During rush hour in major Chinese cities, the streets are filled with cyclists. In the Netherlands, there are as many bicycles as there are people. Separate bike paths are found in many parts of the country.

Assessment Note
Ex. 20 This assessment exercise is appropriate for peer evaluation.

Practice 72 for Section 11.1

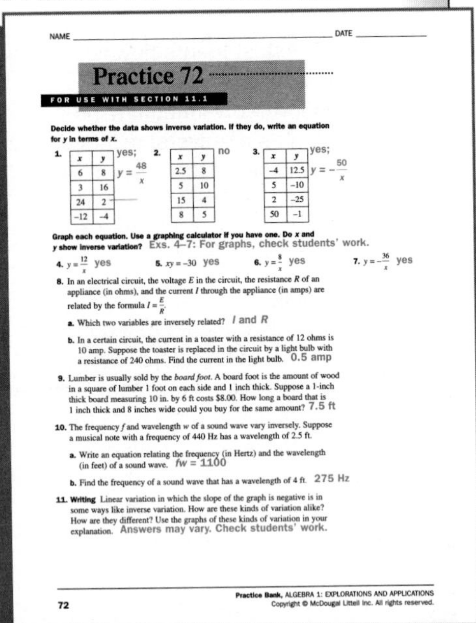

NAME ______ DATE ______

Practice 72

FOR USE WITH SECTION 11.1

Decide whether the data shows inverse variation. If they do, write an equation for y in terms of x.

1.

x	y
6	8
3	16
24	2
−12	−4

yes; $y = \frac{48}{x}$

2.

x	y
2.5	8
5	10
15	4
8	5

no

3.

x	y
−4	12.5
5	−10
2	−25
50	−1

yes; $y = -\frac{50}{x}$

Graph each equation. Use a graphing calculator if you have one. Do x and y show inverse variation? Exs. 4–7: For graphs, check students' work.

4. $y = \frac{12}{x}$ yes
5. $xy = -30$ yes
6. $y = \frac{8}{x}$ yes
7. $y = -\frac{36}{x}$ yes

8. In an electrical circuit, the voltage E in the circuit, the resistance R of an appliance (in ohms), and the current I through the appliance (in amps) are related by the formula $I = \frac{E}{R}$.
   a. Which two variables are inversely related? I and R
   b. In a certain circuit, the current in a toaster with a resistance of 12 ohms is 10 amp. Suppose the toaster is replaced in the circuit by a light bulb with a resistance of 240 ohms. Find the current in the light bulb. 0.5 amp

9. Lumber is usually sold by the *board foot*. A board foot is the amount of wood in a square of lumber 1 foot on each side and 1 inch thick. Suppose a 1-inch thick board measuring 10 in. by 6 ft costs $8.00. How long a board that is 1 inch thick and 8 inches wide could you buy for the same amount? 7.5 ft

10. The frequency f and wavelength w of a sound wave vary inversely. Suppose a musical note with a frequency of 440 Hz has a wavelength of 2.5 ft.
   a. Write an equation relating the frequency (in Hertz) and the wavelength (in feet) of a sound wave. $fw = 1100$
   b. Find the frequency of a sound wave that has a wavelength of 4 ft. 275 Hz

11. **Writing** Linear variation in which the slope of the graph is negative is in some ways like inverse variation. How are these kinds of variation alike? How are they different? Use the graphs of these kinds of variation in your explanation. Answers may vary. Check students' work.

72

Practice Bank, ALGEBRA 1: EXPLORATIONS AND APPLICATIONS

Connection RECREATION

Bicycles in the late 1800s often had very large front wheels and no chains. The larger the wheel, the farther it traveled in one rotation and the faster a bicyclist could go.

Modern bicycles use a gear system to achieve similar speeds with much smaller wheels. The 24-speed bicycle below has gear numbers as large as 99. This means it can go as far with one turn of the pedals as an old-fashioned bicycle with a front wheel with diameter 99 in.

15. The bike at the right has a gear number of 40 when the chain is on the front chainwheel with 32 teeth and the rear sprocket with 21 teeth. If the chain remains on the same front chainwheel, then the gear number g is inversely related to the number of teeth n on the rear sprocket. Write an equation relating g and n.

16. What is the gear number when the chain is on the rear sprocket with 12 teeth?

17. What is the gear number when the chain is on the rear sprocket with 28 teeth?

18. Challenge If you always pedal at a constant rate, how does the number of teeth in the rear sprocket affect the speed of the bicycle? Explain.

19. Investigation How is the graph of an equation in the form $xy = k$ different when k is negative than when k is positive? Explain how you found your answer.

ONGOING ASSESSMENT

20. Writing List at least three things that all graphs of inverse variation functions have in common.

SPIRAL REVIEW

Find the mean, the median, and the mode(s) of each set of data. *(Section 1.3)*

21. 83, 88, 81, 87, 81, 85, 87, 87, 77

22. 101.25, 103.75, 102.25, 102.25

23. 11, 11, 11, 12, 12, 13, 13, 13, 13, 14, 14, 14, 15, 23

Solve each system of equations by adding or subtracting. *(Section 7.3)*

24. $3a - 2b = 2$
$4a + 2b = -16$

25. $x + y = 4$
$\frac{2}{3}x - y = 6$

15. $g = \frac{840}{n}$

16. 70 17. 30

18. Answers may vary. An example is given. Assuming that you pedal at a constant rate, as the number of teeth in the rear sprocket increases, the speed of the bicycle increases. A higher gear number means the bicycle goes farther with each turn of the pedals. If you pedal at a constant rate and go farther with each turn of the pedals, you travel faster.

19. Answers may vary. An example is given. When k is negative, the two parts of the graph lie in Quadrants II and IV, rather than in Quadrants I and III. I found this answer by graphing a few equations such as $xy = -1$ and $xy = -4$.

20. Answers may vary. An example is given. All graphs of inverse variation functions have two separate curved parts, have no x-intercept or y-intercept, and approach an axis as one of the variables approaches zero.

21. 84; 85; 87

22. 102.375; 102.25; 102.25

23. 13.5; 13; 13

24. (−2, −4)

25. (6, −2)

SECTION

11.2 Using Weighted Averages

Learn how to...
- find weighted averages

So you can...
- summarize information such as grades or game scores

Sometimes when you find an average, you may want to give more weight to one number than to the others. For example, your teacher may use a *weighted average* to calculate your semester grade average, so that your final exam grade counts more than each quiz grade. In the Exploration, you will see another use for weighted averages.

EXPLORATION
COOPERATIVE LEARNING

Comparing Game Scores

Work in a group of four.
You will need:
- a bowl
- a narrow cup or glass
- about two cups of dried beans

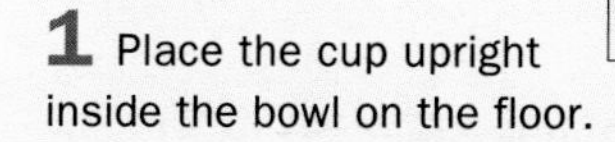

1 Place the cup upright inside the bowl on the floor.

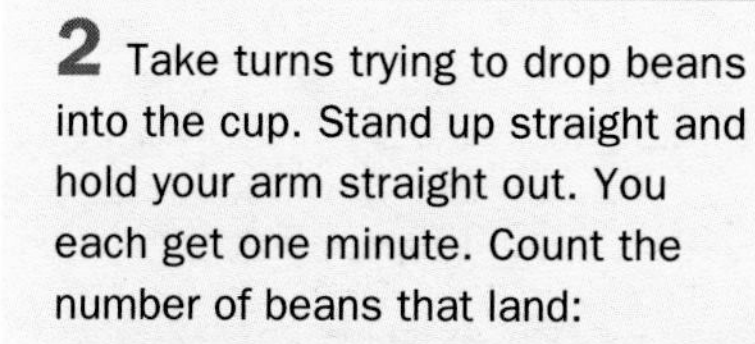

2 Take turns trying to drop beans into the cup. Stand up straight and hold your arm straight out. You each get one minute. Count the number of beans that land:
- inside the cup (10 points)
- inside the bowl but outside the cup (5 points)
- outside the bowl (0 points)

Record your results in a table.

Name	Number in cup	× 10 points	Number in bowl	× 5 points	Number outside bowl	× 0 points
Lisa	18	180	6	30	3	0
?	?	?	?	?	?	?

3 Compute the total number of points for each person in your group. Can you tell who had the best aim by comparing total numbers of points? Explain.

4 Compute the average number of points per drop for each person in your group. Can you use this information to decide who had the best aim? Explain.

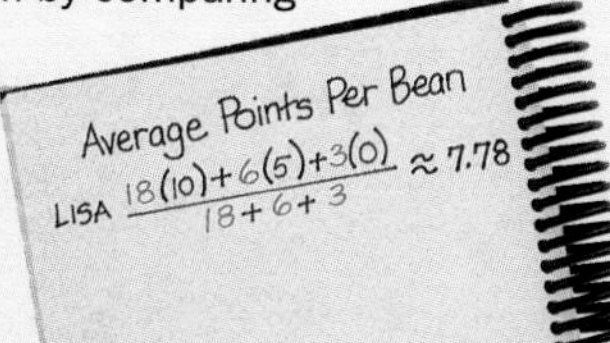

Exploration Note

Purpose
The purpose of this Exploration is to introduce students to the concept of a weighted average.

Materials/Preparation
Plastic cups and bowls can eliminate the possibility of accidental breakage. If more than one group is doing the Exploration at the same time, it would be desirable for each group to have the same size cups and bowls to facilitate a discussion and comparison of results.

Procedure
One student in each group should be assigned the task of recording the results. All students should compute their own average.

Closure
Discuss why the computation in Step 4 gives a better way of comparing the players' aims than the computation in Step 3. Be sure students understand that a higher average means a better aim.

Explorations Lab Manual
See the Manual for more commentary on this Exploration.
Diagram Master 19

Plan⇔Support

Objectives
- Find weighted averages.
- Use weighted averages to solve real-world problems.

Recommended Pacing
Core and Extended Courses
Section 11.2: 1 day
Block Schedule
Section 11.2: $\frac{1}{2}$ block
(with Section 11.1)

Resource Materials
Lesson Support
Lesson Plan 11.2
Warm-Up Transparency 11.2
Practice Bank: Practice 73
Study Guide: Section 11.2
Explorations Lab Manual: Diagram Master 19
Assessment Book: Test 47

Technology
Technology Book: Spreadsheet Activity 11
Graphing Calculator
McDougal Littell Software *Function Investigator* *Stats!*
Internet: http://www.hmco.com

Warm-Up Exercises
Find the mean of each group of numbers.

1. 78, 59, 84, 65, 90 75.2
2. 800, 800, 350, 350, 350, 350 500
3. 8, 7, 9, 9, 8, 7, 5, 13 8.25

Solve each equation for x.

4. $\frac{86 + 2x}{3} = 84$ 83
5. $\frac{4(88) + 3x}{7} = 91$ 95
6. $\frac{30 + 2x}{5} = 12$ 15

Teach⇔Interact

Section Notes

Communication: Discussion
Discuss the meaning of the word *weight*. Ask students what they think this word means as used here. Lead students to understand that weighted means to give more importance or significance to something.

Reasoning
Ask students to suppose that some questions on a test count more than others. The test grade is determined by adding all the points for the questions answered correctly. Is this an example of a weighted average? Why or why not? (No, because the points are added but not then divided by the number of questions. No type of average has been computed.)

Additional Example 1

Refer to Example 1. Suppose Luke's teacher decides to let the midterm count as 3 tests and the final exam count as 4 tests. What must Luke earn on the final exam to get an average grade of at least 81 for the semester?

Write and solve an inequality. Let x = Luke's grade on the final.

$$\frac{75 + 80 + 3(89) + 62 + 85 + 4x}{1 + 1 + 3 + 1 + 1 + 4} \geq 81$$

$$\frac{569 + 4x}{11} \geq 81$$

$$569 + 4x \geq 891$$

$$4x \geq 322$$

$$x \geq 80.5$$

Luke needs a grade of 81 or better on the final to earn an average of 81 for the semester.

Think and Communicate

When discussing question 1, you may wish to have students compute both the mean and the weighted average for the first five tests. (78.2, 80)

Luke's Test Grades	
Test 1	75
Test 2	80
Midterm (worth 2 tests)	89
Test 3	62
Test 4	85
Final Exam (worth 3 tests)	?

To determine your final score for the semester, your teacher may look at many things, such as test grades, homework grades, and midterm grades. Some of these grades may carry more weight than others.

For example, the table at the left shows Luke's test grades so far in his algebra class. Luke found his average grade so far.

The red numbers are called weights.

The midterm counts as **two** tests.

$$\frac{\text{point total}}{\text{number of tests}} = \frac{1(75) + 1(80) + 2(89) + 1(62) + 1(85)}{1 + 1 + 2 + 1 + 1} = 80$$

The average is a **weighted average**.

EXAMPLE 1 Application: Test Scores

To earn a B for the semester, Luke's average grade must be at least 81. What must Luke earn on the final exam to get a B for the semester? The final exam counts as **three** tests.

SOLUTION

Write and solve an inequality.
Let x = Luke's grade on the final.

The final counts as 3 tests.

$$\frac{75 + 80 + 2(89) + 62 + 85 + 3x}{1 + 1 + 2 + 1 + 1 + 3} \geq 81$$

$$\frac{480 + 3x}{9} \geq 81$$

Luke's average grade must be at least 81.

$$480 + 3x \geq 729$$

$$3x \geq 249$$

$$x \geq 83$$

Luke needs a grade of 83 or better on the final to earn a B for the semester.

THINK AND COMMUNICATE

1. Find the mean of Luke's scores on the first five tests, including the midterm. How is a weighted average different from a mean?
2. You found a weighted average in Step 4 of the Exploration. What did your average describe? What were the weights?

ANSWERS Section 11.2

Exploration

1–4. Answers may vary. Students should recognize that comparing total numbers of points does not tell you who has the best aim. It is better to compare average numbers of points per drop.

Think and Communicate

1. 78.2; A mean counts the data values equally, while a weighted average does not.
2. the number of points per drop for each person; the number of beans in each category

EXAMPLE 2 Application: Business

Karen Arndt works for a development company that is designing a new shopping mall. For the project to be approved, Karen needs to make sure that the mall will bring in an average monthly rent of **$8/ft²**.

Two large department stores have each agreed to rent **100,000 ft²** at **$1.25/ft²**. Karen predicts that she can rent space to smaller stores for **$18.00/ft²**. How much space for small stores should be included in the mall's design to bring in an average monthly rent of $8/ft²?

SOLUTION

Step 1 Model the situation with an equation.

Let x = the amount of space (in square feet) designed for small stores.

$$\text{Average monthly rent} = \frac{\text{Total monthly rent}}{\text{Total area}}$$

2 large stores are each renting 100,000 ft².

$$\$8/\text{ft}^2 = \frac{2 \bullet 100{,}000 \text{ ft}^2 \bullet \$1.25/\text{ft}^2 + x \bullet \$18.00}{2 \bullet 100{,}000 \text{ ft}^2 + x}$$

She needs to include x ft² for small stores.

$$8 = \frac{250{,}000 + 18x}{200{,}000 + x}$$

Step 2 Use a graph to solve the equation.

Graph the equations $y = 8$ and $y = \frac{250{,}000 + 18x}{200{,}000 + x}$ in the same coordinate plane.

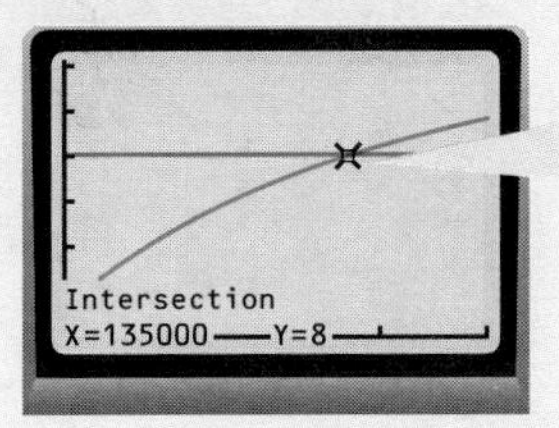

Find the x-coordinate of the point where the graphs intersect.

The mall's design should include 135,000 ft² of space for small stores.

Mall of America in Bloomington, Minnesota, is so large that seven Yankee Stadiums could fit inside. It houses over 400 stores, a miniature golf course, an amusement park, and over 400 trees up to 35 ft tall.

THINK AND COMMUNICATE

3. In Example 2, explain why the x-coordinate of the intersection of the two graphs is the solution of the equation $8 = \frac{250{,}000 + 18x}{200{,}000 + x}$.
4. If Karen provides more than 135,000 ft² of space for small stores, will the average rent go up or down? Explain. What problems might Karen run into if she provides more space for small stores than is necessary?
5. In Examples 1 and 2, what is being averaged? What are the weights?

Teach⇔Interact

About Example 2

Visual Thinking

You may wish to have students work in teams to create a sketch of the floor plan of the mall described in this Example. Have them display their floor plans and explain their space allocation between large stores and small stores. This activity involves the visual skills of *exploration* and *self-expression*.

Additional Example 2

A company that rents office space plans to put new carpeting in two buildings having spaces of 90,000 ft² each. The cost for carpet and installation is $9/yd². A much larger building of 250,000 ft² does not require carpeting throughout and has carpeting costs of $10.80/yd². How many square feet can the company have carpeted in the larger building in order to have the average carpeting cost come out to $9.90/yd²?

Step 1 Model the situation with an equation. Since office spaces are rented in square feet, use the cost per square foot for each of the carpeting costs. ($1/ft² for the smaller buildings, $1.20/ft² for the larger building, and $1.10/ft² for the average cost) Let x = the number of square feet of space to be carpeted in the larger building.

$$\$1.10/\text{ft}^2 = \frac{2 \cdot 90{,}000 \text{ ft}^2 \cdot \$1/\text{ft}^2 + x \text{ ft}^2 \cdot \$1.20/\text{ft}^2}{2 \cdot 90{,}000 \text{ ft}^2 + x \text{ ft}^2}$$

$$1.1 = \frac{180{,}000 + 1.2x}{180{,}000 + x}$$

Step 2 Use a graph to solve the equation. Graph the equations $y = 1.1$ and $y = \frac{180{,}000 + 1.2x}{180{,}000 + x}$ in the same coordinate plane.

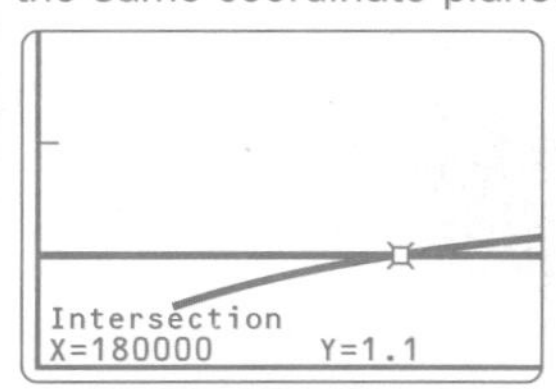

Find the x-coordinate of the point where the graphs intersect.

The company can have 180,000 ft² of the larger building carpeted.

Think and Communicate

3. The intersection of the graphs is the point for which $y = 8$ and $y = \frac{250{,}000 + 18x}{200{,}000 + x}$. Thus, the x-coordinate of the intersection point is the solution of the equation $8 = \frac{250{,}000 + 18x}{200{,}000 + x}$.
4. The average rent will go up because the rental cost for small stores is much higher than the cost for the department stores. Answers may vary. Examples are given. She might exceed the amount of money budgeted for the mall construction or the amount of land available. Also, the management for the department stores might object to an increase in the amount of space provided for small stores. Finally, Karen may have difficulty renting all the space if she provides more small-store space than necessary.
5. In Example 1, Luke's grades are being averaged and the weights are the numbers 1, 2, and 3. In Example 2, the rent per square foot is being averaged and weights are the amounts of area in each category.

Teach⇔Interact

Closure Question

How is a weighted average for a set of numbers computed?

Multiply each number by its weight. Add all the products and divide their sum by the sum of the weights.

Apply⇔Assess

Suggested Assignment

Core Course
Exs. 1–3, 5–7, 9–12, 14, 15, 19–25, AYP

Extended Course
Exs. 1–12, 14–25, AYP

Block Schedule
Day 70 Exs. 1–3, 5–7, 9–12, 14, 15, 19–25, AYP

Exercise Notes

Communication: Discussion
Ex. 1 When discussing this exercise, ask students to explain how it is related to the concept of a weighted average.

Application
Exs. 1–8 Weighted averages have many uses in real-world situations. These exercises demonstrate uses in business, education, sports, and building design. Students may wish to research other uses in these and other areas.

CHECKING KEY CONCEPTS

	Test 1	Test 2	Term paper
Daniel	85	85	75
Peter	75	85	75
Sarah	75	75	85

1. The term paper in an English class counts as two tests. The table at the right shows three students' grades. Find each student's weighted average.
2. At the yearly Weld College clam bake, you can order either the lobster platter for $10 or the hamburger platter for $3. This year 124 people ordered the lobster platter and 95 ordered the hamburger platter. What was the average cost of a meal per person?
3. In Questions 1 and 2, what is being averaged? What are the weights?
4. **Technology** Look back at the Exploration on page 469. Instead of taking careful aim with each bean, Saul decided to pour all of the beans at once. He counted 66 beans in the cup and 83 in the bowl. Instead of counting the beans that spilled onto the floor, Saul estimated that his average score was about 4. About how many beans are on the floor if his estimate is correct?

11.2 Exercises and Applications

Extra Practice exercises on page 579

1. **BUSINESS** On a business trip, Michael Simms recorded information about his gasoline purchases.

EXPENSE REPORT
Name: *Michael Simms*

Date	Price/gallon	Amount (gal)
3/12	$1.20	18
3/14	$1.14	15
3/15	$1.16	10
3/18	$1.22	19

   a. What is the total cost of the gasoline?
   b. How many gallons of gasoline did Michael buy in all?
   c. Find the average cost per gallon.
2. **Test Scores** Look back at Example 1 on page 470. The minimum grade for an A average is 91. Is there any way Luke's final exam grade can bring his average up to an A? Explain.
3. **Test Scores** To earn an A for the semester in her Spanish class, Josie must have an average grade of at least 91. She earned grades of 87, 95, 92, and 88 on the tests during the semester. The final counts as two tests. What score does she need on the final to earn an A?
4. **Writing** A project in a math class involved comparison shopping. Students checked the prices of a gallon of low-fat milk at 12 stores. The results are given in the table at the left.

Price of milk (dollars)	Number of stores
1.80	‖ ②
1.95	‖‖ ④
2.35	𝍸 ⑤
2.65	| ①

   a. Without calculating, estimate whether the average price will be closer to $1.80 or $2.65. Explain your reasoning.
   b. Calculate the average price per gallon.

Checking Key Concepts

1. Daniel, 80; Peter, 77.5; Sarah, 80
2. about $6.96
3. In Question 1, the students' grades are being averaged, and the weights are 1 (for the tests) and 2 (for the term paper). In Question 2, the meal costs are being averaged, and the weights are the numbers of people who ordered each meal choice.
4. about 120 beans

Exercises and Applications

1. a. $73.48
   b. 62 gal
   c. about $1.19 per gal
2. The only way Luke could bring his average up to an A would be if the teacher added bonus questions to the final exam. Luke would need to score at least 113 on the final.
3. 92 or better

Connection SPORTS

Olympic figure skating judges use weighted scoring to rank skaters. Each skater is judged twice, in the technical program and in the free skating program. To find the final ranking, each skater's rank in the technical program is multiplied by 0.5 and added to the rank in the free skate. The skater with the lowest total wins the gold medal.

The table shows the rankings in both events of the 1994 Olympic Ladies' Figure Skating Championship in Hamar, Norway.

Skater COUNTRY	Ranking: Technical Program	Ranking: Free Skate
Oksana Baiul UKR	2	1
Surya Bonaly FRA	3	4
Lu Chen CHN	4	3
Tonya Harding USA	10	7
Nancy Kerrigan USA	1	2
Yuka Sato JPN	7	5
Tanja Szewczenko GER	5	6
Katarina Witt GER	6	8

5. Find the final ranking.

6. Explain how weights are involved in the scoring.

7. How would the final ranking be different if the scores were not weighted?

8. 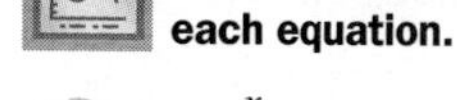 **Technology** Look back at Example 2 on page 471. Suppose one of the larger stores decides not to rent space in the mall. The other agrees to a larger area of 150,000 ft^2. How much space should Karen Arndt provide for small stores as she revises her plans?

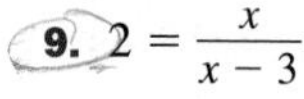 **Technology For Exercises 9–11, use a graphing calculator to solve each equation.**

9. $2 = \dfrac{x}{x - 3}$ **10.** $2 = \dfrac{x + 4}{x + 3}$ **11.** $3 = \dfrac{4x - 10}{x - 3}$

12. BUSINESS Martin Dionne sells peanuts for \$1.70 per pound and cashews for \$6 per pound. He decides to make a mixture of 70 lb of peanuts and 35 lb of cashews.

a. How much should he charge per pound for the mixture if he wants to earn the same amount as if he sold the nuts separately?

b. Write a variable expression to show how much Martin Dionne should charge if he adds x pounds of almonds to the mixture. He usually charges \$4 per pound for almonds.

Apply⇔Assess

Exercise Notes

Multicultural Note

Exs. 5–7 Born in 1977, Lu Chen lives in Beijing, China and frequently trains in Los Angeles. As a young girl, Chen did not see much figure skating on television, but she found a role model in Kristi Yamaguchi, the 1992 Olympic gold-medalist from the United States. After finishing third at the 1992 and 1993 World Championships, and third at the 1994 Olympics, Lu Chen won first place in the 1995 World Championships. She is the first world figure skating champion from China.

 **Using Technology**

Exs. 9–11 Students may use graphing calculators or the *Function Investigator* software to solve these equations.

 **Communication: Discussion**

Ex. 12 Have students discuss how weighted averages are used in this situation. Ask if they can think of other examples that use weighted averages to set prices.

4. a. Answers may vary. For example, I think the average price will be closer to \$1.65 than to \$2.50 because half the prices are within \$.15 of \$1.65.

b. \$2.15

5. Baiul, Kerrigan, Chen, Bonaly, Sato and Szewczenko, Witt, Harding

6. The ranking in the technical program counts half as much as the ranking in the free skate.

7. Baiul and Kerrigan tied, Bonaly and Chen tied, Szewczenko, Sato, Witt, Harding

8. Estimates may vary; about 101,000 ft^2.

9–11. Check students' work.

9. 6

10. −2

11. 1

12. a. about \$3.13

b. $\dfrac{70(1.7) + 35(6) + 4x}{105 + x}$

Apply⇔Assess

Exercise Notes

Problem Solving

Ex. 13 This exercise, like several others in this section, can be approached in more than one way. Ask students to think of different ways to use available technology to answer the questions. (spreadsheets with the *Stats!* software, tables with the TI-82 graphing calculator, or graphs with a graphing calculator or the *Function Investigator* software.)

Communication: Listening

Ex. 17 Students may find it interesting to hear their classmates answers to this exercise. Hearing different problems will most likely broaden their awareness of situations that can involve the use of weighted averages.

Using Manipulatives

Ex. 18 This exercise provides students with a physical model of weighted averages and demonstrates the physical principles used in making mobiles. The groups may enjoy displaying their mobiles in class.

13. **Technology** Look back at the Exploration on page 469. In his first 30 seconds, Soneth drops 5 beans into the cup and 3 beans into the bowl.

a. What is Soneth's average score after the first 30 seconds?

b. Soneth suspects that if he drops the next few beans into the cup and then stops, he will get a higher average than his friend Tim. Tim's average score was 8.4. How many beans must Soneth drop into the cup to beat Tim's score?

Visual Learning For Exercises 14 and 15, graph each equation and compare the graphs. How can you predict what the graph of a rational equation will look like?

14. a. $y = \frac{1}{x}$ b. $y = \frac{1}{x+1}$ c. $y = \frac{1}{x+2}$ d. $y = \frac{1}{x-2}$

15. a. $y = \frac{x}{x+1}$ b. $y = \frac{2x}{x+1}$ c. $y = \frac{3x}{x+1}$ d. $y = \frac{-3x}{x+1}$

16. **Challenge** Eiji Hirai and his friend Josh Smith share a two-bedroom apartment in a suburb of Boston. The rent is \$950 per month. Both friends use common areas such as the kitchen and living room, so they each pay \$225 toward those areas. They use weights to split the rest of the rent based on the sizes of their rooms. Eiji's room is 160 ft^2 and Josh's room is only 90 ft^2. How much rent should each pay? Explain how you found your answer.

17. a. **Open-ended Problem** Write a problem that can be solved using the weighted average $\frac{200(12) + x(20)}{200 + x} = 15$.

b. Solve the equation and tell what the answer means in terms of the situation you described in part (a).

18. **Cooperative Learning** Work with a partner to design and build a mobile. You need a lightweight dowel, string, and a collection of identical objects that are heavier than the dowel.

Step 1 Make a scale on the dowel as shown.

Step 2 Tie strings to several numbers. If you hang a single object from each string, the mobile will balance when it is hung from the average of the numbers. If you hang multiple objects from each string, the mobile will balance when it is hung from the weighted average of the numbers.

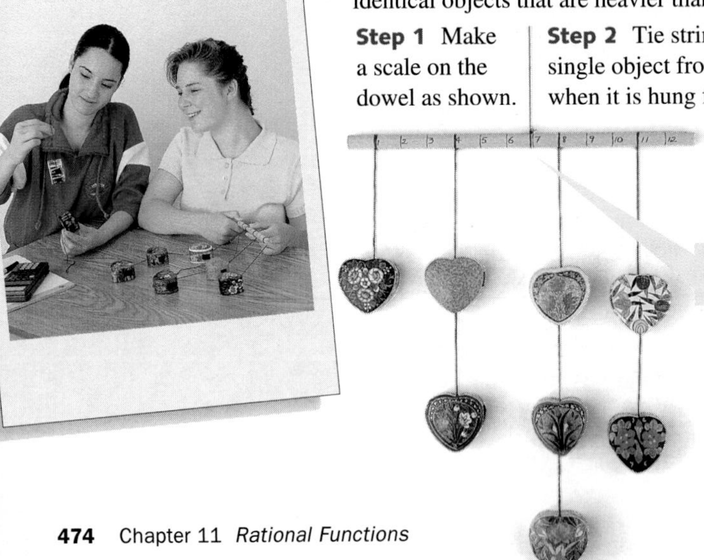

$$6\frac{7}{8} = \frac{1(1) + 2(4) + 3(8) + 2(11)}{1 + 2 + 3 + 2}$$

Step 3 Design your mobile first, using multiple objects on at least one string. Then test your design by making the mobile.

13. a. 8.125

b. at least 2 beans

14. a.

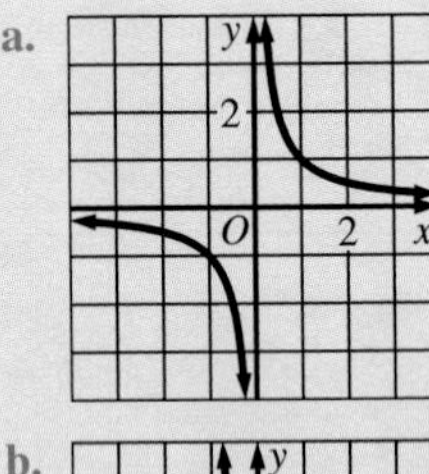

b.

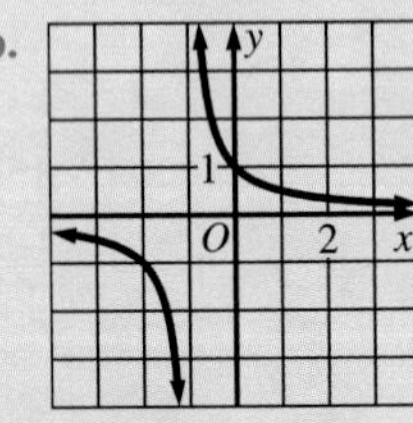

c.

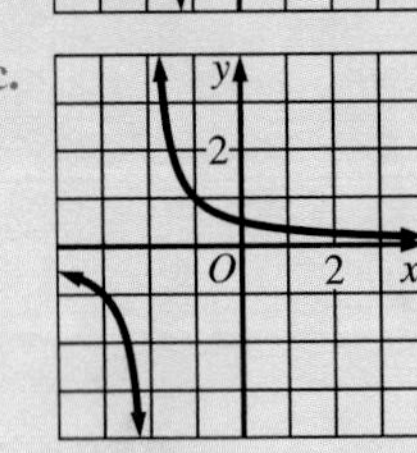

d.

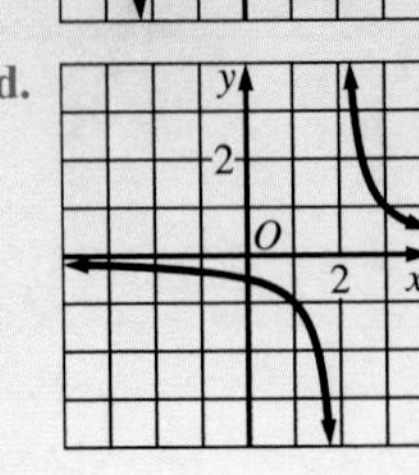

Answers may vary. An example is given. The graph of $y = \frac{1}{x+k}$ is the graph of $y = \frac{1}{x}$ shifted k units to the left if $k > 0$ and $|k|$ units to the right if $k < 0$. The graph of $y = \frac{1}{x+k}$ never crosses the x-axis or the line $x = -k$.

15. a.

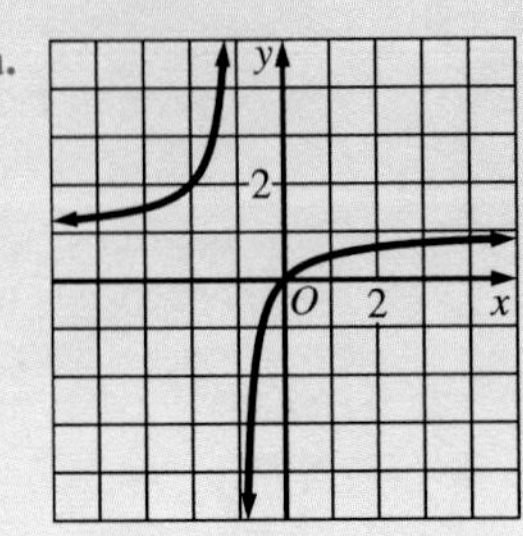

b.

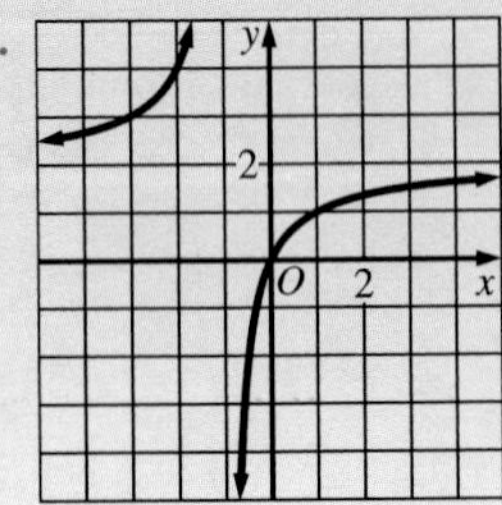

ONGOING ASSESSMENT

19. **Writing** Explain how weighted averages can be used in determining grades. When do you think it is to your advantage to have weighted averages used in calculating a grade? When is it to your disadvantage?

SPIRAL REVIEW

Solve each equation. *(Section 4.4)*

20. $\frac{x}{5} + \frac{x}{6} = 7$

21. $5 - \frac{2}{3}y = -\frac{7}{15}y$

22. $\frac{b}{3} - \frac{b}{9} = -\frac{1}{4}$

Tell whether each expression is a polynomial. Write *Yes* or *No*. If not, explain why not. *(Section 10.1)*

23. $2 + 3x^2 - 5x$

24. -5

25. $\frac{2}{x^3} - \frac{1}{x^2} + \frac{2}{5}$

ASSESS YOUR PROGRESS

VOCABULARY

inverse variation (p. 464)
constant of variation (p. 464)
weighted average (p. 470)

Do the data show inverse variation? If they do, write an equation for *y* in terms of *x*. *(Section 11.1)*

1.

x	y
12	12
16	8
18	6

2.

x	y
5	24
6	20
8	15

3. **GEOMETRY** For a triangle with fixed area, the height and the length of the base vary inversely. One triangle has base length 32 cm and height 12 cm. A triangle with the same area has base length 24 cm. Find its height. *(Section 11.1)*

4. In Mark's statistics class the final is worth twice as much as each of the other tests. Mark earned the following grades on his tests: 80, 78, 96. Mark got a 90 on the final. What is his average grade? *(Section 11.2)*

5. **Technology** Jay Graydon is designing a hotel. He will include 50 economy rooms and at least 20 executive suites. If he plans to charge \$80 per night for the economy rooms and \$110 per night for the executive suites, how many executive suites must he include to bring in an average income per room of \$90 per night? *(Section 11.2)*

6. **Journal** How are inverse variation and direct variation alike? How are they different? Include both equations and graphs in your answer.

Apply⇔Assess

Assess Your Progress

Journal Entry
Urge students to include real-world examples in their responses. A class discussion of the journal entries of several students can help all students to consolidate their understanding of concepts from this chapter and Chapter 3.

Progress Check 11.1–11.2

See page 502.

Practice 73 for Section 11.2

NAME ______________ DATE ______

Practice 73
FOR USE WITH SECTION 11.2

In Exercises 1–3, find each average.

1. The average fuel economy of a car that went 200 highway miles on 6 gal of gas and 100 city miles on 4 gal of gas. 30 $\frac{\text{mi}}{\text{gal}}$
2. The average yards-per-carry of a running back who had 3 carries of 8 yd each and 5 carries of 6 yd each in one football game. 6.75 $\frac{\text{yd}}{\text{carry}}$
3. Your average cost for 1 megabyte (MB) of RAM (read-only memory) for a computer, if you bought one 4 MB chip for \$210 and one 8 MB chip for \$330. \$45
4. Juanita has had 2 grades of 83, 4 grades of 86, and 3 grades of 92 on the 9 math tests she has taken so far. The final exam will count as much as 3 tests.
   a. Using x for the grade Juanita gets on the final exam, write an equation that states that Juanita's average for the term is 88. $\frac{166 + 344 + 276 + 3x}{12} = 88$
   b. What grade does Juanita need to get on the final exam in order to have an 88 average for the term? 90
5. In baseball, you find a player's *slugging average* by dividing the player's number of total bases by his or her number of times at bat. In calculating total bases, each single counts as 1, each double counts as 2, and so on.
   a. So far this season, Ozzie has 7 singles, 4 doubles, 1 triple, and 3 home runs in 45 times at bat. Suppose he doubles in each of his next n at-bats. Write an equation that states that his slugging average is .750 after those at-bats. $\frac{7 + 8 + 3 + 12 + 2n}{45 + n} = 0.75$
   b. How many straight doubles would Ozzie need to bring his slugging average up to .750? 3
6. On a camping trip, Ming Liang's friend has twisted her ankle, and Ming Liang needs to take the pair's canoe to go for help. Their campsite is at a point where a river flows into a lake. Ming Liang has a choice of paddling 1.2 mi up and 1.2 mi back down the river, or the same distance across the lake and back, in order to bring help to her friend.
   a. Suppose Ming Liang paddles at 8 mi/h, and the current in the river flows at 2 mi/h. Find the time it will take Ming Liang to go up the river and back to the campsite. 0.32 h
   b. Find the time it would take Ming Liang to paddle across the lake and back. Which choice will get help to her friend faster? 0.3 h; Going across the lake is slightly faster.

Practice Bank, ALGEBRA 1: EXPLORATIONS AND APPLICATIONS

73

c.

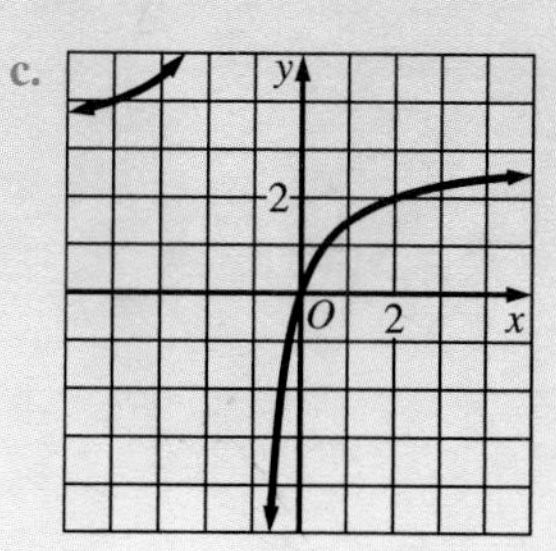

d.

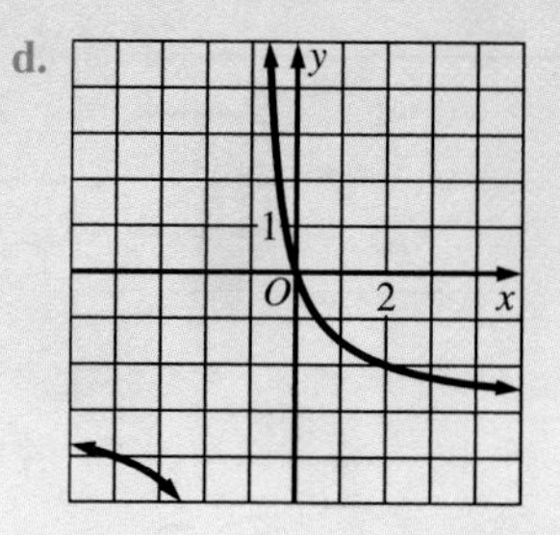

Answers may vary. An example is given. The graph of $y = \frac{kx}{x+1}$ is the graph of $y = \frac{x}{x+1}$ stretched vertically by a factor of $|k|$. If $k < 0$, the graph is also reflected in the x-axis.

16–19. See answers in back of book.

20. $19\frac{1}{11}$
21. 25
22. $-\frac{9}{8}$
23. Yes.
24. Yes.
25. No; the first two terms cannot be written as monomials with exponents that are whole numbers.

Assess Your Progress

1. No.
2. Yes; $y = \frac{120}{x}$.
3. 16 cm
4. 86.8
5. 25 executive suites
6. See answers in back of book.

Plan⇔Support

Objectives

- Solve rational equations.
- Solve real-world problems involving rational equations.

Recommended Pacing

Core and Extended Courses
Section 11.3: 1 day

Block Schedule
Section 11.3: $\frac{1}{2}$ block
(with Section 11.4)

Resource Materials

Lesson Support
Lesson Plan 11.3
Warm-Up Transparency 11.3
Practice Bank: Practice 74
Study Guide: Section 11.3
Explorations Lab Manual: Diagram Master 2

Technology
McDougal Littell Software *Function Investigator*

Internet: http://www.hmco.com

Warm-Up Exercises

1. Simplify $2(x + 5) \cdot \frac{9}{x + 5}$. 18
2. Solve the equation $3x + 9 = 39$. 10
3. Find the least common denominator for the fractions $\frac{1}{3}, \frac{1}{25}$, and $\frac{1}{45}$. 225

Tell whether the given number is a solution of the equation. Write *Yes* or *No.*

4. $\frac{7x - 7}{x - 1} = 7$; $x = 5$ Yes.
5. $\frac{7x - 7}{x - 1} = 7$; $x = 1$ No.
6. $\frac{2}{3x} + \frac{5}{2x} = 2$; $x = 2$ No.

SECTION

11.3 Solving Rational Equations

Learn how to...

- solve rational equations

So you can...

- solve problems involving rational equations such as designing a mall, finding hiking speeds, and exploring weight at different altitudes

Look back at Example 2 on page 471 about designing a shopping mall. It showed how to solve the equation $8 = \frac{250{,}000 + 18x}{200{,}000 + x}$ by graphing. You can also solve this equation algebraically.

EXAMPLE 1

Solve the equation $8 = \frac{250{,}000 + 18x}{200{,}000 + x}$.

SOLUTION

$$8 = \frac{250{,}000 + 18x}{200{,}000 + x}$$

Eliminate the fraction by multiplying both sides of the equation by $(200{,}000 + x)$.

$$(200{,}000 + x)8 = (200{,}000 + x)\frac{250{,}000 + 18x}{200{,}000 + x}$$

$$1{,}600{,}000 + 8x = 250{,}000 + 18x$$

$$1{,}350{,}000 = 10x$$

$$135{,}000 = x$$

The solution is 135,000.

Check

$$8 \stackrel{?}{=} \frac{250{,}000 + 18(135{,}000)}{200{,}000 + 135{,}000}$$

$$8 \stackrel{?}{=} \frac{2{,}680{,}000}{335{,}000}$$

$$8 = 8 \checkmark$$

Examples of Rational Expressions

$$\frac{300}{s}$$

$$\frac{1(75) + 2(89) + 1(x)}{1 + 2 + 1}$$

$$\frac{x - y}{x^2 - 2xy + y^2}$$

$$\frac{b}{b - \sqrt{2}}$$

The variable expression in Example 1 is an example of a *rational expression.* A **rational expression** is an expression that can be written as a fraction. The numerator and denominator must be polynomials.

A **rational equation** is an equation that contains only rational expressions.

Numerator is a polynomial. → $\frac{250{,}000 + 18x}{200{,}000 + x} = 8$ ← Can be written as $\frac{8}{1}$.
Denominator is a polynomial. →

THINK AND COMMUNICATE

1. Decide whether each expression is a rational expression. Explain.

a. $\frac{a + 2}{b - 1}$ **b.** $\frac{1}{\sqrt{y}}$ **c.** $x - 1$

ANSWERS Section 11.3

Think and Communicate

1. a. Yes; the expression is a fraction in which both the numerator and the denominator are polynomials.
   b. No; the denominator is not a polynomial.
   c. Yes; the expression can be written as a fraction in which both the numerator and the denominator are polynomials, that is, $\frac{x - 1}{1}$.

EXAMPLE 2

Solve the rational equation $\frac{3}{2(t+1)} + \frac{5}{4t} = \frac{2}{t}$.

SOLUTION

Step 1 Find the least common denominator (LCD) of the fractions.

Look at the factors of the denominators.

$$\frac{3}{2 \cdot (t+1)} + \frac{5}{2 \cdot 2 \cdot t} = \frac{2}{t}$$

Remember that the LCD is the least common multiple of the denominators.

The LCD is $2 \cdot 2 \cdot t \cdot (t+1) = 4t(t+1)$.

Step 2 Multiply both sides of the equation by the LCD and solve.

$$4t(t+1)\left(\frac{3}{2(t+1)}\right) + 4t(t+1)\left(\frac{5}{4t}\right) = 4t(t+1)\left(\frac{2}{t}\right)$$

Each denominator is a **factor** of the **LCD**.

$$6t + (5t + 5) = 8t + 8$$

$$11t + 5 = 8t + 8$$

$$3t = 3$$

$$t = 1$$

The solution is 1.

Check

$$\frac{3}{2(1+1)} + \frac{5}{4} \stackrel{?}{=} \frac{2}{1}$$

$$\frac{8}{4} = 2 \checkmark$$

EXAMPLE 3

Solve the equation $\frac{5x-4}{2(x-4)} + \frac{9}{2} = \frac{8}{x-4}$.

SOLUTION

Multiply both sides of the equation by the LCD $2(x-4)$.

$$2(x-4)\frac{5x-4}{2(x-4)} + 2(x-4)\left(\frac{9}{2}\right) = 2(x-4)\left(\frac{8}{x-4}\right)$$

$$(5x-4) + (9x-36) = 16$$

$$14x = 56$$

$$x = 4$$

The equation has no solution.

Check

$$\frac{5 \cdot 2 - 4}{2(4-4)} + \frac{9}{2} \stackrel{?}{=} \frac{8}{4-4}$$

$$\frac{6}{0} + \frac{9}{2} \neq \frac{8}{0}$$

You can't have zero in the denominator, so 4 is not a solution.

WATCH OUT!

Always check your answer! Sometimes you will get answers that are not solutions of the original equation.

Teach⇔Interact

Additional Example 1

Solve the equation $\frac{480 + 7x}{100 - x} = 3$.

Eliminate the fraction by multiplying both sides of the equation by $(100 - x)$.

$$(100 - x)\frac{480 + 7x}{100 - x} = (100 - x)3$$

$$480 + 7x = 300 - 3x$$

$$10x = -180$$

$$x = -18$$

Check

$$\frac{480 + 7(-18)}{100 - (-18)} \stackrel{?}{=} 3$$

$$\frac{354}{118} \stackrel{?}{=} 3$$

$$3 = 3 \checkmark$$

Additional Example 2

Solve the rational equation $\frac{3}{2(x-1)} + \frac{2}{x} = \frac{31}{8x}$.

Step 1 Find the LCD of the fractions. Look at the factors of the denominators. Determine their least common multiple. The LCD is $2 \cdot 2 \cdot 2 \cdot x \cdot (x-1) = 8x(x-1)$.

Step 2 Multiply both sides of the equation by the LCD and solve.

$$8x(x-1)\left(\frac{3}{2(x-1)}\right) + 8x(x-1)\left(\frac{2}{x}\right) = 8x(x-1)\left(\frac{31}{8x}\right)$$

$$12x + (16x - 16) = 31x - 31$$

$$28x - 16 = 31x - 31$$

$$-3x = -15$$

$$x = 5$$

Check

$$\frac{3}{2(5-1)} + \frac{2}{5} \stackrel{?}{=} \frac{31}{8 \cdot 5}$$

$$\frac{3}{8} + \frac{2}{5} = \frac{31}{40} \checkmark$$

Additional Example 3

Solve the equation $\frac{-3x + 19}{7(x+3)} - \frac{3}{7} = \frac{4}{x+3}$.

Multiple both sides of the equation by the LCD $7(x + 3)$.

$$7(x+3)\frac{-3x+19}{7(x+3)} - 7(x+3)\left(\frac{3}{7}\right) = 7(x+3)\left(\frac{4}{x+3}\right)$$

$$(-3x + 19) - (3x + 9) = 28$$

$$-6x + 10 = 28$$

$$-6x = 18$$

$$x = -3$$

Check

$$\frac{-3 \cdot -3 + 19}{7(-3+3)} - \frac{3}{7} \stackrel{?}{=} \frac{4}{-3+3}$$

$$\frac{10}{0} - \frac{3}{7} \neq \frac{4}{0}$$

You cannot have zero in the denominator, so -3 is not a solution. The equation has no solution.

Teach⇔Interact

Additional Example 4

Margie's speed driving home from a rock concert was 1.25 times her speed driving to the concert. She lives 28 mi from where the concert was given. Her total travel time was 1.5 h. What was Margie's average speed driving to the concert? her average speed driving home from the concert?

Step 1 Organize the information in a table. Let s = Margie's speed driving to the concert in miles per hour. Then $1.25s$ = Margie's speed from the concert in miles per hour.

Use the formula Time $= \frac{\text{Distance}}{\text{Rate}}$ to find how long each part of the trip took.

	Distance	Rate	Time
To Concert	28	s	$\frac{28}{s}$
From Concert	28	$1.25s$	$\frac{28}{1.25s}$

Step 2 Model the situation with an equation. The entire trip took 1.5 h.

Time to Concert + Time from Concert = 1.5

$$\frac{28}{s} + \frac{28}{1.25s} = 1.5$$

Step 3 Use a graph to solve the equation. Graph the equations $y = \frac{28}{x} + \frac{28}{1.25x}$ and $y = 1.5$ on a graphing calculator. Find the x-coordinate of the point where the graphs intersect.

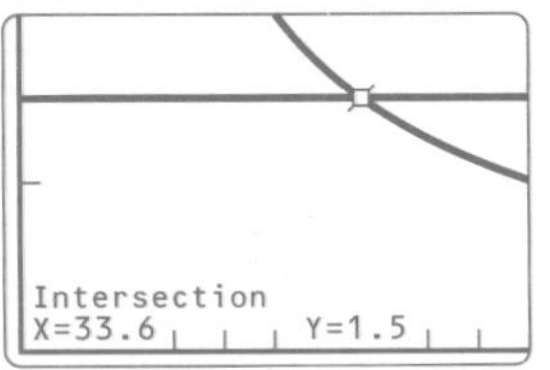

Margie's average speed driving to the concert was about 33.6 mi/h. Her average speed driving home from the concert was about 1.25(33.6) = 42 mi/h.

EXAMPLE 4 Application: Recreation

Dan Patterson hikes downhill **twice as fast** as he hikes uphill. It took Dan **5.5 h** to hike from Lake Louise to the top of Saddleback and back. The trail up Saddleback is about **3.7 mi** long. What is Dan's average uphill hiking speed? his average downhill speed?

SOLUTION

Step 1 Organize the information in a table.

Let s = Dan's **uphill hiking speed** in miles per hour.

Then $2s$ = Dan's **downhill hiking speed** in miles per hour.

Use the formula **Time** $= \frac{\text{Distance}}{\text{Rate}}$ to find how long each part of the trip takes.

	Distance	*Rate*	*Time*
Uphill hike	3.7	s	$\frac{3.7}{s}$
Downhill hike	3.7	$2s$	$\frac{3.7}{2s}$

Distance traveled one way.

Step 2 Model the situation with an equation. The entire trip takes **5.5 h**.

Time uphill + Time downhill = 5.5

$$\frac{3.7}{s} + \frac{3.7}{2s} = 5.5$$

Step 3 Use a graph to solve the equation.

Graph the equations $y = \frac{3.7}{x} + \frac{3.7}{2x}$ and $y = 5.5$ on a graphing calculator.

Find the x-coordinate of the point where the graphs intersect.

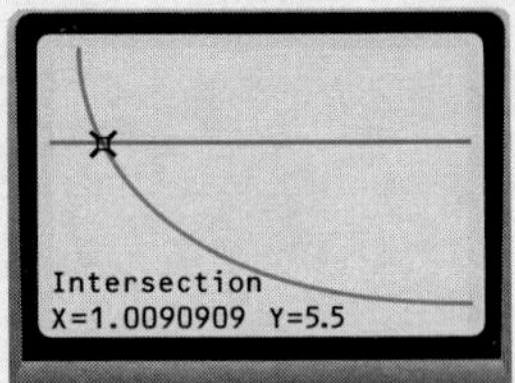

Dan's average uphill hiking speed is about 1 mi/h. His average downhill speed is about 2(1) = 2 mi/h.

THINK AND COMMUNICATE

2. Suppose Beth hikes uphill for 2 h and downhill twice as fast for 1 h, covering a total of 5 mi. Organize this information in a table like the one in Example 4 and write an equation. Let s = uphill speed.
3. Show how you can solve the equation in Example 4 algebraically.

2.

	Distance	Rate	Time
Uphill hike	$2s$	s	2 h
Downhill hike	$2s$	$2s$	1 h

$2s + 2s = 5$; $s = 1.25$

3. If you multiply both sides of the equation $\frac{3.7}{s} + \frac{3.7}{2s} = 5.5$ by the LCD, $2s$, you get $7.4 + 3.7 = 11s$; $s = \frac{11.1}{11} \approx 1.01$.

Checking Key Concepts

1. Yes; both sides are rational expressions.
2. Yes; both sides are rational expressions.
3. Yes; both sides are rational expressions.
4. No; the numerator of the right side is not a polynomial.
5. 24
6. $4x$
7. $10(x + 1)$
8. 3
9. no solution
10. $\frac{7}{13}$

CHECKING KEY CONCEPTS

Tell whether each equation is a rational equation. Explain your answer.

1. $y = \frac{36}{x}$

2. $xy = 36$

3. $\frac{16(0) + 13(5) + x(10)}{16 + 13 + x} = 7.5$

4. $x = \frac{-b \pm \sqrt{b^2 - 4ac}}{2a}$

Find the least common denominator of the fractions in each equation.

5. $\frac{x-2}{3} + \frac{3x}{8} = \frac{7}{6}$

6. $\frac{x}{4} + \frac{3}{x} = \frac{10x + 3}{4x}$

7. $\frac{6}{5x} = \frac{1}{10(x+1)}$

Solve each equation. If the equation has no solution, write *no solution*.

8. $\frac{1}{3} + \frac{2}{x} = \frac{3}{x}$

9. $\frac{9}{n} = 3 - \frac{4n - 9}{n}$

10. $\frac{2}{6x} - \frac{1}{x-1} = \frac{3}{2x}$

11.3 Exercises and Applications

Extra Practice *exercises on page 579*

Tell whether each equation is a rational equation. If it is not, explain why not.

1. $\frac{3}{x+1} = \frac{2}{x}$

2. $P = 0.2\sqrt{L}$

3. $\frac{26.4 + 10x}{4.7 + x} = 7.3$

4.  **Technology** Graph the equations $y = \frac{3}{x+1}$ and $y = \frac{2}{x}$ in the same coordinate plane. Use the graphs to solve the equation $\frac{3}{x+1} = \frac{2}{x}$. Check your answer by solving the equation algebraically.

Solve each equation. If the equation has no solution, write *no solution*.

5. $\frac{5}{x-1} = \frac{2}{3}$

6. $\frac{4}{x} + 2 = \frac{12}{x}$

7. $\frac{x}{4} + \frac{x}{6} = 2$

8. $\frac{6}{x+2} + 10 = \frac{1}{x+2}$

9. $\frac{6}{y} + \frac{3}{4} = 2$

10. $\frac{a+5}{a-6} = \frac{11}{a-6} + 4$

11. PHYSICS On or near Earth, the weight w of an object depends on its altitude.

Let h = height in miles above sea level.
Let s = weight in pounds at sea level.

$$w = \frac{(4000)^2 s}{(4000 + h)^2}$$

a. On long flights, passenger airplanes often fly at an altitude of about 6 mi. How much does a 150 lb passenger weigh 6 mi above sea level?

b. How high above sea level must a plane fly for a 120 lb person to weigh 119 lb?

BY THE WAY

The Condor, a robotic spy plane, set an altitude record of 12.7 mi for piston-powered aircraft. It could fly for more than two and a half days without refueling.

Exercises and Applications

1. Yes.
2. No; the right side is not a rational expression.
3. Yes.
4. 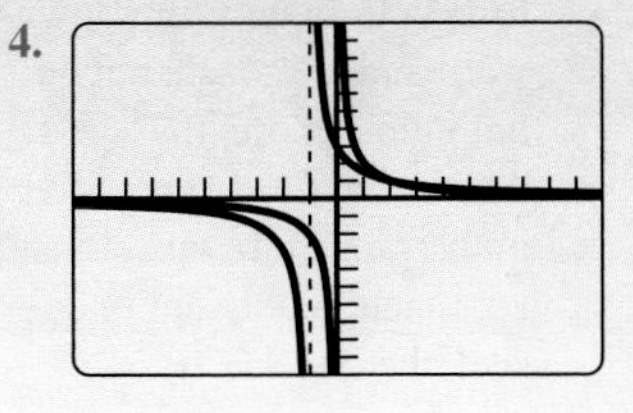

The x-coordinate of the intersection point, 2, is the solution. Algebraic solution:
$x(x+1) \cdot \frac{3}{x+1} = x(x+1) \cdot \frac{2}{x}$;
$3x = 2x + 2$; $x = 2$

5. 8.5
6. 4
7. 4.8
8. –2.5
9. 4.8
10. no solution
11. a. about 149.6 lb
    b. about 16.8 mi above sea level

Teach⇔Interact

Checking Key Concepts

Reasoning
Question 4 may be difficult for some students to understand. They are accustomed to thinking of *a*, *b*, and *c* in the quadratic formula as representing the coefficients of a quadratic equation which are constants. This would lead to the conclusion that the equation is rational. If they think of *a*, *b*, and *c* simply as variables, however, then the equation is not rational because the numerator is not a polynomial.

Closure Question

Why is it always necessary to check your answers when solving rational equations?
In some cases, the values found for the variables after eliminating fractions may result in a denominator of zero when substituted into the original rational equation.

Apply⇔Assess

Suggested Assignment

Core Course
Exs. 1–10, 12, 13, 15–24, 27, 31–37

Extended Course
Exs. 1–37

Block Schedule
Day 71 Exs. 1–10, 12, 13, 15–24, 27, 31–37

Exercise Notes

Using Technology
Ex. 4 Students can work on this exercise by using a graphing calculator or the *Function Investigator* software.

Common Error
Exs. 5–10 When solving rational equations, students may make errors in eliminating the fractions or in doing other algebraic steps to find the solution. Or they may state that a number is a solution when in fact it is not. Urge students to double-check their algebraic steps and to check their solution by substituting it into the original equation.

Apply⇔Assess

Exercise Notes

Mathematical Procedures
Ex. 13 Point out to students that using the means-extremes property is a very efficient way to solve rational equations that are proportions. Students may wish to use this property to solve Exs. 15 and 21.

Integrating the Strands
Ex. 20 This exercise integrates ideas from algebra, geometry, and probability.

Using Technology
Ex. 24 This exercise can be approached in two ways. The approach used in this section is to graph two equations (one for the left side of the equation and another for the right side) and then to find the point of intersection of the two graphs. Another approach is to get 0 on the right side by rewriting the equation as

$\frac{x+10}{x-1} - \frac{12-x}{x-1} - 5 = 0.$

You can then graph

$y = \frac{x+10}{x-1} - \frac{12-x}{x-1} - 5$

(or $y = \frac{2x-2}{x-1} - 7$) and find where the graph intersects the x-axis.

Either approach can be used with a graphing calculator or the *Function Investigator* software.

Communication: Writing
Ex. 26 Suggest that students use specific examples to illustrate their responses to this exercise.

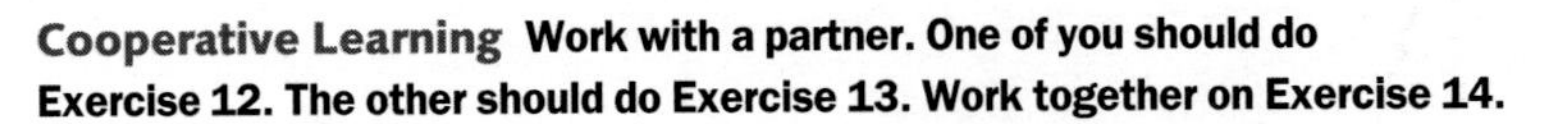

Cooperative Learning **Work with a partner. One of you should do Exercise 12. The other should do Exercise 13. Work together on Exercise 14.**

12. a. What is the LCD of the fractions in the equation $\frac{5}{x+3} = \frac{4}{x-1}$?

b. Solve the equation in part (a) by multiplying both sides by the LCD.

13. a. Is the equation $\frac{5}{x+3} = \frac{4}{x-1}$ a proportion? Explain.

b. Solve the equation in part (a) by using the means-extremes property.

14. Writing Compare the methods you used in Exercises 12 and 13. How are they alike? How are they different? Which do you prefer? Why?

Solve each equation. If the equation has no solution, write *no solution*.

15. $\frac{c-2}{5} = \frac{c-1}{4}$

16. $\frac{2}{x} + \frac{3}{5} = \frac{1}{4x}$

17. $\frac{t-1}{t} + \frac{3}{2t} = \frac{1}{3}$

18. $\frac{2(t-2)}{t} - 1 = \frac{-4}{t}$

19. SPORTS In his last NBA season, Larry Bird of the Boston Celtics made 52 of the 128 three-point shots he attempted, for an average of 40.6%. The Celtics lost in the semifinal round of the Eastern Conference Championship. If they had continued playing, how many consecutive three-point shots would Larry Bird have had to make to raise his average to 50%?

20. A game manufacturer is designing a dart board, as shown at the left.

a. Write an expression for the area of a bull's-eye with radius r. Write an expression for the area of the entire dart board.

b. The manufacturer wants the probability of a randomly thrown dart hitting the bull's-eye to be $\frac{1}{9}$. (Assume the dart hits the board.) What should the radius of the bull's-eye be?

Solve each equation. If the equation has no solution, write *no solution*.

21. $\frac{5}{z+2} = \frac{5}{z-4}$

22. $\frac{11}{5-b} = 4 + \frac{b+6}{5-b}$

23. $\frac{3}{4} + \frac{1}{6(t-5)} = \frac{5}{2(t-5)}$

24. a. **Technology** Solve the equation $\frac{x+10}{x-1} = \frac{12-x}{x-1} + 5$ algebraically.

b. Use a graphing calculator to solve the equation in part (a). Explain how the graph supports the answer you found in part (a).

25. Challenge The loudness of a sound decreases as the listener moves away from the sound. Suppose you are playing a radio at the beach with the volume very high. The intensity of the sound in Watts/m^2 can be modeled by the formula $I = \frac{0.01}{d^2}$, where d is the distance from the radio in meters. If your parents want to hear a sound only as intense as normal conversation (10^{-6} Watts/m^2), how far from the radio should they be?

26. Writing How are rational expressions like rational numbers? How are they different?

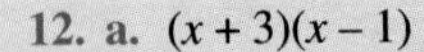

12. a. $(x+3)(x-1)$
b. 17
13. a. Yes, both sides of the equation are ratios.
b. 17
14. Answers may vary. An example is given. The methods are similar because both lead to the equation $5(x-1) = 4(x+3)$. In Ex. 12, both sides of the equation are multiplied by the product of two binomials. In Ex. 13, the numerator of each side is multiplied by a binomial. I prefer the method in Ex. 13 because it requires fewer steps and I do not need to find the LCD in order to solve the equation.

15. −3 16. $-\frac{35}{12}$
17. −0.75 18. no solution
19. 24 shots

20. a. πr^2; $\pi(r+2)^2$
b. 1 in.
21. no solution 22. no solution
23. $\frac{73}{9}$
24. a. no solution
b.

The graphs of $y = \frac{x+10}{x-1}$ and $y = \frac{12-x}{x-1} + 5$ do not intersect, so there is no solution. As x increases from 0 to 1, both graphs turn downward and approach, but do not intersect, the line $x = 1$. As x decreases from 2 to 1, both graphs turn upward and approach, but do not intersect, the line $x = 1$.

Connection RECREATION

In New Hampshire's White Mountains, there is a range of mountains named after presidents of the United States. Occasionally experienced hikers hike the 23 mi from one end of the range to the other.

27. Tanja and Adam Cook are hiking from Route 302 to Dolly Copp Campground along the path shown below. It took them 9 h to hike from Route 302 to the top of Mt. Washington. They usually hike downhill twice as fast as they hike uphill. What was their average uphill rate? their average downhill rate?

28. Predict how long it will take Tanja and Adam to hike from the top of Mt. Washington to Dolly Copp Campground. Explain your answer.

29. RESEARCH Which of the peaks do not have names of former presidents?

30. Open-ended Problem Write a rational equation that has no solution. Explain how you came up with your equation.

BY THE WAY

Normal summer temperatures on Mt. Washington are only about 47°F. Winds can reach up to 231 mi/h.

ONGOING ASSESSMENT

31. Writing How is finding the least common denominator of the fractions $\frac{3}{2(x-3)}$, $\frac{5}{4x}$, and $\frac{2}{2x}$ like finding the least common denominator of the fractions $\frac{3}{8}$, $\frac{5}{12}$, and $\frac{2}{3}$? How is it different?

SPIRAL REVIEW

Graph each equation. *(Section 3.4)*

32. $y = 2x - 5$ **33.** $y = -4x$ **34.** $y = \frac{2}{3}x + 3$

Solve each equation by factoring. *(Section 10.5)*

35. $2x^2 + 5x - 3 = 0$ **36.** $3y^2 - 15y + 12 = 0$ **37.** $3a^2 + 11a - 4 = 0$

Apply⇔Assess

Exercise Notes

Reasoning

Ex. 25 Ask students if they can express the relationship shown by the equation using the language of variation. (The quantities I and d^2 vary inversely.)

Problem Solving

Ex. 30 You may wish to extend this problem by asking students how to write an equation that is true for all but a specified group of numbers. For example, find an equation that is true for all real numbers *except* 3, 8, and –1.5. (Sample answer: $\frac{(x-3)(x-8)(x+1.5)}{(x-3)(x-8)(x+1.5)} = 1$)

Topic Spiraling: Preview

Ex. 31 Students who have difficulty with this exercise should be led to see the connection presented here as it will help them with the concepts that will be presented in Section 11.6.

Practice 74 for Section 11.3

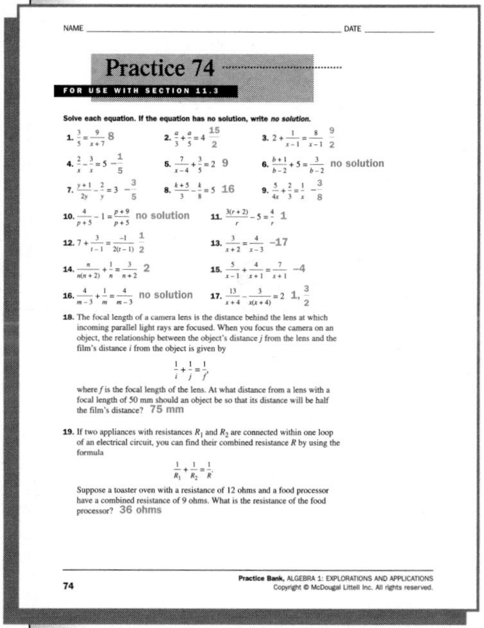
NAME ______ DATE ______

Practice 74

FOR USE WITH SECTION 11.3

Solve each equation. If the equation has no solution, write *no solution*.

1. $\frac{3}{5} = \frac{9}{x+7}$ 8
2. $\frac{a}{3} + \frac{a}{5} = 4$ $\frac{15}{2}$
3. $2 + \frac{1}{x-1} = \frac{8}{x-1}$ $\frac{9}{2}$
4. $\frac{2}{x} - \frac{3}{x} = 5$ $-\frac{1}{5}$
5. $\frac{7}{x-4} + \frac{3}{5} = 2$ 9
6. $\frac{b+1}{b-2} + 5 = \frac{3}{b-2}$ no solution
7. $\frac{y+1}{2y} - \frac{2}{y} = 3$ $-\frac{3}{5}$
8. $\frac{k+5}{3} - \frac{k}{8} = 5$ 16
9. $\frac{5}{4x} + \frac{2}{3} = \frac{1}{x}$ $-\frac{3}{8}$
10. $\frac{4}{p+5} - 1 = \frac{p+9}{p+5}$ no solution
11. $\frac{3(r+2)}{r} - 5 = \frac{4}{r}$ 1
12. $7 + \frac{3}{t-1} = \frac{-1}{2(t-1)}$ $\frac{1}{2}$
13. $\frac{3}{x+2} = \frac{4}{x-3}$ –17
14. $\frac{n}{n(n+2)} + \frac{1}{n} = \frac{3}{n+2}$ 2
15. $\frac{5}{x-1} + \frac{4}{x+1} = \frac{7}{x+1}$ –4
16. $\frac{4}{m-3} + \frac{1}{m} = \frac{4}{m-3}$ no solution
17. $\frac{13}{x+4} - \frac{3}{x(x+4)} = 2$ 1, $\frac{3}{2}$
18. The focal length of a camera lens is the distance behind the lens at which incoming parallel light rays are focused. When you focus the camera on an object, the relationship between the object's distance j from the lens and the film's distance i from the object is given by

$$\frac{1}{i} + \frac{1}{j} = \frac{1}{f},$$

where f is the focal length of the lens. At what distance from a lens with a focal length of 50 mm should an object be so that its distance will be half the film's distance? 75 mm

19. If two appliances with resistances R_1 and R_2 are connected within one loop of an electrical circuit, you can find their combined resistance R by using the formula

$$\frac{1}{R_1} + \frac{1}{R_2} = \frac{1}{R}.$$

Suppose a toaster oven with a resistance of 12 ohms and a food processor have a combined resistance of 9 ohms. What is the resistance of the food processor? 36 ohms

74 **Practice Bank,** ALGEBRA 1: EXPLORATIONS AND APPLICATIONS
Copyright © McDougal Littell Inc. All rights reserved.

25. 100 m
26. Answers may vary. An example is given. Rational expressions are like rational numbers in that both can be written as fractions with nonzero denominators. However, the numerators and denominators of rational numbers are integers, whereas the numerators and denominators of rational expressions are polynomials.
27. about 1 mi/h; about 2 mi/h

28–31. See answers in back of book.

32.

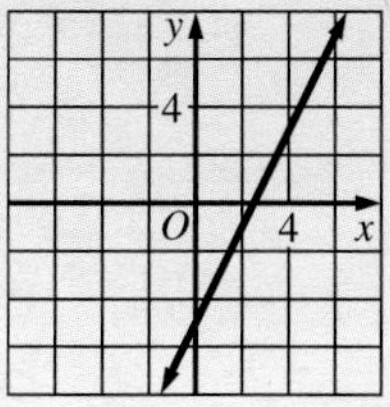

33.

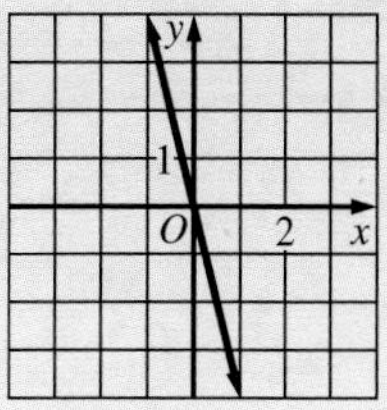

34.

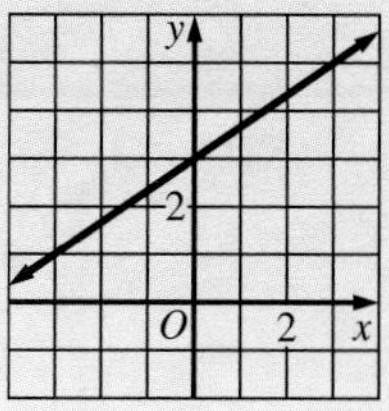

35. $-3; \frac{1}{2}$
36. 1; 4
37. $-4; \frac{1}{3}$

Plan⇔Support

Objectives

- Solve a formula for one of its variables.
- Perform multiple calculations with the same formula.

Recommended Pacing

Core and Extended Courses
Section 11.4: 1 day

Block Schedule
Section 11.4: $\frac{1}{2}$ block
(with Section 11.3)

Resource Materials

Lesson Support
Lesson Plan 11.4
Warm-Up Transparency 11.4
Practice Bank: Practice 75
Study Guide: Section 11.4
Assessment Book: Test 48

Technology
Graphing Calculator
Spreadsheet Software
McDougal Littell Software *Stats!*
Internet: http://www.hmco.com

Warm-Up Exercises

Solve each equation.

1. $3x + 8 = 29$ 7
2. $8x - 3 = 4x + 7$ $\frac{5}{2}$
3. $\frac{x+4}{3} = 10$ 26

Simplify.

4. $\frac{a+b}{a-b}(a-b)$ $a + b$
5. $\frac{cd}{c^2 - d^2} \cdot \frac{1}{3c}$ $\frac{d}{3c^2 - 3d^2}$

SECTION

11.4 Applying Formulas

Learn how to...

- **solve a formula for one of its variables**

So you can...

- **perform multiple calculations with the same formula**
- **graph equations on graphing calculators**
- **make calculations with spreadsheets**

You may think you never use formulas outside math class. But most people use formulas all the time. For example, you use a formula if you leave a 15% tip in a restaurant or estimate how long it will take to travel a given distance at a given speed.

EXAMPLE 1 Application: Recreation

Some hiking guidebooks estimate the time a hiker needs to hike a trail by using the formula $t = \frac{1}{2}m + v$, where t is the **time** in hours, m is the **length** of the trail in miles, and v is the **vertical rise** in thousands of feet. Jonathan wants to choose a trail which is not very strenuous. His guidebook says that the Tumbledown Mountain loop trail is **2 mi long** and takes about $2\frac{3}{4}$ **h** to hike. Use the formula to estimate the vertical rise of the trail.

SOLUTION

Method 1 Substitute 2.75 for t and 2 for m. Then solve for v.

$$t = \frac{1}{2}m + v$$

$$2.75 = \frac{1}{2} \cdot 2 + v$$

$$2.75 = 1 + v$$

$$1.75 = v$$

The vertical rise of the trail is about 1750 ft.

Method 2 Solve the formula for v. Then substitute 2.75 for t and 2 for m and simplify.

$$t = \frac{1}{2}m + v$$

Subtract $\frac{1}{2}m$ from both sides.

$$t - \frac{1}{2}m = v$$

$$2.75 - \frac{1}{2} \cdot 2 = v$$

$$2.75 - 1 = v$$

$$1.75 = v$$

The vertical rise of the trail is about 1750 ft.

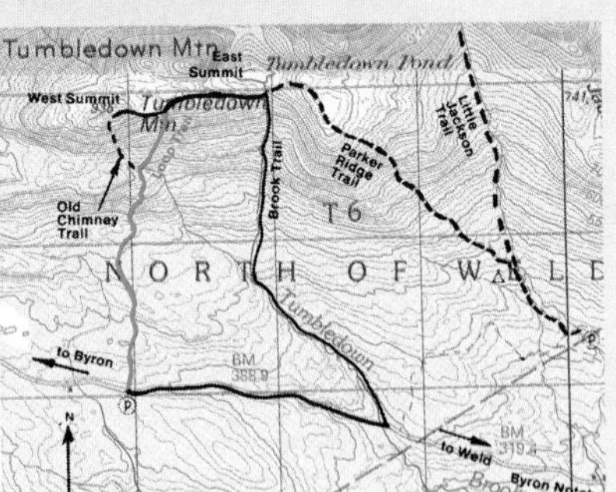

THINK AND COMMUNICATE

1. If you want to find the vertical rises of several trails, would you rather use the formula $t = \frac{1}{2}m + v$ or $t - \frac{1}{2}m = v$? Explain.
2. Do you think the Tumbledown Mountain loop trail is likely to be a strenuous hike? Explain.

If you need to use a formula many times, you may want to solve the formula for one of its variables first. You also solve an equation for a variable to graph it on a graphing calculator.

EXAMPLE 2

Graph the equation $x = \frac{y}{y+2}$ on a graphing calculator.

SOLUTION

Solve the equation for y. Get y alone on one side of the equation.

$$x = \frac{y}{y+2}$$

$$(y+2)x = \frac{y}{y+2}(y+2)$$

Multiply both sides by $(y + 2)$ and simplify. Note that $y \neq -2$.

$$(y+2)x = y$$

Move all terms containing y to the same side and factor out y.

$$yx + 2x = y$$

$$yx - y = -2x$$

$$y(x-1) = -2x$$

Divide both sides by $x - 1$ so y is alone. Note that $x \neq 1$.

$$y = \frac{-2x}{x-1}$$

Now you can graph the equation.

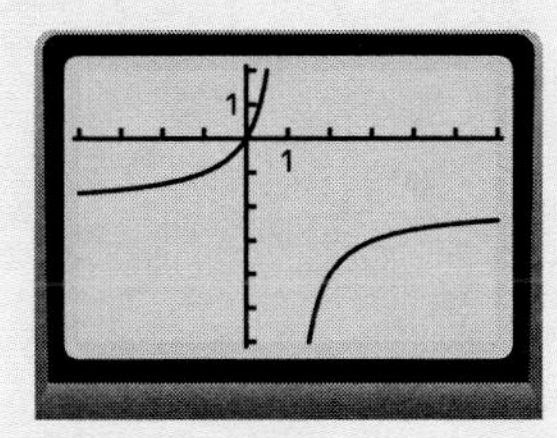

WATCH OUT!

The graph of your final equation may include a point that gives the original equation a zero denominator. If this happens, use an open dot to indicate that the point is not a part of the graph.

THINK AND COMMUNICATE

3. In the solution of Example 2, why must you factor out y?

Teach⇔Interact

Additional Example 1

Suppose Jonathan's guidebook gives the time for hiking the Black Horse Mountain trail as $3\frac{1}{2}$ h and the vertical rise as 2800 ft. Use the formula in Example 1 to estimate the length of the trail.

Since the formula requires the value of v to be in thousands of feet, use 2.8 for v. Use 3.5 for t.

Method 1 Substitute 2.8 for v and 3.5 for t and solve for m.

$$3.5 = \frac{1}{2}m + 2.8$$

$$0.7 = \frac{1}{2}m$$

$$1.4 = m$$

Method 2 Solve the formula for m. Then substitute 2.8 for v and 3.5 for t and simplify.

$$t = \frac{1}{2}m + v$$

$$t - v = \frac{1}{2}m$$

$$2(t - v) = m$$

$$2(3.5 - 2.8) = m$$

$$2(0.7) = m$$

$$1.4 = m$$

The length of Black Horse Mountain trail is about 1.4 mi.

Additional Example 2

Graph $x + y = \frac{y}{x}$ on a graphing calculator.

Solve the equation for y. Get y alone on one side of the equation.

$$x + y = \frac{y}{x}$$

$$x(x + y) = x\frac{y}{x}$$

$$x^2 + xy = y$$

$$xy - y = -x^2$$

$$y(x - 1) = -x^2$$

$$y = \frac{-x^2}{x-1}$$

Now graph the equation.

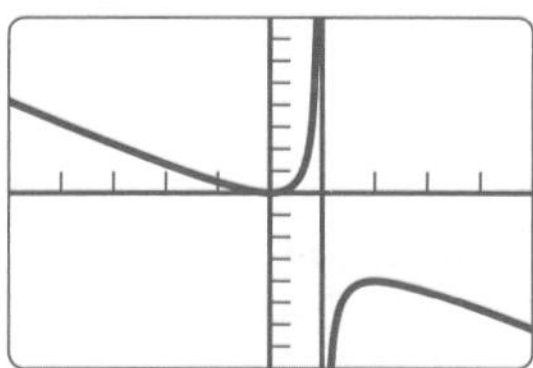

Note that $x \neq 0$ in $x + y = \frac{y}{x}$. However, in $y = \frac{-x^2}{x-1}$, it is possible for $x = 0$. Thus, the graph contains a point, (0, 0), that is not on the graph of the original equation.

ANSWERS Section 11.4

Think and Communicate

1. Answers may vary. An example is given. I would rather use the formula $t - \frac{1}{2}m = v$ because I could calculate v directly by substituting each pair of values of t and m in the formula.
2. Answers may vary. An example is given. I think a vertical rise of 1750 ft over a 2-mile hike would be quite strenuous. The hiker's average rate of speed is only about 0.7 mi/h, which is fairly slow.
3. to separate y from x in the term yx and to separate y from –1 in the term $-y$ so that you can get y alone on one side

Teach⇔Interact

Additional Example 3

Many optical devices use concave circular mirrors. Light rays from an object that travel on paths nearly parallel to the axis of symmetry of such a mirror will be focused at a point in front of the mirror. The focus point will be approximately on the axis.

Let d = the distance from the object to the mirror.

Let i = the distance from the focus point to the mirror.

Let R = the radius of the mirror.

Then d, i, and R are related by the formula

$\frac{1}{d} + \frac{1}{i} \approx \frac{2}{R}$.

a. Solve the formula for i.

$\frac{1}{d} + \frac{1}{i} \approx \frac{2}{R}$

Multiply both sides by the LCD, which is diR.

$diR\left(\frac{1}{d}\right) + diR\left(\frac{1}{i}\right) \approx diR\left(\frac{2}{R}\right)$

Move terms containing i to the same side.

$iR + dR \approx 2di$

$iR - 2di \approx -dR$

Factor out i.

$i(R - 2d) \approx -dR$

Divide both sides by $(R - 2d)$.

$i \approx \frac{-dR}{R - 2d}$ or $i \approx \frac{dR}{2d - R}$

b. Find the approximate distance of the focus point from the mirror for an object located 200 m from a concave circular mirror with radius 1 m.

$i \approx \frac{dR}{2d - R}$

$\approx \frac{(200)(1)}{2(200) - 1}$

$\approx \frac{(200)(1)}{400 - 1}$

≈ 0.501

The focus point is about 0.5 m from the mirror.

Section Note

Historical Connection

In connection with Example 3, students may find it interesting to know that it was not until the 13th century that Europeans discovered the power of glass lenses to magnify objects. Hundreds of years later, when lens grinders had perfected their craft, European scientists discovered that the right combination of lenses could be used to make a telescope or a microscope.

BY THE WAY

If you focus sunlight to make a small, bright dot on the ground, the distance from the dot to the magnifying glass is the focal length of the glass. The focal length depends on the shape of the glass.

EXAMPLE 3 Application: Optics

If you look at a leaf with a magnifying glass, the distances between the leaf, the glass, and your eye are related by the formula below.

Let a = the distance from the magnifying glass to the leaf.

Let b = the distance from your eye to the magnifying glass.

Let f = the focal length of the magnifying glass.

$$\frac{1}{a} + \frac{1}{b} \approx \frac{1}{f}$$

Not drawn to scale

a. Solve the formula for a.

b. How far should you hold the magnifying glass from the leaf if your eye is **18 in.** from the glass and the focal length of the glass is **5 in**?

SOLUTION

a. $\frac{1}{a} + \frac{1}{b} \approx \frac{1}{f}$ — Multiply both sides by the least common denominator abf.

$abf\left(\frac{1}{a}\right) + abf\left(\frac{1}{b}\right) \approx abf\left(\frac{1}{f}\right)$

Move all terms containing a to the same side.

$bf + af \approx ab$

$bf \approx ab - af$ — Factor out a.

$bf \approx a(b - f)$ — Divide both sides by $(b - f)$.

$\frac{bf}{b - f} \approx a$

b. $a \approx \frac{bf}{b-f}$

$\approx \frac{(18)(5)}{18 - 5}$

≈ 6.9

To see the leaf clearly, you should hold the magnifying glass about 7 in. from the leaf.

Checking Key Concepts

1. $z = \frac{2}{5}y$
2. $y = 4 - x$
3. $m = 3 - \frac{1}{2}d$
4. $a = \frac{2d}{t^2}$
5. $b = \frac{1}{2c}$
6. $n = \frac{2}{f+2}$
7. 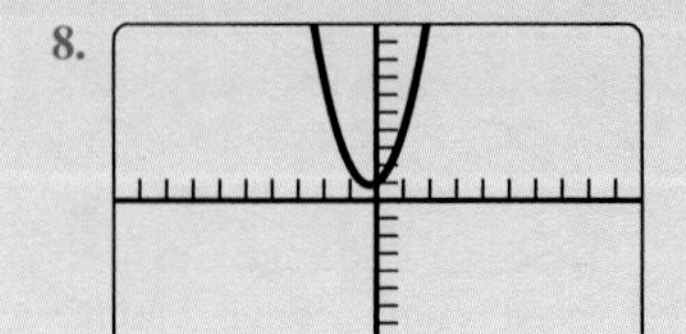
8. 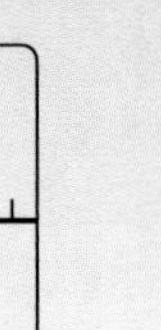
9. $t = \frac{d}{r}$

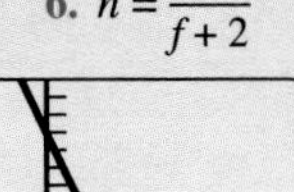

Exercises and Applications

1. a. $h = \frac{3V}{B}$ b. 10 cm
2. 4.5 mi
3. $b = P - a - c$
4. $l = \frac{A}{w}$
5. $C = P + S$
6. $b = \frac{y-a}{x}$
7. $y = 24 - 2x$
8. $y = \frac{45 - 3x}{2}$
9. $y = \frac{c - ax}{b}$
10. $y = \frac{A - 2x^2}{4x}$
11. $x = \frac{1}{y}$

CHECKING KEY CONCEPTS

Solve each equation for the variable in red.

1. $5z = 2y$
2. $x + y = 4$
3. $2m + d = 6$
4. $d = \frac{1}{2}at^2$
5. $\frac{1 - bc}{b} = c$
6. $\frac{f}{2} = \frac{1 - n}{n}$

Technology Graph each equation. Use a graphing calculator if you have one.

7. $3x + y = 7$
8. $y - 2x^2 - 1 = x$
9. **Business** Barbara Andrews travels frequently and needs to determine how long each of her trips will take. Express the formula $d = rt$ in the form she would find most useful.

11.4 Exercises and Applications

1. **GEOMETRY** The formula for the volume of a pyramid is $V = \frac{1}{3}Bh$, where B is the area of the base and h is the height.
   a. Solve the formula for h.
   b. Find h when $V = 140\ \text{cm}^3$ and $B = 42\ \text{cm}^2$.

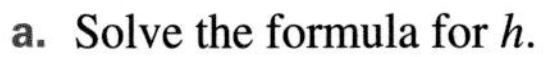

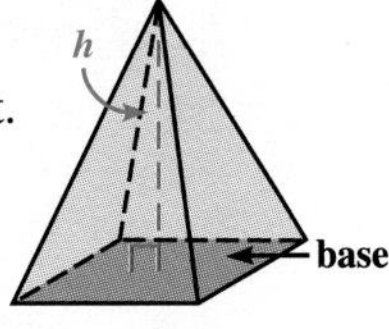

Extra Practice exercises on page 579

2. **RECREATION** It takes 3 h to hike a trail that has a vertical rise of 750 ft. Use the formula from Example 1 to estimate the length of the trail.

Solve each equation for the variable in red.

3. $P = a + b + c$
4. $A = lw$
5. $P = C - S$
6. $y = a + bx$
7. $2x + y = 24$
8. $3x + 2y = 45$
9. $ax + by = c$
10. $A = 2x^2 + 4xy$
11. $y = \frac{1}{x}$

Technology Graph each equation. Use a graphing calculator if you have one.

12. $x^2 + y - 9 = 0$
13. $x + 2y = 3$
14. $x^2 - \frac{1}{2}y = x$
15. $\frac{x - 1}{y} = 3$
16. $2(y - 3) = x^2$
17. $x(y + 1) = y - 1$
18. **Writing** Jen solved the formula $ry = a + y$ for y, as shown at the right. Describe what Jen did. What is wrong with her solution?

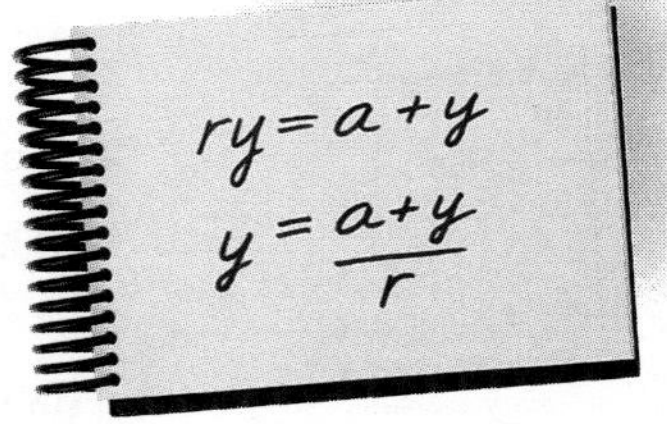

19. a. **OPTICS** Solve the equation in Example 3 on page 484 for f.
    b. Find the focal length of the magnifying glass if you can see the leaf clearly when $a = 40$ in. and $b = 10$ in.

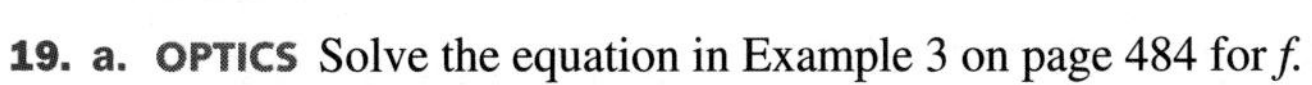

12.

14.

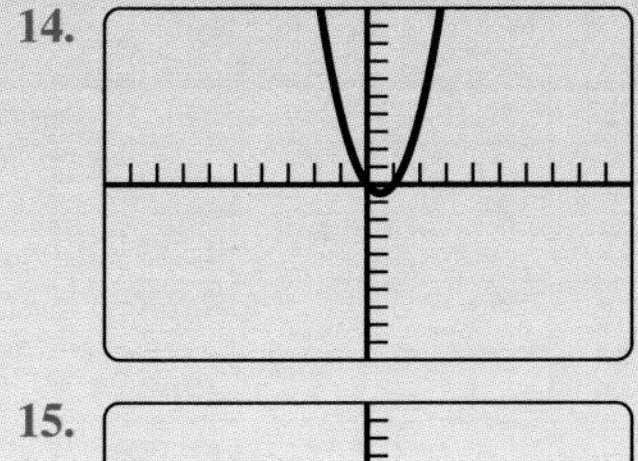

16.

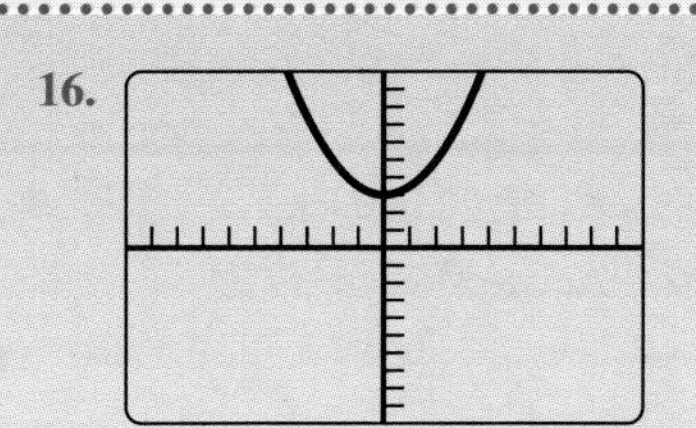

13.

15.

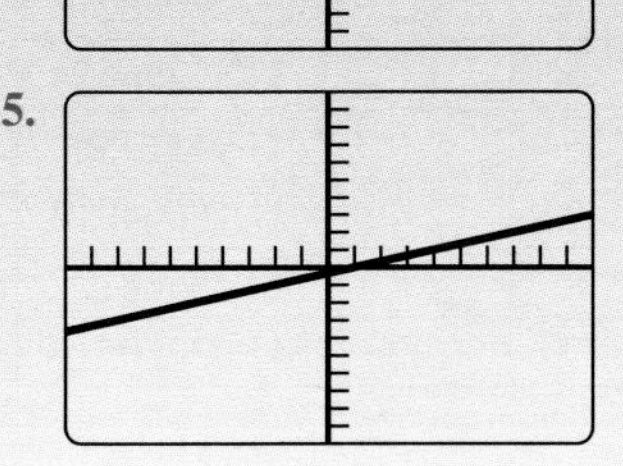

17, 18. See answers in back of book.

19. a. $f \approx \frac{ab}{a + b}$
    b. about 8 in.

Teach⇔Interact

Section Note

Research

Optics and other branches of physics are excellent sources of formulas that students can solve for one variable or another. Have students research other formulas that they can use to solve for one of its variables.

Closure Question

How is solving a formula for a variable similar to or different from solving an equation that contains a single variable?

They are similar in that both require getting all terms involving that variable on the same side of the equation. The same properties of equations are used to do this. The results are different in that solutions for equations with one variable are numbers, while in the case of a formula, the result is an expression containing one or more variables.

Apply⇔Assess

Suggested Assignment

Core Course
Exs. 1–17, 20, 22, 25–31, AYP

Extended Course
Exs. 1–22, 24–31, AYP

Block Schedule
Day 71 Exs. 1–17, 20, 22, 25–31, AYP

Exercise Notes

Integrating the Strands

Ex. 1 Many problems in geometry can be solved by using algebraic formulas. This exercise provides one example of solving a formula for one of its variables.

Alternate Approach

Exs. 3–11 Many students experience some initial difficulties when first learning to solve a formula for one of its variables. They are not accustomed to working with more than two variables in one equation. One approach that can be used to help students with this difficulty is to expose them to different solutions of the same equation. For each of these exercises, call upon these students to write their solutions on the board.

Apply⇔Assess

Exercise Notes

Interview Note

Exs. 20, 21 Aerial photography has a wide variety of applications. Aerial photographs are used to estimate crowd sizes at outdoor events, to study traffic flow in cities, and to survey troop and equipment movement in military situations. Mapmakers and environmentalists make extensive use of aerial photographs in their work. Those who make and use such photographs need a good understanding of mathematics to do their work.

Second-Language Learners

Ex. 22 Some of the terms and concepts related to the engine of a car may be difficult for second-language students to understand. You might read this problem aloud and then have students work cooperatively to solve it.

Reasoning

Ex. 23 Suggest that students think about the equation before they begin solving for y. They should note that y must be restricted to numbers whose absolute values are greater than or equal to 1. Ask why this is so. (If $|y| < 1$, then $y^2 - 1$ will be negative, and negative numbers do not have square roots.)

INTERVIEW Alex MacLean

Look back at the article on pages 460–462.

Focal Lengths of Lenses
24 mm
28 mm
35 mm
50 mm
80–200 mm zoom
300 mm

Alex MacLean can adjust image sizes in aerial photographs by varying his flying height from about 0.2 mi to 2.5 mi or by choosing different lenses. The table at the left shows the focal lengths of the lenses Alex MacLean uses. Focal length f, flying height h, image size i, and actual size a are related by the formula $\frac{i}{a} = \frac{f}{h}$.

20. a. Solve the formula above for f.

b. Alex MacLean took a photograph of a $\frac{1}{2}$ mi long irrigator from a height of 1 mi. The image of the irrigator on 35 mm film is 14 mm long. Which lens did Alex use for this photograph?

A giant pivot irrigator stopping short of erosion gullies along the Teton River.

21. Open-ended Problem Look back at the map of New York's Central Park on page 87. Suppose you ask Alex MacLean to photograph Central Park. The park is about $\frac{1}{2}$ mi wide and $2\frac{1}{2}$ mi long.

a. Alex MacLean uses 35 mm film, which has frames that are 24 mm by 36 mm. Sketch how you would like Central Park to appear within the frame. How long will the image of the park be?

b. Choose a lens and flying height to produce the image you sketched in part (a).

22. CARS Pushing a bicycle pedal causes the wheels of the bicycle to turn. In a car, exploding gas pushes a piston which causes the wheels of the car to turn. The power of the car engine depends on many things.

Let H = the power of a car engine in *horsepower*.
Let p = the mean pressure on the piston in lb/in.^{2}.
Let l = the distance in feet the piston moves.
Let a = the area in square inches of the piston.
Let n = the number of explosions per minute.
Let k = the number of pistons in the engine.

$$H = \frac{plank}{33{,}000}$$

exploding gas
p
a
l
Not drawn to scale

a. Solve the equation for p.

b. Find p for a 300 horsepower engine for which $l = 0.29$ ft, $a = 12.56$ in.^{2}, $n = 2500$/min, and $k = 8$.

BY THE WAY

One horsepower is 33,000 foot-pounds per minute, which is about $1\frac{1}{3}$ times the actual power of a horse.

23. Challenge Solve the equation $\sqrt{y^2 - 1} = x - y$ for y.

20. a. $f = \frac{ih}{a}$

b. the 28 mm lens

21. Answers may vary.

22. a. $p = \frac{33{,}000H}{lank}$

b. about 135.9 lb/in.^{2}

23. $y = \frac{x^2 + 1}{2x}$

24. Answers may vary. An example is given. If $r = 18$ ft, then $v = \sqrt{18ng} = 24\sqrt{n}$.

	A	B
1	n	$v = 24\sqrt{n}$
2	0.4	15.17893277
3	0.8	21.46625258
4	1.2	26.29068276
5	1.6	30.35786554
6	2.0	33.9411255
7	2.4	37.18064012
8	2.8	40.15968127
9	3.2	42.93250517

25. Answers may vary. An example is given. It is helpful to solve a formula for one of its variables when you need to find the value of that variable for various values of the other variable(s). It is also helpful when you need to input the equation into a calculator or a computer and the given formula does not express the required variable in terms of the

24. **Spreadsheets** Suppose you are designing a spinning circular room for an amusement park. The faster the room spins, the more the riders are pressed against the wall. If the ride spins quickly enough, the riders are pressed into the wall more than they are pulled to the ground by gravity.

Let g = how much you feel "pulled" to the ground by gravity. (32 ft/s^2)
Let n = the number of g's with which you are pressed against the wall.
Let r = the radius of the circular room in feet.
Let v = the speed of the riders in feet per second.

Choose a reasonable value for r. Then use the formula $ng = \frac{v^2}{r}$ and a spreadsheet to find the speed v that will result in each value of n: 0.4, 0.8, 1.2, 1.6, 2.0, 2.4, 2.8, 3.2

ONGOING ASSESSMENT

25. **Writing** When is it helpful to solve a formula for one of its variables? Give some examples to support your answer.

SPIRAL REVIEW

Find the reciprocal of each number. *(Section 4.2)*

26. 4
27. $\frac{1}{3}$
28. $-\frac{2}{3}$

Simplify each expression. *(Section 9.7)*

29. $(4x^3)^3$
30. $\left(\frac{a}{b^2}\right)^5$
31. $\left(\frac{m^5}{n^2p^3}\right)^2$

ASSESS YOUR PROGRESS

VOCABULARY

rational expression (p. 476)   **rational equation** (p. 476)

Solve. *(Section 11.3)*

1. $\frac{x-1}{3} = \frac{x+1}{4}$
2. $\frac{5}{4x} + \frac{7}{8} = \frac{3}{x}$
3. $\frac{2}{x} + \frac{1}{3} = \frac{11}{3x}$

Solve each equation for the variable in red. *(Section 11.4)*

4. $x - 2y = 8$
5. $2(x - 3) = y^2 + 1$
6. $\frac{a+b}{2c} = b + d$
7. The formula $C = \frac{5}{9}(F - 32)$ gives the temperature C in degrees Celsius for a given temperature F in degrees Fahrenheit. Solve for F. Find F when $C = 20°$. *(Section 11.4)*
8. **Journal** What do you find most difficult about solving rational equations? How do you think you could make this less difficult?

Apply⇔Assess

Exercise Notes

Using Technology
Ex. 24 Students can use the *Stats!* software for this exercise.

Assessment Note
Ex. 25 The examples that students give will provide useful information on how well they understand the role of formulas in real-world and mathematical applications of algebra.

Assess Your Progress

Journal Entry
Students may find it helpful to discuss their ideas in small groups before putting their journal entries in final form.

Progress Check 11.3–11.4

See page 503.

Practice 75 for Section 11.4

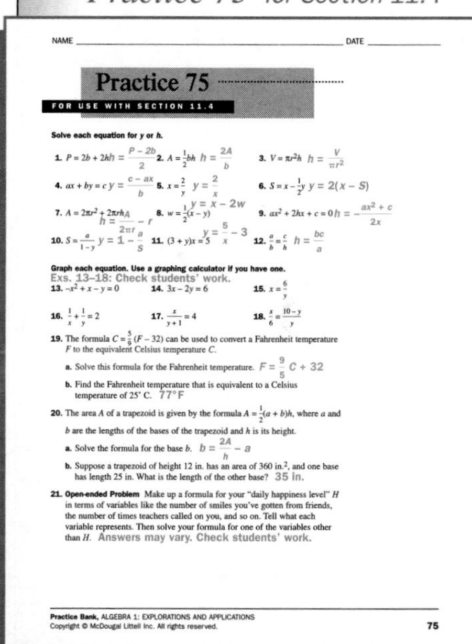
NAME ______________________ DATE __________

Practice 75

FOR USE WITH SECTION 11.4

Solve each equation for y or h.

1. $P = 2b + 2h$ $h = \frac{P-2b}{2}$
2. $A = \frac{1}{2}bh$ $h = \frac{2A}{b}$
3. $V = \pi r^2 h$ $h = \frac{V}{\pi r^2}$
4. $ax + by = c$ $y = \frac{c - ax}{b}$
5. $x = \frac{2}{y}$ $y = \frac{2}{x}$
6. $S = x - \frac{1}{2}y$ $y = 2(x - S)$
7. $A = 2\pi r^2 + 2\pi rh$ $h = \frac{A}{2\pi r} - r$
8. $w = \frac{1}{2}(x - y)$ $y = x - 2w$
9. $ax^2 + 2hx + c = 0$ $h = -\frac{ax^2 + c}{2x}$
10. $S = \frac{a}{1-y}$ $y = 1 - \frac{a}{S}$
11. $(3 + y)x = 5$ $y = \frac{5}{x} - 3$
12. $\frac{a}{b} = \frac{c}{h}$ $h = \frac{bc}{a}$

Graph each equation. Use a graphing calculator if you have one.
Exs. 13–18: Check students' work.

13. $-x^2 + x - y = 0$
14. $3x - 2y = 6$
15. $x = \frac{6}{y}$
16. $\frac{1}{x} + \frac{1}{y} = 2$
17. $\frac{x}{y+1} = 4$
18. $\frac{x}{6} = \frac{10-y}{y}$
19. The formula $C = \frac{5}{9}(F - 32)$ can be used to convert a Fahrenheit temperature F to the equivalent Celsius temperature C.
   a. Solve this formula for the Fahrenheit temperature. $F = \frac{9}{5}C + 32$
   b. Find the Fahrenheit temperature that is equivalent to a Celsius temperature of 25° C. 77°F
20. The area A of a trapezoid is given by the formula $A = \frac{1}{2}(a + b)h$, where a and b are the lengths of the bases of the trapezoid and h is its height.
   a. Solve the formula for the base b. $b = \frac{2A}{h} - a$
   b. Suppose a trapezoid of height 12 in. has an area of 360 in.^{2}, and one base has length 25 in. What is the length of the other base? 35 in.
21. **Open-ended Problem** Make up a formula for your "daily happiness level" H in terms of variables like the number of smiles you've gotten from friends, the number of times teachers called on you, and so on. Tell what each variable represents. Then solve your formula for one of the variables other than H. Answers may vary. Check students' work.

Practice Bank, ALGEBRA 1: EXPLORATIONS AND APPLICATIONS

75

other variable(s). For example, if you know the lengths of the hypotenuse and one leg for several right triangles and you want to find the lengths of the other legs, you might solve the equation $a^2 + b^2 = c^2$ for a or b. If you want to graph the equation $2x + 5y = -20$ on a graphing calculator, you must first solve the equation for y.

26. $\frac{1}{4}$
27. 3
28. $-\frac{3}{2}$
29. $64x^9$
30. $\frac{a^5}{b^{10}}$
31. $\frac{m^{10}}{n^4p^6}$

Assess Your Progress

1. 7
2. 2
3. 5
4. $y = \frac{1}{2}x - 4$
5. $x = \frac{y^2+7}{2}$
6. $b = \frac{a - 2cd}{2c - 1}$
7. $F = \frac{9}{5}C + 32$; $F = 68°$
8. Answers may vary. An example is given. I think that the most difficult part is simplifying the equation after multiplying both sides by the least common denominator. I hope that writing neatly and getting more practice with this skill will make this part easier for me.

Plan⇔Support

Objectives

- Multiply and divide rational expressions.
- Use multiplication and division of rational expressions to solve mathematical and real-world problems.

Recommended Pacing

Core and Extended Courses
Section 11.5: 1 day

Block Schedule
Section 11.5: $\frac{1}{2}$ block
(with Section 11.6)

Resource Materials

Lesson Support
Lesson Plan 11.5
Warm-Up Transparency 11.5
Practice Bank: Practice 76
Study Guide: Section 11.5
Explorations Lab Manual:
Additional Exploration 17

Technology
Graphing Calculator
Internet:
http://www.hmco.com

Warm-Up Exercises

Write each number as a product of powers of prime factors.

1. 72 $2^3 \cdot 3^2$

2. 150 $2 \cdot 3 \cdot 5^2$

3. 4900 $2^2 \cdot 5^2 \cdot 7^2$

4. 392 $2^3 \cdot 7^2$

Simplify.

5. $\frac{2^5 \cdot 7^3}{11^2} \cdot \frac{11^2}{2^3 \cdot 7}$ 196

6. $5^{-3} \cdot \frac{2^2}{5}$ 0.0064

SECTION

11.5 Working with Rational Expressions

Learn how to...

- **multiply and divide rational expressions**

So you can...

- **solve problems such as finding the area of an ice cream cone or the turning radius of a train**

Joline Jacques wants to make 500 ice cream cones out of cookie dough for her ice cream shop. How can she tell how much dough to make?

Imagine cutting along the side of a cone and flattening it. The resulting shape is a sector of a circle. The area of the sector is the area of the dough needed to make the cone.

EXAMPLE 1 Application: Cooking

What is the area of the piece of dough Joline cuts out for each cone?

Let A = the area of the sector of dough as described above.
Let C = the circumference of the top of the cone.
Let s = the slant height of the cone.

SOLUTION

$$\frac{\text{area of sector of circle}}{\text{area of entire circle}} = \frac{\text{circumference of sector of circle}}{\text{circumference of entire circle}}$$

$$\frac{A}{\pi s^2} = \frac{C}{2\pi s}$$

To solve for A, multiply both sides of the equation by πs^2.

$$\pi s^2 \cdot \frac{A}{\pi s^2} = \pi s^2 \cdot \frac{C}{2\pi s}$$

Divide out common factors.

$$\cancel{\pi s^2} \cdot \frac{A}{\cancel{\pi s^2}} = \cancel{\pi} \overset{s}{\cancel{s^2}} \cdot \frac{C}{2\cancel{\pi}\cancel{s}}$$

$$A = \frac{sC}{2}$$

You need to cut out a piece of dough with area $A = \frac{sC}{2}$.

THINK AND COMMUNICATE

1. What is the area of the piece of dough Joline cuts out to make a cone with a circumference of 24 cm and a slant height of 10 cm?

2. About how thick should the sheet of dough described in Question 1 be? To make 500 cones, how much batter must Joline make?

ANSWERS Section 11.5

Think and Communicate

1. 120 cm^2
2. Answers may vary. An example is given. 5 mm, or 0.5 cm; 30,000 cm^3

You can multiply rational expressions just as you multiply fractions.

Multiplying fractions

$$\frac{4}{3} \cdot \frac{9}{2} = \frac{4 \cdot 9}{3 \cdot 2} = \frac{\overset{6}{\cancel{36}}}{\cancel{6}} = 6$$

$$\frac{\overset{2}{\cancel{4}}}{\cancel{3}} \cdot \frac{\overset{3}{\cancel{9}}}{\cancel{2}} = 2 \cdot 3 = 6$$

Multiplying rational expressions

$$\frac{x^2}{6y} \cdot \frac{10y}{x} = \frac{x^2 \cdot 10y}{6y \cdot x} = \frac{\overset{5\,x}{\cancel{10x^2y}}}{\underset{3}{\cancel{6xy}}} = \frac{5x}{3}$$

$$\frac{\overset{x}{\cancel{x^2}}}{\underset{3}{\cancel{6y}}} \cdot \frac{\overset{5}{\cancel{10y}}}{\cancel{x}} = \frac{x}{3} \cdot 5 = \frac{5x}{3}$$

Multiply the numerators and **multiply the denominators.**

You can also divide out common factors before you multiply.

A rational expression is in **simplest form** when the greatest common factor of the numerator and denominator is 1.

EXAMPLE 2

Simplify $\frac{4b}{a^2} \cdot \frac{a^3}{14b^2}$.

SOLUTION

$$\frac{4b}{a^2} \cdot \frac{a^3}{14b^2} = \frac{4b \cdot a^3}{a^2 \cdot 14b^2}$$

Multiply numerators and denominators.

$$= \frac{\overset{2}{\cancel{4b}} \cdot \overset{a}{\cancel{a^3}}}{\cancel{a^2} \cdot \underset{7b}{\cancel{14b^2}}}$$

Divide out common factors.

$$= \frac{2a}{7b}$$

EXAMPLE 3

Simplify $\frac{2x+8}{x+1} \cdot \frac{x^2}{3x+12}$.

SOLUTION

$$\frac{2x+8}{x+1} \cdot \frac{x^2}{3x+12} = \frac{2(x+4)}{x+1} \cdot \frac{x^2}{3(x+4)}$$

Factor the numerator and the denominator.

$$= \frac{2\cancel{(x+4)}}{x+1} \cdot \frac{x^2}{3\cancel{(x+4)}}$$

$$= \frac{2x^2}{3(x+1)}$$

Teach⇔Interact

About Example 1

Teaching Tip

When discussing this Example, make sure students understand that the slant height of the cone is also the radius of the circle of which the opened cone is a sector.

Additional Example 1

Suppose Joline bought a set of dough cutters for cutting out dough sectors of different sizes. Each cutter is marked with the number of degrees for the sector it will cut.

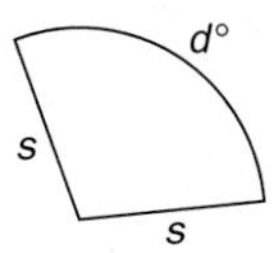

All the cutters have the same radius s. Let A = the area of the sector, and let d = the number of degrees for the sector. What is the area of the sector she will get if she uses the cutter marked $d°$?

$$\frac{\text{degrees in sector}}{\text{degrees in circle}} = \frac{\text{area of sector}}{\text{area of circle}}$$

$$\frac{d}{360} = \frac{A}{\pi s^2}$$

Multiply both sides by $360\pi s^2$. Then divide out common factors.

$$\cancel{360}\pi s^2 \cdot \frac{d}{\cancel{360}} = 360\cancel{\pi s^2} \cdot \frac{A}{\cancel{\pi s^2}}$$

$$\pi s^2 d = 360A$$

$$\frac{\pi s^2 d}{360} = A$$

The area of the sector is $\frac{\pi s^2 d}{360}$.

Additional Example 2

Simplify $\frac{7k^2}{15m^5} \cdot \frac{40m^2}{21k}$.

Multiply numerators and denominators. Then divide out common factors.

$$\frac{7k^2}{15m^5} \cdot \frac{40m^2}{21k} = \frac{7k^2 \cdot 40m^2}{15m^5 \cdot 21k}$$

$$= \frac{\overset{k}{\cancel{7k^2}} \cdot \overset{8}{\cancel{40m^2}}}{\underset{3m^3}{\cancel{15m^5}} \cdot \underset{3}{\cancel{21k}}}$$

$$= \frac{8k}{9m^3}$$

Additional Example 3

Simplify $\frac{3x}{5x+10} \cdot \frac{4x+8}{7x+9}$.

Factor the numerators and denominators.

$$\frac{3x}{5x+10} \cdot \frac{4x+8}{7x+9} = \frac{3x}{5(x+2)} \cdot \frac{4(x+2)}{7x+9}$$

$$= \frac{3x}{5\cancel{(x+2)}} \cdot \frac{4\cancel{(x+2)}}{7x+9}$$

$$= \frac{12x}{5(7x+9)}$$

Teach⇔Interact

Additional Example 4

Simplify $\frac{a^2-6a+9}{a-1} \div (2a-6)$.

$\frac{a^2-6a+9}{a-1} \div (2a-6)$

$= \frac{a^2-6a+9}{a-1} \div \frac{2a-6}{1}$

$= \frac{a^2-6a+9}{a-1} \cdot \frac{1}{2a-6}$

$= \frac{(a-3)^2}{a-1} \cdot \frac{1}{2(a-3)}$

$= \frac{(a-3)(a-3)}{a-1} \cdot \frac{1}{2(a-3)}$

$= \frac{a-3}{2(a-1)}$

Think and Communicate

Students who have difficulty answering question 3 may also have difficulty performing the procedure shown in Example 4. Be certain that students have a clear understanding of why $3m - 3$ was rewritten as $\frac{3m-3}{1}$.

Section Note

Mathematical Procedures

Ask students how they could check to see whether an expression has been simplified correctly. (Go over all the work that was done and check each step. Assign numerical values to each variable and see if the value of the simplified expression equals the value of the original expression.)

Closure Question

Describe how to divide one rational expression by another.

Multiply the first fraction by the reciprocal of the fraction that is the divisor. If the divisor is a polynomial, first write it as a fraction whose denominator is 1. Factor the numerators and denominators where possible. Divide out common factors and then write the final product.

You can divide rational expressions just as you divide fractions. To divide, multiply the first expression by the reciprocal of the second expression.

Dividing fractions	**Dividing rational expressions**
$\frac{2}{3} \div \frac{7}{3} = \frac{2}{3} \cdot \frac{3}{7}$	$\frac{x^2}{2y} \div \frac{2x}{3} = \frac{x^2}{2y} \cdot \frac{3}{2x}$

EXAMPLE 4

Simplify $\frac{m^2-1}{4m} \div (3m-3)$.

SOLUTION

$$\frac{m^2-1}{4m} \div (3m-3) = \frac{m^2-1}{4m} \div \frac{3m-3}{1}$$

$$= \frac{m^2-1}{4m} \cdot \frac{1}{3m-3}$$

$$= \frac{(m+1)(m-1)}{4m} \cdot \frac{1}{3(m-1)}$$

$$= \frac{m+1}{12m}$$

THINK AND COMMUNICATE

3. In Example 4, why was $3m - 3$ written as $\frac{3m-3}{1}$?

4. Explain how to simplify the expression $\frac{b^3}{b^2}$.

CHECKING KEY CONCEPTS

Simplify each expression. If it is already in simplest form, write *simplest form*.

1. $\frac{4}{8x+16}$ **2.** $\frac{3z^2}{4z}$ **3.** $\frac{12x^3}{8x}$

4. $\frac{2y-8}{3y-12}$ **5.** $\frac{x^2+1}{x+1}$ **6.** $\frac{n^2+n}{2n+2}$

7. $\frac{6x^2}{5y} \cdot \frac{10y}{3x}$ **8.** $\frac{x+1}{12} \cdot \frac{8}{2x+2}$ **9.** $\frac{6x}{y+1} \cdot \frac{2y+2}{3}$

10. $\frac{2xy}{3} \div \frac{4x^2}{3}$ **11.** $\frac{n+1}{4} \div \frac{3n+3}{16}$ **12.** $\frac{x+2}{5} \div (x^2+2x)$

Think and Communicate

3. so it would be easier to find the reciprocal of $3m - 3$
4. You use the quotient of powers rule: $\frac{b^3}{b^2} = b^{3-2} = b^1 = b$. You could also divide out common factors: $\frac{b^3}{b^2} = \frac{b \cdot b^2}{b^2} = b$.

Checking Key Concepts

1. $\frac{1}{2x+4}$
2. $\frac{3z}{4}$
3. $\frac{3x^2}{2}$
4. $\frac{2}{3}$
5. simplest form
6. $\frac{n}{2}$
7. $4x$
8. $\frac{1}{3}$
9. $4x$
10. $\frac{y}{2x}$
11. $\frac{4}{3}$
12. $\frac{1}{5x}$

11.5 Exercises and Applications

Simplify each expression. If it is already in simplest form, write *simplest form*.

1. $\frac{b^2}{2b}$ **2.** $\frac{5x^3}{10x^2}$ **3.** $\frac{y^2}{12x}$ **4.** $\frac{4}{4a + 8}$

5. $\frac{c + 1}{c + 2}$ **6.** $\frac{c^2 + c}{c^2}$ **7.** $\frac{4x + 12}{x + 3}$ **8.** $\frac{n^2 + 2n}{2n + 2}$

Extra Practice exercises on page 579

9. Writing Jim and Ted disagree about whether $\frac{3x + 5}{2x + 5}$ can be simplified. Jim says no. Ted says yes. Ted's work is shown at the right. Who is correct? Give an argument to support your answer.

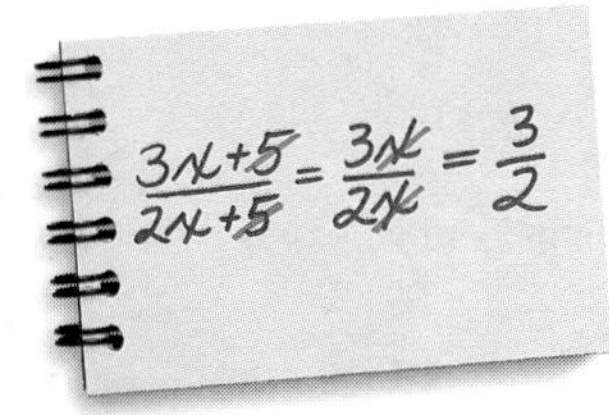

Connection PROBABILITY

The probability of spinning both spinners and having the first pointer land on the red region and the second pointer land on "4" is the product of the individual probabilities: $\frac{1}{3} \cdot \frac{1}{4} = \frac{1}{12}$.

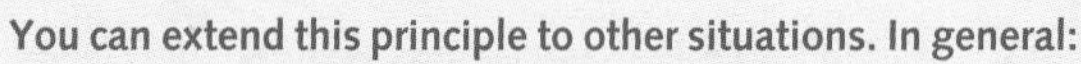

You can extend this principle to other situations. In general:

Let a = the probability that event A will happen.
Let b = the probability that event B will happen.

If events A and B do not depend on each other, then the probability that both events will happen is $a \cdot b$.

10. Suppose that one of two darts thrown randomly hits the green target and the other hits the red target. Write a variable expression for the probability that both darts land in the bonus region of the target. Simplify your expression.

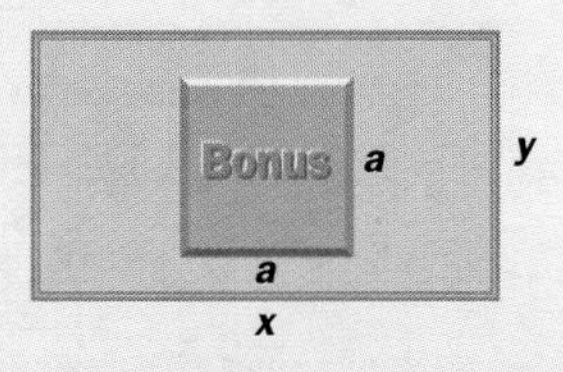

11. When you fly into Mexico City from the United States, you must go through customs. At customs you press a button which randomly activates a light. If the light is green you may pass immediately, but if the light is red you must wait as your bags are searched.

a. If the bags of about one in every n people are searched, what is the probability that your bags will be searched?

b. If you are traveling with a friend, what is the probability that you will both get a green light?

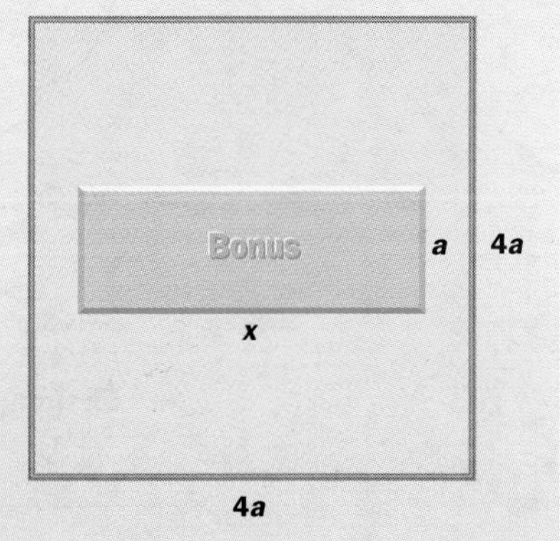

Apply⇔Assess

Suggested Assignment

Core Course
Exs. 1–24, 26, 27, 33–39

Extended Course
Exs. 1–39

Block Schedule
Day 72 Exs. 1–24, 26, 27, 33–39

Exercise Notes

Common Error

Exs. 4–9 Some students may be simplifying rational expressions incorrectly by canceling addends instead of factors. For example, in Ex. 6, some students may write $\frac{c^2 + c}{c^2} = c$. Stress that factors can be canceled but not addends. Also, suggest that students try some values for c. In this case, if $c = 2$, $\frac{c^2 + c}{c^2} = \frac{6}{4}$ or $\frac{3}{2}$. Since $\frac{3}{2} \neq 2$, the expression was not simplified correctly. Also, a class discussion of Ex. 9 should help to correct this problem.

Integrating the Strands

Exs. 10, 11 Many probability questions involve operations with numbers. Ex. 11 shows, however, that rational expressions can be used in probability problems.

Exercises and Applications

1. $\frac{b}{2}$
2. $\frac{x}{2}$
3. simplest form
4. $\frac{1}{a+2}$
5. simplest form
6. $\frac{c+1}{c}$
7. 4
8. simplest form
9. Answers may vary. Jim is correct. Ted made the mistake of canceling a term in the numerator with a term in the denominator. You cannot cancel terms in this way, only factor. Since neither the expression $3x + 5$ nor $2x + 5$ is factorable, the rational expression cannot be simplified.
10. $\frac{a^2}{xy} \cdot \frac{ax}{16a^2}$, or $\frac{a}{16y}$
11. a. $\frac{1}{n}$
    b. The probability that you will get a green light is 1 minus the probability that you will get a red light; $\left(1 - \frac{1}{n}\right)\left(1 - \frac{1}{n}\right) = \frac{n-1}{n} \cdot \frac{n-1}{n} = \left(\frac{n-1}{n}\right)^2$.

Apply⇔Assess

Exercise Notes

Student Progress

Exs. 12–26 Simplifying rational expressions can be difficult for some students. A thorough review of the solutions to these exercises should help all students assess their understanding and skill on these types of problems.

Common Error

Exs. 19–24 Some students may forget to change the divisor to its reciprocal when they rewrite a division problem as a multiplication problem. To correct this error, remind students of the correct procedure as they do each of these exercises.

Challenge

Exs. 27, 28 You may wish to ask students to suggest other examples or formulas from geometry that require simplifying rational expressions.

Using Technology

Ex. 29 Students can use a graphing calculator to graph $y = \frac{x^2 - 2x - 35}{x - 7}$ and $y = x + 5$ on the same screen. Suggest a "friendly" window. Ask what value each expression has when $x = 7$. What does the calculator show?

(The first expression is not defined when $x = 7$. If you try to trace to the point where $x = 7$, the display at the bottom of the screen shows

X = 7 Y =

The value of the second expression is 12, and tracing along the graph will result in a display that shows

X = 7 Y = 12

at the bottom of the screen.)

Simplify each expression.

12. $\frac{6x}{4y^2} \cdot \frac{y^3}{x^2}$

13. $\frac{4m - 8}{10y} \cdot \frac{15y^2}{5m - 10}$

14. $(2a + 14)\frac{1}{3a + 21}$

15. $\frac{9y^2}{x + 1} \cdot \frac{2x + 2}{6y}$

16. $\frac{n^2 - 4}{n + 2} \cdot \frac{n}{3n - 6}$

17. $\frac{a + 1}{a + 2} \cdot \frac{2}{a}$

18. a. Writing Evaluate $\frac{3x + 6}{4x + 8}$ when $x = 5$ and when $x = -3$.

b. Predict the value of $\frac{3x + 6}{4x + 8}$ when $x = 9678$. Explain how you know.

Simplify each expression.

19. $\frac{8z}{9k} \div \frac{4z}{15k^2}$

20. $\frac{5t^2(r - 1)}{4r} \div \frac{r - 1}{tr^2}$

21. $\frac{n}{n + 3} \div \frac{n - 2}{n + 3}$

22. $\frac{ax + bx}{x^2} \div \frac{a + b}{2x}$

23. $\frac{n^2 - 1}{2n} \div \frac{n - 1}{4n^3}$

24. $\frac{6x^2 - 6x}{10} \div (6x - 6)$

25. Open-ended Problem Write three different rational expressions that all equal $\frac{2x}{5}$ when simplified.

26. SAT/ACT Preview Which variable expression is equivalent to $\frac{x^2}{8} \div \frac{2}{x}$?

A. $\frac{2x^2}{8x}$ **B.** $\frac{x}{4}$ **C.** $\frac{x^3}{8}$ **D.** $\frac{x}{2}$ **E.** none of these

GEOMETRY Write a ratio in simplest form comparing the area of the shaded region to the total area of the figure.

27.

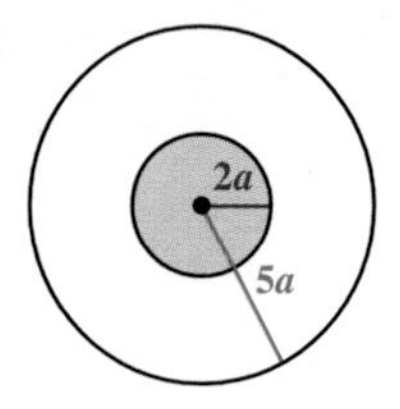

28.

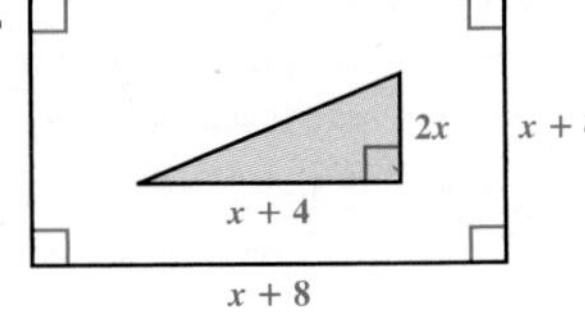

29. Writing Kris and Sandra disagree about whether the two expressions $\frac{x^2 - 2x - 35}{x - 7}$ and $(x + 5)$ are equivalent for all values of x. Kris says they are. Sandra says they are not. Who is correct? Explain.

30. Challenge Complete each simplification.

a. $\frac{15a^2}{7b} \div \underline{\ ?\ } = \frac{3b}{14}$

b. $\underline{\ ?\ } \div \frac{w^2}{a^2 - 4} = \frac{aw}{4(a + 2)}$

12. $\frac{3y}{2x}$
13. $\frac{6y}{5}$
14. $\frac{2}{3}$
15. $3y$
16. $\frac{n}{3}$
17. $\frac{2a + 2}{a^2 + 2a}$
18. a. $\frac{3}{4}; \frac{3}{4}$
   b. $\frac{3}{4}$; $\frac{3x + 6}{4x + 8} = \frac{3(x + 2)}{4(x + 2)} = \frac{3}{4}$ for every value of x except $x = -2$
19. $\frac{10k}{3}$
20. $\frac{5t^3r}{4}$
21. $\frac{n}{n - 2}$
22. 2
23. $2n^2(n + 1)$
24. $\frac{x}{10}$
25. Answers may vary. Examples are given. $\frac{2x^2}{5x}$, $\frac{6xy}{15y}$, and $\frac{2x^2 + 4x}{5x + 10}$
26. E

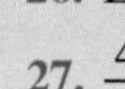

27. $\frac{4}{25}$
28. $\frac{x}{x + 8}$
29. Sandra is correct. The first expression is not defined when $x = 7$. $\frac{x^2 - 2x - 35}{x - 7} = \frac{(x - 7)(x + 5)}{x - 7} = x + 5$ if $x \neq 7$. The two expressions are equivalent for all values of x except $x = 7$.

Connection ENGINEERING

If a train goes around a curve like the one shown, each wheel on the inside of the curve travels a distance of πT, and each outer wheel travels a distance of $\pi(T + t)$. The outer wheels must travel farther, but they cannot turn any faster than the inside wheels because the wheels are connected by solid axles.

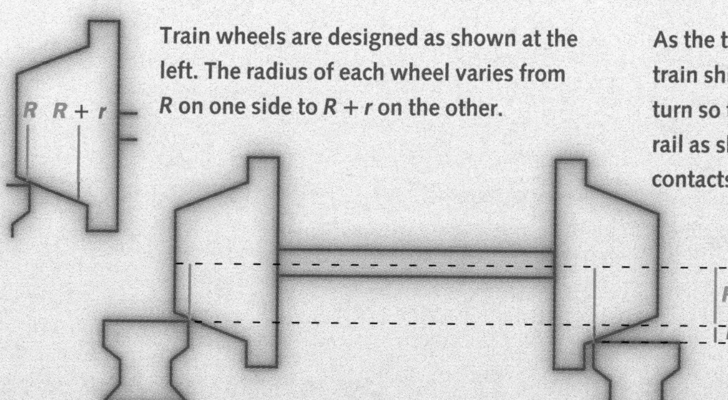

Train wheels are designed as shown at the left. The radius of each wheel varies from R on one side to $R + r$ on the other.

As the train goes around a curve, the train shifts slightly to the outside of the turn so the wheels are off center on the rail as shown. Where the outer wheel contacts the rail, the radius is large. Where the inside wheel contacts the rail, the radius is small. This allows the outer wheels to travel farther with each rotation.

In Exercises 31 and 32, you will find the radius T of the sharpest turn a train can make without its wheels slipping.

31. **a.** Let n = the number of turns the inner wheel makes on the curve shown. With each turn, the wheel travels a distance of $2\pi R$. So $n(2\pi R) = \pi T$. Explain this equation and solve it for n.

b. Write and solve an equation similar to the one in part (a) for the outer wheel.

c. Write an equation relating T, $T + t$, R, and $R + r$. Use the fact that n must be the same for both wheels.

d. Solve the equation from part (c) for T.

32. Suppose a train's wheels vary from a radius of 14 in. to a radius of 14.2 in. On turns, the distance between the rails is 57 in. Find T. Do you think trains ever turn sharply enough for the wheels to slip? Explain.

ONGOING ASSESSMENT

33. Open-ended Problem Find a product that equals $\frac{x+3}{x}$ when simplified. Find a quotient that equals $\frac{x+3}{x}$ when simplified.

SPIRAL REVIEW

Use the distributive property to find each product. *(Section 6.4)*

34. $2x(3 - 5x)$ **35.** $(4x + 2)(5x - 4)$ **36.** $(4\sqrt{3} + 5)(3\sqrt{3} + 1)$

Simplify each expression. *(Section 9.6)*

37. $a^2 \cdot a^3 \cdot a^4$ **38.** $\frac{a^{12}}{a^3}$ **39.** $\frac{c^{-2}}{c^2}$

Apply⇔Assess

Exercise Notes

Interdisciplinary Problems
Exs. 31, 32 Point out to students that many engineering problems that relate to moving objects can be modeled by using formulas and concepts from algebra and geometry.

Cooperative Learning
Exs. 31, 32 These exercises are appropriate for small groups. Students who are not entirely clear on the physical principles involved may understand the mathematics and vice versa. Either type of student can help the other.

Assessment Note
Ex. 33 Constructing problems similar to those in the text can be fun and challenging. Such an approach is an excellent way to assess understanding of mathematical concepts and procedures.

Practice 76 for Section 11.5

NAME ______________________ DATE __________

Practice 76

FOR USE WITH SECTION 11.5

Simplify each expression. If it is already in simplest form, write *simplest form*.

1. $\frac{8y^3}{12y}$ $\frac{2y^2}{3}$
2. $\frac{a^6}{3a^2}$ $\frac{a^4}{3}$
3. $\frac{7x}{x^2+x}$ $\frac{7}{x+1}$
4. $\frac{3b+15}{3}$ $b+5$
5. $\frac{k+3}{2k+6}$ $\frac{1}{2}$
6. $\frac{r-4}{2r-4}$ simplest form
7. $\frac{6x+4}{9x+6}$ $\frac{2}{3}$
8. $\frac{5c+10}{c^2+2c}$ $\frac{5}{c}$
9. $\frac{2x+5}{2x-5}$ simplest form
10. $\frac{n+7}{n^2-49}$ $\frac{1}{n-7}$
11. $\frac{3m^2}{3m+12}$ $\frac{m^2}{m+4}$
12. $\frac{d^2+25}{d+5}$ simplest form

Simplify each expression.

13. $\frac{12a^3}{5b} \cdot \frac{20b^2}{9a^2}$ $\frac{16ab}{3}$
14. $\frac{6x^2}{2x-5} \cdot \frac{4x-10}{21x}$ $\frac{4x}{7}$
15. $\frac{6x^2-2x}{y^4} \cdot \frac{y}{3x-1}$ $\frac{2x}{y^3}$
16. $\frac{1}{35k+7} \cdot (5k+1)$ $\frac{1}{7}$
17. $\frac{3t+9}{t^3} \cdot \frac{2t}{t^2-9}$ $\frac{6}{t^2(t-3)}$
18. $\frac{p}{q(q+4)} \cdot \frac{2q+8}{3p^2}$ $\frac{2}{3pq}$

Simplify each expression.

19. $\frac{c^3}{4d} \div \frac{c^6}{6d^2}$ $\frac{3d}{2c^3}$
20. $\frac{r+1}{3s^2} \div \frac{4r+4}{6s}$ $\frac{1}{2s}$
21. $\frac{x+4}{5y} \div \frac{5x+20}{25y^2}$ y
22. $\frac{n+2}{2n^2} \div \frac{n^2-4}{8n}$ $\frac{4}{n(n-2)}$
23. $\frac{y^3-y^2}{7y} \div (3y-3)$ $\frac{y}{21}$
24. $\frac{ax+4a}{a} \div \frac{x^2-16}{x-4}$ 1

25. The *aspect ratio* of a rectangular computer image is the ratio of the width of the image to its height. Kaneisha generated a computer graphic for her design class of width x in. and height y in. She wanted to increase the width of the graphic by 4 in., while keeping the same aspect ratio.

a. Let t represent the amount Kaneisha must increase the height of her graphic. Write an equation that relates x, y, and t. $\frac{x+4}{y+t} = \frac{x}{y}$

b. Solve the equation you wrote in part (a) for the variable t. $t = \frac{4y}{x}$

76

30. a. $\frac{10a^2}{b^2}$
b. $\frac{aw^2}{4(a+2)^2(a-2)}$

31. a. The number of turns made by the inner wheel, n, times the circumference of the inner wheel, $2\pi R$, should equal the distance around the inner track of the curve, πT; $n = \frac{T}{2R}$.
b. $n(2\pi(R + r)) = \pi(T + t)$; $n = \frac{T+t}{2(R+r)}$
c. $\frac{T}{2R} = \frac{T+t}{2(R+r)}$
d. $T = \frac{tR}{r}$

32. about 59 ft

33. Answers may vary.

34. $6x - 10x^2$

35. $20x^2 - 6x - 8$

36. $41 + 19\sqrt{3}$

37. a^9

38. a^9

39. c^{-4}, or $\frac{1}{c^4}$

Plan⇔Support

Objectives

- Add and subtract rational expressions.
- Explore patterns with fractions.
- Apply addition and subtraction of rational expressions to solve real-world problems.

Recommended Pacing

Core and Extended Courses
Section 11.6: 2 days

Block Schedule
Section 11.6: 2 half-blocks
(with Section 11.5 and Review)

Resource Materials

Lesson Support
Lesson Plan 11.6
Warm-Up Transparency 11.6
Practice Bank: Practice 77
Study Guide: Section 11.6
Explorations Lab Manual:
Diagram Master 1
Assessment Book: Test 49

Technology
Graphing Calculator
Internet:
http://www.hmco.com

Warm-Up Exercises

Add or subtract.

1. $\frac{1}{5}+\frac{1}{7}$ $\frac{12}{35}$
2. $\frac{1}{5}+\frac{3}{14}$ $\frac{29}{70}$
3. $\frac{1}{6}+\frac{5}{12}$ $\frac{7}{12}$
4. $\frac{7}{12}-\frac{1}{3}$ $\frac{1}{4}$
5. $\frac{5}{18}-\frac{1}{9}$ $\frac{1}{6}$
6. $\frac{7}{15}-\frac{1}{18}$ $\frac{37}{90}$

SECTION

11.6 Exploring Rational Expressions

Learn how to...

- **add and subtract rational expressions**

So you can...

- **explore patterns with fractions**
- **solve problems such as investigating car accidents and reducing fuel costs**

Mathematics involves the study of patterns. The Exploration will help you discover a pattern for subtracting rational expressions.

EXPLORATION
COOPERATIVE LEARNING

Subtracting Rational Expressions

Work with a partner.
You will need:

- a ruler
- graph paper

1 Look at the diagram. Explain how it shows that $\frac{1}{2}-\frac{1}{3}=\frac{1}{6}$.

2 Draw a diagram to show that $\frac{1}{3}-\frac{1}{4}=\frac{1}{12}$.

Questions

1. Look for a pattern in the equations from Steps 1 and 2. Use this pattern to complete the following equations.

   a. $\frac{1}{4}-\frac{1}{5}=\underline{?}$ **b.** $\frac{1}{5}-\frac{1}{6}=\underline{?}$ **c.** $\frac{1}{6}-\frac{1}{7}=\underline{?}$

2. Now use the pattern you found to complete this equation. $\frac{1}{x}-\frac{1}{x+1}=\underline{?}$

Exploration Note

Purpose
The purpose of this Exploration is to have students discover a pattern for subtracting successive unit fractions and then to extend that pattern to subtract successive rational expressions with numerators of 1.

Materials/Preparation
Each student should have his or her own ruler and graph paper.

Procedure
Partners should first study the diagram in the text and understand what it shows. Then they can complete Step 2 and answer the questions.

Closure
Students should come to the conclusion that the same procedures used to subtract fractions can be used to subtract rational expressions. You may wish to ask students if they think this applies to addition as well.

Explorations Lab Manual
See the Manual for more commentary on this Exploration.

Diagram Master 1

You can add and subtract rational expressions the same way you add and subtract fractions.

Fractions

$$\frac{5}{8} + \frac{2}{8} = \frac{5+2}{8} = \frac{7}{8}$$

Rational expressions

$$\frac{7}{a+b} - \frac{1}{a+b} = \frac{7-1}{a+b} = \frac{6}{a+b}$$

When the denominators of the expressions are different, use the LCD to rewrite each expression.

Fractions

$$\frac{7}{8} - \frac{3}{10} = \frac{35}{40} - \frac{12}{40} = \frac{35-12}{40} = \frac{23}{40}$$

Rational expressions

$$\frac{1}{x} - \frac{1}{x+1} = \frac{x+1}{x(x+1)} - \frac{x}{x(x+1)} = \frac{x+1-x}{x(x+1)} = \frac{1}{x(x+1)}$$

THINK AND COMMUNICATE

1. Terence simplified the expression $\frac{7}{8} - \frac{3}{10}$ as shown at the right. Compare Terence's method to the method above. Which do you prefer?

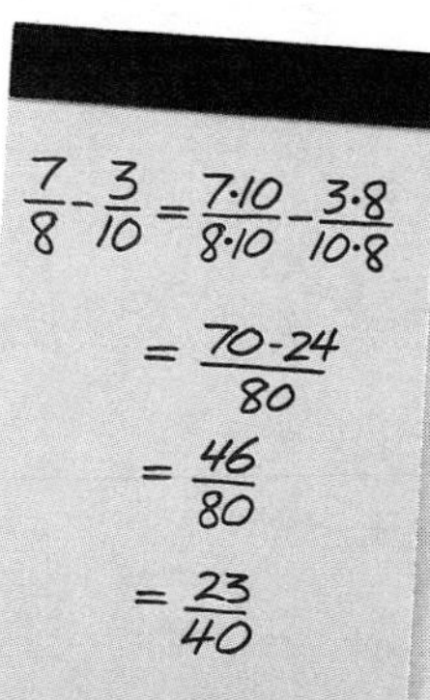

EXAMPLE 1

Simplify each expression.

a. $\frac{7n}{n+2} + \frac{14}{n+2}$ **b.** $\frac{2x+5}{2x} - \frac{x-1}{2x}$

SOLUTION

a. $\frac{7n}{n+2} + \frac{14}{n+2} = \frac{7n+14}{n+2} = \frac{7(n+2)}{n+2} = 7$

Always write your answer in simplest form.

b. $\frac{2x+5}{2x} - \frac{x-1}{2x} = \frac{(2x+5)-(x-1)}{2x} = \frac{2x+5-x+1}{2x} = \frac{x+6}{2x}$

WATCH OUT!

When you subtract expressions, use parentheses to avoid confusion.

Teach⇔Interact

Learning Styles: Visual

Students can learn how to add and subtract rational expressions by seeing many examples that relate operations with rational expressions to previously learned skills with fractions. Since the concepts and procedures used to add and subtract rational expressions are exactly the same when adding and subtracting numerical fractions, this connection should be used to support and build understanding and skill with rational expressions.

Section Note

Communication: Discussion
Discuss the samples at the top of this page thoroughly. Be sure students understand how the LCD is used to rewrite the original fractions.

Additional Example 1

Simplify each expression.

a. $\frac{5x}{x+3} + \frac{15}{x+3}$

$\frac{5x}{x+3} + \frac{15}{x+3} = \frac{5x+15}{x+3} = \frac{5(x+3)}{x+3} = 5$

b. $\frac{5x+7}{12x} - \frac{7x+5}{12x}$

$\frac{5x+7}{12x} - \frac{7x+5}{12x} = \frac{(5x+7)-(7x+5)}{12x} = \frac{5x+7-7x-5}{12x} = \frac{-2x+2}{12x} = \frac{2(-x+1)}{12x} = \frac{-x+1}{6x}$

ANSWERS Section 11.6

Exploration

1. By lining up the end of a bar that represents $\frac{1}{2}$ with the end of a bar that represents $\frac{1}{3}$, you can see that the difference is the length of a bar that represents $\frac{1}{6}$.

2. 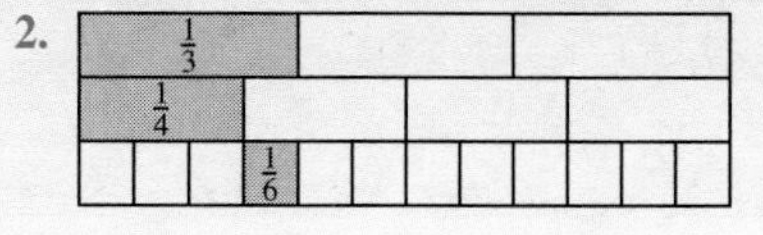

Questions

1. All three numerators in each equation are 1. Also, the first two denominators are consecutive integers, and the denominator of the difference is the product of the integers.

   a. $\frac{1}{20}$ b. $\frac{1}{30}$ c. $\frac{1}{42}$

2. $\frac{1}{x(x+1)}$

Think and Communicate

1. Answers may vary. An example is given. I prefer the method above because the numbers involved are smaller and the result does not have to be simplified as it does with Terence's method.

Teach⇔Interact

Additional Example 2

Simplify the expression $\frac{9}{24x} + \frac{2x-13}{18x^2}$.

First find the LCD. The LCD of $\frac{9}{24x}$ and $\frac{2x-13}{18x^2}$ is $72x^2$.

$$\frac{9}{24x} + \frac{2x-13}{18x^2} = \frac{3x \cdot 9}{3x \cdot 24x} + \frac{4(2x-13)}{4 \cdot 18x^2}$$
$$= \frac{27x}{72x^2} + \frac{8x-52}{72x^2}$$
$$= \frac{27x + 8x - 52}{72x^2}$$
$$= \frac{35x-52}{72x^2}$$

Section Note

Topic Spiraling: Review
The table showing the different types of functions provides an excellent review of these ideas. You may wish to have students work in groups of 3 or 4 to verify the graphs of the types of functions shown. Students should use graphing calculators to do this work.

Closure Question

How are adding and subtracting rational expressions like adding and subtracting numerical fractions?
The procedure for rational expressions and numerical fractions is the same. If the fractions have a common denominator, add or subtract numerators and use the same denominator. If the fractions do not have a common denominator, first find their LCD. Rewrite the fractions as equivalent fractions using the LCD. Perform the addition or subtraction and then simplify the result.

EXAMPLE 2

Simplify the expression $\frac{5}{6x} - \frac{x-3}{4x^2}$.

SOLUTION

First find the LCD. The LCD of $\frac{5}{6x}$ and $\frac{x-3}{4x^2}$ is $12x^2$.

$$\frac{5}{6x} - \frac{x-3}{4x^2} = \frac{2x \cdot 5}{2x \cdot 6x} - \frac{3(x-3)}{3 \cdot 4x^2}$$
$$= \frac{10x}{12x^2} - \frac{3x-9}{12x^2}$$
$$= \frac{10x - 3x + 9}{12x^2}$$
$$= \frac{7x+9}{12x^2}$$

Notice that $\frac{2x}{2x}$ and $\frac{3}{3}$ are both equal to 1. Multiplying by 1 does not affect the value of the fractions.

Types of Functions

The table below summarizes the different kinds of functions that have been described in this book.

Linear	Quadratic	Polynomial
$y = \frac{1}{2}x + 4$ The highest exponent is 1.	$y = \frac{1}{2}x^2 - 4x + 1$ The highest exponent is 2.	$y = x^3$ Exponents can be any whole number.

Exponential	Rational
$y = 2(1.12)^x$ One of the variables is an exponent.	$y = \frac{4}{x-2} + 3$ There can be variables in the denominator.

Checking Key Concepts

1. $\frac{4}{y}$
2. $\frac{3-4x}{x-2}$
3. -2
4. $\frac{x-1}{x^2}$
5. $\frac{12m+5}{15m^2}$
6. $\frac{1}{30p}$
7. rational
8. quadratic
9. linear

CHECKING KEY CONCEPTS

Simplify each expression.

1. $\frac{1}{y} + \frac{3}{y}$
2. $\frac{3}{x-2} - \frac{4x}{x-2}$
3. $\frac{a}{a+1} - \frac{3a+2}{a+1}$
4. $\frac{1}{x} - \frac{1}{x^2}$
5. $\frac{4}{5m} + \frac{1}{3m^2}$
6. $\frac{p}{10p^2} - \frac{1}{15p}$

State whether each equation can be best described as *linear, quadratic, polynomial, rational,* or *exponential.*

7. $p = \frac{1}{q} - \frac{1}{q^2}$
8. $d = 2 + 3t - 16t^2$
9. $\frac{1}{2}a - \frac{1}{3}b = 7$

11.6 Exercises and Applications

Simplify each expression.

Extra Practice exercises on page 579

1. $\frac{2t}{t-1} - \frac{t+2}{t-1}$
2. $\frac{b+a}{ab} + \frac{b-a}{ab}$
3. $\frac{4f+3}{f-1} - \frac{2f+5}{f-1}$
4. $\frac{3}{m} + \frac{2}{m^2}$
5. $\frac{5}{4xy^2} - \frac{3}{10x^2y}$
6. $\frac{a}{a+5} + \frac{2}{a}$

State whether each equation can be best described as *linear, quadratic, polynomial, rational,* or *exponential.*

7. $s = 15t^2 - 4$
8. $m = 4r + 2$
9. $y = 2.9(0.94)^x$
10. $3x + 2y = -7$
11. $b = \frac{1}{a} + \frac{1}{a-5}$
12. $V = d^4 - \frac{1}{2}d^2 + \frac{1}{4}$

Investigation In the Exploration on page 494, you used a diagram to discover a pattern for subtracting fractions. In Exercises 13–16, you will use the diagram to find a pattern for adding fractions.

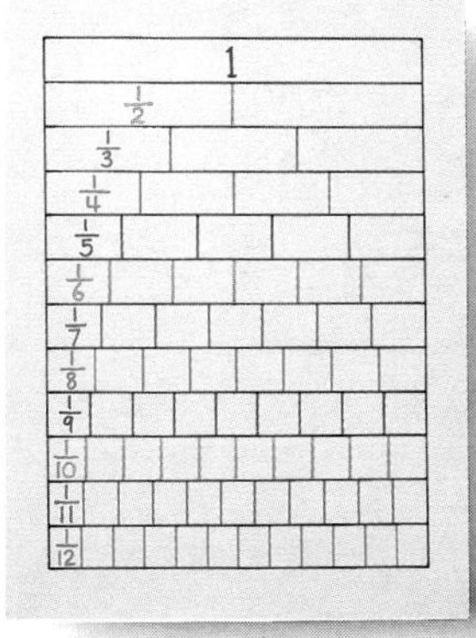

13. Place a blank sheet of paper over the diagram at the right. Trace bars of length $\frac{1}{2}$ and $\frac{1}{3}$ end to end. Find a bar as long as the sum of the two bars you traced. Complete the equation $\frac{1}{2} + \frac{1}{3} = \underline{?}$.

14. Complete the equation $\frac{1}{3} + \frac{1}{4} = \underline{?}$.

15. **Writing** Compare the equations in Exercises 13 and 14. What pattern do you notice? Use the pattern to simplify each expression below.

   a. $\frac{1}{4} + \frac{1}{5}$ b. $\frac{1}{5} + \frac{1}{6}$ c. $\frac{1}{42} + \frac{1}{43}$ d. $\frac{1}{x} + \frac{1}{x+1}$

16. Verify the pattern you discovered in Exercise 15 by simplifying the expression $\frac{1}{x} + \frac{1}{x+1}$.

Apply⇔Assess

Suggested Assignment

Core Course
Day 1 Exs. 1–16, 18–20
Day 2 Exs. 21–26, 28, 30–36, AYP

Extended Course
Day 1 Exs. 1–20
Day 2 Exs. 21–36, AYP

Block Schedule
Day 72 Exs. 1–16, 18–20
Day 73 Exs. 21–26, 28, 30–36, AYP

Exercise Notes

Teaching Tip
Exs. 1–6 If any students have difficulty with these exercises use numerical examples to reinforce the procedures for adding and subtracting fractions. Then apply the same procedures to these algebraic fractions.

Mathematical Procedures
Ex. 6 Students may multiply a and $a + 5$ when they write the denominator of the sum. Point out that it is often preferable to leave the denominator in factored form.

Communication: Drawing
Exs. 13–15 Geometric models are often very useful in helping students develop concepts. Students may wish to make diagrams for the first two parts of Ex. 15.

Challenge
Ex. 16 Ask students if they think the pattern discovered in Ex. 15 is true for the expression $\frac{a}{x} + \frac{a}{x+1}$, where a is any positive integer. (Yes, but the numerator is now multiplied by the value of a.)

Exercises and Applications

1. $\frac{t-2}{t-1}$
2. $\frac{2}{a}$
3. 2
4. $\frac{x+2}{x^2}$
5. $\frac{25x-6y}{20x^2y^2}$
6. $\frac{a^2+2a+10}{a^2+5a}$
7. quadratic
8. linear
9. exponential
10. linear
11. rational
12. polynomial
13. $\frac{5}{6}$
14. $\frac{7}{12}$
15. The numerators of the fractions being added are both 1 and their denominators are consecutive integers. The numerator of the sum is 1 more than twice the first integer. The denominator of the sum is the product of the integers.

   a. $\frac{9}{20}$ b. $\frac{11}{30}$
   c. $\frac{85}{1806}$ d. $\frac{2x+1}{x^2+x}$

Apply⇔Assess

Exercise Notes

Common Errors
Exs. 18–23 Errors in doing these exercises may result from not understanding the concepts or procedures involved or from not performing the computations correctly. Students should review their own work or trade papers with a classmate to discover any errors they may have made.

Integrating the Strands
Ex. 26 This exercise provides a good illustration of how algebra and geometry can be used to solve real-world problems.

Research
Ex. 26 Students may wish to contact highway patrol departments or the local police to inquire how their accident investigators use mathematics when they study an accident scene.

Problem Solving
Ex. 29 Students should be able to use the strategy of working backward to solve this problem.

BY THE WAY

Underinflated tires can cut gas mileage by as much as 2% per pound of pressure below the recommended level.

17. a. **CARS** A car's *gas mileage* tells how far the car goes on one gallon of gasoline. Let x = your car's gasoline mileage in miles per gallon. Write a variable expression for the amount you spend on gasoline each year if gasoline averages \$1.20 per gallon and you drive about 15,000 mi each year.
    b. Write and simplify an expression for the amount of money you will save each year if you can increase your car's gas mileage by 10 mi/gal. (*Hint:* First write an expression for the cost of gasoline if the mileage increases. Then find the difference.)
    c. Maeghan Maloney estimates that she can save \$200 each year on gasoline costs. Use your variable expression from part (b) to estimate her car's present gas mileage to the nearest mile per gallon.

Simplify each expression.

18. $\frac{2}{5cd} - \frac{3}{10cd}$
19. $\frac{2}{x+8} + \frac{6}{x-3}$
20. $\frac{y+6}{9y^2} - \frac{7}{6y}$
21. $\frac{2}{x+1} + 3$
22. $5 - \frac{1}{x}$
23. $\frac{b^2}{4a} - \frac{c}{a}$

24. a. Simplify each expression. Describe any patterns you see.
    $\frac{1}{3} - \frac{1}{6}$ $\quad \frac{1}{4} - \frac{1}{8}$ $\quad \frac{1}{5} - \frac{1}{10}$ $\quad \frac{1}{6} - \frac{1}{12}$
    b. Write and simplify a variable expression to verify your pattern.

25. **TRANSPORTATION** Michael and Danielle Evans are driving 250 mi to the Super Bowl. Michael wants Danielle to drive faster to save time.
    a. Write an expression that gives the time saved if the car's average speed s is increased by 5 mi/h.
    b. How much time will they save if Danielle averages 55 mi/h rather than 50 mi/h?

26. **CARS** If a car goes around a turn too quickly, it will create *yaw marks* on the pavement. A yaw mark looks like an arc of a circle. Accident investigators can use the radius r of the circle and the drag factor f of the road surface to tell how fast the car was going.

    To find the radius r of a yaw mark, investigators pick two points A and B on the arc and measure the distance d between them. Then they measure the perpendicular distance m from the midpoint of segment AB to the arc.

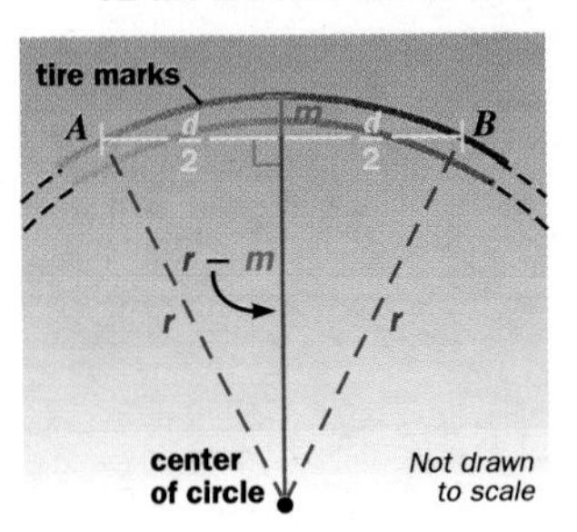

    a. Use the Pythagorean Theorem to write an equation relating d, m, and r. Solve for r.
    b. Find the radius r if d = 35 ft and m = 3.5 ft.
    c. For a level road, the equation $S \approx 3.86\sqrt{rf}$ gives the speed of the car in miles per hour. Find S if f = 0.76. Use your value for r from part (b).

16. $\frac{1}{x} + \frac{1}{x+1} = \frac{x+1}{x(x+1)} + \frac{x}{x(x+1)} = \frac{2x+1}{x^2+x}$

17. a. $\frac{180{,}000}{x}$ dollars
    b. $\frac{180{,}000}{x(x+10)}$ dollars
    c. about 25.4 mi/gal

18. $\frac{1}{\text{[illegible]}}$

19. $\frac{8x+42}{(x+8)(x-3)}$

20. $\frac{12-19y}{18y^2}$

21. $\frac{3x+5}{x+1}$

22. $\frac{5x-1}{x}$

23. $\frac{b^2-4c}{4a}$

24. a. $\frac{1}{6}; \frac{1}{8}; \frac{1}{10}; \frac{1}{12}$; In each case, both numerators are 1. The denominator of the second fraction in the subtraction is twice that of the first. The difference is the second fraction.
    b. $\frac{1}{x} - \frac{1}{2x} = \frac{2}{2x} - \frac{1}{2x} = \frac{1}{2x}$

25. a. $\frac{250}{s} - \frac{250}{s+5} = \frac{1250}{s(s+5)}$
    b. about 27 min

26. a. $r = \frac{m^2 + \frac{1}{4}d^2}{8m}$
    b. 45.5 ft c. about 23 mi/h

27. a. $\frac{5}{4x}$
    b. Answers may vary. Examples are given. $\frac{1}{2x} + \frac{3}{4x}; \frac{1}{2} + \frac{5-2x}{4x}$

28. A

29. Answers may vary. For example, $\frac{2}{x+5} + \frac{5}{x^2+5x} = \frac{2x+5}{x(x+5)}$; To find this answer, I rewrote

27. a. **Open-ended Problem** Simplify the expression $\frac{1}{x} + \frac{1}{4x}$.

b. Find two more rational expressions whose sum is the same as your answer to part (a).

28. **SAT/ACT Preview** If $A = \frac{1}{x} + \frac{1}{y} + 1$ and $B = \frac{x+y}{xy}$ for positive integers x and y, then:

A. $A > B$ B. $B > A$ C. $A = B$ D. relationship cannot be determined

29. **Challenge** Shannon added two rational expressions with different denominators. After simplifying, her answer was $\frac{2x+5}{x^2+5x}$. What might the original expressions have been? How did you get your answer?

ONGOING ASSESSMENT

30. a. **Open-ended Problem** Use a pattern to create several numerical expressions as in Exercises 15 and 24. Simplify each expression and describe any patterns you see.

b. Write and simplify a variable expression to verify your pattern.

SPIRAL REVIEW

Solve each equation. If the equation has no solution, write *no solution*. *(Section 11.3)*

31. $\frac{3}{x} - 8 = \frac{5}{x}$ 32. $\frac{5}{y} + 1 = \frac{1}{2}$ 33. $\frac{5r-4}{2(r-4)} + \frac{9}{2} = \frac{8}{r-4}$

Solve each equation for the variable in red. *(Section 11.4)*

34. $p = 2l + 2w$ 35. $x = 2xy + y$ 36. $\frac{1}{a} + \frac{1}{b} = c$

ASSESS YOUR PROGRESS

VOCABULARY

simplest form of a rational expression (p. 489)

Simplify each expression. *(Sections 11.5 and 11.6)*

1. $\frac{2t}{5t+15} \cdot \frac{t+3}{t}$ 2. $\frac{x+1}{3x^2} \cdot \frac{6x}{5x+5}$ 3. $\frac{3y-3}{y+1} \div \frac{y-1}{y}$

4. $\frac{y}{y+1} + \frac{5}{y+1}$ 5. $\frac{3z}{z+1} - \frac{5}{2(z+1)}$ 6. $\frac{3}{m+2} + \frac{6}{m(m+2)}$

7. **Journal** How is subtracting rational expressions like subtracting fractions? How is dividing rational expressions like dividing fractions? How are they different?

Apply⇔Assess

Exercise Notes

Assessment Note

Ex. 30 This may be a difficult exercise for some students. Students who do well on it are demonstrating a fine knowledge of using patterns to solve problems.

Assess Your Progress

Journal Entry

Students who can write clear explanations to these questions should have a good understanding of the procedures involved in performing operations on rational expressions. You may wish to have students who have written exceptionally clear responses share them with the class.

Progress Check 11.5–11.6

See page 503.

Practice 77 for Section 11.6

NAME ______ DATE ______

Practice 77

FOR USE WITH SECTION 11.6

Simplify each expression. Exs. 4, 5, and 11: See bottom of page.

1. $\frac{2x}{x+3} + \frac{x-1}{x+3}$ $\frac{3x-1}{x+3}$ 2. $\frac{pq-1}{p+4} - \frac{pq+1}{p+4}$ $-\frac{2}{p+4}$ 3. $\frac{b+2c}{b^2c} - \frac{2b-c}{b^2c}$ $\frac{3c-b}{b^2c}$

4. $\frac{a+1}{a^2} + \frac{a-2}{a^3}$ 5. $\frac{x}{3x+1} - \frac{3}{x}$ 6. $\frac{m+2}{5m} + \frac{m-1}{m^2}$ $\frac{m^2+7m-5}{5m^2}$

7. $\frac{2x}{y} + \frac{1}{y^2}$ $\frac{2xy+1}{y^2}$ 8. $\frac{k}{k-1} - \frac{k+1}{k}$ $\frac{1}{k(k-1)}$ 9. $\frac{a+1}{a^2b} + \frac{3}{ab}$ $\frac{4a+1}{a^2b}$

10. $\frac{n}{n+1} - \frac{n}{n-1}$ $-\frac{2n}{n^2-1}$ 11. $\frac{x-2}{4x} + \frac{x+2}{x^2}$ 12. $\frac{3t+1}{t+2} - \frac{2}{t}$ $\frac{3t^2-t-4}{t(t+2)}$

State whether each equation can be best described as linear, quadratic, polynomial, rational, or exponential.

13. $y = 6 - 5x^2$ quadratic 14. $y = 1.5(1.06)^x$ exponential 15. $ax + by = 5$ linear

16. $y = -4x^3 + 2x - 1$ polynomial 17. $y = \frac{x-3}{x} + 1$ rational 18. $y = \frac{x+5}{3}$ linear

19. a. Find the value of each difference:

$\frac{1}{3} - \frac{1}{5}$ $\frac{2}{15}$ $\frac{1}{9} - \frac{1}{11}$ $\frac{2}{99}$ $\frac{1}{97} - \frac{1}{99}$ $\frac{2}{9603}$

b. Verify the pattern you found in part (a) by simplifying the expression $\frac{1}{x-1} - \frac{1}{x+1}$. $\frac{2}{x^2-1}$

20. After Catarina Hernandez had ridden the first 10 mi of a bicycle race at a constant speed, she ran into a headwind that she estimated slowed her down by 4 mi/h during the last 8 mi of the race.

a. Let t = Catarina's speed for the first 10 mi of the race. Write an expression that models her total time for the race. $\frac{10}{t} + \frac{8}{t-4}$

b. Suppose Catarina's total time for the race is 1 h. Find her speed for the first 10 mi. $20\ \frac{\text{mi}}{\text{h}}$

21. **Open-ended Problem** Make up a situation involving a quantity that can be modeled by the expression $\frac{1}{x} + \frac{1}{2x}$. Tell what the variable x represents. Answers may vary. Check students' work

4. $\frac{a^2+2a-2}{a^3}$ 5. $\frac{x^2-9x-3}{x(3x+1)}$ 11. $\frac{x^2+2x+8}{4x^2}$

Practice Bank, ALGEBRA 1: EXPLORATIONS AND APPLICATIONS

77

$\frac{2x+5}{x^2+5x}$ as $\frac{2x}{x^2+5x} + \frac{5}{x^2+5x}$. Then I simplified the first fraction by dividing out an x. $\frac{2x}{x(x+5)} + \frac{5}{x(x+5)} = \frac{2}{x+5} + \frac{5}{x^2+5x}$

30. Answers may vary. Examples are given.

a. $\frac{1}{3} - \frac{1}{5} = \frac{2}{15}$; $\frac{1}{5} - \frac{1}{7} = \frac{2}{35}$; $\frac{1}{6} - \frac{1}{8} = \frac{2}{48}$; In each case, both numerators are 1. The denominator of the first fraction in the subtraction is a positive integer and the denominator of the second fraction is 2 more than that of the first. The numerator of the difference is 2 and the denominator is the product of the two integers.

b. $\frac{1}{x} - \frac{1}{x+2} = \frac{x+2}{x(x+2)} - \frac{x}{x(x+2)} = \frac{x+2-x}{x(x+2)} = \frac{2}{x(x+2)}$

31. $-\frac{1}{4}$ 32. -10

33. no solution 34. $w = \frac{p-2l}{2}$

35. $y = \frac{x}{2x+1}$ 36. $a = \frac{b}{bc-1}$

Assess Your Progress

1. $\frac{2}{5}$ 2. $\frac{2}{5x}$ 3. $\frac{3y}{y+1}$

4. $\frac{y+5}{y+1}$ 5. $\frac{6z-5}{2(z+1)}$ 6. $\frac{3}{m}$

7. Answers may vary. An example is given. You subtract rational expressions just as you subtract numerical fractions. Write the fractions using the least common denominator, subtract, and simplify the result. You divide rational expressions just as you divide numerical fractions. Rewrite the division as the multiplication of the first expression by the reciprocal of the second, multiply the numerators, multiply the denominators, and simplify the result. In rational expressions, the numerators and denominators may be variable expressions. Therefore, the work is more complicated than it is for numerical fractions and may involve factoring and rewriting expressions.

Project Notes

Mathematical Goals

- Collect real-world data and make a scatter plot of the data.
- Write an equation that relates distance as a function of the number of beans in a bag.
- Explore rational expressions that involve real-world data.
- Write and simplify a variable expression.

Planning

Materials
- String
- Plastic bags
- Yardstick
- Dried beans
- Binder clip
- Tape

Project Teams
Students can choose partners to work with and then discuss how they wish to proceed. Partners can share in the collection of the materials needed to do the project.

Guiding Students' Work

Students will need a place from which to hang their balance. You can set up the working areas prior to the beginning of the project. Try to have some extra materials available if students need them. These steps will facilitate the project work by all groups.

Second-Language Learners
Before writing a summary of their results, students learning English may benefit from explaining their experiment orally to an aide or peer tutor.

PORTFOLIO PROJECT

Balancing Weights

Have you ever ridden a seesaw or teeter-totter? Often you must adjust your position on the board to balance with the person on the other end. In the picture, which way should the person on the left move to make the seesaw balance?

PROJECT GOAL **Explore the relationship between weights and their distances from the center of a balance.**

Making a balance

Work with a partner to make a balance, as shown.

1. USE loops of string to hang plastic bags from a yardstick.

2. HANG the balance using string and a binder clip attached to the middle of the yardstick.

3. PUT different numbers of marbles in the bags.

4. SLIDE the bags until they balance.

Exploring Inverse Variation

1. Let b = the distance from bag B to the center of the balance. Record the distance b to the nearest $\frac{1}{4}$ in., and record the number of marbles in bag B.

2. Use tape to hold bag A in place. Put a different number of marbles in bag B and move it until both bags are balanced again. Record the distance b and the number of marbles in bag B. Repeat this step at least six more times.

3. Make a scatter plot of your data. Write an equation for the distance b as a function of the number of marbles in bag B.

Exploring Rational Expressions

4. Add bag C to the balance as shown at the right.

5. Adjust the three bags until they balance. Record the number of marbles in each bag and measure the distances a, b, and c. Repeat Step 4 at least 7 times, putting different numbers of marbles in the bags for each trial.

6. For each trial, make the calculations shown at the right. What do you notice?

7. Suppose the bags are balanced and contain different numbers of marbles. If you add x marbles to each bag, how far must you move bag A so they balance again? Write and simplify a variable expression.

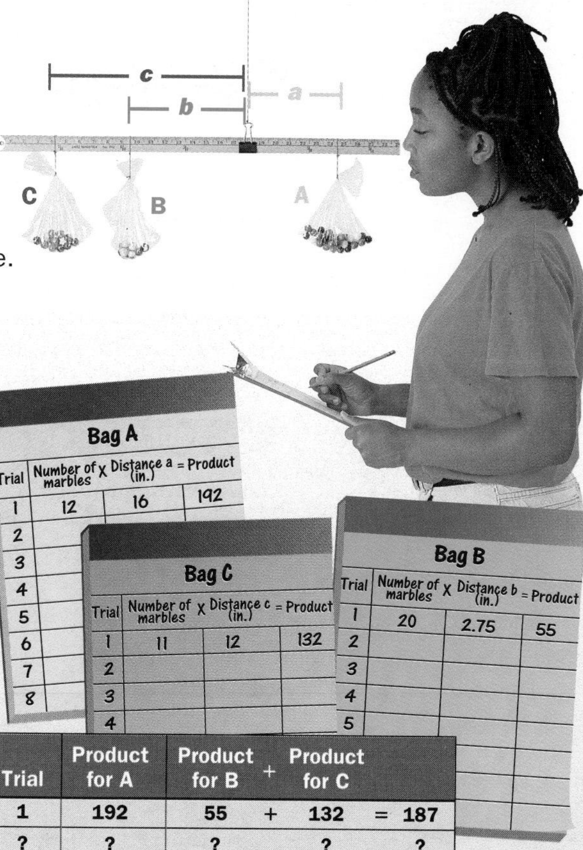
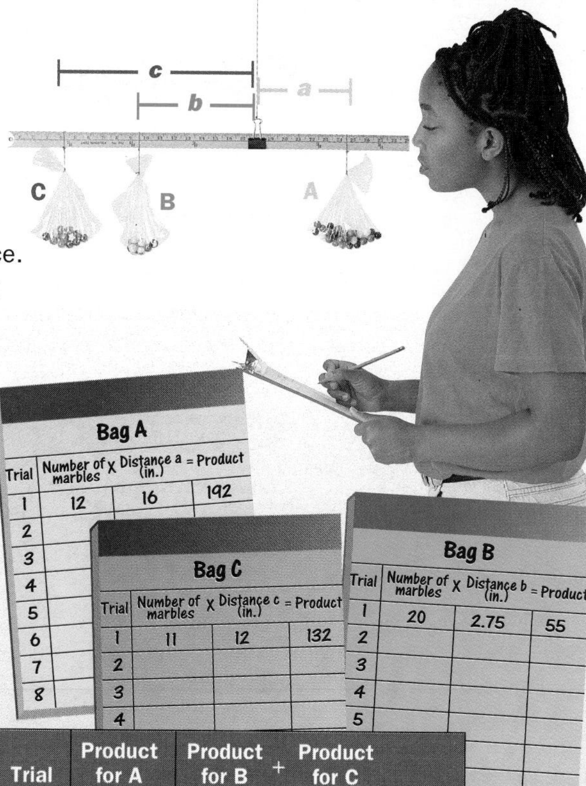

Trial	Product for A	Product for B	+	Product for C	=	
1	192	55	+	132	=	187
?	?	?		?		?

Summarizing Your Results

Write a report about your experiments. Include your data, calculations, and results.

Remember to:

- Explain how the number of marbles in bag B affects the distance at which it balances.
- Explain how you constructed the variable expression in Step 7, and show how you simplified it.

You may want to extend your report to include some of the ideas below:

- Find the weight of an object by balancing it with available objects of known weight.
- Ask your science teacher to show you a beam scale. Explain how it works.

Self-Assessment

In your report, describe what you learned while completing this project. You may want to mention things you learned about balances as well as things you learned about how you approach problems and how you work with others.

Project Notes

Guiding Students' Work

Rubric for Chapter Project

4 Students make a balance, collect their data, make a scatter plot of the data, and write an equation that shows the distance b as a function of the number of beans in bag B. The work is well organized and correctly done. Students also complete Steps 4–7 and draw the correct conclusions from the calculations in their table. They summarize their results in a report that is complete, well written, and mathematically correct.

3 Students complete all steps of the experiment. The final report is complete but has a few mathematical errors. The explanations are correct but could have been stated more clearly. Diagrams are not used to clarify the explanations.

2 Students make a balance and collect their data to explore inverse variation. The balance and data collection are done correctly, but the work to explore rational expressions has some serious problems involving the data collection, the calculations, and the variable expression written. Students attempt to write a report, but it is incomplete and the explanations given are confusing.

1 Students have difficulty making a balance and collecting data for exploring inverse variation. They do not proceed beyond this point in the experiment. Students should be encouraged to speak with the teacher as soon as possible to review their work and to make a new start on the project.

Chapter Support

Course Guide: Chapter 11
Lesson Plans: Chapter 11
Practice Bank:
Cumulative Practice 78
Study Guide: Chapter 11 Review
Assessment Book:
Chapter Tests 50 and 51
Chapter 11 Alternative Assessment
Test Generator software
Portfolio Project Book
Preparing for College Entrance Tests
Professional Handbook

Progress Check 11.1–11.2

For Exs. 1 and 2, tell whether the data show inverse variation. If they do, write an equation for y in terms of x. *(Section 11.1)*

1.

x	y
2	−12
6	4
8	−3
24	1

No.

2.

x	y
2	18
3	12
4	9
12	3

Yes; $y = \frac{36}{x}$.

3. Marco said that filling a container with water shows inverse variation, because as the amount of water added increases, the amount that can still be added decreases. Do you agree or disagree? Explain. *(Section 11.1)* disagree; Suppose the container holds 10 gal. If 1 gal is poured in, it can still hold 9 gal more. If 6 gal are poured in, it can still hold 4 more, and $1 \cdot 9 \neq 6 \cdot 4$.

4. Woh Yan has test grades of 85 and 78. His homework average is 89, which counts as one test. His final exam grade is 92, which counts as two tests. What is his semester average? *(Section 11.2)* about 87

5. Jamie has three test scores of 86 and a final exam score of 86. Is there any way to give extra weight to the final so that his semester average will be more than 86? *(Section 11.2)* No.

6. A bag of gem stones contains 8 blue stones and 3 yellow stones. Each blue stone costs $15, and each yellow stone costs $20. What is the average cost of a stone? *(Section 11.2)* about $16.36

CHAPTER

11 Review

STUDY TECHNIQUE

Look over the sections in the chapter. List the key skills presented in each section. For each skill, solve a related exercise from the text. If possible, choose an exercise that you have not already worked on.

VOCABULARY

inverse variation (p. 462)
constant of variation (p. 462)
weighted average (p. 470)
rational expression (p. 476)
rational equation (p. 476)
simplest form of a rational expression (p. 489)

SECTION 11.1

If two variables have a constant, nonzero product, they show **inverse variation**.

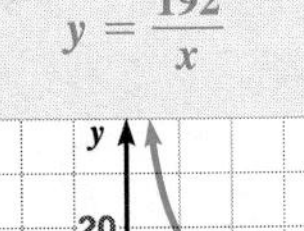
$y = \frac{192}{x}$

x	y
12	16
32	6
4	48
−24	−8

$12 \cdot 16 = 192$

$32 \cdot 6 = 192$

$4 \cdot 48 = 192$

$-24(-8) = 192$

$xy = 192,\ y = \frac{192}{x}$

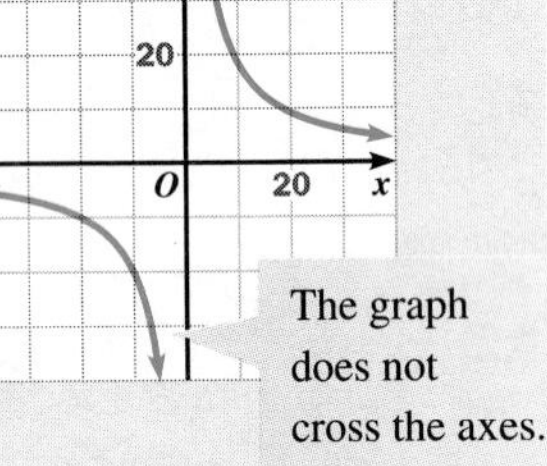

The graph does not cross the axes.

SECTIONS 11.2 and 11.3

In a **weighted average**, the data values have different weights.

Quiz	82
Midterm	90
Quiz	70
Final	80

The midterm and final count **twice** as much as each quiz.
Weighted average:

$$\frac{1(82) + 2(90) + 1(70) + 2(80)}{1 + 2 + 1 + 2} = 82$$

You can solve a **rational equation** by multiplying both sides of the equation by the **least common denominator (LCD)**.

$$\frac{5}{x} = \frac{1}{2-x} + \frac{3}{2(2-x)}$$

$$2x(2-x)\frac{5}{x} = 2x(2-x)\frac{1}{2-x} + 2x(2-x)\frac{3}{2(2-x)}$$

$$20 - 10x = 2x + 3x$$

$$x = \frac{4}{3}$$

Remember to check your answer.

The solution is about 1.33.

You can also use a graph to solve the rational equation.

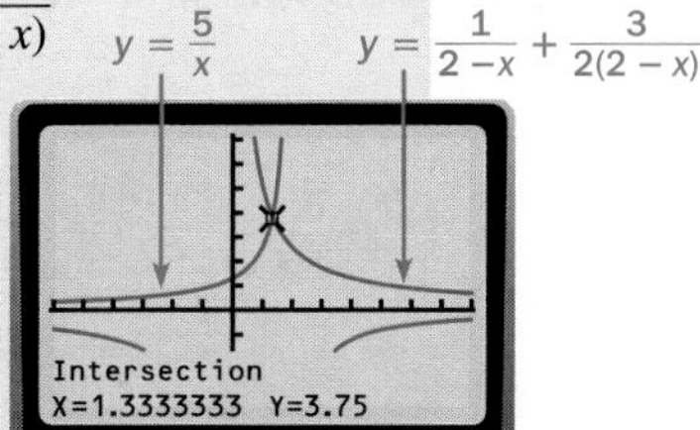

SECTION 11.4

To solve a formula for a variable, you need to get the variable alone on one side of the equation. Perform the same operations on both sides.

$$\frac{1}{x} + \frac{1}{y} = \frac{1}{2}$$

To solve for y, first multiply both sides by the **LCD, $2xy$**, and simplify.

$$2y + 2x = xy$$

Move all of the y-terms to one side.

$$2y - xy = -2x$$

$$y(2-x) = -2x$$

Factor out y.

$$y = \frac{-2x}{2-x}$$

SECTIONS 11.5 *and* 11.6

1. To divide by a rational expression, multiply by the reciprocal.

2. Multiply the numerators and multiply the denominators.

3. Simplify if possible.

$$\frac{2(n+1)}{3} \div \frac{n+1}{2} = \frac{2(n+1)}{3} \cdot \frac{2}{n+1} = \frac{2 \cdot 2(n+1)}{3(n+1)} = \frac{4}{3}$$

1. To add or subtract rational expressions, rewrite each using the **LCD**.

2. Then add or subtract the numerators. Simplify if possible.

$$\frac{1}{3r^2} + \frac{1}{2r} = \frac{2}{6r^2} + \frac{3r}{6r^2} = \frac{2+3r}{6r^2}$$

Progress Check 11.3–11.4

Solve each equation. *(Section 11.3)*

1. $\frac{1}{6x} + \frac{1}{2x} = \frac{4}{9}$ 1.5

2. $\frac{2}{5(x-1)} - \frac{3}{15x(x-1)} = \frac{8}{15x}$ 2.5

3. Is the equation $\frac{1}{\sqrt{x^2-1}} + \frac{4}{3x} = 7$ a rational equation? Explain. *(Section 11.3)*
No; the denominator of the first fraction is not a polynomial.

4. Solve the formula $A = \frac{1}{2}(a+b)h$ for b. *(Section 11.4)*
$b = \frac{2A - ah}{h}$

5. Solve the formula $d = \frac{1}{2}gt^2$ for t. (Assume that d, g, and t are all positive.) *(Section 11.4)*
$t = \sqrt{\frac{2d}{g}}$

6. Solve the equation $3k + \frac{m}{n} = 5p$ for n. *(Section 11.4)*
$n = \frac{m}{5p - 3k}$

Progress Check 11.5–11.6

Simplify. *(Section 11.5)*

1. $\frac{2x+3}{7x^2} \cdot \frac{x}{2x+3}$ $\frac{1}{7x}$

2. $\frac{6x-4}{8x} \cdot \frac{x}{3x-2}$ $\frac{1}{4}$

3. $\frac{3x^2}{4x-1} \div \frac{5x^3}{4x-1}$ $\frac{3}{5x}$

Simplify. *(Section 11.6)*

4. $\frac{7}{x+3} - \frac{7-x}{x+3}$ $\frac{x}{x+3}$

5. $\frac{8a^2}{10b^3} - \frac{a^2}{5b}$ $\frac{8a^2 - 2a^2b^2}{10b^3}$

6. $9 + \frac{3x}{x-6}$ $\frac{12x-54}{x-6}$

Chapter 11 Assessment
Form A Chapter Test

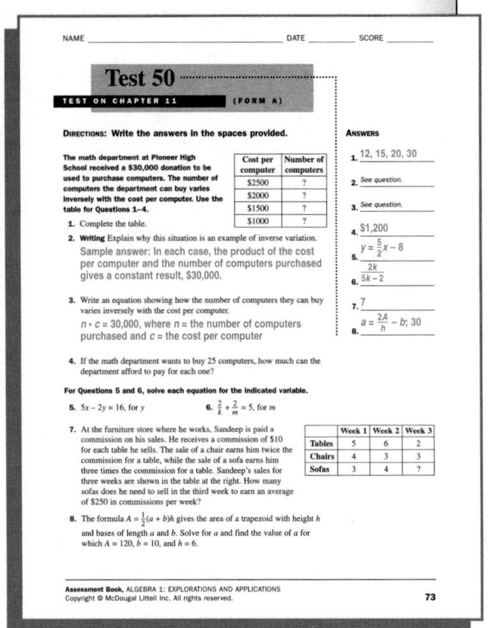

NAME ____________ DATE ______ SCORE ______

Test 50

TEST ON CHAPTER 11 (FORM A)

DIRECTIONS: Write the answers in the spaces provided.

The math department at Pioneer High School received a $30,000 donation to be used to purchase computers. The number of computers the department can buy varies inversely with the cost per computer. Use the table for Questions 1–4.

Cost per computer	Number of computers
$2500	?
$2000	?
$1500	?
$1000	?

1. Complete the table.
2. **Writing** Explain why this situation is an example of inverse variation.
   Sample answer: In each case, the product of the cost per computer and the number of computers purchased gives a constant result, $30,000.
3. Write an equation showing how the number of computers they can buy varies inversely with the cost per computer.
   $n \cdot c = 30{,}000$, where n = the number of computers purchased and c = the cost per computer
4. If the math department wants to buy 25 computers, how much can the department afford to pay for each one?

For Questions 5 and 6, solve each equation for the indicated variable.

5. $5x - 2y = 16$, for y
6. $\frac{2}{k} + \frac{2}{m} = 5$, for m
7. At the furniture store where he works, Sandeep is paid a commission on his sales. He receives a commission of $10 for each table he sells. The sale of a chair earns him twice the commission for a table, while the sale of a sofa earns him three times the commission for a table. Sandeep's sales for three weeks are shown in the table at the right. How many sofas does he need to sell in the third week to earn an average of $250 in commissions per week?

	Week 1	Week 2	Week 3
Tables	5	6	2
Chairs	4	3	3
Sofas	3	4	?

8. The formula $A = \frac{1}{2}(a + b)h$ gives the area of a trapezoid with height h and bases of length a and b. Solve for a and find the value of a for which $A = 120$, $b = 10$, and $h = 6$.

ANSWERS

1. 12, 15, 20, 30
2. *See question.*
3. *See question.*
4. $1,200
5. $y = \frac{5}{2}x - 8$
6. $\frac{2k}{5k - 2}$
7. 7
8. $a = \frac{2A}{h} - b$; 30

Assessment Book, ALGEBRA 1: EXPLORATIONS AND APPLICATIONS

73

Chapter 11 Assessment
Form B Chapter Test

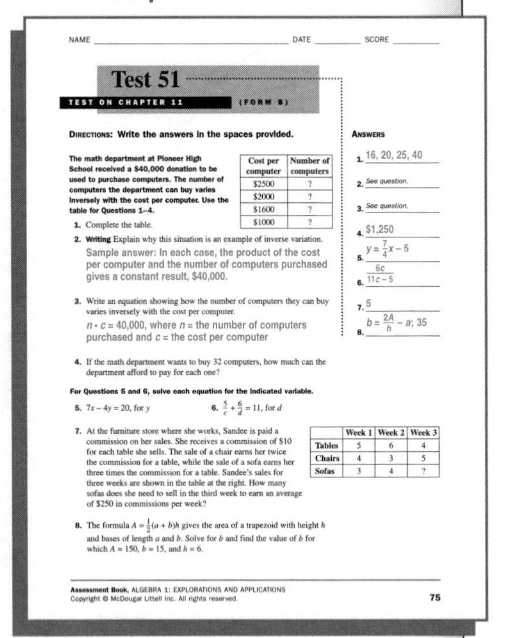

NAME ____________ DATE ______ SCORE ______

Test 51

TEST ON CHAPTER 11 (FORM B)

DIRECTIONS: Write the answers in the spaces provided.

The math department at Pioneer High School received a $40,000 donation to be used to purchase computers. The number of computers the department can buy varies inversely with the cost per computer. Use the table for Questions 1–4.

Cost per computer	Number of computers
$2500	?
$2000	?
$1600	?
$1000	?

1. Complete the table.
2. **Writing** Explain why this situation is an example of inverse variation.
   Sample answer: In each case, the product of the cost per computer and the number of computers purchased gives a constant result, $40,000.
3. Write an equation showing how the number of computers they can buy varies inversely with the cost per computer.
   $n \cdot c = 40{,}000$, where n = the number of computers purchased and c = the cost per computer
4. If the math department wants to buy 32 computers, how much can the department afford to pay for each one?

For Questions 5 and 6, solve each equation for the indicated variable.

5. $7x - 4y = 20$, for y
6. $\frac{5}{c} + \frac{6}{d} = 11$, for d
7. At the furniture store where she works, Sandee is paid a commission on her sales. She receives a commission of $10 for each table she sells. The sale of a chair earns her twice the commission for a table, while the sale of a sofa earns her three times the commission for a table. Sandee's sales for three weeks are shown in the table at the right. How many sofas does she need to sell in the third week to earn an average of $250 in commissions per week?

	Week 1	Week 2	Week 3
Tables	5	6	4
Chairs	4	3	5
Sofas	3	4	?

8. The formula $A = \frac{1}{2}(a + b)h$ gives the area of a trapezoid with height h and bases of length a and b. Solve for b and find the value of b for which $A = 150$, $b = 15$, and $h = 6$.

ANSWERS

1. 16, 20, 25, 40
2. *See question.*
3. *See question.*
4. $1,250
5. $y = \frac{7}{4}x - 5$
6. $\frac{6c}{11c - 5}$
7. 5
8. $b = \frac{2A}{h} - a$; 35

Assessment Book, ALGEBRA 1: EXPLORATIONS AND APPLICATIONS

75

CHAPTER

11 Assessment

VOCABULARY REVIEW

1. Explain what is meant by a weighted average. Give an example of a situation in which you might use a weighted average.
2. Write an equation that is an example of inverse variation.
3. Give an example of a rational expression that is not in simplest form.

SECTIONS 11.1 *and* 11.2

CONSUMER ECONOMICS The math department at Stanley High School received a $6000 grant to buy calculators. The number of calculators the department can buy varies inversely with the cost per calculator, as shown in the table. Use the table for Questions 4–8.

Cost per calculator	Number of calculators
$100	?
$60	?
$50	?
$30	?

4. Complete the table.
5. **Writing** Explain why this situation is an example of inverse variation.
6. Write an equation showing how the number of calculators the math department can buy varies inversely with the cost per calculator.
7. Graph the equation from Question 6. Be sure to show all quadrants. Which parts of the graph make sense in the situation described above?
8. If the math department wants to buy 250 graphing calculators, how much can the department afford to pay for each one?

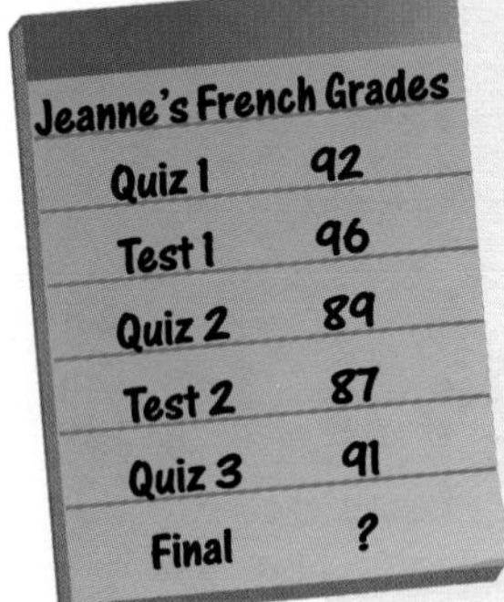

Jeanne's French Grades
Quiz 1 92
Test 1 96
Quiz 2 89
Test 2 87
Quiz 3 91
Final ?

9. Jeanne's grades in French class are shown at the left. Each test is worth two quizzes, and the final is worth three quizzes. What must Jeanne earn on the final to get at least a 92 average for the semester?
10. **Technology** Peder Anderson is designing an apartment complex. He plans to devote 11,500 ft^2 to two-bedroom apartments that bring in $.93/$ft^2$ in rent each month. Peder will also include one-bedroom apartments that have a monthly rent of $1.03/$ft^2$. How many square feet of space should Peder allow for one-bedroom apartments if he wants to receive an average rent of $1.00/$ft^2$?

ANSWERS Chapter 11

Assessment

1–3. Answers may vary. Examples are given.

1. A weighted average is one in which some numbers are given more weight than others. Examples: calculating roommates' shares of utility bills when one roommate is home more than others; calculating grades when certain tests or test items count for more than others.
2. $y = \frac{12}{x}$ or $xy = 12$
3. $\frac{2x + 4}{6x^2 + 24x + 24}$
4.

Cost per calculator	Number of calculators
$100	60
$60	100
$50	120
$30	200

5. The two variables, the cost per calculator and the number of calculators, have a constant product, 6000.
6. Let y = the number of calculators the department can buy and x = the cost per calculator; $y = \frac{6000}{x}$.
7.

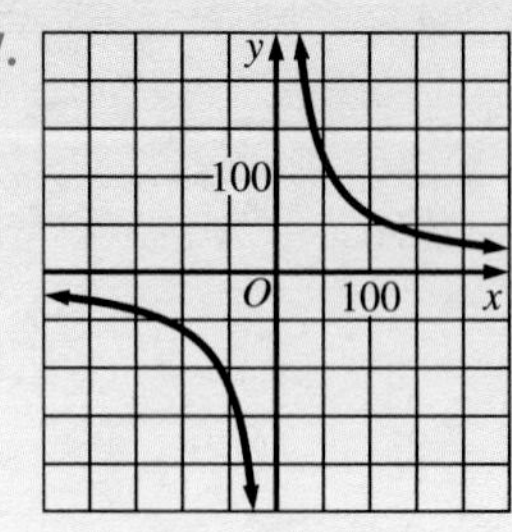

Only the portion of the graph in Quadrant I makes sense, since both the number of calculators and the cost per calculator must be positive.

SECTIONS 11.3 *and* 11.4

Solve each equation. If the equation has no solution, write *no solution*.

11. $7 = \frac{4 - x}{2 + 3x}$

12. $\frac{3}{2} - \frac{x + 4}{x + 2} = \frac{5}{2(x + 2)}$

13. $\frac{4}{x - 5} - \frac{3}{x} = \frac{20}{x(x - 5)}$

14. $\frac{2}{r - 1} - \frac{3}{2r} = \frac{5}{4(r - 1)}$

Solve each equation for the variable in red.

15. $3x + 4y = 12$

16. $\frac{1}{m} + \frac{1}{n} = 2$

17. **Technology** Rewrite the equation $2x = \frac{4y}{y + 2}$ so it can be graphed on a graphing calculator. Then graph it.

18. **GEOMETRY** Triangles and squares are examples of *polygons*. The figure at the right is a 5-sided polygon. The formula $S = (n - 2)180$ gives the sum, S, of the measures of the angles inside any polygon with n sides. Solve for n and find the value of n for which $S = 1260$.

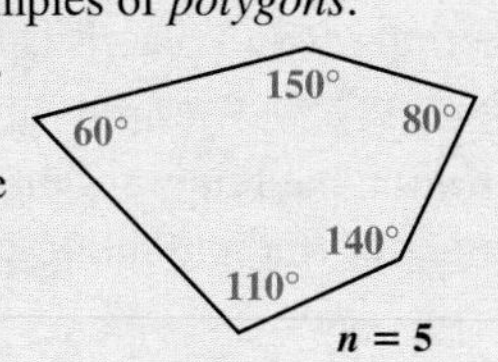

$n = 5$

SECTIONS 11.5 *and* 11.6

Simplify each expression.

19. $\frac{15ab}{4} \cdot \frac{2b^2}{3a^2}$

20. $\frac{8c}{6c + 2} \cdot \frac{3c + 1}{5}$

21. $\frac{2t + 7}{9t} \div (6t + 21)$

22. $\frac{p^2 - 4}{8} \div \frac{p + 2}{p - 2}$

23. $\frac{5t - 1}{3t} + \frac{t + 1}{3t}$

24. $\frac{3k + 4}{k + 2} - \frac{2k - 1}{k + 2}$

25. $\frac{y + 1}{3y^2} + \frac{1}{6y}$

26. $\frac{3}{z + 1} - \frac{5}{z}$

27. $\frac{n}{n - 7} - \frac{n + 7}{n}$

28. **Open-ended Problem** Give an example of each type of equation: linear, quadratic, polynomial, rational, and exponential.

PERFORMANCE TASK

29. Ask your teacher how he or she assigns final grades. Determine what your final grade will be as a function of your grade on the final exam. You may need to make some assumptions about your grades for class participation, homework, etc.

Chapter 11 Assessment
Form C Alternative Assessment

Chapter 11

ALTERNATIVE ASSESSMENT

1. a. If you were given a table of x- and y-values for a function, how could you tell if the function was a model of direct variation or inverse variation?
   b. If you were given a graph of points, how could you tell if the function was a model of direct variation or inverse variation?
   c. Give an example of a real-life function that is a model of direct variation and an example of a real-life function that is a model of inverse variation.

2. **Open-ended Problem** The table at the right gives the grades for the last quiz in an Algebra class. To find the mean of the grades, Laura found the sum of the grades (33) and divided by the number of grades (14). She got an average of 8.25. The correct average is 9. What did she do wrong?

Quiz grade	Frequency
10	6
9	4
8	3
6	1

3. **Open-ended Problem** Compare rational expressions to polynomials. Are there rational expressions that are also polynomials? If so, give examples.

4. **Open-ended Problem** Write three different rational expressions that simplify to the same expression.

5. **Group Activity** Simplify the expression $\frac{x^2 - 6x}{x^2 - 2x - 24}$. Verify that your answer is equivalent to the original expression. Using several values of x, make a table of values for the expression $\frac{x^2 - 6x}{x^2 - 2x - 24}$ and for your simplified expression. Set each expression equal to y and graph the resulting rational equations. Discuss your findings.

6. **Performance Task** Is $\frac{x^2 + 3x + 2}{x + 1} = \frac{(x + 2)(x + 1)}{x + 1} = x + 2$ true for all values of x? Explain why or why not.

7. **Open-ended Problem** Compare simplifying rational expressions with simplifying rational numbers. Use examples in your comparison.

108

Assessment Book, ALGEBRA 1: EXPLORATIONS AND APPLICATIONS

8. \$24
9. ~~98 or better~~ 7/94
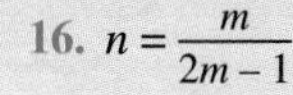
10. Estimates may vary. An example is given. about 26,800 ft²
11. $-\frac{5}{11}$
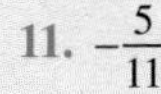
12. 7
13. no solution
14. 2
15. $x = -\frac{4}{3}y + 4$
16. $n = \frac{m}{2m - 1}$
17. $y = \frac{2x}{2 - x}$

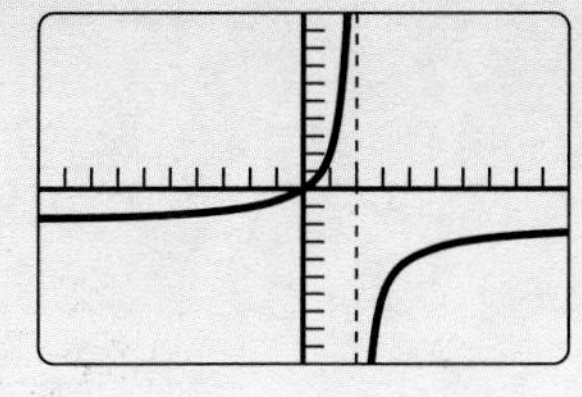

18. $n = \frac{S}{180} + 2$ or $n = \frac{S + 360}{180}$; 9
19. $\frac{5b^3}{2a}$
20. $\frac{4c}{5}$
21. $\frac{1}{27t}$
22. $\frac{(p - 2)^2}{8}$
23. 2
24. $\frac{k + 5}{k + 2}$
25. $\frac{3y + 2}{6y^2}$
26. $\frac{-2z - 5}{z(z + 1)}$
27. $\frac{49}{n(n - 7)}$
28. Answers may vary. Examples are given. linear: $y = \frac{1}{2}x - 5$; quadratic: $y = x^2 - 2x + 1$; polynomial: $y = x^3 + \frac{3}{4}x^2 - 11$; rational: $y = \frac{1}{x + 1} + x(x - 2)$; exponential: $y = 2.5(1.08)^x$
29. Answers may vary.

CHAPTER

12 Discrete Mathematics

OVERVIEW

Connecting to Prior and Future Learning

⇔ Chapter 12 begins with a presentation of algorithms. Learning to write algorithms helps students develop their ability to think logically. This ability is developed further in Section 12.2 as students find paths and trees.

⇔ Recognizing fairness in elections and divisions is the focus of Section 12.3. Students use the skill of developing algorithms as they work through the examples and exercises in this section.

⇔ Students study the multiplication counting principle, permutations, and combinations and learn to use counting strategies to find probabilities. These topics will be expanded upon in Algebra 2.

Chapter Highlights

Interview with Phil Roman: Creating animated figures requires the use of discrete mathematics. This connection is discussed in the interview, with related exercises on page 513.

Explorations in Chapter 12 focus on writing an algorithm and writing an equation for the number of edges in a complete graph.

The Portfolio Project: Game theory is the focus of this project. Students work with a partner to develop strategies for playing and winning a game from Ghana called *wari.*

Technology: Students write an algorithm to use the subtraction key on a calculator. Scientific or graphing calculators are used to find factorials and to find permutations or combinations. Spreadsheets are used to simulate and solve the handshake problem. Graphing calculators are used to make scatter plots. The programming capabilities of graphing calculators are used to analyze a program to model the roll of a 6-sided die.

OBJECTIVES

Section	Objectives	NCTM Standards
12.1	• Write and use algorithms. • Use algorithms to understand and solve problems.	1, 2, 3, 4, 5, 12
12.2	• Find the best path or tree for a graph.	1, 2, 3, 4, 5, 12
12.3	• Recognize fairness in elections and divisions.	1, 2, 3, 4, 5, 10, 12
12.4	• Count all the ways groups of objects can be arranged.	1, 2, 3, 4, 5, 11, 12
12.5	• Count the number of possible pairs of objects in a group. • Find the number of edges in a complete graph.	1, 2, 3, 4, 5, 11, 12
12.6	• Apply counting strategies to find probabilities. • Find the probability of winning a random drawing.	1, 2, 3, 4, 5, 11, 12

INTEGRATION

Mathematical Connections	12.1	12.2	12.3	12.4	12.5	12.6
algebra	**509–515***	**516–522**	**523–529**	**530–535**	**536–541**	**542–547**
geometry	514	**516–522**			538, 540	
data analysis, probability, discrete math	**509–515**	**516–522**	**523–529**	**530–535**	**536–541**	**542–547**
patterns and functions	514					
logic and reasoning	**509–515**	517, 519–522	524–528	531, 533–535	**536–541**	543–547
Interdisciplinary Connections and Applications						
reading and language arts						546
business and economics	515	521, 522	527		540	
arts and entertainment	512, 514					547
sports and recreation				535	541	544
politics and government			528		539	544
fundraising, cars, architecture, volunteering, fashion design, Braille, social studies, photography	510, 513	520	527, 529	530, 531, 533, 534	539	546

****Bold page numbers*** *indicate that a topic is used throughout the section.*

TECHNOLOGY

	opportunities for use with	
Section	**Student Book**	**Support Material**
12.1	graphing calculator	**Technology Book:** Calculator Activity 12
12.2	scientific calculator	
12.3	graphing calculator	
12.4	scientific calculator graphing calculator	
12.5	graphing calculator McDougal Littell Software *Stats!*	
12.6	scientific calculator graphing calculator	**Technology Book:** Spreadsheet Activity 12

PLANNING GUIDE

Regular Scheduling (45 min)

Section	Materials Needed	Core Assignment	Extended Assignment	*exercises that feature* Applications	Communication	Technology
12.1	graph paper, graphing calculator	1–3, 6–9, 17–19, 21, 23–32	1–32	6, 12, 17–22	1–5, 10, 11, 13–20, 22	
12.2		1–9, 11–17, 22–31	1–31	5–9, 16–21	10, 17, 18, 21, 22	
12.3		1–3, 5, 6, 8–10, 14–21, AYP*	1–21, AYP	4–12	3, 8, 9, 11–13	
12.4	scientific calculator	1–3, 7, 8, 11–25	1–25	5–11, 16	3, 8–10, 14–16	13–15
12.5	graphing calculator, spreadsheet software	1–13, 15–17, 19–30	1–30	1–6, 11, 12, 16, 19	10, 11, 13, 18, 20, 21	13, 14, 17
12.6	scientific calculator	1–10, 14–28, AYP	1–28, AYP	6, 14–17	5, 8–13, 18	9, 13
Review/ Assess		**Day 1:** 1–9 **Day 2:** 10–17 **Day 3:** Ch. 12 Test	**Day 1:** 1–9 **Day 2:** 10–17 **Day 3:** Ch. 12 Test	12–14, 16	1–3 10, 11, 17	
Portfolio Project		Allow 2 days.	Allow 2 days.			

Yearly Pacing (with Portfolio Project)	**Chapter 12 Total** 11 days	**Chapters 1–12 Total** 160 days	**Remaining** 0 days	**Total** 160 days

Block Scheduling (90 min)

	Day 76	Day 77	Day 78	Day 79	Day 80	Day 81
Teach/Interact	12.1: Exploration, page 409	12.2 12.3	12.4 12.5: Exploration, page 536	12.6 Review	Review Port. Proj.	Port. Proj. Ch. 12 Test
Apply/Assess	**12.1:** 1–10, 12, 14–19, 21, 23–32	**12.2:** 1–9, 11–17, 22–31 **12.3:** 1–3, 5, 6, 8–10, 14–21, AYP*	**12.4:** 1–4, 7, 8, 11–25 **12.5:** 1–13, 15–17, 19–30	**12.6:** 1–10, 14–28, AYP **Review:** 1–9	**Review:** 10–17 **Port. Proj.**	**Port. Proj.** **Ch. 12 Test**

NOTE: A one-day block has been added for the Portfolio Project—timing and placement to be determined by teacher.

Yearly Pacing (with Portfolio Project)	**Chapter 12 Total** 6 days	**Chapters 1–12 Total** 81 days	**Remaining** 0 days	**Total** 81 days

* ***AYP*** *is Assess Your Progress.*

Lesson Support

Section	Practice Bank	Study Guide*	Assessment Book*	Visuals	Explorations Lab Manual	Lesson Plans	Technology Book
12.1	79	12.1		Warm-Up 12.1	Master 2 Add. Expl. 18	12.1	Calculator Act. 12
12.2	80	12.2		Warm-Up 12.2 Folder 12		12.2	
12.3	81	12.3	Test 52	Warm-Up 12.3		12.3	
12.4	82	12.4		Warm-Up 12.4		12.4	
12.5	83	12.5		Warm-Up 12.5 Folder 12		12.5	
12.6	84	12.6	Test 53	Warm-Up 12.6 Folder 5		12.6	Spreadsheet Act. 12
Review Test	85	Chapter Review	Tests 54, 55, 56 Alternative Assessment			Review Test	

***Spanish versions** of *Study Guide* and *Assessment Book* are available.

Chapter Support

- Course Guide
- Lesson Plans
- Portfolio Project Book
- Preparing for College Entrance Tests
- Multi-Language Glossary
- *Test Generator* Software
- Professional Handbook

Software Support

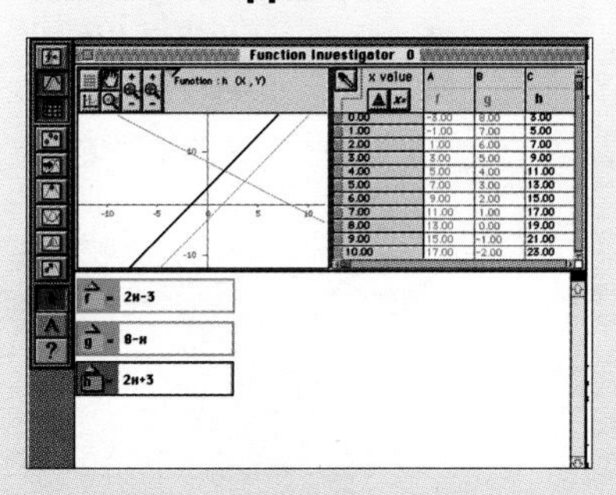

McDougal Littell Software
Stats!

Internet Support

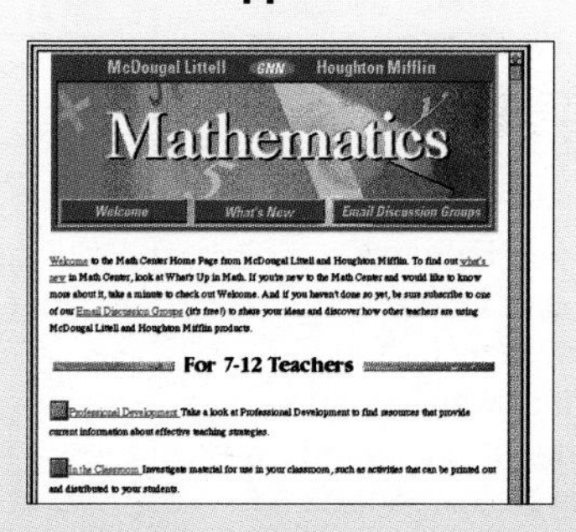

http://www.hmco.com
Next go to McDougal Littell; then the Education Center; then Secondary Math.

Outside Resources

Books, Periodicals

Williams, John. "Graph Coloring Used to Model Traffic Lights." *Mathematics Teacher* (March 1992): pp. 212–214.

Perham, Arnold E. and Bernadette H. Perham. "Discrete Mathematics and Historical Analysis: A Study of Magellan." *Mathematics Teacher* (February 1995): pp. 106–112.

Discrete Mathematics Across the Curriculum, K–12. Margaret Kenney ed. 1991 Yearbook. Reston, VA: NCTM.

Copes, Wayne, William Sacco, Clifford Sloyer, and Robert Stark. "Reachability." *Contemporary Applied Mathematics, Graph Theory.* pp. 47–51. Janson Publications.

Activities, Manipulatives

The Madison Project. "Matrices and Space Capsules." *Explorations in Math.* pp. 372–380.

Copes, Wayne, William Sacco, Clifford Sloyer, and Robert Stark. "Queues." *Contemporary Applied Mathematics, Graph Theory.* Part IV, "Variable Arrivals and Service Times—Simulation." pp. 16–34. Janson Publications.

Software

Geometric Probability. IBM PC or compatible. Accompanied by teaching materials including 19 problems or activities. Reston, VA: NCTM.

Internet

To search for economical local access to the information super highway, investigate the nonprofit National Public Telecommunications Network by sending e-mail to:
info@nptn.org or call 216-498-4050.

Search for mathematics-related software, teaching materials, other gophers, through
gopher archives.math.utk.edu

Interview Notes

Background

Motion Pictures
Animated films are based upon the same technology as all motion picture photography. Film showing was invented in 1895, when the Lumiere brothers developed the first movie projector. Moving pictures were an instant success and within a decade a new industry was born. A motion picture camera works like a still camera, but instead of taking just one picture, it takes a series of discrete pictures as frames. The usual rate for taking and projecting professional movies is 24 frames per second. Older silent movies were shot and projected at 16 frames per second. Each picture, when taken either of actors or drawings created by animators, is a static view of a scene. Anything that moves is in a slightly different position in each succeeding frame. When the pictures are projected in sequence, movement is created.

Phil Roman
Phil Roman grew up in Fresno, California. Today, he is CEO and president of the company he founded called Film Roman, which is now the largest independent animation production house in Hollywood. As a young person, Phil Roman was strongly influenced by animated movies, such as *Bambi* and *Pinocchio.* He attended the Art Center School in Los Angeles to learn animation and eventually worked for Disney Studios and Warner Brothers. After many years of successful work, he decided to launch his own company. His first major project won an Emmy Award.

CHAPTER

12 Discrete Mathematics

Animating the Imagination

INTERVIEW Phil Roman

Growing up in Fresno, California, during the Great Depression of the 1930s, Phil Roman picked grapes with his family to earn money for food. His life changed when he turned eleven and saw the Disney movie *Bambi* for the first time. "I was captured by those images on the screen," Roman recalls. "Here were drawn animals that you could relate to. It really struck a chord in me." After he left the movie theater, he began to draw figures of Bambi everywhere–on school notebooks and any scrap of paper he could lay his hands on. "That was the start of it all," he says. "I read books on animation and gradually pointed myself in that direction."

"I love to draw, to bring characters to life, to animate."

A Dream Come True

Roman arrived in Hollywood at the age of 18 with $60 in his pocket and an urge to draw. He took a job as an usher at a movie theater, using half his life savings as a down payment for art school. His dream came true a year later when he landed a job at Disney. In 1984, after almost three decades working at several different studios, he started his own company. Initially just "three people in a small room," the company is now the largest independent animation studio in Hollywood. Its productions include popular television shows, such as *Garfield and Friends* and *The Simpsons.*

"Animation takes ... an understanding of math to make it look natural."

From Script to Film

As studio head, Roman oversees every aspect of producing a film. The starting point is a script that describes the action and characters. Next to be created is a *storyboard*—an elaborate comic strip that tells the story in a visual way. Animators then create thousands of drawings that are photographed by a movie camera, frame by frame. Music and sound effects are added during *post-production*, with some last-minute cutting and trimming done in the editing stage. Finally, the lights go dim, the projector rolls, and the characters come to life.

507

Interview Notes

Background

Modern Animation

Animation has come a long way since the early days of *Bambi* and other Disney movies. Today, animation studios are equipped with advanced technology and graphics, capable of extremely life-like and versatile animation. One such studio is Colossal Pictures, a leading animation company based in San Francisco, which was founded by Gary Gutierrez, a graduate of the San Francisco Art Institute and Drew Takahashi, a graduate of UCLA Film School. Half the company's staff consists of talented animators, many of whom have come to the United States from other nations. The company designs and animates commercials, CD-ROM titles, music videos, and live-action interactive films.

Second-Language Learners

Students learning English may be unfamiliar with the expression *struck a chord.* Explain that when something *strikes a chord* with a person, that person is left with a deep impression about it.

Interview Notes

Mathematical Connection

The story board for an animated film is essentially an algorithm for creating the thousands of drawings that make up the film. In Section 12.1, students learn how to write and use algorithms so that they can solve real-world problems, such as how to animate an action.

Explore and Connect

Research
Each student should select one technique for his or her report. Since it may be somewhat difficult for some students to find reference materials, students should share their sources.

Project
You may wish to assign students to small groups to work on this project. Since not all students are good at making drawings, each group should assign an *artist* to create the pages. The group as a whole can contribute to the algorithm for the story.

Writing
The differences students have noticed should be discussed in class.

Creating an Illusion

When you watch an animated film, the action on the screen looks smooth and seamless. But if you examine the film itself, you will see a string of pictures that are each slightly different. The frames are *discrete*. When the film is projected at a speed of 24 frames per second, your eye is tricked into seeing the images as continuous movement. The challenge for an animator is to choose the series of pictures that will model continuous motion. In *discrete mathematics*, you will use such techniques as choosing the correct series of steps for solving a problem and counting the possibilities at each stage in a process. "When I got into animation, I had no idea that the stuff I learned in high school would be so important," Roman says. "But the math comes in very handy. I use it every day." Although the movement in animated films seems natural, that doesn't just happen, he explains. "It takes hard work, a lot of planning, and an understanding of math to make it look natural."

EXPLORE AND CONNECT

Phil Roman with two animated characters.

1. Research There are many different techniques used to create animated films. Some animators create thousands of drawings. Some use puppets, clay figures, and other three-dimensional objects. Some use computer animation. Find out more about one of these or another animation technique. Report your findings.

2. Project Create an animated flip book. Each page of a flip book shows a picture that is slightly different from the one before it. When you flip through the pages quickly, you see a short piece of animation.

3. Writing Have you noticed any differences between animation in movies and animation in television shows? Describe the differences.

Mathematics & Phil Roman

In this chapter, you will learn more about how mathematics is related to animation.

Related Explorations and Exercises

Section 12.1
- **Exploration**
- **Exercises 14–16**

SECTION

12.1 Exploring Algorithms

Learn how to...
- write and use algorithms

So you can...
- understand real-world problems such as how to animate an action; or
- assign shelf keeping units

Animators like Phil Roman break simple motions into discrete steps creating sequences of pictures that give the appearance of natural motion on film. You can write an *algorithm* to break a motion into steps. An **algorithm** is a step-by-step method for accomplishing a goal.

EXPLORATION
COOPERATIVE LEARNING

Writing an Algorithm

Work in a group of 3–5 students.

1 Work together to write an algorithm that explains how to sit down in a chair. Make each step as clear and precise as possible.

2 Have one of the members of your group follow your algorithm exactly. Did each step work the way you planned?

3 If necessary, revise your algorithm. Have someone in your group test your new algorithm.

4 Compare your algorithm with those of other groups. Are there different steps you could use, or are all of the algorithms basically the same?

Exploration Note

Purpose
The purpose of this Exploration is to have students learn how to write an algorithm, in this case, one that explains how to sit down in a chair.

Materials/Preparation
Students should first choose a chair. The algorithms for sitting in an armchair, a straight-back chair without arms, or a chair that is attached to a desk would be slightly different.

Procedure
Students should number each step in their algorithm. As one student in the group follows the algorithm, have another student check off each step completed successfully. Students should begin revising their algorithm at the first step that did not receive a checkmark.

Closure
Discuss the similarities and differences in the different algorithms. Ask students which algorithm they think gives the best instructions in the least number of steps.

Explorations Lab Manual
See the Manual for more commentary on this Exploration.

For answers to the Exploration, see following page.

Plan⇔Support

Objectives
- Write and use algorithms.
- Use algorithms to understand and solve problems.

Recommended Pacing
Core and Extended Courses
Section 12.1: 1 day
Block Schedule
Section 12.1: 1 block

Resource Materials
Lesson Support
Lesson Plan 12.1
Warm-Up Transparency 12.1
Practice Bank: Practice 79
Study Guide: Section 12.1
Explorations Lab Manual: Additional Exploration 18, Diagram Master 2

Technology
Technology Book: Calculator Activity 12
Graphing Calculator
Internet: http://www.hmco.com

Warm-Up Exercises
Simplify each expression.

1. 403×15 6045
2. 74×123 9102
3. 2^8 256
4. 3^4 81
5. $348 \div 24$ $14\frac{1}{2}$
6. $1072 \div 36$ $29\frac{7}{9}$

Teach⇔Interact

Section Note

Communication: Discussion Ask students to suggest examples from their own experiences when they have had to follow a series of discrete steps to perform a task.

Additional Example 1

The freshman class deposited $500 in a savings account that pays 7% annual interest. The class treasurer wants to project the class's account balance in three years if they make no additional deposits or withdrawals to the account. The treasurer discovered that the exponent on her calculator was not working. Write an algorithm she could use to find a solution without using the exponent key. Use your algorithm to find a solution.

Method 1 Write an algorithm to use the percent key.
Step 1 Enter 500.
Step 2 Enter + 7%.
Step 3 Repeat Step 2 two more times.
The account balance will be about $612.52.

Method 2 Write an algorithm to use the multiplication key.
Step 1 Enter 500.
Step 2 Enter × 1.07
Step 3 Repeat Step 2 two more times.
The account balance will be about $612.52.

Algorithms break a complicated task into discrete steps. If you have ever followed a recipe, you have used an algorithm. If you have ever programmed a calculator or computer, you have written an algorithm.

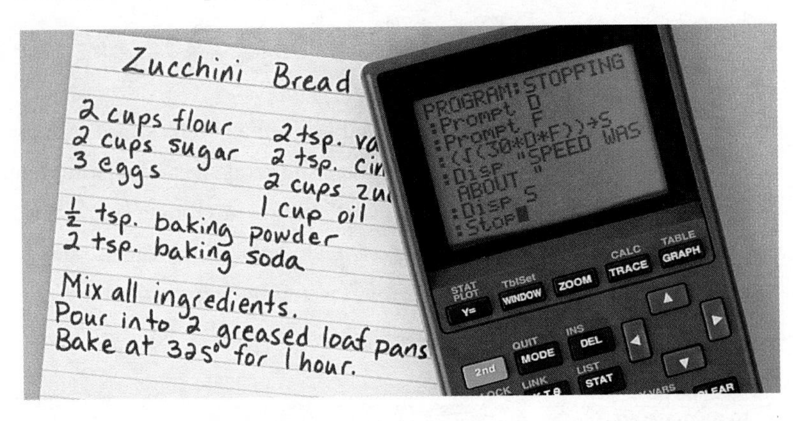

EXAMPLE Application: Fundraising

The drama club took orders for 204 plants to raise money. Danielle needed to determine how many plants each of the 17 club members should deliver. She tried to use her calculator, but found that the division key was broken. Write an algorithm Danielle could use to find a solution without using the division key. Use your algorithm to find a solution.

SOLUTION

Method 1 Write an algorithm to use the subtraction key.

Step 1 Subtract 17 from 204.

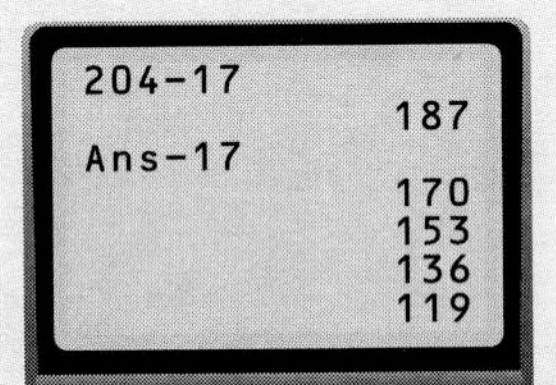

Step 2 Subtract 17 from the remainder.

Step 3 Repeat Step 2 until the number left is less than 17.

Step 4 Count how many times 17 was subtracted.

```
102
 85
 68
 51
 34
 17
  0
```

You need to subtract 12 times. There is no remainder. Each student should deliver 12 plants.

ANSWERS Section 12.1

Exploration

1. Answers may vary. An example is given. Step 1: Stand in front of the chair with your back to it. Step 2: Step back until your legs are near or actually touching the chair. Step 3: Bend your knees, leaning slightly forward above the waist until you are seated on the chair.

2–4. Answers may vary.

Think and Communicate

1–3. Answers may vary. Examples are given.

1. I prefer to use the subtraction algorithm because the guess-and-check algorithm often takes me longer.
2. I would use the guess-and-check algorithm because Step 3 of Method 1 would have to be repeated 71 times. On the other hand, I could use the solution to the example to choose a reasonable starting point for guessing and checking.
3. Modify Step 4: Is the result too small? Subtract the product from 204. Is the difference less than the guess? If so, the guess is correct and the difference is the remainder. If the difference is larger than the guess, start again with a larger guess.

Checking Key Concepts

1–3. Answers may vary. Examples are given.

1. You might be trying to cook the eggs in their shells. You might end up eating uncooked eggs.
2. In Step 1, it should be made clear that the eggs must be broken. In Step 2, it should be noted that if the pan does not have a non-stick surface, it must be oiled. A step must be

Method 2 Write an algorithm using multiplication and guess and check.

In Steps 3–5, if your answer is *no*, skip to the next step.

Algorithm		First try	Second try	Third try
Step 1	Make a guess.	Guess 15.	Guess 10.	Guess 12.
Step 2	Multiply your guess by 17.	$15 \cdot 17 = 255$	$10 \cdot 17 = 170$	$12 \cdot 17 = 204$
Step 3	Is the result correct? You are done.	No	No	12 is correct. Done.
Step 4	Is the result too big? Try again with a smaller number.	15 is too big. Try again.	No	
Step 5	Is the result too small? Try again with a larger number.		10 is too small. Try again.	

Each student should deliver 12 plants.

BY THE WAY

The word *algorithm* comes from Persia. It is derived from the name of the mathematician Muhammad ibn Mūsā al-Khwārizmī, who lived around 780–850 A.D.

THINK AND COMMUNICATE

For Questions 1–3, look at the Example on pages 510 and 511.

1. Which of the two algorithms do you prefer? Why?
2. Suppose there were 1230 plants to distribute. Which algorithm would you prefer? Why?
3. How could you change the algorithm in Method 2 to make it work for a problem with a remainder?

CHECKING KEY CONCEPTS

For Questions 1–3, refer to the algorithm for making scrambled eggs.

Step 1 Take the eggs out of the refrigerator and put them in a bowl.
Step 2 Mix them all up together and pour them in a pan.
Step 3 Stir them around as they cook.

1. What could go wrong if you follow this algorithm exactly?
2. Which steps in this algorithm need to be clarified? What steps would you add to this algorithm?
3. **Cooperative Learning** Work with a partner to write a clear, concise, and complete algorithm for making scrambled eggs.
4. Look back at the Example on page 510. Write an algorithm that uses the calculator's addition key to solve Danielle's problem.

Teach⇔Interact

Think and Communicate

These questions lead students to understand that one algorithm may not be the best for every situation and that a good algorithm is clear and concise. Students should also see that when an algorithm does not work in a given situation, it may be possible to adapt it so that it can be applied to the situation.

Checking Key Concepts

Mathematical Procedures

These questions illustrate an important feature of algorithms, namely that all steps must be stated and stated clearly. An algorithm should be thought of as a mathematical procedure that will accomplish a task in exactly the same way each time it is used.

Closure Question

How is an algorithm like a recipe?

An algorithm is a set of detailed instructions as is a recipe. The instructions in an algorithm and a recipe should be clear and concise and contain all necessary steps. No unnecessary steps should be given.

added to make it clear that the pan must be applied to a cooking source, such as a hot burner or a microwave oven. Step 3 should indicate that the eggs must be cooked until they are done. An additional step might suggest removing the eggs from the pan to a dish, or adding water, milk, or some other ingedient to the eggs. A step might also be added to specify the number of eggs per serving.

3. The following is an example of an algorithm for cooking scrambled eggs on a conventional stovetop. Step 1: Take the eggs out of the refrigerator and put them in a bowl. Step 2: Break the eggs into a bowl and stir until the yolks are broken and the eggs are well mixed. Step 3: Put a frying pan on a stove burner over medium heat. If the pan does not have a non-stick surface, oil the pan before putting it on the stove. Step 4: When the pan is heated, pour the eggs into the pan. Step 5: Stir the eggs as they cook. When the eggs are cooked to the degree you prefer, remove the pan from the heat.

4. Step 1: Add 17 to the number showing on the calculator. (Initially, this number is 0.) Step 2: Repeat Step 1 until the sum is greater than or equal to 204. Step 3: Count the number of times 17 was added. Since the final sum is 204, this number is the solution.

Apply⇔Assess

Suggested Assignment

Core Course
Exs. 1–3, 6–9, 17–19, 21, 23–32

Extended Course
Exs. 1–32

Block Schedule
Day 76 Exs. 1–10, 12, 14–19, 21, 23–32

Exercise Notes

Second-Language Learners
Exs. 1–3 Since an incorrect placement of words can make an algorithm incorrect, second-language learners may feel more comfortable writing algorithms if they are offered the assistance of a peer tutor or an aide.

Cooperative Learning
Exs. 1–3 These exercises provide an opportunity for students to learn cooperatively. By writing the algorithms as a group, students can share their ideas and then test the algorithms to see how well they work.

Multicultural Note
Ex. 6 The pottery made by the Pueblo for cooking is characterized by a series of pinched coils that produced a corrugated effect. Their more decorative pottery is characterized by elaborate patterns in black on a white background. During the period from 1100–1300, when the Pueblos were constructing their elaborate cliff dwellings of stone and adobe on ledges at Mesa Verde in southwest Colorado, the white backgrounds on their decorative pottery were often replaced by red or orange.

12.1 Exercises and Applications

Extra Practice exercises on page 581

For Exercises 1–3, write an algorithm that could be used to program a robot or computer to complete each task.

1. Make your favorite sandwich.
2. Use two given points to write an equation of a line.
3. Graph the equation $y = \frac{1}{x + 2}$ on a graphing calculator.
4. Look back at Exercises 20–23 on page 163. Write an algorithm for the method described by the Ahmes papyrus.
5. **Writing** How would you change the two algorithms in the Example on pages 510 and 511 if 19 students were going to help deliver plants?
6. **FINE ARTS** Santana Martinez makes coil pots in the traditional Pueblo style. The photographs show some of the key steps she follows. Put the steps in the correct order. Write an algorithm for the process.

BY THE WAY

Santana Martinez's mother-in-law, Maria Martinez, rediscovered techniques used by the Pueblos to make black-on-black pots eight centuries earlier. Maria Martinez was the first Pueblo potter to sign her work.

A. 

B. 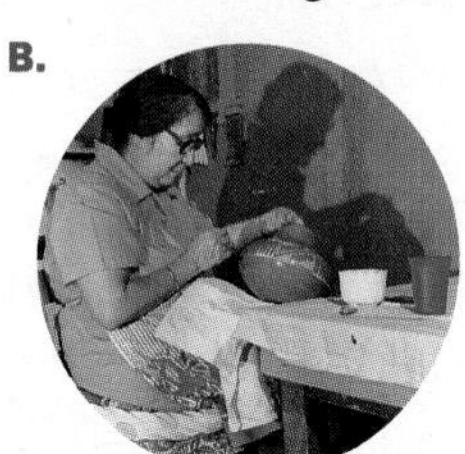

C. 

D. 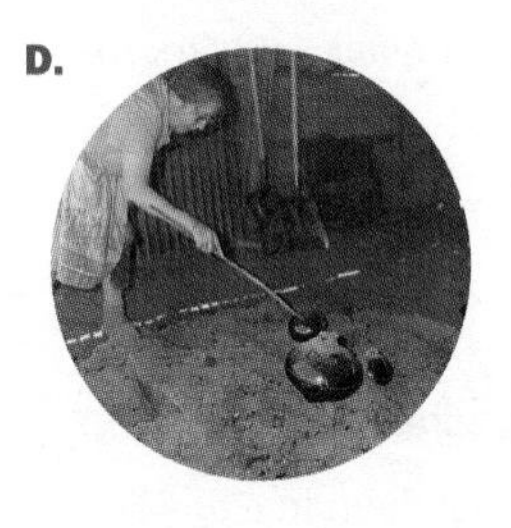

E. 

F. 

For Exercises 7–9, write an algorithm that explains how to use a calculator to solve each problem.

7. Divide 326 by 52 without using the division key.
8. Multiply 13 by 34 without using the multiplication key.
9. Compute 4^8 without using the exponent key.
10. **Open-ended Problem** Write an algorithm for a common activity, such as tying your shoe or climbing a flight of stairs.
11. **Challenge** Write an algorithm for playing tic-tac-toe.

Exercises and Applications

1–3. Algorithms may vary. Examples are given.

1. Step 1: Take out two slices of bread. Step 2: Spread peanut butter on one slice. Step 3: Spread jelly on the other slice. Step 4: Put the slices together with unspread sides out.
2. Step 1: Subtract the y-coordinates of the points. Step 2: Subtract the x-coordinates of the points (in the same order as in Step 1). Step 3: Write the ratio of the answer from Step 1 to the answer from Step 2 as a fraction in simplest form. Step 4: Choose the coordinates of one of the two points and the fraction from Step 3 to write an equation of the form $y = mx + b$. Step 5: Solve the equation from Step 4 for b. Step 6: Use the value of m from Step 3 and the value of b from Step 5 to write the equation for the line.
3. Step 1: If the range is not set on standard, enter [ZOOM] 6. Step 2: Enter Y=1/(X + 2). Step 3: Enter [GRAPH]. Step 4: If necessary, adjust the range.

4–11. See answers in back of

12. **CARS** The algorithm below describes a method for computing fuel mileage.

    Step 1 Drive the car until it runs out of fuel.

    Step 2 Put exactly one gallon of fuel in the tank.

    Step 3 Record the mileage on the odometer.

    Step 4 Drive until you run out of fuel again.

    Step 5 Record the mileage on the odometer. Subtract the old figure from the new figure. This is the rate of fuel consumption in miles per gallon.

    a. Comment on this algorithm. Will it work? Is it practical?

    b. Write a better algorithm to compute a car's fuel mileage.

13. **Open-ended Problem** Suppose you have a computer program that can factor numbers into primes. Write an algorithm that uses this computer program to find the least common multiple of two numbers.

INTERVIEW Phil Roman

Look back at the interview on pages 506–508.

Cooperative Learning *Suppose an animation studio wishes to include a sequence on making a cake in an upcoming film. You are part of the team of animators working on this segment. Work in a group of three students.*

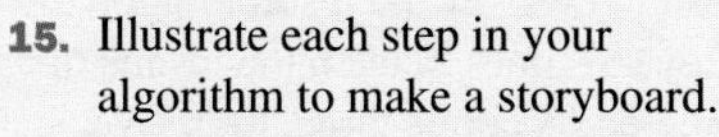

14. Write an algorithm for making a cake. Think about how the process appears to someone watching the action.

15. Illustrate each step in your algorithm to make a storyboard.

16. Estimate the time each step will take in the film. At a rate of 24 frames per second, how many frames will be required for each step? for the entire sequence?

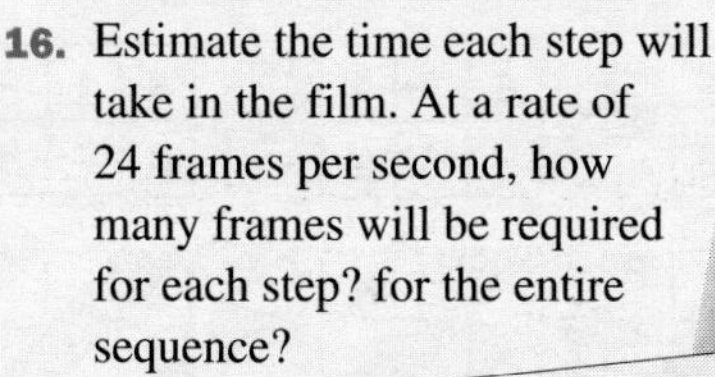

Phil Roman reviews the storyboards for an animated film.

Your storyboard can be made by drawing pictures, or you might want to use photos from magazines.

Apply⇔Assess

Exercise Notes

Student Study Tip

Exs. 7–9 These exercises can be used to point out a fundamental feature of algorithms written for a computer. An algorithm must have a finite number of steps so the computer will not be caught in an endless series of steps. In Ex. 7, for example, if successive subtraction is used to find the quotient, the algorithm must provide for the subtraction to stop when the difference is less than or equal to the divisor.

Reasoning

Ex. 12 Analyzing an algorithm and writing a better algorithm helps students to develop their critical thinking skills. To further develop these skills, you may wish to have students work in groups to analyze each other's algorithms. This can be done not only for this exercise, but also for any exercise where students have written algorithms.

Alternate Approach

Ex. 13 You may wish to suggest that students assume the program finds the GCF of a pair of numbers, a and b. Use the formula LCM = $(ab) \div$ GCF of a and b.

Research

Exs. 14–16 Some students may wish to research the kinds of jobs that are available in an animation studio.

book.

12. a. The algorithm will work, but it is totally impractical. For example, you would want to be sure that in both instances you ran out of fuel at a filling station. This would be unlikely.

    b. This sample algorithm assumes the car does not have a trip odometer which can be set to 0. Step 1: Wait until the gauge shows that the fuel tank is, say, one-quarter full. Record the mileage on the odometer. Step 2: Fill the fuel tank. Record the number of gallons bought. Step 3: Repeat Steps 1 and 2 several times, ending with a repetition of Step 1. Step 4: Subtract the first odometer reading from the last. Divide by the total number of gallons of fuel purchased. This is an average fuel mileage in miles per gallon.

13. Step 1: Use the computer to factor the first number. Step 2: Use the computer to factor the second number. Step 3: Compare the prime factorizations. List any prime number of which a power appears in both of the given numbers. Step 4: Determine the greater power of each base described in Step 3. Step 5: List any number that appears in only one of the prime factorizations. Step 6: Write the product of all the numbers found in Steps 4 and 5. This product is the least common multiple of the two numbers.

14. The algorithm in the example given is based on using a packaged mix. Step 1: Preheat the oven to the required temperature. Step 2: Oil and flour the pans. Step 3: Put the mix in a bowl. Add ingredients as directed on the package and stir. Step 4: Pour the mix into the pans, dividing it equally among them. Step 5: Put the pans in the preheated oven and bake the required length of time. Step 6: Remove the pans from the oven and let them cool. Step 7: Remove the cake from the pans and frost.

15. Check students' work.

16. Answers may vary. Check students' work.

Apply⇔Assess

Exercise Notes

Using Manipulatives

Exs. 17, 18 Some students may find it easier to complete these exercises if they make copies of the original image and then cut out the six squares of the rectangle. This will allow them to manipulate the cutout figures to determine the new pattern each time the algorithm is applied. If students are working in pairs, you might suggest that one person cut out the original image with only the numbered squares and that the other person copy and cut out the original image with the pattern drawn on it. For each application of the algorithm, one student can manipulate the numbered squares, while the other person manipulates the patterned squares.

Interdisciplinary Problems

Exs. 19, 20 These exercises allow students with an interest in graphic arts to show their creativity and originality in addition to providing them with the opportunity to apply a mathematical concept to their art. If students work in groups, you might want to assign a student with artistic ability to each of the groups. These students can use their artistic and visual abilities to help the members of the group produce interesting designs.

Connection ART

Bob Brill is an algorithmic artist. Instead of drawing, he writes computer programs to create art. Brill repeatedly applied the P2 algorithm to transform the original image (shown at the right) into the images below. Use this algorithm for Exercises 17–19.

Bob Brill's original image

P2 Algorithm

Step 1 Divide the *starting* rectangle into rows and columns of squares.

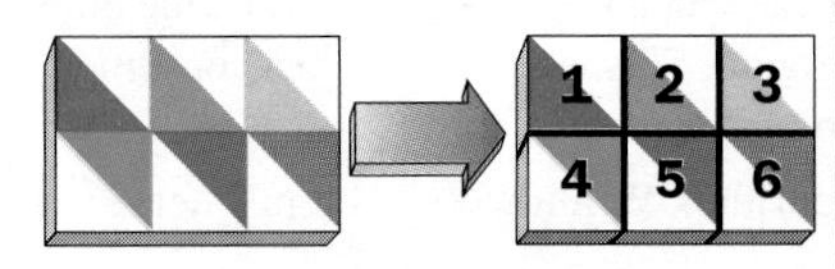

Step 2 Use the rules below to move the squares from the *starting* rectangle to a *new* rectangle with the same dimensions as the starting rectangle.

Rule 1 Start at the top left of the *starting* rectangle. Remove the pieces of the top row from left to right. When a row is empty, start at the left end of the next row.

Rule 2 Start at the top right of the *new* rectangle. Place the squares by column from top to bottom. When a column is full, start filling the column to its left.

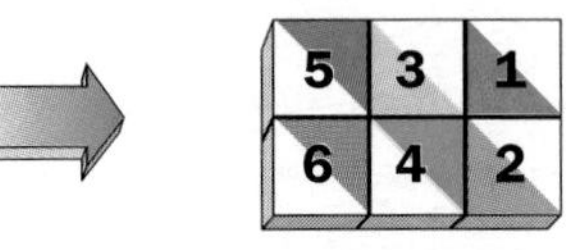

Algorithm applied 254 times.

Algorithm applied 341 times.

17. The images below show the result of applying the P2 algorithm to a rectangle several times. Sketch the results of applying the P2 algorithm to the numbered rectangle three, four, five, and six times.

1 2 3 / 4 5 6 → P2 → 5 3 1 / 6 4 2 → P2 → 4 1 5 / 2 6 3 → P2 → ? ? ? / ? ? ?

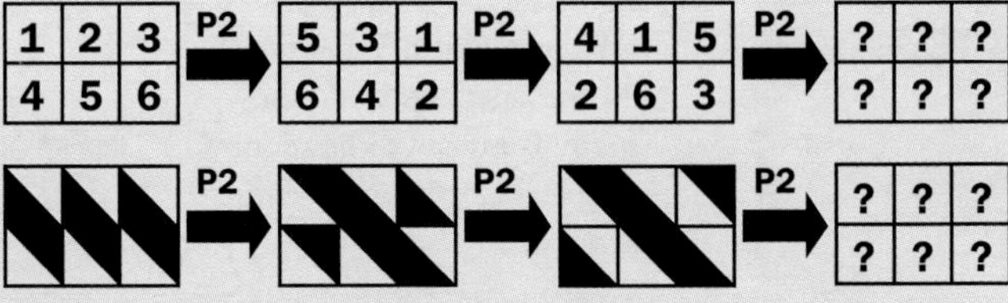

18. Use the patterned rectangle. Sketch the results of applying the P2 algorithm three, four, five, and six times.

19. Create your own pattern in a rectangle. Cut it into equally sized squares. Apply the P2 algorithm to your pattern until it repeats. Sketch the original image and the result of several applications of the algorithm.

20. **Open-ended Problem** Create your own algorithm for changing a pattern. Explain your algorithm. Does the pattern return to its original form if you repeat the algorithm enough times?

By the Way **shows the algorithm applied 512 times.**

BY THE WAY

The images on this page are all rectangles. If you apply the P2 algorithm to a square, the image will simply rotate around the center of the square.

17. 6 5 4 / 3 2 1 → 2 4 6 / 1 3 5 → 3 6 2 / 5 1 4 → 1 2 3 / 4 5 6

18.

19. Patterns may vary. Check students' results.

20. Answers may vary.

21. A: 03–b–01; B: 03–b–03; C: 03–c–03; D: 03–c–02

22. Answers may vary. SKUs should take into account that items that sell very well should be easy to find and to reach. Also, packages that are heavy or large should be on lower shelves for ease in lifting, whereas lighter, smaller packages could be placed on higher shelves.

23. Answers may vary. Examples are given.

a. Step 1: Substitute any number for x and see if the value makes the equation true. If so, you are done. If the value you tried was too big, go to Step 2. If the value you tried was too small, go to Step 3. Step 2: Substitute a smaller number for x and see if the value makes the equation true. If so, you are done. If the value you tried was too big, repeat Step 2. If the value you tried was too small, go to Step 3. Step 3: Substitute a larger number for x and see if the value makes the equation true. If so, you are done. If the value you tried was too big, go to Step 2. If the value you tried was too small, repeat Step 3. I do not think this is a very good algorithm. It might take quite a while to get the right solution this way.

BUSINESS A mail order company is organized to fill customers' orders quickly. Each item is assigned a shelf keeping unit (SKU). The letters and digits of the SKU tell the packer where to find the item.

21. In one company, the first two digits of the SKU indicate a shelving unit. The lower case letter indicates a shelf. The last two digits indicate the position on the shelf. Match each SKU with a lettered package on shelving unit 03 in the photo at the right.

03-b-03	03-c-02
03-a-03	03-b-01

22. Writing When assigning SKUs, the company considers how well the item is expected to sell, as well as the weight and dimensions of the package. Explain how each of these would affect where you would shelve an item. Write an algorithm for assigning SKUs.

ONGOING ASSESSMENT

23. Open-ended Problem The results of using an algorithm to solve the equation $12x + 9 = 45$ are given below:

Try $x = 10$.	$12(10) + 9 = 129$	Too big!
Try $x = 1$.	$12(1) + 9 = 21$	Too small!
Try $x = 5$.	$12(5) + 9 = 69$	Too big!
Try $x = 3$.	$12(3) + 9 = 45$	That's it!

a. Write an algorithm that could have given these results. Do you think this is a good algorithm to use?

b. Write two different algorithms that you could use to solve the same equation.

SPIRAL REVIEW

Graph each inequality. *(Section 7.5)*

24. $y > 5$ **25.** $y \leq 2x - 4$ **26.** $4x - 5y < -10$

Simplify each expression. *(Section 11.5)*

27. $-\frac{3x}{2y} \cdot \frac{2}{5x}$ **28.** $\frac{x}{y} \div 2x$ **29.** $\frac{21r}{20s^2} \cdot \frac{15s^3}{7r}$

30. $\frac{x^2 + 2}{x} \div \frac{x + 2}{3x}$ **31.** $\frac{xy}{4x - 4y} \cdot \frac{8x + 2y}{x^2}$ **32.** $\frac{3x + 1}{x - 2} \div \frac{x^2}{x^2 - 4}$

Apply⇔Assess

Exercise Notes

Application
Exs. 21, 22 Discuss with students the fact that problem-solving situations like those in Exs. 21 and 22 are common in many workplaces. Employees must be able to work together to solve problems.

Assessment Note
Ex. 23 This exercise can be used to discuss the merits of the guess-and-check method of problem solving. Students should understand that the pattern of guessing should provide the fewest number of steps possible to the solution.

Practice 79 for Section 12.1

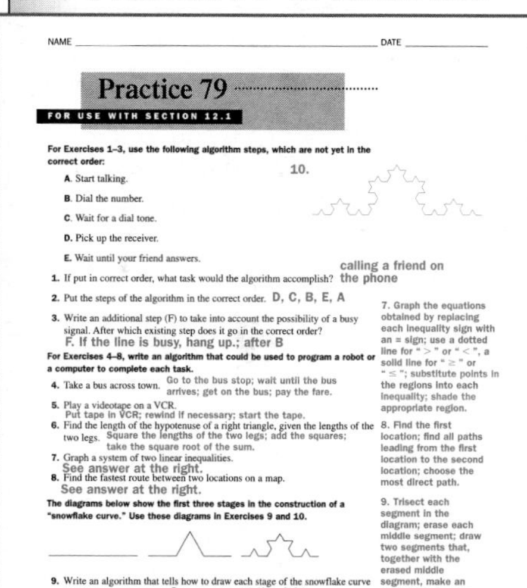
NAME ______________________ DATE __________

Practice 79
FOR USE WITH SECTION 12.1

For Exercises 1–3, use the following algorithm steps, which are not yet in the correct order:

A. Start talking.
B. Dial the number.
C. Wait for a dial tone.
D. Pick up the receiver.
E. Wait until your friend answers.

1. If put in correct order, what task would the algorithm accomplish? calling a friend on the phone

2. Put the steps of the algorithm in the correct order. D, C, B, E, A

3. Write an additional step (F) to take into account the possibility of a busy signal. After which existing step does it go in the correct order? F. If the line is busy, hang up.; after B

For Exercises 4–8, write an algorithm that could be used to program a robot or a computer to complete each task.

4. Take a bus across town. Go to the bus stop; wait until the bus arrives; get on the bus; pay the fare.

5. Play a videotape on a VCR. Put tape in VCR; rewind if necessary; start the tape.

6. Find the length of the hypotenuse of a right triangle, given the lengths of the two legs. Square the lengths of the two legs; add the squares; take the square root of the sum.

7. Graph a system of two linear inequalities. See answer at the right.

8. Find the fastest route between two locations on a map. See answer at the right.

The diagrams below show the first three stages in the construction of a "snowflake curve." Use these diagrams in Exercises 9 and 10.

9. Write an algorithm that tells how to draw each stage of the snowflake curve from the preceding stage. See answer at the right.

10. Use your algorithm to draw the next stage of the snowflake curve. See top of page.

11. Open-ended Problem Write an algorithm that could be used to program a computer or a robot to get to school the same way you get to school each day. Answers may vary. Check students' work.

10.

7. Graph the equations obtained by replacing each inequality sign with an = sign; use a dotted line for ">" or "<", a solid line for "≥" or "≤"; substitute points in the regions into each inequality; shade the appropriate region.

8. Find the first location; find all paths leading from the first location to the second location; choose the most direct path.

9. Trisect each segment in the diagram; erase each middle segment; draw two segments that, together with the erased middle segment, make an equilateral triangle.

Practice Bank, ALGEBRA 1: EXPLORATIONS AND APPLICATIONS

79

---

b. Two examples are given. (1) Step 1: Subtract 9 from both sides of the equation. Step 2: Divide both sides of the equation by 12. (2) Step 1: Divide both sides of the equation by 12. Step 2: Subtract $\frac{3}{4}$ from both sides of the equation.

24.
y, 2, O, 2, x

25.
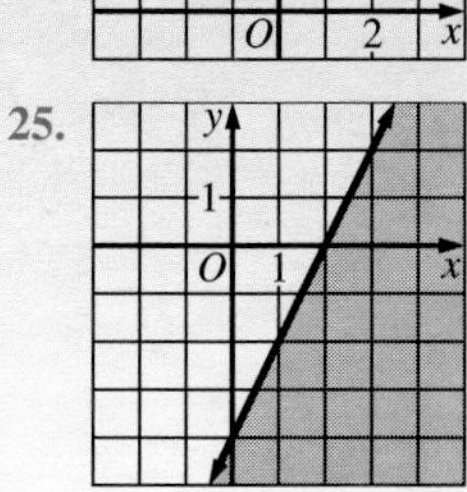

26.
y, 3, O, 2, x

27. $-\frac{3xw}{2yz}$

28. $\frac{1}{2y}$

29. $-\frac{3r^2}{5s(s-2)}$

30. $\frac{3(a-2)}{(a+1)(a-1)}$

31. $\frac{y(x+y)}{4}$

32. $\frac{x+2}{3x-1}$

Plan⇔Support

Objective

- Find the best path or tree for a graph.

Recommended Pacing

Core and Extended Courses
Section 12.2: 1 day

Block Schedule
Section 12.2: $\frac{1}{2}$ block
(with Section 12.3)

Resource Materials

Lesson Support
Lesson Plan 12.2
Warm-Up Transparency 12.2
Overhead Visuals:
Folder 12: Connecting Cities with Graphs
Practice Bank: Practice 80
Study Guide: Section 12.2

Technology
Internet:
http://www.hmco.com

Warm-Up Exercises

1. Name the vertices of triangle *ABC*. *A*, *B*, and *C*
2. Name the two sides of quadrilateral *WXYZ* that share vertex *X*. $\overline{WX}$ and $\overline{XY}$

For each of the figures listed below, state the number of distinct segments that can be drawn connecting all possible pairs of vertices, including the sides of the figure.

3. quadrilateral 6
4. pentagon 10
5. hexagon 15

SECTION

12.2 Finding Paths and Trees

Learn how to...
- find the best path or tree for a graph

So you can...
- minimize the distance you must travel on a trip
- find the way to connect a group of computers that takes the least amount of cable

How do you decide where to go first when you have several places to go? You might want to arrange your trip so that you travel the shortest distance. You can use discrete math to find a good *path.*

A group of students is planning a trip to visit five Anasazi sites. The students found information about the sites and located them on a map. They want to find a good route to use to visit all of the sites.

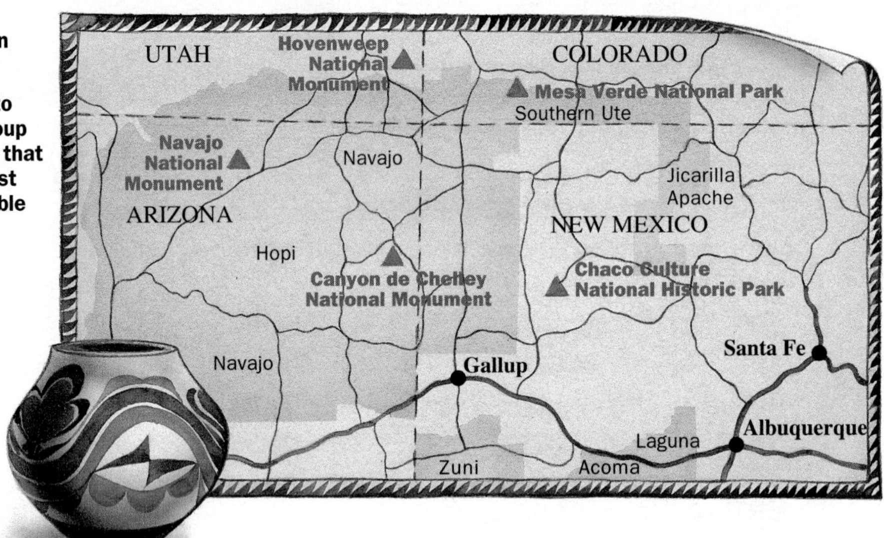

The students decided to simplify the map by modeling it with a **graph** consisting of points, called **vertices**, and lines, called **edges**. The graph below models the Anasazi sites and the distances between them.

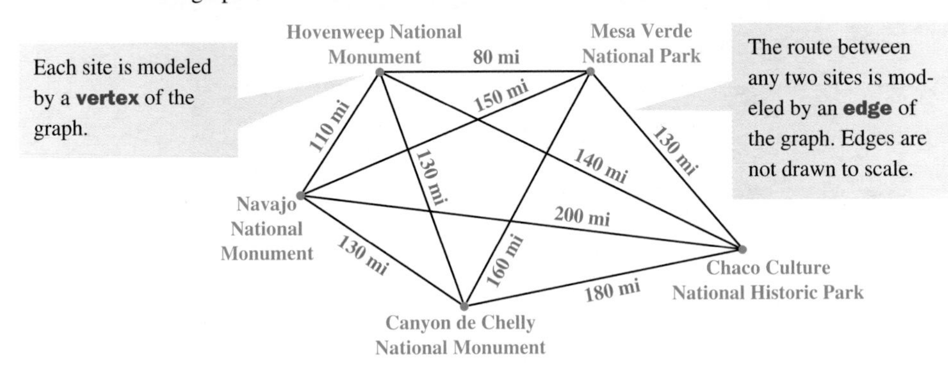

A **path** among the vertices on a graph is a collection of edges that connect one vertex to another without repeating any vertex or edge. Example 1 uses a *greedy algorithm* to find a short path. A **greedy algorithm** always chooses the best edge from the ones currently available.

EXAMPLE 1

Use a greedy algorithm to find a short path for the students to use to visit each of the sites in the graph on page 516. How far will the students travel if they use this path?

SOLUTION

Step 1 Choose a site from which to start.

Step 2 Go to the closest site that you have not visited.

Step 3 Continue choosing the next closest site until all of the sites have been visited.

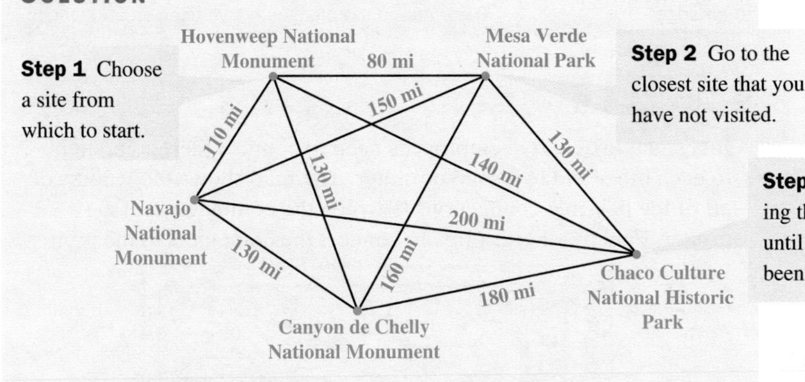

The algorithm suggests visiting the sites in this order: Hovenweep, Mesa Verde, Chaco Culture, Canyon de Chelly, Navajo.

Find the sum of the lengths of the edges used.

$$80 + 130 + 180 + 130 = 520$$

The students will travel 520 miles if they use this path.

BY THE WAY

The Anasazi civilization built many of their villages in sheltered recesses in the faces of cliffs. They also built dwellings, some containing as many as 1000 rooms, along canyons or mesa walls.

The only known way to find the *best* path is to try all the possible paths. A greedy algorithm will usually find a good path.

THINK AND COMMUNICATE

1. Use the graph in Example 1. If you start at Canyon de Chelly, there are two sites that are the same distance from your starting point. Use a greedy algorithm to determine which of the two choices is better.
2. Why is *greedy algorithm* a good name for the steps used to find a good path through a graph?
3. Why is the algorithm used in Example 1 a greedy algorithm?

ANSWERS Section 12.2

Think and Communicate

1. Hovenweep National Monument; From that site you can choose an edge 80 mi long next, while from the other possible choice, Navajo National Monument, you must choose among edges 130 mi long, 150 mi long, and 200 mi long.
2. A greedy algorithm always chooses the best edge from the ones currently available, just as a greedy person might take the best of everything.
3. Using this algorithm, you always choose the best (in this case, the shortest) edge available.

Teach⇔Interact

Section Note

Common Error

Some students may mistakenly identify the intersection of any two edges in a graph as a vertex of the graph. Point out that vertices of a graph are always labeled.

Additional Example 1

Use the graph in Example 1. Use a greedy algorithm to find a short path for the students to use to visit each of the sites in the graph if they begin at Chaco Culture National Historic Park. How far will the students travel if they use this path?

Step 1 Start at Chaco Culture National Historic Park.

Step 2 Go to the closest site that you have not visited.

Step 3 Continue choosing the next closest site until all of the sites have been visited.

The algorithm suggests visiting the sites in this order: Chaco Culture, Mesa Verde, Hovenweep, Navajo, Canyon de Chelly.

Find the sum of the lengths of the edges used.

$130 + 80 + 110 + 130 = 450$

The students will travel 450 miles if they use this path.

Section Note

Visual Thinking

After discussing Additional Example 1, you may wish to assign students to work in teams and invent a location with 8 to 10 sites, or to select a location with which they are familiar. Examples may include malls, parks, or carnivals. Ask students to create a map of their location and its sites. Then have them create a *greedy algorithm* for their location. Encourage them to present and discuss their solutions with the class. This activity involves the visual skills of *identification* and *self-expression.*

Think and Communicate

These questions focus students' attention on the meaning of *greedy algorithm* to ensure that they understand its relation to the shortest path on a graph. Students may need to be reminded that no vertex or edge is repeated in a path. For question 2, students could make a list of synonyms for *greedy.*

Teach⇔Interact

Section Note

Teaching Tip

Make sure students understand that although each vertex in a tree must be connected to an edge, not all edges of a graph need to be drawn in a tree. Point out to students that if they have correctly drawn a tree, there will be exactly one path between any two vertices and the number of edges will be one less than the number of vertices.

Additional Example 2

A vice president of a software supply company in Houston, Texas wants to have a teleconference with his four regional managers. The map shows the locations of the regional offices and the length of all the possible telephone connections between the regional offices and the vice president. Find the shortest tree to connect the regional offices by telephone to the vice president.

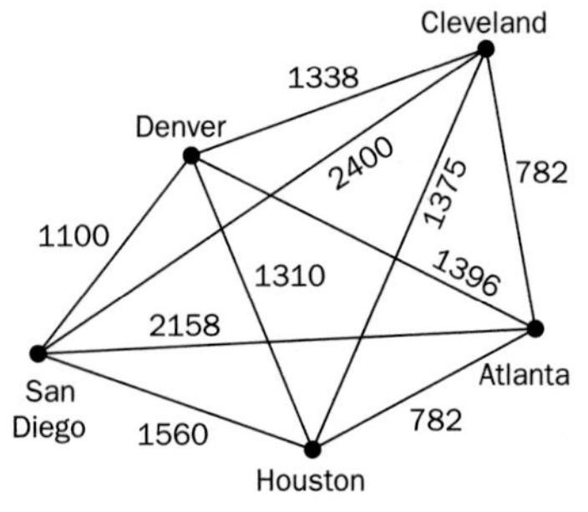

Model the map with a graph. Use a greedy algorithm.

Step 1 Choose the shortest edge.

Step 2 Choose the next shortest edge. Do not make any loops.

Step 3 Repeat Step 2 until all the vertices have been connected.

The length of the shortest tree is $782 + 782 + 1310 + 1100 = 3974$ mi.

Trees

A **tree** connects all the vertices in a graph with the smallest possible number of edges. You can use a greedy algorithm to find the best tree for a graph.

This is not a tree. One vertex is not connected to an edge.

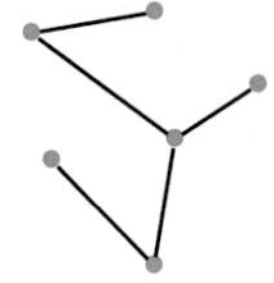

This is a tree. All of the vertices are connected. There are no extra edges.

This is not a tree. The extra edge between the vertices creates a loop.

EXAMPLE 2

In a small office, five employees need to connect their computers to each other and to a shared printer. The map shows the lengths of all of the possible connections between the computers and the printer. Find the shortest tree to connect the computers to the printer.

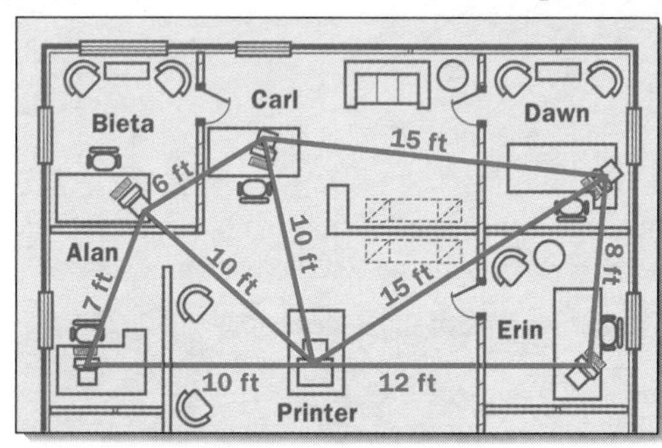

SOLUTION

Model the map with a graph. Use a greedy algorithm.

Step 1 Choose the shortest edge. If there is a tie, choose one.

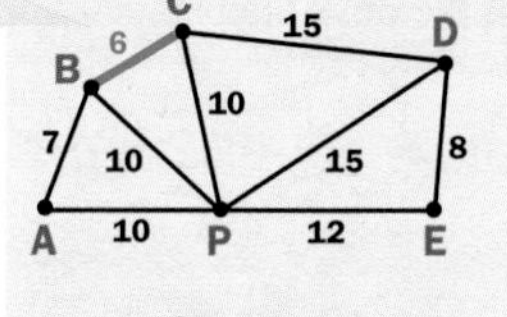

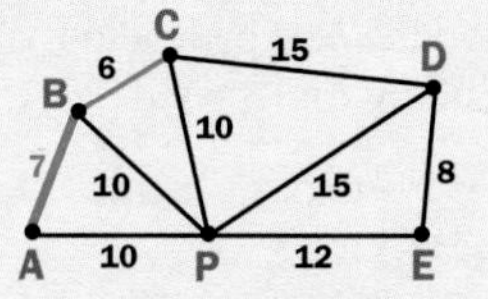

Step 2 Choose the next shortest edge. Do not make any loops.

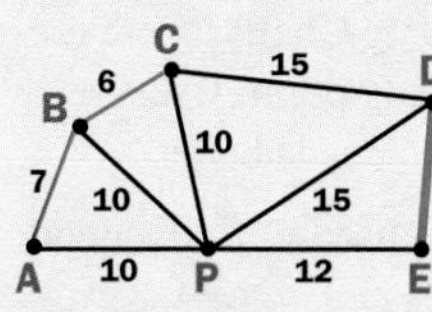

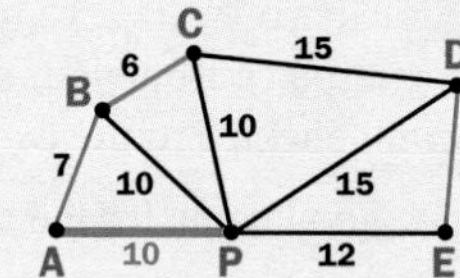

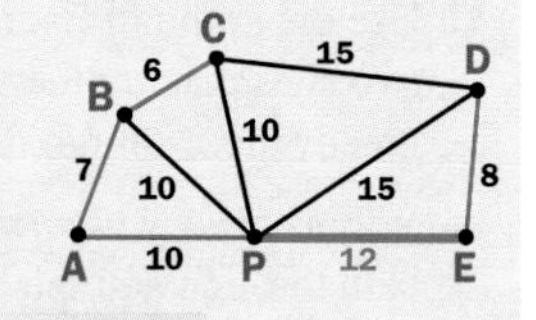

Step 3 Repeat Step 2 until all the vertices have been connected.

The length of the shortest tree is $6 + 7 + 8 + 10 + 12 = 43$ ft.

Think and Communicate

4. Example 1 is solved using a path, because the students want to travel from one point to another without repeating any vertex or edge. Example 2 is solved using a tree because what is needed is a collection of edges that connects all the vertices, without the need to travel from one to another. A path has a beginning and an end, while a tree does not. Also, a path has no breaks other than between the beginning and end. It has no "forks" as a tree might.
5. The algorithm in Example 1 begins by choosing a vertex, while the algorithm in Example 2 begins by choosing an edge. Both, however, continue by choosing the shortest edge at each repetition of Step 2. Also, in Example 1, each vertex is connected to a single other vertex, while in Example 2, a vertex may be connected to more than one other vertex.

Checking Key Concepts

1. a tree
2. both
3. The shortest path is 9 units long.
4. Paths may vary. Yes.

THINK AND COMMUNICATE

4. Compare Example 1 to Example 2. How is a path different from a tree? Explain why the graph at the right is both a path *and* a tree.

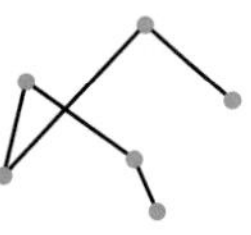

5. How is the algorithm in Example 1 different from the one given in Example 2? How are they alike?

CHECKING KEY CONCEPTS

Tell whether each graph is a *path*, a *tree*, *both*, or *neither*.

1.

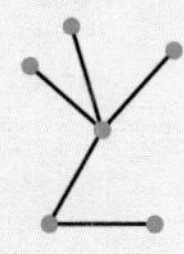

2.

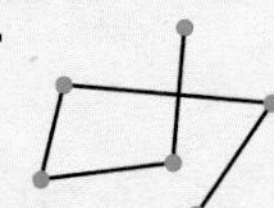

For Questions 3–5, copy the graph at the right.

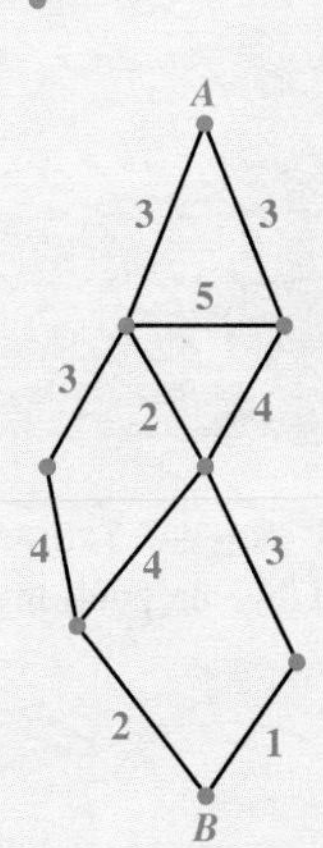

3. Use a greedy algorithm to find a short path from vertex A to vertex B. Your path does not need to visit all the vertices. How long is the path?
4. Start at vertex A. Find a path through all of the vertices. How long is it? Try starting at other vertices. Can you always find a path?
5. Find a tree for this graph. How long is it? Is it the shortest tree? How do you know?

12.2 Exercises and Applications

Tell whether each graph is a *path*, a *tree*, *both*, or *neither*. Explain why.

Extra Practice exercises on page 581

1.

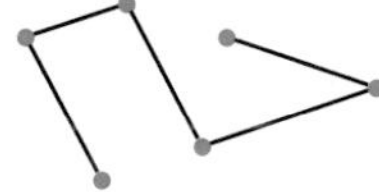

2.

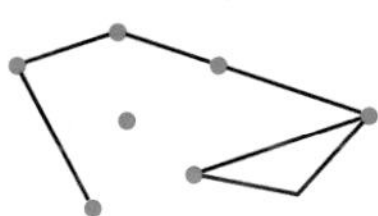

3.

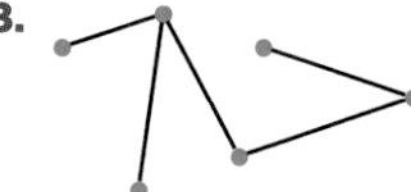

4.

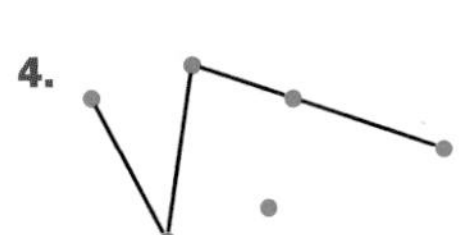

5.

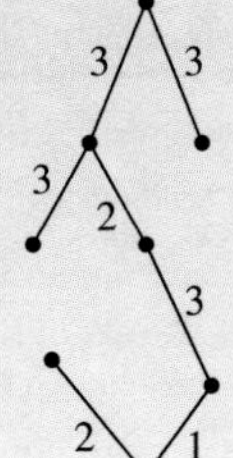

17 units; Yes; the seven edges are the seven shortest edges of the graph.

Exercises and Applications

1. both; The edges allow you to travel from one vertex to the other without repeating any vertex or edge, so the graph is a path. The vertices are connected with the smallest possible number of edges, so the graph is a tree.
2. neither; One of the vertices is not connected to an edge.
3. The graph is a tree because the vertices are connected with the smallest possible number of edges. The graph is not a path because you cannot travel the graph without repeating any edges or vertices.
4. neither; One of the vertices is not connected to an edge.

Teach⇔Interact

Think and Communicate

These questions help students to evaluate their understanding of the difference between a path and a tree. You may wish to have some students draw examples of each on the board to help those students with visual learning styles.

Checking Key Concepts

Reasoning
Question 3 reminds students that a path does not need to include all the vertices of a graph. Students should understand that this is an important distinction between a path and a tree because a tree must include all the vertices of a graph.

Closure Question

When is a graph both a path and a tree?
A graph is both a path and a tree when all the vertices in the graph are connected with the smallest possible number of edges.

Apply⇔Assess

Suggested Assignment

Core Course
Exs. 1–9, 11–17, 22–31

Extended Course
Exs. 1–31

Block Schedule
Day 77 Exs. 1–9, 11–17, 22–31

Exercise Notes

Teaching Tip
Exs. 1–4 Students who have difficulty with these exercises may benefit from drawing their own examples of trees and paths.

Apply⇔Assess

Exercise Notes

Career Connection

Ex. 5 The interiors of almost all public buildings are designed by an interior designer with college-level or art-school training and considerable experience. Generally, an architect plans the shell of a public building and consults with the interior designer while the building is still in the planning stage. The architect and the interior designer often work together to decide on the basic layout of the building, including the location of stairs, elevators, plumbing, and lighting in a way that meets the needs of their client. Once a client decides on the floor plan, the interior designer then helps choose colors, fabrics, furniture, and even art that will make the interior space of the building useful, pleasant, and comfortable.

Multicultural Note

Exs. 6, 7 In 1984, the country of Upper Volta changed its name to Burkina Faso, meaning "land of the honest people." Formerly a French colony, Burkina Faso gained its independence in 1960. Fifty percent of the country's population is made up of the Mossi people. Most of the people of Burkina Faso rely on farming or cattle rearing for their livelihood. Crops such as peanuts, sugar cane, sorghum, and sesame are grown in this land-locked country. Sheep, cattle, and goats are raised and exported. The capital, Ouagadougou (pronounced WAH guh Doo goo), is also the largest city.

Using Manipulatives

Exs. 6–9 Students may enjoy making scale models of the plans in these exercises. This activity will also give students a better visual image of these buildings. Students might enjoy displaying their models in a central area of the school such as the library.

Connection ARCHITECTURE

Architects may use an *access graph* to model the floor plan of a building. In an access graph, each vertex represents a room. An edge is drawn between two vertices if you can walk directly from one room to the other.

5. A client gives an architect a list of preferences for the design of a restaurant.

a. Copy and complete the access graph for the restaurant.

b. Is there more than one possible graph? Explain your reasoning.

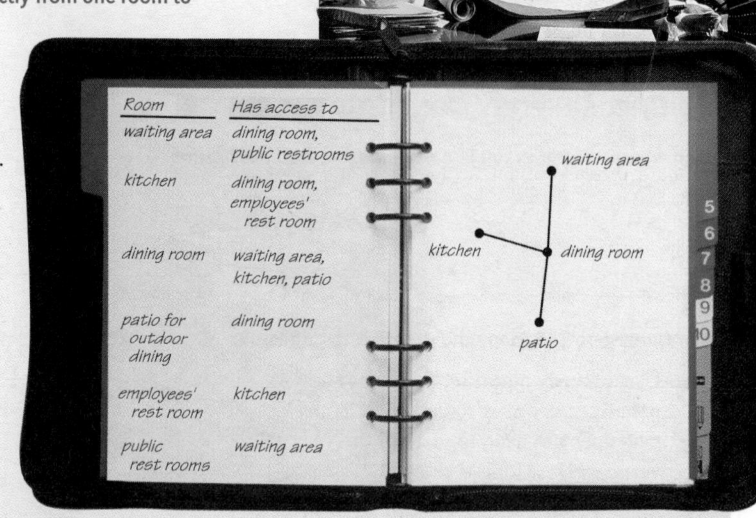

a traditional home in Segenega, Burkina Faso

For Exercises 6 and 7, draw an access graph to model each plan of a traditional dwelling unit in Burkina Faso in west Africa.

6. **7.**

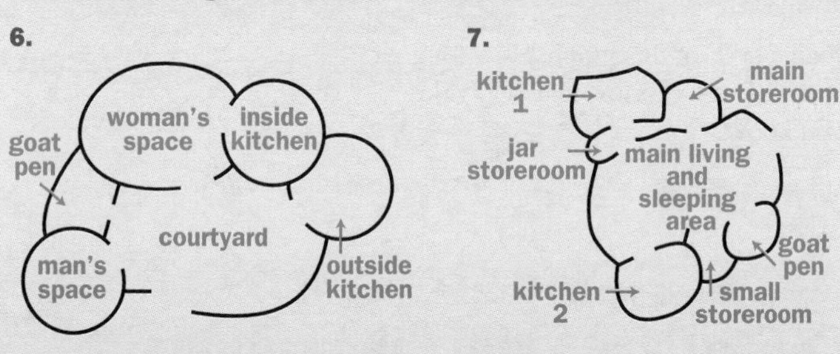

For Exercises 8 and 9, draw an access graph to model each house plan.

8.

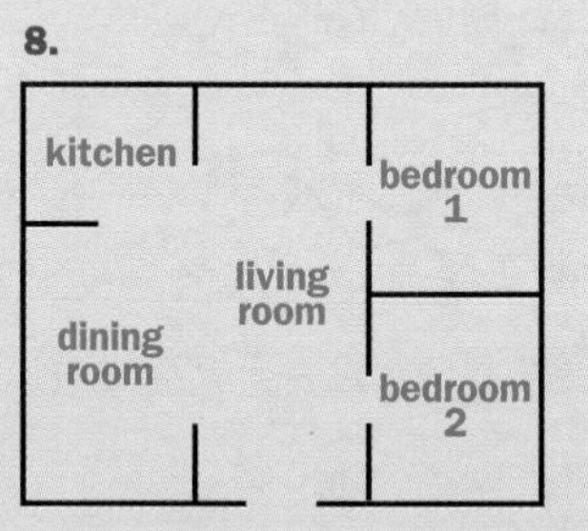

9.

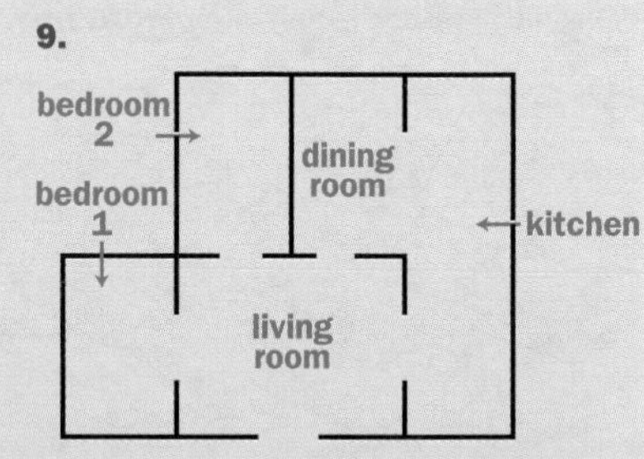

10. Writing How are the access graphs for Exercises 8 and 9 alike? Do you see any patterns? How are the house plans different?

5. a. Answers may vary. An example is given.

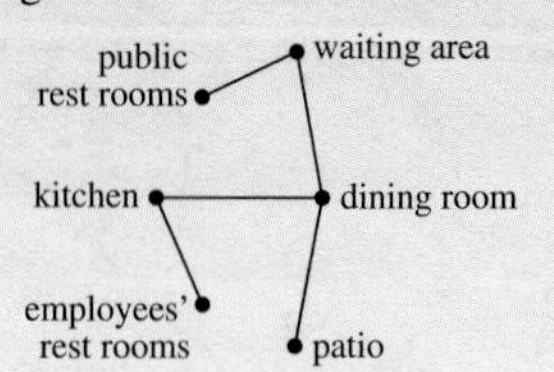

b. Other graphs are possible, but all would have 6 vertices and 5 edges and would be essentially the same except for the position of the vertices.

6–9. Drawings may vary. Examples are given.

6.

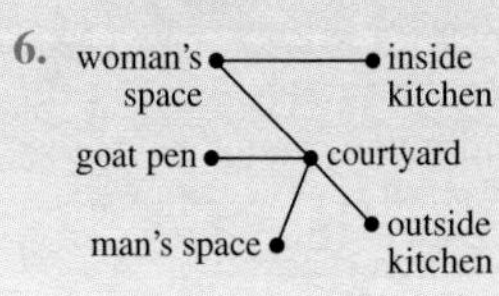

7.

kitchen 1
jar storeroom
main storeroom
main living and sleeping area
goat pen
kitchen 2
small storeroom

8, 9.

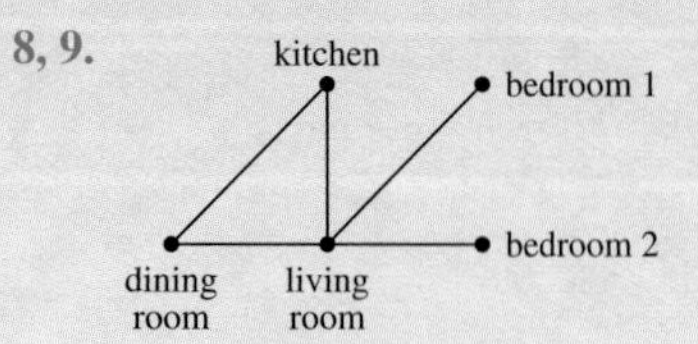

10. Answers may vary. As shown in the answer for Exs. 8 and 9 the access graphs may be identical. Both house plans have the same number of rooms and the same access between rooms. The houses are just laid out differently.

For Exercises 11–14, copy the graph below.

11. Find a short path from vertex A to vertex B. How long is this path?

12. Start at vertex A. Find a path through all of the points. How long is this path?

13. Find the shortest tree for the graph.

14. Write a greedy algorithm to find the *longest* tree for a graph. Use your algorithm to find the longest tree for the graph.

15. SAT/ACT Preview Evaluate the expression $7^2 + 2(7)(3) + 3^2$.

A. 42 **B.** 100 **C.** 1764 **D.** 1000 **E.** 84

BUSINESS For Exercises 16–18, use the map of the camping area.

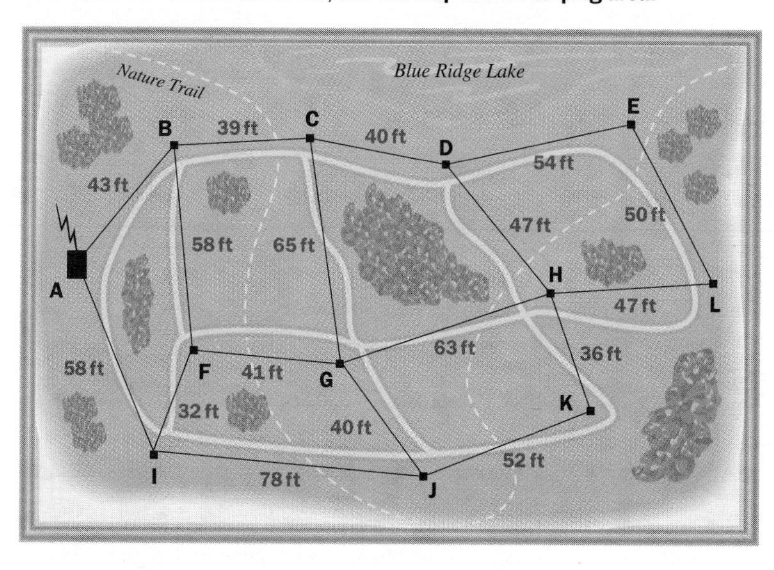

16. Dave Stapp owns a camping area. He is installing power lines to provide electricity to each site. He plans to locate the power source at A. The power lines will run along existing roads.

a. Does Dave need to find a *path* or a *tree* to connect the sites?

b. Use a greedy algorithm to decide where Dave should put his power lines. How many feet of power line does Dave need?

17. Writing The power source could also be located at site I. Would changing the location of the power source affect the amount of power line Dave needs? Explain your answer.

18. Challenge Dave checks each camp site every day. Find a short path through all of the sites. Does it matter where the path begins and ends?

Apply⇔Assess

Exercise Notes

Challenge

Ex. 12 Students should notice that a path through all the vertices does not include each edge. Ask students what must be true of the number of edges that flow from each vertex for a path to include each edge exactly once and return to the starting vertex. (The number of edges must be even.)

Problem Solving

Ex. 14 An important problem-solving technique is to base a solution to a problem on a similar problem. By asking students to consider adapting the algorithm for finding the shortest tree, students should have an intuitive sense of how to approach this exercise.

Journal Entry

Exs. 16, 17 Before students begin these exercises, make sure they can distinguish between a path and a tree. Students can use their journals to make a list of the features of each and then refer to the lists to verify their answer to Ex. 16(b). They can then continue writing their explanations for Ex. 17 in the same entry. Some students may wish to include a sketch of the power line in their journals to strengthen their argument.

Application

Ex. 18 This exercise provides an opportunity for students to connect the skills they have learned in this section to a real-world problem. Most students probably understand that an efficient use of time is an important consideration in business.

---

11. The edges along the bottom of the graph form a path 17 units long.

12. Paths may vary. An example is given. It is 36 units long.

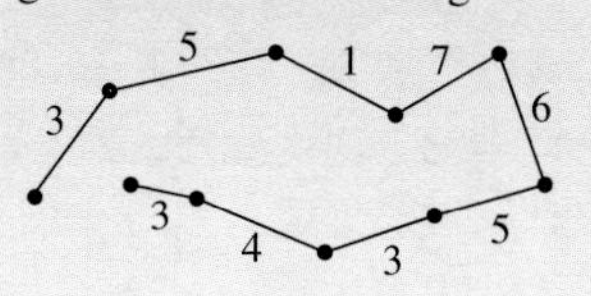

13.

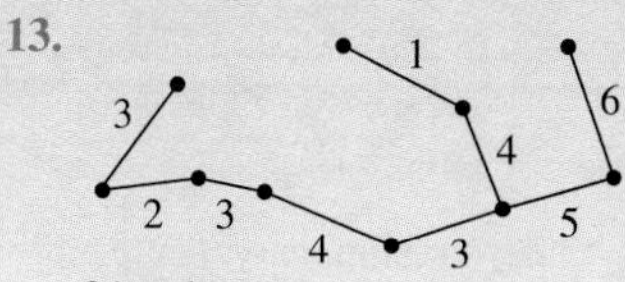

31 units

14. Step 1: Choose a vertex from which to start. Step 2: Choose the longest edge from that vertex to another. Step 3: Continue choosing the next longest edge until all the vertices have been reached.

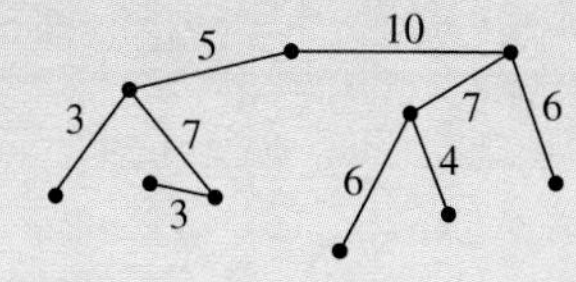

51 units

15. B

16. a. a tree

b. Answers may vary. An example is given. Dave should string lines from A to B to C to D to H to K to J to G to F to I, with additional lines from H to L to E. Dave needs 467 ft of power line.

17. No. The shortest tree is determined by starting with the shortest edge and then connecting all the vertices. It does not matter where the power source is located.

18. Answers may vary. An example is given. Dave could walk from A to I to F to G to J to K to H to L to E to D to C and, finally, to B. Yes; it does matter where he begins and ends.

Apply⇔Assess

Exercise Notes

Topic Spiraling: Preview

Exs. 19, 20 Students have learned that a greedy algorithm finds a good path, but the only way to find the best path is to try all possibilities. Students should have a sense that there are a number of paths to try in these exercises. In Section 12.4, students will learn to use the multiplication counting principle to determine the total number of possible choices.

Practice 80 for Section 12.2

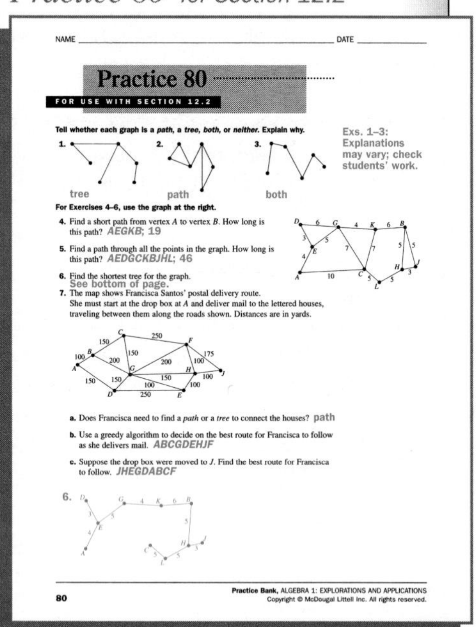

NAME ______________________ DATE __________

Practice 80

FOR USE WITH SECTION 12.2

Tell whether each graph is a *path*, a *tree*, *both*, or *neither*. Explain why.

Exs. 1–3: Explanations may vary; check students' work.

1. tree
2. path
3. both

For Exercises 4–6, use the graph at the right.

4. Find a short path from vertex *A* to vertex *B*. How long is this path? *AEGKB*; 19
5. Find a path through all the points in the graph. How long is this path? *AEDGCKBJHL*; 46
6. Find the shortest tree for the graph. See bottom of page.
7. The map shows Francisca Santos' postal delivery route. She must start at the drop box at *A* and deliver mail to the lettered houses, traveling between them along the roads shown. Distances are in yards.

   a. Does Francisca need to find a *path* or a *tree* to connect the houses? path

   b. Use a greedy algorithm to decide on the best route for Francisca to follow as she delivers mail. *ABCGDEHJF*

   c. Suppose the drop box were moved to *J*. Find the best route for Francisca to follow. *JHEGDABCF*

6.

80

Practice Bank, ALGEBRA 1: EXPLORATIONS AND APPLICATIONS

Connection BUSINESS

Mary Longacre is installing a new computer system at six regional offices of her company. Her travel agent gave her prices for routes between each of the cities she will visit.

Airfare Between Cities (dollars)

	Bos	Dal	Phnx	Atl	Chi	S F
Boston	0	654	671	461	304	734
Dallas		0	456	361	461	576
Phoenix			0	556	565	644
Atlanta				0	356	669
Chicago					0	712
San Francisco						0

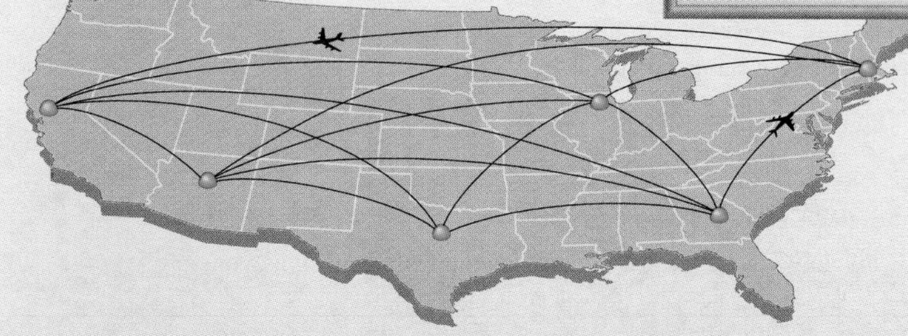

BY THE WAY

People in sales and industry are often concerned with finding the best route through a city or group of cities. Mathematicians often call this type of problem a *Traveling Salesman Problem.*

19. Draw a graph that models the cost of traveling between the cities.

20. a. Does Mary need to find a *path* or a *tree* for her trip?

b. Use a greedy algorithm to find a low-cost route through all six of the cities.

c. What is the total cost for the trip?

21. Open-ended Problem Do you think that the route with the lowest cost will always be the same as the route that takes the shortest time? Explain your reasoning.

ONGOING ASSESSMENT

22. Open-ended Problem Think of a situation where you have a choice of paths between several places. Model the situation with a graph. How do you decide which is the shortest route? Does a greedy algorithm give you the same route?

SPIRAL REVIEW

Solve each equation. *(Section 4.4)*

23. $11.5x - 3.8 = 12.3$ **24.** $\frac{3}{0.7}x + \frac{4}{1.4}x = 20$ **25.** $\frac{8}{11} - \frac{5}{3}y = y$

Solve each equation by factoring. *(Section 10.5)*

26. $x^2 + 2x - 35 = 0$ **27.** $2x^2 - 5x - 3 = 0$ **28.** $3a^2 + 5a = 2$

29. $4y^2 - 25 = 0$ **30.** $6a^2 - 11a + 3 = 0$ **31.** $4y^2 - 4y = 3$

19. Answers may vary. An example is given.

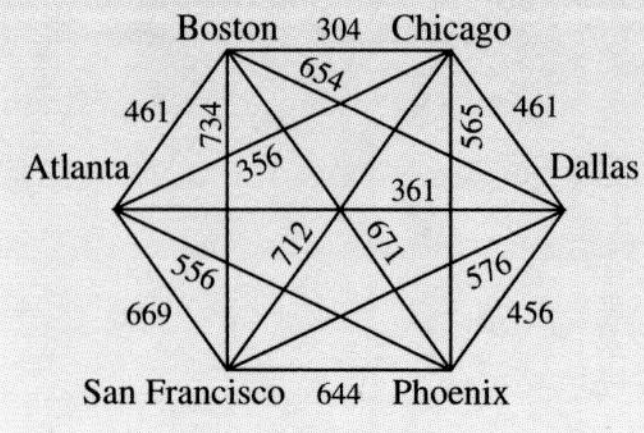

20. a. a path

b. Answers may vary. An example is given. Boston-Chicago-Atlanta-Dallas-Phoenix-San Francisco

c. \$2121

21. Answers may vary. An example is given. The costs and the distances are not necessarily related. The cost of a flight usually depends more on how popular the route is. Consider, for example, that the difference between the distances from Boston to Phoenix and Boston to Dallas is about 750 mi, while the cost difference is less than \$20.

22. Answers may vary.

23. 1.4

24. 2.8

25. $\frac{3}{11}$

26. −7, 5

27. $-\frac{1}{2}$, 3

28. $\frac{1}{3}$, −2

29. $\frac{5}{2}, -\frac{5}{2}$

30. $\frac{1}{3}, \frac{3}{2}$

31. $\frac{3}{2}, -\frac{1}{2}$

12.3 Voting and Fair Division

Learn how to...

- recognize fairness in elections and divisions

So you can...

- choose the best voting method for school elections

You are constantly faced with deciding how to be fair. You may need to decide how to divide something up among several people. You may be part of a group that has to make a choice.

Rocky Mount School is having a student council election. The dress code is the biggest topic in the campaign. Two of the candidates, Jack and Denzel, conducted a poll using the questionnaire shown below.

What kind of dress code do you think the school should have?
(Check one.)
1. No dress code
2. Minimal dress code
3. Moderate dress code
4. Detailed dress code
5. Strict dress code
6. School uniform

Jack and Denzel made a bar graph of the poll results. They asked the other candidates which type of school dress code they would support and added this information to their bar graph. Then they predicted the election results.

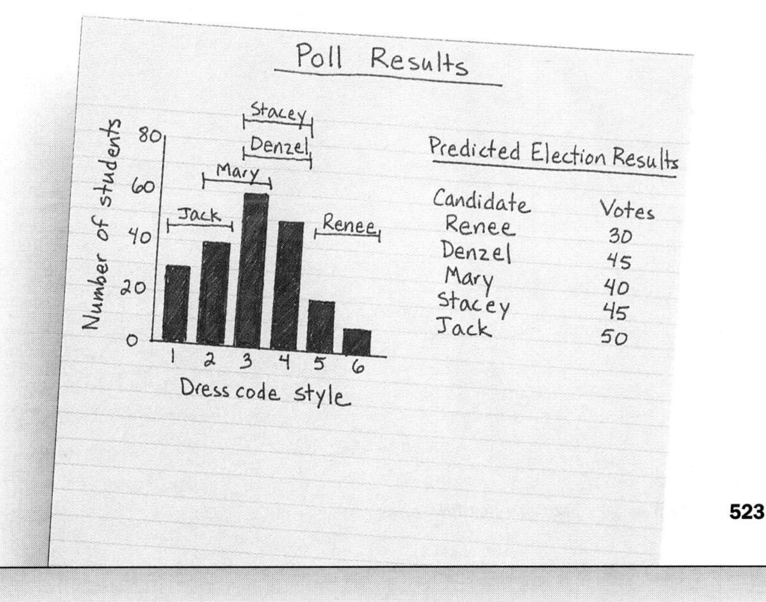

Plan⇔Support

Objective

- Recognize fairness in elections and divisions.

Recommended Pacing

Core and Extended Courses
Section 12.3: 1 day

Block Schedule
Section 12.1: $\frac{1}{2}$ block
(with Section 12.2)

Resource Materials

Lesson Support
Lesson Plan 12.3
Warm-Up Transparency 12.3
Practice Bank: Practice 81
Study Guide: Section 12.3
Assessment Book: Test 52

Technology
Graphing Calculator

Internet:
http://www.hmco.com

Warm-Up Exercises

1. If Sue received 31.2% of the votes in a school election and there were 538 votes, how many votes did she receive? Round your answer to the nearest whole number. 168 votes

Write an algorithm to accomplish each task.

2. Answer Ex. 1 using a calculator.
Step 1: Type 0.312 [×] 538 [=]
Step 2: Round the displayed number to the nearest whole number.

3. Sweep a floor.
Step 1: Get out the broom and dustpan.
Step 2: Sweep the dirt into a pile.
Step 3: Sweep the pile into the dustpan.
Step 4: Empty dustpan in the trash.
Step 5: Put away broom and dustpan.

Think and Communicate

For Questions 1–3, use the results of Jack and Denzel's poll on page 523.

1. Explain why Jack and Denzel predicted that Renee will receive 30 votes. Why do they predict that Mary will receive 40 votes?
2. Based on the bar graph, which candidates represent the views of the most students? Explain your reasoning.
3. Jack and Denzel predict that Jack will win the election. Why might he get more votes than candidates who represent the views of more students?

In a **one person-one vote** election, each person can vote for one candidate. In an **approval voting** election, each person can vote for any number of candidates. In the event of a tie, there is a tiebreaking election.

Example 1

Writing Use the results from Jack and Denzel's poll to predict the results of an approval voting election. Do you think the results of an approval voting election would better reflect the opinion of the voters?

Solution

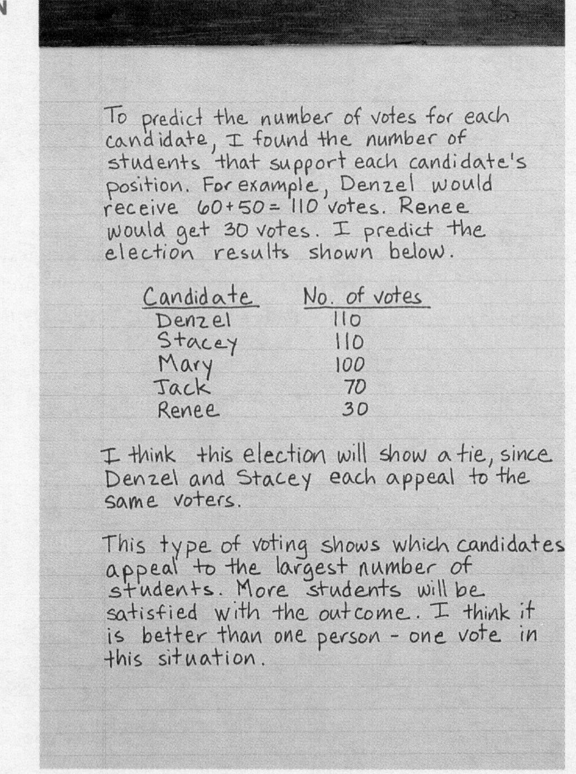

To predict the number of votes for each candidate, I found the number of students that support each candidate's position. For example, Denzel would receive 60+50= 110 votes. Renee would get 30 votes. I predict the election results shown below.

Candidate	No. of votes
Denzel	110
Stacey	110
Mary	100
Jack	70
Renee	30

I think this election will show a tie, since Denzel and Stacey each appeal to the same voters.

This type of voting shows which candidates appeal to the largest number of students. More students will be satisfied with the outcome. I think it is better than one person - one vote in this situation.

Teach⇔Interact

Example 1

The biggest topic in a local school committee election is where to make some necessary spending cuts. Two candidates, Jeannette Swanson and Eduardo Hernandez, conducted a poll using the questionnaire below.

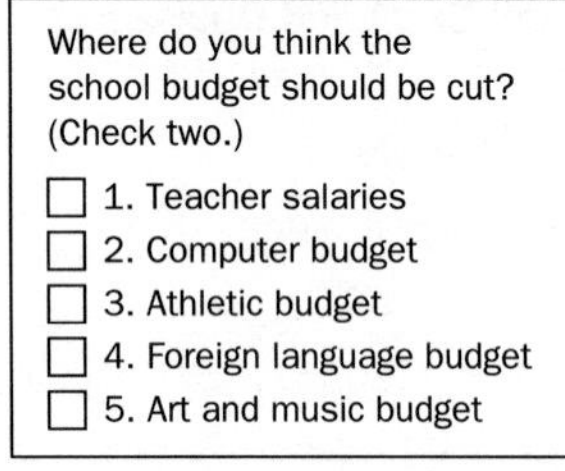

Where do you think the school budget should be cut? (Check two.)

- ☐ 1. Teacher salaries
- ☐ 2. Computer budget
- ☐ 3. Athletic budget
- ☐ 4. Foreign language budget
- ☐ 5. Art and music budget

They asked the other candidates which areas of the budget they thought should be cut and added this information to their bar graph.

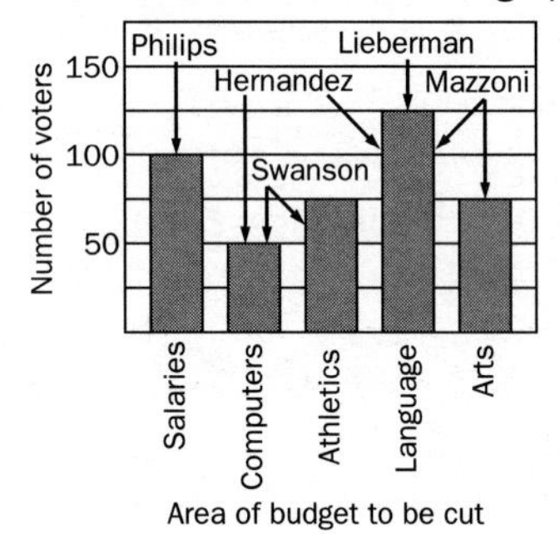

Use the results from the poll to predict the results of an approval voting election for the school board. Do you think the results of an approval voting election would reflect the opinion of voters better than a one person-one vote election?

To predict the number of votes for each candidate, find the number of voters that support each candidate's position. For example, Swanson would receive 50 + 75 = 125 votes. The predicted election results follow.

Candidate	No. of votes
Philips	100
Hernandez	175
Swanson	125
Lieberman	125
Mazzoni	200

According to the prediction, Mazzoni will win the election.

This type of election allows people to voice more than one opinion, which is important in a situation like this where more than one area of the budget can be cut.

ANSWERS Section 12.3

Think and Communicate

1. Renee is the only candidate that supports positions 5 and 6, so Jack and Denzel assume she will get the votes of all students supporting one of those positions.
2. Jack and Denzel assume that if candidates share a position, they will share the votes from those supporting those positions equally. Mary and Jack support position 2 and Stacey, Mary, and Denzel support position 3. So Denzel and Jack assume that Mary will get half the votes of those supporting position 2 and one-third of the votes of those supporting position 3.
3. The candidates who represent the views of more students must share their votes. For example, 60 students voted for position 3, but those 60 students may vote for any one of the three candidates supporting that position. On the other hand, the 30 students supporting position 1 may well all vote for Jack.
4. In both cases, it is predicted that Renee will get the votes of all the people who support positions 5 and 6, since she is the only candidate who sup-

Think and Communicate

4. Explain why Renee is predicted to receive 30 votes in both an approval voting election and a one person-one vote election.
5. The student who wrote the solution in Example 1 suggests that an approval voting election would result in a tie. What led the student to make this conclusion? In this situation, do you think that a tie is likely to occur in an approval voting election? Explain why or why not.
6. Discuss which candidate you think *should* win the election.

Voting is a way for a large group of people to decide among a few choices. In a **fair division** problem a group of objects needs to be divided among a group of people. Example 2 shows one type of fair division strategy.

Example 2

Amy decided not to take all of her CDs with her to college. She gave the rest of her collection to two of her friends to split. Pam and Maria like all of the CDs. They want to be friendly about the division. How can they split the collection fairly?

Solution

First, Pam should divide the CDs into two groups so that she will be equally happy with either group.

Second, Maria picks the group of CDs that she likes best.

Third, Pam gets the remaining group.

Pam and Maria used the **cut-and-choose** algorithm.

Step 1 The first person decides on a fair way to divide the item(s) into two parts.

Step 2 The second person chooses one of the two parts.

Step 3 The first person gets the remaining part.

By the Way

Dividing things fairly in a large group is much harder than dividing into two pieces. In 1994, mathematicians Steven J. Brams and Alan D. Taylor solved the problem of *envy-free division* for any number of people.

Think and Communicate

7. Why is this way of dividing something fair for the first person? Why is it fair for the second person?
8. In Example 2, Pam might choose to split the CDs so that one group has fewer CDs than the other. Explain why she might call this a fair division of the CDs.
9. Will the cut-and-choose algorithm always work for two people? Explain why or why not.

Teach⇔Interact

About Example 2

Mathematical Procedures
Make sure students understand that the cut-and-choose algorithm requires that one party divides the items into two groups and that the other party then chooses the group he or she wants. If this sequence is not followed, the division may not be considered fair.

Additional Example 2

Louise and Ron Swiftwater are moving into a smaller house and cannot take all their furniture with them. They are giving the rest of their furniture to their sons, Travis and Peter, to split. Travis and Peter both like all of the furniture and want to split it fairly. How can they do this?
First, Travis should divide the furniture into two groups so that he will be equally happy with either group. Second, Peter picks the group of furniture that he likes best. Third, Travis gets the remaining group.

Think and Communicate

For questions 7 and 8, students examine the assumptions implicit in the cut-and-choose algorithm. Students should understand that before this algorithm can be applied to a situation, both parties must desire a fair division and both parties must be interested in most or all of the items being divided. Students may wish to examine the meaning of what constitutes a fair division for both parties involved. For question 9, students can write scenarios involving a division in which the procedures of the cut-and-choose algorithm do not work.

ports those, and only those, positions.

5. There are two people, Denzel and Stacey, with the same number of votes. No; I think there are probably other issues that will result in either Denzel or Stacey getting more votes.
6. Answers may vary. An example is given. I think there are probably other issues in the campaign and the outcome may not be determined by the candidates' positions on a single issue. If the dress code is the only issue, I think Mary, Stacey, or Denzel should win the election since they represent the views of the majority of the student. To choose one of the three, I would have to know more about them.
7. The first person is the one who divides the group into two smaller groups, either of which would satisfy him or her, and so should be happy with either. The second person gets his or her choice of the two groups.
8. Answers may vary. An example is given. Some of the CDs may be considered to be more valuable than others.
9. It should work as long as the items in question can be divided evenly into two groups.

Teach⇔Interact

Checking Key Concepts

Teaching Tip
Since these questions require students to be able to distinguish between one person-one vote elections and an approval voting election, you may wish to have students review these two voting methods before answering the questions.

Closure Question

How can tie votes result in both a one person-one vote election and an approval voting election?
In a one person-one vote election, two or more candidates could receive the same number of votes from individual voters. In an approval voting election, two or more candidates could receive the same number of total votes from different groups of voters.

Apply⇔Assess

Suggested Assignment

Core Course
Exs. 1–3, 5, 6, 8–10, 14–21, AYP

Extended Course
Exs. 1–21, AYP

Block Schedule
Day 77 Exs. 1–3, 5, 6, 8–10, 14–21, AYP

Exercise Notes

Reasoning
Ex. 4 This exercise encourages students to examine the effects of weighting a choice. Students should see similarities between the results of assigning different weights to a vote and the results of assigning different weights to the edges of a path.

CHECKING KEY CONCEPTS

Jack and Denzel discovered that they missed a stack of questionnaires when they tallied the results shown on page 523. For Questions 1–3, combine the results from these questionnaires with those used on page 523.

Dress code	1	2	3	4	5	6
Number of votes	0	2	9	8	3	0

1. Predict the results of a one person-one vote election.
2. Predict the results of an approval voting election.
3. Suppose a new candidate who supports the same type of dress code as Jack enters the campaign. Predict the results of both a one person-one vote election and an approval voting election with this new candidate.

12.3 Exercises and Applications

Extra Practice exercises on page 582

For Exercises 1 and 2, tell how to apply the cut-and-choose algorithm in each situation.

1. There is one piece of pie left. You and your sister both want some.
2. You and a friend are being paid to weed a neighbor's flower garden.
3. **Writing** Explain the difference between a one person-one vote election and an approval voting election.
4. In major league baseball, the Cy Young Award is given to the best pitcher in each league. The winner is chosen by 28 members of the Baseball Writers Association of America. The electors rank their top three choices.
   a. Give each pitcher 5 points for each first-place vote, 3 for each second-place vote, and 1 for each third-place vote. Rank the pitchers based on total points.
   b. Suppose the number of points awarded is changed to 3 points for first-place, 2 for second-place, and 1 for third-place. Explain how this affects the order of the ranking of the pitchers.

BY THE WAY

The Cy Young Award winner is chosen using a Borda Count. In a Borda Count, voters rank their choices. The Borda Count is named after Jean-Charles de Borda (1733–1799), the French naval captain who invented it.

Cy Young Award

Pitcher, Team	1	2	3
John Burkett, San Francisco	—	3	—
Tom Glavine, Atlanta	4	7	8
Tommy Greene, Philadelphia	—	—	2
Bryan Harvey, Florida	—	—	1
Greg Maddux, Atlanta	22	2	3
Randy Myers, Chicago	—	—	1
Mark Portugal, Houston	—	—	2
Jose Rijo, Cincinnati	—	1	5
Bill Swift, San Francisco	2	15	6

Checking Key Concepts

1, 2.

	Number of votes	
Candidate	One person-one vote	Approval
Jack	54	76
Denzel	52	127
Stacey	52	127
Mary	45	113
Renee	33	33

3. Denzel and Stacey (tie); Jack, with Denzel and Stacey in a tie for second

Exercises and Applications

1. You should divide the piece into two smaller pieces, either of which would satisfy you. Your sister should choose one of the pieces. You get the remaining piece.
2. You should divide the garden into two sections, either of which you would be willing to weed. Your friend should choose one of the sections. You get the remaining section.
3. In a one person-one vote election, each voter can vote for only one candidate. In an approval voting election, each voter can vote for any number of candidates.

BUSINESS For Exercises 5–7, the diagram shows the layout of a small restaurant. The number of seats at each table is indicated.

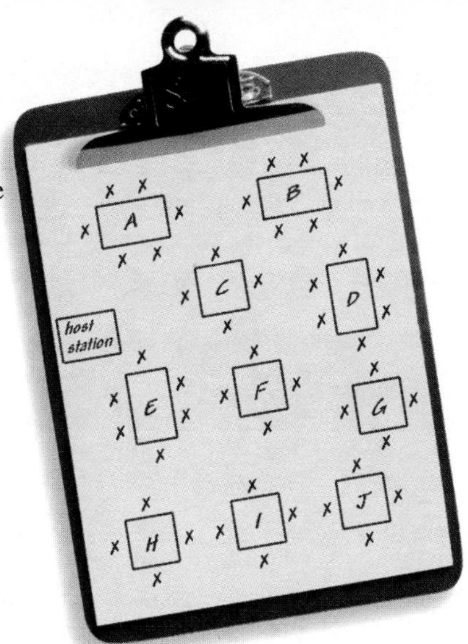

5. **Open-ended Problem** How do you think the manager should decide on a fair distribution of tables among the people waiting on the tables?
6. a. How would you divide up the tables fairly between two people?

   b. What is a fair division of the tables among three people?
7. **Challenge** Suppose twelve people want to sit together. How can the restaurant seat them? Does this change the division of the tables among three people? Justify the fairness of the division.

Connection VOLUNTEERING

Hospital

Food Bank

Animal Shelter

Local Cleanup

8. **Cooperative Learning** Consider each volunteer opportunity.

   a. Rank each type of volunteer work in order of your preference. Give your first choice 4 points, your second choice 3 points, your third choice 2 points, and your fourth choice 1 point.

   b. Predict the ranking of each type of volunteer work in order of your class's preference. Explain your reasoning.

   c. Collect the rankings of other students in your class. Add up the scores. Compare the class ranking to your prediction.
9. **Open-ended Problem** Three students decide to work as a group to do volunteer work. They each rank their interest in the four volunteer opportunities shown in the table. They rank their first choice with a 1. Create a schedule for them to divide their time fairly. Explain your reasoning.

Volunteer Opportunity	Tomoko	Samuel	Rita
Hospital	2	2	4
Food Bank	3	3	1
Animal Shelter	1	4	3
Local Cleanup	4	1	2

Each student can volunteer four hours per week.

Each volunteer opportunity needs students to help three hours per week.

Apply⇔Assess

Exercise Notes

Communication: Discussion
Exs. 5–7 Since some high school students wait on tables as a part-time job, these exercises pose a relevant situation. Through discussion of the exercises, students should see that the problem involves more than simply dividing up the number of tables equally. Factors such as the amount of seating at each table, the closeness of the tables to the kitchen, and whether the customers seat themselves or are assigned to tables by a hostess all affect the fair division of the tables.

Second-Language Learners
Ex. 8 Some second-language learners may not be familiar with the types of volunteer jobs young people do in one or more of these institutions. It will be helpful to them if class members briefly discuss their knowledge of and experiences with volunteer work.

Cooperative Learning
Ex. 9 Students can work in groups of three on this exercise. Each student should do the exercises and create a schedule for the volunteers. The group should then confer and create what they feel is the best possible schedule for the volunteers.

4. See answers in back of book.

5–7. Answers may vary. Examples are given.

5. The tables should be distributed so that equal numbers of people are being served by each server.

6. a. Assign each server two of the larger tables and three of the smaller tables.

   b. There are 48 people to be served, so each server should have 16. Two servers could be assigned two large tables and one small table. The third could be assigned four small tables.

7. Answers may vary. An example is given. Two large tables and one small table could be put together to make a table for 12 people. One server could wait on this table. Another server could wait on the two remaining large tables and one small table. The third server could weait on the four remaining small tables.

8. Answers may vary.

9. Answers may vary. An example is given.

Volunteer opportunity	Tomoko	Samuel	Rita
Hospital	1	1	1
Food Bank	1	—	2
Animal Shelter	2	1	—
Local Cleanup	—	2	1

I assigned each student two hours at his or her first choice and one hour at the second choice. I assigned the remaining three hours to meet the organizations' needs.

Apply⇔Assess

Exercise Notes

Reasoning

Ex. 11 Before students can make an informed judgment of which vote better represents the opinion of the voters, they should understand that presidential electors are elected through the political parties. Each voter must choose an entire slate of electoral candidates. Consequently, one party in each state wins all of that state's electoral votes. During the entire history of the United States, there have only been a few instances in which presidential electors did not vote for their party's candidate.

Research

Ex. 12 Students who choose Topic A may wish to do further research on the percent of the popular vote and the number of electoral college votes a presidential candidate received in each of the last five elections in their state. Students who chose Topic B may include in their research the way in which congressional districts in their state are divided up.

Assessment Note

Ex. 13 Once students have worked out their explanations in writing, they can share their suggestions for applying one method or the other to their own student council elections. After hearing other points of view, some students may wish to reconsider their own position on the issue.

Connection POLITICS

The President of the United States is not elected directly by the votes of citizens. The final vote is made by the Electoral College. In most states, the candidate who gets the most votes in the state receives all of the Electoral College votes for that state. For Exercises 10 and 11, use the table of election results.

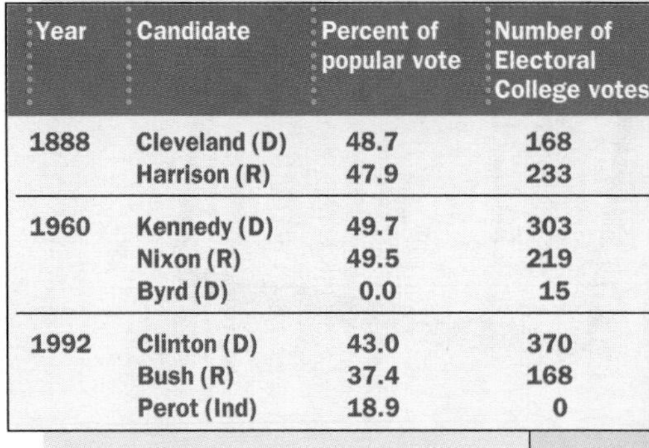

Year	Candidate	Percent of popular vote	Number of Electoral College votes
1888	Cleveland (D)	48.7	168
	Harrison (R)	47.9	233
1960	Kennedy (D)	49.7	303
	Nixon (R)	49.5	219
	Byrd (D)	0.0	15
1992	Clinton (D)	43.0	370
	Bush (R)	37.4	168
	Perot (Ind)	18.9	0

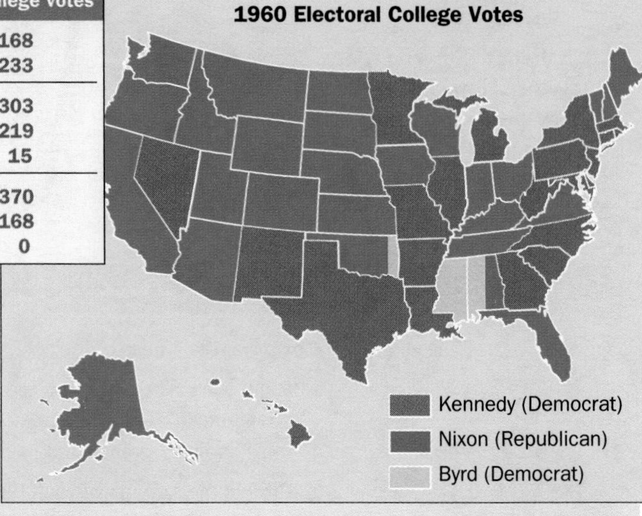

10. For each of the elections in the table, calculate the percent of the Electoral College votes each candidate received. In each election, how does the percent of the Electoral College vote compare to the percent of the popular vote?

11. Open-ended Problem Do you think the popular vote or the Electoral College vote better indicates the opinion of the voters? Why?

12. Research Find out more about one of the topics below.

Topic A
How is the number of Electoral College votes decided for each state?

Topic B
Dividing areas into voting districts is a complicated fair division problem. What factors are considered in making this division?

ONGOING ASSESSMENT

13. Writing Write a letter to your student council explaining the different voting methods you have learned about in this section. Which method do you think your school should use in student council elections? Explain why the method you suggest is best for your student council elections.

10. 1888: Cleveland, 41.9%, Harrison, 58.1%; 1960: Kennedy, 56.4%, Nixon, 40.8%, Byrd, 2.8%; 1992: Clinton, 68.8%, Bush, 31.2%, Perot, 0%; In the 1960 and 1992 elections, the candidate with the greater percentage of the popular vote had the greater percentage of the Electoral College votes. In the 1988 election, the candidate with the greater percentage of the popular vote had the smaller percentage of the Electoral College votes, and lost the election.

11. Answers may vary. Examples are given. I think the popular vote better indicates the opinion of the voters, because their opinions are voiced directly.

12. Topic A: The number of electoral votes each state has is equal to the total number of Senators and Representatives it has. The District of Columbia also has 3 electoral votes. Topic B: Summaries may vary. In the United States, the primary factor is population (except in the U.S. Senate, where districts are geographical). However, other factors are also considered.

13. Answers may vary. An example is given. Dear student council members, In one person-one vote elections, each person can vote for one candidate. In an approval voting election, each person can vote for any number of candidates. In the event of a tie, there is a tiebreaking election. I think we should use the one person-one vote method. With this method, I feel each voter has to make a better thought-out choice. Also, in an approval election, all voters may not necessarily cast the same num-

SPIRAL REVIEW

Find the theoretical probability that a randomly thrown dart that hits each target hits the shaded area. *(Section 5.6)*

14. 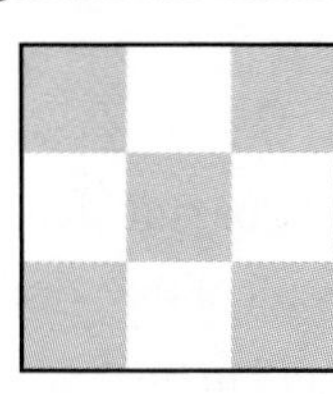

15. 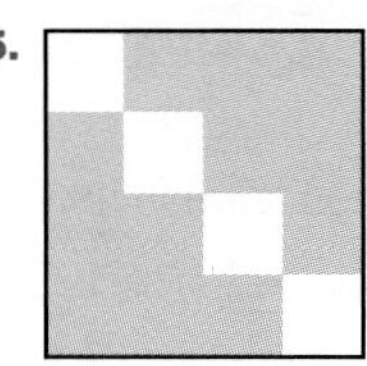

Write each number in scientific notation. *(Section 9.5)*

16. 10,025 **17.** 0.0034 **18.** 54,000,000

19. 0.0010078 **20.** 9.6 million **21.** 3 ten-thousandths

ASSESS YOUR PROGRESS

VOCABULARY

algorithm (p. 509)
graph (p. 516)
vertex (p. 516)
edge (p. 516)
path (p. 517)
greedy algorithm (p. 517)
tree (p. 518)
one person-one vote (p. 524)
approval voting (p. 524)
fair division (p. 525)
cut-and-choose (p. 525)

1. Write an algorithm for adding two trinomials. *(Section 12.1)*

VOLUNTEERING **For Exercises 2 and 3, use the graph below. The graph models the locations where donations for a food bank need to be picked up today.**

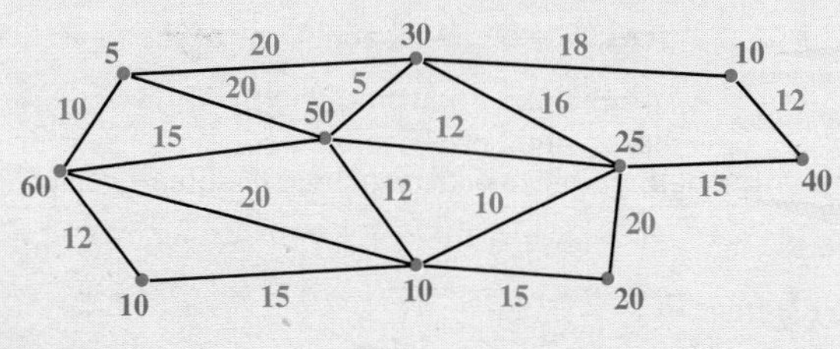

weight in pounds of the donation at this location

time in minutes to drive between two locations

2. Open-ended Problem Find a short path for one person to use to visit each of the locations. *(Section 12.2)*

3. Suppose two people will each pick up some of the donations. Copy the graph and suggest a fair division of the locations. Explain why your division is fair. *(Section 12.3)*

4. Journal Explain what is *discrete* about each of the topics covered in Sections 12.1–12.3.

Apply⇔Assess

Exercises Notes

Using Technology

Exs. 16–21 Students can use a TI-81 or TI-82 graphing calculator to check their answers to these exercises. They should set their calculators to scientific notation mode by pressing MODE SCI ENTER. Then type in a number as it is given in an exercise and press ENTER. For example, in Ex. 19, students can type 0.0010078 ENTER and the calculator will display 1.0078E–3, which confirms that the answer is 1.0078×10^{-3}.

Progress Check 12.1–12.3

See page 550.

Practice 81 for Section 12.3

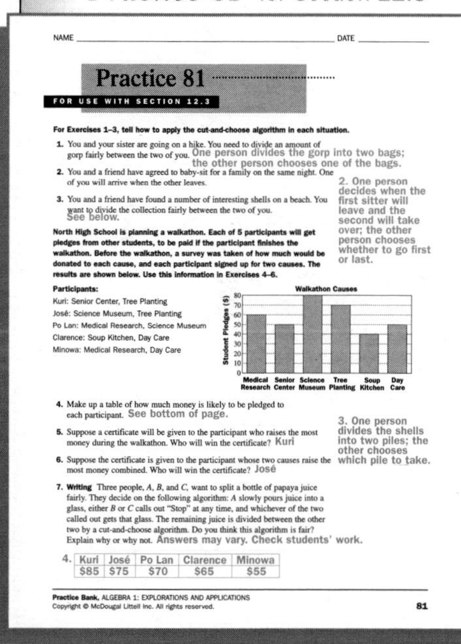

NAME ________________ DATE ________

Practice 81

FOR USE WITH SECTION 12.3

For Exercises 1–3, tell how to apply the cut-and-choose algorithm in each situation.

1. You and your sister are going on a hike. You need to divide an amount of gorp fairly between the two of you. One person divides the gorp into two bags; the other person chooses one of the bags.
2. You and a friend have agreed to baby-sit for a family on the same night. One of you will arrive when the other leaves. 2. One person decides when the first sitter will leave and the second will take over; the other person chooses whether to go first or last.
3. You and a friend have found a number of interesting shells on a beach. You want to divide the collection fairly between the two of you. See below.

North High School is planning a walkathon. Each of 5 participants will get pledges from other students, to be paid if the participant finishes the walkathon. Before the walkathon, a survey was taken of how much would be donated to each cause, and each participant signed up for two causes. The results are shown below. Use this information in Exercises 4–6.

Participants:
Kuri: Senior Center, Tree Planting
José: Science Museum, Tree Planting
Po Lan: Medical Research, Science Museum
Clarence: Soup Kitchen, Day Care
Minowa: Medical Research, Day Care

Walkathon Causes
Student Pledges ($): 0, 10, 20, 30, 40, 50, 60, 70, 80
Medical Research, Senior Center, Science Museum, Tree Planting, Soup Kitchen, Day Care

4. Make up a table of how much money is likely to be pledged to each participant. See bottom of page.
5. Suppose a certificate will be given to the participant who raises the most money during the walkathon. Who will win the certificate? Kuri
6. Suppose the certificate is given to the participant whose two causes raise the most money combined. Who will win the certificate? José
7. **Writing** Three people, *A*, *B*, and *C*, want to split a bottle of papaya juice fairly. They decide on the following algorithm: *A* slowly pours juice into a glass, either *B* or *C* calls out "Stop" at any time, and whichever of the two called out gets that glass. The remaining juice is divided between the other two by a cut-and-choose algorithm. Do you think this algorithm is fair? Explain why or why not. Answers may vary. Check students' work.

3. One person divides the shells into two piles; the other chooses which pile to take.

4.

Kuri	José	Po Lan	Clarence	Minowa
$85	$75	$70	$65	$55

Practice Bank, ALGEBRA 1: EXPLORATIONS AND APPLICATIONS

81

---

ber of votes and I do not think that is fair. Sincerely, Estelle Lambert

14. $\frac{5}{9}$ **15.** $\frac{3}{4}$

16. 1.0025×10^4

17. 3.4×10^{-3}

18. 5.4×10^7

19. 1.0078×10^{-3}

20. 9.6×10^6

21. 3×10^{-4}

Assess Your Progress

1–3. Answers may vary. Examples are given.

1. Step 1: If necessary, write both trinomials with the square term first, the linear term next, and the constant term last. Step 2: Combine similar terms.

2.

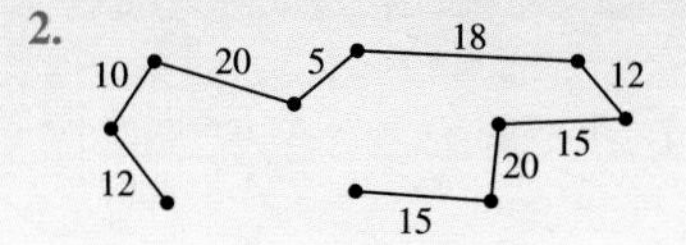

3. Check students' graphs. The first person should pick up donations weighing 5 lb, 10 lb, 25 lb, 30 lb, and 60 lb. The second person should pick up donations weighing 10 lb, 10 lb, 20 lb, 40 lb, and 50 lb. I think this is fair because each person has the same number of stops and nearly the same weight to pick up, with the weights distributed fairly evenly.

4. The number of steps in an algorithm can be counted. Greedy algorithms can be used to determine the shortest path using a number of vertices that can be counted. The number of vertices and edges of paths and trees can be counted. Voting and fair division are both based on counting choices.

Plan⇔Support

Objective

- Count all the ways groups of objects can be arranged.

Recommended Pacing

Core and Extended Courses
Section 12.4: 1 day

Block Schedule
Section 12.4: $\frac{1}{2}$ block
(with Section 12.5)

Resource Materials

Lesson Support
Lesson Plan 12.4
Warm-Up Transparency 12.4
Practice Bank: Practice 82
Study Guide: Section 12.4

Technology
Scientific Calculator
Graphing Calculator
Internet:
http://www.hmco.com

Warm-Up Exercises

1. Maria is planning a trip from Atlanta to Boston and Chicago. How many different paths can she take? 2
2. How many ways can you arrange the letters P and Q? 2
3. Write out all the different arrangements of the letters P, E, and T. How many are there? PET, PTE, TEP, TPE, EPT, ETP; 6

Evaluate when $n = 5$.

4. $n(n-1)(n-2)(n-3)$ 120
5. $\frac{n(n-1)(n-2)(n-3)}{(n-2)(n-3)}$ 20

SECTION

12.4 Permutations

Learn how to...

- count all the ways groups of objects can be arranged

So you can...

- find all the possible outfits of clothing; or
- all possible computer passwords; or
- all possible Braille characters

In Section 12.2, you learned that the only known way to be sure that you have found the best path between two vertices is to try all of the possibilities. How do you know that you have found them all? In this section, you will learn a way of counting all the possibilities.

A fashion designer is planning a new collection of clothing that will look good together in any combination. The collection includes two pairs of pants, four shirts, and three sweaters.

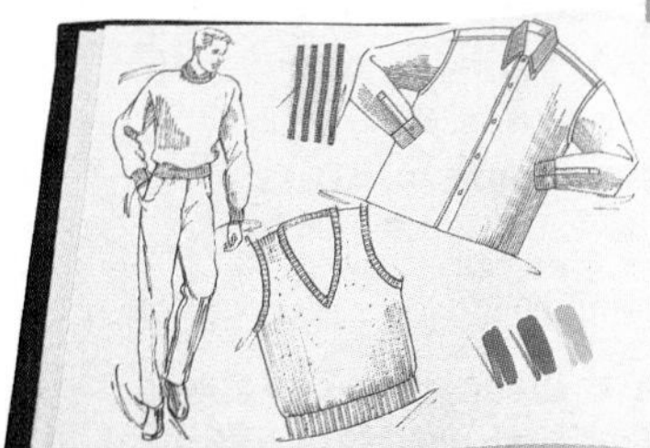

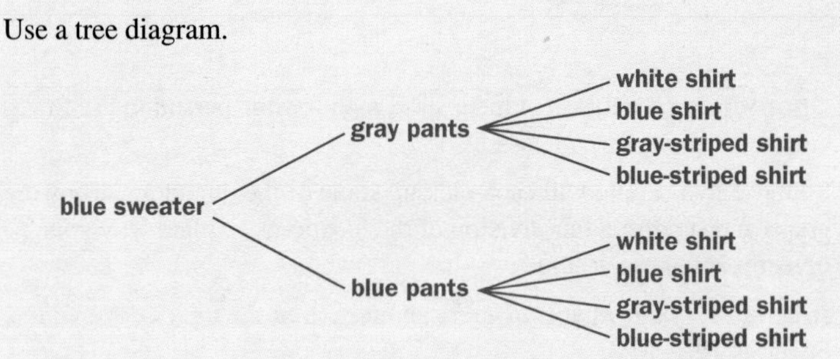

EXAMPLE 1 Application: Fashion Design

How many different three-piece outfits include the blue sweater?

SOLUTION

Use a tree diagram.

blue sweater	gray pants	white shirt
		blue shirt
		gray-striped shirt
		blue-striped shirt
	blue pants	white shirt
		blue shirt
		gray-striped shirt
		blue-striped shirt

There are eight different outfits that include the blue sweater.

THINK AND COMMUNICATE

For Questions 1–3, use the information about the clothing collection on page 530.

1. How would the tree diagram in Example 1 look if you chose a shirt first and then a pair of pants to go with the blue sweater?
2. How many different three-piece outfits include the gray sweater?
3. How many different three-piece outfits are possible if you use all the items planned for the collection?

A tree diagram models all the possible ways to choose items so that you have one item from each of several categories. You can use a tree diagram to count the total number of choices, as you did in Example 1. Another way to find the number of possible choices is to use the *multiplication counting principle.*

Multiplication Counting Principle

When you have m choices in one category and n choices in a second category, you have $m \cdot n$ possible choices.

When you have r choices in a third category, you have a total of $m \cdot n \cdot r$ possible choices.

EXAMPLE 2 Application: Fashion Design

The fashion designer in Example 1 decides to add one pair of pants and two shirts to the collection in colors that coordinate well with those already included. There are now 3 sweaters, 3 pairs of pants, and 6 shirts. What is the total number of different three-piece outfits in the collection?

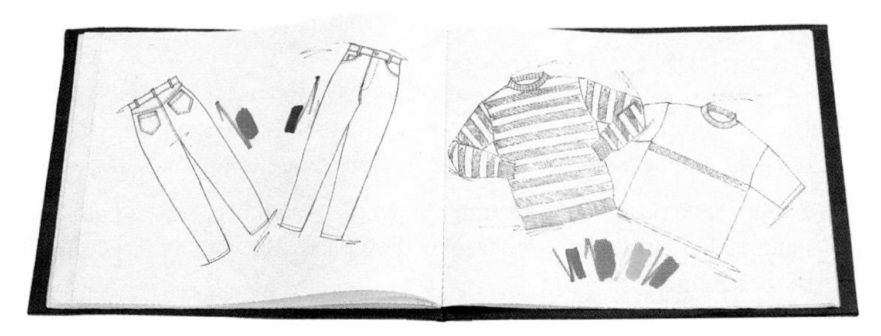

SOLUTION

Use the multiplication principle.

Sweaters	×	Pants	×	Shirts	=	Number of outfits
3	•	3	•	6	=	54

There are 54 different three-piece outfits.

Teach⇔Interact

Additional Example 1

Use the information about the clothing collection. How many different three-piece outfits include the white shirt?

Use a tree diagram.

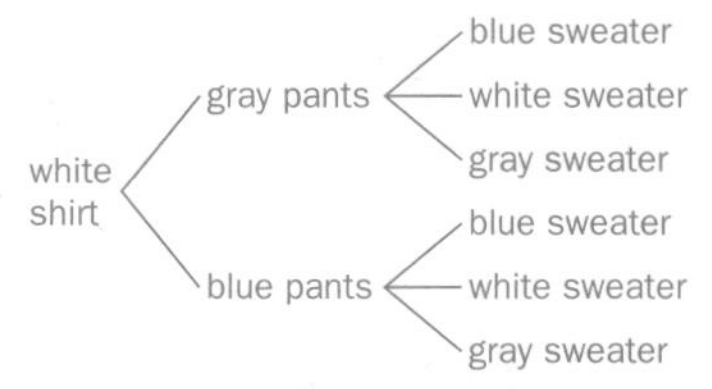

There are six different outfits that include the white shirt.

Think and Communicate

By drawing several tree diagrams that represent the different orders in which the outfit can be chosen, students should see that changing the design of the tree does not affect the total number of choices represented by the tree. Students can also see that the order of selection does not matter. For example, choosing a white shirt, blue pants, and a blue sweater gives the same outfit as choosing a blue sweater, a white shirt, and blue pants.

Additional Example 2

Mary Ann is buying a new pickup truck. There are two models. Her choices of color are forest green, metallic blue, or white. She can buy either a two-wheel drive truck or a four-wheel drive truck. Considering these three choices, how many different pickup trucks could Mary Ann buy?

Use the multiplication principle.

pickup models	×	colors	×	type of drive	=	number of trucks
2	•	3	•	2	=	12

She could buy 12 different pickup trucks.

ANSWERS Section 12.4

Think and Communicate

1. There would be four choices for the second branch, each of which would in turn have two branches.
2. 8 outfits
3. 24 outfits

Teach⇔Interact

Additional Example 3

Kyle bought 4 rose bushes of different colors to plant in a row in his rose bed. How many different arrangements of the 4 new rose bushes can he make?

1st bush	2nd bush	3rd bush	4th bush
4 choices	3 choices	2 choices	1 choice

$4 \cdot 3 \cdot 2 \cdot 1 = 24$

Kyle can make 24 different arrangements of the rose bushes.

Section Note

Visual Thinking
Ask students to create a tree diagram about a real-world situation with which they are familiar. Examples might include colors of paint for a house, trim, and shutters; their own outfit combinations; or password names. Have students present their diagrams to the class and discuss them in terms of *factorials* and *permutations.* This activity involves the visual skills of *correlation* and *communication.*

Checking Key Concepts

Topic Spiraling: Review
For question 3, students have to relate a problem involving paths to the permutation of n items. They should realize that the formula for $n!$ can be used because each of the sites is connected by an edge to each of the other sites. Thus, a visitor has the choice of beginning at any of the 5 sites. The visitor then has 4 choices for his next site, 3 choices for the third site, and so on until only one site remains to be visited.

Closure Question

How is the formula for finding the permutations of n objects related to the multiplication counting principle? The formula for the permutation of n items allows you to find the number of possible choices for any number of categories when you have m items in the first category, $m - 1$ items in a second category, $m - 2$ items in a third category, and so on until you have 1 item in the final category.

EXAMPLE 3

Kevin decides to rearrange the letters of his name to use as a computer password. How many different arrangements can he use?

SOLUTION

1st letter	2nd letter	3rd letter	4th letter	5th letter
5 choices	4 choices	3 choices	2 choices	1 choice

$5 \cdot 4 \cdot 3 \cdot 2 \cdot 1 = 120$

There are 120 different arrangements of the letters in *Kevin*. One of them is *Kevin*, so he can use 119 of them as passwords.

In Example 3, the product $5 \cdot 4 \cdot 3 \cdot 2 \cdot 1$ can be written 5!, which is read as *five factorial*. The symbol ! indicates a **factorial**. The number $n!$ is the product of all the integers from 1 to n. By definition, $0! = 1$.

In Example 3, each password is a *permutation* of the letters in Kevin's name. A **permutation** is an arrangement of a group of items in a definite order. The order of the items is important. In Example 3, *ekniv* is different from *vnkei*.

For more information on using a calculator to find factorials, see the *Technology Handbook*, p. 602.

Permutations

A group of n objects has $n!$ permutations.

$n! = n(n-1)(n-2)\ldots(3)(2)(1)$

A group of 5 objects has 5! permutations.

$5! = 5 \cdot 4 \cdot 3 \cdot 2 \cdot 1$

CHECKING KEY CONCEPTS

1. Use the multiplication counting principle to solve Example 1. How does the multiplication counting principle relate to the tree diagram?
2. Suppose the designer in Example 2 decides to add two more pairs of pants to her collection for a total of five pairs. How many three-piece outfits are now possible?
3. How many different paths are possible through all five Anasazi sites in Example 1 on page 517?
4. a. Evaluate 4!, 7!, and 10!. How can you use 7! to help find 10!?
   b. Try finding other factorials with your calculator. How large a factorial can your calculator evaluate?
5. How many different passwords can be made by rearranging the letters in *Claire*?

WATCH OUT!
Your calculator cannot evaluate large factorials. If you get an error message when you try to evaluate a large factorial, write your answer as a factorial.

Checking Key Concepts

1. $1 \cdot 2 \cdot 4 = 8$; The product is the product of the number of branches at each stage of the tree diagram.
2. 90 outfits 3. 5! or 120 paths
4. a. 24; 5040; 3,628,800; $10! = 10 \cdot 9 \cdot 8 \cdot 7! = 720 \cdot 7!$
   b. Answers may vary. The largest factorial many calculators will evaluate is 69!.
5. $6! = 720$ passwords, if you include Claire

Exercises and Applications

1.

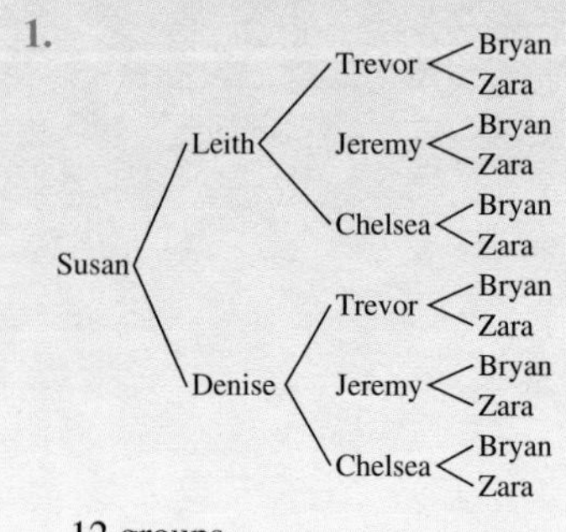

12 groups

2. 48 groups

3. a.

n	$n!$
1	1
2	2
3	6
4	24
5	120
6	720
7	5040
8	40,320
9	362,880
10	3,628,800
11	39,916,800
12	479,001,600

12.4 Exercises and Applications

Extra Practice exercises on page 582

The ninth grade at Central High School is about to elect class officers for next year. For Exercises 1 and 2, use the list of candidates.

President	Vice President	Secretary	Treasurer
Susan	Leith	Trevor	Bryan
Naomi	Denise	Jeremy	Zara
Anna		Chelsea	
Miguel			

1. Make a tree diagram to show all the possible groups of officers if Susan is elected president. How many different groups are possible?
2. How many possible groups of officers are there in all, including all the presidential candidates?
3. a. **Writing** Make a table giving the value of $n!$ for $n = 1$ to $n = 12$.
   b. Do you think the last digit of every factorial larger than 4! is 0? Explain your reasoning.
4. **Challenge** Show why $6! \times 7! = 10!$.

Connection BRAILLE

Each character in Braille is a grouping of raised dots. The dots are arranged in a 3-by-2 rectangle. In the photo, the large black circles represent raised dots. The small black circles represent positions in the rectangle that do not contain raised dots. Characters are used to represent letters, words, and punctuation marks.

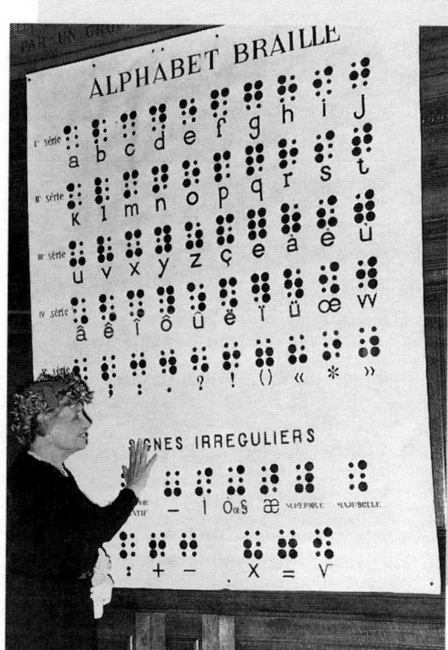

5. a. How many different ways can the raised and unraised dots be grouped to make a Braille character? Explain how you know.
   b. One of these groupings is not used as a Braille character. Why?
6. Braille uses some pairs of characters for numbers and commonly used words. How many different pairs of Braille characters are possible? Explain how you know.

BY THE WAY

Louis Braille was blinded in an accident when he was 3. He invented a method of reading with raised dots in 1824, when he was 15 years old.

The picture at the left shows Helen Keller during a presentation she gave in Paris, France.

Apply⇔Assess

Suggested Assignment

Core Course
Exs. 1–3, 7, 8, 11–25

Extended Course
Exs. 1–25

Block Schedule
Day 78 Exs. 1–4, 7, 8, 11–25

Exercise Notes

Problem Solving

Exs. 1, 2 These exercises should confirm students' intuition that although tree diagrams provide a graphical method for solving problems like these, they can become quite cumbersome as the number of categories and choices increases. As this occurs, the multiplication counting principle becomes a better method of solution.

Multicultural Note

Exs. 5, 6 Braille has been widely used throughout the world ever since it was developed in France by Louis Braille. A cell of six dots used in different arrangements is the basis of Braille alphabets, numbers, punctuation, and even music. Braille can be both read and written. A "braillewriter," which looks much like a manual typewriter, is used to print Braille onto paper, and smaller, more portable instruments like the "Braille slate" are used for writing quick notes in Braille. Today many textbooks, works of literature, and magazines are available in Braille.

b. Yes; every factorial greater than 4! has 2 and 5 as factors, and so is a multiple of 10.

4. $6! = 6 \cdot 5 \cdot 4 \cdot 3 \cdot 2 \cdot 1 = 3 \cdot 2 \cdot 5 \cdot 4 \cdot 3 \cdot 2 \cdot 1$; $6! = (5 \cdot 2)(3 \cdot 3)(4 \cdot 2) = 10 \cdot 9 \cdot 8$. Then $6! \cdot 7! = 10 \cdot 9 \cdot 8 \cdot 7 \cdot 6 \cdot 5 \cdot 4 \cdot 3 \cdot 2 \cdot 1 = 10!$.

5. a. 64 ways
   b. The one in which none of the 6 dots is raised would not be readable.

6. 3969 pairs; There are 63 choices for each member of the pair, so there are $63 \cdot 63 = 3969$ pairs.

Apply⇔Assess

Exercise Notes

Challenge

Ex. 8 Students can relate this exercise to their study of theoretical probability by considering the probability that any country not chosen for a position on the economic council will be chosen for a position on the security council. They can then consider the probability that any country not chosen for a position on the security council will be chosen for a position on the development council. Students should understand that as the pool of remaining countries decreases, the probability that a country not yet chosen for a council position will be chosen increases. Students can then use this information to make a conjecture about the relative importance of each committee appointment.

Using Technology

Ex. 9 One means of selecting the remaining members is to assign numbers to each of the countries and then generate a list of random numbers that represents the countries. For example, each country can be assigned a number from 1 to 30. Students can then use a graphics calculator or software to generate a list of 70 numbers between 1 and 30 inclusive. On the TI-82, in the MODE menu, the floating decimal should be set to zero so that only integers will be randomly generated. Then students should perform the following procedures to generate the 70 random numbers.

1. Enter 1 + 29.
2. Press [MATH].
3. Highlight PRB in the top row.
4. Select 1:rand.
5. Press [ENTER].

Students can discuss whether the resulting distribution is fair.

Problem Solving

Ex. 11 This exercise requires that students adapt the formula for the permutation of *n* objects if the order of players is limited. Ask them how the formula can be adjusted if the pitcher always bats last in the lineup. (The first eight positions will have one less choice than the number of the position.)

Connection SOCIAL STUDIES

King High School is hosting a Model United Nations (MUN). Each participating school will represent one of the countries listed below.

North and Central America	South America	Africa	Europe	Asia
Canada	Brazil	Algeria	Estonia	China
Costa Rica	Chile	Botswana	France	Indonesia
Cuba	Colombia	Chad	Germany	Japan
El Salvador	Peru	Egypt	Spain	Pakistan
Guatemala	Uruguay	Somalia		Thailand
Mexico		Togo		Vietnam
United States		Zaire		
		Zimbabwe		

BY THE WAY

The United Nations was created after World War II. Countries that had cooperated during the war hoped that they could maintain international peace by continuing to work together.

7. The MUN will have an economic council consisting of one representative from each continent. The representative can come from any country. How many different councils are possible?

8. **Writing** After the economic council is chosen, the security and development councils will be picked. Each council will consist of one representative from each continent. No country can be on more than one council. How many different ways can the security council be chosen? the development council? Explain your reasoning.

9. a. **Open-ended Problem** The MUN will have an assembly of 100 members. At least one representative from each country will be in the assembly. The remaining seats must be distributed among the countries. Write an algorithm for fairly distributing assembly seats in any MUN.

   b. Discuss your algorithm with other students. What do the algorithms have in common? How are they different?

10. **Research** Find out about one of the programs run by the United Nations.

7. 6720 councils
8. 2520 ways; 720 ways; There will be one fewer possible choice from each continent for the second council and two fewer for the third.
9. a. Answers may vary. An example is given based on representing all countries as fairly as possible without regard to population.
   Step 1: Since there are 30 countries and $\frac{100}{30} = 3\frac{1}{3}$, choose three representatives from each country.
   Step 2: Put the name of each country on a slip of paper. Place the slips of paper in a container.
   Step 3: Draw one slip and name a representative from the country indicated. Discard the slip of paper.
   Step 4: Repeat Step 3 until all 100 seats are filled.
   b. Answers may vary.
10. Answers may vary.
11. 362,880 ways; Answers may vary. Examples might include the abilities of the players or the use of designated hitters.
12. C
13. 2730 orders

11. SPORTS There are nine players on an intramural softball team. In how many different orders can the players go to bat? What factors might limit the number of orders the team actually uses?

12. SAT/ACT Preview Pauline read 24 pages of an assigned book in 20 min before lunch. How long will it take her to read the remaining 60 pages of her assignment if she continues reading at the same rate?

A. 30 min **B.** 40 min **C.** 50 min **D.** 60 min **E.** 70 min

Technology Your calculator has a function ${}_nP_r$ that will find the number of permutations of *r* items chosen from a group of *n* items. For Exercises 13–15, use the function ${}_nP_r$.

13. The *Mind Rockers* have 15 songs that they perform in concert. At the upcoming Battle of the Bands, they will play three songs. In how many different orders could the band perform three of their songs?

14. Challenge On a sales trip, Jane Gupta needs to visit 10 locations in 2 days. She decides to visit 5 locations each day. How many possible paths can she take the first day? How many possible paths remain on the second day? Explain your reasoning.

15. Open-ended Problem How many seven-digit numbers are possible if no digit is repeated? if digits are repeated? How many seven-digit numbers could be telephone numbers in the United States? Explain how you decided which numbers could *not* be telephone numbers.

ONGOING ASSESSMENT

16. Open-ended Problem Automatic teller machines (ATMs) usually require customers to use a four-digit password. Many computer systems require a password that is at least six characters long and can include letters and numbers.

a. How many passwords are possible in each system?

b. What are the advantages of requiring a longer password? What are the disadvantages?

c. Why do you think banks choose to require a password with only four digits for ATM use?

SPIRAL REVIEW

Solve each system of equations by adding or subtracting. *(Section 7.3)*

17. $3x - y = -3$
$4x + y = -11$

18. $4p - 2q = -8$
$4p - 5q = -17$

19. $5m - n = 17$
$m + n = 1$

Solve each equation using the quadratic formula. *(Section 8.5)*

20. $x^2 + 2x - 9 = 0$ **21.** $y^2 - 5y - 1 = 0$ **22.** $3z^2 + 5z + 1 = 0$

23. $5x^2 + 3x = 5$ **24.** $11n^2 - 3n + 4 = 0$ **25.** $2a^2 - 7a = 8$

Apply⇔Assess

Exercise Notes

Using Technology

Ex. 13 In writing, the number of permutations of 15 objects taken 3 at a time is denoted ${}_{15}P_3$. To use a TI-82 or TI-81 graphing calculator to evaluate such an expression, go to the home screen and use 15 [MATH] [▶] [▶] [▶] <nPr>3. The home screen will show 15nPr3. Next, press [ENTER]. The calculator will display the answer 2730.

Common Error

Ex. 15 Some students may use the multiplication counting principle to find 7 • 6 • 5 • 4 • 3 • 2 • 1. Remind these students that the number in each position must reflect the number of choices for that position. If all ten digits are allowed in any position of the phone number, then there are 10 choices for the first position.

Practice 82 for Section 12.4

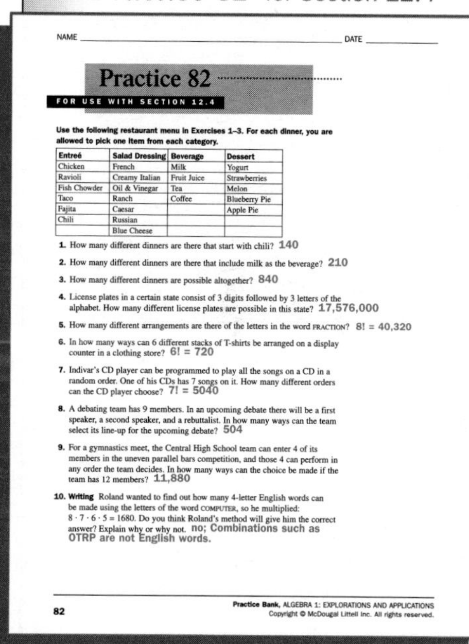

NAME ______________ DATE ________

Practice 82

FOR USE WITH SECTION 12.4

Use the following restaurant menu in Exercises 1–3. For each dinner, you are allowed to pick one item from each category.

Entreé	Salad Dressing	Beverage	Dessert
Chicken	French	Milk	Yogurt
Ravioli	Creamy Italian	Fruit Juice	Strawberries
Fish Chowder	Oil & Vinegar	Tea	Melon
Taco	Ranch	Coffee	Blueberry Pie
Fajita	Caesar		Apple Pie
Chili	Russian		
	Blue Cheese		

1. How many different dinners are there that start with chili? 140
2. How many different dinners are there that include milk as the beverage? 210
3. How many different dinners are possible altogether? 840
4. License plates in a certain state consist of 3 digits followed by 3 letters of the alphabet. How many different license plates are possible in this state? 17,576,000
5. How many different arrangements are there of the letters in the word FRACTION? 8! = 40,320
6. In how many ways can 6 different stacks of T-shirts be arranged on a display counter in a clothing store? 6! = 720
7. Indivar's CD player can be programmed to play all the songs on a CD in a random order. One of his CDs has 7 songs on it. How many different orders can the CD player choose? 7! = 5040
8. A debating team has 9 members. In an upcoming debate there will be a first speaker, a second speaker, and a rebuttalist. In how many ways can the team select its line-up for the upcoming debate? 504
9. For a gymnastics meet, the Central High School team can enter 4 of its members in the uneven parallel bars competition, and those 4 can perform in any order the team decides. In how many ways can the choice be made if the team has 12 members? 11,880
10. **Writing** Roland wanted to find out how many 4-letter English words can be made using the letters of the word COMPUTER, so he multiplied: $8 \cdot 7 \cdot 6 \cdot 5 = 1680$. Do you think Roland's method will give him the correct answer? Explain why or why not. no; Combinations such as OTRP are not English words.

82 **Practice Bank,** ALGEBRA 1: EXPLORATIONS AND APPLICATIONS

---

14. 30,240 paths; 120 paths; The number of paths for five locations chosen from among ten is ${}_{10}P_5 = 30{,}240$. The second day there are five locations left, so there are 5! = 120 paths.

15. 604,800; 10,000,000; Assume the first digit cannot be 0 and that digits can repeat; the 9,000,000 numbers are possible.

16. a. ATM: 10,000 passwords; computer system: 1,000,000 passwords

b. Answers may vary. Examples are given. Requiring a longer password allows for a greater number of passwords; however, it also means a password may be difficult to remember.

c. A short password is easier to remember than a long one, which makes it less likely that a customer would write the password down, making it less secure. The short password is still reasonably safe. Most ATMs give you only three chances to get the code correct without destroying the card. It is nearly impossible to guess at or chance upon a number with four digits in three tries.

17. $(-2, -3)$ **18.** $\left(-\frac{1}{2}, 3\right)$ **19.** $(3, -2)$

20. $-1 + \sqrt{10}, -1 - \sqrt{10}$

21. $\frac{5}{2} - \frac{\sqrt{29}}{2}; \frac{5}{2} + \frac{\sqrt{29}}{2}$

22. $-\frac{5}{6} - \frac{\sqrt{13}}{6}; -\frac{5}{6} + \frac{\sqrt{13}}{6}$

23. $-\frac{3}{10} - \frac{\sqrt{109}}{10}; -\frac{3}{10} + \frac{\sqrt{109}}{10}$

24. no solution

25. $\frac{7}{4} - \frac{\sqrt{113}}{4}; \frac{7}{4} + \frac{\sqrt{113}}{4}$

Plan⇔Support

Objectives

- Count the number of possible pairs of objects in a group.
- Find the number of edges in a complete graph.

Recommended Pacing

Core and Extended Courses
Section 12.5: 1 day

Block Schedule
Section 12.5: $\frac{1}{2}$ block
(with Section 12.4)

Resource Materials

Lesson Support
Lesson Plan 12.5
Warm-Up Transparency 12.5
Overhead Visuals:
Folder 12: Connecting Cities with Graphs
Practice Bank: Practice 83
Study Guide: Section 12.5

Technology
Graphing Calculator
McDougal Littell Software
Stats!
Internet:
http://www.hmco.com

Warm-Up Exercises

Simplify.

1. 7! 5040
2. 4! · 2! 48
3. $\frac{8!}{5!}$ 336
4. $\frac{7 \cdot 6 \cdot 5!}{5!}$ 42

How many edges are there in a graph with the given number of vertices if there is an edge between each pair of vertices?

5. 3 vertices 3
6. 4 vertices 6

SECTION

12.5 Combinations

Learn how to...

- count the number of possible pairs of objects in a group

So you can...

- find the number of edges in a complete graph
- find the number of games in a tournament

How many different ways can you choose two people from a group? How many distances do you need to know to find the best path through a group of cities? In the Exploration, you will investigate a related problem.

EXPLORATION
COOPERATIVE LEARNING

Making Connections

Work in a group of 2–4 students.
You will need:

- a graphing calculator

SET UP A **complete graph** has a single edge between every pair of vertices.

1 Draw a complete graph with 2 vertices. How many edges does it have? Draw complete graphs with 3, 4, 5, 6, and 7 vertices. How many edges does each graph have? Organize your results in a table.

Put the number of vertices on the horizontal axis.

2 Use a graphing calculator to make a scatter plot showing how the number of edges is related to the number of vertices. Choose a good viewing window.

3 Predict the number of edges in a complete graph with 8 vertices. Explain your reasoning. Check your prediction.

4 Write an equation for the number of edges in a complete graph as a function of the number of vertices. Explain your reasoning.

5 Graph the function from Step 4 on your scatter plot from Step 2. Does it seem to fit the data?

Exploration Note

Purpose
The purpose of this Exploration is to have students discover the relationship between the number of edges in a complete graph and the number of vertices in the graph.

Materials/Preparation
Students should know how to graph a scatter plot on their graphing calculators. They should set their viewing window so that $0 < x \le 10$ and $0 < y \le 35$.

Procedure
Students should record the number of edges for each complete graph. Check to see that each group of students graphs the number of edges as a function of the number of vertices.

Closure
Groups can share their answers and results with each other. All students should recognize that the number of edges in a complete graph can be found by using the expression $\frac{n(n-1)}{2}$, where n represents the number of vertices in the graph.

Explorations Lab Manual
See the Manual for more commentary on this Exploration.

If you choose three students from your class in a particular order, you are creating a permutation. You can also choose the students without paying any attention to order. A selection of items chosen from a group in which order is not important is called a **combination**.

In the Exploration, you wrote an equation that allows you to count combinations of two items chosen from a group.

Combinations

The number of **combinations** of **2** items chosen from a group of ***n*** items is given by the formula:

$${}_nC_2 = \frac{n(n-1)}{2!}$$

For example:

$${}_5C_2 = \frac{5 \cdot 4}{2 \cdot 1}$$

The number of **combinations** of **3** items chosen from a group of ***n*** items is given by the formula:

$${}_nC_3 = \frac{n(n-1)(n-2)}{3!}$$

For example:

$${}_5C_3 = \frac{5 \cdot 4 \cdot 3}{3 \cdot 2 \cdot 1}$$

EXAMPLE 1

Paul Green is leading a workshop for 20 people. He wants each person to meet every other person. If each pair of people shake hands when they meet, how many handshakes will occur?

SOLUTION

There is one handshake for each pair in a group of 20 people. Count the number of combinations of 2. Use the formula for ${}_nC_2$ and let $n = 20$.

$${}_{20}C_2 = \frac{n(n-1)}{2}$$

$$= \frac{20(20-1)}{2}$$

$$= 190$$

There will be 190 handshakes.

THINK AND COMMUNICATE

1. Explain the difference between a *combination* and a *permutation*. Give an example of each.
2. Explain how you can use a complete graph to model the number of handshakes in a group of people.

Teach⇔Interact

Learning Styles: Verbal

Some students have difficulty distinguishing between permutations and combinations. Students with verbal learning styles can benefit from paying attention to whether they are considering an *arrangement* of a set of items or a *group* of items. Because an arrangement implies ordering, a permutation is involved. When a group of equal members is created order is *not* a consideration, and therefore a combination is involved.

Additional Example 1

The Student Council is forming a 2-member advisory committee from among 10 people. How many different committees are possible?

There is one committee for each group of 2 people from among the 10 people. Count the number of combinations of 2. Use the formula ${}_nC_2$ and let $n = 10$.

$${}_{10}C_2 = \frac{n(n-1)}{2}$$

$$= \frac{10(10-1)}{2}$$

$$= 45$$

There are 45 possible committees.

Think and Communicate

For question 2, students have to relate the idea of creating groups to the number of edges in a graph. Emphasize that counting the number of edges involves creating pairs of vertices. Students should see that the pairing of vertices *A* and *B* is the same as pairing vertices *B* and *A*. In other words, there is only one distinct edge between the two vertices and order is not important when forming pairs of vertices.

ANSWERS Section 12.5

Exploration

1.

number of vertices	number of edges
2	1
3	3
4	6
5	10
6	15
7	21

2.

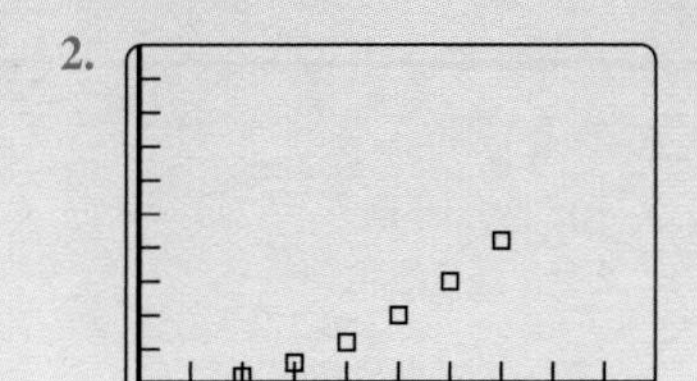

3. 28 edges; For each added vertex, you add a number of edges equal to the original number of vertices. Adding an eighth vertex adds seven edges to the previous number, for a total of 28 edges.

4. Let n = the number of vertices and $f(n)$ = the number of edges; $f(n) = \frac{n(n-1)}{2}$. Each vertex can be connected to all the other vertices, so there are $n(n-1)$ edges. Only half of them are unique since each edge is counted twice in that number.

5. Yes.

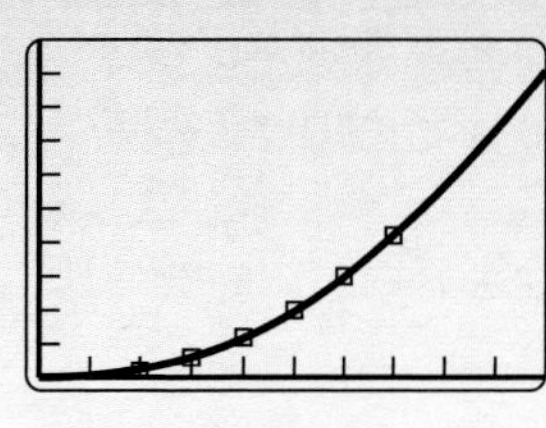

Think and Communicate

1. In a combination, order is not important in selecting items; in a permutation, order is important. Answers may vary.
2. Let each vertex in a complete graph represent a person and each edge represent a handshake between two people.

Teach⇔Interact

Additional Example 2

Suppose Cecilia Jordan decided to use only groups of two different students each day to lead the reading discussions. Can she still choose a different combination for each day of the school year?

Find the number of ways to choose 2 students from a group of 25.

$$_{25}C_2 = \frac{n(n-1)}{2!}$$
$$= \frac{25(25-1)}{2 \cdot 1}$$
$$= 300$$

There are 300 possible groups of two. She can still choose a different group for each day of the school year.

Think and Communicate

Questions 3 and 4 relate geometry to graph theory. Before answering question 4, students may wish to check their conjectures by experimenting with hexagons, heptagons, octagons, and so on.

Closure Question

What is a combination of n items selected from a group?

A combination is a selection of items from a group in which the order of the selection is not important.

EXAMPLE 2

Cecilia Jordan plans to choose a different group of three students each day to lead the discussion of the assigned reading from the previous day. Can she choose a different combination of three students from her class of 25 students each day of the 180-day school year?

SOLUTION

Find the number of ways to choose 3 students from a group of 25.

$$_{25}C_3 = \frac{n(n-1)(n-2)}{3!}$$
$$= \frac{25(25-1)(25-2)}{3 \cdot 2 \cdot 1}$$
$$= 2300$$

There are 2300 possible groups of three. Not only could Cecilia Jordan choose a different group each day of the school year, she could choose a different group each school day for more than twelve years.

THINK AND COMMUNICATE

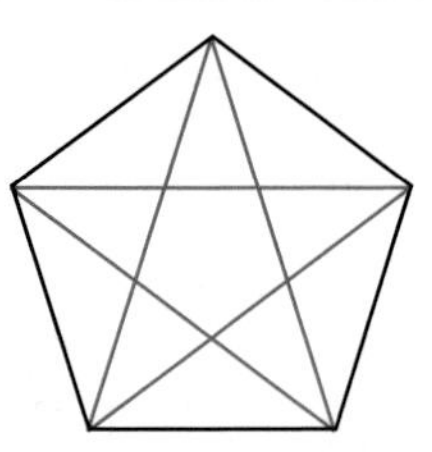

3. How many diagonals are there in a polygon with 5 sides? How is this related to the number of edges in a complete graph with 5 vertices?

4. Explain how to change the formula for the number of edges in a complete graph to get a formula for the number of diagonals of a polygon as a function of the number of sides.

CHECKING KEY CONCEPTS

Tell how many handshakes there would be in a group of each size if each pair of people shook hands.

1. 8 **2.** 15 **3.** 100 **4.** 1000

Evaluate.

5. $_7C_2$ **6.** $_6C_3$ **7.** $_{11}C_2$ **8.** $_{15}C_3$

9. Maya must read two of the six books on the summer reading list at her school. She can pick any two of the books. How many different combinations of two books can she choose?

10. How many edges does a complete graph with 18 vertices have?

11. How many ways can you choose a combination of 3 students from a class of 28 students?

Think and Communicate

3. 5 diagonals; it is five fewer, because the number of edges in the complete graph would include the five sides of the polygon.
4. Subtract the number of sides of the polygon: $f(n) = \frac{n(n-1)}{2} - n$.

Checking Key Concepts

1. 28 handshakes
2. 105 handshakes
3. 4950 handshakes
4. 499,500 handshakes
5. 21
6. 20
7. 55
8. 455
9. 15 combinations
10. 153 edges
11. 3276 ways

12.5 Exercises and Applications

GOVERNMENT For Exercises 1–4, use the 12 members of the student council listed below.

Julie	Karen	Nathan	Susan
Booker	Eric	Tamai	Ivan
Mariah	Paula	Isabel	Greg

Extra Practice exercises on page 583

1. The council is sending a two-person delegation to the state model legislature. How many two-person delegations are possible?
2. How many of the possible delegations have either Nathan or Tamai on them? (*Hint:* Only count the Nathan-Tamai delegation once.)
3. How many possible delegations will not have Isabel, Karen, or Booker? Explain your reasoning.
4. How many three-person delegations are possible?

Connection SOCIAL STUDIES

Many countries have flags with three horizontal stripes. The colors of the stripes have meanings for each country. For Exercises 5 and 6, use the colors red, blue, orange, white, green, black, and yellow.

5. Both Austria and Belarus have flags with red and white horizontal stripes. How many different pairs of colors could you choose for a two-color flag? How many different two-color flags with three horizontal stripes of alternating colors could be made?

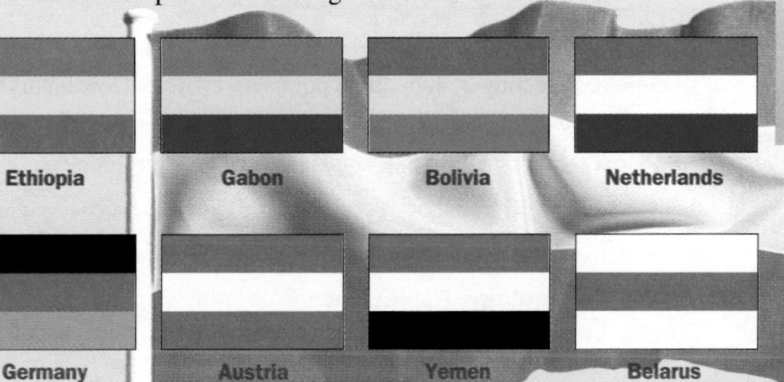

BY THE WAY

When the Olympic flag was first flown in 1920, every nation's flag contained at least one of the five colors on the Olympic flag.

6. a. How many different combinations of three colors could you choose for a three-color flag?
   b. How many different flags with three horizontal stripes can be made with green, red, and blue? How many different flags with three horizontal stripes can be made with three colors?

Apply⇔Assess

Suggested Assignment

Core Course
Exs. 1–13, 15–17, 19–30

Extended Course
Exs. 1–30

Block Schedule
Day 78 Exs. 1–13, 15–17, 19–30

Exercise Notes

Communication: Drawing
Exs. 2, 3 Students can begin with a related but simpler problem. They might first consider choosing from only a group of 4 members or a group of 5 members. This will allow them to draw diagrams and consider how the 12 members of the student council are related to combinations that include either Nathan or Tamai or exclude either Isabel, Karen, or Booker.

Multicultural Note
Exs. 5, 6 In 1990, the formerly separate nations of Yemen and Southern Yemen united to create one country, Yemen. Located on the southern end of the Arabian peninsula between the Red Sea and the Arabian Sea, Yemen is largely desert but also includes temperate regions that receive enough rainfall to support agriculture. Fruits, nuts, sorghum, millet, coffee, and cotton are some of the crops farmed. Livestock are raised for milk, hides, and meat. Recent discoveries of petroleum give the promise of greater prosperity and industrial development in the coming years. A large percentage of the population live in San'a, the capital, or in Aden (once the capital of Southern Yemen). Arabic is the principal language.

Exercises and Applications

1. 66 delegations
2. 21 delegations
3. 36 delegations; The number of such delegations is the number of two-person delegations that can be chosen from a group of 9, or ${}_9C_2$.
4. 220 delegations
5. 21 pairs; If the colors alternate, there are 42 possible flags; if the stripes appear in any order, there are 126 possible flags.
6. a. 35 different combinations
   b. 6 different flags; 210 different flags

Apply⇔Assess

Exercise Notes

Mathematical Procedures
Exs. 7–10 Students should realize that each factor in the denominator of the fractions generated by these exercises can be divided out with a factor in the numerator. This technique should allow students to evaluate these expressions quickly and without the aid of a calculator.

Using Technology
Exs. 13, 14, 17 Students can use the *Stats!* software to work on Exs. 13 and 14. For Ex. 17, the function $_nC_r$ is evaluated on the TI-82 and TI-81 exactly as $_nP_r$. (See the Using Technology note on page 535.) A quick way to get the table for Ex. 17 is to enter the equation Y1 = 9nCr X on the Y= list. Then press [2nd][TblSet] and use the following: TblMin = 1, △Tbl = 1. Finally, press [2nd][TABLE] to display the table. Scroll down through the table by using the [▼] key. Note that when there are 10 or more items, the table shows zeros. Ask students how they interpret this. (Possible answer: There are no ways to select 10 or more items from a group that contains only 9 items to start with.)

Integrating the Strands
Ex. 15 This exercise relates a newly learned algebraic procedure to a topic in geometry. Students should see that the number of triangles formed is equal to the combination of *n* points taken 3 at a time.

Evaluate.

7. $_{13}C_2$ **8.** $_9C_3$ **9.** $_{17}C_3$ **10.** $_{21}C_2$

11. a. Jill Keyes asked her students to answer two of the five essay questions on an English test. How many different choices do her students have?

b. For extra credit the students can answer a third essay question. How many combinations of three questions can the students answer?

12. MANUFACTURING A book publisher selects two books at random from each lot for a check of production quality. How many different combinations of two books can be selected from a lot of 25 books?

Spreadsheet Virginia set up the spreadsheet below to calculate the number of different handshakes possible for a group of people. She assumed that each pair of people shakes hands once.

B8		=A7+B7
	A	B
1	People	Handshakes
2	1	
3	2	1
4	3	3
5	4	6
6	5	10
7	6	15
8	7	21

13. Cooperative Learning Work in a group of 2 or 3 students. Discuss how the formula for Column B works. Explain in words what the formula A7 + B7 in cell B8 means. Why does it give the correct number of handshakes?

14. Make a spreadsheet of your own. Use it to find the number of handshakes in a group of 35 people.

15. GEOMETRY Draw three points on a circle. How many triangles can you draw connecting the points? How many triangles can you draw connecting four points on a circle? Make a table for the number of triangles connecting 3, 4, 5, and 6 points on a circle. How is this related to the formula for $_nC_3$?

16. At a local restaurant, you can choose from nine pizza toppings.

a. How many different two topping pizzas can you choose?

b. How many different three topping pizzas can you choose?

17. Technology The $_nC_r$ function on your calculator will calculate the number of ways you can choose *r* items from a list of *n* possibilities. Let $n = 9$ pizza toppings. Make a table of the number of pizzas with *r* toppings for values of *r* from 1 to 9. Explain any patterns you see.

18. a. Challenge Write a formula for $_nC_4$. You may want to think about how many ways there are to choose 4 of the pizza toppings in Exercise 17. Explain your reasoning.

b. Write a general formula for $_nC_r$. Test your formula by using the $_nC_r$ function on your calculator and choosing values for *n* and *r*.

7. 78 8. 84
9. 680 10. 210
11. a. 10 choices
b. 10 combinations
12. 300 combinations
13. The entry in column B is the sum of the entries in columns A and B of the preceding row. Since the new person does not shake hands with himself or herself, the total number of added handshakes each time is the number of people in the group before the new person was added. Then the total number of handshakes is the previous total plus the number of people before the new person was added.
14. Check students' work. 595 handshakes
15. Check students' work.

number of points	number of triangles
3	1
4	4
5	10
6	20

The number of triangles that can be drawn by connecting *n* points is $_nC_3$.

19. SPORTS North Central High is planning a tournament for the city's intramural volleyball league. Each team will play a three-game match against each of the other teams. There are enough volleyball courts to play about ten matches each day of the tournament. If nine teams participate, how many days will the tournament take to complete? Explain your reasoning.

20. Open-ended Problem Write a counting problem that could be solved by the equation $\frac{15 \cdot 14}{2} = 105$.

ONGOING ASSESSMENT

21. Writing Melissa tried to find how many ways she could choose two cassette tapes to listen to in the car. What do the numbers 1190 and 595 represent?

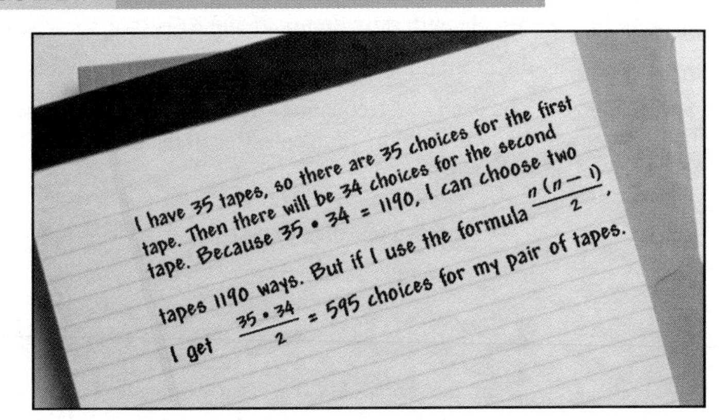

I have 35 tapes, so there are 35 choices for the first
tape. Then there will be 34 choices for the second
tape. Because 35 • 34 = 1190, I can choose two
tapes 1190 ways. But if I use the formula n(n − 1)/2,
I get 35 • 34/2 = 595 choices for my pair of tapes.

SPIRAL REVIEW

The equations in Exercises 22–27 describe quantities that grow or decay exponentially. For each equation, tell whether the quantity described *grows* or *decays*. Then give the growth or decay factor for the quantity. *(Sections 9.2 and 9.3)*

22. $y = 2.1(1.32)^x$ **23.** $y = 0.15(0.52)^x$ **24.** $y = 3.2(0.91)^x$

25. $y = 0.02(2)^x$ **26.** $y = A(1 + r)^x$ **27.** $y = ab^x$

Decide whether the data show inverse variation. If they do, write an equation for *y* in terms of *x*. *(Section 11.1)*

28.

x	y
−2.5	6.00
2.5	−6.00
6.5	−2.31
10.5	−1.43

29.

x	y
−1.5	−2.79
−0.5	−0.93
1.0	1.86
3.5	6.51

30.

x	y
−14	−7.2
6	16.7
8	12.5
12	8.4

Apply⇔Assess

Exercise Notes

Alternate Approach

Ex. 20 The given formula should help students see that the multiplication counting principle is involved because items are taken two at a time. But since the number of permutations of two items is divided out, the equation can be used to find the combination of pairs of 15 items.

Assessment Note

Ex. 21 You may wish to ask students to consider in their writing assignment whether they think Melissa is concerned with the order in which she listens to the two tapes.

Practice 83 for Section 12.5

NAME ______________________ DATE __________

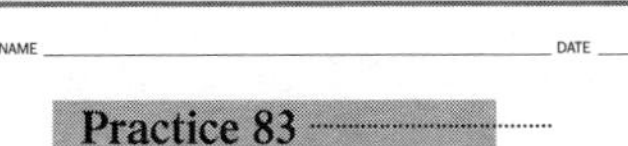

Practice 83

FOR USE WITH SECTION 12.5

Aretha Jethroe raises dogs. Use the following list of the names of her dogs in Exercises 1–4.

Rusty	**Henrietta**
Elmer	**Slim**
Minnie	**Sylvester**
Millicent	**Sapphire**
Harpo	**Oona**

1. Aretha must choose two dogs to bring to an upcoming show. How many selections of 2 dogs are possible? 45
2. How many of the possible selections do not include Elmer or Oona? 28
3. How many of the possible selections include either Harpo or Henrietta or Sylvester? (*Hint*: Find the number of selections that include none of these dogs. Then subtract.) 24
4. Suppose Aretha is allowed to bring 3 dogs to the show. In how many ways can the selection be made? 120
5. Suppose Aretha chooses 3 dogs and must specify the order in which she wants them to appear. How many choices does she have? 720
6. Six girls want to choose sides for a half-court basketball game. In how many ways can they be divided into two teams? 120
7. In how many ways can you choose 3 CDs, out of a collection of 25 CDs, to bring to a party? 2300
8. A local telephone company wants to build a communication network among 7 cities, with one line joining each pair of cities. How many lines will they need to build? 21
9. In the diagram points A, B, C, and D lie on the diameter of the semicircle. Points E, F, and G are on the semicircle. How many different triangles have 3 of these points as vertices? (*Hint*: Any combination of 3 points determines a triangle *unless* the 3 points all lie on diameter $\overline{AD}$. Use subtraction.) 31
10. The Outing Club, consisting of 12 members, wants to choose a president and a vice-president. In how many ways can the choice be made? Did you calculate your answer using permutations or combinations? Explain your choice. 132; permutations; order matters

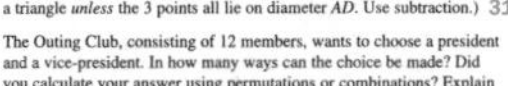

11. After a soccer game ended in a tie, each team had to select one player for a "shoot-out" of free kicks to determine a winner. If there are 11 players on each team, in how many ways can the two players be selected? Why won't the combination formula work in this situation? 121; There is a restriction on the choice of 2 out of 22 players.

Practice Bank, ALGEBRA 1: EXPLORATIONS AND APPLICATIONS

83

---

16. a. 36 pizzas
 b. 84 pizzas

17, 18. See answers in back of book.

19. Yes; last year's tournament involved ${}_6C_2$ or 15 games and with one game played at a time, all 15 were completed in two days. Since three games can be played at once, there would be time to play 45 games in this year's tournament. However, only ${}_9C_2$ or 36 games are necessary.

20. Answers may vary. An example is given. A two-member executive council is to be chosen from the 15 members of the student government. How many different councils may be chosen?

21. the number of permutations of 35 items taken two at a time; the number of combinations of 35 items taken two at a time

22. grows; 1.32
23. decays; 0.52
24. decays; 0.91
25. grows; 2
26. grows; $1 + r$
27. grows if $b > 1$ and decays if $0 < b < 1$; b
28. Yes; $y = -\frac{15}{x}$.
29. No.
30. Yes; $y = \frac{100}{x}$.

Plan⇔Support

Objectives

- Apply counting strategies to find probabilities.
- Find the probability of winning a random drawing.

Recommended Pacing

Core and Extended Courses
Section 12.6: 1 day

Block Schedule
Section 12.6: $\frac{1}{2}$ block
(with Review)

Resource Materials

Lesson Support
Lesson Plan 12.6
Warm-Up Transparency 12.6
Overhead Visuals:
Folder 5: Probability Spinners
Practice Bank: Practice 84
Study Guide: Section 12.6
Assessment Book: Test 53

Technology
Technology Book:
Spreadsheet Activity 12
Scientific Calculator
Graphing Calculator
Internet:
http://www.hmco.com

Warm-Up Exercises

Find the probability of rolling each number on a single roll of a six-sided die.

1. a 5 $\frac{1}{6}$

2. a 2 or a 3 $\frac{1}{3}$

3. a number greater than 3 $\frac{1}{2}$

Simplify.

4. ${}_5C_2$ 10

5. ${}_{15}C_3$ 455

SECTION

12.6 Connecting Probability and Counting

Learn how to...

- **use counting strategies to find probabilities**

So you can...

- **find the probability of winning a random drawing**

Nancy Greenleaf has a deal for her algebra class. She agrees to cancel all of the tests for the rest of the year and give everyone an A for the class. But she will only do this if she rolls five 1's when she rolls five dice with different numbers of sides. How likely is that to happen?

You can use the counting strategies you have learned in this chapter to solve many probability problems. Recall the meaning of *theoretical probability* from Chapter 5.

$$\text{Theoretical probability} = \frac{\text{Number of favorable outcomes}}{\text{Number of possible outcomes}}$$

EXAMPLE 1

Find the probability that Nancy Greenleaf will roll five 1's when she rolls a 6-sided die, an 8-sided die, a 10-sided die, a 12-sided die, and a 20-sided die.

SOLUTION

The number of favorable outcomes is the number of ways to roll five 1's. There is one favorable outcome.

Use the multiplication counting principle to find the number of possible outcomes.

$$6 \cdot 8 \cdot 10 \cdot 12 \cdot 20 = 115{,}200$$

$$\frac{\text{Favorable outcomes}}{\text{Possible outcomes}} = \frac{1}{115{,}200}$$

The probability of rolling five 1's is $\frac{1}{115{,}200}$.

EXAMPLE 2

Deb and Kyle are two of the 15 people who submitted correct solutions to a puzzle contest. Two prize winners will be chosen in a random drawing.

a. What is the probability that both Deb and Kyle will win a prize?

b. What is the probability that Deb will be one of the prize winners?

SOLUTION

a. The number of favorable outcomes is the number of ways to choose both Deb and Kyle. There is one favorable outcome.

The number of possible outcomes is the number of combinations of two winners. Use the formula for ${}_nC_2$ and let $n = 15$.

$$ {}_{15}C_2 = \frac{n(n-1)}{2} $$

$$ = \frac{15 \cdot 14}{2} $$

$$ = 105 $$

There are 105 ways to choose 2 winners out of 15 people.

$$ \frac{\text{Favorable outcomes}}{\text{Possible outcomes}} = \frac{1}{105} $$

The probability that both Deb and Kyle will be chosen is $\frac{1}{105}$.

b. Use the multiplication counting principle. If Deb is one winner, there are $15 - 1 = 14$ choices for the second winner. There are $1 \cdot 14 = 14$ favorable outcomes.

$$ \frac{\text{Favorable outcomes}}{\text{Possible outcomes}} = \frac{14}{105} = \frac{2}{15} $$

The probability that Deb will be a winner is $\frac{2}{15}$.

THINK AND COMMUNICATE

1. How do you decide when to use the multiplication counting principle and when to use the number of combinations to determine the number of possible results?
2. What is the probability of Kyle being selected as one of the two prize winners in Example 2?
3. What is the probability of getting four heads if you flip one coin four times? if you flip four coins one time?
4. When Robyn put her new license plate on her car, she was surprised to notice that her license plate had the same four digits in a different order as her sister's license plate. Explain how you would find the probability of this happening.

ANSWERS Section 12.6

Think and Communicate

1. Use the multiplication counting principles when the outcomes involve numbers of choices from different categories. Use the number of combinations when the outcomes involve choosing a number of items from a group of items.

2. $\frac{1}{15}$

3. $\frac{1}{16}, \frac{1}{16}$

4. There are 10,000 four-digit numbers. There are 23 other permutations of the four digits on Robyn's sister's license plate that are in a different order. Then the probability that Robyn's license plate has the same four digits in a different order is $\frac{23}{10{,}000}$.

Teach⇔Interact

Additional Example 1

Find the probability that Jennifer Hartman will roll a 1, then a 2, then a 3 on three successive rolls of a 6-sided die.

The number of favorable outcomes is the number of ways to roll a 1, then a 2, then a 3. There is one favorable outcome. Use the multiplication counting principle to find the number of possible outcomes.

$6 \cdot 6 \cdot 6 = 216$

$$ \frac{\text{Favorable outcomes}}{\text{Possible outcomes}} = \frac{1}{216} $$

The probability of rolling a 1, then a 2, then a 3 is $\frac{1}{216}$.

Additional Example 2

Suppose that in Example 2, 20 people submitted correct solutions to the puzzle contest.

a. What is the probability that both Deb and Kyle will win a prize?

The number of favorable outcomes is the number of ways to choose both Deb and Kyle. There is still one favorable outcome. The number of possible outcomes is the number of combinations of two winners. Use the formula for ${}_nC_2$ and let $n = 20$.

$$ {}_{20}C_2 = \frac{n(n-1)}{2} $$

$$ = \frac{20 \cdot 19}{2} $$

$$ = 190 $$

There are 190 ways to choose 2 winners out of 20 people.

$$ \frac{\text{Favorable outcomes}}{\text{Possible outcomes}} = \frac{1}{190} $$

The probability that both Deb and Kyle will be chosen is $\frac{1}{190}$.

b. What is the probability that Deb will be one of the prize winners?

Use the multiplication counting principle. If Deb is one winner, there are $20 - 1 = 19$ choices for the second winner. There are $1 \cdot 19 = 19$ favorable outcomes.

$$ \frac{\text{Favorable outcomes}}{\text{Possible outcomes}} = \frac{19}{190} $$

The probability that Deb will be a winner is $\frac{19}{190}$, or $\frac{1}{10}$.

Teach⇔Interact

Checking Key Concepts

Topic Spiraling: Review
Students may find it helpful to draw tree diagrams to answer questions 1–4. They can use the diagrams to count outcomes and determine the probability of each toss.

Closure Question

How can you use counting strategies to find the probability that you and your friend both have the same first and last initials but in a different order?
Use the multiplication counting principle to find the number of favorable outcomes. There are two ways to choose the first initial and one way to choose the second, $2 \cdot 1 = 2$. Use the formula for combinations to find the number of possible outcomes. The number of possible outcomes is the number of combinations of two letters. Use the formula ${}_nC_2$ and let $n = 26$. Write the probability fraction with the number of favorable outcomes in the numerator and the number of possible outcomes in the denominator.

Apply⇔Assess

Suggested Assignment

Core Course
Exs. 1–10, 14–28, AYP
Extended Course
Exs. 1–28, AYP
Block Schedule
Day 79 Exs. 1–10, 14–28, AYP

Exercise Notes

Teaching Tip
Ex. 4 Remind students that this exercise asks for the theoretical probability. Students may need to review the coin-tossing experiments from Chapter 5.

CHECKING KEY CONCEPTS

For Questions 1–4, suppose that you toss a penny four times.

1. List all of the possible results.
2. What is the probability that you will get four heads?
3. What is the probability that you will get four tails?
4. What is the probability that you will get two heads and two tails?
5. **POLITICS** Ben and Lara want to be the delegates for a local political convention. Two people will be chosen at random from a group of 11 students. What is the probability of Ben and Lara being chosen?

12.6 Exercises and Applications

Extra Practice exercises on page 583

For Exercises 1 and 2, find the probability of getting the result shown when each group of coins is tossed.

1. 

2. 

For Exercises 3–5, suppose you toss six pennies.

3. What is the total number of possible results?
4. What is the probability that the pennies will all come up heads?
5. **Writing** Explain why the probability of getting one tail and five heads is the same as the probability of getting five tails and one head.
6. a. **SPORTS** There are 12 players on Elisa's basketball team. Each team member plays a one-on-one game with each of the other members of the team. How many games are played?
   b. The school's video club is going to pick one game at random to film for the video yearbook. What is the probability that Elisa and Bridget's game will be filmed? What is the probability that one of Elisa's games will be filmed?
7. Suppose you throw a 4-sided die, a 20-sided die, and a 100-sided die at the same time. What's the probability of rolling three 4's?
8. Suppose you roll three 4-sided dice. What is the probability that you will roll three of the same number? Explain your reasoning.

Checking Key Concepts

1. HHHH, HHHT, HHTH, HHTT, HTHH, HTHT, HTTH, HTTT, THHH, THHT, THTH, THTT, TTHH, TTHT, TTTH, TTTT
2. $\frac{1}{16}$
3. $\frac{1}{16}$
4. $\frac{3}{8}$
5. $\frac{1}{55}$

Exercises And Applications

1. $\frac{3}{8}$
2. $\frac{1}{4}$
3. 64 possible results
4. $\frac{1}{64}$
5. Both situations involve choosing one item from among six.
6. a. 66 games
   b. $\frac{1}{66}$; $\frac{1}{6}$
7. $\frac{1}{8000}$
8. $\frac{1}{16}$; There are 64 possible outcomes, four of which are positive: 1, 1, 1; 2, 2, 2; 3, 3, 3; and 4, 4, 4; the probability is $\frac{4}{64} = \frac{1}{16}$.
9. Answers may vary.
10. $\frac{5}{16}$; This game is the same as tossing a coin 5 times. (In the case of the coin, you assume the coin does not land on an edge. In the case of the random

Cooperative Learning For Exercises 9–12, work with a partner. Use the game explained below. You will need a calculator with a random number generator.

Take turns until one player wins. On each turn:

Step 1 Start with a score of zero.

Step 2 Generate a random number, n.
If $n \le 0.5$, subtract 1 from your score.
If $n > 0.5$, add 1 to your score.

Step 3 Repeat Step 2 five times.

a. If your score = 3 or −3, you win the game.

b. If your score is anything else, it is your partner's turn.

9. Play at least six games. Record how many turns it takes to win the game. Record the ending score after each turn during the games. You may want to use bar graphs to organize this information.

10. What is the probability of winning the game on any given turn? Explain your reasoning.

11. Open-ended Problem Choose a way to change the rules in the game described above. Explain how to play and win your new game.

12. Challenge Play the game you described in Exercise 11 several times. Can you determine the probability of winning the game on any given turn? Explain why or why not.

13. **Technology** You can program your calculator to model the roll of a 6-sided die. See your calculator's manual for information on how to program your calculator.

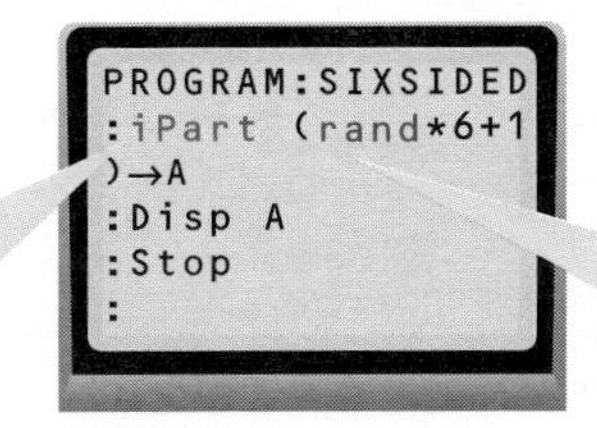

PROGRAM:SIXSIDED
:iPart (rand*6+1
)→A
:Disp A
:Stop
:

`iPart` tells the calculator to ignore everything to the right of the decimal point.

`rand` generates a random number between 0 and 1.

a. Writing Explain how this model works. How can you use it without the `iPart` function? Why do you need to add 1 to the random number?

b. How can you model the roll of a 20-sided die?

c. Program your calculator to model the roll of a 6-sided die. Run your program 36 times. Do you get the results you would expect from a fair die?

Apply⇔Assess

Exercise Notes

Integrating the Strands

Ex. 9 This exercise relates an idea from statistics to both counting and probability theory. Students may wish to draw their bar graphs on a graphiing calculator or to use computer software and then create a bulletin board display of their graphs.

Reasoning

Exs. 11, 12 These exercises require students to modify the game described at the top of this page. Students can compare their answers to these exercises with those of Exs. 9 and 10 and then determine if their new game is fair.

Using Technology

Ex. 13 If you press MATH ▶ on the TI-82, you will see a menu of special functions. Item 2 is iPart, which gives the integer part of any decimal that you enter. For example, if you enter iPart 27.043 and press ENTER, the calculator displays the result 27, the integer part of 27.043.

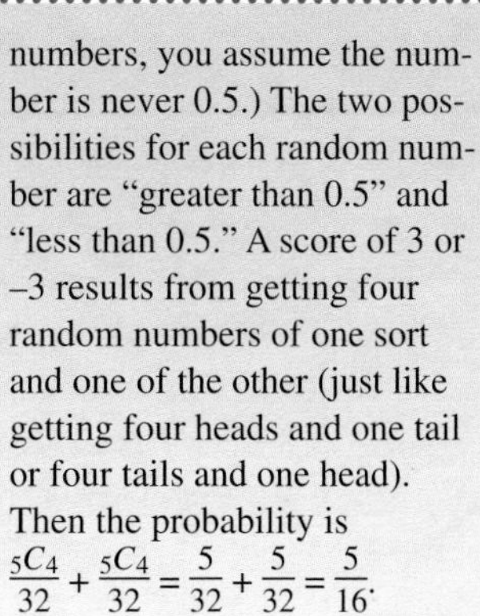

numbers, you assume the number is never 0.5.) The two possibilities for each random number are "greater than 0.5" and "less than 0.5." A score of 3 or −3 results from getting four random numbers of one sort and one of the other (just like getting four heads and one tail or four tails and one head). Then the probability is $\frac{{}_5C_4}{32} + \frac{{}_5C_4}{32} = \frac{5}{32} + \frac{5}{32} = \frac{5}{16}$.

11. Answers may vary.

12. Answers may vary.

13. a. The program produces random numbers between 1 and 6, modeling the results of rolling the die. Without the iPart function, you could use the int function: int(rand*6 + 1), or you could ignore everything to the right of the decimal point. If you do not add 1, the random numbers produced would be between 0 and 5.

b. Use the same function, but replace 6 with 20.

c. Answers may vary. Results should generally be what you would expect from a fair die, with each of the six outcomes represented fairly equally.

Apply⇔Assess

Exercise Notes

Interdisciplinary Problems

Exs. 15–17 Tom Stoppard's play is a retelling of William Shakespeare's play *Hamlet* through the eyes of the court's couriers, Rosencrantz and Guildenstern. These exercises illustrate how the author was able to incorporate his knowledge of mathematics into the play in an interesting and lively manner.

Second-Language Learners

Exs. 15–17 The complexity of the language in Guildenstern's last speech in this passage will make it difficult for second-language learners to read independently. You might read the passage aloud to the class and then call on an English-proficient student to paraphrase it.

Communication: Discussion

Ex. 18 Students will undoubtedly generate a number of suggestions for Ex. 18(b). A class discussion of these suggestions will benefit all students by allowing them to see methods different from their own.

14. PHOTOGRAPHY Luisa removed the used batteries from her camera. She accidently dropped the two used batteries into her pile of eight new ones. The batteries all look the same. What is the probability that she will pick up at least one used battery when she chooses two to put into her camera? What is the probability that she will pick up two new batteries?

Connection LITERATURE

Rosencrantz and Guildenstern are Dead **by Tom Stoppard is a twentieth century play about two characters from Shakespeare's *Hamlet*. When the play begins, Rosencrantz and Guildenstern are carrying large leather money bags. Guildenstern's (Guil) is almost empty. Rosencrantz's (Ros) is almost full.**

15. Guildenstern tosses 92 coins, all of which come up heads. What is the number of possible outcomes of tossing 92 coins? Explain how you know.

16. What is the probability of getting 92 heads in a row?

17. Guildenstern makes another wager later in the play. "*Bet* me then. . . . Year of your birth. Double it. Even numbers I win, odd numbers I lose." What is the probability that Guildenstern will win this wager?

ROSENCRANTZ AND GUILDENSTERN ARE DEAD

Guil: *takes a coin out of his bag, spins it, letting it fall.*
Ros: Heads. *He picks it up and puts it in his bag. The process is repeated.*

Heads. *Again.*
Heads. *Again.*
Heads. *Again.*
Heads.

Guil: . . . *flips over two more coins. . . . Ros announces each of them as heads."*
Ros: . . . Eighty-five in a row—beaten the record!
Guil: . . . And if you'd lost? If they'd come down against you, eighty-five times, one after another, just like that?
Ros: . . . Eighty-five in a row? *Tails?*
Guil: Yes! What would you think?
Ros: *(doubtfully)*: Well . . . . Well, I'd have a good look at your coins for a start! . . .
Guil: . . . It must be indicative of something, besides the redistribution of wealth. . . . each individual coin spun individually *(he spins one)* is as likely to come down heads as tails and therefore should cause no surprise each individual time it does. . . . The equanimity of your average tosser of coins depends upon a law, or rather a tendency, or let us say a probability, or at any rate a mathematically calculable chance, which ensures that he will not upset himself by losing too much or upset his opponent by winning too often.

14. $\frac{16}{45}$; $\frac{28}{45}$

15. 2^{92}; On each toss, there are two possible outcomes, so the total number of outcomes is $2 \cdot 2 \cdot 2 \cdot \cdots \cdot 2$ (92 times) $= 2^{92} \approx 4.95 \times 10^{27}$.

16. $\frac{1}{2^{92}} \approx 2.02 \times 10^{-28}$

17. 1; Doubling any number produces an even number.

18. a. The four possible outcomes are HH, HT, TH, and TT. The probability of getting two heads and the probability of getting two tails are the same, $\frac{1}{4}$. The probability of getting one head and one tail is $\frac{1}{2}$.

b. Answers may vary. An example is given. Roll a die. Assign two possible outcomes to each person. For example, if the result is a 1 or 2, George will go first; if it is a 3 or 4, Helen will go first; and if it is a 5 or 6, Inez will go first. Since the probability of winning is the same for each person, this method is fair to all three.

19. $5x - 10$

20. $a^2 - 10a + 10$

21. $4 - 2n - 8n^2$

22. $4x - 11y - 5z + 5$

ONGOING ASSESSMENT

18. a. Writing George, Helen, and Inez agree to toss a coin twice to decide who will go first at batting practice. George goes first if the result is two heads. Helen does if the result is two tails. Inez does if the result is one head and one tail. Explain why Inez is most likely to go first.

b. Suggest a fair way to decide who goes first using coins or dice. Why is your suggestion fair to all three people?

SPIRAL REVIEW

Add. *(Section 10.1)*

19. $(x + 2) + (4x - 12)$

20. $(4a^2 - 3a + 1) - (3a^2 + 7a - 9)$

21. $(4 - 3n - 5n^2) + (n - 3n^2)$

22. $(3x - 7y - 5z) - (x - 4y - 5)$

Simplify each equation. *(Section 11.6)*

23. $\frac{3}{2ab} - \frac{5}{3ab}$

24. $\frac{6}{m+1} + 7$

25. $\frac{2t+3}{5} + \frac{5-t}{4}$

26. $\frac{m^2}{7n} - \frac{5}{n}$

27. $\frac{3}{2a} - \frac{1+a}{b}$

28. $\frac{a+b}{3} + \frac{a+b}{5} + \frac{a+b}{10}$

ASSESS YOUR PROGRESS

VOCABULARY

multiplication counting principle (p. 531)
factorial (p. 532)
permutation (p. 532)
complete graph (p. 536)
combination (p. 537)

1. How many ways can you select 1 shirt, 1 sweater, and 1 hat from 3 shirts, 5 sweaters, and 4 hats? *(Section 12.4)*

2. A company has 713 employees. Explain why at least two of them must have the same pair of first and last initials. For example, Ann Martin has the same initials as Alan Monchick. *(Section 12.4)*

3. a. ENTERTAINMENT Martha and Gwen have 150 movies to choose from at the video store. They plan to rent 2 movies. How many different pairs of movies could they choose? *(Section 12.5)*

b. They decide to rent 1 of 36 comedies and 1 of 23 new releases. How many different pairs could they choose? *(Section 12.4)*

4. If you throw two six-sided dice, what is the probability that you will roll at least one 6? *(Section 12.6)*

5. Journal Give examples of probability problems that help you to remember the difference between permutations and combinations. Explain how permutations or combinations are used in each problem.

Apply⇔Assess

Assess Your Progress

Journal Entry

Students may wish to begin this exercise by writing definitions for a permutation and a combination. Their journal entries will be even more valuable if they include explanations of why each example helps them remember the difference between the two counting methods.

Progress Check 12.4–12.6

See page 551.

Practice 84 for Section 12.6

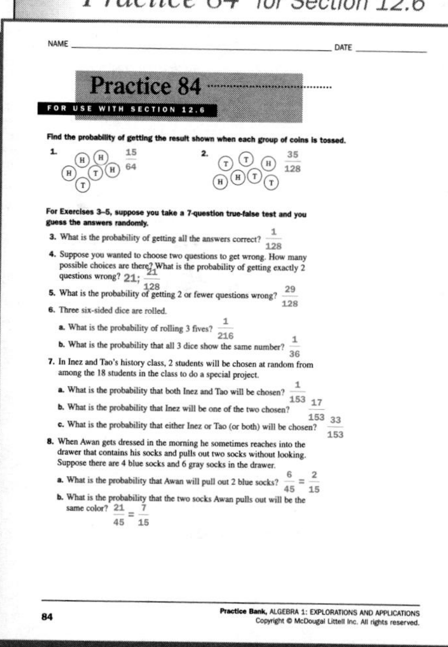
NAME ______ DATE ______

Practice 84

FOR USE WITH SECTION 12.6

Find the probability of getting the result shown when each group of coins is tossed.

1. (H)(H)(H)(T)(H)(T) $\frac{15}{64}$

2. (T)(T)(H)(H)(H)(T)(T) $\frac{35}{128}$

For Exercises 3–5, suppose you take a 7-question true-false test and you guess the answers randomly.

3. What is the probability of getting all the answers correct? $\frac{1}{128}$

4. Suppose you wanted to choose two questions to get wrong. How many possible choices are there? What is the probability of getting exactly 2 questions wrong? 21; $\frac{21}{128}$

5. What is the probability of getting 2 or fewer questions wrong? $\frac{29}{128}$

6. Three six-sided dice are rolled.

a. What is the probability of rolling 3 fives? $\frac{1}{216}$

b. What is the probability that all 3 dice show the same number? $\frac{1}{36}$

7. In Inez and Tao's history class, 2 students will be chosen at random from among the 18 students in the class to do a special project.

a. What is the probability that both Inez and Tao will be chosen? $\frac{1}{153}$

b. What is the probability that Inez will be one of the two chosen? $\frac{17}{153}$

c. What is the probability that either Inez or Tao (or both) will be chosen? $\frac{33}{153}$

8. When Awan gets dressed in the morning he sometimes reaches into the drawer that contains his socks and pulls out two socks without looking. Suppose there are 4 blue socks and 6 gray socks in the drawer.

a. What is the probability that Awan will pull out 2 blue socks? $\frac{6}{45} = \frac{2}{15}$

b. What is the probability that the two socks Awan pulls out will be the same color? $\frac{21}{45} = \frac{7}{15}$

84

Practice Bank, ALGEBRA 1: EXPLORATIONS AND APPLICATIONS

23. $-\frac{1}{6ab}$

24. $\frac{7m+13}{m+1}$

25. $\frac{3t+37}{20}$

26. $\frac{m^2-35}{7n}$

27. $\frac{3b-2a-2a^2}{2ab}$

28. $\frac{19a+19b}{30}$

Assess Your Progress

1. 60 ways

2. The number of pairs of initials (in the same order) is the number of two-letter permutations of the 26 letters of the alphabet, which is $26 \cdot 26$, or 676. There are not enough pairs to allow all 713 employees to have unique pairs of initials.

3. a. 11,175 pairs

   b. 828 pairs

4. $\frac{11}{36} \approx 0.31$

5. Answers may vary. Examples are given. Suppose in one club of 20 members a president and vice president are elected, while in another 20-member club, a two-member executive committee is chosen. In the first case, order matters and permutations should be used to count the possible outcomes. $(20 \cdot 19 = 380)$ In the second case, order does not matter and combinations should be used to count the possible outcomes. $({}_{20}C_2 = 190)$

Project Notes

Mathematical Goal

- Develop a strategy for winning a game.

Planning

Materials
- Cups
- Counters
- Seeds

Project Teams
Students can choose a partner to play the game and then collect the materials needed. They can follow Steps 1–3 to begin playing.

Guiding Students' Work

As students set up their boards and start to play, you can circulate around the room to make sure each group is playing the game according to the rules. Encourage students to play several games and to discuss their strategies for winning.

Second-Language Learners
Students learning English should have no difficulties understanding any of the vocabulary used in this project.

PORTFOLIO PROJECT

Playing the Game

This mancala board is from the Neolithic Age (7000–5000 B.C.).

For centuries, people in Africa and Asia have been developing strategies for winning games that involve moving counters around a board. Some versions of this ancient game are *mancala, adi, kala,* and *oware.* Finding strategies for winning many different kinds of games is an important part of a field of mathematics known as *game theory.*

PROJECT GOAL **Develop strategies for winning a game of wari from Ghana, a country in Africa.**

How to Play Wari

MATERIALS The Wari board has 12 playing cups, in two rows of six, and two scoring cups. The game is played with 48 counters—stones or seeds are often used.

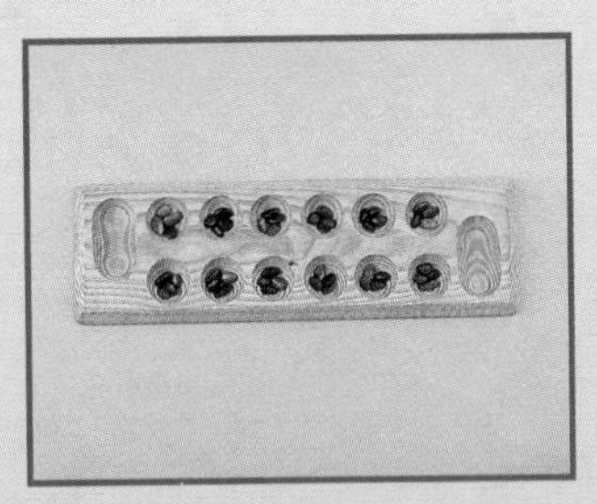

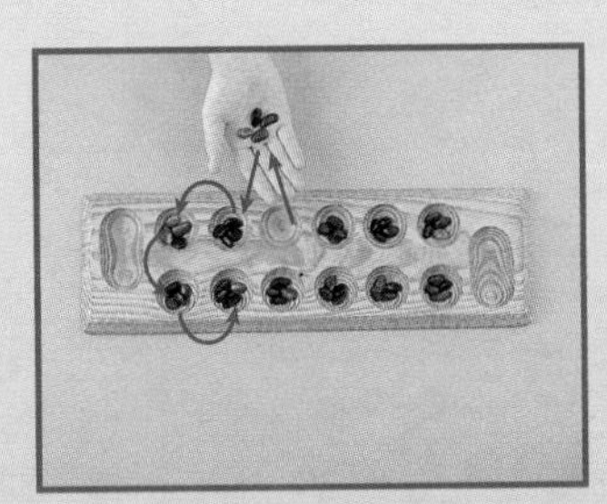

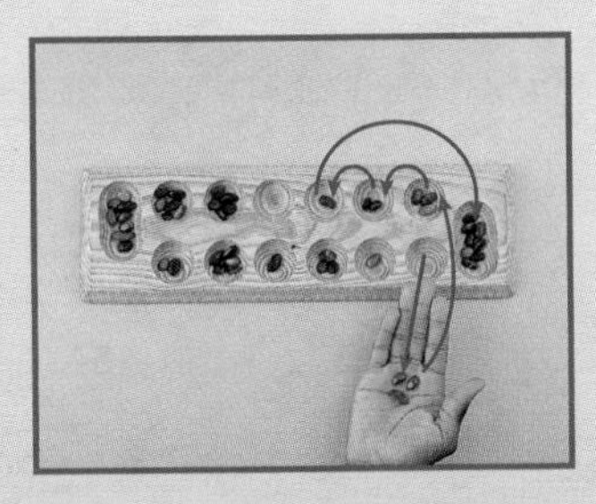

1. SET UP Work with a partner. Each of you should choose one side of the board. Your scoring cup is on your left. Place 4 stones in every cup but the scoring cups.

2. PLAY On your turn, choose any cup on your side of the board. Take all the stones out of this cup. Moving counter-clockwise drop one stone in each cup (not including the scoring cups) until you run out of stones.

3. CAPTURE If the last stone is placed in one of your partner's cups and makes the total number of stones in that cup 2 or 3, capture those stones and put them in your scoring cup.

If your partner's side of the board is empty, you must play so that you drop *at least one stone* into your partner's cups. If you cannot, you capture all remaining stones.

WINNING The game ends when no captures are possible. The player who has captured the most stones is the winner.

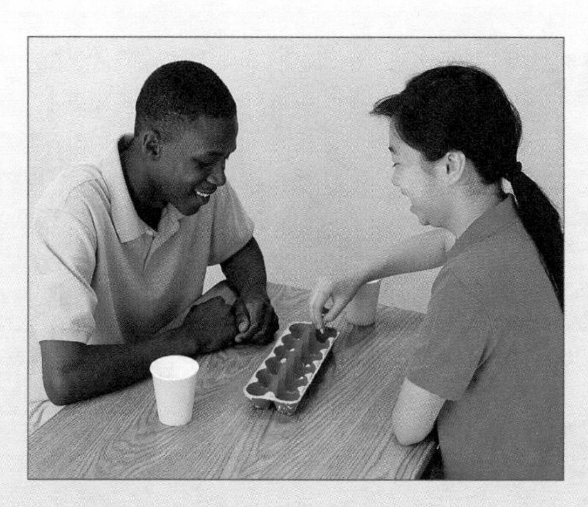

Developing Strategies

Play several games of *wari*. Take turns going first. While you are playing try different strategies for collecting as many stones as possible. Discuss your strategies. Which seem to work best? Do any of your strategies depend on whether you play first?

You may want to use an egg carton for your board. You may want to use dried beans or pumpkin seeds instead of stones.

Presenting Your Results

Give an oral report on this project to your class. Demonstrate the strategies you developed. Explain how they work and how you developed them.

Senet

You may want to extend your report to explore other ideas:

• Research and explore some of the variations of this game. Develop a strategy for winning a different version of the game. How does this new strategy compare with the strategy that you developed for *wari?*

• Research the history and directions for another game. Where and when was the game developed? Explore the game and develop a strategy for winning that game. Present your findings to your classmates. Some games you might research are *Go, Nim, Senet,* and *Nyout.*

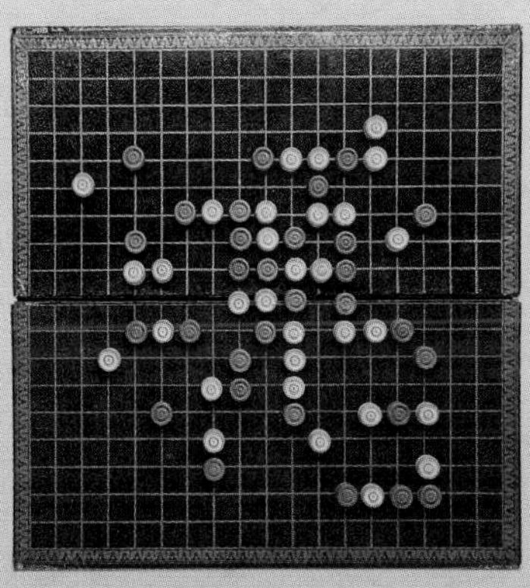

Go

Self-Assessment

What general advice would you give someone who is trying to develop strategies for winning a new game? How did you and your partner work together to develop strategies? Based on your experience playing wari, does the same strategy for winning always work? How do your partner's moves affect your strategy?

Project Notes

Guiding Students' Work

Rubric for Chapter Project

4 Students play the game and develop strategies for winning. They give a clear and interesting oral report to the class and demonstrate successful strategies for winning. They explain how the strategies work and how they were developed.

3 Students play the game and develop strategies for winning. Their oral report includes a demonstration of the strategies, but the explanation of how they were developed is not entirely clear.

2 Students play the game but cannot seem to develop any strategies for winning. They give an oral report and attempt to explain their strategies, but the class is not convinced they work.

1 Students start to play the game but are not sure about how to proceed. They have difficulty understanding the rules and do not develop any strategies for winning. Students should be encouraged to speak with the teacher as soon as possible to review their work and to make a new start on the project.

Chapter Support

Course Guide: Chapter 12
Lesson Plans: Chapter 12
Practice Bank:
Cumulative Practice 85
Study Guide: Chapter 12 Review
Assessment Book:
Chapter Tests 54 and 55
Chapter 12 Alternative Assessment
Cumulative Test 56
Test Generator software
Portfolio Project Book
Preparing for College Entrance Tests
Professional Handbook

Progress Check 12.1–12.3

1. Write an algorithm to add a positive number and a negative number. *(Section 12.1)*
   Step 1: Find the absolute value of each number.
   Step 2: Subtract the smaller absolute value from the larger absolute value.
   Step 3: Attach the sign of the number with the greater absolute value to the difference found in Step 2.

At *The Stone Shed Gardens*, there are seven display areas: 3 flower gardens, 2 herb gardens, and 2 ponds. The graph below models the locations of the display areas. The numbers show the distance in feet between display areas.

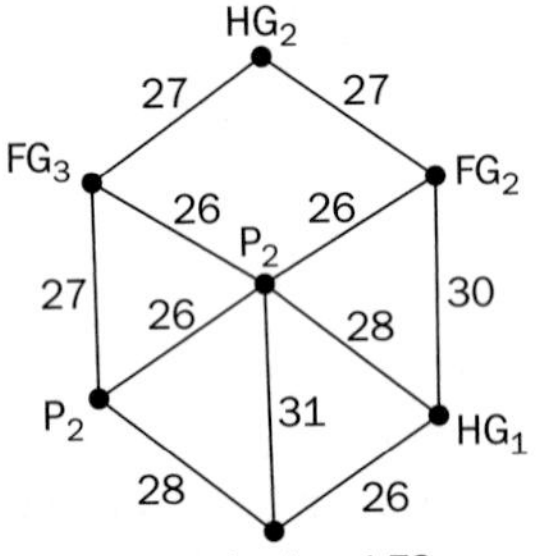

2. Use a greedy algorithm to find a short path starting from the stone shed for a person to use to visit each of the display areas. How far will a person walk if he or she uses this path? *(Section 12.2)*
   26 + 28 + 26 + 27 + 27 + 27 = 161; 161 ft
3. All the paths between the display areas are now dirt. The owner wants to install stone walks so that people can get from one location to another without

CHAPTER

12 Review

Study Technique

Look back at the study techniques you used in other Review and Assessment sections. Decide which technique(s) helped you most. Why was it helpful? Choose a technique you think would be most helpful in studying Chapter 12.

Vocabulary

algorithm (p. 509)
graph (p. 516)
vertices (p. 516)
edges (p. 516)
path (p. 517)
greedy algorithm (p. 517)
tree (p. 518)
one person-one vote (p. 524)
approval voting (p. 524)
fair division (p. 525)
cut-and-choose (p. 525)
multiplication counting principle (p. 531)
factorial (p. 532)
permutation (p. 532)
complete graph (p. 536)
combination (p. 537)

Sections 12.1 *and* 12.2

A **graph** consists of points called **vertices** and lines called **edges**. A **path** is a group of edges that connect one vertex to another without repeating any vertices or edges. A **tree** uses as few edges as possible to connect all the vertices in a graph.

This is a graph.

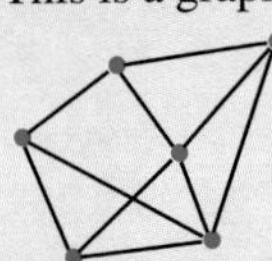

This is a path from A to E.

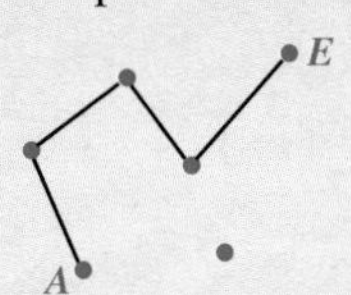

This is a tree.

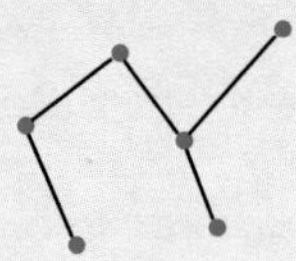

An **algorithm** is a step-by-step method for accomplishing a goal.

You can use a **greedy algorithm** to find a short path.

Step 1 Choose a vertex.

Step 2 Choose the nearest vertex you have not already chosen.

Step 3 Repeat Step 2 until you've connected all the vertices.

SECTION 12.3

In a **one person-one vote** election, each person can vote for one candidate. In an **approval voting** election, each person can vote for any number of candidates. The table shows poll results based on both types of voting.

	Type of Election	
Candidate	One person-one vote	Approval voting
Phuong	48	73
Gioberti	62	68
Wallach	41	82

The poll results show that Gioberti would probably win a one person-one vote election. Wallach would probably win an approval voting election.

You can use a **cut-and-choose** algorithm to create a **fair division** of a collection into two groups.

SECTIONS 12.4, 12.5, *and* 12.6

A **permutation** is an arrangement of a group of items in a definite order. A **combination** is a selection of items from a group in which the order of selection does not matter.

Problem	Type	Number
Hang 4 flags in a row.	permutation	$4! = 4 \cdot 3 \cdot 2 \cdot 1 = 24$
Hang 2 of 4 flags in a row.	permutation	$4 \cdot 3 = 12$
Choose 2 of 4 flags.	combination	${}_4C_2 = \frac{4 \cdot 3}{2!} = 6$

To find the probability of getting exactly 2 heads when 5 coins are tossed, you can use the formula for theoretical probability.

The number of favorable outcomes is the number of combinations of 2 heads.

$${}_5C_2 = \frac{5 \cdot 4}{2!} = 10$$

There are two possible outcomes for each coin toss. By the multiplication counting principle, the number of possible outcomes when 5 coins are tossed is $2 \cdot 2 \cdot 2 \cdot 2 \cdot 2 = 32$.

$$\text{Theoretical probability} = \frac{\text{Number of favorable outcomes}}{\text{Number of possible outcomes}} = \frac{10}{32}.$$

The probability of getting exactly 2 heads when 5 coins are tossed is $\frac{5}{16}$.

walking on dirt. What is the most economical way to put in the stone walks? *(Section 12.2)*

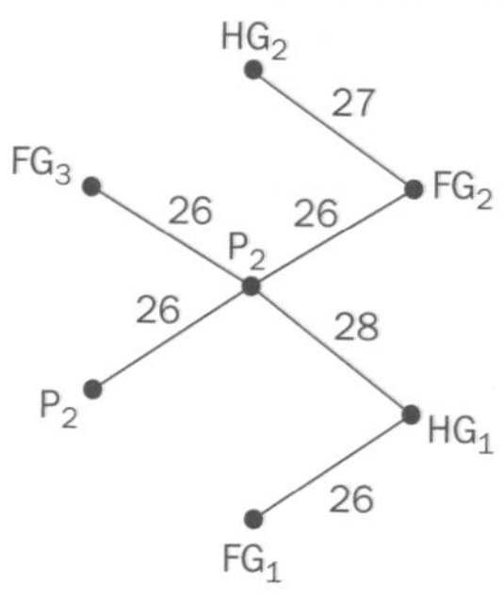

4. The dirt paths have to be hoed by 2 workers. How can they split the hoeing fairly? *(Section 12.3)*
Step 1: Considering the length of the paths and the distance of the paths from each other and from the shed, one worker divides the number of feet into 2 groups.
Step 2: The second worker chooses the group he likes.
Step 3: The other worker gets the remaining group.

5. In which kind of voting, one person-one vote or approval voting, can a person vote for all the candidates with a platform he or she likes? *(Section 12.3)*
approval voting

Progress Check 12.4–12.6

1. Evaluate 6! *(Section 12.4)* 720

2. In how many ways can 9 students be seated in 9 chairs in the front row of an auditorium? *(Section 12.4)* 362,880 ways

Write the formula to calculate each combination. *(Section 12.5)*

3. ${}_nC_2$ ${}_nC_2 = \frac{n(n-1)}{2!}$

4. ${}_nC_3$ ${}_nC_3 = \frac{n(n-1)(n-2)}{3!}$

5. An office is hiring 3 data entry personnel. If they have 8 applications to choose from, how many different groups of data entry personnel can they hire? *(Section 12.5)* 56 groups

6. Are the probabilities of rolling at least 3 heads or at least 3 tails on 3 consecutive rolls of a regular die equal? Explain. *(Section 12.6)* Yes; the number of favorable outcomes are the same for each as are the number of possible outcomes.

7. Using the letters M, A, T, H, what is the probability that a two-letter arrangement of these letters will begin with M? *(Section 12.6)*
$\frac{1}{4}$

Chapter 12 Assessment
Form A Chapter Test

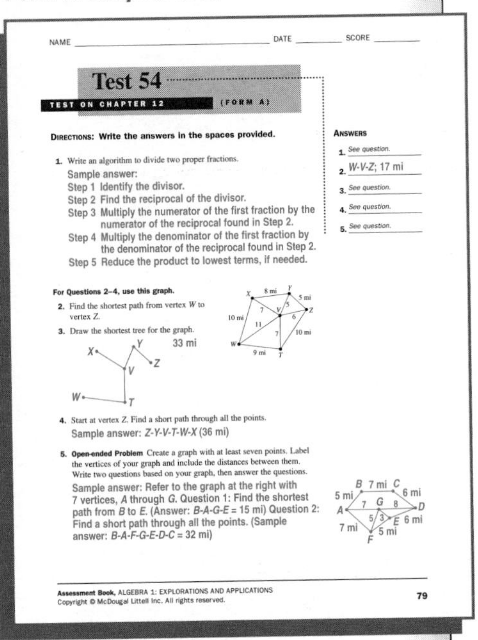
NAME ______ DATE ______ SCORE ______

Test 54
TEST ON CHAPTER 12 (FORM A)

DIRECTIONS: **Write the answers in the spaces provided.**

1. Write an algorithm to divide two proper fractions.
   Sample answer:
   Step 1 Identify the divisor.
   Step 2 Find the reciprocal of the divisor.
   Step 3 Multiply the numerator of the first fraction by the numerator of the reciprocal found in Step 2.
   Step 4 Multiply the denominator of the first fraction by the denominator of the reciprocal found in Step 2.
   Step 5 Reduce the product to lowest terms, if needed.

For Questions 2–4, use this graph.

2. Find the shortest path from vertex W to vertex Z.
3. Draw the shortest tree for the graph. 33 mi
4. Start at vertex Z. Find a short path through all the points.
   Sample answer: Z-Y-V-T-W-X (36 mi)
5. **Open-ended Problem** Create a graph with at least seven points. Label the vertices of your graph and include the distances between them. Write two questions based on your graph, then answer the questions.
   Sample answer: Refer to the graph at the right with 7 vertices, A through G. Question 1: Find the shortest path from B to E. (Answer: B-A-G-E = 15 mi) Question 2: Find a short path through all the points. (Sample answer: B-A-F-G-E-D-C = 32 mi)

ANSWERS
1. See question.
2. W-V-Z; 17 mi
3. See question.
4. See question.
5. See question.

Assessment Book, ALGEBRA 1: EXPLORATIONS AND APPLICATIONS

79

Chapter 12 Assessment
Form B Chapter Test

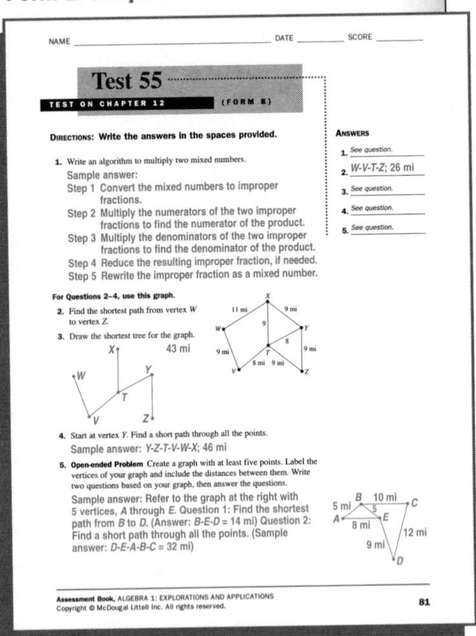
NAME ______ DATE ______ SCORE ______

Test 55
TEST ON CHAPTER 12 (FORM B)

DIRECTIONS: **Write the answers in the spaces provided.**

1. Write an algorithm to multiply two mixed numbers.
   Sample answer:
   Step 1 Convert the mixed numbers to improper fractions.
   Step 2 Multiply the numerators of the two improper fractions to find the numerator of the product.
   Step 3 Multiply the denominators of the two improper fractions to find the denominator of the product.
   Step 4 Reduce the resulting improper fraction, if needed.
   Step 5 Rewrite the improper fraction as a mixed number.

For Questions 2–4, use this graph.

2. Find the shortest path from vertex W to vertex Z.
3. Draw the shortest tree for the graph. 43 mi
4. Start at vertex Y. Find a short path through all the points.
   Sample answer: Y-Z-T-V-W-X; 46 mi
5. **Open-ended Problem** Create a graph with at least five points. Label the vertices of your graph and include the distances between them. Write two questions based on your graph, then answer the questions.
   Sample answer: Refer to the graph at the right with 5 vertices, A through E. Question 1: Find the shortest path from B to D. (Answer: B-E-D = 14 mi) Question 2: Find a short path through all the points. (Sample answer: D-E-A-B-C = 32 mi)

ANSWERS
1. See question.
2. W-V-T-Z; 26 mi
3. See question.
4. See question.
5. See question.

Assessment Book, ALGEBRA 1: EXPLORATIONS AND APPLICATIONS

81

CHAPTER

12 Assessment

VOCABULARY REVIEW

For Questions 1–3, complete each paragraph.

1. Describe the difference between a permutation and a combination.
2. Give an example of a greedy algorithm.
3. Give an example of a situation in which you would find a *tree* to solve a problem. Give an example of a situation in which you would find a *path*.

SECTION 12.1

For Questions 4 and 5, refer to the algorithm for doing laundry.

Step 1 Divide the laundry evenly into loads.
Step 2 Put one load in the washing machine.
Step 3 Turn the machine on.
Step 4 When the cycle is done, remove the laundry from the machine.
Step 5 Repeat Steps 2 through 4 until the laundry is finished.

4. What could go wrong if you follow this algorithm exactly?
5. How could you clarify this algorithm?
6. Write an algorithm to add two fractions with different denominators.

SECTION 12.2

For Questions 7–9, use the graph at the right.

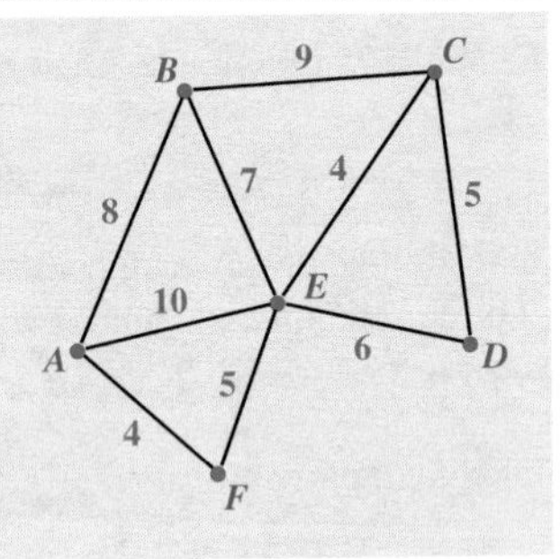

7. Find the shortest path from vertex A to vertex D.
8. Start at vertex D. Find a short path through all the points.
9. Find the shortest tree for the graph.

ANSWERS Chapter 12

Assessment

1. A permutation is an arrangement of a group of items in a definite order. A combination is a selection of items chosen from a group in which order does not matter.
2. Answers may vary. An example is given. In the diagram. a greedy algorithm would choose the path from A to B to C as the best path from A to C.

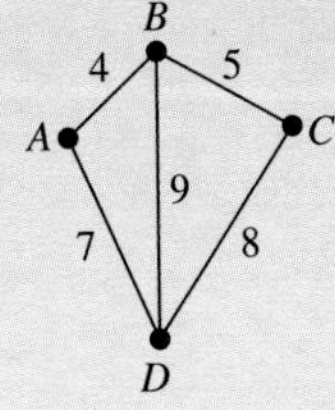

3. Answers may vary. Examples are given. A tree could be used to join a series of computers together with the least amount of cable. A path could be used to determine a good way to visit a collection of sites, such as pickups or drops on a delivery route, or vacation destinations.
4. Answers may vary. Examples are given. (1) You might not add detergent or fabric rinse. (2) You might not dry the laundry.
5. Answers may vary. Examples, based on the answers to Ex. 4 above, are given. (1) Add "Add detergent and fabric conditioner to the washing machine" to Step 2. (2) Add "Put the clean laundry into the dryer or hang it on a line until it is dry" to Step 4.
6. Step 1: Find the least common denominator (LCD) of the fractions. Step 2: Divide the LCD by the denominator of the first fraction. Multiply the

SECTION 12.3

10. Writing Ballots are being printed for an election. Explain how the instructions for a one person-one vote election would be different from those for an approval voting election.

11. Open-ended Problem Describe a situation in which the cut-and-choose algorithm could be used to create a fair division. Explain how to apply the algorithm in this situation.

SECTIONS 12.4 *and* 12.5

12. In a class election, there are four candidates for president, five candidates for vice-president, and six candidates for secretary. How many possible groups of officers are there?

13. In how many ways can a three-member finance committee be chosen from among ten students?

14. There are 14 baseball teams in a community league. Suppose the league champion is determined by having each team in the league play every other team. The team with the best record wins. How many games are necessary to determine the league champion?

SECTION 12.6

15. Three coins are tossed. Find the probability of getting exactly two heads.

16. Luisa and Keith each prepared an individual project for their history class. Each day, 2 of the 17 students in the class will present their projects. Find the probability that both Luisa and Keith will be chosen to present their projects on the first day.

PERFORMANCE TASK

17. Choose one story of a building to model with a graph. Let each vertex represent a room. Connect two vertices with an edge if you can walk from one room to the other without going through another room. Give an example of a path through all the rooms that does not visit any room twice, or explain why such a path is not possible. In how many different orders can you visit all of the rooms in the building? (You may pass through the same room more than once. Only count the first time you visit each room.)

Chapter 12 Assessment
Form C Alternative Assessment

Chapter 12

ALTERNATIVE ASSESSMENT

1. **Group Activity** Write an algorithm for finding a path through a maze. Draw a diagram of a maze of your own design. Trade your maze with another group. Use your algorithm to solve the maze from the other group. Write an evaluation describing how well your algorithm worked.

2. **Open-ended Problem** When you program a VCR to tape a movie, you are using an algorithm. Describe the algorithm.

3. **Project** During homecoming week at Pierce High School, an honorary court consisting of four female students and four male students is chosen. The following method has traditionally been used to select this court.

   On Monday of homecoming week, one female and one male representative are chosen from each homeroom. From these representatives, all the high school students will choose the court. On Wednesday, each student receivets a ballot listing all the homeroom representatives. The student circles one female name and one male name as their choices. The four female students with the most votes and the four males with the most votes are on the honorary court.

   Design another way to organize the selection of the court. Compare your method to the method above. Write a letter to the student council trying to convince them that your method is better. Consider the fairness of both methods in your letter.

4. **Performance Task** Plan a travel wardrobe of several coordinated outfits. Include at least 2 skirts or pairs of slacks, 3 sweaters or jackets, and 4 blouses or shirts in your wardrobe. Draw or cut out pictures from magazines of your clothing pieces. Using a tree diagram, make a poster display of the different outfits that can be created from your wardrobe. Can you think of a way to use permutations or combinations to count the number of possible outfits?

5. a. What is the difference between a permutation problem and a combination problem?
   b. Describe two situations that could be counted using permutations.
   c. Describe two situations that could be counted using combinations.

Assessment Book, ALGEBRA 1: EXPLORATIONS AND APPLICATIONS

109

6. **Performance Task** Mom, Dad, Laura, and Tyler are going to a play performed by a regional theatre company. They have tickets for four seats in the sixth row.
   a. How many different ways can the family sit together for the play? List all the possible seating arrangements. Use the permutation formula to confirm that you have the correct number of arrangements.
   b. Due to occasional misbehavior, it is best if Laura and Tyler do not sit next to each other. On the list from part (a), cross out any undesirable seating arrangements. How many seating arrangements are now available to the family? Use the permutation formula and subtraction or division to confirm that you have the correct number of seating arrangements.

110
Assessment Book, ALGEBRA 1: EXPLORATIONS AND APPLICATIONS

numerator and denominator of the first fraction by the quotient. Step 3: Repeat Step 2 for the second fraction. Step 4: Add the numerators of the fractions obtained in Steps 2 and 3. Step 5: Write the fraction whose numerator is the result from Step 4 and whose denominator is the LCD. This is the sum of the original fractions.

7. *A* to *F* to *E* to *D*
8. *F* to *A* to *B* to *E* to *C* to *D*
9. *D* to *C* to *E* to *B* to *A* to *F*
10. The ballots for a one person-one vote election should instruct the voter to vote for only one candidate. The ballots for an approval election should instruct the voter to vote for as many candidates as the voter chooses.
11. Answers may vary. An example is given. Two sisters want to share a small pizza. One should divide the pizza into two pieces, either of which would satisfy her. The second sister should choose one of the pieces. The first sister would get the remaining piece.
12. 120 groups
13. 120 ways
14. 91 games
15. $\frac{3}{8}$
16. $\frac{1}{136}$
17. Answers may vary.

Cumulative Assessment

CHAPTERS 10–12

CHAPTER 10

Add, subtract, or multiply each pair of polynomials as indicated. Write each answer in standard form and give the degree of the polynomial.

1. $(4x - 3) + (-2x^2 - x + 1)$
2. $(z^2 - 6z + 5) - (3z^2 + 10z + 7)$
3. $(4y^3 - 7y^2 + 3) + (5y^2 - 2y + 9)$
4. $(7s - 2)(6s - 4)$
5. $8x(10x^2 - 4x + 7)$
6. $(2y - 9)(4y^2 - 7y + 5)$

Factor, if possible. If not, write *not factorable*.

7. $x^2 + 4x - 21$
8. $x^2 - 49$
9. $a^2 - 9a + 18$
10. $y^2 - 10y - 56$
11. $3z^2 + 25z - 18$
12. $4x^2 - 16x + 15$

Solve each equation.

13. $x^2 + x - 12 = 0$
14. $r^2 + 16r + 64 = 0$
15. $81y^2 - 25 = 0$
16. $9z^2 + 9z + 5 = 3$
17. $3s^2 - 8s - 1 = 2$
18. $6x^2 = 11x - 5$

19. If $4x^2 + bx + 9$ is a perfect square trinomial, what is the value of b?

20. Writing When can you combine terms in a polynomial? Write a few sentences explaining your answer.

21. Open-ended Problem Write a quadratic equation that cannot be solved by factoring. Show how to solve it using another method. Explain why you chose that method.

CHAPTER 11

Tell whether the data show *direct variation, inverse variation,* or *neither.* If the data show direct or inverse variation, write an equation describing the situation.

22.

x	y
1	3
3	1
6	$\frac{1}{2}$

23.

x	y
6.2	1.55
7.0	1.75
8.4	2.10

24.

x	y
1	2
3	6
6	12

Solve each equation.

25. $\frac{3}{a+3} + 2 = \frac{7}{a+3}$
26. $\frac{5}{6y} + \frac{2}{3} = \frac{1}{y}$

ANSWERS Chapters 10–12

Cumulative Assessment

1. $-2x^2 + 3x - 2$; 2
2. $-2z^2 - 16z - 2$; 2
3. $4y^3 - 2y^2 - 2y + 12$; 3
4. $42s^2 - 40s + 8$; 2
5. $80x^3 - 32x^2 + 56x$; 3
6. $8y^3 - 50y^2 + 73y - 45$; 3
7. $(x + 7)(x - 3)$
8. $(x + 7)(x - 7)$
9. $(a - 6)(a - 3)$
10. $(y - 14)(y + 4)$
11. $(3z - 2)(z + 9)$
12. $(2x - 5)(2x - 3)$
13. $-4, 3$
14. -8
15. $\pm\frac{5}{9}$
16. $-\frac{2}{3}, -\frac{1}{3}$
17. $-\frac{1}{3}, 3$
18. $\frac{5}{6}, 1$
19. 12
20. Answers may vary. An example is given. Only like terms can be added or subtracted. These are terms with the same variables to the same degree. Add only the coefficients of the terms; leave the variable portions the same.
21. Answers may vary. An example is given. $2x^2 - 3x - 4 = 0$; Use the quadratic formula because it gives exact solutions. The solutions are $\frac{3 + \sqrt{41}}{4}$ and $\frac{3 - \sqrt{41}}{4}$.
22. inverse variation; $y = \frac{3}{x}$
23. direct variation; $y = \frac{1}{4}x$
24. direct variation; $y = 2x$
25. -1
26. $\frac{1}{4}$
27. 85

27. Here are scores for nine students in a math contest. Find the mean score.

Score	75	83	87	88
Number of students	1	2	4	2

28. a. In a trapezoid with height h and bases b_1 and b_2, the area A can be calculated using the formula $A = \frac{1}{2}h(b_1 + b_2)$. Solve the formula for b_1.

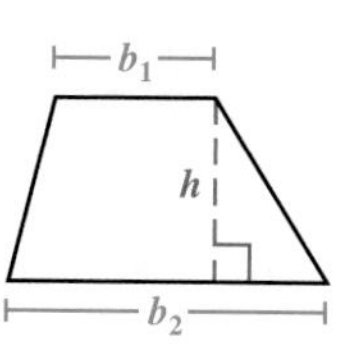

b. Find b_1 for a trapezoid with an area of 60 cm^2, a height of 4 cm, and $b_2 = 17$ cm.

Simplify each expression.

29. $\frac{7}{3ab} - \frac{5}{9a}$ **30.** $\frac{5y}{3y^2} \div \frac{15}{9y^4}$ **31.** $\frac{z-1}{2(z+1)} + \frac{3z+5}{2(z+1)}$

32. Writing Describe how to solve a rational expression.

33. Open-ended Problem Write two rational expressions. Use the same variable(s) in each expression. Multiply your expressions and simplify the resulting expression.

CHAPTER 12

34. Write an algorithm that could be used to program a computer to follow the order of operations. Be sure to include all of the operations you have learned this year.

35. Use a greedy algorithm to find the shortest tree for the graph at the right.

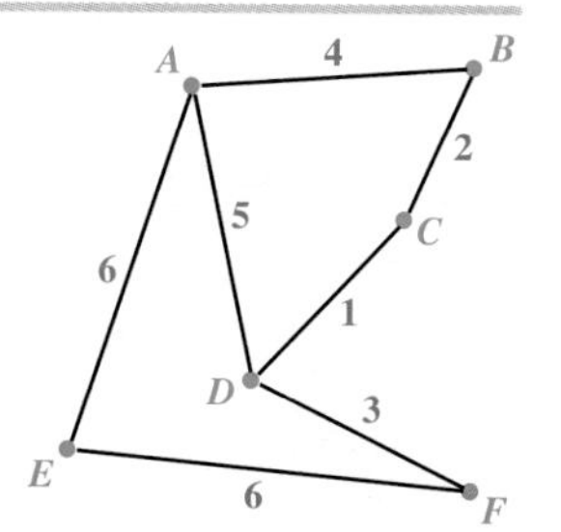

Calculate.

36. 7! **37.** 9! **38.** ${}_8C_2$ **39.** ${}_9C_3$ **40.** ${}_{13}C_2$

41. Each day, the cafeteria offers a lunch special that includes a sandwich, a salad, and a dessert. The menu includes 10 sandwiches, 4 salads, and 6 desserts. How many different lunch specials can the cafeteria offer?

42. a. Suppose you toss a penny and a nickel. What is the probability that the penny shows heads and the nickel shows tails?

b. When you toss a penny and a nickel, what is the probability of getting exactly one head and one tail?

43. Writing Suppose every pair of people in a group shakes hands once. Explain why the number of possible handshakes in a group of people is related to the number of combinations of two items chosen from a group.

44. Open-ended Problem Give an example of a situation where you might want to find the number of combinations. Give an example of a situation where you might want to find the number of permutations.

28. a. $b_1 = \frac{2A}{h} - b_2$
b. 13 cm

29. $\frac{21 - 5b}{9ab}$

30. y^3 31. 2

32. Answers may vary. An example is given. You need to find the least common denominator and rewrite each term with this denominator. Then add the numerators and reduce if possible.

33. Answers may vary. An example is given. $\frac{2xy}{5} \cdot \frac{3x}{y} = \frac{6x^2}{5}$

34. Answers may vary. Step 1: Simplify all expressions in parentheses. Step 2: Raise bases to their powers. Step 3: Multiply and divide from left to right. Step 4: Add and subtract from left to right.

35. $\overline{EA}$ or $\overline{EF}$, and $\overline{AB}$, $\overline{BC}$, $\overline{CD}$, and $\overline{DF}$

36. 5040 37. 362,880

38. 28 39. 84

40. 78 41. 240

42. a. $\frac{1}{4}$ b. $\frac{1}{2}$

43. Answers may vary. An example is given. Each handshake involves two people so the number of possible handshakes is related to the total number of people involved, taken two at a time.

44. Answers may vary. Examples are given. Use combinations to find the total number of possible committees of 4 from a group of 20 people. Use permutations to find the total possible class officer rosters containing 4 names from a group of 20 names.

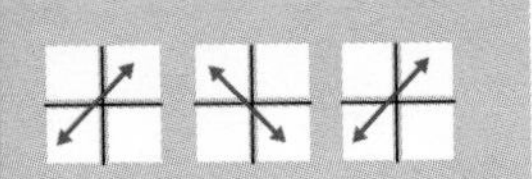

Contents of Student Resources

556

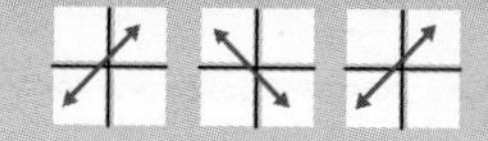

Extra Practice

CHAPTER 1

For Exercises 1–6, use the histogram at the right. *Section 1.1*

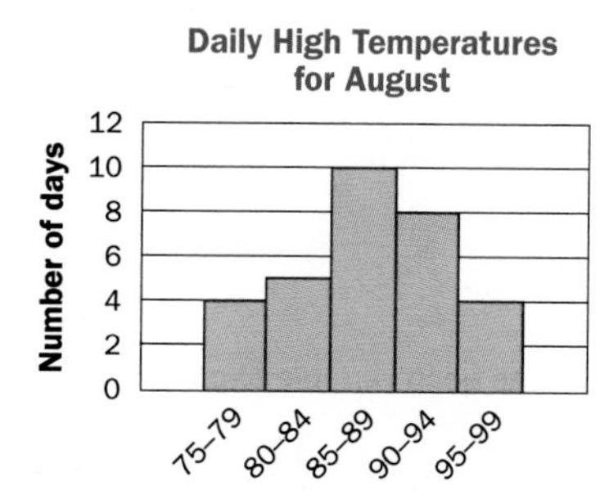

1. How many days does the histogram represent?
2. For how many days was the high temperature 90°F to 94°F?
3. a. If a high temperature of 80°F to 89°F is considered normal, how many days had normal temperatures?
   b. How many days had above-normal temperatures?
4. For how many days was the temperature $t \leq 89$?
5. Write two inequalities that describe a daily high temperature of at least 90°F.
6. Write two inequalities that describe the range of daily high temperatures for the month of August.

Simplify each expression. *Section 1.2*

7. $4(10 - 8) + 3$
8. $15 - 3(2 - 1.5)$
9. $28 - 15 \div 3 \cdot 5$
10. $\frac{16 + 8}{2 \cdot 6}$
11. $\frac{4(8)}{2(6)}$
12. $2 + \frac{7}{20} \cdot 4$
13. $0.8(25 - 3.7)$
14. $3 + (82 - 8) \div 7$
15. $\frac{8.9 - 5}{6 + 7}$
16. $\frac{3}{2} + 4(10 - 2)$
17. $\frac{5 + 7 \cdot 32}{17 \cdot 4}$
18. $\frac{(19 + 18 \div 2) \div 2}{5 + 2}$

Evaluate each variable expression when $r = 2$, $s = 5$, and $t = 6$. *Section 1.2*

19. $s^2 - 4 + t$
20. $r^2 + t^2$
21. $6t - r$
22. $3s + 2r$
23. $\frac{7t}{r + s}$
24. $7(r + s) + t$
25. $\frac{t - s}{5r}$
26. $3(t - r) + 7$
27. $\frac{t + s - r}{(s - r)^2}$
28. $s^2 - rt + r^2 - s$
29. $t^2 + 7t - st$
30. $\frac{3r^2 - t}{2s}$

Find the mean, the median, and the mode(s) of each set of data. *Section 1.3*

31. Kilometers hiked each day
    10, 18, 15, 13, 19, 6, 13
32. Number of people in computer classes
    22, 18, 25, 21, 17, 17, 18, 20
33.

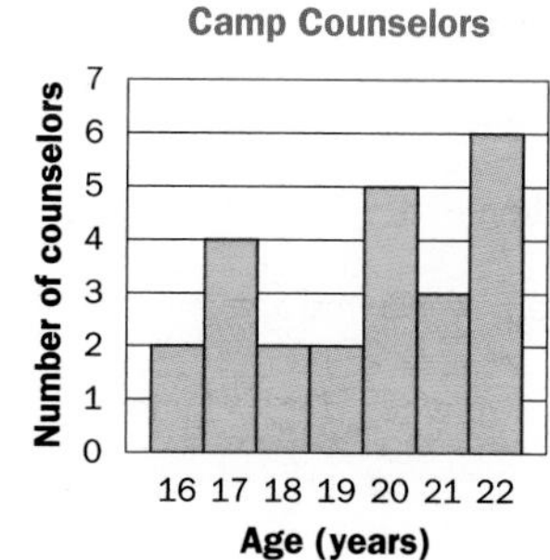

ANSWERS Extra Practice

Chapter 1

1. 31 days
2. 8 days
3. a. 15 days
   b. 12 days
4. 19 days
5. Answers may vary. Example: Let t = the daily high temperature; $t \geq 90$ and $t \leq 99$.
6. Answers may vary. Example: Let t = the daily high temperature; $t \geq 75$ and $t \leq 99$.
7. 11
8. 13.5
9. 3
10. 2
11. $2\frac{2}{3}$
12. $3\frac{2}{5}$
13. 17.04
14. $13\frac{4}{7}$
15. 0.3
16. $33\frac{1}{2}$
17. $3\frac{25}{68}$
18. 2
19. 27
20. 40
21. 34
22. 19
23. 6
24. 55
25. $\frac{1}{10}$
26. 19
27. 1
28. 12
29. 48
30. $\frac{3}{5}$
31. $13\frac{3}{7}$; 13; 13
32. 19.75; 19; 17 and 18
33. $19\frac{13}{24}$; 20; 22

34. Terri's quiz scores
99, 85, 94, 100, 82, 76, 94, 88, 95, 81

35. Number of people who swim laps each day
12, 8, 5, 9, 8, 15, 8, 8, 8, 9, 9, 10, 14, 10, 11, 12, 1

Simplify each expression. *Section 1.4*

36. $-3 + 10$

37. $9 + (-14)$

38. $5 - 17$

39. $-5 + (-16)$

40. $4 - (-2)$

41. $-8 - (-3)$

42. $15 - 19$

43. $-2.1 + 5.3$

44. $|-2| + 5$

45. $|-6| - |-5|$

46. $3|-7|$

47. $|-3 - 2|$

48. $4 - |10| + 2$

49. $-8 + |5 - 9|$

50. $|4 - 1| + |3 - 10|$

51. $\frac{|-5 - 2|}{14}$

52. $\frac{4|2 - 5|}{3}$

53. $|7.4 - 6| - (-7)$

54. $\frac{|-25|}{|-15|}$

55. $\frac{2(-1 + 5)}{12|-2|}$

Evaluate each variable expression when $a = 3$, $b = -2$, and $c = -4$. *Section 1.4*

56. $b + a$

57. $b + (-5)$

58. $a - c$

59. $4 - c + 1$

60. $a^2 + b + c$

61. $\frac{2a^2}{b + 8}$

62. $|c - b|$

63. $\frac{|b + c|}{2}$

64. $|b| - |a + c|$

65. $a|b| - (-10)$

66. $2|a + b|$

67. $7|3 - a + c|$

Simplify each expression. *Section 1.5*

68. $-7(8)$

69. $\frac{-18}{2}$

70. $(-9)(-8)$

71. $\frac{-42}{-7}$

72. $(-1)(3 - 7)$

73. $-8(5) + (-8)$

74. $|6 - 4 - (-4)|$

75. $\left(\frac{1}{2} \cdot 5\right) \cdot (-2)$

76. $\frac{|(-4)(6)|}{-3}$

77. $-6 + (-3) \cdot 7 + 4$

78. $0.5 \cdot (-15) \cdot 10$

79. $\frac{-1 + (-3)(-9)}{-7}$

80. $\frac{7}{8}(-5)(16)$

81. $4 - (5 - 8) \cdot (-3)$

82. $(-72 \div 6) - 25(-2)$

Evaluate each variable expression for the given values. *Section 1.5*

83. $4x$, when $x = -3$

84. $-5a + 3$, when $a = 4$

85. $-3p - 7$, when $p = -5$

86. $\frac{|x|}{2}$, when $x = -6$

87. $4r + s$, when $r = -5$ and $s = -4$

88. $5xyz$, when $x = -4$, $y = 2$, and $z = -1$

89. $-2|d - e| + d$, when $d = -3$ and $e = 7$

90. $\frac{-16 - 9}{2n - p}$, when $n = -2$ and $p = 1$

34. 89.4; 91; 94
35. $9\frac{4}{17}$; 9; 8
36. 7
37. –5
38. –12
39. –21
40. 6
41. –5
42. –4
43. 3.2
44. 7
45. 1
46. 21
47. 5
48. –4
49. –4
50. 10
51. $\frac{1}{2}$
52. 4
53. 8.4
54. $\frac{5}{3}$
55. $\frac{1}{3}$
56. 1
57. –7
58. 7
59. 9
60. 3
61. 3
62. 2
63. 3
64. 1
65. 16
66. 2
67. 28
68. –56
69. –9
70. 72
71. 6
72. 4
73. –48
74. 6
75. –5
76. –8
77. –23
78. –75
79. $-\frac{26}{7}$
80. –70
81. –5
82. 38
83. –12
84. –17
85. 8
86. 3
87. –24
88. 40
89. –23
90. 5

Write a variable expression for each situation. *Section 1.6*

91. The rate for a long-distance call is \$.35 for the first minute and \$.10 for each additional minute. Write a variable expression that represents the total cost of a call.

92. The cost of a basic platter at a fast-food restaurant in the mall is \$3.75. Each additional item costs \$.75. Write a variable expression that represents the cost of any platter that is ordered.

93. The estimated deer population in a certain area is 2200. It is estimated that the population increases by 130 deer each year. Write a variable expression that represents the predicted population for each future year.

Simplify each variable expression. *Section 1.7*

94. $5(2a + 3b)$ **95.** $-8(-4 + 2x)$ **96.** $-(7y - z)$

97. $7a + 3b - 3a$ **98.** $3m^2 - 4 + 9m^2 + 6$ **99.** $18n - 3n^2 - 5n^2 + n$

100. $5c + 2cd + d$ **101.** $9 - (6f + 8) + 1$ **102.** $-4(g + 8) - (5g + 3)$

103. $3(y - 4y) - 4(2 + y)$ **104.** $4t + 9t^2 - (2t^2 - t)$

105. $2(a^2 - 5a + 2a) + 9a^2 + 6a$ **106.** $-9x - 7y + 3(-x + 5y)$

For Exercises 107–109, use matrices *E*, *F*, *G*, and *H*. *Section 1.8*

$$E = \begin{bmatrix} -4 & 2 & 1 & 9 \\ 3 & -8 & 7 & -9 \\ 0 & 1 & -2 & 6 \end{bmatrix} \qquad F = \begin{bmatrix} -2 & 5 & -15 & 3 \\ 10 & -4 & 13 & -9 \\ 3 & 7 & -1 & -1 \end{bmatrix}$$

$$G = \begin{bmatrix} -1 & 9 & 3 \\ 0 & -11 & 1 \\ -6 & 14 & -4 \\ 12 & 30 & -6 \end{bmatrix} \qquad H = \begin{bmatrix} 16 & -12 & 8 \\ -2 & 8 & -2 \\ -3 & 4 & 0 \\ -13 & -20 & -1 \end{bmatrix}$$

107. Give the dimensions of each matrix.

108. Add each pair of matrices.

a. *E* and *F* b. *G* and *H*

109. Which pairs of matrices cannot be added? Explain.

CHAPTER 2

Solve each equation. *Section 2.1*

1. $3.1 + a = 5$ **2.** $n - 15 = 40$ **3.** $0.9 - b = 2.2$ **4.** $-4 = x + (-9)$

5. $4s = -64$ **6.** $22 = \frac{v}{4}$ **7.** $15 = 5 - q$ **8.** $\frac{r}{-2} = 9$

9. $-4 = -b$ **10.** $-7w = -3.5$ **11.** $\frac{m}{3} = 0.12$ **12.** $s + 9 = -45$

13. $-8a = 0.56$ **14.** $k + \frac{2}{3} = \frac{5}{6}$ **15.** $5 - x = \frac{1}{2}$ **16.** $\frac{n}{-9} = -7$

91. Let x = the number of minutes a call lasts and y = the cost in dollars; $y = 0.35 + 0.10(x - 1)$.

92. Let x = the number of additional items and y = the cost in dollars; $y = 3.75 + 0.75x$.

93. Let x = the number of years from now and y = the deer population; $y = 2200 + 130x$.

94. $10a + 15b$ 95. $32 - 16x$

96. $-7y + z$ 97. $4a + 3b$

98. $12m^2 + 2$

99. $19n - 8n^2$

100. $5c + 2cd + d$

101. $-6f + 2$

102. $-9g - 35$

103. $-13y - 8$

104. $5t + 7t^2$

105. $11a^2$

106. $-12x + 8y$

107. E: 3 × 4; F: 3 × 4; G: 4 × 3; H: 4 × 3

108. a. $\begin{bmatrix} -6 & 7 & -14 & 12 \\ 13 & -12 & 20 & -18 \\ 3 & 8 & -3 & 5 \end{bmatrix}$

b. $\begin{bmatrix} 15 & -3 & 11 \\ -2 & -3 & -1 \\ -9 & 18 & -4 \\ -1 & 10 & -7 \end{bmatrix}$

109. *E* and *G*, *E* and *H*, *F* and *G*, *F* and *H*; They do not have the same dimensions.

Chapter 2

1. 1.9 2. 55
3. –1.3 4. 5
5. –16 6. 88
7. –10 8. –18
9. 4 10. 0.5
11. 0.36 12. –54
13. –0.07 14. $\frac{1}{6}$
15. $4\frac{1}{2}$ 16. 63

Solve each equation. *Section 2.2*

17. $3x + 5 = 23$ **18.** $-22 = 4n + 2$ **19.** $12 - \frac{x}{2} = -3$ **20.** $9 + 4b = -7$

21. $13 - 6c = -17$ **22.** $-17 = -13 - 6c$ **23.** $-1 = 9 + \frac{c}{6}$ **24.** $\frac{s}{3} - 8 = 7$

25. $2m + 0.5 = 6.5$ **26.** $-42 = 3v - 6$ **27.** $27 = -9 - \frac{n}{3}$ **28.** $-3 = 42 - a$

29. $-10 + \frac{a}{6} = -2.7$ **30.** $-82 + 9b = -19$ **31.** $32 = -5f - 6$ **32.** $\frac{c}{21} + 2 = 0$

Write and solve an equation for each problem. *Section 2.2*

33. Chad bought a bottle of fruit juice that cost \$2.29 and 9 juice boxes. He paid a total of \$5.17. How much did each juice box cost?

34. The perimeter of a rectangle is 22 in. If the length of the rectangle is 7 in., what is the width of the rectangle?

For Exercises 35–37:

a. Write an equation for the function.

b. Describe the domain and the range of the function.

c. Solve the problem. *Section 2.3*

35. Connie Nardelli rents a car for \$27 per day. The total cost is a function of the number of days she rents the car. How much will she pay for an 8-day rental?

36. A city parking lot charges \$4.50 for the first hour and \$1.50 for each additional hour. The total cost is a function of the amount of time spent in the lot. How much does it cost to park for 12 hours?

37. Calligraphy pens cost \$3 each. A package of special paper costs \$5.75. The total cost of a package of paper and some pens is a function of the number of pens purchased. How much will the paper and 6 pens cost?

For Exercises 38 and 39, use the coordinate graph. *Section 2.4*

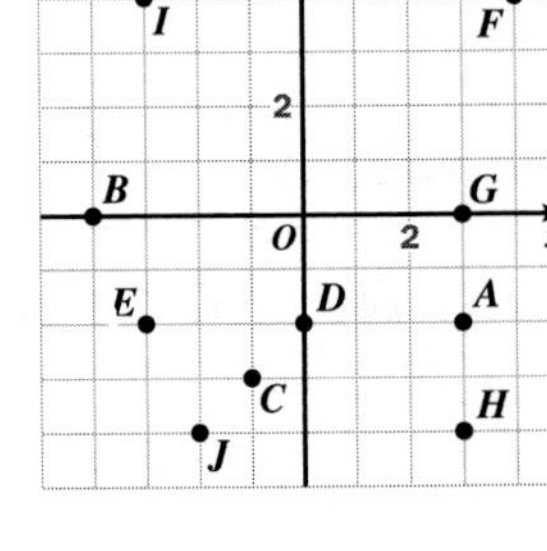

38. Write an ordered pair for each labeled point.

39. Name the quadrant (if any) in which each point lies.

Graph each point in a coordinate plane. Label each point with its letter. *Section 2.4*

40. $A(2, 0)$ **41.** $B(1, -2)$ **42.** $C(3, 3)$

43. $D(-3, 2)$ **44.** $E(4, 2)$ **45.** $F(-1, 4)$

Graph each equation. Name an ordered pair that is a solution of the equation. *Section 2.5*

46. $y = x + 5$ **47.** $y = -2x + 2$ **48.** $y = 4 - 3x$ **49.** $y = 3x$

50. $y = -x + 4$ **51.** $y = x + 2$ **52.** $y = 3 + 3x$ **53.** $y = -3x - 1$

17. 6 18. −6
19. 30 20. −4
21. 5 22. $\frac{2}{3}$
23. −60 24. 45
25. 3 26. −12
27. −108 28. 45
29. 43.8 30. 7
31. −7.6 32. −42

33. Let x = the cost in dollars of a juice box; $9x + 2.29 = 5.17$; $x = 0.32$; 32¢.

34. Let w = the width in inches; $2w + 2(7) = 22$; $w = 4$; 4 in.

35. a. Let C = the total cost in dollars and n = the number of days the car is rented ; $C = 27n$.
   b. The domain consists of the positive whole numbers and the range consists of the nonnegative multiples of 27.
   c. \$216

36. a. Let C = the total cost in dollars and n = the number of hours in the lot; $C = 4.50 + 1.5(x - 1)$.
   b. The domain consists of the positive whole numbers and the range consists of the numbers 4.50, 6.00, 7.50, … .
   c. \$21

37. a. Let C = the total cost in dollars and n = the number of pens purchased; $C = 3n + 5.75$.
   b. The domain consists of the positive whole numbers and the range consists of the numbers 5.75, 8.75, 11.75, … .
   c. \$23.75

38. $A(3, -2)$; $B(-4, 0)$; $C(-1, -3)$; $D(0, -2)$; $E(-3, -2)$; $F(4, 4)$; $G(3, 0)$; $H(3, -4)$; $I(-3, 4)$; $J(-2, -4)$

39. Point F lies in Quadrant I, point I lies in Quadrant II, points C, E, and J lie in Quadrant III, and points A and H lie in Quadrant IV. Points B, D, and G do not lie in any quadrant; they lie on axes.

40–45.

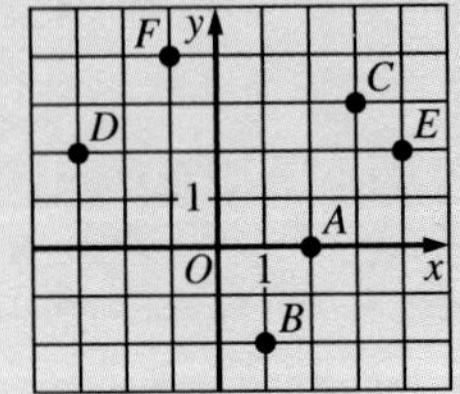

46–53. Solutions may vary. Examples are given.

46. (4, 9)

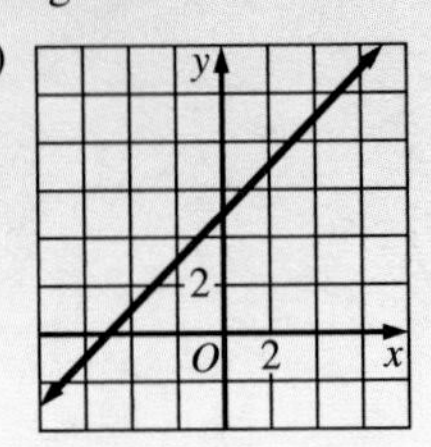

47. (−1, 4)

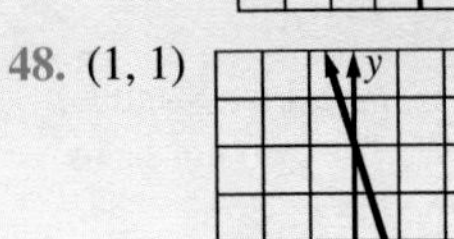

48. (1, 1)

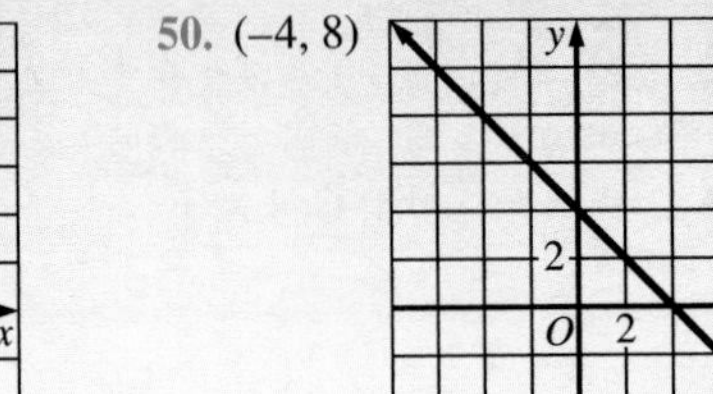

49. (−1, −3)

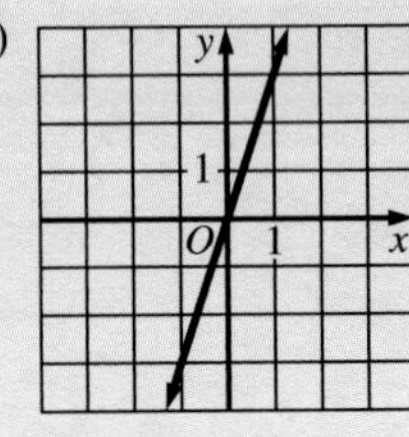

50. (−4, 8)

51. (1, 3)

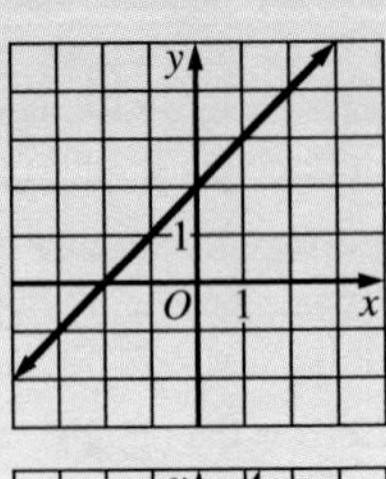

52. (0, 3)

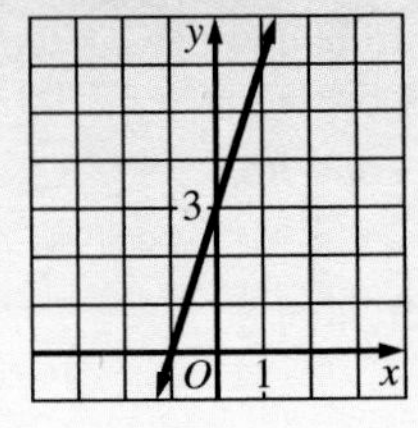

For Exercises 54 and 55, show how to use a graph to solve the problem. Then show how to use an equation. *Section 2.6*

54. Alison is trying to raise money for her soccer team by selling 100 raffle tickets. She has already sold 12 tickets. She estimates that she can sell an average of 5 tickets per day. About how many days will it take her to sell all 100 tickets?

55. Mike Ellis is paying back a loan from his parents. The original amount of the loan was \$7500. He pays \$125 per month, and has already paid \$1625. How many more payments must he make?

CHAPTER 3

Copy and complete each equation. *Section 3.1*

1. $\frac{50 \text{ cycles}}{1 \text{ s}} \cdot \frac{60 \text{ s}}{1 \text{ min}} = \underline{?}$

2. $\frac{\$450}{1 \text{ week}} \cdot \underline{?} = \frac{\$23{,}400}{1 \text{ year}}$

3. $\frac{\$1.53}{1 \text{ gal}} \cdot \frac{18 \text{ gal}}{1 \text{ tank}} = \underline{?}$

4. $\frac{\$200}{1 \text{ h}} \cdot \frac{1 \text{ h}}{60 \text{ min}} \approx \underline{?}$

5. $\frac{8 \text{ pages}}{15 \text{ min}} \cdot \underline{?} = \frac{32 \text{ pages}}{1 \text{ h}}$

6. $\frac{22{,}000 \text{ rotations}}{1 \text{ h}} \cdot \frac{1 \text{ h}}{60 \text{ min}} \approx \underline{?}$

7. $\frac{1170 \text{ mi}}{3 \text{ h}} \cdot \frac{1 \text{ h}}{60 \text{ min}} = \underline{?}$

8. $\frac{2 \text{ ft}}{1 \text{ min}} \cdot \underline{?} \cdot \frac{1 \text{ min}}{5280 \text{ ft}} = \frac{1 \text{ mi}}{44 \text{ h}}$

Express each rate in the given units. For conversion rates, see the Table of Measures on page 612. *Section 3.1*

9. 88 km/h ≈ __?__ m/s

10. 1270 ft/h ≈ __?__ mi/day

11. 2 km/h = __?__ cm/h

12. \$30,000/min = __?__ per second

13. \$290/month = __?__/year

14. 14,700 papers/week = __?__ papers/day

Tell whether the data show direct variation. If they do, give the constant of variation and write an equation that describes each situation. *Section 3.2*

15.

Tickets sold	Money received (dollars)
20	150.00
50	375.00
75	562.50
100	750.00
120	900.00

16.

Old height	New height
15 cm	31.5 cm
24 cm	50.4 cm
30 cm	63.0 cm
45 cm	94.5 cm
50 cm	105.0 cm

17.

Hours bicycling	Distance traveled
3	45 km
4	64 km
5	70 km
6	95 km
7	105 km

53. (−1, 2)

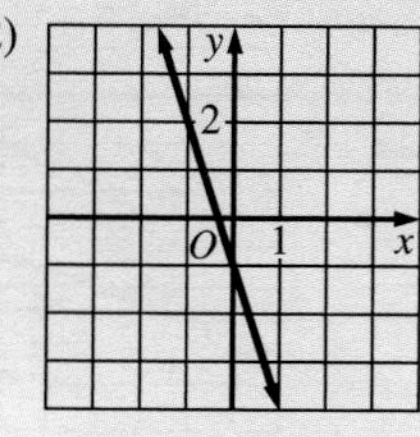

54. Let d = the number of days she sells tickets and t = the number of tickets sold; then the equation $t = 5d + 12$ describes the situation. To solve graphically, graph the equation and find 100 on the vertical axis. Then move across to the graph of the equation. Move down to the horizontal axis to find the number of days, about 18.

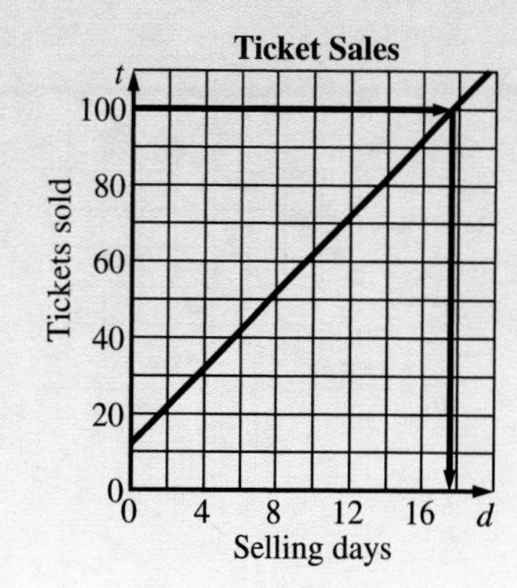

To solve using an equation, substitute 100 for t in the given equation and solve for d: $5d + 12 = 100$; $d = \frac{88}{5} = 17.6$, or about 18 days.

55. Let n = the number of payments he makes and A = the amount in dollars of the loan; then the equation $A = 125n + 1625$ describes the situation. To solve graphically, graph the equation and find 7500 on the vertical axis. Then move across to the graph of the equation. Move down to the horizontal axis to find the number of payments, 47.

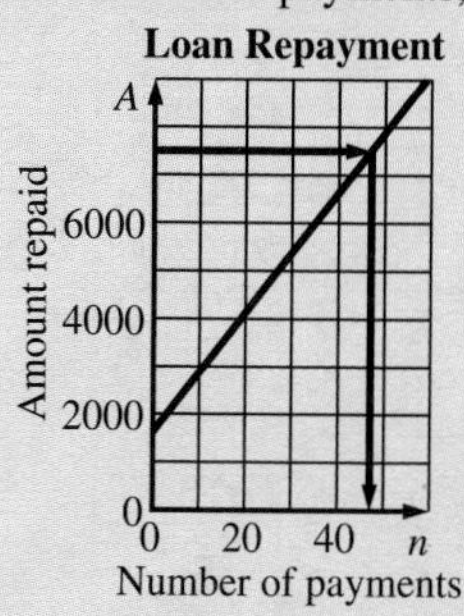

To solve using an equation, substitute 7500 for A in the given equation and solve for n: $7500 = 125n + 1625$; $n = 47$; 47 payments.

Chapter 3

1. 3000 cycles/min
2. 52 weeks/year
3. \$27.54/tank
4. \$3.33/min
5. 60 min/h
6. 366.67 rotations/min
7. 6.5 mi/min
8. 60 min/h
9. 24.44
10. 5.77
11. 200,000
12. \$500
13. \$3480
14. 2100
15. Yes; 7.5; $y = 7.5x$.
16. Yes; 2.1; $y = 2.1x$.
17. No.

18. Yes; 0.45; $y = 0.45x$

19. No.

20. No.

21–32. See answers in back of book.

33. 1 34. –2

35. 5 36. –2

37. $\frac{6}{5}$ 38. 1

39. $\frac{1}{4}$ 40. 3

41. $\frac{5}{4}$ 42. $-\frac{3}{22}$

43. $-\frac{15}{19}$ 44. 1

45. $-\frac{10}{7}$ 46. –3

47. $\frac{18}{19}$

48. $-\frac{4}{3}; -\frac{7}{3}; y = -\frac{4}{3}x - \frac{7}{3}$

49. –2; 8; $y = -2x + 8$

50. $\frac{2}{3}; -\frac{11}{3}; y = \frac{2}{3}x - \frac{11}{3}$

51. 1; –2

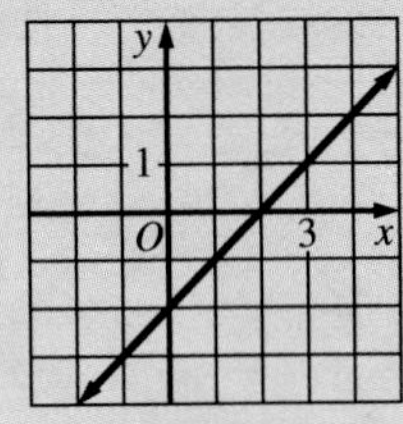

52. 4; –2

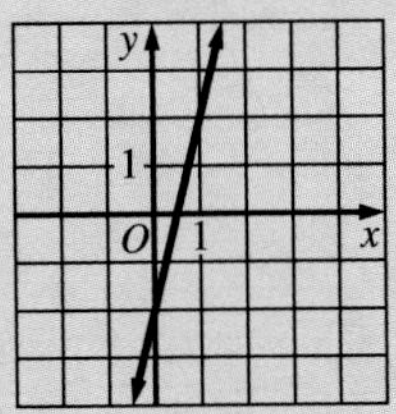

53. –3; 1

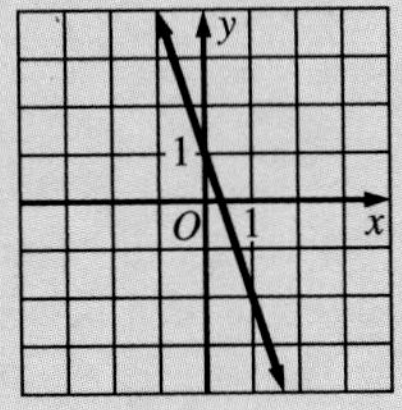

54. –1; 5

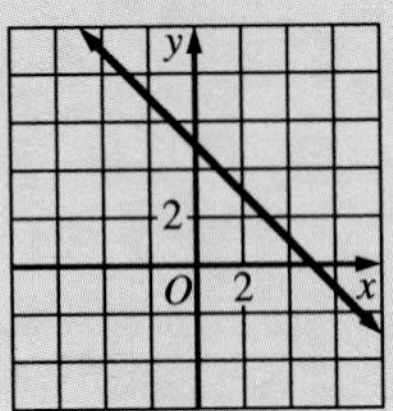

55. 8; 0

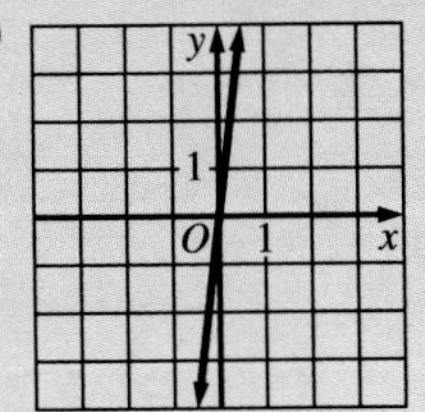

56. not defined; none

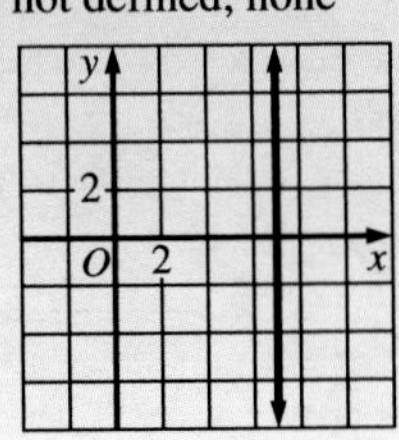

18.

Number of bagels	Cost (dollars)
1	0.45
2	0.90
3	1.35
6	2.70
8	3.60

19.

Number of counties	State population (millions)
4	1.1
8	3.3
16	1.2
58	29.8

20.

Building height (ft)	Number of stories
400	33
550	40
600	44
900	64
1050	77

Graph each equation. Find the slope of the line. *Section 3.3*

21. $y = -2x + 3$ **22.** $y = \frac{1}{3}x - 1$ **23.** $y = 3x$

24. $y = \frac{1}{4}x + 5$ **25.** $y = 2$ **26.** $y = x - 6$

27. $y = 4 - x$ **28.** $y = -4x - 3$ **29.** $y = 9 + 2x$

30. $y = 2 + x$ **31.** $y = 5x - 2$ **32.** $y = -3 - x$

Find the slope of the line through each pair of points. *Section 3.3*

33. (2, 3) and (5, 6) **34.** (−1, 5) and (2, −1) **35.** (3, −3) and (5, 7)

36. (5, −2) and (6, −4) **37.** (8, 2) and (3, −4) **38.** (−4, −4) and (3, 3)

39. (−6, 2) and (−2, 3) **40.** (10, −7) and (15, 8) **41.** (−7, −9) and (1, 1)

42. (14, 6) and (−8, 9) **43.** (8, −12) and (−11, 3) **44.** (20, 2) and (4, −14)

45. (4, −13) and (−3, −3) **46.** (25, 35) and (20, 50) **47.** (−21, −7) and (−2, 11)

Give the slope and *y*-intercept of each graph. Then write an equation in slope-intercept form. *Section 3.4*

48.

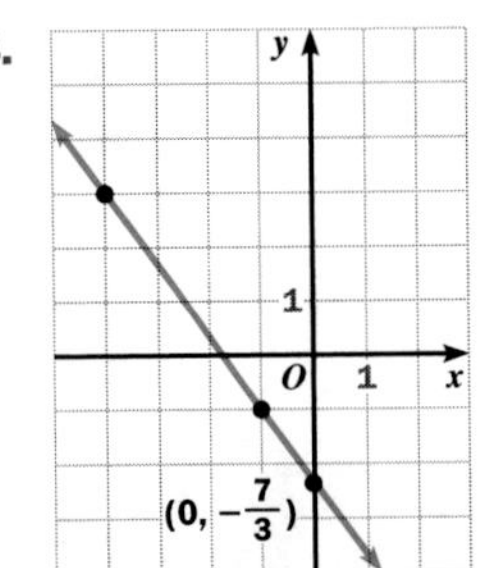

49.

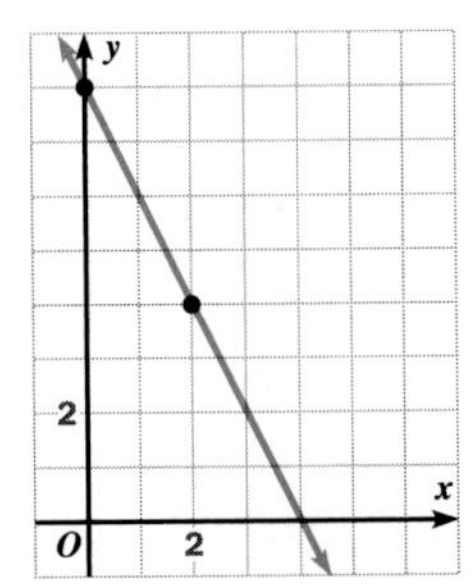

50.

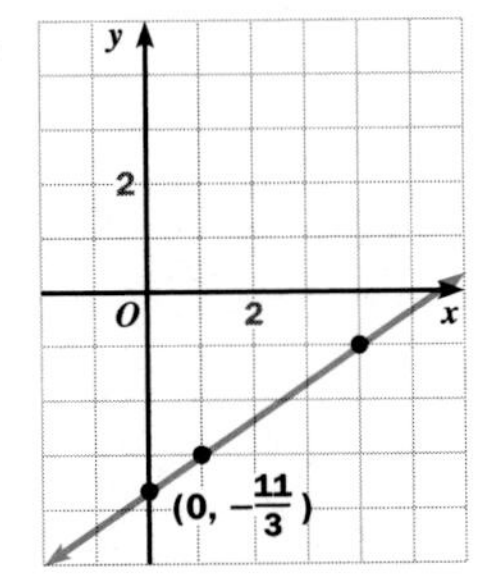

Graph each equation. Give the slope and *y*-intercept of the graph if they exist. *Section 3.4*

51. $y = x - 2$ **52.** $y = 4x - 2$ **53.** $y = -3x + 1$ **54.** $y = -x + 5$

55. $y = 8x$ **56.** $x = 7$ **57.** $y = \frac{1}{2}x + 6$ **58.** $y = x + 1$

57. $\frac{1}{2}$; 6

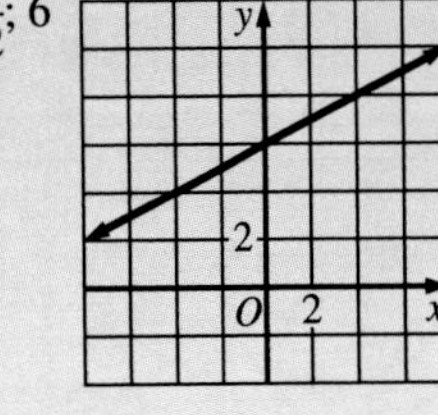

58. 1; 1

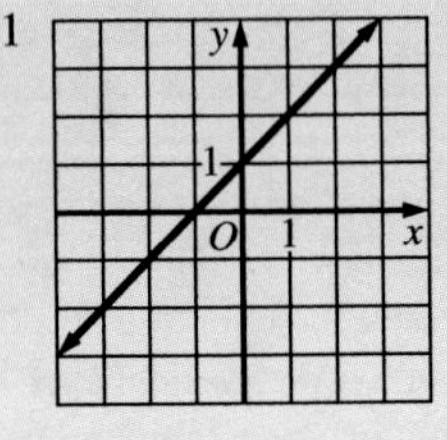

59. $\frac{1}{3}$; –2

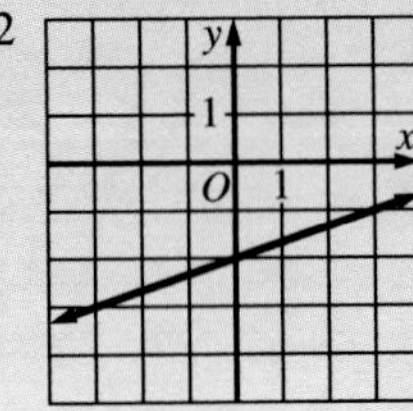

59. $y = \frac{1}{3}x - 2$ 60. $y = \frac{2}{3}x + 4$ 61. $y = \frac{3}{2}x$ 62. $y = 5x - 10$

63. $y = -5x - (-1)$ 64. $y = -1$ 65. $y = -7x - 8$ 66. $y = \frac{5}{2} - \frac{3}{2}x$

Find an equation of the line with the given slope and through the given point.
Section 3.5

67. slope $= 2$; $(1, 3)$ 68. slope $= 5$; $(-2, 1)$ 69. slope $= -3$; $(4, -5)$

70. slope $= \frac{1}{3}$; $(-8, 2)$ 71. slope $= 0$; $(-4, 2)$ 72. slope $= 6$; $(8, -12)$

73. slope is undefined; $(5, -14)$ 74. slope $= -\frac{2}{3}$; $(9, 4)$ 75. slope $= \frac{1}{4}$; $(3, -2)$

76. slope $= \frac{7}{2}$; $(6, 1)$ 77. slope $= -11$; $\left(\frac{1}{2}, 0\right)$ 78. slope $= -4$; $\left(-\frac{1}{5}, \frac{3}{10}\right)$

Find an equation of the line through the given points. *Section 3.5*

79. $(1, 5), (2, 4)$ 80. $(-2, 3), (4, 9)$ 81. $(4, -3), (-5, -5)$ 82. $(-1, -7), (2, 2)$

83. $(2, -1), (-1, -9)$ 84. $(0, 0), (3, -6)$ 85. $(3, 4), (6, 0)$ 86. $(-4, 10), (-5, 7)$

Find an equation that models the data in each scatter plot. *Section 3.6*

87.
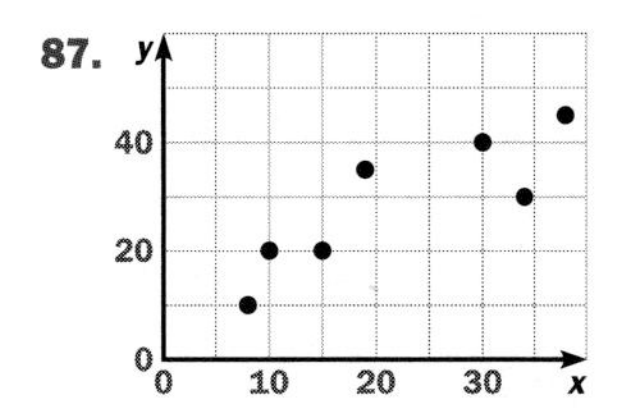

88.
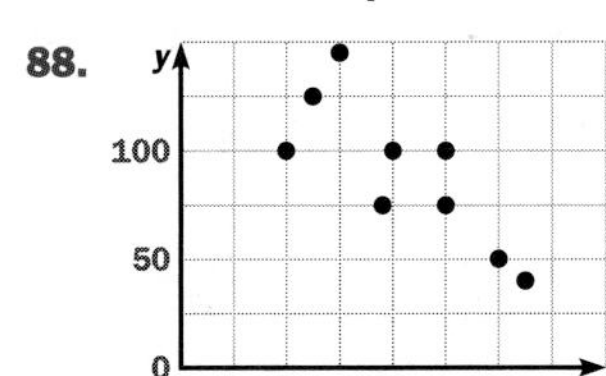

89.
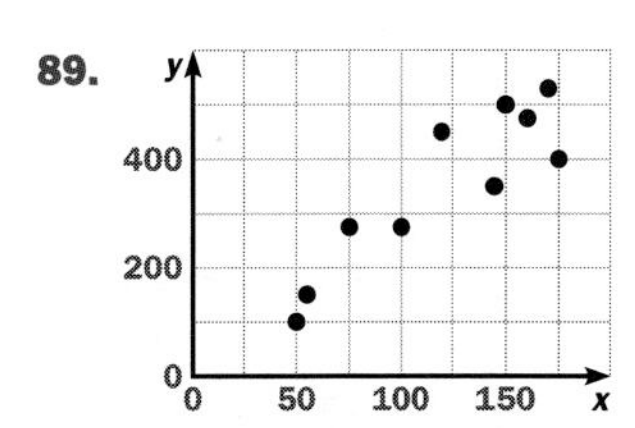

CHAPTER 4

For Exercises 1–4, use the graph at the right. *Section 4.1*

1. a. What is the cost of one round-trip ride?
   b. Write an expression for the cost of x round-trip rides.
2. Suppose you want to spend as little as possible on bus fares. If you plan to make 15 round-trip rides this month, should you buy a monthly pass or pay each day? Why?
3. Which costs more, 23 daily rides or a monthly pass? How much more?
4. When does it become more economical to buy a monthly bus pass? Explain.

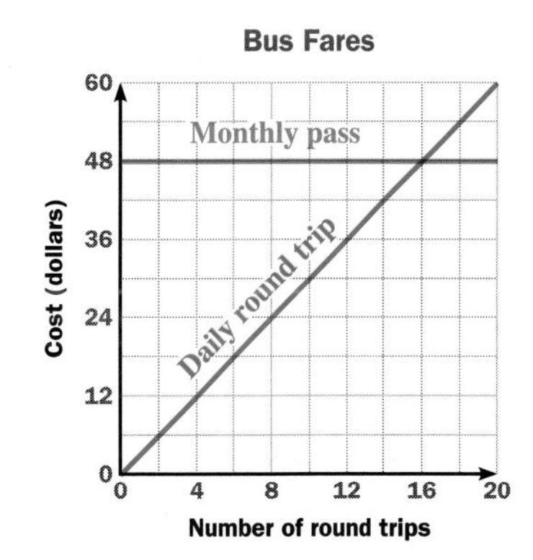

63. –5; 1
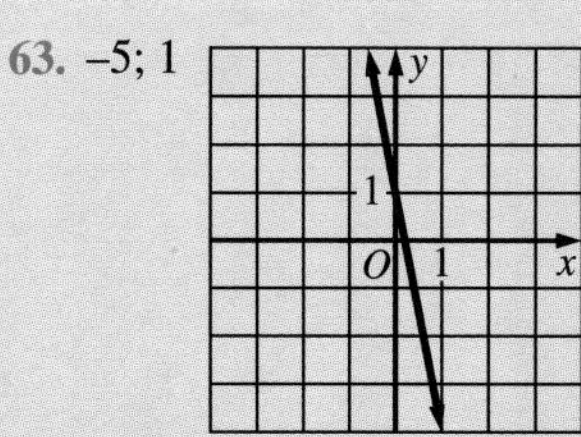

64. 0; –1
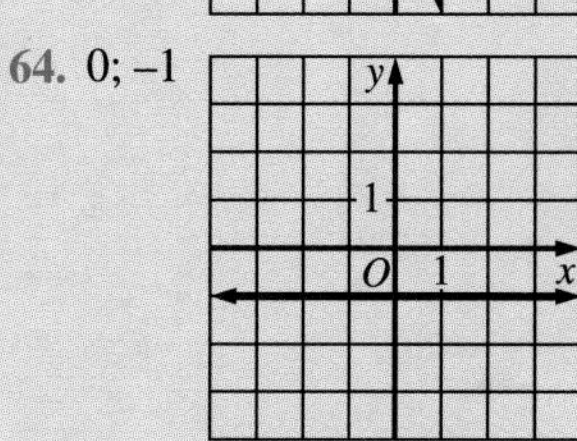

65. –7; –8
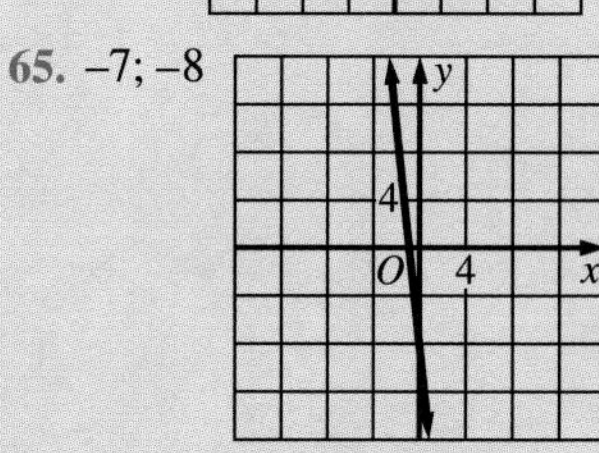

66. $-\frac{3}{2}; \frac{5}{2}$
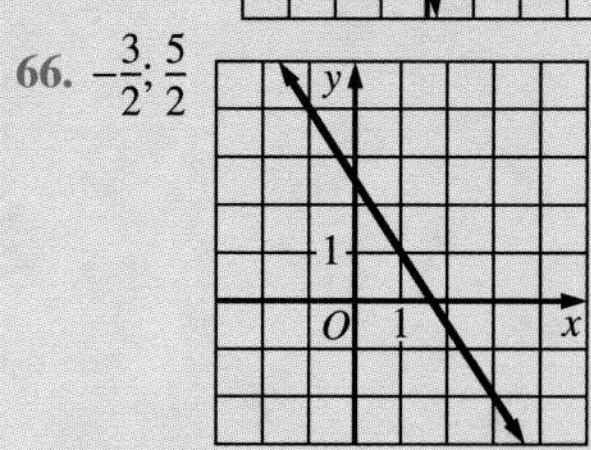

67. $y = 2x + 1$ 68. $y = 5x - 9$
69. $y = -3x + 7$ 70. $y = \frac{1}{3}x + \frac{14}{3}$
71. $y = 2$ 72. $y = 6x - 60$
73. $x = 5$ 74. $y = -\frac{2}{3}x + 10$
75. $y = \frac{1}{4}x - \frac{11}{4}$ 76. $y = \frac{7}{2}x - 20$
77. $y = -11x + \frac{11}{2}$ 78. $y = -4x - \frac{1}{2}$
79. $y = -x + 6$ 80. $y = x + 5$
81. $y = \frac{2}{9}x - \frac{35}{9}$ 82. $y = 3x - 4$
83. $y = \frac{8}{3}x - \frac{19}{3}$ 84. $y = -2x$
85. $y = -\frac{4}{3}x + 8$ 86. $y = 3x + 22$
87–89. Answers may vary. Examples are given.
87. $y = x + 8$ 88. $y = -50x + 200$
89. $y = 2.7x + 26.4$

60. $\frac{2}{3}$; 4
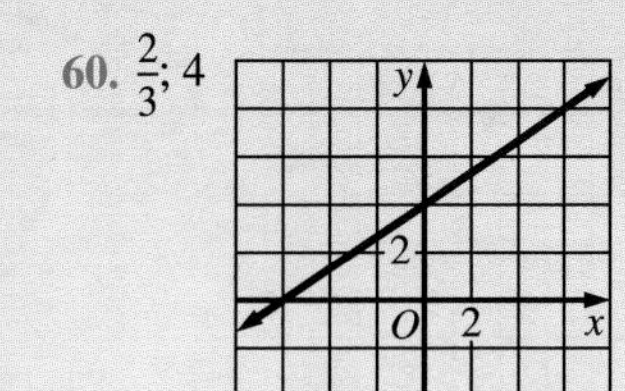

61. $\frac{3}{2}$; 0
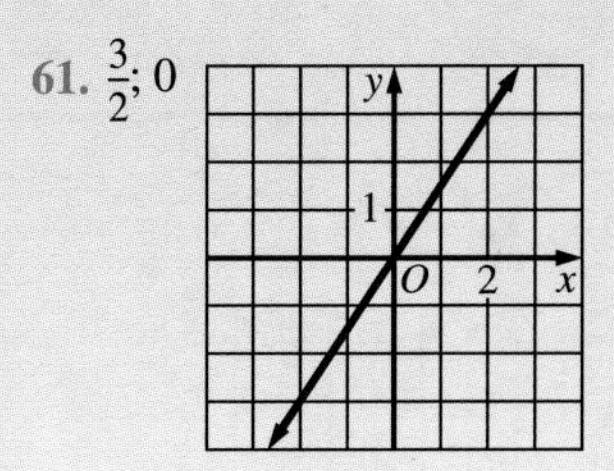

62. 5; –10
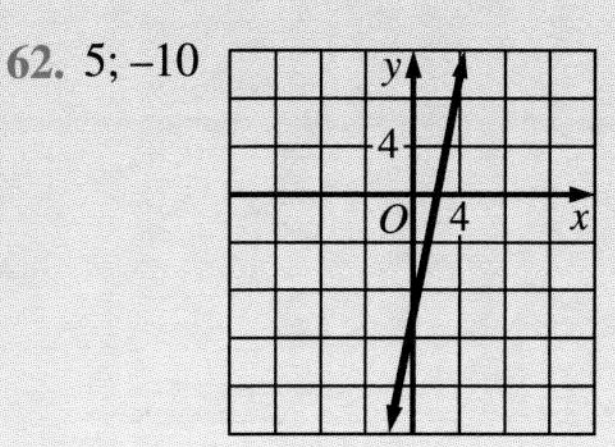

Chapter 4

1. a. \$3 b. $3x$
2. Answers may vary. An example is given. I think you should pay each day because the total cost is less than the cost of a pass. However, since the cost difference is only \$3, if it is more convenient to use the pass, it might be worthwhile to buy one.
3. 23 daily rides; \$21
4. when you take more than 16 round trips per month; When x is greater than 16, the graph for the daily round trip is above the graph for the monthly pass, indicating that the monthly pass is cheaper.

Rewrite each expression with a fraction as the coefficient. Then give the reciprocal of the coefficient. *Section 4.2*

5. $\frac{3a}{-2}$ **6.** $\frac{h}{12}$ **7.** $\frac{-5b}{8}$ **8.** $\frac{-r}{9}$

Solve each equation. *Section 4.2*

9. $\frac{2}{3}n = 14$ **10.** $\frac{-1}{8}a = -3$ **11.** $\frac{2x}{5} = 6$ **12.** $15 = -\frac{1}{4}h$

13. $\frac{-7c}{3} = -14$ **14.** $-9 = \frac{2}{7}y$ **15.** $\frac{-z}{6} = 5$ **16.** $-\frac{3t}{8} = -24$

17. $\frac{-4x}{3} = -8$ **18.** $\frac{3}{4}x + 1 = 16$ **19.** $8 = \frac{2r}{3} - 4$ **20.** $\frac{3}{5}n + 4 = -5$

21. $7 = \frac{1}{2}x - 5$ **22.** $-\frac{3}{2}y + 2 = -13$ **23.** $-3 + \frac{1}{4}s = 19$ **24.** $-12 = -\frac{5n}{2} + 3$

25. $\frac{-a}{7} - 6 = 2$ **26.** $-\frac{5}{6}n + 1 = 0$ **27.** $\frac{-b}{3} - 1 = -5$ **28.** $3 = \frac{2x}{5} + 6$

Solve each equation. *Section 4.3*

29. $2x + 7 = 3x$ **30.** $9a = 6a - 6$ **31.** $5b - 2 = 8b - 14$

32. $-8n + 3 = -11 - n$ **33.** $3r - 10 = -5r - 14$ **34.** $2n + 6 = 5n$

35. $8y = 3y - 10$ **36.** $5(a + 3) - 6 = 2a - 6$ **37.** $\frac{1}{2}(4n + 6) = 3n + 2(9 - 2n)$

Solve each equation. If an equation is an *identity* or there is *no solution*, say so. *Section 4.3*

38. $5n - 4 = 10n + 6$ **39.** $3(x - 7) = 15 + 3x$ **40.** $7a + 3(a - 9) = 3$

41. $2(n - 3) = 2n - 6$ **42.** $4(x + 4) = 3(x - 1)$ **43.** $-7(m + 2) = 5(m + 3)$

44. $3(2x - 1) = 5x + 18$ **45.** $7(x - 2) = 3 + 7x$ **46.** $3d + 7 = 6 - d$

47. $\frac{4}{5}x - 1 = \frac{4}{5}x + 1$ **48.** $-5 - \frac{1}{2}c = 6 + \frac{1}{2}c$ **49.** $2n + 6 - 6n = 2(3 - 2n)$

50. $2(-5 - 3a) = 7a + 3$ **51.** $4(x - 5) = 3(6 - 2x)$ **52.** $8g - 12 = 3(4g - 4)$

53. $-1 - t = -t + 1$ **54.** $6n + 3 = 2n + 1$ **55.** $4(h - 3) = h - 11 + 3h$

56. $44 - 5x = 4(-3x - 3)$ **57.** $\frac{3}{2}(6b - 10) = 4b$ **58.** $9c - 15 = -3(5 - 3c)$

Find the least common denominator of the fractions in each equation. Then solve each equation. *Section 4.4*

59. $\frac{r}{4} - \frac{r}{5} = 6$ **60.** $\frac{1}{2}a + 6 = \frac{1}{3}a$ **61.** $\frac{1}{5}n - 2n = 1$ **62.** $2 - \frac{2}{3}s = \frac{5}{6}s$

63. $\frac{x}{3} - \frac{x}{5} = \frac{4}{3}$ **64.** $\frac{3}{4}b - \frac{2}{3}b = \frac{1}{6}$ **65.** $\frac{1}{8}n - \frac{1}{2} = \frac{1}{4}n$ **66.** $\frac{3c}{5} - \frac{c}{3} = -24$

5. $-\frac{3}{2}a; -\frac{2}{3}$ 6. $\frac{1}{12}h$; 12
7. $-\frac{5}{8}b; -\frac{8}{5}$ 8. $-\frac{1}{9}r$; –9
9. 21 10. 24
11. 15 12. –60
13. 6 14. $-\frac{63}{2}$
15. –30 16. 64
17. 6 18. 20
19. 18 20. –15
21. 24 22. 10
23. 88 24. 6
25. –56 26. $\frac{6}{5}$
27. 12 28. $-\frac{15}{2}$
29. 7 30. –2
31. 4 32. 2
33. $-\frac{1}{2}$ 34. 2
35. –2 36. –5
37. 5 38. –2
39. no solution 40. 3
41. identity 42. –19
43. $-\frac{29}{12}$ 44. 21
45. no solution 46. $-\frac{1}{4}$
47. no solution 48. –11
49. identity 50. –1
51. $3\frac{4}{5}$ 52. 0
53. no solution 54. $-\frac{1}{2}$
55. no solution 56. –8
57. 3 58. identity
59. 20; 120 60. 6; –36
61. 5; $-\frac{5}{9}$ 62. 6; $\frac{4}{3}$
63. 15; 10 64. 12; 2
65. 8; –4 66. 15; –90
67. 10; 18 68. 10; 3
69. 100; 3 70. 100; 1
71. 10; 3 72. 10; $\frac{1}{3}$
73. 1000; 359 74. 100; 161
75. 10; 5 76. 100; 1
77. 100; $\frac{1}{60}$ 78. 10; 0.8

79. $y < -1$

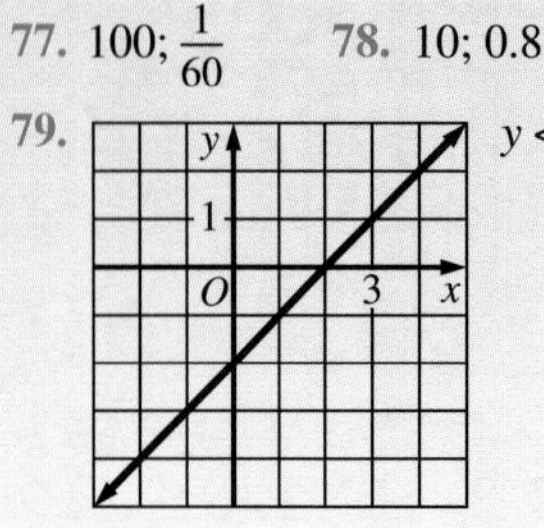

80. $y > -3$

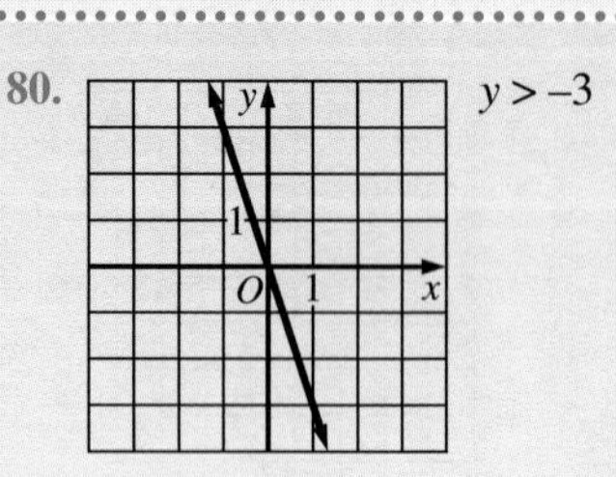

Tell what power of ten you would multiply by to make each equation easier to solve. Then solve each equation. *Section 4.4*

67. $3 = 0.1n + 1.2$ **68.** $5 - 1.5m = 0.5$ **69.** $0.5x - 0.02 = 1.48$

70. $0.15 + 1.8c = 1.95$ **71.** $2.1 = 0.5x + 0.2x$ **72.** $1.5 = 6n - 1.5n$

73. $0.002y - 0.018 = 0.7$ **74.** $0.02t = 4.83 - 0.01t$ **75.** $1.8 - 0.3b = 0.3$

76. $0.09d + 0.11 = 0.2d$ **77.** $3x - 0.02 = 1.8x$ **78.** $0.2 - 0.3a = 1.2a - 1$

For Exercises 79–81, graph each equation. Then write an inequality for each graph that represents the *y*-values when $x < 1$. *Section 4.5*

79. $y = x - 2$ **80.** $y = -3x$ **81.** $y = -x + 3$

For Exercises 82–84, use the graph. *Section 4.5*

82. The graph shows the costs of buying videos through two video clubs. Which club charges a membership fee? What is the fee?

83. Write an inequality that shows when it is more economical to join Club B. Let v = the number of videos purchased.

84. Write an inequality that shows when it is more economical to join Club A. Let v = the number of videos purchased.

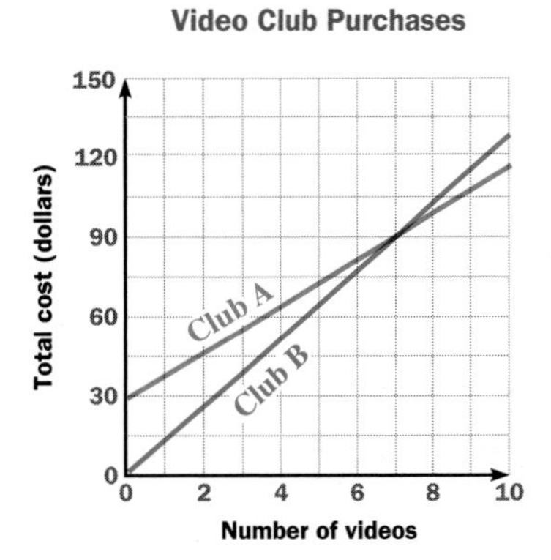

Solve each inequality. Graph the solution on a number line. If there is no solution, write *no solution*. *Section 4.6*

85. $x - 3 > 1$ **86.** $-4a > -20$ **87.** $\frac{1}{2}x < 5$

88. $6 \le -3d$ **89.** $-3t + 4 < -8$ **90.** $3x - 1 \ge 2$

91. $7n + 6 > 6n$ **92.** $2t + 4 \le -7 + 2t$ **93.** $32 - 5a > -22$

94. $1 - 3s < -8 - s$ **95.** $4r + 3 \le 5r - (-3)$ **96.** $5c \ge 6(c - 1)$

97. $\frac{2n}{3} > \frac{n}{4} - 5$ **98.** $-4k + 3.5 \le 2.3 - 4k$ **99.** $4(3x + 4) < -2(-2 + x)$

Write and solve an inequality for each problem. *Section 4.6*

100. Dawn's neighbor will pay her \$13 to mow the front lawn and \$4 per hour to take care of the garden. If she mows the front lawn, what is the least number of hours she has to work on the garden in order to earn at least \$25?

101. The weight of a carton of books to be shipped must be under 44 lb. If each book weighs 0.6 lb and the carton weighs 2 lb, what is the greatest number of books that can go in the carton?

81. $y > 2$

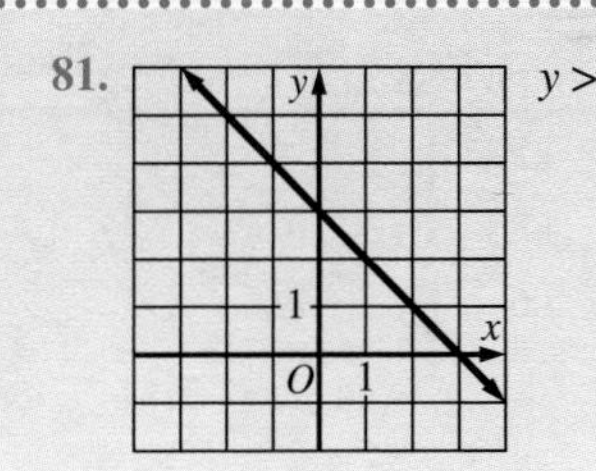

82. Club A; Estimates may vary; about \$30.

83. $v < 7$

84. $v > 7$

85. $x > 4$

86. $a < 5$

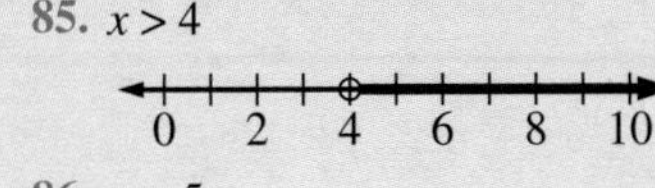

87. $x < 10$

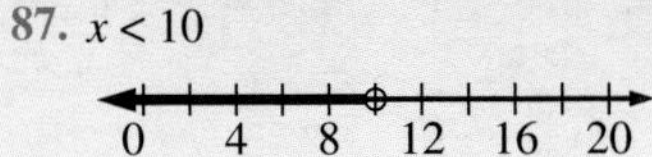

88. $d \le -2$

89. $t > 4$

90. $x \ge 1$

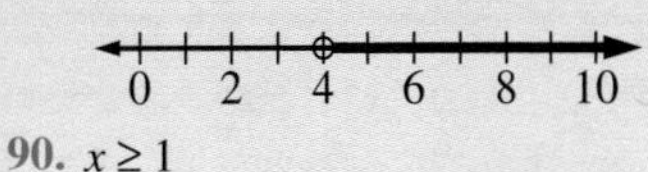

91. $n > -\frac{6}{13}$

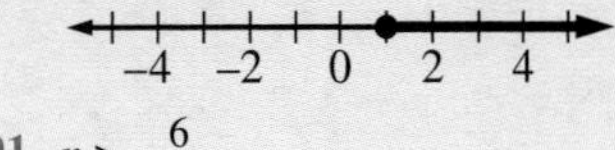

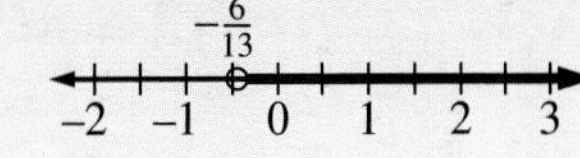

92. no solution

93. $a < 10\frac{4}{5}$

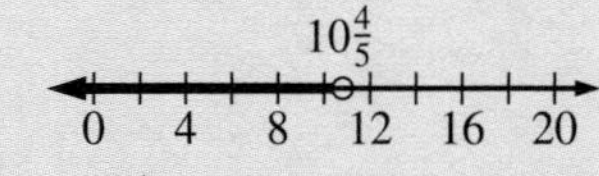

94. $s > 4\frac{1}{2}$

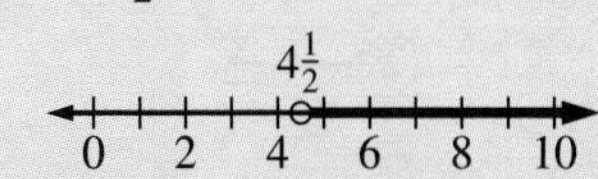

95. $r \ge 0$

96. $c \le 6$

97. $n > -12$

98. no solution

99. $x < -\frac{6}{7}$

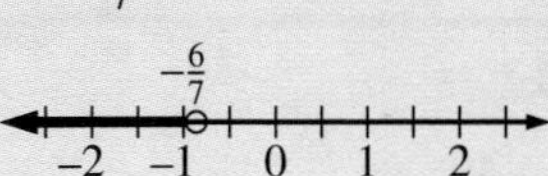

100. Let n = the number of hours she has to work; $4n + 13 \ge 25$; $n \ge 3$; 3 h.

101. Let n = the number of books in a carton; $0.6n + 2 < 44$; $n < 70$; at most, 69 books can go in the carton.

CHAPTER 5

Find the ratio of the first quantity to the second quantity. Write your answer as a fraction in lowest terms. *Section 5.1*

1. 150 m; 300 m
2. 6 in.; 2 ft
3. 1 m; 1 km
4. 200 mL; 2 kL
5. 1 mi; 2000 ft
6. 1 mm; 2 m
7. 192 oz; 3 lb
8. 20 g; 4 kg
9. 0.72 m; 40 cm

Solve each proportion. *Section 5.1*

10. $\frac{5}{12} = \frac{10}{n}$
11. $\frac{1}{4} = \frac{b}{10}$
12. $\frac{v}{8} = \frac{63}{18}$
13. $\frac{12}{6} = \frac{a}{18}$
14. $\frac{1.2}{4.8} = \frac{0.7}{k}$
15. $\frac{2}{5.6} = \frac{3}{n}$
16. $\frac{2}{3} = \frac{12}{2a}$
17. $\frac{x}{3.5} = \frac{2}{7}$
18. $\frac{21}{1.5} = \frac{1.4}{v}$
19. $\frac{x+2}{3} = \frac{9}{2}$
20. $\frac{9}{4a} = \frac{1}{6}$
21. $\frac{1}{x+3} = \frac{6}{7}$
22. $\frac{c}{1+c} = \frac{5}{9}$
23. $\frac{3p-2}{4p} = \frac{3}{8}$
24. $\frac{d+3}{6} = \frac{d-1}{2}$

25. A floor plan of a house is drawn to a scale of 1 in. = $2\frac{1}{2}$ ft. Find the dimensions of these rooms in the floor plan. *Section 5.2*
  a. living room: 15 ft by 20 ft
  b. bathroom: 8 ft by 10 ft
  c. kitchen: 12 ft by 13 ft
  d. bedroom: 14 ft by 16 ft

26. Give the scale factor used in each situation.
  a. enlarge a 24 cm long picture to 64 cm long
  b. reduce a 3 ft high poster to 15 in. high
  c. reduce a 9 ft long banner to 9 in. long
  d. enlarge a 7.5 ft wide billboard photo to 4 yd wide

27. Fred is using a copy machine. Give the scale factor he should use to make each change.
  a. enlarge a 3 in. wide photo to $4\frac{1}{2}$ in. wide
  b. enlarge a 9 cm wide chart to 21 cm wide
  c. reduce an 11 in. by 14 in. paper to approximately $8\frac{1}{2}$ in. by 11 in.
  d. reduce a 7 in. wide diagram to $3\frac{1}{2}$ in. wide

Chapter 5

1. $\frac{1}{2}$
2. $\frac{1}{4}$
3. $\frac{1}{1000}$
4. $\frac{1}{10,000}$
5. $\frac{66}{25}$
6. $\frac{1}{2000}$
7. $\frac{4}{1}$
8. $\frac{1}{200}$
9. $\frac{9}{5}$
10. 24
11. $2\frac{1}{2}$
12. 28
13. 36
14. 2.8
15. 8.4
16. 9
17. 1
18. 0.1
19. $\frac{23}{2}$
20. $13\frac{1}{2}$
21. $-\frac{11}{6}$
22. $\frac{5}{4}$
23. $\frac{4}{3}$
24. 3
25. a. 6 in. by 8 in.
  b. 3.2 in. by 4 in.
  c. 4.8 in. by 5.2 in.
  d. 5.6 in. by 6.4 in.
26. a. $\frac{8}{3}$
  b. $\frac{5}{12}$
  c. $\frac{1}{12}$
  d. $\frac{8}{5}$

For Exercises 28–33, use figures *ABCD* and *PQRS*. Figure *ABCD* ~ figure *PQRS*. *Section 5.3*

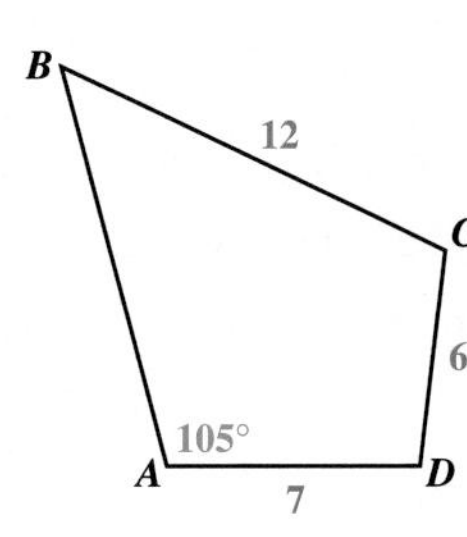

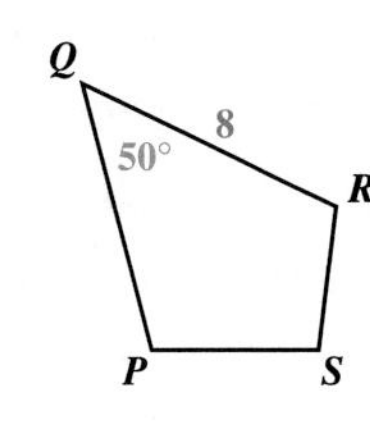

28. Find the measure of $\angle B$.

29. Find the measure of $\angle P$.

30. $\overline{DC}$ and $\underline{\ ?\ }$ are corresponding sides.

31. What is the ratio of BC to QR?

32. Find RS.

33. Find PS.

For Exercises 34–36, tell whether the triangles are similar. If they are, list each pair of corresponding angles and sides. If they are not similar, explain why. *Section 5.3*

34.

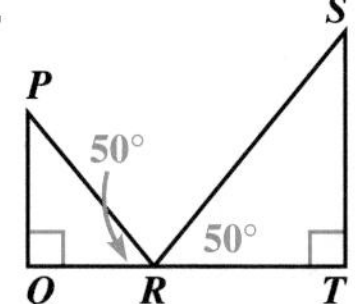

35.

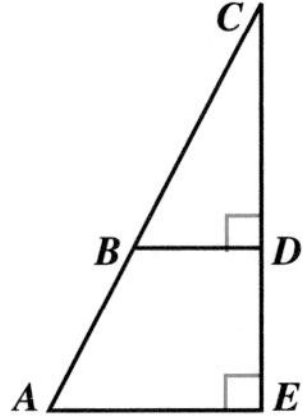

36.

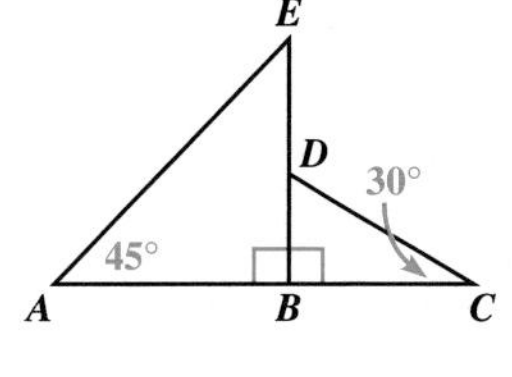

37. Suppose quadrilateral *DEFG* is similar to quadrilateral *HIJK*, and $\angle F = 120°$. What is the measure of $\angle J$?

38. a. Suppose $\triangle ABC \sim \triangle DEF$ and $AB = 20$, $DE = 5$, and $AC = 16$. How long is $\overline{DF}$?

b. Suppose $EF = 3$. How long is $\overline{BC}$?

Find the missing values for each pair of similar figures. *Section 5.4*

39.

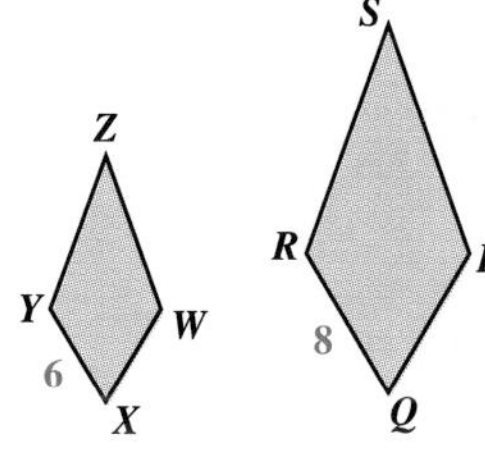

$P = \underline{\ ?\ }$
$A = 42$

$P = 48$
$A = \underline{\ ?\ }$

40.

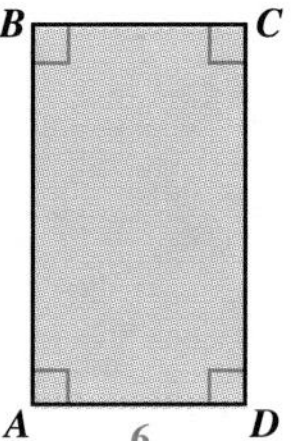

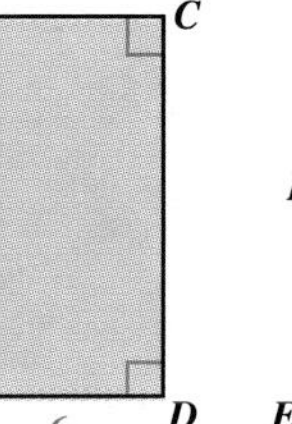

$P = 40$
$AB = \underline{\ ?\ }$
$A = \underline{\ ?\ }$

$P = \underline{\ ?\ }$
$EF = \underline{\ ?\ }$

41.

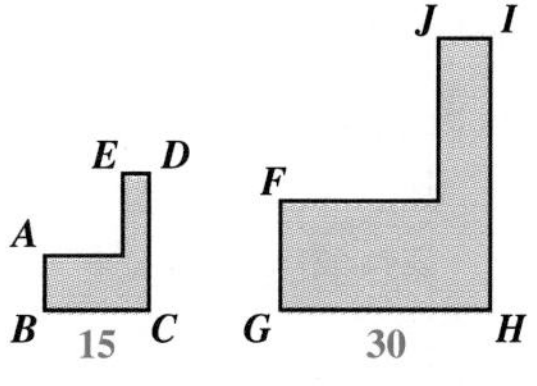

$P = \underline{\ ?\ }$
$A = 200$

$P = 140$
$A = \underline{\ ?\ }$

27. a. $\frac{3}{2}$

b. $\frac{7}{3}$

c. either $\frac{17}{22}$ or $\frac{11}{14}$

d. $\frac{1}{2}$

28. 50°

29. 105°

30. $\overline{SR}$

31. 3 to 2

32. 4

33. $4\frac{2}{3}$

34. Yes; $\angle P$ and $\angle S$, $\angle Q$ and $\angle T$, $\angle PRQ$ and $\angle SRT$; $\overline{PQ}$ and $\overline{ST}$; $\overline{PR}$ and $\overline{SR}$; $\overline{QR}$ and $\overline{TR}$.

35. Yes; $\angle C$ and $\angle C$, $\angle CBD$ and $\angle A$, $\angle CDB$ and $\angle E$; $\overline{CB}$ and $\overline{CA}$; $\overline{CD}$ and $\overline{CE}$; $\overline{BD}$ and $\overline{AE}$.

36. No.

37. 120°

38. a. 4

b. 12

39. $P = 36$; $A = 74\frac{2}{3}$

40. $AB = 14$; $A = 84$; $P = 20$; $EF = 7$

41. $P = 70$; $A = 800$

For Exercises 42–46, use the spinner shown. Write each probability as a fraction and as a percent. *Section 5.5*

42. a. What is the probability of spinning a 6?

b. What is the probability of spinning a number other than 6?

43. What is the probability of spinning an even number?

44. What is the probability of spinning a number greater than 8?

45. What is the probability of spinning a number less than 5?

46. a. Suppose that for 120 spins, an odd number came up 40 times. What is the experimental probability of spinning an odd number?

b. What is the theoretical probability of spinning an odd number?

For Exercises 47–52, the targets are a regular hexagon, squares, an equilateral triangle, and a circle. Find the probability that a randomly thrown dart that hits each target hits the shaded area. *Section 5.6*

47.

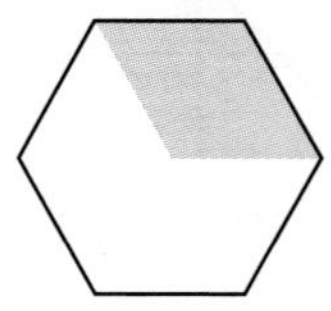

48.

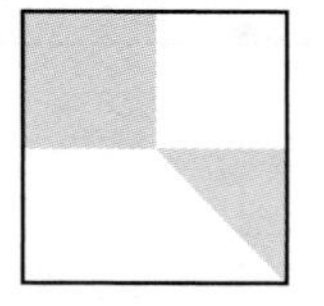

49.

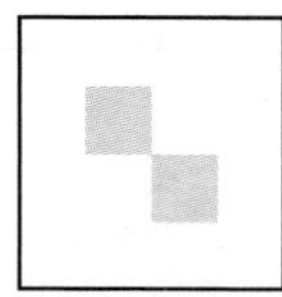

50.

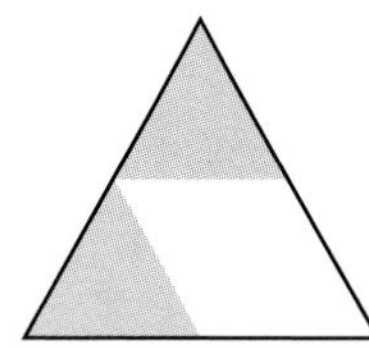

51.

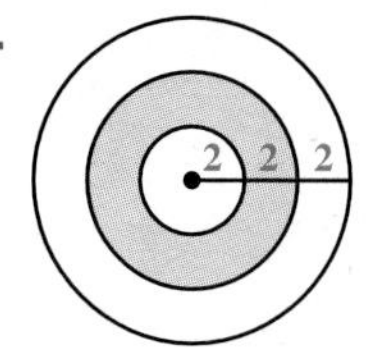

52.

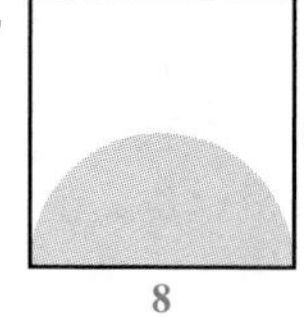

CHAPTER 6

Find the unknown length for each right triangle. *Section 6.1*

1.

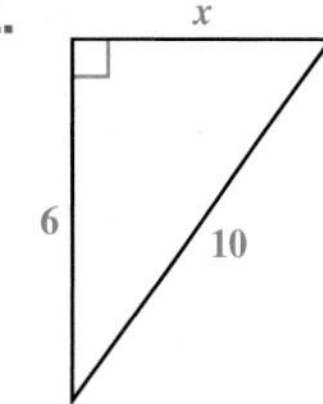

2.

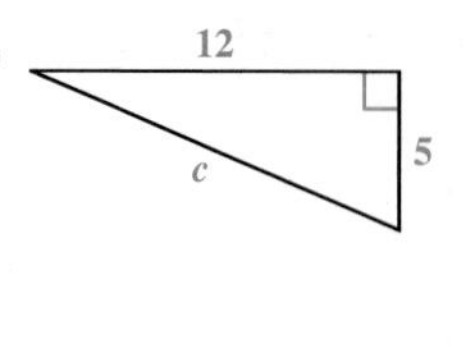

3.

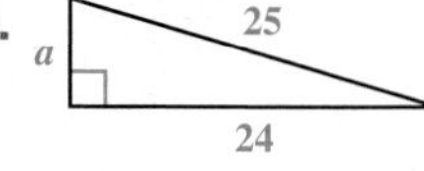

Tell whether the given lengths can be the sides of a right triangle. *Section 6.1*

4. 8 cm, 15 cm, 17 cm
5. 5 in., 8 in., 9 in.
6. 20 ft, 48 ft, 52 ft
7. 1 m, 2.4 m, 2.6 m
8. 7 yd, 7 yd, 13 yd
9. 0.3 cm, 0.4 cm, 0.5 cm
10. 33 cm, 544 cm, 545 cm
11. 3.5 m, 8.4 m, 9.1 m
12. 56 ft, 105 ft, 119 ft

42. a. $\frac{1}{8} = 12.5\%$
b. $\frac{7}{8} = 87.5\%$
43. $\frac{1}{2} = 50\%$
44. 0
45. $\frac{1}{2} = 50\%$
46. a. $\frac{1}{3} \approx 33.33\%$
b. $\frac{1}{2} = 50\%$

47. $\frac{1}{3} \approx 33.33\%$
48. $\frac{3}{8} = 37.5\%$
49. $\frac{1}{8} = 12.5\%$
50. $\frac{1}{2} = 50\%$
51. $\frac{1}{3} \approx 33.33\%$
52. $\frac{\pi}{8} \approx 39.27\%$

Chapter 6

1. 8
2. 13
3. 7
4. Yes.
5. No.
6. Yes.
7. Yes.
8. No.
9. Yes.
10. Yes.
11. Yes.
12. Yes.
13. $c = 15$
14. $b = 20$
15. $a = 15$
16. $b = 24$
17. $c = 34$
18. $a = 11$

The legs of each right triangle have lengths *a* and *b*. The hypotenuse has length *c*. Find the unknown length for each right triangle. *Section 6.1*

13. $a = 9, b = 12$ **14.** $a = 15, c = 25$ **15.** $b = 36, c = 39$

16. $a = 7, c = 25$ **17.** $a = 16, b = 30$ **18.** $b = 60, c = 61$

Write each decimal as a fraction in lowest terms. *Section 6.2*

19. $0.111\ldots$ **20.** 0.02 **21.** $0.7272\ldots$

22. $0.012012\ldots$ **23.** $0.\overline{6}$ **24.** $0.\overline{396}$

25. $1.\overline{7}$ **26.** $0.011011\ldots$ **27.** $0.\overline{006}$

Tell whether each number is *rational* or *irrational*. *Section 6.2*

28. $1.666\ldots$ **29.** $-\sqrt{25}$ **30.** $\sqrt{10}$

31. $\frac{22}{7}$ **32.** $0.0202202220\ldots$ **33.** $\sqrt{121}$

Estimate each square root within a range of two integers. *Section 6.3*

34. $\sqrt{30}$ **35.** $\sqrt{95}$ **36.** $\sqrt{41}$ **37.** $\sqrt{135}$

Simplify each expression. *Section 6.3*

38. $\sqrt{32}$ **39.** $\sqrt{45}$ **40.** $\sqrt{63}$ **41.** $\sqrt{80}$

42. $\sqrt{28}$ **43.** $\sqrt{90}$ **44.** $\sqrt{144}$ **45.** $\sqrt{126}$

46. $\sqrt{68}$ **47.** $3\sqrt{75}$ **48.** $3\sqrt{500}$ **49.** $\sqrt{\frac{3}{4}}$

50. $\sqrt{\frac{9}{49}}$ **51.** $\sqrt{\frac{1}{36}}$ **52.** $9\sqrt{3} + 3\sqrt{3}$ **53.** $2\sqrt{27} + \sqrt{12}$

54. $\sqrt{50} - 2\sqrt{2}$ **55.** $3\sqrt{20} + 3\sqrt{45}$ **56.** $\sqrt{48} - \sqrt{75}$ **57.** $\sqrt{5} \cdot \sqrt{5}$

58. $2\sqrt{2} \cdot \sqrt{8}$ **59.** $\sqrt{12} \cdot 2\sqrt{6}$ **60.** $2\sqrt{6^2}$ **61.** $3\sqrt{10} \cdot 2\sqrt{9}$

For Exercises 62–64, write an equation for each tile model. *Section 6.4*

62.

63.

64.

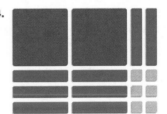

Find each product. *Section 6.4*

65. $2(3x - 6)$ **66.** $6(x + 1)$ **67.** $5x(x)$ **68.** $x(9x)$

69. $4x(2x + 7)$ **70.** $(4 - 3x)7$ **71.** $5x(2 + 4x)$ **72.** $(x - 2)(x + 3)$

73. $(x + 8)(5x - 6)$ **74.** $(12x - 15)(-3x)$ **75.** $(5x - 3)(2x + 8)$ **76.** $(-4)(5 + \sqrt{5})$

77. $4(6 + \sqrt{5})$ **78.** $\sqrt{2}(1 + \sqrt{2})$ **79.** $(5 - \sqrt{14})(3 - \sqrt{2})$ **80.** $(4\sqrt{2} + 2)(3\sqrt{3} - 1)$

19. $\frac{1}{9}$ 20. $\frac{1}{50}$
21. $\frac{8}{11}$ 22. $\frac{4}{333}$
23. $\frac{2}{3}$ 24. $\frac{44}{111}$
25. $\frac{16}{9}$ 26. $\frac{11}{999}$
27. $\frac{2}{333}$ 28. rational
29. rational 30. irrational
31. rational 32. irrational
33. rational

34. $5 < \sqrt{30} < 6$
35. $9 < \sqrt{95} < 10$
36. $6 < \sqrt{41} < 7$
37. $11 < \sqrt{135} < 12$
38. $4\sqrt{2}$ 39. $3\sqrt{5}$
40. $3\sqrt{7}$ 41. $4\sqrt{5}$
42. $2\sqrt{7}$ 43. $3\sqrt{10}$
44. 12 45. $3\sqrt{14}$
46. $2\sqrt{17}$ 47. $15\sqrt{3}$
48. $30\sqrt{5}$ 49. $\frac{\sqrt{3}}{2}$

50. $\frac{3}{7}$ 51. $\frac{1}{6}$
52. $12\sqrt{3}$ 53. $8\sqrt{3}$
54. $3\sqrt{2}$ 55. $15\sqrt{5}$
56. $-\sqrt{3}$ 57. 5
58. 8 59. $12\sqrt{2}$
60. 12 61. $18\sqrt{10}$
62. $(x + 3)(2x + 1) = 2x^2 + 7x + 3$
63. $(x + 3)(x + 2) = x^2 + 5x + 6$
64. $(x + 3)(2x + 2) = 2x^2 + 8x + 6$
65. $6x - 12$ 66. $6x + 6$

67. $5x^2$ 68. $9x^2$
69. $8x^2 + 28x$ 70. $28 - 21x$
71. $10x + 20x^2$ 72. $x^2 + x - 6$
73. $5x^2 + 34x - 48$
74. $-36x^2 + 45x$
75. $10x^2 + 34x - 24$
76. $-20 - 4\sqrt{5}$ 77. $24 + 4\sqrt{5}$
78. $\sqrt{2} + 2$
79. $15 - 3\sqrt{14} - 5\sqrt{2} + 2\sqrt{7}$
80. $12\sqrt{6} + 6\sqrt{3} - 4\sqrt{2} - 2$

81. $x^2 + 8x + 16$ 82. $y^2 - 36$
83. $a^2 - 16a + 64$ 84. $g^2 - 16$
85. $4w^2 - 1$ 86. $16x^2 - 16x + 4$
87. $49k^2 + 5.6k + 0.16$
88. $r^2 - 2rs + s^2$
89. $36b^2 - 24bc + 4c^2$
90. $x^2 - 16y^2$ 91. $11 - 4\sqrt{7}$
92. 44 93. 961
94. 891 95. 27.04
96. $20\frac{1}{4}$
97. $14 + 6\sqrt{5}$; $12 + 4\sqrt{5}$
98. $n^2 - 49$

Chapter 7

1–12. Tables may vary.

1.
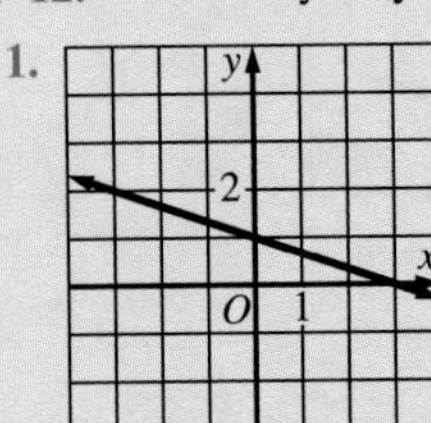

2.
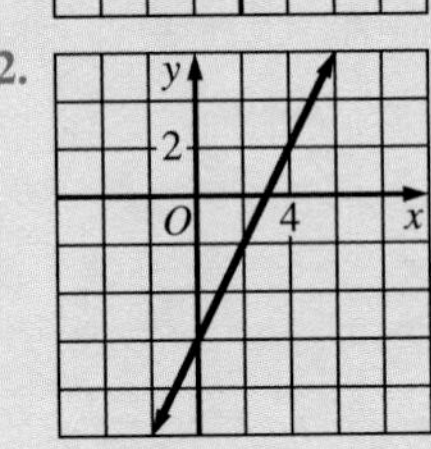

3.
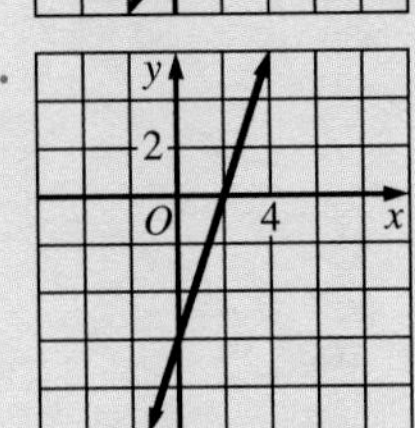

4.
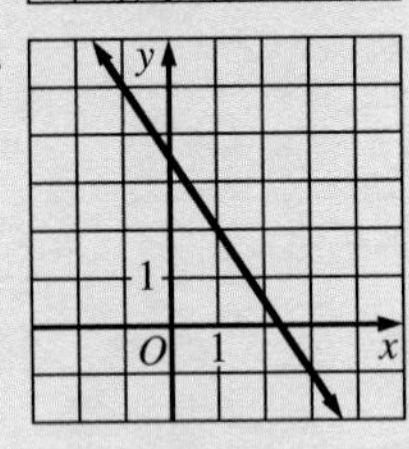

5.
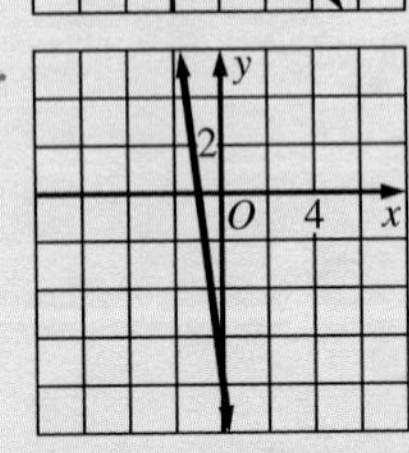

6.
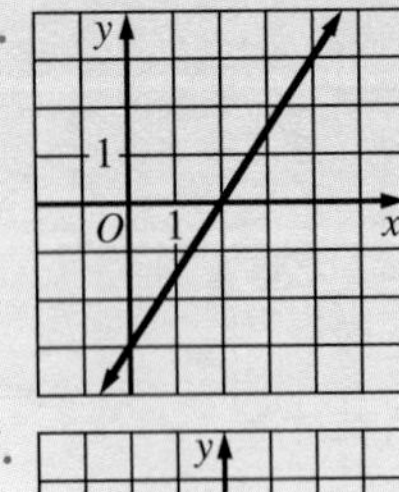

7.
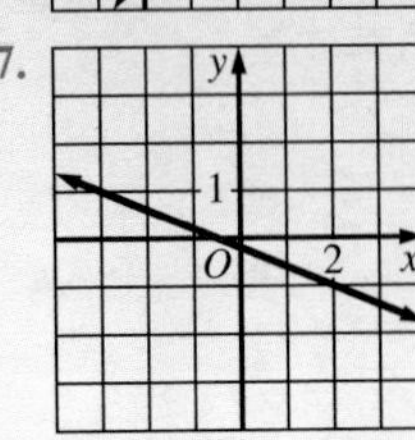

Find each product. *Section 6.5*

81. $(x + 4)^2$ **82.** $(y + 6)(y - 6)$ **83.** $(a - 8)^2$ **84.** $(g - 4)(g + 4)$

85. $(2w + 1)(2w - 1)$ **86.** $(4x - 2)^2$ **87.** $(7k + 0.4)^2$ **88.** $(r - s)^2$

89. $(6b - 2c)^2$ **90.** $(x + 4y)(x - 4y)$ **91.** $(2 - \sqrt{7})^2$ **92.** $(7 + \sqrt{5})(7 - \sqrt{5})$

Find each product using mental math. *Section 6.5*

93. 31^2 **94.** $27 \cdot 33$ **95.** $(5.2)^2$ **96.** $\left(4\frac{1}{2}\right)^2$

Solve. *Section 6.5*

97. The side of a square is $3 + \sqrt{5}$ units long. What is the area of the square? What is the perimeter?

98. The length of a rectangle is $n + 7$ units and the width is $n - 7$ units. What expression represents the area of the rectangle?

CHAPTER 7

Graph each equation by making a table and graphing each point. *Section 7.1*

1. $x + 3y = 3$ **2.** $4x - 2y = 12$ **3.** $-3x + y = -6$

4. $2y = -3x + 7$ **5.** $8x + y = -8$ **6.** $-6 + 3x = 2y$

7. $-1 - 2x = 5y$ **8.** $-x + 7y = 14$ **9.** $0.6x + 0.1y = 2$

10. $\frac{1}{2}x + \frac{2}{3}y = 2$ **11.** $\frac{2}{5}x = \frac{1}{3}y - 3$ **12.** $0.6x - 1.5y = 3$

For Exercises 13–15, match each equation with its *x*- and *y*-intercepts. *Section 7.1*

13. $y = \frac{1}{2}x - 3$ **A.** The x-intercept is 2. The y-intercept is 2.

14. $x + y = 2$ **B.** The x-intercept is 6. The y-intercept is 4.

15. $y = -\frac{2}{3}x + 4$ **C.** The x-intercept is 6. The y-intercept is -3.

Use substitution to solve each system of equations. *Section 7.2*

16. $x + y = 3$, $x - y = 1$ **17.** $2x - y = 1$, $x - 2y = 1$ **18.** $x - y = 2$, $3x + y = 6$ **19.** $y = -2x + 3$, $y = 3x - 1$

20. $x = y$, $x = 3$ **21.** $3x + y = 2$, $x + 2y = -1$ **22.** $x + y = 8$, $x - y = -5$ **23.** $y = -x + 3$, $y = 2x - 3$

Technology **Solve each system of equations by graphing. Use a graphing calculator if you have one.**

24. $2x + y = -2$, $-5x + 2y = 3$ **25.** $4x = y$, $\frac{1}{2}x = y$ **26.** $2x - 2y = 1$, $x + 3y = 6$ **27.** $5 - 2y = x$, $6 + y = 2x$

8.
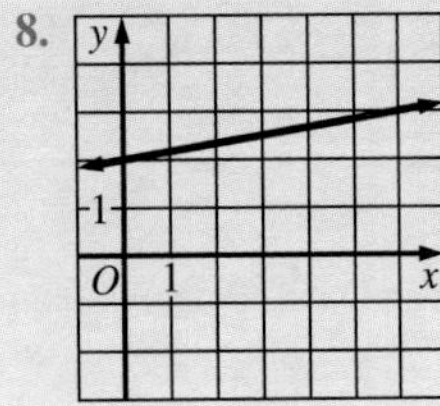

9.
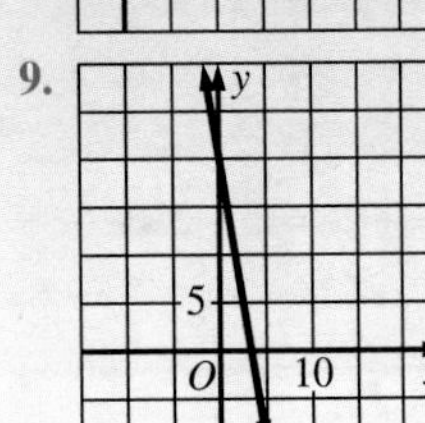

10.
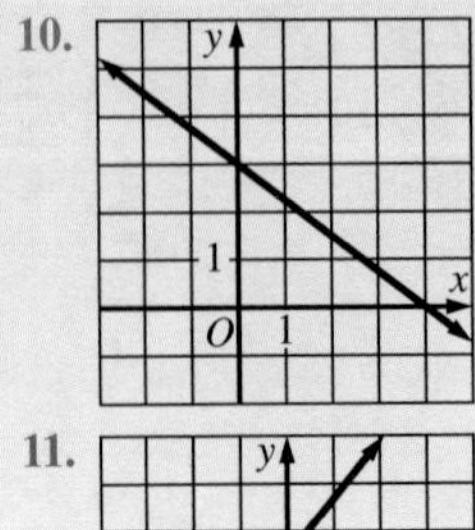

11.
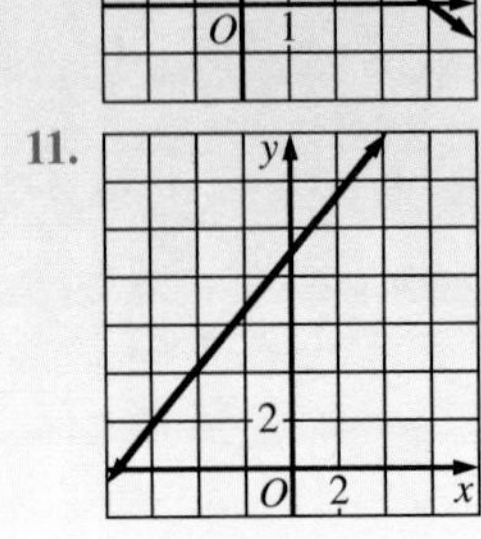

12.
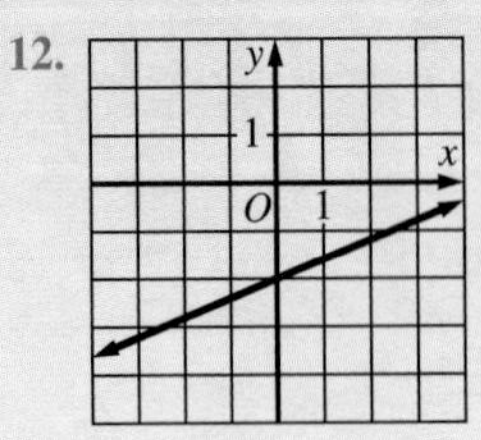

13. C 14. A
15. B 16. (2, 1)
17. $\left(\frac{1}{3}, -\frac{1}{3}\right)$ 18. (2, 0)
19. $\left(\frac{4}{5}, \frac{7}{5}\right)$ 20. (3, 3)

Solve each system of equations by adding or subtracting. *Section 7.3*

28. $2x + y = 8$
 $x - y = 1$

29. $4a - 6b = 2$
 $-4a + b = 5$

30. $3n - 2p = 6$
 $3n - 7p = -4$

31. $3 = -2x + 2y$
 $-7 = 3x - 2y$

32. $6x - 2 = -4y$
 $-2x + 10 = 4y$

33. $a + 2b = -6$
 $7a + 2b = 6$

34. $a + b = 6$
 $a - b = 4$

35. $x - y = -4$
 $x = 6 - y$

36. $2n + p = 8$
 $-2n + 3p = 0$

37. $3x + y = 13$
 $2x = y + 2$

38. $5y = 3x + 12$
 $7x - 5y = 8$

39. $-8 = 3a - b$
 $-6 = -3a - b$

For each system of equations, give the solution or state whether there are *no solutions* or *many solutions*. *Section 7.4*

40. $x + y = 3$
 $2x + 2y = 2$

41. $x + 2y = 5$
 $2x + y = 7$

42. $3a + 2b = 8$
 $6a + 4b = 12$

43. $-3 = r - 2s$
 $9 = 3r - 3s$

44. $2x + y = 2$
 $2y = -4x + 4$

45. $2a + b = 7$
 $3a - 2b = 7$

46. $4 = 2x + 3y$
 $2 = 3x + 4y$

47. $b = -3a - 6$
 $2b = a + 2$

48. $y = -3x + 4$
 $2y = -x + 3$

49. $y = 4x + 2$
 $8x + 2y = 4$

50. $3r + 4s = -7$
 $2r + s = -3$

51. $2y = 8x + 6$
 $y = 2x + 9$

52. $x + y = -5$
 $2x + 2y = -10$

53. $3x - 2y = 0$
 $3x - 2y = 5$

54. $2a + 3b = -7$
 $3a - 2b = -4$

55. $2x + 5y = 10$
 $15y = -6x + 30$

Match each inequality with its graph. *Section 7.5*

56. $y \le 2$

57. $x \ge 2$

58. $y \ge -2$

59. $x \le 2$

A.

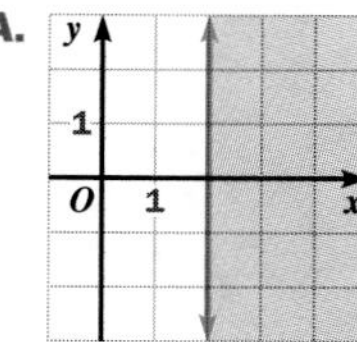

B.

C.

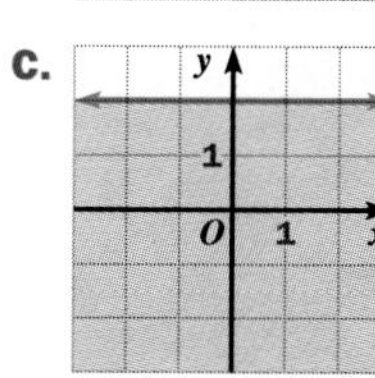

D.

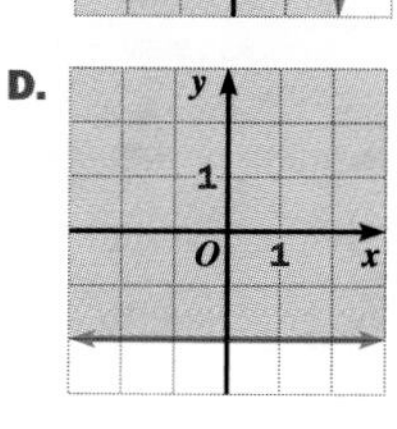

Graph each inequality. *Section 7.5*

60. $y \le x - 1$

61. $y > 2x + 6$

62. $x + y \ge -2$

63. $y \le 4x$

64. $x - y < -1$

65. $2x - 3y < 9$

66. $y \ge -x + 3$

67. $y > 3x$

68. $x \ge 5$

69. $x < y$

70. $\frac{1}{2}x + 2y \ge 3$

71. $3x - 4y > 12$

72. $6 < y - 2x$

73. $0.4x - y < 3$

74. $-\frac{x}{3} + \frac{y}{5} \le \frac{2}{5}$

75. $y \le 1.6x + 2$

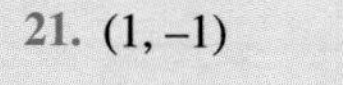

21. (1, –1)
22. $\left(\frac{3}{2}, \frac{13}{2}\right)$
23. (2, 1)
24. $\left(-\frac{7}{9}, -\frac{4}{9}\right)$
25. (0, 0)
26. $\left(\frac{15}{8}, \frac{11}{8}\right)$
27. $\left(\frac{17}{5}, \frac{4}{5}\right)$
28. (3, 2)
29. $\left(-\frac{8}{5}, -\frac{7}{5}\right)$
30. $\left(\frac{10}{3}, 2\right)$
31. $\left(-4, -\frac{5}{2}\right)$
32. $\left(-2, \frac{7}{2}\right)$
33. (2, –4)
34. (5, 1)
35. (1, 5)
36. (3, 2)
37. (3, 4)
38. $\left(5, \frac{27}{5}\right)$
39. $\left(-\frac{1}{3}, 7\right)$
40. no solution
41. (3, 1)
42. no solution
43. (9, 6)
44. many solutions
45. (3, 1)
46. (–10, 8)
47. (–2, 0)
48. (1, 1)
49. (0, 2)
50. (–1, –1)
51. (3, 15)
52. many solutions
53. no solution
54. (–2, –1)
55. many solutions
56. C
57. A
58. D
59. B
60.

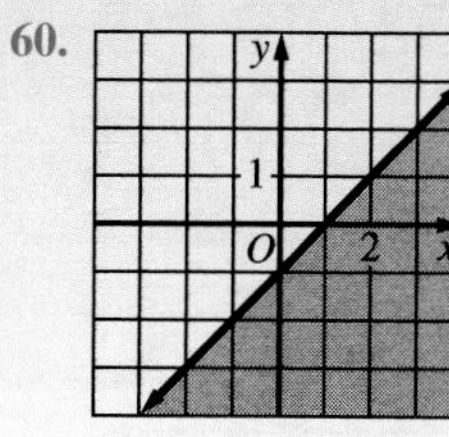

61.

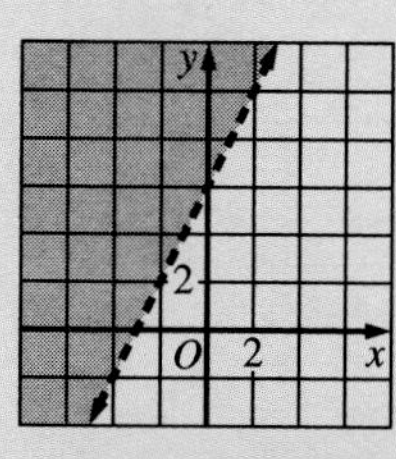

62.

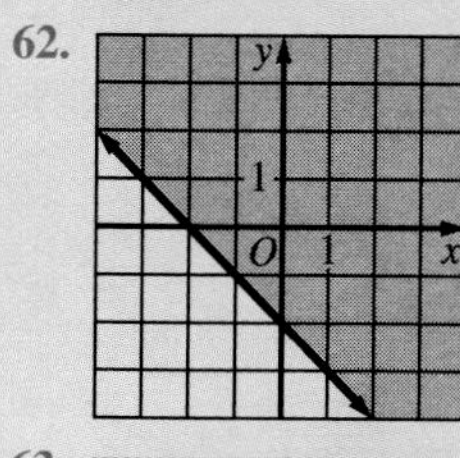

63.

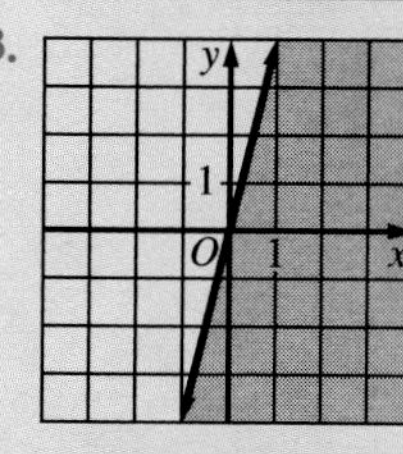

64.

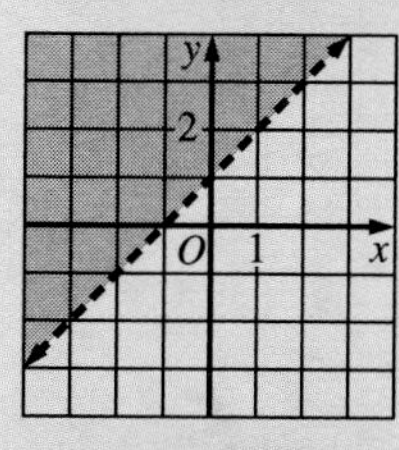

65.

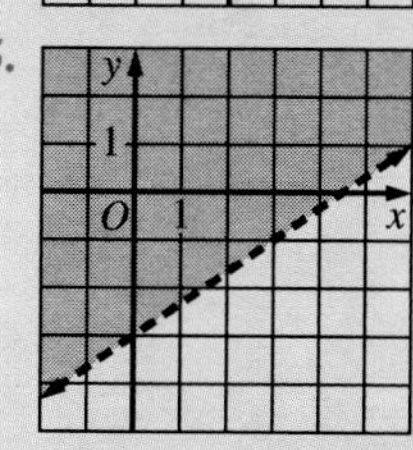

66.

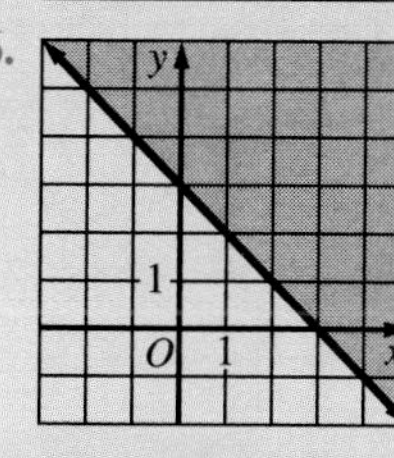

67.

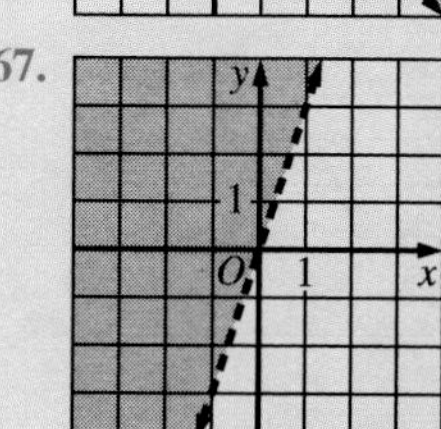

68.

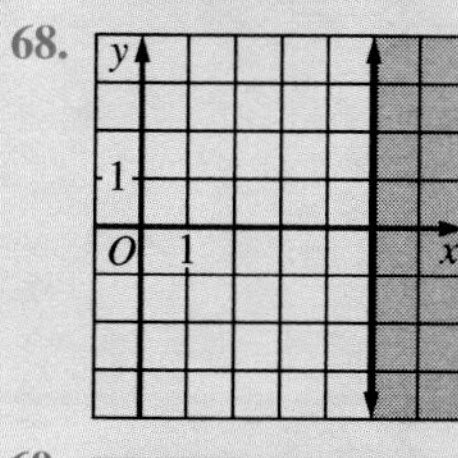

69.

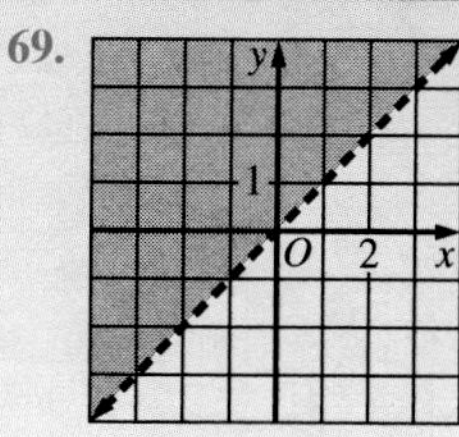

70–75. See answers in back of book.

76. C 77. D 78. A 79. B

80.

81.

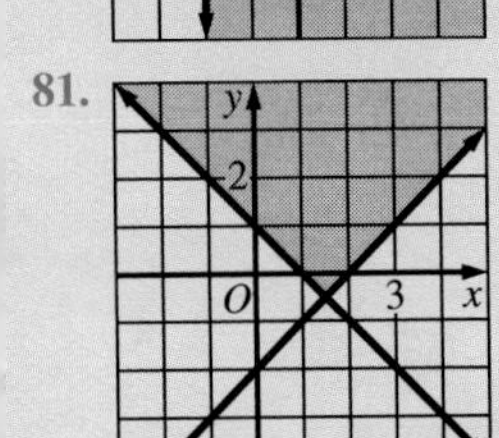

82.

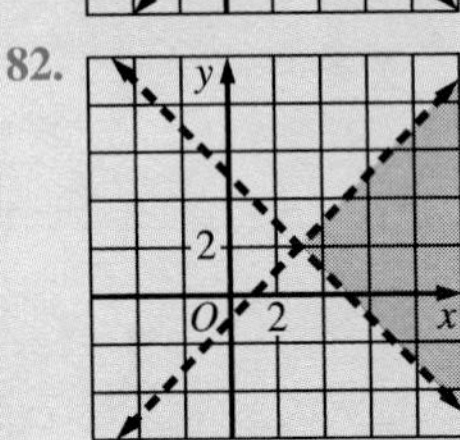

83.

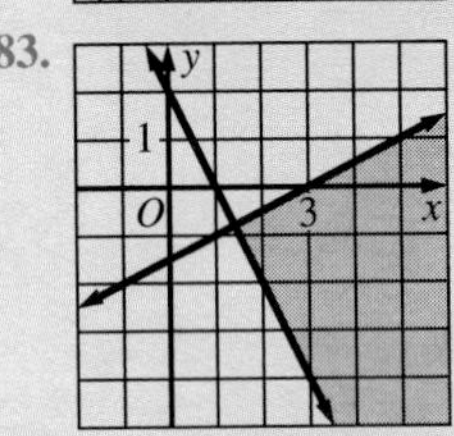

84.

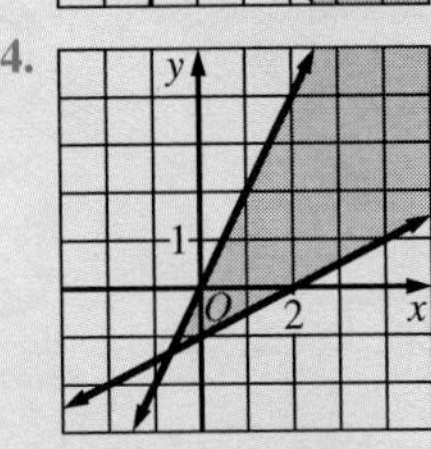

85.

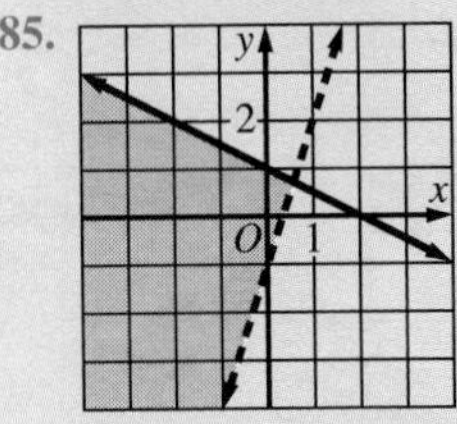

86.

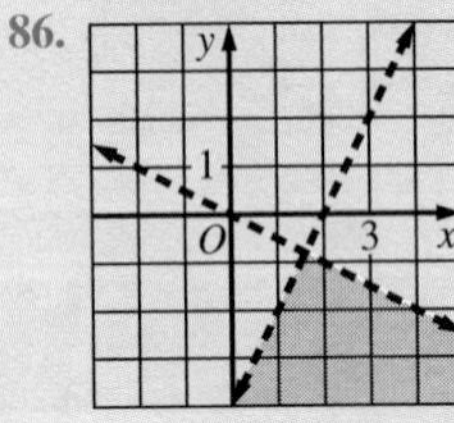

87.

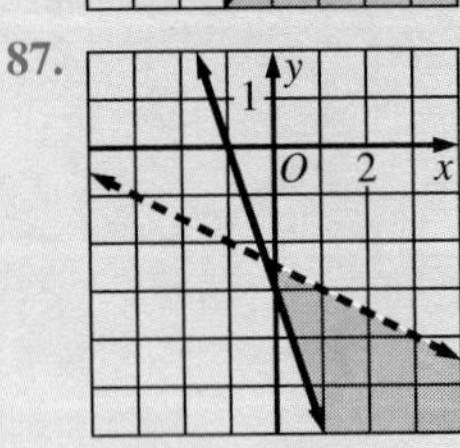

For Exercises 76–79, tell which region, *A*, *B*, *C*, or *D*, is the solution of each system of inequalities. *Section 7.6*

76. $y < -2x + 2$
$y > x - 1$

77. $y < x - 1$
$y < -2x + 2$

78. $y < x - 1$
$y > -2x + 2$

79. $y > -2x + 2$
$y > x - 1$

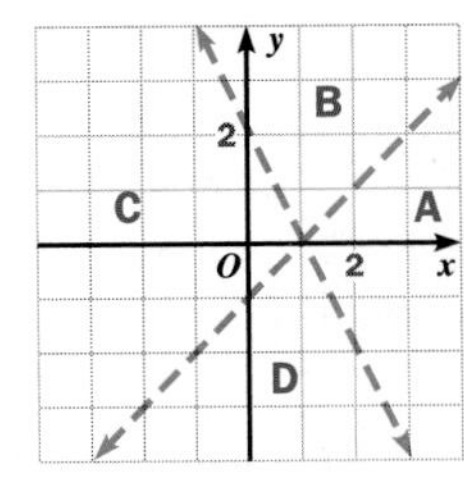

Graph each system of inequalities. *Section 7.6*

80. $y < 3$
$x > -2$

81. $y \geq x - 2$
$y \geq -x + 1$

82. $x + y > 5$
$y < x - 1$

83. $2x + y \geq 2$
$x - 2y \geq 3$

84. $y \geq \frac{1}{2}x - 1$
$y \leq 2x$

85. $3x - y < 1$
$x + 2y \leq 2$

86. $4 < 2x - y$
$-2y > x$

87. $2y < -x - 5$
$3x + y \geq -3$

CHAPTER 8

Graph each equation. Tell whether each equation is *linear* or *nonlinear*. *Section 8.1*

1. $y = 4x - 2$

2. $y = |x| - 2$

3. $y = 2x^2$

4. $y = \frac{1}{2}x$

5. $y = -2x^2 + 1$

6. $y = -2|x|$

7. $y = 2x + 1$

8. $x^2 - 2 = y$

9. $y = \frac{1}{3}x^3$

10. $y = 6x$

11. $y = -x + 3$

12. $y = -4x$

Tell whether each relationship is *linear* or *nonlinear*. *Section 8.1*

13. the relationship between any positive number x and its reciprocal, $\frac{1}{x}$

14. the relationship between the edge length s of a cube and its surface area, $6s^2$

15. the relationship between any number x and its opposite, $-x$

Technology For Exercises 16–23, use a graphing calculator to check your predictions. *Section 8.2*

a. Predict how the graph of the equation will compare with the graph of $y = x^2$.

b. Sketch the graph of each equation.

16. $y = -2x^2$

17. $y = \frac{1}{3}x^2$

18. $y = -x^2$

19. $y = -\frac{1}{4}x^2$

20. $y = 7x^2$

21. $y = 0.1x^2$

22. $y = \frac{5x^2}{2}$

23. $y = -1.75x^2$

Chapter 8

1. linear

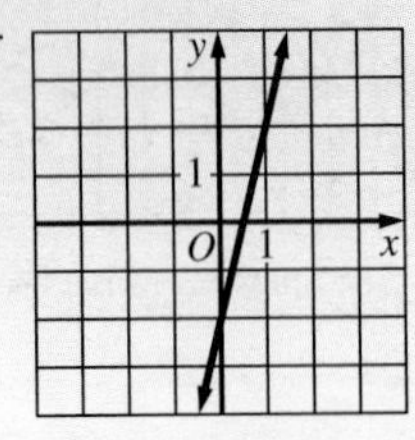

2. nonlinear

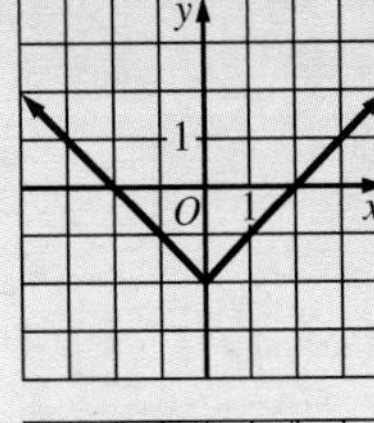

3. nonlinear

4. linear

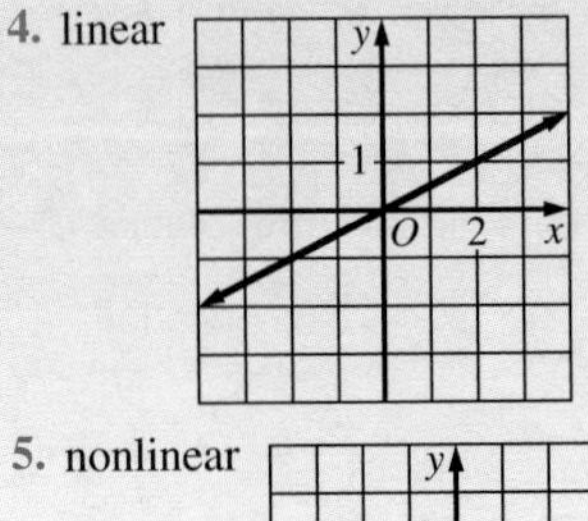

5. nonlinear

6. nonlinear

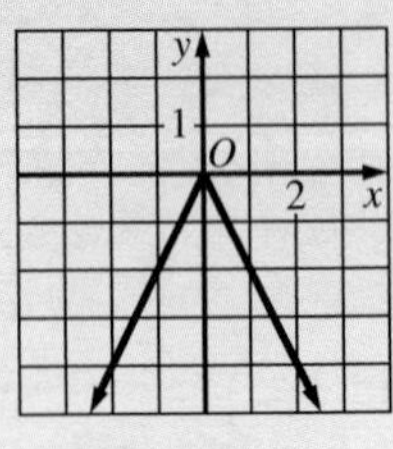

7. linear

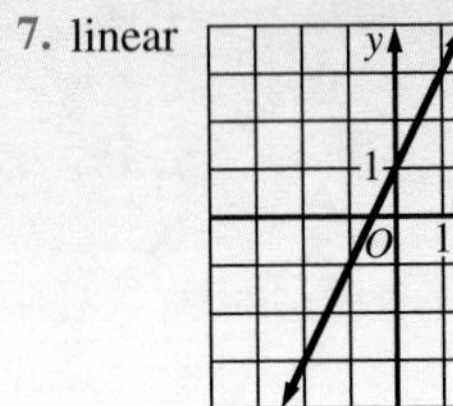

For Exercises 24 and 25, match each graph with its equation. *Section 8.2*

24.

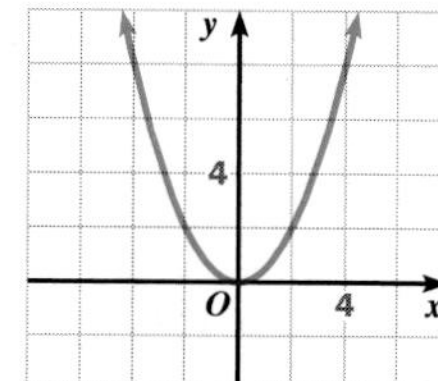

25.

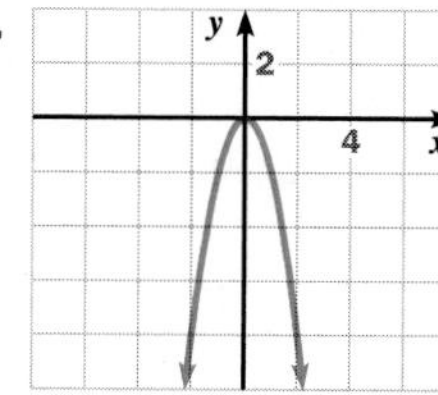

A. $y = -\frac{1}{4}x^2$

B. $y = -2x^2$

C. $y = \frac{1}{2}x^2$

D. $y = 4x^2$

Solve each equation algebraically or by using a graph. *Section 8.3*

26. $x^2 - 25 = 0$ **27.** $144 = n^2$ **28.** $4a^2 = 64$ **29.** $-3x^2 = -27$

30. $\frac{1}{3}p^2 = 12$ **31.** $42 = n^2 + 6$ **32.** $b^2 - 15 = 1$ **33.** $4x^2 = 196$

34. $196 = x^2$ **35.** $3x^2 + 2 = 245$ **36.** $11 = 2n^2 + 7$ **37.** $g^2 = 0.01$

38. $\frac{2}{5}x^2 - 3 = 1$ **39.** $\frac{1}{2}c^2 - 2 = -1$ **40.** $12a^2 - 5 = 55$ **41.** $\frac{3}{4}x^2 - 4 = 2$

Solve each equation using the given graph. *Section 8.4*

42. $x^2 + x - 6 = 0$ **43.** $x^2 + 6x + 9 = 0$ **44.** $-x^2 + 5x - 4 = 0$

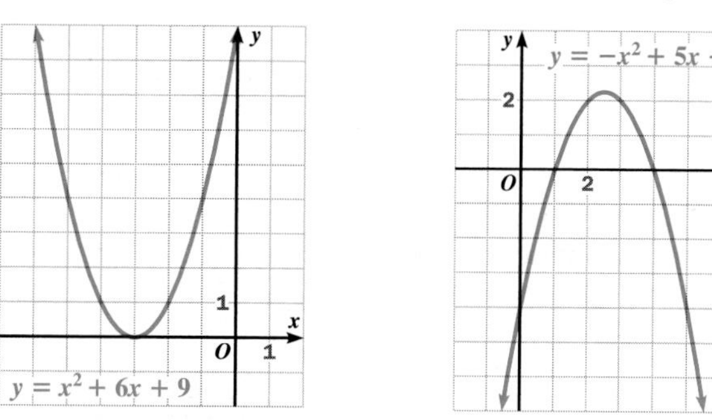

Technology Use a graphing calculator to solve each equation. If an equation has no solution, write *no solution*. *Section 8.4*

45. $x^2 + 3x - 8 = 0$ **46.** $x^2 - 8x = -15$ **47.** $2x^2 - 8x + 6 = 0$ **48.** $-x^2 + x - 3 = 0$

49. $2x^2 - 3x = 5$ **50.** $-x^2 + x = 2$ **51.** $42x^2 = 10$ **52.** $-5x^2 - 2x + 18 = 0$

53. $3x - 5x^2 = 6$ **54.** $3x^2 + 6x + 4 = 0$ **55.** $-x^2 + x = 5$ **56.** $2x^2 - 2x + 6 = 0$

Solve each equation using the quadratic formula. *Section 8.5*

57. $x^2 - 9x + 20 = 0$ **58.** $3x^2 - 3x - 6 = 0$ **59.** $x^2 - 2x - 15 = 0$ **60.** $-2x^2 + 4x = 0$

61. $-n^2 + 5n - 6 = 0$ **62.** $a^2 - 6a + 9 = 0$ **63.** $2k^2 + 9k - 5 = 0$ **64.** $p^2 + 2p + 1 = 0$

65. $-x^2 + 6x = -16$ **66.** $x^2 + 2x + 1 = 0$ **67.** $9x^2 - 18x + 9 = 0$ **68.** $5x^2 + 10x - 7 = 0$

69. $-2m^2 + 4m = -6$ **70.** $16g^2 + 16g = 32$ **71.** $x^2 - x = 42$ **72.** $3x^2 - 23x = -30$

8. nonlinear

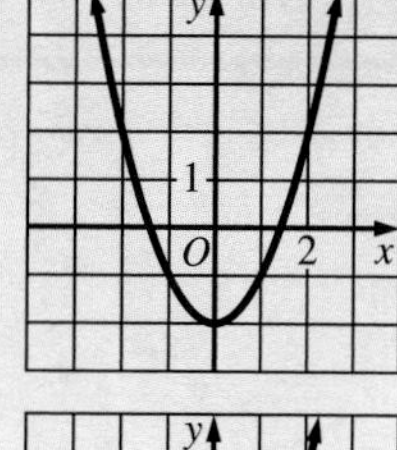

9. nonlinear

10. linear

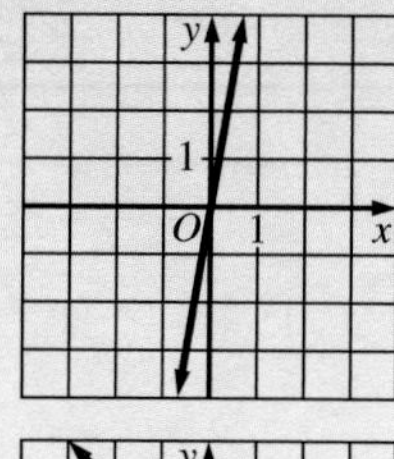

11. linear

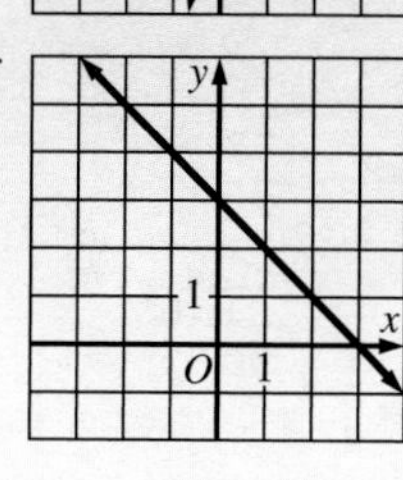

12. linear

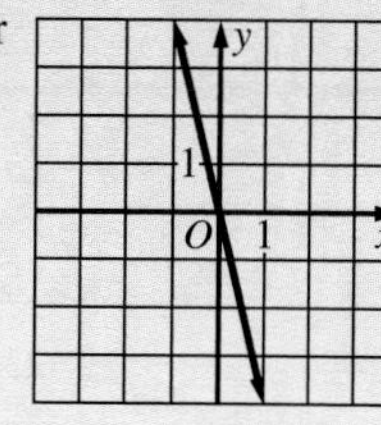

13. nonlinear 14. nonlinear

15. linear

16. a. narrower b. down

17. a. wider b. up

18. a. as wide b. down

19. a. wider b. down

20. a. narrower b. up

21. a. wider b. up

22. a. narrower b. up

23. a. narrower b. down

24. C 25. B

26. ± 5 27. ± 12

28. ± 4 29. ± 3

30. ± 6 31. ± 6

32. ± 4 33. ± 7

34. ± 14 35. ± 9

36. $\pm\sqrt{2}$ 37. ± 0.1

38. $\pm\sqrt{10}$ 39. $\pm\sqrt{2}$

40. $\pm\sqrt{5}$ 41. $\pm 2\sqrt{2}$

42. –3; 2 43. –3

44. 1; 4

45. about 1.70; about –4.70

46. 3; 5 47. 1; 3

48. no solution 49. –1; 2.5

50. no solution 51. about ± 0.49

52. about –2.11; about 1.71

53. no solution 54. no solution

55. no solution 56. no solution

57. 4; 5 58. –1; 2

59. –3; 5 60. 0; 2

61. 3; 2 62. 3

63. $-5; \frac{1}{2}$ 64. –1

65. 8; –2

66. –1

67. 1

68. $-1 \pm \frac{2\sqrt{15}}{5}$ (or about –2.55; about 0.55)

69. –1; 3 70. –2; 1

71. –6; 7 72. $\frac{5}{3}$; 6

For each equation, find the number of solutions. *Section 8.6*

73. $x^2 + 7x + 12 = 0$ **74.** $y^2 + y + 1 = 0$ **75.** $2c^2 + 5 = 0$ **76.** $6x^2 - x - 2 = 0$

77. $-2x^2 = 2x + 12$ **78.** $3x^2 + 6x + 3 = 0$ **79.** $x^2 + 8 = 7x$ **80.** $3x^2 + x + 2 = 0$

For each equation, find the number of solutions. Then solve using the quadratic formula. *Section 8.6*

81. $x^2 + 4x + 4 = 0$ **82.** $x^2 - 3x = 18$ **83.** $x^2 - 5x - 24 = 0$ **84.** $3x^2 + 13x - 10 = 0$

85. $-x^2 + x + 20 = 0$ **86.** $x^2 - 4x = 0$ **87.** $x^2 + 10x = -25$ **88.** $-x^2 + 14x - 48 = 0$

89. $5x^2 + 3x + 2 = 0$ **90.** $4x^2 - 17x = -4$ **91.** $x^2 + 16x + 64 = 0$ **92.** $x^2 + 4x = 1$

CHAPTER 9

Rewrite each product using exponents. *Section 9.1*

1. $ab \cdot ab \cdot ab \cdot ab \cdot ab \cdot ab$

2. $-4(6 \cdot 6 \cdot 6 \cdot 6 \cdot 6 \cdot 6 \cdot 6 \cdot 6)$

3. $\left(-\frac{1}{2}\right)\left(-\frac{1}{2}\right)\left(-\frac{1}{2}\right)\left(-\frac{1}{2}\right)\left(-\frac{1}{2}\right)$

4. $-b \cdot b \cdot b$

5. $3\left(\frac{2}{5}\right)\left(\frac{2}{5}\right)$

6. $cde \cdot cde \cdot cde \cdot cde$

Evaluate each power. *Section 9.1*

7. 7^3 **8.** 3^5 **9.** $(-5)^2$ **10.** $(1.1)^4$

11. $\left(\frac{1}{3}\right)^3$ **12.** $\left(\frac{1}{2}\right)^6$ **13.** $(-1.5)^3$ **14.** $-\left(\frac{1}{6}\right)^2$

Evaluate each expression when $x = 2$ and $y = 4$. *Section 9.1*

15. x^2y **16.** $2y^2$ **17.** $-5x^2y$ **18.** $x^2 - y^3$

19. $(y - x)^3$ **20.** y^x **21.** $4(x + y)^3$ **22.** y^y

The equations in Exercises 23–28 model the growth of a quantity y over a certain number of intervals. For each equation:

a. Give the initial amount of the quantity.

b. Give the growth factor and the growth rate.

c. Give the number of intervals. *Section 9.2*

23. $y = 26(1.25)^3$ **24.** $y = 2(1.5)^7$ **25.** $y = 300(1.35)^{10}$

26. $y = 750(1.01)^5$ **27.** $y = 1500(1.8)^4$ **28.** $y = 50(2.1)^9$

For Exercises 29–34, find the value of y when $x = 6$. Round decimal answers to the nearest hundredth. *Section 9.2*

29. $y = 30(1.05)^x$ **30.** $y = 33(2.01)^x$ **31.** $y = 52(12)^x$

32. $y = 2.5(1.04)^x$ **33.** $y = 600(1.15)^x$ **34.** $y = 14(2.5)^x$

35. A population of 250 fish grows at the rate of 20% per year. What is the expected population in five years?

73. two solutions
74. no solution
75. no solution
76. two solutions
77. no solution
78. one solution
79. two solutions
80. no solution
81. one solution; –2
82. two solutions; –3; 6
83. two solutions; –3; 8
84. two solutions; –5; $\frac{2}{3}$
85. two solutions; –4; 5
86. two solutions; 0; 4
87. one solution; –5
88. two solutions; 6; 8
89. no solution
90. two solutions; $\frac{1}{4}$; 4
91. one solution; –8
92. two solutions; $-2 \pm \sqrt{5}$ (about –4.24; about 0.24)

Chapter 9

1. $(ab)^6$ 2. $-4(6)^8$
3. $\left(-\frac{1}{2}\right)^5$ 4. $-(b)^3$
5. $3\left(\frac{2}{5}\right)^2$ 6. $(cde)^4$
7. 343 8. 243
9. 25 10. 1.4641
11. $\frac{1}{27}$ 12. $\frac{1}{64}$
13. –3.375 14. $-\frac{1}{36}$
15. 16 16. 32
17. –80 18. –60
19. 8 20. 16
21. 864 22. 256
23. a. 26 b. 1.25; 0.25 c. 3
24. a. 2 b. 1.5; 0.5 c. 7
25. a. 300 b. 1.35; 0.35 c. 10
26. a. 750 b. 1.01; 0.01 c. 5
27. a. 1500 b. 1.8; 0.8 c. 4
28. a. 50 b. 2.1; 0.1 c. 9
29. 40.20 30. 2176.16
31. 155,271,168
32. 3.16 33. 1387.84
34. 3417.97 35. about 622
36. 1.56 37. 0.02
38. 0.20 39. 1.69
40. 72.9 41. 12.8
42. 12.15 43. 40.63
44. 266.67 45. 0.60
46. 343.28
47. a. grows b. 455 c. 1.15

Find the value of y when $x = 4$. Round decimal answers to the nearest hundredth. *Section 9.3*

36. $y = 25(0.5)^x$ **37.** $y = 200(0.1)^x$ **38.** $y = 50\left(\frac{1}{4}\right)^x$

Find the value of y when $x = 3$. Round decimal answers to the nearest hundredth. *Section 9.3*

39. $y = 4(0.75)^x$ **40.** $y = 100(0.9)^x$ **41.** $y = 25(0.8)^x$ **42.** $y = 450(0.3)^x$

43. $y = 325\left(\frac{1}{2}\right)^x$ **44.** $y = 900\left(\frac{2}{3}\right)^x$ **45.** $y = 0.7(0.95)^x$ **46.** $y = 1250(0.65)^x$

The equations in Exercises 47–52 describe quantities that grow or decay exponentially. For each equation:

a. Tell whether the quantity described *grows* or *decays*.

b. Give the initial amount of the quantity.

c. Give the growth or decay factor for the quantity. *Section 9.3*

47. $y = 455(1.15)^x$ **48.** $y = 20(0.6)^x$ **49.** $y = 335(1.01)^x$

50. $y = 75(0.87)^x$ **51.** $y = 100(2.3)^x$ **52.** $y = 575(0.1)^x$

Rewrite each expression using only positive exponents. *Section 9.4*

53. x^{-3} **54.** a^{-10} **55.** $b^{-2}c^{-5}$ **56.** $7m^{-4}$

57. $y^{-4}z^7$ **58.** $6p^4q^{-2}$ **59.** $w^2x^{-1}z^3$ **60.** $\left(\frac{x}{y}\right)^{-6}$

61. $\frac{b^{-5}}{4}$ **62.** $10a^{-2}b^{-6}$ **63.** $\frac{m^{-5}}{n^6}$ **64.** $\frac{a^7}{b^{-3}}$

65. $(3c)^{-2}$ **66.** $(ab)^{-4}$ **67.** $2x^{-5}y^{-4}z^8$ **68.** $(pqr)^{-3}$

Evaluate each expression when $a = 2$, $b = 3$, and $c = 4$. *Section 9.4*

69. a^{-5} **70.** bc^{-1} **71.** $a^{-2}c^0$ **72.** $\left(\frac{1}{c}\right)^{-3}$

73. a^4b^{-2} **74.** $(a + c)^{-4}$ **75.** $b^{-1} + c^{-2}$ **76.** $a^{-1}b^3c^{-3}$

77. $\left(\frac{a}{c}\right)^{-3}$ **78.** 4^{c-b} **79.** $a^{-5} - b^{-1}$ **80.** $b^{-a}c^a$

Write each number in decimal notation. *Section 9.5*

81. 3×10^5 **82.** 1.4×10^{-4} **83.** 5.341×10^0 **84.** 3.97×10^6

85. 4.503×10^{-5} **86.** 5.7×10^{-8} **87.** 3.001×10^{10} **88.** 6.517×10^{-7}

89. 9.9×10^{-6} **90.** 6.5×10^{13} **91.** 6.0032×10^{11} **92.** 7.03×10^{-9}

Write each number in scientific notation. *Section 9.5*

93. 9675 **94.** 3,000,000 **95.** 0.0058 **96.** 0.000045

97. 2.7 million **98.** 0.000000862 **99.** 87,492,000,000 **100.** 0.0004

48. a. decays b. 20 c. 0.6
49. a. grows b. 335 c. 1.01
50. a. decays b. 75 c. 0.87
51. a. grows b. 100 c. 2.3
52. a. decays b. 575 c. 0.1

53. $\frac{1}{x^3}$
54. $\frac{1}{a^{10}}$
55. $\frac{1}{b^2c^5}$
56. $\frac{7}{m^4}$
57. $\frac{z^7}{y^4}$
58. $\frac{6p^4}{q^2}$
59. $\frac{w^2z^3}{x}$
60. $\frac{y^6}{x^6}$
61. $\frac{1}{4b^5}$
62. $\frac{10}{a^2b^2}$
63. $\frac{1}{m^5n^6}$
64. a^7b^3
65. $\frac{1}{9c^2}$
66. $\frac{1}{a^4b^4}$
67. $\frac{2z^8}{x^5y^4}$
68. $\frac{1}{p^3q^3r^3}$
69. $\frac{1}{32}$
70. $\frac{3}{4}$
71. $\frac{1}{4}$
72. 64
73. $\frac{16}{9}$
74. $\frac{1}{1296}$
75. $\frac{19}{48}$
76. $\frac{27}{128}$
77. 8
78. 4
79. $-\frac{29}{96}$
80. $\frac{16}{9}$

81. 300,000
82. 0.00014
83. 5.341
84. 3,970,000
85. 0.00004503
86. 0.000000057
87. 30,010,000,000
88. 0.0000006517
89. 0.0000099
90. 65,000,000,000,000
91. 600,320,000,000
92. 0.00000000703
93. 9.675×10^3
94. 3×10^6
95. 5.8×10^{-3}
96. 4.5×10^{-5}
97. 2.7×10^6
98. 8.62×10^{-7}
99. 8.7492×10^{10}
100. 4×10^{-4}

101. 9 millionths | 102. 4.92 hundredths | 103. 0.0004000562 | 104. 455 hundredths
105. 4506 millionths | 106. 475 million | 107. 6.8 million | 108. 543,570,000,000,000

Write each expression as a single power. *Section 9.6*

109. $10^6 \cdot 10^3$
110. $2^8 \cdot 2^3$
111. $4^4 \cdot 4^{-1}$
112. $\frac{5^8}{5^4}$
113. $\frac{9^{-4}}{9^3}$
114. $7^{-5} \cdot 7^{-7}$
115. $\frac{3^5}{3^{-6}}$
116. $\frac{6^3 \cdot 6^{-2}}{6^4}$
117. $\frac{2^{-2}}{2^{-6}}$
118. $8^{-3} \cdot 8^6 \cdot 8^{-5}$
119. $\frac{2^9}{2^5 \cdot 2^{-2}}$
120. $\frac{3^{-4}}{3^6 \cdot 3^{-2}}$

Simplify each expression. Write your answers using positive exponents. *Section 9.6*

121. $a^3 \cdot a^6$
122. $x^{-4} \cdot x^{-3}$
123. $a^2 \cdot b^{-3}$
124. $\frac{x^8}{x^6}$
125. $\frac{c^{-2}}{c^5}$
126. $\frac{y^5}{3y^3}$
127. $3b^5 \cdot 4b^{-2}$
128. $(5a^{-7}b)(3a^{-2}b^{-4})$
129. $(7c^8)(2c^{-3})$
130. $(2x^3y)(4x^{-6}y^4)$
131. $\frac{5bc^5}{b^2c^{-3}}$
132. $(5 \times 10^2)(1 \times 10^8)$
133. $\frac{6 \times 10^9}{2 \times 10^6}$
134. $\frac{4 \times 10^6}{8 \times 10^3}$
135. $\frac{4x^{-1}y^7}{2x^5y^{-3}}$
136. $(3 \times 10^6)(4 \times 10^5)$

Evaluate each power. Write your answers in scientific notation. *Section 9.7*

137. $(4 \times 10^6)^2$
138. $(2 \times 10^7)^3$
139. $(1.5 \times 10^{-4})^2$
140. $(1.0 \times 10^4)^{-2}$
141. $(3 \times 10^{-3})^4$
142. $(2.21 \times 10^{-8})^2$

Simplify each expression. *Section 9.7*

143. $(b^5)^3$
144. $(a^2c^3)^4$
145. $(3x)^3$
146. $\left(\frac{x}{y}\right)^8$
147. $\left(\frac{a^2}{b^4}\right)^3$
148. $(2c^4)^3$
149. $4(y^3)^4$
150. $(3a^3)^2$
151. $\left(\frac{b^3}{c^4}\right)^4$
152. $(x^3y^6)^5$
153. $(7a^2b^3c^5)^2$
154. $(2xy^4)^4$

CHAPTER 10

Write each polynomial in standard form. Then classify the polynomial as *linear*, *quadratic*, *cubic*, or *none of these*. *Section 10.1*

1. $2x - 3x^2 + 7$
2. $4a^2 + 2a - 3 - 3a^3$
3. $5c - 4c^2$
4. $2 - 3y^4 + 6y$
5. $\frac{1}{2}b^3 - 4 + 2b$
6. $1.6x + 4.5x^2 - 7.2$
7. $22 + \frac{1}{2}y$
8. $r^2 - 3r^4 + 7r + 1 - r^3$
9. $5t^2 - 8 + 9t$

101. 9×10^{-6}
102. 4.92×10^{-2}
103. 4.000562×10^{-4}
104. 4.55×10^0
105. 4.506×10^{-3}
106. 4.75×10^8
107. 6.8×10^6
108. 5.4357×10^{14}
109. 10^9
110. 2^{11}
111. 4^3
112. 5^4
113. $\frac{1}{9^7}$
114. $\frac{1}{7^{12}}$
115. 3^{11}
116. $\frac{1}{6^3}$
117. 2^4
118. $\frac{1}{8^2}$
119. 2^6
120. $\frac{1}{3^8}$
121. a^9
122. $\frac{1}{x^7}$
123. $\frac{a^2}{b^3}$
124. x^2
125. $\frac{1}{c^7}$
126. $\frac{y^2}{3}$
127. $12b^3$
128. $\frac{15}{a^9b^3}$
129. $14c^5$
130. $\frac{8y^5}{x^3}$
131. $\frac{5c^8}{b}$
132. 5×10^{10}
133. 3×10^3
134. 5×10^2
135. $\frac{2y^{10}}{x^6}$
136. 1.2×10^{12}
137. 1.6×10^{13}
138. 8×10^{21}
139. 2.25×10^{-8}
140. 1.0×10^{-8}
141. 8.1×10^{-11}
142. 4.8841×10^{-16}
143. b^{15}
144. a^8c^{12}
145. $27x^3$
146. $\frac{x^8}{y^8}$
147. $\frac{a^6}{b^{12}}$
148. $8c^{12}$
149. $4y^{12}$
150. $9a^6$
151. $\frac{b^{12}}{c^{16}}$
152. $x^{15}y^{30}$
153. $49a^4b^6c^{10}$
154. $16x^4y^{16}$

Chapter 10

1. $-3x^2 + 2x + 7$; quadratic
2. $-3a^3 + 4a^2 + 2a - 3$; cubic
3. $-4c^2 + 5c$; quadratic
4. $-3y^4 + 6y + 2$; none of these
5. $\frac{1}{2}b^3 + 2b - 4$; cubic
6. $4.5x^2 + 1.6x - 7.2$; quadratic
7. $\frac{1}{2}y + 22$; linear
8. $-3r^4 - r^3 + r^2 + 7r + 1$; none of these
9. $5t^2 + 9t - 8$; quadratic
10. $8x^2 + 2x + 4$
11. $2a^2 - 8a - 2$
12. $5x^2 - 9xy - y^2$
13. $-b^2 - 4b - 10$
14. $9a^2 - 3ab - 4b^2$
15. $5x + 4y - 11$
16. $3st^2 + st - 2$
17. $3a^2 - a + 3b - 6$
18. $9s^3 - 9$

Add or subtract. *Section 10.1*

10. $(5x^2 - 4x + 6) + (3x^2 + 6x - 2)$
11. $(4a^2 - 3a + 5) - (2a^2 + 5a + 7)$
12. $(6x^2 - 7xy + 5y^2) - (x^2 + 2xy + 6y^2)$
13. $(3b^2 - 4b + 5) - (4b^2 + 15)$
14. $(a^2 - 2b^2) + (8a^2 - 3ab - 2b^2)$
15. $(7x + y - 4) - (2x + 7 - 3y)$
16. $(3st^2 - 8st + 12) + (9st - 14)$
17. $(-a - 4b + 2) + (3a^2 + 7b - 8)$
18. $(8s^3 + 3s^2 - 5) + (s^3 - 3s^2 - 4)$
19. $(-r^3 - 3r - 7) - (-7r^3 + 2r^2 - 8r - 9)$

Multiply. *Section 10.2*

20. $2a(2a^2 - 3a)$
21. $6n(3n + 8)$
22. $-s(2s^2 + 5s + 4)$
23. $(x + 3)(x + 5)$
24. $(t + 9)(t - 4)$
25. $(4b - 3)(b - 7)$
26. $(3c - 8)(4c + 6)$
27. $(-4x + 5)(-4x - 5)$
28. $5n(2n^2 + 3n - 9)$
29. $0.2a(4a^2 - 3a - 5)$
30. $(n - 1)(n^2 + 3n)$
31. $(3d + 10)(-2d + 7)$
32. $(c - 3)(c^2 + 5c + 2)$
33. $(s + 4)(3s^2 - 5s - 12)$
34. $(7r^2 + 2r - 4)(3r - 5)$
35. $(-3y - 3)(6y^2 + y - 4)$
36. $(3a + 2b)(4a - 7b + 5)$
37. $(x + 5y)(x + 3y - 1)$
38. $(2x^2 - 4x + 5)(x - 5)$
39. $(2t - 3)(t^2 + 4t + 2)$
40. $(7x - 4)(x^2 - x - 8)$

Write a polynomial in standard form to represent the volume of each figure.
Section 10.2

41.

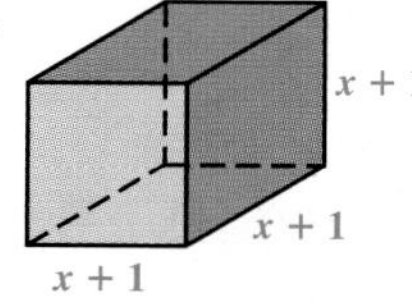

42.

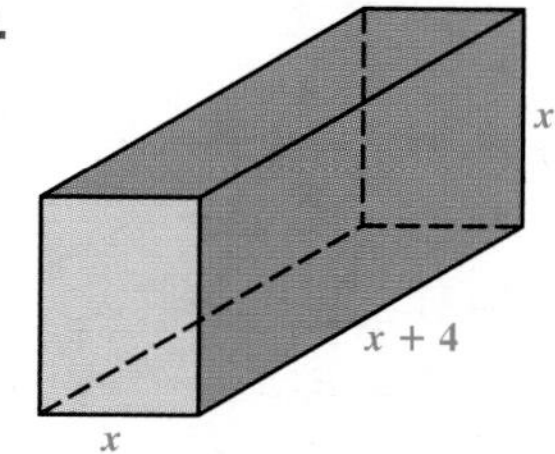

43.

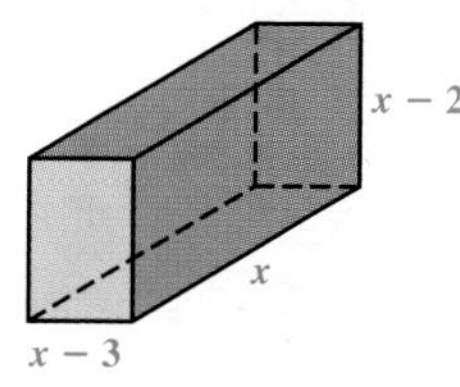

Factor each polynomial. *Section 10.3*

44. $a^2 + 8a$
45. $7n + 14$
46. $16 + 4b^2$
47. $x^3 - 5x$
48. $4c^2 + 8c$
49. $24t^2 - 6t$
50. $12y^2 + 8y$
51. $3v^2 - 6v$
52. $30b^2 - 25b$
53. $15a^2 + 9a - 9$
54. $16b^2 - 2b + 4$
55. $8z^3 - 4z$
56. $y^3 - 3y^2 - 3y$
57. $18x^3 + 6x^2 + 12x$
58. $2p^2 - 4p + 4$
59. $18x^3 - 14x^2 + 9x$

Solve each equation. *Section 10.3*

60. $b(b - 5) = 0$
61. $0 = 3x(x + 9)$
62. $8a(3a - 12) = 0$
63. $0 = 4x(6x + 3)$
64. $7y(5y - 8) = 0$
65. $0 = 0.9c(4c - 24)$
66. $(x - 4)(x - 9) = 0$
67. $(s - 8)(s + 3) = 0$
68. $(2t + 8)(t - 6) = 0$
69. $0 = (6x + 6)(3x + 4)$
70. $(-r + 5)(2r + 7) = 0$
71. $0 = (4k + 3)(7k - 5)$
72. $6t^2 - 2t = 0$
73. $0 = 4x^2 - 12x$
74. $15y^2 = -y$
75. $9a = 18a^2$
76. $20d^2 + 15d = 0$
77. $x^2 - 16x = 0$

48. $4c(c + 2)$
49. $6t(4t - 1)$
50. $4y(3y + 2)$
51. $3v(v - 2)$
52. $5b(6b - 5)$
53. $3(5a^2 + 3a - 3)$
54. $2(8b^2 - b + 2)$
55. $4z(2z^2 - 1)$
56. $y(y^2 - 3y - 3)$
57. $6x(3x^2 + x + 2)$
58. $2(p^2 - 2p + 2)$
59. $x(18x^2 - 14x + 9)$
60. 0; 5
61. 0; –9
62. 0; 4
63. $0; -\frac{1}{2}$
64. $0; \frac{8}{5}$
65. 0; 6
66. 4; 9
67. 8; –3
68. –4; 6
69. $-1; -\frac{4}{3}$
70. $5; -\frac{7}{2}$
71. $-\frac{3}{4}; \frac{5}{7}$
72. $0; \frac{1}{3}$
73. 0; 3
74. $0; -\frac{1}{15}$
75. $0; \frac{1}{2}$
76. $0; -\frac{3}{4}$
77. 0; 16

19. $6r^3 - 2r^2 + 5r + 2$
20. $4a^3 - 6a^2$
21. $18n^2 + 48n$
22. $-2s^3 - 5s^2 - 4s$
23. $x^2 + 8x + 15$
24. $t^2 + 5t - 36$
25. $4b^2 - 31b + 21$
26. $12c^2 - 14c - 48$
27. $16x^2 - 25$
28. $10n^3 + 15n^2 - 45n$
29. $0.8a^3 - 0.6a^2 - a$
30. $n^3 + 2n^2 - 3n$
31. $-6d^2 + d + 70$
32. $c^3 + 2c^2 - 13c - 6$
33. $3s^3 + 7s^2 - 32s - 48$
34. $21r^3 - 29r^2 - 22r + 20$
35. $-18y^3 - 21y^2 + 9y + 12$
36. $12a^2 - 13ab + 15a - 14b^2 + 10b$
37. $x^2 + 8xy - x + 15y^2 - 5y$
38. $2x^3 - 14x^2 + 25x - 25$
39. $2t^3 + 5t^2 - 8t - 6$
40. $7x^3 - 11x^2 - 52x + 32$
41. $x^3 + 3x^2 + 3x + 1$
42. $x^3 + 5x^2 + 4x$
43. $x^3 - 5x^2 + 6x$
44. $a(a + 8)$
45. $7(n + 2)$
46. $4(4 + b^2)$
47. $x(x^2 - 5)$

For each set of algebra tiles, write a polynomial that represents the area. Then express the area as the product of two binomials. *Section 10.4*

78. 

79. 

80. 

Factor, if possible. If not, write *not factorable*. *Section 10.4*

81. $x^2 + 4x + 4$
82. $a^2 + 7a + 12$
83. $b^2 - 8b + 7$
84. $y^2 + 13y + 13$
85. $c^2 - 7c + 10$
86. $n^2 + 11n + 28$
87. $b^2 - 14b + 49$
88. $t^2 + 13t + 30$
89. $g^2 - 12g + 24$
90. $3n^2 + 7n + 3$
91. $5n^2 - 7n + 2$
92. $7n^2 + 8n + 2$
93. $4c^2 + 8c + 3$
94. $4y^2 + 13y + 3$
95. $6n^2 - 5n + 1$
96. $9x^2 + 9x + 2$
97. $9t^2 - 19t + 2$
98. $10n^2 - 19n + 6$

Factor each polynomial. *Section 10.5*

99. $x^2 + 3x - 4$
100. $n^2 - n - 12$
101. $y^2 + y - 6$
102. $a^2 - 64$
103. $t^2 + 3t - 28$
104. $b^2 - 15b + 36$
105. $3x^2 + x - 2$
106. $4x^2 - 5x - 6$
107. $7d^2 - 25d + 12$
108. $8b^2 - 2b - 15$
109. $3x^2 - 12x - 15$
110. $6x^2 - 19x + 3$
111. $15c^2 + c - 2$
112. $10n^2 - 3n - 4$
113. $3r^2 - 8r - 3$

Solve each equation by factoring. *Section 10.5*

114. $a^2 - a - 12 = 0$
115. $x^2 - 11x + 30 = 0$
116. $0 = y^2 - 7y - 8$
117. $n^2 + 10n + 21 = 0$
118. $c^2 - 3c - 10 = 0$
119. $b^2 + 13b = -42$
120. $x^2 + 5x - 30 = 6$
121. $3n^2 - n - 4 = 0$
122. $4z^2 - 9 = 0$
123. $2t^2 - 5t = 3$
124. $8y^2 = 32$
125. $12x^2 - 4x - 21 = 0$

Solve each equation using whichever method you prefer. *Section 10.5*

126. $y^2 - 81 = 0$
127. $(3a - 2)(a + 3) = 0$
128. $0 = 9b^2 - 6b$
129. $12x^2 - x - 6 = 0$
130. $0 = x^2 + 16x + 64$
131. $(5x - 7)(2x + 9) = 0$
132. $8 = 3b^2 - 2b$
133. $9x^2 + 7x = 14 - 8x$
134. $10n^2 - 20n = 14n + 24$

78. $x^2 + 4x$; $x(x + 4)$
79. $2x^2 + 4x$; $2x(x + 2)$
80. $x^2 + 6x + 8$; $(x + 2)(x + 4)$
81. $(x + 2)^2$
82. $(a + 3)(a + 4)$
83. $(b - 7)(b - 1)$
84. not factorable
85. $(c - 5)(c - 2)$
86. $(n + 4)(n + 7)$
87. $(b - 7)^2$
88. $(t + 10)(t + 3)$
89. not factorable
90. not factorable
91. $(5n - 2)(n - 1)$
92. not factorable
93. $(2c + 1)(2c + 3)$
94. $(4y + 1)(y + 3)$
95. $(3n - 1)(2n - 1)$
96. $(3x + 1)(3x + 2)$
97. $(9t - 1)(t - 2)$
98. $(5n - 2)(2n - 3)$
99. $(x + 4)(x - 1)$
100. $(n - 4)(n + 3)$
101. $(y + 3)(y - 2)$
102. $(a + 8)(a - 8)$
103. $(t + 7)(t - 4)$
104. $(b - 12)(b - 3)$
105. $(3x - 2)(x + 1)$
106. $(4x + 3)(x - 2)$
107. $(7d - 4)(d - 3)$
108. $(4b + 5)(2b - 3)$
109. $3(x + 1)(x - 5)$
110. $(6x - 1)(x - 3)$
111. $(5c + 2)(3c - 1)$
112. $(5n - 4)(2n + 1)$
113. $(3r + 1)(r - 3)$
114. 4; –3
115. 5; 6
116. 8; –1
117. –3; –7
118. 5; –2
119. –6; –7
120. –9; 4
121. $\frac{4}{3}$; –1

CHAPTER 11

Decide whether the data show inverse variation. If they do, find the missing value. *Section 11.1*

1.

x	y
6	18
12	9
?	72
36	3

2.

x	y
14	12
4	42
−6	?
−7	−24

3.

x	y
$\frac{2}{3}$	?
3	112
4	84
8	42

4. It takes Roy four hours to travel to North Lake when he drives at an average speed of 45 mi/h. The time it takes him to drive to the lake varies inversely with his average speed.
   a. How fast must he drive in order to return home in $3\frac{1}{2}$ hours?
   b. If he averages 42 mi/h on the return trip, how long will it take him to get home?

Solve. *Section 11.2*

5. The average scores for the five teams in the Eastern Division of a mathematics competition are shown at the right. What is the average score for the students in the Eastern Division?

Team	Average score	Number of students
A	46	5
B	47	4
C	42	6
D	49	4
E	44	6

6. Gina's test grades in history are 92, 89, 89, 85, and 90. Her final exam for the marking period counts as two tests. What score does she need on the exam in order to get an average of at least 90 for the marking period?

7. The drama club has 350 adult tickets and 150 children's tickets for a performance. The club plans to sell the children's tickets for $3 and wants to earn an average price of $6 per ticket.
   a. What is the minimum amount the club needs to charge for the adult tickets?
   b. Suppose the club decides to charge $7 for an adult. What must it charge for a child's ticket in order to keep the average price at $6 per ticket?

Find the least common denominator of the fractions in each equation. *Section 11.3*

8. $\frac{1}{4x} + \frac{1}{8} = \frac{1}{2x}$

9. $\frac{1}{3x^2} + \frac{1}{4x^2} = \frac{1}{x+1}$

10. $\frac{2}{3(a+1)} = \frac{a}{3(a-1)}$

11. $\frac{4t}{2(t+1)} - \frac{1}{3t} = \frac{5}{5(t+1)}$

12. $\frac{8}{b^2} = \frac{3}{b(b+1)} + \frac{b}{b+1}$

13. $\frac{1}{4y} - \frac{3y}{2(y-2)} = \frac{7}{6(y+2)}$

122. $\pm\frac{3}{2}$
123. $-\frac{1}{2}; 3$
124. ± 2
125. $\frac{3}{2}; -\frac{7}{6}$
126. ± 9
127. $\frac{2}{3}; -3$
128. $0; \frac{2}{3}$
129. $\frac{3}{4}; -\frac{2}{3}$
130. −8
131. $\frac{7}{5}; -\frac{9}{2}$
132. $-\frac{4}{3}; 2$
133. $-\frac{7}{3}; \frac{2}{3}$
134. $4; -\frac{3}{5}$

Chapter 11

1. Yes; 1.5.
2. Yes; –28.
3. Yes; 504.
4. a. about 51.4 mi/h
   b. about 4.3 h
5. 45.2
6. at least 92.5
7. a. about $7.29
   b. about $3.67
8. $8x$
9. $12x^2(x+1)$
10. $3(a+1)(a-1)$
11. $30t(t+1)$
12. $b^2(b+1)$
13. $12y(y-2)(y+2)$

Solve each equation. If the equation has no solution, write *no solution*.

Section 11.3

14. $\frac{5}{x} + 3 = \frac{2}{x}$ **15.** $\frac{2}{x} + \frac{2}{3} = \frac{2}{3}$ **16.** $\frac{a}{5} + \frac{a}{2} = 3$

17. $\frac{1}{x-2} + \frac{1}{2} = \frac{2}{x-2}$ **18.** $\frac{b-2}{b+3} = 2 + \frac{5}{b+3}$ **19.** $\frac{x+1}{6} = \frac{x+1}{4}$

20. $\frac{a-2}{a} + \frac{4}{3a} = \frac{1}{3}$ **21.** $\frac{3}{y} + \frac{2}{3} = \frac{1}{3y}$ **22.** $\frac{3}{x+2} = \frac{2}{x-2}$

23. $\frac{4}{x+3} = \frac{4}{x-6}$ **24.** $\frac{4}{a^2} + \frac{3}{a} = \frac{5}{a^2}$ **25.** $\frac{2}{x-2} + \frac{3}{(x+2)(x-2)} = \frac{1}{x+2}$

26. $\frac{3}{4(c-2)} + \frac{1}{3(c-2)} = \frac{2}{c+3}$ **27.** $\frac{1}{x(x-1)} - \frac{2}{x^2} = \frac{1}{x-1}$ **28.** $\frac{3}{b^3} + \frac{3}{b^2} = \frac{6}{b}$

Solve each equation for the variable in red. *Section 11.4*

29. $t = v + w + x$ **30.** $2a + b = c$ **31.** $cd = fg$

32. $4m - 5n = 25$ **33.** $y = cx + b$ **34.** $2m + 3n = 4p$

35. $t + st = 2$ **36.** $3 - 5m = 4n$ **37.** $a = \frac{b+2}{a}$

38. $\frac{2}{x} - \frac{1}{y} = z$ **39.** $4 - 6a = 15$ **40.** $\frac{2 - ab}{2} = c$

41. $\frac{y}{3} = \frac{2+x}{7}$ **42.** $a(b+2) = b - 2 + c$ **43.** $2ax + 4ay = 5$

44. $4(a-5) = b^2 - 2$ **45.** $\frac{a}{b} + \frac{c}{d} = a$ **46.** $a + 3b = 4a - 8$

Simplify each expression. If it is already in simplest form, write *simplest form*.

Section 11.5

47. $\frac{x^3}{5x}$ **48.** $\frac{7a^2}{3a}$ **49.** $\frac{a^2b}{a+b}$

50. $\frac{5y+10}{5}$ **51.** $\frac{9t^2}{12y}$ **52.** $\frac{c^2-4}{c+2}$

53. $\frac{n-6}{4n^2-24n}$ **54.** $\frac{(x+1)^2}{x-1}$ **55.** $\frac{4b^2+2b}{2b}$

Simplify each expression. *Section 11.5*

56. $\frac{4a^3}{3b} \cdot \frac{9b^2}{2a}$ **57.** $\frac{6x}{5y} \div \frac{2x^2}{15y^2}$ **58.** $(3n+12) \cdot \frac{2}{2n+8}$

59. $\frac{4}{n+1} \div \frac{2}{3n+3}$ **60.** $\frac{3x+6}{x-3} \cdot \frac{x^2}{x+2}$ **61.** $\frac{6a+9}{9b^2} \cdot \frac{3b}{4a+6}$

62. $\frac{1}{3x^2} \div \frac{x-2}{12x}$ **63.** $\frac{y}{y-4} \div \frac{2y^2+y}{y-4}$ **64.** $\frac{x^2-9}{2(x+1)} \cdot \frac{6x+1}{x+3}$

14. -1
15. no solution
16. $\frac{30}{7}$
17. 4
18. -13
19. -1
20. 1
21. -4
22. 10
23. no solution
24. $\frac{1}{3}$
25. -9
26. $\frac{87}{11}$
27. -2
28. $-\frac{1}{2}$; 1
29. $t - v - x$
30. $\frac{c-b}{2}$
31. $\frac{cd}{f}$
32. $\frac{4}{5}m - 5$
33. $\frac{y-b}{c}$
34. $-\frac{3}{2}n + 2p$
35. $-\frac{4}{5}n + \frac{3}{5}$
36. $\frac{2}{1+s}$
37. $a^2 - 2$
38. $\frac{2y}{1+zy}$
39. $-\frac{11}{6}$
40. $\frac{2-2c}{a}$
41. $\frac{7y-6}{3}$
42. $\frac{-2a-2+c}{a-1}$
43. $\frac{-2ax+5}{4a}$
44. $\frac{b^2+18}{4}$
45. $\frac{ad}{ad-c}$
46. $b + \frac{8}{3}$
47. $\frac{x^2}{5}$
48. $\frac{7a}{3}$
49. simplest form
50. $y + 2$
51. $\frac{3t^2}{4y}$
52. $c - 2$
53. $\frac{1}{4n}$
54. simplest form
55. $2b + 1$
56. $6a^2b$
57. $\frac{9y}{x}$
58. 3
59. 6
60. $\frac{3x^2}{x-3}$
61. $\frac{1}{2b}$
62. $\frac{4}{x(x-2)}$
63. $\frac{1}{2y+1}$
64. $\frac{(x-3)(6x+1)}{2(x+1)}$
65. $\frac{2}{y}$
66. $\frac{3a-4}{a-1}$
67. $\frac{1-y}{xy}$
68. $\frac{4a^2+1}{a^2}$
69. $\frac{y+6}{y(y+2)}$
70. $\frac{6b+5a}{3a^2b^2}$
71. $\frac{(x+2)(x-1)}{x(x-2)}$
72. $\frac{8x-11}{(x+3)(x-4)}$
73. $\frac{2x+1}{x+1}$
74. $\frac{y-6}{3x}$
75. $\frac{5(x-2)}{x(x-5)}$
76. $\frac{4}{5a}$
77. $\frac{x-2}{x+1}$
78. $\frac{3y-8}{4y^2}$
79. $\frac{x-y}{x^2y^2}$
80. $\frac{-8n+9}{4n(4n+3)}$

Simplify each expression. *Section 11.6*

65. $\frac{1+x}{xy}+\frac{x-1}{xy}$ **66.** $\frac{2a}{a-1}+\frac{a-4}{a-1}$ **67.** $\frac{3x}{3x^2y}-\frac{3xy}{3x^2y}$ **68.** $\frac{4a^2}{a^2}+\frac{1}{a^2}$

69. $\frac{3}{y}-\frac{2}{y+2}$ **70.** $\frac{4}{2a^2b}+\frac{5}{3ab^2}$ **71.** $\frac{x}{x-2}+\frac{1}{x}$ **72.** $\frac{5}{x+3}+\frac{3}{x-4}$

73. $2-\frac{1}{x+1}$ **74.** $\frac{y}{3x}-\frac{2}{x}$ **75.** $\frac{3}{x-5}+\frac{2}{x}$ **76.** $\frac{1}{2a}+\frac{3}{10a}$

77. $\frac{x}{x+1}-\frac{2}{x+1}$ **78.** $\frac{6}{8y}-\frac{2}{y^2}$ **79.** $\frac{1}{xy^2}-\frac{1}{x^2y}$ **80.** $\frac{3}{4n}-\frac{5}{4n+3}$

CHAPTER 12

For Exercises 1–3, write an algorithm that explains how to use a calculator to solve each problem. *Section 12.1*

1. Multiply 15 by 52 without using the multiplication key.

2. Divide 451 by 79 without using the division key.

3. Estimate the value of $\sqrt{30}$ without using the square root key.

For Exercises 4–7, write an algorithm for each task. *Section 12.1*

4. Make change from a \$5 bill for a \$3.52 purchase.

5. Find 30% of 132 without using a calculator.

6. Make soup from a can of condensed soup.

7. Graph the equation $y = 3x + 2$ in a coordinate plane.

For Exercises 8–12, copy the graph at the right. *Section 12.2*

8. Use a greedy algorithm to find a short path from vertex A to vertex B.

9. Find another path from vertex A to vertex B through all the vertices. How long is it? How does it compare to the path you found in Exercise 8?

10. Find a path from vertex C to vertex B through all the vertices. How long is the path?

11. Find the shortest tree for this graph. How long is it?

12. Suppose each vertex represents a home in a new neighborhood. Vertex C represents the main entry for the telephone lines for the neighborhood. Would you use a path or a tree to bring phone lines to all the homes in the neighborhood? Explain.

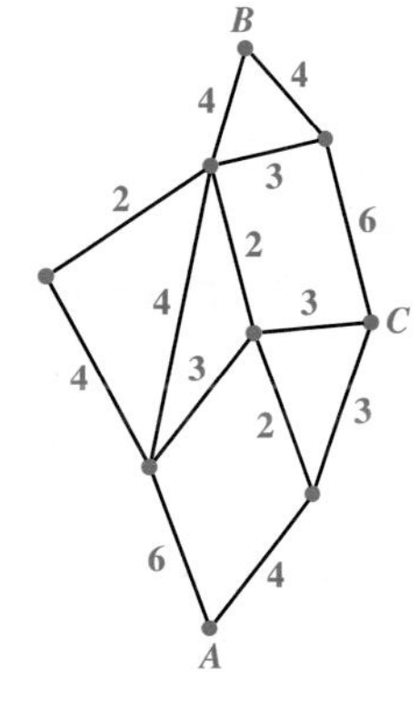

Chapter 12

1–7. Algorithms may vary. Examples are given.

1. Step 1: Enter 52. Step 2: Add 52. Step 3: Repeat Step 2 thirteen times.

2. Step 1: Enter 451. Step 2: Subtract 79. Step 3: Repeat Step 2 until the result is less than 79. The number of times you subtracted is the quotient. The number on the screen is the remainder.

3. Step 1: Choose a number and square it. If the result is less than 30, go to Step 2. If the result is greater than 30, go to Step 3. Step 2: Choose a larger number and square it. If the result is less than 30, repeat Step 2. If the result is greater than 30, go to Step 3. Step 3: Choose a smaller number and square it. If the result is less than 30, go to Step 2. If the result is greater than 30, repeat Step 3. Step 4: Continue until the number you find is satisfactorily close to $\sqrt{30}$.

4. This algorithm calculates change by adding to the charges until the amount of money offered is reached. Step 1: Count out 3 pennies to bring the amount to \$3.55. Step 2: Count out 2 dimes to bring the amount to \$3.75. Step 3: Add 1 quarter to bring the amount to \$4. Step 4: Add 1 \$1 bill to bring the amount to \$5.

5. Step 1: Move the decimal point one place to the left to find 10% of 132, 13.2. Step 2: Multiply the result by 3 to find 30% of 132, 39.6.

6. Step 1: Open the can and pour the contents into a pan. Step 2: Add one soup can of water to the pan. Step 3: Put the pan on a burner on a stovetop and turn the heat to medium. Step 4: Stir the soup occasionally as it heats. Step 5: When the soup is heated to your taste, remove the pan from the stove and turn off the heat. Step 6: Pour the contents of the pan into a bowl and serve.

7. Step 1: Choose three values for x. Step 2: Substitute each of the chosen values into the equation and find the corresponding values of y. Step 3: Draw coordinate axes. Graph the points corresponding to the ordered pairs you found in Steps 1 and 2. Step 4: Draw a line through the three points.

8. $4 + 2 + 2 + 4 = 12$

9. Answers may vary. One possible path is $4 + 3 + 3 + 3 + 4 + 2 + 3 + 4 = 26$. It is longer.

10. Answers may vary. One possible path is $3 + 2 + 4 + 6 + 4 + 2 + 3 + 4 = 28$.

11. $3 + 2 + 4 = 9$

12. a tree; All the vertices must be connected.

Four students voted for their favorite after-school activities, with 1 being their first choice. The results are shown in the table below. *Section 12.3*

13. Rank the activities in order of popularity. Explain how you decided on your ranking.

14. If two additional students both chose newspaper, community service, yearbook, and computer club, in that order, how would the group's preferences change?

After-school Activities

	Anna	Glenn	Justin	Jenna
Newspaper	2	4	3	2
Yearbook	1	3	1	4
Community Service	4	1	2	3
Computer Club	3	2	4	1

For Exercises 15–17, tell how to apply the cut-and-choose algorithm in each situation. *Section 12.3*

15. You and your sister both want the remaining ear of corn-on-the-cob.

16. You and a friend share a newspaper delivery route.

17. You and a friend volunteer to pick up litter in front of your school.

Find the number of possibilities for each situation. *Section 12.4*

18. ordering a three-course meal from a menu with 6 appetizers, 23 main dishes, and 8 desserts

19. sandwiches made with a choice of 8 different fillings on a choice of 4 types of bread

20. fabric for a chair in 7 designs, 2 fabric weights, and 6 colors

21. phones with or without built-in answering machines, with or without speaker-phone capability, and available in 7 different colors

For Exercises 22–26, find the number of permutations. *Section 12.4*

22. the number of ways four students can arrange themselves in a line

23. the number of ways you can arrange the letters in the word *pencil*

24. the number of different codes that can be programmed into a lock that has 4 digits, ranging from 0 through 9

25. the number of orders in which 8 different award winners can be announced

26. the number of different license plates that can be made using 3 different letters followed by 3 different digits

13. Yearbook; Community Service and Computer Club (tie); Newspaper. The Yearbook had the most 1's, Community Service and Computer Club had one each 1, 2, 3, and 4, and the Newspaper had zero 1's.

14. Newspaper; Community Service; Yearbook; Computer Club

15. One of you should break or cut the ear of corn into two pieces, either of which you would be willing to take. The other should choose one of the pieces.

16. One of you should divide the route into two routes, either of which you would be willing to do. The other should choose one of the routes.

17. One of you should divide the task into two tasks, either of which you would be willing to do. (The task may be divided by area or by time, for example.) The other should choose one of the tasks.

18. 1104 meals

19. 32 sandwiches

20. 84 chairs

21. 28 phones

For Exercises 27 and 28, use the cards at the right. *Section 12.5*

1 4 2 5 3 6

27. a. How many pairs of cards can be made from the pictured cards?

b. List the pairs that can be made.

c. How many of the possible pairs have either a 2 or a 3 in them?

28. a. How many sets of three cards can be made from the pictured cards?

b. List the possible sets.

c. How many sets will not have a 4, a 5, or a 6 in them?

29. How many different ways can a committee of 3 students be selected from a class of 22 students?

30. For a history test, students must select 3 out of 8 questions to answer. How many different sets of questions can be chosen?

31. a. A jukebox at a diner has 50 songs. If you pay $.50, you can choose any 3 different songs. How many different combinations of 3 songs are available?

b. In how many different orders could you choose 3 songs?

For Exercises 32 and 33, suppose three spinners are spun at the same time. *Section 12.6*

32. a. If the spinners have equally sized sections numbered 1–8, 1–5, and 1–6, respectively, what is the probability of spinning three 4's?

b. What is the probability of spinning three 6's?

33. If the spinners each have 10 equally sized sections numbered 1–10, what is the probability of spinning three 7's?

34. a. Suppose you are among 200 people who have entered a contest given by the local supermarket. If one name is drawn at random to be the winner, what is the probability that you will win?

b. Suppose two names will be drawn at random as winners, and your friend also has entered the contest. What is the probability that you both will win?

22. 24 ways
23. 720 ways
24. 10,000 codes
25. 40,320 orders
26. 11,232,000 license plates
27. a. 15 pairs
    b. 1, 2; 1, 3; 1, 4; 1, 5; 1, 6; 2, 3; 2, 4; 2, 5; 2, 6; 3, 4; 3, 5; 3, 6; 4, 5; 4, 6; 5, 6
    c. 9 pairs
28. a. 20 sets
    b. 1, 2, 3; 1, 2, 4; 1, 2, 5; 1, 2, 6; 1, 3, 4; 1, 3, 5; 1, 3, 6; 1, 4, 5; 1, 4, 6; 1, 5, 6; 2, 3, 4; 2, 3, 5; 2, 3, 6; 2, 4, 5; 2, 4, 6; 2, 5, 6; 3, 4, 5; 3, 4, 6; 3, 5, 6; 4, 5, 6
    c. one set: 1, 2, 3
29. 1540 ways
30. 56 sets
31. a. 19,600 combinations
    b. 117,600 ways
32. a. $\frac{1}{240}$
    b. 0
33. $\frac{1}{1000}$
34. a. $\frac{1}{200}$
    b. $\frac{1}{19,900}$

Toolbox

FACTORS AND MULTIPLES

A common factor of two numbers is a number that is a factor of both numbers. The *greatest common factor* (GCF) of two numbers is the *largest* number that is a common factor of both numbers.

EXAMPLE 1

Find the GCF of 24 and 180.

SOLUTION

Find the prime factorization of each number.

$$24 = 2 \cdot 2 \cdot 2 \cdot 3$$
$$180 = 2 \cdot 2 \cdot 3 \cdot 3 \cdot 5$$

Find the factors that appear in both lists. These are the **common factors**.

The GCF is the product of the common factors.

$$\text{GCF} = 2 \cdot 2 \cdot 3 = 12$$

A common multiple of two numbers is a number that is a multiple of both numbers. For example: $8 \cdot 10 = 80$ and $20 \cdot 4 = 80$, so 80 is a common multiple of 8 and 20. The *least common multiple* (LCM) of two numbers is the *smallest* number that is a common multiple of both numbers.

EXAMPLE 2

Find the LCM of 8 and 20.

SOLUTION

First find the GCF of the numbers.

$$8 = 2 \cdot 2 \cdot 2$$
$$20 = 2 \cdot 2 \cdot 5$$

The LCM is the product of the GCF and the other factors.

$$\text{LCM} = 4 \cdot 2 \cdot 5 = 40$$

The GCF is 4.

PRACTICE

Find the GCF and LCM of each set of numbers.

1. 10, 45 **2.** 28, 49 **3.** 15, 70 **4.** 20, 24

5. 54, 72 **6.** 18, 25 **7.** 16, 64 **8.** 14, 42, 70

9. Pencils are sold in packages of 10, and pencil-top erasers are sold in packages of 6. Find the smallest number of each type of package you can buy so that you have the same number of pencils as erasers.

ANSWERS Toolbox

Practice

1. GCF: 5; LCM: 90
2. GCF: 7; LCM: 196
3. GCF: 5; LCM: 210
4. GCF: 4; LCM: 120
5. GCF: 18; LCM: 216
6. GCF: 1; LCM: 450
7. GCF: 16; LCM: 64
8. GCF: 14; LCM: 210
9. You can buy 3 packages of pencils and 5 packages of pencil-top erasers.

POWERS OF TEN

Numbers like the ones in the table below are **powers of ten**.

$10 = 10$	$\frac{1}{10} = 0.1$
$10 \cdot 10 = 100$	$\frac{1}{10} \cdot \frac{1}{10} = \frac{1}{100} = 0.01$
$10 \cdot 10 \cdot 10 = 1000$	$\frac{1}{10} \cdot \frac{1}{10} \cdot \frac{1}{10} = \frac{1}{1000} = 0.001$

The examples below show you some shortcuts to use when multiplying or dividing with powers of ten.

EXAMPLE 1

Multiply.

a. 1.2×100

b. 37×0.001

SOLUTION

a. $1.2 \times 100 = 1.20 = 120$

There are 2 zeros in 100. Move the decimal point 2 places to the right.

b. $37 \times 0.001 = .037 = 0.037$

There are 3 decimal places in 0.001. Move the decimal point 3 places to the left.

Insert a zero as a place holder.

Divide.

a. $23 \div 10{,}000$

b. $9.45 \div 10$

SOLUTION

a. $23 \div 10{,}000 = .0023 = 0.0023$

There are 4 zeros in 10,000. Move the decimal point 4 places to the left.

Insert zeros as place holders.

b. $9.45 \div 10 = 9.45 = 0.945$

There is 1 zero in 10. Move the decimal point 1 place to the left.

PRACTICE

Multiply.

1. 25×1000
2. 18.4×0.01
3. $1.94 \times 100{,}000$
4. 0.003×100
5. 3.68×0.001
6. 0.05×10
7. 0.781×0.0001
8. 9346×0.00001

Divide.

9. $29.3 \div 100$
10. $108.3 \div 1000$
11. $468 \div 10$
12. $413 \div 1000$
13. $0.06 \div 100$
14. $0.09 \div 10$
15. $1995 \div 10{,}000$
16. $21.973 \div 100{,}000$

Practice

1. 25,000
2. 0.184
3. 194,000
4. 0.3
5. 0.00368
6. 0.5
7. 0.0000781
8. 0.09346
9. 0.293
10. 0.1083
11. 46.8
12. 0.413
13. 0.0006
14. 0.009
15. 0.1995
16. 0.00021973

Ratios and Rates

You can use *ratios* to compare two numbers using division. A ratio can be written in three ways:

a to b $\quad a : b \quad \frac{a}{b}$

Example 1

Write the ratio 6 to 8 in lowest terms.

Solution

To write a ratio in lowest terms, write it as a fraction in lowest terms.

$$\frac{6}{8} = \frac{3}{4} \quad \text{or} \quad 3 \text{ to } 4 \quad \text{or} \quad 3:4$$

Example 2

Write the ratio $\frac{6\text{ h}}{3\text{ days}}$ in lowest terms.

Solution

When the quantities being compared are in different units, first write the quantities in the same unit and then write the ratio.

Convert days to hours.

$$\frac{6\text{ h}}{3\text{ days}} = \frac{6}{72} = \frac{1}{12} \quad \text{or} \quad 1 \text{ to } 12 \quad \text{or} \quad 1:12$$

A *rate* is a ratio that compares two different quantities, such as dollars and hours. A *unit rate* is a rate per one unit of a given quantity.

Example 3

What is the unit rate if you earn \$420 in 21 hours?

Solution

Write the comparison as a fraction: $\frac{\text{dollars} \longrightarrow}{\text{hours} \longrightarrow} \frac{420}{21} = \frac{20}{1}$

The rate is \$20 per hour.

Practice

Write each ratio in lowest terms.

1. 12 to 4
2. 62 : 279
3. $\frac{42}{64}$
4. 85 to 15
5. 20 min to 3 h
6. 75 mg : 65 g
7. 9 in. : 4 ft
8. 12 oz to 1 lb

Write each unit rate.

9. 750 mi in 15 h
10. \$12.50 for 5 hats
11. 840 vibrations in 10 s
12. 63 hours in 7 days
13. \$522 in 12 months
14. \$3.60 for 8 pens

Practice

1–8. Forms of ratios may vary.

1. 3 to 1
2. 2:9
3. $\frac{21}{32}$
4. 17 to 3
5. 1 to 9
6. 3:2600
7. 3:16
8. 3 to 4
9. 50 miles per hour
10. \$2.50 per hat
11. 84 vibrations per second
12. 9 hours per day
13. \$43.50 per month
14. \$.45 per pen

FRACTIONS, DECIMALS, AND PERCENTS

A percent is a ratio that compares a number to 100. Remembering that *percent* means *per 100* can help you convert fractions, decimals, and percents.

EXAMPLE 1

a. Write 28% as a decimal and as a fraction in lowest terms.

b. Write 0.375 as a percent and as a fraction in lowest terms.

SOLUTION

a. $28\% = \frac{28}{100} = 0.28$ and $28\% = \frac{28}{100} = \frac{7}{25}$

b. $0.375 = 37.5\%$ and $0.375 = \frac{375}{1000} = \frac{375 \div 125}{1000 \div 125} = \frac{3}{8}$

Move the decimal point two places to the right. Include the % symbol.

EXAMPLE 2

Write each fraction as a decimal and as a percent.

a. $\frac{5}{4}$ **b.** $\frac{1}{12}$

SOLUTION

a. $\frac{5}{4} = \frac{5 \times 25}{4 \times 25} = \frac{125}{100} = 1.25 = 125\%$

b. $\frac{1}{12} = 1 \div 12 = 0.8333\ldots = 83.\overline{3}\% \approx 83\%$

Equivalent Percents, Decimals, and Fractions		
$1\% = 0.01 = \frac{1}{10}$	$33\frac{1}{3}\% = 0.\overline{3} = \frac{1}{3}$	$60\% = 0.6 = \frac{3}{5}$
$10\% = 0.1 = \frac{10}{100}$	$40\% = 0.4 = \frac{2}{5}$	$66\frac{2}{3}\% = 0.\overline{6} = \frac{2}{3}$
$25\% = 0.25 = \frac{1}{4}$	$50\% = 0.5 = \frac{1}{2}$	$100\% = 1$

You may want to memorize the information in the chart.

PRACTICE

Write each percent as a decimal and as a fraction in lowest terms.

1. 4% **2.** 87% **3.** 175% **4.** 82.4% **5.** $\frac{1}{2}\%$

Write each decimal as a percent and as a fraction in lowest terms.

6. 0.75 **7.** $0.\overline{6}$ **8.** 0.025 **9.** 1.6 **10.** 0.513

Write each fraction as a decimal and as a percent.

11. $\frac{7}{20}$ **12.** $\frac{4}{5}$ **13.** $\frac{27}{80}$ **14.** $\frac{14}{25}$ **15.** $\frac{8}{15}$

Practice

1. $0.04; \frac{1}{25}$
2. $0.87; \frac{87}{100}$
3. $1.75; \frac{7}{4}$
4. $0.824; \frac{103}{125}$
5. $0.005; \frac{1}{200}$
6. $75\%; \frac{3}{4}$
7. $66\frac{2}{3}\%; \frac{2}{3}$
8. $2.5\%; \frac{1}{40}$
9. $160\%; \frac{8}{5}$
10. $51.3\%; \frac{513}{1000}$
11. 0.35; 35%
12. 0.8; 80%
13. 0.3375; 33.75%
14. 0.56; 56%
15. $0.5\overline{3}; 53\frac{1}{3}\%$

INTEGERS

The *integers* are the numbers . . . , −3, −2, −1, 0, 1, 2, 3, ... You can use integer chips to model integers.

+1, or 1 **−1** **0** **−2** **0 + 3, or 3**

Since $1 + (-1) = 0$, a positive chip and a negative chip form a *zero pair*. The value of a zero pair is 0.

EXAMPLE 1

Add.

a. $-2 + (-5)$

b. $4 + (-3)$

SOLUTION

Model the first number with chips and add chips for the second number. Count the chips.

a.

$-2 + (-5) = -7$

b.

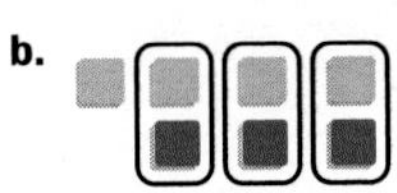

Add 3 negative chips to 3 positive chips to form zero pairs. 1 chip remains.

$4 + (-3) = 1$

EXAMPLE 2

Subtract.

a. $-3 - (-1)$

b. $-3 - 1$

c. $-2 - (-4)$

SOLUTION

Model the first number with chips. You may need to add zero pairs so you are able to subtract.

a.

Subtract 1 negative chip. 2 negative chips remain.

$-3 - (-1) = -2$

b.

Add a zero pair so you have a positive chip to subtract. Subtract 1 positive chip. 4 negative chips remain.

$-3 - 1 = -4$

c.

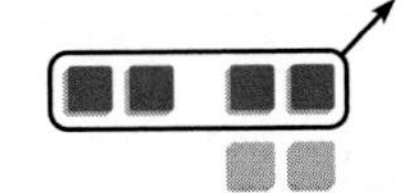

Add zero pairs until you have 4 negative chips to subtract. Subtract 4 negative chips. 2 positive chips remain.

$-2 - (-4) = 2$

PRACTICE

Add or subtract.

1. $7 + (-2)$ **2.** $-7 + (-2)$ **3.** $-7 - (-2)$

4. $-2 - (-7)$ **5.** $0 - 8$ **6.** $-3 + (-3)$

7. $-10 + (-3)$ **8.** $-7 - 4$ **9.** $2 - (-9)$

10. $-6 - (-1)$ **11.** $5 - (-5)$ **12.** $-6 - (-15)$

13. $-6 - (-8)$ **14.** $2 - (-4)$ **15.** $-10 - 2$

EXAMPLE 3

Multiply -4×2.

SOLUTION

Two groups of -4 chips are the same as one group of -8 chips.

$$-4 \times 2 = -8$$

EXAMPLE 4

Divide $-9 \div 3$.

SOLUTION

One group of -9 chips is the same as three groups of -3 chips.

$$-9 \div 3 = -3$$

PRACTICE

Multiply or divide.

16. $6 \times (-3)$ **17.** $-6 \div 3$ **18.** -6×3

19. -6×0 **20.** 4×5 **21.** $1 \times (-1)$

22. $-15 \div 5$ **23.** -1×3 **24.** $4 \times (-4)$

25. $4 \div 4$ **26.** $0 \div 2$ **27.** $-8 \div 8$

28. $2 \times (-5)$ **29.** $7 \times (-3)$ **30.** $-18 \div 9$

Practice

1. 5
2. –9
3. –5
4. 5
5. –8
6. –6
7. –13
8. –11
9. 11
10. –5
11. 10
12. 9
13. 2
14. 6
15. –12
16. –18
17. –2
18. –18
19. 0
20. 20
21. –1
22. –3
23. –3
24. –16
25. 1
26. 0
27. –1
28. –10
29. –21
30. –2

INEQUALITIES

A mathematical statement that has an *inequality symbol* between two numbers or quantities is called an *inequality*.

You can use a *variable* such as n to represent numbers or quantities that vary. The inequality $n < 7$ means "all numbers less than 7."

Inequality symbol	Meaning
$<$	less than
$\leq$	less than or equal to
$>$	greater than
$\geq$	greater than or equal to

EXAMPLE 1

Write an inequality to describe the numbers less than or equal to 50.

SOLUTION

Choose a variable and tell what it represents. Let n = any number.

$n \leq 50$ — Use the $\leq$ symbol.

EXAMPLE 2

Write an inequality to describe the phrase "at least $10."

SOLUTION

Choose a variable and tell what it represents. Let d = number of dollars.

$d \geq 10$ — Use the $\geq$ symbol.

EXAMPLE 3

What integers make the inequality $x < 2$ a true statement?

SOLUTION

The inequality $x < 2$ describes all the integers that are less than 2.

The integers $1, 0, -1, -2, -3, \ldots$ make $x < 2$ a true statement.

PRACTICE

Write a phrase to describe each inequality.

1. $a > 6$ **2.** $x \leq 9$ **3.** $h < -2$ **4.** $p \geq 0$

Write an inequality for each statement. Tell what the variable represents.

5. numbers greater than or equal to 7

6. numbers less than 20

7. numbers greater than 3

8. numbers greater than or equal to 24

9. a price less than $30

10. a hike less than or equal to 12 km

Describe the integers that make each inequality true.

11. $n > 7$ **12.** $r \geq -1$ **13.** $v < 5$ **14.** $t \leq 3$

Practice

1. all numbers greater than 6
2. all numbers less than or equal to 9
3. all numbers less than –2
4. all numbers greater than or equal to 0

5–8. Let n = any number.

5. $n \geq 7$
6. $n < 20$
7. $n > 3$
8. $n \geq 24$
9. Let p = a price; $p < 30$.
10. Let h = a hiking distance; $h \leq 12$.
11. The integers 8, 9, 10, … make $n > 7$ a true statement.
12. The integers –1, 0, 1, 2, 3, … make $r \geq -1$ a true statement.
13. The integers 4, 3, 2, 0, –1, –2, … make $v < 5$ a true statement.
14. The integers 3, 2, 1, 0, –1, –2, …. make $t \leq 3$ a true statement.

GRAPHING INEQUALITIES

You can use a number line to *graph* an inequality.

EXAMPLE 1

Graph each inequality on a number line.

a. $x > -2$

b. $x \leq 5$

SOLUTION

Step 1 Draw a number line.

Step 2 Draw an open or closed circle at the number. Use an open circle for $<$ and $>$, and a closed circle for $\leq$ and $\geq$.

Step 3 Decide which direction the arrow should point, to the left for $<$ and $\leq$, or to the right for $>$ and $\geq$. Then draw the line and the arrow.

The open circle means -2 is not included in the graph.

a. $x > -2$

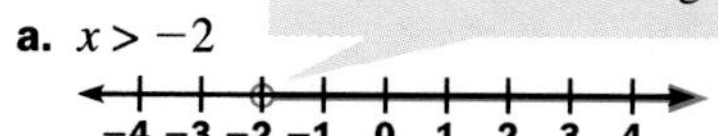

The closed circle means that 5 is included in the graph.

b. $x \leq 5$

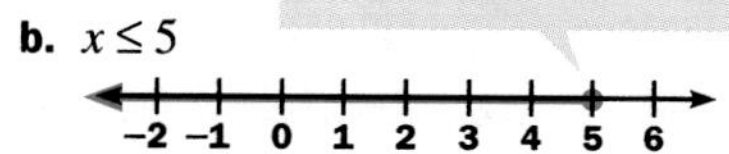

EXAMPLE 2

Write an inequality for each graph.

a.

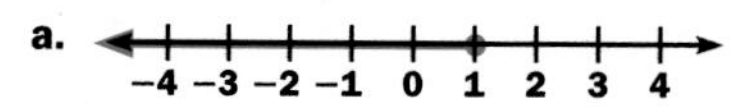

b. –4 –3 –2 –1 0 1 2 3 4

SOLUTION

The arrow is pointing to the left. The inequality is either $<$ or $\leq$.

The closed circle includes the value 1. Choose $\leq$.

a.

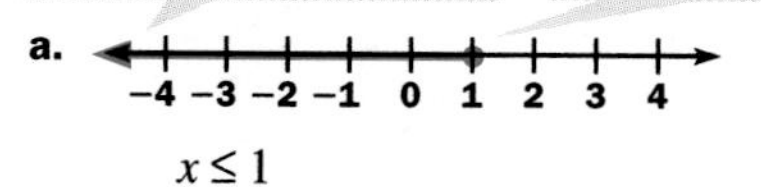

$x \leq 1$

The open circle does not include the value -3. The inequality is either $>$ or $<$.

The arrow is pointing to the right. Choose $>$.

b.

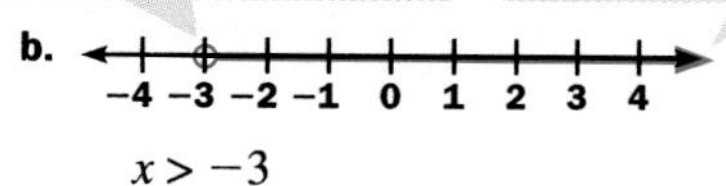

$x > -3$

PRACTICE

Graph each inequality on a number line.

1. $n \leq 6$

2. $r > -1$

3. $a < 0$

4. $t \geq 4$

5. $x \geq -5$

6. $k < 7$

7. $h > 3$

8. $b \leq -5$

Write an inequality for each graph.

9.

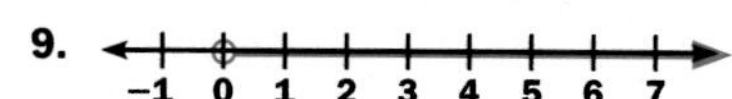

10.

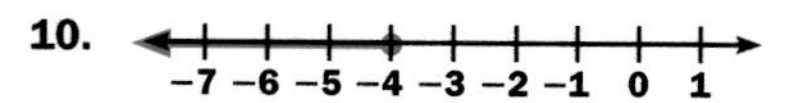

Practice

1.

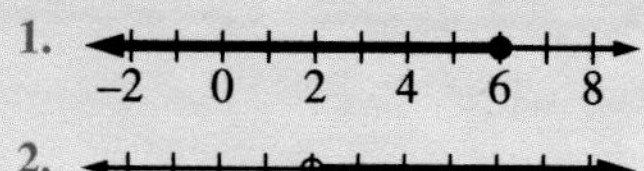

2.

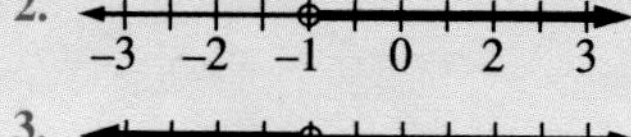

3.

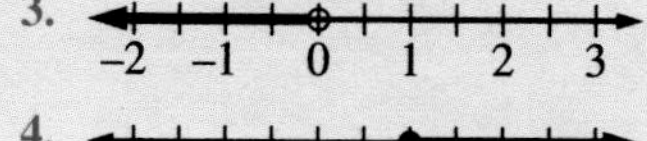

4.

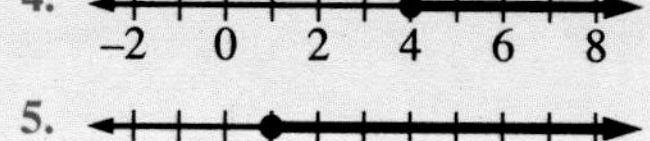

5. –8 –6 –4 –2 0 2

6.

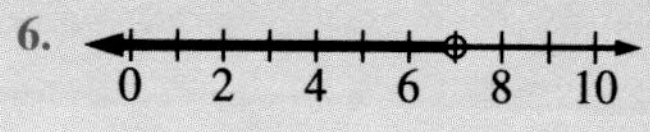

7.

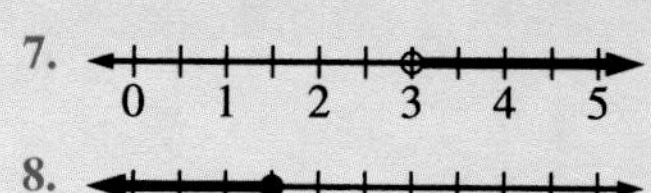

8. –8 –6 –4 –2 0 2

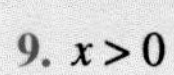

9. $x > 0$

10. $x \leq -4$

SYSTEMS OF MEASUREMENT

The **United States customary system** of measurement uses these units for length, capacity, and weight.

Length	Capacity	Weight
12 in. (inches) = 1 ft (foot)	8 fl oz (fluid ounces) = 1 c (cup)	16 oz = 1 lb (pound)
3 ft = 1 yd (yard)	2 c = 1 pt (pint)	2000 lb = 1 T (ton)
5280 ft = 1 mi (mile)	2 pt = 1 qt (quart)	
	4 qt = 1 gal (gallon)	

The **metric system** of measurement uses these units for length, capacity, and weight. The prefixes *milli-* (0.001), *centi-* (0.01), and *kilo-* (1000) are added to indicate different amounts.

Length (meter)	Capacity (liter)	Mass (gram)
1 mm = 0.001 m	1 mL = 0.001 L	1 mg = 0.001 g
1 cm = 0.01 m	1 cL = 0.01 L	1 cg = 0.01 g
1 m = 0.001 km	1 L = 0.001 kL	1 g = 0.001 kg
1 km = 1000 m	1 kL = 1000 L	1 kg = 1000 g

EXAMPLE

Convert each measurement.

a. 10 quarts to gallons

b. 0.7 kg to grams

SOLUTION

a. 10 qt = ? gal

Since 4 qt = 1 gal, divide by 4.

$\frac{10}{4} = 2\frac{2}{4}$

10 qt = $2\frac{1}{2}$ gal

When you convert to a larger unit (gal), you will have fewer gallons than quarts.

b. 0.7 kg = ? g

Since 1 kg = 1000 g, multiply by 1000.

(0.7)(1000) = 700

0.7 kg = 700 g

When you convert to a smaller unit (g), you will have more grams than kilograms.

PRACTICE

Convert each measurement.

1. 87 in. = ? ft ? in.

2. 2.5 T = ? lb

3. 1 qt = ? fl oz

4. 550 m = ? cm

5. 3400 g = ? kg

6. $7\frac{1}{4}$ yd = ? ft

7. $5\frac{1}{2}$ qt = ? pt

8. 250 L = ? mL

Practice

1. 7 ft 3 in.
2. 5000 lb
3. 32 fl oz
4. 55,000 cm
5. 3.4 kg
6. $21\frac{3}{4}$ ft
7. 11 pt
8. 250,000 mL

PERIMETER AND CIRCUMFERENCE

The *perimeter* of a figure is the distance around it.

The *circumference* of a circle is the distance around it.

Formulas for finding perimeter and circumference are listed on page 615.

EXAMPLE

Find the perimeter or circumference of each figure.

a.

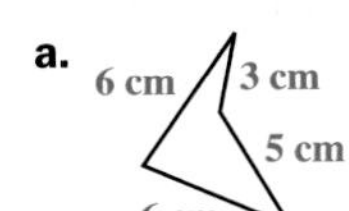

b.

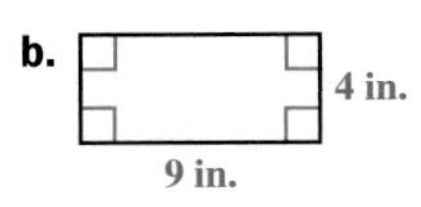

c.

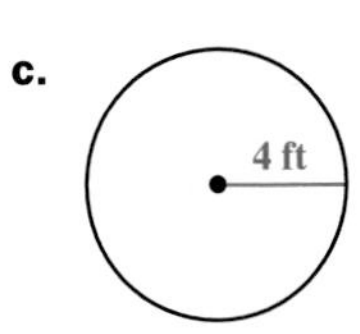

SOLUTION

a. Add the measures of each side.

$6 + 6 + 5 + 3 = 20$

The perimeter is 20 cm.

b. Use a formula.

$P = 2l + 2w$

$P = 2(9) + 2(4)$

$= 18 + 8$

$= 26$

The perimeter is 26 in.

c. Use a formula.

$C = 2\pi r$ Use 3.14 for π.

$\approx 2(3.14)(4)$

≈ 25.12

The circumference is about 25.12 ft.

PRACTICE

Find the perimeter or circumference of each figure.

1.

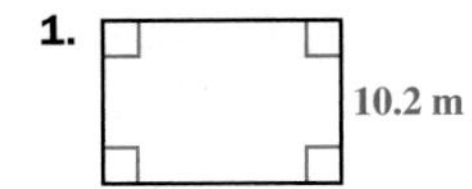

2.

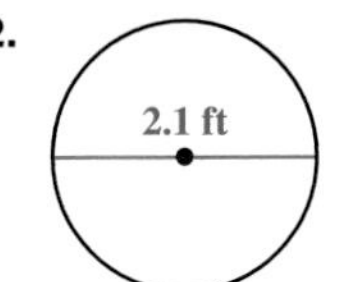

3.

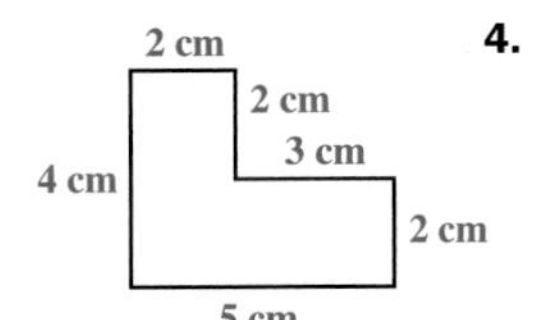

4. 8 in.

5. a square with a side of 4.2 cm

6. a circle with radius of 3.3 mm

7. an equilateral triangle with sides of $5\frac{1}{2}$ in.

8. a circle with diameter of 14 yd

9. The diameter of Josh's bicycle wheel is 22 inches. About how far will the wheel travel in one turn?

Practice

1. 51 m
2. about 6.59 ft
3. 18 cm
4. about 50.24 in.
5. 16.8 cm
6. about 20.72 mm
7. $16\frac{1}{2}$ in.
8. about 43.96 yd
9. about 69 in.

Area and Volume

The *area* of a figure is the number of square units it encloses.

The *volume* of a space figure is the amount of space it encloses. Volume is measured in cubic units.

Formulas for finding area and volume are listed on page 615.

Example

a. Find the area.

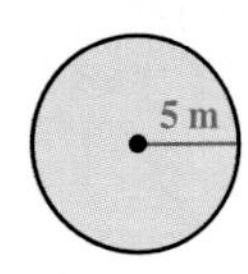

b. Find the volume.

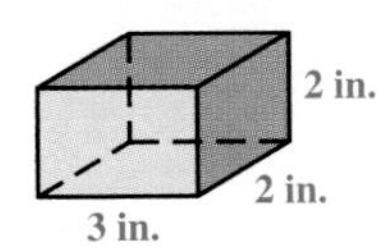

Solution

Use 3.14 for π.

a. $A = \pi r^2$

$\approx (3.14)(5)^2$

$\approx (3.14)(5)(5)$

≈ 78.5

The area is about 78.5 m^2.

b. $V = lwh$

$= (3)(2)(2)$

$= 12$

The volume is 12 in.^{3}.

Practice

Find the area of each figure.

1. circle

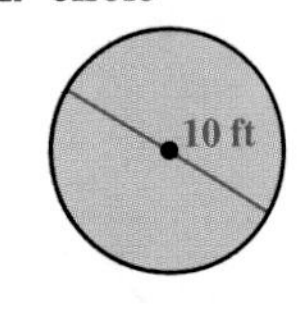

2. rectangle

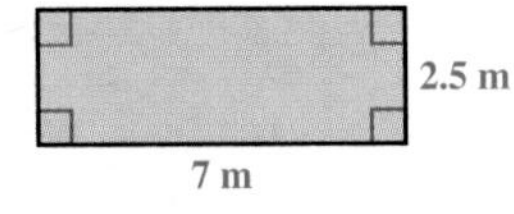

3. parallelogram

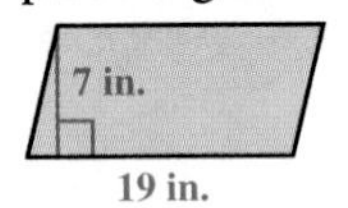

4. triangle

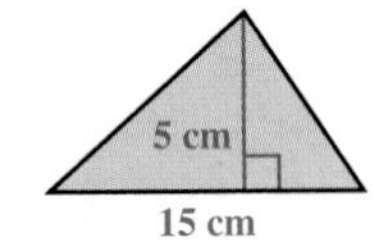

Find the volume of each rectangular prism.

5.

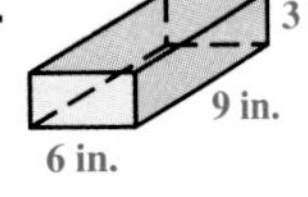

6.

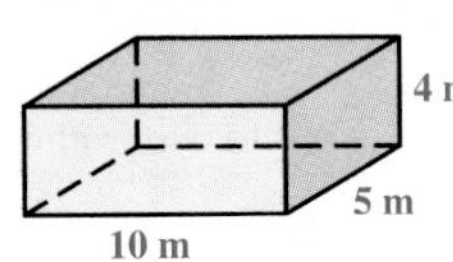

7. A cube with sides of length 25 in.

8. A box with length 7.5 cm, width 9.5 cm, and height 3.5 cm

9. A rectangular hole 7 yd long, 4 yd wide, and 2 yd deep was dug for a swimming pool. How many cubic yards of dirt were removed from the hole?

10. A pizza restaurant charges $9 for a large round pizza that has a diameter of 14 in. This week the rectangular pizza, which is 15 in. by 12 in., also costs $9. Which pizza has the greater area?

Practice

1. 113.04 ft^2
2. 17.5 m^2
3. 133 in.^{2}
4. 37.5 cm^2
5. 162 in.^{3}
6. 200 m^3
7. 15,625 in.^{3}
8. 249.375 cm^3
9. 56 yd^3
10. The rectangular pizza has the greater area.

ANGLES

An angle is formed by two rays (called *sides*) with a common endpoint (called a *vertex*).

An angle is measured in degrees.

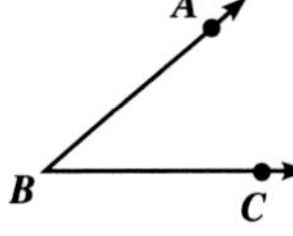

$\angle$ is the symbol for angle.

This angle is named $\angle ABC$, $\angle CBA$, or $\angle B$.

Some angles have special names based on their measures.

right angle: 90°

acute angle: less than 90°

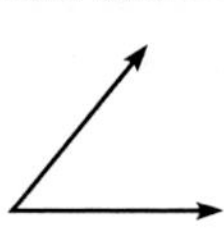

obtuse angle: greater than 90° and less than 180°

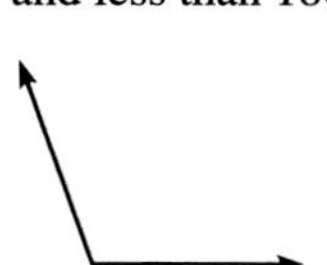

straight angle: 180°

EXAMPLE 1

Measure the angle and tell whether it is *right*, *acute*, *obtuse*, or *straight.*

SOLUTION

Read the number of degrees from zero that the other side passes through.

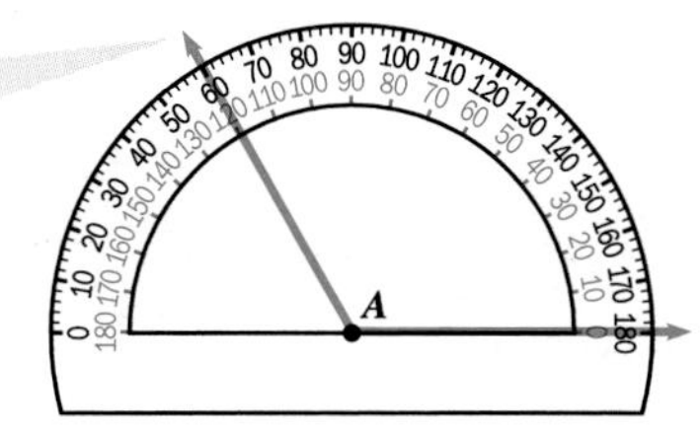

Line up one side of the angle with 0°.

The measure of the angle is 120°. It is obtuse.

PRACTICE

Find the measure of each angle. Use a protractor if necessary. Then tell whether the angle is *right*, *acute*, *obtuse*, or *straight*.

1.

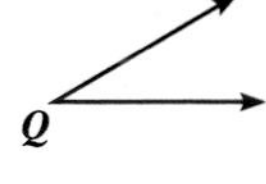

2.

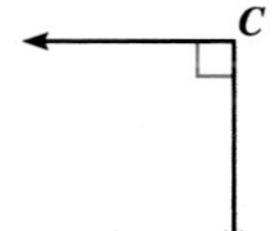

3.

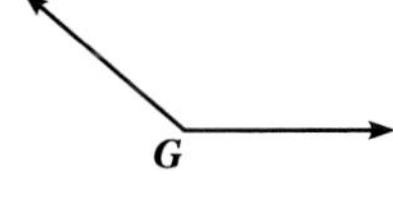

4.

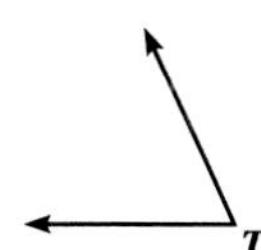

5. Draw an angle that has a measure of 85°.

6. Draw an angle that has a measure of 160°.

7. Draw a figure showing two intersecting lines forming angles of 75° and 105°.

Practice

1. about 30°; acute
2. about 90°; right
3. about 140°; obtuse
4. about 65°; acute
5.

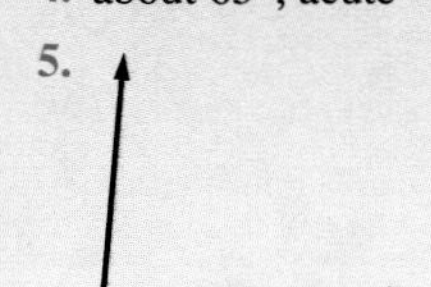

6.

7.

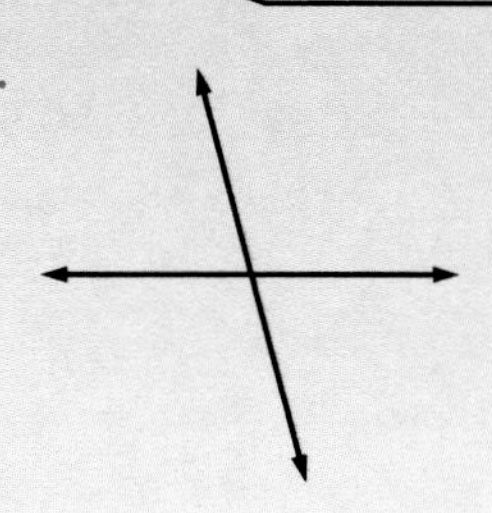

STATISTICAL GRAPHS

A *bar graph* is a visual display of data that fall into distinct categories.

EXAMPLE 1

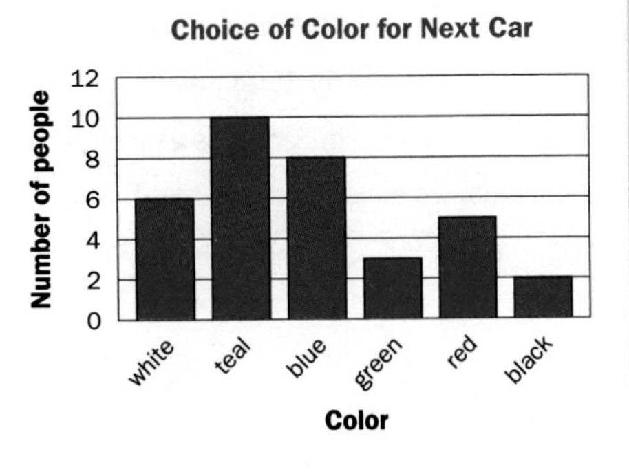

Use the bar graph at the right.

a. What is the most popular car color among the people surveyed?

b. How many people in the survey plan to buy a red car?

SOLUTION

a. The teal bar is the tallest. The most popular color is teal.

b. The bar for red is 5 units high. Five people plan to buy a red car.

PRACTICE

1. How many people are represented by the bar graph?

2. What is the least popular color?

3. How many more people choose blue than choose white?

A *line graph* is a graph that shows the amount and the direction of change in data over a period of time.

EXAMPLE 2

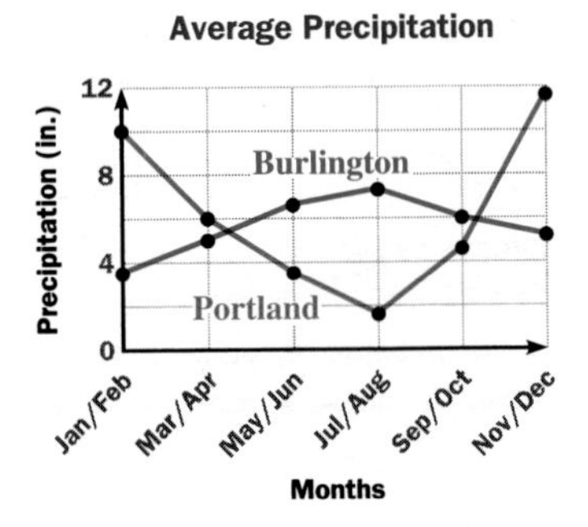

The line graph shows the average precipitation in Portland, Oregon, and Burlington, Vermont.

a. Estimate the average precipitation during May and June in Burlington.

b. During January and February, about how much more precipitation is there in Portland than in Burlington?

SOLUTION

a. Locate May/June on the horizontal axis. Move vertically up the grid line to the bullet and then move horizontally to the vertical axis. Estimate the number.

The average precipitation during May and June is about 7 inches.

b. Estimate the average precipitation for **Portland** and **Burlington** during January and February. Then subtract.

$$10 - 4 = 6$$

During January and February, Portland gets about 6 more inches of precipitation than Burlington.

Practice

1. 34 people
2. black
3. 2 people

PRACTICE

Use the line graph in Example 2.

4. Estimate the average precipitation for Burlington and the average precipitation for Portland during November and December.
5. When does Burlington get the most precipitation?
6. During July and August about how much more precipitation is there in Burlington than in Portland?

A *pictograph* can be used to display data that fall into distinct categories. A symbol is used to represent a given number of items.

EXAMPLE 3

Use the pictograph below.

a. About how many Super Trekker bicycles does Super Cycles manufacture?

b. About how many more Mountain Blazers are manufactured than Super Trekkers?

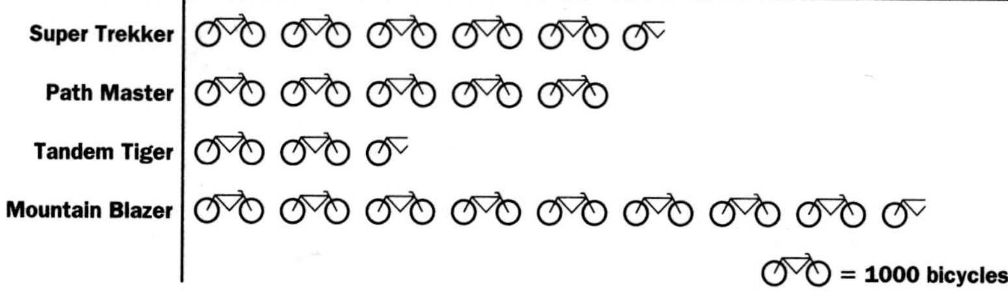

SOLUTION

a. Each symbol represents 1000 bicycles.

There are $5\frac{1}{2}$ symbols for Super Trekker.

Multiply.
$(5.5)(1000) = 5500$

Super Cycle manufactures about 5500 Super Trekkers.

b. Compare the number of symbols. The Mountain Blazer has about 3 more symbols than the Super Trekker.

Multiply.
$3(1000) = 3000$

There are about 3000 more Mountain Blazers manufactured than Super Trekkers.

PRACTICE

Use the pictograph in Example 3.

7. About how many bicycles are manufactured altogether?
8. For which bicycle is the amount manufactured the smallest?
9. How many symbols would you use to represent 9500 bicycles?

Practice

4. Burlington: about 5 in.; Portland: about 12 in.
5. July/August
6. about 5 in.
7. about 21,500 bicycles
8. Tandem Tiger
9. 9.5 symbols

A *circle graph* shows how parts relate to a whole and to each other.

The conductor of a community orchestra analyzed the ideal makeup of the musicians in the orchestra. The results are shown in the *circle graph* at the right.

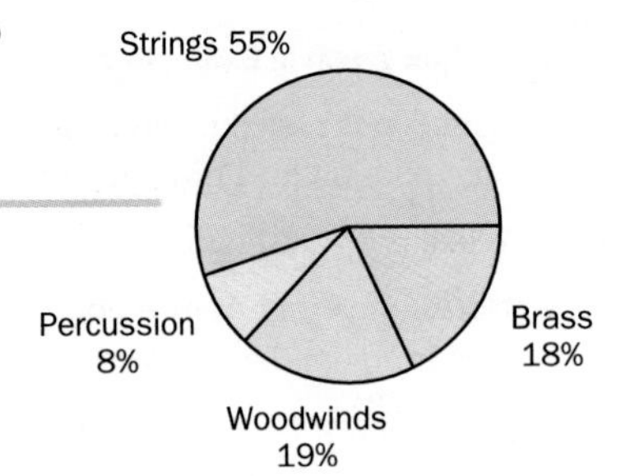

EXAMPLE 4

a. Suppose there are 65 musicians in the orchestra. About how many should be string players?

b. Suppose the conductor would like to have 22 woodwind players. About how big should the orchestra be to maintain an ideal balance?

SOLUTION

a. 55% of the orchestra's musicians are string players. Let s = the number of string players.

$s = 0.55(65)$

$= 35.75$

There should be about 36 string players.

b. 19% of the orchestra are woodwind players. Let N = the number of orchestra members.

$0.19N = 22$

$\frac{0.19N}{0.19} = \frac{22}{0.19} \approx 115.8$

There should be about 116 members in the orchestra.

PRACTICE

Use the circle graph in Example 4.

10. The conductor decides to expand the orchestra for a performance. To maintain balance, about how many players should be in each section in a 140 piece orchestra?

11. The conductor decides to have 65 string players. About how big should the orchestra be to maintain an ideal balance?

A *stem-and-leaf plot* displays data grouped into intervals. For example, the stem-and-leaf plot at the right groups daily protein intake of health club members for one day into intervals of 10. Each data item consists of two parts, a *stem* and a *leaf.* For example the number 57 is written

5 | 7

5 is the stem. 7 is the leaf.

Protein Consumed (grams)

Stem	Leaf
4	3 4
5	2 6 7
6	3 3 4 8 8 9
7	0 1 2 5 6
8	2 4 4 7

EXAMPLE 5

Use the stem-and-leaf plot showing protein consumed.

a. How many different amounts of protein are represented?

b. What is the highest amount of protein? the lowest?

Practice

10. Strings: about 77; Woodwinds: about 27; Brass: about 25; Percussion: about 11

11. about 118 players

SOLUTION

a. Count each leaf. Do not count duplicates. There are 17 different amounts represented.

b. Find the highest stem, then find the highest leaf in that row. The highest amount of protein consumed is 87 grams.

Find the smallest stem, then find the smallest leaf in that row. The lowest amount of protein consumed is 43 grams.

PRACTICE

Use the stem-and-leaf plot showing finishing times of racers.

12. How many racers finished in under 50 minutes?

13. What was the fastest time? the slowest?

Finishing Times (minutes)

Stem	Leaves
3	2 4
4	1 3 4 5 5 6 7
5	0 1 1 2 2 8
6	2 3 5 6
7	1

A *histogram* is a bar graph that shows how many data items occur in each of one or more intervals. The height of each bar shows the frequency, or number of items in each interval. Note that there are no spaces between the bars of a histogram.

EXAMPLE 6

Use the histogram at the right.

a. How many students walked from 80 to 89 laps?

b. How many students in all are represented by the histogram?

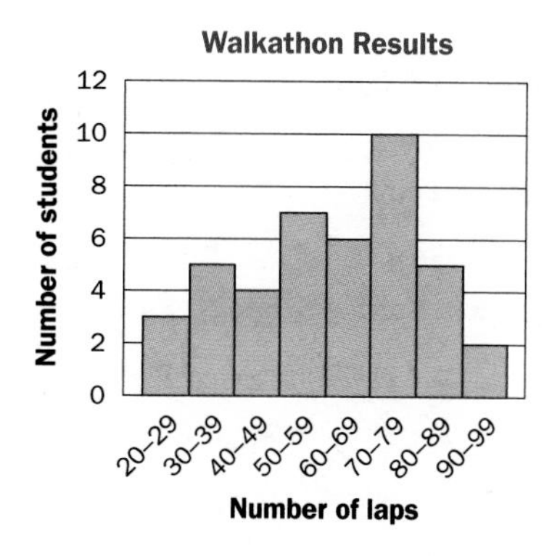

SOLUTION

a. Find the height of the bar for the interval 80–89. About 5 students walked from 80 to 89 laps.

b. Find the height of each bar. The sum of these heights is the total number of students.

$$3 + 5 + 4 + 7 + 6 + 10 + 5 + 2 = 42$$

The histogram represents 42 students.

PRACTICE

Use the histogram in Example 6.

14. How many students walked 40 or more laps?

15. How many students walked fewer than 40 laps?

16. Which interval has the smallest frequency? What is the frequency in this interval?

17. Change the intervals to 20–39, 40–59, 60–79 and 80–99. Redraw the histogram using these intervals.

Practice

12. 9 racers
13. 32 min; 71 min
14. 34 students
15. 8 students
16. 90–99 laps; 2 students

17.

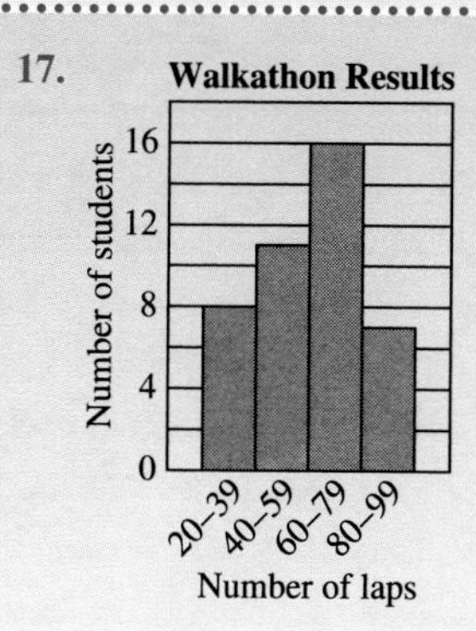

PROBLEM SOLVING

Often there is more than one way to solve a problem. If you are not sure how to solve a problem, consider using one or more of the following strategies.

Strategy	Application
Break the problem into parts	When intermediate steps are needed
Guess, check, and revise	When you do not seem to have all the needed information
Identify a pattern	When you can identify several cases
Make an organized list	When organizing information helps you solve the problem
Make a diagram	When words describe a picture
Solve a simpler problem	When simpler numbers help you solve the problem
Use a formula	When you recognize a known relationship
Use logical reasoning	When you use the facts to reach a conclusion
Work backward	When you are looking for a fact leading to a known result

EXAMPLE 1

How many segments can be drawn to connect ten points equally spaced around a circle?

SOLUTION

Problem Solving Strategy: Find a pattern.

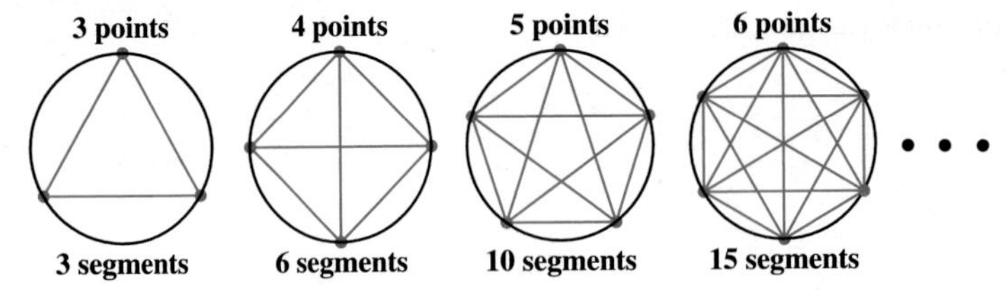

Compare the numbers of points and segments in one circle to the numbers in the previous circle to find a pattern. Then continue the pattern until the number of points equals 10.

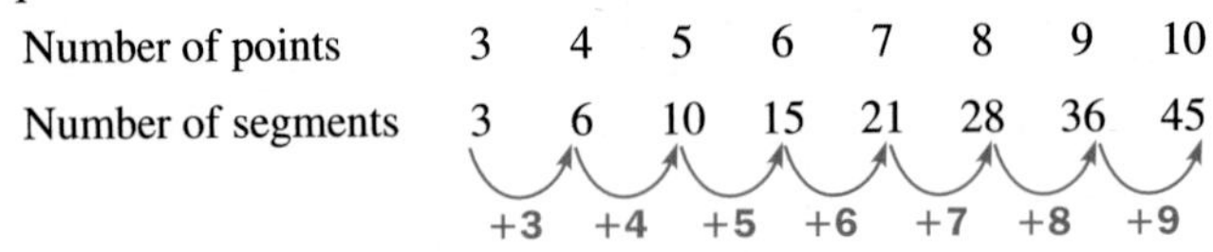

Number of points	3	4	5	6	7	8	9	10
Number of segments	3	6	10	15	21	28	36	45

A circle with 10 evenly spaced points can have 45 drawn segments.

EXAMPLE 2

Marva's savings account statement for May showed a deposit of $250, two withdrawals of $50 and $80, an interest payment of $1.29, and an ending balance of $637.52. Find the starting balance.

SOLUTION

Problem Solving Strategy: Work backward.

$637.52	Ending balance
− 1.29	Subtract the interest paid.
$636.23	
+ 80.00	Add back the withdrawals.
+ 50.00	
$766.23	
− 250.00	Subtract the deposit.
$516.23	

Marva's starting balance was $516.23.

PRACTICE

Solve each problem.

1. Each digit from 1 to 9 is used only once in the problem at the right. Find the missing digits.

	1	2	?
+	?	?	7
	4	?	?

2. List all the ways you can pay a bridge toll of $1.50 if you have only quarters and dimes.
3. Complete the pattern.

Days	1	2	3	4	5	6	7
Length of vine (inches)	12	24	36	?	?	?	?

4. The perimeter of a rectangular garden is 20 ft and the width is 4 ft. Find the area of the garden.
5. The Changs drove for 2 h at a speed of 50 mi/h, and then drove for 45 min at 40 mi/h. If their car gets about 25 miles per gallon of gasoline, about how many gallons of gasoline did the trip require?
6. A stack of cans in a grocery store has 1 can on top, 3 cans in the second row, 5 in the third, and so on. If there are 20 rows of cans altogether, how many cans are there in the stack?
7. Lois wants to buy a compact disk player that costs $145 plus 5% sales tax. She has already saved $35. Lois earns $4.75 an hour working at a department store. How many more hours does Lois need to work to be able to buy the compact disk player?
8. The sum of the measures of the angles of a triangle is 180°. Suppose the three angles of a triangle are equal in measure. Find the measure of each angle.

Practice

1. 128 + 367 = 495 or 129 + 357 = 486
2. 15 dimes; 10 dimes and 2 quarters; 5 dimes and 4 quarters; 6 quarters
3. 48; 60; 72; 84
4. 24 ft^2
5. about 5.2 gal
6. 400 cans
7. about 25 hours
8. 60°

Technology Handbook

This handbook introduces you to the basic features of most graphing calculators. Check your calculator's instruction manual for specific keystrokes and any details not provided here.

PERFORMING CALCULATIONS

The Keyboard

Look closely at your calculator's keyboard. Notice that most keys serve more than one purpose. Each key is labeled with its primary purpose, and labels for any secondary purposes appear somewhere near the key. You may need to press **2nd**, **SHIFT**, or **ALPHA** to use a key for a secondary purpose.

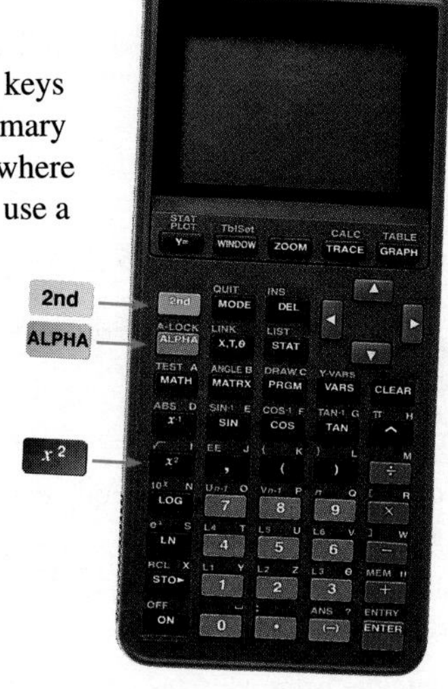

Examples of using the x^2 key:

Press x^2 to square a number.

Press **2nd** and then x^2 to take a square root.

Press **ALPHA** and then x^2 to get the letter I.

The Home Screen

Your calculator has a *home screen* where you can do calculations. You can usually enter a calculation on a graphing calculator just as you would write it on a piece of paper.

Shown below are some things to remember as you enter calculations on your graphing calculator.

The calculator has a subtraction key, **−**, and a negation key, **(−)**. If you use these incorrectly, you will get an error message.

You may need to use parentheses to avoid confusion. For example, enter $|5-9|$ as **ABS (5 − 9)** so the calculator does not interpret the expression as $|5|-9$.

To perform an operation that is not often used, such as to calculate 8! or ${}_{20}C_2$, you may need to press **MATH** to display a menu of operations.

```
3--2
                5
abs (5-9)
                4
20 nCr 2
              190
```

CALCULATOR PRACTICE

Use your calculator to find the value of each expression.

1. $-5 - 3$ **2.** $|-30|$ **3.** 2π **4.** $\sqrt{12.25}$

5. 2^5 **6.** 2.34×5^{-3} **7.** $8!$ **8.** ${}_7C_2$

DISPLAYING GRAPHS

Entering and Graphing a Function

To graph a function, enter its equation in the form $y = \ldots$. Rewrite the equation with x and y if necessary. For example, rewrite $t = 3 - s$ as $y = 3 - x$ and rewrite $y - 3 = \frac{1}{2}x$ as $y = \frac{1}{2}x + 3$.

The graph of $y = \frac{1}{2}x + 3$ is shown.

Use parentheses. If you enter $y = 1/2x + 3$, the calculator may interpret the equation as $y = \frac{1}{2x} + 3$.

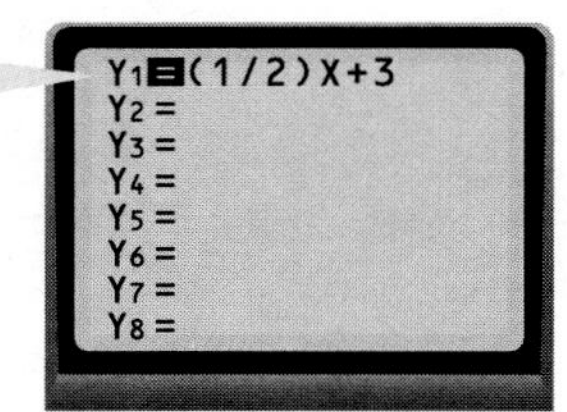

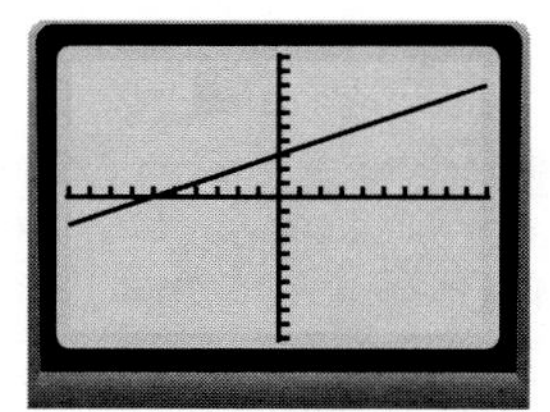

The Viewing Window

Think of the calculator screen as a *viewing window* that lets you look at a portion of the coordinate plane.

On many calculators, the *standard window* shows values from -10 to 10 on both the x- and y-axes. You can adjust the viewing window by pressing WINDOW or RANGE and entering new values for the window variables.

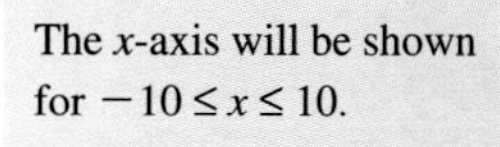

The x-axis will be shown for $-10 \le x \le 10$.

The y-axis will be shown for $-10 \le y \le 10$.

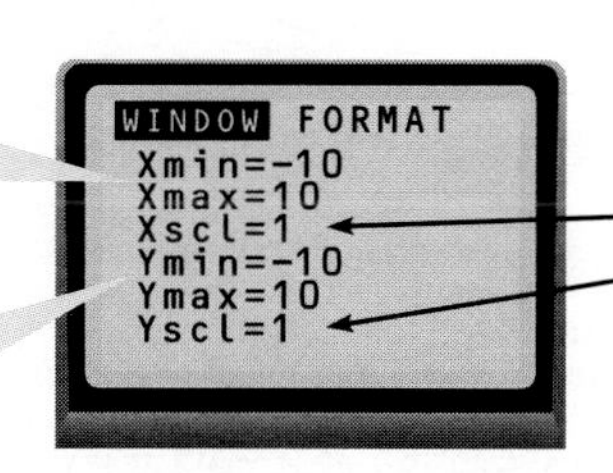

With scale variables set to equal 1, tick marks will be 1 unit apart on both axes.

ANSWERS Technology Handbook

Calculator Practice

1. –8 2. 30
3. about 6.28 4. 3.5
5. 32 6. 0.01872
7. 40,320 8. 21

Squaring the Screen

A *square screen* is a viewing window with equal unit spacing on the two axes. For example, the graph of $y = x$ is shown for two different windows.

Standard Viewing Window

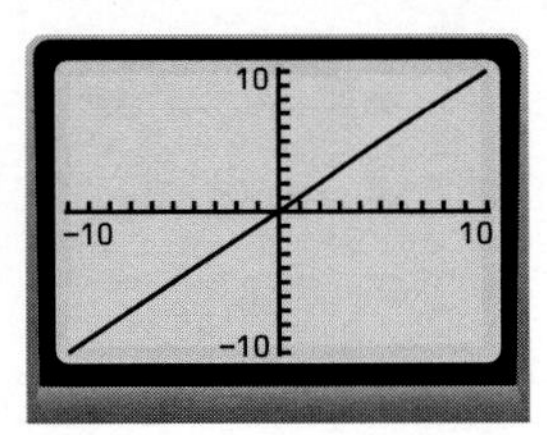

Square Screen Window

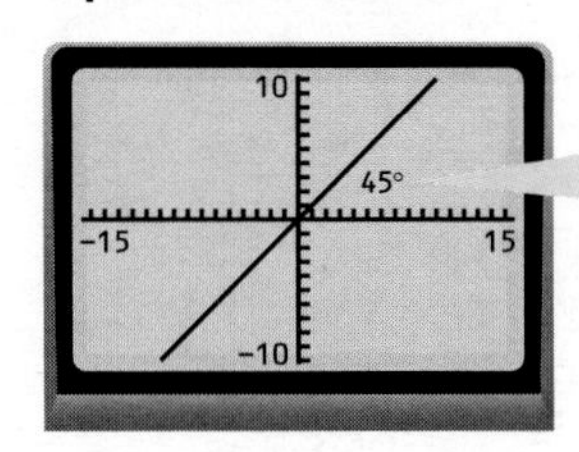

On a square screen, the line $y = x$ makes a 45° angle with the x-axis.

On many graphing calculators, the ratio of the screen's height to its width is about 2 to 3. Your calculator may have a feature that gives you a square screen. If not, choose values for the window variables that make the "length" of the y-axis about two-thirds the "length" of the x-axis:

$$(\text{Ymax} - \text{Ymin}) \approx \frac{2}{3}(\text{Xmax} - \text{Xmin})$$

The *Zoom* Feature

To get a closer look at a point of interest on a graph, you can use the *zoom* feature. Your calculator may have more than one way to zoom. A common way is to put a *zoom box* around the point.

Place one corner of the zoom box. Then place the opposite corner.

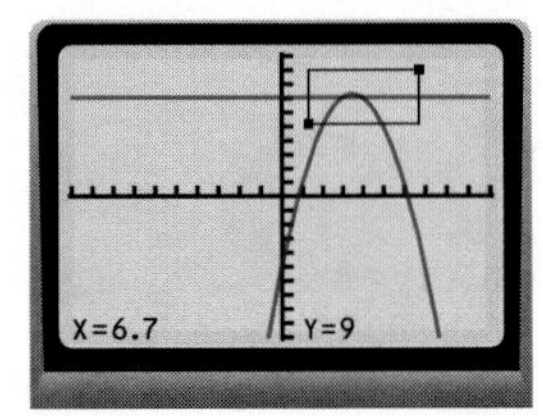

The calculator draws what is inside the box at full-screen size.

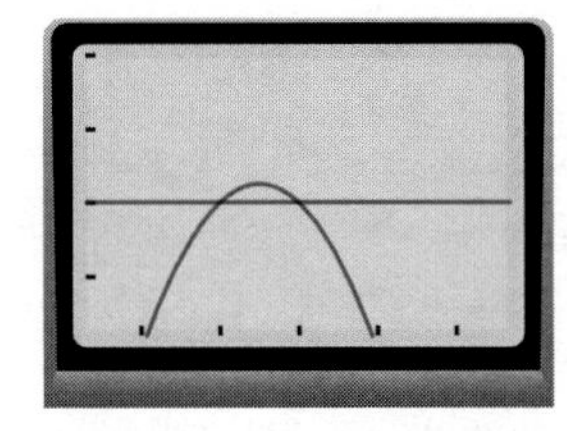

CALCULATOR PRACTICE

9. Enter and graph each equation separately. Use the standard viewing window.

a. $y = 3x + 1$ **b.** $y = x^2 - 4x + 2$ **c.** $y = \frac{5}{x}$ **d.** $y = 2^x$

e. Zoom in on part of your graph from part (d) and sketch it.

10. Find a good viewing window for the graph of $y = 65 - 3x$. Be sure your window shows where the graph crosses both axes. Be sure the window is square.

Calculator Practice

9. a.

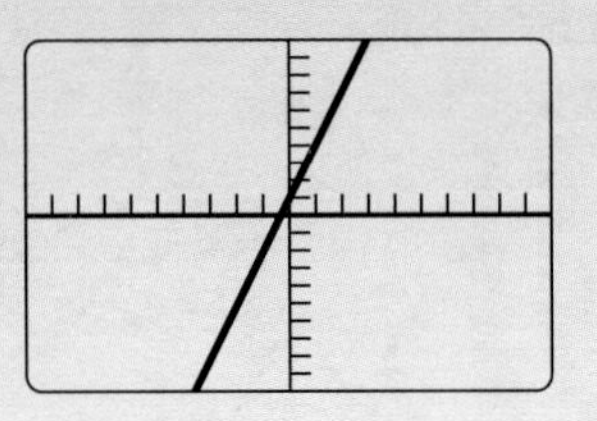

b.

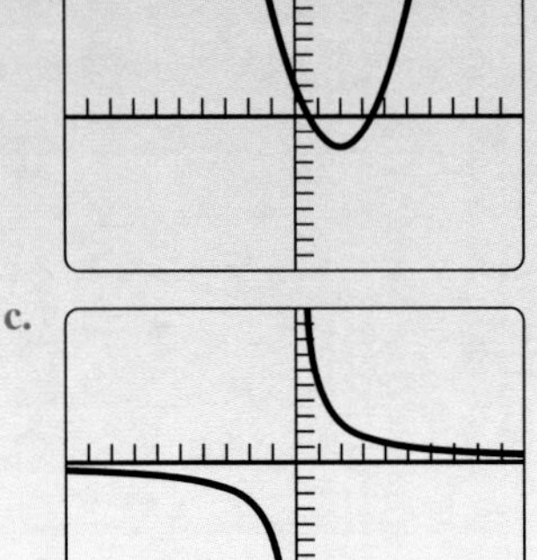

c.

d.

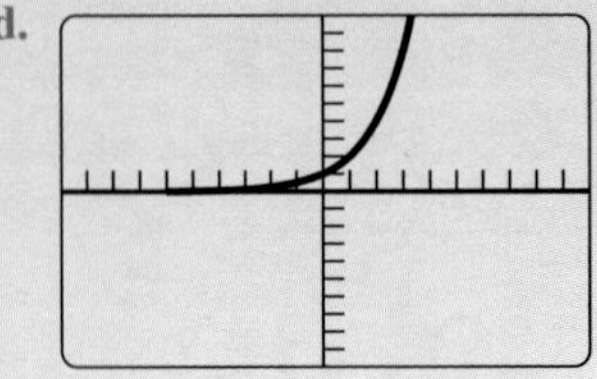

e. Zoom windows may vary. An example is given.

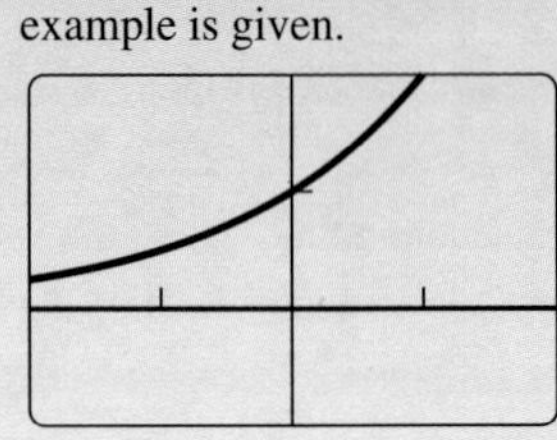

10. Answers may vary. An example is given; $-75 \le x \le 75$, $-20 \le x \le 80$

READING A GRAPH

The *Trace* Feature

After a graph is displayed, you can use the calculator's *trace* feature. When you press TRACE, a flashing cursor appears on the graph. The x- and y-coordinates of the cursor's location are shown at the bottom of the screen. Press ◄ and ► to move the *trace* cursor along the graph.

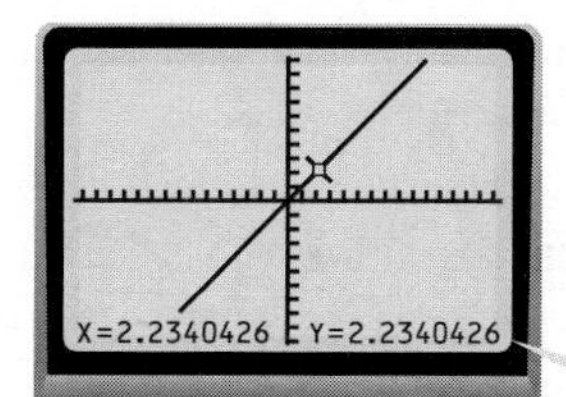

The *trace* cursor is at the point (2.2340426, 2.2340426) on the graph of $y = x$.

Suppose the equation $y = 7x - 70$ gives your profit y from selling x T-shirts. You can use the *trace* feature to find how many T-shirts you must sell to make a profit of \$800.

Graph the equation $y = 7x - 70$. Using TRACE, move the cursor along the graph until the y-value is approximately equal to 800. This gives you the value of x, the number of T-shirts you must sell.

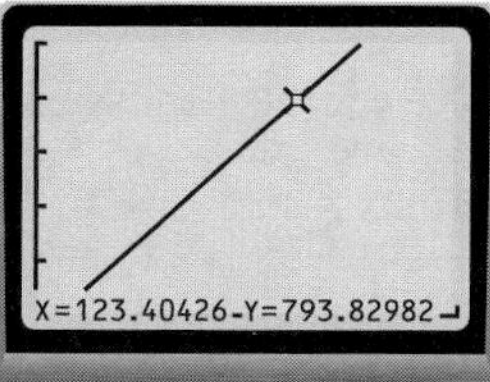

Another way to solve the problem is to graph two equations on the same screen: $y = 7x - 70$ and $y = 800$. Use *trace* to move the cursor to the intersection of the two lines to find your answer. On many graphing calculators you can use ▲ and ▼ to move between graphs. This helps you find the intersection.

Trace along one line until you are near the intersection.

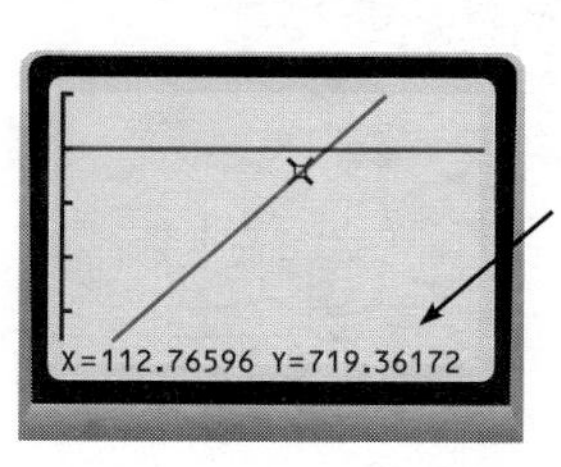

The x-coordinates are the same. The y-coordinates are close.

Press ▲. Now you are on the other line.

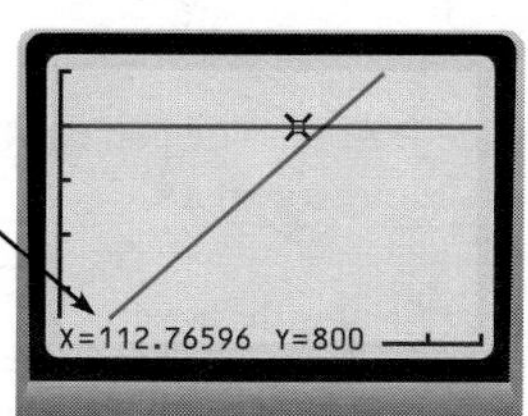

You can *zoom* in and repeat the process described above until the y-coordinates are the same to the nearest tenth, hundredth, or any other decimal place.

605

Friendly Windows

As you press ▸ while tracing a graph, the x-coordinate may increase by "unfriendly" increments, giving "unfriendly" x-values.

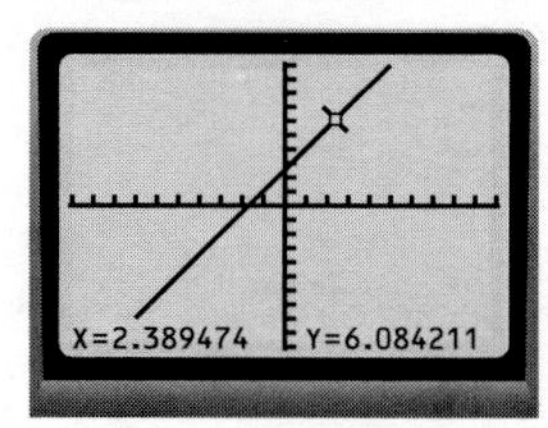

Your calculator may allow you to control the x-increment, ΔX, directly. If not, you can control it indirectly by choosing an appropriate Xmax for a given Xmin.

For example, on a TI-82 graphing calculator, choose Xmin and Xmax so that

$$\text{Xmax} - \text{Xmin} = 94\Delta X \qquad \text{or} \qquad \text{Xmax} = \text{Xmin} + 94\Delta X.$$

This number depends upon the calculator you are using.

Suppose you want ΔX to equal 0.1 and Xmin to equal -5. Then:

$$\begin{aligned} \text{Xmax} &= -5 + 94(0.1) \\ &= 4.4 \end{aligned}$$

Set Xmax to 4.4 for a "friendly window" in which the *trace* cursor's x-coordinate will increase by 0.1 each time you press ▸.

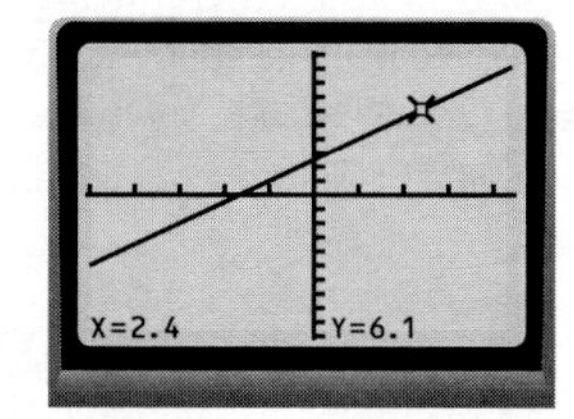

Calculating Coordinates

Some calculators have a *calculate* feature. You can use this feature to find the coordinates of a point of interest automatically.

For example, on a TI-82 use *value* to find the value of y for a given value of x.

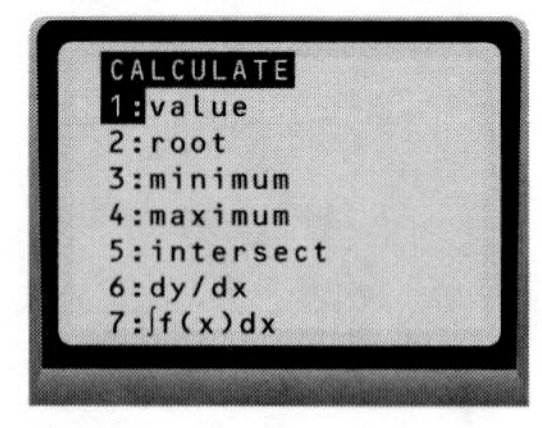

Enter the x-value.

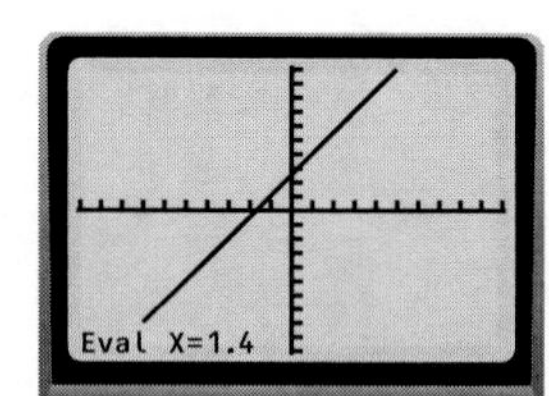

Press ENTER.

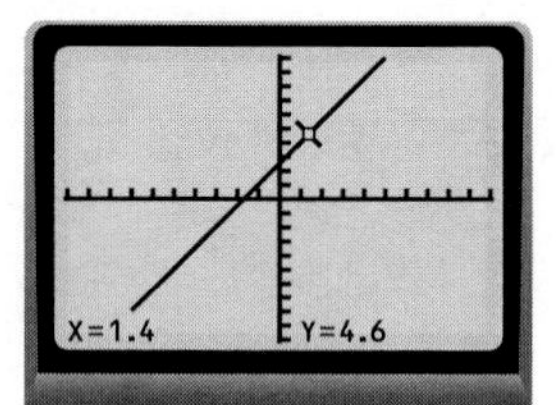

You can use the *intersection* feature to find the intersection of two graphs.

Use ▲ and ▼ to choose the two graphs.

If there is more than one point of intersection, use ▶ and ◀ to choose one of them.

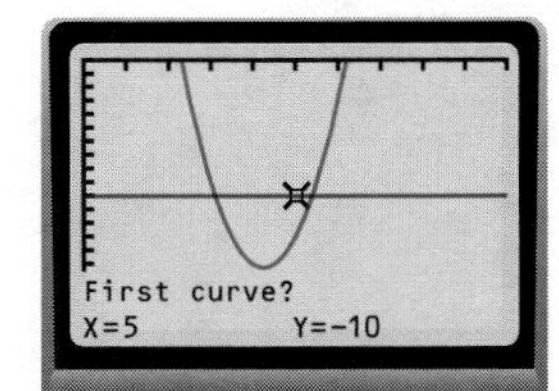

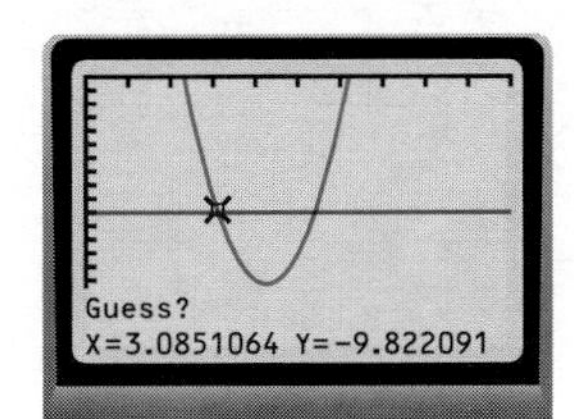

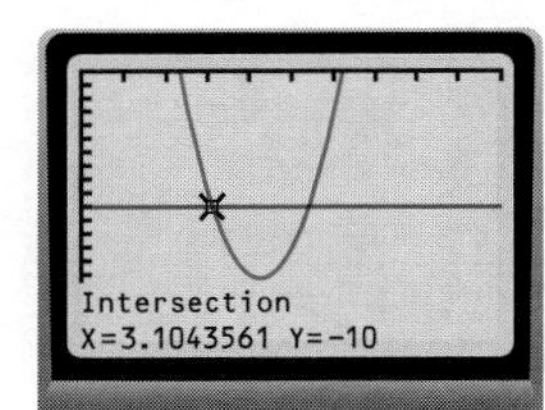

You can use the *root* feature to find the point where a graph crosses the x-axis.

Choose a root by marking a point to the left of it and a point to the right of it.

Then guess the location of the root to help the calculator find it more quickly.

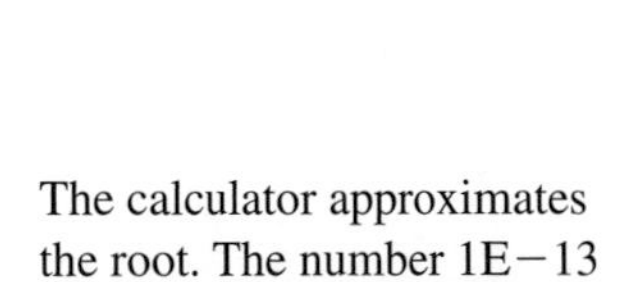

The calculator approximates the root. The number 1E−13 means 10^{-13} or about zero.

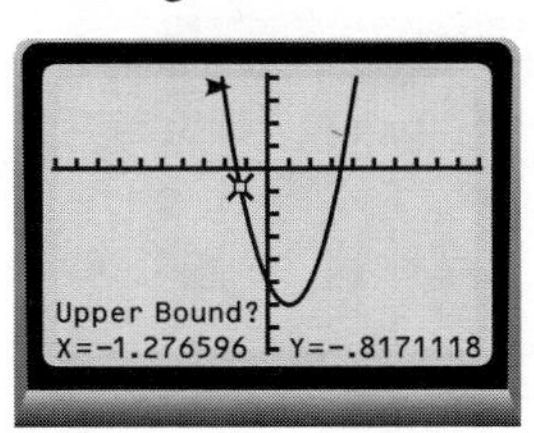

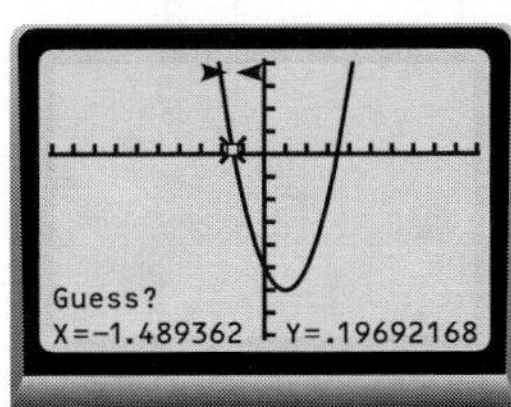

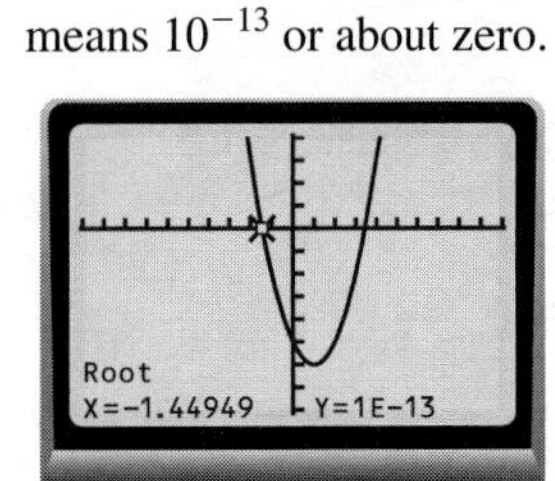

CALCULATOR PRACTICE

11. Graph the equation $y = x + \frac{1}{7}$.

a. Choose a friendly window where $\Delta X = 0.1$. Use the *trace* feature to determine, to the nearest tenth, the value of x for which $y = 5$.

b. Try zooming in on the point where $y = 5$. Between what two values, to the nearest hundredth, does the x-coordinate of the point lie?

12. Graph $y = 2\pi x$. Use the *calculate* feature to find an approximate value of y when $x = 9$.

13. Graph the equations below. Use the *calculate* feature to find the coordinates of the point of intersection.

$$9x + 3y = 14$$
$$-3x + 2y = 8$$

14. Use the *calculate* feature to find the roots of the equation $y = -4x^2 + 2x + 15$.

Calculator Practice

11. a. about 4.9
    b. between 4.85 and 4.86
12. about 56.55
13. x is about 0.15 and y is about 4.22.
14. about –1.70 and about 2.20

WORKING WITH STATISTICS

Adding Matrices

Suppose you want to use matrices to find each baseball team's total wins and losses for the 1992 season. Use your calculator's *matrix* feature to enter (or *edit*) the data into matrices.

AMERICAN LEAGUE

Games Before the All-Star Game		
EAST	W	L
Toronto	53	34
Baltimore	49	38
Milwaukee	45	41
Boston	42	43
New York	42	45
Detroit	41	48
Cleveland	36	52
WEST	W	

AMERICAN LEAGUE

Games After the All-Star Game		
EAST	W	L
Toronto	43	32
Baltimore	40	35
Milwaukee	47	29
Boston	31	46
New York	34	41
Detroit	34	39
Cleveland	40	34
WEST	W	

Give the dimensions of matrix A and enter the data for games before the All-Star Game.

```
MATRIX[A] 7 ×2
[ 53   34      ]
[ 49   38      ]
[ 45   41      ]
[ 42   43      ]
[ 42   45      ]
[ 41   48      ]
[ 36   52      ]
7,2=52
```

Enter the data for games after the All-Star Game in matrix B.

```
MATRIX[B] 7 ×2
[ 43   0       ]
[ 0    0       ]
[ 0    0       ]
[ 0    0       ]
[ 0    0       ]
[ 0    0       ]
[ 0    0       ]
1,1=43
```

From the home screen, add the two matrices. To type [A] you may need to press MATRX 1 or 2nd 1.

The answer is a matrix with the same dimensions as the original matrices. Its elements are the sums of elements in the same positions in [A] and in [B].

Scatter Plots and Finding a Line of Fit

Some calculators use linear regression to find a line of fit for a set of ordered pairs. First enter the two lists of data and then graph the paired data in a scatter plot.

On a TI-82 graphing calculator, press STAT to edit, or enter, the data.

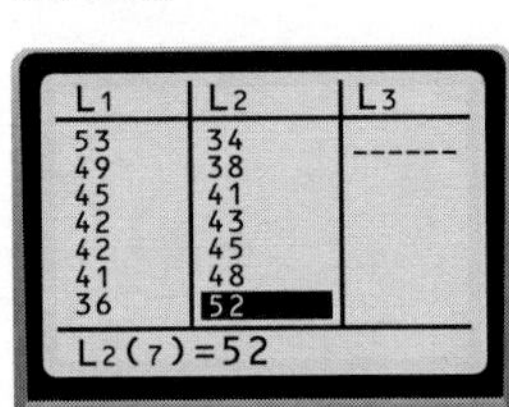

Turn STAT PLOT on and GRAPH the data. See your manual or your teacher if you need help.

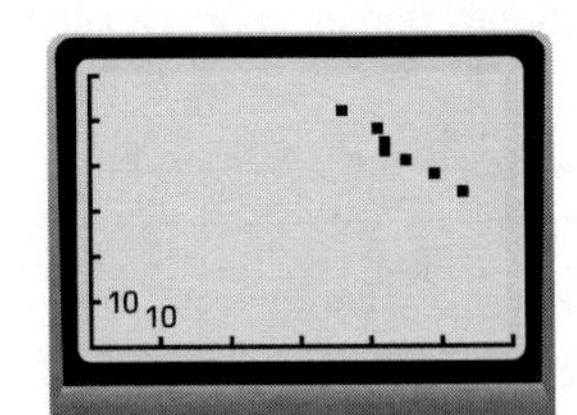

Use linear regression in the statistics calculations menu to fit a line to your data.

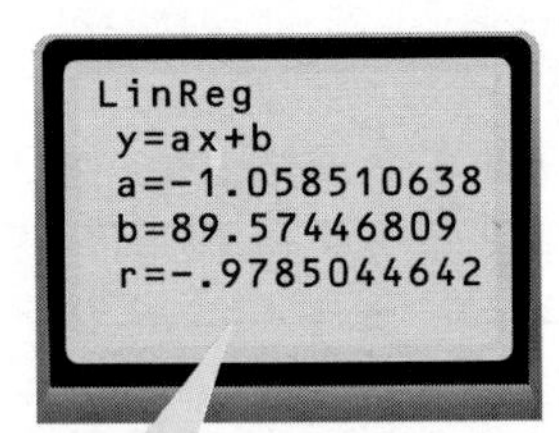

The equation $y = -x + 90$ models the data.

The *Table* Feature

Some calculators have a *table* feature. You can use this feature to examine the values of a function. The screen shows a table of values for $y = x^2 - 6x + 7$.

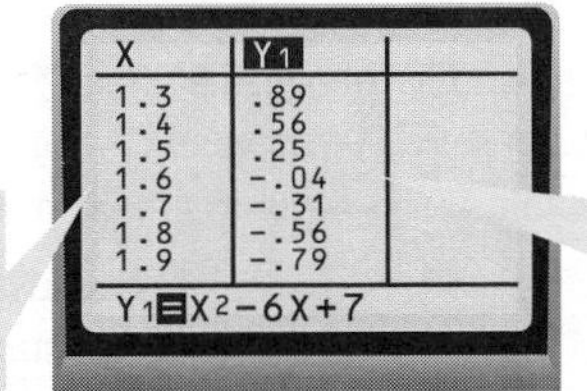

Some calculators have a table set-up feature that allows you to change the *x*-values.

Notice that the *y*-values change sign between $x = 1.5$ and $x = 1.6$.

You can also use the table feature like a spreadsheet. Suppose you want to change the units of the data at the right from centimeters to inches.

Heights of Children (cm)				
102	106	96	106	102
101	100	106	112	102
105	106	100	100	100
100	94	94	106	97

Use **TblSet** to instruct the calculator to ask for *x*-values.

```
TABLE SETUP
 TblMin=0
 ΔTbl=.1
Indpnt: Auto Ask
Depend: Auto Ask
```

Enter the *x*-values in the **TABLE**.

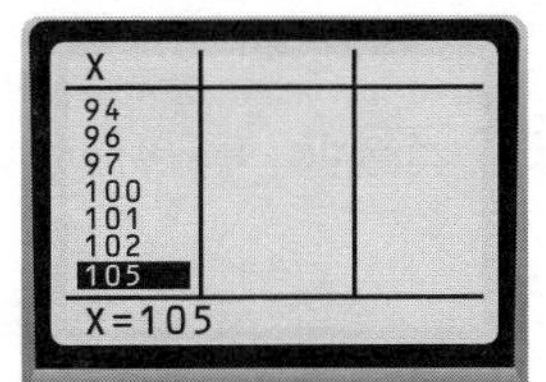

Using Y=, enter a formula to change the units of the data to inches.

```
X     Y1
94    37.008
96    37.795
97    38.189
100   39.37
101   39.764
102   40.157
105   41.339
Y1=0.3937X
```

Now you can use **Y-VARS** to perform calculations on Y_1.

CALCULATOR PRACTICE

15. Add matrices *A* and *B*.

$$A = \begin{bmatrix} 1 & -2 & 7 \\ -1 & 5 & 0 \end{bmatrix} \quad B = \begin{bmatrix} -3 & 1 & -2 \\ 0 & -1 & 5 \end{bmatrix}$$

16. Graph the data about can sizes. Find a line of fit. If a can has a diameter of 3.1 in., what will the circumference be?

17. Create a table of values for the equation $y = x^2 - 6x + 7$. As the *x*-values increase, at approximately what *x*-value do the *y*-values stop decreasing and begin increasing?

18. Use the *table* feature to transform the data at the right from hours to minutes.

Sizes of Cans	
Diameter (inches)	Circumference (inches)
2.1	6.7
2.7	8.4
3.0	9.4
3.2	10.0
3.4	10.8
4.1	12.8
4.3	13.4
6.2	19.4

Time Spent on Homework	
Day	Time (hours)
Monday	1
Tuesday	1.5
Wednesday	2
Thursday	1
Friday	0
Saturday	4.75
Sunday	0

Calculator Practice

15. $\begin{bmatrix} -2 & -1 & 5 \\ -1 & 4 & 5 \end{bmatrix}$

16. Equations may vary. An example is given; $y = 3.11x + 0.1$; about 9.7 in.

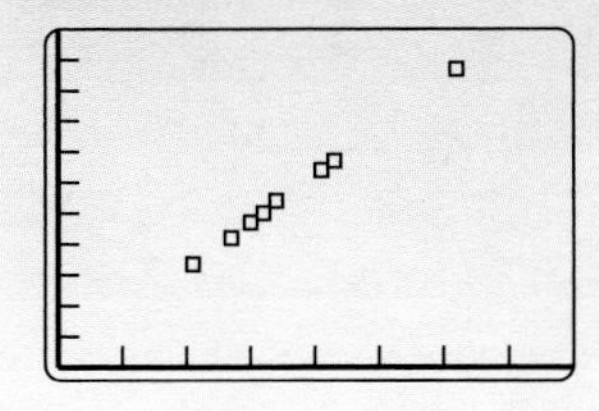

17. Possible table:

```
X    Y1
0    7
1    2
2    -1
3    -2
4    -1
5    2
6    7
X=3
```

The *y*-values stop decreasing at $x = 3$.

18. Monday: 60 min; Tuesday: 90 min; Wednesday: 120 min; Thursday: 60 min; Friday: 0 min; Saturday: 285 min; Sunday: 0 min

USING A SPREADSHEET

Suppose you want to buy a CD player that costs $195, including tax. You already have $88 and can save $9 per week. After how many weeks can you buy the player? You can use a computer spreadsheet to solve this problem.

row numbers

column letters

CD Savings

	A	B
1	Week number	Total saved
2	0	88
3	= A2 + 1	= B2 + 9
4	= A3 + 1	= B3 + 9
5	= A4 + 1	= B4 + 9
6	= A5 + 1	= B5 + 9
7	= A6 + 1	= B6 + 9
8	= A7 + 1	= B7 + 9
9	= A8 + 1	= B8 + 9
10	= A9 + 1	= B9 + 9
11	= A10 + 1	= B10 + 9
12	= A11 + 1	= B11 + 9
13	= A12 + 1	= B12 + 9
14	= A13 + 1	= B13 + 9
15	= A14 + 1	= B14 + 9

A spreadsheet is made up of cells named by a column letter and a row number, like A3 or B4. Cells can contain labels, numbers, or formulas.

Cell B1 contains the label "Total saved."

Cell B2 contains the number 88.

Cell B3 contains the formula " = B2 + 9." This formula tells the computer to take the number in cell B2, add 9 to it, and put the result in cell B3.

Instead of typing a formula into each cell individually, you can use the spreadsheet's *series*, *fill down*, and *copy* commands.

CD Savings

	A	B
1	Week number	Total saved
2	0	88
3	1	97
4	2	106
5	3	115
6	4	124
7	5	133
8	6	142
9	7	151
10	8	160
11	9	169
12	10	178
13	11	187
14	12	196
15	13	205

The computer replaces all the formulas with calculated values. You can have the computer draw a scatter plot with a line connecting the plotted points.

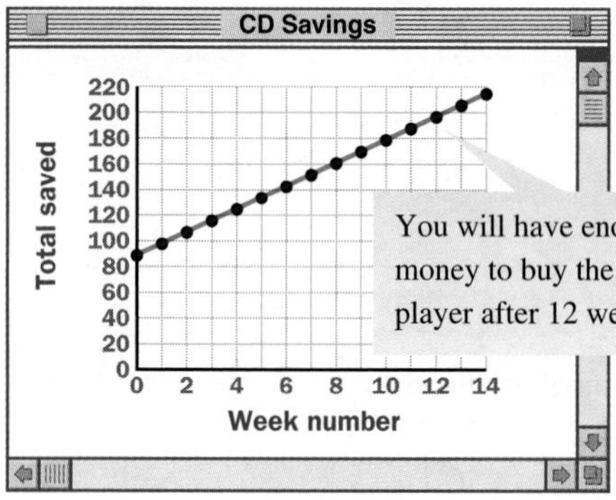

CD Savings

Total saved

220
200
180
160
140
120
100
80
60
40
20
0

0 2 4 6 8 10 12 14

Week number

You will have enough money to buy the CD player after 12 weeks.

See pages 41–42 of Section 1.8 for more information about spreadsheets.

SPREADSHEET PRACTICE

19. Suppose you borrowed $150 from your parents and are paying back $7 each week. Use a spreadsheet to find out when you will owe only $45. How long will it take to pay back the loan completely?

Calculator Practice

19. You will owe $45 at 15 weeks. The loan will be paid back in 22 weeks.

Table of Measures

Time

60 seconds (s)	=	1 minute (min)
60 minutes	=	1 hour (h)
24 hours	=	1 day
7 days	=	1 week
4 weeks (approx.)	=	1 month
365 days 52 weeks (approx.) 12 months	=	1 year
10 years	=	1 decade
100 years	=	1 century

Metric

Length

10 millimeters (mm)	=	1 centimeter (cm)
100 cm 1000 mm	=	1 meter (m)
1000 m	=	1 kilometer (km)

Area

100 square millimeters (mm^2)	=	1 square centimeter (cm^2)
10,000 cm^2	=	1 square meter (m^2)
10,000 m^2	=	1 hectare (ha)

Volume

1000 cubic millimeters (mm^3)	=	1 cubic centimeter (cm^3)
1,000,000 cm^3	=	1 cubic meter (m^3)

Liquid Capacity

1000 milliliters (mL)	=	1 liter (L)
1000 L	=	1 kiloliter (kL)

Mass

1000 milligrams (mg)	=	1 gram (g)
1000 g	=	1 kilogram (kg)
1000 kg	=	1 metric ton (t)

Temperature

Degrees Celsius (°C)

0°C	=	freezing point of water
37°C	=	normal body temperature
100°C	=	boiling point of water

United States Customary

Length

12 inches (in.)	=	1 foot (ft)
36 in. 3 ft	=	1 yard (yd)
5280 ft 1760 yd	=	1 mile (mi)

Area

144 square inches ($in.^2$)	=	1 square foot (ft^2)
9 ft^2	=	1 square yard (yd^2)
43,560 ft^2 4840 yd^2	=	1 acre (A)

Volume

1728 cubic inches ($in.^3$)	=	1 cubic foot (ft^3)
27 ft^3	=	1 cubic yard (yd^3)

Liquid Capacity

8 fluid ounces (fl oz)	=	1 cup (c)
2 c	=	1 pint (pt)
2 pt	=	1 quart (qt)
4 qt	=	1 gallon (gal)

Weight

16 ounces (oz)	=	1 pound (lb)
2000 lb	=	1 ton (t)

Temperature

Degrees Fahrenheit (°F)

32°F	=	freezing point of water
98.6°F	=	normal body temperature
212°F	=	boiling point of water

Table of Symbols

Symbol		Page	Symbol		Page
$\geq$	is greater than or equal to	3	$^\circ$	degree(s)	172
$\leq$	is less than or equal to	3	$a : b$	the ratio of a to b	192
$>$	is greater than	3	$\triangle ABC$	triangle ABC	205
$<$	is less than	3	$\overline{AB}$	segment AB	205
$=$	equals, is equal to	3	AB	the length of $\overline{AB}$	205
( )	parentheses—a grouping symbol	8	$\angle A$	angle A	205
$\cdot$	multiplication, times ($\times$)	9	$\sim$	is similar to	206
$\approx$	is approximately equal to	14	π	pi, an irrational number approximately equal to 3.14	215
. . .	and so on	19	$\sqrt{a}$	the nonnegative square root of a	240
$\lvert x \rvert$	absolute value of x	19	$\pm$	plus or minus	240
$-a$	opposite of a	20	$0.\overline{27}$	the repeating decimal 0.272727. . .	246
[ ]	brackets—a grouping symbol	26	a^n	nth power of a	370
$\stackrel{?}{=}$	is this statement true?	28	a^{-n}	$\frac{1}{a^n}, a \neq 0$	391
$\begin{bmatrix} 1 & 0 \\ 0 & 1 \end{bmatrix}$	matrix	40	!	factorial	532
(x, y)	ordered pair	73	$_nP_r$	permutation	535
m	slope	120	$_nC_r$	combination	537
$\frac{1}{a}$	reciprocal of a, $a \neq 0$	153			
$\neq$	is not equal to	154			

Table of Squares and Square Roots

No.	Square	Sq. Root
1	1	1.000
2	4	1.414
3	9	1.732
4	16	2.000
5	25	2.236
6	36	2.449
7	49	2.646
8	64	2.828
9	81	3.000
10	100	3.162
11	121	3.317
12	144	3.464
13	169	3.606
14	196	3.742
15	225	3.873
16	256	4.000
17	289	4.123
18	324	4.243
19	361	4.359
20	400	4.472
21	441	4.583
22	484	4.690
23	529	4.796
24	576	4.899
25	625	5.000
26	676	5.099
27	729	5.196
28	784	5.292
29	841	5.385
30	900	5.477
31	961	5.568
32	1,024	5.657
33	1,089	5.745
34	1,156	5.831
35	1,225	5.916
36	1,296	6.000
37	1,369	6.083
38	1,444	6.164
39	1,521	6.245
40	1,600	6.325
41	1,681	6.403
42	1,764	6.481
43	1,849	6.557
44	1,936	6.633
45	2,025	6.708
46	2,116	6.782
47	2,209	6.856
48	2,304	6.928
49	2,401	7.000
50	2,500	7.071

No.	Square	Sq. Root
51	2,601	7.141
52	2,704	7.211
53	2,809	7.280
54	2,916	7.348
55	3,025	7.416
56	3,136	7.483
57	3,249	7.550
58	3,364	7.616
59	3,481	7.681
60	3,600	7.746
61	3,721	7.810
62	3,844	7.874
63	3,969	7.937
64	4,096	8.000
65	4,225	8.062
66	4,356	8.124
67	4,489	8.185
68	4,624	8.246
69	4,761	8.307
70	4,900	8.367
71	5,041	8.426
72	5,184	8.485
73	5,329	8.544
74	5,476	8.602
75	5,625	8.660
76	5,776	8.718
77	5,929	8.775
78	6,084	8.832
79	6,241	8.888
80	6,400	8.944
81	6,561	9.000
82	6,724	9.055
83	6,889	9.110
84	7,056	9.165
85	7,225	9.220
86	7,396	9.274
87	7,569	9.327
88	7,744	9.381
89	7,921	9.434
90	8,100	9.487
91	8,281	9.539
92	8,464	9.592
93	8,649	9.644
94	8,836	9.695
95	9,025	9.747
96	9,216	9.798
97	9,409	9.849
98	9,604	9.899
99	9,801	9.950
100	10,000	10.000

No.	Square	Sq. Root
101	10,201	10.050
102	10,404	10.100
103	10,609	10.149
104	10,816	10.198
105	11,025	10.247
106	11,236	10.296
107	11,449	10.344
108	11,664	10.392
109	11,881	10.440
110	12,100	10.488
111	12,321	10.536
112	12,544	10.583
113	12,769	10.630
114	12,996	10.677
115	13,225	10.724
116	13,456	10.770
117	13,689	10.817
118	13,924	10.863
119	14,161	10.909
120	14,400	10.954
121	14,641	11.000
122	14,884	11.045
123	15,129	11.091
124	15,376	11.136
125	15,625	11.180
126	15,876	11.225
127	16,129	11.269
128	16,384	11.314
129	16,641	11.358
130	16,900	11.402
131	17,161	11.446
132	17,424	11.489
133	17,689	11.533
134	17,956	11.576
135	18,225	11.619
136	18,496	11.662
137	18,769	11.705
138	19,044	11.747
139	19,321	11.790
140	19,600	11.832
141	19,881	11.874
142	20,164	11.916
143	20,449	11.958
144	20,736	12.000
145	21,025	12.042
146	21,316	12.083
147	21,609	12.124
148	21,904	12.166
149	22,201	12.207
150	22,500	12.247

Properties of Algebra

Identity Property of Multiplication, p. 24

For every number a:

$1 \cdot a = a \qquad 1 \cdot 3 = 3$

Multiplicative Property of −1, p. 24

For every number a:

$-1 \cdot a = -a \qquad -1 \cdot 3 = 3$

Commutative Property, p. 26

You can add or multiply numbers in any order.

$a + b = b + a \qquad -3 + 7 = 7 + (-3)$

$ab = ba \qquad (-3)(7) = (7)(-3)$

Associative Property, p. 26

When you add or multiply three or more numbers, you can group the numbers without affecting the result,

For example: $(7 + 8) + 2 = 7 + (8 + 2)$

$(7 \cdot 8) \cdot 2 = 7 \cdot (8 \cdot 2)$

Distributive Property, p. 35

For all numbers a, b, and c:

$a(b + c) = ab + ac$

For example: $3(10 + 2) = 3(10) + 3(2)$

Property of Reciprocals, p. 153

For every nonzero number a, there is exactly one number $\frac{1}{a}$ such that

$a \cdot \frac{1}{a} = 1$ and $\frac{1}{a} \cdot a = 1$

For example: $2 \cdot \frac{1}{2} = 1$ and $\frac{1}{2} \cdot 2 = 1$

Means-Extremes Property, p. 193

For any proportion, the product of the extremes is equal to the product of the means.

If $\frac{a}{b} = \frac{c}{d}$, then $ad = bc$.

Properties of Square Roots, p. 252

For all nonnegative numbers a and b:

$\sqrt{a^2} = a$

$\sqrt{ab} = \sqrt{a} \cdot \sqrt{b}$

$\sqrt{\frac{a}{b}} = \frac{\sqrt{a}}{\sqrt{b}}, b \neq 0$

Zero and Negative Exponents, p. 391

For any nonzero number a and any integer n:

$a^0 = 1$ and $a^{-n} = \frac{1}{a^n}$

Rules for Products and Quotients of Powers, p. 402

For any nonzero number a and any integers m and n:

$a^m \cdot a^n = a^{m+n}$ and $\frac{a^m}{a^n} = a^{m-n}$

Rules for Powers of Products and Quotients, p. 408

For any nonzero numbers a and b and any integer m:

$(ab)^m = a^m b^m$ and $\left(\frac{a}{b}\right)^m = \frac{a^m}{b^m}$

Power of a Power Rule, p. 408

For any nonzero number a and any integers m and n:

$(a^m)^n = a^{m \cdot n}$

Zero-Product Property, p. 437

If the product of factors is equal to zero, then one or more of the factors is equal to zero.

For example: If $(x - 3)(x - 4) = 0$,

then $(x - 3) = 0$ or $(x - 4) = 0$

$x = 3$ or $x = 4$

Formulas from Geometry

To find the perimeter P of any plane figure, find the sum of the lengths of each side of the figure.

For the formulas below, let A = area, C = circumference, V = volume, and B = the area of the base of a space figure. Let $\pi \approx 3.14$.

Parallelogram

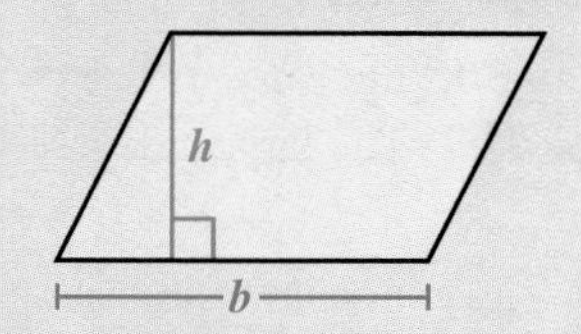

$A = \text{base} \times \text{height}$

$A = bh$

Triangle

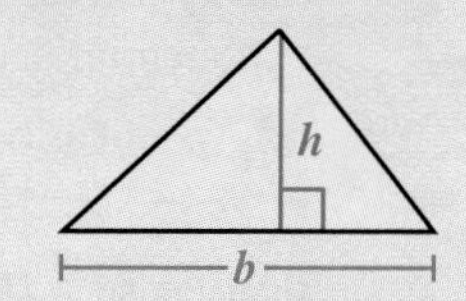

$A = \frac{1}{2} \times \text{base} \times \text{height}$

$A = \frac{1}{2}bh$

Trapezoid

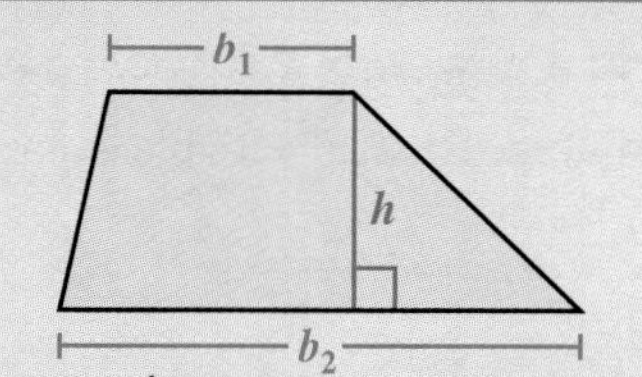

$A = \frac{1}{2} \times \text{sum of bases} \times \text{height}$

$A = \frac{1}{2}(b_1 + b_2)h$

Circle

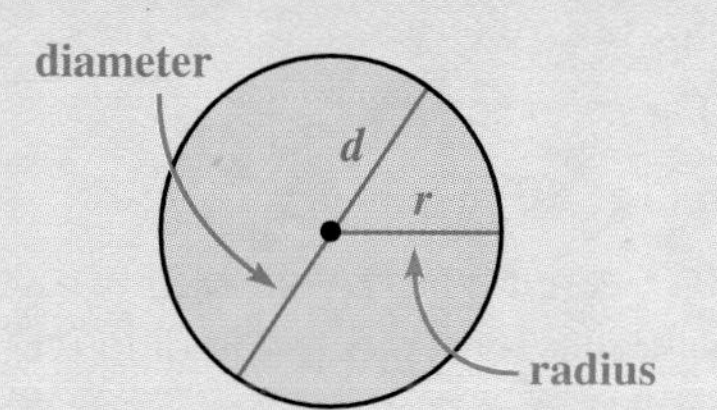

$C = 2\pi r$ or $C = \pi d$

$A = \pi r^2$

Rectangular Prism

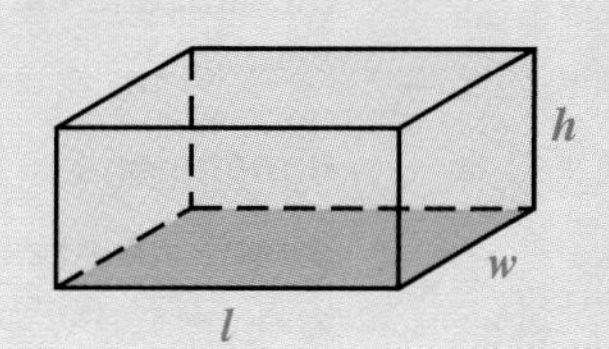

$V = \text{area of base} \times \text{height}$

$V = Bh = lwh$

Cylinder

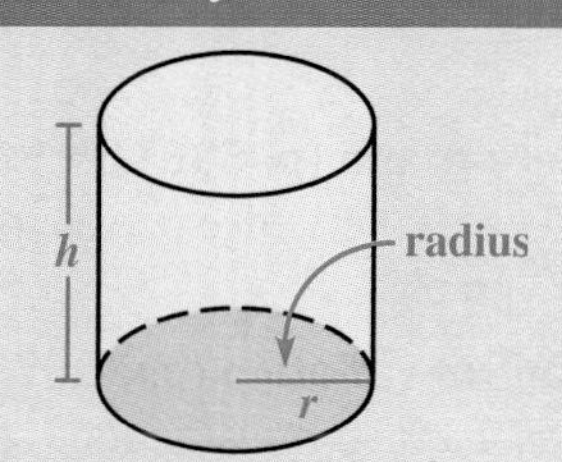

Area of base $B = \pi r^2$

$V = \text{area of base} \times \text{height}$

$V = Bh = \pi r^2 h$

Pyramid

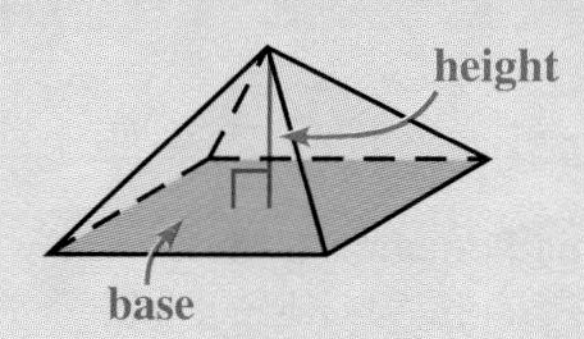

$V = \frac{1}{3} \times \text{area of base} \times \text{height}$

$V = \frac{1}{3}Bh$

Cone

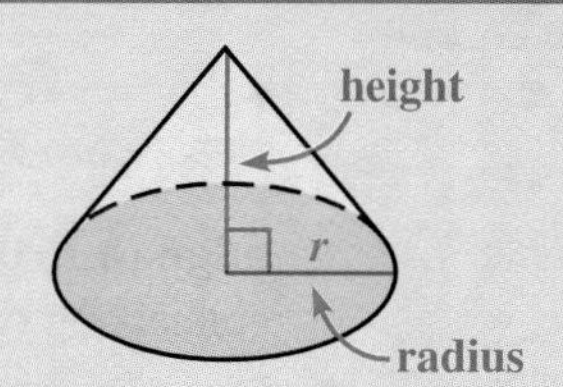

Area of base $B = \pi r^2$

$V = \frac{1}{3} \times \text{area of base} \times \text{height}$

$V = \frac{1}{3}Bh = \frac{1}{3}\pi r^2 h$

Sphere

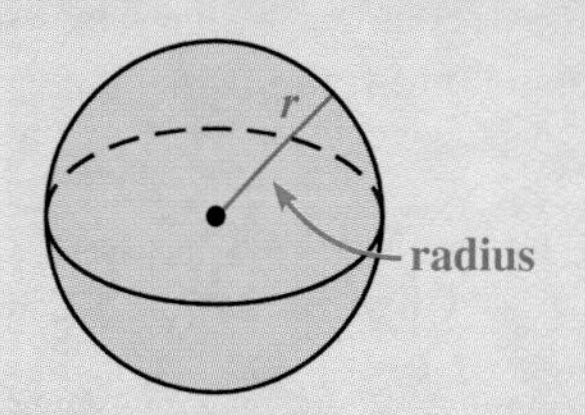

$V = \frac{4}{3}\pi r^3$

Glossary

absolute value (p. 19) The distance that a number x is from zero on a number line, written $|x|$.

algorithm (p. 509) A step-by-step method for accomplishing a goal.

approval voting (p. 524) A voting method in which each person can vote for any number of candidates.

base of a power (p. 370) *See* power.

binomial (p. 260) A variable expression with two terms.

coefficient (p. 36) A number multiplied by a variable in a term of a variable expression. For example, the coefficient of $6x^2$ is 6.

combination (p. 537) A selection made from a group of items. The order of the selected items does not matter.

complete graph (p. 536) In discrete mathematics, a graph with a single edge between every pair of vertices. *See also* graph.

constant of variation (pp. 107, 464) *See* direct variation *and* inverse variation.

converse of the Pythagorean theorem (p. 241) If the lengths of a triangle's sides have the relationship $a^2 + b^2 = c^2$, then the triangle is a right triangle. *See also* Pythagorean theorem.

coordinate plane (p. 73) A grid formed by two *axes* that intersect at the *origin.* The axes divide the coordinate plane into four *quadrants.*

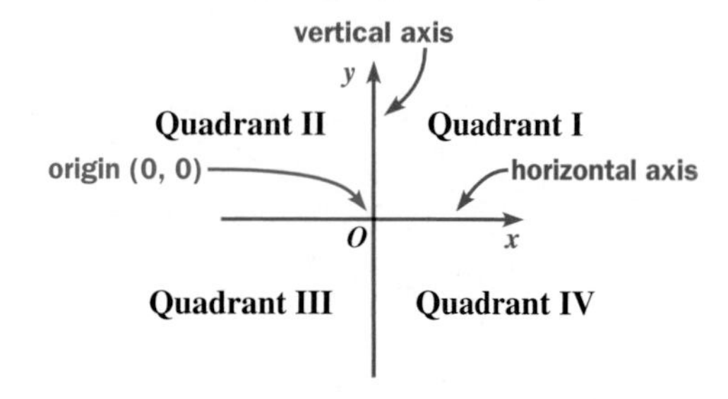

cubic polynomial (p. 423) A polynomial of degree three.

cut-and-choose (p. 525) An algorithm used for fair division in which one person makes a division, another person chooses the part he or she wants, and the first person gets the remaining part.

data (p. 3) A collection of numerical facts.

decay factor (p. 384) The expression $1 - r$ in an exponential decay function. *See also* exponential decay.

degree of a polynomial (p. 423) The largest degree of all the terms.

degree of a term (p. 423) The exponent of the variable part of a term. If there is more than one variable, the degree is the sum of the exponents of the variables.

dense (p. 246) A group of numbers is dense if between any two numbers there is always a third number. For example, the real numbers are dense.

difference of two squares (p. 266) The product of the sum and the difference of two terms:

$$(a + b)(a - b) = a^2 - b^2$$

dimensions of a matrix (p. 40) *See* matrix.

direct variation (p. 107) A relationship between two variables in which the variables have a constant ratio k, $k \neq 0$. If x and y are the variables, then $\frac{y}{x} = k$, or $y = kx$. The number k is called the *constant of variation.*

discriminant (p. 355) The value of the expression $b^2 - 4ac$ in the quadratic formula.

domain of a function (p. 67) *See* function.

edge (p. 516) In discrete mathematics, a route connecting two vertices of a graph. *See also* graph.

element of a matrix (p. 40) *See* matrix.

equation (p. 55) A mathematical sentence that says that two numbers or expressions are equal.

equivalent expressions (p. 35) Two expressions that are equal.

evaluate a variable expression (p. 9) Finding the value of a variable expression for given values of the variable(s).

event (p. 218) *See* probability.

experimental probability (p. 219) The ratio of the number of times an event actually occurs to the number of times an experiment is performed.

exponent (pp. 10, 370) *See* power.

exponential decay (p. 384) A decreasing exponential function modeled by the equation $y = a(1 - r)^x$, where r is the fixed rate of decrease and a represents the initial amount before any decrease.

exponential growth (p. 377) An increasing exponential function modeled by the equation $y = a(1 + r)^x$, where r is the fixed rate of growth and a represents the initial amount before any growth.

extremes of a proportion (p. 193) In the proportion $\frac{a}{b} = \frac{c}{d}$, a and d are the extremes.

factorial (p. 532) The product of all integers from 1 to n, represented by the symbol $n!$.

factoring a polynomial (p. 437) Rewriting a polynomial as a product of polynomial factors.

fair division (p. 525) A way of dividing something among a group of people so that everyone feels he or she has received a fair portion.

favorable outcome (p. 220) *See* probability.

function (p. 66) A relationship between input and output. For each input, there is exactly one output. The *domain* of a function consists of all possible input values. The *range* consists of all possible output values. The range depends on the domain.

geometric probability (p. 224) A probability based on area.

graph (p. 516) In discrete mathematics, a collection of points, called *vertices*, and lines, called *edges*, that connect the vertices.

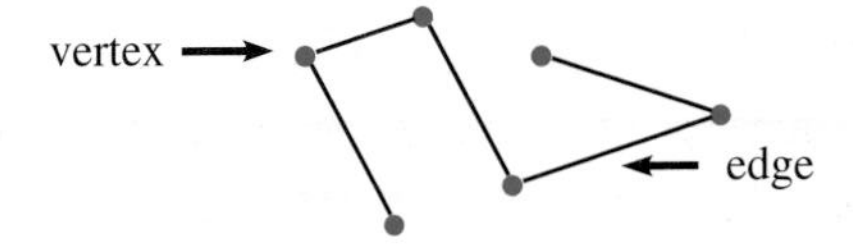

greedy algorithm (p. 517) A method for finding a good path or tree by choosing the best edge from the ones currently available on a graph. *See also* graph, path, and tree.

growth factor (p. 377) The expression $1 + r$ in an exponential growth function. *See also* exponential growth.

growth rate (p. 377) The constant rate at which an exponential function increases. *See also* exponential growth.

histogram (p. 599) A bar graph that shows how many data items occur in each of one or more intervals. The height of each bar shows the frequency, or number of items, in each interval.

horizontal axis (p. 72) *See* coordinate plane.

horizontal coordinate (p. 73) *See* ordered pair.

horizontal intercept (p. 282) *See* x-intercept.

hypotenuse (p. 239) In a right triangle, the side opposite the right angle. *See also* Pythagorean theorem.

identity (p. 161) An equation that is true for all numbers.

inequality (p. 171) A mathematical statement formed by placing an inequality symbol between numerical or variable expressions.

integers (p. 19) The numbers $\ldots, -3, -2, -1, 0, 1, 2, 3, \ldots$.

inverse operations (p. 56) Operations that undo each other. Addition and subtraction are inverse operations. Multiplication and division are inverse operations.

inverse variation (p. 464) A relationship between variables in which the variables have a constant product k, $k \neq 0$. If x and y are the variables, then $xy = k$, or $y = \frac{k}{x}$, $x \neq 0$. The number k is called the *constant of variation*.

irrational number (p. 246) A real number that cannot be written as a ratio of two integers; an irrational number written as a decimal never repeats or terminates.

leg (p. 239) In a right triangle, one of two sides that are shorter than the hypotenuse. *See also* Pythagorean theorem.

like terms (p. 36) In a variable expression, terms that have identical variable parts.

line of fit (p. 130) A line on a scatter plot that is as close as possible to all the data points on the scatter plot. A way to model linear data.

line of symmetry (p. 333) *See* symmetric.

linear data (p. 131) Paired data that come close to forming a line when displayed in a scatter plot.

linear equation (p. 119) An equation whose graph is a line.

linear inequality (p. 307) An inequality whose graph on a coordinate plane is a region bounded by a line.

linear polynomial (p. 423) A polynomial of degree one.

mathematical model (p. 130) A table, a graph, an equation, a function, or an inequality that describes a real-world situation.

matrix, matrices (p. 40) A group of numbers arranged in rows and columns. Each number is an *element* of the matrix. The number of rows and columns in a matrix are its *dimensions*. Matrix A has dimensions 2×3.

$$A = \begin{bmatrix} 3 & 11 & 0 \\ -1 & 4 & 7 \end{bmatrix}$$

mean (p. 14) In a data set, the sum of all the data divided by the number of data items.

means of a proportion (p. 193) In the proportion $\frac{a}{b} = \frac{c}{d}$, b and c are the means.

median (p. 14) In a data set, the middle number when you put the data in order from smallest to largest. When the number of data items is even, the median is the mean of the two middle numbers.

mode (p. 14) In a data set, the most frequently appearing data item, or items.

modeling (p. 130) Using a mathematical model.

monomial (p. 260) An expression with a single term.

multiplication counting principle (p. 531) When you have m choices in one category and n choices in a second category, you have $m \cdot n$ possible choices. When you have r choices in a third category, then you have a total of $m \cdot n \cdot r$ possible choices.

nonlinear equation (p. 326) An equation whose graph is not a line.

one person-one vote (p. 524) A voting method in which each person can vote for one candidate.

opposites (p. 20) Two numbers whose sum is zero. For example, the numbers 3 and -3 are opposites.

order of operations (pp. 8, 371) A set of rules people agree to use so that an expression has only one value.

ordered pair (p. 73) A pair of numbers, such as $(5, -3)$, that can be graphed on a coordinate plane. The first number is the *horizontal coordinate*, or *x-coordinate*, and the second number is the *vertical coordinate*, or *y-coordinate*.

origin (p. 73) On a coordinate plane, the point with coordinates (0, 0). *See also* coordinate plane.

outcome (p. 218) One possible result of a probability experiment. *See also* probability.

parabola (p. 331) A nonlinear graph that is U-shaped. *See also* quadratic equation.

path (p. 517) In discrete mathematics, a collection of edges on a graph that allow you to travel from one vertex to another without repeating any vertex or edge. *See also* graph.

perfect square (p. 240) A number that is the square of an integer.

perfect square trinomial (p. 265) A trinomial that is the product of squaring a binomial. *See also* trinomial.

permutation (p. 532) An arrangement of a group of items in a definite order.

polynomial (p. 423) An expression that can be written as a sum of monomials with exponents that are whole numbers.

polynomial equation (p. 436) An equation in which both sides are polynomials.

power (p. 370) The product when a number or expression is used as a factor a given number of times. In the power 3^5, 3 is the *base* and 5 is the *exponent*. $3^5 = 3 \cdot 3 \cdot 3 \cdot 3 \cdot 3$.

probability (p. 218) A ratio that measures how an *event* is to happen. For example, suppose you want to get an even number when you roll a six-sided die. There are six possible *outcomes* when you roll the die, only three of which are *favorable* to making this event happen. The probability of rolling an even number is $\frac{3}{6}$ or $\frac{1}{2}$.

proportion (p. 193) An equation that shows two ratios to be equal.

Pythagorean theorem (p. 239) If the length of the hypotenuse of a right triangle is c and the lengths of the legs are a and b, then $a^2 + b^2 = c^2$.

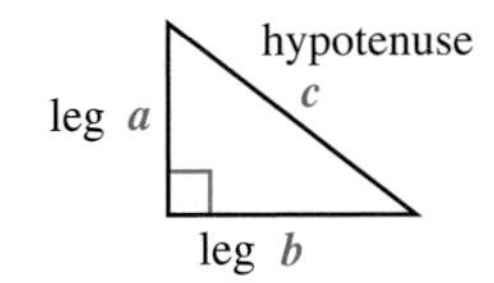

quadrant (p. 73) *See* coordinate plane.

quadratic equation (p. 345) An equation that can be written in the form $ax^2 + bx + c = 0$, when $a \neq 0$. The graph of a quadratic equation is a parabola.

quadratic formula (p. 350) The formula $x = \frac{-b \pm \sqrt{b^2 - 4ac}}{2a}$, for the solutions of the equation $ax^2 + bx + c = 0$, when $a \neq 0$.

quadratic function (p. 332) A function or an equation whose graph is a parabola.

quadratic polynomial (p. 423) A polynomial of degree two.

radical form (p. 252) When an expression is written using the symbol $\sqrt{\ }$, it is in radical form.

range of a function (p. 67) *See* function.

rate (p. 99) A ratio that compares the amounts of two unlike quantities, for example, miles per hour.

rate of decrease (p. 384) The constant rate at which an exponential function decreases. *See also* exponential decay.

ratio (p. 192) A quotient that compares two quantities.

rational equation (p. 476) An equation that contains only rational expressions.

rational expression (p. 476) A variable expression that can be written as a fraction, the numerator and denominator of which are polynomials.

rational number (p. 246) A real number that can be written as a ratio of two integers; when written as a decimal, a rational number repeats or terminates.

real number (p. 246) Any number that is either rational or irrational.

reciprocals (p. 153) Two numbers whose product is 1.

right triangle (p. 239) A triangle that has a right, or 90°, angle.

scale (p. 198) The ratio of the size of a representation of an object to the actual size of the object.

scale factor (p. 199) The ratio of a length on an enlarged or reduced image to the corresponding length on the original figure.

scatter plot (p. 106) A graph that shows a relationship between two sets of data.

scientific notation (p. 396) A number expressed in the form $a \times 10^n$, where $1 \leq a < 10$ and n is an integer.

similar figures (p. 206) Two figures are similar if corresponding angles have equal measure and the ratios of the lengths of corresponding sides are equal.

simplest form of an expression (p. 36) A variable expression with no parentheses and with all like terms combined.

simplest form of a rational expression (p. 489) A rational expression that consists of a single fraction whose numerator and denominator have no common factors other than 1.

slope (pp. 113, 115) The measure of the steepness of a line.

slope-intercept form (p. 120) The equation of a line in the form $y = mx + b$, where m is the slope of the line and b is the vertical intercept.

solution of an equation (p. 56) Values of variables that make an equation true.

solution of an equation in two variables (p. 79) An ordered pair that makes the equation true.

solution of an inequality (p. 177) The set of values of a variable that make an inequality true.

solution of a system of inequalities (p. 312) Any ordered pair of numbers that satisfy all the inequalities of the system.

solution of a system of linear equations (p. 288) Any ordered pair of numbers that satisfy all the equations of the system.

solving algebraically (p. 62) Using symbols to solve an equation.

solving an equation (p. 55) Finding all values of variables that make the equation true.

spreadsheet (p. 41) A computer-generated arrangement of data in rows and columns. A spreadsheet can be used to perform calculations on large groups of data.

square root (p. 240) One of two equal factors of a number.

standard form of a linear equation (p. 282) A linear equation written in the form $ax + by = c$, where a, b, and c are integers and x and y are variables.

standard form of a polynomial (p. 423) A polynomial that has been simplified and whose terms are written in order from greatest exponent to smallest exponent.

symmetric (p. 333) When a graph or shape can be folded so that one side fits exactly over the other side, the graph or shape is symmetric about the fold line, called the *line of symmetry.*

system of inequalities (p. 312) Two or more inequalities that state relationships between the same variables.

system of linear equations (p. 288) Two or more linear equations that state relationships between the same variables.

terms (p. 36) The parts of a variable expression that are added together.

theoretical probability (p. 220) When all possible outcomes of an experiment are equally likely, the ratio of the number of favorable outcomes to the total number of possible outcomes.

tree (p. 518) In discrete mathematics, a path that connects all the vertices in a graph with the smallest possible number of edges. *See also* graph.

trinomial (p. 260) A variable expression with three terms.

unit rate (p. 99) A rate per one given unit.

value of a variable (p. 3) A numeric value of a variable.

variable (p. 3) A letter used to stand for a quantity that varies.

variable expression (p. 9) An expression made up of variables, numbers, and operations.

vertex, vertices (p. 516) In discrete mathematics, a point used to model a location on a graph. *See also* graph.

vertex of a parabola (p. 332) The point where the function reaches a minimum or maximum value.

vertical axis (p. 72) *See* coordinate plane.

vertical coordinate (p. 73) *See* ordered pair.

vertical intercept (pp. 120, 282) *See* y-intercept.

weighted average (p. 470) An average of data values in which each value is multiplied by a weight, so that some values count more than others.

***x*-axis (p. 73)** The horizontal axis of a coordinate plane. *See also* coordinate plane.

***x*-coordinate (p. 73)** *See* ordered pair.

***x*-intercept (p. 282)** The x-value of the point where a graph crosses the x-axis. *Also called* horizontal intercept.

***y*-axis (p. 73)** The vertical axis of a coordinate plane. *See also* coordinate plane.

***y*-coordinate (p. 73)** *See* ordered pair.

***y*-intercept (pp. 120, 282)** The y-value of the point where a graph crosses the y-axis. *Also called* vertical intercept.

Index

E

F

630

Credits

Cover

Front cover Earth Imaging/Tony Stone Images/Chicago Inc.(tl) James Andrew Bareham/Tony Stone Images/Chicago Inc.(r)
Back cover Magnus Rietz/The Image Bank

Chapter Opener Writers

Linda Borcover interviewed Carrie Teegardin (pp. xxx–2), Mae Jemison (pp. 96–98), Bill Pinkney (pp. 322–324), Linda Pei (pp. 420–422); Laird Harrison interviewed Daniel Galvez (pp. 188–190); Yleana Martinez interviewed Lori Arviso Alvord (pp. 366–368); Steve Nadis interviewed Bob Hall (pp. 52–54), Barbara Romanowicz (pp. 144–146), Nathan Shigemura (pp. 236–238), Nancy Clark (pp. 278–280), Alex MacLean (pp. 460–462), Phil Roman (pp. 506–508).

Acknowledgements

xxx–2 From "Where Children Live" (map) and masthead logo from the June 6, 1993, issue of *The Atlanta Journal/The Atlanta Constitution*. Reprinted with permission from The Atlanta Journal and The Atlanta Constitution. **96** From "Music of the Spheres" from the July 10, 1993, issue of *The Boston Globe*. Reprinted by permission of The Boston Globe. **210** From *Was Pythagoras Chinese?: An Examination of Right Triangle Theory in Ancient China* by Frank J. Swetz and T.I. Kao. University Park: The Pennsylvania State University Press, 1977. **285** From "Counting Carbohydrates" by Nancy Clark from the January 1992 issue of *American Fitness*. Copyright ©1992 by Aerobics and Fitness Association of America. Reprinted by permission of the author. **303** From *The Crest of the Peacock: Non-European Roots of Mathematics*, by George Gheverghese Joseph. London: I.B. Tauris & Co. Ltd., 1991. **304** "The Two Parallels" from *Songs From the Gallows*, by Christian Morgenstern. Copyright ©1993 by Yale University. Reprinted by permission of Yale University Press. **404** From *The Rolling Stones*, by Robert A. Heinlein. New York: Charles Scribner's Sons, 1952, 1980. **546** From *Rosencrantz and Guildenstern Are Dead*, by Tom Stoppard. Copyright ©1967 by Tom Stoppard. Used by permission of Grove/Atlantic, Inc.

Stock Photography

i James Andrew Bareham/Tony Stone Images/Chicago Inc.; **iv** Courtesy of Arlene Blum/Expedition photographer unknown(t); **iv** David Weintraub/Stock Boston(m); **vi** Manoj Shah/Tony Stone Images/Chicago Inc.(t); **vi** NASA(b); **viii** Jeffrey Dunn(t); **viii** Jeffrey Dunn(b); **ix** Superstock(t); **ix** Clifford R. Mueller/Illinois State Police. Courtesy of Nathan S. Shigemura(b); **xi** M. & E. Bernheim/Woodfin Camp and Associates(t); **xi** Paul Elledge (b); **xii** Clive Brunskill/Allsport(t); **xii** Courtesy of *The Chicago Tribune*(b); **xiii** Courtesy of Linda Pei/Photo by Gus Bower; **xiv** Alex S. MacLean/Landslides(b); **xv** Bob Brill(t); **xv** Bob Brill(m); **xv** Courtesy of Film Roman(b); **xvii** Craig T. Mathew/Courtesy of Film Roman(tl); **xvii** Michael Heller(tr); **xvii** Clifford R. Mueller/Illinois State Police. Courtesy of Nathan S. Shigemura(br); **xxi** David Ulmer/Stock Boston(bl); **xxi** Mark Segal/Tony Stone Images/Chicago Inc.(br); **xxiii** Michel Hans/Allsport; **xxiv** Lonny Kalfus/Tony Stone Images/Chicago Inc.(l); **xxv** Tom McHugh/Photo Researchers, Inc.(t); **xxv** Warren Morgan/Westlight(b); **xxix** Jeffrey Dunn(bl); **6** Superstock(br); **7** Negative # 314993, Courtesy of Department of Library Services, American Museum of Natural History; **12** Norm Thomas/Photo Researchers, Inc.(tl); **12** Terry Raines(m); **12** Ian Murphy/Tony Stone Images/Chicago Inc.(bl); **12** Terry Raines(br); **15** Tony Duffy/Allsport; **16** Paul Kenward/Tony Stone Images/Chicago Inc.; **17** ©Tom Van Sant/Geosphere Project, Santa Monica/Science Photo Library/Photo Researchers, Inc.(tr); **17** Gregory Silber(ml); **17** C. Faesi/Projecto Vaquita/Marine Mammal Images(bl); **19** David Weintraub/Stock Boston(tr); **19** Courtesy of Arlene Blum/Expedition photographer unknown(l); **19** David Weintraub/Stock Boston(br); **22** Mike Yamashita/Woodfin Camp and Associates(tl); **22** Robert Frerck/Odyssey Productions/Chicago(lower ml); **22** Loren McIntyre/Woodfin Camp and Associates(upper ml); **22** Michael J. Howell/The Picture Cube(mr); **22** Giraudon/Art Resource(bl); **23** Paul J. Sutton/Duomo(t); **23** Duomo(b); **26** David Harding/Tony Stone Images/Chicago Inc.; **28** Craig Newbauer/Peter Arnold, Inc.(tl); **28** Dennis MacDonald/PhotoEdit(ml); **28** Stephen Frisch/Stock Boston(bl); **32** Runk/Schoenberger/Grant Heilman Photography(bl); **32** Runk/Schoenberger/Grant Heilman Photography(bm); **32** Barry L. Runk/Grant Heilman Photography(br); **34** Cuisine Studio, 400(t); **34** Felicia Martinez/PhotoEdit(tm); **34** Cuisine Studio, 400(b); **34** Felicia Martinez/PhotoEdit(bm); **38** Miro Vintoniv/Stock Boston(l); **41** Peter Tarry/Action-Plus; **44** Phil McCarten/PhotoEdit(t); **44** Richard Hutchings/Photo Researchers, Inc.(b); **46** Warren Morgan/Westlight(tl); **53** Courtesy of Boston Athletic Association/Fay Foto(b); **55** Neal Boenzi/New York Times Pictures **58** Focus on Sports, Inc.; **59** UPI/Bettmann; **60** Courtesy of Bob Hall(r); **63** The Bettmann Archive; **64** Ferguson & Katzman/Tony Stone Images/Chicago Inc.; **70** Lawrence Migdale/Stock Boston; **72** Kevin Schafer/Tony Stone Images/Chicago Inc.; **75** Mark Segal/Stock Boston(l); **75** David Burnett/Contact Press Images(r); **76** Michael J. Howell/The Stock Shop(t); **76** *"The Region of The Great Wall of China"*. #15261e2. By permission of The British Library(b); **80** Letraset(bkgrnd); **81**Gabe Palmer/The Stock Market; **87** Jack Gescheidt; **88** Joe McDonald/Animals Animals(inset); **96-98** NASA; **100** Mark Wagner/Tony Stone Images/Chicago Inc.; **102** David Nunuk/Science Photo Library/Photo Researchers, Inc.; **104** NASA; **105** ©Tom Van Sant/The Geosphere Project/The Stock Market(tr); **105** Dorothy Littell/Stock Boston(br); **110** Brooks Air Force Base(bl); **110** NASA(br); **111** Art Resource; **113** Zefa Germany/The Stock Market; **116** Mark Antman/Stock Boston; **117** Michael Newman/PhotoEdit(tl); **117** Terry Qing/FPG International(br); **123** The Bettmann Archive(tl); **123** Photographs courtesy of The Time Museum, Rockford, Illinois(r, bl); **125-126** Ferne Saltzman/Albuquerque International Balloon Fiesta; **128** William Johnson/Stock Boston; **130** David Ulmer/Stock Boston(t); **145** U.S. Geological Survey/Albuquerque Seismological Laboratory(t); **145** AP/Wide World Photos(b); **146** AP/Wide World Photos(t); **146** AP/Wide World Photos(l); **146** Natural History Museum, London(r); **156** Bob Daemmrich/Uniphoto; **162** Mark C. Burnett/Stock Boston; **163** Michael Holford; **165** AP/Wide World Photos; **165** U.S. Geological Survey/Albuquerque Seismological Laboratory(t); **171** George Hunter/Tony Stone Images/Chicago Inc.; **172** Ulrike Welsch/Photo Researchers, Inc.;

179 Don Smetzer/Tony Stone Images/Chicago Inc.(tl); **179** Pete Seaward/Tony Stone Images/Chicago Inc.(tr); **179** Dennis O'Clair/Tony Stone Images/Chicago Inc.(bl); **187** Courtesy of Plimouth Plantation, Inc., Plymouth, Massachusetts USA. Photos by Ted Curtin.; **188** Jeffrey Dunn; **189** Jeffrey Dunn(t); **189** Frank Siteman/Stock Boston(b); **190** Jeffrey Dunn; **192** John S. Flannery/Bruce Coleman Inc.(tl inset); **192** Michael J. Howell/The Picture Cube(tr inset); **194** Courtesy of Willa Mae Jones **195** Jeffrey Dunn(tl); **195** Leonard Lee Rue III/Animals Animals(tr); **195** Jeffrey Dunn(r); **195** John and Karen Hollingsworth(mr); **195** Jeffrey Dunn(bl); **196** Mark Segal/Tony Stone Images/Chicago Inc.(tl); **196** The Bettmann Archive(mr); **197** David Young-Wolff/PhotoEdit; **198** Renee Lynn/Photo Researchers, Inc.(b); **201** Lawrence Migdale/Tony Stone Images/Chicago Inc.(tl); **201** NASA(tm); **201** Will McIntyre/Photo Researchers, Inc.(m); **201** David Lissy/The Picture Cube(r); **201** Mark Burnett/Stock Boston(bl); **201** Ken Lax/Photo Researchers, Inc.(bm); **202** Map from the Road Atlas ©1995 by Rand McNally, R.L. 95-S-161; **207** Barry Herem/Tony Stone Images/Chicago Inc.; **209** Uesugi Shrine, Yamagata Prefecture; **210** Reproduced by permission of The Commercial Press (Hong Kong)Limited, from the publication *Chinese Mathematics: A Concise History*. ; **217** Rhoda Sidney/PhotoEdit(tr) **217** Harry Scull/Allsport(br); **218** Amy C. Etra/PhotoEdit(t); **219** Jerry Wachter/Focus on Sports, Inc.; **221** Catalogue # 221399, Department of Anthropology, Smithsonian Institution; **222** Superstock(tl); **227** Stephen E. Sutton/Duomo; **232** NASA; **236** Brian G. Miller/Illinois State Police. Courtesy of Nathan S. Shigemura(b); **237** Clifford R. Mueller/Illinois State Police. Courtesy of Nathan S. Shigemura(t); **237** Brian G. Miller/Illinois State Police. Courtesy of Nathan S. Shigemura(m); **238** Clifford R. Mueller/Illinois State Police. Courtesy of Nathan S. Shigemura(t); **238** Brian G. Miller/Illinois State Police. Courtesy of Nathan S. Shigemura(b); **240** Superstock; **241** Brian G. Miller/Illinois State Police. Courtesy of Nathan S. Shigemura; **243** Superstock; **245** Bob Daemmrich(ml); **248** Fred Kahn/Envision(m); **249** Clifford R. Mueller/Illinois State Police. Courtesy of Nathan S. Shigemura(t); **249** William Johnson/Stock Boston(l); **249** Bill Horsman/Stock Boston(r); **249** William Johnson/Stock Boston(ml); **249** E. Sparks/The Stock Market(mr); **256** World Perspectives/Tony Stone Images/Chicago Inc.(l); **262** By permission of The British Library; **263** Mark C. Burnett/Stock Boston; **266** Comstock; **268** Michael Schimpf(r); **268** Spencer Grant/Photo Researchers, Inc.(ml); **268** Richard Kolar/Animals Animals(mr); **268** Michael Holford(bl); **279** Courtesy of Boston Athletic Association/Fay Foto; **281** Courtesy of Boston Athletic Association/Fay Foto; **287** Greg Pease/Tony Stone Images/Chicago Inc.(l); **287** Douglas T. Mesney/The Stock Market(r); **288** Pat Wellenbach/AP/Wide World Photos; **291** NASA; **292** Jeff Zaruba/Tony Stone Images/Chicago Inc.(tr); **292** Superstock(b); **292** Chris Sorensen(bl); **292** Gerrit Buntrock/The Anthony Blake Photo Library(br); **293** Roy Garner/Rex USA; **297** David Young-Wolff/PhotoEdit(br); **298** Jerry Wachter/Photo Researchers, Inc. **301** "*Pyramid*". Pieced by Gussie Wells, Oakland, California, 1984. Quilted by Bessie Moore, Oakland, California, 1985, from *Who'd A Thought It*. Courtesy of The San Francisco Craft and Folk Art Museum; **302** "*Links and Lineage*" Commissioned by the Middlesex County Chapter of the Links, Inc. Paul Goodnight©1987; **303** Reproduced by permission of The Commercial Press (Hong Kong) Limited from the publication *Chinese Mathematics: A Concise History* (br); **304** Reproduced by permission of Insel Verlag(tr); **310** Richard Pasley/Viesti Associates, Inc.(inset); **311** Phil Schermeister/Tony Stone Images/Chicago Inc.(tl); **314** Graham Watson(b); **316** Cathlyn Melloan/Tony Stone Images/Chicago Inc.(tr); **316** John Elk III/Stock Boston(l); **316** G. Brad Lewis/Tony Stone Images/Chicago Inc.(br); **317** ©1995 M.C. Escher/Cordon Art - Baarn - Holland. All rights reserved(t); **317** Phillip Bailey/The Stock Market(m); **322** Paul Elledge (l); **324** John Curtis/The Stock Market(l); **328** Union-Tribune/Jerry Rife; **329** Courtesy of Wilton Industries, Inc.; **333** Ray Coleman/Photo Researchers, Inc.; **333** M. & E. Bernheim/Woodfin Camp and Associates(br); **335** Robin Scagell/Science Photo Library/Photo Researchers, Inc.(tr); **335** W.E. Carter/National Oceanic and Atmospheric Administration/Stocktrek(l); **335** R. Gaede/Hartebeesthoek Radio Astronomy Observatory(br); **337** Richard B. Levine; **342** Catherine Pichon/Vandystadt/Allsport; **344** Michel Hans/Vandystadt/Allsport; **346** Brown Brothers; **347** Tony Stone Images/Chicago Inc.; **348** Joe Bator/The Stock Market(t); **348** Rick Strange/The Picture Cube(l); **348** P. Quittemelle/Stock Boston(r); **351** Paul J. Sutton/Duomo; **353** Larry Lefever/Grant Heilman Photography; **354** Bettmann Newsphotos(t); **354** Mike Hewitt/Allsport (l); **354** S. Botterill/Allsport (r); **357** Ben Osborne/Tony Stone Images/Chicago Inc.; **358** Peter Pearson/Tony Stone Images/Chicago Inc.; **367** Courtesy of *The Chicago Tribune*; **368** Michael Heller; **370** The Granger Collection(tl); **374** Superstock(l); **374** Tom Tracy/Tony Stone Images/Chicago Inc.(r); **376** Tom McHugh/Photo Researchers, Inc.; **377** F.J. Alsop III/Bruce Coleman Inc.; **379** Zig Leszczynski/Animals Animals(l); **379** Robert Lubeck/Animals Animals(r); **380** Michael Heller(tl); **380** J.J. Cardamone, Jr. and B.K. Pugashetti, University of Pittsburgh/Biological Photo Service(m); **380** E.R. Degginger/Animals Animals/Earth Scenes(br); **382** Tim Holt/Photo Researchers, Inc.; **386** Clive Brunskill/Allsport; **388** Fred Conrad/Discover Magazine; **394** Charles McNulty/Tony Stone Images/Chicago Inc.; **399** Bill Longcore/Science Source/Photo Researchers, Inc.; **403** Chris Butler/Science Photo Library/Photo Researchers, Inc.; **410** Gregory G. Dimijian/Photo Researchers, Inc.; **412** Dr. Jeremy Burgess/Science Photo Library/Photo Researchers, Inc.(t); **412** Kathleen Campbell/Tony Stone Images/Chicago Inc.(m); **420-421** David Ladd Nelson(m); **427** Courtesy of Linda Pei/Photo by Gus Bower(l); **433** "*Origins*", made by the quilters of Humbolt County, CA; coordinated by Mary Ann Spencer; photo by Lindsay Olsen. ; **435** Doranne Jacobson; **438** Michael S. Yamashita/Westlight; **440** Bruce Ayers/Tony Stone Images/Chicago Inc.(t); **449** Robert Ginn/PhotoEdit(l); **452** Duomo(t); **452** Ron Vesely/Focus on Sports, Inc.(b); **460** Alex S. MacLean/Landslides(b); **461** Alex S. MacLean/Landslides(m); **462** Alex S. MacLean/Landslides; **465** Brian Parker/Tom Stack & Associates; **467** Alex S. MacLean/Landslides; **468** Archive Photos(t); **468** Courtesy of TREK USA(b); **471** Courtesy of the Mall of America(l); **471** Owen Franken/Stock Boston(r); **473** Richard Martin/Vandystadt/Allsport(t); **473** Clive Brunskill/Allsport(m); **473** Richard Martin/Vandystadt/Allsport(b); **479** Courtesy of Boeing Defense and Space Group; **481** Dan Bouchard/Appalachian Mountain Club(tr); **481** Robert J. Kozlow/Appalachian Mountain Club(l,r); **481** Paul Rezendes(m); **482** Dwight Kuhn(t); **482** Copyright©1989, The Countryman Press, Inc. Reprinted with permission from *Fifty Hikes in Northern Maine*, by Cloe Caputo(b); **484** Mark Antman/The Image Works; **486** Alex S. MacLean/Landslides; **498** John Parent Stock; **500** Kingston Collection/ProFiles West; **506-7** ©PAWS (m); **506** Courtesy of Film Roman(b); **507** Craig T. Mathew/Courtesy of Film Roman(m); **508** Craig T. Mathew/Courtesy of Film Roman(tr); **508** ©Turner Entertainment Co. and WMG. All rights reserved. Courtesy of Film Roman(bl); **509** ©PAWS (t); **512** Photos by Susan Peterson, from her book *The Living Tradition of Maria Martinez*, Kodansha International; **512** Jerry Jacka Photography(l); **513** Craig T. Mathew/Courtesy of Film Roman(l); **514** Bob Brill; **515** Cindy Charles/PhotoEdit; **517** Ed Simpson/Tony Stone Images/Chicago Inc.; **520** Ron Chapple/FPG International(tr);

520 Beryl Goldberg(bl); **523** Mary Kate Denny/PhotoEdit(tr); **527** Myrleen Ferguson Cate/PhotoEdit(t); **527** Jeff Greenberg/The Picture Cube(ml); **527** Bob Daemmrich/Stock Boston(mr); **527** Bob Daemmrich(b); **528** The Bettmann Archive(tr); **528** Bob Daemmrich/The Image Works(bl); **530** Bruce Ayres/Tony Stone Images/Chicago Inc.(t); **533** David R. Frazier/Photo Researchers, Inc.(mr); **533** UPI/Bettmann(bl); **534** Wesley Bocxe/Photo Researchers, Inc.(tr); **534** Courtesy of the Dallas Chapter of the United Nations Association(ml, bl); **539** Zefa/The Stock Market(bkgrnd); **546** Martha Swope/©Time Inc.; **548** Bill Lyons(t); **549** MFA - Harvard University Expedition/Courtesy Museum of Fine Arts, Boston. #11.3095-97(m);

Assignment Photographers;

Kindra Clineff **148** (l),**322-323** (t), **323** (r), **341** (tl).;
Jeffrey Dunn **v** (b), **xxix**(tl), **52**, **53** (t), **54**, **57**, **60** (l), **82**.;
Richard Haynes **x** (t), **x** (m), **x** (b), **xiv** (t),**xx**(l), **xxii**(t), **xxii**(b), **xxiii**(bl), **xxiv**(l), **xxvi**(b), **xxvii**(t),**1**, **2**, **3**, **4**, **5**, **8**, **14**, **28** (br), **30**, **34**, **38** (tr), **38** (mr), **38** (br), **44** (tl), **44** (bl), **46** (t), **46** (m), **51**, **69**, **88**, **91** (b), **103** (mr), **106**, **107**, **118**, **119**, **122**, **127**, **130** (b), **132**, **136**, **137**, **147**, **148** (t), **155**, **158**, **159**, **166**, **167**, **170**, **176**, **177**, **182**, **183**, **191**, **192**, **200**, **203**, **205**, **211**, **212**, **218** (m), **223**, **225**, **226**, **230**, **231**, **245** (mr), **246** (bl), **251**, **253**, **255**, **256** (tr), **256** (mr), **256** (br), **259**, **271**, **278**, **282**, **283** (l), **285**, **297** (t), **297** (m), **299**, **304** (b), **306**, **309**, **310**, **311** (r), **314** (tl), **316**, **317**, **325**, **330**, **331**, **332**, **333** (tl), **333** (tr), **336**, **341** (br), **355**, **359**, **360**, **361**, **366**, **369**, **370** (tr), **383**, **387**, **391**, **393**, **401**, **406**, **408**, **413**, **428**, **429**, **430**, **431**, **434**, **436**, **440** (l), **440** (r), **441**, **442**, **449** (r), **454**, **455**, **463**, **469**, **472** (b), **474**, **485**, **488**, **491**, **494**, **495**, **497**, **500**, **501**, **509** (t), **509** (b), **510** (t), **513** (bl), **513** (br), **523** (tl), **530** (t), **531** (m), **536** (t), **536** (bl), **542**, **544** (m), **545**, **548** (m), **548** (ml), **548** (mr), **548** (bl), **549** (t).
Ken Karp **270**.;
Erik S. Lesser iv (b), **xxix**(br), **xxx**, **1** (t), **6** (bl).
Lawrence Migdale **vii**, **144**, **169**, **421** (t), **422** (b).
Tony Scarpetta **90** (l), **222** (bl).
Ellen Silverman **549** (b).;
Tracey Wheeler **v** (t), **61**, **78**, **84**, **86**, **90** (t), **90** (r), **91** (t).

Illustrations

Chris Costello **104** (bl), **117** (tr, m), **152**, **213**(tr), **215**(br); Steve Cowden **160**, **198**(t), **199**(t), **207**, **210**(t), **250**, **409**; Christine Czernota **230**(b), **369**, **383**, **391**, **393**, **401**; DLF Group **59**, **75**(tr), **128**, **144**, **169**, **199**(b), **224**, **230**(tl), **244**(t), **396**, **407**, **498**, **528** (tr), **534** (tr); Dartmouth Publishing, Inc. **87**, **100**, **104** (m), **110** (tl,tr), **125**, **202**(tl), **294** (r), **296**, **313**
Bob Doucet **516**; Piotr Kaczmarek **294** (l); Joe Klim **239**(tr)
Heidi Lutts **191**, **200**(tl), **203**(bl), **413**; Precision Graphics **5**, **7**, **17** (chart), **18**, **44**, **82**, **113**(tr), **120**(m), **133** (m , b), **134**(b), **151**, **174**, **181**, **197**, **203**(t), **216**, **222**, **286**, **368**, **465**, **467** (m, b), **473** (l), **480**, **484**, **486**, **491**, (b), **521** (m), **522**, **526** (b); Lisa Rahon **17**, **34**, **493**, Photo manipulation by Lisa Rahon **17**, **297**, **311**, **468**, **473**, **488**, **500–501** (background); Ann Raymond **6** (t), **33** (t), **231**, **267**, **280** (r), **372**, **374**, **388**, **390**, **394**, **395**, **404**, **410**, **462**, **472**, **514** (tl, m, br), **518** (m, b); Patrice Rossi **88**(inset); D.A. Shepherd Design **13** (l), **19**, **22**, **23**, **31**, **32**, **33** (b), **70**(tl), **72**, **75**(tl), **129** (m), **213**(t), **218**, **219**, **243** (b), **248**, **262**(t), **284**, **293**, **303**, **375**(t), **422**, **463** (t); Nancy Smith-Evers **62**(bl), **64**(t), **69**(b), **74**(b), **75**(br); Jeremy Spiegel **46** (b), **228**(tl), **278**, **279**, **298**, **301**, **353**, **354**, **366**, **367**, **373**, **375**(b), **404**(bl), **405**(b), **434**, **438**, **449**; Doug Stevens **16**, **40**, **173**, **255**, **227**, **324** (r); Richard Vyse: **530** (m), **531**

Selected Answers

CHAPTER 1

Page 5 Checking Key Concepts

1. Answers may vary. **3.** Answers may vary. An example is given. I disagree. It is true that $h > 0$ and $h < 100$, but these inequalities apply to the heights of all human beings and give no useful description of this particular group. **5.** 17 students **7.** that you spend at most 6 h visiting

Pages 5–7 Exercises and Applications

1. 0 **3.** 13 states **5.** 20%; 40% **7.** Let e = weekly earnings; $e \leq 600$. **9.** Let l = price of lunch; $l \geq 3.95$. Let d = price of dinner; $d \geq 4.95$. **13.** about 15% **19.** If w = the weight in grams, then $w \geq 2$ and $w \leq 30{,}000{,}000$; if w = the weight in kilograms, then $w \geq 0.002$ and $w \leq 30{,}000$. **21.** Answers may vary. An example is given. 29.95, 49.99, 119.99 **27.** United States **29.** 2.67 **31.** 5.56 **33.** 0.13

Page 11 Checking Key Concepts

1. 9 **3.** 5 **5.** 5.1 **7.** $\frac{1}{2}$ **9.** 0 **11.** **13.** 10.7 **15.** $\frac{7}{17}$

Pages 11–13 Exercises and Applications

1. a. 0 **b.** 4 **3. a.** 42 **b.** 42 **5. a.** 8 **b.** 11.5 **9.** 3.5 **11.** 5.88 **13.** $6\frac{1}{5}$ **15.** $\frac{2}{11}$ **17.** $32\frac{1}{2}$ **19.** $46\frac{1}{8}$ **23.** $25g$ mi **25.** 36 in. **27.** 53 **29.** $4\frac{3}{8}$ **31.** $\frac{11}{6}$ **33.** 1 **35.** 0 **37.** about 256 ft **41.** Answers may vary. Example: [( ] 2 [×] 5 [−] 7 [ )] [÷] 15 [=] **43.** C **47.** Choices of variables may vary. An example is given. Let s = the number of species counted; $s > 80$.

Page 16 Checking Key Concepts

1. about 3.2; 3; 2 **3.** about 21.9, 22, 22

Pages 16–18 Exercises and Applications

1. 2; 2; 0 **3.** about 234.9; 240; 244; Answers may vary. An example is given. The three values are fairly close, so I think no one average describes the data better than the others. **11.** $\frac{1}{20}$; 5% **13.** $\frac{1}{4}$; 25% **15.** $\frac{7}{4}$; 175%

Page 18 Assess Your Progress

1. 16 **2.** 2 **3.** $\frac{1}{5}$ **4.** $18\frac{2}{3}$ **5.** 10 **6.** $\frac{1}{5}$ **7.** 24 students **8.** 15 students **9.** For example, $n \geq 0$ and $n \leq 6$; the variable n represents the number of movies seen by each student. **10.** 2.75; 3, 3 and 4

Page 21 Checking Key Concepts

1. –24 **3.** $-\frac{1}{4}$ **5.** $-n$ **7. a.** 6 **b.** 6 **9. a.** –6 **b.** –2 **11. a.** 2 **b.** 8

Pages 21–23 Exercises and Applications

1. –10 **3.** –13 **5.** –8 **7.** –4 **9.** 3 **13.** 1050 years **15.** 507 years **17.** 1100 years **19.** –12 **21.** 78 **23.** 4.3 **25.** –3 **27.** $\frac{9}{4}$ **29.** Answers may vary. Examples are given. **a.** $x = 5, y = 2$ **b.** $x = 2, y = 5$ **c.** $x = 5, y = 5$ **31.** –1 **33.** 8 **35.** 0 **37.** 12 **41.** D **47.** 8; 4 **49.** $4\pi \approx 12.56$; $4\pi \approx 12.56$ **51.** 7; 2.5

Page 27 Checking Key Concepts

1. –6 **3.** –6 **5.** 18 **7.** $\frac{13}{16}$; The quotient of two numbers is positive if both numbers are negative. **9.** $-\frac{12}{7}$; The quotient of two numbers is positive if both numbers are negative, so $\frac{-12}{-7} = \frac{12}{7}$. The opposite of a positive number is negative, so $-\left(\frac{12}{7}\right) = -\frac{12}{7}$. **11.** Answers may vary. An example is given. You can multiply 7 and 8 and then multiply the product by 0, or you can multiply 8 and 0 and then multiply the product by 7. I prefer the second way because the product of 8 and 0 is 0, and the product of 0 and 7 is 0; this is easier than remembering the product of 7 and 8.

Pages 27–29 Exercises and Applications

3. –4 **5.** 70 **7.** 7 **9.** 7 **15.** 5°F **17.** –12 **19.** –30 **21.** 6 **23.** 12 **27.** equal; associative property **29.** equal; associative property **31.** not equal **33.** equal; both properties **39.** 8; 120 **41.** 1.2; 120% **43.** 0.888; 88.8%

Page 29 Assess Your Progress

1. 9 **2.** 0 **3.** –3 **4.** 5 **5.** 12 **6.** 8 **9.** –8 **10.** 5 **11.** 7 **12.** $-\frac{8}{3}$ **13.** 12

Page 32 Checking Key Concepts

1. $0.75(g - 5)$ dollars **3.** Let f = the number of friends she invites to the party; $5 + 2f$ people. The expression represents the total number of people she invites to the party, which is 5 more than twice the number of friends she invites.

Pages 32–34 Exercises and Applications

1. $10 + 1.5t$ dollars; Answers may vary. An example is given. whole numbers from 0 to 4 **3.** Let n = the number of ballerinas; $200n$ pairs of toe shoes are ordered.

7. Let d = the number of days Chris is on vacation; $2d + 2$ books (if "a couple" of books means two books); the number of books Chris wants to take along is two more than twice the number of vacation days. **15.** Let n = the number of people to be served; $1.5n$ pounds. **19.** 1 **21.** 195 **23.** 48 **25.** 9

Page 37 Checking Key Concepts

1. $2(2x + 1) = 4x + 2$ **3.** $2x - 8$ **5.** $40x - 16$ **7.** $13 + 8x$ **9.** $2x - 7$ **11.** $-3x - 20$

Pages 37–39 Exercises and Applications

1. $2x + 2y$ **3.** $-10t - 14$ **5.** $-\frac{5}{3}a + 2b$ **7.** $24 - 12y$ **9.** $15z + 20$ **11.** $-1.75 + x$ **13.** Answers may vary. An example is given. The teacher can give each student a book and a calculator at the same time (which is like adding, then multiplying). The result will be the same if the teacher gives a book to each student and then a calculator to each student (which is like multiplying first). In symbols, $(1 + 1)n = n + n$, where n = number of students. **17.** $3m + 4$ **19.** $3x + 4y^2$ **21.** 15 **23.** $-21u + 21uv - 2v$ **25.** If he walks m miles, then the amount of money he raises is $m + 1.5m + 2m + m + m$, or $6.5m$ dollars. I prefer to evaluate the expression $6.5m$ because it requires multiplying once. Evaluating the other expression requires multiplying twice and then adding five numbers. **27.** $36 + 4x$ **29.** $13 - 3x$ **31.** $6x + 15y - 7$ **33.** 0 **35. a.** evaluating: $-5(6 + 12(5)) + 3(20(5) + 10) = -5(66) + 3(110) = -330 + 330 = 0$; simplifying first: $-30 - 60x + 60x + 30 = 0$; It is easier to simplify the expression first because you can see that the expression is equal to 0. **b.** Answers may vary. An example is given. $4(3x - 6y) - 3(-8y + 4x) = 12x - 24y + 24y - 12x = 0$ **41.** 20 **43.** -2 **45.** 22

Page 42 Checking Key Concepts

1. A: 1×3; B: 3×1; C: 1×3; D: 3×1
3. a. 12 * B3 + C3; 88 **b.** 82.25; The number is the mean height of the four athletes in inches.

Pages 43–45 Exercises and Applications

1. A: 2×3; B: 2×3: C: 3×2; D: 3×2 **5. a.** The sales manager used the fill down command to divide each person's expenses for the month by that person's mileage for the month. **b.** Miner, \$.04; Dubois, \$.05; Perez, \$.04; Chen, \$.04 **9.** The number of males in the work force and the number of females in the work force increased steadily from 1960 to 1980 and then decreased from 1980 to 1990.
11. Answers may vary. An example is given. The number and percent of 16- and 17-year-olds in the work force increased between 1960 and 1980 and then declined. I think the percents give a better sense of how many teenagers were in the work force. They tell how likely it is a teenager was in the work force in a given year. **17.** 0 **19.** 7 **21.** $\frac{1}{2}$

Page 45 Assess Your Progress

1. $60h$ min **2.** $m + 250$ mi **3.** $3x - 6$ **4.** $5x^2 + 2y - x$ **5.** $x - 1$ **6.** X: 3×3; Y: 3×3; Z: 2×2;

$$X + Y = \begin{bmatrix} 6.6 & 2.25 & -2 \\ -2 & 18 & 3 \\ 5 & 3 & 11 \end{bmatrix}$$

7. B6 = B2 + B3 + B4 + B5

Pages 50–51 Chapter 1 Assessment

1. Answers may vary. An example is given. $3x + 2y - 5x + 6 = -2x + 2y + 6$ **2.** The absolute value of a number is its distance from zero on the number line. For example, $|-7| = 7$, $|3| = 3$, and $|0| = 0$. **3.** The mean is the sum of the data divided by the number of data items. The median is the middle number when you put the data in order from smallest to largest. When the number of data items is even, the median is the mean of the two middle numbers. **4.** 10 **5.** 25 **6.** $g \geq 76$ and $g \leq 100$ **7.** 24 **8.** 8 **9.** 1 **10. a.** 19.6; 19; 19 **11.** -3 **12.** 1 **13.** -7 **14.** 30 **15.** -5 **16.** 5 **17.** 1 **18.** $-11\frac{2}{3}$ **19.** -11 **20.** -20 **21.** -10 **22.** $24d$; 168 h **23.** Let f = the number of friends; $2.90(f + 1)$ dollars; \$23.20. **24.** $11x + 11$ **25.** $5 + 5x + 7y$ **26.** y **27.** $-6x^2 + 6x$ **28.** amount; price per item; total cost; D2 = B2*C2 **29.** A: 3×2; B: 2×2; C: 3×1; D: 3×2

30. $A + D = \begin{bmatrix} 2 & 8 \\ 7 & 0 \\ 0 & 6 \end{bmatrix}$

CHAPTER 2

Page 58 Checking Key Concepts

1. -23 **3.** 0.75 **5.** -16 **7.** 140 **9.** -10 **11.** 4.1 **13.** about 10.31 m/s

Pages 58–60 Exercises and Applications

1. -415 **3.** 6.5 **5.** $-\frac{3}{8}$ **7.** -5 **9.** 0 **11.** 0 **13.** 6.25 **15.** $-\frac{1}{2}$ **17.** -4.5 **21.** A **25.** about 135.97 mi/h **27.** $29{,}887 = 100d$; about 299 days **29.** A; The rate per day, 15 mi, times the number of days d equals the distance traveled, 2000 mi; $d \approx 133.33$; about 133.3 days. **31.** about 0.22 mi/min **35.** 26 **37.** $\frac{2}{3}$ **39.** 6 **41.** $6x + 40$; 100

Page 63 Checking Key Concepts

1. $4x + 3 = 11$; $x = 2$ **3.** $\frac{7}{5} = 1.4$ **5.** 9 **7.** $-\frac{78}{11} \approx -7.09$ **9.** -9 **11.** 10

Pages 64–65 Exercises and Applications

1. 31 **3.** –1 **5.** –6 **7.** 2 **9.** –4 **11.** –42 **13.** 72 **15.** –67.5 **17. a.** subtraction of 13; division by 6 **19.** 18 mi **23.** 2.5 **25.** –3 **27.** 30 **29.** –49 **31.** 130 **33.** –100 **35.** $x = 1$; 5 in. **37.** $x = 7.5$; 7.5 mm, 10.5 mm, 12.5 mm, and 12.5 mm **41.** $s + 5$; Rob's sister's height in inches **43.** $25t - 7$ **45.** $-8m + 28$ **47.** $-2x$

Page 68 Checking Key Concepts

1. The steering wheel is like a function because the angle through which the car turns depends on how far you turn the steering wheel. For each amount you turn the wheel, the car will turn in a predictable way. A TV channel selector is like a function. For each number you select, one channel will be chosen.

Pages 69–71 Exercises and Applications

1. No; for some values in the domain (13 and 14), there is more than one value in the range. **3.** Yes; for each value in the domain, there is exactly one value in the range. **5.** C; p = the price of the book in dollars and C = her change in dollars. **7.** B; n = the number of friends and p = the price in dollars each friend pays. **9.** Ex. 5: the domain is 0.01, 0.02, 0.03, …, 10.00 and the range is 0, 0.01, 0.02, …, 9.99, since she must pay at least \$.01 for the book and can spend no more than \$10. Ex. 6: the domain is 1, 2, 3, … and the range is 10, 20, 30, …, since the bike is rented by the day. Ex. 7: the domain is 1, 2, 3, 4, … and the range is 10, 5, 3.33, 2.50, …, since the number of friends is a whole number greater than 0. Ex. 8: the domain is 10.01, 10.02, 10.03, … and the range is 0.01, 0.02, 0.03, since the original price of the shoes must be at least \$10.01. **11.** 4 tickets **13. a.** $p = \frac{n}{5{,}021{,}000}$, where n = the number of teens associated with a portion of the circle graph and p = the related percent. **b.** Yes; for each value of n, there is exactly one value of p. **15. a.** \$12.50 **b.** \$18.75 **c.** 8 h **17.** $P = 2s + 16$; the domain is all positive numbers and the range is all numbers greater than 16. **25.** 52 **27.** about 6.64 **29.** 2.5 **31.** 15 **33.** –51 **35.** 21

Page 71 Assess Your Progress

1. $-\frac{1}{6}$ **2.** –8 **3.** –41 **4.** 8 **5.** –20 **6.** 14 **7.** –48 **8.** 78 **9.** 4.5 min **10. a.** $C = 3 + 2p$, where p is the number of plants bought and C is the total cost in dollars. The domain is the positive whole numbers and the range consists of these amounts in dollars: 5, 7, 9, …. **b.** 21 plants

Page 74 Checking Key Concepts

1. about 400 ft below the surface **3.** C; I **5.** A; II **7.** G; none **9.** D; III

Pages 74–77 Exercises and Applications

1–15.

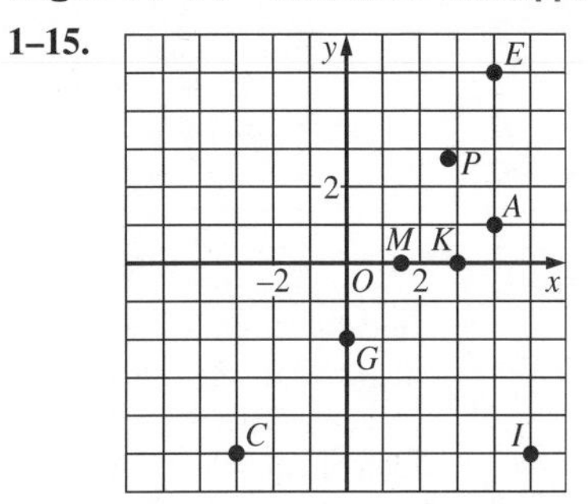

1. I **3.** III **5.** I **7.** none **9.** IV **11.** none **13.** none **15.** I **17.** the number of cars washed; the amount of money raised **19.** 40 cars **21.** March and September; about 12 h per day **25. a.** \$100; The point (0, 100) is on the graph, so 0 weeks after the money is borrowed, \$100 is owed. **b.** \$40 **c.** 10 weeks; The point (10, 0) is on the graph, so 10 weeks after the money is borrowed, the amount owed is \$0. **27.** Estimates may vary; about 1820 mi. The band covers about 13 grids horizontally. Each grid is about 140 mi on a side. **31.** $A(-5, -5)$; $B(-2, -3)$, $C(3, -3)$; $D(5, 5)$; $E(1, 3)$; $F(-2, 2)$; $G(-5, -1)$ **33.** B and C **35.** C and F **37.** about 5.66; about 5.88; 6.75 **39.** A: 3 × 2; B: 2 × 3; C: 3 × 3; D: 1 × 4; E: 2 × 2; F: 1 × 3; G: 3 × 2; H: 1 × 4; I: 3 × 3; A and G, C and I, D and H;

$$A + G = \begin{bmatrix} 5 & 3 \\ 13 & 15 \\ -11 & 5 \end{bmatrix}; C + I = \begin{bmatrix} 2 & 1 & 13 \\ 1 & 0 & -4 \\ 5 & 7 & 1 \end{bmatrix}; D + H = [16\ \ 5\ \ 3\ \ 7]$$

Page 81 Checking Key Concepts

1. B; $1 = \frac{1}{2}(2)$ and $-1 = \frac{1}{2}(-2)$ **3.** A; $-2 = -(2)$ and $2 = -(-2)$ **5.** (1) an equation: $C = 8 + 0.32n$

(2) a table of values:

number of invitations sent	total cost (dollars)
0	8
1	8.32
2	8.64
…	…
15	12.80

(3) a graph:

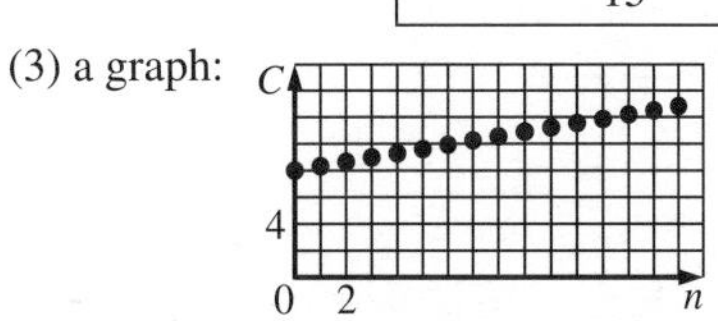

Pages 81–83 Exercises and Applications

1. Yes. **3.** No.

5. No.

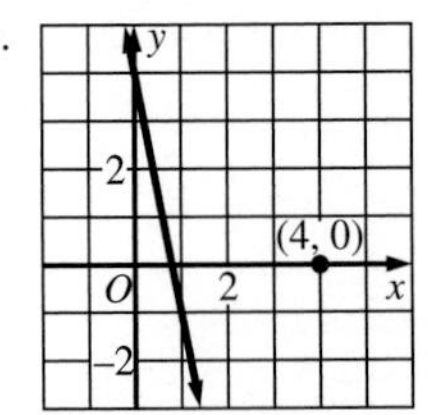

7. Yes.

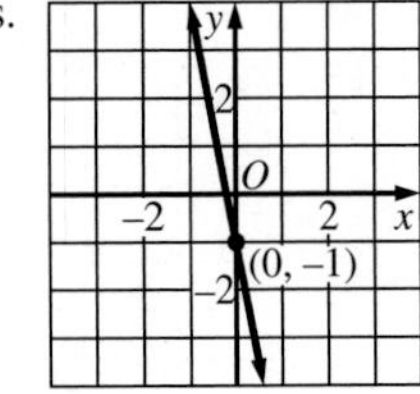

11. a.

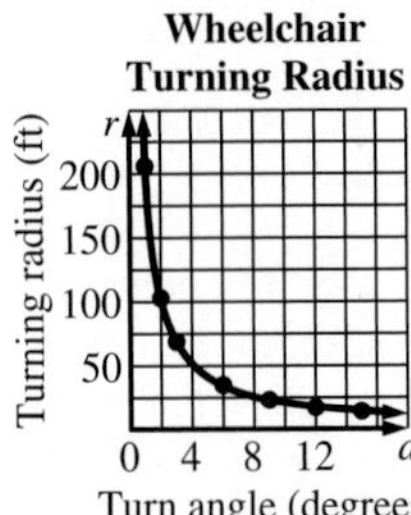

b. Estimates may vary; about 50 ft **c.** The domain is the positive numbers less than or equal to 15 and the range is the positive numbers greater than 14.

13. Estimates and methods may vary. Examples are given. I estimate there were about 15 male wheelchair athletes and about 3 female wheelchair athletes in 1983. I estimated by using the values in the table, giving special attention to the years 1984 to 1993, and following the trends backward from 1984 to 1983. **15.** D; –57 **17.** B; –28 **19.** A; B

21.

	A	B
1	–5	–24
2	–4	–23
3	–3	–22
4	–0.5	–19.5
5	0	–19
6	1	–18
7	2	–17
8	5	–14

23.

	A	B
1	1	0.5
2	2	1
3	3	1.5
4	4	2
5	5	2.5
6	6	3
7	7	3.5
8	8	4
...	...	...

29. $\frac{17}{12}$ **31.** $19s + 4t - 7$ **33.** $-6(2m - 3n) + 12m = -12m + 18n + 12m = 18n + 0 = 18n$

Page 86 Checking Key Concepts

1. a. $19 + 9d = 126$; $d \approx 12$; about 12 days

b.

Pages 86–89 Exercises and Applications

1. 10 T-shirts **3.** 2.5 mi, 0.5 mi, 0.5 mi **5.** Estimates may vary. **a.** about 30°F **b.** about 70°F **c.** about –5°F **d.** about –20°F **7.** Ex. 5: 32°C; 68°C; –4°C; –22°C; Ex. 6: 0°F; about 32.2°F; about –17.8°F; –40°F **11.** 63°F

13. a. 45°F **b.** 50°F **c.** 60°F **d.** 65°F

17, 19. Answers may vary. Examples are given.

17. $x = 5$

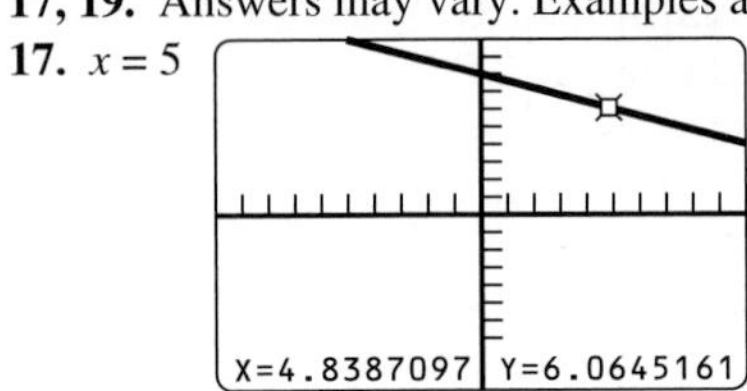

19. $x = 18$

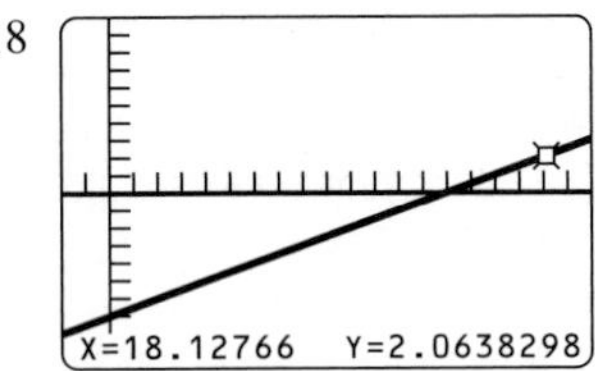

21. B **23.** No. **25.** No.

27. (1) an equation: $y = 2x + 3$

(2) a table of values:

x	y
–1	1
0	3
1	5
...	...

(3) a graph:

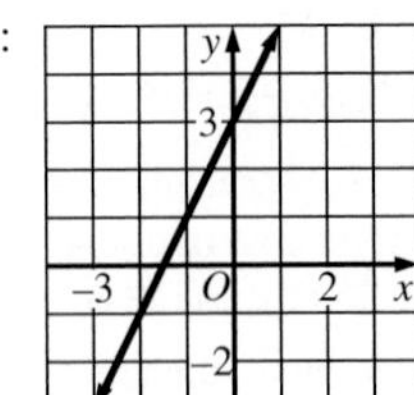

29. (1) an equation: $y = 4 - x$

(2) a table of values:

x	y
–1	5
0	4
1	3
...	...

(3) a graph: 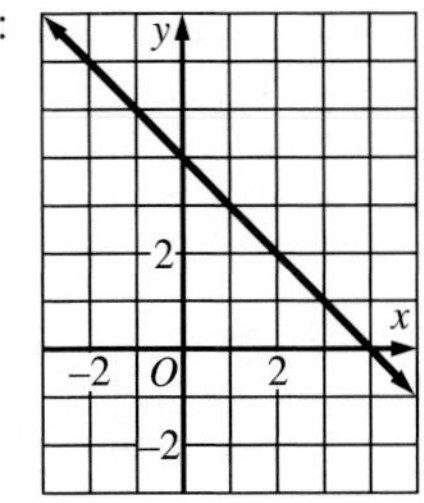

Page 89 Assess Your Progress

1–8. 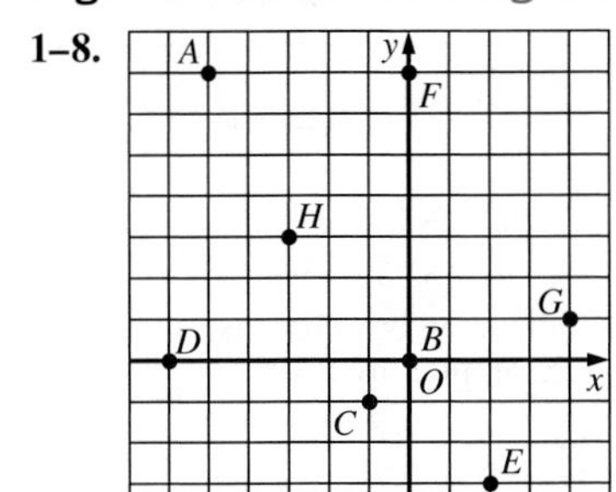

1. II **2.** none **3.** III **4.** none **5.** IV **6.** none **7.** I
8. II

9. 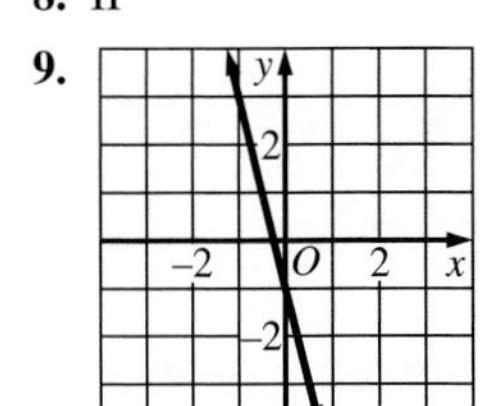

10.

11. 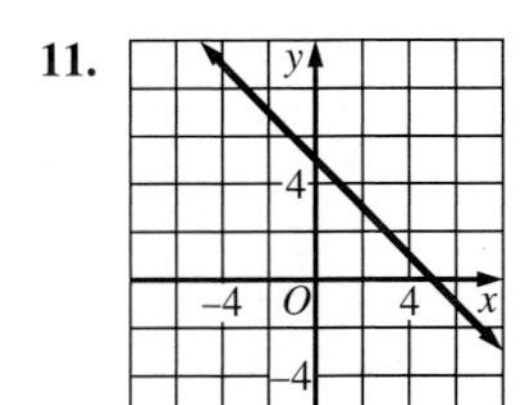

12. $P = 8n - 780$

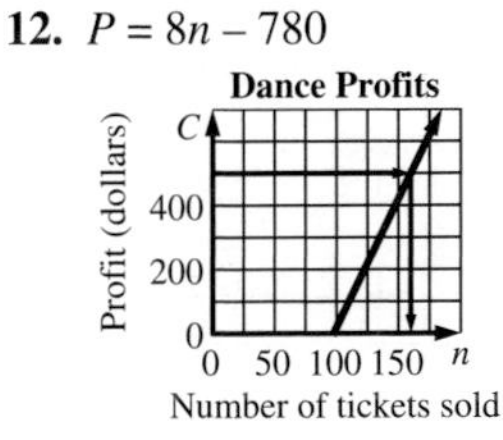

Estimates may vary; the actual number of tickets is 160.

Pages 94–95 Chapter 2 Assessment

1. domain; range **2.** multiplication or division; inverse operations **3.** 26 **4.** $-\frac{5}{16}$ **5.** –23 **6.** 32 **7.** –4 **8.** 5
9. 104 **10.** 54 **11.** –3 **12.** about 16,456.5 mi/h
13. a. $C = 50 + 12n$, where n is the number of times she plays golf and C is the total cost in dollars **b.** The domain is 0, 1, 2, 3, … and the range is 50, 62, 74, 86, …. **c.** 16 times
14. a. Estimates may vary. highest temperature: about –30°C after 15 h; lowest temperature: about –85°C after 5 h **b.** IV

15–18. 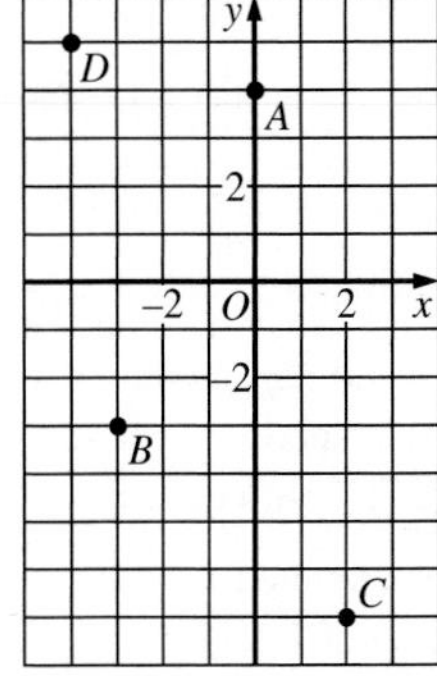 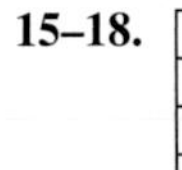

15. none **16.** III **17.** IV **18.** II

19. 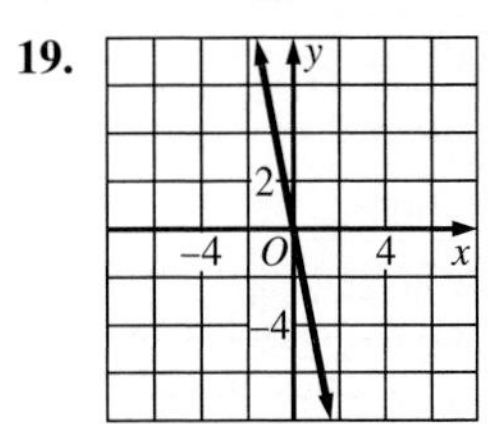

20. 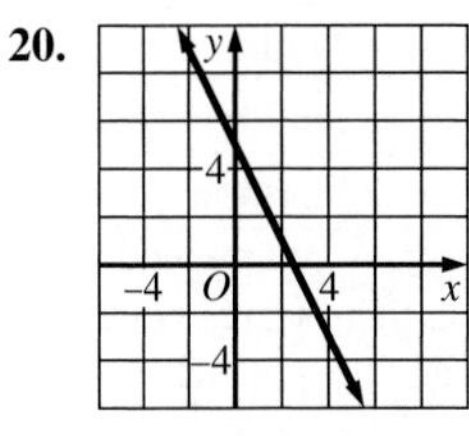

21. 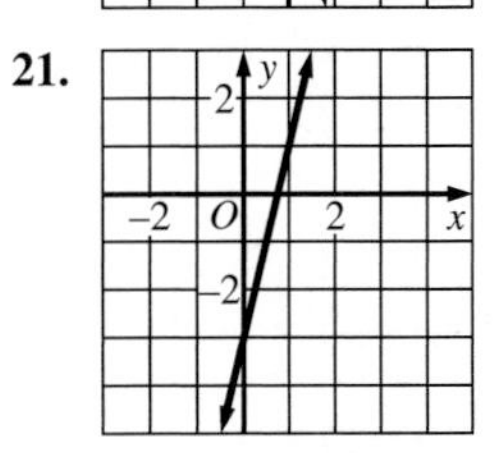

22. 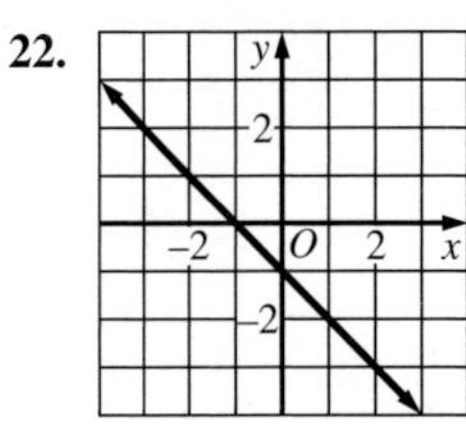

23. 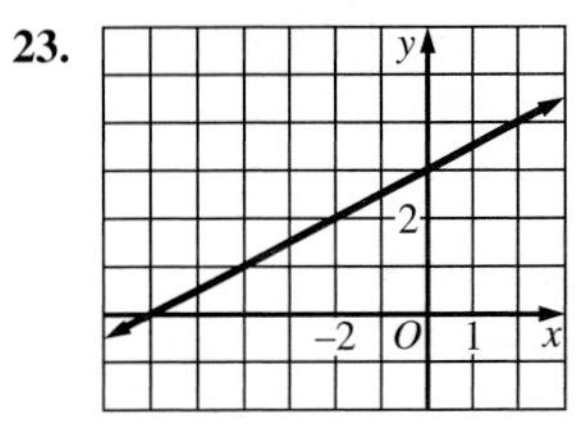

24. 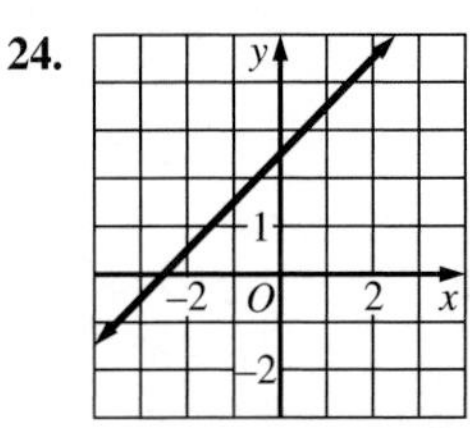

25. the equation $y = 4x - 3$ in Ex. 21 **27. a.** $C = 450 + 20n$, where n is the number of people at the party and C is the total cost in dollars

b. 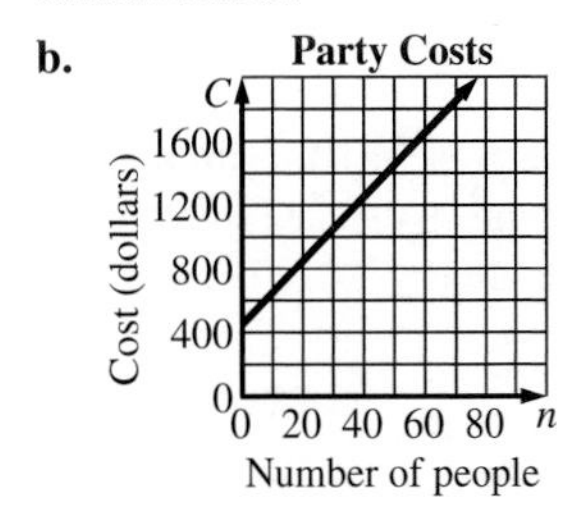

c. 77 people

CHAPTER 3

Page 102 Checking Key Concepts

1. $\frac{\$54}{1 \text{ day}}$ **3.** 315 mi **5.** about 34.2 km/h; about 21.2 mi/h
7. Estimates may vary. about 2800 ft/min

Pages 102–105 Exercises and Applications

1. $\frac{36 \text{ students}}{1 \text{ dormitory}}$ **3.** $\frac{60 \text{ s}}{1 \text{ min}}$ **5.** 0.9 **7.** 1760 **9.** 369.6

11. He multiplied by $\frac{1 \text{ h}}{60 \text{ min}}$ when he should have multiplied by $\frac{60 \text{ min}}{1 \text{ h}}$ to get 36 as the average number of shooting stars per hour. **13.** Answers may vary. Examples are given. 4 wins out of the first 5 games, 1 touchdown per half **15.** about 84 yards per game **17.** Estimates may vary.

	Rate of change for LP sales (LPs/y)	Rate of change for CD sales (CDs/y)
a.	about –40 million	about 20 million
b.	about –40 million	about 30 million
c.	about –20 million	about 50 million
d.	about –30 million	about 50 million

23. Answers may vary. An example is given using centimeters per minute. Unionville: 3.1 cm/min; Curtea-de-Arges: 1.025 cm/min; Holt: about 0.73 cm/min; Cilaos: about 0.13 cm/min; Cherrapunji: about 0.02 cm/min **25.** 300 min, or 5 h; 930 cm $\cdot \frac{1 \text{ min}}{3.1 \text{ cm}}$ = 300 min **27.** C **29.** 0.38

31. 1.25 **33, 35.**

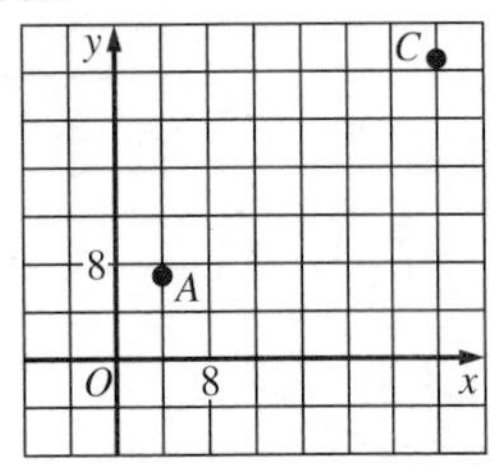

33. I **35.** I

Page 109 Checking Key Concepts

1. No; the graph is not a line through the origin. **3.** about 0.8; If x = drop height in inches and y = bounce height in inches, then $y = 0.8x$.

Pages 109–112 Exercises and Applications

1. No. **3.** No. **5.** Yes; the points lie on a line that goes through the origin.

7.

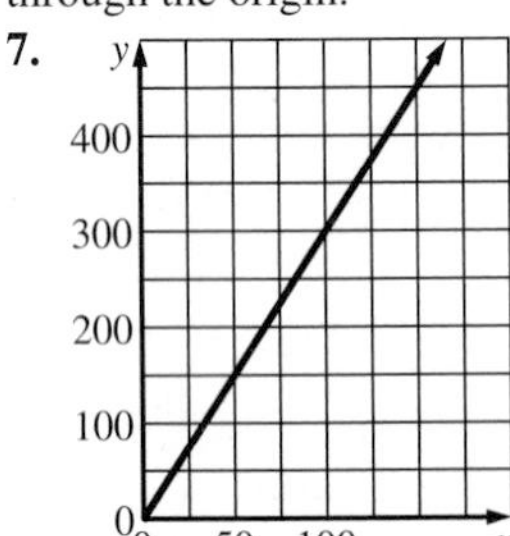

The graph is a line through the origin, so the data show direct variation.

9. $l = \frac{1}{6}f$ **13.** the graph on the right; This graph shows that Mina receives no commissions on sales at or below $1000. **15.** square: $P = 4x$; rectangle: $P = 2x + 20$; The perimeter of the square varies directly with x because the equation $P = 4x$ is in the form of a direct variation equation. The perimeter of the rectangle does not vary directly with x because the equation for the perimeter cannot be written in the form $P = kx$ and because the graph of $P = 2x + 20$ does not pass through the origin. **17.** mean, $60,000; median, $42,500; mode, $25,000

19. No.

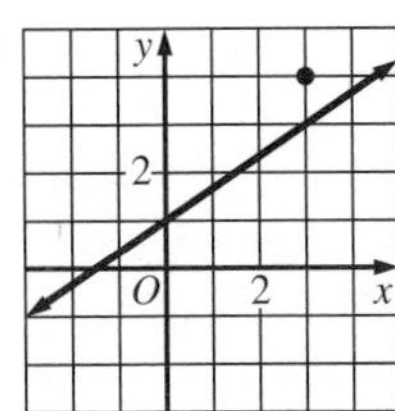

21. Yes.

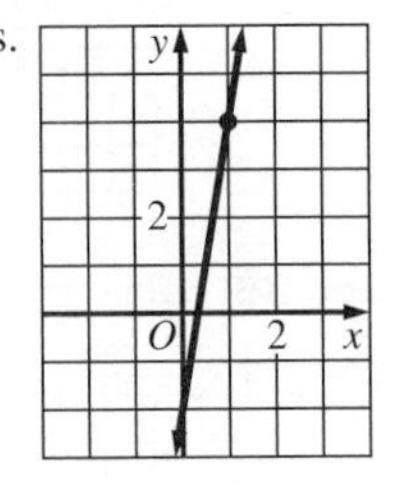

Page 112 Assess Your Progress

1. 21 km/h **2.** about 4.17 km/h **3.** 3.6 km/h **4.** No. **5.** Yes; 3; $y = 3x$, where x = the number of tickets sold and y = profit in dollars.

Page 115 Checking Key Concepts

1. E and F **3.** A **5.** E

Pages 116–118 Exercises and Applications

1. $\frac{1}{2}$ **3.** $-\frac{1}{3}$

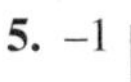

5. –1

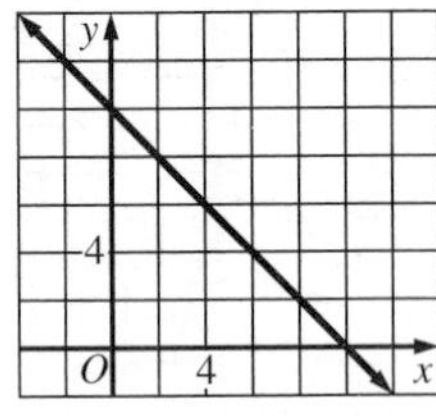

7. $\frac{1}{2}$

9. –2

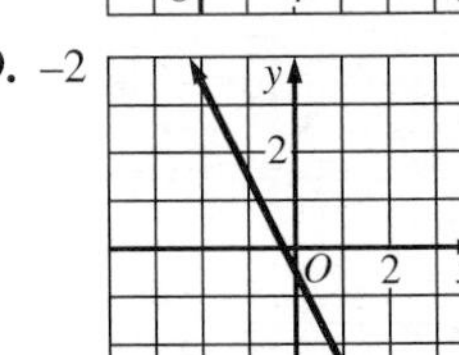

11. $-\frac{3}{5}$ **13.** undefined **15.** $-\frac{32}{15}$ **17.** $-\frac{15}{121}$

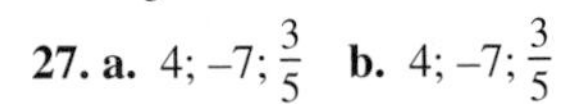

27. a. 4; –7; $\frac{3}{5}$ **b.** 4; –7; $\frac{3}{5}$

c. The constant of variation in each direct variation equation is equal to the slope of the graph of the equation.

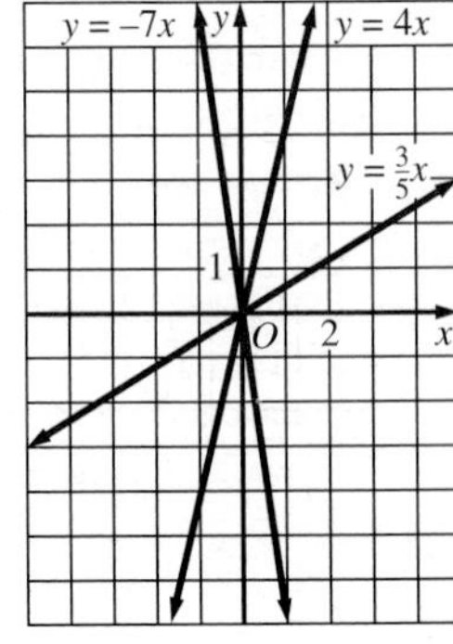

31. No; for each value in the domain, there are two values in the range. **33.** 1.8 **35.** 74.4

Page 122 Checking Key Concepts

1. 1; 4

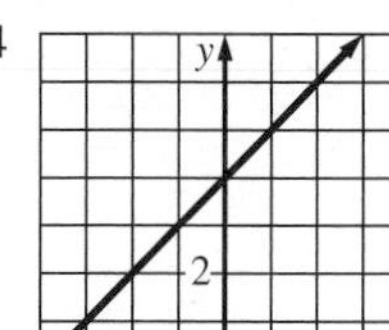

3. 1; 0

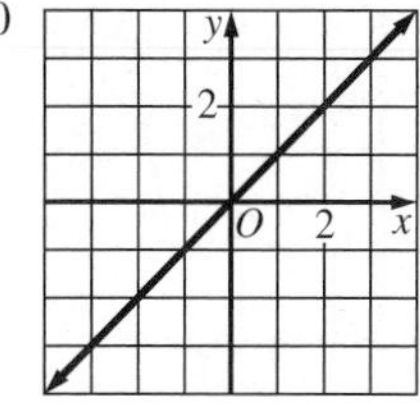

Pages 122–124 Exercises and Applications

1. $y = \frac{1}{3}x + 1$ **3.** $y = -\frac{2}{3}x - 1$ **5.** $y = -3x$

7.

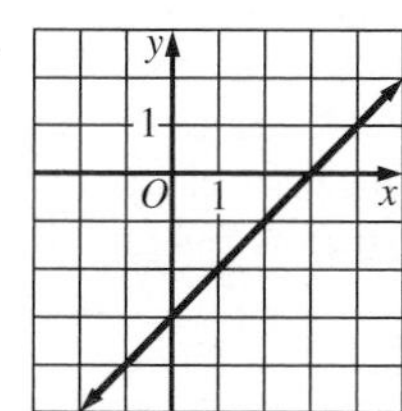

9.

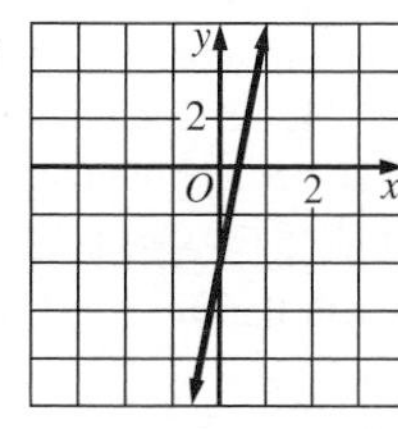

11.

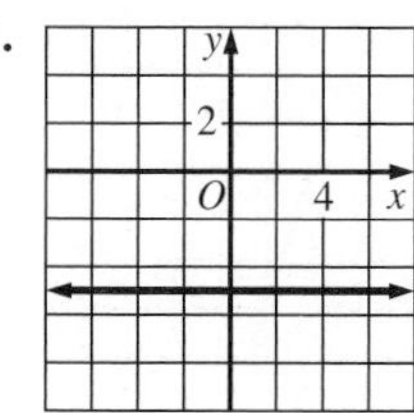

13.

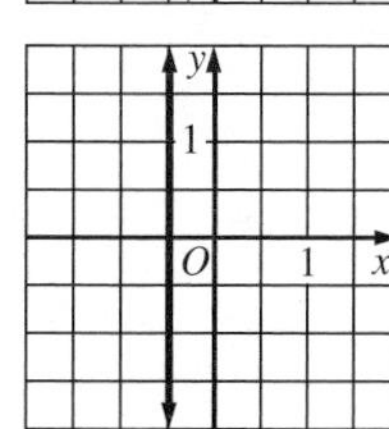

15.

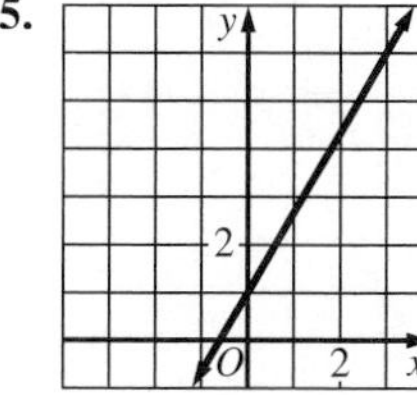

21. –3; the rate at which the height of the candle changes, –3 in./h; 12; the original height of the candle, 12 in.

23. $h = -3t + 12$ **27.** $\frac{8}{3}$ **29.** 15 **31.** 2 **33.** –6

Page 124 Assess Your Progress

1. 2 **2.** undefined **3.** –1 **4. a.** $y = -50x + 350$ **b.** The slope, –50, is the rate of change of Sue's distance from Tampa, –50 mi/h. The vertical intercept, 350, is Sue's distance from Tampa at $t = 0$, that is, the distance from her home to Tampa.

5. 3; –4

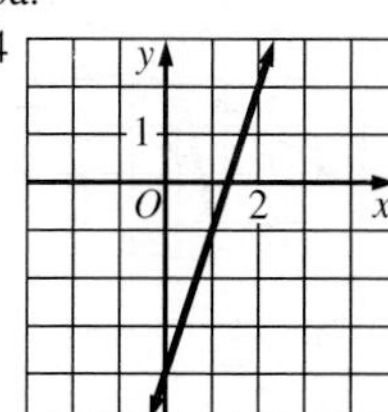

6. –1, 2

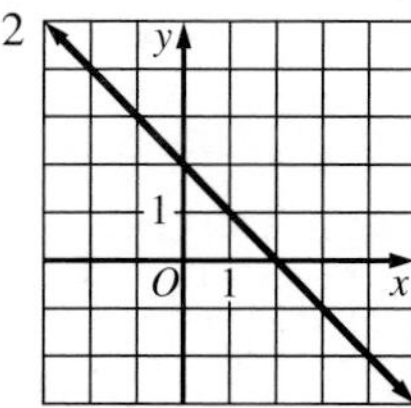

7. $\frac{1}{3}, -\frac{4}{3}$

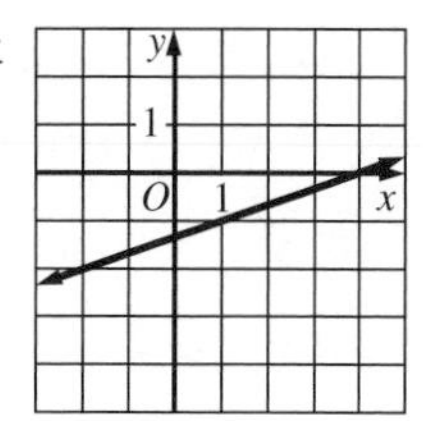

Page 127 Checking Key Concepts

1. $y = 2x - 3$ **3.** $y = 4$ **5.** $y = -x + 2$

Pages 127–129 Exercises and Applications

1. $y = x + 3$ **3.** $y = -3x + 31$ **5.** $y = -2x$ **7.** $y = 5$

9. $x = 1$ **11. a.** $y = -\frac{4}{5}x + 4$

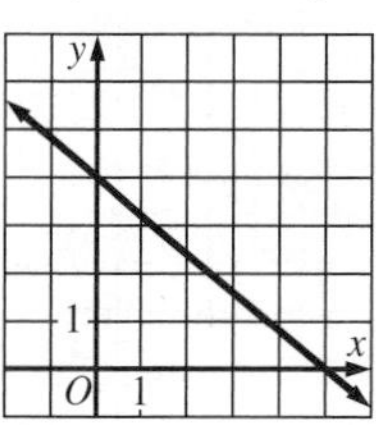

b. $y = -\frac{4}{5}x + 6$

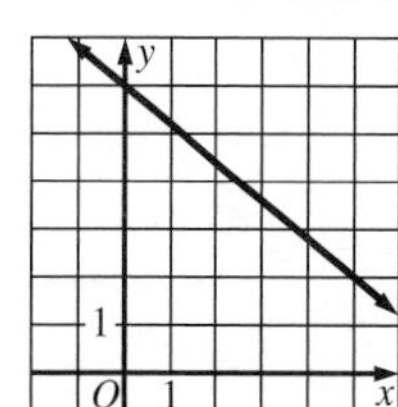

13. $y = 18$ **15.** $y = 2x - 3$ **17.** $y = -\frac{5}{3}x + \frac{5}{3}$

19. $y = -\frac{19}{7}x + \frac{65}{7}$ **23.** Answers may vary, depending on rounding. $t = -0.004e + 109.2$; about 76° **25.** Answers may vary. Examples are given. about 91°, about 60°; about 88°, about 56° **29.** 22 students **31.** about 14.7%

33. $15p + 5a$, where p is the cost of a postcard stamp and a is the cost of an airmail stamp

Page 132 Checking Key Concepts

1. Yes. Equations may vary. An example is given. $y = 5x$

3. about 102.6 min **5.** about 106.2 min

Pages 133–135 Exercises and Applications

Equations based on line of fit may vary throughout. Examples are given. **1.** $y = \frac{3}{16}x$ **5.** Answers may vary. Examples are given. **a.** about 73.4 years **b.** about 78.2 years **c.** about 62.6 years **9.** D **11.** $14b + 3$ **13.** $9 - 9z$ **15.** $-7n + 24$

17.

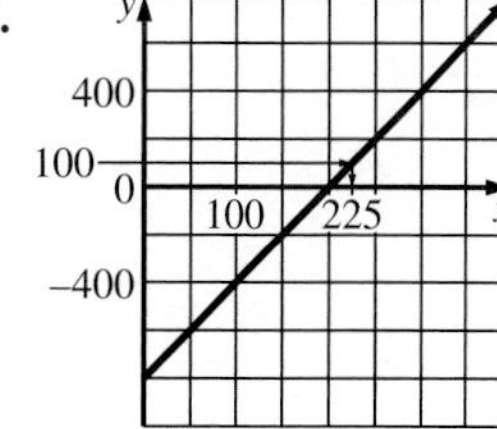

at least 225 tickets

19. 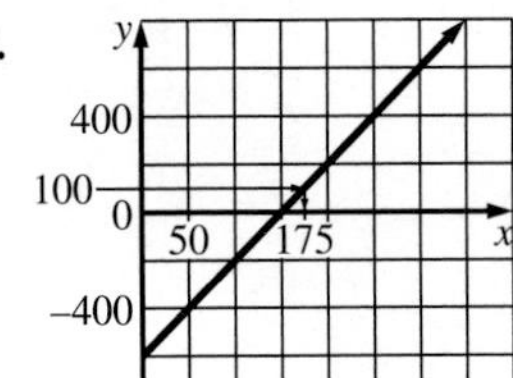

at least 175 tickets

Page 135 Assess Your Progress

1. $y = x - 5$ **2.** $y = 3$ **3.** $x = 6$ **4.** $y = -2x - 2$

5. $y = \frac{2}{3}x + 5$ **6.** $y = 5.5x - 30.5$

7. a. 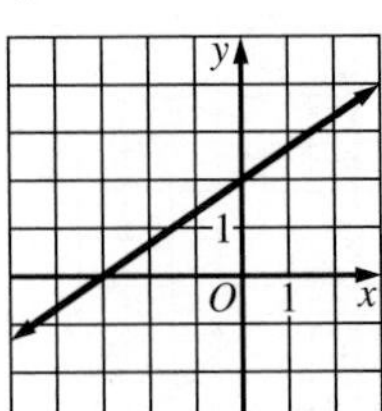
$y = \frac{2}{3}x + 2$

b. 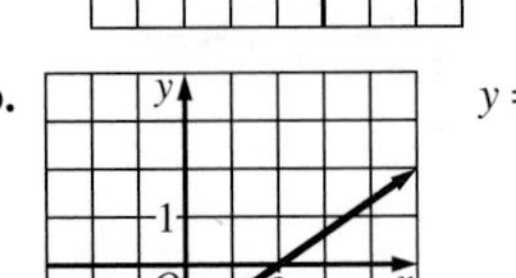 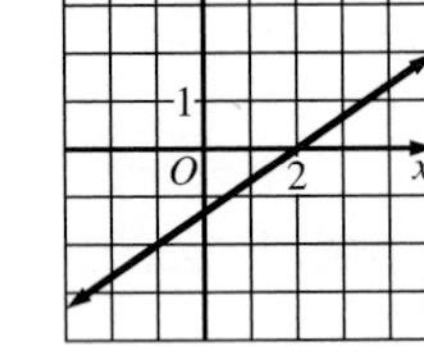
$y = \frac{2}{3}x - \frac{4}{3}$

8–10. Answers may vary. Examples are given.

8, 9. 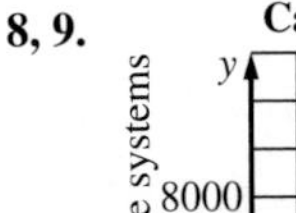

$y = 560x + 4000$, where x is the number of years since 1980 and y is the number of cable systems

10. about 12,400 cable systems

Pages 140–141 Chapter 3 Assessment

1. Answers may vary. An example is given. The data are from Example 1 on page 107. The constant variation is 5.75.

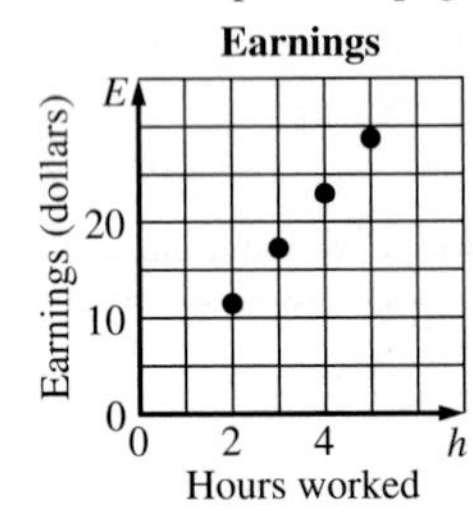

2. Answers may vary. Examples are given.

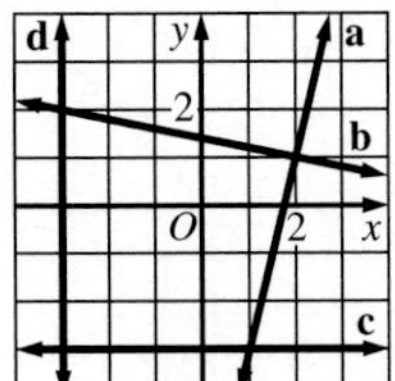

3. $\frac{7 \text{ days}}{1 \text{ week}}$ **4.** $\frac{\$58}{1 \text{ day}}$ **5.** 134 **6.** 22.5 **7.** B

8. about 25,000 tons of garbage **9.** Yes; the constant of variation is about 0.159; $r = 0.159C$, where C is the circumference in centimeters and r is the radius in centimeters.

11. Let m = number of miles towed and C = cost in dollars; $C = 2m + 34$; the slope indicates the rate of change of the cost, \$2/mi. The vertical intercept indicates the basic charge of the trip excluding the cost per mile. The tow costs \$34 plus the cost per mile. **12.** –1 **13.** $-\frac{3}{4}$ **14.** $y = -x + 1$

15. $y = \frac{3}{4}x - \frac{1}{4}$

16. 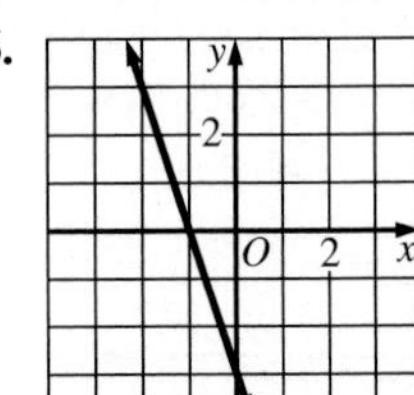

17. 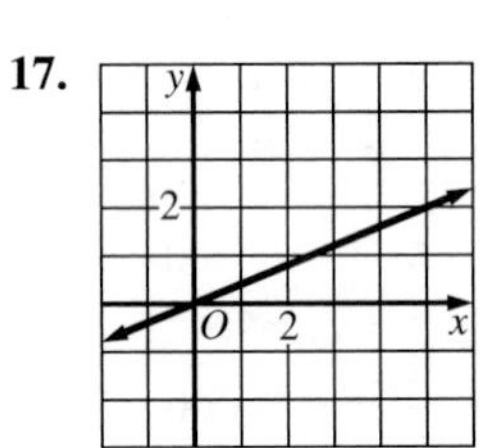

18. 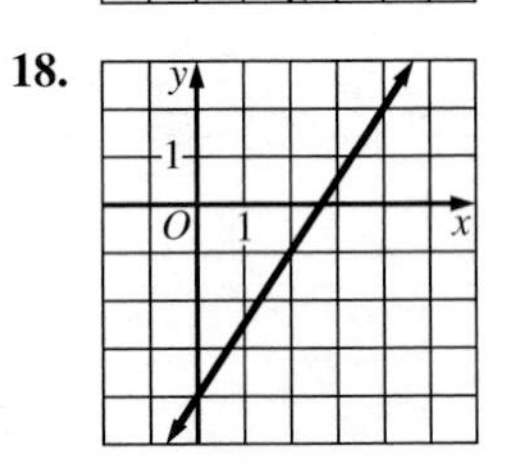

19. $y = \frac{3}{2}x$ **20.** $y = x - 1$ **21.** $y = 7$ **22.** $y = 3x - 5$

23. Answers may vary. Examples are given.

a, b. 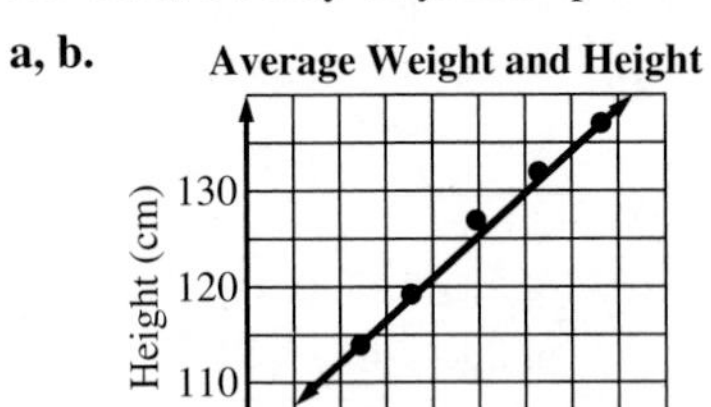

c. Answers may vary. $h = 2.2w + 68$
d. about 23.6 kg

CHAPTERS 1–3

Pages 142–143 Cumulative Assessment

1. 3500 **3.** 20–24 **5.** –5 **7.** –10 **9.** $\frac{22}{5}$, or 4.4

11. mean 23.4; median: 21; mode: 21 **17.** –208 **19.** 11

21. 32

23, 25. 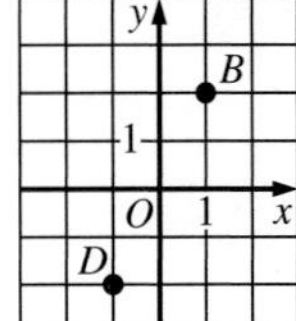

23. B: I **25.** D: III

27. about 2 s **31.** $\frac{8 \text{ h}}{1 \text{ day}}$ **33. a.** $\frac{3}{2}$ **b.** $y = \frac{3}{2}x + 6$ **c.** $y = 6$

CHAPTER 4

Page 150 Checking Key Concepts

1. $y = 60$; The cost is constant no matter how many visits a person makes to the pool. **3.** No; the cost for a pass is \$60 and the cost for 10 visits without a pass is \$40.

Pages 150–152 Exercises and Applications

1. Club A, since at 0 months Club A has a cost of \$100 and Club B has a cost of \$0; Club B, since the cost of the first month is 50 – 0 or \$50 and the cost at Club A is 140 – 100 or \$40 **3.** during the tenth month of membership

5.

No. of hours of overtime per year	Total earnings from Job A (dollars)	Total earnings from Job B (dollars)
50	30,000	24,600
100	30,000	25,200
150	30,000	25,800
200	30,000	26,400
250	30,000	27,000
300	30,000	27,600
350	30,000	28,200
400	30,000	28,800
450	30,000	29,400
500	**30,000**	**30,000**
550	30,000	30,600

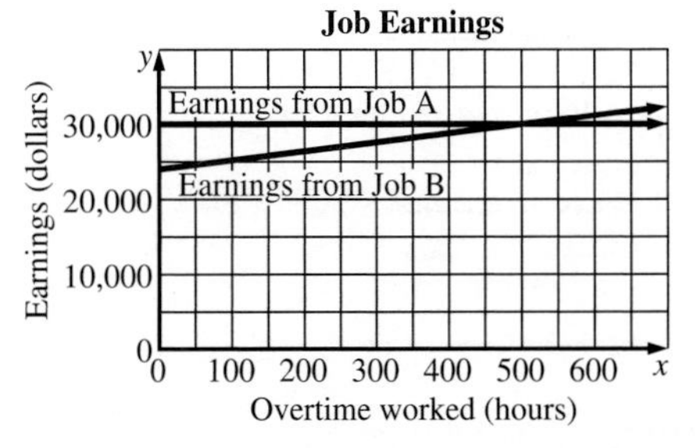

7. If you work no more than 500 h of overtime per year, then Job A would pay more than Job B. If you work more than 500 h of overtime per year, then Job B would pay more. **9.** The cost of using a fluorescent bulb increases sharply after each 9000 hour period; about \$20. **11.** Answers may vary.

13. If a customer plans to attend up to 15 shows, then it is less expensive to pay the nonmember rate. If a customer plans to attend more than 16 shows, buying a membership is the better choice. At 16 shows, the costs are equal. **17.** 10 **19.** 2 **21.** –5 **23.** No.

Page 155 Checking Key Concepts

1. $\frac{10}{9}$ **3.** $\frac{1}{8}$ **5.** $\frac{2}{3}n$ **7.** $-\frac{6}{5}c$ **9.** 20 **11.** 50

Pages 156–157 Exercises and Applications

1. $\frac{9}{2}$; 36 **3.** $\frac{6}{5}$; $\frac{72}{5} = 14\frac{2}{5}$ **5.** $\frac{7}{16}c$ **7.** $-\frac{5}{11}x$ **9.** –45 **11.** $\frac{21}{2} = 10\frac{1}{2}$ **13.** $\frac{18}{5} = 3\frac{3}{5}$ **15.** –80 **17.** E **19. a.** $l = \frac{3}{4}h$, where l = the frequency of the lower note and h = the frequency of the higher note **b.** $586\frac{2}{3}$ vibrations per second **c.** 330 vibrations per second **21.** $2\frac{2}{3}$ yd **23.** $1\frac{1}{3}$ ft **25.** 12 **27.** 55 **29.** –36 **31.** 10 **33.** $-10\frac{4}{5}$ **35.** 350 mi **39.** 2 **41.** 5 **43.** $16\frac{2}{5}$ **45.** $5a^2 + 5a - 2$ **47.** $-8m + 12n$ **49.** $8p + 56$

Page 161 Checking Key Concepts

1. $2x$; 3; 3; 2 **3.** Subtract $2x$ from both sides; $x = \frac{9}{2}$. **5.** Subtract $5n$ from both sides; $n = -10$. **7.** Rewrite $2(6x + 4)$ as $12x + 8$; no solution.

Pages 162–164 Exercises and Applications

3. –1 **5.** 3 **7.** 5 **9.** $\frac{1}{3}$ **11.** –1 **13.** 3 **15.** Let m = the number of miles they must drive; $30 + 0.6m = 55 + 0.35m$; $m = 100$; after 100 miles. **17.** about 50 m **19.** $8(t - 2) = 6t$; $t = 8$; The faster runner, Pam, caught up to the slower runner, Beth, 8 s after the race began. At that time, they were 48 m from the starting line. **21.** Assume the quantity is 3 because $\frac{1}{3}$ of 3 is an integer: $3 + \frac{1}{3} \cdot 3 = 4$, not 36. If you multiply both sides by 9, you get $3 \cdot 9 + \frac{1}{3} \cdot 3 \cdot 9 = 4 \cdot 9$, or $27 + 9 = 36$. Since $9 = \frac{1}{3} \cdot 27$, the unknown quantity is 27. **23.** Ex. 20: $x + \frac{1}{4}x = 25$; $\frac{5}{4}x = 25$; $x = \frac{4}{5} \cdot 25 = 20$. Ex. 21: $x + \frac{1}{3}x = 36$; $\frac{4}{3}x = 36$; $x = \frac{3}{4}x \cdot 36 = 27$. Ex. 22: $x + \frac{1}{12}x = 65$; $\frac{13}{12}x = 65$; $x = \frac{12}{13} \cdot 65 = 60$ **25.** –4 **27.** identity **29.** 11 **31.** $\frac{1}{4}$ **37.** 30 **39.** 60 **41.** The domain consists of the nonnegative real numbers. The range consists of the numbers 0, 0.20, 0.40, 0.60, … .

Page 164 Assess Your Progress

1. a.

No. of Checks	Minimum Account Cost ($)	Maximum Account Cost ($)
10	2.50	7.00
11	3.25	7.00
12	4.00	7.00
13	4.75	7.00
14	5.50	7.00
15	6.25	7.00
16	7.00	7.00

b. The monthly cost will be the same when 16 checks are written in a month. **2.** 9 **3.** –7 **4.** –12 **5.** –65 **6.** 8 **7.** –1 **8.** 3 **9.** no solution **10.** identity

Page 167 Checking Key Concepts

1. 30 **3.** 72 **5.** $10^2 = 100$ **7.** $\frac{6}{5}$ **9.** 16 **11.** 3

Pages 168–170 Exercises and Applications

1. 6 **3.** 90 **5.** 10 **7.** 21 **9.** –7 **11.** 20

13.

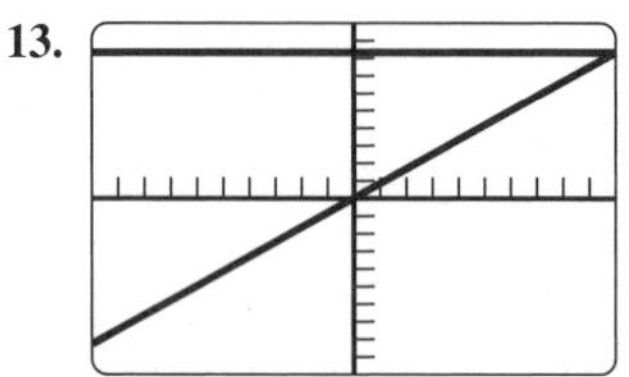

15. The equations in Ex. 14 are obtained by multiplying the right-hand side of the equations in Ex. 13 by the LCD, 6, of the fractions in the equations in Ex. 13. The intersections of both graphs have the same *x*-coordinate, 10.

17. a.

	Distance across (km)	Rate (km/h)	Time (hours)
Wind at his back	d	14	$\frac{d}{14}$
Into the wind	d	12	$\frac{d}{12}$

b. $\frac{d}{14} + \frac{d}{12} = \frac{1}{2}$; $84\left(\frac{d}{14} + \frac{d}{12}\right) = 84\left(\frac{1}{2}\right)$; $6d + 7d = 42$; $d \approx 3.2$; The distance across the lake is about 3.2 km. **19.** –25 **21.** 0.2 **23.** –2000 **25.** 2 **27.** E **29.** about 380 km

33.

	A	B	C
1	Number of pages	Cost (ordinary)	Cost (special)
2	1,000	1278.82	646.14
3	10,000	1322.20	863.43
4	100,000	1756.00	3036.29

37.

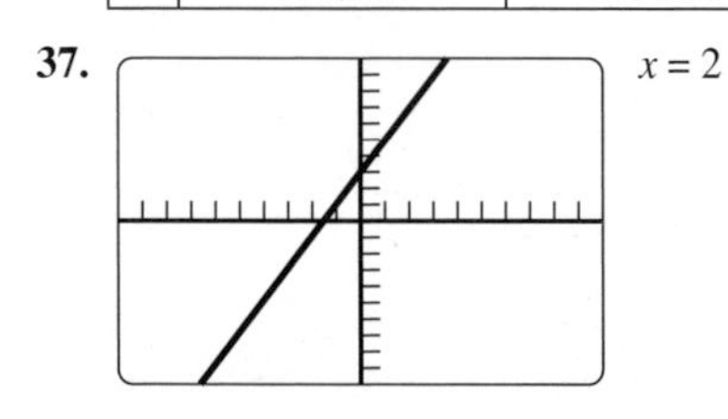

$x = 2$

39.

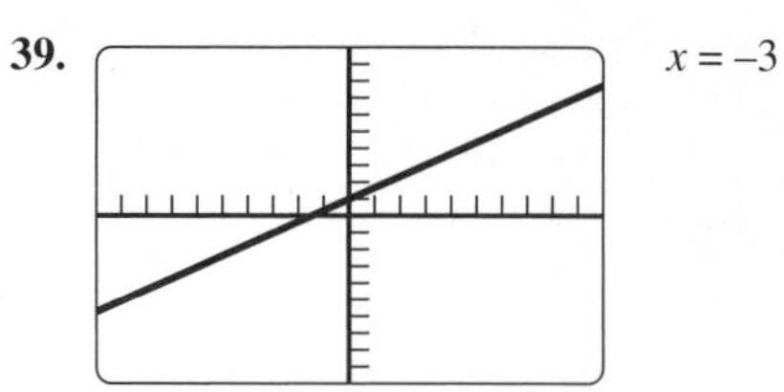

$x = -3$

Page 173 Checking Key Concepts

1. $x < 8$ **3.** $x > 2$ **5.** $x > 60$, where x = the number of dances attended in a year

Pages 173–175 Exercises and Applications

1. a. $20 < s < 65$

3. a.

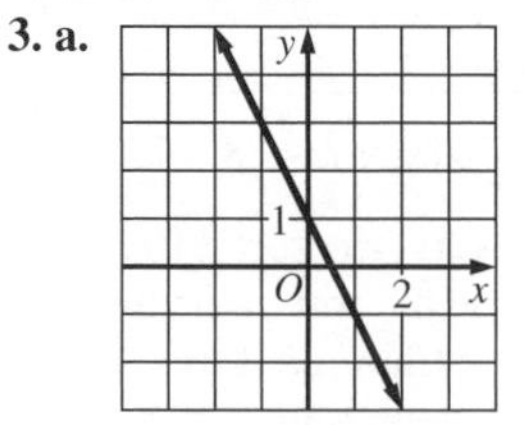

b. $x < 0$

5. Estimates may vary. $-55 \le t \le 15$; $-85 \le t \le 0$

7. Estimates may vary. $78 < a < 102$ **11. a.** Company A charges $12 for the first 60 min and $.20 for each additional minute. Company B charges $10 for the first 60 min and $.25 for each additional minute. **b.** $t > 100$, where t = the number of minutes of calling per month **13.** $d > 18$ or $d \ge 19$, where d is the number of day **15.** $y = 2x - 7$ **17.** $y = -12$

19. –6 –4 –2 0 2

21. –4 –2 0 2 2

Page 178 Checking Key Concepts

1. Yes; $x < -2$. **3.** No; $a \ge 1$. **5.** Yes; $x \le -20$.

7. $x < -5$ –10 –8 –6 –4 –2 0

9. $t < -8$ –16 –12 –8 –4 0 4

11. $x < 3$ 0 1 2 3 4 5

13. if he rents skates 16 or more times

Pages 179–180 Exercises and Applications

1. $x > 5$ 0 2 4 6 8 10

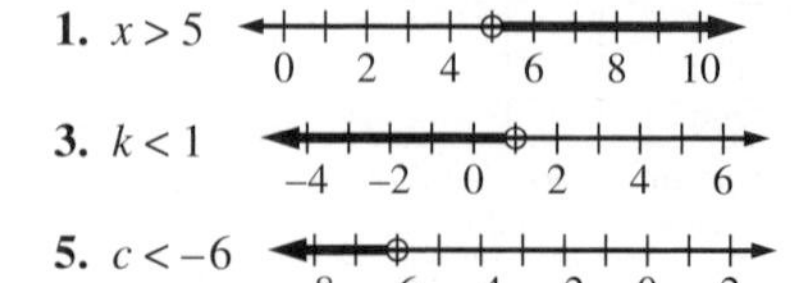

3. $k < 1$ –4 –2 0 2 4 6

5. $c < -6$ –8 –6 –4 –2 0 2

7. $x < 0$ –4 –2 0 2 4 6

9. $n \le -4$ –6 –4 –2 0 2 4

11. $a > \frac{20}{3}$ $\frac{20}{3}$ 0 2 4 6 8 10

13. Let g be the number of guests already invited; $g + 8 > 40$; $g > 32$; more than 32 people have already been invited. **15.** 9 or more days **17.** never **19.** B **21.** A

23. $n < 1$ 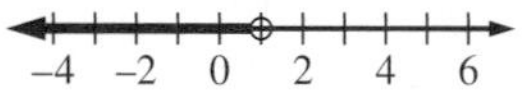

25. $c \leq -\frac{3}{4}$ 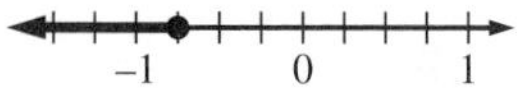

27. $t \geq -4$ 

29. $a < -6$ 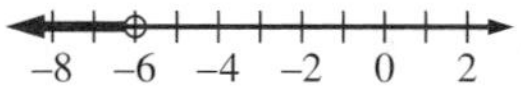

31. $x \leq \frac{1}{2}$ 

33. identity 

35. Let s be Veronica's annual sales. $19{,}500 + 0.05s > 43{,}000$; $s > 470{,}000$; Her annual sales last year were more than \$470,000. **39.** 28 trips **41.** 5280 ft/mi; 88 **43.** $-\frac{4}{9}$

Page 181 Assess Your Progress

1. $\frac{15}{2}$ **2.** 2 **3.** $-\frac{5}{4}$ **4.** 2 **5.** 1.25 **6.** 3.2 **7.** Let d be the distance of the ball from the node. **a.** $-4 \leq d \leq 7$ **b.** $-2.8 \leq d \leq 4.8$

8. $n > -3$ 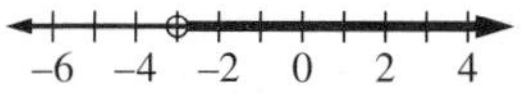

9. $x \leq -40$ 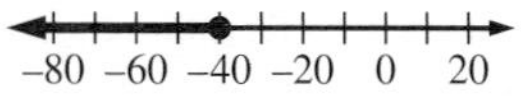

10. $a > -2$ 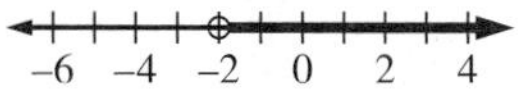

11. $t \leq \frac{1}{2}$ 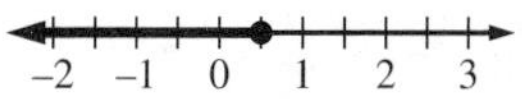

12. $k \geq -2$ 

13. identity 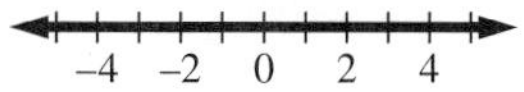

Pages 186–187 Chapter 4 Assessment

1. reciprocal **2.** identity

3. 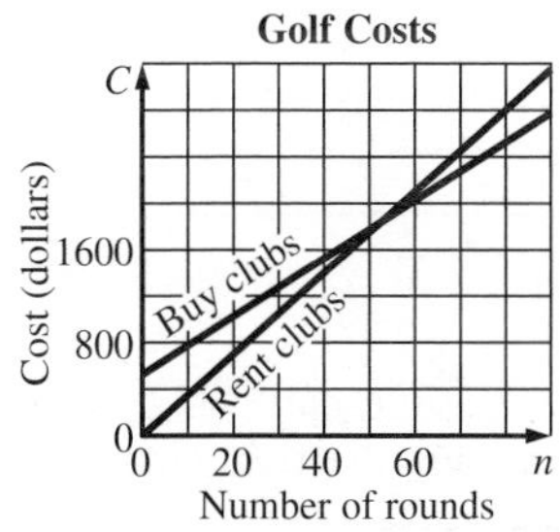

Estimates may vary; after about 55 rounds.

4. Let n = the number of rounds and C = the cost in dollars; $35n = 25n + 525$. **5.** $\frac{5}{3}$ **6.** $-\frac{8}{7}$ **7.** $\frac{2}{3}$ **8.** 25 **9.** 18 **10.** 42 **11.** 9 **12.** −5 **13.** $\frac{1}{3}$ **14.** A; D **15.** Answers may vary. An example is given. First use the distributive property to simplify both sides: $3x + 6 = 5x - 15 + 13$; $3x + 6 = 5x - 2$. Next subtract $3x$ from both sides to get the variables on one side of the equation: $3x + 6 - 3x = 5x - 2 - 3x$; $6 = 2x - 2$. Add 2 to both sides: $6 + 2 = 2x - 2 + 2$; $8 = 2x$. Finally, divide both sides by 2: $\frac{8}{2} = \frac{2x}{2}$; $x = 4$. **16.** $-\frac{2}{3}$

17. identity **18.** no solution **19.** 2 **20.** Let n = the number of performances; $35n = 140$; $n = 4$. **21.** $10^3 = 1000$ **22.** 90 **23.** $-\frac{5}{9}$ **24.** −5 **25.** 2 **26.** 0.5 **27.** Let s = the speed of the car in mi/h. Estimates may vary; $15 \leq s < 20$ and $65 < s \leq 75$. **28.** Let s = the speed of the car in mi/h. Estimates may vary; $25 < s < 55$. **29.** Yes; No; Yes; No; if both sides of a true inequality are multiplied by the same positive number, the result is a true inequality. If both sides of a true inequality are multiplied by the same negative number, the result is not a true inequality. The order of the inequality sign must be reversed to make the resulting inequality true.

30. B **31.** 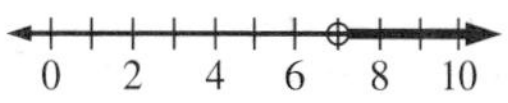

32. 

33. 

34. $x \geq 3$ 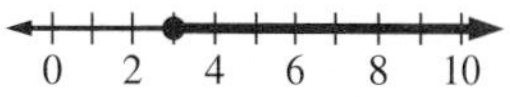

35. $x \leq -15$ 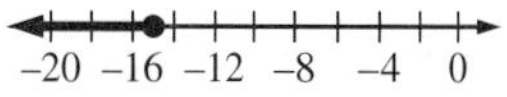

36. $x < -3$ 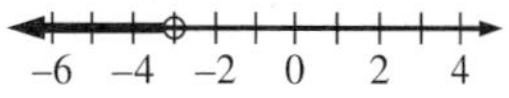

CHAPTER 5

Page 194 Checking Key Concepts

1. $\frac{1}{3}$ **3.** $\frac{4}{1}$ **5.** 24 **7.** about 387.62 Cal

Pages 194–197 Exercises and Applications

1. $\frac{9}{4}$ **3.** $\frac{1}{375{,}000}$ **5.** 2 **7.** 12 **9.** 24 **11.** 4 **13.** $\frac{33}{7} \approx 4.7$ **17.** Estimates may vary; about $\frac{4}{1}$. **19.** The portrait is about 4 times as large as the girl. The ratio in the photo is about 4 to 1, so the ratio of the actual sizes is about 4 to 1. **23.** $\frac{13}{18}$ **25.** 7 **27.** $-\frac{14}{33}$ **29.** 2 **31.** 5 **33. a.** Estimates may vary due to rounding; about 61 home runs. **35.** Answers may vary. An example is given for a school with a population of 625 students. Mandarin: about 95; Hindustani: about 38; English: about 37; Spanish: about 38; Bengali: about 21; Arabic: about 21 **37.** $y = 3x - 7$ **39.** $y = \frac{3}{4}x$ **41.** $x = 0$ **43.** −0.94 **45.** 1.13

Page 201 Checking Key Concepts

1. 11.25 in. by 25 in. **3.** $\frac{5}{24}$ **5.** Answers may vary. An example is given. Let each of the given intervals on the horizontal axis represent 10 years instead of 5.

Pages 201–204 Exercises and Applications

1. Estimates may vary; about 120 ft. **3.** $\frac{1}{2}$; $\frac{3}{4}$ in. **5.** $\frac{2}{3}$; $1\frac{1}{3}$ in.

11. 5 to 1; The scale factor is $\frac{\text{distance on Montgomery map}}{\text{distance on Alabama map}}$. Consider a distance of 4 mi. The corresponding distance on the Montgomery map is 1 in. and on the Alabama map is $\frac{4}{20}$ in. $= \frac{1}{5}$ in.; $\frac{1}{\frac{1}{5}} = 1 \cdot \frac{5}{1} = \frac{5}{1}$ **13.** Answers may vary.

15. Answers may vary. Moving the projector forward will make an image smaller. Moving the projector back will make the image larger. **17.** C

19. $x < 3$

−2 0 2 4 6

21. $x \geq -1.2$

−1.2 −0.6 0

23. $x \leq -\frac{5}{7}$

$-\frac{5}{7}$

−2 −1 0 1 2

Page 204 Assess Your Progress

1. $\frac{1}{24}$ **2.** $\frac{15}{1}$ **3.** 10 **4.** $33.\overline{3}$ **5.** 8 **6.** Estimates may vary due to rounding; about 9347 votes. **7. a.** $1\frac{7}{8}$ in. **b.** 20 ft by $29\frac{1}{3}$ ft

Page 208 Checking Key Concepts

1. a. $\angle F$ **b.** BC **3.** 155° **5.** 3

Pages 208–211 Exercises and Applications

1. a. $\angle Q$ **b.** $\overline{MO}$ **c.** PQ; QR **3.** Answers may vary. Because the two figures are rectangles, there are several ways to write the similarity. For example, $MNOP \sim QRST$ and $MNOP \sim RSTQ$. The examples are based on the similarity $MNOP \sim QRST$; $\angle M$ and $\angle Q$, $\angle N$ and $\angle R$, $\angle O$ and $\angle S$, $\angle P$ and $\angle T$, $\overline{MN}$ and $\overline{QR}$, $\overline{NO}$ and $\overline{RS}$, $\overline{PO}$ and $\overline{TS}$, $\overline{MP}$ and $\overline{QT}$. **5.** $\angle V$ and $\angle V$, $\angle VWX$ and $\angle Y$, and $\angle VXW$ and $\angle Z$, and $\overline{VW}$ and $\overline{VY}$, $\overline{WX}$ and $\overline{YZ}$, $\overline{VX}$ and $\overline{VZ}$ **7. a.** Yes; the corresponding angles have equal measure and the ratios of the lengths of corresponding sides are equal. **b.** Yes; the corresponding angles have equal measure and the ratios of the lengths of corresponding sides are equal. **c.** No; the corresponding angles do not have equal measure and the ratios of the lengths of corresponding sides are not equal. **9. a.** Yes; all four angles of a square are right angles, so corresponding angles of the two squares have equal measure. Let x be the length of a side of one square and y the length of a side of the other. The ratio of the length of any side of the first square to any side of the second is $\frac{x}{y}$. **b.** Yes; the measure of each angle of an equilateral triangle is 60°, so corresponding angles of the two triangles have equal measure. Let x be the length of a side of one equilateral triangle and y the length of a side of the other. The ratio of the length of any side of the first triangle to any side of the second is $\frac{x}{y}$. **c.** Yes; Yes. Corresponding angles of the two pentagons have equal measure. Let x be the length of a side of one pentagon and y the length of a side of the other. The ratio of the length of any side of the first pentagon to any side of the second is $\frac{x}{y}$. **d.** Regular polygons with the same number of sides are similar. **13.** $\triangle AFG$ and $\triangle ADE$ share $\angle A$ and $\overline{ED}$ and $\overline{GF}$ are both perpendicular to the ground, so $\angle D$ and $\angle F$ are both right angles. Two angles of $\triangle AFG$ have the same measures as two angles of $\triangle ADE$, so the triangles are similar. **17.** about 1635 ch'ih; Use the fact that $\triangle AFG \sim \triangle ABC$ to write the proportion $\frac{y}{y + 56} = \frac{7}{x}$. Substitute the value of y found in Ex. 14 and solve for x. **19.** A **23.** 3840 **25.** No. **27.** No.

Page 214 Checking Key Concepts

1. 35.2; 66.56

Pages 215–217 Exercises and Applications

1. $\frac{4}{9}$; $\frac{16}{81}$ **3.** 132.3; 36 **5.** 225π; 15; 5 **7. a.** $\frac{25}{49}$ **b.** \$14.70; The area of the large pizza is $\frac{49}{25}$ times that of the small pizza, so the large pizza should cost $\frac{49}{25}$ times as much as the small pizza. **11.** Answers may vary. An example is given. In the upper pictograph, I would choose the 22¢ stamp. In the lower pictograph, I would choose the 29¢ stamp. Both appear to have area about twice that of the May 29, 1978 stamp. **13.** about 64 mm^2; the 22¢ stamp (upper pictograph), 32¢ stamp (lower pictograph); No, Yes; No, No. **15.** D **17.** 10 **19.** $-\frac{14}{5}$ **21.** 45

Page 217 Assess Your Progress

2. 7.5 cm **3.** 3.3 cm **4.** $\frac{5}{3}$ **5.** $\frac{25}{9}$ **6.** The ratio of the lengths of the sides should be about $\frac{3}{2}$.

Page 221 Checking Key Concepts

1. a. 0 **b.** 1 **c.** $\frac{1}{3}$ **3.** theoretical

Pages 221–223 Exercises and Applications

1, 3. Answers may vary. Examples are given.

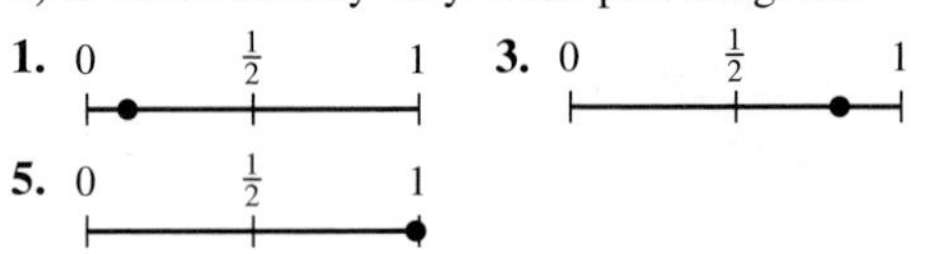

9. about 9.9%; Divide the number of people with less than a ninth-grade education (9,060,000) by the total number (91,367,000). **13.** The domain is the positive even numbers greater than 2, and the range is the reciprocals of numbers in the domain. Let x = the number of faces on the die and let y = the probability of rolling a 1; $y = \frac{1}{x}$; for any number x of faces, there are x possible outcomes, one of which is the desired outcome. **15.** No; for any number of faces, the probability or rolling a 1 is the same as the probability of rolling a 2. **17.** D

21, 23, 25. **21.** none **23.** IV **25.** I

27. $-\frac{5}{7}$ **29.** $\frac{6}{11}$ **31.** no solution

Page 226 Checking Key Concepts

1. $\frac{5}{9} \approx 56\%$ **3.** 75%

Pages 226–229 Exercises and Applications

1. 50% **3.** 50% **5.** 50% **9.** $\frac{4}{9} \approx 44\%$

15. a.

Stage	Probability of hitting the target's unshaded area
0	1
1	$\frac{8}{9}$
2	$\frac{64}{81}$

b. $\frac{512}{729}$ **17. a.** $x \le 0.5$ and $y \le 0.5$ **b.** 25% **c.** Answers may vary. **19.** 104 **21.** 1636 **23.** $7\frac{3}{4}$

Page 229 Assess Your Progress

1. $\frac{2}{3}$ **2.** $\frac{1}{6}$ **1, 2.** Answers may vary. An example is given. A die is rolled 30 times. It lands with a number less than 5 showing 25 times. The experimental probability of rolling a die and having it land with a number less than 5 showing is $\frac{5}{6}$. **3.** about 21.5% **4.** 37.5% **5.** 50%

Pages 234–235 Chapter 5 Assessment

1. proportion; means; extremes **2.** similar; angle measures; sides **3.** theoretical; experimental **4.** 1; 0 **5.** $\frac{1}{11}$ **6.** $\frac{17}{2}$ **7. a.** $\frac{5}{28}$ **b.** $\frac{5}{16}$ **c.** $\frac{3}{28}$ **8.** 192.5 **9.** 1 **10.** 30 **11.** 18 **12.** 56 **13.** $4\frac{4}{9}$ **15.** 11 in. = 20 ft, or 1 in. ≈ 1 ft 10 in.; $\frac{11}{240}$ **18.** 6.8; 5 **19.** 90°; 127°; 53°; 90° **20.** 24 **21.** 20.16 **22. a.** True. **b.** False. **23.** $\frac{1}{3}; \frac{11}{25}$ **24.** $\frac{\pi}{16} \approx 0.2$

CHAPTER 6

Page 242 Checking Key Concepts

1. 17 **3.** 2 **5.** No. **7.** Yes.

Pages 242–244 Exercises and Applications

1. 5 **3.** ±10 **5.** ±20 **7.** 29 **9.** 24 **11.** $a = 1$ **13.** $b = 35$ **15.** No. **17.** Yes.

27. $y = 1.22x$

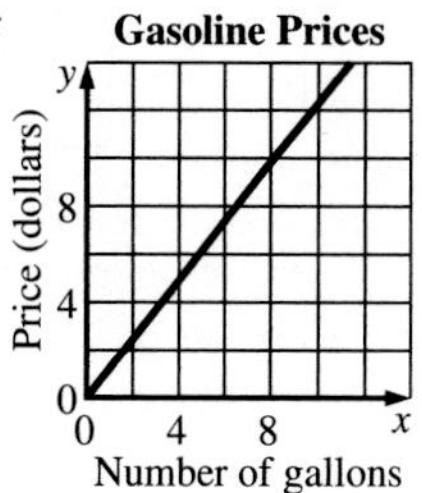

29. $-\frac{36}{13} \approx -2.77$ **31.** $-\frac{1}{11} \approx 0.09$

Page 247 Checking Key Concepts

1. $-5, -\sqrt{16}, 0, 3$ **3.** $-\sqrt{29}, \sqrt{11}$ **5.** $\frac{2}{5}$ **7.** $\frac{63}{50}$ **9.** about 44.90 mi/h

Pages 248–250 Exercises and Applications

1. $-6, -\sqrt{25}, 0, 8, \sqrt{961}$ **3.** $-\sqrt{29}, \sqrt{19}$ **5.** $\frac{7}{20}$ **7.** $\frac{7}{11}$ **9.** $\frac{5}{11}$ **11, 13.** Estimates may vary. Answers are given to the nearest hundredth. **11.** about 211.31 mi **13.** about 40.96 mi **15.** Answers may vary. An example is given. 1.07 **17.** not possible **19.** 0.2, $0.\overline{2}$, 0.26, $0.\overline{26}$, 0.265, $0.\overline{265}$, $0.2\overline{6}$ **21.** $a \approx 10.95$ **23.** $b \approx 17.66$ **25.** $b \approx 73.97$ **35.** 5:8; 25:64 **37.** $c = 13$ cm **39.** $c = 12$ ft

Page 254 Checking Key Concepts

1. $2\sqrt{3}$ **3.** $\frac{\sqrt{13}}{6}$ **5.** $3\sqrt{5}$ **7.** $2 < \sqrt{5} < 3$ **9.** $9 < \sqrt{84} < 10$

Pages 254–257 Exercises and Applications

1. $2\sqrt{10}$ **3.** $4 + 2\sqrt{10}$ **5.** $6\sqrt{5}$ **7.** $3\sqrt{7}$ **9.** 75 **11.** $\frac{\sqrt{5}}{4}$ **13.** Nina; Answers may vary. An example is given. You could evaluate $11 + 6\sqrt{2}$ and $17\sqrt{2}$ using a calculator to show they are not equal, or you could point out that $11 + 6\sqrt{2}$ cannot be simplified because only one term contains $\sqrt{2}$. **15.** $20 + 6\sqrt{2}$; $30\sqrt{2}$ **17.** about 1.81 m^2 **19. a.** Answers may vary. **b.** Answers may vary. **21. a.** The length of the hypotenuse of the second triangle is $\sqrt{3}$ units. **b.** The length of the hypotenuse of the third triangle is $\sqrt{4} = 2$ units. **c.** The lengths of the hypotenuses of the succeeding triangles are $\sqrt{5}$ units, $\sqrt{6}$ units, $\sqrt{7}$ units, and so on. The pattern is that the nth triangle has a hypotenuse that is $\sqrt{n+1}$ units long. **23.** C **25.** 7 **27.** 4 **29.** 5 **31.** $\angle B = 101°$; $\angle C = 24°$; $\angle P = 55°$; $PQ = 4.5$ m; $AC = 6$ m

Page 257 Assess Your Progress

1. $b = 12$ **2.** $c = \sqrt{261} \approx 16.16$ **3.** $b = \sqrt{259} \approx 16.09$ **4.** $\frac{73}{20}$ **5.** $\frac{118}{33}$ **6.** $\frac{23}{9}$ **7.** $3\sqrt{5}$ **8.** $5 + 4\sqrt{5}$ **9.** $\frac{2\sqrt{7}}{11}$

Page 261 Checking Key Concepts

1. $x(2x) = 2x^2$ **3.** $(x + 2)x = x^2 + 2x$

5. 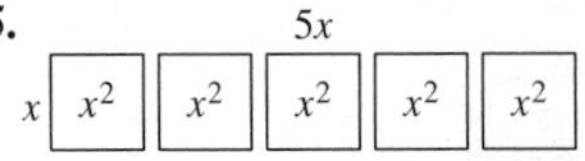

7. 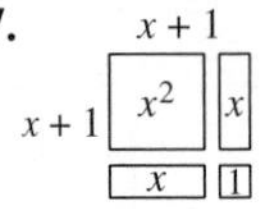

9. binomial **11.** monomial

Pages 261–263 Exercises and Applications

1. $(x + 1)2x = 2x^2 + 2x$ **3.** $(x + 2)(x + 4) = x^2 + 6x + 8$

5. 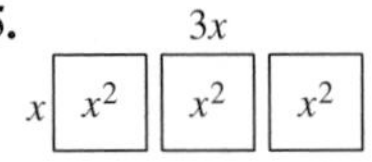

7. 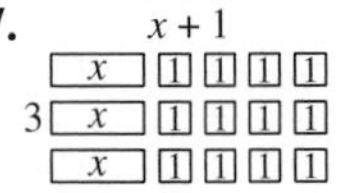

9. 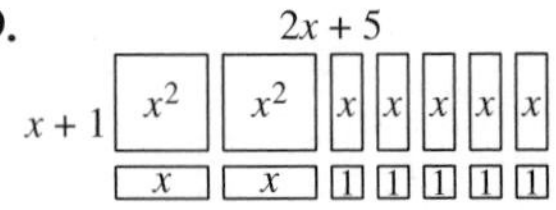

11. $4x + 4$ **13.** $4x^2 + 12x$ **15.** $2x^2 + 7x - 4$
17. $15 + 3\sqrt{2}$ **19.** $3\sqrt{6} + 2\sqrt{2} + 3\sqrt{3} + 2$
21. a. $(a + b)(a + b) = a^2 + 2ab + b^2$ **b.** $4\left(\frac{1}{2}ab\right) + c^2 = 2ab + c^2$ **27.** $\frac{1}{2}$ **29.** Answers may vary. An example is given. $y = 0.72x + 0.75$

Page 267 Checking Key Concepts

1. $x^2 + 16x + 64$ **3.** $x^2 - 81$ **5.** $52 - 14\sqrt{3}$
7. $(50 + 4)^2 = 50^2 + 2(50)(4) + 4^2 = 2500 + 400 + 16 = 2916$

Pages 267–269 Exercises and Applications

1. $w^2 - 4w + 4$ **3.** $x^2 - 6x + 9$ **5.** $y^2 - 16$
7. $4r^2 + 24r + 36$ **9.** $100t^2 - 36$ **11.** The greater the width, the greater the value of x and the greater the area of the path; the cost depends on the area. The area of the path is equal to the combined area of the path and garden minus the area of the garden alone. Then the area of the path = $(2x + 20)^2 - 20^2 = 4x^2 + 80x$. **13.** $(40 - 1)^2 = 40^2 - 2(40)(1) + 1^2 = 1600 - 80 + 1 = 1521$ **15.** $(2 + 0.1)^2 = 2^2 + 2(2)(0.1) + (0.1)^2 = 4 + 0.4 + 0.01 = 4.41$
17. $\left(10 - \frac{1}{2}\right)\left(10 + \frac{1}{2}\right) = 10^2 - \left(\frac{1}{2}\right)^2 = 100 - \frac{1}{4} = 99\frac{3}{4}$
19. $x^2 - 2xy + y^2$ **21.** $x^2 - y^2$ **23.** $16x^2 - 49$
25. $28 + 16\sqrt{3}$ **27.** B **33.** $y = 4x$ **35.** $y = \frac{1}{2}x - 4$
37. $y = \frac{2}{3}x + 4$ **39.** $6m^2 - 15m$ **41.** $-25x^2 + 10x - 1$
43. $20s - 15s^2$

Page 269 Assess Your Progress

1. $5x^2 + 15x$ **2.** $10 + 5\sqrt{3}$ **3.** $x^2 - 25$ **4.** $46 + 12\sqrt{10}$
5. $6x^2 + 9x$ **6.** $x^2 - 8x + 16$ **7.** $15x^2$ **8.** $2x^2 + 11x + 5$
9. $x^2 + 4\sqrt{2}x + 8$ **10.** $(n + 8)^2 = n^2 + 16n + 64$
11. $4x$; $x^2 - 49$

Pages 274–275 Chapter 6 Assessment

1. 90; hypotenuse; legs; $a^2 + b^2 = c^2$, that is, the square of the length of the hypotenuse is equal to the sum of the squares of the lengths of the legs. **2.** a number whose square root is an integer; examples: 100, 12^2, 0 **3.** any number that can be written as a ratio of two integers $\frac{a}{b}$, where $b \neq 0$; examples: 1.5, $\frac{7}{6}$, 2; any number that cannot be written as a ratio of two integers, and whose decimal form never terminates or repeats; examples: $\sqrt{7}$, 1.010010001..., $\sqrt{3}$ **4.** $m = 120$ **5.** $p = 25$
6. $r = 7.5$ **7.** Yes. **8.** No. **9.** 13.75 in. **10.** $\frac{171}{250}$ **11.** $\frac{7}{9}$
12. $\frac{95}{99}$ **13.** $4\sqrt{3}$ **14.** $3 + 10\sqrt{2}$ **15.** $8\sqrt{2}$ **16.** $\frac{3}{5}$
17. $24\sqrt{5}$ **18.** $\frac{8\sqrt{3}}{11}$ **19.** $7 < \sqrt{54} < 8$
20. $10 < \sqrt{107} < 11$ **21.** $8 < \sqrt{72} < 9$ **22.** $16\sqrt{5}$; 75; $\sqrt{170}$ **23.** $14x + 35$ **24.** $3y^2 - 27y$ **25.** $6m^2 - 11m - 35$
26. $8\sqrt{3} + 3$ **27.** 13 **28.** $195 + 47\sqrt{5}$
29. a. $(2x - 6)(x - 5) = 2x^2 - 16x + 30$ **b.** 6
30. $n^2 - 16n + 64$ **31.** $4a^2 + 12a + 9$ **32.** $61 + 28\sqrt{3}$
33. $c^2 - 81$ **34.** $36x^2 - \frac{1}{25}$ **35.** 20 **36.** $24\frac{3}{4}$ **37.** 851
38. 784

CHAPTERS 4–6

Pages 276–277 Cumulative Assessment

1. –20 **3.** 1.5 **5.** 2.5

7. $x < 4$ 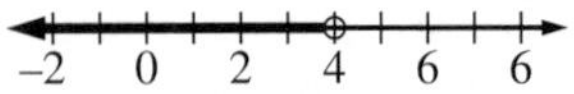

9. $p \leq \frac{1}{3}$ 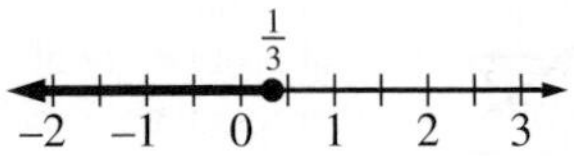

11. $m > 100$ **15.** 15 **17.** 9 **19.** $\frac{3}{4}$ **21.** 22.5 **23.** 4:5
29. Yes; $15^2 + 36^2 = 39^2$. **31.** No; $8.2^2 + 9.1^2 \neq 11.9^2$.
33. $\frac{1}{33}$ **35.** $\frac{961}{999}$ **37.** $19\sqrt{2}$ **39.** $20\sqrt{3}$ **41.** $8\sqrt{7}$
43. $86 + 14\sqrt{3}$ **45.** $36x^2 - 5$

CHAPTER 7

Page 283 Checking Key Concepts

1. a. $8a + 5c$ **b.** $8a + 5c = 40$ **3.** No; the variable terms are not on one side. **5.** 10; 6 **7.** Answers may vary. Examples are given. One method is to graph the x- and y-intercepts and then draw the line through the points. Another method is to rewrite the equation in slope-intercept form and enter this equation into a graphing calculator.

Pages 284–286 Exercises and Applications

1. 15 points **3.** 5 points **5. a.** $2w + t = 13$

b. 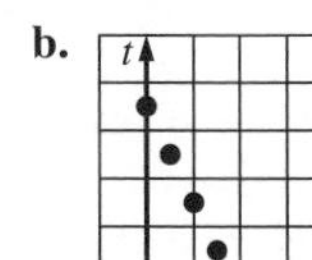

The t-intercept 13 represents the number of ties it would take to have 13 points without any wins. The graph has no w-intercept since the domain and range of this function include only whole numbers. If the graph did intersect the w-axis, the w-intercept would represent the number of wins it would take to have 13 points without any ties.

7. \$123 **9.** $15c + 9t$

11. 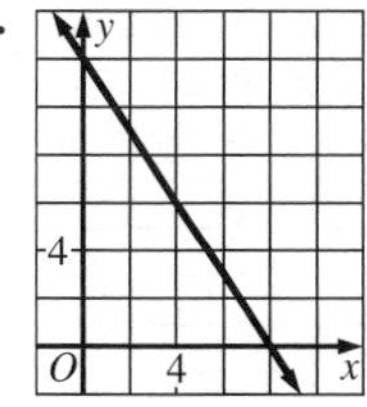

13. 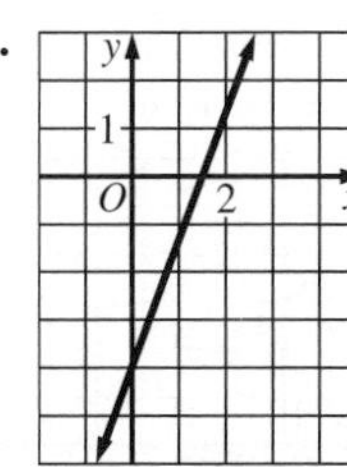

15. 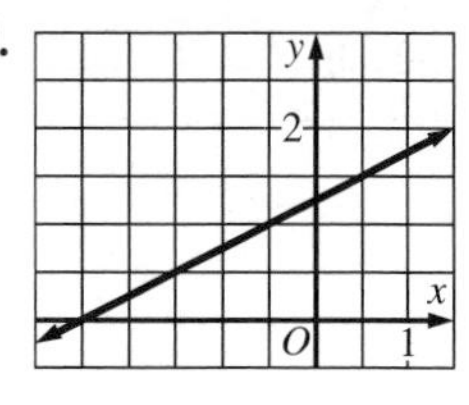

17. 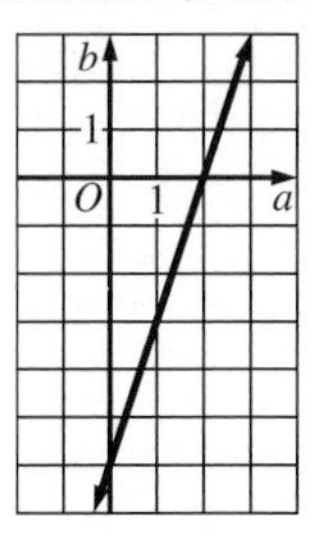

19. 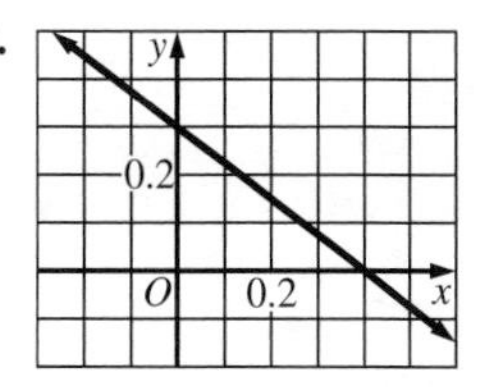

21. $y = -1.5x + 5$ 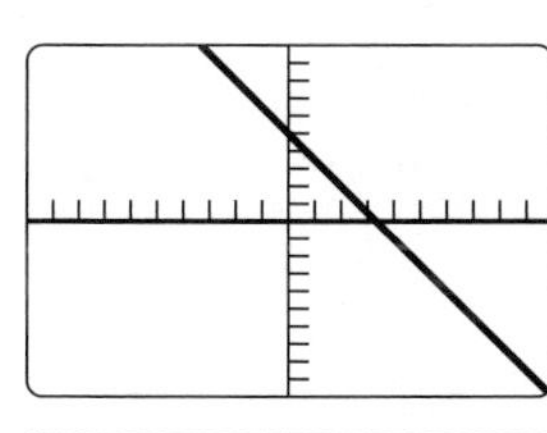

23. $y = 0.8x + 2.4$ 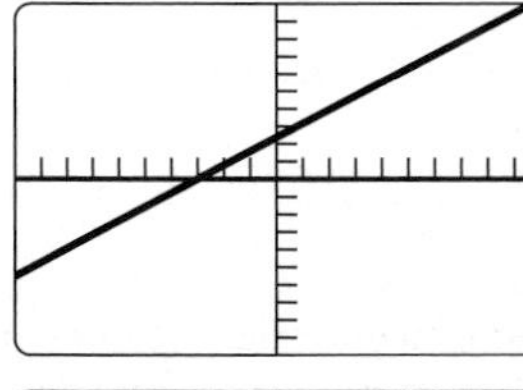

25. $y = -\frac{4}{3}x + \frac{8}{3}$ 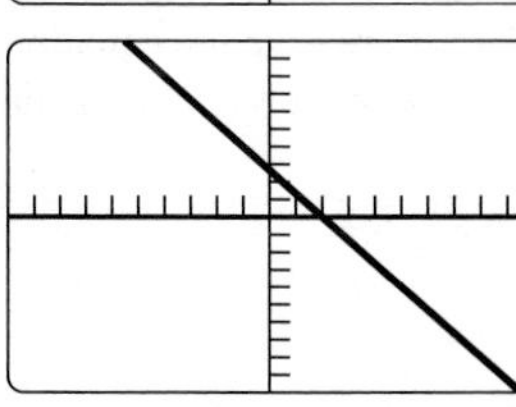

27. $y = -1.5x + 6$ 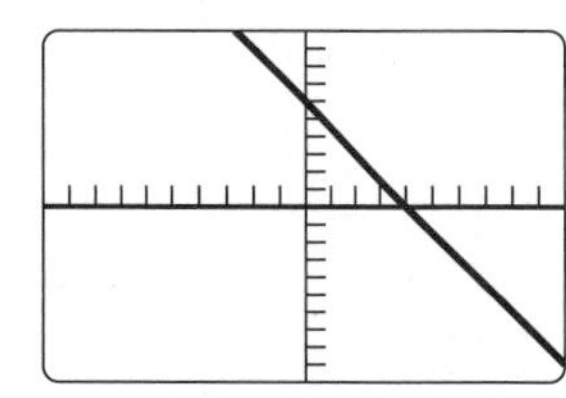

29. the equations in Exs. 17, 18, 19, 27, and 28 because a, b, and c are not all integers and the equation in Ex. 25 because it does not have both variable terms on one side

31. $1.5x + y = 10$ **35.** $54c + 100r = 100$, where c = the number of cups of oatmeal and r = the number of cups of raisins; Solutions may vary. Examples are given. 1 cup of raisins alone, about 1.85 cups of oatmeal alone, and 1 cup of oatmeal with 0.46 cup of raisins

41. 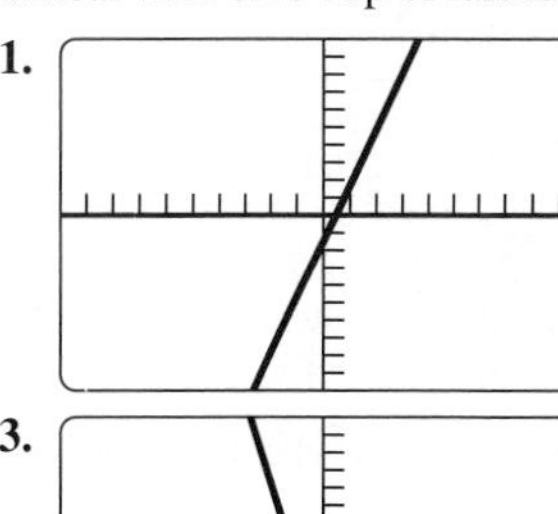 0.46875

43. 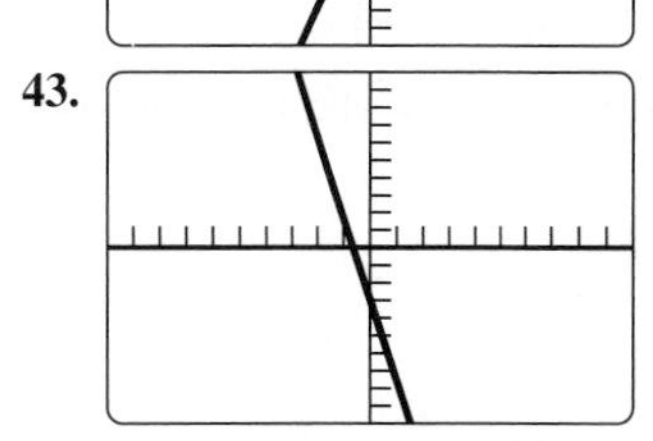 about −0.21

45. 2.5 **47.** 0.5 **49.** $-\frac{16}{7}$

Page 290 Checking Key Concepts

1. $\left(\frac{2}{7}, \frac{24}{7}\right)$; $\left(7, \frac{8}{3}\right)$; $(3, -2)$

Pages 290–293 Exercises and Applications

1. 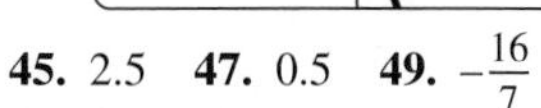 $(6, -4)$

3. 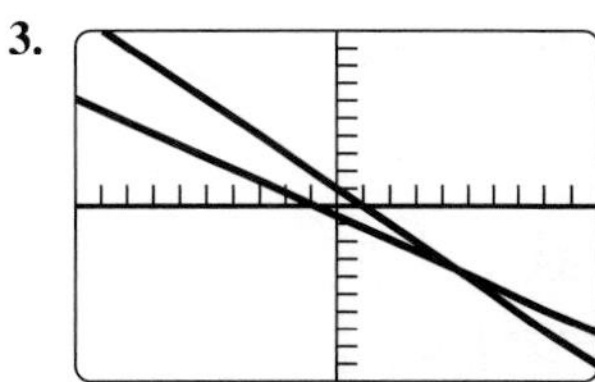 Estimates may vary; about $(4.7, -3.7)$.

5. $(4, -2)$ **7.** $(-3, 1)$ **9.** $(2, -2)$ **11.** $(4, -1)$ **13.** $(2, 4)$ **17.** $L = 2N$ and $L = N + 10$; The normal rate is 10 pints/min and the launch rate is 20 pints/min. **19.** Let b = the cost in dollars of a beef meal and c = the cost in dollars of a chicken meal. $98b + 66c = 1000$; $c = 0.8b$; about \$6.63 for a beef meal and \$5.31 for a chicken meal **23. a.** about 1030 ft **b.** about 11.7 mi/h

Page 293 **Assess Your Progress**

1. –3, –4

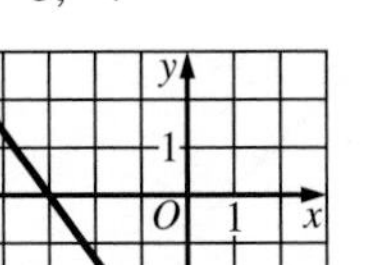

2. $-4, \frac{3}{2}$

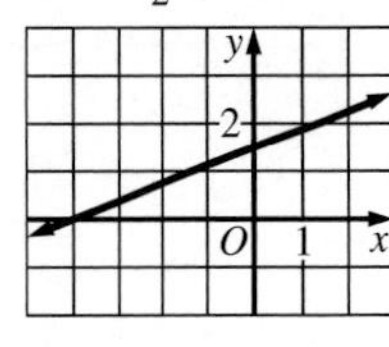

3. $y = 5x - 5$

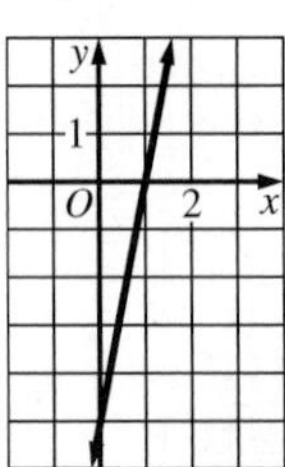

4. $y = -\frac{2}{3}x + 3$

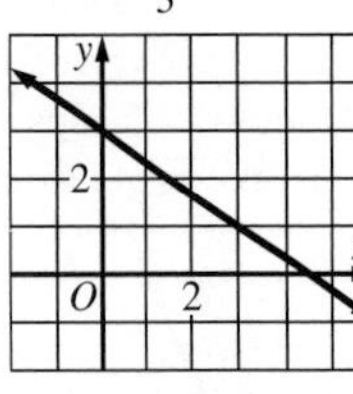

5. Let c = the number of correct answers and w = the number of wrong answers. **a.** $4c - w = 30$ **b.** $c + w = 20$

c.

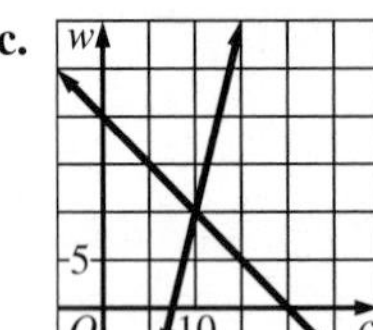

10 correct answers and 10 wrong ones

6. (1, –1) **7.** (5, 2)

Page 296 **Checking Key Concepts**

1. addition; (3, 1) **3.** subtraction; (–14, –25)
5. addition; (–3, –2)

Pages 296–298 **Exercises and Applications**

3. (2, 1) **5.** (4, –1) **7.** (–2, 3) **9.** (–1, –2) **11.** $495
13. $\left(-3, -\frac{2}{5}\right)$ **15.** (1, –2) **17.** (7, 5)

19.

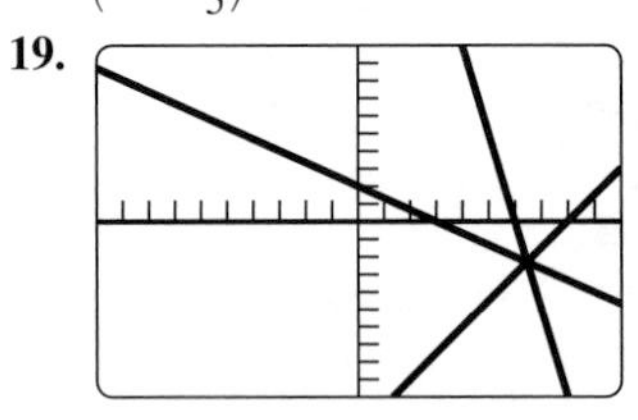

The graph of the third equation contains the intersection point of the system.

21. 2 cups **27.** –2 **29.** 11

31. $x \leq 3$

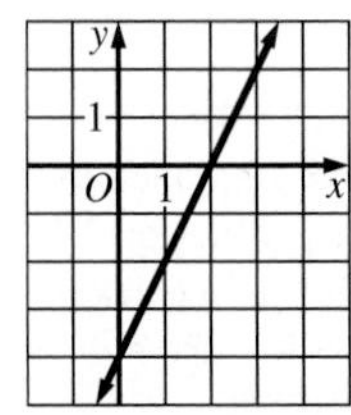

33. $x \geq \frac{1}{2}$

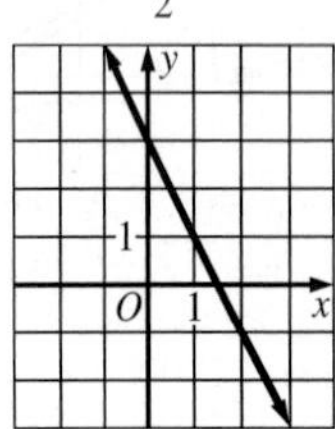

Page 303 **Checking Key Concepts**

1, 3. Answers may vary. Examples are given. **1.** addition or subtraction **3.** graphing or multiplication **5.** (5, –1)

Pages 303–305 **Exercises and Applications**

1. (3, –1) **3.** (1, 2) **5.** many solutions (all (x, y) such that $3x + 4y = 5$) **7.** $\left(6\frac{1}{3}, 2\right)$ **9.** (2, –1) **11.** $\left(-\frac{22}{29}, \frac{32}{29}\right)$

13. (6, 0) **15.** $\left(\frac{17}{8}, \frac{1}{2}\right)$ **21.** She can buy any number of Freesia bulbs from 0 to 15 in combination with any number of dahlia bulbs from 15 to 0, just so the number of bulbs totals 15.

25. No.

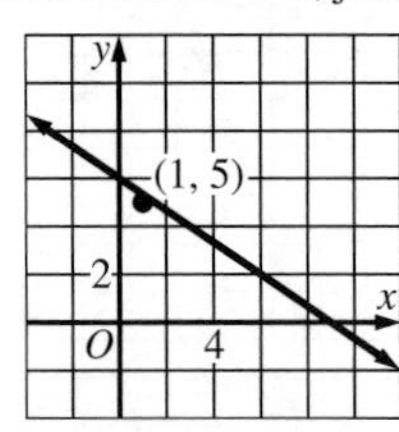

27.

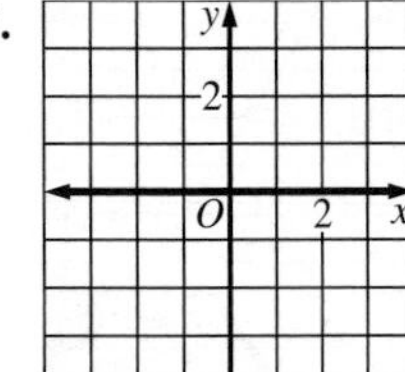

29.

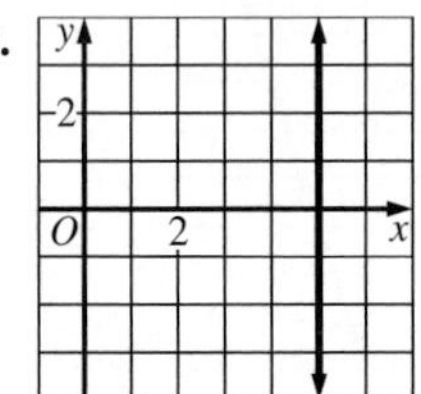

Page 305 **Assess Your Progress**

1. (–2, 3) **2.** (–4, 7) **3.** (–8, –1) **4.** all (s, t) such that $5 = 2t - s$ **5.** all (s, t) such that $4t - 6s = 2$ **6.** no solution
7. Emily: 8 visits; Adam: 20 visits

Page 308 **Checking Key Concepts**

1. C **3.** B **5.**

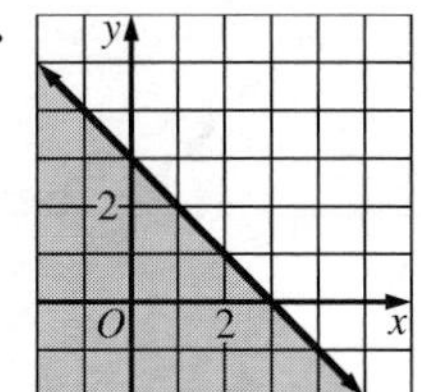

Pages 308–310 **Exercises and Applications**

1. D **3.** B

5.

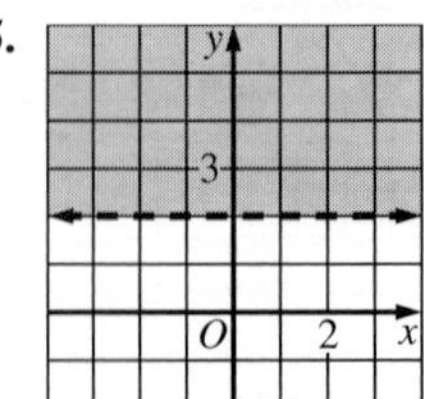

7.

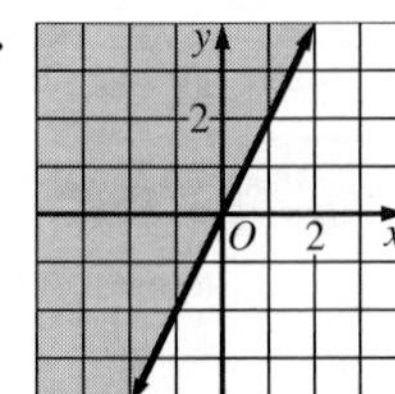

9.

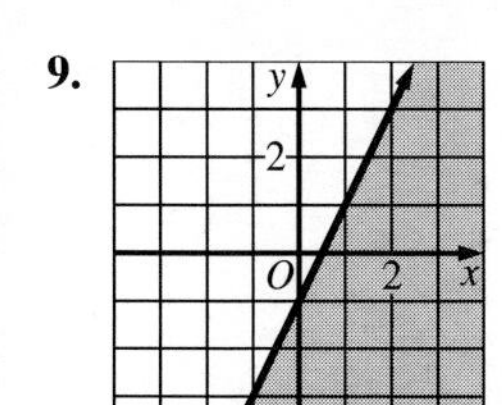

11.

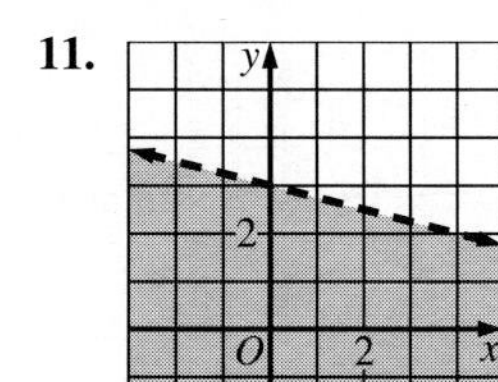

13.

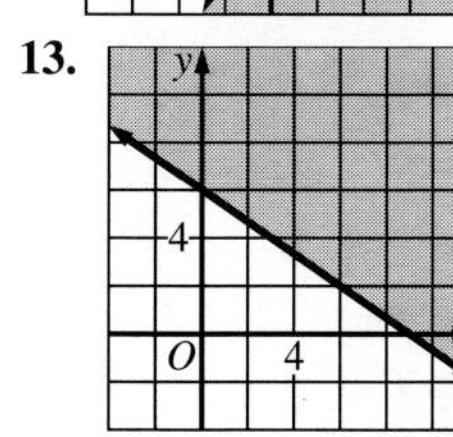

15.

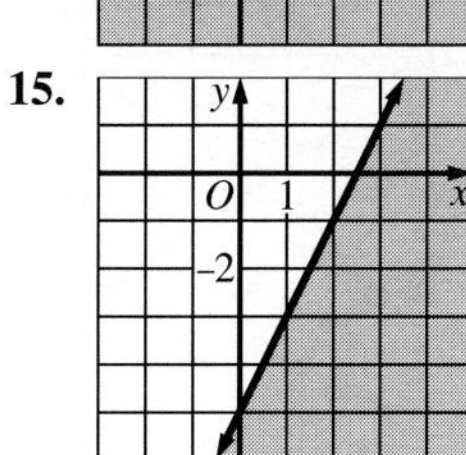

17.

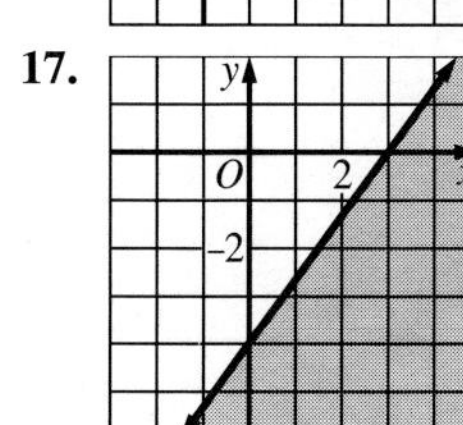

19. Answers may vary. An example is given. (1) I solved for y to get $y \leq \frac{4}{3}x - 4$. Then I graphed $y = \frac{4}{3}x - 4$, using a solid line. Since y is less than or equal to $\frac{4}{3}x - 4$, I shaded the points below the line. (2) I graphed the solid line $4x - 3y = 12$ by using the intercepts. I chose (3, –2) as a test point that does not lie on the line and checked to see if the coordinates make the inequality $4x - 3y \geq 12$ true or false. Since the coordinates make the inequality true, I shaded the part of the graph that contains the point (3, –2). I prefer to use the method in Ex. 18 because it is hard to solve for y and sometimes I forget when to reverse the inequality sign. Using a test point seems easier.

21.

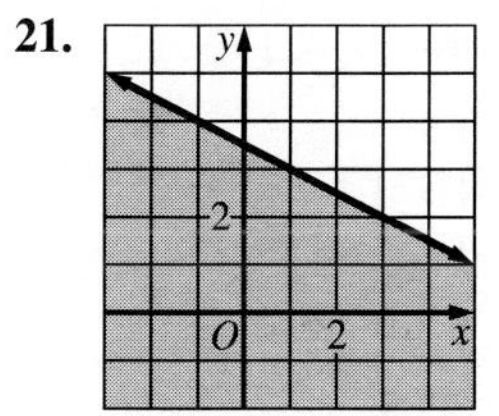

23.

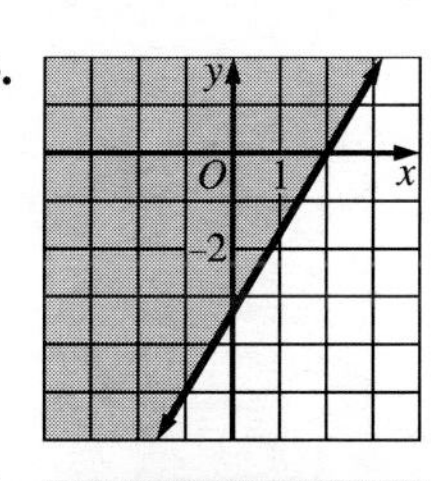

25.

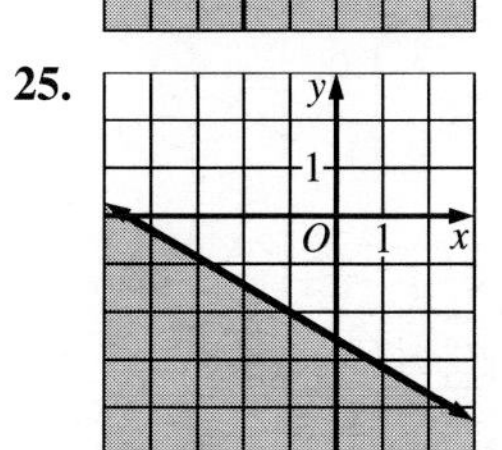

27.

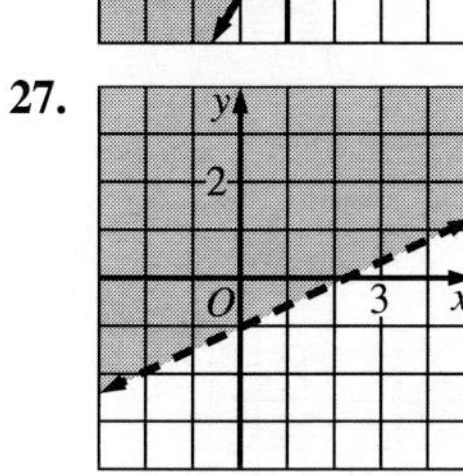

29. a. $300x + 180y > 1200$

b.

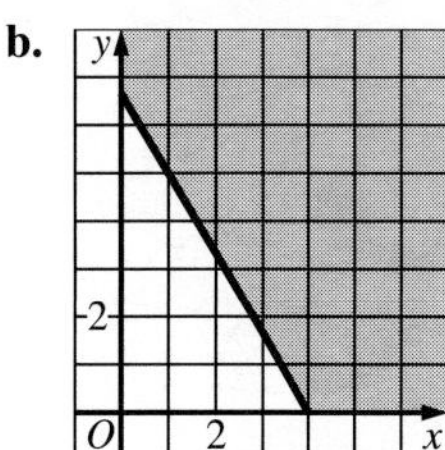

Answers may vary. Examples are given. (8, 0) and (4, 4)

33. B **37.** $2x^2 + 14x$ **39.** $a^2 + 2a - 35$ **41.** $-6 + 2\sqrt{5}$ **43.** (1, 1) **45.** (4, 11)

Page 313 Checking Key Concepts

1.

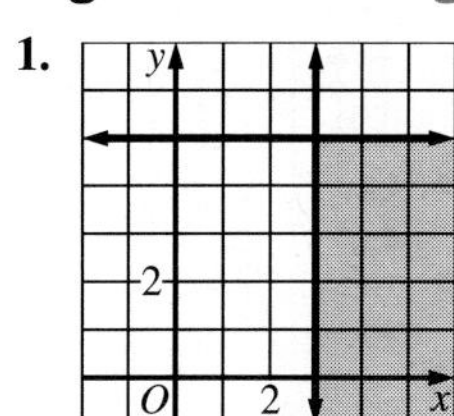

3.

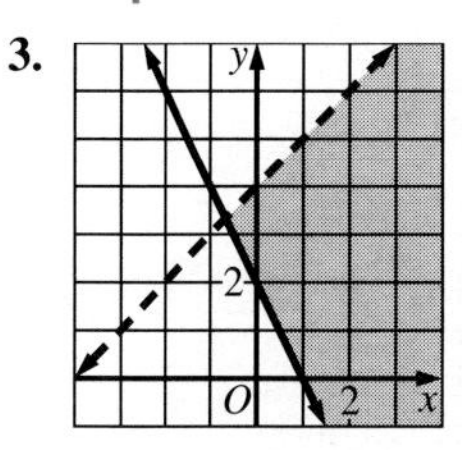

5. B: $y > 7 - x$ and $y > x - 3$; C: $y > 7 - x$ and $y < x - 3$; D: $y < 7 - x$ and $y < x - 3$

Pages 313–315 Exercises and Applications

1.

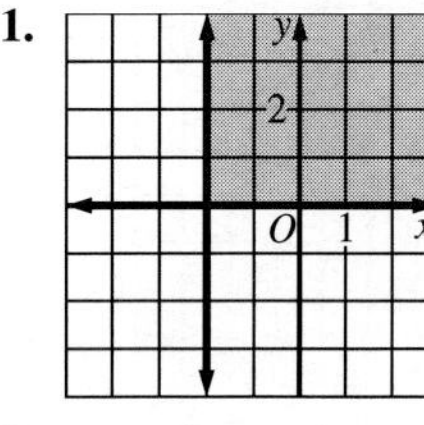

3.

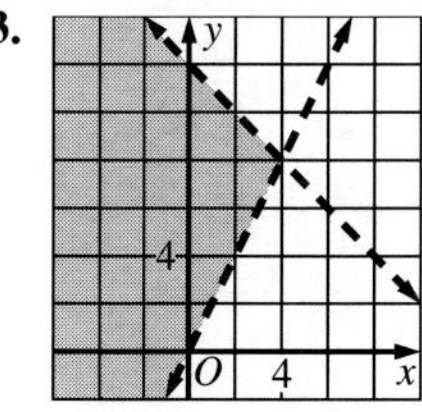

5.

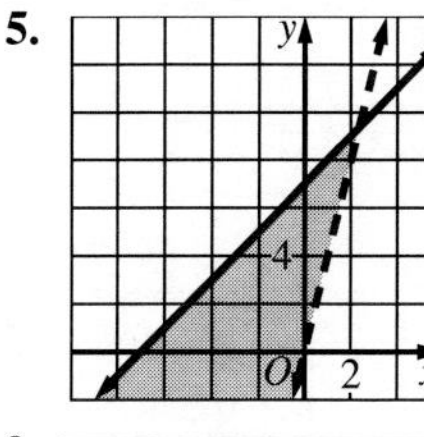

7.

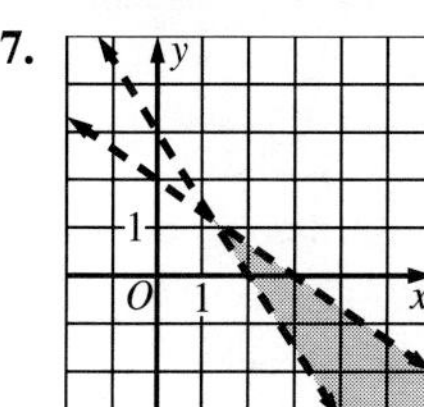

9.

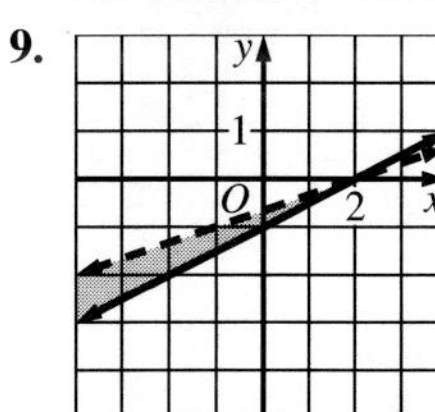

11.

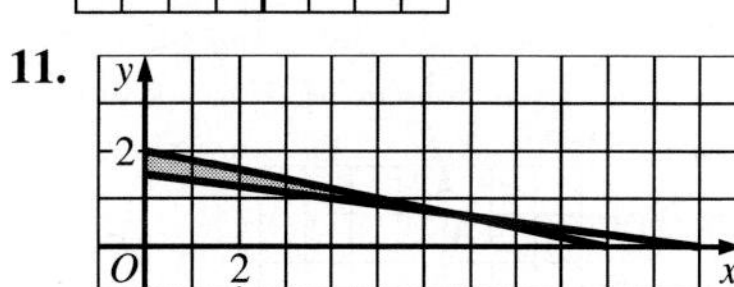

While the system has many solutions, there are only three with whole-number coordinates: (0, 2), (4, 1), and (5, 1).

13. 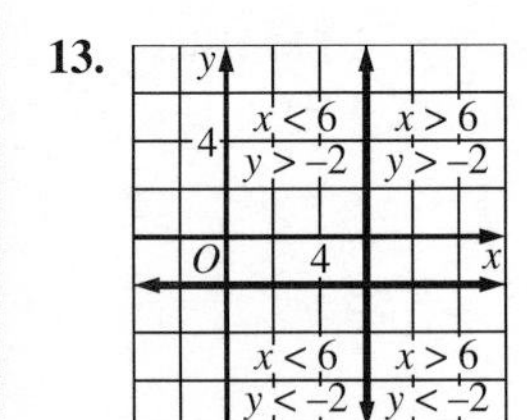

15. a. $c \geq 0.6t$ **b.** $c \leq 0.7t$ **21.** B **25.** $\frac{6}{7}$ **27.** 20
29. $\frac{14}{3}$ **31.** $x^2 + 4x + 4$ **33.** $x^2 - 25$

Page 315 Assess Your Progress

1. 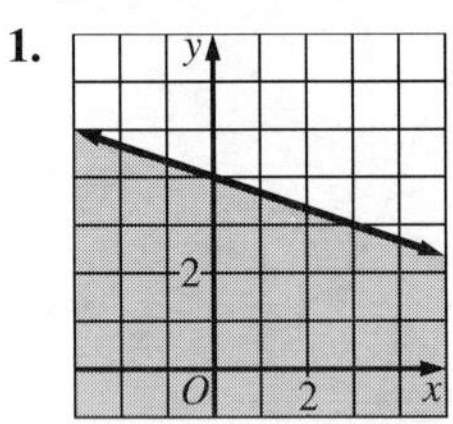

2. 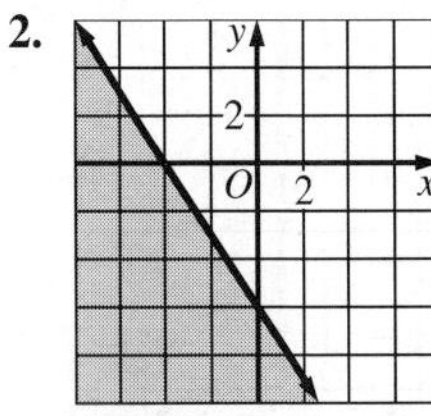

3. 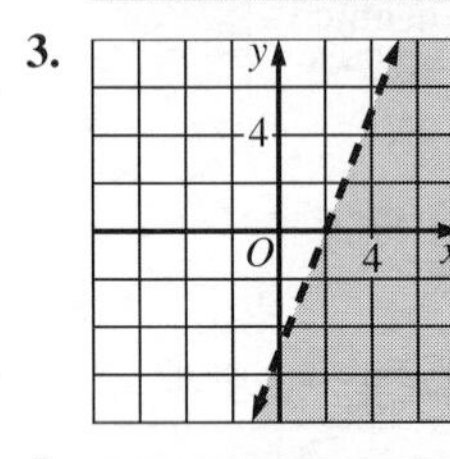

4. 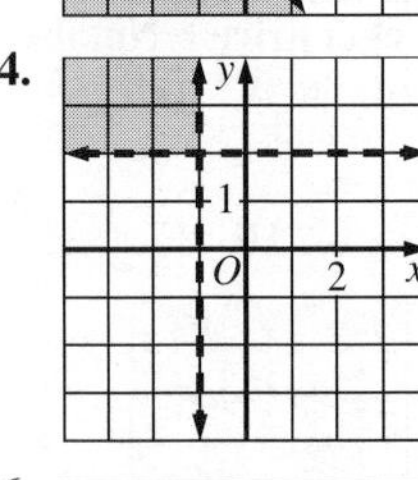

5. 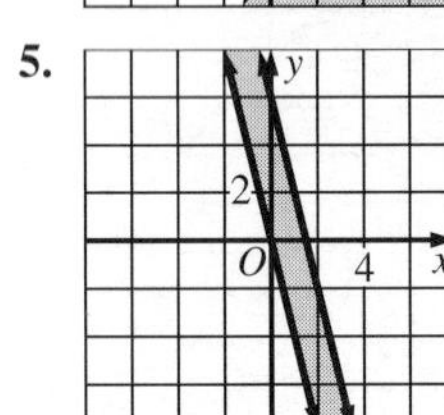

6. 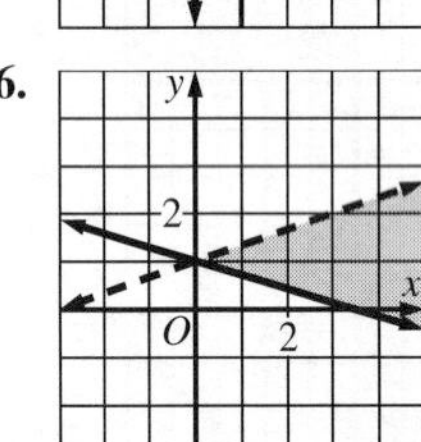

7. Let t = the number of T-shirts and s = the number of sweatshirts; $t \geq 10$; $s \geq 10$; $5t + 10s \geq 200$.

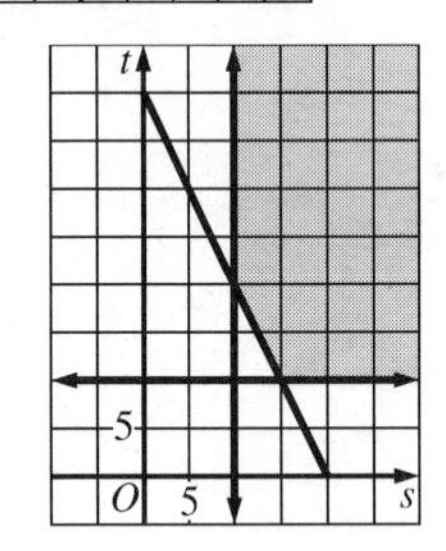

Pages 320–321 Chapter 7 Assessment

1. Any equation $ax + bx = c$, where a, b, and c are integers and a and b are not both zero is in standard form. For example, $2x + 5y = 7$ is in standard form. Any equation $y = mx + b$ is in slope-intercept form. For example, $y = \frac{3}{4}x - \frac{1}{8}$ is in slope-intercept form.

2. Answers may vary. Examples are given. $y \leq 2x - 1$; $y < 7$; $y \geq 3x + \frac{1}{5}$; $x > 5$ **3.** A horizontal intercept is on the x-axis and has coordinates $(x, 0)$ for some number x. A vertical intercept is on the y-axis and has coordinates $(0, y)$ for some number y.

4. 6; –8 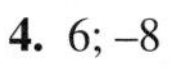

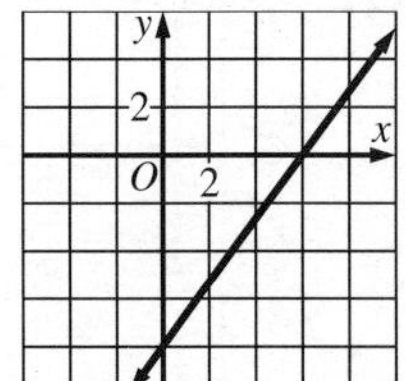

5. $8\frac{1}{2}$; $5\frac{2}{3}$

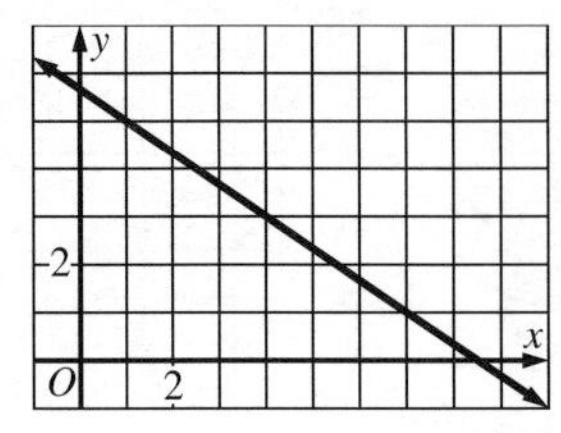

6. $y = \frac{5}{2}x - \frac{11}{2}$

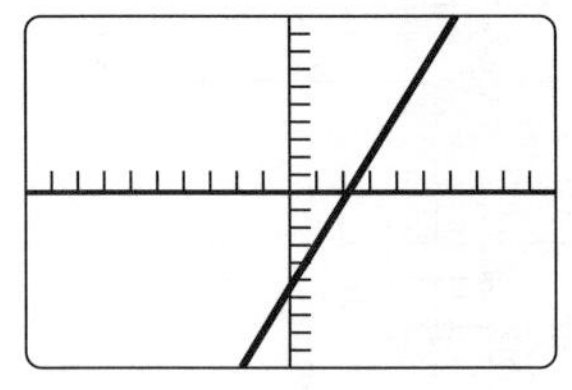

7. $y = \frac{2}{3}x - 4$

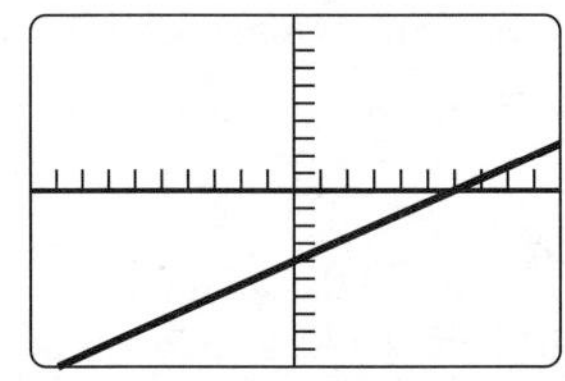

8. Let x = the number of pounds of Heavenly Brew and y = the number of pounds of Great Taste Tea.
a. $6x + 4y = 20$ **b.** $x + y = 4$ **c.** (2, 2); Anita buys two pounds of each kind of tea. **9.** (3, 4) **10.** $\left(1\frac{2}{3}, \frac{2}{3}\right)$
11. (3, 1) **12.** (1, –2) **13.** (1, 2) **14.** $\left(0, 3\frac{1}{3}\right)$
15. all (x, y) such that $2x - 3y = 11$ **16.** (–14, 11)
17. 250 children's tickets

19. 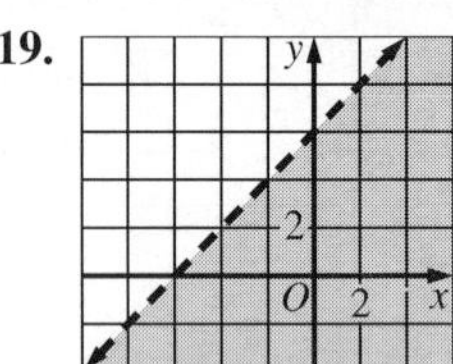

20.

21.

22. 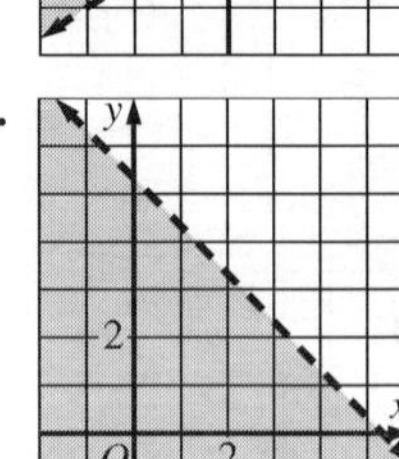

23. 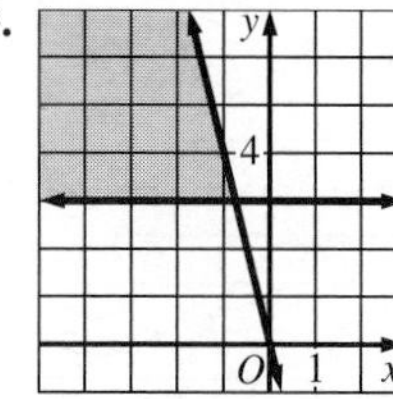

24. no solution

25. 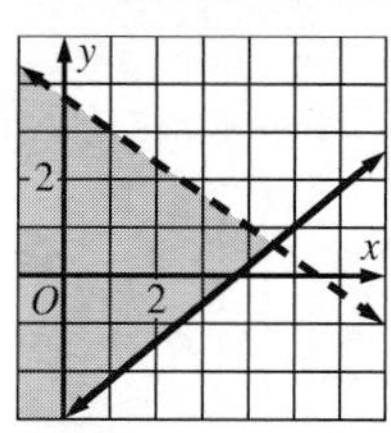

26. 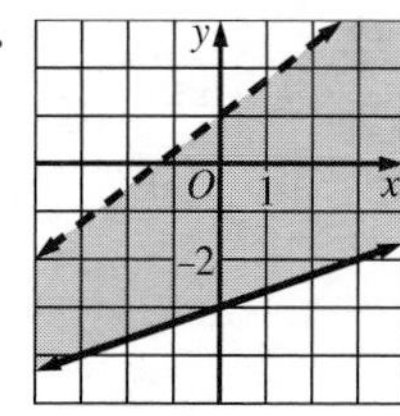

27. 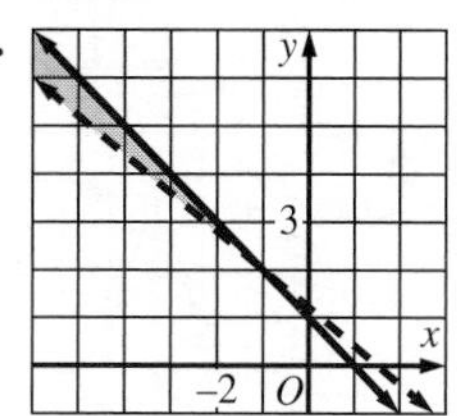

28. a. $x + y \geq 10$ **b.** $2x + 5y \leq 25$

c. 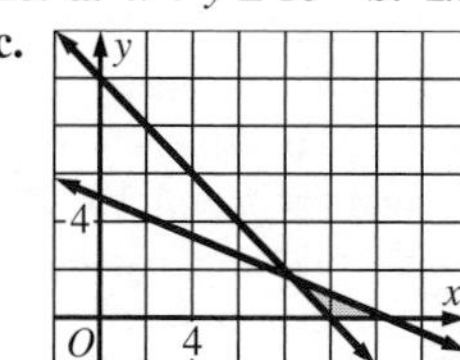

d. $1\frac{2}{3}$ lb; The greatest amount is the maximum y-value, which occurs where the graphs of $x + y = 10$ and $2x + 5y = 25$ intersect.

CHAPTER 8

Page 327 Checking Key Concepts

1. nonlinear **3.** linear

5. nonlinear

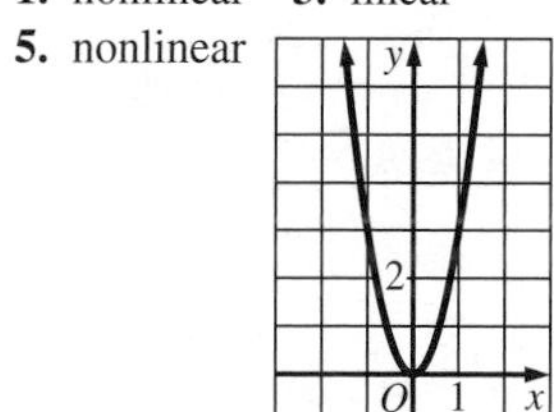

7. nonlinear; The graph is not a line.

Pages 328–330 Exercises and Applications

1. linear

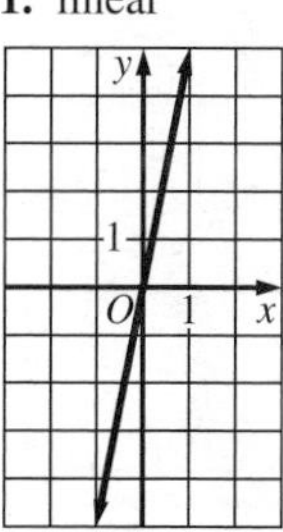

3. nonlinear

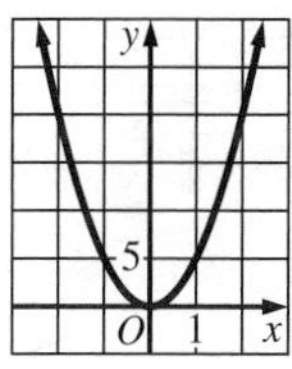

5. linear 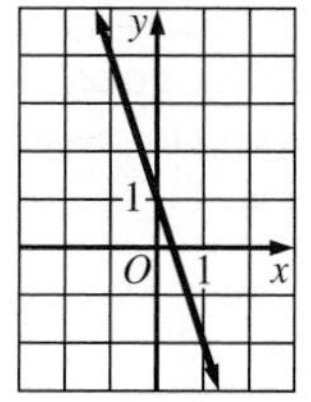

7. linear 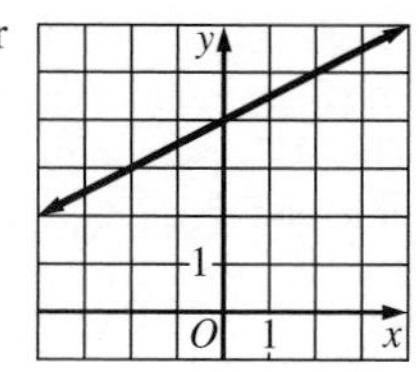

9. nonlinear 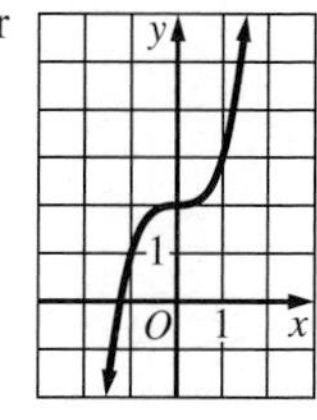

11. linear; The graph of the equation is a line; also, the equation is of the form $y = mx + b$, where $m = 2\pi$ and $b = 0$.

13. nonlinear; The graph of the equation is not a line; also, the equation cannot be written in the form $y = mx + b$.

15.

Ticket price (dollars)	Number of tickets sold	Income (dollars)
3.00	500	1500
3.50	450	1575
4.00	400	1600
4.50	350	1575
5.00	300	1500

17. nonlinear; The points on the scatter plot do not lie on a line.

19. 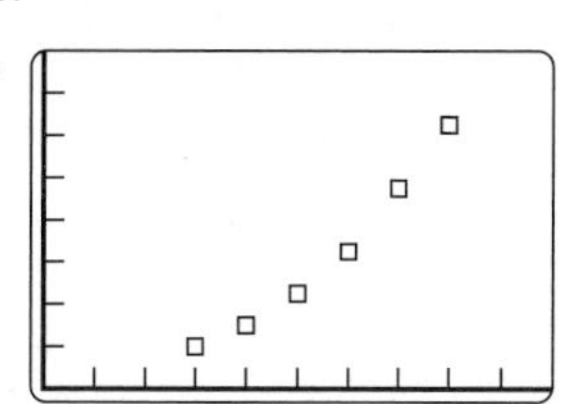

21. linear; Answers may vary. An example is given. In the scatter plot, numbers of servings is on the horizontal axis.

23. a. 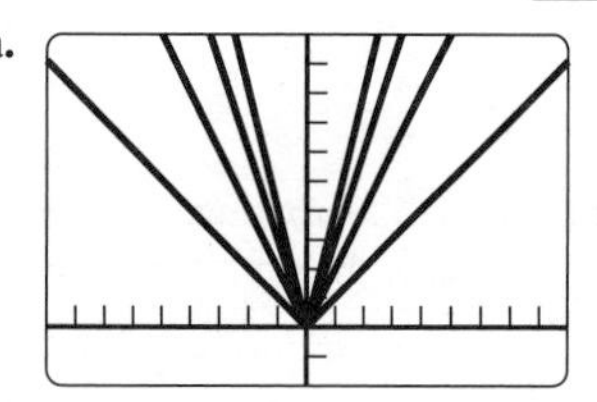

b. 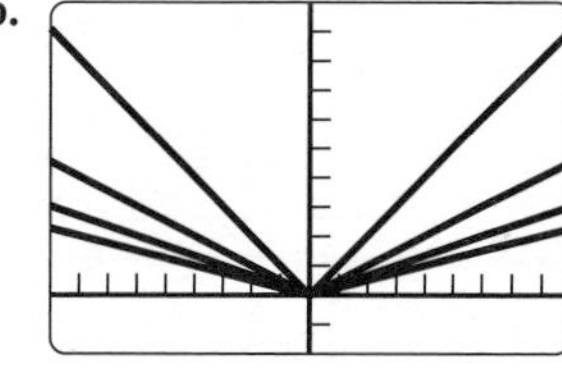

c. Answers may vary. An example is given. As the coefficient increases, the sides of the "V" become steeper. As the coefficient decreases toward 0, the sides of the "V" flatten out.

25. Graph $y = -2|x|$, $y = -3|x|$, $y = -0.5|x|$, and $y = -0.25|x|$ on your graphing calculator along with the graph of $y = |x|$. The "V" opens in the downward direction for negative values of a, and as $|a|$ increases, the sides of the "V" become steeper. **29.** $p^2 + 4p - 21$ **31.** $6r^2 + 21r + 15$
33. $2 + 2\sqrt{3} + \sqrt{5} + \sqrt{15}$

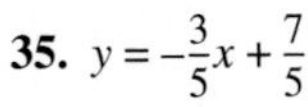

35. $y = -\frac{3}{5}x + \frac{7}{5}$ **37.** $y = -\frac{5}{7}x + \frac{3}{7}$

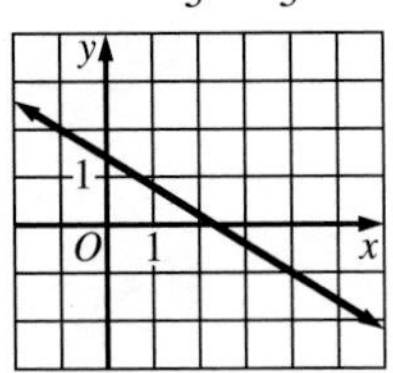

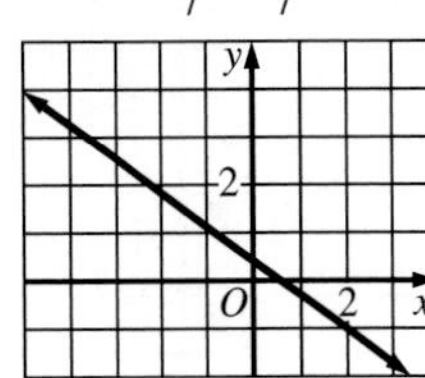

39. $y = \frac{3}{7}x - \frac{5}{7}$

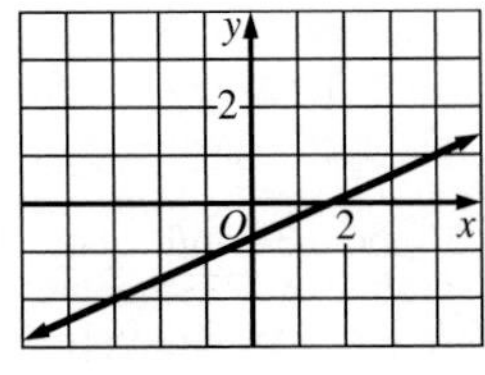

Page 333 Checking Key Concepts

1. narrower; up; The function reaches a minimum at the vertex. **3.** wider; up; The function reaches a minimum at the vertex. **5.** wider; up; The function reaches a minimum at the vertex.

7.

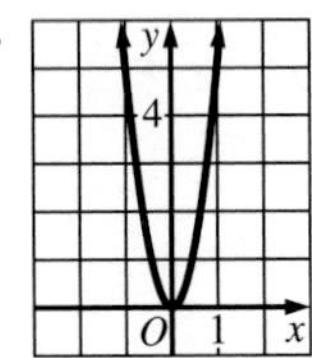

9.

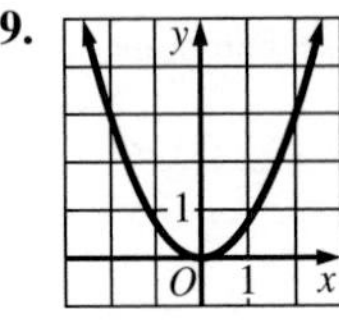

Pages 334–336 Exercises and Applications

1. C **3.** B **5.** Ex. 1: minimum; Ex. 2: maximum; Ex. 3: minimum; Ex. 4: minimum **7. a.** Since $6 > 1$, the graph will be narrower. Since $6 > 0$, the graph will also open up. **b.**

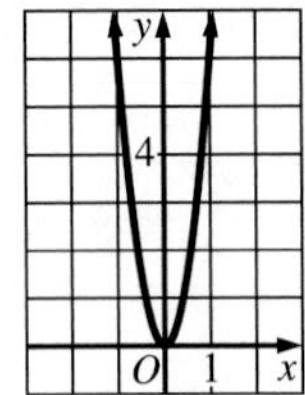

9. a. Since $|-10| > 1$, the graph will be narrower. Since $-10 < 0$, the graph will open down. **b.**

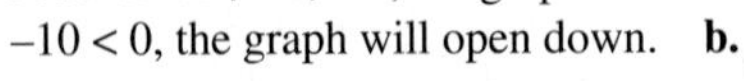

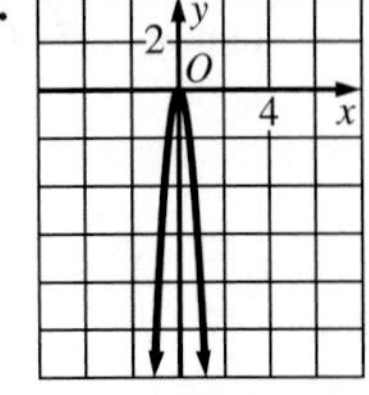

11. a. Since $\left|-\frac{1}{3}\right| < 1$, the graph will be wider. Since $-\frac{1}{3} < 0$, the graph will open down. **b.**

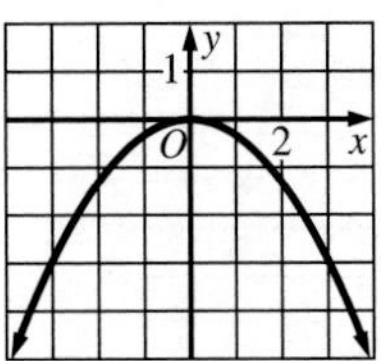

13. a. Since $0.5 < 1$, the graph will be wider. Since $0.5 > 0$, the graph will also open up. **b.**

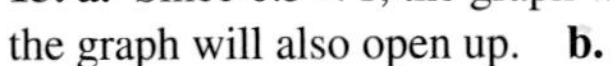

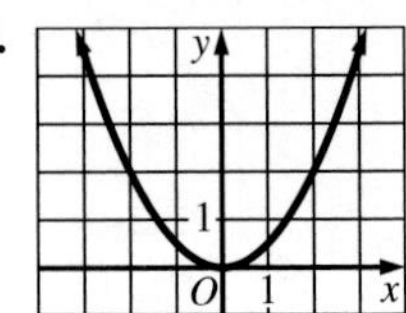

15. a. Since $|-2.5| > 1$, the graph will be narrower. Since $-1.25 < 0$, the graph will open down. **b.**

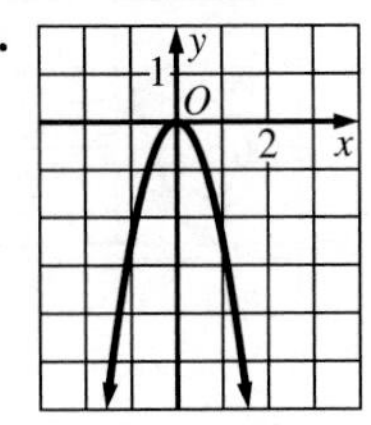

19. No; if the graph were folded along the y-axis, the two halves would not match exactly. **21.** C; The coefficient of x^2 is less than the coefficients in the equations in both Ex. 22 and Ex. 23, so the parabola is wider than both those parabolas. **23.** B; The coefficient of x^2 is greater than that in the equation in Ex. 21 and less than that in the equation in Ex. 22, so it is narrower than the first parabola and wider than the second. **25.** The area of each circle should be a rough approximation of πr^2. **27.** The graphs should be similar. **31.** $y^2 + 6y + 9$ **33.** $4p^2 + 4p + 1$ **35.** $x^2 - 36$ **37.** 500 **39.** 36

Page 340 Checking Key Concepts

1. 0 **3.** ± 4 **5.** ± 4 **7.** $\pm\sqrt{41}$, or about 6.4 and about -6.4 **9.** $\pm 2\sqrt{3}$, or about 3.46 and about -3.46

Pages 340–343 Exercises and Applications

1. ± 9 **3.** ± 3 **5.** ± 2 **7.** ± 12 **9.** ± 0.5 **11, 13.** Estimates may vary. Examples are given. **11.** about -1.75, about 1.75 **13.** about -4.5, about 4.5 **15.** about 30 knots
17. a. Vertices: (0, 0), (0, 1), (0, 2), (0, 3)

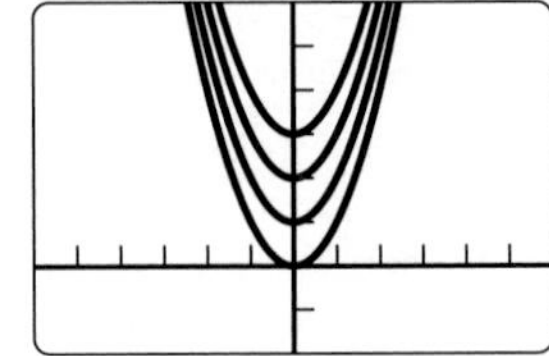

b. Vertices: (0, 0), (0, –1), (0, –2), (0, –3)

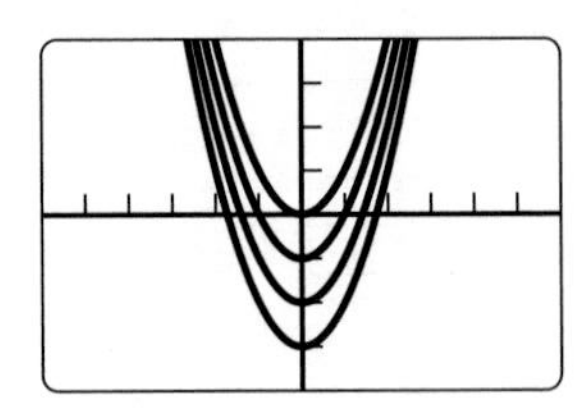

c. The graphs of $y = x^2 + 1$, $y = x^2 + 2$, and $y = x^2 + 3$ are the graph of $y = x^2$ shifted up 1, 2, and 3 units, respectively. The graphs of $y = x^2 - 1$, $y = x^2 - 2$, and $y = x^2 - 3$ are the graph of $y = x^2$ shifted down 1, 2, and 3 units, respectively.
19. Answers may vary. Examples are given. Graph equations using different values for c and compare the graphs to the original graph. Use both positive and negative values. The graph of $y = |x| + c$ is the graph of $y = |x|$ shifted up c units if c is positive, and shifted down $|c|$ units if c is negative.
21. ± 2 **23.** $\pm\sqrt{3}$, or about 1.73 and about -1.73 **25.** ± 2
27. $\pm\frac{4}{3}$ **29.** $\pm\sqrt{39}$, or about 6.24 and about -6.24

33.

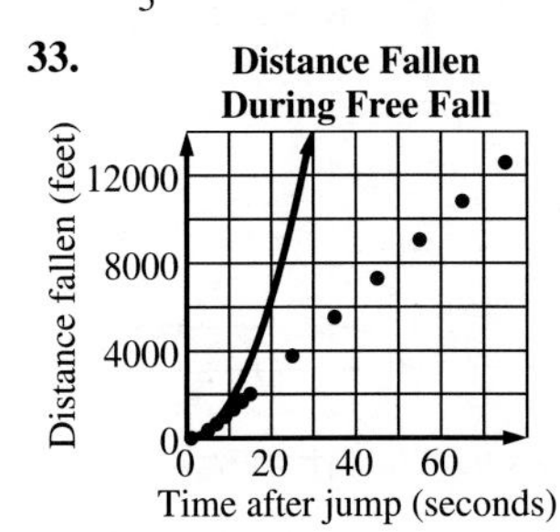

The scatter plot falls below the graph of the equation. That means the skydiver does not fall as far during each time interval as the equation would indicate.

37. $13\frac{1}{3}$ in. **39.** 128; 36

Page 343 Assess Your Progress

1. linear

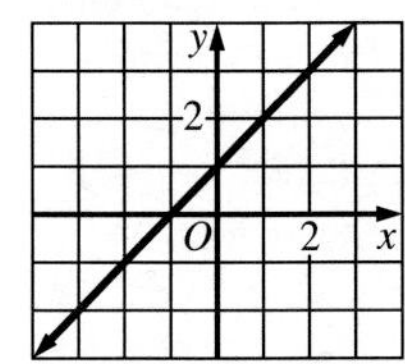

2. nonlinear

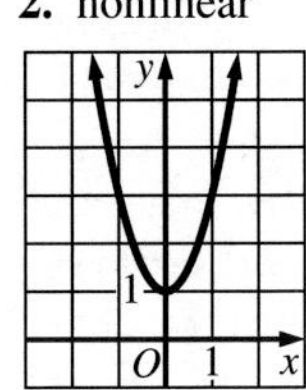

3. nonlinear

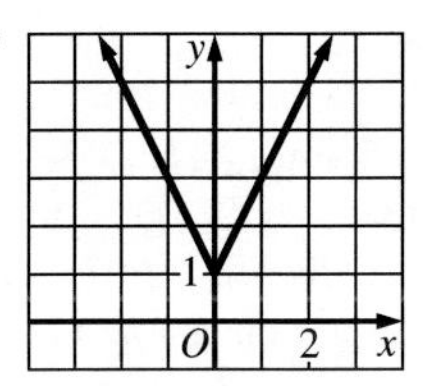

4. Since $7(0) = 0$, $7 > 0$, and $7 > 1$, the graph will have the same vertex, will open in the same direction but will be narrower and steeper.

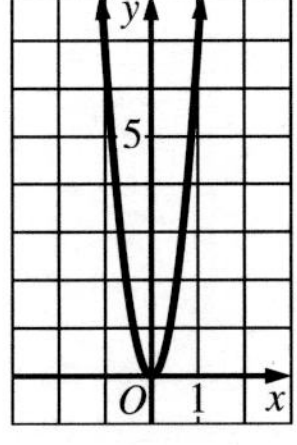

5. Since $-7(0) = 0$, $-7 < 0$, and $|-7| > 1$, the graph will have the same vertex, will open down, and will be narrower and steeper.

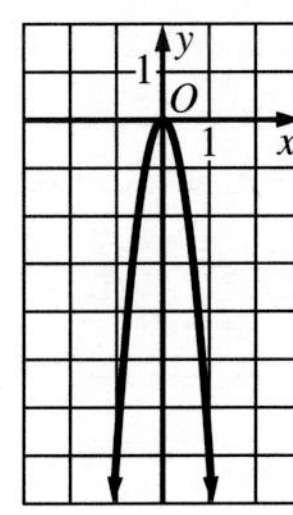

6. Since $0.7(0) = 0$, $0.7 < 1$, and $0.7 > 0$, the graph will have the same vertex, will open up but will be wider and flatter.

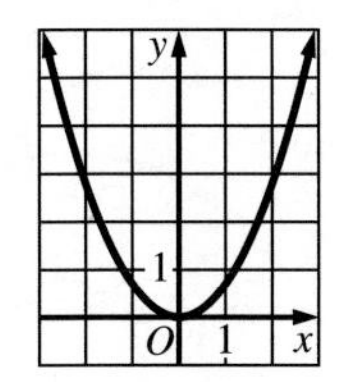

7. ± 12 **8.** ± 3 **9.** ± 2 **10.** ± 2 **11.** $\pm 2\sqrt{2}$, or about 2.83 and about -2.83 **12.** $\pm\sqrt{6}$, or about 2.45 and about -2.45

Page 346 Checking Key Concepts

1. $x^2 + 3x - 8 = 0$ **3.** $5x^2 + x - 14 = 0$ **5.** no solution
7. $1, -\frac{2}{3}$ **9.** 2

Pages 347–349 Exercises and Applications

1. 1, 3 **3.** about $-4\frac{1}{4}$, about $\frac{1}{4}$ **5, 7.** Estimates may vary. Examples are given. **5.** about 34 mi/h, about 44 mi/h
7. $\sqrt{\frac{15}{0.013}}$ or about 33.97 mi/h, $\sqrt{\frac{25}{0.013}}$ or about 43.85 mi/h
9. $-4, 2$ **11.** ± 3 **13.** 4 **15.** $-\frac{3}{2}$ **17, 19.** Estimates may vary. **17.** about -3.68, about 0.68 **19.** about -1.27, about 6.27 **21.** A **23.** about 122 km/h; Germany and some places in France **33.** $\frac{2}{3}$ **35.** $\frac{1}{999}$ **37.** $\frac{7}{3}$ **39.** $4\sqrt{3}$
41. $5\sqrt{2}$

Page 352 Checking Key Concepts

1. There is only one x-intercept so the equation has only one solution. **3.** $1, -3, -10$ **5.** 35, 3, 0 **7.** $4x^2 - 2x - 1 = 0$; $\frac{1 \pm \sqrt{5}}{4}$, or about 0.81 and about -0.31 **9.** $4.9t^2 - 7t = 0$; $0, \frac{10}{7}$

Pages 353–354 Exercises and Applications

1. $-3, 6$ **3.** $\frac{1}{3}, 3$ **5.** $2, -\frac{1}{5}$ **9.** 0, 4 **11.** 1, 2
13. $\frac{1 \pm \sqrt{19}}{6}$, or about 0.89 and about -0.56 **15.** $1, \frac{1}{4}$
17. about 20 ft **21.** 180 ft **23.** The ball achieves a height of 8 ft twice, once about 10 ft before reaching the goal, and again about 118 ft on the far side of the goal. **25.** irrational
27. irrational **29.** $(1, -1)$ **31.** $\left(\frac{1}{3}, \frac{1}{4}\right)$ **33.** $(11, 23)$

Page 357 Checking Key Concepts

1. 1; The equation has two solutions. **3.** 0; The equation has one solution. **5.** 0; The equation has one solution.
7. $a = -0.0069$, $b = 0.0549$, $c = 999.8266$
9. 999.8266 mg/cm^3

Pages 357–359 Exercises and Applications

1. two solutions **3.** two solutions **5.** two solutions
7. no solutions **9.** two solutions **11.** two solutions; about -8, about 208 (Note that while both are solutions of the equation and can be used to sketch its graph, only the positive solution makes sense as a horizontal distance.)

13. 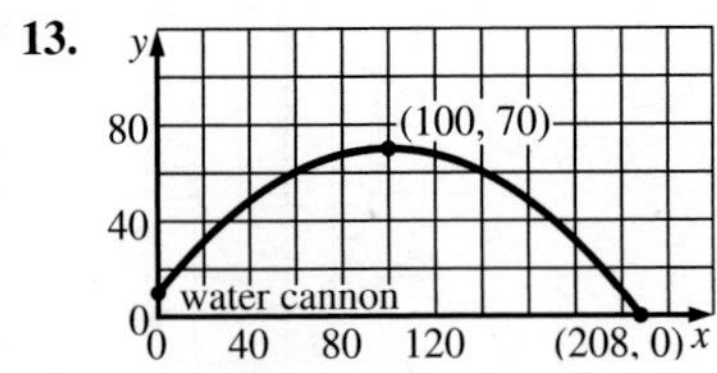

15. two solutions; –6, –1 **17.** one solution; 1
19. two solutions; $1 \pm \sqrt{3}$, or about –0.73 and about 2.73
21. one solution; 4 **23.** two solutions; $\frac{3 \pm 3\sqrt{2}}{2}$, or about –0.62 and about 3.62 **27.** B **31.** (–1, –4) **33.** ±2
35. ±1 **37.** $\pm\sqrt{5}$, or about –2.24 and about 2.24

Page 359 Assess Your Progress

1. about –0.7, about 5.7 **2.** 1, 2 **3.** no solutions **4.** –3, 8
5. $-3, \frac{5}{2}$ **6.** $\frac{1 \pm \sqrt{17}}{4}$, or about 1.28 and about –0.7807
7. two solutions **8.** no solutions **9.** one solution

Pages 364–365 Chapter 8 Assessment

1. Answers may vary. Examples are given; $y = 3x$; $y = 3x^2$.
2. Sketches may vary. An example is given; $y = x^2$.

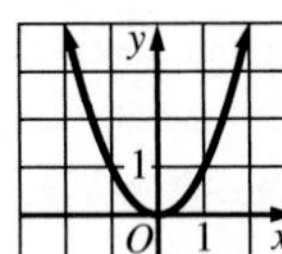

Symmetry is a correspondence in size, shape, and position of parts on opposite sides of a dividing line, plane, or point.

3. All quadratic equations must contain an x^2-term whose coefficient is not equal to zero. All graphs of quadratic equations are parabolas with a vertex and an axis of symmetry.
4. The discriminant is the value of $b^2 - 4ac$ in the quadratic formula. The discriminant gives you information about the number of solutions when you solve a quadratic equation.
5. nonlinear **6.** linear **7.** nonlinear **8.** nonlinear
9. linear **10.** The graph opens in the same direction, but it is narrower and steeper. **11.** The graph opens in the opposite direction and is narrower and steeper. **12.** The graph opens in the opposite direction and is wider and flatter.
13. C **14.** A **15.** B **16.** –4, 4 **17.** –5, 5 **18.** –3, 3
19. –4, 4 **20.** –6, 6 **21.** $\pm\sqrt{26}$, or about –5.1 and about 5.1 **22.** about –1.15, about 1.15 **23.** about –1.6, about 1.6 **24.** –0.5, 3 **25.** about –2.76, about –0.24
26. no solutions **27.** –11, –1 **28.** $\frac{1}{4}$ **29.** $\frac{3 \pm \sqrt{29}}{10}$, or about 0.84 and about –0.24 **30.** two solutions
31. no solutions **32.** one solution **33. a.** $4x^2 = 4x + 24$
b. two solutions

CHAPTER 9

Page 373 Checking Key Concepts

1. 5^4 **3.** $2^1 \cdot y^3$ **5.** $\frac{1}{343}$ **7.** 54 **9.** 512

Pages 373–375 Exercises and Applications

1. $(4.1)^5$ **3.** $\left(-\frac{3}{8}\right)^3$ **5.** $pq(4r)^4$ **7.** 64 **9.** 1331 **11.** $\frac{1}{32}$
13. 7.84 **15.** 6^7 **17.** y^{10} **19. a.** 390,625
b. 1829.19 **c.** –40,353,607 **25.** 106 **27.** –108
29. 250 **31.** 641 **33.** 25 **35.** –270

37.

Radius	Surface area	Volume
1	4π	$\frac{4}{3}\pi$
2	16π	$\frac{32}{3}\pi$
4	64π	$\frac{256}{3}\pi$
8	256π	$\frac{2048}{3}\pi$
n	$4\pi n^2$	$\frac{4}{3}\pi n^3$
$2n$	$16\pi n^2$	$\frac{32}{3}\pi n^3$

45. Answers may vary. An example is given. the total price of 4.5 lb of a commodity that cost p dollars per pound; $9
47. $(r + 5)^2 = r^2 + 10r + 25$; 9 = 9
49. $(7 - 3r)^2 = 49 - 42r + 9r^2$; 169 = 169

Page 378 Checking Key Concepts

1. Answer is given to the nearest hundredth. 773.18
3. 14,592 **5.** the growth rate of the number of cellular telephone subscribers after 1983

Pages 379–382 Exercises and Applications

1, 3, 5, 7, 9. When necessary, answers are rounded to the nearest hundredth. **1.** 2430 **3.** 32,659.2 **5.** 29.48
7. 722.11 **9.** 7,100,000 **11, 13, 15.** Answers are obtained from the equation $y = a(1 + r)^x$, where y = the population in thousands, x = the number of years after 1980, a is the initial population, and r is the growth rate. **11. a.** 7,549,000
b. 0.002 **c.** 7,686,000 **13. a.** 4,780,000 **b.** 0.004
c. 4,955,000 **15. a.** 56,451,000 **b.** 0.002 **c.** 57,475,000
17. a. $y = 200(1.035)^x$, where y = the amount in dollars and x = the number of years from now. **b.** $237.54

23. a–c. The graphs all show exponential growth with the same growth factor but different initial values.

29. a. $y = 1653(1.04)^x$, where y is the number of vehicle-miles in billions x years after 1983. **b.** about 2092 billion vehicle-miles to the nearest billion; $x = 6$ since 1989 is 6 years after 1983. **33.** $\left(0, -\frac{11}{7}\right)$ **35.** (1, 3) **37.** $\left(1, -\frac{1}{2}\right)$
39. $\frac{1}{2} \pm \frac{\sqrt{5}}{2}$ **41.** $-1; 3\frac{1}{2}$ **43.** $\frac{1}{4} \pm \frac{\sqrt{5}}{4}$

Page 385 Checking Key Concepts

Where necessary, answers are rounded to the nearest hundredth. **1.** 82.71 **3.** $\frac{4}{3}$ **5.** 5%; The decay factor, $1 - r$, is 0.95, so $r = 0.05$, or 5%.

Pages 386–389 Exercises and Applications

1. 0.29 **3.** 5970.74 **5.** 1351.77 **7.** 0.75 **9.** 262.14 **11.** $y = 100(0.83)^x$, where y = the number of grams remaining and x = the number of hours after 3:00 P.M. **15.** exponential growth; The value of y increases exponentially as x increases. **17. a.** decays **b.** 14 **c.** 0.23 **19. a.** grows **b.** 0.25 **c.** 1.88 **21. a.** grows **b.** 0.99 **c.** 1.01 **23.** 133.12 mg **27. a.** about 0.46 h **b.** about 1.83 h **c.** about 7.37 h **29.** $n = 2^x$ **33.** $\frac{3}{5}$ **35.** $5\frac{3}{4}$ **37.** $-\frac{3}{2}$; -6 **39.** $-\frac{7}{3}$; 49 **41.** 8; 64

Page 389 Assess Your Progress

1. 64 **2.** 64 **3.** $\frac{1}{625}$ **4.** 80 **5.** 343 **6.** 42 **7.** -35 **8.** -59 **9.** 0 **10. a.** $y = 250(1.05)^x$ **b.** \$289.41

11. a. $y = 591(0.98)^x$, where y = fuel consumption in gallons and x = the number of years since 1980.

b.

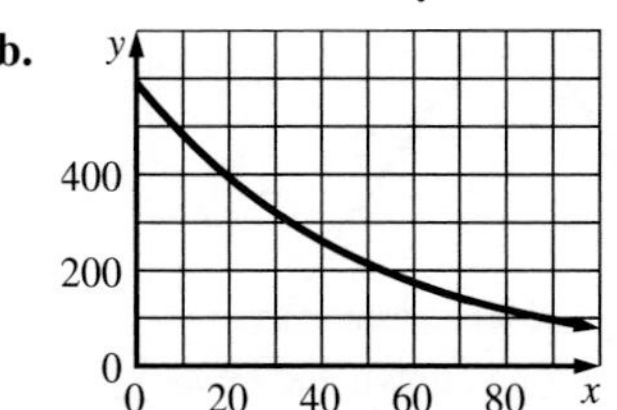

about 523.5 gal per car

Page 393 Checking Key Concepts

1. $\frac{1}{121}$ **3.** $\frac{1}{49}$ **5.** $\frac{7}{m^3}$ **7.** $\frac{1}{a^4c^6}$ **9.** 1 **11.** $-\frac{1}{8}$

Pages 393–395 Exercises and Applications

1. 1 **3.** $\frac{1}{32}$ **5.** 1 **7.** $\frac{81}{16}$ **9.** $\frac{1}{a^2}$ **11.** $\frac{4}{b^6}$ **13.** $\frac{b^3}{a^3}$ **15.** $\frac{2}{w^8}$ **17.** $\frac{1}{y^2z^2}$ **19.** $\frac{1}{4}$ **21.** $\frac{1}{16}$ **23.** $\frac{53}{27}$ **25.** $\frac{1}{9}$ **31. a.** \$502.53 **b.** \$349.80 **c.** \$273.74 **d.** \$228.34 **35.** A **39.** $m \geq -3$ **41.** $q > -\frac{14}{3}$ **43.** $y \geq \frac{1}{6}$

Page 398 Checking Key Concepts

1. 600,000 **3.** 5,030,000 **5.** 0.000772 **7.** 8×10^6 **9.** 3.40056×10^{10} **11.** 7.4×10^{-4}

Pages 398–400 Exercises and Applications

1. 2000 **3.** 4,900,000 **5.** 676,100,000 **7.** 7.8×10^2 **9.** 2.1×10^4 **11.** 9.7×10^{12} **13.** 3.2×10^6 **15.** 8.023×10^{-2} **19.** 7 to 12 microns; $0.000007 = 7 \times 10^{-6}$ and $0.000012 = 1.2 \times 10^{-5} = 12 \times 10^{-6}$ **23.** 9.54×10^{-7} **25.** 3.6288×10^{11} ft³ **29.** B **33.** 1296 **35.** -125 **37.** 16

Page 400 Assess Your Progress

1. 1 **2.** $\frac{1}{1024}$ **3.** $\frac{27}{64}$ **4.** $\frac{1}{32}$ **5.** $\frac{9}{16}$ **6.** 4 **7. a.** about 13.81 million **b.** 14.6 million **c.** about 13.06 million **8.** 7,003,000,000 **9.** 0.0000421 **10.** 0.000008 **11.** 5.4×10^{10} **12.** 3.28×10^0 **13.** 7.09×10^{-6}

Page 403 Checking Key Concepts

1. 4^8 **3.** 2^0 **5.** t^4 **7.** $\frac{2}{3}$ **9.** $15p^6q^6$

Pages 404–405 Exercises and Applications

1. 2^7 **3.** 3^6 **5.** 9^{13} **7.** a^5 **9.** $\frac{1}{c^7}$ **11.** $40d^{10}$ **13.** p^2q^2 **15.** $\frac{b}{3a^6}$ **17.** 6750 people

19.

Country	Population	Area (square miles)
China	1.192×10^9	3.696×10^6
France	5.8×10^7	2.10×10^5
Iran	6.0×10^7	6.32×10^5
United States	2.61×10^8	3.679×10^6

21. a. 1 light-year = $\frac{186{,}282 \text{ mi}}{\text{s}} \cdot \frac{60 \text{ s}}{\text{min}} \cdot \frac{60 \text{ min}}{\text{h}} \cdot \frac{24 \text{ h}}{\text{d}} \cdot \frac{365.25 \text{ d}}{\text{y}} \cdot 1 \text{ y} \approx 5.88 \times 10^{12}$ mi **b.** 2.352×10^{16} mi **23.** 1 **25.** $\frac{343}{64}$ **27.** $-\frac{2048}{177{,}147}$

Page 408 Checking Key Concepts

1. 1296 **3.** 64 **5.** $\frac{49}{64}$ **7.** a^9b^9 **9.** z^{16} **11.** $25c^2$

Pages 409–411 Exercises and Applications

1. 9×10^8 **3.** 6.25×10^{30} **5.** $\frac{64}{343}$ **7.** a^5b^5 **9.** $\frac{16a^6c^{12}d^8}{b^6}$ **11.** $81a^4b^4$ **13.** $\frac{q^{35}}{p^{14}}$ **15.** $\frac{b^{30}c^5}{d^{15}}$

21. a. 28.544 g **b.** Let x = length of short snake. Then $2x$ = length of longer snake. Weight of longer snake = $446(2x)^3 = 446 \cdot 8x^3 = 8(446x^3) = 8$(weight of shorter snake). **23.** B **25.** $-14u - 9v$ **27.** $6 - m$ **29.** $3q + 5$ **31.** $f = 36$ **33.** $p = 42\frac{6}{7}$ **35.** $x = 2$

Page 411 Assess Your Progress

1. a^{16} **2.** b^4 **3.** $\frac{a^7}{b^4}$ **4.** a^4b^4 **5.** c^{56} **6.** $\frac{bc^{10}}{d^{10}}$ **7.** $\frac{p^6q^{18}}{r^{12}}$ **8.** $\frac{x^8}{y^8}$ **9.** about 1.53×10^{12} mi **10.** 4; Let n_A = the number of seconds that Rock *A* has been falling; then $2n_A$ = the time in seconds that Rock *B* has been falling. Then the distance traveled by Rock $B = 16(2n_A)^2 = 16(2^2)(n_A^2) = 16(4)n_A^2 = 4(16n_A^2) = 4$ times the distance traveled by Rock *A*.

Pages 416–417 Chapter 9 Assessment

1. base; exponent; power **2.** exponential growth; growth rate; growth factor **3.** exponential decay; decay factor **5.** 729 **6.** 81 **7.** $\frac{1}{1024}$ **8.** 0.125 **9.** 2^3; 8 **10.** 3^2; 9

11. -79 **12.** -200 **13.** 8 **14. a.** $y = 2000(1.075)^x$ **b.** \$36,088.48 **15. a.** $y = 13{,}000(0.88)^x$ **b.** \$3620.51

c. 

(0, 13,000); the value of the car when it was new

16. 1 **17.** $\frac{1}{32}$ **18.** $\frac{1}{729}$ **19.** $\frac{125}{8}$ **20.** $-\frac{1}{125}$ **21.** $\frac{25}{8}$ **22.** $-\frac{5}{6}$ **23.** $\frac{103}{4}$ **24. a.** about 1.15 g **b.** 2.1 g **c.** about 0.057 g **25.** 45,700,000 **26.** 0.0000038 **27.** 70,230 **28.** 6.78×10^{-9} **29.** 2.3×10^{10} **30.** 5.0004×10^{0} **32.** $\frac{m^2}{d^3}$ **33.** $16b^8$ **34.** $-6c^7d^{15}$ **35.** $\frac{yz}{x^3}$ **36.** $-\frac{8a^3}{27b^3}$ **37.** $\frac{s^{12}}{t^4}$ **38.** about \$10.44

CHAPTERS 7–9

Pages 418–419 Cumulative Assessment

1. horizontal intercept: -2; vertical intercept: 10; $y = 5x + 10$ **3.** many solutions **5.** Answers may vary. Methods may be graphical or algebraic. **7.** $2.5i + 2m = 20$

9. 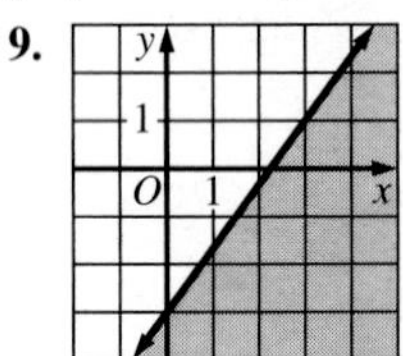

11. 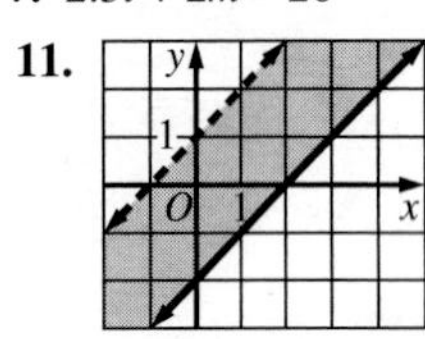

13. 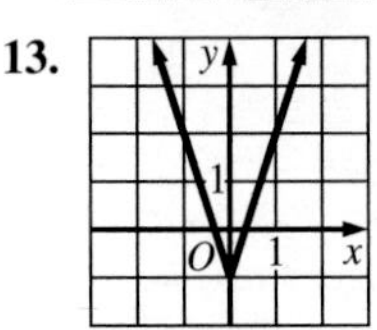

nonlinear

15. ± 20 **17.** C **19.** B **21.** about -0.47 and about 2.14; $\frac{-5 \pm \sqrt{61}}{-6}$ **25.** -8 **27.** $a^{14}b^6$ **29.** $128m^{11}$ **31.** $\frac{32r^5}{243s^5}$ **33.** $y = 14{,}952{,}000(1.006)^x$ **35.** 15,874,000

CHAPTER 10

Page 426 Checking Key Concepts

1. Yes. **3.** No; the exponent of n^{-3} is not a whole number. **5.** Yes. **7.** $5x^2 - 2x + 1$; quadratic **9.** $-x^4 + 1$; none of these **11.** $-\frac{3}{4}h + 14$; linear **13.** $6x^2 - 9$ **15.** $8x - 11xy + 2y$

Pages 426–428 Exercises and Applications

1. E **3.** A **5.** B **7.** $6x^2 - x - 4$; 2 **9.** $2r^3 - 3r^2 - 2r + 3$; 3 **11.** $17n^2 + 2n - 3$ **13.** $y^2 - 4y - 7$ **15.** $-a - 4b + 7$ **17.** $2b^3 + b^2 - 6b + 1$ **19.** $x^2 + 8x - 8$ **21.** 30 cm, 40 cm, and 50 cm **25.** $c^2 + 3c + 4$ **27.** $4z^3 - 8z^2 - 8z + 8$ **29.** $-8m^2 - 5m + 4$ **31.** $-2x^2 + x + 7$ **33.** $-3y^3 - y^2 + 9$ **35.** $-a^2 + 12ab - 18b^2$ **39.** $x^2 - 1$ **41, 43.** Estimates may vary. Examples are given. **41.** about 1.4, about -1.4 **43.** about 2.1, about -2.1

Page 432 Checking Key Concepts

1. $x, x; x, 8; 4, x; 4, 8$ **3.** $2n, n; 2n, -8; 4, n; 4, -8$ **5.** $4ab$ **7.** $m^2 + 9m + 20$ **9.** $k^3 + 2k^2 - 24k$ **11.** $3y^3 - 11y^2 + 4y + 4$

Pages 433–435 Exercises and Applications

1. $c^2 + 12c + 35$ **3.** $s^2 - 6s + 8$ **5.** $12a^2 + 38a + 6$ **7.** $n^3 + 10n$ **9.** $45b^3 + 30b$ **11.** $-z^3 + 5z^2 - 4z$ **15. a.** The width is $(5x + 6y + 20)$ in. The length is $(6x + 7y + 20)$ in. **b.** The perimeter is $(22x + 26y + 80)$ in. The area is $(30x^2 + 220x + 71xy + 260y + 42y^2 + 400)$ in^2. **17.** Substituting $x = 12$ and $y = 3$ in the expressions for perimeter and area gives $P = 422$ and $A = 11{,}074$. Since $P = 2(98) + 2(113) = 422$ and $A = (98)(113) = 11{,}074$, the polynomials check. **19.** $18x^2 + 15x + 2$ **21.** $3y^2 + 10y + 3$ **23.** $6n^3 - 10n^2 - 15n - 27$ **25.** $5x^3 - 16x^2 - 47x + 10$ **27.** $42j^3 - 13j^2 + 25j - 4$ **29.** $6x^3 + 23x^2 - 30x + 8$ **33.** length $= (4 - 2x)$ in.; width $= (3 - 2x)$ in.; height $= x$ in. **35.** $y = 4x^3 - 14x^2 + 12x$; The maximum volume, y, is created when x is about 0.57 in.

39. (2, 3) **41.** The graph will open down and be narrower than the graph of $y = x^2$. The prediction results from the fact that $-4 < 0$ and $|-4| > 1$. **43.** The graph will open down and be narrower than the graph of $y = x^2$. The prediction results from the fact that $-1.2 < 0$ and $|-1.2| > 1$.

Page 435 Assess Your Progress

1. No. The term $\frac{-1}{x}$, or $-1x^{-1}$, is not a monomial with a whole-number exponent. **2.** $5x^2 - 5x - 3$; 2 **3.** $-3y^2 + 7y - 3$; 2 **4.** $2a$; 1 **5.** $z^3 - 2z^2 + 7z + 3$; 3 **6.** $8y^2 - 34y + 35$ **7.** $7n^3 + 21n^2 - 56n$ **8.** $5j^3 - 17j^2 + 18j + 18$ **9.** Methods may vary. Examples are given. Subtract the area of the center square from the total area: $(x + 24)^2 - x^2 = 48x + 576$. Divide the border into eight parts—four that are 12 by x and four that are 12 by 12. The add the areas: $4(12x) + 4(12^2) = 48x + 576$.

Page 439 Checking Key Concepts

1. x **3.** $2y$ **5.** $7x(2x + 1)$ **7.** $15n(2n - 3)$ **9.** $3t(t^2 - 2t + 3)$ **11.** $\frac{2}{5}, -\frac{5}{2}$ **13.** $0, -\frac{3}{4}$

Pages 439–441 Exercises and Applications

1. $x(x - 4)$ **3.** $4(y^2 + 1)$ **5.** $2x(x + 3)$ **7.** $5n(3n + 2)$ **9.** $a(a^2 - 4a + 1)$ **11.** $2x(27x^2 + 9x + 1)$ **15.** 0, 2 **17.** 0, 4 **19.** $0, \frac{1}{3}$ **21.** $-4, 7$ **23.** $-\frac{3}{4}, -\frac{1}{2}$

25. As the price increases from \$0 to \$1500, net sales increase from \$0 to \$225 million. As the price continues to increase to \$3000, net sales decrease back to \$0. **29.** $0, -\frac{7}{3}$ **31.** 0, –1 **33.** 0, 5 **35.** 0, 2 **39. a.** 2 **b.** –4 **c.** –5 **d.** 3 **41.** The graph is above the x-axis except that its vertex is $(-k, 0)$. The new equation is shifted $|k|$ units to the left if k is positive and $|k|$ units to the right if k is negative. **43.** The graph of $y = |x - 2|$ is obtained by shifting the graph of $y = |x|$ two units to the right. The graph of $y = |x + 2|$ is obtained by shifting the graph of $y = |x|$ two units to the left. The graph of $y = |x + k|$ is the graph of $y = |x|$ shifted horizontally so that its vertex is $(-k, 0)$. **45.** 1 **47.** 1 **49.** $\frac{121}{25}$, or $4\frac{21}{25}$ **51.** $3x^2 + 7x + 2$ **53.** $2z^2 - 17z + 21$ **55.** $4a^2 - 1$

Page 445 Checking Key Concepts

1. 10, 1; –10, –1; 2, 5; –2, –5; 2 + 5 = 7 **3.** 12, 1; –12, –1; 6, 2; –6, –2; 4, 3; –4, –3; 4 + 3 = 7 **5.** $(y - 2)^2$ **7.** $(a + 9)(a + 2)$ **9.** $(3x - 2)(x - 4)$ **11.** $(5p + 3)^2$ **13.** The factors of 4 are 1 and 4, –1 and –4, 2 and 2, or –2 and –2. None of these pairs add to –3.

Pages 445–446 Exercises and Applications

1. $(x + 3)(x + 1)$ **3.** $(k - 4)(k - 2)$ **5.** $(x + 2)(x + 10)$ **7.** $(y - 9)(y - 4)$ **9.** $(s - 5)(s - 4)$ **11.** $(t - 8)(t - 3)$ **15.** Area $= 2x^2 + 9x + 10 = (x + 2)(2x + 5)$ **17.** $(5k + 1)(k + 1)$ **19.** $(3n + 2)(n + 2)$ **21.** $(4a - 9)(a - 1)$ **23.** $(5x - 8)(x - 1)$ **25.** $(2g - 1)(5g - 3)$ **27.** $(3p - 4)(3p - 2)$ **31.** \$1000 is the amount invested. The growth factor is $(1 + r)$. At the end of year one, the value is $1000(1 + r)$. At the end of year two, the value is $1000(1 + r) \cdot (1 + r)$ or $1000(1 + r)^2$. **33. a.** Answers may vary. The population of persons 65 and over will increase substantially between 1995 and 2040. **b.** Estimates may vary. An example is given: In 50 years, the investment will be worth $5000(1 + 0.07)^{50}$, or about \$147,285. **41.** ±7 **43.** ±15 **45.** $\pm\frac{3}{5}$ **47.** –3, –4 **49.** $\frac{5 \pm \sqrt{33}}{4}$, or about 2.69 and about –0.19 **51.** 3, –1

Page 450 Checking Key Concepts

1. No. The factors of $2x^2$ are $2x$ and x; the factors of 1 are 1 and 1. To get the right quadratic and constant terms, the product would have to be $(2x + 1)(x + 1)$. But this product gives the wrong linear term. **3.** B; $(3y - 1)(y - 2)$ **5.** A; $(2r + 3)^2$ **7.** C; $(p + 10)(p - 10)$ **9.** Methods may vary. An example is given. $2 \pm \sqrt{3}$, or about 3.73 and about 0.27; The polynomial is not factorable, so use the quadratic formula.

Pages 451–453 Exercises and Applications

1. +; – **3.** –; – **5.** $(t - 8)(t + 1)$ **7.** $(z - 9)(z + 4)$ **9.** $(y + 6)(y - 6)$ **11.** $(2d + 11)(d - 1)$ **13.** $(2x + 3)(x - 2)$ **15.** $(2y - 1)(3y - 4)$ **17.** C **19.** 1, –1, 11, –11, 19, –19, 41, –41 **23.** 2, –1 **25.** 4, –1.5 **27.** $\frac{2}{3}, -3$ **29.** $\frac{2}{3}, -\frac{2}{3}$ **31. a.** 0.75 s after release **b.** No. Explanations may vary. An example is given. The graph of the equation shows that 10 ft is the maximum height. **33.** $\frac{3 \pm \sqrt{7}}{3}$, or about 2.82 and about 0.18 **35.** $\frac{1}{2}, -3$ **37.** $3, -\frac{2}{3}$ **39.** about 4.43 s; 17.30 ft; Yes, it could clear the centerfield wall or the right-field wall.

47, 49, 51.

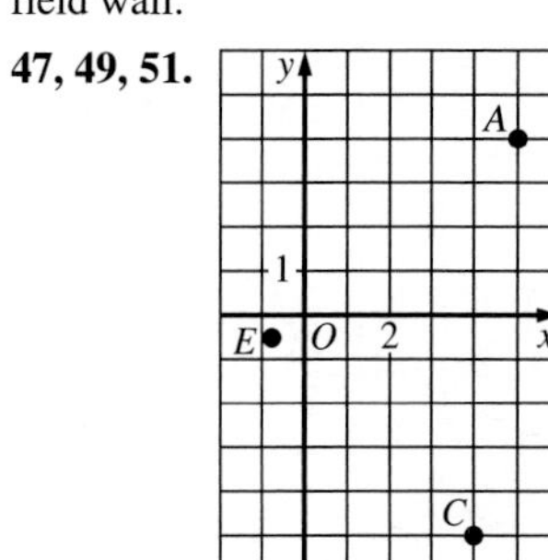

47. I **49.** IV **51.** III

53. Yes; the points lie on a straight line that passes through the origin.

Page 453 Assess Your Progress

1. $\frac{3}{2}, -\frac{5}{4}$ **2.** 0, 3 **3.** 0, –4 **4.** 0, –20; The car must be traveling at 0 mi/h in order to stop in 0 ft. The negative solution does not make sense in this real-world problem. Speed cannot be negative. **5.** $(x + 2)(x + 5)$ **6.** $(k + 7)(k - 4)$ **7.** $(k - 14)(k + 2)$ **8.** $(2a - 1)(a - 5)$ **9.** $(5z + 1)^2$ **10.** $(2m - 1)(2m - 49)$ **11.** $(3c - 2)(3c - 5)$ **12.** $(9t + 8)(9t - 8)$ **13.** $(9d - 10)(d + 1)$ **14.** 4, –7 **15.** $-2, -\frac{7}{4}$ **16.** $\frac{1}{3}, -\frac{1}{8}$ **17.** 10 mi/h

Pages 458–459 Chapter 10 Assessment

1. Answers may vary. An example is given. $2x^3 - 4x + 6$ **2.** Answers may vary. Examples are given. linear: $4x - 6$; quadratic: $x^2 - 2$; cubic: $4x^3 + 8x^2 + x - 5$ **3.** To factor a polynomial means to write it as a product of two or more polynomials. **4.** Yes. **5.** Yes. **6.** Yes. **7.** Yes. **8.** No; the power of x is not a whole number. **9.** No; the power of x in the term $-\frac{2}{x^2} = -2x^{-2}$ is not a whole number. **10.** $8x + 10$; 1 **11.** $7a^2 - a + 2$; 2 **12.** $n^3 + 6n^2 - 4n + 4$; 3 **13.** $8x^2 + 2x + 11$ **14.** $4r^2 - 14r + 2$ **15.** $6x^3 - 14x^2 - 10x + 14$ **16.** $3b - 2c - 5$ **17.** $x^2 - 7x + 12$ **18.** $b^2 - 4b - 45$ **19.** $5r^2 - 21r + 18$ **20.** $6c^2 + 22c + 16$ **21.** $6a^3 + 27a$ **22.** $-d^3 + 3d^2 + 6d$ **23.** $2h^3 - h^2 - 16h + 42$ **24.** $4x^3 - 29x^2 - 22x - 16$ **25.** $5x^2 - 4x$ **26.** $y(13y - 4)$ **27.** $2a(3a - 5)$ **28.** $2x(3x^2 + 2x - 4)$ **29.** 0, 7 **30.** $0, -\frac{1}{4}$ **31.** $1, -\frac{8}{5}$ **32.** 0, 5 **33.** 0, –1 **34.** 0, 2 **35.** 2 in. **36.** $(x + 10)(x + 1)$ **37.** $(y - 8)(y - 3)$ **38.** not factorable **39.** $(q + 7)(q - 1)$ **40.** $(y - 8)(y - 2)$ **41.** $(a + 4)(a - 3)$ **42.** $(m + 3)(m - 3)$ **43.** $(x - 1)(2x - 1)$ **44.** $(2n + 1)(2n + 5)$ **45.** $-1, \frac{11}{3}$ **46.** $-\frac{8}{7}, 1$ **47.** $-\frac{5}{4}, 2$

CHAPTER 11

Page 466 Checking Key Concepts

1. inverse variation; $y = \frac{60}{x}$ **3.** neither

5. $y = -\frac{150}{x}$ or $xy = -150$

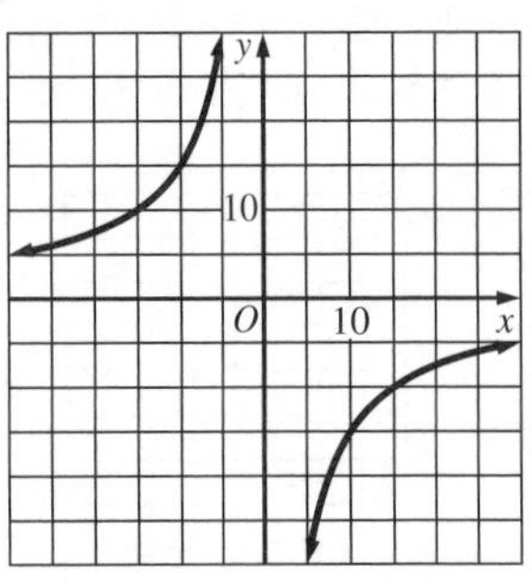

Pages 466–468 Exercises and Applications

1. No **3.** Yes. 12

5.

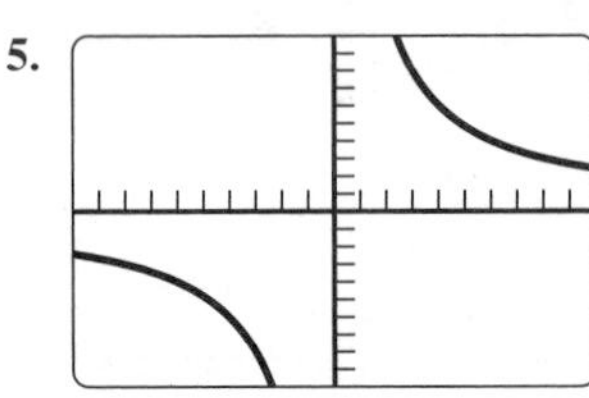

Yes; the product of x and y is constant.

7.

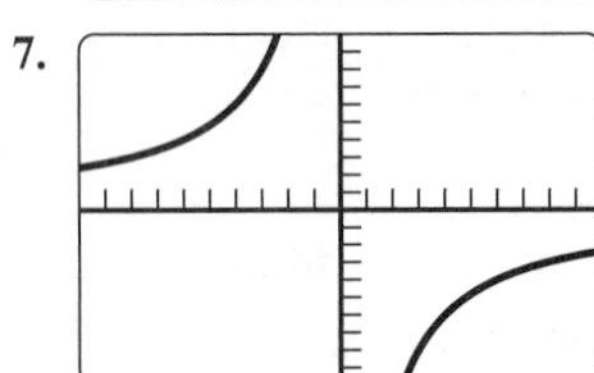

Yes; the equation has the form $y = \frac{k}{x}$.

9. B **11.** 500 ft; 2000 ft **15.** $g = \frac{1404.2}{n}$ **17.** 50.15

19. Answers may vary. An example is given. When k is negative, the two parts of the graph lie in Quadrants II and IV, rather than in Quadrants I and III. I found this answer by graphing a few equations such as $xy = -1$ and $xy = -4$.

21. 84; 85; 87 **23.** 13.5; 13; 13 **25.** (6, –2)

Page 472 Checking Key Concepts

1. Daniel, 80; Peter, 77.5; Sarah, 80 **3.** In Question 1, the students' grades are being averaged, and the weights are 1 (for the tests) and 2 (for the term paper). In Question 2, the meal costs are being averaged, and the weights are the numbers of people who ordered each meal choice.

Pages 472–475 Exercises and Applications

1. a. \$73.48 **b.** 62 gal **c.** about \$1.19 per gal **3.** 92 or better **5.** The final score is a weighted average of the scores in the technical program and the free skating program with weights $\frac{1}{2}$ and 1 respectively. **7.** Baiul and Kerrigan tied, Bonaly and Chen tied, Szewczenko, Sato, Witt, Harding **9.** 6 **11.** 1

15. a.

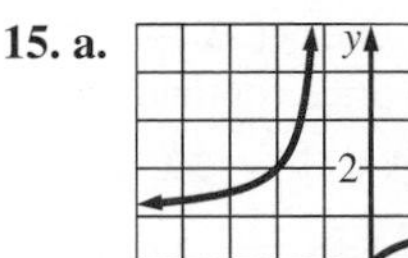

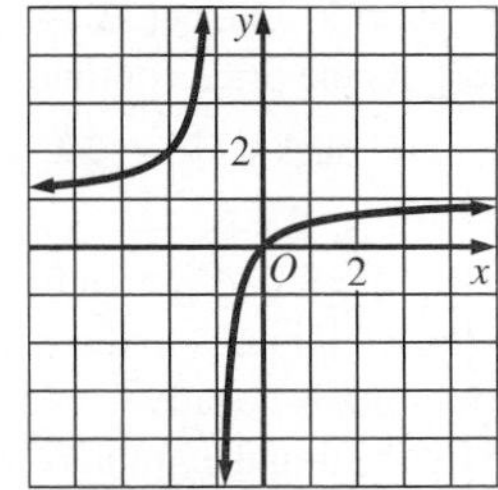

b.

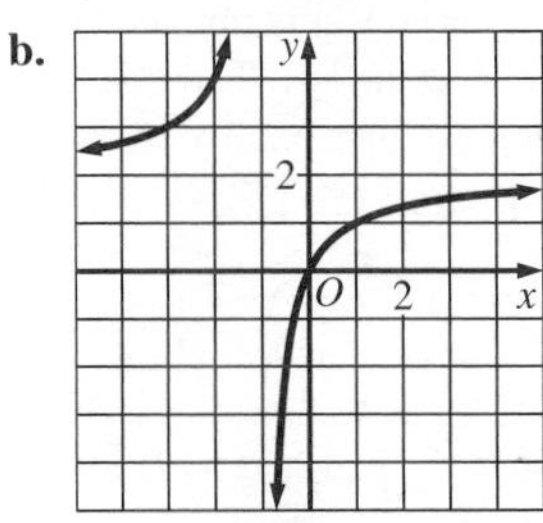

c.

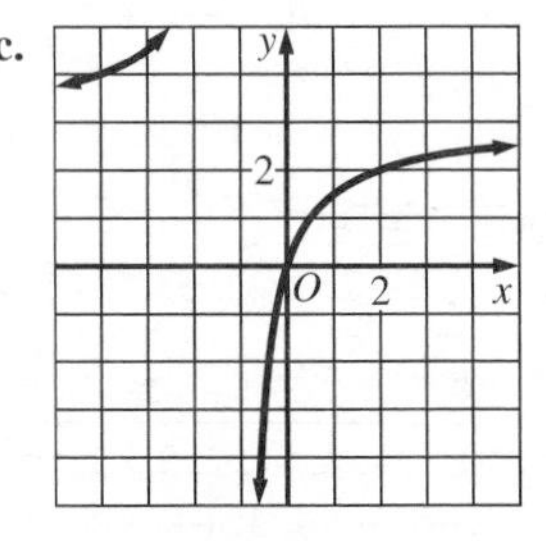

d.

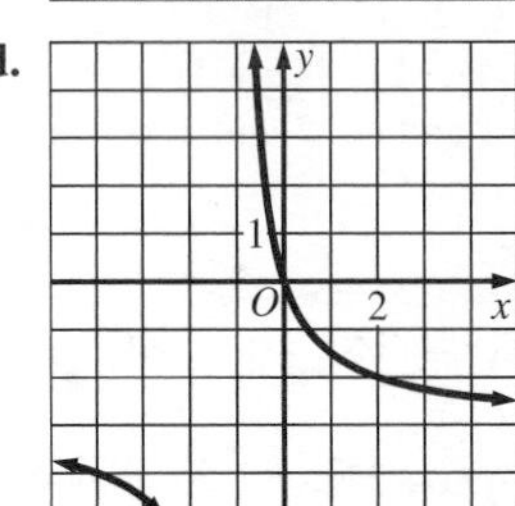

Answers may vary. An example is given. The graph of $y = \frac{kx}{x+1}$ is the graph of $y = \frac{x}{x+1}$ stretched vertically by a factor of $|k|$. If $k < 0$, the graph is also reflected in the x-axis.

21. 25 **23.** Yes. **25.** No; the first two terms cannot be written as monomials with exponents that are whole numbers.

Page 475 Assess Your Progress

1. No. **2.** Yes; $y = \frac{120}{x}$. **3.** 16 cm **4.** 86.8

5. 25 executive suites

Page 479 Checking Key Concepts

1. Yes; both sides are rational expressions. **3.** Yes; both sides are rational expressions. **5.** 24 **7.** $10x(x + 1)$

9. no solution

Pages 479–481 Exercises and Applications

1. Yes. **3.** Yes. **5.** 8.5 **7.** 4.8 **9.** 4.8 **13. a.** Yes, both sides of the equation are ratios. **b.** 17 **15.** –3

17. –0.75 **19.** 24 shots **21.** no solution **23.** $\frac{73}{9}$

27. about 1 mi/h; about 2 mi/h

33.

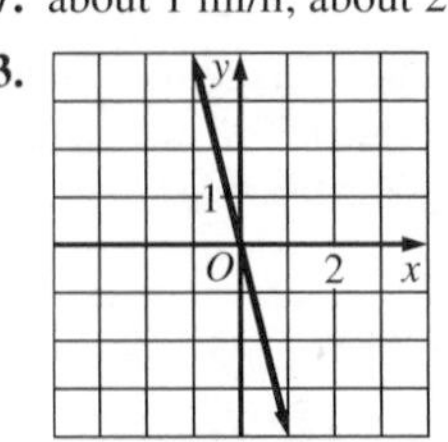

35. $-3; \frac{1}{2}$ **37.** $-4; \frac{1}{3}$

Page 485 Checking Key Concepts

1. $z = \frac{2}{5}y$ **3.** $m = 3 - \frac{1}{2}d$ **5.** $b = \frac{1}{2c}$

7.

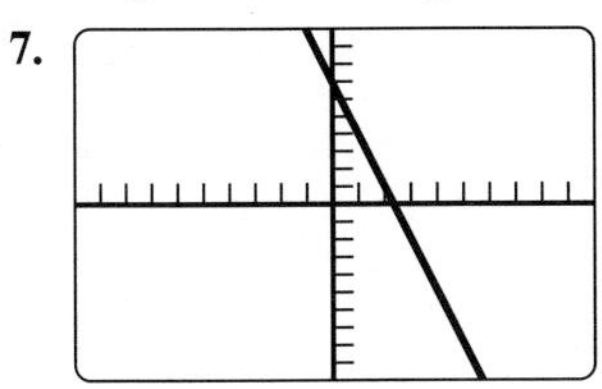

9. $t = \frac{d}{r}$

Pages 485–487 Exercises and Applications

1. a. $h = \frac{3V}{B}$ **b.** 10 cm **3.** $b = P - a - c$ **5.** $C = P + S$

7. $y = 24 - 2x$ **9.** $y = \frac{c - ax}{b}$ **11.** $x = \frac{1}{y}$

13.

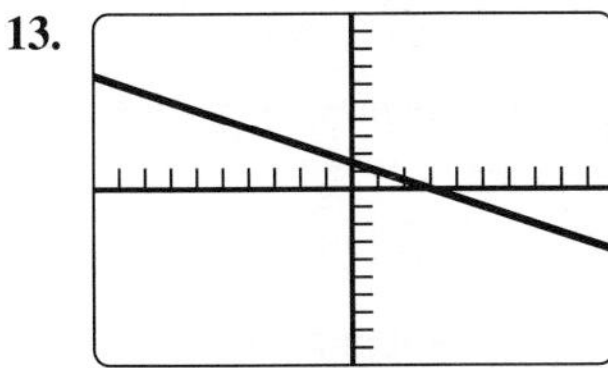

15.

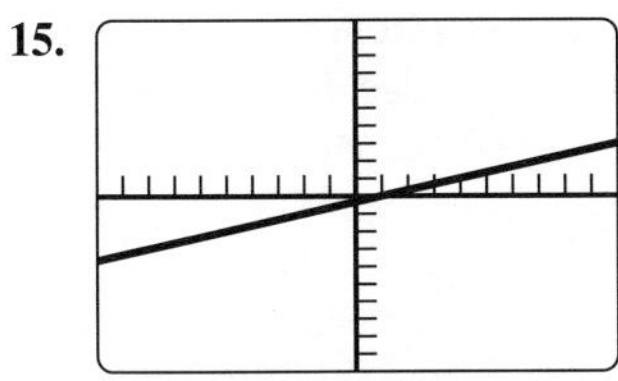

17.

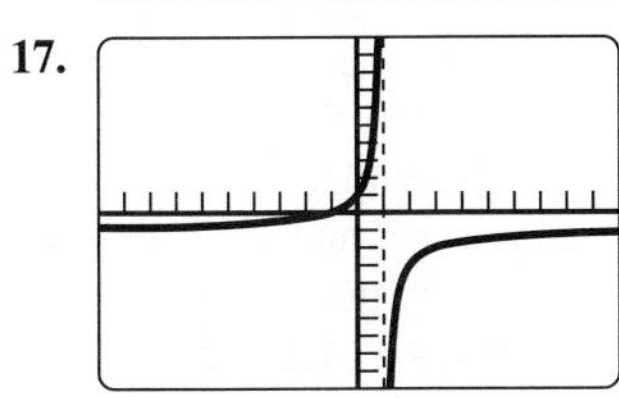

27. 3 **29.** $64x^9$ **31.** $\frac{m^{10}}{n^4p^6}$

Page 487 Assess Your Progress

1. 7 **2.** 2 **3.** 5 **4.** $y = \frac{1}{2}x - 4$ **5.** $x = \frac{y^2 + 7}{2}$

6. $b = \frac{a - 2cd}{2c - 1}$ **7.** $F = \frac{9}{5}C + 32$; $F = 68°$

Page 490 Checking Key Concepts

1. $\frac{1}{2x + 4}$ **3.** $\frac{3x^2}{2}$ **5.** simplest form **7.** $4x$ **9.** $4x$ **11.** $\frac{4}{3}$

Pages 491–493 Exercises and Applications

1. $\frac{b}{2}$ **3.** simplest form **5.** simplest form **7.** 4 **11. a.** $\frac{1}{n}$ **b.** The probability that you will get a green light is 1 minus the probability that you will get a red light; $\left(1 - \frac{1}{n}\right)\left(1 - \frac{1}{n}\right) = \frac{n-1}{n} \cdot \frac{n-1}{n} = \left(\frac{n-1}{n}\right)^2$. **13.** $\frac{6y}{5}$ **15.** $3y$

17. $\frac{2a + 2}{a^2 + 2a}$ **19.** $\frac{10k}{3}$ **21.** $\frac{n}{n - 2}$ **23.** $2n^2(n + 1)$ **27.** $\frac{4}{25}$

35. $20x^2 - 6x - 8$ **37.** a^9 **39.** c^{-4}, or $\frac{1}{c^4}$

Page 497 Checking Key Concepts

1. $\frac{4}{y}$ **3.** -2 **5.** $\frac{12m + 5}{15m^2}$ **7.** rational **9.** linear

Pages 497–499 Exercises and Applications

1. $\frac{t - 2}{t - 1}$ **3.** 2 **5.** $\frac{25x - 6y}{20x^2y^2}$ **7.** quadratic **9.** exponential

11. rational **13.** $\frac{5}{6}$ **19.** $\frac{8x + 42}{(x + 8)(x - 3)}$ **21.** $\frac{3x + 5}{x + 1}$

23. $\frac{b^2 - 4c}{4a}$ **25. a.** $\frac{250}{s} - \frac{250}{s + 5} = \frac{1250}{s(s + 5)}$ **b.** about 27 min

31. $-\frac{1}{4}$ **33.** no solution **35.** $y = \frac{x}{2x + 1}$

Page 499 Assess Your Progress

1. $\frac{2}{5}$ **2.** $\frac{2}{5x}$ **3.** $\frac{3y}{y + 1}$ **4.** $\frac{y + 5}{y + 1}$ **5.** $\frac{6z - 5}{2(z + 1)}$ **6.** $\frac{3}{m}$

Pages 504–505 Chapter 11 Assessment

1–3. Answers may vary. Examples are given.

1. A weighted average is one in which some numbers are given more weight than others. Examples: calculating roommates' shares of utility bills when one roommate is home more than others; calculating grades when certain tests or test items count for more than others. **2.** $y = \frac{12}{x}$ or $xy = 12$

3. $\frac{2x + 4}{6x^2 + 24x + 24}$

4.

Cost per calculator	Number of calculators
\$100	60
\$60	100
\$50	120
\$30	200

6. Let y = the number of calculators the department can buy and x = the cost per calculator; $y = \frac{6000}{x}$.

7.

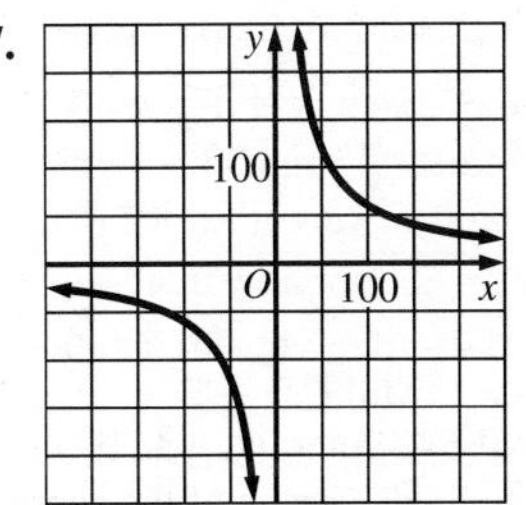

Only the portion of the graph in Quadrant I makes sense, since both the number of calculators and the cost per calculator must be positive.

8. \$24 **9.** 94 or better **10.** Estimates may vary. An example is given. about 26,800 ft^2 **11.** $-\frac{5}{11}$ **12.** 7 **13.** no solution **14.** 2 **15.** $x = -\frac{4}{3}y + 4$ **16.** $n = \frac{m}{2m - 1}$

17. $y = \frac{2x}{2 - x}$

18. $n = \frac{S}{180} + 2$ or $n = \frac{S + 360}{180}$; 9 **19.** $\frac{5b^3}{2a}$ **20.** $\frac{4c}{5}$ **21.** $\frac{1}{27t}$

22. $\frac{(p-2)^2}{8}$ **23.** 2 **24.** $\frac{k+5}{k+2}$ **25.** $\frac{3y+2}{6y^2}$ **26.** $\frac{-2z-5}{z(z+1)}$

27. $\frac{49}{n(n-7)}$ **28.** Answers may vary. Examples are given.

linear: $y = \frac{1}{2}x - 5$; quadratic: $y = x^2 - 2x + 1$;

polynomial: $y = x^3 + \frac{3}{4}x^2 - 11$; rational: $y = \frac{1}{x+1} + x(x-2)$;

exponential: $y = 2.5(1.08)^x$

CHAPTER 12

Page 511 Checking Key Concepts

1, 3. Answers may vary. Examples are given. **1.** You might be trying to cook the eggs in their shells. You might end up eating uncooked eggs. **3.** The following is an example of an algorithm for cooking scrambled eggs on a conventional stovetop. Step 1: Take the eggs out of the refrigerator and put them in a bowl. Step 2: Break the eggs into a bowl and stir until the yolks are broken and the eggs are well mixed. Step 3: Put a frying pan on a stove burner over medium heat. If the pan does not have a non-stick surface, oil the pan before putting it on the stove. Step 4: When the pan is heated, pour the eggs into the pan. Step 5: Stir the eggs as they cook. When the eggs are cooked to the degree you prefer, remove the pan from the heat.

Pages 512–515 Exercises and Applications

1, 3. Answers may vary. Examples are given. **1.** Step 1: Take out two slices of bread. Step 2: Spread peanut butter on one slice. Step 3: Spread jelly on the other slice. Step 4: Put the slices together with unspread sides out. **3.** Step 1: If the range is not set on standard, enter ZOOM 6. Step 2: Enter Y=1/(X + 2). Step 3: Enter GRAPH. Step 4: If necessary, adjust the range. **7.** Step 1: Subtract 52 from 326. Step 2: Subtract 52 from the remainder. Step 3: Repeat Step 2 until the remainder is less than 52. Step 4: Count how many times 52 was subtracted. That number is the quotient and the number left is the remainder. **9.** Step 1: Multiply 4 by 4. Step 2: Multiply the product by 4. Step 3: Repeat Step 2 five more times.

17.

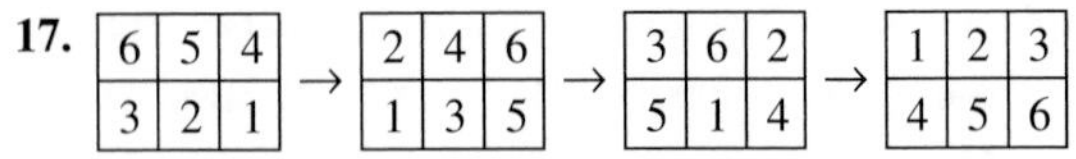

6	5	4
3	2	1

→

2	4	6
1	3	5

→

3	6	2
5	1	4

→

1	2	3
4	5	6

21. A: 03–b–01; B: 03–b–03; C: 03–a–03; D: 03–c–02

25.

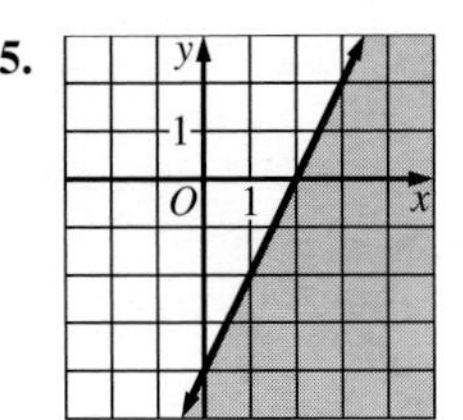

27. $-\frac{3xw}{2yz}$ **29.** $\frac{9s}{4}$ **31.** $\frac{y(4x+y)}{2x(x-y)}$

Page 519 Checking Key Concepts

1. a tree **3.** The shortest path is 9 units long.

5.

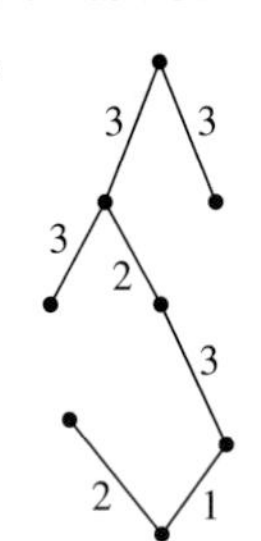

17 units; Yes; the seven edges are the seven shortest edges of the graph.

Pages 519–522 Exercises and Applications

1. both; The edges allow you to travel from one vertex to the other without repeating any vertex or edge, so the graph is a path. The vertices are connected with the smallest possible number of edges, so the graph is a tree. **3.** The graph is a tree because the vertices are connected with the smallest possible number of edges. The graph is not a path because you cannot travel the graph without repeating any edges or vertices. **5. a.** Answers may vary. An example is given.

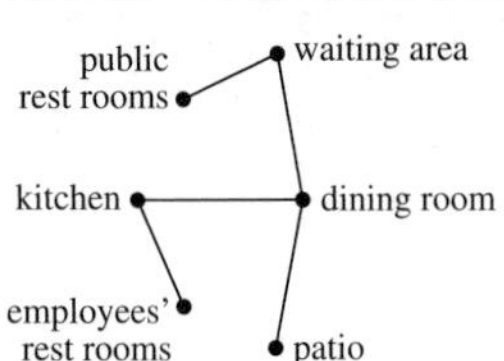

b. Other graphs are possible, but all would have 6 vertices and 5 edges and would be essentially the same except for the position of the vertices.

7, 9. Drawings may vary. Examples are given.

7.

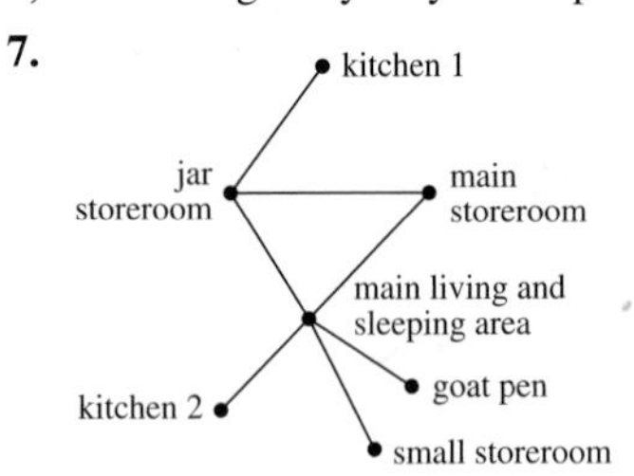

9.

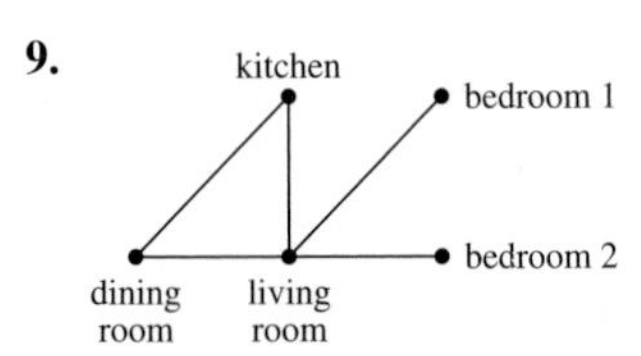

11. The edges along the bottom of the graph form a path 17 units long.

13.

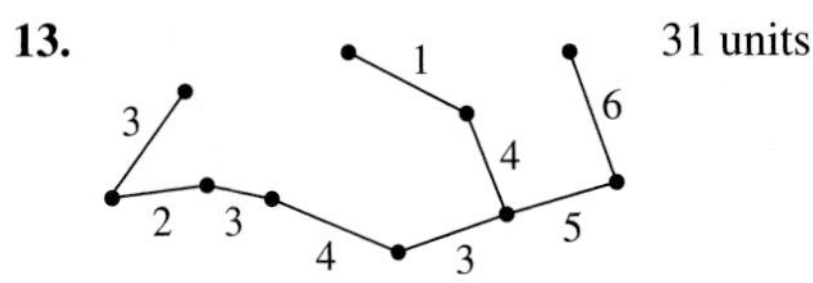

31 units

15. B

19. Answers may vary. An example is given.

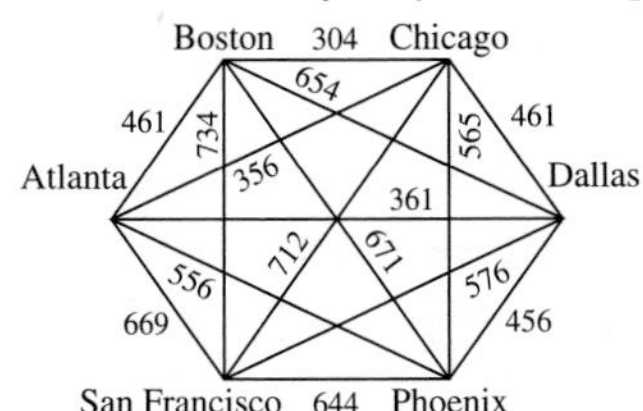

23. 1.4 **25.** $\frac{3}{11}$ **27.** $-\frac{1}{2}, 3$ **29.** $\frac{5}{2}, -\frac{5}{2}$ **31.** $\frac{3}{2}, -\frac{1}{2}$

Page 526 Checking Key Concepts

1.

Candidate	Number of votes
Jack	51
Denzel	52
Stacey	52
Mary	44
Renee	33

Denzel and Stacey (tie); Jack; Mary; Renee
3. For both elections: Denzel and Stacey (tie); Mary; Jack and new person (tie); Renee

Pages 526–529 Exercises and Applications

1. You should divide the piece into two smaller pieces, either of which would satisfy you. Your sister should choose one of the pieces. You get the remaining piece. **15.** $\frac{3}{4}$
17. 3.4×10^{-3} **19.** 1.0078×10^{-3} **21.** 3×10^{-4}

Page 529 Assess Your Progress

1, 3. Answers may vary. Examples are given. **1.** Step 1: If necessary, write both trinomials with the square term first, the linear term next, and the constant term last. Step 2: Combine similar terms. **3.** The first person should pick up donations weighing 5 lb, 10 lb, 25 lb, 30 lb, and 60 lb. The second person should pick up donations weighing 10 lb, 10 lb, 20 lb, 40 lb, and 50 lb. I think this is fair because each person has the same number of stops and nearly the same weight to pick up, with the weights distributed fairly evenly.

Page 532 Checking Key Concepts

1. $1 \cdot 2 \cdot 4 = 8$; The number of branches at each stage of the tree diagram is equal to the product of the number of items at that stage. **3.** 5! or 120 paths **5.** 6! = 720 passwords, if you include Claire

Pages 533–535 Exercises and Applications

1.

Susan
- Leith
  - Trevor: Bryan, Zara
  - Jeremy: Bryan, Zara
  - Chelsea: Bryan, Zara
- Denise
  - Trevor: Bryan, Zara
  - Jeremy: Bryan, Zara
  - Chelsea: Bryan, Zara

12 groups

7. 6720 councils **11.** 362,880 ways; Answers may vary. Examples might include the abilities of the players or the use of designated hitters. **13.** 2730 orders **17.** (–2, –3)
19. (3, –2) **21.** $\frac{5}{2} - \frac{\sqrt{29}}{2}; \frac{5}{2} + \frac{\sqrt{29}}{2}$ **23.** $-\frac{3}{10} - \frac{\sqrt{109}}{10}; -\frac{3}{10} + \frac{\sqrt{109}}{10}$ **25.** $\frac{7}{4} - \frac{\sqrt{113}}{4}; \frac{7}{4} + \frac{\sqrt{113}}{4}$

Page 538 Checking Key Concepts

1. 28 handshakes **3.** 4950 handshakes **5.** 21 **7.** 55
9. 15 combinations **11.** 3276 ways

Pages 539–541 Exercises and Applications

1. 66 delegations **3.** 36 delegations; The number of such delegations is the number of two-person delegations that can be chosen from a group of 9, or ${}_9C_2$. **5.** 21 pairs; If the colors alternate, there are 42 possible flags; if the stripes appear in any order, there are 126 possible flags. **7.** 78 **9.** 680
11. a. 10 choices **b.** 10 combinations **13.** The entry in column B is the sum of the entries in columns A and B of the preceding row. Since the new person does not shake hands with himself or herself, the total number of added handshakes each time is the number of people in the group before the new person was added. Then the total number of handshakes is the previous total plus the number of people before the new person was added.

15.

number of points	number of triangles
3	1
4	4
5	10
6	20

The number of triangles that can be drawn by connecting n points is ${}_nC_3$.

17.

number of toppings	number of pizzas
1	9
2	36
3	84
4	126
5	126
6	84
7	36
8	9
9	1

${}_9C_1 = {}_9C_8$; ${}_9C_2 = {}_9C_7$; In fact, for any whole number r between 1 and 9, ${}_9C_r = {}_9C_{9-r}$. For example, the number of ways of choosing three things from among nine is the same as the number of ways of choosing six things from among nine. When you choose r items from among n items, you automatically chose the other $n - r$ items as well. **19.** Yes; last year's tournament involved ${}_6C_2$ or 15 games and with one game played at a time, all 15 were completed in two days. Since three games can be played at once, there would be time to play 45 games in this year's tournament. However, only ${}_9C_2$ or 36 games are necessary. **23.** decays; 0.52 **25.** grows; 2 **27.** grows if $b > 1$ and decays if $0 < b < 1$; b **29.** No.

Page 544 Checking Key Concepts

1. HHHH, HHHT, HHTH, HHTT, HTHH, HTHT, HTTH, HTTT, THHH, THHT, THTH, THTT, TTHH, TTHT, TTTH, TTTT **3.** $\frac{1}{16}$ **5.** $\frac{1}{55}$

Pages 544–547 Exercises and Applications

1. $\frac{3}{8}$ **3.** 64 possible results **7.** $\frac{1}{8000}$ **9.** Answers may vary. **15.** 2^{92}; On each toss, there are two possible outcomes, so the total number of outcomes is $2 \cdot 2 \cdot 2 \cdot \cdots \cdot 2$ (92 times) $= 2^{92} \approx 4.95 \times 10^{27}$. **17.** 1; Doubling any number produces an even number. **19.** $5x - 10$ **21.** $4 - 2n - 8n^2$ **23.** $-\frac{1}{6ab}$ **25.** $\frac{3t + 37}{20}$ **27.** $\frac{3b - 2a - 2a^2}{2ab}$

Page 547 Assess Your Progress

1. 60 ways **2.** The number of pairs of initials (in the same order) is the number of two-letter permutations of the 26 letters of the alphabet, which is $26 \cdot 26$, or 676. There are not enough pairs to allow all 713 employees to have unique pairs of initials. **3. a.** 11,175 pairs **b.** 828 pairs **4.** $\frac{11}{36} \approx 0.31$

Pages 552–553 Chapter 12 Assessment

1. A permutation is an arrangement of a group of items in a definite order. A combination is a selection of items chosen from a group in which order does not matter.

2. Answers may vary. An example is given. In the diagram, a greedy algorithm would choose the path from A to B to C as the best path from A to C.

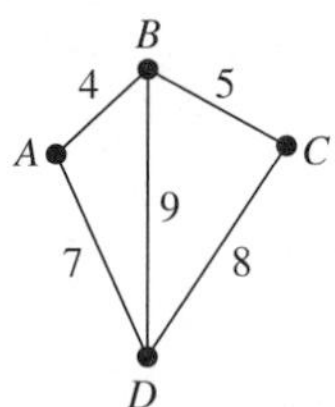

3. Answers may vary. Examples are given. A tree could be used to join a series of computers together with the least amount of cable. A path could be used to determine a good way to visit a collection of sites, such as pickups or drops on a delivery route, or vacation destinations. **4.** Answers may vary. Examples are given. (1) You might not add detergent or fabric rinse. (2) You might not dry the laundry. **5.** Answers may vary. Examples, based on the answers to Ex. 4 above, are given. (1) Add "Add detergent and fabric conditioner to the washing machine" to Step 2. (2) Add "Put the clean laundry into the dryer or hang it on a line until it is dry" to Step 4. **6.** Step 1: Find the least common denominator (LCD) of the fractions. Step 2: Divide the LCD by the denominator of the first fraction. Multiply the numerator and denominator of the first fraction by the quotient. Step 3: Repeat Step 2 for the second fraction. Step 4: Add the numerators of the fractions obtained in Steps 2 and 3. Step 5: Write the fraction whose numerator is the result from Step 4 and whose denominator is the LCD. This is the sum of the original fractions. **7.** A to F to E to D **8.** F to A to B to E to C to D **9.** D to C to E to B to A to F **11.** Answers may vary. An example is given. Two sisters want to share a small pizza. One should divide the pizza into two pieces, either of which would satisfy her. The second sister should choose one of the pieces. The first sister would get the remaining piece. **12.** 120 groups **13.** 120 ways **14.** 91 games **15.** $\frac{3}{8}$ **16.** $\frac{1}{136}$

CHAPTERS 10–12

Pages 554–555 Cumulative Assessment

1. $-2x^2 + 3x - 2$; 2 **3.** $4y^3 - 2y^2 - 2y + 12$; 3 **5.** $80x^3 - 32x^2 + 56x$; 3 **7.** $(x + 7)(x - 3)$ **9.** $(a - 6)(a - 3)$ **11.** $(3z - 2)(z + 9)$ **13.** $-4, 3$ **15.** $\pm\frac{5}{9}$ **17.** $-\frac{1}{3}, 3$ **19.** 12 **23.** direct variation; $y = \frac{1}{4}x$ **25.** -1 **27.** 85 **29.** $\frac{21 - 5b}{9ab}$ **31.** 2 **35.** $\overline{EA}$ or $\overline{EF}$, and $\overline{AB}$, $\overline{BC}$, $\overline{CD}$, and $\overline{DF}$ **37.** 362,880 **39.** 84 **41.** 240 **43.** Answers may vary. An example is given. Each handshake involves two people so the number of possible handshakes is related to the total number of people involved, taken two at a time.

EXTRA PRACTICE

Pages 557–559 Chapter 1

1. 31 days **3. a.** 15 days **b.** 12 days **5.** Answers may vary. Example: Let t = the daily high temperature; $t \geq 90$ and $t \leq 99$. **7.** 11 **9.** 3 **11.** $2\frac{2}{3}$ **13.** 17.04 **15.** 0.3 **17.** $3\frac{25}{68}$ **19.** 27 **21.** 34 **23.** 6 **25.** $\frac{1}{10}$ **27.** 1 **29.** 48 **31.** $13\frac{3}{7}$; 13; 13 **33.** $19\frac{13}{24}$; 20; 22 **35.** $9\frac{4}{17}$; 9; 8 **37.** –5 **39.** –21 **41.** –5 **43.** 3.2 **45.** 1 **47.** 5 **49.** –4 **51.** $\frac{1}{2}$ **53.** 8.4 **55.** $\frac{1}{3}$ **57.** –7 **59.** 9 **61.** 3 **63.** 3 **65.** 16 **67.** 28 **69.** –9 **71.** 6 **73.** –48 **75.** –5 **77.** –23 **79.** $-\frac{26}{7}$ **81.** –5 **83.** –12 **85.** 8 **87.** –24 **89.** –23 **91.** Let x = the number of minutes a call lasts and y = the cost in dollars; $y = 0.35 + 0.10(x - 1)$. **93.** Let x = the number of years from now and y = the deer population; $y = 2200 + 130x$. **95.** $32 - 16x$ **97.** $4a + 3b$ **99.** $19n - 8n^2$ **101.** $-6f + 2$ **103.** $-13y - 8$ **105.** $11a^2$ **107.** E: 3 × 4; F: 3 × 4; G: 4 × 3; H: 4 × 3 **109.** E and G, E and H, F and G, F and H; They do not have the same dimensions.

Pages 559–561 Chapter 2

1. 1.9 **3.** –1.3 **5.** –16 **7.** –10 **9.** 4 **11.** 0.36 **13.** –0.07 **15.** $4\frac{1}{2}$ **17.** 6 **19.** 30 **21.** 5 **23.** –60 **25.** 3 **27.** –108 **29.** 43.8 **31.** –7.6 **33.** Let x = the cost in dollars of a juice box; $9x + 2.29 = 5.17$; $x = 0.32$; 32¢. **35. a.** Let C = the total cost in dollars and n = the number of days the car is rented ; $C = 27n$. **b.** The domain consists of the positive whole numbers and the range consists of the non-negative multiples of 27. **c.** \$216 **37. a.** Let C = the total cost in dollars and n = the number of pens purchased; $C = 3n + 5.75$. **b.** The domain consists of the positive whole numbers and the range consists of the numbers 5.75, 8.75, 11.75, ... **c.** \$23.75 **39.** Point F lies in Quadrant I, point I lies in Quadrant II, points C, E, and J lie in Quadrant III, and points A and H lie in Quadrant IV. Points B, D, and G do not lie in any quadrant; they lie on axes.

41, 43, 45.

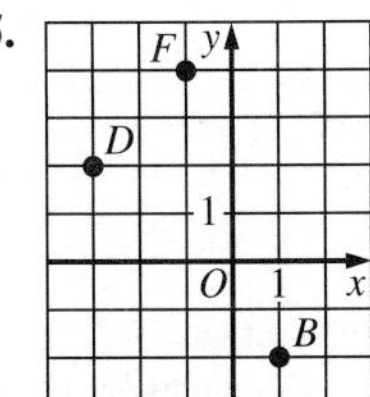

47, 49, 51, 53. Solutions may vary. Examples are given.

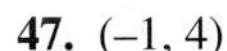

47. (–1, 4)

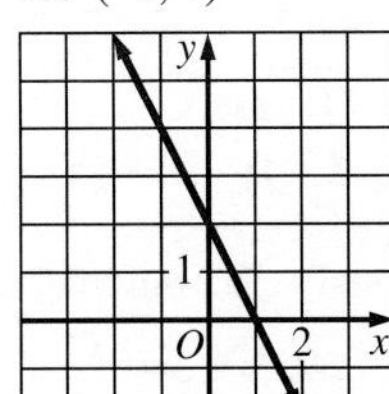

49. (–1, –3)

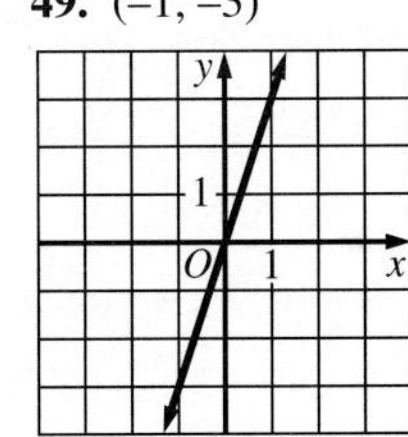

51. (1, 3)

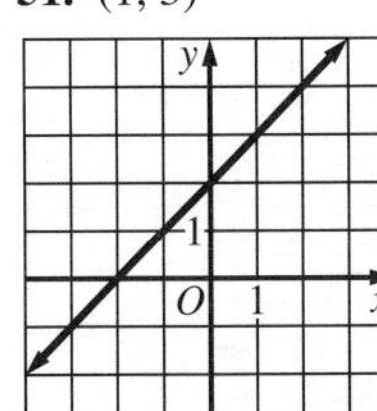

53. (–1, 2)

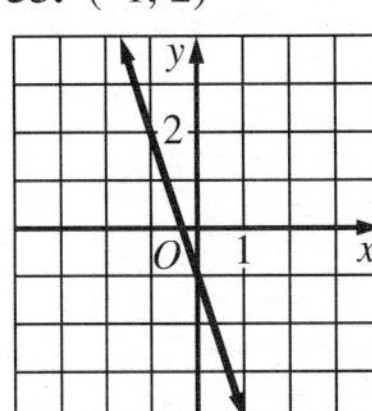

55. Let n = the number of payments he makes and A = the amount in dollars of the loan; then the equation $A = 125n + 1625$ describes the situation. To solve graphically, graph the equation and find 7500 on the vertical axis. Then move across to the graph of the equation. Move down to the horizontal axis to find the number of payments, 47.

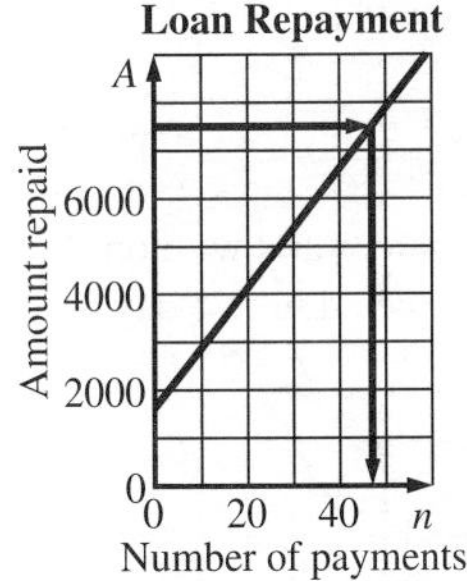

To solve using an equation, substitute 7500 for A in the given equation and solve for n: $7500 = 125n + 1625$; $n = 47$; 47 payments.

Pages 561–563 Chapter 3

1. 3000 cycles/min **3.** \$27.54/tank **5.** 60 min/h **7.** 6.5 mi/min **9.** 24.44 **11.** 200,000 **13.** \$3480 **15.** Yes; 7.5; $y = 7.5x$. **17.** No. **19.** No.

21. –2

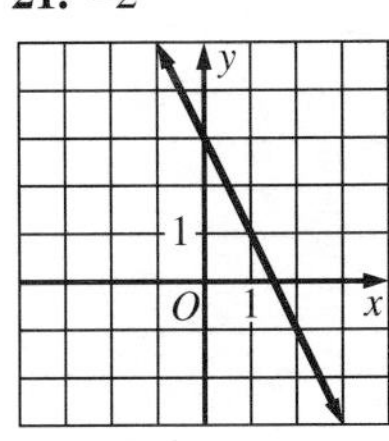

23. 3

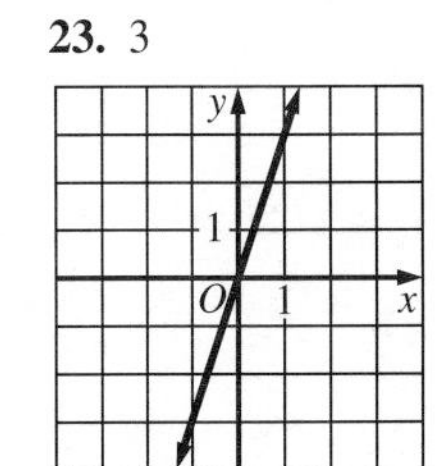

25. 0

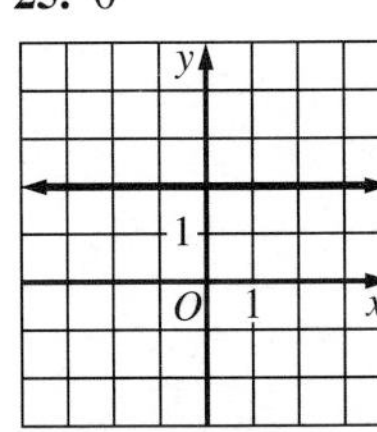

27. –1

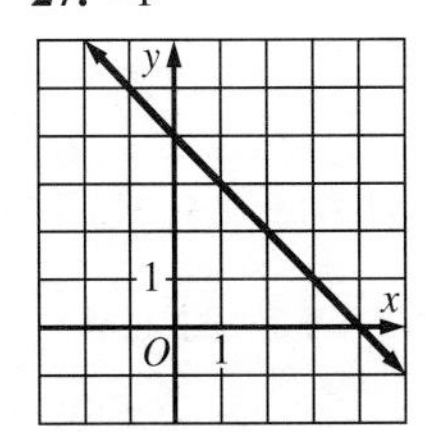

29. 2

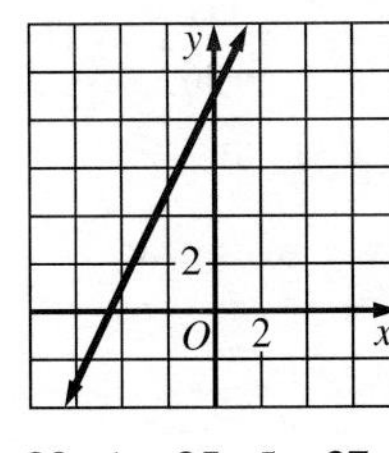

31. 5

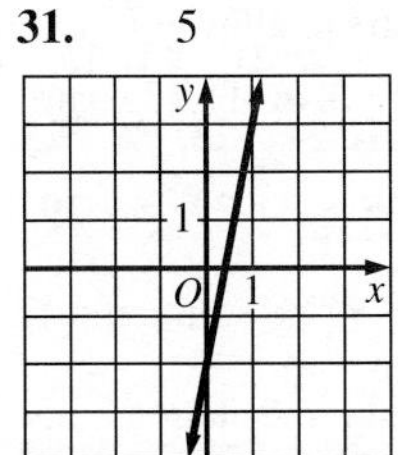

33. 1 **35.** 5 **37.** $\frac{6}{5}$ **39.** $\frac{1}{4}$ **41.** $\frac{5}{4}$ **43.** $-\frac{15}{19}$ **45.** $-\frac{10}{7}$

47. $\frac{18}{19}$ **49.** -2; 8; $y = -2x + 8$

51. 1; -2

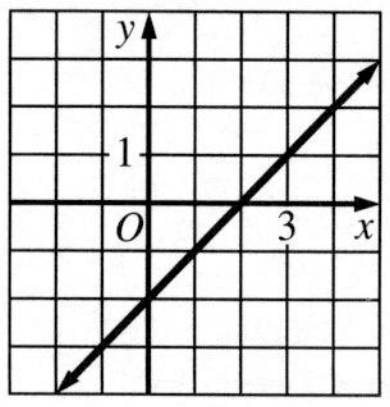

53. -3; 1

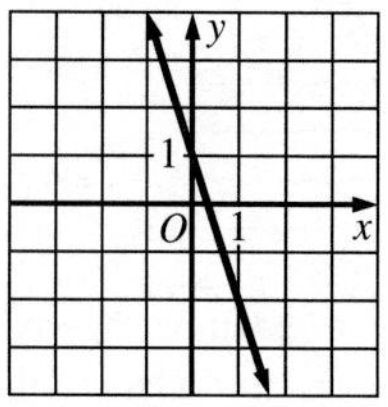

55. 8; 0

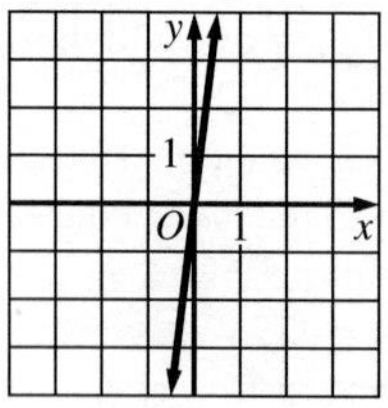

57. $\frac{1}{2}$; 6

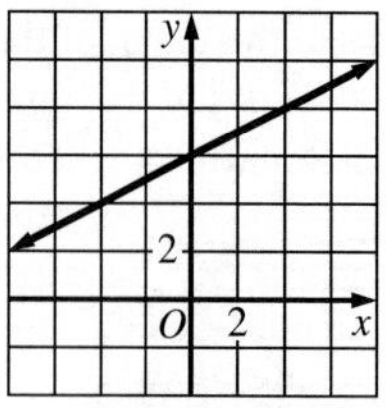

59. $\frac{1}{3}$; -2

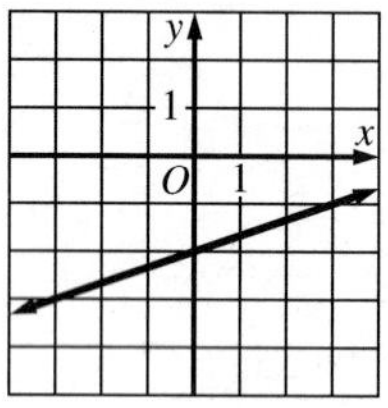

61. $\frac{3}{2}$; 0

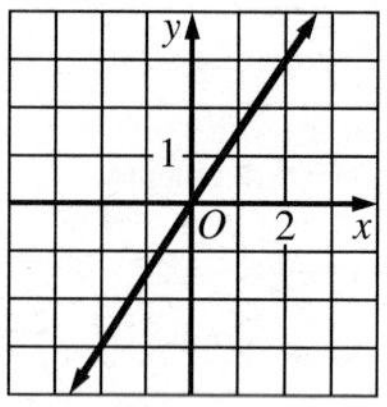

63. -5; 1

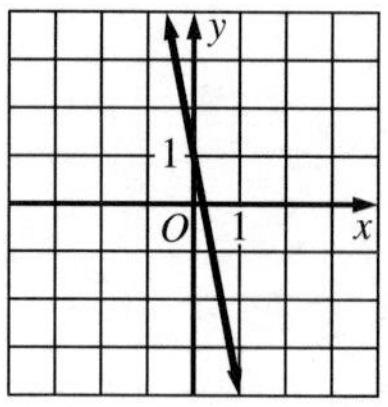

65. -7; -8

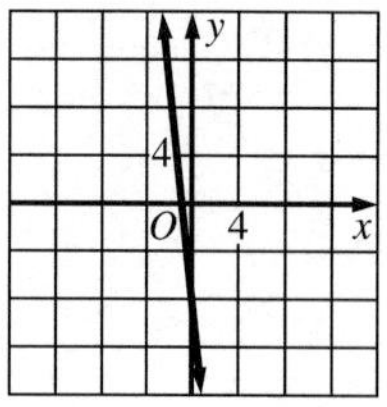

67. $y = 2x + 1$ **69.** $y = -3x + 7$ **71.** $y = 2$ **73.** $x = 5$
75. $y = \frac{1}{4}x - \frac{11}{4}$ **77.** $y = -11x + \frac{11}{2}$ **79.** $y = -x + 6$
81. $y = \frac{2}{9}x - \frac{35}{9}$ **83.** $y = \frac{8}{3}x - \frac{19}{3}$ **85.** $y = -\frac{4}{3}x + 8$
87, 89. Answers may vary. Examples are given.
87. $y = x + 8$ **89.** $y = 2.7x + 26.4$

Pages 563–565 Chapter 4

1. a. \$3 **b.** $3x$ **3.** 23 daily rides; \$21 **5.** $-\frac{3}{2}a$; $-\frac{2}{3}$
7. $-\frac{5}{8}b$; $-\frac{8}{5}$ **9.** 21 **11.** 15 **13.** 6 **15.** -30 **17.** 6
19. 18 **21.** 24 **23.** 88 **25.** -56 **27.** 12 **29.** 7 **31.** 4
33. $-\frac{1}{2}$ **35.** -2 **37.** 5 **39.** no solution **41.** identity
43. $-\frac{29}{12}$ **45.** no solution **47.** no solution **49.** identity
51. $3\frac{4}{5}$ **53.** no solution **55.** no solution **57.** 3
59. 20; 120 **61.** 5; $-\frac{5}{9}$ **63.** 15; 10 **65.** 8; -4
67. 10; 18 **69.** 100; 3 **71.** 10; 3 **73.** 1000; 359

75. 10; 5 **77.** 100; $\frac{1}{60}$

79.

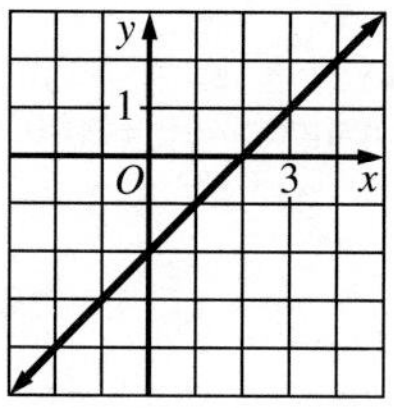

$y < -1$

81.

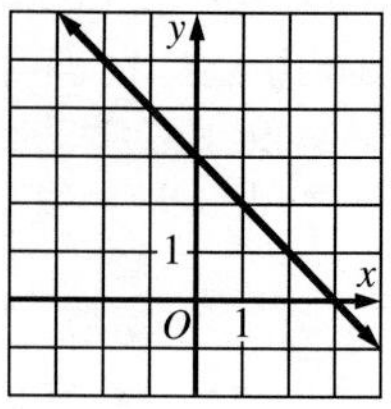

$y > 2$

83. $v < 7$

85. $x > 4$ 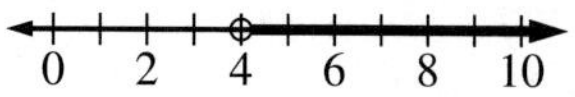

87. $x < 10$ 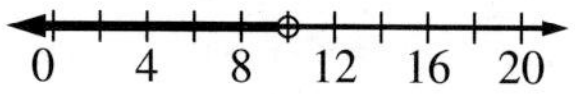

89. $t > 4$ 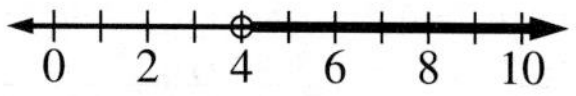

91. $n > -\frac{6}{13}$ 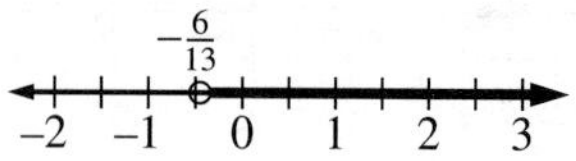

93. $a < 10\frac{4}{5}$ 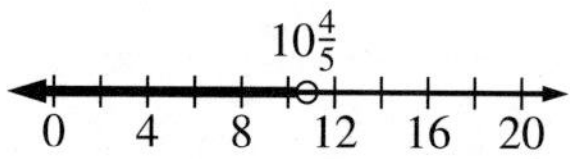

95. $r \geq 0$ 

97. $n > -12$ 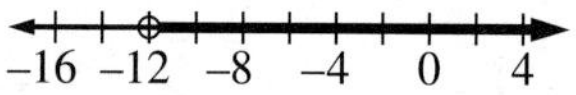

99. $x < -\frac{6}{7}$ 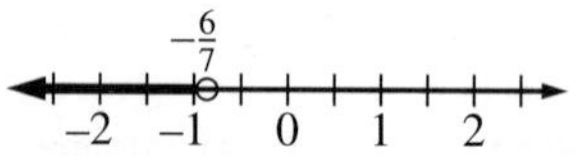

101. Let n = the number of books in a carton; $0.6n + 2 < 44$; $n < 70$; at most, 69 books can go in the carton.

Pages 566–568 Chapter 5

1. $\frac{1}{2}$ **3.** $\frac{1}{1000}$ **5.** $\frac{66}{25}$ **7.** $\frac{4}{1}$ **9.** $\frac{9}{5}$ **11.** $2\frac{1}{2}$ **13.** 36
15. 8.4 **17.** 1 **19.** $\frac{23}{2}$ **21.** $-\frac{11}{6}$ **23.** $\frac{4}{3}$
25. a. 6 in. by 8 in. **b.** 3.2 in. by 4 in. **c.** 4.8 in. by 5.2 in. **d.** 5.6 in. by 6.4 in. **27. a.** $\frac{3}{2}$ **b.** $\frac{7}{3}$ **c.** either $\frac{17}{22}$ or $\frac{11}{14}$
d. $\frac{1}{2}$ **29.** 105° **31.** 3 to 2 **33.** $4\frac{2}{3}$ **35.** Yes; $\angle C$ and $\angle C$, $\angle CBD$ and $\angle A$, $\angle CDB$ and $\angle E$; $\overline{CB}$ and $\overline{CA}$; $\overline{CD}$ and $\overline{CE}$; $\overline{BD}$ and $\overline{AE}$. **37.** 120° **39.** $P = 36$; $A = 74\frac{2}{3}$
41. $P = 70$; $A = 800$ **43.** $\frac{1}{2} = 50\%$ **45.** $\frac{1}{2} = 50\%$
47. $\frac{1}{3} \approx 33.33\%$ **49.** $\frac{1}{8} = 12.5\%$ **51.** $\frac{1}{3} \approx 33.33\%$

Pages 568–570 Chapter 6

1. 8 **3.** 7 **5.** No. **7.** Yes. **9.** Yes. **11.** Yes.
13. $c = 15$ **15.** $a = 15$ **17.** $c = 34$ **19.** $\frac{1}{9}$ **21.** $\frac{8}{11}$ **23.** $\frac{2}{3}$
25. $\frac{16}{9}$ **27.** $\frac{2}{333}$ **29.** rational **31.** rational **33.** rational
35. $9 < \sqrt{95} < 10$ **37.** $11 < \sqrt{135} < 12$ **39.** $3\sqrt{5}$

41. $4\sqrt{5}$ **43.** $3\sqrt{10}$ **45.** $3\sqrt{14}$ **47.** $15\sqrt{3}$ **49.** $\frac{\sqrt{3}}{2}$
51. $\frac{1}{6}$ **53.** $8\sqrt{3}$ **55.** $15\sqrt{5}$ **57.** 5 **59.** $12\sqrt{2}$
61. $18\sqrt{10}$ **63.** $(x + 3)(x + 2) = x^2 + 5x + 6$ **65.** $6x - 12$
67. $5x^2$ **69.** $8x^2 + 28x$ **71.** $10x + 20x^2$
73. $5x^2 + 34x - 48$ **75.** $10x^2 + 34x - 24$ **77.** $24 + 4\sqrt{5}$
79. $15 - 3\sqrt{14} - 5\sqrt{2} + 2\sqrt{7}$ **81.** $x^2 + 8x + 16$
83. $a^2 - 16a + 64$ **85.** $4w^2 - 1$ **87.** $49k^2 + 5.6k + 0.16$
89. $36b^2 - 24bc + 4c^2$ **91.** $11 - 4\sqrt{7}$ **93.** 961
95. 27.04 **97.** $14 + 6\sqrt{5}$; $12 + 4\sqrt{5}$

Pages 570–572 Chapter 7

1, 3, 5, 7, 9, 11. Tables may vary.

1.

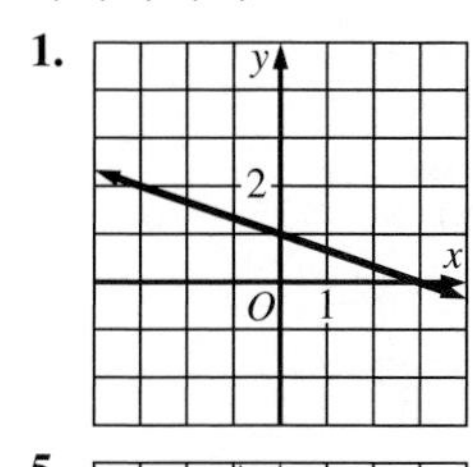

3.

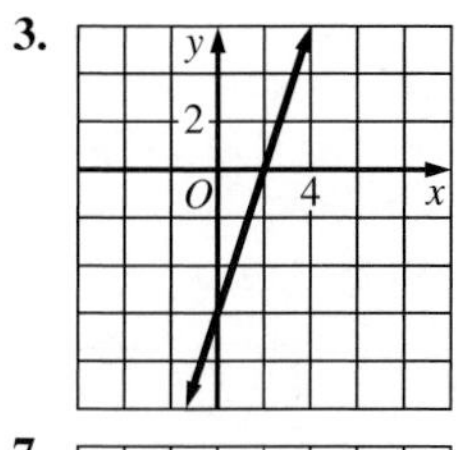

5.

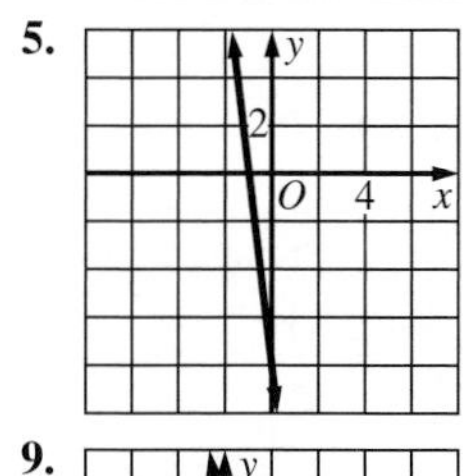

7.

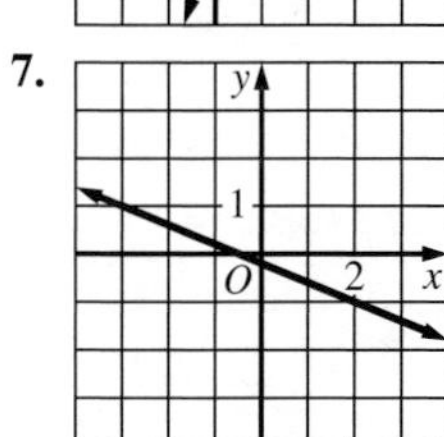

9.

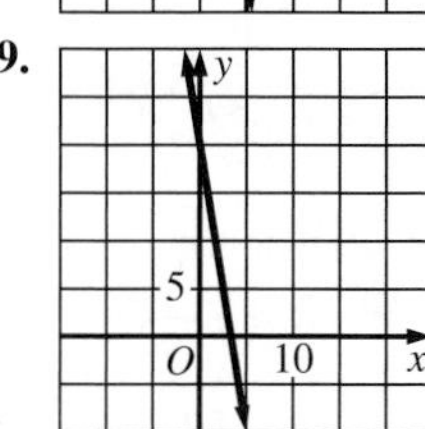

11.

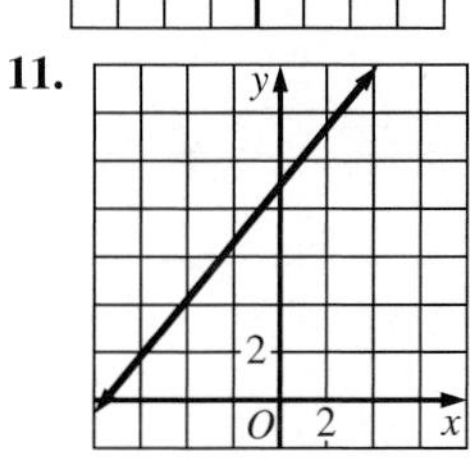

13. C **15.** B **17.** $\left(\frac{1}{3}, -\frac{1}{3}\right)$ **19.** $\left(\frac{4}{5}, \frac{7}{5}\right)$ **21.** $(1, -1)$

23. $(2, 1)$ **25.** $(0, 0)$ **27.** $\left(\frac{17}{5}, \frac{4}{5}\right)$ **29.** $\left(-\frac{8}{5}, -\frac{7}{5}\right)$

31. $\left(-4, -\frac{5}{2}\right)$ **33.** $(2, -4)$ **35.** $(1, 5)$ **37.** $(3, 4)$

39. $\left(-\frac{1}{3}, 7\right)$ **41.** $(3, 1)$ **43.** $(9, 6)$ **45.** $(3, 1)$ **47.** $(-2, 0)$
49. $(0, 2)$ **51.** $(3, 15)$ **53.** no solution
55. many solutions **57.** A **59.** B

61.

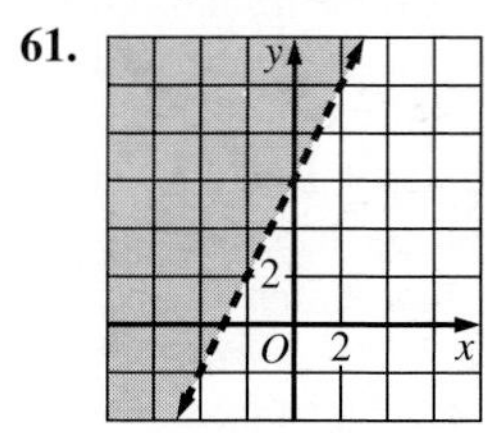

63.

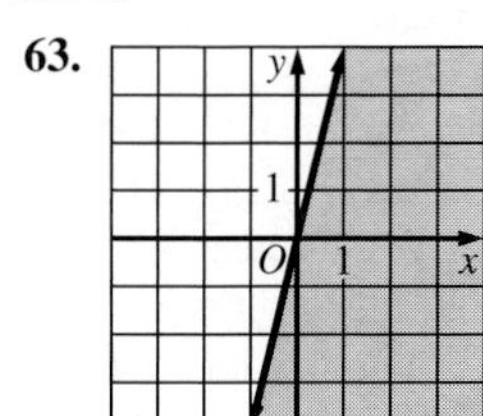

65.

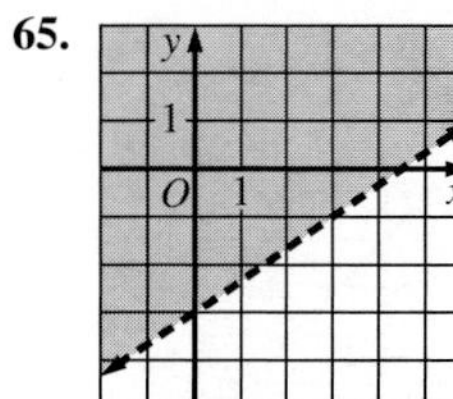

67.

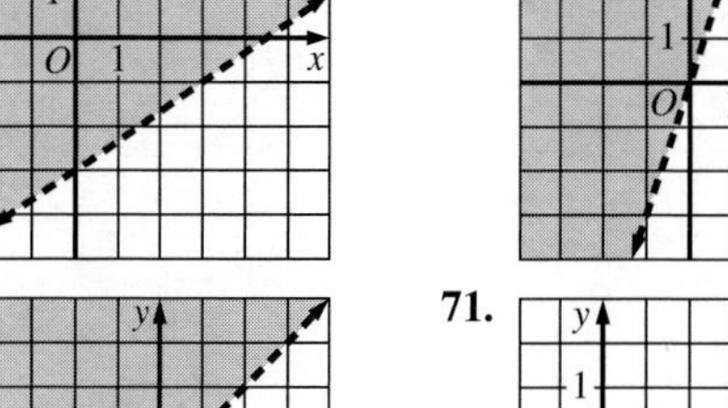

69.

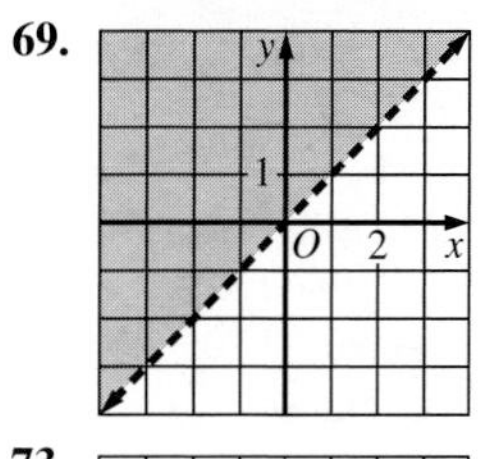

71.

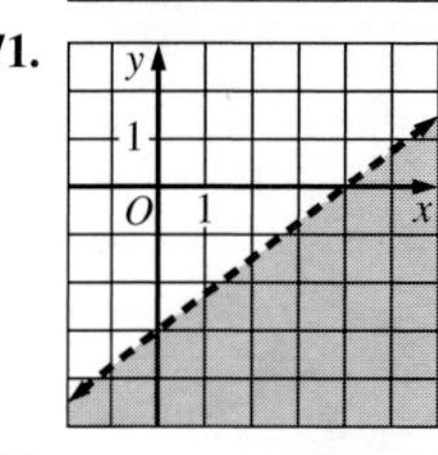

73.

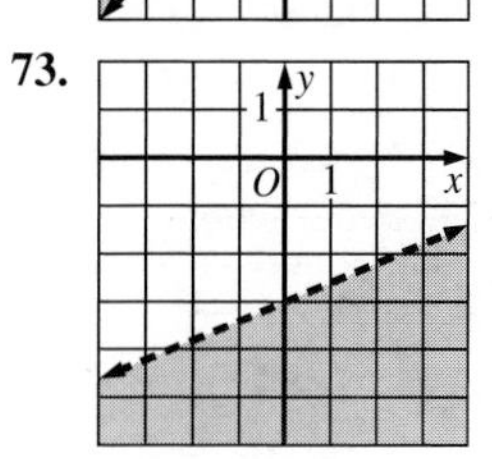

75.

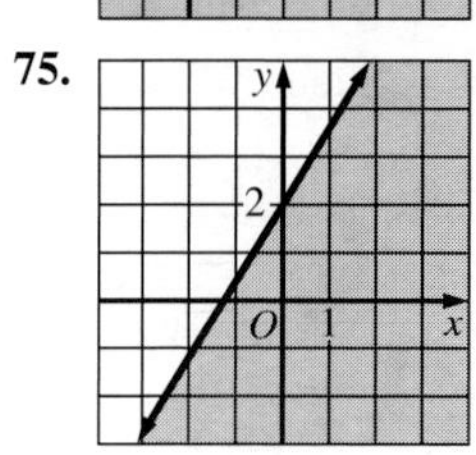

77. D **79.** B

81.

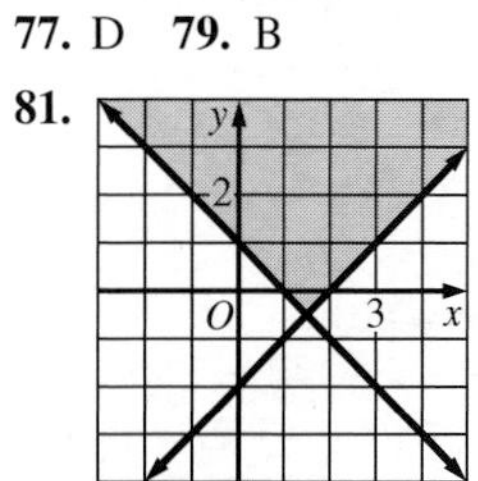

83.

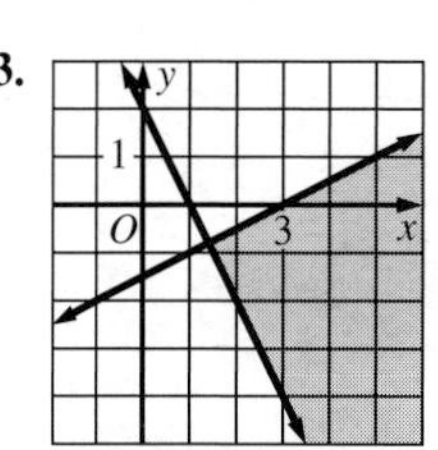

85.

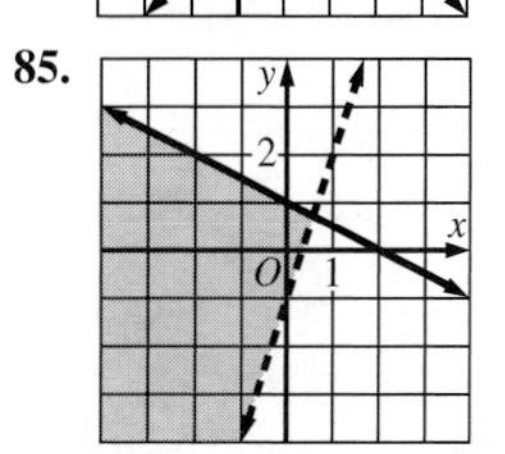

87.

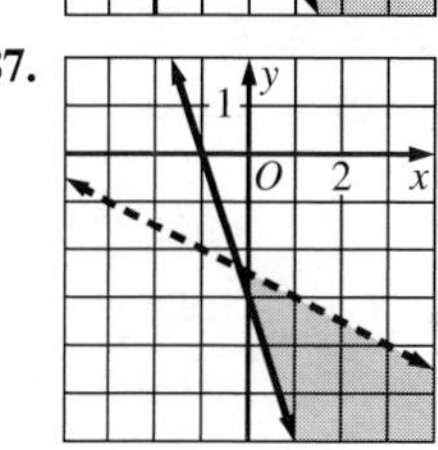

Pages 572–574 Chapter 8

1. linear

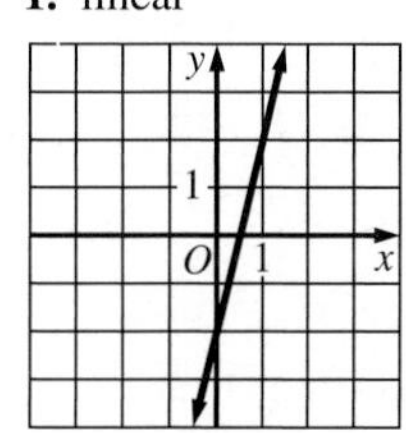

3. nonlinear

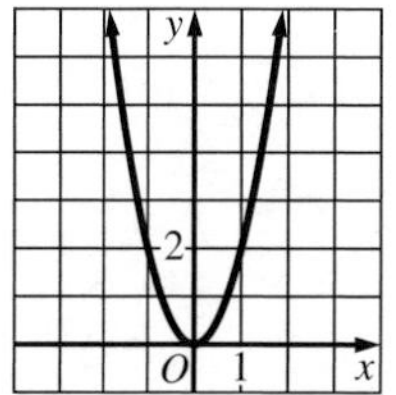

5. nonlinear

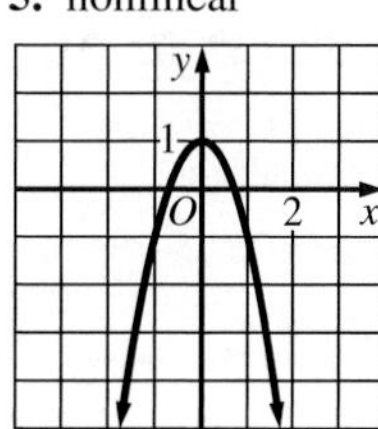

7. linear

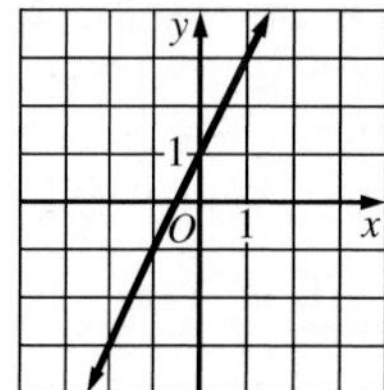

9. nonlinear

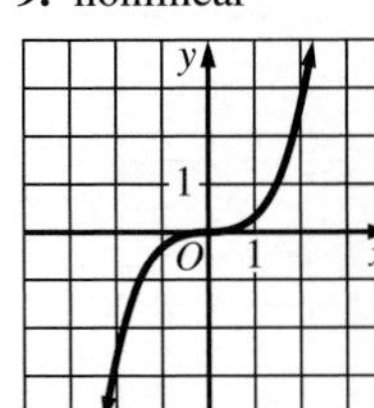

11. linear

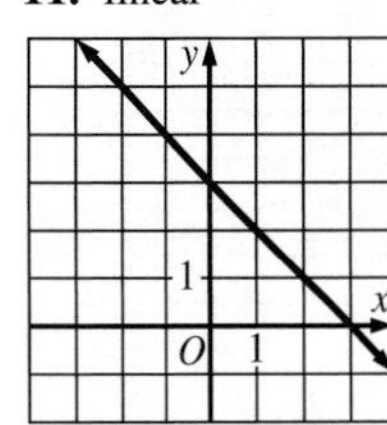

13. nonlinear **15.** linear **17. a.** wider **b.** up
19. a. wider **b.** down **21. a.** wider **b.** up
23. a. narrower **b.** down **25.** B **27.** ± 12 **29.** ± 3
31. ± 6 **33.** ± 7 **35.** ± 9 **37.** ± 0.1 **39.** $\pm\sqrt{2}$
41. $\pm 2\sqrt{2}$ **43.** –3 **45.** about 1.70; about –4.70 **47.** 1; 3
49. –1; 2.5 **51.** about ± 0.49 **53.** no solution
55. no solution **57.** 4; 5 **59.** –3; 5 **61.** 3; 2 **63.** $-5; \frac{1}{2}$
65. 8; –2 **67.** 1 **69.** –1; 3 **71.** –6; 7 **73.** two solutions
75. no solution **77.** no solution **79.** two solutions
81. one solution; –2 **83.** two solutions; –3; 8
85. two solutions; –4; 5 **87.** one solution; –5
89. no solution **91.** one solution; –8

Pages 574–576 Chapter 9

1. $(ab)^6$ **3.** $\left(-\frac{1}{2}\right)^5$ **5.** $3\left(\frac{2}{5}\right)^2$ **7.** 343 **9.** 25 **11.** $\frac{1}{27}$
13. –3.375 **15.** 16 **17.** –80 **19.** 8 **21.** 864 **23. a.** 26
b. 1.25; 0.25 **c.** 3 **25. a.** 300 **b.** 1.35; 0.35 **c.** 10
27. a. 1500 **b.** 1.8; 0.8 **c.** 4 **29.** 40.20
31. 155,271,168 **33.** 1387.84 **35.** about 622 **37.** 0.02
39. 1.69 **41.** 12.8 **43.** 40.63 **45.** 0.60 **47. a.** grows
b. 455 **c.** 1.15 **49. a.** grows **b.** 335 **c.** 1.01
51. a. grows **b.** 100 **c.** 2.3 **53.** $\frac{1}{x^3}$ **55.** $\frac{1}{b^2c^5}$ **57.** $\frac{z^7}{y^4}$
59. $\frac{w^2z^3}{x}$ **61.** $\frac{1}{4b^5}$ **63.** $\frac{1}{m^5n^6}$ **65.** $\frac{1}{9c^2}$ **67.** $\frac{2z^8}{x^5y^4}$ **69.** $\frac{1}{32}$
71. $\frac{1}{4}$ **73.** $\frac{16}{9}$ **75.** $\frac{19}{48}$ **77.** 8 **79.** $-\frac{29}{96}$ **81.** 300,000
83. 5.341 **85.** 0.00004503 **87.** 30,010,000,000
89. 0.0000099 **91.** 600,320,000,000 **93.** 9.675×10^3
95. 5.8×10^{-3} **97.** 2.7×10^6 **99.** 8.7492×10^{10}
101. 9×10^{-6} **103.** 4.000562×10^{-4} **105.** 4.506×10^{-3}
107. 6.8×10^6 **109.** 10^9 **111.** 4^3 **113.** $\frac{1}{9^7}$ **115.** 3^{11}
117. 2^4 **119.** 2^6 **121.** a^9 **123.** $\frac{a^2}{b^3}$ **125.** $\frac{1}{c^7}$ **127.** $12b^3$
129. $14c^5$ **131.** $\frac{5c^8}{b}$ **133.** 3×10^3 **135.** $\frac{2y^{10}}{x^6}$
137. 1.6×10^{13} **139.** 2.25×10^{-8} **141.** 8.1×10^{-11}
143. b^{15} **145.** $27x^3$ **147.** $\frac{a^6}{b^{12}}$ **149.** $4y^{12}$ **151.** $\frac{b^{12}}{c^{16}}$
153. $49a^4b^6c^{10}$

Pages 576–578 Chapter 10

1. $-3x^2 + 2x + 7$; quadratic **3.** $-4c^2 + 5c$; quadratic
5. $\frac{1}{2}b^3 + 2b - 4$; cubic **7.** $\frac{1}{2}y + 22$; linear **9.** $5t^2 + 9t - 8$; quadratic **11.** $2a^2 - 8a - 2$ **13.** $-b^2 - 4b - 10$
15. $5x + 4y - 11$ **17.** $3a^2 - a + 3b - 6$
19. $6r^3 - 2r^2 + 5r + 2$ **21.** $18n^2 + 48n$ **23.** $x^2 + 8x + 15$
25. $4b^2 - 31b + 21$ **27.** $16x^2 - 25$ **29.** $0.8a^3 - 0.6a^2 - a$
31. $-6d^2 + d + 70$ **33.** $3s^3 + 7s^2 - 32s - 48$
35. $-18y^3 - 21y^2 + 9y + 12$ **37.** $x^2 + 8xy - x + 15y^2 - 5y$
39. $2t^3 + 5t^2 - 8t - 6$ **41.** $x^3 + 3x^2 + 3x + 1$
43. $x^3 - 5x^2 + 6x$ **45.** $7(n + 2)$ **47.** $x(x^2 - 5)$
49. $6t(4t - 1)$ **51.** $3v(v - 2)$ **53.** $3(5a^2 + 3a - 3)$
55. $4z(2z^2 - 1)$ **57.** $6x(3x^2 + x + 2)$ **59.** $x(18x^2 - 14x + 9)$
61. 0; –9 **63.** $0; -\frac{1}{2}$ **65.** 0; 6 **67.** 8; –3 **69.** $-1; -\frac{4}{3}$
71. $-\frac{3}{4}; \frac{5}{7}$ **73.** 0; 3 **75.** $0; \frac{1}{2}$ **77.** 0; 16
79. $2x^2 + 4x$; $2x(x + 2)$ **81.** $(x + 2)^2$ **83.** $(b - 7)(b - 1)$
85. $(c - 5)(c - 2)$ **87.** $(b - 7)^2$ **89.** not factorable
91. $(5n - 2)(n - 1)$ **93.** $(2c + 1)(2c + 3)$
95. $(3n - 1)(2n - 1)$ **97.** $(9t - 1)(t - 2)$ **99.** $(x + 4)(x - 1)$
101. $(y + 3)(y - 2)$ **103.** $(t + 7)(t - 4)$ **105.** $(3x - 2)(x + 1)$
107. $(7d - 4)(d - 3)$ **109.** $3(x + 1)(x - 5)$
111. $(5c + 2)(3c - 1)$ **113.** $(3r + 1)(r - 3)$ **115.** 5; 6
117. –3; –7 **119.** –6; –7 **121.** $\frac{4}{3}; -1$ **123.** $-\frac{1}{2}; 3$
125. $\frac{3}{2}; -\frac{7}{6}$ **127.** $\frac{2}{3}; -3$ **129.** $\frac{3}{4}; -\frac{2}{3}$ **131.** $\frac{7}{5}; -\frac{9}{2}$
133. $-\frac{7}{3}; \frac{2}{3}$

Pages 579–581 Chapter 11

1. Yes; 1.5. **3.** Yes; 504. **5.** 45.2 **7. a.** about \$7.29
b. about \$3.67 **9.** $12x^2(x + 1)$ **11.** $30t(t + 1)$
13. $12y(y - 2)(y + 2)$ **15.** no solution **17.** 4 **19.** –1
21. –4 **23.** no solution **25.** –9 **27.** –2 **29.** $t - v - x$
31. $\frac{cd}{f}$ **33.** $\frac{y - b}{c}$ **35.** $-\frac{4}{5}n + \frac{3}{5}$ **37.** $a^2 - 2$ **39.** $-\frac{11}{6}$
41. $\frac{7y - 6}{3}$ **43.** $\frac{-2ax + 5}{4a}$ **45.** $\frac{ad}{ad - c}$ **47.** $\frac{x^2}{5}$
49. simplest form **51.** $\frac{3t^2}{4y}$ **53.** $\frac{1}{4n}$ **55.** $2b + 1$ **57.** $\frac{9y}{x}$
59. 6 **61.** $\frac{1}{2b}$ **63.** $\frac{1}{2y + 1}$ **65.** $\frac{2}{y}$ **67.** $\frac{1 - y}{xy}$ **69.** $\frac{y + 6}{y(y + 2)}$
71. $\frac{(x + 2)(x - 1)}{x(x - 2)}$ **73.** $\frac{2x + 1}{x + 1}$ **75.** $\frac{5(x - 2)}{x(x - 5)}$ **77.** $\frac{x - 2}{x + 1}$
79. $\frac{x - y}{x^2y^2}$

Pages 581–583 Chapter 12

1, 3, 5, 7. Algorithms may vary. Examples are given.
1. Step 1: Enter 52. Step 2: Add 52. Step 3: Repeat Step 2 thirteen times. **3.** Step 1: Choose a number and square it. If the result is less than 30, go to Step 2. If the result is greater than 30, go to Step 3. Step 2: Choose a larger number and square it. If the result is less than 30, repeat Step 2. If the result is greater than 30, go to Step 3. Step 3: Choose a smaller number and square it. If the result is less than 30, go to Step 2. If the result is greater than 30, repeat Step 3. Step 4: Continue until the number you find is satisfactorily close to $\sqrt{30}$. **5.** Step 1: Move the decimal point one place to the left to find 10% of 132, 13.2. Step 2: Multiply the result by 3 to find 30% of 132, 39.6. **7.** Step 1: Choose three values for x. Step 2: Substitute each of the chosen values into the equation and find the corresponding values of y. Step 3: Draw

coordinate axes. Graph the points corresponding to the ordered pairs you found in Steps 1 and 2. Step 4: Draw a line through the three points. **9.** Answers may vary. One possible path is $4 + 3 + 3 + 3 + 4 + 2 + 3 + 4 = 26$. It is longer. **11.** $3 + 2 + 4 = 9$ **13.** Yearbook; Community Service and Computer Club (tie); Newspaper. The Yearbook had the most 1's, Community Service and Computer Club had one each 1, 2, 3, and 4, and the Newspaper had zero 1's. **15.** One of you should break or cut the ear of corn into two pieces, either of which you would be willing to take. The other should choose one of the pieces. **17.** One of you should divide the task into two tasks, either of which you would be willing to do. (The task may be divided by area or by time, for example.) The other should choose one of the tasks.
19. 32 sandwiches **21.** 28 phones **23.** 720 ways **25.** 40,320 orders **27. a.** 15 pairs **b.** 1, 2; 1, 3; 1, 4; 1, 5; 1, 6; 2, 3; 2, 4; 2, 5; 2, 6; 3, 4; 3, 5; 3, 6; 4, 5; 4, 6; 5, 6 **c.** 9 pairs **29.** 1540 ways **31. a.** 19,600 combinations **b.** 117,600 ways **33.** $\frac{1}{1000}$

TOOLBOX

Page 584 Practice

1. GCF: 5; LCM: 90 **2.** GCF: 7; LCM: 196 **3.** GCF: 5; LCM: 210 **4.** GCF: 4; LCM: 120 **5.** GCF: 18; LCM: 216 **6.** GCF: 1; LCM: 450 **7.** GCF: 16; LCM: 64 **8.** GCF: 14; LCM: 210 **9.** You can buy 3 packages of pencils and 5 packages of pencil-top erasers.

Page 585 Practice

1. 25,000 **2.** 0.184 **3.** 194,000 **4.** 0.3 **5.** 0.00368 **6.** 0.5 **7.** 0.0000781 **8.** 0.09346 **9.** 0.293 **10.** 0.1083 **11.** 46.8 **12.** 0.413 **13.** 0.0006 **14.** 0.009 **15.** 0.1995 **16.** 0.00021973

Page 586 Practice

1–8. Forms of ratios may vary. **1.** 3 to 1 **2.** 2 : 9 **3.** $\frac{21}{32}$ **4.** 17 to 3 **5.** 1 to 9 **6.** 3 : 2600 **7.** 3 : 16 **8.** 3 to 4 **9.** 50 miles per hour **10.** $2.50 per hat **11.** 84 vibrations per second **12.** 9 hours per day **13.** $43.50 per month **14.** $.45 per pen

Page 587 Practice

1. 0.04; $\frac{1}{25}$ **2.** 0.87; $\frac{87}{100}$ **3.** 1.75; $\frac{7}{4}$ **4.** 0.824; $\frac{103}{125}$ **5.** 0.005; $\frac{1}{200}$ **6.** 75%; $\frac{3}{4}$ **7.** $66\frac{2}{3}$%; $\frac{2}{3}$ **8.** 2.5%; $\frac{1}{40}$ **9.** 160%; $\frac{8}{5}$ **10.** 51.3%; $\frac{513}{1000}$ **11.** 0.35; 35% **12.** 0.8; 80% **13.** 0.3375; 33.75% **14.** 0.56; 56% **15.** $0.5\overline{3}$; $53\frac{1}{3}$%

Page 589 Practice

1. 5 **2.** –9 **3.** –5 **4.** 5 **5.** –8 **6.** –6 **7.** –13 **8.** –11 **9.** 11 **10.** –5 **11.** 10 **12.** 9 **13.** 2 **14.** 6 **15.** –12 **16.** –18 **17.** –2 **18.** –18 **19.** 0 **20.** 20 **21.** –1 **22.** –3 **23.** –3 **24.** –16 **25.** 1 **26.** 0 **27.** –1 **28.** –10 **29.** –21 **30.** –2

Page 590 Practice

1. all numbers greater than 6 **2.** all numbers less than or equal to 9 **3.** all numbers less than –2 **4.** all numbers greater than or equal to 0 **5–8.** Let n = any number. **5.** $n \geq 7$ **6.** $n < 20$ **7.** $n > 3$ **8.** $n \geq 24$ **9.** Let p = a price; $p < 30$. **10.** Let h = a hiking distance; $h \leq 12$. **11.** The integers 8, 9, 10, ... make $n > 7$ a true statement. **12.** The integers –1, 0, 1, 2, 3, ... make $r \geq -1$ a true statement. **13.** The integers 4, 3, 2, 0, –1, –2, ... make $v < 5$ a true statement. **14.** The integers 3, 2, 1, 0, –1, –2, ... make $t \leq 3$ a true statement.

Page 591 Practice

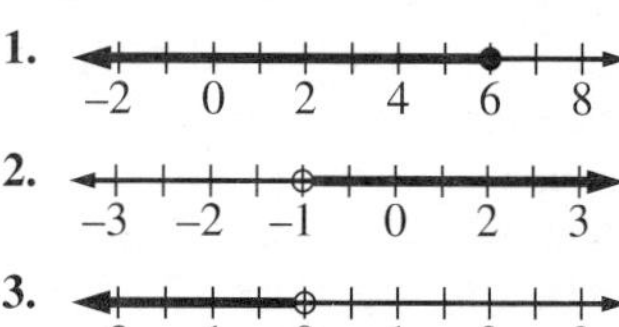

4. –2 0 2 4 6 8

5. –8 –6 –4 –2 0 2

6. 0 2 4 6 8 10

7. 0 1 2 3 4 5

8. –8 –6 –4 –2 0 2

9. $x > 0$ **10.** $x \leq -4$

Page 592 Practice

1. 7 ft 3 in. **2.** 5000 lb **3.** 32 fl oz **4.** 55,000 cm **5.** 3.4 kg **6.** $21\frac{3}{4}$ ft **7.** 11 pt **8.** 250,000 mL

Page 593 Practice

1. 51 m **2.** about 6.59 ft **3.** 18 cm **4.** about 50.24 in. **5.** 16.8 cm **6.** about 20.72 mm **7.** $16\frac{1}{2}$ in. **8.** about 43.96 yd **9.** about 69 in.

Page 594 Practice

1. 113.04 ft^2 **2.** 17.5 m^2 **3.** 133 in.^{2} **4.** 37.5 cm^2 **5.** 162 in.^{3} **6.** 200 m^3 **7.** 15,625 in.^{3} **8.** 249.375 cm^3 **9.** 56 yd^3 **10.** The rectangular pizza has the greater area.

Page 595 Practice

1. about 30°; acute **2.** about 90°; right **3.** about 140°; obtuse **4.** about 65°; acute

5. **6.**

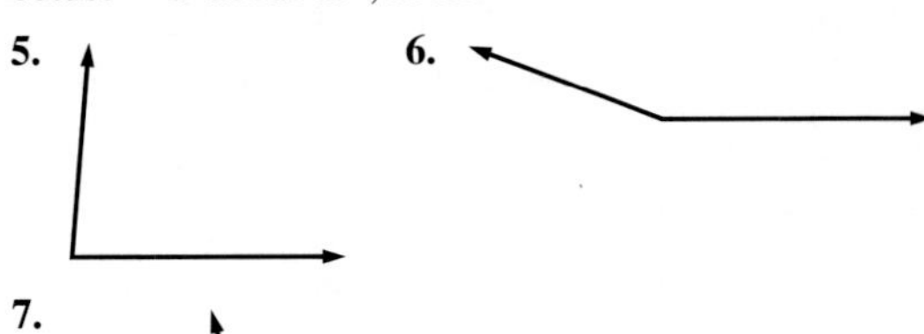

7.

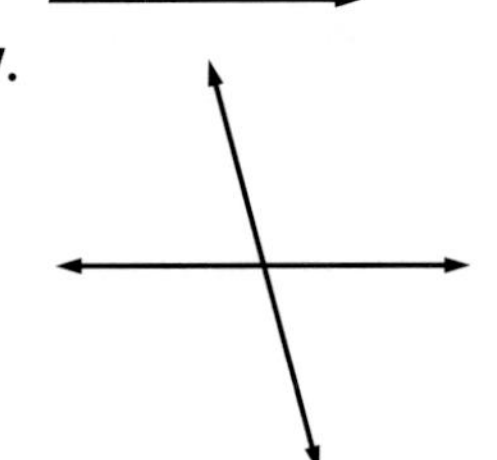

Pages 596–599 Practice

1. 34 people **2.** black **3.** 2 people **4.** Burlington: about 5 in.; Portland: about 12 in. **5.** July/August **6.** about 5 in. **7.** about 21,500 bicycles **8.** Tandem Tiger **9.** 9.5 symbols **10.** Strings: about 77; Woodwinds: about 27; Brass: about 25; Percussion: about 11 **11.** about 118 players **12.** 9 racers **13.** 32 min; 71 min **14.** 34 students **15.** 8 students **16.** 90–99 laps; 2 students

17.

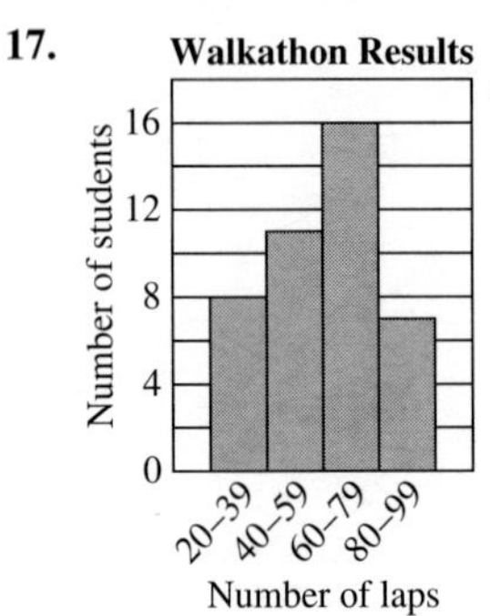

Page 601 Practice

1. 128 + 367 = 495 or 129 + 357 = 486 **2.** 15 dimes; 10 dimes and 2 quarters; 5 dimes and 4 quarters; 6 quarters **3.** 48; 60; 72; 84 **4.** 24 ft² **5.** about 5.2 gal **6.** 400 cans **7.** about 25 hours **8.** 60°

TECHNOLOGY HANDBOOK

Pages 603–610 Calculator Practice

1. –8 **2.** 30 **3.** about 6.28 **4.** 3.5 **5.** 32 **6.** 0.01872 **7.** 40,320 **8.** 21

9. a.

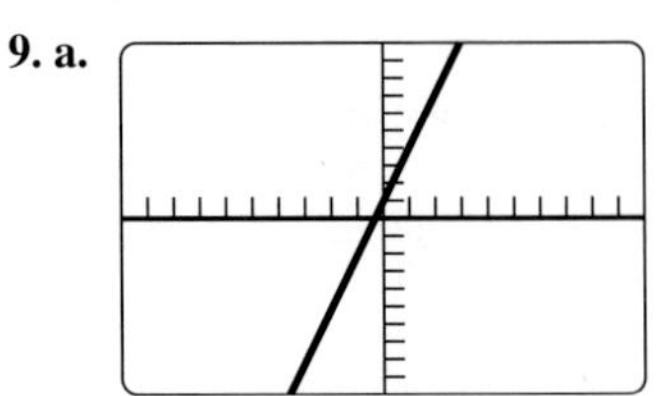

b.

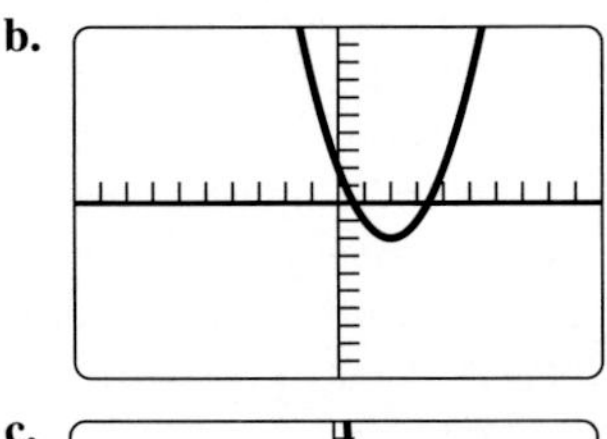

c.

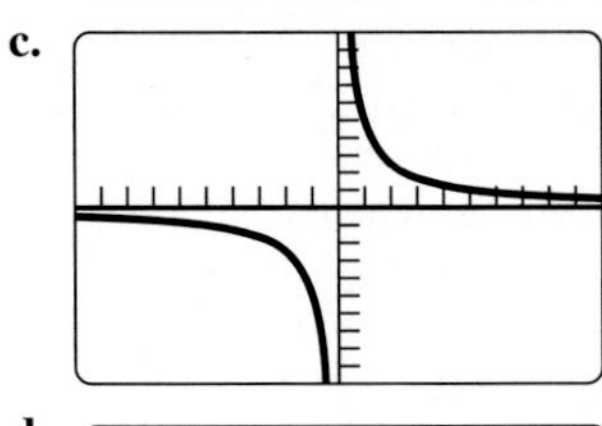

d.

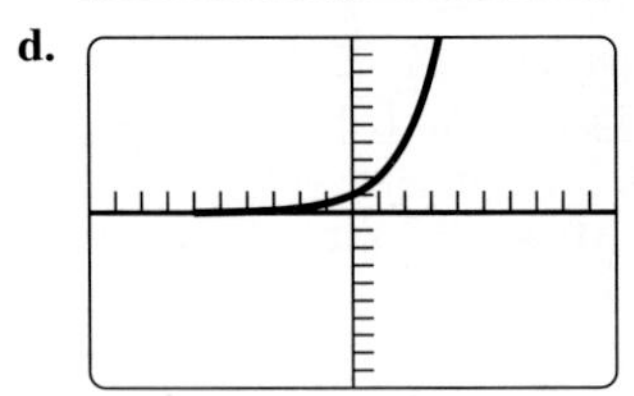

e. Zoom windows may vary. An example is given.

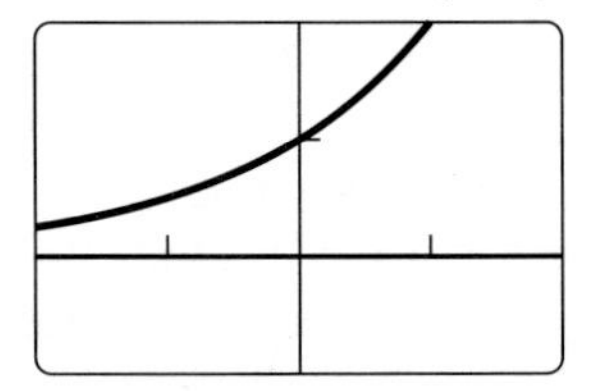

10. Answers may vary. An example is given; $-75 \le x \le 75$, $-20 \le x \le 80$ **11. a.** about 4.9 **b.** between 4.85 and 4.86 **12.** about 56.55 **13.** x is about 0.15 and y is about 4.22.

14. about –1.70 and about 2.20 **15.** $\begin{bmatrix} -2 & -1 & 5 \\ -1 & 4 & 5 \end{bmatrix}$

16. Equations may vary. An example is given; $y = 3.11x + 0.1$; about 9.7 in.

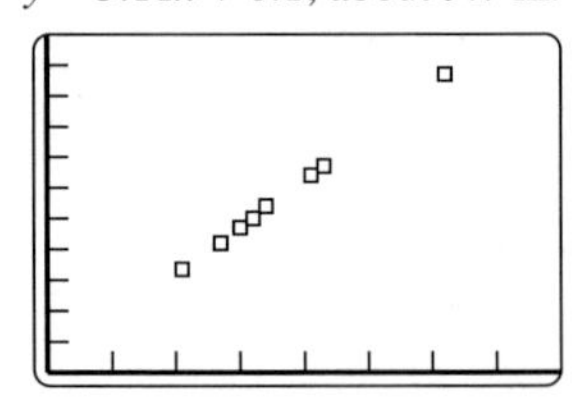

17. Possible table:

X	Y1	
0	7	
1	2	
2	−1	
3	−2	
4	−1	
5	2	
6	7	
X=3		

The y-values stop decreasing at $x = 3$.

18. Monday: 60 min; Tuesday: 90 min; Wednesday: 120 min; Thursday: 60 min; Friday: 0 min; Saturday: 285 min; Sunday: 0 min **19.** You will owe $45 at 15 weeks. The loan will be paid back in 22 weeks.

Additional Answers

CHAPTER 1

Page 7 Exercises and Applications

28. Answers may vary. Examples are given. The bars could be arranged in order of height from greatest to least or in alphabetical order. The bars of a histogram are arranged in numerical order of the categories, from least to greatest. Rearranging the bars of a histogram would make it very difficult to interpret.
29. 2.67 **30.** 0.43 **31.** 5.56 **32.** 5.78 **33.** 0.13 **34.** 3.33

Page 17 Exercises and Applications

7. a. Answers may vary. An example is given. I think the typical age is best described by the median, 2, which is in the middle of the data values. The mean is influenced by the atypical age of the two oldest vaquitas, and the mode is clearly too low to be typical. **b.** Answers may vary. Sample questions: How old must vaquitas be before they are able to reproduce? How many offspring does a typical vaquita produce?

Page 22 Exercises and Applications

13. 1050 years **14.** 644 years **15.** 507 years **16.** 93 years
17. 1100 years **18.** Answers may vary. **19.** –12 **20.** –1 **21.** 78 **22.** 9
23. 4.3 **24.** 2 **25.** –3 **26.** 17 **27.** $\frac{9}{4}$

Page 27 Checking Key Concepts

7. $\frac{13}{16}$; The quotient of two numbers is positive if both numbers are negative.

8. $-\frac{4}{15}$; The quotient of two numbers is negative if one number is negative.

9. $-\frac{12}{7}$; The quotient of two numbers is positive if both numbers are negative, so $\frac{-12}{-7} = \frac{12}{7}$. The opposite of a positive number is negative, so $-\left(\frac{12}{7}\right) = -\frac{12}{7}$.

Page 33 Exercises and Applications

12. the total cost of renting a car for a day at \$20 a day plus \$.15 per mile when you drive m miles **13.** the amount you pay for a month of long-distance telephone service if the cost is \$.15 a minute, the first 60 minutes are free, and you spend m minutes on long-distance calls in the month **14. a.** Let f = the number of grams of fat, p = the number of grams of protein, and c = the number of grams of carbohydrates in the food; $9f + 4p + 4c$. **b.** Answers may vary. An example is given. A serving of canned pasta with tomato and cheese sauce contains 2 g of fat, 5 g of protein, and 36 g of carbohydrates. To find the number of calories in the serving of pasta, substitute 2 for f, 5 for p, and 36 for c in the expression for part (a): $9(2) + 4(5) + 4(36) = 182$; there are about 182 calories per serving.

Page 34 Exercises and Applications

18. Answers may vary. Examples are given. **a.** You need $3n + 1$ toothpicks to make a row of n squares.

Number of squares	Number of toothpicks
1	4
2	7
3	10
4	13
n	$3n + 1$

b. You need $4n + 1$ toothpicks to make a row of n pentagons.

Number of pentagons	Number of toothpicks
1	5
2	9
3	13
4	17
n	$4n + 1$

c. The expressions all have the form (constant $\times$ n) + 1. In each expression, the value of the constant is 1 less than the number of sides of the figures formed. The constant is 2 for triangles, 3 for squares, and 4 for pentagons.

Page 39 Exercises and Applications

39. Use the distributive property to simplify $10(d - 6)$ and $3(2d + 8)$. Subtract the second expression from the first and combine like terms to get $4d - 84$. Answers may vary. An example is given. The two expressions are equivalent because they have the same value no matter what number you substitute for d.

CHAPTER 2

Page 65 Exercises and Applications

39. Answers will vary. An example is given. If an equation involves both (1) multiplication or division and (2) addition or subtraction, undo the addition or subtraction first. Use addition to undo subtraction, and subtraction to undo addition. Then undo the multiplication or division. Use multiplication to undo division, and division to undo multiplication. For example, to solve $17 - 2x = -5$, subtract 17 from both sides to get $-2x = -22$. Then divide both sides by –2 to get $x = 11$. To solve $\frac{k}{3} - 19 = 8$, add 19 to both sides to get $\frac{k}{3} = 27$. Then multiply both sides by 3 to get $k = 81$.

Page 81 Exercises and Applications

4. No.

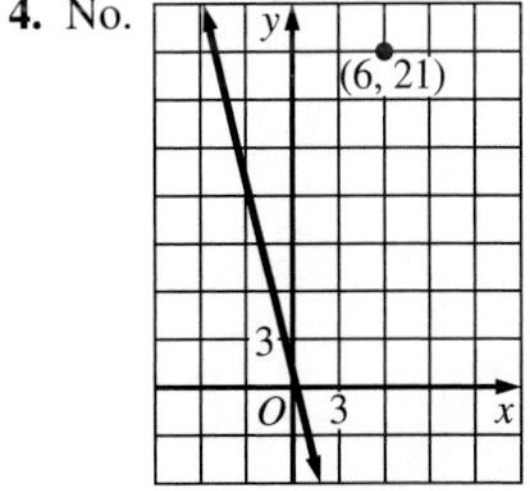

5. No.

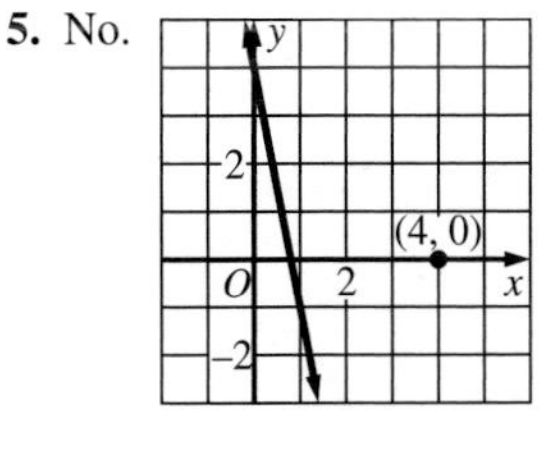

6. No.

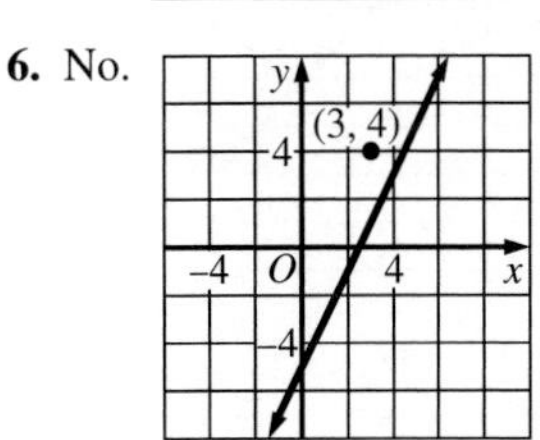

7. Yes.

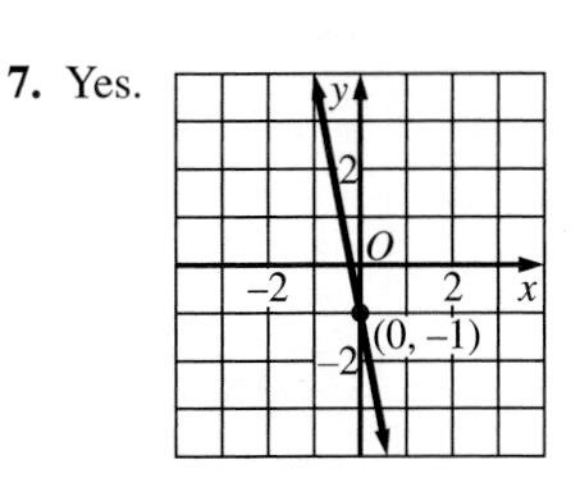

8. Yes.

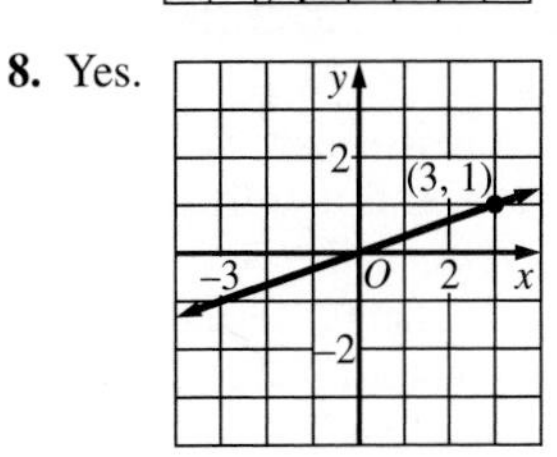

9. The third point serves as a check on your work.

10. a. $C = 1000 + 68n$

b.

n	C
0	1000
1	1068
2	1136
3	1204
…	…

c. Refrigerator Costs

Total cost (dollars) · 1000 · 1200 · 1400 · 1600 · C · 0 2 4 6 8 n · Years of use

Page 89 Exercises and Applications

20. 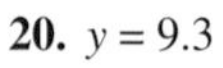 $y = 9.3$

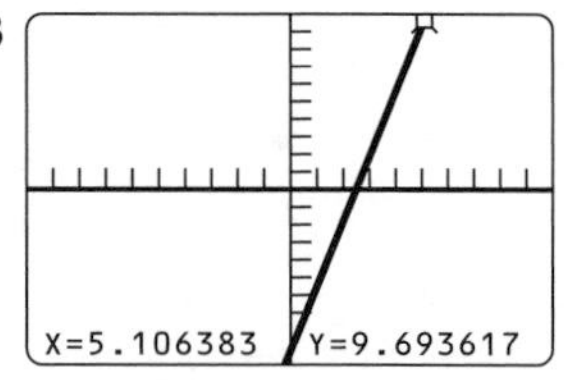

21. B

22. a.

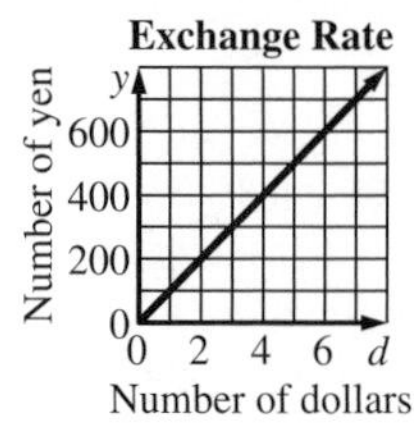

b. Extend the graph and use the process described in Example 1 on page 84 to find the dollar values that correspond to 100,000 yen and to 150,000 yen. You should exchange between \$1000 and \$1500.

23. No. **24.** Yes. **25.** No. **26.** Yes.

27. (1) an equation: $y = 2x + 3$

(2) a table of values:

x	y
–1	1
0	3
1	5
…	…

(3) a graph:

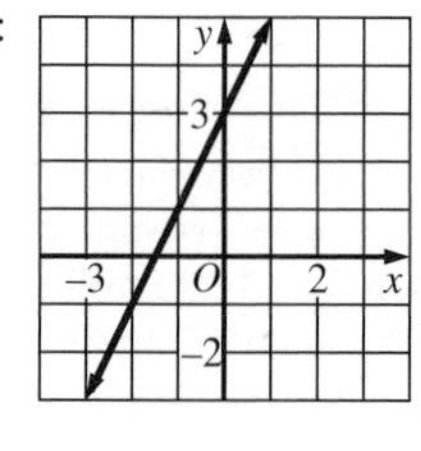

28. (1) an equation: $y = 5x$

(2) a table of values:

x	y
–1	–5
0	0
1	5
…	…

(3) a graph:

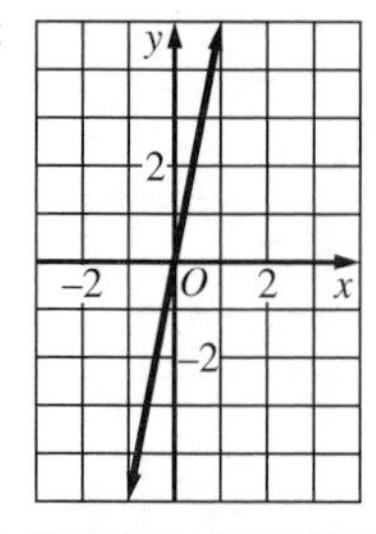

29. (1) an equation: $y = 4 - x$

(2) a table of values:

x	y
–1	5
0	4
1	3
…	…

(3) a graph:

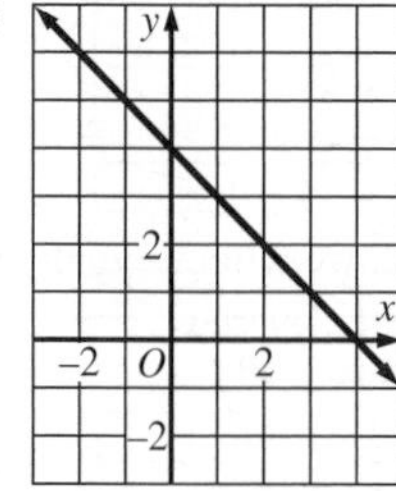

Page 95 Chapter 2 Assessment

27. a. $C = 450 + 20n$, where n is the number of people at the party and C is the total cost in dollars

b.

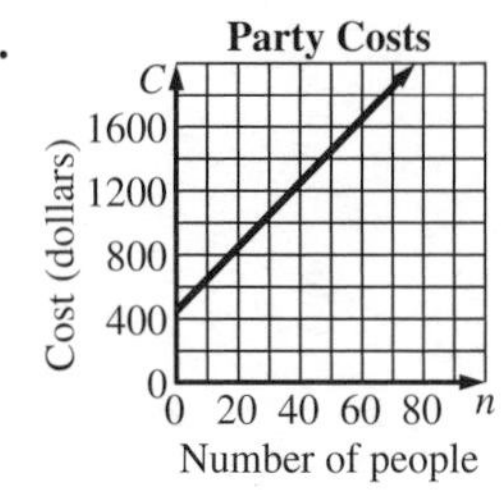

c. 77 people **d.** Answers may vary. One way to solve the problem is to solve the equation $2000 = 450 + 20n$. Another is to graph $y = 450 + 20x$ on a graphing calculator and use the trace feature to find the value of x when the value of y is close to 2000.

CHAPTER 3

Page 112 Exercises and Applications

19. No.

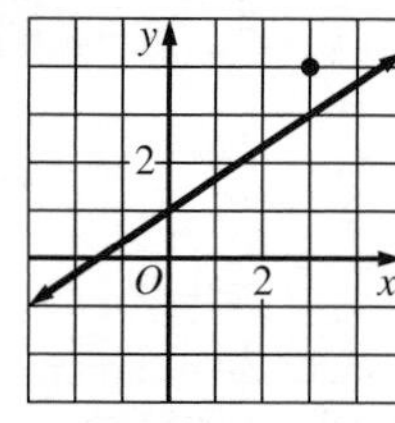

20. No.

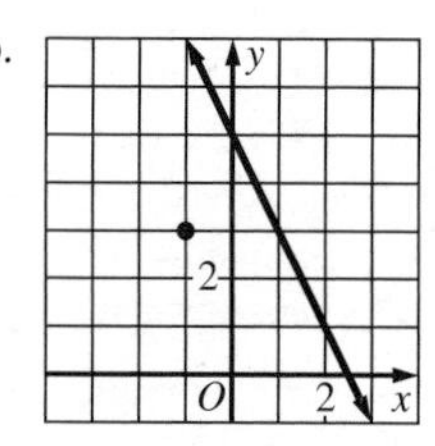

21. Yes.

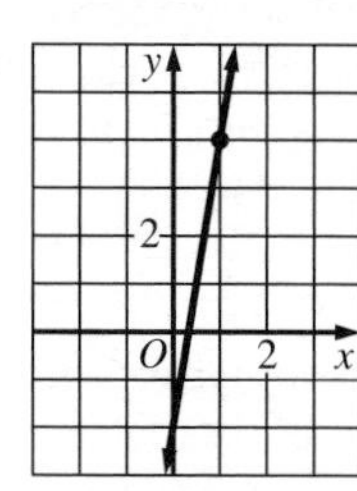

Page 112 Assess Your Progress

6. The data in a table show direct variation if the data pairs have a constant ratio. For example, in the table below, the distance you can bike at 10 mi/h varies directly with the time spent biking. An equation is a direct variation equation if it can be written in the form $y = kx$ or $\frac{y}{x} = k$, where k is a constant. For example, the equation $y = 10x$ is a direct variation equation. A scatter plot or other graph shows direct variation if the points of the graph lie on or near a line through the origin.

Time (h)	Distance (mi)	$\frac{\text{Distance}}{\text{Time}}$
0.5	5	10
1	10	10
2	20	10
2.5	25	10

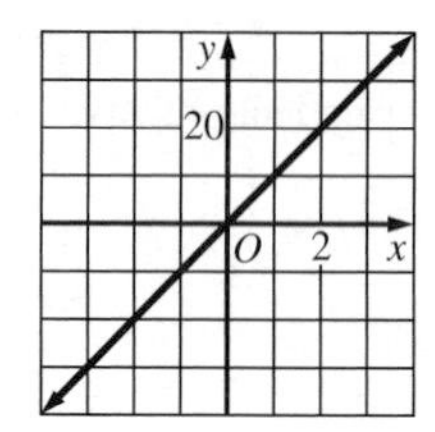

Page 117 Exercises and Applications

21. Answers may vary. Examples are given. If each step in both the new and old designs is x inches across, then the old design requires $14(8.25)x + 13(9)x = 232.5x$ in.^{2} of wood and the new design requires $16(7)x + 15(11)x = 277x$ in.^{2} of wood. The old design requires less wood, so it is less costly to build. With the new design, the tread is deeper and more of a person's foot fits on the step. Also, the new design has a smaller rise, so it is less steep than the old design. The new stairs are less tiring to use, and so more safe. **22. a.** the new design; $\frac{9}{14} \approx 0.64 \approx \frac{7}{11}$ and $\frac{8.25}{9} \approx 0.92$ **b.** about 9.8 ft; With the old design, the ratio of the height of the staircase to the length would be about $\frac{8.25}{9} \approx 0.92$; $\frac{9}{l} \approx 0.92$; $l \approx 9.8$ ft.

Page 123 Exercises and Applications

18. Answers may vary. Check students' work. **19.** Any line through (0, –2) has vertical intercept –2, so its equation in slope-intercept form is $y = mx - 2$. Any line with slope 3 has an equation of the form $y = 3x + b$. Exactly one line has slope 3 and passes through the point (0, –2), namely, the line with equation $y = 3x - 2$.

20.

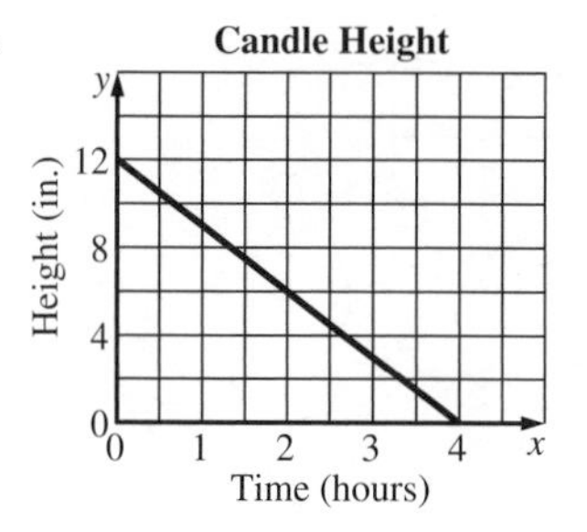

21. –3; the rate at which the height of the candle changes, –3 in./h; 12; the original height of the candle, 12 in. **22.** 4 h; Answers may vary. For example, the horizontal intercept of the graph in Ex. 20 is 4, indicating that the candle's height was 0 when 4 h had elapsed. **23.** $h = -3t + 12$ **24.** Answers may vary. Samples of clepsydra from as early as A.D. 1400 have been found. The earliest such devices were simple containers that measured time by the gradual flow of water. The Romans later developed a more complicated clepsydra that consisted of a cylinder into which water dripped from a reservoir. Readings were taken against a scale with a float in the cylinder. Since the water level increased over time, the slope of a graph for this type of clepsydra would be positive. At some point, the cylinder would be emptied and the process begun again.

Page 124 Assess Your Progress

1. 2 **2.** undefined **3.** –1 **4. a.** $y = -50x + 350$ **b.** The slope, –50, is the rate of change of Sue's distance from Tampa, –50 mi/h. The vertical intercept, 350, is Sue's distance from Tampa at $t = 0$, that is, the distance from her home to Tampa.

5. 3; –4

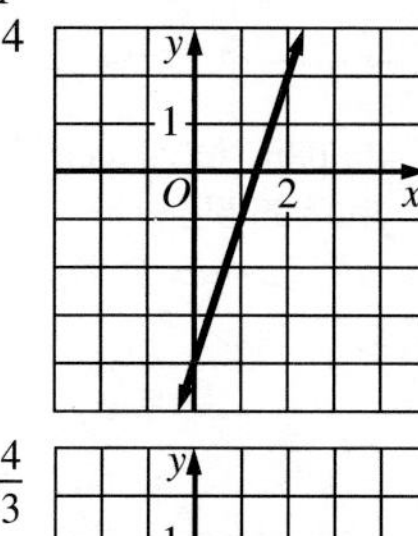

6. –1, 2

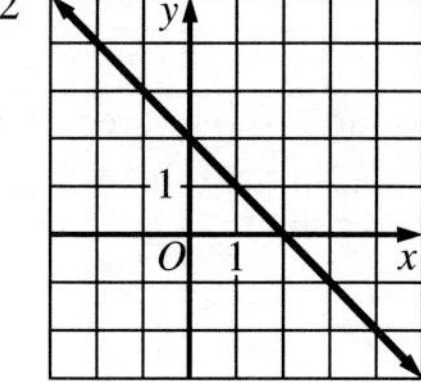

7. $\frac{1}{3}, -\frac{4}{3}$

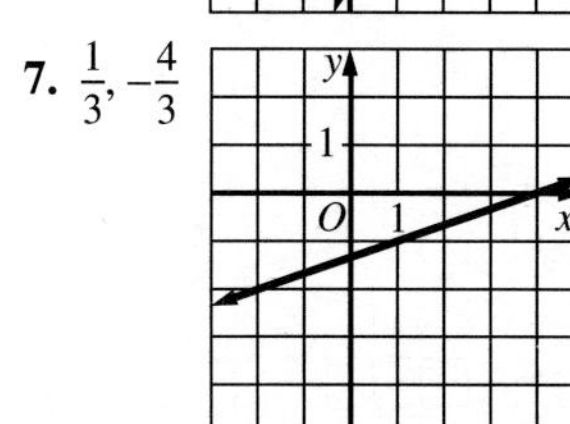

8. Answers may vary. An example is given. One way to find the slope is to locate two points on the line and then use the formula $m = \frac{y_2 - y_1}{x_2 - x_1}$. Another way is to locate two points on the line, count the number of units in the vertical change and in the horizontal change, and compute the ratio of these values. I prefer to count units, when possible, because it is faster than the formula, I think.

Page 129 Exercises and Applications

27. Answers may vary. An example is given. Substitute the desired temperature for y in the equation you found in Ex. 26 and solve for x. **28.** Let n = the number of visits and C = the cost in dollars. The slope indicates the cost per visit and the y-intercept indicates the basic cost, the cost independent of the number of visits.

(a) $C = 20n$; The cost per visit is \$20 and the basic cost is 0.

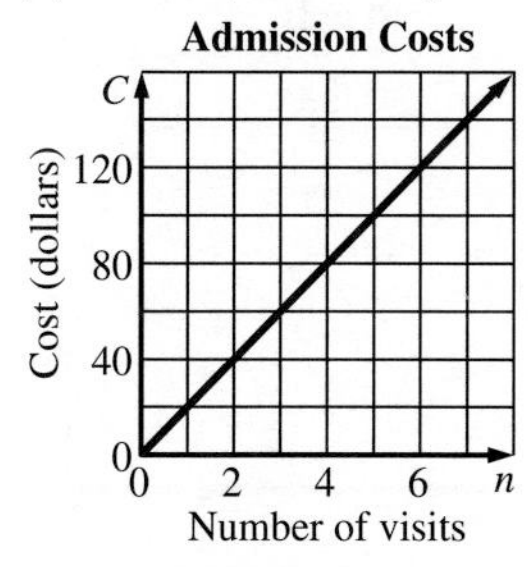

(b) $C = 80$; The cost per visit is 0 and the basic cost is \$80.

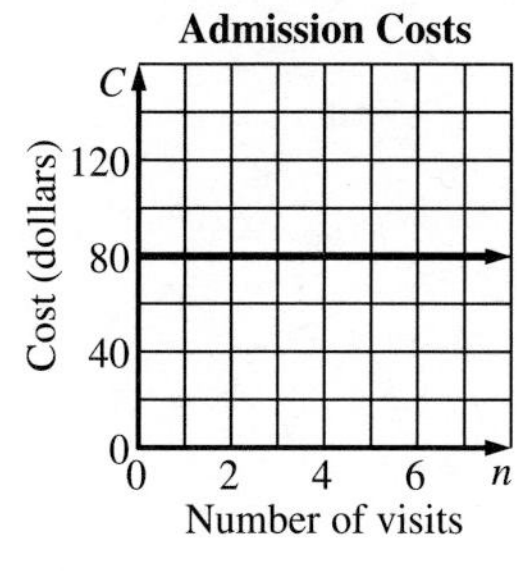

(c) $C = 50 + 12n$; The cost per visit is \$12 and the basic cost is \$50.

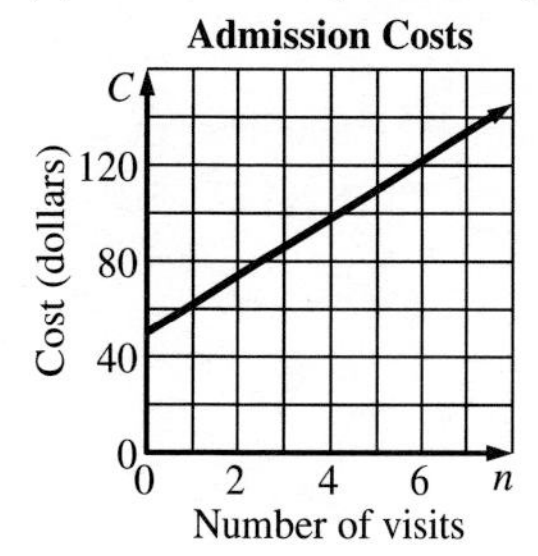

Page 132 Exploration

Answers will vary. Examples are given.

1.

Number of pieces of spaghetti	Number of marbles needed to break spaghetti
1	3
2	7
3	11
4	16
5	20

2. Let n = the number of marbles and m = the number of pieces of spaghetti; $n = 4m$.

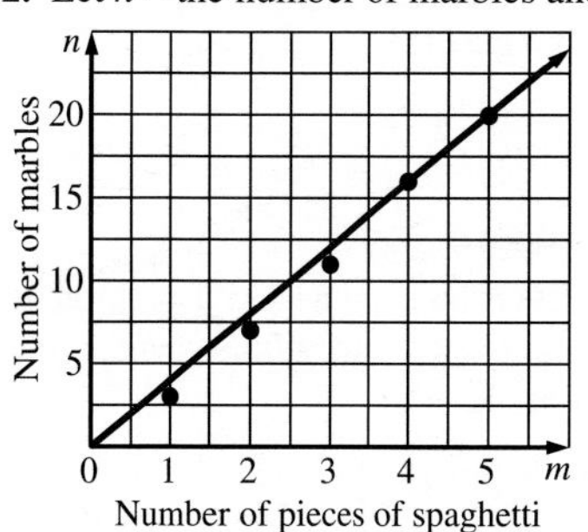

3. If $m = 6$, $n \approx 24$. **4.** Moving the chairs does affect the results. If the chairs are moved farther apart, fewer marbles are needed to break the spaghetti; if the chairs are moved closer together, more marbles are needed.

Page 133 Exercises and Applications

3. Answers may vary. For example, the scatter plots below suggest that garbage generated and garbage recycled have both increased steadily from 1960 to 1990. By fitting a line to the points in each scatter plot, you can find equations such as $y = 0.05x + 2.7$ for garbage generated and $y = 0.01x + 0.1$ for garbage recycled. (y is the number of pounds of garbage per person per day and x is the number of years since 1960.) This means that the rate of increase in recycling garbage, about 0.01 lb/person/day, is only about one-fifth the rate of increase in generating garbage, about 0.05 lb/person/day. Unless these trends are reversed, we will eventually run out of landfill space.

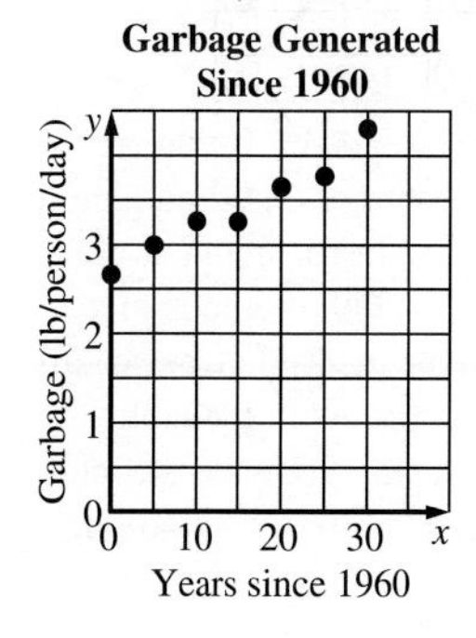

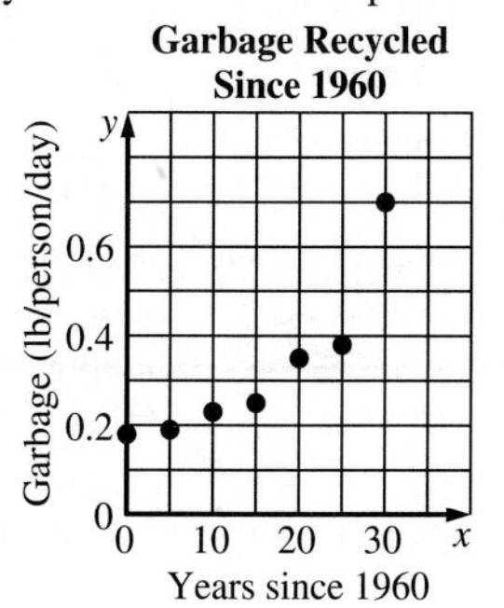

Page 135 Assess Your Progress

8–11. Answers may vary. Examples are given.

8, 9.

Cable Systems

Number of cable systems; 8000; 6000; 4000; 0; 4; 8; Years since 1980

$y = 560x + 4000$, where x is the number of years since 1980 and y is the number of cable systems

10. about 12,400 cable systems **11.** I have seen physical representations of real objects, such as model cars, trains, and airplanes, whereas mathematical modeling uses abstractions such as tables, graphs, functions, equations, or inequalities to represent real-life situations. Physical models and mathematical models are alike because both represent real-world ideas and both are tools for exploring these real-world ideas.

CHAPTER 4

Pages 147–148 Exploration

1.

No. of CDs	Cost at store ($)	CD club cost ($)
14	154	90
15	165	105
16	176	120

2. the CD club; No, since the cost for each CD is less at the store than through the club, at some point the savings per CD will overtake the initial savings from the 8 free CDs and the total cost will be less at the store.

3.

No. of CDs	Cost at store ($)	CD club cost ($)
20	220	180
30	330	330
40	440	480
50	550	630

4. The costs are equal for 30 CDs. The entries in the the second and third columns of the table are the same for 30 CDs. If you are choosing based only on costs, choose the club if the number of CDs you plan to buy is greater than or equal to 14 and less than 30; choose the discount store otherwise. If you plan to buy 30 CDs, choose either. **5.** $y = 11x$ **6.** $y = 15(x - 8)$

7.

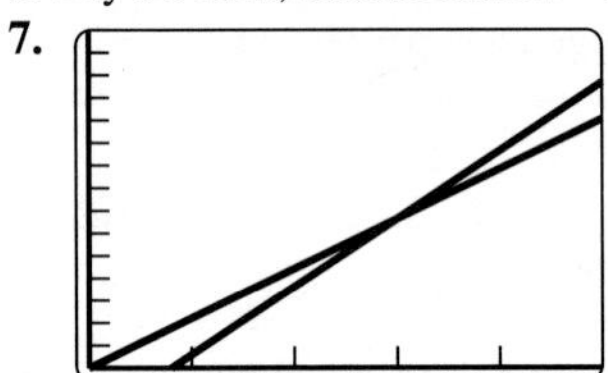

8. The costs are equal for 30 CDs. The graphs intersect at $x = 30$. If you are choosing based only on costs, choose the club if the number of CDs you plan to buy is greater than or equal to 14 and less than 30; choose the discount store otherwise. If you plan to buy 30 CDs, choose either. **9.** Both methods give the same answer. Answers may vary. Examples: The table method gives specific values. This can be helpful when the value you need appears in the table. However, if the value you need does not appear in the table, it may be difficult to find an exact solution. A graph gives a good visual picture of the situation, but may take more time than a table to create.

Page 150 Exercises and Applications

2.

No. of months	Cost at Club A ($)	Cost at Club B ($)
4	260	200
5	300	250
6	340	300
7	380	350
8	420	400
9	460	450
10	**500**	**500**
11	540	550
12	580	600
13	620	650
14	660	700
15	700	750
16	740	800
17	780	850
18	820	900

Graph $y = 100 + 40x$ and $y = 50x$ on the same coordinate plane.

Fitness Club Costs

Cost (dollars); 800; 600; 400; 200; 0; Cost for Club A; Cost for Club B; 4; 8; 12; 16; Months of membership

5.

No. of hours of overtime per year	Total earnings from Job A (dollars)	Total earnings from Job B (dollars)
50	30,000	24,600
100	30,000	25,200
150	30,000	25,800
200	30,000	26,400
250	30,000	27,000
300	30,000	27,600
350	30,000	28,200
400	30,000	28,800
450	30,000	29,400
500	**30,000**	**30,000**
550	30,000	30,600

Job Earnings

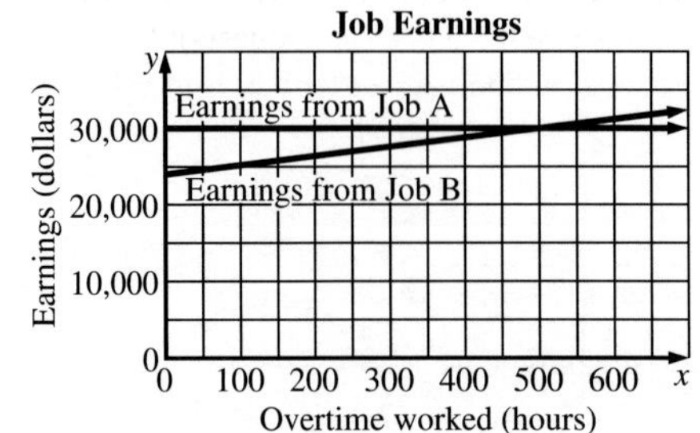

Page 152 Exercises and Applications

14. a. They pass each other 1 h and 20 min after each begins driving. At that time, they are 120 km from Montreal. Tables and graphs may vary. Examples are given.

Traveling time (h)	Sylvie's distance from Montreal (km)	Anne's distance from Montreal (km)
$\frac{1}{6}$	15	242.5
$\frac{1}{3}$	30	225
$\frac{1}{2}$	45	207.5
$\frac{2}{3}$	60	190
$\frac{5}{6}$	75	172.5
1	90	155
$\frac{7}{6}$	105	137.5
$\frac{4}{3}$	**120**	**120**

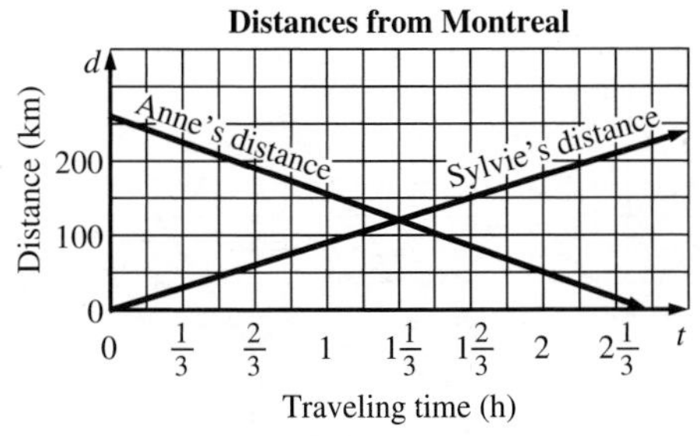

b. $90t = 260 - 105t$; $195t = 260$; $t = \frac{4}{3}$, that is, $\frac{4}{3}$ h, or 1 h 20 min

Page 157 Exercises and Applications

36. This equation has a variable in the denominator and the other equations in this section do not. Methods may vary. An example is given. Subtract 2 from both sides to get $\frac{3}{5x} = 6$. Multiply both sides by $5x$ to get $3 = 30x$. Then divide both sides by 30 to get $x = \frac{1}{10}$.

37. Answers may vary. An example is given.
1. What number has no reciprocal?
Solve each equation by using reciprocals.
2. $\frac{1}{3}x = 2$ 3. $-\frac{3}{5}y = 15$ 4. $-\frac{k}{2} + 4 = 3$ 5. $\frac{3v}{7} - 4 = 11$
Answers: 0; 6; –25; 2; 35

Page 159 Think and Communicate

3. Use the distributive property to simplify $3(x + 10)$. Then subtract $3x$ from both sides. Finally, divide both sides by 2. **4. a.** Subtract $3x$ from both sides. Then divide both sides by 2. **b.** No; to solve the equation in question 3 you need to use the distributive property first. You do not need to do this to solve the equation in part (a).

Page 164 Assess Your Progress

1. a.

No. of Checks	Minimum Account Cost ($)	Maximum Account Cost ($)
10	2.50	7.00
11	3.25	7.00
12	4.00	7.00
13	4.75	7.00
14	5.50	7.00
15	6.25	7.00
16	7.00	7.00

b. The monthly cost will be the same when 16 checks are written in a month.

Page 168 Exercises and Applications

17. a.

	Distance across (km)	Rate (km/h)	Time (hours)
Wind at his back	d	14	$\frac{d}{14}$
Into the wind	d	12	$\frac{d}{12}$

b. $\frac{d}{14} + \frac{d}{12} = \frac{1}{2}$; $84\left(\frac{d}{14} + \frac{d}{12}\right) = 84\left(\frac{1}{2}\right)$; $6d + 7d = 42$; $d \approx 3.2$; The distance across the lake is about 3.2 km.

Page 170 Exercises and Applications

33.

	A	B	C
1	Number of pages	Cost (ordinary)	Cost (special)
2	1,000	1278.82	646.14
3	10,000	1322.20	863.43
4	100,000	1756.00	3036.29

34. For the fax machine that uses ordinary paper, $C = 1274 + \left(\frac{2.41}{500}\right)p$ and for the fax machine that uses special paper, $C = 622 + \left(\frac{8.45}{350}\right)p$, where C is the total cost and p is the number of pages. $1274 + \left(\frac{2.41}{500}\right)p = 622 + \left(\frac{8.45}{350}\right)p$; $p \approx 33{,}742.4$; about 33,742 pages **35.** Answers may vary. An example is given. Questions you might ask include: About how many pages do you expect to use per year? How essential is it that the copies be very easy to read? How much money is available for the purchase of a fax machine and for paper for the machine? The cheaper machine will be less expensive if the company's total paper usage per year is less than about 33,750 sheets. Otherwise, the more expensive machine will have the lower total cost. If clear copies are essential and can prevent costly errors, the more expensive machine may be the wiser purchase, regardless of the number of pages used.

Page 176 Exploration

1. a.

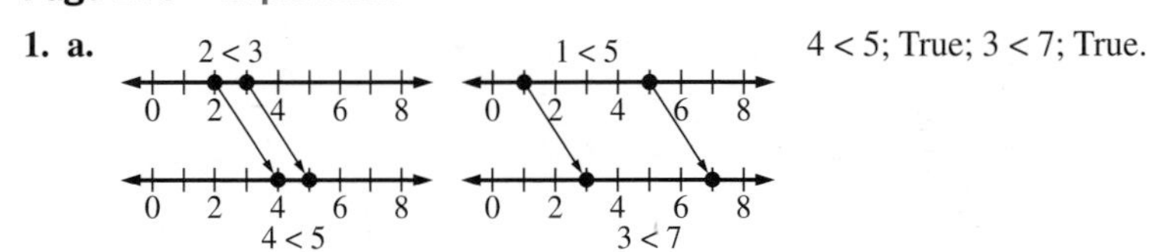

$4 < 5$; True; $3 < 7$; True.

b.

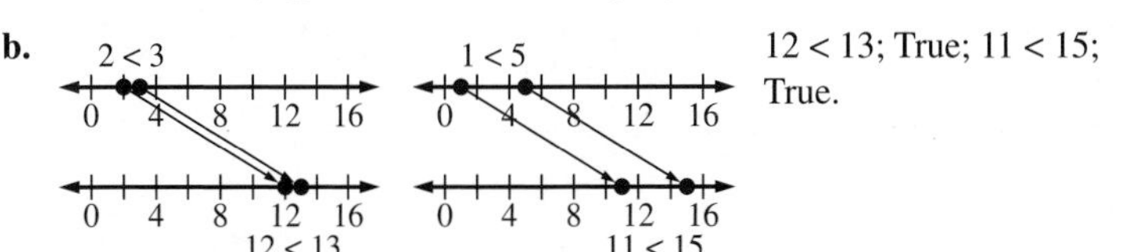

$12 < 13$; True; $11 < 15$; True.

c.

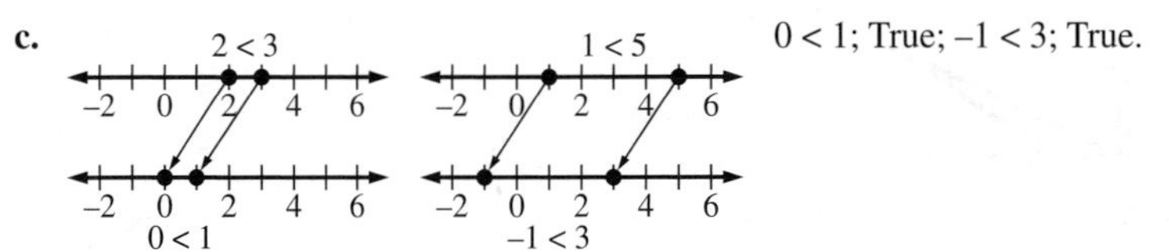

$0 < 1$; True; $-1 < 3$; True.

d.

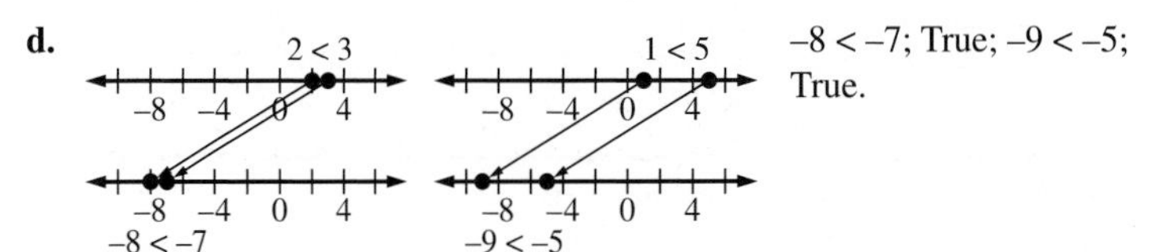

$-8 < -7$; True; $-9 < -5$; True.

2. a.

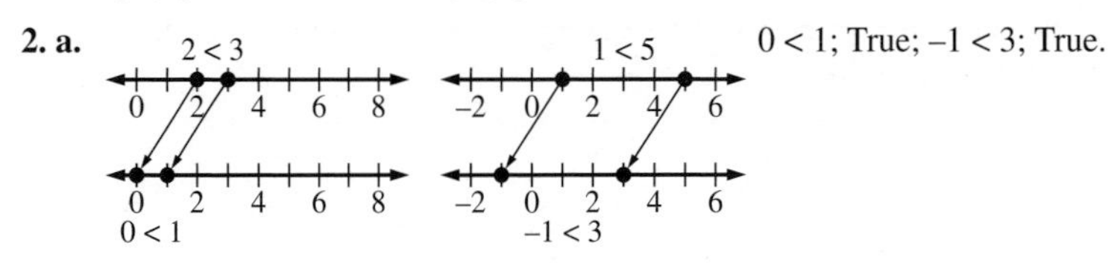

$0 < 1$; True; $-1 < 3$; True.

b. 2 < 3; 1 < 5

$-8 < -7$; $-9 < -5$

$-8 < -7$; True; $-9 < -5$; True.

c. 2 < 3; 1 < 5

$4 < 5$; $3 < 7$

$4 < 5$; True; $3 < 7$; True.

d. 2 < 3; 1 < 5

$12 < 13$; $11 < 15$

$12 < 13$: True; $11 < 15$; True.

3. a. 2 < 3; 1 < 5

$4 < 6$; $2 < 10$

$4 < 6$; True; $2 < 10$; True.

b. 2 < 3; 1 < 5

$20 < 30$; $10 < 50$

$20 < 30$; True; $10 < 50$; True.

c. 2 < 3; 1 < 5

$-4 > -6$; $-2 > -10$

$-4 < -6$; False; $-2 < -10$; False.

d. 2 < 3; 1 < 5

$-20 > -30$; $-10 > -50$

$-20 < -30$; False; $-10 < -50$; False.

4. a. 2 < 3; 1 < 5

$1 < 1.5$; $0.5 < 2.5$

$1 < 1.5$; True; $0.5 < 2.5$; True.

b. 2 < 3; 1 < 5

$0.2 < 0.3$; $0.1 < 0.5$

$0.2 < 0.3$; True; $0.1 < 0.5$; True.

c. 2 < 3; 1 < 5

$-1 > -1.5$; $-0.5 > -2.5$

$-1 < -1.5$; False; $-0.5 < -2.5$; False.

d. 2 < 3; 1 < 5

$-0.2 > -0.3$; $-0.1 > -0.5$

$-0.2 < -0.3$; False; $-0.1 < -0.5$; False.

Page 179 Exercises and Applications

18. No. Answers may vary. An example is given. You need to consider your work schedule and the schedule for each commuting option, the distance from each commuting site to your work location, the time required by each commuting option, and the comfort provided by each option.

CHAPTER 5

Page 209 Exercises and Applications

12. Check students' work. An example is given.

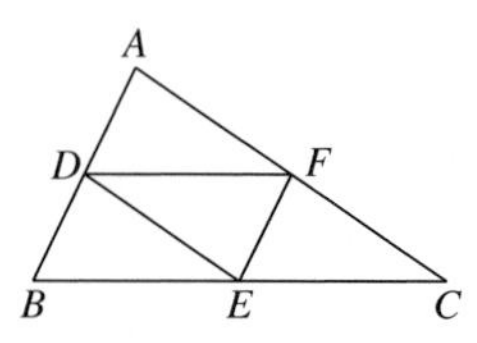

a. 4 triangles **b.** $\triangle ABC$, $\triangle ADF$, $\triangle DBE$, $\triangle FEC$, and $\triangle EFD$; For $\triangle ABC$ and each of the smaller triangles, the ratio is 2:1. For any two of the smaller triangles, the ratio is 1:1. **c.** Check students' work. The method will work for any triangle.

CHAPTER 6

Page 243 Exercises and Applications

22. Drawings may vary. Examples are shown for an acute triangle, a right triangle, and an obtuse triangle.

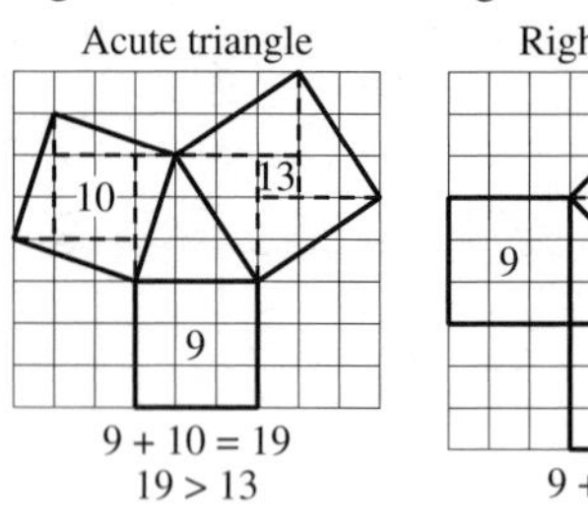

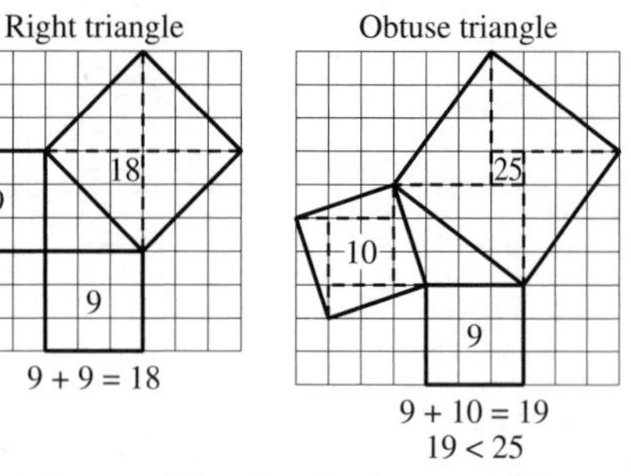

Students should conclude that for any right triangle, the sum of the area of the two smaller squares is equal to the area of the larger square. For an acute triangle, the sum of the areas of the two smaller squares is greater than the area of the larger square. For an obtuse triangle, the sum of the areas of the two smaller squares is less than the area of the larger square.

Page 244 Exercises and Applications

24. Answers may vary. An example is given. The route from location 2 to location 8 along Av. Brasil is about 5 units, while the route along Av. Joao Pinheiro is about 7 units. Since a diagonal route is usually shorter, city planners might include diagonal routes between major locations in the city.

Page 249 Exercises and Applications

29. The greater the value of f, the greater the calculated speed of the car; because the calculated minimum speed of the car depends on both the slipperiness of the road, measured by the friction coefficient, and the length of the skid marks. Consider the accidents described in Rows 4 and 6 of the table in Ex. 28. Those cars were estimated to have about the same minimum speed, but one skidded 30 ft more than the other. If you did not take the road conditions into account, you would have to guess that the car that skidded farther was going faster.

30. Answers may vary. An example is given. A road with a high coefficient of friction is less slippery than a road with a low coefficient of friction. A dry road will have a high coefficient of friction. A wet or icy road will have a low coefficient of friction.

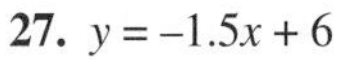

Page 261 Exercises and Applications

8.

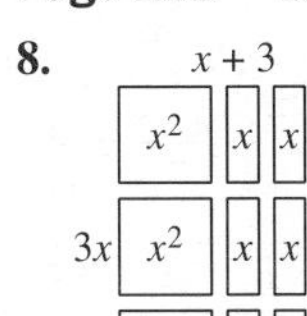

9.

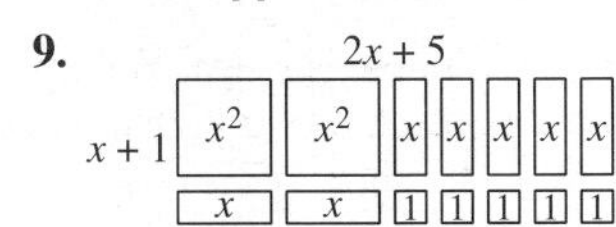

10.

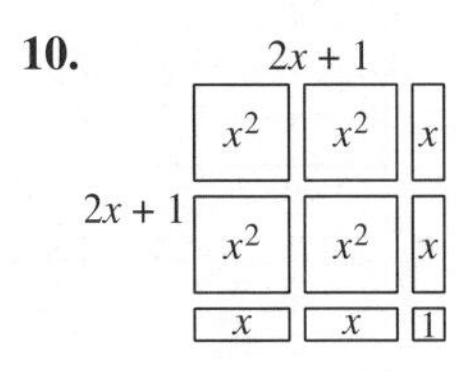

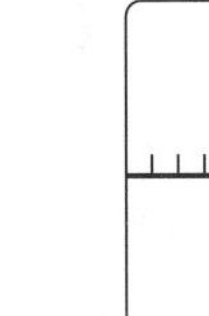

CHAPTER 7

Pages 284–285 Exercises and Applications

13.

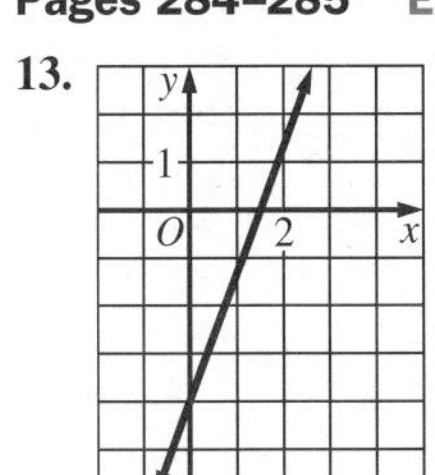

14.

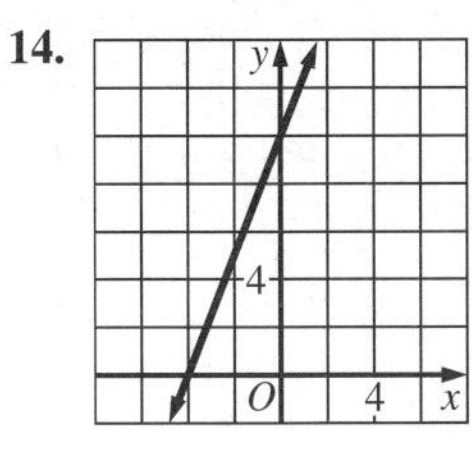

15.

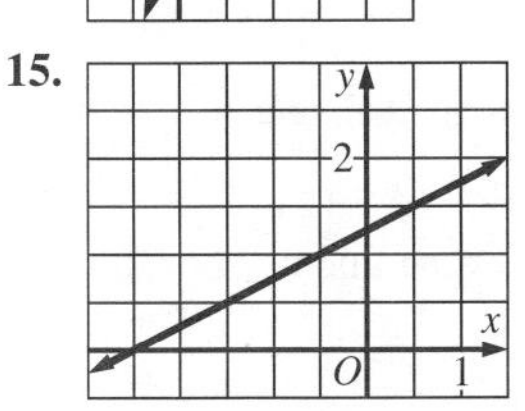

16.

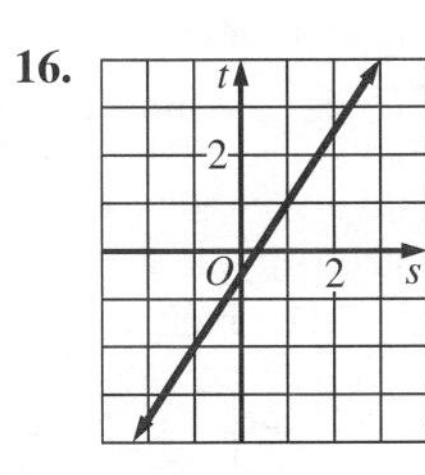

17.

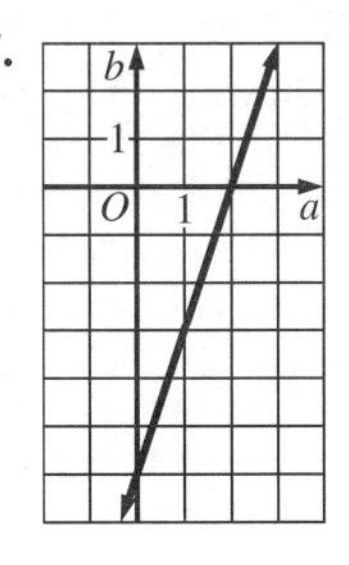

18.

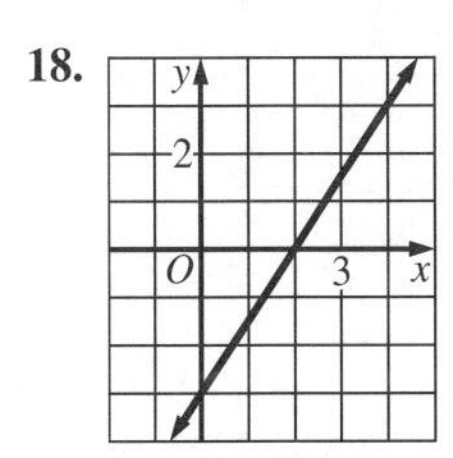

19.

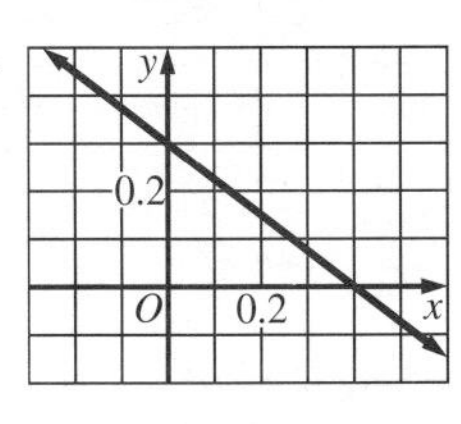

20. $y = -2x + 8$

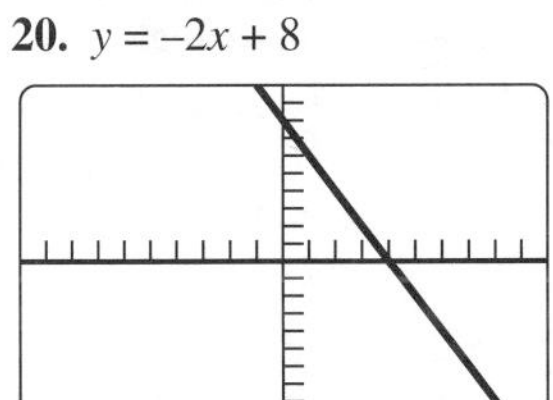

21. $y = -1.5x + 5$

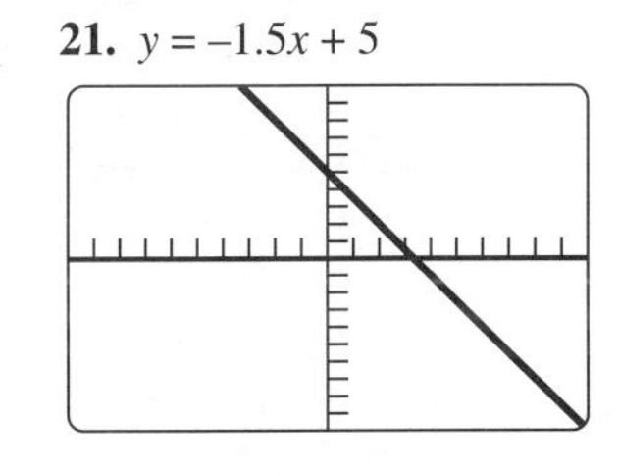

22. $y = \frac{2}{7}x - 3$

23. $y = 0.8x + 2.4$

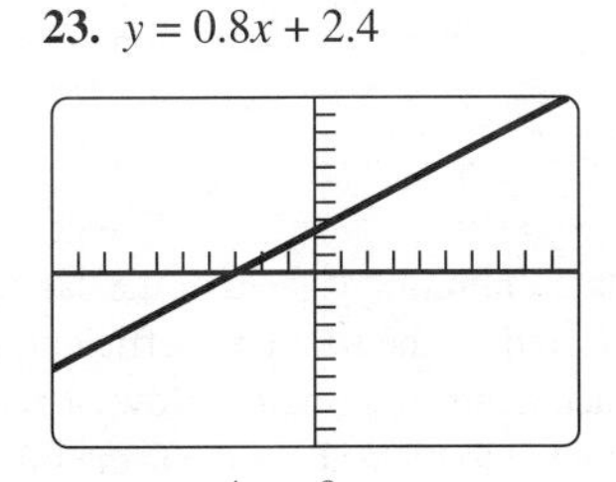

24. $y = -x + 3.5$

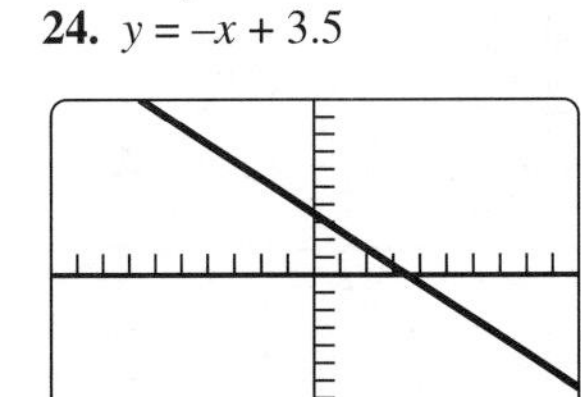

25. $y = -\frac{4}{3}x + \frac{8}{3}$

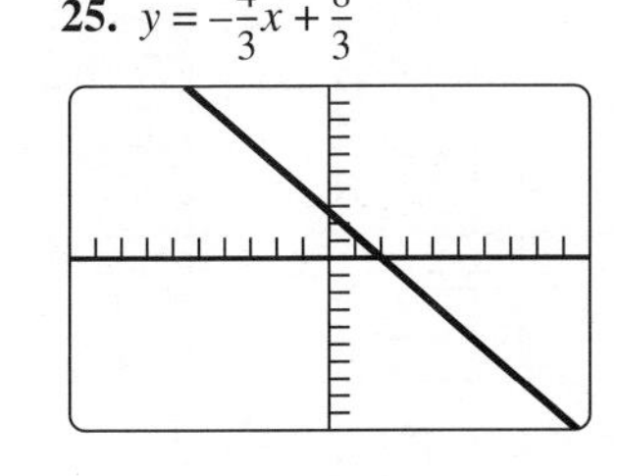

26. $y = -1.25x + 12.5$

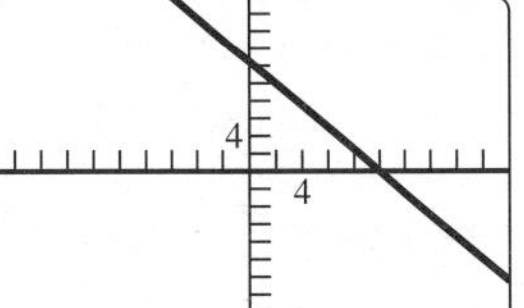

27. $y = -1.5x + 6$

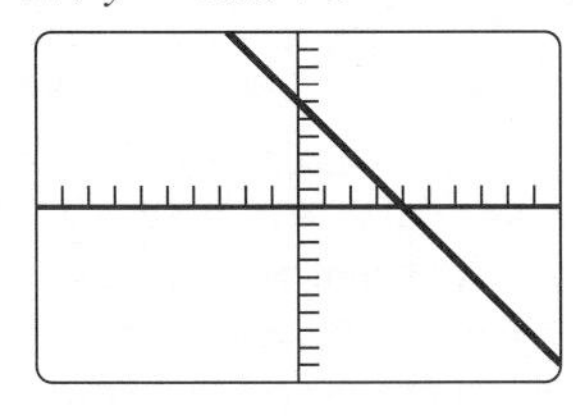

28. $y = 1.5x - 3$

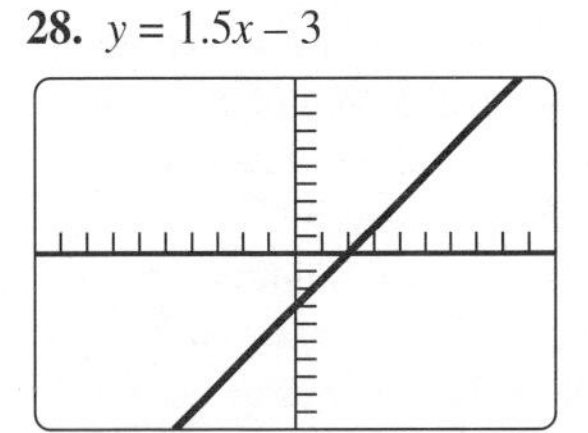

29. the equations in Exs. 17, 18, 19, 27, and 28 because a, b, and c are not all integers and the equation in Ex. 25 because it does not have both variable terms on one side **30.** D **31.** $1.5x + y = 10$ **32.** $1.5x + y = 10$ **33.** Since each roller coaster ride (pound of grapes) costs \$1.50, the cost of x rides (pounds) is represented by $1.5x$. Since each bumper car ride (pound of mushrooms) costs \$1.00, the cost of y rides (pounds) is represented by y. The graph on the left is the graph of $1.5x + y = 10$ for all positive values of x and y. It represents the situation in Ex. 32. The graph on the right is the graph of the same equation for whole-number values of x and y. It represents the situation in Ex. 31. The graphs are different because the number of rides you take must be a whole number, but you can buy any positive number of pounds of grapes or mushrooms. (For weighing and pricing purposes, the number of pounds would be rounded to tenths or hundredths and the price to cents.)

Page 286 Exercises and Applications

40. Answers may vary. An example is given. My brother and I are saving to buy a gift for our grandmother. It will cost \$25. If x = the amount I contribute and y = the amount my brother contributes, the equation $x + y = 25$ describes the situation. Some solutions (x, y) are (\$10, \$15), (\$13, \$12), and (\$12.50, \$12.50).

41. **42.** **43.** **44.**

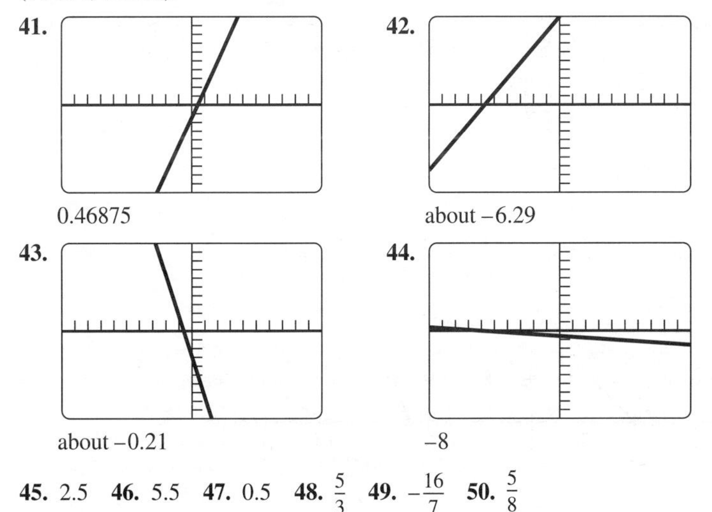

41. 0.46875

42. about –6.29

43. about –0.21

44. –8

45. 2.5 **46.** 5.5 **47.** 0.5 **48.** $\frac{5}{3}$ **49.** $-\frac{16}{7}$ **50.** $\frac{5}{8}$

Page 291 Exercises and Applications

15. Let f = the number of French francs, m = the number of German marks, and d = the number of dollars. **a.** $f = 0.19d$ and $m = 0.68d$ **b.** $f = \frac{19m}{68}$ or $m = \frac{68f}{19}$; Substitute 2000 for f in either equation and solve for m; about 7158 marks.
16. No, there is no common solution. **17.** $L = 2N$ and $L = N + 10$; The normal rate is 10 pints/min and the launch rate is 20 pints/min.

Page 298 Exercises and Applications

25. a. slope $= \frac{10-2}{5-3} = \frac{8}{2} = 4$; Choose one of the two points, say (3, 2), and substitute in the equation $y = 4x + b$; $2 = 4(3) + b$; $b = -10$; an equation of the line is $y = 4x - 10$. **b.** $2 = 3m + b$ and $10 = 5m + b$; $m = 4$ and $b = -10$; $y = 4x - 10$ **c.** Preferences may vary. Method 1: Use the coordinates of the two given points and the definition of slope to find the slope of the equation. Next use the fact that the coordinates of each point make the equation true. Substitute the x- and y-coordinates in the equation $y = mx + b$ (for which you have already found m) and solve for b. Method 2: Write a system of equations using the general slope-intercept form and the coordinates of both pairs of points. Every line with equation $2 = 3m + b$ contains the point (3, 2). Every line with equation $10 = 5m + b$ contains the point (5, 10). The solution of the system indicates the slope and y-intercept of the line that contains both.
26. Answers may vary. Examples are given. Graphing can always be used and is an easy method when a graphing calculator is available. Substitution is a good method when one equation expresses one variable in terms of another or can easily be written in this form. Addition is a good method when the coefficients of one of the variables are opposites. Subtraction is a good method when the coefficients of one of the variables are the same.

Page 305 Exercises and Applications

25. No.

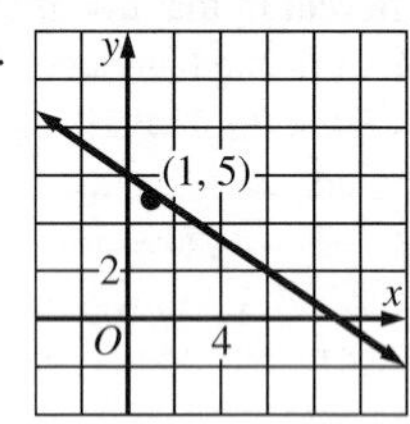

26. Yes.

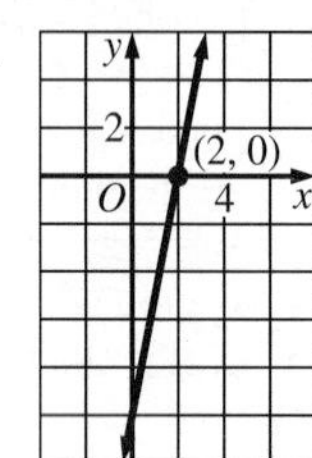

27.

28.

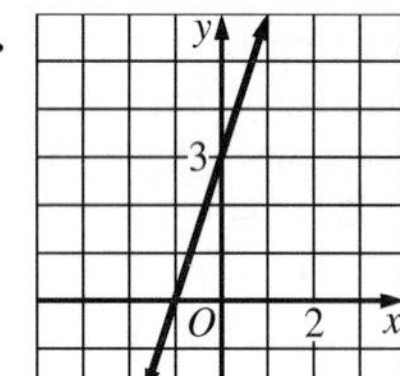

29.

Page 305 Assess Your Progress

1. (–2, 3) **2.** (–4, 7) **3.** (–8, –1) **4.** all (s, t) such that $5 = 2t - s$ **5.** all (s, t) such that $4t - 6s = 2$ **6.** no solution **7.** Emily: 8 visits; Adam: 20 visits **8.** By multiplying one or both equations by a number, you can get a system of linear equations for which one of the variables has either the same coefficients or opposite coefficients. The resulting system can then be solved by addition or subtraction.

Pages 308–309 Exercises and Applications

12.

13.

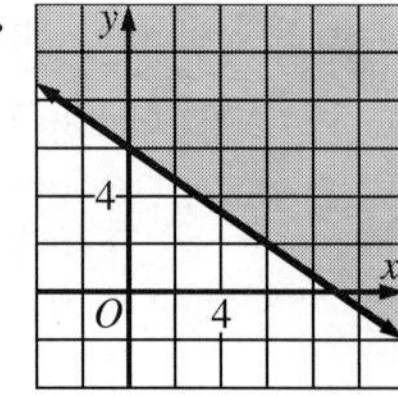

14.

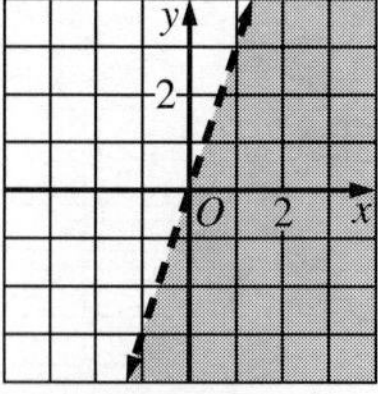

15.

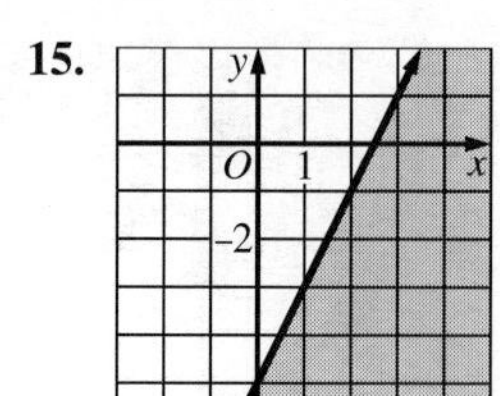

16.

17.

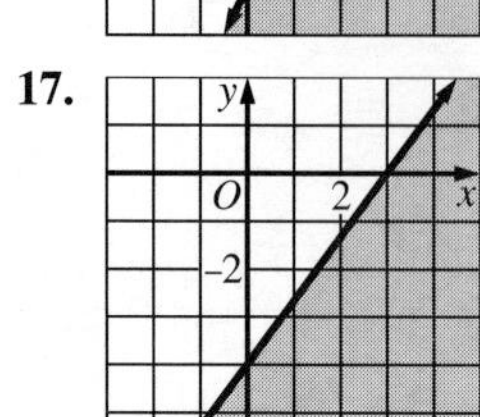

18. a, b.

b. Answers may vary. An example is given. The point (3, –2) does not lie on the line, but the coordinates do make the inequality $4x - 3y \geq 12$ true. This tells me that the solutions of the inequality lie on the side of the line that contains (3, –2), so I know which part of the graph to shade.

c.

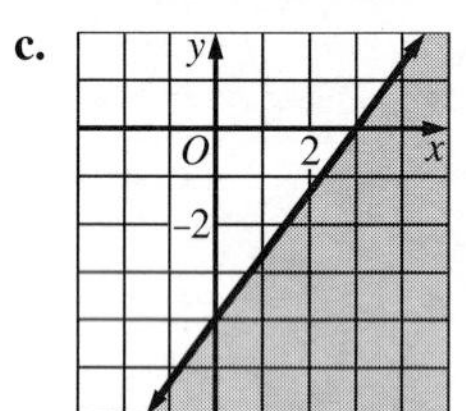

19. Answers may vary. An example is given. (1) I solved for y to get $y \leq \frac{4}{3}x - 4$. Then I graphed $y = \frac{4}{3}x - 4$, using a solid line. Since y is less than or equal to $\frac{4}{3}x - 4$, I shaded the points below the line. (2) I graphed the solid line $4x - 3y = 12$ by using the intercepts. I chose (3, –2) as a test point that does not lie on the line and checked to see if the coordinates make the inequality $4x - 3y \geq 12$ true or false. Since the coordinates make the inequality true, I shaded the part of the graph that contains the point (3, –2). I prefer to use the method in Ex. 18 because it is hard to solve for y and sometimes I forget when to reverse the inequality sign. Using a test point seems easier.

20.

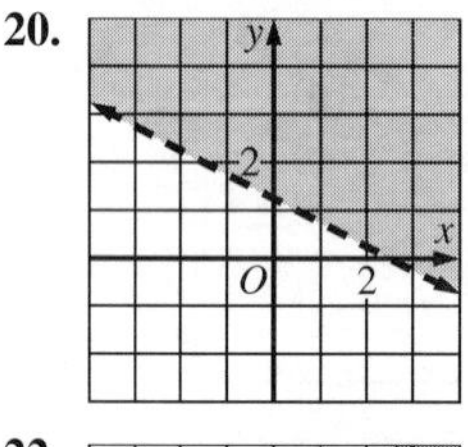

21.

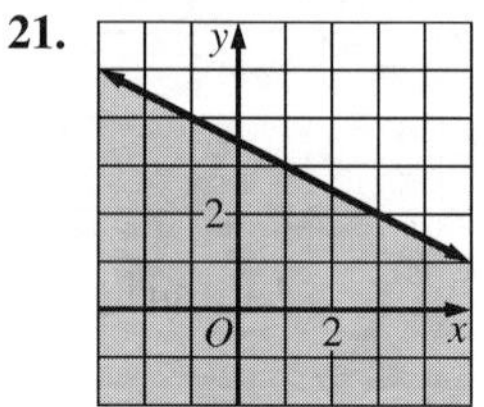

22.

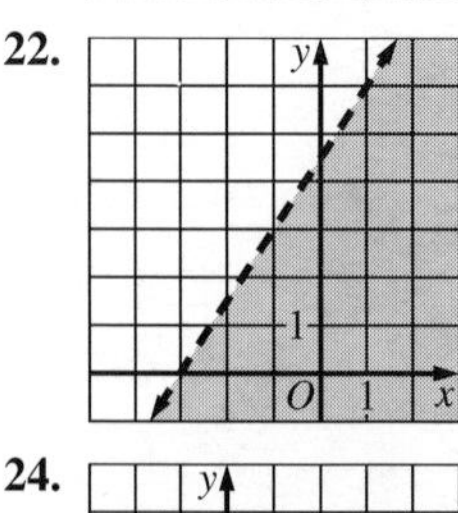

23.

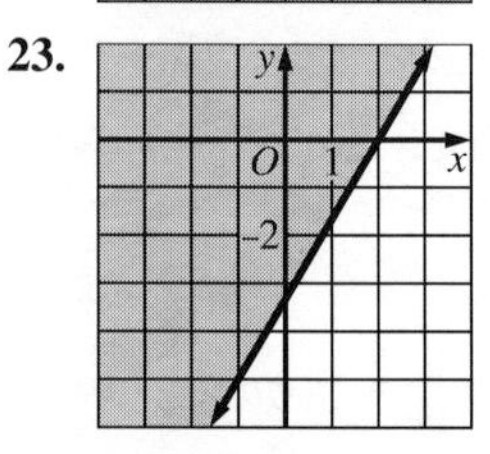

24.

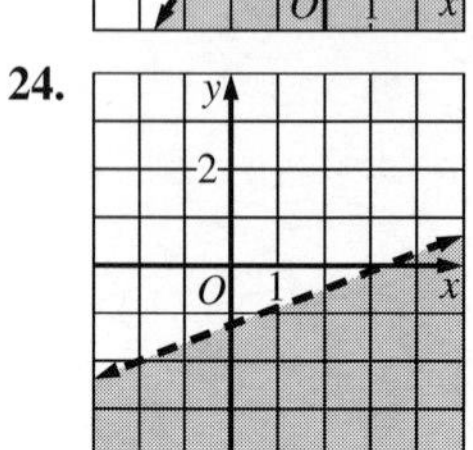

25.

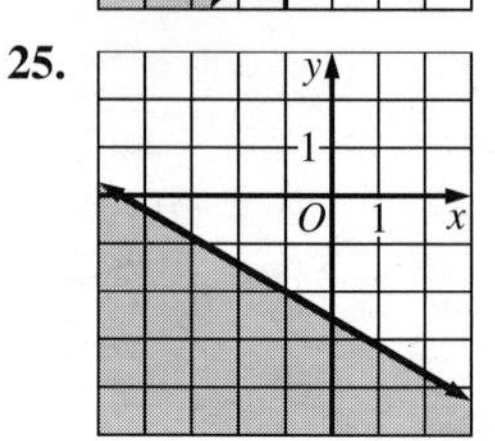

26.
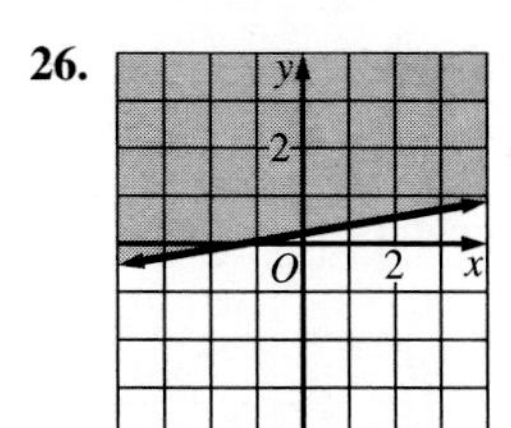

27.
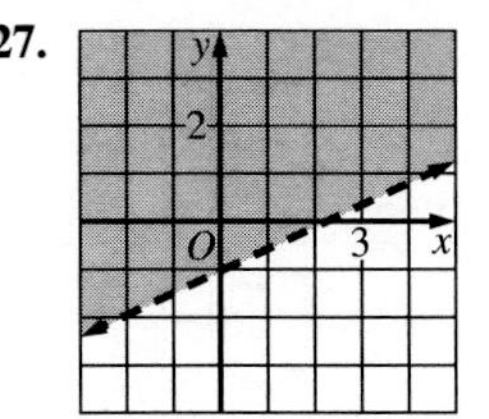

28.
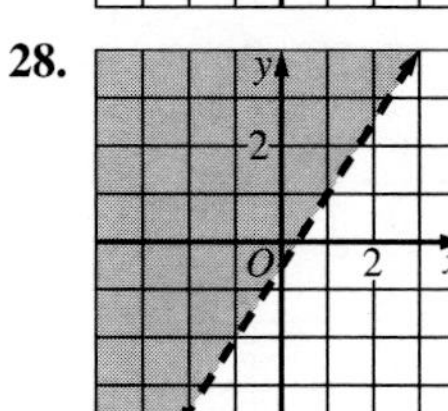

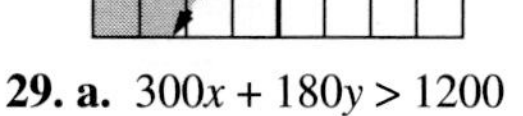

29. a. $300x + 180y > 1200$ **b.**
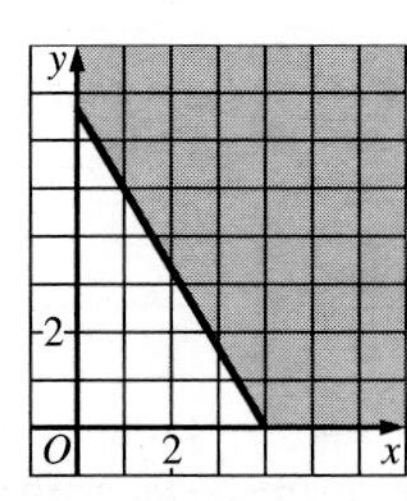

Answers may vary. Examples are given. (8, 0) and (4, 4)

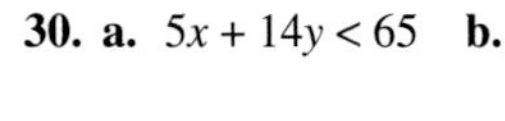

30. a. $5x + 14y < 65$ **b.**
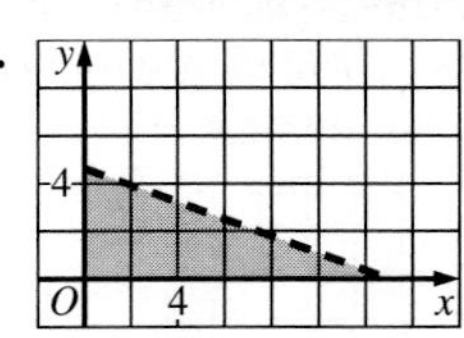

Answers may vary. Examples are given. (8, 0) and (4, 2)

31. Yes. Answers may vary. An example is given. If you have 6 cups of low-fat milk and 1 cup of ice cream, you get 1980 mg of calcium and 44 g of fat. (If you graph the inequalities from Ex. 29 and Ex. 30 on the same axes, any point inside the doubly shaded region is a solution.)

Page 313 Checking Key Concepts

1.
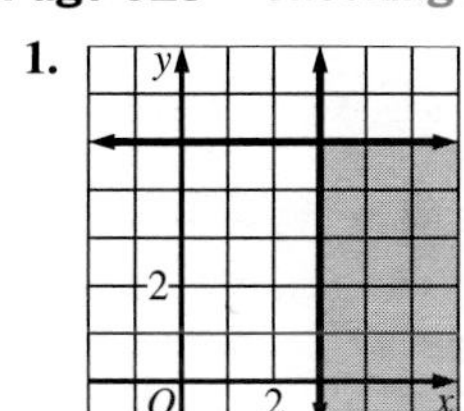

2.
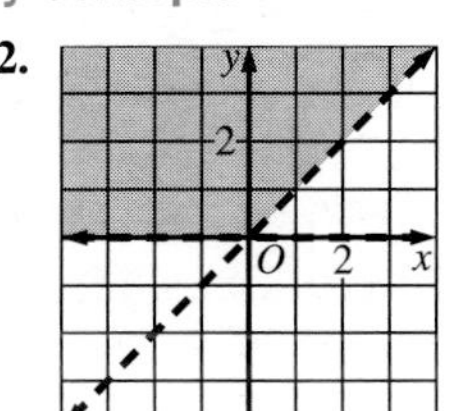

3.
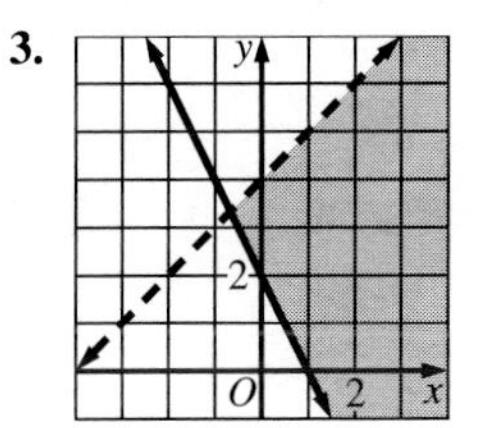

4.
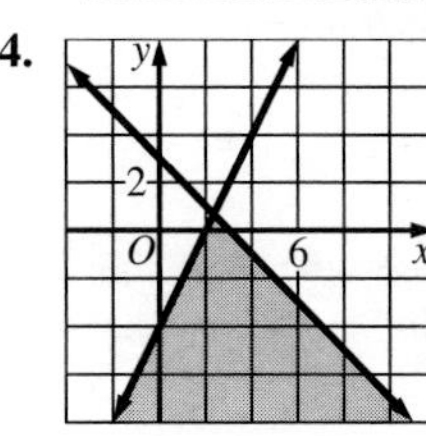

5. B: $y > 7 - x$ and $y > x - 3$; C: $y > 7 - x$ and $y < x - 3$; D: $y < 7 - x$ and $y < x - 3$

Page 313 Exercises and Applications

9.
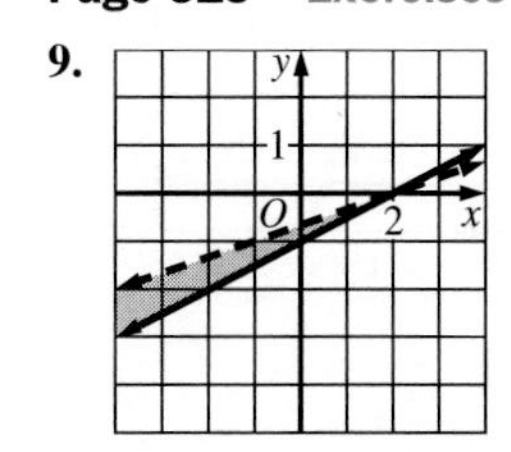

10. Let x = the number of cans of cat food and y = the number of containers of dry food. **a.** $0.5x + 2.5y \leq 5$ **b.** $6x + 48y \geq 72$

11.
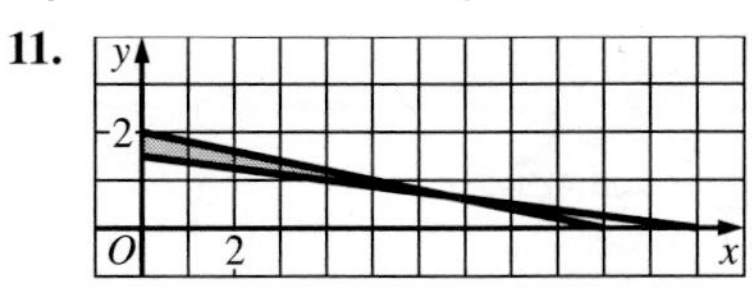

While the system has many solutions, there are only three with whole-number coordinates: (0, 2), (4, 1), and (5, 1).

12. $6x > 48y$
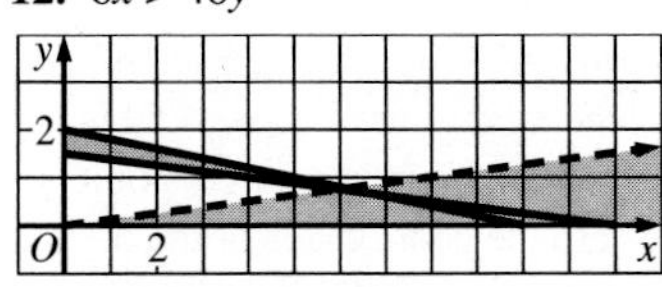

There is no way for Erik to meet all three requirements. Of the three possible solutions given in the answer to Ex. 11, note that none satisfies the inequality $6x > 48y$.

13.
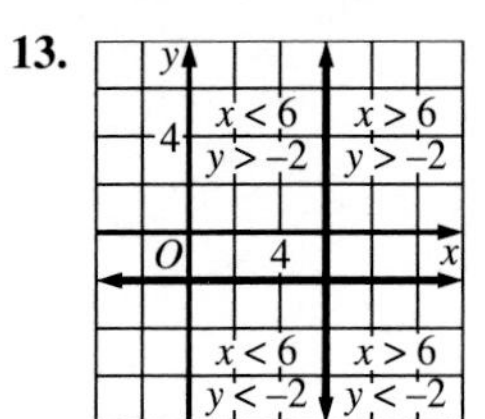

14.
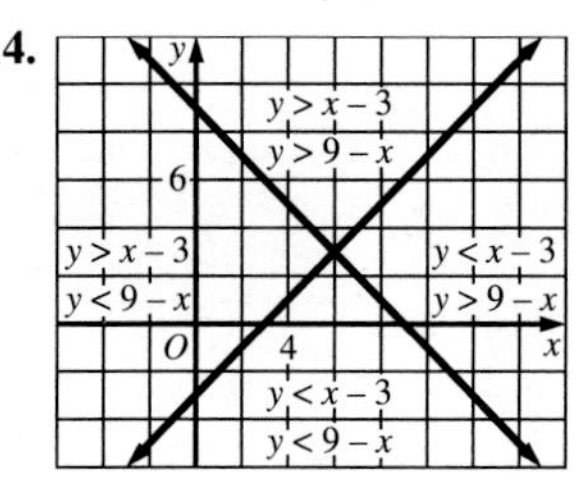

Page 314 Exercises and Applications

16.
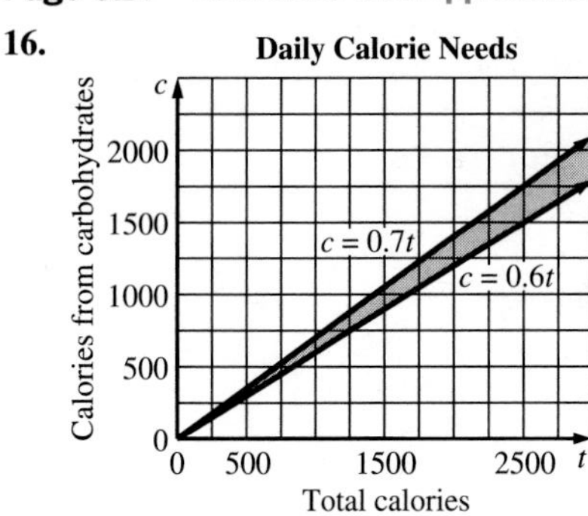

Answers may vary based on scales used. An example is given. The graphs are alike in that the same inequalities are graphed, and the solution region is the same. They are different in that my graph uses a different scale than the graph shown on page 280. That graph uses a scale of $2000 \leq t \leq 3000$ for the horizontal axis and $1000 \leq c \leq 2200$ for the vertical axis. My graph uses scales of $0 \leq t \leq 3000$ and $0 \leq c \leq 2500$.

Page 315 Exercises and Applications

24. Answers may vary. An example is given. The methods of solving are alike in that both involve finding one or more ordered pairs that satisfy all the requirements in the system at one time. Also, graphing can be used to solve both types of systems. The solution (if it exists) of a system of equations is a single point or a line. The solution (if it exists) of a system of inequalities is a region of the coordinate plane. **25.** $\frac{6}{7}$ **26.** $\frac{21}{8}$ **27.** 20 **28.** ± 6 **29.** $\frac{14}{3}$ **30.** 1 **31.** $x^2 + 4x + 4$ **32.** $19 + 8\sqrt{3}$ **33.** $x^2 - 25$ **34.** $3 + 2\sqrt{2}$

Page 315 Assess Your Progress

1.
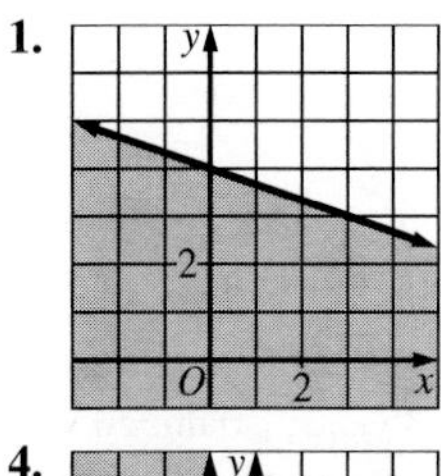

2.
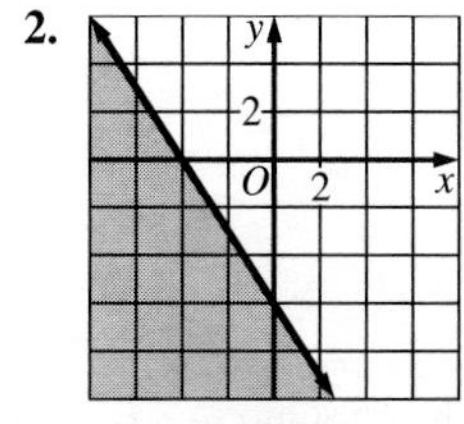

3.
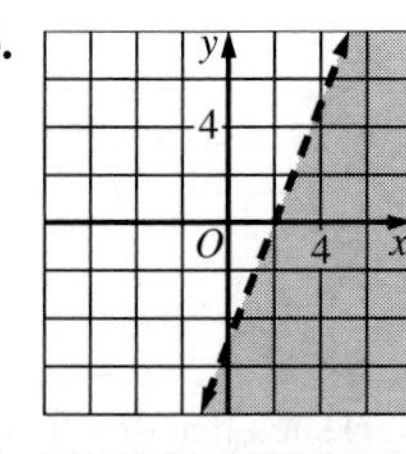

4.
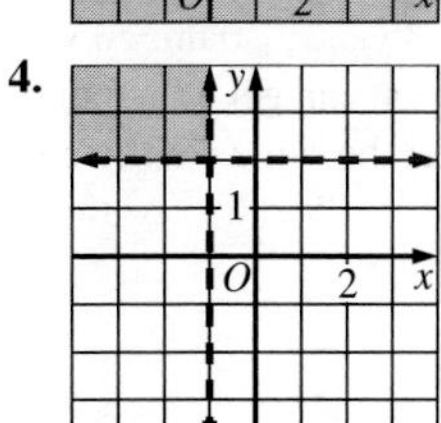

5.
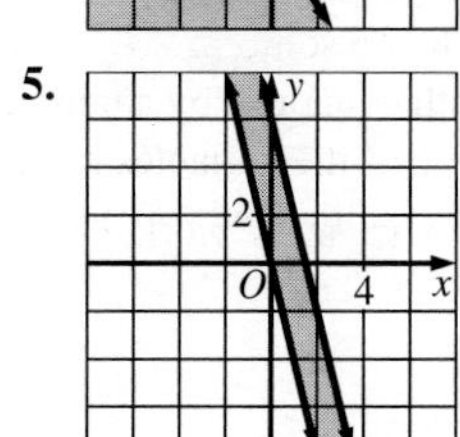

6.
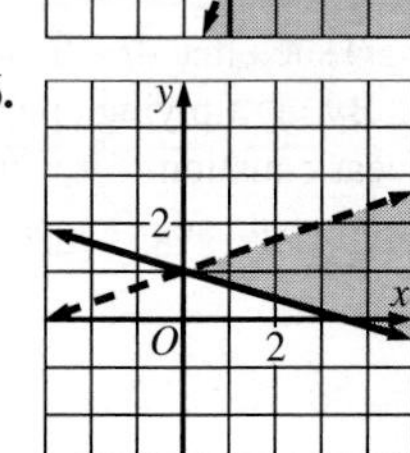

7. Let t = the number of T-shirts and s = the number of sweatshirts; $t \geq 10$; $s \geq 10$; $5t + 10s \geq 200$.
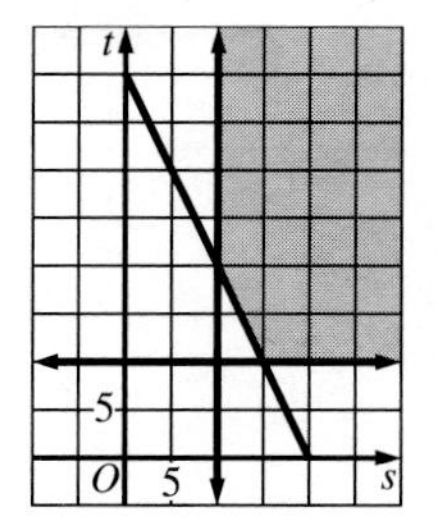

Pages 320–321 Chapter 7 Assessment

18. Summaries may vary. In each case, let $y = m_1x + b_1$ be the equation of the upper line and $y = m_2x + b_2$ be the equation of the lower line. **a.** m_1 and m_2 are both negative and $m_1 < m_2$; $b_1 > b_2$; the graphs intersect in a single point, so the system has one solution whose coordinates are the coordinates of the intersection point. **b.** m_1 and m_2 are both positive and $m_1 = m_2$; $b_1 > b_2$; the graphs do not intersect and are not the same line, so the system has no solution.

19. **20.** **21.** **22.**

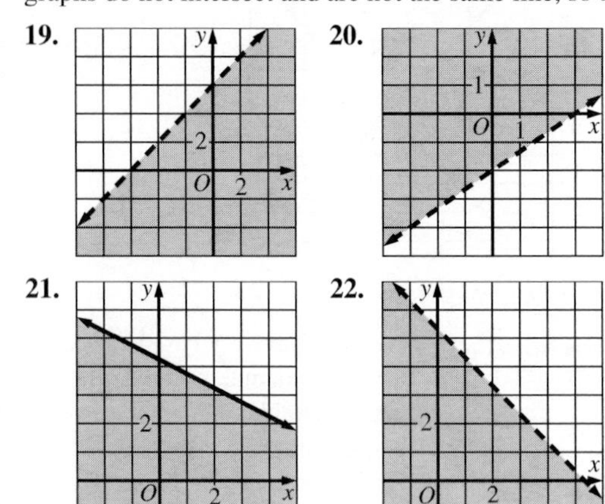

23.

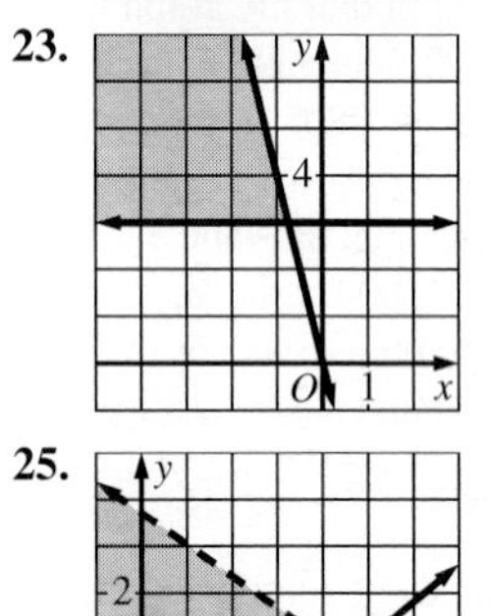

24. no solution

25. **26.**

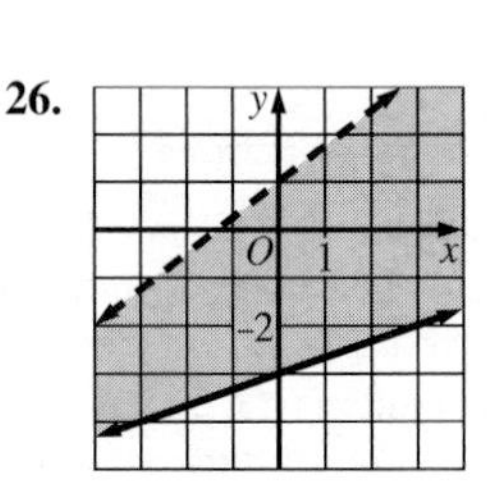

27.

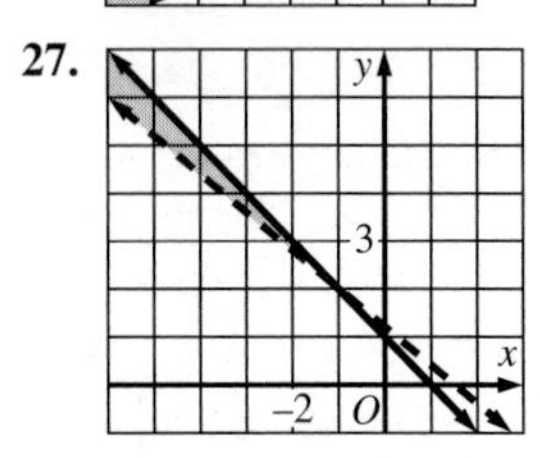

28. a. $x + y \geq 10$ **b.** $2x + 5y \leq 25$

c.

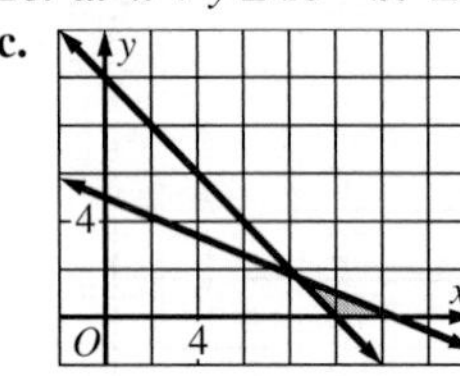

d. $1\frac{2}{3}$ lb; The greatest amount is the maximum y-value, which occurs where the graphs of $x + y = 10$ and $2x + 5y = 25$ intersect.

29. Answers may vary. Check students' work.

CHAPTER 8

Page 325 Exploration

1.

Length of Side 1 (ft)	Length of Side 2 (ft)	Perimeter (ft)	Area (ft^2)
1	14	30	14
14	1	30	14
2	13	30	26
13	2	30	26
3	12	30	36
12	3	30	36
4	11	30	44
11	4	30	44
5	10	30	50
10	5	30	50
6	9	30	54
9	6	30	54
7	8	30	56
8	7	30	56

2. 7 ft by 8 ft or 8 ft by 7 ft; 1 ft by 14 ft or 14 ft by 1 ft; Answers may vary. An example is given. I would recommend the 7 ft by 8 ft or 8 ft by 7 ft pen because its area is 4 times as large as the area of the 1 ft by 14 ft pen.

3. The graph is shaped like an upside-down U.

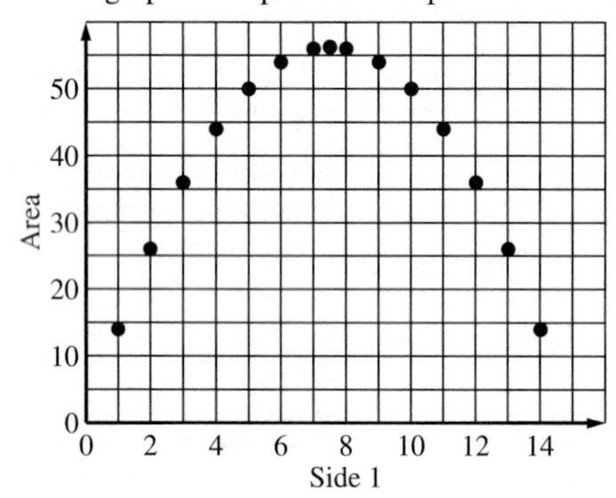

4. length of Side 2 $= \frac{1}{2}(30 - 2x) = 15 - x$; Let y be the area of the pen; $y = x(15 - x) = 15x - x^2$.

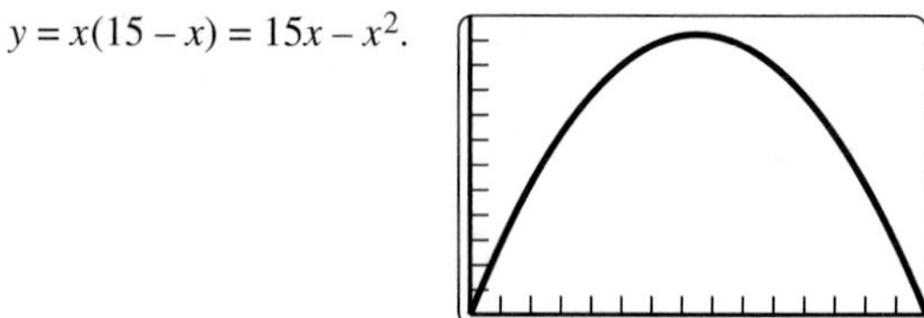

5. Answers may vary. An example is given. The graphs have the same shape; the graph in Step 3 consists of those points on the graph in Step 4 for which x and y are both integers.

Pages 328–329 Exercises and Applications

15.

Ticket price (dollars)	Number of tickets sold	Income (dollars)
3.00	500	1500
3.50	450	1575
4.00	400	1600
4.50	350	1575
5.00	300	1500

16.

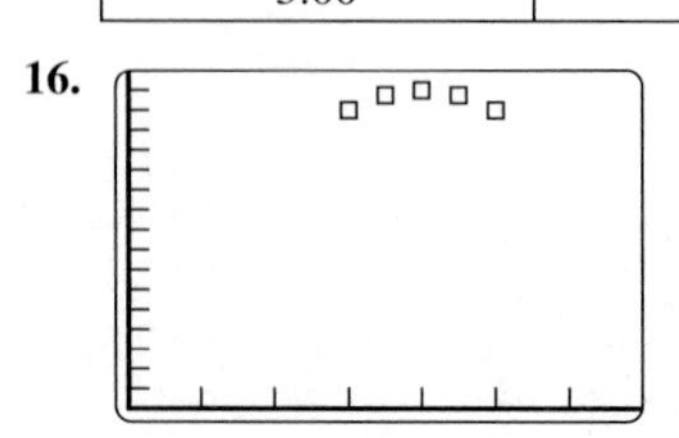

17. nonlinear; The points on the scatter plot do not lie on a line.

18. \$4.00; That price yields the greatest income.

19.

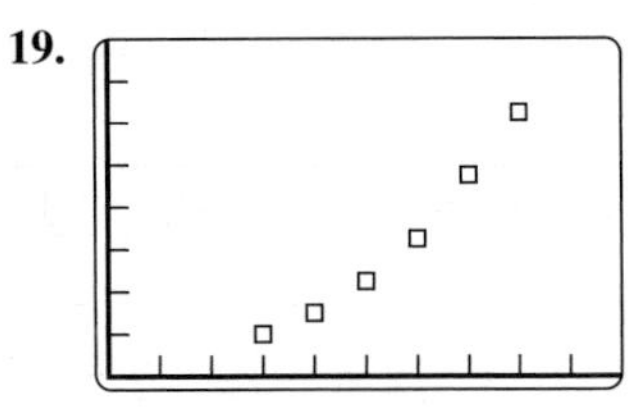

20. nonlinear; The points on the scatter plot do not lie on a line.

21. linear; Answers may vary. An example is given. In the scatter plot, numbers of servings is on the horizontal axis.

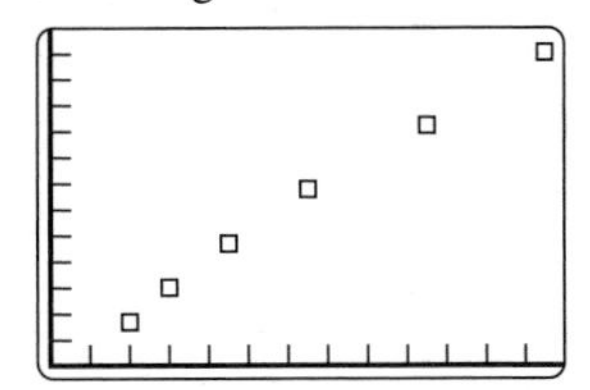

Page 330 Exercises and Applications

27.

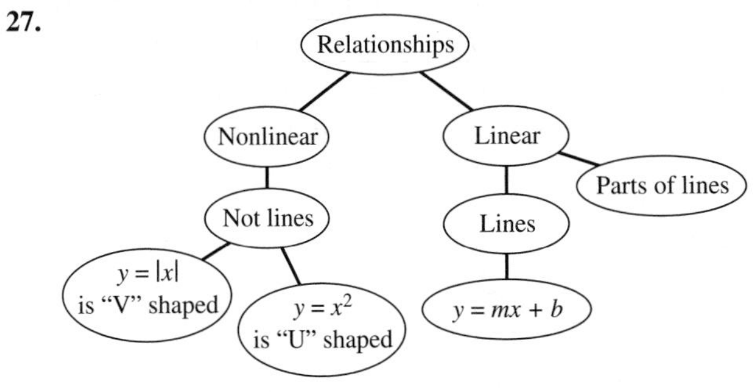

28. $75t^2 - 10t$ **29.** $p^2 + 4p - 21$ **30.** $2u^2 + 5u - 12$ **31.** $6r^2 + 21r + 15$
32. $17 - \sqrt{3}$ **33.** $2 + 2\sqrt{3} + \sqrt{5} + \sqrt{15}$

34. $y = \frac{3}{5}x - \frac{7}{5}$

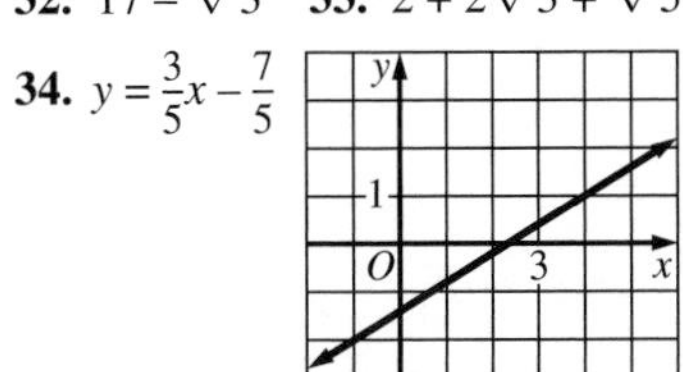

35. $y = -\frac{3}{5}x + \frac{7}{5}$

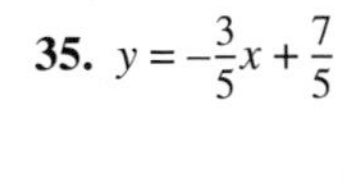

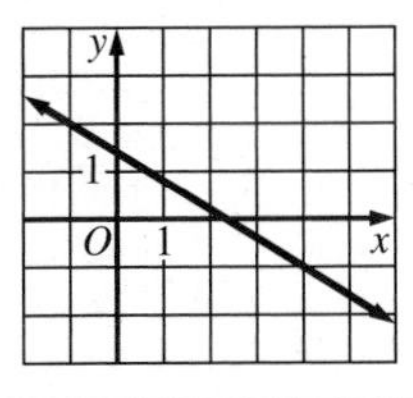

36. $y = \frac{3}{5}x + \frac{7}{5}$

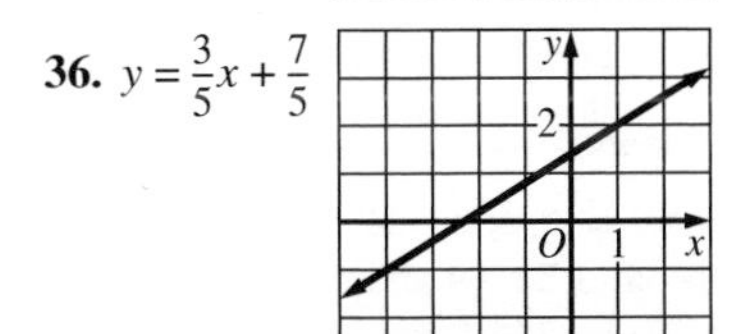

37. $y = -\frac{5}{7}x + \frac{3}{7}$

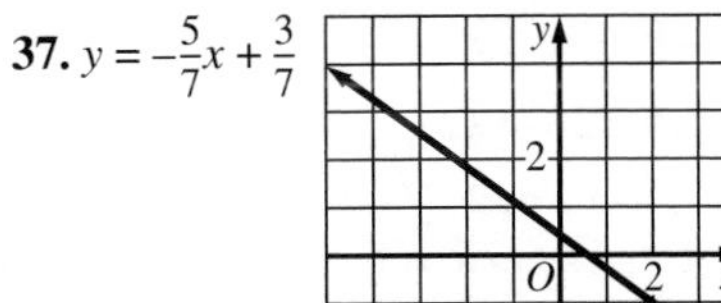

38. $y = -\frac{3}{7}x + \frac{5}{7}$

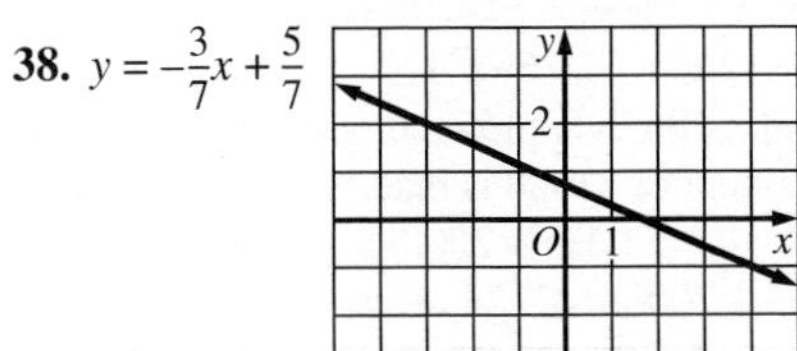

39. $y = \frac{3}{7}x - \frac{5}{7}$

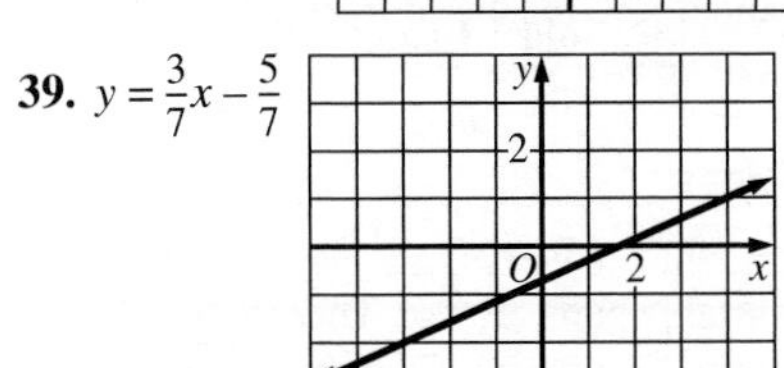

Pages 334–335 Exercises and Applications

17. No. The expression x^2 is always positive or zero. If $a > 0$, then $ax^2 \geq 0$, and it has a minimum value of 0 at $x = 0$, the vertex. If $a < 0$, then $ax^2 \leq 0$, and it has a maximum value of 0 at $x = 0$, the vertex. **18.** Yes; if the graph were folded along the y-axis, the two halves would match exactly. **19.** No; if the graph were folded along the y-axis, the two halves would not match exactly. **20.** Yes; if the graph were folded along the y-axis, the two halves would overlap exactly. **21.** C; The coefficient of x^2 is less than the coefficients in the equations in both Ex. 22 and Ex. 23, so the parabola is wider than both those parabolas. **22.** A; The coefficient of x^2 is greater than the ones in the equations in both Ex. 21 and Ex. 23, so the parabola is narrower than both those parabolas. **23.** B; The coefficient of x^2 is greater than that in the equation in Ex. 21 and less than that in the equation in Ex. 22, so it is narrower than the first parabola and wider than the second.

Page 341 Exercises and Applications

20. a. The graph will be the graph of $y = 2x^2$ shifted down 4 units.

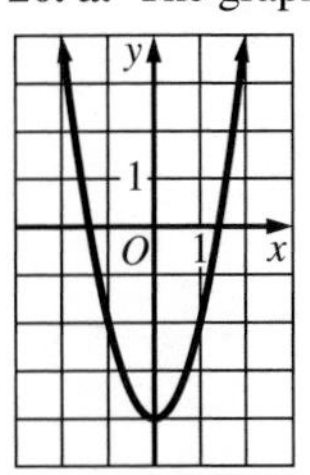

b. The graph will be the graph of $y = 2x^2$ shifted up 4 units.

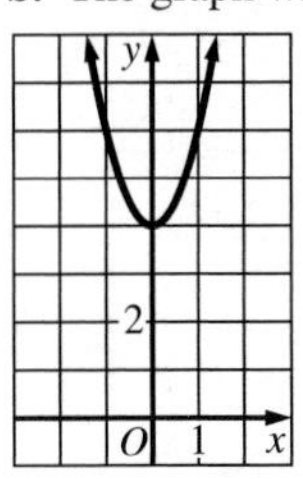

c. The graph will be the graph of $y = 2x^2$ shifted up 0.25 unit.

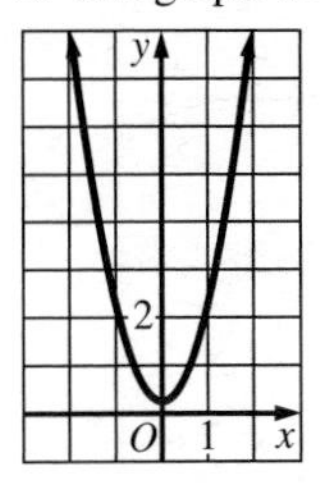

Page 342 Exercises and Applications

34. between 15 and 25 s; 176 ft/s; The table shows the average velocity for each time period. Average velocity is the change in distance fallen from one row to the next divided by the change in "time after jump" from one row to the next. The velocities increase as shown in the table to a terminal velocity of 176 ft/s.

Time after jump (s)	Distance fallen (ft)	Average velocity (ft/s)
1	16	16
3	138	61
5	366	114
7	652	143
9	970	159
11	1,310	170
13	1,660	175
15	2,010	175
25	3,770	176
35	5,530	176
45	7,290	176
55	9,050	176
65	10,810	176
75	12,570	176

Page 357 Checking Key Concepts

1. 1; The equation has two solutions. **2.** –39; The equation has no solutions. **3.** 0; The equation has one solution. **4.** –407; The equation has no solutions. **5.** 0; The equation has one solution. **6.** 81; The equation has two solutions. **7.** $a = -0.0069$, $b = 0.0549$, $c = 999.8266$ **8.** two solutions; no solutions **9.** 999.8266 mg/cm³

Page 357 Exercises and Applications

10. a. $A = s^2 + 8s$ **b.** 256; There are two possible solutions. **c.** –12, 4; The negative solution does not make sense because length cannot be negative; the solution is 4.

Page 359 Assess Your Progress

7. two solutions **8.** no solutions **9.** one solution **10.** Answers may vary. The methods include solving algebraically, using a graph, and using the quadratic formula.

CHAPTER 9

Page 381 Exercises and Applications

28. a.

Comparing Growth				
B2	=2*A2			
	A	B	C	D
1	x	2x	x^2	2^x
2	1	2	1	2
3	2	4	4	4
4	3	6	9	8
5	4	8	16	16
6	5	10	25	32
7	6	12	36	64
8	7	14	49	128
9	8	16	64	256
10	9	18	81	512
11	10	20	100	1024
12	11	22	121	2048
13	12	24	144	4096
14	13	26	169	8192
15	14	28	196	16,384
16	15	30	225	32,768
17	16	32	256	65,536
18	17	34	289	131,072
19	18	36	324	262,144
20	19	38	361	524,288
21	20	40	400	1,048,576

b. $y = 2^x$; Comparing the table values appears to justify the claim. For example, consider $x = 2$ and $x = 10$. The three functions have the same value at $x = 2$, while $2(10) = 20$, $10^2 = 100$, and $2^{10} = 1024$.

c. 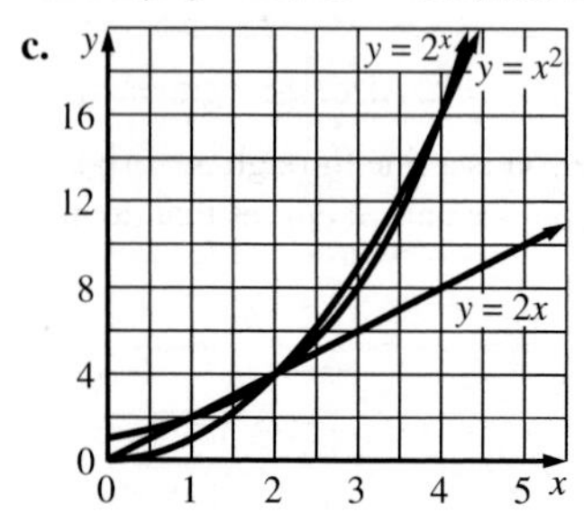

Depending on your graph, it may be difficult to tell which function is the fastest growing and which is slowest. You can determine which is fastest growing by which is steepest. For $x \geq 2$, a graph will support the answers to part (b). For $0 \leq x \leq 1$, $y = 2x$ is the fastest growing, with the other two functions growing at about the same rate. For $1 \leq x \leq 2$, $y = x^2$ is the fastest growing and $y = 2x$ the slowest.

Pages 398–399 Exercises and Applications

16. a. Pluto; Mercury

b.

Planet	Distance from Sun (miles)
Earth	93,000,000
Mars	140,000,000
Jupiter	480,000,000
Uranus	1,800,000,000
Venus	67,000,000
Mercury	36,000,000
Pluto	3,700,000,000
Saturn	890,000,000
Neptune	2,800,000,000

17. Answers may vary. **18.** 7.2 microns; $0.0000072 = 7.2 \times 10^{-6}$ **19.** 7 to 12 microns; $0.000007 = 7 \times 10^{-6}$ and $0.000012 = 1.2 \times 10^{-5} = 12 \times 10^{-6}$ **20.** 3 microns; $0.000003 = 3 \times 10^{-6}$
21. Answers may vary. In the examples shown, the scale is $\frac{1}{16}$ in.:1 micron.

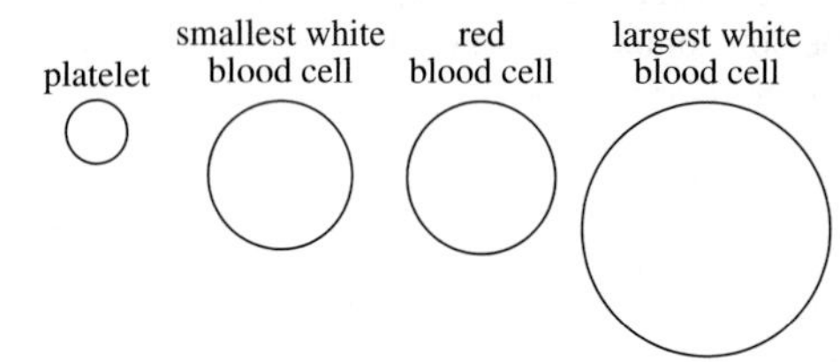

Page 400 Exercises and Applications

31. To compare two numbers in scientific notation, first compare the values of x in the expressions $a \times 10^x$. The number for which x is larger is the larger number. If the exponents are the same, compare the values of a. The number for which a is larger is the larger number. Examples:
$2.7 \times 10^{15} > 7.6 \times 10^{12}$ because $15 > 12$.
$3.1 \times 10^9 < 5.2 \times 10^9$ because $3.1 < 5.2$.
$8.1 \times 10^{-11} < 9.3 \times 10^{-8}$ because $-11 < -8$.
$1.2 \times 10^{-8} < 1.6 \times 10^{-8}$ because $1.2 < 1.6$.

CHAPTER 10

Pages 426–427 Exercises and Applications

7. $6x^2 - x - 4$; 2 **8.** $5y - 10$; 1 **9.** $2r^3 - 3r^2 - 2r + 3$; 3 **10.** Answers may vary. An example is given. The polynomial is $3x^2 + 2x - 4$ and it can be represented as a sum in many ways. For example, $(3x^2 - 2) + (2x - 2)$.
11. $17n^2 + 2n - 3$ **12.** $10x^2 - 5$ **13.** $y^2 - 4y - 7$ **14.** $k^2 - 5k - 5$
15. $-a - 4b + 7$ **16.** $10x^2 + 6xy$ **17.** $2b^3 + b^2 - 6b + 1$
18. $a^3 + 6a^2 - 3a + 1$ **19.** $x^2 + 8x - 8$ **20.** $3x^2 + 15x - 6$ **21.** 30 cm, 40 cm, and 50 cm **22.** No. An investment of $1000 is made each year for a period of three years. **23. a.** $2000g^4$ **b.** $2000g^4 + 2000g^3 + 2000g^2 + 2000g$ **c.** $4000g^4 + 2000g^3 + 2000g^2 + 2000g$; 4 **24.** Descriptions may vary. The percent of women in management jobs increased from 1983 to 1991, but the rate of increase slowed down from year to year.

Page 433 Exercises and Applications

14. a. $x^2 + 5x + 6$ **b.** $x^2 + 5xy + 6y^2$ **c.** $4a^2 - 23a - 35$ **d.** $4a^2 - 23ab - 35b^2$ The coefficients are the same, but the variable parts are not. There is a "y" in the middle term, and a "y^2" in the last term. Yes; the coefficients are the same, but the middle term has a "b" and the last term has a "b^2." In both cases, the coefficients of the binomial factors are the same, so the coefficents in the products are the same. **15. a.** The width is $(5x + 6y + 20)$ in. The length is $(6x + 7y + 20)$ in. **b.** The perimeter is $(22x + 26y + 80)$ in. The area is $(30x^2 + 220x + 71xy + 260y + 42y^2 + 400)$ in^2.
16. $5x + 6y + 20 = 98$
$6x + 7y + 20 = 113$; $x = 12$; $y = 3$

Page 436 Exploration

1. 1 and B, 2 and D, 3 and A, 4 and C

Equation 1

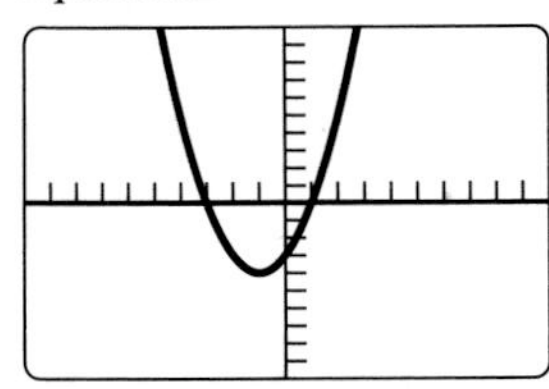

Equation 2

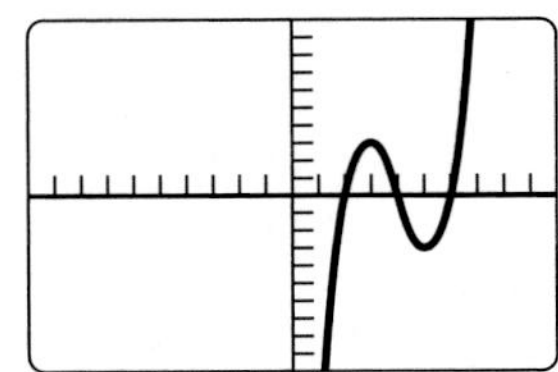

Equation 3

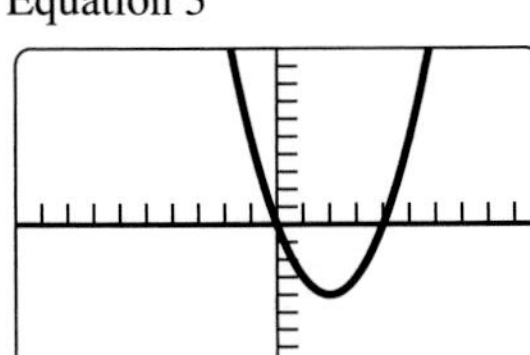

Equation 4

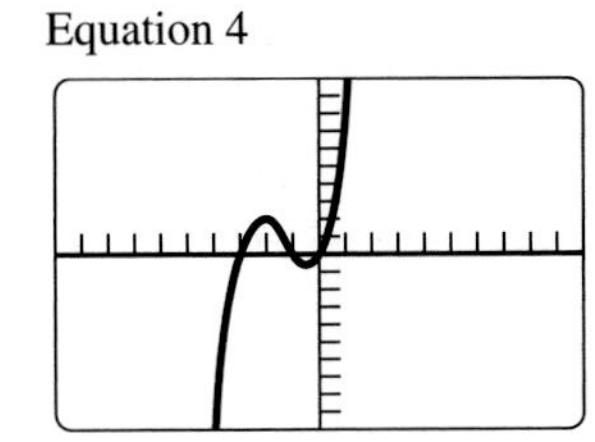

2. 1: –3 and 1; 2: 2, 4, and 6; 3: 0 and 4; 4: –3, –1, and 0; Each x-intercept makes one of the factors in factored form equal to 0. **3.** The x-intercepts of the graph are the solutions to the polynomial equation; the solutions are where either of the factors x and $x - 6$ is zero, or at 0 and 6.

Page 445 Exercises and Applications

13. Examples may vary. If a quadratic trinomial has integral coefficients and the discriminant is a perfect square, then the polynomial is factorable. If the discriminant is not a perfect square, then the polynomial is not factorable.
14. A **15.** Area $= 2x^2 + 9x + 10 = (x + 2)(2x + 5)$

16. 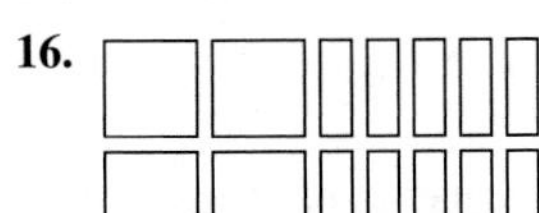 $4x^2 + 12x + 5 = (2x + 1)(2x + 5)$

17. $(5k + 1)(k + 1)$ **18.** $(2d - 3)(d - 2)$ **19.** $(3n + 2)(n + 2)$ **20.** $(2a + 3)^2$ **21.** $(4a - 9)(a - 1)$ **22.** $(2m + 3)(m + 6)$ **23.** $(5x - 8)(x - 1)$ **24.** $(2n + 1)(3n + 7)$ **25.** $(2g - 1)(5g - 3)$ **26.** $(2q - 1)(3q - 4)$ **27.** $(3p - 4)(3p - 2)$ **28.** $(3x + 5)(5x - 2)$ **29.** Answers may vary. Examples are given. $x^2 + 5x + 6 = (x + 3)(x + 2)$; $x^2 + 5x + 4 = (x + 4)(x + 1)$; $2x^2 + 5x + 2 = (2x + 1)(x + 2)$; $2x^2 + 5x + 3 = (2x + 3)(x + 1)$; $3x^2 + 5x + 2 = (3x + 2)(x + 1)$

CHAPTER 11

Page 465 Think and Communicate

2. t decreases; t increases; In direct variation, both variables increase together and decrease together, but in $t = \frac{300}{s}$, as one variable increases, the other decreases. **3.** No; the product of water depth and bubble volume is not constant. **4.** No. Explanations may vary. As the length of a side of a square increases, the area of the square also increases. If the length of the side and the area were inversely related, as the length of a side increased, the area would decrease. **5.** Yes. The number of people sharing a pizza equally multiplied by the number of slices per person is always equal to the number of slices in the pizza. For example, if the pizza has 8 slices, then 4 people could each have 2 slices, or 16 people could each have $\frac{1}{2}$ slice.

Page 466 Exercises and Applications

5. 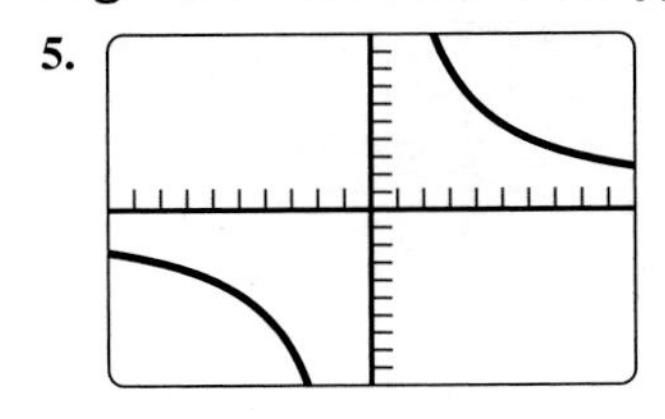 Yes; the product of x and y is constant.

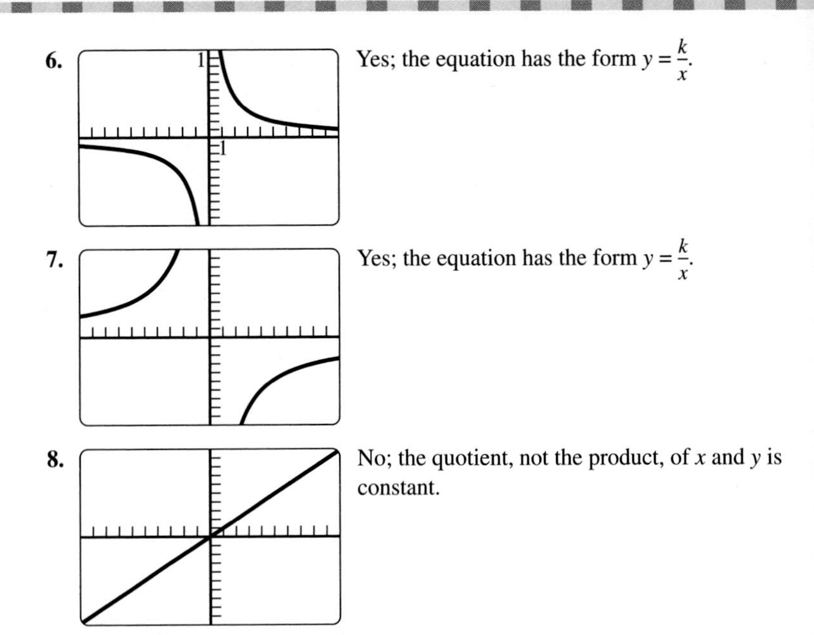

6. Yes; the equation has the form $y = \frac{k}{x}$.

7. Yes; the equation has the form $y = \frac{k}{x}$.

8. No; the quotient, not the product, of x and y is constant.

Page 475 Exercises and Applications

16. Eiji should pay \$545, and Josh should pay \$405. Explanations may vary. An example is given. After each pays \$225, the remaining rent is \$500. Eiji's room is $\frac{160}{160 + 90}$ or 0.64 of the bedroom area, so Eiji's rent should be $\$225 + 0.64(500) = \$545$. Josh's room is 0.36 of the area, so Josh's rent should be $225 + 0.36(500) = \$405$. **17.** Answers may vary. An example is given. **a.** The varsity club has sold 200 T-shirts at a profit of \$12 each. If the club plans to sell 120 sweatshirts, how much profit must they make on each to achieve an average profit of \$15 per item? **b.** $x = 120$; They must make a profit of \$20 per sweatshirt to achieve an average profit of \$15 per item.
18. Check students' work. **19.** Answers may vary. An example is given. Weighted averages can be used by giving some grades more importance than others. For example, a teacher might consider a term paper as worth two test grades and the final exam as worth three test grades. It would be to your advantage if you tended to do better on big projects. You might tend to prepare better for final exams, or put more effort into writing term papers than preparing for tests. It would be to your disadvantage if you feel more pressured by big projects and do not do as well as you do on regular tests.

Page 475 Assess Your Progress

6. Answers may vary. An example is given. Both direct variation and inverse variation are relationships in which two variables are related by a nonzero constant. However, in direct variation, the quotient of the variables is constant, while in inverse variation, the product of the variables is constant. The form of the equations is different. A direct variation equation has the form $y = kx$ and an inverse variation equation has the form $y = \frac{k}{x}$ or $xy = k$. Also, their graphs are different. The graph of a direct variation equation is a line through the origin, while the graph of an inverse variation equation is a pair of curves that do not cross either axis.

Page 481 Exercises and Applications

28. about 5 h; From the top of Mt. Washington to the base of Mt. Madison is about 2.8 mi uphill and about 3.5 mi downhill. From Ex. 27, they hike at an average uphill rate of about 1 mi/h and an average downhill rate of about 2 mi/h, so they need about $\frac{2.8}{1} + \frac{3.5}{2} \approx 4.6$ h for the hike. **29.** Mt. Webster, Mt. Franklin, and Mt. Clay **30.** Answers may vary. For example, the equation $\frac{x}{x - 5} = \frac{5}{x - 5}$ has no solution. To find this equation, I began with the equation $x = 5$. I rewrote this equation by dividing both sides by $x - 5$ so the "solution" $x = 5$ would cause a zero in the denominator. The equation $\frac{x}{x - 5} = \frac{5}{x - 5}$ has no solution.

31. Answers may vary. An example is given. To find the least common denominator in either problem, you need to find the least common multiple of the denominators by factoring. However, to find the least common denominator of the numerical fractions, you just factor integers. To find the least common denominator of the rational expressions, you need to consider both the numerical factors and the algebraic factors of the denominators.

Page 485 Exercises and Applications

17.

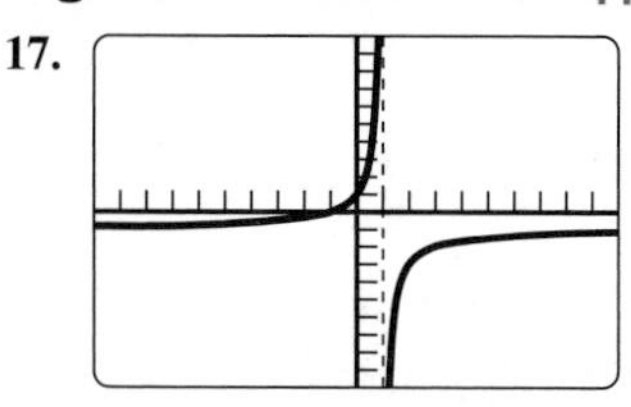

18. Jen divided both sides of the equation by r. The variable y is still on both sides of the equation.

CHAPTER 12

Page 512 Checking Key Concepts

4. Step 1: Use the denominator of the fraction as your first guess. Add this number to 1. Step 2: Compare your result from Step 1 to the desired sum. If they are the same, you are done. If not, go to Step 3. Step 3: Divide the desired sum by your result from Step 1. Multiply your first guess by this quotient. The product is the unknown quantity. Step 4: Check your work. Multiply your first guess, 1, and their sum by the quotient from Step 2. The sum of the first two products should equal the third. **5.** In Method 1, each occurrence of 17 should be changed to 19. The algorithm must also be changed to allow for a remainder. (Example: Step 4: Count how many times 19 was subtracted. That number is the quotient and the number left is the remainder.) In Method 2, each guess should be multiplied by 19 rather than 17. The algorithm must also be changed to allow for a remainder. (Example: Step 4: Is the result too small? Subtract the product from 204. Is the difference 0 or is it less than the guess? If so, the guess is correct and the difference is the remainder. If the difference is too big, start again with a larger number.) **6.** E, C, F, A, D, B; Step 1: Make the base of the pot by patting the clay into a circular disk; Step 2: Form coils and stack them on the base; Step 3: Smooth the coils to form the sides; Step 4: Finish the lip of the pot; Step 5: Fire the pot; Step 6: Decorate the pot. **7.** Step 1; Subtract 52 from 326. Step 2: Subtract 52 from the remainder. Step 3: Repeat Step 2 until the remainder is less than 52. Step 4: Count how many times 52 was subtracted. That number is the quotient and the number left is the remainder. **8.** Step 1: Add 34 and 34. Step 2: Add 34 to the sum. Step 3: Repeat Step 2 ten more times. **9.** Step 1: Multiply 4 by 4. Step 2: Multiply the product by 4. Step 3: Repeat Step 2 five more times. **10.** Answers may vary. **11.** Step 1: Draw a three-by-three square grid. Step 2: Player 1 makes an X in one of the nine boxes. Step 3: Player 2 makes an O in one of the remaining boxes. Step 4: Each player in turn makes a mark (either an X or an O, accordingly) in one of the remaining boxes in such a way as to prevent the other player from putting three of his or her marks in a line horizontally, vertically, or diagonally. Step 5: Repeat Step 4 until one of the players has managed to make three of his or her marks in a line horizontally, vertically, or diagonally and is declared the winner or until all the squares are filled with no three marks in a line as described and the game is declared a draw.

Page 526 Exercises and Applications

4. a..

Pitcher	Point total	Rank
Maddux	119	1
Swift	61	2
Glavine	49	3
Burkett	9	4
Rijo	8	5
Greene	2	6 (tie)
Portugal	2	6 (tie)
Harvey	1	8 (tie)
Myers	1	8 (tie)

b. The pitchers' ranks would be unaffected, except for John Burkett, who would move from fourth to fifth, and Jose Rijo, who would move from fifth to fourth.

Page 540 Exercises and Applications

17.

number of toppings	number of pizzas
1	9
2	36
3	84
4	126
5	126
6	84
7	36
8	9
9	1

${}_9C_1 = {}_9C_8$; ${}_9C_2 = {}_9C_7$; In fact, for any whole number r between 1 and 9, ${}_9C_r = {}_9C_{9-r}$. For example, the number of ways of choosing three things from among nine is the same as the number of ways of choosing six things from among nine. When you choose r items from among n items, you automatically chose the other $n - r$ items as well.

18. a. ${}_nC_4 = \frac{n(n-1)(n-2)(n-3)}{4!}$; There are n ways to choose the first topping, $n - 1$ ways to choose the second topping, $n - 2$ ways to choose the third topping, and $n - 3$ ways to choose the fourth topping. Since the order of the toppings does not matter, the product is divided by 4! to eliminate repetitions.

b. ${}_nC_r = \frac{n(n-1)(n-2)\cdots(n-(r-1))}{r!}$

EXTRA PRACTICE

Page 562 Chapter 3

21. –2

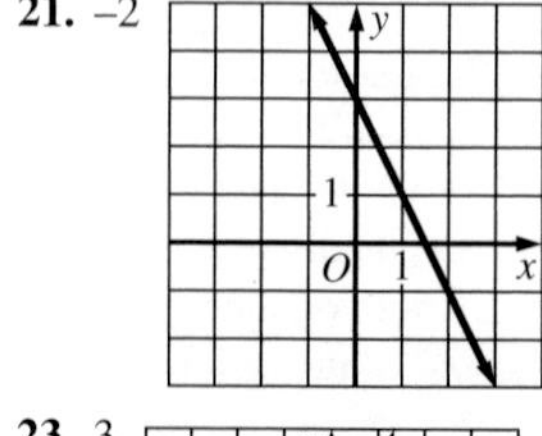

22. $\frac{1}{3}$

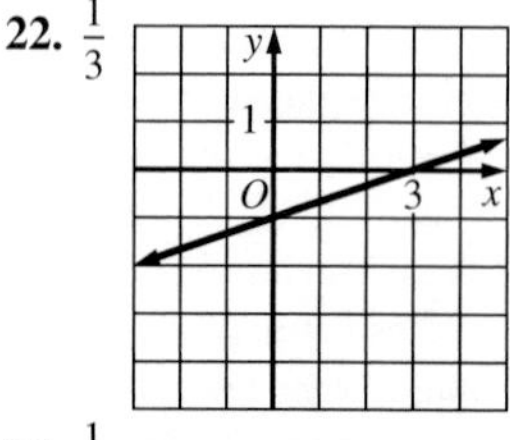

23. 3

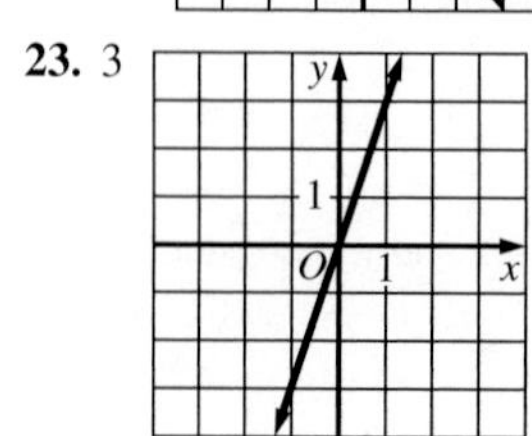

24. $\frac{1}{4}$

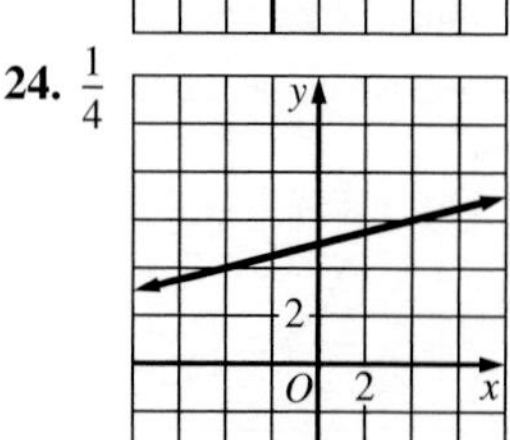

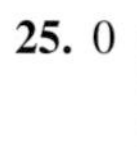

25. 0

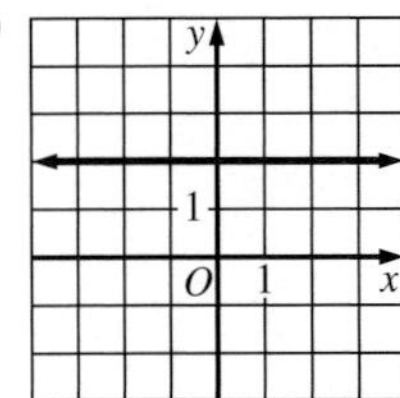

26. 1

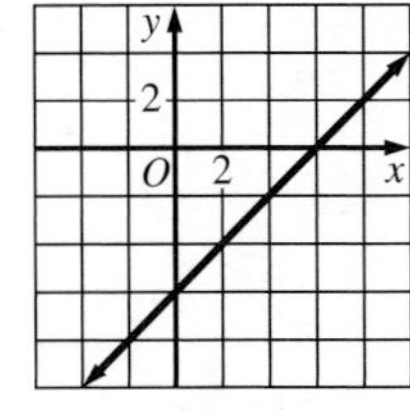

27. –1

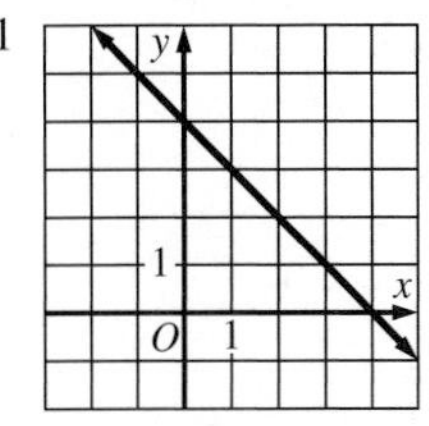

28. –4

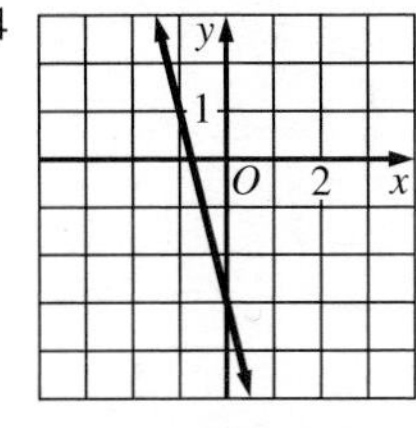

29. 2

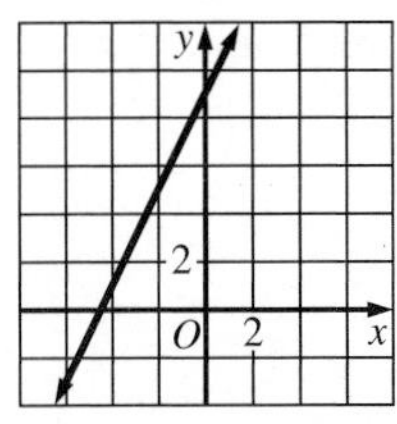

30. 1

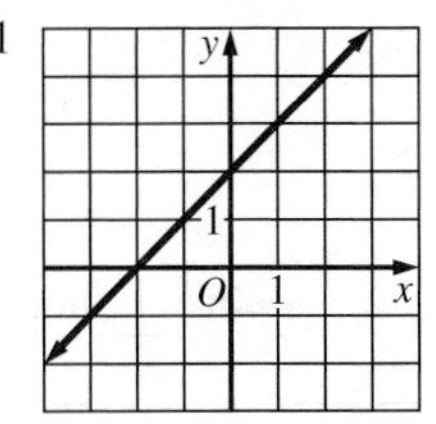

31. 5

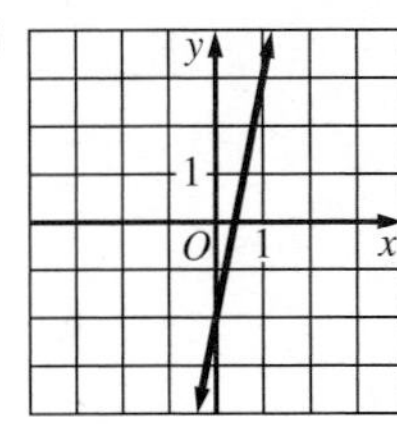

32. –1

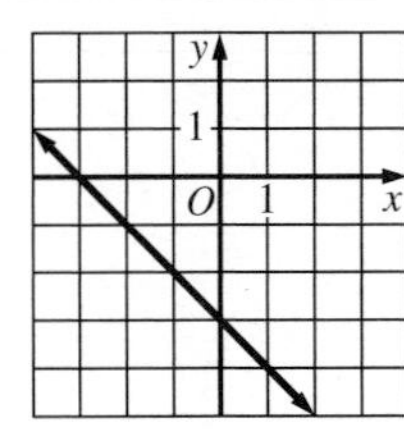

Page 571 **Chapter 7**

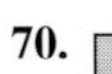

70.

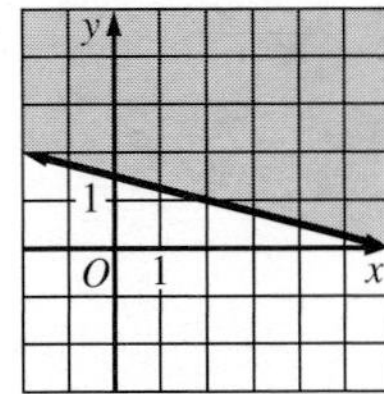

71.

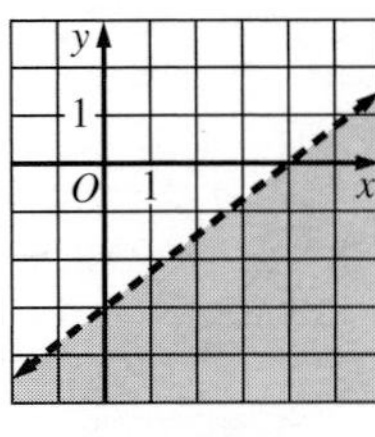

72.

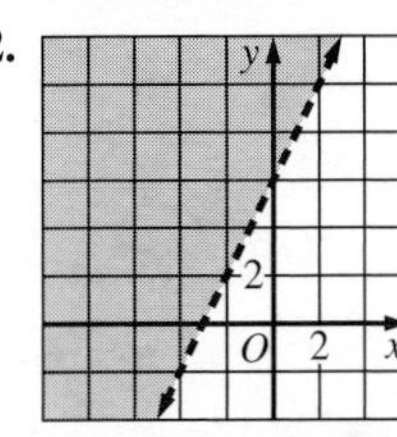

73.

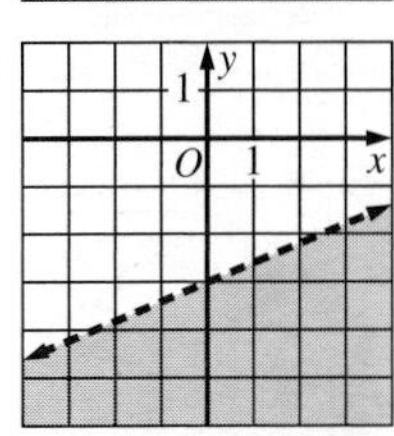

74.

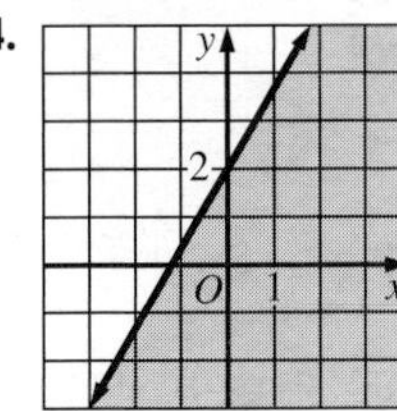

75.

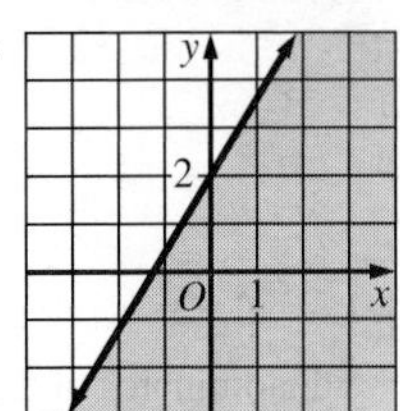